INTRODUCTORY & INTERMEDIATE ALGEBRA

Second Edition

D. FRANKLIN WRIGHT
CERRITOS COLLEGE

HAWKES LEARNING

Editor: Nina Waldron
Vice President, Development: Marcel Prevuznak
Production Editor: Kara Roché
Editorial Assistants: Jessica Ballance, Kaitlin Marley
Layout: E. Jeevan Kumar, D. Kanthi, U. Nagesh, B. Syamprasad
Copy Editors: A. Bhaskar Rao, Nicholas Boyer, B. Padmavathi, G. Ram Jagadish, Taylor Hamrick, Rebecca Hughes, K. Ramesh Narasimha, K.V.S. Anil, M. Prasanna Lakshmi, N. Narasimha Rao, U. Madhavi, U. Ravindra Kumar, Claudia Vance
Contributors: Thomas Clark, Elizabeth Thomas, Melissa Turowski, Barry Wright, III
Answer Key Editors: Vidya Bachina, Kirk Boyer, Peter Bull, Melissa Hecht, Caroline Holbrook, Taylor Jones, Saurabh Joshi, Eric Ketcham, Michael Lane, Joseph Miller, Pradeep Nagalla, Hemanth Nallagatla, Jacob Stauch, Yinliang Tan, Joseph Tracy, James von der Lieth, Colin Williams
Art: Anushree C Bhattacharya, Ayvin Samonte
Cover Art and Design: Jessica Cokins, Johnson Design

Photograph Credits:
iStockphoto.com and Digital Vision with the exception of:
71: Berners, edited by A. Winterberger; 126: NASA/JPL-Caltech; 195: Mark Wagner.

HAWKES LEARNING

A division of Quant Systems, Inc.

546 Long Point Road
Mount Pleasant, SC 29464

Library of Congress Control Number: 2010901560

Printed in the United States of America

10 9 8 7 6 5 4 3 2

ISBN: 978-1-932628-77-7

AIE ISBN: 978-1-64277-192-3

This book is dedicated to the memory of

Greg Hill

TABLE OF CONTENTS

Chapter 6 Factoring Polynomials and Solving Quadratic Equations 429

Chapter 7 Rational Expressions 507

Chapter 8 Systems of Linear Equations 587

PREFACE

Purpose and Style

Introductory & Intermediate Algebra is a comprehensive and versatile teaching tool. For the beginning student, this book develops the more abstract skills and reasoning abilities needed to master algebra. For the more experienced student, it provides a solid base for further studies in mathematics.

Based on this text's feedback from users, insightful comments from reviewers, and skillful editing and design by the editorial staff at Hawkes Learning Systems, we have confidence that students and instructors alike will find that this text is a superior teaching and learning tool. The text may be used independently or in conjunction with the software package ***Hawkes Learning Systems: Introductory & Intermediate Algebra***.

The introductory portion of the text is written assuming that students have basic arithmetic knowledge and no previous experience with algebra. In the second half of the text, students will find that the pace of coverage is somewhat faster and in more depth than they have seen in previous courses. As with any text in mathematics, students should read the text carefully and thoroughly.

The style of the text is informal and nontechnical while maintaining mathematical accuracy. Each topic is developed in a straightforward, step-by-step manner. Each section contains carefully developed and worked out examples to lead the students successfully through the exercises and prepare them for examinations. Whenever appropriate, information is presented in list form for organized learning and easy reference. Common errors are highlighted and explained so that students can avoid such pitfalls and better understand the correct corresponding techniques. Practice problems with answers are provided in nearly every section to allow the students to "warm up" and to provide the instructor with immediate classroom feedback.

A special feature in many sections is "Writing and Thinking About Mathematics." These questions are placed at the end of the exercises for the section and ask the students to delve deeper into mathematical concepts and to become accustomed to organizing their thoughts and writing about mathematics in their own words.

The NCTM and AMATYC curriculum standards have been taken into consideration in the development of the topics throughout the text. In particular:

- there is an emphasis on reading and writing skills as they relate to mathematics,
- techniques for using a graphing calculator are discussed early,
- a special effort has been made to make the exercises motivating and interesting,
- geometric concepts are integrated throughout, and
- statistical concepts, such as interpreting bar graphs and calculating elementary statistics, are included where appropriate.

CHAPTER 1

Real Numbers

Did You Know?

Arithmetic operations defined on the set of positive integers, negative integers, and zero are studied in this chapter. The integer zero will be shown to have interesting properties under the operations of addition, subtraction, multiplication, and division.

Curiously, zero was not recognized as a number by early Greek mathematicians. When Hindu scientists developed the place-value numeration system we currently use, the zero symbol was initially a place holder but not a number. The spread of Islam transmitted the Hindu number system to Europe where it became known as the Hindu-Arabic system and replaced Roman numerals. The word "zero" comes from the Hindu word meaning "void," which was translated into Arabic as "sifr" and later into Latin as "zephirum," hence the derivation of our English words "zero" and "cipher."

Almost all of the operational properties of zero were known to the Hindus. However, the Hindu mathematician Bhaskara the Learned (1114 – 1185?) asserted that a number divided by zero was zero, or possibly infinite. Bhaskara did not seem to understand the role of zero as a divisor since division by zero is undefined and hence, an impossible operation in mathematics.

Albert Einstein, in his development of a proof that the universe was stable and unchangeable in time, divided both sides of one of his intermediate equations by a complicated expression that under certain circumstances could become zero. When the expression became zero, Einstein's proof did not hold and the possibility of a pulsating, expanding, or contracting universe had to be considered. This error was pointed out to Einstein, and he was forced to withdraw his proof that the universe was stable. The moral of this story is that although zero seems like a "harmless" number, ... positive

"How ca... all a pro... of experi... objects o...

Albert Ei...

Did You Know?

This feature draws students into each new chapter's topics through an interesting bit of math history.

Introduction

Presented before the first section of every chapter, the introduction gives students insight into the topics to be covered and the purpose of the chapter.

102 CHAPTER 2 Algebraic Expre...

Chapter 2 provides an integrated introduction to elementary algebra concepts using variables, equations, and positive and negative numbers. You will learn to simplify algebraic expressions and then translate English phrases into algebraic expressions and vice versa. Next you will develop one of the most important skills needed in mathematics: solving equations. You will begin by learning to solve the simplest forms of equations and then expand the related skills to solve more complex and challenging equations. In any case, the basic skills and techniques developed here are fundamental to the ideas related to solving a variety of types of equations. You will see that equations can be used in understanding number problems and geometry problems and in determining percents of profit, gas mileage, taxes, and discounts. So have patience and learn to follow the recommended steps in solving even the basic forms of equations.

2.1 Simplifying and Evaluating Algebraic Expressions

- *Simplify algebraic expressions by combining **like terms**.*
- *Evaluate expressions for given values of the variables.*

Simplifying Algebraic Expressions

A single number is called a **constant**. Any constant or variable or the indicated product and/or quotient of constants and variables is called a **term**. Examples of terms are

$$16,\ 3x,\ -5.2,\ 1.3xy,\ -5x^2,\ 14b^3,\ \text{and}\ -\frac{x}{y}.$$

Note: As discussed in Section 1.5, an expression with 2 as the exponent is read "squared" and an expression with 3 as the exponent is read "cubed." Thus $7x^2$ is read "seven x squared" and $-4y^3$ is read "negative four y cubed."

The number written next to a variable is called the **coefficient** (or the **numerical coefficient**) of the variable. For example, in

$5x^2$, 5 is the coefficient of x^2.

Similarly, in

$1.3xy$, 1.3 is the coefficient of xy.

NOTES

If no number is written next to a variable, the coefficient is understood to be 1. If a negative sign (−) is next to a variable, the coefficient is understood to be −1. For example,

$$x = 1 \cdot x,\ a^3 = 1 \cdot a^3,\ -x = -1 \cdot x,\ \text{and}\ -y^5 = -1 \cdot y^5.$$

Objectives

The objectives provide students with a clear and concise list of the concepts taught in each section, enabling students to focus their time and effort on any less known skill sets.

Every counting number, except the number 1, has **at least two** factors, as illustrated in the following list. Note that in this list, every number has at least two factors, but 5, 19, and 23 have **exactly two** factors.

Examples of Counting Numbers	Factors
5	1, 5
6	1, 2, 3, 6
19	1, 19
23	1, 23
33	1, 3, 11, 33
42	1, 2, 3, 6, 7, 14, 21, 42
96	1, 2, 3, 4, 6, 8, 12, 16, 24, 32, 48, 96

In the list above 5, 19, and 23 have exactly two different factors. Such numbers are called **prime numbers**. The other numbers in the list (6, 33, 42, and 96) are called **composite numbers**.

Prime Numbers

A **prime number** is a counting number greater than 1 that has exactly two different factors (or divisors), namely 1 and itself.

Composite Numbers

A **composite number** is a counting number with more than two different factors (or divisors).

Example 3: Prime Numbers

Some prime numbers:

2 2 has exactly two different factors, 1 and 2
3 3 has exactly two different factors, 1 and 3
11 11 has exactly two different factors, 1 and
29 29 has exactly two different factors, 1 and

Definition Boxes

Straightforward definitions are presented in highly-visible boxes for easy reference.

Notes

Notes highlight common mistakes and how to avoid them as well as offer additional information to students on subtle details.

Examples

Examples are denoted with titled headers indicating the problem solving skill being presented. Each section contains many carefully explained examples with appropriate tables, diagrams, and graphs. Examples are presented in an easy to understand, step-by-step fashion and annotated with notes for additional clarification.

SECTION 1

Exponent and Base

A whole number n is an **exponent** if it is used to tell how many times another whole number a is used as a factor. The repeated factor a is called the **base** of the exponent. Symbolically,

$$\underbrace{a \cdot a \cdot a \cdot \ldots \cdot a \cdot a}_{n \text{ factors}} = a^n$$

exponent
base

NOTES

COMMON ERROR

Do not multiply the base and the exponent.

$10^2 = 10 \cdot 2$ INCORRECT

$6^3 = 6 \cdot 3$ INCORRECT

Do multiply the base by itself.

$10^2 = 10 \cdot 10$ CORRECT

$6^3 = 6 \cdot 6 \cdot 6$ CORRECT

In expressions with exponent 2, the base is said to be **squared**. In expressions with exponent 3, the base is said to be **cubed**.

Example 2: Translating Expressions with Exponents

a. $8^2 = 64$ is read "eight squared is equal to sixty-four."

b. $5^3 = 125$ is read "five cubed is equal to one hundred twenty-five."

Expressions with exponents other than 2 or 3 are read as the base "to the ____ power." For example,

$2^5 = 32$ is read "two to the fifth power is equal to thirty-two."

Prime Numbers and Composite Numbers

Because of the relationship between multiplication and division, **factors** are also called **divisors** of the product. Division by a factor of a number gives a remainder of 0. For example,

factors

$7 \cdot 6 = 42$ and $42 \div 7 = 6$, remainder 0

Thus 7 and 6 are divisors as well as factors of 42.

SECTION 4.6 Graphing Linear Inequalities in Two Variables 331

b. Graph the linear inequality $-5x + 4y > -8$.

Solution:

Step 1: Solving the inequality for y gives: $y > \frac{5}{4}x - 2$.

Step 2: Press the key [Y=] and enter the function: \Y₁ = (5/4)X − 2.

Step 3: Go to the \ and hit [ENTER] two times so that the display appears as follows:

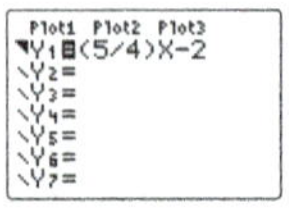

Step 4: Press [GRAPH] and (using the standard WINDOW settings) the following graph should appear on the display. (**Note:** The boundary line should actually be dotted.)

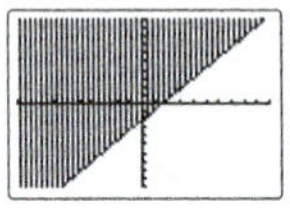

Practice Problems

1. Which of the following points satisfy the inequality $x + y < 3$?

a. $(2, 1)$ **b.** $\left(\frac{1}{2}, 3\right)$ **c.** $(0, 5)$ **d.** $(-5, 2)$

...ty $x - 2y \geq 0$?

Calculator Instruction

As many students are visual learners, screen shots of a TI-84 Plus are provided for visual reference and for step-by-step instructions introducing students to basic graphing skills with a TI-84 Plus calculator.

Practice Problems

Practice Problems with answers are presented at the end of almost every section giving the students an opportunity to practice their newly acquired skills. The answers to the problems are provided at the bottom of the page so students can immediately assess their understanding of the topic at hand.

SECTION 2.4 Solving Linear Equations: $ax + b = c$ 131

Practice Problems

Solve the following linear equations.

1. $x + 14 - 8x = -7$
2. $2.4 = 2.6y - 5.9y - 0.9$
3. $n - \frac{2n}{3} - \frac{1}{2} = \frac{1}{6}$
4. $\frac{3x}{14} + \frac{1}{2} - \frac{x}{7} = 0$

2.4 Exercises

Solve each of the following linear equations.

1. $3x + 11 = 2$
2. $3x + 10 = -5$
3. $5x - 4 = 6$
4. $4y - 8 = -12$
5. $6x + 10 = 22$
6. $3n + 7 = 19$
7. $9x - 5 = 13$
8. $2x - 4 = 12$
9. $1 - 3y = 4$
10. $5 - 2x = 9$
11. $14 + 9t = 5$
12. $5 + 2x = -7$
13. $-5x + 2.9 = 3.5$
14. $3x + 2.7 = -2.7$
15. $10 + 3x - 4 = 18$
16. $5 + 5x - 6 = 9$
17. $15 = 7x + 7 + 8$
18. $14 = 9x + 5 + 8$
19. $5y - 3y + 2 = 2$
20. $6y + 8y - 7 = -7$
21. $x - 4x + 25 = 31$
22. $3y + 9y - 13 = 11$
23. $-20 = 7y - 3y + 4$
24. $-20 = 5y + y + 16$
25. $4n - 10n + 35 = 1 - 2$
26. $-5n - 3n + 2 = 34$
27. $3n - 15 - n = 1$
28. $2n + 12 + n = 0$
29. $5.4x - 0.2x = 0$
30. $0 = 5.1x + 0.3x$
31. $\frac{1}{2}x + 7 = \frac{7}{2}$
32. $\frac{3}{5}x + 4 = \frac{9}{5}$
33. $\frac{1}{2} - \frac{8}{3}x = \frac{5}{6}$
34. $\frac{2}{5} - \frac{1}{2}x = \frac{7}{4}$
35. $\frac{3}{2} = \frac{1}{3}x + \frac{11}{3}$
36. $\frac{11}{8} = \frac{1}{5}x + \frac{4}{5}$
37. $\frac{7}{2} - 5 - \frac{5}{2}x = 9$
38. $\frac{8}{3} + 2 - \frac{7}{3}x = 6$
39. $\frac{5}{8}x - \frac{1}{4}x + \frac{1}{2} = \frac{3}{10}$

...s to Practice Problems: **1.** $x = 3$ **2.** $y = -1$ **3.** $n = 2$ **4.** $x = -7$

Exercises

Each section includes a variety of paired and graded exercises to give the students much needed practice applying and reinforcing the skills learned in the section. More than 4600 carefully selected exercises are provided in the sections. The exercises progress from relatively easy problems to more difficult problems. New for this edition, each chapter has a set of review exercises labeled to match the corresponding section in the chapter.

164 CHAPTER 2 Algebraic Expressions, Linear Equations, and Applications

Chapter 2 Index of Key Ideas and Terms

Section 2.1 Simplifying and Evaluating Algebraic Expressions

Algebraic Vocabulary pages 102-103
Constant – A single number
Term – Any constant or variable or the indicated product and/or quotient of constants and variables
Numerical coefficient – The number written next to a variable
Algebraic expression – A combination of variables and numbers using any of the operations of addition, subtraction, multiplication, or division, as well as exponents

Like Terms page 103
Like terms (or **similar terms**) are terms that are constants or terms that contain the same variables that are raised to the same exponents.

Combining Like Terms page 104
To **combine like terms**, add (or subtract) the coefficients and keep the common variable expression.

Evaluating Algebraic Expressions page 105-106
1. Combine like terms, if possible.
2. Substitute the values given for any variables.
3. Follow the rules for order of operations.

Section 2.2 Translating English Phrases and Algebraic

Translating English Phrases into Algebraic Expressions

Translating Algebraic Expressions into English Phrases

Section 2.3 Solving Linear Equations: $x + b = c$ and ax

Solution
A **solution** to an equation is a number that gives a
statement when substituted for the variable in the

Index of Key Ideas and Terms

Each chapter contains an index highlighting the main concepts within the chapter. Page numbers are given for easy reference as well as complete definitions and concise steps to solve particular types of problems.

80 CHAPTER 1 Real Numbers

Use a graphing calculator to evaluate each of the expressions.

45. $3.4 \div 4 + 5 \cdot 8.32$

46. $8.1 \div 5 + 16.3 \cdot 7$

47. $0.75 \div 1.5 + 7 \cdot 3.1^2$

48. $1.05 \div (-3) \cdot 3.7 - 1.1^2$

49. $6.32 \cdot 8.4 \div 16.8 + 3.5^2$

50. $(82.7 + 16.2) \div (14.83 - 19.83)^2$

Use the rules for order of operations to simplify each of the following expressions.

51. $\frac{3}{4a} \div \frac{3}{16a} - \frac{2b}{3} \cdot \frac{3}{4b}$

52. $\frac{3}{4x} - \frac{5}{6} \div \frac{5x}{8} - \frac{1}{3x}$

53. $-\frac{5}{7y} - \frac{1}{2y} \cdot \frac{2}{3} - \frac{1}{21y}$

54. $\frac{6}{5y} + \frac{3}{2y} - \frac{5}{8} + \frac{y}{12}$

55. $\left(\frac{7}{10x} + \frac{3}{5x}\right) \div \left(\frac{1}{2x} + \frac{3}{7x}\right)$

56. $\left(\frac{1}{5a} - \frac{2}{3a}\right) \div \left(\frac{7}{7b} - \frac{1}{3b}\right)$

57. $\frac{3}{4a} \cdot \left(\frac{1}{3}\right)^2 \div \frac{5}{2} \cdot \frac{1}{9a}$

58. $\frac{1}{24y} \cdot \left(\frac{4}{3}\right)^2 + \left(\frac{2}{3}\right)^2 + \frac{y}{3}$

Writing and Thinking About Mathematics

59. Explain, in your own words, why the following expression cannot be evaluated.

$$(24 - 2^4) \div 6(3 - 5) \div (3^2 - 9)$$

60. Consider any number between 0 and 1. If you square this number, will the result be larger or smaller than the original number? Is this always the case? Explain.

61. Consider any number between −1 and 0. If you square this number, will the result be larger or smaller than the original number? Is this always the case? Explain.

HAWKES LEARNING SYSTEMS: INTRODUCTORY & INTERMEDIATE ALGEBRA SOFTWARE

- 1.8 Order of Operations

Writing and Thinking About Mathematics

In this feature, students are given an opportunity to independently explore and expand on concepts presented in the chapter. These questions will foster a better understanding of the concepts learned within each section.

Additional Features

Calculator Problems: Denoted with a calculator icon, each calculator problem is designed to highlight the usefulness of a graphing calculator in solving certain complex problems, while still emphasizing the importance of understanding the concepts behind the problem.

Chapter Review: At the end of each chapter, a chapter review provides extra problems organized by section. These problems give students an opportunity to review concepts presented throughout the chapter and to identify strengths and potential weaknesses before taking an exam.

Chapter Test: Each chapter also includes a chapter test that provides an opportunity for the students to practice the skills presented in the chapter in a test format.

Cumulative Review: As new concepts build on previous concepts, the cumulative review provides students with an opportunity to continually reinforce existing skills while practicing newer skills.

Answers: Answers are provided for odd-numbered section exercises and for all exercises in the Chapter Reviews, Chapter Tests, and Cumulative Reviews.

Review Chapter: A special chapter reviewing the first seven chapters of the text is available on our website **hawkeslearning.com** to help ease students' transition from 1st semester material to 2nd semester material.

Instructor's Annotated Edition:

Answers: Answers to all the exercises are conveniently located in the margins next to the problems.

Teaching Notes: Suggestions for more in-depth classroom discussions and alternate methods and techniques are located in the margins.

Changes included in the new edition:

- New, reader-friendly layout and improved use of color
- Rearrangement of chapters for better flow, continuity, and progression
- Progressive Chapter Review problems at the end of each chapter
- Calculator Instructions (new emphasis on the TI-84 Plus graphing calculator)

Content

Calculators: The TI-84 Plus graphing calculator has been made an integral part of many of the presentations in this textbook. To get maximum benefits from the use of this text, students are encouraged to have one of these calculators (or a calculator with similar features). Directions are given for using the related calculator commands as they are needed throughout. Generally, calculator discussions and exercises are placed at the end of a section to give the instructor flexibility in what specific calculator usage to include in the course.

Chapter 1, Real Numbers, develops the algebraic concept of integers and the basic skills of operations with integers. Real, rational, and irrational numbers are discussed in conjunction with the real number line. Variables, absolute values, and exponents are defined and expressions are evaluated by using the rules for order of operations. Rational numbers (or fractions) are introduced. Two sections cover basic operations with fractions. The chapter closes with a discussion of the properties of addition and multiplication with real numbers.

Chapter 2, Algebraic Expressions, Linear Equations, and Applications, shows how arithmetic concepts, through the use of variables and signed numbers, can be generalized with algebraic expressions. Algebraic expressions are simplified by combining like terms and evaluated by using the rules for order of operations. As a lead-in to interpreting and understanding word problems, a section involving translating English phrases and algebraic expressions is included. The chapter then goes on to develop the techniques for solving linear (or first-degree) equations in a step-by-step manner over three sections. Techniques include combining like terms and use of the distributive property. Applications in this section relate to number problems, consecutive integers, and percent.

Chapter 3, Formulas and Linear Inequalities, begins by applying the techniques acquired in the previous chapter to work with formulas. A variety of word problems relating to distance, interest, and average as well as formulas related to the geometric concepts of perimeter, area, and volume are included. Other topics include sets and set-builder notation, interval notation, solving absolute value equations, solving linear and absolute value inequalities, and graphing the solution sets of inequalities.

Chapter 4, Linear Equations and Functions, allows for the early introduction of a graphing calculator and the ideas and notation related to functions. Included are complete discussions on the three basic forms for equations of straight lines in a plane: the standard form, the slope-intercept form, and the point-slope form. Slope is discussed for parallel and perpendicular lines and treated as a rate of change. Functions are introduced and the vertical line test is used to tell whether or not a graph represents a function. Use of a TI-84 Plus graphing calculator is an integral part of this introduction to functions as well as part of graphing linear inequalities in the last section.

Chapter 5, Exponents and Polynomials, studies the properties of exponents in depth and shows how to read and write scientific notation. The remainder of the chapter is concerned with definitions and operations related to polynomials. Included are the FOIL method of multiplication with two binomials, special products of binomials, and the division algorithm.

Chapter 6, Factoring Polynomials and Solving Quadratic Equations, discusses methods of factoring polynomials, including finding common monomial factors, factoring by grouping, factoring trinomials by grouping and by trial-and-error, and factoring special products. A special section provides students with tips on determining which type of factoring to use and provides extra exercises to allow them to practice their skills. The topic of solving quadratic equations is introduced, and quadratic equations are solved by factoring only. Applications with quadratic equations are included and involve topics such as the use of function notation to represent area, the Pythagorean theorem, and consecutive integers.

Chapter 7, Rational Expressions, provides still more practice with factoring and shows how to use factoring to operate with rational expressions. Included are the topics of multiplication, division, addition, and subtraction with rational expressions, simplifying complex fractions, and solving equations containing rational expressions. Applications are related to work, distance-rate-time, and variation.

Chapter 8, Systems of Linear Equations, covers solving systems of two equations in two variables and systems of three equations in three variables. The basic methods of graphing, substitution, and addition are included along with matrices and Gaussian elimination. Double subscript notation is now used with matrices for an easy transition to the use of matrices in solving systems of equations with the TI-84 Plus calculator. Applications involve mixture, interest, work, algebra, and geometry. The last section discusses half-planes and graphing systems of linear inequalities, again including the use of a graphing calculator.

Chapter 9, Roots, Radicals, and Complex Numbers, introduces roots and fractional exponents and the use of a calculator to find approximate values. Arithmetic with radicals includes simplifying radical expressions, addition, subtraction, and rationalizing denominators. A section on functions with radicals shows how to analyze the domain and range of radical functions and how to graph these functions by using a graphing calculator. Complex numbers are introduced along with the basic operations of addition, subtraction, multiplication, and division. These are skills needed for the work with quadratic equations and quadratic functions in Chapter 10.

Chapter 10, Quadratic Equations, reviews solving quadratic equations by factoring and introduces the methods of using the square root property and completing the square. The quadratic formula is developed by completing the square and students are encouraged to use the most efficient method for solving any particular quadratic equation. Applications are related to the Pythagorean theorem, projectiles, geometry, and cost. The fourth section covers solving equations in quadratic form and the last two sections deal with quadratic functions (graphing parabolas) and solving quadratic inequalities.

Chapter 11, Exponential and Logarithmic Functions, begins with a section on the algebra of functions and leads to the development of the composition of functions and methods for finding the inverses of one-to-one functions. This introduction lays the groundwork for understanding the relationship between exponential functions and logarithmic functions. While the properties of real exponents and logarithms are presented completely, most numerical calculations are performed with the aid of a calculator. Special emphasis is placed on the number *e* and applications with natural logarithms. Students will find the applications with exponential and logarithmic functions among the most interesting and useful in their mathematical studies. Those students who plan to take a course in calculus should be aware that many of the applications found in calculus involve exponential and logarithmic expressions in some form.

Chapter 12, Conic Sections, provides a basic understanding of conic sections (parabolas, circles, ellipses, and hyperbolas) and their graphs. The first section gives detailed analyses of translations involving horizontal and vertical shifting. Function notation is used in discussing reflections and translations of a variety of types of functions. Vertical and horizontal parabolas are developed as conic sections. The distance formula is developed and used to find equations of circles. The thorough development of ellipses and hyperbolas includes graphs with centers not at the origin. Solving systems with nonlinear equations is the final topic of the chapter.

Chapter 13, Sequences, Series, and the Binomial Theorem, introduces the concept of sequences, their basic properties, and sigma notation. The final section includes factorials and the binomial theorem. The topics presented here are likely to appear in courses in probability and statistics, finite mathematics, and higher level courses in mathematics. Any of these topics covered at this time will give students additional mathematical experience and insight for future studies.

I recommend that the topics be covered in the order presented because most sections assume knowledge of the material in previous sections. This is particularly true of the cumulative review sections at the end of each chapter. Of course, time and other circumstances may dictate another sequence of topics. For example, in some programs, Chapters 1 and 2 might be considered review or Chapters 12 and 13 might be considered part of a college algebra course.

About the annotations in the Instructor's Edition

The annotations in the margin entitled "Teaching Notes:" are ideas for on-the-spot helpful classroom suggestions. I am sure you are familiar with most of them; but many of us, myself included, have thought just after a class has been dismissed, "Oh, I wish I had mentioned _____ for interest or motivation or variety." These annotations include extra examples, historical comments, ideas for possible class discussions, and other comments that you might find interesting or helpful. Obviously, they are for use at your own discretion.

Thank you for using my text, and I look forward to receiving any suggestions, comments, or errata that you might bring to my attention for future printings and editions.

Frank

Acknowledgements

I would like to thank Editor Nina Waldron, Production Editor Kara Roché, and Vice President of Development Marcel Prevuznak for their hard work and invaluable assistance in the development and production of this text.

Many thanks go to the following manuscript reviewers who offered constructive and critical comments:

Elaine Arrington, *University of Montana*
Darcee Bex, *South Louisiana Community College*
Cindy Bond, *Butler Community College*
Randall Dorman, *Cochise College*
Debra Gupton, *New River Community College*
Kimberly Haughee, *University of Montana*
Rebecca Heiskell, *Mountain View College*
Bobbie Jo Hill, *Coastal Bend College*
Sandee House, *Georgia Perimeter College*
Marjorie Hunter, *Butler Community College*
Joanne Koratich, *Muskegon Community College*
Jim Martin, *Cochise College*
Dr. Carol Okigbo, *Minnesota State University*
Gabriel Perrow, *Eastern Maine Community College*
Amy Rexrode, *Navarro College*
Harriete Roadman, *New River Community College*
Connie Rost, *South Louisiana Community College*
Jim Sheff, *Spoon River College*
Dr. Melanie Smith, *Bishop State Community College*
Nan Strebeck, *Navarro College*
Emily Whaley, *Georgia Perimeter College*

Finally, special thanks go to James Hawkes for his faith in this second edition and his willingness to commit so many resources to guarantee a top-quality product for students and teachers.

D. Franklin Wright

TO THE STUDENT

The goal of this text and of your instructor is for you to succeed in Introductory & Intermediate Algebra. Certainly, you should make this your goal as well. What follows is a brief discussion about developing good work habits and using the features of this text to your best advantage. For you to achieve the greatest return on your investment of time and energy you should practice the following three rules of learning:

1. Reserve a block of time for study every day.
2. Study what you don't know.
3. Don't be afraid to make mistakes.

How to Use this Book

The following seven-step guide will not only make using this book a more worthwhile and efficient task, but it will also help you benefit more from classroom lectures or the assistance that you receive in a math lab.

1. Try to look over the assigned section(s) before attending class or lab. In this way, you will be more comfortable with new ideas presented in class or lab. This will also help you anticipate where you need to ask questions about material that is difficult for you to understand.

2. Read examples carefully. They have been chosen and written to show all of the problem-solving steps that you need to be familiar with. You might even try to solve example problems independently before studying the solutions that are given.

3. Work the section exercises as they are assigned. Problem-solving practice is the single most important element in achieving success in any math class, and there is no good substitute for actually doing this work yourself. Demonstrating that you can think independently through each step of each type of problem will also build your confidence in your ability to answer questions on quizzes and exams. Check the Answer Key periodically while working section exercises to be sure that you have the right ideas and are proceeding in the right manner. Identify and correct your mistakes as you work.

4. Use the Writing and Thinking About Mathematics questions as an opportunity to explore the way that you think about math. A big part of learning and understanding mathematics is being able to communicate mathematical ideas and the thinking that occurs to you as you approach new concepts and problems. These questions can help you analyze your own approach to mathematics and, in class or group discussions, learn from ideas expressed by your fellow students.

5. Use the Chapter Index of Key Ideas and Terms as a recap when you begin to prepare for a Chapter Test. It will reference all the major ideas that you should be familiar with from that chapter and indicate where you can turn if review is needed. You can also use the Chapter Index as a final checklist once you feel you have completed your review and are prepared for the Chapter Test.

6. Chapter Tests are provided so that you can practice for the tests that are actually given in class or lab. To simulate a test situation, block out a one-hour, uninterrupted period in a quiet place where your only focus is on accurately completing the Chapter Test. Use the Answer Key at the back of the book as a self-check only after you have completed all of the questions on the test.

7. Chapter Reviews and Cumulative Reviews will help you retain the skills that you acquired in studying earlier material. Approach them in much the same manner as you would the Chapter Tests in order to keep all of your skills sharp throughout the entire course.

How to Prepare for an Exam

Gaining Skill and Confidence

The stress that many students feel while trying to succeed in mathematics is what you have probably heard called "math anxiety." It is a real-life phenomenon, and many students experience such a high level of anxiety during mathematics exams in particular that they struggle to perform to the best of their abilities. It is possible to overcome this stress by building your confidence in your ability to do mathematics and by minimizing your fears of making mistakes.

No matter how much it may seem that in mathematics you must either be right or wrong, with no middle ground, you should realize that you can be learning just as much from the times that you make mistakes as you can from the times that your work is correct. Success will come. Don't think that making mistakes at first means that you'll never be any good at mathematics. Learning mathematics requires lots of practice. Most importantly, it requires a true confidence in yourself and in the fact that, with practice and persistence, the mistakes will become fewer, the successes will become greater, and you will be able to say, "I can do this."

Showing What You Know

If you have attended class or lab regularly, taken good notes, read your textbook, kept up with homework exercises, and asked for help when it was needed, then you have already made significant progress in preparing for an exam and conquering any anxiety. Here are a few other suggestions to maximize your preparedness and minimize your stress.

1. Give yourself enough time to review. You will generally have several days notice before an exam. Set aside a block of time each day with the goal of reviewing a manageable portion of the material that the test will cover. Don't cram!

2. Work lots of problems to refresh your memory and sharpen you skills. Go back to rework selected exercises from all of your homework assignments.

3. Reread your text and your notes and use the Chapter Index of Key Ideas and Terms, the Chapter Review, and the Chapter Test to recap major ideas and do a self-evaluated test simulation.

4. Study with a friend or classmate. Peer tutoring almost always helps in gaining better understanding of concepts and developing better problem solving skills.

5. Be sure that you are well-rested so that you can be alert and focused during the exam.

6. Don't study up to the last minute. Give yourself some time to wind down before the exam. This will help you to organize your thoughts and feel more calm as the test begins.

7. As you take the test, realize that its purpose is not to trick you, but to give you and your instructor an accurate idea of what you have learned. Good study habits, a positive attitude, and confidence in your own ability will be reflected in your performance on any exam.

8. Finally, you should realize that your responsibility does not end with taking the exam. When your instructor returns your corrected exam, you should review your instructor's comments and any mistakes that you might have made. Take the opportunity to learn from this important feedback about what you have accomplished, where you could work harder, and how you can best prepare for future exams.

HAWKES LEARNING SYSTEMS: Introductory & Intermediate Algebra

Overview

This multimedia courseware allows students to become better problem-solvers by creating a mastery level of learning in the classroom. The software includes an "Instruct," "Practice," "Tutor," and "Certify" mode in each lesson, allowing students to learn through step-by-step interactions with the software. The automated homework system's tutorial and assessment modes extend instructional influence beyond the classroom. Intelligence is what makes the tutorials so unique. By offering intelligent tutoring and mastery level testing to measure what has been learned, the software extends the instructor's ability to influence students to solve problems. This courseware can be ordered either separately or bundled together with this text.

Minimum Requirements

In order to run ***HLS: Introductory & Intermediate Algebra***, you will need:

1 GHz or faster processor
Windows® XP (with Service Pack 3) or later
256 MB RAM
200 MB hard drive space (compact install), or up to 1.5 GB (complete install)
800x600 resolution (1024x768 recommended)
Internet Explorer 6.0 (or higher), Mozilla Firefox 2.0 (or higher), or Google Chrome 2.0 (or higher)
CD-ROM drive

Getting Started

Before you can run ***HLS: Introductory & Intermediate Algebra***, you will need an access code. This 30 character code is your personal access code. To obtain an access code, go to **hawkeslearning.com** and follow the links to the access code request page (unless directed otherwise by your instructor).

Installation

Insert the ***HLS: Introductory & Intermediate Algebra*** installation CD-ROM into the CD-ROM drive. Select the Start>Run command, type in the CD-ROM drive letter followed by :\setup.exe. (For example, d:\setup.exe where d is the CD-ROM drive letter.)

The compact installation may use over 200 MB of hard drive space and will install the entire product, except the multimedia files, on your hard drive.

After selecting the desired installation option, follow the on-screen instructions to complete your installation of ***HLS: Introductory & Intermediate Algebra***.

Starting the Courseware

After you install ***HLS: Introductory & Intermediate Algebra*** on your computer, to run the courseware select Start>Programs>Hawkes Learning Systems>Introductory & Intermediate Algebra.

You will be prompted to enter your access code with a message box similar to the following:

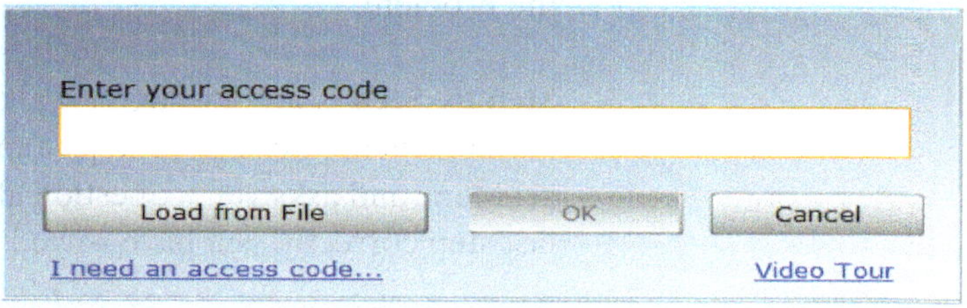

Type your entire access code in the box. When you are finished, press OK.

If you typed in your access code correctly, you will be prompted to save the code to a disk. If you choose to save your code to a disk, typing in the access code each time you run ***HLS: Introductory & Intermediate Algebra*** will not be necessary. Instead, select the 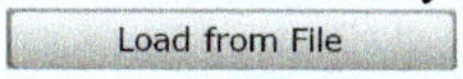

button when prompted to enter your access code and choose the path to your saved access code.

Now that you have entered your access code and saved it, you are ready to run a lesson. From the table of contents screen, choose the appropriate chapter and then choose the lesson you wish to run.

Features

Each lesson in ***HLS: Introductory & Intermediate Algebra*** has four modes: Instruct, Practice, Tutor, and Certify.

Instruct: Instruct provides an exposition on the material covered in the lesson in a multimedia environment. This same instruct mode can be accessed via the tutor mode.

Practice: Practice allows you to hone your problem-solving skills. It provides an unlimited number of randomly generated problems. Practice also provides access to the Tutor mode by selecting the Tutor button located next to the Submit button.

Tutor: Tutor mode is broken up into several parts: Instruct, Explain Error, Step by Step, and Solution.

1. **Instruct**, which can also be selected directly from Practice mode, contains a multimedia lecture of the material covered in a lesson.

2. **Explain Error** is active whenever a problem is incorrectly answered. It will attempt to explain the error that caused you to incorrectly answer the problem.

3. **Step by Step** is an interactive walkthrough of the problem. It breaks each problem into several steps, explains to you each step in solving the problem, and asks you a question about the step. After you answer the last step correctly, you have solved the problem.

4. **Solution** will provide you with a detailed "worked-out" solution to the problem.

Throughout the Tutor, you will see words or phrases colored green with a dashed underline. These are called Hot Words. Clicking on a Hot Word will provide you with more information on these words or phrases.

Certify: Certify is the testing mode. You are given a finite number of problems and a certain number of strikes (problems you can get wrong). If you answer the required number of questions correctly, you will receive a certification code and a certificate. Write down your certification code and/or print out your certificate. The certification code will be used to update your records in your progress report. Note that the Tutor is not available in Certify.

Integration of Courseware and Textbook

Throughout the text you will find references that will help you integrate the *Introductory & Intermediate Algebra* textbook and the ***HLS: Introductory & Intermediate Algebra*** courseware. At the end of each section and again at the end of each chapter you will find a list of which ***HLS: Introductory & Intermediate Algebra*** lessons you should use in order to test yourself on the subject material and to review the contents of a chapter.

Support

If you have questions about ***HLS: Introductory & Intermediate Algebra*** or are having technical difficulties, we can be contacted as follows:

Phone: (843) 571-2825
Email: support@hawkeslearning.com
Web: hawkeslearning.com

Our support hours are 8:30 a.m. to 5:30 p.m., EST, Monday through Friday.

[illegible]

CHAPTER 1

Real Numbers

Did You Know?

Arithmetic operations defined on the set of positive integers, negative integers, and zero are studied in this chapter. The integer zero will be shown to have interesting properties under the operations of addition, subtraction, multiplication, and division.

Curiously, zero was not recognized as a number by early Greek mathematicians. When Hindu scientists developed the place-value numeration system we currently use, the zero symbol was initially a place holder but not a number. The spread of Islam transmitted the Hindu number system to Europe where it became known as the Hindu-Arabic system and replaced Roman numerals. The word "zero" comes from the Hindu word meaning "void," which was translated into Arabic as "sifr" and later into Latin as "zephirum," hence the derivation of our English words "zero" and "cipher."

Almost all of the operational properties of zero were known to the Hindus. However, the Hindu mathematician Bhaskara the Learned (1114 – 1185?) asserted that a number divided by zero was zero, or possibly infinite. Bhaskara did not seem to understand the role of zero as a divisor since division by zero is undefined and hence, an impossible operation in mathematics.

Albert Einstein, in his development of a proof that the universe was stable and unchangeable in time, divided both sides of one of his intermediate equations by a complicated expression that under certain circumstances could become zero. When the expression became zero, Einstein's proof did not hold and the possibility of a pulsating, expanding, or contracting universe had to be considered. This error was pointed out to Einstein, and he was forced to withdraw his proof that the universe was stable. The moral of this story is that although zero seems like a "harmless" number, its operational properties are different from those of the positive and negative integers.

"How can it be that mathematics, being after all a product of human thought independent of experience, is so admirably adapted to the objects of reality?"

Albert Einstein (1879 – 1955)

Numbers and number concepts form the foundation for the study of algebra. In this chapter, you will learn about positive and negative numbers and how to operate with these numbers. Believe it or not, the key number is 0. Pay particularly close attention to the idea of the magnitude of a number, called its absolute value, and the terminology used to represent different types of numbers.

1.1 The Real Number Line and Absolute Value

- *Identify types of numbers.*
- *Graph sets of numbers on a real number line.*
- *Determine if given numbers are greater than, less than, or equal to other given numbers.*
- *Determine **absolute values**.*

We begin with a development of the terminology of numbers that form the foundation for the study of algebra. The following kinds of numbers are studied in some detail in beginning algebra courses.

Types of Numbers

A **set** is a collection of objects or numbers. The set of numbers

$$\mathbb{N} = \{1, 2, 3, 4, 5, 6, 7, 8, 9, 10, 11, \ldots\}$$

is called the **counting numbers** or **natural numbers**. The three dots ... are called an ellipsis and are used to indicate that the pattern is to continue without end. Putting 0 with the set of natural numbers gives the set of **whole numbers**.

$$\mathbb{W} = \{0, 1, 2, 3, 4, 5, 6, 7, 8, 9, 10, 11, \ldots\}$$

Thus mathematicians make the distinction that 0 is a whole number but not a natural number.

To help in understanding different types of numbers and their relationships to each other, we begin with a "picture" called a **number line**. For example, choose some point on a horizontal line and label it with the number 0 (Figure 1).

Teaching Notes:
Point out that this simple action represents the initial connection between numbers and geometry; and it is the basis for graphing.

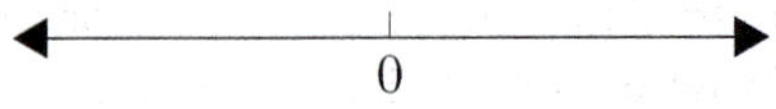

Figure 1

Now choose another point on the line to the right of 0 and label it with the number 1 (Figure 2). This point is arbitrary and once it is chosen a units scale is determined for the remainder of the line.

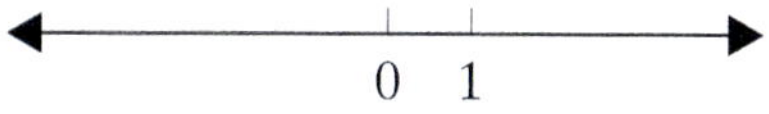

Figure 2

We now have a number line. Points corresponding to all the whole numbers are determined. The point corresponding to 2 is the same distance from 1 as 1 is from 0, and so on (Figure 3).

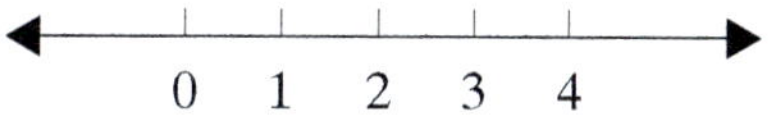

Figure 3

The **graph** of a number is the point that corresponds to the number and the number is called the **coordinate** of the point. We will follow the convention of using the terms "number" and "point" interchangeably. For example, one point might be "seven" and another point "two." The graph of 7 is indicated by marking the point corresponding to 7 with a large dot (Figure 4).

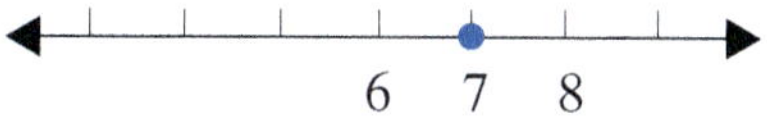

Figure 4

The graph of the set $A = \{2, 4, 6\}$ is shown in Figure 5.

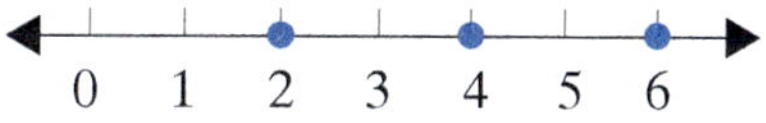

Figure 5

On a horizontal number line, the point one unit to the left of 0 is the **opposite** of 1. It is called **negative** 1 and is symbolized –1. Similarly, the point two units to the left of 0 is the opposite of 2, called negative 2, and symbolized –2, and so on (Figure 6).

The opposite of 1 is –1;	The opposite of –1 is $-(-1) = +1$;
The opposite of 2 is –2;	The opposite of –2 is $-(-2) = +2$;
The opposite of 3 is –3;	The opposite of –3 is $-(-3) = +3$;
and so on.	and so on.

NOTES

The negative sign (–) indicates the opposite of a number which we call a negative number. It is also used, as we will see in Section 1.3, to indicate subtraction. To avoid confusion, you must learn (by practice) just how the – sign is used in each particular situation.

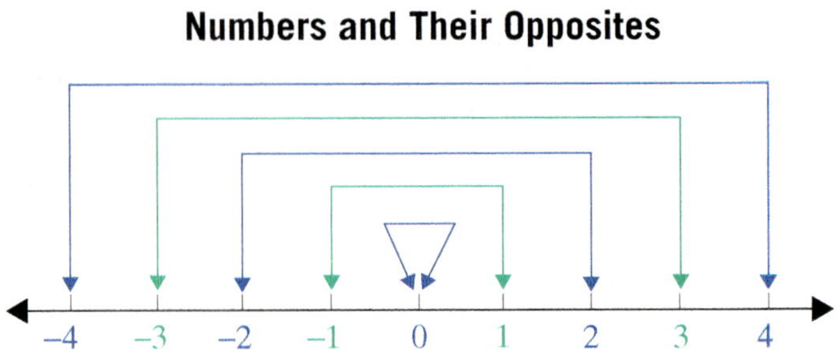

Figure 6

Integers

The set of numbers consisting of the whole numbers and their opposites is called the set of **integers**: $\mathbb{Z} = \{\ldots, -3, -2, -1, 0, 1, 2, 3, \ldots\}$.

The natural numbers are also called **positive integers**. Their opposites are called **negative integers**. **Zero is its own opposite and is neither positive nor negative** (Figure 7)**.** Note that the opposite of a positive integer is a negative integer, and the opposite of a negative integer is a positive integer.

Integers:	$\{\ldots, -3, -2, -1, 0, 1, 2, 3, \ldots\}$
Positive integers:	$\{1, 2, 3, 4, 5, \ldots\}$
Negative integers:	$\{\ldots, -4, -3, -2, -1\}$

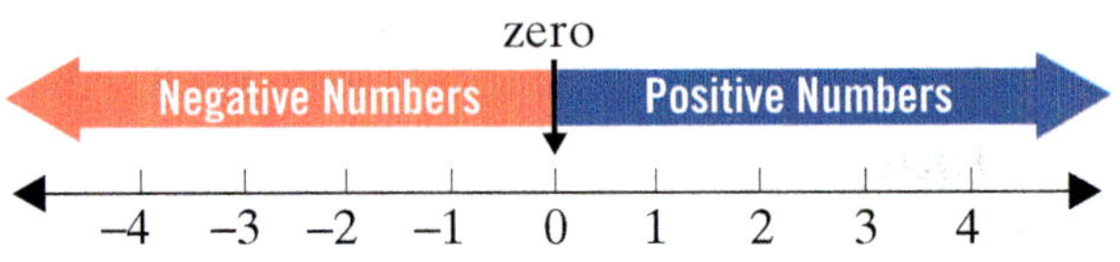

Figure 7

Example 1: Opposites

a. State the opposite of 7.

Solution: -7

b. State the opposite of -3.

Solution: $-(-3)$ or $+3$

In words, the opposite of -3 is $+3$.

Example 2: Number Line

a. Graph the set of integers $\{-3, -1, 1, 3\}$.

Solution:

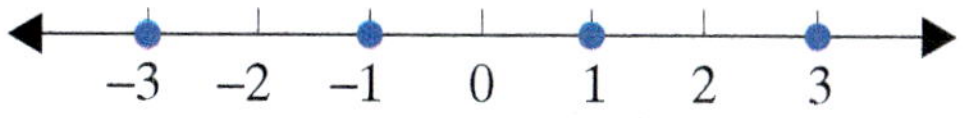

b. Graph the set of integers $\{\ldots, -5, -4, -3\}$.

Solution:

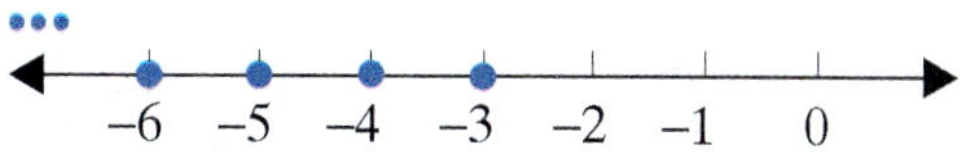

The three dots above the number line indicate that the pattern in the graph continues without end.

The integers are not the only numbers that can be represented on a number line. Fractions and decimal numbers such as $\frac{1}{2}, -\frac{4}{3}, \frac{3}{4},$ and -2.3, as well as numbers such as π and $\sqrt{2}$, can also be represented (Figure 8).

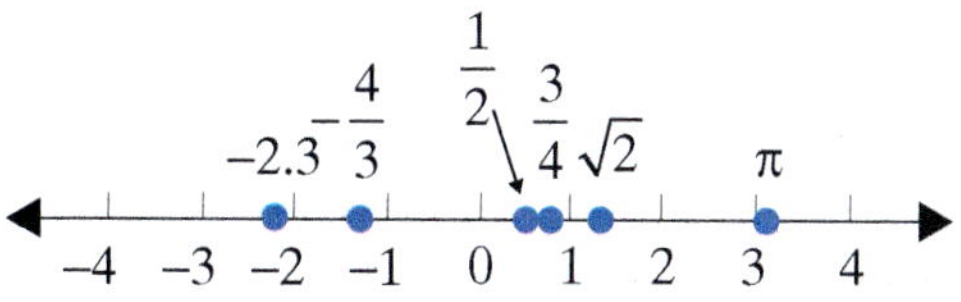

Figure 8

Teaching Notes: You might mention that the word "rational," in this case, comes from "ratio."

Numbers that can be written as fractions and whose numerators and denominators are integers have the technical name **rational numbers** ($\mathbb{Q}$). Positive and negative decimal numbers and the integers themselves can also be classified as rational numbers. For example, the following numbers are all rational numbers:

$$1.3 = \frac{13}{10}, \quad 5 = \frac{5}{1}, \quad -4 = \frac{-4}{1}, \quad \frac{3}{8}, \quad \text{and} \quad \frac{17}{6}.$$

Variables are needed to be able to state rules and definitions in general forms.

Variable

A **variable** is a symbol (generally a letter of the alphabet) that is used to represent an unknown number (or numbers).

The variables a and b are used to represent integers in the following definition of a **rational number**.

Rational Numbers

A **rational number** is a number that can be written in the form of $\frac{a}{b}$ where a and b are integers and $b \neq 0$. ($\neq$ is read "is not equal to.")

OR

A **rational number** is a number that can be written in decimal form as a terminating decimal or as an infinite repeating decimal.

Other numbers on a number line, such as $\sqrt{2}, \sqrt{3}, \pi,$ and $\sqrt[3]{5}$, are called **irrational numbers**. These numbers can be written as **infinite nonrepeating decimal numbers**. All rational numbers and irrational numbers are classified as **real numbers** ($\mathbb{R}$) and can be written in some decimal form. The number line is called the **real number line**.

We will discuss rational numbers later in this chapter and irrational numbers in Chapter 9. For now, we are only interested in recognizing various types of numbers and locating their positions on the real number line.

With a calculator, you can find the following decimal values and approximations:

Examples of Rational Numbers:

$\frac{3}{4} = 0.75$ — This decimal number is terminating.

$\frac{1}{3} = 0.33333333...$ — There is an infinite number of 3's in this repeating pattern.

$\frac{3}{11} = 0.27272727...$ — This repeating pattern shows that there may be more than one digit in the pattern.

Teaching Notes:
Point out that any truncation of repeating decimals gives an approximation. Often, in algebra, we prefer the fractional form for a rational number because it remains exact.

Examples of Irrational Numbers:

$\sqrt{2} = 1.414213562...$ — There is an infinite number of digits with no repeating pattern.

$\pi = 3.141592653...$ — There is an infinite number of digits with no repeating pattern. (A TI-84 Plus graphing calculator will show a 4 in place of the ninth place digit 3 because it rounds off decimal numbers.)

1.414414441444441... — There is an infinite number of digits and a pattern of sorts. However, the pattern is nonrepeating.

The following diagram (Figure 9) illustrates the relationships among the various categories of real numbers.

Teaching Notes:
You might want to show your students a Venn diagram such as

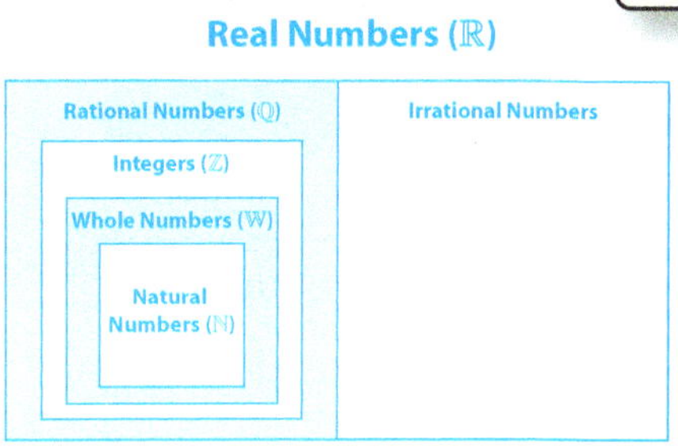

as another illustration of these relationships.

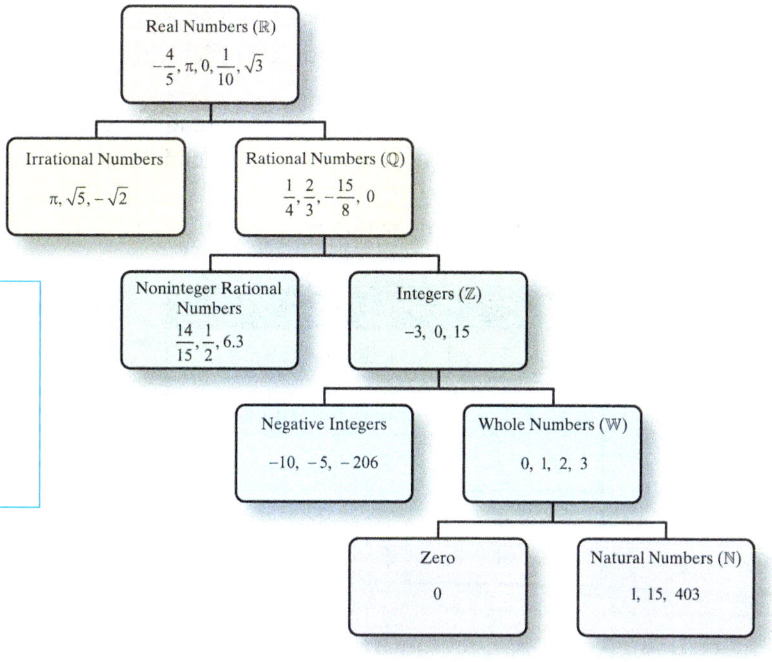

Figure 9

Example 3: Identifying Types of Numbers

List the numbers in the set $S = \left\{-5, -\frac{3}{4}, 0, \sqrt{2}, 17\right\}$ that are:

a. Whole numbers

Solution: 0 and 17 are whole numbers.

b. Integers

Solution: −5, 0, and 17 are integers.

c. Rational numbers

Solution: $-5, -\frac{3}{4}, 0,$ and 17 are rational numbers.

d. Real numbers

Solution: All numbers in S are real numbers.

Inequality Symbols

On a horizontal number line, **smaller numbers are always to the left of larger numbers.** Each number is smaller than any number to its right and larger than any number to its left. Two symbols used to indicate order are

$<$, read "is less than"
and $>$, read "is greater than."

Using the real number line in Figure 10, you can see the following relationships.

Using <	
$0 < 3$	0 is less than 3
$-2 < 1$	-2 is less than 1
$-7 < -4$	-7 is less than -4
$\frac{1}{4} < \frac{9}{8}$	$\frac{1}{4}$ is less than $\frac{9}{8}$

Using >	
$3 > 0$	3 is greater than 0
$1 > -2$	1 is greater than -2
$-4 > -7$	-4 is greater than -7
$\frac{9}{8} > \frac{1}{4}$	$\frac{9}{8}$ is greater than $\frac{1}{4}$

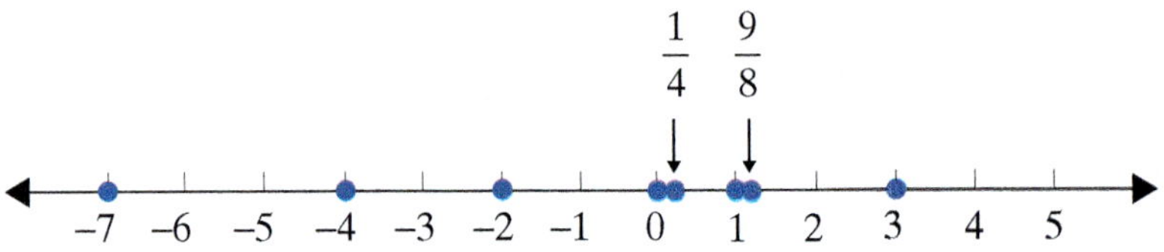

Figure 10

Two other symbols commonly used are

$\leq$, read "is less than or equal to"
and $\geq$, read "is greater than or equal to."

For example, $5 \geq -10$ is true since 5 is greater than -10. Also, $5 \geq 5$ is true since 5 does equal 5.

Symbols of Equality and Inequality

$=$	is equal to	$\neq$	is not equal to
$<$	is less than	$>$	is greater than
$\leq$	is less than or equal to	$\geq$	is greater than or equal to

NOTES

Special Note About the Inequality Symbols

Each symbol can be read from left to right as was just indicated in "Symbols of Equality and Inequality." However, each symbol can also be read from right to left. Thus any inequality can be read in two ways. For example, $6 < 10$ can be read from left to right as "6 is less than 10," but also from right to left as "10 is greater than 6." We will see that this flexibility is particularly useful when reading expressions with variables in Section 3.4.

Example 4: Inequalities

a. Determine whether each of the following statements is true or false.

$7 < 15$	True, since 7 is less than 15. 7 is to the left of 15 on the number line.
$3 > -1$	True, since 3 is greater than –1. 3 is to the right of –1 on the number line.
$4 \geq -4$	True, since 4 is greater than –4. 4 is to the right of –4 on the number line.
$2.7 \geq 2.7$	True, since 2.7 is equal to 2.7.
$-5 < -6$	False, since –5 is greater than –6.

Note: $7 < 15$ can be read as "7 is less than 15" or as "15 is greater than 7."
$3 > -1$ can be read as "3 is greater than –1" or as "–1 is less than 3."
$4 \geq -4$ can be read as "4 is greater than or equal to –4" or as "–4 is less than or equal to 4."

b. Graph the set of **real numbers** $\left\{-\frac{3}{4}, 0, 1, 1.5, 3\right\}$.

Solution:

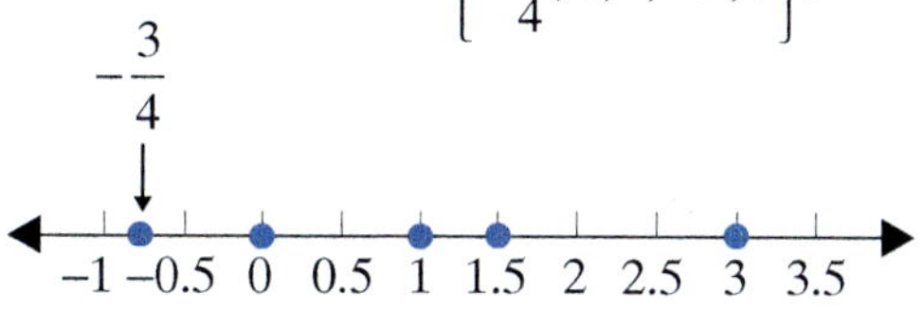

c. Graph all **natural numbers** less than or equal to 3.

Solution:

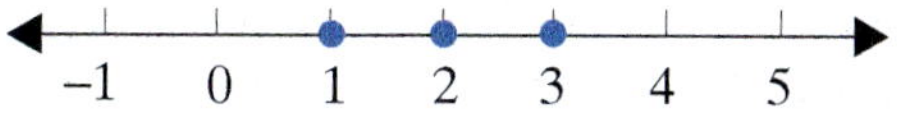

Remember that the natural numbers are 1, 2, 3, 4, ...

d. Graph all **integers** less than 0.

Solution:

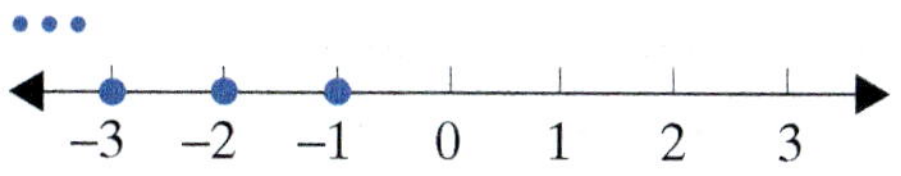

Remember, the three dots above the number line indicate that the pattern in the graph continues without end.

Absolute Value

In working with the real number line, you may have noticed that any integer and its opposite lie the same number of units from 0 on the number line. For example, both +7 and −7 are seven units from 0 (Figure 11). The + and − signs indicate direction and the 7 indicates distance.

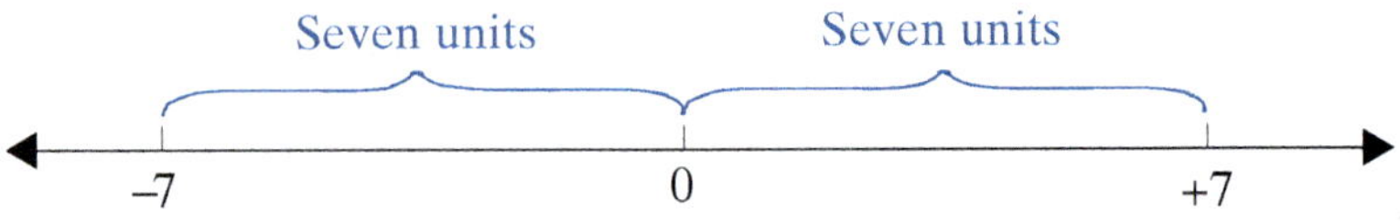

Figure 11

The **distance a number is from 0 on a number line** is called its **absolute value** and is symbolized by two vertical bars, $|\quad|$. Thus $|+7| = 7$ and $|-7| = 7$. Similarly,

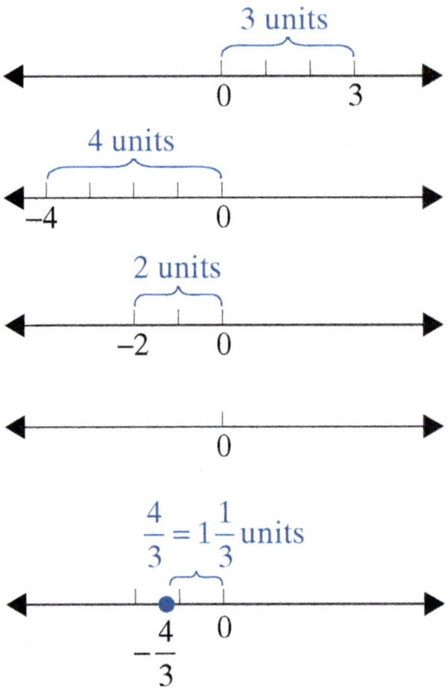

$|3| = 3$ — 3 is 3 units from 0.

$|-4| = 4$ — −4 is 4 units from 0.

$|-2| = 2$ — −2 is 2 units from 0.

$|0| = 0$

$\left|-\frac{4}{3}\right| = \frac{4}{3}$ — $-\frac{4}{3}$ is $\frac{4}{3} = 1\frac{1}{3}$ units from 0.

Since distance (similar to length) is never negative, the absolute value of a number is never negative. Or, the absolute value of a nonzero number is always positive.

Absolute Value

The **absolute value** of a real number is its distance from 0. Note that the absolute value of a real number is never negative.

$|\boldsymbol{a}| = \boldsymbol{a}$ if a is a positive number or 0.

$|\boldsymbol{a}| = -\boldsymbol{a}$ if a is a negative number.

NOTES

The symbol $-a$ should be thought of as the "opposite of a." Since a is a variable, a might represent a positive number, a negative number, or 0. This use of symbols can make the definition of absolute value difficult to understand at first. As an aid to understanding the use of the negative sign, consider the following examples.

$$\text{If } a = -6, \text{ then } -a = -(-6) = 6.$$

Similarly,

$$\text{If } x = -1, \text{ then } -x = -(-1) = 1.$$
$$\text{If } y = -10, \text{ then } -y = -(-10) = 10.$$

Remember that $-a$ (the opposite of a) represents a positive number whenever a represents a negative number.

Example 5: Absolute Value

Teaching Notes:
You may want to show your students how to use the absolute value key and the negative key on the graphing calculator at this time. These ideas are included in the exercises.

a. $|6.3| = 6.3$

The number 6.3 is 6.3 units from 0. Also, 6.3 is positive so its absolute value is the same as the number itself.

b. $|-5.1| = -(-5.1) = 5.1$

The number –5.1 is 5.1 units from 0.

c. $-|-2.9| = -(2.9) = -2.9$

The opposite of the absolute value of –2.9.

d. If $|x| = 7$, what are the possible values for x?

Solution: $x = 7$ or $x = -7$ since $|7| = 7$ and $|-7| = 7$.

e. True or false: $|-4| \geq 4$

Solution: True, since $|-4| = 4$ and $4 \geq 4$.

f. True or false: $\left|-5\frac{1}{2}\right| < 5\frac{1}{2}$

Solution: False, since $\left|-5\frac{1}{2}\right| = 5\frac{1}{2}$ and $5\frac{1}{2} \not< 5\frac{1}{2}$.

($\not<$ is read "is not less than")

g. If $|x| = -3$, what are the possible values for x?

Continued on the next page...

Solution: There are no values of x for which $|x| = -3$. The absolute value can never be negative. There is **no solution.**

h. If $|x| < 3$, what are the possible integer values for x? Graph these numbers on a number line.

Solution: The integers are within 3 units of 0: $-2, -1, 0, 1, 2$.

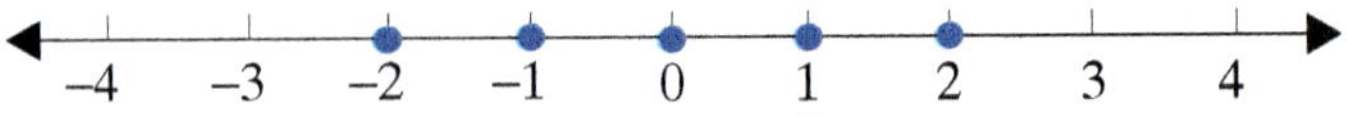

i. If $|x| \geq 4$, what are the possible integer values for x? Graph these numbers on a number line.

Solution: The integers must be 4 or more units from 0: $\ldots, -7, -6, -5, -4, 4, 5, 6, 7, \ldots$

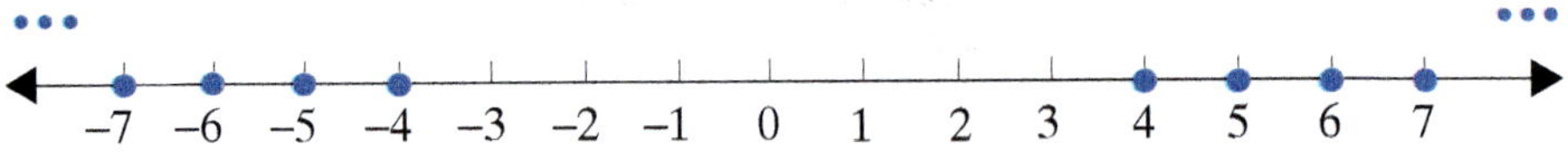

Practice Problems

Fill in the blank with the appropriate symbol: <, >, or =.

1. -2 ____ 1 **2.** $1\frac{6}{10}$ ____ 1.6 **3.** $-(-4.1)$____-7.2

4. Graph the set of all negative integers on a number line.

5. True or false: $3.6 \leq |-3.6|$

6. If $|x| = 8$, what are the possible values for x?

7. If $|x| = -6$, what are the possible values for x?

1. 4, 8
2. 0, 4, 8
3. $-7, -2, 0, 4, 8$
4. $-\sqrt{6}, \sqrt{5}, \sqrt{11}$

1.1 Exercises

List the numbers in the set $A = \left\{-7, -\sqrt{6}, -2, -\frac{5}{3}, -1.4, 0, \frac{3}{5}, \sqrt{5}, \sqrt{11}, 4, 5.9, 8\right\}$ *that are described in each exercise.*

1. Natural numbers **2.** Whole numbers **3.** Integers

4. Irrational numbers **5.** Rational numbers **6.** Real numbers

5. $-7, -2, -\frac{5}{3}, -1.4, 0, \frac{3}{5}, 4, 5.9, 8$
6. All numbers in A are real numbers.

Answers to Practice Problems: **1.** < **2.** = **3.** > **4.** 

5. True **6.** 8, –8 **7.** None

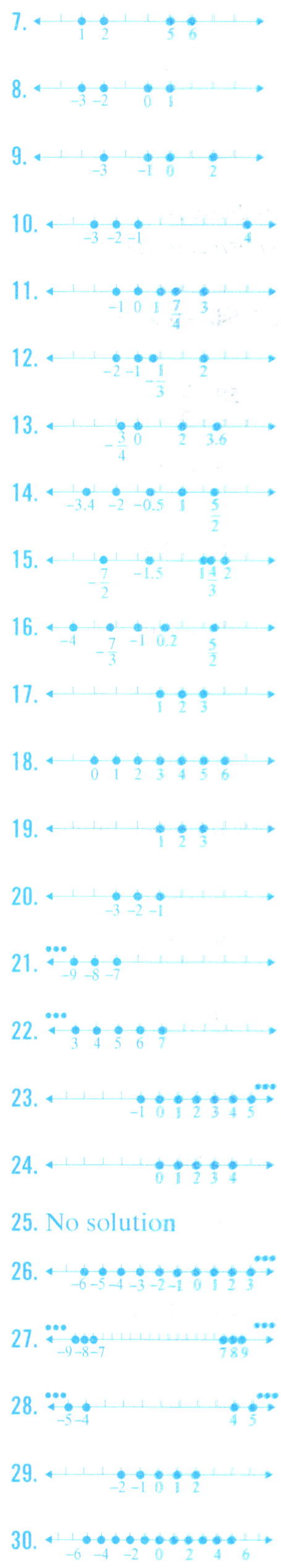

Graph each set of real numbers on a real number line. For decimal representations of fractions, use a calculator.

7. $\{1, 2, 5, 6\}$

8. $\{-3, -2, 0, 1\}$

9. $\{2, -3, 0, -1\}$

10. $\{-2, -1, 4, -3\}$

11. $\left\{0, -1, \frac{7}{4}, 3, 1\right\}$

12. $\left\{-2, -1, -\frac{1}{3}, 2\right\}$

13. $\left\{-\frac{3}{4}, 0, 2, 3.6\right\}$

14. $\left\{-3.4, -2, -0.5, 1, \frac{5}{2}\right\}$

15. $\left\{-\frac{7}{2}, -1.5, 1, \frac{4}{3}, 2\right\}$

16. $\left\{-4, -\frac{7}{3}, -1, 0.2, \frac{5}{2}\right\}$

Graph each set of integers on a real number line.

17. All positive integers less than 4

18. All whole numbers less than 7

19. All positive integers less than or equal to 3

20. All negative integers greater than or equal to –3

21. All integers less than –6

22. All integers less than 8

23. All integers greater than or equal to –1

24. All whole numbers less than or equal to 4

25. All negative integers greater than or equal to 2

26. All integers greater than or equal to –6

27. All integers more than 6 units from 0

28. All integers more than 3 units from 0

29. All integers less than 3 units from 0

30. All integers less than 6 units from 0

Find the value of each of the following absolute value expressions.

31. $|-10|$

32. $|-5.6|$

33. $-|-4|$

34. $-|-3.4|$

35. $-|11.3|$

36. $-|15|$

Fill in the blanks with the appropriate symbol: <, >, or =.

37. 4 ____ 6

38. –3 ____ 1

39. –2 ____ –4

40. –8 ____ 0

41. 5 ____ –(–5)

42. –(–4.3) ____ 4.3

43. 5.6 ____ –(–8.7)

44. 2.3 ____ 1.6

45. $-\frac{3}{4}$ ____ –1

31. 10 32. 5.6
33. -4 34. -3.4
35. -11.3 36. -15
37. $<$ 38. $<$
39. $>$ 40. $<$
41. $=$ 42. $=$
43. $<$ 44. $>$
45. $>$ 46. $=$
47. $<$ 48. $<$
49. $=$ 50. $<$
51. True 52. True
53. False; $-22 < -16$
54. True 55. True
56. True 57. True
58. False; $\frac{3}{5} > \frac{1}{4}$
59. True 60. True
61. True 62. True
63. False; $-|-3| > -|4|$
64. False; $-|73| \le |-73|$
65. False; $\left|-\frac{5}{2}\right| > 2$
66. False; $|-3.4| > 0$
67. True 68. True
69. False; $-|5| < -|3.1|$
70. False $|-1.6| < |-2.1|$
71. True 72. True
73. 

74. 

75. 

76. 

77. 

78. 

79. No solution
80. No solution
81. 

82. 

83. 

84. 

85. 

86. 

87. 

46. -2.3 ____ $-2\frac{3}{10}$ **47.** $\frac{1}{3}$ ____ $\frac{1}{2}$ **48.** $-\frac{1}{2}$ ____ $-\frac{1}{3}$

49. $-\frac{2}{8}$ ____ $-\frac{1}{4}$ **50.** $\frac{9}{16}$ ____ $\frac{3}{4}$

Determine whether each statement is true or false. If a statement is false, rewrite it in a form that is a true statement. (There may be more than one way to correct a statement.)

51. $0 = -0$ **52.** $11 = -(-11)$ **53.** $-22 > -16$ **54.** $-6 > -8$

55. $-17 \le 17$ **56.** $4.7 \ge 3.5$ **57.** $-\frac{1}{3} \le 0$ **58.** $\frac{3}{5} < \frac{1}{4}$

59. $|-5| = 5$ **60.** $-|-6.2| = -6.2$ **61.** $-|-7| \ge -|7|$ **62.** $|-6| \ge 6$

63. $-|-3| < -|4|$ **64.** $-|73| \ge |-73|$ **65.** $\left|-\frac{5}{2}\right| < 2$ **66.** $|-3.4| < 0$

67. $\frac{2}{3} < |-1|$ **68.** $3 > \left|-\frac{4}{3}\right|$ **69.** $-|5| > -|3.1|$ **70.** $|-1.6| > |-2.1|$

71. $|2.5| = \left|-\frac{5}{2}\right|$ **72.** $|-1.75| = \left|\frac{7}{4}\right|$

List the possible values for x for each statement. Graph the numbers on a real number line.

73. $|x| = 4$ **74.** $|x| = 6$ **75.** $|x| = 9$ **76.** $13 = |x|$

77. $0 = |x|$ **78.** $|x| = 0$ **79.** $-2 = |x|$ **80.** $|x| = -3$

81. $|x| = 3.5$ **82.** $|x| = 4.7$

On a real number line, graph the integers that satisfy the conditions stated.

83. $|x| \le 4$ **84.** $|x| \le 2$ **85.** $|x| > 4$ **86.** $|x| > 2$

87. $|x| < 7$ **88.** $|x| > 6$ **89.** $|x| \le x$ **90.** $|x| > x$

91. $|x| = x$ **92.** $|x| = -x$

88. 

89. 

90. 

91. 

92. 

93. 46 94. 36

95. –6 96. –21

97. 61.4 98. 24.2

99. If y is a negative number then $-y$ represents a positive number. For example, if $y = -2$, then $-y = -(-2) = 2$.

Graphing Calculator Problems

A TI-84 Plus graphing calculator has the absolute value command built in. To access the absolute command press MATH *, go to* NUM *at the top of the screen and press* 1 *or* ENTER*. The command* `abs(` *will appear on the display screen. Then enter any arithmetic expression you wish and the calculator will print the absolute value of that expression. (**Note:** The negative sign is on the key marked* (–) *next to the* ENTER *key.)*

Follow the directions above and use your calculator to find the value of each of the following expressions.

93. $|34-80|$ **94.** $|-62+26|$ **95.** $-|10-16|$

96. $-|-10-11|$ **97.** $|17.5+16.3-95.2|$ **98.** $|-52.1+41.7-13.8|$

Writing and Thinking About Mathematics

100. The absolute value is the distance the number is from 0 on a number line.

99. Explain, in your own words, how an expression such as $-y$ might represent a positive number.

100. Explain, in your own words, the meaning of absolute value.

HAWKES LEARNING SYSTEMS: INTRODUCTORY & INTERMEDIATE ALGEBRA SOFTWARE

- 1.1a The Real Number Line and Inequalities
- 1.1b Introduction to Absolute Values

1.2 Addition with Real Numbers

- *Add real numbers.*

Picture a straight line in an open field and numbers marked on a number line. An archer stands at 0 and shoots an arrow to +3, then stands at 3 and shoots the arrow 5 more units in the positive direction (to the right). Where will the arrow land (Figure 1)?

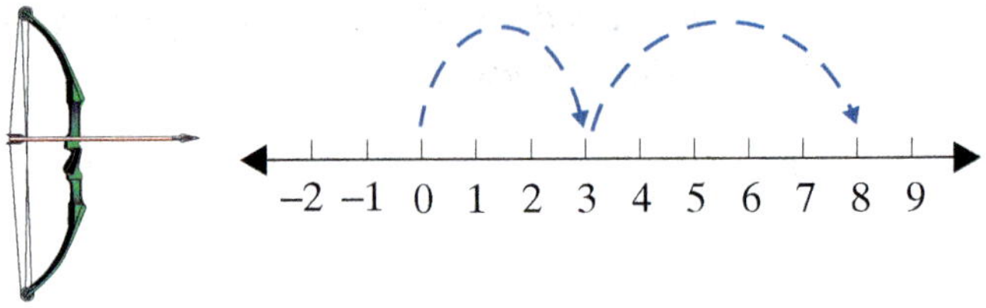

Figure 1

Naturally, you have figured out that the answer is +8. What you have done is added the two positive integers, +3 and +5.

$$(+3)+(+5)=+8 \qquad \text{or} \qquad 3+5=8$$

Suppose another archer shoots an arrow in the same manner as the first but in the opposite direction. Where would his arrow land? The arrow lands at −8. You have just added −3 and −5 (Figure 2).

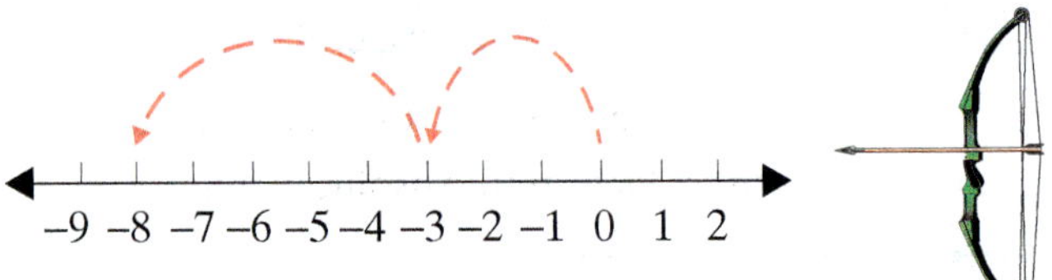

$$(-3)+(-5)=-8$$

Figure 2

If the archer stands at 0 and shoots an arrow to +3 and then the archer goes to +3 and turns around and shoots an arrow 5 units in the opposite direction, where will the arrow stick? Would you believe at −2 (Figure 3)?

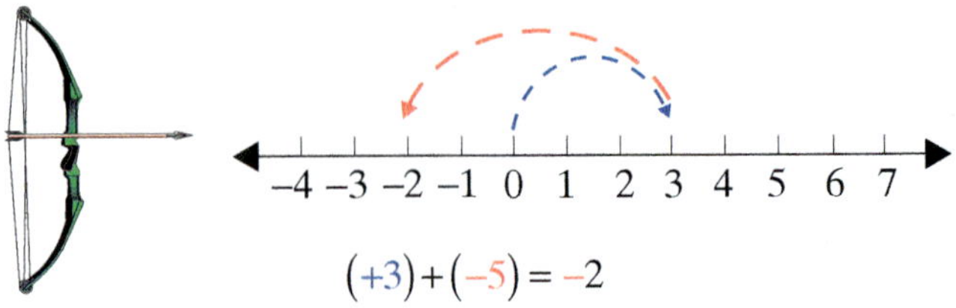

$$(+3)+(-5)=-2$$

Figure 3

For our final archer, the first shot is to –3. Then after going to –3, he turns around and shoots 5 units in the opposite direction. Where is the arrow? It is at +2 (Figure 4).

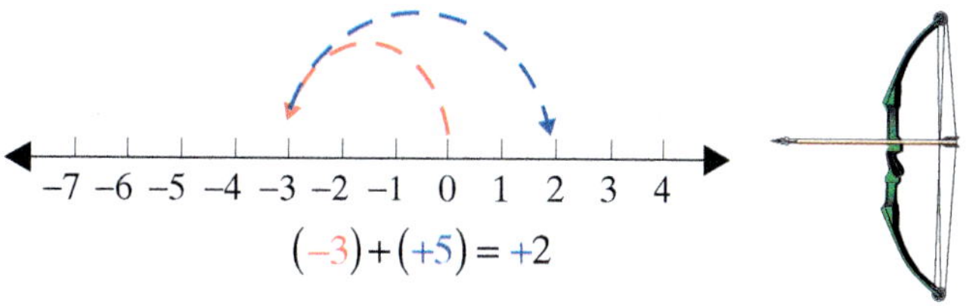

Figure 4

In summary:

1. The sum of two positive real numbers is positive.

$$(+3) + (+5) = +8$$

positive plus positive is **positive**

2. The sum of two negative real numbers is negative.

$$(-3) + (-5) = -8$$

negative plus negative is **negative**

3. The sum of a positive real number and a negative real number may be negative or positive (or zero) depending on which number is farther from 0.

$$(+3) + (-5) = -2$$

positive plus negative is **negative**

$$(+5) + (-3) = +2$$

positive plus negative is **positive**

You probably did quite well and understand how to add real numbers. The rules for addition with positive and negative real numbers can be summarized in the following manner.

Rules for Addition with Real Numbers

1. To add two real numbers with **like signs**,
   a. add their absolute values and
   b. use the common sign.
2. To add two real numbers with **unlike signs**,
   a. subtract their absolute values (the smaller from the larger), and
   b. use the sign of the number with the larger absolute value.

Study the following examples carefully to help in understanding the rules for addition.

Example 1: Addition with Like Signs

a. $(+10)+(+3)$
$= +(|+10|+|+3|)$
$= +(10+3)$
$= 13$

b. $(-10)+(-3)$
$= -(|-10|+|-3|)$
$= -(10+3)$
$= -13$

c. $(-1.4)+(-2.5)$
$= -(|-1.4|+|-2.5|)$
$= -(1.4+2.5)$
$= -3.9$

Note: A more detailed discussion of decimal numbers and operations with decimal numbers is included in Appendix A.1.

Example 2: Addition with Unlike Signs

a. $(-10)+(+3)$
$= -(|-10|-|+3|)$
$= -(10-3)$
$= -7$

b. $(+10)+(-3)$
$= +(|+10|-|-3|)$
$= +(10-3)$
$= 7$

c. $(-9.5)+(8.7)$
$= -(|-9.5|-|8.7|)$
$= -(9.5-8.7)$
$= -0.8$

NOTES

In adding real numbers, if a negative number occurs after an addition symbol then we must place the number in parentheses so that the two operation symbols are not next to each other.

Because equations in algebra are almost always written horizontally, you should become used to working with sums written horizontally. However, there are situations (as in long division) where sums (and differences) are written vertically with one number directly under another. Example 3 illustrates this technique.

Example 3: Vertical Addition

Find each sum.

a. $\begin{array}{r} -7 \\ \underline{10} \\ +3 \end{array}$ **b.** $\begin{array}{r} 6 \\ \underline{-15} \\ -9 \end{array}$ **c.** $\begin{array}{r} -5 \\ \underline{-9} \\ -14 \end{array}$ **d.** $\begin{array}{r} -10.8 \\ \underline{7.6} \\ -3.2 \end{array}$

Note: Remember that when adding decimal numbers in a vertical format, decimal points must be aligned vertically.

Practice Problems

Find each sum. Add from left to right if there are more than two numbers.

1. $-14+(-6)$	**2.** $(16)+(-10)$
3. $-12+8$	**4.** $11+7$
5. $-13+13$	**6.** $-11+(-8)$
7. $-8.4+(-5.3)$	**8.** $-10.6+(15.4)$
9. $6+(-7)+(-1)$	**10.** $100+(-100)+10$

1.2 Exercises

Simplify the expressions.

1. $4+9$	**2.** $8+(-3)$	**3.** $-9+5$
4. $-7+(-3)$	**5.** $-9+9$	**6.** $2+(-8)$
7. $11+(-6)$	**8.** $-12+3$	**9.** $-18+5$
10. $26+(-26)$	**11.** $-5+(-3)$	**12.** $11+(-2)$
13. $-2+(-8)$	**14.** $10+(-3)$	**15.** $17+(-17)$
16. $-7+20$	**17.** $2.1+(-4.6)$	**18.** $-1.5+(-3.1)$
19. $-12+(-17)$	**20.** $24+(-16)$	**21.** $-4.3+(-5.8)$
22. $-6.9+(-8.5)$	**23.** $9.7+(-12.2)$	**24.** $-12.2+9.7$
25. $38+(-16)$	**26.** $-20+(-11)$	**27.** $-33+(-21)$
28. $-21+18$	**29.** $60+(-20)$	**30.** $-35+75$
31. $-3+4+(-8)$	**32.** $-9+(-6)+5$	**33.** $-9+(-2)+(-5)$
34. $-21+6+15$	**35.** $-13+(-1)+(-12)$	**36.** $-19+(-2)+(-4)$
37. $27+(-14)+(-13)$	**38.** $-33+29+2$	**39.** $-43+(-16)+27$
40. $-68+(-3)+42$	**41.** $-3.8+4.9+(-6.2)$	**42.** $10.2+(-9.3)+(-6.6)$

1. 13 2. 5
3. −4 4. −10
5. 0 6. −6
7. 5 8. −9
9. −13 10. 0
11. −8 12. 9
13. −10 14. 7
15. 0 16. 13
17. −2.5 18. −4.6
19. −29 20. 8
21. −10.1 22. −15.4
23. −2.5 24. −2.5
25. 22 26. −31
27. −54 28. −3
29. 40 30. 40
31. −7 32. −10
33. −16 34. 0
35. −26 36. −25
37. 0 38. −2
39. −32 40. −29
41. −5.1 42. −5.7

Answers to Practice Problems: **1.** −20 **2.** 6 **3.** −4 **4.** 18 **5.** 0 **6.** −19 **7.** −13.7 **8.** 4.8 **9.** −2 **10.** 10

Find each sum indicated in a vertical format.

43.	44.	45.	46.	47.
−21	−12	−15	−7	20.4
−62	17	19	−9	−7.3

48.	49.	50.	51.	52.
−8.4	−16	31	−33	−29
14.7	−31	−15	−9	−17

53. Add −17 and 5.

54. Add 9 and 11.

55. Add −8 and −7.

56. Add 20 and −14.

57. Find the sum of 6 and −12.

58. Find the sum of −18 and 5.

59. Find the sum of 2, −6, and 10.

60. Find the sum of −3, −7, and 5.

Calculator Problems

Use your graphing calculator to find the value of each of the following expressions. (Remember that the key marked **(−)** *next to the* ENTER *key is used to indicate negative numbers.)*

61. $47+(-29)+66$

62. $56+(-41)+(-28)$

63. $2932+4751+(-3876)$

64. $(-8154)+2147+(-136)$

65. $(-16,945)+(-27,302)+(-53,467)$

66. $(-12,299)+15,631+(-47,558)$

Writing and Thinking About Mathematics

67. Show how the sum of the absolute values of two numbers might be 0. (Is this even possible?)

HAWKES LEARNING SYSTEMS: INTRODUCTORY & INTERMEDIATE ALGEBRA SOFTWARE

- 1.2 Addition with Real Numbers

43. −83
44. 5
45. 4
46. −16
47. 13.1
48. 6.3
49. −47
50. 16
51. −42
52. −46
53. −12
54. 20
55. −15
56. 6
57. −6
58. −13
59. 6
60. −5
61. 84
62. −13
63. 3807
64. −6143
65. −97,714
66. −44,226
67. $|0|+|0|=0$

1.3 Subtraction with Real Numbers

- *Find the additive inverse (opposite) of a real number.*
- *Subtract real numbers.*
- *Find the change in value between two numbers.*
- *Find the net change for a set of numbers..*

In basic arithmetic, subtraction is defined in terms of addition. For example, we know that the difference $32 - 25$ is equal to 7 because $25 + 7 = 32$. A beginning student in arithmetic does not know how to find a difference such as $15 - 20$, where a larger number is subtracted from a smaller number, because negative numbers are not yet defined and there is no way to add a positive number to 20 and get 15. Now, with our knowledge of negative numbers, we will define subtraction in such a way that larger numbers may be subtracted from smaller numbers. We will still define subtraction in terms of addition, but we will apply our new rules for addition with real numbers.

Before we proceed to develop the techniques for subtraction with real numbers, we will state and illustrate an important relationship between any real number and its opposite.

Additive Inverse

The **opposite** of a real number is called its **additive inverse**. The sum of a number and its additive inverse is zero. Symbolically, for any real number a,

$$\boldsymbol{a + (-a) = 0}.$$

Example 1: Additive Inverse

a. Find the additive inverse (opposite) of 3.

Solution: The additive inverse of 3 is –3. $3 + (-3) = 0$

b. Find the additive inverse (opposite) of –7.

Solution: The additive inverse of –7 is $-(-7) = +7$. $(-7) + (+7) = 0$

c. Find the additive inverse (opposite) of 0.

Solution: The additive inverse of 0 is $-0 = 0$.
That is, 0 is its own opposite. $(0) + (-0) = 0 + 0 = 0$

d. Find the additive inverse (opposite) of 13.67.

Solution: The additive inverse of 13.67 is –13.67. $13.67 + (-13.67) = 0$

Subtraction (Change in Value)

Intuitively, we can think of addition of numbers (positive or negative or both) as "*piling on*" or "*accumulating*" numbers. For example, when we add positive numbers (or negative numbers), the sum is a number more positive (or more negative). Figure 1 illustrates these ideas.

a. Adding positive numbers "piles on" or "accumulates" the numbers in a positive direction.

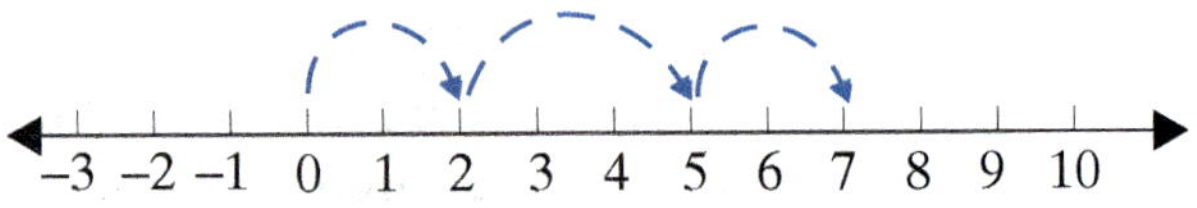

$2+3+2=7$

b. Adding negative numbers "piles on" or "accumulates" the numbers in a negative direction.

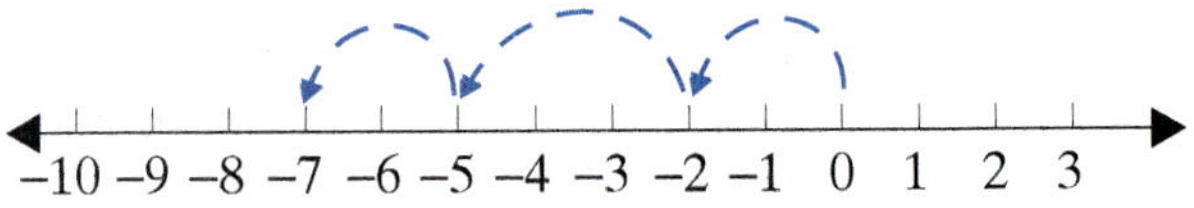

$(-2)+(-3)+(-2)=-7$

Figure 1

In **subtraction** we want to find the "*difference between*" two numbers. On a number line this translates as the "*distance between*" the two numbers **with direction considered**. As illustrated in Figure 2, the *distance between* 6 and 1 is five units. To find this *distance*, we subtract: $6-1=6+(-1)=5$. Note that to subtract 1, we add the opposite of 1 (which is –1).

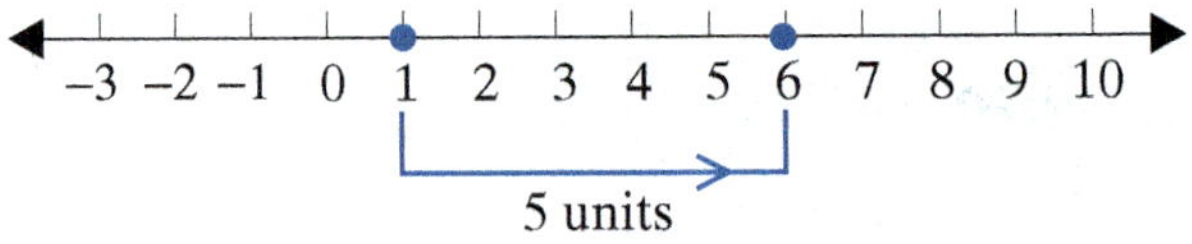

$6-1=6+(-1)=5$

Figure 2

In Figure 3 we see that the distance between 6 and –4 is ten units. Again, to find this distance, we subtract $6-(-4)$. But we know that we must have 10 as an answer. To get 10, we add the opposite of –4 (which is +4) as follows: $6-(-4)=6+(+4)=10$.

To understand the "direction" in subtraction, think of subtraction on the number line as:

$$(\text{end value}) - (\text{beginning value}).$$

This means that for $6-(-4)=6+(+4)=10$,

we have $6 \quad - \quad (-4) \quad = \quad 6 \quad + \quad (+4) \quad = \quad 10$

↑ ↑

(end value) – (beginning value)

and, as illustrated in Figure 3, we would start at –4 and move 10 units in the **positive direction** to end at 6.

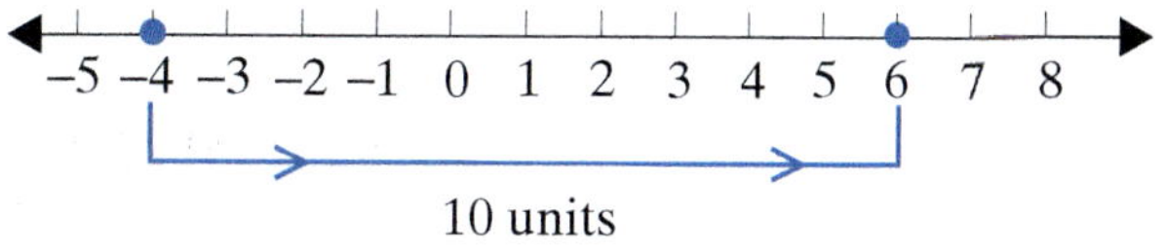

$6-(-4)=6+(+4)=10$

Figure 3

As illustrated in Figures 1, 2, and 3, **subtraction is defined in terms of addition.**

To subtract, add the opposite of the number being subtracted.

If we reverse the order of subtraction, then the answer must indicate the opposite direction.

This means that for $-4-6=-4+(-6)=-10$,

we have $-4 \quad - \quad 6 \quad = \quad -4 \quad + \quad (-6) \quad = \quad -10$

↑ ↑

(end value) – (beginning value)

and, as illustrated in Figure 4, we would start at 6 and move 10 units in the **negative direction** to end at –4. We see that –4 and 6 are still ten units apart but subtraction now indicates a negative direction.

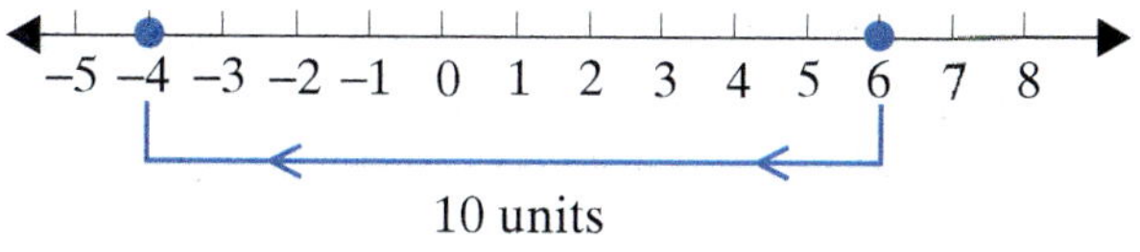

$-4-6=-4+(-6)=-10$

Figure 4

The formal definition of subtraction is as follows.

Subtraction

For any real numbers a and b,

$$\boldsymbol{a - b = a + (-b)}.$$

In words,

to subtract b from a, **add** the **opposite** of b to a.

We change the sign of the second number and follow the rules for addition.

$$(+1)-(-4)=(+1)+[-(-4)]=(+1)+(+4)=+5$$

add opposite

$$(-10)-(-3)=(-10)+[-(-3)]=(-10)+(+3)=-7$$

add opposite

Example 2: Subtraction

a. $(-1)-(+4)=(-1)+(-4)=-5$

b. $(-1.00)-(-8.23)=(-1.00)+(+8.23)=7.23$

c. $(10)-(+2)=(10)+(-2)=8$

d. $(-10)-(-5.5)=(-10)+(+5.5)=-4.5$

Teaching Notes:
The following ideas might help students understand that the "–" is being used in two different ways. One way is as a verb ("subtract"), as in $a - b$. The other way is as an adjective ("negative"), as in $a+(-b)$. We tie the two uses together with $a+(-b)=a-b$.

In practice, the notation $a - b$ is thought of as addition of signed numbers. That is, because $a-b=a+(-b)$, we think of the plus sign, +, as being present in $a - b$. In fact, an expression such as 4 – 19 can be thought of as "four plus negative nineteen." We have

$$4-19=4+(-19)=-15$$
$$-25-30=-25+(-30)=-55$$
$$-3-(-17)=-3+(+17)=14$$
$$24-11-6=24+(-11)+(-6)=7.$$

Generally, the second step is omitted and we go directly to the answer by computing the sum mentally.

$$4-19=-15$$
$$-25-30=-55$$
$$-3-(-17)=14$$
$$24-11-6=7$$

The numbers may also be written vertically. In this case, the sign of the number being subtracted (the bottom number) is changed and addition is performed.

Example 3: Vertical Subtraction

a. **Subtract** 43 and $-(-25)$ — sign change → **Add** 43 and $+25$, giving 68

b. **Subtract** -38 and $-(+11)$ — sign change → **Add** -38 and -11, giving -49

c. **Subtract** -7.3 and $-(-3.2)$ — sign change → **Add** -7.3 and $+3.2$, giving -4.1

d. **Subtract** 17 and $-(+69)$ — sign change → **Add** 17 and -69, giving -52

Change in Value

To find the **change in value** between two numbers, take the end value and subtract the beginning value. Symbolically,

Change in Value = (End Value) – (Beginning Value).

Example 4: Change in Value

a. At noon on Tuesday the temperature was 34°F. By noon on Thursday the temperature had changed to –5° F. How much did the temperature change between Tuesday and Thursday?

Continued on the next page...

Solution: For change in value:

$$(\text{end value}) - (\text{beginning value}) = (-5) - (+34)$$

$$= -5 + (-34)$$

$$= -39$$

Between Tuesday and Thursday the temperature changed –39° F (or dropped 39° F).

b. A jet pilot flew her plane from an altitude of 30,000 ft to an altitude of 12,000 ft. What was the change in altitude?

Solution: end altitude – beginning altitude = change in altitude

12,000 – 30,000 = –18,000 ft

(This means that the plane *descended* 18,000 ft.)

The **net change** in a measure is the algebraic sum of several numbers. Example 5 illustrates how positive and negative numbers can be used to find the net change of weight (gain or loss) and sales over a period of time.

Example 5: Net Change

a. Susan is a salesperson for a shoe store. Last week her sales of pairs of shoes were as follows:

Day	Sales	Returns	Daily Net Sales
Monday	7	1	6
Tuesday	3	0	3
Wednesday	2	4	–2
Thursday	6	1	5
Friday	8	3	5

What were Susan's net sales for last week?

Solution: $6+3+(-2)+5+5=17$

Susan's net sales for the week were 17 pairs of shoes.

b. Robert weighed 230 lb when he started to diet. The first month he lost 7 lb, the second month he gained 2 lb, and the third month he lost 5 lb. What was his weight after 3 months of dieting?

Solution: $230+(-7)+(+2)+(-5)$

$$= 223+(+2)+(-5)$$
$$= 225+(-5)$$
$$= 220\text{ lb}$$

Practice Problems

1. What is the additive inverse of 85?
2. Find the difference: $-6-(-5)$
3. Simplify: $-6-4-(-2)$
4. True or false: $-5+(-3)<-5-(-3)$
5. Perform the indicated subtraction: $\begin{array}{r} 63.1 \\ \underline{-(-27.8)} \end{array}$

1.3 Exercises

Find the additive inverse for each integer.

1. 11 **2.** 17 **3.** −6 **4.** −23 **5.** 47

6. −34 **7.** 0 **8.** 10.1 **9.** −5.2 **10.** −257

Simplify the expressions.

11. $8-3$ **12.** $5-7$ **13.** $-4-6$ **14.** $-18-17$

15. $3-(-4)$ **16.** $5-(-7)$ **17.** $-8-(-11)$ **18.** $0-(-12)$

19. $2-7$ **20.** $4-12$ **21.** $6-(-10)$ **22.** $3-(-8)$

23. $-9-4$ **24.** $-11-5$ **25.** $2.8-(-3.1)$ **26.** $5.3-(-1.7)$

27. $-1.4-2.6$ **28.** $-8.5-7.1$ **29.** $1.6-(-8.4)$ **30.** $1.5-2.3$

1. −11 2. −17
3. 6 4. 23
5. −47 6. 34
7. 0 8. −10.1
9. 5.2 10. 257
11. 5 12. −2
13. −10 14. −35
15. 7 16. 12
17. 3 18. 12
19. −5 20. −8
21. 16 22. 11
23. −13 24. −16
25. 5.9 26. 7
27. −4 28. −15.6
29. 10 30. −0.8

Answers to Practice Problems: **1.** −85 **2.** −1 **3.** −8 **4.** True **5.** 90.9

31. -15
32. -7
33. -16
34. -33
35. -57
36. -60
37. -5.4
38. -4.5
39. 1
40. 42
41. -7
42. 8
43. -26
44. -40
45. 11
46. -27
47. -15
48. 7
49. -1
50. -8
51. -3
52. -10
53. -6
54. 1
55. 0
56. -13.9
57. $-8 < -5$
58. $-1 > -7$
59. $-3 < 3$
60. $10 > -10$
61. $8 = 8$
62. $-6 < 6$
63. $0 > -27$
64. $-4 > -5$
65. $-23.7 > -24.8$
66. $-3.7 < -3.4$

Subtract as indicated.

31. $\begin{array}{r} 27 \\ \underline{-(+42)} \end{array}$ **32.** $\begin{array}{r} 19 \\ \underline{-(+26)} \end{array}$ **33.** $\begin{array}{r} -23 \\ \underline{-(-7)} \end{array}$ **34.** $\begin{array}{r} -41 \\ \underline{-(-8)} \end{array}$

35. $\begin{array}{r} -21 \\ \underline{-(+36)} \end{array}$ **36.** $\begin{array}{r} -47 \\ \underline{-(+13)} \end{array}$ **37.** $\begin{array}{r} -2.7 \\ \underline{-(+2.7)} \end{array}$ **38.** $\begin{array}{r} -1.9 \\ \underline{-(+2.6)} \end{array}$

39. Find the difference between –5 and –6. (**Hint:** Subtract the numbers in the order given.)

40. Find the difference between 30 and –12. (**Hint:** Subtract the numbers in the order given.)

41. Subtract –3 from –10.

42. Subtract –2 from 6.

43. Subtract 13 from –13.

44. Subtract 20 from –20.

45. Find the sum of –12 and 6. Then subtract –17.

46. Find the sum of 11 and –13. Then subtract 25.

Find the net change in value of each expression by performing the indicated operations.

47. $-6+(-4)-5$ **48.** $-2-2+11$ **49.** $6+(-3)+(-4)$ **50.** $-3+(-7)+2$

51. $-5-2-(-4)$ **52.** $-8-5-(-3)$ **53.** $-3-(-3)+(-6)$ **54.** $-7-(-2)+6$

55. $9.7-1.6-(8.1)$ **56.** $-11.3+5.3-7.9$

Perform the operations on each side of the blank and then fill in the blank with the proper symbol: <, >, or =.

57. $-6+(-2)$_______$3+(-8)$ **58.** $-4-(-3)$_______$-4+(-3)$

59. $5-8$_______$8-5$ **60.** $7-(-3)$_______$-3-7$

61. $11+(-3)$_______$11-3$ **62.** $0-6$_______$0-(-6)$

63. $-8-(-8)$_______$-14-13$ **64.** $-7-(-3)$_______$4-9$

65. $-15.1-8.6$_______$-10.7-14.1$ **66.** $2.5-6.2$_______$-1.1-2.3$

67. −18°F
68. −540°F
69. $8
70. −$750
71. −14,777 ft
72. 30,280 ft
73. 85 years old
74. 50 years old
75. 1044 > −39
76. −43,241 < 27,180
77. −15,254 > −35,090
78. −37,402 < −31,985

67. Temperature: At 2 p.m. the temperature was 76°F. At 8 p.m. the temperature was 58°F. What was the change in temperature?

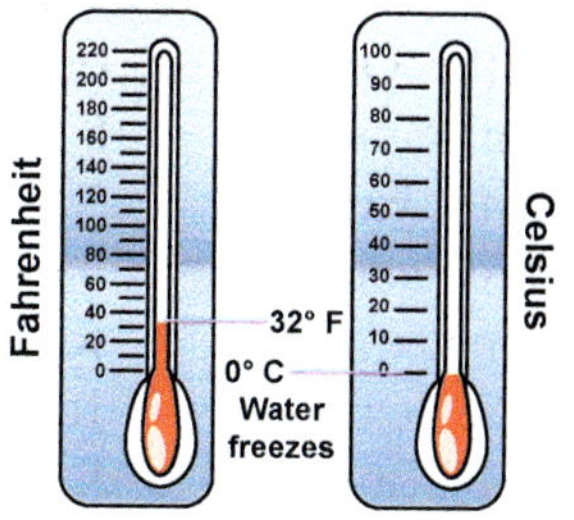

68. Astronomy: According to NASA, the temperature on the earth's moon during the day is 260°F and during the night it is −280°F. What is the daily temperature change on the moon?
Source: NASA.gov

69. Stock market: On Monday, February 8, 2010 NIKE stock opened at $61 per share. A month later, on Monday, March 8, 2010, the stock opened at $69 per share. Find the change in price of the stock.
Source: NYSE

70. Used cars: Mr. Meade is having a hard time selling his old car. He just slashed the price from $3500 to $2750. By how much did he change the price?

71. Travel: If you travel from the top of Mt. Whitney, elevation 14,495 ft, to the floor of Death Valley, elevation 282 ft below sea level, what is the change in elevation?

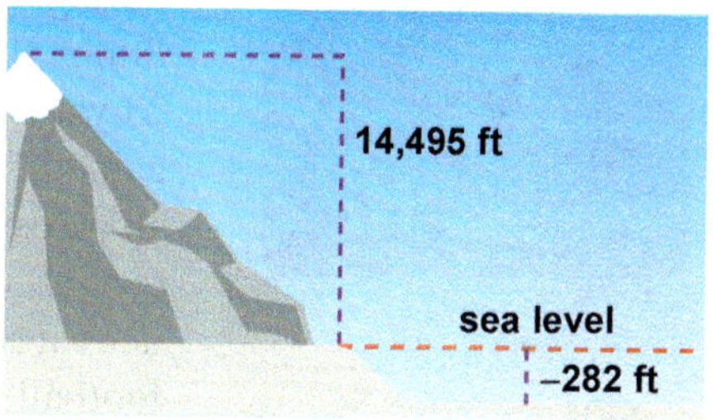

72. Military: A submarine submerged 280 ft below the surface of the sea and fired a rocket that reached an altitude of 30,000 ft. What was the change in altitude of the rocket?

73. History: The great English mathematician and scientist Isaac Newton was born in 1642 and died in 1727. How old was he when he died? (Assume he lived past his birthday.)

74. History: The famous pop culture icon Michael Jackson was born in August 1958 and died in June 2009. How old was he when he died? (Notice that he did not reach his birthday in 2009.)

Calculator Problems

Use your graphing calculator to find the value indicated on each side of the blank and then fill in the blank with the proper symbol: <, >, or =.

75. $648-(-396)$ ________ $124-163$

76. $-19{,}824-23{,}417$ ________ $12{,}793-(-14{,}387)$

77. $-43{,}931-(-28{,}677)$ ________ $-(13{,}665+21{,}425)$

78. $-(24{,}295+13{,}107)$ ________ $-48{,}261-(-16{,}276)$

79. Gained 35 yards
80. 6° below 0°C (or −6°C)

Temperatures above 0° and below 0° as well as gains and losses can be thought of in terms of positive and negative numbers. Use positive and negative numbers to answer the following problems.

81. Lost 7 pounds; 203 pounds
82. 2316 points
83. Add the opposite of the second number to the first number.
84. 0; Because −0 = 0.

79. Football: At the 2010 Rose Bowl, during one possession for the Ohio State Buckeyes, the team gained 28 yards, gained 9 yards, lost 2 yards, gained 5 yards, gained 1 yard, gained 10 yards, lost 5 yards on a penalty, and lost 11 yards. What was the team's net yardage on this possession of the football?

Source: ESPN

80. Temperature: Beginning at a temperature of 10°C, the temperature in a scientific experiment was measured hourly for four hours. It dropped 5°, dropped 8°, dropped 6°, then rose 3°. What was the final temperature recorded?

81. Dieting: Harry and his wife went on a diet for 5 weeks. During those 5 weeks, Harry lost 5 pounds, gained 3 pounds, lost 2 pounds, lost 4 pounds, and gained 1 pound. What was his total loss (or gain) for the 5 weeks? If he weighed 210 pounds when he started the diet plan, what did he weigh at the end of the 5-week period? During the same time, his wife lost 10 pounds.

82. Stock market: In a 5-day week the NASDAQ stock posted a gain of 38 points, a loss of 65 points, a loss of 32 points, a gain of 10 points, and a gain of 15 points. If the NASDAQ started the week at 2350 points, what was the market at the end of the week?

Writing and Thinking About Mathematics

83. Explain, in your own words, how to find the difference between two numbers.

84. What is the additive inverse of 0? Why?

HAWKES LEARNING SYSTEMS: INTRODUCTORY & INTERMEDIATE ALGEBRA SOFTWARE

- 1.3 Subtraction with Real Numbers

1.4 Multiplication and Division with Real Numbers

- *Multiply real numbers.*
- *Divide real numbers.*
- *Calculate the average (mean) of a set of numbers.*

Multiplication with Real Numbers

Multiplication can be represented by any of the following symbols.

Symbols for Multiplication

Symbol	Description	Example
$\cdot$	raised dot	$4 \cdot 7$
( )	numbers inside or next to parentheses	$5(10)$ or $(5)10$ or $(5)(10)$
$\times$	cross sign	6×12 or $\begin{array}{r} 12 \\ \times 6 \\ \hline \end{array}$
	number written next to variable variable written next to variable	$8x$ xy

Table 1

Multiplication is shorthand for repeated addition. That is, repeatedly adding a positive real number such as 7, we have

$$7+7+7+7+7=5 \cdot 7=35$$

which leads to the following familiar statement:

The product of two positive real numbers is positive.

Now repeatedly adding the number –6 gives the following

$$(-6)+(-6)+(-6)=3(-6)=-18$$

and similarly,

$$5(-3)=(-3)+(-3)+(-3)+(-3)+(-3)=-15.$$

We see that repeated addition with a negative number results in a product of a positive number and a negative number. Because the sum of negative numbers is negative, we can reasonably conclude that:

The product of a positive real number and a negative real number is negative.

Example 1: Products of Positive and Negative Real Numbers

a. $3(-2.1) = (-2.1)+(-2.1)+(-2.1) = -6.3$

b. $7(-10) = -70$

c. $42(-1) = -42$

The product of two negative real numbers can be explained in terms of **opposites**. We also need the fact that for any real number a, we can think of the opposite of a, $(-\boldsymbol{a})$, as the product of -1 and a. That is, we have

$$-a = -1 \cdot a.$$

Thus

$$-4 = -1 \cdot 4 = -1(4)$$

and in the product $-4(-7)$ we have

$$-4(-7) = -1 \cdot 4(-7) = -1[4(-7)] = -1[-28] = \underset{\text{opposite}}{-}(-28) = 28.$$

Although one example does not prove a rule, this process can be used in general to arrive at the following correct conclusion:

The product of two negative real numbers is positive.

Example 2: Products of Negative Real Numbers

a. $(-5)(-9) = 45$

b. $-7(-8) = 56$

c. $-5.2(-4) = 20.8$

d. $(-1)(-5)(-3)(-2) = +5(-3)(-2) = -15(-2) = 30$

(Here we multiplied from left to right. Later, we will see that multiplication can be done in any order.)

What happens if a number is multiplied by 0? For example, $3(0) = 0 + 0 + 0 = 0$. In fact, **multiplication by 0 always gives a product of 0.**

Example 3: Multiplication by 0

a. $6 \cdot 0 = 0$

b. $-13(0) = 0$

Rules for Multiplication with Positive and Negative Real Numbers

For positive real numbers a and b,

1. The product of two positives is positive: $(\boldsymbol{a})(\boldsymbol{b}) = +\boldsymbol{ab}$
2. The product of two negatives is positive: $(-\boldsymbol{a})(-\boldsymbol{b}) = +\boldsymbol{ab}$
3. The product of a positive and a negative is negative: $\boldsymbol{a}(-\boldsymbol{b}) = -\boldsymbol{ab}$
4. The product of 0 and any number is 0: $\boldsymbol{a} \cdot \mathbf{0} = \mathbf{0}$

In summary:

The product of real numbers with like signs is positive.
The product of real numbers with unlike signs is negative.
The product of any number and 0 is 0.

Example 4: Multiplication with Real Numbers

a. $9(-4)(2) = -36(2) = -72$

b. $(-2.1)(-0.03) = 0.063$

c. $-6.2(0.6) = -3.72$

d. $-17(0)(+13) = 0(+13) = 0$

NOTES

Review Note about Multiplication with Decimal Numbers

When multiplying decimal numbers:

a. count the number of places to the right of each decimal point

b. find the sum of these counts

c. use the sum as the total number of places to the right of the decimal point in the product.

Appendix A.1 contains a review of decimal numbers and operations with decimal numbers.

Practice Problems

Find the following products.

1. $5(-3)$
2. $-6(-4)$
3. $-8(4)$
4. $-12(0)$
5. $-9(-2)(-1)$
6. $3(-20)(5)$
7. $5(-3.7)$
8. $-4.1(-4.5)$

Division with Real Numbers

The rules for multiplication lead directly to the rules for division because division is defined in terms of multiplication. For convenience, division is indicated in fraction form.

Division with Real Numbers

For real numbers a, b, and x (where $b \neq 0$),

$$\frac{\boldsymbol{a}}{\boldsymbol{b}} = \boldsymbol{x} \text{ means that } \boldsymbol{a} = \boldsymbol{b} \cdot \boldsymbol{x}.$$

For real numbers a and b (where $b \neq 0$),

$$\frac{\boldsymbol{a}}{\mathbf{0}} \text{ is } \textbf{undefined} \text{ and } \frac{\mathbf{0}}{\boldsymbol{b}} = \mathbf{0}.$$

The following discussion explains why division by 0 is undefined.

Division by 0 is Undefined

1. Suppose that $a \neq 0$ and $\frac{a}{0} = x$. Then since division is related to multiplication, we must have $a = 0 \cdot x$. But this is not possible because $0 \cdot x = 0$ for any value of x and we stated that $a \neq 0$.
2. Suppose that $\frac{0}{0} = x$. Then $0 = 0 \cdot x$ which is true for all values of x. But we must have a unique answer for x.

Therefore, in any case, we conclude that **division by 0 is undefined**.

Answers to Practice Problems: **1.** -15 **2.** 24 **3.** -32 **4.** 0 **5.** -18 **6.** -300 **7.** -18.5 **8.** 18.45

Example 5 illustrates several division exercises with positive and negative real numbers.

Example 5: Division with Real Numbers

a. $\frac{-36}{9} = -4$ because $-36 = 9(-4)$.

b. $\frac{-36}{-9} = 4$ because $-36 = -9(4)$.

c. $\frac{30.6}{-2} = -15.3$

d. $\frac{-18}{-6} = 3$

e. $-\frac{51}{3} = -17$

f. $\frac{-23}{0}$ is undefined.

The rules for division with real numbers can be stated as follows.

Rules for Division with Positive and Negative Real Numbers

If a and b are positive real numbers,

1. The quotient of two positive numbers is positive: $\frac{a}{b} = +\frac{a}{b}$
2. The quotient of two negative numbers is positive: $\frac{-a}{-b} = +\frac{a}{b}$
3. The quotient of a positive number and a negative number is negative:

$$\frac{-a}{b} = -\frac{a}{b} \text{ and } \frac{a}{-b} = -\frac{a}{b}$$

In summary,

The quotient of numbers with like signs is positive.
The quotient of numbers with unlike signs is negative.

NOTES

The following common rules about multiplication and division with two nonzero real numbers are helpful in remembering the signs of answers.

1. If the numbers have the same sign, both the product and quotient will be positive.
2. If the numbers have different signs, both the product and quotient will be negative.

Average (or Mean)

You may already be familiar with the concept of the **average** of a set of numbers. The average is also called the **arithmetic average** or **mean**. Your grade in this course may be based on the average of your exam scores. Newspapers and magazines report average income, average sales, average attendance at sporting events, and so on. The mean of a set of numbers is particularly important in the study of statistics. For example, scientists might be interested in the mean IQ of the students attending a certain university or the mean height of students in the fourth grade.

Average

The **average** (or **mean**) of a set of numbers is the value found by adding the numbers in the set and then dividing the sum by the number of numbers in the set.

Example 6: Average

Teaching Notes:
Students may find additional "averages" meaningful to them - e.g., the average stopping distance for a Taurus at 50 mph, mean temperature for freezing of an automobile engine coolant, mean local temperature in January, average price of new homes, etc.

a. At noon on five consecutive days in Aspen, Colorado the temperatures were –5°, 7°, 6°, –7°, and 14° (in degrees Fahrenheit). (Negative numbers represent temperatures below zero.) Find the average of these noonday temperatures.

Solution: First, add the five temperatures.

$$(-5)+7+6+(-7)+14=15$$

Now divide the sum, 15, by the number of temperatures, 5.

$$\frac{15}{5}=3$$

The average noon temperature was 3° F.

b. In a placement exam for mathematics, a group of ten students had the following scores: 3 students scored 75, 2 students scored 80, 1 student scored 82, 3 students scored 85, and 1 student scored 88. What was the mean score for this group of students?

Solution: To find the total of all the scores, we multiply and then add. This is more efficient than adding all ten scores.

$$\left.\begin{aligned} 75\cdot 3 &= 225 \\ 80\cdot 2 &= 160 \\ 82\cdot 1 &= 82 \\ 85\cdot 3 &= 255 \\ 88\cdot 1 &= 88 \end{aligned}\right\}\text{ Multiply.}$$

$225+160+82+255+88=810$ Add.

$810\div 10=81$ Divide by the number of scores.

The mean score on the placement test for this group of students was 81.

c. The following speeds (in miles per hour) of fifteen cars were recorded at a certain point on a freeway.

70	75	65	60	61
64	68	72	59	68
82	76	70	68	50

Find the average speed of these cars. (One car received a speeding ticket, while another had a broken muffler.)

Solution: Using a calculator, the sum of the speeds is 1008 mph.

Dividing by 15 gives the average speed:

$1008\div 15=67.2$ mph

Practice Problems

Find the quotients.

1. $\frac{-30}{10}$ **2.** $\frac{40}{-10}$ **3.** $\frac{-20}{-10}$ **4.** $\frac{-7}{0}$

5. $\frac{0}{13}$ **6.** $\frac{7.5}{-3}$ **7.** $\frac{-4.32}{-4}$ **8.** $\frac{-5.2}{-2.6}$

9. Find the mean of the set of integers −16, 20, 32, and 92.

Answers to Practice Problems: **1.** −3 **2.** −4 **3.** 2 **4.** Undefined **5.** 0 **6.** −2.5 **7.** 1.08 **8.** 2 **9.** 32

1.4 Exercises

1. -12
2. -30
3. 48
4. 57
5. 56
6. 22
7. -21
8. -35
9. 56
10. 66
11. 26
12. 72
13. -70
14. -60
15. -24
16. -162
17. -288
18. 18
19. 0
20. 0
21. -9
22. 18.5
23. -7.31
24. 0.52
25. 4
26. 2
27. -6
28. -17
29. 2
30. 13
31. 0
32. Undefined
33. -3
34. -11
35. -17
36. -4
37. Undefined
38. 0
39. 5
40. 3
41. -0.6
42. -0.8
43. 20

Find the products.

1. $4(-3)$ **2.** $6(-5)$ **3.** $12 \cdot 4$ **4.** $19 \cdot 3$

5. $(-8)(-7)$ **6.** $(-11)(-2)$ **7.** $-3 \cdot 7$ **8.** $-7 \cdot 5$

9. $(-14)(-4)$ **10.** $(-11)(-6)$ **11.** $(-13)(-2)$ **12.** $(-8)(-9)$

13. $10(-7)$ **14.** $(-5)(12)$ **15.** $(-2)(-3)(-4)$ **16.** $(-6)(-3)(-9)$

17. $-8 \cdot 4 \cdot 9$ **18.** $(-3)(2)(-3)$ **19.** $(-7)(-16)(0)$ **20.** $-9 \cdot 0 \cdot 4$

21. $(-2)(4.5)$ **22.** $(-5)(-3.7)$ **23.** $4.3(-1.7)$ **24.** $(-2.6)(-0.2)$

Find the quotients.

25. $\frac{-8}{-2}$ **26.** $\frac{-20}{-10}$ **27.** $\frac{-30}{5}$ **28.** $\frac{-51}{3}$

29. $\frac{-26}{-13}$ **30.** $\frac{-91}{-7}$ **31.** $\frac{0}{6}$ **32.** $\frac{16}{0}$

33. $\frac{39}{-13}$ **34.** $\frac{44}{-4}$ **35.** $\frac{-34}{2}$ **36.** $\frac{-36}{9}$

37. $\frac{-3}{0}$ **38.** $\frac{0}{-7}$ **39.** $\frac{-60}{-12}$ **40.** $\frac{-48}{-16}$

41. $\frac{-4.8}{8}$ **42.** $\frac{-5.6}{7}$ **43.** $\frac{-4}{-0.2}$ **44.** $\frac{-3}{-8}$

45. $\frac{2.99}{-1.3}$ **46.** $\frac{2.8}{-1.4}$

Determine whether each statement is true or false. If a statement is false, rewrite it in a form that is true. (There may be more than one correct new form.)

47. $(-4)(6) \geq 3 \cdot 8$ **48.** $(-7)(-9) \leq 3 \cdot 21$

49. $(-12)(6) = 9(-8)$ **50.** $(-6)(9) = (18)(-2)$

51. $6(-3) > (-14) + (-4)$ **52.** $7 + 8 > (-10) + (-5)$

44. 0.375
45. −2.3
46. −2
47. False; $(-4)(6) < 3 \cdot 8$
48. True
49. True
50. False; $(-6)(9) < (18)(-2)$
51. False; $6(-3) = (-14) + (-4)$
52. True
53. True
54. False; $17 + (-3) > (-14) + (-4)$
55. True
56. True
57. −12
58. −48
59. $163
60. 37,673 accidents
61. 26.6 calls
62. 9.3 milligrams per deciliter
63. 80
64. 78.57

53. $(-7) + 0 \le (-7)(0)$

54. $17 + (-3) < (-14) + (-4)$

55. $(-4)(9) = (-24) + (-12)$

56. $14 + 6 \le (-2)(-10)$

57. Find the mean of the following set of integers: −10, 15, 16, −17, −34, and −42.

58. Find the mean of the following set of integers: −72, −100, −54, 82, and −96.

59. Airline travel: The costs of a one-way flight from Baltimore, MD to Orlando, FL on seven different airlines are as follows: $189, $134, $131, $231, $134, $213, $109. What is the average cost of a flight from Baltimore to Orlando?
Source: Expedia.com

60. Car accidents: The numbers of fatal motor-vehicle accidents in the United States each year from 1998 to 2008 are as follows: 37,107; 37,140; 37,526; 37,862; 38,491; 38,477; 38,444; 39,252; 38,648; 37,435; and 34,017. What was the average number of fatal accidents in the United States per year from 1998 to 2008? (Round your answer to the nearest integer.)
Source: NHTSA

61. Business: Twenty business executives made the following numbers of telephone calls during one week. Find the mean number of calls (to the nearest tenth) made by these executives.

20	16	14	11	51
40	36	28	52	25
18	16	42	49	12
18	22	33	9	19

62. Health: The blood calcium level (in milligrams per deciliter) for 20 patients was reported as follows:

8.2	10.2	9.3	8.5	7.3
9.7	9.6	8.3	9.8	9.1
9.4	11.1	10.0	8.5	9.9
8.6	10.2	9.4	9.1	9.2

Find the mean blood calcium level (to the nearest tenth) for these patients.

63. Exam scores: Fifteen students scored the following scores on an exam in accounting: 1 scored 67, 4 scored 73, 3 scored 77, 2 scored 80, 3 scored 88, and 2 scored 93. What was the average score for these students?

64. Exam scores: On an exam in history, a class of twenty-one students had the following test scores: 4 scored 65, 3 scored 70, 6 scored 78, 2 scored 82, 1 scored 85, 3 scored 91, and 2 scored 95. What was the mean score (to the nearest hundredth) on this test for the class?

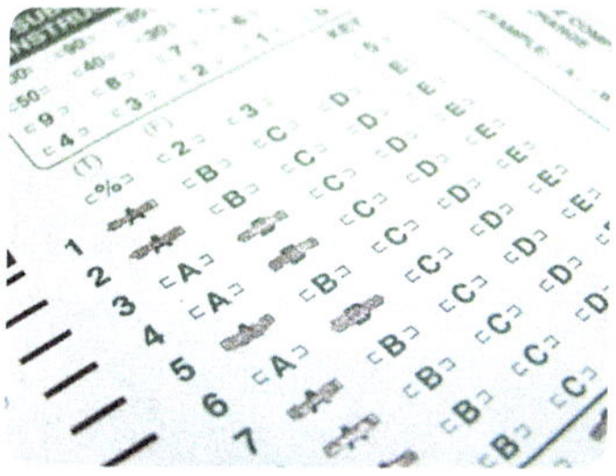

65. 78.6 inches
66. 2.2 books
67. 31.5 hours
68. 42 calls

The ***frequency*** *of a number is simply* ***a count of how many times that number appears.*** *In statistics, data is commonly given in the table form of a frequency distribution as illustrated. To find the mean, multiply each number by its frequency, add these products, and divide the sum by the sum of the frequencies.*

65. Height: The heights of the top 30 NBA scorers for the 2009-2010 season are listed in the frequency table. Find the mean height (to the nearest tenth of an inch) for these men.

Source: NBA

Height (in inches)	72	73	74	75	76	78	79	80	81	82	83	84
Frequency	1	1	1	5	1	5	4	3	3	1	3	2

66. Books read: The students in a psychology class were asked the number of books that they had read in the last month. The following frequency distribution indicates the results. Find the mean number of books read (to the nearest tenth) by these students.

Number of Books	0	1	2	3	4	5
Frequency	3	2	6	4	2	1

67. Watching TV: The bar graph to the right shows the approximate amounts of time per week spent watching TV for six groups (by age and sex) of people 18 years of age and older. What is the average amount of time per week people over the age of 18 spend watching TV? (Assume each group has the same number of people.)

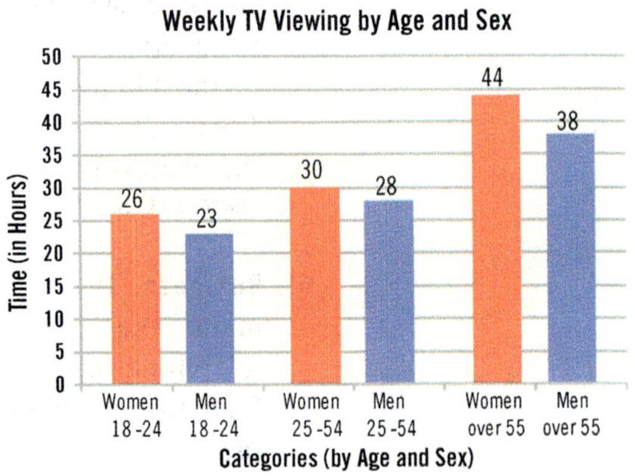

68. Phone calls: The following pictograph shows the number of phone calls received by a 1-hour radio talk show in Seattle during one week. (Not all calls actually get on the air.) What was the mean number of calls per show received that week?

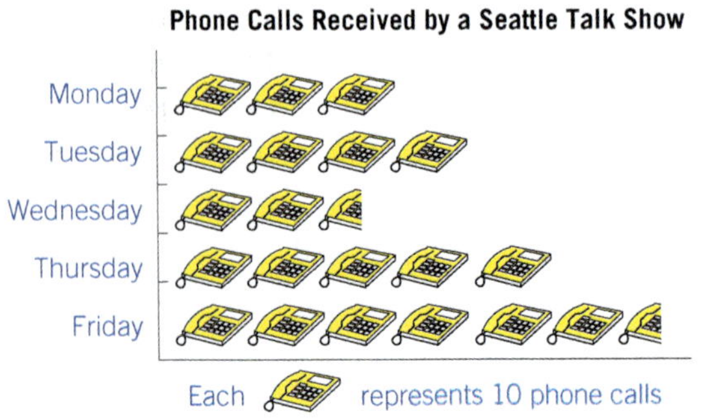

69. 18,902 square miles
70. 1,179,360
71. −34,459,110
72. 22,032
73. −2671
74. 1682
75. −45.33
76. −45.5
77. −10.29
78. −9.5
79. 2.83
80. −54.605
81. −90.365
82. Dividing 0 by any number other than 0 gives a quotient of 0.
83. See page 34.

69. Geography: The bar graph on the right shows the area of each of the five Great Lakes. Lake Superior, with an area of about 31,800 square miles, is the world's largest fresh water lake. What is the mean size of these lakes?

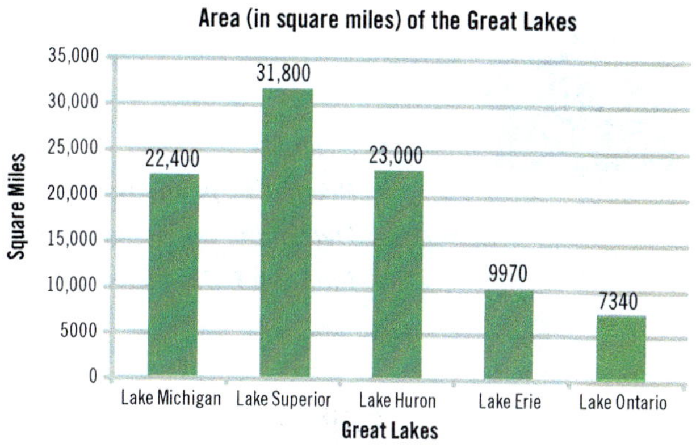

Calculator Problems

Use a graphing calculator to find the value of each expression. Round quotients to the nearest hundredth, if necessary. Remember the negative sign **(−)** *is next to* ENTER. *For example,* $-14.8 \div (-5)$ *would appear as follows:*

```
-14.8/(-5)
                2.96
```

70. $(273)(-24)(-180)$ **71.** $(-4613)(-45)(-166)$ **72.** $(54)(-17)(-24)$

73. $(-77,459) \div 29$ **74.** $(-62,234) \div (-37)$ **75.** $(-35 - 45 - 56) \div 3$

76. $(-52 - 30 - 40 - 60) \div 4$ **77.** $72 \div (15 - 22)$ **78.** $95 \div (-3 - 7)$

79. $-15.3 \div (-5.4)$ **80.** $(-13.4)(-2.5)(-1.63)$ **81.** $(-2.5)(-3.41)(-10.6)$

Writing and Thinking About Mathematics

82. Explain the conditions under which the quotient of two numbers is 0.

83. Explain, in your own words, why division by 0 is not a valid arithmetic operation.

HAWKES LEARNING SYSTEMS: INTRODUCTORY & INTERMEDIATE ALGEBRA SOFTWARE

- 1.4 Multiplication and Division with Real Numbers

1.5 Exponents, Prime Numbers, and LCM

- *Evaluate expressions with exponents.*
- *Recognize* ***prime numbers*** *less than 50.*
- *Determine the* ***prime factorization*** *of a* ***composite number****.*
- *Find the* ***LCM*** *(least common multiple) of a set of counting numbers.*

Exponents

We know that repeated addition with whole numbers can be shortened using multiplication. The result of multiplication is called the **product**, and the numbers being multiplied are called **factors** of the product. For example,

$$3+3+3+3=4\cdot 3=12 \quad \text{and} \quad 10+10+10+10+10=5(10)=50.$$

factors (4 · 3; 5(10)) product (12; 50)

In a similar manner, repeated multiplication by the same number can be shortened using **exponents**. For example, if 3 is used as a factor 4 times, we can write:

$$3\cdot 3\cdot 3\cdot 3=3^4=81.$$

4 is the exponent

3 is the base

81 is the 4th power of 3

Teaching Notes:
To help students understand the meaning of the word "power," consider explaining that the numbers 1, 10, 100, 1000, and so on are called the powers of 10.

In the equation $3^4=81$, 3 is the **base**, and 4 is the **exponent**. (Exponents are written slightly to the right and above the base.) The expression 3^4 is an **exponential expression** and is read "3 to the fourth power." We can also say that "81 is the fourth power of 3."

Example 1: Writing Exponents

	With Repeated Multiplication	With Exponents
a.	$5\cdot 5=25$	$5^2=25$
b.	$2\cdot 2\cdot 2=8$	$2^3=8$
c.	$5\cdot 5\cdot 5=125$	$5^3=125$
d.	$10\cdot 10\cdot 10\cdot 10=10{,}000$	$10^4=10{,}000$

Exponent and Base

A whole number n is an **exponent** if it is used to tell how many times another whole number a is used as a factor. The repeated factor a is called the **base** of the exponent. Symbolically,

$$\underbrace{a \cdot a \cdot a \cdot \ldots \cdot a \cdot a}_{n \text{ factors}} = a^n.$$

(In a^n: n is the exponent, a is the base.)

NOTES

COMMON ERROR

Do not multiply the base and the exponent.

$10^2 = 10 \cdot 2$ INCORRECT

$6^3 = 6 \cdot 3$ INCORRECT

Do multiply the base by itself.

$10^2 = 10 \cdot 10$ CORRECT

$6^3 = 6 \cdot 6 \cdot 6$ CORRECT

In expressions with exponent 2, the base is said to be **squared**. In expressions with exponent 3, the base is said to be **cubed**.

Example 2: Translating Expressions with Exponents

a. $8^2 = 64$ is read "eight squared is equal to sixty-four."

b. $5^3 = 125$ is read "five cubed is equal to one hundred twenty-five."

Expressions with exponents other than 2 or 3 are read as the base "to the ____ power." For example,

$2^5 = 32$ is read "two to the fifth power is equal to thirty-two."

Prime Numbers and Composite Numbers

Because of the relationship between multiplication and division, **factors** are also called **divisors** of the product. Division by a factor of a number gives a remainder of 0. For example,

factors ↓ ↓

$7 \cdot 6 = 42$ and

$$\begin{array}{r} 6 \\ 7\overline{)42} \\ \underline{42} \\ 0 \end{array} \leftarrow \text{remainder}$$

Thus 7 and 6 are divisors as well as factors of 42.

Every counting number, except the number 1, has **at least two** factors, as illustrated in the following list. Note that in this list, every number has at least two factors, but 5, 19, and 23 have **exactly two** factors.

Examples of Counting Numbers	Factors
5	1, 5
6	1, 2, 3, 6
19	1, 19
23	1, 23
33	1, 3, 11, 33
42	1, 2, 3, 6, 7, 14, 21, 42
96	1, 2, 3, 4, 6, 8, 12, 16, 24, 32, 48, 96

In the list above 5, 19, and 23 have exactly two different factors. Such numbers are called **prime numbers**. The other numbers in the list (6, 33, 42, and 96) are called **composite numbers**.

Prime Numbers

A **prime number** is a counting number greater than 1 that has exactly two different factors (or divisors), namely 1 and itself.

Composite Numbers

A **composite number** is a counting number with more than two different factors (or divisors).

Example 3: Prime Numbers

Some prime numbers:

2	2 has exactly two different factors, 1 and 2.
3	3 has exactly two different factors, 1 and 3.
11	11 has exactly two different factors, 1 and 11.
29	29 has exactly two different factors, 1 and 29.

Example 4: Composite Numbers

Some composite numbers:

15	1, 3, 5, and 15 are all factors of 15.
39	1, 3, 13, and 39 are all factors of 39.
49	1, 7, and 49 are all factors of 49.
51	1, 3, 17, and 51 are all factors of 51.

Because we will be looking for factors in algebra, particularly in dealing with fractions, we need to be aware of the concept of a number being **exactly divisible** by another number. A number is exactly divisible by (or just **divisible by**) another number if the remainder in the division process is 0. For example,

$$\begin{array}{r} 57 \\ 5\overline{)285} \\ \underline{25} \\ 35 \\ \underline{35} \\ \mathbf{0} \end{array} \quad \text{remainder} \qquad\qquad \begin{array}{r} 142 \\ 2\overline{)285} \\ \underline{2} \\ 8 \\ \underline{8} \\ 5 \\ \underline{4} \\ \mathbf{1} \end{array} \quad \text{remainder}$$

Thus 5 and 57 are both **factors** of 285 and both **divide exactly** into 285. The number 2 does not divide exactly into 285 and is not a factor of 285.

The related concepts of even and odd whole numbers are also useful.

Even and Odd Whole Numbers

If a whole number is divisible by 2, it is **even**.
If a whole number is not divisible by 2, it is **odd**.

The **even** whole numbers are

2, 4, 6, 8, 10, 12, ...

The **odd** whole numbers are

1, 3, 5, 7, 9, 11, 13, ...

The prime numbers less than 50 are

2, 3, 5, 7, 11, 13, 17, 19, 23, 29, 31, 37, 41, 43, 47.

For convenience and ease in working with fractions, you should memorize, or at least recognize, these primes. Also, you should note the following two facts about prime numbers:

1. 2 is the only even prime number; and
2. All other prime numbers are odd, but not all odd numbers are prime. (For example, 9, 15, and 33 are odd, but not prime.)

Prime Factorization of Composite Numbers

Using basic knowledge of multiplication and division, we can find factors of relatively small counting numbers. For example, basic multiplication facts give

$$63 = 9 \cdot 7.$$

However, for use in dealing with fractions, we need to find a factorization of 63 in which all of the factors are prime numbers. In this example, 9 is not a prime number. By factoring 9, we can write

$$63 = 3 \cdot 3 \cdot 7.$$

This last product $(3 \cdot 3 \cdot 7)$ contains all prime factors and is called the **prime factorization** of 63. Note that, because multiplication is a commutative operation (See Section 1.9 for a detailed discussion of the properties of real numbers.), the order of the factors is not important. Thus we could write $63 = 3 \cdot 7 \cdot 3$ as the prime factorization. However, for consistency, we will generally write the factors in order, from smallest to largest. Also, using exponents, we can write $63 = 3^2 \cdot 7$.

Regardless of the method used to find the prime factorization, **there is only one prime factorization for any composite number**. The following procedure can be used to find prime factorizations.

To Find the Prime Factorization of a Composite Number

1. Factor the composite number into any two factors.
2. Factor each factor that is not prime into two more factors.
3. Continue this process until all factors are prime.

The **prime factorization** is the product of all the prime factors.

NOTES

You may have studied quick tests for divisibility by 2, 3, 5, 6, 9, and 10 in a previous course in mathematics. For example, a number is divisible by 2, and therefore even, if the units digit is 0, 2, 4, 6, or 8. We will make reference to some of these tests for divisibility in the examples. See the end of the section for a brief review of this helpful topic.

Example 5: Prime Factorization

Find the prime factorization of 90.

Solution: $90 = 9 \cdot 10$ Since the units digit is 0, we know that 10 is a factor.

$= 3 \cdot 3 \cdot 2 \cdot 5$ 9 and 10 can both be factored so that each factor is a prime number. This is the prime factorization.

OR

$90 = 3 \cdot 30$ 3 is prime, but 30 is not.

$= 3 \cdot 10 \cdot 3$ 10 is not prime.

$= 3 \cdot 2 \cdot 5 \cdot 3$ All factors are prime.

Note that the final prime factorization was the same in both factor trees even though the first pair of factors was different.

Since multiplication is commutative, the order of the factors is not important. What is important is that **all the factors are prime**. Writing the factors in order, we can write

$$90 = 2 \cdot 3 \cdot 3 \cdot 5$$

or, with exponents,

$$90 = 2 \cdot 3^2 \cdot 5.$$

Example 6: Prime Factorization

Find the prime factorizations of each number:

a. 65

Solution: $65 = 5 \cdot 13$ 5 is a factor because the units digit is 5. Since both 5 and 13 are prime, $5 \cdot 13$ is the prime factorization.

b. 72

Solution: $72 = 8 \cdot 9$

$= 2 \cdot 4 \cdot 3 \cdot 3$

$= 2 \cdot 2 \cdot 2 \cdot 3 \cdot 3$

$= 2^3 \cdot 3^2$

Continued on the next page...

c. 294

Solution:

$$294 = 2 \cdot 147$$

2 is a factor because the units digits is even.

$$= 2 \cdot 3 \cdot 49$$

3 is a factor of 147 because the sum of the digits is divisible by 3.

$$= 2 \cdot 3 \cdot 7 \cdot 7$$

$$= 2 \cdot 3 \cdot 7^2$$

Using exponents

If we begin with the product $294 = 6 \cdot 49$, we see that the prime factorization is the same.

$$294 = 6 \cdot 49$$

$$= 2 \cdot 3 \cdot 7 \cdot 7$$

$$= 2 \cdot 3 \cdot 7^2$$

Least Common Multiple (LCM)

The **multiples** of a number are the products of that number with the counting numbers. Thus the first multiple of any number is the number itself. All other multiples are larger than that number. We are interested in finding common multiples and more particularly the **least common multiple (LCM)** for a set of counting numbers. For example, consider the lists of multiples of 8 and 12 shown here.

Counting Numbers:	**1,**	**2,**	**3,**	**4,**	**5,**	**6,**	**7,**	**8,**	**9,**	**10,**	**11,**	**12...**
Multiples of 8:	8,	16,	(24,)	32,	40,	(48,)	56,	64,	(72,)	80,	88,	(96.)..
Multiples of 12:	12,	(24,)	36,	(48,)	60,	(72,)	84,	(96,)	108,	(120,)	132,	(144,)..

The common multiples of 8 and 12 are 24, 48, 72, 96, 120, …. The **least common multiple (LCM)** is 24.

Listing all the multiples, as we just did for 8 and 12, and then choosing the least common multiple (LCM) is not very efficient. The following technique involving prime factorizations is generally much easier to use.

To Find the LCM of a Set of Counting Numbers

1. Find the prime factorization of each number.
2. List the prime factors that appear in any one of the prime factorizations.
3. Find the product of these primes using each prime the greatest number of times it appears in any one of the prime factorizations.

Example 7: Least Common Multiple (LCM)

Find the least common multiple (LCM) of 8, 10, and 30.

Solution:

Step 1: Prime factorizations:

$8 = 2 \cdot 2 \cdot 2$ three 2's
$10 = 2 \cdot 5$ one 2, one 5
$30 = 2 \cdot 3 \cdot 5$ one 2, one 3, one 5

Step 2: Prime factors that are present are 2, 3, and 5.

The **most** number of times each prime factor is used in **any one** factorization:

Three 2's (in 8)
One 3 (in 30)
One 5 (in 10 and in 30)

Step 3: Find the product of these primes.

$$\text{LCM} = 2 \cdot 2 \cdot 2 \cdot 3 \cdot 5 = 2^3 \cdot 3 \cdot 5 = 120$$

120 is the LCM and therefore the smallest number divisible by 8, 10, and 30.

Example 8: Least Common Multiple (LCM)

Find the LCM of 27, 30, 35, and 42.

Solution:

Step 1: Prime factorizations:

$27 = 3 \cdot 3 \cdot 3$ three 3's
$30 = 2 \cdot 3 \cdot 5$ one 2, one 3, one 5
$35 = 5 \cdot 7$ one 5, one 7
$42 = 2 \cdot 3 \cdot 7$ one 2, one 3, one 7

Continued on the next page...

Step 2: Prime factors present are 2, 3, 5, and 7.

The most number of times each prime factor is used in any one factorization:

One 2	(in 30 and in 42)
Three 3's	(in 27)
One 5	(in 30 and in 35)
One 7	(in 35 and in 42)

Step 3: Find the product of these primes.

$$\begin{aligned} \text{LCM} &= 2\cdot 3\cdot 3\cdot 3\cdot 5\cdot 7 \\ &= 2\cdot 3^3\cdot 5\cdot 7 \\ &= 1890 \end{aligned}$$

1890 is the smallest number divisible by all four of the numbers 27, 30, 35, and 42.

Example 9: LCM with Variables

Find the LCM for $4x, x^2y,$ and $6x^2$.

Solution: We treat each variable in the same manner as a prime number;

$$\left.\begin{aligned} 4x &= 2^2\cdot x \\ x^2y &= x^2\cdot y \\ 6x^2 &= 2\cdot 3\cdot x^2 \end{aligned}\right\} \text{LCM} = 2^2\cdot 3\cdot x^2\cdot y = 12x^2y$$

Tests for Divisibility

As mentioned in the special note on page 46, here are the quick tests for divisibility.

An integer is divisible:

By 2: if the units digit is 0, 2, 4, 6, or 8.

By 3: if the sum of the digits is divisible by 3.

By 5: if the units digit is 0 or 5.

By 6: if the number is divisible by both 2 and 3.

By 9: if the sum of the digits is divisible by 9.

By 10: if the units digit is 0.

Practice Problems

1. Determine which numbers, if any, are prime: $\{13, 17, 29, 36, 37, 49\}$.

Find the prime factorization of each counting number.

2. 70 **3.** 240 **4.** 507

Find the LCM of each set of counting numbers and expressions.

5. $\{3, 5, 11\}$ **6.** $\{25, 50, 60\}$

7. $\{88, 99, 121\}$ **8.** $\{x^2, 36xy, 20xy^3\}$

1.5 Exercises

1. 7^4 2. 11^3
3. $2 \cdot 3^2 \cdot 11^2$
4. $5^3 \cdot 7^2$ 5. $3^3 \cdot 7^3$
6. $2^4 \cdot 11^2 \cdot 13^2$
7. 2^4 or 4^2
8. 2^2 9. 6^2
10. 10^2
11. 10^3
12. 9^2 or 3^4
13. A counting number greater than 1 that has exactly two different factors (or divisors): itself and 1.
14. 2, 3, 5, 7, 11, 13, 17, 19, 23, 29, 31, 37, 41, 43, 47
15. A counting number with more than two different factors (or divisors).

Rewrite the following products using exponents.

1. $7 \cdot 7 \cdot 7 \cdot 7$ **2.** $11 \cdot 11 \cdot 11$ **3.** $2 \cdot 3 \cdot 3 \cdot 11 \cdot 11$

4. $5 \cdot 5 \cdot 5 \cdot 7 \cdot 7$ **5.** $3 \cdot 3 \cdot 3 \cdot 7 \cdot 7 \cdot 7$ **6.** $2 \cdot 2 \cdot 2 \cdot 2 \cdot 11 \cdot 11 \cdot 13 \cdot 13$

Find an exponential expression for each of the following numbers without using the exponent 1.

7. 16 **8.** 4 **9.** 36

10. 100 **11.** 1000 **12.** 81

13. Define *prime number*.

14. List the prime numbers less than 50.

15. Define *composite number*.

16. Are the following statements true or false? If a statement is false, explain why.

a. All prime numbers are even.

b. All prime numbers are odd.

c. All odd numbers are prime.

d. 2 is the only even prime number.

Answers to Practice Problems: **1.** 13, 17, 29, 37 **2.** $2 \cdot 5 \cdot 7$ **3.** $2^4 \cdot 3 \cdot 5$ **4.** $3 \cdot 13^2$ **5.** 165 **6.** 300 **7.** 8712 **8.** $180x^2y^3$

Determine which numbers, if any, in each set of counting numbers are prime.

17. $\{13, 15, 17, 21\}$ **18.** $\{11, 19, 23, 51\}$

19. $\{2, 4, 6, 8, 10, 12, 14\}$ **20.** $\{7, 16, 25, 36, 47, 49\}$

Find two factors of each number (other than 1 and the number itself) to determine that the number is composite. (Answers will vary.)

21. 72 **22.** 63 **23.** 68 **24.** 39 **25.** 502

26. 417 **27.** 170 **28.** 99 **29.** 444 **30.** 230

Find the prime factorization of each of the numbers. If a number is prime, write "prime."

31. 52 **32.** 60 **33.** 616 **34.** 460

35. 308 **36.** 155 **37.** 79 **38.** 43

39. 289 **40.** 361 **41.** 125 **42.** 343

43. 400 **44.** 500 **45.** 120 **46.** 196

47. 231 **48.** 675 **49.** 1692 **50.** 1716

51. List the first ten **multiples** of each number.

a. 5 **b.** 6 **c.** 10 **d.** 15

52. From the lists you made in Exercise 51, find the least common multiple for each of the following sets of numbers.

a. $\{5, 6\}$ **b.** $\{6, 10\}$ **c.** $\{5, 10, 15\}$ **d.** $\{6, 10, 15\}$

Find the LCM of each of the following sets of counting numbers.

53. $\{3, 5, 7\}$ **54.** $\{2, 7, 11\}$ **55.** $\{8, 10\}$

56. $\{9, 12\}$ **57.** $\{2, 3, 11\}$ **58.** $\{3, 5, 13\}$

59. $\{4, 14, 35\}$ **60.** $\{10, 12, 20\}$ **61.** $\{50, 75\}$

62. $\{30, 70\}$ **63.** $\{20, 90\}$ **64.** $\{50, 80\}$

65. $\{28, 98\}$ **66.** $\{45, 75\}$ **67.** $\{10, 15, 35\}$

68. $\{6, 24, 30\}$ **69.** $\{15, 45, 90\}$ **70.** $\{14, 28, 56\}$

16. a. False; All even numbers are divisible by 2. Thus all even numbers other than 2 have at least 3 different factors: 1, 2, and themselves.
b. False; The number 2 is both even and prime. It is divisible by 2 and it has exactly two different factors: 1 and 2.
c. False; Numbers such as 9 and 15 are odd. However both numbers have more than two factors. (9 is divisible by 1, 3, and 9 and 15 is divisible by 1, 3, 5, and 15.)
d. True
17. 13, 17
18. 11, 19, 23
19. 2 20. 7, 47
21. 8, 9 22. 7, 9
23. 4, 17 24. 3, 13
25. 2, 251 26. 3, 139
27. 10, 17 28. 9, 11
29. 4, 111 30. 10, 23
31. $2^2 \cdot 13$
32. $2^2 \cdot 3 \cdot 5$
33. $2^3 \cdot 7 \cdot 11$
34. $2^2 \cdot 5 \cdot 23$
35. $2^2 \cdot 7 \cdot 11$
36. $5 \cdot 31$
37. 79 is prime.
38. 43 is prime.
39. 17^2 40. 19^2
41. 5^3 42. 7^3
43. $2^4 \cdot 5^2$

44. $2^2 \cdot 5^3$
45. $2^3 \cdot 3 \cdot 5$
46. $2^2 \cdot 7^2$
47. $3 \cdot 7 \cdot 11$
48. $3^3 \cdot 5^2$
49. $2^2 \cdot 3^2 \cdot 47$
50. $2^2 \cdot 3 \cdot 11 \cdot 13$
51. a. 5, 10, 15, 20, 25, 30, 35, 40, 45, 50 b. 6, 12, 18, 24, 30, 36, 42, 48, 54, 60 c. 10, 20, 30, 40, 50, 60, 70, 80, 90, 100 d. 15, 30, 45, 60, 75, 90, 105, 120, 135, 150
52. a. 30 b. 30 c. 30 d. 30
53. 105 54. 154

71. $\{20, 50, 100\}$ **72.** $\{30, 60, 120\}$ **73.** $\{10, 15, 25\}$

74. $\{22, 44, 121\}$ **75.** $\{26, 28, 91\}$ **76.** $\{34, 51, 54\}$

77. $\{35, 40, 72\}$ **78.** $\{30, 35, 63\}$ **79.** $\{12, 21, 44\}$

80. $\{20, 28, 45\}$ **81.** $\{99, 121, 231\}$ **82.** $\{81, 225, 324\}$

83. $\{48, 120, 144, 192\}$ **84.** $\{125, 135, 225, 250\}$ **85.** $\{40, 56, 160, 196\}$

86. $\{35, 49, 63, 126\}$

Find the LCM for each set of expressions. Treat each variable as you would a prime number.

87. $\{8x, 10y, 20xy\}$ **88.** $\{20xz, 24xy, 32yz\}$

89. $\{14x^2, 21xy, 35xy^2\}$ **90.** $\{60a, 105a^2b, 120ab\}$

Writing and Thinking About Mathematics

91. List five prime numbers larger than 50.

92. Describe, in your own words, how to find the LCM of a set of counting numbers.

93. **a.** Explain why 1 is not a prime number.
b. Explain why 1 is not a composite number.

55. 40 56. 36
57. 66 58. 195
59. 140 60. 60
61. 150 62. 210
63. 180 64. 400
65. 196 66. 225
67. 210 68. 120
69. 90 70. 56
71. 100 72. 120
73. 150 74. 484

HAWKES LEARNING SYSTEMS: INTRODUCTORY & INTERMEDIATE ALGEBRA SOFTWARE

- 1.5 Exponents, Prime Numbers, and LCM

75. 364 76. 918 77. 2520 78. 630 79. 924 80. 1260 81. 7623 82. 8100 83. 2880 84. 6750 85. 7840 86. 4410
87. $40xy$ 88. $480xyz$ 89. $210x^2y^2$ 90. $840a^2b$ 91. 53, 59, 61, 67, 71 Answers will vary.
92. First, find the prime factorization of each number. Next, identify the prime factors that appear in any one of the prime factorizations. Finally, form the product of these primes using each prime the most number of times it appears in any one of the prime factorizations.
93. a. To be prime, a number has to have exactly two different factors. The number 1 only has one unique factor, itself.
b. A composite number has more than two different factors. The number 1 only has one unique factor, itself.

1.6 Multiplication and Division with Fractions

- *Reduce fractions to lowest terms.*
- *Write fractions as equivalent fractions with specified denominators.*
- *Multiply and divide fractions.*

Rational Numbers Defined

In Section 1.4 we used the fraction form to indicate division with real numbers. Now we develop the knowledge and skills related to operating (multiplying, dividing, adding, and subtracting) with fractions. Technically, fractions of the form

$$\frac{\boldsymbol{a}}{\boldsymbol{b}} \quad \frac{\leftarrow \text{numerator}}{\leftarrow \text{denominator}}$$

where a and b are **integers** and $\boldsymbol{b \neq 0}$ are called **rational numbers** as discussed in Section 1.1.

Rational Number

A **rational number** is a number that can be written in the form $\frac{\boldsymbol{a}}{\boldsymbol{b}}$ where a and b are **integers** and $\boldsymbol{b \neq 0}$.

NOTES

There are fractions that cannot be written with the numerator and denominator as integers. For example, $\frac{\pi}{3}$ is a fraction (very useful in trigonometry) but also an irrational number.

For now, when we deal with the fractional form $\frac{a}{b}$ we will assume the fraction represents a rational number and **use the terms fraction and rational number interchangeably**.

Examples of rational numbers (fractions):

$$\frac{1}{2}, \frac{3}{10}, -\frac{11}{7}, -\frac{13}{2}, -10, \text{ and } 15.$$

Note that every integer is also a rational number because integers can be written in fraction form with a denominator of 1. Thus

$$0=\frac{0}{1},\ 1=\frac{1}{1},\ 2=\frac{2}{1},\ 3=\frac{3}{1},\ \text{and so on.}$$

Also,

$$-1=\frac{-1}{1},\ -2=\frac{-2}{1},\ -3=\frac{-3}{1},\ -4=\frac{-4}{1},\ \text{and so on.}$$

In general, fractions can be used to indicate:

1. Equal parts of a whole, or
2. Division.

Example 1: Fractions

a. $\frac{1}{2}$ can mean 1 of 2 equal parts.

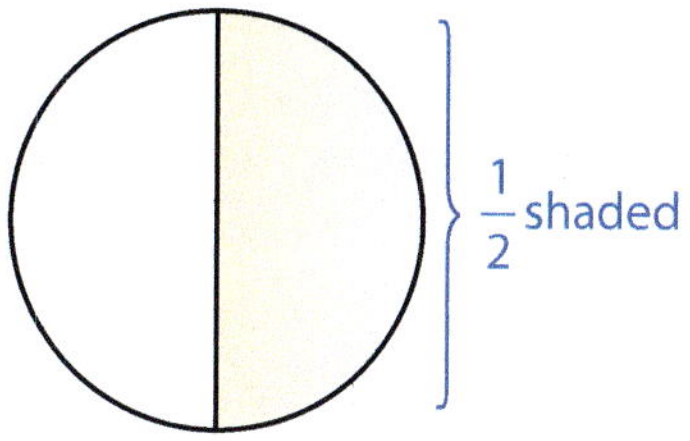

If you read $\frac{1}{2}$ of a book, then you can view this as having read one of two equal parts of the book. If the book has 50 pages, then you read 25 pages, since, as we will see, $\frac{25}{50}=\frac{1}{2}$.

b. $\frac{-45}{9}$ can mean to divide: $(-45)\div 9$.

Teaching Notes:
You might point out that the "÷" sign is a fraction bar with dots indicating numerator and denominator. Also, whether one actually interprets $\frac{a}{b}$ as an implied division depends on the context.

Before actually operating with fractions, we need to understand the use and placement of negative signs in fractions. Consider the following three results involving fractions, negative signs, and the meaning of division.

$$-\frac{16}{8}=-2,\ \frac{-16}{8}=-2,\ \text{and}\ \frac{16}{-8}=-2$$

Thus the three different placements of a negative sign give the same result, as noted in the following general statement or rule.

Placement of Negative Signs in Fractions

If a and b are real numbers, and $b\neq 0$, then

$$-\frac{\boldsymbol{a}}{\boldsymbol{b}}=\frac{-\boldsymbol{a}}{\boldsymbol{b}}=\frac{\boldsymbol{a}}{-\boldsymbol{b}}.$$

For example:

$$-\frac{3}{4b}=\frac{-3}{4b}=\frac{3}{-4b}$$

Example 2: Negative Signs in Fractions

a. $-\frac{12}{4}=\frac{-12}{4}=\frac{12}{-4}=-3$

b. $-\frac{1}{5}=\frac{-1}{5}=\frac{1}{-5}$

Multiplication

As stated below, multiplication with fractions is accomplished by multiplying the numerators and multiplying the denominators.

Multiplication with Fractions

To **multiply** two fractions, multiply the numerators and multiply the denominators.

$$\frac{a}{b}\cdot\frac{c}{d}=\frac{a\cdot c}{b\cdot d} \text{ where } b, d \neq 0$$

For example:

$$\frac{2}{3}\cdot\frac{7}{5}=\frac{2\cdot 7}{3\cdot 5}=\frac{14}{15}.$$

Teaching Notes:
Emphasize that multiplication by 1 in the form of $\frac{k}{k}$ changes the form of a fraction, but not its value. For example, $\frac{6}{8}$ of a pie is the same as $\frac{3}{4}$ of the same pie.

The number 1 is called the **multiplicative identity** since the product of 1 with any number is that number. That is,

$$\frac{a}{b}\cdot 1=\frac{a}{b}.$$

Thus, if $k \neq 0$, we have

$$\frac{a}{b}=\frac{a}{b}\cdot 1=\frac{a}{b}\cdot\frac{k}{k}=\frac{a\cdot k}{b\cdot k}.$$

This relationship is called the **fundamental principle of fractions.**

The Fundamental Principle of Fractions

$$\frac{a}{b}=\frac{a\cdot k}{b\cdot k} \text{ where } b, k \neq 0$$

We can use the fundamental principle to find equal fractions with larger denominators (**build to higher terms**) or find equal fractions with smaller denominators (**reduce to lower terms**). Equal fractions are said to be **equivalent**.

To Build a Fraction to Higher Terms

To build a fraction to higher terms, use the fundamental principle and multiply both the numerator and denominator by the same nonzero number.

Example 3: Building Fractions to Higher Terms

a. Write $-\frac{3}{7}$ in the form of an equivalent fraction with denominator 28.

Solution: Because we know $7 \cdot 4 = 28$, use $k = 4$ and

write $-\frac{3}{7} = -\frac{3}{7} \cdot \frac{4}{4} = -\frac{12}{28}$.

b. Build $-\frac{5}{8}$ to higher terms with denominator of $16a$.

Solution: Because we know that $8 \cdot 2a = 16a$, use $k = 2a$ and

write $-\frac{5}{8} = -\frac{5}{8} \cdot \frac{2a}{2a} = -\frac{10a}{16a}$.

To Reduce a Fraction

To reduce a fraction, factor both the numerator and denominator and use the fundamental principle to "divide out" any common factors. If the numerator and the denominator have no common prime factors, the fraction has been reduced to **lowest terms**.

Example 4: Reducing Fractions

a. Reduce $\frac{-12}{20}$ to lowest terms using prime factorization.

Solution: $\frac{-12}{20} = \frac{-1 \cdot \cancel{2} \cdot \cancel{2} \cdot 3}{\cancel{2} \cdot \cancel{2} \cdot 5} = \frac{-3}{5} = -\frac{3}{5}$

Note that the negative sign can be placed in the numerator or in front of the fraction. We seldom use the negative sign in the denominator.

Continued on the next page...

b. Find the product $\frac{15ac}{28b}\cdot\frac{4bc}{9a}$ in lowest terms. (Do not find the product directly. Factor and reduce as you multiply. Treat the variables as you would numbers.)

Solution: $\frac{15ac}{28b}\cdot\frac{4bc}{9a}=\frac{\cancel{3}\cdot 5\cdot\cancel{a}\cdot c\cdot\cancel{2}\cdot\cancel{2}\cdot\cancel{b}\cdot c}{\cancel{2}\cdot\cancel{2}\cdot 7\cdot\cancel{b}\cdot\cancel{3}\cdot 3\cdot\cancel{a}}=\frac{5c^2}{21}$

c. Find the product $\frac{4a}{12b}\cdot\frac{3b}{7a}$ in lowest terms. (**Note:** The number 1 is implied to be a factor even if it is not written.)

Solution: $\frac{4a}{12b}\cdot\frac{3b}{7a}=\frac{\cancel{4}\cdot\cancel{a}\cdot\cancel{3}\cdot\cancel{b}\cdot 1}{\cancel{4}\cdot\cancel{3}\cdot\cancel{b}\cdot 7\cdot\cancel{a}}=\frac{1}{7}$

Remember that if all factors are divided out (in the numerator or denominator), you must use 1 even if it is not written.

Finding the product of two fractions can be thought of as finding one fractional part **of** the other. Thus to find a fraction **of** a number means to multiply the fraction and the number.

Example 5: Fraction of a Number

Find $\frac{2}{3}$ of $\frac{5}{7}$.

Solution: $\frac{2}{3}\cdot\frac{5}{7}=\frac{2\cdot 5}{3\cdot 7}=\frac{10}{21}$

Division

We will now see that division with fractions is accomplished by multiplication. That is, if we know how to multiply fractions, we automatically know how to divide them.

Reciprocal

If $a\neq 0$ and $b\neq 0$, the **reciprocal** of $\frac{a}{b}$ is $\frac{b}{a}$, and $\frac{a}{b}\cdot\frac{b}{a}=1$.

0 has no reciprocal, since $\frac{1}{0}$ is **undefined**.

To divide by a number, we multiply by its reciprocal. For example,

$$\frac{2}{3} \div \frac{5}{6} = \frac{2}{3} \cdot \frac{6}{5}.$$

This gives the result (reduced) as

$$\frac{2}{3} \div \frac{5}{6} = \frac{2}{3} \cdot \frac{6}{5} = \frac{2 \cdot \cancel{3} \cdot 2}{\cancel{3} \cdot 5} = \frac{4}{5}.$$

Division with Fractions

To **divide** by a nonzero fraction, multiply by its reciprocal.

$$\frac{\boldsymbol{a}}{\boldsymbol{b}} \div \frac{\boldsymbol{c}}{\boldsymbol{d}} = \frac{\boldsymbol{a}}{\boldsymbol{b}} \cdot \frac{\boldsymbol{d}}{\boldsymbol{c}} \text{ where } b, c, d \neq 0$$

Example 6: Dividing Fractions with Negative Numbers

$$\left(-\frac{3}{4}\right) \div \left(-\frac{2}{5}\right)$$

Solution: $\left(-\frac{3}{4}\right) \div \left(-\frac{2}{5}\right) = \left(-\frac{3}{4}\right) \cdot \left(-\frac{5}{2}\right) = \frac{15}{8}$

Note that, just as with integers, the product of two negative fractions is positive. In algebra, $\frac{15}{8}$, an **improper fraction** (a fraction with the numerator greater than the denominator), is preferred to the mixed number $1\frac{7}{8}$.

Improper fractions are perfectly acceptable as long as they are reduced, meaning the numerator and denominator have no common factors.

Example 7: Dividing Fractions with Variables

$$\frac{-21a}{5b} \div 3a$$

Solution: $\frac{-21a}{5b} \div 3a = \frac{-21a}{5b} \div \frac{3a}{1} = \frac{-21a}{5b} \cdot \frac{1}{3a} = \frac{-1 \cdot \cancel{3} \cdot 7 \cdot \cancel{a}}{5 \cdot b \cdot \cancel{3} \cdot \cancel{a}} = \frac{-7}{5b} = -\frac{7}{5b}$

Note: The reciprocal of $3a$ is $\frac{1}{3a}$, because $3a = \frac{3a}{1}$.

Example 8: Multiplication and Division of Fractions

Multiply the quotient of $\frac{3}{4}$ and $\frac{15}{8}$ by $\frac{7}{2}$.

Solution: First find the quotient: $\frac{3}{4} \div \frac{15}{8} = \frac{3}{4} \cdot \frac{8}{15} = \frac{\cancel{3} \cdot \cancel{4} \cdot 2}{\cancel{4} \cdot \cancel{3} \cdot 5} = \frac{2}{5}$

Now multiply: $\frac{\cancel{2}}{5} \cdot \frac{7}{\cancel{2}} = \frac{7}{5}$

The answer is $\frac{7}{5}$.

Example 9: Multiplication and Division of Fractions

If $\frac{3}{8}$ of a number is 99, what is the number?

Solution: Because we know the product, 99, the missing number can be found by dividing the product by $\frac{3}{8}$.

$$99 \div \frac{3}{8} = \frac{\overset{33}{\cancel{99}}}{1} \cdot \frac{8}{\cancel{3}} = 264$$

The other number is 264. That is $\frac{3}{8}$ **of** 264 is 99. (or $\frac{3}{8} \cdot 264 = 99$)

NOTES

Important Note About Expressions Involving Fractions in Algebra

An expression with a fraction as the coefficient can be written in two different forms. For example, $\frac{9}{8}x$ and $\frac{9x}{8}$ have the same meaning. That is, for all values of x, we have $\frac{9}{8}x = \frac{9}{8} \cdot \frac{x}{1} = \frac{9x}{8}$.

Practice Problems

Multiply or divide as indicated and reduce each answer to lowest terms.

1. $\frac{-2}{3}\cdot\frac{3}{4}$

2. $\frac{12x}{18}\div\frac{4}{6x}$

3. $0\div\frac{7y}{14}$

4. $\frac{-15a}{20b}\cdot\frac{6ab}{21}\cdot\frac{40}{28b}$

5. Multiply the quotient of $\frac{3}{8}$ and $\frac{9}{10}$ by $\frac{7}{2}$.

6. The product of $\frac{2}{3}$ with another number is $\frac{-6}{11}$. What is the other number?

1.6 Exercises

Supply the missing numbers (or terms) so that each fraction will be built to higher terms as indicated.

1. $\frac{5}{6}=\frac{5}{6}\cdot\frac{?}{?}=\frac{?}{48}$

2. $\frac{3}{13}=\frac{3}{13}\cdot\frac{?}{?}=\frac{?}{52}$

3. $\frac{2}{9}=\frac{2}{9}\cdot\frac{?}{?}=\frac{?}{63b}$

4. $\frac{8}{3}=\frac{8}{3}\cdot\frac{?}{?}=\frac{?}{36x}$

5. $-\frac{7}{24}=-\frac{7}{24}\cdot\frac{?}{?}=-\frac{?}{72x}$

6. $-\frac{3}{5}=-\frac{3}{5}\cdot\frac{?}{?}=-\frac{?}{35y}$

Reduce each fraction to lowest terms.

7. $\frac{18}{45}$

8. $\frac{35}{63}$

9. $\frac{150xy}{350y}$

10. $\frac{60ab}{75a}$

11. $\frac{-12a}{-100ab}$

12. $\frac{-6x}{-51xy}$

13. $\frac{-30y}{45y}$

14. $\frac{-66ab}{88ab}$

15. $\frac{-28}{56x}$

16. $\frac{34}{-51y}$

17. $\frac{12xyz}{-35xy}$

18. $\frac{-8abc}{15ab}$

Find the fraction of a number by multiplying.

19. Find $\frac{1}{2}$ of $\frac{3}{4}$.

20. Find $\frac{2}{7}$ of $\frac{5}{7}$.

21. Find $\frac{7}{8}$ of 40.

22. Find $\frac{1}{3}$ of $\frac{1}{3}$.

1. $\frac{5}{6}=\frac{5}{6}\cdot\frac{8}{8}=\frac{40}{48}$
2. $\frac{3}{13}=\frac{3}{13}\cdot\frac{4}{4}=\frac{12}{52}$
3. $\frac{2}{9}=\frac{2}{9}\cdot\frac{7b}{7b}=\frac{14b}{63b}$
4. $\frac{8}{3}=\frac{8}{3}\cdot\frac{12x}{12x}=\frac{96x}{36x}$
5. $-\frac{7}{24}=-\frac{7}{24}\cdot\frac{3x}{3x}=-\frac{21x}{72x}$
6. $-\frac{3}{5}=-\frac{3}{5}\cdot\frac{7y}{7y}=-\frac{21y}{35y}$
7. $\frac{2}{5}$ 8. $\frac{5}{9}$
9. $\frac{3x}{7}$ 10. $\frac{4b}{5}$
11. $\frac{3}{25b}$ 12. $\frac{2}{17y}$
13. $-\frac{2}{3}$ 14. $-\frac{3}{4}$
15. $-\frac{1}{2x}$ 16. $-\frac{2}{3y}$
17. $-\frac{12z}{35}$ 18. $-\frac{8c}{15}$
19. $\frac{3}{8}$ 20. $\frac{10}{49}$
21. 35 22. $\frac{1}{9}$

Answers to Practice Problems: **1.** $-\frac{1}{2}$ **2.** x^2 **3.** 0 **4.** $-\frac{15a^2}{49b}$ **5.** $\frac{35}{24}$ **6.** $-\frac{9}{11}$

23. $-\frac{1}{6}$ 24. $-\frac{6}{35}$
25. $\frac{81}{100}$ 26. 4
27. $-\frac{3y}{7}$ 28. $-\frac{14y}{9}$
29. $-\frac{9x}{5}$ 30. $-\frac{3a}{2}$
31. $\frac{2}{3}$ 32. $\frac{49}{12}$
33. $-\frac{50}{9}$ 34. -7
35. $\frac{2}{5b}$ 36. $\frac{168}{x}$
37. $\frac{5y}{3}$ 38. 16
39. 0 40. 0
41. Undefined
42. Undefined
43. 22 44. $\frac{3}{2}$
45. $-\frac{1}{10}$ 46. $-\frac{1}{10}$
47. $\frac{9a}{10b}$ 48. $\frac{25}{98n}$
49. $\frac{81}{320n}$
50. $\frac{16}{5ab}$
51. $-\frac{7x}{100z}$
52. $-\frac{21}{8xy}$
53. $\frac{38}{3}$ 54. $\frac{9}{16}$
55. $\frac{14}{5}$ 56. $\frac{9}{5}$
57. $-\frac{25}{24}$ 58. $\frac{6}{7}$

Multiply or divide as indicated and reduce each answer to lowest terms.

23. $\frac{-3}{8}\cdot\frac{4}{9}$ **24.** $\frac{4}{5}\cdot\frac{-3}{14}$ **25.** $\frac{9}{10x}\div\frac{10}{9x}$ **26.** $\frac{4}{5a}\div\frac{1}{5a}$

27. $\frac{-16y}{7}\div\frac{16}{3}$ **28.** $\frac{-20y}{9}\div\frac{10}{7}$ **29.** $\frac{-15}{2}\cdot\frac{6x}{25}$ **30.** $\frac{-30}{10}\cdot\frac{4a}{8}$

31. $\frac{26b}{51a}\cdot\frac{17a}{13b}$ **32.** $\frac{35y}{15x}\cdot\frac{14x}{8y}$ **33.** $\frac{-15}{4}\cdot\frac{5}{6}\cdot\frac{16}{9}$ **34.** $\frac{-20}{3}\cdot\frac{7}{6}\cdot\frac{9}{10}$

35. $\frac{9a}{15}\cdot\frac{10}{3b}\cdot\frac{1}{5a}$ **36.** $\frac{20}{3x}\cdot\frac{18}{5x}\cdot\frac{14x}{2}$ **37.** $\frac{-35y}{12}\cdot\frac{-6}{14y}\cdot\frac{4y}{3}$

38. $\frac{-4ab}{18a}\cdot\frac{-24a}{13b}\cdot\frac{39}{a}$ **39.** $0\div\frac{37y}{21x}$ **40.** $0\div\frac{5a}{3b}$

41. $\frac{6xy}{15x}\div 0$ **42.** $\frac{28x}{7xy}\div 0$ **43.** $\frac{92}{7a}\div\frac{46}{77a}$

44. $\frac{57x}{32}\div\frac{19x}{16}$ **45.** $\frac{21}{30}\div(-7)$ **46.** $\frac{18}{20}\div(-9)$

47. $\frac{45a}{30b}\cdot\frac{12ab}{18b}\cdot\frac{9c}{10ac}$ **48.** $\frac{15m}{3m}\cdot\frac{2mn}{28n}\cdot\frac{10}{14mn}$ **49.** $\frac{-35m}{21n}\cdot\frac{27n}{40mn}\cdot\frac{-9m}{40m}$

50. $\frac{-72a}{4b}\cdot\frac{-16b}{36ab}\cdot\frac{8c}{20ac}$ **51.** $\frac{-9x}{40y}\cdot\frac{35xy}{15x}\cdot\frac{5}{10}\cdot\frac{8xy}{30xyz}$ **52.** $\frac{-3}{2x}\cdot\frac{7x}{44y}\cdot\frac{17}{xy}\cdot\frac{22xy}{34x}$

53. Find the quotient of $\frac{19}{2}$ and $\frac{3}{4}$.

54. Find the product of $\frac{5}{8}$ and $\frac{9}{10}$.

55. Multiply $\frac{21}{4}$ by $\frac{8}{15}$.

56. Divide $\frac{36}{15}$ by $\frac{4}{3}$.

57. The product of $\frac{3}{5}$ with another number is $-\frac{5}{8}$. What is the other number?

58. The product of $-\frac{1}{2}$ with another number is $-\frac{3}{7}$. What is the other number?

59. 61st floor
60. 450 students are over 4 feet tall; 50 students are less than or equal to 4 feet tall.
61. 98 pounds
62. 58,535,000 square miles
63. 344 yards
64. $\frac{7}{20}$, $\frac{13}{20}$
65. $\frac{5}{8}$ of the student body
66. $\frac{6}{17}$ of the land
67. a. More than 60
 b. Less than 60
 c. 72 passengers
68. a. More than 372
 b. 465 passengers

59. Building height: The Empire State Building is 102 stories tall. If you work $\frac{3}{5}$ of the way up the building, what floor do you work on? (Round your answer to the nearest whole number)?

60. Height: In one elementary school $\frac{9}{10}$ of the students are over 4 feet tall. If the school had an enrollment of 500 students, how many are over 4 feet tall? How many are less than or equal to 4 feet tall?

61. Anatomy: A study showed that $\frac{7}{10}$ of the human body weight is water. If a person weighs 140 pounds, how many pounds is water?

62. Geography: The continent of Africa covers approximately 11,707,000 square miles. This is $\frac{1}{5}$ of the land area in the world. What is the approximate total land area in the world?

63. Golf: The seventeenth hole at the local golf course is a par 4 hole. Ralph drove his ball 258 yards. If this distance was $\frac{3}{4}$ the length of the hole, how long is the seventeenth hole?

64. Restaurants: If you have \$20 and you spend \$7 on a glass of milk and a piece of pie, what fraction of your money did you spend? What fraction of your money do you still have?

65. Education: At Harvard University, a college that enrolls approximately 6400 students, 2400 students have social science majors. What fraction of the student body have majors that are not social science?

66. National parks: The United States National Park Service manages 85 million acres of land in the United States. 55 million acres managed by the National Park Service are located in Alaska. What fraction of the land managed by the National Park Service is not located in Alaska?

67. Transportation: A bus is carrying 60 passengers. This is $\frac{5}{6}$ of the capacity of the bus.

a. Is the capacity of the bus more or less than 60?

b. If you were to multiply 60 by $\frac{5}{6}$, would the product be more or less than 60?

c. What is the capacity of the bus?

68. Airplanes: A Boeing 747 jet is carrying 372 passengers from New York City to Los Angeles. This is $\frac{4}{5}$ of the plane's capacity.

a. Is the capacity more or less than 372?

b. What is the plane's full capacity?

69. $\dfrac{1}{18,009,460}$

70. $\dfrac{3}{4}x = \dfrac{3}{4} \cdot \dfrac{x}{1} = \dfrac{3x}{4}$

69. Lottery: The California lottery is based on choosing any six of the integers from 1 to 51. The probability of winning the lottery can be found by multiplying the fractions

$$\frac{6}{51} \cdot \frac{5}{50} \cdot \frac{4}{49} \cdot \frac{3}{48} \cdot \frac{2}{47} \cdot \frac{1}{46}.$$

Multiply and reduce the product of these fractions to find the probability of winning the lottery in the form of a fraction with numerator 1.

Writing and Thinking About Mathematics

70. Show why $\dfrac{3}{4}x$ and $\dfrac{3x}{4}$ have the same meaning.

71. Explain, in your own words, why 0 does not have a reciprocal.

HAWKES LEARNING SYSTEMS: INTRODUCTORY & INTERMEDIATE ALGEBRA SOFTWARE

- 1.6a Reducing Fractions
- 1.6b Multiplication and Division with Fractions

71. Division by zero is undefined. For example we could write $0 = \dfrac{0}{1}$. Then the reciprocal would be $\dfrac{1}{0}$, but this reciprocal is undefined since division by zero is undefined. Thus 0 does not have a reciprocal.

1.7 Addition and Subtraction with Fractions

- *Add and subtract fractions with like denominators.*
- *Add and subtract fractions with unlike denominators.*

Addition with Like Denominators

Finding the sum of two or more fractions with the same denominator is similar to adding whole numbers of some particular item. For example, the sum of 5 *apples* and 6 *apples* is 11 *apples*. Similarly, the sum of 5 *seventeenths* and 6 *seventeenths* is 11 *seventeenths*.

$$\frac{5}{17}+\frac{6}{17}=\frac{11}{17}$$

Adding Fractions with Like Denominators

To **add** two (or more) fractions with like denominators, add the numerators and use the common denominator.

$$\frac{\boldsymbol{a}}{\boldsymbol{b}}+\frac{\boldsymbol{c}}{\boldsymbol{b}}=\frac{\boldsymbol{a}+\boldsymbol{c}}{\boldsymbol{b}} \quad \text{where } b \text{ is a nonzero number or expression}$$

Example 1: Adding Fractions with Like Denominators

a. $\frac{3}{8}+\frac{4}{8}=\frac{3+4}{8}=\frac{7}{8}$ Here the common denominator is the number 8.

b. $\frac{3}{8x}+\frac{4}{8x}=\frac{3+4}{8x}=\frac{7}{8x}$ Here the common denominator is the algebraic term $8x$.

c. $\frac{2}{15}+\frac{3}{15}+\frac{1}{15}+\frac{6}{15}=\frac{2+3+1+6}{15}=\frac{12}{15}=\frac{\cancel{3}\cdot 4}{\cancel{3}\cdot 5}=\frac{4}{5}$ Reduce when possible.

Addition with Unlike Denominators

To find the sum of fractions with different denominators, we need the concepts of **multiples** and **least common multiple (LCM)** that we developed in Section 1.5. For example, to find the sum

$$\frac{5}{6}+\frac{1}{8}$$

we need the LCM of the denominators 6 and 8. This will be the **least common denominator (LCD)**.

$$\left.\begin{aligned} 6 &= 2\cdot 3 \\ 8 &= 2^3 \end{aligned}\right\} \text{LCM} = \text{LCD} = 2^3\cdot 3 = 24$$

Now change each fraction to an equivalent fraction with 24 as its denominator and add these new fractions.

$$\begin{aligned} \frac{5}{6}+\frac{1}{8} &= \left(\frac{5}{6}\cdot\frac{4}{4}\right)+\left(\frac{1}{8}\cdot\frac{3}{3}\right) \\ &= \frac{20}{24}+\frac{3}{24} \\ &= \frac{23}{24} \end{aligned}$$

This procedure can be generalized in the following way.

Adding Fractions with Unlike Denominators

1. Find the LCM of the denominators.
   (This is called the **least common denominator** (**LCD**) and may involve variables.)
2. Change each fraction to an equivalent fraction with the LCM as the denominator.
3. Add the new fractions. Reduce if possible.

Example 2: Adding Fractions with Unlike Denominators

a. $\frac{1}{6}+\frac{5}{4}$

Solution: $\left.\begin{aligned} 6 &= 2\cdot 3 \\ 4 &= 2\cdot 2 \end{aligned}\right\} \text{LCM} = \text{LCD} = 2^2\cdot 3 = 12 = 6\cdot 2 = 4\cdot 3$

Thus,

$$\frac{1}{6}+\frac{5}{4}=\left(\frac{1}{6}\cdot\frac{2}{2}\right)+\left(\frac{5}{4}\cdot\frac{3}{3}\right)$$ Multiply each fraction by 1 in the form $\frac{k}{k}$.

$$=\frac{2}{12}+\frac{15}{12}$$ Each fraction has the same denominator.

$$=\frac{17}{12}$$ Add the fractions.

b. $\frac{1}{4}+\frac{3}{8}+\frac{3}{10}$

Solution:

$$\left.\begin{aligned}4&=2^2\\8&=2^3\\10&=2\cdot5\end{aligned}\right\}\text{ LCM} = \text{LCD} = 2^3\cdot5=40$$

To get the common denominator in each fraction, we see

$$40=4\cdot10=8\cdot5=10\cdot4.$$

$$\frac{1}{4}+\frac{3}{8}+\frac{3}{10}=\left(\frac{1}{4}\cdot\frac{10}{10}\right)+\left(\frac{3}{8}\cdot\frac{5}{5}\right)+\left(\frac{3}{10}\cdot\frac{4}{4}\right)$$ Multiply each fraction by 1 in the form $\frac{k}{k}$.

$$=\frac{10}{40}+\frac{15}{40}+\frac{12}{40}$$

$$=\frac{37}{40}$$

c. $\frac{5}{21a}+\frac{5}{28a}$

Solution:

$$\left.\begin{aligned}21a&=3\cdot7\cdot a\\28a&=2^2\cdot7\cdot a\end{aligned}\right\}\text{ LCM} = \text{LCD} = 2^2\cdot3\cdot7\cdot a=84a=21a\cdot4=28a\cdot3$$

In each fraction, the numerator and denominator are multiplied by the same number to get $84a$ as the denominator.

$$\frac{5}{21a}+\frac{5}{28a}=\left(\frac{5}{21a}\cdot\frac{4}{4}\right)+\left(\frac{5}{28a}\cdot\frac{3}{3}\right)$$ Multiply each fraction by 1 in the form $\frac{k}{k}$.

$$=\frac{20}{84a}+\frac{15}{84a}$$

$$=\frac{35}{84a}$$

$$=\frac{\cancel{7}\cdot5}{\cancel{7}\cdot12a}$$ Reduce.

$$=\frac{5}{12a}$$

Subtraction

When subtracting fractions with like denominators, we subtract the numerators and use the common denominator.

Subtracting Fractions with Like Denominators

To **subtract** two fractions with like denominators, subtract the numerators and use the common denominator.

$$\frac{a}{b}-\frac{c}{b}=\frac{a-c}{b} \quad \text{where } b \text{ is a nonzero number or expression}$$

Example 3: Subtracting Fractions with Like Denominators

a. $\frac{3}{16}-\frac{1}{16}$

Solution: $\frac{3}{16}-\frac{1}{16}=\frac{3-1}{16}=\frac{2}{16}=\frac{\cancel{2}\cdot 1}{\cancel{2}\cdot 8}=\frac{1}{8}$

b. $\frac{2}{3}-\frac{17}{3}$

Solution: $\frac{2}{3}-\frac{17}{3}=\frac{2-17}{3}=\frac{-15}{3}=-5$

c. $\frac{1}{8a}-\frac{5}{8a}$

Solution: $\frac{1}{8a}-\frac{5}{8a}=\frac{1-5}{8a}=\frac{-4}{8a}=\frac{-1\cdot\cancel{4}}{\cancel{4}\cdot 2\cdot a}=\frac{-1}{2a}$ $\left(\text{or } -\frac{1}{2a}\right)$

Subtracting Fractions with Unlike Denominators

Just as with addition, if the two fractions do not have the same denominator, find equivalent fractions with the LCD and subtract the new fractions.

Subtracting Fractions with Unlike Denominators

1. Find the LCM of the denominators.
2. Change each fraction to an equivalent fraction with the LCM as the denominator.
3. Subtract the new fractions. Reduce if possible.

Example 4: Subtracting Fractions with Unlike Denominators

a. $\frac{1}{45}-\frac{1}{72}$

Solution: $\left.\begin{array}{l}45=3^2\cdot 5\\72=2^3\cdot 3^2\end{array}\right\}$ LCM = LCD = $2^3\cdot 3^2\cdot 5=360=45\cdot 8=72\cdot 5$

$$\frac{1}{45}-\frac{1}{72}=\left(\frac{1}{45}\cdot\frac{8}{8}\right)-\left(\frac{1}{72}\cdot\frac{5}{5}\right)=\frac{8}{360}-\frac{5}{360}=\frac{3}{360}=\frac{\cancel{3}\cdot 1}{\cancel{3}\cdot 120}=\frac{1}{120}$$

b. $\frac{3}{x}-\frac{2}{5x}$

Solution: In this case the LCD = $5x$.

$$\frac{3}{x}-\frac{2}{5x}=\left(\frac{3}{x}\cdot\frac{5}{5}\right)-\frac{2}{5x}=\frac{15}{5x}-\frac{2}{5x}=\frac{13}{5x}$$

This fraction cannot be reduced.

c. $\frac{1}{4a}-\frac{1}{6a}$

Solution: The numbers are small and we can see that the LCD is $12a$.

$$\frac{1}{4a}-\frac{1}{6a}=\frac{1}{4a}\cdot\frac{3}{3}-\frac{1}{6a}\cdot\frac{2}{2}=\frac{3}{12a}-\frac{2}{12a}=\frac{1}{12a}$$

This fraction cannot be reduced.

Practice Problems

Perform the indicated operations and reduce all answers to lowest terms.

1. $\frac{7}{4}-\frac{5}{4}$
2. $\frac{1}{2a}+\frac{1}{2a}$
3. $\frac{3}{20}-\frac{1}{2}$
4. $-\frac{5}{2}+\frac{3}{10}$
5. $\frac{4}{x}-\frac{3}{5x}$
6. $\frac{5}{12x}+\frac{7}{36x}$

Answers to Practice Problems: **1.** $\frac{1}{2}$ **2.** $\frac{1}{a}$ **3.** $-\frac{7}{20}$ **4.** $-\frac{11}{5}$ **5.** $\frac{17}{5x}$ **6.** $\frac{11}{18x}$

1.7 Exercises

1. $\frac{7}{9}$ 2. $\frac{10}{7}$
3. $\frac{3}{7}$ 4. $\frac{2}{5}$
5. $\frac{7}{10}$ 6. $\frac{17}{24}$
7. $-\frac{7}{20}$ 8. $\frac{53}{40}$
9. $\frac{5}{8}$ 10. $-\frac{5}{34}$
11. 1 12. $\frac{1}{3}$
13. $-\frac{1}{6}$ 14. $-\frac{1}{4}$
15. $\frac{2}{15}$ 16. $\frac{3}{40}$
17. $\frac{17}{36}$ 18. $-\frac{13}{21}$
19. $\frac{17}{21}$ 20. $\frac{9}{13}$
21. $\frac{63}{50}$ 22. $\frac{89}{200}$
23. $\frac{4}{7}$ 24. $\frac{4}{9}$
25. $\frac{4}{y}$ 26. $\frac{7}{y}$
27. $\frac{5}{x}$ 28. $\frac{1}{b}$
29. $\frac{16}{15a}$ 30. $\frac{7}{3x}$
31. $-\frac{2}{5x}$ 32. $-\frac{1}{x}$
33. $\frac{1}{2y}$ 34. $\frac{1}{2c}$
35. $\frac{43}{72x}$ 36. $\frac{19}{36a}$
37. $\frac{5}{6a}$ 38. $-\frac{37}{80b}$
39. $-\frac{11}{12x}$ 40. $\frac{3}{8y}$
41. 0 42. 0

Perform the indicated operations and reduce each answer to lowest terms.

1. $\frac{2}{9}+\frac{5}{9}$ **2.** $\frac{2}{7}+\frac{8}{7}$ **3.** $\frac{3}{14}+\frac{3}{14}$ **4.** $\frac{1}{10}+\frac{3}{10}$

5. $\frac{2}{5}+\frac{3}{10}$ **6.** $\frac{5}{8}+\frac{1}{12}$ **7.** $\frac{2}{5}-\frac{3}{4}$ **8.** $\frac{7}{10}+\frac{5}{8}$

9. $\frac{3}{4}+\left(-\frac{1}{8}\right)$ **10.** $\frac{5}{17}+\left(-\frac{15}{34}\right)$ **11.** $\frac{24}{23}-\frac{1}{23}$ **12.** $\frac{7}{15}-\frac{2}{15}$

13. $\frac{11}{12}-\frac{13}{12}$ **14.** $\frac{5}{16}-\frac{9}{16}$ **15.** $\frac{5}{6}-\frac{7}{10}$ **16.** $\frac{9}{20}-\frac{3}{8}$

17. $\frac{8}{9}-\frac{5}{12}$ **18.** $\frac{3}{14}-\frac{5}{6}$ **19.** $\frac{2}{7}+\frac{4}{21}+\frac{1}{3}$ **20.** $\frac{2}{39}+\frac{1}{3}+\frac{4}{13}$

21. $\frac{1}{2}+\frac{3}{4}+\frac{1}{100}$ **22.** $\frac{1}{4}+\frac{1}{8}+\frac{7}{100}$ **23.** $1-\frac{3}{7}$ **24.** $1-\frac{5}{9}$

25. $\frac{5}{y}-\frac{1}{y}$ **26.** $\frac{6}{y}+\frac{1}{y}$ **27.** $\frac{4}{x}+\frac{1}{x}$ **28.** $\frac{3}{b}-\frac{2}{b}$

29. $\frac{11}{15a}+\frac{5}{15a}$ **30.** $\frac{2}{3x}+\frac{5}{3x}$ **31.** $\frac{-19}{25x}+\frac{9}{25x}$ **32.** $\frac{-5}{13x}-\frac{8}{13x}$

33. $\frac{7}{10y}-\frac{1}{5y}$ **34.** $\frac{5}{6c}-\frac{1}{3c}$ **35.** $\frac{11}{24x}+\frac{5}{36x}$ **36.** $\frac{11}{12a}-\frac{7}{18a}$

37. $\frac{2}{3a}+\frac{1}{6a}$ **38.** $\frac{7}{16b}-\frac{9}{10b}$ **39.** $\frac{1}{3x}+\left(-\frac{5}{4x}\right)$ **40.** $\frac{7}{8y}+\left(-\frac{3}{6y}\right)$

41. $\frac{3}{4a}-\frac{6}{8a}$ **42.** $\frac{5}{6x}-\frac{10}{12x}$

43. $\frac{25}{24}$
44. $\frac{35}{36}$
45. $\frac{41}{48}$
46. $\frac{69}{35}$
47. $\frac{11}{18}$ of the play
48. $\frac{8}{21}$ of the panels
49. 3 hours
50. $\frac{9}{2}$ miles
51. $\frac{9}{10}$ lb
52. $\frac{25}{8}$ or $3\frac{1}{8}$ measures

43. Find the sum of $\frac{5}{6}$ and $\frac{7}{12}$. Then subtract $\frac{3}{8}$. What is the difference?

44. Subtract $\frac{2}{3}$ from $\frac{8}{9}$. Then add $\frac{3}{4}$. What is the sum?

45. Find the product of $\frac{9}{16}$ and $\frac{1}{3}$. Then add $\frac{2}{3}$. What is the sum?

46. Find the sum of $\frac{4}{5}$ and $\frac{11}{15}$. Then divide by $\frac{7}{9}$. What is the quotient?

47. Theater: The microphone of the leading actor in the play *Peter Pan* did not work the first $\frac{2}{9}$ of the play nor the last $\frac{1}{6}$ of the play. What fraction of the time was the microphone working?

48. Recycled glass: A new window is to be made of panels of recycled glass bottles. If $\frac{2}{7}$ of the panels are made from blue bottles and $\frac{1}{3}$ are made from green bottles, how many of the panels are made from other recycled glass?

49. Parades: In preparation for the 2009 Macy's Thanksgiving Day Parade it took $\frac{3}{2}$ hours to blow up the Shrek balloon, $\frac{4}{5}$ hours to blow up Kermit the Frog, and $\frac{7}{10}$ hours to blow up the Pillsbury Dough Boy. How long did it take to blow up all three balloons?

50. Hiking: Part of a $\frac{24}{5}$ mile hiking trail was washed away by heavy rains leaving hikers to hike a trail that is $\frac{3}{10}$ of a mile shorter. How long is the new trail?

51. Fudge: Leila bought 1 lb of Rocky Road fudge and 1 lb of Peanut Butter fudge. Her roommates ate $\frac{3}{5}$ of the Rocky Road and $\frac{1}{2}$ of the Peanut Butter. How much of the fudge was left?

52. Music: In musical notation, a half note is $\frac{1}{2}$ a measure; a quarter note is $\frac{1}{4}$ a measure, and an eighth note is $\frac{1}{8}$ a measure. If there are 8 measures in a piece of music and 5 eighth notes, 3 quarter notes, and 7 half notes are erased, how many measures are left?

53. a. $\frac{13}{30}$
b. \$1170
54. a. Vote for
b. Pass by 40 votes
55. Multiplying the numerator and denominator by the same number is equivalent to multiplying the fraction by 1 and any number multiplied by 1 equals the original number.
56. This occurs when the fraction is multiplied by a negative number.

53. Budgeting: Delia's income is \$2700 a month and she plans to budget $\frac{1}{3}$ of her income for rent and $\frac{1}{10}$ of her income for food.

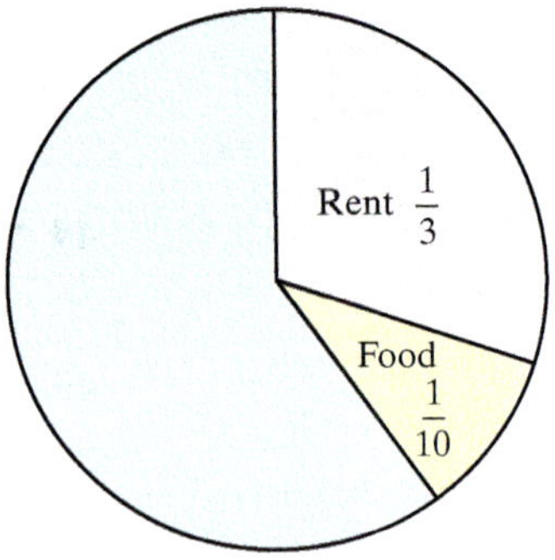

Monthly Budget

a. What fraction of her income does she plan to spend each month on these two items?
b. What amount of money does she plan to spend each month on these two items?

54. Tennis: The tennis club has 400 members, and they are considering putting in a new tennis court. The cost of the new court is going to involve a charge of \$250 for each member. Of the six-tenths of the members who live near the club, $\frac{4}{5}$ are in favor of the charge. However, of the other members who live further away, only $\frac{3}{10}$ are in favor of the charge.

a. If a vote were taken today, would more than one-half of the members vote for or against the charge?
b. By how many votes would the charge pass or fail if more than one-half of the members must vote in favor for the question to pass?

Writing and Thinking About Mathematics

55. Explain why, when a fraction is changed to higher terms, both the numerator and denominator are multiplied by the same number.

56. Explain under what circumstances a fraction between 0 and 1 can be multiplied by a number and the result be larger than the number.

HAWKES LEARNING SYSTEMS: INTRODUCTORY & INTERMEDIATE ALGEBRA SOFTWARE

- 1.7 Addition and Subtraction with Fractions

1.8 Order of Operations

- *Follow the rules for order of operations to evaluate expressions.*

Mathematicians have agreed on a set of rules for the order of performing operations when evaluating numerical expressions. These expressions may contain grouping symbols, exponents, and the operations of addition, subtraction, multiplication, and division. The rules are called the **rules for order of operations** and are used throughout all levels of mathematics and science (as well as calculators) so that there is only one correct answer for the value of an expression. For example, evaluate the following expression:

$$36 \div 4 + 6 \cdot 2^2.$$

Did you get 33 or 60? The correct answer is 33. In this section we will see how to apply the **rules for order of operations** with integers, decimals, and fractions.

Rules for Order of Operations

1. Simplify within grouping symbols, such as parentheses ( ), brackets [ ], and braces { }, working from the innermost grouping outward.
2. Find any powers indicated by exponents.
3. Moving from **left to right**, perform any multiplications **or** divisions **in the order they appear**.
4. Moving from **left to right**, perform any additions **or** subtractions **in the order they appear**.

NOTES

Other grouping symbols are the absolute value bars (such as $|3+5|$), the fraction bar (as in $\frac{4+7}{10+1}$), and the square root symbol (such as $\sqrt{5+11}$).

These rules are very explicit and should be studied carefully. Note that in rule 3, neither multiplication nor division has priority over the other. Whichever of these operations occurs first, **moving left to right**, is done first. In rule 4, addition and subtraction are handled in the same way. Unless they occur within grouping symbols, **addition and subtraction are the last operations to be performed**.

A well-known mnemonic device for remembering the rules for order of operations is the following:

Please	Excuse	My	Dear	Aunt	Sally
↓	↓	↓	↓	↓	↓
Parentheses	**Exponents**	**Multiplication**	**Division**	**Addition**	**Subtraction**

NOTES

Even though the mnemonic **PEMDAS** is helpful, remember that multiplication and division are performed as they appear, left to right. Also, addition and subtraction are performed as they appear, left to right.

For example

$$12 \div 3 \cdot 4 = 4 \cdot 4 = 16,$$

but

$$12 \cdot 3 \div 4 = 36 \div 4 = 9.$$

A negative sign in front of a variable indicates that the variable is to be multiplied by -1. For example,

$$-x^2 = -1 \cdot x^2.$$

This is consistent with the rules for order of operations (indicating that exponents come before multiplication) and is particularly useful in determining the values of expressions involving negative numbers and exponents. For example, each of the expressions

$$-7^2 \quad \text{and} \quad (-7)^2$$

has a different value. By the order of operations,

$$-7^2 = -1 \cdot 7^2 = -1 \cdot 49 = -49,$$

but

$$(-7)^2 = (-7)(-7) = 49.$$

In the second expression the base is a negative number. Remember that if the base is a negative number, then the negative number must be placed in parentheses.

The following examples show how to apply the rules. In some cases, more than one step can be performed at the same time. This is possible when parts are separated by + or – signs or are within separate symbols of inclusion. **Work through each of the following examples step by step, and rewrite the examples on a separate sheet of paper.**

Example 1: Order of Operations with Integers

Use the rules for order of operations to evaluate each of the following expressions.

a. $36 \div 4 - 6 \cdot 2^2$

Solution: $36 \div 4 - 6 \cdot 2^2$

$= 36 \div 4 - 6 \cdot 4$ Exponents

$= 9 - 24$ Divide and multiply, left to right.

$= -15$ Subtract.

b. $2(3^2 - 1) - 3 \cdot 2^3$

Solution: $2(3^2 - 1) - 3 \cdot 2^3$

$= 2(9 - 1) - 3 \cdot 8$ Exponents

$= 2(8) - 3 \cdot 8$ Subtract inside the parentheses.

$= 16 - 24$ Multiply.

$= -8$ Subtract (or add algebraically).

c. $9 - 2[(3 \cdot 5 - 7^2) \div 2 + 2^2]$

Solution: $9 - 2[(3 \cdot 5 - 7^2) \div 2 + 2^2]$

$= 9 - 2[(3 \cdot 5 - 49) \div 2 + 4]$ Exponents

$= 9 - 2[(15 - 49) \div 2 + 4]$ Multiply inside the parentheses.

$= 9 - 2[(-34) \div 2 + 4]$ Subtract inside the parentheses.

$= 9 - 2[-17 + 4]$ Divide inside the brackets.

$= 9 - 2[-13]$ Add inside the brackets.

$= 9 + 26$ Multiply.

$= 35$ Add.

Note: Because of the rules for order of operations at no time did we even subtract $9 - 2$.

Example 2: Order of Operations with Fractions

Use the rules for order of operations to evaluate each of the following expressions.

a. $2\frac{1}{3} \div \left(\frac{1}{4} + \frac{1}{3}\right)$

Solution: $2\frac{1}{3} \div \left(\frac{1}{4} + \frac{1}{3}\right)$

$= \frac{7}{3} \div \left(\frac{3}{12} + \frac{4}{12}\right)$ Change the mixed number to an improper fraction and find the LCD in the parentheses.

$= \frac{7}{3} \div \left(\frac{7}{12}\right)$ Add within parentheses.

$= \frac{\cancel{7}}{\cancel{3}} \cdot \frac{\overset{4}{\cancel{12}}}{\cancel{7}}$ Divide and reduce.

$= 4$

b. $\frac{3}{5} \cdot \frac{5}{6} + \frac{1}{4} \div \left(\frac{5}{2}\right)^2$

Solution: $\frac{3}{5} \cdot \frac{5}{6} + \frac{1}{4} \div \left(\frac{5}{2}\right)^2$

$= \frac{3}{5} \cdot \frac{5}{6} + \frac{1}{4} \div \left(\frac{25}{4}\right)$ Exponents

$= \frac{\cancel{3}}{\cancel{5}} \cdot \frac{\cancel{5}}{\underset{2}{\cancel{6}}} + \frac{1}{\cancel{4}} \cdot \frac{\cancel{4}}{25}$ Divide, then multiply and reduce.

$= \frac{1}{2} + \frac{1}{25}$

$= \frac{25}{50} + \frac{2}{50}$ Find the LCD.

$= \frac{27}{50}$ Add.

c. $\frac{1}{10} \div \frac{3}{4} \cdot \frac{1}{2} - \frac{3}{5a} \cdot \frac{7a}{30}$

Solution: $\frac{1}{10} \div \frac{3}{4} \cdot \frac{1}{2} - \frac{3}{5a} \cdot \frac{7a}{30}$

$= \frac{1}{10} \cdot \frac{4}{3} \cdot \frac{1}{2} - \frac{3}{5a} \cdot \frac{7a}{30}$ Divide.

$= \frac{1}{\cancel{10}_{5}} \cdot \frac{\cancel{4}}{3} \cdot \frac{1}{\cancel{2}} - \frac{\cancel{3}}{5\cancel{a}} \cdot \frac{7\cancel{a}}{\cancel{30}_{10}}$ Multiply and reduce.

$= \frac{1}{15} - \frac{7}{50}$ Simplify.

$= \frac{10}{150} - \frac{21}{150}$ Find the LCD.

$= -\frac{11}{150}$ Subtract.

Example 3: Order of Operations with Decimals (using Calculators)

Graphing calculators are programmed to use the rules for order of operations. Use your graphing calculator to find the value of each of the following expressions. Note that the x^2 key gives the exponent 2.

a. $1.5 \div 6 - 4 \cdot 3.2^2$

Solution:

```
1.5/6-4*3.2²
              ⁻40.71
```

b. $2.1\left(4.5^2 - 5^2\right) - 3(-1.6)^2$

Solution:

```
2.1(4.5²-5²)-3(-
1.6)²
             ⁻17.655
```

Continued on the next page...

c. $15.8-3.1\left(7^2-6.3^2\right)+5\cdot 2.5$

Solution:

```
15.8-3.1(7²-6.3²
)+5*2.5
              -.561
```

Practice Problems

Use the rules for order of operations to evaluate each expression.

1. $36\div 2\cdot 6$

2. $8\cdot 2+4(-2)-4^2$

3. $9\cdot 3\div\left(2^2-5\right)$

4. $\left(\frac{1}{6}\right)^2\div\frac{7}{12}-\frac{5}{8}$

5. $\left(-\frac{3}{4}\right)+\frac{3}{8}\cdot\frac{4}{5}\div\frac{2}{5}+\frac{2}{3}$

6. $-15\div\left(\frac{1}{3}+\frac{1}{10}\right)$

Simplify.

7. $\frac{2}{5a}\div\frac{16}{3a}-\frac{2b}{7}\cdot\frac{7}{5b}$

1.8 Exercises

Use the rules for order of operations to evaluate the expressions. (Change mixed numbers to improper fractions before evaluating.)

1. a. 36 b. 16
2. a. 8 b. 50
3. −25
4. 0
5. −10
6. −59
7. −45
8. 45
9. −137
10. −48
11. 152
12. −3

1. a. $24\div 4\cdot 6$

b. $24\cdot 4\div 6$

2. a. $20\div 5\cdot 2$

b. $20\cdot 5\div 2$

3. $15\div(-3)\cdot 3-10$

4. $20\cdot 2\div 2^2+5(-2)$

5. $3^2\div(-9)\cdot\left(4-2^2\right)+5(-2)$

6. $4^2\div(-8)(-2)+3\left(2^2-5^2\right)$

7. $14\cdot 3\div(-2)-6(4)$

8. $6(13-15)^2\cdot 8\div 2^2+3(-1)$

9. $-10+15\div(-5)\cdot 3^2-10^2$

10. $16\cdot 3\div\left(2^2-5\right)$

11. $2-5\left[(-20)\div(-4)\cdot 2-40\right]$

12. $9-6\left[(-21)\div 7\cdot 2-(-8)\right]$

Answers to Practice Problems: **1.** 108 **2.** −8 **3.** −27 **4.** $-\frac{97}{168}$ **5.** $\frac{2}{3}$ **6.** $-\frac{450}{13}$ **7.** $-\frac{13}{40}$

13. -189
14. -530
15. 143
16. 426
17. -36
18. 260
19. -10
20. 129

21. $\frac{11}{30}$ 22. $\frac{17}{35}$
23. $\frac{1}{24}$ 24. $-\frac{1}{21}$
25. $\frac{31}{24}$ 26. $\frac{49}{150}$
27. $\frac{107}{24}$ 28. $-\frac{107}{12}$
29. $\frac{41}{32}$ 30. $-\frac{33}{28}$
31. $\frac{3}{5}$ 32. $\frac{379}{60}$
33. $-\frac{341}{30}$ 34. $-\frac{37}{9}$
35. $-\frac{15}{17}$ 36. $\frac{23}{6}$
37. 24 38. -20
39. $\frac{7}{10}$ 40. $\frac{229}{540}$
41. $5\frac{1}{12}$ or $\frac{61}{12}$
42. $-\frac{5}{8}$
43. $-\frac{13}{64}$
44. $-\frac{112}{165}$

13. $(7-10)[49\div(-7)+20\cdot3-(-10)]$

14. $(9-11)[(-10)^2\cdot2+6(-5)^2-10^2+3\cdot5]$

15. $8-9[(-39)\div(-13)+7(-2)-(-2)^2]$

16. $6-20[(-15)\div3\cdot5+6\cdot2\div3]$

17. $|16-20|[32\div|3-5|-5^2]$

18. $|10-30|[4^2\cdot|5-8|\div(-2)^2+|17-18|]$

19. $(-10)+(-2)+|2-4|$

20. $|16-20|+(-10)^2+5^2$

21. $\frac{3}{8}\cdot\frac{4}{5}+\frac{1}{15}$

22. $\frac{1}{4}\cdot\frac{12}{15}+\frac{2}{7}$

23. $\frac{1}{3}\div\frac{1}{2}-\frac{5}{6}\cdot\frac{3}{4}$

24. $\frac{2}{9}\div\frac{14}{3}-\frac{1}{6}\cdot\frac{4}{7}$

25. $\left(\frac{5}{6}\right)^2\div\frac{5}{12}-\frac{3}{8}$

26. $\left(\frac{2}{5}\right)^2\cdot\frac{3}{8}+\frac{1}{5}\div\frac{3}{4}$

27. $\frac{7}{6}\cdot2^2-\frac{2}{3}\div3\frac{1}{5}$

28. $\frac{3}{4}\div3^2-4\left(\frac{3}{2}\right)^2$

29. $\left(-\frac{3}{4}\right)\div\left(-\frac{3}{5}\right)\cdot\frac{7}{8}+\frac{3}{16}$

30. $\left(-\frac{2}{3}\right)\div\frac{7}{12}-\frac{2}{7}+\left(-\frac{1}{2}\right)^2$

31. $\left(-\frac{9}{10}\right)+\frac{5}{8}\cdot\frac{4}{5}\div\frac{6}{10}+\frac{2}{3}$

32. $\frac{5}{8}\div\frac{1}{10}+\left(-\frac{1}{3}\right)^2\cdot\frac{3}{5}$

33. $-2\frac{1}{2}\cdot3\frac{1}{5}\div\frac{3}{4}-\frac{7}{10}$

34. $-\frac{5}{9}-\frac{1}{3}\cdot\frac{2}{3}-3\frac{1}{3}$

35. $\left(\frac{1}{2}-1\frac{3}{4}\right)\div\left(\frac{2}{3}+\frac{3}{4}\right)$

36. $\left(\frac{5}{8}+2\frac{1}{4}\right)\div\left(1-\frac{1}{4}\right)$

37. $-15\div\left(\frac{1}{4}-\frac{7}{8}\right)$

38. $-12\div\left(\frac{1}{2}+\frac{1}{10}\right)$

39. $\left(1\frac{1}{10}-3\frac{1}{5}\right)\div3\frac{1}{2}+1\frac{3}{10}$

40. $4\left(\frac{1}{2}\right)^2+\frac{7}{10}\div\frac{3}{5}\cdot\frac{5}{18}-\frac{9}{10}$

41. Find the average of the three numbers: $3\frac{2}{5}$, $5\frac{3}{4}$, and $6\frac{1}{10}$.

42. Find the average of the four numbers: $1\frac{1}{8}$, $2\frac{1}{2}$, $-5\frac{1}{2}$, and $-\frac{5}{8}$.

43. If the square of $\frac{7}{8}$ is subtracted from the square of $\frac{3}{4}$, what is the difference?

44. Find the quotient if the sum of $\frac{4}{5}$ and $\frac{2}{15}$ is divided by the difference between $\frac{7}{8}$ and $2\frac{1}{4}$.

45. 42.45
46. 115.72
47. 67.77
48. –2.505
49. 15.41
50. 3.956
51. $\frac{7}{2}$ 52. $-\frac{11}{12x}$
53. $-\frac{23}{21y}$ 54. $-\frac{24}{5y}$
55. $\frac{7}{5}$ 56. $-\frac{7b}{10a}$
57. $\frac{13}{36a}$ 58. $\frac{38}{27y}$
59. $(3^2-9)=0$ and division by 0 is undefined.

Use a graphing calculator to evaluate each of the expressions.

45. $3.4 \div 4 + 5 \cdot 8.32$

46. $8.1 \div 5 + 16.3 \cdot 7$

47. $0.75 \div 1.5 + 7 \cdot 3.1^2$

48. $1.05 \div (-3) \cdot 3.7 - 1.1^2$

49. $6.32 \cdot 8.4 \div 16.8 + 3.5^2$

50. $(82.7 + 16.2) \div (14.83 - 19.83)^2$

Use the rules for order of operations to simplify each of the following expressions.

51. $\frac{3}{4a} \div \frac{3}{16a} - \frac{2b}{3} \cdot \frac{3}{4b}$

52. $\frac{3}{4x} - \frac{5}{6} \div \frac{5x}{8} - \frac{1}{3x}$

53. $-\frac{5}{7y} - \frac{1}{2y} \cdot \frac{2}{3} - \frac{1}{21y}$

54. $\frac{6}{5y} + \frac{3}{2y} - \frac{5}{8} \div \frac{y}{12}$

55. $\left(\frac{7}{10x} + \frac{3}{5x}\right) \div \left(\frac{1}{2x} + \frac{3}{7x}\right)$

56. $\left(\frac{1}{5a} - \frac{2}{3a}\right) \div \left(\frac{7}{7b} - \frac{1}{3b}\right)$

57. $\frac{3}{4a} \cdot \left(\frac{1}{3}\right)^2 + \frac{5}{2} \cdot \frac{1}{9a}$

58. $\frac{1}{24y} \cdot \left(\frac{4}{3}\right)^2 + \left(\frac{2}{3}\right)^2 \div \frac{y}{3}$

Writing and Thinking About Mathematics

60. Smaller; When any positive number is multiplied by a fraction (or decimal) between 0 and 1, the result will be smaller. This is what is happening when a number between 0 and 1 is squared. Answers will vary.

59. Explain, in your own words, why the following expression cannot be evaluated.

$$(24 - 2^4) + 6(3 - 5) \div (3^2 - 9)$$

60. Consider any number between 0 and 1. If you square this number, will the result be larger or smaller than the original number? Is this always the case? Explain.

61. Consider any number between –1 and 0. If you square this number, will the result be larger or smaller than the original number? Is this always the case? Explain.

HAWKES LEARNING SYSTEMS: INTRODUCTORY & INTERMEDIATE ALGEBRA SOFTWARE

- 1.8 Order of Operations

61. It will always be larger. If you square a negative number you get a positive number. All positive numbers are larger than all negative numbers.

1.9 Properties of Real Numbers

- *Apply the properties of real numbers to complete statements.*
- *Name the real number properties that justify given statements.*

Teaching Notes:
At some point, it may be important to show how this property of addition helps overcome the failure of subtraction to be commutative - e.g., that

$$5-3=5+(-3)$$
$$=-3+5.$$

In this chapter we have discussed real numbers, the order of real numbers, and the placement of real numbers on a real number line as well as operations with real numbers. As we continue to work with all types of real numbers and algebraic expressions, we will need to understand that while certain properties are true for addition and multiplication these same properties are not true for subtraction and division. For example,

the order of the numbers in addition ***does not*** change the result:

$$17+5=5+17=22$$

but, the order of the numbers in subtraction ***does*** change the result:

$$17-5=12 \quad \text{but} \quad 5-17=-12.$$

We say that addition is **commutative** while subtraction is **not commutative**. The various properties of real numbers under the operations of addition and multiplication are summarized here. These properties are used throughout algebra and mathematics in developing formulas and general concepts.

Properties of Addition and Multiplication

In this table a, b, and c are real numbers.

For Addition	Name of Property	For Multiplication
$\boldsymbol{a+b=b+a}$ $3+6=6+3$	Commutative property	$\boldsymbol{ab=ba}$ $4\cdot 9=9\cdot 4$
$\boldsymbol{(a+b)+c=a+(b+c)}$ $(2+5)+4=2+(5+4)$	Associative property	$\boldsymbol{a(bc)=(ab)c}$ $6\cdot(2\cdot 7)=(6\cdot 2)\cdot 7$
$\boldsymbol{a+0=0+a=a}$ $20+0=0+20=20$	Identity	$\boldsymbol{a\cdot 1=1\cdot a=a}$ $-2\cdot 1=1\cdot(-2)=-2$

Continued on the next page...

Properties of Addition and Multiplication (Cont.)

Name of Property

For Addition		**For Multiplication**
$\boldsymbol{a+(-a)=0}$	Inverse	$\boldsymbol{a\cdot\frac{1}{a}=1}$ **(for** $\boldsymbol{a\neq 0}$**)**
$10+(-10)=0$		$3\cdot\frac{1}{3}=1$

Zero-Factor Law

$\boldsymbol{a\cdot 0=0\cdot a=0}$ $\qquad -5\cdot 0=0\cdot(-5)=0$

Distributive Property of Multiplication over Addition

$\boldsymbol{a(b+c)=ab+ac}$ $\qquad 3(x+5)=3\cdot x+3\cdot 5$

NOTES

The number 0 is called the **additive identity** because when 0 is added to a number the result is the same number. Likewise, the number 1 is called the **multiplicative identity** because when a number is multiplied by 1 the result is the same number. Also, the **additive inverse** of a number is its **opposite** and the **multiplicative inverse** of a number is its **reciprocal**.

The raised dot is optional when indicating multiplication between two variables or a number and a variable. For example,

$$x\cdot y=xy \quad \text{and} \quad 5\cdot x=5x.$$

In the case of a number and a variable, the number is called the **coefficient** of the variable. So in

$5x$ the number 5 is the **coefficient** of x.

Thus, in an illustration of the distributive property involving a variable, we can write

$$5(x-6)=5\cdot x-5\cdot 6=5x-30.$$

Example 1: Properties of Addition and Multiplication

State the name of each property being illustrated.

a. $(-7)+13=13+(-7)$

Solution: Commutative property of addition

b. $8+(9+1)=(8+9)+1$

Solution: Associative property of addition

c. $(-25)\cdot 1=-25$

Solution: Multiplicative identity

d. $3(x+y)=3x+3y$

Solution: Distributive property

e. $4\cdot(3\cdot 2)=(4\cdot 3)\cdot 2$

Solution: Associative property of multiplication

Example 2: Properties of Addition and Multiplication

In each of the following equations, state the property illustrated and show that the statement is true for the value given for the variable by substituting the value in the equation and evaluating.

a. $x+14=14+x$ given that $x=-4$

Solution: The commutative property of addition is illustrated.

$$(-4)+14=10 \text{ and } 14+(-4)=10$$

b. $(3\cdot 6)x=3(6x)$ given that $x=5$

Solution: The associative property of multiplication is illustrated.

$$(3\cdot 6)\cdot 5=18\cdot 5=90 \text{ and } 3\cdot(6\cdot 5)=3\cdot 30=90$$

c. $12(y+3)=12y+36$ given that $y=-2$

Solution: The distributive property is illustrated.

$$12(-2+3)=12(1)=12 \text{ and } 12(-2)+36=-24+36=12$$

Practice Problems

Determine the property being illustrated.

1. $(-2\cdot5)\cdot2=-2\cdot(5\cdot2)$
2. $15\cdot0=0\cdot15=0$
3. $2+7=7+2$
4. $2(y+5)=2y+10$

1.9 Exercises

1. $3+7$
2. $6\cdot(9\cdot3)$
3. $4\cdot19$
4. $5+18$
5. $30+48$
6. $(16+9)+11$
7. $(2\cdot3)\cdot x$
8. $3x+15$
9. $(3+x)+7$
10. $9x+45$
11. 0
12. 6
13. $x+7$
14. 0
15. $2x-24$
16. 0
17. 0
18. 1
19. Commutative property of addition
20. Commutative property of multiplication
21. Multiplicative identity
22. Additive identity
23. Associative property of addition

Complete the expressions using the given property.

1. $7+3=$ ______ Commutative property of addition
2. $(6\cdot9)\cdot3=$ ______ Associative property of multiplication
3. $19\cdot4=$ ______ Commutative property of multiplication
4. $18+5=$ ______ Commutative property of addition
5. $6(5+8)=$ ______ Distributive property
6. $16+(9+11)=$ ______ Associative property of addition
7. $2\cdot(3x)=$ ______ Associative property of multiplication
8. $3(x+5)=$ ______ Distributive property
9. $3+(x+7)=$ ______ Associative property of addition
10. $9(x+5)=$ ______ Distributive property
11. $6\cdot0=$ ______ Zero-factor law
12. $6\cdot1=$ ______ Multiplicative identity
13. $0+(x+7)=$ ______ Additive identity
14. $0\cdot(-13)=$ ______ Zero-factor law

Answers to Practice Problems: **1.** Associative property of multiplication **2.** Zero-factor law **3.** Commutative property of addition **4.** Distributive property

24. Commutative property of addition
25. Commutative property of multiplication
26. Distributive property
27. Commutative property of multiplication
28. Associative property of multiplication
29. Multiplicative inverse
30. Multiplicative inverse
31. Additive inverse
32. Additive inverse
33. Multiplicative identity
34. Associative property of multiplication
35. Zero-factor law
36. Additive identity
37. Associative property of addition
38. Distributive property
39. $6(11)=66$ and $6\cdot 3+6\cdot 8=66$
40. $7(3)=21$ and $7\cdot 8-7\cdot 5=21$
41. $10(-7)=-70$ and $10\cdot 2-10\cdot 9=-70$
42. $13(8)=104$ and $13\cdot 5+13\cdot 3=104$
43. Commutative property of multiplication; $6\cdot 4=4\cdot 6=24$

15. $2(x-12)=$ ______ Distributive property

16. $(-5)+5=$ ______ Additive inverse

17. $6.3+(-6.3)=$ ______ Additive inverse

18. $3\cdot\frac{1}{3}=$ ______ Multiplicative inverse

Name the property of real numbers illustrated.

19. $5+16=16+5$

20. $5\cdot 16=16\cdot 5$

21. $32\cdot 1=32$

22. $32+0=32$

23. $5+(3+1)=(5+3)+1$

24. $5+(3+1)=(3+1)+5$

25. $13(y+2)=(y+2)\cdot 13$

26. $13(y+2)=13y+26$

27. $6(2\cdot 9)=(2\cdot 9)\cdot 6$

28. $6(2\cdot 9)=(6\cdot 2)\cdot 9$

29. $5\cdot\frac{1}{5}=1$

30. $14\cdot\frac{1}{14}=1$

31. $7.1+(-7.1)=0$

32. $(-9)+9=0$

33. $1\cdot 14.2=14.2$

34. $(5\cdot 3)\cdot -7=5(3\cdot -7)$

35. $5.68\cdot 0=0\cdot 5.68=0$

36. $0+5.68=5.68$

37. $2+(x+6)=(2+x)+6$

38. $2(x+6)=2x+12$

First evaluate each expression using the rules for order of operations and then use the distributive property to evaluate the same expression. The value must be the same.

39. $6(3+8)$

40. $7(8-5)$

41. $10(2-9)$

42. $13(5+3)$

In each of the following equations, state the property illustrated and show that the statement is true for the value of $x=4$, $y=-2$, or $z=3$ by substituting the corresponding value in the equation and evaluating.

43. $6\cdot x=x\cdot 6$

44. $19+z=z+19$

45. $8+(5+y)=(8+5)+y$

46. $(2\cdot 7)\cdot x=2\cdot(7x)$

47. $5(x+18)=5x+90$

48. $(2z+14)+3=2z+(14+3)$

44. Commutative property of addition; $19 + 3 = 3 + 19 = 22$

49. $(6\cdot y)\cdot 9 = 6\cdot(y\cdot 9)$

50. $11\cdot x = x\cdot 11$

51. $z+(-34) = -34+z$

52. $3(y+15) = 3y+45$

53. $2(3+x) = 2(x+3)$

54. $(y+2)(y-4) = (y-4)(y+2)$

55. $5+(x-15) = (x-15)+5$

56. $z+(4+x) = (4+x)+z$

57. $(3x)\cdot 5 = 3\cdot(x\cdot 5)$

58. $(x+y)+z = x+(y+z)$

Writing and Thinking About Mathematics

45. Associative property of addition; $8+(5+(-2)) = (8+5)+(-2) = 11$

59. a. The distributive property illustrated as $a(b+c) = ab+ac$ is said to "distribute multiplication over addition." Explain, in your own words, the meaning of this phrase.

b. What would an expression that "distributes addition over multiplication" look like? Explain why this would or would not make sense.

HAWKES LEARNING SYSTEMS: INTRODUCTORY & INTERMEDIATE ALGEBRA SOFTWARE

- 1.9 Properties of Real Numbers

46. Associative property of multiplication; $(2\cdot 7)\cdot 4 = 2\cdot(7\cdot 4) = 56$
47. Distributive property; $5(4+18) = 5(4)+90 = 110$
48. Associative property of addition; $(2(3)+14)+3 = 2(3)+(14+3) = 23$
49. Associative property of multiplication; $(6\cdot(-2))\cdot 9 = 6\cdot(-2\cdot 9) = -108$
50. Commutative property of multiplication; $11\cdot 4 = 4\cdot 11 = 44$
51. Commutative property of addition; $3+(-34) = (-34)+3 = -31$
52. Distributive property; $3(-2+15) = 3(-2)+45 = 39$
53. Commutative property of addition; $2(3+4) = 2(4+3) = 14$
54. Commutative property of multiplication; $(-2+2)(-2-4) = (-2-4)(-2+2) = 0$
55. Commutative property of addition; $5+(4-15) = (4-15)+5 = -6$
56. Commutative property of addition; $3+(4+4) = (4+4)+3 = 11$
57. Associative property of multiplication; $(3\cdot 4)\cdot 5 = 3\cdot(4\cdot 5) = 60$
58. Associative property of addition; $(4+(-2))+3 = 4+(-2+3) = 5$
59. a. Multiply each term that is part of the sum (in the parentheses) by a. b. An expression that "distributes addition over multiplication" would look like $a+(bc) = (a+b)(a+c)$, however this is not a true statement. Answers will vary.

Chapter 1 Index of Key Ideas and Terms

Section 1.1 The Real Number Line and Absolute Value

Types of Numbers

Counting numbers (or **natural numbers**) page 2

$\mathbb{N} = \{1, 2, 3, 4, 5, 6, 7, 8, 9, 10, 11, \ldots\}$

Whole numbers page 2

$\mathbb{W} = \{0, 1, 2, 3, 4, 5, 6, 7, 8, 9, 10, 11, \ldots\}$

Integers page 4

Integers: $\mathbb{Z} = \{\ldots, -3, -2, -1, 0, 1, 2, 3, \ldots\}$

Positive integers: $\{1, 2, 3, 4, 5, \ldots\}$

Negative integers: $\{\ldots, -4, -3, -2, -1\}$

The integer 0 is neither positive nor negative.

Rational numbers pages 5-6

A **rational number** is a number that can be written in the form $\frac{a}{b}$ where a and b are integers and $b \neq 0$.

OR

A **rational number** is a number that can be written in decimal form as a terminating decimal or as an infinite repeating decimal.

Irrational numbers page 6

Irrational numbers are numbers that can be written as infinite nonrepeating decimals.

Real numbers page 6

The **real numbers** consist of all rational and irrational numbers.

Diagram of Types of Numbers page 7

Variables page 5

A **variable** is a symbol (generally a letter of the alphabet) that is used to represent an unknown number (or numbers).

Equality and Inequality Symbols pages 8-9

Read from left to right:

$=$ "is equal to"

$\neq$ "is not equal to"

$<$ "is less than"

$>$ "is greater than"

$\leq$ "is less than or equal to"

$\geq$ "is greater than or equal to"

Note: Inequality symbols may be read from left to right or from right to left.

Continued on the next page...

Section 1.1 The Real Number Line and Absolute Value (cont.)

Absolute Value page 10

The **absolute value** of a real number is its distance from 0. Symbolically,

$|a| = a$ if a is a positive number or 0.

$|a| = -a$ if a is a negative number.

Note: The absolute value of a number is never negative.

Section 1.2 Addition with Real Numbers

Addition with Real Numbers pages 16-17

1. To add two real numbers with **like signs**,
   a. add their absolute values and
   b. use the common sign.
2. To add two real numbers with **unlike signs**,
   a. subtract their absolute values (the smaller from the larger), and
   b. use the sign of the number with the larger absolute value.

Section 1.3 Subtraction with Real Numbers

Additive Inverse page 21

The **opposite** of a real number is called its **additive inverse**. The sum of a number and its additive inverse is zero. Symbolically, for any real number a, $a + (-a) = 0$.

Subtraction with Real Numbers pages 22-24

For any real numbers a and b, $a - b = a + (-b)$.
In words,

to subtract b from a, add the opposite of b to a.

Change in Value and Net Change pages 25-26

To find the **change in value** between two numbers, take the end value and subtract the beginnning value. The **net change** in a measure is the algebraic sum of several numbers.

Section 1.4 Multiplication and Division with Real Numbers

For positive real numbers a and b,

1. The product of two positives is positive:
$$(a)(b) = +ab$$

2. The product of two negatives is positive:
$$(-a)(-b) = +ab$$

3. The product of a positive and a negative is negative:
$$a(-b) = -ab$$

4. The product of 0 and any number is 0:
$$a \cdot 0 = 0$$

For real numbers a, b, and x $(\text{where } b \neq 0)$,

$\frac{a}{b} = x$ means that $a = b \cdot x$.

For real numbers a and b $(\text{where } b \neq 0)$,

$\frac{a}{0}$ is undefined and $\frac{0}{b} = 0$.

If a and b are positive real numbers

1. The quotient of two positive numbers is positive:
$$\frac{a}{b} = +\frac{a}{b}$$

2. The quotient of two negative numbers is positive:
$$\frac{-a}{-b} = +\frac{a}{b}$$

3. The quotient of a positive number and a negative number is negative: $\frac{-a}{b} = -\frac{a}{b}$ and $\frac{a}{-b} = -\frac{a}{b}$

The **average** (or **mean**) of a set of numbers is the value found by adding the numbers in the set and then dividing the sum by the number of numbers in the set.

Section 1.5 Exponents, Prime Numbers, and LCM

Exponent and Base pages 42-43

A whole number n is an **exponent** if it is used to tell how many times another whole number a is used as a factor. The repeated factor a is called the **base** of the exponent.

$$\underbrace{a \cdot a \cdot a \cdot \ldots \cdot a \cdot a}_{n \text{ factors}} = a^n \quad \leftarrow \text{exponent } (n), \text{ base } (a)$$

Prime Numbers page 44

A **prime number** is a counting number greater than 1 that has exactly two different factors (or divisors), namely 1 and itself.

Composite Numbers page 44

A **composite number** is a counting number with more than two different factors (or divisors).

Even and Odd Whole Numbers page 45

If a whole number is divisible by 2, it is **even**.

If a whole number is not divisible by 2, it is **odd**.

Prime Factorization of Composite Numbers page 46

To Find the Prime Factorization of a Composite Number

1. Factor the composite number into any two factors.
2. Factor each factor that is not prime into two more factors.
3. Continue this process until all factors are prime.

The **prime factorization** is the product of all the prime factors.

Note: There is only one prime factorization for any composite number.

Least Common Multiple (LCM) pages 48-49

To Find the LCM of a Set of Counting Numbers

1. Find the prime factorization of each number.
2. List the prime factors that appear in any one of the prime factorizations.
3. Find the product of these primes using each prime the greatest number of times it appears in any one of the prime factorizations.

Section 1.6 Multiplication and Division with Fractions

Rational Numbers page 54

Uses for Fractions page 55

In general, fractions can be used to indicate:

1. Equal parts of a whole, or
2. Division

Placement of Negative Signs in Fractions page 55

If a and b are real numbers, and $b \neq 0$, then

$$-\frac{a}{b} = \frac{-a}{b} = \frac{a}{-b}.$$

Multiplication with Fractions page 56

To **multiply** two fractions, multiply the numerators and multiply the denominators.

$$\frac{a}{b} \cdot \frac{c}{d} = \frac{a \cdot c}{b \cdot d} \text{ where } b, d \neq 0$$

The Fundamental Principle of Fractions page 56

$$\frac{a}{b} = \frac{a \cdot k}{b \cdot k} \text{ where } b, k \neq 0$$

To Build a Fraction to Higher Terms page 57

To build a fraction to higher terms, use the fundamental principle and multiply both the numerator and denominator by the same nonzero number.

To Reduce a Fraction page 57

To reduce a fraction, factor both the numerator and denominator and use the fundamental principle to "divide out" any common factors. If the numerator and the denominator have no common prime factors, the fraction has been reduced to **lowest terms**.

Reciprocal page 58

If $a \neq 0$ and $b \neq 0$, the **reciprocal** of $\frac{a}{b}$ is $\frac{b}{a}$ and $\frac{a}{b} \cdot \frac{b}{a} = 1$.

Division with Fractions page 59

To **divide** by a nonzero fraction, multiply by its reciprocal:

$$\frac{a}{b} \div \frac{c}{d} = \frac{a}{b} \cdot \frac{d}{c} \text{ where } b, c, d \neq 0$$

Section 1.7 Addition and Subtraction with Fractions

Adding (or Subtracting) Fractions with Like Denominators pages 65, 68

To **add** (or **subtract**) two (or more) fractions with like denominators, add (or subtract) the numerators and use the common denominator.

$$\frac{a}{b}+\frac{c}{b}=\frac{a+c}{b}$$ where b is a nonzero number or algebraic expression

or

$$\frac{a}{b}-\frac{c}{b}=\frac{a-c}{b}$$ where b is a nonzero number or algebraic expression

Adding (or Subtracting) Fractions with Unlike Denominators pages 66, 68

1. Find the LCM of the denominators. (This is called the **least common denominator** (**LCD**) and may involve variables.)
2. Change each fraction to an equivalent fraction with the LCM as the denominator.
3. Add (or subtract) the new fractions. Reduce if possible.

Section 1.8 Order of Operations

Rules for Order of Operations page 73

1. Simplify within grouping symbols, such as parentheses ( ), brackets [ ], and braces { }, working from the inner most grouping outward.
2. Find any powers indicated by exponents.
3. Moving from **left to right**, perform any multiplications **or** divisions **in the order they appear.**
4. Moving from **left to right**, perform any additions **or** subtractions **in the order they appear.**

Section 1.9 Properties of Real Numbers

Properties of Addition and Multiplication

pages 81-82

For Addition	Name of Property	For Multiplication
$a+b=b+a$	Commutative property	$ab=ba$
$(a+b)+c=a+(b+c)$	Associative property	$a(bc)=(ab)c$
$a+0=0+a=a$	Identity	$a\cdot 1=1\cdot a=a$
$a+(-a)=0$	Inverse	$a\cdot\frac{1}{a}=1$ (for $a\neq 0$)

Zero-Factor Law

$$a\cdot 0=0\cdot a=0$$

Distributive Property of Multiplication over Addition

$$a(b+c)=ab+ac$$

HAWKES LEARNING SYSTEMS: INTRODUCTORY & INTERMEDIATE ALGEBRA SOFTWARE

- 1.1a The Real Number Line and Inequalities
- 1.1b Introduction to Absolute Values
- 1.2 Addition with Real Numbers
- 1.3 Subtraction with Real Numbers
- 1.4 Multiplication and Division with Real Numbers
- 1.5 Exponents, Prime Numbers, and LCM
- 1.6a Reducing Fractions
- 1.6b Multiplication and Division with Fractions
- 1.7 Addition and Subtraction with Fractions
- 1.8 Order of Operations
- 1.9 Properties of Real Numbers

Chapter 1 Review

1.1 The Real Number Line and Absolute Value

1. Given the set of numbers $\left\{-9.3, -2, -1.343434..., -\frac{3}{4}, 0, \sqrt{2}, \frac{2}{1}, \frac{10}{3}\right\}$, tell which numbers are

a. Natural numbers **b.** Integers

c. Rational numbers **d.** Irrational numbers

Graph each set of real numbers on a real number line.

2. $\{-3, 0, 3, 4\}$ **3.** $\left\{-4, -1, -\frac{1}{4}, 2.5\right\}$

4. All positive integers less than or equal to 5.

5. All integers greater than –2.

Find the value of each absolute value expression.

6. $|-6|$ **7.** $-|7.3|$ **8.** $-|-4.1|$

Determine whether each statement is true or false. If a statement is false, rewrite it in a form that is a true statement. (There may be more than one way to correct a statement.)

9. $-14 \le 14$ **10.** $|-1.5| \ge 1.5$ **11.** $|-2.3| < 0$ **12.** $7 = |-7|$

List the possible values for x for each statement. Graph the numbers on a real number line.

13. $|x| = 5$ **14.** $|x| = -5$ **15.** $3.5 = |x|$

On a real number line, graph the integers that satisfy the conditions stated.

16. $|x| \le 3$ **17.** $|x| > 3$ **18.** $x = |x|$

1.2 Addition with Real Numbers

Simplify the expressions.

19. $6+9$ **20.** $15+7$ **21.** $8+(-3)$ **22.** $(-9)+6$

23. $18+(-18)$ **24.** $75+(-75)$ **25.** $-2.3+(-1.1)$ **26.** $-0.7+1.4$

27. $-12+(-1)+(-5)$ **28.** $-6.5+(-4.2)+4.7$

1. a. $\frac{2}{1}$
b. $-2, 0, \frac{2}{1}$
c. $-9.3, -2, -1.343434..., -\frac{3}{4}, 0, \frac{2}{1}, \frac{10}{3}$
d. $\sqrt{2}$

2. 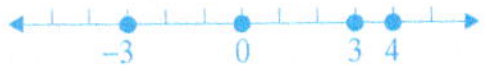

3. 

4. 

5. 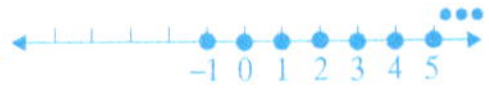

6. 6
7. –7.3
8. –4.1
9. True
10. True
11. False; $|-2.3| > 0$
12. True

13. 

14. No solution

15. 

16. 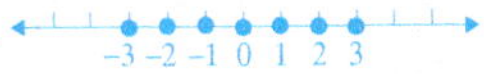

17. 

18. 

19. 15
20. 22
21. 5
22. –3
23. 0
24. 0
25. –3.4
26. 0.7
27. –18
28. –6

29. 15
30. –5
31. –35
32. –38
33. –18
34. 4
35. 6
36. –5.4
37. –7
38. –2
39. –19
40. –8
41. –8
42. –17
43. 0.9
44. 0.7
45. –4
46. –31
47. 2.1
48. –23.8
49. –1
50. –41
51. –33°F
52. –6000 feet
53. –35
54. –25.6
55. 90
56. 91
57. 24
58. 0
59. 5
60. 17

29. $\begin{array}{r} -8 \\ \underline{23} \end{array}$ **30.** $\begin{array}{r} 13 \\ \underline{-18} \end{array}$ **31.** $\begin{array}{r} -22 \\ \underline{-13} \end{array}$ **32.** $\begin{array}{r} -6 \\ \underline{-32} \end{array}$

33. Add –6 and –12. **34.** Find the sum of 1, –4, and 7.

1.3 Subtraction with Real Numbers

Find the additive inverse for each integer.

35. –6 **36.** 5.4

Simplify the expressions.

37. –4 – 3 **38.** 6 – 8 **39.** –15 – 4 **40.** 10 – 18

Subtract as indicated.

41. $\begin{array}{r} 32 \\ \underline{-(40)} \end{array}$ **42.** $\begin{array}{r} -24 \\ \underline{-(-7)} \end{array}$ **43.** $\begin{array}{r} -1.6 \\ \underline{-(-2.5)} \end{array}$ **44.** $\begin{array}{r} -2.9 \\ \underline{-(-3.6)} \end{array}$

Find the net change in value of each expression by performing the indicated operations.

45. $-7+(-2)-(-5)$ **46.** $-10+(-7)-14$ **47.** $2.5-(1.7)-(-1.3)$ **48.** $9.8-1.6-(32)$

49. Find the sum of 16 and –30. Then subtract –13.

50. Find the sum of –13 and –20. Then subtract 8.

51. Temperature: Find the change in temperature if the temperature drops from 30°F at noon to –3°F at midnight.

52. Airplanes: An airplane drops in altitude from 30,000 feet to 24,000 feet. What is the change in altitude in terms of signed numbers?

1.4 Multiplication and Division with Real Numbers

Find the value of each of the following expressions.

53. $(-5)\cdot 7$ **54.** $8(-3.2)$ **55.** $(-9)(-10)$ **56.** $(-7)(-13)$

57. $-4\cdot 3\cdot(-2)$ **58.** $8(-9)\cdot 0$ **59.** $\frac{-10}{-2}$ **60.** $\frac{-51}{-3}$

61. Undefined
62. 0
63. −0.4
64. −40
65. 62
66. 85.6
67. 65.8 inches
68. 8 hurricanes
69. $3^3 \cdot 5^2$
70. $2^4 \cdot 11^2 \cdot 13$
71. $5^2 \cdot 7$
72. $2 \cdot 7 \cdot 11$
73. $3 \cdot 5 \cdot 13$
74. $3^2 \cdot 5^2 \cdot 11$
75. 1890
76. 120
77. 120
78. 462
79. $270x^2y$
80. $468a^2bc^2$

61. $\frac{-16}{0}$ **62.** $\frac{0}{-16}$ **63.** $\frac{3.2}{-8}$ **64.** $\frac{100}{-2.5}$

65. Find the mean of the following set of integers: 30, 25, 70, 85, and 100.

66. Exam scores: In a statistics class, two students scored 82 and three students scored 88. What was the average score for these five students?

67. Height: The heights of twenty women were recorded as shown in the following frequency distribution. Find the mean height (to the nearest tenth of an inch) for these women.

Height (in inches)	60	61	63	66	70	71
Frequency	2	1	6	4	4	3

68. Hurricanes: The bar graph shows the number of hurricanes during the Atlantic Hurricane Seasons from 2004 – 2009. What was the average number of hurricanes per season in these years? (Round your answer to the nearest integer.)

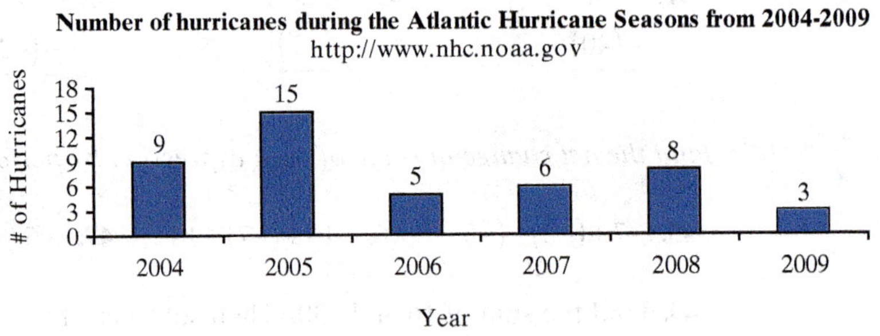

1.5 Exponents, Prime Numbers, and LCM

Rewrite the following products using exponents.

69. $3 \cdot 3 \cdot 3 \cdot 5 \cdot 5$ **70.** $2 \cdot 2 \cdot 2 \cdot 2 \cdot 11 \cdot 11 \cdot 13$

Find the prime factorization of each of the numbers.

71. 175 **72.** 154 **73.** 195 **74.** 2475

Find the LCM of each of the following sets of numbers and/or expressions.

75. $\{27, 30, 35\}$ **76.** $\{15, 24, 40\}$

77. $\{8, 10, 20, 60\}$ **78.** $\{11, 22, 66, 77\}$

79. $\{10x, 27x^2, 30xy\}$ **80.** $\{18abc, 36abc^2, 78a^2bc\}$

81. $-\frac{3}{16}$ 82. $-\frac{1}{12}$

83. $\frac{2}{5}$ 84. 1

85. $-\frac{1}{5}$ 86. $\frac{3}{10}$

87. $\frac{4}{11}$ 88. $\frac{7}{12}$

89. 12

90. $\frac{3}{5}$ was spent; $\frac{2}{5}$ remains

91. $\frac{9}{7}$

92. $-\frac{3}{16}$

93. $\frac{53}{60}$

94. $\frac{7}{24}$

95. $\frac{2}{5}$

96. $\frac{5}{16}$

97. $\frac{7}{8}$

98. 0

99. $\frac{17}{3a}$

100. $-\frac{49}{5x}$

101. $\frac{3}{14x}$

102. $\frac{5}{12x}$

103. a. $\frac{28}{135}$ b. 1120 species

104. 2790 miles

1.6 Multiplication and Division with Fractions

Multiply or divide as indicated and reduce each answer to lowest terms.

81. $\frac{-7}{8}\cdot\frac{3}{14}$ **82.** $\frac{-5}{16}\cdot\frac{4}{15}$ **83.** $\frac{6}{31}\cdot\frac{62}{9}\cdot\frac{3}{10}$

84. $\frac{15}{27}\div\frac{5}{9}$ **85.** $\frac{9}{15}\div(-3)$ **86.** $\frac{6x}{32}\div\frac{5x}{8}$

87. Find $\frac{1}{2}$ of $\frac{8}{11}$. **88.** Find $\frac{2}{3}$ of $\frac{21}{24}$.

89. Presidents: $\frac{3}{11}$ of the U.S. presidents have served two full terms as president. If there have been 44 presidents, how many presidents have served two full terms?

90. Fast food: If you have $20 and spend $12 on a soft drink and hamburger, what fraction of your money did you spend? What fraction of your money do you still have?

1.7 Addition and Subtraction with Fractions

Perform the indicated operations and reduce each answer to lowest terms.

91. $\frac{3}{7}+\frac{6}{7}$ **92.** $\frac{9}{32}-\frac{15}{32}$ **93.** $\frac{5}{12}+\frac{7}{15}$ **94.** $\frac{1}{6}+\frac{1}{8}$

95. $1-\frac{3}{5}$ **96.** $1-\frac{11}{16}$ **97.** $\frac{1}{5}+\frac{3}{8}+\frac{3}{10}$ **98.** $\frac{1}{28}+\frac{8}{21}-\frac{5}{12}$

99. $\frac{5}{a}+\frac{2}{3a}$ **100.** $\frac{1}{5x}-\frac{10}{x}$ **101.** $\frac{2}{7x}-\frac{1}{14x}$ **102.** $\frac{3}{4x}-\frac{1}{3x}$

103. Zoology: There are 5400 species of mammals on earth. $\frac{5}{27}$ of these mammals are aerial mammals and $\frac{1}{45}$ of them are marine mammals.

a. What fraction of mammals are aerial mammals or marine mammals?

b. How many species of mammals fit into these categories?

104. Travel: Adam is driving from New York City to Los Angeles. He has driven 2232 miles over the past week and is $\frac{4}{5}$ of the way to his final destination. How far is Los Angeles from New York City?

105. 88 106. 16
107. 18 108. −4
109. 26 110. 32
111. $\frac{49}{36}$ 112. $-\frac{9}{2}$
113. $-\frac{1}{360}$
114. $\frac{31}{28}$
115. $\frac{5}{4x}$
116. $\frac{1}{7y}$
117. $-\frac{11}{81}$
118. $-\frac{9}{8}$
119. Commutative property of addition
120. Associative property of addition
121. Additive identity
122. Additive inverse
123. Associative property of multiplication
124. Commutative property of multiplication
125. Multiplicative identity
126. Multiplicative inverse
127. Distributive property
128. Distributive property
129. 238.43
130. −74.65
131. −40,473.75
132. 5.1252

1.8 Order of Operations

Use the rules for order of operations to evaluate (and simplify) each expression.

105. $16 \div 2 \cdot 8 + 24$

106. $32 - 8 \cdot 14 \div 7$

107. $-36 \div (-2)^2 + 25 - 2(16 - 17)$

108. $9(-1 + 2^2) - 7 - 6 \cdot 2^2$

109. $2(-1 + 5^2) - 4 - 2 \cdot 3^2$

110. $10 - 2[(19 - 14) \div 5 + 3(-4)]$

111. $\frac{1}{9} + \frac{1}{2} \div \frac{1}{6} \cdot \frac{5}{12}$

112. $\left(\frac{1}{2} - 2\right) \div \left(1 - \frac{2}{3}\right)$

113. $\left(\frac{1}{15} - \frac{1}{12}\right) \div 6$

114. $\frac{2}{3} \div \frac{7}{12} - \frac{2}{7} + \left(\frac{1}{2}\right)^2$

115. $\frac{3}{2x} - \frac{1}{4x}$

116. $\frac{4}{7y} - \frac{3}{14y} \div \frac{1}{2}$

117. If the square of $\frac{2}{3}$ is subtracted from the square of $-\frac{5}{9}$, what is the difference?

118. If the product of $\frac{3}{5}$ and $\frac{3}{4}$ is divided by the product of $\frac{4}{5}$ and $-\frac{1}{2}$, what is the quotient?

1.9 Properties of Real Numbers

Name the property of real numbers illustrated.

119. $9 + 15 = 15 + 9$

120. $8 + (3 + 2) = (8 + 3) + 2$

121. $7 + 0 = 7$

122. $6 + (-6) = 0$

123. $2(3 \cdot 4) = (2 \cdot 3) \cdot 4$

124. $8(-7) = -7(8)$

125. $9.2 \cdot 1 = 9.2$

126. $19 \cdot \frac{1}{19} = 1$

127. $3(2 + x) = 6 + 3x$

128. $5(x - 7) = 5x - 35$

Calculator Problems

Use a graphing calculator to find the value of each of the following expressions (if necessary, round answers to four decimal places).

129. $100.9 - (-32.8) + 104.73$

130. $-76.35 - (-14.7) + (-13)$

131. $93.75(17.2)(-25.1)$

132. $83.54 \div 16.3$

Chapter 1 Test

1. a. $-5, -1, 0$
   b. $-5, -1, -\frac{1}{3}, 0, \frac{1}{2}, 3\frac{1}{4}, 7.121212...$
   c. $-\pi$
   d. All are real numbers
2. a. True, since for any integer n, $n = \frac{n}{1}$ which is a rational number since the numerator and the denominator are both integers.
   b. False, because rational numbers also include non-integers, such as $\frac{2}{3}$.
3. a. < b. > c. =
4. a. $\{-2, -1, 0, 1, 2\}$
   b. $\{..., -11, -10, -9, 9, 10, 11, ...\}$
5. -13
6. 23
7. 112
8. 0
9. -3
10. -0.3
11. -60
12. -22.5
13. 21.7 mph
14. $3^3 \cdot 5 \cdot 7^2$
15. $2^4 \cdot 3^2$
16. $90x^2y^2$
17. $-\frac{9}{2}$ 18. 8
19. $\frac{9}{10}$ 20. $-\frac{2}{y}$

1. Given the set of numbers $\left\{-5, -\pi, -1, -\frac{1}{3}, 0, \frac{1}{2}, 3\frac{1}{4}, 7.121212...\right\}$, tell which numbers are

a. Integers **b.** Rational numbers
c. Irrational numbers **d.** Real numbers

2. a. True or false: All integers are rational numbers. Explain your answer.
b. True or false: All rational numbers are integers. Explain your answer.

3. Fill in the blanks with the proper symbol: <, >, or =.

a. -4 _____ -2 **b.** $-(-2)$ _____ 0 **c.** $|-9|$ _____ $|9|$

4. List the integers that satisfy the conditions stated.

a. $|x| < 3$ **b.** $|x| \geq 9$

Perform the indicated operations.

5. $15 - 45 + 17$ **6.** $13 - (-16) + (-6)$ **7.** $(-7)(-16)$

8. $41(-5)(-3)(0)$ **9.** $\frac{-57}{19}$ **10.** $\frac{4.5}{-15}$

11. From the sum of -14 and -16, subtract the product of -6 and -5.

12. Find the average of the following set of numbers: $\{-12, -15, -32, -31\}$.

13. Find the mean (to the nearest tenth) of the data shown in the following frequency distribution.

Bicycle Speeds (mph)	19	20	22	25	28
Frequency	4	5	7	3	1

14. Rewrite the following product using exponents: $3 \cdot 3 \cdot 3 \cdot 5 \cdot 7 \cdot 7$.

15. Find the prime factorization of the following number: 144.

16. Find the LCM of $10xy^2$, $18x^2y$, and $15xy$.

Perform the indicated operations. Simplify any fractional answers.

17. $\frac{-15x}{7} \cdot \frac{42}{20x}$ **18.** $\frac{6}{7a} \div \frac{3}{28a}$ **19.** $\frac{4}{15} + \frac{1}{3} + \frac{3}{10}$ **20.** $\frac{7}{2y} - \frac{8}{2y} - \frac{3}{2y}$

21. -44
22. 19
23. $\frac{29}{40}$
24. $\frac{17}{90}$
25. $-\frac{33}{16}$
26. a. 15 gallons
b. No, you need \$46.50 to fill the tank so you are short \$6.50.
27. Additive identity
28. Commutative property of multiplication
29. Associative property of addition
30. Zero-factor law

Use the rules for order of operations to evaluate (or simplify) each expression.

21. $36 \div (-9) \cdot 2 - 16 + 4(-5)$

22. $4^2 + [24 + 3(4-5)] \div 7$

23. $\frac{7}{8} - \frac{1}{3} \div \frac{5}{6} + \frac{1}{4}$

24. $\frac{7}{12} \cdot \frac{9}{56} \div \frac{5}{16} - \left(\frac{1}{3}\right)^2$

25. Find the sum of $\frac{3}{4}$ and $\frac{9}{10}$. Then multiply by $-\frac{5}{4}$. What is the product?

26. Gasoline: You know that the gas tank in your car holds 20 gallons of gas and the gauge reads $\frac{1}{4}$ full.

a. How many gallons will be needed to fill the tank?

b. If the price of gas is \$3.10 per gallon and you have \$40, do you have enough cash to fill the tank? How much extra money do you have or how much are you short?

Name each property of real numbers illustrated.

27. $-3 + 0 = -3$

28. $8 \cdot 2 = 2 \cdot 8$

29. $13 + (9 + 1) = (13 + 9) + 1$

30. $-5 \cdot 0 = 0$

CHAPTER 2

Algebraic Expressions, Linear Equations, and Applications

Did You Know?

Traditionally, algebra has been defined to mean generalized arithmetic where letters represent numbers. For example, $3+3+3+3=4\cdot 3$ is a special case of the more general algebraic statement that $x+x+x+x=4\cdot x$. The name algebra comes from an Arabic word, al-jabr, which means "to restore."

A notorious Muslim ruler Harun al-Rashid, the caliph made famous in *Tales of the Arabian Nights,* and his son Al-Mamun brought to their Baghdad court many renowned Muslim scholars. One of these scholars was the mathematician Muhammad ibn-Musa al-Khwarizmi, who wrote a text (c. A.D. 800) entitled *Al-Jabr wa- al-Muqabilah.* The text included instructions for solving equations by adding terms to both sides of the equation, thus "restoring" equality. The abbreviated title of Al-Khwarizmi's text, *Al-Jabr,* became our word for equation solving and operations on letters standing for numbers, **algebra**.

Al-Khwarizmi

Al-Khwarizmi's algebra was brought to western Europe through Moorish Spain in a Latin translation done by Robert of Chester (c. A.D. 1140). Al-Kwarizmi's name may have sounded familiar to you. It eventually was translated as "algorithm," which came to mean any series of steps used to solve a problem. Thus we speak of the division algorithm used to divide one number by another. Al-Khwarizmi is known as the father of algebra just as Euclid is known as the father of geometry.

One of the hallmarks of algebra is its use of specialized notation that enables complicated statements to be expressed using compact notation. Al-Khwarizmi's algebra used only a few symbols with most statements written out in words. He did not even use symbols for numbers! The development of algebraic symbols and notation occurred over the next 1000 years.

"Algebra is generous. She often gives more than is asked of her."

Jean le Rond d'Alembert (1717 – 1783)

Chapter 2 provides an integrated introduction to elementary algebra concepts using variables, equations, and positive and negative numbers. You will learn to simplify algebraic expressions and then translate English phrases into algebraic expressions and vice versa. Next you will develop one of the most important skills needed in mathematics: solving equations. You will begin by learning to solve the simplest forms of equations and then expand the related skills to solve more complex and challenging equations. In any case, the basic skills and techniques developed here are fundamental to the ideas related to solving a variety of types of equations. You will see that equations can be used in understanding number problems and geometry problems and in determining percents of profit, gas mileage, taxes, and discounts. So have patience and learn to follow the recommended steps in solving even the basic forms of equations.

2.1 Simplifying and Evaluating Algebraic Expressions

- *Simplify algebraic expressions by combining **like terms**.*
- *Evaluate expressions for given values of the variables.*

Simplifying Algebraic Expressions

A single number is called a **constant**. Any constant or variable or the indicated product and/or quotient of constants and variables is called a **term**. Examples of terms are

$$16,\ 3x,\ -5.2,\ 1.3xy,\ -5x^2,\ 14b^3,\ \text{and}\ -\frac{x}{y}.$$

Note: As discussed in Section 1.5, an expression with 2 as the exponent is read "squared" and an expression with 3 as the exponent is read "cubed." Thus $7x^2$ is read "seven x squared" and $-4y^3$ is read "negative four y cubed."

The number written next to a variable is called the **coefficient** (or the **numerical coefficient**) of the variable. For example, in

$$5x^2,\ 5 \text{ is the coefficient of } x^2.$$

Similarly, in

$$1.3xy,\ 1.3 \text{ is the coefficient of } xy.$$

NOTES

If no number is written next to a variable, the coefficient is understood to be 1. If a negative sign $(-)$ is next to a variable, the coefficient is understood to be -1. For example,

$$x = 1 \cdot x,\ a^3 = 1 \cdot a^3,\ -x = -1 \cdot x,\ \text{and}\ -y^5 = -1 \cdot y^5.$$

Like Terms

Like terms (or **similar terms**) are terms that are constants or terms that contain the same variables raised to the same exponents.

Note: The sum of the exponents on the variables is the **degree** of the term.

Like Terms

$-6,\ 1.84,\ 145,\ \frac{3}{4}$	are like terms because each term is a constant.
$-3a,\ 15a,\ 2.6a,\ \frac{2}{3}a$	are like terms because each term contains the same variable a, raised to the same exponent, 1. (Remember that $a = a^1$.) These terms are first-degree in a.
$5xy^2$ and $-3.2xy^2$	are like terms because each term contains the same two variables, x and y, with x first-degree in both terms and y second-degree in both terms.

Unlike Terms

$8x$ and $-9x^2$	are unlike terms (**not** like terms) because the variable x is not of the same power in both terms. $8x$ is first-degree in x and $-9x^2$ is second-degree in x.

Example 1: Like Terms

From the following list of terms, pick out the like terms.

$$-7,\ 2x,\ 4.1,\ -x,\ 3x^2y,\ 5x,\ -6x^2y, \text{ and } 0$$

Solution: $-7, 4.1$, and 0 are like terms. (All are constants.)

$2x, -x$, and $5x$ are like terms.

$3x^2y$ and $-6x^2y$ are like terms.

An **algebraic expression** is a combination of variables and numbers using any of the operations of addition, subtraction, multiplication, or division, as well as exponents. Examples of algebraic expressions are

$$x^2 - 14,\quad \frac{2xy}{3z^3} + y^2,\quad \frac{12x-9}{3x-4},\quad \text{and}\quad 10x^3 - 4x^2 - 11x + 1.$$

To simplify expressions that contain like terms we want to **combine like terms**.

Combining Like Terms

To **combine like terms**, add (or subtract) the coefficients and keep the common variable expression.

The procedure for combining like terms uses the distributive property in the form

$$ba + ca = (b + c)a.$$

In particular, with numerical coefficients, we can combine like terms as follows.

$9x + 6x = (9 + 6)x$	By the distributive property
$= 15x$	Add the coefficients algebraically.
$3x^2 - 5x^2 = (3 - 5)x^2$	By the distributive property
$= -2x^2$	Add the coefficients algebraically.
$xy - 1.6xy = (1.0 - 1.6)xy$	By the distributive property **Note:** $xy = 1xy = 1.0xy$
$= -0.6xy$	Add the coefficients algebraically.

Example 2: Combining Like Terms

Combine like terms whenever possible.

a. $8x + 10x$

Solution: $8x + 10x = (8 + 10)x = 18x$ By the distributive property

b. $6.5y - 2.3y$

Solution: $6.5y - 2.3y = (6.5 - 2.3)y = 4.2y$ By the distributive property

c. $2x^2 + 3a + x^2 - a$

Solution: $2x^2 + 3a + x^2 - a = 2x^2 + x^2 + 3a - a$ Use the commutative property of addition.

$= (2 + 1)x^2 + (3 - 1)a$ **Note:** $+x^2 = +1x^2$ and $-a = -1a$

$= 3x^2 + 2a$

d. $4(n-7)+5(n+1)$

Solution: $4(n-7)+5(n+1)=4n-28+5n+5$ Simplify by using the distributive property twice.

$=4n+5n-28+5$ Use the commutative property of addition.

$=(4+5)n+(-28+5)$ By the distributive property

$=9n-23$ Combine like terms.

e. $\frac{x+3x}{2}+5x$

Solution: A fraction bar is a grouping symbol, similar to parentheses. So combine like terms in the numerator first.

$\frac{x+3x}{2}+5x=\frac{4x}{2}+5x$

$=\frac{4}{2}\cdot x+5x$

$=2x+5x$ Reduce the fraction.

$=7x$ Combine like terms.

Evaluating Algebraic Expressions

In most cases, if an expression is to be evaluated, like terms should be combined first and then the resulting expression evaluated by following the rules for order of operations.

Parentheses must be used around negative numbers when substituting.

Without parentheses, an evaluation can be dramatically changed and lead to wrong answers, particularly when even exponents are involved. We analyze with the exponent 2 as follows.

In general, except for $x=0$,

1. $-x^2$ **is negative** $[-6^2=-1\cdot 6^2=-1\cdot 36=-36]$
2. $(-x)^2$ **is positive** $[(-6)^2=(-6)(-6)=36]$
3. $-x^2\neq(-x)^2$ $[-36\neq 36]$

To Evaluate an Algebraic Expression

1. Combine like terms, if possible.
2. Substitute the values given for any variables.
3. Follow the rules for order of operations. (See Section 1.8.)

Example 3: Evaluating Algebraic Expressions

a. Evaluate x^2 for $x = 3$ and for $x = -4$.

Solution: For $x = 3$, $x^2 = (3)^2 = 9$.

For $x = -4$, $x^2 = (-4)^2 = 16$.

b. Evaluate $-x^2$ for $x = 3$ and for $x = -4$.

Solution: For $x = 3$, $-x^2 = -(3)^2 = -1(9) = -9$.

For $x = -4$, $-x^2 = -(-4)^2 = -1(16) = -16$.

Example 4: Simplifying and Evaluating Algebraic Expressions

Simplify each expression below by combining like terms. Then evaluate the resulting expression using the given values for the variables.

a. Simplify and evaluate $2x + 5 + 7x$ for $x = -3$.

Solution: Simplify first.

$$\begin{aligned} 2x + 5 + 7x &= 2x + 7x + 5 \\ &= 9x + 5 \end{aligned}$$

Now evaluate.

$$\begin{aligned} 9x + 5 &= 9(-3) + 5 \\ &= -27 + 5 \\ &= -22 \end{aligned}$$

b. Simplify and evaluate $3ab - 4ab + 6a - a$ for $a = 2, b = -1$.

Solution: Simplify first.

$$3ab - 4ab + 6a - a = -ab + 5a$$

Now evaluate.

$$\begin{aligned} -ab + 5a &= -1(2)(-1) + 5(2) \\ &= 2 + 10 \\ &= 12 \end{aligned}$$

Note: $-ab = -1ab$

c. Simplify and evaluate $\frac{5x+3x}{4}+2(x+1)$ for $x=5$.

Solution: Simplify first.

$$\frac{5x+3x}{4}+2(x+1)=\frac{8x}{4}+2x+2$$
$$=2x+2x+2$$
$$=4x+2$$

Now evaluate.

$$4x+2=4(5)+2$$
$$=20+2$$
$$=22$$

1. -5, $\frac{1}{6}$, and 8 are like terms; $7x$ and $9x$ are like terms.
2. $-2x^2$, $5x^2$, and $14x^2$ are like terms; $-13x^3$ and $10x^3$ are like terms.
3. $-x^2$ and $2x^2$ are like terms; $5xy$ and $-6xy$ are like terms; $3x^2y$ and $5x^2y$ are like terms.
4. $3ab^2$ and $-ab^2$ are like terms; $8ab$ and ab are like terms; $9a^2b$ and $-10a^2b$ are like terms.

Practice Problems

Simplify the following expressions by combining like terms.

1. $-2x-5x$
2. $12y+6-y+10$
3. $5(x-1)+4x$
4. $2b^2-a+b^2+a$

Simplify the expression. Then evaluate the resulting expression for $x=3$ and $y=-2$.

5. $2(x+3y)+4(x-y)$

2.1 Exercises

5. 24, 8.3, and -6 are like terms; $1.5xyz$, $-1.4xyz$, and xyz are like terms.
6. $-35y$, $-y$, and $75y$ are like terms; 1.62 and $\frac{1}{2}$ are like terms. $-y^2$, $3y^2$, and $2.5y^2$ are like terms.

Pick out the like terms in each list of terms.

1. $-5,\ \frac{1}{6},\ 7x,\ 8,\ 9x,\ 3y$
2. $-2x^2,\ -13x^3,\ 5x^2,\ 14x^2,\ 10x^3$
3. $5xy,\ -x^2,\ -6xy,\ 3x^2y,\ 5x^2y,\ 2x^2$
4. $3ab^2,\ -ab^2,\ 8ab,\ 9a^2b,\ -10a^2b,\ ab,\ 12a^2$
5. $24,\ 8.3,\ 1.5xyz,\ -1.4xyz,\ -6,\ xyz,\ 5xy^2z,\ 2xyz^2$
6. $-35y,\ 1.62,\ -y^2,\ -y,\ 3y^2,\ \frac{1}{2},\ 75y,\ 2.5y^2$

Answers to Practice Problems: **1.** $-7x$ **2.** $11y+16$ **3.** $9x-5$ **4.** $3b^2$ **5.** $6x+2y$; 14

7. 64 8. −64
9. −121 10. 36
11. $15x$ 12. $11y$
13. $3x$ 14. $4x$
15. $-2n$ 16. $-2x$
17. $5y^2$ 18. $11z^2$
19. $34x^2$ 20. $25x^3$
21. $7x+2$ 22. $4x-1$
23. $x-4y$
24. $2x-y$
25. $13x^2-2y$
26. $5a-3b$
27. $2n+3$
28. $3n-7$
29. $7a-8b$
30. $6a+b$
31. $8x+y$
32. $10x-y$
33. $2x^2-x$
34. y^2+y
35. $-2n^2+2n-2$
36. $2n^2+3n-9$
37. $3x^2-xy+y^2$
38. $2x^2+6xy+3$
39. $2x$
40. $-y$
41. $-y$
42. $-z$
43. a. $3x+4$ b. 16
44. a. $6x-17$ b. 7
45. a. $-2x-8$ b. −16
46. a. $10a+13$ b. −7
47. a. $5y+1$ b. 16
48. a. $9y+2$ b. 29
49. a. $-3x-7y$ b. −33
50. a. $-5a+b$ b. 9
51. a. $3.6x^2+2$ b. 59.6
52. a. $-3.1a^2+4a$ b. −20.4
53. a. $3ab+b^2+b^3$ b. 6
54. a. $8a-ab^2$ b. −14

Find the value of each numerical expression.

7. $(-8)^2$ **8.** -8^2 **9.** -11^2 **10.** $(-6)^2$

Simplify each expression by combining like terms.

11. $8x+7x$ **12.** $3y+8y$ **13.** $5x+(-2x)$ **14.** $7x+(-3x)$

15. $-n-n$ **16.** $-x-x$ **17.** $6y^2-y^2$ **18.** $16z^2-5z^2$

19. $23x^2+11x^2$ **20.** $18x^3+7x^3$ **21.** $4x+2+3x$ **22.** $3x-1+x$

23. $2x-3y-x-y$ **24.** $x+y+x-2y$ **25.** $2x^2-2y+5x^2+6x^2$

26. $4a+2a-3b-a$ **27.** $3(n+1)-n$ **28.** $2(n-4)+n+1$

29. $5(a-b)+2a-3b$ **30.** $4a-3b+2(a+2b)$

31. $3(2x+y)+2(x-y)$ **32.** $4(x+5y)+3(2x-7y)$

33. $2x+3x^2-3x-x^2$ **34.** $2y^2+4y-y^2-3y$

35. $2n^2-6n+1-4n^2+8n-3$ **36.** $3n^2+2n-5-n^2+n-4$

37. $3x^2+4xy-5xy+y^2$ **38.** $2x^2-5xy+11xy+3$

39. $\dfrac{x+5x}{6}+x$ **40.** $\dfrac{2y+3y}{5}-2y$

41. $y-\dfrac{2y+4y}{3}$ **42.** $z-\dfrac{3z+5z}{4}$

For the following expressions, **a.** *simplify, and* **b.** *evaluate the simplified expression for $x=4$, $y=3$, $a=-2$, and $b=-1$.*

43. $5x+4-2x$ **44.** $7x-17-x$

45. $x-10-3x+2$ **46.** $6a+5a-a+13$

47. $3(y-1)+2(y+2)$ **48.** $4(y+3)+5(y-2)$

49. $-5(x+y)+2(x-y)$ **50.** $-2(a+b)+3(b-a)$

51. $8.3x^2-5.7x^2+x^2+2$ **52.** $3.1a^2-0.9a^2+4a-5.3a^2$

53. $5ab+b^2-2ab+b^3$ **54.** $5a+ab^2-2ab^2+3a$

55. a. $3.7x + 1.1$ b. 15.9
56. a. $3.9y - 18.2$ b. −6.5
57. a. $8a$ b. −16
58. a. $9a$ b. −18
59. a. $3b$ b. −3
60. a. $3b$ b. −3
61. a. $-x - 54$ b. −58
62. a. $-7x + 30$ b. 2

55. $2.4(x+1)+1.3(x-1)$

56. $1.3(y+2)-2.6(8-y)$

57. $\frac{3a+5a}{-2}+12a$

58. $8a+\frac{5a+4a}{9}$

59. $\frac{-4b-2b}{-3}+\frac{2b+5b}{7}$

60. $\frac{5b+3b}{4}+\frac{-4b-b}{-5}$

61. $2x+3[x-2(9+x)]$

62. $5x-2[x+5(x-3)]$

Writing and Thinking About Mathematics

63. -5^2 is the square of 5 multiplied by −1 by the order of operations; while $(-5)^2$ is the square of −5. ($-5^2 = -25$ and $(-5)^2 = 25$)

63. Explain the difference between -5^2 and $(-5)^2$.

64. The text recommends simplifying an expression (combining like terms) before evaluating. Do you think this is necessary? Evaluate the expression $4x^2-5(x+2)+3x+10+2x$ for $x=3$:
a. by substituting and then evaluating.
b. by first simplifying and then evaluating.
Which method would you recommend? Why?

HAWKES LEARNING SYSTEMS: INTRODUCTORY & INTERMEDIATE ALGEBRA SOFTWARE

- 2.1a Variables and Algebraic Expressions
- 2.1b Simplifying Expressions
- 2.1c Evaluating Algebraic Expressions

64. a. For $x = 3$

$$\begin{aligned}4x^2-5(x+2)+3x+10+2x &= 4(3)^2-5((3)+2)+3(3)+10+2(3)\\ &= 4(9)-5(5)+9+10+6\\ &= 36-25+9+10+6\\ &= 36\end{aligned}$$

b. $$\begin{aligned}4x^2-5(x+2)+3x+10+2x &= 4x^2-5x-10+3x+10+2x\\ &= 4x^2\end{aligned}$$

For $x = 3$ $$\begin{aligned}4x^2 &= 4(3)^2\\ &= 4(9)\\ &= 36\end{aligned}$$

The second method, as it requires less calculations.

2.2 Translating English Phrases and Algebraic Expressions

- *Write the meaning of algebraic expressions in words.*
- *Write algebraic expressions for word phrases.*

Translating English Phrases into Algebraic Expressions

Algebra is a language of mathematicians, and to understand mathematics, you must understand the language. We want to be able to change English phrases into their "algebraic" equivalents and vice versa. So if a problem is stated in English, we can translate the phrases into algebraic symbols and proceed to solve the problem according to the rules and methods developed for algebra.

Certain words are the keys to the basic operations. Some of these words are listed here and highlighted in boldface in Example 1.

Key Words to Look For when Translating Phrases				
Addition	**Subtraction**	**Multiplication**	**Division**	**Exponent (Powers)**
add	subtract (from)	multiply	divide	square of
sum	difference	product	quotient	cube of
plus	minus	times		
more than	less than	twice		
increased by	decreased by	of (with fractions and percent)		
	less			

The following examples illustrate how these key words used in English phrases can be translated into algebraic expressions. **Note that in each case "a number" or "the number" implies the use of a variable (an unknown quantity).**

Example 1: Translating English Phrases

English Phrase	Algebraic Expression
a. the **product** of 3 and x 3 **times** x 3 **multiplied by** the number represented by x	$3x$
b. 3 **added to** a number the **sum** of z and 3 z **plus** 3 3 **more than** z z **increased by** 3	$z+3$
c. **twice** the **sum** of x and 1 the **product** of 2 with the **sum** of x and 1 2 **times** the quantity found by **adding** a number to 1	$2(x+1)$
d. **twice** x **plus** 1 the **sum** of **twice** x and 1 2 **times** x **increased** by 1 1 **more than** the **product** of 2 and a number	$2x+1$
e. the **difference** between 5 **times** a number and 3 3 **less than** the **product** of a number and 5 5 **times** a number **minus** 3 3 **subtracted from** $5n$ 5 **multiplied by** a number, **less** 3	$5n-3$
f. the **square of** a number a number **squared**	x^2
g. the **cube of** a number a number **cubed**	n^3

NOTES

In Example 1b, the phrase "the sum of z and 3" was translated as $z + 3$. If the expression had been translated as $3 + z$, there would have been no mathematical error because addition is commutative. That is, $z + 3 = 3 + z$. However, in part **e.**, the phrase "3 less than the product of a number and 5" must be translated as it was because subtraction is **not** commutative. Thus

"3 less than 5 times a number" means $5n - 3$
while "5 times a number less than 3" means $3 - 5n$
and "3 less 5 times a number" means $3 - 5n$.

Therefore, be very careful when writing and/or interpreting expressions indicating subtraction. Be sure that the subtraction is in the order indicated by the wording in the problem. The same is true with expressions involving division.

The words **quotient** and **difference** deserve special mention because their use implies that the numbers given are to be operated on in the order given. That is, division and subtraction are done with the values in the same order that they are given in the problem. For example:

the quotient of y and 5 → $\frac{y}{5}$

the quotient of 5 and y → $\frac{5}{y}$

the difference between 6 and x → $6 - x$

the difference between x and 6 → $x - 6$

If we did not have these agreements concerning subtraction and division, then the phrases just illustrated might have more than one interpretation and be considered **ambiguous**.

An **ambiguous phrase** is one whose meaning is not clear or for which there may be two or more interpretations. This is a common occurrence in ordinary everyday language, and misunderstandings occur frequently. Imagine the difficulties diplomats have in communicating ideas from one language to another trying to avoid ambiguities. Even the order of subjects, verbs, and adjectives may not be the same from one language to another. Translating grammatical phrases in any language into mathematical expressions is quite similar. To avoid ambiguous phrases in mathematics, we try to be precise in the use of terminology, to be careful with grammatical construction, and to follow the rules for order of operations.

Translating Algebraic Expressions into English Phrases

Consider the three expressions to be translated into English:

$$7(n+1),\ 6(n-3),\ \text{and}\ 7n+1.$$

In the first two expressions, we indicate the parentheses with a phrase such as "the quantity" or "the sum of" or "the difference between." **Without the parentheses, we agree that the operations used in the expression are to be indicated in the order given.** Thus

	$7(n+1)$	can be translated as "seven times the sum of a number and 1,"
	$6(n-3)$	can be translated as "six times the difference between a number and 3,"
while	$7n+1$	can be translated as "seven times a number plus 1."

Example 2: Translating Algebraic Expressions to Phrases

Write an English phrase that indicates the meaning of each algebraic expression.

	Algebraic Expression	Possible English Phrase
a.	$5x$	The product of 5 and a number
b.	$2n+8$	Twice a number increased by 8
c.	$3(a-2)$	Three times the difference between a number and 2

Example 3: Translating Phrases to Algebraic Expressions

Change each phrase into an equivalent algebraic expression.

	Phrase	Algebraic Expression
a.	The quotient of a number and –4	$\frac{x}{-4}$
b.	6 less than 5 times a number	$5y-6$
c.	Twice the sum of 3 and a number	$2(3+n)$
d.	The number of minutes in h hours	$60h$
e.	The cost of renting a truck for one day and driving x miles if the rate is \$30 per day plus \$0.25 per mile	$30+0.25x$

Practice Problems

Change the following phrases to algebraic expressions.

1. 7 less than a number

2. The quotient of y and 5

3. 14 more than 3 times a number

Change the following algebraic expressions into English phrases. (There may be more than one correct translation.)

4. $10 - x$ **5.** $2(y-3)$ **6.** $5n + 3n$

2.2 Exercises

Translate each algebraic expression into an equivalent English phrase. (There may be more than one correct translation.)

1. $4x$ **2.** $-9x$ **3.** $x + 5$ **4.** $4x - 7$

5. $7(x+1.1)$ **6.** $3.2(x+2.5)$ **7.** $-2(x-8)$ **8.** $10(x+4)$

9. $\frac{6}{(x-1)}$ **10.** $\frac{9}{(x+3)}$ **11.** $5(2x+3)$ **12.** $3(4x-5)$

Write each pair of expressions in words. Notice the differences between the algebraic expressions and the corresponding English phrases.

13. $3x+7;\ 3(x+7)$ **14.** $4x-1;\ 4(x-1)$

15. $7x-3;\ 7(x-3)$ **16.** $5(x+6);\ 5x+6$

Write the algebraic expressions described by the English phrases. Choose your own variable.

17. 6 added to a number

18. 7 more than a number

19. 4 less than a number

20. A number decreased by 13

21. The quotient of twice a number and 10

1. 4 times a number
2. The product of a number and −9
3. 5 more than a number
4. 7 less than 4 times a number
5. 7 times the sum of a number and 1.1
6. 3.2 times the sum of a number and 2.5
7. −2 times the difference between a number and 8
8. 10 times the sum of a number and 4
9. 6 divided by the difference between a number and 1
10. 9 divided by the sum of a number and 3
11. 5 times the sum of twice a number and 3
12. 3 times the difference of 4 times a number and 5

Answers to Practice Problems: **1.** $x - 7$ **2.** $\frac{y}{5}$ **3.** $3y + 14$ **4.** 10 decreased by a number **5.** Twice the difference between a number and 3 **6.** 5 times a number plus 3 times the same number

13. 3 times a number plus 7; 3 times the sum of a number and 7
14. 1 less than 4 times a number; 4 times the difference of a number and 1
15. The product of 7 and a number minus 3; 7 times the difference between a number and 3
16. 5 times the sum of a number and 6; the sum of 5 times a number and 6
17. $x + 6$
18. $x + 7$
19. $x - 4$
20. $x - 13$
21. $\frac{2x}{10}$
22. $\frac{x-3}{7}$
23. $3x - 5$
24. $2x + 4x$
25. $8 - 2x$
26. $x + 9x$
27. $20 - 4.8x$
28. $3x - 5x$
29. $9(x+2)$
30. $3(x-8)$
31. $4(x+1)-13$
32. $8(x-6)+4$
33. $3(x+6)+8$
34. $2(x-7)-6$
35. $3(7-x)-4$
36. $2(17+x)+9$
37. a. $x - 6$
b. $6 - x$

22. The difference between a number and 3, all divided by 7

23. 5 subtracted from three times a number

24. The sum of twice a number and four times the number

25. 8 minus twice a number

26. The sum of a number and 9 times the number

27. Twenty decreased by 4.8 times a number

28. The difference between three times a number and five times the same number

29. 9 times the sum of a number and 2

30. 3 times the difference between a number and 8

31. 13 less than the product of 4 with the sum of a number and 1

32. 4 more than the product of 8 with the difference between a number and 6

33. Eight more than the product of 3 and the sum of a number and 6

34. Six less than twice the difference between a number and 7

35. Four less than 3 times the difference between 7 and a number

36. Nine more than twice the sum of 17 and a number

37. **a.** 6 less than a number
b. 6 less a number

38. **a.** 20 less than a number
b. 20 less a number

39. **a.** 5 less than 3 times a number
b. 5 less 3 times a number

40. **a.** 6 less than 4 times a number
b. 6 less 4 times a number

Write the algebraic expressions described by the English phrases.

41. Time: The number of hours in d days

42. Graphing calculators: The cost of x graphing calculators if one calculator costs \$115

43. Gas prices: The cost of x gallons of gasoline if the cost of one gallon is \$3.15

44. Time: The number of seconds in m minutes

45. Time: The number of days in y years (Assume 365 days in a year.)

46. Candy: The cost of x pounds of candy at \$4.95 a pound

47. Time: The number of days in t weeks and 3 days

48. Time: The number of minutes in h hours and 20 minutes

38. a. $x - 20$
b. $20 - x$
39. a. $3x - 5$
b. $5 - 3x$
40. a. $4x - 6$
b. $6 - 4x$
41. $24d$
42. $115x$
43. $3.15x$
44. $60m$
45. $365y$
46. $4.95x$
47. $7t + 3$
48. $60h + 20$
49. $7t + 3$
50. $0.2w$
51. $20 + 0.15m$
52. $x + (2x + 8)$
$= 3x + 8$
53. $2w + 2(2w - 3)$
$= 6w - 6$

49. Football: The points scored by a football team on t touchdowns (7 points) and 1 field goal (3 points)

50. Vacation time: The amount of vacation days an employee has after w weeks if she gets 0.2 vacation days for every week she works

51. Car rentals: The cost of renting a car for one day and driving m miles if the rate is \$20 per day plus 15 cents per mile

52. Fishing: The cost of purchasing a fishing rod and reel if the rod costs x dollars and the reel costs \$8 more than twice the cost of the rod

53. Rectangles: The perimeter of a rectangle if the width is w centimeters and the length is 3 cm less than twice the width

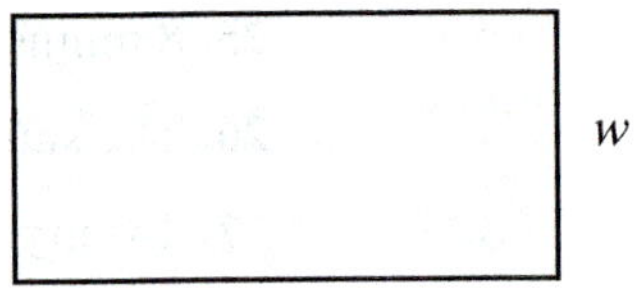

3 cm less than twice the width

54. Squares: The area of a square with side c centimeters

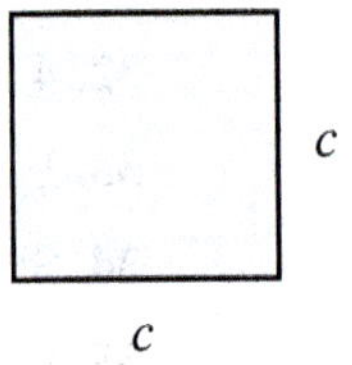

Writing and Thinking About Mathematics

55. Discuss the meaning of the term "ambiguous phrase."

HAWKES LEARNING SYSTEMS: INTRODUCTORY & INTERMEDIATE ALGEBRA SOFTWARE

- 2.2 Translating Phrases into Algebraic Expressions

54. c^2 55. A phrase whose meaning is not clear or for which there may be two or more interpretations.

2.3 Solving Linear Equations: $x + b = c$ and $ax = c$

- *Define the term* ***linear equation***.
- *Solve equations of the form* $\boldsymbol{x+b=c}$.
- *Solve equations of the form* $\boldsymbol{ax=c}$.

In this section, we will discuss solving linear equations in the following two forms:

$$x+b=c \quad \text{and} \quad ax=c.$$

In these equations we treat a, b, and c as constants and x as the unknown quantity.

In the following sections, we will combine the techniques developed here and discuss solving linear equations in the forms:

$$ax+b=c \quad \text{and} \quad ax+b=cx+d.$$

An **equation** is a statement that two algebraic expressions are equal. That is, both expressions represent the same number. If an equation contains a variable, any number that gives a true statement when substituted for the variable is called a **solution** to the equation. The solutions to an equation form a **solution set**. The process of finding the solution set is called **solving the equation**.

Linear Equation in x

If a, b, and c are **constants** and $a \neq 0$ then a **linear equation in x** is an equation that can be written in the form

$$ax+b=c.$$

Note: A linear equation in x is also called a **first-degree equation in x** because the variable x can be written with the exponent 1. That is, $x = x^1$.

Determining Possible Solutions

We begin by determining whether or not a particular number satisfies an equation. A number is said to be a **solution** or to **satisfy an equation** if it gives a true statement when substituted for the variable.

Example 1: Determining a Possible Solution

Determine whether or not the given real number is a solution to the given equation by substituting for the variable and checking to see if the resulting equation is true or false.

a. $x+5=-2$ given that $x=-7$

Solution: $(-7)+5=-2$ is true, so -7 **is** a solution.

b. $1.4+z=0.5$ given that $z=-1.1$

Solution: $1.4+(-1.1)=0.5$ is false because $1.4+(-1.1)=0.3\neq 0.5$.
So -1.1 **is not** a solution.

c. $5.6-y=2.9$ given that $y=2.7$

Solution: $5.6-2.7=2.9$ is true. So, 2.7 **is** a solution.

d. $|z|-14=-3$ given that $z=-10$

Solution: $|(-10)|-14=-3$ is false because $|-10|-14=10-14=-4\neq -3$. So, -10 **is not** a solution.

Note: Notice that parentheses were used around negative numbers in the substitutions. This should be done to keep operations properly separated, particularly when negative numbers are involved.

Solving Equations of the Form $x + b = c$

To begin, we need the **addition principle of equality**.

Addition Principle of Equality

If the same algebraic expression is added to both sides of an equation, the new equation has the same solutions as the original equation. Symbolically, if A, B, and C are algebraic expressions, then the equations

$$\mathbf{A=B}$$

and

$$\mathbf{A+C=B+C}$$

have the same solutions.

Equations with the same solutions are said to be **equivalent equations**.

The objective of solving linear (or first-degree) equations is to get the variable by itself (with a coefficient of +1) on one side of the equation and any constants on the other side. The following procedure will help in solving linear equations such as

$$x-3=7, \quad -11=y+5, \quad 3z-2z+2.5=3.2+0.8, \quad \text{and} \quad x-\frac{2}{5}=\frac{3}{10}.$$

Procedure for Solving Linear Equations that Simplify to the Form $x + b = c$

1. Combine any like terms on each side of the equation.
2. Use the **addition principle of equality** and add the opposite of the constant b to both sides. The objective is to isolate the variable on one side of the equation (either the left side or the right side) with a coefficient of +1.
3. Check your answer by substituting it for the variable in the original equation.

Every linear equation has exactly one solution. This means that once a solution has been found, there is no need to search for another solution.

Example 2: Solving $x + b = c$

Solve each of the following linear equations.

a. $x-3=7$

Solution:

$x-3=7$ Write the equation.

$x-3+3=7+3$ Add 3 (the opposite of −3) to both sides.

$x=10$ Simplify.

Check:

$x-3=7$

$(10)-3\stackrel{?}{=}7$ Substitute $x=10$.

$7=7$ True statement

b. $-11=y+5$

Solution:

$-11=y+5$ Write the equation. Note that the variable can be on the right side.

$-11-5=y+5-5$ Add −5 (the opposite of +5) to both sides.

$-16=y$ Simplify.

Check:

$-11=y+5$

$-11\stackrel{?}{=}(-16)+5$ Substitute $y=-16$.

$-11=-11$ True statement

Continued on the next page...

c. $x-\frac{2}{5}=\frac{3}{10}$

Solution:

$x-\frac{2}{5}=\frac{3}{10}$ Write the equation.

$x-\frac{2}{5}+\frac{2}{5}=\frac{3}{10}+\frac{2}{5}$ Add $\frac{2}{5}$, the opposite of $-\frac{2}{5}$, to both sides.

$x=\frac{3}{10}+\frac{4}{10}$ Simplify. (The common denominator is 10.)

$x=\frac{7}{10}$ Simplify.

Check:

$x-\frac{2}{5}=\frac{3}{10}$

$\left(\frac{7}{10}\right)-\frac{2}{5}\stackrel{?}{=}\frac{3}{10}$ Substitute $x=\frac{7}{10}$.

$\frac{7}{10}-\frac{4}{10}\stackrel{?}{=}\frac{3}{10}$ The common denominator is 10.

$\frac{3}{10}=\frac{3}{10}$ True statement

Example 3: Simplifying and Solving Equations

$3z-2z+2.5=3.2+0.8$

Solution:

$3z-2z+2.5=3.2+0.8$ Write the equation.

$z+2.5=4.0$ Combine like terms on both sides of the equation.

$z+2.5-2.5=4.0-2.5$ Add -2.5 (the opposite of $+2.5$) to both sides.

$z=1.5$ Simplify.

Check:

$3z-2z+2.5=3.2+0.8$

$3(1.5)-2(1.5)+2.5\stackrel{?}{=}3.2+0.8$ Substitute $z=1.5$.

$4.5-3.0+2.5\stackrel{?}{=}3.2+0.8$ Simplify.

$4.0=4.0$ True statement

Note that, as illustrated in Examples 1-3, variables other than x may be used.

Solving Equations of the Form $ax = c$

To solve equations of the form $ax = c$, where $a \neq 0$, we can use the idea of the reciprocal of the coefficient a. For example, as we studied in Section 1.6, the reciprocal of $\frac{3}{4}$ is $\frac{4}{3}$ and $\frac{3}{4} \cdot \frac{4}{3} = 1$. Also, we need the **multiplication** (or **division**) **principle of equality** as stated below.

Multiplication (or Division) Principle of Equality

If both sides of an equation are multiplied by (or divided by) the same nonzero constant, the new equation has the same solutions as the original equation. Symbolically, if A and B are algebraic expressions and C is any nonzero constant, then the equations

$$\mathbf{A = B}$$

$$\text{and} \quad \mathbf{AC = BC} \quad \text{where } C \neq 0$$

$$\text{and} \quad \mathbf{\frac{A}{C} = \frac{B}{C}} \quad \text{where } C \neq 0$$

have the same solutions. We say that the equations are equivalent.

Remember that the objective is to get the variable by itself on one side of the equation. That is, we want the variable to have +1 as its coefficient. The following procedure will accomplish this.

Procedure for Solving Linear Equations that Simplify to the Form $ax = c$

1. Combine any like terms on each side of the equation.
2. Use the **multiplication** (or **division**) **principle of equality** and multiply both sides of the equation by the reciprocal of the coefficient of the variable. (**Note:** This is the same as dividing both sides of the equation by the coefficient.) Thus the coefficient of the variable will become +1.
3. Check your answer by substituting it for the variable in the original equation.

Example 4: Solving $ax = c$

Solve the following equation.

a. $5x = 20$

Solution:

$5x = 20$	Write the equation.
$\frac{1}{5} \cdot (5x) = \frac{1}{5} \cdot 20$	Multiply by $\frac{1}{5}$, the reciprocal of 5.
$\left(\frac{1}{5} \cdot 5\right)x = \frac{1}{5} \cdot \frac{20}{1}$	Use the associative property of multiplication.
$1 \cdot x = 4$	Simplify.
$x = 4$	

Check:

$5x = 20$	
$5 \cdot (4) \stackrel{?}{=} 20$	Substitute $x = 4$.
$20 = 20$	True statement

Multiplying by the reciprocal of the coefficient is the same as **dividing** by the coefficient itself. So, we can multiply both sides by $\frac{1}{5}$, as we did, or we can divide both sides by 5. In either case, the coefficient of x becomes +1.

$5x = 20$	
$\frac{5x}{5} = \frac{20}{5}$	Divide both sides by 5.
$x = 4$	Simplify.

b. $1.1x + 0.2x = 12.2 - 3.1$

Solution: When decimal coefficients or constants are involved, you might want to use a calculator to perform some of the arithmetic.

$1.1x + 0.2x = 12.2 - 3.1$	Write the equation.
$1.3x = 9.1$	Combine like terms.
$\frac{1.3x}{1.3} = \frac{9.1}{1.3}$	Use a calculator or pencil and paper to divide.
$x = 7.0$	

Check:

$1.1x + 0.2x = 12.2 - 3.1$	
$1.1(7) + 0.2(7) \stackrel{?}{=} 12.2 - 3.1$	Substitute $x = 7$.
$7.7 + 1.4 \stackrel{?}{=} 9.1$	Simplify.
$9.1 = 9.1$	True statement

c. $\frac{4x}{5} = \frac{3}{10}$ (This could be written $\frac{4}{5}x = \frac{3}{10}$ because $\frac{4}{5}x$ is the same as $\frac{4x}{5}$.)

Solution:

$\frac{4x}{5} = \frac{3}{10}$ Write the equation.

$\frac{5}{4} \cdot \frac{4}{5}x = \frac{5}{4} \cdot \frac{3}{10}$ Multiply both sides by $\frac{5}{4}$.

$1 \cdot x = \frac{1 \cdot \cancel{5}}{4} \cdot \frac{3}{2 \cdot \cancel{5}}$ Simplify.

$x = \frac{3}{8}$

Check:

$\frac{4x}{5} = \frac{3}{10}$

$\frac{4}{5} \cdot \frac{3}{8} \stackrel{?}{=} \frac{3}{10}$ Substitute $x = \frac{3}{8}$.

$\frac{3}{10} = \frac{3}{10}$ True statement

d. $-x = 4$

Solution:

$-x = 4$ Write the equation.

$-1x = 4$ -1 is the coefficient of x.

$\frac{-1x}{-1} = \frac{4}{-1}$ Divide by -1 so that the coefficient will become $+1$.

$x = -4$

Check:

$-x = 4$

$-(-4) \stackrel{?}{=} 4$ Substitute $x = -4$.

$4 = 4$ True statement

Example 5: Application

The original price of a Blu-Ray player was reduced by \$45.50. The sale price was \$165.90. Solve the equation $y - 45.50 = 165.90$ to determine the original price of the Blu-Ray player.

Solution:

$y - 45.50 = 165.90$

$y - 45.50 + 45.50 = 165.90 + 45.50$ Use the addition principle by adding 45.50 to both sides.

$y = 211.40$ Simplify.

The original price of the Blu-Ray player was \$211.40.

Practice Problems

Solve the following equations.

1. $-16 = x + 5$ **2.** $6y - 1.5 = 7.5$ **3.** $4x = -20$

4. $\frac{3y}{5} = 33$ **5.** $1.7z + 2.4z = 8.2$ **6.** $-x = -8$

7. $3x = 10$ **8.** $5x - 4x + 1.6 = -2.7$ **9.** $\frac{4}{5} = 3x$

2.3 Exercises

1. –2 is a solution
2. 18 is a solution
3. 4 is not a solution
4. 1 is not a solution
5. –4 is a solution
6. 5 is a solution
7. –18 is a solution
8. –72 is not a solution
9. –28 is a solution
10. –11 is a solution
11. $x = 7$
12. $x = 19$
13. $y = -4$
14. $y = -7$
15. $x = -19$
16. $x = -27$
17. $n = 37$
18. $n = 56$
19. $z = -6$
20. $z = 17$
21. $x = 5$
22. $x = -1$
23. $y = -5.9$
24. $y = -5.3$
25. $x = -1.2$
26. $x = 1.1$
27. $y = \frac{11}{20}$

Determine whether or not the given number is a solution to the given equation by substituting and then evaluating.

1. $x + 4 = 2$ given that $x = -2$

2. $z + (-12) = 6$ given that $z = 18$

3. $x - 3 = -7$ given that $x = 4$

4. $x - 2 = -3$ given that $x = 1$

5. $-10 + x = -14$ given that $x = -4$

6. $-9 - x = -14$ given that $x = 5$

7. $-26 + |x| = -8$ given that $x = -18$

8. $42 + |z| = -30$ given that $z = -72$

9. $|x| - |-3| = 25$ given that $x = -28$

10. $|-2| + |x| = 13$ given that $x = -11$

Solve each of the following linear equations. (Remember that the objective is to isolate the variable on one side of the equation with a coefficient of +1.)

11. $x - 6 = 1$ **12.** $x - 10 = 9$ **13.** $y + 7 = 3$ **14.** $y + 12 = 5$

15. $x + 15 = -4$ **16.** $x + 17 = -10$ **17.** $22 = n - 15$ **18.** $36 = n - 20$

19. $6 = z + 12$ **20.** $18 = z + 1$ **21.** $x - 20 = -15$ **22.** $x - 10 = -11$

23. $y + 3.4 = -2.5$ **24.** $y + 1.6 = -3.7$ **25.** $x + 3.6 = 2.4$ **26.** $x + 2.7 = 3.8$

27. $x + \frac{1}{20} = \frac{3}{5}$ **28.** $n - \frac{2}{7} = \frac{3}{14}$ **29.** $5x = 45$ **30.** $9x = 108$

31. $32 = 4y$ **32.** $51 = 17y$ **33.** $\frac{3x}{4} = 15$ **34.** $\frac{5x}{7} = 65$

Answers to Practice Problems: **1.** $x = -21$ **2.** $y = 1.5$ **3.** $x = -5$ **4.** $y = 55$ **5.** $z = 2$ **6.** $x = 8$ **7.** $x = \frac{10}{3}$ **8.** $x = -4.3$ **9.** $\frac{4}{15} = x$

35. $4 = \frac{2y}{5}$

36. $46 = \frac{46y}{5}$

37. $4x - 3x = 10 - 12$

38. $7x - 6x = 13 + 15$

39. $7x - 8x = 13 - 25$

40. $10n - 11n = 20 - 14$

41. $3n - 2n + 6 = 14$

42. $7n - 6n + 13 = 22$

43. $1.7y + 1.3y = 6.3$

44. $2.5y + 7.5y = 4.2$

45. $\frac{3}{4}x = \frac{5}{3}$

46. $\frac{5}{6}x = \frac{5}{3}$

47. $7.5x = -99.75$

48. $-14 = 0.7x$

49. $1.5y - 0.5y + 6.7 = -5.3$

50. $2.6y - 1.6y - 5.1 = -2.9$

51. $10x - 9x - \frac{1}{2} = -\frac{9}{10}$

52. $6x - 5x + \frac{3}{4} = -\frac{1}{12}$

53. $1.4x - 0.4x + 2.7 = -1.3$

54. $3.5y - 2.5y - 6.3 = -1.0 - 2.5$

55. $\frac{7x}{4} - \frac{3x}{4} + \frac{7}{8} = \frac{3}{2}$

56. $\frac{5n}{2} - \frac{3n}{2} + \frac{4}{5} = \frac{7}{5} - \frac{1}{10}$

57. $6.2 = -3.5 + 7n - 6n$

58. $-7.2 = 1.3n - 0.3n - 1.0$

59. $1.7x = -5.1 - 1.7$

60. $3.2x = 2.8 - 9.2$

61. World Languages: The Japanese writing system consists of three sets of characters, two with 81 characters (which all Japanese students must know), and a third, kanji, with over 50,000 characters (of which only some are used in everyday writing). If a Japanese student knows 2107 total characters, solve the equation

$$x + 2(81) = 2107$$

to determine the number of kanji characters the student knows.

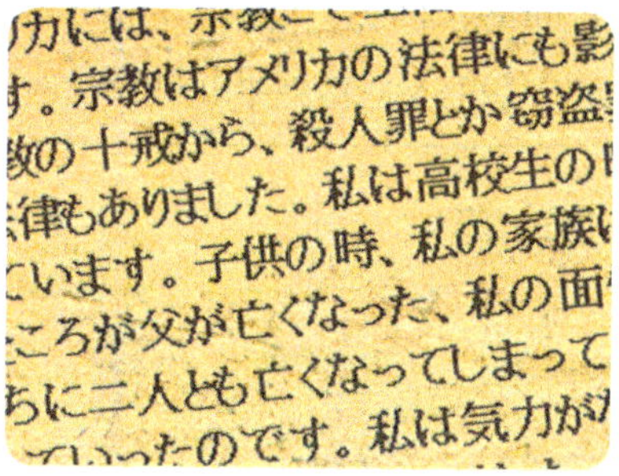

62. Writing: An author is determined to have his first novel published by the publisher of George Orwell's *1984*, his favorite book. However, his contract with the publisher requires his novel to be at least 75,000 words, and he has only written 63,500. Solve the following equation to determine how many more words he must write.

$$63,500 + x = 75,000$$

63. Sculpture: A sculptor has decided to begin a project to make scale models of famous landmarks out of stone. His first model will be of one of the moai, giant human figures carved from stone on Easter Island. If his model is to be 1/12 scale and the original moai weighs 75 tons, solve the equation

$$12x = 75$$

to determine how many tons his completed sculpture will weigh.

28. $n = \frac{1}{2}$
29. $x = 9$
30. $x = 12$
31. $y = 8$
32. $y = 3$
33. $x = 20$
34. $x = 91$
35. $y = 10$
36. $y = 5$
37. $x = -2$
38. $x = 28$
39. $x = 12$
40. $n = -6$
41. $n = 8$
42. $n = 9$
43. $y = 2.1$
44. $y = 0.42$
45. $x = \frac{20}{9}$
46. $x = 2$
47. $x = -13.3$
48. $x = -20$
49. $y = -12$
50. $y = 2.2$
51. $x = -\frac{2}{5}$
52. $x = -\frac{5}{6}$
53. $x = -4$
54. $y = 2.8$
55. $x = \frac{5}{8}$
56. $n = \frac{1}{2}$
57. $n = 9.7$
58. $n = -6.2$
59. $x = -4$
60. $x = -2$
61. 1945 kanji characters
62. 11,500 words
63. 6.25 tons

64. 4.24 light years
65. $x = -50.753$
66. $x = 100.979$
67. $x = -17.214$
68. $y = 27.087$
69. $x = 246$
70. $y = -1036$
71. $x = -153.17$
72. $y = -121.962$
73. a. Yes; It is stating that 6 + 3 is equal to 9.
b. No; If we substitute 4 for x, we get the statement 9 = 10, which is not true.

64. Astronomy: The diameter of the Milky Way, the galaxy our solar system is in, is approximately 23,585 times the distance from the sun to the nearest star, Proxima Centauri. Considering that the Milky Way is roughly 100,000 light years across, solve the following equation to find the number of light years from the sun to this star. (Round your answer to two decimal places.)

$$23{,}585x = 100{,}000$$

Calculator Problems

Use a graphing calculator to help solve the following equations.

65. $y + 32.861 = -17.892$

66. $x - 41.625 = 59.354$

67. $17.61x - 16.61x + 27.059 = 9.845$

68. $14.83y - 8.65 - 13.83y = 17.437 + 1.0$

69. $2.637x = 648.702$

70. $-0.3057y = 316.7052$

71. $-x = 145.6 + 17.89 - 10.32$

72. $-y = 143.5 + 178.462 - 200$

Writing and Thinking About Mathematics

73. a. Is the expression $6 + 3 = 9$ an equation? Explain.
b. Is 4 a solution to the equation $5 + x = 10$? Explain.

HAWKES LEARNING SYSTEMS: INTRODUCTORY & INTERMEDIATE ALGEBRA SOFTWARE

- 2.3a Solving Linear Equations Using Addition and Subtraction
- 2.3b Solving Linear Equations Using Multiplication and Division

2.4 Solving Linear Equations: $ax + b = c$

- *Solve equations of the form* $\boldsymbol{ax + b = c}$.

Solving Equations of the Form $ax + b = c$

In Section 2.3 the equations to be solved were in one of two forms: $\boldsymbol{x + b = c}$ or $\boldsymbol{ax = c}$. In the first type the coefficient of the variable x was always +1 and we had to add or subtract constants to solve the equation. (We used the addition principle.) In the second type, we had to multiply both sides by the reciprocal of the coefficient of the variable (or divide both sides by the coefficient itself). Now we will apply both of these techniques in solving the same equation. That is, in the form $\boldsymbol{ax + b = c}$ the coefficient $\boldsymbol{a}$ may be a number other than 1.

The **general procedure for solving linear equations** is now a combination of the procedures stated in Section 2.3.

Procedure for Solving Linear Equations that Simplify to the Form $ax + b = c$

1. Combine like terms on both sides of the equation.
2. Use the **addition principle of equality** and add the opposite of the constant b to both sides.
3. Use the **multiplication** (or **division**) **principle of equality** to multiply both sides by the reciprocal of the coefficient of the variable (or divide both sides by the coefficient itself). The coefficient of the variable will become +1.
4. Check your answer by substituting it for the variable in the original equation.

Example 1: Solving Linear Equations

Solve each of the following equations.

a. $3x + 3 = -18$

Solution:

$3x + 3 = -18$	Write the equation.
$3x + 3 - 3 = -18 - 3$	Add -3 to both sides.
$3x = -21$	Simplify.
$\frac{3x}{3} = \frac{-21}{3}$	Divide both sides by 3.
$x = -7$	Simplify.

Continued on the next page...

Check:

$3x+3=-18$

$3(-7)+3\overset{?}{=}-18$ Substitute $x=-7$.

$-21+3\overset{?}{=}-18$ Simplify.

$-18=-18$ True statement

b. $-26=2y-14-4y$

Solution:

$-26=2y-14-4y$ Write the equation.

$-26=-2y-14$ Combine like terms.

$-26+14=-2y-14+14$ Add 14 to both sides.

$-12=-2y$ Simplify.

$\frac{-12}{-2}=\frac{-2y}{-2}$ Divide both sides by -2.

$6=y$ Simplify.

Check:

$-26=2y-14-4y$

$-26\overset{?}{=}2(6)-14-4(6)$ Substitute $y=6$.

$-26\overset{?}{=}12-14-24$ Simplify.

$-26=-26$ True statement

Examples 2a and 2b illustrate solving equations with decimal coefficients. You may choose to work with the decimal coefficients as they are. However, another approach, shown here, is to multiply both sides in such a way to give integer coefficients. Generally, integers are easier to work with than decimals.

Example 2: Solving Linear Equations Involving Decimals

Solve each of the following equations.

a. $16.53-18.2z-7.43=0$

Solution:

$16.53-18.2z-7.43=0$ Write the equation.

$100(16.53-18.2z-7.43)=100(0)$ Multiply both sides by 100. (This results in integer coefficients.)

$1653-1820z-743=0$ Simplify.

$910-1820z=0$ Combine like terms.

$910-1820z-910=0-910$ Add -910 to both sides.

$-1820z=-910$ Simplify.

$$\frac{-1820z}{-1820} = \frac{-910}{-1820}$$ Divide both sides by −1820.

$$z = 0.5\left(\text{or } z = \frac{1}{2}\right)$$ Simplify.

Check: $16.53 - 18.2z - 7.43 = 0$

$$16.53 - 18.2(0.5) - 7.43 \stackrel{?}{=} 0$$ Substitute $z = 0.5$.

$$16.53 - 9.10 - 7.43 \stackrel{?}{=} 0$$ Simplify.

$$0 = 0$$ True statement

b. $5.1x + 7.4 - 1.8x = -9.1$

Solution: $5.1x + 7.4 - 1.8x = -9.1$ Write the equation.

$$10(5.1x + 7.4 - 1.8x) = 10(-9.1)$$ Multiply both sides by 10. (This results in integer coefficients.)

$$51x + 74 - 18x = -91$$ Simplify.

$$33x + 74 = -91$$ Combine like terms.

$$33x + 74 - 74 = -91 - 74$$ Add −74 to both sides.

$$33x = -165$$ Simplify.

$$\frac{33x}{33} = \frac{-165}{33}$$ Divide both sides by the coefficient 33.

$$x = -5$$ Simplify.

Examples 3a and 3b illustrate solving equations with coefficients that are fractions. You may choose to work with the the coefficients as they are. However, another approach, shown here, is to multiply both sides by the LCM of the denominators which will give integer coefficients Generally, integers are easier to work with than fractions.

Example 3: Solving Linear Equations with Fractional Coefficients

Solve each of the following equations.

a. $\frac{5}{6}x - \frac{5}{2} = -\frac{10}{9}$

Solution: $\frac{5}{6}x - \frac{5}{2} = -\frac{10}{9}$ Write the equation.

$$18\left(\frac{5}{6}x - \frac{5}{2}\right) = 18\left(-\frac{10}{9}\right)$$ Multiply both sides by 18 (the LCM of the denominators).

$$18\left(\frac{5}{6}x\right) - 18\left(\frac{5}{2}\right) = 18\left(-\frac{10}{9}\right)$$ Apply the distributive property.

$$15x - 45 = -20$$ Simplify.

Continued on the next page...

$$15x - 45 + 45 = -20 + 45$$ Add 45 to both sides.

$$15x = 25$$ Simplify.

$$\frac{15x}{15} = \frac{25}{15}$$ Divide both sides by 15.

$$x = \frac{5}{3}$$ Simplify.

Check: $$\frac{5}{6}x - \frac{5}{2} = -\frac{10}{9}$$

$$\frac{5}{6}\left(\frac{5}{3}\right) - \frac{5}{2} \stackrel{?}{=} -\frac{10}{9}$$ Substitute $y = \frac{5}{3}$.

$$\frac{25}{18} - \frac{45}{18} \stackrel{?}{=} -\frac{20}{18}$$ Simplify.

$$-\frac{20}{18} = -\frac{20}{18}$$ True statement

b. $\frac{1}{2}x + \frac{3}{4}x + \frac{7}{2} - \frac{2}{3}x = 0$

Solution: $$\frac{1}{2}x + \frac{3}{4}x + \frac{7}{2} - \frac{2}{3}x = 0$$ Write the equation.

$$12\left(\frac{1}{2}x + \frac{3}{4}x + \frac{7}{2} - \frac{2}{3}x\right) = 12(0)$$ Multiply both sides by 12 (the LCM of the denominators).

$$12\left(\frac{1}{2}x\right) + 12\left(\frac{3}{4}x\right) + 12\left(\frac{7}{2}\right) - 12\left(\frac{2}{3}x\right) = 12(0)$$ Apply the distributive property.

$$6x + 9x + 42 - 8x = 0$$ Simplify.

$$7x + 42 = 0$$ Combine like terms.

$$7x + 42 - 42 = 0 - 42$$ Add −42 to both sides.

$$7x = -42$$ Simplify.

$$\frac{7x}{7} = \frac{-42}{7}$$ Divide both sides by 7.

$$x = -6$$ Simplify. Checking will show that −6 is the solution.

NOTES

ABOUT CHECKING

Checking can be quite time-consuming and need not be done for every problem. This is particularly important on exams. You should check only if you have time after the entire exam is completed.

Practice Problems

Solve the following linear equations.

1. $x + 14 - 8x = -7$
2. $2.4 = 2.6y - 5.9y - 0.9$
3. $n - \frac{2n}{3} - \frac{1}{2} = \frac{1}{6}$
4. $\frac{3x}{14} + \frac{1}{2} - \frac{x}{7} = 0$

2.4 Exercises

1. $x = -3$
2. $x = -5$
3. $x = 2$
4. $y = -1$
5. $x = 2$
6. $n = 4$
7. $x = 2$
8. $x = 8$
9. $y = -1$
10. $x = -2$
11. $t = -1$
12. $x = -6$
13. $x = -0.12$
14. $x = -1.8$
15. $x = 4$
16. $x = 2$
17. $x = 0$
18. $x = \frac{1}{9}$
19. $y = 0$
20. $y = 0$
21. $x = -2$
22. $y = 2$
23. $y = -6$
24. $y = -6$
25. $n = 6$
26. $n = -4$
27. $n = 8$
28. $n = -4$
29. $x = 0$
30. $x = 0$

Solve each of the following linear equations.

1. $3x + 11 = 2$
2. $3x + 10 = -5$
3. $5x - 4 = 6$
4. $4y - 8 = -12$
5. $6x + 10 = 22$
6. $3n + 7 = 19$
7. $9x - 5 = 13$
8. $2x - 4 = 12$
9. $1 - 3y = 4$
10. $5 - 2x = 9$
11. $14 + 9t = 5$
12. $5 + 2x = -7$
13. $-5x + 2.9 = 3.5$
14. $3x + 2.7 = -2.7$
15. $10 + 3x - 4 = 18$
16. $5 + 5x - 6 = 9$
17. $15 = 7x + 7 + 8$
18. $14 = 9x + 5 + 8$
19. $5y - 3y + 2 = 2$
20. $6y + 8y - 7 = -7$
21. $x - 4x + 25 = 31$
22. $3y + 9y - 13 = 11$
23. $-20 = 7y - 3y + 4$
24. $-20 = 5y + y + 16$
25. $4n - 10n + 35 = 1 - 2$
26. $-5n - 3n + 2 = 34$
27. $3n - 15 - n = 1$
28. $2n + 12 + n = 0$
29. $5.4x - 0.2x = 0$
30. $0 = 5.1x + 0.3x$
31. $\frac{1}{2}x + 7 = \frac{7}{2}$
32. $\frac{3}{5}x + 4 = \frac{9}{5}$
33. $\frac{1}{2} - \frac{8}{3}x = \frac{5}{6}$
34. $\frac{2}{5} - \frac{1}{2}x = \frac{7}{4}$
35. $\frac{3}{2} = \frac{1}{3}x + \frac{11}{3}$
36. $\frac{11}{8} = \frac{1}{5}x + \frac{4}{5}$
37. $\frac{7}{2} - 5 - \frac{5}{2}x = 9$
38. $\frac{8}{3} + 2 - \frac{7}{3}x = 6$
39. $\frac{5}{8}x - \frac{1}{4}x + \frac{1}{2} = \frac{3}{10}$

Answers to Practice Problems: **1.** $x = 3$ **2.** $y = -1$ **3.** $n = 2$ **4.** $x = -7$

31. $x=-7$
32. $x=-\frac{11}{3}$
33. $x=-\frac{1}{8}$
34. $x=-\frac{27}{10}$
35. $x=-\frac{13}{2}$
36. $x=\frac{23}{8}$
37. $x=-\frac{21}{5}$
38. $x=-\frac{4}{7}$
39. $x=-\frac{8}{15}$
40. $x=2$
41. $y=\frac{28}{5}$
42. $y=23$
43. $x=2$
44. $x=-\frac{1}{4}$
45. $y=\frac{7}{5}$
46. $y=-10$
47. $x=-4.5$
48. $x=2.2$
49. $x=-44$
50. $y=50$
51. $x=2$
52. $x=6$
53. $x=-4$
54. $x=2$
55. $y=0.5$
56. $x=0.3$
57. $x=1.5$
58. $y=3.6$
59. $x=0.2$
60. $x=-0.03$
61. 36.5°C
62. 14,000 tickets per hour
63. 135 yards
64. 367.5 feet

40. $\frac{1}{2}x+\frac{3}{4}x-\frac{5}{3}=\frac{5}{6}$

41. $\frac{y}{2}+\frac{1}{5}=3$

42. $\frac{y}{3}-\frac{2}{3}=7$

43. $\frac{7}{8}=\frac{3}{4}x-\frac{5}{8}$

44. $\frac{1}{10}=\frac{4}{5}x+\frac{3}{10}$

45. $\frac{y}{7}+\frac{y}{28}+\frac{1}{2}=\frac{3}{4}$

46. $\frac{5y}{6}-\frac{7y}{8}-\frac{1}{12}=\frac{1}{3}$

47. $x+1.2x+6.9=-3.0$

48. $3x-0.75x-1.72=3.23$

49. $10=x-0.5x+32$

50. $33=y+3-0.4y$

51. $2.5x+0.5x-3.5=2.5$

52. $4.7-0.5x-0.3x=-0.1$

53. $6.4+1.2x+0.3x=0.4$

54. $5.2-1.3x-1.5x=-0.4$

55. $-12.13=2.42y+0.6y-13.64$

56. $-7.01=1.75x+3.05x-8.45$

57. $-0.4x+x+17.2=18.1$

58. $y-0.75y+13.76=14.66$

59. $0=17.3x-15.02x-0.456$

60. $0=20.5x-16.35x+0.1245$

61. Temperature: Jeff, who lives in England, is reading a letter from his pen pal in the United States. His pen pal says that the temperature in his city was 97.7° Fahrenheit one day, so it was too hot to play soccer outside. Jeff doesn't know how hot this is, because he is used to temperatures in Celsius. Help Jeff solve the equation below to determine the temperature in degrees Celsius.

$$1.8C+32=97.7$$

62. Music: The tickets for a concert featuring the new hit band, Flying Sailor, sold out in 2.5 hours. If there were 35,000 tickets sold, solve the equation $35{,}000-2.5x=0$ to find the number of tickets sold per hour.

63. Parking lots: A rectangular shaped parking lot is to have a perimeter of 450 yards. If the width must be 90 yards because of a building code, solve the equation $2l+2(90)=450$ to determine the length of the parking lot.

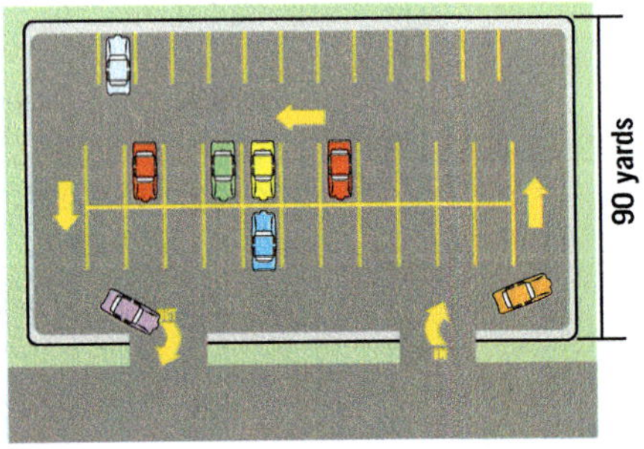

64. Height: The tallest man-made structure in the world is the Burj Khalifa in Dubai, which stands at 2717 feet tall. The tallest tree in the world is a Mendocino tree in California. If 7 of these trees were stacked on top of each other, they would still be 144.5 feet shorter than the Burj Khalifa. Solve the equation below to determine the height of the tree.

$$7x+144.5=2717$$

65. $x = 6.1$
66. $y = -2.3$
67. $x = 1.12$
68. $y = 5.79$

Calculator Problems

Use a graphing calculator to help solve the linear equations.

65. $0.15x + 5.23x - 17.815 = 15.003$

66. $15.97y - 12.34y + 16.95 = 8.601$

67. $13.45x - 20x - 17.36 = -24.696$

68. $26.75y - 30y + 23.28 = 4.4625$

HAWKES LEARNING SYSTEMS: INTRODUCTORY & INTERMEDIATE ALGEBRA SOFTWARE

- 2.4 Solving Linear Equations

2.5 Solving Linear Equations: $ax + b = cx + d$

- *Solve equations of the form* $\boldsymbol{ax + b = cx + d}$.
- *Understand the terms* ***conditional equations***, ***identities***, *and* ***contradictions***.

Now we are ready to solve linear equations of the most general form $\boldsymbol{ax + b = cx + d}$ where constants and variables may be on both sides. There may also be parentheses or other symbols of inclusion. **Remember that the objective is to get the variable on one side of the equation by itself with a coefficient of +1.**

General Procedure for Solving Linear Equations that Simplify to the Form $ax + b = cx + d$

1. Simplify each side of the equation by removing any grouping symbols and combining like terms on both sides of the equation.
2. Use the **addition principle of equality** and add the opposite of a constant term and/or variable term to both sides so that variables are on one side and constants are on the other side.
3. Use the **multiplication** (or **division**) **principle of equality** to multiply both sides by the reciprocal of the coefficient of the variable (or divide both sides by the coefficient itself). The coefficient of the variable will become +1.
4. Check your answer by substituting it for the variable in the original equation.

Example 1: Solving Equations with Variables on Both Sides

Teaching Notes:
You might want to emphasize to your students that there is more than one possible first step. For example in Example 1a, the text shows adding -3 to both sides. You can show them that adding $+18$, $-2x$, or $-5x$ as a first step will give the same solution.

Solve the following equations.

a. $5x + 3 = 2x - 18$

Solution:

$5x + 3 = 2x - 18$	Write the equation.
$5x + 3 - 3 = 2x - 18 - 3$	Add -3 to both sides.
$5x = 2x - 21$	Simplify.
$5x - 2x = 2x - 21 - 2x$	Add $-2x$ to both sides.
$3x = -21$	Simplify.
$\frac{3x}{3} = \frac{-21}{3}$	Divide both sides by 3.
$x = -7$	Simplify.

Check:

$$5x+3=2x-18$$

$$5(-7)+3\stackrel{?}{=}2(-7)-18 \quad \text{Substitute } x=-7.$$

$$-35+3\stackrel{?}{=}-14-18 \quad \text{Simplify.}$$

$$-32=-32 \quad \text{True statement}$$

b. $4x+1-x=2x-13+5$

Solution:

$$4x+1-x=2x-13+5 \quad \text{Write the equation.}$$

$$3x+1=2x-8 \quad \text{Combine like terms.}$$

$$3x+1-1=2x-8-1 \quad \text{Add } -1 \text{ to both sides.}$$

$$3x=2x-9 \quad \text{Simplify.}$$

$$3x-2x=2x-9-2x \quad \text{Add } -2x \text{ to both sides.}$$

$$x=-9 \quad \text{Simplify.}$$

Check:

$$4x+1-x=2x-13+5$$

$$4(-9)+1-(-9)\stackrel{?}{=}2(-9)-13+5 \quad \text{Substitute } x=-9.$$

$$-36+1+9\stackrel{?}{=}-18-13+5 \quad \text{Simplify.}$$

$$-26=-26 \quad \text{True statement}$$

Example 2: Solving Linear Equations Involving Decimals

Teaching Notes:
Another good approach in Example 2 is to first multiply each term by 10 to clear the decimals. You might want to go further and give an example where they could multiply by 100 to clear decimals.

$6y+2.5=7y-3.6$

Solution:

$$6y+2.5=7y-3.6 \quad \text{Write the equation.}$$

$$6y+2.5+3.6=7y-3.6+3.6 \quad \text{Add 3.6 to both sides.}$$

$$6y+6.1=7y \quad \text{Simplify.}$$

$$6y+6.1-6y=7y-6y \quad \text{Add } -6y \text{ to both sides.}$$

$$6.1=y \quad \text{Simplify.}$$

Check:

$$6y+2.5=7y-3.6$$

$$6(6.1)+2.5\stackrel{?}{=}7(6.1)-3.6 \quad \text{Substitute } y=6.1.$$

$$36.6+2.5\stackrel{?}{=}42.7-3.6 \quad \text{Simplify.}$$

$$39.1=39.1 \quad \text{True statement}$$

Example 3: Solving Linear Equations with Fractional Coefficients

$\frac{1}{3}x+\frac{1}{6}=\frac{2}{5}x-\frac{7}{10}$

Solution:

$\frac{1}{3}x+\frac{1}{6}=\frac{2}{5}x-\frac{7}{10}$	Write the equation.
$30\left(\frac{1}{3}x+\frac{1}{6}\right)=30\left(\frac{2}{5}x-\frac{7}{10}\right)$	Multiply both sides by the LCM, 30.
$30\left(\frac{1}{3}x\right)+30\left(\frac{1}{6}\right)=30\left(\frac{2}{5}x\right)-30\left(\frac{7}{10}\right)$	Apply the distributive property.
$10x+5=12x-21$	Simplify.
$10x+5-5=12x-21-5$	Add -5 to both sides.
$10x=12x-26$	Simplify.
$10x-12x=12x-26-12x$	Add $-12x$ to both sides.
$-2x=-26$	Simplify.
$\frac{-2x}{-2}=\frac{-26}{-2}$	Divide both sides by -2.
$x=13$	Simplify.

Example 4: Solving Equations with Parentheses

Solve the following equations.

a. $2(y-7)=4(y+1)-26$

Solution:

$2(y-7)=4(y+1)-26$	Write the equation.
$2y-14=4y+4-26$	Use the distributive property.
$2y-14=4y-22$	Combine like terms.
$2y-14+22=4y-22+22$	Add 22 to both sides. Here we will put the variable on the right side to get a positive coefficient of y.
$2y+8=4y$	Simplify.
$2y+8-2y=4y-2y$	Add $-2y$ to both sides.
$8=2y$	Simplify.
$\frac{8}{2}=\frac{2y}{2}$	Divide both sides by 2.
$4=y$	Simplify.

b. $-2(5x+13)-2=-6(3x-2)-41$

Solution:

$-2(5x+13)-2=-6(3x-2)-41$	Write the equation.
$-10x-26-2=-18x+12-41$	Use the distributive property. Be careful with the signs.
$-10x-28=-18x-29$	Combine like terms.
$-10x-28+18x=-18x-29+18x$	Add $18x$ to both sides.
$8x-28=-29$	Simplify.
$8x-28+28=-29+28$	Add 28 to both sides.
$8x=-1$	Simplify.
$\frac{8x}{8}=\frac{-1}{8}$	Divide both sides by 8.
$x=-\frac{1}{8}$	Simplify.

Conditional Equations, Identities, and Contradictions

When solving equations, there are times that we are concerned with the number of solutions that an equation has. If an equation has a finite number of solutions (the number of solutions is a countable number), the equation is said to be a **conditional equation**. As stated earlier, every linear equation has exactly one solution. Thus **every linear equation is a conditional equation**. However, in some cases, simplifying an equation will lead to a statement that is always true, such as $0 = 0$. In these cases the original equation is called an **identity** and has an infinite number of solutions which can be written as all real numbers or $\mathbb{R}$. If the equation simplifies to a statement that is never true, such as $0 = 2$, then the original equation is called a **contradiction** and there is no solution. Table 1 summarizes these ideas.

Type of Equation	Number of Solutions
Conditional	Finite number of solutions
Identity	Infinite number of solutions
Contradiction	No solution

Table 1

Example 5: Solutions of Equations

Determine whether each of the following equations is a conditional equation, an identity, or a contradiction.

a. $3x + 16 = -11$

Solution:

$3x + 16 = -11$	Write the equation.
$3x + 16 - 16 = -11 - 16$	Add -16 to both sides.
$3x = -27$	Simplify.
$\frac{3x}{3} = \frac{-27}{3}$	Divide both sides by 3.
$x = -9$	Simplify.

The equation has one solution and it is a conditional equation.

b. $3(x - 25) + 3x = 6(x + 10)$

Solution:

$3(x - 25) + 3x = 6(x + 10)$	Write the equation.
$3x - 75 + 3x = 6x + 60$	Use the distributive property.
$6x - 75 = 6x + 60$	Combine like terms.
$6x - 75 - 6x = 6x + 60 - 6x$	Add $-6x$ to both sides.
$-75 = 60$	Simplify.

The last equation is never true. Therefore, the original equation is a contradiction and has no solution.

c. $-2(x - 7) + x = 14 - x$

Solution:

$-2(x - 7) + x = 14 - x$	Write the equation.
$-2x + 14 + x = 14 - x$	Use the distributive property.
$14 - x = 14 - x$	Combine like terms.
$14 - x + x = 14 - x + x$	Add x to both sides.
$14 = 14$	Simplify.

The last equation is always true. Therefore, the original equation is an identity and has an infinite number of solutions. Every real number is a solution.

Practice Problems

Solve the following linear equations.

1. $x+14-6x=2x-7$
2. $6.4x+2.1=3.1x-1.2$
3. $\frac{2x}{3}-\frac{1}{2}=x+\frac{1}{6}$
4. $\frac{3}{14}n+\frac{1}{4}=\frac{1}{7}n-\frac{1}{4}$
5. $5-(y-3)=14-4(y+2)$

Determine whether each of the following equations is a conditional equation, an identity, or a contradiction.

6. $7(x-3)+42=7x+21$
7. $-2x+14=-2(x+1)+10$

2.5 Exercises

1. $x=-5$
2. $x=-2$
3. $n=3$
4. $y=-2$
5. $y=6$
6. $y=-\frac{5}{6}$
7. $x=3$
8. $x=-3$
9. $n=0$
10. $x=0$
11. $y=0$
12. $x=0$
13. $z=-1$
14. $x=3$
15. $y=\frac{1}{5}$
16. $x=5$
17. $x=-3$
18. $y=-3$
19. $x=-4$
20. $x=-1$
21. $x=-21$
22. $n=-15$
23. $y=0$
24. $n=0$
25. $y=1$

Solve each of the following linear equations.

1. $3x+2=x-8$
2. $5x+1=2x-5$
3. $4n-3=n+6$
4. $6y+3=y-7$
5. $3y+18=7y-6$
6. $2y+5=8y+10$
7. $3x+11=8x-4$
8. $9x+3=5x-9$
9. $14n=3n$
10. $1.6x=0.8x$
11. $6y-2.1=y-2.1$
12. $13x+5=2x+5$
13. $2(z+1)=3z+3$
14. $6x-3=3(x+2)$
15. $16y+23y-3=16y-2y+2$
16. $5x-2x+4=3x+x-1$
17. $0.25+3x+6.5=0.75x$
18. $0.9y+3=0.4y+1.5$
19. $6.5+1.2x=0.5-0.3x$
20. $x-0.1x+0.8=0.2x+0.1$
21. $\frac{2}{3}x+1=\frac{1}{3}x-6$
22. $\frac{4}{5}n+2=\frac{2}{5}n-4$
23. $\frac{y}{5}+\frac{3}{4}=\frac{y}{2}+\frac{3}{4}$
24. $\frac{5n}{6}+\frac{1}{9}=\frac{3n}{2}+\frac{1}{9}$
25. $\frac{3}{8}\left(y-\frac{1}{2}\right)=\frac{1}{8}\left(y+\frac{1}{2}\right)$
26. $\frac{1}{2}\left(\frac{x}{2}+1\right)=\frac{1}{3}\left(\frac{x}{2}-1\right)$

Answers to Practice Problems: **1.** $x=3$ **2.** $x=-1$ **3.** $x=-2$ **4.** $n=-7$ **5.** $y=-\frac{2}{3}$ **6.** Identity **7.** Contradiction

26. $x=-10$
27. $x=-\frac{3}{2}$
28. $x=-\frac{2}{3}$
29. $x=\frac{1}{4}$
30. $x=\frac{7}{60}$
31. $x=\frac{3}{17}$
32. $x=-\frac{15}{7}$
33. $x=-\frac{1}{4}$
34. $x=\frac{8}{3}$
35. $x=\frac{8}{5}$
36. $y=-\frac{1}{2}$
37. $x=\frac{2}{3}$
38. $x=-\frac{7}{3}$
39. $x=6$
40. $x=20$
41. $x=-11$
42. $x=-30$
43. $x=\frac{1}{2}$
44. $x=\frac{24}{19}$
45. $x=-5$
46. $y=4.8$
47. $n=-1.5$
48. $n=0.97$
49. $x=0$
50. $x=0$
51. Conditional
52. Contradiction
53. Conditional
54. Identity
55. Contradiction

27. $\frac{2x}{3}+\frac{x}{3}=-\frac{3}{4}+\frac{x}{2}$

28. $\frac{3}{4}x+\frac{1}{5}x=\frac{1}{2}x-\frac{3}{10}$

29. $x+\frac{2}{3}x-2x=\frac{x}{6}-\frac{1}{8}$

30. $3x+\frac{1}{2}x-\frac{2}{5}x=\frac{x}{10}+\frac{7}{20}$

31. $3(1+9x)=6(2-4x)$

32. $4(5-x)=8(3x+10)$

33. $3(4x-1)=4(2x-3)+8$

34. $7(2x-1)=5(x+6)-13$

35. $5-3(2x+1)=4(x-5)+6$

36. $-2(y+5)-4=6(y-2)+2$

37. $8+4(2x-3)=5-(x+3)$

38. $8(3x+5)-9=9(x-2)+14$

39. $4.7-0.3x=0.5x-0.1$

40. $5.8-0.1x=0.2x-0.2$

41. $0.2(x+3)=0.1(x-5)$

42. $0.4(x+3)=0.3(x-6)$

43. $\frac{1}{2}(4-8x)=\frac{1}{3}(4x+7)-3$

44. $3+\frac{1}{4}(x-4)=\frac{2}{5}(2+3x)$

45. $0.6x-22.9=1.5x-18.4$

46. $0.1y+3.8=5.72-0.3y$

47. $0.12n+0.25n-5.895=4.3n$

48. $0.15n+32n-21.0005=10.5n$

49. $0.7(x+14.1)=0.3(x+32.9)$

50. $0.8(x-6.21)=0.2(x-24.84)$

Determine whether each of the following equations is a conditional equation, an identity, or a contradiction.

51. $2(3x-1)+5=3$

52. $-2x+13=-2(x-7)$

53. $5x+13=-2(x-7)+3$

54. $3x+9=-3(x-3)+6x$

55. $7(x-1)=-3(3-x)+4x$

56. $3(x-2)+4x=6(x-1)+x$

57. $5(x+1)=3(x+1)+2(x+1)$

58. $8x-20+x=-3(5-2x)+3(x-4)$

59. $2x+3x=5.2(3-x)$

60. $5.2x+3.4x=0.2(x-0.42)$

61. Construction: A farmer is putting a shed on his property. He has two designs. One uses wood and would cost \$2 per square foot plus an extra \$8400 in materials. The other design is metal and would cost \$4 per square foot plus an additional \$8800. Both sheds are the same size, and the wood shed costs $\frac{3}{4}$ what the metal shed costs. Solve the following equation to determine how many square feet the shed will be.

$$2x+8400=\frac{3}{4}(4x+8800)$$

56. Identity
57. Identity
58. Contradiction
59. Conditional
60. Conditional
61. 1800 square feet
62. 150 boxes
63. 240 sundaes
64. 3 hours
65. $x = -50.21$
66. $x = -9.293$
67. $x = 1.067$
68. $x = 4.58$

62. **Renting buildings:** Two rival shoe companies want to rent the same empty building for their office and shipping space. Schulster's Shoes needs 1000 square feet for offices, 600 for shipping, and another 6 square feet for every packaged box of shoes. Shoes, Shoes, Shoes! needs 750 square feet for offices, 400 for shipping, and 9 square feet per packaged box of shoes. If only one company will fit exactly into the empty building and they both plan to have the same amount of inventory, solve the following equation to determine how many boxes of shoes both companies hope to have at any given time.

$$1000 + 600 + 6x = 750 + 400 + 9x$$

63. **Ice cream:** An ice cream shop is having a special "Ice Cream Sunday" event in which they are giving away giant mixed sundaes of 3 scoops of vanilla ice cream and 2 scoops of chocolate. If they have 24 gallons of chocolate and 36 gallons of vanilla to start with, solve the given equation to determine how many sundaes they will have made when they run out of ice cream. (For this problem, we assume a gallon = 20 scoops.)

$$36 - \frac{1}{20}(3x) = 24 - \frac{1}{20}(2x)$$

64. **Music:** A guitarist and a drummer are getting ready for a gig. The length of the gig will depend on how much material they have prepared. For every hour of the show, the guitarist must practice for 5 days and the drummer for 3 days. Since the guitarist knows some of the songs, he saves 3 days of practice time. The drummer hurts his hand and loses 3 days of practice. If they plan to start and finish practicing at the same time, solve the following equation to determine how long the show will be.

$$5x - 3 = 3x + 3$$

Calculator Problems

Use a graphing calculator to help solve the linear equations.

65. $0.17x - 23.0138 = 1.35x + 36.234$

66. $48.512 - 1.63x = 2.58x + 87.63553$

67. $0.32(x + 14.1) = 2.47x + 2.21795$

68. $1.6(9.3 + 2x) = 0.2(3x + 133.94)$

HAWKES LEARNING SYSTEMS: INTRODUCTORY & INTERMEDIATE ALGEBRA SOFTWARE

- 2.5 More Linear Equations: $ax + b = cx + d$

2.6 Introduction to Problem Solving

- *Understand the four basic steps in solving applications.*
- *Solve word problems involving translating number phrases, consecutive integers, and other applications.*

George Pólya (1877 – 1985), a famous professor at Stanford University, studied the process of discovery learning. Among his many accomplishments, he developed the following four-step process as an approach to problem solving.

1. **Understand the problem.**
2. **Devise a plan.**
3. **Carry out the plan.**
4. **Look back over the results.**

For a complete discussion of these ideas, see *How To Solve It by Pólya* (Princeton University Press, 2nd edition, 1957). The following quote by Pólya illustrates his sense of humor and his understanding of students' dilemmas: "The traditional mathematics professor of the popular legend is absent-minded. He usually appears in public with a lost umbrella in each hand. He prefers to face a blackboard and to turn his back on the class. He writes a, he says b, he means c, but it should be d. Some of his sayings are handed down from generation to generation."

There are a variety of types of applications discussed throughout this text and subsequent courses in mathematics, and you will find these four steps helpful as guidelines for understanding and solving all of them. Applying the necessary skills to solve exercises, such as combining like terms or solving equations, is not the same as accumulating the knowledge to solve problems. **Problem solving can involve careful reading, reflection, and some original or independent thought.**

Basic Steps for Solving Applications

1. Understand the problem. For example,
  - **a.** Read the problem carefully, maybe several times.
  - **b.** Understand all the words.
  - **c.** If it helps, restate the problem in your own words.
  - **d.** Be sure that there is enough information.

2. Devise a plan. For example,
  - **a.** Guess, estimate, or make a list of possibilities.
  - **b.** Draw a picture or diagram.
  - **c.** Represent the unknown quantity with a variable and form an equation.

Continued on the next page...

Basic Steps for Solving Applications (cont.)

3. Carry out the plan. For example,
   a. Try all the possibilities you have listed.
   b. Study your picture or diagram for insight into the solution.
   c. Solve any equation that you may have set up.
4. Look back over the results. For example,
   a. Can you see an easier way to solve the problem?
   b. Does your solution actually work? Does it make sense in terms of the wording of the problem? Is it reasonable?
   c. If there is an equation, check your answer in the equation.

NOTES

You may find that many of the applications in this section can be solved by "reasoning," and there is nothing wrong with that approach. Reasoning is a fundamental part of all of mathematics. However, keep in mind that the algebraic techniques you are learning are important. They also involve reasoning and will prove very useful in solving more complicated problems in later sections and in later courses.

Number Problems

In Section 2.2, we discussed translating English phrases into algebraic expressions. Now we will use those skills to read number problems and translate the sentences and phrases in the problem into a related equation. The solution of this equation will be the solution to the problem.

Example 1: Number Problems

a. If a number is decreased by 36 and the result is 76 less than twice the number, what is the number?

Solution: Let n = the unknown number.

$$\underbrace{n-36}_{\text{a number decreased by 36}} \quad \underbrace{=}_{\text{the result is}} \quad \underbrace{2n-76}_{\text{76 less than twice the number}}$$

$$n-36-n=2n-76-n$$

$$-36=n-76$$

$$-36+76=n-76+76$$

$$40=n$$

The number is 40.

Continued on the next page...

b. Three times the sum of a number and 5 is equal to twice the number plus 5. Find the number.

Solution: Let x = the unknown number.

$$\underbrace{3(x+5)}_{\text{3 times the sum of a number and 5}} \quad \underbrace{=}_{\text{is equal to}} \quad \underbrace{2x+5}_{\text{twice the number plus 5}}$$

$$3x+15=2x+5$$
$$3x+15-2x=2x+5-2x$$
$$x+15=5$$
$$x+15-15=5-15$$
$$x=-10$$

The number is –10.

c. One integer is 4 more than three times a second integer. Their sum is 24. What are the two integers?

Solution: Let n = the second integer,
then $3n+4$ = the first integer.

$$(\text{1st integer})+(\text{2nd integer})=24$$
$$(3n+4)+n=24 \qquad \text{Their sum is 24.}$$
$$4n+4=24$$
$$4n+4-4=24-4$$
$$4n=20$$
$$\frac{4n}{4}=\frac{20}{4}$$
$$n=5$$
$$3n+4=19$$

The two integers are 5 and 19.

Consecutive Integers

Remember that the set of **integers** consists of the whole numbers and their opposites.

$$\mathbb{Z}=\{\ldots,-4,-3,-2,-1,0,1,2,3,4,\ldots\}$$

Even integers are integers that are divisible by 2. The even integers are

$$\{\ldots, -6, -4, -2, 0, 2, 4, 6, \ldots\}.$$

Odd integers are integers that are not even. If an odd integer is divided by 2 the remainder will be 1. The odd integers are

$$\{\ldots, -5, -3, -1, 1, 3, 5, \ldots\}.$$

NOTES

In this discussion we will be dealing only with integers. Therefore, if you get a result that has a fraction or decimal number (not an integer), you will know that an error has been made and you should correct some part of your work.

The following terms and the ways of representing the integers must be understood before attempting the problems.

Consecutive Integers

Integers are **consecutive** if each is 1 more than the previous integer. Three consecutive integers can be represented as

$$\boldsymbol{n}, \ \boldsymbol{n+1}, \ \text{and} \ \boldsymbol{n+2}.$$

Consecutive Even Integers

Even integers are **consecutive** if each is 2 more than the previous even integer. Three consecutive even integers can be represented as

$$\boldsymbol{n}, \ \boldsymbol{n+2}, \ \text{and} \ \boldsymbol{n+4}$$

where n is an **even** integer.

Consecutive Odd Integers

Odd integers are **consecutive** if each is 2 more than the previous odd integer. Three consecutive odd integers can be represented as

$$\boldsymbol{n}, \ \boldsymbol{n+2}, \ \text{and} \ \boldsymbol{n+4}$$

where n is an **odd** integer.

Note that consecutive even and consecutive odd integers are represented in the same way:

$$\boldsymbol{n},\ \boldsymbol{n+2},\ \text{and}\ \ \boldsymbol{n+4}.$$

The value of the first integer, n, determines whether the remaining integers are odd or even. For example,

<u>n is odd</u>		**<u>n is even</u>**
If $n = 11$,	**or**	If $n = 36$,
then $n + 2 = 13$		then $n + 2 = 38$
and $n + 4 = 15$.		and $n + 4 = 40$.

Example 2: Consecutive Integers

a. Three consecutive **odd** integers are such that their sum is –3. What are the integers?

Solution: Let $n =$ the first odd integer,

then $n + 2$ = the second odd integer

and $n + 4$ = the third odd integer.

Set up and solve the related equation.

$$(\text{1st integer}) + (\text{2nd integer}) + (\text{3rd integer}) = -3$$

$$n + (n+2) + (n+4) = -3$$

$$3n + 6 = -3$$

$$3n + 6 - 6 = -3 - 6$$

$$3n = -9$$

$$\frac{3n}{3} = \frac{-9}{3}$$

$$n = -3$$

$$n + 2 = -1$$

$$n + 4 = 1$$

The three consecutive odd integers are –3, –1, and 1.

Check: $(-3) + (-1) + (1) = -3$

b. Find three consecutive integers such that the sum of the first and third is 76 less than three times the second.

Solution: Let $n =$ the first integer,

then $n + 1$ = the second integer

and $n + 2$ = the third integer.

Set up and solve the related equation.

$$(\text{1st integer})+(\text{3rd integer})=3(\text{2nd integer})-76$$
$$n+(n+2)=3(n+1)-76$$
$$2n+2=3n+3-76$$
$$2n+2=3n-73$$
$$2n+2+73=3n-73+73$$
$$2n+75=3n$$
$$2n+75-2n=3n-2n$$
$$75=n$$
$$76=n+1$$
$$77=n+2$$

The three consecutive integers are 75, 76, and 77.

Check: $75+77=152$ and $3(76)-76=228-76=152$

Other applications

As you learn more abstract mathematical ideas, you will find that you will use these ideas and the related processes to solve a variety of everyday problems as well as problems in specialized fields of study. Generally, you may not even be aware of the fact that you are using your mathematical knowledge. However, these skills and ideas will be part of your thinking and problem solving techniques for the rest of your life.

Example 3: Applications

a. Joe pays \$800 per month to rent an apartment. If this is $\frac{2}{5}$ of his monthly income, what is his monthly income?

Solution: Let $x=$ Joe's monthly income, then $\frac{2}{5}x=$ rent.

$$\frac{2}{5}x=800$$
$$\frac{5}{2}\cdot\frac{2}{5}x=\frac{5}{2}\cdot\frac{800}{1}$$
$$x=2000$$

Joe's monthly income is \$2000.

b. A student bought a calculator and a textbook for a total of \$200.80 (including tax). If the textbook cost \$20.50 more than the calculator, what was the cost of each item?

Solution: Let $x=$ cost of the calculator,
then $x+20.50=$ cost of the textbook.

Continued on the next page...

The equation to be solved is:

$$\underbrace{x+20.50}_{\text{cost of textbook}} + \underbrace{x}_{\text{cost of calculator}} = \underbrace{200.80}_{\text{total spent}}$$

$$2x+20.50 = 200.80$$

$$2x+20.50-20.50 = 200.80-20.50$$

$$2x = 180.30$$

$$\frac{2x}{2} = \frac{180.30}{2}$$

$$x = 90.15 \quad \text{Cost of calculator}$$

$$x+20.50 = 110.65 \quad \text{Cost of textbook}$$

The calculator costs \$90.15 and the textbook costs \$90.15 + \$20.50 = \$110.65, with tax included in each price.

2.6 Exercises

Read each problem carefully, translate the various phrases into algebraic expressions, set up an equation, and solve the equation.

1. Five less than a number is equal to 13 decreased by the number. Find the number.
2. Three less than twice a number is equal to the number. What is the number?
3. Thirty-six is 4 more than twice a certain number. Find the number.
4. Fifteen decreased by twice a number is 27. Find the number.
5. Seven times a certain number is equal to the sum of twice the number and 35. What is the number?
6. The difference between twice a number and 3 is equal to 6 decreased by the number. Find the number.
7. Fourteen more than three times a number is equal to 6 decreased by the number. Find the number.
8. Two added to the quotient of a number and 7 is equal to –3. What is the number?
9. The quotient of twice a number and 5 is equal to the number increased by 6. What is the number?
10. Three times the sum of a number and 4 is equal to –9. Find the number.
11. Four times the difference between a number and 5 is equal to the number increased by 4. What is the number?
12. When 17 is added to six times a number, the result is equal to 1 plus twice the number. What is the number?

Answers:

1. $x-5=13-x$; 9
2. $2x-3=x$; 3
3. $36=2x+4$; 16
4. $15-2x=27$; -6
5. $7x=2x+35$; 7
6. $2x-3=6-x$; 3
7. $3x+14=6-x$; -2
8. $\frac{x}{7}+2=-3$; -35
9. $\frac{2x}{5}=x+6$; -10
10. $3(x+4)=-9$; -7
11. $4(x-5)=x+4$; 8
12. $6x+17=1+2x$; -4

13. $\frac{2x+5}{11}=4-x$; 3

14. $\frac{x-21}{2}=4x$; -3

15. $2x+3x=4(x+3)$; 12

16. $2(x-10)=6x+16$; -9

17. $n+(n+2)=60$; 29, 31

18. $n+(n+2)=78$; 38, 40

19. $n+(n+1)+(n+2)=69$; 22, 23, 24

20. $n+(n+1)+(n+2)=93$; 30, 31, 32

21. $n+(n+1)+(n+2)+(n+3)=74$; 17, 18, 19, 20

22. $n+(n+1)+(n+2)+(n+3)=90$; 21, 22, 23, 24

23. $171-n=(n+1)+(n+2)$; 56, 57, 58

24. $120-n=(n+1)+(n+2)$; 39, 40, 41

25. $208-3n=(n+1)+(n+2)+(n+3)-50$; 42, 43, 44, 45

26. $2n+3(n+1)=83$; 16, 17

27. $n+2(n+2)=4(n+4)-54$; 42, 44, 46

28. $(n+2)+(n+4)-n=66$; 60, 62, 64

29. $4n=(n+2)+(n+4)+44$; 25, 27, 29

30. $n+(n+2)+(n+4)=(n+2)+168$; 82, 84, 86

13. If the sum of twice a number and 5 is divided by 11, the result is equal to the difference between 4 and the number. Find the number.

14. If the difference between a number and 21 is divided by 2, the result is 4 times the number. What is the number?

15. Twice a number increased by three times the number is equal to 4 times the sum of the number and 3. Find the number.

16. Twice the difference between a number and 10 is equal to 6 times the number plus 16. What is the number?

17. The sum of two consecutive odd integers is 60. What are the integers?

18. The sum of two consecutive even integers is 78. What are the integers?

19. Find three consecutive integers whose sum is 69.

20. Find three consecutive integers whose sum is 93.

21. The sum of four consecutive integers is 74. What are the integers?

22. Find four consecutive integers whose sum is 90.

23. 171 minus the first of three consecutive integers is equal to the sum of the second and third. What are the integers?

24. If the first of three consecutive integers is subtracted from 120, the result is the sum of the second and third. What are the integers?

25. Four consecutive integers are such that if 3 times the first is subtracted from 208, the result is 50 less than the sum of the other three. What are the integers?

26. Find two consecutive integers such that twice the first plus three times the second equals 83.

27. Find three consecutive even integers such that the first plus twice the second is 54 less than four times the third.

28. Find three consecutive even integers such that if the first is subtracted from the sum of the second and third, the result is 66.

29. Find three consecutive odd integers such that 4 times the first is 44 more than the sum of the second and third.

30. Find three consecutive even integers such that their sum is 168 more than the second.

31. Find three consecutive odd integers such that the sum of twice the first and three times the second is 7 more than twice the third.

32. Find three consecutive even integers such that the sum of three times the first and twice the third is twenty less than six times the second.

33. School supplies: A mathematics student bought a graphing calculator and a textbook for a course in statistics. If the text costs \$49.50 more than the calculator, and the total cost for both was \$125.74, what was the cost of each item?

31. $2n+3(n+2) = 2(n+4)+7$; 3, 5, 7

32. $3n+2(n+4) = 6(n+2)-20$; 16, 18, 20

33. $c+(c+49.50) = 125.74$; Calculator: \$38.12, Textbook: \$87.62

34. $(c-170.70)+c = 225.50$; Flash drive: \$27.40; Printer: \$198.10

35. $2x + 90{,}000 = 310{,}000$; \$110,000

36. $3x + 1500 = 12{,}000$; \$3500

37. $(x+56)+x = 542$; 243 boys

38. $x+x+(x+3) +(x+3+5) = 19$; 2 magazines

39. $50 - 2x = 10.50$; \$19.75

40. $7x = 91{,}399$; 13,057

41. $(x+68)+x = 158$; \$45 million on electric guitars

42. $2x+0.28(250) = 140$; \$35

43. $20 + 0.22x = 66.20$; 210 miles

44. $0.35(3)+0.15x = 12.30$; 75 minutes

34. **Electronics:** The total cost of a computer flash drive and a color printer was \$225.50, including tax. If the cost of the flash drive was \$170.70 less than the printer, what was the cost of each item?

35. **Real estate:** A real estate agent says that the current value of a home is \$90,000 more than twice its value when it was new. If the current value is \$310,000, what was the value of the home when it was new? The home is 25 years old.

36. **Classic cars:** A classic car is now selling for \$1500 more than three times its original price. If the selling price is now \$12,000, what was the car's original price?

37. **Class size:** A high school graduating class is made up of 542 students. If there are 56 more girls than boys, how many boys are in the class?

38. **Mail:** On August 24, the Fernandez family received 19 pieces of mail, consisting of magazines, bills, letters, and ads. If they received the same number of magazines as letters, three more bills than letters, and five more ads than bills, how many magazines did they receive?

39. **Golfing:** Lucinda bought two boxes of golf balls. She gave the pro-shop clerk a 50-dollar bill and received \$10.50 in change. What was the cost of one box of golf balls? (Tax was included.)

40. **Guitars:** On average the number of electric guitars sold in Texas each year is 91,399, which is seven times the average number of guitars sold each year in Wyoming. How many electric guitars, on average, are sold each year in Wyoming?

41. **Guitars:** A guitar manufacturer spent \$158 million on the production of acoustic and electric guitars last year. If the amount the company spent producing acoustic guitars was \$68 million more than it spent on producing electric guitars, how much did the company spend producing electric guitars?

42. **Rental cars:** If the rental price on a car is a fixed price per day plus \$0.28 per mile, what was the fixed price per day if the total paid was \$140 and the car was driven for 250 miles in two days?

43. **Rental cars:** The U-Drive Company charges \$20 per day plus 22¢ per mile driven. For a one-day trip, Louis paid a rental fee of \$66.20. How many miles did he drive?

44. **Telephone calls:** For a long-distance call, the telephone company charges 35¢ for each of the first three minutes and 15¢ for each additional minute. If the cost of a call was \$12.30, how many minutes did the call last?

45. $x + x + 25{,}000 = 275{,}000$; Lot: \$125,000, House: \$150,000
46. $x + x + 2 + x + 6 = 29$; 7 feet, 9 feet, 13 feet
47. $x + 3x - 1 + 2x + 5 = 64$; 10 inches, 29 inches, 25 inches

45. Construction: Joe Johnson decided to buy a lot and build a house on the lot. He knew that the cost of constructing the house was going to be \$25,000 more than the cost of the lot. He told a friend that the total cost was going to be \$275,000. As a test to see if his friend remembered the algebra they had together in school, he challenged his friend to calculate what he paid for the lot and what he was going to pay for the house. What was the cost of the lot and the cost of the house?

46. Carpentry: A 29 foot board is cut into three pieces at a sawmill. The second piece is 2 feet longer than the first and the third piece is 4 feet longer than the second. What are the lengths of the three pieces?

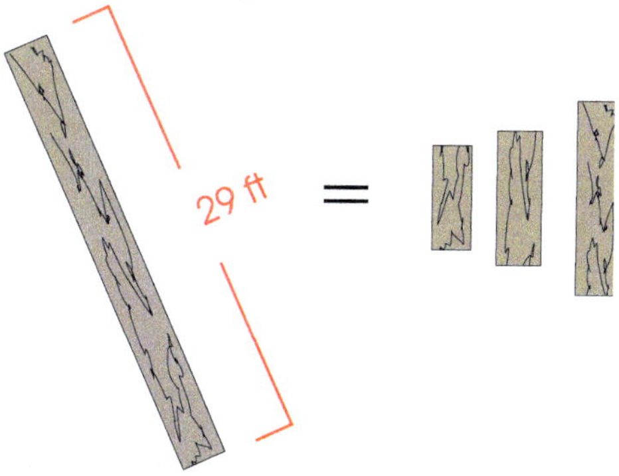

47. Triangles: The three sides of a triangle are x, $3x - 1$, and $2x + 5$ (as shown in the figure below). If the perimeter of the triangle is 64 inches, what is the length of each side? (**Reminder:** The perimeter of a triangle is the sum of the lengths of the sides.)

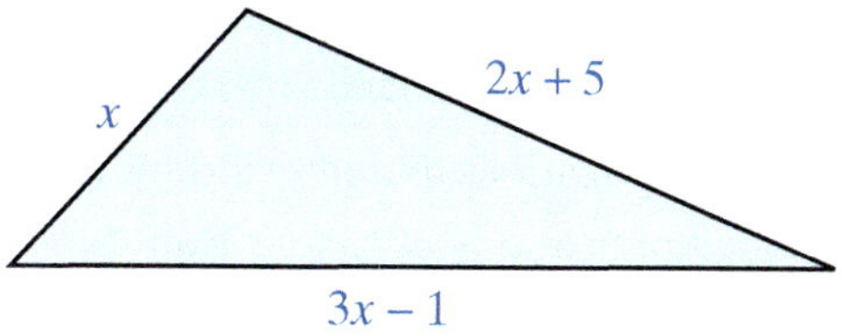

48. $n + 4n - 4 + 2n + 7 = 59$; 8 cm, 28 cm, 23 cm

Note: Answers for 49 – 54 will vary.

49. The difference between five times a number and the number is equal to 8; 2

50. Twice a number increased by 3 equals 9; 3

51. Find two consecutive integers whose sum is 33; 16, 17

48. Triangles: The three sides of a triangle are n, $4n - 4$, $2n + 7$ (as shown in the figure below). If the perimeter of the triangle is 59 cm, what is the length of each side?

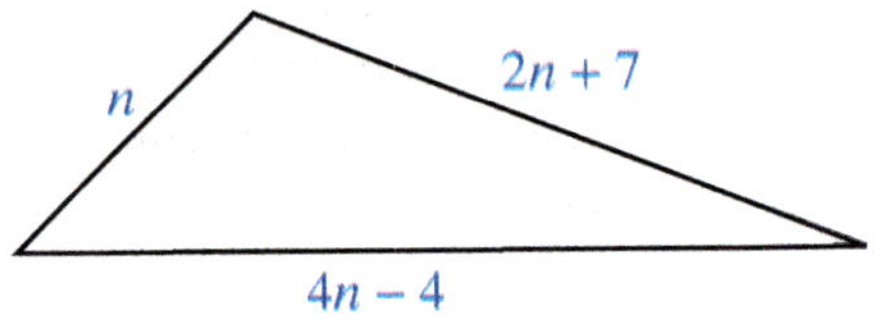

Make up your own word problem that might use the given equation in its solution. Be creative! Then solve the equation and check to see that the answer is reasonable.

49. $5x - x = 8$

50. $2x + 3 = 9$

51. $n + (n+1) = 33$

52. $n + (n+4) = 3(n+2)$

53. $3(n+1) = n + 53$

54. $2(n+2) - 1 = n + 4 - n$

Writing and Thinking About Mathematics

52. Find 3 consecutive even integers such that the sum of the first and the third is 3 times the second; –2, 0, 2

53. Find two consecutive integers such that 3 times the second is 53 more than the first; 25, 26

55. a. How would you represent four consecutive odd integers?

b. How would you represent four consecutive even integers?

c. Are these representations the same? Explain.

56. Discuss, briefly, how you would apply Pólya's four-step problem-solving process to a problem that you faced today. (For example, what route to take to school, what time to spend studying algebra, what movie to see, etc.)

a. What was the problem?

b. What was the plan?

c. How did you carry out the plan?

d. Did your solution make sense? Could you have solved the problem in a different way?

HAWKES LEARNING SYSTEMS: INTRODUCTORY & INTERMEDIATE ALGEBRA SOFTWARE

- 2.6a Applications: Number Problems and Consecutive Integers
- 2.6b Applications of Linear Equations Addition and Subtraction
- 2.6c Applications of Linear Equations: Multiplication and Division

54. Find a number such that twice the sum of the number and 2, decreased by 1 is equal to the sum of the number and 4 minus the number; $\frac{1}{2}$

55. **a.** $n, n + 2, n + 4, n + 6$ **b.** $n, n + 2, n + 4, n + 6$ **c.** Yes; Answers will vary.

56. See pages 142-143. Answers will vary.

2.7 Applications with Percent

- *Find percents of numbers.*
- *Find fractional parts of numbers.*
- *Solve word problems involving decimals, fractions, and percents.*

Our daily lives are filled with decimal numbers and percents: stock market reports, batting averages, won-lost records, salary raises, measures of pollution, taxes, interest on savings, home loans, discounts on clothes and cars, and on and on. Since decimal numbers and percents play such a prominent role in everyday life, we need to understand how to operate with them and apply them correctly in a variety of practical situations.

Brief Review of Decimals and Percents

Note: A more detailed review of decimals and percents is included in Appendix 1.

The word percent comes from the Latin *per centum,* meaning "per hundred." So, **percent means hundredths**. Thus

$$72\%,\quad \frac{72}{100},\quad \text{and } 0.72 \text{ all have the same meaning.}$$

Change a decimal to percent:

$0.036 = 3.6\%$ Move the decimal point two places to the right and add the % sign.

Change a percent to a decimal:

$56\% = 0.56$ Move the decimal point two places to the left and drop the % sign.

Change a percent to a fraction (or mixed number):

$30\% = \frac{30}{100} = \frac{3}{10}$ Write the percent in hundredths form and reduce.

Change a fraction (or mixed number) to a percent:

$\frac{1}{4} = 0.25 = 25\%$ Write the fraction in decimal form and then change the decimal to a percent.

The Basic Percent Equation

Consider the statement

"15% of 80 is 12."

This statement has three numbers in it. In general, solving a percent problem involves knowing two of these numbers and trying to find the third. That is, **there are three basic types of percent problems**. We can break down the sentence in the following way.

Sentence →	15%	of	80	is	12
Equation →	0.15	·	80	=	12
	percent written as decimal	"of" changed to times		"is" changed to =	
Basic Formula →	**Rate**	·	**Base**	=	**Amount**

The terms that we have just discussed are explained in detail in the following box.

The Basic Formula $R \cdot B = A$

R = **RATE** or percent (as a decimal or fraction)
B = **BASE** (number we are finding the percent of)
A = **AMOUNT** (a part of the base)
"**of**" means to multiply.
"**is**" means equal (=).
The relationship among R, B, and A is given in the equation

$$\boldsymbol{R \cdot B = A} \quad (\text{or } \boldsymbol{A = R \cdot B}).$$

Teaching Notes:
Some students might find the following diagram helpful as a mnemonic device:

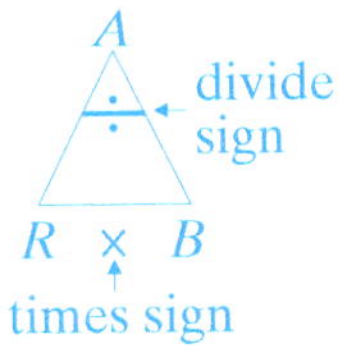

Even though there are just three basic types of percent problems, many people have difficulty deciding whether to multiply or divide in a particular problem. Using the equation $R \cdot B = A$ helps to avoid these difficulties. If the values of any two of the quantities in the formula are known, they can be substituted in the equation and then the missing number can be found by solving the equation. **The process of solving the equation for the unknown quantity determines whether multiplication or division is needed.**

The following examples illustrate how to substitute into the equation and how to solve the resulting equations.

Example 1: Percent of a Number

a. What is 72% of 800?

Solution: $R = 0.72$ and $B = 800$ and A is unknown.

$$R \cdot B = A$$
$$\downarrow \quad \downarrow \quad \downarrow$$
$$0.72 \cdot 800 = A \qquad \text{Here, simply multiply to find } A.$$
$$576 = A$$

So **576** is 72% of 800.

b. 57% of what number is 163.191?

Solution: $R = 0.57$ and B is unknown and A is 163.191.

$$R \cdot B = A$$
$$\downarrow \quad \downarrow \quad \downarrow$$
$$0.57 \cdot B = 163.191$$
$$\frac{0.57 \cdot B}{0.57} = \frac{163.191}{0.57} \qquad \text{Now divide both sides by 0.57 to find } B.$$
$$B = 286.3$$

So 57% of **286.3** is 163.191.

c. What percent of 180 is 45?

Solution: R is unknown and B is 180 and A is 45.

$$R \cdot B = A$$
$$\downarrow \quad \downarrow \quad \downarrow$$
$$R \cdot 180 = 45$$
$$\frac{R \cdot 180}{180} = \frac{45}{180} \qquad \text{Now divide both sides by 180 to find } R.$$
$$R = 0.25$$
$$R = 25\% \qquad \text{Write the decimal in percent form.}$$

So **25%** of 180 is 45.

Applications with Percent

To sell goods that have been in stock for some time or simply to attract new customers, retailers and manufacturers sometimes offer a **discount**, a reduction in the selling price usually stated as a percent of the original price. The new, reduced price is called the **sale price**.

Example 2: Discounts

A bicycle was purchased at a discount of 25% of its original price of \$1600. What was the sale price?

Solution: There are two ways to approach this problem. One way is to find the discount and then subtract this amount from \$1600. Another way is to subtract 25% from 100% to get 75% and then find 75% of \$1600.

Here, we will calculate the discount and then subtract from the original price.

$$R \cdot B = A$$
$$\downarrow \quad \downarrow \quad \downarrow$$
$$0.25 \cdot 1600 = 400 \qquad \text{Discount}$$

Original Price – Discount = \$1600 – \$400 = \$1200 Sale Price

The sale price for the bicycle was \$1200.

Sales tax is a tax charged on goods sold by retailers, and it is assessed by states and cities for income to operate various services. The **rate of sales tax** (a percent) varies by location.

Example 3: Sales Tax

Suppose 6% sales tax was added to the sale price \$1200 of the bicycle in Example 2. What was the total paid for the bicycle?

Solution: The sales tax must be calculated on the sale price and then added to the sale price.

$$R \cdot B = A$$
$$\downarrow \quad \downarrow \quad \downarrow$$
$$0.06 \cdot 1200 = 72 \qquad \text{Sales Tax}$$

Total Paid = Sale Price + Sales Tax = \$1200 + \$72 = \$1272

The total paid for the bicycle, including sales tax, was \$1272.

A **commission** is a fee paid to an agent or salesperson for a service. Commissions are usually a percent of a negotiated contract (as to a real estate agent) or a percent of sales.

Example 4: Commission

Teaching Notes:
Examples from real estate can be effective here. Some of your students may even be working in real estate and can relate their experiences in buying and selling real estate.

A saleswoman earns a salary of $1200 a month plus a commission of 8% on whatever she sells after she has sold $8000 in furniture. What did she earn the month she sold $25,000 worth of furniture?

Solution: First subtract $8000 from $25,000 to find the amount on which the commission is based.

$25,000 – $8000 = $17,000

Now find the amount of the commission.

$$R \cdot B = A$$
$$\downarrow \quad \downarrow \quad \downarrow$$
$$0.08 \cdot 17{,}000 = 1360 \text{ commission}$$

Now add the commission to her salary to find what she earned that month.

$1200 + $1360 = $2560 earned for the month

Problems involving percent come in a variety of forms. The next example illustrates another situation you may encounter in your own life.

Example 5: Commission

The Berrys sold their house. After paying the real estate agent a commission of 6% of the selling price and then paying $1486 in other costs and $90,000 on the mortgage, they received $49,514. What was the selling price of the house?

Solution: Use the relationship:

selling price – cost = profit ($S - C = P$).

Let S = selling price

cost = $0.06S + 1486 + 90{,}000$

Continued on the next page...

$$\underbrace{\text{selling price}} - \underbrace{\text{cost}} = \underbrace{\text{profit}}$$

$$\begin{aligned} S - (0.06S + 1486 + 90{,}000) &= 49{,}514 \\ S - 0.06S - 1486 - 90{,}000 &= 49{,}514 \\ 0.94S &= 141{,}000 \\ S &= 150{,}000 \end{aligned}$$

Check:

\$150,000	selling price	\$9000	commission	\$150,000	selling price
× 0.06	commission %	1486	other costs	−100,486	cost
\$9000	commission	+ 90,000	mortgage	\$49,514	profit
		\$100,486	cost		

The selling price was \$150,000.

Percent of profit for an investment is the ratio (a fraction) that compares the money made to the money invested. If you make two investments of different amounts of money, then the amount of money you make on each investment is not a fair comparison. In comparing such investments, the investment with the greater percent of profit is considered the better investment. **To find the percent of profit, form the fraction (or ratio) of profit divided by the investment and change the fraction to a percent.**

Example 6: Percent of Profit

Calculate the percent of profit for both **a.** and **b.** and tell which is the better investment.

a. \$300 profit on an investment of \$2400

b. \$500 profit on an investment of \$5000

Solution: Set up ratios and find the corresponding percents.

a. $\dfrac{\$300 \text{ profit}}{\$2400 \text{ invested}} = \dfrac{300 \cdot 1}{300 \cdot 8} = \dfrac{1}{8} = 0.125 = 12.5\%$ Percent of profit

b. $\dfrac{\$500 \text{ profit}}{\$5000 \text{ invested}} = \dfrac{500 \cdot 1}{500 \cdot 10} = \dfrac{1}{10} = 0.1 = 10\%$ Percent of profit

Clearly, \$500 is more than \$300, but 12.5% is greater than 10% and investment **a.** is the better investment.

Practice Problems

1. Write 0.3 as a percent.
2. Write 6.4% as a decimal.
3. Write $\frac{1}{5}$ as a percent.
4. Find 12% of 200.
5. Find the percent of profit if \$480 is made on an investment of \$1500.

2.7 Exercises

1. 91%
2. 62.5%
3. 137%
4. 0.75%
5. 62.5%
6. 87%
7. 75%
8. 150%
9. 0.69
10. 0.82
11. 1.62
12. 2.35
13. 0.075
14. 0.113
15. 0.005
16. 0.314
17. $\frac{7}{20}$
18. $\frac{18}{25}$
19. $\frac{13}{10}$
20. $\frac{2}{5}$
21. a. 32%
b. 28%
c. 12%

Write each of the following numbers in percent form.

1. 0.91 **2.** 0.625 **3.** 1.37 **4.** 0.0075

5. $\frac{5}{8}$ **6.** $\frac{87}{100}$ **7.** $\frac{3}{4}$ **8.** $1\frac{1}{2}$

Write each of the following percents in decimal form.

9. 69% **10.** 82% **11.** 162% **12.** 235%

13. 7.5% **14.** 11.3% **15.** 0.5% **16.** 31.4%

Write each of the following percents in fraction form (reduced).

17. 35% **18.** 72% **19.** 130% **20.** 40%

21. Budgeting: Heather's monthly income is \$2500. Using the given circle graph answer the following questions.

a. What percent of her income does Heather spend on rent each month?

b. What percent of her income is spent on food and entertaiment each month?

c. What percent of her income does Heather save each month?

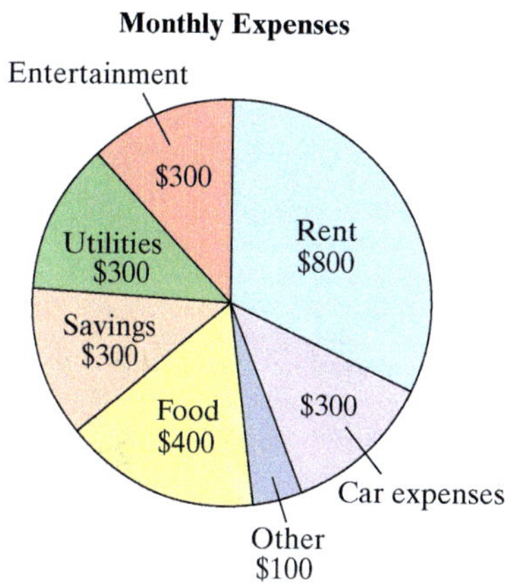

Answers to Practice Problems: **1.** 30% **2.** 0.064 **3.** 20% **4.** 24 **5.** 32%

22. a. 56.7%
b. 35.0%
c. 43.3%
23. a. 38.9%
b. 19.4%
c. 66.7%
24. a. 12%
b. 15%
c. 36%
25. 61.56
26. 88.74
27. 40
28. 150
29. 2180
30. 225

22. Earthquakes: There were 4257 registered earthquakes in the U.S. in 2009. Using the given graph answer the following questions. (Round answers to the nearest tenth of a percent.)

a. What percent of U.S. earthquakes in 2009 were less than a magnitude of 3.0?

b. What percent of earthquakes were of magnitude 3.0 – 3.9?

c. What percent of earthquakes were 3.0 and above?

Magnitude of US Earthquakes in 2009

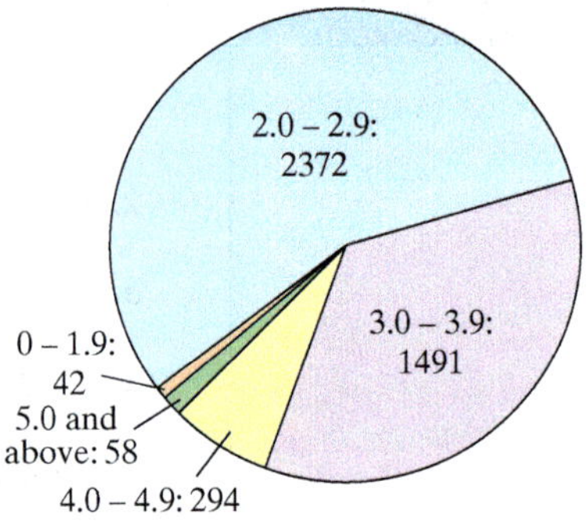

[Data from the U.S. National Earthquake Information Center]

23. Apple's revenue: In 2009, Apple's revenue was \$36 billion. (Round answers to the nearest tenth of a percent.)

a. What percent of Apple's revenue came from computer sales?

b. What percent of Apple's revenue came from mobile phones?

c. What percent of Apple's revenue did not come from music related products?

Apple's Revenue in 2009
(Data is in millions of dollars)

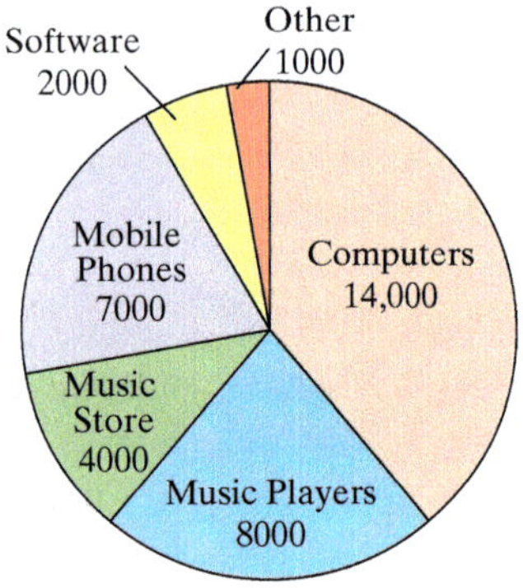

[Data from http://tech.fortune.cnn.com]

24. Music sales: In July, Daddy Fat's Record Shop sold 3900 CDs.

a. What percent of their sales came from classical CDs?

b. What percent of their sales came from R&B and rap CDs?

c. What percent of their sales was neither rock nor country CDs?

Daddy Fat's CD Sales for July

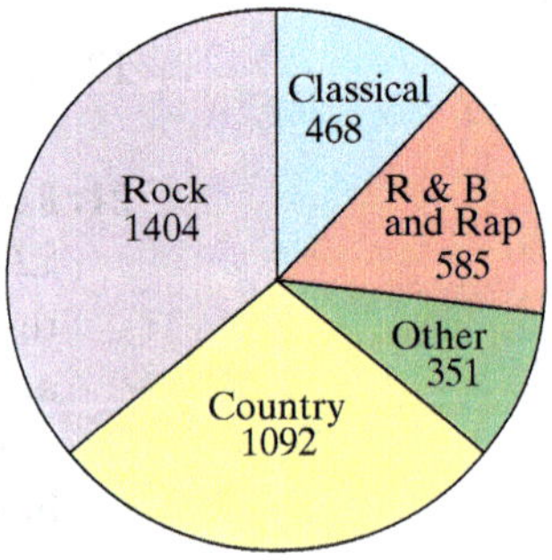

Find the missing rate, base, or amount.

25. 81% of 76 is ______.

26. 102% of 87 is _______.

27. _______% of 150 is 60.

28. _______% of 160 is 240.

29. 3% of _______ is 65.4.

30. 20% of _______ is 45.

31. 125%
32. 2500
33. 80
34. 58.56
35. a. 8%
b. 10%
c. b
36. a. 20%
b. 17%
c. a
37. $85.33
38. $128.81
39. $1952.90
40. $20,000
41. 1.5%
42. 129
43. 40
44. a. $22.50
b. $9

31. What percent of 32 is 40?

32. 1250 is 50% of what number?

33. 100 is 125% of what number?

34. Find 24% of 244.

35. What is the percent of profit if
a. $400 is made on an investment of $5000?
b. $350 is made on an investment of $3500?
c. Which was the better investment?

36. What is the percent of profit if
a. $400 is made on an investment of $2000?
b. $510 is made on an investment of $3000?
c. Which was the better investment?

37. Tipping: Patrick took his parents to dinner and the bill was $70 plus 6% tax. If he left a tip of 15% of the total (food plus tax), what was the total cost of the dinner?

38. Sales tax: In North Carolina sales tax on food is 4% and sales tax on all other products is 5.75%. At a superstore, Nevaeh buys $72 of food and $51 of other products. What will be the total cost with tax?

39. Salary: A calculator salesman's monthly income is $1500 salary plus a commission of 7% of his sales over $5000. What was his income the month that he sold $11,470 in calculators?

40. Salary bonus: A computer programmer was told that he would be given a bonus of 5% of any money his programs could save the company. How much would he have to save the company to earn a bonus of $1000?

41. Property tax: The property tax on a home was $2550. What was the tax rate if the home was valued at $170,000?

42. Basketball: In one season a basketball player missed 14% of her free throws. How many free throws did she make if she attempted 150 free throws?

43. Test taking: A student missed 6 problems on a statistics test and received a grade of 85%. If all the problems were of equal value, how many problems were on the test?

44. Shopping: At a department store sale, sheets and pillowcases were discounted 25%. Sheets were originally marked $30.00 and pillowcases were originally marked $12.00.
a. What was the sale price of the sheets?
b. What was the sale price of the pillowcases?

45. a. $700
b. 33.33%
46. $11.60; $171.60
47. a. $3125
b. $3375
48. a. 10%
b. 11.11%
c. The first percent is a percent of his original weight, while the second percent is a percent of his weight after he lost weight.
49. a. more than $27
b. $30
50. a. $7134.29
b. $6559.23
51. 2000 baskets
52. 62,500 pounds

45. Golfing: The Golf Pro Shop had a set of golf clubs that were marked on sale for $560. This was a discount of 20% off the original selling price.

a. What was the original selling price?

b. If the clubs cost the shop $420, what was the percent of profit based on cost?

46. Buying textbooks: If sales tax is figured at 7.25%, how much tax will be added to the total purchase price of three textbooks, priced at $35.00, $55.00, and $70.00? What will be the total amount paid for the books?

47. Buying a car: A car dealer bought a used car for $2500. He marked up the price so that he would make a profit of 25% of his cost.

a. What was the selling price?

b. If the customer paid 8% of the selling price in taxes and fees, what was the total cost of the car to the customer? The car was 6 years old.

48. Losing weight: A man weighed 200 pounds. He lost 20 pounds in 3 months. Then he gained back 20 pounds 2 months later.

a. What percent of his weight did he lose in the first 3 months?

b. What percent of his weight did he gain back?

c. The loss and the gain are the same amount, but the two percents are different. Explain.

49. Technology: Data cartridges for computer backup were on sale for $27.00 (a package of two cartridges). This price was a discount of 10% off the original price.

a. Was the original price more than $27 or less than $27?

b. What was the original price?

50. Buying a motorcycle: The dealer's discount to the buyer of a new motorcycle was $499.40. The manufacturer gave a rebate of $1000 on this model.

a. What was the original price if the dealer discount was 7% of the original price?

b. What would a customer pay for the motorcycle if taxes were 5% of the final selling price (after the rebate) and license fees were $642.60?

51. Farming: Farmer McGregor raises strawberries. They cost him $0.80 a basket to produce. He is able to sell only 85% of those he produces. If he sells his strawberries at $2.40 a basket, how many baskets must he produce to make a profit of $2480?

52. Farming: A citrus farmer figures that his fruit costs 96¢ a pound to grow. If he lost 20% of the crop he produced due to a frost and he sold the remaining 80% at $1.80 a pound, how many pounds did he produce to make a profit of $30,000?

53. a. $8800
b. Sam
c. Maria

Collaborative Learning Exercise

With the class separated into teams of two to four students, each team is to analyze the following problem and decide how to answer the related questions. Then each team leader is to present the team's answers and related ideas to the class for general discussion.

53. Business: Sam works at a picture framing store and gets a salary of $500 per month plus a commission of 3% on sales he makes over $2000. Maria works at the same store, but she has decided to work on a straight commission of 8% of her sales.
a. At what amount of sales will Sam and Maria make the same amount of money?
b. Up to that point, who would be making more?
c. After that point, who would be making more? Explain briefly. (If you were offered a job at that store, which method of payment would you choose?)

Writing and Thinking About Mathematics

54. a. Because the 6% commission is against the selling price, not the $141,000 the couple wanted.
b. The selling price is 100%. Therefore, $141,000 is 94% of the selling price (selling price (100%) – realtor fee (6%) = amount for couple(94%)). The selling price should be $150,000.

54. A man and his wife wanted to sell their house and contacted a realtor. The realtor said that the price for the services (listing, advertising, selling, and paperwork) was 6% of the selling price. The realtor asked the couple how much cash they wanted after the realtor fee was deducted from the selling price. They said $141,000 and that therefore, the asking price should be 106% of $141,000 or a total of $149,460. The realtor said that this was the wrong figure and that the percent used should be 94% and not 106%.
a. Explain why 106% is the wrong percent and $149,460 is not the right selling price.
b. Explain why and how 94% is the correct percent and just what the selling price should be for the couple to receive $141,000 after the realtor's fee is deducted.

HAWKES LEARNING SYSTEMS: INTRODUCTORY & INTERMEDIATE ALGEBRA SOFTWARE

- 2.7 Applications with Percent

Chapter 2 Index of Key Ideas and Terms

Section 2.1 Simplifying and Evaluating Algebraic Expressions

Algebraic Vocabulary pages 102-103

Constant – A single number

Term – Any constant or variable or the indicated product and/or quotient of constants and variables

Numerical coefficient – The number written next to a variable

Algebraic expression – A combination of variables and numbers using any of the operations of addition, subtraction, multiplication, or division, as well as exponents

Like Terms page 103

Like terms (or **similar terms**) are terms that are constants or terms that contain the same variables that are raised to the same exponents.

Combining Like Terms page 104

To **combine like terms**, add (or subtract) the coefficients and keep the common variable expression.

Evaluating Algebraic Expressions pages 105-106

1. Combine like terms, if possible.
2. Substitute the values given for any variables.
3. Follow the rules for order of operations.

Section 2.2 Translating English Phrases and Algebraic Expressions

Translating English Phrases into Algebraic Expressions pages 110-112

Translating Algebraic Expressions into English Phrases page 113

Section 2.3 Solving Linear Equations: $x + b = c$ and $ax = c$

Solution page 117

A **solution** to an equation is a number that gives a true statement when substituted for the variable in the equation.

Continued on the next page...

Section 2.3 Solving Linear Equations: $x + b = c$ and $ax = c$ (cont.)

Section 2.4 Solving Linear Equations: $ax + b = c$

Section 2.5 Solving Linear Equations: $ax + b = cx + d$

General Procedure for Solving Linear Equations that Simplify to the Form $ax + b = cx + d$ page 134

1. Simplify each side of the equation by removing any grouping symbols and combining like terms on both sides of the equation.
2. Use the **addition principle of equality** and add the opposite of a constant term and/or variable term to both sides so that variables are on one side and constants are on the other side.
3. Use the **multiplication** (or **division**) **principle of equality** to multiply both sides by the reciprocal of the coefficient of the variable (or divide both sides by the coefficient itself). The coefficient of the variable will become +1.
4. Check your answer by substituting it for the variable in the original equation.

Types of Equations and their Solutions page 137

Type of Equation	Number of Solutions
Conditional	Finite number of solutions
Identity	Infinite number of solutions
Contradiction	No solutions

Section 2.6 Introduction to Problem Solving

Procedure for Solving Applications pages 142-143

George Pólya's four-step process for problem solving:

1. Understand the problem.
2. Devise a plan.
3. Carry out the plan.
4. Look back over the results.

Number Problems pages 143-144

Consecutive Integers pages 144-145

Consecutive integers are two integers that differ by 1. Three consecutive integers can be represented as $n, n + 1$, and $n + 2$.

Consecutive Even Integers page 145

Consecutive even integers are two even integers that differ by 2. Two consecutive even integers can be represented as n and $n + 2$ where n is even.

Continued on the next page...

Section 2.6 Introduction to Problem Solving (cont.)

Consecutive Odd Integers page 145

Consecutive odd integers are two odd integers that differ by 2. Two consecutive odd integers can be represented as n and $n + 2$ where n is odd.

Section 2.7 Applications with Percent

Review of Decimals and Percents page 153

The Basic Percent Formula page 154

$\boldsymbol{R \cdot B = A}$ (or $\boldsymbol{A = R \cdot B}$)
$\boldsymbol{R}$ = **RATE** or percent (as a decimal or fraction)
$\boldsymbol{B}$ = **BASE** (number we are finding the percent of)
$\boldsymbol{A}$ = **AMOUNT** (a part of the base)
"**of**" means to multiply.
"**is**" means equal $(=)$.

Applications with Percent pages 156-157

Discounts, Sales Tax, and Commission

Percent of Profit page 158

To find the **percent of profit**, form the fraction (or ratio) of profit divided by the investment and change the fraction to a percent.

HAWKES LEARNING SYSTEMS: INTRODUCTORY & INTERMEDIATE ALGEBRA SOFTWARE

- 2.1a Variables and Algebraic Expressions
- 2.1b Simplifying Expressions
- 2.1c Evaluating Algebraic Expressions
- 2.2 Translating Phrases into Algebraic Expressions
- 2.3a Solving Linear Equations Using Addition and Subtraction
- 2.3b Solving Linear Equations Using Multiplication and Division
- 2.4 Solving Linear Equations
- 2.5 More Linear Equations: $ax + b = cx + d$
- 2.6a Applications: Number Problems and Consecutive Integers
- 2.6b Applications of Linear Equations: Addition and Subtraction
- 2.6c Applications of Linear Equations: Multiplication and Division
- 2.7 Applications with Percent

Chapter 2 Review

2.1 Simplifying and Evaluating Algebraic Expressions

1. $13x$
2. $6y$
3. $25x^2$
4. $-3a^2$
5. $-3x+2$
6. $3n+4$
7. $5x-34$
8. y^2+7y
9. $7a^2+4a$
10. $9y$
11. a. $-a^2+4a+7$ b. -5
12. a. $17x-17$ b. -34
13. a. $3.2y^2-0.4y-13.6$ b. 36
14. a. $5x+5$ b. 0
15. a. $-2y-26$ b. -34
16. a. $7y$ b. 28
17. 3 more than 5 times a number
18. 5 less than 6 times a number
19. −3 multiplied by the sum of a number and 2
20. −2 multiplied by the sum of a number and 6
21. 5 times the difference between a number and 3
22. 4 times the difference between a number and 10
23. 7 times a number divided by 33
24. 50 divided by the product of 6 and a number

Simplify each expression by combining like terms.

1. $9x+3x+x$ **2.** $3y+4y-y$ **3.** $18x^2+7x^2$

4. $-a^2-5a^2+3a^2$ **5.** $4x+2-7x$ **6.** $4(n+1)-n$

7. $5(x-6)-4$ **8.** $2y^2+5y-y^2+2y$ **9.** $12a+3a^2-8a+4a^2$

10. $\dfrac{-6y+2y}{4}+10y$

For the following expressions, ***a.*** *simplify, and* ***b.*** *evaluate the simplified expression for $x=-1$, $y=4$, and $a=-2$.*

11. $5a^2+3a-6a^2+a+7$ **12.** $20-17+13x-20+4x$

13. $2.7y+2.2y^2+y^2-13.6-3.1y$ **14.** $3(x+5)+2(x-5)$

15. $-4(y+6)+2(y-1)$ **16.** $3y+\dfrac{10y-2y}{2}$

2.2 Translating English Phrases and Algebraic Expressions

Translate each algebraic expression into an equivalent English phrase. (There may be more than one correct translation.)

17. $5n+3$ **18.** $6x-5$ **19.** $-3(x+2)$ **20.** $-2(n+6)$

21. $5(x-3)$ **22.** $4(x-10)$ **23.** $\dfrac{7n}{33}$ **24.** $\dfrac{50}{6y}$

Write the algebraic expressions described by the English phrases. Choose your own variable.

25. 5 times a number increased by 3 times the same number

26. 28 decreased by 6 times a number

27. 72 plus 8 times the sum of a number and 2

28. Twice a number plus 3

29. 22 divided by the sum of a number and 9

30. 32 less than the product of a number and 10

31. Time: The number of hours in x days and 5 hours

32. Frozen yogurt: The cost of y ounces of yogurt at 49¢ an ounce

25. $5x+3x$
26. $28-6x$
27. $72+8(x+2)$
28. $2x+3$
29. $\frac{22}{x+9}$
30. $10x-32$
31. $24x+5$
32. $0.49y$
33. -11 is a solution
34. -0.28 is a solution
35. 4.05 is not a solution
36. -10 is a solution
37. $x=7$
38. $x=14$
39. $y=-11$
40. $y=-8$
41. $n=-8$
42. $x=-4$
43. $n=10$
44. $n=6$
45. $y=-18$
46. $y=-30$
47. $x=18$
48. $y=0.4$
49. $x=-5$
50. $x=-8$
51. $y=9$
52. $y=-7$
53. $x=12$
54. $x=-25$
55. $y=6$
56. $y=-3$
57. $n=\frac{17}{9}$
58. $n=\frac{2}{5}$
59. $x=7$
60. $x=28$
61. $x=-12.35$
62. $x=-8.9$
63. $x=12$
64. $x=8$
65. $y=1.02$

2.3 Solving Linear Equations: $x+b=c$ and $ax=c$

Determine whether or not the given number is a solution to the given equation by substituting and then evaluating.

33. $y+(-6)=-17$ given that $y=-11$

34. $5y=-1.4$ given that $y=-0.28$

35. $-4x=16.2$ given that $x=4.05$

36. $-36+|z|=-26$ given that $z=-10$

Solve each of the following linear equations.

37. $x-6=1$

38. $x-4=10$

39. $y+5=-6$

40. $y+7=-1$

41. $0=n+8$

42. $13=x+17$

43. $6n=60$

44. $7n=42$

45. $\frac{2}{3}y=-12$

46. $\frac{5}{6}y=-25$

47. $2.9x-1.9x=7+11$

48. $3.2y-2.2y=1.3-0.9$

2.4 Solving Linear Equations: $ax+b=c$

Solve each of the following linear equations.

49. $2x+3=-7$

50. $3x+5=-19$

51. $2y+2y-6=30$

52. $3y+10+2y=-25$

53. $20=5x-3x-4$

54. $-30=6x-4x+20$

55. $9y+3y-50=22$

56. $5y+6y+17=-16$

57. $\frac{1}{5}n+\frac{2}{5}n-\frac{1}{3}=\frac{4}{5}$

58. $\frac{3}{8}n+\frac{1}{4}n+\frac{1}{2}=\frac{3}{4}$

59. $\frac{2}{3}x-\frac{1}{2}x+\frac{5}{6}=2$

60. $\frac{3}{4}x-\frac{7}{12}x-\frac{2}{3}=4$

61. $4.78-0.3x+0.5x=2.31$

62. $5.32+0.4x-0.6x=7.1$

63. $8.62=0.45x-0.15x+5.02$

64. $1.21=1.83x-0.86x-6.55$

2.5 Solving Linear Equations: $ax+b=cx+d$

Solve each of the following linear equations.

65. $8y-2.1=y+5.04$

66. $6y-4.5=y+4.5$

67. $2.4n=3.6n$

68. $8n=-1.3n$

69. $2(x+3)=3(x-7)$

70. $4(x-2)=3(x+6)$

71. $\frac{1}{3}n+6=\frac{1}{4}n+5$

72. $\frac{2}{5}n+7=\frac{1}{5}n+4$

66. $y = 1.8$
67. $n = 0$
68. $n = 0$
69. $x = 27$
70. $x = 26$
71. $n = -12$
72. $n = -15$
73. $x = 4$
74. $x = \frac{27}{5}$
75. $y = -2$
76. $x = 13$
77. $x = -\frac{1}{9}$
78. $x = \frac{2}{15}$
79. $y = -35.6$
80. Contradiction
81. Contradiction
82. Identity
83. Identity
84. Conditional
85. 19
86. 6
87. 9
88. −9
89. 24, 25, 26
90. 91, 93, 95
91. 10, 12, 14
92. 46, 47, 48, 49
93. $245,000
94. 12 months

73. $6(2x-1)=4(x+3)+14$

74. $8+4(x-4)=5-(x-14)$

75. $-2(y+2)-4=6-(y+12)$

76. $7-3(x+5)=-2(x+3)-15$

77. $\frac{5x}{6}+\frac{1}{18}=\frac{x}{3}$

78. $\frac{x}{6}+\frac{1}{15}-\frac{2x}{3}=0$

79. $2(y+8.3)+1.2=y-16.6-1.2$

Determine whether each of the following equations is a conditional equation, an identity, or a contradiction.

80. $3(x-1)=3-x+4x$

81. $5(x+5)+10=3(x-4)+2x$

82. $7+2(3x+1)=3(2x-5)+24$

83. $8x+3(4-x)=5(x+2)+2$

84. $\frac{1}{2}(x-24)=\frac{1}{3}(x-24)$

2.6 Introduction to Problem Solving

85. Forty-two is 4 more than twice a certain number. What is the number?

86. The difference between three times a number and 7 is equal to 17 decreased by the number. Find the number.

87. The quotient of twice a number and 9 is seven less than the number. What is the number?

88. Twice the difference between a number and 8 is equal to 5 times the number plus 11. What is the number?

89. Find three consecutive integers whose sum is 75.

90. Find three consecutive odd integers whose sum is 279.

91. Three consecutive even integers are such that if the second is added to twice the first the result is 18 more than the third. Find these integers.

92. Find four consecutive integers whose sum is 190.

93. Real estate: A real estate agent says that the current value of a home is $150,000 more than twice its value when it was new. If the current value is $640,000, what was the value of the home when it was new?

94. Buying a car: Karen wants to pay cash for a new car that costs $15,325. So far, she has saved $7621 for the car, and has determined that she can afford to put away $642 a month towards it. How many months will it take her to save up for the car?

95. 25 houses
96. 17 packs
97. 80%
98. $\frac{18}{25}$
99. 78.2
100. 70%
101. a. 8.4%
b. 6.25%
c. a
102. 1.2%
103. 18
104. $60.30
105. $21.45; $281.45
106. 6.25%

95. Candy: Jeremy has devised a plan to get the most candy possible this Halloween. He intends to visit a neighborhood that gives away 5 pieces of candy per house and then finish at his neighbor's house where he will get 36 pieces at the end of the night. If he ends the night with 161 pieces of candy, how many houses did he visit in the other neighborhood?

96. Collecting: Anferny loves collecting pens of all types. His favorites are rare, old pens, but he also likes regular store-bought ones. His pen collection currently has 16 rare pens, and 82 regular ones. If his mother wants to bring his collection up to an even 200 pens for his birthday, how many packs of pens must she get him if each pack has 6 pens?

2.7 Applications with Percent

97. Write $\frac{4}{5}$ in percent form.

98. Write 72% in fraction form (reduced).

Find the missing rate, base or amount.

99. 92% of 85 is ________.

100. ________% of 150 is 105.

101. Investing: What is the percent of profit if
a. $420 is made on an investment of $5000?
b. $225 is made on an investment of $3600?
c. Which was the better investment?

102. Property tax: The property tax on a home was $4440. What was the tax rate if the home was valued at $370,000?

103. Basketball: In one season a basketball player made 85% of her free throws. How many free throws did she miss if she attempted 120 free throws?

104. Shopping: At a department store sale shoes were discounted 33%. What was the sale price of shoes that were originally marked $90.00?

105. Sales tax: If sales tax is figured at 8.25%, how much tax will be added to a purchase of $260? What will be the total amount paid?

106. Pets: A golden retriever weighed 80 pounds and was put on a diet so she lost 5 pounds. What percent of her weight did she lose?

Chapter 2 Test

*For the following expressions, **a.** simplify, and **b.** evaluate the simplified expression for $x = -2$ and $y = 3$.*

1. $7x + 8x^2 - 3x^2$

2. $5y - y - 6 + 2y$

3. $2(x-5) + 3(x+4)$

4. $3y^2 + 6y - y^2 + y$

5. Translate each algebraic expression into an equivalent English phrase. (There may be more than one correct translation.)

a. $5n + 18$ **b.** $3(x+6)$ **c.** $42 - 7y$ **d.** $6x - 11$

6. Write the algebraic expression described by each English phrase.

a. The product of a number and 6 decreased by 3

b. Twice the sum of a number and 5

c. 4 less than twice the sum of a number and 15

d. The quotient of a number and 10 increased by twice the number

Solve each of the following linear equations.

7. $0 = 2x + 8$

8. $y + 13 = -16$

9. $\frac{5}{3}x + 1 = -4$

10. $3 + 6x - 8x = 15$

11. $4x - 5 - x = 2x + 5 - x$

12. $8(3+y) = -4(2y-6)$

13. $\frac{3}{2}a + \frac{1}{2}a = -3 + \frac{3}{4}$

14. $6 + \frac{3}{2}x = 9 + \frac{1}{3}x - \frac{5}{6}x$

15. $0.7x + 2 = 0.4x + 8$

16. $4(2x-1) + 3 = 2(x-4) - 5$

Determine whether each of the following equations is a conditional equation, an identity, or a contradiction.

17. $9x + x - 4(x+3) = 6(x-2)$

18. $\frac{1}{4}(4x+1) = -\frac{1}{2}(2x-1)$

Set up an equation for each word problem and solve the equation.

19. One number is 5 more than twice another number. Their sum is –22. Find the numbers.

20. Find three consecutive odd integers such that three times the second is equal to 27 more than the sum of the first and the third.

21. Find two consecutive integers such that twice the first added to three times the second is equal to 83.

22. **Basketball:** Brandon scored 21 points in his basketball game last night. If he made two free throws (1 point each) and one 3-point shot, how many 2-point shots did he make?

---

1. a. $5x^2 + 7x$ b. 6
2. a. $6y - 6$ b. 12
3. a. $5x + 2$ b. –8
4. a. $2y^2 + 7y$ b. 39
5. a. The product of 5 and a number increased by 18
   b. 3 multiplied by the sum of a number and 6
   c. 42 less the product of 7 and a number
   d. The product of 6 and a number decreased by 11
6. a. $6x - 3$
   b. $2(x+5)$
   c. $2(x+15) - 4$
   d. $\frac{x}{10} + 2x$
7. $x = -4$
8. $y = -29$
9. $x = -3$
10. $x = -6$
11. $x = 5$
12. $y = 0$
13. $a = -\frac{9}{8}$
14. $x = \frac{3}{2}$
15. $x = 20$
16. $x = -2$
17. Identity
18. Conditional
19. $(2y+5) + y = -22$; –9, –13
20. $3(n+2) = n + (n+4) + 27$; $n = 25, 27, 29$
21. $2n + 3(n+1) = 83$; 16, 17

22. $2(1)+1(3)+2x = 21$; 8 2-pt shots
23. 111.6
24. 32%
25. The \$6000 investment is the better investment since it has a higher percent of profit, 6%, than the \$10,000 investment, 5%.
26. 3 meters, 4 meters, and 5 meters
27. \$1875
28. a. \$730
b. \$605.35

Find the missing rate, base, or amount.

23. 62% of 180 is _____.

24. 48 is _____% of 150.

Solve each of the following application problems.

25. Investing: Which is the better investment, an investment of \$6000 that earns a profit of \$360 or an investment of \$10,000 that earns a profit of \$500? Explain your answer in terms of percent.

26. Triangles: The triangle shown here indicates that the sides can be represented as x, $x+1$, and $2x-1$. What is the length of each side if the perimeter is 12 meters?

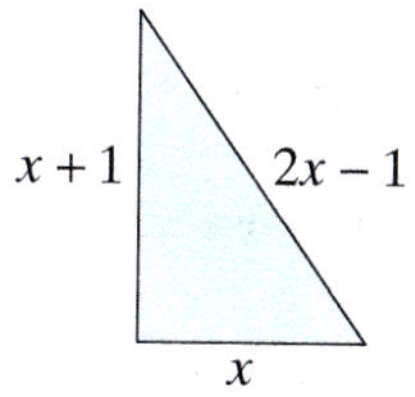

27. Investing: Julie made an investment of \$2000 and lost 25% in one year. In the next year she made 25%. How much money did she have at the end of the two years?

28. Shopping: A man's suit was on sale for \$547.50.

a. If this price was a discount of 25% from the original price, what was the original price?

b. What would the suit cost a buyer if sales tax was 6% and the alterations were \$25?

Cumulative Review: Chapters 1 – 2

1. 20
2. −14
3. 9
4. −15
5. −126
6. 459
7. −39
8. 124
9. 67
10. 23.3
11. See page 34.
12. 108
13. $72xy^2$
14. Associative property of multiplication
15. Commutative property of addition
16. Multiplicative inverse
17. Additive identity
18. a. $\{-4, 4\}$ b. No solution
19. True
20. True
21. False; $\frac{3}{4} \le |-1|$
22. 
23. 
24. 7
25. 18
26. 189
27. 13
28. 35
29. $\frac{5}{6}$
30. $\frac{2}{9}$
31. $\frac{6}{5y}$
32. $-\frac{1}{36}$

Perform the indicated operations. Reduce each answer to lowest terms.

1. $18 + 7 - 8 + 3$ **2.** $9 - 16 - 11 + 4$ **3.** $26 - 17$ **4.** $18 - 33$

5. $(14)(-9)$ **6.** $(-27)(-17)$ **7.** $\frac{273}{-7}$ **8.** $\frac{-744}{-6}$

Find the average of each set of numbers.

9. $\{32, 56, 70, 86, 91\}$ **10.** $\{13.62, 20.15, 26.24, 33.19\}$

11. Explain, in your own words, why division by 0 is undefined.

Find the LCM for each set of numbers or expressions.

12. $\{18, 27, 36\}$ **13.** $\{12x, 9xy, 24xy^2\}$

Name the property of real numbers illustrated.

14. $(3x)y = 3(xy)$ **15.** $x + 17 = 17 + x$ **16.** $17 \cdot \frac{1}{17} = 1$ **17.** $16 + 0 = 16$

18. Find the values of x that satisfy each equation:

a. $|x| = 4$ **b.** $|x| = -5$

Determine whether each statement is true or false. If a statement is false, rewrite it in a form that is a true statement. (There may be more than one way to correct a statement.)

19. $|-5| \le 5$ **20.** $2^3 - 8 > 8 - 3 \cdot 4$ **21.** $\frac{3}{4} \ge |-1|$

On a real number line, graph the integers that satisfy the conditions stated.

22. $|x| \le 6$ **23.** $|x| > 4$

Use the rules for order of operations to evaluate each expression.

24. $3 \cdot 2^2 - 5$ **25.** $(18 + 2 \cdot 3) \div 4 \cdot 3$

26. $5^2 \cdot 3^2 - 24 \div 2 \cdot 3$ **27.** $-36 \div (-2)^2 + 20 - 2(16 - 17)$

28. $(7 - 10)[49 \div (-7) + 20 \cdot 3 - 5 \cdot 15 - (-10)] - 1$

Perform the indicated operations. Reduce each answer to lowest terms.

29. $\frac{11}{9} \div \frac{22}{15}$ **30.** $\frac{8x}{21} \cdot \frac{7}{12x}$ **31.** $\frac{4}{5y} + \frac{2}{5y}$ **32.** $\frac{5}{12} - \frac{4}{9}$

33. $\frac{1}{6}$
34. 0
35. $\frac{49}{40}$
36. $-\frac{31}{5}$
37. a. $-2y-12$ b. -18
38. a. $4x+7$ b. -1
39. a. x^2+13x b. -22
40. a. $x-7$ b. -9
41. $13x-5$
42. $\frac{a}{6}$
43. a. $2x-3$
b. $\frac{x}{6}+3x$
c. $2(x+10)-13$
44. a. The sum of 6 and 4 times a number
b. 18 times the difference of a number and 5
45. $x=-9.8$
46. $x=18$
47. $y=-3$
48. $y=6$
49. $x=-1$
50. $x=3$
51. $x=-11$
52. $x=-4$
53. $x=\frac{31}{8}$
54. $y=\frac{24}{5}$
55. $y=\frac{15}{2}$
56. $y=0$
57. Conditional
58. Contradiction
59. Contradiction
60. Identity
61. Identity

Simplify the expressions using the rules for order of operations. Reduce all answers to lowest terms.

33. $\frac{1}{2}+\frac{3}{8}\div\frac{3}{4}-\frac{5}{6}$

34. $\frac{1}{3}+\frac{2}{5}\cdot\frac{5}{8}\div\frac{3}{2}-\frac{1}{2}$

35. $\left(\frac{2}{3}\right)^2\div\frac{5}{18}-\frac{3}{8}$

36. $\left(\frac{9}{10}+\frac{2}{15}\right)\div\left(\frac{1}{2}-\frac{2}{3}\right)$

For the following expressions, **a.** *simplify, and* **b.** *evaluate the simplified expression for* $x=-2$ *and* $y=3$.

37. $-4(y+3)+2y$

38. $3(x+4)-5+x$

39. $2(x^2+4x)-(x^2-5x)$

40. $\frac{3(5x-x)}{4}-2x-7$

Simplify each expression.

41. $6x+3x-5+4x$

42. $\frac{a}{7}+\frac{a}{14}-\frac{a}{21}$

43. Translate each English phrase into an algebraic expression:

a. Twice a number decreased by 3

b. The quotient of a number and 6 increased by 3 times the number

c. Thirteen less than twice the sum of a number and 10

44. Translate each algebraic expression into an English phrase:

a. $6+4n$ **b.** $18(n-5)$

Solve each of the following linear equations.

45. $x+4.2=-5.6$

46. $-5+x=13$

47. $17y=-51$

48. $-16y=-96$

49. $9x-8=4x-13$

50. $6-(x-2)=5$

51. $10+4x=5x-3(x+4)$

52. $4.4+0.6x=1.2-0.2x$

53. $-2(5x+12)-5=-6(3x-2)-10$

54. $\frac{3}{4}y-\frac{1}{2}=\frac{5}{8}y+\frac{1}{10}$

55. $\frac{1}{2}+\frac{1}{5}y=\frac{2}{15}y+1$

56. $-\frac{3}{4}y+\frac{3}{8}y=\frac{1}{6}y$

Determine whether each of the following equations is a conditional equation, an identity, or a contradiction.

57. $0.2x+25=-1.6+0.1x$

58. $3(x-5)+4=x+2(x-5)$

59. $4x-3(x+6)=x-12$

60. $-x+7(x+3)=3x+3(x+7)$

61. $8-4(3-x)+2x=6(x+2)-16$

62. $-\frac{5}{51}$
63. -5
64. $\frac{10}{3}$ cups (or $3\frac{1}{3}$ cups)
65. 12%
66. a. 1000 Republicans are in favor
b. 2500 registered voters are in favor
67. 11
68. 18, 20, 22
69. −2, −1, 0, 1
70. Hard drive: $84.60 Battery: $46.15
71. $506.94
72. $555.80
73. These are equally good investments since the percent of profit, 8%, is the same for each investment.

62. If the difference between $\frac{1}{2}$ and $\frac{11}{18}$ is divided by the sum of $\frac{2}{3}$ and $\frac{7}{15}$, what is the quotient?

63. From the sum of −2 and −11, subtract the product of −4 and 2.

64. Entertaining: Lucia is making punch for a party. She is making 5 gallons. The recipe calls for $\frac{2}{3}$ cup of sugar per gallon of punch. How many cups of sugar does she need?

65. College admission: The percentages of applicants accepted into the eight Ivy League colleges are as follows: 8%, 19%, 10%, 7%, 18%, 11%, 13%, and 10%. What is the average percentage of applicants accepted to an Ivy League college?
Source: College Board

66. Politics: There are 4000 registered voters in Greenville, and $\frac{3}{8}$ of these voters are registered Republicans. A survey indicates that $\frac{2}{3}$ of the registered Republicans are in favor of a certain Proposition and $\frac{3}{5}$ of the other registered voters are in favor of the Proposition.

a. How many of the registered Republicans are in favor of the Proposition?

b. How many of the registered voters are in favor of the Proposition?

Set up an equation for each problem and solve for the unknown quantity.

67. If twice a certain number is increased by 3, the result is 8 less than three times the number. Find the number.

68. Find three consecutive even integers such that the sum of the first and twice the second is equal to 14 more than twice the third.

69. Find four consecutive integers such that the sum of the first, second, and fourth integers is 2 less than the third.

70. Electronics: The total cost of a portable hard drive and a spare laptop battery was $130.75. If the hard drive cost $38.45 more than the battery, what was the cost of each item?

71. Computers: Computers are on sale at a 30% discount. If you know that you will pay 6.5% in sales taxes, what will you pay for a computer that is priced at $680?

72. Commission: Barry makes $500 a month plus commission working at a clothing store on the weekends. If he receives a 3% commission on all clothes he sells, what was his income the month he sold $1860 worth of clothes?

73. Investing: Which is the better investment: **a.** a profit of $800 on an investment of $10,000 or **b.** a profit of $1200 on an investment of $15,000? Explain in terms of percents.

Formulas and Linear Inequalities

CHAPTER 3

Did You Know?

In Chapter 3, you will find a great many symbols defined, as well as rules of manipulation for these symbolic expressions. Most people think that algebra has always existed, complete with all the common symbols in use today. That is not the case, since modern symbols did not appear consistently until the beginning of the sixteenth century. Prior to that, algebra was rhetorical. That is, all problems were written out in words using either Latin, Arabic, or Greek, and some nonstandard abbreviations. Numbers were written out. The common use of Hindu-Arabic numerals did not begin until the sixteenth century, although these numerals had been introduced into Europe in the twelfth century.

The sign for addition, +, was a contraction of the Latin *et*, which means "and." Gradually, the *e* was contracted and the crossed *t* became the plus sign. The minus sign or bar, –, is thought to be derived from the habit of early scribes of using a bar to represent the letter *m*. Thus the word *summa* was often written *sūma*. The bar came to represent the missing *m*, the first letter of the word *minus*. The radical symbol, $\sqrt{\ }$, is derived from a small printed *r*, which stood for the Latin word *radix*, or root. The symbol for times, a cross, ×, was developed from cross multiplication or for the purpose of indicating products in proportions.
Thus

$$\begin{matrix} 2 & & 6 \\ & \times & \\ 3 & & 9 \end{matrix} \text{ stood for } \frac{2}{3} = \frac{6}{9}.$$

The cross is not well suited for algebra, since it resembles the symbol *x*, which is used for variables. Therefore, a dot is usually used to indicate multiplication in algebra. The dot seemed to have developed from an Italian practice of separating columns in multiplication tables with a dot. Exponents were used as early as the fourteenth century by the mathematician Oresme (1320? – 1382), who gave the first known use of the rules for fractional exponents in a text book he wrote. The equal sign is attributed to Robert Recorde (1510? – 1558), who wrote, "I will sette as I doe often in woorke use, a paire of paralleles, or Gemowe [twin] lines of one lengthe, thus: = , because noe .2. thynges, can be moare equalle." As you can tell, the development of algebraic symbols occurred over a long period of time, and symbols became standardized through usage and convenience. If you are interested in the history of numerical symbolism, you will find more information in D.E. Smith's *History of Mathematics*, Volume II.

Recorde

"The Mathematician, carried along on his flood of symbols, dealing apparently with purely formal truths, may still reach results of endless importance for our description of the physical universe."

Karl Pearson (1857 – 1936)

3.1 Working with Formulas

- *Evaluate formulas for given values of the variables.*
- *Solve formulas for specified variables in terms of the other variables.*
- *Use formulas to solve a variety of applications.*

Formulas are general rules or principles stated mathematically. There are many formulas in such fields of study as business, economics, medicine, physics, and chemistry as well as mathematics. Some of these formulas and their meanings are shown here.

NOTES

SPECIAL COMMENT: Be sure to use the letters just as they are given in the formulas. In mathematics, there is little or no flexibility between capital and small letters as they are used in formulas. In general, capital letters have special meanings that are different from corresponding small letters. For example, capital **A** may mean the area of a triangle and small **a** may mean the length of one side, two completely different ideas.

Listed here are a few formulas with a description of the meaning of each formula. There are of course many more formulas. Some you will see in the exercises here and more (related to geometry) in Section 3.2.

Teaching Notes:
The formulas shown in this section may represent for students, in familiar contexts, the notation of "modeling". Modeling with mathematics is a central theme and formulas are a primary example of modeling.

Formula	Meaning
$I = Prt$	The **simple interest** I earned by investing money is equal to the product of the principal P times the rate of interest r times the time t in one year or part of a year.
$C = \frac{5}{9}(F-32)$	**Temperature** in degrees Celsius C equals $\frac{5}{9}$ times the difference between the Fahrenheit temperature F and 32.
$d = rt$	The **distance traveled** d equals the product of the rate of speed r and the time t.
$P = 2l + 2w$	The **perimeter** P **of a rectangle** is equal to twice the length l plus twice the width w.
$L = 2\pi rh$	The **lateral surface area** L (top and bottom not included) of a cylinder is equal to 2π times the radius r of the base times the height h.
$F = ma$	In physics, the **force** F acting on an object is equal to its mass m times its acceleration a.

Formula	Meaning
$\alpha + \beta + \gamma = 180°$	The **sum of the angles** (α, β, and γ) **of a triangle** is 180°. **Note:** α, β, and γ are the Greek lowercase letters alpha, beta, and gamma, respectively.
$P = a + b + c$	The **perimeter** P **of a triangle** is equal to the sum of the lengths of the three sides a, b, and c.

Table 1

Evaluating Formulas

If you know values for all but one variable in a formula, you can substitute those values and find the value of the unknown variable using the techniques for solving equations discussed in Chapter 2. This section presents a variety of formulas from real-life situations, with complete descriptions of the meanings of these formulas. Working with these formulas will help you become familiar with a wide range of applications of algebra and provide practice in solving equations.

Example 1: Evaluating Formulas

Teaching Notes:
You might want to discuss various situations involving interest. For simple interest with $t > 1$, the interest must be paid at the end of the year because the interest rate is an annual rate. For compound interest, interest is paid on interest and the formula involves exponents: $A = P\left(1 + \frac{r}{n}\right)^{nt}$.

a. A **note** is a loan for a period of 1 year or less, and the interest earned (or paid) is called **simple interest**. A note involves only one payment at the end of the term of the note and includes both principal and interest. The formula for calculating simple interest is:

$I = Prt$

where I = **interest** (earned or paid)
P = **principal** (the amount invested or borrowed)
r = **rate** of interest (stated as an annual or yearly rate)
t = **time** (one year or part of a year)

Note: The rate of interest is usually given in percent form and converted to decimal or fraction form for calculations. **For the purpose of calculations, we will use 360 days in one year and 30 days in a month.** Before the use of computers, this was common practice in business and banking.

Maribel loaned \$5000 to a friend for 90 days at an annual interest rate of 8%. How much will her friend pay her at the end of the 90 days?

Solution: Here, $P = \$5000$,
$r = 8\% = 0.08$,
$t = 90 \text{ days} = \frac{90}{360} \text{ year} = \frac{1}{4} \text{ year}$.

Find the interest by substituting in the formula $I = Prt$ and evaluating.

Continued on the next page...

$$I = 5000 \cdot \overset{0.02}{\cancel{0.08}} \cdot \frac{1}{\cancel{4}}$$
$$= 5000 \cdot 0.02$$
$$= \$100.00$$

The interest is \$100 and the amount to be paid at the end of 90 days is principal + interest = \$5000 + \$100 = \$5100.

b. Given the formula $C = \frac{5}{9}(F - 32)$, first find C if $F = 212°$ (212 degrees Fahrenheit) and then find F if $C = 20°$ (20 degrees Celsius).

Solution: $F = 212°$, so substitute 212 for F in the formula.

$$C = \frac{5}{9}(212 - 32)$$
$$= \frac{5}{9}(180)$$
$$= 100$$

That is, 212°F is the same as 100°C. Water will boil at 212°F at sea level. This means that if the temperature is measured in degrees Celsius instead of degrees Fahrenheit, water will boil at 100°C at sea level.

$C = 20°$, so substitute 20 for C in the formula.

$20 = \frac{5}{9}(F - 32)$	Now solve for F.
$\frac{9}{5} \cdot 20 = \frac{9}{5} \cdot \frac{5}{9}(F - 32)$	Multiply both sides by $\frac{9}{5}$.
$36 = F - 32$	Simplify.
$36 + 32 = F - 32 + 32$	Add 32 to both sides.
$68 = F$	Simplify.

That is, a temperature of 20°C is the same as a comfortable spring day temperature of 68°F.

c. The lifting force F exerted on an airplane wing is found by multiplying some constant k by the area A of the wing's surface and by the square of the plane's velocity v. The formula is $F = kAv^2$. Find the force on a plane's wing during take off if the area of the wing is 120 ft², k is $\frac{4}{3}$, and the plane is traveling 80 miles per hour during take off.

Solution: We know that $k = \frac{4}{3}$, $A = 120$, and $v = 80$. Substitution gives

$$F = \frac{4}{3} \cdot 120 \cdot 80^2$$
$$= \frac{4}{3} \cdot 120 \cdot 6400$$
$$= 160 \cdot 6400$$

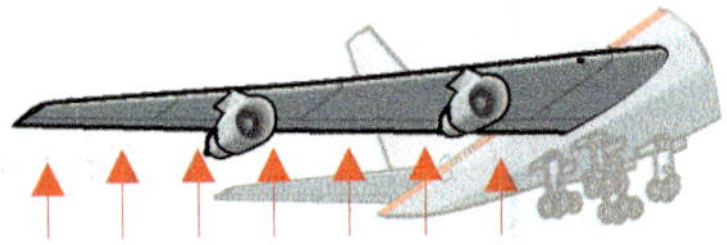

$F = 1{,}024{,}000$ lb   The force is measured in pounds.

d. The perimeter of a triangle is 38 feet. One side is 5 feet long and a second side is 18 feet long. How long is the third side?

Solution: Using the formula $P = a + b + c$, substitute $P = 38$, $a = 5$, and $b = 18$.

Then solve for the third side.

$$\begin{aligned} P &= a + b + c \\ 38 &= 5 + 18 + c \\ 38 &= 23 + c \\ 38 - 23 &= 23 + c - 23 \\ 15 &= c \end{aligned}$$

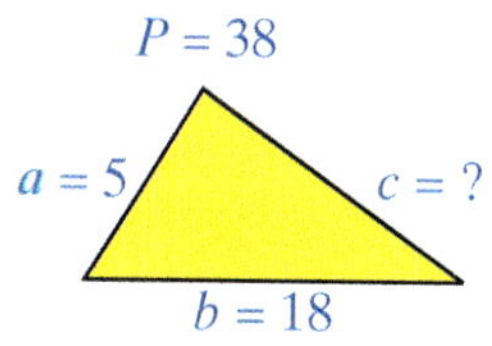

The third side is 15 feet long.

Solving Formulas for Different Variables

We say that the formula $d = rt$ is "solved for" d in terms of r and t. Similarly, the formula $A = \frac{1}{2}bh$ is solved for A in terms of b and h, and the formula $P = R - C$ (profit is equal to revenue minus cost) is solved for P in terms of R and C. Many times we want to use a certain formula in another form. We want the formula "solved for" some variable other than the one given in terms of the remaining variables. **Treat the variables just as you would constants in solving linear equations.** Study the following examples carefully.

Example 2: Solving for Different Variables

a. Given $d = rt$, solve for t in terms of d and r. We want to represent the time in terms of distance and rate. We will use this concept later in word problems.

Solution:

$d = rt$ — Treat r and d as if they were constants.

$\frac{d}{r} = \frac{rt}{r}$ — Divide both sides by r.

$\frac{d}{r} = t$ — Simplify.

b. Given $V = \frac{k}{P}$, solve for P in terms of V and k.

Solution:

$V = \frac{k}{P}$

$P \cdot V = P \cdot \frac{k}{P}$ — Multiply both sides by P.

$PV = k$ — Simplify.

$\frac{PV}{V} = \frac{k}{V}$ — Divide both sides by V.

$P = \frac{k}{V}$ — Simplify.

Continued on the next page...

c. Given $C=\frac{5}{9}(F-32)$ as in Example 1b, solve for F in terms of C. This would give a formula for finding Fahrenheit temperature given a Celsius temperature value.

Solution:

$$C=\frac{5}{9}(F-32) \quad \text{Treat } C \text{ as a constant.}$$

$$\frac{9}{5}\cdot C=\frac{9}{5}\cdot\frac{5}{9}(F-32) \quad \text{Multiply both sides by } \frac{9}{5}.$$

$$\frac{9}{5}C=F-32 \quad \text{Simplify.}$$

$$\frac{9}{5}C+32=F-32+32 \quad \text{Add 32 to both sides.}$$

$$\frac{9}{5}C+32=F \quad \text{Simplify.}$$

Thus $F=\frac{9}{5}C+32$ is solved for F, and $C=\frac{5}{9}(F-32)$ is solved for C. These are two forms of the same formula.

Teaching Notes:
You might ask the students if anyone observed that they could divide by 2 first. Explain that looking for a common factor (or divisor) can make the problem easier to solve later.

d. Given the equation $2x + 4y = 10$,
1. solve first for x in terms of y, and then
2. solve for y in terms of x.
This equation is typical of the algebraic equations that we will discuss in Chapter 4.

Solution: **1.** Solving for x yields

$$2x+4y=10 \quad \text{Treat } 4y \text{ as a constant.}$$

$$2x+4y-4y=10-4y \quad \text{Subtract } 4y \text{ from both sides. (This is the same as adding } -4y.\text{)}$$

$$2x=10-4y \quad \text{Simplify.}$$

$$\frac{2x}{2}=\frac{10-4y}{2} \quad \text{Divide both sides by 2.}$$

$$x=\frac{10}{2}-\frac{4y}{2}$$

$$x=5-2y \quad \text{Simplify.}$$

2. Solving for y yields

$$2x+4y=10 \quad \text{Treat } 2x \text{ as a constant.}$$

$$2x+4y-2x=10-2x \quad \text{Subtract } 2x \text{ from both sides.}$$

$$4y=10-2x \quad \text{Simplify.}$$

$$\frac{4y}{4}=\frac{10-2x}{4} \quad \text{Divide both sides by 4.}$$

$$y=\frac{10}{4}-\frac{2x}{4}$$

$$y=\frac{5}{2}-\frac{x}{2} \quad \text{Simplify.}$$

or we can write

$$y = \frac{5-x}{2} \quad \text{or} \quad y = -\frac{1}{2}x + \frac{5}{2}. \qquad \text{All forms are correct.}$$

e. Given $3x - y = 15$, solve for y in terms of x.

Solution: Solving for y gives

$$3x - y = 15$$

$$3x - y - 3x = 15 - 3x \qquad \text{Subtract } 3x \text{ from both sides.}$$

$$-y = 15 - 3x$$

$$-1(-y) = -1(15 - 3x) \qquad \text{Multiply both sides by } -1 \text{ (or divide both sides by } -1\text{).}$$

$$y = -15 + 3x \qquad \text{Simplify using the distributive property.}$$

or $$y = 3x - 15$$

Practice Problems

1. $2x - y = 5$; solve for y.
2. $2x - y = 5$; solve for x.
3. $A = \frac{1}{2}bh$; solve for h.
4. $L = 2\pi rh$; solve for r.
5. $P = 2l + 2w$; solve for w.

3.1 Exercises

*Refer to Example 1a for information concerning simple interest and the related formula $I = Prt$. (**Note:** For ease of calculation, assume 360 days in a year and 30 days in each month.)*

Simple Interest

1. $120
2. 72 days
3. $10,000
4. 2.5%

1. You want to borrow $4000 at 12% for only 90 days. How much interest would you pay?
2. For how many days must you leave $1000 in a savings account at 5.5% to earn $11.00 in interest?
3. What principal would you need to invest to earn $450 in simple interest in 6 months if the interest rate was 9%?
4. After 30 days, Gustav received $25 in simple interest on his savings account of $12,000. What was the interest rate?

Answers to Practice Problems: **1.** $y = 2x - 5$ **2.** $x = \frac{y+5}{2}$ **3.** $h = \frac{2A}{b}$ **4.** $r = \frac{L}{2\pi h}$ **5.** $w = \frac{P-2l}{2}$

5. a. $183.75
b. $3683.75
6. a. $62.50
b. 10%
7. 176 ft/sec
8. 2 seconds
9. 4 milliliters
10. 100 milligrams
11. $1030
12. 0.5 years

5. A savings account of $3500 is left for 9 months and draws simple interest at a rate of 7%.
a. How much interest is earned?
b. What is the balance in the account at the end of the 9 months?

6. Tim just deposited $2562.50 to pay off a 3 month loan of $2500.
a. How much of what he deposited was interest on the loan?
b. What rate of interest was he charged?

In the following application problems, read the descriptive information carefully and then substitute the values given in the problem for the corresponding variables in the formulas. Evaluate the resulting expression for the unknown variable.

Velocity

If an object is shot upward with an initial velocity v_0 in feet per second, the velocity v in feet per second is given by the formula $v = v_0 - 32t$, where t is time in seconds. (v_0 is read "v sub zero." The 0 is called a ***subscript****.)*

7. Find the initial velocity of an object if the velocity after 4 seconds is 48 feet per second.

8. An object projected upward with an initial velocity of 106 feet per second has a velocity of 42 feet per second. How many seconds have passed?

Medicine

In nursing, one procedure for determining the dosage for a child is

$$\text{child's dosage} = \frac{\text{age of child in years}}{\text{age of child} + 12} \cdot \text{adult dosage}$$

9. If the adult dosage of a drug is 20 milliliters, how much should a 3-year-old child receive?

10. If the adult dosage of a drug is 340 milligrams, how much should a 5-year-old child receive?

Investment

The total amount of money in an account with P dollars invested in it is given by the formula $A = P + Prt$, where r is the rate expressed as a decimal and t is time (one year or part of a year).

11. If $1000 is invested at 6% interest, find the total amount in the account after 6 months.

12. How long will it take an investment of $600, at an annual rate of 5%, to be worth $615?

Carpentry

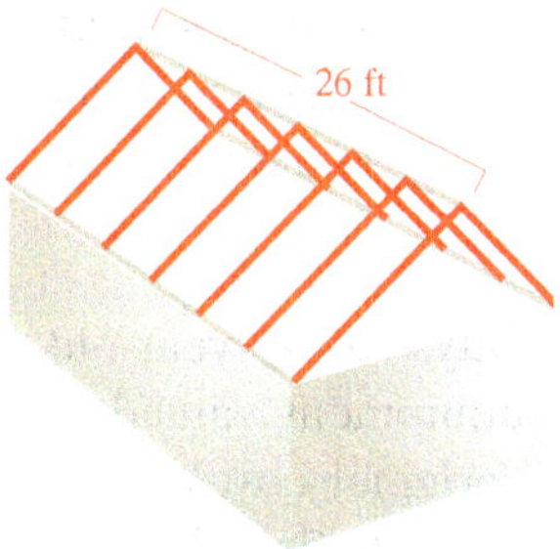

The number N of rafters in a roof or studs in a wall can be found by the formula $N = \frac{L}{d} + 1$, where L is the length of the roof or wall and d is the center-to-center distance from one rafter or stud to the next. Note that l and d must be in the same units.

13. 14 rafters
14. 16 studs
15. 336 in. or 28 ft
16. 3 ft
17. $1030
18. 230 calculators
19. $76.50
20. $337.50
21. $5 million
22. $2.5 million

13. How many rafters will be needed to build a roof 26 ft long if they are placed 2 ft on center?

14. A wall has studs placed 16 in. on center. If the wall is 20 ft long, how many studs are in the wall?

15. How long is a wall if it requires 22 studs placed 16 in. on center?

16. What should the center-to-center distance be if you are building a 33 ft long roof using 12 rafters?

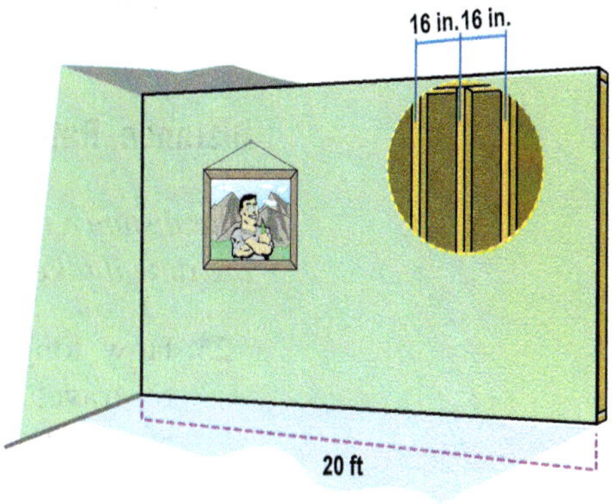

Cost

The total cost C of producing x items can be found by the formula $C = ax + k$, where a is the cost per item and k is the fixed costs (rent, utilities, and so on).

17. Find the total cost of producing 30 items if each costs $15 and the fixed costs are $580.

18. It costs a company $3.60 to produce a calculator. Last week the total costs were $1308. If the fixed costs are $480 weekly, how many calculators were produced last week?

19. Each week an electronics company builds 60 mp3 players for a total cost of $5340. If the fixed costs for a week are $750, what is the cost to produce each mp3 player?

20. The total cost to produce 80 dolls is $1097.50. If each doll costs $9.50 to produce, find the fixed costs.

Profit

The profit P is given by the formula $P = R - C$, where R is the revenue and C is the cost.

21. Find the revenue (income) of a company that shows a profit of $3.2 million and costs of $1.8 million.

22. Find the revenue of a company that shows a profit of $3.2 million and costs of $5.7 million.

23. \$2400
24. 5%
25. 7 hours
26. 0.32 miles per minute
27. 1.625 miles per hour
28. 3.5 hours
29. $b = P - a - c$
30. $s = \frac{P}{3}$
31. $m = \frac{F}{a}$
32. $d = \frac{C}{\pi}$
33. $w = \frac{A}{l}$
34. $C = R - P$
35. $n = \frac{R}{p}$
36. $k = v - gt$
37. $P = A - I$
38. $h = \frac{L}{2\pi r}$
39. $m = 2A - n$
40. $a = P - 2b$
41. $t = \frac{I}{Pr}$
42. $E = RI$
43. $b = \frac{P - a}{2}$
44. $b^2 = c^2 - a^2$
45. $\beta = 180° - \alpha - \gamma$
46. $x = \frac{y - b}{m}$

Depreciation

Many items decrease in value as time passes. This decrease in value is called **depreciation**. *One type of depreciation is called* **linear depreciation**. *The value V of an item after t years is given by $V = C - Crt$, where C is the original cost and r is the rate of depreciation expressed as a decimal.*

23. If you buy a car for \$6000 and it depreciates linearly at a rate of 10% per year, what will be its value after 6 years?

24. A contractor buys a 4 year-old piece of heavy equipment valued at \$20,000. If the original cost of this equipment was \$25,000, find the rate of depreciation.

Distance, Rate, and Time

The distance traveled d is given by the formula $d = rt$, where r is the rate of speed and t is the time it takes.

25. How long will a truck driver take to travel 350 miles if he averages 50 mph?

26. What is the average rate of speed of a biker who bikes 21.92 miles in 68.5 min?

27. What is Jonathan's average rate of speed if he hikes 10.4 miles in 6.4 hours?

28. How long will it take a train traveling at 40 mph to go 140 miles?

Solve each formula for the indicated variable.

29. $P = a + b + c$; solve for b.

30. $P = 3s$; solve for s.

31. $F = ma$; solve for m.

32. $C = \pi d$; solve for d.

33. $A = lw$; solve for w.

34. $P = R - C$; solve for C.

35. $R = np$; solve for n.

36. $v = k + gt$; solve for k.

37. $I = A - P$; solve for P.

38. $L = 2\pi rh$; solve for h.

39. $A = \frac{m + n}{2}$; solve for m.

40. $P = a + 2b$; solve for a.

41. $I = Prt$; solve for t.

42. $R = \frac{E}{I}$; solve for E.

43. $P = a + 2b$; solve for b.

44. $c^2 = a^2 + b^2$; solve for b^2.

45. $\alpha + \beta + \gamma = 180°$; solve for β.

46. $y = mx + b$; solve for x.

47. $h = \frac{V}{lw}$
48. $t = \frac{v_0 - v}{g}$
49. $b = \frac{2A}{h}$
50. $I = \frac{E}{R}$
51. $\pi = \frac{A}{r^2}$
52. $L = \frac{R}{2A}$
53. $g = \frac{mv^2}{2K}$
54. $y = \frac{4-x}{4}$
55. $y = \frac{6-2x}{3}$
56. $y = 3x - 14$
57. $x = \frac{11-2y}{5}$
58. $x = \frac{2y-5}{2}$
59. $b = \frac{2A-hc}{h}$ or $b = \frac{2A}{h} - c$
60. $r = \frac{A-P}{Pt}$
61. $x = \frac{8R+36}{3}$ or $x = \frac{8R}{3} + 12$
62. $x = 5 - 3y$
63. $y = -x - 12$
64. $h = \frac{3V}{\pi r^2}$
65. $C = nt + 9$
66. $C = 0.12x + 25$
67. $C = 325n + 5400$
68. $P = 358n - 5400$

47. $V = lwh$; solve for h.

48. $v = -gt + v_0$; solve for t.

49. $A = \frac{1}{2}bh$; solve for b.

50. $R = \frac{E}{I}$; solve for I.

51. $A = \pi r^2$; solve for π.

52. $A = \frac{R}{2L}$; solve for L.

53. $K = \frac{mv^2}{2g}$; solve for g.

54. $x + 4y = 4$; solve for y.

55. $2x + 3y = 6$; solve for y.

56. $3x - y = 14$; solve for y.

57. $5x + 2y = 11$; solve for x.

58. $-2x + 2y = 5$; solve for x.

59. $A = \frac{1}{2}h(b+c)$; solve for b.

60. $A = P(1+rt)$; solve for r.

61. $R = \frac{3(x-12)}{8}$; solve for x.

62. $-2x - 5 = -3(x+y)$; solve for x.

63. $3y - 2 = x + 4y + 10$; solve for y.

64. $V = \frac{1}{3}\pi r^2 h$; solve for h.

Make up a formula for each of the following situations.

65. Concert tickets: Each ticket for a concert costs $\$t$ per person and parking costs \$9.00. What is the total cost per car C if there are n people in a car?

66. Car rentals: ABC Car Rental charges \$25 per day plus \$0.12 per mile. What would you pay per day for renting a car from ABC if you were to drive the car x miles in one day?

67. Computers: Top-of-The-Line computer company knows that the cost (labor and materials) of producing a computer is \$325 per computer per week and the fixed overhead costs (lighting, rent, etc.) are \$5400 per week. What are the company's weekly costs of producing n computers per week?

68. Computers: If the Top-of-The-Line computer company (see Exercise 67) sells its computers for \$683 each, what is its profit per week if it sells the same number n that it produces? (Remember that profit is equal to revenue minus costs, or $P = R - C$.)

Writing and Thinking About Mathematics

69. a. 0; No, because the numerator will be zero and thus the whole fraction will be equal to zero for all values of s.
b. $x < 70$
c. Answers will vary.

70. a. 1
b. −1
c. 1.5
d. −2

69. The formula $z = \frac{x - \overline{x}}{s}$ is used extensively in statistics. In this formula, x represents one of a set of numbers, $\overline{x}$ represents the average (or mean) of those numbers in the set, and s represents a value called the standard deviation of the numbers. (The standard deviation is a positive number and is a measure of how "spread out" the numbers are.) The values for z are called z-scores, and they measure the number of standard deviation units a number x is from the mean $\overline{x}$.

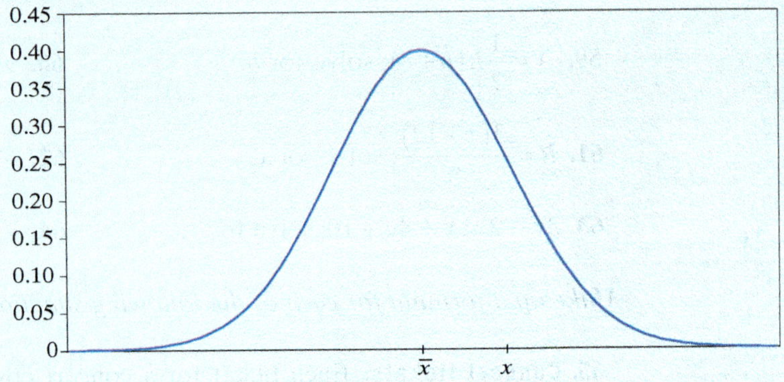

a. If $\overline{x} = 70$, what will be the z-score for $x = 70$? Does this z-score depend on the value of s? Explain.
b. For what values of x will the corresponding z-scores be negative?
c. Calculate your z-score on each of the last two test scores in this class. (Your instructor will give you the mean and standard deviation for each test.) What do these scores tell you about your performance on the two exams?

70. Suppose that, for a particular set of exam scores, $\overline{x} = 72$ and $s = 6$. Find the z-score that corresponds to a score of
a. 78 **b.** 66 **c.** 81 **d.** 60

HAWKES LEARNING SYSTEMS: INTRODUCTORY & INTERMEDIATE ALGEBRA SOFTWARE

- 3.1 Working with Formulas

3.2 Formulas in Geometry

- *Recognize and use appropriate geometric formulas for computations.*

A **formula** is an equation that represents a general relationship between two or more quantities or measurements. Several variables may appear in a formula. And, as was illustrated by the variety of formulas presented in Section 3.1, formulas are useful in mathematics, economics, chemistry, medicine, physics, and many other fields of study. In this section, we will discuss some formulas related to simple geometric figures such as rectangles, circles, and triangles.

Perimeter

The **perimeter** P of a geometric figure is the total distance around the figure. (For circles, the perimeter is called the **circumference** C.) Perimeters are measured in units of length such as inches, feet, yards, miles, centimeters, and meters. The formulas for the perimeters of various geometric figures are shown in Table 1.

Formulas for Perimeter		
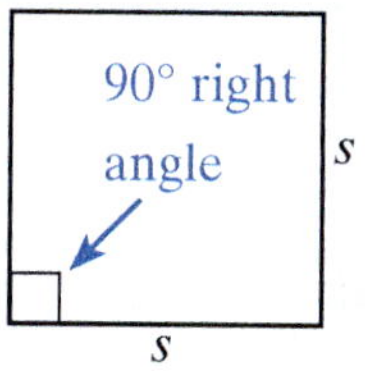 Square $P = 4s$	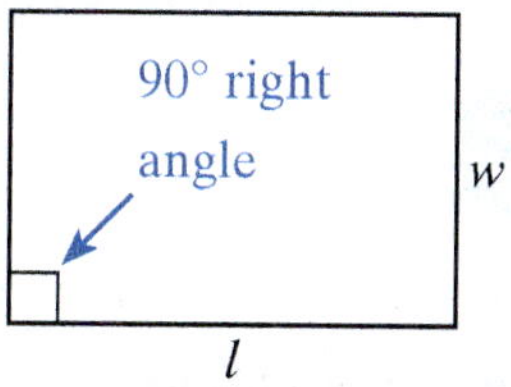 Rectangle $P = 2l + 2w$	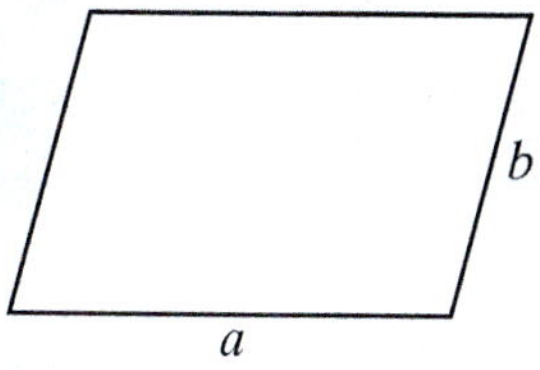 Parallelogram $P = 2a + 2b$
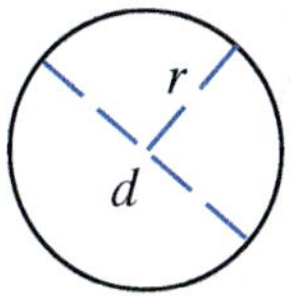 Circle $C = 2\pi r$ $C = \pi d$	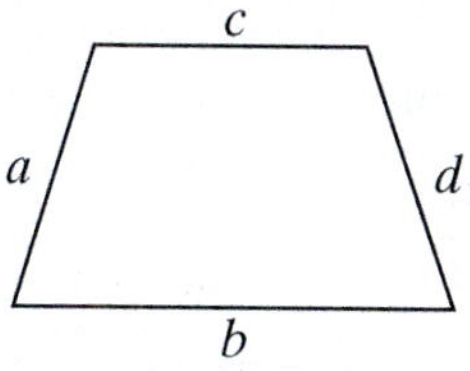 Trapezoid $P = a + b + c + d$	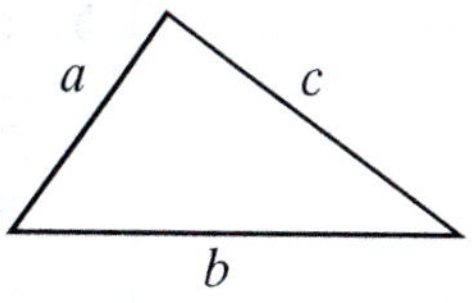 Triangle $P = a + b + c$

Table 1

Terminology and Notation Related to Circles		
Term	**Notation**	**Definition**
Circumference	C	The perimeter of a circle is called its **circumference**.
Radius	r	The distance from the center of a circle to any point on the circle is called the **radius** of the circle.
Diameter	d	The distance from one point on a circle to another point on the circle measured through its center is called the **diameter** of the circle. **Note:** The diameter is twice the radius: $\boldsymbol{d = 2r}$.
Pi	π	π is a Greek letter used for the constant 3.1415926535… (π is an irrational number.)

Table 2

For the calculations involving π in this text we will round off to two decimal places and use $\pi = 3.14$. Remember, however, that π is an irrational number and is an infinite nonrepeating decimal. Historically, π was discovered by mathematicians trying to find a relationship between the circumference and diameter of any circle. **The result is that π is the ratio of the circumference to the diameter of any circle** $\left(\frac{C}{d} = \pi\right)$, a rather amazing fact.

Example 1: Perimeter

a. Find the perimeter of the triangle with sides labeled as in the figure.

Solution: $P = a + b + c$

$P = 3 + 6.2 + 8.1$

$= 17.3 \text{ in.}$

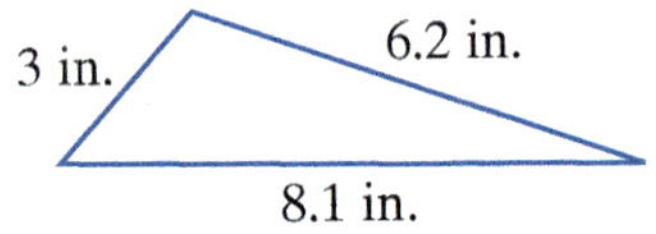

The perimeter is 17.3 inches.

b. Find the circumference of a circle with a diameter of 3 centimeters.

Solution:

Method 1: Sketch the figure.

$C = \pi d$

$C = 3.14(3)$

$= 9.42 \text{ cm}$

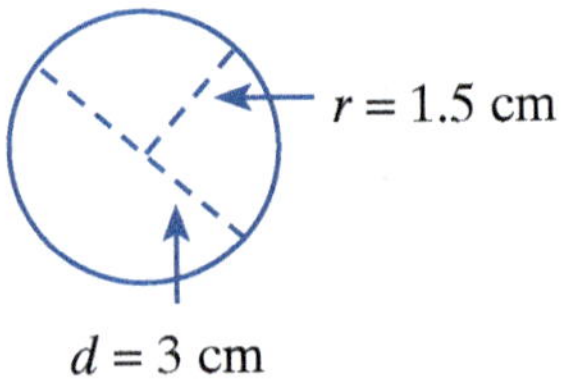

Method 2: Sketch the figure.

$$C = 2\pi r$$
$$C = 2(3.14)(1.5)$$
$$= 9.42\text{ cm}$$

Note that the result is the same with either formula. The circumference is **approximately** 9.42 centimeters.

Example 2: Perimeter

The perimeter of a football field is 1040 feet. If the length is 360 feet, what is the width?

Solution:

$$P = 2l + 2w$$

$1040 = 2(360) + 2w$ Substitute 1040 for P and 360 for l.

$1040 = 720 + 2w$ Simplify.

$1040 - 720 = 720 + 2w - 720$ Subtract 720 from both sides.

$320 = 2w$ Simplify.

$\frac{320}{2} = \frac{2w}{2}$ Divide both sides by 2.

$160 = w$ Simplify.

The width of the field is 160 feet.

Area

Area is a measure of the interior, or enclosure, of a surface. Area is measured in square units such as square feet, square inches, square meters, or square miles. For example, the area enclosed by a square of 1 inch on each side is 1 sq in. (or 1 in.^{2}) (Figure 1a), while the area enclosed by a square of 1 centimeter on each side is 1 sq cm (or 1 cm^2) (Figure 1b).

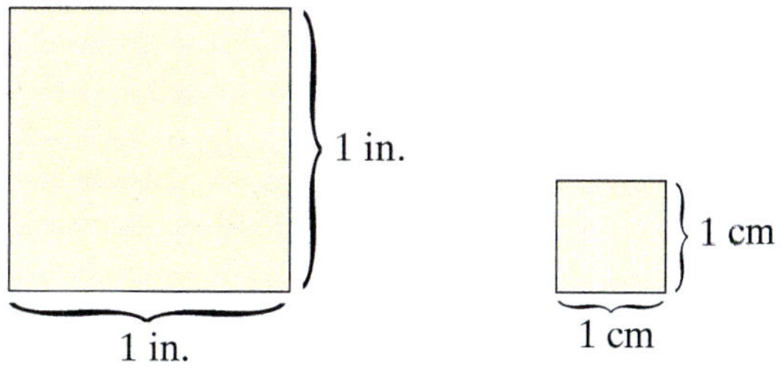

Area = 1 sq in. = 1 in.^{2} **a.**

Area = 1 sq cm = 1 cm^2 **b.**

Figure 1

The formulas for finding the areas of several geometric figures are shown in Table 3.

Formulas for Area		
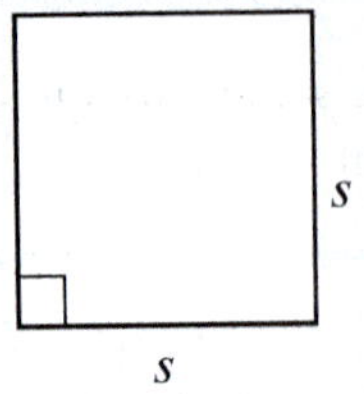 Square $A = s^2$	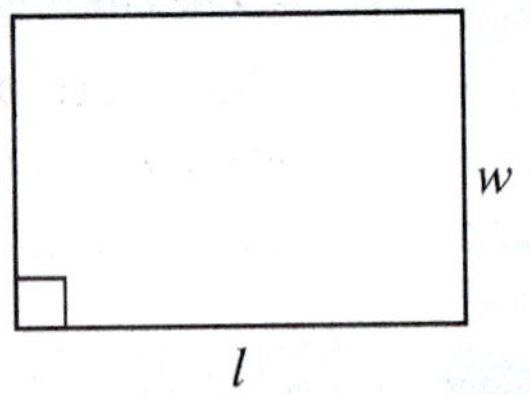 Rectangle $A = lw$	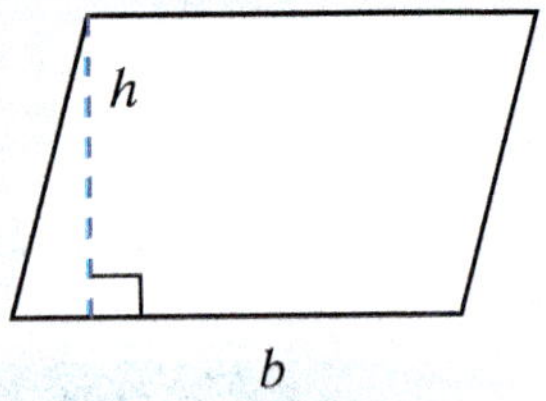 Parallelogram $A = bh$
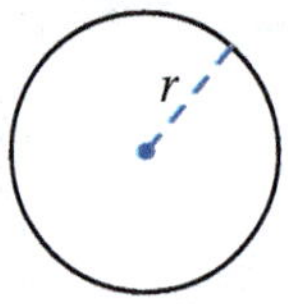 Circle $A = \pi r^2$	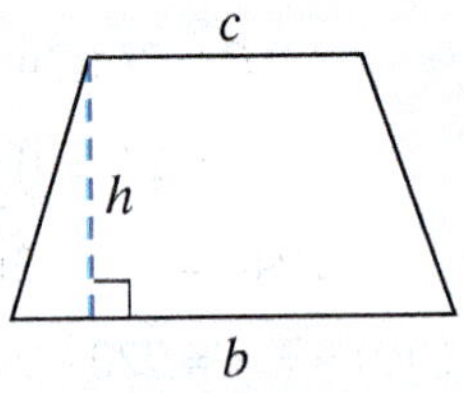 Trapezoid $A = \frac{1}{2}h(b+c)$	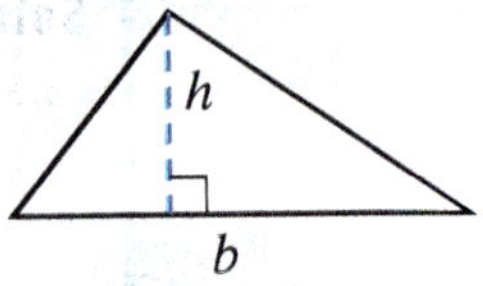 Triangle $A = \frac{1}{2}bh$

Table 3

Example 3: Area

a. Find the area of a triangle with a height of 3 cm and a base of 4 cm.

Solution: Sketch the figure.

$$A = \frac{1}{2} \cdot b \cdot h$$

$$A = \frac{1}{2} \cdot 4 \cdot 3$$

$$= 2 \cdot 3$$

$$= 6$$

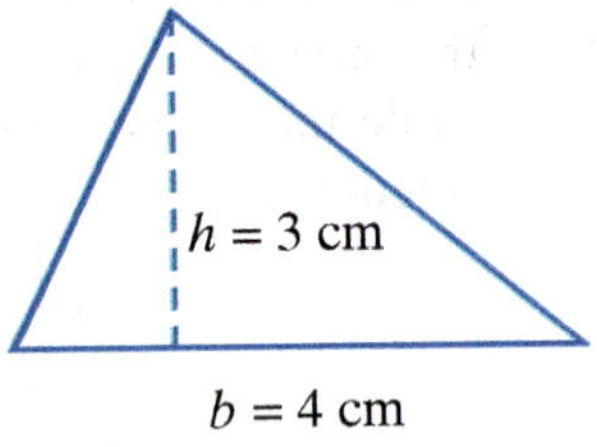

The area is 6 cm^2.

Teaching Notes:
You might mention that 36π is also an acceptable and exact answer.

b. Find the area of a circle with a radius of 6 inches. (Use $\pi = 3.14$.)

Solution: Sketch the figure.

$A = \pi r^2$

$A = 3.14(6)^2$

$= 3.14(36)$

$= 113.04$

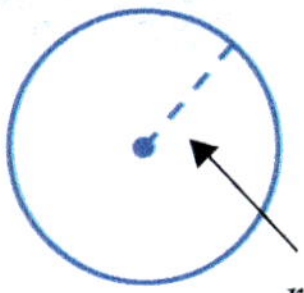

The area is approximately 113.04 square inches.

Example 4: Area

Central Park in New York City is rectangular in shape and has an area of approximately 1.25 square miles. If it is 2.5 miles long, how wide is it?

Solution: $A = lw$

$1.25 = 2.5w$ Substitute 1.25 for A and 2.5 for l.

$\frac{1.25}{2.5} = \frac{2.5w}{2.5}$ Divide both sides by 2.5.

$0.5 = w$

The width of the park is 0.5 miles.

Volume

Teaching Notes:
For meaningful applications, relate to the student that the economics of package design and the analysis of material requirements can be determined from the formulas for area and volume.

Volume is a measure of the space enclosed by a three-dimensional figure. Volume is measured in cubic units, such as cubic inches, cubic centimeters, and cubic feet. For example, the volume enclosed by a cube of 1 inch on each edge is 1 cu in. (or 1 in.^{3}) (Figure 2a), while the volume enclosed by a cube of 1 centimeter on each edge is 1 cu cm (or 1 cm^3) (Figure 2b).

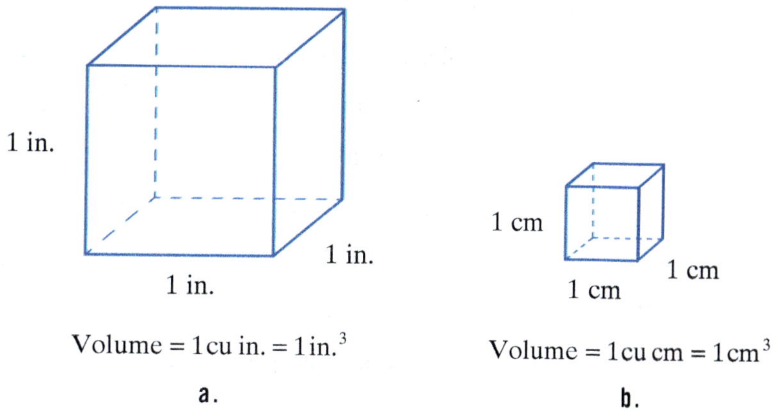

Figure 2

The formulas for finding the volume of several geometric figures are shown in Table 4.

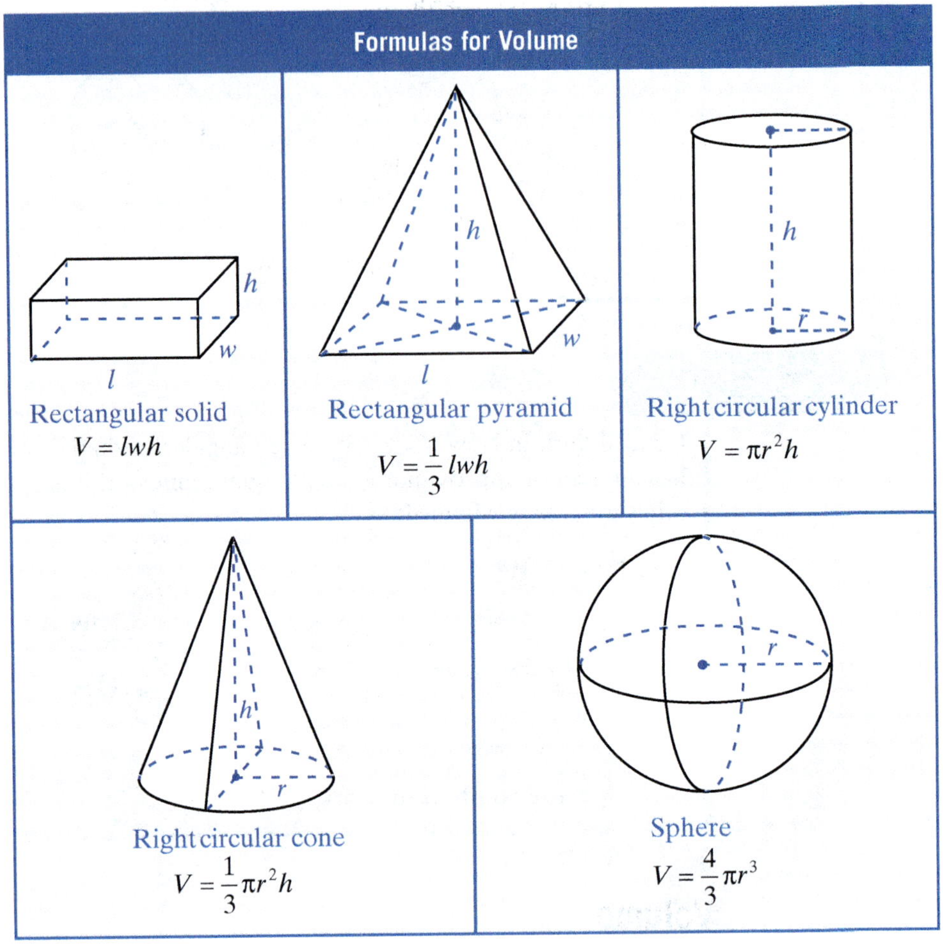

Table 4

Example 5: Volume

a. Find the volume of a right circular cylinder with a radius of 2 centimeters and a height of 5 centimeters.

Solution: Sketch the figure.

$$V = \pi r^2 h$$

$$V = 3.14(2)^2 \cdot 5$$

$$= 3.14(4) \cdot 5$$

$$= 3.14(20)$$

$$= 62.80 \text{ cm}^3$$

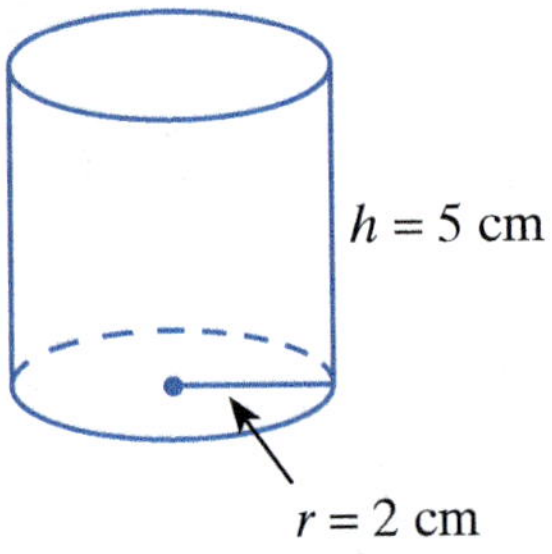

The volume is approximately 62.80 cubic centimeters.

b. Find the volume of a sphere with a radius of 4 feet.

Solution: Sketch the figure.

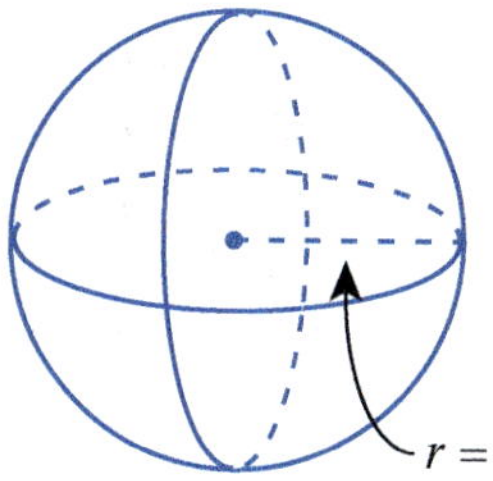

$$V = \frac{4}{3}\pi r^3$$

$$V = \frac{4}{3}(3.14)\cdot 4^3$$

$$= \frac{4(3.14)\cdot 64}{3}$$

$$= \frac{803.84}{3}\text{ ft}^3 \text{ or about } 267.95 \text{ ft}^3$$

The volume is approximately 267.95 cubic feet.

Example 6: Volume

The Luxor Hotel in Las Vegas is 30 stories tall and shaped like a pyramid. Its volume is approximately 36 million cubic feet. If the base of the hotel is a square with sides 556 ft long, what is the approximate height of the pyramid? (Round your answer to the nearest integer.)

Solution:

$$V = \frac{1}{3}lwh$$

$$36{,}000{,}000 = \frac{1}{3}(556)(556)h$$ Substitute 36,000,000 for V and 556 for l and w.

$$36{,}000{,}000 = \frac{309{,}136}{3}h$$ Simplify.

$$\frac{3}{309{,}136}\cdot 36{,}000{,}000 = \frac{3}{309{,}136}\cdot\frac{309{,}136}{3}h$$ Multiply both sides by $\frac{3}{309{,}136}$.

$$h = \frac{6{,}750{,}000}{19{,}321}\text{ ft or about 349 ft}$$

The height is approximately 349 feet.

Practice Problems

1. Find the area of a square with sides 4 cm long.
2. Find the area of a circle with a diameter 6 in.
3. Find the perimeter of a rectangle 3.5 m long and 1.6 m wide.
4. Find the volume of a rectangular solid with a length of 16 ft, a width of 10 ft, and a height of 1.4 ft.

3.2 Exercises

Select the answer from the right-hand column that correctly matches the statement.

1. e
2. j
3. k
4. f
5. c
6. b
7. d
8. o
9. p
10. a
11. l
12. n
13. m
14. i
15. h
16. g

______	**1.** The formula for the perimeter of a square	**a.** $V = \frac{1}{3}lwh$
______	**2.** The formula for the circumference of a circle	**b.** $A = \frac{1}{2}h(b+c)$
______	**3.** The formula for the perimeter of a triangle	**c.** $A = s^2$
______	**4.** The formula for the area of a rectangle	**d.** $A = \frac{1}{2}bh$
______	**5.** The formula for the area of a square	**e.** $P = 4s$
______	**6.** The formula for the area of a trapezoid	**f.** $A = lw$
______	**7.** The formula for the area of a triangle	**g.** $P = a+b+c+d$
______	**8.** The formula for the area of a parallelogram	**h.** $V = \frac{1}{3}\pi r^2 h$
______	**9.** The formula for the perimeter of a rectangle	**i.** $V = \frac{4}{3}\pi r^3$
______	**10.** The formula for the volume of a rectangular pyramid	**j.** $C = 2\pi r$
______	**11.** The formula for the volume of a rectangular solid	**k.** $P = a+b+c$
______	**12.** The formula for the area of a circle	**l.** $V = lwh$
______	**13.** The formula for the volume of a right circular cylinder	**m.** $V = \pi r^2 h$
______	**14.** The formula for the volume of a sphere	**n.** $A = \pi r^2$
______	**15.** The formula for the volume of a right circular cone	**o.** $A = bh$
______	**16.** The formula for the perimeter of a trapezoid	**p.** $P = 2l + 2w$

Answers to Practice Problems: **1.** 16 cm^2 **2.** 28.26 $in.^2$ (or exactly 9π $in.^2$) **3.** 10.2 m **4.** 224 ft^3

17. a. 72 mm; b. 324 mm^2
18. a. 70 cm; b. 300 cm^2
19. a. 38 in.; b. 72 in.^{2}
20. a. 54 mm; b. 126 mm^2
21. a. 32 cm; b. 44 cm^2
22. a. 31.4 in.; b. 78.5 in.^{2}
23. 108.85 cm^3
24. 1436.03 in.^{3}
25. 470.4 ft^3
26. 729 m^3
27. 203.47 in.^{3}

*Find **a.** the perimeter P (or circumference C) and **b.** the area A of each figure.*

17. $P =$ _____ $A =$ _____

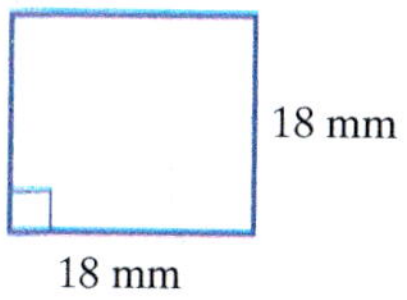

18. $P =$ _____ $A =$ _____

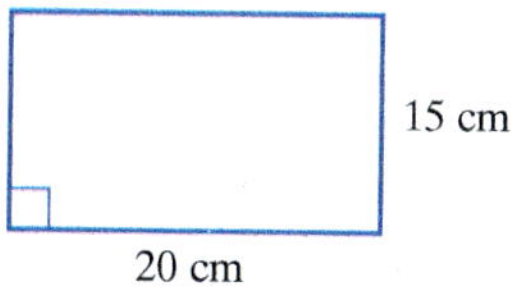

19. $P =$ _____ $A =$ _____

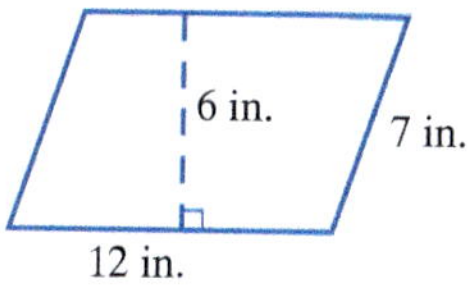

20. $P =$ _____ $A =$ _____

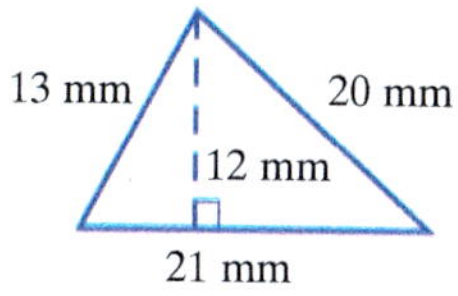

21. $P =$ _____ $A =$ _____

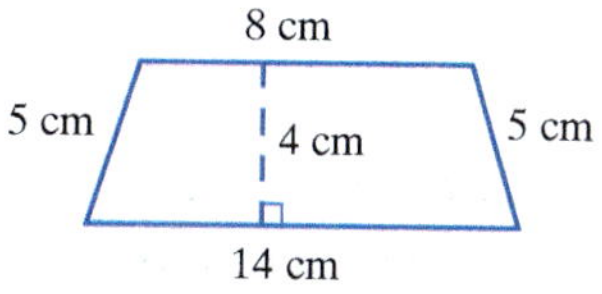

22. $C =$ _____ $A =$ _____

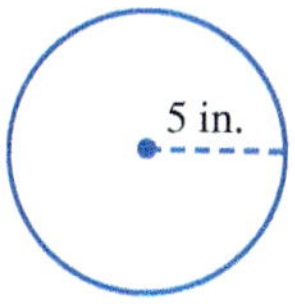

Find the volume of each figure. (Use $\pi = 3.14$ and round to the nearest hundredth.)

23. $V =$ _____

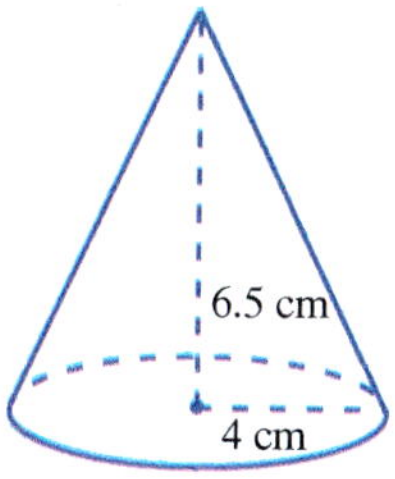

24. $V =$ _____

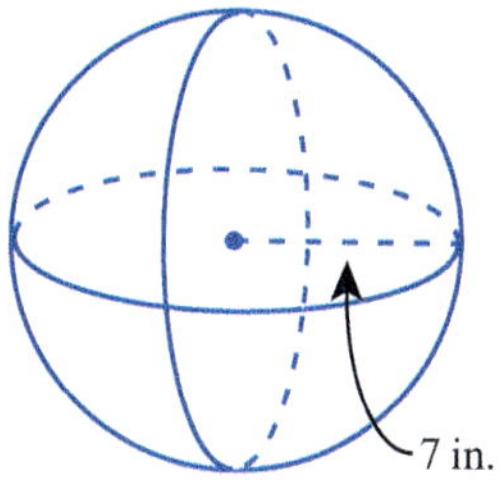

25. $V =$ _____

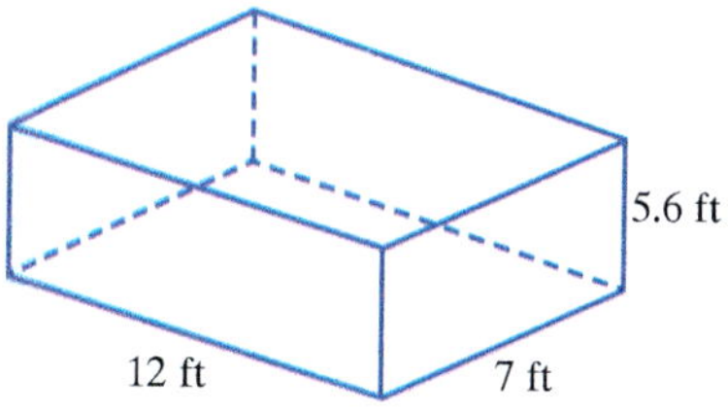

26. $V =$ _____

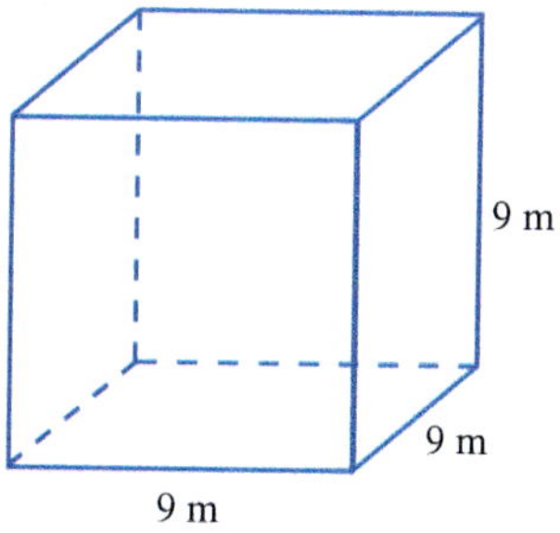

27. $V =$ _____

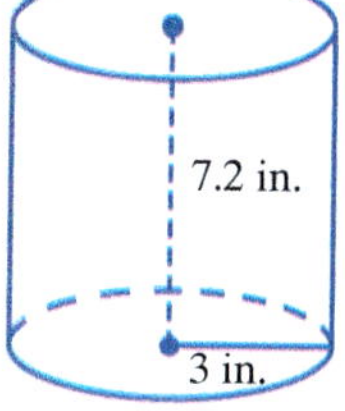

28. 56 in.
29. 81 ft^2
30. 63 cm^2
31. 50.24 m
32. 6154.4 ft^3
33. 24.2 m
34. 125 ft^3
35. 2520 ft^2
36. 272 cm^2
37. 1483.2 in.^{3}
38. 529.875 in.^{3}
39. 4186.67 cm^3
40. 28 cm; 48 cm^2
41. 54.84 in.; 200.52 in.^{2}

Solve the following problems. (Use $\pi = 3.14$.)

28. Rectangles: What is the perimeter of a rectangle with length of 17 in. and width of 11 in.?

29. Squares: Find the area of a square with sides that are 9 ft long.

30. Triangles: The base of a triangle is 14 cm and the height is 9 cm. Find the area.

31. Circles: Find the circumference of a circle with a radius of 8 m.

32. Cylinders: The radius of the base of a cylindrical tank is 14 ft. If the tank is 10 ft high, find the volume.

33. Triangles: The sides of a triangle are 6.2 m, 8.6 m, and 9.4 m. Find the perimeter.

34. Cubes: What is the volume of a cube with edges of 5 ft?

35. Rectangles: A rectangular garden plot is 60 ft long and 42 ft wide. Find the area.

36. Parallelograms: A parallelogram has a base of 20 cm and a height of 13.6 cm. Find the area.

37. Rectangular solids: A rectangular box is 18 in. long, 10.3 in. wide, and 8 in. high. Find the volume.

38. Cones: The diameter of the base of a right circular cone is 15 in. If the height of the cone is 9 in., find the volume.

39. Spheres: Find the volume of a sphere whose radius is 10 cm. (Round to the nearest hundredth.)

Find the perimeter and area for each figure. (Use $\pi = 3.14$.)

40. $P =$ _____ $A =$ _____

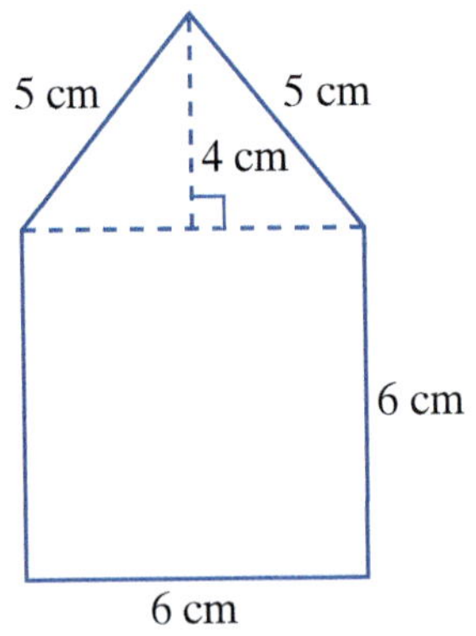

41. $P =$ _____ $A =$ _____

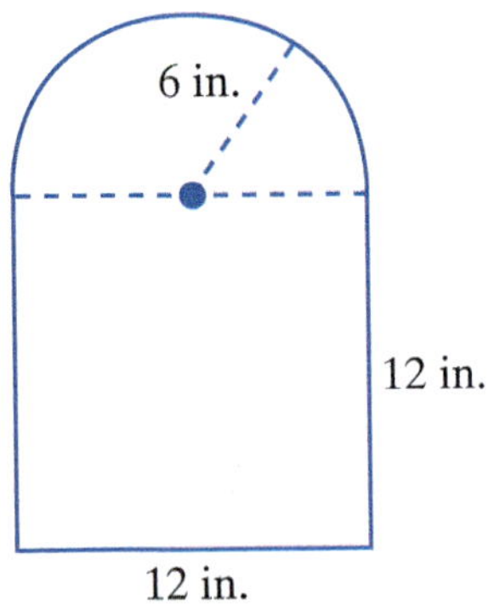

42. a. $P = 4s$

b. $s = \frac{P}{4}$

c. $2\frac{2}{3}$ meters

43. a. $P = 2l + 2w$

b. $w = \frac{P - 2l}{2}$

c. 13 feet

44. a. $C = 2\pi r$

b. $r = \frac{C}{2\pi}$

c. 13 centimeters

45. a. $A = bh$

b. $b = \frac{A}{h}$

c. 47 inches

46. a. $A = \frac{1}{2}bh$

b. $b = \frac{2A}{h}$

c. 64 cm

47. a. $P = a + b + c$

b. $a = P - b - c$

c. 61 inches

48. a. $A = \frac{1}{2}h(b + c)$

b. $h = \frac{2A}{b + c}$

c. 6 meters

49. a. $A = \frac{1}{2}h(b + c)$

b. $b = \frac{2A}{h} - c$

c. 5 feet

For the following exercises, ***a.*** *state the formula that relates to the given information, then* ***b.*** *solve the formula for the unknown quantity, and* ***c.*** *substitute the given values in the formula to determine the value of the unknown quantity. (Use $\pi = 3.14$.)*

42. Squares: The perimeter of a square is $10\frac{2}{3}$ meters. Find the length of the sides.

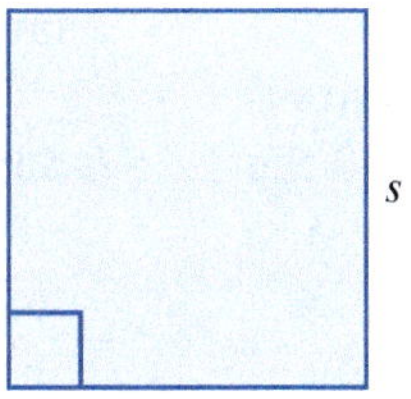

43. Rectangles: The perimeter of a rectangle is 88 feet. If the length is 31 feet, find the width.

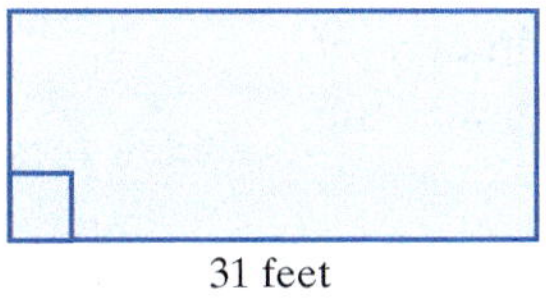

44. Circles: The circumference of a circle is 26π centimeters. Find the radius.

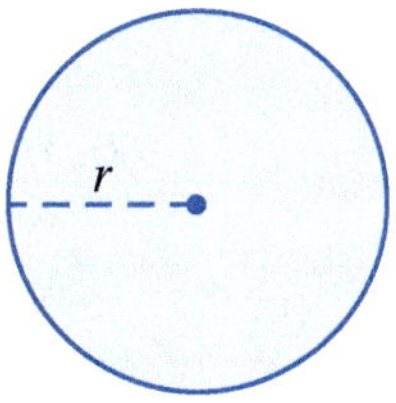

45. Parallelograms: The area of a parallelogram is 1081 square inches. If the height is 23 inches, find the length of the base.

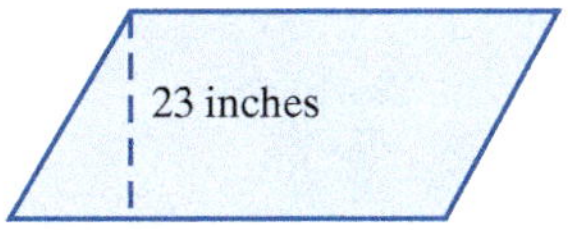

46. Rectangles: The area of a triangle is 800 cm^2. If the height is 25 cm, find the length of the base.

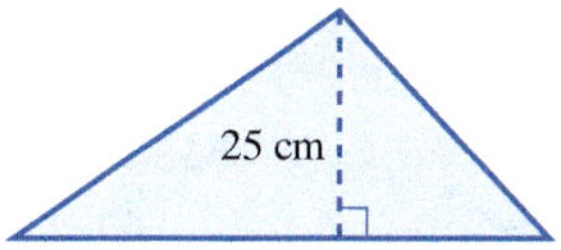

47. Triangles: The perimeter of a triangle is 147 inches. Two of the sides measure 38 inches and 48 inches. Find the length of the third side.

48. Trapezoids: The area of a trapezoid is 51 square meters. One base is 7 meters long, and the other is 10 meters long. Find the height of the trapezoid.

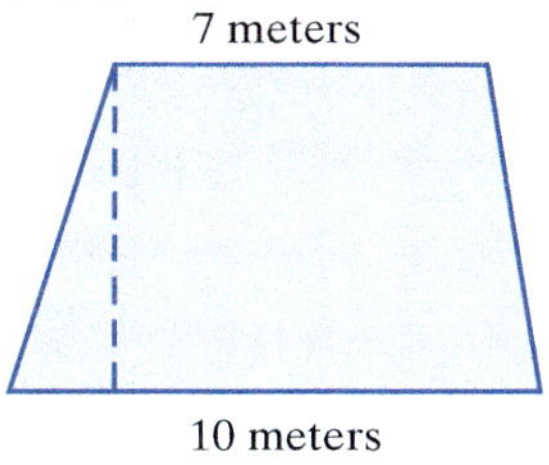

49. Trapezoids: The area of a trapezoid is 76 square feet. One base is 14 feet long, and the height is 8 feet. Find the height of the trapezoid.

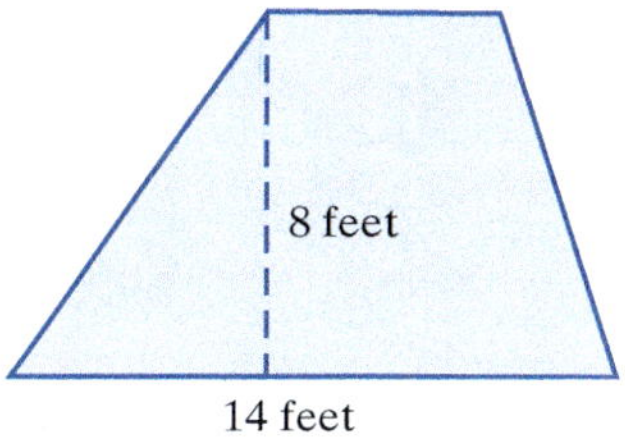

50. 3.75 inches
51. $16\frac{1}{8}$ or 16.125 meters
52. 12 cm, 12 cm, 21 cm
53. Side $a = 17$ m; Side $b = 11$ m
54. 64 feet
55. 15 ft
56. 15 inches
57. 40 inches
58. 19 yards
59. 5 in.
60. 12 ft, 38 ft
61. 9 m, 15 m
62. 12 ft

Use the formulas for perimeter, area, and volume to find the unknown values. (Use $\pi = 3.14$.)

50. Triangles: One side of a triangle is 11.5 inches long. A second side is 8.75 inches long. If the perimeter is 24 inches, find the length of the third side.

51. Squares: The perimeter of a square is 4 times the length of a side. Find the length of a side if the perimeter of a square is $64\frac{1}{2}$ meters.

52. Triangles: Two sides of a triangle have the same length and the third side is 3 cm less than the sum of the other two sides. If the perimeter of the triangle is 45 cm, what are the lengths of the three sides?

53. Parallelograms: The perimeter of a parallelogram is 56 m. If side a is 6 m more than side b, what are the lengths of the sides?

54. Lawn care: One hundred eighty-four feet of fencing is needed to enclose a rectangular-shaped garden plot. If the plot is 28 feet wide, what is the length?

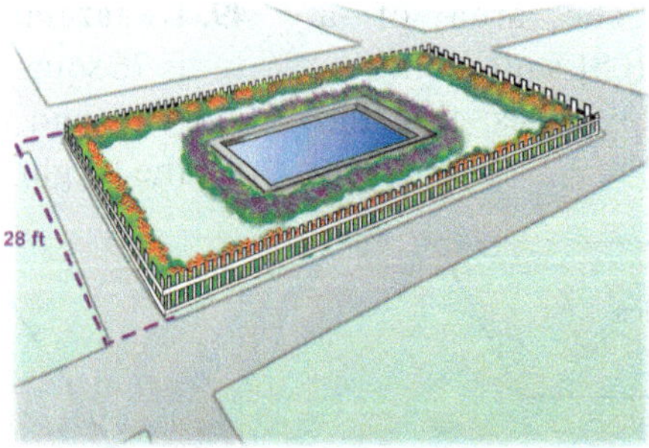

55. Laying carpet: Maria is laying down carpet in her home. She knows that her living room has a length of 25 ft and a perimeter of 80 ft. How wide is her living room?

56. Railroad crossing: The smallest railroad crossing sign U.S. Barricades offers has a circumference of 47.1 inches. What is the diameter of a railroad crossing sign with that circumference?

57. Sun Temple of Qosqo: The perimeter of one of the trapezoidal shaped doors of the Sun Temple of Qosqo in Peru is 194 inches. If the top of the doorway is 28 inches and both sides of the door measure 63 inches, what does the bottom of the doorway measure?

58. Rectangles: Find the width of a rectangle with a length of 21 yards and an area of 399 square yards.

59. Triangles: Find the length of the base of a triangle if the height is 18 in. and the area is 45 $in.^2$

60. Trapezoids: The height of a trapezoid is 10 ft and its area is 250 ft^2. If one of the bases is 2 feet more than 3 times the other base, what are the lengths of the bases?

61. Trapezoids: The area of a trapezoid is 192 m^2. If one of the bases is 3 m less than twice the other base, and the height is 16 m, what are the lengths of the bases?

62. Flags: An American flag flying on a 50 ft flagpole has a hoist (width) of 8 ft and an area of 96 ft^2. What is the fly (length) of the flag?

63. 10.5 cm
64. 30 ft
65. 5.5 ft
66. 15 mm
67. 21 in.
68. 17 ft
69. 28 meters
70. 25 in.
71. 4.5 cm
72. 26 in.
73. 36 cm

63. Kites: A diamond kite from Kitty Hawk Kites is made up from two identical triangles. If the area of the kite is 393.75 cm^2 and the length of the kite is 75 cm what is the width of the kite?

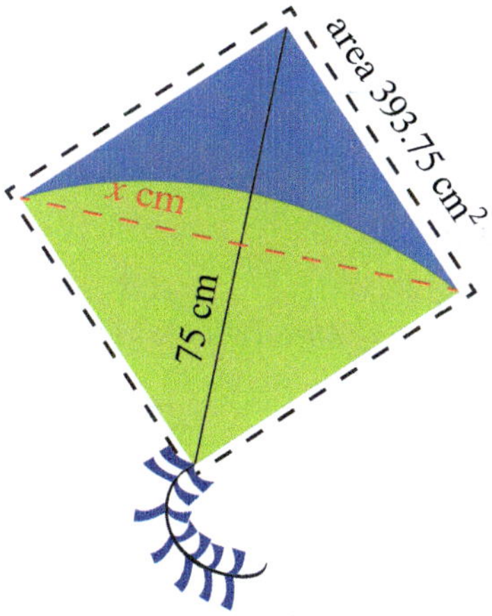

64. Sailing: A sail is in the shape of a triangle. Find the height of the sail if its base is 20 ft and its area is 300 ft^2.

65. Architecture: Each of the balconies of an apartment building in New York City are in the shape of a parallelogram. If the area of a balcony is 33 ft^2 and the distance from the wall to the edge is 6 ft, how long is the balcony?

66. Circular cylinders: The volume of a right circular cylinder is 1695.6 mm^3. If the radius is 6 mm, what is the height?

67. Rectangular pyramids: The volume of a rectangular pyramid is 2520 $in.^3$ If the length is 20 in. and the height is 18 in., what is the width?

68. Rectangular solid: The width of a rectangular solid is 10 ft. The height is 22 ft. What is the length of the solid if the volume is 3740 ft^3?

69. Circular cones: The radius of a right circular cone is 12 meters. Find the height if the volume of the cone is 4220.16 cubic meters.

70. Computers: A custom PC mod case in the form of a glass pyramid has a square base with sides that measure 18 in. If the volume of the case is 2700 $in.^3$ how tall is the case?

71. Rectangular boxes: Find the width of a cereal box with volume 1260 cm^3 if its length is 14 cm and its height is 20 cm.

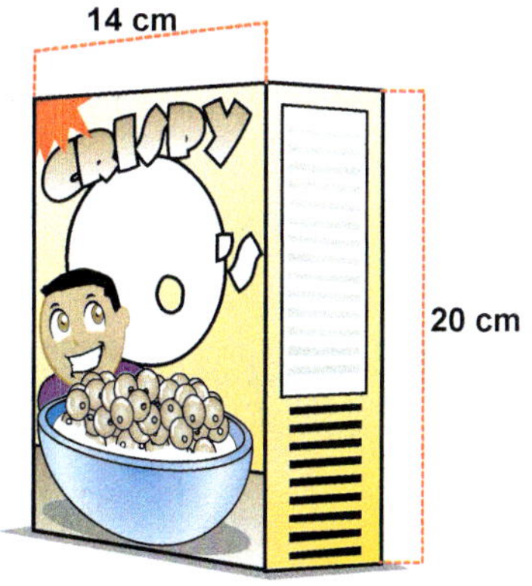

72. Telescopes: The volume of a telescope optical tube is 1306.24 $in.^3$ If the diameter of the tube is 8 in., what is the length of the tube?

73. Baking: A pastry bag filled with icing is shaped like a right circular cone. If the radius of the pastry bag is 7 cm and it contains 1846.32 cm^3 worth of icing, what is the length of the pastry bag (the height of the cone)?

74. $x = 2$
75. $x = 5$
76. $212.04\ \text{m}^2$
77. 52.25 in.
78. $130.08\ \text{in.}^2$
79. $1326.93\ \text{cm}^2$
80. $6{,}334{,}233.21\ \text{cm}^3$
81. $510.46\ \text{ft}^3$
82.

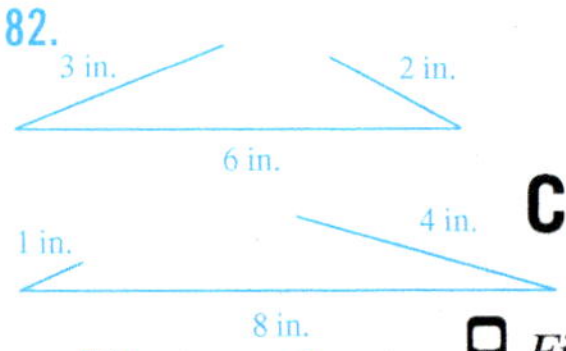

The two shorter sides are not long enough to touch and form a triangle. From this we can guess that the sum of the lengths of any two sides of a triangle must be greater than the length of the third side.

74. Area: Find the value of x in the figure if the area of the shaded portion is 95 square inches.

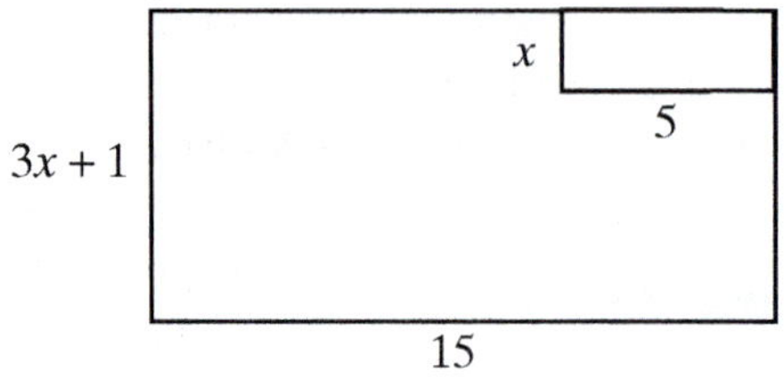

75. Area: Find the value of x in the figure if the area of the shaded portion is $193\ \text{cm}^2$.

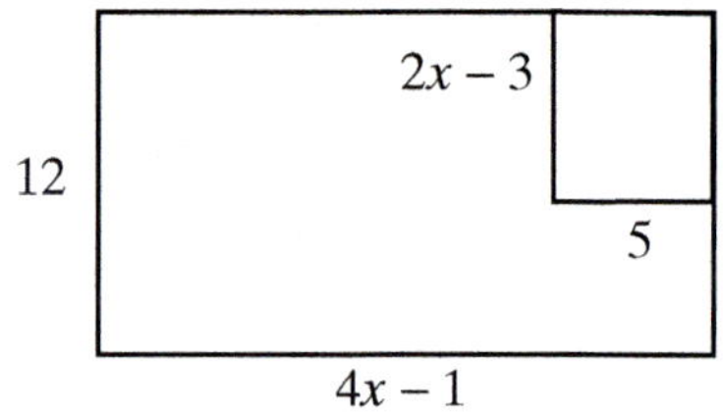

Calculator Problems

Find the answers accurate to two decimal places. (Use $\pi = 3.14$.)

76. Rectangles: Find the area of a rectangle that is 16.54 m long and 12.82 m wide.

77. Circles: The radius of a circle is 8.32 in. Find the circumference.

78. Trapezoids: Find the area of a trapezoid with bases of 22.36 in. and 17.48 in. and a height of 6.53 in.

79. Triangles: Find the area of a triangle with a base of 63.52 cm and a height of 41.78 cm.

80. Spheres: The radius of a sphere is 114.8 cm. Find the volume.

81. Cones: The radius of a right circular cone is 6.32 ft and its height is 12.21 ft. Find the volume.

Writing and Thinking About Mathematics

82. Triangles: Sketch a triangle with sides of length 3 inches, 6 inches, and 2 inches. Now sketch another triangle with sides of length 1 inch, 8 inches, and 4 inches. What is wrong with your triangles? Discuss what you think must be true concerning the lengths of the sides of any triangle.

HAWKES LEARNING SYSTEMS: INTRODUCTORY & INTERMEDIATE ALGEBRA SOFTWARE

- 3.2 Formulas in Geometry

3.3 Applications: Distance-Rate-Time, Interest, Average

Solve the following types of problems using linear equations:

- *Distance-rate-time problems,*
- *Simple interest problems,*
- *Average problems, and*
- *Cost problems.*

Word problems (or applications) are designed to teach you to read carefully, to organize, and to think clearly. Whether or not a particular problem is easy for you depends a great deal on your personal experiences and general reasoning abilities. The problems generally do not give specific directions to add, subtract, multiply, or divide. You must decide what relationships are indicated through careful analysis of the problem.

Distance-Rate-Time

Problems involving distance usually make use of the relationship indicated by the formula $d = rt$, where $d =$ distance, $r =$ rate, and $t =$ time. A chart or table showing the known and unknown values is quite helpful and is illustrated in the next example.

Example 1: Distance-Rate-Time

A motorist averaged 45 mph for the first part of a trip and 54 mph for the last part of the trip. If the total trip of 303 miles took 6 hours, what was the time for each part?

Solution: **Analysis of strategy:**

What is being asked for?
Total time minus time for 1st part of trip gives time for 2nd part of trip.

Let $t =$ time for 1st part of trip
$6 - t =$ time for 2nd part of trip

	rate	· time	= distance
1st Part	45	t	$45t$
2nd Part	54	$6 - t$	$54(6 - t)$

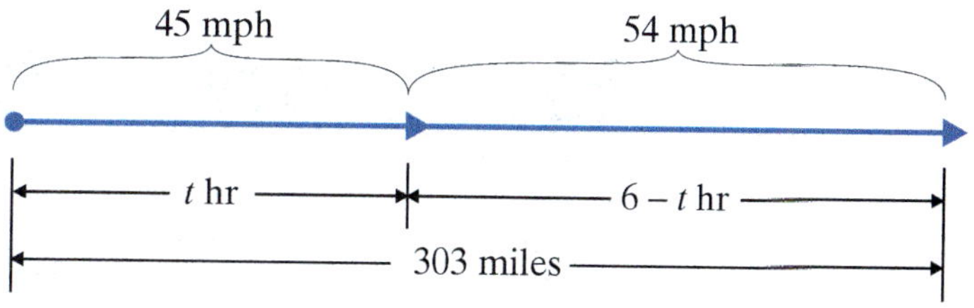

Continued on the next page...

$$\underbrace{\text{1st part distance}} + \underbrace{\text{2nd part distance}} = \underbrace{\text{total distance}}$$

Form an equation relating the given information.

$$45t + 54(6-t) = 303$$

Solve the equation.

$$45t + 324 - 54t = 303$$
$$324 - 9t = 303$$
$$-9t = -21$$
$$t = \frac{-21}{-9} = \frac{7}{3} \text{ hr}$$

1st part of the trip

$$6 - t = 6 - \frac{7}{3} = \frac{11}{3} \text{ hr}$$

2nd part of the trip

Check: $45 \cdot \frac{7}{3} = 15 \cdot 7 = 105$ miles (1st part)

$54 \cdot \frac{11}{3} = 18 \cdot 11 = 198$ miles (2nd part)

$105 + 198 = 303$ miles in total

The first part took $\frac{7}{3}$ hr or $2\frac{1}{3}$ hr. The second part took $\frac{11}{3}$ hr or $3\frac{2}{3}$ hr.

Interest

To solve problems related to interest on money invested for one year, you need to know the basic relationship among the principal P (amount invested), the annual rate of interest r, and the amount of interest I (money earned). This relationship is described in the formula $P \cdot r = I$. (This is the formula for simple interest, $I = Prt$, with $t = 1$.) We use this relationship in Example 2.

Example 2: Interest

Kara has had \$40,000 invested for one year, some in a savings account which paid 7% and the rest in a high-risk stock which yielded 12% for the year. If her interest income last year was \$3550, how much did she have in the savings account and how much did she invest in the stock?

Solution: Let x = amount invested at 7%

$40{,}000 - x$ = amount invested at 12%

Total amount invested minus amount invested at 7% represents amount invested at 12%.

	principal ·	rate =	interest
Savings Account	x	0.07	$0.07(x)$
Stock	$40{,}000 - x$	0.12	$0.12(40{,}000 - x)$

$$\underbrace{\text{interest at }7\%}_{0.07(x)} + \underbrace{\text{interest at }12\%}_{0.12(40{,}000-x)} = \underbrace{\text{total interest}}_{3550}$$

$$0.07(x) + 0.12(40{,}000 - x) = 3550$$

$$
\begin{aligned}
7x + 12(40{,}000 - x) &= 355{,}000 && \text{Multiply both sides of the equation by 100 to eliminate the decimal.}\\
7x + 480{,}000 - 12x &= 355{,}000\\
-5x &= -125{,}000\\
x &= 25{,}000 && \text{Amount invested at 7\%}\\
40{,}000 - x &= 15{,}000 && \text{Amount invested at 12\%}
\end{aligned}
$$

Check: $25{,}000(0.07) = 1750$ and $15{,}000(0.12) = 1800$
and $\$1750 + \$1800 = \$3550$.
Kara had \$25,000 in the savings account at 7% interest and invested \$15,000 in the stock at 12% interest.

Average (or Mean)

The **average** (or **mean**) of a set of numbers was defined and discussed in Section 1.4. In this section we use the concept of average to find unknown numbers and to read and understand graphs in more detail.

Example 3: Average (or Mean)

Suppose that you have scores of 85, 92, 82 and 88 on four exams in your English class. What score will you need on the fifth exam to have an average of 90?

Solution: Let x = your score on the fifth exam.
The sum of all the scores, including the unknown fifth exam, divided by 5 must equal 90.

$$\frac{85 + 92 + 82 + 88 + x}{5} = 90$$

$$\frac{347 + x}{5} = 90$$

$$5 \cdot \frac{347 + x}{5} = 5 \cdot 90$$

$$347 + x = 450$$

$$x = 103$$

Assuming that each exam is worth 100 points, you cannot attain an average of 90 on the five exams.

Example 4: Bar Graphs

The bar graph shows the enrollment at the main campuses at six Big Ten universities. Use the graph to find the following (note that the units on the graph are in thousands):

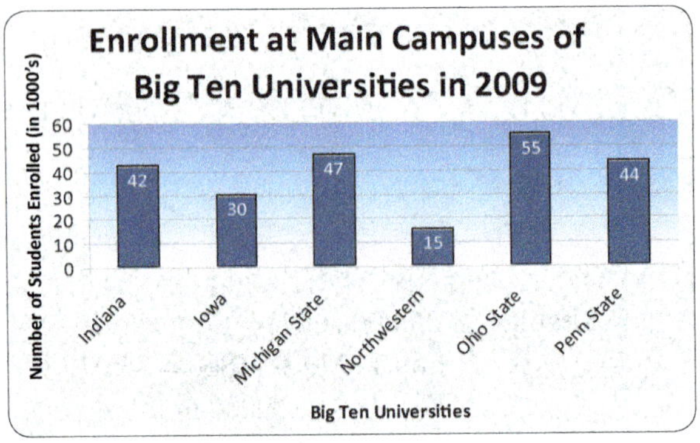

a. Find the average enrollment over the six schools. (Round to the nearest thousand.)

Solution: Find the sum: $42+30+47+15+55+44=233$.

Divide by 6: $233 \div 6 \approx 39$.

The average enrollment is approximately 39,000 students.

b. Which university has the lowest enrollment?

Solution: Northwestern has the lowest enrollment: 15,000 students.

c. Find the difference in enrollment between Ohio State and Penn State.

Solution: The difference is $55{,}000 - 44{,}000 = 11{,}000$ students.

Cost

Example 5: Cost

A jeweler paid $350 for a ring. He wants to price the ring for sale so that he can give a 30% discount on the marked selling price and still make a profit of 20% on his cost. What should be the marked selling price of the ring?

Solution: Again, we make use of the relationship $S - C = P$ (selling price – cost = profit).

Let x = marked selling price,

then $x - 0.30x$ = actual selling price

and 350 = cost.

actual selling price	−	cost	=	profit
$x - 0.30x$	−	350	=	$0.20(350)$
$0.70x$	−	350	=	70
		$0.70x$	=	420
		x	=	600

The actual selling price is the marked selling price minus the 30% discount on the ring.

The profit is 20% of what he paid originally.

Check: **Step 1:**

$600 marked selling price
× 0.30 discount %
$180 discount

Step 2:

$600 marked selling price
− $180 discount
$420 actual selling price

Step 3:

$420 actual selling price
− $350 cost
$70 profit

As a double check, $350 cost
× 0.20 profit %
$70 profit.

The jeweler should set the selling price at $600.

3.3 Exercises

Refer to the formulas listed in Section 3.1 as necessary.

1. 1.68 mph
2. 18 mph
3. 8.75 hours
4. 4.75 hours
5. 3 hours

1. Hiking: What is Nathan's average rate of speed if he hikes 12.6 miles in 7.5 hours?

2. Biking: What is Amy's average rate of speed if she bikes 39.6 miles in 2.2 hours?

3. Traveling by car: Jamie plans to take the scenic route from Los Angeles to San Francisco. Her GPS tells her it is a 420 mile trip. If she figures she'll average 48 mph, how long will the trip take her?

4. Traveling by car: Scott's averaging 60 mph on his drive from St. Louis, MO to Memphis, TN. If the total trip is 285 miles, how long should he expect the drive to take?

5. Biking: Jane rides her bike to Lake Junaluska. Going to the lake, she averages 12 mph. On the return trip, she averages 10 mph. If the round trip takes a total of 5.5 hours, how long does the return trip take?

	rate	· time	= distance
Going	12	$5.5 - t$	
Returning	10	t	

6. 375 mph; 450 mph
7. 60 mph; 300 miles
8. 4.8 hours
9. 7 hours
10. 54 mph
11. 36 mph; 60 mph
12. 12 miles
13. Day = 56 mph; Night = 69 mph

6. **Plane speeds:** Two planes which are 2475 miles apart fly toward each other. Their speeds differ by 75 mph. If they pass each other in 3 hours, what is the speed of each?

	rate	· time	= distance
1st plane	r	3	
2nd plane	$r+75$	3	

7. **Traveling by car:** Marcus drives from Chicago to Detroit in 6 hours. On the return trip, his speed is increased by 10 mph and the trip takes 5 hours. Find his rate on the return trip. How far apart are the towns?

	rate	· time	= distance
Going	r	6	
Returning	$r+10$	5	

8. **Hiking:** Tim and Barb have 8 hours to spend on a mountain hike. They can walk up the trail at an average of 2 mph and can walk down at an average of 3 mph. How long should they plan to hike uphill before turning around?

	rate	· time	= distance
Uphill	2	x	
Downhill	3	$8-x$	

9. **Traveling by car:** The Reeds are moving across Texas. Mr. Reed leaves $3\frac{1}{2}$ hours before Mrs. Reed. If he averages 40 mph and she averages 60 mph, how long will Mrs. Reed have to drive before she overtakes Mr. Reed?

10. **Traveling by car:** After traveling for 40 minutes, Mr. Koole had to slow to $\frac{2}{3}$ his original speed for the rest of the trip due to heavy traffic. The total trip of 84 miles took 2 hours. Find his original speed.

11. **Train speeds:** A train leaves Cincinnati at 2:00 PM. A second train leaves the same station in the same direction at 4:00 PM. The second train travels 24 mph faster than the first. If the second train overtakes the first at 7:00 PM, what is the speed of each of the two trains?

12. **Running:** Maria runs through the countryside at a rate of 10 mph. She returns along the same route at 6 mph. If the total trip took 1 hour 36 minutes, how far did she run in total?

13. **Traveling by car:** The distance from Atlanta, Georgia to Washington, DC is 620 miles. Driving in the middle of the night it takes about 9 hours to get to Washington. Due to higher traffic volume, it takes 2 more hours to travel there during the day. What is the average rate of the driver during the day and during the night? (Round to the nearest whole number.)

14. 1st half = 40 mph; 2nd half = 60 mph
15. 4.5 miles
16. Walks for 0.6 hours or 36 minutes; Runs for 0.3 hours or 18 minutes
17. $14,000 at 5%; $11,000 at 6%
18. $7400 at 5.5%; $2600 at 6%
19. 4.5% on $10,000; 5.5% on $6000
20. 6.5% on $4000; 6% on $3000
21. $7000 at 6%; $9000 at 8%
22. $5500 at 8%; $6500 at 10%
23. $24,000 at 4.5%; $18,000 at 6%
24. $6720 at 5.5%; $5280 at 7%

14. Traveling by car: Mr. Kent drove to a conference. The first half of the trip took 3 hours due to traffic. Traffic let up for the second half of the trip and he was able to increase his speed by 20 mph to make sure he got there on time. Find his rates of speed if he traveled 2 hours at the second rate.

15. Exercising: Jayden walked to his friend's house at a rate of 4 mph to borrow his friend's bicycle. Coming back home, he rode the bicycle at an average rate of 12 mph. The total time for the round trip was 1 hour 30 minutes. How far away does Jayden's friend live?

16. Exercising: Once a week Felicia walks/runs for a total of 6 miles. Felicia spends twice as much time walking as she does running. If she walks at a rate of 4 mph and runs three times faster than she walks, what is the time for each part?

17. Investing: Amanda invests $25,000, part at 5% and the rest at 6%. The annual return on the 5% investment exceeds the annual return on the 6% investment by $40. How much did she invest at each rate?

18. Investing: Mr. Hill invests $10,000, part at 5.5% and part at 6%. The annual interest from the 5.5% investment exceeds the annual interest from the 6% investment by $251. How much did he invest at each rate?

19. Investing: The annual interest earned on a $6000 investment was $120 less than the interest earned on $10,000 invested at 1% less interest per year. What was the rate of interest on each amount?

20. Investing: The annual interest on a $4000 investment exceeds the interest earned on a $3000 investment by $80. The $4000 is invested at a 0.5% higher rate of interest than the $3000. What is the interest rate of each investment?

21. Investing: Two investments totaling $16,000 produce an annual income of $1140. One investment yields 6% a year, while the other yields 8% per year. How much is invested at each rate?

22. Investing: D'Andra makes two investments that total $12,000. One investment yields 8% per year and the other 10% per year. The total interest for one year is $1090. Find the amount invested at each rate.

23. Investing: A company invested $42,000 into two simple interest accounts. The annual interest rate on one of the accounts is 4.5% while the rate on the other is 6%. How much should the company invest in each account so that the two accounts will produce an equal annual interest income?

24. Investing: Sebastian would like both of his investments, a total of $12,000, to bring him the same annual interest income. One of his investments is at 5.5% annual interest rate and the other at 7%. Find the amount of money that Sebastian should invest in each account.

25. $600 at 2.5%; $800 at 4%
26. $9200 at 5%; $7000 at 6%
27. $11,000 at 4%; $9500 at 5%
28. $12,000 at 1.5%; $6000 at 4%
29. $112.50
30. $4200
31. $60.50
32. $113.75
33 $40.50
34. $220

25. Investing: Gabriella got some money from her grandparents as a graduation present. She decided to invest all of it. Part of the money was invested at a 2.5% interest rate, and the rest at a 4% interest rate. She invested $200 more in the 4% account than the 2.5% account. If her annual interest income was $47, how much did she invest at each rate?

26. Investing: Last year an individual invested some money at a 5% interest rate and $2200 less than that amount at a 6% interest rate. If his interest income was $880, how much did he invested at each rate?

27. Investing: A small company invested $20,500 such that a part of the money is in an account with a 4% interest rate and the rest at a 5% rate. The annual interest from the 5% account is $35 more than the interest earned from the 4% account. Find the amount of money the company invested at each rate.

28. Investing: Jordan is earning 1.5% interest from money invested in a savings account and 4% interest on a mutual bond fund. If the total of his investments is $18,000 and the annual interest from the savings account is less than the annual interest from the bond by $60, how much has Jordan invested at each rate?

29. Shopping: A particular style of shoe costs the dealer $81 per pair. At what price should the dealer mark them so he can sell them at a 10% discount off the selling price and still make a 25% profit?

30. Used car sales: A car dealer paid $2850 for a used car. For his upcoming Labor Day sale, he wants to offer a 5% discount off the posted selling price, but would still like to make a 40% profit. What price should he advertise for that car?

31. Shopping: During the month of January, a department store would like to have a sale of 40% off of women's long boots. The store purchased the boots for $33 per pair. How much should the selling price be, if the manager of the store wants to make a 10% profit per pair?

32. Shopping: A store purchased a certain style of leather jacket at $70 per jacket. If the store wants to sell the jackets at a 20% discount and still make a profit of 30%, what should be the marked selling price for each jacket?

33. Shopping: Carla plans on buying a new pair of sandals for the summer. They are on sale for 20% off of the original price. What was the original price of the sandals, if she pays $34.83, with 7.5% sales tax?

34. Electronics: After receiving a 10% off coupon in the mail, Mark decided that it was time to buy a Blu-Ray player. Using the coupon and paying 8% sales tax, the final price came to $213.84. What was the listed price of the Blu-Ray player?

35. 62 min
36. $20
37. 6 hours
38. 4 phone calls
39. 71
40. 94
41. a. Not possible
b. 192
42. 67

35. Biking: While riding her bike to the park and back home five times, Stacey timed herself at 60 min, 62 min, 55 min (the wind was helping), 58 min, and 63 min. She had set a goal of averaging 60 minutes for her rides. How many minutes will she need on her sixth ride to attain her goal?

36. Budgeting: A college student realized that he was spending too much money on video games. For the remaining 5 months of the year his goal is to spend an average of $50 a month towards his hobby. How much can he spend in December, taking into consideration that the other 4 months he spent $70, $25, $105, $30 respectively?

37. TV watching: While growing up, Jason was allowed to watch TV an average of 3 hours a day over a one week period. One particular week he watched 1 hour, 2 hours, 1 hour, 3 hours, 3 hours, and 5 hours. How many hours could Jason watch the seventh and last day of the week and still obey his parents?

38. Cell phones: For every 4 week period, Lauren wants to make an average of 6 phone calls per week from her prepaid cell phone. The first week she made 9 phone calls, the second week 6 phone calls, and the third week 5 phone calls. How many phone calls does Lauren need to make in the fourth week to make sure she stays on track with her goal?

39. Exam scores: Marissa has five exam scores of 75, 82, 90, 85, and 77 in her chemistry class. What score does she need on the final exam to have an average of 80 (and thus earn a grade of B)? (All exams have a maximum of 100 points.)

40. Exam scores: Gerald had scores of 80, 92, 89, and 95 on four exams in his algebra class. What score will he need on his fifth exam to have an overall average of 90? (All exams have a maximum of 100 points.)

41. Exam scores: A statistics student has grades of 86, 91, 95, and 76 on four hour-long exams. What score must he receive on the final exam to have an average of 90 if:

a. the final is equivalent to a single hour-long exam (100 points maximum)?

b. the final is equivalent to two hour-long exams (200 points maximum)?

42. Exam scores: Wade has scores of 59, 68, 76, 84 and 69 on the first five tests in his social studies class. He knows that the final exam counts as two tests. What score will he need on the final to have an average of 70? (All tests and exams have a maximum of 100 points.)

43. a. 53.8°F
b. 44°F
c. 17°F
44. a. 3.2 inches
b. 7.1 inches
c. 6.9 inches
45. a. $24.2 billion
b. $26 billion
c. $24 billion

43. Temperature: Given the monthly temperatures over a year for Christchurch, New Zealand:

a. Find the average temperature for the year. (Round to the nearest tenth.)

b. Find the minimum average temperature for the year.

c. Find the difference in average temperature between the months of June and December.

Source: http://www.weather.com

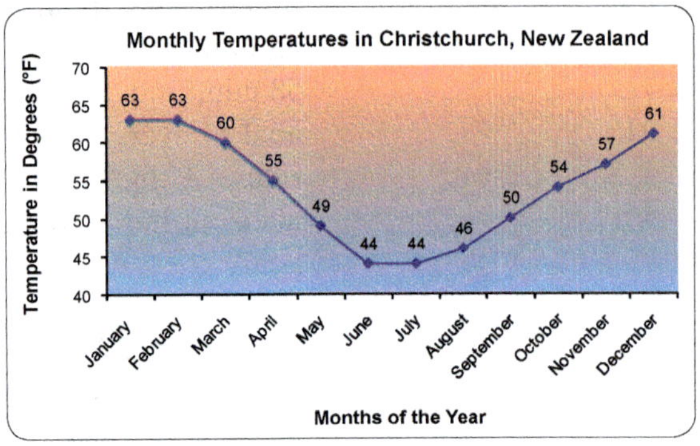

44. Average rainfall: Given the monthly rainfall averages over a year for Visakhapatnam, India:

a. Find the average rainfall for the year. (Round to the nearest tenth.)

b. Find the maximum average rainfall for the year.

c. Find the difference in average rainfall between the months of October and December.

Source: http://www.weather.com

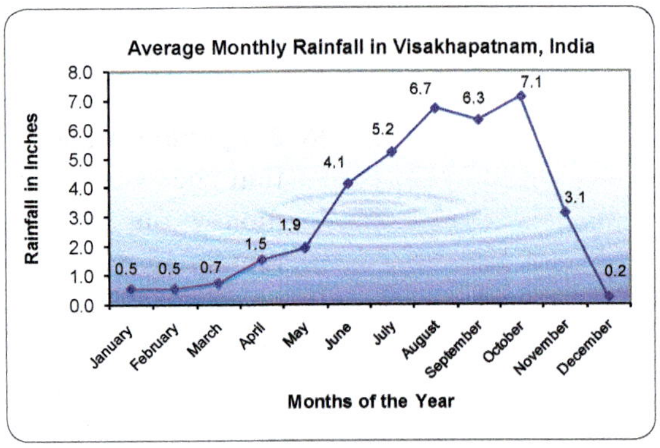

45. Company profits: Given the yearly profits of the world's largest companies in 2009:

a. Find the average profits of the companies in the year 2009. (Round to the nearest tenth.)

b. Find the yearly profits of Royal Dutch Shell in the year 2009.

c. What was the difference in profits between Exxon Mobil and BP?

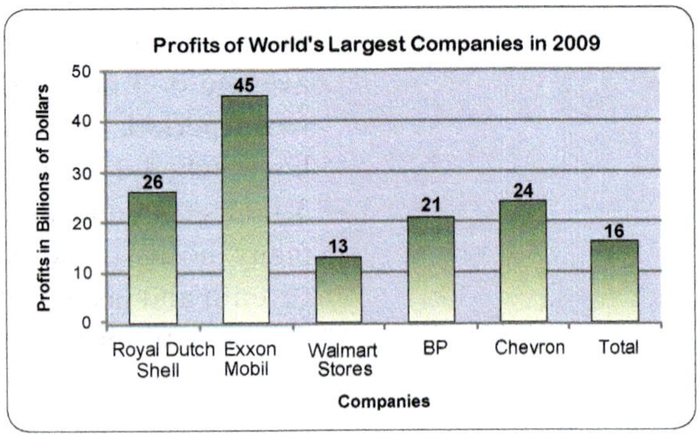

Source: http://money.cnn.com/magazines/fortune/global500/2009/full_list/

46. a. 67.2 million
b. 27 million
c. 66 million

46. Airport usage: Given the number of passengers at the following airports:

a. Find the average number of passengers.

b. Find the difference in passengers between Atlanta, GA and Tokyo.

c. What was the total number of passengers to go through London?

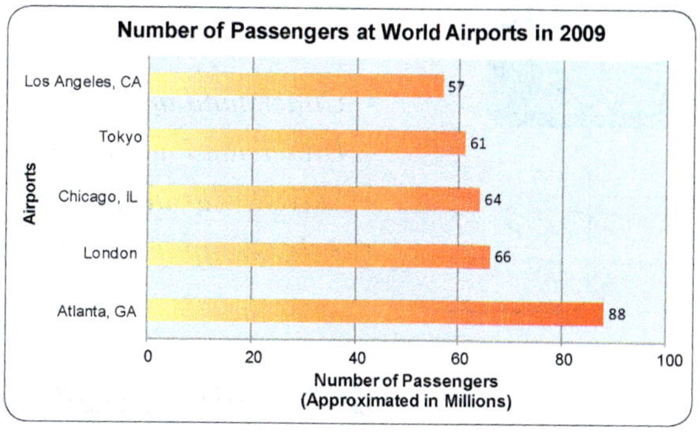

Source: Airports Council International – North America

HAWKES LEARNING SYSTEMS: INTRODUCTORY & INTERMEDIATE ALGEBRA SOFTWARE

- 3.3 Applications

3.4 Linear Inequalities

- *Understand and use* ***set-builder notation.***
- *Understand and use* ***interval notation.***
- *Solve linear inequalities.*
- *Solve compound inequalities.*

Sets and Set-Builder Notation

A **set** is a collection of objects or numbers. The items in the set are called **elements**, and sets are indicated with braces $\{\ \}$, and named with capital letters. If the elements are listed within the braces the set is said to be in **roster form**. For example,

$$A = \left\{\frac{3}{4}, 1.7, 2, 4, 6, \sqrt{89}\right\}$$

$$\mathbb{W} = \{0, 1, 2, 3, 4, 5, \ldots\}$$

$$\mathbb{Z} = \{\ldots, -4, -3, -2, -1, 0, 1, 2, 3, 4, \ldots\}$$

are all in roster form.

The symbol $\in$ is read "is an element of" and is used to indicate that a particular number belongs to a set. For example, in the previous illustrations, $1.7 \in A$, $0 \in \mathbb{W}$, and $-3 \in \mathbb{Z}$.

If the elements in a set can be counted, as in A above, the set is said to be **finite**. If the elements cannot be counted, as in $\mathbb{W}$ and $\mathbb{Z}$ above, the set is said to be **infinite**. If a set has absolutely no elements, it is called the **empty set** or **null set** and is written in the form $\{\ \}$ or with the special symbol $\varnothing$. For example, the set of all people over 15 feet tall is the empty set, $\varnothing$.

The notation $\{x|\quad\}$ is read "the set of all x such that ..." and is called **set-builder notation**. The vertical bar $(\,|\,)$ is read "such that." A statement following the bar gives a condition (or restriction) for the variable x. For example,

$\{x|x \text{ is an even integer}\}$ is read "the set of all x such that x is an even integer."

The following note highlights the two important set concepts **union** and **intersection**.

NOTES

Special Comments about Union and Intersection

The concepts of union and intersection are part of set theory which is very useful in a variety of courses including abstract algebra, probability, and statistics. These concepts are also used in analyzing inequalities and analyzing relationships among sets in general.

The **union** (symbolized $\cup$, as in $A \cup B$) of two (or more) sets is the set of all elements that belong to either one set or the other set or to both sets. The **intersection** (symbolized $\cap$, as in $A \cap B$) of two (or more) sets is the set of all elements that belong to both sets. The word **or** is used to indicate union and the word **and** is used to indicate intersection. For example, if $A = \{1, 2, 3\}$ and $B = \{2, 3, 4\}$, then the numbers that belong to A **or** B is the set $A \cup B = \{1, 2, 3, 4\}$. The set of numbers that belong to A **and** B is the set $A \cap B = \{2, 3\}$. These relationships can be illustrated using the following Venn diagram.

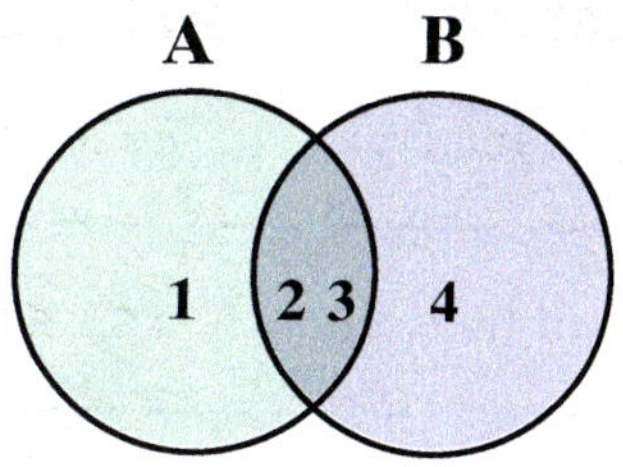

Similarly, union and intersection notation can be used for sets with inequalities.

For example, $\{x | x < a \textbf{ or } x > b\}$ can be written in the form

$$\{x | x < a\} \cup \{x | x > b\}.$$

Also, $\{x | x > a \textbf{ and } x < b\}$ can be written in the form

$$\{x | x > a\} \cap \{x | x < b\} \text{ or } \{x | a < x < b\}.$$

Intervals of Real Numbers

Suppose that a and b are two real numbers and that $a < b$. The set of all real numbers between a and b is called an **interval of real numbers**. Intervals of real numbers are used in relationship to everyday concepts such as the length of time you talk on your cell phone, the shelf life of chemicals or medicines, and the speed of an airplane.

Teaching Notes:
Inequalities seem to be underplayed in importance in the classroom. The "real world" uses inequality concepts at every turn. Cite several examples to illustrate common thinking in inequality terms: "...made to fit *within certain limits*..., ...manufactured to a specific *tolerance*..., ...must have an annual income *greater than*..., ...the ten year financial goal is to have grown *at least* by..., ...we need *a minimum* of..."

Teaching Notes:
The alternate notation of open and closed circles may be more familiar to your students, however we chose to graph using parentheses and brackets in order to correspond to interval notation. Also, this notation is common in higher level mathematics.

As an aid in reading inequalities and graphing inequalities correctly, note that an inequality may be read either from left to right or right to left. Because we are concerned about which numbers satisfy an inequality, we **read the variable first**:

$x > 7$ is read from **left to right** as "**x is greater than 7.**"
and $7 < x$ is read from **right to left** as "**x is greater than 7.**"

A compound interval such as $-3 < x < 6$ is read

"**x is greater than -3 and x is less than 6.**"

Again, note that *the variable is read first.*

Various types of intervals and their corresponding **interval notation** are listed in Table 1.

Types of Intervals and Interval Notation

Type of Interval	Algebraic Notation	Interval Notation	Graph
Open Interval	$a < x < b$	(a, b)	
Closed Interval	$a \le x \le b$	$[a, b]$	
Half-open Interval	$\begin{cases} a \le x < b \\ a < x \le b \end{cases}$	$[a, b)$ $(a, b]$	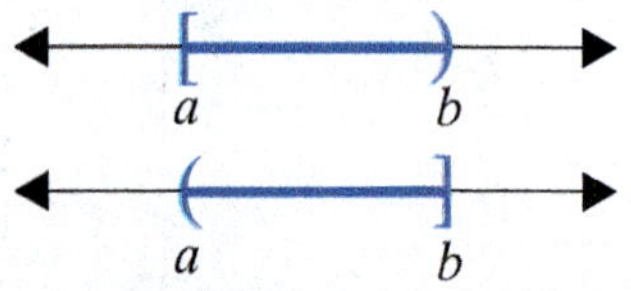
Open Interval	$\begin{cases} x > a \\ x < b \end{cases}$	(a, ∞) $(-\infty, b)$	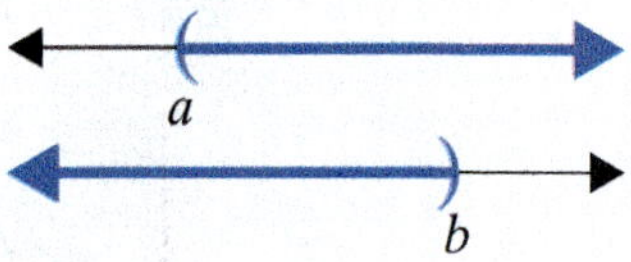
Half-open Interval	$\begin{cases} x \ge a \\ x \le b \end{cases}$	$[a, \infty)$ $(-\infty, b]$	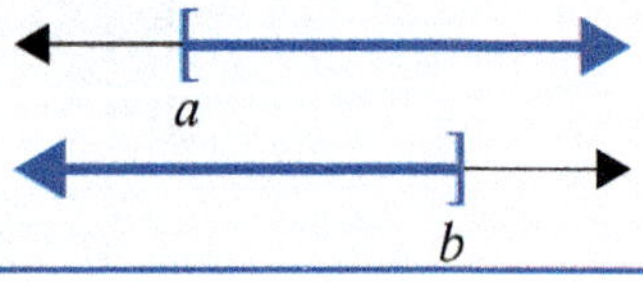

Table 1

NOTES
The symbol for infinity ∞ (or $-\infty$) is not a number. It is used to indicate that the interval is to include all real numbers from some point on (either in the positive direction or the negative direction) without end.

Example 1: Graphing Intervals

a. Graph the open interval $(3, \infty)$.

Solution:

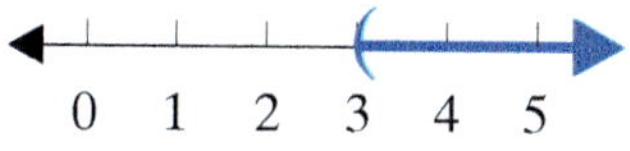

b. Graph the half-open interval $0 < x \le 4$.

Solution:

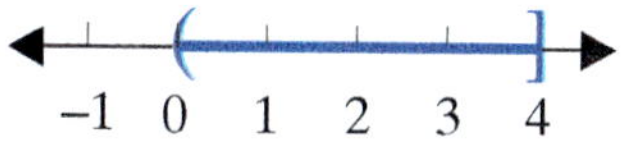

c. Represent the following graph using algebraic notation, and state what kind of interval it is.

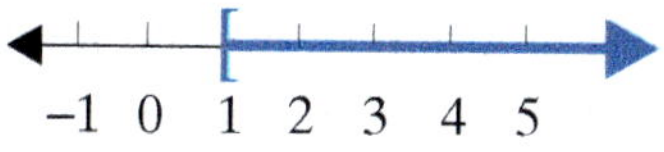

Solution: $x \ge 1$ is a half-open interval.

d. Represent the following graph using interval notation, and state what kind of interval it is.

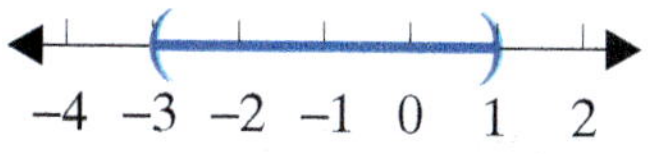

Solution: $(-3, 1)$ is an open interval.

Example 2 and Example 3 illustrate how the concepts of **union** (indicated by **or**) and **intersection** (indicated by **and**) are related to the graphs of intervals of real numbers on a real number line.

Example 2: Sets of Real Numbers Illustrating Union

Graph the set $\{x | x > 5 \textbf{ or } x \le 4\}$. The word **or** implies those values of x that satisfy **at least one** of the inequalities.

Solution: $x > 5$

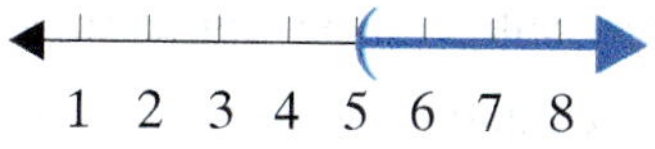

$x \le 4$

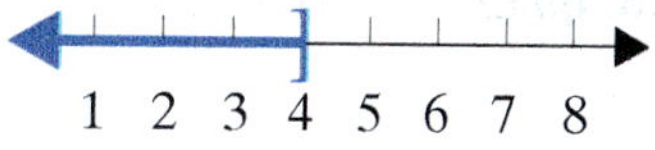

$x > 5$ **or** $x \le 4$

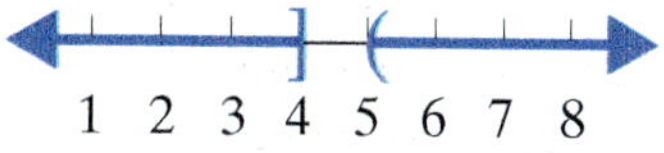

The solution graph shows the union $\cup$ of the first two graphs

Example 3: Sets of Real Numbers Illustrating Intersection

Graph the set $\{x \mid x \le 2 \textbf{ and } x \ge 0\}$. The word **and** implies those values of x that satisfy **both** inequalities.

Solution: $x \le 2$

$x \ge 0$

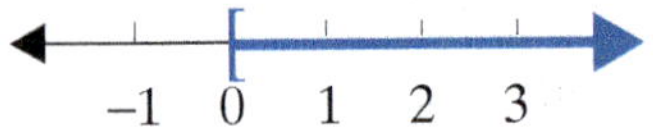

$x \le 2$ **and** $x \ge 0$

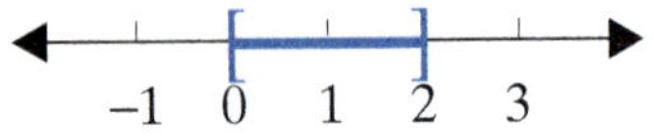

The solution graph shows the intersection $\cap$ of the first two graphs. In other words, the third graph shows the points in common between the first two graphs in this example.

This set can also be indicated as $\{x \mid 0 \le x \le 2\}$.

Solving Linear Inequalities

In this section we will solve **linear inequalities**, such as $6x + 5 \le -7$, and write the solution in **interval notation**, such as $(-\infty, -2]$, to indicate all real numbers x less than or equal to -2. Note that this can also be written in set-builder notation as $\{x \mid x \in (-\infty, -2]\}$ or $\{x \mid x \le -2\}$.

Linear Inequalities

Inequalities of the given form, where a, b, and c are real numbers and $a \ne 0$,

$$\boldsymbol{ax + b < c} \quad \text{and} \quad \boldsymbol{ax + b \le c}$$

$$\boldsymbol{ax + b > c} \quad \text{and} \quad \boldsymbol{ax + b \ge c}$$

are called **linear inequalities.**

The inequalities $\boldsymbol{c < ax + b < d}$ and $\boldsymbol{c \le ax + b \le d}$ are called **compound linear inequalities**. (This includes $\boldsymbol{c < ax + b \le d}$ and $\boldsymbol{c \le ax + b < d}$ as well.)

The solutions to linear inequalities are intervals of real numbers, and the methods for solving linear inequalities are similar to those used to solve linear equations. There is only one important exception.

Multiplying or dividing both sides of an inequality by a negative number causes the "sense" of the inequality to be reversed.

By the sense of the inequality, we mean "less than" or "greater than." Consider the following examples.

We know that $6 < 10$.

Add 5 to both sides	**Add –7 to both sides**	**Multiply both sides by 3**
$6 < 10$	$6 < 10$	$6 < 10$
$6+5 \ ? \ 10+5$	$6+(-7) \ ? \ 10+(-7)$	$3\cdot 6 \ ? \ 3\cdot 10$
$11 < 15$	$-1 < 3$	$18 < 30$

In the three cases just illustrated, addition, subtraction, and multiplication by a positive number, the sense of the inequality stayed the same. It remained <. Now we will see that multiplying or dividing each side by a negative number will **reverse the sense** of the inequality, from < to > or from > to <. This concept also applies to ≤ and ≥.

Teaching Notes:
You will probably need to give several more examples of division and multiplication by negative numbers to help students remember this important rule.

Multiply both sides by –3	**Divide both sides by –2**
$6 < 10$	$6 < 10$
$-3\cdot 6 \ ? \ -3\cdot 10$	$\frac{6}{-2} \ ? \ \frac{10}{-2}$
$-18 > -30$	$-3 > -5$

In each of these last two examples, the sense of the inequality is changed from < to >. While two examples do not prove a rule to be true, this particular rule is true and is included in the following rules for solving inequalities.

Rules for Solving Linear Inequalities

1. Simplify each side of the inequality by removing any grouping symbols and combining like terms.
2. Use the addition property of equality to add the opposites of constants or variable expressions so that variable expressions are on one side of the inequality and constants are on the other.
3. Use the multiplication property of equality to multiply both sides by the reciprocal of the coefficient of the variable (that is, divide both sides by the coefficient) so that the new coefficient is 1. **If this coefficient is negative, reverse the sense of the inequality.**
4. A quick (and generally satisfactory) check is to select any one number in your solution and substitute it into the original inequality.

As with solving equations, the object of solving an inequality is to find equivalent inequalities of simpler form that have the same solution set. We want the variable with a coefficient of +1 on one side of the inequality and any constants on the other side. One key difference between solving linear equations and solving linear inequalities is that linear equations have only one solution while linear inequalities generally have an infinite number of solutions.

Example 4: Solving Linear Inequalities

Solve the following linear inequalities and graph the solution set. Write the solution set using interval notation. Assume that x is a real number.

a. $6x+5\le -1$

Solution:

$6x+5\le -1$ — Write the inequality.

$6x+5-5\le -1-5$ — Add -5 to both sides.

$6x\le -6$ — Simplify.

$\frac{6x}{6}\le\frac{-6}{6}$ — Divide both sides by 6.

$x\le -1$ — Simplify.

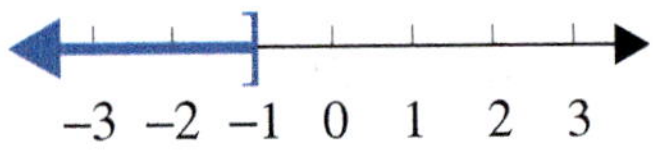

x is in $(-\infty,-1]$ — Use interval notation. Note that the interval $(-\infty,-1]$ is a half-open interval.

b. $x-3>3x+4$

Solution:

$x-3>3x+4$ — Write the inequality.

$x-3-x>3x+4-x$ — Add $-x$ to both sides.

$-3>2x+4$ — Simplify.

$-3-4>2x+4-4$ — Add -4 to both sides.

$-7>2x$ — Simplify.

$\frac{-7}{2}>\frac{2x}{2}$ — Divide both sides by 2.

$\frac{-7}{2}>x$ — Simplify.

or $x<-\frac{7}{2}$

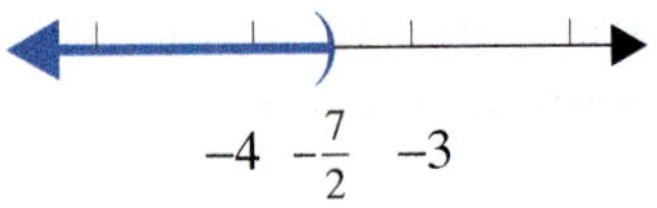

x is in $\left(-\infty,-\frac{7}{2}\right)$ — Use interval notation. Note that the interval $\left(-\infty,-\frac{7}{2}\right)$ is an open interval.

c. $6-4x \le x+1$

Solution:

$6-4x \le x+1$	Write the inequality.
$6-4x-x \le x+1-x$	Add $-x$ to both sides.
$6-5x \le 1$	Simplify.
$6-5x-6 \le 1-6$	Add -6 to both sides.
$-5x \le -5$	Simplify.
$\frac{-5x}{-5} \ge \frac{-5}{-5}$	Divide both sides by -5. **Note the reversal of the inequality sign!**
$x \ge 1$	Simplify.

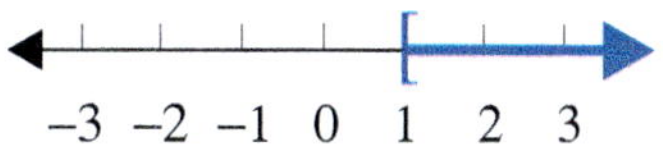

x is in $[1, \infty)$	Use interval notation. Note that the interval $[1, \infty)$ is a half-open interval.

d. $2x+5 < 3x-(7-x)$

Solution:

$2x+5 < 3x-(7-x)$	Write the inequality.
$2x+5 < 3x-7+x$	Distribute the negative sign.
$2x+5 < 4x-7$	Combine like terms.
$2x+5-2x < 4x-7-2x$	Add $-2x$ to both sides.
$5 < 2x-7$	Simplify.
$5+7 < 2x-7+7$	Add 7 to both sides.
$12 < 2x$	Simplify.
$\frac{12}{2} < \frac{2x}{2}$	Divide both sides by 2.
$6 < x$	Simplify.

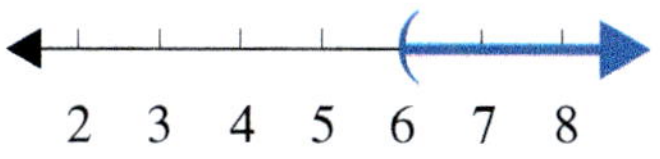

x is in $(6, \infty)$	Use interval notation. Note that the interval $(6, \infty)$ is an open interval.

Solving Compound Inequalities

Compound inequalities have three parts and can arise when a variable or variable expression is to be between two numbers. For example, the inequality

$$5 < x + 3 < 10$$

indicates that the values for the expression $x + 3$ are to be between 5 and 10. To solve this inequality, subtract 3 (or add -3) from each part.

Teaching Notes:
You may want to discuss how the compound inequality $5 < x + 3 < 10$ is a way of writing the intersection of two inequalities $5 < x + 3$ and $x + 3 < 10$.

$$5 < x + 3 < 10$$
$$5 - 3 < x + 3 - 3 < 10 - 3$$
$$2 < x < 7$$

Thus the variable x is isolated with coefficient +1, and we see that the solution set is the interval of real numbers $(2, 7)$. The graph of the solution set is the following.

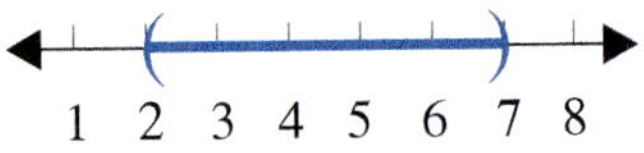

Example 5: Solving Compound Inequalities

a. Solve the compound inequality $-5 \le 4x - 1 < 11$ and graph the solution set. Write the solution set using interval notation. Assume that x is a real number.

Solution:

$-5 \le 4x - 1 < 11$	Write the inequality.
$-5 + 1 \le 4x - 1 + 1 < 11 + 1$	Add 1 to each part.
$-4 \le 4x < 12$	Simplify.
$\frac{-4}{4} \le \frac{4x}{4} < \frac{12}{4}$	Divide each part by 4.
$-1 \le x < 3$	Simplify.

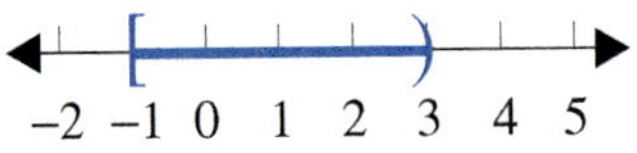

The solution set is the half-open interval $[-1, 3)$.

b. Solve the compound inequality $5 \le -3 - 2x \le 13$ and graph the solution set. Write the solution set using interval notation. Assume that x is a real number.

Solution:

$5 \le -3 - 2x \le 13$	Write the inequality.
$5 + 3 \le -3 - 2x + 3 \le 13 + 3$	Add 3 to each part.
$8 \le -2x \le 16$	Simplify.

$$\frac{8}{-2} \geq \frac{-2x}{-2} \geq \frac{16}{-2}$$ Divide each part by -2. **Note that the inequalities change sense.**

$$-4 \geq x \geq -8$$ Simplify.

$$(\text{or } -8 \leq x \leq -4)$$

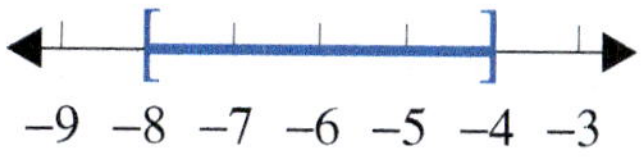

The solution set is the closed interval $[-8, -4]$.

c. Solve the compound inequality $0 < \frac{3x-5}{4} < 3$ and graph the solution set. Write the solution set using interval notation. Assume that x is a real number.

Solution:

$$0 < \frac{3x-5}{4} < 3$$ Write the inequality.

$$0 \cdot 4 < \frac{3x-5}{4} \cdot 4 < 3 \cdot 4$$ Multiply each part by 4.

$$0 < 3x - 5 < 12$$ Simplify each part.

$$0 + 5 < 3x - 5 + 5 < 12 + 5$$ Add 5 to each part.

$$5 < 3x < 17$$ Simplify.

$$\frac{5}{3} < \frac{3x}{3} < \frac{17}{3}$$ Divide each part by 3.

$$\frac{5}{3} < x < \frac{17}{3}$$ Simplify.

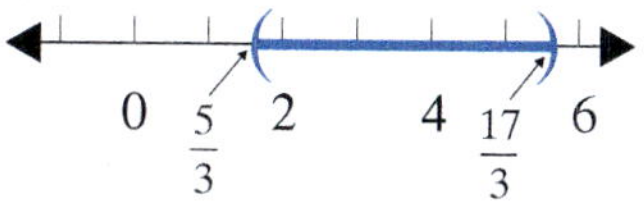

The solution set is the open interval $\left(\frac{5}{3}, \frac{17}{3}\right)$.

Applications of Linear Inequalities

In the following two examples, we show how inequalities can be related to real-world problems.

Example 6: Application with an Inequality

A math student has grades of 85, 98, 93, and 90 on four examinations. If he must average 90 or better to receive an A for the course, what scores can he receive on the final exam and earn an A? (Assume that the final exam counts the same as the other exams.)

Solution: Let x = score on final exam.

The average is found by adding the scores and dividing by 5.

$$\frac{85+98+93+90+x}{5} \geq 90$$

$$\frac{366+x}{5} \geq 90 \qquad \text{Simplify the numerator.}$$

$$5\left(\frac{366+x}{5}\right) \geq 5 \cdot 90 \qquad \text{Multiply both sides by 5.}$$

$$366+x \geq 450 \qquad \text{Simplify.}$$

$$366+x-366 \geq 450-366 \qquad \text{Add } -366 \text{ to each side.}$$

$$x \geq 84$$

If the student scores 84 or more on the final exam, he will average 90 or more and receive an A in math.

Example 7: Application with an Inequality

Ellen is going to buy 30 stamps, some 28-cent and some 44-cent. If she has \$9.68, what is the maximum number of 44-cent stamps she can buy?

Solution: Let x = number of 44-cent stamps,

then $30 - x$ = number of 28-cent stamps.

Ellen cannot spend more than \$9.68.

$$0.44x + 0.28(30-x) \leq 9.68$$

$$0.44x + 8.40 - 0.28x \leq 9.68$$

$$0.16x + 8.40 \leq 9.68$$

$$0.16x + 8.40 - 8.40 \leq 9.68 - 8.40$$

$$0.16x \leq 1.28$$

$$\frac{0.16x}{0.16} \leq \frac{1.28}{0.16}$$

$$x \leq 8$$

Ellen can buy at most eight 44-cent stamps if she buys a total of 30 stamps.

Practice Problems

Graph each set of real numbers on a real number line.

1. $\{x|x \le 3 \text{ and } x > 0\}$ **2.** $\{x|x \ge -1.5\}$ **3.** $\{x|x > 2 \text{ or } x < -4\}$

Solve each of the following inequalities and graph the solution sets. Write each solution set in interval notation. Assume that x is a real number.

4. $7 + x < 3$ **5.** $\frac{x}{2} + 1 \le \frac{1}{3}$ **6.** $-5 \le 2x + 1 < 9$

3.4 Exercises

Graph each set of indicated numbers on a real number line.

1. $\{x|x \text{ is a whole number less than } 3\}$

2. $\{x|x \text{ is an integer with } |x| \le 3\}$

3. $\{x|x \text{ is a prime number less than } 20\}$

4. $\{x|x \text{ is a positive whole number divisible by } 0\}$

5. $\{x|x \text{ is a composite number between 2 and } 10\}$

6. $\{x|x \ge 4, \ x \text{ is an integer}\}$

7. $\{x|-8 < x < 0, \ x \text{ is a whole number}\}$

8. $\{x|-2 < x < 12, \ x \text{ is an integer}\}$

Use set-builder notation to indicate each set of numbers as described.

9. The set of all real numbers between 3 and 5, including 3

10. The set of all real numbers between –4 and 4

11. The set of all real numbers greater than or equal to –2.5

12. The set of all real numbers between –1.8 and 5, including both of these numbers

1. 
2. 
3. 
4. No solution
5. 
6. 
7. No solution
8. 
9. $\{x|3 \le x < 5\}$
10. $\{x|-4 < x < 4\}$
11. $\{x|x \ge -2.5\}$
12. $\{x|-1.8 \le x \le 5\}$
13. a. $\{x|-3 \le x \le 1\}$
    b. $[-3, 1]$
    c. Closed interval
14. a. $\{x|x \ge 2\}$
    b. $[2, \infty)$
    c. Half-open interval

Answers to Practice Problems: **1.**  **2.**  **3.** 
4. $(-\infty, -4)$  **5.** $\left(-\infty, -\frac{4}{3}\right]$ 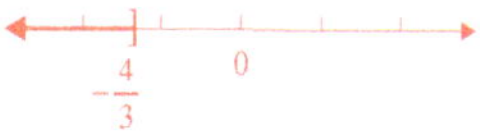
6. $[-3, 4)$ 

15. a. $\{x \mid -8 < x \le -2\}$
b. $(-8, -2]$
c. Half-open interval

16. a. $\{x \mid -2 \le x < 6\}$
b. $[-2, 6)$
c. Half-open interval

17. a. $\{x \mid x \le 1\}$
b. $(-\infty, 1]$
c. Half-open interval

18. a. $\{x \mid x > 4\}$
b. $(4, \infty)$
c. Open interval

19. a. $\{x \mid -4 < x < 4\}$
b. $(-4, 4)$
c. Open interval

20. a. $\{x \mid -6 \le x \le 0\}$
b. $[-6, 0]$
c. Closed interval

21. 

Half-open interval

22. 

Half-open interval

23. 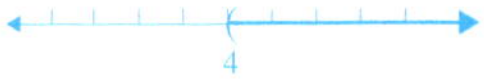

Open interval

24. 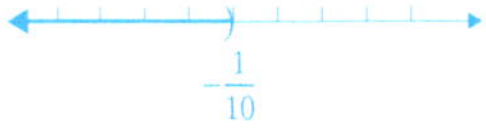

Open interval

25. 

Half-open interval

26. 

Half-open interval

27. 

Closed interval

28. 

Closed interval

29. 

Half-open interval

30. 

Half-open interval

The graphs of sets of real numbers are given. ***a.*** *Use set-builder notation to indicate the set of numbers shown in each graph.* ***b.*** *Use interval notation to represent the graph.* ***c.*** *Tell what type of interval is illustrated.*

13. 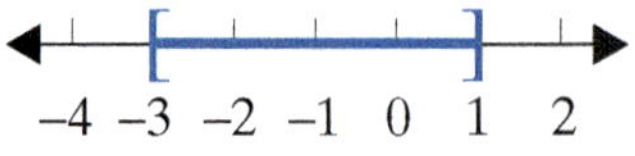

14. 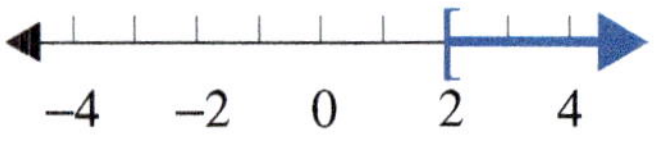

15. 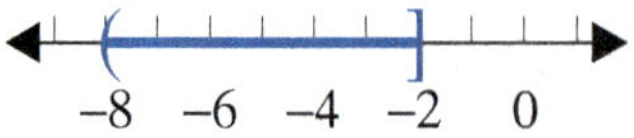

16. 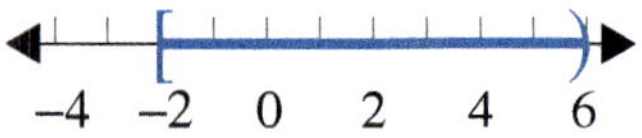

17. 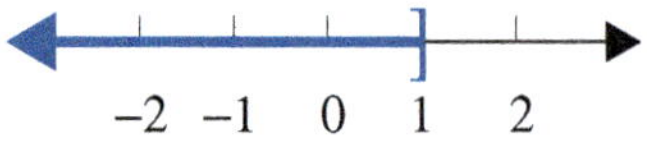

18. 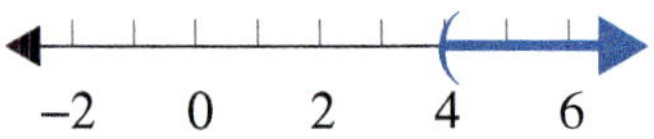

19. 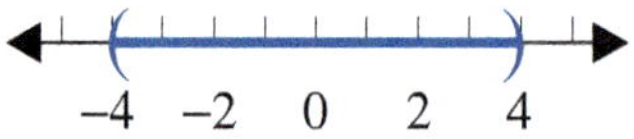

20. 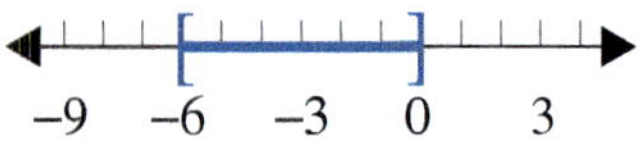

Graph each interval on a real number line and tell what type of interval it is. Assume that x is a real number.

21. $x \le -3$ **22.** $x \ge -0.5$ **23.** $x > 4$ **24.** $x < -\frac{1}{10}$

25. $0 < x \le 2.5$ **26.** $-1.5 \le x < 3.2$ **27.** $-2 \le x \le 0$ **28.** $-1 \le x \le 1$

29. $4 > x \ge 2$ **30.** $0 > x \ge -5$

Solve the inequalities and graph the solution sets. Write each solution using interval notation. Assume that x is a real number.

31. $2x + 3 < 5$ **32.** $4x - 7 \ge 9$ **33.** $14 - 5x < 4$

34. $23 < 7x - 5$ **35.** $6x - 15 > 1$ **36.** $9 - 2x < 8$

37. $5.6 + 3x \ge 4.4$ **38.** $12x - 8.3 < 6.1$ **39.** $1.5x + 9.6 < 12.6$

40. $0.8x - 2.1 \ge 1.1$ **41.** $2 + 3x \ge x + 8$ **42.** $x - 6 \le 4 - x$

43. $3x - 1 \le 11 - 3x$ **44.** $5x + 6 \ge 2x - 2$ **45.** $4 - 2x < 5 + x$

46. $4 + x > 1 - x$ **47.** $x - 6 > 3x + 5$ **48.** $4 + 7x \le 4x - 8$

49. $\frac{x}{2} - 1 \le \frac{5x}{2} - 3$ **50.** $\frac{x}{4} + 1 \le 5 - \frac{x}{4}$ **51.** $\frac{x}{3} - 2 > 1 - \frac{x}{3}$

52. $\frac{5x}{3} + 2 > \frac{x}{3} - 1$ **53.** $6x + 5.91 < 1.11 - 2x$ **54.** $4.3x + 21.5 \ge 1.7x + 0.7$

31. 
$(-\infty, 1)$

32. 
$[4, \infty)$

33. 
$(2, \infty)$

34. 
$(4, \infty)$

35. 
$\left(\frac{8}{3}, \infty\right)$

36. 
$\left(\frac{1}{2}, \infty\right)$

37. 
$[-0.4, \infty)$

38. 
$(-\infty, 1.2)$

39. 
$(-\infty, 2)$

40. 
$[4, \infty)$

41. 
$[3, \infty)$

42. 
$(-\infty, 5]$

43. 
$(-\infty, 2]$

44. 
$\left[-\frac{8}{3}, \infty\right)$

45. 
$\left(-\frac{1}{3}, \infty\right)$

46. 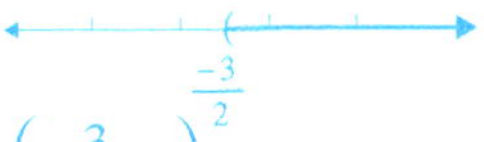
$\left(-\frac{3}{2}, \infty\right)$

47. 
$\left(-\infty, -\frac{11}{2}\right)$

55. $6.2x - 5.9 > 4.8x + 3.2$ **56.** $0.9x - 11.3 < 3.1 - 0.7x$ **57.** $4(6-x) < -2(3x+1)$

58. $-3(2x-5) \le 3(x-1)$ **59.** $-(3x+8) \ge 2(3x+1)$ **60.** $6(3x+1) < 5(1-2x)$

61. $11x+8-5x \ge 2x-(4-x)$ **62.** $1-(2x+8) < (9+x)-4x$

63. $5-3(4-x)+x \le -2(3-2x)-x$ **64.** $x-2(x+3) \ge 7-(4-x)+11$

65. $\frac{2(x-1)}{3} < \frac{3(x+1)}{4}$ **66.** $\frac{3(x-2)}{2} \ge \frac{4(x-1)}{3}$

67. $\frac{x-2}{4} > \frac{x+2}{2} + 6$ **68.** $\frac{x+4}{9} \le \frac{x}{3} - 2$

69. $\frac{2x+7}{4} \le \frac{x+1}{3} - 1$ **70.** $\frac{4x}{7} - 3 > \frac{x-6}{2} - 4$

Solve the compound inequalities and graph the solution sets. Write each solution using interval notation. Assume that x is a real number.

71. $-4 < x + 5 < 6$ **72.** $2 \le -x + 2 \le 6$ **73.** $3 \ge 4x - 3 \ge -1$

74. $13 > 3x + 4 > -2$ **75.** $1 \le \frac{2}{3}x - 1 \le 9$ **76.** $-2 \le \frac{1}{2}x - 5 \le -1$

77. $14 > -2x - 6 > 4$ **78.** $-11 \ge -3x + 2 > -20$ **79.** $-1.5 < 2x + 4.1 < 3.5$

80. $0.9 < 3x + 2.4 < 6.9$

81. Test scores: A statistics student has grades of 82, 95, 93, and 78 on four hour-long exams. He must average 90 or higher to receive an A for the course. What scores can he receive on the final exam and earn an A if:
a. The final is equivalent to a single hour-long exam (100 points maximum)?
b. The final is equivalent to two hourly exams (200 points maximum)?

82. Test scores: To receive a grade of B in a chemistry class, Melissa must average 80 or more but less than 90. If her five hour-long exam scores were 75, 82, 90, 85, and 77, what score does she need on the final exam (100 points maximum) to earn a grade of B?

83. Car sales: A car salesman makes \$1000 each day that he works and makes approximately \$250 commission for each car he sells. If a car salesman wants to make at least \$3500 in one day, how many cars does he need to sell?

48. $(-\infty, -4]$

49. $[1, \infty)$

50. $(-\infty, 8]$

84. Postage: Allison is going to the post office to buy 38¢ stamps and 2¢ adjustment stamps. Since the current postage rate is 44¢, she will need 3 times as many 2¢ adjustment stamps as 38¢ stamps. If she has $11 to spend, what is the largest number of 38¢ stamps she can buy?

Writing and Thinking About Mathematics

51. $\left(\frac{9}{2}, \infty\right)$

85. a. Write a list of three situations where inequalities might be used in your daily life.

b. Illustrate these situations with algebraic inequalities and appropriate numbers.

HAWKES LEARNING SYSTEMS: INTRODUCTORY & INTERMEDIATE ALGEBRA SOFTWARE

- 3.4 Solving Linear Inequalities

52. $\left(-\frac{9}{4}, \infty\right)$ 53. $(-\infty, -0.6)$ 54. $[-8, \infty)$

55. $(6.5, \infty)$ 56. $(-\infty, 9)$ 57. $(-\infty, -13)$

58. $[2, \infty)$ 59. $\left(-\infty, -\frac{10}{9}\right]$ 60. $\left(-\infty, -\frac{1}{28}\right)$

61. $[-4, \infty)$ 62. $(-\infty, 16)$ 63. $(-\infty, 1]$

64. $(-\infty, -10]$ 65. $(-17, \infty)$ 66. $[10, \infty)$

67. $(-\infty, -30)$ 68. $[11, \infty)$ 69. $\left(-\infty, -\frac{29}{2}\right]$

70. $(-56, \infty)$ 71. $(-9, 1)$ 72. $[-4, 0]$

73. $\left[\frac{1}{2}, \frac{3}{2}\right]$ 74. $(-2, 3)$ 75. $[3, 15]$

76. $[6, 8]$ 77. $(-10, -5)$ 78. $\left[\frac{13}{3}, \frac{22}{3}\right)$

79. $(-2.8, -0.3)$ 80. $(-0.5, 1.5)$

81. a. The student would need a score higher than 102 points, which is not possible. Thus he cannot earn an A in the course. b. The student must score at least 192 points to earn an A in the course.

82. Melissa must score at least a 71 to earn a B for the course. 83. He must sell at least 10 cars.

84. Allison can buy at most 25 of the 38¢ stamps. 85. Answers will vary.

3.5 Absolute Value Equations and Inequalities

- *Solve absolute value equations.*
- *Solve equations with two absolute value expressions.*
- *Solve absolute value inequalities.*

Absolute Value Equations

The definition of **absolute value** was given in Section 1.1 and is stated again here for easy reference. Remember, the absolute value of a number is its distance from 0 on a number line.

Absolute Value

For any real number x,
$$|x| = \begin{cases} x & \text{if } x \geq 0 \\ -x & \text{if } x < 0 \end{cases}$$

Equations involving absolute value may have more than one solution (all of which must be included when giving an answer). For example, suppose that $|x| = 3$. Since $|3| = 3$ and $|-3| = -(-3) = 3$, we have either $x = 3$ or $x = -3$. We can say that the solution set is $\{3, -3\}$. In general, **any number and its opposite have the same absolute value**.

Solving Absolute Value Equations

For $c > 0$:

a. If $|x| = c$, then $x = c$ or $x = -c$.

b. If $|ax + b| = c$, then $ax + b = c$ or $ax + b = -c$.

Note: If the absolute value expression is isolated on one side of the equation, we say that the equation is in **standard form**. You may need to manipulate the absolute value equation to get it into standard form before you can solve it. (See Example 1d.)

Example 1: Solving Absolute Value Equations

Solve the following equations involving absolute value.

a. $|x| = 5$

Solution: $x = 5$ or $x = -5$

b. $|3x - 4| = 5$

Solution:
$$\begin{aligned} 3x - 4 &= 5 & \text{or} \quad 3x - 4 &= -5 \\ 3x &= 9 & 3x &= -1 \\ x &= 3 & x &= -\frac{1}{3} \end{aligned}$$

c. $|4x - 1| = -8$

Solution: There is no number that has a negative absolute value. Therefore, this equation has no solution. (The solution is $\varnothing$ and the equation is a contradiction.)

d. $5|3x + 17| - 4 = 51$

Solution:

$5|3x + 17| - 4 = 51$ Write the equation.

$5|3x + 17| = 55$ Add 4 to both sides.

$|3x + 17| = 11$ Divide both sides by 5. **We must put the equation in standard form.**

$$\begin{aligned} 3x + 17 &= 11 & \text{or} \quad 3x + 17 &= -11 \\ 3x &= -6 & 3x &= -28 \\ x &= -2 & x &= -\frac{28}{3} \end{aligned}$$

Equations with Two Absolute Value Expressions

If two numbers have the same absolute value, then either they are equal or they are opposites of each other. This fact can be used to solve equations that involve two absolute value expressions.

Solving Equations with Two Absolute Value Expressions

If $|a| = |b|$, then either $\boldsymbol{a = b}$ or $\boldsymbol{a = -b}$.

More generally,

if $|\boldsymbol{ax + b}| = |\boldsymbol{cx + d}|$, then either $\boldsymbol{ax + b = cx + d}$ or $\boldsymbol{ax + b = -(cx + d)}$.

Example 2: Solving Equations with Two Absolute Value Expressions

Solve $|x+5| = |2x+1|$.

Solution: In this case, the two expressions $(x+5)$ and $(2x+1)$ are equal to each other or are opposites of each other.

$$|x+5| = |2x+1|$$

$x+5=2x+1$	or	$x+5=-(2x+1)$	Note the use of parentheses. We want the opposite of the entire expression $(2x+1)$.
$5=x+1$		$x+5=-2x-1$	
$4=x$		$3x+5=-1$	
		$3x=-6$	
		$x=-2$	Make sure to check that both 4 and −2 satisfy the original equation.

Absolute Value Inequalities

Now consider an inequality with absolute value such as $|x| < 3$. For a number to have an absolute value less than 3, it must be within 3 units of 0. That is, the numbers between −3 and 3 have their absolute values less than 3 because they are within 3 units of 0. Thus for $|x| < 3$.

Algebraic Notation	Graph	Interval Notation
$|x| < 3$ $-3 < x < 3$ **(the intersection)**	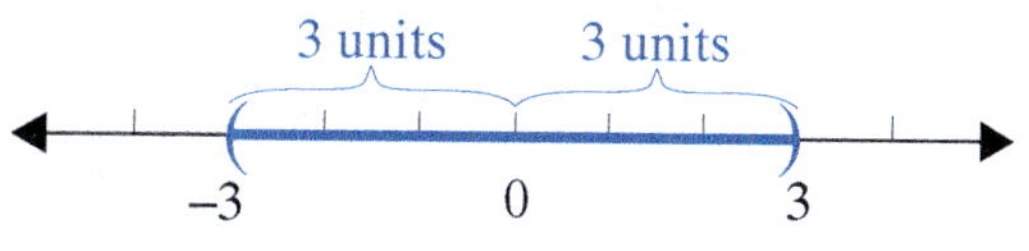	$(-3, 3)$

Table 1

The inequality $|x-5| < 3$ means that the distance between x and 5 is less than 3. That is, we want all the values of x that are within 3 units of 5. The inequality is solved algebraically as follows.

$|x-5| < 3$	
$-3 < x-5 < 3$	$x-5$ is between −3 and 3.
$-3+5 < x-5+5 < 3+5$	Add 5 to each part of the expression, just as in solving linear inequalities.
$2 < x < 8$	Simplify each expression.
x is in $(2, 8)$	Use interval notation.

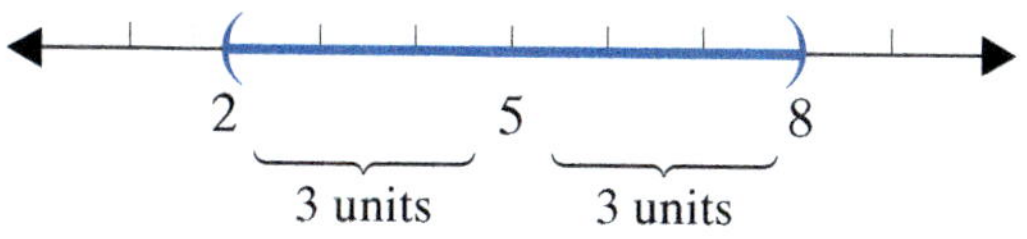

The values for x are between 2 and 8 and are within 3 units of 5.

Solving Absolute Value Inequalities with < (or ≤)

For $c > 0$:

a. If $|x| < c$, then $-c < x < c$.

b. If $|ax+b| < c$, then $-c < ax + b < c$.

The inequalities in **a.** and **b.** are also true if $<$ is replaced by $\leq$.

Note: If the absolute value expression is isolated on one side of the inequality, we say that the inequality is in **standard form**. You may need to manipulate the absolute value inequality to get it into standard form before you can solve it. (See Example 3c and 3e.)

Example 3: Solving Absolute Value Inequalities

Solve the following absolute value inequalities and graph the solution sets.

a. $|x| \leq 6$

Solution:

$$|x| \leq 6$$
$$-6 \leq x \leq 6$$

or x is in $[-6, 6]$

b. $|x+3| < 2$

Solution:

$$|x+3| < 2$$
$$-2 < x+3 < 2$$
$$-2-3 < x+3-3 < 2-3$$
$$-5 < x < -1$$

or x is in $(-5, -1)$

c. $3|2x-7| < 15$

Solution:

$$3|2x-7| < 15$$
$$|2x-7| < 5$$
$$-5 < 2x-7 < 5$$
$$-5+7 < 2x-7+7 < 5+7$$
$$2 < 2x < 12$$
$$1 < x < 6$$

or x is in $(1, 6)$

Divide both sides by 3 in order to get the inequality in **standard form**. Remember, we must get the expression in standard form before solving any absolute value inequality.

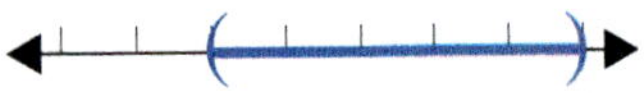

d. $|x+9|<-\frac{1}{2}$

Solution: Since absolute value is always nonnegative (greater than or equal to 0), no number has an absolute value less than $-\frac{1}{2}$. Thus there is **no solution**, $\varnothing$.

e. $|2x+4|+4<7$

Solution:

$$|2x+4|+4<7$$

$$|2x+4|<3$$

Add −4 to both sides in order to get the expression in **standard form**.

$$-3<2x+4<3$$

$$-7<2x<-1$$

$$-\frac{7}{2}<x<-\frac{1}{2}$$

So x is in $\left(-\frac{7}{2},-\frac{1}{2}\right)$.

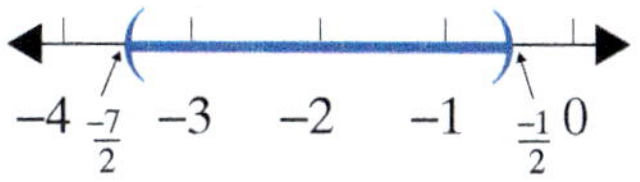

We have been discussing inequalities in which the absolute value is less than some positive number. Now consider an inequality where the absolute value is greater than some positive number, such as $|x|>3$. For a number to have an absolute value greater than 3, its distance from 0 must be greater than 3. That is, numbers that are greater than 3 **or** less than –3 will have absolute values greater than 3. Thus for $|x|>3$:

Algebraic Notation	Graph	Interval Notation
$\lvert x\rvert>3$ $x>3$ **or** $x<-3$ **(the union)**	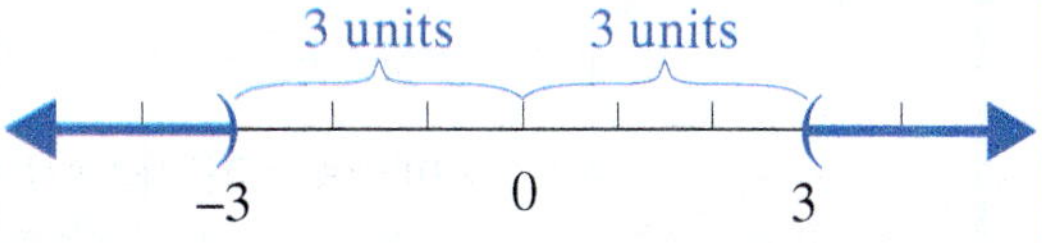	$(-\infty,-3)\cup(3,\infty)$

Table 2

NOTES

The expression $x>3$ or $x<-3$ **cannot** be combined into one inequality expression. The word **or** must separate the inequalities since any number that satisfies one **or** the other is a solution to the absolute value inequality. There are **no** numbers that satisfy **both** inequalities.

The inequality $|x-5|>6$ means that the distance between x and 5 is more than 6. That is, we want all values of x that are more than 6 units from 5. The inequality is solved algebraically as follows.

$|x-5|>6$ indicates that

$x-5<-6$ **or** $x-5>6$. $\quad$ $x-5$ is less than –6 or greater than 6

Solving both inequalities gives

$$x-5+5<-6+5 \quad \textbf{or} \quad x-5+5>6+5,$$ Add 5 to each side, just as in solving linear inequalities.

$$x<-1 \quad \textbf{or} \quad x>11.$$ Simplify.

So x is in $(-\infty,-1)\cup(11,\infty)$.

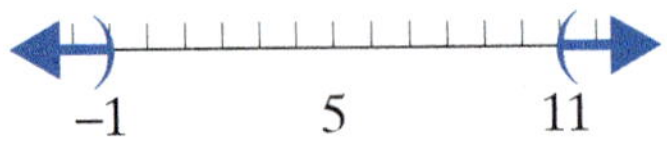

Note: The values for x less than –1 or greater than 11 are more than 6 units from 5. Thus we can interpret the inequality $|x-5|>6$ to mean that the distance from x to 5 is greater than 6.

Solving Absolute Value Inequalities with > (or ≥)

For $c>0$:

a. If $|x|>c$, then $x<-c$ **or** $x>c$.

b. If $|ax+b|>c$, then $ax+b<-c$ **or** $ax+b>c$.

The inequalities in **a.** and **b.** are true if > is replaced by ≥.

Example 4: Solving Absolute Value Inequalities

Solve the following absolute value inequalities and graph the solution set.

a. $|x|\geq 5$

Solution: $|x|\geq 5$

$x\leq -5$ or $x\geq 5$

So x is in $(-\infty,-5]\cup[5,\infty)$.

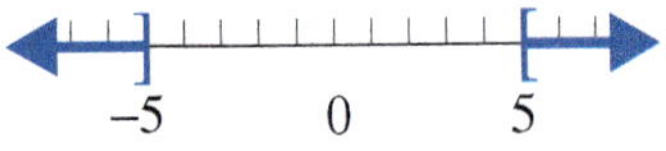

b. $|4x-3|>2$

Solution: $|4x-3|>2$

$4x-3<-2$ or $4x-3>2$

$4x<1$ or $4x>5$

$x<\frac{1}{4}$ or $x>\frac{5}{4}$

So x is in $\left(-\infty,\frac{1}{4}\right)\cup\left(\frac{5}{4},\infty\right)$.

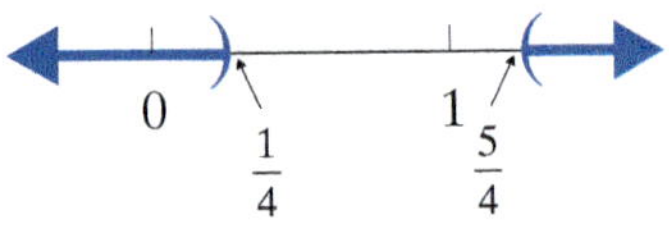

1. $x = -8, 8$
2. $x = -6, 6$
3. No solution
4. $z = -\frac{1}{5}, \frac{1}{5}$
5. $x = -5, -1$
6. No solution
7. $x = -\frac{4}{3}, \frac{5}{3}$
8. $x = -3, \frac{7}{3}$
9. $n = -2, \frac{2}{3}$
10. $x = -\frac{5}{3}, 5$
11. No solution
12. No solution
13. $x = 2$
14. $y = -\frac{2}{3}$
15. $x = -\frac{3}{2}, 2$
16. $x = -1, \frac{11}{3}$
17. $x = -\frac{1}{5}, 1$
18. $x = 3, 4$
19. $x = -1, \frac{11}{3}$
20. $x = -3, 13$
21. $x = -22, 26$
22. $y = 0, \frac{20}{3}$
23. $x = -8, 4$
24. $x = 5, 15$
25. $x = -6, 0$
26. $x = 4, 20$

c. $|3x-8| > -6$

Solution: There is nothing to do here except observe that no matter what is substituted for x, the absolute value will be greater than -6. Absolute value is always nonnegative (greater than or equal to 0). The solution to the inequality is **all real numbers**, so shade the entire number line. In interval notation, x is in $(-\infty, \infty)$.

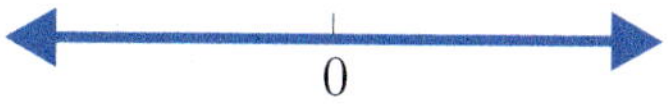

d. $|2x-5|-5 \geq 4$

Solution: $|2x-5|-5 \geq 4$

$|2x-5| \geq 9$

Add 5 to both sides in order to get the inequality in **standard form**.

$2x-5 \leq -9$ or $2x-5 \geq 9$

$2x \leq -4$ or $2x \geq 14$

$x \leq -2$ or $x \geq 7$

So x is in $(-\infty, -2] \cup [7, \infty)$.

Practice Problems

Solve the following equations.

1. $|x| = 10$ **2.** $|2x-1| = 8.2$ **3.** $|4x+1| = -5$ **4.** $|x+4| = |3x-2|$

Solve the following absolute value inequalities and graph the solution sets.

5. $|x-6| \leq 7$ **6.** $|4x+1| < 13$ **7.** $3|2x+6| \geq 20-2$

3.5 Exercises

Solve each of the absolute value equations.

1. $|x| = 8$ **2.** $|x| = 6$ **3.** $|z| = -\frac{1}{5}$

4. $|z| = \frac{1}{5}$ **5.** $|x+3| = 2$ **6.** $|y+5| = -7$

Answers to Practice Problems: **1.** $x = -10$ or $x = 10$ **2.** $x = -3.6$ or $x = 4.6$ **3.** No solution **4.** $x = -\frac{1}{2}$ or $x = 3$ **5.** $[-1, 13]$

6. $\left(-\frac{7}{2}, 3\right)$

7. $(-\infty, -6] \cup [0, \infty)$

27. $x = -\frac{1}{3}, 3$

28. $x = 2, \frac{8}{3}$

29. $x = 1$

30. $x = 2$

31. $x = -\frac{5}{2}, \frac{3}{4}$

32. $x = -\frac{5}{2}, -\frac{3}{8}$

33. $x = -\frac{20}{7}, \frac{4}{5}$

34. $x = -10, \frac{18}{7}$

35. $x = -20, \frac{40}{9}$

36. $x = -\frac{108}{19}, -\frac{60}{13}$

37. 

$(-\infty, \infty)$

38. 

$(-\infty, -3] \cup [3, \infty)$

39. 

$\left[-\frac{4}{5}, \frac{4}{5}\right]$

40. 

$\left(-\infty, -\frac{7}{2}\right] \cup \left[\frac{7}{2}, \infty\right)$

41. 

$(-\infty, 1) \cup (5, \infty)$

42. 

$[-1, 9]$

43. 

$[-10, -2]$

44. 

No solution

45. 

$(-\infty, -8] \cup [-2, \infty)$

46. 

$(-5, 7)$

7. $|6x-1| = 9$ **8.** $|3x+1| = 8$ **9.** $|6n+4| = 8$

10. $|3x-5| = 10$ **11.** $|3x+4| = -9$ **12.** $|-2x+1| = -3$

13. $|-5x+10| = 0$ **14.** $|6y+4| = 0$ **15.** $|-4x+1| = 7$

16. $|-3x+4| = 7$ **17.** $|5x-2|+4 = 7$ **18.** $|2x-7|-1 = 0$

19. $|-3x+4|-2 = 5$ **20.** $|-x+5|+1 = 9$ **21.** $\left|\frac{1}{4}x - \frac{1}{2}\right| = 6$

22. $\left|\frac{1}{5}y - \frac{2}{3}\right| = \frac{2}{3}$ **23.** $5\left|\frac{x}{2}+1\right| - 7 = 8$ **24.** $6\left|\frac{x}{5}-2\right| + 5 = 11$

25. $3\left|\frac{x}{3}+1\right| - 5 = -2$ **26.** $2\left|\frac{x}{4}-3\right| + 6 = 10$ **27.** $|2x-1| = |x+2|$

28. $|2x-5| = |x-3|$ **29.** $|x+3| = |x-5|$ **30.** $|x-8| = |x+4|$

31. $|3x+1| = |4-x|$ **32.** $|5x+4| = |1-3x|$ **33.** $\left|\frac{3x}{2}+2\right| = \left|\frac{x}{4}+3\right|$

34. $\left|\frac{x}{3}-4\right| = \left|\frac{5x}{6}+1\right|$ **35.** $\left|\frac{2x}{5}-3\right| = \left|\frac{x}{2}-1\right|$ **36.** $\left|\frac{4x}{3}+7\right| = \left|\frac{x}{4}+2\right|$

Solve each of the absolute value inequalities and graph the solution sets. Write each solution using interval notation. Assume that x is a real number.

37. $|x| \geq -2$ **38.** $|x| \geq 3$ **39.** $|x| \leq \frac{4}{5}$ **40.** $|x| \geq \frac{7}{2}$

41. $|x-3| > 2$ **42.** $|y-4| \leq 5$ **43.** $|x+6| \leq 4$ **44.** $|x+2| \leq -4$

45. $|x+5| \geq 3$ **46.** $|x-1| < 6$ **47.** $|2x-1| \geq 2$ **48.** $|3x+4| > -8$

49. $|3-2x| < -2$ **50.** $|4+3x| > 5$ **51.** $|5+4x| \leq 3$ **52.** $|5x-2| < 8$

53. $|3x+4|-1 < 0$ **54.** $|2x-3|-3 \leq 0$ **55.** $\left|\frac{3x}{2}-4\right| \geq 5$ **56.** $\left|\frac{3}{7}y+\frac{1}{2}\right| > 2$

57. $|2x-9|-7 \leq 4$ **58.** $|3x-7|+4 \leq 4$ **59.** $-4 < |6x-1|+4$ **60.** $4 \leq |3x+1|-6$

47. 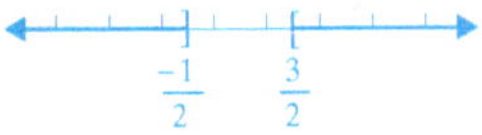

$\left(-\infty, -\frac{1}{2}\right] \cup \left[\frac{3}{2}, \infty\right)$

48. 

$(-\infty, \infty)$

49. 

No solution

61. $5>|4-2x|+2$ **62.** $7>|8-5x|+3$ **63.** $3|4x+5|-5>10$

64. $6|4x-7|+7>19$ **65.** $4|7x+9|-3<17$ **66.** $2|7x-3|+4\geq 12$

Writing and Thinking About Mathematics

A set of real numbers is described. ***a.*** *Sketch a graph of the set on a real number line.* ***b.*** *Represent each set using absolute value notation.* ***c.*** *Represent each set using interval notation. If the set is one interval, state what type of interval it is.*

67. The set of real numbers between –10 and 10, inclusive

68. The set of real numbers within 7 units of 4

69. The set of real numbers more than 6 units from 8

70. The set of real numbers greater than or equal to 3 units from –1

71. The set of real numbers within 2 units of –5

50. 

$(-\infty,-3)\cup\left(\frac{1}{3},\infty\right)$

51. 

$\left[-2,-\frac{1}{2}\right]$

52. 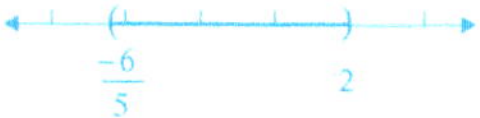

$\left(-\frac{6}{5},2\right)$

HAWKES LEARNING SYSTEMS: INTRODUCTORY & INTERMEDIATE ALGEBRA SOFTWARE

- 3.5a Solving Absolute Value Equations
- 3.5b Solving Absolute Value Inequalities

53. 

$\left(-\frac{5}{3},-1\right)$

54. 

$[0,3]$

55. 

$\left(-\infty,-\frac{2}{3}\right]\cup[6,\infty)$

56. 

$\left(-\infty,-\frac{35}{6}\right)\cup\left(\frac{7}{2},\infty\right)$

57. 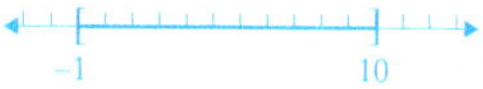

$[-1,10]$

58. 

$\left\{\frac{7}{3}\right\}$

59. 

$(-\infty,\infty)$

60. 

$\left(-\infty,-\frac{11}{3}\right]\cup[3,\infty)$

61. 

$\left(\frac{1}{2},\frac{7}{2}\right)$

62. 

$\left(\frac{4}{5},\frac{12}{5}\right)$

63. 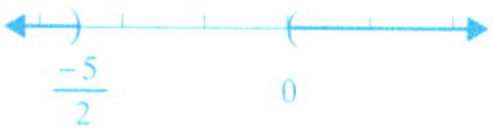

$\left(-\infty,-\frac{5}{2}\right)\cup(0,\infty)$

64. 

$\left(-\infty,\frac{5}{4}\right]\cup\left[\frac{9}{4},\infty\right)$

65. 

$\left(-2,-\frac{4}{7}\right)$

66. 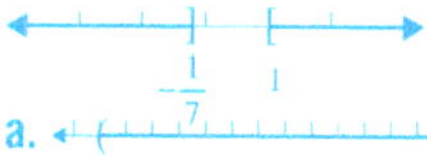

$\left(-\infty,-\frac{1}{7}\right]\cup[1,\infty)$

67. a. 

b. $|x|\leq 10$ c. $[-10,10]$, Closed interval

68. a. 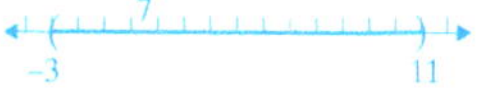

b. $|x-4|<7$ c. $(-3,11)$, Open interval

69. a. 

b. $|x-8|>6$ c. $(-\infty,2)\cup(14,\infty)$

70. a. 

b. $|x+1|\geq 3$ c. $(-\infty,-4]\cup[2,\infty)$

71. a. 

b. $|x+5|<2$ c. $(-7,-3)$, Open interval

Chapter 3 Index of Key Ideas and Terms

Section 3.1 Working with Formulas

Section 3.2 Formulas in Geometry

Section 3.3 Applications: Distance-Rate-Time, Interest, Average

Section 3.4 Linear Inequalities

Terminology Regarding Sets page 214
Set
Elements
Finite and infinite sets
The empty set (or null set), $\varnothing$
Set-builder notation

Union page 215
The **union** (symbolized $\cup$, as in $A \cup B$) of two (or more) sets is the set of all elements that belong to either one set or the other set or to both sets.

Intersection page 215
The **intersection** (symbolized $\cap$, as in $A \cap B$) of two (or more) sets is the set of all elements that belong to both sets.

Interval Notation page 216

Type of Interval	Interval Notation
Open	(a, b), (a, ∞), $(-\infty, b)$
Closed	$[a, b]$
Half-open	$[a, \infty)$, $(-\infty, b]$, $[a, b)$, $(a, b]$

Linear Inequalities pages 218-220
Inequalities of the given form where a, b, and c are real numbers and $a \neq 0$,

$ax + b < c$ and $ax + b \leq c$
$ax + b > c$ and $ax + b \geq c$

Continued on the next page...

Section 3.4 Linear Inequalities (cont.)

Rules for Solving Linear Inequalities page 219

1. Simplify each side of the inequality by removing any grouping symbols and combining like terms.
2. Use the addition property of equality to add the opposites of constants or variable expressions so that variable expressions are on one side of the inequality and constants are on the other.
3. Use the multiplication property of equality to multiply both sides by the reciprocal of the coefficient of the variable (that is, divide both sides by the coefficient) so that the new coefficient is 1. **If this coefficient is negative, reverse the sense of the inequality.**
4. A quick check is to select any one number in your solution and substitute it into the original inequality.

Compound Inequalities pages 218, 222

The inequalities $c < ax + b < d$ and $c \leq ax + b \leq d$ are called **compound linear inequalities**. (This includes $c < ax+b \leq d$ and $c \leq ax+b < d$ as well.)

Applications of Linear Inequalities pages 223-224

Section 3.5 Absolute Value Equations and Inequalities

Solving Absolute Value Equations page 229

For $c > 0$:

a. If $|x| = c$, then $x = c$ **or** $x = -c$.

b. If $|ax + b| = c$, then $ax + b = c$ **or** $ax + b = -c$.

Solving Equations with Two Absolute Value Expressions page 230

If $|a| = |b|$, then either $a = b$ or $a = -b$.

If $|ax+b| = |cx+d|$, then either $ax+b = cx+d$ or $ax+b = -(cx+d)$.

Continued on the next page....

Section 3.5 Absolute Value Equations and Inequalities (cont.)

Absolute Value Inequalities pages 231-234

For $c > 0$:

a. If $|x| < c$, then $-c < x < c$.

b. If $|ax + b| < c$, then $-c < ax + b < c$.

The inequalities in **a.** and **b.** are also true if $<$ is replaced by $\leq$.

For $c > 0$:

a. If $|x| > c$, then $x < -c$ **or** $x > c$.

b. If $|ax + b| > c$, then $ax + b < -c$ **or** $ax + b > c$.

The inequalities in **a.** and **b.** are also true if $>$ is replaced by $\geq$.

HAWKES LEARNING SYSTEMS: INTRODUCTORY & INTERMEDIATE ALGEBRA SOFTWARE

- 3.1 Working with Formulas
- 3.2 Formulas in Geometry
- 3.3 Applications
- 3.4 Solving Linear Inequalities
- 3.5a Solving Absolute Value Equations
- 3.5b Solving Absolute Value Inequalities

Chapter 3 Review

1. $C = 0°$
2. 47.5 mph
3. $y = -2$
4. 40 meters
5. $P = 0.16$
6. $\pi = \dfrac{L}{2rh}$
7. $R = P + C$
8. $b = \dfrac{2A}{h}$
9. $\alpha = 180° - \beta - \gamma$
10. $g = -\dfrac{v - v_0}{t}$
11. $m = \dfrac{2gK}{v^2}$
12. $y = 6 - 3x$
13. $y = 6x - 3$
14. $x = \dfrac{10 + 2y}{5}$
15. $x = \dfrac{4y + 7}{3}$
16. a. $P = 2a + 2b$
    b. $A = \pi r^2$
    c. $V = \dfrac{4}{3}\pi r^3$
17. a. 50 cm
    b. 144 cm^2
18. a. 38 in.
    b. 80 $in.^2$

3.1 Working with Formulas

Solve the following word problems.

1. Temperature: Given the formula $C = \frac{5}{9}(F - 32)$, find the value of C if $F = 32°$.

2. Driving speed: Using the formula $d = rt$, find Karl's average rate of speed if he drove 190 miles in 4 hours.

3. Equation of a line: Given the equation $4x + y = 10$, find the value of y that corresponds to a value of 3 for x.

4. Soccer: If the perimeter of a rectangular soccer field is planned to be 150 meters and because of building restrictions the width has to be 35 meters, what is the planned length of the field? $(P = 2l + 2w)$

5. Pressure of a gas: Volume and pressure of gas in a container are known to be related by the formula $V = \frac{k}{P}$ where k is a constant that depends on the type of gas involved. Find P if $k = 3.2$ and V = 20 cubic centimeters.

Solve each formula for the indicated variable.

6. $L = 2\pi rh$; solve for π.

7. $P = R - C$; solve for R.

8. $A = \frac{1}{2}bh$; solve for b.

9. $\alpha + \beta + \gamma = 180°$; solve for α.

10. $v = v_0 - gt$; solve for g.

11. $K = \frac{mv^2}{2g}$; solve for m.

12. $3x + y = 6$; solve for y.

13. $6x - y = 3$; solve for y.

14. $5x - 2y = 10$; solve for x.

15. $-3x + 4y = -7$; solve for x.

3.2 Formulas in Geometry

16. Write the formula for **a.** the perimeter of a parallelogram, **b.** the area of a circle, and **c.** the volume of a sphere.

*Find **a.** the perimeter P and **b.** the area A of each figure.*

17. $P =$ _____ $A =$ _____

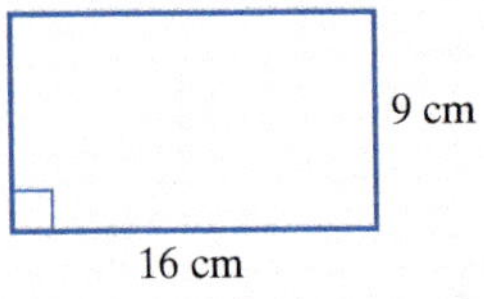

18. $P =$ _____ $A =$ _____

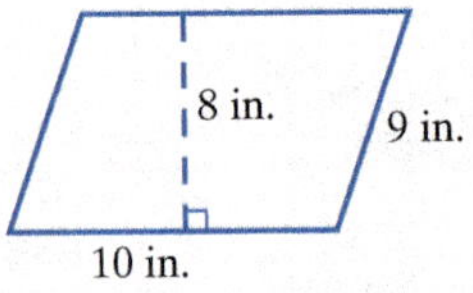

19. 20.2 in.
20. 530.66 cm^2
21. 27 ft^3
22. 468 m^2
23. 78.5 $in.^2$
24. 686.2 cm^3
25. $2\frac{2}{3}$ meters
26. 47 inches
27. 61 inches
28. 84 feet by 46 feet
29. 13 cm
30. 5 yd
31. 6 meters
32. $P = 48$ in.; $A = 120$ $in.^2$

19. Perimeter: Find the perimeter of the triangle with sides labeled as in the figure below.

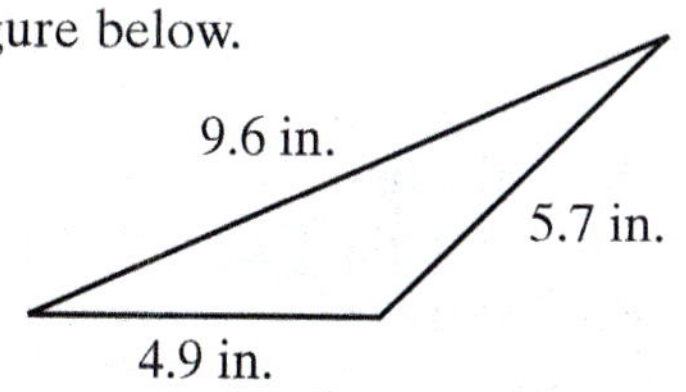

20. Circles: Find the area of a circle with radius 13 cm. (Use $\pi = 3.14$.)

21. Cubes: What is the volume of a cube with sides 3 ft long?

22. Gardening: A flower garden is in the shape of a parallelogram with a base of 30 m and a height of 15.6 m. What is the area of the flower garden?

23. Area of a pizza: One size of pizza served at Joe's Pizza Parlor is 10 in. in diameter. What is the area of this particular (circular) size pizza? (Use $\pi = 3.14$.)

24. Volume of a can: A can of beans has a diameter of 8.2 cm and a height of 13 cm. What is the volume of the can of beans (to the nearest tenth of a cubic centimeter)?

25. Squares: The perimeter of a square is $10\frac{2}{3}$ meters. Find the length of the sides.

26. Parallelograms: The area of a parallelogram is 1081 square inches. If the height is 23 inches, find the length of the base.

27. Triangles: The perimeter of a triangle is 147 inches. Two of the sides measure 38 inches and 48 inches. Find the length of the third side.

28. Fencing a yard: The length of a rectangular-shaped backyard is 8 feet less than twice the width. If 260 feet of fencing is needed to enclose the yard, find the dimensions of the yard.

29. Circles: The circumference of a circle is 26π cm. Find the radius.

30. Circular cylinders: The volume of a right circular cylinder is 7598.8 yd^3. What is the height of the cylinder if the radius is 22 yd? (Use $\pi = 3.14$.)

31. Trapezoids: The area of a trapezoid is 51 square meters. One base is 7 meters long, and the other is 10 meters long. Find the height of the trapezoid.

32. Geometry: Find the perimeter and area for the figure shown below.

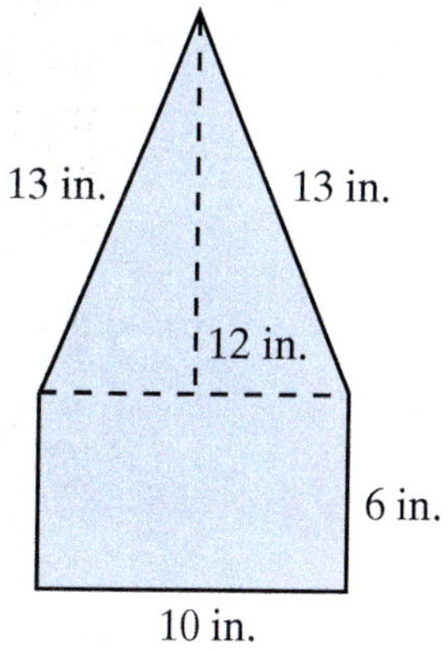

33. $\frac{11}{3}$ hr; $\frac{7}{3}$ hr
34. 55 mph, 247.5 miles
35. George: 55 mph; Joshua: 50 mph
36. \$30,000 at 4%, \$10,000 at 6.5%
37. \$4000 at 6%, \$16,000 at 8%
38. \$80
39. \$220
40. \$220
41. 77 points

3.3 Applications: Distance-Rate-Time, Interest, Average

33. Traveling by car: Alicia drove to her parents' home in 6 hours. She figured that she averaged 50 mph for the first part of the trip and 62 mph for the second part. The trip was 328 miles. As a math quiz for her sister, she asked her sister how much time she took for each part of the trip. Her sister, being a math whiz, told her what two numbers?

34. Traveling by car: Jerry drove from St. Louis, MO to Kansas City, MO in 5.5 hours. On the return trip he decided to increase his speed by 10 mph and the return took 4.5 hours. Find the rate on the return trip. What is the distance from St. Louis to Kansas City?

35. Traveling by car: Jacksonville and Tampa are about 210 miles apart. George leaves Tampa at the same time that Joshua leaves Jacksonville. They drive towards each other and meet in 2 hours. What was each person's speed, knowing that George traveled 5 mph faster than Joshua?

36. Investing: Reginald has had \$40,000 invested for one year, some in a savings account which paid 4%, the rest in stocks which earned 6.5% for the year. If his interest income for the year was \$1850, how much did he have in each investment?

37. Investing: Ms. Clark has two investments totaling \$20,000. One investment yields 6% a year and the other yields 8%. If her annual income from these investments is \$1520, how much did she invest at each rate?

38. Shopping: A men's store pays \$50 per pair for a certain style of men's pants. The store wants to be able to advertise a sale of 15% off the marked price and still make a profit of 36%. What should be the marked price?

39. Shopping: Maria used a coupon for 20% off her purchase to buy a new digital camera. If sales tax is 8% and she paid \$190.08, what was the original price of the camera?

40. Saving: For the next six months Sarah is planning to put money in her savings account for an average of \$200 a month. The first five months she managed to save the following amounts: \$250, \$100, \$180, \$260, \$190 respectively. How much will she need to save the last month to keep up with her goal?

41. Test scores: Sundar received scores of 71, 85, 83, 62, and 70 on his five Physics tests this semester. The final exam counts as two tests. What score will he need on the final to have an average of 75 for the semester? (All tests and exams have a maximum of 100 points.)

42. a. 30.85 million
b. Aspro Group
c. 2.7 million

43. 

44. 

45. 

46. 

47. 

48. $\{x|3 \le x < 6\}$

49. $\{x|-4 < x < 4\}$

50. $\{x|x \ge -4.6\}$

51. a. $\{x|-4 < x < 2\}$
b. $(-4, 2)$
c. Open interval

52. a. $\{x|x \ge 3\}$
b. $[3, \infty)$
c. Half-open interval

53. a. $\{x|x < 1.5\}$
b. $(-\infty, 1.5)$
c. Open interval

54. a. $\{x|-2 \le x \le 0\}$
b. $[-2, 0]$
c. Closed interval

55. 

$[1, \infty)$

56. 

$(-\infty, 7]$

57. 

$(-\infty, -1)$

58. 

$(-\infty, -5)$

59. 

$(-\infty, -6]$

60. 

$\left(\frac{36}{7}, \infty\right)$

42. Amusement parks: Given the attendance at the top ten amusement park chains:

a. Find the average attendance over the ten amusement park chains.

b. Find the amusement park chain with the lowest attendance.

c. Find the difference in attendance between Six Flags and Cedar Fair.

Source: http://www.teaconnect.org/etea/2009report.pdf

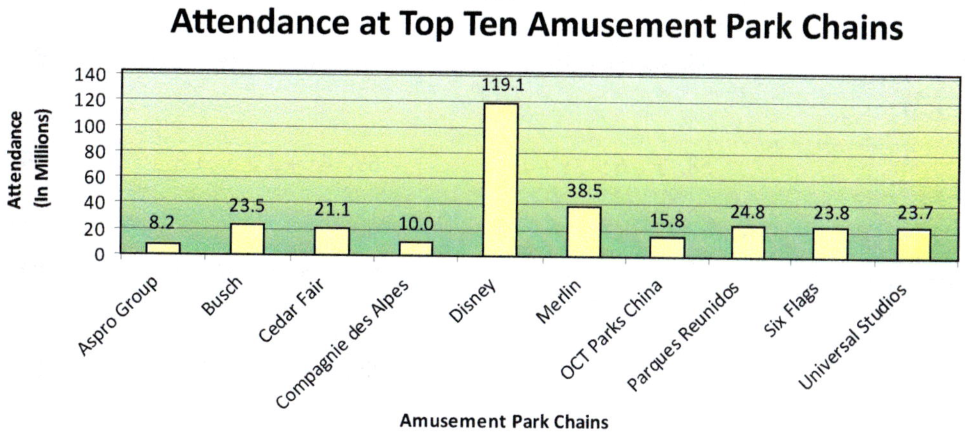

3.4 Linear Inequalities

Graph each set of indicated numbers on a real number line.

43. $\{x|x \text{ is an integer greater than } -3\}$

44. $\left\{x|x \text{ is a whole number with } |x| \le \frac{3}{4}\right\}$

45. $\{x|x \text{ is an even prime number}\}$

46. $\{x|x \text{ is any real number divisible by } 0\}$

47. $\{x|x \text{ is a composite number between 2 and 20}\}$

Use set-builder notation to indicate each set of numbers as described.

48. The set of all real numbers between 3 and 6, including 3

49. The set of all real numbers between –4 and 4

50. The set of all real numbers greater than or equal to –4.6

The graphs of sets of real numbers are given. **a.** *Use set-builder notation to indicate the set of numbers shown in each graph.* **b.** *Use interval notation to represent the graph.* **c.** *Tell what type of interval is illustrated.*

51. 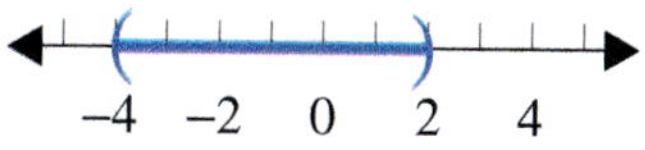

52. 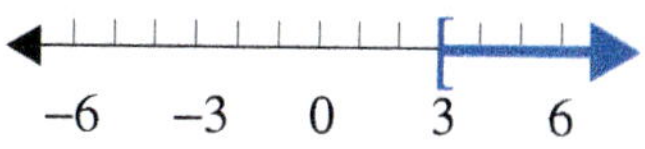

61. 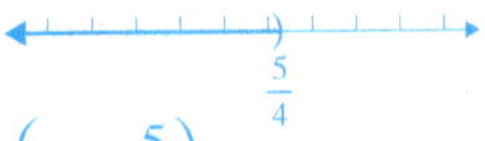
$\left(-\infty, \frac{5}{4}\right)$

62. 
$[2,5]$

63. 
$(-1,3)$

64. 
$(-1, 2]$

65. 
$\left[-4, \frac{4}{3}\right)$

66. 
$[2,6)$

67. $x = -3, \frac{7}{3}$

68. $x = -10, 4$

69. No solution

70. $x = 3, 11$

71. $x = -5$

72. $x = 0, 2$

73. 
$(-2, 5)$

74. 
$[-24, -10]$

75. 
$(0, 4)$

53. 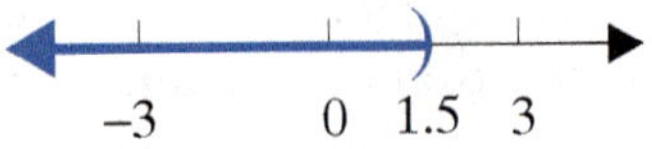

54. 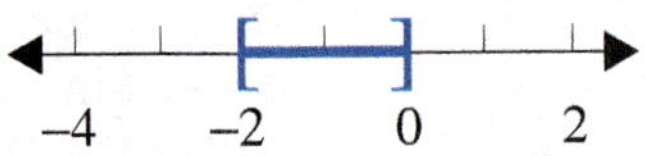

Solve the inequalities and graph the solution sets. Write each solution using interval notation. Assume that x is a real number.

55. $4x + 6 \geq 10$ **56.** $3x - 7 \leq 14$ **57.** $5x + 8 < 2x + 5$

58. $x - 6 > 3x + 4$ **59.** $4x + 2 \geq 5x + 8$ **60.** $\frac{x}{3} - 1 > 2 - \frac{x}{4}$

61. $\frac{1}{2} + \frac{x}{5} < 1 - \frac{x}{5}$ **62.** $1 \leq 3x - 5 \leq 10$ **63.** $-4 < 5x + 1 < 16$

64. $-6 < 7x + 1 \leq 15$ **65.** $0 < -\frac{3}{4}x + 1 \leq 4$ **66.** $0 < -\frac{1}{2}x + 3 \leq 2$

3.5 Absolute Value Equations and Inequalities

Solve each of the absolute value equations.

67. $|3x + 1| = 8$ **68.** $2|x + 3| - 4 = 10$ **69.** $|5x - 2| = -10$

70. $3|x - 7| - 15 = -3$ **71.** $4|x + 5| - 2 = -2$ **72.** $|2x + 1| = |3x - 1|$

Solve each of the absolute value inequalities and graph the solution sets. Write each solution using interval notation. Assume that x is a real number.

73. $|2x - 3| < 7$ **74.** $2|x + 17| - 5 \leq 9$ **75.** $10 > |4 - 2x| + 6$

76. $|x - 8| \geq -4$ **77.** $|2 - 5x| < -3$ **78.** $2|3x + 2| < 22$

76.  $(-\infty, \infty)$ 77.  No solution 78.  $\left(-\frac{13}{3}, 3\right)$

Chapter 3 Test

1. $m = \frac{N-p}{rt}$
2. $y = \frac{7-5x}{3}$
3. \$29.00
4. a. 18 months
   b. \$490
5. a. 27.42 ft
   b. 50.13 ft^2
6. 4186.67 cm^3
7. 200 ft^3
8. 3 meters, 4 meters, and 5 meters
9. 10 cm
10. a. $w = 24$ in.
    b. $A = 888$ in.^{2}

Solve each formula for the indicated variable.

1. $N = mrt + p$; solve for m.

2. $5x + 3y - 7 = 0$; solve for y.

Solving the following application problems.

3. Text messaging: The cost to send/receive t text messages in a given month on AT&T's Messaging 1500 plan is $C = 0.05(t-1500) + 15.00$ when $t \geq 1500$. How much did Jena spend the month she sent/received 1780 text messages.

4. Payment plans: When purchasing an item on an installment plan, the total cost C equals the down payment d plus the product of the monthly payment p and the number of months t. ($C = d + pt$) Use this information to answer the following questions.

a. A refrigerator costs \$857.60 if purchased on the installment plan. If the monthly payments are \$42.50 and the down payment is \$92.60, how long will it take to pay for the refrigerator?

b. A used car will cost \$3250 if purchased on an installment plan. If the monthly payments are \$115 for 24 months, what will be the down payment?

5. Geometry: Find **a.** the perimeter and **b.** the area of the figure shown here. (Use $\pi = 3.14$.)

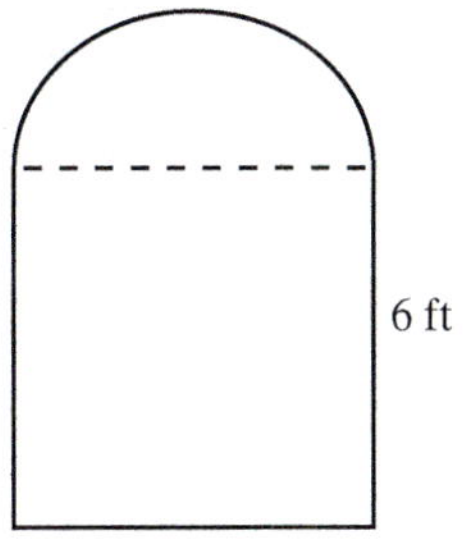

6. Volume: Find the volume of a sphere with diameter 20 cm. (Use $\pi = 3.14$ and round to the nearest hundreth.)

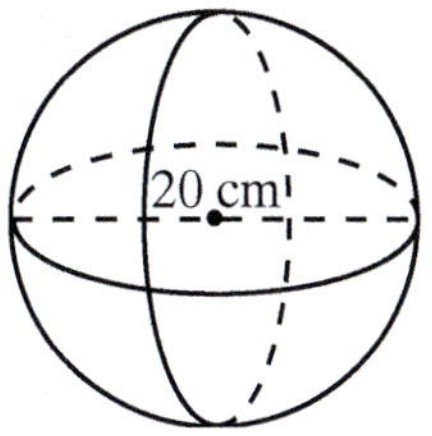

7. Volume: Find the volume (in cubic feet) of concrete needed for a sidewalk that is 6 feet by 100 feet by 4 inches.

8. Triangles: The triangle shown here indicates that the sides can be represented as x, $x + 1$, and $2x - 1$. What is the length of each side, if the perimeter is 12 meters?

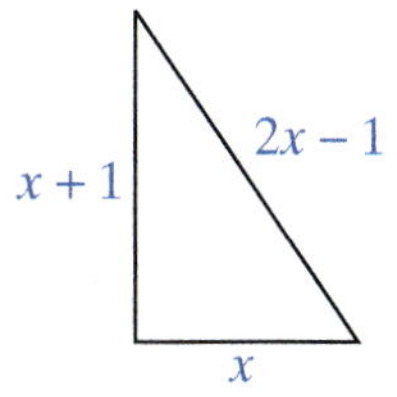

9. Trapezoids: The area of a trapezoid is 108 cm^2. The height is 12 cm and the length of one of the parallel sides is 8 cm. Find the length of the second of the parallel sides.

10. Rectangles: A rectangle is 37 in. long and has a perimeter of 122 in.

a. Find the width of the rectangle.

b. Find the area of the rectangle.

11. \$12,000 at 5%, \$13,000 at 3.5%
12. 4 hours
13. \$332.50
14. 88
15. a. 175.05 million
 b. 184.5 million
 c. 128.5 million
16. a. $\{x|x<5\}$
 b. $(-\infty,5)$
 c. Open interval
17. a. $\{x|-1\le x\le 4\}$
 b. $[-1,4]$
 c. Closed interval

11. Investing: George and Marie have investments in two accounts totaling \$25,000. One investment yields 5% a year and the other yields 3.5%. If their annual income from the 5% account yields \$145 more each year than the other account, what amount do they have in each account?

12. Traveling by bus: A bus leaves Kansas City headed for Phoenix traveling at a rate of 48 mph. Thirty minutes later, a second bus follows, traveling at 54 mph. How long will it take the second bus to overtake the first?

13. Shopping: The manager of a jewelry store purchased a selection of watches from the manufacturer for \$190 each. Next weekend he plans to have a 20% off sale. What price should he mark on each watch so that he will still make a profit of 40% on his cost?

14. Test scores: In his calculus class a student has the following test scores: 77, 82, and 73. If his last test will have a maximum of 100 points, what score does he need to finish the class with an average of 80?

15. Cell phone usage: Given the amount of U.S. cellular telephone subscribers:

a. Find the average number of subscribers from 2002-2005, inclusive.

b. Find the difference in the number of subscribers from 1999 and 2008.

c. How many subscribers were there in 2001?

Source: International Telecommunications Union

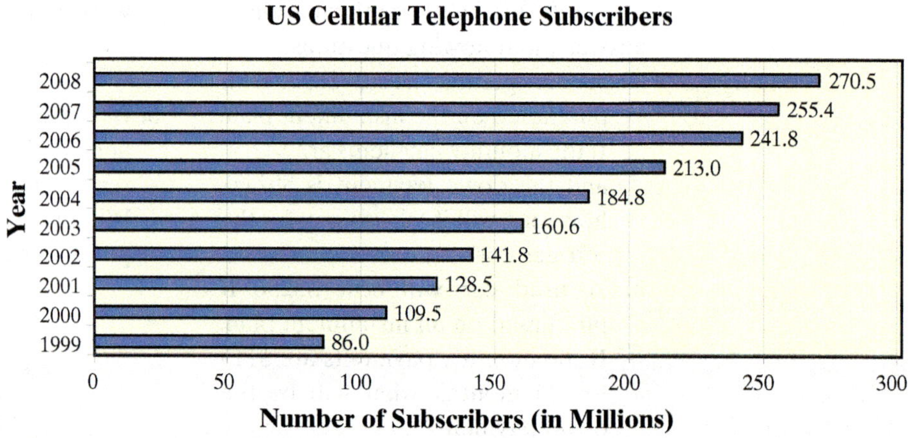

The graphs of sets of real numbers are given. ***a.*** *Use set-builder notation to indicate the set of numbers shown in each graph.* ***b.*** *Use interval notation to represent the graph.* ***c.*** *Tell what type of interval is illustrated.*

16.

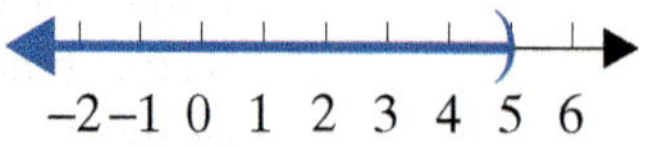

17.

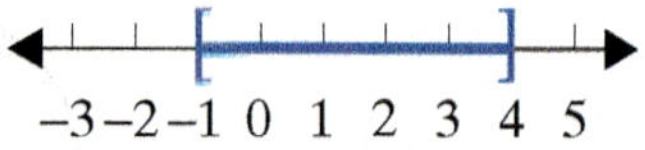

18. $x = -1.9, 0.9$

19. $x = \frac{2}{3}, \frac{8}{3}$

20. $x = -\frac{1}{3}, -13$

21. 

$(-5, \infty)$

22. 

$\left(-\infty, -\frac{7}{2}\right)$

23. 

$(-1, 5)$

24. 

$(2, 5)$

25. 

$(-\infty, -0.85) \cup (1.85, \infty)$

Solve each of the absolute value equations.

18. $|2x+1| = 2.8$ **19.** $|5-3x|+1=4$ **20.** $|x-6| = |2x+7|$

Solve the inequalities and graph the solution sets. Write each solution in interval notation. Assume that x is a real number.

21. $3x+7 < 4(x+3)$ **22.** $\frac{2x+5}{4} > x+3$ **23.** $-1 < 3x+2 < 17$

24. $|7-2x| < 3$ **25.** $|2(x-3)+5| > 2.7$

Cumulative Review: Chapters 1 – 3

1. a. 17
   b. –26
2. 21.7
3. $x + 15$
4. $3x + 8$
5. 540
6. $120a^2b^3$
7. False, $-15 < 5$
8. True
9. False, $\frac{7}{8} \geq \frac{7}{10}$
10. True
11. $x = 3$
12. $x = 4$
13. $x = \frac{8}{5}$
14. $x = -\frac{1}{2}$
15. $x = \frac{3}{8}$
16. $x = -3$
17. Conditional
18. Identity
19. Identity
20. Conditional

1. Find the average of each set of numbers.

a. $\{17, 14, 20\}$ **b.** $\{-18, -22, -27, -37\}$

2. Find the average (to the nearest tenth) of the data in the following frequency distribution.

Clocked Speeds of Bicycles (to nearest mph)	Frequency
19	4
20	5
22	7
25	3
28	1

Use the distributive property to complete the expression.

3. $3x + 45 = 3(\quad)$ **4.** $6x + 16 = 2(\quad)$

Find the LCM for each set of numbers or terms.

5. $\{12, 15, 54\}$ **6.** $\{6a^2, 24ab^3, 30ab, 40a^2b^2\}$

Determine whether the inequality is true or false. If the inequality is false, rewrite it in a form that is true using the same numbers or expressions.

7. $-15 > 5$ **8.** $|-6| \geq 6$ **9.** $\frac{7}{8} \leq \frac{7}{10}$ **10.** $-10 < 0$

Solve the equations.

11. $5x - 4 = 11$ **12.** $3x = 7x - 16$

13. $7(4 - x) = 3(x + 4)$ **14.** $-2(5x + 1) + 2x = 4(x + 1)$

15. $\frac{2}{3}x + \frac{1}{2} = \frac{3}{4}$ **16.** $1.5x - 3.7 = 3.6x + 2.6$

Determine whether each of the following equations is conditional, an identity, or a contradiction.

17. $5x + 17 = 3(2x - 5) + x - 2$ **18.** $14 + 2(x + 1) = 3(x + 5) - x + 1$

19. $5(x + 2) - 10 = 3x + 2x$ **20.** $9(x - 2) + 15 = 3x + 12$

21. $v = \frac{h+16t^2}{t}$

22. $r = \frac{A-P}{Pt}$

23. $y = -\frac{14-5x}{3}$

24. $h = \frac{3V}{\pi r^2}$

25. $d = \frac{C}{\pi}$

26. $y = \frac{10-3x}{5}$

27. $\{x | 2 < x < 5\}$

28. $\{x | -3 \le x \le 3\}$

29. $\{x | x \ge -8.3\}$

30. 

31. 

32. 

33. 

34. 

35. 
$(-\infty, 2]$

36. 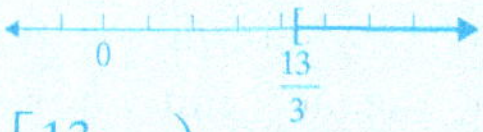
$\left[\frac{13}{3}, \infty\right)$

37. 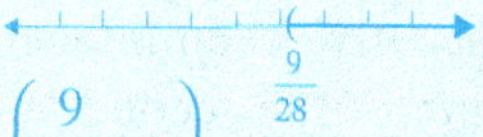
$\left(\frac{9}{28}, \infty\right)$

38. 
$[-1.1, 7.4]$

39. 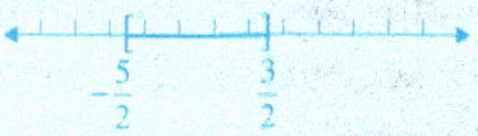
$\left[-\frac{5}{2}, \frac{3}{2}\right]$

40. 
$(-6, 2)$

41. $x = -3, 3$

42. No solution

Solve for the indicated variable.

21. $h = vt - 16t^2$; solve for v.

22. $A = P + Prt$; solve for r.

23. $5x - 3y = 14$; solve for y.

24. $V = \frac{1}{3}\pi r^2 h$; solve for h.

25. $C = \pi d$; solve for d.

26. $3x + 5y = 10$; solve for y.

Use set-builder notation to indicate each set of numbers as described.

27. The set of all real numbers between 2 and 5

28. The set of all real numbers between –3 and 3, inclusive

29. The set of all real numbers greater than or equal to –8.3

Graph each set of indicated numbers on a real number line.

30. $\{x | x \text{ is an integer less than } -1\}$

31. $\{x | x \text{ is a whole number with } |x| = 0\}$

32. $\{x | x \text{ is a prime number less than } 15\}$

33. $\{x | x \text{ is an integer divisible by } 0\}$

34. $\{x | x \text{ is a positive composite number less than or equal to } 25\}$

Solve each of the inequalities. Graph each solution set on a real number line and write the answer using interval notation.

35. $2x + 5 - 3 \le 6$

36. $-3(7 - 2x) \ge 2 + 3x - 10$

37. $-\frac{1}{2}x + \frac{7}{8} < \frac{2}{3}x + \frac{1}{2}$

38. $-6.2 \le 2x - 4 \le 10.8$

39. $-11 \le 4x - 1 \le 5$

40. $0 < \frac{1}{2}x + 3 < 4$

Solve each of the absolute value equations.

41. $|x| = 3$

42. $|y| = -1$

43. $|x + 9| = 10$

44. $|a + 0.5| = 1.4$

45. $|2x + 9| - 7 = 5$

46. $\left|\frac{2x}{5} - 6\right| = 18$

47. $|3x - 2| = |x + 2|$

48. $|4 - x| = |4 + x|$

43. $x = -19, 1$
44. $a = -1.9, 0.9$
45. $x = \frac{-21}{2}, \frac{3}{2}$
46. $x = -30, 60$
47. $x = 0, 2$
48. $x = 0$
49. a. $[-3, 3]$
b. $(6, 8)$
c. $[0.3, 2.7)$
50. 

$(-6, 6)$
51. 

No solution
52. 

$[-3, 2]$
53. 

$[0, 7]$
54. 

$(-\infty, -1.2) \cup (2.2, \infty)$
55. 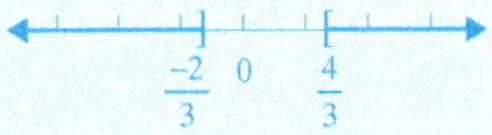

$\left(-\infty, \frac{-2}{3}\right] \cup \left[\frac{4}{3}, \infty\right)$
56. a. 45.7 cm
b. 139.25 cm^2
57. 1177.5 in.^{3}
58. 10.71 in.
59. \$3360
60. 20, 22, 24
61. 15, 17, 19

The graphs of sets of real numbers are given. Use interval notation to represent the graph.

49. a. 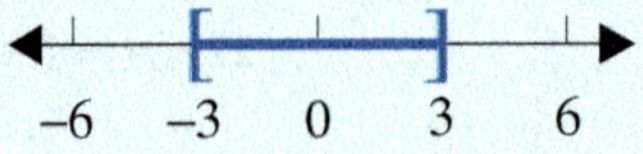

b. 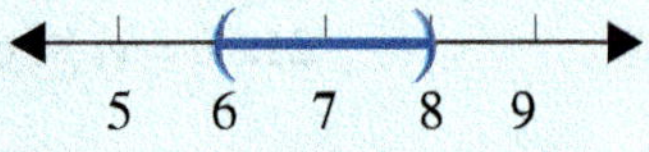

c. 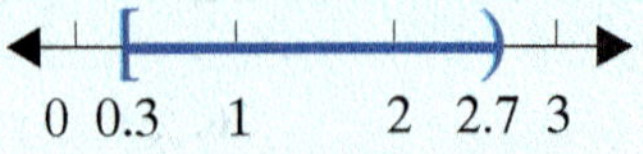

Solve the following absolute value inequalities and graph each solution set on a real number line. Write each solution set in interval notation.

50. $|x| < 6$

51. $|x+3| < -2$

52. $|2x+1| \le 5$

53. $|7-2x| \le 7$

54. $|2(x-3)+5| > 3.4$

55. $2|3x-1|+3 \ge 9$

56. Find **a.** the perimeter and **b.** area of the figure shown here. (Use $\pi = 3.14$.)

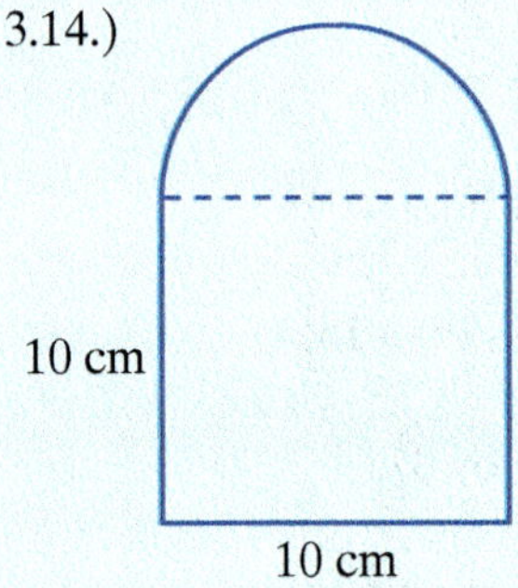

57. Find the volume of the circular cylinder with dimensions as shown in the figure. (Use $\pi = 3.14$.)

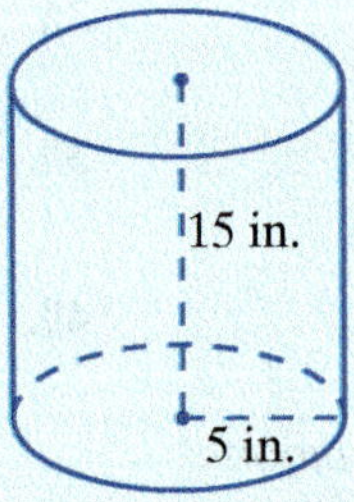

58. Find the height of a cone that has a volume of 175π in.^{3} and a radius of 7 in. (Use $\pi = 3.14$.)

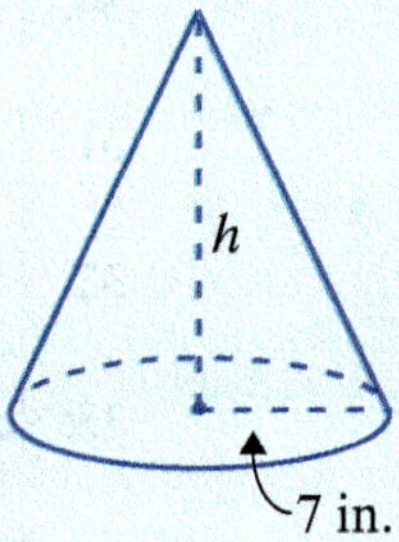

59. Investing: If \$3000 is invested at a yearly rate of 6%, what will be the total amount in the account after 2 years? Use the formula $A = P + Prt$.

60. Find three consecutive even integers such that the sum of the second and third is equal to three times the first decreased by 14.

61. Find three consecutive odd integers such that the sum of the first and three times the second is 28 more than twice the third.

62. Length = 25 cm; Width = 16 cm
63. 5 inches, 12 inches, 13 inches
64. –19
65. 1.8 hours (or 1 hour 48 minutes)
66. $16,500 at 3%; $5500 at 4.5%
67. Less than 140 miles
68. 87.25
69. $86 \le x \le 100$

62. Rectangles: The length of a rectangle is 9 cm more than its width. The perimeter is 82 cm. Find the dimensions of the rectangle.

63. Triangles: An artist has a piece of wire 30 inches long and wants to bend it into the shape of a triangle so that the sides are of length x, $2x + 2$, and $2x + 3$. What will be the length of each side of the triangle?

64. Twice the difference between a number and 16 is equal to 4 times the number increased by 6. What is the number?

65. Traveling by train: Diego took the train into New York City to see a concert. On the way to the concert the train averaged 90 miles per hour. On the way home from the concert the train made a lot more stops and averaged only 50 miles per hour. If Diego spent a total of 2.8 hours on the train, how long did the return trip take?

66. Investing: Two investments totaling $22,000 produce an annual income of $742.50. One investment yields 4.5% a year, while the other yields 3% per year. How much is invested at each rate?

67. Car rentals: Acme Car Rental charges $0.25 for each mile driven plus $15 a day. Zenith Car Rental charges $50 a day with no extra charge for mileage. How many miles a day can you drive an Acme car and still keep the cost less than a Zenith car?

68. Final grades: From the syllabus in Alesha's English class, the final grade consists of 20% homework, 10% quiz average, 30% semester paper, and 40% final exam. Going into the final exam, Alesha has an 87 for her homework average, an 89 for her quiz average, and a 96 on her semester paper. What is the minimum grade Alesha can receive on the final exam to earn an A (90 to 100) in the class?

69. Exam scores: The range for a grade of B is an average of 75 or more but less than 85 in an English class. If your grades on the first four exams were 73, 65, 77, and 74, what possible grades can you get on the final exam to earn a grade of B? (Assume that 100 is the maximum number of points on the final exam.)

CHAPTER 4

Linear Equations and Functions

Did You Know?

In Chapter 4, you will be introduced to the idea of a graph of an algebraic equation. A graph is simply a picture of an algebraic relationship. This topic is more formally called **analytic geometry**. It is a combination of algebra (the equation) and geometry (the picture).

Descartes

The idea of combining algebra and geometry was not thought of until René Descartes wrote his famous *Discourse on the Method of Reasoning* in 1637. The third appendix in this book, "La Geometrie," made Descartes' system of analytic geometry known to the world. In fact, you will find that analytic geometry is sometimes called **Cartesian geometry**.

René Descartes is perhaps better known as a philosopher than as a mathematician; he is often referred to as the father of modern philosophy. His method of reasoning was to apply the same logical structure to philosophy that had been developed in mathematics, especially geometry.

In the Middle Ages, the highest forms of knowledge were believed to be mathematics, philosophy, and theology. Many famous people in history who have reputations as poets, artists, philosophers, and theologians were also creative mathematicians. Almost every royal court had mathematicians whose work reflected glory on the royal sponsor who paid the mathematician for his research and court presence.

Descartes, in fact, died in 1650 after accepting a position at the court of the young warrior-queen, Christina of Sweden. Apparently, the frail French philosopher-mathematician, who spent his mornings in bed doing mathematics, could not stand the climate of Sweden and the hardships imposed by Christina in her demand that Descartes tutor her in mathematics each morning at 5 o'clock in an unheated castle library.

"Divide each difficulty into as many parts as is feasible and necessary to resolve it."

René Descartes (1596 – 1650)

This chapter uses the **Cartesian coordinate system** to develop relationships between algebra and geometry. These relationships have proven extremely important to the development of mathematics and mathematical problem-solving. In particular, you will develop graphing skills for linear equations and linear inequalities in two variables. The graphs of linear equations are lines, hence the term linear. The graphs of linear inequalities are half-planes separated by lines.

Three useful forms of linear equations are the **standard form**, the **slope-intercept form**, and the **point-slope form**. Each form is equally important and useful depending on the information given and the application of the equation. Your algebraic skills should allow you to change from one form to another and to recognize when the same equation is in a different form.

The concept of **slope** underlies all our work with lines. Horizontal lines and vertical lines can be related to slope. Relationships between two or more lines can be discussed in terms of their slopes. Parallel lines have the same slope, and perpendicular lines have slopes that are negative reciprocals of each other.

Slope occurs as a part of our daily lives in many ways. The slope of a roof is particularly important in areas with lots of snow. Large trucks must be very careful on mountain roads with steep slopes. Airplanes must deal with their rate of descent or the slope of their paths in landing and taking off. We will see that slope can be related to the rate of change in prices of products that we purchase.

4.1 The Cartesian Coordinate System

- *Graph and label ordered pairs of real numbers as points on a plane.*
- *Find ordered pairs of real numbers that satisfy a given equation.*
- *Locate points on a given graph of a line.*

René Descartes (1596 – 1650), a famous French mathematician, developed a system for solving geometric problems using algebra. This system is called the **Cartesian coordinate system** in his honor. Descartes based his system on a relationship between points in a plane and **ordered pairs** of real numbers. This section begins by relating algebraic formulas with ordered pairs and then shows how these ideas can be related to geometry.

Equations in Two Variables

Equations such as $d = 60t$, $I = 0.05P$, and $y = 2x + 3$ represent relationships between pairs of variables. For example, in the first equation, if $t = 3$, then $d = 60 \cdot 3 = 180$. With the understanding that t is first and d is second, we can represent $t = 3$ and $d = 180$ in the form of an ordered pair $(3, 180)$. In general, if t is the first number and d is the second number, then solutions to the equation $d = 60t$ can be written in the form of ordered pairs (t, d). Thus we see that $(180, 3)$ is different from $(3, 180)$. **The order of the numbers in an ordered pair is critical.**

We say that $(3, 180)$ **is a solution of** (or **satisfies**) the equation $d = 60t$. Similarly, $(5, 300)$ represents $t = 5$ and $d = 300$ and satisfies the equation $d = 60t$. In the same way, $(100, 5)$ satisfies $I = 0.05P$ where $P = 100$ and $I = 0.05 \cdot 100 = 5$. In this equation, solutions are ordered pairs in the form (P, I).

For the equation $y = 2x + 3$, ordered pairs are in the form (x, y), and $(2, 7)$ satisfies the equation. If $x = 2$, then substituting in the equation gives $y = 2 \cdot 2 + 3 = 7$. In the ordered pair (x, y), x is called the **first coordinate** and y is called the **second coordinate**. To find ordered pairs that satisfy an equation in two variables, we can **choose any value** for one variable and find the corresponding value for the other variable by substituting into the equation. For example:

For the equation $y = 2x + 3$:

Choices for x:	**Substitution:**	**Ordered Pairs:**
$x = 1$	$y = 2(1) + 3 = 5$	$(1, 5)$
$x = -2$	$y = 2(-2) + 3 = -1$	$(-2, -1)$
$x = \frac{1}{2}$	$y = 2\left(\frac{1}{2}\right) + 3 = 4$	$\left(\frac{1}{2}, 4\right)$

All the ordered pairs $(1, 5), (-2, -1)$, and $\left(\frac{1}{2}, 4\right)$ satisfy the equation $y = 2x + 3$. **There is an infinite number of such ordered pairs. Any real number could have been chosen for x and the corresponding value for y calculated.**

Since the equation $y = 2x + 3$ is solved for y, we say that the value of y "depends" on the choice of x. Thus in an ordered pair of the form (x, y), the first coordinate x is called the **independent variable** and the second coordinate y is called the **dependent variable**.

In the following table the first variable in each case is the independent variable and the second variable is the dependent variable. Corresponding ordered pairs would be of the form $(t, d), (P, I)$, and (x, y). The choices for the values of the independent variables are arbitrary. There are an infinite number of other values that could have just as easily been chosen.

$d = 60t$			$I = 0.05P$			$y = 2x + 3$		
t	d	(t, d)	P	I	(P, I)	x	y	(x, y)
5	$60(5) = 300$	$(5, 300)$	100	$0.05(100) = 5$	$(100, 5)$	-2	$2(-2) + 3 = -1$	$(-2, -1)$
10	$60(10) = 600$	$(10, 600)$	200	$0.05(200) = 10$	$(200, 10)$	-1	$2(-1) + 3 = 1$	$(-1, 1)$
12	$60(12) = 720$	$(12, 720)$	500	$0.05(500) = 25$	$(500, 25)$	0	$2(0) + 3 = 3$	$(0, 3)$
15	$60(15) = 900$	$(15, 900)$	1000	$0.05(1000) = 50$	$(1000, 50)$	3	$2(3) + 3 = 9$	$(3, 9)$

Table 1

Graphing Ordered Pairs

The Cartesian coordinate system relates algebraic equations and ordered pairs to geometry. In this system, two number lines intersect at right angles and separate the plane into four **quadrants**. The **origin**, designated by the ordered pair $(0, 0)$, is the point of intersection of the two lines. The horizontal number line is called the **horizontal axis** or ***x*-axis**. The vertical number line is called the **vertical axis** or ***y*-axis**. Points that lie on either axis are not in any quadrant. They are simply on an axis (Figure 1).

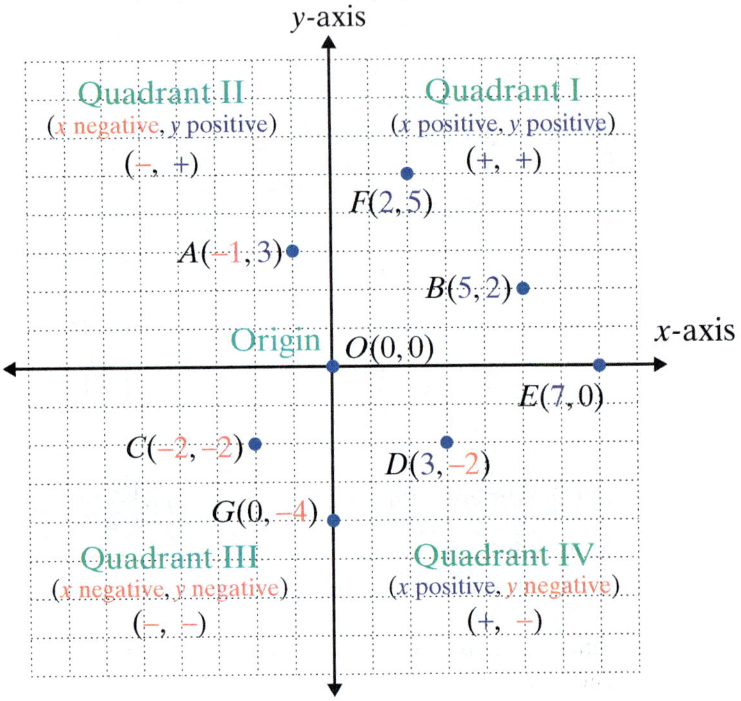

Figure 1

The following important relationship between ordered pairs of real numbers and points in a plane is the cornerstone of the Cartesian coordinate system.

One-to-One Correspondence

There is a **one-to-one correspondence** between points in a plane and ordered pairs of real numbers.

In other words, for each point in a plane there is one and only one corresponding ordered pair of real numbers, and for each ordered pair of real numbers there is one and only one corresponding point in the plane.

The **graphs of the points** $A(2,1)$, $B(-2,3)$, $C(-3,-2)$, $D(1,-2)$, and $E(3,0)$ are shown in Figure 2. (**Note:** An ordered pair of real numbers and the corresponding point on the graph are frequently used to refer to each other. Thus the ordered pair $(2,1)$ and the point $(2,1)$ are interchangeable ideas.)

POINT	QUADRANT
$A(2,1)$	I
$B(-2,3)$	II
$C(-3,-2)$	III
$D(1,-2)$	IV
$E(3,0)$	x-axis

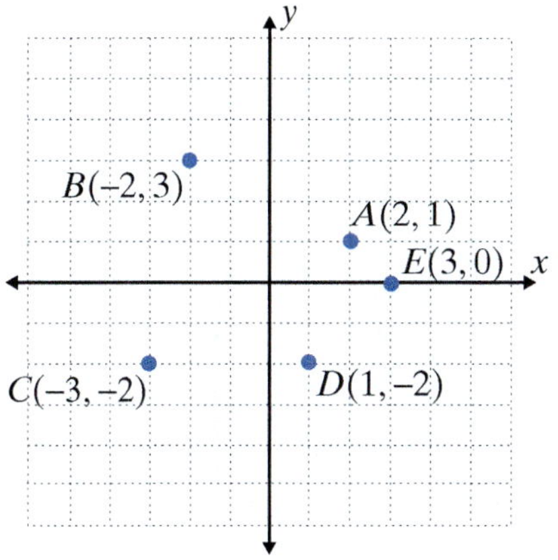

Figure 2

NOTES Unless otherwise stated, assume that the grid lines (as shown here in Figure 2) are one unit apart.

Example 1: Graphing Ordered Pairs

Graph the sets of ordered pairs.

a. $\{A(-2,1), B(-1,-4), C(0,2), D(1,3), E(2,-3)\}$

Note: The listing of ordered pairs within the braces can be in any order.

Solution: To locate points: start at the **origin** $(0,0)$, move left or right for the x-coordinate and up or down for the y-coordinate.

For $A(-2,1)$, move 2 units left and 1 unit up.

For $B(-1,-4)$, move 1 unit left and 4 units down.

For $C(0,2)$, move no units left or right and 2 units up.

For $D(1,3)$, move 1 unit right and 3 units up.

For $E(2,-3)$, move 2 units right and 3 units down.

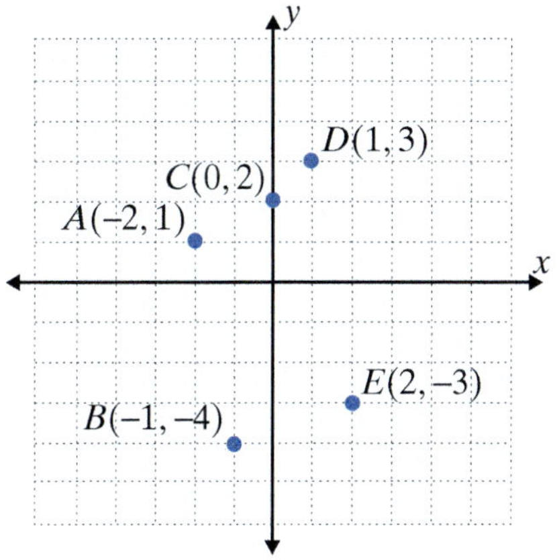

Continued on the next page...

Teaching Notes:
You may need to remind your students that there is no significance in the order of the elements in a set. However, the order of each pair is important.

b. $\{A(-1,3), B(0,1), C(1,-1), D(2,-3), E(3,-5)\}$

Solution: To locate each point, start at the **origin**, and:

For $A(-1,3)$, move 1 unit left and 3 units up.

For $B(0,1)$, move no units left or right and 1 unit up.

For $C(1,-1)$, move 1 unit right and 1 unit down.

For $D(2,-3)$, move 2 units right and 3 units down.

For $E(3,-5)$, move 3 units right and 5 units down.

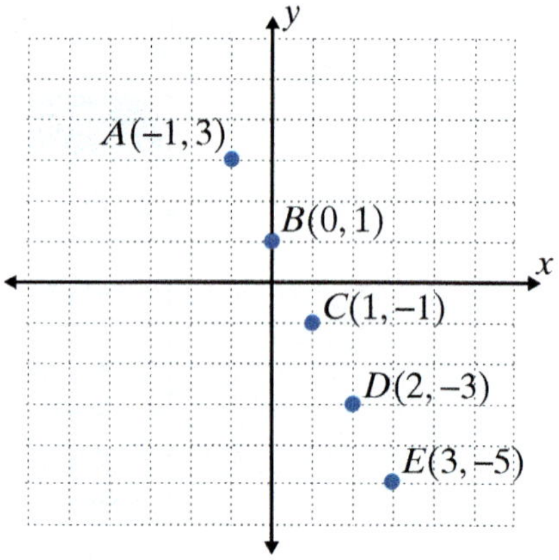

The points (ordered pairs) in Example 1b can be shown to satisfy the equation $y = -2x + 1$. For example, using $x = -1$ in the equation yields,

$$y = -2(-1) + 1 = 2 + 1 = 3$$

and the ordered pair $(-1, 3)$ satisfies the equation. Similarly, letting $y = 1$ gives,

$$1 = -2x + 1$$
$$0 = -2x$$
$$0 = x$$

and the ordered pair $(0, 1)$ satisfies the equation.

We can write all the ordered pairs in Example 1b in table form.

x	$-2x + 1 = y$	(x, y)
−1	$-2(-1)+1=3$	$(-1,3)$
0	$-2(0)+1=1$	$(0,1)$
1	$-2(1)+1=-1$	$(1,-1)$
2	$-2(2)+1=-3$	$(2,-3)$
3	$-2(3)+1=-5$	$(3,-5)$

Table 2

Example 2: Determining Ordered Pairs

a. Determine which, if any, of the ordered pairs $(0, -2)$, $\left(\frac{2}{3}, 0\right)$, and $(2, 5)$ satisfy the equation $y = 3x - 2$.

Solution: We will substitute 0, $\frac{2}{3}$, and 2 for x in the equation $y = 3x - 2$ and see if the corresponding y-values match those in the given ordered pairs.

$x = 0$: $\quad y = 3(0) - 2 = -2$ $\quad$ so, $(0, -2)$ satisfies the equation.

$x = \frac{2}{3}$: $\quad y = 3\left(\frac{2}{3}\right) - 2 = 0$ $\quad$ so, $\left(\frac{2}{3}, 0\right)$ satisfies the equation.

$x = 2$: $\quad y = 3(2) - 2 = 4$ $\quad$ so, $(2, 4)$ satisfies the equation.

The point $(2, 5)$ does not satisfy the equation $y = 3x - 2$ because, as just illustrated, $y = 4$ when $x = 2$, not 5.

b. Determine the missing coordinate in each of the following ordered pairs so that the points will satisfy the equation $2x + 3y = 12$:

$$(0, \ ), (3, \ ), (\ , 0), (\ , -2).$$

Solution: The missing values can be found by substituting the given values for x (or for y) into the equation $2x + 3y = 12$ and solving for the other variable.

For $(0, \ )$, let $x = 0$:

$$2(0) + 3y = 12$$
$$3y = 12$$
$$y = 4$$

The ordered pair is $(0, 4)$.

For $(3, \ )$, let $x = 3$:

$$2(3) + 3y = 12$$
$$6 + 3y = 12$$
$$3y = 6$$
$$y = 2$$

The ordered pair is $(3, 2)$.

For $(\ , 0)$, let $y = 0$:

$$2x + 3(0) = 12$$
$$2x = 12$$
$$x = 6$$

The ordered pair is $(6, 0)$.

For $(\ , -2)$, let $y = -2$:

$$2x + 3(-2) = 12$$
$$2x - 6 = 12$$
$$2x = 18$$
$$x = 9$$

The ordered pair is $(9, -2)$.

Continued on the next page...

c. Complete the table below so that each ordered pair will satisfy the equation $y = 1 - 2x$.

x	$y = 1 - 2x$	y	(x, y)
0			
		3	
$\frac{1}{2}$			
5			

Solution: Substituting each given value for x or y into the equation $y = 1 - 2x$ gives the following table of ordered pairs.

x	$y = 1 - 2x$	y	(x, y)
$\mathbf{0}$	$1 = 1 - 2(\mathbf{0})$	1	$(\mathbf{0}, 1)$
-1	$\mathbf{3} = 1 - 2(-1)$	$\mathbf{3}$	$(-1, \mathbf{3})$
$\mathbf{\frac{1}{2}}$	$0 = 1 - 2\left(\mathbf{\frac{1}{2}}\right)$	0	$\left(\mathbf{\frac{1}{2}}, 0\right)$
$\mathbf{5}$	$-9 = 1 - 2(\mathbf{5})$	-9	$(\mathbf{5}, -9)$

For $x = \mathbf{0}$:

$$y = 1 - 2(\mathbf{0}) = 1$$

For $y = \mathbf{3}$:

$$(\mathbf{3}) = 1 - 2x$$
$$2 = -2x$$
$$-1 = x$$

For $x = \mathbf{\frac{1}{2}}$:

$$y = 1 - 2\left(\mathbf{\frac{1}{2}}\right) = 0$$

For $x = \mathbf{5}$:

$$y = 1 - 2(\mathbf{5}) = -9$$

NOTES

Although this discussion is related to ordered pairs of real numbers, most of the examples use ordered pairs of **integers**. This is because ordered pairs of integers are relatively easy to locate on a graph and relatively easy to read from a graph. Ordered pairs with fractions, decimals, or radicals must be located by estimating the positions of the points. The precise coordinates intended for such points can be difficult or impossible to read because large dots must be used so the points can be seen. **Even with these difficulties, you should understand that we are discussing ordered pairs of real numbers and that points with fractions, decimals, and radicals as coordinates do exist and should be plotted by estimating their positions.**

Example 3: Reading Points on a Graph

The graphs of two lines are given. Each line contains an infinite number of points. Use the grid to help you locate (or estimate) three points on each line.

a.

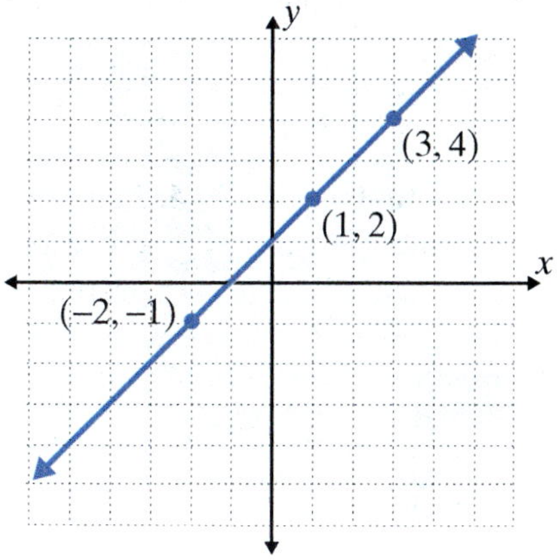

Solution:

Three points on this graph are $(-2,-1)$, $(1,2)$, and $(3,4)$. (Of course there is more than one correct answer to this type of question. Use your own judgment.)

b.

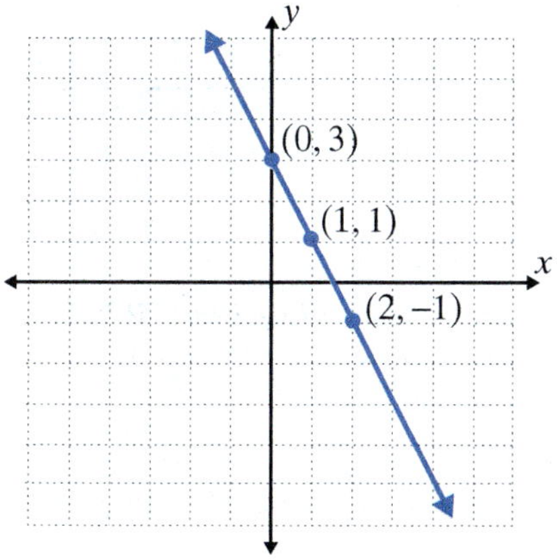

Solution:

Three points on this graph are $(0,3)$, $(1,1)$, and $(2,-1)$. (You may also estimate with fractions. For example, one point appears to be approximately $\left(\frac{1}{2},2\right)$.)

Practice Problems

1. Determine which ordered pairs satisfy the equation $3x + y = 14$.

a. $(5, -1)$ **b.** $(4, 2)$ **c.** $(-1, 17)$

2. Given $3x + y = 5$, find the missing coordinate of each ordered pair so that it will satisfy the equation.

a. $(0, \ )$ **b.** $\left(\frac{1}{3}, \ \right)$ **c.** $(\ , 2)$

3. Complete the table so that each ordered pair will satisfy the equation $y = \frac{2}{3}x + 1$.

x	y
0	
	−2
−3	
6	

4. List the sets of ordered pairs corresponding to the points on the graph.

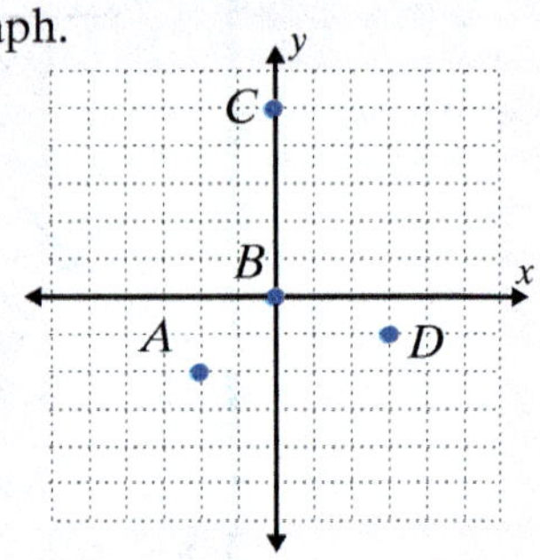

4.1 Exercises

List the sets of ordered pairs corresponding to the points on the graphs. Assume that the grid lines are marked one unit apart.

1. $\{A(-5, 1), B(-3, 3), C(-1, 1), D(1, 2), E(2, -2)\}$

2. $\{M(-3, 1), N(-1, 5), O(0, 3), P(1, 5), Q(3, -2)\}$

1.

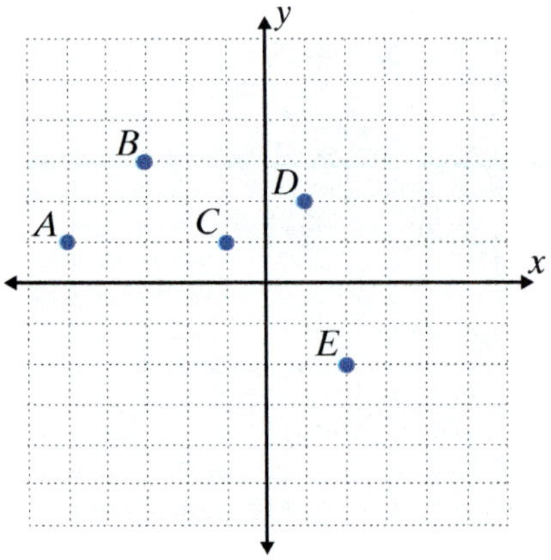

2.

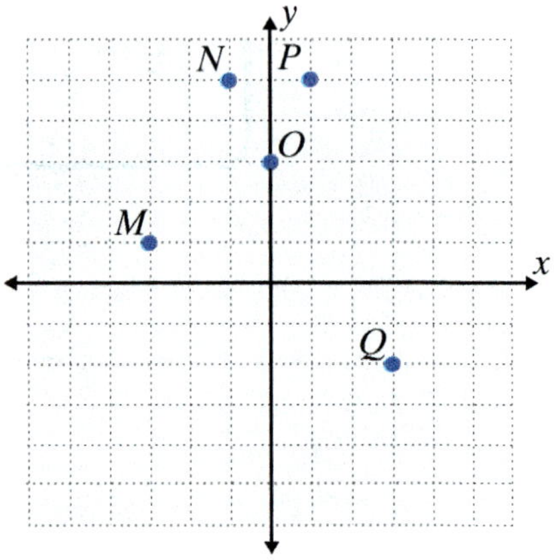

Answers to Practice Problems: **1.** All satisfy the equation. **2. a.** $(0, 5)$ **b.** $\left(\frac{1}{3}, 4\right)$ **c.** $(1, 2)$

3. $(0, 1), \left(-\frac{9}{2}, -2\right), (-3, -1), (6, 5)$ **4.** $\{A(-2, -2), B(0, 0), C(0, 5), D(3, -1)\}$

3. $\{A(-3,-2), B(-1,-3), C(-1,3), D(0,0), E(2,1)\}$

4. $\{M(-3,2), N(-1,-1), O(1,5), P(3,-2), Q(6,5)\}$

5. $\{A(-4,4), B(-3,-4), C(0,-4), D(0,3), E(4,1)\}$

6. $\{M(-4,-5), N(-3,1), O(0,-2), P(1,1), Q(3,-1)\}$

7. $\{A(-3,-5), B(-1,4), C(0,-1), D(3,1), E(6,0)\}$

8. $\{M(-6,2), N(-1,6), O(0,0), P(1,-6), Q(6,3)\}$

9. $\{A(-5,0), B(-2,2), C(-1,-4), D(0,6), E(2,0)\}$

10. $\{M(-5,-2), N(-3,4), O(2,-2), P(2,6), Q(4,3)\}$

3.

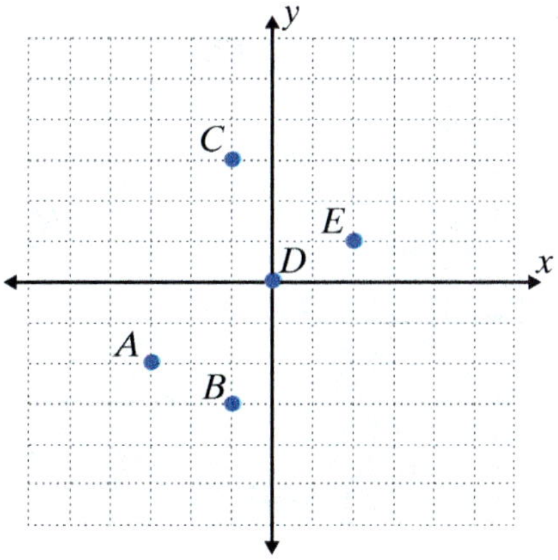

4.

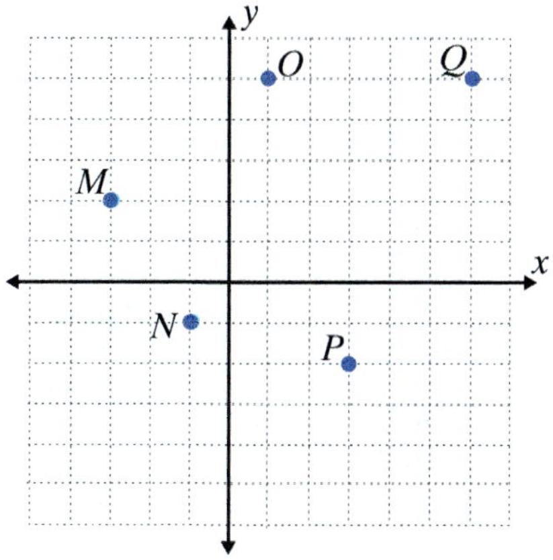

5.

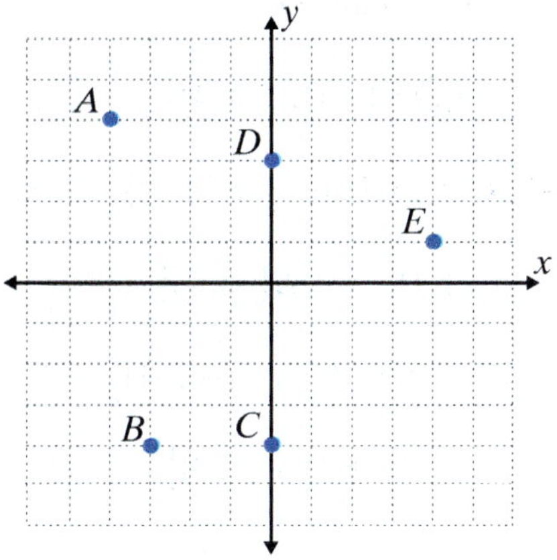

6.

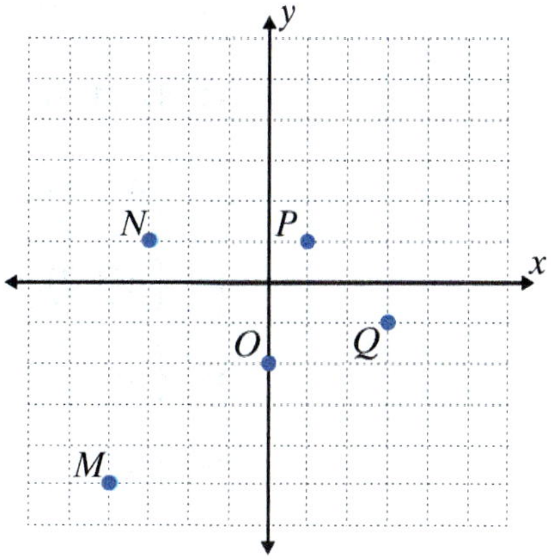

7.

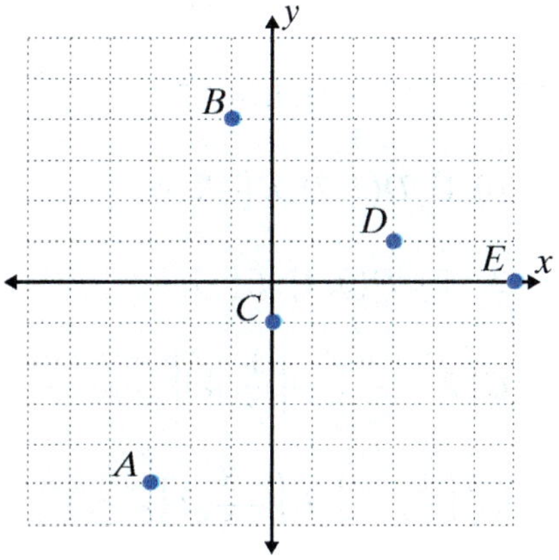

8.

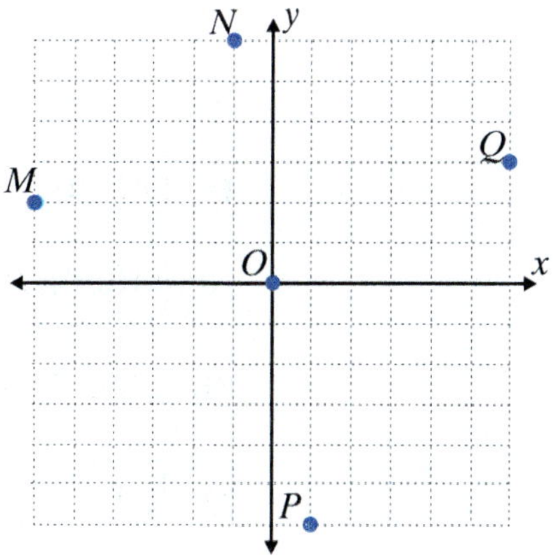

9.

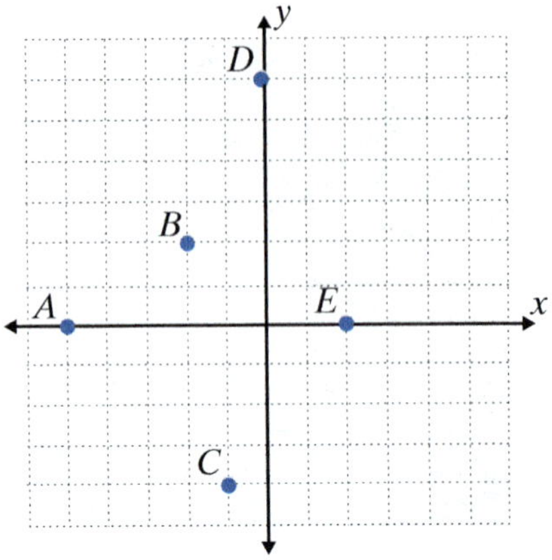

10.

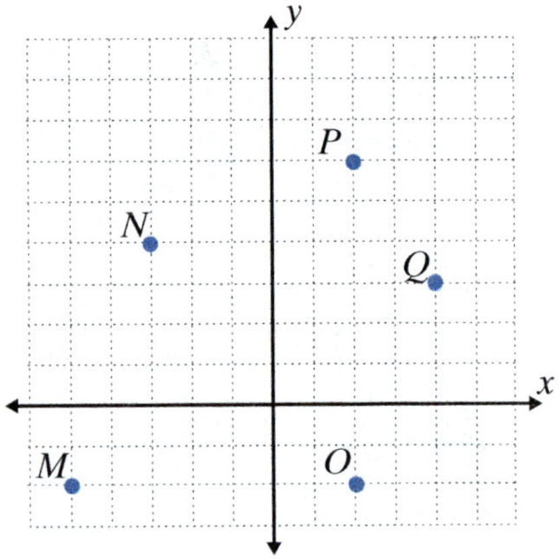

Graph the sets of ordered pairs and label the points.

11. $\{A(4,-1), B(3,2), C(0,5), D(1,-1), E(1,4)\}$

12. $\{A(-1,-1), B(-3,-2), C(1,3), D(0,0), E(2,5)\}$

13. $\{A(1,2), B(0,2), C(-1,2), D(2,2), E(-3,2)\}$

14. $\{A(-1,4), B(0,-3), C(2,-1), D(4,1), E(-1,-1)\}$

15. $\{A(1,0), B(3,0), C(-2,1), D(-1,1), E(0,0)\}$

16. $\{A(-1,-1), B(0,1), C(1,3), D(2,5), E(3,10)\}$

17. $\{A(4,1), B(0,-3), C(1,-2), D(2,-1), E(-4,2)\}$

18. $\{A(0,1), B(1,0), C(2,-1), D(3,-2), E(4,-3)\}$

19. $\{A(1,4), B(-1,-2), C(0,1), D(2,7), E(-2,-5)\}$

20. $\{A(0,0), B(-1,3), C(3,-2), D(0,4), E(-7,0)\}$

21. $\left\{A(1,-3), B\left(-4, \frac{3}{4}\right), C\left(2, -2\frac{1}{2}\right), D\left(\frac{1}{2}, 4\right)\right\}$

22. $\left\{A\left(\frac{3}{4}, \frac{1}{2}\right), B\left(2, -\frac{5}{4}\right), C\left(\frac{1}{3}, -2\right), D\left(-\frac{5}{3}, 2\right)\right\}$

23. $\{A(1.6,-2), B(3,2.5), C(-1,1.5), D(0,-2.3)\}$

24. $\{A(-2,2), B(-3,1.6), C(3,0.5), D(1.4,0)\}$

11.

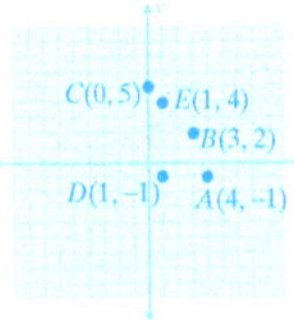

12.

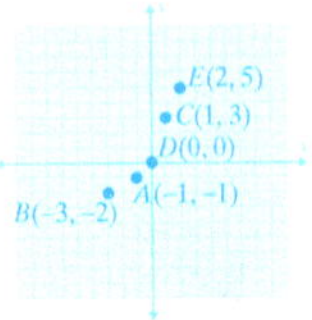

13.

14.

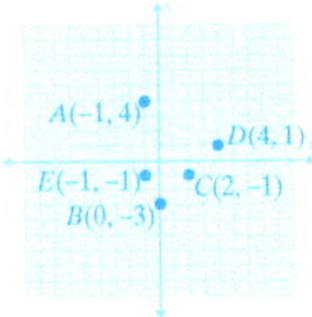

15.

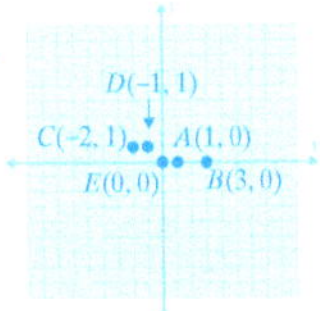

16.

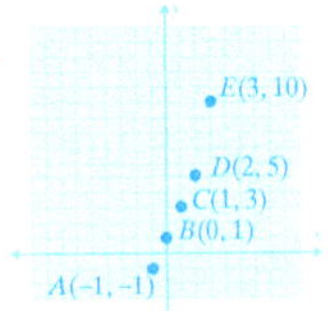

17.

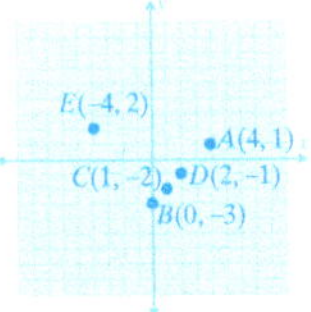

18.

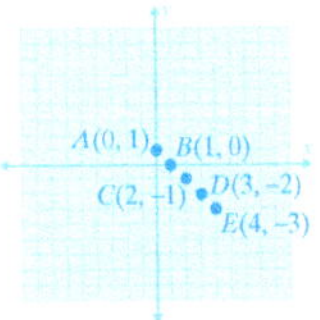

19.

20.

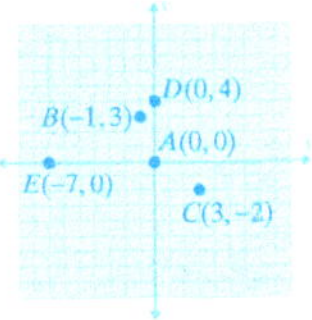

21.

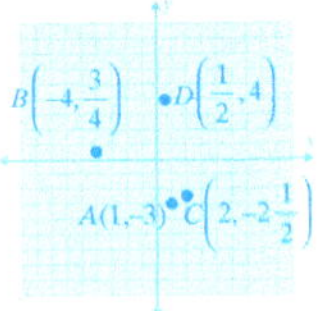

22.

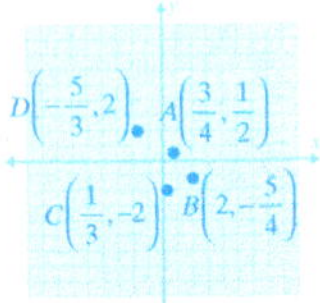

23.

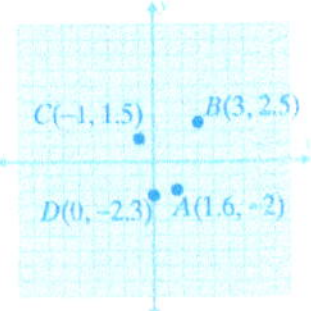

24.

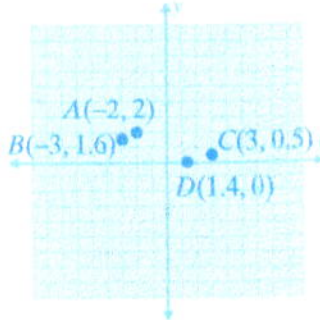

25. b, c, d
26. a, d
27. a, c
28. b, c
29. a, c, d
30. b, d
31. a. $(0, -4)$ b. $(2, -2)$ c. $(4, 0)$ d. $(1, -3)$
32. a. $(0, 7)$ b. $(-1, 8)$ c. $(7, 0)$ d. $(4, 3)$
33. a. $(0, 3)$ b. $(2, 2)$ c. $(6, 0)$ d. $(-2, 4)$
34. a. $(0, 9)$ b. $(4, -3)$ c. $(3, 0)$ d. $(2, 3)$
35. a. $(0, -8)$ b. $(1, -4)$ c. $(2, 0)$ d. $(3, 4)$
36. a. $(0, -1)$ b. $(4, 1)$ c. $(2, 0)$ d. $(8, 3)$
37. a. $(0, 2)$ b. $\left(-1, \frac{8}{3}\right)$ c. $(3, 0)$ d. $(6, -2)$
38. a. $(0, 5)$ b. $\left(2, \frac{5}{3}\right)$ c. $(3, 0)$ d. $\left(\frac{3}{5}, 4\right)$
39. a. $\left(0, -\frac{7}{4}\right)$ b. $(1, -1)$ c. $\left(\frac{7}{3}, 0\right)$ d. $\left(3, \frac{1}{2}\right)$
40. a. $\left(0, \frac{6}{5}\right)$ b. $\left(\frac{1}{2}, 1\right)$ c. $(3, 0)$ d. $(-2, 2)$

Determine which, if any, of the ordered pairs satisfy the given equations.

25. $2x - y = 4$

a. $(1, 1)$ **b.** $(2, 0)$ **c.** $(1, -2)$ **d.** $(3, 2)$

26. $x + 2y = -1$

a. $(1, -1)$ **b.** $(1, 0)$ **c.** $(2, 1)$ **d.** $(3, -2)$

27. $4x + y = 5$

a. $\left(\frac{3}{4}, 2\right)$ **b.** $(4, 0)$ **c.** $(1, 1)$ **d.** $(0, 3)$

28. $2x - 3y = 7$

a. $(1, 3)$ **b.** $\left(\frac{1}{2}, -2\right)$ **c.** $\left(\frac{7}{2}, 0\right)$ **d.** $(2, 1)$

29. $2x + 5y = 8$

a. $(4, 0)$ **b.** $(2, 1)$ **c.** $(1, 1.2)$ **d.** $(1.5, 1)$

30. $3x + 4y = 10$

a. $(-2, 3)$ **b.** $(0, 2.5)$ **c.** $(4, -2)$ **d.** $(1.2, 1.6)$

Determine the missing coordinate in each of the ordered pairs so that the point will satisfy the equation given.

31. $x - y = 4$

a. $(0, \quad)$ **b.** $(2, \quad)$ **c.** $(\quad, 0)$ **d.** $(\quad, -3)$

32. $x + y = 7$

a. $(0, \quad)$ **b.** $(-1, \quad)$ **c.** $(\quad, 0)$ **d.** $(\quad, 3)$

33. $x + 2y = 6$

a. $(0, \quad)$ **b.** $(2, \quad)$ **c.** $(\quad, 0)$ **d.** $(\quad, 4)$

34. $3x + y = 9$

a. $(0, \quad)$ **b.** $(4, \quad)$ **c.** $(\quad, 0)$ **d.** $(\quad, 3)$

35. $4x - y = 8$

a. $(0, \quad)$ **b.** $(1, \quad)$ **c.** $(\quad, 0)$ **d.** $(\quad, 4)$

36. $x - 2y = 2$

a. $(0, \quad)$ **b.** $(4, \quad)$ **c.** $(\quad, 0)$ **d.** $(\quad, 3)$

37. $2x + 3y = 6$

a. $(0, \quad)$ **b.** $(-1, \quad)$ **c.** $(\quad, 0)$ **d.** $(\quad, -2)$

38. $5x + 3y = 15$

a. $(0, \quad)$ **b.** $(2, \quad)$ **c.** $(\quad, 0)$ **d.** $(\quad, 4)$

39. $3x - 4y = 7$

a. $(0, \quad)$ **b.** $(1, \quad)$ **c.** $(\quad, 0)$ **d.** $\left(\quad, \frac{1}{2}\right)$

40. $2x + 5y = 6$

a. $(0, \quad)$ **b.** $\left(\frac{1}{2}, \quad\right)$ **c.** $(\quad, 0)$ **d.** $(\quad, 2)$

41. $(0, 0), (-1, -3), (-2, -6), (2, 6)$

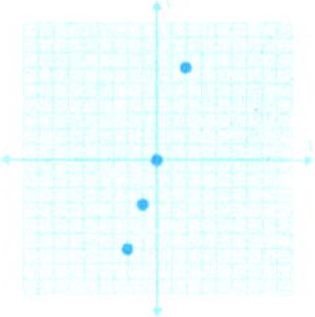

42. $(0, 0), (-2, 4), (3, -6), (1, -2)$

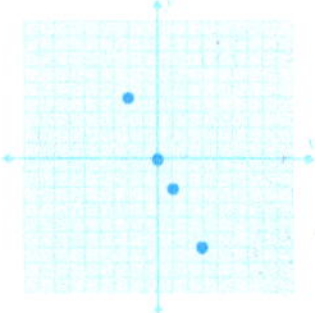

43. $(0, -3), (1, -1), (-2, -7), (3, 3)$

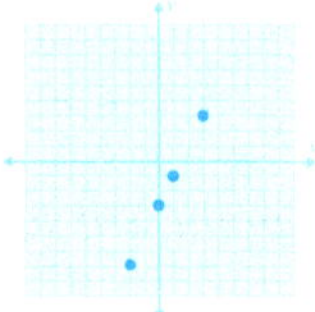

44. $(0, 5), (-3, -4), (-2, -1), (-1, 2)$

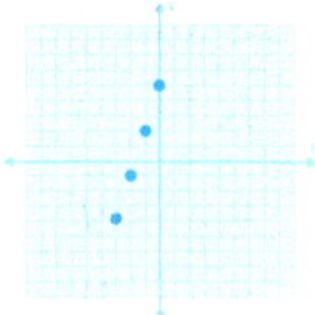

45. $(0, 9), (3, 0), (1, 6), (4, -3)$

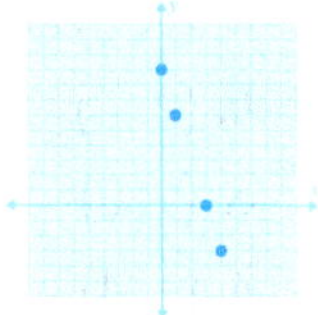

46. $(0, 6), (3, 0), (-2, 10), (4, -2)$

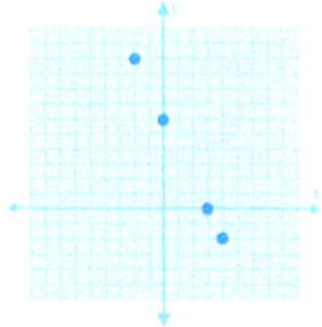

Complete the tables so that each ordered pair will satisfy the given equation. Graph the resulting sets of ordered pairs.

41. $y = 3x$

x	y
0	
	–3
–2	
	6

42. $y = -2x$

x	y
0	
	4
3	
	–2

43. $y = 2x - 3$

x	y
0	
	–1
–2	
	3

44. $y = 3x + 5$

x	y
0	
	–4
–2	
	2

45. $y = 9 - 3x$

x	y
0	
	0
1	
	–3

46. $y = 6 - 2x$

x	y
0	
	0
–2	
	–2

47. $y = \frac{3}{4}x + 2$

x	y
0	
	5
–4	
	$\frac{5}{4}$

48. $y = \frac{3}{2}x - 1$

x	y
0	
	2
–2	
	$-\frac{5}{2}$

49. $3x - 5y = 9$

x	y
0	
	0
–2	
	–1

50. $4x + 3y = 6$

x	y
0	
	0
3	
	–1

51. $5x - 2y = 10$

x	y
0	
	0
–1	
	5

52. $3x - 2y = 12$

x	y
	0
0	
	–3
6	

53. $2x + 3.2y = 6.4$

x	y
0	
3.2	
	0.8
	–0.2

54. $3x + y = -2.4$

x	y
	0
0	
	0.6
1.6	

47. $(0, 2), (4, 5), (-4, -1), \left(-1, \frac{5}{4}\right)$

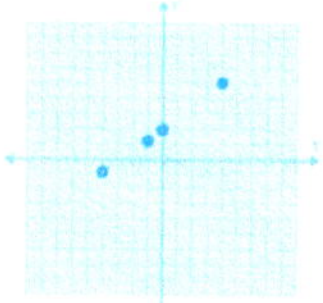

48. $(0, -1), (2, 2), (-2, -4), \left(-1, -\frac{5}{2}\right)$

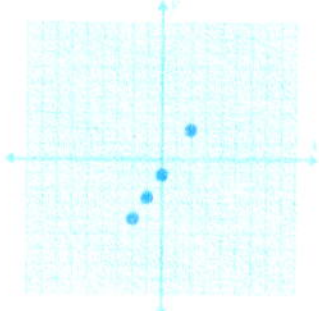

49. $\left(0, -\frac{9}{5}\right), (3, 0), (-2, -3), \left(\frac{4}{3}, -1\right)$

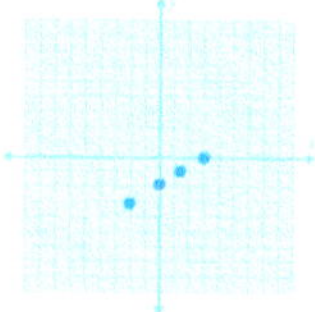

50. $(0, 2), \left(\frac{3}{2}, 0\right), (3, -2), \left(\frac{9}{4}, -1\right)$

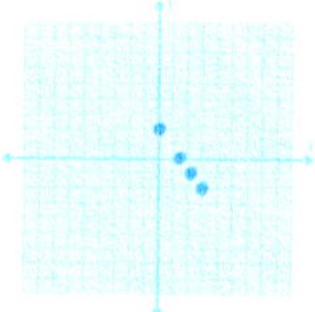

51. $(0, -5), (2, 0), \left(-1, -\frac{15}{2}\right), (4, 5)$

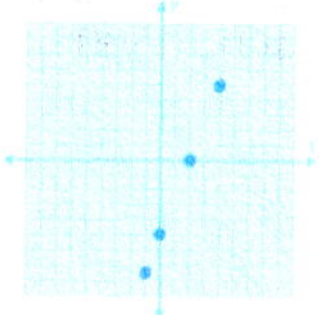

The graph of a line is shown. Each line contains an infinite number of points. List any three points on each line. (There is more than one correct answer.)

55.

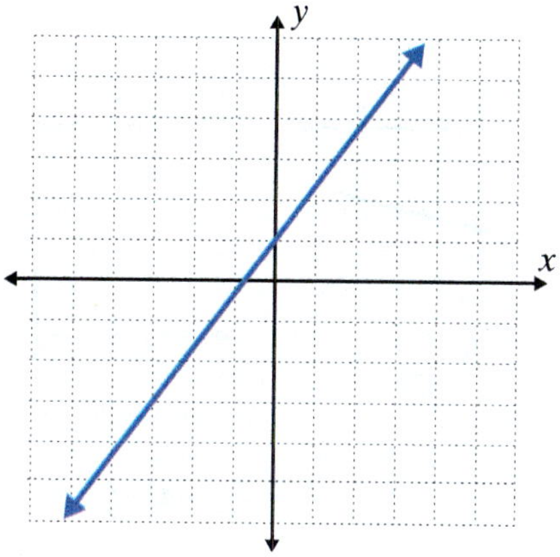

56.

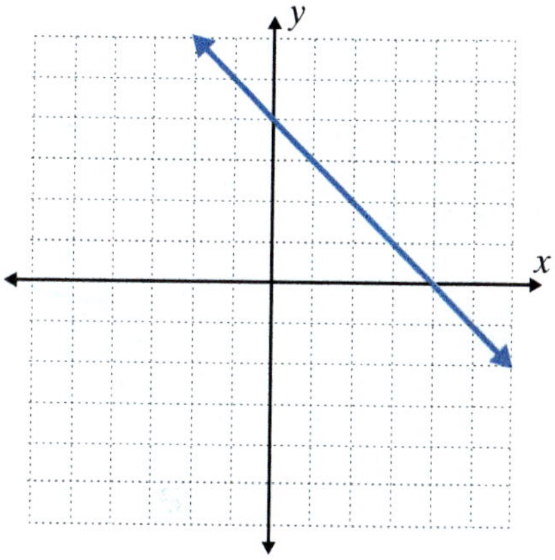

57.

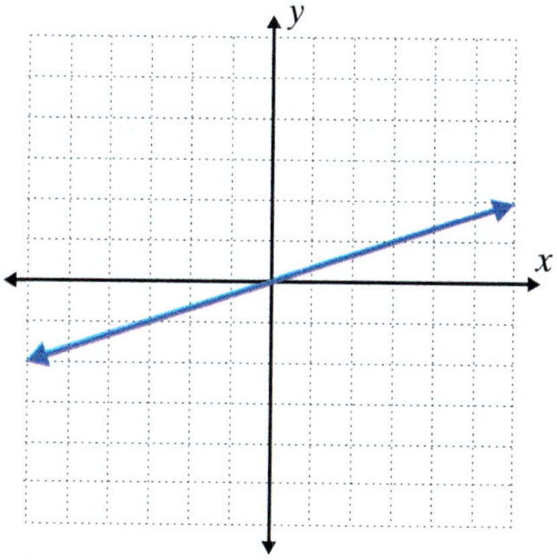

58.

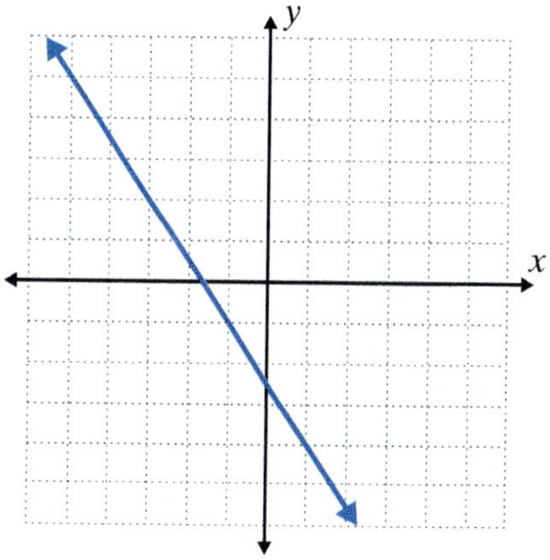

59.

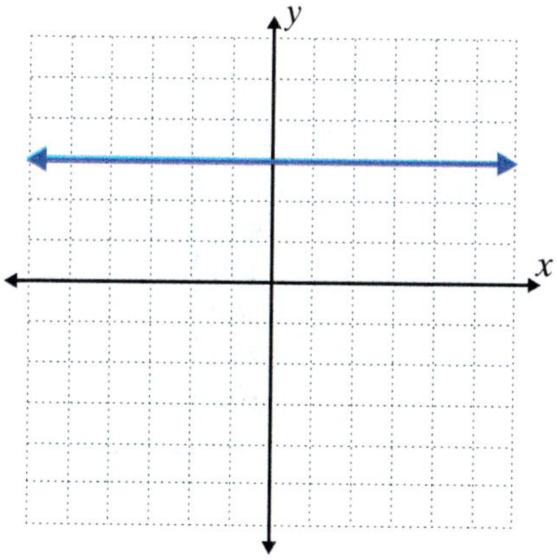

60.

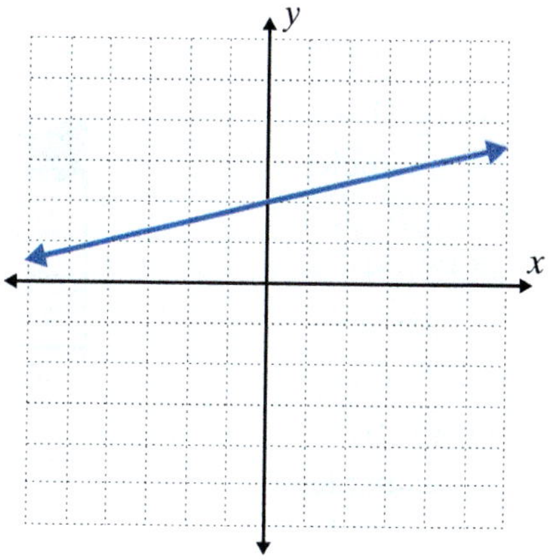

52. $(4, 0), (0, -6), (2, -3), (6, 3)$

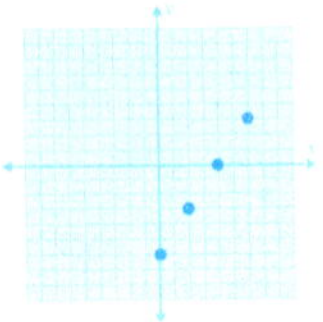

53. $(0, 2), (3.2, 0), (1.92, 0.8), (3.52, -0.2)$

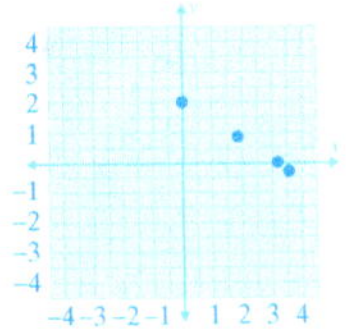

54. $(-0.8, 0), (0, -2.4), (-1, 0.6), (1.6, -7.2)$

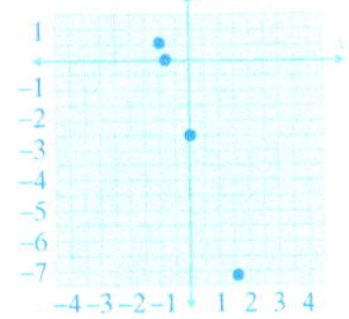

55. For example, $(-3, -3), (0, 1)$, and $(3, 5)$

56. For example, $(0, 4), (2, 2)$, and $(4, 0)$

57. For example, $(-3, -1), (0, 0)$, and $(3, 1)$

58. For example, $(-5, 5), (-3, 2)$, and $(1, -4)$

59. For example, $(-2, 3), (0, 3)$, and $(4, 3)$

60. For example, $(-4, 1), (0, 2)$, and $(4, 3)$

61.

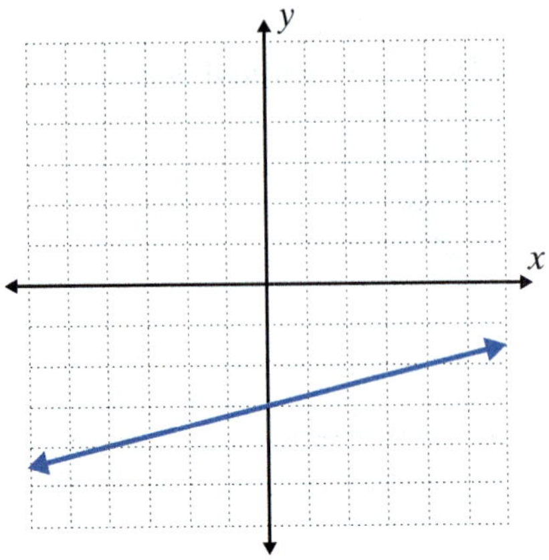

62.

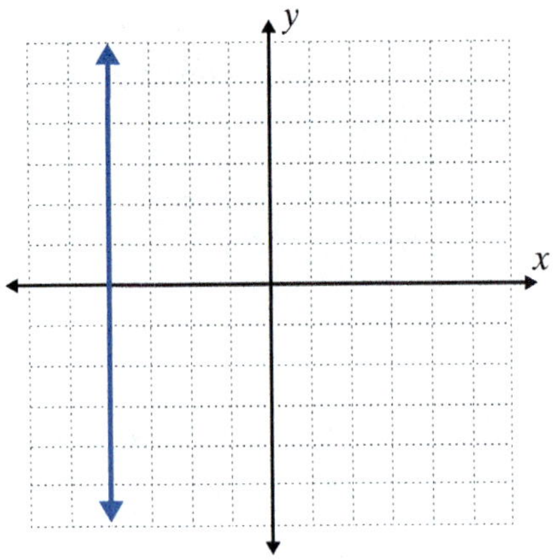

63.

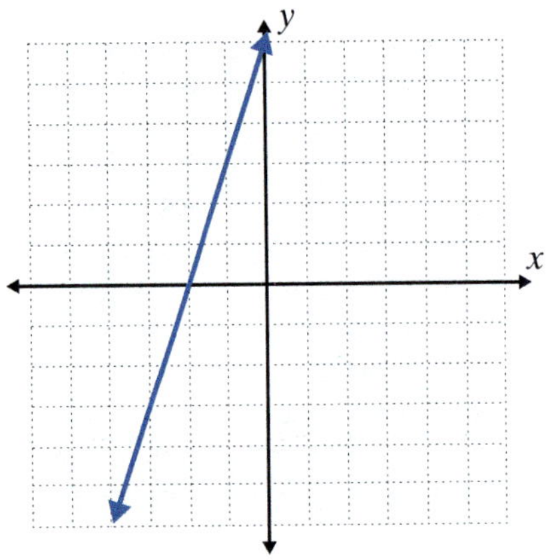

64.

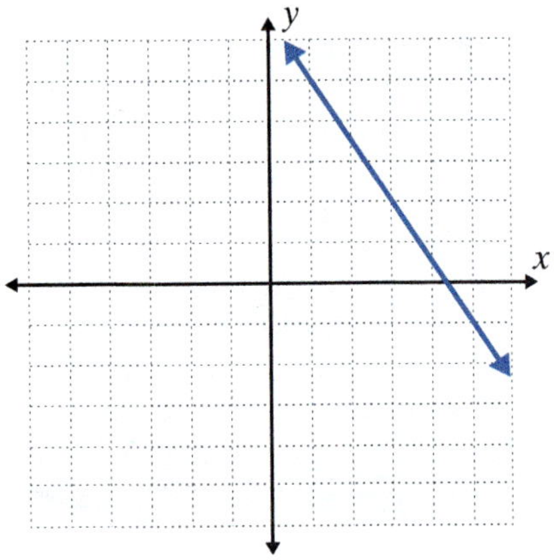

65. Exchange rate: At one point in 2010, the current exchange rate from U.S. dollars to Euros was $E = 0.7802D$ where E is Euros and D is dollars.

a. Make a table of ordered pairs for the values of D and E if D has the values \$100, \$200, \$300, \$400, and \$500.

b. Graph the points corresponding to the ordered pairs.

D	E
100	
200	
300	
400	
500	

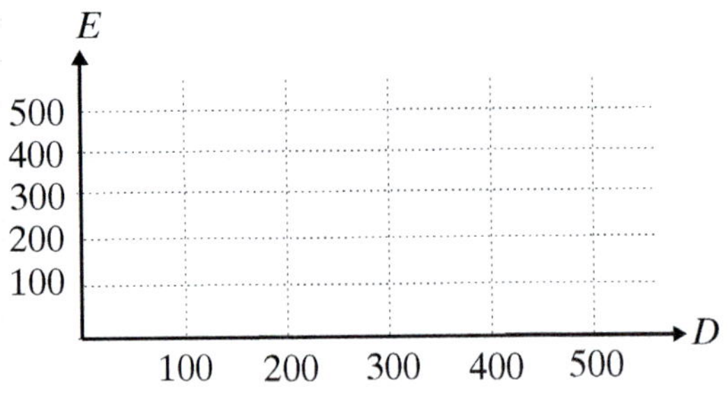

66. Temperature: Given the equation $F = \frac{9}{5}C + 32$ where C is temperature in degrees Celsius and F is the corresponding temperature in degrees Fahrenheit:

a. Make a table of ordered pairs for the values of C and F if C has the values $-20°$, $-10°$, $-5°$, $0°$, $5°$, $10°$, and $15°$.

b. Graph the points corresponding to the ordered pairs.

C	F
–20	
–10	
–5	
0	
5	
10	
15	

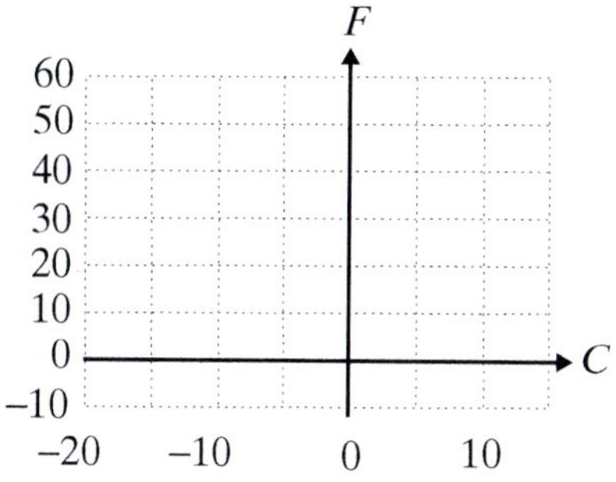

67. Falling objects: Given the equation $d = 16t^2$, where d is the distance an object falls in feet and t is the time in seconds that the object falls:

a. Make a table of ordered pairs for the values of t and d with the values of 1, 2, 3.5, 4, 4.5, and 5 for t seconds.

b. Graph the points corresponding to the ordered pairs.

c. These points do not lie on a straight line. What feature of the equation might indicate to you that the graph is not a straight line?

t	d
1	
2	
3.5	
4	
4.5	
5	

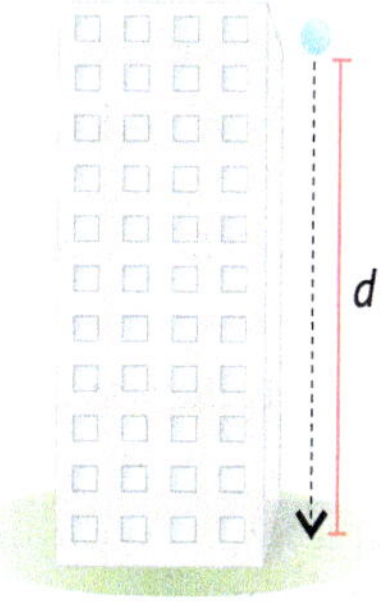

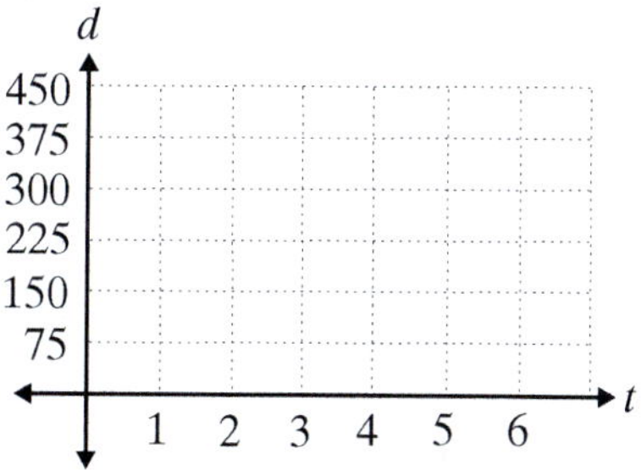

68. Volume: Given the equation $V = 9h$, where V is the volume (in cubic centimeters) of a box with a variable height h in centimeters and a fixed base of area 9 cm^2.

a. Make a table of ordered pairs for the values of h and V with h as the values 2 cm, 3 cm, 5 cm, 8 cm, 9 cm, and 10 cm.

b. Graph the points corresponding to the ordered pairs.

h	V
2	
3	
5	
8	
9	
10	

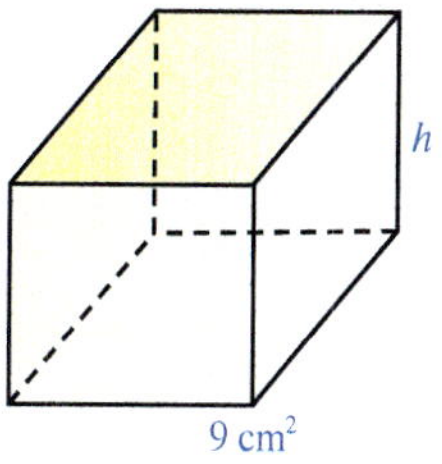

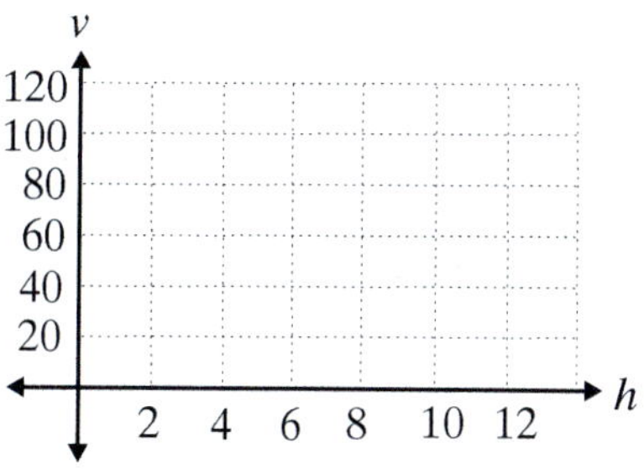

61. For example, $(-4,-4), (0,-3)$, and $(4,-2)$

62. For example, $(-4,-3), (-4,0)$, and $(-4,2)$

63. For example, $(-3,-3), (-2,0)$, and $(-1,3)$

64. For example, $(1,5), (3,2)$, and $(5,-1)$

65. a. $(100, 78.02), (200, 156.04), (300, 234.06), (400, 312.08), (500, 390.1)$

b.

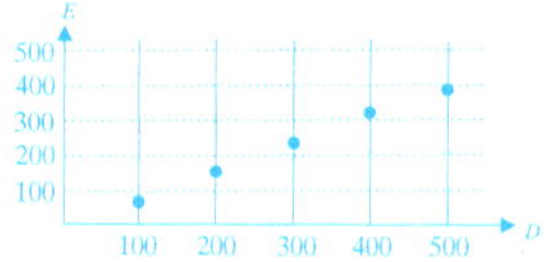

66. a. $(-20,-4), (-10,14), (-5,23), (0,32), (5,41), (10,50), (15,59)$

b.

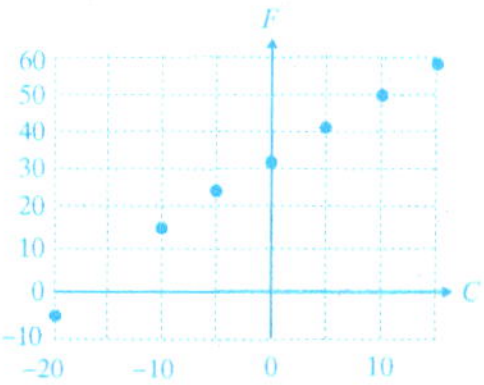

67. a. $(1,16), (2,64), (3.5,196), (4,256), (4.5,324), (5,400)$

b.

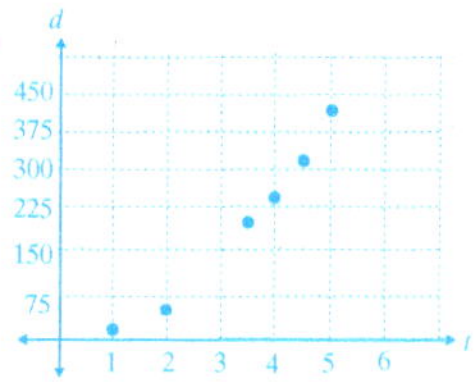

c. The time t is squared in the equation.

68. a. $(2, 18), (3, 27), (5, 45), (8, 72), (9, 81), (10, 90)$

b.

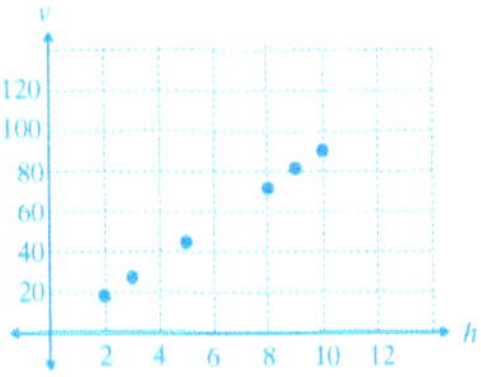

69. a.

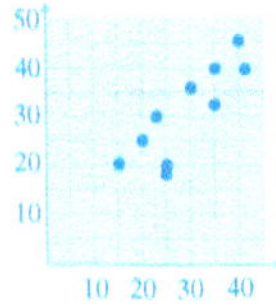

b. Yes, the more sit-ups a person can do it appears the more push-ups he/she can do.

c. 24 sit-ups, 33 sit-ups, 36 sit-ups, 45 sit-ups; Answers will vary.

70. Answers will vary.

71. Answers will vary. Not all scatterplots can be used to predict information related to the two variables graphed because not all variables are related.

Writing and Thinking About Mathematics

In statistics, data is sometimes given in the form of ordered pairs where each ordered pair represents two pieces of information about one person. For example, ordered pairs might represent the height and weight of a person or the person's number of years of education and that person's annual income. The ordered pairs are plotted on a graph and the graph is called a **scatter diagram** (or **scatter plot**). Such scatter diagrams are used to see if there is any pattern to the data and, if there is, then the diagram is used to predict the value for one of the variables if the value of the other is known. For example, if you know that a person's height is 5 ft 6 in., then his or her weight might be predicted from information indicated in a scatter diagram that has several points of known information about height and weight.

69. a. The following table of values indicates the number of push-ups and the number of sit-ups that ten students did in a physical education class. Plot these points in a scatter diagram.

Person	#1	#2	#3	#4	#5	#6	#7	#8	#9	#10
x (push-ups)	20	15	25	23	35	30	42	40	25	35
y (sit-ups)	25	20	20	30	32	36	40	45	18	40

b. Does there seem to be a pattern in the relationship between push-ups and sit-ups? What is this pattern?

c. Using the scatter diagram in part **a.**, predict the number of sit-ups that a student might be able to do if he or she has just done each of the following numbers of push-ups: 22, 32, 35, and 45. (**Note:** In each case, there is no one correct answer. The answers are only estimates based on the diagram.)

70. Ask ten friends or fellow students what their height and shoe size is. (You may want to ask all boys or all girls since the scale for boys and girls' shoe sizes is different.) Organize the data in table form and then plot the corresponding scatter diagram. Knowing your own height, does the pattern indicated in the scatter diagram seem to predict your shoe size?

71. Ask ten friends or fellow students what their height and age is. Organize the data in table form and then plot the corresponding scatter diagram. Knowing your own height, does the pattern indicated in the scatter diagram seem to predict your age? Do you think that all scatter diagrams can be used to predict information related to the two variables graphed? Explain.

HAWKES LEARNING SYSTEMS: INTRODUCTORY & INTERMEDIATE ALGEBRA SOFTWARE

- 4.1 Introduction to the Cartesian Coordinate System

4.2 Graphing Linear Equations in Two Variables: $Ax + By = C$

- *Plot points that satisfy a linear equation and draw the corresponding line.*
- *Recognize the standard form of a linear equation in two variables: $Ax + By = C$*
- *Find the x-intercept and y-intercept of a line and graph the corresponding line.*

The Standard Form: $Ax + By = C$

In Section 4.1, we discussed ordered pairs of real numbers and graphed a few points (ordered pairs) that satisfied particular equations. Now suppose we want to graph all the points that satisfy an equation such as

$$3 - 3x = y.$$

The **solution set** for equations of this type (in the two variables x and y) consists of an infinite set of ordered pairs in the form (x, y) that satisfy the equation.

To find some of the solutions of the equation $3 - 3x = y$, we form a table (as we did in Section 4.1) by:

1. choosing arbitrary values for x and
2. finding the corresponding values for y by substituting into the equation.

In Figure 1, we have found five ordered pairs that satisfy the equation and graphed the corresponding points.

Choices	Substitutions	Results
x	$3 - 3x = y$	(x, y)
-1	$3-3(-1)=6$	$(-1, 6)$
0	$3-3(0)=3$	$(0, 3)$
$\frac{2}{3}$	$3-3\left(\frac{2}{3}\right)=1$	$\left(\frac{2}{3}, 1\right)$
2	$3-3(2)=-3$	$(2, -3)$
3	$3-3(3)=-6$	$(3, -6)$

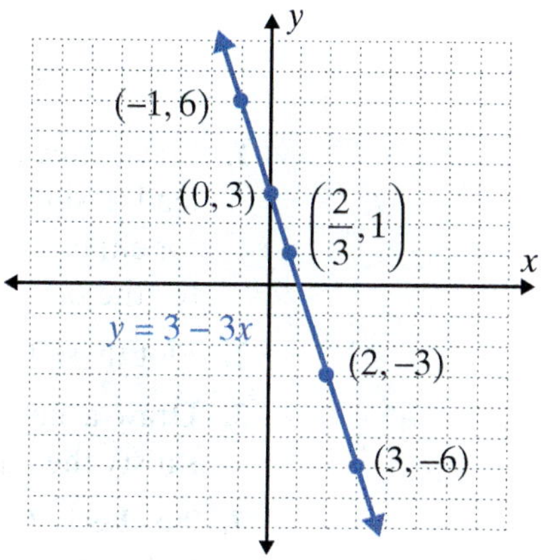

Figure 1

The five points in Figure 1 appear to lie on a line. They, in fact, do lie on a line, and any ordered pair that satisfies the equation $y = 3 - 3x$ will also lie on that same line.

Just as we use the terms **ordered pair** and **point** (the graph of an ordered pair) interchangeably, we use the terms **equation** and **graph of an equation** interchangeably. The equations

$$2x + 3y = 4, \quad y = -5, \quad x = 1.4, \quad \text{and} \quad y = 3x + 2$$

are called **linear equations**, and their graphs are lines on the Cartesian plane.

Standard Form of a Linear Equation

Any equation of the form

$$\boldsymbol{Ax + By = C,}$$

where A, B, and C are real numbers and A and B are not both equal to 0, is called the **standard form** of a **linear equation**.

NOTES

Note that in the standard form $Ax + By = C$, A and B may be positive, negative, or 0, but A and B cannot **both** equal 0.

Every line corresponds to some linear equation, and the graph of every linear equation is a line. We know from geometry that **two points determine a line**. This means that the graph of a linear equation can be found by locating any two points that satisfy the equation.

To Graph a Linear Equation in Two Variables

1. Locate any two points that satisfy the equation. (Choose values for x and y that lead to simple solutions. Remember that there is an infinite number of choices for either x or y. But, once a value for x or y is chosen, the corresponding value for the other variable is found by substituting into the equation.)
2. Plot these two points on a Cartesian coordinate system.
3. Draw a line through these two points. (**Note:** Every point on that line will satisfy the equation.)
4. **To check:** Locate a third point that satisfies the equation and check to see that it does indeed lie on the line.

Example 1: Graphing a Linear Equation in Two Variables

Teaching Notes:
Emphasize that choosing 0 for x and then 0 for y usually leads to easy calculations for the other variable.

Graph each of the following linear equations.

a. $2x + 3y = 6$

Solution: Make a table with headings x and y and, whenever possible, **choose values for x or y that lead to simple solutions for the other variable.** (Values chosen for x and y are shown in red.)

x	$2x + 3y = 6$	y
0	$2(0) + 3y = 6$	2
−3	$2(-3) + 3y = 6$	4
3	$2x + 3(0) = 6$	0
$\frac{5}{2}$	$2x + 3\left(\frac{1}{3}\right) = 6$	$\frac{1}{3}$

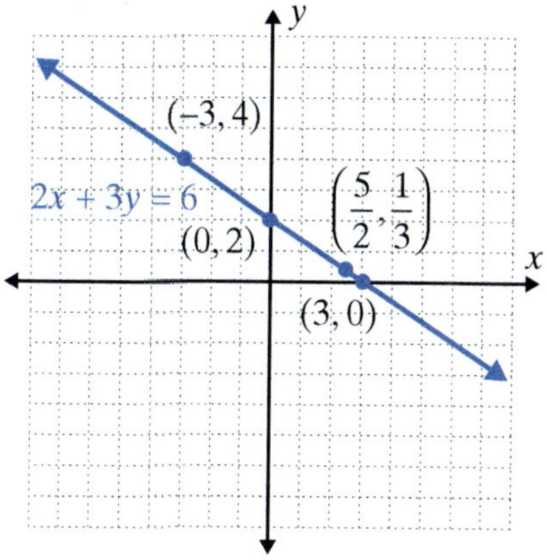

b. $x - 2y = 1$

Solution: Solve the equation for x $(x = 2y + 1)$ and substitute 0, 1, and 2 for y.

Results	Substitutions	Choices
x	$x = 2y + 1$	y
1	$x = 2(0) + 1$	0
3	$x = 2(1) + 1$	1
5	$x = 2(2) + 1$	2

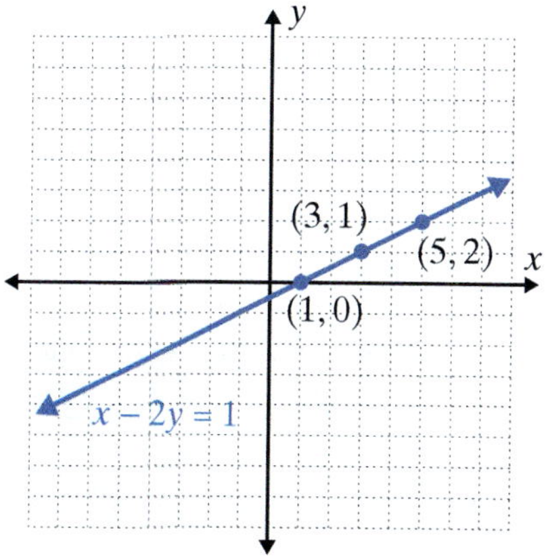

c. $y = 2x$

Solution: Substitute −1, 0, and 1 for x.

Choices	Substitutions	Results
x	$y = 2x$	y
−1	$y = 2(-1)$	−2
0	$y = 2(0)$	0
1	$y = 2(1)$	2

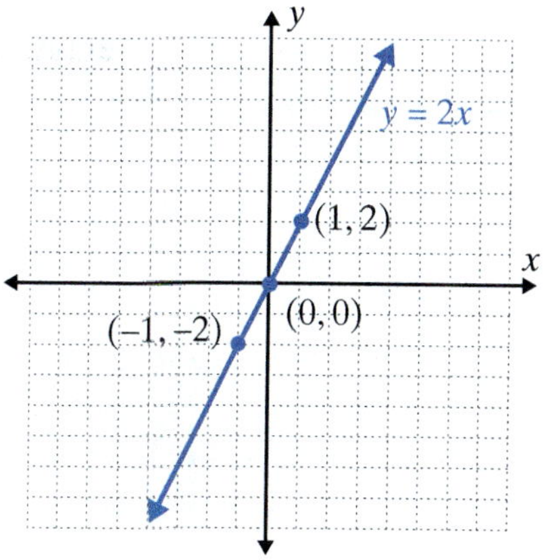

Locating the *y*-intercept and *x*-intercept

While the choice of the values for x or y can be arbitrary, letting $x = 0$ will locate the point on the graph where the line crosses (or intercepts) the y-axis. This point is called the **y-intercept** and is of the form $(0, y)$. The **x-intercept** is the point found by letting $y = 0$. This is the point where the line crosses (or intercepts) the x-axis and is of the form $(x, 0)$. These two points are generally easy to locate and are frequently used as the two points for drawing the graph of a linear equation. If the line passes through the point $(0, 0)$, then the y-intercept and the x-intercept are the same point, namely the origin. In this case you will need to locate some other point to draw the graph.

Intercepts

1. To find the **y-intercept** (where the line crosses the y-axis), substitute $x = 0$ and solve for y.
2. To find the **x-intercept** (where the line crosses the x-axis), substitute $y = 0$ and solve for x.

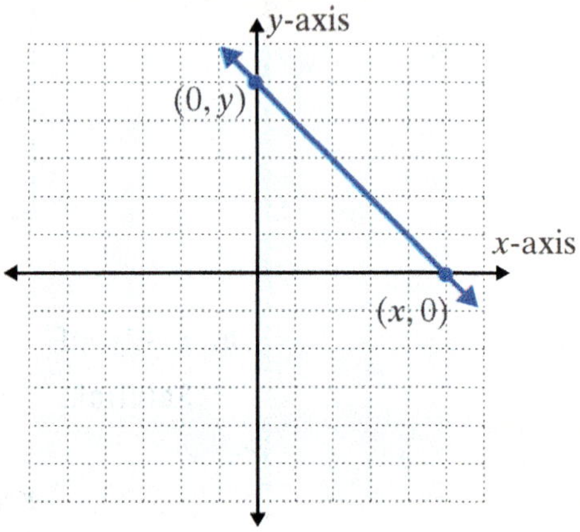

Example 2: *x*- and *y*-Intercepts

Teaching Notes:
You may want to remind students to find a third point as a check even when graphing using the intercepts.

Graph the following linear equations by locating the y-intercept and the x-intercept.

a. $x + 3y = 9$

Solution: $x = 0 \longrightarrow (0) + 3y = 9$

$$3y = 9$$

$$y = 3$$

$(0, 3)$ is the y-intercept.

$y = 0 \longrightarrow x + 3(0) = 9$

$$x = 9$$

$(9, 0)$ is the x-intercept.

Plot the two intercepts and draw the line that contains them.

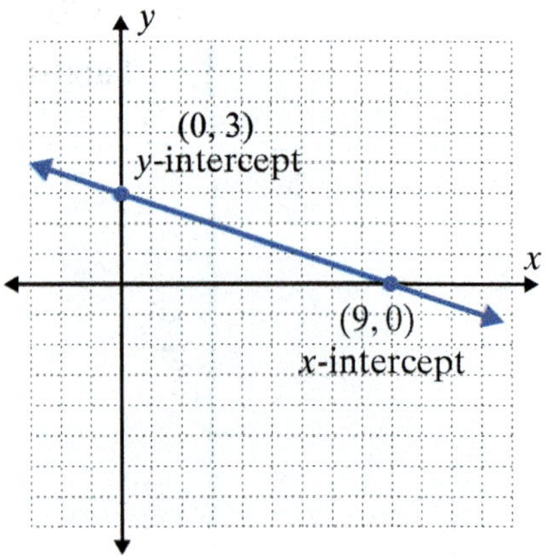

1. a 2. e
3. d 4. c
5. f 6. b

7.
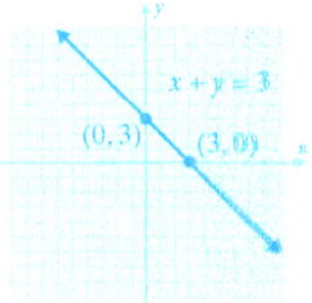

8.
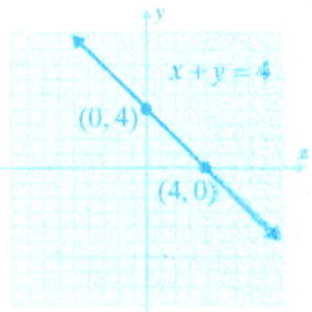

9.
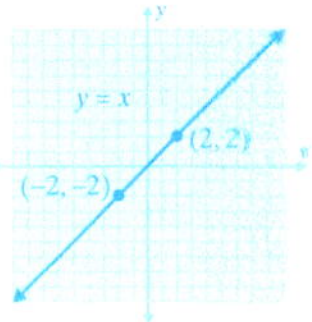

10.
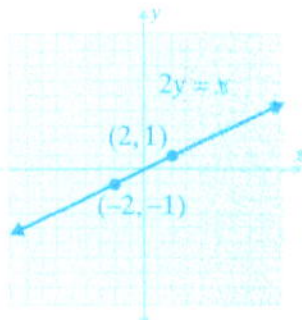

b. $3x - 2y = 12$

Solution:

$$x = 0 \longrightarrow 3(0) - 2y = 12$$
$$-2y = 12$$
$$y = -6$$

$(0, -6)$ is the y-intercept.

$$y = 0 \longrightarrow 3x - 2(0) = 12$$
$$3x = 12$$
$$x = 4$$

$(4, 0)$ is the x-intercept.

Plot the two intercepts and draw the line that contains them.

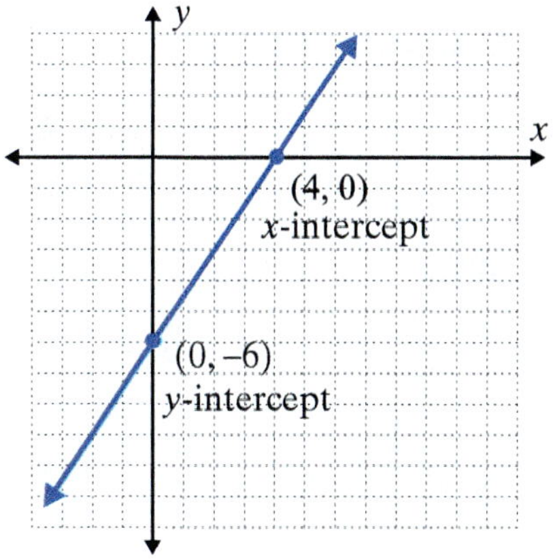

NOTES

In general, the intercepts are easy to find because substituting 0 for x or y leads to an easy solution for the other variable. However, when the intercepts result in a point with fractional (or decimal) coordinates and estimation is involved, then a third point that satisfies the equation should be found to verify that the line is graphed correctly.

Practice Problems

1. Find the missing coordinate of each ordered pair so that it belongs to the solution set of the equation $2x + y = 4$:
$$(0,\ \ ),\ (\ \ ,0),\ (\ \ ,8),\ (-1,\ \ ).$$
2. Does the ordered pair $\left(1, \frac{3}{2}\right)$ satisfy the equation $3x + 2y = 6$?
3. Find the x-intercept and y-intercept of the equation $-3x + y = 9$.
4. Graph the linear equation $x - 2y = 3$.

11.
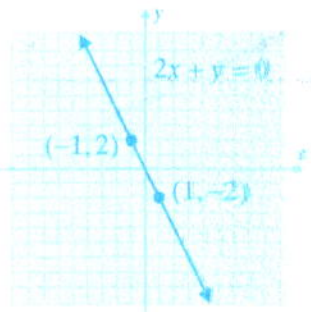

12.
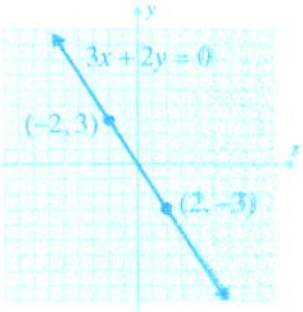

13.
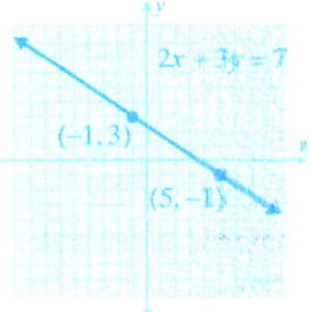

14.
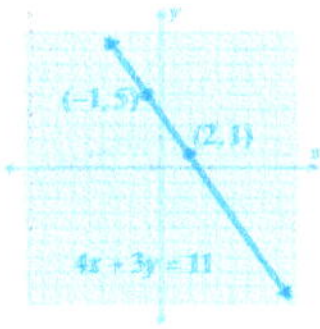

15.
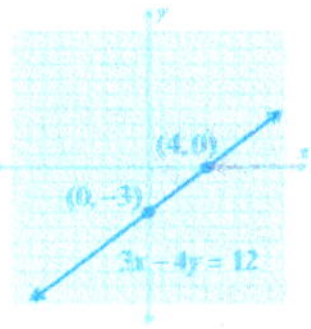

16.
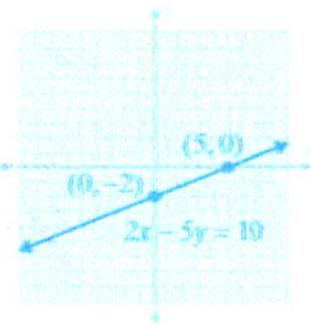

17.
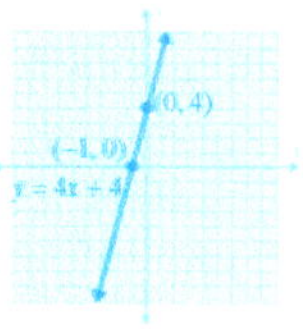

18.
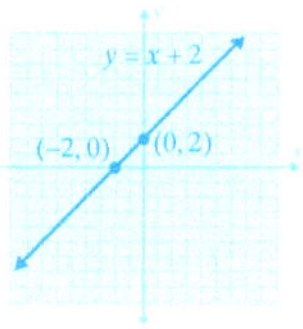

Answers to Practice Problems: **1.** $(0, 4), (2, 0), (-2, 8), (-1, 6)$ **2.** Yes

3. x-intercept $= (-3, 0)$, y-intercept $= (0, 9)$ **4.**

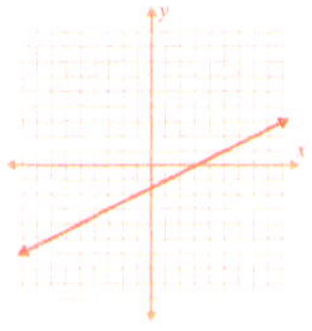

4.2 Exercises

19.
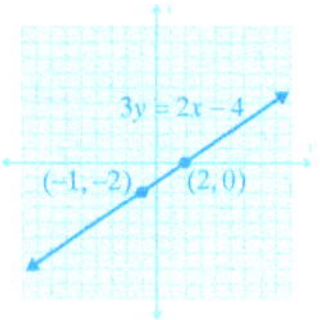

20.
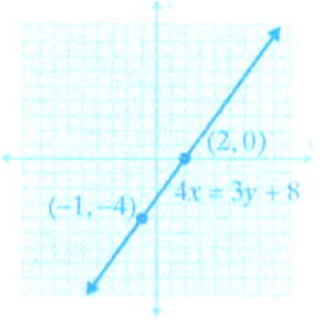

21.
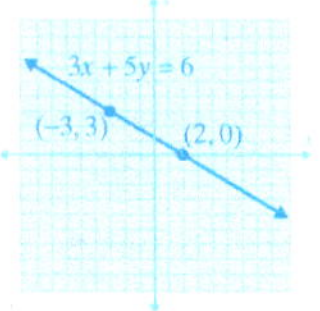

22.
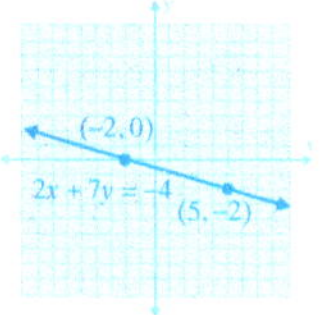

23.
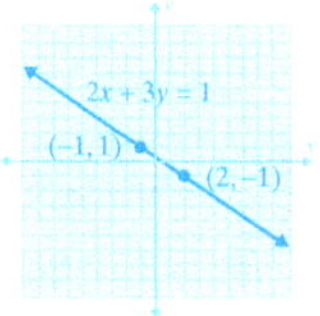

24.
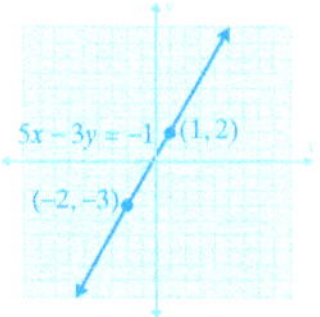

25.
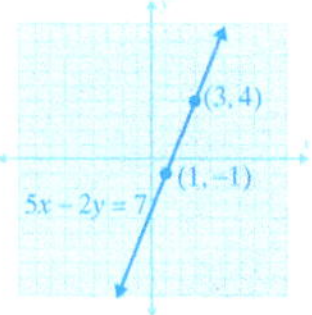

26.
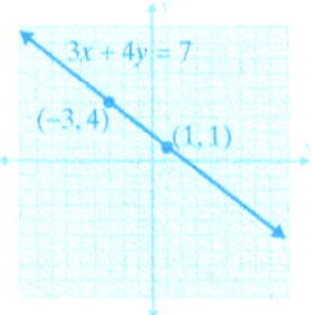

27.
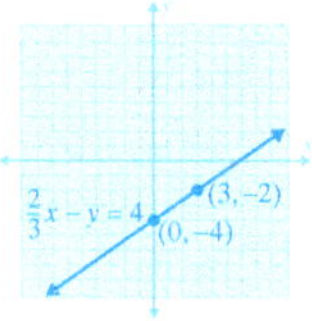

Use your knowledge of x-intercepts and y-intercepts to match each of the following equations with its graph.

1. $4x + 3y = 12$ **2.** $4x - 3y = 12$ **3.** $x + 2y = 8$

4. $-x + 2y = 8$ **5.** $x + 4y = 0$ **6.** $5x - y = 10$

a.
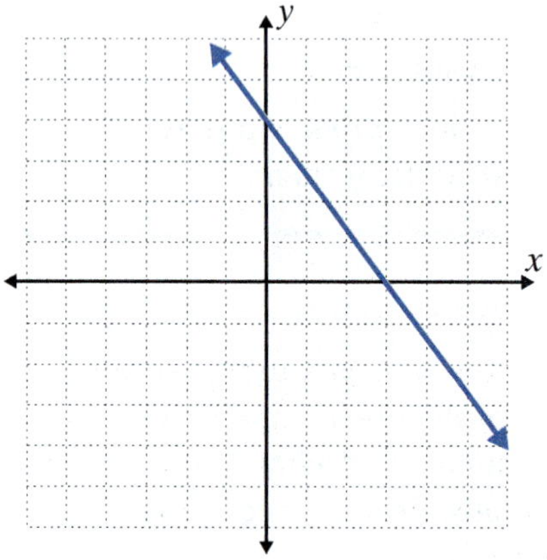

b.
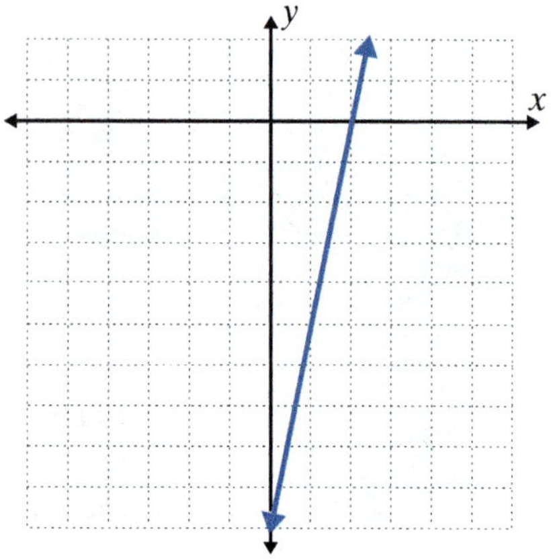

c.
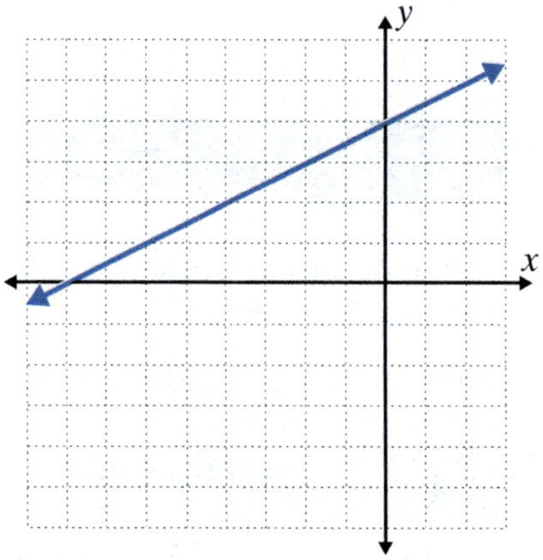

d.
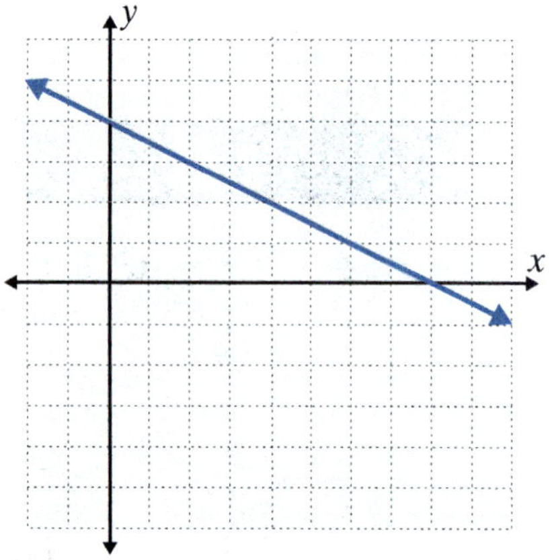

e.
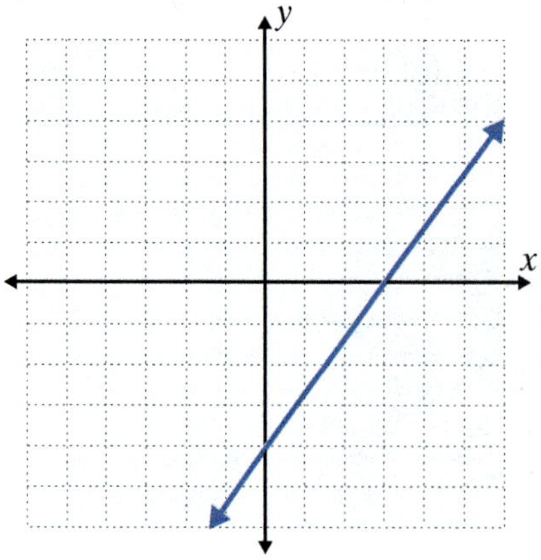

f.
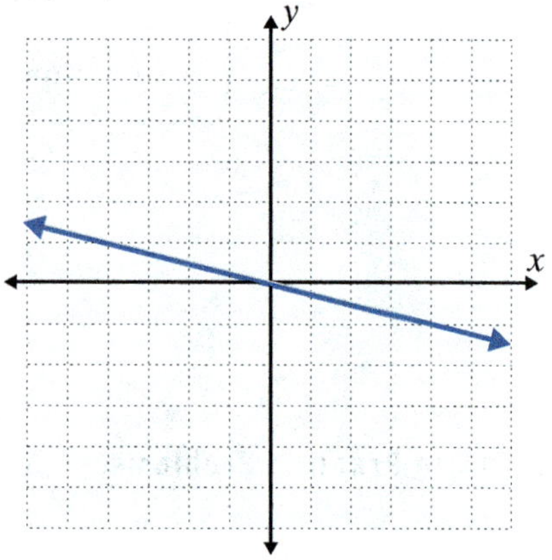

Locate at least two ordered pairs of real numbers that satisfy each of the linear equations and graph the corresponding line in the Cartesian coordinate system.

7. $x + y = 3$ **8.** $x + y = 4$ **9.** $y = x$

10. $2y = x$ **11.** $2x + y = 0$ **12.** $3x + 2y = 0$

13. $2x + 3y = 7$ **14.** $4x + 3y = 11$ **15.** $3x - 4y = 12$

16. $2x - 5y = 10$ **17.** $y = 4x + 4$ **18.** $y = x + 2$

19. $3y = 2x - 4$ **20.** $4x = 3y + 8$ **21.** $3x + 5y = 6$

22. $2x + 7y = -4$ **23.** $2x + 3y = 1$ **24.** $5x - 3y = -1$

25. $5x - 2y = 7$ **26.** $3x + 4y = 7$ **27.** $\frac{2}{3}x - y = 4$

28. $x + \frac{3}{4}y = 6$ **29.** $2x + \frac{1}{2}y = 3$ **30.** $\frac{2}{5}x - 3y = 5$

31. $5x = y + 2$ **32.** $4x = 3y - 5$

Graph the following linear equations by locating the x-intercept and the y-intercept.

33. $x + y = 6$ **34.** $x + y = 4$ **35.** $x - 2y = 8$

36. $x - 3y = 6$ **37.** $4x + y = 8$ **38.** $x + 3y = 9$

39. $x - 4y = -6$ **40.** $x - 6y = 3$ **41.** $y = 4x - 10$

42. $y = 2x - 9$ **43.** $3x - 2y = 6$ **44.** $5x + 2y = 10$

45. $2x + 3y = 12$ **46.** $3x + 7y = -21$ **47.** $3x - 7y = -21$

48. $3x + 2y = 15$ **49.** $5x + 3y = 7$ **50.** $2x + 3y = 5$

51. $y = \frac{1}{2}x - 4$ **52.** $y = -\frac{1}{3}x + 3$ **53.** $\frac{2}{3}x - 3y = 4$

54. $\frac{1}{2}x + 2y = 3$ **55.** $\frac{1}{2}x - \frac{3}{4}y = 6$ **56.** $\frac{2}{3}x + \frac{4}{3}y = 8$

28.
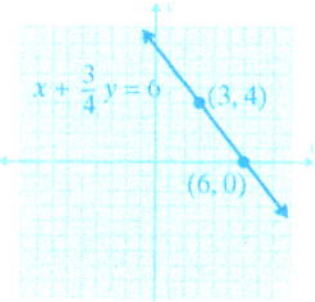

29.
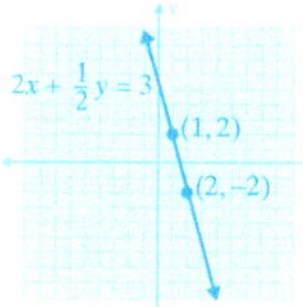

30.
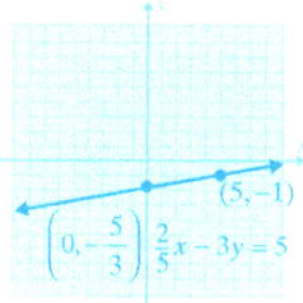

31.
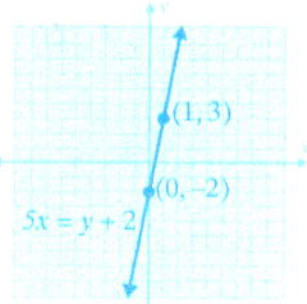

32.
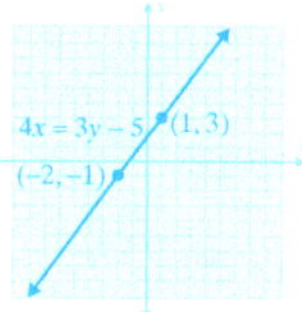

33.
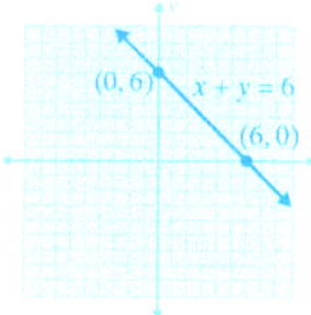

34.
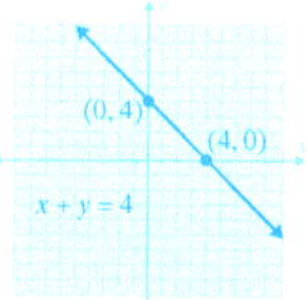

35.
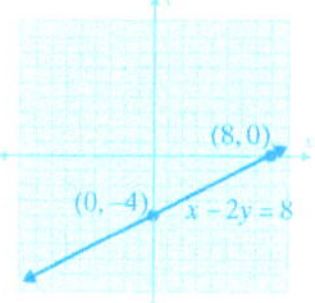

36.
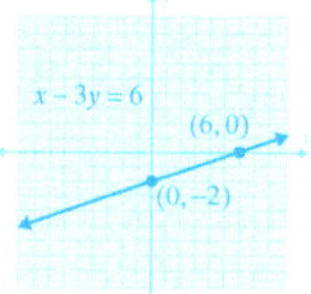

37.
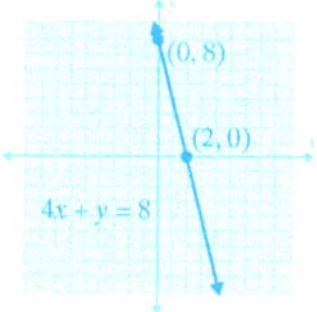

38.
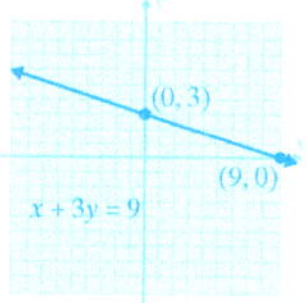

39.
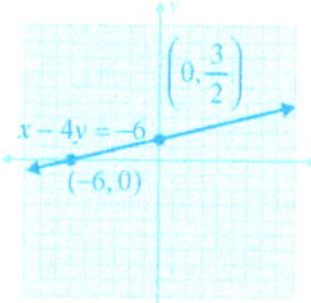

40.
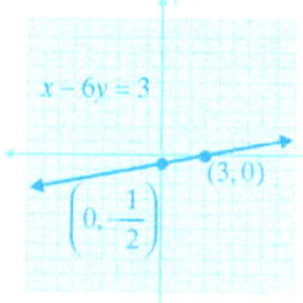

41.
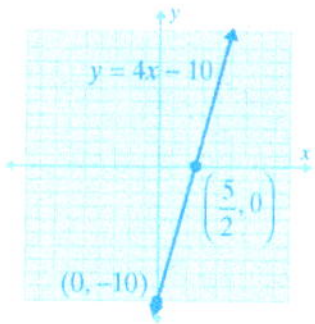

42.
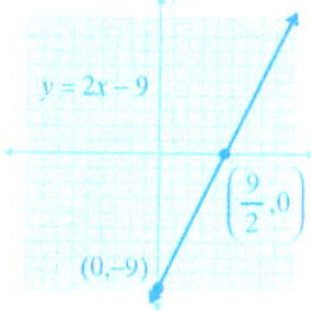

Writing and Thinking About Mathematics

57. Explain, in your own words, why it is sufficient to find the x-intercept and y-intercept to graph a line (assuming that they are not the same point).

58. Explain, in your own words, how you can determine if an ordered pair is a solution to an equation.

HAWKES LEARNING SYSTEMS: INTRODUCTORY & INTERMEDIATE ALGEBRA SOFTWARE

- 4.2 Graphing Linear Equations by Plotting Points

43. 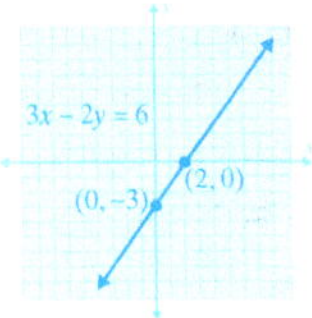

44. 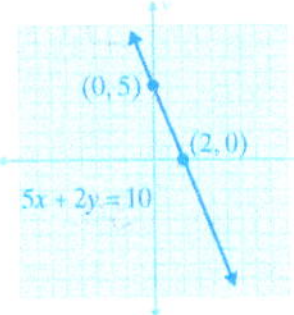

45. 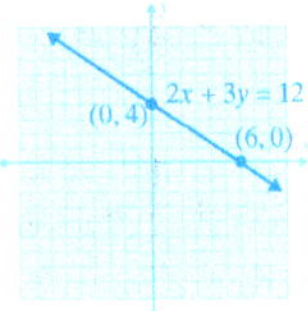

46. 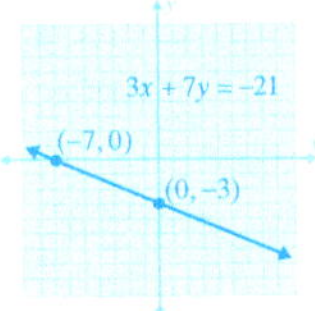

47. 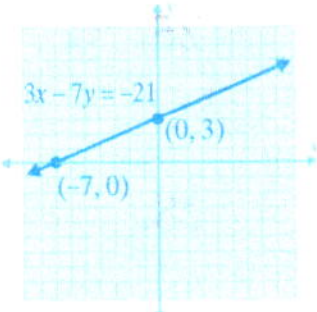

48. 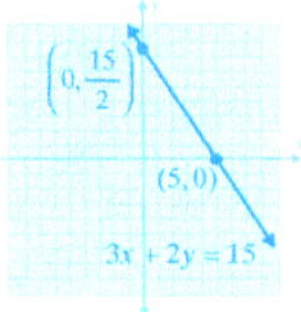

49. 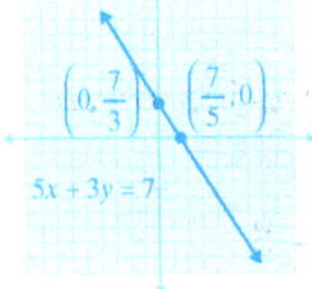

50. 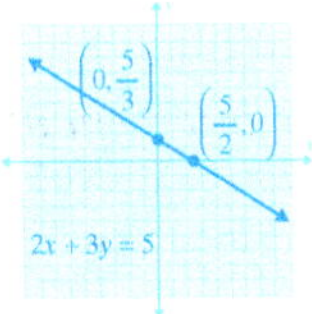

51. 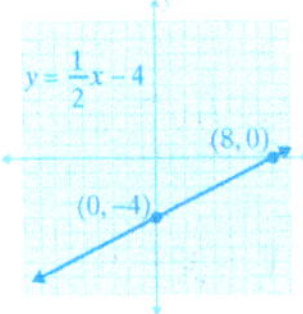

52. 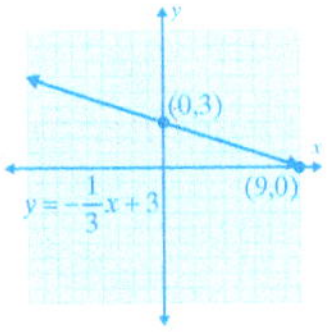

53. 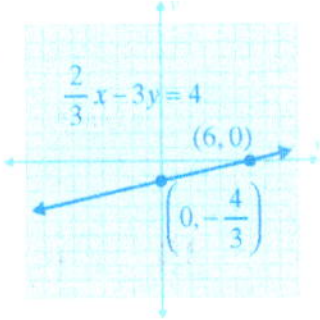

54. 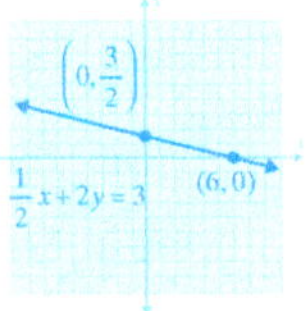

55. 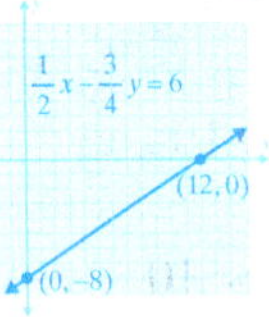

56. 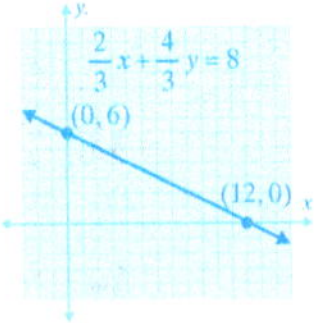

57. Two (unique) points determine a line.

58. Substitute the x and y values into the equation. Then evaluate both sides to see if the equation is true.

4.3 The Slope-Intercept Form: $y = mx + b$

- *Interpret the slope of a line as a rate of change.*
- *Find the slope of a line given two points.*
- *Find the slopes of and graph horizontal and vertical lines.*
- *Find the slopes and y-intercepts of lines and then graph the lines.*
- *Write the equations of lines given the slopes and y-intercepts.*

The Meaning of Slope

If you ride a bicycle up a mountain road, you certainly know when the **slope** (a measure of steepness called the **grade** for roads) increases because you have to pedal harder. The contractor who built the road was aware of the **slope** because trucks traveling the road must be able to control their downhill speed and be able to stop in a safe manner. A carpenter given a set of house plans calling for a roof with a **pitch** of 7 : 12 knows that for every 7 feet of rise (vertical distance) there are 12 feet of run (horizontal distance). That is, the ratio of rise to run is $\frac{rise}{run} = \frac{7}{12}$.

Figure 1

Note that this ratio can be in units other than feet, such as inches or meters. (See Figure 2.)

$$\frac{rise}{run} = \frac{7 \text{ inches}}{12 \text{ inches}} = \frac{3.5 \text{ feet}}{6 \text{ feet}} = \frac{14 \text{ feet}}{24 \text{ feet}}$$

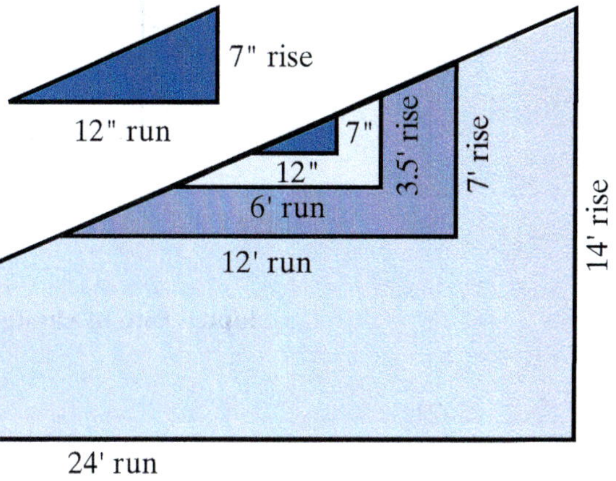

Figure 2

For a line, the **ratio of rise to run** is called the **slope of the line**. The graph of the linear equation $y=\frac{1}{3}x+2$ is shown in Figure 3. What do you think is the slope of the line? Do you think that the slope is positive or negative? Do you think the slope might be $\frac{1}{3}$? $\frac{3}{1}$? 2?

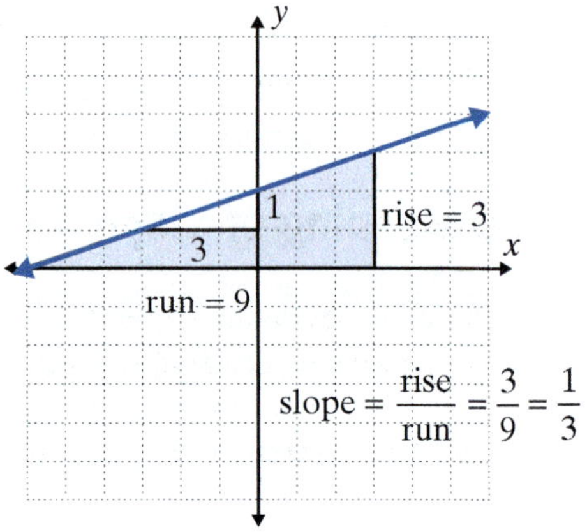

Figure 3

The concept of slope also relates to situations that involve **rate of change**. For example, the graphs in Figure 4 illustrate slope as miles per hour that a car travels and as pages per minute that a printer prints.

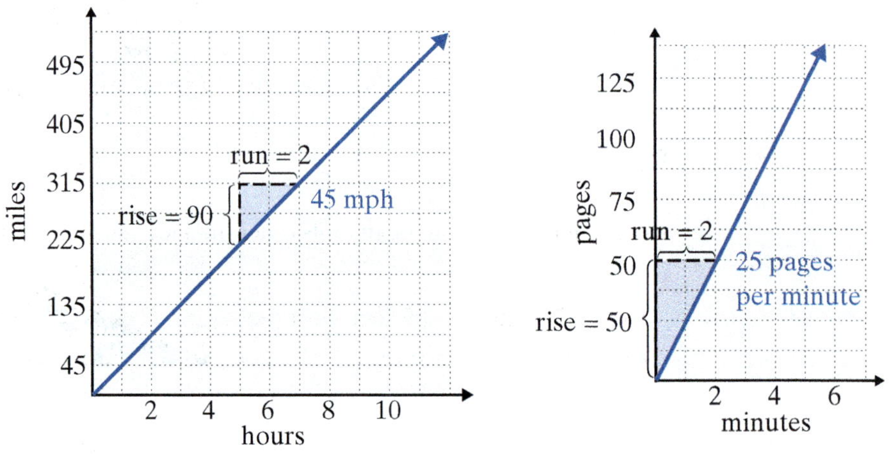

Figure 4

slope = rate of change $= \dfrac{90 \text{ miles}}{2 \text{ hours}}$ $= 45$ mph

slope = rate of change $= \dfrac{50 \text{ pages}}{2 \text{ minutes}}$ $= 25$ pages per minute

In general, the ratio of a change in one variable (say y) to a change in another variable (say x) is called the **rate of change of y with respect to x.** Figure 5 shows how the rate of change (the slope) can change over periods of time.

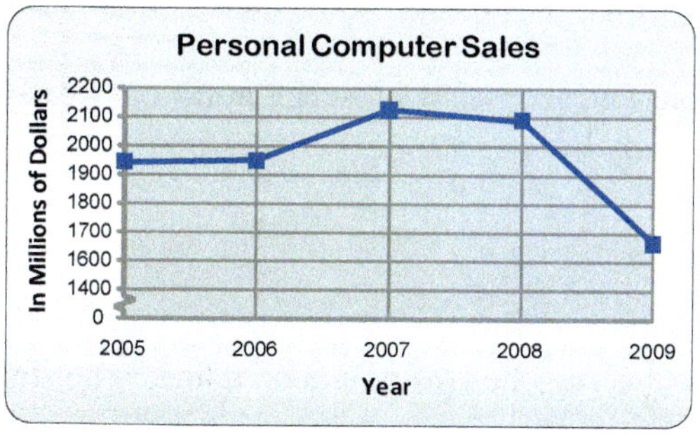

Source: Consumer Electonics Association

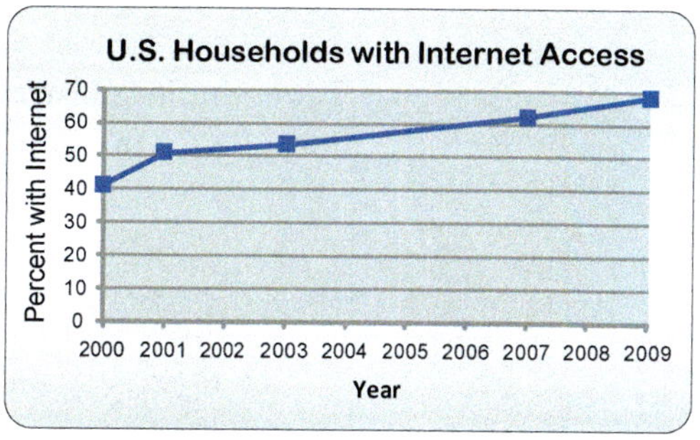

Source: U.S. Dept. of Commerce

Figure 5

Calculating the Slope: $m = \dfrac{y_2 - y_1}{x_2 - x_1}$

Consider the line $y = 2x + 3$ and two points on the line $P_1(-2, -1)$ and $P_2(2, 7)$ as shown in Figure 6. (**Note:** In the notation P_1, 1 is called a **subscript** and P_1 is read "P sub 1". Similarly, P_2 is read "P sub 2." Subscripts are used in "labeling" and are not used in calculations.)

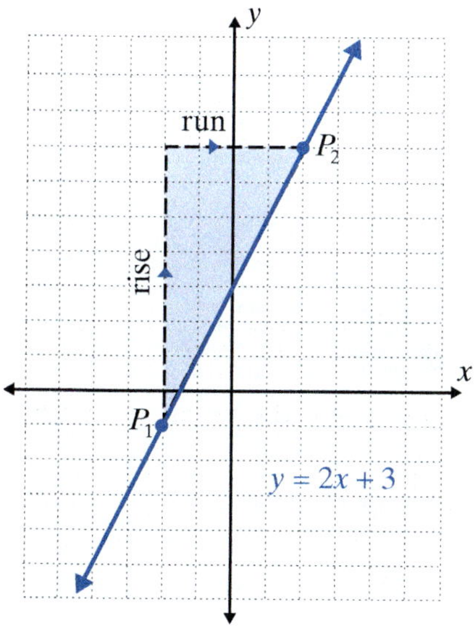

a.

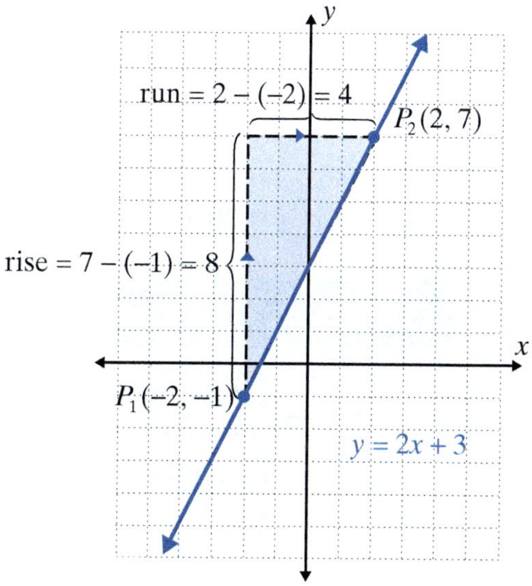

b.

Figure 6

For the line $y = 2x + 3$ and using the points $(-2, -1)$ and $(2, 7)$ that are on the line,

$$\textbf{slope} = \frac{\text{rise}}{\text{run}} = \frac{\text{difference in } y\text{-values}}{\text{difference in } x\text{-values}} = \frac{7-(-1)}{2-(-2)} = \frac{8}{4} = 2.$$

From similar illustrations and the use of subscript notation, we can develop the following formula for the slope of any line.

Slope

Let $P_1(x_1, y_1)$ and $P_2(x_2, y_2)$ be two points on a line. The **slope** can be calculated as follows:

$$\textbf{slope} = m = \frac{\textbf{rise}}{\textbf{run}} = \frac{y_2 - y_1}{x_2 - x_1}.$$

Note: The letter m is standard notation for representing the slope of a line.

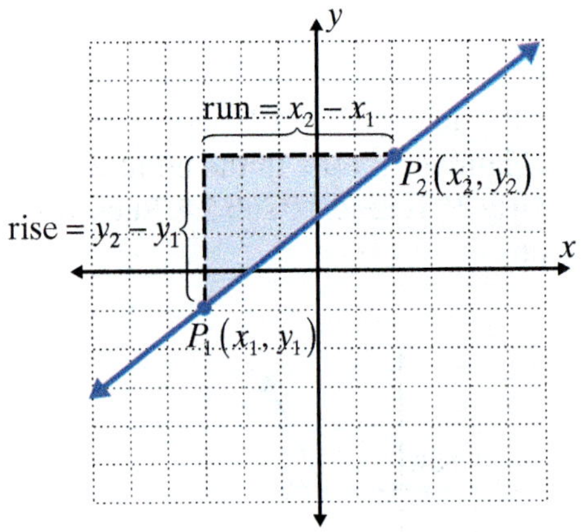

$$\text{slope} = m = \frac{\text{rise}}{\text{run}} = \frac{y_2 - y_1}{x_2 - x_1}$$

Figure 7

Example 1: Finding the Slope of a Line

Find the slope of the line that contains the points $(-1, 2)$ and $(3, 5)$, and then graph the line.

Solution: Using $(-1, 2)$ and $(3, 5)$,

$(-1, 2)$ ↑↑ (x_1, y_1) and $(3, 5)$ ↑↑ (x_2, y_2)

$$\text{slope} = m = \frac{y_2 - y_1}{x_2 - x_1} = \frac{5-2}{3-(-1)} = \frac{3}{4}$$

Or, using $(3, 5)$ and $(-1, 2)$,

$$(x_1, y_1) \quad (x_2, y_2)$$

$$\text{slope} = m = \frac{y_2 - y_1}{x_2 - x_1} = \frac{2-5}{-1-3} = \frac{-3}{-4} = \frac{3}{4}$$

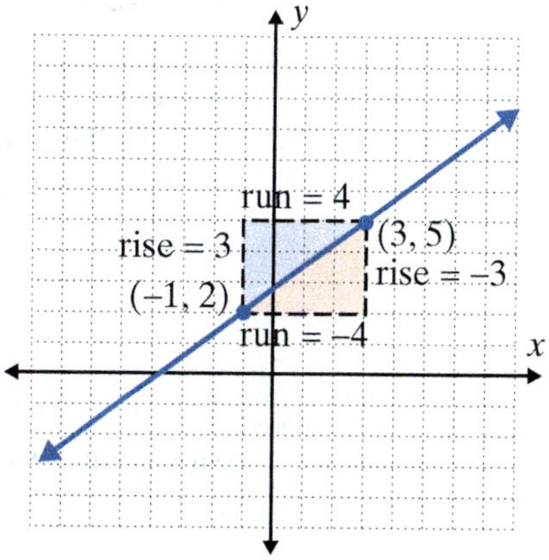

As we see in Example 1, **the slope is the same even if the order of the points is reversed.** The important part of the procedure is that **the coordinates must be subtracted in the same order in both the numerator and the denominator.**

In general,

$$\text{slope} = \frac{y_2 - y_1}{x_2 - x_1} = \frac{y_1 - y_2}{x_1 - x_2}.$$

Example 2: Finding the Slope of a Line

Find the slope of the line that contains the points $(1, 3)$ and $(5, 1)$, and then graph the line.

Solution: Using $(1, 3)$ and $(5, 1)$,

$$(x_1, y_1) \quad (x_2, y_2)$$

$$\text{slope} = m = \frac{1-3}{5-1} = \frac{-2}{4} = -\frac{1}{2}$$

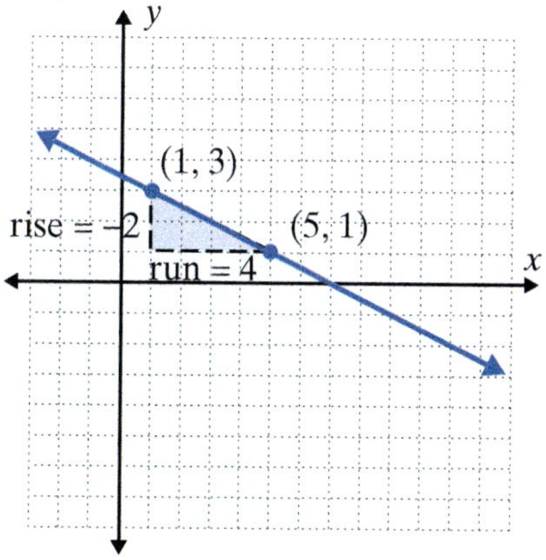

NOTES

Lines with **positive slope go up** (increase) as we move along the line from left to right.

Lines with **negative slope go down** (decrease) as we move along the line from left to right.

Slopes of Horizontal and Vertical Lines

Suppose that two points on a line have the same y-coordinate, such as $(-2, 3)$ and $(5, 3)$. Then the line through these two points will be **horizontal** as shown in Figure 8. In this case, the y-coordinates of the horizontal line are all 3, and the equation of the line is simply $y = 3$. The slope is

$$m = \frac{3-3}{5-(-2)} = \frac{0}{7} = 0.$$

For any horizontal line, all of the y-values will be the same. Consequently, the formula for slope will always have 0 in the numerator. Therefore, **the slope of every horizontal line is 0.**

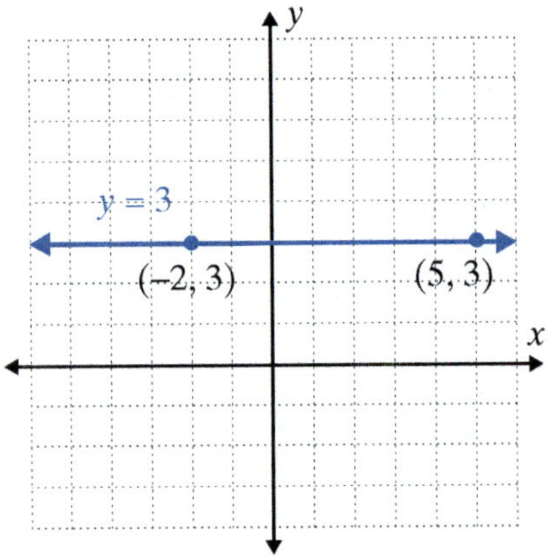

Figure 8

If two points have the same x-coordinates, such as $(1, 3)$ and $(1, -2)$, then the line through these two points will be **vertical** as in Figure 9. The x-coordinates for every point on the vertical line are all 1, and the equation of the line is simply $x = 1$. The slope is

$$m = \frac{-2-3}{1-1} = \frac{-5}{0},$$ which is **undefined**.

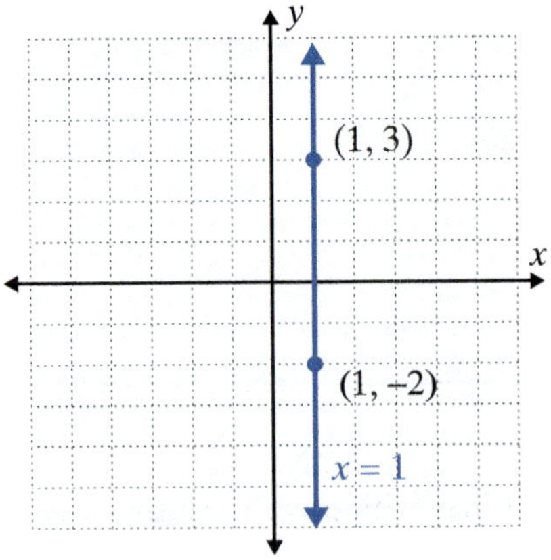

Figure 9

Horizontal and Vertical Lines

The following two general statements are true for horizontal and vertical lines:

1. For **horizontal lines** (of the form $y = b$), the **slope is 0**.
2. For **vertical lines** (of the form $x = a$), the **slope is undefined**.

Example 3: Slopes of Horizontal and Vertical Lines

a. Find the equation and slope of the horizontal line through the point $(-2, 5)$.

Solution: The equation is $y = 5$ and the slope is 0.

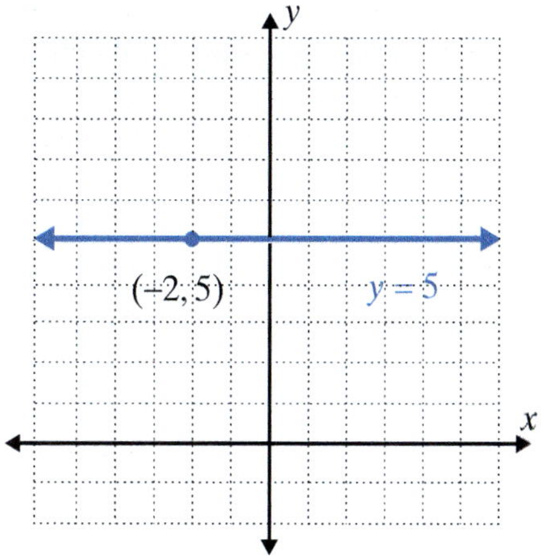

b. Find the equation and slope of the vertical line through the point $(3, 2)$.

Solution: The equation is $x = 3$ and the slope is undefined.

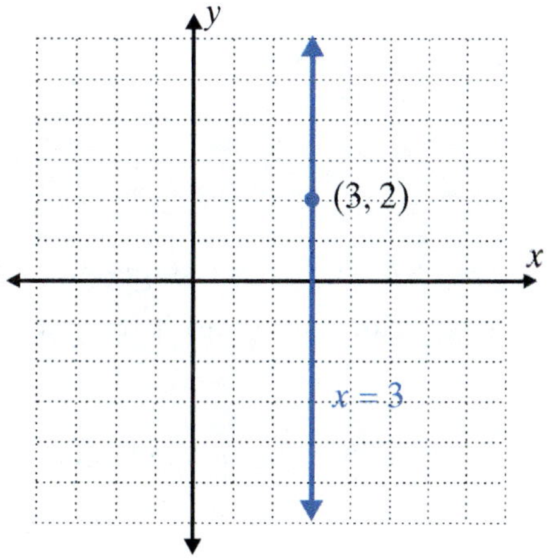

Teaching Notes:
A good idea is to have students practice (again) solving for y from the standard form of an equation. This skill needs frequent reinforcement.

Slope-Intercept Form: $y = mx + b$

There are certain relationships between the coefficients in the equation of a line and the graph of that line. For example, consider the equation

$$y = 5x - 7.$$

First, find two points on the line and calculate the slope. $(0, -7)$ and $(2, 3)$ both satisfy the equation.

$$\text{slope} = m = \frac{3-(-7)}{2-0} = \frac{10}{2} = 5$$

Observe that the slope, $m = 5$, is the same as the coefficient of x in the equation $y = 5x - 7$. This is not just a coincidence. In fact, if a linear equation is solved for y, then the coefficient of x will always be the slope of the line.

For $y = mx + b$, m is the Slope

For an equation in the form $y = mx + b$, the slope of the line is m.

For the line $y = mx + b$, the point where $x = 0$ is the point where the line crosses the y-axis. Recall that this point is called the **y-intercept**. By letting $x = 0$, we get

$$y = mx + b$$
$$y = m \cdot 0 + b$$
$$y = b.$$

Thus the point $(0, b)$ is the y-intercept. The concepts of slope and y-intercept lead to the following definition.

Slope-Intercept Form

$\boldsymbol{y = mx + b}$ is called the **slope-intercept form** for the equation of a line, where m is the **slope** and $(0, b)$ is the **y-intercept**.

As illustrated in Example 4, an equation in the **standard form**

$$\boldsymbol{Ax + By = C} \quad \text{with } B \neq 0$$

can be written in the slope-intercept form by solving for y.

Example 4: Using the Form $y = mx + b$

a. Find the slope and y-intercept of $-2x+3y=6$ and graph the line.

Solution: Solve for y.

$$-2x+3y=6$$
$$3y=2x+6$$
$$\frac{3y}{3}=\frac{2x}{3}+\frac{6}{3}$$
$$y=\frac{2}{3}x+2$$

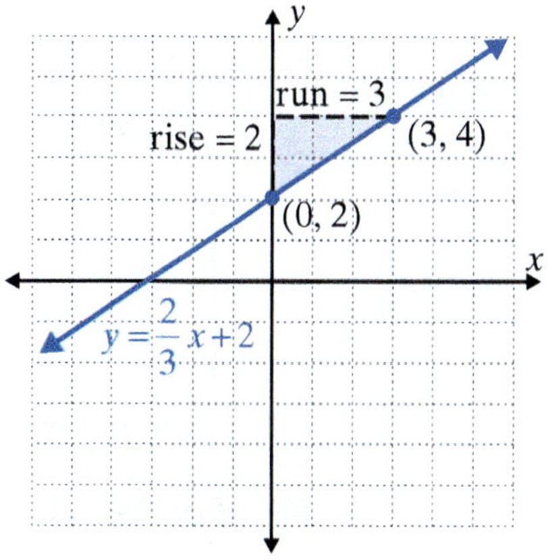

Thus $m=\frac{2}{3}$, which is the slope, and b is 2 making the y-intercept equal $(0, 2)$.

As shown in the graph, if we "rise" 2 units up and "run" 3 units to the right **from the *y*-intercept $(0, 2)$**, we locate another point $(3, 4)$. The line can be drawn through these two points.

Note: As shown in the graph on the right, we could also first "run" 3 units right and "rise" 2 units up from the y-intercept to locate the point $(3, 4)$ on the graph.

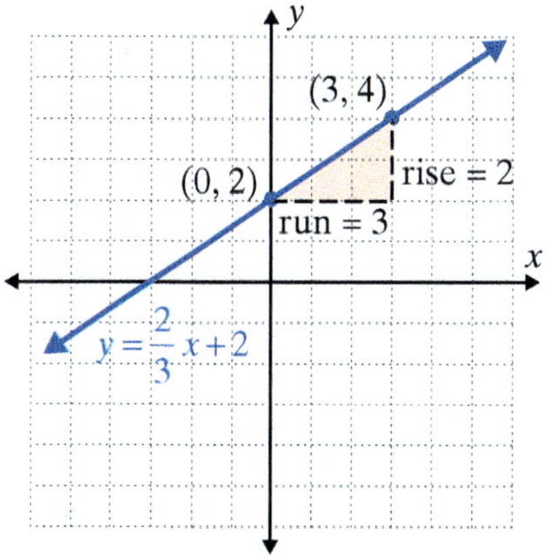

b. Find the slope and y-intercept of $x+2y=-6$ and graph the line.

Solution: Solve for y.

$$x+2y=-6$$
$$2y=-x-6$$
$$\frac{2y}{2}=\frac{-x}{2}-\frac{6}{2}$$
$$y=-\frac{1}{2}x-3$$

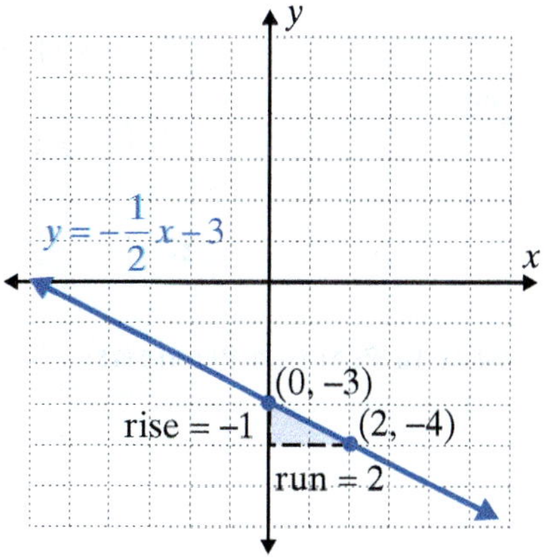

Thus $m=-\frac{1}{2}$, which is the slope, and b is -3 making the y-intercept equal $(0, -3)$.

Continued on the next page...

1. $m = 5$
2. m is undefined
3. $m = -\frac{1}{7}$
4. $m = -\frac{8}{7}$ 5. $m = 0$
6. m is undefined
7. $m = \frac{1}{2}$ 8. $m = \frac{3}{2}$
9. $m = 2$
10. $m = -\frac{3}{10}$
11. $m = \frac{1}{5}$ 12. $m = \frac{5}{4}$

We can treat $m = -\frac{1}{2}$ as $m = \frac{-1}{2}$ and the "rise" as –1 and the "run" as 2. Moving from $(0, -3)$ as shown in the figure, we locate another point $(2, -4)$ on the graph and draw the line.

c. Find the equation of the line through the point $(0, -2)$ with slope $\frac{1}{2}$.

Solution: Because the x-coordinate is 0, we know that the point $(0, -2)$ is the y-intercept. So $b = -2$. The slope is $\frac{1}{2}$. So $m = \frac{1}{2}$. Substituting in slope-intercept form $y = mx + b$ gives the result:

$$y = \frac{1}{2}x - 2.$$

Practice Problems

13. Horizontal line; $m = 0$

1. Find the slope of the line through the two points $(1, 3)$ and $(4, 6)$. Graph the line.
2. Find the equation of the line through the point $(0, 5)$ with slope $-\frac{1}{3}$.
3. Find the slope and y-intercept for the line $2x + y = 7$.
4. Write the equation for the horizontal line through the point $(-1, 3)$. What is the slope of this line?
5. Write the equation for the vertical line through the point $(-1, 3)$. What is the slope of this line?

4.3 Exercises

Find the slope of the line determined by each pair of points.

14. Horizontal line; $m = 0$

1. $(2, 4); (1, -1)$ **2.** $(1, -2); (1, 4)$ **3.** $(-6, 3); (1, 2)$

4. $(-3, 7); (4, -1)$ **5.** $(-5, 8); (3, 8)$ **6.** $(-2, 3); (-2, -1)$

7. $(5, 1); (3, 0)$ **8.** $(0, 0); (-2, -3)$ **9.** $\left(\frac{3}{4}, \frac{3}{2}\right); (1, 2)$

10. $\left(4, \frac{1}{2}\right); (-1, 2)$ **11.** $\left(\frac{3}{2}, \frac{4}{5}\right); \left(-2, \frac{1}{10}\right)$ **12.** $\left(\frac{7}{2}, \frac{3}{4}\right); \left(\frac{1}{2}, -3\right)$

Answers to Practice Problems: **1.** $m = 1$

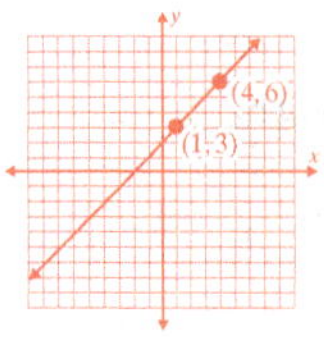

2. $y = -\frac{1}{3}x + 5$ **3.** $m = -2$; y-intercept $= (0, 7)$
4. $y = 3$; slope is 0
5. $x = -1$; slope is undefined

15. Vertical line; m is undefined.

16. Vertical line; m is undefined.

17. Horizontal line; $m = 0$

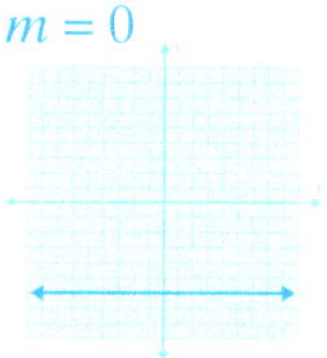

18. Vertical line; m is undefined.

19. Vertical line; m is undefined.

20. Horizontal line; $m = 0$

21.

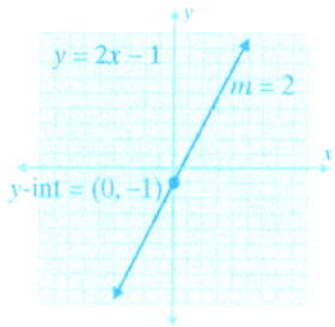

Tell whether each equation represents a horizontal line or vertical line and give its slope. Graph the line.

13. $y = 5$ **14.** $y = -2$ **15.** $x = -3$ **16.** $x = 1.7$

17. $3y = -18$ **18.** $4x = 2.4$ **19.** $-3x + 21 = 0$ **20.** $2y + 5 = 0$

Write each equation in slope-intercept form. Find the slope and the y-intercept, and then use them to draw the graph.

21. $y = 2x - 1$ **22.** $y = 3x - 4$ **23.** $y = 5 - 4x$ **24.** $y = 4 - x$

25. $y = \frac{2}{3}x - 3$ **26.** $y = \frac{2}{5}x + 2$ **27.** $x + y = 5$ **28.** $x - 2y = 6$

29. $x + 5y = 10$ **30.** $4x + y = 0$ **31.** $4x + y + 3 = 0$ **32.** $2x + 7y + 7 = 0$

33. $2y - 8 = 0$ **34.** $3y - 9 = 0$ **35.** $2x = 3y$ **36.** $4x = y$

37. $3x + 9 = 0$ **38.** $4x + 7 = 0$ **39.** $5x - 6y = 18$ **40.** $3x + 6 = 6y$

41. $5 - 3x = 4y$ **42.** $5x = 11 - 2y$ **43.** $6x + 4y = -8$ **44.** $7x + 2y = 4$

45. $6y = -6 + 3x$ **46.** $4x = 3y - 7$ **47.** $5x - 2y + 5 = 0$ **48.** $6x + 5y = -15$

In reference to the equation $y = mx + b$, sketch the graph of three lines for each of the two characteristics listed below.

49. $m > 0$ and $b > 0$ **50.** $m < 0$ and $b > 0$

51. $m > 0$ and $b < 0$ **52.** $m < 0$ and $b < 0$

Find an equation in slope-intercept form for the line passing through the given point with the given slope.

53. $(0, 3);\ m = -\frac{1}{2}$ **54.** $(0, 2);\ m = \frac{1}{3}$ **55.** $(0, -3);\ m = \frac{2}{5}$

56. $(0, -6);\ m = \frac{4}{3}$ **57.** $(0, -5);\ m = 4$ **58.** $(0, 9);\ m = -1$

59. $(0, -4);\ m = 1$ **60.** $(0, 6);\ m = -5$ **61.** $(0, -3);\ m = -\frac{5}{6}$

62. $(0, -1);\ m = -\frac{3}{2}$

22.

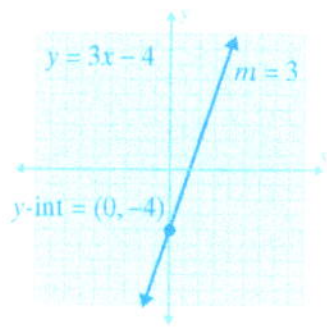

23.

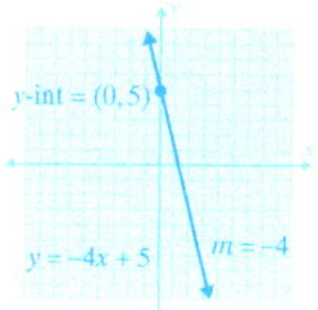

24.

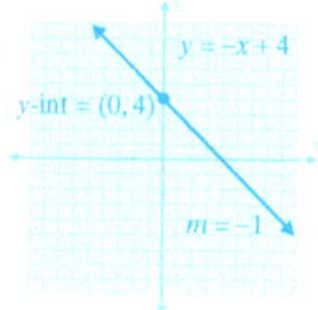

*The graph of a line is shown with two points highlighted. Find **a.** the slope, **b.** the y-intercept (if there is one), and **c.** the equation of the line in slope-intercept form.*

63.

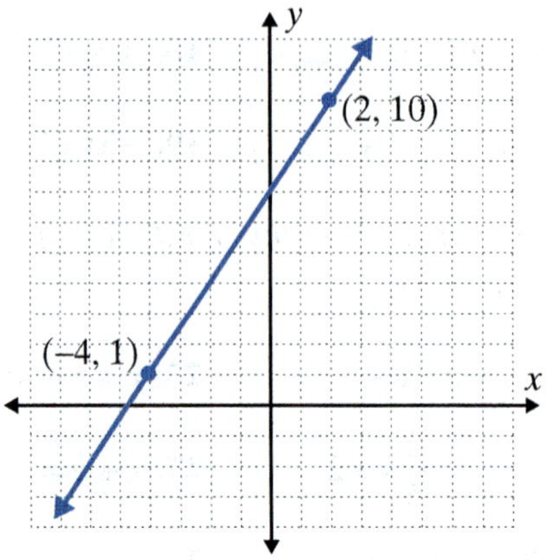

64.

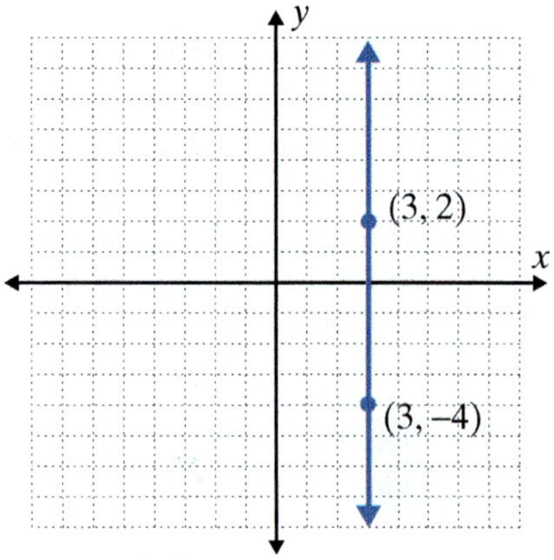

65.

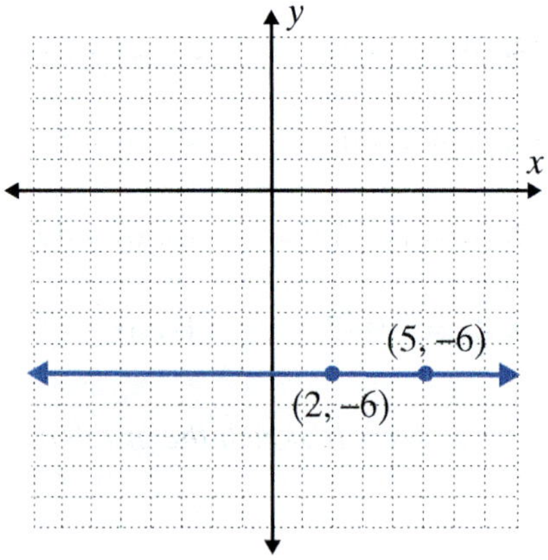

66.

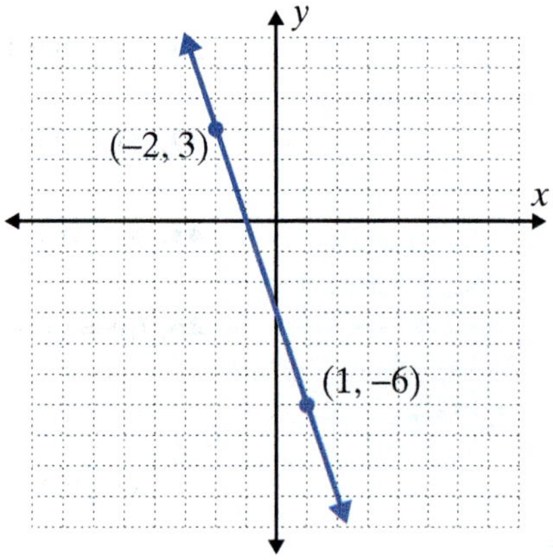

67.

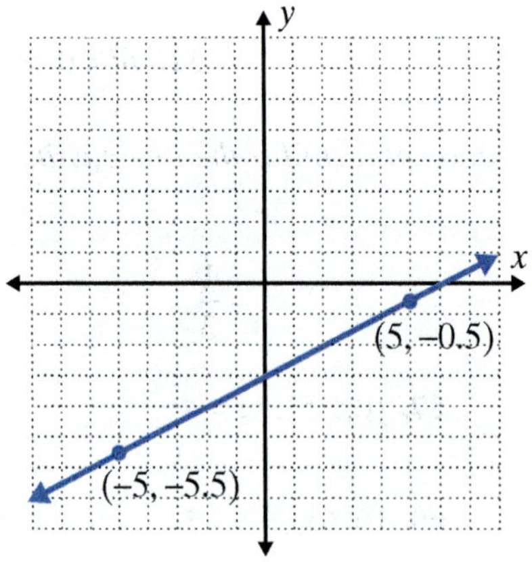

68.

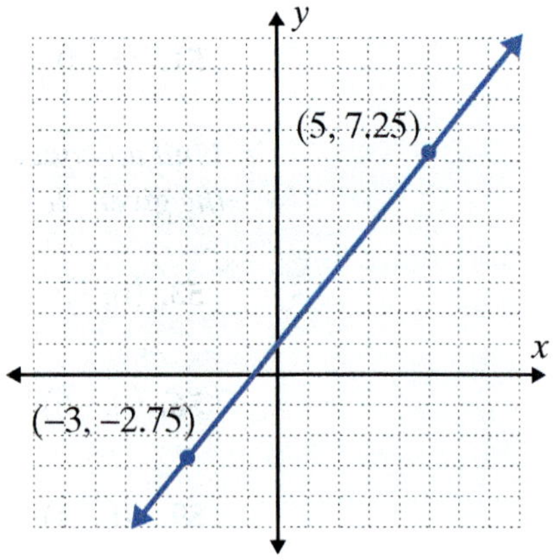

25.

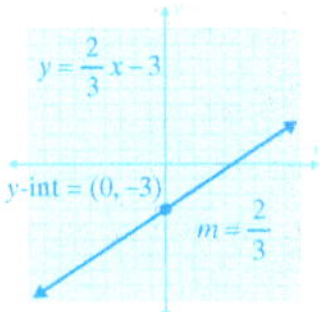

26.

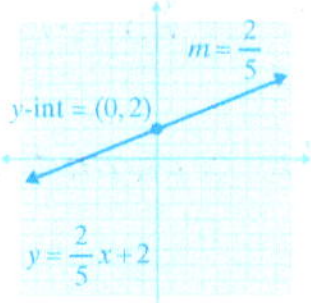

27.

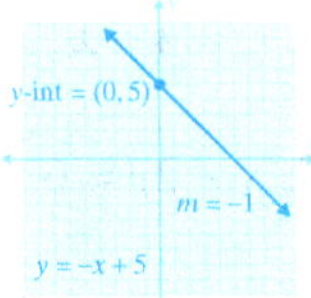

28.

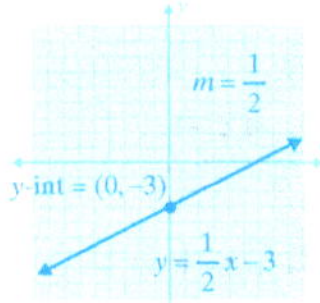

29.

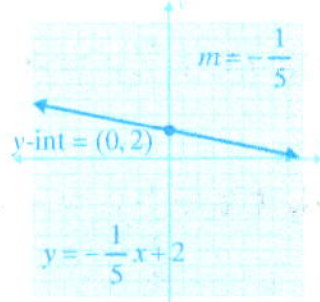

30.

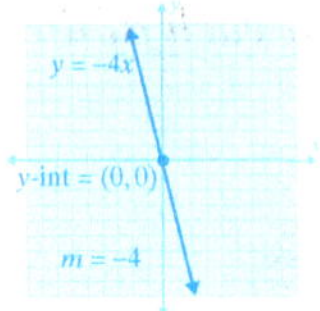

31.

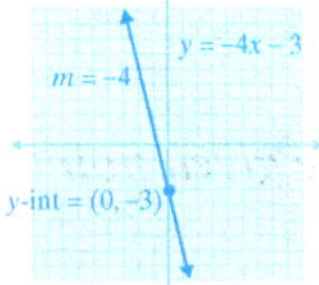

32.

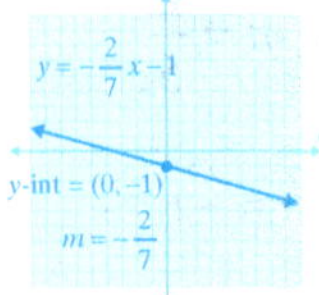

33.

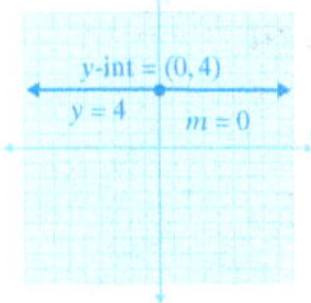

34.

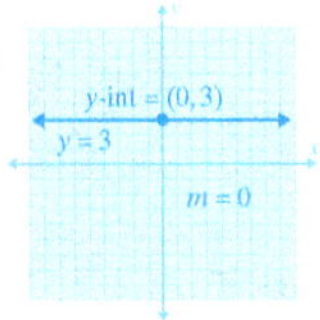

35.

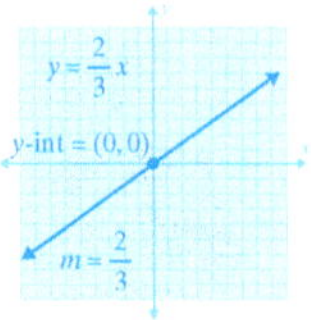

36.

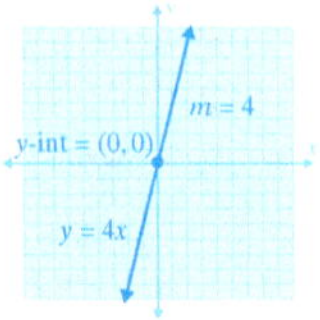

37.

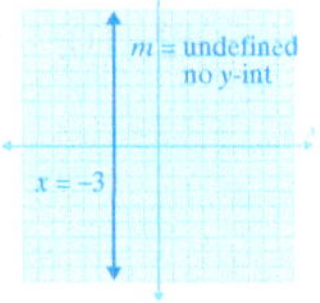

38.

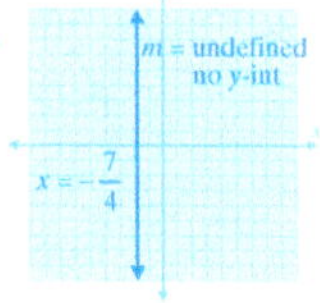

69.

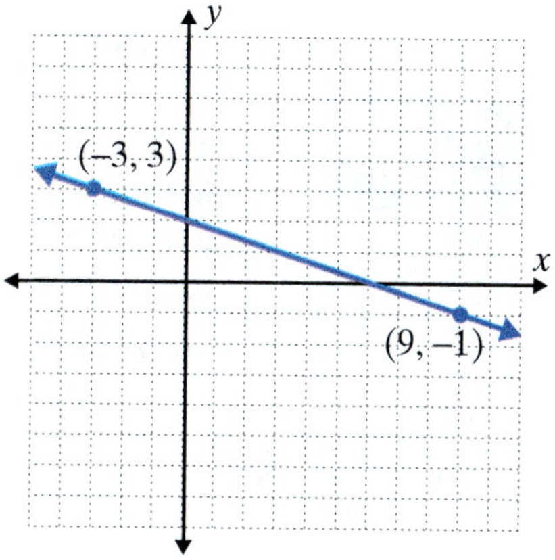

70.

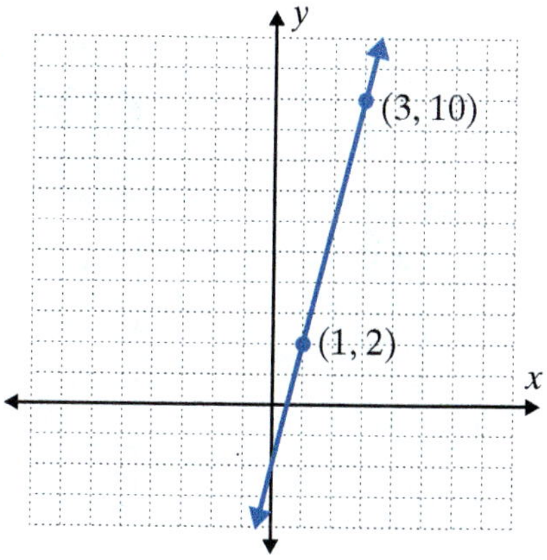

Points are said to be ***collinear*** *if they are on a straight line. If points are collinear, then the slope of the line through any two of them must be the same (because the line is the same line). Use this idea to determine whether or not the three points in each of the sets are collinear.*

71. $\{(-1, 3), (0, 1), (5, -9)\}$

72. $\{(-2, -4), (0, 2), (3, 11)\}$

73. $\{(-2, 0), (0, 30), (1.5, 5.25)\}$

74. $\{(-1, -7), (1, 1), (2.5, 7)\}$

75. $\left\{\left(\frac{2}{3}, \frac{1}{2}\right), \left(0, \frac{5}{6}\right), \left(-\frac{3}{4}, \frac{29}{24}\right)\right\}$

76. $\left\{\left(\frac{3}{2}, -\frac{1}{3}\right), \left(0, \frac{1}{6}\right), \left(-\frac{1}{2}, \frac{3}{4}\right)\right\}$

77. Buying a new car: John bought his new car for \$35,000 in the year 2007. He knows that the value of his car has depreciated linearly. If the value of the car in 2010 was \$23,000, what was the annual rate of depreciation of his car? Show this information on a graph. (When graphing, use years as the x-coordinates and the corresponding values of the car as the y-coordinates.)

78. Cell phone usage: The number of people in the United States with mobile cellular phones was about 180 million in 2004 and about 286 million in 2009. If the growth in mobile cellular phones was linear, what was the approximate rate of growth per year from 2004 to 2009. Show this information on a graph (When graphing, use years as the x-coordinates and the corresponding number of users as the y-coordinates.)

39.

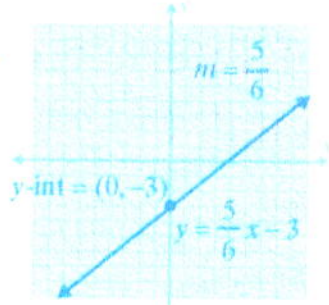

40.

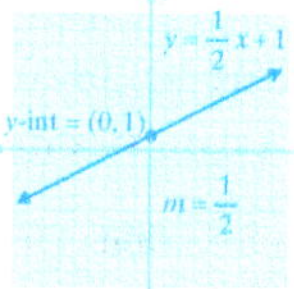

41.

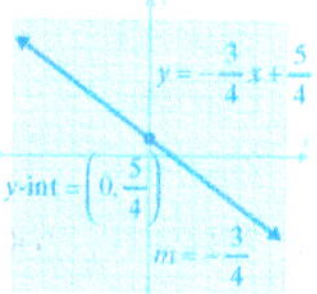

42.

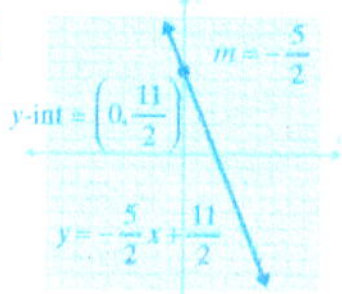

43.

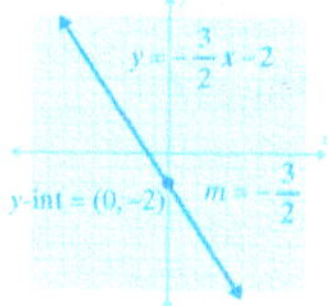

44.

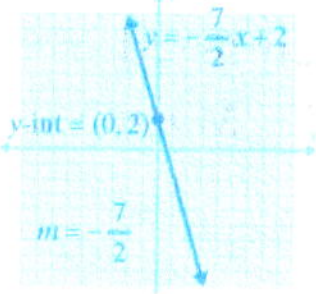

45.

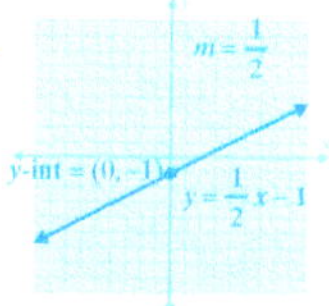

46.

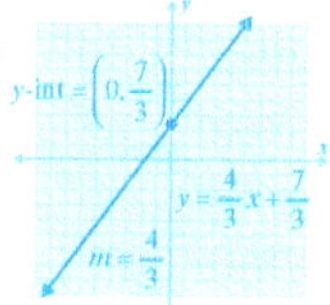

47.

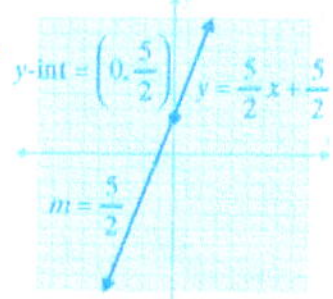

48.

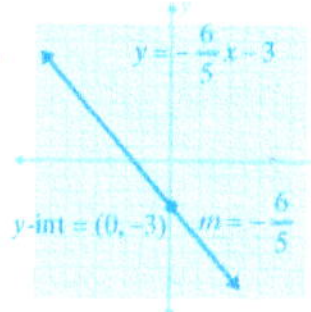

49-52. Answers will vary.

49.

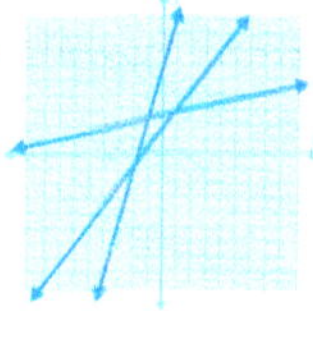

50.

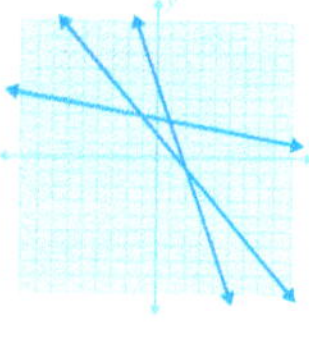

51.

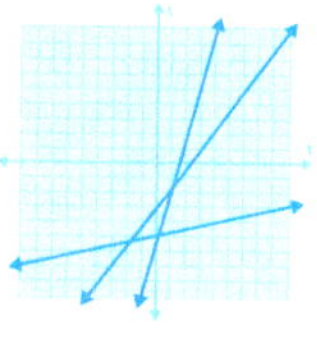

52.

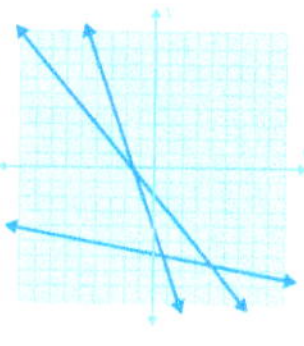

53. $y = -\frac{1}{2}x + 3$

54. $y = \frac{1}{3}x + 2$

55. $y = \frac{2}{5}x - 3$

56. $y = \frac{4}{3}x - 6$

57. $y = 4x - 5$

58. $y = -x + 9$

59. $y = x - 4$

60. $y = -5x + 6$

61. $y = -\frac{5}{6}x - 3$

62. $y = -\frac{3}{2}x - 1$

63. a. $m = \frac{3}{2}$ b. $(0, 7)$
c. $y = \frac{3}{2}x + 7$

64. a. m is undefined.
b. No y-intercept
c. $x = 3$

65. a. $m = 0$ b. $(0, -6)$
c. $y = -6$

66. a. $m = -3$ b. $(0, -3)$
c. $y = -3x - 3$

67. a. $m = \frac{1}{2}$ b. $(0, -3)$
c. $y = \frac{1}{2}x - 3$

68. a. $m = \frac{5}{4}$ b. $(0, 1)$
c. $y = \frac{5}{4}x + 1$

69. a. $m = -\frac{1}{3}$ b. $(0, 2)$
c. $y = -\frac{1}{3}x + 2$

70. a. $m = 4$ b. $(0, -2)$
c. $y = 4x - 2$

71. Yes 72. Yes

73. No 74. Yes

75. Yes 76. No

77. \$4000/year

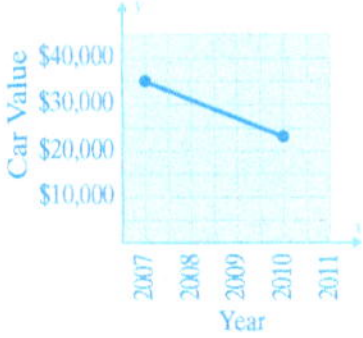

79. Internet usage: The given table shows the estimated number of internet users from 2004 to 2008. The number of users for each year is shown in millions.

Year	Internet Users (in millions)
2004	185
2005	198
2006	210
2007	220
2008	231

a. Plot these points on a graph.
b. Connect the points with line segments.
c. Find the slope of each line segment.
d. Interpret each slope as a rate of change.

Source: International Telecommunications Union Yearbook of Statistics

80. Urban growth: The following table shows the urban growth from 1850 to 2000 in New York, NY.

Year	Population
1850	515,547
1900	3,437,202
1950	7,891,957
2000	8,008,278

a. Plot these points on a graph.
b. Connect the points with line segments.
c. Find the slope of each line segment.
d. Interpret each slope as a rate of change.

Source: U.S. Census Bureau

81. Military: The following graph shows the number of female active duty military personnel over a span from 1945 to 2009. The number of women listed includes both officers and enlisted personnel from the Army, the Navy, the Marine Corps, and the Air Force.

a. Plot these points on a graph.
b. Connect the points with line segments.
c. Find the slope of each line segment.
d. Interpret each slope as a rate of change.

Source: U.S. Dept. of Defense

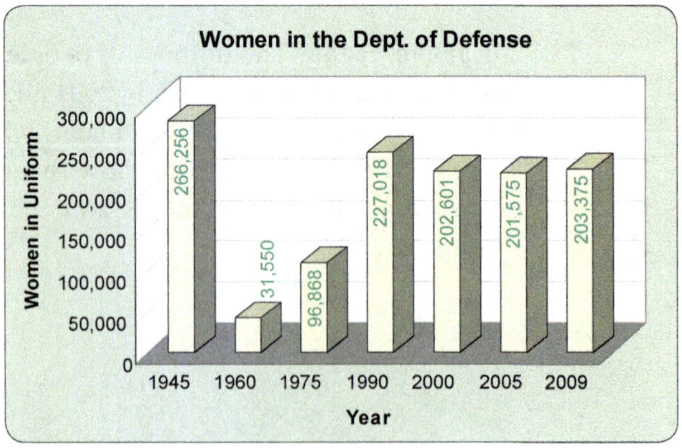

78. 21.2 million/year

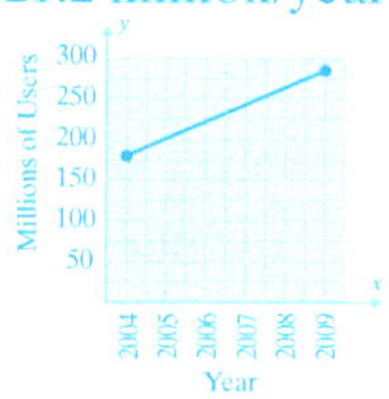

82. Marriage: The following graph shows the rates of marriage per 1000 people in the U.S., over a span from 1920 to 2008.

a. Plot these points on a graph.
b. Connect the points with line segments.
c. Find the slope of each line segment.
d. Interpret each slope as a rate of change.

Source: U.S. National Center for Health Statistics

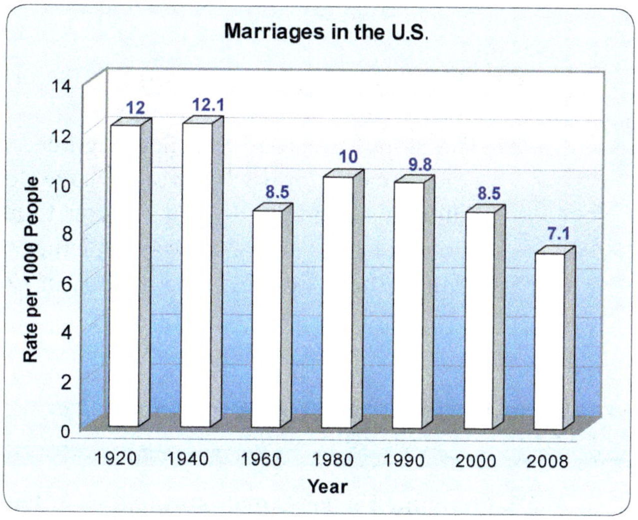

79. a. and **b.**

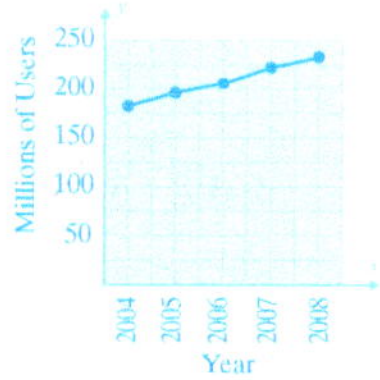

c. 13, 12, 10, 11
d. The number of internet users increased by 13 million people/year from '04-'05, 12 mppy from '05-'06, 10 mppy from '06-'07, and 11 mppy from '07-'08.

80. a. and **b.**

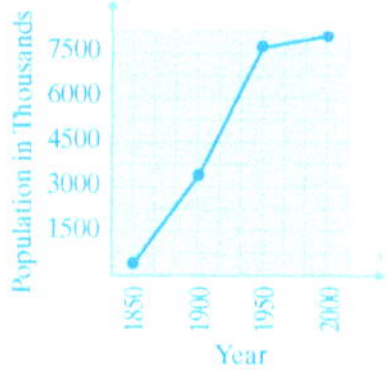

c. 58,433.1; 89,095.1; 2326.42
d. The population of New York increased 58,433 people/year from 1850-1900; 89,095 ppy from 1900-1950; and 2326 ppy from 1950-2000.

Collaborative Learning Exercise

83. The class should be divided into teams of 2 or 3 students. Each team will need access to a digital camera, a printer, and a ruler.

a. Take pictures of 8 things with a defined slope. (**Suggestions:** A roof, a stair railing, a beach umbrella, a crooked tree, etc. Be creative!)
b. Print each picture.
c. Use a ruler to draw a coordinate system on top of each picture. You will probably want to use increments of in. or cm, depending on the size of your picture.
d. Identify the line in each picture whose slope you are calculating and then use the coordinate systems you created to identify the coordinates of two points on each line .
e. Use the points you just found to calculate the slope of the line in each picture.
f. Share your findings with the class.

81. a. and **b.**

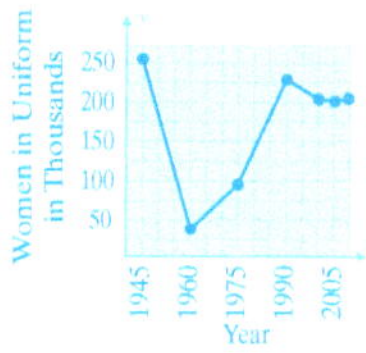

c. –15,647.07; 4354.53; 8676.67; –2441.7; –205.2; 450
d. Number of female active duty military personnel decreased by 15,647.07 women/year from 1945-1960; increased by 4354.53 wpy from '60-'75; increased by 8676.67 wpy from '75-'90; decreased by 2441.7 wpy from '90-'00; decreased by 205.2 wpy from '00-'05; and increased by 450 wpy from '05-'09.

Writing and Thinking About Mathematics

84. a. Explain in your own words why the slope of a horizontal line must be 0.

b. Explain in your own words why the slope of a vertical line must be undefined.

85. a. Describe the graph of the line $y = 0$.

b. Describe the graph of the line $x = 0$.

86. In the formula $y = mx + b$ explain the meaning of m and the meaning of b.

87. The slope of a road is called a **grade**. A steep grade is cause for truck drivers to have slow speed limits in mountains. What do you think that a "grade of 12%" means? Draw a picture of a right triangle that would indicate a grade of 12%.

HAWKES LEARNING SYSTEMS: INTRODUCTORY & INTERMEDIATE ALGEBRA SOFTWARE

- 4.3 Graphing Linear Equations in Slope-Intercept Form

82. a. and **b.**

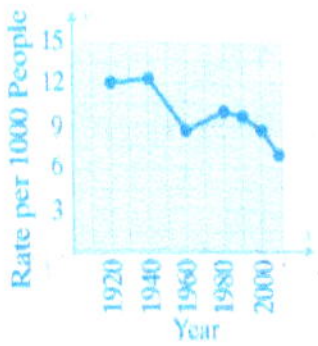

c. 0.005, −0.18, 0.075, −0.02, −0.13, −0.175 **d.** # of marriages increased 0.005 marriages/1000 people from 1920-1940, decreased 0.18 marriages/1000 people people from '40-'60, increased 0.075 marriages/1000 people from '60-'80, decreased 0.02 marriages/1000 people from '80-'90, decreased 0.13 marriages/1000 people from '90-'00, and decreased marriages/1000 people 0.175 from '00-'08.

83. Answers will vary.

84. a. For any horizontal line, all of the y values will be the same. Thus the formula for slope will always have 0 in the numerator making the slope of every horizontal line 0.

b. For any vertical line, all of the x values will be the same. Thus the formula for slope will always have 0 in the denominator making the slope of every vertical line undefined.

85. a. The x-axis **b.** The y-axis **86.** $y = mx + b$, m = slope of the line, b = y-intercept.

87. A grade of 12% means the slope of the road is 0.12. For every 100 feet of horizontal distance (run) there is 12 feet of vertical distance (rise).

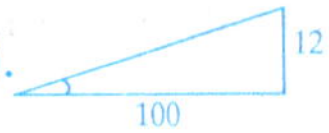

4.4 The Point-Slope Form: $y - y_1 = m(x - x_1)$

- *Graph a line given its slope and one point on the line.*
- *Find the equation of a line given its slope and one point on the line using the formula* $y - y_1 = m(x - x_1)$.
- *Find the equation of a line given two points on the line.*
- *Recognize and know how to find lines that are parallel and perpendicular.*

Graphing a Line Given a Point and the Slope

Lines represented by equations in the **standard form** $Ax + By = C$ and in the **slope-intercept form** $y = mx + b$ have been discussed in Sections 4.2 and 4.3. In Section 4.3 we graphed lines using the y-intercept $(0, b)$ and the slope by moving vertically and then horizontally (or by moving horizontally and then vertically) from the y-intercept. This same technique can be used to graph lines if the given point is on the line but is not the y-intercept. Consider the following example.

Example 1: Graph a Line Given a Point and the Slope

Graph the line with slope $m = -\frac{3}{4}$ and which passes through the point $(2, 5)$.

Solution: Start from the point $(2, 5)$ and locate another point on the line using the slope as $\frac{rise}{run} = \frac{-3}{4}$ or $\frac{3}{-4}$. There are four ways to proceed. Here are two:

1. Move 4 units right and 3 units down, or
2. Move 3 units down and 4 units right.

Either way, you arrive at the same point $(6, 2)$.

This means that we can move from the given point either with the rise first or the run first.

Note: Any numbers in the ratio of –3 to 4 can be used for the moves, such as –6 to 8 or 9 to –12.

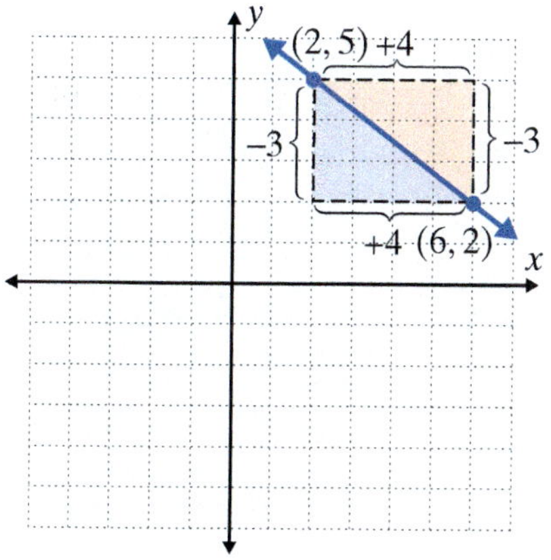

Point-Slope Form: $y - y_1 = m(x - x_1)$

Now consider finding the equation of the line given the point (x_1, y_1) on the line and the slope m. If (x, y) is **any other point** on the line, then the slope formula gives the equation

$$\frac{y - y_1}{x - x_1} = m.$$

Multiplying both sides of this equation by the denominator (assuming the denominator is not 0) gives

$$y - y_1 = m(x - x_1) \qquad \text{which is called the } \textbf{point-slope form}.$$

For example, suppose that a point $(x_1, y_1) = (8, 3)$ and the slope $m = -\frac{3}{4}$ are given.

If (x, y) represents any point on the line other than $(8, 3)$, then substituting into the formula for slope gives

$$\frac{y - y_1}{x - x_1} = m \qquad \text{Formula for slope}$$

$$\frac{y - 3}{x - 8} = -\frac{3}{4} \qquad \text{Substitute the given information.}$$

$$(x - 8)\left(\frac{y - 3}{x - 8}\right) = -\frac{3}{4}(x - 8) \qquad \text{Multiply both sides by } (x - 8).$$

$$y - 3 = -\frac{3}{4}(x - 8). \qquad \text{Point-slope form: } y - y_1 = m(x - x_1)$$

From this point-slope form, we can manipulate the equation to get the other two forms:

$$y - 3 = -\frac{3}{4}(x - 8)$$

$$y - 3 = -\frac{3}{4}x + 6$$

or $$y = -\frac{3}{4}x + 9 \qquad \text{Slope-intercept form: } y = mx + b$$

or $$4(y) = 4\left(-\frac{3}{4}x + 9\right)$$

$$4y = -3x + 36$$

$$3x + 4y = 36. \qquad \text{Standard form: } Ax + By = C$$

Point-Slope Form

An equation of the form

$$\mathbf{y - y_1 = m(x - x_1)}$$

is called the **point-slope form** for the equation of a line that contains the point (x_1, y_1) and has slope m.

Example 2: Finding Equations of Lines Using the Slope and a Point

Find the equation of the line with a slope of $-\frac{1}{2}$ and passing through the point $(2, 3)$. Graph the line using the point and slope.

Solution: Substitute the values into the point-slope form.

$$y - y_1 = m(x - x_1)$$

$$y - 3 = -\frac{1}{2}(x - 2) \qquad \text{Point-slope form}$$

$$y - 3 = -\frac{1}{2}x + 1$$

or $$y = -\frac{1}{2}x + 4 \qquad \text{Slope-intercept form}$$

or $$2y = -x + 8$$

$$x + 2y = 8 \qquad \text{Standard form}$$

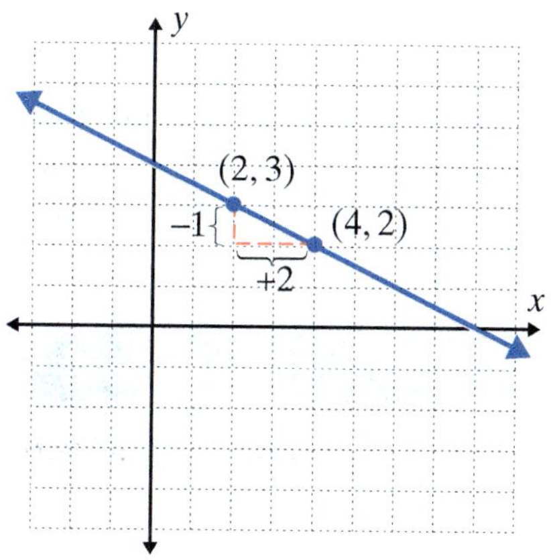

The point one unit down and two units right from $(2, 3)$ will be on the line because the slope is $m = \frac{\text{rise}}{\text{run}} = \frac{-1}{2} = -\frac{1}{2}$.

With a negative slope, either the rise is negative and the run is positive, or the rise is positive and the run is negative. In either case, as the previous figure and the following figure illustrate, the line is the same.

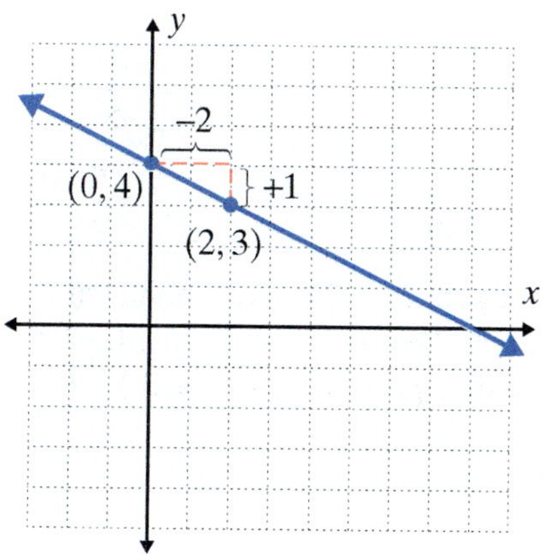

The point one unit up and two units to the left from $(2, 3)$ is on the line because the slope is $m = \frac{\text{rise}}{\text{run}} = \frac{1}{-2} = -\frac{1}{2}$.

In Example 2, the equation of the line is written in all three forms: point-slope form, slope-intercept form, and standard form. Generally any one of these forms is sufficient. However, there are situations in which one form is preferred over the others. Therefore, manipulation among the forms is an important skill. Also, if the answer in the text is in one form and your answer is in another form, you should be able to recognize that the answers are equivalent.

Finding the Equations of Lines Given Two Points

Given two points that lie on a line, the equation of the line can be found using the following method.

Finding the Equation of a Line Given Two Points

To find the equation of a line given two points on the line

1. Use the formula $m = \frac{y_2 - y_1}{x_2 - x_1}$ to find the slope.

2. Use this slope, m, and either point in the point-slope formula $y - y_1 = m(x - x_1)$ to find the equation.

Example 3: Using Two Points to Find the Equation of a Line

Teaching Notes:
Emphasize, using several examples, that either point may be used after the slope is determined, and have students demonstrate to their own satisfaction that either point produces the same, or an equivalent, equation for the line.

Find the equation of the line containing the two points $(-1, 2)$ and $(4, -2)$.

Solution: First, find the slope.

$$\begin{aligned} m &= \frac{y_2 - y_1}{x_2 - x_1} \\ &= \frac{-2-2}{4-(-1)} \\ &= \frac{-4}{5} \\ &= -\frac{4}{5} \end{aligned}$$

Now use one of the given points and the point-slope form for the equation of a line. (**Note:** $(-1, 2)$ and $(4, -2)$ are used on the next page to illustrate that either point may be used.)

Using $(-1, 2)$		Using $(4, -2)$
$y - y_1 = m(x - x_1)$	Point-slope form	$y - y_2 = m(x - x_2)$
$y - 2 = -\frac{4}{5}[x - (-1)]$	Substitute.	$y - (-2) = -\frac{4}{5}(x - 4)$
$y - 2 = -\frac{4}{5}x - \frac{4}{5}$	Distributive property	$y + 2 = -\frac{4}{5}x + \frac{16}{5}$
$y = -\frac{4}{5}x - \frac{4}{5} + 2$		$y = -\frac{4}{5}x + \frac{16}{5} - 2$
$y = -\frac{4}{5}x + \frac{6}{5}$	Slope-intercept form	$y = -\frac{4}{5}x + \frac{6}{5}$
or $4x + 5y = 6$	Standard form	$4x + 5y = 6$

Parallel Lines and Perpendicular Lines

Parallel and Perpendicular Lines

Parallel lines are lines that never intersect (never cross each other) and these lines have the **same slope**. (**Note:** All vertical lines (undefined slopes) are parallel to one another.)

Perpendicular lines are lines that intersect at 90º (right) angles and whose slopes are **negative reciprocals** of each other. Horizontal lines are perpendicular to vertical lines.

As illustrated in Figure 1, the lines $y = 2x + 1$ and $y = 2x - 3$ are **parallel**. They have the same slope, 2. The lines $y = \frac{2}{3}x + 1$ and $y = -\frac{3}{2}x - 2$ are **perpendicular**. Their slopes are negative reciprocals of each other, $\frac{2}{3}$ and $-\frac{3}{2}$.

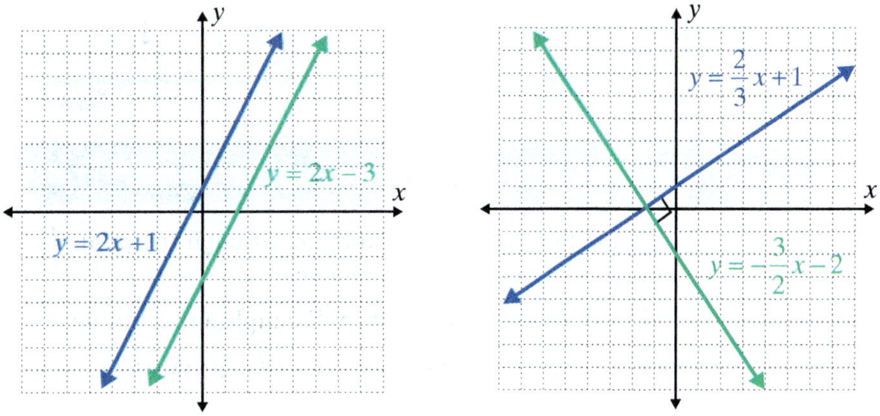

Figure 1 Parallel lines Perpendicular lines

Example 4: Finding the Equations of Parallel Lines

Find the equation of the line through the point $(2, 3)$ and parallel to the line $5x + 3y = 1$. Graph both lines.

Solution: First, solve for y to find the slope of the given line.

$$5x + 3y = 1$$
$$3y = -5x + 1$$
$$y = -\frac{5}{3}x + \frac{1}{3}$$ Thus any line parallel to this line has slope $-\frac{5}{3}$.

Now use the point-slope form $y - y_1 = m(x - x_1)$ with $m = -\frac{5}{3}$ and $(x_1, y_1) = (2, 3)$.

$y - 3 = -\frac{5}{3}(x - 2)$ Point-slope form

$3(y - 3) = -5(x - 2)$ Multiply both sides by the LCD, 3.

$3y - 9 = -5x + 10$ Simplify.

$5x + 3y = 19$ Standard form

or $y = -\frac{5}{3}x + \frac{19}{3}$ Slope-intercept form

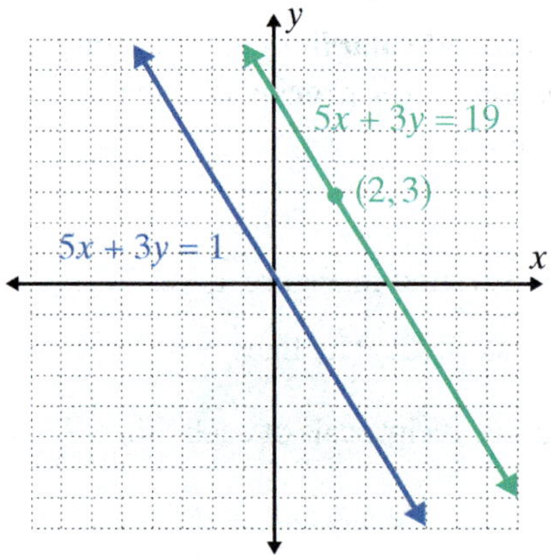

Example 5: Finding the Equations of Perpendicular Lines

Find the equation of the line through the point $(2, 3)$ and perpendicular to the line $5x + 3y = 1$. Graph both lines.

Solution: We know from Example 4 that the slope of the line $5x + 3y = 1$ is $-\frac{5}{3}$. Thus any line perpendicular to this line must have slope $m = \frac{3}{5}$ (the negative reciprocal of $-\frac{5}{3}$).

Now using the point-slope form $y - y_1 = m(x - x_1)$ with $m = \frac{3}{5}$, and $(x_1, y_1) = (2, 3)$, we have

	$y - 3 = \frac{3}{5}(x - 2)$	Point-slope form
	$5(y - 3) = 3(x - 2)$	Multiply both sides by the LCD, 5.
	$5y - 15 = 3x - 6$	Simplify.
	$3x - 5y = -9$	Standard form
or	$y = \frac{3}{5}x + \frac{9}{5}$	Slope-intercept form

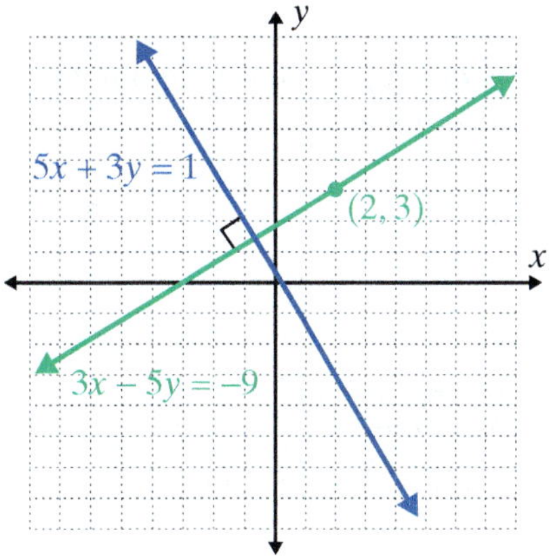

For easy reference, the following table summarizes what we know about straight lines.

Summary of Formulas and Properties of Straight Lines

1. $Ax + By = C$	Standard form
2. $m = \frac{y_2 - y_1}{x_2 - x_1}$	Slope of a line
3. $y = mx + b$	Slope-intercept form
4. $y - y_1 = m(x - x_1)$	Point-slope form
5. $y = b$	Horizontal line, slope 0
6. $x = a$	Vertical line, undefined slope

7. Parallel lines have the same slope.

8. Perpendicular lines have slopes that are negative reciprocals of each other.

1. a. $m = 2$ b. $(3, 1)$

c.

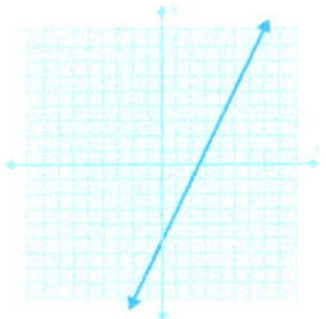

2. a. $m = \frac{1}{2}$ b. $(1, 4)$

c.

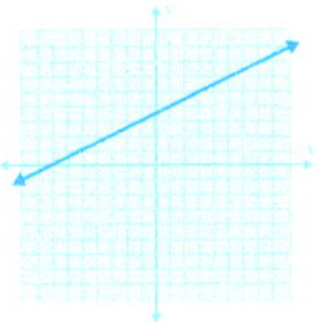

3. a. $m = -5$ b. $(0, -2)$

c.

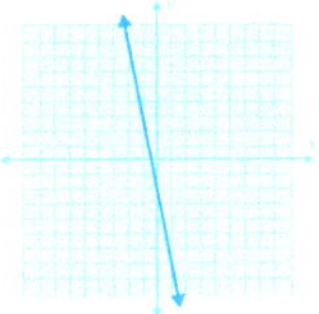

4. a. $m = -1$ b. $(-8, 0)$

c.

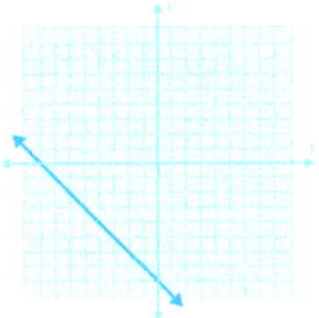

5. a. $m = -\frac{1}{4}$ b. $(-2, 3)$

c.

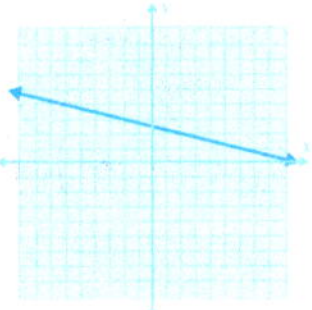

6. a. $m = \frac{1}{3}$ b. $(7, -6)$

c.

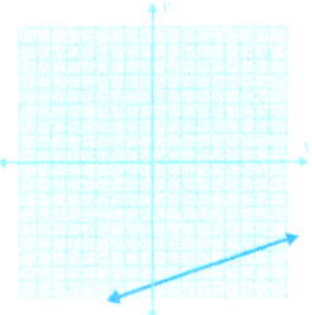

7. $2x + y = -3$

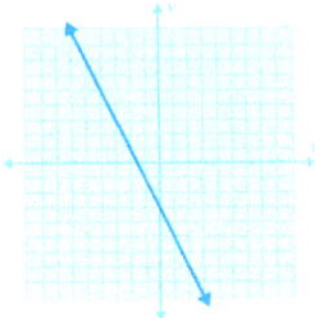

Practice Problems

Find a linear equation in standard form whose graph satisfies the given conditions.

1. Passes through the point $(4, -1)$ with $m = 2$
2. Parallel to $y = -3x + 4$ and contains the point $(-1, 5)$
3. Perpendicular to $2x + y = 1$ and passes through the origin $(0, 0)$
4. Contains the two points $(6, -2)$ and $(2, 0)$

4.4 Exercises

8. $3x - y = 5$

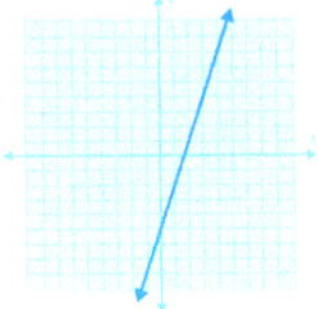

9. $y = -2$

10. $3x + y = 0$

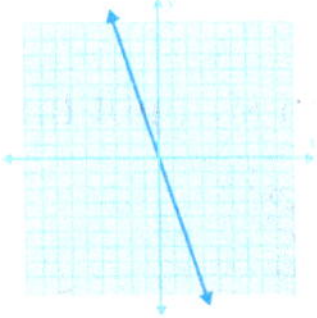

11. $x - 2y = -15$

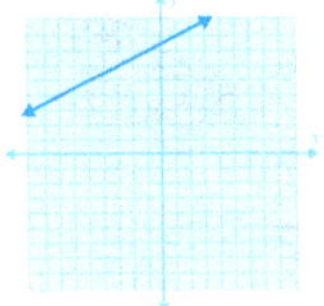

12. $x = -3$

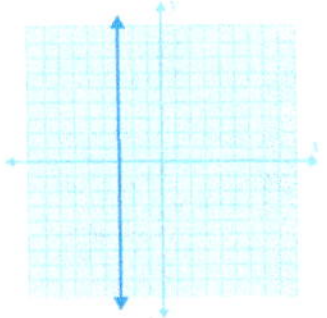

*Find **a.** the slope, **b.** a point on the line, and **c.** the graph of the line for the given equations in point-slope form.*

1. $y-1=2(x-3)$ **2.** $y-4=\frac{1}{2}(x-1)$ **3.** $y+2=-5(x)$

4. $y=-(x+8)$ **5.** $y-3=-\frac{1}{4}(x+2)$ **6.** $y+6=\frac{1}{3}(x-7)$

Find an equation in standard form for the line passing through the given point with the given slope. Graph the line.

7. $(-2,1)$; $m=-2$ **8.** $(3,4)$; $m=3$ **9.** $(5,-2)$; $m=0$

10. $(0,0)$; $m=-3$ **11.** $(-3,6)$; $m=\frac{1}{2}$

12. $(-3,-1)$; m is undefined **13.** $(7,10)$; $m=\frac{3}{5}$

14. $(-1,-1)$; $m=-\frac{1}{4}$ **15.** $\left(-2,\frac{1}{3}\right)$; $m=\frac{2}{3}$ **16.** $\left(\frac{5}{2},\frac{1}{2}\right)$; $m=-\frac{4}{3}$

Find an equation in slope-intercept form for the line passing through the two given points.

17. $(-5,2)$; $(3,6)$ **18.** $(-3,4)$; $(2,1)$ **19.** $(-5,1)$; $(2,0)$ **20.** $(-4,-4)$; $(3,1)$

21. $(0,2)$; $\left(1,\frac{3}{4}\right)$ **22.** $\left(\frac{5}{2},0\right)$; $\left(2,-\frac{1}{3}\right)$ **23.** $(2,-5)$; $(4,-5)$

24. $(0,4)$; $\left(1,\frac{1}{2}\right)$ **25.** $(-2,6)$; $(3,1)$ **26.** $(8,2)$; $(0,0)$

Answers to Practice Problems: **1.** $2x-y=9$ **2.** $3x+y=2$ **3.** $x-2y=0$ **4.** $x+2y=2$

13. $3x - 5y = -29$

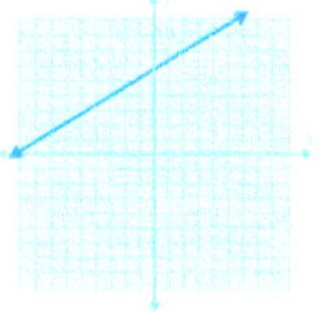

14. $x + 4y = -5$

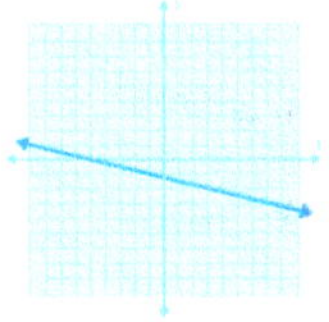

15. $2x - 3y = -5$

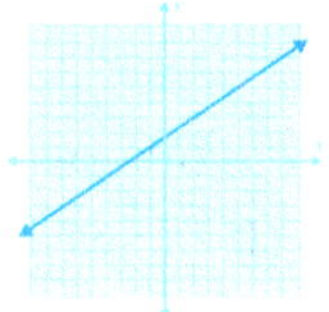

16. $8x + 6y = 23$

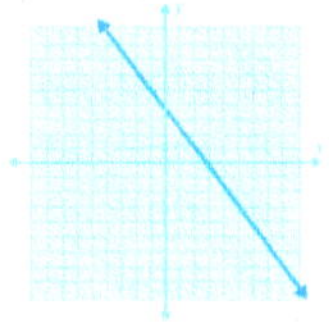

17. $y = \frac{1}{2}x + \frac{9}{2}$

18. $y = -\frac{3}{5}x + \frac{11}{5}$

19. $y = -\frac{1}{7}x + \frac{2}{7}$

20. $y = \frac{5}{7}x - \frac{8}{7}$

21. $y = -\frac{5}{4}x + 2$

22. $y = \frac{2}{3}x - \frac{5}{3}$

23. $y = -5$

24. $y = -\frac{7}{2}x + 4$

25. $y = -x + 4$

26. $y = \frac{1}{4}x$

27. $y = 6$

28. $x = -1$

29. $y = 7$

30. $x = 2$

31. $y = 7$

32. $y = 2$

Find the slope-intercept form of the equations described.

27. Find an equation for the horizontal line through the point $(-2, 6)$.

28. Find an equation for the vertical line through the point $(-1, -4)$.

29. Write an equation for the line parallel to the x-axis and containing the point $(2, 7)$.

30. Find an equation for the line parallel to the y-axis and containing the point $(2, -4)$.

31. Find an equation for the line perpendicular to $x = 4$ and that passes through $(-1, 7)$.

32. Find an equation for the line parallel to the line $-6y = 1$ and containing the point $(-3, 2)$.

33. Write an equation for the line parallel to the line $2x - y = 4$ and containing the origin. Graph both lines.

34. Find an equation for the line parallel to $7x - 3y = 1$ and containing the point $(1, 0)$. Graph both lines.

35. Write an equation for the line parallel to $5x = 7 + y$ and through the point $(-1, -3)$. Graph both lines.

36. Write an equation for the line that contains the point $(2, 2)$ and is perpendicular to the line $4x + 3y = 4$. Graph both lines.

37. Find an equation for the line that passes through the point $(4, -1)$ and is perpendicular to the line $5x - 3y + 4 = 0$. Graph both lines.

38. Write an equation for the line that is perpendicular to $8 - 3x - 2y = 0$ and passes through the point $(-4, -2)$.

39. Write an equation for the line through the origin that is perpendicular to $3x - y = 4$.

40. Find an equation for the line that is perpendicular to $2x + y = 5$ and that passes through $(6, -1)$.

41. Write an equation for the line that is perpendicular to $2x - y = 7$ and has the same y-intercept as $x - 3y = 6$.

42. Find an equation for the line with the same y-intercept as $5x + 4y = 12$ and that is perpendicular to $3x - 2y = 4$.

43. Show that the points $A(-2, 4)$, $B(0, 0)$, $C(6, 3)$, and $D(4, 7)$ are the vertices of a rectangle. (Plot the points and show that opposite sides are parallel and that adjacent sides are perpendicular.)

44. Show that the points $A(0, -1)$, $B(3, -4)$, $C(6, 3)$, and $D(9, 0)$ are the vertices of a parallelogram. (Plot the points and show that opposite sides are parallel.)

33. $y = 2x$

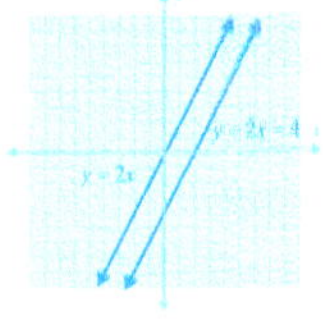

34. $y = \frac{7}{3}x - \frac{7}{3}$

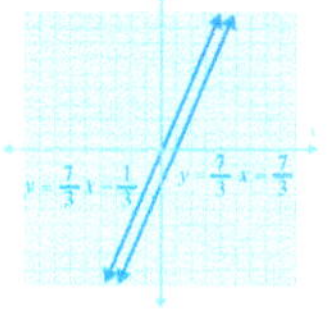

35. $y = 5x + 2$

36. $y = \frac{3}{4}x + \frac{1}{2}$

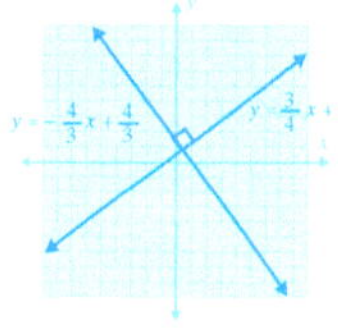

37. $y = -\frac{3}{5}x + \frac{7}{5}$

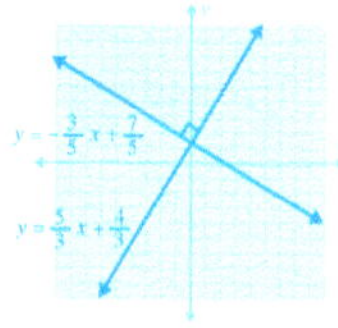

38. $y = \frac{2}{3}x + \frac{2}{3}$

39. $y = -\frac{1}{3}x$

40. $y = \frac{1}{2}x - 4$

41. $y = -\frac{1}{2}x - 2$

42. $y = -\frac{2}{3}x + 3$

*Determine whether each pair of lines is **a.** parallel, **b.** perpendicular, or **c.** neither. Graph both lines. (**Hint:** Write the equations in slope-intercept form and then compare slopes and y-intercepts.)*

45. $\begin{cases} y = -2x + 3 \\ y = -2x - 1 \end{cases}$

46. $\begin{cases} y = 3x + 2 \\ y = -\frac{1}{3}x + 6 \end{cases}$

47. $\begin{cases} 4x + y = 4 \\ x - 4y = 8 \end{cases}$

48. $\begin{cases} 2x + 3y = 5 \\ 3x + 2y = 10 \end{cases}$

49. $\begin{cases} 2x + 2y = 9 \\ 2x - y = 6 \end{cases}$

50. $\begin{cases} 3x - 4y = 16 \\ 4x + 3y = 15 \end{cases}$

Writing and Thinking About Mathematics

43. 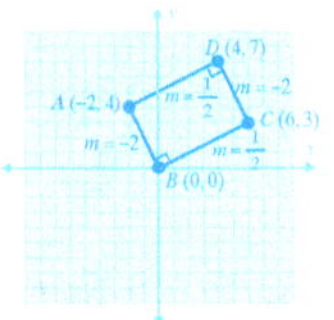

44. 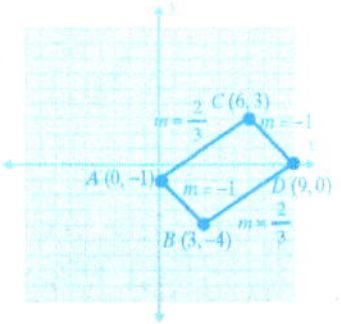

51. Handicapped access: Ramps for persons in wheelchairs or otherwise handicapped are now built into most buildings and walkways. (If ramps are not present in a building, then there must be elevators.) What do you think that the slope of a ramp should be for handicapped access? Look in your library or contact your local building permit office to find the recommended slope for such ramps.

52. Discuss the difference between the concepts of a line having slope 0 and a line having undefined slope.

HAWKES LEARNING SYSTEMS: INTRODUCTORY & INTERMEDIATE ALGEBRA SOFTWARE

- 4.4a Finding the Equation of a Line
- 4.4b Graphing Linear Equations in Point-Slope Form

45. 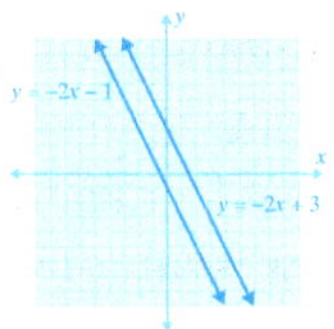
Parallel

46. 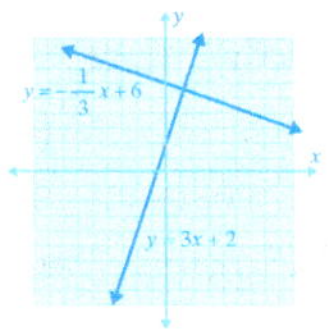
Perpendicular

47. 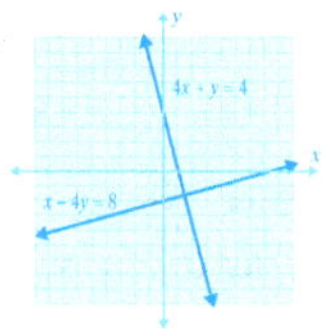
Perpendicular

48. 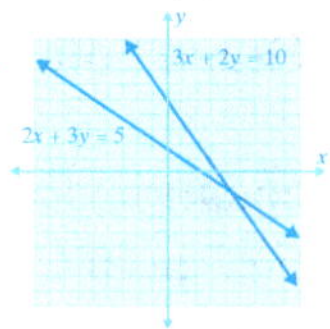
Neither

49. 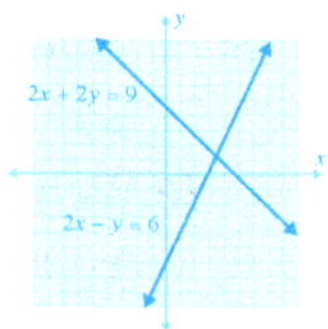
Neither

50. 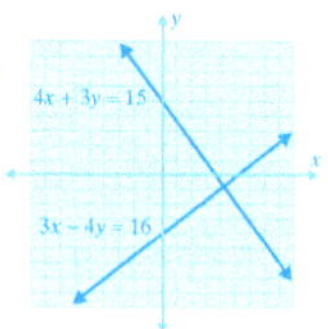
Perpendicular

51. The Americans with Disabilities Act recommends a slope of $\frac{1}{12}$ for business use and a slope of $\frac{1}{6}$ for residential use. Answers will vary.

52. A line with slope 0 has no rise and is horizontal. A line with undefined slope has no run and is vertical. Answers will vary.

4.5 Introduction to Functions and Function Notation

- *Find the domain and range of a relation or function.*
- *Determine whether a relation is or is not a function.*
- *Use the vertical line test to determine whether a graph is or is not the graph of a function.*
- *Write a function using function notation.*
- *Use a graphing calculator to graph functions.*

Everyday use of the term **function** is not far from the technical use in mathematics. For example, distance traveled is a function of time; profit is a function of sales; heart rate is a function of exertion; and interest earned is a function of principal invested. In this sense, one variable "depends on" (or "is a function of") another.

Mathematicians distinguish between graphs of ordered pairs of real numbers as those that represent **functions** and those that do not. For example, every equation of the form $y = mx + b$ represents a function and we say that y "is a function of" x. Thus lines that are not vertical are the graphs of functions. As the following discussion indicates, vertical lines do not represent functions.

NOTES

The ordered pairs discussed in this text are ordered pairs of real numbers. However, more generally, ordered pairs might be other types of pairs such as (child, mother), (city, state), or (name, batting average).

Relations and Functions

Relation, Domain, and Range

A **relation** is a set of ordered pairs of real numbers.

The **domain**, ***D***, of a relation is the set of all first coordinates in the relation.

The **range**, ***R***, of a relation is the set of all second coordinates in the relation.

In the graph of a relation, the horizontal axis (the x-axis) is called the **domain axis**, and the vertical axis (the y-axis) is called the **range axis**.

Example 1: Finding the Domain and Range

Find the domain and range for each of the following relations.

a. $g = \{(5, 7), (6, 2), (6, 3), (-1, 2)\}$

Solution: $D = \{5, 6, -1\}$ The set of all the first coordinates in g

$R = \{7, 2, 3\}$ The set of all the second coordinates in g

Note that 6 is written only once in the domain and 2 is written only once in the range, even though each appears more than once in the relation.

b. $f = \{(-1, 1), (1, 5), (0, 3)\}$

Solution: $D = \{-1, 1, 0\}$ The set of all the first coordinates in f

$R = \{1, 5, 3\}$ The set of all the second coordinates in f

Example 2: Reading the Domain and Range from the Graph of a Relation

Identify the domain and range from the graph of each relation.

a.

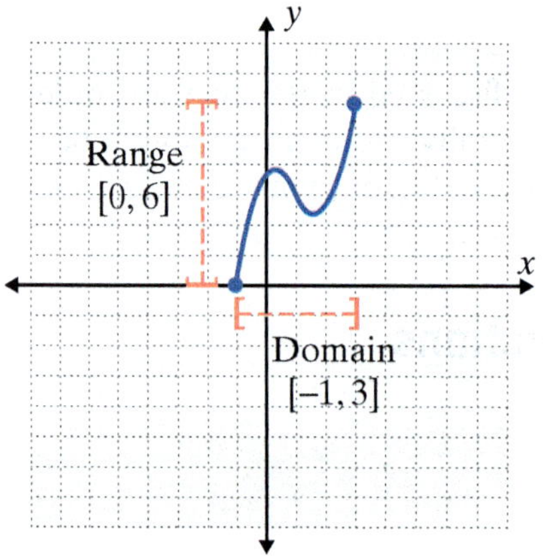

Solution:

The domain consists of the set of x-values for all points on the graph. In this case the domain is the interval $[-1, 3]$. The range consists of the set of y-values for all points on the graph. In this case, the range is the interval $[0, 6]$.

b.

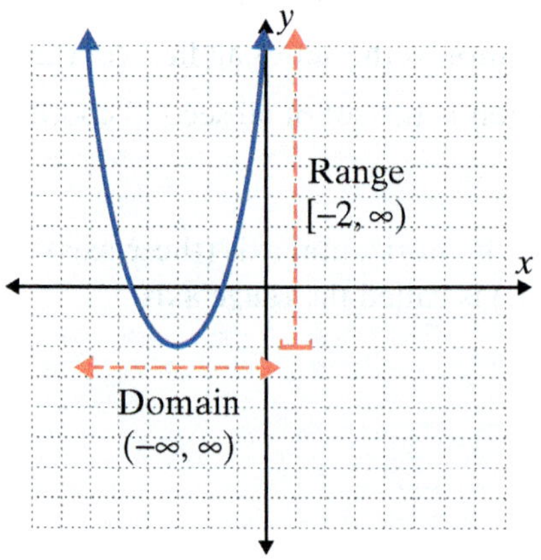

Solution:

There is no restriction on the x-values which means that for every real number there is a point on the graph with that number as its x-value. Thus the domain is the interval $(-\infty, \infty)$. The y-values begin at -2 and then increase to infinity. The range is the interval $[-2, \infty)$.

The relation $f = \{(-1, 1), (1, 5), (0, 3)\}$, used in Example 1b, meets a particular condition in that each first coordinate has a unique corresponding second coordinate. Such a relation is called a **function**. Notice that g in Example 1a is **not** a function because the first coordinate 6 has two corresponding second coordinates, 2 and 3. Also, for ease in discussion and understanding, the relations illustrated in Examples 1 and 3 have only a finite number of ordered pairs. The graphs of these relations are isolated dots or points. As we will see, the graphs of most relations and functions have an infinite number of points, and their graphs are smooth curves. (**Note:** Lines are also deemed to be curves in mathematics.)

Function

A **function** is a relation in which each domain element has exactly one corresponding range element.

The definition can also be stated in the following ways:

1. A function is a relation in which each first coordinate appears only once.
2. A function is a relation in which no two ordered pairs have the same first coordinate.

Example 3: Functions

Determine whether or not each of the following relations is a function.

a. $s = \{(2, 3), (1, 6), (2, \sqrt{5}), (0, -1)\}$

Solution: s is not a function. The number 2 appears as a first coordinate more than once.

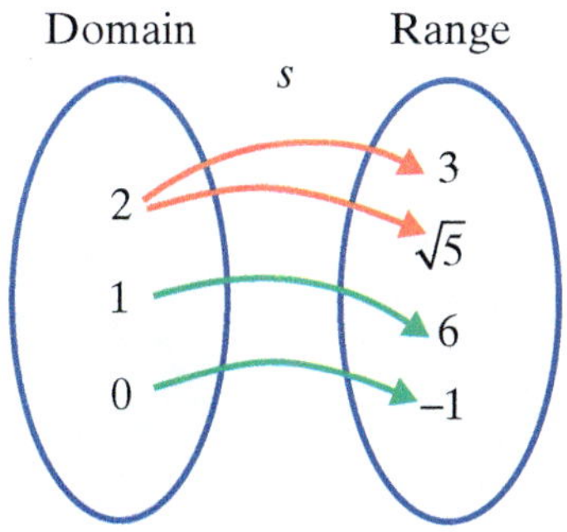

b. $t = \{(1, 5), (3, 5), (\sqrt{2}, 5), (-1, 5), (-4, 5)\}$

Solution: t is a function. Each first coordinate appears only once. The fact that the second coordinates are all the same has no effect on the concept of a function.

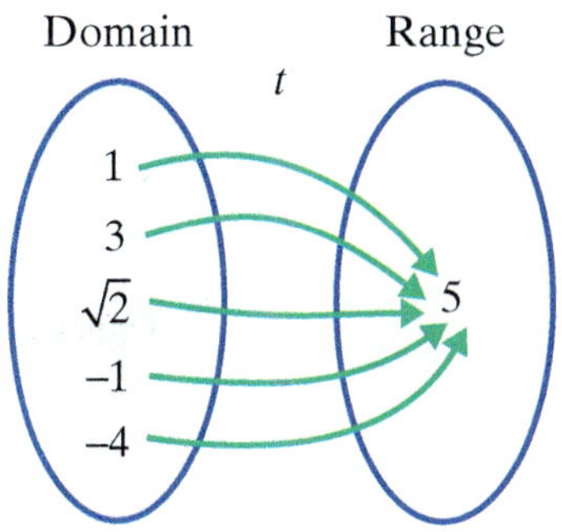

Vertical Line Test

If one point on the graph of a relation is directly above or below another point on the graph, then these points have the same first coordinate (or x-coordinate). Such a relation is **not** a function. Therefore, the **vertical line test** can be used to tell whether or not a graph represents a function (See Figures 1 and 2).

Vertical Line Test

If **any** vertical line intersects the graph of a relation at more than one point, then the relation is **not** a function.

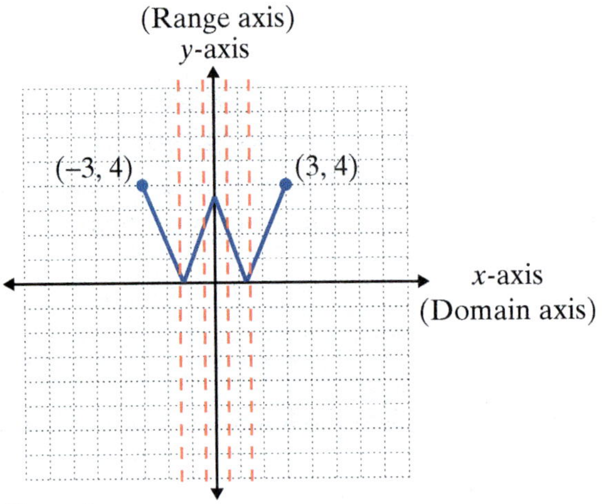

Figure 1

The vertical lines in Figure 1 indicate that this graph represents a function. From the graph, we see that the domain of the function is the interval of real numbers $[-3, 3]$ and the range of the function is the interval of real numbers $[0, 4]$.

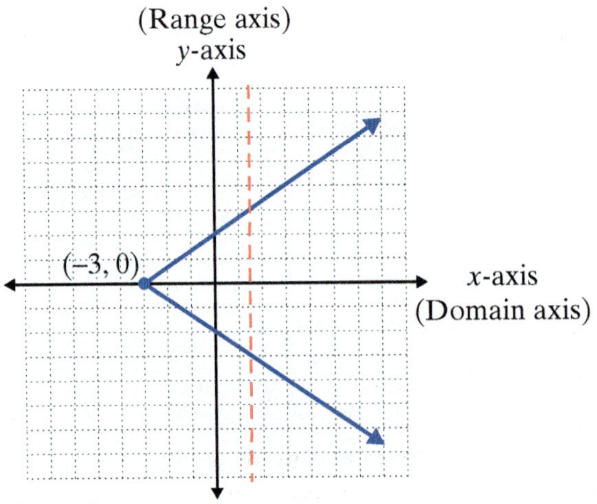

Figure 2

The relation in Figure 2 is **not** a function because the vertical line drawn intersects the graph at more than one point. Thus for that x-value, there is more than one corresponding y-value. Here $D = [-3, \infty)$ and $R = (-\infty, \infty)$.

Example 4: Vertical Line Test

Use the vertical line test to determine whether or not each of the following graphs represents a function. Then list the domain and range of each graph.

a.

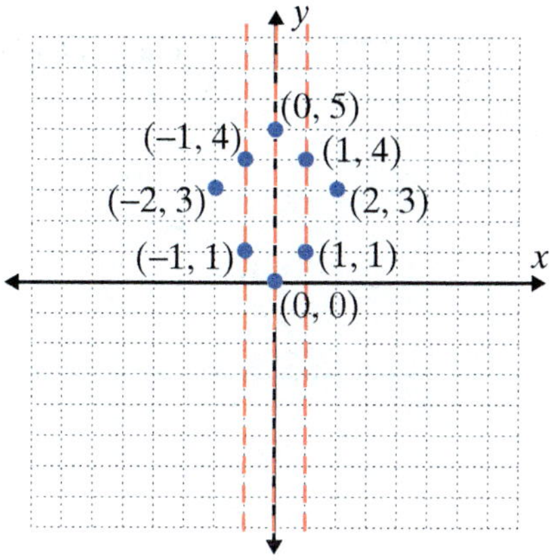

Solution:

The relation is **not** a function since a vertical line can be drawn that intersects the graph at more than one point. Listing the ordered pairs shows that several x-coordinates appear more than once.

$$r = \begin{Bmatrix} (-2,3),(-1,1),(-1,4), \\ (0,0),(0,5),(1,1),(1,4),(2,3) \end{Bmatrix}$$

Here $D = \{-2, -1, 0, 1, 2\}$ and $R = \{0, 1, 3, 4, 5\}$.

b.

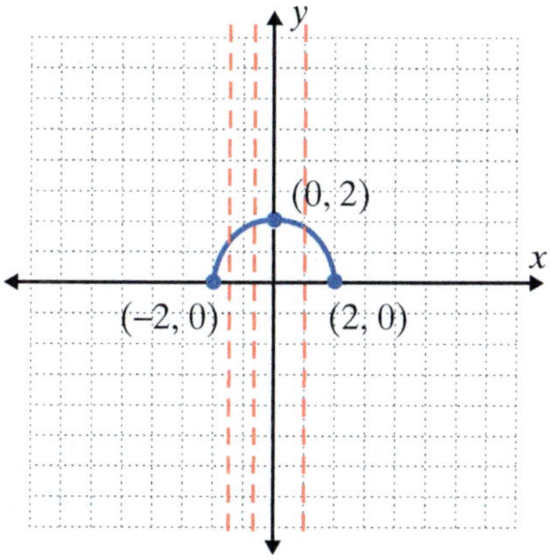

Solution:

The relation is a **function**. No vertical line will intersect the graph at more than one point. Several vertical lines are drawn to illustrate this.

For this function, we see from the graph that $D = [-2,2]$ and $R = [0,2]$.

c.

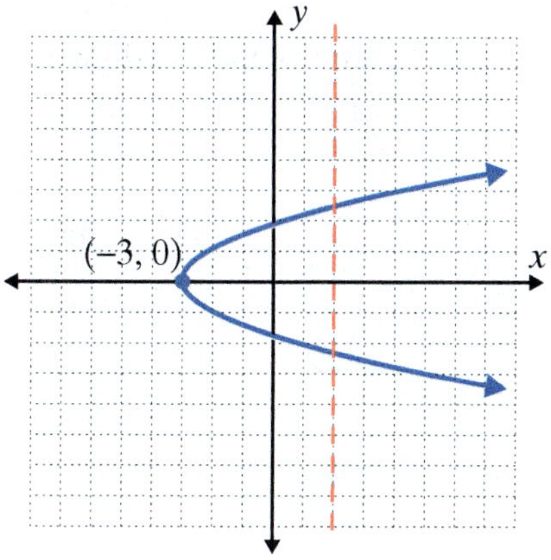

Solution:

The relation is **not** a function. At least one vertical line (drawn) intersects the graph at more than one point.

Here $D = [-3,\infty)$ and $R = (-\infty,\infty)$

Continued on the next page...

d. 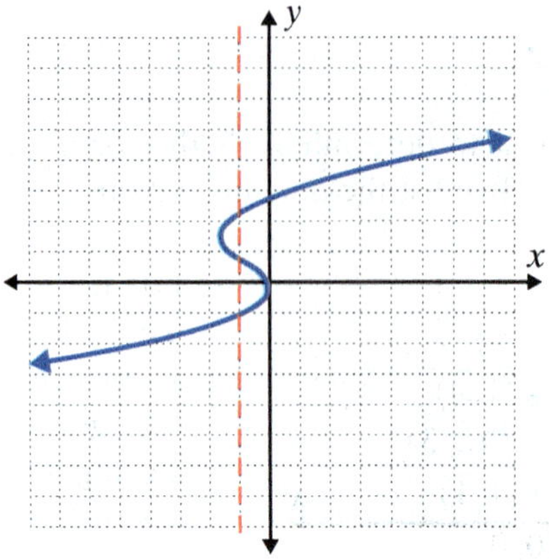

Solution:

The relation is **not** a function. At least one vertical line intersects the graph at more than one point.

Here $D=(-\infty,\infty)$ and $R=(-\infty,\infty)$.

Linear Functions

All non-vertical lines represent functions. Thus we have the following definition for a linear function.

Linear Function

A **linear function** is a function represented by an equation of the form

$$y = mx + b.$$

The domain of a linear function is the set of all real numbers: $D=(-\infty,\infty)$.

If the graph of a linear function is not a horizontal line, then the range is also the set of all real numbers. If the line is horizontal, then the domain is still the set of all real numbers; however, the range is a set containing just a single number. For example, the graph of the linear equation $y=5$ is a horizontal line. The domain of the function is the set of all real numbers and the range is the set containing only the number 5 (written $\{5\}$). Figures 3 and 4 show two linear functions and the domain and range of each function.

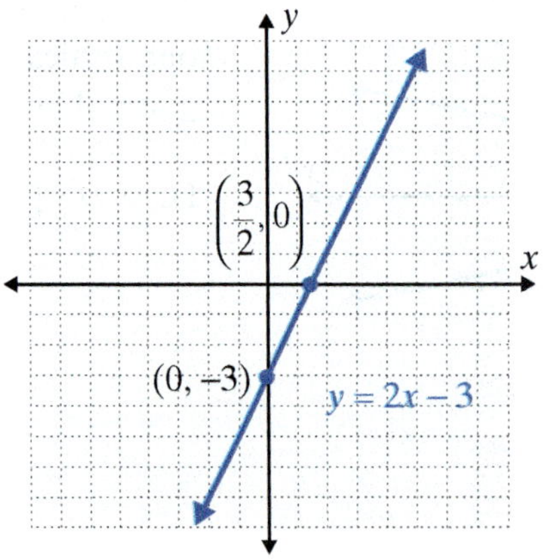

Figure 3

The graph of the linear function $y=2x-3$ is shown here.

$D=\{\text{all real numbers}\}=(-\infty,\infty)$

$R=\{\text{all real numbers}\}=(-\infty,\infty)$

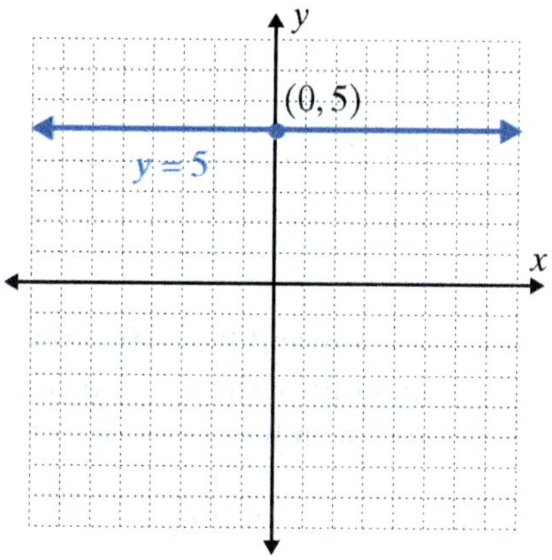

The graph of the linear function $y = 5$ is shown here.

$D = \{\text{all real numbers}\} = (-\infty, \infty)$

$R = \{5\}$

Figure 4

Domains of Nonlinear Functions

As we have seen the domain for a linear function is the set of all real numbers. Now for the nonlinear function

$$y = \frac{2}{x-1}$$

we say that the domain (all possible values for x) is every real number for which the expression $\frac{2}{x-1}$ is **defined**. Because the denominator cannot be 0, the domain consists of all real numbers except 1. That is, $D = (-\infty, 1) \cup (1, \infty)$ or simply $x \neq 1$. We adopt the following rule concerning equations and domains:

Unless a finite domain is explicitly stated, the domain will be implied to be the set of all real *x*-values for which the given function is defined. That is, the domain consists of all values of *x* that give real values for *y*.

NOTES

In determining the domain of a function, one fact to remember at this stage is that **no denominator can equal 0**. In future chapters we will discuss other nonlinear functions with limited domains.

Example 5: Domain

Find the domain for the function $y = \frac{2x+1}{x-5}$.

Solution: The domain is all real numbers for which the expression $\frac{2x+1}{x-5}$ is defined. Thus $D = (-\infty, 5) \cup (5, \infty)$ or $x \neq 5$, because the denominator is 0 when $x = 5$.

Note: Here interval notation tells us that x can be any real number except 5.

Function Notation

We have used the ordered pair notation (x, y) to represent points in relations and functions. As the vertical line test will show, linear equations of the form

$$y = mx + b$$

where the equation is solved for y, represent **linear functions**. Another notation, called **function notation**, is more convenient for indicating calculations of values of a function and indicating operations performed with functions. In function notation,

instead of writing $\boldsymbol{y}$, write $\boldsymbol{f(x)}$, read "***f*** **of** ***x***."

The letter f is the name of the function. The letters f, g, h, F, G, and H are commonly used in mathematics, but any letter other than x will do. We have used r, s, and t in previous examples.

The linear equation $y = -3x + 2$ represents a linear function and we can replace y with $f(x)$ as follows:

$$f(x) = -3x + 2.$$

Now in function notation, $f(4)$ means to replace x with 4 in the function.

$$f(4) = -3 \cdot (4) + 2 = -12 + 2 = -10$$

Thus the ordered pair $(4, -10)$ can be written as $(4, f(4))$.

Example 6: Function Evaluation

For the function $g(x) = 4x + 5$, find:

a. $g(2)$

Solution: $g(2) = 4(2) + 5 = 13$

b. $g(-1)$

Solution: $g(-1) = 4(-1) + 5 = 1$

c. $g(0)$

Solution: $g(0) = 4(0) + 5 = 5$

Function notation is valid for a wide variety of types of functions. Example 7 illustrates the use of function notation with a nonlinear function.

Example 7: Nonlinear Function Evaluation

For the function $h(x) = x^2 - 3x + 2$, find:

a. $h(4)$

Solution: $h(4) = (4)^2 - 3(4) + 2 = 16 - 12 + 2 = 6$

b. $h(0)$

Solution: $h(0) = (0)^2 - 3(0) + 2 = 0 - 0 + 2 = 2$

c. $h(-3)$

Solution: $h(-3) = (-3)^2 - 3(-3) + 2 = 9 + 9 + 2 = 20$

Using a TI-84 Plus Graphing Calculator to Graph Functions

There are many types and brands of graphing calculators available. For convenience and so that directions can be specific, only the TI-84 Plus graphing calculator is used in the related discussions in this text. Other graphing calculators may be used, but the steps required may be different from those indicated in the text. If you choose to use another calculator, be sure to read the manual for your calculator and follow the relevant directions.

In any case, remember that a calculator is just a tool to allow for fast calculations and to help in understanding some abstract concepts. A calculator does not replace your ability to think and reason or the need for algebraic knowledge and skills.

You should practice and experiment with your calculator until you feel comfortable with the results. **Do not be afraid of making mistakes. Note that** CLEAR **or** 2ND QUIT **will get you out of most trouble and allow you to start over.**

Some Basics about the TI-84 Plus

1. MODE: Turn the calculator ON and press the MODE key. The screen should be highlighted as shown below. If it is not, use the arrow keys in the upper right corner of the keyboard to highlight the correct words and press ENTER. It is particularly important that Func is highlighted. This stands for function. See the manual for the meanings of the rest of the terms.

2. WINDOW: Press the WINDOW key and the standard window will be displayed. By default, the standard window displays a graph with x-values and y-values ranging from –10 to 10 with tic marks on the axis every 1 unit.

```
WINDOW
 Xmin=-10
 Xmax=10
 Xscl=1
 Ymin=-10
 Ymax=10
 Yscl=1
 Xres=1
```

This window can be changed at any time by changing the individual numbers or pressing the ZOOM key and selecting an option from the menu displayed. Because of the shape of the display screen, the standard screen is not a square screen (one unit along the x-axis looks longer than one unit along the y-axis). Be aware that the slopes of lines are not truly depicted unless the screen is in a scale of about 3 : 2. A square screen can be attained by pressing zoom and `5: ZSquare` or by pressing the window key and setting `Xmin` = –15 and `Xmax` = 15 to give the x-axis a length of 30 and the y-axis a length of 20 (a ratio of 3:2).

3. Y=: The Y= key is in the upper left corner of the keyboard. This key will allow ten different functions to be entered. These functions are labeled as $Y_1, \ldots, Y_{10}$. The variable x may be entered using the X,T,θ,n key. The ^ key is used to indicate exponents. (Also note that the negative sign (–) is next to the ENTER key.) For example, the equation $y = x^2 + 3x$ would be entered as:

$$Y_1 = X\text{^}2 + 3X.$$

To change an entry, practice with the keys DEL (delete), CLEAR, and 2ND INS (insert).

```
Plot1 Plot2 Plot3
\Y1=X^2+3X
\Y2=
\Y3=
\Y4=
\Y5=
\Y6=
\Y7=
```

4. GRAPH: If this key is pressed, then the screen will display the graph of whatever functions are indicated in the Y= list with the = sign highlighted using the minimum and maximum values indicated in the current `WINDOW`. In many cases the `WINDOW` must be changed to accommodate the domain and range of the function or to show a point where two functions intersect.

5. TRACE: The TRACE key will display the current graph even if it is not already displayed and give the x- and y- coordinates of a point highlighted on the graph. The curve may be traced by pressing the left and right arrow keys. At each point on the graph, the corresponding x- and y- coordinates are indicated at the bottom of the screen. **(Remember that because of the limitations of the pixels (lighted dots) on the screen, these x- and y- coordinates are generally only approximations.)**

6. CALC: The CALC key (press 2ND TRACE) gives a menu with seven items. Items 1 – 5 are used with graphs.

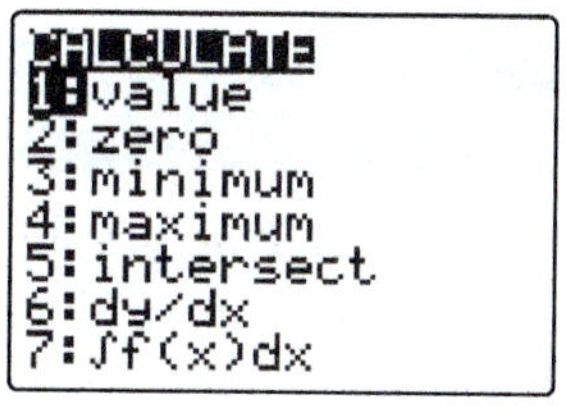

After displaying a graph, select CALC. Then press **2** and follow the steps outlined below to locate the point where the graph crosses the x-axis (the x-intercept). The graph must actually cross the axis. The x-value of this point is called a **zero** of the function because the corresponding y-value will be 0.

Step 1: With the left arrow, move the cursor to the left of the x-intercept on the graph. Press ENTER in response to the question "LeftBound?".

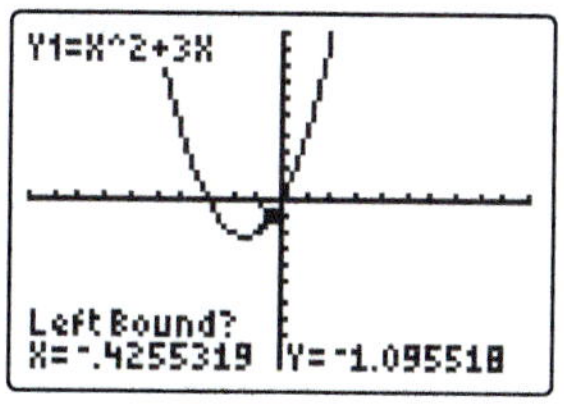

Step 2: With the right arrow, move the cursor to the right of the x-intercept on the graph. Press ENTER in response to the question "RightBound?".

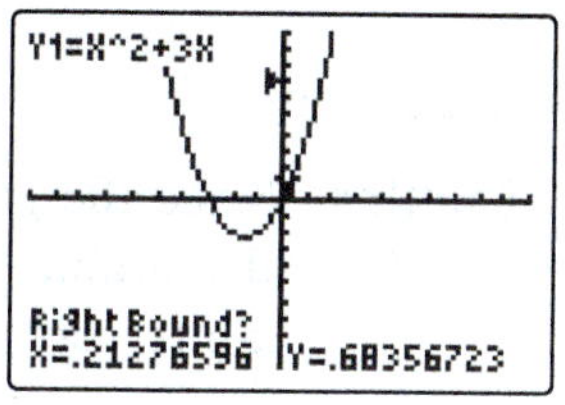

Step 3: With the left arrow, move the cursor near the x-intercept. Press ENTER in response to the question "Guess?". The calculator's estimate of the zero will appear at the bottom of the display.

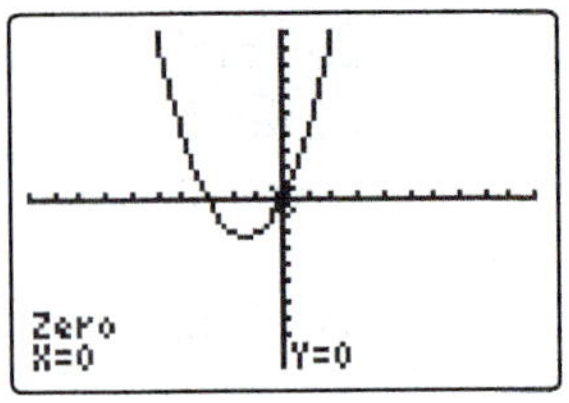

Example 8: Graphing Functions with a TI-84 Plus

Use a TI-84 Plus graphing calculator to find the graphs of each of the following functions. Use the CALC key to find the point where each graph intersects the x-axis. Changing the WINDOW may help you get a "better" or "more complete" picture of the function. This is a judgment call on your part.

Teaching Notes:
Tell your students that if the message ERR: SYNTAX 1: Quit 2: Goto appears on the display, the error will be highlighted if they press "2: Goto".

a. $3x + y = -1$

Solution: To have the calculator graph a nonvertical straight line, you must first solve the equation for y. Solving for y gives,

$$y = -3x - 1.$$

(It is important that the (–) key be used to indicate the negative sign in front of $3x$. This is not the same as the subtraction key.)

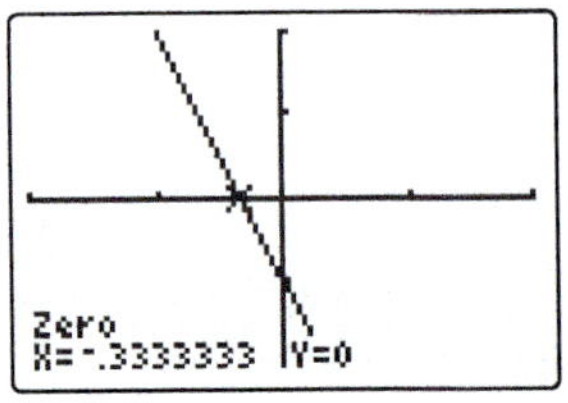

Note: Vertical lines are not functions and cannot be graphed by the calculator in function mode.

b. $y = x^2 + 3x$

Solution: Since the graph of this function has two x-intercepts, we have shown the graph twice. Each graph shows the coordinates of a distinct x-intercept.

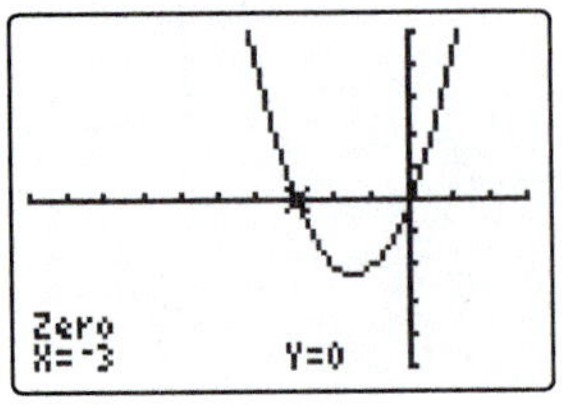

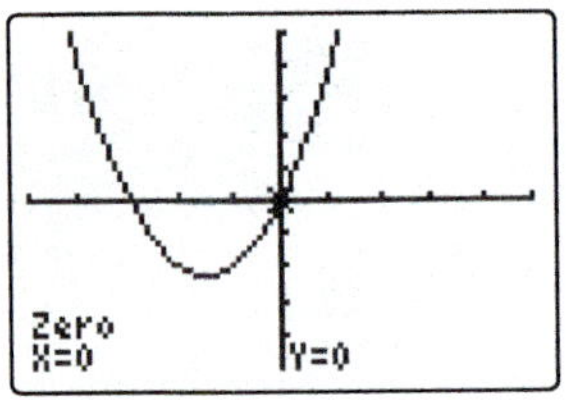

c. $y = 2x - 1$; $y = 2x + 1$; $y = 2x + 3$

Solution:

$y = 2x - 1$

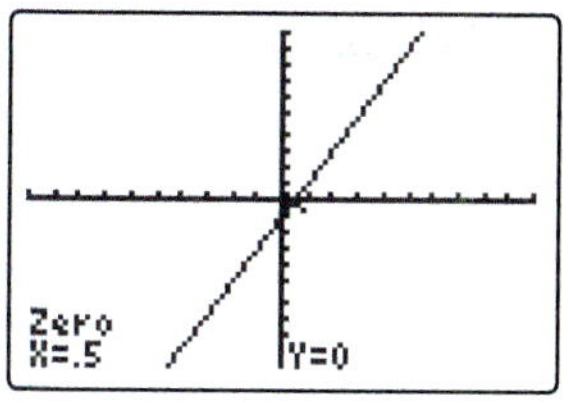

$y = 2x + 1$

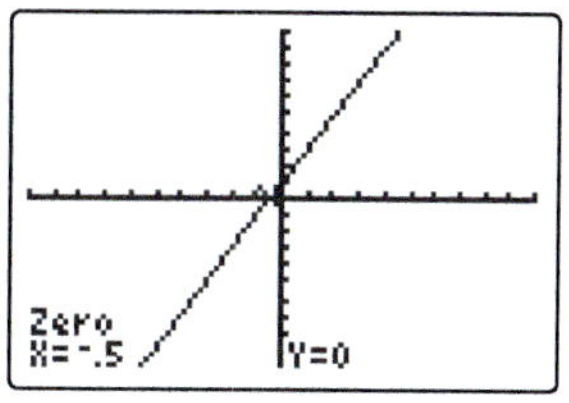

$y = 2x + 3$

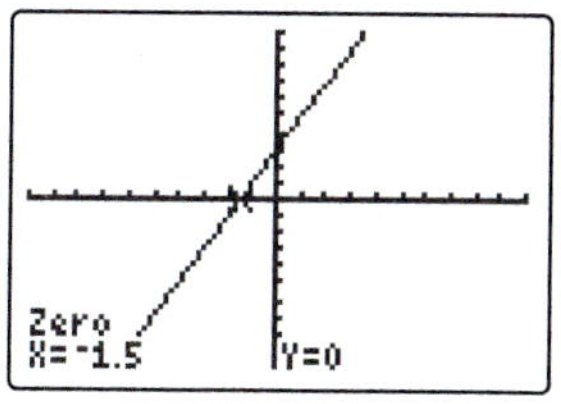

NOTES

The standard window shows 96 pixels across the window and 64 pixels up and down the window. This gives a ratio of 3 to 2 and can give a slightly distorted view of the actual graph because the vertical pixels are squeezed into a smaller space. For Example 8c, the graphs of all three functions are in the standard window. Experiment by changing the window to a square window, say –9 to 9 for x and –6 to 6 for y. Then graph the functions and notice the slight differences (and better representation) in the appearances on the display.

Practice Problems

1. State the domain and range of the relation $\{(5,6),(7,8),(9,0.5),(11,0.3)\}$. Is the relation a function? Explain briefly.

2. Use the vertical line text to determine whether the graph on the right represents a function.

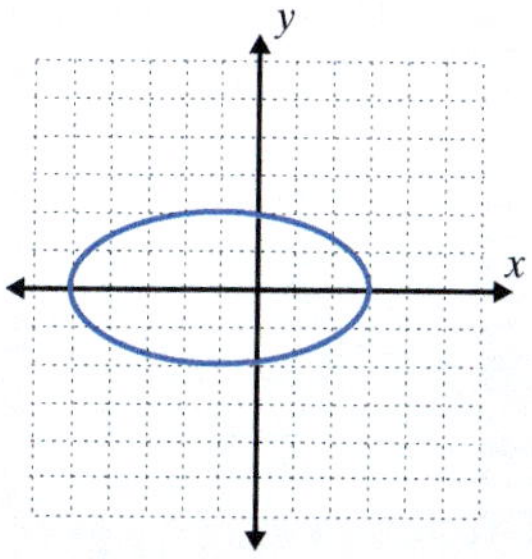

3. For the function $f(x) = 3x^2 + x - 4$, find **a.** $f(2)$ **b.** $f(0)$ **c.** $f(-1)$.

Answers to Practice Problems: **1.** $D = \{5, 7, 9, 11\}$; $R = \{0.3, 0.5, 6, 8\}$ Yes, the relation is a function because each x-coordinate appears only once.
2. Not a function **3. a.** 10 **b.** –4 **c.** –2

4.5 Exercises

1. $\begin{Bmatrix} (-4,0),(-1,4), \\ (1,2),(2,5), \\ (6,-3) \end{Bmatrix}$; $D=\{-4,-1,1,2,6\}$; $R=\{-3,0,2,4,5\}$; Function

2. $\begin{Bmatrix} (-4,-1),(0,0), \\ (0,3),(2,-2),(3,1) \end{Bmatrix}$; $D=\{-4,0,2,3\}$; $R=\{-2,-1,0,1,3\}$; Not a function

3. $\begin{Bmatrix} (-5,-4),(-4,-2), \\ (-2,-2),(1,-2),(2,1) \end{Bmatrix}$; $D=\{-5,-4,-2,1,2\}$; $R=\{-4,-2,1\}$; Function

4. $\begin{Bmatrix} (-2,-1),(0,3), \\ (3,0),(4,-2),(6,2) \end{Bmatrix}$; $D=\{-2,0,3,4,6\}$; $R=\{-2,-1,0,2,3\}$; Function

5. $\begin{Bmatrix} (-4,-3),(-4,1), \\ (-1,-1),(-1,3),(3,-4) \end{Bmatrix}$; $D=\{-4,-1,3\}$; $R=\{-4,-3,-1,1,3\}$; Not a function

6. $\begin{Bmatrix} (-2,6),(-1,-5), \\ (0,-7),(3,2),(4,6) \end{Bmatrix}$; $D=\{-2,-1,0,3,4\}$; $R=\{-7,-5,2,6\}$; Function

7. $\begin{Bmatrix} (-5,-5),(-5,3),(0,5), \\ (1,-2),(1,2) \end{Bmatrix}$; $D=\{-5,0,1\}$; $R=\{-5,-2,2,3,5\}$; Not a function

List the sets of ordered pairs that correspond to the points. State the domain and range and indicate which of the relations are also functions.

1.

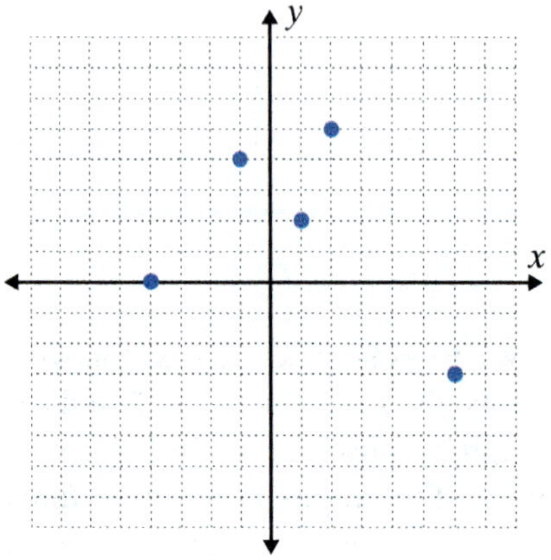

2.

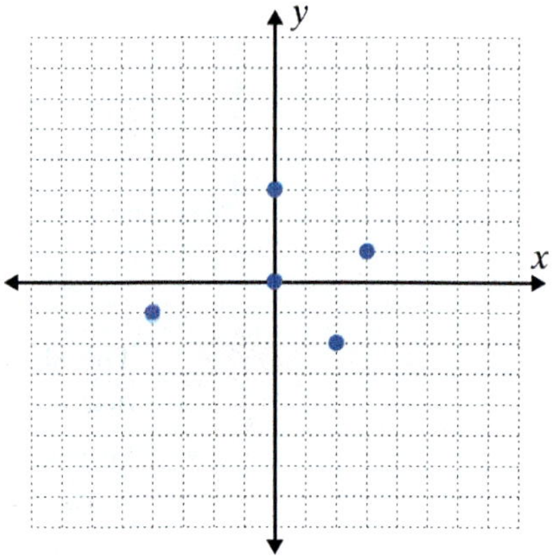

3.

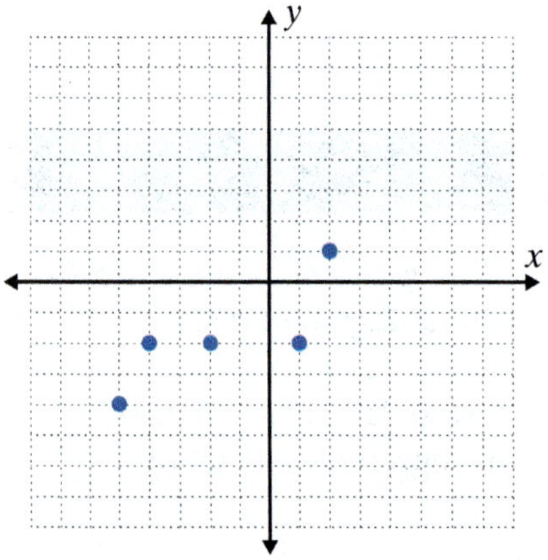

4.

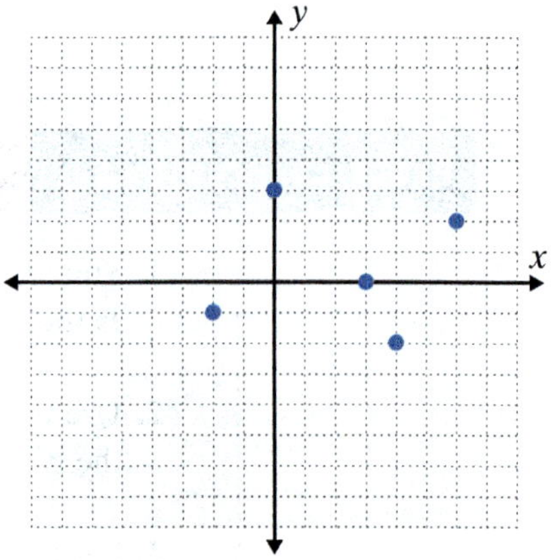

5.

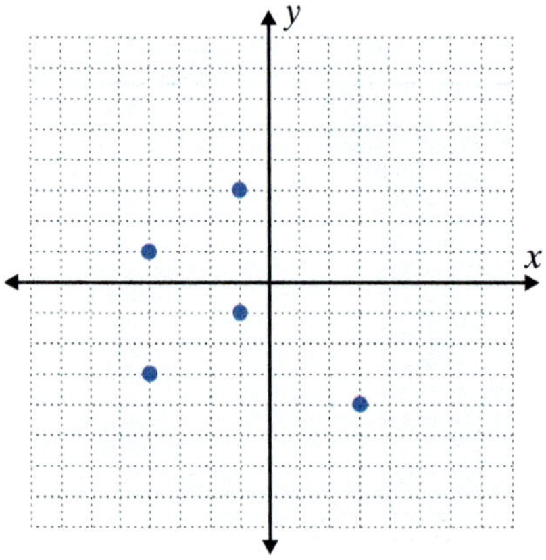

6.

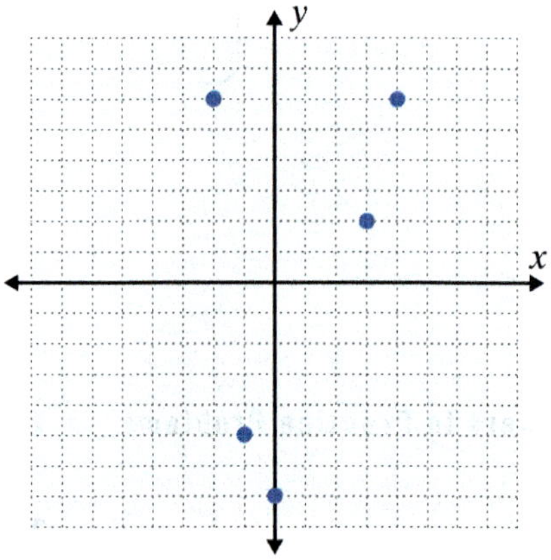

8. $\left\{\begin{matrix}(-3, 3), (-1, 2), \\ (1, 1), (3, 0), (5, -1)\end{matrix}\right\}$; $D = \{-3, -1, 1, 3, 5\}$; $R = \{-1, 0, 1, 2, 3\}$; Function

9. $D = \{-3, 0, 1, 2, 4\}$; $R = \{-2, -1, 0, 5, 6\}$; Function

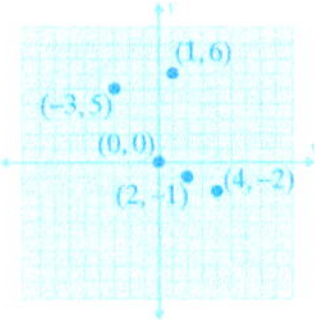

7.

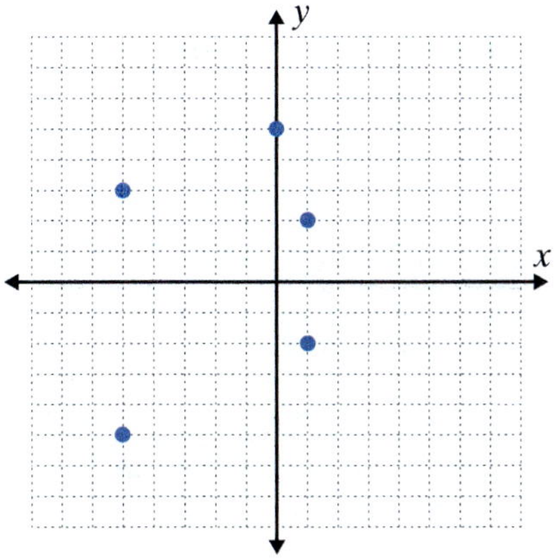

8.

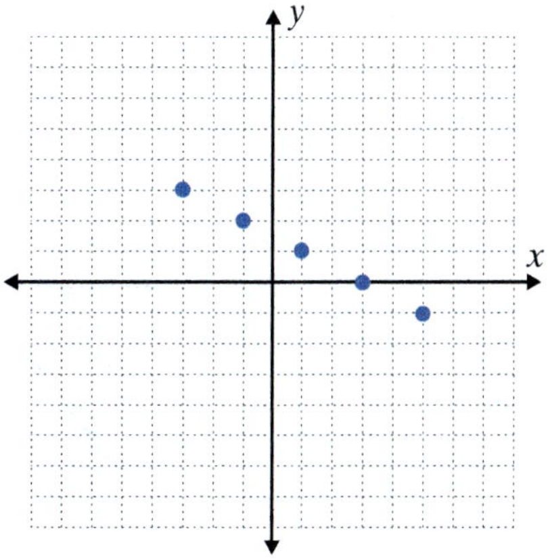

Graph the relations. State the domain and range and indicate which of the relations are functions.

10. $D = \{-1, 0, 1, 2, 4\}$; $R = \{-5, -3, 2, 3\}$; Function

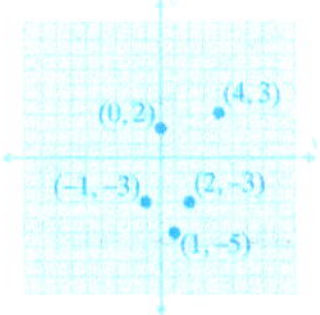

9. $f = \{(0, 0), (1, 6), (4, -2), (-3, 5), (2, -1)\}$

10. $h = \{(1, -5), (2, -3), (-1, -3), (0, 2), (4, 3)\}$

11. $g = \{(-4, 4), (-3, 4), (1, 4), (2, 4), (3, 4)\}$

12. $f = \{(-3, -3), (0, 1), (-2, 1), (3, 1), (5, 1)\}$

13. $s = \{(0, 2), (-1, 1), (2, 4), (3, 5), (-3, 5)\}$

14. $t = \{(-1, -4), (0, -3), (2, -1), (4, 1), (1, 1)\}$

15. $f = \{(-1, 4), (-1, 2), (-1, 0), (-1, 6), (-1, -2)\}$

16. $g = \{(0, 0), (-2, -5), (2, 0), (4, -6), (5, 2)\}$

11. $D = \{-4, -3, 1, 2, 3\}$; $R = \{4\}$; Function

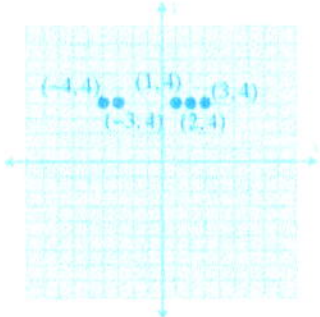

12. $D = \{-3, -2, 0, 3, 5\}$; $R = \{-3, 1\}$; Function

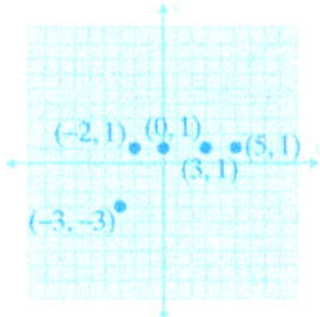

13. $D = \{-3, -1, 0, 2, 3\}$; $R = \{1, 2, 4, 5\}$; Function

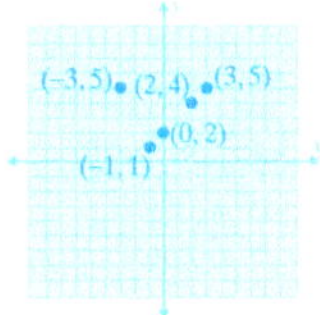

14. $D = \{-1, 0, 1, 2, 4\}$; $R = \{-4, -3, -1, 1\}$; Function

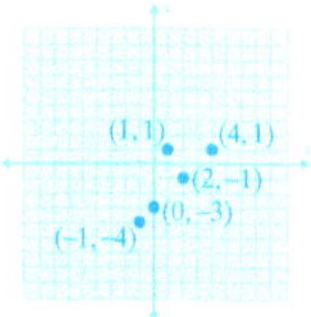

15. $D = \{-1\}$; $R = \{-2, 0, 2, 4, 6\}$; Not a function

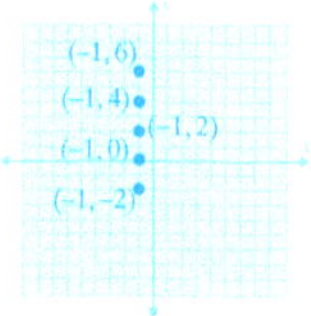

Use the vertical line test to determine whether or not each graph represents a function. State the domain and range using interval notation.

17.

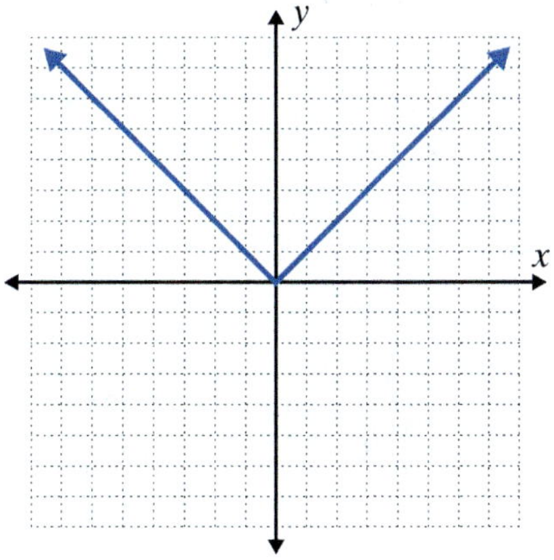

18.

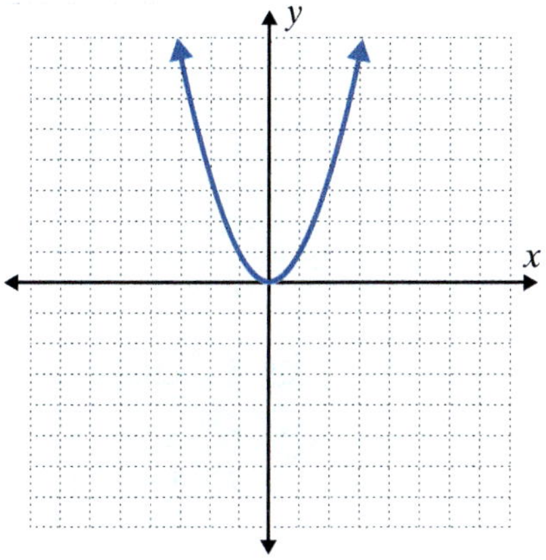

16. $D=\{-2,0,2,4,5\}$; $R=\{-6,-5,0,2\}$; Function

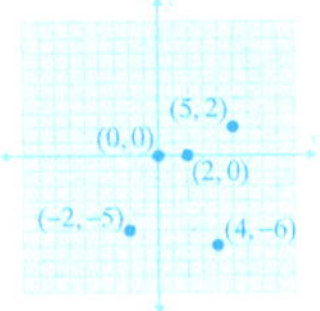

17. Function; $D=(-\infty,\infty)$; $R=[0,\infty)$

18. Function; $D=(-\infty,\infty)$; $R=[0,\infty)$

19. Function; $D=(-\infty,\infty)$; $R=(-\infty,\infty)$

20. Function; $D=(-\infty,\infty)$; $R=(-\infty,3]$

21. Not a function; $D=(-\infty,\infty)$; $R=(-\infty,\infty)$

22. Function; $D=(-\infty,\infty)$; $R=(-\infty,5]$

23. Not a function; $D=(-\infty,\infty)$; $R=(-\infty,\infty)$

24. Not a function; $D=[-6,4]$; $R=[-3,3]$

25. Function; $D=[-5,5]$; $R=[-2,2]$

26. Not a function; $D=[-4,4]$; $R=[-3,3]$

27. Not a function; $D=\{-3\}$; $R=(-\infty,\infty)$

19.

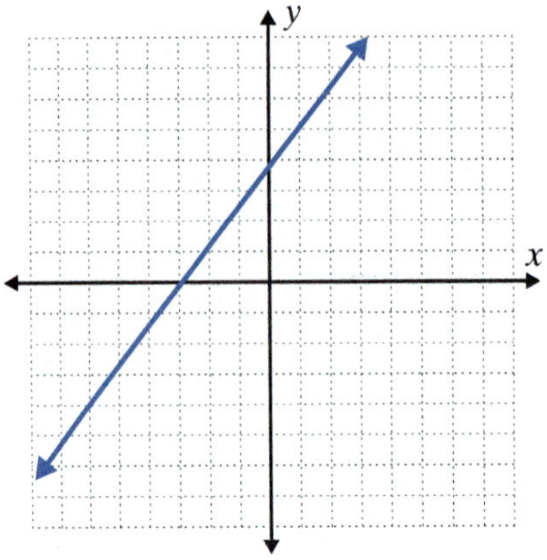

20.

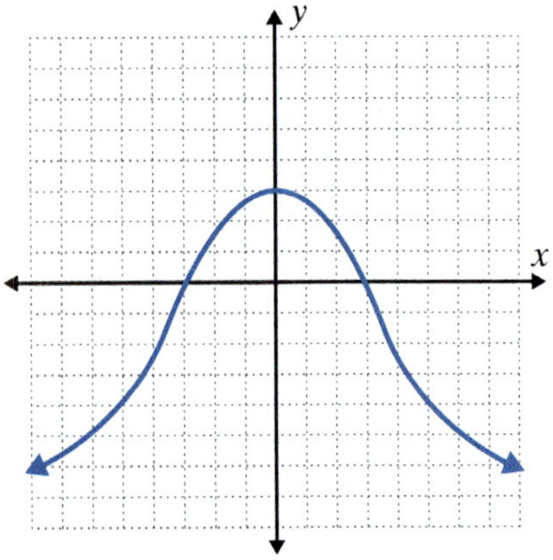

21.

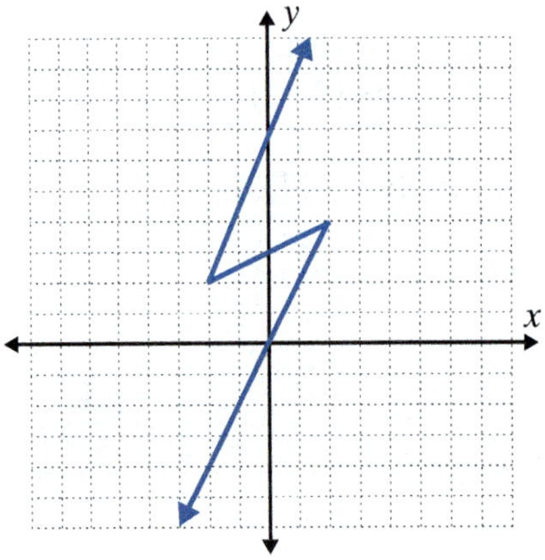

22.

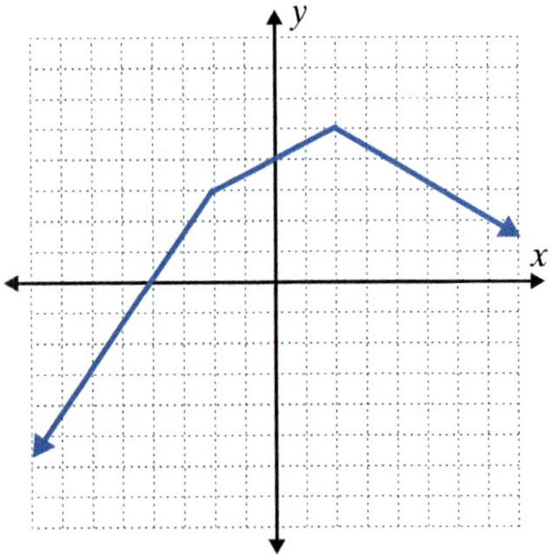

23.

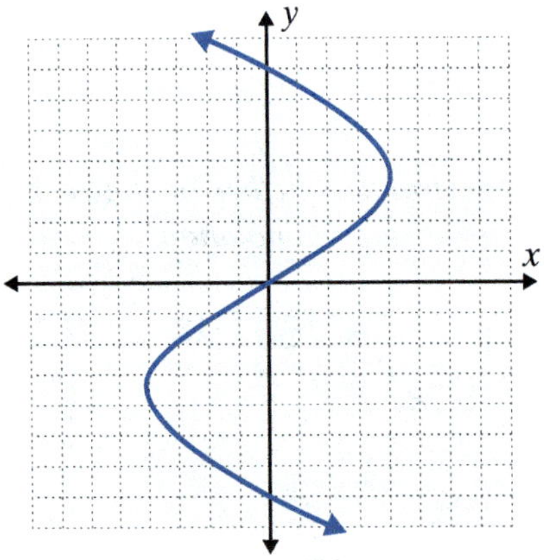

24.

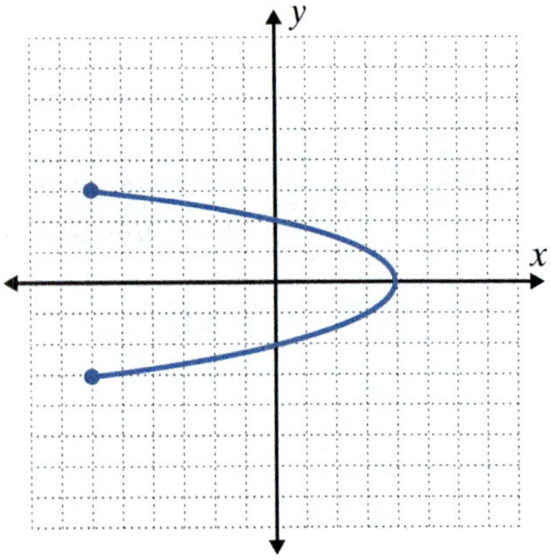

28. Function; $D=(-\infty,-2]\cup[-1,\infty)$; $R=(-\infty,\infty)$

29. $\left\{(-9,-26),\left(-\frac{1}{3},0\right),(0,1),\left(\frac{4}{3},5\right),(2,7)\right\}$

30. $\left\{(-4,5),\left(-2,\frac{7}{2}\right),(0,2),\left(3,-\frac{1}{4}\right),(4,-1)\right\}$

31. $\{(-2,-11),(-1,-2),(0,1),(1,-2),(2,-11)\}$

32. $\left\{(-1,3),(0,0),\left(\frac{1}{2},-\frac{15}{8}\right),(1,-3),(2,0)\right\}$
33. $D=(-\infty,\infty)$
34. $D=(-\infty,\infty)$
35. $D=(-\infty,0)\cup(0,\infty)$ or $x\neq 0$
36. $D=(-\infty,0)\cup(0,\infty)$ or $x\neq 0$
37. $D=(-\infty,3)\cup(3,\infty)$ or $x\neq 3$
38. $D=(-\infty,6)\cup(6,\infty)$ or $x\neq 6$
39. a. -4 b. -16 c. -10
40. a. 19 b. -17 c. 7
41. a. 0 b. 12 c. 56
42. a. -10 b. 86 c. 86
43. a. -3 b. 0 c. 3
44. a. 0 b. 144 c. 9

45. 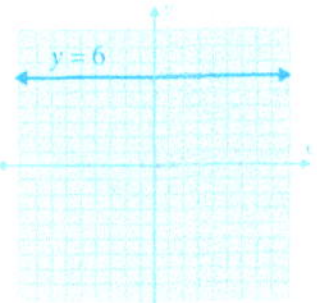

46. 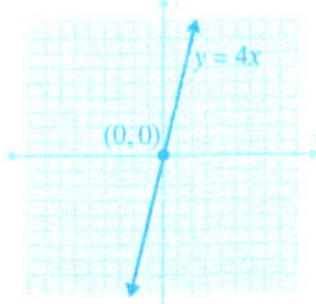

47. 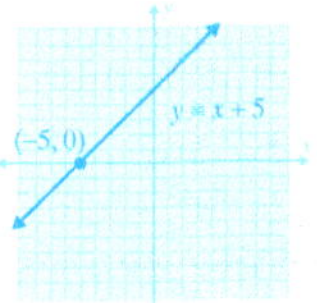

48. 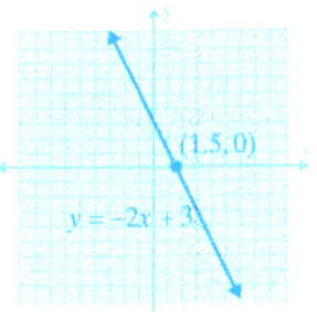

49. 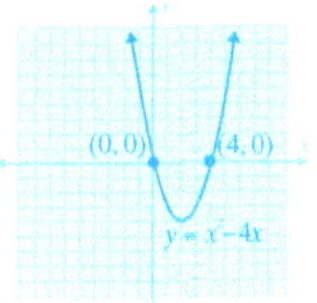

25. 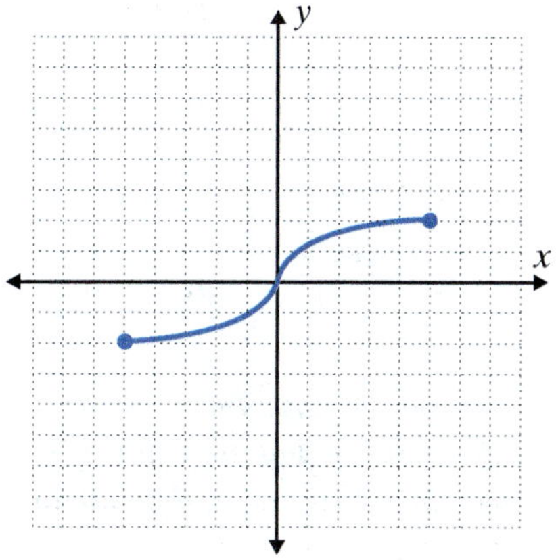

26. 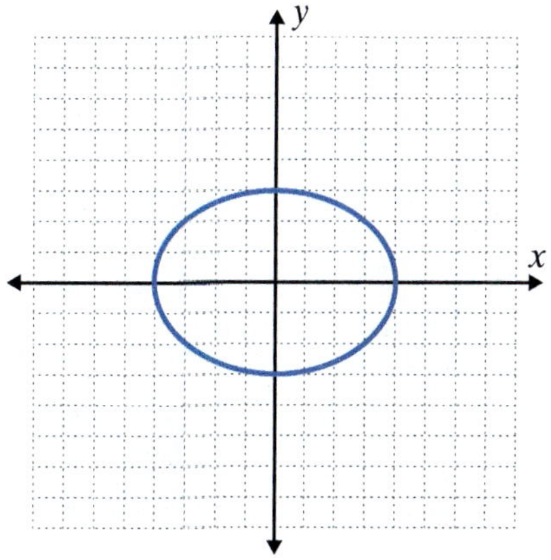

27. 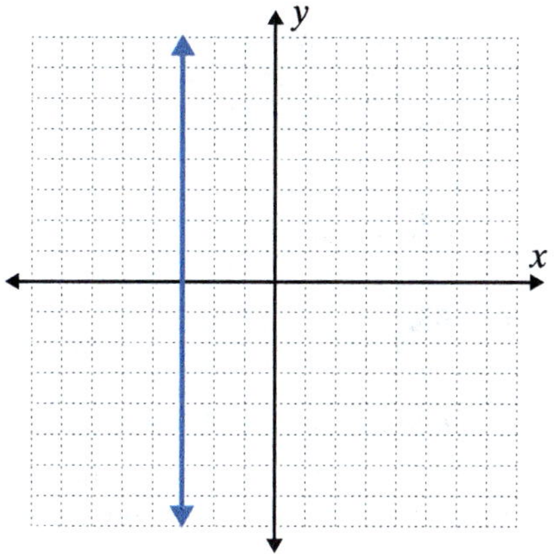

28. 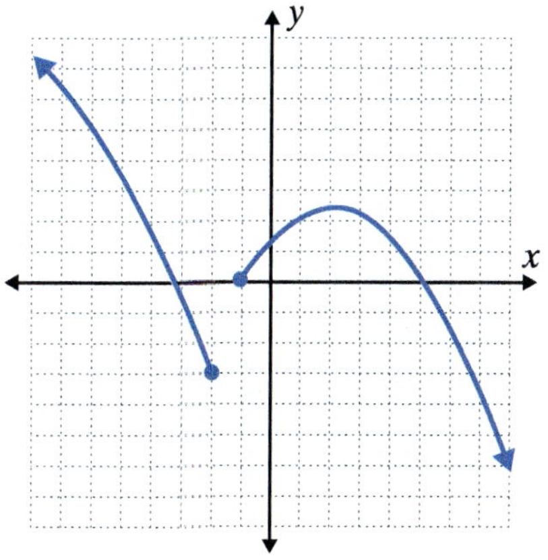

*Express the function as a set of ordered pairs for the given equation and given domain. (**Hint:** Substitute each domain element for x and find the corresponding y-coordinate.)*

29. $y=3x+1;\ D=\left\{-9,-\frac{1}{3},0,\frac{4}{3},2\right\}$

30. $y=-\frac{3}{4}x+2;\ D=\{-4,-2,0,3,4\}$

31. $y=1-3x^2;\ D=\{-2,-1,0,1,2\}$

32. $y=x^3-4x;\ D=\left\{-1,0,\frac{1}{2},1,2\right\}$

State the domains of the functions.

33. $y=-5x+10$

34. $2x+y=14$

35. $g(x)=\frac{8}{x}$

36. $h(x)=\frac{7}{3x}$

37. $y=\frac{13x^2-5x+8}{x-3}$

38. $f(x)=\frac{35}{x-6}$

Find the values of the functions as indicated.

39. $f(x)=3x-10$ Find **a.** $f(2)$ **b.** $f(-2)$ **c.** $f(0)$

40. $g(x)=-4x+7$ Find **a.** $g(-3)$ **b.** $g(6)$ **c.** $g(0)$

50.

51.
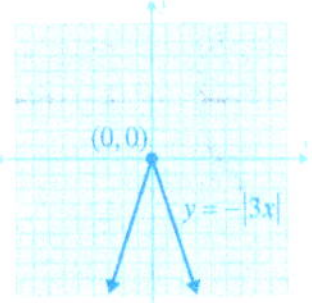

52.
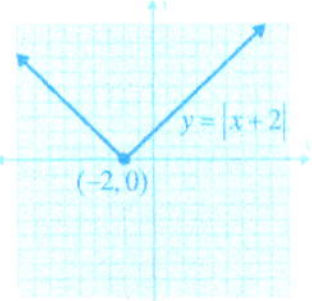

53.

54.
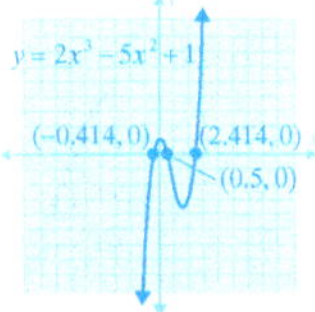

55.
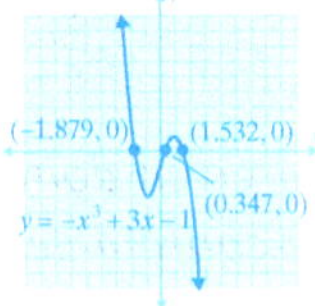

56.
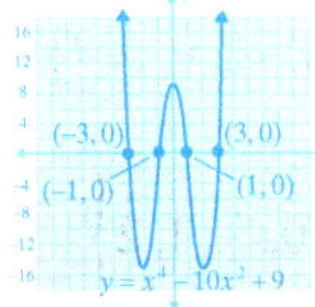

57.
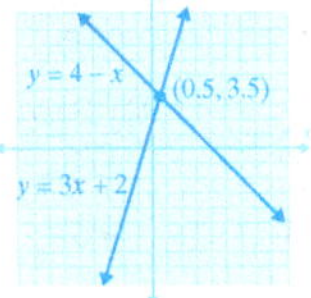

58.
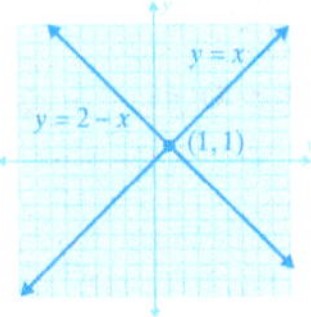

41. $G(x) = x^2 + 5x + 6$ Find **a.** $G(-2)$ **b.** $G(1)$ **c.** $G(5)$

42. $F(x) = 6x^2 - 10$ Find **a.** $F(0)$ **b.** $F(-4)$ **c.** $F(4)$

43. $h(x) = x^3 - 8x$ Find **a.** $h(-3)$ **b.** $h(0)$ **c.** $h(3)$

44. $P(x) = x^2 + 4x + 4$ Find **a.** $P(-2)$ **b.** $P(10)$ **c.** $P(-5)$

Use a graphing calculator to graph the functions. Use the CALC *features of the calculator to find x-intercepts, if any. (Remember that the value of y will be 0 at those points.) For absolute value functions, select the* MATH *menu, then the* NUM *menu, and then* 1: abs(. *Remember to press*) *after entering the absolute value.*

45. $y = 6$ **46.** $y = 4x$ **47.** $y = x + 5$

48. $y = -2x + 3$ **49.** $y = x^2 - 4x$ **50.** $y = 1 + 2x - x^2$

51. $y = -|3x|$ **52.** $y = |x + 2|$ **53.** $y = |x^2 - 3x|$

54. $y = 2x^3 - 5x^2 + 1$ **55.** $y = -x^3 + 3x - 1$ **56.** $y = x^4 - 10x^2 + 9$

Use the CALC *features of the calculator to find the coordinates of any points of intersection of the graphs. (**Hint:** Item 5 on the* CALC *menu* 5: intersect *will help in finding the point (or points) of intersection of two functions, if there is one.) In the* Y = *menu use both* Y_1 = *and* Y_2 = *to be able to graph both functions at the same time.*

57. $y = 3x + 2$, $y = 4 - x$ **58.** $y = 2 - x$, $y = x$ **59.** $y = 2x - 1$, $y = x^2$ **60.** $y = x + 3$, $y = -x^2 + x + 7$

The calculator display shows an incorrect graph for the corresponding equation. Explain how you know, just by looking at the graph, that a mistake has been made.

61. $y = 2x + 5$ **62.** $y = -3x + 4$ **63.** $y = \frac{2}{3}x - 2$

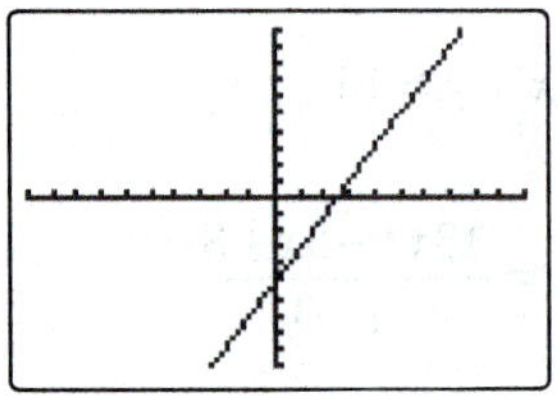
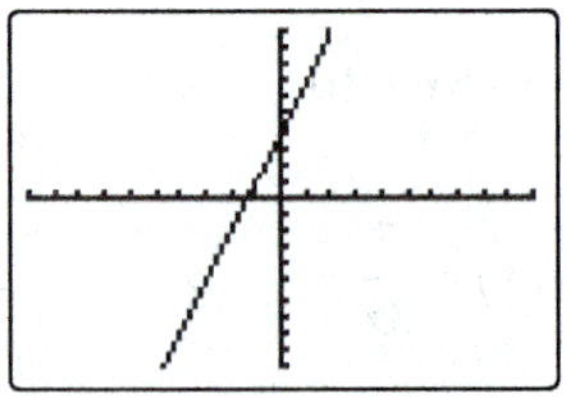
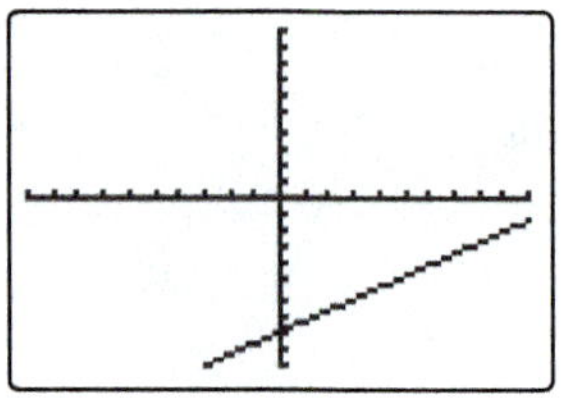

59. 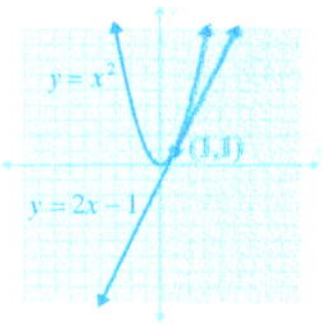

64. $y = -4x$

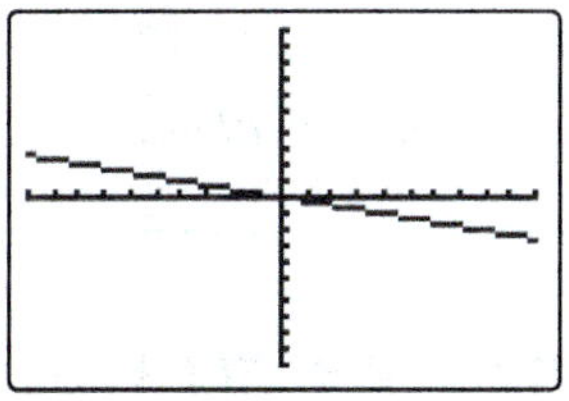

65. $y = -\frac{1}{3}x$

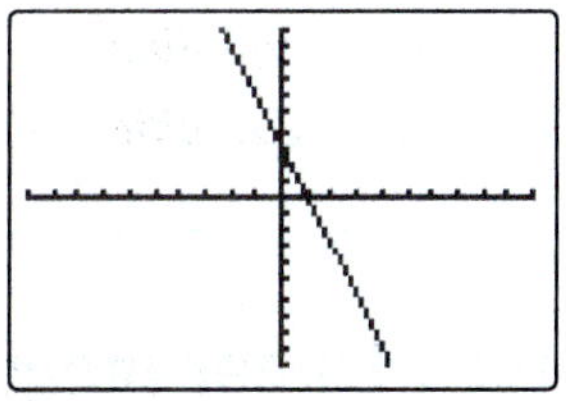

Writing and Thinking About Mathematics

60. 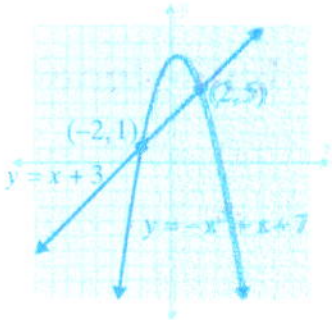

66. JUST FOR FUN: Enter a variety of functions in your calculator, investigate your findings, and report these to your class. Certainly, interesting discussions will follow!

HAWKES LEARNING SYSTEMS: INTRODUCTORY & INTERMEDIATE ALGEBRA SOFTWARE

- 4.5 Introduction to Functions and Function Notation

61. y-intercept = $(0,-5)$ (should be $(0,5)$) 62. slope = 4 (should be −3)

63. y-intercept = $(0,-8)$ (should be $(0,-2)$) 64. slope = $-\frac{1}{4}$ (should be the reciprocal, −4)

65. slope = −3 (should be the reciprocal, $-\frac{1}{3}$) and y-intercept = $(0,2)$ (should be $(0,0)$)

66. Answers will vary.

4.6 Graphing Linear Inequalities in Two Variables

- *Graph linear inequalities.*
- *Graph linear inequalities using a graphing calculator.*

Graphing Linear Inequalities: $y < mx + b$

In this section, we will develop techniques for analyzing and graphing linear inequalities. We need the following terminology:

Half-plane A straight line separates a plane into two **half-planes.** The points on one side of the line are in one of the half-planes, and the points on the other side of the line are in the other half-plane.

Boundary line The line itself is called the **boundary line.**

Closed half-plane If the boundary line is included in the solution set, then the half-plane is said to be **closed.**

Open half-plane If the boundary line is not included in the solution set, then the half-plane is said to be **open.**

Figure 1 shows both **a.** an open half-plane and **b.** a closed half-plane with the line $2x - 3y = 10$ as the boundary line. The solution set is the shaded region.

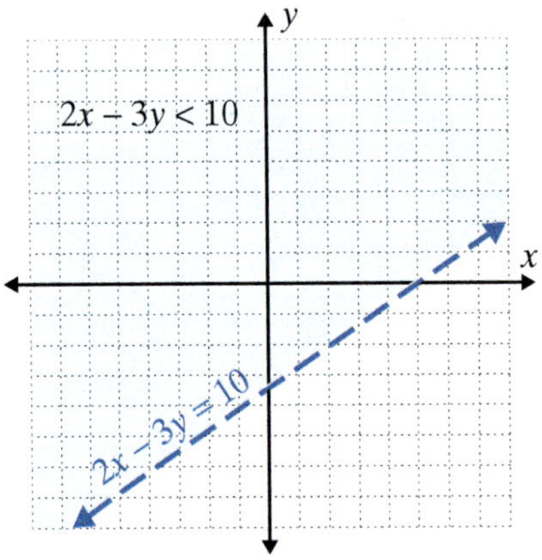

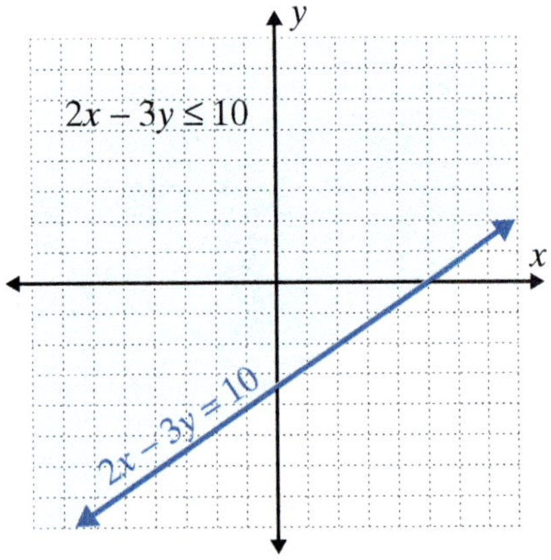

a. The points on the line $2x - 3y = 10$ are not included in the solution set so the line is dashed. The half-plane is open.

b. The points on the line $2x - 3y = 10$ are included in the solution set so the line is solid. The half-plane is closed.

Figure 1

Graphing Linear Inequalities

1. First, graph the boundary line (dashed if the inequality is $<$ or $>$, solid if the inequality is $\le$ or $\ge$).
2. Next, determine which side of the line to shade using one of the following methods.

Method 1

a. Test any one point obviously on one side of the line.

b. If the test-point satisfies the inequality, shade the half-plane on that side of the line. Otherwise, shade the other half-plane.

(**Note:** The point $(0, 0)$, if it is not on the boundary line, is usually the easiest point to test.)

Method 2

a. Solve the inequality for y (assuming that the line is not vertical).

b. If the solution shows $y <$ or $y \le$, then shade the half-plane below the line.

c. If the solution shows $y >$ or $y \ge$, then shade the half-plane above the line.

(**Note:** If the boundary line is vertical, then solve for x. If the solution shows $x >$ or $x \ge$, then shade the half-plane to the right. If the solution shows $x <$ or $x \le$, then shade the half-plane to the left.)

3. The shaded half-plane (and the line if it is solid) is the solution to the inequality.

Example 1: Graphing Linear Inequalities

Teaching Notes:
While some students will recognize that the origin $(0, 0)$ is the easiest point to use as a test point, it is a good idea to have students "test" some points on both sides of the boundary line upon first introduction, so that they develop confidence in the test (and solution set).

a. Graph the half-plane that satisfies the inequality $2x + y \le 6$.

Solution: Method 1 is used in this example.

Step 1: Graph the boundary line $2x + y = 6$ as a solid line because the inequality is $\le$ (less than or **equal to**).

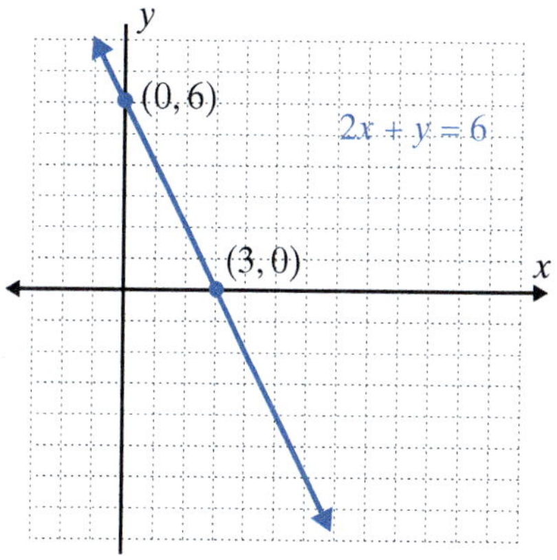

Continued on the next page...

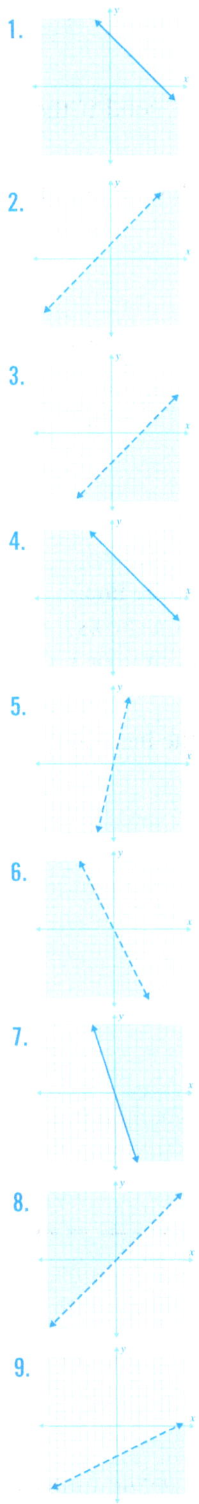

Step 2: Test any point on one side of the line. In this example, we have chosen $(0, 0)$.

$$2(0)+(0)\le 6$$
$$0\le 6$$

This is a true statement.

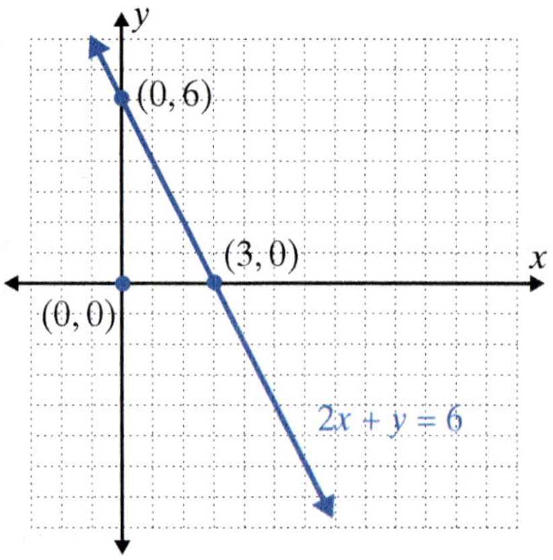

Step 3: Shade the half-plane on the same side of the line as the point $(0, 0)$. (The shaded half-plane and the boundary line is the solution set to the inequality.)

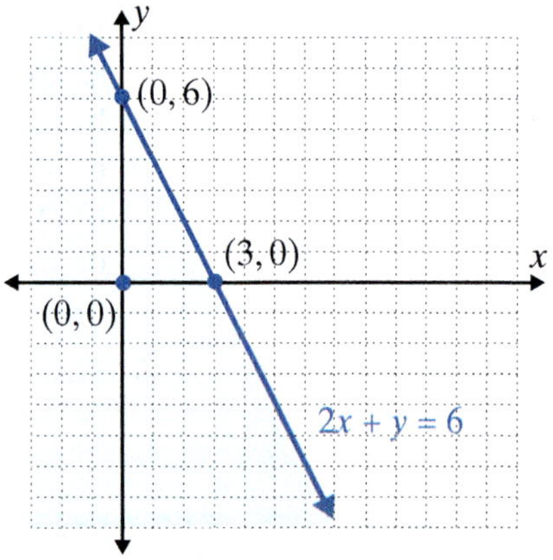

b. Graph the solution set to the inequality $y > 2x$.

Solution: Since the inequality is already solved for y, Method 2 is easy to apply.

Step 1: Graph the boundary line $y = 2x$ as a dashed line because the inequality is $>$.

Step 2: The solution shows $>$, so by Method 2, the graph consists of those points above the line. Shade the half-plane above the line.

Note: As a check, we see that the point $(-3, 0)$ gives $0 > -6$, a true statement. Thus we know we have shaded the correct half-plane.

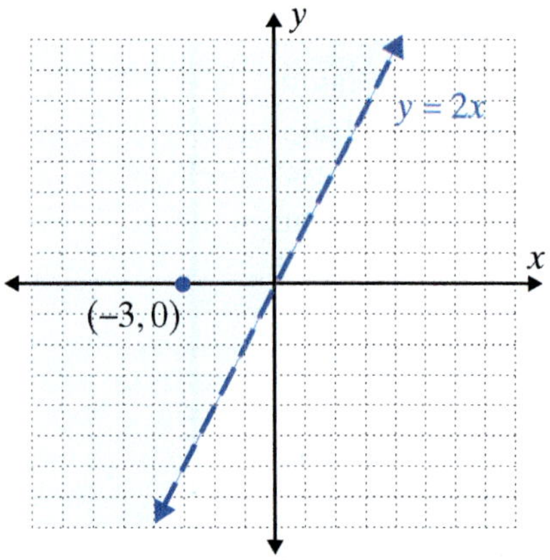

10.
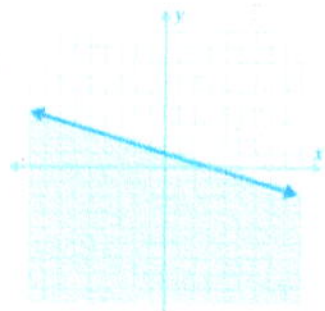

11.
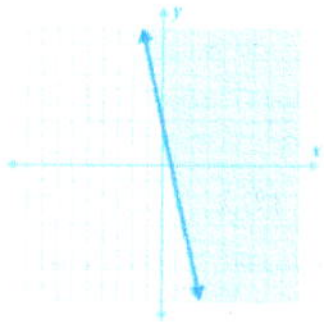

12.
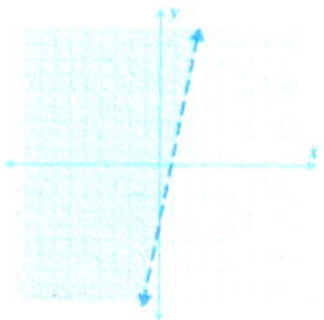

13.
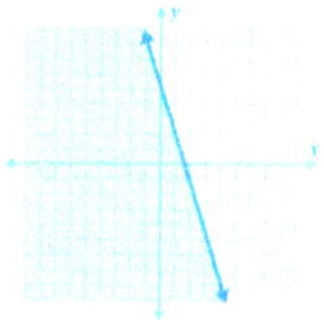

14.
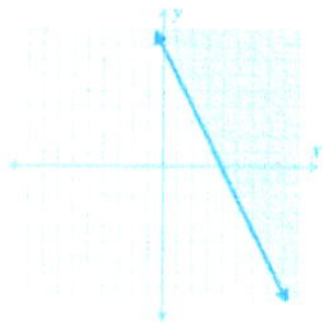

15.
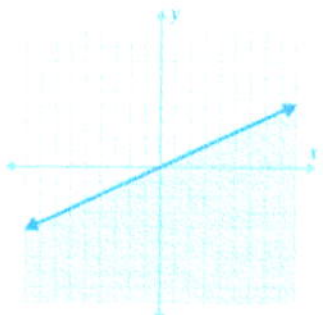

16.
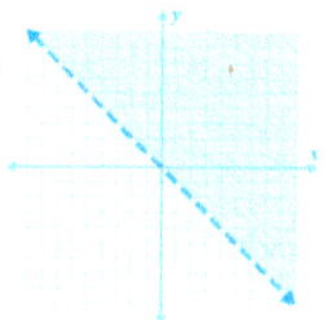

17.
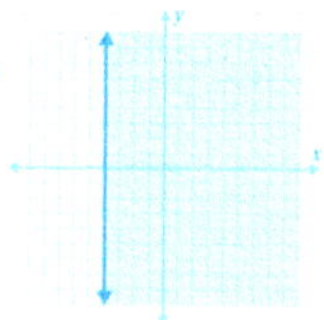

18.
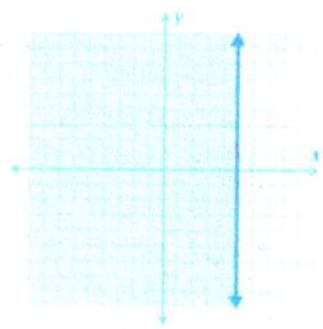

c. Graph the half-plane that satisfies the inequality $y > 1$.

Solution: Again, the inequality is already solved for y and Method 2 is used.

Step 1: Graph the boundary line $y = 1$ as a dashed line because the inequality is $>$. (The boundary line is a horizontal line.)

Step 2: By Method 2, shade the half-plane above the line.

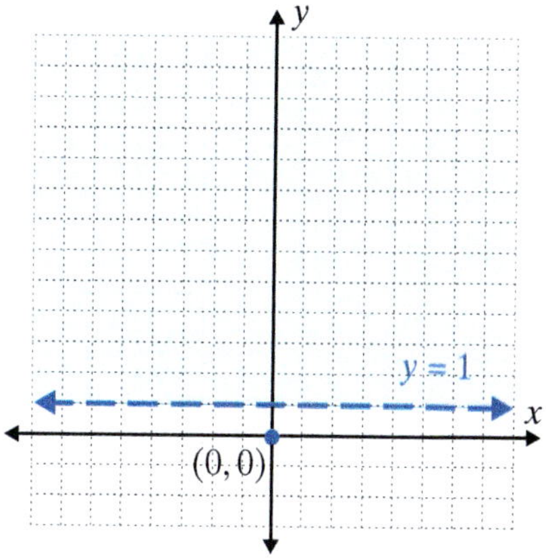

d. Graph the solution set to the inequality $x \le 0$.

Solution: The boundary line is a vertical line and Method 1 is used.

Step 1: Graph the boundary line $x = 0$ as a solid line because the inequality is $\le$ (less than or **equal to**). Note that this is the y-axis.

Step 2: Test the point $(-2, 1)$.

$$-2 \le 0$$

This statement is true.

Step 3: Shade the half-plane on the same side of the line as $(-2, 1)$. This half-plane consists of the points with x-coordinate 0 or negative.

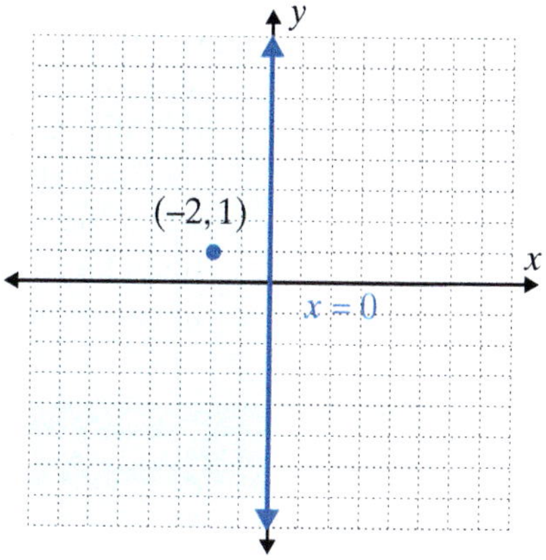

Using a TI-84 Plus Graphing Calculator to Graph Linear Inequalities

The first step in using the TI-84 Plus (or any other graphing calculator) to graph a linear inequality is to solve the inequality for y. This is necessary because this is the way that the boundary line equation can be graphed as a function. Thus Method 2 for graphing the correct half-plane is appropriate.

Note that when you press the Y= key on the calculator, a slash (\ ) appears to the left of the Y expression as in $\backslash Y_1 =$. This slash is actually a command to the calculator to graph the corresponding function as a solid line or curve. If you move the cursor to position it over the slash and hit ENTER repeatedly, the following options will appear.

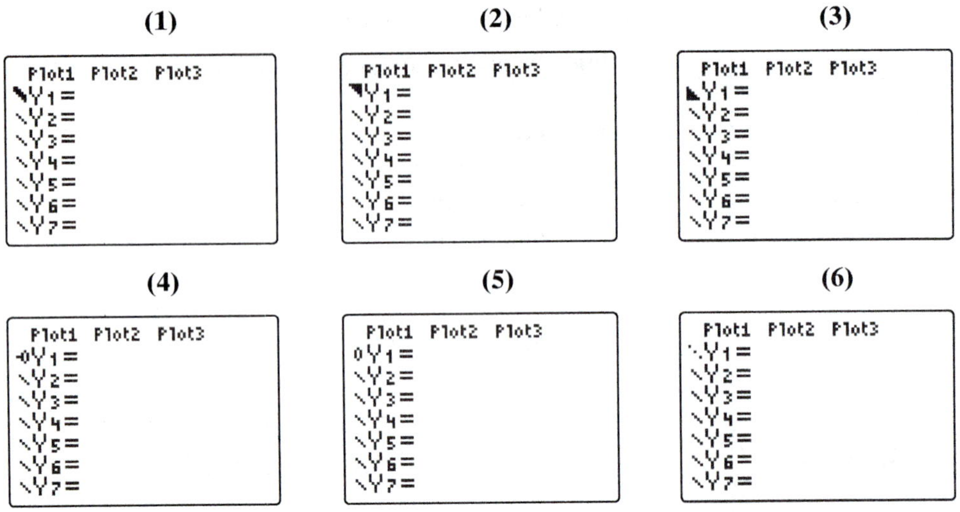

Figure 2

If the slash (which is actually four dots if you look closely) becomes a set of three dots **(6)**, then the corresponding graph of the function will be dotted. By setting the shading above the slash **(2)**, the corresponding graph on the display will show shading above the line or curve. By setting the shading below the slash **(3)**, the corresponding graph on the display will show shading below the line or curve. (The solid line occurs only when the slash is four dots, so the calculator is not good for determining whether the boundary curve is included or not.) Options 1, 4, and 5 are not used when graphing linear inequalities. The following examples illustrate two situations.

Example 2: Graphing Linear Inequalities Using a Calculator

a. Graph the linear inequality $2x + y \leq 7$.

Solution:

Step 1: Solving the inequality for y gives: $y \leq -2x + 7$.

Step 2: Press the key Y= and enter the function: $\backslash Y_1 = -2X + 7$.

Step 3: Go to the \ and hit ENTER three times so that the display appears as follows:

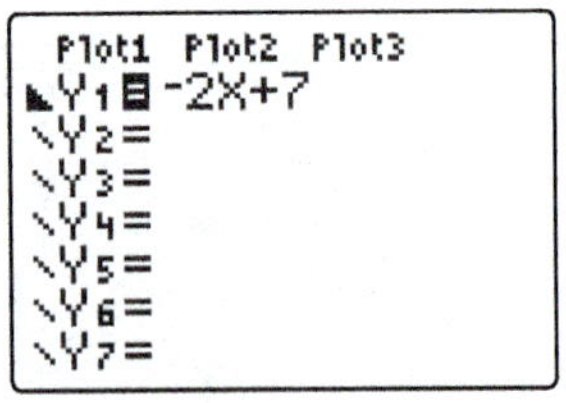

Step 4: Press GRAPH and (using the standard WINDOW settings) the following graph should appear on the display.

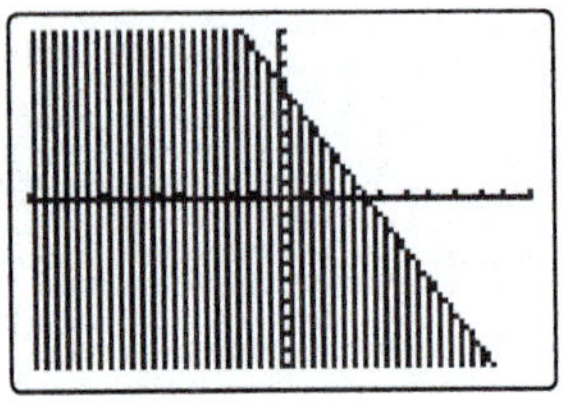

27.
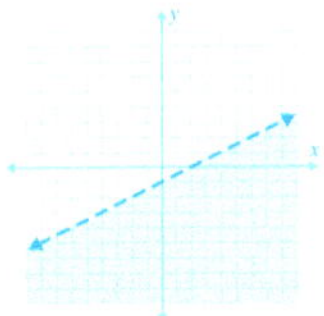

28.
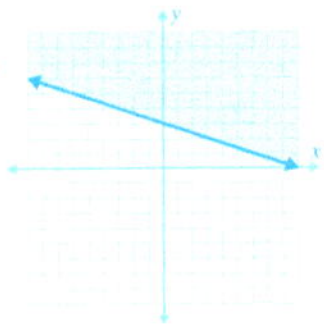

29.
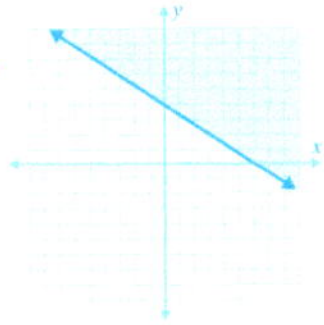

30.
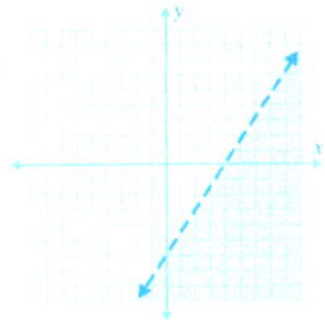

b. Graph the linear inequality $-5x + 4y > -8$.

Solution:

Step 1: Solving the inequality for y gives: $y > \frac{5}{4}x - 2$.

Step 2: Press the key Y= and enter the function: \Y_1 = (5/4)X − 2.

Step 3: Go to the \ and hit ENTER two times so that the display appears as follows:

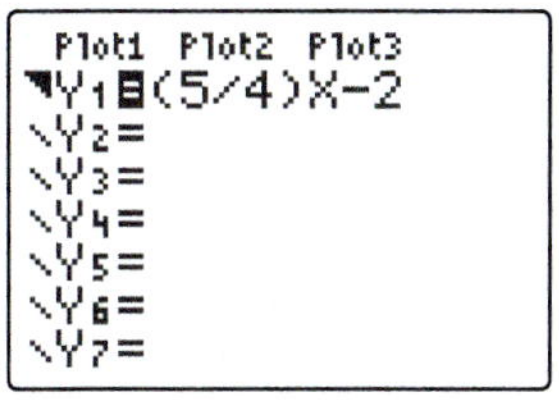

Step 4: Press GRAPH and (using the standard WINDOW settings) the following graph should appear on the display. (**Note:** The boundary line should actually be dotted.)

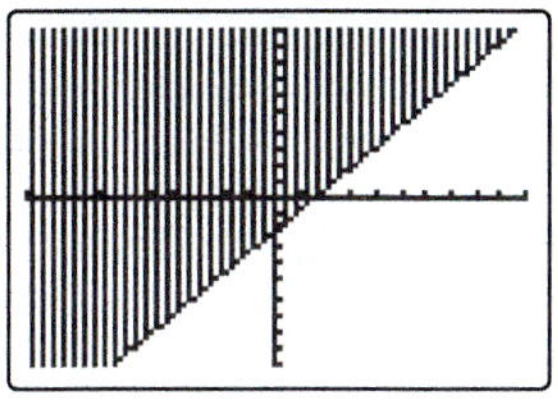

Practice Problems

31.
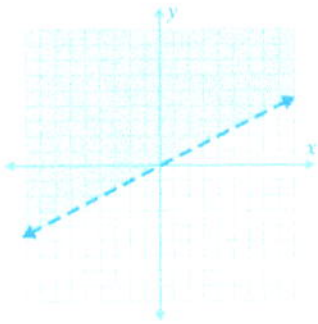

32.
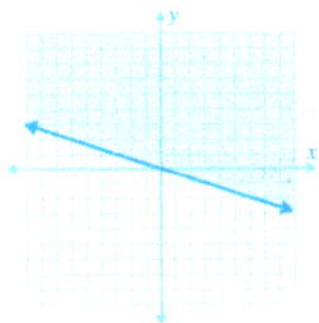

1. Which of the following points satisfy the inequality $x + y < 3$?

a. $(2, 1)$ **b.** $\left(\frac{1}{2}, 3\right)$ **c.** $(0, 5)$ **d.** $(-5, 2)$

2. Which of the following points satisfy the inequality $x - 2y \geq 0$?

a. $(2, 1)$ **b.** $(1, 3)$ **c.** $(4, 2)$ **d.** $(3, 1)$

3. Which of the following points satisfy the inequality $x < 3$?

a. $(1, 0)$ **b.** $(0, 1)$ **c.** $(4, -1)$ **d.** $(2, 3)$

4.6 Exercises

33.
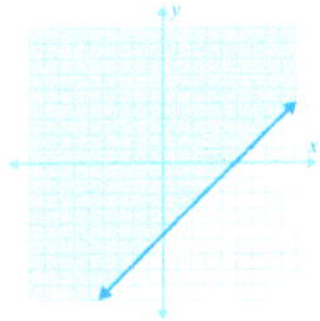

Graph the solution set of each of the linear inequalities.

1. $x + y \leq 7$ **2.** $x - y > -2$ **3.** $x - y > 4$ **4.** $x + y \leq 6$

5. $y < 4x$ **6.** $y < -2x$ **7.** $y \geq -3x$ **8.** $y > x$

Answers to Practice Problems: **1.** d **2.** a, c, d **3.** a, b, d

9. $x - 2y > 8$ **10.** $x + 3y \le 3$ **11.** $4x + y \ge 2$ **12.** $5x - y < 4$

13. $y \le 5 - 3x$ **14.** $y \ge 8 - 2x$ **15.** $2y - x \le 0$ **16.** $x + y > 0$

17. $x + 4 \ge 0$ **18.** $x - 5 \le 0$ **19.** $y \ge -2$ **20.** $y + 3 < 0$

21. $4x < -3y + 9$ **22.** $3x < 2y - 4$ **23.** $3y > 4x + 6$ **24.** $5x < 2y - 6$

25. $x + 3y < 7$ **26.** $3x + 4y > 11$ **27.** $\frac{1}{2}x - y > 1$ **28.** $\frac{1}{3}x + y \ge 3$

29. $\frac{2}{3}x + y \ge 4$ **30.** $2x - \frac{4}{3}y > 8$

Use a graphing calculator to graph each of the linear inequalities.

31. $y > \frac{1}{2}x$ **32.** $2x \ge -6y$ **33.** $x - y \le 5$ **34.** $x + 2y > 8$

35. $y \ge -3$ **36.** $y \le -4$ **37.** $2x + y \le 6$ **38.** $x - 3y \ge 9$

39. $3x + 2y \ge 12$ **40.** $3x - 4y > 15$

Writing and Thinking About Mathematics

41. Explain in your own words how to test to determine which side of the graph of an inequality should be shaded.

42. Describe the difference between a closed and an open half-plane.

HAWKES LEARNING SYSTEMS: INTRODUCTORY & INTERMEDIATE ALGEBRA SOFTWARE

- 4.6 Graphing Linear Inequalities

34.
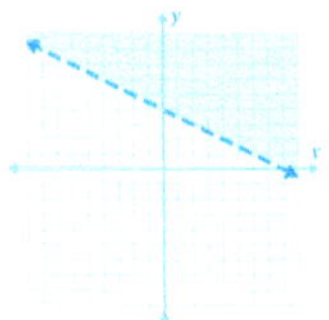

35.
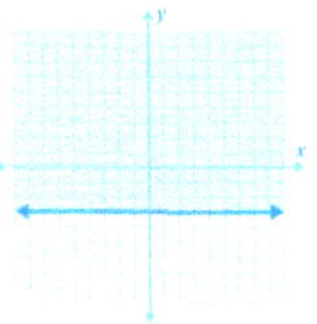

36.
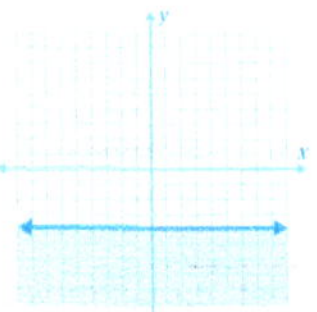

37.
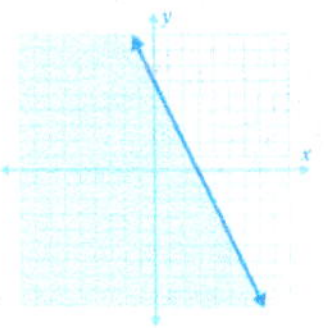

38.
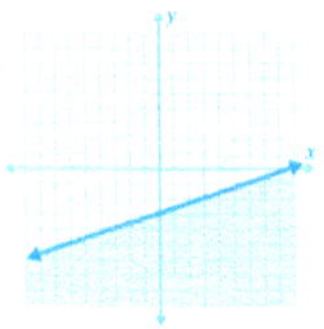

39.
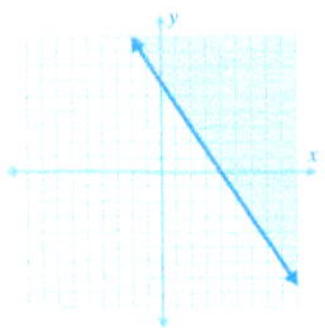

40.
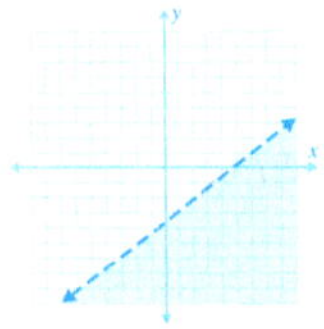

41. Test any point not on the line. If the test point satisfies the inequality, shade the half-plane on that side of the line. Otherwise, shade the other half-plane.

42. A closed half-plane includes points on the line and is symbolized by a solid boundary line. An open half-plane does not include points on the line and is symbolized by a dashed boundary line.

Chapter 4 Index of Key Ideas and Terms

Section 4.1 The Cartesian Coordinate System

Cartesian Coordinate System pages 256-257

Ordered Pairs (x, y) pages 256-257

As solutions to linear equations

First coordinate (the independent variable)

Second coordinate (the dependent variable)

Quadrants page 258

Origin $(0, 0)$ page 258

Horizontal Axis (x-axis) page 258

Vertical Axis (y-axis) page 258

One-To-One Correspondence page 258

There is a **one-to-one correspondence** between points in a plane and ordered pairs of real numbers.

Graphs of Points pages 258-259

Section 4.2 Graphing Linear Equations in Two Variables: $Ax + By = C$

Standard Form of a Linear Equation page 274

Any equation of the form $Ax + By = C$, where A, B, and C are real numbers and where A and B are not both 0, is called the **standard form** of a **linear equation**.

To Graph a Linear Equation in Two Variables page 274

1. Locate any two points that satisfy the equation.
2. Plot these two points on a Cartesian coordinate system.
3. Draw a line through these two points.
   Note: Every point on that line will satisfy the equation.
4. To check, locate a third point that satisfies the equation and check to see that it does indeed lie on the line

***y*-intercept** page 276

The ***y*-intercept** is the point on the graph where the line crosses the y-axis. The x-coordinate will be 0, which makes the y-intercept of the form $(0, y)$.

***x*-intercept** page 276

The ***x*-intercept** is the point on the graph where the line crosses the x-axis. The y-coordinate will be 0, which makes the x-intercept of the form $(x, 0)$.

Section 4.3 The Slope-Intercept Form: $y = mx + b$

Slope as a Rate of Change page 283

Slope of a Line pages 281-284

Let $P_1(x_1, y_1)$ and $P_2(x_2, y_2)$ be two points on a line. The **slope** can be calculated as follows:

$$\text{slope} = m = \frac{rise}{run} = \frac{y_2 - y_1}{x_2 - x_1}.$$

Positive and Negative Slopes page 285

Lines with **positive slope** go up as we move along the line from left to right.
Lines with **negative slope** go down as we move along the line from left to right.

Horizontal and Vertical Lines pages 286-287

The following two general statements are true for horizontal and vertical lines:

1. For **horizontal lines** (of the form $y = b$), the **slope is 0.**
2. For **vertical lines** (of the form $x = a$), the **slope is undefined.**

Slope-Intercept Form page 288

Any equation of the form $y = mx + b$ is called the **slope-intercept form** for the equation of a line. m is the slope and $(0, b)$ is the y-intercept.

Section 4.4 The Point-Slope Form: $y - y_1 = m(x - x_1)$

Graphing a Line Given a Point and the Slope page 297

Point-Slope Form page 298

An equation of the form $y - y_1 = m(x - x_1)$ is called the **point-slope form** for the equation of a line that contains the point (x_1, y_1) and has slope m.

Finding the Equation of a Line Given Two Points page 300

Parallel Lines page 301

Parallel lines are lines that never intersect (cross each other) and these lines have the same slope. (**Note:** All vertical lines (undefined slopes) are parallel to one another.)

Continued on the next page....

Section 4.4 The Point-Slope Form: $y - y_1 = m(x - x_1)$ (cont.)

Perpendicular Lines page 301

Perpendicular lines are lines that intersect at 90° (right) angles and whose slopes are negative reciprocals of each other. Horizontal lines are perpendicular to vertical lines.

Section 4.5 Introduction to Functions and Function Notation

Relation, Domain, and Range page 307

A **relation** is a set of ordered pairs of real numbers. The **domain**, ***D***, of a relation is the set of all first coordinates in the relation. The **range**, ***R***, of a relation is the set of all second coordinates in the relation.

Functions page 309

A **function** is a relation in which each domain element has exactly one corresponding range element. The definition can also be stated in the following ways:

1. A function is a relation in which each first coordinate appears only once.
2. A function is a relation in which no two ordered pairs have the same first coordinate.

Vertical Line Test page 310

If **any** vertical line intersects the graph of a relation at more than one point, then the relation graphed is **not** a function.

Linear Functions page 312

A **linear function** is a function represented by an equation of the form $y = mx + b$. The domain of a linear function is the set of all real numbers, $D = (-\infty, \infty)$.

Domains of Non-Linear Functions page 313

Function Notation page 314

In **function notation**, instead of writing y, write $f(x)$, read "f of x." The letter f is the name of the function. The notation $f(4)$ means to replace x with 4 in the function.

Using a Graphing Calculator to Graph Functions pages 315-318

Section 4.6 Graphing Linear Inequalities in Two Variables

Terminology for Graphing Linear Inequalities page 326

Half-plane
Boundary line
Closed Half-plane
Open Half-plane

Graphing Linear Inequalities page 327

1. First, graph the boundary line (dashed if the inequality is < or >, solid if the inequality is ≤ or ≥).
2. Next, determine which side of the line to shade using one of the following methods.

   Method 1

   a. Test any one point obviously on one side of the line.

   b. If the test-point satisfies the inequality, shade the half-plane on that side of the line. Otherwise, shade the other half-plane.

   Method 2

   a. Solve the inequality for y (assuming that the line is not vertical).

   b. If the solution shows $y <$ or $y \leq$, then shade the half-plane below the line.

   c. If the solution shows $y >$ or $y \geq$, then shade the half-plane above the line.
3. The shaded half-plane (and the line if it is solid) is the solution to the inequality.

Using a Graphing Calculator to Graph Linear Inequalities pages 329-330

HAWKES LEARNING SYSTEMS: INTRODUCTORY & INTERMEDIATE ALGEBRA SOFTWARE

- 4.1 Introduction to the Cartesian Coordinate System
- 4.2 Graphing Linear Equations by Plotting Points
- 4.3 Graphing Linear Equations in Slope-Intercept Form
- 4.4a Finding the Equation of a Line
- 4.4b Graphing Linear Equations in Point-Slope Form
- 4.5 Introduction to Functions and Function Notation
- 4.6 Graphing Linear Inequalities

Chapter 4 Review

4.1 The Cartesian Coordinate System

List the sets of ordered pairs corresponding to the points on the graphs. Assume that the grid lines are marked one unit apart.

1.

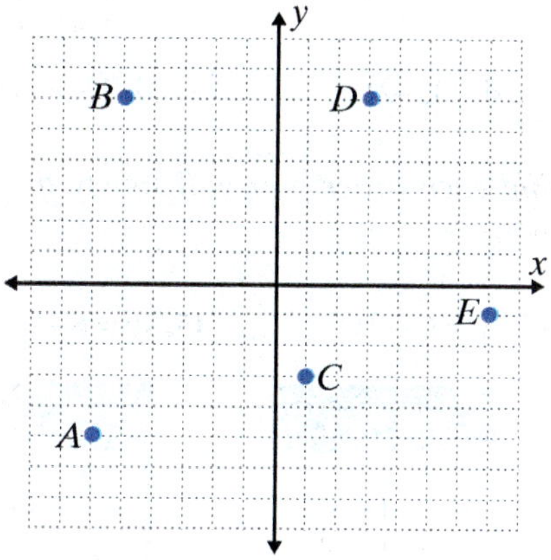

2.

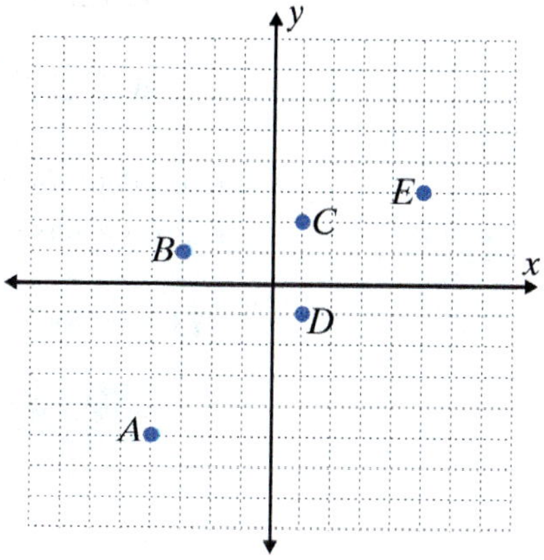

3.

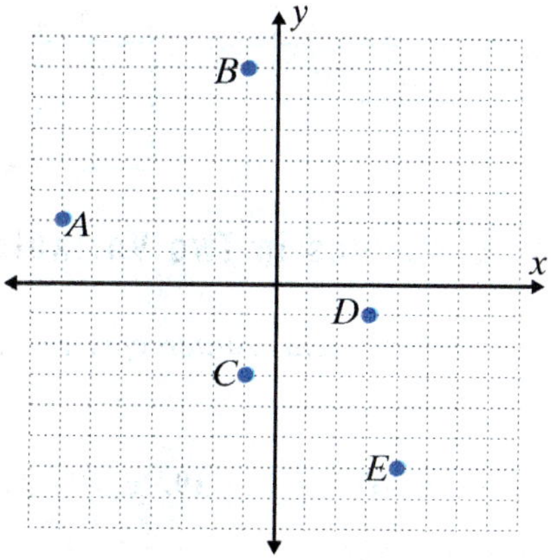

4.

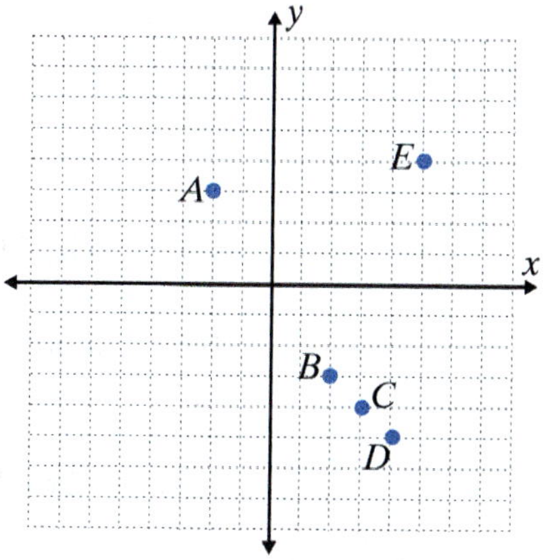

Graph the sets of ordered pairs and label the points.

5. $\{A(-3,1), B(-2,-2), C(0,-5), D(1,4)\}$

6. $\{E(-3,-4), F(0,0), G(5,-1), H(1.5,-3)\}$

7. $\{M(6,2), N(3,-1), O(0,-2), P(-4,-5)\}$

8. $\{Q(-4,-2), R(-4,0), S(-4,3), T(-4,5)\}$

1. $\{A(-6,-5), B(-5,6), C(1,-3), D(3,6), E(7,-1)\}$
2. $\{A(-4,-5), B(-3,1), C(1,2), D(1,-1), E(5,3)\}$
3. $\{A(-7,2), B(-1,7), C(-1,-3), D(3,-1), E(4,-6)\}$
4. $\{A(-2,3), B(2,-3), C(3,-4), D(4,-5), E(5,4)\}$
5.

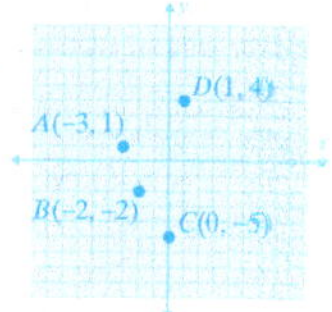

6.

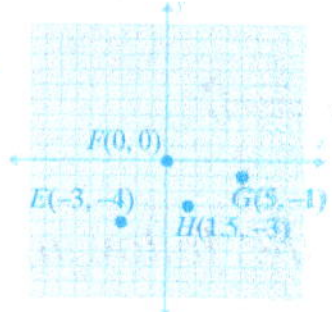

7.

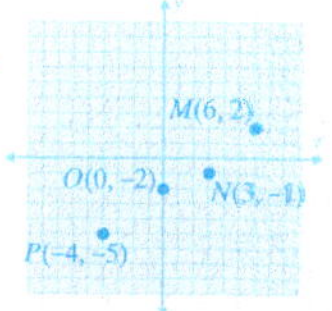

8.

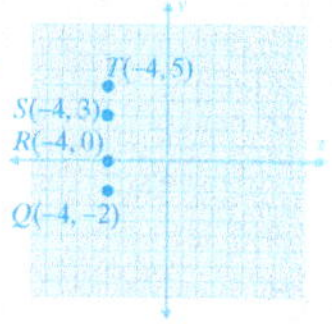

9. None
10. c
11. a, c, d
12. a, b, c
13. $(0,0), (-1,-5), (-2,-10), (1,5)$

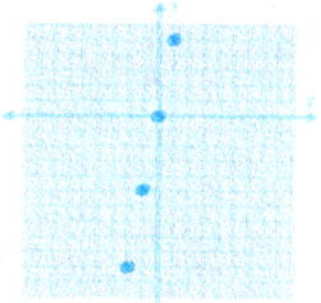

14. $(0,6), (2,0), (-1,9), (4,-6)$

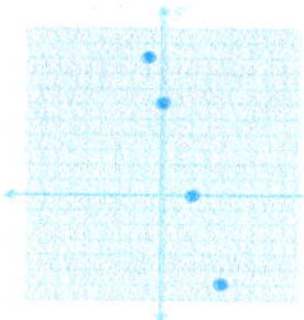

15. $(0,8), (4,0), (3,2), \left(\frac{17}{4}, -\frac{1}{2}\right)$

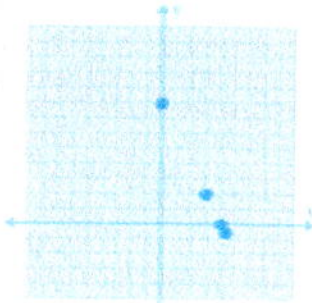

16. $(0,-5), \left(-\frac{4}{3}, -1\right), (-3,4), (-1.5,-0.5)$

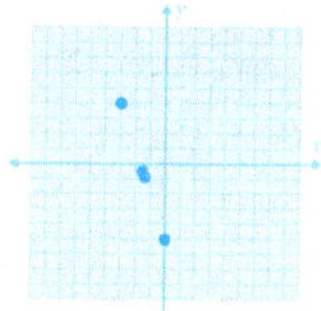

17. c 18. b
19. d 20. a

21.
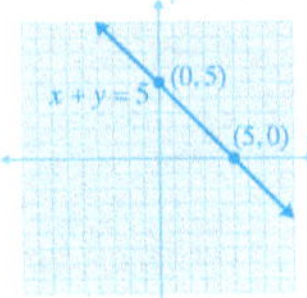

22.
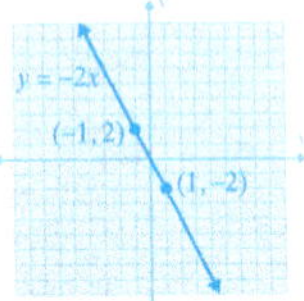

23.
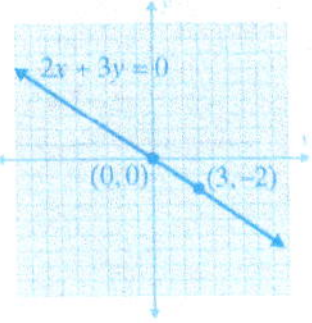

24.
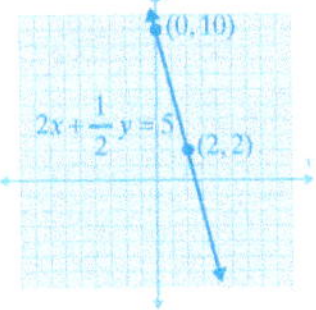

25.
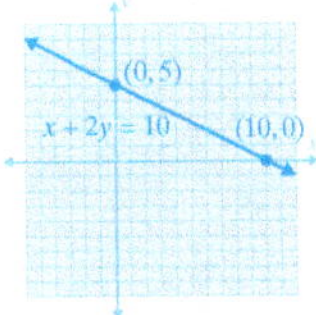

26.
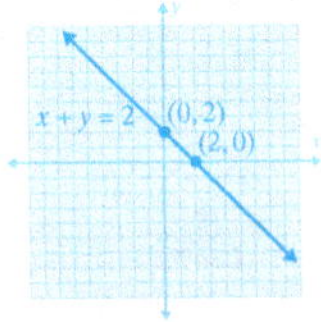

27.
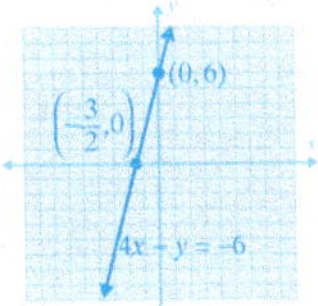

28.
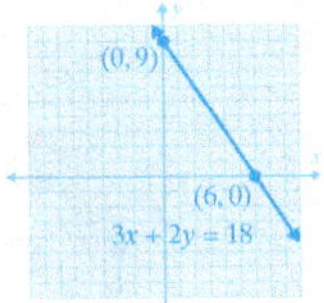

29. $m = -2$
30. $m = 3$
31. $m = 4$
32. $m = -\frac{1}{2}$

Determine which, if any, of the ordered pairs satisfy the given equations.

9. $3x - y = -2$
- **a.** $(1, 1)$
- **b.** $(-1, 1)$
- **c.** $(2, 2)$
- **d.** $(-2, 2)$

10. $x + 2y = 3$
- **a.** $(-3, 1)$
- **b.** $(2, 1)$
- **c.** $\left(2, \frac{1}{2}\right)$
- **d.** $(1, -2)$

11. $y = 4x$
- **a.** $(1, 4)$
- **b.** $(-1, 4)$
- **c.** $\left(\frac{1}{4}, 1\right)$
- **d.** $(-3, -12)$

12. $2x - 3y = 5$
- **a.** $(1, -1)$
- **b.** $\left(2, -\frac{1}{3}\right)$
- **c.** $(2.5, 0)$
- **d.** $(0, -2)$

Complete the tables so that each order pair will satisfy the given equation. Graph the resulting sets of ordered pairs.

13. $y = 5x$

x	y
0	
	−5
−2	
	5

14. $y = 6 - 3x$

x	y
0	
	0
−1	
	−6

15. $y = 8 - 2x$

x	y
0	
	0
3	
	$-\frac{1}{2}$

16. $3x + y = -5$

x	y
0	
	−1
−3	
	−0.5

4.2 Graphing Linear Equations in Two Variables: *Ax* + *By* = *C*

Use your knowledge of x-intercepts and y-intercepts to match each of the following equations with its graph.

17. $y = 2x + 4$

18. $y = 6 - x$

19. $y = 3 - 3x$

20. $-5x + y = 10$

a.
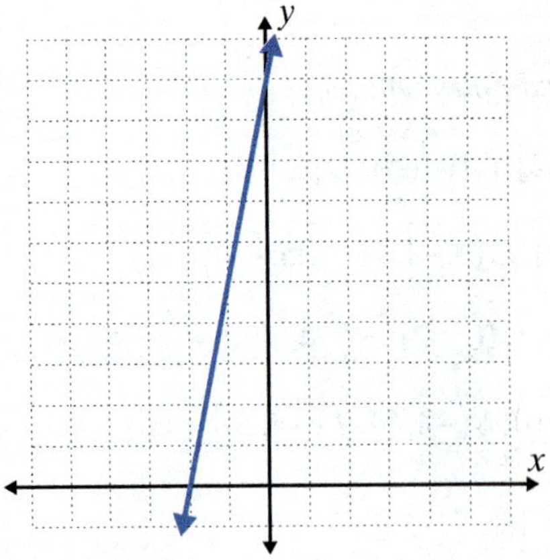

b.
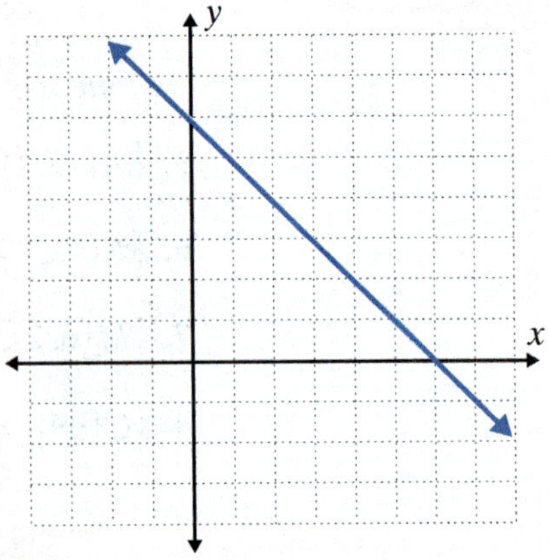

c.

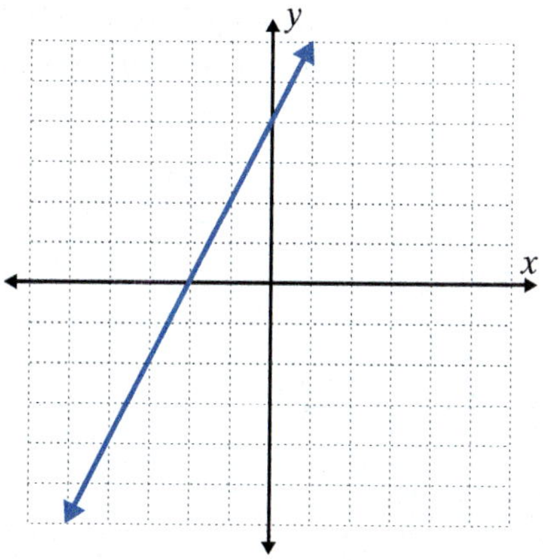

d.

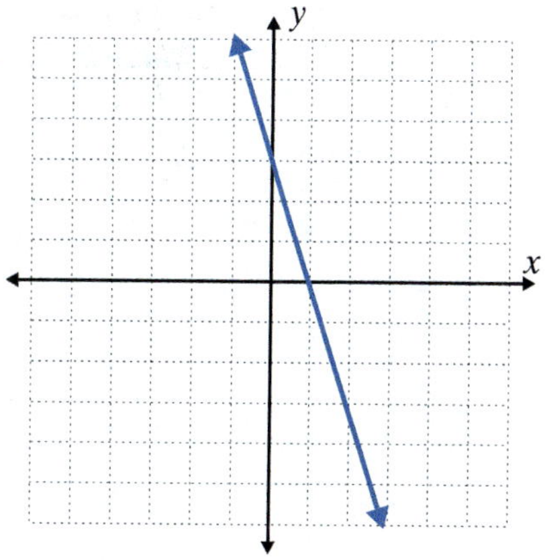

Locate at least two ordered pairs of real numbers that satisfy each of the linear equations and graph the corresponding line in the Cartesian coordinate system.

21. $x+y=5$ **22.** $y=-2x$ **23.** $2x+3y=0$ **24.** $2x+\frac{1}{2}y=5$

Graph the following linear equations by locating the x-intercept and the y-intercept.

25. $x+2y=10$ **26.** $x+y=2$ **27.** $4x-y=-6$ **28.** $3x+2y=18$

4.3 The Slope-Intercept Form: $y=mx+b$

Find the slope of the line determined by each pair of points.

29. $(-1,3);(0,1)$ **30.** $(-2,-4);(0,2)$ **31.** $(-1,-7);(1,1)$ **32.** $\left(\frac{2}{3},\frac{1}{2}\right);\left(0,\frac{5}{6}\right)$

Tell whether each equation represents a horizontal line or vertical line and give its slope. Graph the line.

33. $3x=-5.4$ **34.** $3y-4=0$

Write each equation in slope-intercept form. Find the slope and the y-intercept, and then use them to draw the graph.

35. $y=-\frac{2}{5}x+1$ **36.** $y=\frac{4}{3}x-2$ **37.** $x-2y=4$ **38.** $2x+y+8=0$

39. $2y-6=0$ **40.** $3x=-9$ **41.** $2x+6=3y$ **42.** $4x=-2y+7$

33. Vertical line; m is undefined.

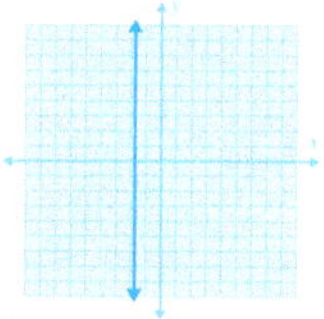

34. Horizontal line; $m=0$

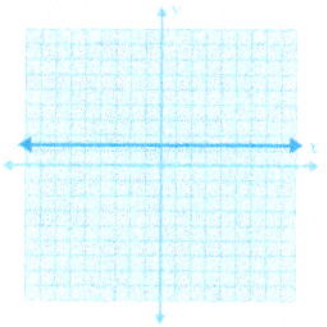

35.

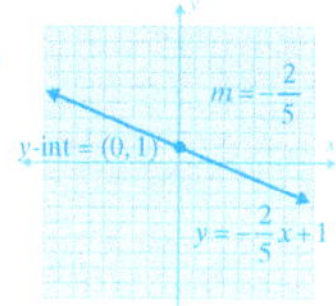

36.

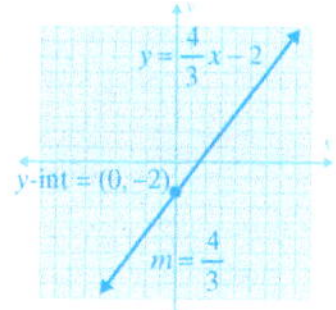

37.

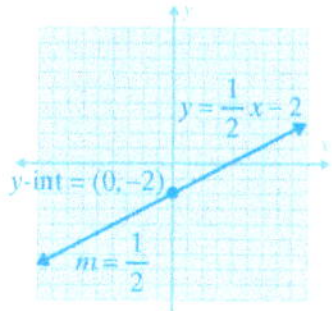

38.

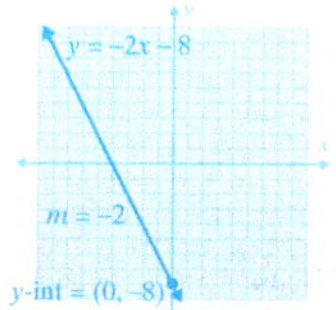

39.

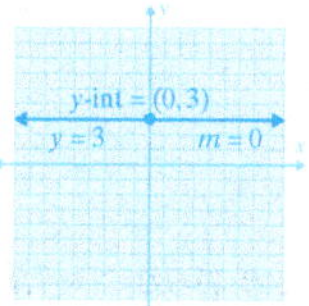

40.

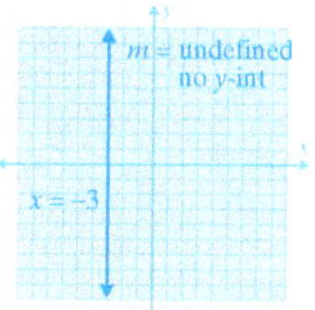

41.

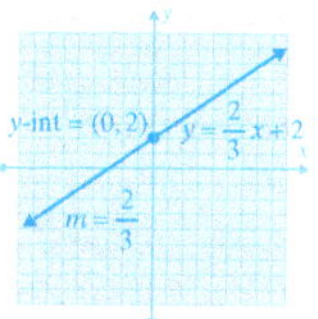

42.

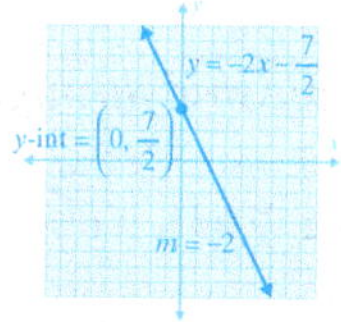

43. a. $m=-\frac{3}{2}$
b. $(0,5)$
c. $y=-\frac{3}{2}x+5$

44. a. m is undefined
b. No y-intercept
c. $x=-2$

45. a. $m=0$
b. $(0,4)$
c. $y=4$

*The graph of a line is shown with two points highlighted. Find **a.** the slope, **b.** the y-intercept (if there is one), and **c.** the equation of the line in slope-intercept form.*

43.

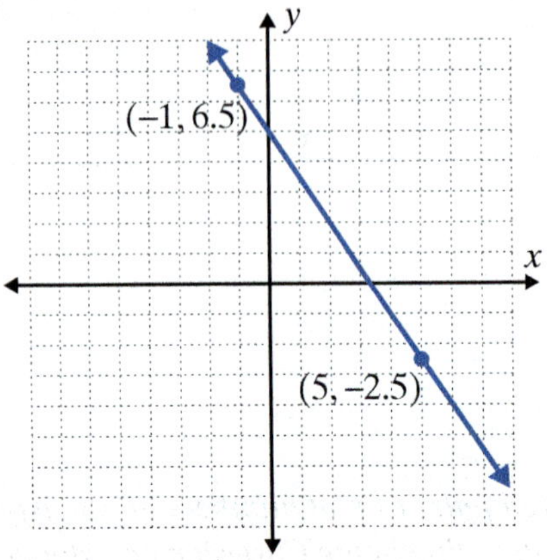

44.

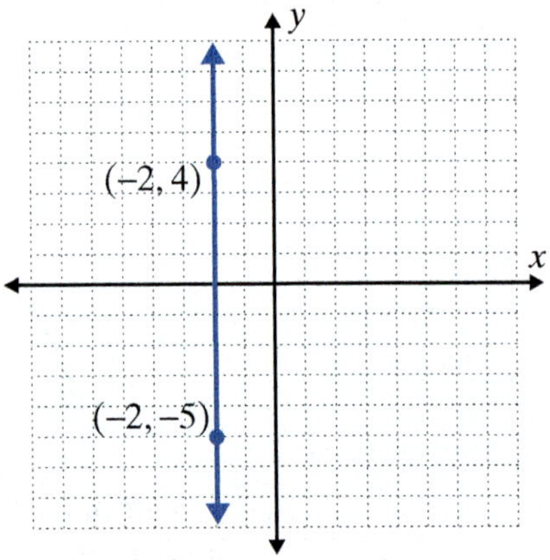

45.

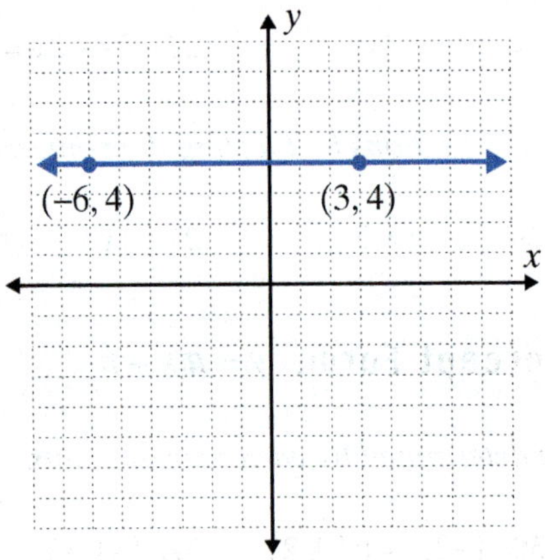

46.

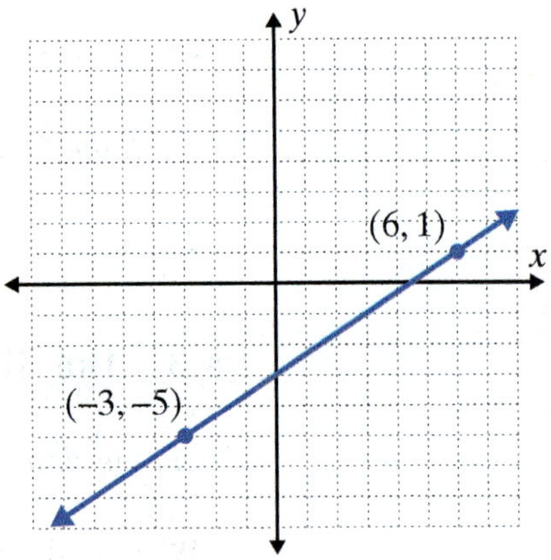

4.4 The Point-Slope Form: $y - y_1 = m(x - x_1)$

*Find **a.** the slope, **b.** a point on the line, and **c.** the graph of the line for the given equations in point-slope form.*

47. $y - 7 = 3(x + 1)$

48. $y + 3 = -(x - 1)$

Find an equation in standard form for the line passing through the given point with the given slope. Graph the line.

49. $m = -3; (-4, 1)$

50. $m = \frac{5}{2}; (3, -2)$

51. m is undefined; $(-4, -7)$

52. $m = 0; \left(-6, -\frac{3}{2}\right)$

Find an equation in slope-intercept form for the line passing through the two given points.

53. $(-3, 5); (1, -3)$

54. $(4, 2); (6, 1)$

46. a. $m = \frac{2}{3}$ b. $(0, -3)$

c. $y = \frac{2}{3}x - 3$

47. a. $m = 3$ b. $(-1, 7)$

c.

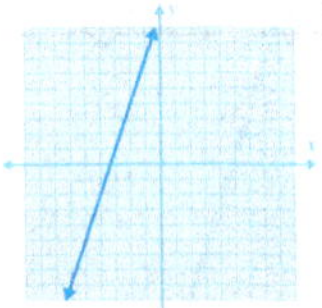

48. a. $m = -1$ b. $(1, -3)$

c.

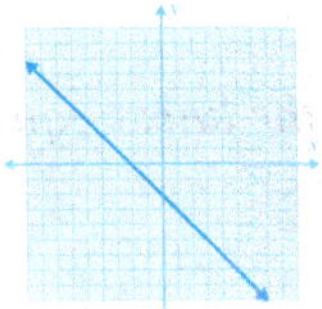

49. $3x + y = -11$

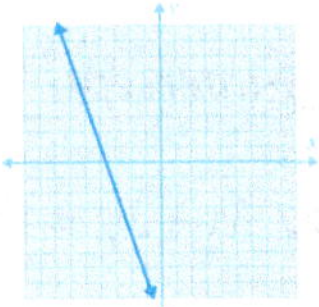

50. $5x - 2y = 19$

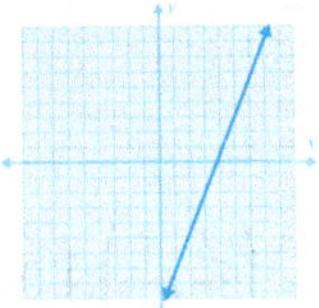

51. $x = -4$

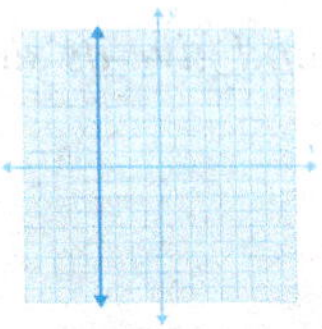

52. $y = -\frac{3}{2}$

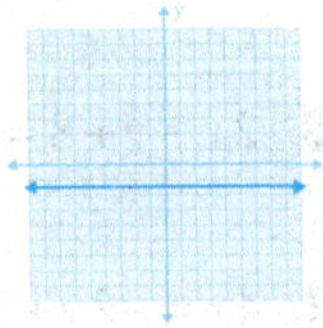

53. $y = -2x - 1$

54. $y = -\frac{1}{2}x + 4$

55. $y = -\frac{1}{8}x + \frac{1}{12}$

55. $\left(\frac{2}{3}, 0\right); \left(-2, \frac{1}{3}\right)$

56. $\left(0, -\frac{1}{2}\right); \left(\frac{3}{4}, \frac{1}{5}\right)$

Find the slope-intercept form of the equations described.

57. Write an equation for the horizontal line through the point $(-3, 5)$.

58. Write an equation for the vertical line through the point $(5, 7)$.

59. Write an equation for the line through the point $(-6, -2)$ and parallel to the line $x - 4y = 4$. Graph both lines.

60. Write an equation for the line through the point $(-4, 1)$ and parallel to the line $x + y = -6$. Graph both lines.

61. Write an equation for the line through the point $(1, 1)$ and perpendicular to the line $2x + y = 9$. Graph both lines.

62. Write an equation for the line perpendicular to the line $x + 2y = 10$ and passing through $(-3, 0)$.

63. Write an equation for the line parallel to the line $2x + 3y = 5$ with the same y-intercept as the line $x - 2y = 8$.

64. Write an equation for the line perpendicular to the line $3x - y = 2$ with the same y-intercept as the line $2x + y = 5$.

Determine whether each pair of lines is **a.** *parallel,* **b.** *perpendicular, or* **c.** *neither. Graph both lines. (**Hint:** Write the equations in slope-intercept form and then compare slopes and y-intercepts.)*

65. $\begin{cases} 5x + y = 5 \\ x - 5y = 10 \end{cases}$

66. $\begin{cases} 2x + y = 8 \\ y = -\frac{1}{2}x \end{cases}$

4.5 Introduction to Functions and Function Notation

List the sets of ordered pairs that correspond to the points. State the domain and range and indicate which of the relations are also functions.

67.

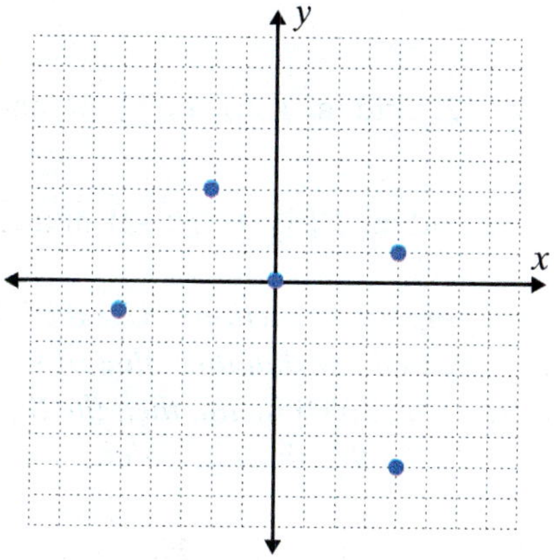

68.

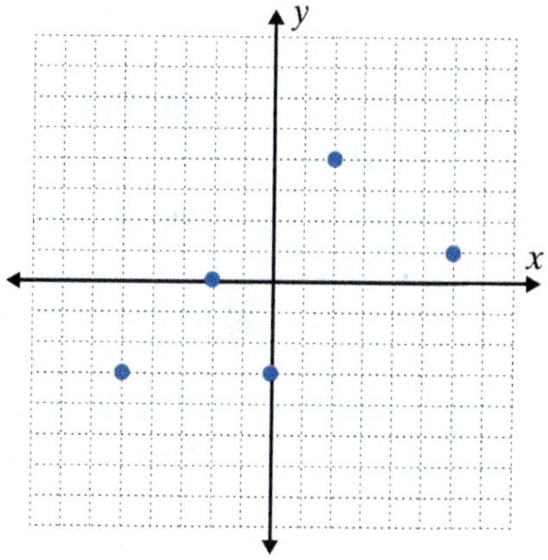

---

56. $y = \frac{14}{15}x - \frac{1}{2}$

57. $y = 5$

58. $x = 5$

59. $y = \frac{1}{4}x - \frac{1}{2}$

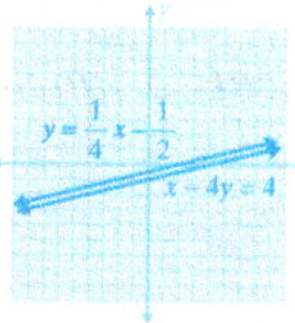

60. $y = -x - 3$

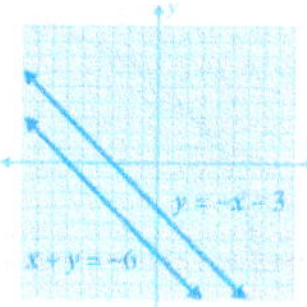

61. $y = \frac{1}{2}x + \frac{1}{2}$

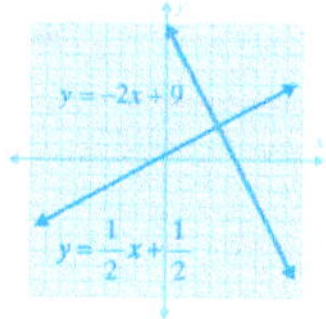

62. $y = 2x + 6$

63. $y = -\frac{2}{3}x - 4$

64. $y = -\frac{1}{3}x + 5$

65. Perpendicular

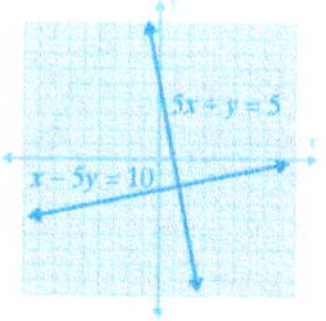

66. Neither

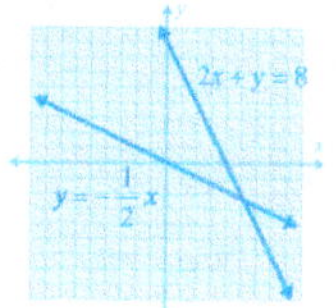

67. $\left\{\begin{matrix} (-5, -1), (-2, 3), \\ (0, 0), (4, 1), \\ (4, -6) \end{matrix}\right\}$

$D = \{-5, -2, 0, 4\}$;

$R = \{-6, -1, 0, 1, 3\}$;

Not a function

Graph the relations. State the domain and range and indicate which of the relations are functions.

68. $\begin{Bmatrix} (-5,-3),(-2,0), \\ (0,-3),(2,4), \\ (6,1) \end{Bmatrix}$

69. $g=\{(-5,1),(-3,-2),(-3,2),(0,0),(1,1)\}$

70. $f=\{(-4,3.2),(-3,0.1),(0,0.1),(0,-2),(2,3.2)\}$

Use the vertical line test to determine whether or not each graph represents a function. State the domain and range.

71.

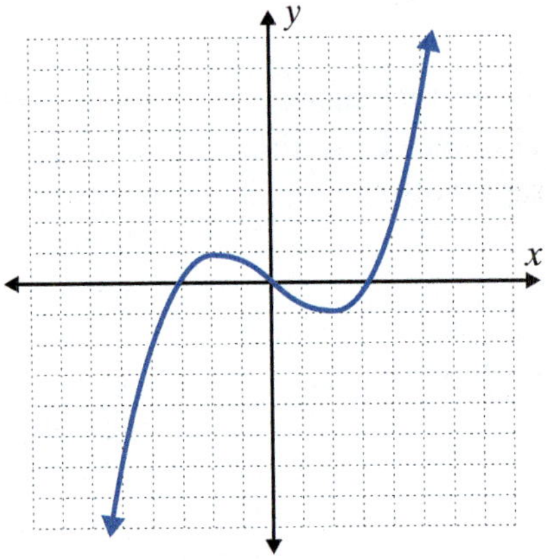

72.

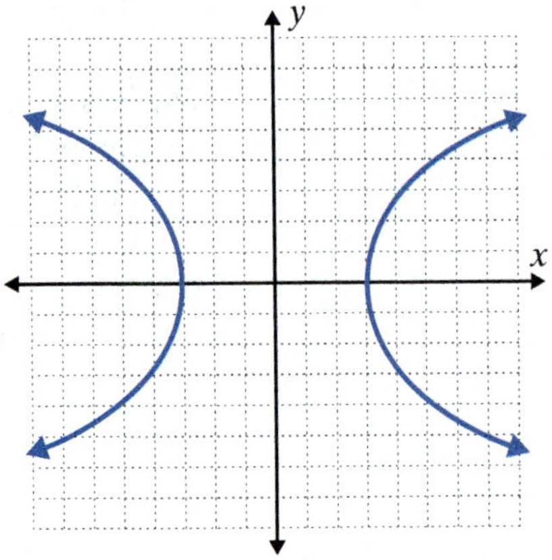

73.

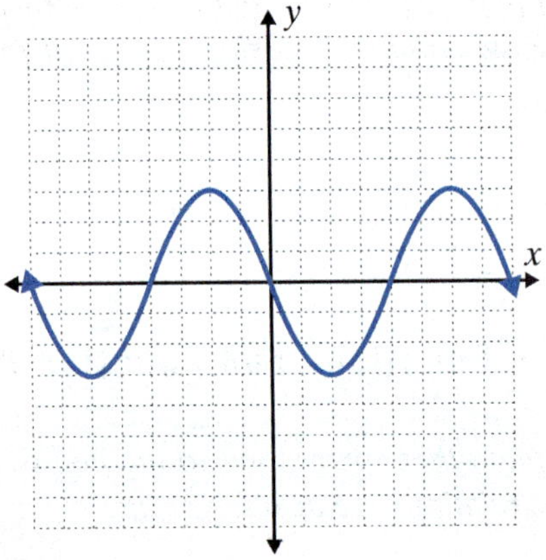

74.

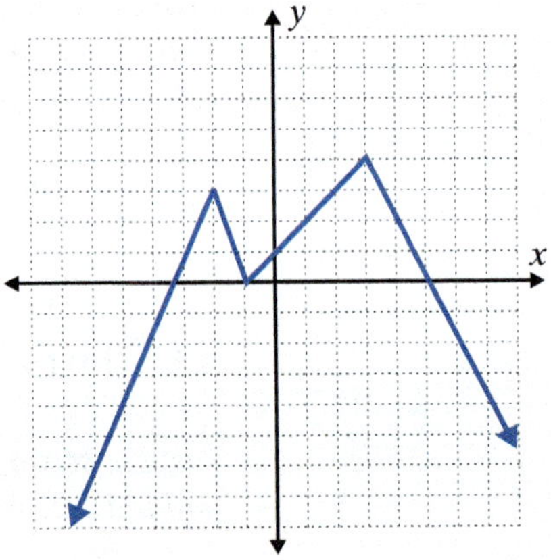

75. Given that $g(x)=5x-12$, find **a.** $g(6)$ **b.** $g(-1)$ and **c.** $g\left(\frac{3}{5}\right)$.

76. For $R(x)=x^2-5x+6$, find **a.** $R(5)$ **b.** $R(-2)$ and **c.** $R(0)$.

Use a graphing calculator to graph the functions. Use the CALC *features of the calculator to find x-intercepts, if any. (Remember that the value of y will be 0 at those points.) For absolute value functions, select the* MATH *menu, then the* NUM *menu, and then* 1: abs(*. Remember to press*) *after entering the absolute value.*

77. $y=-3x$ **78.** $y=3x+1$ **79.** $y=|x^2-2x|$ **80.** $y=x^3-4x-1$

---

68. $D=\{-5,-2,0,2,6\}$; $R=\{-3,0,1,4\}$; Function

69. $D=\{-5,-3,0,1\}$; $R=\{-2,0,1,2\}$; Not a function

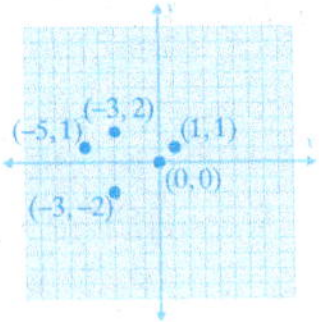

70. $D=\{-4,-3,0,2\}$; $R=\{-2,0.1,3.2\}$; Not a function

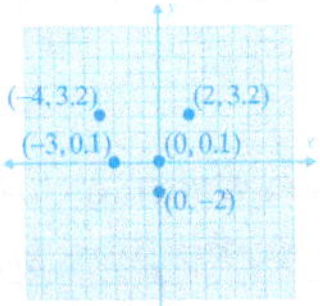

71. Function; $D=(-\infty,\infty)$; $R=(-\infty,\infty)$

72. Not a function; $D=(-\infty,-3]\cup[3,\infty)$; $R=(-\infty,\infty)$

73. Function; $D=(-\infty,\infty)$; $R=[-3,3]$

74. Function; $D=(-\infty,\infty)$; $R=(-\infty,4]$

75. a. 18 **b.** −17 **c.** −9

76. a. 6 **b.** 20 **c.** 6

77.

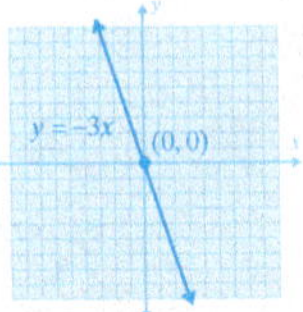

78.
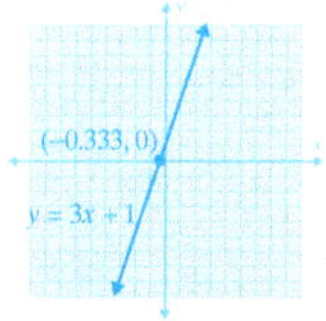

79.
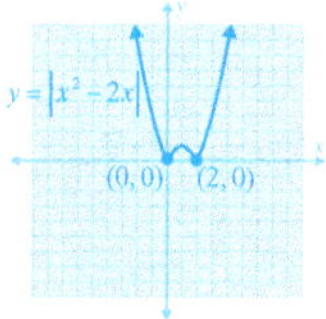

80.
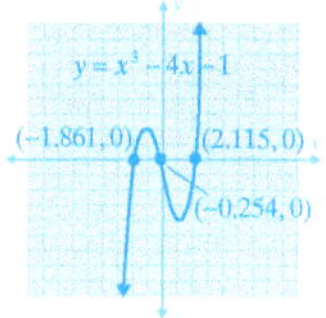

4.6 Graphing Linear Inequalities in Two Variables

Graph the solution set of each of the linear inequalities.

81. $y \le 2x - 1$

82. $y > -5$

83. $y < 3x$

84. $y - 3 > 6x$

85. $2x + 4 \ge 2y$

86. $3x + 2y > 12$

87. $\frac{3}{5}x + 6 \le y$

88. $y > \frac{1}{4}x - 3$

Use a graphing calculator to graph each of the linear inequalities.

89. $y > 2x$

90. $2x + 6y > 0$

91. $x + y \le 5$

92. $x - y > -2$

81.
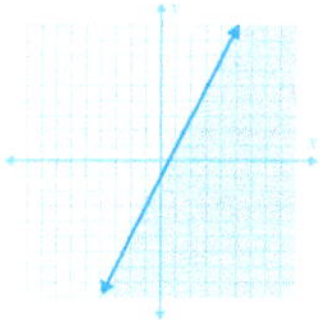

82.
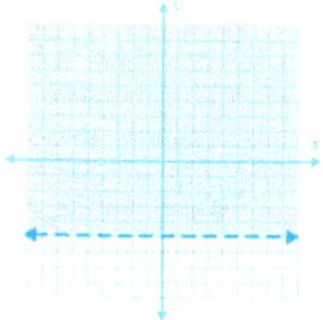

83.
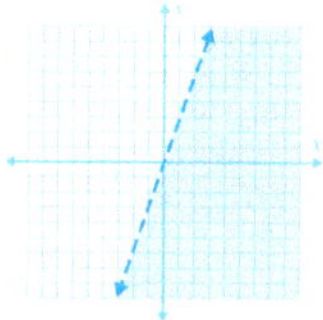

84.
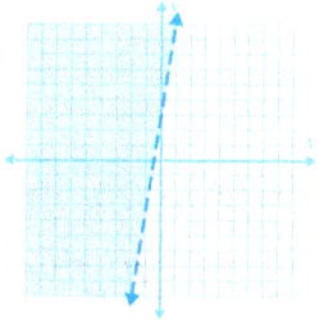

85.
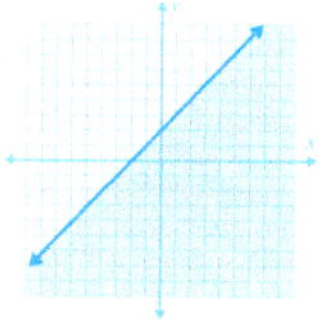

86.
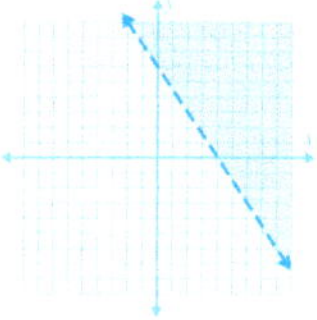

87.
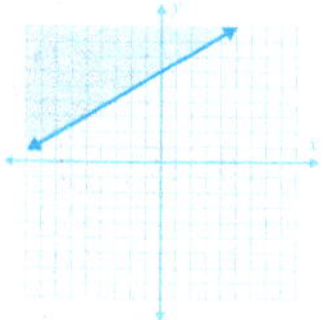

88.
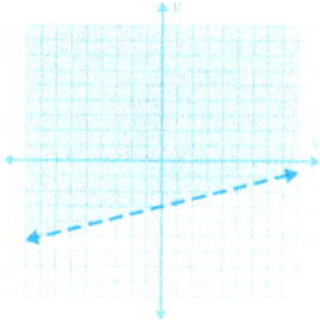

89.
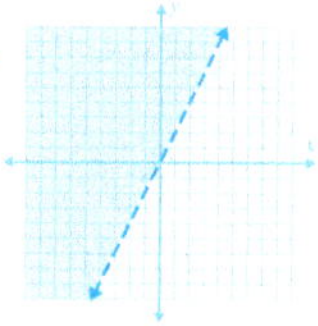

90.
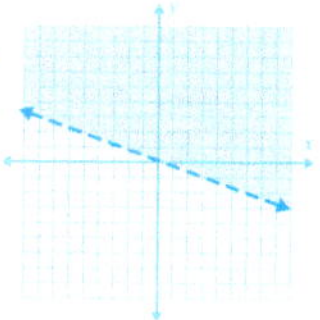

91.
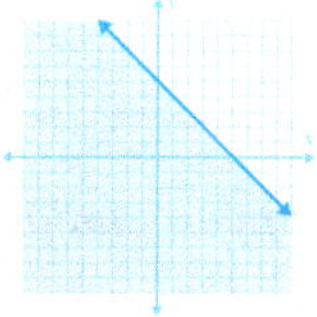

92.
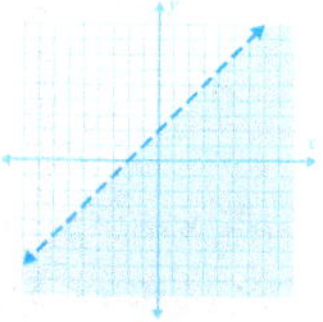

Chapter 4 Test

1. $\left\{\begin{array}{l} A(-3,2), B(-2,-1), \\ C(0,-3), D(1,2), \\ E(3,4), F(5,0), \\ G(4,-2) \end{array}\right\}$

2. 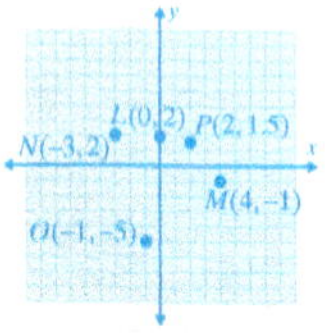

3. a. $(0,2)$ b. $\left(\frac{2}{3},0\right)$ c. $(-2,8)$ d. $(3,-7)$

4. a. $\left(0,-\frac{6}{5}\right)$ b. $(6,0)$ c. $(11,1)$ d. $(-4,-2)$

5. 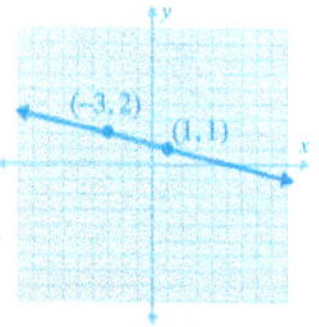

6. 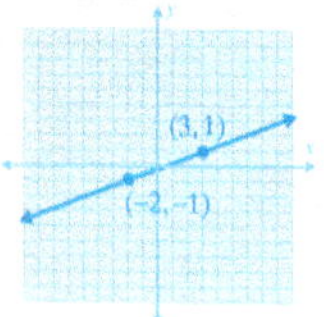

7. 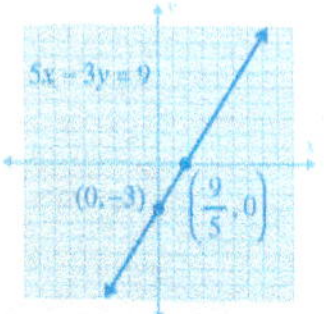

8. 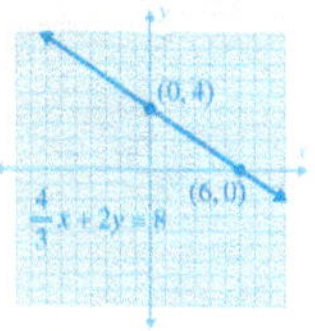

9. $m=\frac{9}{8}$

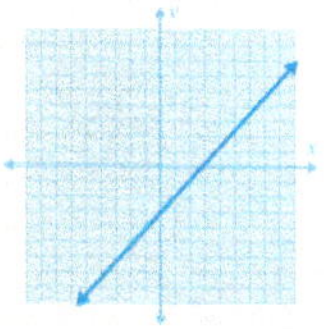

1. List the set of ordered pairs corresponding to the points on the graph. Assume that the grid lines are marked one unit apart.

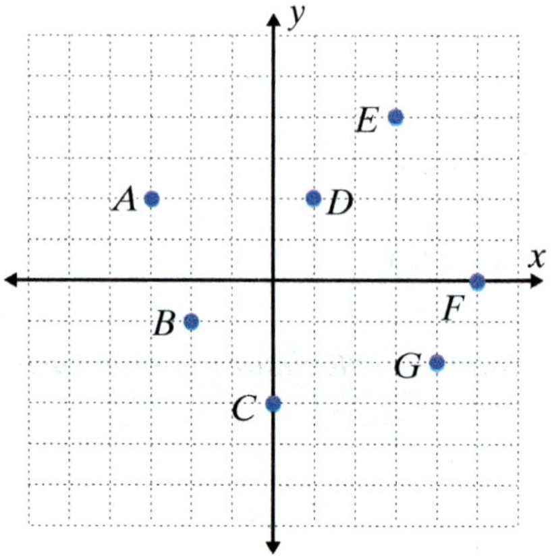

2. Graph the following set of ordered pairs and label the points.

$$\{L(0,2), M(4,-1), N(-3,2), O(-1,-5), P(2,1.5)\}$$

Determine the missing coordinate in each of the ordered pairs so that the point will satisfy the equation given.

3. $3x+y=2$

a. $(0,\ \ )$

b. $(\ \ ,0)$

c. $(-2,\ \ )$

d. $(\ \ ,-7)$

4. $x-5y=6$

a. $(0,\ \ )$

b. $(\ \ ,0)$

c. $(11,\ \ )$

d. $(\ \ ,-2)$

Locate at least two ordered pairs of real numbers that satisfy each of the linear equations and graph the corresponding line in the Cartesian coordinate system.

5. $x+4y=5$

6. $2x-5y=1$

Graph the following linear equations by locating the x-intercept and the y-intercept.

7. $5x-3y=9$

8. $\frac{4}{3}x+2y=8$

Find the slope of the line determined by each pair of points. Graph the line.

9. $(1,-2),(9,7)$

10. $(-2,5),(8,3)$

Write each equation in slope-intercept form. Find the slope and y-intercept, and then use them to draw the graph.

11. $x - 3y = 4$

12. $4x + 3y = 3$

13. Bicycling: The following table shows the number of miles Sam rode his bicycle from one hour to another.

a. Plot these points on a graph.
b. Connect the points with line segments.
c. Find the slope of each line segment.
d. Interpret the slope as a rate of change.

Hour	Miles
First	20
Second	31
Third	17

14. Find an equation in standard form for the line passing through the point $(3, 7)$ and with the slope $m = -\frac{5}{3}$. Graph the line.

15. Find an equation in standard form for the line passing through the points $(-4, 6)$ and $(3, -2)$. Graph the line.

Find an equation in slope-intercept form for each of the equations described.

16. Horizontal and passing through $(-1, 6)$

17. Parallel to $3x + 2y = -1$ and passing through $(2, 4)$

18. Perpendicular to the y-axis and passing through the point $(3, -2)$

19. Write the definition of a function.

20. List the set of ordered pairs that corresponds to the points. State the domain and range and indicate whether or not the relation is a function.

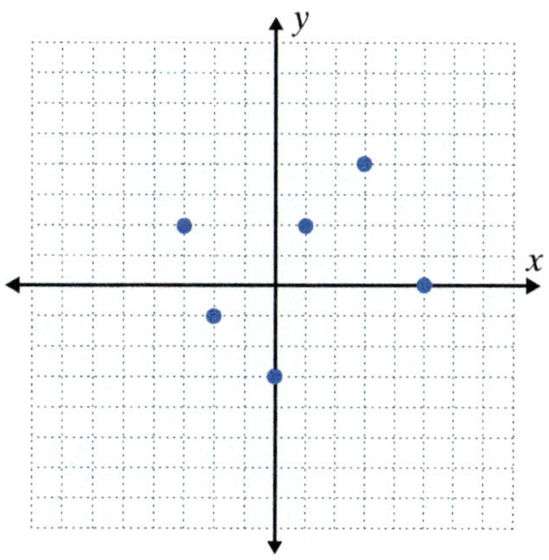

10. $m = -\frac{1}{5}$

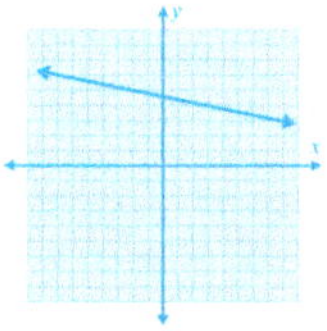

11.

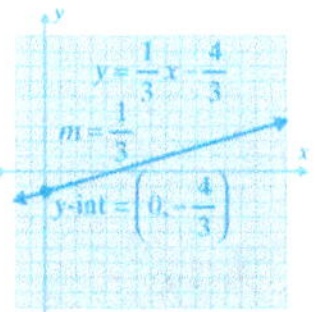

12.

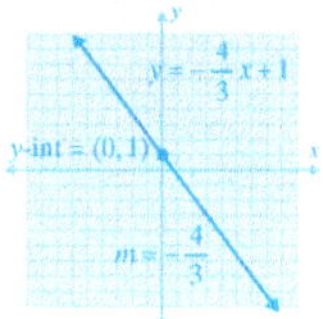

13. a. and b.

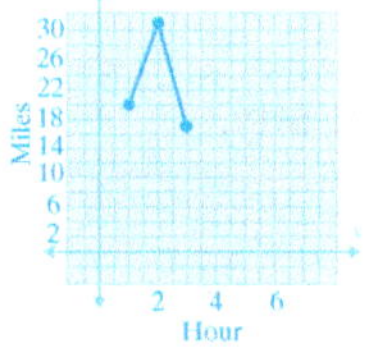

c. 11, –14
d. Sam's speed increased 11 mph from the 1st to the 2nd hour and decreased 14 mph from the 2nd to the 3rd hour.

14. $5x + 3y = 36$

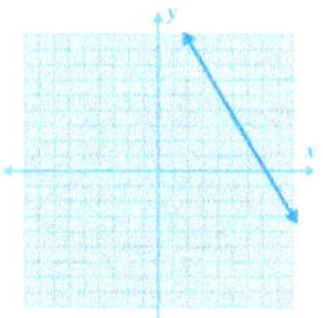

15. $8x + 7y = 10$

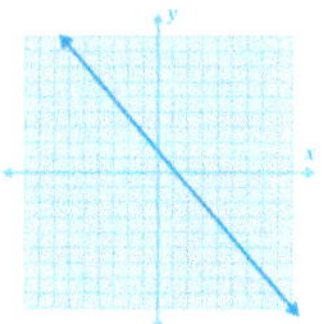

16. $y = 6$

17. $y = -\frac{3}{2}x + 7$

18. $y = -2$

19. See page 309.

20. $f = \left\{ \begin{array}{l} (-3, 2), (-2, -1), \\ (0, -3), (1, 2), \\ (3, 4), (5, 0) \end{array} \right\}$;
$D = \{-3, -2, 0, 1, 3, 5\}$;
$R = \{-3, -1, 0, 2, 4\}$;
Function

21. Function;
$D = (-\infty, \infty)$;
$R = [0, \infty)$

22. **a.** 13 **b.** 5 **c.** 4

23. **a.**

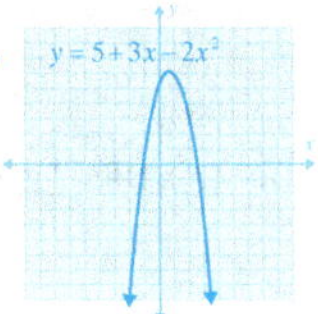

b. $(-1, 0), (2.5, 0)$

24.

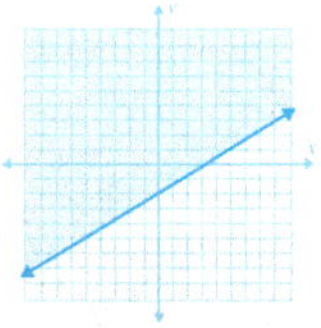

25.

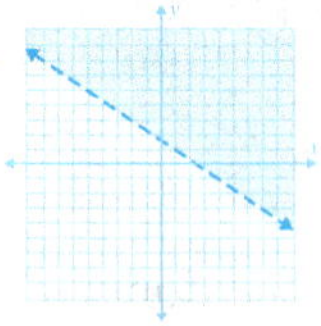

26.

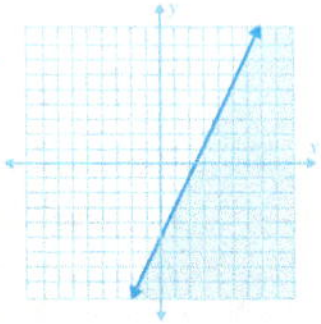

21. Use the vertical line test to determine whether or not the following graph represents a function. State the domain and range.

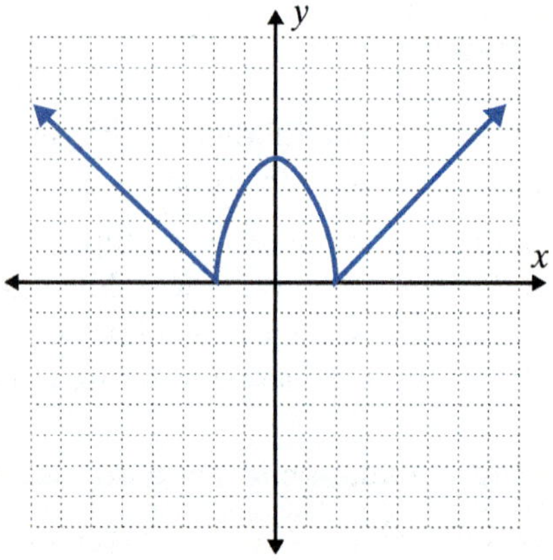

22. Given that $f(x) = x^2 - 2x + 5$, find **a.** $f(-2)$ **b.** $f(0)$ and **c.** $f(1)$.

23. **a.** Use a graphing calculator to graph the function $y = 5 + 3x - 2x^2$.
b. Use the CALC features of the calculator to find x-intercepts, if any.

Graph the solution set of each of the linear inequalities.

24. $3x - 5y \leq 10$

25. $3x + 4y > 7$

26. Use a graphing calculator to graph the following linear inequality: $y - 2x \leq -5$.

Cumulative Review: Chapters 1 – 4

1. $93 = 3 \cdot 31$
2. $300 = 2^2 \cdot 3 \cdot 5^2$
3. $245 = 5 \cdot 7^2$
4. 180
5. 2100
6. $120a^2b^2$
7. 13.125
8. 83.3
9. a. $\{\sqrt{9}\}$
   b. $\{0, \sqrt{9}\}$
   c. $\left\{-10, -\sqrt{25}, -1.6, 0, \frac{1}{5}, \sqrt{9}\right\}$
   d. $\left\{-10, -\sqrt{25}, 0, \sqrt{9}\right\}$
   e. $\left\{-\sqrt{7}, \pi, \sqrt{12}\right\}$
   f. $\left\{-10, -\sqrt{25}, -1.6, -\sqrt{7}, 0, \frac{1}{5}, \sqrt{9}, \pi, \sqrt{12}\right\}$
10. 

11. 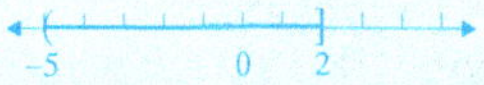

12. Associative property of addition
13. Additive inverse
14. Multiplicative identity
15. $x + 15$
16. $3x + 8$
17. 11
18. 22
19. –0.8
20. $-\frac{1}{12}$
21. –40
22. $\frac{2}{3}$
23. Undefined
24. 0

Find the prime factorization of each of the composite numbers.

1. 93 **2.** 300 **3.** 245

Find the LCM of each set of numbers or algebraic expressions.

4. $\{12, 15, 45\}$ **5.** $\{6, 20, 25, 35\}$ **6.** $\{6a^2, 24ab^2, 30ab, 40a^2b\}$

7. Find the average of the set of numbers: $\{3.6, 8.9, 14.7, 25.3\}$

8. Fernando scored 80, 88, and 82 on three exams in statistics. What was his average score on these exams (to the nearest tenth)?

9. Given the set of numbers $\left\{-10, -\sqrt{25}, -1.6, -\sqrt{7}, 0, \frac{1}{5}, \sqrt{9}, \pi, \sqrt{12}\right\}$, list those numbers that belong to each of the following sets:

a. $\{x \mid x \text{ is a natural number}\}$ **b.** $\{x \mid x \text{ is a whole number}\}$

c. $\{x \mid x \text{ is a rational number}\}$ **d.** $\{x \mid x \text{ is an integer}\}$

e. $\{x \mid x \text{ is an irrational number}\}$ **f.** $\{x \mid x \text{ is a real number}\}$

Graph each of the sets of real numbers described on a real number line.

10. $\{x \mid x \text{ is an integer between } -1.8 \text{ and } 3.1\}$

11. $\{x \mid x \text{ is a real number greater than } -5 \text{ but less or equal to } 2\}$

State the properties of real numbers that are illustrated. All variables represent real numbers.

12. $7 + (x + 3) = (7 + x) + 3$ **13.** $\frac{4}{5} + \left(-\frac{4}{5}\right) = 0$ **14.** $x \cdot 1 = x$

Use the distributive property to complete the expression.

15. $3x + 45 = 3(\quad)$ **16.** $6x + 16 = 2(\quad)$

Perform the indicated operations.

17. $|-9| + |-2|$ **18.** $17 - (-5)$ **19.** $6.5 + (-4.2) - 3.1$

20. $\frac{3}{4} - \frac{2}{3} + \left(-\frac{1}{6}\right)$ **21.** $8(-5)$ **22.** $(-4) \div (-6)$

23. $8 \div 0$ **24.** $0 \div \frac{3}{5}$

25. -2 26. -35
27. 26 28. -59
29. $-2x-12$
30. 0
31. $3x^3-x^2+4x-1$
32. $-10x+4$
33. $x=2$ 34. $x=-4$
35. $x=-4$ 36. $x=-8$
37. $x=-3.3, 2.3$
38. $x=2, \frac{10}{3}$
39. Conditional
40. Identity
41. a. $n=2A-m$
b. $f=\frac{\omega}{2\pi}$
42. 

$(4, \infty)$
43. 

$[-12.8, \infty)$
44. 

$(-\infty, 2]$
45. 

$(-\infty, -5]$
46. 

$\left(-\infty, -\frac{9}{5}\right)$ or $(1, \infty)$
47. 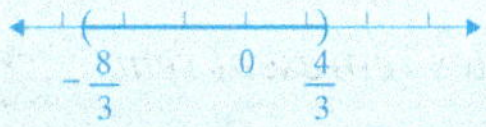

$\left(-\frac{8}{3}, \frac{4}{3}\right)$
48. a. $(0, -4)$ b. $(2, 0)$
c. $(1, -2)$ d. $(3, 2)$
49. a. $(0, 2)$ b. $(6, 0)$
c. $\left(2, \frac{4}{3}\right)$ d. $(9, -1)$
50. 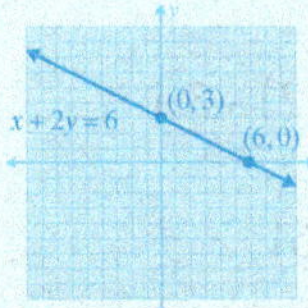

Find the value of each expression by using the rules for order of operations.

25. $12-6\div 2\cdot 3-5$

26. $5-(13\cdot 5-5)\div 3\cdot 2$

27. $3^2+5\cdot 4-10+|7|$

28. $6\cdot 4-2^3-(5\cdot 10)-5^2$

Simplify each expression by combining like terms.

29. $-4(x+3)+2x$

30. $x+\frac{x-5x}{4}$

31. $(x^3+4x-1)-(-2x^3+x^2)$

32. $-2[7x-(2x+5)+3]$

Solve each of the equations.

33. $9x-11=x+5$

34. $5(1-2x)=3x+57$

35. $5(2x+3)=3(x-4)-1$

36. $\frac{7x}{8}+5=\frac{x}{4}$

37. $|2x+1|=5.6$

38. $|2(x-4)+x|=2$

Determine whether each of the following equations is a conditional equation, an identity, or a contradiction.

39. $12x-7=-4(4x-1)-3$

40. $2(4z+5)+16=8z+26$

41. Solve each equation for the indicated variable.

a. Solve for n: $A=\frac{m+n}{2}$

b. Solve for f: $\omega=2\pi f$

Solve each inequality and graph the solution set on a real number line. Write each solution in interval notation. Assume that x is a real number.

42. $5x-7>x+9$

43. $5x+10\le 6(x+3.8)$

44. $x+8-5x\ge 2(x-2)$

45. $\frac{2x+1}{3}\le\frac{3x}{5}$

46. $|5x+2|>7$

47. $|3x+2|+4<10$

Determine the missing coordinate in each of the ordered pairs so that the point will satisfy the equation given.

48. $2x-y=4$

a. (0,)

b. (,0)

c. (1,)

d. (,2)

49. $x+3y=6$

a. (0,)

b. (,0)

c. (2,)

d. (,−1)

Graph the following linear equations by locating the x-intercept and the y-intercept.

50. $x + 2y = 6$ **51.** $2x - 5y = 10$ **52.** $3x - 4y = 6$

Write each equation in slope-intercept form. Find the slope and the y-intercept, and use them to draw the graph.

53. $x + 5y = 10$ **54.** $3x + y = 1$ **55.** $3x - 7y = 7$

Find an equation in standard form for the line determined by the given point and slope or two points.

56. $(6, -1), m = \frac{2}{5}$ **57.** $(-1, 2), m = \frac{4}{3}$ **58.** $(0, 0), m = 2$

59. $(5, 2)$, m is undefined **60.** $(0, 3), (5, -1)$ **61.** $(5, -2), (1, 6)$

Find an equation in slope-intercept form for each of the equations described.

62. Parallel to $3x + 2y - 6 = 0$ and passing through $(2, 3)$

63. Parallel to the y-axis and passing through $(1, -7)$

64. Perpendicular to $4x + 3y = 5$ and passing through $(4, 0)$

65. Perpendicular to $3x - 5y = 1$ and passing through $(6, -2)$

List the ordered pairs corresponding to the points in the given graph. State the domain and range and whether or not the relation is a function.

66.

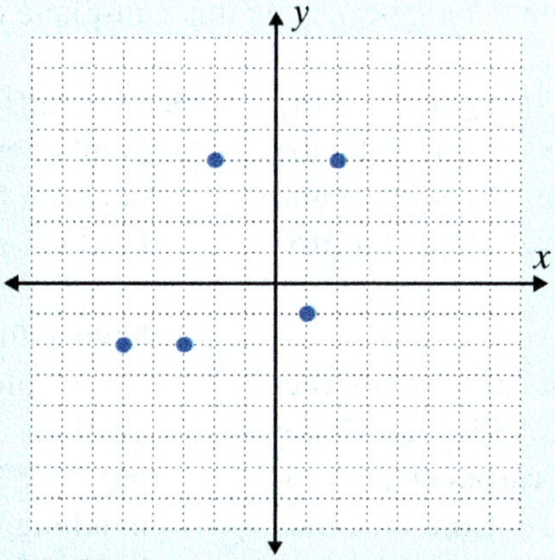

67.

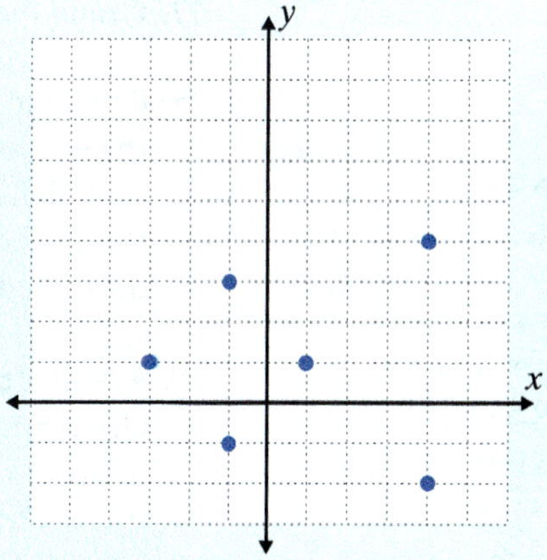

51.

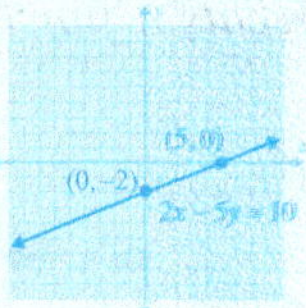

52.

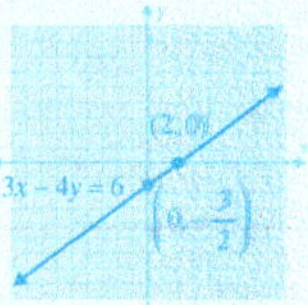

53.

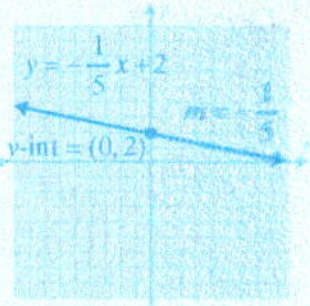

54.

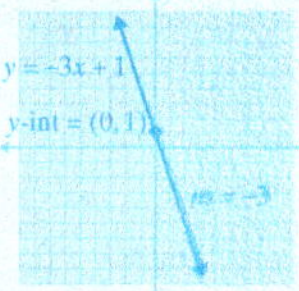

55.

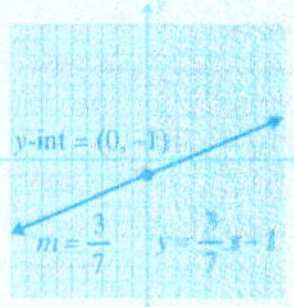

56. $2x - 5y = 17$

57. $4x - 3y = -10$

58. $2x - y = 0$

59. $x = 5$

60. $4x + 5y = 15$

61. $2x + y = 8$

62. $y = -\frac{3}{2}x + 6$

63. $x = 1$

64. $y = \frac{3}{4}x - 3$

65. $y = -\frac{5}{3}x + 8$

66. $\left\{\begin{matrix}(-5, -2), \\ (-3, -2), (-2, 4), \\ (1, -1), (2, 4)\end{matrix}\right\}$; $D = \{-5, -3, -2, 1, 2\}$; $R = \{-2, -1, 4\}$; Function

67. $\left\{\begin{matrix}(-3, 1), (-1, -1), (-1, 3), \\ (1, 1), (4, -2), (4, 4)\end{matrix}\right\}$; $D = \{-3, -1, 1, 4\}$; $R = \{-2, -1, 1, 3, 4\}$; Not a function

68. Not a function; $D = (-\infty, 0]$; $R = (-\infty, \infty)$

Use the vertical line test to determine whether or not each graph represents a function. State the domain and range.

68.

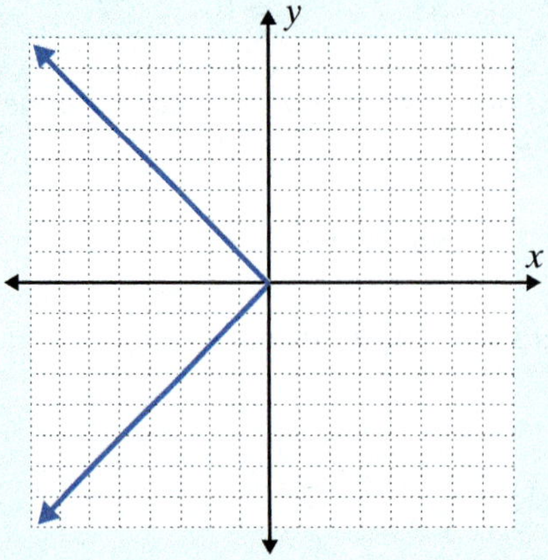

69.

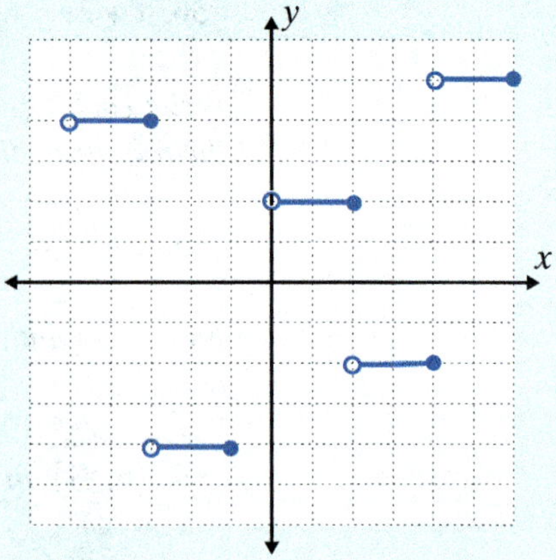

70. For $f(x) = -3x + 14$, find $f(5)$.

71. For $g(x) = x^2 - 4x + 7$, find $g(3)$.

72. For $F(x) = 3x^2 - 8x - 10$, find $F(0)$.

73. For $G(x) = 5x^3 - 4x$, find $G(-2)$.

Graph the solution set of each of the linear inequalities.

74. $y \geq 4x$

75. $3x + y < 2$

76. Graph the half-plane $y \geq -2x + 3$. Is this half-plane open or closed?

77. Graph the half-plane $-3x + y < 6.3$. Is this half-plane open or closed?

78. Consecutive even integers: Find three consecutive even integers such that the sum of the second and third is equal to three times the first decreased by 14.

79. Rental cost: The local supermarket charges a flat rate of \$5, plus \$3 per hour for rental of a carpet cleaner. If it cost Ron \$26 to rent the machine, how many hours did he keep it?

80. The difference between twice a number and 3 is equal to the difference between five times the number and 2. Find the number.

81. Buying Online: Sarah found a purse she would like to buy at an online auction. If she submits a winning bid of \$72 then with shipping and handling, the total cost of the purchase is \$76.80. What percent of the total cost is the shipping and handling?

69. Function; $D = (-5, -1] \cup (0, 6]$; $R = \{-4, -2, 2, 4, 5\}$

70. -1

71. 4

72. -10

73. -32

74.

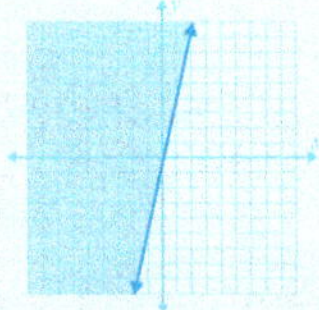

75.

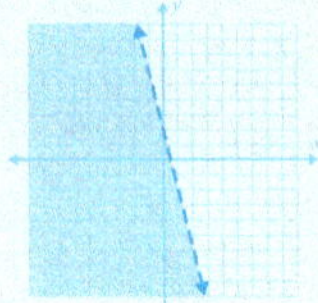

76. Closed

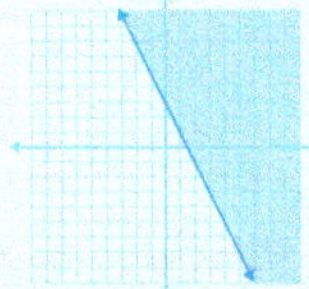

77. Open

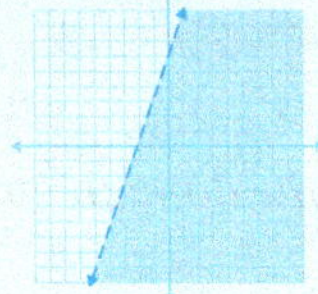

78. 20, 22, 24

79. 7 hours

80. $-\frac{1}{3}$

81. 6.25%

82. Length = 25 cm; Width = 16 cm

83. 84

84. 10 miles

85. \$3400 at 4.5% \$1100 at 7%

82. Rectangles: The length of a rectangle is 9 cm more than its width. The perimeter is 82 cm. Find the dimensions of the rectangle.

83. Entrance exams: The mathematics component of the entrance exam at a certain Midwestern college consists of three parts: one part on geometry, one part on algebra, and one part on trigonometry. Prospective students must score at least 50 on each part and average at least 70 on the three parts. Beth learned that she had scored 60 and 66 on the first two parts of the exam. What minimum score did she need on the third part for her to pass this portion of the exam to gain entrance to the college?

84. Distance traveled: Stephanie rode her new moped to Rod's house. Traveling the side streets, she averaged 20 mph. To save time on the return trip, they loaded the bike into Rod's truck and took the freeway, averaging 50 mph. The freeway distance is 2 miles less than the distance on the side streets and saves 0.4 hours. Find the distance traveled on the return trip.

85. Interest: Christopher opened a sandwich shop and wants to invest some of his earnings. He decides to invest a total of \$4500, part at 4.5% and the other at 7%. How much money does he need to invest at each rate so that the interest made is \$230?

86. World Series of Poker: The following table shows the number of entrants in the main event of the World Series of Poker from 2003 to 2008.

Year	Number of Entrants
2003	839
2004	2576
2005	5619
2006	8773
2007	6358
2008	6844

a. Plot these points on a graph.
b. Connect the points with line segments.
c. Find the slope of each line segment.
d. Interpret the slope as a rate of change.

Source: Harrah's License Company

87. a. Use a graphing calculator to graph the linear functions $y = -2x + 7$ and $y = 3x$.
b. Use the CALC feature of the calculator to find the point of intersection of these two linear functions.

88. a. Use a graphing calculator to graph the function $y = x^4 - 3x^3 + 1$.
b. Use the CALC feature of the calculator to find the values of the x-intercepts.

86. a. and b.

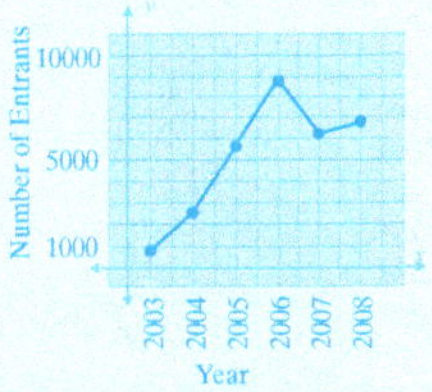

c. 1737; 3043; 3154; –2415; 486

d. Number of entrants increased by 1737 people/year from 2003-2004; increased by 3043 people/year from 2004-2005; increased by 3154 people/year from 2005-2006; decreased by 2415 people/year from 2006-2007; increased by 486 people/year from 2007-2008.

87. a.

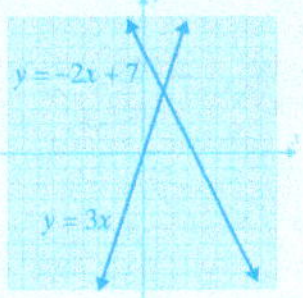

b. (1.4, 4.2)

88. a.

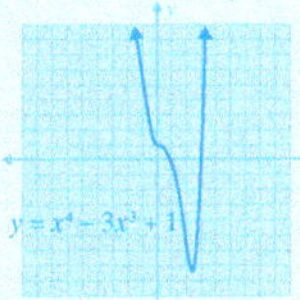

b. (0.7648, 0), (2.9615, 0)

CHAPTER 5

Exponents and Polynomials

Did You Know?

One of the most difficult challenges for students in beginning algebra is to become comfortable with the idea that letters or symbols can be manipulated just like numbers in arithmetic. These symbols may be the cause of "math anxiety." A great deal of publicity has recently been given to the concept that a large number of people suffer from math anxiety, a painful uneasiness caused by mathematical symbols or problem-solving situations. Persons affected by math anxiety feel that it is almost impossible to learn mathematics, or they may be able to learn but be unable to apply their knowledge or do well on tests. Persons suffering from math anxiety often develop math avoidance, so they avoid careers, majors, or classes that will require mathematics courses or skills. The sociologist Lucy Sells has determined that mathematics is a critical filter in the job market. Persons who lack quantitative skills are channeled into high unemployment, low-paying, non-technical areas.

What causes math anxiety? Researchers are investigating the following hypotheses:

1. A lack of skills that leads to a lack of confidence and, therefore, to anxiety;
2. An attitude that mathematics is not useful or significant to society;
3. Career goals that seem to preclude mathematics;
4. A self-concept that differs radically from the stereotype of a mathematician;
5. Perceptions that parents, peers, or teachers have low expectations for the person in mathematics;
6. Social conditioning to avoid mathematics.

We hope that you are finding your present experience with algebra successful and that the skills you are acquiring now will enable you to approach mathematical problems with confidence.

"Mathematics is the queen of the sciences and arithmetic the queen of mathematics."

Carl F. Gauss (1777 – 1855)

In Chapter 5, you will learn the rules of exponents and how exponents can be used to simplify very large and very small numbers. Astronomers, chemists, and biologists frequently make use of these ideas and the notation used is appropriately called scientific notation. This notation is also used in scientific calculators. (Multiply 5,000,000 by 5,000,000 on your calculator and read the results. What do you think the answer means?)

Polynomials and operations with polynomials (the topics in Chapters 5, 6, and 7) are used throughout all levels of mathematics from elementary and intermediate algebra through statistics, calculus, and beyond. **Be aware that the knowledge and skills you learn about polynomials will be needed again and again and may be one of the determining factors in successfully completing this course and future courses, so work particularly hard at this early stage.**

5.1 Exponents

- *Simplify expressions by using properties of integer exponents.*
- *Recognize which property of exponents is used to simplify an expression.*

The Product Rule

In Section 1.5, an **exponent** was defined as a number that tells how many times a number (called the **base**) is used in multiplication. This definition is limited because it is valid only if the exponents are positive integers. In this section, we will develop four properties of exponents that will help in simplifying algebraic expressions and expand your understanding of exponents to include variable bases, negative exponents, and the exponent 0. (In later sections you will study fractional exponents.)

From Section 1.5, we know that

Exponent

$$6^2 = 6 \cdot 6 = 36$$

Base

and

$$6^3 = 6 \cdot 6 \cdot 6 = 216.$$

Also, the base may be a variable so that

$$x^3 = x \cdot x \cdot x \quad \text{and} \quad x^5 = x \cdot x \cdot x \cdot x \cdot x.$$

Teaching Notes:
Be sure to emphasize that the exponent **counts** the number of factors in these expressions. This is especially important when students are dealing with expressions such as $2^3 \cdot 2^4$, because there is a tendency for them to multiply the base numbers and erroneously write 4^7.

Now to find the products of expressions such as $6^2 \cdot 6^3$ or $x^3 \cdot x^5$ and to simplify these products, we can write down all the factors as follows:

$$6^2 \cdot 6^3 = (6 \cdot 6) \cdot (6 \cdot 6 \cdot 6) = 6^5$$

and

$$x^3 \cdot x^5 = (x \cdot x \cdot x) \cdot (x \cdot x \cdot x \cdot x \cdot x) = x^8.$$

With these examples in mind, what do you think would be a simplified form for the product $3^4 \cdot 3^3$? You were right if you thought 3^7. That is, $3^4 \cdot 3^3 = 3^7$. Notice that in each case, **the base stays the same**.

The preceding discussion, along with the basic concept of whole-number exponents, leads to the following **product rule for exponents**.

The Product Rule for Exponents

If a is a nonzero real number and m and n are integers, then

$$a^m \cdot a^n = a^{m+n}.$$

In words, to multiply two powers with the same base, keep the base and add the exponents.

NOTES

Remember about the **exponent 1**. If a variable or constant has no exponent written, the exponent is understood to be 1.

For example,

$$y = y^1$$

and

$$7 = 7^1.$$

In general, for any real number a,

$$a = a^1.$$

Example 1: The Product Rule for Exponents

Use the product rule for exponents to simplify the following expressions.

a. $x^2 \cdot x^4$

Solution: $x^2 \cdot x^4 = x^{2+4} = x^6$

Continued on the next page...

Teaching Notes:
You might want to show your students an expression such as $x^3 \cdot y^4 = x^3y^4$ where the bases are different and the exponents are **not** added.

b. $y \cdot y^6$

Solution: $y \cdot y^6 = y^1 \cdot y^6 = y^{1+6} = y^7$

c. $4^2 \cdot 4$

Solution: $4^2 \cdot 4 = 4^{2+1} = 4^3 = 64$

Note that the base stays 4. That is, the bases are not multiplied.

d. $2^3 \cdot 2^2$

Solution: $2^3 \cdot 2^2 = 2^{3+2} = 2^5 = 32$

Note that the base stays 2. That is, the bases are not multiplied.

e. $(-2)^4(-2)^3$

Solution: $(-2)^4(-2)^3 = (-2)^{4+3} = (-2)^7 = -128$

To multiply terms that have numerical coefficients and variables with exponents, **the coefficients are multiplied** as usual and **the exponents are added by using the product rule**. Generally variables and constants may need to be rearranged using the commutative and associative properties of multiplication. Example 2 illustrates these concepts.

Example 2: Multiplying Terms with Coefficients

Use the product rule for exponents when simplifying the following expressions.

a. $2y^2 \cdot 3y^9$

Solution:
$$2y^2 \cdot 3y^9 = 2 \cdot 3 \cdot y^2 \cdot y^9$$
$$= 6y^{2+9}$$
$$= 6y^{11}$$

Coefficients 2 and 3 are multiplied and exponents 2 and 9 are added.

b. $(-3x^3)(-4x^3)$

Solution:
$$(-3x^3)(-4x^3) = (-3)(-4) \cdot x^3 \cdot x^3$$
$$= 12x^{3+3}$$
$$= 12x^6$$

Coefficients –3 and –4 are multiplied and exponents 3 and 3 are added.

c. $(-6ab^2)(8ab^3)$

Solution:
$$(-6ab^2)(8ab^3) = (-6) \cdot 8 \cdot a^1 \cdot a^1 \cdot b^2 \cdot b^3$$
$$= -48 \cdot a^{1+1} \cdot b^{2+3}$$
$$= -48a^2b^5$$

Coefficients –6 and 8 are multiplied and exponents on each variable are added.

The Exponent 0

The product rule is stated for m and n as **integer** exponents. This means that the rule is also valid for 0 and for negative exponents. As an aid for understanding 0 as an exponent, consider the following patterns of exponents for powers of 2, 3, and 10.

Powers of 2	Powers of 3	Powers of 10
$2^5 = 32$	$3^5 = 243$	$10^5 = 100{,}000$
$2^4 = 16$	$3^4 = 81$	$10^4 = 10{,}000$
$2^3 = 8$	$3^3 = 27$	$10^3 = 1000$
$2^2 = 4$	$3^2 = 9$	$10^2 = 100$
$2^1 = 2$	$3^1 = 3$	$10^1 = 10$
$2^0 = ?$	$3^0 = ?$	$10^0 = ?$

Do you notice that the patterns indicate that the exponent 0 gives the same value for the last number in each column? That is, $2^0 = 1$, $3^0 = 1$, and $10^0 = 1$.

Another approach to understanding 0 as an exponent is to consider the product rule. Remember that the product rule is stated for **integer exponents**. Applying this rule with the exponent 0 and the fact that 1 is the multiplicative identity gives results such as the following:

$$5^0 \cdot 5^2 = 5^{0+2} = 5^2 \qquad \text{This implies that } 5^0 = 1.$$

$$4^0 \cdot 4^3 = 4^{0+3} = 4^3 \qquad \text{This implies that } 4^0 = 1.$$

This discussion leads directly to the **rule for 0 as an exponent**.

The Exponent 0

If a is a nonzero real number, then

$$a^0 = 1.$$

The expression 0^0 is undefined.

NOTES

Throughout this text, unless specifically stated otherwise, we will assume that the bases of exponents are nonzero.

Example 3: The Exponent 0

Simplify the following expressions using the rule for 0 as an exponent.

a. 10^0

Solution: $10^0 = 1$

b. $x^0 \cdot x^3$

Solution: $x^0 \cdot x^3 = x^{0+3} = x^3$ or $x^0 \cdot x^3 = 1 \cdot x^3 = x^3$

c. $(-6)^0$

Solution: $(-6)^0 = 1$

The Quotient Rule

Teaching Notes:
While we are emphasizing the quotient rule for exponents here, you might want to discuss the fact that $\frac{5^4}{5^2} = \frac{625}{25} = 25$, and, as long as the base is numerical, we can always check this way.

Now consider a fraction in which the numerator and denominator are powers with the same base, such as $\frac{5^4}{5^2}$ or $\frac{x^5}{x^2}$. We can write,

$$\frac{5^4}{5^2} = \frac{\cancel{5} \cdot \cancel{5} \cdot 5 \cdot 5}{\cancel{5} \cdot \cancel{5} \cdot 1} = \frac{5^2}{1} = 25 \quad \text{or} \quad \frac{5^4}{5^2} = 5^{4-2} = 5^2 = 25$$

and

$$\frac{x^5}{x^2} = \frac{\cancel{x} \cdot \cancel{x} \cdot x \cdot x \cdot x}{\cancel{x} \cdot \cancel{x} \cdot 1} = \frac{x^3}{1} = x^3 \quad \text{or} \quad \frac{x^5}{x^2} = x^{5-2} = x^3.$$

In fractions, as just illustrated, the exponents can be subtracted. Again, the base remains the same. We now have the following **quotient rule for exponents**.

The Quotient Rule for Exponents

If a is a nonzero real number and m and n are integers, then

$$\frac{a^m}{a^n} = a^{m-n}.$$

In words, to divide two powers with the same base, keep the base and subtract the exponents. (Subtract the denominator exponent from the numerator exponent.)

Example 4: The Quotient Rule for Exponents

Use the quotient rule for exponents to simplify the following expressions.

a. $\frac{x^6}{x}$

Solution: $\frac{x^6}{x} = x^{6-1} = x^5$

b. $\frac{y^8}{y^2}$

Solution: $\frac{y^8}{y^2} = y^{8-2} = y^6$

c. $\frac{x^2}{x^2}$

Solution: $\frac{x^2}{x^2} = x^{2-2} = x^0 = 1$

Note how this example shows another way to justify the idea that $a^0 = 1$. Since the numerator and denominator are the same and not 0, it makes sense that the fraction is equal to 1.

In division with terms that have numerical coefficients the **coefficients are divided** as usual and any **exponents are subtracted** by using the quotient rule. These ideas are illustrated in Example 5.

Example 5: Dividing Terms with Coefficients

Use the quotient rule for exponents when simplifying the following expressions.

a. $\frac{15x^{15}}{3x^3}$

Solution:

$$\frac{15x^{15}}{3x^3} = \frac{15}{3} \cdot \frac{x^{15}}{x^3}$$

$$= 5 \cdot x^{15-3}$$

Coefficients 15 and 3 are divided and exponents 15 and 3 are subtracted.

$$= 5x^{12}$$

b. $\frac{20x^{10}y^6}{2x^2y^3}$

Solution:

$$\frac{20x^{10}y^6}{2x^2y^3} = \frac{20}{2} \cdot \frac{x^{10}}{x^2} \cdot \frac{y^6}{y^3}$$

$$= 10 \cdot x^{10-2} \cdot y^{6-3}$$

Coefficients 20 and 2 are divided and exponents on each variable are subtracted.

$$= 10x^8y^3$$

Negative Exponents

The quotient rule for exponents leads directly to the development of an understanding of negative exponents. In Examples 4 and 5, for each base, the exponent in the numerator was larger than or equal to the exponent in the denominator. Therefore, when the exponents were subtracted using the quotient rule, the result was either a positive exponent or the exponent 0. But, what if the larger exponent is in the denominator and we still apply the quotient rule?

Teaching Notes:
You might want to refer back to the patterns discussed on page 357 and help the students to logically decide that the next steps in the respective columns are $2^{-1} = \frac{1}{2}$, $3^{-1} = \frac{1}{3}$, and $5^{-1} = \frac{1}{5}$.

For example, applying the quotient rule to $\frac{4^3}{4^5}$ gives

$$\frac{4^3}{4^5} = 4^{3-5} = 4^{-2} \text{ which results in a negative exponent.}$$

But, simply reducing $\frac{4^3}{4^5}$ gives

$$\frac{4^3}{4^5} = \frac{\cancel{4}\cdot\cancel{4}\cdot\cancel{4}\cdot 1}{\cancel{4}\cdot\cancel{4}\cdot\cancel{4}\cdot 4\cdot 4} = \frac{1}{4\cdot 4} = \frac{1}{4^2}.$$

This means that $4^{-2} = \frac{1}{4^2}$.

Similar discussions will show that $2^{-1} = \frac{1}{2}$, $5^{-2} = \frac{1}{5^2}$, and $x^{-3} = \frac{1}{x^3}$.

The **rule for negative exponents** follows.

Rule for Negative Exponents

If a is a nonzero real number and n is an integer, then

$$\boldsymbol{a^{-n} = \frac{1}{a^n}}.$$

Example 6: Negative Exponents

Use the rule for negative exponents to simplify each expression so that it contains only positive exponents.

a. 5^{-1}

Solution: $5^{-1} = \frac{1}{5^1} = \frac{1}{5}$ Using the rule for negative exponents

b. x^{-3}

Solution: $x^{-3} = \frac{1}{x^3}$ Using the rule for negative exponents

c. $x^{-9} \cdot x^7$

Solution: Here we use the product rule first and then the rule for negative exponents.

$$x^{-9} \cdot x^7 = x^{-9+7} = x^{-2} = \frac{1}{x^2}$$

Teaching Notes: Scientific notation (as discussed in Section 5.2) is an example of preference for the use of negative exponents.

Each of the expressions in Example 7 is simplified by using the appropriate rules for exponents. Study each example carefully. In each case, **the expression is considered simplified if each base appears only once and each base has only positive exponents.**

NOTES

There is nothing wrong with negative exponents. In fact, negative exponents are preferred in some courses in mathematics and science. However, so that all answers are the same, in this course we will consider expressions to be simplified if:

1. all exponents are positive, and
2. each base appears only once.

Example 7: Combining the Rules for Exponents

Simplify each expression so that it contains only positive exponents.

a. $2^{-5} \cdot 2^8$

Solution: $2^{-5} \cdot 2^8 = 2^{-5+8}$ Using the product rule with positive and negative exponents

$= 2^3$

$= 8$

b. $\frac{x^6}{x^{-1}}$

Solution: $\frac{x^6}{x^{-1}} = x^{6-(-1)}$ Using the quotient rule with positive and negative exponents

$= x^{6+1}$

$= x^7$

Continued on the next page...

c. $\dfrac{10^{-5}}{10^{-2}}$

Solution: $\dfrac{10^{-5}}{10^{-2}} = 10^{-5-(-2)}$ — Using the quotient rule with negative exponents

$= 10^{-5+2}$

$= 10^{-3}$

$= \dfrac{1}{10^3}$ or $\dfrac{1}{1000}$ — Using the rule for negative exponents

d. $\dfrac{x^6y^3}{x^2y^5}$

Solution: $\dfrac{x^6y^3}{x^2y^5} = x^{6-2}y^{3-5}$ — Using the quotient rule with two variables

$= x^4y^{-2}$

$= \dfrac{x^4}{y^2}$ — Using the rule for negative exponents

e. $\dfrac{15x^{10}\cdot 2x^2}{3x^{15}}$

Solution: $\dfrac{15x^{10}\cdot 2x^2}{3x^{15}} = \dfrac{(15\cdot 2)x^{10+2}}{3x^{15}}$ — Using the product rule

$= \dfrac{30x^{12}}{3x^{15}}$

$= 10x^{12-15}$ — Using the quotient rule

$= 10x^{-3}$

$= \dfrac{10}{x^3}$ — Using the rule for negative exponents

NOTES

Special Note About Using the Quotient Rule

Regardless of the size of the exponents or whether they are positive or negative, the following single subtraction rule can be used with the quotient rule.

(numerator exponent – denominator exponent)

This subtraction will always lead to the correct answer.

Summary of the Rules for Exponents

For any nonzero real number a and integers m and n:

1. The exponent 1: $a = a^1$
2. The exponent 0: $a^0 = 1$
3. The product rule: $a^m \cdot a^n = a^{m+n}$
4. The quotient rule: $\dfrac{a^m}{a^n} = a^{m-n}$
5. Negative exponents: $a^{-n} = \dfrac{1}{a^n}$

Using a TI-84 Plus Graphing Calculator to Evaluate Expressions with Exponents

On a TI-84 Plus graphing calculator the caret key ^ is used to indicate an exponent. Example 8 illustrates the use of a graphing calculator in evaluating expressions with exponents.

Example 8: Evaluating Expressions with Exponents

Use a graphing calculator to evaluate each expression.

a. 2^{-3} **b.** 23.18^0 **c.** $(-3.2)^3(1.5)^2$

Solutions: The following solutions show how the caret key ^ is used to indicate exponents. Be careful to use the negative sign key **(–)** (and not the minus sign key) for negative numbers and negative exponents.

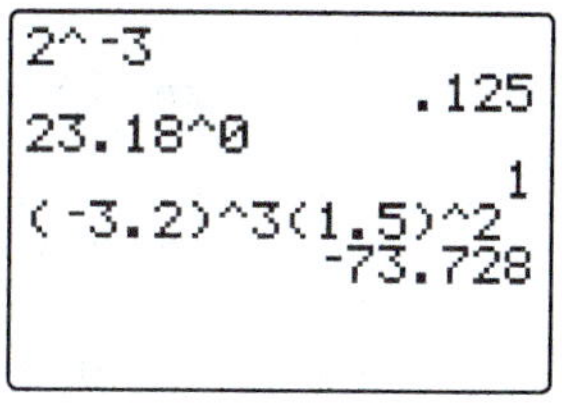

Practice Problems

Simplify each expression.

1. $2^3 \cdot 2^4$
2. $\dfrac{2^3}{2^4}$
3. $\dfrac{x^7 \cdot x^{-3}}{x^{-2}}$
4. $\dfrac{10^{-8} \cdot 10^2}{10^{-7}}$
5. $\dfrac{14x^{-3}y^2}{2x^{-3}y^{-2}}$
6. $(9x^4)^0$

Use a graphing calculator to evaluate each expression.

7. 72^4
8. $(19.3)^0(15)^3$

5.1 Exercises

1. 27 2. 16,807
3. 512 4. 25
5. $\frac{1}{3}$ 6. $\frac{1}{16}$
7. $\frac{1}{25}$ 8. $\frac{1}{216}$
9. 16 10. −64
11. 24 12. 54
13. −500 14. −54
15. $\frac{3}{8}$ 16. $\frac{4}{9}$
17. $-\frac{3}{25}$ 18. $-\frac{5}{4}$
19. x^5 20. x^4
21. y^2 22. y^{11}
23. $\frac{1}{x^3}$ 24. $\frac{1}{y^2}$
25. $\frac{2}{x}$ 26. $\frac{5}{y^4}$
27. $-\frac{8}{y^2}$ 28. $-\frac{10}{x^3}$
29. $\frac{5x^6}{y^4}$ 30. $\frac{1}{y^2}$
31. 4 32. 2
33. 49 34. 729

Simplify each expression. The final form of the expressions with variables should contain only positive exponents. Assume that all variables represent nonzero numbers.

1. $3^2 \cdot 3$ **2.** $7^2 \cdot 7^3$ **3.** $8^3 \cdot 8^0$ **4.** $5^0 \cdot 5^2$

5. 3^{-1} **6.** 4^{-2} **7.** 5^{-2} **8.** 6^{-3}

9. $(-2)^4(-2)^0$ **10.** $(-4)^3(-4)^0$ **11.** $3(2^3)$ **12.** $6(3^2)$

13. $-4(5^3)$ **14.** $-2(3^3)$ **15.** $3(2^{-3})$ **16.** $4(3^{-2})$

17. $-3(5^{-2})$ **18.** $-5(2^{-2})$ **19.** $x^2 \cdot x^3$ **20.** $x^3 \cdot x$

21. $y^2 \cdot y^0$ **22.** $y^3 \cdot y^8$ **23.** x^{-3} **24.** y^{-2}

25. $2x^{-1}$ **26.** $5y^{-4}$ **27.** $-8y^{-2}$ **28.** $-10x^{-3}$

29. $5x^6y^{-4}$ **30.** x^0y^{-2} **31.** $3x^0 + y^0$ **32.** $5y^0 - 3x^0$

33. $\dfrac{7^3}{7}$ **34.** $\dfrac{9^5}{9^2}$ **35.** $\dfrac{10^3}{10^4}$ **36.** $\dfrac{10}{10^5}$

37. $\dfrac{2^3}{2^6}$ **38.** $\dfrac{5^7}{5^4}$ **39.** $\dfrac{x^4}{x^2}$ **40.** $\dfrac{x^6}{x^3}$

Answers to Practice Problems: **1.** $2^7 = 128$ **2.** $\frac{1}{2}$ **3.** x^6 **4.** 10 **5.** $7y^4$ **6.** 1 **7.** 26,873,856 **8.** 3375

35. $\frac{1}{10}$ 36. $\frac{1}{10{,}000}$
37. $\frac{1}{8}$ 38. 125
39. x^2 40. x^3
41. x^2 42. y^5
43. x^4 44. x^5
45. $\frac{1}{x^4}$ 46. $\frac{1}{x^4}$
47. x^6 48. x^6
49. x^2 50. $\frac{1}{x^3}$
51. y^2 52. y^6
53. $3x^3$ 54. $3y^5$
55. x^4 56. $\frac{1}{x}$
57. $36x^3$ 58. $15x^6$
59. $-14x^5$ 60. $-18y^5$
61. $-12x^6$ 62. $30y^9$
63. $4y$ 64. $4x^3$
65. $3y^2$ 66. $-5x^4$
67. $-2y^2$ 68. $-4x^3$
69. $\frac{1}{x^2}$ 70. x^7
71. 1000 72. $\frac{1}{100}$
73. 1 74. 1
75. $-18x^5y^7$
76. $\frac{15x^3}{y^2}$
77. $-\frac{2y^2}{x}$
78. $-\frac{2y^6}{x^4}$
79. $12a^3b^9c$
80. $-24ab^{12}$
81. $-4a^{10}b^3c$
82. $\frac{x^2z^2}{4y^3}$ 83. $\frac{5}{x}$
84. $6b^5$ 85. 1
86. 25.6036
87. 0.390625
88. 0.10798
89. 8875.147264
90. 99,376.2576

41. $\frac{x^3}{x}$ **42.** $\frac{y^7}{y^2}$ **43.** $\frac{x^7}{x^3}$ **44.** $\frac{x^8}{x^3}$

45. $\frac{x^{-2}}{x^2}$ **46.** $\frac{x^{-3}}{x}$ **47.** $\frac{x^4}{x^{-2}}$ **48.** $\frac{x^5}{x^{-1}}$

49. $\frac{x^{-3}}{x^{-5}}$ **50.** $\frac{x^{-4}}{x^{-1}}$ **51.** $\frac{y^{-2}}{y^{-4}}$ **52.** $\frac{y^3}{y^{-3}}$

53. $3x^3 \cdot x^0$ **54.** $3y \cdot y^4$ **55.** $x^3 \cdot x^2 \cdot x^{-1}$ **56.** $x^{-3} \cdot x^0 \cdot x^2$

57. $(4x^3)(9x^0)$ **58.** $(5x^2)(3x^4)$ **59.** $(-2x^2)(7x^3)$ **60.** $(3y^3)(-6y^2)$

61. $(-4x^5)(3x)$ **62.** $(6y^4)(5y^5)$ **63.** $\frac{8y^3}{2y^2}$ **64.** $\frac{12x^4}{3x}$

65. $\frac{9y^5}{3y^3}$ **66.** $\frac{-10x^5}{2x}$ **67.** $\frac{-8y^4}{4y^2}$ **68.** $\frac{12x^6}{-3x^3}$

69. $\frac{x^{-1}x^2}{x^3}$ **70.** $\frac{x \cdot x^3}{x^{-3}}$ **71.** $\frac{10^4 \cdot 10^{-3}}{10^{-2}}$ **72.** $\frac{10 \cdot 10^{-1}}{10^2}$

73. $(9x^2)^0$ **74.** $(-2x^{-3}y^5)^0$ **75.** $(9x^2y^3)(-2x^3y^4)$

76. $(-3xy)(-5x^2y^{-3})$ **77.** $\frac{-8x^2y^4}{4x^3y^2}$ **78.** $\frac{-8x^{-2}y^4}{4x^2y^{-2}}$

79. $(3a^2b^4)(4ab^5c)$ **80.** $(-6a^3b^4)(4a^{-2}b^8)$ **81.** $\frac{36a^5b^0c}{-9a^{-5}b^{-3}}$

82. $\frac{7x^2y^{-2}}{28x^0yz^{-2}}$ **83.** $\frac{25y^6 \cdot 3y^{-2}}{15xy^4}$ **84.** $\frac{12a^{-2} \cdot 18a^4}{36a^2b^{-5}}$

Calculator Problems

Use a graphing calculator to evaluate each expression.

85. $(2.16)^0$ **86.** $(-5.06)^2$ **87.** $(1.6)^{-2}$

88. $(2.1)^{-3}$ **89.** $(6.4)^4(2.3)^2$ **90.** $(-14.8)^2(21.3)^2$

Writing and Thinking About Mathematics

91. Show why each of the following statements is WRONG.

a. $3^2 \cdot 3^2 = 6^2$ **b.** $3^2 \cdot 2^2 = 6^4$ **c.** $3^2 \cdot 3^2 = 9^4$

HAWKES LEARNING SYSTEMS: INTRODUCTORY & INTERMEDIATE ALGEBRA SOFTWARE

- 5.1 Simplifying Integer Exponents I

91. a. $3^2 \cdot 3^2 = 3^4 = 81$, but $6^2 = 36$. b. $3^2 \cdot 2^2 = 3 \cdot 3 \cdot 2 \cdot 2 = 6^2$ not 6^4 c. $3^2 \cdot 3^2 = 3^4$ not 9^4

5.2 Exponents and Scientific Notation

- *Simplify powers of expressions by using the properties of integer exponents.*
- *Write decimal numbers in scientific notation.*
- *Operate with decimal numbers by using scientific notation.*

The summary of the rules for exponents given in Section 5.1 is repeated here for easy reference.

Summary of the Rules for Exponents

For any nonzero real number a and integers m and n:

1. The exponent 1: $\boldsymbol{a = a^1}$
2. The exponent 0: $\boldsymbol{a^0 = 1}$
3. The product rule: $\boldsymbol{a^m \cdot a^n = a^{m+n}}$
4. The quotient rule: $\boldsymbol{\dfrac{a^m}{a^n} = a^{m-n}}$
5. Negative exponents: $\boldsymbol{a^{-n} = \dfrac{1}{a^n}}$

Power Rule

Now consider what happens when a power is raised to a power. For example, to simplify the expressions $\left(x^2\right)^3$ and $\left(2^5\right)^2$, we can write

$$\left(x^2\right)^3 = x^2 \cdot x^2 \cdot x^2 = x^{2+2+2} = x^6$$

and

$$\left(2^5\right)^2 = 2^5 \cdot 2^5 = 2^{5+5} = 2^{10}.$$

However, this technique can be quite time-consuming when the exponent is large such as in $\left(3y^3\right)^{17}$. The **power rule for exponents** gives a convenient way to handle powers raised to powers.

Power Rule for Exponents

If a is a nonzero real number and m and n are integers, then

$$\left(a^m\right)^n = a^{mn}.$$

In words, the value of a power raised to a power can be found by multiplying the exponents and keeping the base.

Example 1: Power Rule for Exponents

Simplify each expression by using the power rule for exponents.

a. $\left(x^2\right)^4$

Solution: $\left(x^2\right)^4 = x^{2(4)} = x^8$

b. $\left(x^5\right)^{-2}$

Solution: $\left(x^5\right)^{-2} = x^{5(-2)} = x^{-10} = \dfrac{1}{x^{10}}$

or $\left(x^5\right)^{-2} = \dfrac{1}{\left(x^5\right)^2} = \dfrac{1}{x^{5(2)}} = \dfrac{1}{x^{10}}$

c. $\left(y^{-7}\right)^2$

Solution: $\left(y^{-7}\right)^2 = y^{(-7)(2)} = y^{-14} = \dfrac{1}{y^{14}}$

d. $\left(2^3\right)^{-2}$

Solution: $\left(2^3\right)^{-2} = 2^{3(-2)} = 2^{-6} = \dfrac{1}{2^6}$ $\left(\text{Evaluating gives } \dfrac{1}{2^6} = \dfrac{1}{64}.\right)$

Another approach, because we have a numerical base, would be

$\left(2^3\right)^{-2} = (8)^{-2} = \dfrac{1}{8^2}.$ $\left(\text{Evaluating gives } \dfrac{1}{8^2} = \dfrac{1}{64}.\right)$

We see that while the base and exponent may be different, the value is the same.

Rule for Power of a Product

If the base of an exponent is a product, we will see that each factor in the product can be raised to the power indicated by the exponent. For example, $(10x)^3$ indicates that the product of 10 and x is to be raised to the 3rd power and $\left(-2x^2y\right)^5$ indicates that the product of $-2, x^2,$ and y is to be raised to the 5th power. We can simplify these expressions as follows.

$$\begin{aligned}(10x)^3 &= 10x \cdot 10x \cdot 10x \\ &= 10 \cdot 10 \cdot 10 \cdot x \cdot x \cdot x \\ &= 10^3 \cdot x^3 \\ &= 1000x^3\end{aligned}$$

and

$$\begin{aligned}(-2x^2y)^5 &= (-2x^2y)(-2x^2y)(-2x^2y)(-2x^2y)(-2x^2y) \\ &= (-2)(-2)(-2)(-2)(-2) \cdot x^2 \cdot x^2 \cdot x^2 \cdot x^2 \cdot x^2 \cdot y \cdot y \cdot y \cdot y \cdot y \\ &= (-2)^5 (x^2)^5 \cdot y^5 \\ &= -32x^{10}y^5\end{aligned}$$

We can simplify expressions such as these in a much easier fashion by using the following **power of a product rule for exponents**.

Rule for Power of a Product

If a and b are nonzero real numbers and n is an integer then

$$(ab)^n = a^n b^n.$$

In words, a power of a product is found by raising each factor to that power.

Example 2: Rule for Power of a Product

Simplify each expression by using the rule for power of a product.

a. $(5x)^2$

Solution: $(5x)^2 = 5^2 \cdot x^2 = 25x^2$

b. $(xy)^3$

Solution: $(xy)^3 = x^3 \cdot y^3 = x^3y^3$

c. $(-7ab)^2$

Solution: $(-7ab)^2 = (-7)^2 a^2b^2 = 49a^2b^2$

d. $(ab)^{-5}$

Solution: $(ab)^{-5} = a^{-5} \cdot b^{-5} = \frac{1}{a^5} \cdot \frac{1}{b^5} = \frac{1}{a^5b^5}$

or, using the rule of negative exponents first and then the rule for the power of a product,

$$(ab)^{-5} = \frac{1}{(ab)^5} = \frac{1}{a^5b^5}$$

Continued on the next page...

e. $\left(x^2y^{-3}\right)^4$

Solution: $\left(x^2y^{-3}\right)^4 = \left(x^2\right)^4\left(y^{-3}\right)^4 = x^8 \cdot y^{-12} = \dfrac{x^8}{y^{12}}$

NOTES

Special Note about Negative Numbers and Exponents

In an expression such as $-x^2$, we know that -1 is understood to be the coefficient of x^2. That is,

$$-x^2 = -1 \cdot x^2.$$

The same is true for expressions with numbers such as -7^2. That is,

$$-7^2 = -1 \cdot 7^2 = -1 \cdot 49 = -49.$$

We see that the exponent refers to 7 and **not** to -7. For the exponent to refer to -7 as the base, -7 **must be in parentheses** as follows:

$$(-7)^2 = (-7)(-7) = +49.$$

As another example,

$$-2^0 = -1 \cdot 2^0 = -1 \cdot 1 = -1 \quad \text{and} \quad (-2)^0 = 1.$$

Rule for Power of a Quotient

If the base of an exponent is a quotient (in fraction form), we will see that both the numerator and denominator are raised to the power indicated by the exponent. For example, in the expression $\left(\dfrac{2}{x}\right)^3$ the quotient (or fraction) $\dfrac{2}{x}$ is raised to the 3rd power.

We can simplify this expression as follows:

$$\left(\frac{2}{x}\right)^3 = \frac{2}{x} \cdot \frac{2}{x} \cdot \frac{2}{x} = \frac{2 \cdot 2 \cdot 2}{x \cdot x \cdot x} = \frac{2^3}{x^3} = \frac{8}{x^3}.$$

Or, we can simplify the expression in a much easier manner by applying the following **rule for the power of a quotient**.

Rule for Power of a Quotient

If a and b are nonzero real numbers and n is an integer, then

$$\left(\frac{a}{b}\right)^n = \frac{a^n}{b^n}.$$

In words, a power of a quotient (in fraction form) is found by raising both the numerator and the denominator to that power.

Example 3: Rule for the Power of a Quotient

Simplify each expression by using the rule for the power of a quotient.

a. $\left(\frac{y}{x}\right)^5$

Solution: $\left(\frac{y}{x}\right)^5 = \frac{y^5}{x^5}$

b. $\left(\frac{3}{4}\right)^3$

Solution: $\left(\frac{3}{4}\right)^3 = \frac{3^3}{4^3} = \frac{27}{64}$

c. $\left(\frac{2}{a}\right)^4$

Solution: $\left(\frac{2}{a}\right)^4 = \frac{2^4}{a^4} = \frac{16}{a^4}$

d. $\left(\frac{x}{7}\right)^2$

Solution: $\left(\frac{x}{7}\right)^2 = \frac{x^2}{7^2} = \frac{x^2}{49}$

Using Combinations of Rules for Exponents

There may be more than one way to apply the various rules for exponents. As illustrated in the following examples, if you apply the rules correctly (even in a different sequence), the answer will be the same in every case.

Example 4: Applying Combinations of Rules for Exponents

Simplify each expression by using the appropriate rules for exponents.

a. $\left(\frac{-2x}{y^2}\right)^3$

Solution: $\left(\frac{-2x}{y^2}\right)^3 = \frac{(-2x)^3}{(y^2)^3} = \frac{(-2)^3 x^3}{y^6} = \frac{-8x^3}{y^6}$

b. $\left(\frac{3a^2b}{a^3b^2}\right)^2$

Solution:

Method 1: Simplify inside the parentheses first.

$$\left(\frac{3a^2b}{a^3b^2}\right)^2 = \left(3a^{2-3}b^{1-2}\right)^2 = \left(3a^{-1}b^{-1}\right)^2 = 3^2a^{-2}b^{-2} = \frac{9}{a^2b^2}$$

Method 2: Apply the power of a quotient rule first.

$$\left(\frac{3a^2b}{a^3b^2}\right)^2 = \frac{(3a^2b)^2}{(a^3b^2)^2} = \frac{3^2a^{2(2)}b^2}{a^{3(2)}b^{2(2)}} = \frac{9a^4b^2}{a^6b^4} = 9a^{4-6}b^{2-4} = 9a^{-2}b^{-2} = \frac{9}{a^2b^2}$$

Note that the answer is the same even though the rules were applied in a different order.

Teaching Notes:
Example 4b illustrates two procedures for simplifying an expression. You should emphasize the idea that all of these rules for exponents allow for flexibility. Of course, the challenge is for the student to use the rules correctly regardless of the order of usage.

Another general approach with fractions involving negative exponents is to note that

$$\left(\frac{a}{b}\right)^{-n} = \frac{a^{-n}}{b^{-n}} = \frac{b^n}{a^n} = \left(\frac{b}{a}\right)^n.$$

NOTES

In effect, there are two basic shortcuts with negative exponents and fractions:

1. Taking the reciprocal of a fraction changes the sign of any exponent in the fraction.
2. Moving any term from numerator to denominator or vice versa, changes the sign of the corresponding exponent.

Example 5: Two Approaches with Fractional Expressions and Negative Exponents

Simplify: $\left(\frac{x^3}{y^5}\right)^{-4}$

Solution:

Method 1: Use the ideas of reciprocals first.

$$\left(\frac{x^3}{y^5}\right)^{-4} = \left(\frac{y^5}{x^3}\right)^4 = \frac{y^{5(4)}}{x^{3(4)}} = \frac{y^{20}}{x^{12}}$$

Method 2: Apply the power of a quotient rule first.

$$\left(\frac{x^3}{y^5}\right)^{-4} = \frac{\left(x^3\right)^{-4}}{\left(y^5\right)^{-4}} = \frac{x^{3(-4)}}{y^{5(-4)}} = \frac{x^{-12}}{y^{-20}} = \frac{y^{20}}{x^{12}}$$

Example 6: A More Complex Example

This example involves the application of a variety of steps. Study it carefully and see if you can get the same result by following a different sequence of steps.

Simplify: $\left(\frac{2x^2y^3}{3xy^{-2}}\right)^{-2}\left(\frac{4x^2y^{-1}}{3x^{-5}y^3}\right)^{-1}$

Solution:

$$\left(\frac{2x^2y^3}{3xy^{-2}}\right)^{-2}\left(\frac{4x^2y^{-1}}{3x^{-5}y^3}\right)^{-1} = \frac{2^{-2}\cdot x^{2(-2)}\cdot y^{3(-2)}}{3^{-2}\cdot x^{-2}y^{-2(-2)}}\cdot\frac{4^{-1}\cdot x^{2(-1)}\cdot y^{-1(-1)}}{3^{-1}\cdot x^{-5(-1)}\cdot y^{3(-1)}}$$

$$= \frac{3^2}{2^2}\cdot\frac{x^{-4}}{x^{-2}}\cdot\frac{y^{-6}}{y^4}\cdot\frac{3}{4}\cdot\frac{x^{-2}}{x^5}\cdot\frac{y^1}{y^{-3}}$$

$$= \frac{9\cdot3\cdot x^{-4-2}y^{-6+1}}{4\cdot4\cdot x^{-2+5}y^{4-3}} = \frac{27x^{-6}y^{-5}}{16x^3y^1}$$

$$= \frac{27x^{-6-3}y^{-5-1}}{16} = \frac{27x^{-9}y^{-6}}{16} = \frac{27}{16x^9y^6}$$

A complete summary of the rules for exponents includes the following eight rules.

Summary of the Rules for Exponents

If a and b are nonzero real numbers and m and n are integers:

1. The exponent 1: $\boldsymbol{a = a^1}$
2. The exponent 0: $\boldsymbol{a^0 = 1}$
3. The product rule: $\boldsymbol{a^m \cdot a^n = a^{m+n}}$
4. The quotient rule: $\boldsymbol{\dfrac{a^m}{a^n} = a^{m-n}}$
5. Negative exponents: $\boldsymbol{a^{-n} = \dfrac{1}{a^n}}$
6. Power rule: $\boldsymbol{\left(a^m\right)^n = a^{mn}}$
7. Power of a product: $\boldsymbol{(ab)^n = a^n b^n}$
8. Power of a quotient: $\boldsymbol{\left(\dfrac{a}{b}\right)^n = \dfrac{a^n}{b^n}}$

Scientific Notation and Calculators

A basic application of integer exponents occurs in scientific disciplines, such as astronomy and biology, when very large and very small numbers are involved. For example, the distance from the earth to the sun is approximately 93,000,000 miles, and the approximate radius of a carbon atom is 0.0000000077 centimeters.

In **scientific notation** (an option in all scientific and graphing calculators), **decimal numbers are written as the product of a number greater than or equal to 1 and less than 10, and an integer power of 10**. In scientific notation there is just one digit to the left of the decimal point. For example,

$$250{,}000 = 2.5 \times 10^5 \quad \text{and} \quad 0.000000345 = 3.45 \times 10^{-7}.$$

The exponent tells how many places the decimal point is to be moved and in what direction. If the exponent is positive, the decimal point is moved to the right.

$$5.6 \times 10^4 = 56{,}000. \quad \text{4 places right}$$

A negative exponent indicates that the decimal point should move to the left.

$$4.9 \times 10^{-3} = 0.004.9 \quad \text{3 places left}$$

Scientific Notation

If N is a decimal number, then in **scientific notation**

$$N = a \times 10^n \text{ where } 1 \le a < 10 \text{ and } n \text{ is an integer.}$$

Example 7: Decimals in Scientific Notation

Write the following decimal numbers in scientific notation.

a. 8,720,000

Solution: $8{,}720{,}000 = 8.72 \times 10^6$ 8.72 is between 1 and 10

To check, move the decimal point 6 places to the right and get the original number:

$$8.72 \times 10^6 = 8.720000. = 8{,}720{,}000$$

1 2 3 4 5 6

b. 0.000000376

Solution: $0.000000376 = 3.76 \times 10^{-7}$ 3.76 is between 1 and 10

To check, move the decimal point 7 places to the left and get the original number:

$$3.76 \times 10^{-7} = 0.0000003.76 = 0.000000376$$

7 6 5 4 3 2 1

Example 8: Scientific Notation and Properties of Exponents

Simplify the following expressions by first writing the decimal numbers in scientific notation and then using the properties of exponents.

a. $\dfrac{(0.085)(41{,}000)}{0.00017}$

Solution:

$$\frac{(0.085)(41{,}000)}{0.00017} = \frac{(8.5\times10^{-2})(4.1\times10^{4})}{1.7\times10^{-4}}$$

$$= \frac{(\overset{5}{\cancel{8.5}})(4.1)}{\cancel{1.7}} \times \frac{(10^{-2})(10^{4})}{10^{-4}}$$

$$= 20.5 \times \frac{10^2}{10^{-4}}$$

$$= 2.05 \times 10^1 \times \frac{10^2}{10^{-4}}$$

$$= 2.05 \times 10^{1+2-(-4)}$$

$$= 2.05 \times 10^7$$

b. $\dfrac{(11{,}100)(0.064)}{(8{,}000{,}000)(370)}$

Solution: $\dfrac{(11{,}100)(0.064)}{(8{,}000{,}000)(370)} = \dfrac{(1.11\times10^{4})(6.4\times10^{-2})}{(8.0\times10^{6})(3.7\times10^{2})}$

$$= \frac{(\overset{0.3}{\cancel{1.11}})\,(\overset{0.8}{\cancel{6.4}})}{(\cancel{8.0})\,(\cancel{3.7})} \times \frac{10^{2}}{10^{8}}$$

$$= 0.24\times10^{2-8}$$

$$= 2.4\times10^{-1}\times10^{-6}$$

$$= 2.4\times10^{-1-6}$$

$$= 2.4\times10^{-7}$$

c. Light travels approximately 3×10^{8} meters per second. How many meters per minute does light travel?

Solution: Since there are 60 seconds in one minute, multiply by 60.

$$3\times10^{8}\times60 = 180\times10^{8}$$

$$= 1.8\times10^{2}\times10^{8}$$

$$= 1.8\times10^{10}$$

Thus light travels 1.8×10^{10} meters per minute.

Example 9: Scientific Notation and Calculators

a. Use a graphing calculator to evaluate the expression $\dfrac{0.0042\cdot3{,}000{,}000}{0.21}$. Leave the answer in scientific notation.

Note: You can press the MODE key and select SCI on the first line to have all decimal calculations in scientific notation.

Solution: With a TI-84 Plus calculator (set in scientific notation mode) the display should appear as shown below. Note that the E in the display indicates an exponent with base 10.

```
(0.0042*3000000)
/(0.21)
                6E4
```

Continued on the next page...

1. -81 2. -25
3. -16 4. -400
5. 1,000,000
6. 4096 7. $36x^6$
8. $9x^8$ 9. $-108x^6$
10. $\frac{112}{y^8}$ 11. $\frac{5x^2}{y}$
12. -3
13. $-\frac{2y^6}{27x^{15}}$
14. $-\frac{4x^3}{5y}$
15. $\frac{27x^3}{y^3}$
16. $\frac{16x^2}{y^4}$
17. 1
18. $\frac{9x^4}{y^6}$
19. $4x^4y^4$
20. $\frac{y^{10}}{4x^2}$
21. $\frac{y^2}{x^2}$
22. $\frac{b}{2a}$
23. $\frac{1}{3xy^2}$
24. $\frac{1}{64a^6b^9}$

b. A light-year is the distance light travels in one year. Use a graphing calculator to find the length of a light-year in scientific notation if light travels 186,000 miles per second.

Solution: 60 sec = 1 minute
60 min = 1 hour
24 hr = 1 day
365 days = 1 year

Multiplication gives the following display on your calculator.

```
186000*60*60*24*
365
        5.865696E12
```

Thus a light-year is 5,865,696,000,000 miles (5 trillion, 865 billion, 696 million miles).

c. Use a calculator to evaluate the expression $\frac{8600\left(3.0\times10^5\right)}{1.5\times10^{-6}}$. Leave the answer in scientific notation.

Note: Remember, the caret key ^ is used to indicate an exponent.

Solution: With a graphing calculator (set in scientific notation mode) the display should appear as shown below.

```
(8600(3.0*10^5))
/(1.5*10^-6)
            1.72E15
```

Note: The numerator and denominator must be set in parentheses.

Practice Problems

25. $-\frac{x^3y^6}{27}$
26. $25x^2y^4$
27. m^2n^4
28. $16a^4b^4$
29. $-\frac{y^2}{49x^2}$

Simplify each expression.

1. $\frac{x^{-2}x^5}{x^{-7}}$ **2.** $\left(\frac{a^2b^3}{4}\right)^0$ **3.** $\frac{-3^2\cdot5}{2\cdot(-3)^2}$ **4.** $\left(\frac{5x}{3b}\right)^{-2}$

5. Write the number in scientific notation: 186,000 miles per second (speed of light in miles per second).

Answers to Practice Problems: **1.** x^{10} **2.** 1 **3.** $-\frac{5}{2}$ **4.** $\frac{9b^2}{25x^2}$ **5.** 1.86×10^5 miles per second

5.2 Exercises

30. $\frac{1}{8a^3b^6}$
31. $\frac{25x^6}{y^2}$
32. $\frac{y^8}{16x^8}$
33. $\frac{x^6}{y^6}$
34. $\frac{8a^6}{b^9}$
35. $\frac{36y^{14}}{x^4}$
36. $27x^{18}y^{12}$
37. $\frac{49y^4}{x^6}$
38. $\frac{25x^{12}y^6}{9}$
39. $\frac{24y^5}{x^7}$
40. $\frac{25x^9}{8y^4}$
41. $\frac{4x^3}{3}$
42. $\frac{3}{10y}$
43. $\frac{16x^{13}}{243y^6}$
44. $\frac{11{,}520b^{28}}{a}$
45. $96x^2y^4z^4$
46. $\frac{729}{49a^6b^4c^4}$
47. 8.6×10^4
48. 9.27×10^5
49. 3.62×10^{-2}
50. 6.1×10^{-3}
51. 1.83×10^7
52. 3.76×10^8
53. 0.042
54. 0.00835

Use the rules for exponents to simplify each of the expressions. Assume that all variables represent nonzero real numbers.

1. -3^4 **2.** -5^2 **3.** -2^4 **4.** -20^2

5. $(-10)^6$ **6.** $(-4)^6$ **7.** $(6x^3)^2$ **8.** $(-3x^4)^2$

9. $4(-3x^2)^3$ **10.** $7(2y^{-2})^4$ **11.** $5(x^2y^{-1})$ **12.** $-3(7xy^2)^0$

13. $-2(3x^5y^{-2})^{-3}$ **14.** $-4(5x^{-3}y)^{-1}$ **15.** $\left(\frac{3x}{y}\right)^3$ **16.** $\left(\frac{-4x}{y^2}\right)^2$

17. $\left(\frac{6m^3}{n^5}\right)^0$ **18.** $\left(\frac{3x^2}{y^3}\right)^2$ **19.** $\left(\frac{-2x^2}{y^{-2}}\right)^2$ **20.** $\left(\frac{2x}{y^5}\right)^{-2}$

21. $\left(\frac{x}{y}\right)^{-2}$ **22.** $\left(\frac{2a}{b}\right)^{-1}$ **23.** $\left(\frac{3x}{y^{-2}}\right)^{-1}$ **24.** $\left(\frac{4a^2}{b^{-3}}\right)^{-3}$

25. $\left(\frac{-3}{xy^2}\right)^{-3}$ **26.** $\left(\frac{5xy^3}{y}\right)^2$ **27.** $\left(\frac{m^2n^3}{mn}\right)^2$ **28.** $\left(\frac{2ab^3}{b^2}\right)^4$

29. $\left(\frac{-7^2x^2y}{y^3}\right)^{-1}$ **30.** $\left(\frac{2ab^4}{b^2}\right)^{-3}$ **31.** $\left(\frac{5x^3y}{y^2}\right)^2$ **32.** $\left(\frac{2x^2y}{y^3}\right)^{-4}$

33. $\left(\frac{x^3y^{-1}}{y^2}\right)^2$ **34.** $\left(\frac{2a^2b^{-1}}{b^2}\right)^3$ **35.** $\left(\frac{6y^5}{x^2y^{-2}}\right)^2$ **36.** $\left(\frac{3x^4}{x^{-2}y^{-4}}\right)^3$

37. $\frac{(7x^{-2}y)^2}{(xy^{-1})^2}$ **38.** $\frac{(-5x^3y^4)^2}{(3x^{-3}y)^2}$ **39.** $\frac{(3x^2y^{-1})^{-2}}{(6x^{-1}y)^{-3}}$ **40.** $\frac{(2x^{-3})^{-3}}{(5y^{-2})^{-2}}$

41. $\frac{(4x^{-2})(6x^5)}{(9y)(2y^{-1})}$ **42.** $\frac{(5x^2)(3x^{-1})^2}{(25y^3)(6y^{-2})}$

43. $\left(\frac{3xy^3}{4x^2y^{-3}}\right)^{-1}\left(\frac{2x^3y^{-1}}{9x^{-3}y^{-1}}\right)^2$ **44.** $\left(\frac{5a^4b^{-2}}{6a^{-4}b^3}\right)^{-2}\left(\frac{5a^3b^4}{2^{-2}a^{-2}b^{-2}}\right)^3$

45. $\left(\frac{6x^{-4}yz^{-2}}{4^{-1}x^{-4}y^3z^{-2}}\right)^{-1}\left(\frac{2^{-2}xyz^{-3}}{12x^2y^2z^{-1}}\right)^{-2}$ **46.** $\left(\frac{3^{-5}a^5b^3c^{-1}}{3^{-2}abc}\right)^{-2}\left(\frac{7^{-1}a^{-4}bc^2}{7^{-2}a^{-3}bc^{-2}}\right)^{-2}$

Write the following numbers in scientific notation.

47. 86,000 **48.** 927,000 **49.** 0.0362

50. 0.0061 **51.** 18,300,000 **52.** 376,000,000

Write the following numbers in decimal form.

53. 4.2×10^{-2} **54.** 8.35×10^{-3} **55.** 7.56×10^{6}

56. 6.132×10^{-5} **57.** 8.515×10^{8} **58.** 9.374×10^{7}

First write each of the numbers in scientific notation. Then perform the indicated operations and leave your answer in scientific notation.

59. $300\cdot0.00015$ **60.** $0.000024\cdot40{,}000$ **61.** $0.0003\cdot0.0000025$

62. $0.00005\cdot0.00013$ **63.** $\dfrac{3900}{0.003}$ **64.** $\dfrac{4800}{12{,}000}$

65. $\dfrac{125}{50{,}000}$ **66.** $\dfrac{0.0046}{230}$ **67.** $\dfrac{0.02\cdot3900}{0.013}$

68. $\dfrac{0.0084\cdot0.003}{0.21\cdot60}$ **69.** $\dfrac{0.005\cdot650\cdot3.3}{0.0011\cdot2500}$ **70.** $\dfrac{5.4\cdot0.003\cdot50}{15\cdot0.0027\cdot200}$

71. $\dfrac{(1.4\times10^{-2})(922)}{(3.5\times10^{3})(2.0\times10^{6})}$ **72.** $\dfrac{(4300)(3.0\times10^{2})}{(1.5\times10^{-3})(860\times10^{-2})}$

73. **Atomic Mass:** The mass of a hydrogen atom is approximately

0.00000000000000000000000167 grams.

Write this number in scientific notation.

74. **Circumference of the earth:** The circumference of the earth is approximately

1,580,000,000 inches.

Express this circumference in scientific notation.

75. **Anatomy:** There are approximately 6×10^{13} cells in an adult human body. Express this number in decimal form.

76. **World population:** The world population is approximately 6.7×10^{9} people. Write this number in decimal form.

77. **Mass of the earth:** The mass of the earth is about

5,980,000,000,000,000,000,000,000,000 grams.

Write this number in scientific notation.

55. 7,560,000
56. 0.00006132
57. 851,500,000
58. 93,740,000
59. $(3\times10^{2})(1.5\times10^{-4})$; 4.5×10^{-2}
60. $(2.4\times10^{-5})(4\times10^{4})$; 9.6×10^{-1}
61. $(3\times10^{-4})(2.5\times10^{-6})$; 7.5×10^{-10}
62. $(5\times10^{-5})(1.3\times10^{-4})$; 6.5×10^{-9}
63. $\dfrac{3.9\times10^{3}}{3\times10^{-3}}$; 1.3×10^{6}
64. $\dfrac{4.8\times10^{3}}{1.2\times10^{4}}$; 4×10^{-1}
65. $\dfrac{1.25\times10^{2}}{5\times10^{4}}$; 2.5×10^{-3}
66. $\dfrac{4.6\times10^{-3}}{2.3\times10^{2}}$; 2×10^{-5}
67. $\dfrac{(2\times10^{-2})(3.9\times10^{3})}{1.3\times10^{-2}}$; 6×10^{3}
68. $\dfrac{(8.4\times10^{-3})(3\times10^{-3})}{(2.1\times10^{-1})(6\times10)}$; 2×10^{-6}
69. $\dfrac{(5\times10^{-3})(6.5\times10^{2})(3.3\times10^{0})}{(1.1\times10^{-3})(2.5\times10^{3})}$; 3.9×10^{0}
70. $\dfrac{(5.4\times10^{0})(3\times10^{-3})(5\times10)}{(1.5\times10)(2.7\times10^{-3})(2\times10^{2})}$; 1×10^{-1}
71. $\dfrac{(1.4\times10^{-2})(9.22\times10^{2})}{(3.5\times10^{3})(2.0\times10^{6})}$; 1.844×10^{-9}
72. $\dfrac{(4.3\times10^{3})(3.0\times10^{2})}{(1.5\times10^{-3})(8.60\times10^{0})}$; 1×10^{8}

73. 1.67×10^{-24} grams
74. 1.58×10^{9} inches
75. 60,000,000,000,000 cells
76. 6,700,000,000 people
77. 5.98×10^{27} grams

78. Time: One year is approximately 31,500,000 seconds. Express this time in scientific notation.

79. Speed of light: One light-year is approximately 9.46×10^{15} meters. The distance to a certain star is 4.3 light-years. How many meters is this?

80. Speed of light: Light travels approximately 3×10^{10} centimeters per second. How many centimeters would this be per minute? Per hour? Express your answers in scientific notation.

81. Atomic weight: An atom of gold weighs approximately 3.25×10^{-22} grams. What would be the weight of 2000 atoms of gold? Express your answer in scientific notation.

82. Atomic weight: An ounce of gold contains 5×10^{22} atoms. All the gold ever taken out of the earth is estimated to be 3.0×10^{31} atoms. How many ounces of gold is this?

78. 3.15×10^{7} seconds
79. 4.0678×10^{16} m
80. 1.8×10^{12} cm / min; 1.08×10^{14} cm / hr
81. 6.5×10^{-19} grams
82. 6×10^{8} ounces
83. 3E8
84. 9E−7
85. 8.5E7
86. 1.32E8
87. 1E2

Use your calculator (set in scientific notation mode) to evaluate each expression. Leave the answer in scientific notation.

83. $90{,}000\div0.0003$

84. $0.0081\div9000$

85. $400\times175{,}000+5000\times3000$

86. $7000\times6000+200\times450{,}000$

87. $\dfrac{5.6\cdot0.003\cdot5000}{15\cdot0.0028\cdot20}$

88. $\dfrac{0.0006\cdot660\cdot40.4}{0.00011\cdot3600}$

89. $\dfrac{\left(1.8\times10^{-3}\right)(932)}{\left(4.5\times10^{3}\right)\left(2.0\times10^{-6}\right)}$

90. $\dfrac{(86{,}000)\left(3.0\times10^{4}\right)}{\left(4.3\times10^{-2}\right)\left(1.5\times10^{-3}\right)}$

Writing and Thinking About Mathematics

88. 4.04E1
89. 1.864E2
90. 4E13
91. See page 370.

91. Without looking at the text, show that $\left(\dfrac{x}{y}\right)^{-n}=\left(\dfrac{y}{x}\right)^{n}$ by using the power rules and the rule for negative exponents. (Check to see if your method is similar to that on page 370.)

HAWKES LEARNING SYSTEMS: INTRODUCTORY & INTERMEDIATE ALGEBRA SOFTWARE

- 5.2a Simplifying Integer Exponents II
- 5.2b Scientific Notation

5.3 Introduction to Polynomials

- *Define a polynomial.*
- *Classify a polynomial as a monomial, binomial, trinomial, or a polynomial with more than three terms.*
- *Evaluate a polynomial for given values of the variable.*

Definition of a Polynomial

A **term** is an expression that involves only multiplication and/or division with constants and/or variables. Remember that a number written next to a variable indicates multiplication, and the number is called the **numerical coefficient** (or **coefficient**) of the variable. For example,

$$3x, \quad -5y^2, \quad 17, \quad \text{and} \quad \frac{x}{y}$$

are all algebraic terms.

In the term

$3x$, 3 is the coefficient of x,

$-5y^2$, -5 is the coefficient of y^2, and

$\frac{x}{y}$, 1 is the coefficient.

A term that consists of only a number, such as 17, is also called a **constant** or a **constant term**.

Monomial

A **monomial in *x*** is a term of the form

$$kx^n$$

where k is a real number and n is a whole number.
n is called the **degree** of the term, and k is called the **coefficient**.

A monomial may have more than one variable, and **the degree of such a monomial is the sum of the degrees of its variables**. For example,

$$4x^2y^3 \text{ is a 5}^{\text{th}} \text{ degree monomial in } x \text{ and } y.$$

However, in this chapter, only monomials of one variable (note that any variable may be used in place of x) will be discussed.

Since $x^0 = 1$, a nonzero constant can be multiplied by x^0 without changing its value. Thus we say that a **nonzero constant is a monomial of degree 0.** For example,

$$17 = 17x^0 \quad \text{and} \quad -6 = -6x^0$$

which means that the constants 17 and –6 are monomials of degree 0. However, for the special number 0, we can write

$$0 = 0x^2 = 0x^5 = 0x^{13}$$

and we say that **the constant 0 is a monomial of no degree.**

Monomials may have fractional or negative coefficients; however, **monomials may not have fractional or negative exponents.** These facts are part of the definition since in the expression kx^n, k (the coefficient) can be any real number, but n (the exponent) must be a whole number.

Expressions that **are not** monomials: $3\sqrt{x},\ -15x^{\frac{2}{3}},\ 4a^{-2}$

Expressions that **are** monomials: $17,\ 3x,\ 5y^2,\ \frac{2}{7}a^4,\ \pi x^2,\ \sqrt{2}x^3$

A **polynomial** is a monomial or the indicated sum or difference of monomials. Examples of polynomials are

$$3x, \quad y+5, \quad 4x^2-7x+1, \quad \text{and} \quad a^{10}+5a^3-2a^2+6.$$

Polynomial

A **polynomial** is a monomial or the indicated sum or difference of monomials.

The **degree of a polynomial** is the largest of the degrees of its terms.

The coefficient of the term of the largest degree is called the **leading coefficient**.

Special Terminology for Some Polynomials

		Examples
Monomial:	polynomial with one term	$-2x^3$ and $4a^5$
Binomial:	polynomial with two terms	$3x+5$ and a^2+3
Trinomial:	polynomial with three terms	x^2+6x-7 and a^3-8a^2+12a

Polynomials with four or more terms are simply referred to as **polynomials.**

Examples of polynomials are

$-1.4x^5$ a fifth-degree monomial in x (leading coefficient is -1.4),

$4z^3 - 7.5z^2 - 5z$ a third-degree trinomial in z (leading coefficient is 4),

$\frac{3}{4}y^4 - 2y^3 + 4y - 6$ a fourth-degree polynomial in y (leading coefficient is $\frac{3}{4}$).

In each of these examples, the terms have been written so that the exponents on the variables decrease in order from left to right. We say that the terms are written in **descending order**. If the exponents on the terms increase in order from left to right, we say that the terms are written in **ascending order**. **As a general rule, for consistency and style in operating with polynomials, the polynomials in this chapter will be written in descending order.**

Example 1: Simplifying Polynomials

Simplify each of the following polynomials by combining **like terms** (To review the definition of like terms, see section 2.1). Write the polynomial in descending order and state the degree and type of the polynomial.

a. $5x^3 + 7x^3$

Solution: $5x^3 + 7x^3 = (5+7)x^3 = 12x^3$ Third-degree monomial

b. $5x^3 + 7x^3 - 2x$

Solution: $5x^3 + 7x^3 - 2x = 12x^3 - 2x$ Third-degree binomial

c. $\frac{1}{2}y + 3y - \frac{2}{3}y^2 - 7$

Solution: $\frac{1}{2}y + 3y - \frac{2}{3}y^2 - 7 = -\frac{2}{3}y^2 + \frac{7}{2}y - 7$ Second-degree trinomial

d. $x^2 + 8x - 15 - x^2$

Solution: $x^2 + 8x - 15 - x^2 = 8x - 15$ First-degree binomial

e. $-3y^4 + 2y^2 + y^{-1}$

Solution: This expression is not a polynomial since y has a negative exponent.

Evaluating Polynomials

To evaluate a polynomial for a given value of the variable, substitute the value for the variable wherever it occurs in the polynomial and follow the rules for order of operations. A convenient notation for evaluating polynomials is the function notation, $p(x)$ (read "p of x") discussed in Section 4.5.

For example,

$$\text{if} \quad p(x)=x^2-4x+13,$$

$$\text{then} \quad p(5)=(5)^2-4(5)+13=25-20+13=18.$$

Example 2: Evaluating Polynomials

Teaching Notes:
Some students may be perplexed by the notation "$p(x)$" and continue confusing it as meaning multiplication. Remind the students of the context for functions and be aware of the dual interpretations. Also, mention that the use of "p" is purely arbitrary and other letters may be used.

a. Given $p(x)=4x^2+5x-15$, find $p(3)$.

Solution: For $p(x)=4x^2+5x-15$,

$$\begin{aligned} p(3)&=4(3)^2+5(3)-15 \quad \text{Substitute 3 for } x.\\ &=4(9)+15-15\\ &=36+15-15\\ &=36 \end{aligned}$$

b. Given $p(y)=5y^3+y^2-3y+8$, find $p(-2)$.

Solution: For $p(y)=5y^3+y^2-3y+8$,

$$\begin{aligned} p(-2)&=5(-2)^3+(-2)^2-3(-2)+8 \quad \text{Note the use of parentheses around } -2.\\ &=5(-8)+4-3(-2)+8\\ &=-40+4+6+8\\ &=-22 \end{aligned}$$

c. Rewrite the polynomial expression $f(x)=6x+13$ by substituting for x as indicated by the function notation $f(2a+1)$.

Solution: $f(2a+1)=6(2a+1)+13=12a+6+13=12a+19$

Practice Problems

Combine like terms and state the degree and type of the polynomial.

1. $8x^3-3x^2-x^3+5+3x^2$

2. $5y^4+y^4-3y^2+2y^2+4$

3. For the polynomial $p(x)=x^2-5x-5$, find **a.** $p(3)$ and **b.** $p(-1)$.

Answers to Practice Problems: **1.** $7x^3+5$; Third-degree binomial **2.** $6y^4-y^2+4$; Fourth-degree trinomial **3. a.** -11 **b.** 1

5.3 Exercises

Identify the expression as a monomial, binomial, trinomial, or not a polynomial.

1. $3x^4$

2. $5y^2-2y+1$

3. $8x^3-7$

4. $-2x^{-2}$

5. $14a^7-2a-6$

6. $17x^{\frac{2}{3}}+5x^2$

7. $6a^3+5a^2-a^{-3}$

8. $-3y^4+2y^2-9$

9. $\frac{1}{2}x^3-\frac{2}{5}x$

10. $\frac{5}{8}x^5+\frac{2}{3}x^4$

Simplify the polynomials. Write the polynomial in descending order and state the degree and type of the simplified polynomial. For each polynomial, also state the leading coefficient.

11. $y+3y$

12. $4x^2-x+x^2$

13. x^3+3x^2-2x

14. $3x^2-8x+8x$

15. $x^4-4x^2+2x^2-x^4$

16. $2-6y+5y-2$

17. $-x^3+6x+x^3-6x$

18. $11x^2-3x+2-7x^2$

19. $6a^5+2a^2-7a^3-3a^2$

20. $2x^2-3x^2+2-4x^2-2+5x^2$

21. $4y-8y^2+2y^3+8y^2$

22. $2x+9-x+1-2x$

23. $5y^2+3-2y^2+1-3y^2$

24. $13x^2-6x-9x^2-4x$

25. $7x^3+3x^2-2x+x-5x^3+1$

26. $-3y^5+7y-2y^3-5+4y^2+y^2$

27. $x^4+3x^4-2x+5x-10-x^2+x$

28. $a^3+2a^2-6a+3a^3+2a^2+7a+3$

29. $2x+4x^2+6x+9x^3$

30. $15y-y^3+2y^2-10y^2+2y-16$

Evaluate the given polynomials as indicated.

31. Given $p(x)=x^2+14x-3$, find $p(-1)$.

32. Given $p(x)=-5x^2-8x+7$, find $p(-3)$.

33. Given $p(x)=3x^3-9x^2-10x-11$, find $p(3)$.

34. Given $p(y)=y^3-5y^2+6y+2$, find $p(2)$.

1. Monomial
2. Trinomial
3. Binomial
4. Not a polynomial
5. Trinomial
6. Not a polynomial
7. Not a polynomial
8. Trinomial
9. Binomial
10. Binomial
11. $4y$; First-degree monomial; Leading coefficient 4
12. $5x^2-x$; Second-degree binomial; Leading coefficient 5
13. x^3+3x^2-2x; Third-degree trinomial; Leading coefficient 1
14. $3x^2$; Second-degree monomial; Leading coefficient 3
15. $-2x^2$; Second-degree monomial; Leading coefficient -2
16. $-y$; First-degree monomial; Leading coefficient -1
17. 0; Monomial of no degree; Leading coefficient 0
18. $4x^2-3x+2$; Second-degree trinomial; Leading coefficient 4
19. $6a^5-7a^3-a^2$; Fifth-degree trinomial; Leading coefficient 6
20. 0; Monomial of no degree; Leading coefficient 0
21. $2y^3+4y$; Third-degree binomial; Leading coefficient 2

22. $-x+10$; First-degree binomial; Leading coefficient -1
23. 4; Monomial of degree 0; Leading coefficient 4
24. $4x^2-10x$; Second-degree binomial; Leading coefficient 4
25. $2x^3+3x^2-x+1$; Third-degree polynomial; Leading coefficient 2
26. $-3y^5-2y^3+5y^2+7y-5$; Fifth-degree polynomial; Leading coefficient -3
27. $4x^4-x^2+4x-10$; Fourth-degree polynomial; Leading coefficient 4
28. $4a^3+4a^2+a+3$; Third-degree polynomial; Leading coefficient 4

35. Given $p(y)=-4y^3+5y^2+12y-1$, find $p(-10)$.

36. Given $p(a)=a^3+4a^2+a+2$, find $p(-5)$.

37. Given $p(a)=2a^4+3a^2-8a$, find $p(-1)$.

38. Given $p(x)=8x^4+2x^3-6x^2-7$, find $p(-2)$.

39. Given $p(x)=x^5-x^3+x-2$, find $p(-2)$.

40. Given $p(x)=3x^6-2x^5+x^4-x^3-3x^2+2x-1$, find $p(1)$.

A polynomial is given. Rewrite the polynomial by substituting for the variable as indicated with the function notation.

41. Given $p(x)=3x^4+5x^3-8x^2-9x$, find $p(a)$.

42. Given $p(x)=6x^5+5x^2-10x+3$, find $p(c)$.

43. Given $f(x)=3x+5$, find $f(a+2)$.

44. Given $f(x)=-4x+6$, find $f(a-2)$.

45. Given $g(x)=5x-10$, find $g(2a+7)$.

46. Given $g(x)=-4x-8$, find $g(3a+1)$.

29. $9x^3+4x^2+8x$; Third-degree trinomial; Leading coefficient 9
30. $-y^3-8y^2+17y-16$; Third-degree polynomial; Leading coefficient -1
31. -16 32. -14
33. -41 34. 2
35. 4379 36. -28
37. 13 38. 81
39. -28 40. -1
41. $3a^4+5a^3-8a^2-9a$
42. $6c^5+5c^2-10c+3$

Writing and Thinking About Mathematics

First-degree polynomials are also called **linear polynomials**, second-degree polynomials are called **quadratic polynomials**, and third-degree polynomials are called **cubic polynomials**. The related functions are called linear functions, quadratic functions, and cubic functions, respectively.

47. Use a graphing calculator to graph the following linear functions.

a. $p(x)=2x+3$ **b.** $p(x)=-3x+1$ **c.** $p(x)=\frac{1}{2}x$

48. Use a graphing calculator to graph the following quadratic functions.

a. $p(x)=x^2$ **b.** $p(x)=x^2+6x+9$ **c.** $p(x)=-x^2+2$

49. Use a graphing calculator to graph the following cubic functions.

a. $p(x)=x^3$ **b.** $p(x)=x^3-4x$ **c.** $p(x)=x^3+2x^2-5$

Continued on the next page...

Writing and Thinking About Mathematics (cont.)

43. $3a + 11$
44. $-4a + 14$
45. $10a + 25$
46. $-12a - 12$

50. Make up a few of your own linear, quadratic, and cubic functions and graph these functions with your calculator. Using the results from Exercises 47, 48, and 49, and your own functions, describe in your own words:

a. the general shape of the graphs of linear functions.
b. the general shape of the graphs of quadratic functions.
c. the general shape of the graphs of cubic functions.

HAWKES LEARNING SYSTEMS: INTRODUCTORY & INTERMEDIATE ALGEBRA SOFTWARE

- 5.3 Identifying and Evaluating Polynomials

47. a. 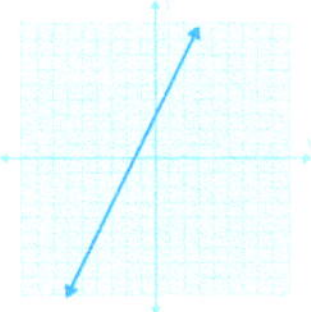 b. 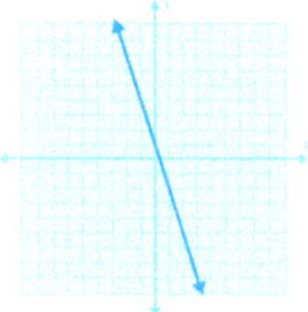 c. 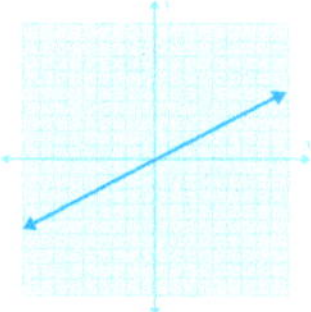

48. a. 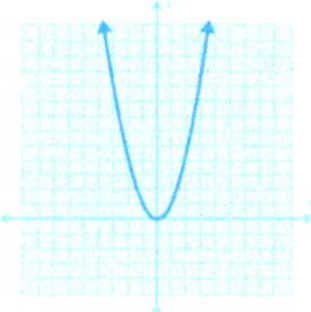 b. 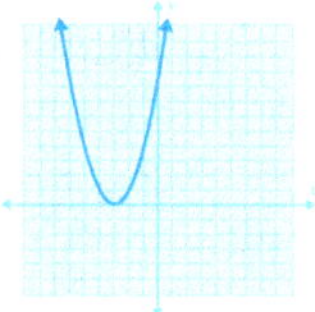 c. 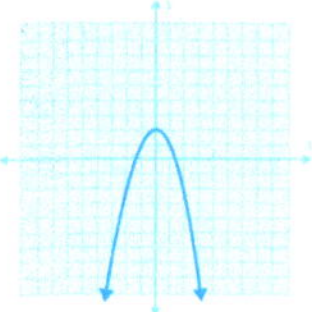

49. a. 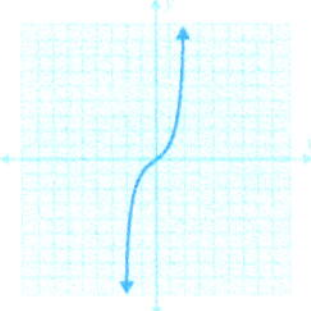 b. 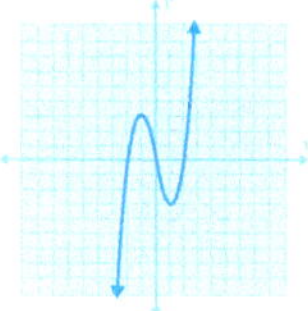 c. 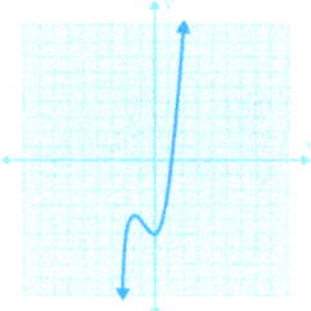

50. Answers will vary.

5.4 Addition and Subtraction with Polynomials

- *Add polynomials.*
- *Subtract polynomials.*
- *Simplify expressions by removing grouping symbols and combining like terms.*

Addition with Polynomials

The **sum** of two or more polynomials is found by combining **like terms**. Remember that like terms (or similar terms) are constants or terms that contain the same variables raised to the same powers. (See Section 2.1.) For example,

> $3x^2$, $-10x^2$, and $1.4x^2$ are all **like terms**. Each has the same variable raised to the same power.
>
> $5a^3$ and $-7a$ are **not** like terms. Each has the same variable, but the powers are different.

When adding polynomials, the polynomials may be written horizontally or vertically. For example,

$$\begin{aligned}&\left(x^2-5x+3\right)+\left(2x^2-8x-4\right)+\left(3x^3+x^2-5\right)\\&=3x^3+\left(x^2+2x^2+x^2\right)+\left(-5x-8x\right)+\left(3-4-5\right)\\&=3x^3+4x^2-13x-6.\end{aligned}$$

Teaching Notes: For simplifying expressions like these, the roles of the commutative and associative properties are important and students need to be reminded frequently that these properties are being used.

If the polynomials are written in a vertical format, we align like terms, one beneath the other, in a column format and combine like terms in each column.

$$\begin{array}{rrrr} & x^2 & -\ 5x & +3\\ & 2x^2 & -\ 8x & -4\\ 3x^3 + & x^2 & & -5\\ \hline 3x^3 & +4x^2 & -13x & -6\end{array}$$

Example 1: Adding Polynomials

a. Add as indicated: $\left(5x^3-8x^2+12x+13\right)+\left(-2x^2-8\right)+\left(4x^3-5x+14\right)$

Solution: $\left(5x^3-8x^2+12x+13\right)+\left(-2x^2-8\right)+\left(4x^3-5x+14\right)$

$$\begin{aligned}&=\left(5x^3+4x^3\right)+\left(-8x^2-2x^2\right)+\left(12x-5x\right)+\left(13-8+14\right)\\&=9x^3-10x^2+7x+19\end{aligned}$$

Continued on the next page...

b. Find the sum: $\left(x^3 - x^2 + 5x\right) + \left(4x^3 + 5x^2 - 8x + 9\right)$

Solution:
$$\begin{array}{r} x^3 - x^2 + 5x \\ \underline{4x^3 + 5x^2 - 8x + 9} \\ 5x^3 + 4x^2 - 3x + 9 \end{array}$$

Subtraction with Polynomials

A negative sign written in front of a polynomial in parentheses indicates the **opposite of the entire polynomial**. The opposite can be found by changing the sign of every term in the polynomial.

$$-\left(2x^2 + 3x - 7\right) = -2x^2 - 3x + 7$$

We can also think of the opposite of a polynomial as –1 times the polynomial, then applying the distibutive property as follows:

$$\begin{aligned} -\left(2x^2 + 3x - 7\right) &= -1\left(2x^2 + 3x - 7\right) \\ &= -1\left(2x^2\right) - 1(3x) - 1(-7) \\ &= -2x^2 - 3x + 7 \end{aligned}$$

Teaching Notes:
Some students may need to be reminded that "combining" like terms is just a use of the distributive property.

The result is the same with either approach. So the **difference** between two polynomials can be found by changing the sign of each term of the second polynomial and then combining like terms.

$$\begin{aligned} \left(5x^2 - 3x - 7\right) - \left(2x^2 + 5x - 8\right) &= 5x^2 - 3x - 7 - 2x^2 - 5x + 8 \\ &= \left(5x^2 - 2x^2\right) + (-3x - 5x) + (-7 + 8) \\ &= 3x^2 - 8x + 1 \end{aligned}$$

If the polynomials are written in a vertical format, one beneath the other, we change the signs of the terms of the polynomial being subtracted and then combine like terms.

Subtract:

$$\begin{array}{r} 5x^2 - 3x - 7 \\ \underline{-\left(2x^2 + 5x - 8\right)} \end{array} \longrightarrow \begin{array}{r} 5x^2 - 3x - 7 \\ \underline{-2x^2 - 5x + 8} \\ 3x^2 - 8x + 1 \end{array}$$

Example 2: Subtracting Polynomials

a. Subtract as indicated: $(9x^4 - 22x^3 + 3x^2 + 10) - (5x^4 - 2x^3 - 5x^2 + x)$

Solution: $(9x^4 - 22x^3 + 3x^2 + 10) - (5x^4 - 2x^3 - 5x^2 + x)$

$$= 9x^4 - 22x^3 + 3x^2 + 10 - 5x^4 + 2x^3 + 5x^2 - x$$
$$= (9x^4 - 5x^4) + (-22x^3 + 2x^3) + (3x^2 + 5x^2) - x + 10$$
$$= 4x^4 - 20x^3 + 8x^2 - x + 10$$

b. Find the difference:

$$\begin{array}{r} 8x^3 + 5x^2 - 14 \\ \underline{-(-2x^3 + \ x^2 + 6x)} \end{array}$$

Solution:

$$\begin{array}{r} 8x^3 + 5x^2 - 14 \\ \underline{-(-2x^3 + \ x^2 + 6x)} \end{array} \longrightarrow \begin{array}{r} 8x^3 + 5x^2 + 0x - 14 \\ \underline{2x^3 - \ x^2 - 6x + \ 0} \\ 10x^3 + 4x^2 - 6x - 14 \end{array}$$

Write in 0's for missing powers to help with alignment of like terms.

Simplifying Algebraic Expressions

If an algebraic expression contains more than one pair of grouping symbols, such as parentheses ( ), brackets [ ], or braces { }, simplify by working to remove the innermost pair of symbols first. **Apply the rules for order of operations** just as if the variables were numbers and proceed to combine like terms.

Example 3: Simplifying Algebraic Expressions

Simplify each of the following expressions.

a. $5x - [2x + 3(4 - x) + 1] - 9$

Solution: $5x - [2x + 3(4 - x) + 1] - 9$

$$= 5x - [2x + 12 - 3x + 1] - 9$$
$$= 5x - [-x + 13] - 9$$
$$= 5x + x - 13 - 9$$
$$= 6x - 22$$

Work with the parentheses first since they are included inside the brackets.

Continued on the next page...

b. $-3(x-4)+2[x+3(x-3)]$

Solution: $-3(x-4)+2[x+3(x-3)]$

$= -3(x-4)+2[x+3x-9]$

$= -3(x-4)+2[4x-9]$

$= -3x+12+8x-18$

$= 5x-6$

Work with the parentheses first since they are included inside the brackets.

Practice Problems

1. Add: $(15x+4)+(3x^2-9x-5)$

2. Subtract: $(-5x^3-3x+4)-(3x^3-x^2+4x-7)$

3. Simplify: $2-[3a-(4-7a)+2a]$

5.4 Exercises

Find the indicated sum.

1. $(2x^2+5x-1)+(x^2+2x+3)$

2. $(x^2+3x-8)+(3x^2-2x+4)$

3. $(x^2+7x-7)+(x^2+4x)$

4. $(x^2+2x-3)+(x^2+5)$

5. $(2x^2-x-1)+(x^2+x+1)$

6. $(3x^2+5x-4)+(2x^2+x-6)$

7. $(-2x^2-3x+9)+(3x^2-2x+8)$

8. $(x^2+6x-7)+(3x^2+x-1)$

9. $(-4x^2+2x-1)+(3x^2-x+2)+(x-8)$

10. $(8x^2+5x+2)+(-3x^2+9x-4)+(2x^2+6)$

11. $(x^2+2x-1)+(3x^2-x+2)+(2x^3-4x-8)$

12. $(x^3+2x-9)+(x^2-5x+2)+(x^3-4x^2+1)$

1. $3x^2+7x+2$
2. $4x^2+x-4$
3. $2x^2+11x-7$
4. $2x^2+2x+2$
5. $3x^2$
6. $5x^2+6x-10$
7. $x^2-5x+17$
8. $4x^2+7x-8$
9. $-x^2+2x-7$
10. $7x^2+14x+4$
11. $2x^3+4x^2-3x-7$
12. $2x^3-3x^2-3x-6$

Answers to Practice Problems: **1.** $3x^2+6x-1$ **2.** $-8x^3+x^2-7x+11$ **3.** $-12a+6$

13. $-x^2+7x-3$
14. $5x^2-5x-1$
15. $-x^3+2x^2+3x-4$
16. $6x^3+3x^2-3x+5$
17. $4x^3+x^2+7$
18. $5x^3+11x^2+10x-14$
19. $2x^3+11x^2-4x-3$
20. x
21. x^2+x+6
22. $2x^2+5x-11$
23. $-3x^2-6x-2$
24. $2x^2+13x+9$
25. $3x^2-4x-8$
26. $9x^2+3x$
27. $-x^4-2x^3-16$
28. $-2x^3+3x^2-5x-8$
29. $-4x^4-7x^3-11x^2$
$+5x+13$
30. $-x^5+5x^3-7x^2$
$-5x-4$
31. $-4x^2+6x+1$
32. $4x^2+3x+11$
33. $3x^4-7x^3+3x^2$
$-5x+9$
34. $-x^5-2x^3-3x^2$
$-5x-22$

13. $$\begin{array}{r} x^2+4x-4 \\ \underline{-2x^2+3x+1} \end{array}$$

14. $$\begin{array}{r} 2x^2+4x-3 \\ \underline{3x^2-9x+2} \end{array}$$

15. $$\begin{array}{r} x^3+3x^2+x \\ \underline{-2x^3-x^2+2x-4} \end{array}$$

16. $$\begin{array}{r} 4x^3+5x^2+11 \\ \underline{2x^3-2x^2-3x-6} \end{array}$$

17. $$\begin{array}{r} 7x^3+5x^2+x-6 \\ -3x^2+4x+11 \\ \underline{-3x^3-x^2-5x+2} \end{array}$$

18. $$\begin{array}{r} x^3+5x^2+7x-3 \\ 4x^2+3x-9 \\ \underline{4x^3+2x^2-2} \end{array}$$

19. $$\begin{array}{r} x^3+3x^2-4 \\ 7x^2+2x+1 \\ \underline{x^3+x^2-6x} \end{array}$$

20. $$\begin{array}{r} x^3+2x^2-5 \\ -2x^3+x-9 \\ \underline{x^3-2x^2+14} \end{array}$$

Find the indicated difference.

21. $(2x^2+4x+8)-(x^2+3x+2)$

22. $(3x^2+7x-6)-(x^2+2x+5)$

23. $(x^2-9x+2)-(4x^2-3x+4)$

24. $(6x^2+11x+2)-(4x^2-2x-7)$

25. $(2x^2-x-10)-(-x^2+3x-2)$

26. $(7x^2+4x-9)-(-2x^2+x-9)$

27. $(x^4+8x^3-2x^2-5)-(2x^4+10x^3-2x^2+11)$

28. $(x^3+4x^2-3x-7)-(3x^3+x^2+2x+1)$

29. $(-3x^4+2x^3-7x^2+6x+12)-(x^4+9x^3+4x^2+x-1)$

30. $(2x^5+3x^3-2x^2+x-5)-(3x^5-2x^3+5x^2+6x-1)$

31. $(9x^2-5)-(13x^2-6x-6)$

32. $(8x^2+9)-(4x^2-3x-2)$

33. $(3x^4-2x^3-8x-1)-(5x^3-3x^2-3x-10)$

34. $(x^5+6x^3-3x^2-5)-(2x^5+8x^3+5x+17)$

35. $6x^2-7x+18$
36. $5x^2+2x+3$
37. $8x^4+6x^2+15$
38. $14x^2-15$
39. $2x^3+4x^2+3x-10$
40. $2x^3-5x^2+11x-11$
41. $4x-13$
42. $-15x+22$
43. $-7x+9$
44. $-3x-6$
45. $2x+17$
46. $-9x+26$
47. $3x^3+13x^2+9$
48. $10x^3-3x^2+7$
49. $8x^2-x-2$
50. $-7x^2+6x-2$
51. $3x-19$
52. $5x-7$
53. $x^2-2x+13$
54. $-x^2+6x-20$
55. $7x-3$
56. $x+5$
57. $4x^2+3x+5$
58. $7x^3-2x^2+9x$
59. $7x-18$
60. $7x-6$
61. $x^2+11x-8$
62. $5x^2-16x-11$
63. $7x^3-x^2-7$
64. x^3+x^2+6x-2

35. $\begin{array}{r} 14x^2-6x+9 \\ \underline{-(8x^2+x-9)} \end{array}$

36. $\begin{array}{r} 9x^2-3x+2 \\ \underline{-(4x^2-5x-1)} \end{array}$

37. $\begin{array}{r} 5x^4+8x^2+11 \\ \underline{-(-3x^4+2x^2-4)} \end{array}$

38. $\begin{array}{r} 11x^2+5x-13 \\ \underline{-(-3x^2+5x+2)} \end{array}$

39. $\begin{array}{r} x^3+6x^2\qquad-3 \\ \underline{-(-x^3+2x^2-3x+7)} \end{array}$

40. $\begin{array}{r} 3x^3\qquad+9x-17 \\ \underline{-(x^3+5x^2-2x-6)} \end{array}$

Simplify each of the algebraic expressions, and write the polynomials in descending order.

41. $5x+2(x-3)-(3x+7)$

42. $-4(x-6)-(8x+2)-3x$

43. $11+[3x-2(1+5x)]$

44. $2+[9x-4(3x+2)]$

45. $8x-[2x+4(x-3)-5]$

46. $17-[-3x+6(2x-3)+9]$

47. $3x^3-[5-7(x^2+2)-6x^2]$

48. $10x^3-[8-5(3-2x^2)-7x^2]$

49. $(2x^2+4)-[-8+2(7-3x^2)+x]$

50. $-[6x^2-3(4+2x)+9]-(x^2+5)$

51. $2[3x+(x-8)-(2x+5)]-(x-7)$

52. $-3[-x+(10-3x)-(8-3x)]+(2x-1)$

53. $(x^2-1)+2[4+(3-x)]$

54. $(4-x^2)+3[(2x-3)-5]$

55. $-(x-5)+[6x-2(4-x)]$

56. $2(2x+1)-[5x-(2x+3)]$

57. Find the sum of $4x^2-3x$ and $6x+5$.

58. Subtract $2x^2-4x$ from $7x^3+5x$.

59. Subtract $3(x+1)$ from $5(2x-3)$.

60. Find the sum of $10x-2(3x+5)$ and $3(x-4)+16$.

61. Subtract $3x^2-4x+2$ from the sum of $4x^2+x-1$ and $6x-5$.

62. Subtract $-2x^2+6x+12$ from the sum of $2x^2+3x-1$ and $x^2-13x+2$.

63. Add $5x^3-8x+1$ to the difference between $2x^3+14x-3$ and x^2+6x+5.

64. Add $2x^3+4x^2+1$ to the difference between $-x^2+10x-3$ and x^3+2x^2+4x.

Writing and Thinking About Mathematics

65. Write the definition of a polynomial.

66. Explain, in your own words, how to subtract one polynomial from another.

67. Describe what is meant by the degree of a polynomial in x.

68. Give two examples that show how the sum of two binomials might not be a binomial.

65. Any monomial or algebraic sum of monomials

66. Add the opposite of each term in the polynomial being subtracted to the original polynomial and combine like terms.

HAWKES LEARNING SYSTEMS: INTRODUCTORY & INTERMEDIATE ALGEBRA SOFTWARE

- 5.4 Adding and Subtracting Polynomials

67. The largest of the degrees of its terms after like terms have been combined.

68. $(2x+4)+(x^2+2x)=x^2+4x+4$; $(x^9+x^3)+(x^7+x^5)=x^9+x^7+x^5+x^3$; Answers will vary.

5.5 Multiplication with Polynomials

- *Multiply a polynomial by a monomial.*
- *Multiply two polynomials.*

Up to this point, we have multiplied terms such as $5x^2 \cdot 3x^4 = 15x^6$ by using the product rule for exponents. Also, we have applied the distributive property to expressions such as $5(2x+3) = 10x+15$.

Now we will use both the product rule for exponents and the distributive property to multiply polynomials. We discuss three cases here:

a. the product of a monomial with a polynomial of two or more terms,
b. the product of two binomials, and
c. the product of a binomial with a polynomial of more than two terms.

Teaching Notes: While we continue to emphasize the use of the distributive property in so many instances of simplifying, it is also important to point out our continued use of the commutative and associative properties toward this end. For example, in this demonstration we reach a point of having "$5x \cdot 3$" and we simplify this as "$15x$" (by the commutative and associative properties of multiplication).

Multiplying a Polynomial by a Monomial

Using the distributive property $a(b+c) = ab+ac$ with multiplication indicated on the left, we can find the product of a monomial with a polynomial of two or more terms as follows:

$$5x(2x+3) = 5x \cdot 2x + 5x \cdot 3 = 10x^2 + 15x$$

$$3x^2(4x-1) = 3x^2 \cdot 4x + 3x^2(-1) = 12x^3 - 3x^2$$

$$-4a^5(a^2 - 8a + 5) = -4a^5 \cdot a^2 - 4a^5(-8a) - 4a^5(5) = -4a^7 + 32a^6 - 20a^5$$

Multiplying Two Polynomials

Now suppose that we want to multiply two binomials, say, $(x+3)(x+7)$. We will apply the distributive property in the following way with multiplication indicated on the right of the parentheses.

Compare $(x+3)(x+7)$ to $(a+b)c = ac+bc$.

Think of $(x+7)$ as taking the place of c. Thus

$$(a+b)c \quad = \quad ac \quad + \quad bc$$

takes the form $(x+3)(x+7) = x(x+7) + 3(x+7)$.

Completing the products on the right, using the distributive property twice again, gives

$$\begin{aligned}(x+3)(x+7) &= x(x+7)+3(x+7)\\ &= x\cdot x+x\cdot 7+3\cdot x+3\cdot 7\\ &= x^2+7x+3x+21\\ &= x^2+10x+21.\end{aligned}$$

In the same manner,

$$\begin{aligned}(x+2)(3x+4) &= x(3x+4)+2(3x+4)\\ &= x\cdot 3x+x\cdot 4+2\cdot 3x+2\cdot 4\\ &= 3x^2+4x+6x+8\\ &= 3x^2+10x+8.\end{aligned}$$

Similarly,

$$\begin{aligned}(2x-1)(x^2+x-5) &= 2x(x^2+x-5)-1(x^2+x-5)\\ &= 2x\cdot x^2+2x\cdot x+2x\cdot(-5)-1\cdot x^2-1\cdot x-1(-5)\\ &= 2x^3+2x^2-10x-x^2-x+5\\ &= 2x^3+x^2-11x+5.\end{aligned}$$

The product of two polynomials can also be found by writing one polynomial under the other. **The distributive property is applied by multiplying each term of one polynomial by each term of the other.** Consider the product $(2x^2+3x-4)(3x+7)$. Now writing one polynomial under the other and applying the distributive property, we obtain the following:

Multiply by $+7$:

$$\begin{array}{rrr} 2x^2+ & 3x & -4\\ & 3x & +7\\ \hline 14x^2+ & 21x & -28\end{array}$$

Multiply by $3x$:

$$\begin{array}{rrrr} & 2x^2+ & 3x & -\ 4\\ & & 3x & +\ 7\\ \hline & 14x^2 & +21x & -28\\ 6x^3+ & 9x^2 & -12x & \\ \hline \end{array}$$

Align the like terms so that they can be easily combined.

Finally, combine like terms:

$$\begin{array}{rrrr} & 2x^2 & +\ 3x & -\ 4\\ & & 3x & +\ 7\\ \hline & 14x^2 & +\ 21x & -\ 28\\ 6x^3+ & 9x^2 & -12x & \\ \hline 6x^3+ & 23x^2 & +\ 9x & -\ 28\end{array}$$

Combine like terms.

Example 1: Multiplying Polynomials

Find each product.

a. $-4x(x^2-3x+12)$

Solution: $-4x(x^2-3x+12) = -4x\cdot x^2 - 4x(-3x) - 4x\cdot 12$

$= -4x^3 + 12x^2 - 48x$

b. $(2x-4)(5x+3)$

Solution: $(2x-4)(5x+3) = 2x(5x+3) - 4(5x+3)$

$= 2x\cdot 5x + 2x\cdot 3 - 4\cdot 5x - 4\cdot 3$

$= 10x^2 + 6x - 20x - 12$

$= 10x^2 - 14x - 12$

c. $7y^2 - 3y + 2$

$\underline{\quad 2y + 3}$

Solution:

$$\begin{array}{rrrrl} & 7y^2 & -\ 3y & +\ 2 & \\ & & 2y & +\ 3 & \\ \hline & 21y^2 & -9y & +6 & \text{Multiply by } 3. \\ 14y^3 & -\ 6y^2 & +4y & & \text{Multiply by } 2y. \\ \hline 14y^3 & +15y^2 & -\ 5y & +6 & \text{Combine like terms.} \end{array}$$

d. $x^2 + 3x - 1$

$\underline{x^2 - 3x + 1}$

Solution:

$$\begin{array}{rrrrrl} & & x^2 & +3x & -1 & \\ & & x^2 & -3x & +1 & \\ \hline & & x^2 & +3x & -1 & \text{Multiply by } 1. \\ & -3x^3 & -9x^2 & +3x & & \text{Multiply by } -3x. \\ x^4 & +3x^3 & -\ x^2 & & & \text{Multiply by } x^2. \\ \hline x^4 & & -9x^2 & +6x & -1 & \text{Combine like terms.} \end{array}$$

e. $(x-5)(x+2)(x-1)$

Solution: First multiply $(x-5)(x+2)$, then multiply this result by $(x-1)$.

$$(x-5)(x+2) = x(x+2) - 5(x+2)$$
$$= x^2 + 2x - 5x - 10$$
$$= x^2 - 3x - 10$$

$$(x^2-3x-10)(x-1) = (x^2-3x-10)x + (x^2-3x-10)(-1)$$
$$= x^3 - 3x^2 - 10x - x^2 + 3x + 10$$
$$= x^3 - 4x^2 - 7x + 10$$

Practice Problems

Find each product.

1. $2x(3x^2+x-1)$
2. $(x+3)(x-7)$
3. $(x-1)(x^2+x-4)$
4. $(x+1)(x-2)(x+2)$

1. $-6x^5-15x^3$
2. $-20x^4+30x^2$
3. $4x^7-12x^6+4x^5$
4. $18x^6-9x^5+45x^4$
5. $-y^5+8y-2$
6. $-14y^4-21y^2-7$
7. $-4x^8+8x^7-12x^4$
8. $-2x^7+2x^6-4x^5$
9. $25x^5-5x^4+10x^3$
10. $-2x^5-10x^3+8x^2$
11. $a^7+2a^6-5a^3+a^2$
12. $-7t^6+35t^5+14t^4+7t^3$
13. $6x^2-x-2$
14. $3x^2+25x+28$
15. $9a^2-25$
16. $6x^2-x-5$
17. $-10x^2+39x-14$
18. y^3-y^2+y-1

5.5 Exercises

Multiply as indicated and simplify if possible.

1. $-3x^2(2x^3+5x)$
2. $5x^2(-4x^2+6)$
3. $4x^5(x^2-3x+1)$
4. $9x^3(2x^3-x^2+5x)$
5. $-1(y^5-8y+2)$
6. $-7(2y^4+3y^2+1)$
7. $-4x^3(x^5-2x^4+3x)$
8. $-2x^4(x^3-x^2+2x)$
9. $5x^3(5x^2-x+2)$
10. $-2x^2(x^3+5x-4)$
11. $a^2(a^5+2a^4-5a+1)$
12. $7t^3(-t^3+5t^2+2t+1)$
13. $3x(2x+1)-2(2x+1)$
14. $x(3x+4)+7(3x+4)$
15. $3a(3a-5)+5(3a-5)$
16. $6x(x-1)+5(x-1)$
17. $5x(-2x+7)-2(-2x+7)$
18. $y(y^2+1)-1(y^2+1)$

Answers to Practice Problems: **1.** $6x^3+2x^2-2x$ **2.** $x^2-4x-21$ **3.** x^3-5x+4 **4.** x^3+x^2-4x-4

19. x^3+5x^2+8x+4
20. $4x^3-x^2+x+3$
21. x^2+x-12
22. $x^2+2x-35$
23. $a^2-2a-48$
24. x^2-2x-8
25. x^2-3x+2
26. $x^2-15x+56$
27. $3t^2-3t-60$
28. $-4x^2+4x+168$
29. x^3+11x^2+24x
30. t^3-11t^2+28t
31. $2x^2-7x-4$
32. $3x^2+11x-4$
33. $6x^2+17x-3$
34. $8x^2+23x+15$
35. $4x^2-9$
36. $9t^2-25$
37. $16x^2+8x+1$
38. $25x^2-20x+4$
39. y^3+2y^2+y+12
40. $2x^3-13x^2-3x+2$
41. $3x^2-8x-35$
42. $2x^2+12x+18$
43. $5x^3+6x^2-22x-9$
44. $-16x^3+50x^2+25x-14$
45. $2x^4+7x^3+5x^2+x-15$
46. $12x^4+28x^3+47x^2+34x+48$
47. $3x^2+2x-8$
48. $4t^2+17t-42$
49. $2x^2+3x-5$
50. $5a^2+17a-12$
51. $7x^2-13x-2$
52. $3x^2+2x-16$
53. $6x^2-13x-8$

19. $x(x^2+3x+2)+2(x^2+3x+2)$ **20.** $4x(x^2-x+1)+3(x^2-x+1)$

21. $(x+4)(x-3)$ **22.** $(x+7)(x-5)$ **23.** $(a+6)(a-8)$

24. $(x+2)(x-4)$ **25.** $(x-2)(x-1)$ **26.** $(x-7)(x-8)$

27. $3(t+4)(t-5)$ **28.** $-4(x+6)(x-7)$ **29.** $x(x+3)(x+8)$

30. $t(t-4)(t-7)$ **31.** $(2x+1)(x-4)$ **32.** $(3x-1)(x+4)$

33. $(6x-1)(x+3)$ **34.** $(8x+15)(x+1)$ **35.** $(2x+3)(2x-3)$

36. $(3t+5)(3t-5)$ **37.** $(4x+1)(4x+1)$ **38.** $(5x-2)(5x-2)$

39. $(y+3)(y^2-y+4)$ **40.** $(2x+1)(x^2-7x+2)$

Multiply as indicated and simplify.

41. $\begin{array}{r} 3x+7 \\ \underline{x-5} \end{array}$ **42.** $\begin{array}{r} 2x+6 \\ \underline{x+3} \end{array}$ **43.** $\begin{array}{r} x^2+3x+1 \\ \underline{5x-9} \end{array}$

44. $\begin{array}{r} 8x^2+3x-2 \\ \underline{-2x+7} \end{array}$ **45.** $\begin{array}{r} 2x^2+3x+5 \\ \underline{x^2+2x-3} \end{array}$ **46.** $\begin{array}{r} 6x^2-x+8 \\ \underline{2x^2+5x+6} \end{array}$

Find the product and simplify if possible.

47. $(3x-4)(x+2)$ **48.** $(t+6)(4t-7)$ **49.** $(2x+5)(x-1)$

50. $(5a-3)(a+4)$ **51.** $(7x+1)(x-2)$ **52.** $(x-2)(3x+8)$

53. $(2x+1)(3x-8)$ **54.** $(3x+7)(2x-5)$ **55.** $(2x+3)(2x+3)$

56. $(5y+2)(5y+2)$ **57.** $(x+3)(x^2-4)$ **58.** $(y^2+2)(y-4)$

59. $(2x+7)(2x-7)$ **60.** $(3x-4)(3x+4)$ **61.** $(x+1)(x^2-x+1)$

62. $(x-2)(x^2+2x+4)$ **63.** $(7a-2)(7a-2)$ **64.** $(5a-6)(5a-6)$

54. $6x^2 - x - 35$
55. $4x^2 + 12x + 9$
56. $25y^2 + 20y + 4$
57. $x^3 + 3x^2 - 4x - 12$
58. $y^3 - 4y^2 + 2y - 8$
59. $4x^2 - 49$
60. $9x^2 - 16$
61. $x^3 + 1$
62. $x^3 - 8$
63. $49a^2 - 28a + 4$
64. $25a^2 - 60a + 36$
65. $2x^3 + x^2 - 5x - 3$
66. $3x^3 - 2x^2 + 26x + 9$
67. $x^3 + 6x^2 + 11x + 6$
68. $t^3 - 6t^2 + 11t - 6$
69. $a^4 - a^2 + 2a - 1$
70. $y^4 + 2y^3 + y^2 - 4$
71. $t^4 + 6t^3 + 13t^2 + 12t + 4$
72. $a^4 - 8a^3 + 18a^2 - 8a + 1$
73. $2y^2 - 61$
74. $2y^2 - 5$
75. $3a^2 - 17a + 11$
76. $3x^2 + 4x + 10$
77. $-3x - 21$
78. $10t + 5$
79. $a^2 + 9a - 1$
80. $t^2 + 18t - 10$
81. Answers will vary.

65. $(2x+3)(x^2-x-1)$

66. $(3x+1)(x^2-x+9)$

67. $(x+1)(x+2)(x+3)$

68. $(t-1)(t-2)(t-3)$

69. $(a^2+a-1)(a^2-a+1)$

70. $(y^2+y+2)(y^2+y-2)$

71. $(t^2+3t+2)^2$

72. $(a^2-4a+1)^2$

Simplify.

73. $(y+6)(y-6)+(y+5)(y-5)$

74. $(y-2)(y+2)+(y-1)(y+1)$

75. $(2a+1)(a-5)+(a-4)(a-4)$

76. $(x+4)(2x+1)+(x-3)(x-2)$

77. $(x-3)(x+5)-(x+3)(x+2)$

78. $(t+3)(t+3)-(t-2)(t-2)$

79. $(2a+3)(a+1)-(a-2)(a-2)$

80. $(4t-3)(t+4)-(t-2)(3t+1)$

Writing and Thinking About Mathematics

81. We have seen how the distributive property is used to multiply polynomials.

a. Show how the distributive property can be used to find the product $\begin{array}{r} 75 \\ \times 93 \\ \hline \end{array}$ (**Hint:** 75 = 70 + 5 and 93 = 90 + 3)

b. In the multiplication algorithm for multiplying whole numbers (as in the product above), we are told to "move to the left" when multiplying. For example

$$\begin{array}{r} 75 \\ \times 93 \\ \hline 15 \\ 21 \\ 45 \\ 63 \\ \hline \end{array}$$

Why are the 21 and 45 moved one place to the left in the alignment? When 9 and 7 are multiplied, we move the 63 two places left. Why?

HAWKES LEARNING SYSTEMS: INTRODUCTORY & INTERMEDIATE ALGEBRA SOFTWARE

- 5.5 Multiplying Polynomials

5.6 Special Products of Binomials

- *Multiply binomials using the FOIL method.*
- *Multiply binomials, finding products that are the difference of squares.*
- *Square binomials, finding products that are perfect square trinomials.*
- *Identify the difference of two squares and perfect square trinomials.*

Teaching Notes:
You may want to explain to students that they could multiply all the polynomials in this section using the methods already discussed in Section 5.5. The new methods shown here in Section 5.6 are simply a tool to make multiplying certain types of polynomials faster and easier.

In the case of the **product of two binomials** such as $(2x+5)(3x-7)$, the **FOIL** method is useful. **F-O-I-L** is a mnemonic device (memory aid) to help in remembering which terms of the binomials to multiply together. First, by using the distributive property we can see how the terms are multiplied.

$$(2x+5)(3x-7) = 2x(3x-7)+5(3x-7)$$
$$= 2x \cdot 3x + 2x \cdot (-7) + 5 \cdot 3x + 5 \cdot (-7)$$

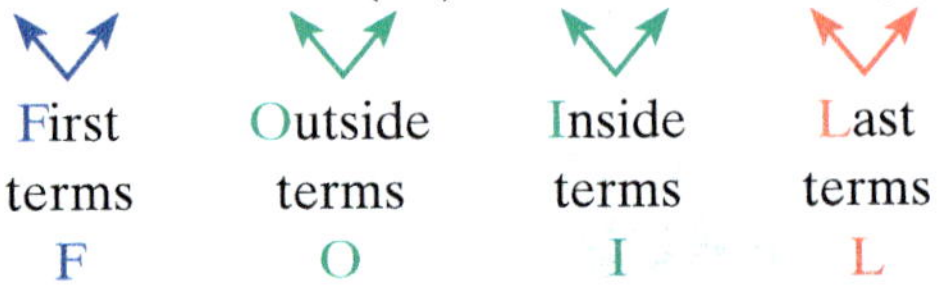

Now we can use the FOIL method and then combine like terms to go directly to the answer.

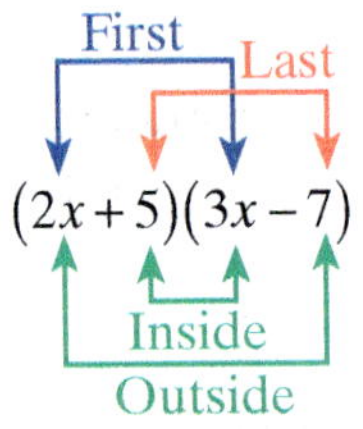

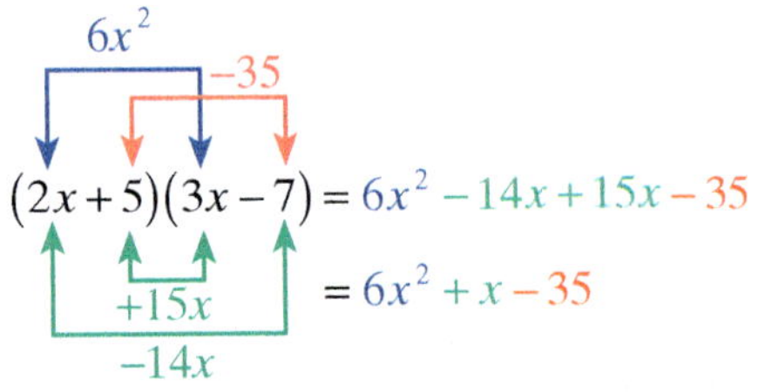

Example 1: FOIL Method

Use the FOIL method to find the products of the given binomials.

a. $(x+3)(2x+8)$

Solution:

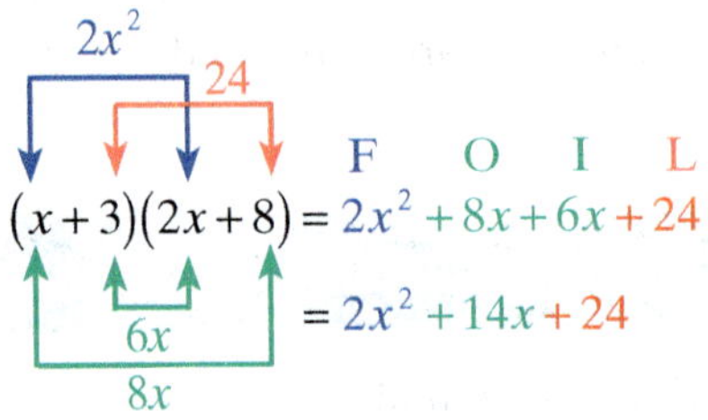

b. $(2x-3)(3x-5)$

Solution:

$6x^2$ 15

F O I L

$$(2x-3)(3x-5) = 6x^2 - 10x - 9x + 15$$

$$= 6x^2 - 19x + 15$$

$-9x$

$-10x$

c. $(x+7)(x-7)$

Solution: $(x+7)(x-7) = x^2 - 7x + 7x - 49$ Apply the FOIL method mentally.

$= x^2 - 49$ Note that in this special case the two middle terms are opposites and their sum is 0.

The Difference of Two Squares: $(x+a)(x-a) = x^2 - a^2$

In Example 1c, the middle terms, $-7x$ and $+7x$, are opposites of each other and their sum is 0. Therefore the resulting product has only two terms and each term is a square.

$$(x+7)(x-7) = x^2 - 49$$

In fact, when two binomials are in the form of the sum and difference of the same two terms, the product will always be the difference of the squares of the terms. The product is called the **difference of two squares**.

Difference of Two Squares

$$(x+a)(x-a) = x^2 - a^2$$

In order to quickly recognize the difference of two squares, you should memorize the following squares of the positive integers from 1 to 20. The squares of integers are called **perfect squares**.

Perfect Squares From 1 to 400

1, 4, 9, 16, 25, 36, 49, 64, 81, 100, 121, 144, 169, 196, 225, 256, 289, 324, 361, 400

Example 2: Difference of Two Squares

Find the following products.

a. $(x+4)(x-4)$

Solution: The two binomials represent the sum and difference of x and 4. So, the product is the difference of their squares.

$(x+4)(x-4)=x^2-4^2=x^2-16$ Difference of two squares

b. $(3y+7)(3y-7)$

Solution: $(3y+7)(3y-7)=(3y)^2-7^2=9y^2-49$ Difference of two squares

c. $(x^3-6)(x^3+6)$

Solution: $(x^3-6)(x^3+6)=(x^3)^2-6^2=x^6-36$ Difference of two squares

Squares of Binomials: $\begin{cases}(x+a)^2 = x^2+2ax+a^2 \\ (x-a)^2 = x^2-2ax+a^2\end{cases}$

Now we consider the case where the two binomials being multiplied are **the same**. That is, we want to consider the **square of a binomial**. The following examples, using the distributive property, illustrate two patterns that, after some practice, allow us to go directly to the products.

$$(x+3)^2=(x+3)(x+3)=x^2+3x+3x+9$$

$$=x^2+2\cdot 3x+9$$ The middle term is doubled. $3x+3x=2\cdot 3x$

$$=x^2+6x+9$$ Perfect square trinomial

$$(x-11)^2=(x-11)(x-11)=x^2-11x-11x+121$$

$$=x^2-2\cdot 11x+121$$ The middle term is doubled. $-11x-11x=2(-11x)=-2\cdot 11x$

$$=x^2-22x+121$$ Perfect square trinomial

Note that in each case **the result of squaring the binomial is a trinomial**. These trinomials are called **perfect square trinomials**.

Squares of Binomials (Perfect Square Trinomials)

$(x+a)^2 = x^2 + 2ax + a^2$ **Square of a Binomial Sum**

$(x-a)^2 = x^2 - 2ax + a^2$ **Square of a Binomial Difference**

Example 3: Squares of Binomials

Find the following products.

a. $(2x+3)^2$

Solution: The pattern for squaring a binomial gives

$$(2x+3)^2 = (2x)^2 + 2 \cdot 3 \cdot 2x + (3)^2 \qquad \textbf{Note: } (2x)^2 = 2x \cdot 2x = 4x^2$$

$$= 4x^2 + 12x + 9$$

b. $(5x-1)^2$

Solution: $(5x-1)^2 = (5x)^2 - 2(1)(5x) + (1)^2$

$$= 25x^2 - 10x + 1$$

c. $(9-x)^2$

Solution: $(9-x)^2 = (9)^2 - 2(x)(9) + x^2$

$$= 81 - 18x + x^2$$

$$= x^2 - 18x + 81$$

d. $(y^3+1)^2$

Solution: $(y^3+1)^2 = (y^3)^2 + 2(1)(y^3) + 1^2 \qquad \textbf{Note: } (y^3)^2 = y^3 \cdot y^3 = y^{3+3} = y^6$

$$= y^6 + 2y^3 + 1$$

NOTES

COMMON ERROR

Many algebra students make the following error.

$$(x+a)^2 = x^2 + a^2$$

$$(x+6)^2 = x^2 + 36$$

INCORRECT

Avoid this error by remembering that **the square of a binomial is a trinomial**.

$$(x+6)^2 = x^2 + 2 \cdot 6x + 36$$

$$= x^2 + 12x + 36$$

CORRECT

Practice Problems

Find the indicated products.

1. $(x+10)(x-10)$
2. $(x+3)^2$
3. $(2x-1)(x+3)$
4. $(2x-5)^2$
5. $(x^2+4)(x^2-3)$

1. $x^2-14x+49$; Perfect square trinomial
2. $x^2-10x+25$; Perfect square trinomial
3. $x^2+8x+16$; Perfect square trinomial
4. $x^2+16x+64$; Perfect square trinomial
5. x^2-9; Difference of two squares
6. x^2-36; Difference of two squares
7. x^2-81; Difference of two squares
8. x^2-144; Difference of two squares
9. $2x^2+x-3$
10. $6x^2+17x+5$
11. $9x^2-24x+16$; Perfect square trinomial
12. $9x^2+6x+1$; Perfect square trinomial
13. $25x^2-4$; Difference of two squares
14. $4x^2-1$; Difference of two squares

5.6 Exercises

Find the product and identify those that are the difference of two squares or perfect square trinomials.

1. $(x-7)^2$
2. $(x-5)^2$
3. $(x+4)(x+4)$
4. $(x+8)(x+8)$
5. $(x+3)(x-3)$
6. $(x-6)(x+6)$
7. $(x+9)(x-9)$
8. $(x+12)(x-12)$
9. $(2x+3)(x-1)$
10. $(3x+1)(2x+5)$
11. $(3x-4)^2$
12. $(3x+1)^2$
13. $(5x+2)(5x-2)$
14. $(2x+1)(2x-1)$
15. $(3x-2)(3x-2)$
16. $(3+x)^2$
17. $(8-x)(8-x)$
18. $(5-x)(5-x)$
19. $(4x+5)(4x-5)$
20. $(11-x)(11+x)$
21. $(5x-9)(5x+9)$
22. $(9x+2)(9x-2)$
23. $(4-x)^2$
24. $(3x+2)^2$
25. $(2x+7)(2x-7)$
26. $(6x+5)(6x-5)$
27. $(5x^2+2)(2x^2-3)$
28. $(4x^2+7)(2x^2+1)$
29. $(1+7x)^2$
30. $(2-5x)^2$

Write the indicated products.

31. $(x+2)(5x+1)$
32. $(7x-2)(x-3)$
33. $(4x-3)(x+4)$
34. $(x+11)(x-8)$
35. $(3x-7)(x-6)$
36. $(x+7)(2x+9)$

Answers to Practice Problems: **1.** x^2-100 **2.** x^2+6x+9 **3.** $2x^2+5x-3$ **4.** $4x^2-20x+25$ **5.** x^4+x^2-12

37. $(5+x)(5+x)$

38. $(3-x)(3-x)$

39. $(x^2+1)(x^2-1)$

40. $(x^2+5)(x^2-5)$

41. $(x^2+3)(x^2+3)$

42. $(x^3+8)(x^3+8)$

43. $(x^3-2)^2$

44. $(x^2-4)^2$

45. $(x^2-6)(x^2+9)$

46. $(x^2+3)(x^2-5)$

47. $\left(x+\frac{2}{3}\right)\left(x-\frac{2}{3}\right)$

48. $\left(x-\frac{1}{2}\right)\left(x+\frac{1}{2}\right)$

49. $\left(x+\frac{3}{4}\right)\left(x-\frac{3}{4}\right)$

50. $\left(x+\frac{3}{8}\right)\left(x-\frac{3}{8}\right)$

51. $\left(x+\frac{3}{5}\right)\left(x+\frac{3}{5}\right)$

52. $\left(x+\frac{4}{3}\right)\left(x+\frac{4}{3}\right)$

53. $\left(x-\frac{5}{6}\right)^2$

54. $\left(x-\frac{2}{7}\right)^2$

55. $\left(x+\frac{1}{4}\right)\left(x-\frac{1}{2}\right)$

56. $\left(x-\frac{1}{5}\right)\left(x+\frac{2}{3}\right)$

57. $\left(x+\frac{1}{3}\right)\left(x+\frac{1}{2}\right)$

58. $\left(x-\frac{4}{5}\right)\left(x-\frac{3}{10}\right)$

Calculator Problems

Use a graphing calculator as an aid in multiplying the binomials.

59. $(x+1.4)(x-1.4)$

60. $(x-2.1)(x+2.1)$

61. $(x-2.5)^2$

62. $(x+1.7)^2$

63. $(x+2.15)(x-2.15)$

64. $(x+1.36)(x-1.36)$

65. $(x+1.24)^2$

66. $(x-1.45)^2$

67. $(1.42x+9.6)^2$

68. $(0.46x-0.71)^2$

69. $(11.4x+3.5)(11.4x-3.5)$

70. $(2.5x+11.4)(1.3x-16.9)$

71. $(12.6x-6.8)(7.4x+15.3)$

72. $(3.4x+6)(3.4x-6)$

15. $9x^2-12x+4$; Perfect square trinomial
16. x^2+6x+9; Perfect square trinomial
17. $x^2-16x+64$; Perfect square trinomial
18. $x^2-10x+25$; Perfect square trinomial
19. $16x^2-25$; Difference of two squares
20. $-x^2+121$; Difference of two squares
21. $25x^2-81$; Difference of two squares
22. $81x^2-4$; Difference of two squares
23. $x^2-8x+16$; Perfect square trinomial
24. $9x^2+12x+4$; Perfect square trinomial
25. $4x^2-49$; Difference of two squares
26. $36x^2-25$; Difference of two squares
27. $10x^4-11x^2-6$
28. $8x^4+18x^2+7$
29. $49x^2+14x+1$; Perfect square trinomial
30. $25x^2-20x+4$; Perfect square trinomial
31. $5x^2+11x+2$
32. $7x^2-23x+6$
33. $4x^2+13x-12$
34. $x^2+3x-88$
35. $3x^2-25x+42$
36. $2x^2+23x+63$
37. $x^2+10x+25$
38. x^2-6x+9
39. x^4-1
40. x^4-25
41. x^4+6x^2+9
42. x^6+16x^3+64
43. x^6-4x^3+4
44. x^4-8x^2+16

45. x^4+3x^2-54
46. x^4-2x^2-15
47. $x^2-\frac{4}{9}$
48. $x^2-\frac{1}{4}$
49. $x^2-\frac{9}{16}$
50. $x^2-\frac{9}{64}$
51. $x^2+\frac{6}{5}x+\frac{9}{25}$
52. $x^2+\frac{8}{3}x+\frac{16}{9}$
53. $x^2-\frac{5}{3}x+\frac{25}{36}$
54. $x^2-\frac{4}{7}x+\frac{4}{49}$
55. $x^2-\frac{1}{4}x-\frac{1}{8}$
56. $x^2+\frac{7}{15}x-\frac{2}{15}$
57. $x^2+\frac{5}{6}x+\frac{1}{6}$
58. $x^2-\frac{11}{10}x+\frac{6}{25}$
59. $x^2-1.96$
60. $x^2-4.41$
61. $x^2-5x+6.25$
62. $x^2+3.4x+2.89$
63. $x^2-4.6225$
64. $x^2-1.8496$
65. $x^2+2.48x+1.5376$
66. $x^2-2.9x+2.1025$
67. $2.0164x^2+27.264x+92.16$
68. $0.2116x^2-0.6532x+0.5041$
69. $129.96x^2-12.25$
70. $3.25x^2-27.43x-192.66$
71. $93.24x^2+142.46x-104.04$
72. $11.56x^2-36$

73. A square is 20 inches on each side. A square x inches on each side is cut from each corner of the square.

a. Represent the area of the remaining portion of the square in the form of a polynomial function $A(x)$.

b. Represent the perimeter of the remaining portion of the square in the form of a polynomial function $P(x)$.

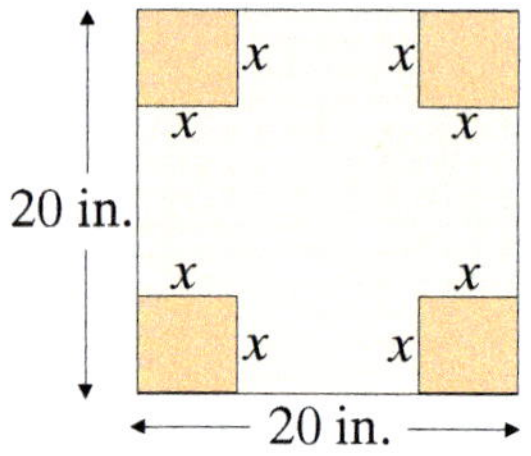

74. In the case of binomial probabilities, if x is the probability of success in one trial of an event, then the expression $f(x)=15x^4(1-x)^2$ is the probability of 4 successes in 6 trials where $0 \le x \le 1$.

a. Represent the expression $f(x)$ as a single polynomial by multiplying the polynomials.

b. If a fair coin is tossed, the probability of heads occurring is $\frac{1}{2}$. That is, $x=\frac{1}{2}$. Find the probability of 4 heads occurring in 6 tosses.

75. A rectangle has sides $(x+3)$ ft and $(x+5)$ ft. If a square x ft on a side is cut from the rectangle, represent the remaining area in the form of a polynomial function $A(x)$.

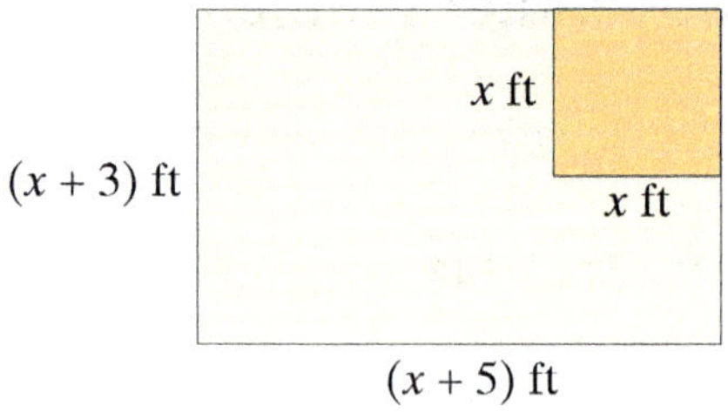

76. Architecture: The Americans with Disabilities Act requires sidewalks to be x feet wide in order for wheelchairs to fit on them. At the bottom, the Empire State Building is 425 feet long and 190 feet wide and a regulation sidewalk surrounds the building.

a. Represent the area covered by the building and the sidewalk in the form of a polynomial function.

b. Represent the area covered by the sidewalk only in the form of a polynomial function.

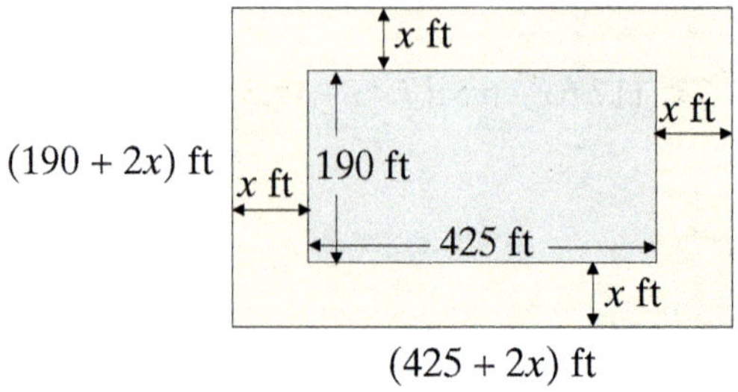

73. a. $A(x)=400-4x^2$
b. $P(x)=4(20-2x)+8x$ $=80$

74. a. $f(x)=15x^6-30x^5+15x^4$
b. 0.234375

75. $A(x)=8x+15$

76. a. $A(x)=4x^2+1230x+80,750$
b. $A(x)=4x^2+1230x$

77. a. $A(x)=150-4x^2$
b. $P(x)=2(10-2x)+2(15-2x)+8x$ $=50$
c. $V(x)=150x-50x^2+4x^3$

78. $V(x)=x^3+440x^2+48,084x+1,073,520$

77. A rectangular piece of cardboard that is 10 inches by 15 inches has squares of length x inches on a side cut from each corner. (Assume that $0<x<5$.)

a. Represent the remaining area in the form of a polynomial function $A(x)$.

b. Represent the perimeter of the remaining figure in the form of a polynomial function $P(x)$.

c. If the flaps of the figure are folded up, an open box is formed. Represent the volume of this box in the form of a polynomial function $V(x)$.

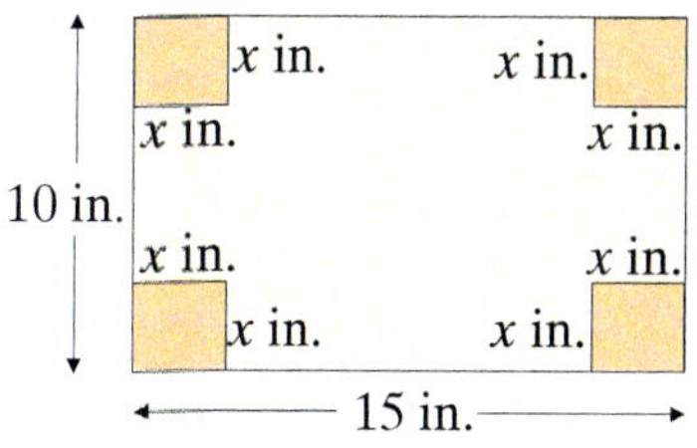

78. **Museums:** The world's largest single aquarium habitat at the Georgia Aquarium is 284 feet long, 126 feet wide, and 30 feet deep. Another aquarium is attempting to make a tank that is x feet longer, wider, and deeper. Represent the volume of the new tank as a polynomial function $V(x)$.

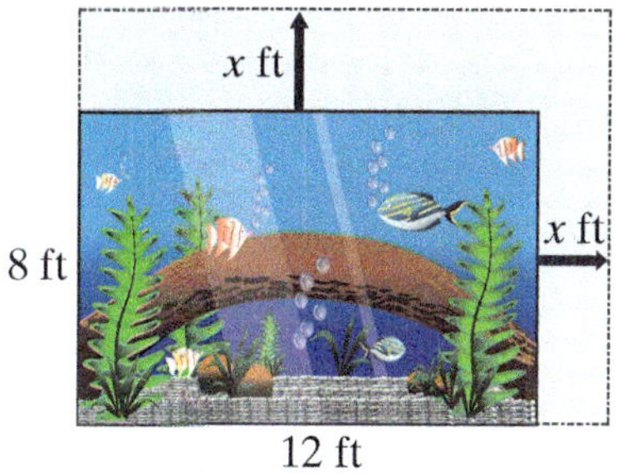

Writing and Thinking About Mathematics

79. As indicated in the diagram $(x+5)^2=x^2+2(5x)+5^2$ Answers will vary.

79. A square with sides of length $(x+5)$ can be broken up as shown in the diagram. The sums of the areas of the interior rectangles and squares is equal to the total area of the square: $(x+5)^2$. Show how this fits with the formula for the square of a sum.

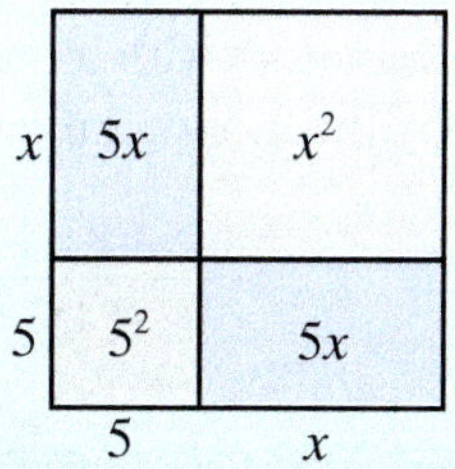

HAWKES LEARNING SYSTEMS: INTRODUCTORY & INTERMEDIATE ALGEBRA SOFTWARE

- 5.6a The FOIL Method
- 5.6b Special Products

5.7 Division with Polynomials

- *Divide a polynomial by a monomial.*
- *Divide polynomials using the division algorithm.*

Fractions, such as $\frac{127}{2}$ and $\frac{1}{8}$, in which the numerator and denominator are integers are called rational numbers. Fractions in which the numerator and denominator are polynomials are called **rational expressions**. (No denominator can be 0.)

Rational Expressions: $\frac{x^3-6x^2+2x}{3x}$, $\frac{6x^2-7x-2}{2x-1}$, $\frac{5x^3-3x+1}{x+3}$, and $\frac{1}{10x}$

In this section, we will treat a rational expression as a division problem. With this basis, there are two situations to consider:

1. the denominator (divisor) is a monomial, or
2. the denominator (divisor) is not a monomial.

Dividing by a Monomial

We know that the sum of fractions with the same denominator can be written as a single fraction by adding the numerators and using the common denominator. For example,

$$\frac{5}{a}+\frac{3b}{a}+\frac{2c}{a}=\frac{5+3b+2c}{a}.$$

If instead of adding the fractions, we start with the sum and want to divide the numerator by the denominator (with a monomial in the denominator), we divide each term in the numerator by the monomial denominator and simplify each fraction.

$$\frac{4x^3+8x^2-12x}{4x}=\frac{4x^3}{4x}+\frac{8x^2}{4x}-\frac{12x}{4x}=x^2+2x-3$$

Example 1: Dividing by a Monomial

Divide each polynomial by the **monomial denominator** by writing each fraction as the sum (or difference) of fractions. Simplify each fraction, if possible.

a. $\frac{x^3-6x^2+2x}{3x}$

Solution: $\frac{x^3-6x^2+2x}{3x}=\frac{x^3}{3x}-\frac{6x^2}{3x}+\frac{2x}{3x}=\frac{x^2}{3}-2x+\frac{2}{3}$

b. $\dfrac{15y^4 - 20y^3 + 5y^2}{5y^2}$

Solution: $\dfrac{15y^4 - 20y^3 + 5y^2}{5y^2} = \dfrac{15y^4}{5y^2} - \dfrac{20y^3}{5y^2} + \dfrac{5y^2}{5y^2} = 3y^2 - 4y + 1$

The Division Algorithm

Teaching Notes: You might want to remind your students of the long division process so that they can see the similarities. Expect many students to have forgotten the process.

In arithmetic, the **division algorithm** (called **long division**) is the process (or series of steps) that we follow when dividing two numbers. By this division algorithm, we can find $64 \div 5$ as follows.

	12	← Quotient
Divisor →	5)64	← Dividend
	5	← Subtract
	14	
	10	← Subtract
	4	← 4 is the remainder (The remainder is always smaller than the divisor.)

Check: $5 \cdot 12 + 4 = 60 + 4 = 64$ (Multiply the divisor times the quotient and add the remainder. The result should be the original dividend.)

We can also write the division in fraction form with the remainder over the divisor, giving a mixed number.

$$64 \div 5 = \frac{64}{5} = 12 + \frac{4}{5} = 12\frac{4}{5}$$

In algebra, **the division algorithm with polynomials** is quite similar. In dividing one polynomial by another, with the degree of the divisor smaller than the degree of the dividend, the quotient will be another polynomial with a remainder.

	For $64 \div 5$	For $P \div D$	
	$64 = 12 \cdot 5 + 4$	$P = Q \cdot D + R$	where
or			P is the dividend,
	$\dfrac{64}{5} = 12 + \dfrac{4}{5}$	$\dfrac{P}{D} = Q + \dfrac{R}{D}$	D is the divisor,
			Q is the quotient, and
			R is the remainder.

The remainder must be of smaller degree than the divisor. If the remainder is 0, then the divisor and quotient are factors of the dividend.

The Division Algorithm

For polynomials P and D, the division algorithm gives

$$\frac{P}{D} = Q + \frac{R}{D},\ D \neq 0$$

where Q and R are polynomials and the **degree of *R* < degree of *D***.

The actual process of long division is not clear from this abstract definition. Although you are familiar with the process of long division with decimal numbers, this same procedure appears more complicated with polynomials. The **division algorithm** (**long division**) is illustrated in a step-by-step form in the following example. Study it carefully.

Example 2: The Division Algorithm

Simplify $\dfrac{6x^2 - 7x - 2}{2x - 1}$ by using long division.

Solution:

Calculation	Explanation
Step 1: $2x-1\overline{)6x^2-7x-2}$	Write both polynomials in order of descending powers. **If any powers are missing, fill in with 0's.**
Step 2: $\begin{array}{r} 3x \\ 2x-1\overline{)\,6x^2-7x-2} \end{array}$	Mentally divide $6x^2$ by $2x$: $\frac{6x^2}{2x} = 3x$. Write $3x$ above $6x^2$.
Step 3: $\begin{array}{r} 3x \\ 2x-1\overline{)\,6x^2-7x-2} \\ -(6x^2-3x) \end{array}$	Multiply $3x$ times $(2x-1)$ and write the terms of the product, $6x^2-3x$, under the like terms in the dividend. Use a '–' sign to indicate that the product is to be subtracted.
Step 4: $\begin{array}{r} 3x \\ 2x-1\overline{)\,6x^2-7x-2} \\ -6x^2+3x \\ \hline -4x \end{array}$	Subtract $6x^2-3x$ by changing signs and adding.
Step 5: $\begin{array}{r} 3x \\ 2x-1\overline{)\,6x^2-7x-2} \\ -6x^2+3x \downarrow \\ \hline -4x-2 \end{array}$	Bring down the -2.

Step 6:
$$\begin{array}{r} 3x - 2 \\ 2x-1 \overline{)\, 6x^2 - 7x - 2} \\ \underline{-6x^2 + 3x} \\ -4x - 2 \end{array}$$

Mentally divide $-4x$ by $2x$: $\frac{-4x}{2x} = -2$.

Write -2 in the quotient.

Step 7:
$$\begin{array}{r} 3x - 2 \\ 2x-1 \overline{)\, 6x^2 - 7x - 2} \\ \underline{-6x^2 + 3x} \\ -4x - 2 \\ \underline{-(-4x + 2)} \end{array}$$

Multiply -2 times $(2x-1)$ and write the terms of the product, $-4x+2$, under the like terms in the expression $-4x-2$. Use a '–' sign to indicate that the product is to be subtracted.

Step 8:
$$\begin{array}{r} 3x - 2 \\ 2x-1 \overline{)\, 6x^2 - 7x - 2} \\ \underline{-6x^2 + 3x} \\ -4x - 2 \\ \underline{+4x - 2} \\ -4 \end{array}$$

Subtract $-4x+2$ by changing signs and adding.

Thus the quotient is $3x-2$ and the remainder is -4.

In the form $Q + \frac{R}{D}$ we can write $\frac{6x^2 - 7x - 2}{2x - 1} = 3x - 2 - \frac{4}{2x-1}$.

Check: Show $Q \cdot D + R = P$.

$$(3x-2)(2x-1) - 4 = 6x^2 - 3x - 4x + 2 - 4 = 6x^2 - 7x - 2$$

Example 3: Long Division (Remainder 0)

Divide $(25x^3 - 5x^2 + 3x + 1) \div (5x + 1)$ by using long division.

Solution:
$$\begin{array}{r} 5x^2 - 2x + 1 \\ 5x+1 \overline{)\, 25x^3 - 5x^2 + 3x + 1} \\ \underline{-(25x^3 + 5x^2)} \\ -10x^2 + 3x \\ \underline{-(-10x^2 - 2x)} \\ 5x + 1 \\ \underline{-(5x+1)} \\ 0 \end{array}$$

There is no remainder, thus the quotient is simply $5x^2 - 2x + 1$.

In Example 3, because the remainder is 0, both $(5x+1)$ and $(5x^2-2x+1)$ are **factors** of $25x^3-5x^2+3x+1$. That is,

$$(5x^2-2x+1)(5x+1)=25x^3-5x^2+3x+1 \quad \text{or} \quad Q\cdot D=P.$$

Factoring polynomials will be discussed in detail in Sections 6.2, 6.3, and 6.4.

Example 4: Long Division (Terms Missing)

Simplify $\dfrac{x^4+9x^2-3x+5}{x^2-x+2}$ using long division.

Solution: Note that 0 is written as a placeholder for any missing powers of the variable. In this way, like terms are easily aligned vertically.

$$\begin{array}{r} x^2+\ x+\ 8 \\ x^2-x+2\overline{)x^4+0x^3+9x^2-3x+5} \\ \underline{-(x^4-\ x^3+2x^2)} \\ x^3+7x^2-3x \\ \underline{-(x^3-\ x^2+2x)} \\ 8x^2-5x+5 \\ \underline{-(8x^2-8x+16)} \\ 3x-11 \end{array}$$

Note that the remainder is of smaller degree than the divisor.

Thus the quotient is x^2+x+8 and the remainder is $3x-11$.

In the form $Q+\dfrac{R}{D}$ we can write $x^2+x+8+\dfrac{3x-11}{x^2-x+2}$.

Practice Problems

1. Express the quotient as a sum of fractions in simplified form: $\dfrac{8x^2+6x+1}{2x}$.

Use the division algorithm to divide. Write the answer in the form $Q+\dfrac{R}{D}$.

2. $\dfrac{3x^2-8x+5}{x+2}$

3. $(x^3+4x^2-10)\div(x^2+x-1)$

Answers to Practice Problems: 1. $4x+3+\dfrac{1}{2x}$ 2. $3x-14+\dfrac{33}{x+2}$ 3. $x+3+\dfrac{-2x-7}{x^2+x-1}$

5.7 Exercises

1. $y^2 - 2y + 3$
2. $3x^2 + 4x + 6$
3. $2x^2 - 3x + 1$
4. $7y^2 + 14y + 6$
5. $10x^3 - 11x^2 + x$
6. $x^4 + 2x^3 - 3$
7. $-4x^2 + 7x - \frac{5}{2}$
8. $6x^2 - 4x + \frac{7}{2}$
9. $y^3 - \frac{7}{2}y^2 - \frac{15}{2}y + 4$
10. $5y^3 - \frac{7}{2}y^2 + \frac{21}{4}y + \frac{21}{2}$
11. $x - 6 + \frac{4}{x+4}$
12. $x + 10 + \frac{5}{x-1}$
13. $3x - 4 - \frac{7}{2x-1}$
14. $2x + 2 - \frac{1}{5x+3}$
15. $3x + 4 + \frac{1}{7x-1}$
16. $5x + 2 - \frac{3}{3x-4}$
17. $x - 9$
18. $x - 7$
19. $x^2 - x - \frac{3}{x-8}$
20. $x^2 - 4x - \frac{5}{x-2}$
21. $4x^2 - 6x + 9 - \frac{17}{x+2}$
22. $3x^2 + 3x + 5 - \frac{10}{x+1}$
23. $x^2 + 7x + 55 + \frac{388}{x-7}$

Express each quotient as a sum (or difference) of fractions and simplify if possible.

1. $\dfrac{8y^3 - 16y^2 + 24y}{8y}$ **2.** $\dfrac{18x^4 + 24x^3 + 36x^2}{6x^2}$ **3.** $\dfrac{34x^5 - 51x^4 + 17x^3}{17x^3}$

4. $\dfrac{14y^4 + 28y^3 + 12y^2}{2y^2}$ **5.** $\dfrac{110x^4 - 121x^3 + 11x^2}{11x}$ **6.** $\dfrac{15x^7 + 30x^6 - 45x^3}{15x^3}$

7. $\dfrac{-56x^4 + 98x^3 - 35x^2}{14x^2}$ **8.** $\dfrac{108x^6 - 72x^5 + 63x^4}{18x^4}$

9. $\dfrac{16y^6 - 56y^5 - 120y^4 + 64y^3}{16y^3}$ **10.** $\dfrac{20y^5 - 14y^4 + 21y^3 + 42y^2}{4y^2}$

Divide by using the division algorithm. Write the answers in the form $Q + \frac{R}{D}$*, where the degree of R < the degree of D.*

11. $\dfrac{x^2 - 2x - 20}{x+4}$ **12.** $\dfrac{x^2 + 9x - 5}{x-1}$ **13.** $\dfrac{6x^2 - 11x - 3}{2x-1}$

14. $\dfrac{10x^2 + 16x + 5}{5x+3}$ **15.** $\dfrac{21x^2 + 25x - 3}{7x-1}$ **16.** $\dfrac{15x^2 - 14x - 11}{3x-4}$

17. $\dfrac{x^2 - 12x + 27}{x-3}$ **18.** $\dfrac{x^2 - 12x + 35}{x-5}$ **19.** $\dfrac{x^3 - 9x^2 + 8x - 3}{x-8}$

20. $\dfrac{x^3 - 6x^2 + 8x - 5}{x-2}$ **21.** $\dfrac{4x^3 + 2x^2 - 3x + 1}{x+2}$ **22.** $\dfrac{3x^3 + 6x^2 + 8x - 5}{x+1}$

23. $\dfrac{x^3 + 6x + 3}{x-7}$ **24.** $\dfrac{2x^3 + 3x - 2}{x-1}$ **25.** $\dfrac{2x^3 - 5x^2 + 6}{x+2}$

26. $\dfrac{4x^3 - x^2 + 13}{x-1}$ **27.** $\dfrac{21x^3 + 41x^2 + 13x + 5}{3x+5}$ **28.** $\dfrac{6x^3 - 7x^2 + 14x - 8}{3x-2}$

29. $\dfrac{2x^3 + 7x^2 + 10x - 6}{2x+3}$ **30.** $\dfrac{6x^3 - 4x^2 + 5x - 7}{x-2}$ **31.** $\dfrac{x^3 - x^2 - 10x - 10}{x-4}$

32. $\dfrac{2x^3 - 3x^2 + 7x + 4}{2x-1}$ **33.** $\dfrac{10x^3 + 11x^2 - 12x + 9}{5x+3}$ **34.** $\dfrac{6x^3 + 19x^2 - 3x - 7}{6x+1}$

24. $2x^2+2x+5+\frac{3}{x-1}$
25. $2x^2-9x+18-\frac{30}{x+2}$
26. $4x^2+3x+3+\frac{16}{x-1}$
27. $7x^2+2x+1$
28. $2x^2-x+4$
29. $x^2+2x+2-\frac{12}{2x+3}$
30. $6x^2+8x+21+\frac{35}{x-2}$
31. $x^2+3x+2-\frac{2}{x-4}$
32. $x^2-x+3+\frac{7}{2x-1}$
33. $2x^2+x-3+\frac{18}{5x+3}$
34. $x^2+3x-1-\frac{6}{6x+1}$
35. $2x^2-8x+25-\frac{98}{x+4}$
36. $2x^2-2x+6-\frac{27}{x+3}$
37. $3x^2+2x-5-\frac{1}{3x-2}$

35. $\frac{2x^3-7x+2}{x+4}$ **36.** $\frac{2x^3+4x^2-9}{x+3}$ **37.** $\frac{9x^3-19x+9}{3x-2}$

38. $\frac{4x^3-8x^2-9x}{2x-3}$ **39.** $\frac{6x^3+11x^2+25}{2x+5}$ **40.** $\frac{16x^3+7x+12}{4x+3}$

41. $\frac{x^4-3x^3+2x^2-x+2}{x-3}$ **42.** $\frac{x^4+x^3-4x^2+x-3}{x+6}$ **43.** $\frac{x^4+2x^2-3x+5}{x-2}$

44. $\frac{3x^4+2x^3-2x^2-1}{x+1}$ **45.** $\frac{x^4-x^2+3}{x-\frac{1}{2}}$ **46.** $\frac{x^3+2x^2+1}{x-\frac{2}{3}}$

47. $\frac{3x^3+5x^2+7x+9}{x^2+2}$ **48.** $\frac{2x^4+2x^3+3x^2+6x-1}{2x^2+3}$ **49.** $\frac{x^4+x^3-4x+1}{x^2+4}$

50. $\frac{2x^4+x^3-8x^2+3x-2}{x^2-5}$ **51.** $\frac{6x^3+5x^2-8x+3}{3x^2-2x-1}$ **52.** $\frac{x^3-9x^2+20x-38}{x^2-3x+5}$

53. $\frac{3x^4-7x^3+5x^2+x-2}{x^2+x+1}$ **54.** $\frac{2x^4-x^3-10x^2-3x-1}{x^2-3x+1}$ **55.** $\frac{x^4+3x-7}{x^2+2x-3}$

56. $\frac{3x^4-4x^2+3}{x^2+x-1}$ **57.** $\frac{x^3-27}{x-3}$ **58.** $\frac{x^3+125}{x+5}$

59. $\frac{x^6-1}{x+1}$ **60.** $\frac{x^6-1}{x-1}$ **61.** $\frac{x^5+1}{x-1}$

62. $\frac{x^6+1}{x+1}$ **63.** $\frac{x^5-x^3+x}{x+\frac{1}{2}}$ **64.** $\frac{x^4-2x^3+4}{x+\frac{4}{5}}$

Writing and Thinking About Mathematics

38. $2x^2-x-6-\frac{18}{2x-3}$
39. $3x^2-2x+5$
40. $4x^2-3x+4$
41. $x^3+2x+5+\frac{17}{x-3}$
42. $x^3-5x^2+26x-155+\frac{927}{x+6}$

65. Suppose that a polynomial is divided by $(3x-2)$ and the answer is given as $x^2+2x+4+\frac{20}{3x-2}$. What was the original polynomial? Explain how you arrived at this conclusion.

66. Suppose that a polynomial is divided by $(x+5)$ and the answer is given as $x^2-3x+2-\frac{6}{x+5}$. What was the original polynomial? Explain how you arrived at this conclusion.

Writing and Thinking About Mathematics (cont.)

67. Given that $P(x) = 2x^3 - 8x^2 + 10x + 15$.

a. Find $P(2)$ then divide $P(x)$ by $x - 2$.

b. Find $P(-1)$ then divide $P(x)$ by $x + 1$.

c. Find $P(4)$ then divide $P(x)$ by $x - 4$.

Do you see any pattern in the values of $P(a)$ for $x = a$ and the remainders you found in the division process? (**Hint:** Check the appendix section on Synthetic Division and the Remainder Theorem.)

HAWKES LEARNING SYSTEMS: INTRODUCTORY & INTERMEDIATE ALGEBRA SOFTWARE

- 5.7a Division by a Monomial
- 5.7b The Division Algorithm

43. $x^3 + 2x^2 + 6x + 9 + \frac{23}{x-2}$

44. $3x^3 - x^2 - x + 1 - \frac{2}{x+1}$

45. $x^3 + \frac{1}{2}x^2 - \frac{3}{4}x - \frac{3}{8} + \frac{45}{16\left(x - \frac{1}{2}\right)}$

46. $x^2 + \frac{8}{3}x + \frac{16}{9} + \frac{59}{27\left(x - \frac{2}{3}\right)}$ **47.** $3x + 5 + \frac{x-1}{x^2+2}$ **48.** $x^2 + x + \frac{3x-1}{2x^2+3}$ **49.** $x^2 + x - 4 + \frac{-8x+17}{x^2+4}$

50. $2x^2 + x + 2 + \frac{8x+8}{x^2-5}$ **51.** $2x + 3 + \frac{6}{3x^2-2x-1}$ **52.** $x - 6 + \frac{-3x-8}{x^2-3x+5}$ **53.** $3x^2 - 10x + 12 + \frac{-x-14}{x^2+x+1}$

54. $2x^2 + 5x + 3 + \frac{x-4}{x^2-3x+1}$ **55.** $x^2 - 2x + 7 + \frac{-17x+14}{x^2+2x-3}$ **56.** $3x^2 - 3x + 2 + \frac{-5x+5}{x^2+x-1}$ **57.** $x^2 + 3x + 9$

58. $x^2 - 5x + 25$ **59.** $x^5 - x^4 + x^3 - x^2 + x - 1$ **60.** $x^5 + x^4 + x^3 + x^2 + x + 1$ **61.** $x^4 + x^3 + x^2 + x + 1 + \frac{2}{x-1}$

62. $x^5 - x^4 + x^3 - x^2 + x - 1 + \frac{2}{x+1}$ **63.** $x^4 - \frac{1}{2}x^3 - \frac{3}{4}x^2 + \frac{3}{8}x + \frac{13}{16} - \frac{13}{32\left(x + \frac{1}{2}\right)}$

64. $x^3 - \frac{14}{5}x^2 + \frac{56}{25}x - \frac{224}{125} + \frac{3396}{625\left(x + \frac{4}{5}\right)}$ **65.** $3x^3 + 4x^2 + 8x + 12$ **66.** $x^3 + 2x^2 - 13x + 4$

67. a. 19; $2x^2 - 4x + 2 + \frac{19}{x-2}$ **b.** −5; $2x^2 - 10x + 20 - \frac{5}{x+1}$ **c.** 55; $2x^2 + 10 + \frac{55}{x-4}$ Answers will vary.

Chapter 5 Index of Key Ideas and Terms

Section 5.1 Exponents

Section 5.2 Exponents and Scientific Notation

Section 5.3 Introduction to Polynomials

Section 5.3 Introduction to Polynomials (cont.)

Polynomial page 381

A **polynomial** is a monomial or the indicated sum or difference of monomials. The **degree of a polynomial** is the largest of the degrees of its terms. The coefficient of the term of the largest degree is called the **leading coefficient**.

Special Terminology for Some Polynomials page 381

Monomial: polynomial with one term
Binomial: polynomial with two terms
Trinomial: polynomial with three terms

Evaluation of Polynomials pages 382-383

A polynomial can be treated as a function and the notation $p(x)$ can be used.

Section 5.4 Addition and Subtraction with Polynomials

Addition with Polynomials page 387

Subtraction with Polynomials page 388

Simplifying Algebraic Expressions page 389

Section 5.5 Multiplication with Polynomials

Multiplying a Polynomial by a Monomial page 394

Multiplying Two Polynomials pages 394-395

Section 5.6 Special Products of Binomials

The FOIL Method page 400

Multiply **F**irst Terms, **O**utside Terms, **I**nside Terms, **L**ast Terms

The Difference of Two Squares page 401

$$(x+a)(x-a)=x^2-a^2$$

Continued on the next page...

Section 5.6 Special Products of Binomials (cont.)

Square of Binomials pages 402-403

$$(x+a)^2 = (x+a)(x+a) = x^2 + 2ax + a^2$$

$$(x-a)^2 = (x-a)(x-a) = x^2 - 2ax + a^2$$

Section 5.7 Division with Polynomials

Dividing by a Monomial page 408

The Division Algorithm page 409-410

For polynomials P and D, the division algorithm gives

$$\frac{P}{D} = Q + \frac{R}{D}, \quad D \neq 0$$

where Q and R are polynomials and the **degree of R < degree of D**.

HAWKES LEARNING SYSTEMS: INTRODUCTORY & INTERMEDIATE ALGEBRA SOFTWARE

- 5.1 Simplifying Integer Exponents I
- 5.2a Simplifying Integer Exponents II
- 5.2b Scientific Notation
- 5.3 Identifying and Evaluating Polynomials
- 5.4 Adding and Subtracting Polynomials
- 5.5 Multiplying Polynomials
- 5.6a The FOIL Method
- 5.6b Special Products
- 5.7a Division by a Monomial
- 5.7b The Division Algorithm

Chapter 5 Review

1. 125
2. 64
3. $-\frac{3}{8}$
4. $\frac{5}{9}$
5. $\frac{1}{y^3}$
6. x^7
7. $\frac{1}{y^2}$
8. $\frac{4}{x}$
9. $-12x^4$
10. $-6x^4$
11. 1
12. $-6a^4b^4c^2$
13. $\frac{3a^7}{b^3}$
14. $\frac{5}{xy}$
15. $-8x^{12}$
16. $\frac{17}{x^6}$
17. $\frac{x}{y}$
18. $9x^2y^4$
19. $\frac{9x^6}{y^4}$
20. m^6n^9
21. $-10x^4y^6$
22. $\frac{5}{x^3}$
23. $\frac{4}{27a^{14}b^3}$
24. $\frac{9}{a^4b^6}$
25. 2.93×10^7
26. 7.5×10^{-3}
27. 0.000724
28. $94{,}850{,}000$
29. $\frac{5.8\times10^{-3}}{2.9\times10^2}$; 2×10^{-5}
30. $\frac{2.7\times10^0\cdot2\times10^{-3}\cdot2.5\times10^1}{5.4\times10^1\cdot5\times10^{-4}}$; 5.0×10^0
31. $\frac{1.5\times10^{-3}\cdot8.22\times10^2}{4.11\times10^3\cdot3\times10^2}$; 1×10^{-6}

5.1 Exponents

Simplify each expression using the rules for exponents. The final form of the expressions with variables should contain only positive exponents. Assume that all variables represent nonzero numbers.

1. $5^2\cdot5$ **2.** $4^3\cdot4^0$ **3.** $-3\cdot2^{-3}$ **4.** $5\cdot3^{-2}$

5. y^{-3} **6.** $x^3\cdot x^4$ **7.** $y^{-2}\cdot y^0$ **8.** $4x^{-1}$

9. $(6x^2)(-2x^2)$ **10.** $\frac{-12x^7}{2x^3}$ **11.** $(-3a^3)^0$

12. $(3a^3b^2c)(-2ab^2c)$ **13.** $\frac{36a^6b^0}{12a^{-1}b^3}$ **14.** $\frac{35y^5\cdot2y^{-2}}{14xy^4}$

5.2 Exponents and Scientific Notation

Use the rules for exponents to simplify each of the expressions. Assume that all variables represent nonzero real numbers.

15. $(-2x^4)^3$ **16.** $17(x^{-3})^2$ **17.** $\left(\frac{y}{x}\right)^{-1}$ **18.** $\left(\frac{3x}{y^{-2}}\right)^2$

19. $\left(\frac{3x^3y}{y^3}\right)^2$ **20.** $\left(\frac{mn^{-1}}{m^3n^2}\right)^{-3}$ **21.** $\frac{(5x^{-2})(-2x^6)}{(y^2)^{-3}}$ **22.** $\frac{(6x^{-2})(5x^{-1})}{(3y)(2y^{-1})}$

23. $\left(\frac{15a^3b^{-1}}{20ab^2}\right)^{-1}\left(\frac{-3a^2b}{a^{-4}b^{-2}}\right)^{-2}$ **24.** $\left(\frac{3a^4bc^{-1}}{7ab^{-1}c^2}\right)^{-2}\left(\frac{9abc^{-1}}{7b^2c^2}\right)^2$

Write the following numbers in scientific notation.

25. 29,300,000 **26.** 0.0075

Write the following numbers in decimal form.

27. 7.24×10^{-4} **28.** 9.485×10^7

First write each of the numbers in scientific notation. Then perform the indicated operations and leave your answer in scientific notation.

29. $\frac{0.0058}{290}$ **30.** $\frac{2.7\cdot0.002\cdot25}{54\cdot0.0005}$ **31.** $\frac{(1.5\times10^{-3})(822)}{(4.11\times10^3)(300)}$

32. 3.5×10^4 ft
33. Not a polynomial
34. Trinomial
35. Binomial
36. $6x$; First-degree monomial; Leading coefficent 6
37. $6x^2 - x$; Second-degree binomial; Leading coefficient 6
38. $-10a^3 + 4a^2 + 9a$; Third-degree trinomial; Leading coefficient -10
39. 5; Monomial of 0 degree; Leading coefficient 5
40. $8x^3 - 6x^2 - 5$; Third-degree trinomial; Leading coefficient 8
41. $4a^3 - 4a^2 - 2a + 9$; Third-degree polynomial; Leading coefficient 4
42. 19 43. 14
44. 154 45. 78
46. $a^2 + 6a + 9$
47. $8a - 9$
48. $4x^2 + 9x + 3$
49. $-x^2 + x + 3$
50. $2x^2 + 19x + 4$
51. $-5x^2 - 3x - 4$
52. $-2x^3 + 3x^2 + 3x + 7$
53. $2x^3 + 13x^2 - x - 7$

32. Altitude: A commercial airplane has a cruising speed of 555 mph at an altitude of 35,000 ft. Express the altitude of the plane in scientific notation.

5.3 Introduction to Polynomials

Identify the expression as a monomial, binomial, trinomial, or not a polynomial.

33. $2x^2 + 3y^{\frac{1}{2}} + 3$ **34.** $3 + 5y^3 + 20y$ **35.** $10y - \frac{4}{7}y^3$

Simplify the polynomials. Write the polynomial in descending order and state the degree and type of the simplified polynomial. For each polynomial, state the leading coefficient.

36. $x + 5x$ **37.** $5x^2 - x + x^2$

38. $3a + 4a^2 + 6a - 10a^3$ **39.** $6x^2 + 4 - 2x^2 + 1 - 4x^2$

40. $11x^3 - 6x^2 - 9x^3 - 5 + 6x^3$ **41.** $a^3 - 2a^2 - 8a + 3a^3 - 2a^2 + 6a + 9$

Evaluate the given polynomials as indicated.

42. Given $p(x) = x^2 - 13x + 5$, find $p(-1)$.

43. Given $p(x) = 3x^2 + 4x - 6$, find $p(2)$.

44. Given $p(y) = 3y^3 - 3y^2 + 10$, find $p(4)$.

45. Given $p(y) = y^2 - 8y + 13$, find $p(-5)$.

46. Given $p(x) = x^2 + 6x + 9$, find $p(a)$.

47. Given $p(x) = 2x - 7$, find $p(4a - 1)$.

5.4 Addition and Subtraction with Polynomials

Find the indicated sum.

48. $(3x^2 + 6x - 1) + (x^2 + 3x + 4)$ **49.** $(-5x^2 + 2x - 3) + (4x^2 - x + 6)$

50. $(x^2 + 7x + 12) + (2x - 8) + (x^2 + 10x)$

51. $(-8x^2 + 5x + 2) + (3x^2 - 9x + 5) + (x - 11)$

52.
$$\begin{array}{rrrr} x^3 & +4x^2 & +x & \\ \underline{-3x^3} & \underline{-x^2} & \underline{+2x} & \underline{+7} \end{array}$$

53.
$$\begin{array}{rrrr} x^3 & +4x^2 & & -6 \\ & 8x^2 & +3x & -1 \\ \underline{x^3} & \underline{+x^2} & \underline{-4x} & \end{array}$$

54. $2x^2 - x - 10$
55. $2x^4 + 7x^3 - 6x^2 + 10$
56. $6x^2 - x$
57. $3x^2 + 2x - 2$
58. $4x^2 - 7x + 2$
59. $10x^3 - 5x^2 + 20x + 2$
60. $5x - 18$
61. $-x - 13$
62. $5x^2 - x + 4$
63. $17x - 91$
64. $5x^2 - 8x - 1$
65. $-x^2 - 11x + 12$
66. $-6x^5 - 8x^3$
67. $-35a^5 + 21a^4 - 7a^3$
68. $-6a^2 + 8a + 30$
69. $x^3 + 3x^2 - 6x - 8$
70. $x^2 - 5x - 6$
71. $7t^2 - 11t - 6$
72. $9a^2 - 18a + 9$
73. $4x^2 + 19x - 5$
74. $-4x^2 + 8x + 60$
75. $2y^3 - 5y^2 + y + 2$
76. $3y^3 + 4y^2 - 29y - 8$
77. $x^4 + 6x^3 + 12x^2 + 9x - 10$
78. $4x^2 - 14x - 3$
79. $7x + 46$

Find the indicated difference.

54. $(3x^2 + 4x - 4) - (x^2 + 5x + 6)$

55. $(x^4 + 7x^3 - x + 13) - (-x^4 + 6x^2 - x + 3)$

56. $(5x^2 - x - 10) - (-x^2 - 10)$

57. $(7x^2 - 14) - (4x^2 - 2x - 12)$

58.
$$\begin{array}{r} 13x^2 - 6x + 7 \\ \underline{-(9x^2 + x + 5)} \end{array}$$

59.
$$\begin{array}{r} 11x^3 + 10x - 20 \\ \underline{-(x^3 + 5x^2 - 10x - 22)} \end{array}$$

Simplify each of the algebraic expressions and write the polynomials in descending order.

60. $6x + 2(x - 7) - (3x + 4)$

61. $3x + [8x - 4(3x + 1) - 9]$

62. $(x^2 + 4) - [-6 + 2(3 - 2x^2) + x]$

63. $2[4x + 5(x - 8) - (2x + 3)] + (3x - 5)$

64. Subtract $3x^2 - 2x + 4$ from $8x^2 - 10x + 3$.

65. Subtract $4x^2 - 8x + 1$ from the sum of $2x^2 - 5x + 2$ and $x^2 - 14x + 11$.

5.5 Multiplication with Polynomials

Multiply as indicated and simplify if possible.

66. $-2x^2(3x^3 + 4x)$

67. $7a^3(-5a^2 + 3a - 1)$

68. $3a(-2a + 6) + 5(-2a + 6)$

69. $x(x^2 + 5x + 4) - 2(x^2 + 5x + 4)$

70. $(x - 6)(x + 1)$

71. $(7t + 3)(t - 2)$

72. $(3a - 3)(3a - 3)$

73. $(4x - 1)(x + 5)$

74. $-4(x + 3)(x - 5)$

75. $(2y + 1)(y^2 - 3y + 2)$

76.
$$\begin{array}{r} y^2 + 4y + 1 \\ \underline{3y - 8} \end{array}$$

77.
$$\begin{array}{r} x^2 + 3x + 5 \\ \underline{x^2 + 3x - 2} \end{array}$$

Simplify.

78. $(x + 4)(x - 2) + (3x - 1)(x - 5)$

79. $(x + 2)(x + 5) - (x - 6)(x + 6)$

5.6 Special Products of Binomials

Find the product and identify those that are the difference of two squares or perfect square trinomials.

80. $(2x+3)(x+1)$ **81.** $(2x+9)^2$ **82.** $(x+13)(x-13)$

83. $(y+8)(y-8)$ **84.** $(3x+5)(2x+5)$ **85.** $(3x-1)^2$

86. $(y^2+6)(y^2-4)$ **87.** $(3x-7)(3x+7)$ **88.** $(7-3x)^2$

89. $(x^3-10)(x^3+10)$ **90.** $\left(t+\frac{1}{4}\right)\left(t-\frac{1}{4}\right)$ **91.** $\left(y+\frac{2}{3}\right)\left(y-\frac{1}{2}\right)$

5.7 Division with Polynomials

Express each quotient as a sum (or difference) of fractions and simplify if possible.

92. $\dfrac{4y^2+16y+20}{2y^2}$ **93.** $\dfrac{90x^6+72x^5-108x^3}{18x^4}$

94. $\dfrac{2x^4+8x^3-12x^2+4x}{4x}$ **95.** $\dfrac{12y^4+6y^3-3y^2+11y}{3y^2}$

Divide by using the division algorithm. Write the answers in the form $Q+\frac{R}{D}$, where the degree of $R <$ the degree of D.

96. $\dfrac{x^2-2x-15}{x-5}$ **97.** $\dfrac{8y^2+10y+5}{y+3}$ **98.** $\dfrac{x^3+9x^2+20x}{x+5}$

99. $\dfrac{4x^3-20x+3}{x-2}$ **100.** $\dfrac{64x^3-125}{4x-5}$ **101.** $\dfrac{8x^3+27}{2x+3}$

102. $\dfrac{3x^3+4x^2-7x+1}{x^2+2}$ **103.** $\dfrac{10x^3+17x^2-2x+45}{2x+5}$ **104.** $\dfrac{x^3-8x^2+20x-50}{x^2-2x+3}$

105. $\dfrac{x^4-3x^3+4x^2-10x-6}{x-4}$

80. $2x^2+5x+3$
81. $4x^2+36x+81$; Perfect square trinomial
82. x^2-169; Difference of two squares
83. y^2-64; Difference of two squares
84. $6x^2+25x+25$
85. $9x^2-6x+1$; Perfect square trinomial
86. y^4+2y^2-24
87. $9x^2-49$; Difference of two squares
88. $9x^2-42x+49$; Perfect square trinomial
89. x^6-100; Difference of two squares
90. $t^2-\frac{1}{16}$; Difference of two squares
91. $y^2+\frac{1}{6}y-\frac{1}{3}$
92. $2+\frac{8}{y}+\frac{10}{y^2}$
93. $5x^2+4x-\frac{6}{x}$
94. $\frac{1}{2}x^3+2x^2-3x+1$
95. $4y^2+2y-1+\frac{11}{3y}$
96. $x+3$
97. $8y-14+\frac{47}{y+3}$
98. x^2+4x
99. $4x^2+8x-4-\frac{5}{x-2}$
100. $16x^2+20x+25$
101. $4x^2-6x+9$
102. $3x+4+\frac{-13x-7}{x^2+2}$
103. $5x^2-4x+9$
104. $x-6+\frac{5x-32}{x^2-2x+3}$
105. $x^3+x^2+8x+22+\frac{82}{x-4}$

Chapter 5 Test

1. $-10a^5$ 2. 1
3. $\frac{4x^3}{y^7}$ 4. $\frac{x}{3y^2}$
5. $\frac{x^2}{4y^2}$ 6. $4x^2y^4$
7. a. 135,000
b. 0.0000027
8. a. $2.5\times10^2\cdot5\times10^5$; 1.25×10^8
b. $\frac{6.5\times10^1\cdot1.2\times10^{-2}}{1.5\times10^3}$; 5.2×10^{-4}
9. $8x^2+3x$; Second-degree binomial; Leading coefficient 8
10. $-x^3+3x^2+3x-1$; Third-degree polynomial; Leading coefficient -1
11. $5x^5+2x^4-11x+3$; Fifth-degree polynomial; Leading coefficient 5
12. a. 20 b. -110
13. $-3x+2$
14. $20x-8$
15. $5x^3-x^2+6x+5$
16. $7x^4+14x^2+4$
17. $15x^7-20x^6+15x^5-40x^4-10x^2$
18. $49x^2-9$; Difference of two squares
19. $16x^2+8x+1$; Perfect square trinomial

Use the rules for exponents to simplify each expression. Each answer should have only positive exponents. Assume that all variables represent nonzero numbers.

1. $(5a^2b^5)(-2a^3b^{-5})$ **2.** $(-7x^4y^{-3})^0$ **3.** $\frac{(-8x^2y^{-3})^2}{16xy}$

4. $\frac{3x^{-2}}{9x^{-3}y^2}$ **5.** $\left(\frac{4xy^2}{x^3}\right)^{-1}$ **6.** $\left(\frac{2x^0y^3}{x^{-1}y}\right)^2$

7. Write each of the following numbers in decimal notation.

a. 1.35×10^5 **b.** 2.7×10^{-6}

8. First write each of the numbers in scientific notation. Then perform the indicated operations and leave your answer in scientific notation.

a. $250\cdot500{,}000$ **b.** $\frac{65\cdot0.012}{1500}$

Simplify the polynomials. Write the polynomial in descending order and state the degree and type of the simplified polynomial. For each polynomial, state the leading coefficient.

9. $3x+4x^2-x^3+4x^2+x^3$ **10.** $2x^2+3x-x^3+x^2-1$

11. $17-14+4x^5-3x+2x^4+x^5-8x$

12. For the polynomial $P(x)=2x^3-5x^2+7x+10$, find **a.** $P(2)$ and **b.** $P(-3)$.

Simplify the following expressions.

13. $7x+[2x-3(4x+1)+5]$ **14.** $12x-2[5-(7x+1)+3x]$

Perform the indicated operations and simplify each expression. Tell which, if any, answers are the difference of two squares and which, if any, are perfect square trinomials.

15. $(5x^3-2x+7)+(-x^2+8x-2)$ **16.** $(x^4+3x^2+9)-(-6x^4-11x^2+5)$

17. $5x^2(3x^5-4x^4+3x^3-8x^2-2)$ **18.** $(7x+3)(7x-3)$

19. $(4x+1)^2$ **20.** $(6x-5)(6x-5)$

21. $(2x+5)(6x-3)$ **22.** $3x(x-7)(2x-9)$

20. $36x^2 - 60x + 25$; Perfect square trinomial
21. $12x^2 + 24x - 15$
22. $6x^3 - 69x^2 + 189x$
23. $7x^2 + 7x + 14$
24. $10x^4 + 4x^3 - 15x^2 - 41x - 14$
25. $2x + \frac{3}{2} - \frac{3}{x}$
26. $\frac{5a}{3} + 2a^2 + \frac{1}{a}$
27. $x - 6 - \frac{2}{2x+3}$
28. $x - 9 + \frac{15x - 12}{x^2 + x - 3}$
29. $2x^2 - 3x - 13$
30. a. $A(x) = 240 - (x^2 + 3x) = -(x^2 + 3x - 240)$
b. $P(x) = 64$

23. $(3x+1)(3x-1)-(2x+3)(x-5)$

24. $\begin{array}{r} 2x^3 - 3x - 7 \\ \times \quad 5x + 2 \\ \hline \end{array}$

Express each quotient as a sum (or difference) of fractions and simplify if possible.

25. $\dfrac{4x^3 + 3x^2 - 6x}{2x^2}$

26. $\dfrac{5a^3 + 6a^4 + 3a}{3a^2}$

Divide by using the division algorithm. Write the answers in the form $Q + \frac{R}{D}$ where the degree of $R <$ the degree of D.

27. $(2x^2 - 9x - 20) \div (2x + 3)$

28. $\dfrac{x^3 - 8x^2 + 3x + 15}{x^2 + x - 3}$

29. Subtract $5x^2 - 3x + 4$ from the sum of $4x^2 + 2x + 1$ and $3x^2 - 8x - 10$.

30. Rectangles: A sheet of metal is in the shape of a rectangle with width 12 inches and length 20 inches. A slot (see the figure) of width x inches and length $x + 3$ inches is cut from the top of the rectangle.

a. Write a polynomial function $A(x)$ that represents the area of the remaining figure.

b. Write a polynomial function $P(x)$ that represents the perimeter of the remaining figure.

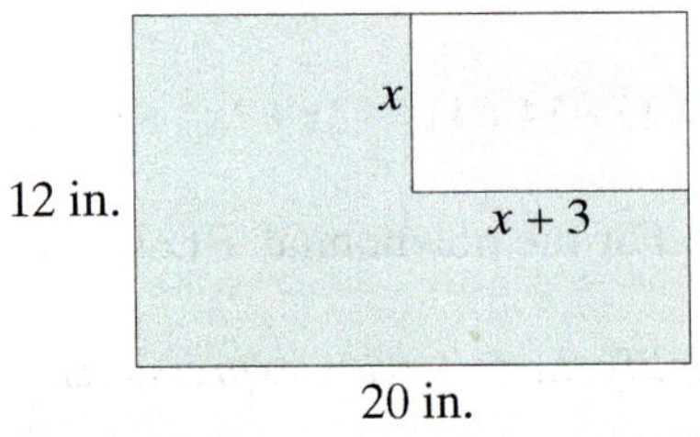

Cumulative Review: Chapters 1 – 5

1. $\frac{34}{9}$ 2. $-\frac{7}{22}$
3. $x+15$
4. $3x+8$
5. 540 6. $120a^2b^3$
7. $x=-3$
8. $x=\frac{8}{5}$
9. $x=-\frac{1}{2}$
10. $x=\frac{3}{8}$
11. $x=-3$
12. $x=-\frac{20}{7}, 4$
13. $x=\frac{y-b}{m}$
14. $y=\frac{10-3x}{5}$
15. 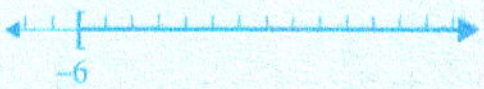

$[-6, \infty)$
16. 

$(-7, 4)$
17. 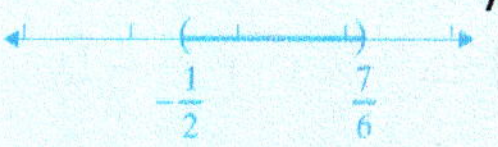

$\left(-\frac{1}{2}, \frac{7}{6}\right)$
18. 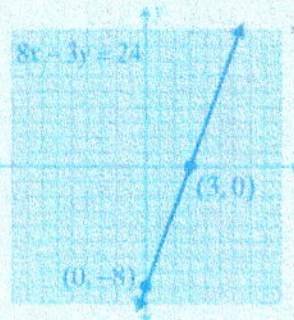

19. $-2x+3y=8$
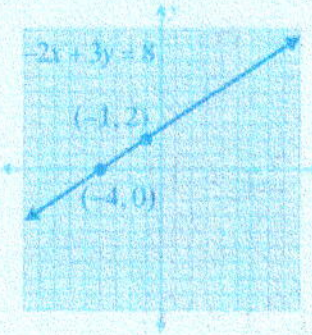

Use the rules for order of operations to evaluate the expressions.

1. $\frac{2}{3} \div \frac{1}{5} + \frac{4}{9}$ **2.** $\frac{8}{11} \div \left(-\frac{4}{5}\right) \cdot \frac{1}{2} + \frac{3}{22}$

Use the distributive property to complete the expressions.

3. $3x + 45 = 3(\qquad)$ **4.** $6x + 16 = 2(\qquad)$

Find the LCM for each set of numbers or terms.

5. 12, 15, 54 **6.** $6a^2, 24ab^3, 30ab, 40a^2b^2$

Solve the following equations.

7. $13x - 5x + 7 = 2 - 19$ **8.** $7(4 - x) = 3(x + 4)$

9. $-2(5x+1) + 2x = 4(x+1)$ **10.** $\frac{2}{3}x + \frac{1}{2} = \frac{3}{4}$

11. $1.5x - 3.7 = 3.6x + 2.6$ **12.** $|7x - 4| = 24$

Solve each formula for the indicated variable.

13. $y = mx + b$; solve for x. **14.** $3x + 5y = 10$; solve for y.

Solve the following inequalities and graph the solution sets. Write each solution in interval notation. Assume that x is a real number.

15. $5x + 3 \ge 2x - 15$ **16.** $-16 < 3x + 5 < 17$ **17.** $3|6y - 2| - 11 < 4$

18. Graph the linear equation $8x - 3y = 24$ by locating the x-intercept and the y-intercept.

19. Write an equation in standard form for the line that has slope $m = \frac{2}{3}$ and contains the point $(-1, 2)$. Graph the line.

20. Write an equation in slope-intercept form for the line passing through the two points $(-5, 1)$ and $(3, -4)$. Graph the line.

21. Write an equation in standard form for the line that is parallel to the line $x + 6y = 6$ and passes through the point $(0, 7)$. Graph both lines.

22. For the function $f(x) = 3x^2 - 4x + 2$, find **a.** $f(-2)$ and **b.** $f(5)$.

20. $y = -\frac{5}{8}x - \frac{17}{8}$

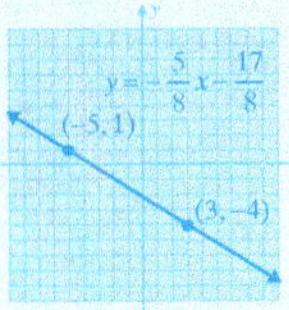

21. $x + 6y = 42$

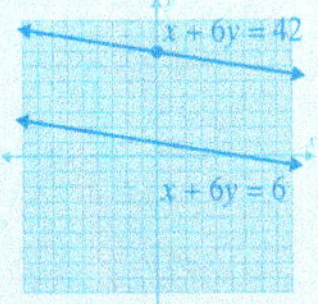

22. a. 22 b. 57

23. Open;

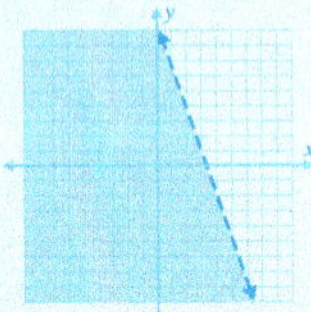

24. Closed;

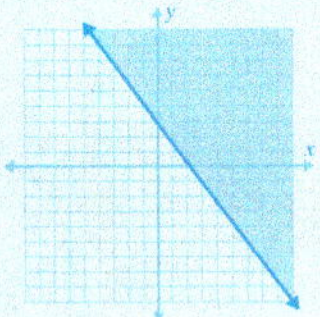

25. $2x$

26. $64x^6y^3$

27. $\frac{49x^{10}}{y^4}$

28. $\frac{36x^4}{y^{10}}$

29. $\frac{a^6}{b^4}$ 30. $\frac{x}{3y^4}$

31. 1 32. $\frac{x^8}{9y^6}$

33. a. 0.00000028
b. 35,100

34. $1.5\times10^{-3}\cdot4.2\times10^3$; 6.3×10^0

35. $\frac{8.4\times10^2}{2.1\times10^{-4}}$; 4×10^6

36. $\frac{5\times10^{-3}\times7.7\times10}{1.1\times10^{-2}\times3.5\times10^3}$; 1×10^{-2}

Graph the following inequalities and tell what type of half-plane is graphed.

23. $3x + y < 10$ **24.** $4x + 3y \geq 9$

Simplify each expression so that it has no exponents or only positive exponents.

25. $\frac{4x^3}{2x^{-2}x^4}$ **26.** $(4x^2y)^3$ **27.** $(7x^5y^{-2})^2$ **28.** $\left(\frac{6x^2}{y^5}\right)^2$

29. $(a^{-3}b^2)^{-2}$ **30.** $\left(\frac{3xy^4}{x^2}\right)^{-1}$ **31.** $\left(\frac{8x^{-3}y^2}{xy^{-1}}\right)^0$ **32.** $\left(\frac{3^{-1}x^3y^{-1}}{x^{-1}y^2}\right)^2$

33. Write each number in decimal notation.

a. 2.8×10^{-7} **b.** 3.51×10^4

First write each of the numbers in scientific notation. Then perform the indicated operations and leave your answer in scientific notation.

34. $0.0015\cdot4200$ **35.** $\frac{840}{0.00021}$ **36.** $\frac{0.005\cdot77}{0.011\cdot3500}$

For each of the following functions, **a.** *simplify each polynomial,* **b.** *find $P(3)$ and* **c.** *find $P(a)$.*

37. $P(x) = 9x - x^3 + 3x^2 - x + x^3$ **38.** $P(x) = 8x - 7x^2 + x^2 - 6x - x^3 - 4$

Simplify the polynomials. Write the polynomial in descending order and state the degree and type of the simplified polynomial. For each polynomial, state the leading coefficient.

39. $4x - [(6x+7) - (3x+4)] - 6$ **40.** $3x^2 - 8(x-5) + x^4 - 2x^3 + x^2 - 2x$

Perform the indicated operations and simplify each expression.

41. $(x^3 + 4x^2 - x) - (-2x^3 + 6x + 3)$ **42.** $(6x^2 + x - 10) - (x^3 - x^2 + x - 4)$

43. $(x^2 + 2x + 6) + (5x^2 - x - 2) - (8x + 3)$

44. $(2x^2 - 5x - 7) - (3x^2 - 4x + 1) + (x^2 - 9)$

45. $-3x(x^2 - 4x + 1)$ **46.** $5x^3(x^2 + 2x)$ **47.** $(x+6)(x-6)$

48. $(x+4)(x-3)$ **49.** $(3x+7)(3x+7)$ **50.** $(2x-1)(2x+1)$

51. $(x^2 - 2)^2$ **52.** $(2x-9)(x+4)$ **53.** $(3x-4)(2x+3)$

54. $(4x+1)(x^2 - x)$ **55.** $(y-5)(y^2 + 5y + 25)$ **56.** $(x+3y)(x^2 - 3xy + 9y^2)$

37. a. $3x^2+8x$
b. 51
c. $3a^2+8a$
38. a. $-x^3-6x^2+2x-4$
b. −79
c. $-a^3-6a^2+2a-4$
39. $x-9$; First-degree binomial; Leading coefficient 1
40. $x^4-2x^3+4x^2-10x+40$; Fourth-degree polynomial; Leading coefficient 1
41. $3x^3+4x^2-7x-3$
42. $-x^3+7x^2-6$
43. $6x^2-7x+1$
44. $-x-17$
45. $-3x^3+12x^2-3x$
46. $5x^5+10x^4$
47. x^2-36
48. x^2+x-12
49. $9x^2+42x+49$
50. $4x^2-1$
51. x^4-4x^2+4
52. $2x^2-x-36$
53. $6x^2+x-12$
54. $4x^3-3x^2-x$
55. y^3-125
56. x^3+27y^3
57. $4x-7+\frac{3}{x}$
58. $\frac{y^3}{7}-3y+\frac{1}{y}-4y^2$
59. $x-2$
60. $2x^2-x+3-\frac{2}{x+3}$
61. $2x^3-x^2-5x+8-\frac{48}{x^2+2x+3}$

Express each quotient as a sum (or difference) of fractions and simplify if possible.

57. $\dfrac{8x^2-14x+6}{2x}$

58. $\dfrac{y^5-21y^3+7y-28y^4}{7y^2}$

Divide by using the division algorithm and express each answer in the form $Q+\frac{R}{D}$, where the degree of $R <$ the degree of D.

59. $\dfrac{x^2+7x-18}{x+9}$

60. $\dfrac{2x^3+5x^2+7}{x+3}$

61. $\dfrac{2x^5+3x^4-x^3-5x^2+x-24}{x^2+2x+3}$

Use the vertical line test to determine whether or not each graph represents a function. State the domain and range.

62.

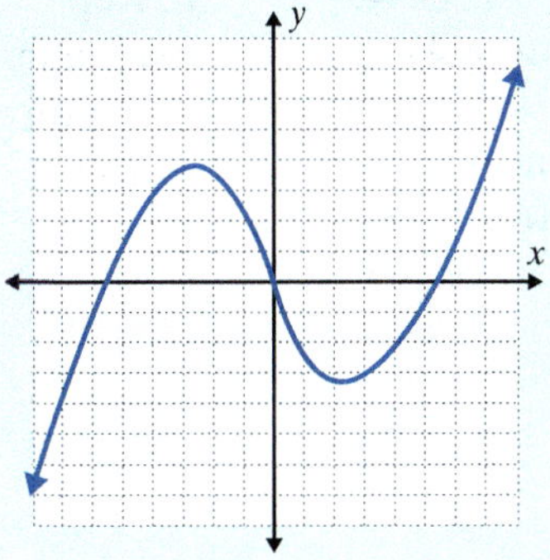

63.

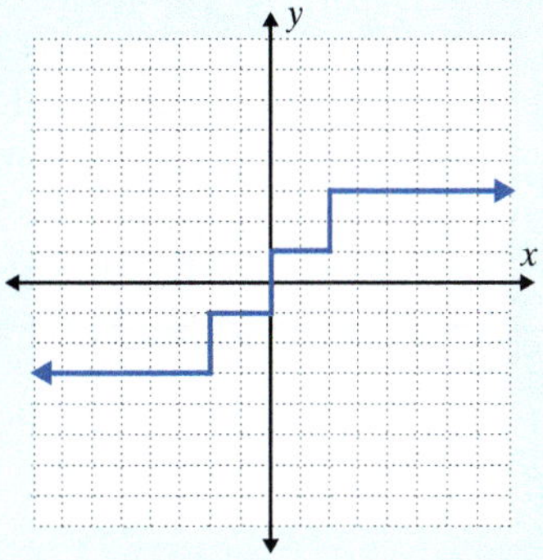

64. Find three consecutive even integers such that the sum of the second and third is equal to three times the first decreased by 14.

65. The length of a rectangle is 9 centimeters more than its width. The perimeter is 82 centimeters. Find the dimensions of the rectangle.

66. Twice the difference between a number and 16 is equal to 4 times the number increased by 6. What is the number?

67. Triangles: An isosceles triangle has two equal sides. The third side is 4 feet more than the length of the two equal sides. If each side is increased by 3 feet in length, the perimeter of the triangle will be 127 feet. What were the lengths of each side of the original triangle?

68. Investing: Sylvester has two investments that total $100,000. One investment earns interest at 6% and the other investment earns interest at 8%. If the total amount of interest from the two investments is $6700 in one year, how much money does he have invested at each rate?

69. Landscaping: Stan has decided to fence in his back yard using his house as the fourth side of a rectangular shape. He wants to use 300 feet of fencing (costs are a factor) and he wants the length to be 15 feet longer than the width of his house. What will be the dimensions of his fenced-in area?

62. Function;
$D = (-\infty, \infty)$;
$R = (-\infty, \infty)$

63. Not a function;
$D = (-\infty, \infty)$;
$R = [-3, 3]$

64. 20, 22, and 24

65. Length = 25 centimeters; Width = 16 centimeters

66. −19

67. 38 feet, 38 feet, 42 feet

68. \$35,000 at 8%; \$65,000 at 6%

69. Length = 105 ft; Width = 90 ft

70. \$337.50

71. 10 miles

72. a. $A(x) = 900 - 4x^2$
b. $P(x) = 4(30 - 2x) + 8x$
$= 120$

70. Farming: A farmer sells her award-winning triple-berry jam at the local farmer's market. She figures that it costs \$0.52 for the blueberries, \$0.65 for the blackberries, \$0.65 for the raspberries, and \$1.10 for other materials per jar. This week she brought 150 jars to the market but 9 jars broke on the way. At the end of the day, she had sold the remaining 141 jars. How much profit did she make if each jar sold for \$5.50?

71. Biking: Rosa went bike riding. She biked from her home to the beach at a rate of 10 mph and returned at a rate 5 mph faster. (She had a tailwind coming back.) If her return trip took 20 minutes less than the trip to the beach, how far is her home from the beach?

72. A square is 30 inches on each side. A square x inches on each side is cut from each corner of the square.

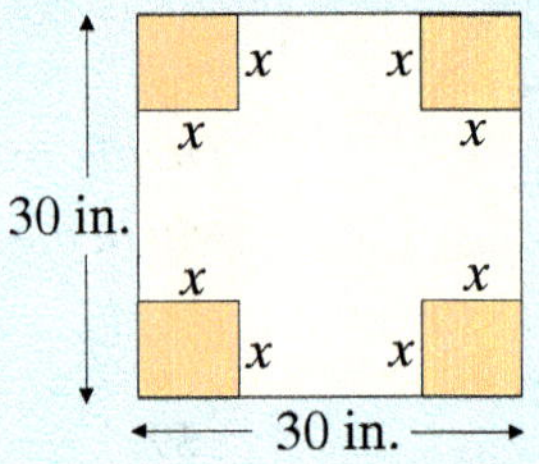

a. Represent the area of the remaining portion of the square in the form of a polynomial function, $A(x)$.

b. Represent the perimeter of the remaining portion of the square in the form of a polynomial function, $P(x)$.

CHAPTER 6

Factoring Polynomials and Solving Quadratic Equations

Did You Know?

You have noticed by now that almost every algebraic skill somehow relates to equation-solving and applied problems. The emphasis on equation-solving has always been a part of classical algebra.

In Italy, during the Renaissance, it was the custom for one mathematician to challenge another mathematician to an equation-solving contest. A large amount of money, often in gold, was supplied by patrons or sponsoring cities as the prize. At that time, it was important not to publish equation-solving methods, since mathematicians could earn large amounts of money if they could solve problems that their competitors could not. Equation-solving techniques were passed down from a mathematician to an apprentice, but they were never shared.

A Venetian mathematician, Niccolo Fontana (1500? – 1557) known as Tartaglia, "the stammerer," discovered how to solve third-degree or cubic equations. At that time, everyone could solve first- and second-degree equations and special kinds of equations of higher degree. Tartaglia easily won equation-solving contests simply by giving his opponents third-degree equations to solve.

Tartaglia planned to keep his method secret, but after receiving a pledge of secrecy, he gave his method to Girolamo Cardano (1501 – 1576). Cardano broke his promise by publishing one of the first successful Latin algebra texts, *Ars Magna*, "The Great Art." In it, he included not only Tartaglia's solution to the third-degree equations but also a pupil's (Ferrari) discovery of the general solution to fourth-degree equations. Until recently, Cardano received credit for discovering both methods.

It was not until 300 years later that it was shown that there are no general algebraic methods for solving fifth- or higher-degree equations. As you can see, a great deal of time and energy has gone into developing the methods of equation-solving that you are learning.

"Algebra is the intellectual instrument which has been created for rendering clear the quantitative aspects of the world."

Alfred North Whitehead (1861 – 1947)

Factoring is the reverse of multiplication. That is, to factor polynomials, you need to remember how you multiplied them. In this way, the concept of factoring is built on your previous knowledge and skills with multiplication. For example, if you are given a product, such as $x^2 - a^2$ (the difference of two squares), you must recall the factors as $(x+a)$ and $(x-a)$ from your work in multiplying polynomials in Chapter 5.

Studying mathematics is a building process with each topic dependent on previous topics with a few new ideas added each time. The equations and applications in Chapter 6 involve many of the concepts studied earlier, yet you will find them more interesting and more challenging.

6.1 Greatest Common Factor and Factoring by Grouping

- *Factor polynomials by finding the* ***greatest common factor****.*
- *Factor polynomials by* ***grouping****.*

The result of multiplication is called the **product** and the numbers or expressions being multiplied are called **factors** of the product. The reverse of multiplication is called **factoring**. That is, given a product, we want to find the factors.

Multiplying Polynomials	**Factoring Polynomials**
$3x(x+5) = 3x^2 + 15x$	$3x^2 + 15x = 3x(x+5)$
factors ↑ *product*	*product* ↑ *factors*

Factoring polynomials relies heavily on the multiplication techniques developed in Section 5.5. You must remember how to multiply in order to be able to factor. Furthermore, you will find that the skills used in factoring polynomials are necessary when simplifying rational expressions (Chapter 7) and when solving equations. In other words, study this section and Sections 6.2, 6.3, and 6.4 with extra care.

Teaching Notes:
You may want to carefully explain the distinction between finding the LCM and the GCF of a set of expressions.

Greatest Common Factor of a Set of Terms

The **greatest common factor (GCF)** of two or more integers is the largest integer that is a factor (or divisor) of all of the integers. For example, the GCF of 30 and 40 is 10. Note that 5 is also a common factor of 30 and 40, but 5 is not the **greatest** common factor. The number 10 is the largest number that will divide into both 30 and 40.

One way of finding the GCF is to use the prime factorization of each number. For example, to find the GCF for 36 and 60, we can write

$$36 = 4 \cdot 9 = 2 \cdot 2 \cdot 3 \cdot 3$$

$$60 = 4 \cdot 15 = 2 \cdot 2 \cdot 3 \cdot 5$$

The common factors are 2, 2, and 3 and their product is the GCF.

$$\text{GCF} = 2 \cdot 2 \cdot 3 = 12.$$

Writing the prime factorizations using exponents gives

$$36 = 2^2 \cdot 3^2 \quad \text{and} \quad 60 = 2^2 \cdot 3 \cdot 5.$$

We can see that the GCF is the product of the greatest power of each prime factor that is common to both numbers. That is,

$$\text{GCF} = 2^2 \cdot 3 = 12.$$

This procedure can be used to find the GCF for any set of integers or algebraic terms with integer exponents.

Procedure for Finding the GCF of a Set of Terms

1. Find the prime factorization of all integers and integer coefficients.
2. List all the factors that are common to all terms, including variables.
3. Choose the greatest power of each factor that is common to all terms.
4. Multiply these powers to find the GCF.

Note: If there is no common prime factor or variable, then the GCF is 1.

Example 1: Finding the GCF

Find the GCF for each of the following sets of algebraic terms.

a. $\{30, 45, 75\}$

Solution: Find the prime factorization of each number:

$$30 = 2 \cdot 3 \cdot 5, \; 45 = 3^2 \cdot 5, \; \text{and} \; 75 = 3 \cdot 5^2.$$

The common factors are 3 and 5 and the greatest power of each common to all numbers is 3^1 and 5^1.

Thus $\text{GCF} = 3^1 \cdot 5^1 = 15$.

Continued on the next page...

b. $\{20x^4y, 15x^3y, 10x^5y^2\}$

Solution: Writing each integer coefficient in prime factored form gives:

$$20x^4y = 2^2 \cdot 5 \cdot x^4 \cdot y,$$
$$15x^3y = 3 \cdot 5 \cdot x^3 \cdot y,$$
$$10x^5y^2 = 2 \cdot 5 \cdot x^5 \cdot y^2.$$

The common factors are $5, x,$ and y and after finding the greatest power of each common to all three terms, we have 5^1, x^3, and y^1.

Thus $\text{GCF} = 5^1 \cdot x^3 \cdot y^1 = 5x^3y.$

Factoring Out the Greatest Common Monomial Factor

Teaching Notes:
Emphasize the use of the distributive property as the springboard to factoring the common monomial factor from binomials and other polynomials. This will help students understand the need to factor out common binomial factors, as well.

Now consider the polynomial $3n + 15$. We want to write this polynomial as a product of two factors. Since

$$3n + 15 = 3 \cdot n + 3 \cdot 5$$

we see that 3 is a common factor of the two terms in the polynomial. By using the distributive property, we can write

$$3n + 15 = 3 \cdot n + 3 \cdot 5 = 3(n + 5).$$

In this way, the polynomial $3n + 15$ has been **factored** into the product of 3 and $(n + 5)$.

For a more general approach to finding factors of a polynomial, we can use the quotient rule for exponents.

$$\frac{a^m}{a^n} = a^{m-n}$$

This property is used when dividing terms. For example,

$$\frac{35x^8}{5x^2} = 7x^6 \quad \text{and} \quad \frac{16a^5}{-8a} = -2a^4.$$

To divide a polynomial by a monomial (a procedure that we discussed in Chapter 5) each term in the polynomial is divided by the monomial. For example,

$$\frac{8x^3 - 14x^2 + 10x}{2x} = \frac{8x^3}{2x} - \frac{14x^2}{2x} + \frac{10x}{2x} = 4x^2 - 7x + 5.$$

Note: With practice, this division can be done mentally.

Teaching Notes:
You may want to remind your students that they can check their factoring by multiplying the factors. While they may not have the greatest common factor, the product must be the expression being factored.

This concept of dividing each term by a monomial is part of finding a monomial factor of a polynomial. Finding the greatest common monomial factor (GCF) of a polynomial means to **find the GCF of the terms of the polynomial**. This monomial will be one factor, and the sum of the various quotients found by dividing each term by the GCF will be the other factor. Thus using the quotients just found and the fact that $2x$ is the GCF of the terms,

$$8x^3-14x^2+10x=2x\left(4x^2-7x+5\right).$$

We say that $2x$ is **factored out** and $2x$ and $\left(4x^2-7x+5\right)$ are the factors of $8x^3-14x^2+10x$.

Factoring Out the GCF

To find a monomial that is the greatest common factor (GCF) of a polynomial:

1. Find the variable(s) of highest degree and the largest integer coefficient that is a factor of each term of the polynomial. (This is one factor.)
2. Divide this monomial factor into each term of the polynomial, resulting in another polynomial factor.

Now factor $24x^6-12x^4-18x^3$. The GCF is $6x^3$ and is factored out as follows:

$$\begin{aligned}24x^6-12x^4-18x^3&=6x^3\cdot 4x^3+6x^3(-2x)+6x^3(-3)\\&=6x^3\left(4x^3-2x-3\right).\end{aligned}$$

The factoring can be checked by multiplying $6x^3$ and $4x^3-2x-3$. The product should be the expression we started with:

$$6x^3\left(4x^3-2x-3\right)=24x^6-12x^4-18x^3.$$

By definition, the GCF of a polynomial will have a positive coefficient. **However, if the leading coefficient is negative, we may choose to factor out the negative of the GCF (or $-1\cdot$GCF).** This technique will leave a positive coefficient for the first term of the other polynomial factor. For example,

the GCF for $-10a^4b+15a^4$ is $5a^4$ and we can factor as follows:

$$-10a^4b+15a^4=5a^4(-2b+3)$$

or

$$-10a^4b+15a^4=-5a^4(2b-3)$$

Both answers are correct.

Suppose we factor $8n + 16$ as follows.

$$8n + 16 = 4(2n + 4)$$

But note that $2n+4$ has 2 as a common factor. Therefore, we have not factored **completely**. **An expression is factored completely if none of its factors can be factored.**

Factoring polynomials means to find factors that are integers (other than +1) or polynomials with integer coefficients. If this cannot be done, we say that the polynomial is **not factorable**. For example, the polynomials $x^2 + 36$ and $3x + 17$ are not factorable.

Example 2: Factoring Out the GCF of a Polynomial

Factor each polynomial by factoring out the greatest common monomial factor.

a. $6n + 30$

Solution: $6n + 30 = 6 \cdot n + 6 \cdot 5 = 6(n + 5)$

Checking by multiplying gives $6(n+5) = 6n + 30$, the original expression.

b. $x^3 + x$

Solution: $x^3 + x = x \cdot x^2 + x(+1) = x(x^2 + 1)$ +1 is the coefficient of x.

c. $5x^3 - 15x^2$

Solution: $5x^3 - 15x^2 = 5x^2 \cdot x + 5x^2(-3) = 5x^2(x - 3)$

d. $2x^4 - 3x^2 + 2$

Solution: $2x^4 - 3x^2 + 2$

This polynomial has no common monomial factor other than 1. In fact, this polynomial is **not factorable**.

e. $-4a^5 + 2a^3 - 6a^2$

Solution: The GCF is $2a^2$ and we can factor as follows.

$$-4a^5 + 2a^3 - 6a^2 = 2a^2(-2a^3 + a - 3)$$

However, the leading coefficient is negative and we can also factor as follows.

$$-4a^5 + 2a^3 - 6a^2 = -2a^2(2a^3 - a + 3)$$

Both answers are correct. However, we will see later that having a positive leading coefficient for the polynomial in parentheses may make that polynomial easier to factor.

A polynomial may be in more than one variable. For example, $5x^2y+10xy^2$ is in the two variables x and y. Thus the GCF may have more than one variable.

$$5x^2y+10xy^2 = 5xy\cdot x+5xy\cdot 2y$$
$$= 5xy(x+2y)$$

Similarly,

$$4xy^3-2x^2y^2+8xy^2 = 2xy^2\cdot 2y+2xy^2(-x)+2xy^2\cdot 4$$
$$= 2xy^2(2y-x+4).$$

Example 3: Factoring out the GCF of a Polynomial

Factor each polynomial by finding the GCF (or $-1\cdot$GCF).

a. $4ax^3+4ax$

Solution: $4ax^3+4ax=4ax(x^2+1)$ Note that $4ax=1\cdot 4ax$.

Checking by multiplying gives $4ax(x^2+1)=4ax^3+4ax$, the original expression.

b. $3x^2y^2-6xy^2$

Solution: $3x^2y^2-6xy^2=3xy^2(x-2)$

c. $-14by^3-7b^2y+21by^2$

Solution: $-14by^3-7b^2y+21by^2=-7by(2y^2+b-3y)$ Note that by factoring out a negative term, in this case $-7by$, the leading coefficient in parentheses is positive.

d. $13a^4b^5-3a^2b^9-4a^6b^3$

Solution: $13a^4b^5-3a^2b^9-4a^6b^3=a^2b^3(13a^2b^2-3b^6-4a^4)$

Factoring by Grouping

Consider the expression

$$y(x+4)+2(x+4)$$

as the sum of two terms, $y(x+4)$ and $2(x+4)$. Each of these "terms" has the common **binomial** factor $(x+4)$. Factoring out this common binomial factor by using the distributive property gives

$$y(x+4)+2(x+4)=(x+4)(y+2).$$

Similarly,

$$3(x-2)-a(x-2)=(x-2)(3-a).$$

Now consider the product

$$\begin{aligned}(x+3)(y+5) &= (x+3)y+(x+3)5 \\ &= xy+3y+5x+15\end{aligned}$$

which has four terms and no like terms. Yet the product has two factors, namely $(x+3)$ and $(y+5)$. Factoring polynomials with four or more terms can sometimes be accomplished by grouping the terms and using the distributive property, as in the above discussion and the following examples. Keep in mind that the common factor can be a binomial or other polynomial.

Example 4: Factoring Out a Common Binomial Factor

Factor each polynomial.

a. $3x^2(5x+1)-2(5x+1)$

Solution: $3x^2(5x+1)-2(5x+1)=(5x+1)(3x^2-2)$

b. $7a(2x-3)+(2x-3)$

Solution: $7a(2x-3)+(2x-3)=7a(2x-3)+1\cdot(2x-3)$ 1 is the understood coefficient of $(2x-3)$.

$=(2x-3)(7a+1)$

In the examples just discussed, the common binomial factor was in parentheses. However, many times the expression to be factored is in a form with four or more terms. For example, by multiplying in Example 4a, we get the following expression.

$$3x^2(5x+1)-2(5x+1)=15x^3+3x^2-10x-2$$

The expression $15x^3+3x^2-10x-2$ has four terms with no common monomial factor; yet, we know that it has the two binomial factors $(5x+1)$ and $(3x^2-2)$. We can find the binomial factors by **grouping**. This means looking for common factors in each group and then looking for common binomial factors. The process is illustrated in Example 5.

Example 5: Factoring by Grouping

Factor each polynomial by grouping.

a. $xy+5x+3y+15$

Solution: $xy+5x+3y+15=(xy+5x)+(3y+15)$ Group terms that have a common monomial factor.

$=x(y+5)+3(y+5)$ Using the distributive property

$=(y+5)(x+3)$ $(y+5)$ is a common binomial factor.

Checking gives $(y+5)(x+3)=xy+5x+3y+15$.

Teaching Notes:
In Example 5a, you might want to show your students that another grouping will result in the same factors. For Example 5a we could have grouped as follows:
$xy+5x+3y+15$
$=(xy+3y)+(5x+15)$
$=y(x+3)+5(x+3)$
$=(x+3)(y+5).$

b. $x^2-xy-5x+5y$

Solution: $x^2-xy-5x+5y=(x^2-xy)+(-5x+5y)$

$$=x(x-y)+5(-x+y)$$

This does not work because $(x-y)\neq(-x+y)$. However, these two expressions are **opposites**. Thus we can find a common factor by factoring –5 instead of +5 from the last two terms.

$$x^2-xy-5x+5y=(x^2-xy)+(-5x+5y)$$
$$=x(x-y)-5(x-y)$$
$$=(x-y)(x-5) \quad \text{Success!}$$

c. $x^2+ax+3x+3y$

Solution: $x^2+ax+3x+3y=(x^2+ax)+(3x+3y)$

$$=x(x+a)+3(x+y)$$

But $x+a\neq x+y$ and there is no common factor.

So $x^2+ax+3x+3y$ is **not factorable**.

d. $xy+5x+y+5$

Solution: $xy+5x+y+5=(xy+5x)+(y+5)$

$$=x(y+5)+1\cdot(y+5)$$

1 is the understood coefficient of $(y+5)$.

$$=(y+5)(x+1)$$

e. $5xy+6uv-3vy-10ux$

Solution: In the expression $5xy+6uv-3vy-10ux$ there is no common factor in the first two terms. However, the first and third terms have a common factor so we rearrange the terms as follows.

$$5xy+6uv-3vy-10ux=(5xy-3vy)+(6uv-10ux)$$
$$=y(5x-3v)+2u(3v-5x)$$

Now we see that $5x-3v$ and $3v-5x$ are opposites and we factor out $-2u$ from the last two terms. The result is as follows.

$$5xy+6uv-3vy-10xu=(5xy-3vy)+(6uv-10ux)$$
$$=y(5x-3v)-2u(-3v+5x)$$
$$=y(5x-3v)-2u(5x-3v)$$
$$=(5x-3v)(y-2u)$$

Note that $5x-3v=-3v+5x$.

Practice Problems

Factor each expression completely.

1. $2x - 16$
2. $-5x^2 - 5x$
3. $7ax^2 - 7ax$
4. $a^4b^5 + 2a^3b^2 - a^3b^3$
5. $9x^2y^2 + 12x^2y - 6x^3$
6. $6a^3(x+3) - (x+3)$
7. $5x + 35 - xy - 7y$

6.1 Exercises

1. 5 2. 5
3. 8 4. 6
5. 1 6. 22
7. $10x^3$ 8. $5y$
9. $4a^2$ 10. $12xy$
11. $13ab$ 12. $14cd^2$
13. $15xy^2z^2$
14. $7a^3b^4c^2$
15. x^4 16. x^5
17. $-4y$ 18. $6x$
19. $3x^3$ 20. $-5x^4$
21. $2x^2y$ 22. $-7x^3y$
23. $m + 9$ 24. $x + 9$
25. $x - 6$ 26. $y - 4$
27. $b + 1$ 28. $2x - 1$
29. $3y + 4x + 1$
30. $3m^2 + m + 2$
31. $11(x-11)$
32. $7(2x+3)$
33. $4y(4y^2+3)$
34. $-3x(x-2)$
35. $-3a(2x-3y)$
36. $4a(x-2y)$
37. $5xy(2x-5)$
38. $2x^2y(8x^2-7)$

Find the GCF for each set of terms.

1. $\{10, 15, 20\}$
2. $\{25, 30, 75\}$
3. $\{16, 40, 56\}$
4. $\{30, 42, 54\}$
5. $\{9, 14, 22\}$
6. $\{44, 66, 88\}$
7. $\{30x^3, 40x^5\}$
8. $\{15y^4, 25y\}$
9. $\{8a^3, 16a^4, 20a^2\}$
10. $\{36xy, 48xy, 60xy\}$
11. $\{26ab^2, 39a^2b, 52a^2b^2\}$
12. $\{28c^2d^3, 14c^3d^2, 42cd^2\}$
13. $\{45x^2y^2z^2, 75xy^2z^3\}$
14. $\{21a^5b^4c^3, 28a^3b^4c^3, 35a^3b^4c^2\}$

Simplify the expressions.

15. $\dfrac{x^7}{x^3}$
16. $\dfrac{x^8}{x^3}$
17. $\dfrac{-8y^3}{2y^2}$
18. $\dfrac{12x^2}{2x}$
19. $\dfrac{9x^5}{3x^2}$
20. $\dfrac{-10x^5}{2x}$
21. $\dfrac{4x^3y^2}{2xy}$
22. $\dfrac{21x^4y^3}{-3xy^2}$

Complete the factoring of the polynomial as indicated.

23. $3m + 27 = 3(\quad)$
24. $2x + 18 = 2(\quad)$
25. $5x^2 - 30x = 5x(\quad)$
26. $6y^3 - 24y^2 = 6y^2(\quad)$
27. $13ab^2 + 13ab = 13ab(\quad)$
28. $8x^2y - 4xy = 4xy(\quad)$
29. $-15xy^2 - 20x^2y - 5xy = -5xy(\quad)$
30. $-9m^3 - 3m^2 - 6m = -3m(\quad)$

Answers to Practice Problems: **1.** $2(x-8)$ **2.** $-5x(x+1)$ **3.** $7ax(x-1)$ **4.** $a^3b^2(ab^3+2-b)$ **5.** $3x^2(3y^2+4y-2x)$ **6.** $(6a^3-1)(x+3)$ **7.** $(5-y)(x+7)$

39. $-2yz(9yz-1)$
40. $-14x^2y(y^2+1)$
41. $8(y^2-4y+1)$
42. $5(x^2-3x-1)$
43. $x(2y^2-3y-1)$
44. $a(d^2+10d+25)$
45. $4m^2(2x^3-3y+z)$
46. $3t^2x^2(12x^2-15x+8)$
47. $-7x^2z^3(8x^2+14xz+5z^2)$
48. $17x^3y^4(2xy^2-3y+x^2)$
49. $x^4y^2(15+24x^2y^4-32x^3y)$
50. $-3x^2y^3(y+2xy+3)$
51. $(y+3)(7y^2+2)$
52. $(a-7)(6a-5)$
53. $(x-4)(3x+1)$
54. $(x+5)(2x^2+1)$
55. $(x-2)(4x^3-1)$
56. $(x+1)(9a-1)$
57. $(2y+3)(10y-7)$
58. $(x+5)(a+b)$
59. $(x-2)(a-b)$
60. $(x-10)(3a+5b)$
61. $(b+c)(x+1)$
62. $(3+a)(x+y)$
63. $(x^2+6)(x+3)$
64. $(2z^2+3)(z-7)$
65. Not factorable
66. $(x+6y)(x-4)$
67. $(3-b)(x+y)$
68. $(a+3)(x+5y)$
69. $(y-4)(5x+z)$
70. $(x+2z)(1-3y)$
71. $(z^2+3)(a+1)$
72. Not factorable
73. $(6x+1)(a+2)$

Factor each of the polynomials by finding the GCF (or $-1 \cdot GCF$).

31. $11x-121$ **32.** $14x+21$ **33.** $16y^3+12y$

34. $-3x^2+6x$ **35.** $-6ax+9ay$ **36.** $4ax-8ay$

37. $10x^2y-25xy$ **38.** $16x^4y-14x^2y$ **39.** $-18y^2z^2+2yz$

40. $-14x^2y^3-14x^2y$ **41.** $8y^2-32y+8$ **42.** $5x^2-15x-5$

43. $2xy^2-3xy-x$ **44.** $ad^2+10ad+25a$

45. $8m^2x^3-12m^2y+4m^2z$ **46.** $36t^2x^4-45t^2x^3+24t^2x^2$

47. $-56x^4z^3-98x^3z^4-35x^2z^5$ **48.** $34x^4y^6-51x^3y^5+17x^5y^4$

49. $15x^4y^2+24x^6y^6-32x^7y^3$ **50.** $-3x^2y^4-6x^3y^4-9x^2y^3$

Factor each expression by factoring out the common binomial factor.

51. $7y^2(y+3)+2(y+3)$ **52.** $6a(a-7)-5(a-7)$ **53.** $3x(x-4)+(x-4)$

54. $2x^2(x+5)+(x+5)$ **55.** $4x^3(x-2)-(x-2)$ **56.** $9a(x+1)-(x+1)$

57. $10y(2y+3)-7(2y+3)$ **58.** $a(x+5)+b(x+5)$ **59.** $a(x-2)-b(x-2)$

60. $3a(x-10)+5b(x-10)$

Factor each of the polynomials by grouping. If a polynomial cannot be factored, write "not factorable."

61. $bx+b+cx+c$ **62.** $3x+3y+ax+ay$ **63.** $x^3+3x^2+6x+18$

64. $2z^3-14z^2+3z-21$ **65.** $10a^2-5az+2a+z$ **66.** $x^2-4x+6xy-24y$

67. $3x+3y-bx-by$ **68.** $ax+5ay+3x+15y$ **69.** $5xy+yz-20x-4z$

70. $x-3xy+2z-6zy$ **71.** z^2+3+az^2+3a **72.** x^2-5+x^2y+5y

73. $6ax+12x+a+2$ **74.** $4xy+3x-4y-3$ **75.** $xy+x+y+1$

76. $xy+x-y-1$ **77.** $10xy-2y^2+7yz-35xz$ **78.** $7xy-3y+2x^2-3x$

74. $(4y+3)(x-1)$
75. $(x+1)(y+1)$
76. $(x-1)(y+1)$

79. $3xy - 4uy - 6vx + 8uv$ **80.** $xy + 5vy + 6ux + 30uv$ **81.** $3ab + 4ac + 2b + 6c$

82. $24y - 3yz + 2xz - 16x$ **83.** $6ac - 9ad + 2bc - 3bd$ **84.** $2ac - 3bc + 6ad - 9bd$

Writing and Thinking About Mathematics

77. $(2y-7z)(5x-y)$
78. Not factorable

85. Explain why the GCF of $-3x^2 + 3$ is 3 and not -3.

HAWKES LEARNING SYSTEMS: INTRODUCTORY & INTERMEDIATE ALGEBRA SOFTWARE

- 6.1a Greatest Common Factor of Two or More Terms
- 6.1b Greatest Common Factor of a Polynomial
- 6.1c Factoring Expressions by Grouping

79. $(3x-4u)(y-2v)$ 80. $(x+5v)(y+6u)$ 81. Not factorable 82. $(2x-3y)(z-8)$ 83. $(2c-3d)(3a+b)$
84. $(2a-3b)(3d+c)$
85. Although both can be factored out of $-3x^2+3$, 3 is greater than -3 making it the greatest common factor.

6.2 Factoring Trinomials: $x^2 + bx + c$

- *Factor trinomials with leading coefficient 1 (of the form $x^2 + bx + c$).*
- *Factor out a common monomial factor and then factor the remaining factor, a trinomial with leading coefficient 1.*

In Chapter 5 we learned to use the FOIL method to multiply two binomials. In many cases the simplified form of the product was a trinomial. In this section we will learn to factor trinomials by reversing the FOIL method. In particular, we will focus on only factoring trinomials in one variable with leading coefficient 1.

Factoring Trinomials with Leading Coefficient 1

Teaching Notes:
You might want to show your students a rectangular format such as the following for finding the product $(x+5)(x+3)$. Some students will find this arrangement helpful in keeping track of all the terms.

	x	$+5$
x	x^2	$5x$
$+3$	$3x$	15

$= x^2 + 8x + 15$

Using the FOIL method to multiply $(x+5)$ and $(x+3)$, we find

F O I L

$$(x+5)(x+3) = x^2 + 3x + 5x + 3 \cdot 5 = x^2 + 8x + 15$$

$8x$ ↑ $3+5$ 15 ↑ $3 \cdot 5$

We see that the leading coefficient (coefficient of x^2) is 1, the coefficient 8 is the sum of 5 and 3, and the constant term 15 is the product of 5 and 3.

More generally,

$$(x+a)(x+b) = x^2 + ax + bx + ab$$
$$= x^2 + (a+b)x + ab$$

$(a+b)$ ↑ sum of constants a and b ab ↑ product of constants a and b

Now given a trinomial with leading coefficient 1, we want to find the binomial factors, if any. Reversing the relationship between a and b, as shown above, we can proceed as follows:

To factor a trinomial with leading coefficient 1, find two factors of the constant term whose sum is the coefficient of the middle term. (If these factors do not exist, the trinomial is **not factorable**.)

For example, to factor $x^2 + 11x + 30$, we need positive factors of +30 whose sum is +11:

Positive Factors of 30		Sums of These Factors
$1 \cdot 30$	→	$1 + 30 = 31$
$2 \cdot 15$	→	$2 + 15 = 17$
$3 \cdot 10$	→	$3 + 10 = 13$
$5 \cdot 6$	→	$5 + 6 = 11$

Now because $5 \cdot 6 = 30$ and $5 + 6 = 11$, we have

$$x^2 + 11x + 30 = (x + 5)(x + 6).$$

Example 1: Factoring Trinomials with Leading Coefficient 1

Factor the following trinomials.

a. $x^2 + 8x + 12$

Solution: 12 has three pairs of positive integer factors as illustrated in the following table. Of these 3 pairs, only 2 + 6 is equal to 8:

Factors of 12		
$1 \cdot 12$		
$2 \cdot 6$	→	$2 + 6 = 8$ Thus $x^2 + 8x + 12 = (x + 2)(x + 6)$.
$3 \cdot 4$		

Note: If the middle term had been $-8x$, then we would have wanted pairs of negative integer factors to find a sum of -8.

b. $y^2 - 8y - 20$

Solution: We want a pair of integer factors of -20 whose sum is -8. In this case, because the product is negative, one of the factors must be positive and the other negative.

Factors of –20		
$-1 \cdot 20$		
$1 \cdot -20$		
$-2 \cdot 10$		
$2 \cdot -10$	→	$2 + (-10) = -8$
$-4 \cdot 5$		
$4 \cdot -5$		

We have listed all the pairs of integer factors of –20. You can see that 2 and –10 are the only two whose sum is –8. Thus listing all the pairs is not necessary. This stage is called the **trial-and-error stage**. That is, you can **try** different pairs (mentally or by making a list) until you find the correct pair. If such a pair does not exist, the polynomial is **not factorable**.

In this case, we have

$$y^2 - 8y - 20 = (y+2)(y-10).$$

Note that by the commutative property of multiplication, the order of the factors does not matter. That is, we can also write

$$y^2 - 8y - 20 = (y-10)(y+2)$$

To Factor Trinomials of the Form $x^2 + bx + c$

To factor $\boldsymbol{x^2 + bx + c}$, if possible, find an integer pair of factors of c whose sum is b.

1. If c is positive, then both factors must have the same sign.

- **a.** Both will be positive if b is positive.
  Example: $x^2 + 5x + 4 = (x+4)(x+1)$
- **b.** Both will be negative if b is negative.
  Example: $x^2 - 5x + 4 = (x-4)(x-1)$

2. If c is negative, then one factor must be positive and the other negative.
Example: $x^2 + 6x - 7 = (x+7)(x-1)$ and $x^2 - 6x - 7 = (x-7)(x+1)$

Finding a Common Monomial Factor First

If a trinomial does not have a leading coefficient of 1, then we look for a common monomial factor. **If there is a common monomial factor, factor out this common monomial factor first** and then factor the remaining trinomial factor, if possible. (We will discuss factoring trinomials with leading coefficients other than 1 in the next section.) A polynomial is **completely factored** if none of its factors can be factored.

Example 2: Finding a Common Monomial Factor

Teaching Notes:
Reinforce the need for students not to overlook the common monomial factor, and that once they have found one, they need to maintain it. Often students "lose" it or discard it in the next step.

Completely factor the following trinomials by first factoring out the GCF in the form of a common monomial factor.

a. $5x^3 - 15x^2 + 10x$

Solution: First factor out the GCF, $5x$.

$$5x^3 - 15x^2 + 10x = 5x(x^2 - 3x + 2)$$ Factored, but not completely factored.

Now factor the trinomial $x^2 - 3x + 2$. Look for factors of +2 that add up to −3. Because $(-1)(-2) = +2$ and $(-1)+(-2) = -3$, we have

$$\begin{aligned} 5x^3 - 15x^2 + 10x &= 5x(x^2 - 3x + 2) \\ &= 5x(x-1)(x-2). \end{aligned}$$ Completely factored.

Check: $$\begin{aligned} 5x(x-1)(x-2) &= 5x(x^2 - 2x - x + 2) \\ &= 5x(x^2 - 3x + 2) \\ &= 5x^3 - 15x^2 + 10x \end{aligned}$$ The original expression

b. $10y^5 - 20y^4 - 80y^3$

Solution: First factor out the GCF, $10y^3$.

$$10y^5 - 20y^4 - 80y^3 = 10y^3(y^2 - 2y - 8)$$

Now factor the trinomial $y^2 - 2y - 8$. Look for factors of −8 that add up to −2. Because $(-4)(+2) = -8$ and $(-4)+(+2) = -2$, we have

$$10y^5 - 20y^4 - 80y^3 = 10y^3(y^2 - 2y - 8) = 10y^3(y-4)(y+2).$$

The factoring can be checked by multiplying the factors.

NOTES

When factoring polynomials, always look for a common monomial factor first. Then, if there is one, remember to include this common monomial factor as part of the answer. Not all polynomials are factorable. For example, no matter what combinations are tried, $x^2 + 3x + 4$ does not have two binomial factors with integer coefficients. (There are no factors of +4 that will add to +3.) We say that the polynomial is **not factorable** (or **prime**). **A polynomial is not factorable if it cannot be factored as the product of polynomials with integer coefficients.**

Practice Problems

Completely factor each polynomial. If the polynomial cannot be factored, write "not factorable."

1. $x^2 - 3x - 28$ **2.** $x^2 + 4x + 3$ **3.** $y^2 - 12y + 35$

4. $a^2 - a + 1$ **5.** $2y^3 + 24y^2 + 64y$ **6.** $4a^4 + 48a^3 + 108a^2$

6.2 Exercises

List all pairs of integer factors for the given integer. Remember to include negative integers as well as positive integers.

1. 15 **2.** 12 **3.** 20 **4.** 30 **5.** –6

6. –7 **7.** 16 **8.** 18 **9.** –10 **10.** –25

Find the pair of integers whose product is the first integer and whose sum is the second integer.

11. 12, 7 **12.** 25, 26 **13.** –14, –5 **14.** –30, –1 **15.** –8, 7

16. –40, 6 **17.** 36, –12 **18.** 16, –10 **19.** 20, –9 **20.** 4, –5

Complete the factorization.

21. $x^2 + 6x + 5 = (x+5)(\quad)$ **22.** $y^2 - 7y + 6 = (y-1)(\quad)$

23. $p^2 - 9p - 10 = (p+1)(\quad)$ **24.** $m^2 + 4m - 45 = (m-5)(\quad)$

25. $a^2 + 12a + 36 = (a+6)(\quad)$ **26.** $n^2 - 2n - 3 = (n-3)(\quad)$

Completely factor the trinomials. If the trinomial cannot be factored, write "not factorable."

27. $x^2 - x - 12$ **28.** $x^2 - 6x - 27$ **29.** $y^2 + y - 30$

30. $x^2 + 6x - 36$ **31.** $m^2 + 3m - 1$ **32.** $x^2 + 3x - 18$

33. $x^2 - 8x + 16$ **34.** $a^2 + 10a + 25$ **35.** $x^2 + 7x + 12$

Answers to Practice Problems: **1.** $(x-7)(x+4)$ **2.** $(x+3)(x+1)$ **3.** $(y-5)(y-7)$ **4.** Not factorable **5.** $2y(y+8)(y+4)$ **6.** $4a^2(a+3)(a+9)$

1. $\{1, 15\}, \{-1, -15\}, \{3, 5\}, \{-3, -5\}$
2. $\{1, 12\}, \{-1, -12\}, \{2, 6\}, \{-2, -6\}, \{4, 3\}, \{-4, -3\}$
3. $\{1, 20\}, \{-1, -20\}, \{4, 5\}, \{-4, -5\}, \{2, 10\}, \{-2, -10\}$
4. $\{1, 30\}, \{-1, -30\}, \{6, 5\}, \{-6, -5\}, \{15, 2\}, \{-15, -2\}, \{10, 3\}, \{-10, -3\}$
5. $\{1, -6\}, \{6, -1\}, \{2, -3\}, \{3, -2\}$
6. $\{1, -7\}, \{7, -1\}$
7. $\{1, 16\}, \{-1, -16\}, \{4, 4\}, \{-4, -4\}, \{8, 2\}, \{-8, -2\}$
8. $\{1, 18\}, \{-1, -18\}, \{2, 9\}, \{-2, -9\}, \{3, 6\}, \{-3, -6\}$
9. $\{1, -10\}, \{10, -1\}, \{5, -2\}, \{2, -5\}$
10. $\{1, -25\}, \{-1, 25\}, \{5, -5\}$
11. 4, 3
12. 25, 1
13. –7, 2
14. –6, 5
15. 8, –1
16. 10, –4
17. –6, –6

18. $-8, -2$
19. $-5, -4$
20. $-1, -4$
21. $x+1$ 22. $y-6$
23. $p-10$ 24. $m+9$
25. $a+6$ 26. $n+1$
27. $(x-4)(x+3)$
28. $(x-9)(x+3)$
29. $(y+6)(y-5)$
30. Not factorable
31. Not factorable
32. $(x+6)(x-3)$
33. $(x-4)(x-4)$
34. $(a+5)(a+5)$
35. $(x+4)(x+3)$
36. Not factorable
37. $(y-1)(y-2)$
38. $(y-12)(y-2)$
39. Not factorable
40. $(y+7)(y+5)$
41. $(x+8)(x-9)$
42. $(y+7)(y+1)$
43. $(z-6)(z-9)$
44. $(a-3)(a+7)$
45. $x(x+7)(x+3)$
46. $x(x+5)(x+3)$
47. $5(x-4)(x+3)$
48. $6(x+3)(x+1)$
49. $10y(y-3)(y+2)$
50. $7y(y-6)(y-4)$
51. $4p^2(p+1)(p+8)$
52. $15m^3(m-1)(m-1)$
53. $2x^2(x-9)(x+2)$
54. $3y^4(y+6)(y+5)$
55. $2(x^2-x-36)$
56. $3(x^2-6x+10)$
57. $2a^2(a-10)(a+6)$
58. $2a^2(a+9)(a+3)$
59. $3y^3(y-8)(y+1)$

36. a^2+a+2 **37.** y^2-3y+2 **38.** $y^2-14y+24$

39. x^2+3x+5 **40.** $y^2+12y+35$ **41.** x^2-x-72

42. y^2+8y+7 **43.** $z^2-15z+54$ **44.** $a^2+4a-21$

Completely factor the given polynomials.

45. x^3+10x^2+21x **46.** x^3+8x^2+15x **47.** $5x^2-5x-60$

48. $6x^2+24x+18$ **49.** $10y^3-10y^2-60y$ **50.** $7y^3-70y^2+168y$

51. $4p^4+36p^3+32p^2$ **52.** $15m^5-30m^4+15m^3$ **53.** $2x^4-14x^3-36x^2$

54. $3y^6+33y^5+90y^4$ **55.** $2x^2-2x-72$ **56.** $3x^2-18x+30$

57. $2a^4-8a^3-120a^2$ **58.** $2a^4+24a^3+54a^2$ **59.** $3y^5-21y^4-24y^3$

60. $4y^5+28y^4+24y^3$ **61.** x^3-10x^2+16x **62.** x^3-2x^2-3x

63. $5a^2+10a-30$ **64.** $6a^2+24a+12$ **65.** $20a^4+40a^3+20a^2$

66. $6x^4-12x^3+6x^2$

67. Triangles: The area of a triangle is $\frac{1}{2}$ the product of its base and its height. If the area of the triangle shown is given by the function $A(x)=\frac{1}{2}x^2+24x$, find representations for the lengths of its base and its height (where the base is longer than the height).

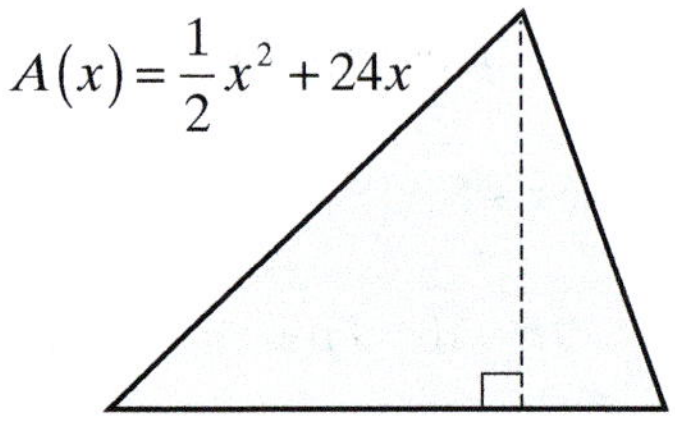

68. Triangles: The area of a triangle is $\frac{1}{2}$ the product of its base and its height. If the area of the triangle shown is given by the function $A(x)=\frac{1}{2}x^2+14x$, find representations for the lengths of its base and its height (where the height is longer than the base).

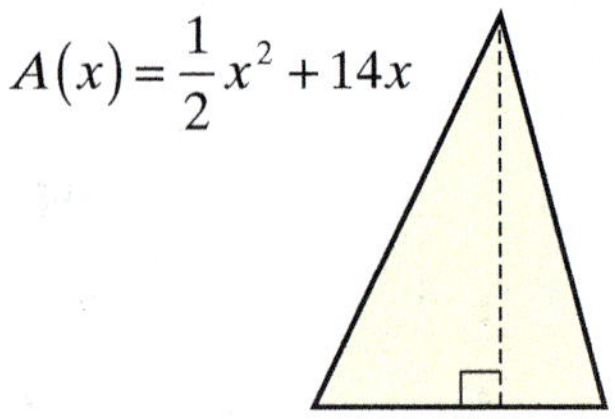

60. $4y^3(y+6)(y+1)$
61. $x(x-2)(x-8)$
62. $x(x-3)(x+1)$
63. $5(a^2+2a-6)$
64. $6(a^2+4a+2)$
65. $20a^2(a+1)(a+1)$
66. $6x^2(x-1)(x-1)$
67. Base = $x + 48$; Height = x
68. Base = x; Height = $x + 28$
69. $x + 5$
70. $x + 3$

69. Rectangles: The area of the rectangle shown is given by the polynomial function $A(x) = 4x^2 + 20x$. If the width of the rectangle is $4x$, what is the length?

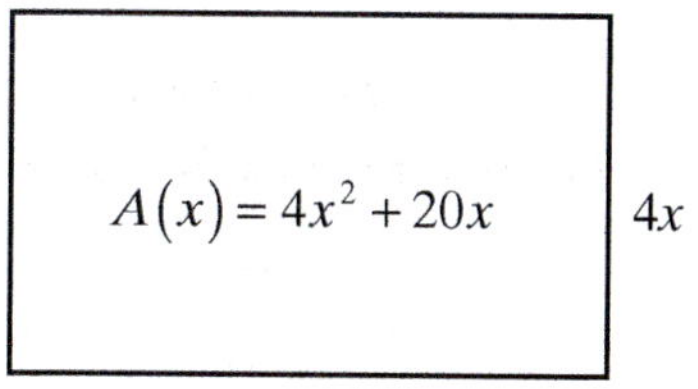

70. Rectangles: The area of the rectangle shown is given by the polynomial function $A(x) = x^2 + 11x + 24$. If the length of the rectangle is $(x+8)$, what is the width?

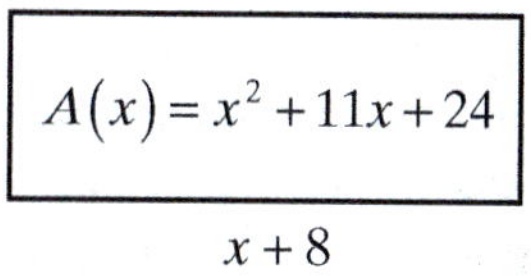

Writing and Thinking About Mathematics

71. It is true that $2x^2 + 10x + 12 = (2x+6)(x+2) = (2x+4)(x+3)$. Explain how the trinomial can be factored in two ways. Is there some kind of error?

72. It is true that $5x^2 - 5x - 30 = (5x-15)(x+2)$. Explain why this is not the completely factored form of the trinomial.

HAWKES LEARNING SYSTEMS: INTRODUCTORY & INTERMEDIATE ALGEBRA SOFTWARE

- 6.2 Factoring Trinomials: Leading Coefficient 1

71. This is not an error, but the trinomial is not completely factored. The completely factored form of this trinomial is $2(x+2)(x+3)$.
72. This is not the completely factored form because $(5x-15)$ can be factored into $5(x-3)$.

6.3 Factoring Trinomials: $ax^2 + bx + c$

- *Factor trinomials using the* ***trial-and-error method****.*
- *Factor trinomials using the* ***ac-method****.*

Second-degree trinomials in the variable x are of the general form

$\boldsymbol{ax^2+bx+c}$ where the coefficients a, b, and c are real numbers and $a \neq 0$.

In this section we will discuss two methods of factoring trinomials of the form ax^2+bx+c in which the coefficients are restricted to integers. These two methods are the **trial-and-error method** (a reverse of the **FOIL** method of multiplication) and the ***ac*-method** (a form of **grouping**).

The Trial-and-Error Method of Factoring

To help in understanding this method we review the FOIL method of multiplying two binomials as follows:

$$(2x+5)(3x+1)=6x^2+17x+5$$

$$\begin{aligned}(2x+5)(3x+1) &= \overset{F}{2x\cdot 3x} + \overset{O}{2x\cdot 1} + \overset{I}{5\cdot 3x} + \overset{L}{5\cdot 1}\\ &= 6x^2 \quad + \quad 17x \quad + 5\end{aligned}$$

F The product of the **first** two terms is $6x^2$.

O, **I** The sum of **inner** and **outer** products is $17x$.

L The product of the **last** two terms is 5.

Now consider the problem of factoring

$$6x^2+23x+7$$

as the product of two binomials.

$\mathbf{F} = 6x^2$ $\quad \mathbf{L} = +7$

$$6x^2 + 23x + 7 = (\quad)(\quad)$$

For $\mathbf{F} = 6x^2$, we know that $6x^2 = 6x \cdot x$ and $6x^2 = 3x \cdot 2x$.

For $\mathbf{L} = +7$, we know that $+7 = (+1)(+7)$ and $+7 = (-1)(-7)$.

Now we use various combinations for **F** and **L** in the **trial-and-error method** as follows:

1. List all the possible combinations of factors of $6x^2$ and $+7$ in their respective **F** and **L** positions. (See the following list.)
2. Check the sum of the products in the **O** and **I** positions until you find the sum to be $+23x$.
3. If none of these sums is $+23x$, the trinomial is not factorable.

F **L**

a. $(6x+1)(x+7)$

b. $(6x+7)(x+1)$

c. $(3x+1)(2x+7)$

d. $(3x+7)(2x+1)$

e. $(6x-1)(x-7)$

f. $(6x-7)(x-1)$

g. $(3x-1)(2x-7)$

h. $(3x-7)(2x-1)$

We really don't need to check these last four because the **O** and **I** would have to be negative, and we are looking for $+23x$. In this manner, the trial-and-error method is more efficient than it first appears to be.

Now investigating only the possibilities in the list with positive constants, we need to check the sums of the outer (**O**) and inner (**I**) products to find $+23x$.

a. $(6x+1)(x+7)$: $\mathbf{O} + \mathbf{I} = 42x + x = 43x$

x $\quad 42x$

b. $(6x+7)(x+1)$: $\mathbf{O} + \mathbf{I} = 6x + 7x = 13x$

$7x$ $\quad 6x$

c. $(3x+1)(2x+7)$: $\mathbf{O}+\mathbf{I} = 21x + 2x = 23x$ ← We found $23x$! Yeah! With a little luck we could have found this first.

$2x$
$21x$

The correct factors, $(3x+1)$ and $(2x+7)$, have been found so we need not take the time to try the next product in the list, $(3x+7)(2x+1)$. Thus even though the list of possibilities of factors may be long, the actual time involved may be quite short if the correct factors are found early in the trial-and-error method. So we have

$$6x^2 + 23x + 7 = (3x+1)(2x+7).$$

Look at the constant term to determine what signs to use for the constants in the factors. The following guidelines will help limit the trial-and-error search.

Guidelines for the Trial-and-Error Method

1. If the sign of the constant term is positive (+), the signs in both factors will be the same, either both positive or both negative.
2. If the sign of the constant term is negative (–), the signs in the factors will be different, one positive and one negative.

Example 1: Using the Trial-and-Error Method

Factor by using the trial-and-error method.

a. $x^2 + 6x + 5$

Solution: Since the middle term is $+6x$ and the constant is 5, we know that the two factors of 5 must both be positive, +5 and +1.

$$x^2 + 6x + 5 = (x+5)(x+1)$$

$5x$
x

b. $4x^2 - 4x - 15$

Solution: For **F**: $4x^2 = 4x \cdot x$ and $4x^2 = 2x \cdot 2x$.
For **L**: $-15 = -15 \cdot 1$, $-15 = -1 \cdot 15$, $-15 = -3 \cdot 5$, and $-15 = -5 \cdot 3$.

Trials: $(2x-15)(2x+1)$ — $2x - 30x = -28x$ is the wrong middle term.

$-30x$
$2x$

$(2x-3)(2x+5)$ — inner product: $-6x$; outer product: $10x$

$10x - 6x = +4x$ is the wrong middle term only because the sign is wrong. So just switch the signs and the factors will be right.

$(2x+3)(2x-5)$ — inner product: $6x$; outer product: $-10x$

$-10x + 6x = -4x$ is the right middle term.

Thus $4x^2 - 4x - 15 = (2x+3)(2x-5)$.

Now that we have the answer, there is no need to try all the possibilities with $(4x \quad)(x \quad)$.

c. $6a^2 - 31a + 5$

Solution: Since the middle term is $-31a$ and the constant is $+5$, we know that the two factors of 5 must both be negative, -5 and -1. We try, $\mathbf{F} = 6a^2 = 6a \cdot a$.

$6a^2 - 31a + 5 = (6a-1)(a-5)$ — inner product: $-a$; outer product: $-30a$

$-30a - a = -31a$

So we found the correct factors on the first try.

NOTES

Reminder: To factor completely means to find factors of the polynomial, none of which are themselves factorable. Thus

$$2x^2 + 12x + 10 = (2x+10)(x+1)$$

is not factored completely because $2x + 10 = 2(x+5)$.

We could write

$$\begin{aligned} 2x^2 + 12x + 10 &= (2x+10)(x+1) \\ &= 2(x+5)(x+1). \end{aligned}$$

This problem can be avoided by first factoring out the GCF (in this case, 2).

Example 2: Factoring Completely

Factor completely. Be sure to look first for the greatest common monomial factor.

a. $6x^3 - 8x^2 + 2x$

Solution: $6x^3 - 8x^2 + 2x = 2x(3x^2 - 4x + 1) = 2x(3x - 1)(x - 1)$

Check: Check the factorization by multiplying.

$$2x(3x-1)(x-1) = 2x(3x^2 - 3x - x + 1)$$
$$= 2x(3x^2 - 4x + 1)$$
$$= 6x^3 - 8x^2 + 2x \quad \text{The original polynomial}$$

b. $-2x^2 - x + 6$

Solution: $-2x^2 - x + 6 = -1(2x^2 + x - 6) = -1(2x - 3)(x + 2)$

Note that factoring out -1 gives a positive leading coefficient for the trinomial.

c. $10x^3 + 5x^2 + 5x$

Solution: $10x^3 + 5x^2 + 5x = 5x(2x^2 + x + 1)$

Now consider the trinomial:

$$2x^2 + x + 1 = (2x + ?)(x + ?)$$

The factors of $+1$ need to be $+1$ and $+1$, but

$$(2x+1)(x+1) = 2x^3 + \underline{3x} + 1$$

So there is no way to factor and get a middle term of $+x$ for the product. This trinomial, $2x^2 + x + 1$, is **not factorable**.

We have

$$10x^3 + 5x^2 + 5x = 5x(2x^2 + x + 1) \quad \text{Factored completely}$$

Teaching Notes:
There is some debate about which method, the *ac*-method or the trial-and-error method, is better (or easier) for beginning algebra students. You may choose to teach one method or the other, or both. The *ac*-method is more "organized" and appeals to some students because of this. However, if students become accustomed to the trial-and-error method, they will probably find it easier and faster in the long run, and they will be better able to handle rational expressions. Basically, the choice depends on the size of the numbers involved and the factoring skills of the students.

The *ac*-Method of Factoring

The ***ac*-method** of factoring is in reference to the coefficients a and c in the general form $ax^2 + bx + c$ and involves the method of factoring by grouping discussed in Section 6.1. The method is best explained by analyzing an example and explaining each step as follows.

We consider the problem of factoring the trinomial

$$2x^2 + 9x + 10 \text{ where } a = 2, b = 9, \text{ and } c = 10.$$

Analysis of Factoring by the *ac*-Method

	General Method	Example
	$ax^2 + bx + c$	$2x^2 + 9x + 10$
Step 1:	Multiply $a \cdot c$.	Multiply $2 \cdot 10 = 20$.
Step 2:	Find two integers whose product is *ac* and whose sum is *b*. If this is not possible, then the trinomial is **not factorable**.	Find two integers whose product is 20 and whose sum is 9. (In this case, $4 \cdot 5 = 20$ and $4 + 5 = 9$.)
Step 3:	Rewrite the middle term (bx) using the two numbers found in Step 2 as coefficients.	Rewrite the middle term $(+9x)$ using +4 and +5 as coefficients. $2x^2 + 9x + 10 = 2x^2 + 4x + 5x + 10$
Step 4:	Factor by grouping the first two terms and the last two terms.	Factor by grouping the first two terms and the last two terms. $2x^2 + 4x + 5x + 10 = (2x^2 + 4x) + (5x + 10)$ $= 2x(x+2) + 5(x+2)$
Step 5:	Factor out the common binomial factor. This will give two binomial factors of the trinomial $ax^2 + bx + c$.	Factor out the common binomial factor $(x+2)$. Thus $2x^2 + 9x + 10 = 2x^2 + 4x + 5x + 10$ $= (2x^2 + 4x) + (5x + 10)$ $= 2x(x+2) + 5(x+2)$ $= (x+2)(2x+5)$.

Example 3: Using the *ac*-Method

Teaching Notes:
While "*ac*" suggests to students "multiplication," it will be helpful to remind them that they will be interested in the combination of the factors within "*ac*," rather than the actual product.

a. Factor $4x^2 + 33x + 35$ using the *ac*-method.

Solution: $a = 4, b = 33, c = 35$

Step 1: Find the product *ac*: $4 \cdot 35 = 140$.

Step 2: Find two integers whose product is 140 and whose sum is 33.

$$(+5)(+28) = 140 \text{ and } (+5) + (+28) = +33$$

Note: this step may take some experimenting with factors. You might try prime factoring. For example

$$140 = 10 \cdot 14 = 2 \cdot 5 \cdot 2 \cdot 7.$$

Continued on the next page...

With combinations of these prime factors we can write

Factors of 140	Sum
$1 \cdot 140$	$1 + 140 = 141$
$2 \cdot 70$	$2 + 70 = 72$
$4 \cdot 35$	$4 + 35 = 39$
$5 \cdot 28$	$5 + 28 = 33$

We can stop here!

5 and 28 are the desired coefficients.

Step 3: Rewrite $+33x$ as $+5x + 28x$, giving

$$4x^2 + 33x + 35 = 4x^2 + 5x + 28x + 35.$$

Step 4: Factor by grouping.

$$\begin{aligned} 4x^2 + 33x + 35 &= 4x^2 + 5x + 28x + 35 \\ &= x(4x+5) + 7(4x+5) \end{aligned}$$

Step 5: Factor out the common binomial factor $(4x+5)$.

$$\begin{aligned} 4x^2 + 33x + 35 &= 4x^2 + 5x + 28x + 35 \\ &= x(4x+5) + 7(4x+5) \\ &= (4x+5)(x+7) \end{aligned}$$

Note that in Step 3 we could have written $+33x$ as $+28x + 5x$. Try this to convince yourself that the result will be the same two factors.

b. Factor $12y^3 - 26y^2 + 12y$ using the *ac*-method.

Solution: First factor out the greatest common factor $2y$.

$$12y^3 - 26y^2 + 12y = 2y\left(6y^2 - 13y + 6\right)$$

Now factor the trinomial $6y^2 - 13y + 6$ with $a = 6$, $b = -13$, and $c = 6$.

Step 1: Find the product *ac*: $6(6) = 36$.

Step 2: Find two integers whose product is 36 and whose sum is -13.

Note: This may take some time and experimentation. We do know that both numbers must be negative because the product is positive and the sum is negative.

$$(-9)(-4) = +36 \quad \text{and} \quad -9 + (-4) = -13$$

Steps 3 and 4:

Factor by grouping.

$$6y^2 - 13y + 6 = 6y^2 - 9y - 4y + 6$$
$$= 3y(2y-3) - 2(2y-3)$$

Note: –2 is factored from the last two terms so that there will be a common binomial factor $(2y-3)$.

Step 5: Factor out the common binomial factor $(2y-3)$.

$$6y^2 - 13y + 6 = 6y^2 - 9y - 4y + 6$$
$$= 3y(2y-3) - 2(2y-3)$$
$$= (2y-3)(3y-2)$$

Thus for the original expression,

$$12y^3 - 26y^2 + 12y = 2y(6y^2 - 13y + 6)$$
$$= 2y(2y-3)(3y-2)$$

Do not forget to write the common monomial factor, $2y$, in the answer.

c. Factor $4x^2 - 5x - 6$ using the *ac*-method.

Solution: $a = 4, b = -5, c = -6$

Step 1: Find the product *ac*: $4(-6) = -24$.

Step 2: Find two integers whose product is –24 and whose sum is –5. (**Note:** We know that one number must be positive and the other negative because the product is negative.) In this example we have

$$(+3)(-8) = -24 \text{ and } (+3) + (-8) = -5.$$

Steps 3 and 4:

Factor by grouping.

$$4x^2 - 5x - 6 = 4x^2 + 3x - 8x - 6$$
$$= x(4x+3) - 2(4x+3)$$

Step 5: Factor out the common binomial factor $(4x+3)$.

$$4x^2 - 5x - 6 = 4x^2 + 3x - 8x - 6$$
$$= x(4x+3) - 2(4x+3)$$
$$= (4x+3)(x-2).$$

1. $(x+2)(x+3)$
2. $(x-2)(x-4)$
3. $(2x-5)(x+1)$
4. $(3x-7)(x+1)$
5. $(6x+5)(x+1)$
6. $(4x-3)(x-2)$
7. $-(x-2)(x-1)$
8. $-(x+2)(x+3)$
9. $(x-5)(x+2)$
10. $(x-10)(x-1)$
11. $-(x-14)(x+1)$
12. $-(x-6)(x-6)$
13. Not Factorable
14. Not Factorable
15. $-x(2x+1)(x-1)$
16. $-y(2y+1)(y+1)$
17. $(t-1)(4t+1)$
18. $(2x+1)(x-2)$
19. $(5a-6)(a+1)$
20. $(3a+1)(a+1)$
21. $(7x-2)(x+1)$

Tips to Keep in Mind while Factoring

a. When factoring polynomials, always look for the greatest common factor first. Then, if there is one, remember to include this common factor as part of the answer.

b. **To factor completely** means to find factors of the polynomial such that none of the factors are themselves factorable.

c. Not all polynomials are factorable. (See $2x^2+x+1$ in Example 2c.) **Any polynomial that cannot be factored as the product of polynomials with integer coefficients is not factorable.**

d. Factoring can be checked by multiplying the factors. The product should be the original expression.

NOTES

No matter which method you use (the *ac*-method or the trial-and-error method), factoring trinomials takes time. With practice you will become more efficient with either method. Make sure to be patient and observant.

Practice Problems

Factor completely.

1. $3x^2+7x-6$ **2.** $2x^2+6x-8$ **3.** $3x^2+15x+18$

4. $10x^2-41x-18$ **5.** $x^2+11x+28$ **6.** $-3x^3+6x^2-3x$

7. $3x^2+4x-8$

6.3 Exercises

22. $(4x+3)(x+5)$
23. $(2x-3)(4x+1)$
24. $(2x+3)(3x+7)$
25. $(3x+4)(3x-5)$
26. Not factorable
27. $2(2x-5)(3x-2)$

Completely factor each of the given polynomials. If a polynomial cannot be factored, write "not factorable."

1. x^2+5x+6 **2.** x^2-6x+8 **3.** $2x^2-3x-5$

4. $3x^2-4x-7$ **5.** $6x^2+11x+5$ **6.** $4x^2-11x+6$

Answers to Practice Problems: **1.** $(3x-2)(x+3)$ **2.** $2(x-1)(x+4)$ **3.** $3(x+2)(x+3)$
4. $(5x+2)(2x-9)$ **5.** $(x+7)(x+4)$ **6.** $-3x(x-1)(x-1)$
7. Not factorable

28. $3(2b-1)(2b-1)$
29. $(3x-1)(x-2)$
30. $(7x+3)(x-2)$
31. $(3x-1)(3x-1)$
32. $(2x+1)(2x+1)$
33. $(3y+2)(2y+1)$
34. $(3y-4)(4y+3)$
35. $(x-1)(x-45)$
36. $(x-2)(x+8)$
37. Not factorable
38. $(5a-2)(a-1)$
39. $2b(4a-3)(a-2)$
40. $2mn(6m-1)(m-4)$
41. Not factorable
42. Not factorable
43. $(4x-1)(4x-1)$
44. $(3x+1)(x-4)$
45. $(8x-3)(8x-3)$
46. $(3x-2)(3x-2)$
47. $2(3x-5)(x+2)$
48. $3(4y-1)(y-1)$
49. $5(2x+3)(x+2)$
50. $4(3y-1)(2y+1)$
51. $-2(9x^2-36x+4)$
52. $x^2(7x^2-5x+3)$
53. $-15(3y+4)(y-2)$
54. $-2(m-2)(6m+1)$
55. $3(2x-5)(2x-5)$
56. $2(16y^2+25)$
57. $3x(2x-1)(x+2)$
58. $-5(y-2)(y-6)$
59. $3x(2x-9)(2x-9)$
60. $3a(5a+1)(2a+3)$
61. $9xy^3(x^2+x+1)$
62. $6y(8x-3)(x-7)$
63. $4xy(3y-4)(4y-3)$
64. $3x(8a^2x+24a^2+81)$
65. $7y^2(y-4)(3y-2)$
66. $9a(4a-3)(2a-7)$

7. $-x^2+3x-2$
8. $-x^2-5x-6$
9. $x^2-3x-10$
10. $x^2-11x+10$
11. $-x^2+13x+14$
12. $-x^2+12x-36$
13. $x^2+8x+64$
14. x^2+2x+3
15. $-2x^3+x^2+x$
16. $-2y^3-3y^2-y$
17. $4t^2-3t-1$
18. $2x^2-3x-2$
19. $5a^2-a-6$
20. $3a^2+4a+1$
21. $7x^2+5x-2$
22. $4x^2+23x+15$
23. $8x^2-10x-3$
24. $6x^2+23x+21$
25. $9x^2-3x-20$
26. $4x^2+40x+25$
27. $12x^2-38x+20$
28. $12b^2-12b+3$
29. $3x^2-7x+2$
30. $7x^2-11x-6$
31. $9x^2-6x+1$
32. $4x^2+4x+1$
33. $6y^2+7y+2$
34. $12y^2-7y-12$
35. $x^2-46x+45$
36. $x^2+6x-16$
37. $3x^2+9x+5$
38. $5a^2-7a+2$
39. $8a^2b-22ab+12b$
40. $12m^3n-50m^2n+8mn$
41. x^2+x+1
42. x^2+2x+2
43. $16x^2-8x+1$
44. $3x^2-11x-4$
45. $64x^2-48x+9$
46. $9x^2-12x+4$
47. $6x^2+2x-20$
48. $12y^2-15y+3$
49. $10x^2+35x+30$
50. $24y^2+4y-4$
51. $-18x^2+72x-8$
52. $7x^4-5x^3+3x^2$
53. $-45y^2+30y+120$
54. $-12m^2+22m+4$
55. $12x^2-60x+75$
56. $32y^2+50$
57. $6x^3+9x^2-6x$
58. $-5y^2+40y-60$
59. $12x^3-108x^2+243x$
60. $30a^3+51a^2+9a$
61. $9x^3y^3+9x^2y^3+9xy^3$
62. $48x^2y-354xy+126y$
63. $48xy^3-100xy^2+48xy$
64. $24a^2x^2+72a^2x+243x$
65. $21y^4-98y^3+56y^2$
66. $72a^3-306a^2+189a$

Writing and Thinking About Mathematics

67. Discuss, in your own words, how the sign of the constant term determines what signs will be used in the factors when factoring trinomials.

68. Volume of a box: The volume of an open box is found by cutting equal squares (x inches on a side) from a sheet of cardboard that is 5 inches by 25 inches. The function representing this volume is $V(x) = 4x^3 - 60x^2 + 125x$, where $0 < x < 2.5$. Factor this function and use the factors to explain, in your own words, how the function represents the volume.

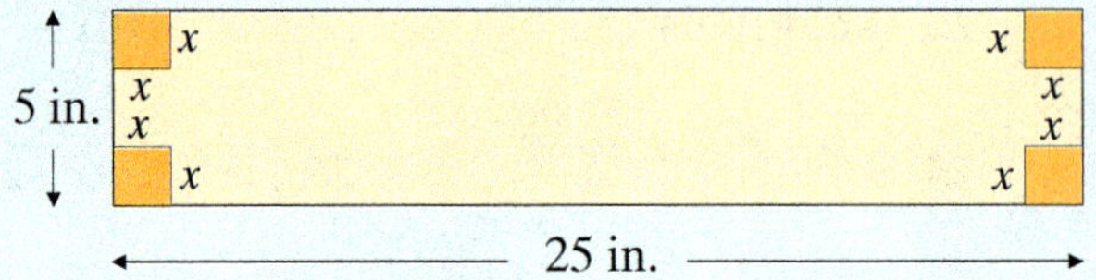

HAWKES LEARNING SYSTEMS: INTRODUCTORY & INTERMEDIATE ALGEBRA SOFTWARE

- 6.3a Factoring Trinomials by Trial and Error
- 6.3b Factoring Trinomials by Grouping

67. If the sign of the constant term is positive, the signs in the factors will both be positive or both be negative. If the sign of the constant term is negative, the sign in one factor will be positive and the sign in the other factor will be negative.

68. $x(5-2x)(25-2x)$
The height is x, the width is $(5-2x)$, and the length is $(25-2x)$.

6.4 Special Factoring Techniques

- *Factor the difference of two squares.*
- *Factor perfect square trinomials.*
- *Factor the sums and differences of two cubes.*

In Section 5.6 we discussed the following three products.

I. $(x+a)(x-a)=x^2-a^2$ Difference of Two Squares

II. $(x+a)^2=x^2+2ax+a^2$ Square of a Binomial Sum

III. $(x-a)^2=x^2-2ax+a^2$ Square of a Binomial Difference

Two objectives in this section are to learn to factor products of these types (difference of two squares and squares of binomials) without referring to the trial-and-error method. That is, with practice you will learn to recognize the "form" of these products and go directly to the factors. **Memorize the products and their names listed above.**

For easy reference to squares, Table 1 lists the squares of the integers from 1 to 20.

Perfect Squares from 1 to 400										
Integer n	1	2	3	4	5	6	7	8	9	10
Square n^2	1	4	9	16	25	36	49	64	81	100
Integer n	11	12	13	14	15	16	17	18	19	20
Square n^2	121	144	169	196	225	256	289	324	361	400

Table 1

Difference of Two Squares: $x^2-a^2=(x+a)(x-a)$

Consider the polynomial x^2-25. By recognizing this expression as the **difference of two squares**, we can go directly to the factors:

$$x^2-25=(x)^2-(5)^2=(x+5)(x-5).$$

Similarly, we have

$$9-y^2=(3)^2-(y)^2=(3+y)(3-y) \quad \text{and} \quad 49x^2-36=(7x)^2-(6)^2=(7x+6)(7x-6).$$

Remember to **look for a common monomial factor first**. For example,

$$6x^2y-24y=6y(x^2-4)=6y(x+2)(x-2).$$

Example 1: Factoring the Difference of Two Squares

Factor completely.

a. $3a^2b - 3b$

Solution: $3a^2b - 3b = 3b(a^2 - 1)$ Factor out the GCF, $3b$.

$= 3b(a+1)(a-1)$ Difference of two squares

Don't forget that $3b$ is a factor.

b. $x^6 - 400$ Even powers, such as x^6, can always be treated as squares: $x^6 = (x^3)^2$.

Solution: $x^6 - 400 = (x^3)^2 - 20^2$

$= (x^3 + 20)(x^3 - 20)$ Difference of two squares

Sum of Two Squares

The **sum of two squares** is an expression of the form $x^2 + a^2$ and is **not factorable**. For example, $x^2 + 36$ is the sum of two squares and is not factorable. There are no factors with integer coefficients whose product is $x^2 + 36$. To understand this situation, write

$$x^2 + 36 = x^2 + 0x + 36$$

and note that there are no factors of +36 that will add to 0.

Example 2: Using The Sum of Two Squares

Factor completely. Be sure to look first for the greatest common monomial factor.

a. $y^2 + 64$

Solution: $y^2 + 64$ is the sum of two squares and is not factorable.

b. $4x^2 + 100$

Solution: $4x^2 + 100 = 4(x^2 + 25)$ Factored completely

We see that 4 is the greatest common monomial factor and $x^2 + 25$ is the sum of two squares and not factorable.

Perfect Square Trinomials: $\begin{cases} x^2 + 2ax + a^2 = (x+a)^2 \\ x^2 - 2ax + a^2 = (x-a)^2 \end{cases}$

We know that **squaring a binomial** leads to a **perfect square trinomial**. Therefore, factoring a perfect square trinomial gives the square of a binomial. We simply need to recognize whether or not a trinomial fits the "form."

In a perfect square trinomial, both the first and last terms of the trinomial must be perfect squares. If the first term is of the form x^2 and the last term is of the form a^2, then the middle term must be of the form $2ax$ or $-2ax$. For example,

$x^2 + 8x + 16 = (x+4)^2$ Here $8x = 2\cdot 4\cdot x = 2ax$ and $16 = 4^2 = a^2$.

$x^2 - 6x + 9 = (x-3)^2$ Here $-6x = -2\cdot 3\cdot x = -2ax$ and $9 = 3^2 = a^2$.

Example 3: Factoring Perfect Square Trinomials

Factor completely.

a. $z^2 - 12z + 36$

Solution: In the form $x^2 - 2ax + a^2$ we have $x = z$ and $a = 6$.

$$\begin{aligned} z^2 - 12z + 36 &= z^2 - 2\cdot 6z + 6^2 \\ &= (z-6)^2 \end{aligned}$$

b. $4y^2 + 12y + 9$

Solution: In the form $x^2 + 2ax + a^2$ we have $x = 2y$ and $a = 3$.

$$\begin{aligned} 4y^2 + 12y + 9 &= (2y)^2 + 2\cdot 3\cdot 2y + 3^2 \\ &= (2y+3)^2 \end{aligned}$$

c. $2x^3 - 8x^2y + 8xy^2$

Solution: Factor out the GCF first. Then factor the **perfect square trinomial**.

$$\begin{aligned} 2x^3 - 8x^2y + 8xy^2 &= 2x\left(x^2 - 4xy + 4y^2\right) \\ &= 2x\left[x^2 - 2\cdot 2y\cdot x + (2y)^2\right] \\ &= 2x(x-2y)^2 \end{aligned}$$

Continued on the next page...

d. $(x^2+6x+9)-y^2$

Solution: Treat x^2+6x+9 as a perfect square trinomial, then factor the **difference of two squares**.

$$\begin{aligned}(x^2+6x+9)-y^2 &= (x+3)^2-y^2\\ &= (x+3+y)(x+3-y)\end{aligned}$$

Sums and Differences of Two Cubes:

$$\begin{cases} x^3+a^3=(x+a)(x^2-ax+a^2)\\ x^3-a^3=(x-a)(x^2+ax+a^2)\end{cases}$$

The formulas for the sums and differences of two cubes are new, and we can proceed to show that they are indeed true as follows:

$$\begin{aligned}(x+a)(x^2-ax+a^2) &= x\cdot x^2-x\cdot ax+x\cdot a^2+a\cdot x^2-a\cdot ax+a\cdot a^2\\ &= x^3-ax^2+a^2x+ax^2-a^2x+a^3\\ &= x^3+a^3 \qquad \text{Sum of two cubes}\end{aligned}$$

$$\begin{aligned}(x-a)(x^2+ax+a^2) &= x\cdot x^2+x\cdot ax+x\cdot a^2-a\cdot x^2-a\cdot ax-a\cdot a^2\\ &= x^3+ax^2+a^2x-ax^2-a^2x-a^3\\ &= x^3-a^3 \qquad \text{Difference of two cubes}\end{aligned}$$

NOTES

Important Notes about the Sum and Difference of Two Cubes

1. In each case, after multiplying the factors the middle terms drop out and only two terms are left.
2. The trinomials in parentheses are not perfect square trinomials. These trinomials are not factorable.
3. The sign in the binomial agrees with the sign in the result.

Because we know that factoring is the reverse of multiplication, we use these formulas to factor the sums and differences of two cubes. For example,

$$\begin{aligned}x^3+27 &= (x)^3+(3)^3\\ &= (x+3)\left((x)^2-(3)(x)+(3)^2\right)\\ &= (x+3)(x^2-3x+9)\end{aligned}$$

and

$$\begin{aligned} x^6 - 125 &= \left(x^2\right)^3 - (5)^3 \\ &= \left(x^2 - 5\right)\left[\left(x^2\right)^2 + (5)\left(x^2\right) + (5)^2\right] \\ &= \left(x^2 - 5\right)\left(x^4 + 5x^2 + 25\right). \end{aligned}$$

As an aid in factoring sums and differences of two cubes, Table 2 contains the cubes of the integers from 1 to 10. These cubes are called perfect cubes.

Perfect Cubes from 1 to 1000										
Integer n	1	2	3	4	5	6	7	8	9	10
Cube n^3	1	8	27	64	125	216	343	512	729	1000

Table 2

Example 4: Factoring Sums and Differences of Two Cubes

Factor completely.

a. $x^3 - 8$

Solution:

$$\begin{aligned} x^3 - 8 &= x^3 - 2^3 \\ &= (x - 2)\left(x^2 + 2 \cdot x + 2^2\right) \\ &= (x - 2)\left(x^2 + 2x + 4\right) \end{aligned}$$

Note: Remember that the second polynomial is not a perfect square trinomial and cannot be factored.

b. $x^6 + 64y^3$

Solution:

$$\begin{aligned} x^6 + 64y^3 &= \left(x^2\right)^3 + (4y)^3 \\ &= \left(x^2 + 4y\right)\left[\left(x^2\right)^2 - 4y \cdot x^2 + (4y)^2\right] \\ &= \left(x^2 + 4y\right)\left(x^4 - 4x^2y + 16y^2\right) \end{aligned}$$

c. $16y^{12} - 250$

Solution: Factor out the GCF first. Then factor the **difference of two cubes**.

$$\begin{aligned} 16y^{12} - 250 &= 2\left(8y^{12} - 125\right) \\ &= 2\left[\left(2y^4\right)^3 - 5^3\right] \\ &= 2\left(2y^4 - 5\right)\left[\left(2y^4\right)^2 + (5)\left(2y^4\right) + 5^2\right] \\ &= 2\left(2y^4 - 5\right)\left(4y^8 + 10y^4 + 25\right) \end{aligned}$$

Practice Problems

Completely factor each of the following polynomials. If a polynomial cannot be factored, write "not factorable."

1. $5x^2-80$ **2.** $9x^2-12x+4$ **3.** $4x^2+20x+25$

4. $2x^3-250$ **5.** $(y^2-4y+4)-x^4$ **6.** x^3+8y^3

6.4 Exercises

Completely factor each of the given polynomials. If a polynomial cannot be factored, write "not factorable."

1. x^2-25	**2.** y^2-121	**3.** $81-y^2$
4. $25-z^2$	**5.** $2x^2-128$	**6.** $3x^2-147$
7. $4x^4-64$	**8.** $5x^4-125$	**9.** y^2+100
10. $4x^2+49$	**11.** $y^2-16y+64$	**12.** $z^2+18z+81$
13. $-4x^2+100$	**14.** $-12x^4+3$	**15.** $9x^2-25$
16. $4x^2-49$	**17.** $y^2-10y+25$	**18.** $x^2+12x+36$
19. $4x^2-4x+1$	**20.** $49x^2-14x+1$	**21.** $25x^2+30x+9$
22. $9y^2+12y+4$	**23.** $16x^2-40x+25$	**24.** $9x^2-12x+4$
25. $4x^3-64x$	**26.** $50x^3-8x$	**27.** $2x^3y+32x^2y+128xy$
28. $3x^2y-30xy+75y$	**29.** y^2+6y+9	**30.** y^2+4y+4
31. $x^2-20x+100$	**32.** $25x^2-10x+1$	**33.** $x^4+10x^2y+25y^2$
34. $16x^4+8x^2y+y^2$	**35.** x^3-125	**36.** x^3-64
37. y^3+216	**38.** y^3+1	**39.** x^3+27y^3

1. $(x-5)(x+5)$
2. $(y-11)(y+11)$
3. $(9-y)(9+y)$
4. $(5-z)(5+z)$
5. $2(x-8)(x+8)$
6. $3(x-7)(x+7)$
7. $4(x-2)(x+2)(x^2+4)$
8. $5(x^2-5)(x^2+5)$
9. Not factorable
10. Not factorable
11. $(y-8)^2$
12. $(z+9)^2$
13. $-4(x-5)(x+5)$
14. $-3(2x^2-1)(2x^2+1)$
15. $(3x-5)(3x+5)$
16. $(2x-7)(2x+7)$
17. $(y-5)^2$
18. $(x+6)^2$
19. $(2x-1)^2$
20. $(7x-1)^2$
21. $(5x+3)^2$
22. $(3y+2)^2$
23. $(4x-5)^2$
24. $(3x-2)^2$
25. $4x(x-4)(x+4)$
26. $2x(5x-2)(5x+2)$
27. $2xy(x+8)^2$
28. $3y(x-5)^2$
29. $(y+3)^2$
30. $(y+2)^2$

Answers to Practice Problems: **1.** $5(x-4)(x+4)$ **2.** $(3x-2)^2$ **3.** $(2x+5)^2$ **4.** $2(x-5)(x^2+5x+25)$ **5.** $(y-2-x^2)(y-2+x^2)$ **6.** $(x+2y)(x^2-2xy+4y^2)$

31. $(x-10)^2$
32. $(5x-1)^2$
33. $(x^2+5y)^2$
34. $(4x^2+y)^2$
35. $(x-5)(x^2+5x+25)$
36. $(x-4)(x^2+4x+16)$
37. $(y+6)(y^2-6y+36)$
38. $(y+1)(y^2-y+1)$
39. $(x+3y)(x^2-3xy+9y^2)$
40. $(2x+1)(4x^2-2x+1)$
41. Not factorable
42. $3(x+3)(x^2-3x+9)$
43. $4(x-2)(x^2+2x+4)$
44. $(4x+3y)(16x^2-12xy+9y^2)$
45. $2(3x-y)(9x^2+3xy+y^2)$
46. $3x(x+5y)(x^2-5xy+25y^2)$
47. $y(x+y)(x^2-xy+y^2)$
48. $x(xy-1)(x^2y^2+xy+1)$
49. $x^2y^2(1-y)(1+y+y^2)$
50. $2x^2(1-2y)(1+2y+4y^2)$
51. $3xy(2x+3y)(4x^2-6xy+9y^2)$
52. $(x^2-4y)(x^4+4x^2y+16y^2)$
53. $(x^2-y^3)(x^4+x^2y^3+y^6)$
54. Not factorable
55. $(3x+y^2)(9x^2-3xy^2+y^4)$
56. $(x+4z)(x^2-4xz+16z^2)$
57. $(2x+y)(4x^2-2xy+y^2)$
58. $(x+5y)(x^2-5xy+25y^2)$
59. $8(y-1)(y^2+y+1)$
60. $36(x+1)(x^2-x+1)$

40. $8x^3+1$ **41.** x^2+64y^2 **42.** $3x^3+81$

43. $4x^3-32$ **44.** $64x^3+27y^3$ **45.** $54x^3-2y^3$

46. $3x^4+375xy^3$ **47.** x^3y+y^4 **48.** x^4y^3-x

49. $x^2y^2-x^2y^5$ **50.** $2x^2-16x^2y^3$ **51.** $24x^4y+81xy^4$

52. x^6-64y^3 **53.** x^6-y^9 **54.** $64x^2+1$

55. $27x^3+y^6$ **56.** x^3+64z^3 **57.** $8x^3+y^3$

58. x^3+125y^3 **59.** $8y^3-8$ **60.** $36x^3+36$

61. $9x^2-y^2$ **62.** x^2-4y^2 **63.** x^4-16y^4

64. $81x^4-1$ **65.** $(x-y)^2-81$ **66.** $(x+2y)^2-25$

67. $(x^2-2xy+y^2)-36$ **68.** $(x^2+4xy+4y^2)-25$ **69.** $(16x^2+8x+1)-y^2$

70. $x^2-(y^2+6y+9)$

71. a. Represent the area of the shaded region of the square shown below as the difference of two squares.

b. Use the factors of the expression in part **a.** to draw (and label the sides of) a rectangle that has the same area as the shaded region.

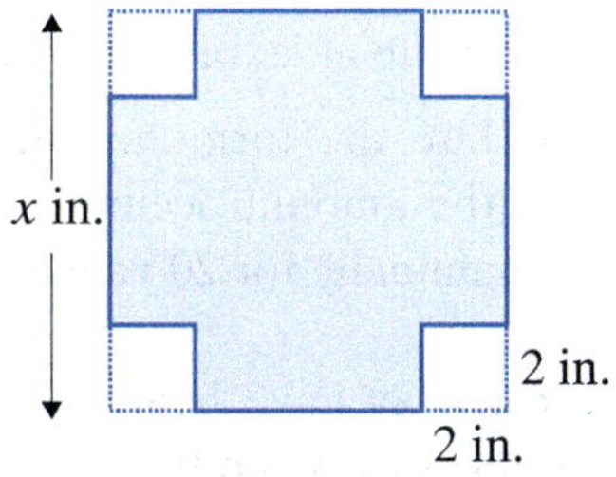

72. a. Use a polynomial function to represent the area of the shaded region of the square.

b. Use a polynomial function to represent the perimeter of the shaded figure.

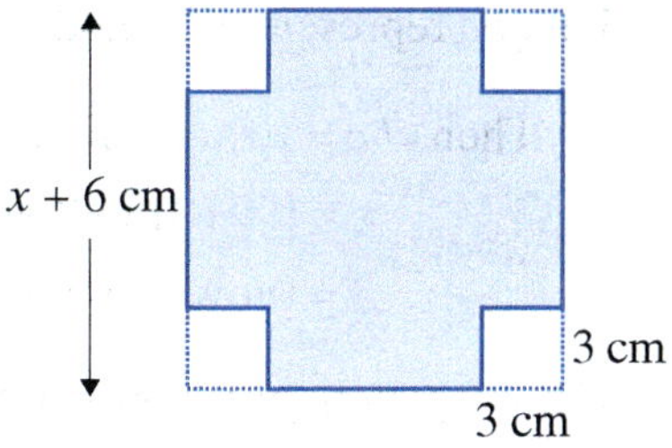

Writing and Thinking About Mathematics

73. a. Show that the sum of the areas of the rectangles and squares in the figure is a perfect square trinomial.

b. Rearrange the rectangles and squares in the form of a square and represent its area as the square of a binomial.

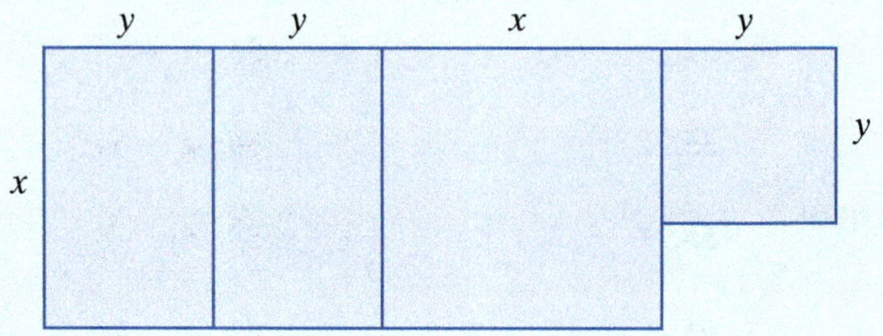

74. Compound interest is interest earned on interest. If a principal P is invested and compounded annually (once a year) at a rate of r, then the amount A_1 accumulated in one year is $A_1 = P + Pr$.
In factored form, we have: $A_1 = P + Pr = P(1+r)$.
At the end of the second year the amount accumulated is

$$A_2 = (P + Pr) + (P + Pr)r.$$

a. Write the expression for A_2 in factored form similar to that for A_1.

b. Write an expression for the amount accumulated in three years, A_3, in factored form.

c. Write an expression for A_n the amount accumulated in n years.

d. Use the formula you developed in part **c.** and your calculator to find the amount accumulated if $10,000 is invested at 6% and compounded annually for 20 years.

75. You may have heard of (or studied) the following rules for division of an integer by 3 and 9:

1. An integer is divisible by 3 if the sum of its digits is divisible by 3.

2. An integer is divisible by 9 if the sum of its digits is divisible by 9.

The proofs of both **1.** and **2.** can be started as follows:

Let abc represent a three-digit integer.

$$\begin{aligned} \text{Then } abc &= 100a + 10b + c \\ &= (99+1)a + (9+1)b + c \\ &= (\text{now you finish the proofs}) \end{aligned}$$

Use the pattern just shown and prove both **1.** and **2.** for a four-digit integer.

HAWKES LEARNING SYSTEMS: INTRODUCTORY & INTERMEDIATE ALGEBRA SOFTWARE

- 6.4a Special Factorizations - Squares
- 6.4b Special Factorizations - Cubes

61. $(3x-y)(3x+y)$ **62.** $(x-2y)(x+2y)$ **63.** $(x-2y)(x+2y)(x^2+4y^2)$ **64.** $(3x-1)(3x+1)(9x^2+1)$
65. $(x-y-9)(x-y+9)$ **66.** $(x+2y-5)(x+2y+5)$ **67.** $(x-y-6)(x-y+6)$ **68.** $(x+2y-5)(x+2y+5)$
69. $(4x+1-y)(4x+1+y)$ **70.** $(x+y+3)(x-y-3)$ **71. a.** x^2-16 **b.**

72. a. x^2+12x **b.** $4x+24$ **73. a.** $xy+xy+x^2+y^2=x^2+2xy+y^2=(x+y)^2$ **b.**

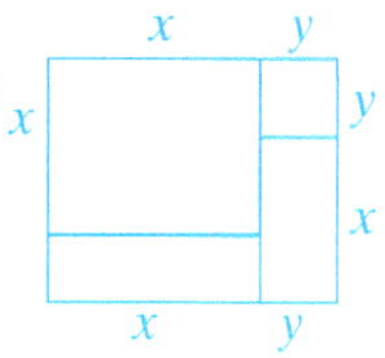

$(x+y)(x+y)=(x+y)^2$

74. a. $A_2=P(1+r)^2$ **b.** $A_3=P(1+r)^3$ **c.** $A_n=P(1+r)^n$ **d.** \$32,071.35

75. For a 3-digit integer:

$abc=100a+10b+c=(99+1)a+(9+1)b+c=9(11a+b)+a+b+c$

So, if the sum $(a+b+c)$ is divisible by 3 (or 9), then the number *abc* will be divisible by 3 (or 9).

For a 4-digit integer:

$abcd=1000a+100b+10c+d=(999+1)a+(99+1)b+(9+1)c+d=9(111a+11b+c)+a+b+c+d$

So, if the sum $(a+b+c+d)$ is divisible by 3 (or 9), then the number *abcd* will be divisible by 3 (or 9).

6.5 Additional Factoring Practice

- *General guidelines for factoring polynomials.*

As we have seen so far in Chapter 6, factoring is a basic algebraic skill. Factoring is useful in solving equations and, as we will see in Chapter 7, necessary for simplifying algebraic fractions. This section is provided as extra practice in factoring. The following guidelines are provided for easy reference.

General Guidelines for Factoring Polynomials

1. **Always look for a common monomial factor first.** If the leading coefficient is negative, factor out a negative monomial even if it is just –1.
2. **Check the number of terms.**
   a. Two terms:
      1. difference of two squares? – factorable
      2. sum of two squares? – not factorable
      3. difference of two cubes? – factorable
      4. sum of two cubes? – factorable
   b. Three terms:
      1. perfect square trinomial?
      2. use trial-and-error method?
      Guidelines for the trial-and-error method
         a. If the sign of the constant term is positive, the signs in both factors will be the same, either both positive or both negative.
         b. If the sign of the constant term is negative, the signs in the factors will be different, one positive and one negative.
      3. use *ac*-method?
      Guidelines for the *ac*-method
         a. Multiply $a \cdot c$.
         b. Find two integers whose product is ac and whose sum is b.
         If this is not possible, the trinomial is not factorable.
         c. Rewrite the middle term (bx) using the two numbers found in step **b.** as coefficients.
         d. Factor by grouping.
   c. Four terms:
      Group terms with a common factor and factor out any common binomial factor.
3. **Check the possibility of factoring any of the factors.**

Checking: Factoring can be checked by multiplying the factors. The product should be the original expression.

6.5 Exercises

1. $(m+6)(m+1)$
2. $(a-3)(a-1)$
3. $(x+9)(x+2)$
4. $(y+5)(y+3)$
5. $(x-10)(x+10)$
6. $(n-6)(n-2)$
7. $(m-3)(m+2)$
8. $(y-7)(y+7)$
9. Not factorable
10. $-(x+7)(x+5)$
11. $(8a-1)(8a+1)$
12. Not factorable
13. $(x+5)^2$
14. $(x+5)(x-2)$
15. $(x+12)(x-3)$
16. $(x+8)^2$
17. $3(a+6)(a-2)$
18. $-2(y-5)(y-7)$
19. $-5(x-6)(x-8)$
20. $7(t-4)(t+6)$
21. Not factorable
22. $3(x-7)(x+7)$
23. $x(x-6)(x+2)$
24. $3n(n+3)(n+2)$
25. $-2a(a+8)(a-7)$
26. $-4x(x-10)(x+5)$
27. $4x(2x-5)(2x+5)$
28. $3x(4x-3)(4x+3)$
29. $-(x-5)(3x-2)$
30. $(2x+1)(x+3)$
31. $(2x-1)(3x-4)$
32. $(2x-5)(6x-1)$
33. $(4m+3)(3m-2)$
34. $(3t-7)(2t+5)$
35. $2(2x-1)(x-3)$

Completely factor each of the given polynomials. If a polynomial cannot be factored, write "not factorable."

1. m^2+7m+6 **2.** a^2-4a+3 **3.** $x^2+11x+18$

4. $y^2+8y+15$ **5.** x^2-100 **6.** $n^2-8n+12$

7. m^2-m-6 **8.** y^2-49 **9.** $a^2+2a+24$

10. $-x^2-12x-35$ **11.** $64a^2-1$ **12.** $49x^2+4$

13. $x^2+10x+25$ **14.** $x^2+3x-10$ **15.** $x^2+9x-36$

16. $x^2+16x+64$ **17.** $3a^2+12a-36$ **18.** $-2y^2+24y-70$

19. $-5x^2+70x-240$ **20.** $7t^2+14t-168$ **21.** $64+49t^2$

22. $3x^2-147$ **23.** x^3-4x^2-12x **24.** $3n^3+15n^2+18n$

25. $112a-2a^2-2a^3$ **26.** $200x+20x^2-4x^3$ **27.** $16x^3-100x$

28. $48x^3-27x$ **29.** $-3x^2+17x-10$ **30.** $2x^2+7x+3$

31. $6x^2-11x+4$ **32.** $12x^2-32x+5$ **33.** $12m^2+m-6$

34. $6t^2+t-35$ **35.** $4x^2-14x+6$ **36.** $-4x^2+18x-20$

37. $8x^2+6x-35$ **38.** $12x^2+5x-3$ **39.** $20x^2-21x-54$

40. $21x^2-x-10$ **41.** $14+11x-15x^2$ **42.** $24+x-3x^2$

43. $-8a^2+22a-15$ **44.** $63x^2-40x-12$ **45.** $20y^2+9y-20$

46. $35x^2-x-6$ **47.** $18x^2-15x+2$ **48.** $12x^2-47x+11$

49. $-150x^2+96$ **50.** $252x-175x^3$ **51.** $12n^2-60n-75$

52. $-12x^3-2x^2-70x$ **53.** $21a^3-13a^2-2a$ **54.** $13x^3+120x^2+100x$

55. $36x^3+21x^2-30x$ **56.** $63x-3x^2-30x^3$ **57.** $16x^3-52x^2+22x$

58. $24y^3-4y^2-160y$ **59.** $75+10m+120m^2$ **60.** $144x^3-10x^2-50x$

61. $xy+3y-4x-12$ **62.** $2xz+10x+z+5$ **63.** $x^2+2xy-6x-12y$

36. $-2(2x-5)(x-2)$
37. $(4x-7)(2x+5)$
38. $(4x+3)(3x-1)$
39. $(5x+6)(4x-9)$
40. $(7x-5)(3x+2)$
41. $-(5x-7)(3x+2)$
42. $-(x-3)(3x+8)$
43. $-(2a-3)(4a-5)$
44. $(7x-6)(9x+2)$
45. $(4y+5)(5y-4)$

64. $2y^2+6yz+5y+15z$ **65.** $-x^3+8x^2+5x-40$ **66.** $2x^3-14x^2-3x+21$

67. x^3+125 **68.** y^3-1000 **69.** $x^4y^3-x^4$

70. $x^6y^3-x^3$ **71.** $8a^6+27b^6$ **72.** a^9+64b^3

73. x^6y^3-125 **74.** x^3y^3+216 **75.** $x^3+7x^2-9x-63$

76. $x^5+5x^4-4x-20$ **77.** $9x^2-(y+6)^2$ **78.** $(x+2)^2-25a^2$

79. $(y^2+20y+100)-49x^2$ **80.** $(t^2+22t+121)-16s^2$

HAWKES LEARNING SYSTEMS: INTRODUCTORY & INTERMEDIATE ALGEBRA SOFTWARE

- 6.5 Additional Factoring Practice

46. $(7x-3)(5x+2)$ 47. $(6x-1)(3x-2)$ 48. $(3x-11)(4x-1)$ 49. $-6(5x-4)(5x+4)$
50. $-7x(5x-6)(5x+6)$ 51. $3(4n^2-20n-25)$ 52. $-2x(6x^2+x+35)$ 53. $a(21a^2-13a-2)$
54. $x(13x^2+120x+100)$ 55. $3x(3x-2)(4x+5)$ 56. $-3x(5x-7)(2x+3)$ 57. $2x(2x-1)(4x-11)$
58. $4y(2y+5)(3y-8)$ 59. $5(24m^2+2m+15)$ 60. $2x(9x+5)(8x-5)$ 61. $(y-4)(x+3)$
62. $(2x+1)(z+5)$ 63. $(x+2y)(x-6)$ 64. $(2y+5)(y+3z)$ 65. $-(x^2-5)(x-8)$
66. $(2x^2-3)(x-7)$ 67. $(x+5)(x^2-5x+25)$ 68. $(y-10)(y^2-20y+100)$
69. $x^4(y-1)(y^2+y+1)$ 70. $x^3(xy-1)(x^2y^2+xy+1)$ 71. $(2a^2+3b^2)(4a^4-6a^2b^2+9b^4)$
72. $(a^3+4b)(a^6-4a^3b+16b^2)$ 73. $(x^2y-5)(x^4y^2+5x^2y+25)$ 74. $(xy+6)(x^2y^2-6xy+36)$
75. $(x-3)(x+3)(x+7)$ 76. $(x+5)(x^2-2)(x^2+2)$ 77. $(3x+y+6)(3x-y-6)$
78. $(x+2+5a)(x+2-5a)$ 79. $(y+10+7x)(y+10-7x)$ 80. $(t+11+4s)(t+11-4s)$

6.6 Solving Quadratic Equations by Factoring

- *Solve quadratic equations by factoring.*
- *Write equations given the roots.*

In Chapter 6 the emphasis has been on second-degree polynomials and functions. Such polynomials are particularly evident in physics with the path of thrown objects (or projectiles) affected by gravity, in mathematics involving area (area of a circle, rectangle, square, and triangle), and in many situations in higher level mathematics. The methods of factoring we have studied (*ac*-method, trial-and-error method, difference of two squares, and perfect square trinomials) have been related, in general, to second-degree polynomials. Second-degree polynomials are called **quadratic polynomials** (or **quadratics**) and, as just discussed, they play a major role in many applications of mathematics. In fact, quadratic polynomials and techniques for solving quadratic equations are central topics in the first two courses in algebra.

Solving Quadratic Equations by Factoring

Quadratic Equations

Quadratic equations are equations that can be written in the form

$$\mathbf{ax^2+bx+c=0} \text{ where } a, b, \text{ and } c \text{ are real numbers and } a \neq 0.$$

The form $ax^2+bx+c=0$ is called the **standard form** (or **general form**) of a **quadratic equation**. In the standard form, a quadratic polynomial is on one side of the equation and 0 is on the other side. For example,

$x^2-8x+12=0$ is a quadratic equation in standard form.

While

$3x^2-2x=27$ and $x^2=25x$ are quadratic equations, just not in standard form.

These last two equations can be manipulated algebraically so that 0 is on one side. When solving quadratic equations by factoring, having 0 on one side before we factor is necessary because of the following **zero-factor property**.

Zero-Factor Property

If the product of two (or more) factors is 0, then at least one of the factors must be 0. That is, if a and b are real numbers, then

if $a \cdot b = 0$, then $a = 0$ or $b = 0$ or both.

Example 1: Solving Factored Quadratic Equations

Solve the following quadratic equation.

$$(x-5)(2x-7)=0$$

Solution: Since the quadratic is already factored and the other side of the equation is 0, we use the zero-factor property and set each factor equal to 0. This process yields two linear equations, which can, in turn, be solved.

$$x-5=0 \quad \text{or} \quad 2x-7=0$$
$$x=5 \quad \text{or} \quad 2x=7$$
$$x=\frac{7}{2}$$

Thus the **two solutions** (or **roots**) to the original equation are $x = 5$ and $x=\frac{7}{2}$. Or, we can say that the solution set is $\left\{\frac{7}{2}, 5\right\}$.

The solutions can be **checked** by substituting them one at a time for x in the equation. That is, there will be two "checks."

Substituting $x = 5$ gives

$$\begin{aligned}(x-5)(2x-7)&=(5-5)(2\cdot5-7)\\&=(0)(3)\\&=0.\end{aligned}$$

Substituting $x=\frac{7}{2}$ gives

$$\begin{aligned}(x-5)(2x-7)&=\left(\frac{7}{2}-5\right)\left(2\cdot\frac{7}{2}-7\right)\\&=\left(-\frac{3}{2}\right)(7-7)\\&=\left(-\frac{3}{2}\right)(0)\\&=0.\end{aligned}$$

Therefore, both 5 and $\frac{7}{2}$ are solutions to the original equation.

Teaching Notes:
Point out that solving quadratic equations begins very differently from that of a linear equation, which isolates the variable as soon as possible. Rather, with quadratics, we gather all terms to one side and the other side must be 0.

In Example 1 the polynomial was factored and the other side of the equation was 0. The equation was solved by setting each factor, in turn, equal to 0 and solving the resulting linear equations. These solutions tell us that the original equation has two solutions. **In general, a quadratic equation has two solutions.** In the special cases where the two factors are the same, there is only one solution and it is called a **double solution** (or **double root**). The following examples show how to solve quadratic equations by factoring. Remember that the equation must be in standard form with one side of the equation equal to 0. Study these examples carefully.

Example 2: Solving Quadratic Equations by Factoring

Solve the following equations by writing the equation in standard form with one side 0 and factoring the polynomial. Then set each factor equal to 0 and solve. Checking is left as an exercise for the student.

a. $3x^2 = 6x$

Solution:

$3x^2 = 6x$

$3x^2 - 6x = 0$ — Write the equation in standard form with 0 on one side by subtracting $6x$ from both sides.

$3x(x-2) = 0$ — Factor out the common monomial, $3x$.

$3x = 0$ or $x-2 = 0$ — Set each factor equal to 0.

$x = 0$ or $x = 2$ — Solve each linear equation.

The solutions are 0 and 2. Or, we can say that the solution set is $\{0, 2\}$.

b. $x^2 - 8x + 16 = 0$

Solution:

$x^2 - 8x + 16 = 0$

$(x-4)^2 = 0$ — The trinomial is a perfect square.

$x - 4 = 0$ or $x - 4 = 0$ — Both factors are the same, so there is only one distinct solution.

$x = 4$ or $x = 4$

The only solution is 4, and it is called a **double solution** (or **double root**). The solution set is $\{4\}$.

c. $4x^2 - 4x = 24$

Solution:

$4x^2 - 4x = 24$

$4x^2 - 4x - 24 = 0$ — Add –24 to both sides so that one side is 0.

$4(x^2 - x - 6) = 0$ — Factor out the common monomial, 4.

$4(x-3)(x+2) = 0$

$x - 3 = 0$ or $x + 2 = 0$ — The constant factor 4 can never be 0 and does not affect the solution.

$x = 3$ $\quad$ $x = -2$

The solutions are 3 and –2. Or, we can say the solution set is $\{-2, 3\}$.

Continued on the next page...

d. $(x+5)^2 = 36$

Solution:

$$(x+5)^2 = 36$$

$$x^2 + 10x + 25 = 36 \quad \text{Expand } (x+5)^2.$$

$$x^2 + 10x - 11 = 0 \quad \text{Add } -36 \text{ to both sides so that one side is 0.}$$

$$(x+11)(x-1) = 0 \quad \text{Factor.}$$

$$x + 11 = 0 \quad \text{or} \quad x - 1 = 0$$

$$x = -11 \qquad x = 1$$

The solutions are –11 and 1.

e. $3x(x-1) = 2(5-x)$

Solution:

$$3x(x-1) = 2(5-x)$$

$$3x^2 - 3x = 10 - 2x \quad \text{Use the distributive property.}$$

$$3x^2 - 3x + 2x - 10 = 0 \quad \text{Arrange terms so that 0 is on one side.}$$

$$3x^2 - x - 10 = 0 \quad \text{Simplify.}$$

$$(3x+5)(x-2) = 0 \quad \text{Factor.}$$

$$3x + 5 = 0 \quad \text{or} \quad x - 2 = 0$$

$$3x = -5 \qquad x = 2$$

$$x = -\frac{5}{3}$$

The solutions are $-\frac{5}{3}$ and 2.

f. $\frac{2x^2}{15} - \frac{x}{3} = -\frac{1}{5}$

Solution:

$$\frac{2x^2}{15} - \frac{x}{3} = -\frac{1}{5}$$

$$15 \cdot \frac{2x^2}{15} - \frac{x}{3} \cdot 15 = -\frac{1}{5} \cdot 15 \quad \text{Multiply each term by 15, the LCM of the denominators, to get integer coefficients.}$$

$$2x^2 - 5x = -3 \quad \text{Simplify.}$$

$$2x^2 - 5x + 3 = 0 \quad \text{Add 3 to both sides so that one side is 0.}$$

$$(2x-3)(x-1) = 0 \quad \text{Factor by the trial-and-error method or the } ac\text{-method.}$$

$$2x - 3 = 0 \quad \text{or} \quad x - 1 = 0$$
$$2x = 3 \qquad\qquad x = 1$$
$$x = \frac{3}{2}$$

The solutions are $\frac{3}{2}$ and 1.

In the next example, we show that factoring can be used to solve equations of degrees higher than second-degree. There will be more than two factors, but just as with quadratics, if the product is 0, then the solutions are found by setting each factor equal to 0.

Example 3: Solving Higher Degree Equations

Solve the following equation: $2x^3 - 4x^2 - 6x = 0$.

Solution: $2x^3 - 4x^2 - 6x = 0$

$2x(x^2 - 2x - 3) = 0$ Factor out the common monomial, $2x$.

$2x(x - 3)(x + 1) = 0$ Factor the trinomial.

$2x = 0$ or $x - 3 = 0$ or $x + 1 = 0$ Set each factor equal to 0 and solve.

$x = 0$ $\quad x = 3$ $\quad x = -1$

The solutions are 0, 3, and –1. Or, we can say that the solution set is $\{-1, 0, 3\}$.

The procedure to solve a quadratic equation can be summarized as follows.

To Solve a Quadratic Equation by Factoring

1. Add or subtract terms as necessary so that 0 is on one side of the equation and the equation is in the standard form $ax^2 + bx + c = 0$ where a, b, and c are real numbers and $a \neq 0$.
2. Factor completely. (If there are any fractional coefficients, multiply each term by the least common denominator first so that all coefficients will be integers.)
3. Set each nonconstant factor equal to 0 and solve each linear equation for the unknown.
4. Check each solution, one at a time, in the original equation.

Special Comment: All of the quadratic equations in this section can be solved by factoring. That is, all of the quadratic polynomials are factorable. However, as we have seen in some of the previous sections, not all polynomials are factorable. In Chapter 10 we will develop techniques (other than factoring) for solving quadratic equations whether the quadratic polynomial is factorable or not.

NOTES

COMMON ERROR

A **common error** is to divide both sides of an equation by the variable x. This error can be illustrated by using the equation in Example 2a.

$$3x^2 = 6x$$

$$\frac{3x^2}{x} = \frac{6x}{x}$$ **Do not** divide by x because you lose the solution $x = 0$. INCORRECT

$$3x = 6$$

$$x = 2$$

Factoring is the method to use. By factoring, you will find all solutions as shown in the previous examples.

Finding an Equation Given the Roots

To help develop a complete understanding of the relationships between factors, factoring, and solving equations, we reverse the process of finding solutions. That is, we want to **find an equation that has certain given solutions (or roots)**. For example, to find an equation that has the roots

$$x = 4 \quad \text{and} \quad x = -7$$

we proceed as follows:

1. Rewrite the linear equations with 0 on one side:
$$x - 4 = 0 \quad \text{and} \quad x + 7 = 0.$$
2. Form the product of the factors and set this product equal to 0:
$$(x-4)(x+7) = 0.$$
3. Multiply the factors. The resulting quadratic equation must have the two given roots (because we know the factors):
$$x^2 + 3x - 28 = 0.$$

NOTES

This equation can be multiplied by any nonzero constant and a new equation will be formed, but it will still have the same solutions, namely 4 and –7. Thus technically, there are many equations with these two roots.

The formal reasoning is based on the following theorem called the **factor theorem**.

Factor Theorem

If $x = c$ is a root of a polynomial equation in the form $P(x) = 0$, then $x - c$ is a factor of the polynomial $P(x)$.

Example 4: Using the Factor Theorem to Find Equations with Given Roots

Find a polynomial equation with integer coefficients that has the given roots:

$$x = 3 \text{ and } x = -\frac{2}{3}.$$

Solution: Form the linear equations and then find the product of the factors.

$x = 3$ $\qquad$ $x = -\frac{2}{3}$

$x - 3 = 0$ $\qquad$ $x + \frac{2}{3} = 0$ $\qquad$ Set each equation equal to 0.

$3x + 2 = 0$ $\qquad$ Multiply by 3 to get integer coefficients.

Form the equation by setting the product of the factors equal to 0 and multiplying.

$(x-3)(3x+2) = 0$ $\qquad$ The equation has integer coefficients.

$3x^2 - 7x - 6 = 0$ $\qquad$ This quadratic equation has the two given roots.

Practice Problems

Solve each equation by factoring.

1. $x^2 - 6x = 0$
2. $6x^2 - x - 1 = 0$
3. $(x-2)^2 - 25 = 0$
4. $x^3 - 8x^2 + 16x = 0$
5. Find a polynomial equation with integer coefficients that has the given roots: $x = -4$ and $x = \frac{3}{5}$.

Answers to Practice Problems: **1.** $x = 0, x = 6$ **2.** $x = \frac{1}{2}, x = -\frac{1}{3}$ **3.** $x = 7, x = -3$ **4.** $x = 0, x = 4$ **5.** $5x^2 + 17x - 12 = 0$

6.6 Exercises

1. $x = 2, 3$
2. $x = -5, 2$
3. $x = -2, \frac{9}{2}$
4. $x = -7, \frac{4}{3}$
5. $x = -3$
6. $x = -10, 10$
7. $x = -5$
8. $x = -5, 5$
9. $x = 0, 2$
10. $x = -3, 0$
11. $x = -6$
12. $x = 9$
13. $x = -1, 4$
14. $x = -4, -3$
15. $x = -3, 4$
16. $x = 2, 9$
17. $x = -3, 0$
18. $x = 0, 3$
19. $x = 2, 4$
20. $x = -5, 6$
21. $x = -4, 3$
22. $x = -5, -2$
23. $x = -\frac{1}{2}, 3$
24. $x = -1, \frac{3}{2}$
25. $x = -\frac{2}{3}, 2$
26. $x = 1, \frac{5}{3}$
27. $x = -\frac{1}{2}, 4$
28. $x = -\frac{3}{2}, -\frac{1}{2}$
29. $x = -2, \frac{4}{3}$
30. $x = -\frac{2}{3}, -\frac{1}{2}$
31. $x = \frac{3}{2}$
32. $x = \frac{6}{5}$

Solve the equations.

1. $(x-3)(x-2)=0$ **2.** $(x+5)(x-2)=0$ **3.** $(2x-9)(x+2)=0$

4. $(x+7)(3x-4)=0$ **5.** $0=(x+3)(x+3)$ **6.** $0=(x+10)(x-10)$

7. $(x+5)(x+5)=0$ **8.** $(x+5)(x-5)=0$ **9.** $2x(x-2)=0$

10. $3x(x+3)=0$ **11.** $(x+6)^2=0$ **12.** $5(x-9)^2=0$

Solve the equations by factoring.

13. $x^2-3x-4=0$ **14.** $x^2+7x+12=0$ **15.** $x^2-x-12=0$

16. $x^2-11x+18=0$ **17.** $0=x^2+3x$ **18.** $0=x^2-3x$

19. $x^2+8=6x$ **20.** $x^2=x+30$ **21.** $2x^2+2x-24=0$

22. $9x^2+63x+90=0$ **23.** $0=2x^2-5x-3$ **24.** $0=2x^2-x-3$

25. $3x^2-4x-4=0$ **26.** $3x^2-8x+5=0$ **27.** $2x^2-7x=4$

28. $4x^2+8x=-3$ **29.** $-2x=3x^2-8$ **30.** $6x^2+2=-7x$

31. $4x^2-12x+9=0$ **32.** $25x^2-60x+36=0$ **33.** $8x=5x^2$

34. $15x=3x^2$ **35.** $9x^2-36=0$ **36.** $4x^2-16=0$

37. $5x^2=10x-5$ **38.** $2x^2=4x+6$ **39.** $8x^2+32=32x$

40. $6x^2=18x+24$ **41.** $\frac{x^2}{9}=1$ **42.** $\frac{x^2}{2}=8$

43. $\frac{x^2}{5}-x-10=0$ **44.** $\frac{2}{3}x^2+2x-\frac{20}{3}=0$ **45.** $\frac{x^2}{8}+x+\frac{3}{2}=0$

46. $\frac{x^2}{6}-\frac{1}{2}x-3=0$ **47.** $x^2-x+\frac{1}{4}=0$ **48.** $\frac{x^2}{3}-2x+3=0$

49. $x^3+8x=6x^2$ **50.** $x^3=x^2+30x$ **51.** $6x^3+7x^2=-2x$

52. $3x^3=8x-2x^2$ **53.** $0=x^2-100$ **54.** $0=x^2-121$

55. $3x^2-75=0$ **56.** $5x^2-45=0$ **57.** $x^2+8x+16=0$

33. $x = 0, \frac{8}{5}$
34. $x = 0, 5$
35. $x = -2, 2$
36. $x = -2, 2$
37. $x = 1$
38. $x = -1, 3$
39. $x = 2$
40. $x = -1, 4$
41. $x = -3, 3$
42. $x = -4, 4$
43. $x = -5, 10$
44. $x = -5, 2$
45. $x = -6, -2$
46. $x = -3, 6$
47. $x = \frac{1}{2}$
48. $x = 3$
49. $x = 0, 2, 4$
50. $x = -5, 0, 6$
51. $x = -\frac{2}{3}, -\frac{1}{2}, 0$
52. $x = -2, 0, \frac{4}{3}$
53. $x = -10, 10$

58. $x^2 + 14x + 49 = 0$ **59.** $3x^2 = 18x - 27$ **60.** $5x^2 = 10x - 5$

61. $(x-1)^2 = 4$ **62.** $(x - 3)^2 = 1$ **63.** $(x+5)^2 = 9$

64. $(x+4)^2 = 16$ **65.** $(x+4)(x-1) = 6$ **66.** $(x-5)(x+3) = 9$

67. $27 = (x+2)(x-4)$ **68.** $-1 = (x+2)(x+4)$ **69.** $x(x+7) = 3(x+4)$

70. $x(x+9) = 6(x+3)$ **71.** $3x(x+1) = 2(x+1)$ **72.** $2x(x-1) = 3(x-1)$

73. $x(2x+1) = 6(x+2)$ **74.** $3x(x+3) = 2(2x-1)$

Write a polynomial equation with integer coefficients that has the given roots.

75. $y = 3, y = -2$ **76.** $x = 5, x = 7$ **77.** $x = -5, x = -\frac{1}{2}$

78. $x = \frac{1}{4}, x = -1$ **79.** $x = \frac{1}{2}, x = \frac{3}{4}$ **80.** $y = \frac{2}{3}, y = \frac{1}{6}$

81. $x = 0, x = 3, x = -2$ **82.** $y = 0, y = -4, y = 1$

83. $y = -2, y = 3, y = 3$ (3 is a double root.)

84. $x = -1, x = -1, x = -1$ (-1 is a triple root.)

Writing and Thinking About Mathematics

54. $x = -11, 11$
55. $x = -5, 5$
56. $x = -3, 3$
57. $x = -4$
58. $x = -7$
59. $x = 3$
60. $x = 1$
61. $x = -1, 3$
62. $x = 2, 4$
63. $x = -8, -2$
64. $x = -8, 0$
65. $x = -5, 2$
66. $x = -4, 6$
67. $x = -5, 7$
68. $x = -3$
69. $x = -6, 2$
70. $x = -6, 3$
71. $x = -1, \frac{2}{3}$

85. When solving equations by factoring, one side of the equation must be 0. Explain why this is so.

86. In solving the equation $(x+5)(x-4) = 6$, why can't we just put one factor equal to 3 and the other equal to 2? Certainly $3 \cdot 2 = 6$.

87. A ball is dropped from the top of a building that is 784 feet high. The height of the ball above ground level is given by the polynomial function $h(t) = -16t^2 + 784$ where t is measured in seconds.

a. How high is the ball after 3 seconds? 5 seconds?

b. How far has the ball traveled in 3 seconds? 5 seconds?

c. When will the ball hit the ground? Explain your reasoning in terms of factors.

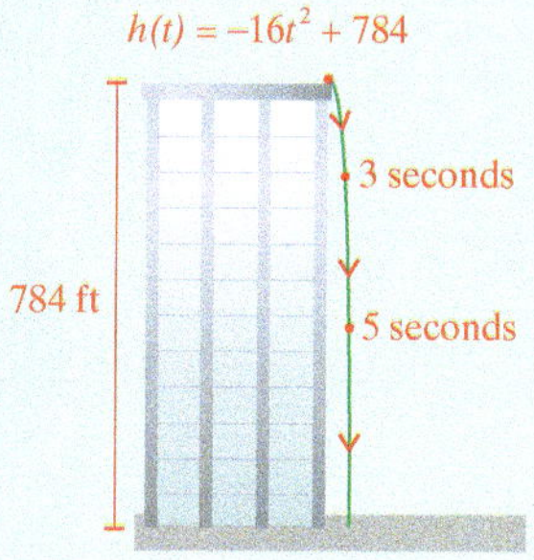

HAWKES LEARNING SYSTEMS: INTRODUCTORY & INTERMEDIATE ALGEBRA SOFTWARE

- 6.6 Solving Quadratic Equations by Factoring

72. $x = 1, \frac{3}{2}$ **73.** $x = -\frac{3}{2}, 4$ **74.** $x = -1, -\frac{2}{3}$ **75.** $y^2 - y - 6 = 0$ **76.** $x^2 - 12x + 35 = 0$ **77.** $2x^2 + 11x + 5 = 0$
78. $4x^2 + 3x - 1 = 0$ **79.** $8x^2 - 10x + 3 = 0$ **80.** $18y^2 - 15y + 2 = 0$ **81.** $x^3 - x^2 - 6x = 0$ **82.** $y^3 + 3y^2 - 4y = 0$
83. $y^3 - 4y^2 - 3y + 18 = 0$ **84.** $x^3 + 3x^2 + 3x + 1 = 0$
85. This allows for use of the zero-factor property which says that for the equation to equal zero one of the factors must equal zero. Answers will vary. **86.** Answers will vary.
87. a. 640 ft; 384 ft **b.** 144 ft; 400 ft **c.** 7 seconds; $0 = -16(t+7)(t-7)$

6.7 Applications of Quadratic Equations

- *Use quadratic equations to solve problems related to numbers, geometry, and consecutive integers.*
- *Solve problems related to the* ***Pythagorean theorem****.*

Whether or not application problems cause you difficulty depends a great deal on your personal experiences and general reasoning abilities. These abilities are developed over a long period of time. A problem that is easy for you, possibly because you have had experience in a particular situation, might be quite difficult for a friend and vice versa.

Most problems do not say specifically to add, subtract, multiply, or divide. You are to know from the nature of the problem what to do. You are to ask yourself, "What information is given? What am I trying to find? What tools, skills, and abilities do I need to use?"

Word problems should be approached in an orderly manner. You should have an "attack plan."

Attack Plan for Application Problems

1. Read the problem carefully at least twice.
2. Decide what is asked for and assign a variable or variable expression to the unknown quantities.
3. Organize a chart, table, or diagram relating all the information provided.
4. Form an equation. (A formula of some type may be necessary.)
5. Solve the equation.
6. Check your solution with the wording of the problem to be sure it makes sense.

Several types of problems lead to quadratic equations. The problems in this section are set up so that the equations can be solved by factoring. More general problems and approaches to solving quadratic equations are discussed in Chapter 10.

Example 1: Applications of Quadratic Equations

a. One number is four more than another and the sum of their squares is 296. What are the numbers?

Solution: Let x = smaller number. Then $x+4$ = larger number.

$$x^2+(x+4)^2=296 \quad \text{Add the squares.}$$

$$x^2+x^2+8x+16=296 \quad \text{Expand } (x+4)^2.$$

$$2x^2+8x-280=0 \quad \text{Write the equation in standard form.}$$

$$2(x^2+4x-140)=0 \quad \text{Factor out the GCF.}$$

$$2(x+14)(x-10)=0 \quad \text{Factor the trinomial.}$$

$$x+14=0 \quad \text{or} \quad x-10=0$$

$$x=-14 \qquad x=10$$

$$x+4=-10 \qquad x+4=14$$

There are two sets of answers to the problem: 10 and 14 or –14 and –10.

Check: $10^2+14^2=100+196=296$

and $(-14)^2+(-10)^2=196+100=296$

b. In an orange grove, there are 10 more trees in each row than there are rows. How many rows are there if there are 96 trees in the grove?

Solution: Let r = number of rows.
Then $r+10$ = number of trees per row.
Set up the equation and solve.

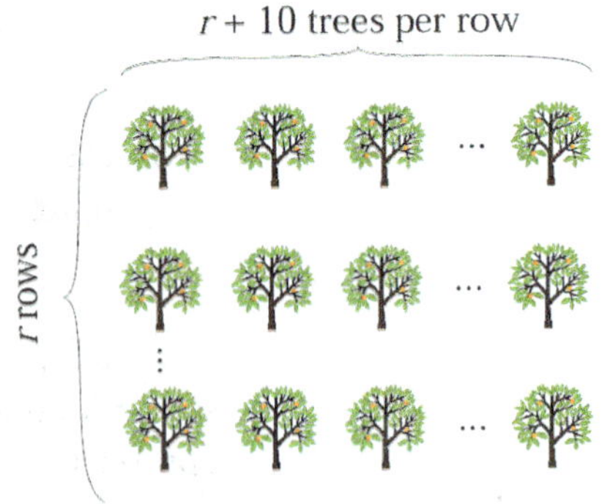

$$r(r+10)=96$$

$$r^2+10r=96$$

$$r^2+10r-96=0$$

$$(r-6)(r+16)=0$$

$$r-6=0 \quad \text{or} \quad r+16=0$$

$$r=6 \qquad r=-16$$

There are 6 rows in the grove ($6\cdot16=96$ trees).

Note: While –16 is a solution to the equation, –16 does not fit the conditions of the problem and is discarded. You cannot have –16 rows.

c. A rectangle has an area of 135 square meters and perimeter of 48 meters. What are the dimensions of the rectangle?

Solution: The area of a rectangle is the product of its length and width ($A = lw$). The perimeter of a rectangle is given by $P = 2l + 2w$.

Since the perimeter is 48 meters, then the length plus the width must be 24 meters (one-half of the perimeter).

Let w = width. Then $24 - w$ = length.

Set up the equation and solve.

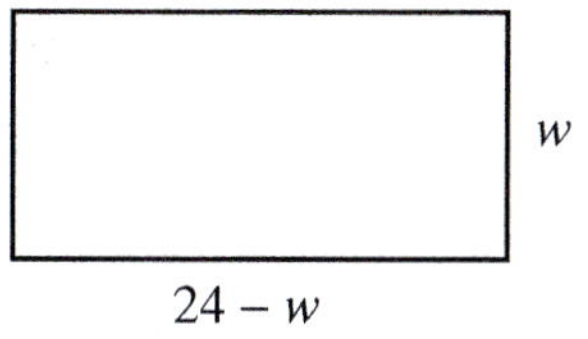

$$(24 - w) \cdot w = 135 \quad \text{Area} = lw = \text{length times width}$$

$$24w - w^2 = 135$$

$$0 = w^2 - 24w + 135$$

$$0 = (w - 9)(w - 15)$$

$$w - 9 = 0 \quad \text{or} \quad w - 15 = 0$$

$$w = 9 \qquad w = 15$$

$$l = 24 - 9 = 15 \qquad l = 24 - 15 = 9$$

The dimensions are 9 meters by 15 meters ($9 \cdot 15 = 135$).

d. A man wants to build a fence on three sides of a rectangular-shaped lot he owns. If 180 feet of fencing is needed and the area of the lot is 4000 square feet, what are the dimensions of the lot?

Solution: Let x = one of two equal sides. Then $180 - 2x$ = third side. Set up the equation and solve.

$$x(180 - 2x) = 4000$$

$$180x - 2x^2 = 4000$$

$$0 = 2x^2 - 180x + 4000$$

$$0 = 2(x^2 - 90x + 2000)$$

$$0 = 2(x - 50)(x - 40)$$

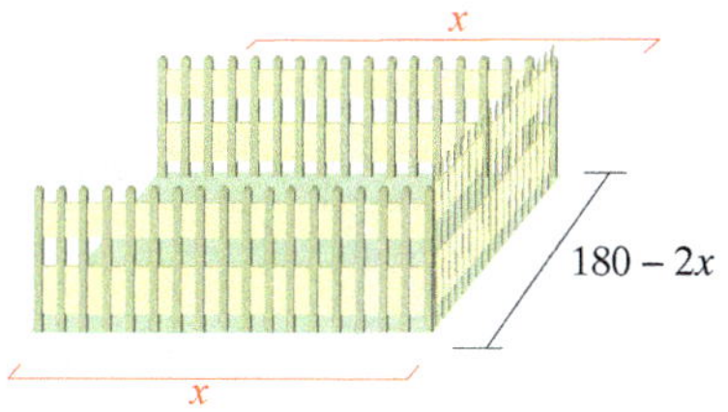

$$x - 50 = 0 \quad \text{or} \quad x - 40 = 0$$

$$x = 50 \qquad x = 40$$

$$180 - 2x = 180 - 2(50) = 80 \qquad 180 - 2x = 180 - 2(40) = 100 \quad \text{Third side}$$

Thus there are two possible answers: the lot is 50 feet by 80 feet or the lot is 40 feet by 100 feet.

Consecutive Integers

In Section 2.6, we discussed applications with **consecutive integers**, **consecutive even integers**, and **consecutive odd integers**. Because the applications involved only addition and subtraction, the related equations were first degree. In this section the applications involve squaring expressions and solving quadratic equations. For convenience, the definitions and representations of the various types of integers are repeated here.

Consecutive Integers

Integers are **consecutive** if each is 1 more than the previous integer. Three consecutive integers can be represented as $\boldsymbol{n}$, $\boldsymbol{n+1}$, and $\boldsymbol{n+2}$.
For example: 5, 6, 7

Consecutive Even Integers

Even integers are consecutive if each is 2 more than the previous even integer. Three consecutive even integers can be represented as $\boldsymbol{n}$, $\boldsymbol{n+2}$, and $\boldsymbol{n+4}$, where n is an even integer.
For example: 24, 26, 28

Consecutive Odd Integers

Odd integers are consecutive if each is 2 more than the previous odd integer. Three consecutive odd integers can be represented as $\boldsymbol{n}$, $\boldsymbol{n+2}$, and $\boldsymbol{n+4}$, where n is an odd integer.
For example: 41, 43, 45

Note that consecutive even and consecutive odd integers are represented in the same way. The value of the first integer *n* determines whether the remaining integers are even or odd.

Example 2: Consecutive Integers

a. Find two consecutive positive integers such that the sum of their squares is 265.

Solution: Let n = first integer
$n+1$ = next consecutive integer
Set up and solve the related equation.

$$n^2 + (n+1)^2 = 265$$

$$n^2 + n^2 + 2n + 1 = 265$$

$$2n^2 + 2n - 264 = 0$$
$$2(n^2 + n - 132) = 0$$
$$2(n+12)(n-11) = 0$$
$$n+12 = 0 \quad \text{or} \quad n-11 = 0$$
$$n = -12 \qquad n = 11$$
$$n+1 = -11 \qquad n+1 = 12$$

Consider the solution $n = -12$. The next consecutive integer, $n + 1$, is -11. While it is true that the sum of their squares is 265, we must remember that the problem calls for **positive** consecutive integers. Therefore, we can only consider positive solutions. Hence, the two integers are 11 and 12.

b. Find three consecutive odd integers such that the product of the first and second is 68 more than the third.

Solution: Let n = first odd integer
and $n + 2$ = second consecutive odd integer
and $n + 4$ = third consecutive odd integer.
Set up and solve the related equation.

$$n(n+2) = (n+4) + 68$$
$$n^2 + 2n = n + 72$$
$$n^2 + n - 72 = 0$$
$$(n+9)(n-8) = 0$$
$$n+9 = 0 \quad \text{or} \quad n-8 = 0$$
$$n = -9 \qquad n = 8$$
$$n+2 = -7 \qquad n+2 = 10$$
$$n+4 = -5 \qquad n+4 = 12$$

The three consecutive odd integers are –9, –7, and –5. Note that 8, 10 and 12 are even and therefore cannot be considered a solution to the problem.

The Pythagorean Theorem

Pythagoras

A geometric topic that often generates quadratic equations is right triangles. In a **right triangle**, one of the angles is a right angle (measures 90°), and the side opposite this angle (the longest side) is called the **hypotenuse**. The other two sides are called **legs**. Pythagoras (c. 585 – 501 B.C.), a famous Greek mathematician, is given credit for proving the following very important and useful theorem (even though history indicates that the Chinese knew of this theorem centuries before Pythagoras). Now there are entire books written that contain only proofs of the Pythagorean theorem developed by mathematicians since the time of Pythagoras. (You might want to visit the library!)

You will see the Pythagorean theorem stated again in Chapter 10 and used throughout your studies in mathematics.

The Pythagorean Theorem

In a right triangle, if c is the length of the hypotenuse and a and b are the lengths of the legs, then

$$c^2 = a^2 + b^2.$$

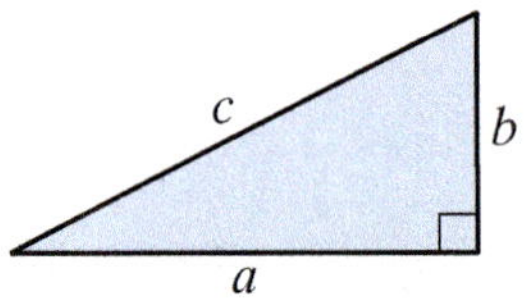

Example 3: The Pythagorean Theorem

A support wire is 25 feet long and stretches from a tree to a point on the ground. The point of attachment on the tree is 5 feet higher than the distance from the base of the tree to the point of attachment on the ground. How far up the tree is the point of attachment?

Solution: Let x = distance from base of tree to point of attachment on ground, then $x + 5$ = height of point of attachment on tree.

By the Pythagorean theorem, we have:

$$(x+5)^2 + x^2 = 25^2$$

$$x^2 + 10x + 25 + x^2 = 625$$

$$2x^2 + 10x - 600 = 0$$

$$2(x^2 + 5x - 300) = 0$$

$$2(x-15)(x+20) = 0$$

$$x = 15 \text{ or } x = -20.$$

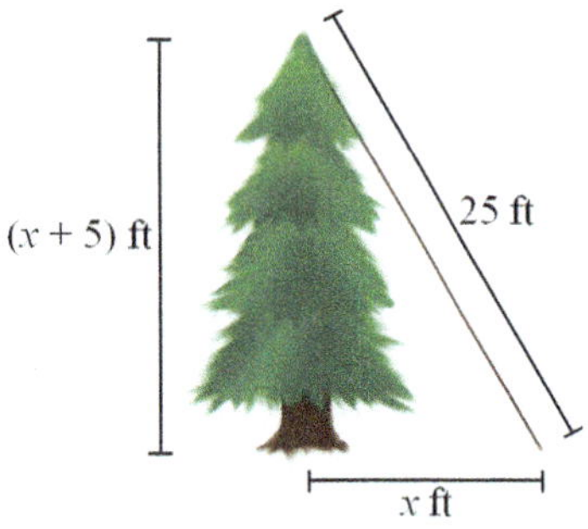

Because distance is positive, –20 is not a possible solution.

The solution is

$$x = 15$$

$$\text{and } x + 5 = 20.$$

Thus the point of attachment is 20 feet up the tree.

6.7 Exercises

Write a quadratic equation for each of the following problems. Then solve the equation.

1. One number is eight more than another. Their product is –16. What are the numbers?

2. One number is 10 more than another. If their product is –25, find the numbers.

3. The square of an integer is equal to seven times the integer. Find the integer.

4. The square of an integer is equal to twice the integer. Find the integer.

5. If the square of a positive integer is added to three times the integer, the result is 28. Find the integer.

6. If the square of a positive integer is added to three times the integer, the result is 54. Find the integer.

7. One number is seven more than another. Their product is 78. Find the numbers.

8. One positive number is three more than twice another. If the product is 27, find the numbers.

9. One positive number is six more than another. The sum of their squares is 260. What are the numbers?

10. One number is five less than another. The sum of their squares is 97. Find the numbers.

11. The difference between two positive integers is 8. If the smaller is added to the square of the larger, the sum is 124. Find the integers.

12. One positive number is 3 more than twice another. If the square of the smaller is added to the larger, the sum is 51. Find the numbers.

13. The product of a negative integer and twice the integer less 5 equals the integer plus 56. Find the integer.

14. Find a positive integer such that the product of the integer with a number three less than the integer is equal to the integer increased by 32.

15. The product of two consecutive positive integers is 72. Find the integers.

16. Find two consecutive integers whose product is 110.

17. Find two consecutive positive integers such that the sum of their squares is 85.

18. Find two consecutive positive integers such that the sum of their squares is 145.

19. The product of two consecutive odd integers is 63. Find the integers.

20. The product of two consecutive even integers is 168. Find the integers.

21. Find two consecutive positive integers such that the square of the second integer added to four times the first is equal to 41.

22. Find two consecutive negative integers such that 6 times the first plus the square of the second equals 34.

1. $x(x+8)=-16$; $x=-4$, so the numbers are –4 and 4
2. $x(x+10)=-25$; $x=-5$, so the numbers are –5 and 5
3. $x^2=7x$; $x=0, 7$
4. $x^2=2x$; $x=0, 2$
5. $x^2+3x=28$; $x=4$
6. $x^2+3x=54$; $x=6$
7. $x(x+7)=78$; $x=-13, 6$, so the numbers are –13 and –6 or 13 and 6
8. $x(2x+3)=27$; $x=3$, so the numbers are 3 and 9
9. $(x+6)^2+x^2=260$; $x=8$, so the numbers are 8 and 14
10. $x^2+(x-5)^2=97$; $x=-4, 9$, so the numbers are –4 and –9 or 9 and 4
11. $x+(x+8)^2=124$; $x=3$, so the integers are 3 and 11
12. $x^2+(3+2x)=51$; $x=-8, 6$, so the numbers are 6 and 15
13. $x(2x-5)=x+56$; $x=-4$
14. $x(x-3)=x+32$; $x=8$
15. $x(x+1)=72$; $x=8$, so the integers are 8 and 9
16. $x(x+1)=110$; $x=-11, 10$, so the integers are –11 and –10 or 10 and 11
17. $x^2+(x+1)^2=85$; $x=6$, so the integers are 6 and 7

18. $(x+1)^2+x^2=145$; $x=8$, so the integers are 8 and 9
19. $x(x+2)=63$; $x=-9, 7$, so the integers are -9 and -7 or 7 and 9
20. $x(x+2)=168$; $x=-14, 12$, so the integers are -14 and -12 or 12 and 14
21. $4x+(x+1)^2=41$; $x=4$, so the integers are 4 and 5
22. $6x+(x+1)^2=34$; $x=-11$, so the integers are -11 and -10
23. $2x(x+1)=(x+1)(x+2)+88$; $x=10$, so the integers are 10, 11, and 12
24. $x(x+4)=10(x+2)+71$; $x=-7, 13$, so the integers are $-7, -5$, and -3 or 13, 15, and 17
25. $6x(x+2)=(x+1)+(x+3)^2$; $x=-2, 1$, so the integers are $-2, -1$, 0, and 1 or 1, 2, 3, and 4
26. $[x+(x+2)]^2=2(x+4)(x+6)+516$; $x=-14, 20$, so the integers are -14, $-12, -10$, and -8 or 20, 22, 24, and 26
27. $w(2w)=72$; $w=6$, so width is 6 in. and length is 12 in.

23. Find three consecutive positive integers such that twice the product of the two smaller integers is 88 more than the product of the two larger integers.

24. Find three consecutive odd integers such that the product of the first and third is 71 more than 10 times the second.

25. Four consecutive integers are such that if the product of the first and third is multiplied by 6, the result is equal to the sum of the second and the square of the fourth. What are the integers?

26. Find four consecutive even integers such that the square of the sum of the first and second is equal to 516 more than twice the product of the third and fourth.

27. **Rectangles:** The length of a rectangle is twice the width. The area is 72 square inches. Find the length and width of the rectangle.

28. **Rectangles:** The length of a rectangle is three times the width. If the area is 147 square centimeters, find the length and width of the rectangle.

29. **Rectangles:** The length of a rectangle is four times the width. If the area is 64 square feet, find the length and width of the rectangle.

30. **Rectangles:** The length of a rectangle is five times the width. If the area is 180 square inches, find the length and width of the rectangle.

31. **Rectangles:** The width of a rectangle is 4 feet less than the length. The area is 117 square feet. Find the length and width of the rectangle.

32. **Rectangles:** The length of a rectangular yard is 12 meters greater than the width. If the area of the yard is 85 square meters, find the length and width of the yard.

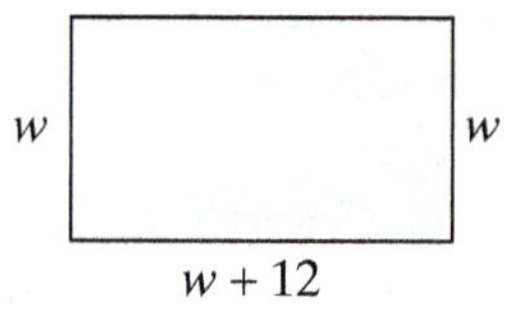

33. **Triangles:** The height of a triangle is 4 feet less than the base. The area of the triangle is 16 square feet. Find the length of the base and height of the triangle.

34. **Triangles:** The base of a triangle exceeds the height by 5 meters. If the area is 42 square meters, find the length of the base and the height of the triangle.

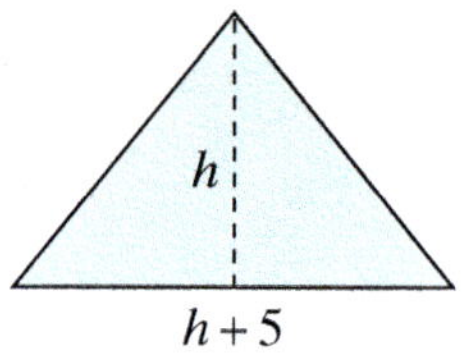

35. **Triangles:** The base of a triangle is 15 inches greater than the height. If the area is 63 square inches, find the length of the base.

36. **Triangles:** The base of a triangle is 6 feet less than the height. The area is 56 square feet. Find the height.

37. **Rectangles:** The perimeter of a rectangle is 32 inches. The area of the rectangle is 48 square inches. Find the dimensions of the rectangle.

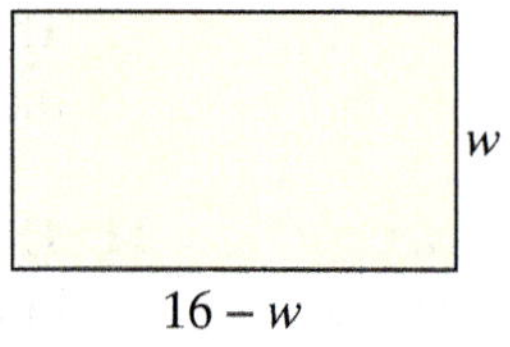

28. $w(3w)=147$; $w=7$, so width is 7 cm and length is 21 cm
29. $w(4w)=64$; $w=4$, so width is 4 ft and length is 16 ft
30. $w(5w)=180$; $w=6$, so width is 6 in. and length is 30 in.
31. $l(l-4)=117$; $l=13$, so width is 9 ft and length is 13 ft
32. $w(w+12)=85$; $w=5$, so width is 5 m and length is 17 m
33. $\frac{1}{2}b(b-4)=16$; $b=8$, so base is 8 ft and height is 4 ft
34. $\frac{1}{2}(h+5)h=42$; $h=7$, so height is 7 m and base is 12 m
35. $\frac{1}{2}(h+15)h=63$; $h=6$, so base is 21 in.
36. $\frac{1}{2}(h-6)h=56$; $h=14$, so height is 14 ft
37. $w(16-w)=48$; $w=4, 12$, so the rectangle is 4 in. by 12 in.
38. $w(10-w)=24$; $w=4, 6$, so the rectangle is 4 cm by 6 cm
39. $r(r+13)=140$; $r=7$, so there are 7 trees in each row
40. $n(n+3)=108$; $n=9$, so there are 9 rows

38. Rectangles: The area of a rectangle is 24 square centimeters. If the perimeter is 20 centimeters, find the length and width of the rectangle.

39. Orchards: An orchard has 140 apple trees. The number of rows exceeds the number of trees per row by 13. How many trees are there in each row?

40. Military: One formation for a drill team is rectangular. The number of members in each row exceeds the number of rows by 3. If there is a total of 108 members in the formation, how many rows are there?

41. Theater: A theater can seat 144 people. The number of rows is 7 less than the number of seats in each row. How many rows of seats are there?

42. College sports: An empty field on a college campus is being used for overflow parking for a football game. It currently has 187 cars in it. If the number of rows of cars is six less than the number of cars in each row, how many rows are there?

43. Parking: The parking garage at Baltimore-Washington International Airport contains 8400 parking spaces. The number of cars that can be parked on each floor exceeds the number of floors by 1675. How many floors are there in the parking garage?

44. Library books: One bookshelf in the public library can hold 175 books. The number of books on each shelf exceeds the number of shelves by 18. How many books are in each row?

45. Rectangles: The length of a rectangle is 7 centimeters greater than the width. If 4 centimeters are added to both the length and width, the new area would be 98 square centimeters. Find the dimensions of the original rectangle.

46. Rectangles: The width of a rectangle is 5 meters less than the length. If 6 meters are added to both the length and width, the new area will be 300 square meters. Find the dimensions of the original rectangle.

47. Gardening: Susan is going to fence a rectangular flower garden in her back yard. She has 50 feet of fencing and she plans to use the house as the fence on one side of the garden. If the area is 300 square feet, what are the dimensions of the flower garden?

48. Ranching: A rancher is going to build a corral with 52 yards of fencing. He is planning to use the barn as one side of the corral. If the area is 320 square yards, what are the dimensions?

49. Communication: A telephone pole is to have a guy wire attached to its top and anchored to the ground at a point that is at a distance 34 feet less than the height of the pole from the base. If the wire is to be 2 feet longer than the height of the pole, what is the height of the pole?

41. $r(r+7)=144$; $r=9$, so there are 9 rows
42. $r(r+6)=187$; $r=11$, so there are 11 rows
43. $n(n+1675)=8400$; $n=5$, so there are 5 floors
44. $b(b-18)=175$; $b=25$, so there are 25 books on each shelf
45. $(w+11)(w+4)=98$; $w=3$, so the rectangle is 3 cm by 10 cm
46. $(l+6)(l+1)=300$; $l=14$, so the rectangle is 14 m by 9 m
47. $w(50-2w)=300$; $w=10, 15$, so width is 10 ft and length is 30 ft or width is 15 ft and length is 20 ft
48. $w(52-2w)=320$; $w=10, 16$, so width is 10 yd and length is 32 yd or width is 16 yd and length is 20 yd
49. $h^2+(h-34)^2=(h+2)^2$; $h=48$, so the height of the pole is 48 ft
50. $h^2+(h-71)^2=(h+1)^2$; $h=84$, so the height of the General Sherman is 84 m
51. $h^2+(h-49)^2=(h+1)^2$; $h=60$, so height is 60 ft

50. Trees: Lucy is standing next to the the General Sherman tree in Sequoia National Park, home of some of the largest trees in the world. The distance from Lucy to the base of the tree is 71 m less than the height of the tree. If the distance from Lucy to the top of the tree is 1 m more than the height of the tree, how tall is the General Sherman?

51. Holiday decorating: A Christmas tree is supported by a wire that is 1 foot longer than the height of the tree. The wire is anchored at a point whose distance from the base of the tree is 49 feet shorter than the height of the tree. What is the height of the tree?

52. Architecture: An architect wants to draw a rectangle with a diagonal of 13 inches. The length of the rectangle is to be 2 inches more than twice the width. What dimensions should she make the rectangle?

53. Gymnastics: Incline mats, or triangle mats, are offered with different levels of incline to help gymnasts learn basic moves. As the name may suggest, two sides of the mat are right triangles. If the height of the mat is 28 inches shorter than the length of the mat and the hypotenuse is 8 inches longer than the length of the mat, what is the length of the mat?

54. Laser show: Bill uses mirrors to augment the "laser experience" at a laser show. At one show he places three mirrors, A, B, C, in a right triangular form. If the distance between A and B is 15 m more than the distance between A and C, and the distance between B and C is 15 m less than the distance between A and C, what is the distance between mirror A and mirror C?

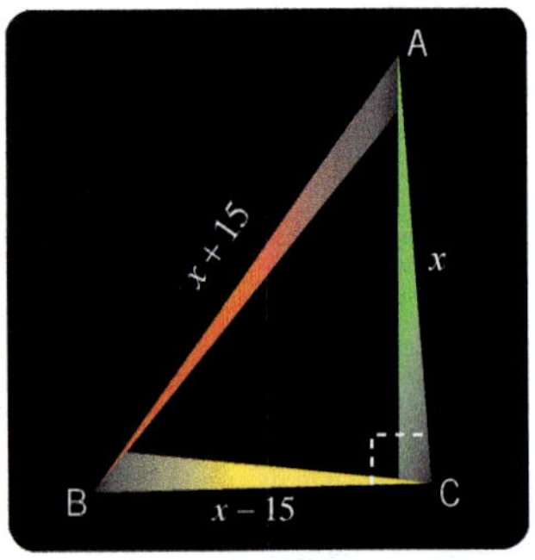

Consumer Demand

The demand for a product is the number of units of the product x that consumers are willing to buy when the market price is p dollars. The consumers' total expenditure for the product S is found by multiplying the price times the demand. $(S=px)$

55. During the summer at a local market, a farmer will sell $8p+588$ pounds of peaches at p dollars per pound. If he sold \$900 worth of peaches this summer, what was the price per pound of the peaches?

52. $w^2+(2w+2)^2=13^2$; $w=5$, so width is 5 in. and length is 12 in.
53. $l^2+(l-28)^2=(l+8)^2$; $l=60$, so the length of the mat is 60 inches

56. On a hot afternoon, fans at a stadium will buy $2000-100p$ drinks for p dollars each. If the total sales after a game were \$4375, what was the price per drink?

57. When fishing reels are priced at p dollars, local consumers will buy $36-p$ fishing reels. What is the price if total sales were \$320?

58. A manufacturer can sell $100-2p$ lamps at p dollars each. If the receipts from the lamps total \$1200, what is the price of the lamps?

Writing and Thinking About Mathematics

Teaching Notes:
You may want to show your students that
$(u^2+v^2)^2=(u^2-v^2)^2+(2uv)^2$.

54. $x^2+(x-15)^2=(x+15)^2$; $x=60$, so the distance between A and C is 60 m
55. \$1.50 per pound
56. \$2.50 or \$17.50 per drink
57. \$16 or \$20 per reel
58. \$20 or \$30 per lamp
59. $12^2+5^2=13^2$; $20^2+21^2=29^2$; $24^2+7^2=25^2$; $14^2+48^2=50^2$; $60^2+11^2=61^2$

59. If three positive integers satisfy the Pythagorean theorem, they are called a **Pythagorean triple**. For example, 3, 4, and 5 are a Pythagorean triple because $3^2+4^2=5^2$. There are an infinite number of such triples. To see how some triples can be found, fill out the following table and verify that the numbers in the rightmost three columns are indeed Pythagorean triples.

u	v	$2uv$	u^2-v^2	u^2+v^2
2	1	4	3	5
3	2			
5	2			
4	3			
7	1			
6	5			

60. The pattern in Kara's linoleum flooring is in the shape of a square 8 inches on a side with right triangles of sides x inches placed on each side of the original square so that a new larger square is formed. What is the area of the new square? Explain why you do not need to find the value of x.

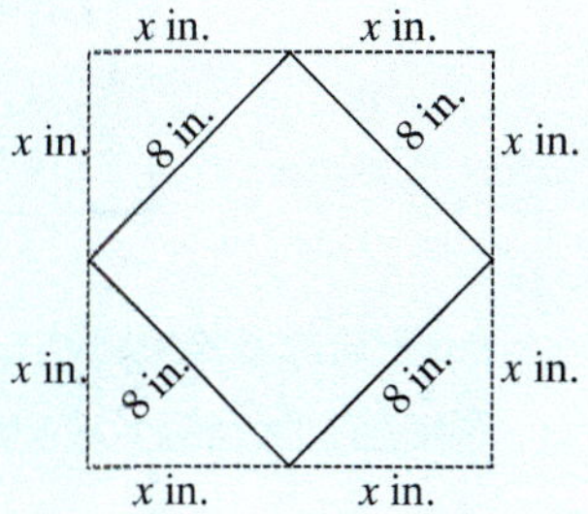

HAWKES LEARNING SYSTEMS: INTRODUCTORY & INTERMEDIATE ALGEBRA SOFTWARE

- 6.7 Applications of Quadratic Equations

60. 128 in.^{2}; We can use the Pythagorean theorem to find the length and the width (by finding the diagonals of the interior square) instead.

Chapter 6 Index of Key Ideas and Terms

Section 6.1 Greatest Common Factor and Factoring by Grouping

Greatest Common Factor (GCF) pages 430-431

The **greatest common factor (GCF)** of two or more integers is the largest integer that is a factor (or divisor) of all of the integers.

Procedure for Finding the GCF page 431

1. Find the prime factorization of all integers and integer coefficients.
2. List all the factors that are common to all terms, including variables.
3. Choose the greatest power of each factor common to all terms.
4. Multiply these powers to find the GCF.
   Note: If there is no common prime factor or variable, then the GCF is 1.

Factoring Out the GCF pages 432-433

To find a monomial that is the GCF of a polynomial:

1. Find the variable(s) of highest degree and the largest integer coefficient that is a factor of each term of the polynomial. (This is one factor.)
2. Divide this monomial factor into each term of the polynomial resulting, in another polynomial factor.

Factoring by Grouping pages 435-436

To **factor by grouping** look for common factors in each group and then look for common binomial factors.

Section 6.2 Factoring Trinomials: $x^2 + bx + c$

Factoring Trinomials with Leading Coefficient 1 ($x^2 + bx + c$) pages 441-443

To factor $x^2 + bx + c$, if possible, find an integer pair of factors of c whose sum is b.

1. If c is positive, then both factors must have the same sign.
   a. Both will be positive if b is positive.
   b. Both will be negative if b is negative.
2. If c is negative, then one factor must be positive and the other negative.

Continued on the next page...

Section 6.2 Factoring Trinomials: $x^2 + bx + c$ (cont.)

Not Factorable (or Prime) page 444

A polynomial is **not factorable** if it cannot be factored as the product of polynomials with integer coefficients.

Section 6.3 Factoring Trinomials: $ax^2 + bx + c$

Trial-and-error method pages 448-450

Guidelines for the Trial-and-Error Method page 450

1. If the sign of the constant term is positive (+), the signs in both factors will be the same, either both positive or both negative.
2. If the sign of the constant term is negative (–), the signs in the factors will be different, one positive and one negative.

The *ac*-method (Grouping) pages 452-453

1. Multiply $a \cdot c$.
2. Find two integers whose product is ac and whose sum is b. If this is not possible, then the trinomial is **not factorable**.
3. Rewrite the middle term (bx) using the two numbers found in Step 2 as coefficients.
4. Factor by grouping the first two terms and the last two terms.
5. Factor out the common binomial factor. This will give two binomial factors of the trinomial $ax^2 + bx + c$.

Section 6.4 Special Factoring Techniques

Special Factoring Techniques

I. $x^2 - a^2 = (x+a)(x-a)$: Difference of two squares page 459

II. $x^2 + 2ax + a^2 = (x+a)^2$: Perfect square trinomial page 461

III. $x^2 - 2ax + a^2 = (x-a)^2$: Perfect square trinomial page 461

IV. $x^3 + a^3 = (x+a)(x^2 - ax + a^2)$: Sum of two cubes pages 462-463

V. $x^3 - a^3 = (x-a)(x^2 + ax + a^2)$: Difference of two cubes page 462-463

Sum of Two Squares page 460

The **sum of two squares** is an expression of the form $x^2 + a^2$ and is **not factorable**.

Section 6.5 Additional Factoring Practice

General Guidelines for Factoring Polynomials page 468

1. **Always look for a common monomial factor first.** If the leading coefficient is negative, factor out a negative monomial even if it is just –1.
2. **Check the number of terms.**
   - **a. Two terms:**
     1. difference of two squares? – factorable
     2. sum of two squares? – not factorable
     3. difference of two cubes? – factorable
     4. sum of two cubes? – factorable
   - **b. Three terms:**
     1. perfect square trinomial?
     2. use trial-and-error method?

        Guidelines for the trial-and-error method

        a. If the sign of the constant term is positive, the signs in both factors will be the same, either both positive or both negative.

        b. If the sign of the constant term is negative, the signs in the factors will be different, one positive and one negative.
     3. use *ac*-method?

        Guidelines for the *ac*-method

        a. Multiply $a \cdot c$.

        b. Find two integers whose product is ac and whose sum is b. If this is not possible, the trinomial is not factorable.

        c. Rewrite the middle term (bx) using the two numbers found in step **b.** as coefficients.

        d. Factor by grouping.
   - **c. Four terms:**

     Group terms with a common factor and factor out any common binomial factor.
3. **Check the possibility of factoring any of the factors.**

Checking: Factoring can be checked by multiplying the factors. The product should be the original expression.

Section 6.6 Solving Quadratic Equations by Factoring

Quadratic Equations page 471

Quadratic equations are equations that can be written in the form $ax^2 + bx + c = 0$ where a, b, and c are real numbers and $a \neq 0$.

Zero-Factor Property page 471

If the product of two (or more) factors is 0, then at least one of the factors must be 0. That is, if a and b are real numbers, then if $a \cdot b = 0$, then $a = 0$ or $b = 0$ or both.

Solving Quadratic Equations by Factoring page 475

1. Add or subtract terms as necessary so that 0 is on one side of the equation and the equation is in the standard form $ax^2 + bx + c = 0$ where a, b, and c are real numbers and $a \neq 0$.
2. Factor completely.
3. Set each nonconstant factor equal to 0 and solve each linear equation for the unknown.
4. Check each solution, one at a time, in the original equation.

Finding an Equation Given the Roots page 476

Factor Theorem page 477

If $x = c$ is a root of a polynomial equation in the form $P(x) = 0$, then $x - c$ is a factor of the polynomial $P(x)$.

Section 6.7 Applications of Quadratic Equations

Attack Plan for Word Problems page 481

1. Read the problem carefully at least twice.
2. Decide what is asked for and assign a variable or variable expression to the unknown quantities.
3. Organize a chart, table, or diagram relating all the information provided.
4. Form an equation. (A formula of some type may be necessary.)
5. Solve the equation.
6. Check your solution with the wording of the problem to be sure it makes sense.

Consecutive Integers page 484

The Pythagorean Theorem pages 485-486

In a right triangle, if c is the length of the hypotenuse and a and b are the lengths of the legs, then $c^2 = a^2 + b^2$.

Hawkes Learning Systems: Introductory & Intermediate Algebra Software

- 6.1a Greatest Common Factor of Two or More Terms
- 6.1b Greatest Common Factor of a Polynomial
- 6.1c Factoring Expressions by Grouping
- 6.2 Factoring Trinomials: Leading Coefficient 1
- 6.3a Factoring Trinomials by Trial and Error
- 6.3b Factoring Trinomials by Grouping
- 6.4a Special Factorizations - Squares
- 6.4b Special Factorizations - Cubes
- 6.5 Additional Factoring Practice
- 6.6 Solving Quadratic Equations by Factoring
- 6.7 Applications of Quadratic Equations

Chapter 6 Review

1. $x-3y+2$
2. $2y+3$
3. $11(x-2)$
4. $-4y(y-7)$
5. $4x^2y(4x-3)$
6. $a(m^2+11m+25)$
7. $-7t^3x(x^3-14tx^2-5t^2)$
8. $-6x^2y^3(xy+y-4)$
9. $(a+7)(3a-2)$
10. $(x-10)(2a+3b)$
11. $(a+c)(x-1)$
12. $(1-4y)(x+2z)$
13. $(z^2+5)(1+c)$
14. Not factorable
15. $(a+b)(7a-3)$
16. $(1-3y)(x-z)$
17. $-7, 6$
18. $3, 9$
19. $(m+6)(m+1)$
20. Not factorable
21. $(n-6)(n-2)$
22. $(x-2)(x+5)$
23. $(a-10)(a+5)$
24. Not factorable
25. $p(p+2)(p-6)$
26. $3(m+2)(m+2)$
27. $2y(y+5)(y+2)$
28. $7x^2(x-1)(x+4)$
29. $9x^3(x+5)(x+4)$
30. $11a(a+1)(a-11)$
31. $(a+5b)(a+6b)$
32. $4(y^2-7xy-x^2)$
33. $(x+7)(x+5)$
34. $-(x-7)(x+3)$
35. $-(6x-1)(2x-5)$
36. $(3x-2)(4x+3)$
37. $(2y-1)(3y-4)$
38. Not factorable

6.1 Greatest Common Factor and Factoring by Grouping

Complete the factoring of the polynomial as indicated.

1. $4x^2y^2-12xy^3+8xy^2=4xy^2(\_\_\_\_\_\_)$ **2.** $14x^2y+21x^2=7x^2(\_\_\_\_\_\_)$

Factor each of the polynomials by finding the GCF (or $-1\cdot$GCF).

3. $11x-22$ **4.** $-4y^2+28y$

5. $16x^3y-12x^2y$ **6.** $am^2+11am+25a$

7. $-7t^3x^4+98t^4x^3+35t^5x$ **8.** $-6x^2y^4-6x^3y^4+24x^2y^3$

Factor each expression by factoring out the common binomial factor.

9. $3a(a+7)-2(a+7)$ **10.** $2a(x-10)+3b(x-10)$

Factor each of the polynomials by grouping. If a polynomial cannot be factored, write "not factorable."

11. $ax-a+cx-c$ **12.** $x-4xy+2z-8zy$ **13.** z^2+5+cz^2+5c

14. $x^2-x^2y+6+6y$ **15.** $7a^2-3a+7ab-3b$ **16.** $x-3xy-z+3zy$

6.2 Factoring Trinomials: x^2+bx+c

Find the pair of integers whose product is the first integer and whose sum is the second integer.

17. $-42, -1$ **18.** $27, 12$

Completely factor each of the given trinomials. If the trinomial cannot be factored, write "not factorable."

19. m^2+7m+6 **20.** $y^2+2y+24$ **21.** $n^2-8n+12$

22. $x^2+3x-10$ **23.** $a^2-5a-50$ **24.** $c^2+3c+10$

25. p^3-4p^2-12p **26.** $3m^2+12m+12$ **27.** $2y^3+14y^2+20y$

28. $7x^4+21x^3-28x^2$ **29.** $9x^5+81x^4+180x^3$ **30.** $11a^3-110a^2-121a$

31. $a^2+11ab+30b^2$ **32.** $4y^2-28xy-4x^2$

39. $3(3x+2)(7x-5)$
40. $2(2x+5)(4x-7)$
41. $-(3x+8)(x-3)$
42. $-(5x-7)(3x+2)$
43. Not factorable
44. Not factorable
45. $-4(x-10)(x+5)$
46. $7x(y-4)(y+6)$
47. $x(3x-2)(6x-1)$
48. $(x^3-10)(x^3+10)$
49. $(4x-5)(4x+5)$
50. Not factorable
51. $4(y+4)^2$
52. Not factorable
53. $2(2x-11)(2x+11)$
54. $3x(x-4)(x+4)$
55. $-3(x+2)^2$
56. $3(4x-y)(4x+y)$
57. $(9-4x)(9+4x)$
58. $(2x^3-y^2)(4x^6+2x^3y^2+y^4)$
59. $3(x+3)(x^2-3x+9)$
60. $4x(x+5)(x^2-5x+25)$
61. $(7y-1)(49y^2+7y+1)$
62. $27(2x-1)(4x^2+2x+1)$
63. $(z+4)(z-9)$
64. $(y+6)(y-7)$
65. Not factorable
66. $4(x+2)(x+3)$
67. $-3(2x-1)(x-5)$
68. $-(3x-2)(x+4)$
69. $5x(x+2)(2x+3)$
70. $(7x+2y)(7x-2y)$
71. $2(x-3y)(x^2+3xy+9y^2)$
72. $(3x^2+1)(x-3)$

6.3 Factoring Trinomials: ax^2+bx+c

Completely factor each of the given polynomials. If the polynomial cannot be factored, write "not factorable."

33. $x^2+12x+35$ **34.** $-x^2+4x+21$ **35.** $-12x^2+32x-5$

36. $12x^2+x-6$ **37.** $6y^2-11y+4$ **38.** $12y^2-11y+4$

39. $63x^2-3x-30$ **40.** $16x^2+12x-70$ **41.** $24+x-3x^2$

42. $14+11x-15x^2$ **43.** $2y^2-13y+5$ **44.** $3y^2+10y+6$

45. $200+20x-4x^2$ **46.** $7xy^2+14xy-168x$ **47.** $18x^3-15x^2+2x$

6.4 Special Factoring Techniques

Completely factor each of the given polynomials. If a polynomial cannot be factored, write "not factorable."

48. x^6-100 **49.** $16x^2-25$ **50.** x^2+144

51. $4y^2+32y+64$ **52.** $4y^2+169$ **53.** $8x^2-242$

54. $3x^3-48x$ **55.** $-3x^2-12x-12$ **56.** $48x^2-3y^2$

57. $81-16x^2$ **58.** $8x^9-y^6$ **59.** $3x^3+81$

60. $4x^4+500x$ **61.** $343y^3-1$ **62.** $216x^3-27$

6.5 Additional Factoring Practice

Completely factor each of the given polynomials. If a polynomial cannot be factored, write "not factorable."

63. $z^2-5z-36$ **64.** y^2-y-42 **65.** $6y^2+23$

66. $4x^2+20x+24$ **67.** $-6x^2+33x-15$ **68.** $-3x^2-10x+8$

69. $10x^3+35x^2+30x$ **70.** $49x^2-4y^2$ **71.** $2x^3-54y^3$

72. $3x^3-9x^2+x-3$

6.6 Solving Quadratic Equations by Factoring

Solve the equations by factoring.

73. $x^2 + 5x = 0$

74. $4x^2 - 24x = 0$

75. $3x^2 - 17x + 10 = 0$

76. $y^2 + 2y - 35 = 0$

77. $16x^3 - 100x = 0$

78. $x^3 - 4x^2 - 12x = 0$

79. $5x^2 = 125$

80. $25y^3 = 36y$

81. $x^2 = 2x + 35$

82. $2a^2 + 28a + 98 = 0$

83. $(x+3)^2 = 9$

84. $(x-5)^2 = 64$

85. $(x-2)(x-6) = 5$

86. $(x+3)(2x+1) = -3$

Write a polynomial equation with integer coefficients that has the given roots.

87. $x = 4, x = -5$

88. $x = -\frac{3}{4}, x = \frac{5}{8}$

89. $x = 0, x = -2, x = \frac{1}{6}$

90. $y = 1,\ y = -4,\ y = -4$ (-4 is a double root.)

6.7 Applications of Quadratic Equations

91. One positive number is three more than another and the sum of their squares is 269. What are the numbers?

92. One number is 20 more than another. Their product is –84. What are the numbers?

93. **Construction:** A tract of new homes is arranged in rectangular fashion with 5 more houses on each street than there are streets. How many streets are there if there are 126 houses in the tract?

94. **Rectangles:** A rectangle has an area of 450 square yards and a perimeter of 90 yards. What are the dimensions of the rectangle?

95. The sum of the squares of two consecutive odd positive integers is 290. Find the integers.

96. The product of two consecutive even integers is 72 more than 4 times the larger integer. Find the integers.

97. Find three consecutive odd integers such that the product of the first and second is 26 more than the third.

98. Find four consecutive integers such that the sum of the squares of the first and second is 65 more than the product of the third and fourth.

99. **Triangles:** The base of a triangle is 12 inches greater than the height. If the area is 80 square inches, find the length of the base.

100. **Rectangles:** The length of a rectangle is 10 meters more than the width. If 3 meters is added to both the length and the width, the new area will be 299 square meters. Find the length and width of the original rectangle.

73. $x = -5, 0$
74. $x = 0, 6$
75. $x = \frac{2}{3}, 5$
76. $y = 5, -7$
77. $x = -\frac{5}{2}, 0, \frac{5}{2}$
78. $x = -2, 0, 6$
79. $x = -5, 5$
80. $y = -\frac{6}{5}, 0, \frac{6}{5}$
81. $x = -5, 7$
82. $a = -7$
83. $x = -6, 0$
84. $x = -3, 13$
85. $x = 1, 7$
86. $x = -2, -\frac{3}{2}$
87. $x^2 + x - 20 = 0$
88. $32x^2 + 4x - 15 = 0$
89. $6x^3 + 11x^2 - 2x = 0$
90. $y^3 + 7y^2 + 8y - 16 = 0$
91. 13, 10
92. 6, –14 or 14, –6
93. 9 streets
94. 15×30 yards
95. 11, 13
96. 10, 12 or –8, –6
97. 5, 7, 9
98. 10, 11, 12, 13 or –7, –6, –5, –4
99. 20 inches
100. $l = 20$ meters; $w = 10$ meters

101. 4 inches
102. 20 rows, 30 seats
103. Length = 21 meters; Width = 20 meters
104. 125 yards

101. Squares: The area of a square (in square inches) is numerically equal to its perimeter (in inches). What is the length of one side of the square?

102. Theater: A small theater can seat 600 people. The number of rows is 10 less than the number of seats in each row. How many rows of seats are there? How many seats are in each row?

103. Rectangles: The diagonal of a rectangle is 29 meters long. The length of the rectangle is 1 meter longer than the width. Find the length and width of the rectangle.

104. Distance: Hector's home is "around the corner" from Paul's home. Hector's home is 100 yards from the corner and Paul's home is 75 yards from the same corner. What is the distance between the two homes?

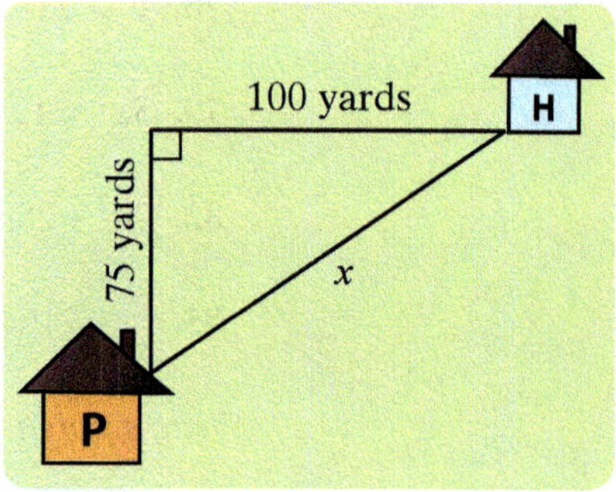

Chapter 6 Test

1. $7ab^2(4x-3y)$
2. $6yz^2(3z-yz+2)$
3. $(x-5)(x-4)$
4. $-(x+7)^2$
5. $(x-5)(y-7)$
6. $6(x+1)(x-1)$
7. $2(6x-5)(x+1)$
8. $(x+3)(3x-8)$
9. $(4x-5y)(4x+5y)$
10. $x(x+1)(2x-3)$
11. $(2x-3)(3x-2)$
12. $(y+7)(2x-3)$
13. Not factorable
14. $-3x(x^2-2x+2)$
15. $10x^3(2y+1)(y+1)$
16. $4y^2(x-3y)(x^2+3xy+9y^2)$
17. $x=-1, 8$
18. $x=-6, 0$
19. $x=-\frac{3}{4}, 5$
20. $x=\frac{3}{2}, 4$
21. $x^2+5x-24=0$
22. $4x^2-4x+1=0$
23. 6, 20 or −30, −4
24. Length = 15 centimeters, Width = 11 centimeters
25. 18, 19
26. 3, 12
27. 18 cm
28. $P(x)=4(3x+5)$

Factor each of the polynomials by finding the GCF (or $-1\cdot GCF$).

1. $28ab^2x-21ab^2y$ **2.** $18yz^3-6y^2z^3+12yz^2$

Completely factor each of the given polynomials. If the polynomial cannot be factored, write "not factorable."

3. $x^2-9x+20$ **4.** $-x^2-14x-49$ **5.** $xy-7x+35-5y$ **6.** $6x^2-6$

7. $12x^2+2x-10$ **8.** $3x^2+x-24$ **9.** $16x^2-25y^2$ **10.** $2x^3-x^2-3x$

11. $6x^2-13x+6$ **12.** $2xy-3y+14x-21$ **13.** $4x^2+25$

14. $-3x^3+6x^2-6x$ **15.** $20x^3y^2+30x^3y+10x^3$ **16.** $4x^3y^2-108y^5$

Solve the equations.

17. $x^2-7x-8=0$ **18.** $-3x^2=18x$

19. $0=4x^2-17x-15$ **20.** $(2x-7)(x+1)=6x-19$

21. Find a polynomial equation with integer coefficients that has $x=3$ and $x=-8$ as solutions.

22. Find a polynomial equation with integer coefficients that has $x=\frac{1}{2}$ as a double root.

23. One number is 10 less than five times another number. Their product is 120. Find the numbers.

24. Rectangles: The length of a rectangle is 7 centimeters less than twice the width. If the area of the rectangle is 165 square centimeters, find the length and width.

25. The product of two consecutive positive integers is 342. Find the two integers.

26. The difference between two positive numbers is 9. If the smaller is added to the square of the larger, the result is 147. Find the numbers.

27. Architecture: The average staircase has steps with a width that is 6 cm more than the height and a diagonal that is 12 cm more than the height. How tall is the average step?

28. Squares: The area of a square can be represented by the function $A(x)=9x^2+30x+25$. Write a polynomial function $P(x)$ that represents the perimeter of the square. (**Hint:** Factor the expression to find the length and the width.)

Cumulative Review: Chapters 1 – 6

1. 120 2. $168x^2y$
3. 6 4. -138
5. $\frac{55}{48}$ 6. $\frac{19}{60a}$
7. $\frac{3}{10}$ 8. $\frac{75x}{23}$
9. $x = \frac{9}{10}$
10. $x = 21$
11. $x = 21$
12. $x = -5, 10$
13. Conditional
14. Identity
15. Contradiction
16. 

$\left[-2, \frac{1}{4}\right]$
17. 

$\left(-\frac{7}{3}, 1\right)$
18. 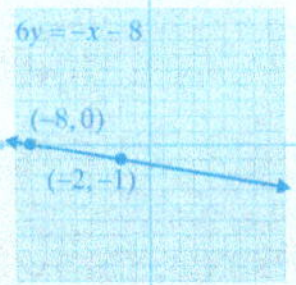

19. 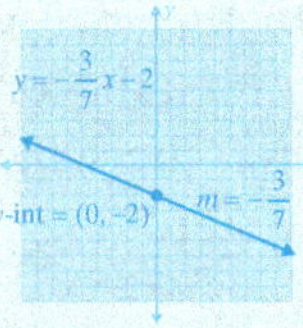

20. $y = 7$
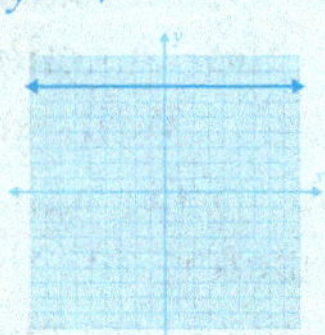
21. $x + y = -2$
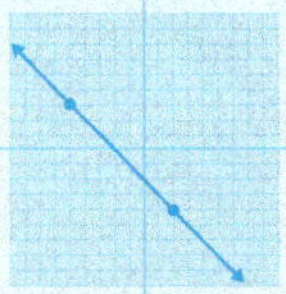

Find the LCM for each set of terms.

1. $\{20, 12, 24\}$

2. $\{8x^2, 14x^2y, 21xy\}$

Find the value of each expression by using the rules for order of operations.

3. $2 \cdot 3^2 \div 6 \cdot 3 - 3$

4. $6 + 3\left[4 - 2\left(3^3 - 1\right)\right]$

Perform the indicated operation. Reduce all answers to lowest terms.

5. $\frac{7}{12} + \frac{9}{16}$

6. $\frac{11}{15a} - \frac{5}{12a}$

7. $\frac{6x}{25} \cdot \frac{5}{4x}$

8. $\frac{40}{92} \div \frac{2}{15x}$

Solve each of the equations.

9. $4(2x - 3) + 2 = 5 - (2x + 6)$

10. $\frac{4x}{7} - 3 = 9$

11. $\frac{2x + 3}{6} - \frac{x + 1}{4} = 2$

12. $\left|\frac{2x}{5} - 1\right| = 3$

Determine whether each of the following equations is a conditional equation, an identity, or a contradiction.

13. $4(x + 2) + 5x - 5 = 2 - (x + 4)$

14. $5(3x - 2) + 1 = 9(x - 1) + 6x$

15. $4(2x + 4) - 3 = 7(x + 2) + x$

Solve the inequalities and graph the solution sets. Write each solution using interval notation. Assume that x is a real number.

16. $-21 \le 8x - 5 \le -3$

17. $|3x + 2| < 5$

18. Locate at least two ordered pairs of real numbers that satisfy the linear equation $6y = -x - 8$ and graph the corresponding line in the Cartesian coordinate system.

19. Write the equation $3x + 7y = -14$ in slope-intercept form. Find the slope and the y-intercept, and then use them to draw the graph.

20. Find the equation for the line passing through the point $(-4, 7)$ with slope 0. Graph the line.

21. Find the equation in standard form of the line determined by the two points $(-5, 3)$ and $(2, -4)$. Graph the line.

22. Find the equation in slope-intercept form of the line parallel to the line $y = 3x - 7$ and passing through the point $(-1, 1)$. Graph both lines.

23. Find the equation in slope-intercept form of the line perpendicular to the line $y = 3x - 7$ and passing through the point $(-1, 1)$. Graph both lines.

24. Given the relation $r = \{(2, -3), (3, -2), (5, 0), (7.1, 3.2)\}$.

a. Graph the relation.

b. State the domain of the relation.

c. State the range of the relation.

d. Is the relation a function? Explain.

Use the vertical line test to determine whether each of the graphs does or does not represent a function. State the domain and range.

25.

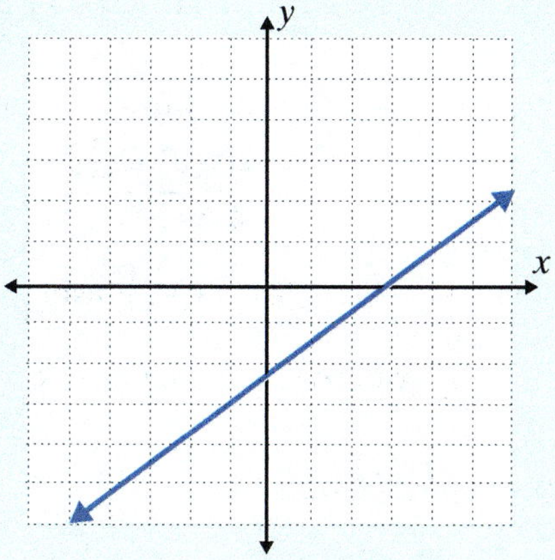

26.

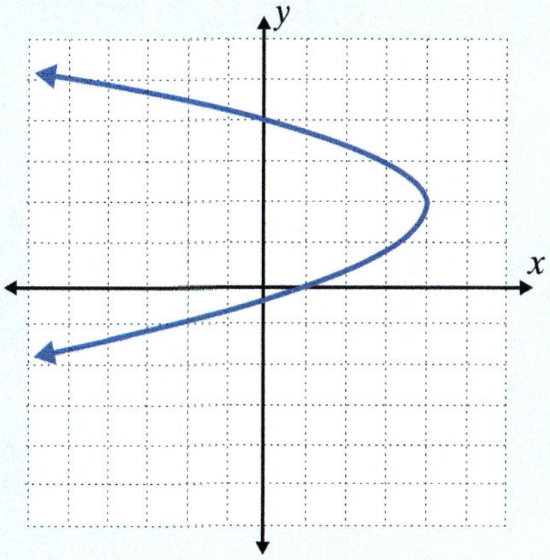

27. For the function $f(x) = x^2 - 3x + 4$, find

a. $f(-6)$ **b.** $f(0)$ **c.** $f\left(\frac{1}{2}\right)$

Graph the inequalities.

28. $y > 6x + 2$ **29.** $3x + 2y \geq 10$

Use the properties of exponents to simplify each of the expressions. Answers should contain only positive exponents. Assume that all variables represent nonzero real numbers.

30. $(4x^2y)^3$ **31.** $(-2x^3y^2)^{-3}$ **32.** $\frac{21xy^2}{3x^{-1}y}$

33. $\left(\frac{6x^2}{y^5}\right)^2$ **34.** $\frac{(xy^0)^3}{(x^3y^{-1})^2}$ **35.** $\left(\frac{x^2y^{-2}}{3y^5}\right)^{-2}$

---

22. $y = 3x + 4$

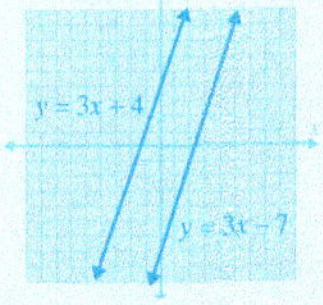

23. $y = -\frac{1}{3}x + \frac{2}{3}$

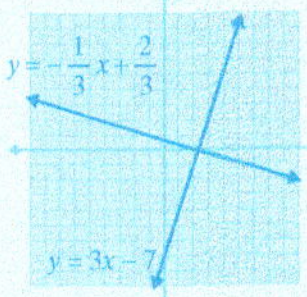

24. a.

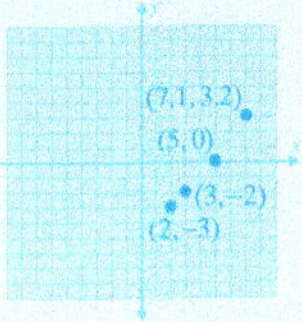

b. $D = \{2, 3, 5, 7.1\}$

c. $R = \{-3, -2, 0, 3.2\}$

d. It is a function because each first coordinate (domain) has only one corresponding second coordinate (range).

25. Function; $D = (-\infty, \infty)$; $R = (-\infty, \infty)$

26. Not a function; $D = (-\infty, 4]$; $R = (-\infty, \infty)$

27. a. 58 b. 4 c. $\frac{11}{4}$

28.

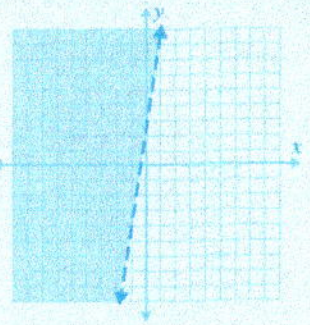

29.

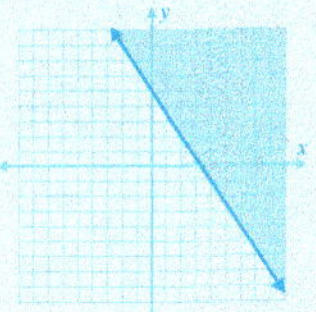

30. $64x^6y^3$
31. $-\dfrac{1}{8x^9y^6}$
32. $7x^2y$
33. $\dfrac{36x^4}{y^{10}}$
34. $\dfrac{y^2}{x^3}$
35. $\dfrac{9y^{14}}{x^4}$
36.
a. $(5.6\times10^{-7})(3\times10^{-4})$; 1.68×10^{-10}
b. $\dfrac{(8.1\times10^4)(6.2\times10^3)}{(3\times10^{-3})(2\times10^{-1})}$; 8.37×10^{11}
37. $13x+1$
38. $x^2+12x+3$
39. $4x^2+5x-8$
40. $-4x^2-2x+3$
41. $2x^2+x-28$
42. $-3x^2-17x+6$
43. $x^2+12x+36$
44. $4x^2-28x+49$
45. $2x-\dfrac{5}{4}+\dfrac{1}{y}$
46. $3x+1+\dfrac{22}{x-2}$
47. $4(2x-5)$
48. $6(x-16)$
49. $(x+2)(y+3)$
50. $(x-2)(a+b)$
51. $(x-6)(x-3)$
52. $(2x-3)(3x+4)$
53. $-4x(3x+4)$
54. $8xy(2x-3)$
55. $5x^2(x-5)(x+8)$
56. $(2x+1)(2x-1)$
57. $3(x+4y)(x-4y)$

36. Write each number in the following expression in scientific notation and simplify. Show the steps you use. Do not use a calculator. Leave the answers in scientific notation.

a. $0.000\,000\,56\times0.0003$ **b.** $\dfrac{81{,}000\times6200}{0.003\times0.2}$

Perform the indicated operations and simplify by combining like terms.

37. $2(4x+3)+5(x-1)$ **38.** $2x(x+5)-(x+1)(x-3)$

39. $(2x^2+6x-7)+(2x^2-x-1)$ **40.** $(x+1)-(4x^2+3x-2)$

41. $(2x-7)(x+4)$ **42.** $-(x+6)(3x-1)$

43. $(x+6)^2$ **44.** $(2x-7)(2x-7)$

45. Express the following quotient as a sum of fractions and simplify, if possible.

$$\frac{8x^2y^2-5xy^2+4xy}{4xy^2}$$

46. Divide the expression $(3x^2-5x+20)\div(x-2)$ using long division and write the answer in the form $Q+\dfrac{R}{D}$, where the degree of $R <$ the degree of D.

Factor each expression as completely as possible.

47. $8x-20$ **48.** $6x-96$ **49.** $xy+3x+2y+6$

50. $ax-2a-2b+bx$ **51.** $x^2-9x+18$ **52.** $6x^2-x-12$

53. $-12x^2-16x$ **54.** $16x^2y-24xy$ **55.** $5x^4+15x^3-200x^2$

56. $4x^2-1$ **57.** $3x^2-48y^2$ **58.** x^2+x+3

59. $3x^2+5x+2$ **60.** $2x^3-20x^2+50x$ **61.** $4x^3+100x$

62. $5x^3-135y^3$

Solve the equations.

63. $(x-7)(x+1)=0$ **64.** $21x-3x^2=0$ **65.** $0=x^2+8x+12$

66. $x^2=3x+28$ **67.** $x^3+14x^2+49x=0$ **68.** $8x=12x+2x^2$

69. $(4x-3)(x+1)=2$ **70.** $2x(x+5)(x-2)=0$

58. Not factorable
59. $(x+1)(3x+2)$
60. $2x(x-5)(x-5)$
61. $4x(x^2+25)$
62. $5(x-3y)(x^2+3xy+9y^2)$
63. $x=-1, 7$
64. $x=0, 7$
65. $x=-6, -2$
66. $x=-4, 7$
67. $x=-7, 0$
68. $x=-2, 0$
69. $x=-\frac{5}{4}, 1$
70. $x=-5, 0, 2$
71. $x^2+15x+50=0$
72. $x^3-17x^2+52x=0$
73. a. 39.7 million tickets
 b. 39.1 million tickets
 c. 12.5 million tickets
74. a. 13%
 b. 11%
 c. a
75. 65 mph
76. \$7500 at 6%; \$2500 at 8%
77. 13 inches by 17 inches
78. 4, 11
79. 8 and 9 or −8 and −9
80. $t=2.5, 3$ seconds

71. Find an equation that has $x=-10$ and $x=-5$ as roots.

72. Find an equation that has $x=0$, $x=4$ and $x=13$ as roots.

73. Given the yearly ticket sales of the top grossing movies in 2009:

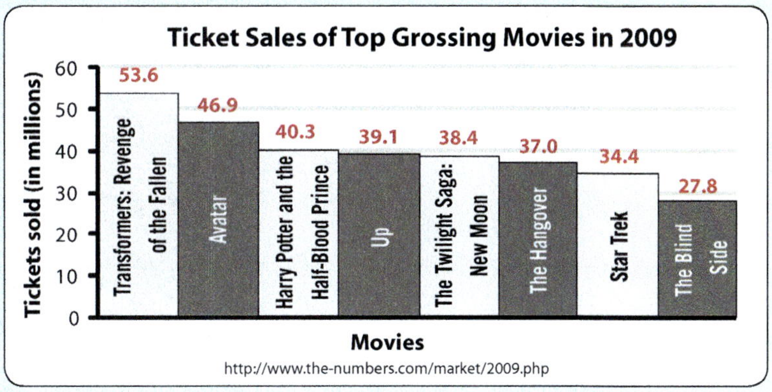

a. Find the average number of tickets sold from the top grossing movies in 2009 (in millions of tickets sold). (Round your answer to the nearest tenth)
b. Find the number of tickets sold for the movie *Up* in 2009.
c. Find the difference in the number of tickets sold for the movies *Avatar* and *Star Trek*.

74. What is the percent of profit if:
a. \$520 is made on an investment of \$4000?
b. \$385 is made on an investment of \$3500?
c. Which was the better investment based on return rate?

75. Traveling by car: For winter break, Lindsey and Sloan decided to make the 510 mile trip from Charlotte, NC to Philadelphia, PA together. Sloan drove first at an average speed of 63 mph for 4.5 hours. Lindsey drove for 3.5 hours. What was Lindsey's average speed? Please round to the nearest mph.

76. Investing: Karl invested a total of \$10,000 in two separate accounts. One account paid 6% interest and the other paid 8% interest. If the annual income from both accounts was \$650, how much did he have invested in each account?

77. Rectangles: The perimeter of a rectangle is 60 inches. The area of the rectangle is 221 square inches. Find the dimensions of the rectangle. (**Hint:** After substituting and simplifying, you will have a quadratic equation.)

78. The difference between two positive numbers is 7. If the square of the smaller is added to the square of the larger, the result is 137. Find the numbers.

79. Find two consecutive integers such that the sum of their squares is equal to 145.

80. Baseball: A baseball is thrown upward with an initial velocity of 88 feet per second $(v_0=88)$. When will the ball be 120 feet above the ground? $[h(t)=-16t^2+v_0t.]$

81. Length = 6 inches; Width = 4 inches
82. 5 m

81. Photography: The length of a photograph is 2 inches longer than its width. If the photo is enlarged and 4 inches are added to both the length and the width, the new area would be 80 square inches. Find the dimensions of the original photograph.

82. Water slides: A pool's straight water slide is 8 m longer than the height of the slide. If the distance from the ladder to the end of the slide is 1 m shorter than the length, what is the height of the slide?

CHAPTER 7

Rational Expressions

Did You Know?

An important property related to rational expressions is the **cross-multiplication property**, which states that two rational expressions are equal if the cross products are equal. Symbolically, this property is stated as $\frac{a}{b}=\frac{c}{d}$ if and only if $a\cdot d=b\cdot c$. This property is the key to understanding how a missing term of a proportion can be found if the remaining three terms are known. For example, using the cross-multiplication property to solve the proportion $\frac{x}{6}=\frac{3}{4}$ yields $4x = 18$, and the solution of this simple linear equation is $x=\frac{9}{2}$, the missing term of the proportion. This type of solution illustrates the so-called Rule of Three, which has been known for over 3000 years. The Hindu mathematician Brahmagupta (c. 628) called the rule by that name in his writings, although problems of this type exist in ancient Egyptian and Chinese writings. Brahmagupta taught and wrote in the town of Ujjain in Central India, a center of Hindu science in the seventh century. Brahmagupta is one of the most famous Hindu mathematicians and he stated the Rule of Three as follows: "In the Rule of Three, argument, fruit and requisition are the names of the terms. Requisition multiplied by fruit and divided by argument is the produce;" or, as in our example, $x=\frac{3\cdot 6}{4}$. In this case, the terms of the proportion have been given very fanciful names, and the cross-multiplication property has been concealed in an arbitrary rule.

The Rule of Three appears in Arabic and Latin works without explanation until the Renaissance. It was used in commercial arithmetic and occasionally was called the Merchant's Key or the Golden Rule. A popular seventeenth-century English arithmetic states: "The Rule of Three is commonly called The Golden Rule; and indeed it might be so termed; for as gold transcends all other metals, so doth this rule all others in Arithmetick." The Rule of Three often appeared in verse as a memory aid.

Arithmetic texts of the sixteenth and seventeenth centuries had pages or chapters called Practice. At that time, the term "practice" was used to mean commercial arithmetic usually involving the Rule of Three and other short processes for solving applied problems. Sometimes such problems were called Italian Practice because the problems often related to the methods developed in Italian commercial arithmetic.

"Multiplication is vexation,
Division is as bad;
The Rule of Three doth puzzle me,
And Practice drives me mad."

Mother Goose Rhyme

In arithmetic, **rational numbers** are numbers that can be written in fraction form with **integers** for the numerator and denominator, the denominator not equal to 0. In algebra, **rational expressions** are expressions that can be written in fraction form with **polynomials** for the numerator and denominator, the denominator not equal to 0. As we will see in this section, all of the rules about operating with fractions in arithmetic can be applied to operating with rational expressions in algebra. For example, to add or subtract rational expressions, we need common denominators. To reduce rational expressions, we need to be able to factor both the numerator and denominator. Thus all of the factoring skills learned in Chapter 6 are now going to be applied in dealing with rational expressions and in solving rational equations.

7.1 Multiplication and Division with Rational Expressions

- *Determine any restrictions on the variable in a rational expression.*
- *Reduce rational expressions to lowest terms.*
- *Multiply rational expressions.*
- *Divide rational expressions.*

Introduction to Rational Expressions

The term **rational number** is the technical name for a fraction in which both the numerator and denominator are integers. Similarly, the term **rational expression** is the technical name for a fraction in which both the numerator and denominator are polynomials.

Rational Expressions

A **rational expression** is an algebraic expression that can be written in the form

$$\frac{\boldsymbol{P}}{\boldsymbol{Q}} \quad \text{where } \boldsymbol{P} \text{ and } \boldsymbol{Q} \text{ are polynomials and } \boldsymbol{Q} \neq \mathbf{0}.$$

Examples of rational expressions are

$$\frac{4x^2}{9}, \quad \frac{y^2-25}{y^2+25}, \quad \text{and} \quad \frac{x^2+7x-6}{x^2-5x-14}.$$

As with rational numbers, the denominators of rational expressions cannot be 0. If a numerical value is substituted for a variable in a rational expression and the denominator assumes a value of 0, we say that the expression is **undefined** for that value of the variable.

NOTES

Remember, the denominator of a rational expression can never be 0. Division by 0 is undefined.

Example 1: Finding Restrictions on the Variable

Determine what values of the variable, if any, will make the rational expression undefined. (These values are called **restrictions** on the variable.)

a. $\frac{5}{3x-1}$

Solution: $3x-1=0$ Set the denominator equal to 0.

$3x=1$ Solve the equation.

$x=\frac{1}{3}$

Thus the expression $\frac{5}{3x-1}$ is undefined for $x=\frac{1}{3}$. Any other real number may be substituted for x in the expression. We write $x \neq \frac{1}{3}$ to indicate the restriction on the variable.

b. $\frac{x^2-4}{x^2-5x-6}$

Solution: $x^2-5x-6=0$ Set the denominator equal to 0.

$(x-6)(x+1)=0$ Solve the equation by factoring.

$x-6=0$ or $x+1=0$

$x=6$ $x=-1$

Thus there are two restrictions on the variable: 6 and −1.
We write $x \neq -1, 6$.

c. $\frac{x+3}{x^2+36}$

Solution: $x^2+36=0$ Set the denominator equal to 0.

$x^2=-36$ Solve the equation.

However there is no real number whose square is −36. Thus there are **no restrictions** on the variable.

NOTES

Comments About the Numerator Being 0

If the numerator of a rational expression has a value of 0 and the denominator is not 0 for that value of the variable, then the expression is defined and has a value of 0. If both numerator and denominator are 0, then the expression is **undefined** just as in the case where only the denominator is 0.

The rules for operating with rational expressions are essentially the same as those for operating with fractions in arithmetic. That is, simplifying, multiplying, and dividing rational expressions involve factoring and reducing. Addition and subtraction of rational expressions require common denominators. The basic rules for fractions were discussed in Sections 1.6 and 1.7 and are summarized here for easy reference.

Summary of Arithmetic Rules for Rational Numbers (or Fractions)

A **fraction** (or **rational number**) is a number that can be written in the form $\frac{a}{b}$ where a and b are integers and $b \neq 0$. (Remember, no denominator can be 0.)

The Fundamental Principle: $\frac{a}{b} = \frac{a \cdot k}{b \cdot k}$ where $b, k \neq 0$

The **reciprocal** of $\frac{a}{b}$ is $\frac{b}{a}$ and $\frac{a}{b} \cdot \frac{b}{a} = 1$ where $a, b \neq 0$.

Multiplication: $\frac{a}{b} \cdot \frac{c}{d} = \frac{a \cdot c}{b \cdot d}$ where $b, d \neq 0$

Division: $\frac{a}{b} \div \frac{c}{d} = \frac{a}{b} \cdot \frac{d}{c}$ where $b, c, d \neq 0$

Addition: $\frac{a}{b} + \frac{c}{b} = \frac{a+c}{b}$ where $b \neq 0$

Subtraction: $\frac{a}{b} - \frac{c}{b} = \frac{a-c}{b}$ where $b \neq 0$

For rational expressions, each rule can be restated by replacing a and b with P and Q where P and Q represent polynomials. In particular, the fundamental principle can be restated as follows:

The Fundamental Principle of Rational Expressions

If $\frac{P}{Q}$ is a rational expression and P, Q, and K are polynomials where $Q, K \neq 0$, then

$$\frac{P}{Q} = \frac{P \cdot K}{Q \cdot K}.$$

Reducing (or Simplifying) Rational Expressions

The fundamental principle can be used to **reduce** (or **simplify**) a rational expression to **lower terms** (for multiplication or division) and to **build** a rational expression to **higher terms** (for addition or subtraction). (Just as with rational numbers, a rational expression is said to be **reduced to lowest terms** if the numerator and denominator have no common factors other than 1 and –1.)

Example 2: Reducing Rational Expressions

Use the fundamental principle to reduce each expression to lowest terms. State any restrictions on the variable by using the fact that no denominator can be 0. This restriction applies to denominators **before and after** a rational expression is reduced.

a. $\dfrac{2x-10}{3x-15}$

Solution: $\dfrac{2x-10}{3x-15}=\dfrac{2\cancel{(x-5)}}{3\cancel{(x-5)}}=\dfrac{2}{3}\quad (x \neq 5)$

Note that $x-5$ is a common **factor**. The key word here is **factor**. We reduce using **factors** only.

b. $\dfrac{x^3-64}{x^2-16}$

Solution: $\dfrac{x^3-64}{x^2-16}=\dfrac{\cancel{(x-4)}(x^2+4x+16)}{(x+4)\cancel{(x-4)}}$

$=\dfrac{x^2+4x+16}{x+4}\quad (x \neq -4, 4)$

Reduce. The common **factor** is $x-4$. Note that x^3-64 is the difference of two cubes. Also, note that $x^2+4x+16$ is not factorable.

c. $\dfrac{y-10}{10-y}$

Solution: $\dfrac{y-10}{10-y}=\dfrac{y-10}{-y+10}$

$=\dfrac{1\cancel{(y-10)}}{-1\cancel{(y-10)}}$

$=\dfrac{1}{-1}=-1\quad (y \neq 10)$

Note that the expression $10-y$ is the opposite of $y-10$. When **nonzero opposites** are divided, the quotient is always -1.

In Example 2c, the result was –1. The expression 10 – *y* is the opposite of *y* – 10 for any value of *y*. That is,

$$10-y=-y+10=-1(y-10)=-(y-10).$$

When nonzero opposites are divided, the quotient is always –1. For example,

$$\frac{-9}{9}=-1,\quad \frac{23}{-23}=-1,\quad \frac{x-5}{5-x}=\frac{\cancel{(x-5)}}{-1\cancel{(x-5)}}=\frac{1}{-1}=-1\quad (x \neq 5).$$

Opposites in Rational Expressions

For a polynomial P, $\frac{-P}{P} = -1$ where $P \neq 0$.

In particular, $\frac{a-x}{x-a} = \frac{-(x-a)}{x-a} = -1$ where $x \neq a$.

Teaching Notes: To help convince your students that −1 is always the result, you might want to have them pick several pairs of unequal values for x and a and evaluate $\frac{a-x}{x-a}$.

Remember that the key word when reducing is **factor**. Many students make mistakes similar to the following when working with rational expressions.

NOTES

COMMON ERROR

"Divide out" only common factors.

INCORRECT	INCORRECT
$\frac{4x+\cancel{8}}{\cancel{8}}$	$\frac{\cancel{x^2}^{\,x}-\cancel{9}^{\,3}}{\cancel{x}-\cancel{3}}$
8 is not a common factor.	3 and x are not common factors.

CORRECT	CORRECT
$\frac{4x+8}{8} = \frac{\cancel{4}(x+2)}{\underset{2}{\cancel{8}}}$	$\frac{x^2-9}{x-3} = \frac{(x+3)\cancel{(x-3)}}{\cancel{(x-3)}}$
4 is a common factor.	$x-3$ is a common factor.

Multiplication with Rational Expressions

To Multiply Rational Expressions

To multiply any two (or more) rational expressions,

1. Completely factor each numerator and denominator.
2. Multiply the numerators and multiply the denominators, keeping the expressions in factored form.
3. "Divide out" any common factors from the numerators and denominators. Remember that no denominator can have a value of 0.

For example, here no factoring is necessary,

$$\frac{2x}{x-6}\cdot\frac{x+5}{x-4}=\frac{2x(x+5)}{(x-6)(x-4)}=\frac{2x^2+10x}{(x-6)(x-4)}. \qquad (x\neq 4,6)$$

However, in this case the numerators and denominators must be factored.

$$\frac{y^2-4}{y^3}\cdot\frac{y^2-3y}{y^2-y-6}=\frac{\cancel{(y+2)}(y-2)\cancel{(y)}\cancel{(y-3)}}{\cancel{y^3}_{y^2}\cancel{(y-3)}\cancel{(y+2)}}=\frac{y-2}{y^2} \qquad (y\neq -2,0,3)$$

Multiplication with Rational Expressions

If $P, Q, R,$ and S are polynomials and $Q, S \neq 0,$ then

$$\frac{\boldsymbol{P}}{\boldsymbol{Q}}\cdot\frac{\boldsymbol{R}}{\boldsymbol{S}}=\frac{\boldsymbol{P\cdot R}}{\boldsymbol{Q\cdot S}}.$$

Example 3: Multiplication with Rational Expressions

Teaching Notes:
Note that in Example 3a I used the concept of negative exponents to simplify the variables for variety. You might want to point this out to your students and compare it with Example 3b.

Multiply and reduce, if possible. Use the rules for exponents when they apply. State any restrictions on the variable(s).

a. $$\frac{5x^2y}{9xy^3}\cdot\frac{6x^3y^2}{15xy^4}=\frac{\cancel{5}\cdot 2\cdot\cancel{3}\cdot x^5\cdot y^3}{3\cdot 3\cdot\cancel{3}\cdot\cancel{5}\cdot x^2\cdot y^7}=\frac{2x^{5-2}y^{3-7}}{9}=\frac{2x^3y^{-4}}{9}=\frac{2x^3}{9y^4} \qquad (x\neq 0, y\neq 0)$$

b. $$\frac{x}{x-2}\cdot\frac{x^2-4}{x^2}=\frac{\cancel{x}(x+2)\cancel{(x-2)}}{\cancel{(x-2)}\cancel{x^2}_{x}}=\frac{x+2}{x} \qquad (x\neq 0,2)$$

c. $$\frac{3x-3}{x^2+x}\cdot\frac{x^2+2x+1}{3x^2-6x+3}=\frac{\cancel{3}\cancel{(x-1)}\cancel{(x+1)^2}^{(x+1)}}{x\cancel{(x+1)}\cdot\cancel{3}\cancel{(x-1)^2}_{(x-1)}}=\frac{x+1}{x(x-1)} \text{ or } \frac{x+1}{x^2-x} \qquad (x\neq -1,0,1)$$

d. $$\frac{x^2-7x+12}{2x+6}\cdot\frac{x^2-4}{x^2-2x-8}=\frac{\cancel{(x-4)}(x-3)\cancel{(x+2)}(x-2)}{2(x+3)\cancel{(x-4)}\cancel{(x+2)}}$$

$$=\frac{(x-3)(x-2)}{2(x+3)} \text{ or } \frac{x^2-5x+6}{2(x+3)} \text{ or } \frac{x^2-5x+6}{2x+6}$$

$(x\neq -3,-2,4)$

NOTES

As shown in Examples 3c and 3d there may be more than one correct form for an answer. After a rational expression has been reduced, the numerator and denominator may be multiplied out or left in factored form. **Generally, the denominator will be left in factored form and the numerator multiplied out.** As we will see in the next section, this form makes the results easier to add and subtract. However, be aware that this form is just an option, and multiplying out the denominator is not an error.

Division with Rational Expressions

To divide any two rational expressions, multiply the first fraction by the **reciprocal** of the second fraction.

Division with Rational Expressions

If P, Q, R, and S are polynomials with $Q, R, S \neq 0$, then

$$\frac{P}{Q} \div \frac{R}{S} = \frac{P}{Q} \cdot \frac{S}{R}.$$

Note that $\frac{S}{R}$ is the reciprocal of $\frac{R}{S}$.

Example 4: Division with Rational Expressions

Divide and reduce, if possible. Assume that no denominator has a value of 0.

a. $\frac{12x^2y}{10xy^2} \div \frac{3x^4y}{xy^3}$

Solution:

$$\frac{12x^2y}{10xy^2} \div \frac{3x^4y}{xy^3} = \frac{12x^2y}{10xy^2} \cdot \frac{xy^3}{3x^4y}$$

$$= \frac{\not{2} \cdot 2 \cdot \not{3} \cdot x^3 \cdot y^4}{\not{2} \cdot 5 \cdot \not{3} \cdot x^5 \cdot y^3}$$

$$= \frac{2x^{3-5}y^{4-3}}{5}$$

$$= \frac{2x^{-2}y}{5} = \frac{2y}{5x^2}$$

Note that in this example we have used the quotient rule for exponents.

b. $\dfrac{x^3-y^3}{x^3} \div \dfrac{y-x}{xy}$

Solution: $\dfrac{x^3-y^3}{x^3} \div \dfrac{y-x}{xy} = \dfrac{x^3-y^3}{x^3} \cdot \dfrac{xy}{y-x}$

$$= \frac{\overset{-1}{\cancel{(x-y)}}\left(x^2+xy+y^2\right)\cancel{x}y}{\underset{x^2}{\cancel{x^3}}\cancel{(y-x)}}$$

Note that $\dfrac{x-y}{y-x} = -1$.

$$= \frac{-y\left(x^2+xy+y^2\right)}{x^2} = \frac{-x^2y-xy^2-y^3}{x^2}$$

c. $\dfrac{x^2-8x+15}{2x^2+11x+5} \div \dfrac{2x^2-5x-3}{4x^2-1}$

Solution: $\dfrac{x^2-8x+15}{2x^2+11x+5} \div \dfrac{2x^2-5x-3}{4x^2-1} = \dfrac{x^2-8x+15}{2x^2+11x+5} \cdot \dfrac{4x^2-1}{2x^2-5x-3}$

$$= \frac{\cancel{(x-3)}(x-5)(2x-1)\cancel{(2x+1)}}{(2x+1)(x+5)\cancel{(x-3)}\cancel{(2x+1)}}$$

$$= \frac{(x-5)(2x-1)}{(2x+1)(x+5)} = \frac{2x^2-11x+5}{(2x+1)(x+5)}$$

Remember that you have the option of leaving the numerator and/or denominator in factored form.

Practice Problems

Reduce to lowest terms. State any restrictions on the variables.

1. $\dfrac{5x+20}{7x+28}$ **2.** $\dfrac{6-3x}{3x-6}$ **3.** $\dfrac{x^2+x-2}{x^2+3x+2}$

Perform the following operations and simplify the results. Assume that no denominator has a value of 0.

4. $\dfrac{x-7}{x^3} \cdot \dfrac{x^2}{49-x^2}$ **5.** $\dfrac{y^2-y-6}{y^2-5y+6} \cdot \dfrac{y^2-4}{y^2+4y+4}$

6. $\dfrac{x^3+3x}{2x+1} \div \dfrac{x^2+3}{x+1}$ **7.** $\dfrac{x^2+2x-3}{x^2-3x-10} \cdot \dfrac{2x^2-9x-5}{x^2-2x+1} \div \dfrac{4x+2}{x^2-x}$

Answers to Practice Problems: **1.** $\dfrac{5}{7}$, $x \neq -4$ **2.** -1, $x \neq 2$ **3.** $\dfrac{x-1}{x+1}$, $x \neq -2, -1$ **4.** $\dfrac{-1}{x(x+7)}$ **5.** 1 **6.** $\dfrac{x^2+x}{2x+1}$ **7.** $\dfrac{x^2+3x}{2(x+2)}$

7.1 Exercises

1. $\frac{3x}{4y}; x \neq 0, y \neq 0$
2. $\frac{2y^3}{3x}; x \neq 0, y \neq 0$
3. $\frac{2x^3}{3y^3}; x \neq 0, y \neq 0$
4. $\frac{3y^2}{4x^3}; x \neq 0, y \neq 0$
5. $\frac{1}{x-3}; x \neq 0, 3$
6. $\frac{3}{x+5}; x \neq -5, 0$
7. $7; x \neq 2$
8. $-1; x \neq 2$
9. $-\frac{3}{4}; x \neq 3$
10. $-\frac{1}{2}; x \neq 4$
11. $\frac{2x}{y}; x \neq -\frac{2}{3}, y \neq 0$
12. $\frac{1}{4x}; x \neq 0, y \neq -\frac{1}{3}$
13. $\frac{x}{x-1}; x \neq -6, 1$
14. $\frac{x-y}{3x}; x \neq 0, x \neq -y$
15. $\frac{x^2-3x+9}{x-3}; x \neq -3, 3$
16. $\frac{x^2+2x+4}{x+2}; x \neq -2, 2$
17. $\frac{x-3}{y-2}; y \neq -2, 2$
18. $-\frac{3x-4}{2x-3}; x \neq -6, \frac{3}{2}$
19. $\frac{x^2+2x+4}{y+5}; x \neq 2, y \neq -5$
20. $\frac{x^2-4x+16}{2x-7}; x \neq -4, \frac{7}{2}$
21. $\frac{ab}{6y}$
22. $\frac{12y}{x}$

Reduce to lowest terms. State any restrictions on the variable(s).

1. $\frac{9x^2y^3}{12xy^4}$ **2.** $\frac{18xy^4}{27x^2y}$ **3.** $\frac{20x^5}{30x^2y^3}$ **4.** $\frac{15y^4}{20x^3y^2}$

5. $\frac{x}{x^2-3x}$ **6.** $\frac{3x}{x^2+5x}$ **7.** $\frac{7x-14}{x-2}$ **8.** $\frac{4-2x}{2x-4}$

9. $\frac{9-3x}{4x-12}$ **10.** $\frac{2x-8}{16-4x}$ **11.** $\frac{6x^2+4x}{3xy+2y}$ **12.** $\frac{1+3y}{4x+12xy}$

13. $\frac{x^2+6x}{x^2+5x-6}$ **14.** $\frac{x^2-y^2}{3x^2+3xy}$ **15.** $\frac{x^3+27}{x^2-9}$ **16.** $\frac{x^3-8}{x^2-4}$

17. $\frac{xy-3y+2x-6}{y^2-4}$ **18.** $\frac{3x^2+14x-24}{18-9x-2x^2}$ **19.** $\frac{x^3-8}{5x-2y+xy-10}$ **20.** $\frac{x^3+64}{2x^2+x-28}$

Perform the indicated operations and reduce to lowest terms. Assume that no denominator has a value of 0.

21. $\frac{3ax^2}{4b} \cdot \frac{6b^2}{27x^2y}$ **22.** $\frac{18x^3}{5y^2} \cdot \frac{30y^3}{9x^4}$ **23.** $\frac{24x^3}{25y^2} \cdot \frac{10y^5}{18x}$

24. $\frac{16x^8}{3y^{11}} \cdot \frac{-21y^9}{10x^7}$ **25.** $\frac{x^2-9}{x^2+2x} \cdot \frac{x+2}{x-3}$ **26.** $\frac{16x^2-9}{3x^2-15x} \cdot \frac{6}{4x+3}$

27. $\frac{x^2+2x-3}{x^2+3x} \cdot \frac{x}{x+1}$ **28.** $\frac{4x+16}{x^2-16} \cdot \frac{x-4}{x}$ **29.** $\frac{x^2+6x-16}{x^2-64} \cdot \frac{1}{2-x}$

30. $\frac{4-x^2}{x^2-4x+4} \cdot \frac{3}{x+2}$ **31.** $\frac{x^2-5x+6}{x^2-4x} \cdot \frac{x-4}{x-3}$ **32.** $\frac{2x^2+x-3}{x^2+4x} \cdot \frac{2x+8}{x-1}$

33. $\frac{2x^2+10x}{3x^2+5x+2} \cdot \frac{6x+4}{x^2}$ **34.** $\frac{x+3}{x^2-16} \cdot \frac{x^2-3x-4}{x^2-1}$

35. $\frac{x}{x^2+7x+12} \cdot \frac{x^2-2x-24}{x^2-7x+6}$ **36.** $\frac{x^2-2x-3}{x+5} \cdot \frac{x^2-5x-14}{x^2-x-6}$

37. $\frac{8-2x-x^2}{x^2-2x} \cdot \frac{x-4}{x^2-3x-4}$ **38.** $\frac{3x^2+21x}{x^2-49} \cdot \frac{x^2-5x+4}{x^2+3x-4}$

39. $\frac{(x-2y)^2}{x^2-5xy+6y^2} \cdot \frac{x+2y}{x^2-4xy+4y^2}$ **40.** $\frac{4x^2+6x}{x^2+3x-10} \cdot \frac{x^2+4x-12}{x^2+5x-6}$

41. $\frac{2x^2+5x+2}{3x^2+8x+4} \cdot \frac{3x^2-x-2}{4x^3-x}$ **42.** $\frac{x^2+5x}{4x^2+12x+9} \cdot \frac{6x^2+7x-3}{x^2+10x+25}$

23. $\frac{8x^2y^3}{15}$

24. $-\frac{56x}{5y^2}$ 25. $\frac{x+3}{x}$

26. $\frac{8x-6}{x(x-5)}$

27. $\frac{x-1}{x+1}$ 28. $\frac{4}{x}$

29. $-\frac{1}{x-8}$

30. $-\frac{3}{x-2}$

31. $\frac{x-2}{x}$ 32. $\frac{4x+6}{x}$

33. $\frac{4x+20}{x(x+1)}$

34. $\frac{x+3}{(x+4)(x-1)}$

35. $\frac{x}{(x+3)(x-1)}$

36. $\frac{x^2-6x-7}{x+5}$

37. $-\frac{x+4}{x(x+1)}$

38. $\frac{3x^2-12x}{(x-7)(x+4)}$

39. $\frac{x+2y}{(x-3y)(x-2y)}$

40. $\frac{4x^2+6x}{(x+5)(x-1)}$

41. $\frac{x-1}{x(2x-1)}$

42. $\frac{3x^2-x}{(2x+3)(x+5)}$

43. $\frac{1}{x+1}$ 44. $\frac{x+3}{2}$

45. $\frac{x+2}{x-2}$ 46. $\frac{2x-1}{2x+1}$

47. $\frac{1}{3xy^6}$ 48. $\frac{49y^3}{6x^5}$

49. $\frac{6y^7}{x^4}$ 50. $\frac{1}{12x}$

51. $\frac{x}{12}$ 52. $\frac{7x}{x+2}$

43. $\frac{x^2+x+1}{x^2-1}\cdot\frac{x^2-2x+1}{x^3-1}$

44. $\frac{x^2-9}{2x^2+4x+8}\cdot\frac{x^3-8}{x^2-5x+6}$

45. $\frac{x-2}{x^2-2x+4}\cdot\frac{x^3+8}{x^2-4x+4}$

46. $\frac{2x^2-7x+3}{x^2-9}\cdot\frac{3x^2+8x-3}{6x^2+x-1}$

47. $\frac{12x^2y}{9xy^9}\div\frac{4x^4y}{x^2y^3}$

48. $\frac{35xy^3}{24x^3y}\div\frac{15x^4y^3}{84xy^4}$

49. $\frac{45xy^4}{21x^2y^2}\div\frac{40x^4}{112xy^5}$

50. $\frac{x-3}{15x}\div\frac{4x-12}{5}$

51. $\frac{x-1}{6x+6}\div\frac{2x-2}{x^2+x}$

52. $\frac{7x-14}{x^2}\div\frac{x^2-4}{x^3}$

53. $\frac{6x^2-54}{x^4}\div\frac{x-3}{x^2}$

54. $\frac{x^2-25}{6x+30}\div\frac{x-5}{x}$

55. $\frac{2x-1}{x^2+2x}\div\frac{10x^2-5x}{6x^2+12x}$

56. $\frac{x+3}{x^2+3x-4}\div\frac{x+2}{x^2+x-2}$

57. $\frac{6x^2-7x-3}{x^2-1}\div\frac{2x-3}{x-1}$

58. $\frac{x^2-9}{2x^2+7x+3}\div\frac{x^2-3x}{2x^2+11x+5}$

59. $\frac{x^2-6x+9}{x^2-4x+3}\div\frac{2x^2-7x+3}{x^2-3x+2}$

60. $\frac{x^3+2x^2}{x^3+64}\div\frac{4x^2}{x^2-4x+16}$

61. $\frac{2x+1}{4x-x^2}\div\frac{4x^2-1}{x^2-16}$

62. $\frac{x^2-4x+4}{x^2+5x+6}\div\frac{x^2+2x-8}{x^2+7x+12}$

63. $\frac{x^2-x-6}{x^2+6x+8}\div\frac{x^2-4x+3}{x^2+5x+4}$

64. $\frac{x^2-x-12}{6x^2+x-9}\div\frac{x^2-6x+8}{3x^2-x-6}$

65. $\frac{6x^2+5x+1}{4x^3-3x^2}\div\frac{3x^2-2x-1}{3x^2-2x+1}$

66. $\frac{8x^2+2x-15}{3x^2+13x+4}\div\frac{2x^2+5x+3}{6x^2-x-1}$

67. $\frac{3x^2+13x+14}{4x^3-3x^2}\div\frac{6x^2-x-35}{4x^2+5x-6}$

68. $\frac{3x^2+2x}{9x^2-4}\div\frac{27x^3-8}{9x^2-6x+4}$

69. $\frac{x^2-8x+15}{x^2-9x+14}\div\frac{x^2+4x-21}{x-1}$

70. $\frac{6-11x-10x^2}{2x^2+x-3}\div\frac{5x^3-2x^2}{3x^2-5x+2}$

71. $\frac{x-6}{x^2-7x+6}\cdot\frac{x^2-3x}{x+3}\cdot\frac{x^2-9}{x^2-4x+3}$

72. $\frac{3x^2+11x+10}{2x^2+x-6}\cdot\frac{x^2+2x-3}{2x-1}\cdot\frac{2x-3}{3x^2+2x-5}$

73. $\frac{x^3+3x^2}{x^2+7x+12}\cdot\frac{2x^2+7x-4}{2x^2-x}\cdot\frac{x^2+4x-5}{2x^2-x-1}$

74. $\frac{x^2+2x-3}{x^2+10x+21}\cdot\frac{x^2+6x+5}{x^2-7x-8}\cdot\frac{x^2-x-56}{x^2-3x-40}$

75. $\frac{2x^2-5x+2}{4xy-2y+6x-3}\div\frac{xy-2y+3x-6}{2y^2+9y+9}$

76. $\frac{2xy-12x+y-6}{y^2-2y-24}\div\frac{2x^2+11x+5}{xy+5y+4x+20}$

53. $\frac{6x+18}{x^2}$
54. $\frac{x}{6}$ 55. $\frac{6}{5x}$
56. $\frac{x+3}{x+4}$ 57. $\frac{3x+1}{x+1}$
58. $\frac{x+5}{x}$ 59. $\frac{x-2}{2x-1}$
60. $\frac{x+2}{4(x+4)}$
61. $-\frac{x+4}{x(2x-1)}$

77. Rectangles: The area of a rectangle (in square feet) is represented by the polynomial function $A(x)=4x^2-4x-15$. If the length of the rectangle is $(2x+3)$ feet, find a representation for the width.

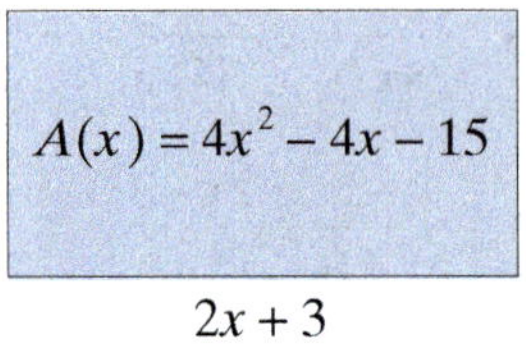

78. Rectangles: The area of a rectangle (in square feet) is represented by the polynomial function $A(x)=3x^2-x-10$. If the length of the rectangle is $(3x+5)$ feet, find a representation for the width.

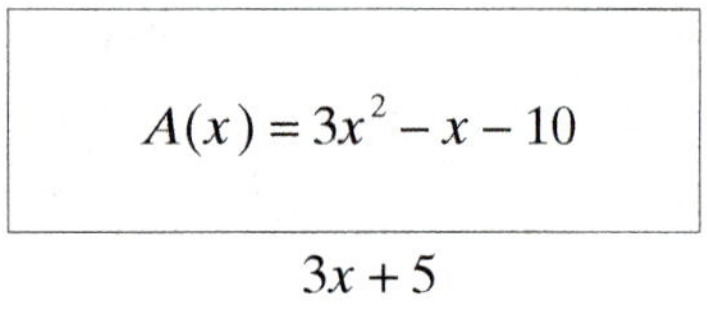

Writing and Thinking About Mathematics

62. $\frac{x-2}{x+2}$ 63. $\frac{x+1}{x-1}$
64. $\frac{3x^3+8x^2-9x-18}{(6x^2+x-9)(x-2)}$
65. $\frac{6x^3-x^2+1}{x^2(4x-3)(x-1)}$
66. $\frac{8x^2-14x+5}{(x+4)(x+1)}$
67. $\frac{x^2+4x+4}{x^2(2x-5)}$
68. $\frac{9x^3-6x^2+4x}{(3x-2)^2(9x^2+6x+4)}$
69. $\frac{x^2-6x+5}{(x-7)(x-2)(x+7)}$

79. a. Define rational expression.

b. Give an example of a rational expression that is undefined for $x = -2$ and $x = 3$ and has a value of 0 for $x = 1$. Explain how you determined this expression.

c. Give an example of a rational expression that is undefined for $x = -5$ and never has a value of 0. Explain how you determined this expression.

80. Write the opposite of each of the following expressions.

a. $3-x$ **b.** $2x-7$ **c.** $x+5$ **d.** $-3x-2$

81. Given the rational function $f(x)=\frac{x-4}{x^2-100}$:

a. For what values, if any, will $f(x)=0$?

b. For what values, if any, is $f(x)$ undefined?

HAWKES LEARNING SYSTEMS: INTRODUCTORY & INTERMEDIATE ALGEBRA SOFTWARE

- 7.1a Defining Rational Expressions
- 7.1b Multiplication and Division with Rational Expressions

70. $-\frac{3x-2}{x^2}$ 71. $\frac{x^2-3x}{(x-1)^2}$ 72. $\frac{x+3}{2x-1}$ 73. $\frac{x^2+5x}{2x+1}$ 74. $\frac{x-1}{x-8}$ 75. 1 76. 1 77. $2x-5$ feet 78. $x-2$ feet
79. a. A rational expression is an algebraic expression that can be written in the form $\frac{P}{Q}$ where P and Q are polynomials and $Q \neq 0$. b. $\frac{x-1}{(x+2)(x-3)}$ Answers will vary. c. $\frac{1}{x+5}$ Answers will vary.
80. a. $x-3$ b. $7-2x$ c. $-x-5$ d. $3x+2$ 81. a. $x=4$ b. $x=-10, 10$

7.2 Addition and Subtraction with Rational Expressions

- *Add rational expressions.*
- *Subtract rational expressions.*

Addition with Rational Expressions

To add rational expressions with a common denominator, proceed just as with fractions: add the numerators and keep the common denominator. For example,

$$\frac{5}{x+2}+\frac{6}{x+2}=\frac{5+6}{x+2}=\frac{11}{x+2}. \qquad (x \neq -2)$$

In some cases the sum can be reduced:

$$\frac{x^2+6}{x+2}+\frac{5x}{x+2}=\frac{x^2+5x+6}{x+2}=\frac{\cancel{(x+2)}(x+3)}{\cancel{x+2}}=x+3. \qquad (x \neq -2)$$

Addition with Rational Expressions

For polynomials P, Q, and R, with $Q \neq 0$,

$$\frac{\boldsymbol{P}}{\boldsymbol{Q}}+\frac{\boldsymbol{R}}{\boldsymbol{Q}}=\frac{\boldsymbol{P}+\boldsymbol{R}}{\boldsymbol{Q}}.$$

Example 1: Adding Rational Expressions with a Common Denominator

Find each sum and reduce if possible. (Note the importance of the factoring techniques we studied in Chapter 6.)

a. $\frac{x}{x^2-1}+\frac{1}{x^2-1}$

Solution:

$$\frac{x}{x^2-1}+\frac{1}{x^2-1}=\frac{x+1}{x^2-1}$$

$$=\frac{\cancel{x+1}}{\cancel{(x+1)}(x-1)}$$

$$=\frac{1}{x-1} \qquad (x \neq -1, 1)$$

Remember, if we use the entire expression in the numerator (or denominator) to reduce, we are left with a factor of 1.

Continued on the next page...

b. $\dfrac{1}{x^2+7x+10}+\dfrac{2x+3}{x^2+7x+10}$

Solution: $\dfrac{1}{x^2+7x+10}+\dfrac{2x+3}{x^2+7x+10}=\dfrac{2x+4}{x^2+7x+10}$

$=\dfrac{2\cancel{(x+2)}}{(x+5)\cancel{(x+2)}}$

$=\dfrac{2}{x+5}$ $\quad (x \neq -5, -2)$

The rational expressions in Example 1 had common denominators. To add expressions with different denominators, we need to find the least common multiple (LCM) of the denominators. The LCM was discussed in Section 1.5. The procedure is stated here for polynomials.

To Find the LCM for a Set of Polynomials

1. Completely factor each polynomial (including prime factors for numerical factors).
2. Form the product of all factors that appear, using each factor the most number of times it appears in any one polynomial.

The LCM of a set of denominators is called the **least common denominator (LCD)**. To add fractions with different denominators, begin by changing each fraction to an equivalent fraction with the LCD as the denominator. This is called **building the fraction to higher terms**.

Use the following procedure when adding rational expressions with different denominators.

Procedure for Adding Rational Expressions with Different Denominators

1. Find the LCD (the LCM of the denominators).
2. Rewrite each fraction in an equivalent form with the LCD as the denominator.
3. Add the numerators and keep the common denominator.
4. Reduce if possible.

Example 2: Adding Rational Expressions with Different Denominators

Find each sum and reduce if possible. Assume that no denominator has a value of 0.

a. $\dfrac{y}{y-3}+\dfrac{6}{y+4}$

Solution: In this case, neither denominator can be factored so the LCD is the product of these factors. That is, $\text{LCD}=(y-3)(y+4)$.

Now, using the fundamental principle, we have

$$\begin{aligned}\frac{y}{y-3}+\frac{6}{y+4}&=\frac{y(y+4)}{(y-3)(y+4)}+\frac{6(y-3)}{(y+4)(y-3)}\\&=\frac{(y^2+4y)+(6y-18)}{(y-3)(y+4)}\\&=\frac{y^2+10y-18}{(y-3)(y+4)}.\end{aligned}$$

The numerator is not factorable and the expression is reduced.

b. $\dfrac{1}{x^2+6x+9}+\dfrac{1}{x^2-9}+\dfrac{1}{2x+6}$

Solution: First, find the LCD.

Step 1: Factor each expression completely.

$$x^2+6x+9=(x+3)^2$$

$$x^2-9=(x+3)(x-3)$$

$$2x+6=2(x+3)$$

Step 2: Form the product of 2, $(x+3)^2$, and $(x-3)$. That is, use each factor the most number of times it appears in any one factorization.

$$\text{LCD}=2(x+3)^2(x-3)$$

Now use the LCD and add as follows.

$$\begin{aligned}&\frac{1}{x^2+6x+9}+\frac{1}{x^2-9}+\frac{1}{2x+6}\\&=\frac{1}{(x+3)^2}+\frac{1}{(x+3)(x-3)}+\frac{1}{2(x+3)}\\&=\frac{1\cdot2(x-3)}{(x+3)^2\cdot2(x-3)}+\frac{1\cdot2(x+3)}{(x+3)(x-3)\cdot2(x+3)}+\frac{1\cdot(x+3)(x-3)}{2(x+3)\cdot(x+3)(x-3)}\\&=\frac{(2x-6)+(2x+6)+(x^2-9)}{2(x+3)^2(x-3)}=\frac{x^2+4x-9}{2(x+3)^2(x-3)}\end{aligned}$$

NOTES

Important Note About the Form of Answers

In Examples 2a and 2b each denominator is left in factored form as a convenience for possibly reducing or adding to some other expression later. You may choose to multiply out these factors. Either form is correct. For consistency, denominators are left in factored form in the answers in the back of the text.

Subtraction with Rational Expressions

When subtracting fractions, the placement of negative signs can be critical. For example, note how –2 can be indicated in three different forms:

$$-\frac{6}{3}=-2,\quad \frac{-6}{3}=-2, \text{ and } \frac{6}{-3}=-2.$$

Thus we have

$$-\frac{6}{3}=\frac{-6}{3}=\frac{6}{-3}.$$

With polynomials we have the following statement about the placement of negative signs which can be **very useful in subtraction**. We seldom leave the negative sign in the denominator.

Placement of Negative Signs

If P and Q are polynomials and $Q \neq 0$, then

$$-\frac{\boldsymbol{P}}{\boldsymbol{Q}}=\frac{\boldsymbol{P}}{-\boldsymbol{Q}}=\frac{-\boldsymbol{P}}{\boldsymbol{Q}}.$$

To subtract rational expressions with a common denominator, proceed just as with fractions: subtract the numerators and keep the common denominator. For example,

$$\frac{17}{x+7}-\frac{23}{x+7}=\frac{17-23}{x+7}=\frac{-6}{x+7}\quad\left(\text{or } -\frac{6}{x+7}\right).$$

Subtraction with Rational Expressions

For polynomials P, Q, and R, with $Q \neq 0$,

$$\frac{\boldsymbol{P}}{\boldsymbol{Q}}-\frac{\boldsymbol{R}}{\boldsymbol{Q}}=\frac{\boldsymbol{P}-\boldsymbol{R}}{\boldsymbol{Q}}.$$

Example 3: Subtracting Rational Expressions with a Common Denominator

Find each difference and reduce if possible. Assume that no denominator has a value of 0.

a. $\frac{2x-5y}{x+y}-\frac{3x-7y}{x+y}$

Solution:

$$\frac{2x-5y}{x+y}-\frac{3x-7y}{x+y}=\frac{2x-5y-(3x-7y)}{x+y}$$ Subtract the entire numerator.

$$=\frac{2x-5y-3x+7y}{x+y}$$

$$=\frac{-x+2y}{x+y}$$

b. $\frac{x^2}{x^2+4x+4}-\frac{2x+8}{x^2+4x+4}$

Solution:

$$\frac{x^2}{x^2+4x+4}-\frac{2x+8}{x^2+4x+4}=\frac{x^2-(2x+8)}{x^2+4x+4}$$ Subtract the entire numerator.

$$=\frac{x^2-2x-8}{x^2+4x+4}$$

$$=\frac{(x-4)\cancel{(x+2)}}{(x+2)\cancel{(x+2)}}$$ Factor and reduce.

$$=\frac{x-4}{x+2}$$

c. $\frac{x}{x-5}-\frac{3}{5-x}$

Solution: Each denominator is the **opposite** of the other. Multiply both the numerator and denominator of the second fraction by -1 so that both denominators will be the same, in this case $x-5$.

$$\frac{x}{x-5}-\frac{3}{5-x}=\frac{x}{x-5}-\frac{3}{(5-x)}\cdot\frac{(-1)}{(-1)}$$

$$=\frac{x}{x-5}-\frac{-3}{x-5}$$

$$=\frac{x-(-3)}{x-5}$$

$$=\frac{x+3}{x-5}$$

NOTES

COMMON ERROR

Many beginning students make a mistake when subtracting rational expressions by not subtracting the entire numerator. They make a mistake similar to the following.

INCORRECT $$\frac{10}{x+5}-\frac{3-x}{x+5}=\frac{10-3-x}{x+5}=\frac{7-x}{x+5}$$

By using parentheses, you can avoid such mistakes.

CORRECT $$\frac{10}{x+5}-\frac{3-x}{x+5}=\frac{10-(3-x)}{x+5}=\frac{10-3+x}{x+5}=\frac{7+x}{x+5}=\frac{x+7}{x+5}$$

As with addition, if the rational expressions do not have the same denominator, find the LCM of the denominators (the LCD) and use the fundamental principle to **build each fraction to higher terms**, if necessary, so that each has the LCD as the denominator.

Example 4: Subtracting Rational Expressions with Different Denominators

Find each difference and reduce if possible. Assume that no denominator has a value of 0.

a. $\frac{x+5}{x-5}-\frac{100}{x^2-25}$

Solution:

$$\left.\begin{aligned}x-5&=x-5\\x^2-25&=(x+5)(x-5)\end{aligned}\right\}\quad \text{LCD}=(x+5)(x-5)$$

$$\begin{aligned}\frac{x+5}{x-5}-\frac{100}{x^2-25}&=\frac{(x+5)(x+5)}{(x-5)(x+5)}-\frac{100}{(x+5)(x-5)}\\&=\frac{(x^2+10x+25)-100}{(x+5)(x-5)}\\&=\frac{x^2+10x+25-100}{(x+5)(x-5)}\\&=\frac{x^2+10x-75}{(x+5)(x-5)}\\&=\frac{(x+15)\cancel{(x-5)}}{(x+5)\cancel{(x-5)}}\\&=\frac{x+15}{x+5}\end{aligned}$$

b. $\dfrac{x+y}{(x-y)^2}-\dfrac{x}{2x^2-2y^2}$

Solution:
$$\left.\begin{aligned}(x-y)^2&=(x-y)^2\\2x^2-2y^2&=2(x-y)(x+y)\end{aligned}\right\rangle\quad \text{LCD}=2(x-y)^2(x+y)$$

$$\begin{aligned}\frac{x+y}{(x-y)^2}-\frac{x}{2x^2-2y^2}&=\frac{(x+y)\cdot 2(x+y)}{(x-y)^2\cdot 2(x+y)}-\frac{x(x-y)}{2(x-y)(x+y)(x-y)}\\&=\frac{2(x^2+2xy+y^2)-(x^2-xy)}{2(x-y)^2(x+y)}\\&=\frac{2x^2+4xy+2y^2-x^2+xy}{2(x-y)^2(x+y)}\\&=\frac{x^2+5xy+2y^2}{2(x-y)^2(x+y)}\end{aligned}$$

c. $\dfrac{3x-12}{x^2+x-20}-\dfrac{x^2+5x}{x^2+9x+20}$

Hint: In this problem, both expressions can be reduced before looking for the LCD.

Solution:
$$\begin{aligned}\frac{3x-12}{x^2+x-20}-\frac{x^2+5x}{x^2+9x+20}&=\frac{3\cancel{(x-4)}}{(x+5)\cancel{(x-4)}}-\frac{x\cancel{(x+5)}}{\cancel{(x+5)}(x+4)}\\&=\frac{3}{x+5}-\frac{x}{x+4}\end{aligned}$$

Now subtract these two expressions with $\text{LCD}=(x+5)(x+4)$.

$$\begin{aligned}\frac{3}{x+5}-\frac{x}{x+4}&=\frac{3(x+4)}{(x+5)(x+4)}-\frac{x(x+5)}{(x+4)(x+5)}\\&=\frac{(3x+12)-(x^2+5x)}{(x+5)(x+4)}\\&=\frac{3x+12-x^2-5x}{(x+5)(x+4)}\\&=\frac{-x^2-2x+12}{(x+5)(x+4)}\end{aligned}$$

Continued on the next page...

d. $\dfrac{x+1}{xy-3y+4x-12}-\dfrac{x-3}{xy+6y+4x+24}$

Solution:

$$\left.\begin{aligned} xy-3y+4x-12 &= y(x-3)+4(x-3) \\ &= (x-3)(y+4) \\ xy+6y+4x+24 &= y(x+6)+4(x+6) \\ &= (x+6)(y+4) \end{aligned}\right\} \text{LCD} = (x-3)(y+4)(x+6)$$

$$\begin{aligned} &\frac{x+1}{xy-3y+4x-12}-\frac{x-3}{xy+6y+4x+24} \\ &= \frac{(x+1)(x+6)}{(y+4)(x-3)(x+6)}-\frac{(x-3)(x-3)}{(x+6)(y+4)(x-3)} \\ &= \frac{x^2+7x+6-(x^2-6x+9)}{(y+4)(x-3)(x+6)} \\ &= \frac{x^2+7x+6-x^2+6x-9}{(y+4)(x-3)(x+6)} \\ &= \frac{13x-3}{(y+4)(x-3)(x+6)} \end{aligned}$$

1. 3 2. 7
3. 2 4. 2
5. 1 6. $x+3$
7. $\dfrac{2}{x-1}$
8. $\dfrac{2x^2+3x+4}{(x-2)(x+2)}$
9. $\dfrac{14}{7-x}$ 10. 6
11. 4 12. $\dfrac{23}{x-10}$
13. $\dfrac{x^2-x+1}{(x+4)(x-3)}$
14. $\dfrac{5}{x-3}$ 15. $\dfrac{x-2}{x+2}$
16. $\dfrac{-2x+3}{(2x+1)(x-1)}$
17. $\dfrac{4x+5}{2(7x-2)}$
18. $\dfrac{x+9}{2(2x+5)}$
19. $\dfrac{6x+15}{(x+3)(x-3)}$
20. $\dfrac{x^2+3x+1}{(x-5)(x+2)}$
21. $\dfrac{x^2-2x+4}{(x+2)(x-1)}$ 22. $\dfrac{-11x+3}{3x-1}$ 23. $\dfrac{-x^2-3x-6}{(x+3)(3-x)}$ 24. $\dfrac{-2x^2+16x-5}{(4-x)(x+5)}$ 25. $\dfrac{8x^2+13x-21}{6(x+3)(x-3)}$
26. $\dfrac{-9x^2+22x+16}{12(x-2)(x+2)}$ 27. $\dfrac{3x^2-20x}{(x+6)(x-6)}$ 28. $\dfrac{2x-7}{(x+5)(x-1)}$ 29. $\dfrac{-4x}{x-7}$ 30. $\dfrac{5x+4}{(x-5)(x+4)}$

Practice Problems

Perform the indicated operations and reduce if possible. Assume that no denominator is 0.

1. $\dfrac{1}{1-y}+\dfrac{2}{y^2-1}$ **2.** $\dfrac{x+3}{x^2+x-6}+\dfrac{x-2}{x^2+4x-12}$

3. $\dfrac{1}{y+2}-\dfrac{1}{y^3+8}$ **4.** $\dfrac{x}{x^2-1}-\dfrac{1}{x-1}$

Answers to Practice Problems: **1.** $\dfrac{-1}{y+1}$ **2.** $\dfrac{2x+4}{(x-2)(x+6)}$ **3.** $\dfrac{y^2-2y+3}{(y+2)(y^2-2y+4)}$ **4.** $\dfrac{-1}{(x+1)(x-1)}$

7.2 Exercises

Perform the indicated operations and reduce if possible. Assume that no denominator has a value of 0.

1. $\frac{3x}{x+4}+\frac{12}{x+4}$

2. $\frac{7x}{x+5}+\frac{35}{x+5}$

3. $\frac{x-1}{x+6}+\frac{x+13}{x+6}$

4. $\frac{3x-1}{2x-6}+\frac{x-11}{2x-6}$

5. $\frac{3x+1}{5x+2}+\frac{2x+1}{5x+2}$

6. $\frac{x^2+3}{x+1}+\frac{4x}{x+1}$

7. $\frac{x-5}{x^2-2x+1}+\frac{x+3}{x^2-2x+1}$

8. $\frac{2x^2+5}{x^2-4}+\frac{3x-1}{x^2-4}$

9. $\frac{13}{7-x}-\frac{1}{x-7}$

10. $\frac{6x}{x-6}+\frac{36}{6-x}$

11. $\frac{3x}{x-4}+\frac{16-x}{4-x}$

12. $\frac{20}{x-10}-\frac{3}{10-x}$

13. $\frac{x^2+2}{x^2+x-12}+\frac{x+1}{12-x-x^2}$

14. $\frac{10}{x^2-x-6}-\frac{5x}{6+x-x^2}$

15. $\frac{x^2+2}{x^2-4}-\frac{4x-2}{x^2-4}$

16. $\frac{2x+5}{2x^2-x-1}-\frac{4x+2}{2x^2-x-1}$

17. $\frac{x+3}{7x-2}+\frac{2x-1}{14x-4}$

18. $\frac{3x+1}{4x+10}+\frac{4-x}{2x+5}$

19. $\frac{5}{x-3}+\frac{x}{x^2-9}$

20. $\frac{x+1}{x^2-3x-10}+\frac{x}{x-5}$

21. $\frac{x}{x-1}-\frac{4}{x+2}$

22. $\frac{x-1}{3x-1}-\frac{8+4x}{x+2}$

23. $\frac{x+2}{x+3}-\frac{4}{3-x}$

24. $\frac{x-1}{4-x}+\frac{3x}{x+5}$

25. $\frac{x+2}{3x+9}+\frac{2x-1}{2x-6}$

26. $\frac{x}{4x-8}-\frac{3x+2}{3x+6}$

27. $\frac{3x}{6+x}-\frac{2x}{x^2-36}$

28. $\frac{4}{x+5}-\frac{2x+3}{x^2+4x-5}$

29. $\frac{4x+1}{7-x}+\frac{x-1}{x^2-8x+7}$

30. $\frac{3x-4}{x^2-x-20}-\frac{2}{5-x}$

31. $\frac{4x}{x^2+3x-28}+\frac{3}{x^2+6x-7}$

32. $\frac{3x}{x^2+2x+1}-\frac{x}{x^2+4x+4}$

33. $\frac{3x+4}{2x^2-23x+30}-\frac{x+5}{2x^2-19x+24}$

34. $\frac{x+1}{x^2-3x+2}+\frac{6}{x^2-6x+8}$

35. $\frac{4x-1}{x^2-5x+4}+\frac{2x+7}{x^2-11x+28}$

36. $\frac{7x+3}{5x^2+27x+36}+\frac{3x-2}{5x^2+22x+24}$

37. $\frac{x-6}{7x^2-3x-4}+\frac{7-x}{7x^2+18x+8}$

38. $\frac{x+10}{x^2+5x+4}-\frac{4}{x^2+6x+8}$

39. $\frac{x-3}{4x^2-5x-6}-\frac{4x+10}{2x^2+x-10}$

40. $\frac{2x+1}{8x^2-37x-15}+\frac{2-x}{8x^2+11x+3}$

31. $\frac{4x^2-x-12}{(x+7)(x-4)(x-1)}$

32. $\frac{2x^3+10x^2+11x}{(x+1)^2(x+2)^2}$

33. $\frac{x-6}{(x-10)(x-8)}$

34. $\frac{x+5}{(x-1)(x-4)}$

35. $\frac{6x}{(x-1)(x-7)}$

36. $\frac{2x}{(x+2)(x+3)}$

37. $\frac{4x-19}{(7x+4)(x-1)(x+2)}$

38. $\frac{x+4}{(x+1)(x+2)}$

39. $\frac{-7x-9}{(4x+3)(x-2)}$

40. $\frac{x^2+10x-9}{(8x+3)(x-5)(x+1)}$

41. $\frac{4x^2-41x+3}{(x+4)(x-4)}$

42. $\frac{x^2-7x+1}{(x-3)(x+3)}$

43. $\frac{x-4}{2(x-2)}$

44. $\frac{-5x^2-11x+47}{(x+6)(x-2)}$

45. $\frac{x^2-4x-6}{(x+2)(x-2)(x-1)}$

46. $\frac{3}{(x+1)(x+2)}$

47. $\frac{3x^2+26x-3}{(x+7)(x-3)(x+1)}$

48. $\dfrac{-20x^2+30x-4}{(3x+2)(9x^2-6x+4)}$

49. $\dfrac{6x+2}{(x-1)(x+3)}$

50. $\dfrac{5x+12}{(x+3)(x+1)}$

51. $\dfrac{2x^2+x-4}{(x-2)(y+1)(x+1)}$

52. $\dfrac{5x+1}{(y-3)(x+1)}$

53. $\dfrac{2x+4xy-15y}{(x+3)(y+2)(x-5)}$

54. $\dfrac{5xy-8y+2}{(y-4)(y+1)(x-2)}$

55. $\dfrac{-2x+2}{x^2+x+1}$

56. $\dfrac{6x^2-12x+5}{2(2x+1)(4x^2-2x+1)}$

57. $\dfrac{2x^2-x-5}{(x-3)(x+3)(x^2+1)}$

58. $\dfrac{14x-43}{(x^2+6)(x-5)(x-2)}$

41. $\dfrac{3x}{4-x}+\dfrac{7x}{x+4}-\dfrac{x-3}{x^2-16}$

42. $\dfrac{x}{x+3}+\dfrac{x+1}{3-x}+\dfrac{x^2+4}{x^2-9}$

43. $-\dfrac{1}{2}+\dfrac{x-5}{x-3}+\dfrac{x-1}{x^2-5x+6}$

44. $-4+\dfrac{1-2x}{x+6}+\dfrac{x^2+1}{x^2+4x-12}$

45. $\dfrac{2}{x^2-4}-\dfrac{3}{x^2-3x+2}+\dfrac{x-1}{x^2+x-2}$

46. $\dfrac{5}{x^2+3x+2}+\dfrac{4}{x^2+6x+8}-\dfrac{6}{x^2+5x+4}$

47. $\dfrac{x}{x^2+4x-21}+\dfrac{1-x}{x^2+8x+7}+\dfrac{3x}{x^2-2x-3}$

48. $\dfrac{x+2}{9x^2-6x+4}+\dfrac{10x-5x^2}{27x^3+8}-\dfrac{2}{3x+2}$

49. $\dfrac{3x+9}{x^2-5x+4}+\dfrac{49}{12+x-x^2}+\dfrac{3x+21}{x^2+2x-3}$

50. $\dfrac{5x+22}{x^2+8x+15}+\dfrac{4}{x^2+4x+3}+\dfrac{6}{x^2+6x+5}$

51. $\dfrac{x}{xy+x-2y-2}+\dfrac{x+2}{xy+x+y+1}$

52. $\dfrac{4x}{xy-3x+y-3}+\dfrac{x+2}{xy+2y-3x-6}$

53. $\dfrac{3y}{xy+2x+3y+6}+\dfrac{x}{x^2-2x-15}$

54. $\dfrac{2}{xy-4x-2y+8}+\dfrac{5y}{y^2-3y-4}$

55. $\dfrac{x+6}{x^2+x+1}-\dfrac{3x^2+x-4}{x^3-1}$

56. $\dfrac{2x-5}{8x^2-4x+2}+\dfrac{x^2-2x+5}{8x^3+1}$

57. $\dfrac{x+1}{x^3-3x^2+x-3}+\dfrac{x^2-5x-8}{x^4-8x^2-9}$

58. $\dfrac{x+4}{x^3-5x^2+6x-30}-\dfrac{x-7}{x^3-2x^2+6x-12}$

59. $\dfrac{x+1}{2x^2-x-1}+\dfrac{2x}{2x^2+5x+2}-\dfrac{2x}{3x^2+4x-4}$

59. $\dfrac{5x^3-x^2+6x-4}{(2x+1)(x-1)(x+2)(3x-2)}$

60. $\dfrac{x-6}{3x^2+10x+3}-\dfrac{2x}{5x^2-3x-2}+\dfrac{2x}{3x^2-2x-1}$

60. $\dfrac{9x^3-19x^2+22x+12}{(3x+1)(x+3)(5x+2)(x-1)}$

Writing and Thinking About Mathematics

61. See page 520. Answers will vary.

61. Discuss the steps in the process you go through when adding two rational expressions with different denominators. That is, discuss how you find the least common denominator when adding rational expressions and how you use this LCD to find equivalent rational expressions that you can add.

HAWKES LEARNING SYSTEMS: INTRODUCTORY & INTERMEDIATE ALGEBRA SOFTWARE

- 7.2 Addition and Subtraction with Rational Expressions

7.3 Complex Fractions

- *Simplify complex fractions.*

Simplifying Complex Fractions (First Method)

A **complex fraction** is a fraction in which the numerator and/or denominator are themselves fractions or the sum or difference of fractions. Examples of complex fractions are

$$\frac{\frac{1}{x+3}-\frac{1}{x}}{1+\frac{3}{x}} \quad \text{and} \quad \frac{x+y}{x^{-1}+y^{-1}}=\frac{x+y}{\frac{1}{x}+\frac{1}{y}}.$$

Teaching Notes:
You will probably need to continually reinforce the concept that the fraction bar serves two important and distinct roles in the symbolism: on the one hand, it can translate as an operation (division), and on the other hand, the bar is a grouping symbol similar to the use of parentheses.

The objective here is to develop techniques for simplifying complex fractions so that they are written in the form of a single reduced rational expression.

In a complex fraction such as $\frac{\frac{1}{x+3}-\frac{1}{x}}{1+\frac{3}{x}}$, the large fraction bar is a symbol of inclusion.

The expression could also be written as follows.

$$\frac{\frac{1}{x+3}-\frac{1}{x}}{1+\frac{3}{x}}=\left(\frac{1}{x+3}-\frac{1}{x}\right)\div\left(1+\frac{3}{x}\right)$$

Thus a complex fraction indicates that the numerator is to be divided by the denominator.

To Simplify Complex Fractions (First Method)

1. Simplify the numerator so that it is a single rational expression.
2. Simplify the denominator so that it is a single rational expression.
3. Divide the numerator by the denominator and reduce to lowest terms.

This method is used to simplify the complex fractions in Examples 1 and 2. Study the examples closely so that you understand what happens at each step.

Example 1: First Method for Simplifying Complex Fractions

Simplify the complex fraction $\dfrac{\frac{3x}{y^2}}{\frac{12x}{7y}}$.

Solution: $\dfrac{\frac{3x}{y^2}}{\frac{12x}{7y}} = \dfrac{\cancel{3}\,\cancel{x}}{\cancel{y^2}_{y}} \cdot \dfrac{7\,\cancel{y}}{\cancel{12}_{4}\,\cancel{x}}$

$= \dfrac{7}{4y}$

To divide, multiply by the reciprocal of the denominator.

Example 2: First Method for Simplifying Complex Fractions

Simplify the following complex fractions.

a. $\dfrac{\frac{1}{x+3}-\frac{1}{x}}{1+\frac{3}{x}}$

Solution: $\dfrac{\frac{1}{x+3}-\frac{1}{x}}{1+\frac{3}{x}} = \dfrac{\frac{1\cdot x}{(x+3)\cdot x}-\frac{1(x+3)}{x(x+3)}}{\frac{x}{x}+\frac{3}{x}}$

Combine the fractions in the numerator and in the denominator separately.

Note that $1=\dfrac{x}{x}$.

$= \dfrac{\frac{x-(x+3)}{x(x+3)}}{\frac{x+3}{x}}$

$= \dfrac{\frac{x-x-3}{x(x+3)}}{\frac{x+3}{x}}$

$= \dfrac{\frac{-3}{x(x+3)}}{\frac{x+3}{x}}$

$= \dfrac{-3}{\cancel{x}(x+3)} \cdot \dfrac{\cancel{x}}{x+3}$

To divide, multiply by the reciprocal of the denominator.

$= \dfrac{-3}{(x+3)^2}$

b. $\dfrac{x+y}{x^{-1}+y^{-1}}$

Solution:

$$\frac{x+y}{x^{-1}+y^{-1}} = \frac{x+y}{\frac{1}{x}+\frac{1}{y}}$$

Recall that $x^{-1} = \frac{1}{x}$ and $y^{-1} = \frac{1}{y}$.

$$= \frac{\frac{x+y}{1}}{\frac{1}{x}\cdot\frac{y}{y}+\frac{1}{y}\cdot\frac{x}{x}}$$

Add the two fractions in the denominator.

$$= \frac{\frac{x+y}{1}}{\frac{y}{xy}+\frac{x}{xy}}$$

$$= \frac{\frac{x+y}{1}}{\frac{y+x}{xy}}$$

$$= \frac{\cancel{x+y}}{1}\cdot\frac{xy}{\cancel{y+x}}$$

Multiply by the reciprocal of the denominator.

$$= \frac{xy}{1}$$

$$= xy$$

Simplifying Complex Fractions (Second Method)

A second method is to find the LCM of the denominators in the fractions in both the original numerator and the original denominator and then multiply **both** the numerator and denominator by this LCM.

To Simplify Complex Fractions (Second Method)

1. Find the LCM of all the denominators in the numerator and denominator of the complex fraction.
2. Multiply both the numerator and denominator of the complex fraction by this LCM.
3. Simplify both the numerator and denominator and reduce to lowest terms.

Example 3: Second Method for Simplifying Complex Fractions

Simplify the following complex fractions.

a. $\dfrac{\dfrac{1}{x+3}-\dfrac{1}{x}}{1+\dfrac{3}{x}}$

Solution:

$$\frac{\frac{1}{x+3}-\frac{1}{x}}{1+\frac{3}{x}}=\frac{\left(\frac{1}{x+3}-\frac{1}{x}\right)\cdot x(x+3)}{\left(1+\frac{3}{x}\right)\cdot x(x+3)}$$

Multiply by $x(x+3)$, the LCM of $\{x, x+3\}$. This multiplication can be done because the net effect is that the fraction is multiplied by 1.

$$=\frac{\frac{1}{\cancel{x+3}}\cdot x\cancel{(x+3)}-\frac{1}{\cancel{x}}\cdot \cancel{x}(x+3)}{1\cdot x(x+3)+\frac{3}{\cancel{x}}\cdot \cancel{x}(x+3)}$$

$$=\frac{x-(x+3)}{x(x+3)+3(x+3)}$$

$$=\frac{x-x-3}{(x+3)(x+3)}$$

$$=\frac{-3}{(x+3)^2}$$

Note that this matches the result found in Example 2a using Method 1.

b. $\dfrac{x+y}{x^{-1}+y^{-1}}$

Solution:

$$\frac{x+y}{x^{-1}+y^{-1}}=\frac{\frac{x+y}{1}}{\frac{1}{x}+\frac{1}{y}}$$

$$=\frac{\left(\frac{x+y}{1}\right)xy}{\left(\frac{1}{x}+\frac{1}{y}\right)xy}$$

Multiply by xy, the LCM of $\{1, x, y\}$.

$$=\frac{(x+y)xy}{\frac{1}{\cancel{x}}\cdot \cancel{x}y+\frac{1}{\cancel{y}}\cdot x\cancel{y}}$$

$$=\frac{\cancel{(x+y)}xy}{\cancel{y+x}}=xy$$

Simplifying Complex Algebraic Expressions

A **complex algebraic expression** is an expression that involves rational expressions and more than one operation. In simplifying such expressions, the rules for order of operations apply. As with complex fractions, the objective is to simplify the expression so that it is written in the form of a single reduced rational expression.

Example 4: Simplifying Complex Algebraic Expressions

Simplify the following expression.

$$\frac{4-x}{x+3}+\frac{x}{x+3}\div\frac{x}{x-3}$$

Solution: In a complex algebraic expression such as

$$\frac{4-x}{x+3}+\frac{x}{x+3}\div\frac{x}{x-3}$$

the rules for order of operations indicate that the division is to be done first.

$$\begin{aligned}\frac{4-x}{x+3}+\frac{x}{x+3}\div\frac{x}{x-3} &= \frac{4-x}{x+3}+\frac{\cancel{x}}{x+3}\cdot\frac{x-3}{\cancel{x}}\\ &= \frac{4-x}{x+3}+\frac{x-3}{x+3}\\ &= \frac{4-x+x-3}{x+3}\\ &= \frac{1}{x+3}\end{aligned}$$

Practice Problems

Simplify each of the following expressions.

1. $\dfrac{\frac{1}{x}}{1+\frac{1}{x}}$

2. $\dfrac{\frac{1}{x+2}-\frac{1}{x}}{1+\frac{2}{x}}$

3. $\dfrac{1+\frac{3}{x-3}}{x-\frac{x^2}{x-3}}$

4. $\dfrac{\frac{1}{x+y}-\frac{1}{x-y}}{\frac{2y}{x^2-y^2}}$

5. $\dfrac{5}{x}-\dfrac{3}{x-2}\div\dfrac{x}{x-2}$

Answers to Practice Problems: **1.** $\frac{1}{x+1}$ **2.** $\frac{-2}{(x+2)^2}$ **3.** $-\frac{1}{3}$ **4.** -1 **5.** $\frac{2}{x}$

7.3 Exercises

Simplify the following complex fractions.

1. $\dfrac{\frac{2x}{3y^2}}{\frac{5x^2}{6y}}$

2. $\dfrac{\frac{6x^2}{5y}}{\frac{x}{10y^2}}$

3. $\dfrac{\frac{12x^3}{7y^2}}{\frac{3x^5}{2y}}$

4. $\dfrac{\frac{9x^2}{7y^3}}{\frac{3xy}{14}}$

5. $\dfrac{\frac{x+3}{2x}}{\frac{2x-1}{4x^2}}$

6. $\dfrac{\frac{x-2}{6x}}{\frac{x+3}{3x^2}}$

7. $\dfrac{\frac{2x-1}{x}}{\frac{2}{x}+3}$

8. $\dfrac{2-\frac{3}{x}}{\frac{x^2-4}{x}}$

9. $\dfrac{\frac{3}{x}+\frac{1}{2x}}{1+\frac{2}{x}}$

10. $\dfrac{\frac{3}{x}+\frac{5}{2x}}{\frac{1}{x}+4}$

11. $\dfrac{1+\frac{1}{x}}{1-\frac{1}{x^2}}$

12. $\dfrac{\frac{2}{y}+1}{\frac{4}{y^2}-1}$

13. $\dfrac{\frac{1}{x}+\frac{1}{3x}}{\frac{x+6}{x^2}}$

14. $\dfrac{\frac{3}{x}-\frac{6}{x^2}}{\frac{x-2}{x^2}}$

15. $\dfrac{\frac{7}{x}-\frac{14}{x^2}}{\frac{1}{x}-\frac{4}{x^3}}$

16. $\dfrac{\frac{3}{x}-\frac{6}{x^2}}{\frac{1}{x}-\frac{2}{x^2}}$

17. $\dfrac{\frac{1}{3}+\frac{1}{x}}{\frac{1}{2}-\frac{1}{x}}$

18. $\dfrac{\frac{x}{y}-\frac{1}{3}}{\frac{6}{y}-\frac{2}{x}}$

19. $\dfrac{\frac{2}{x}+\frac{3}{4y}}{\frac{3}{2x}-\frac{5}{3y}}$

20. $\dfrac{\frac{4}{3x}-\frac{5}{y}}{\frac{1}{3}+\frac{3}{y}}$

21. $\dfrac{1+x^{-1}}{1-x^{-2}}$

22. $\dfrac{x^{-3}+1}{1-x^{-1}}$

23. $\dfrac{1}{x^{-1}+y^{-1}}$

24. $\dfrac{x-y}{x^{-2}-y^{-2}}$

25. $\dfrac{x^{-1}+y^{-1}}{x+y}$

26. $\dfrac{y^{-2}-x^{-2}}{x+y}$

27. $\dfrac{x^{-1}+y^{-1}}{x^{-1}-y^{-1}}$

28. $\dfrac{x^{-1}+y^{-1}}{x^{-2}-y^{-2}}$

29. $\dfrac{\frac{4}{x}-1}{1-\frac{1}{x-3}}$

30. $\dfrac{x+\frac{3}{x-4}}{1-\frac{1}{x}}$

31. $\dfrac{1-\frac{4}{x+3}}{1-\frac{2}{x+1}}$

32. $\dfrac{1+\frac{4}{2x-3}}{1+\frac{x}{x+1}}$

33. $\dfrac{\frac{1}{x+h}-\frac{1}{x}}{h}$

34. $\dfrac{\frac{1}{(x+h)^2}-\frac{1}{x^2}}{h}$

35. $\dfrac{\left(2+\frac{1}{x+h}\right)-\left(2+\frac{1}{x}\right)}{h}$

36. $\dfrac{\left(\frac{1}{(x+h)^2}-3\right)-\left(\frac{1}{x^2}-3\right)}{h}$

37. $\dfrac{x^2-4y^2}{1-\frac{2x+y}{x-y}}$

38. $\dfrac{8x^2-2y^2}{\frac{4x-1}{x-y}-2}$

39. $\dfrac{\frac{x+1}{x-1}-\frac{x-1}{x+1}}{\frac{x+1}{x-1}+\frac{x-1}{x+1}}$

40. $\dfrac{\frac{1}{x^2-1}-\frac{1}{x+1}}{\frac{1}{x-1}+\frac{1}{x^2-1}}$

41. $\dfrac{\frac{x}{x-4}-\frac{1}{x-1}}{\frac{x}{x-1}+\frac{2}{x-3}}$

42. $\dfrac{\frac{1}{x+1}-\frac{x}{x+2}}{\frac{x}{x+2}-\frac{2}{x-1}}$

1. $\frac{4}{5xy}$
2. $12xy$
3. $\frac{8}{7x^2y}$
4. $\frac{6x}{y^4}$
5. $\frac{2x^2+6x}{2x-1}$
6. $\frac{x(x-2)}{2(x+3)}$
7. $\frac{2x-1}{2+3x}$
8. $\frac{2x-3}{(x-2)(x+2)}$
9. $\frac{7}{2(x+2)}$
10. $\frac{11}{2(1+4x)}$
11. $\frac{x}{x-1}$
12. $\frac{y}{2-y}$
13. $\frac{4x}{3(x+6)}$
14. 3
15. $\frac{7x}{x+2}$
16. 3
17. $\frac{2x+6}{3(x-2)}$
18. $\frac{x}{6}$
19. $\frac{24y+9x}{2(9y-10x)}$
20. $\frac{4y-15x}{x(y+9)}$
21. $\frac{x}{x-1}$
22. $\frac{x^3+1}{x^2(x-1)}$

23. $\frac{xy}{x+y}$

24. $\frac{-x^2y^2}{x+y}$

25. $\frac{1}{xy}$

26. $\frac{x-y}{x^2y^2}$

27. $\frac{y+x}{y-x}$

Write each of the expressions as a single fraction reduced to lowest terms.

43. $\frac{1}{x+1}-\frac{3}{2x}\cdot\frac{4x}{x+1}$

44. $\frac{4}{x}-\frac{2}{x^2-2x}\cdot\frac{x-2}{5}$

45. $\left(\frac{8}{x}-\frac{3}{4x}\right)\div\frac{4x+5}{x}$

46. $\left(\frac{2}{x}+\frac{5}{x-3}\right)\div\frac{x}{2x-6}$

47. $\frac{x}{x-1}-\frac{3}{x-1}\cdot\frac{x+2}{x}$

48. $\frac{x+3}{x+2}+\frac{x}{x+2}\cdot\frac{x-3}{x^2}$

49. $\frac{x-1}{x+4}+\frac{x-6}{x^2+3x-4}\div\frac{x-4}{x-1}$

50. $\frac{x}{x+3}-\frac{3}{x+5}\div\frac{x-2}{x^2+3x-10}$

Writing and Thinking About Mathematics

51. Some complex fractions involve the sum (or difference) of complex fractions. Beginning with the "farthest" denominator, simplify each of the following expressions.

a. $1+\cfrac{1}{1+\cfrac{1}{1+\cfrac{1}{1+1}}}$ **b.** $2-\cfrac{1}{2-\cfrac{1}{2-\cfrac{1}{2-1}}}$ **c.** $x+\cfrac{1}{x+\cfrac{1}{x+\cfrac{1}{x+1}}}$

28. $\frac{xy}{y-x}$

29. $\frac{3-x}{x}$

30. $\frac{x^2-3x}{x-4}$

31. $\frac{x+1}{x+3}$

HAWKES LEARNING SYSTEMS: INTRODUCTORY & INTERMEDIATE ALGEBRA SOFTWARE

- 7.3 Complex Fractions

32. $\frac{x+1}{2x-3}$ 33. $\frac{-1}{x(x+h)}$ 34. $-\frac{2x+h}{x^2(x+h)^2}$ 35. $\frac{-1}{x(x+h)}$ 36. $-\frac{2x+h}{x^2(x+h)^2}$ 37. $-(x-2y)(x-y)$

38. $\frac{2(2x-y)(2x+y)(x-y)}{2x+2y-1}$ 39. $\frac{2x}{x^2+1}$ 40. $\frac{2-x}{x+2}$ 41. $\frac{(x-3)(x^2-2x+4)}{(x-4)(x-2)(x+1)}$ 42. $\frac{(2-x^2)(x-1)}{(x-4)(x+1)^2}$ 43. $\frac{-5}{x+1}$

44. $\frac{18}{5x}$ 45. $\frac{29}{4(4x+5)}$ 46. $\frac{14x-12}{x^2}$ 47. $\frac{x^2-3x-6}{x(x-1)}$ 48. $\frac{x^2+4x-3}{x(x+2)}$ 49. $\frac{x^2-4x-2}{(x-4)(x+4)}$ 50. $\frac{-2x-9}{x+3}$

51. a. $\frac{8}{5}$ b. 1 c. $\frac{x^4+x^3+3x^2+2x+1}{x^3+x^2+2x+1}$

7.4 Solving Equations with Rational Expressions

- *Solve proportions.*
- *Solve other equations with rational expressions.*

Proportions

A **ratio** is a comparison of two numbers by division. Ratios are written in the form

$$\boldsymbol{a:b} \quad \text{or} \quad \frac{\boldsymbol{a}}{\boldsymbol{b}} \quad \text{or} \quad \boldsymbol{a} \textbf{ to } \boldsymbol{b}.$$

For example, suppose the ratio of female to male faculty at the local community college is 3 to 2. We can also write this ratio as $\frac{3 \text{ females}}{2 \text{ males}}$ or as 3 females : 2 males. This ratio does not mean that there are only 5 teachers at the college. There are many ratios that reduce to $\frac{3}{2}$. There might be 35 teachers at the college, 21 females and 14 males, because $21 + 14 = 35$ and $\frac{21}{14} = \frac{3}{2}$. Until you know the total number of teachers at the college you cannot know the exact numbers of female and male teachers. But you do know that there are 3 female teachers for every 2 male teachers.

A **proportion** is a special type of equation stating that two ratios are equal. For example,

$$\frac{21}{14} = \frac{3}{2} \quad \text{and} \quad \frac{3.5}{7} = \frac{3.75}{7.5} \quad \text{are proportions.}$$

Proportion

A **proportion** is an equation stating that two ratios are equal.

In symbols, $\frac{a}{b} = \frac{c}{d}$ is a proportion.

Teaching Notes:
Another method of solving proportions is to "cross multiply". However, students often overlook the fact that cross multiplication works only with proportions and not with rational equations in general. In this text, I have tried to avoid this potential confusion by presenting only the method of clearing fractions.

Proportions may involve only numbers as in $\frac{2}{3} = \frac{10}{15}$. However, proportions can also be used to find unknown quantities in applications, and in such cases, will involve variables.

One method of solving proportions with variables is to "clear" the equation of fractions by first multiplying both sides of the equation by the LCM of the denominators, the LCD. This method is illustrated in Example 1.

Example 1: Proportions

Solve the following proportions.

a. $\frac{x-5}{2x}=\frac{6}{3x}$ LCD $=6x$

Solution: $\frac{x-5}{2x}=\frac{6}{3x}$ **Note:** $x \neq 0$

$$6x\cdot\frac{x-5}{2x}=6x\cdot\frac{6}{3x}$$ Multiply both sides by the LCD, $6x$.

$$3(x-5)=2(6)$$
$$3x-15=12$$
$$3x=27$$
$$x=9$$

Check: $\frac{9-5}{2\cdot 9}\stackrel{?}{=}\frac{6}{3\cdot 9}$

$$\frac{4}{18}\stackrel{?}{=}\frac{6}{27}$$
$$\frac{2}{9}=\frac{2}{9}$$ Thus the solution is $x=9$.

b. $\frac{3}{x-6}=\frac{5}{x}$ LCD $=x(x-6)$

Solution: $\frac{3}{x-6}=\frac{5}{x}$ **Note:** $x \neq 0, 6$

$$x(x-6)\cdot\frac{3}{x-6}=x(x-6)\cdot\frac{5}{x}$$ Multiply both sides by the LCD, $x(x-6)$.

$$3x=5x-30$$
$$-2x=-30$$
$$x=15$$

Check: $\frac{3}{15-6}\stackrel{?}{=}\frac{5}{15}$

$$\frac{3}{9}\stackrel{?}{=}\frac{5}{15}$$
$$\frac{1}{3}=\frac{1}{3}$$ Thus the solution is $x=15$.

Proportions can be used to solve many everyday types of word problems. Using the correct ratios of units on both sides of the proportion is critical. One of the following conditions must be true:

1. **The numerators agree in type and the denominators agree in type.**
2. **The numerators correspond and the denominators correspond.**

Example 2 illustrates one situation.

Example 2: Application of Proportions

On an architect's scale drawing of a home, $\frac{1}{2}$ inch represents 10 feet. What length does a measure of $2\frac{1}{2}$ inches represent?

Solution: Set up a proportion representing the information. In this example the numerators are the same type (inches) and the denominators are the same type (feet).

$$\frac{\frac{1}{2}\text{ inch}}{10\text{ feet}} = \frac{2\frac{1}{2}\text{ inches}}{x\text{ feet}} \qquad \text{LCD} = 10x$$

$$\cancel{10}x \cdot \frac{\frac{1}{2}}{\cancel{10}} = 10\cancel{x} \cdot \frac{\frac{5}{2}}{\cancel{x}} \qquad \text{Multiply both sides by } 10x.$$

$$\frac{1}{2}x = 25 \qquad \text{Simplify.}$$

$$2 \cdot \frac{1}{2}x = 2 \cdot 25 \qquad \text{Multiply both sides by } 2.$$

$$x = 50 \qquad \text{Simplify.}$$

On this drawing, $2\frac{1}{2}$ inches represent 50 feet.

NOTES

Any of the following four equations could have been used to solve the problem in Example 2:

Agree in type

$$\frac{\frac{1}{2}\text{ inch}}{10\text{ feet}} = \frac{2\frac{1}{2}\text{ inches}}{x\text{ feet}} \qquad \frac{10\text{ feet}}{\frac{1}{2}\text{ inch}} = \frac{x\text{ feet}}{2\frac{1}{2}\text{ inches}}$$

Correspond:

$$\frac{\frac{1}{2}\text{ inch}}{2\frac{1}{2}\text{ inches}} = \frac{10\text{ feet}}{x\text{ feet}} \qquad \frac{2\frac{1}{2}\text{ inches}}{\frac{1}{2}\text{ inch}} = \frac{x\text{ feet}}{10\text{ feet}}$$

Proportions and Similar Triangles

Proportions are used when working with similar geometric figures. **Similar figures** are figures that meet the following two conditions:

1. **The measures of the corresponding angles are equal.**
2. **The lengths of the corresponding sides are proportional.**

In **similar triangles**, corresponding sides are those sides opposite the equal angles. (See Figure 1.)

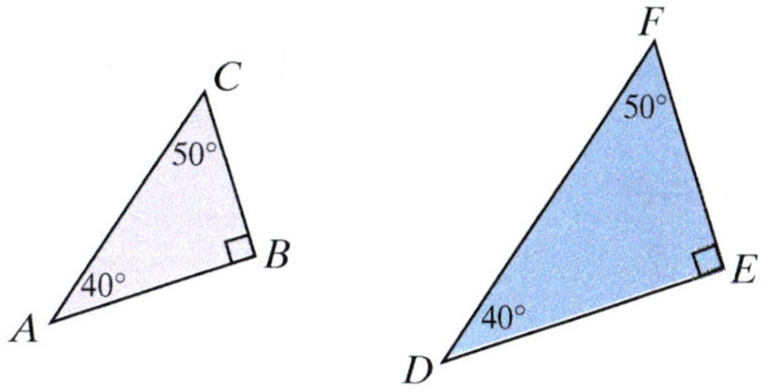

Figure 1

We write $\Delta ABC \sim \Delta DEF$. ($\sim$ is read "**is similar to**".) The lengths of the corresponding sides are proportional. Thus

$$\frac{AB}{DE} = \frac{BC}{EF} \text{ and } \frac{AB}{DE} = \frac{AC}{DF} \text{ and } \frac{BC}{EF} = \frac{AC}{DF}.$$

In a pair of similar triangles, we can often find the length of an unknown side by setting up a proportion and solving. Example 3 illustrates such a situation.

Example 3: Similar Triangles

In the figure shown, $\Delta ABC \sim \Delta PQR$. Find the lengths of the sides $\overline{AB}$ and $\overline{QR}$.

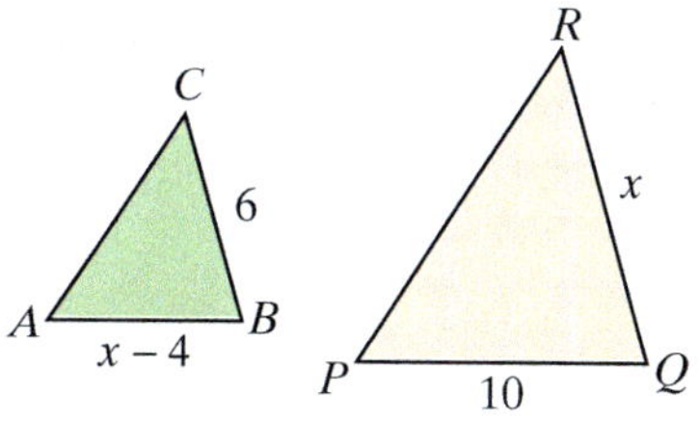

Solution: Set up a proportion involving corresponding sides and solve for x.

$$\frac{x-4}{10} = \frac{6}{x}$$

$$\cancel{10}x \cdot \frac{x-4}{\cancel{10}} = 10\cancel{x} \cdot \frac{6}{\cancel{x}}$$

$$x(x-4) = 10 \cdot 6$$

$$x^2 - 4x = 60$$

Continued on the next page...

$$x^2 - 4x - 60 = 0$$

$$(x-10)(x+6) = 0$$

$$x - 10 = 0 \quad \text{or} \quad x + 6 = 0$$

$$x = 10 \qquad\qquad x = -6$$

Because the length of a side cannot be negative, the only acceptable solution is $x = 10$. Thus $\overline{QR} = 10$. Substituting 10 for x gives $\overline{AB} = 10 - 4 = 6$.

Solving Equations with Rational Expressions

An equation such as

$$\frac{3}{x} + \frac{1}{8} = \frac{13}{4x}$$

that involves the sum of rational expressions is **not a proportion**. However, the method of finding the solution to the equation is similar in that we "clear" the fractions by multiplying both sides of the equation by the LCD of the fractions. In this example the LCD is $8x$ and we can proceed as follows:

$$\frac{3}{x} + \frac{1}{8} = \frac{13}{4x}$$ Note $x \neq 0$.

$$8x\left(\frac{3}{x} + \frac{1}{8}\right) = 8x \cdot \frac{13}{4x}$$ Multiply both sides by $8x$, the LCD.

$$8\cancel{x} \cdot \frac{3}{\cancel{x}} + \cancel{8}x \cdot \frac{1}{\cancel{8}} = \overset{2}{\cancel{8x}} \cdot \frac{13}{\cancel{4x}}$$ Use the distributive property and reduce.

$$24 + x = 26$$ Simplify.

$$x = 2.$$

As we have seen, rational expressions may contain variables in either the numerator or denominator or both. In any case, a general approach to solving equations that contain rational expressions is as follows.

To Solve an Equation Containing Rational Expressions

1. Find the LCD of the fractions.
2. Multiply both sides of the equation by this LCD and simplify.
3. Solve the resulting equation. (This equation will have only polynomials on both sides.)
4. Check each solution in the **original equation**. (Remember that no denominator can be 0 and any solution that gives a 0 denominator is to be discarded.)

Checking is particularly important when equations have rational expressions. Multiplying by the LCD may introduce solutions that are not solutions to the original equation. Such solutions are called **extraneous solutions** or **extraneous roots** and occur because multiplication by a variable expression may, in effect, be multiplying the original equation by 0.

Example 4: Solving Equations Involving Rational Expressions

State any restrictions on the variable, and then solve the equation.

a. $\dfrac{1}{x-4}=\dfrac{3}{x^2-5x}$

Solution: First find the LCD of the fractions, and then multiply both sides of the equation by the LCD.

$$\left.\begin{aligned} x-4 &= x-4 \\ x^2-5x &= x(x-5) \end{aligned}\right\} \quad \text{LCD} = x(x-5)(x-4)$$

$$x(x-5)\cancel{(x-4)}\cdot\frac{1}{\cancel{x-4}}=\cancel{x}\cancel{(x-5)}(x-4)\cdot\frac{3}{\cancel{x}\cancel{(x-5)}} \qquad (x \neq 0, 4, 5)$$

$$x(x-5)=3(x-4)$$
$$x^2-5x=3x-12$$
$$x^2-8x+12=0$$
$$(x-6)(x-2)=0$$
$$x-6=0 \quad \text{or} \quad x-2=0$$
$$x=6 \qquad\qquad x=2$$

Since 6 and 2 are not restrictions, there are two solutions, $x=6$ and $x=2$.

b. $\dfrac{x}{x-2}+\dfrac{x-6}{x(x-2)}=\dfrac{5x}{x-2}-\dfrac{10}{x-2}$

Solution: First find the LCD of the fractions and then multiply each term on both sides of the equation by the LCD.

$$\left.\begin{aligned} &x-2 \\ &x(x-2) \end{aligned}\right\} \quad \text{LCD} = x(x-2)$$

$$x\cancel{(x-2)}\cdot\frac{x}{\cancel{x-2}}+\cancel{x}\cancel{(x-2)}\cdot\frac{x-6}{\cancel{x}\cancel{(x-2)}}=x\cancel{(x-2)}\cdot\frac{5x}{\cancel{x-2}}-x\cancel{(x-2)}\cdot\frac{10}{\cancel{x-2}}$$

$$x^2+x-6=5x^2-10x \qquad (x \neq 0, 2)$$
$$0=4x^2-11x+6$$
$$0=(4x-3)(x-2)$$
$$4x-3=0 \quad \text{or} \quad x-2=0$$
$$4x=3 \qquad\qquad \cancel{x=2}$$
$$x=\frac{3}{4}$$

Continued on the next page...

Teaching Notes:
This section may introduce some students to the fact that multiplying both sides of an equation with a variable term may not strictly result in an equivalent equation – an unsettling realization to these students. You might point out to your students that multiplying by variable terms may introduce extraneous solutions with the increased degree of the resulting polynomial. For example,

$$x=3$$
$$x\cdot x=3\cdot x$$
$$x^2=3x$$

Now $x^2=3x$ has two solutions, whereas $x=3$ had only one.

The only solution is $x = \frac{3}{4}$, since 2 is a restricted value $(x \neq 0, 2)$ and thus **not** a solution. No denominator can be 0.

c. $\frac{2}{x^2 - 9} = \frac{1}{x^2} + \frac{1}{x^2 - 3x}$

Solution: First find the LCD of the fractions and then multiply each term on both sides of the equation by the LCD.

$$\left.\begin{aligned} x^2 - 9 &= (x+3)(x-3) \\ x^2 &= x^2 \\ x^2 - 3x &= x(x-3) \end{aligned}\right\} \text{LCD} = x^2(x+3)(x-3)$$

$$x^2(x+3)(x-3) \cdot \frac{2}{(x+3)(x-3)} = x^2(x+3)(x-3) \cdot \frac{1}{x^2} + x^2(x+3)(x-3) \cdot \frac{1}{x(x-3)}$$

$$2x^2 = (x+3)(x-3) + x(x+3) \qquad (x \neq 0, -3, 3)$$

$$2x^2 = x^2 - 9 + x^2 + 3x$$

$$2x^2 = 2x^2 + 3x - 9$$

$$9 = 3x$$

$$3 = x \qquad \text{Note that 3 is one of the restrictions.}$$

There is no solution. The solution set is the empty set, $\varnothing$. The original equation is a contradiction.

Formulas

As discussed in Section 3.1, many formulas are equations relating more than one variable. The equation is solved for one variable in terms of another variable (maybe more than one). The process of solving equations can be used to manipulate the formula so that it is solved for one of the other variables. As illustrated in Example 5, the resulting equation may involve a rational expression.

Example 5: Solving a Formula for a Specified Variable

The formula $S = 2\pi r^2 + 2\pi rh$ is used to find the surface area (S) of a right circular cylinder, where r is the radius of the cylinder and h is the height of the cylinder. Solve the formula for h.

Solution:

$$S = 2\pi r^2 + 2\pi rh \qquad \text{Write the formula.}$$

$$S - 2\pi r^2 = 2\pi rh \qquad \text{Add } -2\pi r^2 \text{ to both sides of the equation.}$$

$$\frac{S - 2\pi r^2}{2\pi r} = h \qquad \text{Divide both sides by } 2\pi r.$$

Thus the formula solved for h is: $h = \frac{S - 2\pi r^2}{2\pi r}$ $\left(\text{or } h = \frac{S}{2\pi r} - r\right)$.

Practice Problems

Solve each proportion. State any restrictions on the variable.

1. $\frac{5x}{3} = \frac{9x+4}{5}$

2. $\frac{16}{x} = \frac{4}{x-5}$

3. On a road map, each inch represents 50 miles. What distance is represented by 4.5 inches on the map?

Solve each equation. State any restrictions on the variable.

4. $\frac{5}{3x+2} = \frac{4}{3x+1}$

5. $\frac{x}{x-3} - \frac{2x+3}{x^2+x-12} = \frac{x-1}{x+4}$

6. Solve the formula $A = P + Pr$ for r.

7.4 Exercises

State any restrictions on x, and then solve the proportions.

1. $\frac{4x}{7} = \frac{x+5}{3}$

2. $\frac{3x+1}{4} = \frac{2x+1}{3}$

3. $\frac{10}{x} = \frac{5}{x-2}$

4. $\frac{8}{x-3} = \frac{12}{2x-3}$

5. $\frac{4}{x-4} = \frac{2}{x+3}$

6. $\frac{3}{x+5} = \frac{6}{x-2}$

7. $\frac{x+2}{5x} = \frac{x-6}{3x}$

8. $\frac{x-4}{3x} = \frac{x-2}{5x}$

9. $\frac{5x+2}{x-6} = \frac{11}{4}$

10. $\frac{x+9}{3x+2} = \frac{5}{8}$

11. Computers: Making a statistical analysis, Ana found 3 defective computers in a sample of 20 computers. If this ratio is consistent, how many defective computers does she expect to find in a batch of 2400 computers?

12. Manufacturing: At the Bright-As-Day light bulb plant, 3 out of each 100 bulbs produced are defective. If the daily production is 4800 bulbs, how many are defective?

1. $x = 7$
2. $x = 1$
3. $x \neq 0, 2; x = 4$
4. $x \neq \frac{3}{2}, 3;\ x = -3$
5. $x \neq -3, 4; x = -10$
6. $x \neq -5, 2; x = -12$
7. $x \neq 0; x = 18$
8. $x \neq 0; x = 7$
9. $x \neq 6; x = -\frac{74}{9}$
10. $x \neq -\frac{2}{3};\ x = \frac{62}{7}$
11. 360 defective computers
12. 144 defective bulbs

Answers to Practice Problems: **1.** $x = -6$ **2.** $x = \frac{20}{3};\ x \neq 0, 5$ **3.** 225 miles **4.** $x = 1;\ x \neq -\frac{1}{3}, -\frac{2}{3}$ **5.** $x = 1;\ x \neq -4, 3$ **6.** $r = \frac{A-P}{P}$

13. 28,800 students
14. 393 at bats
15. 34 miles
16. 7.5 inches by 10 inches
17. 6.8 cups
18. 8.8 quarts
19. Width = 3 inches; Length = 7.5 inches
20. a. 7.5 miles per hour
b. 55 miles per hour

13. Education: The University of Arizona has a ratio of 1 professor for every 18 students. If there are 1600 faculty members at the university, how many students are enrolled there?

14. Baseball: New York Yankees player Alex Rodriguez averages 14 hits for every 50 times at bat. If he maintains this average, how many at-bats will he have to have to achieve 110 hits? (Round to the nearest whole number.)

15. Cartography: On a map of Maryland, one inch represents 4 miles. If there are 8.5 inches between Baltimore, MD and Washington, DC, how far are the two cities from each other?

16. Architecture: A floor plan is drawn to scale in which 1 inch represents 4 feet. What size will the drawing be for a room that is 30 feet by 40 feet? (**Hint:** Set up two proportions.)

17. Baking: The recipe for Nestle Tollhouse Chocolate Chip Cookies calls for 2 cups of chocolate chips to make 5 dozen cookies. If you want to bake 17 dozen cookies, how many cups of chocolate chips do you need?

18. Car maintenance: Instructions for Never-Ice Antifreeze state that 4 quarts of antifreeze are needed for every 10 quarts of radiator capacity. If Sal's car has a 22-quart radiator, how many quarts of antifreeze will it need?

19. Landscape architecture: An architect is to draw plans for a city park. He intends to use a scale of $\frac{1}{2}$ inch to represent 25 feet. How many inches will be needed to use for the length and width of a rectangular playing field that is 50 yards by 125 yards? (1 yard = 3 feet)

20. Testing cars: A test driver wants to increase the speed of the car he is driving by 3 miles per hour every 2 seconds. But he can only check his speed every 5 seconds because he is busy with other items during the test drive.

a. By how much should he increase his speed in 5 seconds?
b. If he starts checking his speed at 40 miles per hour, how fast should he be going after 10 seconds?

21. $LK = 12, JB = 4$
22. $QP = 3, TU = 12$
23. $AC = 2, ST = 12$
24. $DF = 10, GH = 4$
25. $ST = 8, TU = 12,$
 $QR = 24$
26. $AC = 6, ED = 12$
 $CD = 18$

The following exercises show pairs of similar triangles. Find the lengths of the sides labeled with variables.

21. $\Delta JKL \sim \Delta JTB$

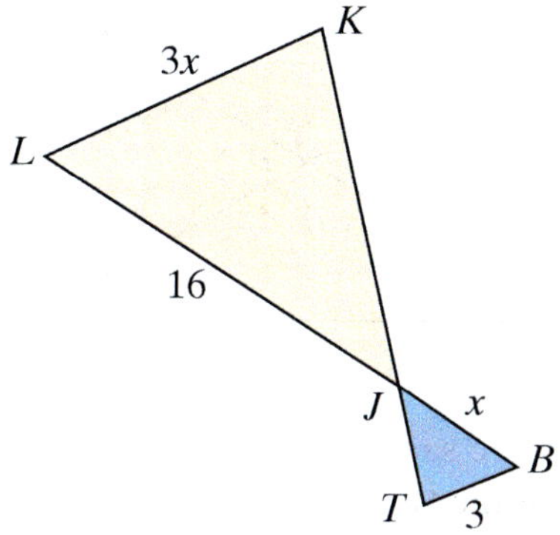

22. $\Delta QRP \sim \Delta TUS$

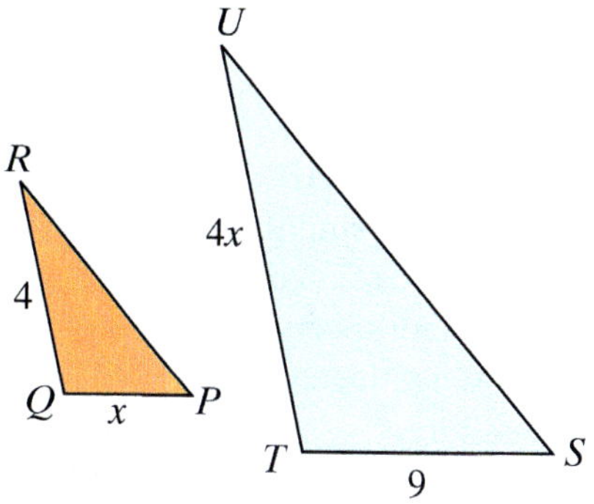

23. $\Delta ABC \sim \Delta RST$

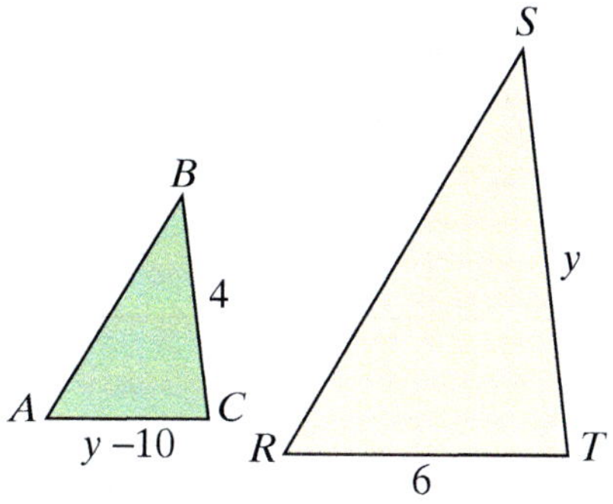

24. $\Delta FED \sim \Delta FGH$

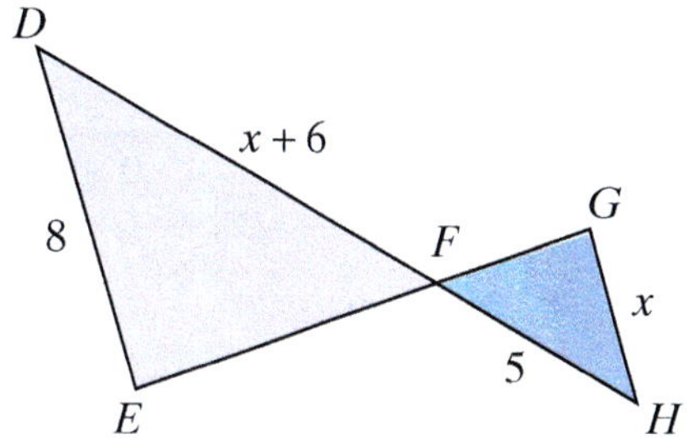

25. $\Delta SUT \sim \Delta PRQ$

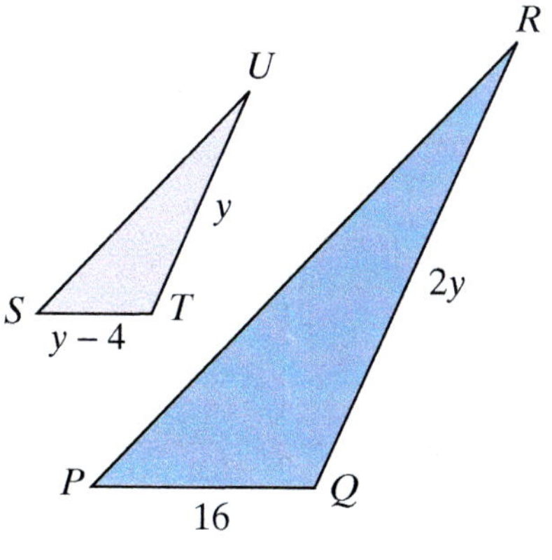

26. $\Delta ABC \sim \Delta DEC$

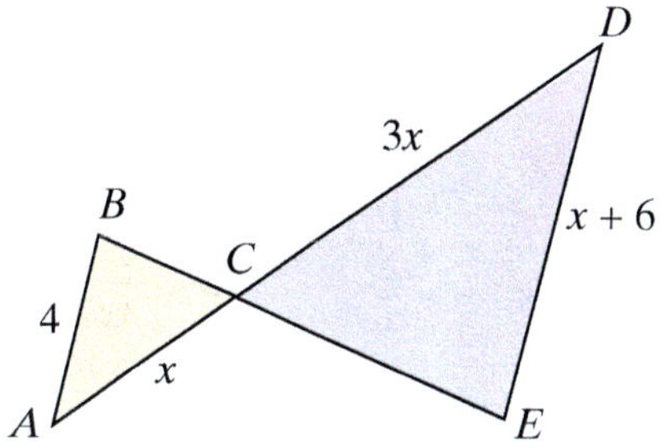

Teaching Notes:
A problem similar to #27 is in the March 2008 issue of *Mathematics Teacher* (#30 in the Calendar problems). The point M can be moved to any position on AB. You could expand this idea and the related geometric properties in an interesting class discussion.

27. $AP = \frac{9}{2}$ in.
$PC = \frac{15}{2}$ in.
28. $AP = PC = 6$ in.
29. $x = \frac{1}{4}$ 30. $x = 1$
31. $x = 6$ 32. $x = 7$
33. $x = 4$ 34. $x = 16$
35. $x \neq 0; x = \frac{10}{3}$
36. $x \neq 0; x = \frac{3}{2}$
37. $x \neq 0; x = -\frac{3}{4}$
38. $x \neq 0; x = \frac{1}{4}$
39. $x \neq 0; x = -\frac{3}{16}$
40. $x \neq 0; x = -\frac{2}{3}$
41. $x \neq -9, -\frac{1}{4}, 0; x = -2, 1$
42. $x \neq -1, 0, \frac{1}{4}; x = \frac{1}{3}, 4$
43. $x \neq \frac{3}{2}, 0, 6; x = \frac{3}{5}, 9$
44. $x \neq -5, 0, \frac{4}{9}; x = \frac{9}{8}, 4$
45. $x \neq \frac{1}{2}, 4; x = -3$

27. Parallelograms: In the parallelogram $ABCD, AB = CD = 10$ in. Diagonal $AC = 12$ in. The point M on AB is 6 in. from A. Point P is the intersection of DM with AC. The triangles APM and CPD are similar. (Symbolically, $\Delta APM \sim \Delta CPD$.) What are the lengths of AP and PC?

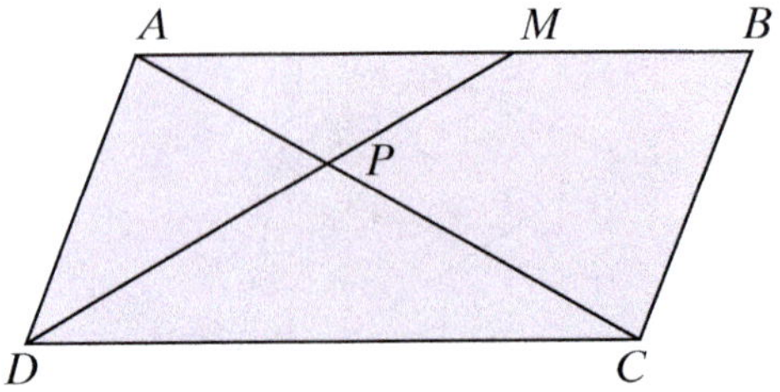

28. Parallelograms: If, in the same figure discussed in Exercise 27, the point P is the point of intersection of the two diagonals, AC and DB, what are the lengths of AP and PC?

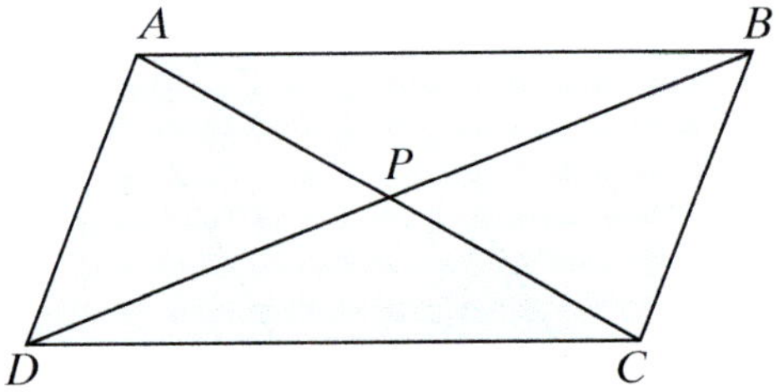

State any restrictions on x, and then solve the equations.

29. $\frac{5x}{4} - \frac{1}{2} = -\frac{3}{16}$

30. $\frac{x}{6} - \frac{1}{42} = \frac{1}{7}$

31. $\frac{3x-1}{6} - \frac{x+3}{4} = \frac{7}{12}$

32. $\frac{x-2}{3} - \frac{x-3}{5} = \frac{13}{15}$

33. $\frac{2+x}{4} - \frac{5x-2}{12} = \frac{8-2x}{5}$

34. $\frac{4x+1}{5} = \frac{2x+3}{2} - \frac{x+2}{4}$

35. $\frac{2}{3x} = \frac{1}{4} - \frac{1}{6x}$

36. $\frac{1}{x} - \frac{8}{21} = \frac{3}{7x}$

37. $\frac{3}{5x} - \frac{1}{5} = \frac{3}{4x}$

38. $\frac{3}{8x} - \frac{7}{10} = \frac{1}{5x}$

39. $\frac{3}{4x} - \frac{1}{2} = \frac{7}{8x} + \frac{1}{6}$

40. $\frac{5}{3x} + \frac{1}{2} = \frac{7}{9x} - \frac{5}{6}$

41. $\frac{2}{4x+1} = \frac{4}{x^2+9x}$

42. $\frac{3}{4x-1} = \frac{4}{x^2+x}$

43. $\frac{9}{x^2-6x} = \frac{5}{2x-3}$

44. $\frac{-9}{x^2+5x} = \frac{8}{4-9x}$

45. $\frac{x}{x-4} - \frac{4}{2x-1} = 1$

46. $\frac{x}{x+3} + \frac{1}{x+2} = 1$

46. $x \neq -3, -2; x = -\frac{3}{2}$
47. $x \neq -4, -1; x = 2$
48. $x \neq -4, 1; x = 6$
49. $x \neq -1, \frac{1}{4}; x = \frac{2}{3}$
50. $x \neq -4, -\frac{1}{2}; x = 1$
51. $x \neq 2, 3; x = \frac{13}{10}$
52. $x \neq -5, 4; x = -2$
53. $x \neq -\frac{2}{3}, 2;$ No solution
54. $x \neq -2, -\frac{2}{3};$ No solution
55. $x \neq -1, \frac{1}{3}, \frac{1}{2}; x = \frac{1}{5}$
56. $x \neq -1, \frac{3}{2}, 5; x = \frac{11}{19}$
57. $r = \frac{S-a}{S}$
58. $x = zs + \overline{x}$
59. $s = \frac{x - \overline{x}}{z}$
60. $d = \frac{a_n - a_1}{n-1}$
61. $y = m(x - x_1) + y_1$
62. $d_2 = v_{avg}(t_2 - t_1) + d_1$
63. $R_{total} = \frac{R_1 R_2}{R_1 + R_2}$
64. $x = \frac{t_1 t_2}{t_1 + t_2}$
65. $P = \frac{A}{1+r}$
66. $x = \frac{b - yd}{yc - a}$

47. $\frac{x+2}{x+1} + \frac{x+2}{x+4} = 2$

48. $\frac{3x-2}{x+4} + \frac{2x+5}{x-1} = 5$

49. $\frac{2}{4x-1} + \frac{1}{x+1} = \frac{3}{x+1}$

50. $\frac{x-2}{x+4} - \frac{3}{2x+1} = \frac{x-7}{x+4}$

51. $\frac{x-2}{x-3} + \frac{x-3}{x-2} = \frac{2x^2}{x^2-5x+6}$

52. $\frac{x}{x-4} - \frac{12x}{x^2+x-20} = \frac{x-1}{x+5}$

53. $\frac{3x+5}{3x+2} + \frac{8x+16}{3x^2-4x-4} = \frac{x+2}{x-2}$

54. $\frac{3x+5}{3x+2} - \frac{4-2x}{3x^2+8x+4} = \frac{x+4}{x+2}$

55. $\frac{3}{3x-1} + \frac{1}{x+1} = \frac{4}{2x-1}$

56. $\frac{2}{x+1} + \frac{4}{2x-3} = \frac{4}{x-5}$

Solve each of the formulas for the specified variables. Assume no denominator has a value of 0.

57. $S = \frac{a}{1-r}$; solve for r (formula for the sum of an infinite geometric sequence)

58. $z = \frac{x - \overline{x}}{s}$; solve for x (formula used in statistics)

59. $z = \frac{x - \overline{x}}{s}$; solve for s (formula used in statistics)

60. $a_n = a_1 + (n-1)d$; solve for d (formula for the nth term in an arithmetic sequence)

61. $m = \frac{y - y_1}{x - x_1}$; solve for y (formula for the slope of a line)

62. $v_{avg} = \frac{d_2 - d_1}{t_2 - t_1}$; solve for d_2 (formula for average velocity)

63. $\frac{1}{R_{total}} = \frac{1}{R_1} + \frac{1}{R_2}$; solve for R_{total} (formula used in electronics)

64. $\frac{1}{x} = \frac{1}{t_1} + \frac{1}{t_2}$; solve for x (formula used in mathematics)

65. $A = P + Pr$; solve for P (formula used for compound interest)

66. $y = \frac{ax+b}{cx+d}$; solve for x (formula used in mathematics)

Writing and Thinking About Mathematics

In simplifying rational expressions, the result is a rational or polynomial expression. However, in solving equations with rational expressions, the goal is to find a value (or values) for the variable that will make the equation a true statement. Many students confuse these two ideas. To avoid confusing the techniques for adding and subtracting rational expressions with the techniques for solving equations, simplify the expression in part **a.** and solve the equation in part **b.** Explain, in your own words, the differences in your procedures. Assume no denominator has a value of 0.

67. a. $\frac{10}{x}+\frac{31}{x-1}+\frac{4x}{x-1}$ **b.** $\frac{10}{x}+\frac{31}{x-1}=\frac{4x}{x-1}$

68. a. $\frac{-4}{x^2-16}+\frac{x}{2x+8}-\frac{1}{4}$ **b.** $\frac{-4}{x^2-16}+\frac{x}{2x+8}=\frac{1}{4}$

69. a. $\frac{3x}{x^2-4}+\frac{5}{x+2}+\frac{2}{x-2}$ **b.** $\frac{3x}{x^2-4}+\frac{5}{x+2}=\frac{2}{x-2}$

70. a. $\frac{7}{5x}+\frac{2}{x-4}-\frac{3}{5x}$ **b.** $\frac{7}{5x}+\frac{2}{x-4}=\frac{3}{5x}$

71. a. $\frac{2}{x+9}-\frac{2}{x-9}+\frac{1}{2}$ **b.** $\frac{2}{x+9}-\frac{2}{x-9}=\frac{1}{2}$

67. a. $\frac{4x^2+41x-10}{x(x-1)}$
b. $x=\frac{1}{4}, 10$ and $x\neq 0,1$

68. a. $\frac{x^2-8x}{4(x-4)(x+4)}$
b. $x=0,8$ and $x\neq 4,-4$

69. a. $\frac{10x-6}{(x-2)(x+2)}$
b. $x=\frac{7}{3}$ and $x\neq 2,-2$

70. a. $\frac{14x-16}{5x(x-4)}$
b. $x=\frac{8}{7}$ and $x\neq 0,4$

71. a. $\frac{x^2-153}{2(x-9)(x+9)}$
b. $x=3,-3$ and $x\neq 9,-9$

HAWKES LEARNING SYSTEMS: INTRODUCTORY & INTERMEDIATE ALGEBRA SOFTWARE

- 7.4a Solving Equations: Ratios and Proportions
- 7.4b Solving Equations with Rational Expressions

7.5 Applications

- *Solve applied problems related to fractions.*
- *Solve applied problems related to work.*
- *Solve applications involving distance, rate, and time.*

The following strategy for solving word problems is valid for all word problems that involve algebraic equations.

Strategy for Solving Word Problems

1. Read the problem carefully. Read it several times if necessary.
2. Decide what is asked for and assign a variable to the unknown quantity.
3. Draw a diagram or set up a chart whenever possible as a visual aid.
4. Form an equation that relates the information provided.
5. Solve the equation.
6. Check your solution with the wording of the problem to be sure it makes sense.

Problems Related to Fractions

We now introduce word problems involving rational expressions with problems relating the numerator and denominator of a fraction.

Example 1: Fractions

The denominator of a fraction is 8 more than the numerator. If both the numerator and denominator are increased by 3, the new fraction is equal to $\frac{1}{2}$. Find the original fraction.

Solution: Reread the problem to be sure that you understand all terminology used. Assign variables to the unknown quantities.

Let n = original numerator

$n+8$ = original denominator

$\frac{n}{n+8}$ = original fraction

Continued on the next page...

$$\frac{n+3}{(n+8)+3}=\frac{1}{2}$$

The numerator and the denominator are each increased by 3, making a new fraction that is equal to $\frac{1}{2}$.

$$\frac{n+3}{n+11}=\frac{1}{2}$$

$$2(n+11)\cdot\frac{n+3}{n+11}=2(n+11)\cdot\frac{1}{2}$$

$$2n+6=n+11$$

$$n=5 \quad \leftarrow \text{Original numerator}$$

$$n+8=13 \quad \leftarrow \text{Original denominator}$$

Check: $\frac{(5)+3}{(13)+3}=\frac{8}{16}=\frac{1}{2}$

The original fraction is $\frac{5}{13}$.

Problems Related to Work

Problems involving work usually translate into equations involving rational expressions. The basic idea is to **represent what part of the work is done in one unit of time**. For example, if a man can dig a ditch in 3 hours, what part (of the ditch-digging job) can he do in one hour? The answer is $\frac{1}{3}$ of the work in one hour. If a fence was painted in 2 days, then $\frac{1}{2}$ of the work of painting the fence was done in 1 day. (These ideas assume a steady working pace.) In general, if the total work took x hours, then $\frac{1}{x}$ of the total work would be done in one hour.

Example 2: Work Problems

a. A carpenter can build a certain type of patio cover in 6 hours. His partner takes 8 hours to build the same cover. How long would it take them working together to build this type of patio cover?

Solution: Let x = number of hours to build the cover working together.

Person(s)	Time of Work (in Hours)	Part of Work Done in 1 Hour
Carpenter	6	$\frac{1}{6}$
Partner	8	$\frac{1}{8}$
Together	x	$\frac{1}{x}$

$$\underbrace{\text{Part done in 1 hr by carpenter}} + \underbrace{\text{Part done in 1 hr by partner}} = \underbrace{\text{Part done in 1 hr together}}$$

$$\frac{1}{6} + \frac{1}{8} = \frac{1}{x}$$

$$\frac{1}{\cancel{6}}\left(\overset{4}{\cancel{24}}x\right)+\frac{1}{\cancel{8}}\left(\overset{3}{\cancel{24}}x\right)=\frac{1}{\cancel{x}}\left(24\cancel{x}\right)$$ Multiply each term on both sides by $24x$, the LCD of the fractions.

$$4x + 3x = 24$$
$$7x = 24$$
$$x = \frac{24}{7}$$

Together, they can build the patio cover in $\frac{24}{7}$ hours, or $3\frac{3}{7}$ hours.

(Note that this answer is reasonable because the time is less than either person would take working alone.)

b. A man can wax his car three times faster than his daughter can. Together they can do the job in 4 hours. How long would it take each of them working alone?

Solution: Let t = number of hours for man alone to wax the car and
$3t$ = number of hours for daughter alone to wax the car.

Person(s)	Time of Work (in Hours)	Part of Work Done in 1 Hour
Man	t	$\frac{1}{t}$
Daughter	$3t$	$\frac{1}{3t}$
Together	4	$\frac{1}{4}$

Continued on the next page...

$$\underbrace{\text{Part done by man alone in 1 hour}} + \underbrace{\text{Part done by daughter alone in 1 hour}} = \underbrace{\text{Part done working together in 1 hour}}$$

$$\frac{1}{t} \qquad + \qquad \frac{1}{3t} \qquad = \qquad \frac{1}{4}$$

$$\frac{1}{\cancel{t}}\left(12\cancel{t}\right)+\frac{1}{\cancel{3t}}\left(\overset{4}{\cancel{12t}}\right)=\frac{1}{\cancel{4}}\left(\overset{3}{\cancel{12}}t\right)$$ Multiply each term on both sides by $12t$, the LCM of the denominators.

$$12+4=3t$$

$$16=3t$$

$$t=\frac{16}{3} \quad \longleftarrow \text{Man's time}$$

$$3t=(3)\left(\frac{16}{3}\right)=16 \quad \longleftarrow \text{Daughter's time}$$

Check: Man's part in 1 hr $=\frac{1}{t}=\frac{1}{\frac{16}{3}}=\frac{3}{16}$

Daughter's part in 1 hr $=\frac{1}{3t}=\frac{1}{3\cdot\frac{16}{3}}=\frac{1}{16}$

Man's part in 4 hr $=\frac{3}{16}\cdot 4=\frac{3}{4}$

Daughter's part in 4 hr $=\frac{1}{16}\cdot 4=\frac{1}{4}$

$\frac{3}{4}+\frac{1}{4}=1$ car waxed in 4 hours.

Working alone, the man takes $\frac{16}{3}$ hours, or 5 hours 20 minutes, and his daughter takes 16 hours.

c. A man was told that his new pool would fill through an inlet valve in 3 hours. He knew something was wrong when the pool took 8 hours to fill. He found he had left the drain valve open. How long would it take to drain the pool once it is completely filled and only the drain valve is open?

Solution: Let t = time to drain pool with only the drain valve open.

Note: We use the information gained when the pool was filled with both valves open. In that situation, the inlet and outlet valves worked against each other.

Valves	Hours to Fill or Drain	Part Filled or Drained in 1 Hour
Inlet	3	$\frac{1}{3}$
Outlet	t	$\frac{1}{t}$
Together	8	$\frac{1}{8}$

$$\underbrace{\text{Part filled by inlet in 1 hour}} - \underbrace{\text{Part emptied by outlet in 1 hour}} = \underbrace{\text{Part filled together in 1 hour}}$$

$$\frac{1}{3} - \frac{1}{t} = \frac{1}{8}$$

$$\frac{1}{\cancel{3}}\left(\overset{8}{\cancel{24}}t\right)-\frac{1}{\cancel{t}}\left(24\cancel{t}\right)=\frac{1}{\cancel{8}}\left(\overset{3}{\cancel{24}}t\right)$$

$$8t - 24 = 3t$$

$$5t = 24$$

$$t = \frac{24}{5}$$

The pool would drain in $\frac{24}{5}$ hours, or 4 hours 48 minutes. (Note that this is more time than the inlet valve would take to fill the pool. If the outlet valve worked faster than the inlet valve, then the pool would never have filled in the first place.)

Problems Related to Distance-Rate-Time: $d = rt$

You may recall that the basic formula involving distance, rate, and time is $d = rt$. This relationship can also be stated in the forms $t = \frac{d}{r}$ and $r = \frac{d}{t}$.

If distance and rate are known or can be represented, then $t = \frac{d}{r}$ is the way to represent time. Similarly, if the distance and time are known or can be represented, then $r = \frac{d}{t}$ is the way to represent rate.

Example 3: Distance-Rate-Time

a. On Lake Itasca a man can row his boat 5 miles per hour. On the nearby Mississippi River it takes him the same time to row 5 miles downstream as it does to row 3 miles upstream. What is the speed of the river current in miles per hour?

Solution: Let $c =$ the speed of the current.

Distance and rate are represented first in the table below. Then the time going downstream and coming back upstream is represented in terms of distance and rate. Since the rate is in miles per hour, the distance is in miles and the time is in hours.

	Distance d	Rate r	Time $t = \frac{d}{r}$
Downstream	5	$5+c$	$\frac{5}{5+c}$
Upstream	3	$5-c$	$\frac{3}{5-c}$

$$\frac{5}{5+c} = \frac{3}{5-c} \qquad \text{The times are equal.}$$

$$(5+c)(5-c)\cdot\frac{5}{5+c} = (5+c)(5-c)\cdot\frac{3}{5-c}$$

$$25-5c = 15+3c$$

$$-8c = -10$$

$$c = \frac{5}{4}$$

Check: $$\text{Time downstream} = \frac{5}{5+\frac{5}{4}} = \frac{5}{\frac{20}{4}+\frac{5}{4}} = \frac{5}{\frac{25}{4}} = 5\cdot\frac{4}{25} = \frac{4}{5}\text{ hours}$$

$$\text{Time upstream} = \frac{3}{5-\frac{5}{4}} = \frac{3}{\frac{20}{4}-\frac{5}{4}} = \frac{3}{\frac{15}{4}} = 3\cdot\frac{4}{15} = \frac{4}{5}\text{ hours}$$

The times are equal. The rate of the river current is $\frac{5}{4}$ mph, or $1\frac{1}{4}$ mph.

b. If a passenger train travels three times as fast as a freight train, and the freight train takes 4 hours longer to travel 210 miles, what is the speed of each train?

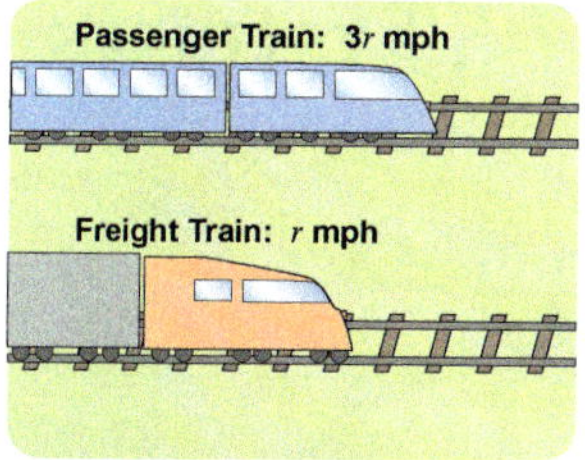

Solution: Let r = rate of freight train in miles per hour
$3r$ = rate of passenger train in miles per hour

	Distance d	Rate r	Time $t = \frac{d}{r}$
Freight	210	r	$\frac{210}{r}$
Passenger	210	$3r$	$\frac{210}{3r}$

Note: If the rate is faster, then the time is shorter. Thus the fraction $\frac{210}{3r}$ is smaller than the fraction $\frac{210}{r}$.

$$\frac{210}{r} - \frac{210}{3r} = 4 \quad \text{The difference between their times is 4 hours.}$$

$$\frac{210}{r} - \frac{70}{r} = 4$$

$$\frac{210}{\cancel{r}} \cdot \cancel{r} - \frac{70}{\cancel{r}} \cdot \cancel{r} = 4 \cdot r$$

$$210 - 70 = 4r$$

$$140 = 4r$$

$$35 = r$$

$$105 = 3r$$

Check: Time for freight train $= \frac{210}{35} = 6$ hours

Time for passenger train $= \frac{210}{105} = 2$ hours

$6 - 2 = 4$ hours difference in time

The freight train travels 35 mph, and the passenger train travels 105 mph.

7.5 Exercises

1. 72, 45
2. 10
3. 9
4. 14
5. $\frac{6}{13}$
6. $\frac{12}{7}$
7. 36, 27
8. 15, 9
9. 7, 12
10. 2, 5
11. 45 shirts
12. \$1500
13. 12 brushes

1. The sum of two numbers is 117 and they are in the ratio of 8 to 5. Find the two numbers.

2. If 4 is subtracted from a certain number and the difference is divided by 2, the result is 1 more than $\frac{1}{5}$ of the original number. Find the original number.

3. What number must be added to both the numerator and denominator of $\frac{16}{21}$ to make the resulting fraction equal to $\frac{5}{6}$?

4. Find the number that can be subtracted from both the numerator and denominator of the fraction $\frac{69}{102}$ so that the result is $\frac{5}{8}$.

5. The denominator of a fraction exceeds the numerator by 7. If the numerator is increased by 3 and the denominator is increased by 5, the resulting fraction is equal to $\frac{1}{2}$. Find the original fraction.

6. The numerator of a fraction exceeds the denominator by 5. If the numerator is decreased by 4 and the denominator is increased by 3, the resulting fraction is equal to $\frac{4}{5}$. Find the original fraction.

7. One number is $\frac{3}{4}$ of another number. Their sum is 63. Find the numbers.

8. The sum of two numbers is 24. If $\frac{2}{5}$ of the larger number is equal to $\frac{2}{3}$ of the smaller number, find the numbers.

9. One number exceeds another by 5. The sum of their reciprocals is equal to 19 divided by the product of the two numbers. Find the numbers.

10. One number is 3 less than another. The sum of their reciprocals is equal to 7 divided by the product of the two numbers. Find the numbers.

11. Shirt sales: A manufacturer sold a group of shirts for \$1026. One-fifth of the shirts were priced at \$18 each and the remainder at \$24 each. How many shirts were sold?

12. Paying bills: Luis spent $\frac{1}{5}$ of his monthly salary for rent and $\frac{1}{6}$ of his monthly salary for his car payment. If \$950 was left, what was his monthly salary?

13. Painting: Suppose that an artist expects that for every 9 special brushes she orders, 7 will be good and 2 will be defective. If she orders 54 brushes, how many will she expect to be defective?

14. 37.5 miles
15. 1875 miles
16. $\frac{12}{5}$ or $2\frac{2}{5}$ hours
17. Person: 505 minutes; Machine: 5.05 minutes
18. $\frac{9}{5}$ or $1\frac{4}{5}$ hours
19. 45 minutes
20. 6 hours
21. Beth: 52 mph; Anna: 48 mph
22. Kirk: 8 mph; Karl: 6 mph
23. Commercial airliner: 300 mph; Private plane: 120 mph
24. 63 mph
25. 6 hours

14. Travel by car: It takes Rosa, traveling at 30 mph, 30 minutes longer to go a certain distance than it takes Melody traveling at 50 mph. Find the distance traveled.

15. Travel by plane: It takes a plane flying at 450 mph 25 minutes longer to travel a certain distance than it takes a second plane to fly the same distance at 500 mph. Find the distance.

16. Landscaping: Toni needs 4 hours to complete the yard work. Her husband, Sonny, needs 6 hours to do the work. How long will the job take if they work together?

17. Manufacturing: In 1921, automated wrapping machines were used to aid in the wrapping of Hershey Kisses in the Hershey chocolate factory. The machine could wrap the candies 100 times faster than a person could. Together they can wrap a crate full of Kisses in 5 minutes. How long would it take each of them working alone?

18. Mass mailings: Ben's secretary can address the weekly newsletters in $4\frac{1}{2}$ hours. Charlie's secretary needs only 3 hours. How long will it take if they both work on the job?

19. Shoveling snow: Working together, Greg and Cindy can clean the snow from the driveway in 20 minutes. It would have taken Cindy, working alone, 36 minutes. How long would it have taken Greg alone?

20. Carpentry: A carpenter and his partner can put up a patio cover in $3\frac{3}{7}$ hours. If the partner needs 8 hours to complete the patio alone, how long would it take the carpenter working alone?

21. Travel: Beth can travel 208 miles in the same length of time it takes Anna to travel 192 miles. If Beth's speed is 4 mph greater than Anna's, find both rates.

22. Biking: Kirk can bike 32 miles in the same amount of time that his twin brother Karl can bike 24 miles. If Kirk bikes 2 mph faster than Karl, how fast does each man bike?

23. Plane speeds: A commercial airliner can travel 750 miles in the same amount of time that it takes a private plane to travel 300 miles. The speed of the airliner is 60 mph more than twice the speed of the private plane. Find the speed of each aircraft.

24. Car speeds: Gabriela drives her car 350 miles and averages a certain speed. If the average speed had been 9 mph less, she could have traveled only 300 miles in the same length of time. What was her average speed?

25. Boating: A family travels 18 miles down river and returns. It takes 8 hours to make the round trip. Their rate in still water is twice the rate of the river's current. How long will the return trip take?

26. Sailboat: 5 mph; Cruise ship: 25 mph
27. 50 mph
28. 20 mph
29. $\frac{9}{4}$ or $2\frac{1}{4}$ hours
30. 120 hours
31. $\frac{9}{2}$ or $4\frac{1}{2}$ days; 9 days
32. John: 11 hours; Ralph: 22 hours; Denny: 33 hours
33. Boat: 14 mph; Current: 2 mph

26. Boat speed: Cruise ships travel 5 times faster than sailboats (in optimal wind conditions). If it takes 16 hours longer for the sailboat to travel 100 miles from Charleston, SC to Savannah, GA, what is the speed of each boat?

27. Wind speed: An airplane can fly 650 mph in still air. If it can travel 2800 miles with the wind in the same time it can travel 2400 miles against the wind, find the wind speed. (**Note:** A tailwind increases the speed of the plane and a headwind decreases the speed of the plane.)

28. Wind speed: A one-engine plane can fly 120 mph in still air. If it can fly 490 miles with a tailwind in the same time that it can fly 350 miles against a headwind, what is the speed of the wind?

29. Filling a pool: Using a small inlet pipe it takes 9 hours to fill a pool. Using a large inlet pipe it only takes 3 hours. If both are used simultaneously, how long will it take to fill the pool?

30. Filling a pool: An inlet pipe on a swimming pool can be used to fill the pool in 36 hours. The drain pipe can be used to empty the pool in 40 hours. If the pool is $\frac{2}{3}$ filled using the inlet pipe and then the drain pipe is accidentally opened, how long from that time will it take to fill the pool?

31. Clearing land: A contractor hires two bulldozers to clear the trees from a 20-acre tract of land. One works twice as fast as the other. It takes them 3 days to clear the tract working together. How long would it take each of them alone?

32. Store maintenance: John, Ralph, and Denny, working together, can clean their bait and tackle store in 6 hours. Working alone, Ralph takes twice as long to clean the store as does John. Denny needs three times as long as does John. How long would it take each man working alone?

33. Boating: Francois rode his jet ski 36 miles downstream and then 36 miles back. The round trip took $5\frac{1}{4}$ hours. Find the speed of the jet ski in still water and the speed of the current if the speed of the current is $\frac{1}{7}$ the speed of the jet ski .

34. 1 mph
35. 14 kilometers per hour

34. Boating: Momence, IL is 12 miles upstream on the same side of the river from Kankakee, IL on the Kankakee River. A motorboat that can travel 8 mph in still water leaves Momence and travels downstream toward Kankakee. At the same time, another boat that can travel 10 mph leaves Kankakee and travels upstream toward Momence. Each boat completes the trip in the same amount of time. Find the rate of the current.

35. Skiing: Samantha rides the ski lift to the top of Blue Mountain, a distance of $1\frac{3}{4}$ kilometers (a little more than 1 mile). She then skis directly down the slope. If she skis five times as fast as the lift travels and the total trip takes 45 minutes, find the rate at which she skis.

Writing and Thinking About Mathematics

36. a. 5 and 7
b. 2 and 4

36. If n is any integer, then $2n$ is an even integer and $2n + 1$ is an odd integer. Use these ideas to solve the following problems.

a. Find two consecutive odd integers such that the sum of their reciprocals is $\frac{12}{35}$.

b. Find two consecutive even integers such that the sum of the first and the reciprocal of the second is $\frac{9}{4}$.

Hawkes Learning Systems: Introductory & Intermediate Algebra Software

- 7.5 Applications Involving Rational Expressions

7.6 Variation

- *Solve problems related to direct variation.*
- *Solve problems related to inverse variation.*
- *Solve problems involving combined variation.*

Teaching Notes:
You might want to make the case for the study of variation (and proportion) as being extremely important to the sciences and technology. Many major relationships in chemistry, physics, and astronomy began by recognizing a variation relationship.

Direct Variation

Suppose that you ride your bicycle at a steady rate of 15 miles per hour (not quite as fast as Lance Armstrong, but you are enjoying yourself). If you ride for 1 hour, the distance you travel would be 15 miles. If you ride for two hours, the distance you travel would be 30 miles. This relationship can be written in the form of the formula $d = 15t$ (or $\frac{d}{t} = 15$) where d is the distance traveled and t is the time in hours. We say that distance and time **vary directly** (or are in **direct variation** or are **directly proportional**). The term proportional implies that the ratio is constant. In this example, 15 is the constant and is called the **constant of variation**. When two variables vary directly, an **increase in the value of one variable indicates an increase in the other**, and the ratio of the two quantities is constant.

Direct Variation

A variable quantity y **varies directly as** (or is **directly proportional to**) a variable x if there is a constant k such that

$$\frac{y}{x} = k \text{ or } y = kx.$$

The constant k is called the **constant of variation**.

Example 1: Direct Variation

If y varies directly as x, and $y = 6$ when $x = 2$, find y if $x = 6$.

Solution:

$y = kx$	General formula for direct variation
$6 = 2k$	Substitute the known values and solve for k.
$3 = k$	Use this value for k in the general formula.

So $y = 3x$. Thus if $x = 6$, then $y = 3 \cdot 6 = 18$.

Example 2: Direct Variation

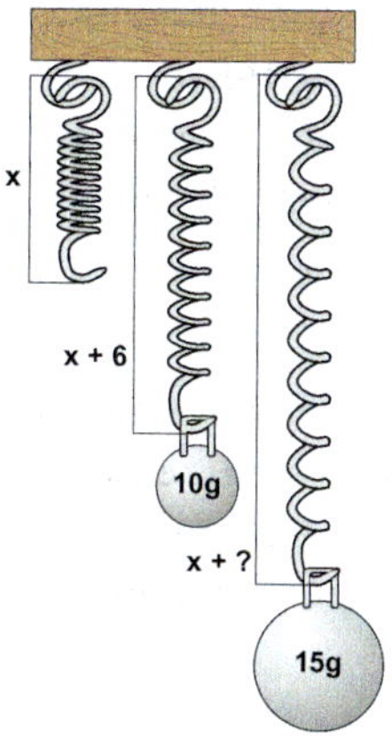

A spring will stretch a greater distance as more weight is placed on the end of the spring. The distance d the spring stretches varies directly as the weight w placed at the end of the spring. This is a property of springs studied in physics and is known as Hooke's Law. If a weight of 10 g stretches a certain spring 6 cm, how far will the spring stretch with a weight of 15 g? (**Note:** We assume that the weight is not so great as to break the spring.)

Solution: Because the two variables are directly proportional, the relationship can be indicated with the formula

$d = k \cdot w$ where d = distance spring stretches in cm, w = weight in g, and k = constant of variation.

First, substitute the given information to find the value for k. (The value of k will depend on the particular spring. Springs made of different material or which are wound more tightly will have different values for k.)

$d = k \cdot w$

$6 = k \cdot 10$ Substitute the known values into the formula.

$\frac{3}{5} = k$ Use this value for k in the general formula.

So $d = \frac{3}{5}w.$ The constant of variation is $\frac{3}{5}$ (or 0.6).

If $w = 15$, we have

$$d = \frac{3}{5} \cdot 15 = 9.$$

The spring will stretch 9 cm if a weight of 15 g is placed at its end.

Listed here are several formulas involving direct variation.

$d = \frac{3}{5}w$ Hooke's Law for a spring where $k = \frac{3}{5}$.

$C = 2\pi r$ The circumference of a circle is directly proportional to the radius.

$A = \pi r^2$ The area of a circle varies directly as the radius squared.

$P = 625d$ Water pressure is proportional to the depth of the water.

Inverse Variation

When two variables vary in such a way that their product is constant, we say that the two variables **vary inversely** (or are **inversely proportional**). For example, if a gas is placed in a container (as in an automobile engine) and pressure is increased on the gas, then the product of the pressure and the volume of gas will remain constant. That is, pressure and volume are related by the formula $V \cdot P = k$ (or $V = \frac{k}{P}$).

Note that if a product of two variables is to remain constant, then **an increase in the value of one variable must be accompanied by a decrease in the other**. Or, in the case of a fraction with a constant numerator, if the denominator increases in value, then the fraction decreases in value. For the gas in an engine, an increase in pressure indicates a decrease in the volume of gas.

Inverse Variation

A variable quantity y **varies inversely as** (or is **inversely proportional to**) a variable x if there is a constant k such that

$$x \cdot y = k \text{ or } y = \frac{k}{x}.$$

The constant k is called the **constant of variation.**

Example 3: Inverse Variation

If y varies inversely as the cube of x, and $y = -1$ when $x = 3$, find y if $x = -3$.

Solution:

$y = \frac{k}{x^3}$ General formula for inverse variation

$-1 = \frac{k}{3^3}$ Substitute the known values and solve for k.

$-1 = \frac{k}{27}$

$-27 = k$ Use this value for k in the general formula.

So $y = \frac{-27}{x^3}$. Thus if $x = -3$, then $y = \frac{-27}{(-3)^3} = \frac{-27}{-27} = 1$.

Example 4: Inverse Variation

The gravitational force F between an object and the Earth is inversely proportional to the square of the distance d from the object to the center of the Earth. Hence we have the formula

$$F \cdot d^2 = k \text{ or } F = \frac{k}{d^2},$$ where F = force, d = distance and k = constant of variation.

(As the distance of an object from the Earth becomes larger, the gravitational force exerted by the Earth on the object becomes smaller.)

If an astronaut weighs 200 pounds on the surface of the Earth, what will he weigh 100 miles above the Earth? Assume that the radius of the Earth is 4000 miles.

Solution: We know $F = 200 = 2 \times 10^2$ pounds when $d = 4000 = 4 \times 10^3$ miles. Use scientific notation to make values simpler to work with in the calculations.

$$2 \times 10^2 = \frac{k}{\left(4 \times 10^3\right)^2}$$ Substitute and solve for k.

$$k = 2 \times 10^2 \times 16 \times 10^6 = 32 \times 10^8 = 3.2 \times 10^9$$

So, $$F = \frac{3.2 \times 10^9}{d^2}.$$

When the astronaut is 100 miles above the Earth, $d = 4100 = 4.1 \times 10^3$ miles. Then

$$F = \frac{3.2 \times 10^9}{16.81 \times 10^6} \approx 0.190 \times 10^3 = 190 \text{ pounds}.$$

That is, 100 miles above the Earth the astronaut will weigh about 190 pounds.

Combined Variation

If a variable varies either directly or inversely with more than one other variable, the variation is said to be **combined variation**. If the combined variation is all direct variation (the variables are multiplied), then it is called **joint variation**. For example, the volume of a cylinder varies jointly as its height and the square of its radius.

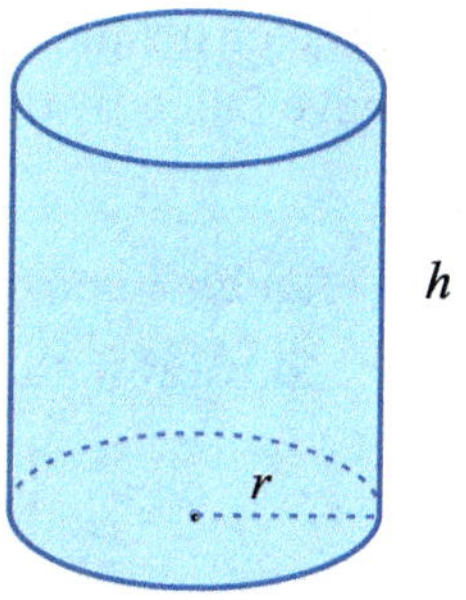

$$V = kr^2h$$

where r = radius, h = height, and k = constant of variation

Using this information, what is the value of k, the constant of variation, if a cylinder has the approximate measurements $V = 198$ cubic feet, $r = 3$ feet, and $h = 7$ feet?

$V = k \cdot r^2 \cdot h$ — V **varies jointly** as r^2 and h.

$198 = k \cdot 3^2 \cdot 7$ — Substitute the known values.

$\frac{198}{9 \cdot 7} = k$

$k = \frac{22}{7} \approx 3.14$ — We know from experience that $k = \pi$. Since the measurements are only approximate, the estimate for k is only approximate.

Substituting the constant of variation, the formula is $V = \pi r^2 h$.

Example 5: Joint Variation

If z varies jointly as x^2 and y, and $z = 18$ when $x = 2$ and $y = 4$, what is z when $x = 4$ and $y = 3$?

Solution: $z = k \cdot x^2 \cdot y$

$18 = k \cdot 2^2 \cdot 4$ — Substitute the known values and solve for k.

$18 = k \cdot 16$

$\frac{9}{8} = k$

So $z = \frac{9}{8}x^2 y.$ — Substitute $\frac{9}{8}$ for k in the general formula.

If $x = 4$ and $y = 3$, then

$z = \frac{9}{8} \cdot 4^2 \cdot 3$

$z = 54.$

Example 6: More Variation

a. The distance an object falls varies directly as the square of the time it falls (until it hits the ground and assuming little or no air resistance). If an object fell 64 feet in two seconds, how far would it have fallen by the end of 3 seconds?

Solution: $d = k \cdot t^2$ — where d = distance, t = time (in seconds), and k = constant of variation

$64 = k \cdot 2^2$ — Substitute the known values and solve for k.

$16 = k$

So $d = 16t^2$. Substitute 16 for k in the general formula.

If $t = 3$, then

$$d = 16 \cdot 3^2$$
$$d = 144.$$

The object would have fallen 144 feet in 3 seconds.

b. The volume of a gas in a container varies inversely as the pressure on the gas. If a gas has a volume of 200 cubic inches under pressure of 5 pounds per square inch, what will be its volume if the pressure is increased to 8 pounds per square inch?

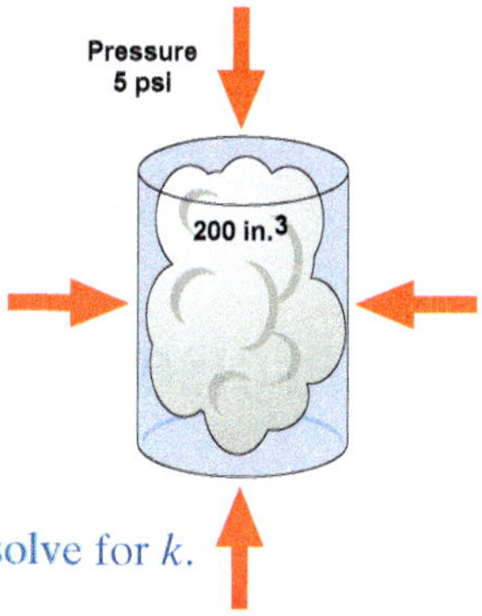

Solution: $V = \dfrac{k}{P}$ where V = volume, P = pressure, and k = constant of variation

$200 = \dfrac{k}{5}$ Substitute the known values and solve for k.

$k = 1000$

So $V = \dfrac{1000}{P}$ Substitute 1000 for k in the general formula.

$$V = \frac{1000}{8} = 125.$$

The volume will be 125 cubic inches.

c. The safe load L of a wooden beam supported at both ends varies jointly as the width w and the square of the depth d and inversely as the length l. A 3 in. wide by 10 in. deep beam that is 8 ft long supports a load of 9600 lb safely. What is the safe load of a beam of the same material that is 4 in. wide, 9 in. deep, and 12 ft long?

Solution: $L = \dfrac{k \cdot w \cdot d^2}{l}$ where L = safe load, w = width, d = depth, and l = length

$9600 = \dfrac{k \cdot 3 \cdot 10^2}{8}$ Substitute the known values and solve for k.

$$9600 = \frac{k \cdot 300}{8}$$
$$k = \frac{9600 \cdot 8}{300}$$
$$k = 256$$

So $L = \dfrac{256 \cdot w \cdot d^2}{l}$ Substitute 256 for k in the general formula.

$$L = \frac{256 \cdot 4 \cdot 9^2}{12} = \frac{256 \cdot 4 \cdot 81}{12} = 6912.$$

The safe load will be 6912 lb.

Practice Problems

1. The length that a hanging spring stretches varies directly as the weight placed on the end of the spring. If a weight of 5 mg stretches a certain spring 3 cm, how far will the spring stretch with a weight of 6 mg?

2. The volume of propane in a container varies inversely to the pressure on the gas. If the propane has a volume of 200 in.^{3} under a pressure of 4 lb per in.^{2}, what will be its volume if the pressure is increased to 5 lb per in.^{2}?

7.6 Exercises

Use the information given to find the unknown value.

1. If y varies directly as x, and $y = 3$ when $x = 9$, find y if $x = 7$.

2. If y is directly proportional to x^2, and $y = 3$ when $x = 2$, what is y when $x = 8$?

3. If y varies inversely as x, and $y = 5$ when $x = 8$, find y if $x = 20$.

4. If y is inversely proportional to x, and $y = 5$ when $x = 4$, what is y when $x = 2$?

5. If y varies inversely as x^2, and $y = -8$ when $x = 2$, find y if $x = 3$.

6. If y is inversely proportional to x^3, and $y = 40$ when $x = \frac{1}{2}$, what is y when $x = \frac{1}{3}$?

7. If y is directly proportional to the square root of x, and $y = 6$ when $x = \frac{1}{4}$, what is y when $x = 9$?

8. If y is directly proportional to the square of x, and $y = 80$ when $x = 4$, what is y when $x = 6$?

9. z varies jointly as x and y, and $z = 60$ when $x = 2$ and $y = 3$. Find z if $x = 3$ and $y = 4$.

10. z varies jointly as x and y, and $z = -6$ when $x = 5$ and $y = 8$. Find z if $x = 12$ and $y = 15$.

11. z varies jointly as x and y^2, and $z = 63$ when $x = 5$ and $y = 3$. Find z if $x = \frac{10}{3}$ and $y = 2$.

12. z varies jointly as x^2 and y, and $z = 20$ when $x = 2$ and $y = 3$. Find z if $x = 4$ and $y = \frac{7}{10}$.

13. z varies directly as x and inversely as y^2. If $z = 5$ when $x = 1$ and $y = 2$, find z if $x = 2$ and $y = 1$.

14. z varies directly as x^3 and inversely as y^2. If $z = 24$ when $x = 2$ and $y = 2$, find z if $x = 3$ and $y = 2$.

1. $\frac{7}{3}$
2. 48
3. 2
4. 10
5. $-\frac{32}{9}$
6. 135
7. 36
8. 180
9. 120
10. −27
11. $\frac{56}{3}$
12. $\frac{56}{3}$
13. 40
14. 81

Answers to Practice Problems: **1.** $\frac{18}{5}$ cm **2.** 160 in.^{3}

15. 54
16. 32
17. $\frac{48}{5}$
18. 54
19. 27
20. 16
21. 400 feet
22. 6 in.
23. \$59.70
24. Rating of 61.36
25. 4.71 feet
26. 254.34 in.^{2}
27. 6 m

15. z varies directly as $\sqrt{x}$ and inversely as y. If $z = 24$ when $x = 4$ and $y = 3$, find z if $x = 9$ and $y = 2$.

16. z varies directly as x^2 and inversely as $\sqrt{y}$. If $z = 108$ when $x = 6$ and $y = 4$, find z if $x = 4$ and $y = 9$.

17. s varies directly as the sum of r and t and inversely as w. If $s = 24$ when $r = 7$ and $t = 8$ and $w = 9$, find s if $r = 9$ and $t = 3$ and $w = 18$.

18. s varies directly as r and inversely as the difference of t and u. If $s = 36$ when $r = 12$ and $t = 9$ and $u = 6$, find s if $r = 18$ and $t = 11$ and $u = 8$.

19. L varies jointly as m and n and inversely as p. If $L = 6$ when $m = 7$ and $n = 8$ and $p = 12$, find L if $m = 15$ and $n = 14$ and $p = 10$.

20. W varies jointly as x and y and inversely as z. If $W = 10$ when $x = 6$ and $y = 5$ and $z = 2$, find W if $x = 12$ and $y = 6$ and $z = 3$.

Solve the following application problems.

21. Free falling object: The distance a free falling object falls is directly proportional to the square of the time it falls (before it hits the ground). If an object fell 256 feet in 4 seconds, how far will it have fallen by the end of 5 seconds?

22. Stretching a spring: The length a hanging spring stretches varies directly with the weight placed on the end. If a spring stretches 5 in. with a weight of 10 lb, how far will the spring stretch if the weight is increased to 12 lb?

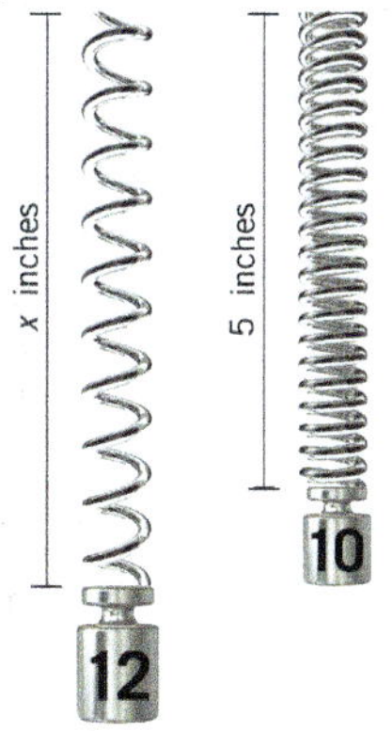

23. Gas prices: The total price P of gasoline purchased varies directly with the number of gallons purchased. If 10 gallons are purchased for \$39.80, what will be the price of 15 gallons?

24. Economics: Research shows that the value of gold and the value of the dollar are inversely proportional. In 2008, gold cost \$900 per ounce and the dollar had a rating of 75 on the US dollar index. In 2010, the cost of gold is \$1100 per ounce. What is the current rating of the dollar? (Round your answer to the nearest hundredth.)

25. Pizza: The circumference of a circle varies directly as the diameter. A circular pizza pie with a diameter of 1 foot has a circumference of 3.14 feet. What will be the circumference of a pizza pie with a diameter of 1.5 feet?

26. Pizza: The area of a circle varies directly as the square of its radius. A circular pizza pie with a radius of 6 in. has an area of 113.04 in.^{2} What will be the area of a pizza pie with a radius of 9 in.?

27. Triangles: Several triangles are to have the same area. In this set of triangles the height and base are inversely proportional. In one such triangle the height is 5 m and the base is 12 m. Find the height of the triangle in this set with a base of 10 m.

28. 232.3 pounds
29. 0.0073 cm
30. 0.008 in.
31. 16,000 lb
32. 1890 lb
33. 9×10^{-11} N
34. 3.6×10^{-10} N

28. Weight in space: If an astronaut weighs 250 pounds on the surface of the earth, what will the astronaut weigh 150 miles above the earth? Assume that the radius of the earth is 4000 miles, and round to the nearest tenth. (See Example 4.)

29. Elongation of a wire: The elongation E in a wire when a mass m is hung at its free end varies jointly as the mass and the length l of the wire and inversely as the cross-sectional area A of the wire. The elongation is 0.0055 cm when a mass of 120 g is attached to a wire 330 cm long with a cross-sectional area of 0.4 cm^2. Find the elongation if a mass of 160 g is attached to the same wire.

30. Elongation of a wire: When a mass of 240 oz is suspended by a wire 49 in. long whose cross-sectional area is 0.035 $in.^2$, the elongation of the wire is 0.016 in. Find the elongation if the same mass is suspended by a 28 in. wire of the same material with a cross-sectional area of 0.04 $in.^2$ (See Exercise 29.)

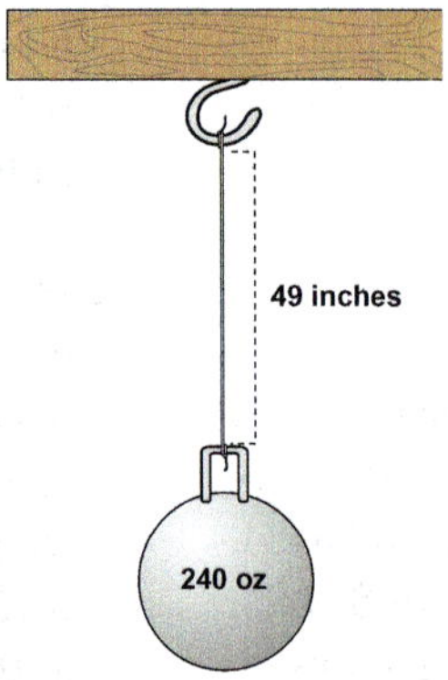

31. Safe load of a wooden beam: The safe load L of a wooden beam supported at both ends varies jointly as the width w and the square of the depth d and inversely as the length l. A 4 in. wide 6 in. deep beam 12 ft long supports a load of 4800 lb safely. What is the safe load of a beam of the same material that is 6 in. wide 10 in. deep and 15 ft long? (See Example 6c.)

32. Safe load of a wooden beam: A wooden beam 2 in. wide, 8 in. deep, and 14 ft long holds up to 2400 lb. What load would a beam 3 in. wide 6 in. deep and 15 ft long, of the same material, support?

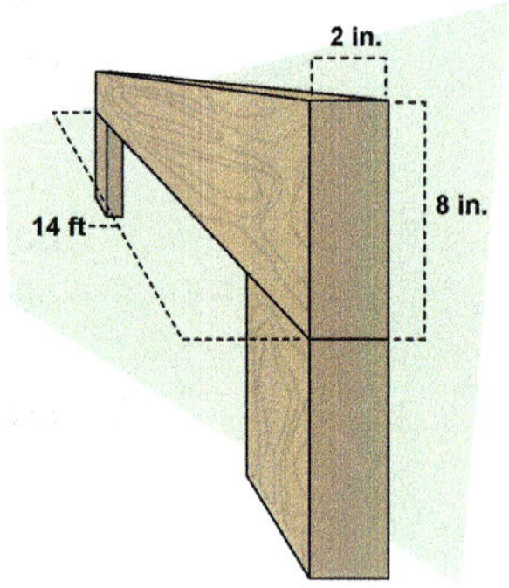

33. Gravitational force: The gravitational force of attraction, F, between two bodies varies directly as the product of their masses, m_1 and m_2, and inversely as the square of the distance, d, between them. The gravitational force between a 5-kg mass and a 2-kg mass 1 m apart is 1.5×10^{-10} N. Find the force between a 24-kg mass and a 9-kg mass that are 6 m apart. (N represents a unit of force called a newton.)

34. Gravitational force: In Exercise 33, what is the force if the distance between the 24-kg mass and the 9-kg mass is cut in half?

35. 15,000 lb
36. 25,200 lb
37. 6400 lb
38. 18,000 lb
39. 1.8 ft^3
40. 200 cm^3
41. 1700 g per $in.^2$
42. 330 bar

Lifting Force

The lifting force (or lift) L in pounds exerted by the atmosphere on the wings of an airplane is related to the area A of the wings in square feet and the speed (or velocity) v of the plane in miles per hour by the formula

$\boldsymbol{L = kAv^2}$, where k is the constant of variation.

35. If the lift is 9600 lb for a wing area of 120 ft^2 and a speed of 80 mph, find the lift of the same airplane at a speed of 100 mph.

36. The lift for a wing of area 280 ft^2 is 34,300 lb when the plane is traveling at 210 mph. What is the lift if the speed is decreased to 180 mph?

37. The lift for a wing with an area of 144 ft^2 is 10,000 lb when the plane is traveling at 150 mph. What is the lift if the speed is decreased to 120 mph?

38. A plane traveling 140 mph with wing area 195 ft^2 has 12,500 lb of lift exerted on the wings. Find the lift for the same plane traveling at 168 mph.

Pressure

Boyle's Law states that if the temperature of a gas sample remains the same, the pressure P of the gas is related to the volume V by the formula

$\boldsymbol{P = \frac{k}{V}}$, where k is the constant of proportionality.

39. A pressure of 1600 lb per ft^2 is exerted by 2 ft^3 of air in a cylinder. If a piston is pushed into the cylinder until the pressure is 1800 lb per ft^2, what will be the volume of the air? Round to the nearest tenth.

40. The volume of gas in a container is 300 cm^3 when the pressure on the gas is 20 g per cm^2. What will be the volume if the pressure is increased to 30 g per cm^2?

41. The pressure in a canister of gas is 1360 g per $in.^2$ when the volume of gas is 5 $in.^3$ If the volume is reduced to 4 $in.^3$, what is the pressure?

42. A scuba diver is using a diving tank that can hold 6 liters of air. If the tank has a pressure rating of 220 bar when full, what is the pressure rating when the volume of gas is 4 liters?

43. 15 ohms
44. 5.2 ohms
45. 2.56 ohms
46. 10 ohms
47. 900 lb
48. $10\frac{2}{3}$ ft from the 120 lb weight (or $1\frac{1}{3}$ ft from the 960 lb weight)
49. 5 ft from the 300 lb weight (or 20 feet from the 75 lb weight)
50. 216 kilograms

Electricity

The resistance, R (in ohms), in a wire is given by the formula

$$R = \frac{kL}{d^2},$$

where k is the constant of variation, L is the length of the wire and d is the diameter.

43. The resistance of a wire 500 ft long with a diameter of 0.01 in. is 20 ohms. What is the resistance of a wire 1500 ft long with a diameter of 0.02 in.?

44. The resistance is 2.6 ohms when the diameter of a wire is 0.02 in. and the wire is 10 ft long. Find the resistance of the same type of wire with a diameter of 0.01 in. and a length of 5 ft.

45. Tristan's car stereo uses a 5 ft audio wire with diameter 0.025 in. and resistance of 1.6 ohms. What is the resistance of 8 ft of the same type of audio wire?

46. Nicole purchased a spool of wire with diameter 0.01 in. for the speakers in her home audio system. If the resistance of 15 ft of this wire is 6 ohms, what is the resistance of 25 ft of the wire?

Levers

If a lever is balanced with weight on opposite sides of its balance point, then the following proportion exists:

$$\frac{W_1}{W_2} = \frac{L_2}{L_1} \text{ or } W_1L_1 = W_2L_2 \qquad \text{where } L_1 + L_2 = L, \text{ the total length of the lever.}$$

47. How much weight can be raised at one end of a bar 8 ft long by the downward force of 60 lb when the balance point is $\frac{1}{2}$ ft from the unknown weight?

48. Where should the balance point of a bar 12 ft long be located if a 120 lb force is to raise a load weighing 960 lb?

49. Find the location of the balance point of a 25 ft board that can raise a 300 lb package with a downward force of 75 lb.

50. How much weight can be raised on one end of a 17 meter board by 90 kilograms, if the balance point is 5 meters from the unknown weight?

Writing and Thinking About Mathematics

51. Explain, in your own words, the meaning of the terms

a. direct variation,
b. inverse variation,
c. joint variation, and
d. combined variation.

Discuss an example of each type of variation that you have observed in your daily life.

HAWKES LEARNING SYSTEMS: INTRODUCTORY & INTERMEDIATE ALGEBRA SOFTWARE

- 7.6 Applications: Variation

51. a. When two variables vary directly, an increase in the value of one variable indicates an increase in the other, and the ratio of the two quantities is constant.
b. When two variables vary inversely, an increase in the value of one variable indicates a decrease in the other, and the product of the two quantities is constant.
c. Joint variation is when a variable varies directly with more than one other variable.
d. Combined variation is when a variable varies directly or inversely with more than one variable.

Chapter 7 Index of Key Ideas and Terms

Section 7.1 Multiplication and Division with Rational Expressions

Section 7.2 Addition and Subtraction with Rational Expressions

Section 7.2 Addition and Subtraction with Rational Expressions (cont.)

To Find the LCM for a Set of Polynomials page 520

1. Completely factor each polynomial.
2. Form the product of all factors that appear, using each factor the most number of times it appears in any one polynomial.

Procedure for Adding (or Subtracting) Rational Expressions with Different Denominators page 520

1. Find the LCD (the LCM of the denominators).
2. Rewrite each fraction in an equivalent form with the LCD as the denominator.
3. Add (or subtract) the numerators and keep the common denominator.
4. Reduce if possible.

Placement of Negative Signs page 522

$$-\frac{P}{Q} = \frac{P}{-Q} = \frac{-P}{Q} \text{ where } Q \neq 0$$

Section 7.3 Complex Fractions

Complex Fractions pages 529-532

A **complex fraction** is a fraction in which the numerator and/or denominator are themselves fractions or the sum or difference of fractions.

To Simplify Complex Fractions

First Method page 529

1. Simplify the numerator so that it is a single rational expression.
2. Simplify the denominator so that it is a single rational expression.
3. Divide the numerator by the denominator and reduce to lowest terms.

Second Method page 531

1. Find the LCM of all the denominators in the numerator and denominator of the complex fraction.
2. Multiply both the numerator and denominator of the complex fraction by this LCM.
3. Simplify both the numerator and denominator and reduce to lowest terms.

Simplifying Complex Algebraic Expressions page 533

Section 7.4 Solving Equations with Rational Expressions

Ratio page 536

A **ratio** is a comparison of two numbers by division.

Proportion pages 536, 538

A **proportion** is an equation stating that two ratios are equal. One of the following conditions must be true:

1. The numerators agree in type and the denominators agree in type.
2. The numerators correspond and the denominators correspond.

Similar Figures (including Similar Triangles) page 539

Similar figures are figures that meet the following two conditions:

1. The measures of the corresponding angles are equal.
2. The lengths of the corresponding sides are proportional.

To Solve an Equation Containing Rational Expressions pages 540-541

1. Find the LCD of the fractions.
2. Multiply both sides of the equation by this LCD and simplify.
3. Solve the resulting equation.
4. Check each solution in the original equation. (Remember that no denominator can equal 0.)

Checking for Extraneous Solutions page 541

Section 7.5 Applications

Strategy for Solving Word Problems page 549

Applications

Number problems related to fractions page 549
Work problems page 550
Distance-Rate-Time problems page 553

Section 7.6 Variation

Direct Variation page 560

A variable quantity y **varies directly as** (or is **directly proportional to**) a variable x if there is a constant k such that

$$\frac{y}{x} = k \text{ or } y = kx.$$

The constant k is called the **constant of variation**.

Inverse Variation page 562

A variable quantity y **varies inversely as** (or is **inversely proportional to**) a variable x if there is a constant k such that

$$x \cdot y = k \text{ or } y = \frac{k}{x}.$$

The constant k is called the **constant of variation**.

Combined Variation pages 563-564

If a variable varies either directly or inversely with more than one other variable, the variation is said to be **combined variation**.

Joint Variation pages 563-564

If the combined variation is all direct variation (the variables are multiplied), then it is called **joint variation**.

HAWKES LEARNING SYSTEMS: INTRODUCTORY & INTERMEDIATE ALGEBRA SOFTWARE

- 7.1a Defining Rational Expressions
- 7.1b Multiplication and Division with Rational Expressions
- 7.2 Addition and Subtraction with Rational Expressions
- 7.3 Complex Fractions
- 7.4a Solving Equations: Ratios and Proportions
- 7.4b Solving Equations with Rational Expressions
- 7.5 Applications Involving Rational Expressions
- 7.6 Applications: Variation

Chapter 7 Review

7.1 Multiplication and Division with Rational Expressions

Reduce to lowest terms. State any restrictions on the variable(s).

1. $\frac{18xy^3}{96x^2y^3}$
2. $\frac{4x-12}{9-3x}$
3. $\frac{3x^2-75}{x^2-10x+25}$
4. $\frac{x^3+64}{x^2-4x+16}$
5. $\frac{6x^2+7x-5}{3x^2-4x-15}$
6. $\frac{2x^2+x-3}{2x^2+13x+15}$

Perform the indicated operations and reduce to lowest terms. Assume that no denominator has a value of 0.

7. $\frac{2x-6}{3x-9}\cdot\frac{2x}{2x-4}$
8. $\frac{25x^2-9}{15x^2-9x}\cdot\frac{6x^2}{5x-3}$
9. $\frac{24x^2y}{9xy^6}\div\frac{4x^4y}{3x^2y^3}$
10. $\frac{x-1}{7x+7}\div\frac{3x-3}{x^2+x}$
11. $\frac{x^2-8x+15}{x^2+5x-14}\div\frac{x^2-9}{x^2-49}$
12. $\frac{x-3}{x^2-3x-4}\div\frac{x^2-x-2}{3x+3}$
13. $\frac{x^2+2x-8}{4x^3}\cdot\frac{5x^2}{2x^2-5x+2}$
14. $\frac{x-1}{x^2+7x+6}\cdot\frac{x^2-4x}{x+4}\cdot\frac{x^2-16}{x^2-5x+4}$
15. $\frac{x^2+3x}{x^2+8x+12}\cdot\frac{2x^2+11x-6}{2x+6}\cdot\frac{x^2+4x+4}{2x^2+5x+2}$
16. $\frac{x^2+2x+3}{x^2+10x+21}\div\frac{2x^2-18}{x^2+14x+49}$

7.2 Addition and Subtraction with Rational Expressions

Perform the indicated operations and reduce if possible. Assume that no denominator has a value of 0.

17. $\frac{6x}{x+2}+\frac{12}{x+2}$
18. $\frac{8x}{x+6}+\frac{48}{x+6}$
19. $\frac{5}{x-2}+\frac{x}{x^2-4}$
20. $\frac{x+1}{x^2-10x-11}+\frac{x}{x-11}$
21. $\frac{4}{x+5}-\frac{2x}{x^2+3x-10}$
22. $\frac{x^2}{x^2+2x+1}-\frac{2x-1}{x^2-2x+1}$
23. $\frac{x}{3x^2+4x+1}+\frac{2x+1}{2x^2+5x+3}$
24. $\frac{6}{2x-3}-\frac{5}{3-2x}$
25. $\frac{6x+24}{x^2+8x+16}+\frac{3x-12}{x^2-16}$
26. $\frac{5x-20}{x^2+x-20}-\frac{x^2+5x}{x^2+9x+20}$
27. $\frac{x+2}{x^2+x+1}-\frac{x-2}{x^3-1}$
28. $\frac{3x+2}{x^2+2x-3}-\frac{x}{3x^2-2x-1}$

Answers

1. $\frac{3}{16x}; x\neq 0, y\neq 0$
2. $-\frac{4}{3}; x\neq 3$
3. $\frac{3x+15}{x-5}; x\neq 5$
4. $x+4$
5. $\frac{2x-1}{x-3};\ x\neq -\frac{5}{3}, 3$
6. $\frac{x-1}{x+5};\ x\neq -5, -\frac{3}{2}$
7. $\frac{2x}{3(x-2)}$
8. $\frac{10x^2+6x}{5x-3}$
9. $\frac{2}{xy^3}$
10. $\frac{x}{21}$
11. $\frac{x^2-12x+35}{(x-2)(x+3)}$
12. $\frac{3x-9}{(x-4)(x+1)(x-2)}$
13. $\frac{5x+20}{4x(2x-1)}$
14. $\frac{x^2-4x}{(x+1)(x+6)}$
15. $\frac{2x^2-x}{2(2x+1)}$
16. $\frac{x^3+9x^2+17x+21}{2(x-3)(x+3)^2}$
17. 6
18. 8
19. $\frac{6x+10}{(x+2)(x-2)}$
20. $\frac{x+1}{x-11}$
21. $\frac{2x-8}{(x+5)(x-2)}$
22. $\frac{x^4-4x^3-2x^2+1}{(x-1)^2(x+1)^2}$

23. $\dfrac{8x^2+8x+1}{(x+1)(3x+1)(2x+3)}$

24. $\dfrac{11}{2x-3}$

25. $\dfrac{9}{x+4}$

26. $\dfrac{20-x^2}{(x+4)(x+5)}$

27. $\dfrac{x^2}{(x-1)(x^2+x+1)}$

28. $\dfrac{8x^2+6x+2}{(x-1)(x+3)(3x+1)}$

29. $\dfrac{14y^2}{5x}$

30. $\dfrac{13}{2(5y-1)}$

31. $\dfrac{y+2}{4}$

32. $\dfrac{-2}{(x+2)^2}$

33. $\dfrac{x-2}{x}$

34. $\dfrac{x}{x^2+x+1}$

35. $\dfrac{2x-2}{x^2}$

36. $\dfrac{4x-12}{x^2-6x-9}$

37. $\dfrac{x}{x-5}$

38. $\dfrac{-2x+10}{x+4}$

39. $-\dfrac{9}{x+2}$

40. $\dfrac{15}{2x}$

41. $\dfrac{20}{3y+2}$

42. $\dfrac{12y+24}{(y+3)^2(y-3)}$

43. $x \neq -3, 3; x = 5$

7.3 Complex Fractions

Simplify the following complex fractions.

29. $\dfrac{\frac{8x^3}{5y}}{\frac{4x^4}{7y^3}}$ **30.** $\dfrac{\frac{4}{y}+\frac{5}{2y}}{5-\frac{1}{y}}$ **31.** $\dfrac{\frac{1}{4}-\frac{1}{y^2}}{\frac{1}{y}-\frac{2}{y^2}}$ **32.** $\dfrac{\frac{1}{x+2}-\frac{1}{x}}{1+\frac{2}{x}}$

33. $\dfrac{1-4x^{-2}}{1+2x^{-1}}$ **34.** $\dfrac{x^{-1}-x^{-2}}{1-x^{-3}}$ **35.** $\dfrac{\frac{2}{x}}{1+\frac{1}{x-1}}$ **36.** $\dfrac{2-\frac{6}{x}}{\frac{x^2-6x-9}{2x}}$

37. $\dfrac{\frac{x^2-25}{x^2-10x+25}}{1+\frac{5}{x}}$ **38.** $\dfrac{\frac{3}{x+2}-\frac{x+4}{x^2-4}}{\frac{x}{x^2-4}-\frac{2}{x-2}}$

Write each of the expressions as a single fraction reduced to lowest terms.

39. $\dfrac{1}{x+2}-\dfrac{5}{2x}\cdot\dfrac{4x}{x+2}$ **40.** $\dfrac{7}{x}+\dfrac{5}{x^2-3x}\cdot\dfrac{x-3}{10}$

41. $\left(\dfrac{4}{y}-\dfrac{2}{3y}\right)\div\dfrac{3y+2}{6y}$ **42.** $\left(\dfrac{4}{y}+\dfrac{2}{y+3}\right)\div\dfrac{y^2-9}{2y}$

7.4 Solving Equations with Rational Expressions

State any restrictions on x, and then solve the proportions.

43. $\dfrac{4}{x+3}=\dfrac{1}{x-3}$ **44.** $\dfrac{x-4}{3x}=\dfrac{2}{x-3}$

45. Antifreeze: In a mixture, there are 3 parts antifreeze for each 8 parts water. If you want a total of 220 gallons of mixture, how many gallons of antifreeze are needed?

46. Classroom size: In an English class the ratio of men to women is 4 to 5. If there are 36 students in the class, how many women are in the class?

47. Baseball: Louis is averaging 13 hits for every 50 times at bat. If he maintains this average, how many "at bats" will he need in order to get 156 hits?

48. Shadows: A flagpole casts a 30 foot shadow at the same time as a $5\frac{1}{2}$ foot woman casts a $4\frac{1}{8}$ foot shadow. How tall is the flagpole?

44. $x \neq 0, 3; x = 1, 12$
45. 60 gallons
46. 20 women
47. 600 times
48. 40 ft
49. $\overline{AB} = 8, \overline{FE} = 5$
50. $\overline{TU} = 6, \overline{VW} = 7$
51. $x \neq 0; x = -\frac{4}{25}$
52. $x \neq -\frac{2}{3}, \frac{4}{7}; x = \frac{8}{53}$
53. $x \neq 0, 1; x = \frac{5}{2}$
54. $x \neq -3, 3; x = 12$
55. $x \neq -3, 4; x = 0, \frac{25}{6}$
56. $x \neq -6; x = -\frac{15}{2}, 4$
57. $x_1 = x - \frac{y - y_1}{m}$
58. $a_1 = \frac{a_n(1-r)}{1-r^n}$
59. 18, 27
60. 15 women
61. $\frac{6}{8}$
62. $\frac{17}{7}$
63. 21, 27
64. 3 hours; 6 hours

The following exercises show pairs of similar triangles. Find the lengths of the sides labeled with variables.

49. $\Delta ABC \sim \Delta DEF$

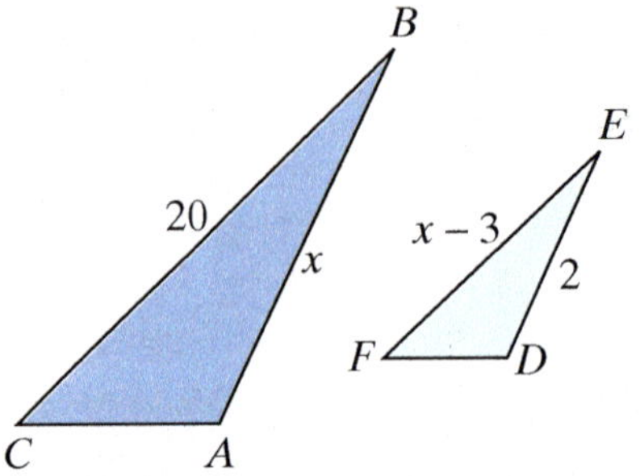

50. $\Delta STU \sim \Delta VWU$

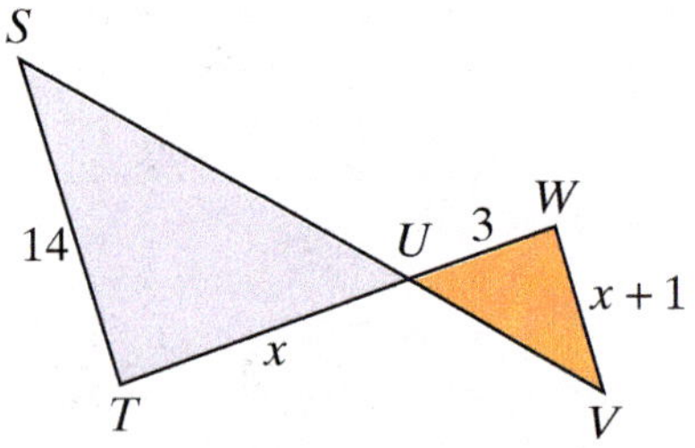

State any restrictions on x and then solve the equations.

51. $\frac{2}{5x} + \frac{3}{8} = \frac{3}{10x} - \frac{1}{4}$

52. $\frac{6}{7x-4} = \frac{-5}{3x+2}$

53. $\frac{5}{x} + \frac{2}{x-1} = \frac{2x}{x-1}$

54. $\frac{9}{x^2-9} + \frac{x}{3x+9} = \frac{1}{3}$

55. $\frac{x}{x-4} - \frac{6x}{x^2-x-12} = \frac{7x}{x+3}$

56. $\frac{2x+7}{15} - \frac{x+2}{x+6} = \frac{2x-6}{5}$

Solve each of the formulas for the specified variable.

57. $\frac{y-y_1}{x-x_1} = m$; solve for x_1

58. $a_n = \frac{a_1(1-r^n)}{1-r}$; solve for a_1

7.5 Applications

59. The ratio of two numbers is 2 to 3. If the sum of the numbers is 45, find the numbers.

60. Education: In a statistics class the ratio of men to women is 6 to 5. If there are 33 students in the class, how many women are in the class?

61. The numerator of a fraction is two less than the denominator. If both the numerator and denominator are decreased by 2, the new fraction will reduce to $\frac{2}{3}$. What was the original fraction? (Do not reduce answer.)

62. The numerator of a fraction is three more than twice the denominator. If both numerator and denominator are increased by five, the simplified result is $\frac{11}{6}$. Find the original fraction.

63. The ratio of two numbers is 7 to 9. If one of the numbers is 6 more than the other, what are the numbers?

64. Swimming pools: One pipe can fill a swimming pool twice as fast as another. When both pipes are used, the pool can be filled in 2 hours. How long would it take each pipe alone to fill the pool?

65. 12 hours
66. Alice takes 6 hrs; Judy takes 4 hrs
67. Tyler takes 2 hrs; His son takes 6 hrs
68. 6 mph
69. 24 mph
70. Initial rate was 37.5 mph and final rate was 62.5 mph
71. Initial rate was 400 mph and final rate was 430 mph
72. Raphael takes 1.5 hrs; Matilde takes 3 hrs
73. 6
74. 5
75. $\frac{490}{9}$
76. 31,250
77. $\frac{100}{7}$ in.

65. Construction: Working together, a bricklayer and his partner can lay a brick sidewalk in 4.8 hours. If he could have completed the job in 8 hours working alone, how long would his partner have needed to complete the job working alone?

66. Accounting: Working together, Alice and Judy can prepare an account report in 2.4 hours. Working alone, Alice would need two hours longer to prepare the same report than Judy would. How long would it take each of them working alone?

67. Painting a room: It takes Tyler four hours less to paint a room than his son would take to paint the same sized room. If they can paint this size room in $1\frac{1}{2}$ hours working together, how long would each of them take to paint such a room alone?

68. Biking: Garrett rides his bike to the top of Mount High, a distance of 12 miles. He then rides back down the mountain four times as fast as he rode up the mountain. If the total ride took 2.5 hours, find the rate at which he rode up the mountain.

69. Boating: A boat can travel 10 miles downstream in the same time it travels 6 miles upstream. If the rate of the current is 6 mph, find the speed of the boat in still water.

70. Car speed: Mr. Lukin had a golf game scheduled with a friend 80 miles from his home. After traveling 30 miles in heavy traffic, he needed to increase his speed by 25 mph to make the scheduled tee off time. If he traveled the same length of time at each rate, find the rates.

71. Plane speed: After flying 760 miles, the pilot increased the speed by 30 mph and continued on for another 817 miles. If the same amount of time was spent at each rate, find the rates.

72. Research projects: It takes Matilde twice as long to complete a certain type of research project as it takes Raphael to do the same type of project. They found that if they worked together they could finish similar projects in 1 hour. How long would it take each of them working alone to finish this type of project?

7.6 Variation

73. If y varies directly as x and $y = 4$ when $x = 20$, find y if $x = 30$.

74. If y is inversely proportional to x and $y = 8$ when $x = 10$, find y when $x = 16$.

75. If y is directly proportional to the square of x, and $y = 10$ when x is 3, what is y when x is 7?

76. If y is inversely proportional to x^4, and $y = 50$ when $x = \frac{1}{2}$, what is y when $x = \frac{1}{10}$?

77. Springs: The length a hanging spring stretches varies directly with the weight placed on the end. If a spring stretches 10 in. when a weight of 14 lb is placed at its end, how far will the spring stretch if the weight is increased to 20 lb?

78. 18.84 in.
79. 113.04 in.^2
80. 162 rpm
81. 576 feet
82. 16.5 ohms
83. 37.68 cubic feet
84. 5 cups of lemon juice

78. Pizza: The circumference of a circle varies directly as the diameter of the circle. A circular pizza with a diameter of 4 in. has a circumference of 12.56 in. What will be the circumference of a pizza with a diameter of 6 in.?

79. Pizza: The area of a circle varies directly as the square of its radius. A circular pizza with a radius of 4 in. has an area of 50.24 in.^2 What is the area of a pizza with a radius of 6 in.?

80. Gears: A gear that has 48 teeth is meshed with a gear that has 16 teeth. Find the speed (in revolutions per minute) of the small gear if the large one is rotating at 54 rpm. (**Hint:** $T_1R_1 = T_2R_2$)

81. Free falling object: The distance a free falling object falls is directly proportional to the square of the time it falls (before it hits the ground). If an object fell 256 feet in 4 seconds, how far will it have fallen by the end of 6 seconds?

82. Resistance in a wire: The resistance, R (in ohms), in a wire is given by the formula $R = \frac{kL}{d^2}$ where k is the constant of variation, L is the length of the wire and d is the diameter of the wire. The resistance of a wire 600 feet long with a diameter of 0.01 in. is 22 ohms. What is the resistance of the same type of wire 1800 feet long with a diameter of 0.02 in.?

83. Volume of a circular cylinder: The volume of a circular cylinder varies jointly as its height and the square of its radius. If a cylinder with $r = 2$ feet and $h = 4$ feet holds 50.24 cubic feet of water, what is the volume of water in the cylinder when $h = 3$ feet?

84. Selling lemonade: Billy realizes that the sale of his homemade lemonade varies directly with the amount of sugar squared and inversely with the amount of lemon juice in the recipe. If he sells 54 glasses when he uses 9 cups of sugar and 6 cups of lemon juice how much lemon juice does he use when he uses 10 cups of sugar and sells 80 glasses of lemonade?

Chapter 7 Test

Reduce to lowest terms. State any restrictions on the variable(s).

1. $\dfrac{x^2+3x}{x^2+7x+12}$ **2.** $\dfrac{2x+5}{4x^2+20x+25}$

Perform the indicated operations and reduce the answers to lowest terms. Assume that no denominator has a value of 0.

3. $\dfrac{x+3}{x^2+3x-4}\cdot\dfrac{x^2+x-2}{x+2}$ **4.** $\dfrac{6x^2-x-2}{12x^2+5x-2}\div\dfrac{4x^2-1}{8x^2-6x+1}$

5. $\dfrac{x}{x^2+3x-10}+\dfrac{3x}{4-x^2}$ **6.** $\dfrac{x-4}{3x^2+5x+2}-\dfrac{x-1}{x^2-3x-4}$

7. $\dfrac{x^2-16}{x^2-4x}\cdot\dfrac{x^2}{x+4}\div\dfrac{x-1}{2x^2-2x}$ **8.** $\dfrac{x}{x+3}-\dfrac{x+1}{x-3}+\dfrac{x^2+4}{x^2-9}$

Simplify the complex fractions.

9. $\dfrac{\frac{5y}{3x^2}}{\frac{10y^4}{9x}}$ **10.** $\dfrac{\frac{4}{3x}+\frac{1}{6x}}{\frac{1}{x^2}-\frac{1}{2x}}$ **11.** $\dfrac{x^{-1}-y^{-1}}{x-y}$ **12.** $\dfrac{\frac{4}{x}+\frac{2}{x+3}}{\frac{x^2-9}{2x}}$

13. Simplify the expression in part **a.** and solve the equation in part **b.**

a. $\dfrac{3}{x}-\dfrac{2}{x+1}+\dfrac{5}{2x}$ **b.** $\dfrac{3}{x}-\dfrac{2}{x+1}=\dfrac{5}{2x}$

State any restrictions on x and then solve the equations.

14. $\dfrac{x-1}{x+4}=\dfrac{4}{5}$ **15.** $\dfrac{4}{7}-\dfrac{1}{2x}=1+\dfrac{1}{x}$

16. $\dfrac{4}{x+4}+\dfrac{3}{x-1}=\dfrac{1}{x^2+3x-4}$ **17.** $\dfrac{x}{x+2}=\dfrac{1}{x+1}-\dfrac{x}{x^2+3x+2}$

Solve each of the formulas for the specified variables.

18. $S=\dfrac{n(a_1+a_n)}{2}$; solve for n **19.** $y=mx+b$; solve for x

---

1. $\dfrac{x}{x+4}$; $x\neq-4,-3$
2. $\dfrac{1}{2x+5}$; $x\neq-\dfrac{5}{2}$
3. $\dfrac{x+3}{x+4}$
4. $\dfrac{3x-2}{3x+2}$
5. $\dfrac{-2x^2-13x}{(x+5)(x+2)(x-2)}$
6. $\dfrac{-2x^2-7x+18}{(3x+2)(x-4)(x+1)}$
7. $2x^2$
8. $\dfrac{x^2-7x+1}{(x+3)(x-3)}$
9. $\dfrac{3}{2xy^3}$
10. $\dfrac{-3x}{x-2}$
11. $-\dfrac{1}{xy}$
12. $\dfrac{12x+24}{(x-3)(x+3)^2}$
13. a. $\dfrac{7x+11}{2x(x+1)}$
    b. $x=\dfrac{1}{3}$
14. $x\neq-4$; $x=21$
15. $x\neq0$; $x=-\dfrac{7}{2}$
16. $x\neq-4,1$; $x=-1$
17. $x\neq-2,-1$; $x=1$
18. $n=\dfrac{2S}{a_1+a_n}$
19. $x=\dfrac{y-b}{m}$

20. $\overline{AC} = 36$, $\overline{DC} = 24$
21. $\frac{2}{7}$
22. 4 hours
23. Carlos: 42 mph; Mario: 57 mph
24. 11 mph
25. $z = 24$
26. $z = \frac{400}{9}$
27. $\frac{15}{2}$ cm
28. 5.13 in.^{3}

20. Triangles ABC and DEC are similar. (Symbolically, $\Delta ABC \sim \Delta DEC$.) Find the lengths of sides AC and DC.

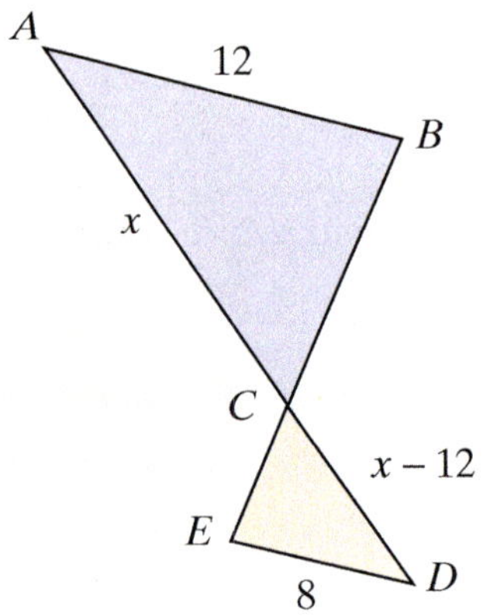

21. The denominator of a fraction is three more than twice the numerator. If eight is added to both the numerator and the denominator, the resulting fraction is equal to $\frac{2}{3}$. Find the original fraction.

22. House cleaning: Sonya can clean the apartment in 6 hours. It takes Lucy 12 hours to clean it. If they work together, how long will it take them?

23. Travel: Mario can travel 228 miles in the same time that Carlos travels 168 miles. If Mario's speed is 15 mph faster than Carlos', find their rates.

24. Boating: Bob travels 4 miles upstream. In the same time, he could have traveled 7 miles downstream. If the speed of the current is 3 mph, find the speed of the boat in still water.

25. z varies directly as x^3. If $z = 3$ when $x = 2$, find z if $x = 4$.

26. z varies directly as x^2 and inversely as $\sqrt{y}$. If $z = 24$ when $x = 3$ and $y = 4$, find z if $x = 5$ and $y = 9$.

27. Stretching a spring: Hooke's Law states that the distance a spring will stretch vertically is directly proportional to the weight placed at its end. If a particular spring will stretch 5 cm when a weight of 4 g is placed at its end, how far will the spring stretch if a weight of 6 g is placed at its end?

28. Candy: The volume of a gumball is directly proportional to the radius cubed. A gumball with a diameter of 1.75 inches has an approximate volume of 2.805 inches cubed. What is the volume of a gumball that has a diameter of 2.14 inches? (Round your answer to two decimal places.)

Cumulative Review: Chapters 1 – 7

1. 

2. a. $\frac{19}{12}$ b. $\frac{-3}{14}$
c. $\frac{1}{4}$ d. $\frac{1}{12}$

3. 5

4. $3x^2-5x-28$

5. $4x^2-20x+25$

6. $6x-23$

7. $10x+4$

8. 

$(6,\infty)$

9. a. 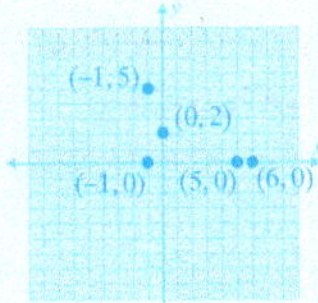

b. $D=\{-1,0,5,6\}$
c. $R=\{0,5,2\}$
d. It is not a function because the x-coordinate -1 has more than one corresponding y-coordinate.

10. a. 8 b. –1 c. $-\frac{59}{27}$

11. 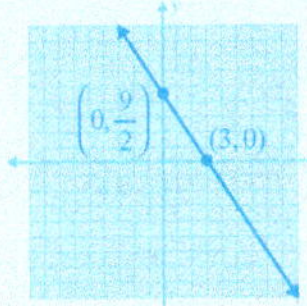

12. $y=\frac{3}{4}x+\frac{11}{2}$;
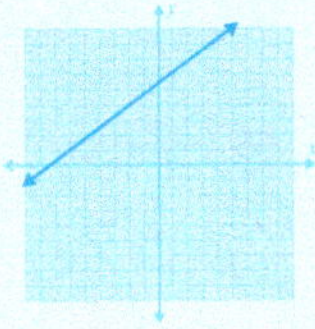

1. Graph the set of integers less than –5 on a number line.

2. Perform the indicated operations and reduce.

a. $\frac{5}{6}+\frac{3}{4}$ **b.** $\frac{1}{2}-\frac{5}{7}$ **c.** $\frac{1}{2}\cdot\frac{1}{2}$ **d.** $\frac{7}{32}\div\frac{21}{8}$

3. Find 12.5% of 40.

Perform the indicated operations and simplify the expressions.

4. $(3x+7)(x-4)$ **5.** $(2x-5)^2$

6. $-4(x+2)+5(2x-3)$ **7.** $x(x+7)-(x+1)(x-4)$

8. Solve the inequality $\frac{x+5}{2}<\frac{3x}{4}+1$ and graph the solution set. Write the solution using interval notation. Assume that x is a real number.

9. Given the relation $r=\{(-1,5),(0,2),(-1,0),(5,0),(6,0)\}$.

a. Graph the relation.
b. What is the domain of the relation?
c. What is the range of the relation?
d. Is the relation a function? Explain.

10. For the function $f(x)=x^3-2x^2-1$, find: **a.** $f(3)$ **b.** $f(0)$ **c.** $f\left(-\frac{2}{3}\right)$

11. Graph the equation $6x+4y=18$ by locating the x-intercept and the y-intercept.

12. Find the equation in slope-intercept form of the line that has slope $\frac{3}{4}$ and passes through the point $(-2,4)$. Graph the line.

13. Find the equation in slope-intercept form of the line passing through the two points $(-5,-1)$ and $(4,2)$. Graph the line.

14. Find the equation of the line passing through the point $(-3,7)$ with slope undefined. Graph the line.

15. Find the equation in standard form of the line perpendicular to the line $x-2y=3$ and passing through the point $(-4,0)$. Graph both lines.

13. $y = \frac{1}{3}x + \frac{2}{3}$;

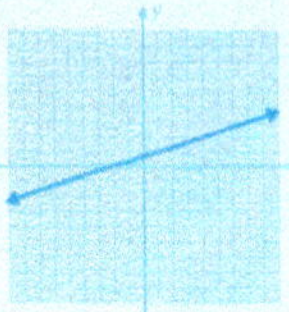

14. $x = -3$;

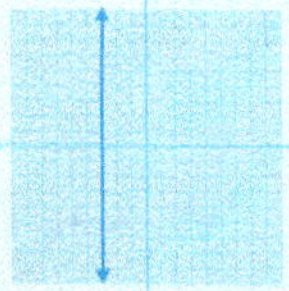

15. $2x + y = -8$;

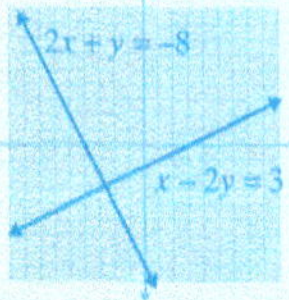

16. Function;
$D = \{-4, -2, 0, 2, 4\}$;
$R = \{-2\}$

17. Function;
$D = (-\infty, \infty)$;
$R = (-\infty, \infty)$

18.

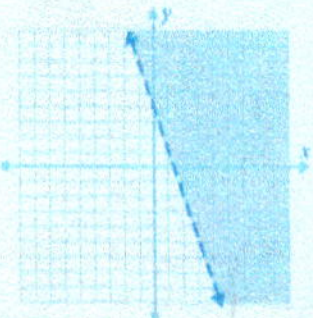

19. $8x^2$

20. $5x^3y^2$

21. $(3x + y)(y + 1)$

22. $(2x - 5)(2x + 3)$

23. $(3x - 2)(2x - 1)$

24. $(2x + 5)(2x + 3)$

25. $2(x + 5)(x - 2)$

26. $2x(3x + 1)(x - 4)$

27. $(3x^3 + 2y)(3x^3 - 2y)$

28. $(2x + 5)(4x^2 - 10x + 25)$

29. $x^2 + 3x - 18 = 0$

Use the vertical line test to determine whether or not each graph represents a function. State the domain and range.

16.

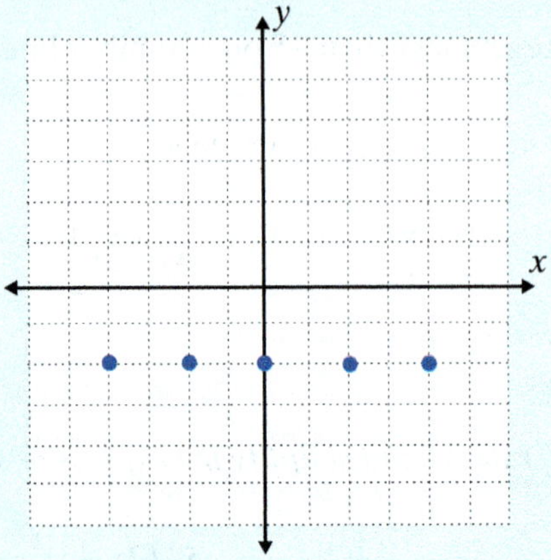

17.

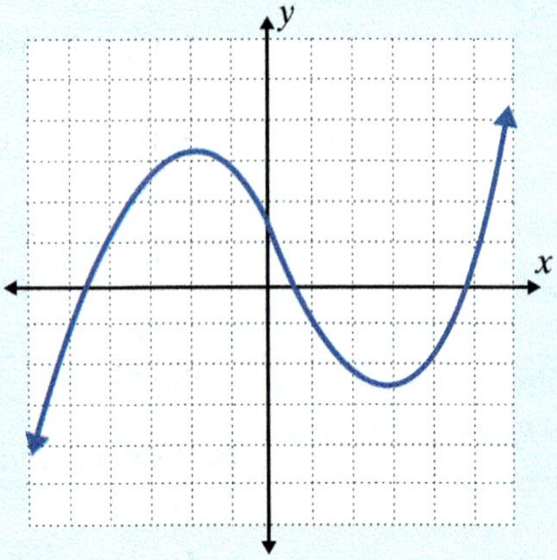

18. Graph the solution to the following inequality: $3x + y > 4$.

Simplify each expression. The final form of the expressions with variables should contain only positive exponents. Assume that all variables represent nonzero numbers.

19. $\frac{32x^3y^2}{4xy^2}$

20. $\frac{15xy^{-1}}{3x^{-2}y^{-3}}$

Completely factor each polynomial.

21. $3xy + y^2 + 3x + y$

22. $4x^2 - 4x - 15$

23. $6x^2 - 7x + 2$

24. $4x^2 + 16x + 15$

25. $2x^2 + 6x - 20$

26. $6x^3 - 22x^2 - 8x$

27. $9x^6 - 4y^2$

28. $8x^3 + 125$

29. Find an equation that has $x = 3$ and $x = -6$ as roots.

30. Find an equation that has $x = -1$, $x = 0$ and $x = -3$ as roots.

Express each quotient as a sum (or difference) of fractions and simplify if possible.

31. $\frac{3x^3y + 10x^2y^2 + 5xy^3}{5x^2y}$

32. $\frac{7x^2y^2 - 21x^2y^3 + 8x^2y^4}{7x^2y^2}$

Divide by using the long division algorithm.

33. $(x^2 - 14x - 16) \div (x + 1)$

34. $\frac{2x^3 + 5x^2 + 7}{x + 3}$

30. $x^3 + 4x^2 + 3x = 0$
31. $\frac{3}{5}x + 2y + \frac{y^2}{x}$
32. $1 - 3y + \frac{8}{7}y^2$
33. $x - 15 - \frac{1}{x+1}$
34. $2x^2 - x + 3 - \frac{2}{x+3}$
35. $\frac{4x^2}{x-3}; x \neq -\frac{1}{2}, 3$
36. $x - 3; x \neq -3$
37. $\frac{1}{x+1}; x \neq -1, 0$
38. $\frac{x+5}{2(x-3)}; x \neq 3$
39. $x - y$
40. $\frac{x}{3(x+1)}$
41. $\frac{x+4}{3x}$
42. $\frac{3x^3+12x^2+12x}{x+3}$
43. $\frac{2x+1}{x-1}$
44. $\frac{2x^2+14x-8}{(x+3)(x-1)(x-2)}$
45. $\frac{2x+4}{(x-1)(x+4)}$
46. $\frac{x-4}{(x+2)(x-2)}$
47. $\frac{19}{14}$
48. $\frac{x+4}{6x}$
49. $x = -\frac{3}{5}$
50. $x = -5, 6$
51. $x = -\frac{5}{2}$
52. $x = -4, 0, 5$
53. $x = 2, 5$

Reduce the rational expressions to lowest terms and indicate any restrictions on the variable.

35. $\frac{8x^3+4x^2}{2x^2-5x-3}$ **36.** $\frac{x^2-9}{x+3}$ **37.** $\frac{x}{x^2+x}$ **38.** $\frac{x^2+2x-15}{2x^2-12x+18}$

Perform the indicated operations and simplify. Assume that no denominator is 0.

39. $\frac{x^2}{x+y} - \frac{y^2}{x+y}$ **40.** $\frac{4x}{3x+3} - \frac{x}{x+1}$ **41.** $\frac{4x}{x-4} \div \frac{12x^2}{x^2-16}$

42. $\frac{x^2+3x+2}{x+3} \cdot \frac{3x^2+6x}{x+1}$ **43.** $\frac{2x+1}{x^2+5x-6} \cdot \frac{x^2+6x}{x}$ **44.** $\frac{8}{x^2+x-6} + \frac{2x}{x^2-3x+2}$

45. $\frac{x}{x^2+3x-4} + \frac{x+1}{x^2-1}$ **46.** $\frac{x+1}{x^2+4x+4} \div \frac{x^2-x-2}{x^2-2x-8}$

Simplify the complex algebraic fractions.

47. $\dfrac{\frac{3}{x}+\frac{1}{6x}}{\frac{7}{3x}}$ **48.** $\dfrac{\frac{1}{4x}+\frac{1}{x^2}}{\frac{1}{2x}+\frac{1}{x}}$

Solve each of the equations.

49. $4(x+2) - 7 = -2(3x+1) - 3$ **50.** $(x+4)(x-5) = 10$

51. $4x^2 + 20x + 25 = 0$ **52.** $x^3 - x^2 = 20x$ **53.** $0 = x^2 - 7x + 10$

54. $\left|x - \frac{9}{2}\right| = |4x+6|$ **55.** $\frac{7}{2x-1} = \frac{3}{x+6}$

56. a. Simplify the following expression: $\frac{3}{x} - \frac{5}{x+3}$.

b. Solve the following equation: $\frac{3}{x} = \frac{5}{x+3}$.

57. a. Simplify the following expression: $\frac{3x}{x+2} + \frac{2}{x-4} + 3$.

b. Solve the following equation: $\frac{3x}{x+2} + \frac{2}{x-4} = 3$.

58. In the formula $A = P + Prt$, solve for t.

54. $x = -\frac{7}{2}, -\frac{3}{10}$

55. $x = -45$

56. a. $\frac{9-2x}{x(x+3)}$

b. $x = \frac{9}{2}$

57. a. $\frac{6x^2-16x-20}{(x-4)(x+2)}$

b. $x = 7$

58. $t = \frac{A-P}{Pr}$

59. $P = 30$ in.; $A = 40$ in.^{2}

60. $\frac{1}{2}$ year

61. She can spend anywhere from 9 to 27 dollars.

62. a. and b.

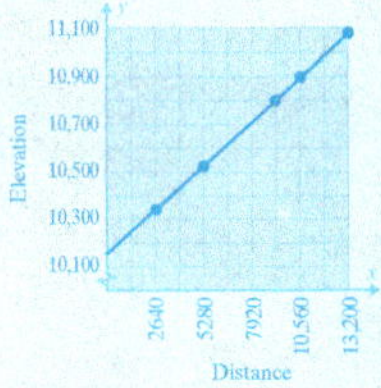

c. Slope = 0.07 for all

d. For every 100 feet the hikers travel there is a 7 ft change in elevation

63. −6, −5, −4 or 7, 8, 9

64. Train's speed is 60 miles per hour; Speed of airplane is 230 mph

65. The father takes $2\frac{2}{3}$ hours and the daughter takes 8 hours

66. 3 mph

67. $7\frac{9}{13}$ cm

68. 5.29 minutes

59. Find **a.** the perimeter P and **b.** the area A of the figure below.

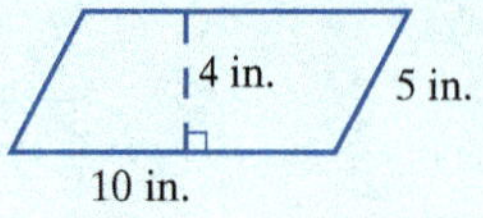

60. Investing: How long will it take for an investment of \$600 at a rate of 5% to be worth \$615? (**Hint:** $I = Prt$)

61. Buying dinner: Kelly wants to buy dinner for her five friends but she wants to stay within a budget of \$17-\$20 per person including a \$20 tip for the server. If her friends order meals that cost \$10, \$14, \$16, \$13, and \$20, what range of money can she spend on her dinner and still stay within her budget?

62. Hiking: A group of hikers decide to hike a trail on Mount Massive in Colorado. They start at the base of the mountain which is 10,152 ft above sea level. Periodically, they see signs that tell them how high and how far they have gone.

Distance	Elevation
2640	10,336.8
5280	10,521.6
9240	10,798.8
10,560	10,891.2
13,200	11,076

a. Plot these points on a graph.
b. Connect the points with line segments.
c. Find the slope of each line segment.
d. Interpret the slope as a rate of change.

63. Find three consecutive integers such that half the product of the two smallest integers is equal to 19 plus the largest.

64. Airplanes: An airplane can travel 1035 miles in the same time that a train travels 270 miles. The speed of the plane is 50 mph more than three times the speed of the train. Find the speed of each.

65. Waxing a car: A man can wax his car three times as fast as his daughter can. Together they can complete the job in 2 hours. How long would it take each of them working alone?

66. Boating: A family travels 18 miles downriver and returns. It takes 8 hours to make the round trip. Their rate in still water is twice the rate of the current. Find the rate of the current.

67. Springs: The weight on a spring varies directly as the length the spring stretches. If a hanging spring stretches 5 cm when a weight of 13 g is placed at its end, how far will the spring stretch if a weight of 20 g is placed at its end?

68. Skydiving: The Airborne Skydiving Company uses rectangular parachutes for their jumps. Assuming that each jumper releases his/her parachute at 2500 feet as instructed, the time it then takes them to reach the ground from that point is jointly proportional to the length and width of the parachute. If it takes 4 minutes for someone to reach the ground using a parachute with a length of 5.2 feet and a width of 3.5 feet, how long will it take for them to reach the ground with a parachute of length 5.6 feet and width 4.3 feet? (Round your answer to the nearest hundredth.)

CHAPTER 8

Systems of Linear Equations

Did You Know?

The subject of systems of linear equations received a great deal of attention in nineteenth-century mathematics. However, problems of this type are very old in the history of mathematics. The solution of simultaneous systems of equations was well known in China and Japan. The great Japanese mathematician Seki Shinsuku Kōwa (1642 – 1708) wrote a book on the subject in 1683 that was well in advance of European work on the subject.

Seki Kōwa is probably the most distinguished of all Japanese mathematicians. Born into a samurai family, he showed great mathematical talent at an early age. He is credited with the independent development of calculus and is sometimes referred to as "the Newton of Japan." There is a traditional story that Seki Kōwa made a journey to the Buddhist shrines at Nara. There, ancient Chinese mathematical manuscripts were preserved, and Seki is supposed to have spent three years learning the contents of the manuscripts that previously no one had been able to understand. As was the custom, much of Seki Kōwa's work was done through the solution of very intricate problems. In 1907, the Emperor of Japan presented a posthumous award to the memory of Seki Kōwa, who did so much to awaken interest in scientific and mathematical research in Japan.

Native Japanese mathematics (the **wasan**) flourished until the nineteenth century when western mathematics and notation were completely adopted. In the 1940's, solution of large systems of equations became a part of the new branch of mathematics called **operations research**. Operations research is concerned with deciding how to best design and operate man-machine systems, usually under conditions requiring the allocation of scarce resources. Although this new science initially had only military applications, recent applications have been in the areas of business, industry, transportation, meteorology, and ecology. Computers now make it possible to solve extremely large systems of linear equations that are used to model the system being studied. Although computers do much of the work involved in solving systems of equations, it is necessary for you to understand the principles involved by studying small systems involving two unknowns as presented in this chapter.

"The advancement and perfection of mathematics are intimately connected with the prosperity of the state."

Napoleon Bonaparte (1769 – 1821)

Many applications involve two (or more) quantities and, by using two (or more) variables, we can form linear equations using the information given. Such a set of equations is called a **system**, and in this chapter we will develop techniques for solving systems of linear equations.

Graphing systems of two equations in two variables is helpful in visualizing the relationships between the equations. However, this approach is somewhat limited in finding solutions since numbers might be quite large, or solutions might involve fractions or decimals that must be estimated on the graph. Therefore, algebraic techniques are necessary to accurately solve systems of linear equations. Graphing in three dimensions will be left to later courses.

Two ideas probably new to you are matrices and determinants. Matrices and determinants provide powerful general approaches to solving large systems of equations with many variables. Discussions in this chapter will be restricted to two linear equations in two variables and three linear equations in three variables.

8.1 Systems of Linear Equations: Solutions by Graphing

- *Determine if given points lie on both lines in a specified system of equations.*
- *Estimate, by graphing, the coordinates of the intersection of a system of linear equations with one solution.*
- *Use a graphing calculator to solve a system of linear equations.*

Systems of Linear Equations

Many applications, as we will see in Sections 8.4 and 8.5, involve solving pairs of linear equations. Such pairs are said to form **systems of equations** or **sets of simultaneous equations**. For example,

$$\begin{cases} x-2y=0 \\ 3x+y=7 \end{cases} \quad \text{and} \quad \begin{cases} y=-x+4 \\ y=2x+1 \end{cases}$$

are two systems of linear equations.

Each individual equation has an infinite number of solutions. That is, there is an infinite number of ordered pairs that satisfy each equation. But, we are interested in finding ordered pairs that satisfy **both** equations.

A **solution of a system** of linear equations is an ordered pair (or point) that satisfies **both** equations. To determine whether or not a particular ordered pair is a solution to a system, substitute the values for x and y in **both** equations. If the results for both equations are true statements, then the ordered pair is a solution to the system.

Example 1: Solution of a System

Show that $(2,1)$ is a solution to the system $\begin{cases} x-2y=0 \\ 3x+y=7 \end{cases}$

Solution: Substitute $x=2$ and $y=1$ into **both** equations.
In the first equation:

$$2-2(1)\stackrel{?}{=}0$$

$$2-2=0 \quad \text{A true statement}$$

In the second equation:

$$3(2)+1\stackrel{?}{=}7$$

$$6+1=7 \quad \text{A true statement}$$

Because $(2,1)$ satisfies both equations, **$(2,1)$ is a solution to the system.**

Teaching Notes:
Consider giving some practical sounding applications in which scientists, business analysts, or engineers might want two conditions to depend on two variables and a simultaneous solution is needed.

Example 2: Not a Solution of a System

Show that $(0,4)$ is not a solution to the system $\begin{cases} y=-x+4 \\ y=2x+1 \end{cases}$

Solution: Substitute $x=0$ and $y=4$ into **both** equations.
In the first equation:

$$4\stackrel{?}{=}-(0)+4$$

$$4=0+4 \quad \text{A true statement}$$

In the second equation:

$$4\stackrel{?}{=}2(0)+1$$

$$4=0+1 \quad \text{A false statement}$$

Because $(0,4)$ does not satisfy **both** equations, **$(0,4)$ is NOT a solution to the system.**

For example, an architect might need a rectangular window that has a perimeter of 140 inches with a length that is 4 more inches than the width. (While this situation may be set up as a single equation, it is also useful for the illustration of a system.)

Solving Systems of Linear Equations by Graphing

The following questions concerning systems of equations need to be addressed:

How do we find the solution, if there is one?
Will there always be a solution to a system of linear equations?
Can there be more than one solution?

In this chapter, we will discuss four methods (graphing, substitution, addition, and using matrices) for solving a system of linear equations. The first method is to **solve by graphing**. In this method, both equations are graphed and the point of intersection (if there is one) is the solution to the system. Table 1 illustrates the three possibilities for a system of two linear equations. Each system will be one of the following:

1. **Consistent** (has exactly one solution)
2. **Inconsistent** (has no solution)
3. **Dependent** (has an infinite number of solutions)

System	Graph	Intersection	Classification
$\begin{cases} 2x+y=5 \\ x-y=1 \end{cases}$	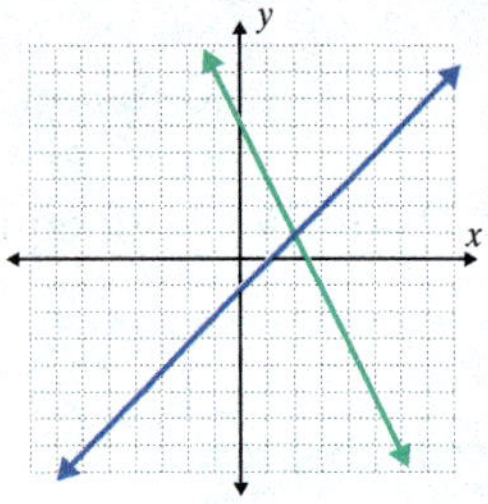	$(2,1)$ (The lines intersect at one point.)	**Consistent**
$\begin{cases} 3x-2y=2 \\ 6x-4y=-4 \end{cases}$	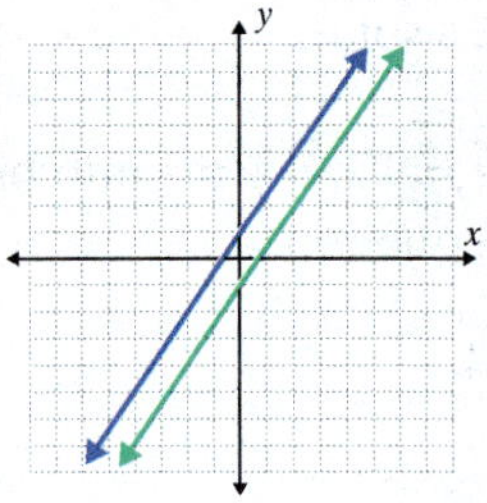	No solution (The lines are parallel.)	**Inconsistent**
$\begin{cases} 2x-4y=6 \\ x-2y=3 \end{cases}$	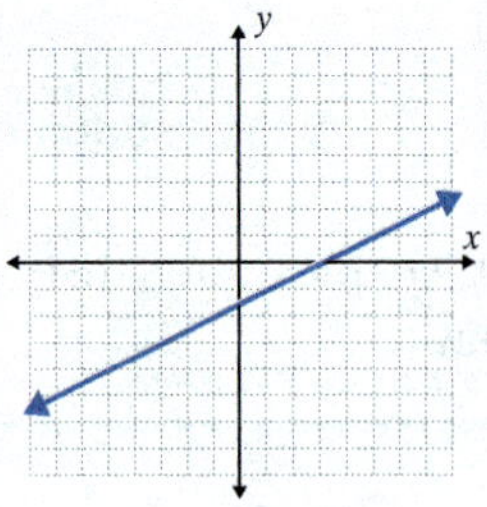	Every ordered pair that satisfies $x-2y=3$. (The lines are the same line.) There is an infinite number of solutions: $(3+2y, y)$, where y is any real number or $\left(x, \dfrac{x-3}{2}\right)$, where x is any real number.	**Dependent**

Table 1

Solutions by Graphing

To Solve a System of Linear Equations by Graphing

Teaching Notes:
It should be made clear to students that graphing is useful for estimating and studying possibilities of solutions to systems of equations, especially prior to determining exact solutions through algebraic techniques.

1. Graph both linear equations on the same set of axes.
2. Observe the point of intersection (if there is one).
   a. If the slopes of the two lines are different, then the lines intersect in one and only one point. The system has a single point as its solution.
   b. If the lines are distinct and have the same slope, then the lines are parallel. The system has no solution.
   c. If the lines are the same line, then all the points on the line constitute the solution. There is an infinite number of solutions.
3. Check the solution (if there is one) in both of the original equations.

Solving by graphing can involve estimating the solutions whenever the intersection of the two lines is at a point not represented by a pair of integers. (There is nothing wrong with this technique. Just be aware that at times it can lack accuracy. See Example 6.)

You might want to review graphing lines in Chapter 4. Recall that lines can be graphed by plotting intercepts (as illustrated in Example 3). Other methods include finding a point on the line and using the slope to find another point (see Examples 4 and 5) or finding two points (see Example 6).

Example 3: A System With One Solution

Solve the following system of linear equations by graphing: $\begin{cases} x+y=6 \\ y=x+4 \end{cases}$

Solution: The two lines intersect at the point $(1,5)$. Thus the solution to the system is $(1,5)$ or $x=1$ and $y=5$.

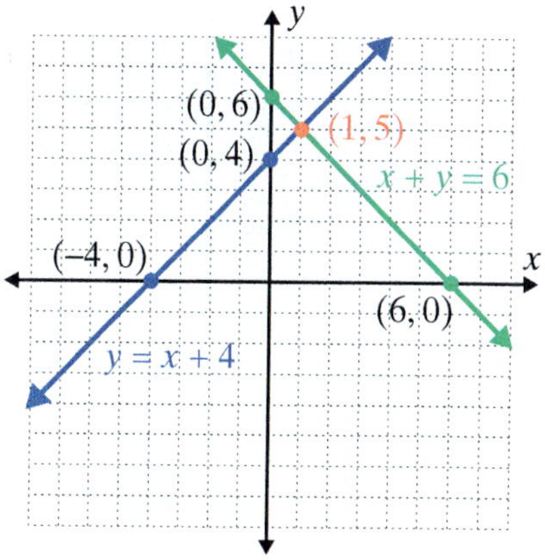

Check: Substitution shows that $(1,5)$ satisfies **both** of the equations in the system.

$$x+y=6 \qquad\qquad y=x+4$$
$$(1)+(5)\stackrel{?}{=}6 \qquad\qquad (5)\stackrel{?}{=}(1)+4$$
$$6=6 \qquad\qquad 5=5$$

Example 4: A System With No Solution

Solve the following system of linear equations by graphing: $\begin{cases} y = 3x \\ y - 3x = -4 \end{cases}$

Solution: The lines are parallel with the same slope, 3, and there are no points of intersection.

There is no solution to the system.

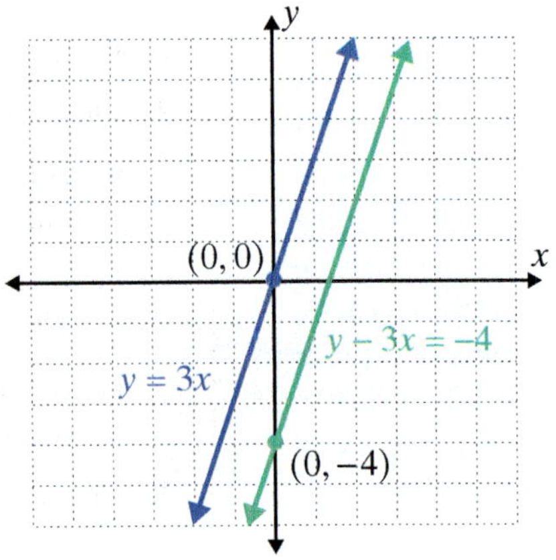

Example 5: A System With Infinitely Many Solutions

Teaching Notes:
The notation $\left(x, -\frac{1}{2}x + 3\right)$ leads nicely to function notation as in $f(x) = -\frac{1}{2}x + 3.$

Solve the following system of linear equations by graphing: $\begin{cases} x + 2y = 6 \\ y = -\frac{1}{2}x + 3 \end{cases}$

Solution: All points that lie on one line also lie on the other line. For example, $(4, 1)$ is a point on the line $x + 2y = 6$ since $4 + 2(1) = 6$. The point $(4, 1)$ is also on the line $y = -\frac{1}{2}x + 3$ since $1 = -\frac{1}{2}(4) + 3$. Therefore, the solution can be stated as all of the points that satisfy the equation $x + 2y = 6$.

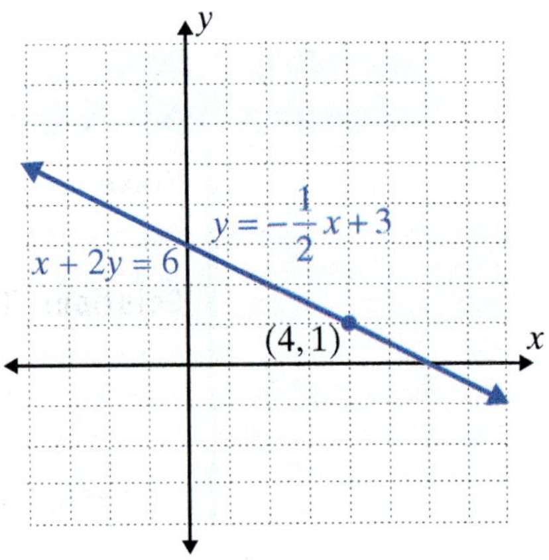

The solution is the set of all points of the form $\left(x, -\frac{1}{2}x + 3\right)$, an infinite set of solutions. (Or, solving the first equation for x we have $(6 - 2y, y)$ for all values of y.)

Example 6: A System That Requires Estimation

Solve the following system of linear equations by graphing: $\begin{cases} x-3y=4 \\ 2x+y=3 \end{cases}$

Solution: The two lines intersect at one point. However, we can only estimate the point of intersection as $\left(2, -\frac{1}{2}\right)$. In this situation be aware that, although graphing gives a good "estimate," finding exact solutions to the system is not likely.

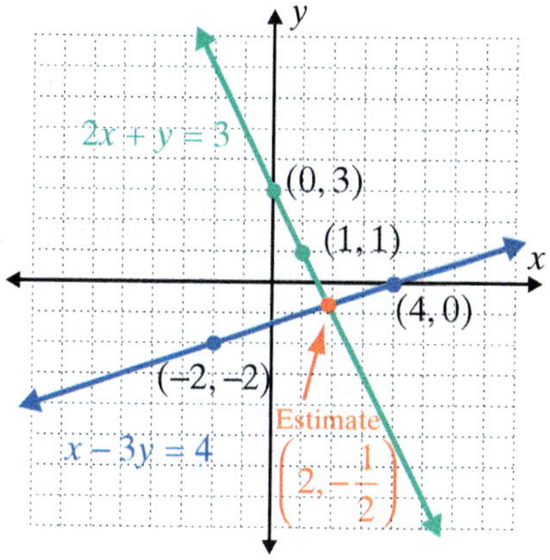

Check: Substitute $x = 2$ and $y = -\frac{1}{2}$.

$$2-3\left(-\frac{1}{2}\right) \stackrel{?}{=} 4 \quad \text{and} \quad 2(2)+\left(-\frac{1}{2}\right) \stackrel{?}{=} 3$$

$$\frac{7}{2} \neq 4 \qquad\qquad \frac{7}{2} \neq 3$$

Thus checking shows that the estimated solution $\left(2, -\frac{1}{2}\right)$ does not satisfy either equation. The estimated point of intersection is just that—an estimate. The following discussion provides a technique based on using of a graphing calculator that would give the exact solution as $\left(\frac{13}{7}, -\frac{5}{7}\right)$.

NOTES

1. To use the graphing method, graph the lines as accurately as you can.
2. Be sure to check your solution by substituting it back into both of the original equations. (Of course, fractional estimates may not check exactly.)

Using a TI-84 Plus Graphing Calculator to Solve a System of Linear Equations

A graphing calculator can be used to locate (or estimate) the point of intersection of two lines (and therefore the solution to the system).

Example 7: Using a Graphing Calculator

Use a graphing calculator to solve the system $\begin{cases} 2x + y = 8 \\ x - y = 1 \end{cases}$ as follows:

Solution:

Step 1: Solve each equation for y. For this system $\begin{cases} y = -2x + 8 \\ y = x - 1 \end{cases}$

Step 2: Press Y= and enter the two expressions for y.

To get the variable X, press X,T,θ,n.
The display screen will appear as follows:

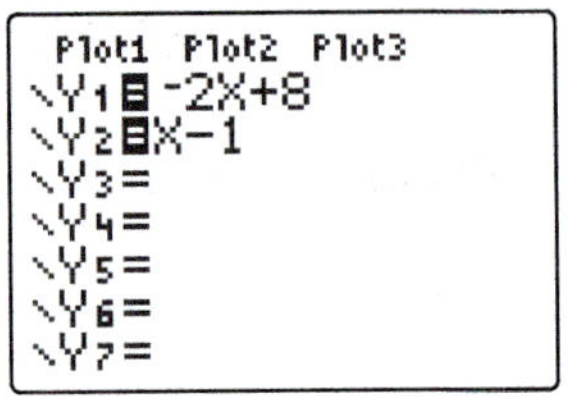

Step 3: Press GRAPH.
(Both lines should appear. If not, you may need to adjust the WINDOW.)

Step 4: Press 2ND and CALC. Select `5: intersect`.
The cursor will appear on one of the lines. Use the right or left arrow to get near the point of intersection and press ENTER. Then move the up or down arrow to get to the other line. Now use the right or left arrow to move closer to the point of intersection on this line and press ENTER. Follow the directions for `Guess?` by moving the cursor to the point of intersection and pressing ENTER.

Step 5: The answer $x = 3$ and $y = 2$ will appear at the bottom of the display screen.

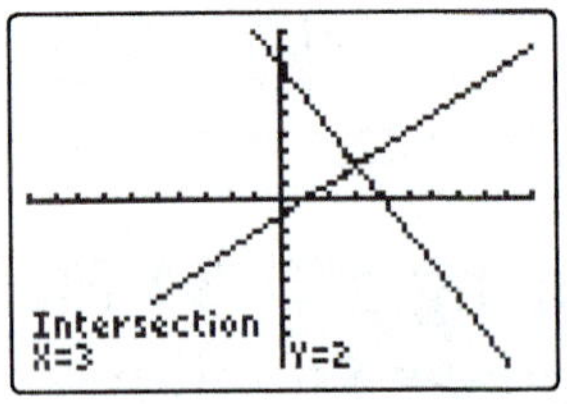

Note: Step 4 may seem somewhat complicated, but TRY IT. It is fun and accurate! (**Note:** If the lines are parallel (an inconsistent system) the calculator will give an error message.)

Teaching Notes:
You may want to show your students how "TRACE" works in locating the intersection. This is basically a one step operation, but the calculator will generally give decimal approximations only.

Practice Problems

1. Determine which of the given points, if any, lie on both of the lines in the given system of equations by substituting each point into both equations.

$$\begin{cases} x+2y=8 \\ 2x-y=1 \end{cases}$$

a. $(8,0)$ **b.** $(2,3)$ **c.** $(0,-1)$ **d.** $(4,2)$

2. Show that the following system of equations is inconsistent by determining the slope of each line and the y-intercept. Explain your reasoning.

$$\begin{cases} 3x+y=1 \\ 6x+2y=7 \end{cases}$$

3. Solve the following system graphically.

$$\begin{cases} y=2x+5 \\ 4x-2y=-10 \end{cases}$$

1. c 2. None
3. a, c 4. b
5. $(0,2)$ 6. $(-1,1)$
7. $(4,2)$ 8. $(1,-3)$
9. $m_1=-2, b_1=3,$ $m_2=-2, b_2=\frac{5}{2}$
10. $m_1=\frac{3}{5}, b_1=-\frac{1}{5},$ $m_2=\frac{3}{5}, b_2=-\frac{2}{5}$
11. $m_1=\frac{1}{2}, b_1=3,$ $m_2=\frac{1}{2}, b_2=-\frac{1}{2}$
12. $m_1=3, b_1=-8,$ $m_2=3, b_2=-6$

8.1 Exercises

13. $(-3,-2)$

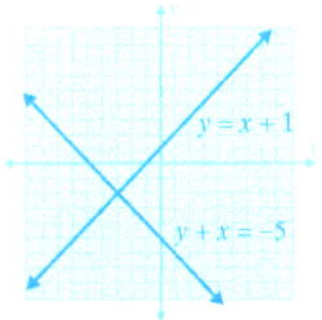

14. No solution

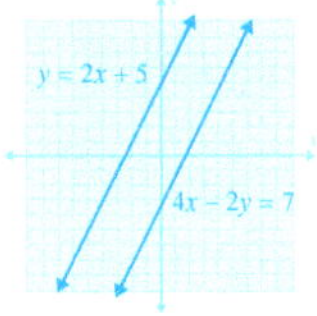

Determine which of the given points, if any, lie on both of the lines in the systems of equations by substituting each point into both equations.

1. $\begin{cases} x-y=6 \\ 2x+y=0 \end{cases}$

a. $(1,-2)$ **b.** $(4,-2)$ **c.** $(2,-4)$ **d.** $(-1,2)$

2. $\begin{cases} x+3y=5 \\ 3y=4-x \end{cases}$

a. $(2,1)$ **b.** $(2,-2)$ **c.** $(-1,2)$ **d.** $(4,0)$

3. $\begin{cases} 2x+4y-6=0 \\ 3x+6y-9=0 \end{cases}$

a. $(1,1)$ **b.** $(2,0)$ **c.** $\left(0,\frac{3}{2}\right)$ **d.** $(-1,3)$

4. $\begin{cases} 5x-2y-5=0 \\ 5x=-3y \end{cases}$

a. $(1,0)$ **b.** $\left(\frac{3}{5},-1\right)$ **c.** $(0,0)$ **d.** $(1,4)$

Answers to Practice Problems: **1.** b **2.** $m_1=-3, b_1=1, m_2=-3, b_2=\frac{7}{2}$; The lines do not intersect because they are parallel, they have the same slope, but different y-intercepts.
3. $(x, 2x+5)$

The graphs of the lines represented by each system of equations are given. Determine the solution of the system by looking at the graph. Check your solution by substituting into both equations.

5. $\begin{cases} x+2y=4 \\ x-y=-2 \end{cases}$ **6.** $\begin{cases} x+2y=1 \\ 2x+y=-1 \end{cases}$ **7.** $\begin{cases} 2x-y=6 \\ 3x+y=14 \end{cases}$ **8.** $\begin{cases} x-y=4 \\ 3x-y=6 \end{cases}$

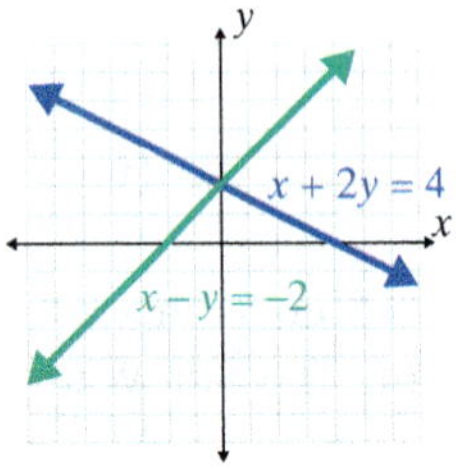

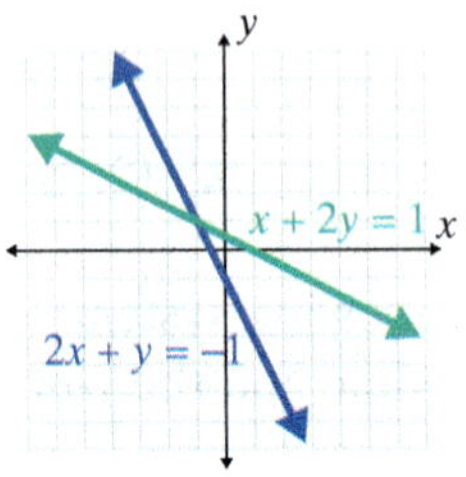

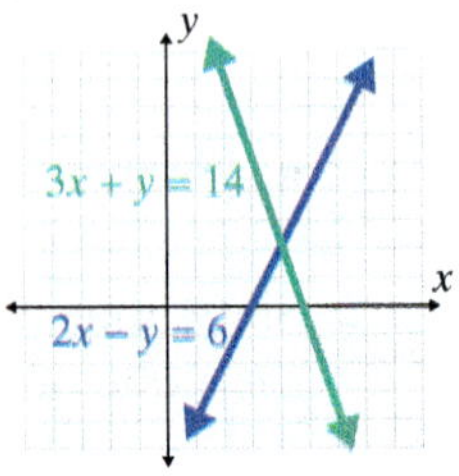

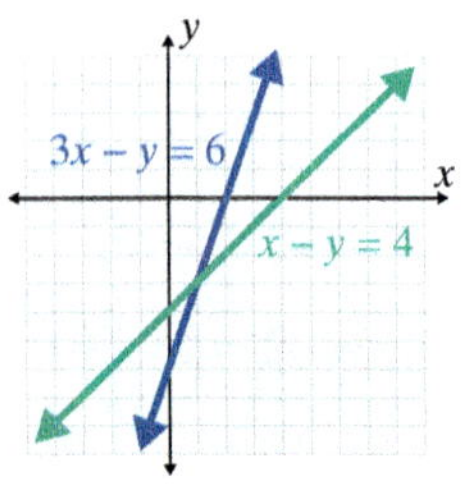

Show that each system of equations is inconsistent by determining the slope of each line and the y-intercept. (That is, show that the lines are parallel and do not intersect.)

9. $\begin{cases} 2x+y=3 \\ 4x+2y=5 \end{cases}$ **10.** $\begin{cases} 3x-5y=1 \\ 6x-10y=4 \end{cases}$ **11.** $\begin{cases} y=\frac{1}{2}x+3 \\ x-2y=1 \end{cases}$ **12.** $\begin{cases} 3x-y=8 \\ x-\frac{1}{3}y=2 \end{cases}$

Solve the following systems graphically.

13. $\begin{cases} y=x+1 \\ y+x=-5 \end{cases}$ **14.** $\begin{cases} y=2x+5 \\ 4x-2y=7 \end{cases}$ **15.** $\begin{cases} 2x-y=4 \\ 3x+y=6 \end{cases}$

16. $\begin{cases} 2x+3y=6 \\ 4x+6y=12 \end{cases}$ **17.** $\begin{cases} x+y-5=0 \\ x=5-y \end{cases}$ **18.** $\begin{cases} y=4x-3 \\ x=2y-8 \end{cases}$

19. $\begin{cases} 3x-y=6 \\ y=3x \end{cases}$ **20.** $\begin{cases} y=2x \\ 2x+y=4 \end{cases}$ **21.** $\begin{cases} x-y=5 \\ x=-3 \end{cases}$

22. $\begin{cases} x-2y=4 \\ x=4 \end{cases}$ **23.** $\begin{cases} x+2y=8 \\ 3x-2y=0 \end{cases}$ **24.** $\begin{cases} 2x+y=0 \\ 4x+2y=-8 \end{cases}$

25. $\begin{cases} 5x-4y=5 \\ 8y=10x-10 \end{cases}$ **26.** $\begin{cases} x+y=8 \\ 5y=2x+5 \end{cases}$ **27.** $\begin{cases} 4x+3y+7=0 \\ 5x-2y+3=0 \end{cases}$

28. $\begin{cases} 4x-2y=10 \\ -6x+3y=-15 \end{cases}$ **29.** $\begin{cases} x=5 \\ y=-1 \end{cases}$ **30.** $\begin{cases} y=7 \\ x=8 \end{cases}$

31. $\begin{cases} \frac{1}{2}x+2y=7 \\ 2x=4-8y \end{cases}$ **32.** $\begin{cases} 4x+y=6 \\ 2x+\frac{1}{2}y=3 \end{cases}$ **33.** $\begin{cases} 7x-2y=1 \\ y=3 \end{cases}$

34. $\begin{cases} x=1.5 \\ x-3y=9 \end{cases}$ **35.** $\begin{cases} y=\frac{1}{2}x+2 \\ x-2y+4=0 \end{cases}$ **36.** $\begin{cases} 2x-5y=6 \\ y=\frac{2}{5}x+1 \end{cases}$

15. $(2, 0)$

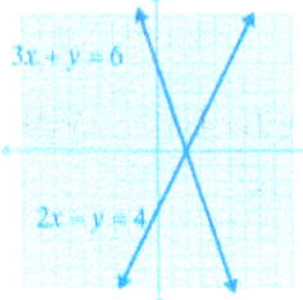

16. $\left(x, -\frac{2}{3}x+2\right)$

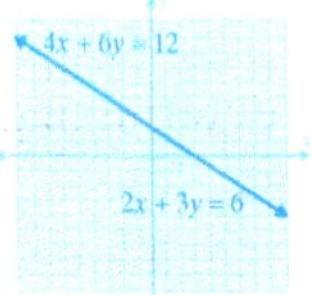

17. $(x, -x+5)$

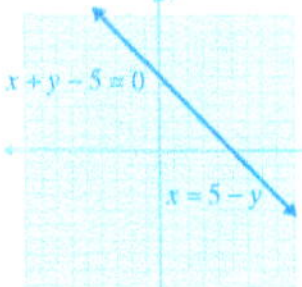

18. $(2, 5)$

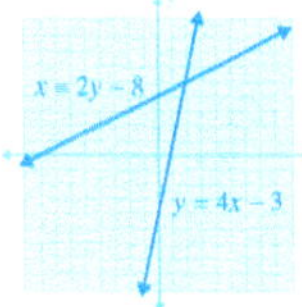

19. No solution;

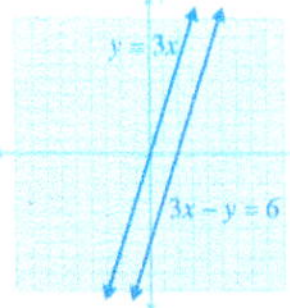

20. $(1, 2)$

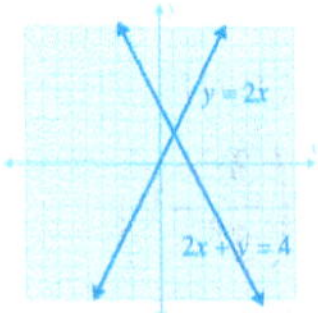

21. $(-3, -8)$

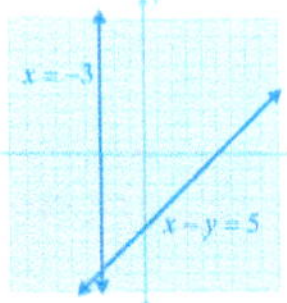

22. $(4, 0)$

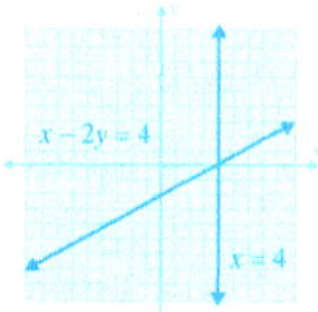

23. $(2, 3)$

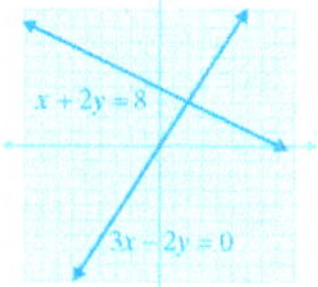

24. No solution

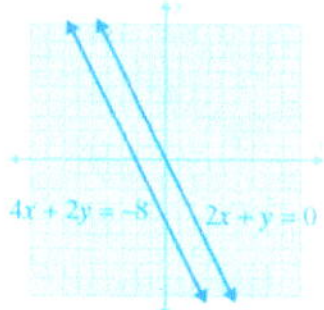

25. $\left(x, \frac{5}{4}x - \frac{5}{4}\right)$

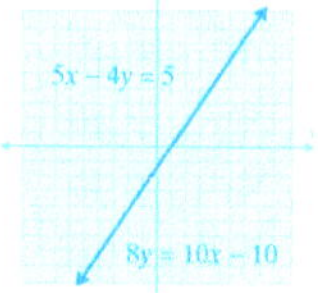

26. $(5, 3)$

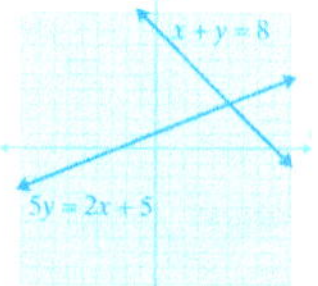

27. $(-1, -1)$

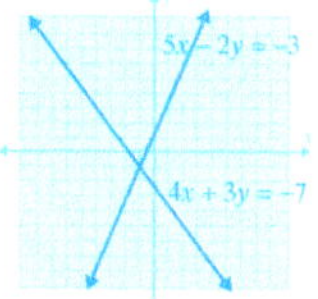

28. $(x, 2x - 5)$

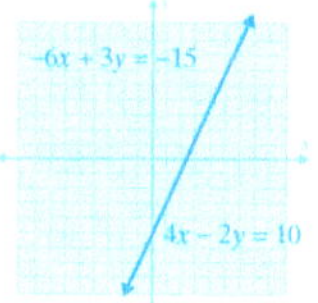

37. $\begin{cases} 2x + 3y = 5 \\ 3x - 2y = 1 \end{cases}$

38. $\begin{cases} \frac{2}{3}x + y = 2 \\ x - 4y = 3 \end{cases}$

39. $\begin{cases} x - y = 4 \\ 2y = 2x - 4 \end{cases}$

40. $\begin{cases} x + y = 4 \\ 2x - 3y = 3 \end{cases}$

41. $\begin{cases} \frac{1}{2}x + \frac{1}{3}y = \frac{1}{6} \\ \frac{1}{4}x + \frac{1}{4}y = 0 \end{cases}$

42. $\begin{cases} \frac{1}{4}x - y = \frac{13}{4} \\ \frac{1}{3}x + \frac{1}{6}y = -\frac{1}{6} \end{cases}$

Each of the following application problems has been modeled using a system of equations. Solve the system graphically.

43. The sum of two numbers is 25 and their difference is 15. What are the two numbers?
Let x = one number and y = the other number.
The corresponding modeling system is $\begin{cases} x + y = 25 \\ x - y = 15 \end{cases}$

44. Rectangles: The perimeter of a rectangle is 50 m and the length is 5 m longer than the width. Find the dimensions of the rectangle.
Let x = the length and y = the width.
The corresponding modeling system is $\begin{cases} 2x + 2y = 50 \\ x - y = 5 \end{cases}$

45. Swimming pools: OSHA recommends that swimming pool owners clean their pool decks with a solvent composed of a 12% chlorine solution and a 3% chlorine solution. Fifteen gallons of the solvent consists of 6% chlorine. How much of each of the mixing solutions were used?
Let x = the number of gallons of the 12% solution
and y = the number of gallons of the 3% solution.
The corresponding modeling system is $\begin{cases} x + y = 15 \\ 0.12x + 0.03y = 0.06(15) \end{cases}$

46. School supplies: A student bought a calculator and a textbook for a course in algebra. He told his friend that the total cost was \$170 (without tax) and that the calculator cost \$20 more than twice the cost of the textbook. What was the cost of each item?
Let x = the cost of the calculator and
y = the cost of the textbook.
The corresponding modeling system is $\begin{cases} x + y = 170 \\ x = 2y + 20 \end{cases}$

29. $(5, -1)$

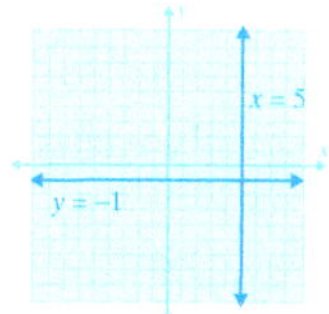

30. $(8, 7)$

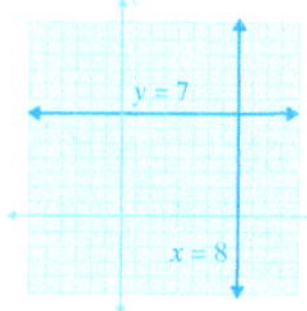

31. No solution

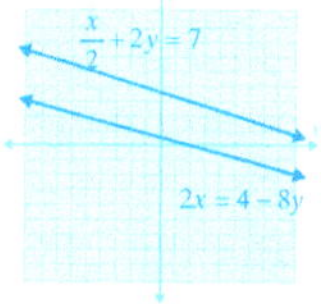

32. $(x, -4x + 6)$

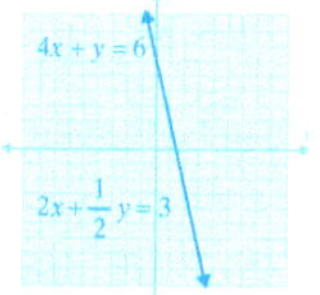

33. $(1, 3)$

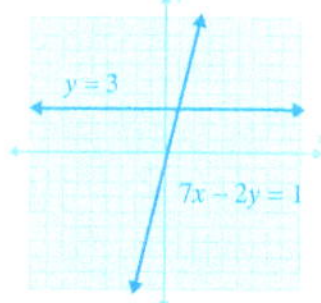

34. (1.5, −2.5)

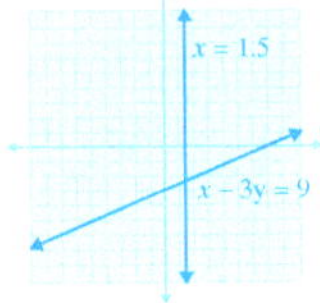

Use a graphing calculator and the CALC *and* `5:intersect` *commands to find the solutions to the given systems of linear equations. If necessary, round values to four decimal places. (Remember to solve each equation for y. Use both* Y_1 *and* Y_2 *in the* Y= *menu.)*

47. $\begin{cases} x+2y=9 \\ x-2y=-7 \end{cases}$

48. $\begin{cases} x-3y=0 \\ 2x+y=7 \end{cases}$

49. $\begin{cases} y=2 \\ 2x-3y=-3 \end{cases}$

50. $\begin{cases} 2x-3y=0 \\ 3x+3y=\frac{5}{2} \end{cases}$

51. $\begin{cases} y=-3 \\ 2x+y=0 \end{cases}$

52. $\begin{cases} 2x-3y=1.25 \\ x+2y=5 \end{cases}$

53. $\begin{cases} x+y=3.5 \\ -2x+5y=7.7 \end{cases}$

54. $\begin{cases} 4x+y=-0.5 \\ x+2y=-8 \end{cases}$

Writing and Thinking About Mathematics

55. Explain, in your own words, why the answer to a consistent system of linear equations can be written as an ordered pair.

HAWKES LEARNING SYSTEMS: INTRODUCTORY & INTERMEDIATE ALGEBRA SOFTWARE

- 8.1 Solving Systems of Linear Equations by Graphing

35. $\left(x, \frac{1}{2}x+2\right)$

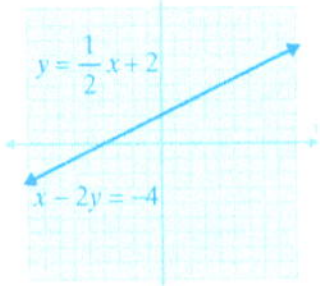

36. No solution

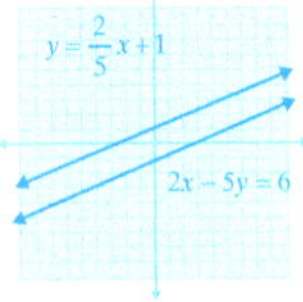

37. (1, 1)

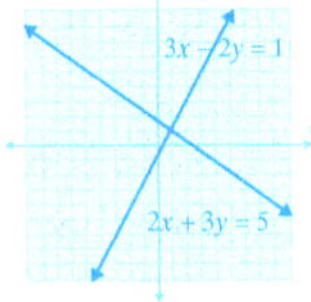

38. (3, 0)

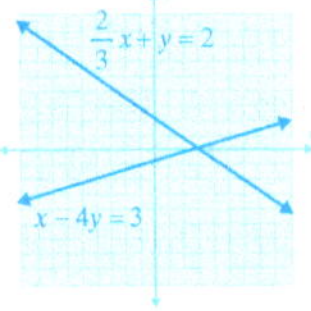

39. No solution

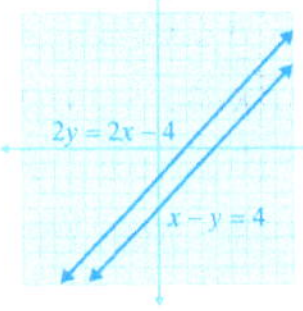

40. (3, 1)

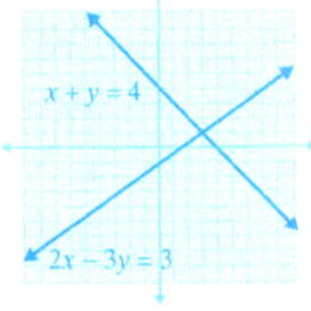

41. (1, −1)

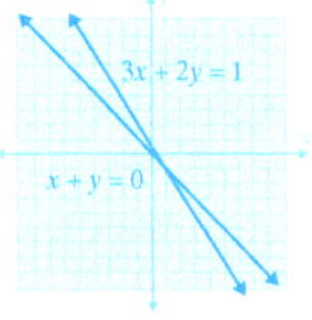

42. (1, −3)

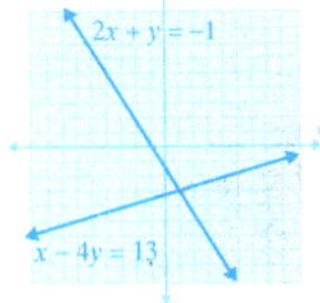

43. (20, 5)

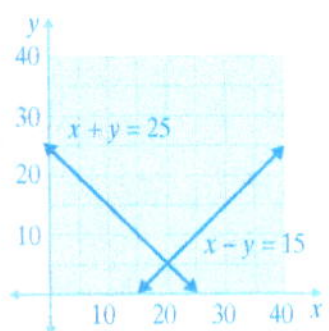

44. 15 m × 10 m

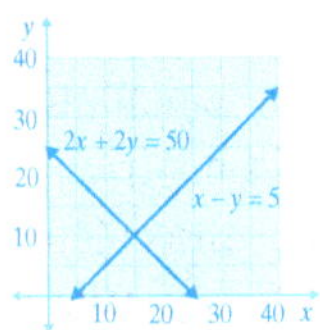

45. 5 gallons of 12%; 10 gallons of 3%

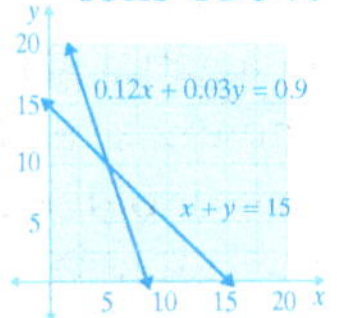

46. Calc. = $120; Book = $50

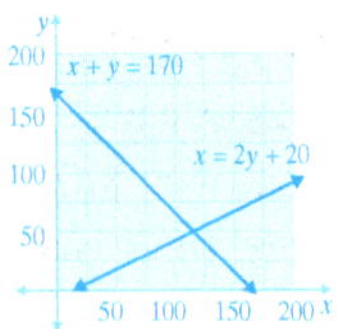

47. (1, 4)

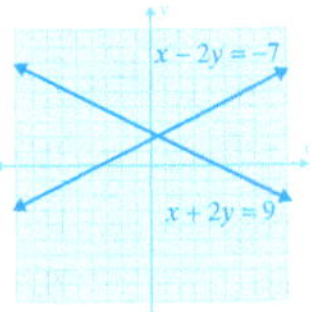

48. (3, 1)

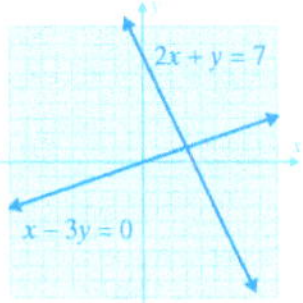

49. (1.5, 2)

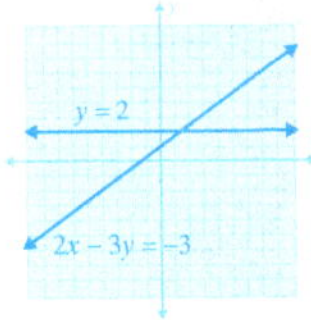

50. (0.5, 0.3333)

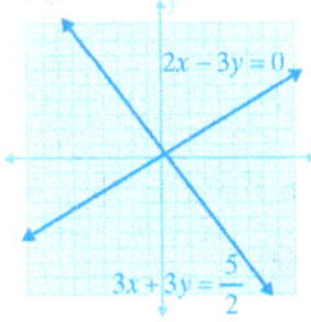

51. (1.5, −3)

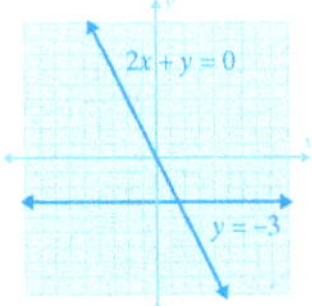

52. (2.5, 1.25)

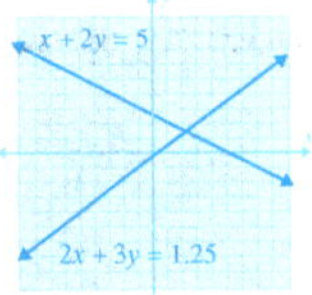

53. (1.4, 2.1)

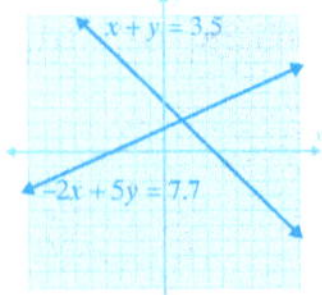

54. (1, −4.5)

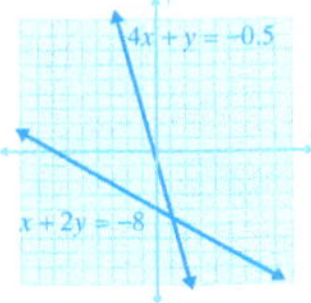

55. The solution to a consistent system of linear equations is a single point, which is easily written as an ordered pair.

8.2 Systems of Linear Equations: Solutions by Substitution

- *Solve systems of linear equations by using the method of substitution.*

Solving Systems of Linear Equations by Substitution

As we discussed in Section 8.1, solving systems of linear equations by graphing is somewhat limited in accuracy. The graphs must be drawn very carefully and even then the points of intersection (if there are any) can be difficult to estimate accurately.

In this section, we will develop an algebraic method called the **method of substitution**. The objective in the substitution method is to eliminate one of the variables so that a new equation is formed with just one variable. If this new equation has one solution then the solution to the system is a single point. If this new equation is never true, then the system has no solution. If this new equation is always true, then the system has an infinite number of solutions.

To Solve a System of Linear Equations by Substitution

1. Solve one of the equations for one of the variables.
2. Substitute the resulting expression into the other equation.
3. Solve this new equation, if possible, and then substitute back into one of the original equations to find the value of the other variable. (This is known as **back substitution**.)
4. Check the solution in both of the original equations.

To illustrate the substitution method, consider the following system:

$$\begin{cases} y = -2x + 5 \\ x + 2y = 1 \end{cases}$$

How would you substitute? Since the first equation is already solved for y, a reasonable substitution would be to put $-2x + 5$ for y in the second equation. Try this and see what happens.

First equation already solved for y: $\quad y = -2x + 5$

Substitute into second equation: $\quad x + 2y = 1$

$$x + 2(-2x + 5) = 1$$

We now have one equation in only one variable, namely x. The problem has been reduced from one of solving two equations in two variables to solving one equation in one variable. Solve this equation for x. Then find the corresponding y-value by substituting this x-value into **either of the two original equations**.

$$\begin{aligned} x+2(-2x+5)&=1 \\ x-4x+10&=1 \\ -3x&=-9 \\ x&=3 \end{aligned}$$

Back substitute $x = 3$ into $y = -2x + 5$.

$$\begin{aligned} y&=-2(3)+5 \\ &=-6+5 \\ &=-1 \end{aligned}$$

Thus the solution to the system is the point $(3, -1)$.

Substitution is not the only algebraic technique for solving a system of linear equations. It does work in all cases but is most often used when one of the equations is easily solved for one variable.

In the following examples, note how the results in Example 2 indicate that the system has no solution and the results in Example 3 indicate that the system has an infinite number of solutions.

Example 1: A System With One Solution

Solve the following system of linear equations by using substitution: $\begin{cases} y=\frac{5}{6}x+2 \\ \frac{1}{6}x+y=8 \end{cases}$

Solution: The first equation is already solved for y. Substituting $\frac{5}{6}x+2$ for y in the second equation gives the following.

$$\begin{aligned} \frac{1}{6}x+\left(\frac{5}{6}x+2\right)&=8 \\ 6\cdot\frac{1}{6}x+6\cdot\left(\frac{5}{6}x+2\right)&=6\cdot 8 \quad \text{Multiply both sides by 6, the LCD.} \\ x+5x+12&=48 \\ 6x+12&=48 \\ 6x&=36 \\ x&=6 \end{aligned}$$

Substituting 6 for x in the first equation gives the corresponding value for y.

$$y=\frac{5}{6}(6)+2=5+2=7$$

The solution to the system is $(6, 7)$.

Check: Substitution shows that $(6,7)$ satisfies **both** of the equations in the system.

$$y=\frac{5}{6}x+2 \qquad \frac{1}{6}x+y=8$$

$$7\stackrel{?}{=}\frac{5}{6}(6)+2 \qquad \frac{1}{6}(6)+7\stackrel{?}{=}8$$

$$7=7 \qquad 8=8$$

Example 2: A System With No Solution

Solve the following system of linear equations by using substitution: $\begin{cases} 3x+y=1 \\ 6x+2y=3 \end{cases}$

Solution: Solving the first equation for y gives $y=1-3x$. Substituting $1-3x$ for y in the second equation gives the following.

$$6x+2(1-3x)=3$$
$$6x+2-6x=3$$
$$2=3$$

This last equation $(2=3)$ is **never true**. This tells us that the system has **no solution**. Graphically, the lines are parallel and there is no intersection.

Example 3: A System With Infinitely Many Solutions

Solve the following system of linear equations by using substitution: $\begin{cases} x-2y=1 \\ 3x-6y=3 \end{cases}$

Solution: Solving the first equation for x gives $x=1+2y$. Substituting $1+2y$ for x in the second equation gives the following.

$$3(1+2y)-6y=3$$
$$3+6y-6y=3$$
$$3=3$$

This last equation $(3=3)$ is **always true**. This tells us that the system has an **infinite number of solutions** and that the solutions are of the form $(1+2y,\ y)$ for all values of y. (Or, solving one of the equations for y we have $\left(x, \frac{1}{2}x-\frac{1}{2}\right)$ for all values of x.)

NOTES

As illustrated in Example 3, the solution can take two forms for a system with an infinite number of solutions. In one form we can solve for x and then use this expression for x in the ordered pair format (x, y). For example, in Example 3, solving either equation for x gives $x = 1 + 2y$ and we can write $(1+2y, y)$ to represent all solutions. Alternatively, we can solve for y which gives $y = \frac{1}{2}x - \frac{1}{2}$ and we have the form $\left(x, \frac{1}{2}x - \frac{1}{2}\right)$ for all solutions. Try substituting various values for x and y in these expressions and you will see that all results satisfy both equations.

Example 4: A System With Decimals

Solve the following system of linear equations by using substitution: $\begin{cases} x + y = 5 \\ 0.2x + 0.3y = 0.9 \end{cases}$

Solution: Solving the first equation for y gives $y = 5 - x$. Substituting $5 - x$ for y in the second equation gives the following.

$$\begin{aligned} 0.2x + 0.3y &= 0.9 \\ 0.2x + 0.3(5 - x) &= 0.9 \\ 0.2x + 1.5 - 0.3x &= 0.9 \\ -0.1x &= -0.6 \\ \frac{-0.1x}{-0.1} &= \frac{-0.6}{-0.1} \\ x &= 6 \\ y = 5 - x &= 5 - 6 = -1 \end{aligned}$$

Note that you could multiply each term by 10 to eliminate the decimal.

The solution to the system is $(6, -1)$.

To check, substitute $x = 6$ and $y = -1$ in both of the original equations.

Practice Problems

Solve the following systems by using the method of substitution.

1. $\begin{cases} x+y=3 \\ y=2x \end{cases}$ **2.** $\begin{cases} y=-3+2x \\ 4x-2y=6 \end{cases}$ **3.** $\begin{cases} x+2y=-1 \\ x-4y=-4 \end{cases}$

4. $\begin{cases} y=4-3x \\ y=-3x+6 \end{cases}$ **5.** $\begin{cases} y=3x-1 \\ 2x+y=4 \end{cases}$

8.2 Exercises

Solve the following systems of linear equations by using the method of substitution.

1. $\begin{cases} x+y=6 \\ y=2x \end{cases}$ **2.** $\begin{cases} 5x+2y=21 \\ x=y \end{cases}$ **3.** $\begin{cases} 3x-7=y \\ 2y=6x-14 \end{cases}$

4. $\begin{cases} y=3x+4 \\ 2y=3x+5 \end{cases}$ **5.** $\begin{cases} x=3y \\ 3y-2x=6 \end{cases}$ **6.** $\begin{cases} 4x=y \\ 4x-y=7 \end{cases}$

7. $\begin{cases} x-5y+1=0 \\ x=7-3y \end{cases}$ **8.** $\begin{cases} 2x+5y=15 \\ x=y-3 \end{cases}$ **9.** $\begin{cases} 7x+y=9 \\ y=4-7x \end{cases}$

10. $\begin{cases} 3y+5x=5 \\ y=3-2x \end{cases}$ **11.** $\begin{cases} 3x-y=7 \\ x+y=5 \end{cases}$ **12.** $\begin{cases} 4x-2y=5 \\ y=2x+3 \end{cases}$

13. $\begin{cases} 3x+5y=-13 \\ y=3-2x \end{cases}$ **14.** $\begin{cases} 15x+5y=20 \\ y=-3x+4 \end{cases}$ **15.** $\begin{cases} x-y=5 \\ 2x+3y=0 \end{cases}$

16. $\begin{cases} 4x=8 \\ 3x+y=8 \end{cases}$ **17.** $\begin{cases} 2y=5 \\ 3x-4y=-4 \end{cases}$ **18.** $\begin{cases} x+y=8 \\ 3x+2y=8 \end{cases}$

19. $\begin{cases} y=2x-5 \\ 2x+y=-3 \end{cases}$ **20.** $\begin{cases} 2x+3y=5 \\ x-6y=0 \end{cases}$ **21.** $\begin{cases} x+5y=1 \\ x-3y=5 \end{cases}$

22. $\begin{cases} 3x+8y=-2 \\ x+2y=-1 \end{cases}$ **23.** $\begin{cases} 9x+3y=6 \\ 3x=2-y \end{cases}$ **24.** $\begin{cases} 5x+2y=-10 \\ 10x=-3-4y \end{cases}$

1. $(2, 4)$
2. $(3, 3)$
3. $(x, 3x-7)$
4. $(-1, 1)$
5. $(-6, -2)$
6. No solution
7. $(4, 1)$
8. $(0, 3)$
9. No solution
10. $(4, -5)$
11. $(3, 2)$
12. No solution
13. $(4, -5)$
14. $(x, -3x+4)$
15. $(3, -2)$
16. $(2, 2)$
17. $\left(2, \frac{5}{2}\right)$
18. $(-8, 16)$
19. $\left(\frac{1}{2}, -4\right)$
20. $\left(2, \frac{1}{3}\right)$
21. $\left(\frac{7}{2}, -\frac{1}{2}\right)$
22. $\left(-2, \frac{1}{2}\right)$
23. $(x, 2-3x)$
24. No solution

Answers to Practice Problems: **1.** $(1, 2)$ **2.** $(x, -3+2x)$ **3.** $\left(-2, \frac{1}{2}\right)$ **4.** No solution **5.** $(1, 2)$

25. $(-2, 1)$
26. $\left(\frac{5}{2}, \frac{1}{8}\right)$
27. $\left(-\frac{4}{5}, -\frac{7}{5}\right)$
28. $(4, -1)$
29. $\left(\frac{11}{7}, \frac{8}{7}\right)$
30. $\left(\frac{13}{5}, -\frac{39}{5}\right)$
31. $\left(x, \frac{1}{6}x + \frac{10}{3}\right)$
32. No solution
33. $(3, 3)$
34. $(9, -3)$
35. $(10, 20)$
36. $(6, -4)$
37. No solution
38. $(-7.5, 0)$
39. $\left(x, -\frac{3}{2}x + 12\right)$
40. $\left(x, -\frac{7}{3}x + 14\right)$
41. $\left(2, \frac{5}{3}\right)$
42. $(10, 4)$
43. $(20, 5)$
44. 15 m × 10 m

25. $\begin{cases} x - 2y = -4 \\ 3x + y = -5 \end{cases}$

26. $\begin{cases} x + 4y = 3 \\ 3x - 4y = 7 \end{cases}$

27. $\begin{cases} 3x - y = -1 \\ 7x - 4y = 0 \end{cases}$

28. $\begin{cases} x + 5y = -1 \\ 2x + 7y = 1 \end{cases}$

29. $\begin{cases} x + 3y = 5 \\ 3x + 2y = 7 \end{cases}$

30. $\begin{cases} 3x - 4y - 39 = 0 \\ 2x - y - 13 = 0 \end{cases}$

31. $\begin{cases} \frac{1}{4}x - \frac{3}{2}y = -5 \\ -x + 6y = 20 \end{cases}$

32. $\begin{cases} \frac{-4}{3}x + 2y = 7 \\ \frac{8}{3}x - 4y = -5 \end{cases}$

33. $\begin{cases} 6x - y = 15 \\ 0.2x + 0.5y = 2.1 \end{cases}$

34. $\begin{cases} x + 2y = 3 \\ 0.4x + y = 0.6 \end{cases}$

35. $\begin{cases} 0.2x - 0.1y = 0 \\ y = x + 10 \end{cases}$

36. $\begin{cases} 0.1x - 0.2y = 1.4 \\ 3x + y = 14 \end{cases}$

37. $\begin{cases} 3x - 2y = 5 \\ y = 1.5x + 2 \end{cases}$

38. $\begin{cases} x = 2y - 7.5 \\ 2x + 4y = -15 \end{cases}$

39. $\begin{cases} \frac{1}{2}x + \frac{1}{3}y = 4 \\ 3x + 2y = 24 \end{cases}$

40. $\begin{cases} \frac{1}{3}x + \frac{1}{7}y = 2 \\ 7x + 3y = 42 \end{cases}$

41. $\begin{cases} \frac{x}{3} + \frac{y}{5} = 1 \\ x + 6y = 12 \end{cases}$

42. $\begin{cases} \frac{x}{5} + \frac{y}{4} - 3 = 0 \\ \frac{x}{10} - \frac{y}{2} + 1 = 0 \end{cases}$

*A word problem is stated with equations given that represent a mathematical model for the problem. Solve the system by using the method of substitution. (**Note:** These exercises were also given in Section 8.1. Check to see that you arrived at the same answers by both methods.)*

43. The sum of two numbers is 25 and their difference is 15. What are the two numbers? Let x = one number and y = the other number.

The corresponding modeling system is $\begin{cases} x + y = 25 \\ x - y = 15 \end{cases}$

44. Rectangles: The perimeter of a rectangle is 50 meters and the length is 5 meters longer than the width. Find the dimensions of the rectangle.
Let x = the length and y = the width.

The corresponding modeling system is $\begin{cases} 2x + 2y = 50 \\ x - y = 5 \end{cases}$

45. 5 gallons of 12%; 10 gallons of 3%
46. Calculator = \$120; Book = \$50
47. Answers will vary.

45. Swimming pools: OSHA recommends that swimming pool owners clean their pool decks with a solvent composed of a 12% chlorine solution and a 3% chlorine solution. Fifteen gallons of the solvent consists of 6% chlorine. How much of each of the mixing solutions were used?
Let x = the number of gallons of the 12% solution
and y = the number of gallons of the 3% solution.

The corresponding modeling system is $\begin{cases} x + y = 15 \\ 0.12x + 0.03y = 0.06(15) \end{cases}$

46. School supplies: A student bought a calculator and a textbook for a course in algebra. He told his friend that the total cost was \$170 (without tax) and that the calculator cost \$20 more than twice the cost of the textbook. What was the cost of each item?

Let x = the cost of the calculator
and y = the cost of the textbook.
The corresponding modeling system is $\begin{cases} x + y = 170 \\ x = 2y + 20 \end{cases}$

Writing and Thinking About Mathematics

47. Explain the advantages of solving a system of linear equations
a. by graphing.
b. by substitution.

HAWKES LEARNING SYSTEMS: INTRODUCTORY & INTERMEDIATE ALGEBRA SOFTWARE

- 8.2 Solving Systems of Linear Equations By Substitution

8.3 Systems of Linear Equations: Solutions by Addition

- *Solve systems of linear equations using the method of addition.*
- *Use systems of equations to find the equation of a line through two given points.*

Solving Systems of Linear Equations by Addition

We have discussed two methods for solving systems of linear equations:

1. **graphing**
2. **substitution**

We know that solutions found by graphing are not necessarily exact (estimation may be involved); and, in some cases, the method of substitution can lead to complicated algebraic steps. In this section, we consider a third method:

3. **addition** (or **method of elimination**)

In the **method of addition**, as with the method of substitution, the objective is to eliminate one of the variables so that a new equation is found with just one variable, if possible. If this new equation:

1. has one solution, the solution to the system is a **single point**.
2. is never true, the system has **no solution**.
3. is always true, the system has **infinite solutions**.

Consider solving the following system where both equations are written in standard form:

$$\begin{cases} x-2y=-9 \\ x+2y=11 \end{cases}$$

In the **method of addition**, we write one equation under the other so that like terms are aligned vertically. (Note that, in this example, the coefficients of y are opposites, namely -2 and $+2$.)

Then add like terms as follows:

$$\begin{aligned} x-2y &= -9 \\ \underline{x+2y} &\underline{= 11} \\ 2x \quad &= 2 \end{aligned}$$ The y terms are eliminated because the coefficients are opposites.

$$x = 1$$

Now substitute $x = 1$ into either of the original equations and solve for y.

$$\begin{aligned} 1 - 2y &= -9 \\ -2y &= -10 \\ y &= 5 \end{aligned} \qquad \text{OR} \qquad \begin{aligned} 1 + 2y &= 11 \\ 2y &= 10 \\ y &= 5 \end{aligned}$$

Therefore the **solution** to the system is $(1, 5)$.

This example was relatively easy because the coefficients for y were opposites. The general procedure can be outlined as follows.

To Solve a System of Linear Equations by Addition

1. Write the equations in **standard form** one under the other so that **like terms are aligned**.
2. Multiply all terms of one equation by a constant (and possibly all terms of the other equation by another constant) so that **two like terms have opposite coefficients**.
3. Add the two equations by **combining like terms** and solve the resulting equation, if possible.
4. **Back substitute into one of the original equations** to find the value of the other variable.
5. Check the solution (if there is one) in both of the original equations.

Example 1: A System with One Solution

Teaching Notes:
You may want to remind your students that when solving a system of equations, two equations may be interchanged. In other words, the order in which the equations are written has no bearing on the solution. If they think that writing one equation before another will make the system easier to solve, they may do so without affecting the solution.

Use the method of addition to solve the following system: $\begin{cases} 3x + 5y = -3 \\ -7x + 2y = 7 \end{cases}$

Solution: Multiply each term in the first equation by 2 and each term in the second equation by -5. This will result in the y-coefficients being opposites. Add the two equations by combining like terms which will eliminate y. Solve for x.

$$\begin{cases} 3x + 5y = -3 \\ -7x + 2y = 7 \end{cases}$$

$$\begin{cases} [2](3x + 5y = -3) \\ [-5](-7x + 2y = 7) \end{cases} \begin{aligned} &\longrightarrow \quad 6x + 10y = -6 \\ &\longrightarrow \quad \underline{35x - 10y = -35} \end{aligned}$$

$$\begin{aligned} 41x &= -41 \qquad y \text{ is eliminated.} \\ x &= -1 \end{aligned}$$

Continued on the next page...

Substitute $x = -1$ into either one of the original equations.

$$\begin{aligned} 3x+5y&=-3 \\ 3(-1)+5y&=-3 \\ -3+5y&=-3 \\ 5y&=0 \\ y&=0 \end{aligned} \qquad \text{OR} \qquad \begin{aligned} -7x+2y&=7 \\ -7(-1)+2y&=7 \\ 7+2y&=7 \\ 2y&=0 \\ y&=0 \end{aligned}$$

The solution is $(-1, 0)$.

Check: Substitution shows that $(-1, 0)$ satisfies **both** of the equations in the system.

$$\begin{aligned} 3x+5y&=-3 \\ 3(-1)+5(0)&\stackrel{?}{=}-3 \\ -3&=-3 \end{aligned} \qquad \begin{aligned} -7x+2y&=7 \\ -7(-1)+2(0)&\stackrel{?}{=}7 \\ 7&=7 \end{aligned}$$

NOTES In Example 1, we could have eliminated x instead of y by multiplying the terms in the first equation by 7 and the terms in the second equation by 3. The solution would be the same. Try this yourself to confirm this method.

Example 2: A System With Infinitely Many Solutions

Solve the following system by using the method of addition: $\begin{cases} 3x-\frac{1}{2}y=6 \\ 6x-y=12 \end{cases}$

Solution: Multiply the first equation by -2 so that the y-coefficients will be opposites.

$$\begin{cases} [-2]\left(3x-\frac{1}{2}y=6\right) \\ (6x-y=12) \end{cases} \begin{aligned} &\longrightarrow & -6x+y&=-12 \\ &\longrightarrow & 6x-y&=12 \\ \hline & & 0&=0 \end{aligned}$$

Because this last equation, $0 = 0$, is **always true**, the system has infinite solutions The solution set consists of all points that satisfy the equation $6x-y=12$. Solving for y gives $y=6x-12$, and we can write the solution in the general form $(x, 6x-12)$. (Or, solving for x, $x=\frac{1}{6}y+2$, and the solution can be written in the form $\left(\frac{1}{6}y+2, y\right)$.)

Example 3: A System with Decimals

Teaching Notes:
You may want to point out to your students that there is more than one set of numbers that they can multiply by to get opposite coefficients.

Solve the following system by using the method of addition: $\begin{cases} x + 0.4y = 3.08 \\ 0.1x - y = 0.1 \end{cases}$

Solution: Multiply the second equation by -10 so that the x-coefficients will be opposites.

$$\begin{cases} (x + 0.4y = 3.08) \longrightarrow & 1.0x + 0.4y = 3.08 \\ [-10](0.1x - y = 0.1) \longrightarrow & \underline{-1.0x + 10.0y = -1.0} \end{cases}$$

$$10.4y = 2.08 \qquad x \text{ is eliminated.}$$

$$y = 0.2$$

Substitute $y = 0.2$ into one of the original equations.

$$x + 0.4y = 3.08$$

$$x + 0.4(0.2) = 3.08$$

$$x + 0.08 = 3.08$$

$$x = 3$$

The solution is $(3, 0.2)$.

To check, substitute $x = 3$ and $y = 0.2$ in both of the original equations.

Example 4: Using Systems to Find the Equation of a Line

Using the formula $y = mx + b$, find the equation of the line determined by the two points $(3, 5)$ and $(-6, 2)$.

Solution: Write two equations in m and b by substituting the coordinates of the points for x and y. This gives the system

$$\begin{cases} 5 = 3m + b \\ 2 = -6m + b \end{cases}$$

Multiply the second equation by -1 so that the b-coefficients will be opposites.

$$\begin{cases} (5 = 3m + b) \longrightarrow & 5 = 3m + b \\ [-1](2 = -6m + b) \longrightarrow & \underline{-2 = 6m - b} \end{cases}$$

$$3 = 9m \qquad b \text{ is eliminated.}$$

$$\frac{1}{3} = m$$

Continued on the next page...

Substitute $m = \frac{1}{3}$ into one of the original equations.

$$5 = 3m + b$$
$$5 = 3\left(\frac{1}{3}\right) + b$$
$$5 = 1 + b$$
$$4 = b$$

The equation of the line is $y = \frac{1}{3}x + 4$.

Now that you know three methods for solving a system of linear equations (graphing, substitution, and addition), which method should you use? Consider the following guidelines when making your decision.

Guidelines for Deciding which Method to Use when Solving a System of Linear Equations

1. The graphing method is helpful in "seeing" the geometric relationship between the lines and finding approximate solutions. A calculator can be very helpful here.
2. Both the substitution method and the addition method give exact solutions.
3. The substitution method may be reasonable and efficient if one of the coefficients of one of the variables is 1.
4. The addition method is particularly efficient if the coefficients for one of the variables are opposites.

Practice Problems

Solve the following systems by using the method of addition.

1. $\begin{cases} 2x + 2y = 4 \\ x - y = -3 \end{cases}$

2. $\begin{cases} 3x + 4y = 12 \\ \frac{1}{3}x - 8y = -5 \end{cases}$

3. $\begin{cases} 2x - 5y = 6 \\ 4x - 10y = -2 \end{cases}$

4. $\begin{cases} 0.02x + 0.06y = 1.48 \\ 0.03x - 0.02y = 0.02 \end{cases}$

5. $\begin{cases} y = 3x + 15 \\ 6x - 2y = -30 \end{cases}$

Answers to Practice Problems: **1.** $\left(-\frac{1}{2}, \frac{5}{2}\right)$ **2.** $\left(3, \frac{3}{4}\right)$ **3.** No solution **4.** $(14, 20)$ **5.** $(x, 3x + 15)$

8.3 Exercises

1. $(4, 3)$
2. $(6, 1)$
3. $\left(1, -\frac{3}{2}\right)$
4. $(x, 4x-7)$
5. No solution
6. $(3, -1)$
7. $(x, 3-x)$
8. $(-4, -3)$
9. $(-2, -3)$
10. $(3, -2)$
11. $(7, 5)$
12. No solution
13. $(1, -5)$
14. $(x, 2x-2)$
15. $\left(x, \frac{1}{2}x-2\right)$
16. No solution
17. $\left(\frac{22}{7}, -\frac{2}{7}\right)$
18. $(-4, 6)$
19. $(2, -2)$
20. $(-3, -1)$
21. $(x, 2x-4)$
22. $(10, -4)$
23. $(2, -1)$
24. $(0, 0)$
25. $(5, -6)$
26. No solution
27. $(3, -1)$
28. $(10, 2)$
29. $(2, 4)$
30. $(7, 0)$
31. $(-6, 2)$
32. $(5, 6)$
33. No solution
34. $(-5, 2)$
35. $(20, 10)$

Solve each system by using the addition method.

1. $\begin{cases} 8x - y = 29 \\ 2x + y = 11 \end{cases}$ **2.** $\begin{cases} x + 3y = 9 \\ x - 7y = -1 \end{cases}$ **3.** $\begin{cases} 3x + 2y = 0 \\ 5x - 2y = 8 \end{cases}$

4. $\begin{cases} 12x - 3y = 21 \\ 4x - y = 7 \end{cases}$ **5.** $\begin{cases} 2x + 2y = 5 \\ x + y = 3 \end{cases}$ **6.** $\begin{cases} 2x - y = 7 \\ x + y = 2 \end{cases}$

7. $\begin{cases} 3x + 3y = 9 \\ x + y = 3 \end{cases}$ **8.** $\begin{cases} 9x + 2y = -42 \\ 5x - 6y = -2 \end{cases}$ **9.** $\begin{cases} \frac{1}{2}x + y = -4 \\ 3x - 4y = 6 \end{cases}$

10. $\begin{cases} x + y = 1 \\ x - \frac{1}{3}y = \frac{11}{3} \end{cases}$ **11.** $\begin{cases} x + y = 12 \\ 0.05x + 0.25y = 1.6 \end{cases}$ **12.** $\begin{cases} x + 0.1y = 8 \\ 0.1x + 0.01y = 0.64 \end{cases}$

Solve each system by using either the substitution method or the addition method (whichever seems better to you).

13. $\begin{cases} x = 11 + 2y \\ 2x - 3y = 17 \end{cases}$ **14.** $\begin{cases} 6x - 3y = 6 \\ y = 2x - 2 \end{cases}$ **15.** $\begin{cases} x - 2y = 4 \\ y = \frac{1}{2}x - 2 \end{cases}$

16. $\begin{cases} 2x + y = 3 \\ 4x + 2y = 7 \end{cases}$ **17.** $\begin{cases} x = 3y + 4 \\ y = 6 - 2x \end{cases}$ **18.** $\begin{cases} y = 2x + 14 \\ x = 14 - 3y \end{cases}$

19. $\begin{cases} 7x - y = 16 \\ 2y = 2 - 3x \end{cases}$ **20.** $\begin{cases} 3x + y = -10 \\ 2y - 1 = x \end{cases}$ **21.** $\begin{cases} 4x - 2y = 8 \\ 2x - y = 4 \end{cases}$

22. $\begin{cases} x + y = 6 \\ 2x + y = 16 \end{cases}$ **23.** $\begin{cases} 3x + 2y = 4 \\ x + 5y = -3 \end{cases}$ **24.** $\begin{cases} x + 2y = 0 \\ 2x = 4y \end{cases}$

25. $\begin{cases} 4x + 3y = 2 \\ 3x + 2y = 3 \end{cases}$ **26.** $\begin{cases} x - 3y = 4 \\ 3x - 9y = 10 \end{cases}$ **27.** $\begin{cases} 5x - 2y = 17 \\ 2x - 3y = 9 \end{cases}$

28. $\begin{cases} \frac{1}{2}x + 2y = 9 \\ 2x - 3y = 14 \end{cases}$ **29.** $\begin{cases} 3x + 2y = 14 \\ 7x + 3y = 26 \end{cases}$ **30.** $\begin{cases} 4x + 3y = 28 \\ 5x + 2y = 35 \end{cases}$

31. $\begin{cases} 2x + 7y = 2 \\ 5x + 3y = -24 \end{cases}$ **32.** $\begin{cases} 7x - 6y = -1 \\ 5x + 2y = 37 \end{cases}$ **33.** $\begin{cases} 10x + 4y = 7 \\ 5x + 2y = 15 \end{cases}$

34. $\begin{cases} 6x - 5y = -40 \\ 8x - 7y = -54 \end{cases}$ **35.** $\begin{cases} 0.5x - 0.3y = 7 \\ 0.3x - 0.4y = 2 \end{cases}$ **36.** $\begin{cases} 0.6x + 0.5y = 5.9 \\ 0.8x + 0.4y = 6 \end{cases}$

37. $\begin{cases} 2.5x + 1.8y = 7 \\ 3.5x - 2.7y = 4 \end{cases}$

38. $\begin{cases} 0.75x - 0.5y = 2 \\ 1.5x - 0.75y = 7.5 \end{cases}$

39. $\begin{cases} \frac{2}{3}x - \frac{1}{2}y = \frac{2}{3} \\ \frac{8}{3}x - 2y = \frac{17}{6} \end{cases}$

40. $\begin{cases} \frac{3}{4}x + \frac{1}{4}y = \frac{3}{8} \\ \frac{3}{2}x + \frac{1}{2}y = \frac{3}{4} \end{cases}$

41. $\begin{cases} \frac{1}{6}x - \frac{1}{12}y = -\frac{13}{6} \\ \frac{1}{5}x + \frac{1}{4}y = 2 \end{cases}$

42. $\begin{cases} \frac{5}{3}x - \frac{2}{3}y = -\frac{29}{30} \\ 2x + 5y = 0 \end{cases}$

Write an equation for the line determined by the two given points using the formula $y = mx + b$ to set up a system of equations with m and b as the unknowns.

43. $(2, 3), (1, -2)$

44. $(0, 6), (-3, -3)$

45. $(1, -3), (5, -3)$

46. $(5, 3), (5, -4)$

47. $(1, 2), (-3, 0)$

48. $(-4, 2), (5, -1)$

A word problem is stated with equations given that represent a mathematical model for the problem. Solve the system by using either the method of substitution or the method of addition.

49. Investing: Georgia had \$10,000 to invest, and she put the money into two accounts. One of the accounts will pay 6% interest and the other will pay 10%. How much did she put in each account if the interest from the 10% account exceeded the interest from the 6% account by \$40?
Let x = amount in 10% account and y = amount in 6% account.

The system that models the problem is $\begin{cases} x + y = 10{,}000 \\ 0.10x - 0.06y = 40 \end{cases}$

50. Baseball: A minor league baseball team has a game attendance of 4500 people. Tickets cost \$5 for children and \$8 for adults. The total revenue made at this game was \$26,100. How many adults and how many children attended the game?
Let x = adults and y = children.

The system that models the problem is $\begin{cases} x + y = 4500 \\ 8x + 5y = 26{,}100 \end{cases}$

51. Acid solutions: How many liters each of a 30% acid solution and a 40% acid solution must be used to produce 100 liters of a 36% acid solution?
Let x = amount of 30% solution and y = amount of 40% solution.

The system that models the problem is $\begin{cases} x + y = 100 \\ 0.30x + 0.40y = 0.36(100) \end{cases}$

---

36. $(4, 7)$
37. $\left(2, \frac{10}{9}\right)$
38. $(12, 14)$
39. No solution
40. $\left(x, -3x + \frac{3}{2}\right)$
41. $\left(-\frac{45}{7}, \frac{92}{7}\right)$
42. $\left(-\frac{1}{2}, \frac{1}{5}\right)$
43. $y = 5x - 7$; $m = 5, b = -7$
44. $y = 3x + 6$; $m = 3, b = 6$
45. $y = -3$; $m = 0$, $b = -3$
46. $x = 5$; m = Undefined b = None
47. $y = \frac{1}{2}x + \frac{3}{2}$; $m = \frac{1}{2}, b = \frac{3}{2}$
48. $y = -\frac{1}{3}x + \frac{2}{3}$; $m = -\frac{1}{3}, b = \frac{2}{3}$
49. \$4000 at 10%; \$6000 at 6%
50. 1200 adults; 3300 children
51. 40 liters of 30% solution (x), 60 liters of 40% solution (y)

52. $x = \frac{7}{2}$ hours, $y = \frac{7}{2}$ hours

53. Answers will vary.

52. Traveling by car: Two cars leave Denver at the same time traveling in opposite directions. One travels at an average speed of 55 mph and the other at 65 mph. In how many hours will they be 420 miles apart?

Let x = time of travel for first car and y = time of travel for second car.

The system that models the problem is $\begin{cases} x = y \\ 55x + 65y = 420 \end{cases}$

Writing and Thinking About Mathematics

53. Explain, in your own words, why the answer to a system with infinite solutions is written as an ordered pair with variables.

HAWKES LEARNING SYSTEMS: INTRODUCTORY & INTERMEDIATE ALGEBRA SOFTWARE

- 8.3 Solving Systems of Linear Equations by Addition

8.4 Applications: Distance-Rate-Time, Number Problems, Amounts and Costs

Use systems of linear equations to solve the following types of applied problems:
- *distance-rate-time,*
- *numbers, and*
- *amounts and costs.*

Solving Systems of Equations as They Relate to Applied Problems

Teaching Notes:
Without getting too mathematical or formal, it might be helpful to have students think of applied situations in which two or more conditions need to be considered when designing or doing something. (For example: the weight of a car for safety versus fuel economy.)

Systems of equations occur in many practical situations such as
supply and demand in business,
velocity and acceleration in engineering,
money and interest in investments, and
mixture in physics and chemistry.

Many of the problems in earlier sections illustrated these ideas. However, in those exercises, the system of equations was given. As you study the applications in Sections 8.4 and 8.5, you may want to refer to some of those exercises, as well as the examples, as guides in solving the given applications. **For these applications you will need to create your own systems of equations.**

Remember that the emphasis in all application problems is to develop your reasoning as well as your reading skills. You must learn how to transfer English phrases into algebraic expressions. This is the THINKING PART.

Example 1: Distance-Rate-Time (Rates Unknown)

A small plane flew 300 miles in 2 hours flying with the wind. Then on the return trip, flying against the wind, it traveled only 200 miles in 2 hours. What were the wind speed and the speed of the plane? (**Note:** The "speed of the plane" means how fast the plane would be flying with no wind.)

Solution: Let s = speed of plane
and w = wind speed

When flying with the wind, the plane's actual rate will increase to $s + w$. When flying against the wind, the plane's actual rate will decrease to $s - w$. (**Note:** If the wind had been strong enough, the plane could actually have been flying backward, away from its destination.)

	Rate ×	Time =	Distance
With the wind	$s+w$	2	$2(s+w)$
Against the wind	$s-w$	2	$2(s-w)$

The system of linear equations is

with the wind → $\begin{cases} 2(s+w)=300 \\ 2(s-w)=200 \end{cases}$ ← against the wind

$$\begin{aligned} 2s+2w&=300 \\ 2s-2w&=200 \\ \hline 4s&=500 \\ s&=125 \end{aligned}$$ Speed of plane

Back substitute $s = 125$ into one of the original equations.

$$2(125+w)=300$$
$$125+w=150$$
$$w=25$$ Wind speed

The speed of the plane was 125 mph, and the wind speed was 25 mph.

Example 2: Distance-Rate-Time (Times Unknown)

Two buses leave a bus station traveling in opposite directions. One leaves at noon and the second leaves at 1 PM. The first one travels at an average speed of 55 mph and the second one at an average speed of 59 mph. At what time will the buses be 226 miles apart?

Solution: Let x = time of travel for first bus
and y = time of travel for second bus

	Rate ×	Time =	Distance
Bus 1	55	x	$55x$
Bus 2	59	y	$59y$

Continued on the next page...

The system of linear equations is

$$\begin{cases} x = y+1 & \text{The first bus travels 1 hour longer than the second.} \\ 55x+59y=226 & \text{The sum of the distances is 226 miles.} \end{cases}$$

Substitution gives

$$\begin{aligned} 55(y+1)+59y &= 226 \\ 55y+55+59y &= 226 \\ 114y &= 171 \\ y &= 1.5 \end{aligned}$$

This gives $x = 1.5 + 1 = 2.5$.

Thus the first bus travels 2.5 hours and the second bus travels 1.5 hours. The buses will be 226 miles apart at 2:30 P.M.

Example 3: Number Problem

Teaching Notes:
Anticipate that some students will recognize many of these problems as solvable with one variable rather than two. Discuss with them that these simple examples can be considered as practice of the method of substitution or addition for more difficult problems. Also, thinking in terms of two or more variables becomes a way of thinking and can be used to solve many practical problems.

The sum of two numbers is 80 and their difference is 10. What are the two numbers?

Solution: Let x = one number
and y = the other number.

The system of linear equations is

$$\begin{cases} x+y=80 & \text{The sum is 80.} \\ x-y=10 & \text{The difference is 10.} \end{cases}$$

Solving by addition gives

$$\begin{aligned} x+y &= 80 \\ x-y &= 10 \\ \hline 2x &= 90 \\ x &= 45 \end{aligned}$$

Back substituting 45 for x in the first equation gives: $45+y=80$
$y = 80-45 = 35$

The two numbers are 45 and 35.

Check: $45 + 35 = 80$ and $45 - 35 = 10$

Example 4: Counting Coins

Mike has \$1.05 worth of change in nickels and quarters. If he has twice as many nickels as quarters, how many of each type of coin does he have?

Solution: We use two equations – one relating the number of coins and the other relating the value of the coins. (The value of each nickel is 5 cents and the value of each quarter is 25 cents.)

Let $n =$ number of nickels
and $q =$ number of quarters.

Coins	# of coins	Value	Total value
Nickels	n	0.05	$0.05n$
Quarters	q	0.25	$0.25q$

The system of linear equations is

$$\begin{cases} n = 2q & \text{There are two times as many nickels as quarters.} \\ 0.05n + 0.25q = 1.05 & \text{The total amount of money is \$1.05.} \end{cases}$$

Note carefully that in the first equation q is multiplied by 2 because the number of nickels is twice the number of quarters. Therefore, n is bigger.

The first equation is already solved for n, so substituting $2q$ for n in the second equation gives

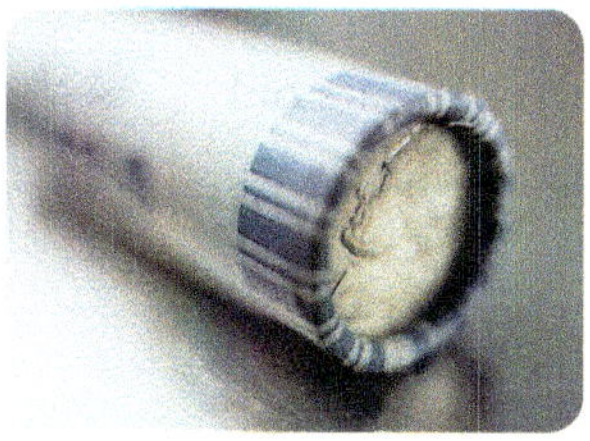

$$0.05(2q) + 0.25q = 1.05$$

$$0.10q + 0.25q = 1.05$$

$$10q + 25q = 105 \qquad \text{Multiply the equation by 100.}$$

$$35q = 105$$

$$q = 3 \qquad \text{Number of quarters}$$

$$n = 2q = 6 \qquad \text{Number of nickels}$$

Mike has 3 quarters and 6 nickels.

Example 5: Calculating Age

Kathy is 6 years older than her sister, Sue. In 3 years, she will be twice as old as Sue. How old is each girl now?

Solution: We use two equations – one relating their ages now and the other relating their ages in 3 years.

Let K = Kathy's age now
and S = Sue's age now.

Then the system of linear equations is

$$\begin{cases} K - S = 6 & \text{The difference in their ages is 6 years.} \\ K + 3 = 2(S + 3) & \text{In 3 years each age is increased by 3 and Kathy is twice as old as Sue.} \end{cases}$$

Rewrite the second equation in standard form and solve by addition.

$$K + 3 = 2(S + 3)$$

$$K + 3 = 2S + 6$$

$$K - 2S = 3 \quad \text{Second equation in standard form}$$

$$\begin{cases} (K - S = 6) \longrightarrow & K - S = 6 \\ [-1](K - 2S = 3) \longrightarrow & \underline{-K + 2S = -3} \end{cases}$$

$$S = 3 \quad \text{Sue's current age}$$

Back substitute $S = 3$ into one of the original equations.

$$K - 3 = 6$$

$$K = 6 + 3$$

$$K = 9 \quad \text{Kathy's current age}$$

Kathy is 9 years old; Sue is 3 years old.

Example 6: Amounts and Costs

Three hot dogs and two orders of French fries cost \$10.30. Four hot dogs and four orders of fries cost \$15.60. What is the cost of a hot dog? What is the cost of an order of fries?

Solution: Let x = cost of one hot dog
and y = cost of one order of fries.

Then the system of linear equations is

$$\begin{cases} 3x+2y=10.30 & \text{Three hot dogs and two orders of French fries cost \$10.30.} \\ 4x+4y=15.60 & \text{Four hot dogs and four orders of fries cost \$15.60.} \end{cases}$$

Both equations are in standard form. Solve using the addition method.

$$\begin{cases} [-2](3x+2y=10.30) \longrightarrow -6x-4y=-20.60 \\ \quad\;\;(4x+4y=15.60) \longrightarrow \;\;\;4x+4y=\;\;15.60 \end{cases}$$

$$-2x = -5.00$$

$$x = 2.50 \quad \text{Cost of one hot dog}$$

Back substitute $x = 2.50$ into one of the original equations.

$$3(2.50)+2y=10.30$$

$$7.50+2y=10.30$$

$$2y=2.80$$

$$y=1.40 \quad \text{Cost of one order of fries}$$

One hot dog costs \$2.50 and one order of fries costs \$1.40.

NOTES

You should consider making tables similar to those illustrated in Examples 1, 2, and 4 when working with applications. These tables can help you organize the information in a more understandable form.

8.4 Exercises

Solve each problem by setting up a system of two equations in two unknowns and solving the system.

1. 23, 33
2. 14, 26
3. 15, 21
4. 23, 79
5. 37, 50

1. The sum of two numbers is 56. Their difference is 10. Find the numbers.

2. The sum of two numbers is 40. The sum of twice the larger and 4 times the smaller is 108. Find the numbers.

3. The sum of two numbers is 36. Three times the smaller plus twice the larger is 87. Find the two numbers.

4. The sum of two integers is 102, and the larger number is 10 more than three times the smaller. Find the two integers.

5. The difference between two integers is 13, and their sum is 87. What are the two integers?

6. 8, 25
7. 80°, 100°
8. 75°, 15°
9. 55°, 55°, 70°
10. 65°, 65°, 50°
11. The rate of the boat is 10 mph and the rate of the current is 2 mph.
12. The speed of the plane is 165 mph and the speed of the wind is 33 mph
13. Bolt's speed was 10.32 meters per second and the wind speed was 0.12 meters per second.
14. The rate of the boat was 38 mph and the rate of the current was 2 mph.
15. He traveled $1\frac{1}{2}$ hours at the first rate and 2 hours at the second rate.

6. The difference between two numbers is 17. Four times the smaller is equal to 7 more than the larger. What are the numbers?

7. Supplementary angles: Two angles are supplementary if the sum of their measures is 180°. Find two supplementary angles such that the smaller is 30° more than one-half of the larger.

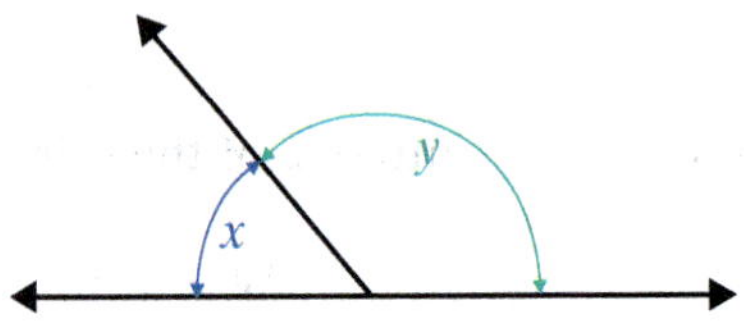

8. Complementary angles: Two angles are complementary if the sum of their measures is 90°. Find two complementary angles such that one is 15° less than six times the other.

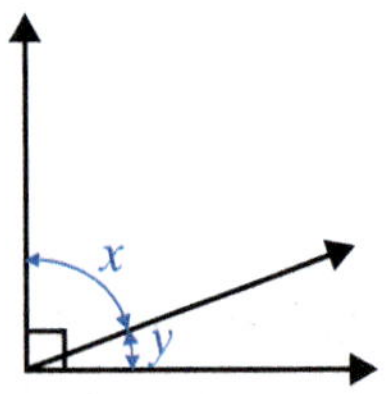

9. Triangles: The sum of the measures of the three angles of a triangle is 180°. In an isosceles triangle, two of the angles have the same measure. What are the measures of the angles of an isosceles triangle in which one angle measures 15° more than each of the other two equal angles?

10. Triangles: The sum of the measures of the three angles of a triangle is 180°. In an isosceles triangle, two of the angles have the same measure. What are the measures of the angles of an isosceles triangle in which each of the two equal angles measures 15° more than the third angle?

11. Boating: Liam makes a 4-mile motorboat trip downstream in 20 minutes $\left(\frac{1}{3}\text{ hr}\right)$. The return trip takes 30 minutes $\left(\frac{1}{2}\text{ hr}\right)$. Find the rate of the boat in still water and the rate of the current.

12. Flying an airplane: Mr. McKelvey finds that flying with the wind he can travel 1188 miles in 6 hours. However, when flying against the wind, he travels only $\frac{2}{3}$ of the distance in the same amount of time. Find the speed of the plane in still air and the wind speed.

13. Running: Usain Bolt, the world-record holder in the 100 meter dash, ran 100 meters in 9.69 seconds with no wind. He later ran the same distance in 9.58 seconds with the wind. What was his speed and what was the wind speed?

14. Boating: Jessica drove her speedboat upriver this morning. It took her 1 hour going upriver and 54 minutes going downriver. If she traveled 36 miles each way, what would have been the rate of the boat in still water and what was the rate of the current (in miles per hour)?

15. Traveling by car: Randy made a business trip of 190 miles. He averaged 52 mph for the first part of the trip and 56 mph for the second part. If the total trip took $3\frac{1}{2}$ hours, how long did he travel at each rate?

16. She traveled $3\frac{3}{4}$ hours at the first rate and 2 hours at the second rate.
17. Marcos traveled at 40 mph and Cana traveled at 51 mph.
18. Naomi's speed was 58 mph and Linda's speed was 50 mph.
19. Steve traveled at 28 mph and Tim traveled at 7 mph.
20. Bella's speed was 45 mph and June's speed was 25 mph
21. The westbound train was traveling 45 mph and the eastbound train was traveling 40 mph.
22. 6:00 pm
23. He jogged about 12 miles.
24. She traveled 45 miles each way.
25. 20 nickels and 10 dimes
26. 18 quarters and 9 dimes
27. 52 nickels and 130 pennies

16. Traveling by car: Marian drove to a resort 335 miles from her home. She averaged 60 mph for the first part of her trip and 55 mph for the second part. If her total driving time was $5\frac{3}{4}$ hours, how long did she travel at each rate?

17. Traveling by car: Marcos lives 364 miles away from his cousin Cana. They start driving at the same time and travel toward each other. Cana's speed is 11 mph faster than Marcos' speed. If they meet in 4 hrs, find their speeds.

18. Traveling by car: Naomi and Linda live 324 miles apart. They start at the same time and travel toward each other. Naomi's speed is 8 mph greater than Linda's. If they meet in 3 hours, find their speeds.

19. Traveling by car: Steve travels 4 times as fast as Tim. Starting at the same point, but traveling in opposite directions, they are 105 miles apart after 3 hours. Find their rates of travel.

20. Traveling by car: Bella travels 5 mph less than twice as fast as June. Starting at the same point and traveling in the same direction, they are 80 miles apart after 4 hours. Find their speeds.

21. Traveling by train: Two trains leave Dallas at the same time. One train travels east and the other travels west. The speed of the westbound train is 5 mph greater than the speed of the eastbound train. After 6 hours, they are 510 miles a part. Find the rate of each train. Assume the trains travel in a straight line in opposite directions.

22. Boating: A boat left Dana Point Marina at 11:00 am traveling at 10 knots (nautical miles per hour). Two hours later, a Coast Guard boat left the same marina traveling at 14 knots trying to catch the first boat. If both boats traveled the same course, at what time did the Coast Guard captain anticipate overtaking the first boat?

23. Jogging: A jogger runs into the countryside at a rate of 10 mph. He returns along the same route at 6 mph. If the total trip took 1 hour 36 minutes, how far did he jog?

24. Biking: A cyclist traveled to her destination at an average rate of 15 mph. By traveling 3 mph faster, she took 30 minutes less to return. What distance did she travel each way?

25. Coin collecting: Sonja has some nickels and dimes. If she has 30 coins worth a total of $2.00, how many of each type of coin does she have?

26. Coin collecting: Conner has a total of 27 coins consisting of quarters and dimes. The total value of the coins is $5.40. How many of each type of coin does he have?

27. Coin collecting: A bag contains pennies and nickels only. If there are 182 coins in all and their value is $3.90, how many pennies and how many nickels are in the bag?

28. 36 quarters and 29 dimes
29. Length is 14 meters and width is 8 meters
30. Length is 33 meters and width is 17 meters
31. Length is 40 feet and width is 30 feet
32. Length is 97.5 meters and width is 32.5 meters
33. 100 yards × 45 yards
34. Side of square is 15 feet, side of hexagon is 10 feet
35. Ava is 12 years old and Curt is 4 years old.

28. Coin collecting: Your friend challenges you to figure out how many dimes and quarters are in a cash register. He tells you that there are 65 coins and that their value is \$11.90. How many dimes and how many quarters are in the register?

29. Rectangles: The length of a rectangle is 10 meters more than one-half the width. If the perimeter is 44 meters, what are the length and width?

30. Rectangles: The length of a rectangle is 1 meter less than twice the width. If each side is increased by 4 meters, the perimeter will be 116 meters. Find the length and the width of the original rectangle.

31. Rectangles: The width of a rectangle is $\frac{3}{4}$ of its length. If the perimeter of the rectangle is 140 feet, what are the dimensions of the rectangle?

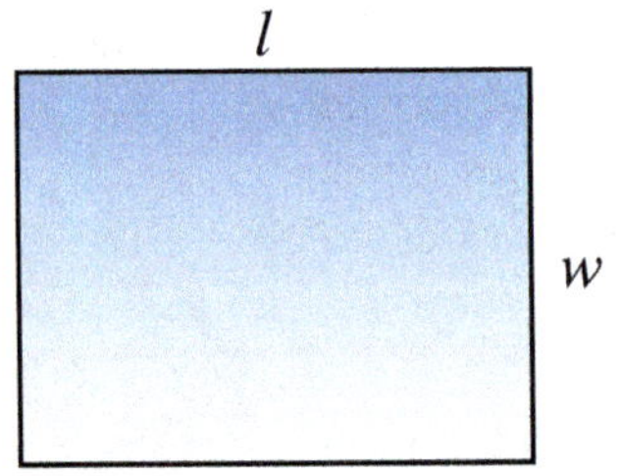

32. Building a fence: A farmer has 260 meters of fencing to build a rectangular corral. He wants the length to be 3 times as long as the width. What dimensions should he make his corral?

33. Soccer: At present, the length of a rectangular soccer field is 55 yards longer than the width. The city council is thinking of rearranging the area containing the soccer field into two square playing fields. A math teacher on the council decided to test the council members' mathematical skills. (You know how math teachers are.) He told them that if the width of the current field were to be increased by 5 yards and the length cut in half, the resulting field would be a square. What are the dimensions of the field currently?

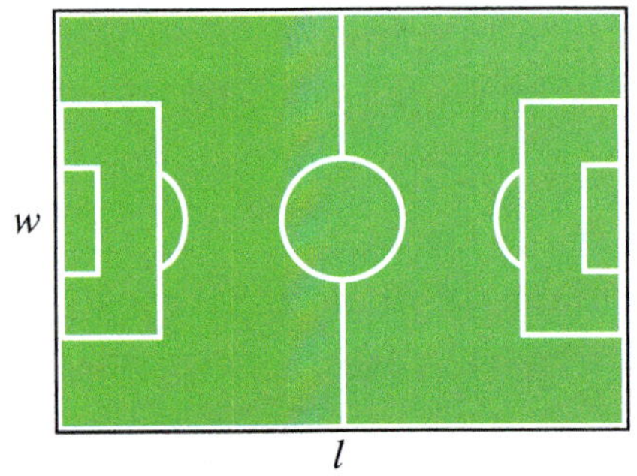

34. Perimeter: Consider a square and a regular hexagon (a six-sided figure with sides of equal length). One side of the square is 5 feet longer than a side of the hexagon, and the two figures have the same perimeter. What are the lengths of the sides of each figure?

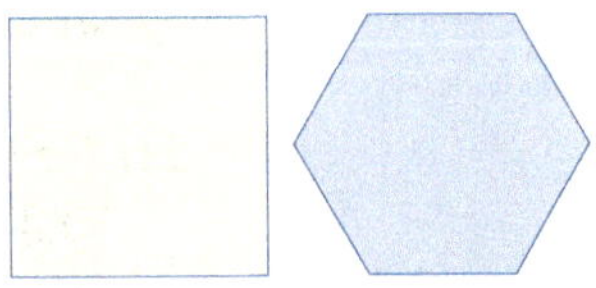

35. Age: Ava is 8 years older than her brother Curt. Four years from now, Ava will be twice as old as Curt. How old is each at the present time?

36. Priscilla was 21 years old and Elvis was 32 years old.
37. 800 adults and 2700 students attended.
38. 250 adult tickets and 70 children's tickets were sold.
39. She bought 10 paperbacks and 5 hardbacks.
40. Morton has 140 pounds of newspapers and 40 pounds of cans.
41. 5000 general admission and 7500 reserved tickets were sold.
42. 1 adult ticket is $3 and 1 child's ticket is $2.
43. 22 at $625 and 25 at $550
44. 16 packages at $1.25 and 26 packages at $1.75

36. Age: When they got married, Elvis Presley was 11 years older than his wife Priscilla. One year later, Priscilla was two-thirds of Elvis' age. How old was each of them when they got married?

37. Basketball admission: Tickets for the local high school basketball game were priced at $3.50 for adults and $2.50 for students. If the income for one game was $9550 and the attendance was 3500, how many adults and how many students attended that game?

38. Charity admission: A Christmas charity party sold tickets for $45.00 for adults and $25.00 for children. The total number of tickets sold was 320 and the total for the ticket sales was $13,000. How many adult and how many children's tickets were sold?

39. Buying books: Joan went to a book sale on campus and bought paperback books for $0.25 each and hardback books for $1.75 each. If she bought a total of 15 books for $11.25, how many of each type of book did she buy?

40. Recycling: Morton took some old newspapers and aluminum cans to the recycling center. Their total weight was 180 pounds. He received 1.5¢ per pound for the newspapers and 30¢ per pound for the cans. The total received was $14.10. How many pounds of each did Morton have?

41. Baseball admission: Admission to the baseball game is $2.00 for general admission and $3.50 for reserved seats. The receipts were $36,250 for 12,500 paid admissions. How many of each ticket, general and reserved, were sold?

42. Going to the theater: Seventy children and 160 adults attended a play. The total receipts were $620. One adult ticket and 2 children's tickets cost $7. Find the price of each type of ticket.

43. Surfing: Last summer, Ernie sold surfboards. One style sold for $625 and the other sold for $550. He sold a total of 47 surfboards. How many of each style did he sell if the sales from each style were equal?

44. Selling Candy: The Candy Shack sells a particular candy in two different size packages. One size sells for $1.25 and the other sells for $1.75. If the store received $65.50 for 42 packages of candy, how many of each size were sold?

45. 360 balls or 30 dozen
46. $140,000 in sales for the improved properties and $80,000 in sales for the unimproved properties
47. The store sold 22 of the $95 jackets and 18 of the $120 jackets.
48. $14 for shirts and $27 for pair of dress pants.

49. One Big Mac costs $3.58 and one order of medium French Fries costs $1.79.
50. 30 of the 42 cent stamps, 50 of the 59 cent stamps
51. They produced 7 of Model X and 10 of Model Y.
52. They built 12 using Style I and 16 using Style II.

45. Golfing: The pro shop at the Divots Country Club ordered two brands of golf balls. Titleless balls cost $1.80 each and the Done Lob balls cost $1.50 each. The total cost of Titleless balls exceeded the total cost of the Done Lob balls by $108. If equal numbers of each brand were ordered, how many dozen of each brand were ordered?

46. Real estate: Sellit Realty Company gets a 6% fee for selling improved properties and 10% for selling unimproved land. Last week, the total sales were $220,000 and their total fees were $16,400. What were the sales from each of the two types of properties?

47. Shopping: A men's clothing store sells two styles of sports jackets, one selling for $95 and one selling for $120. Last month, the store sold 40 jackets, with receipts totaling $4250. How many of each style did the store sell?

48. Shopping: Frank bought 2 shirts and 1 pair of dress pants for a total of $55. If he had bought 1 shirt and 2 pairs of dress pants, he would have paid $68. What was the price of each shirt and each pair of dress pants?

49. Fast food: At McDonalds, 3 Big Macs and 5 orders of medium French Fries cost $19.69. 6 Big Macs and 2 orders of medium French Fries cost $25.06. What is the price of a Big Mac? What is the price of one order of medium French Fries?

50. Stamp collecting: The postal service charges 42¢ for letters that weigh 1 ounce or less and 17¢ more for letters that weigh between 1 and 2 ounces. Jeff, testing his father's math skills, gave his father $42.10 and asked him to purchase 80 stamps for his stamp collection, some 42¢ stamps and some 59¢ stamps. How many of each type of stamp did his dad buy if he used all the money?

51. Manufacturing: A small manufacturer produces two kinds of radios, model X and model Y. Model X takes 4 hours to produce and costs $8 each to make. Model Y takes 3 hours to produce and costs $7 each to make. If the manufacturer decides to allot a total of 58 hours and $126 each week, how many of each model will be produced?

52. Furniture building: A furniture shop makes dining room chairs. Employees can build two styles of chairs. Style I takes 1 day and the materials cost $60. Style II takes $1\frac{1}{2}$ days but the materials only costs $30. If, during the last two months, they spent 36 days and $1200 building chairs, how many chairs of each style did they build?

Writing and Thinking About Mathematics

53. The number is 49.

53. A two digit number can be written as ab, where a and b are the digits. We do **not** mean that the digits are multiplied, but the value of the number is $10a + b$. For example, the two digit number 34 has a value of $10 \cdot 3 + 4$. Set up and solve a system of equations for the following problem:

The sum of the digits of a two digit number is 13. If the digits are reversed, then the value of the number is increased by 45. What is the number?

HAWKES LEARNING SYSTEMS: INTRODUCTORY & INTERMEDIATE ALGEBRA SOFTWARE

- 8.4 Applications: Distance-Rate-Time, Number Problems, Amounts and Costs

8.5 Applications: Interest and Mixture

Use systems of linear equations to solve the following types of applied problems:

- *interest, and*
- *mixture.*

In this section we will study two more types of applications that can be solved using systems of linear equations; interest on money invested and mixture. These applications can be "wordy" and you will need to read carefully and analyze the information thoroughly to be able to translate it into a system involving two variables.

Interest

People in business and banking know several formulas for calculating interest. The formula used depends on the method of payment (monthly or yearly) and the type of interest (simple or compound). Also, penalties for late payments and even penalties for early payments might be involved. In any case, standard notation is the following:

P ⟶ the principal (amount of money invested or borrowed)
r ⟶ the rate of interest (an annual rate)
t ⟶ the time (in one year or part of a year)
I ⟶ the interest (paid or earned)

In this section, we will use only the basic formula for simple interest

$$I = Prt$$

with interest calculated on an annual basis. In this special case, we have $t = 1$ and the formula becomes

$$I = Pr.$$

Example 1: Interest

James has two investment accounts, one pays 6% interest and the other pays 10% interest. He has \$1000 more in the 10% account than he has in the 6% account. Each year, the interest from the 10% account is \$260 more than the interest from the 6% account. How much does he have in each account?

Solution: Careful reading indicates two types of information:

1. He has two accounts.
2. He earns two amounts of interest.

Let x = amount (principal) invested at 6%
y = amount (principal) invested at 10%,

then $0.06x$ = interest earned on first account
$0.10y$ = interest earned on second account.

Now set up two equations.

$$\begin{cases} y = x + 1000 & y \text{ is larger than } x \text{ by \$1000.} \\ 0.10y - 0.06x = 260 & \text{Interest from the 10\% account is \$260 more than interest from the 6\% account.} \end{cases}$$

Because the first equation is already solved for y, we use the substitution method and substitute for y in the second equation.

$$0.10(x + 1000) - 0.06x = 260$$

$$10(x + 1000) - 6x = 26{,}000 \quad \text{Multiply by 100.}$$

$$10x + 10{,}000 - 6x = 26{,}000$$

$$4x = 16{,}000$$

$$x = 4000 \quad \text{Amount at 6\%}$$

Back substitute $x = 4000$ into one of the original equations to find y.

$$y = x + 1000 = 4000 + 1000 = 5000 \quad \text{Amount at 10\%}$$

James has \$4000 invested at 6% and \$5000 invested at 10%.

Example 2: Interest

Lila has \$7000 to invest. She decides to separate her funds into two accounts. One yields interest at the rate of 7% and the other at 12%. If she wants a total annual income from both accounts to be \$690, how should she split the money? (**Note:** The higher interest account is considered more risky. Otherwise, she would put the entire \$7000 into that account.)

Solution: Again, careful reading indicates two types of information:

1. She has two accounts.
2. She earns two amounts of interest.

Let x = amount (principal) invested at 7%
y = amount (principal) invested at 12%,

then $0.07x$ = interest earned on first account
$0.12y$ = interest earned on second account.

Now set up two equations.

$$\begin{cases} x + y = 7000 & \text{The sum of both accounts is \$7000.} \\ 0.07x + 0.12y = 690 & \text{The total interest from both accounts is \$690.} \end{cases}$$

Continued on the next page...

Both equations are in standard form. Solve by addition. Multiply the first equation by –7 and the second by 100 to get opposite coefficients for x as follows:

$$\begin{cases} [-7] \quad (x + \quad y = 7000) \longrightarrow & -7x - 7y = -49{,}000 \\ [100](0.07x + 0.12y = \ 690) \longrightarrow & \underline{7x + 12y = \ 69{,}000} \end{cases}$$

$$5y = 20{,}000$$

$$y = 4000 \quad \text{Amount at 12\%}$$

Substitute $y = 4000$ into one of the original equations.

$$x + 4000 = 7000$$

$$x = 3000 \quad \text{Amount at 7\%}$$

She should invest \$3000 at 7% and \$4000 at 12%.

Mixture

Problems involving mixtures occur in physics and chemistry and in such places as candy stores or coffee shops. Two or more items of a different percentage of concentration of a chemical such as salt, chlorine, or antifreeze are to be mixed; or two or more types of food such as coffee, nuts, or candy are to be mixed to form a final mixture that satisfies certain conditions of percentage of concentration.

The basic plan is to write an equation that deals with only one part of the mixture (such as the salt in the mixture). The following examples explain how this can be accomplished.

Example 3: Mixture

How many ounces each of a 10% salt solution and a 15% salt solution must be used to produce 50 ounces of a 12% salt solution?

Solution: Let x = amount of 10% solution
and y = amount of 15% solution.

	Amount of solution	× Percent of salt	= Amount of salt
10% solution	x	0.10	$0.10x$
15% solution	y	0.15	$0.15y$
12% solution	50	0.12	$0.12(50)$

Then the system of linear equations is

$$\begin{cases} x+y=50 & \text{The sum of the two amounts must be 50 ounces.} \\ 0.10x+0.15y=0.12(50) & \text{The sum of the amounts of salt from the two solutions equals the total amount of salt in the final solution.} \end{cases}$$

Multiplying the first equation by –10 and the second by 100 gives

$$\begin{cases} [-10]\quad(x+\quad y=50) & \longrightarrow & -10x-10y=-500 \\ [100](0.10x+0.15y=0.12(50)) & \longrightarrow & \underline{10x+15y=600} \end{cases}$$

$$5y = 100$$

$$y = 20 \quad \text{Amount of 15\%}$$

Substitute $y = 20$ into one of the original equations.

$$x+20=50$$

$$x=30 \quad \text{Amount of 10\%}$$

Use 30 ounces of the 10% solution and 20 ounces of the 15% solution.

Teaching Notes:
Because "mixture" and "interest" appear to represent different contexts, students may not recognize that the structures of the equations are the same, or that the interest context is a mixture. You might emphasize the fact that both types of problems involve percent.

Example 4: Mixture

How many gallons of a 20% acid solution should be mixed with a 30% acid solution to produce 100 gallons of a 23% solution?

Solution: Let x = amount of 20% solution
and y = amount of 30% solution.

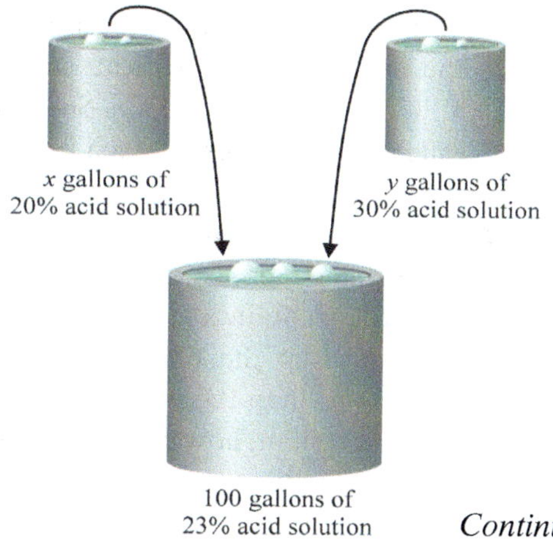

Continued on the next page...

	Amount of solution ×	Percent of acid =	Amount of acid
20% solution	x	0.20	$0.20x$
30% solution	y	0.30	$0.30y$
23% solution	100	0.23	$0.23(100)$

Then the system of linear equations is

$$\begin{cases} x+y=100 & \text{The sum of the two amounts must be 100 gallons.} \\ 0.20x+0.30y=0.23(100) & \text{The sum of the amounts of acid from the two solutions equals the total amount of acid in the final solution.} \end{cases}$$

Multiplying the first equation by –20 and the second by 100 gives,

$$\begin{cases} [-20]\quad (x+\quad y=100) \longrightarrow -20x-20y=-2000 \\ [100](0.20x+0.30y=0.23(100)) \longrightarrow \underline{20x+30y=2300} \end{cases}$$

$$10y=300$$

$$y=30 \quad \text{Amount of 30\%}$$

Substitute $y=30$ into one of the original equations.

$$x+30=100$$

$$x=70 \quad \text{Amount of 20\%}$$

Seventy gallons of the 20% solution should be added to thirty gallons of the 30% solution. This will produce 100 gallons of a 23% solution.

8.5 Exercises

Solve each problem by setting up a system of two equations in two unknowns and solving the system.

1. $5500 at 6%; $3500 at 10%
2. $8500 at 6%; $3500 at 8%
3. $7400 at 5.5%; $2600 at 6%

1. **Investing:** Carmen invested $9000, part in a 6% passbook account and the rest in a 10% certificate account. If her annual interest was $680, how much did she invest at each rate?

2. **Investing:** Mr. Brown has $12,000 invested. Part is invested at 6% and the remainder at 8%. If the interest from the 6% investment is $230 more than the interest from the 8% investment, how much is invested at each rate?

3. **Investing:** Ten thousand dollars is invested, part at 5.5% and part at 6%. The interest from the 5.5% investment is $251 more than the interest from the 6% investment. How much is invested at each rate?

4. **Investing:** On two investments totaling $9500, Darius lost 3% on one and earned 6% on the other. If his net annual receipts were $282, how much was each investment?

4. \$3200 at –3%; \$6300 at 6%
5. \$450 at 8%; \$650 at 10%
6. \$1500 at 9%; \$800 at 13%
7. \$3500 in each or \$7000 total
8. \$900 at 12%; \$2100 at 10%
9. \$20,000 at 24%; \$11,000 at 18%
10. \$750 at 7%; \$2250 at 8%
11. \$800 at 5%; \$2100 at 7%
12. \$2500 at 8%; \$3500 at 12%
13. \$8500 at 9%; \$3500 at 11%
14. \$3800 at 15%; \$4200 at 12%
15. \$87,000 in bonds; \$37,000 in certificates

5. Investing: Merideth has money in two savings accounts. One rate is 8% and the other is 10%. If she has \$200 more in the 10% account, how much is invested at 8% if the total interest is \$101?

6. Investing: Money is invested at two rates. One rate is 9% and the other is 13%. If there is \$700 more invested at 9%, find the amount invested at each rate if the total annual interest is \$239.

7. Investing: Ethan has half of his investments in stock paying an 11% dividend and the other half in a debentured stock paying 13% interest. If his total annual interest is \$840, how much does he have invested?

8. Investing: Betty invested some of her money at 12% interest. She invested \$300 more than twice that amount at 10%. How much is invested at each rate if her interest income is \$318 annually?

9. Investing: GFA invested some money in a development yielding 24% and \$9000 less in a development yielding 18%. If the first investment produces \$2820 more per year than the second, how much is invested in each development?

10. Investing: Victoria invests a certain amount of money at 7% annual interest and three times that amount at 8%. If her annual income is \$232.50, how much does she have invested at each rate?

11. Investing: Jamal has a certain amount of money invested at 5% annual interest and \$500 more than twice that amount invested in bonds yielding 7%. His total income from interest is \$187. How much does he have invested at each rate?

12. Investing: A total of \$6000 is invested, part at 8% and the remainder at 12%. How much is invested at each rate if the annual interest is \$620?

13. Investing: Ms. Merriman has \$12,000 invested. Part is invested at 9% and the remainder at 11%. If the interest from the 9% investment is \$380 more than the interest from the 11% investment, how much is invested at each rate?

14. Investing: Eight thousand dollars is invested, part at 15% and the remainder at 12%. If the annual interest income from the 15% investment is \$66 more than the annual interest income from the 12% investment, how much is invested at each rate?

15. Investing: Morgan inherited \$124,000 from her Uncle Edward. She invested a portion in bonds and the remainder in a long-term certificate account. The amount invested in bonds was \$24,000 less than 3 times the amount invested in certificates. How much was invested in bonds and how much in certificates?

16. \$20,000 at 6%; \$28,000 at 10%
17. 20 pounds of 20%; 30 pounds of 70%
18. 8 tons of 80%; 16 tons of 50%
19. 20 ounces of 30%; 30 ounces of 20%
20. 40 litres of 40%; 20 litres of 55%
21. 450 pounds of 35%; 1350 pounds of 15%
22. 440 pounds of 70%; 200 pounds of 54%
23. 20 pounds of 40%; 30 pounds of 15%
24. Regular costs \$2.70 per gallon; Premium costs \$2.85 per gallon
25. 10 g of acid; 20 g of the 40% solution
26. 12.5 lb of dark beans; 37.5 lb of the 60% mixture

16. Investing: Sang has invested \$48,000, part at 6% and the rest in a higher risk investment at 10%. How much did she invest at each rate to receive \$4000 in interest after one year?

17. Manufacturing: A metallurgist has one alloy containing 20% copper and another containing 70% copper. How many pounds of each alloy must he use to make 50 pounds of a third alloy containing 50% copper?

18. Manufacturing: A manufacturer has received an order for 24 tons of a 60% copper alloy. His stock contains only alloys of 80% copper and 50% copper. How much of each will he need to fill the order?

19. Tobacco: A tobacco shop wants 50 ounces of tobacco that is 24% rare Turkish blend. How much each of a 30% Turkish blend and a 20% Turkish blend will be needed?

20. Acid solutions: How many liters each of a 40% acid solution and a 55% acid solution must be used to produce 60 liters of a 45% acid solution?

21. Dairy production: A dairy man wants to mix a 35% protein supplement and a standard 15% protein ration to make 1800 pounds of a high-grade 20% protein ration. How many pounds of each should he use?

22. Manufacturing: To meet the government's specifications, a certain alloy must be 65% aluminum. How many pounds each of a 70% aluminum alloy and a 54% aluminum alloy will be needed to produce 640 pounds of the 65% aluminum alloy?

23. Food science: A meat market has ground beef that is 40% fat and extra lean ground beef that is only 15% fat. How many pounds of each will be needed to obtain 50 pounds of lean ground beef that is 25% fat?

24. Gasoline: George decides to mix grades of gasoline in his truck. He puts in 8 gallons of regular and 12 gallons of premium for a total cost of \$55.80. If premium gasoline costs \$0.15 more per gallon than regular, what was the price of each grade of gasoline?

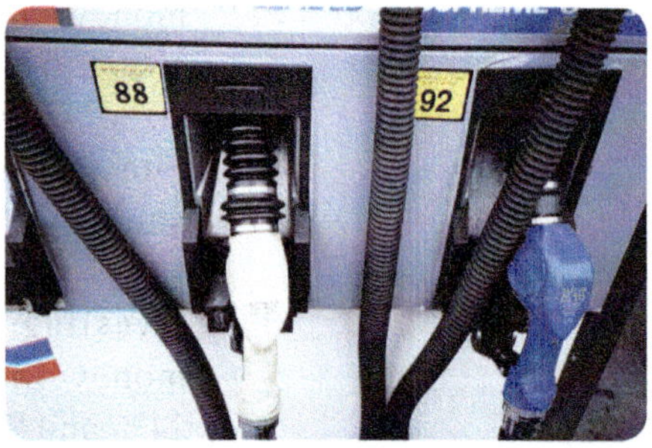

25. Acid solutions: How many grams of pure acid (100% acid) and how many grams of a 40% solution should be mixed together to get a total of 30 grams of a 60% solution?

26. Coffee: Pure dark coffee beans are to be mixed with a mixture that is 60% dark beans. How much of each (pure dark and 60% dark beans) should be used to get a mixture of 50 pounds that contains 70% of the dark beans?

27. 10 oz of salt; 50 oz of the 4% solution
28. 40 liters of 12%; 50 liters of 30%
29. 2 lb of 72%; 4 lb of 42%
30. 240 gallons of 5%; 120 gallons of 2%
31. 27 oz of 25%; 9 oz of 45%
32. 7.11 ounces of 0.05% solution; 0.89 ounces of the 5% solution
33. Answers will vary.

27. Salt solutions: Salt is to be added to a 4% salt solution. How many ounces of salt and how many ounces of the 4% solution should be mixed together to get 60 ounces of a 20% salt solution?

28. Chemistry: How many liters each of a 12% iodine solution and a 30% iodine solution must be used to produce a total mixture of 90 liters of a 22% iodine solution?

29. Food science: A candymaker is making truffles using a mixture of a melted dark chocolate that is 72% cocoa and milk chocolate that is 42% cocoa. If she wants 6 pounds of melted chocolate that is 52% cocoa, how much of each type of chocolate does she need?

30. Dairy farming: A dairy needs 360 gallons of milk containing 4% butterfat. How many gallons each of milk containing 5% butterfat and milk containing 2% butterfat must be used to obtain the desired 360 gallons?

31. Pharmacy: A druggist has two solutions of alcohol. One is 25% alcohol. The other is 45% alcohol. He wants to mix these two solutions to get 36 ounces that will be 30% alcohol. How many ounces of each of these two solutions should he mix together?

32. Body piercings: It is recommended that one cleans new body piercings with a 1% salt solution for the first few weeks with the new piercing. You have a 0.5% solution and a 5% solution and you need to make 8 ounces of the 1% solution. How much of the 0.5% and 5% solutions will you need? (Round your answers to two decimal places.)

Writing and Thinking About Mathematics

33. Your friend has $20,000 to invest and decided to invest part at 4% interest and the rest at 10% interest. Why might you advise him (or her) to invest all of it
a. at 4%?
b. at 10%?

HAWKES LEARNING SYSTEMS: INTRODUCTORY & INTERMEDIATE ALGEBRA SOFTWARE

- 8.5 Applications: Interest and Mixture

8.6 Systems of Linear Equations: Three Variables

- *Solve systems of linear equations in three variables.*
- *Solve applied problems by using systems of linear equations in three variables.*

The equation $2x + 3y - z = 16$ is called a **linear equation in three variables**. The general form is

$$Ax + By + Cz = D \quad \text{where } A, B, \text{ and } C \text{ are not all equal to } 0.$$

The solutions to such equations are called **ordered triples** and are of the form (x_0, y_0, z_0). One ordered triple that satisfies the equation $2x + 3y - z = 16$ is $(1, 4, -2)$. To check this, substitute $x = 1$, $y = 4$, and $z = -2$ into the equation to see if the result is 16:

$$\begin{aligned} 2(1) + 3(4) - (-2) &= 2 + 12 + 2 \\ &= 16. \end{aligned}$$

There are an infinite number of ordered triples that satisfy any linear equation in three variables in which at least two of the coefficients are nonzero. Any two values may be substituted for two of the variables, and then the value for the third variable can be calculated. For example, by letting $x = -1$ and $y = 5$ in the equation $2x + 3y - z = 16$, we find:

$$\begin{aligned} 2(-1) + 3(5) - z &= 16 \\ -2 + 15 - z &= 16 \\ -z &= 3 \\ z &= -3. \end{aligned}$$

Hence, the ordered triple $(-1, 5, -3)$ also satisfies the equation $2x + 3y - z = 16$.

Graphs can be drawn in three dimensions by using a coordinate system involving three mutually perpendicular number lines labeled as the x-axis, y-axis, and z-axis. Three planes are formed: the xy-plane, the xz-plane, and the yz-plane. The three axes separate space into eight regions called **octants**. You can "picture" the first octant as the region bounded by the floor of a room and two walls with the axes meeting in a corner. The floor is the xy-plane. The axes can be ordered in a "right-hand" or "left-hand" format. Figure 1 shows the point represented by the ordered triple $(2, 3, 1)$ in a right-hand system.

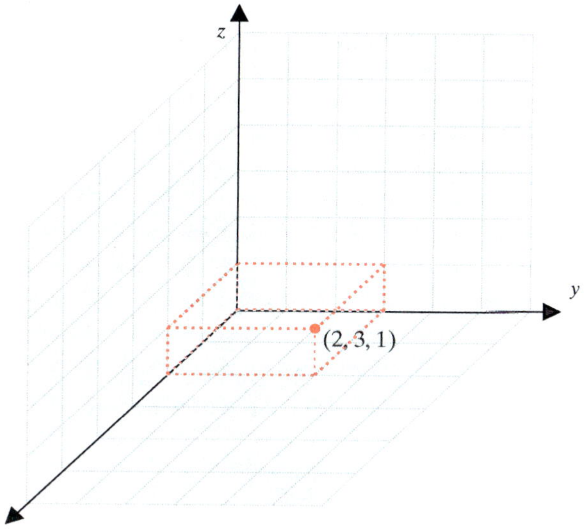

Figure 1

The graphs of linear equations in three variables are planes in three dimensions. A portion of the graph of $2x + 3y - z = 16$ appears in Figure 2.

Teaching Notes:
To help students "see" the three-dimensional aspect, hold three meter sticks perpendicular to each other and discuss coordinates in class.

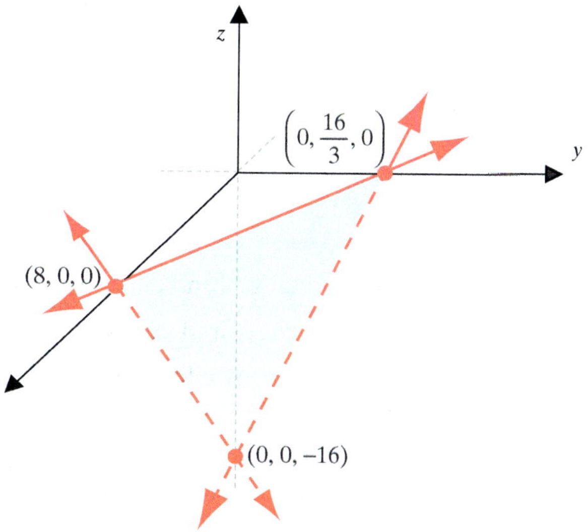

Figure 2

Two distinct planes will either be parallel or they will intersect. If they intersect, their intersection will be a straight line. If three distinct planes intersect, they will intersect in a straight line or in a single point represented by an ordered triple.

The graphs of systems of three linear equations in three variables can be both interesting and informative, but they can be difficult to sketch and points of intersection difficult to estimate. Also, most graphing calculators are limited to graphs in two dimensions, so they are not useful in graphically analyzing systems of linear equations in three variables.

Therefore, in this text, only algebraic techniques for solving these systems will be discussed. Figure 3 illustrates four different possibilities for the relative positions of three planes.

Teaching Notes:
Have students notice commonly created instances of two planes coming together to form a line, (for example, a wall and floor) and three planes coming together to form a point (for example, a wall, floor, and adjacent wall).

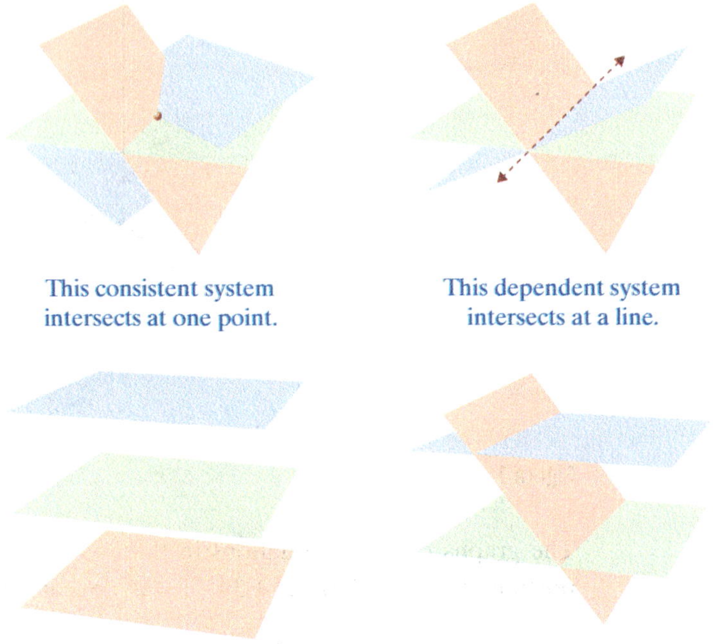

This consistent system intersects at one point.

This dependent system intersects at a line.

These two systems are both inconsistent. All three planes do not have a common intersection.

Figure 3

To Solve a System of Three Linear Equations in Three Variables

1. Select two equations and eliminate one variable by using the addition method.
2. Select a different pair of equations and eliminate the **same** variable.
3. Steps 1 and 2 give **two** linear equations in **two** variables. Solve these equations by either addition or substitution as discussed in Sections 8.2 and 8.3.
4. Back substitute the values found in Step 3 into any one of the original equations to find the value of the third variable.
5. Check the solution (if one exists) in all three of the original equations.

The solution possibilities for a system of three equations in three variables are as follows:

1. There will be exactly one ordered triple solution.
   (Graphically, the three planes intersect at one point.)
2. There will be an infinite number of solutions.
   (Graphically, the three planes intersect in a line or are the same plane.)
3. There will be no solutions.
   (Graphically, there are no points common to all 3 planes.)

The technique is illustrated with the following system.

$$\begin{cases} 2x+3y-z=16 & \text{(I)} \\ x-y+3z=-9 & \text{(II)} \\ 5x+2y-z=15 & \text{(III)} \end{cases}$$

Step 1: Using equations (I) and (II), eliminate y.

Note: We could just as easily have chosen to begin by eliminating x or z. To be sure that you understand the process you might want to solve the system by first eliminating x and then again by first eliminating z. In all cases, the answer will be the same.

$$\begin{array}{ll} \text{(I)} & (2x+3y-z=16) \\ \text{(II)} & [3]\ (x-y+3z=-9) \end{array} \begin{array}{l} \longrightarrow \\ \longrightarrow \end{array} \begin{array}{rcrcrcr} 2x & + & 3y & - & z & = & 16 \\ 3x & - & 3y & + & 9z & = & -27 \\ \hline 5x & & & + & 8z & = & -11 \end{array} \text{ (IV)}$$

Step 2: Using a different pair of equations, (II) and (III), eliminate the **same** variable, y.

$$\begin{array}{ll} \text{(II)} & [2]\ (x-y+3z=-9) \\ \text{(III)} & (5x+2y-z=15) \end{array} \begin{array}{l} \longrightarrow \\ \longrightarrow \end{array} \begin{array}{rcrcrcr} 2x & - & 2y & + & 6z & = & -18 \\ 5x & + & 2y & - & z & = & 15 \\ \hline 7x & & & + & 5z & = & -3 \end{array} \text{ (V)}$$

Step 3: Using the results of Steps 1 and 2, solve the two equations for x and z.

$$\begin{array}{ll} \text{(IV)} & [-7]\ (5x+8z=-11) \\ \text{(V)} & [5]\ (7x+5z=-3) \end{array} \begin{array}{l} \longrightarrow \\ \longrightarrow \end{array} \begin{array}{rcrcr} -35x & - & 56z & = & 77 \\ 35x & + & 25z & = & -15 \\ \hline & & -31z & = & 62 \\ & & z & = & -2 \end{array}$$

Back substitute $z=-2$ into equation (IV) to find x.

$$\begin{aligned} 5x+8(-2)&=-11 \quad \text{Using equation (IV).} \\ 5x-16&=-11 \\ 5x&=5 \\ x&=1 \end{aligned}$$

Step 4: Using $x=1$ and $z=-2$, back substitute to find y.

$$\begin{aligned} 1-y+3(-2)&=-9 \quad \text{Using equation (II).} \\ -y-5&=-9 \\ -y&=-4 \\ y&=4 \end{aligned}$$

The solution is $(1, 4, -2)$. The solution can be checked by substituting the results into **all three** of the original equations.

$$\begin{cases} 2(1)+3(4)-(-2)=16 & \text{(I)} \\ (1)-(4)+3(-2)=-9 & \text{(II)} \\ 5(1)+2(4)-(-2)=15 & \text{(III)} \end{cases}$$

Example 1: Three Variables (A System With One Solution)

Solve the following system of linear equations.

$$\begin{cases} x-y+2z=-4 & \text{(I)} \\ 2x+3y+z=\dfrac{1}{2} & \text{(II)} \\ x+4y-2z=4 & \text{(III)} \end{cases}$$

Solution: Using equations (I) and (III), eliminate z.

$$\begin{array}{rrcrcrcrcr} \text{(I)} & x & - & y & + & 2z & = & -4 & \\ \text{(III)} & x & + & 4y & - & 2z & = & 4 & \\ \hline & 2x & + & 3y & & & = & 0 & \text{(IV)} \end{array}$$

Using equations (I) and (II), eliminate z.

$$\begin{array}{l} \text{(I)} \\ \text{(II)} \end{array} \begin{cases} \ (x - y + 2z = -4) \\ [-2]\ \left(2x + 3y + z = \dfrac{1}{2}\right) \end{cases} \begin{array}{l} \longrightarrow \\ \longrightarrow \end{array} \begin{array}{rcrcrcr} x & - & y & + & 2z & = & -4 \\ -4x & - & 6y & - & 2z & = & -1 \\ \hline -3x & - & 7y & & & = & -5 \quad \text{(V)} \end{array}$$

Eliminate the variable x using the two equations in x and y.

$$\begin{array}{l} \text{(IV)} \\ \text{(V)} \end{array} \begin{cases} [3]\ (2x + 3y = 0) \\ [2]\ (-3x - 7y = -5) \end{cases} \begin{array}{l} \longrightarrow \\ \longrightarrow \end{array} \begin{array}{rcrcr} 6x & + & 9y & = & 0 \\ -6x & - & 14y & = & -10 \\ \hline & & -5y & = & -10 \\ & & y & = & 2 \end{array}$$

Back substituting into equation (IV) to find x yields:

$$2x+3(2)=0$$
$$2x+6=0$$
$$2x=-6$$
$$x=-3.$$

Finally, using $x = -3$ and $y = 2$, back substitute into (I).

$$\begin{aligned}(-3)-(2)+2z &= -4\\ -5+2z &= -4\\ 2z &= 1\\ z &= \frac{1}{2}\end{aligned}$$

The solution is $\left(-3, 2, \frac{1}{2}\right)$.

Check: The solution can be checked by substituting $\left(-3, 2, \frac{1}{2}\right)$ into all three of the original equations.

$$\begin{cases}(-3)-(2)+2\left(\frac{1}{2}\right)=-4 & \text{(I)}\\ 2(-3)+3(2)+\left(\frac{1}{2}\right)=\frac{1}{2} & \text{(II)}\\ (-3)+4(2)-2\left(\frac{1}{2}\right)=4 & \text{(III)}\end{cases}$$

Example 2: Three Variables (A System With No Solution)

Solve the following system of linear equations.

$$\begin{cases}3x-5y+z=6 & \text{(I)}\\ x-y+3z=-1 & \text{(II)}\\ 2x-2y+6z=5 & \text{(III)}\end{cases}$$

Solution: Using equations (I) and (II), eliminate z.

$$\begin{array}{l} \text{(I)} \\ \text{(II)} \end{array}\begin{cases}[-3]\ (3x-5y+z=6) \\ \ (x-y+3z=-1)\end{cases} \begin{array}{l}\longrightarrow \\ \longrightarrow\end{array} \begin{array}{rcrcrcr} -9x & + & 15y & - & 3z & = & -18 \\ x & - & y & + & 3z & = & -1 \\ \hline -8x & + & 14y & & & = & -19 \end{array}\ \text{(IV)}$$

Using equations (II) and (III), eliminate z.

$$\begin{array}{l} \text{(II)} \\ \text{(III)} \end{array}\begin{cases}[-2]\ (x-y+3z=-1) \\ \ (2x-2y+6z=5)\end{cases} \begin{array}{l}\longrightarrow \\ \longrightarrow\end{array} \begin{array}{rcrcrcr} -2x & + & 2y & - & 6z & = & 2 \\ 2x & - & 2y & + & 6z & = & 5 \\ \hline & & & & 0 & = & 7 \end{array}\ \text{(V)}$$

This last equation is **false**. Thus the system has **no solution**.

Example 3: Three Variables (A System with Infinitely Many Solutions)

Solve the following system of linear equations.

$$\begin{cases} 3x-2y+4z=1 & \text{(I)} \\ 3x-y+7z=-1 & \text{(II)} \\ 6x-5y+5z=4 & \text{(III)} \end{cases}$$

Solution: Using equations (I) and (II), eliminate x.

$$\begin{aligned} &\text{(I)} & (3x - 2y + 4z = 1) &\longrightarrow & 3x - 2y + 4z &= 1 \\ &\text{(II)} & [-1](3x - y + 7z = -1) &\longrightarrow & -3x + y - 7z &= 1 \\ & & & & \hline -y - 3z &= 2 \quad \text{(IV)} \end{aligned}$$

Using equations (I) and (III), eliminate x.

$$\begin{aligned} &\text{(I)} & [-2](3x - 2y + 4z = 1) &\longrightarrow & -6x + 4y - 8z &= -2 \\ &\text{(III)} & (6x - 5y + 5z = 4) &\longrightarrow & 6x - 5y + 5z &= 4 \\ & & & & \hline -y - 3z &= 2 \quad \text{(V)} \end{aligned}$$

Using equations (IV) and (V), eliminate y.

$$\begin{aligned} &\text{(IV)} & (-y - 3z = 2) &\longrightarrow & -y - 3z &= 2 \\ &\text{(V)} & [-1](-y - 3z = 2) &\longrightarrow & y + 3z &= -2 \\ & & & & \hline 0 &= 0 \end{aligned}$$

Because this last equation, $0 = 0$, is **always true**, the system has an **infinite number of solutions**.

Example 4: Three Variables (Application)

A cash register contains \$341 in \$20, \$5, and \$2 bills. There are twenty-eight bills in all and three more \$2 bills than \$5 bills. How many bills of each kind are there?

Solution: Let x = number of \$20 bills
y = number of \$5 bills
z = number of \$2 bills

$$\begin{cases} x+y+z=28 & \text{(I)} \\ 20x+5y+2z=341 & \text{(II)} \\ z=y+3 & \text{(III)} \end{cases}$$

(I) There are twenty-eight bills.
(II) The total value is \$341.
(III) There are three more \$2 bills than \$5 bills.

Using equations (I) and (II), eliminate x.

$$\begin{array}{l} \text{(I)} \\ \text{(II)} \end{array}\begin{cases} [-20] & (x + y + z = 28) \longrightarrow \\ & (20x + 5y + 2z = 341) \longrightarrow \end{cases} \begin{array}{rcrcrcr} -20x & - & 20y & - & 20z & = & -560 \\ 20x & + & 5y & + & 2z & = & 341 \\ \hline & - & 15y & - & 18z & = & -219 \end{array} \text{(IV)}$$

As equation (III) already has no x term, we rewrite it in the form $y - z = -3$ and use this equation along with the equation just found.

$$\begin{array}{l} \text{(III)} \\ \text{(IV)} \end{array}\begin{cases} [15] & (y - z = -3) \longrightarrow \\ & (-15y - 18z = -219) \longrightarrow \end{cases} \begin{array}{rcrcr} 15y & - & 15z & = & -45 \\ -15y & - & 18z & = & -219 \\ \hline & & -33z & = & -264 \\ & & z & = & 8 \end{array}$$

Back substituting into equation (III) to solve for y gives:

$$8 = y + 3$$
$$5 = y.$$

Now we can substitute the values $z = 8$ and $y = 5$ into equation (I).

$$x + 5 + 8 = 28$$
$$x + 13 = 28$$
$$x = 15$$

There are fifteen \$20 bills, five \$5 bills, and eight \$2 bills.

Practice Problems

Solve the following system of linear equations: $\begin{cases} 2x + y + z = 4 \\ x + 2y + z = 1 \\ 3x + y - z = -3 \end{cases}$

8.6 Exercises

Solve each system of equations. State which systems, if any, have no solution or an infinite number of solutions.

1. (1, 0, 1)
2. (2, 7, 4)
3. (1, 2, −1)

1. $\begin{cases} x + y - z = 0 \\ 3x + 2y + z = 4 \\ x - 3y + 4z = 5 \end{cases}$ **2.** $\begin{cases} x - y + 2z = 3 \\ -6x + y + 3z = 7 \\ x + 2y - 5z = -4 \end{cases}$ **3.** $\begin{cases} 2x - y - z = 1 \\ 2x - 3y - 4z = 0 \\ x + y - z = 4 \end{cases}$

Answer to Practice Problem: (1, −2, 4)

4. $(1, 3, 3)$
5. Infinitely many solutions
6. $(-2, 9, 1)$
7. $(4, 1, 1)$
8. $(1, -3, 2)$
9. $(1, 2, -1)$
10. Infinitely many solutions
11. $(-2, 3, 1)$
12. $(1, 2, 1)$
13. No solution
14. $(-1, 1, 2)$
15. $(3, -1, 2)$
16. No solution
17. $(2, 1, -3)$
18. $(1, -1, 3)$
19. $\left(\frac{1}{2}, \frac{1}{3}, -1\right)$
20. $\left(\frac{5}{2}, \frac{2}{3}, -\frac{1}{2}\right)$
21. 34, 6, 27
22. 56, 84, 49
23. 18 ones, 16 fives, 12 tens

4. $\begin{cases} y+z=6 \\ x+5y-4z=4 \\ x-3y+5z=7 \end{cases}$

5. $\begin{cases} x+y-2z=4 \\ 2x+y=1 \\ 5x+3y-2z=6 \end{cases}$

6. $\begin{cases} x-y+5z=-6 \\ x+2z=0 \\ 6x+y+3z=0 \end{cases}$

7. $\begin{cases} y+z=2 \\ x+z=5 \\ x+y=5 \end{cases}$

8. $\begin{cases} 2y+z=-4 \\ 3x+4z=11 \\ x+y=-2 \end{cases}$

9. $\begin{cases} x-y+2z=-3 \\ 2x+y-z=5 \\ 3x-2y+2z=-3 \end{cases}$

10. $\begin{cases} x-y-2z=3 \\ x+2y+z=1 \\ 3y+3z=-2 \end{cases}$

11. $\begin{cases} 2x-y+5z=-2 \\ x+3y-z=6 \\ 4x+y+3z=-2 \end{cases}$

12. $\begin{cases} 2x-y+5z=5 \\ x-2y+3z=0 \\ x+y+4z=7 \end{cases}$

13. $\begin{cases} 3x+y+4z=-6 \\ 2x+3y-z=2 \\ 5x+4y+3z=2 \end{cases}$

14. $\begin{cases} 2x+y-z=-3 \\ -x+2y+z=5 \\ 2x+3y-2z=-3 \end{cases}$

15. $\begin{cases} x-2y+z=7 \\ x-y-4z=-4 \\ x+4y-2z=-5 \end{cases}$

16. $\begin{cases} 2x-2y+3z=4 \\ x-3y+2z=2 \\ x+y+z=1 \end{cases}$

17. $\begin{cases} 2x+3y+z=4 \\ 3x-5y+2z=-5 \\ 4x-6y+3z=-7 \end{cases}$

18. $\begin{cases} x+y+z=3 \\ 2x-y-2z=-3 \\ 3x+2y+z=4 \end{cases}$

19. $\begin{cases} 2x-3y+z=-1 \\ 6x-9y-4z=4 \\ 4x+6y-z=5 \end{cases}$

20. $\begin{cases} x+6y+z=6 \\ 2x+3y-2z=8 \\ 2x+4z=3 \end{cases}$

21. The sum of three integers is 67. The sum of the first and second integers is 13 more than the third integer. The third integer is 7 less than the first. Find the three integers.

22. The sum of three integers is 189. The first integer is 28 less than the second. The second integer is 21 less than the sum of the first and third integers. Find the three integers.

23. Money: A wallet contains \$218 in \$10, \$5, and \$1 bills. There are forty-six bills in all and four more fives than tens. How many bills of each kind are there?

24. 11 nickels, 7 dimes, 5 quarters
25. 19 cm, 24 cm, 30 cm
26. 100°, 30°, 50°
27. 300 main floor, 200 mezzanine, and 80 balcony
28. House: $115,000; Lot: $33,000; Improvements: $11,000
29. 3 lillies, 5 roses, and 8 daisies
30. Bananas: $0.60/lb; Apples: $1.60/lb; Grapes: $3.60/lb

24. Money: Sally is trying to get her brother Robert to learn to think algebraically. She tells him that she has 23 coins in her purse, including nickels, dimes, and quarters. She has two more dimes than quarters, and the total value of the coins is $2.50. How many of each kind of coin does she have?

25. Perimeter of a triangle: The perimeter of a triangle is 73 cm. The longest side is 13 cm less than the sum of the other two sides. The shortest side is 11 cm less than the longest side. Find the lengths of the three sides.

26. Triangles: The sum of the measures of the three angles of a triangle is 180°. In one particular triangle, the largest angle is 10° more than three times the smallest angle, and the third angle is one-half the largest angle. What are the measures of the three angles?

27. Theater: A theater has three types of seats for Broadway plays: main floor, balcony, and mezzanine. Main floor tickets are $60, balcony tickets are $45, and mezzanine tickets are $30. Opening night the sales totaled $27,600. Main floor sales were 20 more than the total of balcony and mezzanine sales. Mezzanine sales were 40 more than two times balcony sales. How many of each type of ticket was sold?

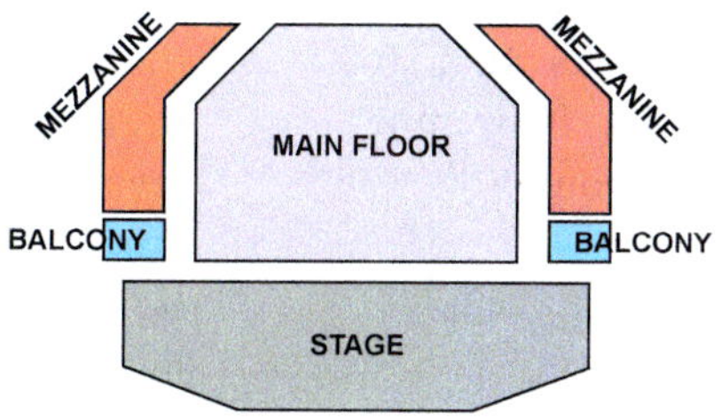

28. Construction: The Tates are having a house built. The cost is split up into three parts: the house, the lot, and the improvements. The cost of building the house is $16,000 more than three times the cost of the lot. The cost of the improvements (the landscaping, sidewalks, and upgrades) is one-third the cost of the lot. If the total cost is $159,000, what is the cost of each part of the construction?

29. Flower arranging: A florist is creating bridesmaids' bouquets for a wedding. Each 16 flower bouquet will have a mixture of lilies, roses, and daisies and cost $92. Lilies cost $10, roses cost $6, and daisies cost $4 a stem. If each bouquet will have as many daisies as it will roses and lilies combined, how many of each type of flower will be in the bouquet?

30. Fruit stand: At Steve's Fruit Stand, 4 pounds of bananas, 2 pounds of apples, and 3 pounds of grapes cost $16.40. Five pounds of bananas, 4 pounds of apples, and 2 pounds of grapes cost $16.60. Two pounds of bananas, 3 pounds of apples, and 1 pound of grapes cost $9.60. Find the price per pound of each kind of fruit.

31. $2.80
32. There were 6 first place finishers, 12 second place finishers and 4 third place finishers.
33. Savings: $30,000; Bonds: $55,000; Stocks: $15,000
34. PepsiCo: $9000; IBM: $5000; Microsoft: $16,000
35. 3 liters of 10%, 4.5 liters of 30%, 1.5 liters of 40%
36. Standard: 70 mixers; Deluxe: 40 mixers; Premium: 30 mixers

31. **Fast food:** At the Happy Burger Drive-In, you can buy 2 hamburgers, 1 chocolate shake, and 2 orders of fries, or 3 hamburgers and 1 order of fries, for $9.50. One hamburger, 2 chocolate shakes, and 1 order of fries cost $7.30. How much does a hamburger cost?

32. **High school sports:** At a high school track meet, the winning team scored 180 points and had 22 athletes score points. A first-place finish earns 10 points, a second-place finish earns 8 points, and a third-place finish earns 6 points. Twice as many athletes came in second place than came in first place. How many athletes finished in each place?

33. **Investing:** Kirk inherited $100,000 from his aunt and decided to invest in three different accounts: savings, bonds, and stocks. The amount in his bond account was $10,000 more than three times the amount in his stock account. At the end of the first year, the savings account returned 5%, the bond 8%, and the stocks 10% for total interest of $7400. How much did he invest in each account?

34. **Stock market:** Melissa has saved a total of $30,000 and wants to invest in three different stocks: PepsiCo, IBM, and Microsoft. She wants the PepsiCo amount to be $1000 less than twice the IBM amount and the Microsoft amount to be $2000 more than the total in the other two stocks. How much should she invest in each stock?

35. **Chemistry:** A chemist wants to mix 9 liters of a 25% acid solution. Because of limited amounts on hand, the mixture is to come from three different solutions, one with 10% acid, another with 30% acid, and a third with 40% acid. The amount of the 10% solution must be twice the amount of the 40% solution, and the amount of the 30% solution must equal the total amount of the other two solutions. How much of each solution must be used?

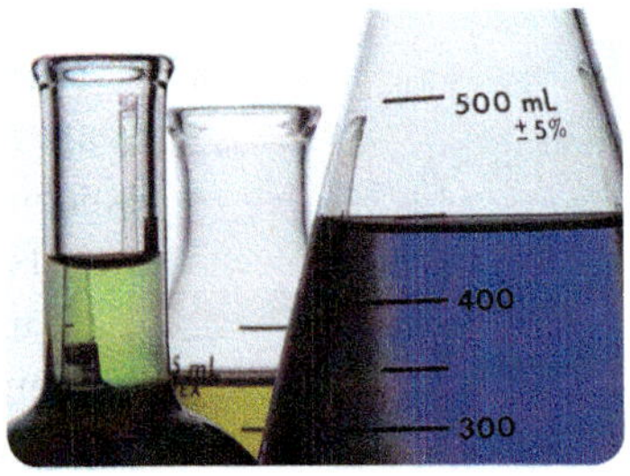

36. **Manufacturing:** An appliance company makes three versions of their popular stand mixer. The standard model has production costs of $100, the deluxe model of $175, and the premium model of $250. On any given day the factory line makes 140 mixers and spends $21,500 on production costs. If the number of standard mixers made equals the sum of the premium and deluxe models made, how many of each kind of mixer are made each day?

Writing and Thinking About Mathematics

37. No. Graphically, the three planes intersect in one point (one solution) or in a line (infinitely many solutions) or they do not have a common intersection (no solution).

37. Is it possible for three linear equations in three unknowns to have exactly two solutions? Explain your reasoning in detail.

38. In geometry, three non-collinear points determine a plane. (That is, if three points are not on a line, then there is a unique plane that contains all three points.) Find the values of A, B, and C (and therefore the equation of the plane) given $Ax + By + Cz = 3$ and the three points on the plane $(0, 3, 2), (0, 0, 1)$ and $(-3, 0, 3)$. Sketch the plane in three dimensions as best you can by locating the three given points.

39. As stated in Exercise 38, three non-collinear points determine a plane. Find the values of A, B, and C (and therefore the equation of the plane) given $Ax + By + Cz = 10$ and the three points on the plane $(2, 0, -2), (3, -1, 0)$ and $(-1, 5, -4)$. Sketch the plane in three dimensions as best you can by locating the three given points.

HAWKES LEARNING SYSTEMS: INTRODUCTORY & INTERMEDIATE ALGEBRA SOFTWARE

- 8.6 Solving Systems of Linear Equations with Three Variables

38. $A = 2, B = -1, C = 3$ 39. $A = 4, B = 2, C = -1$

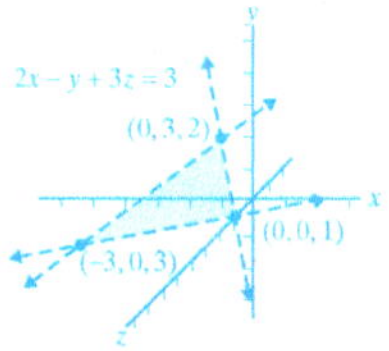

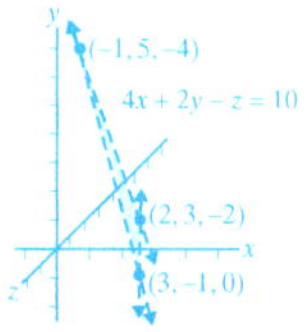

8.7 Matrices and Gaussian Elimination

- *Write a system of equations as a coefficient matrix and an augmented matrix.*
- *Transform a matrix into triangular form by using elementary row operations.*
- *Solve systems of linear equations by using the Gaussian elimination method.*

Matrices

A rectangular array of numbers is called a **matrix** (plural **matrices**). Matrices are usually named with capital letters, and each number in the matrix is called an **entry**. Entries written horizontally are said to form a **row**, and entries written vertically are said to form a **column**. The matrix A shown below has two rows and three columns and is a **2 × 3 matrix** (read "two by three matrix"). We say that the **dimension** of the matrix is the number of rows by the number of columns, which for this matrix is two by three (or 2×3). Similarly, if a matrix has three rows and two columns then its dimension is 3×2.

$$A = \begin{bmatrix} 3 & 4 & 0 \\ 7 & -2 & 5 \end{bmatrix} \qquad \begin{bmatrix} 3 & 4 & 0 \\ 7 & -2 & 5 \end{bmatrix} \begin{matrix} \leftarrow \text{row 1} \\ \leftarrow \text{row 2} \end{matrix} \qquad \begin{matrix} \text{column 1} & \text{column 2} & \text{column 3} \\ \downarrow & \downarrow & \downarrow \end{matrix} \begin{bmatrix} 3 & 4 & 0 \\ 7 & -2 & 5 \end{bmatrix}$$

Three more examples are:

$$B = \begin{bmatrix} 5 & -1 \\ 2 & 3 \end{bmatrix} \qquad C = \begin{bmatrix} 5 & -1 & 0 & 7 \\ 2 & 3 & 2 & 8 \\ 1 & -3 & 0 & 6 \end{bmatrix} \qquad D = \begin{bmatrix} 0 & 4 \\ 1 & 6 \\ -1 & 3 \end{bmatrix}$$

2×2 matrix $\qquad$ 3×4 matrix $\qquad$ 3×2 matrix

A matrix with the same number of rows as columns is called a **square matrix**. Matrix B (shown above) is a square 2×2 matrix.

Elementary Row Operations

Matrices have many uses and are generated from various types of problems because they allow data to be presented in a systematic and orderly manner. (Business majors may want to look up a topic called Markov chains.)

Also, matrices can sometimes be added, subtracted, and multiplied. Some square matrices have inverses, much the same as multiplicative inverses for real numbers. Matrix methods of solving systems of linear equations can be done manually or with graphing calculators and computers. These topics are presented in courses such as finite mathematics and linear algebra.

Teaching Notes:
To generate interest in matrices, you might want to informally discuss and give simple examples of adding, subtracting, and multiplying with matrices and writing entries in the general form a_{ij}. Also, let students know that using row or column matrices to represent vectors is helpful in many physics and calculus applications. Most students find abstract material more worth their time and effort if they can see uses in future courses. Some of your students may even have seen or used Markov chains in a business or economics course and will want to discuss these applications.

In this text we will see that matrices can be used to solve systems of linear equations in which the equations are written in standard form. The two matrices derived from such a system are the **coefficient matrix** (made up of the coefficients of the variables) and the **augmented matrix** (including the coefficients and the constant terms). For example:

System	Coefficient Matrix	Augmented Matrix
$\begin{cases} x - y + z = -6 \\ 2x + 3y = 17 \\ x + 2y + 2z = 7 \end{cases}$	$\begin{bmatrix} 1 & -1 & 1 \\ 2 & 3 & 0 \\ 1 & 2 & 2 \end{bmatrix}$	$\left[\begin{array}{ccc:c} 1 & -1 & 1 & -6 \\ 2 & 3 & 0 & 17 \\ 1 & 2 & 2 & 7 \end{array}\right]$

coefficients constants

Note that 0 is the entry in the second row, third column of both matrices. This 0 corresponds to the missing z-variable in the second equation. The second equation could have been written $2x + 3y + 0z = 17$.

As discussed in Section 3.1, notation with a small number to the right and below a variable is called **subscript** notation. For example, a_1 is read "*a* sub one," b_3 is read "*b* sub three" and R_2 is read "R sub two."

a_1 ← subscript; ↑ variable

b_3 ← subscript; ↑ variable

R_2 ← subscript; ↑ variable

When working with matrices, R_1 represents row 1, R_2 represents row 2, and so on.

In Sections 8.3 and 8.6, systems of linear equations were solved by the addition method and back substitution. In solving these systems, we can make any of the following three manipulations **without changing the solution set of the system.**

The system $\begin{cases} x - y + z = -6 \\ 2x + 3y + 0z = 17 \\ x + 2y + 2z = 7 \end{cases}$ is used here to illustrate some possibilities.

1. Any two equations may be interchanged.

$$\begin{cases} x - y + z = -6 \\ 2x + 3y + 0z = 17 \\ x + 2y + 2z = 7 \end{cases} \xrightarrow{R_1 \leftrightarrow R_2} \begin{cases} 2x + 3y + 0z = 17 \\ x - y + z = -6 \\ x + 2y + 2z = 7 \end{cases}$$

Here we have interchanged the first two equations.

2. All terms of any equation may be multiplied by a constant.

$$\begin{cases} x - y + z = -6 \\ 2x + 3y + 0z = 17 \\ x + 2y + 2z = 7 \end{cases} \xrightarrow{-2R_1} \begin{cases} -2x + 2y - 2z = 12 \\ 2x + 3y + 0z = 17 \\ x + 2y + 2z = 7 \end{cases}$$

Here we have multiplied each term of the first equation by -2.

3. All terms of any equation may be multiplied by a constant and these new terms may be added to like terms of another equation. **(The original equation remains unchanged.)**

$$\begin{cases} x-\ y+\ z=-6 \\ 2x+3y+0z=17 \\ x+2y+2z=7 \end{cases} \xrightarrow{R_3-R_1} \begin{cases} x-\ y+\ z=-6 \\ 2x+3y+0z=17 \\ 0x+3y+\ z=13 \end{cases}$$

Here we have multiplied the first equation by −1 (mentally) and added the results to the third equation.

When dealing with matrices, the three corresponding operations are called **elementary row operations**. These operations are listed below and illustrated in Example 1. Follow the steps outlined in Example 1 carefully, and note how the row operations are indicated, such as $\frac{1}{2}R_3$ to indicate that all numbers in row 3 are multiplied by $\frac{1}{2}$. (Reasons for using these row operations are discussed under **Gaussian Elimination** on page 650.)

Elementary Row Operations

1. Interchange two rows.
2. Multiply a row by a nonzero constant.
3. Add a multiple of a row to another row.

If any elementary row operation is applied to a matrix, the new matrix is said to be **row-equivalent** to the original matrix.

Example 1: Coefficient and Augmented Matrices

a. For the system $\begin{cases} y+\ z=6 \\ x+5y-\ 4z=4 \\ 2x-6y+10z=14 \end{cases}$

write the corresponding coefficient matrix and the corresponding augmented matrix.

Solution: Coefficient Matrix Augmented Matrix

$$\begin{bmatrix} 0 & 1 & 1 \\ 1 & 5 & -4 \\ 2 & -6 & 10 \end{bmatrix} \qquad \left[\begin{array}{ccc|c} 0 & 1 & 1 & 6 \\ 1 & 5 & -4 & 4 \\ 2 & -6 & 10 & 14 \end{array}\right]$$

b. In the augmented matrix in Example 1a, interchange rows 1 and 2 and multiply row 3 by $\frac{1}{2}$.

Solution: $$\left[\begin{array}{ccc|c} 0 & 1 & 1 & 6 \\ 1 & 5 & -4 & 4 \\ 2 & -6 & 10 & 14 \end{array}\right] \xrightarrow[\frac{1}{2}R_3]{R_1 \leftrightarrow R_2} \left[\begin{array}{ccc|c} 1 & 5 & -4 & 4 \\ 0 & 1 & 1 & 6 \\ 1 & -3 & 5 & 7 \end{array}\right]$$

c. For the system $\begin{cases} x - \ \ y = 5 \\ 3x + 4y = 29 \end{cases}$, write the corresponding coefficient matrix and the corresponding augmented matrix.

Solution: Coefficient Matrix Augmented Matrix

$$\begin{bmatrix} 1 & -1 \\ 3 & 4 \end{bmatrix} \qquad \left[\begin{array}{cc|c} 1 & -1 & 5 \\ 3 & 4 & 29 \end{array}\right]$$

d. In the augmented matrix in Example 1c, add -3 times row 1 to row 2. (Note that row 1 is unchanged in the resulting matrix. Only row 2 is changed.)

Solution: $$\left[\begin{array}{cc|c} 1 & -1 & 5 \\ 3 & 4 & 29 \end{array}\right] \xrightarrow{R_2 - 3R_1} \left[\begin{array}{cc|c} 1 & -1 & 5 \\ 0 & 7 & 14 \end{array}\right]$$

General Notation for a Matrix and a System of Equations

With matrices we use capital letters to name a matrix and **double subscript** notation with corresponding lower case letters to indicate both the row and column location of an entry. For example in a matrix A the entries will be designated as described below and as shown on the following page.

a_{11} is read "*a* sub one one" and indicates the entry in the first row and first column;
a_{12} is read "*a* sub one two" and indicates the entry in the first row and second column;
a_{13} is read "*a* sub one three" and indicates the entry in the first row and third column;
a_{21} is read "*a* sub two one" and indicates the entry in the second row and first column;
and so on.

NOTES

We will see in dealing with polynomials later that a_{11} can be read simply as "*a* sub eleven." However, with matrices, we need to indicate the row and column corresponding to the entry. If there are more than nine rows or columns, then commas are used to separate the numbers as $a_{10,10}$. You will see the commas in use on your calculator.

With double subscript notation we can write the general form of a 2×3 matrix A and a 3×3 matrix B as follows.

$$A = \begin{bmatrix} a_{11} & a_{12} & a_{13} \\ a_{21} & a_{22} & a_{23} \end{bmatrix} \qquad B = \begin{bmatrix} b_{11} & b_{12} & b_{13} \\ b_{21} & b_{22} & b_{23} \\ b_{31} & b_{32} & b_{33} \end{bmatrix}$$

We will use this notation when discussing the use of calculators to define matrices and to operate with matrices. The general form of a system of three linear equations might use this notation in the following way.

$$\begin{cases} a_{11}x + a_{12}y + a_{13}z = k_1 \\ a_{21}x + a_{22}y + a_{23}z = k_2 \\ a_{31}x + a_{32}y + a_{33}z = k_3 \end{cases}$$

Teaching Notes:
You might want to illustrate a matrix in lower triangular form to satisfy students' curiosity.

A matrix is in **upper triangular form** (or just **triangular form** for our purposes) if its entries in the lower left triangular region are all 0's. The **main diagonal** consists of the entries in the positions of b_{11}, b_{22}, and b_{33}. Note that in the main diagonal the column and row indices are equal. Thus if all the entries below the main diagonal of a matrix are all 0's, the matrix is in triangular form as shown below.

$$B = \begin{bmatrix} b_{11} & b_{12} & b_{13} \\ 0 & b_{22} & b_{23} \\ 0 & 0 & b_{33} \end{bmatrix}$$

The upper triangular form of a matrix with all 1's in the main diagonal is called the **row echelon form** (or **ref**). We will see that a graphing calculator can be used to change a matrix into the row echelon form.

Gaussian Elimination

Another method of solving a system of linear equations is the **Gaussian elimination** method (named after the famous German mathematician Carl Friedrich Gauss, 1777 – 1855). This method makes use of augmented matrices and elementary row operations. The objective is to transform an augmented matrix into row echelon form and then use back substitution to find the values of the variables. The method is outlined as follows:

Strategy for Gaussian Elimination

1. Write the augmented matrix for the system.
2. Use elementary row operations to transform the matrix into row echelon form.
3. Solve the corresponding system of equations by using back substitution.

The following examples illustrate the method. Study the steps and the corresponding comments carefully.

Example 2: Gaussian Elimination

Solve the following system of linear equations by using the Gaussian elimination method with back substitution.

$$\begin{cases} 2x+4y=-6 \\ 5x-y=7 \end{cases}$$

Solution:

Step 1: Write the augmented matrix.

$$\left[\begin{array}{cc|c} 2 & 4 & -6 \\ 5 & -1 & 7 \end{array}\right]$$

(The following steps show how to use elementary row operations to get the matrix in row echelon form with 0 in the lower left corner.)

Step 2: Multiply row 1 by $\frac{1}{2}$ so that the entry in the upper left corner will be 1. This will help to get 0 below the 1 in the next step.

$$\left[\begin{array}{cc|c} 2 & 4 & -6 \\ 5 & -1 & 7 \end{array}\right] \xrightarrow{\frac{1}{2}R_1} \left[\begin{array}{cc|c} 1 & 2 & -3 \\ 5 & -1 & 7 \end{array}\right]$$

Step 3: To get 0 in the lower left corner, add –5 times row 1 to row 2.

$$\left[\begin{array}{cc|c} 1 & 2 & -3 \\ 5 & -1 & 7 \end{array}\right] \xrightarrow{R_2-5R_1} \left[\begin{array}{cc|c} 1 & 2 & -3 \\ 0 & -11 & 22 \end{array}\right]$$

Step 4: Now multiply row 2 by $-\frac{1}{11}$.

$$\left[\begin{array}{cc|c} 1 & 2 & -3 \\ 0 & -11 & 22 \end{array}\right] \xrightarrow{-\frac{1}{11}R_2} \left[\begin{array}{cc|c} 1 & 2 & -3 \\ 0 & 1 & -2 \end{array}\right]$$

Step 5: The last matrix (in triangular form) in Step 4 represents the following system of linear equations.

$$\begin{cases} x+2y=-3 \\ 0x+y=-2 \end{cases}$$

The last equation gives $y=-2$.

Back substitute to find the value for x.

$$\begin{aligned} x+2(-2)&=-3 \\ x-4&=-3 \\ x&=1 \end{aligned}$$

Thus the solution is $(1,-2)$.

Example 3: Gaussian Elimination

Solve the following system of linear equations by using the Gaussian elimination method with back substitution.

$$\begin{cases} 2x-3y-z=-4 \\ -x+2y+z=6 \\ x-y+2z=14 \end{cases}$$

Solution:

Step 1: Write the augmented matrix.

$$\left[\begin{array}{rrr|r} 2 & -3 & -1 & -4 \\ -1 & 2 & 1 & 6 \\ 1 & -1 & 2 & 14 \end{array}\right]$$

Step 2: Exchange row 1 and row 3 so that the entry in the upper left corner will be 1.

$$\left[\begin{array}{rrr|r} 2 & -3 & -1 & -4 \\ -1 & 2 & 1 & 6 \\ 1 & -1 & 2 & 14 \end{array}\right] \xrightarrow{R_1 \leftrightarrow R_3} \left[\begin{array}{rrr|r} 1 & -1 & 2 & 14 \\ -1 & 2 & 1 & 6 \\ 2 & -3 & -1 & -4 \end{array}\right]$$

Step 3: To get a 0 under the 1 in Column 1, add row 1 to row 2.

$$\left[\begin{array}{rrr|r} 1 & -1 & 2 & 14 \\ -1 & 2 & 1 & 6 \\ 2 & -3 & -1 & -4 \end{array}\right] \xrightarrow{R_2 + R_1} \left[\begin{array}{rrr|r} 1 & -1 & 2 & 14 \\ 0 & 1 & 3 & 20 \\ 2 & -3 & -1 & -4 \end{array}\right]$$

Step 4: To get a_{31} to be 0, add –2 times row 1 to row 3.

$$\left[\begin{array}{rrr|r} 1 & -1 & 2 & 14 \\ 0 & 1 & 3 & 20 \\ 2 & -3 & -1 & -4 \end{array}\right] \xrightarrow{R_3 - 2R_1} \left[\begin{array}{rrr|r} 1 & -1 & 2 & 14 \\ 0 & 1 & 3 & 20 \\ 0 & -1 & -5 & -32 \end{array}\right]$$

Step 5: Add row 2 to row 3 to arrive at triangular form.

$$\left[\begin{array}{rrr|r} 1 & -1 & 2 & 14 \\ 0 & 1 & 3 & 20 \\ 0 & -1 & -5 & -32 \end{array}\right] \xrightarrow{R_3 + R_2} \left[\begin{array}{rrr|r} 1 & -1 & 2 & 14 \\ 0 & 1 & 3 & 20 \\ 0 & 0 & -2 & -12 \end{array}\right]$$

Step 6: Then multiply row 3 by $-\frac{1}{2}$.

$$\left[\begin{array}{rrr|r} 1 & -1 & 2 & 14 \\ 0 & 1 & 3 & 20 \\ 0 & 0 & -2 & -12 \end{array}\right] \xrightarrow{-\frac{1}{2}R_3} \left[\begin{array}{rrr|r} 1 & -1 & 2 & 14 \\ 0 & 1 & 3 & 20 \\ 0 & 0 & 1 & 6 \end{array}\right]$$

Step 7: The triangular matrix in Step 6 represents the following system of linear equations.

$$\begin{cases} x - y + 2z = 14 \\ \quad\; y + 3z = 20 \\ \qquad\quad z = 6 \end{cases}$$

From the last equation, we have $z = 6$.
Back substitution into the equation $y + 3z = 20$ gives:

$$y + 3(6) = 20$$
$$y = 2.$$

Back substitution into the equation $x - y + 2z = 14$ gives:

$$x - 2 + 2(6) = 14$$
$$x = 4.$$

Thus the solution is $(4, 2, 6)$.

If the final matrix, in triangular form, has a row with all entries 0, then the system has an infinite number of solutions.

For example, solving the system $\begin{cases} x + 3y = 8 \\ 2x + 6y = 16 \end{cases}$ will result in the matrix $\left[\begin{array}{cc|c} 1 & 3 & 8 \\ 0 & 0 & 0 \end{array}\right]$

The last line indicates that $0x + 0y = 0$ which is always true. Therefore, the solution to the system is the set of all solutions of the equation $x + 3y = 8$.

If the triangular form of the augmented matrix shows the coefficient entries in one or more rows to be all 0's and the constant not 0, then the system has no solution.

For example, the last row of the augmented matrix $\left[\begin{array}{ccc|c} 1 & 2 & 2 & 7 \\ 0 & 1 & 3 & 6 \\ 0 & 0 & 0 & 15 \end{array}\right]$ indicates that $0x + 0y + 0z = 15$. Since this is never true, the system has no solution.

Using the TI-84 Plus Graphing Calculator to Solve a System of Linear Equations

Teaching Notes:
You may or may not want to allow your students to explore the reduced row echelon form, `rref(` in the MATRIX / MATH menu on the TI-84 Plus graphing calculator. This will lead directly to answers but may take away from their understanding of elementary row operations. You can explain how programmers must understand how to use elementary row operations to arrive at the ref and the rref used in calculators and in computers.

The TI-84 Plus calculator can be used to define a matrix and perform matrix operations. (Matrices can be added, subtracted, and multiplied under special restrictions. These operations are saved for another course.) In particular, the Gaussian elimination method is used by the calculator to solve a system of linear equations by reducing the corresponding augmented matrix to **row echelon form (ref)**.

Pressing the MATRIX key (found by pressing 2ND x^{-1}) will give the menu shown here.

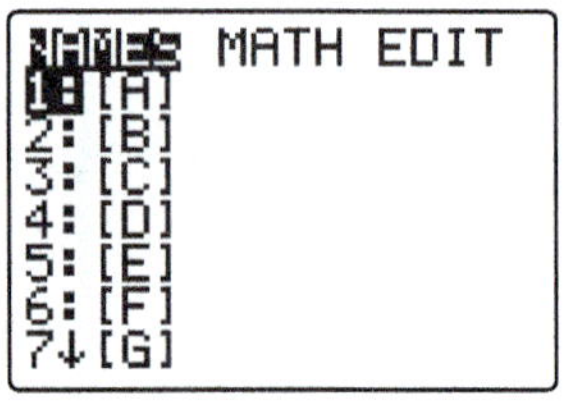

Pressing the right arrow and moving to MATH will give the following choices.

Pressing the right arrow again and moving to EDIT will give the following choices.

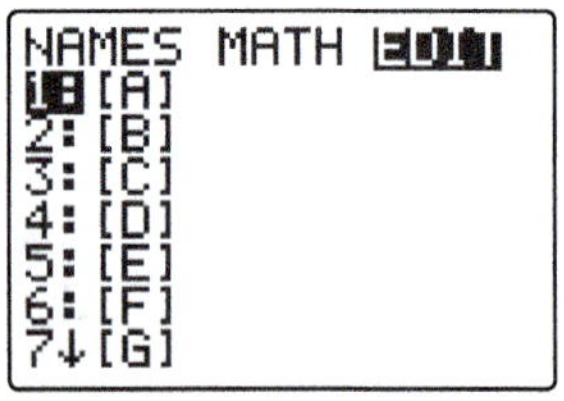

Example 4 shows how to use the calculator to solve a system of three linear equations in three variables. Study each step carefully.

Example 4: Using a Graphing Calculator to Solve Systems of Equations

Use a TI-84 Plus calculator to solve the following system of linear equations.

$$\begin{cases} x+2y+z=1 \\ -x+y+z=-6 \\ 4x-y+3z=-1 \end{cases}$$

Solution:

Step 1: Press 2ND > MATRIX and move to the EDIT menu.

Press ENTER. The following display will appear.

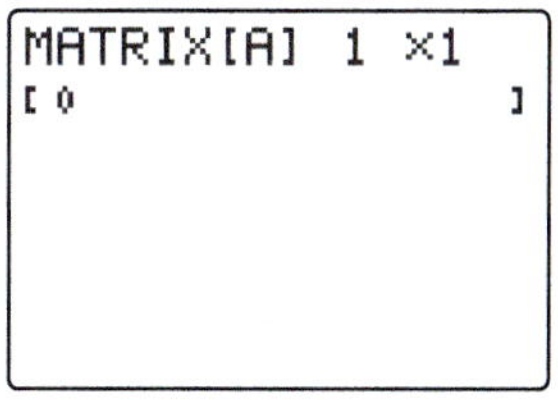

Step 2: The augmented matrix is a 3 × 4 matrix. So, in the top line enter 3, press ENTER, enter 4, press ENTER and the display will appear as shown.

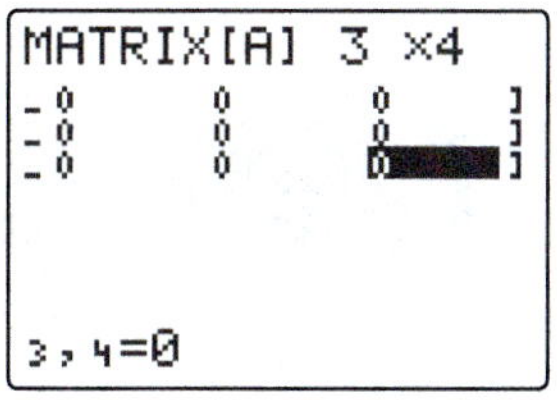

Note: If other numbers are already present on the display, just type over them. The calculator will adjust automatically.

Step 3: Move the cursor to the upper left entry position and enter the coefficients and constants in the matrix. As you enter each number press ENTER and the cursor will automatically move to the next position in the matrix. Note that the double subscripts appear at the bottom of the display as each number is entered. The final display for matrix [A] should appear as shown.

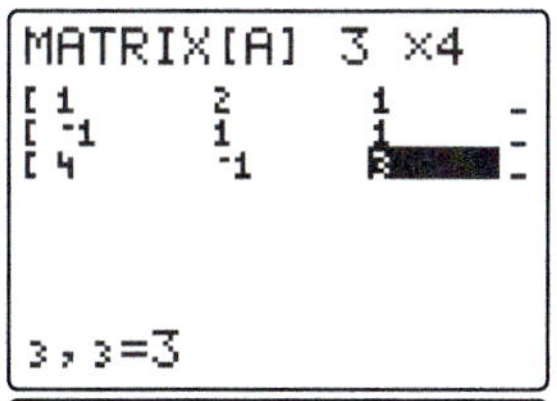

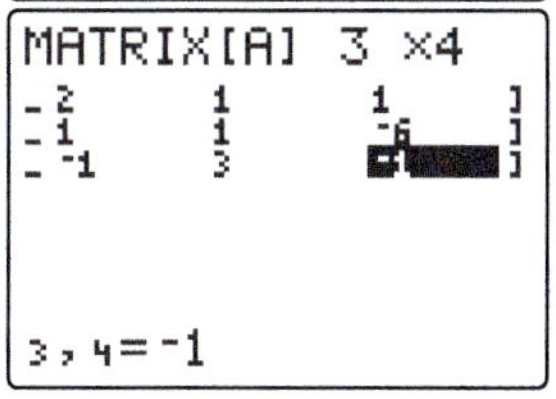

Note: The display only shows three columns at a time.

Step 4: Press 2ND > QUIT, press 2ND > MATRIX again, go to MATH; move the cursor down to A:ref(; press ENTER. The display will appear as shown.

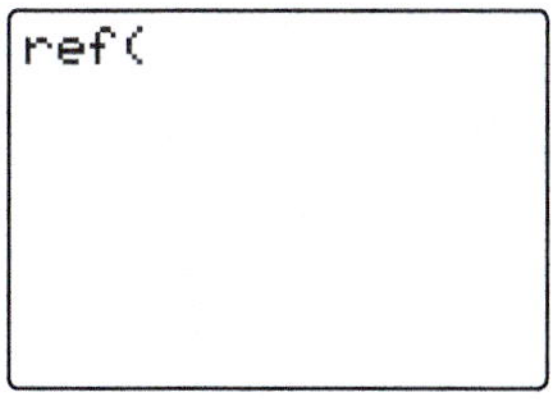

Step 5: Press 2ND > MATRIX again; press ENTER (this selects the matrix A); enter a right parenthesis) ; press the MATH key; and choose 1:>Frac by pressing ENTER. The display will appear as shown.

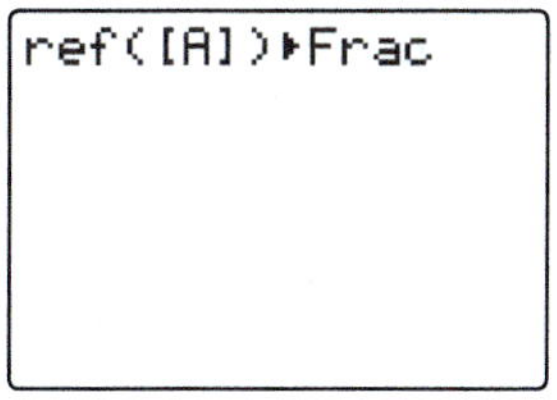

Note: You must select the matrix from the matrix menu. The calculator will not recognize the matrix if you manually type in [A].

Continued on the next page...

Step 6: Press ENTER and the row echelon form of matrix will appear as follows.

```
ref([A])▸Frac
[[1  -1/4 3/4  -1…
 [0  1    1/9  5/…
 [0  0    1    -4…
```

Note: To see the rest of the matrix move the cursor on the screen to the right.

With back substitution we get the following solution $(3, 1, -4)$.

Practice Problems

Solve the following system of linear equations by using the Gaussian elimination method with back substitution.

$$\begin{cases} x - 2y + 3z = 4 \\ 2x + y = 0 \\ 3x + y - z = -4 \end{cases}$$

8.7 Exercises

Write the coefficient matrix and the augmented matrix for the given systems of linear equations.

1. $\begin{cases} 2x + 2y = 13 \\ 5x - y = 10 \end{cases}$

2. $\begin{cases} x + 4y = -1 \\ 2x - 3y = 7 \end{cases}$

3. $\begin{cases} 7x - 2y + 7z = 2 \\ -5x + 3y = 2 \\ 4y + 11z = 8 \end{cases}$

4. $\begin{cases} -8x + 2y - z = 6 \\ 2x + 3z = -3 \\ -4x - 2y + 5z = 13 \end{cases}$

5. $\begin{cases} 3w + x - y + 2z = 6 \\ w - x + 2y - z = -8 \\ 2x + 5y + z = 2 \\ w + 3x + 3z = 14 \end{cases}$

6. $\begin{cases} 4w + x + 3y - 2z = 13 \\ w - 2x + y - 4z = -3 \\ w + x + 4y + 2z = 12 \\ -2w + 3x - y - 3z = 5 \end{cases}$

Write the system of linear equations represented by each of the augmented matrices. Use x, y, and z as the variables.

7. $\left[\begin{array}{cc|c} -3 & 5 & 1 \\ -1 & 3 & 2 \end{array}\right]$

8. $\left[\begin{array}{cc|c} 3 & -1 & 5 \\ -2 & 10 & 9 \end{array}\right]$

9. $\left[\begin{array}{ccc|c} 1 & 3 & 4 & 1 \\ 2 & -3 & -2 & 0 \\ 1 & 1 & 0 & -4 \end{array}\right]$

10. $\left[\begin{array}{ccc|c} 2 & -9 & 14 & 0 \\ -3 & 0 & -8 & 5 \\ 2 & -6 & 1 & 3 \end{array}\right]$

1. $\begin{bmatrix} 2 & 2 \\ 5 & -1 \end{bmatrix}, \left[\begin{array}{cc|c} 2 & 2 & 13 \\ 5 & -1 & 10 \end{array}\right]$

2. $\begin{bmatrix} 1 & 4 \\ 2 & -3 \end{bmatrix}, \left[\begin{array}{cc|c} 1 & 4 & -1 \\ 2 & -3 & 7 \end{array}\right]$

3. $\begin{bmatrix} 7 & -2 & 7 \\ -5 & 3 & 0 \\ 0 & 4 & 11 \end{bmatrix}, \left[\begin{array}{ccc|c} 7 & -2 & 7 & 2 \\ -5 & 3 & 0 & 2 \\ 0 & 4 & 11 & 8 \end{array}\right]$

4. $\begin{bmatrix} -8 & 2 & -1 \\ 2 & 0 & 3 \\ -4 & -2 & 5 \end{bmatrix}, \left[\begin{array}{ccc|c} -8 & 2 & -1 & 6 \\ 2 & 0 & 3 & -3 \\ -4 & -2 & 5 & 13 \end{array}\right]$

5. $\begin{bmatrix} 3 & 1 & -1 & 2 \\ 1 & -1 & 2 & -1 \\ 0 & 2 & 5 & 1 \\ 1 & 3 & 0 & 3 \end{bmatrix}, \left[\begin{array}{cccc|c} 3 & 1 & -1 & 2 & 6 \\ 1 & -1 & 2 & -1 & -8 \\ 0 & 2 & 5 & 1 & 2 \\ 1 & 3 & 0 & 3 & 14 \end{array}\right]$

Answer to Practice Problem: $(-1, 2, 3)$

6. $\begin{bmatrix} 4 & 1 & 3 & -2 \\ 1 & -2 & 1 & -4 \\ 1 & 1 & 4 & 2 \\ -2 & 3 & -1 & -3 \end{bmatrix}$, $\left[\begin{array}{cccc|c} 4 & 1 & 3 & -2 & 13 \\ 1 & -2 & 1 & -4 & -3 \\ 1 & 1 & 4 & 2 & 12 \\ -2 & 3 & -1 & -3 & 5 \end{array}\right]$

7. $\begin{cases} -3x+5y=1 \\ -x+3y=2 \end{cases}$

8. $\begin{cases} 3x-y=5 \\ -2x+10y=9 \end{cases}$

9. $\begin{cases} x+3y+4z=1 \\ 2x-3y-2z=0 \\ x+y=-4 \end{cases}$

10. $\begin{cases} 2x-9y+14z=0 \\ -3x-8z=5 \\ 2x-6y+z=3 \end{cases}$

11. $(-1, 2)$

12. $(2, -1)$

13. $(-1, -1)$

14. $(-2, 2)$

15. $(-1, -2, 3)$

16. No solution

17. $(1, 0, 1)$

18. $(2, -1, 1)$

19. $(2, 1, -1)$

20. Infinitely many solutions

21. $(-2, 9, 1)$

22. $(-2, -1, 5)$

23. No solution

24. $(-2, 0, 1)$

25. Infinitely many solutions

26. $(-2, 3, 1)$

Use the Gaussian elimination method with back substitution to solve the given system of linear equations.

11. $\begin{cases} x+2y=3 \\ 2x-y=-4 \end{cases}$

12. $\begin{cases} 4x+3y=5 \\ -x-2y=0 \end{cases}$

13. $\begin{cases} -8x+2y=6 \\ x-2y=1 \end{cases}$

14. $\begin{cases} 2x+y=-2 \\ 4x+3y=-2 \end{cases}$

15. $\begin{cases} x-3y+2z=11 \\ -2x+4y+z=-3 \\ x-2y+3z=12 \end{cases}$

16. $\begin{cases} x+2y-z=6 \\ x+3y-3z=3 \\ x+y+z=6 \end{cases}$

17. $\begin{cases} x+2y+3z=4 \\ x-y-z=0 \\ 4x-3y+z=5 \end{cases}$

18. $\begin{cases} x+y-2z=-1 \\ 3x+4y-2z=0 \\ x-y+z=4 \end{cases}$

19. $\begin{cases} x-y-2z=3 \\ x+2y-z=5 \\ 2x-3y-2z=3 \end{cases}$

20. $\begin{cases} x+y+3z=2 \\ 2x-y+z=1 \\ 4x+y+7z=5 \end{cases}$

21. $\begin{cases} x-y+5z=-6 \\ x+2z=0 \\ 6x+y+3z=0 \end{cases}$

22. $\begin{cases} x-3y-z=-4 \\ 3x-2y+z=1 \\ -2x+y+2z=13 \end{cases}$

23. $\begin{cases} x-y+2z=5 \\ 2x-2y+4z=5 \\ 3x-3y+6z=8 \end{cases}$

24. $\begin{cases} 2x-y-5z=-9 \\ x-3y+2z=0 \\ 3x+2y+10z=4 \end{cases}$

25. $\begin{cases} x-5y+2z=-7 \\ x+y+4z=7 \\ 2x-y+7z=7 \end{cases}$

26. $\begin{cases} 2x-y+5z=-2 \\ 4x+y+3z=-2 \\ x+3y-z=6 \end{cases}$

27. $\begin{cases} 3x+4z=11 \\ x+y=-2 \\ 2y+z=-4 \end{cases}$

28. $\begin{cases} y+z=2 \\ x+y=5 \\ x+z=5 \end{cases}$

Set up a system of linear equations that represents the information and solve the system using Gaussian elimination.

29. The sum of three integers is 169. The first integer is twelve more than the second integer. The third integer is fifteen less than the sum of the first and second integers. What are the integers?

30. Pizza: A pizzeria sells three sizes of pizzas: small, medium, and large. The pizzas sell for \$6.00, \$8.00, and \$9.50, respectively. One evening they sold 68 pizzas for a total of \$528.00. If they sold twice as many medium-sized pizzas as large-sized pizzas, how many of each size did they sell?

27. $(1, -3, 2)$
28. $(4, 1, 1)$
29. 52, 40, 77
30. 20 small, 32 medium, 16 large
31. Bacon: \$3.09/lb; Eggs: \$4.03/doz; Bread: \$1.40/loaf
32. \$127,500 at 6%; \$62,500 at 8%; \$60,000 at 11%
33. $(0, -4)$
34. $(-4, 3)$
35. $(2, 1, 7)$
36. $(-1, 2, 1)$
37. $(2, -3, 4)$
38. $\left(-\frac{13}{12}, \frac{1}{12}, \frac{8}{3}, -\frac{5}{3}\right)$
39. $\left(\frac{13}{12}, \frac{5}{4}, \frac{8}{3}\right)$
40. $\left(-\frac{1}{3}, 1, \frac{16}{3}\right)$

31. Grocery shopping: Caroline bought a pound of bacon, a dozen eggs, and a loaf of bread. The total cost was \$8.52. The eggs cost \$0.94 more than the bacon. The combined cost of the bread and eggs was \$2.34 more than the cost of the bacon. Find the cost of each item.

32. Investments: An investment firm is responsible for investing \$250,000 from an estate according to three conditions in the will of the deceased. The money is to be invested in three accounts paying 6%, 8%, and 11% interest. The amount invested in the 6% account is to be \$5000 more than the total invested in the other two accounts, and the total annual interest for the first year is to be \$19,250. How much is the firm supposed to invest in each account?

Use your graphing calculator to solve the systems of linear equations.

33. $\begin{cases} x+y=-4 \\ 2x+3y=-12 \end{cases}$

34. $\begin{cases} 2x+3y=1 \\ x-5y=-19 \end{cases}$

35. $\begin{cases} x+y+z=10 \\ 2x-y+z=10 \\ -x+2y+2z=14 \end{cases}$

36. $\begin{cases} 2x+y+2z=2 \\ x-y+4z=1 \\ 3x-y+z=-4 \end{cases}$

37. $\begin{cases} 2y-3z=-18 \\ 4x+5y=-7 \\ 6x-z=8 \end{cases}$

38. $\begin{cases} w+x+y+z=0 \\ w+x-y-z=-2 \\ -w+3x+3y-z=11 \\ y-2z=6 \end{cases}$

39. $\begin{cases} x-3y+z=0 \\ 2x+2y-z=2 \\ x+y+z=5 \end{cases}$

40. $\begin{cases} x-2y-2z=-13 \\ 2x+y-z=-5 \\ x+y+z=6 \end{cases}$

Writing and Thinking About Mathematics

41. Solving the second equation for z, we can back substitute into the first equation, eliminating z. The result is the equation $x + 5y = 6$, which means the system has an infinite number of solutions.

41. Suppose that Gaussian elimination with a system of three linear equations in three unknowns results in the following triangular matrix. Discuss how you can use back substitution to find that the system has an infinite number of solutions and these solutions satisfy the equation $x + 5y = 6$. (**Hint:** Solve the second equation for z.)

$$\begin{bmatrix} 1 & 2 & -1 & 4 \\ 0 & 3 & 1 & 2 \\ 0 & 0 & 0 & 0 \end{bmatrix}$$

HAWKES LEARNING SYSTEMS: INTRODUCTORY & INTERMEDIATE ALGEBRA SOFTWARE

- 8.7 Matrices and Gaussian Elimination

8.8 Graphing Systems of Linear Inequalities

- *Solve systems of linear inequalities graphically.*

In some branches of mathematics, in particular a topic called game theory, the solution to a very sophisticated problem can involve the set of points that satisfies a system of several **linear inequalities**. In business these ideas relate to problems such as minimizing the cost of shipping goods from several warehouses to distribution outlets. In this section we will consider graphing the solution sets to only two inequalities. We will leave the problem solving techniques to another course.

In Section 4.6 we graphed linear inequalities of the form $y < mx + b$ (or $y \le mx + b$ or $y > mx + b$ or $y \ge mx + b$). These graphs are called **half-planes**. The line $y = mx + b$ is called the **boundary line** and the half-planes are **open** (the boundary line is not included) or **closed** (the boundary line is included).

In this section, we will develop techniques for graphing (and therefore solving) **systems of two linear inequalities**. The **solution set** (if there are any solutions) to such a system of linear inequalities consists of the points in the **intersection** of two half-planes and portions of boundary lines indicated by the inequalities. The following procedure may be used to solve a system of linear inequalities.

To Solve a System of Two Linear Inequalities

1. For each inequality, graph the boundary line and shade the appropriate half-plane. (Refer to Section 4.6 to review this process.)
2. Determine the region of the graph that is common to both half-planes (the region where the shading overlaps).

   (This region is called the **intersection** of the two half-planes and is the **solution set** of the system.)
3. To check, pick one test-point in the intersection and verify that it satisfies both inequalities.

Note: If there is no intersection, then the system has no solution.

Example 1: Graphing Systems of Linear Inequalities

a. Graph the points that satisfy the system of inequalities: $\begin{cases} x \le 2 \\ y \ge -x + 1 \end{cases}$

Solution:

Step 1: For $x \le 2$, the points are to the left of and on the line $x = 2$.

Continued on the next page...

Step 2: For $y \geq -x+1$, the points are above and on the line $y = -x+1$.

Step 3: Determine the region that is common to both half-planes. In this case, we test the point $(0, 3)$. On the graph below, the solution is the purple-shaded region and its boundary lines.

$0 \leq 2$ A true statement

$3 \geq -0+1$ A true statement

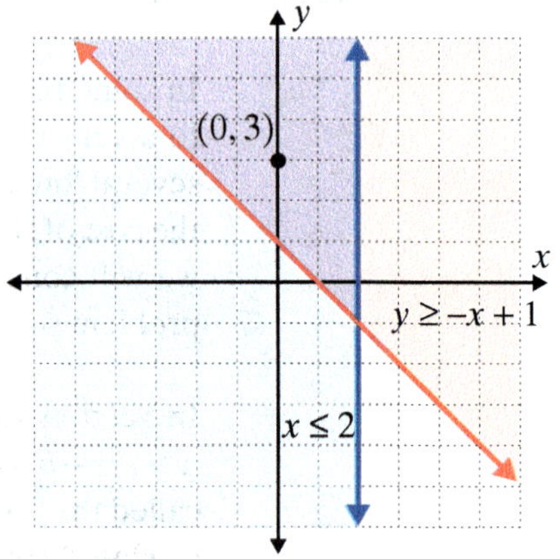

b. Solve the system of linear inequalities graphically: $\begin{cases} 2x+y \leq 6 \\ x+y<4 \end{cases}$

Solution:

Step 1: Solve each inequality for y: $\begin{cases} y \leq -2x+6 \\ y<-x+4 \end{cases}$

Step 2: For $y \leq -2x+6$, the points are below and on the line $y = -2x+6$.

Step 3: For $y < -x+4$, the points are below but not on the line $y = -x+4$.

Step 4: Determine the region that is common to both half-planes. Note that the line $y = -x + 4$ is dashed to indicate that the points on the line are not included. In this case, we test the point $(0, 0)$. On the graph below, the solution is the purple-shaded region and its boundary lines.

$2 \cdot 0+0 \leq 6$ A true statement

$0+0<4$ A true statement

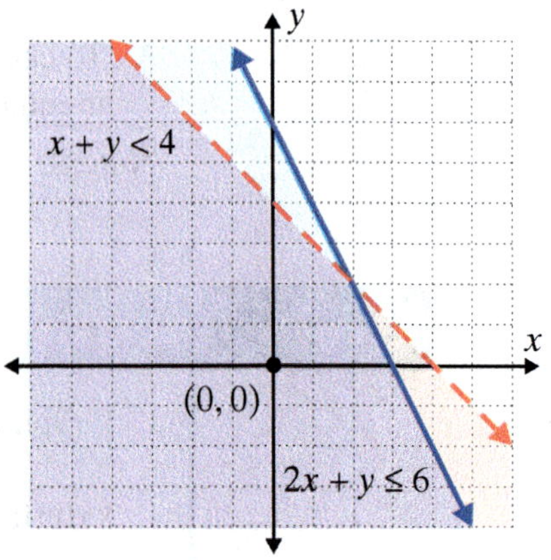

c. Solve the system of linear inequalities graphically: $\begin{cases} y \geq x \\ y \leq x+2 \end{cases}$

Solution:

Step 1: For $y \geq x$, the points are above and on the line $y = x$.

Step 2: For $y \leq x + 2$, the points are below and on the line $y = x + 2$.

Step 3: The solution set consists of the boundary lines and the region between them.

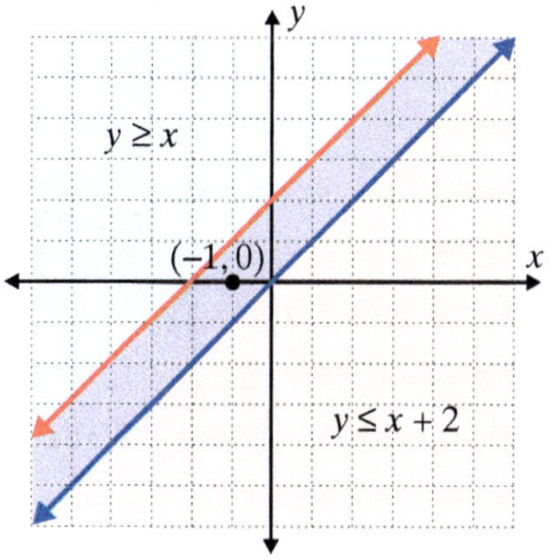

NOTES

When the boundary lines are parallel there are three possibilities:

1. The common region will be in the form of a strip between two lines (as in Example 1c).
2. The common region will be a half-plane, as the solution to one inequality will be entirely contained within the solution of the other inequality.
3. There will be no common region which means there is no solution.

Using a TI-84 Plus Graphing Calculator to Graph Systems of Linear Inequalities

To graph (and therefore solve) a system of linear inequalities (with no vertical line) with a TI-84 Plus graphing calculator, proceed as follows:

1. Solve each inequality for y.
2. Press the Y= key and enter the two expressions for Y_1 and Y_2.
3. Move to the left of Y_1 and Y_2 and press ENTER until the desired graphing symbol appears.
4. Press GRAPH and the desired region will be graphed as a cross-hatched area. (You may need to reset the WINDOW so that both regions appear.)

Example 2 illustrates how this can be done.

Example 2: Using a Graphing Calculator to Solve a System of Linear Inequalities

Use a TI-84 Plus graphing calculator to graph the following system of linear inequalities:

$$\begin{cases} 2x + y < 4 \\ 2x - y \le 0 \end{cases}$$

Solution:

Step 1: Solve each inequality for y: $\begin{cases} y < -2x + 4 \\ y \ge 2x \end{cases}$

Note: Solving $2x - y \le 0$ for y can be written as $2x \le y$ and then as $y \ge 2x$.

Steps 2 and 3: Press the Y= key and enter both functions and the corresponding symbols as they appear here:

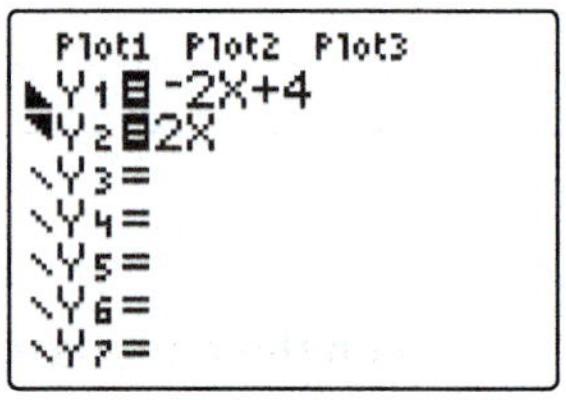

(**Remember:** To shade your graphs, position the cursor over the slash next to Y_1 (or Y_2) and hit ENTER repeatedly until the appropriate shading is displayed.)

Step 4: Press GRAPH. The display should appear as follows. The solution is the cross-hatched region and the points on the line $2x - y = 0$.

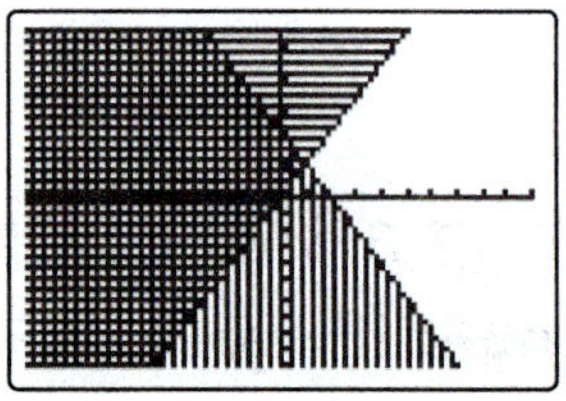

Practice Problems

Solve the systems of two linear inequalities graphically.

1. $\begin{cases} y \le -3 \\ y \ge x - 3 \end{cases}$

2. $\begin{cases} y > 2 \\ x < 2 \end{cases}$

3. $\begin{cases} x + y \ge 0 \\ x - 2y \ge 4 \end{cases}$

Answers to Practice Problems:

1. 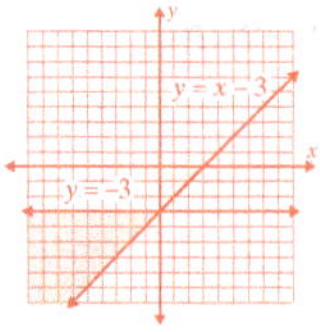

2. 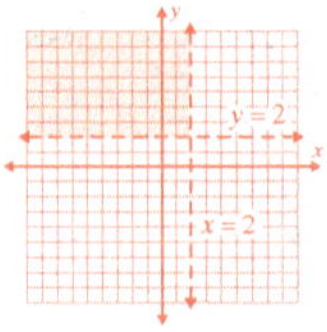

3. 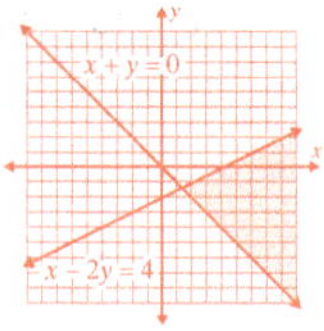

Exercises 8.8

1. 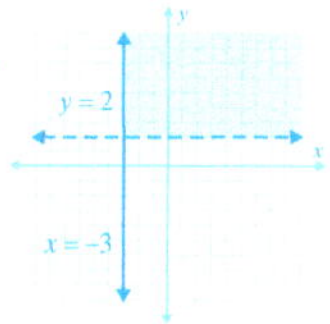

2. 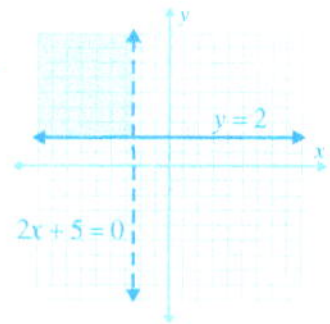

3. 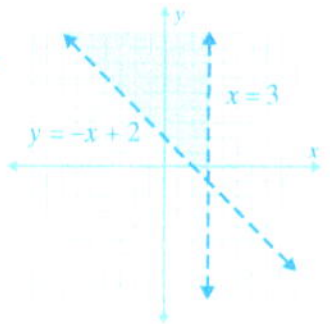

4. 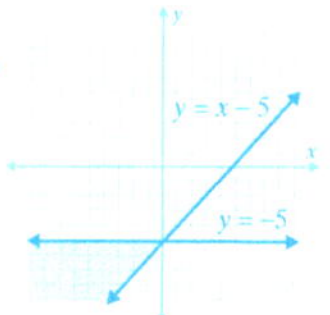

5. 

6. 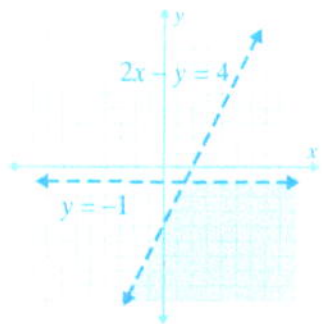

7. 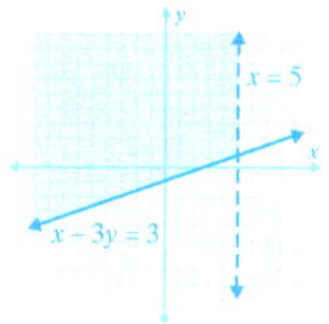

8. 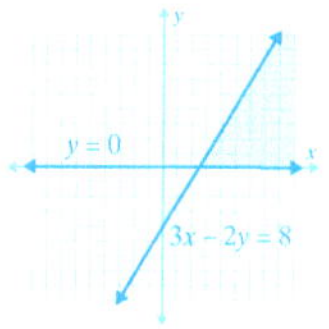

9. 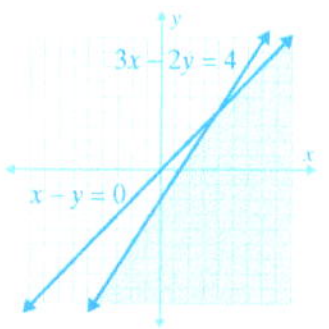

Solve the systems of two linear inequalities graphically.

1. $\begin{cases} y > 2 \\ x \geq -3 \end{cases}$ **2.** $\begin{cases} 2x + 5 < 0 \\ y \geq 2 \end{cases}$ **3.** $\begin{cases} x < 3 \\ y > -x + 2 \end{cases}$

4. $\begin{cases} y \leq -5 \\ y \geq x - 5 \end{cases}$ **5.** $\begin{cases} x \leq 3 \\ 2x + y > 7 \end{cases}$ **6.** $\begin{cases} 2x - y > 4 \\ y < -1 \end{cases}$

7. $\begin{cases} x - 3y \leq 3 \\ x < 5 \end{cases}$ **8.** $\begin{cases} 3x - 2y \geq 8 \\ y \geq 0 \end{cases}$ **9.** $\begin{cases} x - y \geq 0 \\ 3x - 2y \geq 4 \end{cases}$

10. $\begin{cases} y \geq x - 2 \\ x + y \geq -2 \end{cases}$ **11.** $\begin{cases} 3x + y \leq 10 \\ 5x - y \geq 6 \end{cases}$ **12.** $\begin{cases} y > 3x + 1 \\ -3x + y < -1 \end{cases}$

13. $\begin{cases} 3x + 4y \geq -7 \\ y < 2x + 1 \end{cases}$ **14.** $\begin{cases} 2x - 3y \geq 0 \\ 8x - 3y < 36 \end{cases}$ **15.** $\begin{cases} x + y < 4 \\ 2x - 3y < 3 \end{cases}$

16. $\begin{cases} 2x + 3y < 12 \\ 3x + 2y > 13 \end{cases}$ **17.** $\begin{cases} x + y \geq 0 \\ x - 2y \geq 6 \end{cases}$ **18.** $\begin{cases} y \geq 2x + 3 \\ y \leq x - 2 \end{cases}$

19. $\begin{cases} x + 3y \leq 9 \\ x - y \geq 5 \end{cases}$ **20.** $\begin{cases} x - y \geq -2 \\ x + 2y < -1 \end{cases}$ **21.** $\begin{cases} y \leq x + 3 \\ x - y \leq -5 \end{cases}$

22. $\begin{cases} y \geq 2x - 5 \\ 3x + 2y > -3 \end{cases}$ **23.** $\begin{cases} y \leq -2x \\ y > -2x - 6 \end{cases}$ **24.** $\begin{cases} y > x - 4 \\ y < x + 2 \end{cases}$

Use a graphing calculator to solve the systems of linear inequalities.

25. $\begin{cases} y \geq 0 \\ 3x - 5y \leq 10 \end{cases}$ **26.** $\begin{cases} y \leq 0 \\ 3x + y \leq 11 \end{cases}$ **27.** $\begin{cases} 4x - 3y \geq 6 \\ 3x - y \leq 3 \end{cases}$

28. $\begin{cases} 3x + 2y \leq 15 \\ 2x + 5y \geq 10 \end{cases}$ **29.** $\begin{cases} 3x - 4y \geq -6 \\ 3x + 2y \leq 12 \end{cases}$ **30.** $\begin{cases} 3y \leq 2x + 2 \\ x + 2y \leq 11 \end{cases}$

31. $\begin{cases} x + y \leq 8 \\ 3x - 2y \geq -6 \end{cases}$ **32.** $\begin{cases} x + y \leq 7 \\ 2x - y \leq 8 \end{cases}$ **33.** $\begin{cases} y \leq x \\ y < 2x + 1 \end{cases}$

34. $\begin{cases} x - y \geq -2 \\ 4x - y < 16 \end{cases}$

Writing and Thinking About Mathematics

35. Graph the inequalities and explain how you can tell that there is no solution.

$$\begin{cases} y \le 2x - 5 \\ y \ge 2x + 3 \end{cases}$$

HAWKES LEARNING SYSTEMS: INTRODUCTORY & INTERMEDIATE ALGEBRA SOFTWARE

- 8.8 Systems of Linear Inequalities

10. 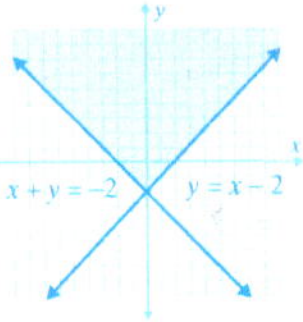

11. 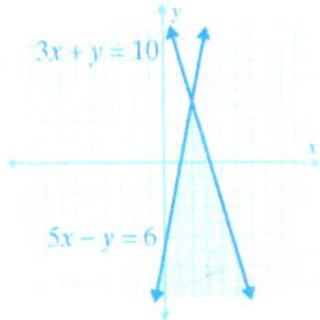

12. No solution

13. 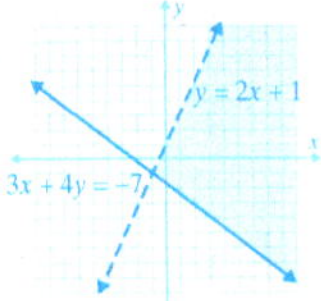

14. 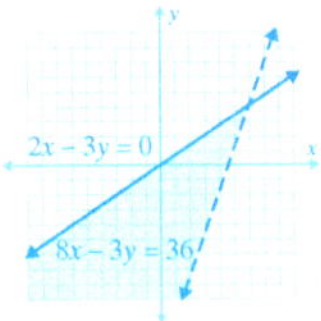

15. 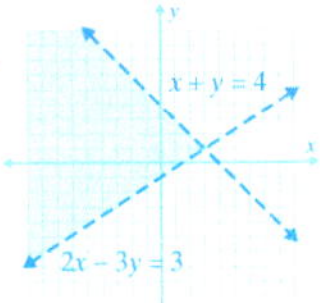

16. 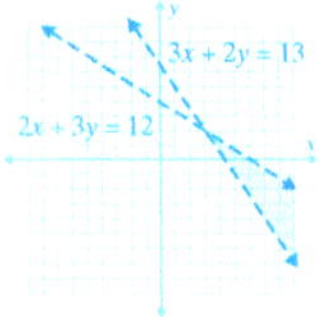

17. 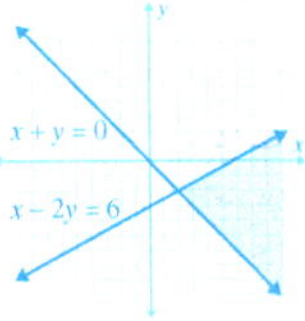

18. 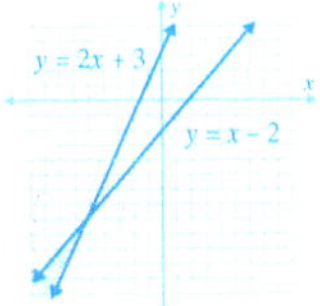

19. 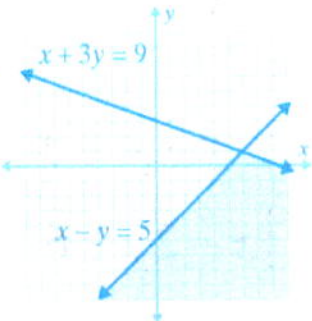

20. 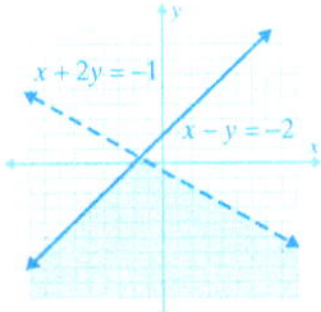

21. No solution

22. 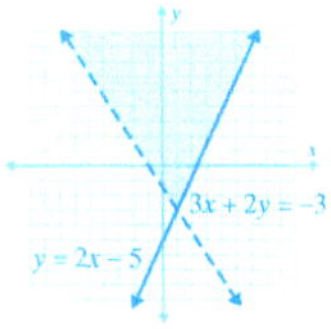

23. 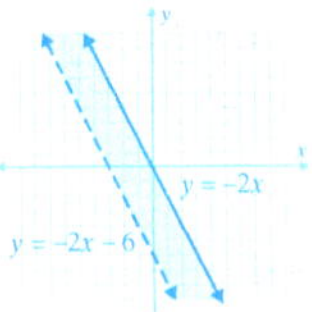

24. 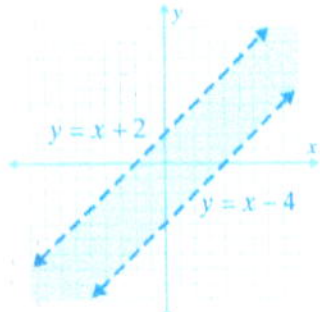

25. 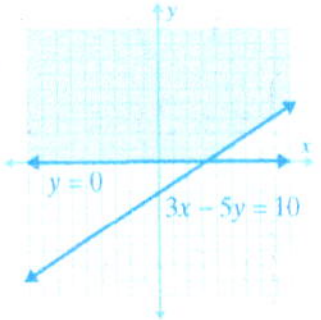

26. 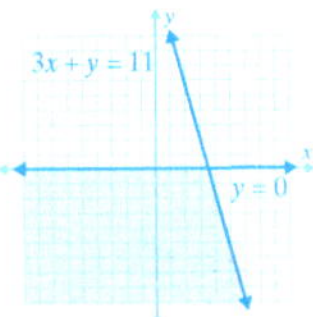

27. 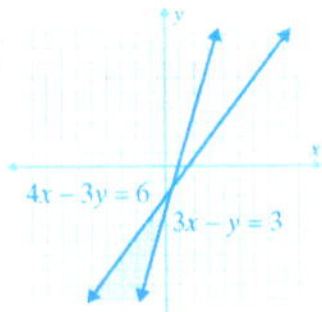

28. 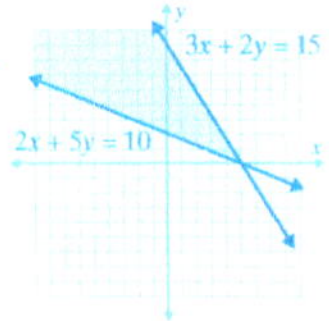

29. 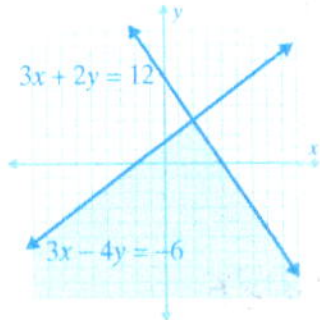

30. 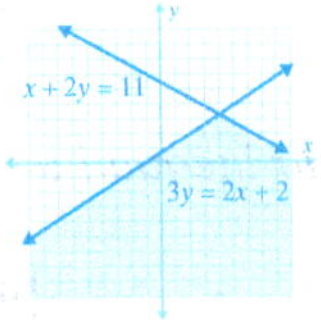

31. 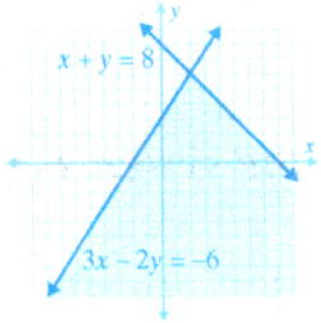

32. 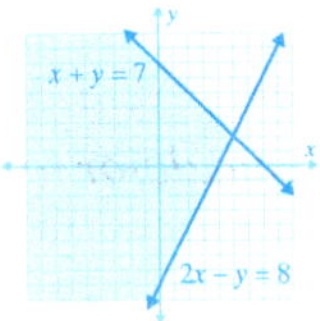

33. 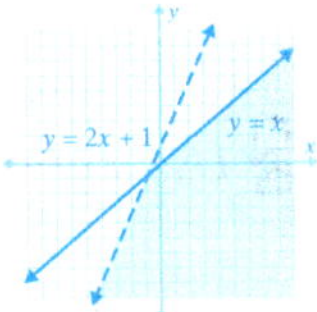

34. 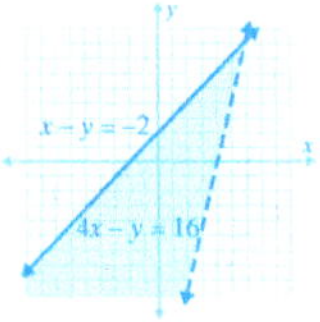

35. 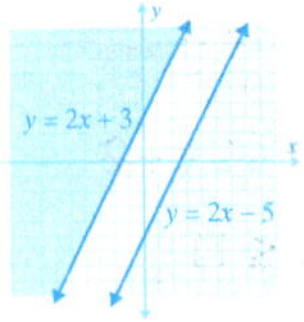

The solutions of the two inequalities do not overlap.

Chapter 8 Index of Key Ideas and Terms

Section 8.1 Systems of Linear Equations: Solutions by Graphing

System of Linear Equations page 588

A pair of linear equations considered together is called a **system of linear equations** (or a **set of simultaneous equations**).

Solution to a System of Equations page 588

A **solution of a system** of linear equations is an ordered pair (or point) that satisfies **both** equations.

Solutions by Graphing page 591

1. Graph both linear equations on the same set of axes.
2. Observe the point of intersection (if there is one).
   - **a.** If the slopes of the two lines are different, then the lines intersect in one and only one point. The system has a single point as its solution.
   - **b.** If the lines are distinct and have the same slope, then the lines are parallel. The system has no solution.
   - **c.** If the lines are the same line, then all the points on the line constitute the solution. There is an infinite number of solutions.
3. Check the solution (if there is one) in both of the original equations.

Using a Graphing Calculator to Solve a System of Linear Equations pages 593-594

Section 8.2 Systems of Linear Equations: Solutions by Substitution

Solutions by Substitution pages 599-600

1. Solve one of the equations for one of the variables.
2. Substitute the resulting expression into the other equation.
3. Solve this new equation, if possible, and then substitute back into one of the original equations to find the value of the other variable. (This is known as **back substitution**.)
4. Check the solution in both of the original equations.

Section 8.3 Systems of Linear Equations: Solutions by Addition

Solutions by Addition pages 606-607

1. Write the equations in **standard form** one under the other so that **like terms are aligned**.
2. Multiply all terms of one equation by a constant (and possibly all terms of the other equation by another constant) so that **two like terms have opposite coefficients**.
3. Add the two equations by **combining like terms** and solve the resulting equation, if possible.
4. **Back substitute into one of the original equations** to find the value of the other variable.
5. Check the solution (if there is one) in both of the original equations.

Guidelines for Deciding which Method to Use when Solving a System of Linear Equations page 610

1. The graphing method is helpful in "seeing" the geometric relationship between the lines and finding approximate solutions. A calculator can be very helpful here.
2. Both the substitution method and the addition method give exact solutions.
3. The substitution method may be reasonable and efficient if one of the coefficients of one of the variables is 1.
4. The addition method is particularly efficient if the coefficients for one of the variables are opposites.

Section 8.4 Applications: Distance-Rate-Time, Number Problems, Amounts and Costs

Applications

Distance-Rate-Time pages 614-616
Number Problems page 616
Amounts and Costs pages 617-619

Section 8.5 Applications: Interest and Mixture

Applications

Interest pages 626-628
Mixture pages 628-630

Section 8.6 Systems of Linear Equations: Three Variables

Linear Equation in Three Variables page 634

The general form of an equation in three variables is $Ax + By + Cz = D$ where A, B, and C are not all equal to 0.

Ordered Triples page 634

An **ordered triple** is an ordering of three real numbers in the form (x_0, y_0, z_0) and is used to represent the solution to a linear equation in three variables.

Graphs in Three Dimensions pages 634-635

Three mutually perpendicular number lines labeled as the x-axis, y-axis, and z-axis are used to separate space into eight regions called **octants**.

To Solve a System of Three Linear Equations in Three Variables pages 636-638

1. Select two equations and eliminate one variable by using the addition method.
2. Select a different pair of equations and eliminate the same variable.
3. Steps 1 and 2 give two linear equations in two variables. Solve these equations by either addition or substitution as discussed in Sections 8.2 and 8.3.
4. Back substitute the values found in Step 3 into any one of the original equations to find the value of the third variable.
5. Check the solution (if there is one) in all three of the original equations.

Section 8.7 Matrices and Gaussian Elimination

Matrices page 646

Matrix – A rectangular array of numbers.
Entry – A number in a matrix
Row – A group of entries written horizontally
Column – A group of entries written vertically
Dimension of a matrix – The number of rows by the number of columns, such as 2×3
Square matrix – A matrix with the same number of rows as columns, such as 3×3

Continued on the next page...

Section 8.7 Matrices and Gaussian Elimination (cont.)

Coefficient Matrix page 647

A matrix formed from the coefficients of the variables in a system of linear equations is called a **coefficient matrix**.

Augmented Matrix page 647

A matrix formed from the coefficients of the variables and the constant terms in a system of linear equations is called an **augmented matrix**.

Elementary Row Operations page 648

There are three elementary row operations with matrices:

1. Interchange two rows.
2. Multiply a row by a nonzero constant.
3. Add a multiple of a row to another row.

If any elementary row operation is applied to a matrix, the new matrix is said to be **row-equivalent** to the original matrix.

Upper Triangular Form page 650

A matrix is said to be in **upper triangular form** if all the entries in the lower left triangular region are 0's.

Row Echelon Form page 650

The upper triangular form of a matrix with all 1's in the main diagonal is called the **row echelon form** (or **ref**).

Gaussian Elimination page 650

To solve a system of linear equations using Gaussian elimination:

1. Write the augmented matrix for the system.
2. Use elementary row operations to transform the matrix into row echelon form.
3. Solve the corresponding system of equations by using back substitution.

Using a Graphing Calculator to Solve a System of Linear Equations pages 654-656

Section 8.8 Graphing Systems of Linear Inequalities

To Solve a System of Two Linear Inequalities page 659

1. For each inequality, graph the boundary line and shade the appropriate half-plane. (Refer to Section 4.6 to review this process.)
2. Determine the region of the graph that is common to both half-planes (the region where the shading overlaps). (This region is called the **intersection** of the two half-planes and is the **solution set** of the system.)
3. To check, pick one test-point in the intersection and verify that it satisfies both inequalities.

Note: If there is no intersection, then the system is inconsistent and has no solution.

HAWKES LEARNING SYSTEMS: INTRODUCTORY & INTERMEDIATE ALGEBRA SOFTWARE

- 8.1 Solving Systems of Linear Equations by Graphing
- 8.2 Solving Systems of Linear Equations by Substitution
- 8.3 Solving Systems of Linear Equations by Addition
- 8.4 Applications: Distance-Rate-Time, Number Problems, Amounts and Costs
- 8.5 Applications: Interest and Mixture
- 8.6 Solving Systems of Linear Equations with Three Variables
- 8.7 Matrices and Gaussian Elimination
- 8.8 Systems of Linear Inequalities

Chapter 8 Review

1. a
2. b, c, d
3. $m_1 = 2,\ b_1 = 0,$ $m_2 = 2,\ b_2 = 5$
4. $m_1 = -3,\ b_1 = 4,$ $m_2 = -3,\ b_2 = -\frac{1}{2}$
5. $(1, 3)$

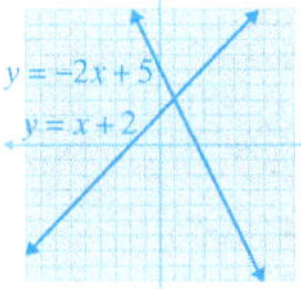

6. No solution

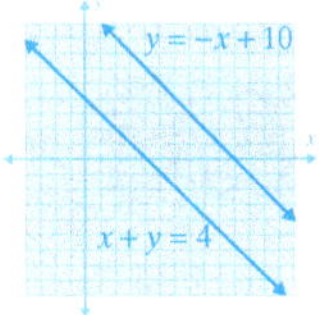

7. $(2, -1)$

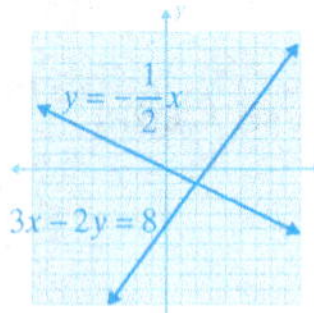

8. $(x, 4-5x)$

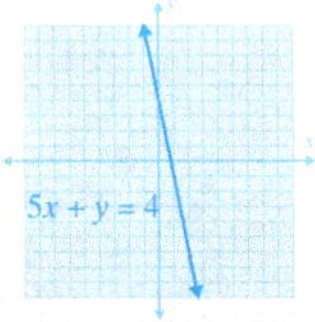

9. No solution

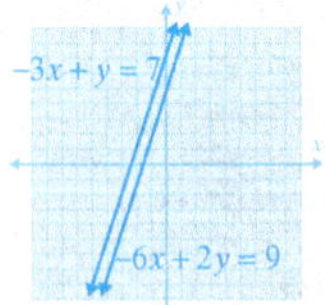

10. $(-1, 4)$

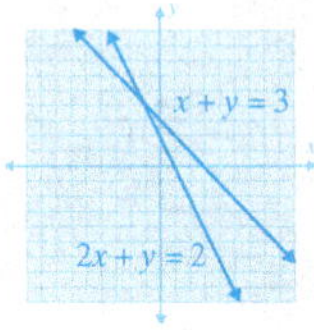

8.1 Systems of Linear Equations: Solutions by Graphing

Determine which of the given points, if any, lie on both of the lines in the systems of equations by substituting each point into both equations.

1. $\begin{cases} y - x = 1 \\ 3y - x = -6 \end{cases}$ **a.** $\left(-\frac{9}{2}, -\frac{7}{2}\right)$ **b.** $(7, 8)$ **c.** $(6, 0)$ **d.** $\left(-5, \frac{1}{4}\right)$

2. $\begin{cases} 6x + 2y - 10 = 0 \\ 12x - 20 = -4y \end{cases}$ **a.** $(7, -5)$ **b.** $(4, -7)$ **c.** $(-1, 8)$ **d.** $(3, -4)$

Show that each system of equations is inconsistent by determining the slope of each line and the y-intercept. (That is, show that the lines are parallel and do not intersect.)

3. $\begin{cases} y = 2x \\ -2x + y = 5 \end{cases}$

4. $\begin{cases} 3x + y = 4 \\ 6x + 2y = -1 \end{cases}$

Solve the following systems graphically.

5. $\begin{cases} y = -2x + 5 \\ y = x + 2 \end{cases}$

6. $\begin{cases} y = -x + 10 \\ x + y = 4 \end{cases}$

7. $\begin{cases} y = -\frac{1}{2}x \\ 3x - 2y = 8 \end{cases}$

8. $\begin{cases} 5x + y = 4 \\ 10x + 2y = 8 \end{cases}$

9. $\begin{cases} -3x + y = 7 \\ -6x + 2y = 9 \end{cases}$

10. $\begin{cases} x + y = 3 \\ 2x + y = 2 \end{cases}$

11. $\begin{cases} x + 2y = 12 \\ 3x - y = -6 \end{cases}$

12. $\begin{cases} x + 4y = 6 \\ y + \frac{1}{4}x = \frac{3}{2} \end{cases}$

Use a graphing calculator to find the solutions to the given systems of linear equations. If necessary, round values to four decimal places.

13. $\begin{cases} 2x + 3y = 10.5 \\ x - y = -1 \end{cases}$

14. $\begin{cases} x - 3y = 2.5 \\ 2x + y = -5.5 \end{cases}$

15. $\begin{cases} y = 7 + x \\ x + 3y = 21 \end{cases}$

16. $\begin{cases} x - 5y = -1.5 \\ x - y = 4.3 \end{cases}$

8.2 Systems of Linear Equations: Solutions by Substitution

Solve the following systems of linear equations by using the method of substitution.

17. $\begin{cases} y = 2x \\ x - y = -1 \end{cases}$

18. $\begin{cases} y = 4x + 1 \\ 8x - 2y = -2 \end{cases}$

19. $\begin{cases} 3x - 2y = 4 \\ y - 6 = \frac{3}{2}x \end{cases}$

20. $\begin{cases} y = -2x + 5 \\ 4x = -2y - 1 \end{cases}$

21. $\begin{cases} 5x - y = 6 \\ 2y = 10x - 12 \end{cases}$

22. $\begin{cases} y = -\frac{1}{8}x \\ x - 4y = 4 \end{cases}$

23. $\begin{cases} 2x - y = 2 \\ 2x + y = -2 \end{cases}$

24. $\begin{cases} x + \frac{1}{2}y = 8 \\ y = x + 1 \end{cases}$

25. $\begin{cases} y = 7x - 5 \\ x - 3y = 0 \end{cases}$

26. $\begin{cases} 6x + y = -4 \\ 3x + \frac{1}{2}y = \frac{1}{2} \end{cases}$

8.3 Systems of Linear Equations: Solutions by Addition

Solve each system by using the addition method.

27. $\begin{cases} 5x + 3y = -11 \\ -5x + 2y = -4 \end{cases}$

28. $\begin{cases} 4x - 5y = 25 \\ 3x - 2y = 17 \end{cases}$

29. $\begin{cases} -4x + 3y = 9 \\ 8x - 2y = -14 \end{cases}$

30. $\begin{cases} 2x - y = 12 \\ 2x + \frac{1}{2}y = 15 \end{cases}$

Solve each system by using either the substitution method or the addition method (whichever seems better to you).

31. $\begin{cases} x + y = 1 \\ x - y = 3 \end{cases}$

32. $\begin{cases} 3x - 2y = 5 \\ x + 3y = 17 \end{cases}$

33. $\begin{cases} x + 2y = 1 \\ 2x - 3y = 0 \end{cases}$

34. $\begin{cases} 5x + y = 1 \\ 3x + y = 3 \end{cases}$

35. $\begin{cases} -3x + y = 6 \\ 6x - 2y = -12 \end{cases}$

36. $\begin{cases} 3x + 4y = 8 \\ -\frac{3}{4}x + 4 = y \end{cases}$

37. $\begin{cases} y = -\frac{1}{2}x + 10 \\ 2y = 6 - x \end{cases}$

38. $\begin{cases} y = \frac{1}{3}x + 10 \\ x = -30 + 3y \end{cases}$

Write an equation for the line determined by the two given points using the formula $y = mx + b$ to set up a system of equations with m and b as the unknowns.

39. $(2, -9), (1, -5)$

40. $(8, -1), (0, 2)$

8.4 Applications: Distance-Rate-Time, Number Problems, Amounts and Costs

41. The sum of two numbers is −12. One number is 3 more than twice the other.

42. The difference between two numbers is 2. Three times the smaller is equal to 8 more than the larger. What are the numbers?

43. **Age:** Alice is 8 years older than her brother John. In 5 years she will be twice as old as John. How old are Alice and John now?

44. **Making change:** Jorge has \$2.10 worth of change in quarters and dimes. If he has eight times as many dimes as quarters how many of each type of coin does he have?

11. $(0, 6)$

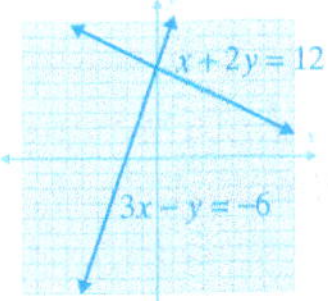

12. $\left(x, -\frac{1}{4}x + \frac{3}{2}\right)$

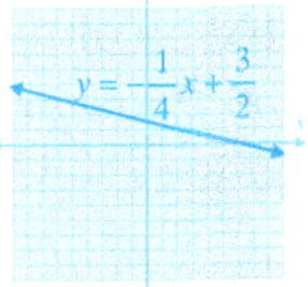

13. $(1.5, 2.5)$

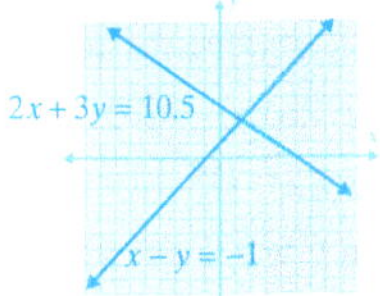

14. $(-2, -1.5)$

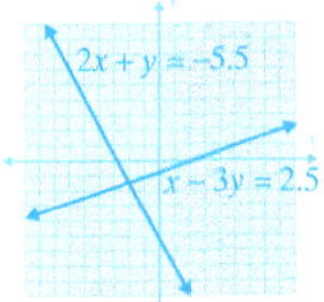

15. $(0, 7)$

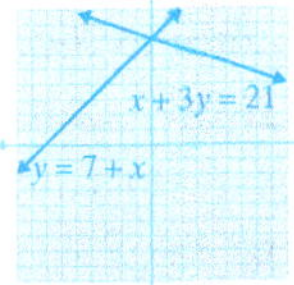

16. $(5.75, 1.45)$

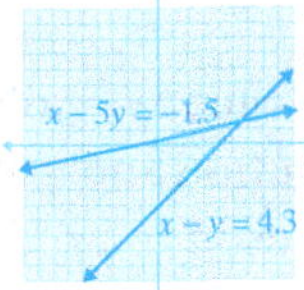

17. $(1, 2)$
18. $(x, 4x + 1)$
19. No solution
20. No solution
21. $(x, 5x - 6)$
22. $\left(\frac{8}{3}, -\frac{1}{3}\right)$
23. $(0, -2)$

24. $(5, 6)$
25. $\left(\frac{3}{4}, \frac{1}{4}\right)$
26. No solution
27. $\left(-\frac{2}{5}, -3\right)$
28. $(5, -1)$
29. $\left(-\frac{3}{2}, 1\right)$
30. $(7, 2)$
31. $(2, -1)$
32. $\left(\frac{49}{11}, \frac{46}{11}\right)$
33. $\left(\frac{3}{7}, \frac{2}{7}\right)$
34. $(-1, 6)$
35. $(x, 3x+6)$
36. No solution
37. No solution
38. $\left(x, \frac{1}{3}x+10\right)$
39. $y = -4x - 1$, $m = -4, b = -1$
40. $y = -\frac{3}{8}x + 2$, $m = -\frac{3}{8}, b = 2$
41. -7 and -5
42. 5 and 7
43. Alice is 11 years old and John is 3 years old.
44. 2 quarters and 16 dimes
45. He averaged 2 mph for the first 3.6 hours and 3 mph for the last 2.4 hours.
46. 36 minutes at 40 mph and 36 minutes at 20 mph

45. Hiking: Brandon decided to hike in the mountains a total distance of 14.4 miles (on a trail he knew). He averaged 2 mph for the first part of the hike and 3 mph for the second part. If the total hike took 6 hours, how many hours did he hike at each rate?

46. Airport Taxi: Sam catches a taxi for a ride to the airport 36 miles away. The taxi maintained an average speed of 40 mph for the first part of the trip, however closer to the airport traffic became heavier and its speed was reduced to 20 mph. How long was spent at each speed if the total trip took 1 hour and 12 minutes?

47. Rectangles: The width of a rectangle is 5 meters less than the length. If the perimeter of the rectangle is 80 meters, what are the length and width?

48. Supplementary angles: Two angles are supplementary if the sum of their measures is 180°. Find two supplementary angles such that one angle is 20° less than three times the other.

49. Shopping: A golf pro shop sells two particular types of shirts, the first for \$110 each and the other for \$65 each. Last month the shop sold 50 of these shirts for a total of \$4600. How many of each type of shirt did they sell?

50. Fair Food: While at the county fair, a family decides to grab a sweet snack. If they buy 5 boxes of candied popcorn and 3 funnel cakes the cost of the snack would be \$26.05. If they buy 2 boxes of popcorn and 4 funnel cakes, the cost would be \$21.90. Find the price of the box of popcorn and the price of the funnel cake.

8.5 Applications: Interest and Mixture

51. Investing: Mr. Smith has \$20,000 invested, part at 6% and the remainder at 8%. How much does he have invested at each rate if the annual interest is \$1300?

52. Investing: Mary Jane invests a certain amount of money at 8% and twice as much at 5%. If her annual income from the two investments is \$720, how much does she have invested at each rate?

53. Investing: If you invest half of your savings at 7% and the other half at 10% and you make \$1275 annually, how much do you have invested at each rate?

54. Investing: A financial advisor advises Jennifer to invest her money equally in stocks and bonds. If the return on bonds is 4.5% and the return on the stocks is anticipated to be 7%, how much should she invest in each type of investment to make \$2530 each year?

55. Salt solutions: How many gallons each of a 25% salt solution and a 40% salt solution must be used to produce 60 gallons of a 35% solution?

56. Grocery shopping: A butcher decides to mix hamburger so it can be sold at a competitive price. If he mixes some that is 22% fat with some that is 10% fat, how many pounds of each type should he mix to get 80 pounds of hamburger that is 16% fat?

57. Acid solutions: How many ounces of pure acid and how many ounces of a 10% acid solution should be mixed to make a total of 50 ounces of a 64% acid solution?

58. Manufacturing: A manufacturer decided to mix two aluminum alloys, one is 20% alloy and the other is 60% alloy. How many tons of each type should he use to get a total of 100 tons of 28% aluminum alloy?

8.6 Systems of Linear Equations: Three Variables

Solve each system of equations. State which systems, if any, have no solution or an infinite number of solutions.

59. $\begin{cases} x+3y=6 \\ 6y+z=12 \\ x-3z=-15 \end{cases}$

60. $\begin{cases} x+y+z=6 \\ -x+y+4z=19 \\ 3x+y-2z=-7 \end{cases}$

61. $\begin{cases} x+y-z=-4 \\ 2x+y+z=6 \\ x+4y-2z=-6 \end{cases}$

62. $\begin{cases} x-2y+3z=0 \\ 3x-y+3z=2 \\ 5x-5y+9z=6 \end{cases}$

63. $\begin{cases} x+y-2z=-2 \\ 2x+y=2 \\ 3y-4z=-14 \end{cases}$

64. $\begin{cases} 2x+3y=4 \\ x-y+2z=1 \\ 5x-5y+2z=9 \end{cases}$

65. The sum of three integers is 65. The sum of the second and third is 25 more than the first. If the third is subtracted from the sum of the first two, the difference is 17. Find the three integers.

66. Investing: Kathleen has three investments for a total of $50,000. She has invested in savings, stocks, and bonds. Her savings account paid 5% interest, stocks earned 6%, and bonds earned 4% for a total interest of $2400 last year. If her investment in bonds is equal to the total invested in the other two accounts, how much does she have invested in each type of investment?

67. Coin collecting: Opal has a small coin collection consisting only of nickels, dimes, and quarters with a total value of $5.40. She has 34 coins in the collection and twice as many quarters as nickels. How many of each kind of coin does she have?

68. Triangles: A triangle has a perimeter of 42 m. The longest side is 2 m less than the sum of the other two sides and four times the shortest side is 32 m more than the difference between the two longer sides. What are the lengths of the three sides of the triangle?

---

47. Length = 22.5 meters; Width = 17.5 meters
48. 130°, 50°
49. They sold 30 of the $110 shirt and 20 of the $65 dollar shirt.
50. Popcorn: $2.75; Funnel Cake: $4.10
51. $15,000 at 6%; $5000 at 8%
52. $4000 at 8% $8000 at 5%
53. $7500 at each rate
54. $22,000 in each type of investment
55. 20 gallons of 25% salt; 40 gallons of 40% salt
56. 40 pounds of 22% fat; 40 pounds of 10% fat
57. 30 ounces of pure acid, 20 ounces of 10% acid
58. 80 tons of 20% alloy, 20 tons of 60% alloy
59. $(3, 1, 6)$
60. Infinitely many solutions
61. $(0, 1, 5)$
62. No solution
63. $\left(3, -4, \frac{1}{2}\right)$
64. $\left(2, 0, -\frac{1}{2}\right)$
65. 20, 21, and 24
66. $10,000 in savings, $15,000 in stocks, and $25,000 in bonds
67. 8 nickels, 10 dimes, and 16 quarters **68.** Longest side = 20 m; Shortest side = 10 m; Other side = 12 m
69. $\begin{bmatrix} 2 & -1 \\ 7 & 1 \end{bmatrix}, \left[\begin{array}{cc|c} 2 & -1 & 2 \\ 7 & 1 & 5 \end{array}\right]$ **70.** $\begin{bmatrix} 2 & -1 & 3 \\ 4 & 0 & -1 \\ 1 & -9 & 2 \end{bmatrix}, \left[\begin{array}{ccc|c} 2 & -1 & 3 & 5 \\ 4 & 0 & -1 & 1 \\ 1 & -9 & 2 & 3 \end{array}\right]$ **71.** $(4, -6)$ **72.** $(-4, 5)$ **73.** $(11, 24, 22)$ **74.** $(6, 1, -1)$

75. $(1, -10, -6)$

76. $(4, -1, -2)$

77. $(5, 7)$

78. $\left(\frac{17}{11}, \frac{10}{11}, -\frac{16}{11}\right)$

79.
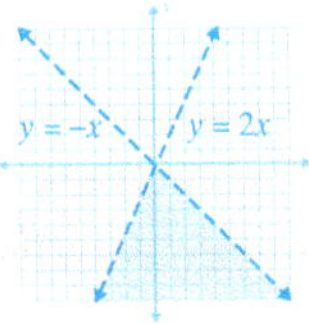

80.
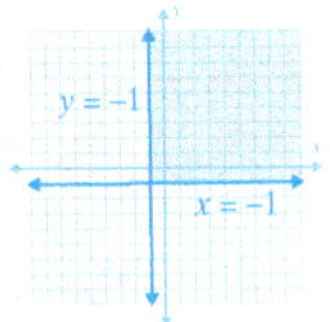

81.
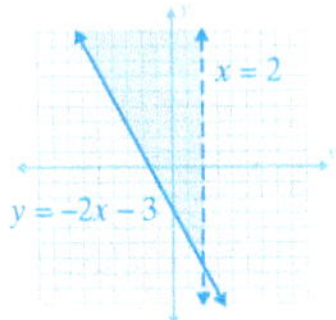

82.
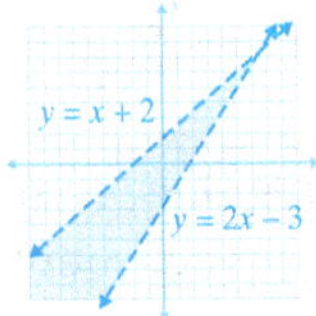

83.
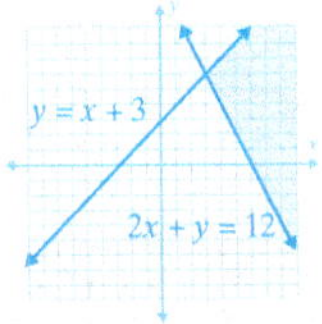

84.
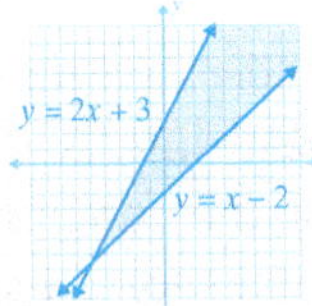

85.
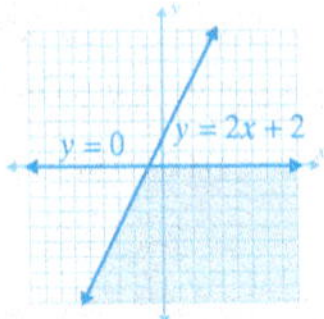

86.
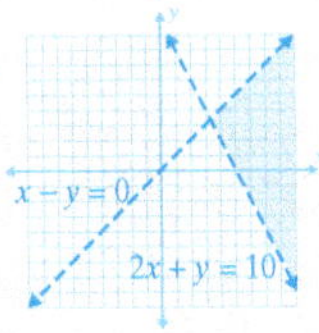

8.7 Matrices and Gaussian Elimination

Write the coefficient matrix and the augmented matrix for the given systems of linear equations.

69. $\begin{cases} 2x - y = 2 \\ 7x + y = 5 \end{cases}$

70. $\begin{cases} 2x - y + 3z = 5 \\ 4x - z = 1 \\ x - 9y + 2z = 3 \end{cases}$

Use the Gaussian elimination method with back substitution to solve the given systems of equations.

71. $\begin{cases} x + y = -2 \\ 3x + 2y = 0 \end{cases}$

72. $\begin{cases} 2x + y = -3 \\ x + 2y = 6 \end{cases}$

73. $\begin{cases} x + y - z = 13 \\ 2x - 2y + z = -4 \\ x + 2y - 2z = 15 \end{cases}$

74. $\begin{cases} 2x - 3y + z = 8 \\ x - 2y + 3z = 1 \\ x + 5y + 4z = 7 \end{cases}$

75. $\begin{cases} 2x + y - z = -2 \\ 5x - 2y + 4z = 1 \\ 3x + y + z = -13 \end{cases}$

76. $\begin{cases} x + y + z = 1 \\ x - 2y + 3z = 0 \\ x + 4y + 2z = -4 \end{cases}$

Use your graphing calculator to solve the systems of linear equations.

77. $\begin{cases} 3x - 2y = 1 \\ 4x - 3y = -1 \end{cases}$

78. $\begin{cases} x + y + z = 1 \\ 3x + 2y + z = 5 \\ x - 4y + 2z = -5 \end{cases}$

8.8 Graphing Systems of Linear Inequalities

Solve the systems of two linear inequalities graphically.

79. $\begin{cases} y < -x \\ y < 2x \end{cases}$

80. $\begin{cases} x \ge -1 \\ y \ge -1 \end{cases}$

81. $\begin{cases} x < 2 \\ y \ge -2x - 3 \end{cases}$

82. $\begin{cases} y > 2x - 3 \\ y < x + 2 \end{cases}$

83. $\begin{cases} 2x + y \ge 12 \\ y \le x + 3 \end{cases}$

84. $\begin{cases} y \ge x - 2 \\ y \le 2x + 3 \end{cases}$

Use a graphing calculator to solve the systems of linear inequalities.

85. $\begin{cases} y \le 0 \\ y \le 2x + 2 \end{cases}$

86. $\begin{cases} 2x + y > 10 \\ x - y > 0 \end{cases}$

Chapter 8 Test

1. Determine which of the given points, if any, lie on both of the lines in the system of equations by substituting each point into both equations.

$$\begin{cases} 3x-7y=5 \\ 5x-2y=-11 \end{cases}$$

a. $(1, 8)$ **b.** $(4, 1)$ **c.** $(-3, -2)$ **d.** $(13, 10)$

Solve the systems by graphing.

2. $\begin{cases} 2x=-3y+9 \\ 4x+6y=18 \end{cases}$

3. $\begin{cases} x-y=3 \\ x+2y=6 \end{cases}$

Solve the systems by using the method of substitution.

4. $\begin{cases} 5x-2y=0 \\ y=3x+4 \end{cases}$

5. $\begin{cases} 3x=y-12 \\ 4x+3y=10 \end{cases}$

Solve the systems by using the method of addition.

6. $\begin{cases} 3x+y=5 \\ -12x-4y=-7 \end{cases}$

7. $\begin{cases} 3x+4y=10 \\ x+6y=1 \end{cases}$

Solve the systems by using any method.

8. $\begin{cases} x+y=2 \\ y=-2x-1 \end{cases}$

9. $\begin{cases} 6x+2y-8=0 \\ y=-3x \end{cases}$

10. $\begin{cases} 7x+5y=-9 \\ 6x+2y=6 \end{cases}$

11. $\begin{cases} x+3y=-2 \\ 2x=-6y-4 \end{cases}$

12. Solve the following system of equations algebraically: $\begin{cases} x-2y-3z=3 \\ x+y-z=2 \\ 2x-3y-5z=5 \end{cases}$

13. Solve the following system of equations algebraically: $\begin{cases} x+2y-2z=0 \\ x-y+z=2 \\ -x+4y-4z=-8 \end{cases}$

1. c
2. $\left(x, -\frac{2}{3}x+3\right)$

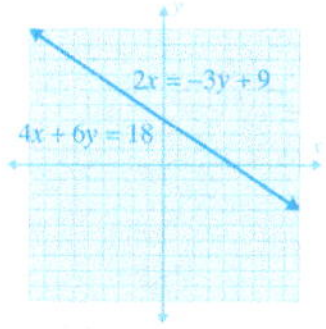

3. $(4, 1)$

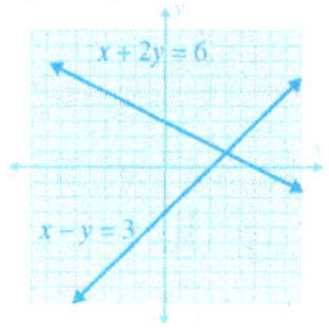

4. $(-8, -20)$
5. $(-2, 6)$
6. No solution
7. $\left(4, -\frac{1}{2}\right)$
8. $(-3, 5)$
9. No solution
10. $(3, -6)$
11. $\left(x, -\frac{1}{3}x-\frac{2}{3}\right)$
12. $(-1, 1, -2)$
13. No solution

14. a. $\begin{bmatrix} 1 & 2 & -3 \\ 1 & -1 & -1 \\ 1 & 3 & 2 \end{bmatrix}$

$\left[\begin{array}{ccc|c} 1 & 2 & -3 & -11 \\ 1 & -1 & -1 & 2 \\ 1 & 3 & 2 & -4 \end{array}\right]$

b. $(1, -3, 2)$

15.

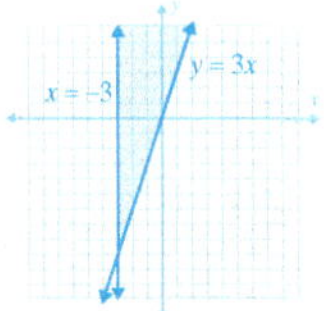

16.

17. Soy latte: \$60.20; Iced coffee: \$18.90

18. Speed of the boat is 14 mph; Speed of the current is 2 mph

19. Pen price: \$0.79; Pencil price: \$0.08

20. \$1600 at 8% and \$960 at 6%

21. 17 inches by 13 inches

22. 1600 lb of 83% and 400 lb of 68%

23. Nickels = 45; Quarters = 60

24. 41¢ stamps: 60; 58¢ stamps: 20; 75¢ stamps: 10

14. For the following system of equations:

a. Write the coefficient matrix and the augmented matrix, and

b. Solve the system using the Gaussian elimination method with back substitution.

$$\begin{cases} x + 2y - 3z = -11 \\ x - y - z = 2 \\ x + 3y + 2z = -4 \end{cases}$$

15. Solve the following system of linear inequalities graphically: $\begin{cases} x \geq -3 \\ y \geq 3x \end{cases}$

16. Solve the following system of linear inequalities graphically: $\begin{cases} y < 3x + 4 \\ 2x + y \geq 1 \end{cases}$

17. Coffee: Every time Tish goes to a coffee shop she either buys a soy latte or an iced coffee. In one month she spent a total of \$79.10. If the money she spent on soy lattes is \$3.50 plus three times the money she spent on iced coffees, how much money did she spend on each drink?

18. Boating: Pete's boat can travel 48 miles upstream in 4 hours. The return trip takes 3 hours. Find the speed of the boat in still water and the speed of the current.

19. School supplies: Eight pencils and two pens cost \$2.22. Three pens and four pencils cost \$2.69. What is the price of each pen and each pencil?

20. Investing: Gary has two investments yielding a total annual interest of \$185.60. The amount invested at 8% is \$320 less than twice the amount invested at 6%. How much is invested at each rate?

21. Rectangles: The perimeter of a rectangle is 60 inches and the length is 4 inches longer than the width. Find the dimensions of the rectangle.

22. Manufacturing: A metallurgist needs 2000 pounds of an alloy that is 80% copper. In stock, he has only alloys of 83% copper and 68% copper. How many pounds of each must be used?

23. Making change: Sonia has a bag of coins with only nickels and quarters. She wants you to figure out how many of each type of coin she has and tells you that she has 105 coins and that the value of the coins is \$17.25. Tell her that you know algebra and determine how many nickels and how many quarters she has.

24. Stamp collecting: Kimberly bought 90 stamps in denominations of 41¢, 58¢, and 75¢. To test her daughter, who is taking an algebra class, she said that she bought three times as many 41¢ stamps as 58¢ stamps and that the total cost of the stamps was \$43.70. How many stamps of each denomination did she buy?

Cumulative Review: Chapters 1 – 8

1. a. $\left\{-9, 3, \sqrt{25}\right\}$

b. $\left\{-9, \frac{1}{4}, 3, \sqrt{25}\right\}$

c. $\left\{-\sqrt{7}, \frac{\pi}{2}\right\}$

d. $\left\{-9, -\sqrt{7}, \frac{1}{4}, \frac{\pi}{2}, 3, \sqrt{25}\right\}$

2. 840

3. $36x^2y^3$

4. $2(x+3)^2(x-3)$

5. $x = -\frac{9}{8}$

6. $x = \frac{29}{11}$

7. $x = -\frac{4}{3}, 4$

8. $\left(-\infty, -\frac{1}{4}\right]$

9. $\left(-\infty, -\frac{50}{3}\right)$

10. $\left[-\frac{8}{3}, 4\right]$

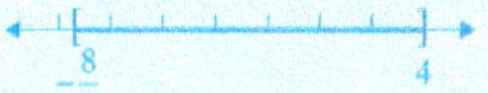

11. m is undefined; No y-intercept; Vertical line

12. $m = \frac{3}{5}$; y-intercept $= \left(0, -\frac{2}{5}\right)$

13. $m = 4$; y-intercept $= (0, -1)$

14. $m = 0$; y-intercept $= \left(0, \frac{3}{2}\right)$; Horizontal line

1. Given the set of numbers $\left\{-9, -\sqrt{7}, \frac{1}{4}, \frac{\pi}{2}, 3, \sqrt{25}\right\}$, list those numbers in the set that are described below.

a. $\{x \mid x \text{ is an integer}\}$ **b.** $\{x \mid x \text{ is a rational number}\}$

c. $\{x \mid x \text{ is an irrational number}\}$ **d.** $\{x \mid x \text{ is a real number}\}$

Find the least common multiple of each set.

2. $\{15, 35, 40\}$ **3.** $\{4x^2, 18x^2y, 12xy^3\}$

4. $\left\{(x+3)^2, 2x+6, 2x^2-18\right\}$

Solve each of the equations.

5. $4(3x-1) = 2(2x-5)-3$ **6.** $\frac{4x-1}{3} + \frac{x-5}{2} = 2$ **7.** $\left|\frac{3x}{4} - 1\right| = 2$

Solve the inequalities and graph the solution sets. Write each solution using interval notation.

8. $x+4-3x \geq 2x+5$ **9.** $\frac{x}{5} - 3.4 > \frac{x}{2} + 1.6$ **10.** $\left|\frac{3x}{2} - 1\right| - 2 \leq 3$

Identify the slope and y-intercept for the following lines. If a line is horizontal or vertical, state it in your answer.

11. $4x = 8$ **12.** $3x - 2 = 5y$ **13.** $4x - y = 1$ **14.** $3 - 2y = 0$

15. Find an equation in standard form for the line that passes through the point $(3, -4)$ and is parallel to the line $2x - y = 5$. Graph both lines.

16. Find an equation in slope-intercept form for the line passing through the points $(1, -4)$ and $(4, 2)$. Graph the line.

17. Three graphs are shown below. Use the vertical line test to determine whether or not each graph represents a function. State each graph's domain and range.

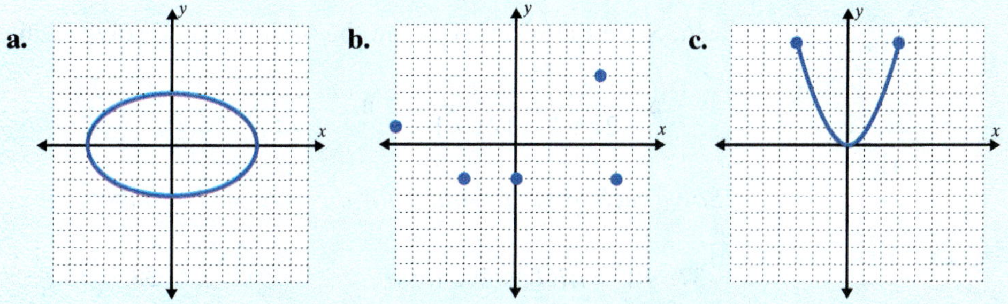

15. $2x - y = 10$

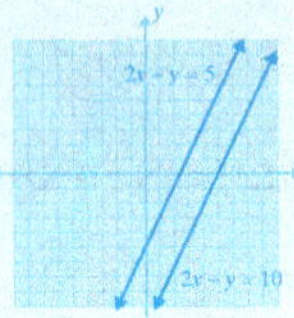

16. $y = 2x - 6$

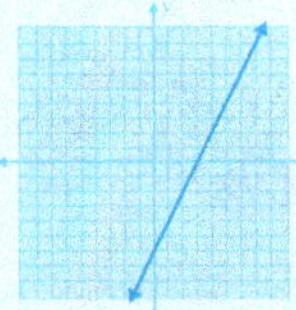

17. a. Not a function; $D = [-5,5]$; $R = [-3,3]$
b. Function; $D = \{-7, -3, 0, 5, 6\}$; $R = \{-2, 1, 4\}$
c. Function; $D = [-3, 3]$; $R = [0, 6]$

18. $\frac{y^6}{9x^4}$ 19. $\frac{x^6}{y^6}$

20. $\frac{3}{2x^4y}$ 21. 6.8

22. 1.05×10^6

23. $(3x+2)(5x+4)$

24. $(x+4)(2x^2+3)$

25. $2(2x-3)(4x^2+6x+9)$

26. $(2x+1)(x-3)$

27. a. -1 b. -12 c. 15

28. $x = -11$

29. $x = -7, 0, 2$

30. $x^2 + x - 20 = 0$

31. $6x^2 - 7x + 2 = 0$

32. $2x^2 + x - 3 + \frac{9}{x+2}$

33. $x - 4 + \frac{6x-15}{x^2-6}$

34. $\frac{2x-1}{(x+1)(x-1)}$

35. $\frac{x^2+15x-26}{(x-4)(x+2)(2x+3)}$

Use the rules of exponents to simplify the expressions so that they contain only positive exponents. Assume that all variables represent nonzero numbers.

18. $\left(3x^2y^{-3}\right)^{-2}$ **19.** $\left(\frac{x^4y}{x^2y^3}\right)^3$ **20.** $\left(\frac{5x^2y^{-1}}{2xy}\right)\left(\frac{3x^{-3}y}{5x^2y^0}\right)$

Write each expression in scientific notation and simplify.

21. $(17{,}000)(0.0004)$ **22.** $\frac{6300}{0.006}$

Factor each expression completely.

23. $15x^2 + 22x + 8$ **24.** $2x^3 + 8x^2 + 3x + 12$

25. $16x^3 - 54$ **26.** $2x^2 - 5x - 3$

27. Given that $P(x) = 4x^3 - 5x^2 + 2x - 1$, find **a.** $P(0)$ **b.** $P(-1)$ **c.** $P(2)$

Solve the following equations by factoring.

28. $x^2 + 22x = -121$ **29.** $5x^2 = 14x - x^3$

30. Find an equation with integer coefficients that has $x = -5$ and $x = 4$ as roots.

31. Find an equation with integer coefficients that has $x = \frac{1}{2}$ and $x = \frac{2}{3}$ as roots.

Use the division algorithm to divide and write the answer in the form $Q + \frac{R}{D}$, where the degree of $R <$ the degree of D.

32. $\frac{2x^3 + 5x^2 - x + 3}{x+2}$ **33.** $\frac{x^3 - 4x^2 + 9}{x^2 - 6}$

Perform the indicated operations.

34. $\frac{x+1}{x^2+3x-4} \cdot \frac{2x^2+7x-4}{x^2-1} \div \frac{x+1}{x-1}$

35. $\frac{x}{x^2-2x-8} + \frac{x-3}{2x^2-5x-12} - \frac{2x-5}{2x^2+7x+6}$

36. State the restrictions on the variables and solve each equation.

a. $\frac{1}{2x+1} = \frac{x+3}{12x+1}$ **b.** $\frac{x}{x+4} - \frac{4}{x^2-16} = \frac{-2x}{x-4}$ **c.** $\frac{x-2}{x-2} + \frac{x-3}{x-3} = \frac{2x^2}{x^2-5x+6}$

Solve each of the formulas for the indicated variables.

37. $PV = nRT$; solve for n **38.** $\frac{1}{S_1} + \frac{1}{S_2} = \frac{1}{f}$; solve for f

36. a. $x \neq -\frac{1}{2}, -\frac{1}{12}$; $x = \frac{1}{2}, 2$
b. $x \neq -4, 4$; $x = -2, \frac{2}{3}$
c. $x \neq 2, 3$; $x = \frac{6}{5}$

37. $n = \frac{PV}{RT}$

38. $f = \frac{S_1 S_2}{S_1 + S_2}$

39. b, c, d

40. (1, 4)

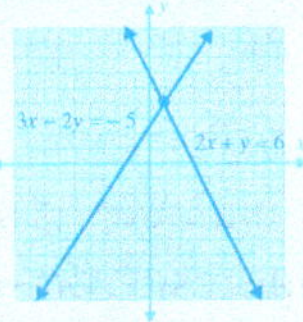

41. (1, 3)

42. (−6, 2)

43. (−1, 4)

44. No solution

45. $\left(x, 2 - \frac{1}{5}x\right)$

46. (2, 3)

47. (4, 3, 2)

48. (1, −5, 2)

49.

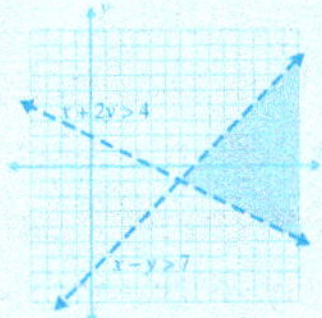

50.

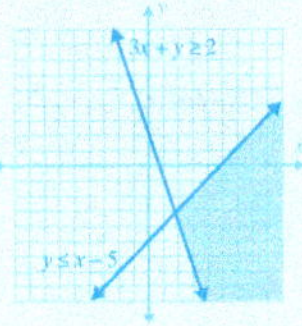

39. Determine which of the given points, if any, lie on both of the lines in the systems of equations by substituting each point into both equations.

$$\begin{cases} x - 2y = 6 \\ y = \frac{1}{2}x - 3 \end{cases}$$

a. (0, 6) **b.** (6, 0) **c.** (2, −2) **d.** $\left(3, -\frac{3}{2}\right)$

40. Solve the system of linear equations by graphing: $\begin{cases} 2x + y = 6 \\ 3x - 2y = -5 \end{cases}$

41. Solve the system of linear equations by the addition method: $\begin{cases} x + 3y = 10 \\ 5x - y = 2 \end{cases}$

42. Solve the system of linear equations by the substitution method: $\begin{cases} x + y = -4 \\ 2x + 7y = 2 \end{cases}$

Solve the systems of linear equations using any method.

43. $\begin{cases} 5x + 2y = 3 \\ y = 4 \end{cases}$ **44.** $\begin{cases} 2x + 4y = 9 \\ 3x + 6y = 8 \end{cases}$ **45.** $\begin{cases} x + 5y = 10 \\ y - 2 = -\frac{1}{5}x \end{cases}$ **46.** $\begin{cases} 2x + y = 7 \\ 2x - y = 1 \end{cases}$

Solve the systems of linear equations using Gaussian elimination.

47. $\begin{cases} x - 3y + 2z = -1 \\ -2x + y + 3z = 1 \\ x - y + 4z = 9 \end{cases}$ **48.** $\begin{cases} x + y + z = -2 \\ 2x - 3y - z = 15 \\ -x + y - 3z = -12 \end{cases}$

Solve the systems of linear inequalities graphically.

49. $\begin{cases} x + 2y > 4 \\ x - y > 7 \end{cases}$ **50.** $\begin{cases} y \leq x - 5 \\ 3x + y \geq 2 \end{cases}$

51. Driving a car: The following table shows the number of feet that Juan drove his car from one minute to another.

a. Plot the points indicated in the table.
b. Calculate the slope of the line segments from point to point.
c. Interpret each slope.

Minutes	Feet
1	1496
2	2024
3	1056
4	440
5	704

51. a.

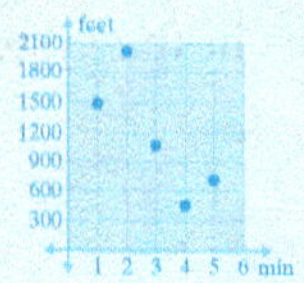

b. 528, −968, −616, 264

c. Juan's average speed increased 528 ft/min^2 from minute 1 to 2; decreased 968 ft/min^2 from minute 2 to 3; decreased 616 ft/min^2 from minute 3 to 4; and increased 264 ft/min^2 from minute 4 to 5

52. 256 feet

53. 24 ft

54. 3 and 11

55. 120 yd^2

56. $\frac{2}{3}$ hr or 40 minutes

57. 5 ft by 9 ft

58. 3 batches of Choc-O-Nut; 2 batches of Chocolate Krunch

59. \$4200 at 7%; \$2800 at 8%

60. Ferris wheel: 5400; Tunnel-of-love: 2250; Tilt-a-whirl: 4350

61. $\left(\frac{8}{5}, \frac{27}{10}\right)$

62. $(-3, 0, 5)$

63.

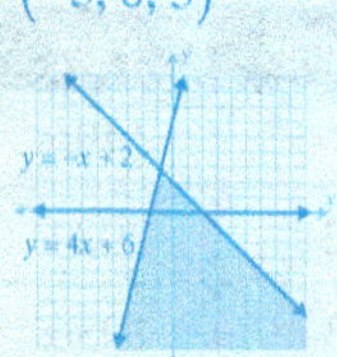

64.

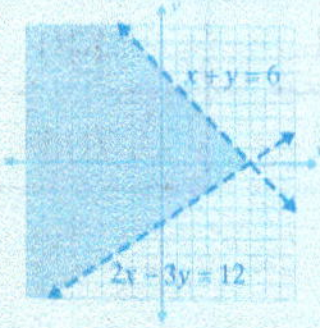

52. Falling objects: The distance an object falls (in feet) varies directly as the square of the time (in seconds) that it falls. If an object falls 16 feet in 1 s, how far will it fall in 4 s?

53. Ladders: A right triangle is formed when a ladder is leaned against a building. If the ladder is 26 ft long and its bottom is placed 10 ft from the base of the building, how far up the building does the ladder reach?

54. The sum of two numbers is 14. Twice the larger number added to three times the smaller is equal to 31. Find both numbers.

55. Carpeting a house: LeeAnne is a carpet layer. She has agreed to carpet a house for \$18 per square yard. The carpet will cost her \$11 per square yard, and she knows that there is approximately 12% waste in cutting and matching. If she plans to make a profit of \$580.80, how many yards of carpet will she buy?

56. Walking: Emily started walking to a town 10 miles away at a rate of 3 mph. After walking part of the way, she got a ride on a bus. The bus traveled at an average rate of 48 mph, and Emily reached the town 50 minutes after she started walking. How long did she walk before catching the bus?

57. Construction: Jeremy wants to build his own plinko board similar to the one found on The Price Is Right. He decides to use a rectangular base with a perimeter of 28 ft. If the length of the board is twice the width minus 1 ft, what are the dimensions of the board?

58. Baking: Karl makes two kinds of cookies. Choc-O-Nut requires 4 oz of peanuts for each 10 oz of chocolate chips. Chocolate Krunch requires 12 oz of peanuts per 8 oz of chocolate chips. How many batches of each can he make if he has 36 oz of peanuts and 46 oz of chocolate chips?

59. Investing: Alicia has \$7000 invested, some at 7% and the remainder at 8%. After one year, the interest from the 7% investment exceeds the interest from the 8% investment by \$70. How much is invested at each rate?

60. State fair: One weekend at the state fair a total of 12,000 tickets were collected from the ferris wheel, tunnel of love, and the tilt-a-whirl rides. The number of ferris wheel tickets plus 3450 equaled twice the amount of tunnel of love tickets plus the tilt-a-whirl tickets. Three times the number of tunnel of love tickets equaled the number of tilt-a-whirl tickets plus 2400. How many tickets were collected for each of the rides?

Solve the systems of equations using a graphing calculator and Gaussian elimination.

61. $\begin{cases} x + 2y = 7 \\ -x + 3y = \dfrac{13}{2} \end{cases}$

62. $\begin{cases} x + y - z = -8 \\ 2x - 3y = -6 \\ x + y + z = 2 \end{cases}$

Use a graphing calculator to solve each system of linear inequalities.

63. $\begin{cases} y \le 4x + 6 \\ y \le -x + 2 \end{cases}$

64. $\begin{cases} 2x - 3y < 12 \\ x + y < 6 \end{cases}$

CHAPTER 9

Roots, Radicals, and Complex Numbers

Did You Know?

In 1545, Girolamo Cardano published *Ars Magna* (*The Great Art*) in which he recognized that solving cubic equations often leads to taking the square root of a negative number. He tried to find two numbers a and b such that $a + b = 10$ and $ab = 40$ which lead to the answer $a = 5 + \sqrt{-15}$ and $b = 5 - \sqrt{-15}$. Mathematicians were still a little wary of negative numbers in general and until this time, it was easy to ignore square roots of negative numbers. John Wallis (1616 – 1703) made his contribution to the acceptance of complex numbers in 1673 by representing complex numbers geometrically. Leonhard Euler was the first to use the symbol i for $\sqrt{-1}$ but Carl Friedrich Gauss' use of complex numbers in several proofs of the fundamental theorem of algebra made them widely accepted.

Unfortunately, there is a stigma with the term **imaginary numbers** which causes a great deal of confusion for most students. Do not despair; imaginary numbers which we now call **complex numbers** can be considered an extension of real numbers and follow the same rules (with a few exceptions). Complex numbers are of the form $a + bi$, where a is the "real" part and b is the "imaginary" part.

Complex numbers play a vital role in the lives of physicists, astronomers, and engineers. Electrical engineers represent the imaginary part of a complex number by j instead of the usual i since, for them, i is used to represent electric current.

Civil engineers use complex numbers when modeling the vibrations of footbridges, skyscrapers, and cars driving on bumpy roads. As we learn more about vibrations of skyscrapers, we are able to improve structural integrity which leads to taller and taller buildings. One such structure will be Burj Mubarak al-Kabir, part of Madinat Al Hareer (City of Silk) in Kuwait.

"Neither the true nor false (negative) roots are always real; sometimes they are imaginary."

René Descartes (1596 – 1650)

In this chapter, we will discuss the meaning of **rational exponents** and their relationship with radical expressions such as square roots $\left(\sqrt{\ }\right)$ and cube roots $\left(\sqrt[3]{\ }\right)$. This relationship allows translation from one type of notation to the other with relative ease and a choice for the form of an answer that best suits the purposes of the problem. For example, in higher level mathematics courses, particularly in calculus, an expression with a square root, such as $\sqrt{x^2+1}$, may be changed into an alternate notation using fractional exponents, such as $\left(x^2+1\right)^{\frac{1}{2}}$. Operations learned in calculus can be performed more easily on expressions with exponents in fractional form than on the equivalent expression in radical form. Then, if desired, answers can be changed back into a form with radical notation.

A new category of numbers, called **complex numbers**, is an expansion of the real number system and includes a group of numbers called **imaginary numbers**. The term "imaginary" is somewhat misleading because these numbers have many practical applications and are no more imaginary than any other type of number. They are particularly useful in electrical engineering and hydrodynamics. Complex numbers arise quite naturally as solutions to quadratic equations.

9.1 Roots and Radicals

- *Evaluate square roots.*
- *Evaluate cube roots.*
- *Use a calculator to evaluate square and cube roots.*

You are probably familiar with the concept of **square roots** and the square root symbol (or **radical sign** $\sqrt{\ }$) from your previous work in algebra and the discussions of real numbers in Chapter 1. For example, $\sqrt{3}$ represents the square root of 3 and is the number whose square is 3.

Perfect Squares and Square Roots

A number is **squared** when it is multiplied by itself. For example,

$$6^2 = 6 \cdot 6 = 36 \qquad \text{and} \qquad (-1.5)^2 = (-1.5)(-1.5) = 2.25.$$

If an integer is squared, the result is called a **perfect square**. The squares for the integers from 1 to 20 are shown in Table 1 for easy reference.

Squares of Integers from 1 to 20 (Perfect Squares)										
Integers (n)	1	2	3	4	5	6	7	8	9	10
Perfect Squares (n^2)	1	4	9	16	25	36	49	64	81	100
Integers (n)	11	12	13	14	15	16	17	18	19	20
Perfect Squares (n^2)	121	144	169	196	225	256	289	324	361	400

Table 1

Now we want to reverse the process of squaring. That is, given a number, we want to find a number that when squared will result in the given number. This is called **finding a square root** of the given number. In general,

if $b^2 = a$, then b is a square root of a.

For example,

- because $5^2 = 25$, 5 is a **square root** of 25 and we write $\sqrt{25} = 5$.
- because $9^2 = 81$, 9 is a **square root** of 81 and we write $\sqrt{81} = 9$.

Radical Terminology

The symbol $\sqrt{\ }$ is called a **radical sign**.

The number under the radical sign is called the **radicand**.

The complete expression, such as $\sqrt{64}$, is called a **radical** or **radical expression**.

Every positive real number has two square roots, one positive and one negative. The positive square root is called the **principal square root**. For example,

- because $(8)^2 = 64, \sqrt{64} = 8.$ ⟵ the **principal square root**
- because $(-8)^2 = 64, -\sqrt{64} = -8.$ ⟵ the **negative square root**

The number 0 has only one square root, namely 0.

Square Root

If a is a nonnegative real number, then

$\sqrt{a}$ is the **principal square root** of a,

and $-\sqrt{a}$. is the **negative square root** of a.

NOTES

Square roots of negative numbers are not real numbers. For example, $\sqrt{-4}$ is not a real number. There is no real number whose square is –4. Numbers of this type will be discussed in Section 9.7.

Example 1: Evaluating Square Roots

a. 64 has two square roots, one positive and one negative. The $\sqrt{\ }$ sign is understood to represent the **positive square root** (or the **principal square root)** and $-\sqrt{\ }$ represents the **negative square root**. Therefore, we have

$$\sqrt{64} = 8 \qquad \text{and} \qquad -\sqrt{64} = -8.$$

b. Because $11^2 = 121$, we have $\sqrt{121} = 11$ and $-\sqrt{121} = -11$.

c. Because $0^2 = 0$, $\sqrt{0^2} = 0$.

d. $\sqrt{-25}$ is not a real number.

Example 2: Evaluating Square Roots

a. Because $\left(\frac{4}{5}\right)^2 = \frac{16}{25}$, we know that $\sqrt{\frac{16}{25}} = \frac{4}{5}$.

b. $-\sqrt{0.0009} = -0.03$ because $(0.03)^2 = 0.0009$.

Recall that **rational numbers** are numbers that can be expressed as the quotient of integers and are of the form $\frac{a}{b}$ where a and b are integers and $b \neq 0$. In decimal form, rational numbers are of the form of terminating decimals or repeating infinite decimals. Square roots of perfect square radicands simplify to rational numbers.

No rational number will square to give 2. So, $\sqrt{2}$ is an **irrational number** and a decimal representation is an infinite nonrepeating decimal. Your calculator will show

$$\sqrt{2} = 1.414213562\ldots \qquad \text{Accurate to 9 decimal places}$$

$$\sqrt{2} = 1.4142 \qquad \text{Rounded to 4 decimal places}$$

To get a better idea of $\sqrt{2}$ we can compare as follows:

$$1 < 2 < 4 \qquad \text{Note that 1 and 4 are perfect squares.}$$

$$\text{and} \quad \sqrt{1} < \sqrt{2} < \sqrt{4}$$

$$\text{which gives} \quad 1 < \sqrt{2} < 2$$

and we see that indeed $\sqrt{2}$ is between 1 and 2 and the value 1.4142 is reasonable.

Example 3: Estimating Square Roots

A calculator will give $\sqrt{30} \approx 5.4772$ accurate to four places. Check that this is a reasonable estimate.

Solution: Because $25 < 30 < 36$, we have $\sqrt{25} < \sqrt{30} < \sqrt{36}$ and $5 < \sqrt{30} < 6$. The approximation 5.4772 is between 5 and 6 and is reasonable.

Another approach is to square as follows:

$(5.4772)^2 = 29.99971984$ which is close to 30.

Cube Roots

A number is **cubed** when it is used as a factor 3 times. For example,

$$5^3 = 5 \cdot 5 \cdot 5 = 125 \qquad \text{and} \qquad (-30)^3 = (-30) \cdot (-30) \cdot (-30) = -27{,}000.$$

If an integer is cubed, the result is called a **perfect cube**. The cubes for the integers from 1 to 10 are shown here for easy reference.

Cubes of Integers from 1 to 10 (Perfect Cubes)				
$1^3 = 1$	$2^3 = 8$	$3^3 = 27$	$4^3 = 64$	$5^3 = 125$
$6^3 = 216$	$7^3 = 343$	$8^3 = 512$	$9^3 = 729$	$10^3 = 1000$

Table 2

The reverse of cubing is finding the **cube root**, symbolized $\sqrt[3]{}$. For example, the cube root of 125 is 5 and we write $\sqrt[3]{125} = 5$.

Cube Root

If a is a real number, then

$$\sqrt[3]{a} \text{ is the } \textbf{cube root} \text{ of } a.$$

NOTES

In the cube root expression $\sqrt[3]{a}$ the number 3 is called the **index**. In a square root expression such as $\sqrt{a}$ the index is understood to be 2 and **is not written**. Expressions with square roots and cube roots (as well as other roots) are called **radical expressions**.

Example 4: Evaluating Cube Roots

a. Because $2^3 = 8$, $\sqrt[3]{8} = 2$.

b. Because $(-6)^3 = -216$, $\sqrt[3]{-216} = -6$.

Note that the cube root of a negative number is a real number and is negative.

c. Because $\left(\frac{1}{3}\right)^3 = \frac{1}{27}$, $\sqrt[3]{\frac{1}{27}} = \frac{1}{3}$.

Using a TI-84 Plus Graphing Calculator to Evaluate Expressions with Radicals

A TI-84 Plus graphing calculator can be used to find decimal approximations for radicals and expressions containing radicals. The displays in Example 5 illustrate the advantage of being able to see the entire expression being evaluated.

Example 5: Evaluating Radical Expressions with a Calculator

The following radical expressions are evaluated by using a TI-84 Plus graphing calculator. In each example the steps (or keys to press) are shown. The TI-84 Plus gives answers accurate up to nine decimal places. You may choose (through the MODE key) to have answers accurate to fewer than nine places.

a. $\sqrt{17}$

Solution: To find $\sqrt{17}$ proceed as follows:

Step 1: Press 2ND x^2 to get the square root symbol $\sqrt{}$. (**Note:** When the $\sqrt{}$ symbol appears, it will appear with a left-hand parenthesis. You should press the right-hand parenthesis to close the square root operation.)

Step 2: Enter 1 7 and the right-hand parenthesis).

Step 3: Press ENTER.

The display will appear as follows:

```
√(17)
      4.123105626
```

b. $3\sqrt{20}$

Note: This expression represents 3 times $\sqrt{20}$.

Solution: To find $3\sqrt{20}$ proceed as follows:

Step 1: Enter 3 .

Step 2: Press 2ND x^2. (This gives the $\sqrt{}$ symbol.)

Step 3: Enter 2 0 and the right-hand parenthesis).

Step 4: Press ENTER.

The display will appear as follows:

```
3√(20)
      13.41640786
```

c. $1+2\sqrt{3}$

Note: The calculator is programmed to follow the rules for order of operations.

Solution: To find $1+2\sqrt{3}$ proceed as follows:

Step 1: Enter 1 .

Step 2: Press the + key.

Step 3: Enter 2 .

Step 4: Press 2ND x^2 (This gives the $\sqrt{}$ symbol.)

Continued on the next page...

Step 5: Enter **3** and the right-hand parenthesis **)**.

Step 6: Press **ENTER**.

The display will appear as follows:

```
1+2√(3)
          4.464101615
```

1. 3 2. 7
3. 9 4. 6
5. 17 6. 11
7. 13 8. 19
9. 1 10. 10
11. 5 12. 7
13. 6 14. 8
15. $\frac{1}{2}$ 16. $\frac{3}{4}$

Practice Problems

Simplify the following square roots and cube roots.

1. $\sqrt{49}$ **2.** $\sqrt[3]{125}$ **3.** $-\sqrt{25}$ **4.** $\sqrt{196}$

Use your knowledge of square roots and cube roots to determine whether each number is rational, irrational, or nonreal.

5. $\sqrt{\frac{9}{16}}$ **6.** $\sqrt{324}$ **7.** $\sqrt{-25}$ **8.** $\sqrt{3}$

Use your calculator to find the value (accurate to four decimal places) of each of the following radical expressions.

9. $5\sqrt{2}$ **10.** $6-4\sqrt{7}$

17. $\frac{3}{4}$ 18. $\frac{1}{2}$
19. 0.2 20. 0.09
21. −10 22. −12
23. −0.04 24. −0.002
25. −3 26. −4
27. −5 28. 10
29. $\frac{3}{5}$ 30. $\frac{5}{9}$
31. $\sqrt{25}<\sqrt{32}<\sqrt{36}$ and $5<\sqrt{32}<6$ because $25<32<36$ or $(5.6569)^2=32.00051761$
32. $\sqrt{16}<\sqrt{18}<\sqrt{25}$ and $4<\sqrt{18}<5$ because $16<18<25$ or $(4.2426)^2=17.99965476$
33. $\sqrt{64}<\sqrt{74}<\sqrt{81}$ and $8<\sqrt{74}<9$ because $64<74<81$ or $(8.6023)^2=73.99956529$

9.1 Exercises

Simplify the following square roots and cube roots.

1. $\sqrt{9}$ **2.** $\sqrt{49}$ **3.** $\sqrt{81}$ **4.** $\sqrt{36}$ **5.** $\sqrt{289}$

6. $\sqrt{121}$ **7.** $\sqrt{169}$ **8.** $\sqrt{361}$ **9.** $\sqrt[3]{1}$ **10.** $\sqrt[3]{1000}$

11. $\sqrt[3]{125}$ **12.** $\sqrt[3]{343}$ **13.** $\sqrt[3]{216}$ **14.** $\sqrt[3]{512}$ **15.** $\sqrt{\frac{1}{4}}$

16. $\sqrt{\frac{9}{16}}$ **17.** $\sqrt[3]{\frac{27}{64}}$ **18.** $\sqrt[3]{\frac{1}{8}}$ **19.** $\sqrt{0.04}$ **20.** $\sqrt{0.0081}$

Answers to Practice Problems: **1.** 7 **2.** 5 **3.** −5 **4.** 14 **5.** Rational **6.** Rational **7.** Nonreal **8.** Irrational **9.** 7.0711 **10.** −4.5830

34. $\sqrt{100} < \sqrt{110} < \sqrt{121}$ and $10 < \sqrt{110} < 11$ because $100 < 110 < 121$ or $(10.4881)^2 = 110.00024161$

35. Rational
36. Irrational
37. Rational
38. Rational
39. Irrational
40. Rational
41. Nonreal
42. Rational
43. Rational
44. Nonreal
45. Irrational
46. Rational
47. 6.2450
48. 12.2474
49. 2.4960 50. 3.0984
51. 0.4472 52. 0.6124
53. 8.9443 54. 10.3923
55. −8.2462 56. −7.3485
57. 7.8990 58. −2.7082
59. −5.4641 60. 14.4868
61. There is no real number that when squared results in a negative number.

62. Cubing a negative real number is equivalent to multiplying a negative number by itself 3 times. Three negative numbers multiplied together results in a negative number.

21. $-\sqrt{100}$ **22.** $-\sqrt{144}$ **23.** $-\sqrt{0.0016}$ **24.** $-\sqrt{0.000004}$ **25.** $\sqrt[3]{-27}$

26. $\sqrt[3]{-64}$ **27.** $\sqrt[3]{-125}$ **28.** $\sqrt[3]{1000}$ **29.** $\sqrt{\frac{9}{25}}$ **30.** $\sqrt{\frac{25}{81}}$

Estimates (accurate to four decimal places) of radicals are given. Show that these are reasonable estimates.

31. $\sqrt{32} \approx 5.6569$ **32.** $\sqrt{18} \approx 4.2426$ **33.** $\sqrt{74} \approx 8.6023$ **34.** $\sqrt{110} \approx 10.4881$

Use your knowledge of square roots and cube roots to determine whether each number is rational, irrational, or nonreal.

35. $\sqrt{4}$ **36.** $\sqrt{17}$ **37.** $\sqrt{169}$ **38.** $\sqrt[3]{8}$

39. $\sqrt{\frac{2}{9}}$ **40.** $-\sqrt{\frac{1}{4}}$ **41.** $\sqrt{-36}$ **42.** $\sqrt[3]{-27}$

43. $-\sqrt[3]{125}$ **44.** $\sqrt{-10}$ **45.** $\sqrt{1.68}$ **46.** $\sqrt{5.29}$

Use a calculator to find the value (accurate to four decimal places) of each radical expression.

47. $\sqrt{39}$ **48.** $\sqrt{150}$ **49.** $\sqrt{6.23}$ **50.** $\sqrt{9.6}$

51. $\sqrt{\frac{1}{5}}$ **52.** $\sqrt{\frac{3}{8}}$ **53.** $4\sqrt{5}$ **54.** $6\sqrt{3}$

55. $-2\sqrt{17}$ **56.** $-3\sqrt{6}$ **57.** $3+2\sqrt{6}$ **58.** $4-3\sqrt{5}$

59. $-2-\sqrt{12}$ **60.** $5+\sqrt{90}$

Writing and Thinking About Mathematics

61. Discuss, in your own words, why the square root of a negative number is not a real number.

62. Discuss, in your own words, why the cube root of a negative number is a negative number.

HAWKES LEARNING SYSTEMS: INTRODUCTORY & INTERMEDIATE ALGEBRA SOFTWARE

- 9.1 Evaluating Radicals

9.2 Simplifying Radicals

- *Simplify radicals, including square roots and cube roots.*

Simplifying Algebraic Expressions with Square Roots

Various roots can be related to solutions of equations, and we want such numbers to be in a **simplified form** for easier calculations and algebraic manipulations. We need the two properties of radicals stated here for square roots.

Properties of Square Roots

If a and b are **positive** real numbers, then

1. $\sqrt{ab} = \sqrt{a}\sqrt{b}$ **2.** $\sqrt{\frac{a}{b}} = \frac{\sqrt{a}}{\sqrt{b}}$.

As an example, we know that $\sqrt{144} = 12$. However, in a situation where you may have forgotten this, you can proceed as follows using property 1 of square roots:

$$\sqrt{144} = \sqrt{36} \cdot \sqrt{4} = 6 \cdot 2 = 12.$$

Similarly, using property 2, we can write

$$\sqrt{\frac{49}{36}} = \frac{\sqrt{49}}{\sqrt{36}} = \frac{7}{6}.$$

Simplest Form for Square Roots

A square root is considered to be in **simplest form** when the radicand has no perfect square as a factor.

The number 200 is not a perfect square and using a calculator we will find $\sqrt{200} \approx 14.1421$. Now to simplify $\sqrt{200}$, we can use property 1 of square roots and any of the following three approaches.

Approach 1: Factor 200 as $4 \cdot 50$ because 4 is a perfect square. This gives,

$$\sqrt{200} = \sqrt{4 \cdot 50} = \sqrt{4} \cdot \sqrt{50} = 2\sqrt{50}.$$

However, $2\sqrt{50}$ is **not in simplest form** because 50 has a perfect square factor, 25. Thus to complete the process, we have

$$\sqrt{200} = 2\sqrt{50} = 2\sqrt{25 \cdot 2} = 2\sqrt{25} \cdot \sqrt{2} = 2 \cdot 5 \cdot \sqrt{2} = 10\sqrt{2}.$$

Approach 2: Note that 100 is a perfect square factor of 200 and $200 = 100 \cdot 2$.

$$\sqrt{200} = \sqrt{100 \cdot 2} = \sqrt{100} \cdot \sqrt{2} = 10\sqrt{2}$$

Approach 3: Use prime factors.

$$\begin{aligned} \sqrt{200} &= \sqrt{2 \cdot 2 \cdot 2 \cdot 5 \cdot 5} \\ &= \sqrt{2 \cdot 2 \cdot 5 \cdot 5} \cdot \sqrt{2} \\ &= \sqrt{2 \cdot 2} \cdot \sqrt{5 \cdot 5} \cdot \sqrt{2} \\ &= 2 \cdot 5 \cdot \sqrt{2} \\ &= 10\sqrt{2} \end{aligned}$$

NOTES

Of these three approaches, the second appears to be the easiest because it has the fewest steps. However, "seeing" the largest perfect square factor may be difficult. If you do not immediately see a perfect square factor, proceed by finding other factors or prime factors as illustrated.

Example 1: Simplifying Numerical Expressions with Square Roots

Simplify each numerical expression so that there are no perfect square numbers in the radicand.

a. $\sqrt{48}$

Solution: $\sqrt{48} = \sqrt{16 \cdot 3} = \sqrt{16} \cdot \sqrt{3} = 4\sqrt{3}$ 16 is the largest perfect square factor.

b. $\sqrt{63}$

Solution: $\sqrt{63} = \sqrt{9 \cdot 7} = \sqrt{9} \cdot \sqrt{7} = 3\sqrt{7}$ 9 is the largest perfect square factor.

c. $\sqrt{\dfrac{75}{16}}$

Solution: $\sqrt{\dfrac{75}{16}} = \dfrac{\sqrt{75}}{\sqrt{16}} = \dfrac{\sqrt{25 \cdot 3}}{\sqrt{16}} = \dfrac{\sqrt{25} \cdot \sqrt{3}}{\sqrt{16}} = \dfrac{5\sqrt{3}}{4}$

Simplifying Radical Expressions with Variables

To simplify square root expressions that contain variables, such as $\sqrt{x^2}$, we must be aware of whether the variable represents a positive number $(x > 0)$, zero $(x = 0)$, or a negative number $(x < 0)$.

For example,

$$\text{if } x = 0, \text{ then } \sqrt{x^2} = \sqrt{0^2} = \sqrt{0} = 0 = x.$$

$$\text{If } x = 5, \text{ then } \sqrt{x^2} = \sqrt{5^2} = \sqrt{25} = 5 = x.$$

$$\text{But, if } x = -5, \text{ then } \sqrt{x^2} = \sqrt{(-5)^2} = \sqrt{25} = 5 \neq x.$$

$$\text{In fact, if } x = -5, \text{ then } \sqrt{x^2} = \sqrt{(-5)^2} = \sqrt{25} = 5 = |-5| = |x|.$$

Thus simplifying radical expressions with variables involves more detailed analysis than simplifying radical expressions with only constants. The following definition indicates the correct way to simplify $\sqrt{x^2}$.

Square Root of x^2

If x is a real number, then $\sqrt{x^2} = |x|$.

Note: If $x \geq 0$ is given, then we can write $\sqrt{x^2} = x$.

Although using the absolute value when simplifying square roots is correct mathematically, we can avoid some confusion by assuming that the variable under the radical sign represents only positive numbers or 0. This eliminates the need for absolute value signs.

Therefore, for the remainder of this text, we will assume that $x > 0$ and write $\sqrt{x^2} = x$, unless specifically stated otherwise, in all square root expressions.

Example 2: Simplifying Square Roots with Variables

Simplify each of the following radical expressions. Assume that all variables represent positive numbers. (Note that, by making this assumption, we need not be concerned about the absolute value sign.)

a. $\sqrt{16y^2}$

Solution: $\sqrt{16y^2} = 4y$

b. $\sqrt{72a^2}$

Solution: $\sqrt{72a^2} = \sqrt{36a^2} \cdot \sqrt{2} = 6a\sqrt{2}$

c. $\sqrt{12x^2y^2}$

Solution: $\sqrt{12x^2y^2} = \sqrt{4x^2y^2} \cdot \sqrt{3} = 2xy\sqrt{3}$

To find the square root of an expression with even exponents, divide the exponents by 2. For example,

$$x^2 \cdot x^2 = x^4 \qquad a^3 \cdot a^3 = a^6 \qquad y^5 \cdot y^5 = y^{10}$$

$$\downarrow \qquad\qquad \downarrow \qquad\qquad \downarrow$$

$$\text{and} \quad \sqrt{x^4} = x^2 \qquad \sqrt{a^6} = a^3 \qquad \sqrt{y^{10}} = y^5.$$

To find the square root of an expression with odd exponents, factor the expression into two terms, one with exponent 1 and the other with an even exponent. For example,

$$x^3 = x^2 \cdot x \quad \text{and} \quad y^9 = y^8 \cdot y$$

which means that

$$\sqrt{x^3} = \sqrt{x^2 \cdot x} = \sqrt{x^2} \cdot \sqrt{x} = x \cdot \sqrt{x} \quad \text{and} \quad \sqrt{y^9} = \sqrt{y^8 \cdot y} = \sqrt{y^8} \cdot \sqrt{y} = y^4\sqrt{y}.$$

Example 3: Simplifying Square Roots

Simplify each of the following radical expressions. Look for perfect square factors and even powers of the variables. Assume that all variables represent positive numbers.

a. $\sqrt{81x^4}$

Solution: $\sqrt{81x^4} = 9x^2$ — The exponent 4 is divided by 2.

b. $\sqrt{64x^5y}$

Solution: $\sqrt{64x^5y} = \sqrt{64x^4} \cdot \sqrt{xy} = 8x^2\sqrt{xy}$

c. $\sqrt{18a^4b^6}$

Solution: $\sqrt{18a^4b^6} = \sqrt{9a^4b^6} \cdot \sqrt{2} = 3a^2b^3\sqrt{2}$ — Each exponent is divided by 2.

d. $\sqrt{\dfrac{9a^{13}}{b^4}}$

Solution: $\sqrt{\dfrac{9a^{13}}{b^4}} = \dfrac{\sqrt{9a^{13}}}{\sqrt{b^4}} = \dfrac{\sqrt{9a^{12}} \cdot \sqrt{a}}{\sqrt{b^4}} = \dfrac{3a^6\sqrt{a}}{b^2}$ — Recall $a, b > 0$.

Simplifying Algebraic Expressions with Cube Roots

When simplifying expressions with cube roots, we need to be aware of perfect cube numbers and variables with exponents that are multiples of 3. (Multiples of 3 are 3, 6, 9, 12, 15, and so on.) **Thus exponents are divided by 3 in simplifying cube root expressions.** For example,

$$x^2 \cdot x^2 \cdot x^2 = x^6 \qquad a^3 \cdot a^3 \cdot a^3 = a^9 \qquad y^5 \cdot y^5 \cdot y^5 = y^{15}$$

$$\downarrow \qquad \downarrow \qquad \downarrow$$

and $\sqrt[3]{x^6} = x^2 \qquad \sqrt[3]{a^9} = a^3 \qquad \sqrt[3]{y^{15}} = y^5$

Simplest Form for Cube Roots

A cube root is considered to be in **simplest form** when the radicand has no perfect cube as a factor.

Example 4: Simplifying Expressions with Cube Roots

Simplify each of the following radical expressions. Look for perfect cube factors and powers of the variables that are multiples of 3.

a. $\sqrt[3]{54x^6}$

Solution: $\sqrt[3]{54x^6} = \sqrt[3]{27x^6} \cdot \sqrt[3]{2} = 3x^2\sqrt[3]{2}$

Note that 27 is a perfect cube and the exponent 6 is divisible by 3.

b. $\sqrt[3]{-40x^4y^{13}}$

Solution: $\sqrt[3]{-40x^4y^{13}} = \sqrt[3]{-8x^3y^{12}} \cdot \sqrt[3]{5xy} = -2xy^4\sqrt[3]{5xy}$

Note that -8 is a perfect cube and the exponents on the variables are separated so that one exponent on each variable is divisible by 3.

c. $\sqrt[3]{250a^8b^{11}}$

Solution: $\sqrt[3]{250a^8b^{11}} = \sqrt[3]{125a^6b^9} \cdot \sqrt[3]{2a^2b^2} = 5a^2b^3\sqrt[3]{2a^2b^2}$

Note that 125 is a perfect cube and the exponents on the variables are separated so that one exponent on each variable is divisible by 3.

Practice Problems

Simplify each of the radical expressions. Assume that all variables represent positive real numbers.

1. $\sqrt{192}$ **2.** $\sqrt{32x^2}$ **3.** $\sqrt{\frac{3}{4}}$

4. $\sqrt{\frac{54y^5}{25x^2}}$ **5.** $\sqrt[3]{120}$ **6.** $\sqrt[3]{32x^5}$

9.2 Exercises

1. $2\sqrt{3}$ 2. $-3\sqrt{5}$
3. $12\sqrt{2}$ 4. $-3\sqrt{7}$
5. $-6\sqrt{2}$ 6. $7\sqrt{2}$
7. $-2\sqrt{14}$ 8. $9\sqrt{2}$
9. $-5\sqrt{5}$ 10. -11
11. $\frac{1}{2}$ 12. $\frac{4\sqrt{2}}{7}$
13. $-\frac{\sqrt{11}}{8}$ 14. $-\frac{\sqrt{5}}{2}$
15. $\frac{2\sqrt{7}}{5}$ 16. $\frac{7\sqrt{3}}{10}$
17. $6x$ 18. $7y$
19. $2x\sqrt{2x}$
20. $3a^2\sqrt{2a}$
21. $2x^5y\sqrt{6x}$
22. $2x^7y\sqrt{5xy}$
23. $5xy^3\sqrt{5x}$
24. $2x^2y^2\sqrt{2x}$
25. $-3xy\sqrt{2}$
26. $-4x^2y^4\sqrt{2}$
27. $2bc\sqrt{3ac}$
28. $3abc^2\sqrt{5b}$
29. $5x^2y^3z^4\sqrt{3}$
30. $10xyz\sqrt{2}$

Simplify each of the radical expressions. Assume that all variables represent positive real numbers.

1. $\sqrt{12}$ **2.** $-\sqrt{45}$ **3.** $\sqrt{288}$ **4.** $-\sqrt{63}$

5. $-\sqrt{72}$ **6.** $\sqrt{98}$ **7.** $-\sqrt{56}$ **8.** $\sqrt{162}$

9. $-\sqrt{125}$ **10.** $-\sqrt{121}$ **11.** $\sqrt{\frac{1}{4}}$ **12.** $\sqrt{\frac{32}{49}}$

13. $-\sqrt{\frac{11}{64}}$ **14.** $-\sqrt{\frac{125}{100}}$ **15.** $\sqrt{\frac{28}{25}}$ **16.** $\sqrt{\frac{147}{100}}$

17. $\sqrt{36x^2}$ **18.** $\sqrt{49y^2}$ **19.** $\sqrt{8x^3}$ **20.** $\sqrt{18a^5}$

21. $\sqrt{24x^{11}y^2}$ **22.** $\sqrt{20x^{15}y^3}$ **23.** $\sqrt{125x^3y^6}$ **24.** $\sqrt{8x^5y^4}$

25. $-\sqrt{18x^2y^2}$ **26.** $-\sqrt{32x^4y^8}$ **27.** $\sqrt{12ab^2c^3}$ **28.** $\sqrt{45a^2b^3c^4}$

29. $\sqrt{75x^4y^6z^8}$ **30.** $\sqrt{200x^2y^2z^2}$ **31.** $\sqrt{\frac{5x^4}{9}}$ **32.** $-\sqrt{\frac{7y^6}{16x^4}}$

33. $\sqrt{\frac{32a^5}{81b^{16}}}$ **34.** $\sqrt{\frac{75x^8}{121y^{12}}}$ **35.** $\sqrt{\frac{200x^8}{289}}$ **36.** $\sqrt{\frac{32x^{15}y^{10}}{169}}$

37. $\sqrt[3]{216}$ **38.** $\sqrt[3]{1}$ **39.** $\sqrt[3]{56}$ **40.** $\sqrt[3]{72}$

41. $\sqrt[3]{-1}$ **42.** $\sqrt[3]{-125}$ **43.** $\sqrt[3]{-128}$ **44.** $\sqrt[3]{-250}$

Answers to Practice Problems: **1.** $8\sqrt{3}$ **2.** $4x\sqrt{2}$ **3.** $\frac{\sqrt{3}}{2}$ **4.** $\frac{3y^2\sqrt{6y}}{5x}$ **5.** $2\sqrt[3]{15}$ **6.** $2x\sqrt[3]{4x^2}$

31. $\frac{x^2\sqrt{5}}{3}$
32. $-\frac{y^3\sqrt{7}}{4x^2}$
33. $\frac{4a^2\sqrt{2a}}{9b^8}$
34. $\frac{5x^4\sqrt{3}}{11y^6}$
35. $\frac{10x^4\sqrt{2}}{17}$
36. $\frac{4x^7y^5\sqrt{2x}}{13}$
37. 6
38. 1
39. $2\sqrt[3]{7}$
40. $2\sqrt[3]{9}$
41. -1
42. -5
43. $-4\sqrt[3]{2}$
44. $-5\sqrt[3]{2}$
45. $5x\sqrt[3]{x}$
46. $4a^4$
47. $-2x^2\sqrt[3]{x^2}$
48. $-8a\sqrt[3]{a^2}$
49. $2a^2b\sqrt[3]{9b}$
50. $3b^3\sqrt[3]{4a}$
51. $6x^2y\sqrt[3]{y^2}$
52. $4x^3\sqrt[3]{y^2}$
53. $2xy^2z^3\sqrt[3]{3x^2y}$
54. $5x^2y^3z^5\sqrt[3]{2}$
55. $\frac{\sqrt[3]{3}}{2}$
56. $\frac{2\sqrt[3]{3}}{5}$
57. $\frac{5\sqrt[3]{3}}{2}$
58. $-\frac{2\sqrt[3]{6}}{5}$
59. $\frac{5y^4}{3x^2}$
60. $\frac{x^2z}{4y^3}$
61. 7 inches
62. 21 centimeters
63. $\sqrt{6} \approx 2.45$ amperes
64. $\sqrt{3} \approx 1.73$ amperes
65. 120 volts
66. $50\sqrt{5} \approx 111.80$ volts

45. $\sqrt[3]{125x^4}$ **46.** $\sqrt[3]{64a^{12}}$ **47.** $\sqrt[3]{-8x^8}$ **48.** $\sqrt[3]{-512a^5}$

49. $\sqrt[3]{72a^6b^4}$ **50.** $\sqrt[3]{108ab^9}$ **51.** $\sqrt[3]{216x^6y^5}$ **52.** $\sqrt[3]{64x^9y^2}$

53. $\sqrt[3]{24x^5y^7z^9}$ **54.** $\sqrt[3]{250x^6y^9z^{15}}$ **55.** $\frac{\sqrt[3]{81}}{6}$ **56.** $\frac{\sqrt[3]{192}}{10}$

57. $\sqrt[3]{\frac{375}{8}}$ **58.** $\sqrt[3]{\frac{-48}{125}}$ **59.** $\sqrt[3]{\frac{125y^{12}}{27x^6}}$ **60.** $\sqrt[3]{\frac{x^6z^3}{64y^9}}$

Use the formula $s = \sqrt[3]{V}$, which relates the length of the side s of a cube and the volume V, to answer the questions below.

61. Puzzle cube: The volume of a puzzle cube is 343 cubic inches. What is the length of one side?

62. Building Blocks: Three cubic blocks of different volumes were stacked on top of each other. The top block was 216 cubic centimeters. The middle block was 343 cubic centimeters, and the bottom block was 512 cubic centimeters. How tall was the stack of blocks?

Use the following two formulas used in electricity to answer the questions below.

$$I = \sqrt{\frac{P}{R}}$$

$$E = \sqrt{PR}$$

P = power (in watts)
I = current (in amperes)
E = voltage (in volts)
R = resistance (in ohms, Ω)

63. Electricity: What is the current in amperes of a light bulb that produces 150 watts of power and has a 25 Ω resistance?

64. Electricity: If a light bulb has a resistance of 30 Ω and produces 90 watts of power, what is its current in amperes?

65. Electricity: How many volts of electricity would Meghan need to produce 48 Ω of resistance from a 300 watt lamp?

66. Electricity: A 5000 Ω resistor is rated at 2.5 watts. What is the maximum voltage of electricity that should be connected across it?

Writing and Thinking About Mathematics

67. When $a < 0$.
68. A cube root has no restrictions as the cube root of a negative number is negative.

67. Under what conditions is the expression $\sqrt{a}$ not a real number?

68. Explain why the expression $\sqrt[3]{y}$ is a real number regardless of whether $y > 0$, $y < 0$, or $y = 0$.

HAWKES LEARNING SYSTEMS: INTRODUCTORY & INTERMEDIATE ALGEBRA SOFTWARE

- 9.2 Simplifying Radicals

9.3 Rational Exponents

- *Understand the meaning of nth root.*
- *Translate expressions using radicals into expressions using rational exponents and translate expressions using rational exponents into expressions using radicals.*
- *Simplify expressions using the properties of rational exponents.*
- *Evaluate expressions of the form $a^{\frac{m}{n}}$ with a calculator.*

n^{th} Roots: $\sqrt[n]{a} = a^{\frac{1}{n}}$

Teaching Notes:
You might want to use the power rule (a little early here) to help justify how fractional exponents relate to roots as in
$\left(\sqrt{a}\right)^2 = a$
and
$\left(a^{\frac{1}{2}}\right)^2 = a^{\frac{1}{2}(2)} = a^1 = a.$
Similarly,
$\left(\sqrt[3]{a}\right)^3 = a$
and
$\left(a^{\frac{1}{3}}\right)^3 = a^{\frac{1}{3}(3)} = a^1 = a.$

In Section 9.1 we restricted our discussions to radicals involving square roots and cube roots. In this section we will expand on those ideas by discussing radicals indicating n^{th} roots in general and how to relate radical expressions to expressions with rational (fractional) exponents. For example, the fifth root of x can be written in radical form as $\sqrt[5]{x}$ and with a fractional exponent as $x^{\frac{1}{5}}$.

To understand roots in general, consider the following notation. (**Note:** In this discussion, we assume that n is a positive integer.)

Type of Root	Radical Notation and Exponential Notation
For square roots,	if $b = \sqrt{a}$, then $b = a^{\frac{1}{2}}$.
For cube roots,	if $b = \sqrt[3]{a}$, then $b = a^{\frac{1}{3}}$.
For fourth roots,	if $b = \sqrt[4]{a}$, then $b = a^{\frac{1}{4}}$.
For n^{th} roots,	if $b = \sqrt[n]{a}$, then $b = a^{\frac{1}{n}}$.

Table 1

For example,

- because $\sqrt[4]{16} = 2$, we can say that $2 = 16^{\frac{1}{4}}$.
- because $\sqrt[5]{243} = 3$, we can say that $3 = 243^{\frac{1}{5}}$.

The following notation is used for all radical expressions.

Radical Notation

If n is a positive integer, then $\sqrt[n]{a} = a^{\frac{1}{n}}$ (assuming $\sqrt[n]{a}$ is a real number).

The expression $\sqrt[n]{a}$ is called a **radical**.

The symbol $\sqrt[n]{}$ is called a **radical sign**.

n is called the **index**.

a is called the **radicand**.

Note: If no index is given, it is understood to be 2. For example, $\sqrt{3} = \sqrt[2]{3} = 3^{\frac{1}{2}}$.

NOTES

Special Notes about the Index n:

For the expression $\sqrt[n]{a}$ (or $a^{\frac{1}{n}}$) to be a real number:

1. when a is nonnegative, n can be any index, and
2. when a is negative, n must be odd.

   (If a is negative and n is even, then $\sqrt[n]{a}$ is nonreal.)

Example 1: Evaluating Principal n^{th} Roots

a. $49^{\frac{1}{2}} = \sqrt{49} = 7$, because $7^2 = 49$.

b. $81^{\frac{1}{4}} = \sqrt[4]{81} = 3$, because $3^4 = 81$.

c. $(-8)^{\frac{1}{3}} = \sqrt[3]{-8} = -2$, because $(-2)^3 = -8$.

d. $(0.00001)^{\frac{1}{5}} = \sqrt[5]{0.00001} = 0.1$, because $(0.1)^5 = 0.00001$.

e. $(-16)^{\frac{1}{2}} = \sqrt{-16}$ is not a real number. Any even root of a negative number is nonreal.

Rational Exponents of the Form $\frac{m}{n}$: $\sqrt[n]{a^m} = a^{\frac{m}{n}}$

Teaching Notes:
Highlight the fact that m and n are not restricted to being positive integers—they may be negative and they may be rational numbers. Another very handy property of exponents (due to the commutative property of multiplication) is $\left(a^m\right)^n = \left(a^n\right)^m$. For example, $\left(81^3\right)^{\frac{1}{4}}$ is much easier to simplify if written as $\left(81^{\frac{1}{4}}\right)^3$.

In Chapter 5 we discussed the rules of exponents using only integer exponents. These same rules of exponents apply to rational exponents (fractional exponents) as well and are repeated here for easy reference.

Summary of the Rules for Exponents

For nonzero real numbers a and b and rational numbers m and n,

The exponent 1: $\boldsymbol{a = a^1}$ (a is any real number.)

The exponent 0: $\boldsymbol{a^0 = 1}$ $(a \neq 0)$

The product rule: $\boldsymbol{a^m \cdot a^n = a^{m+n}}$

The quotient rule: $\boldsymbol{\dfrac{a^m}{a^n} = a^{m-n}}$

Negative exponents: $\boldsymbol{a^{-n} = \dfrac{1}{a^n}, \quad \dfrac{1}{a^{-n}} = a^n}$

Power rule: $\boldsymbol{\left(a^m\right)^n = a^{mn}}$

Power of a product: $\boldsymbol{(ab)^n = a^n b^n}$

Power of a quotient: $\boldsymbol{\left(\dfrac{a}{b}\right)^n = \dfrac{a^n}{b^n}}$

Now consider the problem of evaluating the expression $8^{\frac{2}{3}}$ where the exponent, $\frac{2}{3}$, is of the form $\frac{m}{n}$. Using the power rule for exponents, we can write

$$8^{\frac{2}{3}} = \left(8^{\frac{1}{3}}\right)^2 = (2)^2 = 4$$

or,

$$8^{\frac{2}{3}} = \left(8^2\right)^{\frac{1}{3}} = (64)^{\frac{1}{3}} = 4.$$

The result is the same with either approach. That is, we can take the cube root first and then square the answer. Or, we can square first and then take the cube root. In general, for an exponent of the form $\frac{m}{n}$, taking the nth root first and then raising this root to the exponent m is easier because the numbers are smaller.

For example,

$$81^{\frac{3}{4}} = \left(81^{\frac{1}{4}}\right)^3 = (3)^3 = 27$$

is easier to calculate and work with than

$$81^{\frac{3}{4}} = \left(81^3\right)^{\frac{1}{4}} = (531,441)^{\frac{1}{4}} = 27.$$

The fourth root of 81 is more commonly known than the fourth root of 531,441.

The General Form $a^{\frac{m}{n}}$

If n is a positive integer, m is any integer, and $a^{\frac{1}{n}}$ is a real number, then

$$a^{\frac{m}{n}} = \left(a^{\frac{1}{n}}\right)^m = \left(a^m\right)^{\frac{1}{n}}.$$

In radical notation:

$$a^{\frac{m}{n}} = \left(\sqrt[n]{a}\right)^m = \sqrt[n]{a^m}.$$

Example 2: Conversion from Exponential Notation to Radical Notation

Assume that each variable represents a positive real number. Each expression is changed to an equivalent expression in either radical or exponential notation.

a. $x^{\frac{2}{3}} = \sqrt[3]{x^2}$ Note that the index, 3, is the denominator in the rational exponent.

b. $3x^{\frac{4}{5}} = 3\sqrt[5]{x^4}$ Note that the coefficient, 3, is not affected by the exponent.

c. $-a^{\frac{3}{2}} = -\sqrt{a^3}$ Note that -1 is the understood coefficient.

d. $\sqrt[6]{a^5} = a^{\frac{5}{6}}$ Note that the index, 6, is the denominator of the rational exponent.

e. $5\sqrt{x} = 5x^{\frac{1}{2}}$ Note that in a square root the index is understood to be 2.

f. $-\sqrt[3]{4} = -4^{\frac{1}{3}}$ Note that the coefficient, -1, is not affected by the exponent. Also, we could write $-4^{\frac{1}{3}} = -1 \cdot 4^{\frac{1}{3}}$.

Simplifying Expressions with Rational Exponents

Expressions with rational exponents such as

$$x^{\frac{2}{3}} \cdot x^{\frac{1}{6}}, \quad \frac{x^{\frac{3}{4}}}{x^{\frac{1}{3}}}, \text{ and } \left(2a^{\frac{1}{4}}\right)^3$$

can be simplified using the rules for exponents.

NOTES Unless otherwise stated, we will assume, for the remainder of this chapter, that all variables represent positive real numbers.

Example 3: Simplifying Expressions with Rational Exponents

Each expression is simplified using one or more of the rules for exponents.

a. $x^{\frac{2}{3}} \cdot x^{\frac{1}{6}} = x^{\frac{2}{3}+\frac{1}{6}}$

$= x^{\frac{4}{6}+\frac{1}{6}} = x^{\frac{5}{6}}$ Find a common denominator and add the exponents.

b. $\frac{x^{\frac{3}{4}}}{x^{\frac{1}{3}}} = x^{\frac{3}{4}-\frac{1}{3}}$

$= x^{\frac{9}{12}-\frac{4}{12}} = x^{\frac{5}{12}}$ Find a common denominator and subtract the exponents.

c. $\left(2a^{\frac{1}{4}}\right)^3 = 2^3 \cdot a^{\frac{1}{4}(3)} = 8a^{\frac{3}{4}}$

d. $\left(27y^{-\frac{9}{10}}\right)^{-\frac{1}{3}} = 27^{-\frac{1}{3}} \cdot y^{-\frac{9}{10}\left(-\frac{1}{3}\right)}$

$= \frac{y^{\frac{3}{10}}}{27^{\frac{1}{3}}} = \frac{y^{\frac{3}{10}}}{3}$ Multiply the exponents of y and reduce the fraction to $\frac{3}{10}$.

e. $(-36)^{-\frac{1}{2}} = \frac{1}{(-36)^{\frac{1}{2}}}$ This is not a real number because $(-36)^{\frac{1}{2}} = \sqrt{-36}$ is not real.

f. $9^{\frac{2}{4}} = 9^{\frac{1}{2}} = 3$ The exponent can be reduced as long as the expression is real.

g. $\left(\dfrac{49x^6y^{-2}}{z^{-4}}\right)^{\frac{1}{2}} = \dfrac{\left(49x^6y^{-2}\right)^{\frac{1}{2}}}{\left(z^{-4}\right)^{\frac{1}{2}}}$ **Study this example carefully.**

$= \dfrac{49^{\frac{1}{2}}x^{6\left(\frac{1}{2}\right)}y^{-2\left(\frac{1}{2}\right)}}{z^{-4\left(\frac{1}{2}\right)}}$ Use the power rule four times.

$= \dfrac{7x^3y^{-1}}{z^{-2}}$ Simplify exponents.

$= \dfrac{7x^3z^2}{y}$ Use the properties of negative exponents.

Example 4 shows how to use fractional exponents to simplify rather complicated looking radical expressions. The results may seem surprising at first.

Example 4: Simplifying Radical Notation by Changing to Exponential Notation

Simplify each expression by first changing it into an equivalent expression with rational exponents. Then rewrite the answer in simplified radical form.

a. $\sqrt[4]{\sqrt[3]{x}} = \left(\sqrt[3]{x}\right)^{\frac{1}{4}} = \left(x^{\frac{1}{3}}\right)^{\frac{1}{4}}$

$= x^{\frac{1}{12}} = \sqrt[12]{x}$ Note that $\frac{1}{3}\cdot\frac{1}{4} = \frac{1}{12}$.

b. $\sqrt[3]{a}\sqrt{a} = a^{\frac{1}{3}}\cdot a^{\frac{1}{2}}$

$= a^{\frac{1}{3}+\frac{1}{2}} = a^{\frac{2}{6}+\frac{3}{6}}$

$= a^{\frac{5}{6}} = \sqrt[6]{a^5}$

c. $\dfrac{\sqrt{x^3}\sqrt[3]{x^2}}{\sqrt[5]{x^2}} = \dfrac{x^{\frac{3}{2}}\cdot x^{\frac{2}{3}}}{x^{\frac{2}{5}}} = \dfrac{x^{\frac{3}{2}+\frac{2}{3}}}{x^{\frac{2}{5}}} = \dfrac{x^{\frac{9}{6}+\frac{4}{6}}}{x^{\frac{2}{5}}} = \dfrac{x^{\frac{13}{6}}}{x^{\frac{2}{5}}}$

$= x^{\frac{13}{6}-\frac{2}{5}} = x^{\frac{65}{30}-\frac{12}{30}} = x^{\frac{53}{30}}$

$= x^{\frac{30}{30}}\cdot x^{\frac{23}{30}} = x\cdot x^{\frac{23}{30}} = x\sqrt[30]{x^{23}}$ Note that $\frac{53}{30} = \frac{30}{30}+\frac{23}{30} = 1+\frac{23}{30}$.

Evaluating Roots with a TI-84 Plus Calculator (The ^ key)

The caret key ^ on the TI-84 Plus calculator (and most graphing calculators) is used to indicate exponents. Using this key, roots of real numbers can be calculated with up to nine digit accuracy. To set the number of decimal places you wish in any calculations, press the MODE key and highlight the digit opposite the word FLOAT that indicates the desired accuracy. If no digit is highlighted, then the accuracy will be to nine decimal places (in some cases ten decimal places).

To Find the Value of $a^{\frac{m}{n}}$ with a TI-84 Plus Graphing Calculator

1. Enter the value of the base, a.
2. Press the caret key ^.
3. Enter the fractional exponent enclosed in parentheses. (This exponent may be positive or negative.)
4. Press ENTER.

Example 5: Evaluating Rational Exponents with a Calculator

Evaluate the following expressions using a TI-84 Plus graphing calculator.

a. $125^{\frac{4}{3}}$

Solution: To find $125^{\frac{4}{3}}$ proceed as follows.

Step 1: Enter the base, 125.

Step 2: Press the caret key ^.

Step 3: Enter the exponent in parentheses, $\frac{4}{3}$.

Step 4: Press ENTER.

The display should read as follows.

```
125^(4/3)
                625
```

b. $36^{\frac{3}{5}}$

Solution: To find $36^{\frac{3}{5}}$ proceed as follows.

Step 1: Enter the base, 36.

Step 2: Press the caret key ^.

1. 3 2. 11
3. $\frac{1}{10}$ 4. $\frac{1}{5}$
5. –512 6. Nonreal
7. –4 8. –4
9. Nonreal
10. $-\frac{2}{5}$ 11. $\frac{3}{7}$
12. $\frac{5}{4}$ 13. 16

Step 3: Enter the exponent in parentheses, $\frac{3}{5}$.

Step 4: Press ENTER.

The display should read as follows.

```
36^(3/5)
         8.585814487
```

Practice Problems

14. $\frac{1}{4}$ 15. $-\frac{1}{6}$
16. –5 17. $\frac{5}{2}$
18. $-\frac{27}{8}$ 19. $\frac{1}{4}$
20. $\frac{9}{16}$ 21. $\frac{3}{8}$
22. $\frac{2}{5}$ 23. $-\frac{1}{1000}$
24. $-\frac{1}{16{,}807}$
25. 64 26. $\frac{1}{81}$
27. 8.5499
28. 2187

Simplify each of the following expressions. Leave the answers with rational exponents.

1. $64^{\frac{2}{3}}$ **2.** $x^{\frac{3}{4}} \cdot x^{\frac{1}{5}} \cdot x^{\frac{1}{2}}$ **3.** $\dfrac{x^{\frac{1}{6}} \cdot y^{\frac{1}{2}}}{x^{\frac{1}{3}} \cdot y^{\frac{1}{4}}}$ **4.** $\left(16^{\frac{3}{4}}\right)^{-2}$ **5.** $-81^{\frac{1}{4}}$

Simplify each expression by first changing to an equivalent expression with rational exponents. Rewrite the answer in simplified radical form.

6. $\sqrt[4]{x} \cdot \sqrt{x}$ **7.** $\sqrt[5]{\sqrt[3]{a^2}}$ **8.** $\dfrac{\sqrt{36x^5}}{\sqrt[3]{8x^3}}$

Use a graphing calculator to find the following values accurate to 4 decimal places.

9. $128^{\frac{1}{5}}$ **10.** $100^{-\frac{1}{4}}$

9.3 Exercises

29. 10,000,000
30. 4.6416
31. 99.6055
32. 158.7401
33. 0.0922
34. 0.0131
35. 1.6083
36. 1.6785
37. 0.2236

Simplify each numerical expression.

1. $9^{\frac{1}{2}}$ **2.** $121^{\frac{1}{2}}$ **3.** $100^{-\frac{1}{2}}$ **4.** $25^{-\frac{1}{2}}$ **5.** $-64^{\frac{3}{2}}$

6. $(-64)^{\frac{3}{2}}$ **7.** $(-64)^{\frac{1}{3}}$ **8.** $-(64)^{\frac{1}{3}}$ **9.** $\left(-\frac{4}{25}\right)^{\frac{1}{2}}$ **10.** $-\left(\frac{4}{25}\right)^{\frac{1}{2}}$

11. $\left(\frac{9}{49}\right)^{\frac{1}{2}}$ **12.** $\left(\frac{225}{144}\right)^{\frac{1}{2}}$ **13.** $64^{\frac{2}{3}}$ **14.** $8^{-\frac{2}{3}}$ **15.** $(-216)^{-\frac{1}{3}}$

Answers to Practice Problems: **1.** 16 **2.** $x^{\frac{29}{20}}$ **3.** $\dfrac{y^{\frac{1}{4}}}{x^{\frac{1}{6}}}$ **4.** $\frac{1}{64}$ **5.** –3 **6.** $\sqrt[4]{x^3}$ **7.** $\sqrt[15]{a^2}$ **8.** $3x\sqrt{x}$ **9.** 2.6390 **10.** 0.3162

38. 0.2
39. 7.7460
40. 4.0632
41. 2.0408
42. 1.0605
43. $8x$ 44. $81x^2$
45. $\frac{1}{3a^2}$ 46. $\frac{1}{2a^{\frac{3}{4}}}$
47. $8x^{\frac{5}{2}}$ 48. $3x^{\frac{11}{3}}$
49. $5a^{\frac{13}{6}}$ 50. $a^{\frac{1}{15}}$
51. $x^{\frac{7}{12}}$ 52. $a^{\frac{5}{9}}$
53. $x^{\frac{1}{2}}$ 54. $a^{\frac{7}{6}}$
55. $\frac{1}{a^{\frac{9}{8}}}$ 56. 1
57. $a^{\frac{1}{4}}$ 58. $x^{\frac{7}{6}}$
59. $\frac{a^2}{b^{\frac{6}{5}}}$ 60. $\frac{1}{a^{\frac{3}{4}}b^{\frac{1}{2}}}$
61. $8x^{\frac{3}{2}}y$ 62. a^5
63. $\frac{x^{\frac{3}{2}}}{16y^{\frac{2}{5}}}$ 64. $\frac{a^2}{3b^{\frac{1}{2}}}$
65. $\frac{x^2y^4}{z^4}$ 66. $\frac{27x^3}{y^6z^{\frac{9}{2}}}$
67. $\frac{y^{\frac{3}{2}}z^2}{x}$ 68. $\frac{c^3}{3ab^2}$
69. $\frac{8b^{\frac{9}{4}}}{a^3c^3}$ 70. $-3a^{\frac{2}{3}}bc$
71. $x^{\frac{1}{4}}y^{\frac{5}{4}}$ 72. $\frac{y^{\frac{2}{3}}}{x^{\frac{5}{6}}}$
73. $\frac{y^{\frac{2}{3}}}{50x^{\frac{4}{3}}}$ 74. $\frac{5a^{\frac{7}{5}}}{8b^{\frac{23}{10}}}$
75. $\frac{b^{\frac{5}{12}}}{a^{\frac{11}{12}}}$ 76. $\frac{x^{\frac{7}{4}}}{y^{\frac{3}{4}}}$

16. $(-125)^{\frac{1}{3}}$ **17.** $\left(\frac{8}{125}\right)^{-\frac{1}{3}}$ **18.** $-\left(\frac{16}{81}\right)^{-\frac{3}{4}}$ **19.** $\left(-\frac{1}{32}\right)^{\frac{2}{5}}$ **20.** $\left(\frac{27}{64}\right)^{\frac{2}{3}}$

21. $3\cdot 16^{-\frac{3}{4}}$ **22.** $2\cdot 25^{-\frac{1}{2}}$ **23.** $-100^{-\frac{3}{2}}$ **24.** $-49^{-\frac{5}{2}}$ **25.** $\left[\left(\frac{1}{32}\right)^{\frac{2}{5}}\right]^{-3}$

26. $\left[(-27)^{\frac{2}{3}}\right]^{-2}$

Use a graphing calculator to find the value of each numerical expression accurate to 4 decimal places, if necessary.

27. $25^{\frac{2}{3}}$ **28.** $81^{\frac{7}{4}}$ **29.** $100^{\frac{7}{2}}$ **30.** $100^{\frac{1}{3}}$ **31.** $250^{\frac{5}{6}}$

32. $2000^{\frac{2}{3}}$ **33.** $24^{-\frac{3}{4}}$ **34.** $18^{-\frac{3}{2}}$ **35.** $\sqrt[9]{72}$ **36.** $\sqrt[8]{63}$

37. $\sqrt[4]{0.0025}$ **38.** $\sqrt[5]{0.00032}$ **39.** $\sqrt[4]{3600}$ **40.** $\sqrt[6]{4500}$ **41.** $\sqrt[5]{35.4}$

42. $\sqrt[10]{1.8}$

Simplify each algebraic expression. Assume that all variables represent positive real numbers. Leave the answers in rational exponent form.

43. $\left(2x^{\frac{1}{3}}\right)^3$ **44.** $\left(3x^{\frac{1}{2}}\right)^4$ **45.** $\left(9a^4\right)^{-\frac{1}{2}}$ **46.** $\left(16a^3\right)^{-\frac{1}{4}}$

47. $8x^2\cdot x^{\frac{1}{2}}$ **48.** $3x^3\cdot x^{\frac{2}{3}}$ **49.** $5a^2\cdot a^{-\frac{1}{3}}\cdot a^{\frac{1}{2}}$ **50.** $a^{\frac{2}{3}}\cdot a^{-\frac{3}{5}}\cdot a^0$

51. $\frac{x^{\frac{3}{4}}}{x^{\frac{1}{6}}}$ **52.** $\frac{a^{\frac{2}{3}}}{a^{\frac{1}{9}}}$ **53.** $\frac{x^{\frac{2}{5}}}{x^{-\frac{1}{10}}}$ **54.** $\frac{a^{\frac{1}{2}}}{a^{-\frac{2}{3}}}$

55. $\frac{a^{\frac{3}{4}}\cdot a^{\frac{1}{8}}}{a^2}$ **56.** $\frac{x^{\frac{2}{3}}\cdot x^{\frac{4}{3}}}{x^2}$ **57.** $\frac{a^{\frac{1}{2}}\cdot a^{-\frac{3}{4}}}{a^{-\frac{1}{2}}}$ **58.** $\frac{x^{\frac{2}{3}}x^{-1}}{x^{-\frac{3}{2}}}$

59. $\frac{a^{\frac{3}{2}}b^{\frac{4}{5}}}{a^{-\frac{1}{2}}b^2}$ **60.** $\frac{a^{\frac{3}{4}}b^{-\frac{1}{3}}}{a^{\frac{3}{2}}b^{\frac{1}{6}}}$ **61.** $\left(2x^{\frac{1}{2}}y^{\frac{1}{3}}\right)^3$ **62.** $\left(a^{\frac{1}{2}}a^{\frac{1}{3}}\right)^6$

63. $\left(4x^{-\frac{3}{4}}y^{\frac{1}{5}}\right)^{-2}$ **64.** $\left(81a^{-8}b^2\right)^{-\frac{1}{4}}$ **65.** $\left(-x^3y^6z^{-6}\right)^{\frac{2}{3}}$ **66.** $\left(9x^2y^{-4}z^{-3}\right)^{\frac{3}{2}}$

67. $\left(\frac{x^2y^{-3}}{z^4}\right)^{-\frac{1}{2}}$ **68.** $\left(\frac{27a^3b^6}{c^9}\right)^{-\frac{1}{3}}$ **69.** $\left(\frac{16a^{-4}b^3}{c^4}\right)^{\frac{3}{4}}$ **70.** $\left(\frac{-27a^2b^3}{c^{-3}}\right)^{\frac{1}{3}}$

71. $\frac{\left(x^{\frac{1}{4}}y^{\frac{1}{2}}\right)^3}{x^{\frac{1}{2}}y^{\frac{1}{4}}}$ **72.** $\frac{\left(x^{\frac{1}{2}}y\right)^{-\frac{1}{3}}}{x^{\frac{2}{3}}y^{-1}}$ **73.** $\frac{\left(8x^2y\right)^{-\frac{1}{3}}}{\left(5x^{\frac{1}{3}}y^{-\frac{1}{2}}\right)^2}$ **74.** $\frac{\left(25a^4b^{-1}\right)^{\frac{1}{2}}}{\left(2a^{\frac{1}{5}}b^{\frac{3}{5}}\right)^3}$

77. $\dfrac{3x^{\frac{1}{2}}}{20y^{\frac{2}{3}}}$ 78. $\dfrac{1}{8ab^2}$

79. $\sqrt[6]{a^5}$ 80. $x\sqrt[15]{x^4}$

81. $\sqrt[12]{y^7}$ 82. $x\sqrt[12]{x}$

83. $\sqrt[30]{x^{11}}$ 84. $\sqrt[12]{a^5}$

85. $\sqrt[6]{y}$ 86. $\sqrt[10]{x}$

87. $\sqrt[9]{x}$ 88. $\sqrt[4]{a}$

89. $\sqrt[3]{7a}$ 90. $\sqrt[3]{3x}$

91. $\sqrt[24]{x}$ 92. $\sqrt[60]{x}$

93. $a^{20}b^5c^{10}$

94. $a^9b^{18}c^3$

95. No: $\sqrt[5]{a}\cdot\sqrt{a} = a^{\frac{7}{10}}$

$\sqrt[5]{a^2} = a^{\frac{2}{5}}$

$a^{\frac{7}{10}} \neq a^{\frac{2}{5}}$

75. $\left(\dfrac{a^{-3}b^{\frac{1}{3}}}{a^{\frac{1}{2}}b}\right)^{\frac{1}{2}}\cdot\left(\dfrac{ab^{\frac{1}{2}}}{a^{-\frac{2}{3}}b^{-1}}\right)^{\frac{1}{2}}$

76. $\left(\dfrac{x^2y^{\frac{1}{3}}}{x^{\frac{1}{2}}y^{\frac{3}{2}}}\right)^{\frac{1}{2}}\cdot\left(\dfrac{x^{-\frac{1}{2}}y^{\frac{2}{3}}}{x^{-1}y^{\frac{3}{4}}}\right)^2$

77. $\dfrac{\left(27xy^{\frac{1}{2}}\right)^{\frac{1}{3}}}{\left(25x^{-\frac{1}{2}}y\right)^{\frac{1}{2}}}\cdot\dfrac{\left(x^{\frac{1}{2}}y\right)^{\frac{1}{6}}}{\left(16x^{\frac{1}{3}}y\right)^{\frac{1}{2}}}$

78. $\dfrac{\left(4a^{-6}b\right)^{\frac{1}{2}}}{\left(7a^2b^3\right)^{-1}}\cdot\dfrac{\left(49a^4b^3\right)^{-\frac{1}{2}}}{\left(64a^{-3}b^6\right)^{\frac{2}{3}}}$

Simplify each expression by first changing it into an equivalent expression with rational exponents. Rewrite the answer in simplified radical form.

79. $\sqrt[3]{a}\cdot\sqrt{a}$

80. $\sqrt[3]{x^2}\cdot\sqrt[5]{x^3}$

81. $\dfrac{\sqrt[4]{y^3}}{\sqrt[6]{y}}$

82. $\dfrac{\sqrt[3]{x^4}}{\sqrt[4]{x}}$

83. $\dfrac{\sqrt[3]{x^2}\sqrt[5]{x^6}}{\sqrt{x^3}}$

84. $\dfrac{a\sqrt[4]{a}}{\sqrt[3]{a}\sqrt{a}}$

85. $\sqrt{\sqrt[3]{y}}$

86. $\sqrt[5]{\sqrt{x}}$

87. $\sqrt[3]{\sqrt[3]{x}}$

88. $\sqrt{\sqrt{a}}$

89. $\sqrt[15]{(7a)^5}$

90. $\sqrt[21]{(3x)^7}$

91. $\sqrt[4]{\sqrt[3]{\sqrt{x}}}$

92. $\sqrt[5]{\sqrt[4]{\sqrt[3]{x}}}$

93. $\left(\sqrt[3]{a^4bc^2}\right)^{15}$

94. $\left(\sqrt[4]{a^3b^6c}\right)^{12}$

Writing and Thinking About Mathematics

96. a. n must be an integer greater than 1.
b. n must be a number less than 0.
c. n must equal 0.
d. n must be a rational number that is not an integer.

95. Is $\sqrt[5]{a}\cdot\sqrt{a}$ the same as $\sqrt[5]{a^2}$? Explain why or why not.

96. Assume that x represents a positive real number. Describe what kind of number the exponent n must be for x^n to mean

a. a product **b.** a quotient **c.** 1 **d.** a radical

HAWKES LEARNING SYSTEMS: INTRODUCTORY & INTERMEDIATE ALGEBRA SOFTWARE

- 9.3 Rational Exponents

9.4 Operations with Radicals

- *Perform arithmetic operations with radical expressions.*
- *Rationalize the denominators of radicals.*
- *Use a graphing calculator to evaluate radical expressions.*

Addition and Subtraction with Radical Expressions

Recall that to find the sum $2x^2 + 3x^2 - 8x^2$, you can use the distributive property and write

$$2x^2 + 3x^2 - 8x^2 = (2+3-8)x^2$$
$$= -3x^2.$$

Recall that the terms $2x^2, 3x^2$, and $-8x^2$ are called **like terms** because each term contains the same variable expression, x^2. Similarly,

$$2\sqrt{5} + 3\sqrt{5} - 8\sqrt{5} = (2+3-8)\sqrt{5}$$
$$= -3\sqrt{5}$$

and $2\sqrt{5}$, $3\sqrt{5}$, and $-8\sqrt{5}$ are called **like radicals** because each term contains the same radical expression, $\sqrt{5}$. **Like radicals** have the same index and radicand or they can be simplified so that they have the same index and radicand.

Teaching Notes:
That we cannot combine two exact real numbers into a simple, singular numerical form may be difficult for some students to follow. You may want to emphasize the fact that, for example, $2\sqrt{3}+2\sqrt{7}$ is a single real number.

The terms $2\sqrt{3}$ and $2\sqrt{7}$ are **not** like radicals because the radicands are not the same, and neither expression can be simplified. Therefore, a sum such as

$$2\sqrt{3} + 2\sqrt{7}$$

cannot be simplified. That is, the terms cannot be combined.

In some cases, radicals that are not like radicals can be simplified, and the results may lead to like radicals. For example, $4\sqrt{12}$, $\sqrt{75}$, and $-\sqrt{108}$ are not like radicals. However, simplification of each radical allows the sum of these radicals to be found as follows.

$$4\sqrt{12} + \sqrt{75} - \sqrt{108} = 4\sqrt{4\cdot 3} + \sqrt{25\cdot 3} - \sqrt{36\cdot 3}$$
$$= 4\cdot 2\sqrt{3} + 5\sqrt{3} - 6\sqrt{3}$$
$$= 8\sqrt{3} + 5\sqrt{3} - 6\sqrt{3}$$
$$= (8+5-6)\sqrt{3}$$
$$= 7\sqrt{3}$$

Example 1: Addition and Subtraction with Radicals

Perform the indicated operation and simplify, if possible. Assume that all variables are positive.

a. $\sqrt{32x}+\sqrt{18x}$

Solution:
$$\begin{aligned}\sqrt{32x}+\sqrt{18x}&=\sqrt{16\cdot 2x}+\sqrt{9\cdot 2x}\\&=4\sqrt{2x}+3\sqrt{2x}\\&=(4+3)\sqrt{2x}\\&=7\sqrt{2x}\end{aligned}$$

b. $\sqrt{12}+\sqrt{18}+\sqrt{27}$

Solution:
$$\begin{aligned}\sqrt{12}+\sqrt{18}+\sqrt{27}&=\sqrt{4\cdot 3}+\sqrt{9\cdot 2}+\sqrt{9\cdot 3}\\&=2\sqrt{3}+3\sqrt{2}+3\sqrt{3}\\&=(2+3)\sqrt{3}+3\sqrt{2}\\&=5\sqrt{3}+3\sqrt{2}\end{aligned}$$

Note that $\sqrt{3}$ and $\sqrt{2}$ are **not** like radicals. Therefore, the last expression is already fully simplified.

c. $\sqrt[3]{5x}-\sqrt[3]{40x}$

Solution:
$$\begin{aligned}\sqrt[3]{5x}-\sqrt[3]{40x}&=\sqrt[3]{5x}-\sqrt[3]{8\cdot 5x}\\&=\sqrt[3]{5x}-2\sqrt[3]{5x}\\&=(1-2)\sqrt[3]{5x}\\&=-\sqrt[3]{5x}\end{aligned}$$

d. $x\sqrt{4y^3}-5\sqrt{x^2y^3}$

Solution:
$$\begin{aligned}x\sqrt{4y^3}-5\sqrt{x^2y^3}&=x\sqrt{4y^2}\sqrt{y}-5\sqrt{x^2y^2}\sqrt{y}\\&=2xy\sqrt{y}-5xy\sqrt{y}\\&=-3xy\sqrt{y}\end{aligned}$$

Multiplication with Radical Expressions

To find a product such as $(\sqrt{3}+5)(\sqrt{3}-7)$ treat the two expressions as two binomials and multiply just as with polynomials. For example, using the FOIL method, we get

$$\begin{aligned}(\sqrt{3}+5)(\sqrt{3}-7)&=(\sqrt{3})^2-7\sqrt{3}+5\sqrt{3}+5(-7)\\&=3-7\sqrt{3}+5\sqrt{3}-35\\&=3-35+(-7+5)\sqrt{3}\\&=-32-2\sqrt{3}.\end{aligned}$$

Example 2: Multiplication of Radicals

Multiply and simplify the following expressions.

a. $\sqrt{5}\cdot\sqrt{15}$

Solution: $\sqrt{5}\cdot\sqrt{15}=\sqrt{5}\cdot\sqrt{5}\cdot\sqrt{3}=\left(\sqrt{5}\right)^2\cdot\sqrt{3}=5\sqrt{3}$

b. $\sqrt{7}\left(\sqrt{7}-\sqrt{14}\right)$

Solution:
$$\begin{aligned}\sqrt{7}\left(\sqrt{7}-\sqrt{14}\right)&=\sqrt{7}\cdot\sqrt{7}-\sqrt{7}\cdot\sqrt{14}\\&=\left(\sqrt{7}\right)^2-\sqrt{7}\left(\sqrt{7}\cdot\sqrt{2}\right)\\&=\left(\sqrt{7}\right)^2-\left(\sqrt{7}\right)^2\sqrt{2}\\&=7-7\sqrt{2}\end{aligned}$$

c. $\left(3\sqrt{7}-2\right)\left(\sqrt{7}+3\right)$

Solution:
$$\begin{aligned}\left(3\sqrt{7}-2\right)\left(\sqrt{7}+3\right)&=3\left(\sqrt{7}\right)^2+3\cdot3\sqrt{7}-2\sqrt{7}-2\cdot3\\&=3\cdot7+9\sqrt{7}-2\sqrt{7}-6\\&=21-6+(9-2)\sqrt{7}\\&=15+7\sqrt{7}\end{aligned}$$

d. $\left(\sqrt{6}+\sqrt{2}\right)^2$

Solution:
$$\begin{aligned}\left(\sqrt{6}+\sqrt{2}\right)^2&=\left(\sqrt{6}\right)^2+2\sqrt{6}\sqrt{2}+\left(\sqrt{2}\right)^2 \qquad (a+b)^2=a^2+2ab+b^2\\&=6+2\sqrt{12}+2\\&=8+2\sqrt{4}\cdot\sqrt{3}\\&=8+2\cdot2\sqrt{3}\\&=8+4\sqrt{3}\end{aligned}$$

e. $\left(\sqrt{2x}+5\right)\left(\sqrt{2x}-5\right)$

Solution:
$$\begin{aligned}\left(\sqrt{2x}+5\right)\left(\sqrt{2x}-5\right)&=\left(\sqrt{2x}\right)^2-(5)^2 \qquad (a+b)(a-b)=a^2-b^2\\&=2x-25\end{aligned}$$

Rationalizing Denominators of Rational Expressions

An expression with a radical in the denominator may not be in the most usable form for further algebraic manipulation or operations. In many situations, to make calculations easier, we may want to **rationalize the denominator**. That is, we want to find an equivalent fraction in which the denominator does not have a radical. The numerator may still have a radical in it, but a rational expression with no radicals in the denominator definitely makes arithmetic with radicals much easier. (**Note:** In this section we will deal only with radicals that involve square roots or cube roots, however the ideas discussed here also apply to n^{th} roots in general.)

To Rationalize a Denominator Containing a Square Root or a Cube Root

1. If the denominator contains a square root, multiply both the numerator and denominator by an expression that will give a denominator with no square roots.
2. If the denominator contains a cube root, multiply both the numerator and denominator by an expression that will give a denominator with no cube roots.

Example 3: Rationalizing the Denominator

Teaching Notes:
You may want to illustrate some of these ideas with other purely numerical examples:

$$\sqrt{\frac{3}{7}}=\frac{\sqrt{3}}{\sqrt{7}}=\frac{\sqrt{3}}{\sqrt{7}}\cdot\frac{\sqrt{7}}{\sqrt{7}}=\frac{\sqrt{21}}{7}$$

$$\frac{3}{\sqrt{2}}=\frac{3}{\sqrt{2}}\cdot\frac{\sqrt{2}}{\sqrt{2}}=\frac{3\sqrt{2}}{2}$$

$$\frac{2}{\sqrt[3]{4}}=\frac{2}{\sqrt[3]{4}}\cdot\frac{\sqrt[3]{2}}{\sqrt[3]{2}}=\frac{2\sqrt[3]{2}}{\sqrt[3]{8}}=\frac{\cancel{2}\sqrt[3]{2}}{\cancel{2}}=\sqrt[3]{2}$$

Simplify each radical expression so that the denominator contains no radicals. Assume all variables represent positive numbers.

a. $\dfrac{\sqrt{5}}{\sqrt{2}}$

Solution: Multiply the numerator and denominator by $\sqrt{2}$ because $2\cdot2=4$, a perfect square.

$$\frac{\sqrt{5}}{\sqrt{2}}=\frac{\sqrt{5}}{\sqrt{2}}\cdot\frac{\sqrt{2}}{\sqrt{2}}=\frac{\sqrt{10}}{\sqrt{4}}=\frac{\sqrt{10}}{2}$$

b. $\sqrt{\dfrac{5}{12x}}$

Solution: Multiply the numerator and denominator by $\sqrt{3x}$ because $12x\cdot3x=36x^2$, a perfect square expression.

$$\sqrt{\frac{5}{12x}}=\frac{\sqrt{5}}{\sqrt{12x}}=\frac{\sqrt{5}}{\sqrt{12x}}\cdot\frac{\sqrt{3x}}{\sqrt{3x}}=\frac{\sqrt{15x}}{\sqrt{36x^2}}=\frac{\sqrt{15x}}{6x}$$

Continued on the next page...

c. $\dfrac{7}{\sqrt[3]{32y}}$

Solution: Multiply the numerator and denominator by $\sqrt[3]{2y^2}$ because $32y \cdot 2y^2 = 64y^3$, a perfect cube expression.

$$\frac{7}{\sqrt[3]{32y}} = \frac{7}{\sqrt[3]{32y}} \cdot \frac{\sqrt[3]{2y^2}}{\sqrt[3]{2y^2}} = \frac{7\sqrt[3]{2y^2}}{\sqrt[3]{64y^3}} = \frac{7\sqrt[3]{2y^2}}{4y}$$

If the denominator has a radical expression with **a sum or difference involving square roots** such as

$$\frac{2}{4-\sqrt{2}} \quad \text{or} \quad \frac{12}{3+\sqrt{5}},$$

then a different method is used for rationalizing the denominator. In this method, we think of the denominator in the form of $a - b$ or $a + b$. Thus

$$\text{if } a-b = 4-\sqrt{2}, \text{ then } a+b = 4+\sqrt{2}$$

and

$$\text{if } a+b = 3+\sqrt{5}, \text{ then } a-b = 3-\sqrt{5}.$$

The two expressions $(a-b)$ and $(a+b)$ are called **conjugates** of each other, and, as we know, their product $(a-b)(a+b)$ results in the **difference of two squares**:

$$(a-b)(a+b) = a^2 - b^2.$$

To Rationalize a Denominator Containing a Sum or Difference Involving Square Roots

If the denominator of a fraction contains a sum or difference involving a square root, rationalize the denominator by multiplying both the numerator and denominator by the **conjugate of the denominator**.

1. If the denominator is of the form $a - b$, multiply both the numerator and denominator by $a + b$.
2. If the denominator is of the form $a + b$, multiply both the numerator and denominator by $a - b$.

The new denominator will be the difference of two squares and therefore will not contain a radical term.

Example 4: Rationalizing a Denominator with a Sum or Difference Involving a Square Root

Simplify each expression by rationalizing the denominator.

a. $\frac{2}{4-\sqrt{2}}$

Solution: Multiply the numerator and denominator by $4+\sqrt{2}$.

$$\frac{2}{4-\sqrt{2}} = \frac{2\left(4+\sqrt{2}\right)}{\left(4-\sqrt{2}\right)\left(4+\sqrt{2}\right)}$$

If $a-b=4-\sqrt{2}$, then $a+b=4+\sqrt{2}$.

$$= \frac{2\left(4+\sqrt{2}\right)}{4^2-\left(\sqrt{2}\right)^2}$$

The denominator is the difference of two squares.

$$= \frac{2\left(4+\sqrt{2}\right)}{16-2}$$

$$= \frac{\cancel{2}\left(4+\sqrt{2}\right)}{\cancel{14}_7}$$

$$= \frac{4+\sqrt{2}}{7}$$

The denominator is a rational number. Note that the numerator is now irrational. However, this is generally preferred over having an irrational denominator.

b. $\frac{31}{6+\sqrt{5}}$

Solution: Multiply the numerator and denominator by $6-\sqrt{5}$.

$$\frac{31}{6+\sqrt{5}} = \frac{31\left(6-\sqrt{5}\right)}{\left(6+\sqrt{5}\right)\left(6-\sqrt{5}\right)}$$

$$= \frac{31\left(6-\sqrt{5}\right)}{36-5}$$

$$= \frac{\cancel{31}\left(6-\sqrt{5}\right)}{\cancel{31}}$$

$$= 6-\sqrt{5}$$

c. $\frac{1}{\sqrt{7}-\sqrt{2}}$

Solution: Multiply the numerator and denominator by $\sqrt{7}+\sqrt{2}$.

$$\frac{1}{\sqrt{7}-\sqrt{2}} = \frac{1\left(\sqrt{7}+\sqrt{2}\right)}{\left(\sqrt{7}-\sqrt{2}\right)\left(\sqrt{7}+\sqrt{2}\right)}$$

$$= \frac{\sqrt{7}+\sqrt{2}}{7-2} = \frac{\sqrt{7}+\sqrt{2}}{5}$$

Continued on the next page...

d. $\dfrac{6}{1+\sqrt{x}}$

Solution: Multiply the numerator and denominator by $1-\sqrt{x}$.

$$\frac{6}{1+\sqrt{x}} = \frac{6\left(1-\sqrt{x}\right)}{\left(1+\sqrt{x}\right)\left(1-\sqrt{x}\right)}$$
$$= \frac{6\left(1-\sqrt{x}\right)}{1-x}$$
$$= \frac{6-6\sqrt{x}}{1-x}$$

e. $\dfrac{x-y}{\sqrt{x}-\sqrt{y}}$

Solution: Multiply the numerator and denominator by $\sqrt{x}+\sqrt{y}$.

$$\frac{x-y}{\sqrt{x}-\sqrt{y}} = \frac{(x-y)\left(\sqrt{x}+\sqrt{y}\right)}{\left(\sqrt{x}-\sqrt{y}\right)\left(\sqrt{x}+\sqrt{y}\right)}$$
$$= \frac{\cancel{(x-y)}\left(\sqrt{x}+\sqrt{y}\right)}{\cancel{x-y}}$$
$$= \sqrt{x}+\sqrt{y}$$

Evaluating Radical Expressions with a TI-84 Plus Graphing Calculator

Techniques for using a TI-84 Plus graphing calculator to evaluate radical expressions and expressions with rational exponents were illustrated in Sections 9.1, 9.2, and 9.3. These same basic techniques are used to evaluate numerical expressions that contain sums, differences, products, and quotients of radicals. Be careful to use parentheses to ensure that the rules for order of operations are maintained. **In particular, sums and differences in numerators and denominators of fractions must be enclosed in parentheses.** Study the following examples carefully.

Example 5: Using a TI-84 Graphing Calculator to Evaluate Radical Expressions

Use a TI-84 Plus graphing calculator to evaluate each expression. Round answers to 4 decimal places.

a. $3+2\sqrt{5}$

Solution: The display should appear as follows.

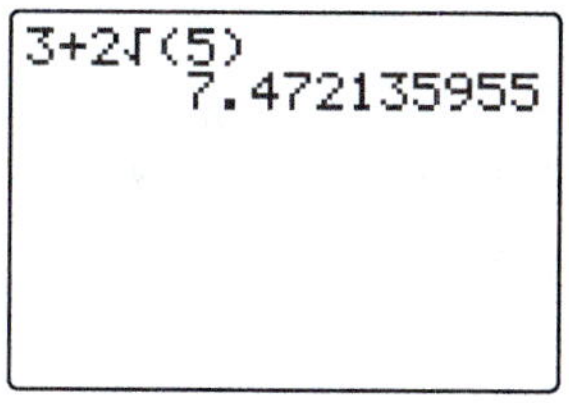

Thus $3+2\sqrt{5}=7.4721359...\approx 7.4721.$ Rounded to 4 decimal places.

b. $(\sqrt{2}+5)(\sqrt{2}-5)$

Solution: The display should appear as follows.

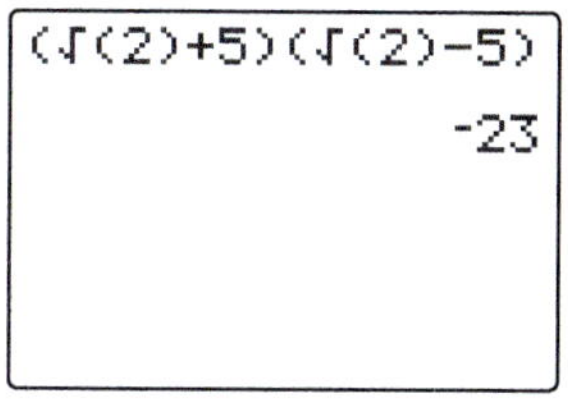

Note that the right parenthesis on 2 must be included. Otherwise, the calculator will interpret the expression as $\sqrt{(2+5)}$ (or $\sqrt{7}$) which is not intended.

Thus $(\sqrt{2}+5)(\sqrt{2}-5)=-23.$

c. $\dfrac{3}{\sqrt{6}-\sqrt{2}}$

Solution: The display should appear as follows.

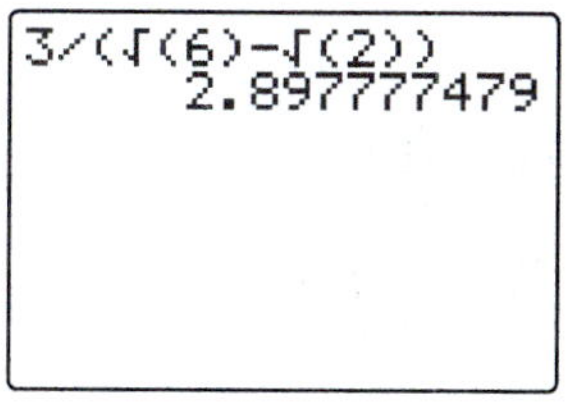

Count the parentheses in pairs.

Thus $\dfrac{3}{\sqrt{6}-\sqrt{2}}=2.8977774...\approx 2.8978.$ Rounded to 4 decimal places.

1. $-6\sqrt{2}$ 2. $-3\sqrt{5}$
3. $7\sqrt{11}$ 4. $5\sqrt{x}$
5. $-3\sqrt[3]{7x^2}$
6. $-4\sqrt[3]{4x}$
7. $10\sqrt{3}$ 8. $-7\sqrt{3}$
9. $-7\sqrt{2}$ 10. $20\sqrt{3}$
11. $6\sqrt{2}-\sqrt{3}$
12. $3\sqrt{7}+8\sqrt{10}$
13. $8\sqrt{3x}$
14. $11\sqrt{2x}+14\sqrt{3x}$
15. $20x\sqrt{5x}$

Practice Problems

Simplify each expression. Assume that all variables represent positive numbers.

1. $2\sqrt{10}-6\sqrt{10}$ **2.** $\sqrt{5}+\sqrt{45}-\sqrt{15}$ **3.** $\sqrt{8x}-3\sqrt{2x}+\sqrt{18x}$

4. $\sqrt[3]{x^5}+x\sqrt[3]{27x^2}$ **5.** $\sqrt{\dfrac{3}{8a^2}}$ **6.** $\dfrac{\sqrt[3]{4ab}}{\sqrt[3]{2a^2b^4}}$

7. $\dfrac{4}{\sqrt{2}+\sqrt{6}}$ **8.** $\dfrac{x-5}{\sqrt{x}-\sqrt{5}}$ **9.** $\left(\sqrt{3}+\sqrt{2}\right)^2$

10. $\left(3+\sqrt{2}\right)^2$ **11.** $\left(\sqrt{3}+\sqrt{8}\right)^2$ **12.** $\dfrac{\sqrt{5}-3\sqrt{2}}{\sqrt{6}+\sqrt{10}}$

9.4 Exercises

16. $-7y\sqrt{2x}$
17. $14\sqrt{5}-3\sqrt{7}$
18. $15\sqrt{3}+4\sqrt{6}$
19. $-7\sqrt[3]{3x^2}-10\sqrt[3]{6x^2}$
20. $4\sqrt[3]{5x}-13\sqrt[3]{2x}$
21. $2x\sqrt{y}-y\sqrt{xy}$
22. $-2xy\sqrt{y}$
23. $x^2y\sqrt{2x}$
24. $4x^2y\sqrt{x}$
25. $13+2\sqrt{2}$
26. 2
27. $3x-9\sqrt{3x}+8$
28. $24+10\sqrt{2x}+2x$
29. $2-2\sqrt{7}$
30. $44-39\sqrt{3}$
31. -3
32. $11+4\sqrt{10}$
33. $13+4\sqrt{10}$
34. $38+12\sqrt{10}$
35. $\sqrt{10}+\sqrt{15}-\sqrt{6}-3$

Perform the indicated operations and simplify. Assume that all variables represent positive numbers.

1. $\sqrt{2}-7\sqrt{2}$ **2.** $3\sqrt{5}-6\sqrt{5}$

3. $6\sqrt{11}+4\sqrt{11}-3\sqrt{11}$ **4.** $2\sqrt{x}+4\sqrt{x}-\sqrt{x}$

5. $9\sqrt[3]{7x^2}-4\sqrt[3]{7x^2}-8\sqrt[3]{7x^2}$ **6.** $12\sqrt[3]{4x}-10\sqrt[3]{4x}-6\sqrt[3]{4x}$

7. $2\sqrt{3}+4\sqrt{12}$ **8.** $2\sqrt{48}-3\sqrt{75}$

9. $2\sqrt{18}+\sqrt{8}-3\sqrt{50}$ **10.** $5\sqrt{48}+2\sqrt{45}-3\sqrt{20}$

11. $2\sqrt{12}+\sqrt{72}-\sqrt{75}$ **12.** $3\sqrt{28}-\sqrt{63}+8\sqrt{10}$

13. $7\sqrt{12x}-4\sqrt{27x}+\sqrt{108x}$ **14.** $\sqrt{32x}+7\sqrt{12x}+\sqrt{98x}$

15. $6x\sqrt{45x}+\sqrt{80x^3}-\sqrt{20x^3}$ **16.** $2\sqrt{18xy^2}+\sqrt{8xy^2}-3y\sqrt{50x}$

17. $\sqrt{125}-\sqrt{63}+3\sqrt{45}$ **18.** $5\sqrt{48}+2\sqrt{24}-\sqrt{75}$

19. $\sqrt[3]{81x^2}-5\sqrt[3]{48x^2}-5\sqrt[3]{24x^2}$ **20.** $\sqrt[3]{16x}-5\sqrt[3]{54x}+2\sqrt[3]{40x}$

Answers to Practice Problems: **1.** $-4\sqrt{10}$ **2.** $4\sqrt{5}-\sqrt{15}$ **3.** $2\sqrt{2x}$ **4.** $4x\sqrt[3]{x^2}$ **5.** $\dfrac{\sqrt{6}}{4a}$ **6.** $\dfrac{\sqrt[3]{2a^2}}{ab}$
7. $\sqrt{6}-\sqrt{2}$ **8.** $\sqrt{x}+\sqrt{5}$ **9.** $5+2\sqrt{6}$ **10.** $11+6\sqrt{2}$ **11.** $11+4\sqrt{6}$
12. $\dfrac{5\sqrt{2}+6\sqrt{3}-6\sqrt{5}-\sqrt{30}}{4}$

36. $6+\sqrt{30}-2\sqrt{3}-\sqrt{10}$
37. $x-2\sqrt{6x}-18$
38. $5-\sqrt{33}$
39. 58 40. $49x-2$
41. $x+10\sqrt{xy}+25y$
42. $9x+6\sqrt{xy}+y$
43. $4x-9\sqrt{xy}-9y$
44. $12x+7\sqrt{2xy}+2y$
45. $\frac{\sqrt{21}}{3}$ 46. $\frac{\sqrt{6}}{2}$
47. $\frac{\sqrt[3]{70}}{2}$ 48. $\frac{\sqrt[3]{30}}{3}$
49. $-\frac{\sqrt{6y}}{3y}$ 50. $-\frac{5\sqrt{x}}{x^2}$
51. $\frac{2\sqrt{10x}}{5y}$
52. $\frac{2\sqrt{3x}}{3y}$ 53. $\frac{2\sqrt{2y}}{y}$
54. $\frac{2\sqrt{b}}{b}$ 55. $\frac{y\sqrt[3]{2x}}{3x}$
56. $\frac{\sqrt[3]{28xy^2}}{2y^2}$
57. $\frac{\sqrt[3]{30a^2b^2}}{5b^2}$
58. $\frac{x\sqrt[3]{3xy^2}}{3y}$
59. $\sqrt{2}-1$ 60. $-\frac{3\sqrt{3}+15}{22}$
61. $-\frac{\sqrt{15}+4\sqrt{3}}{11}$
62. $-\frac{\sqrt{42}-3\sqrt{6}}{2}$
63. $2\sqrt{3}-2\sqrt{2}$
64. $4\sqrt{5}+4\sqrt{3}$
65. $\frac{\sqrt{35}+\sqrt{15}}{4}$
66. $-\frac{5\sqrt{2}+4\sqrt{5}}{3}$
67. $\frac{\sqrt{6}}{3}$

21. $x\sqrt{y}+\sqrt{x^2y}-\sqrt{xy^3}$

22. $x\sqrt{y^3}-2\sqrt{x^2y^3}-y\sqrt{x^2y}$

23. $x\sqrt{2x^3y^2}-3y\sqrt{8x^5}+xy\sqrt{72x^3}$

24. $x\sqrt{9x^3y^2}-5x^2\sqrt{xy^2}+6y\sqrt{x^5}$

25. $(3+\sqrt{2})(5-\sqrt{2})$

26. $(\sqrt{6}+2)(\sqrt{6}-2)$

27. $(\sqrt{3x}-8)(\sqrt{3x}-1)$

28. $(6+\sqrt{2x})(4+\sqrt{2x})$

29. $(2\sqrt{7}+4)(\sqrt{7}-3)$

30. $(5\sqrt{3}-2)(2\sqrt{3}-7)$

31. $(3\sqrt{5}+4\sqrt{3})(3\sqrt{5}-4\sqrt{3})$

32. $(3\sqrt{2}+\sqrt{5})(\sqrt{2}+\sqrt{5})$

33. $(\sqrt{5}+2\sqrt{2})^2$

34. $(2\sqrt{5}+3\sqrt{2})^2$

35. $(\sqrt{2}+\sqrt{3})(\sqrt{5}-\sqrt{3})$

36. $(\sqrt{6}+\sqrt{5})(\sqrt{6}-\sqrt{2})$

37. $(\sqrt{x}+\sqrt{6})(\sqrt{x}-3\sqrt{6})$

38. $(\sqrt{11}+\sqrt{3})(\sqrt{11}-2\sqrt{3})$

39. $(3\sqrt{7}+\sqrt{5})(3\sqrt{7}-\sqrt{5})$

40. $(7\sqrt{x}+\sqrt{2})(7\sqrt{x}-\sqrt{2})$

41. $(\sqrt{x}+5\sqrt{y})^2$

42. $(3\sqrt{x}+\sqrt{y})^2$

43. $(4\sqrt{x}+3\sqrt{y})(\sqrt{x}-3\sqrt{y})$

44. $(2\sqrt{2x}+\sqrt{y})(3\sqrt{2x}+2\sqrt{y})$

Rationalize the denominator and simplify if possible. Assume that all variables represent positive numbers.

45. $\frac{\sqrt{14}}{\sqrt{6}}$

46. $\frac{\sqrt{12}}{\sqrt{8}}$

47. $\frac{\sqrt[3]{35}}{\sqrt[3]{4}}$

48. $\frac{\sqrt[3]{10}}{\sqrt[3]{9}}$

49. $-\sqrt{\frac{2}{3y}}$

50. $-\sqrt{\frac{25}{x^3}}$

51. $\frac{\sqrt{8x}}{\sqrt{5y^2}}$

52. $\frac{\sqrt{4x}}{\sqrt{3y^2}}$

53. $\frac{\sqrt{16y^2}}{\sqrt{2y^3}}$

54. $\frac{\sqrt{24b}}{\sqrt{6b^2}}$

55. $\sqrt[3]{\frac{2y^3}{27x^2}}$

56. $\sqrt[3]{\frac{7x}{2y^4}}$

57. $\frac{\sqrt[3]{6a^4}}{\sqrt[3]{25a^2b^4}}$

58. $\frac{\sqrt[3]{x^5}}{\sqrt[3]{9xy}}$

59. $\frac{1}{\sqrt{2}+1}$

60. $\frac{3}{\sqrt{3}-5}$

61. $\frac{\sqrt{3}}{\sqrt{5}-4}$

62. $\frac{\sqrt{6}}{\sqrt{7}+3}$

63. $\frac{2}{\sqrt{2}+\sqrt{3}}$

64. $\frac{8}{\sqrt{5}-\sqrt{3}}$

65. $\frac{\sqrt{5}}{\sqrt{7}-\sqrt{3}}$

66. $\frac{\sqrt{10}}{\sqrt{5}-2\sqrt{2}}$

67. $\frac{2-\sqrt{6}}{\sqrt{6}-3}$

68. $\frac{\sqrt{10}-5}{2-\sqrt{10}}$

68. $\dfrac{\sqrt{10}}{2}$
69. $-\dfrac{5+\sqrt{21}}{2}$
70. $8+3\sqrt{6}$
71. $\dfrac{2x+y\sqrt{x}-y^2}{x-y^2}$
72. $\dfrac{x^2+4x\sqrt{y}+4y}{x^2-4y}$
73. 4.3397
74. −3.1854
75. 31.6
76. 38.7
77. −57
78. 3
79. −37.3569
80. 47.7788
81. 0.2831
82. 0.4086
83. 0.3820
84. 4.4415
85. $5\sqrt{6}+\sqrt{170}\approx 25.29$ ft
86. $2\sqrt{10}+4\sqrt{5}+\sqrt{30}\approx 20.75$ in.
87. Multiply both the numerator and the denominator by the conjugate of the denominator. This works because multiplying the denominator by its conjugate results in an expression with no square roots. Answers will vary.

69. $\dfrac{\sqrt{3}+\sqrt{7}}{\sqrt{3}-\sqrt{7}}$ **70.** $\dfrac{2\sqrt{3}+\sqrt{2}}{\sqrt{3}-\sqrt{2}}$ **71.** $\dfrac{2\sqrt{x}-y}{\sqrt{x}-y}$ **72.** $\dfrac{x+2\sqrt{y}}{x-2\sqrt{y}}$

Use a graphing calculator to evaluate each expression. Round answers to 4 decimal places, if necessary.

73. $13-\sqrt{75}$ **74.** $5-\sqrt{67}$ **75.** $\sqrt{900}+\sqrt{2.56}$

76. $\sqrt{1600}-\sqrt{1.69}$ **77.** $(\sqrt{7}+8)(\sqrt{7}-8)$ **78.** $(\sqrt{8}-\sqrt{5})(\sqrt{8}+\sqrt{5})$

79. $(2\sqrt{3}+5\sqrt{2})(\sqrt{10}-3\sqrt{5})$ **80.** $(6\sqrt{5}+5\sqrt{7})(3\sqrt{2}-\sqrt{6})$ **81.** $\dfrac{\sqrt{6}}{16-3\sqrt{6}}$

82. $\dfrac{\sqrt{5}}{1+2\sqrt{5}}$ **83.** $\dfrac{\sqrt{10}-\sqrt{2}}{\sqrt{10}+\sqrt{2}}$ **84.** $\dfrac{\sqrt{5}+\sqrt{2}}{\sqrt{5}-\sqrt{2}}$

85. Radio circuits: For a complete radio circuit, $d=\sqrt{2g}+\sqrt{2h}$, where d equals the visual horizon distance and g and h are the heights of the radio antennas at the respective stations. What is d when $g = 75$ ft and $h = 85$ ft?

86. Tile patterns: Mary is making a tile decoration for her wall. Using square tiles of different sizes, Mary created one decoration that is five tiles across, with sides touching. The first tile is 10 in.^{2}, the second is 20 in.^{2}, the third is 30 in.^{2}, the fourth is 20 in.^{2}, and the fifth is 10 in.^{2} What is the length of the decoration?

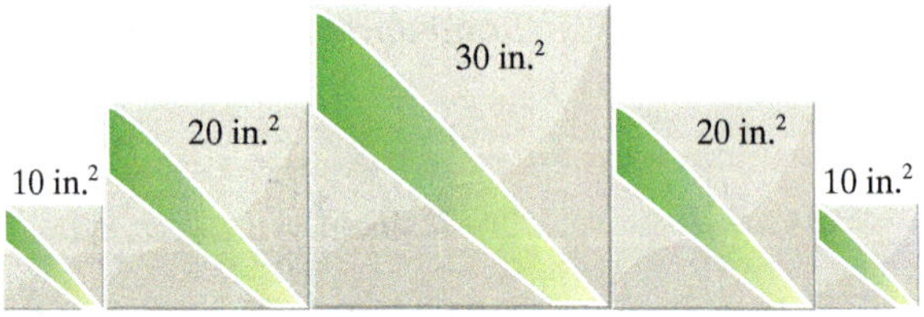

Writing and Thinking About Mathematics

87. In your own words, explain how to rationalize the denominator of a fraction containing the sum or difference of square roots in the denominator. Why does this work?

HAWKES LEARNING SYSTEMS: INTRODUCTORY & INTERMEDIATE ALGEBRA SOFTWARE

- 9.4a Addition and Subtraction with Radicals
- 9.4b Multiplication with Radicals
- 9.4c Rationalizing Denominators

9.5 Equations with Radicals

- *Solve equations that contain one or more radical expressions.*

Teaching Notes:
Up to this point in algebra, students have solved equations by creating a series of equivalent equations. You might want to include a discussion about the procedures of this section that introduce additional variable factors into the series of equations, thereby creating equations that may not be entirely equivalent.

Each of the following equations involves at least one radical expression:

$$x+3=\sqrt{x+5} \qquad \sqrt{x}-\sqrt{2x-14}=1 \qquad \sqrt[3]{x+1}=5.$$

If the radicals are square roots, we solve by squaring both sides of the equations. If the radical is some other root and this root can be isolated on one side of the equation, we solve by raising both sides of the equation to the integer power corresponding to the index of the radical. For example, with a cube root both sides are raised to the third power.

Squaring both sides of an equation may introduce new solutions. For example, the first-degree equation $x=-3$ has only one solution, namely, -3. However, squaring both sides gives the quadratic equation

$$x^2=(-3)^2 \qquad \text{or} \qquad x^2=9.$$

The quadratic equation $x^2=9$ has two solutions, 3 and -3. Thus a new solution that is not a solution to the original equation has been introduced. Such a solution is called an **extraneous solution**.

When both sides of an equation are raised to a power, an extraneous solution may be introduced. Be sure to check all solutions in the original equation.

The following examples illustrate a variety of situations involving radicals. The steps used are related to the following general method.

Method for Solving Equations with Radicals

1. Isolate one of the radicals on one side of the equation. (An equation may have more than one radical.)
2. Raise both sides of the equation to the power corresponding to the index of the radical.
3. If the equation still contains a radical, repeat Steps 1 and 2.
4. Solve the equation after all the radicals have been eliminated.
5. Be sure to check all possible solutions in the original equation and eliminate any extraneous solutions.

Example 1: Equations with One Radical

Solve the following equations.

a. $\sqrt{x^2+13}=7$

Solution: The radical is by itself on one side of the equation, so square both sides.

$$\sqrt{x^2+13}=7$$

$$\left(\sqrt{x^2+13}\right)^2=7^2 \qquad \text{Square both sides.}$$

$$x^2+13=49 \qquad \text{This new equation contains no radical.}$$

$$x^2-36=0$$

$$(x+6)(x-6)=0 \qquad \text{Solve by factoring.}$$

$$x=-6 \quad \text{or} \quad x=6$$

Check: **Check both answers** in the original equation.

$$\sqrt{(-6)^2+13}\stackrel{?}{=}7 \qquad \sqrt{(6)^2+13}\stackrel{?}{=}7$$

$$\sqrt{36+13}\stackrel{?}{=}7 \qquad \sqrt{36+13}\stackrel{?}{=}7$$

$$\sqrt{49}\stackrel{?}{=}7 \qquad \sqrt{49}\stackrel{?}{=}7$$

$$7=7 \qquad 7=7$$

Both –6 and 6 are solutions.

b. $\sqrt{y^2-10y-11}=1+y$

Solution: Since there is only one radical and it is by itself on one side of the equation, square both sides.

$$\sqrt{y^2-10y-11}=1+y$$

$$\left(\sqrt{y^2-10y-11}\right)^2=(1+y)^2 \qquad \text{Square both sides.}$$

$$y^2-10y-11=1+2y+y^2$$

$$-12y-12=0 \qquad \text{Simplifying gives a first-degree equation.}$$

$$-12y=12$$

$$y=-1$$

Check: **Check** in the original equation.

$$\sqrt{(-1)^2-10(-1)-11}\stackrel{?}{=}1+(-1)$$

$$\sqrt{1+10-11}\stackrel{?}{=}0$$

$$\sqrt{0}\stackrel{?}{=}0$$

$$0=0$$

There is one solution, –1.

c. $\sqrt{3x+13}+3=2x$

Solution: $\sqrt{3x+13}+3=2x$

$$\sqrt{3x+13}=2x-3 \quad \text{Isolate the radical.}$$

$$\left(\sqrt{3x+13}\right)^2=(2x-3)^2 \quad \text{Square both sides.}$$

$$3x+13=4x^2-12x+9$$

$$0=4x^2-15x-4$$

$$0=(4x+1)(x-4) \quad \text{Solve by factoring.}$$

$$x=-\frac{1}{4} \quad \text{or} \quad x=4$$

Check: **Check both answers** in the original equation.

$$\sqrt{3\left(-\frac{1}{4}\right)+13}+3\stackrel{?}{=}2\left(-\frac{1}{4}\right) \qquad \sqrt{3(4)+13}+3\stackrel{?}{=}2(4)$$

$$\sqrt{\frac{49}{4}}+3\stackrel{?}{=}-\frac{1}{2} \qquad \sqrt{25}+3\stackrel{?}{=}8$$

$$\frac{7}{2}+3\stackrel{?}{=}-\frac{1}{2} \qquad 5+3\stackrel{?}{=}8$$

$$\frac{13}{2}\neq-\frac{1}{2} \qquad 8=8$$

$-\frac{1}{4}$ is **not** a solution. The only solution is 4.

d. $\sqrt{x+1}=-3$

Solution: We could stop right here. There is **no real solution** to this equation because the radical on the left is nonnegative and cannot possibly equal –3, a negative number. Suppose we did not notice this relationship. Then proceeding as usual, we will find an answer that does not check.

$$\sqrt{x+1}=-3$$

$$\left(\sqrt{x+1}\right)^2=(-3)^2 \quad \text{Square both sides.}$$

$$x+1=9$$

$$x=8$$

Check:

$$\sqrt{(8)+1}\stackrel{?}{=}-3$$

$$\sqrt{9}\stackrel{?}{=}-3$$

$$3\neq-3$$

So 8 does **not** check, and there is **no solution**.

NOTES It is possible that after checking the answers you may find that none of the answers are solutions. In this case the answer is **no solution**.

Example 2: Equations with Two Radicals

Solve the following equations with two radicals. You may need to rearrange terms and to square twice.

a. $\sqrt{x+4}=\sqrt{3x-2}$

Solution: There are two radicals on opposite sides of the equation. Squaring both sides will give a new equation with no radicals.

$$\sqrt{x+4}=\sqrt{3x-2}$$
$$\left(\sqrt{x+4}\right)^2=\left(\sqrt{3x-2}\right)^2$$
$$x+4=3x-2$$
$$6=2x \qquad \text{Simplifying gives a first-degree equation.}$$
$$3=x$$

Check: **Check** in the original equation. $\sqrt{(3)+4}\stackrel{?}{=}\sqrt{3(3)-2}$

$$\sqrt{7}=\sqrt{7}$$

There is one solution, 3.

b. $\sqrt{x}-\sqrt{2x-14}=1$

Solution: Where there is a sum or difference of radicals, squaring is easier if the radicals are on different sides of the equation. Also, squaring both sides of the equation is easier if one of the radicals is by itself on one side of the equation.

$$\sqrt{x}-\sqrt{2x-14}=1$$
$$\sqrt{x}=1+\sqrt{2x-14} \qquad \text{Isolate one of the radicals.}$$
$$\left(\sqrt{x}\right)^2=\left(1+\sqrt{2x-14}\right)^2 \qquad \text{Square both sides.}$$
$$x=1+2\sqrt{2x-14}+(2x-14) \qquad \text{Treat the right-hand side as the square of a binomial.}$$
$$x=2\sqrt{2x-14}+2x-13$$
$$-x+13=2\sqrt{2x-14} \qquad \text{Simplify so that the radical is on one side by itself.}$$

$$(-x+13)^2 = \left(2\sqrt{2x-14}\right)^2 \quad \text{Square both sides again.}$$

$$x^2 - 26x + 169 = 4(2x-14)$$

$$x^2 - 26x + 169 = 8x - 56$$

$$x^2 - 34x + 225 = 0$$

$$(x-9)(x-25) = 0 \quad \text{Solve by factoring.}$$

$$x = 9 \quad \text{or} \quad x = 25$$

Check: **Check both answers** in the original equation.

$$\sqrt{(9)} - \sqrt{2(9)-14} \stackrel{?}{=} 1 \qquad \sqrt{(25)} - \sqrt{2(25)-14} \stackrel{?}{=} 1$$

$$3 - \sqrt{4} \stackrel{?}{=} 1 \qquad 5 - \sqrt{36} \stackrel{?}{=} 1$$

$$3 - 2 \stackrel{?}{=} 1 \qquad 5 - 6 \stackrel{?}{=} 1$$

$$1 = 1 \qquad -1 \neq 1$$

25 is **not** a solution. The only solution is 9.

Example 3: Equation Containing a Cube Root

Solve the following equation containing a cube root: $\sqrt[3]{2x+1} + 1 = 3$

Solution: First, get the radical by itself on one side of the equation. Then since this radical is a cube root, cube both sides of the equation.

$$\sqrt[3]{2x+1} + 1 = 3$$

$$\sqrt[3]{2x+1} = 2 \quad \text{Add } -1 \text{ to both sides.}$$

$$\left(\sqrt[3]{2x+1}\right)^3 = 2^3 \quad \text{Cube both sides.}$$

$$2x + 1 = 8 \quad \text{Solve the equation.}$$

$$x = \frac{7}{2}$$

Check: **Check** in the original equation.

$$\sqrt[3]{2\left(\frac{7}{2}\right)+1} + 1 \stackrel{?}{=} 3$$

$$\sqrt[3]{7+1} + 1 \stackrel{?}{=} 3$$

$$\sqrt[3]{8} + 1 \stackrel{?}{=} 3$$

$$2 + 1 \stackrel{?}{=} 3$$

$$3 = 3$$

There is one solution, $\frac{7}{2}$.

Practice Problems

Solve the following equations.

1. $2\sqrt{x+4} = x+1$
2. $\sqrt{3x+1}+1 = \sqrt{x}$
3. $\sqrt[3]{2x-9}+4 = 3$
4. $\sqrt{2x-5} = -1$

9.5 Exercises

Solve the following equations. Be sure to check your answers in the original equation.

1. $\sqrt{8x+1} = 5$
2. $\sqrt{7x+1} = 6$
3. $\sqrt{3x+4} = -5$
4. $\sqrt{4x-3} = 7$
5. $\sqrt{6-x} = 3$
6. $\sqrt{11-x} = 5$
7. $\sqrt{5x-6} = 8$
8. $\sqrt{2x-5} = -1$
9. $\sqrt{5x+4} = 7$
10. $\sqrt{3x-2} = 4$
11. $\sqrt{x-4}+6 = 2$
12. $\sqrt{6x+4}+2 = 10$
13. $\sqrt{4x+1}+4 = 9$
14. $\sqrt{2x-7}+5 = 3$
15. $\sqrt{x(x+3)} = 2$
16. $\sqrt{x(x-5)} = 6$
17. $\sqrt{x(2x+5)} = 5$
18. $\sqrt{x(3x-14)} = 7$
19. $\sqrt{x+6} = x+4$
20. $\sqrt{x+7} = 2x-1$
21. $\sqrt{x-2} = x-2$
22. $\sqrt{x+3} = x+3$
23. $x-2 = \sqrt{3x-6}$
24. $x+6 = \sqrt{2x+12}$
25. $\sqrt{x^2-16} = 3$
26. $\sqrt{x^2-25} = 12$
27. $5+\sqrt{x+5}-2x = 0$
28. $x-2-\sqrt{x+4} = 0$
29. $2x = \sqrt{7x-3}+3$
30. $x-\sqrt{3x-8} = 4$
31. $\sqrt{2x+5} = \sqrt{4x-1}$
32. $\sqrt{5x-1} = \sqrt{x+7}$
33. $\sqrt{3x+2} = \sqrt{9x-10}$
34. $\sqrt{2-x} = \sqrt{2x-7}$
35. $\sqrt{2x-1} = \sqrt{x+1}$
36. $\sqrt{3x+2} = \sqrt{x+4}$

1. $x=3$ 2. $x=5$
3. No solution
4. $x=13$ 5. $x=-3$
6. $x=-14$ 7. $x=14$
8. No solution
9. $x=9$
10. $x=6$
11. No solution
12. $x=10$ 13. $x=6$
14. No solution
15. $x=-4, 1$
16. $x=-4, 9$
17. $x=-5, \frac{5}{2}$
18. $x=-\frac{7}{3}, 7$
19. $x=-2$
20. $x=2$
21. $x=2, 3$
22. $x=-3, -2$
23. $x=2, 5$
24. $x=-6, -4$
25. $x=-5, 5$
26. $x=-13, 13$
27. $x=4$ 28. $x=5$
29. $x=4$ 30. $x=8$
31. $x=3$ 32. $x=2$
33. $x=2$
34. No solution
35. $x=2$
36. $x=1$

Answers to Practice Problems: **1.** $x=5$ **2.** No solution **3.** $x=4$ **4.** No solution

37. $x=7$ 38. $x=4$
39. $x=4$ 40. $x=5$
41. $x=0$ 42. $x=9$
43. $x=5$ 44. $x=-3$
45. No solution
46. $x=5$ 47. $x=4$
48. $x=4$
49. $x=-1, 3$
50. No solution
51. $x=5$ 52. $x=4$
53. $x=2$ 54. $x=27$
55. $x=1$ 56. $x=11$
57. $x=-4$
58. $x=\frac{727}{9}$
59. $x=12$
60. $x=-18$
61. $x=40$
62. $x=22$
63. $(a+b)^2 = (a+b)(a+b) = a^2+2ab+b^2 \neq a^2+b^2$

37. $\sqrt{x+2}=\sqrt{2x-5}$

38. $\sqrt{2x-5}=\sqrt{3x-9}$

39. $\sqrt{4x-3}=\sqrt{2x+5}$

40. $\sqrt{4x-6}=\sqrt{3x-1}$

41. $\sqrt{3x+1}=1-\sqrt{x}$

42. $\sqrt{x}=\sqrt{x+16}-2$

43. $\sqrt{x+4}=\sqrt{x+11}-1$

44. $\sqrt{1-x}+2=\sqrt{13-x}$

45. $\sqrt{x+1}=\sqrt{x+6}+1$

46. $\sqrt{x+4}=\sqrt{x+20}-2$

47. $\sqrt{x+5}+\sqrt{x}=5$

48. $\sqrt{x}+\sqrt{x-3}=3$

49. $\sqrt{2x+3}=1+\sqrt{x+1}$

50. $\sqrt{5x-18}-4=\sqrt{5x+6}$

51. $\sqrt{3x+1}-\sqrt{x+4}=1$

52. $\sqrt{3x+4}-\sqrt{x+5}=1$

53. $\sqrt{5x-1}=4-\sqrt{x-1}$

54. $\sqrt{2x-5}-2=\sqrt{x-2}$

55. $\sqrt{2x-1}+\sqrt{x+3}=3$

56. $\sqrt{2x+3}-\sqrt{x+5}=1$

57. $\sqrt[3]{4+3x}=-2$

58. $\sqrt[3]{2+9x}=9$

59. $\sqrt[3]{5x+4}=4$

60. $\sqrt[3]{7x+1}=-5$

61. $\sqrt[4]{2x+1}=3$

62. $\sqrt[4]{x-6}=2$

Writing and Thinking About Mathematics

63. Explain, in your own words, why, in general, $(a+b)^2 \neq a^2+b^2$. (Assume $a \neq 0$ and $b \neq 0$.)

HAWKES LEARNING SYSTEMS: INTRODUCTORY & INTERMEDIATE ALGEBRA SOFTWARE

- 9.5 Equations with Radicals

9.6 Functions with Radicals

- *Recognize radical functions.*
- *Evaluate radical functions.*
- *Find the domain and range of radical functions.*
- *Graph radical functions.*

Review of Functions and Function Notation

The concept of functions is among the most important and useful ideas in all of mathematics. Functions were introduced in Chapter 4 along with function notation, such as $f(x)$ (read "*f* of *x*"). In Chapter 4, we discussed **linear functions** and the use of function notation in evaluating functions. Function notation was used again in Chapter 5 where $p(x)$ was used to represent polynomials. In this section the function concept is expanded to include **radical functions** (functions with radicals). The definitions of relations and functions and the vertical line test are restated here for review and easy reference.

Relation, Domain, and Range

A **relation** is a set of ordered pairs of real numbers.

The **domain *D*** of a relation is the set of all first coordinates in the relation.

The **range *R*** of a relation is the set of all second coordinates in the relation.

Function

Teaching Notes: You may want to discuss with your students that both of the characteristics listed (1 and 2) are equivalent to the definition of a function.

A **function** is a relation in which each domain element has exactly one corresponding range element.

The definition can also be stated in the following ways:

1. A function is a relation in which each first coordinate appears only once.
2. A function is a relation in which no two ordered pairs have the same first coordinate.

Vertical Line Test

If **any** vertical line intersects the graph of a relation at more than one point, then the relation is **not** a function.

We have used the ordered pair notation (x, y) to represent points on the graphs of relations and functions. For example, $y = 2x - 5$ represents a linear function and its graph is a line (as shown in the figure below).

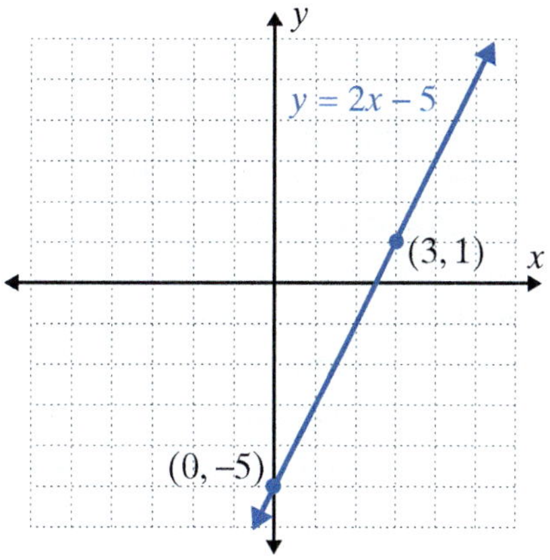

Figure 1

Evaluating Radical Functions

We define **radical functions** (functions with radical expressions) as follows.

Radical Function

A **radical function** is a function of the form $y = \sqrt[n]{g(x)}$ in which the radicand contains a variable expression.
The **domain** of such a function depends on the index, n:

1. If n is an even number, the domain is the set of all x such that $g(x) \geq 0$.
2. If n is an odd number, the domain is the set of all real numbers, $(-\infty, \infty)$.

Examples of radical functions are

$$y = 3\sqrt{x}, \quad f(x) = \sqrt{2x+3}, \quad \text{and} \quad y = \sqrt[3]{x-7}.$$

Example 1: Finding the Domain of a Radical Function

Determine the domain of each radical function.

a. $f(x) = \sqrt{2x+3}$

Solution: Because the index is 2, the radicand must be nonnegative. (That is, the expression under the radical sign cannot be negative.)

Continued on the next page...

Thus we must have

$$2x+3 \geq 0$$
$$2x \geq -3$$
$$x \geq -\frac{3}{2}$$

and the domain is the interval of real numbers $\left[-\frac{3}{2}, \infty\right)$.

b. $y = \sqrt[3]{x-7}$

Solution: Because the index is 3, an odd number, the radicand may be any real number.

Thus the domain is the set of all real numbers $(-\infty, \infty)$.

As illustrated in Chapters 4 and 5, function notation is particularly useful when evaluating functions for specific values of the variable. For example,

if $f(x) = \sqrt{2x+3}$, then $f\left(\frac{1}{2}\right) = \sqrt{2\left(\frac{1}{2}\right)+3} = \sqrt{1+3} = \sqrt{4} = 2$

and $f(0) = \sqrt{2(0)+3} = \sqrt{0+3} = \sqrt{3}$.

A calculator can be used to find decimal approximations. Such approximations are helpful when estimating the locations of points on a graph. For example,

if $f(x) = \sqrt{x-5}$,

then $f(8) = \sqrt{(8)-5} = \sqrt{3} \approx 1.7321$ Accurate to 4 decimal places

and $f(25) = \sqrt{(25)-5} = \sqrt{20} \approx 4.4721$. Accurate to 4 decimal places

Example 2: Evaluating Radical Functions

Complete each table by finding the corresponding $f(x)$ values for the given values of x.

a. $f(x) = 3\sqrt{x}$

x	$f(x)$
0	?
4	?
6	?

b. $f(x) = \sqrt[3]{x-7}$

x	$f(x)$
7	?
6	?
−1	?

Solutions:

a.

x	$f(x)$
0	$3\sqrt{0}=0$
4	$3\sqrt{4}=3\cdot 2=6$
6	$3\sqrt{6}\approx 7.3485$

b.

x	$f(x)$
7	$\sqrt[3]{7-7}=\sqrt[3]{0}=0$
6	$\sqrt[3]{6-7}=\sqrt[3]{-1}=-1$
–1	$\sqrt[3]{-1-7}=\sqrt[3]{-8}=-2$

Graphing Radical Functions

To graph a radical function, we need to be aware of its domain and to plot at least a few points to see the nature of the resulting curve. Example 3 shows how to proceed, at least in the beginning, to graph the radical function $y=\sqrt{x+5}$.

Example 3: Graphing a Radical Function

Graph the function $y=\sqrt{x+5}$.

Solution: For the domain we have

$$x+5\geq 0$$
$$x\geq -5.$$

To see the nature of the graph we select a few values for x in the domain and find the corresponding values of y. Then we plot the points on a graph.

x	y
–5	0
–4	1
–3	$\sqrt{2}\approx 1.41$
0	$\sqrt{5}\approx 2.24$
4	3

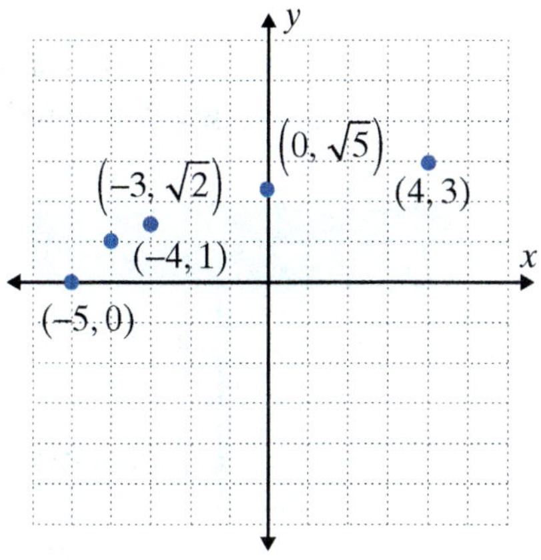

Continued on the next page...

Teaching Notes:
In preparation for the lesson on function notation and translations, you may want to show students the graph of $y = \sqrt{x}$ and point out how the graphs of other radical functions are similar to this basic graph.

Note that $\sqrt{x+5}$ is the principal square root. This means that $y \geq 0$. Thus the point $(-5, 0)$ is on the x-axis and the remaining points on the graph are above the x-axis. So, we can complete the graph by drawing a smooth curve that passes through the selected points. The graph of the function is shown here.

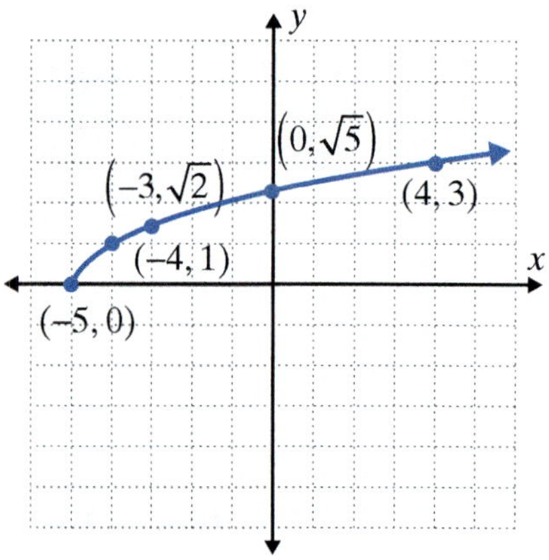

We see that the domain is $[-5, \infty)$ and the range is $[0, \infty)$.

To use a TI-84 Plus graphing calculator to graph this function,

Step 1: Press Y= and enter the function as shown:

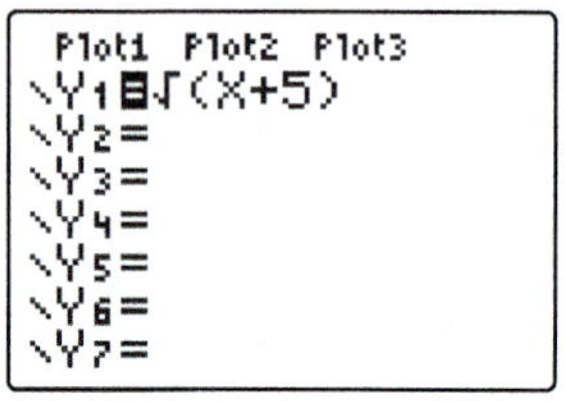

Step 2: Press GRAPH. (You may need to adjust the window.) The result will be the graph as shown here:

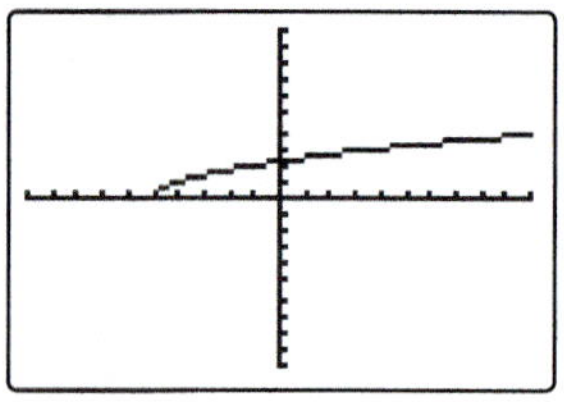

Example 4 shows how to use a TI-84 Plus graphing calculator to find many points on the graph of a radical function and then how to graph the function.

Example 4: Using a TI-84 Plus to Graph a Radical Function

a. Use the TABLE feature of a TI-84 Plus graphing calculator to locate many points on the graph of the function $y = \sqrt[3]{2x-3}$.

Solution: Using the TABLE feature of a TI-84 Plus:

Step 1: Press Y= and enter the function as follows:

1. Press MATH.
2. Choose 4: $\sqrt[3]{\ }$(.
3. Enter 2x – 3) and press ENTER.

Step 2: Press TBLSET (which is 2ND WINDOW) and set the display as shown here:

Step 3: Press TABLE (which is 2ND GRAPH) and the display will appear as follows:

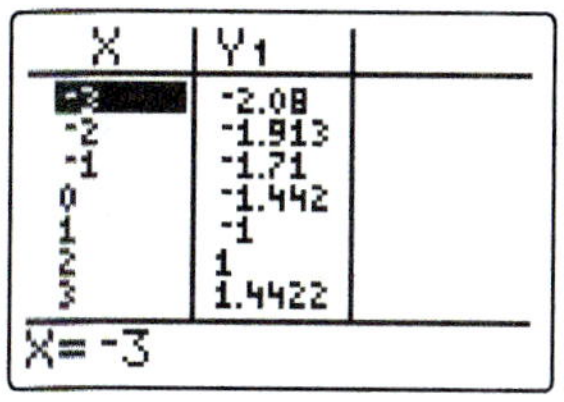

b. Plot several points (approximately) on a graph and then connect them with a smooth curve.

Solution: Once your calculator is displaying the table, you may scroll up and down the display to find as many points as you like. A few are shown here to see the nature of the graph.

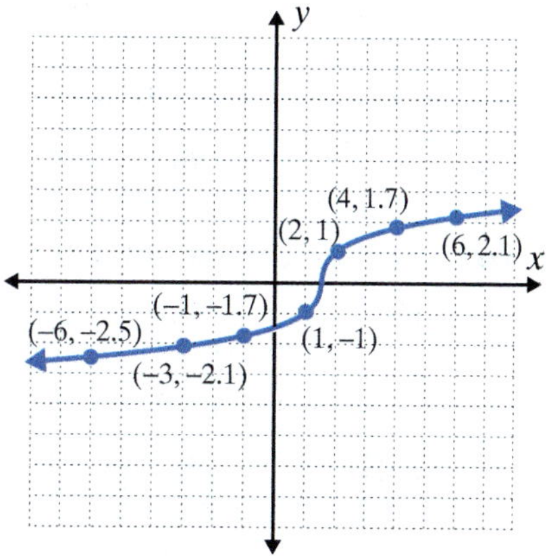

c. Use a TI-84 Plus graphing calculator to graph the function.

Solution: Press GRAPH and the display will appear with the curve as follows:

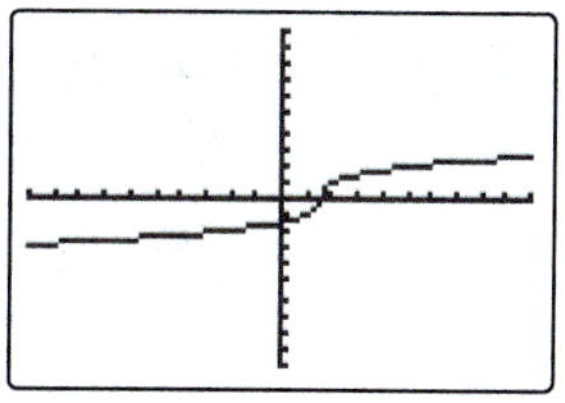

Practice Problems

Find the indicated value of the function.

1. For $f(x)=\sqrt{x+7}$, find $f(5)$.
2. For $g(x)=\sqrt{3x-1}$, find $g\left(\frac{2}{3}\right)$.
3. For $h(x)=\sqrt[3]{x+5}$, find $h(-13)$.

Use a calculator to estimate the value of the function accurate to 4 decimal places.

4. Estimate $f(2)$ for $f(x)=\sqrt{4x-1}$.
5. Estimate $g(-3)$ for $g(x)=\sqrt[3]{1-3x}$.

9.6 Exercises

Find each function value as indicated and write the answers in both radical notation and decimal notation. (If necessary, round decimal values to 4 decimal places.)

1. Given $f(x)=\sqrt{2x+1}$, find
   a. $f(2)$ b. $f(4)$ c. $f(24.5)$ d. $f(1.5)$
2. Given $f(x)=\sqrt{5-3x}$, find
   a. $f(0)$ b. $f(-2)$ c. $f\left(-\frac{20}{3}\right)$ d. $f(-2.4)$
3. Given $g(x)=\sqrt[3]{x+6}$, find
   a. $g(21)$ b. $g(-7)$ c. $g(-14)$ d. $g(18)$
4. Given $h(x)=\sqrt[3]{4-x}$, find
   a. $h(4)$ b. $h(-4)$ c. $h(3.999)$ d. $h(-2.5)$

Use interval notation to indicate the domain of each radical function.

5. $y=\sqrt{x+8}$
6. $y=\sqrt{2x-1}$
7. $y=\sqrt{2.5-5x}$
8. $y=\sqrt{1-3x}$
9. $f(x)=\sqrt[3]{x+4}$
10. $f(x)=\sqrt[3]{6x}$
11. $g(x)=\sqrt[4]{x}$
12. $g(x)=\sqrt[4]{7-x}$
13. $y=\sqrt[5]{4x-1}$
14. $y=\sqrt[5]{8+x}$

1. a. $\sqrt{5}\approx 2.2361$ b. 3 c. $5\sqrt{2}\approx 7.0711$ d. 2
2. a. $\sqrt{5}\approx 2.2361$ b. $\sqrt{11}\approx 3.3166$ c. 5 d. $\sqrt{12.2}\approx 3.4928$
3. a. 3 b. −1 c. −2 d. $2\sqrt[3]{3}\approx 2.8845$
4. a. 0 b. 2 c. 0.1 d. $\sqrt[3]{6.5}\approx 1.8663$
5. $[-8,\infty)$
6. $\left[\frac{1}{2},\infty\right)$
7. $\left(-\infty,\frac{1}{2}\right]$
8. $\left(-\infty,\frac{1}{3}\right]$
9. $(-\infty,\infty)$
10. $(-\infty,\infty)$

Answers to Practice Problems: 1. $2\sqrt{3}$ 2. 1 3. −2 4. 2.6458 5. 2.1544

11. $[0, \infty)$
12. $(-\infty, 7]$
13. $(-\infty, \infty)$
14. $(-\infty, \infty)$
15. E 16. D
17. B 18. F
19. A 20. C

21. 

22. 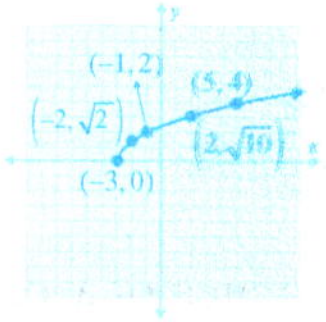

23. 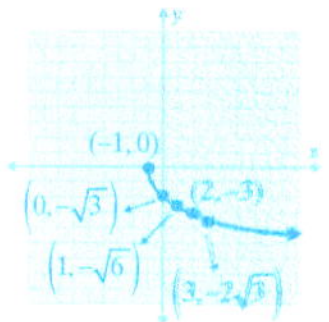

24. 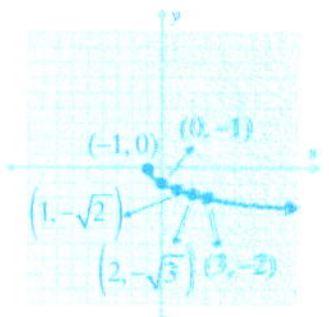

25. 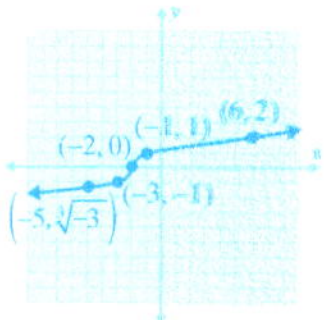

26. 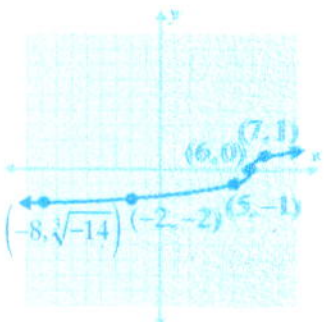

27. 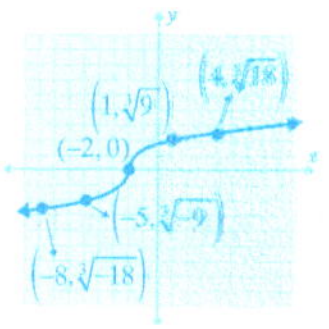

Match the functions given with the graphs of the functions (A) – (F).

15. $y = \sqrt{x-2}$ **16.** $y = \sqrt{2-x}$ **17.** $y = -\sqrt{x-3}$

18. $y = -\sqrt{3-x}$ **19.** $y = \sqrt{x+4}$ **20.** $y = \sqrt{x-4}$

A. 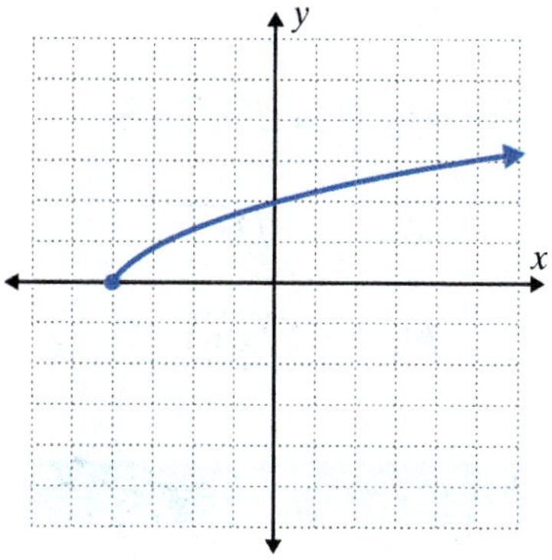

B. 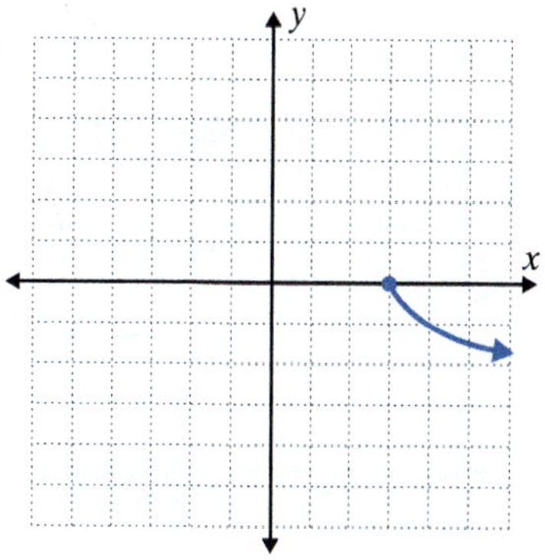

C. 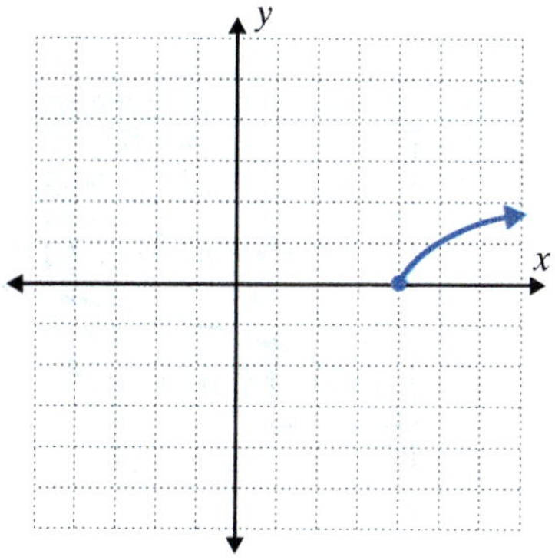

D. 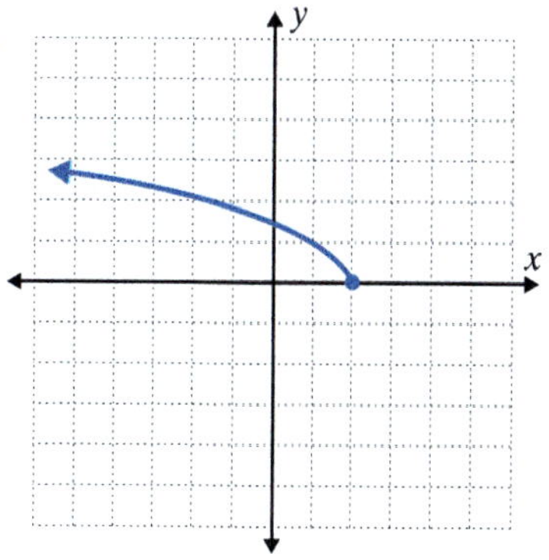

E. 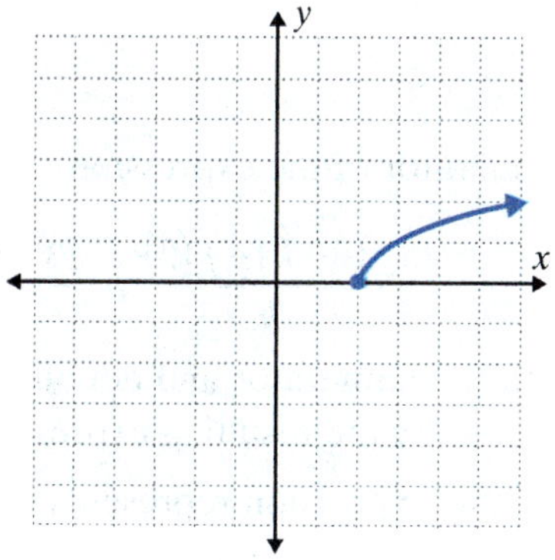

F. 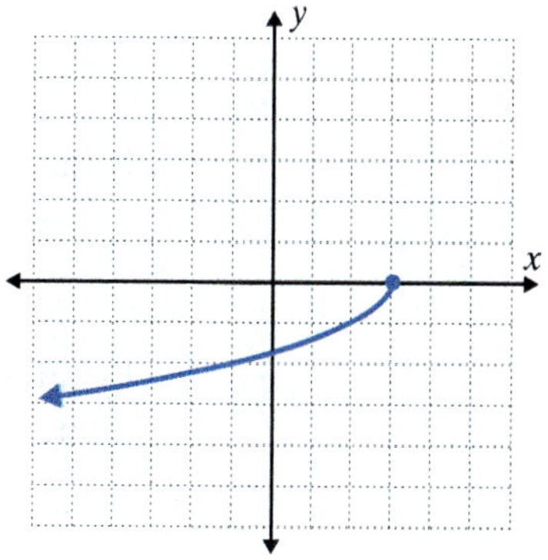

Find and label at least 5 points on the graph of the function and then sketch the graph of the function.

21. $y = \sqrt{x-1}$ **22.** $y = \sqrt{2x+6}$ **23.** $f(x) = -\sqrt{3x+3}$

24. $h(x) = -\sqrt{x+1}$ **25.** $f(x) = \sqrt[3]{x+2}$ **26.** $g(x) = \sqrt[3]{x-6}$

27. $y = \sqrt[3]{3x+6}$ **28.** $y = \sqrt[3]{2x-4}$

Use a graphing calculator to graph each of the functions.

29. $y = 3\sqrt{x+2}$ **30.** $y = 2\sqrt{3-x}$ **31.** $g(x) = -\sqrt{2x}$

32. $f(x) = \sqrt{3x}$ **33.** $f(x) = -\sqrt{x+4}$ **34.** $f(x) = -\sqrt{5-x}$

35. $y = -\sqrt[3]{x+2}$ **36.** $y = -\sqrt[3]{3x+4}$ **37.** $g(x) = -\sqrt[4]{x+5}$

38. $y = \sqrt[4]{2x+6}$ **39.** $y = \sqrt[5]{2x+1}$ **40.** $y = \sqrt[5]{x+7}$

28.
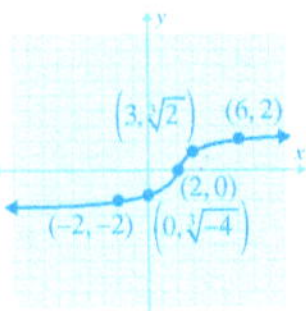

29.
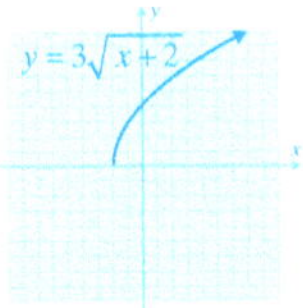

30.
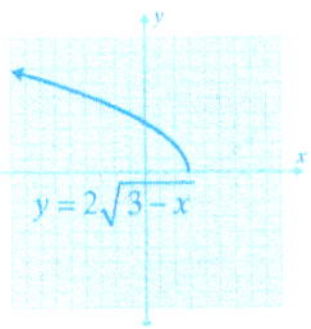

Writing and Thinking About Mathematics

41. The graph of the radical function $f(x) = \sqrt{x}$ is shown with two values of x on the x-axis, 3 and $3 + h$.

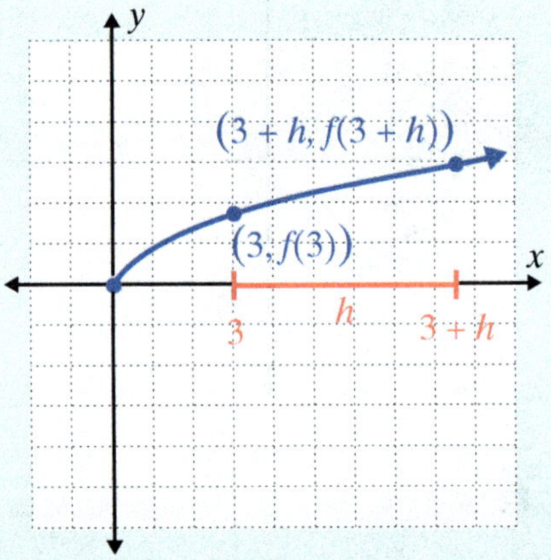

a. Rationalize the numerator of the expression

$$\frac{f(3+h)-f(3)}{h} = \frac{\sqrt{3+h}-\sqrt{3}}{h}$$

by multiplying both the numerator and denominator by the conjugate of the numerator. Then simplify the resulting expression.

b. What do you think this expression represents graphically? (**Hint:** Two points determine a line.)

c. Using your results from parts **a.** and **b.**, what do you see happening on the graph if the value of h shrinks slowly to 0?

d. Using your analysis from part **c.**, what happens to the value of your simplified expression in part **a.** and what do you think this value represents?

31.
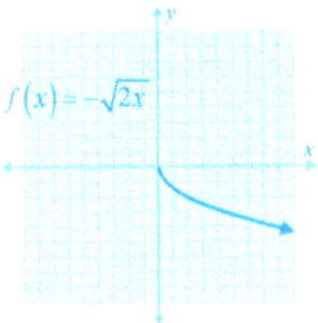

32.

33.
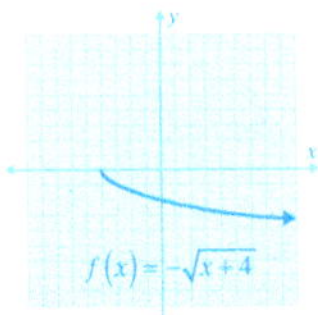

34.

35.
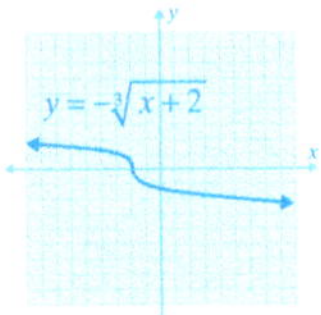

36.
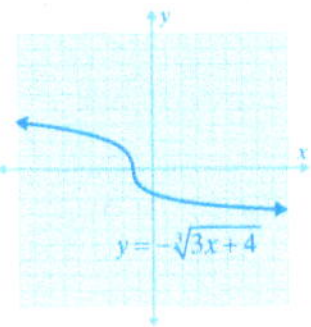

HAWKES LEARNING SYSTEMS: INTRODUCTORY & INTERMEDIATE ALGEBRA SOFTWARE

- 9.6 Functions with Radicals

37. 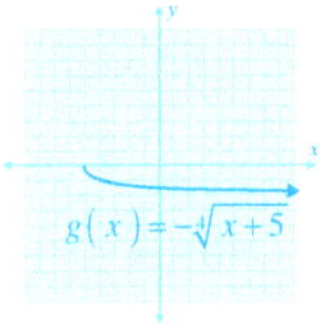

38. 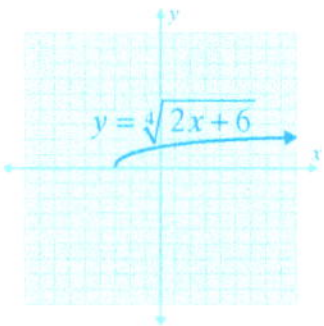

39. 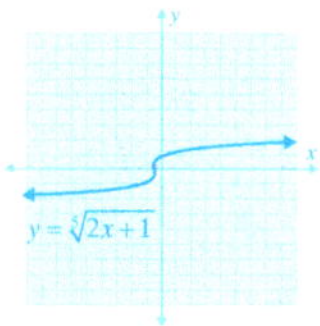

40. 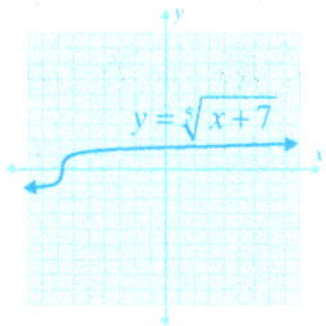

41. a. $\dfrac{1}{\sqrt{3+h}+\sqrt{3}}$ b. Slope of the line connecting $(3+h, f(3+h))$ and $(3, f(3))$

c. A line just touching the curve at one point.

d. $\dfrac{1}{2\sqrt{3}}$; represents the slope of the line tangent to $f(x)$ at $x = 3$.

9.7 Introduction to Complex Numbers

- *Simplify square roots of negative numbers.*
- *Identify the real parts and the imaginary parts of complex numbers.*
- *Solve equations with complex numbers by setting the real parts and the imaginary parts equal to each other.*
- *Add and subtract with complex numbers.*

Introduction to Complex Numbers and the Number *i*

One of the properties of real numbers is that the square of any real number is nonnegative. That is, for any real number x, $x^2 \geq 0$. The square roots of negative numbers, such as $\sqrt{-4}$ and $\sqrt{-5}$, are not real numbers. However, they can be defined as part of the system of **complex numbers**.

Complex numbers include all the real numbers and the even roots of negative numbers. In Chapter 10, we will see how these numbers occur as solutions to quadratic equations. At first such numbers seem to be somewhat impractical because they are difficult to picture in any type of geometric setting and they are not solutions to the types of word problems that are familiar. However, complex numbers do occur quite naturally in trigonometry and higher level mathematics and have practical applications in such fields as electrical engineering.

The first step in the development of complex numbers is to define $\sqrt{-1}$.

i and i^2

$$\boldsymbol{i = \sqrt{-1}} \quad \text{and} \quad \boldsymbol{i^2 = \left(\sqrt{-1}\right)^2 = -1}$$

Using the definition of $\sqrt{-1}$, the following definition for the square root of a negative number can be made.

$\sqrt{-a} = \sqrt{a}\,i$

If a is a positive real number, then

$$\boldsymbol{\sqrt{-a} = \sqrt{a} \cdot \sqrt{-1} = \sqrt{a}\,i}.$$

Note: The number i is not under the radical sign. To avoid confusion, we sometimes write $i\sqrt{a}$.

Example 1: $\sqrt{-a}$

Simplify the following radicals.

a. $\sqrt{-25} = \sqrt{-1}\sqrt{25} = i \cdot 5 = 5i$

$(5i)^2 = 5^2 i^2 = 25(-1) = -25$

b. $\sqrt{-36} = \sqrt{-1}\sqrt{36} = i \cdot 6 = 6i$

c. $\sqrt{-24} = \sqrt{-1}\sqrt{4 \cdot 6} = i \cdot 2 \cdot \sqrt{6} = 2\sqrt{6}\,i$ (or $2i\sqrt{6}$)

d. $\sqrt{-45} = \sqrt{-1}\sqrt{9 \cdot 5} = i \cdot 3 \cdot \sqrt{5} = 3\sqrt{5}\,i$ (or $3i\sqrt{5}$)

We can write $2\sqrt{6}\,i$ and $3\sqrt{5}\,i$ as long as we take care not to include the i under the radical sign.

Complex Numbers

The **standard form** of a **complex number** is $\boldsymbol{a + bi}$, where a and b are real numbers. a is called the **real part** and b is called the **imaginary part**.

If $b = 0$, then $a + bi = a + 0i = a$ is a **real number**.

If $a = 0$, then $a + bi = 0 + bi = bi$ is called a **pure imaginary number** (or an **imaginary number**).

Complex Number: $a + bi$

real part (a) imaginary part (b)

NOTES

The term "imaginary" is somewhat misleading. Complex numbers and imaginary numbers are no more "imaginary" than any other type of number. In fact, all the types of numbers that we have studied (whole numbers, integers, rational numbers, irrational numbers, and real numbers) are products of human imagination.

Example 2: Real and Imaginary Parts

Identify the real and imaginary parts of each complex number.

a. $4 - 2i$ — 4 is the real part; -2 is the imaginary part.

b. $\dfrac{5+2i}{3}$ — $\dfrac{5+2i}{3} = \dfrac{5}{3} + \dfrac{2}{3}i$ in standard form. Thus $\dfrac{5}{3}$ is the real part; $\dfrac{2}{3}$ is the imaginary part.

c. 7 — $7 = 7 + 0i$ in standard form. Thus 7 is the real part; 0 is the imaginary part. (Remember, if $b = 0$, the complex number is a real number.)

Continued on the next page...

d. $-\sqrt{3}i$ $\quad$ $-\sqrt{3}i = 0 - \sqrt{3}i$ in standard form. Thus 0 is the real part; $-\sqrt{3}$ is the imaginary part. (If $a = 0$ and $b \neq 0$, then the complex number is a pure imaginary number.)

In general, if a is a real number, then we can write $a = a + 0i$. This means that a is a complex number. Thus **every real number is a complex number**. Figure 1 illustrates the relationships among the various types of numbers we have studied.

Teaching Notes:
There are alternative models that illustrate the subset relationship with intersecting and non-intersecting sets of numbers. Students may not pick up on the subset relationships shown in Figure 1, and you may know of another diagram that clarifies the subset relationships. Seeing more than one approach helps many students.

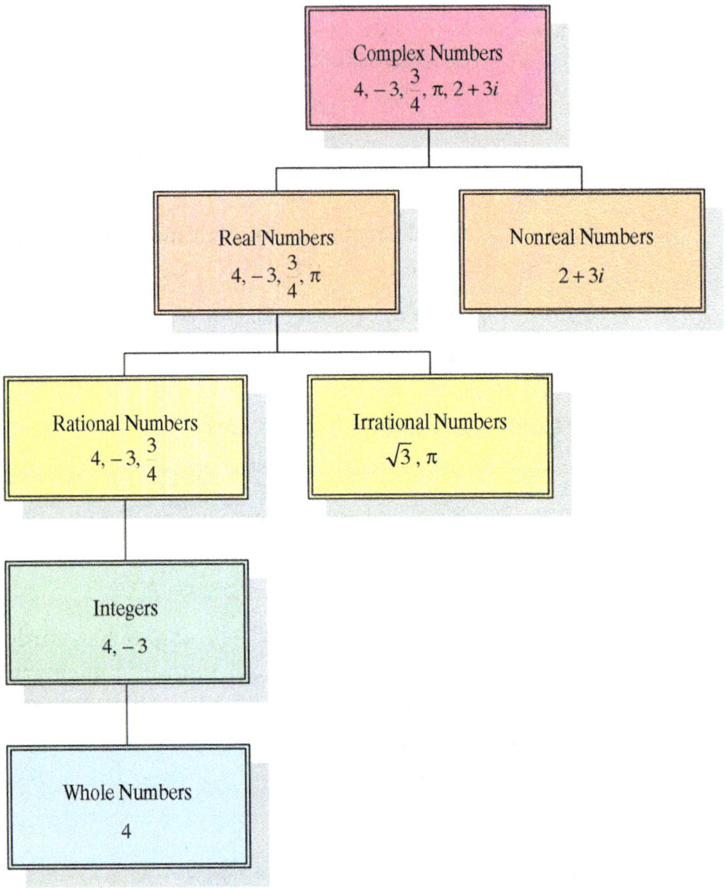

Figure 1

If two complex numbers are equal, then the real parts are equal and the imaginary parts are equal. For example, if

$$x + yi = 7 + 2i$$

then

$$x = 7 \quad \text{and} \quad y = 2.$$

This relationship can be used to solve equations involving complex numbers.

Equality of Complex Numbers

For complex numbers $a + bi$ and $c + di$,

if $\boldsymbol{a + bi = c + di}$, then $\boldsymbol{a = c}$ and $\boldsymbol{b = d}$.

Example 3: Solving Equations

Solve each equation for x and y.

a. $(x+3)+2yi = 7-6i$

Solution: Equate the real parts and the imaginary parts, and solve the resulting equations.

$$x+3=7 \qquad \text{and} \qquad 2y=-6$$
$$x=4 \qquad\qquad\qquad y=-3$$

b. $2y + 3 - 8i = 9 + 4xi$

Solution: Equate the real parts and the imaginary parts, and solve the resulting equations.

$$2y+3=9 \qquad \text{and} \qquad -8=4x$$
$$2y=6 \qquad\qquad\qquad -2=x$$
$$y=3$$

Addition and Subtraction with Complex Numbers

Adding and subtracting complex numbers is similar to adding and subtracting polynomials. Simply combine like terms. For example,

$$\begin{aligned}(2+3i)+(9-8i) &= 2+9+3i-8i \\ &= (2+9)+(3-8)i \\ &= 11-5i.\end{aligned}$$

Similarly,

$$\begin{aligned}(5-2i)-(6+7i) &= 5-2i-6-7i \\ &= (5-6)+(-2-7)i \\ &= -1-9i.\end{aligned}$$

Addition and Subtraction with Complex Numbers

For complex numbers $a + bi$ and $c + di$,

$$(a+bi)+(c+di)=(a+c)+(b+d)i$$

and

$$(a+bi)-(c+di)=(a-c)+(b-d)i.$$

Example 4: Addition and Subtraction with Complex Numbers

Find each sum or difference as indicated.

a. $(6-2i)+(1-2i)$

Solution: $(6-2i)+(1-2i)=(6+1)+(-2-2)i$

$$=7-4i$$

b. $(-8-\sqrt{2}\,i)-(-8+\sqrt{2}\,i)$

Solution: $(-8-\sqrt{2}\,i)-(-8+\sqrt{2}\,i)=(-8-(-8))+(-\sqrt{2}-\sqrt{2})i$

$$=(-8+8)+(-2\sqrt{2})i$$

$$=0-2\sqrt{2}i$$

$$=-2\sqrt{2}\,i\ (\text{or } -2i\sqrt{2})$$

c. $(\sqrt{3}-2i)+(1+\sqrt{5}\,i)$

Solution: $(\sqrt{3}-2i)+(1+\sqrt{5}\,i)=(\sqrt{3}+1)+(-2+\sqrt{5})i$

$$=(\sqrt{3}+1)+(\sqrt{5}-2)i$$

Note: Here, the coefficients do not simplify. This means that the real part is $\sqrt{3}+1$ and the imaginary part is $\sqrt{5}-2$.

1. Real part is 4, imaginary part is –3
2. Real part is $\frac{3}{4}$, imaginary part is 1
3. Real part is –11, imaginary part is $\sqrt{2}$
4. Real part is 6, imaginary part is $\sqrt{3}$
5. Real part is $\frac{3}{8}$, imaginary part is 0
6. Real part is 0, imaginary part is $\frac{4}{7}$
7. Real part is $\frac{4}{5}$, imaginary part is $\frac{7}{5}$
8. Real part is $\frac{1}{2}$, imaginary part is $-\frac{1}{4}$
9. Real part is $\frac{2}{3}$, imaginary part is $\sqrt{17}$
10. Real part is $-\sqrt{5}$, imaginary part is $\frac{\sqrt{2}}{2}$

11. $7i$ 12. $11i$
13. $-8i$ 14. $-13i$
15. $7\sqrt{3}$ 16. $8\sqrt{2}$
17. $10i\sqrt{6}$
18. $12i\sqrt{11}$
19. $-12i\sqrt{3}$
20. $10\sqrt{7}$ 21. $11\sqrt{2}$
22. $8i\sqrt{3}$ 23. $10i\sqrt{10}$
24. $9i\sqrt{3}$
25. $x=6, y=-3$
26. $x=-1, y=-2$
27. $x=-2, y=\sqrt{5}$
28. $x=\frac{\sqrt{2}}{3}, y=-3$
29. $x=\sqrt{2}-3, y=1$
30. $x=-3, y=4+\sqrt{5}$
31. $x=2, y=-8$
32. $x=-4, y=\frac{1}{2}$
33. $x=2, y=-6$
34. $x=1, y=4$
35. $x=3, y=10$
36. $x=-3, y=7$
37. $x=-\frac{4}{3}, y=-3$
38. $x=-2, y=1$
39. $6+2i$ 40. $10+5i$
41. $1+7i$ 42. -9
43. $6-6i$ 44. $13+3i$
45. $14i$ 46. $-8-2i$
47. $(3+\sqrt{5})-6i$
48. $4-\sqrt{2}$
49. $5+(\sqrt{6}+1)i$

Practice Problems

1. Find the real part and the imaginary part of $2-\sqrt{39}i$.

Add or subtract as indicated. Simplify your answers.

2. $(-7+\sqrt{3}i)+(5-2i)$ **3.** $(4+i)-(5+2i)$

Solve for x and y.

4. $x+yi=\sqrt{2}-7i$ **5.** $3y+(x-7)i=-9+2i$

9.7 Exercises

Find the real part and the imaginary part of each of the complex numbers.

1. $4-3i$ **2.** $\frac{3}{4}+i$ **3.** $-11+\sqrt{2}i$ **4.** $6+\sqrt{3}i$ **5.** $\frac{3}{8}$

6. $\frac{4}{7}i$ **7.** $\frac{4+7i}{5}$ **8.** $\frac{2-i}{4}$ **9.** $\frac{2}{3}+\sqrt{17}i$ **10.** $-\sqrt{5}+\frac{\sqrt{2}}{2}i$

Simplify the following radicals.

11. $\sqrt{-49}$ **12.** $\sqrt{-121}$ **13.** $-\sqrt{-64}$ **14.** $-\sqrt{-169}$ **15.** $\sqrt{147}$

16. $\sqrt{128}$ **17.** $2\sqrt{-150}$ **18.** $4\sqrt{-99}$ **19.** $-2\sqrt{-108}$ **20.** $2\sqrt{175}$

21. $\sqrt{242}$ **22.** $\sqrt{-192}$ **23.** $\sqrt{-1000}$ **24.** $\sqrt{-243}$

Solve the equations for x and y.

25. $x+3i=6-yi$ **26.** $2x-8i=-2+4yi$

27. $\sqrt{5}-2i=y+xi$ **28.** $\sqrt{2}-2yi=3x+6i$

29. $\sqrt{2}+i-3=x+yi$ **30.** $\sqrt{5}i-3+4i=x+yi$

31. $2x+3+6i=7-yi-2i$ **32.** $x+yi+8=2i+4-3yi$

33. $x+2i=5-yi-3-4i$ **34.** $3x+2-7i=i-2yi+5$

35. $2+3i+x=5-7i+yi$ **36.** $11i-2x+4=10-3i+2yi$

Answers to Practice Problems: **1.** Real part is 2, imaginary part is $-\sqrt{39}$ **2.** $-2+(\sqrt{3}-2)i$ **3.** $-1-i$
4. $x=\sqrt{2}$ and $y=-7$ **5.** $x=9$ and $y=-3$

50. $\left(\sqrt{11}+5\right)-5i$

51. $\sqrt{3}-5$

52. $\left(\sqrt{5}+1\right)+\left(\sqrt{3}-1\right)i$

53. $-2-5i$ 54. $5+13i$

55. $11-16i$ 56. $4+13i$

57. 2 58. $2-7i$

59. $3+4i$ 60. $-4+10i$

61. a. Yes b. No

62. Answers will vary.

a. Do: $\left\{0, -2, 3, \frac{4}{7}, \sqrt{16}\right\}$; Do not: $\left\{\sqrt{5}, \pi, \sqrt{99}, \sqrt{-4}, i\right\}$

b. Do: $\{-8, -2, 0, 1, 58\}$; Do not: $\left\{2.3, \sqrt{6}, \frac{1}{2}, 3+i, 2\pi\right\}$

c. Do: $\left\{-\sqrt{3}, 0, \frac{5}{6}, 4, \pi^2\right\}$; Do not: $\left\{-i, 2-3i, 4+i, \sqrt{-2}, 9i\right\}$

d. Do: $\left\{-5i, -\sqrt{-4}, i, 3i, \sqrt{-36}\right\}$; Do not: $\{7, 2+i, -3, 4+4i, \pi\}$

e. Do: $\left\{-2, 3+i, 4-\pi i, 55i, \frac{61}{100}\right\}$; Do not: None

f. Do: $\left\{\pi, \sqrt{3}, \sqrt{5}+\sqrt{6}, -\sqrt{11}, 3\pi+4\right\}$; Do not: $\left\{4, 1.1, -\frac{3}{5}, \sqrt{9}, 0\right\}$

37. $2x-2yi+6=6i-x+2$

38. $x+4-3x+i=8+yi$

Find each sum or difference as indicated.

39. $(2+3i)+(4-i)$

40. $(7-i)+(3+6i)$

41. $(4+5i)-(3-2i)$

42. $(-3+2i)-(6+2i)$

43. $(4-3i)+(2-3i)$

44. $(7+5i)+(6-2i)$

45. $(8+9i)-(8-5i)$

46. $(-6+i)-(2+3i)$

47. $\left(\sqrt{5}-2i\right)+(3-4i)$

48. $(4+3i)-\left(\sqrt{2}+3i\right)$

49. $\left(7+\sqrt{6}\,i\right)+(-2+i)$

50. $\left(\sqrt{11}+2i\right)+(5-7i)$

51. $\left(\sqrt{3}+\sqrt{2}\,i\right)-\left(5+\sqrt{2}\,i\right)$

52. $\left(\sqrt{5}+\sqrt{3}\,i\right)+(1-i)$

53. $\left(5+\sqrt{-25}\right)-\left(7+\sqrt{-100}\right)$

54. $\left(1+\sqrt{-36}\right)-\left(-4-\sqrt{-49}\right)$

55. $\left(13-3\sqrt{-16}\right)+\left(-2-4\sqrt{-1}\right)$

56. $\left(7+\sqrt{-9}\right)-\left(3-2\sqrt{-25}\right)$

57. $(4+i)+(-3-2i)-(-1-i)$

58. $(-2-3i)+(6+i)-(2+5i)$

59. $(7+3i)+(2-4i)-(6-5i)$

60. $(-5+7i)+(4-2i)-(3-5i)$

Writing and Thinking About Mathematics

61. Answer the following questions and give a brief explanation of your answer.

a. Is every real number a complex number?

b. Is every complex number a real number?

62. List 5 numbers that do and 5 numbers that do not fit each of the following categories (if possible).

a. rational number
b. integer
c. real number
d. pure imaginary number
e. complex number
f. irrational number

Hawkes Learning Systems: Introductory & Intermediate Algebra Software

- 9.7 Complex Numbers

9.8 Multiplication and Division with Complex Numbers

- *Multiply with complex numbers.*
- *Divide with complex numbers.*
- *Simplify powers of i.*

Multiplication with Complex Numbers

The product of two complex numbers can be found by using the FOIL method for multiplying two binomials. (See Section 5.6.) **Remember that $i^2 = -1$.** For example,

$$(3+5i)(2+i) = 6+3i+10i+5i^2$$
$$= 6+13i-5 \qquad 5i^2 = 5(-1) = -5$$
$$= 1+13i.$$

Example 1: Multiplication with Complex Numbers

Find the following products.

a. $(3i)(2-7i)$

Solution: $(3i)(2-7i) = 6i-21i^2$

$$= 6i-21(-1)$$
$$= 21+6i$$

b. $(5+i)(2+6i)$

Solution: $(5+i)(2+6i) = 5(2+6i)+i(2+6i)$

$$= 10+30i+2i+6i^2$$
$$= 10+32i-6$$
$$= 4+32i$$

c. $(\sqrt{2}-i)(\sqrt{2}-i)$

Solution: $(\sqrt{2}-i)(\sqrt{2}-i) = (\sqrt{2})^2 - \sqrt{2}\cdot i - \sqrt{2}\cdot i + i^2$

$$= 2-2\sqrt{2}\,i-1 \qquad \text{Remember } i^2 = -1.$$
$$= 1-2i\sqrt{2}$$

Continued on the next page...

d. $(-1+i)(2-i)$

Solution: $(-1+i)(2-i) = -2+i+2i-i^2$

$= -2+3i+1$

$= -1+3i$

NOTES

COMMON ERROR

Remember that $\sqrt{a}\cdot\sqrt{b} = \sqrt{ab}$ **only** if a and b are nonnegative real numbers. Applying this rule to negative real numbers can lead to an error. The error can be avoided by first changing the radicals to imaginary form.

INCORRECT	CORRECT
$\sqrt{-6}\cdot\sqrt{-2} = \sqrt{12}$	$\sqrt{-6}\cdot\sqrt{-2} = \sqrt{6}\,i\cdot\sqrt{2}\,i$
$= \sqrt{4}\cdot\sqrt{3}$	$= \sqrt{12}\,i^2$
$= 2\sqrt{3}$	$= 2\sqrt{3}(-1)$
	$= -2\sqrt{3}$

Division with Complex Numbers

The two complex numbers $a + bi$ and $a - bi$ are called **complex conjugates** or simply **conjugates** of each other. As the following steps show, **the product of two complex conjugates will always be a nonnegative real number**.

$$(a+bi)(a-bi) = a^2 - abi + abi - b^2i^2$$
$$= a^2 - b^2i^2$$
$$= a^2 + b^2$$

The resulting product, $a^2 + b^2$, is a real number, and it is nonnegative since it is the sum of the squares of real numbers.

Remember that the form $a + bi$ is called the **standard form** of a complex number. The standard form allows for easy identification of the real and imaginary parts. Thus

$$\frac{1+3i}{5} = \frac{1}{5} + \frac{3}{5}i \text{ in standard form.}$$

The real part is $\frac{1}{5}$ and the imaginary part is $\frac{3}{5}$.

To write the fraction $\frac{1+i}{2-3i}$ in standard form, multiply both the numerator and denominator by $2+3i$ and simplify. This will give a positive real number in the denominator.

$$\frac{1+i}{2-3i}=\frac{(1+i)(2+3i)}{(2-3i)(2+3i)}$$ $2+3i$ is the conjugate of the denominator.

$$=\frac{2+3i+2i+3i^2}{2^2+6i-6i-3^2i^2}$$

$$=\frac{2+5i-3}{4-(9)(-1)}$$ Reminder: $i^2=-1$.

$$=\frac{-1+5i}{13}$$

$$=-\frac{1}{13}+\frac{5}{13}i$$

To Write a Fraction with Complex Numbers in Standard Form

1. Multiply both the numerator and denominator by the complex conjugate of the denominator.
2. Simplify the resulting products in both the numerator and denominator.
3. Write the simplified result in standard form.

Remember the following special product. We restate it here to emphasize its importance.

$$(a+bi)(a-bi)=a^2+b^2$$

Example 2: Division with Complex Numbers

Write the following fractions in standard form.

a. $\frac{4}{-1-5i}$

Solution: $$\frac{4}{-1-5i}=\frac{4(-1+5i)}{(-1-5i)(-1+5i)}$$

$$=\frac{-4+20i}{(-1)^2-(5i)^2}=\frac{-4+20i}{1+25}$$

$$=\frac{-4+20i}{26}=-\frac{4}{26}+\frac{20}{26}i$$

$$=-\frac{2}{13}+\frac{10}{13}i$$

Continued on the next page...

b. $\dfrac{\sqrt{3}+i}{\sqrt{3}-i}$

Solution: $\dfrac{\sqrt{3}+i}{\sqrt{3}-i}=\dfrac{(\sqrt{3}+i)(\sqrt{3}+i)}{(\sqrt{3}-i)(\sqrt{3}+i)}$

$$=\frac{3+2\sqrt{3}\,i+i^2}{(\sqrt{3})^2-i^2}=\frac{2+2\sqrt{3}\,i}{3+1}$$

$$=\frac{2+2\sqrt{3}\,i}{4}=\frac{2}{4}+\frac{2\sqrt{3}}{4}i$$

$$=\frac{1}{2}+\frac{\sqrt{3}}{2}i$$

c. $\dfrac{6+i}{i}$

Solution: $\dfrac{6+i}{i}=\dfrac{(6+i)(-i)}{i(-i)}$

Since $i=0+i$ and $-i=0-i$, the number $-i$ is the conjugate of i.

$$=\frac{-6i-i^2}{-i^2}=\frac{-6i+1}{1}$$

$$=1-6i$$

d. $\dfrac{\sqrt{2}+i}{-\sqrt{2}+i}$

Solution: $\dfrac{\sqrt{2}+i}{-\sqrt{2}+i}=\dfrac{(\sqrt{2}+i)(-\sqrt{2}-i)}{(-\sqrt{2}+i)(-\sqrt{2}-i)}$

$$=\frac{-(\sqrt{2})^2-\sqrt{2}\,i-\sqrt{2}\,i-i^2}{(-\sqrt{2})^2-(i)^2}$$

$$=\frac{-2-2\sqrt{2}\,i+1}{2+1}=\frac{-1-2\sqrt{2}\,i}{3}$$

$$=-\frac{1}{3}-\frac{2\sqrt{2}}{3}i$$

Powers of *i*

The powers of i form an interesting pattern. Regardless of the particular integer exponent, there are only four possible values for any power of i:

$$i,\quad -1,\quad -i,\quad \text{and}\quad 1.$$

The fact that these are the only four possibilities for powers of i becomes apparent from studying the following powers.

$$
\begin{aligned}
i^1 &= i \\
i^2 &= -1 \\
i^3 &= i^2 \cdot i = -1 \cdot i = -i \\
i^4 &= i^2 \cdot i^2 = (-1)(-1) = 1 \\
i^5 &= i^4 \cdot i = 1 \cdot i = i \\
i^6 &= i^4 \cdot i^2 = (1)(-1) = -1 \\
i^7 &= i^4 \cdot i^3 = (1)(-i) = -i \\
i^8 &= i^4 \cdot i^4 = (1)(1) = 1
\end{aligned}
$$

Teaching Notes:
Another approach to evaluating i^n is to divide n by 4 and look at the remainder.
If the remainder is 0, $i^n = i^0 = 1$.
If the remainder is 1, $i^n = i^1 = i$.
If the remainder is 2, $i^n = i^2 = -1$.
If the remainder is 3, $i^n = i^3 = -i$.

Higher powers of i can be simplified using the fact that when i is raised to a power that is a multiple of 4, the result is 1. Thus if n is a positive integer, then

$$
\begin{aligned}
i^{4n} &= \left(i^4\right)^n = 1^n = 1 \\
i^{4n+1} &= i^{4n} \cdot i = 1 \cdot i = i \\
i^{4n+2} &= i^{4n} \cdot i^2 = 1 \cdot (-1) = -1 \\
i^{4n+3} &= i^{4n} \cdot i^3 = 1 \cdot (-i) = -i.
\end{aligned}
$$

Example 3: Powers of i

Simplify each power of i.

a. $i^{45} = i^{44} \cdot i = \left(i^4\right)^{11} \cdot i = 1^{11} \cdot i = i$ — $i = 0 + i$ in standard form.

b. $i^{58} = i^{56} \cdot i^2 = \left(i^4\right)^{14} \cdot i^2 = 1^{14} \cdot (-1) = -1$ — $-1 = -1 + 0i$ in standard form.

c. $i^{-7} = \frac{1}{i^7} = \frac{1}{i^7} \cdot \frac{i}{i} = \frac{i}{i^8} = \frac{i}{1} = i$ — $i = 0 + i$ in standard form.

Practice Problems

Write each of the following numbers in standard form.

1. $-2i(3-i)$ **2.** $(2+4i)(1+i)$ **3.** i^{13}

4. i^{-2} **5.** $\frac{2}{1+5i}$ **6.** $\frac{7+i}{2-i}$

Answers to Practice Problems: **1.** $-2-6i$ **2.** $-2+6i$ **3.** $0+i$ **4.** $-1+0i$ **5.** $\frac{1}{13}-\frac{5}{13}i$ **6.** $\frac{13}{5}+\frac{9}{5}i$

9.8 Exercises

Perform the indicated operations and write each result in standard form.

1. $8(2+3i)$ **2.** $-3(7-4i)$ **3.** $-7(\sqrt{2}-i)$ **4.** $\sqrt{3}(\sqrt{3}+2i)$

5. $3i(4-i)$ **6.** $-4i(6-7i)$ **7.** $-i(\sqrt{3}+i)$ **8.** $2i(\sqrt{5}+2i)$

9. $\sqrt{3}i(2-\sqrt{3}i)$ **10.** $5i(2-\sqrt{2}i)$ **11.** $(5+3i)(1+i)$ **12.** $(2+7i)(6+i)$

13. $(-3+5i)(-1+2i)$ **14.** $(6+2i)(3-i)$ **15.** $(2-3i)(2+3i)$ **16.** $(4+5i)(4-5i)$

17. $(4+3i)(7-2i)$ **18.** $(-2+5i)(i-1)$ **19.** $(5+7i)^2$ **20.** $(3+2i)^2$

21. $(\sqrt{3}+i)(\sqrt{3}-2i)$ **22.** $(2\sqrt{5}+3i)(\sqrt{5}-i)$ **23.** $(5-\sqrt{2}i)(5-\sqrt{2}i)$

24. $(\sqrt{7}+3i)(\sqrt{7}+i)$ **25.** $(4+\sqrt{5}i)(4-\sqrt{5}i)$ **26.** $(7+2\sqrt{3}i)(7-2\sqrt{3}i)$

27. $(\sqrt{5}+2i)(\sqrt{2}-i)$ **28.** $(2\sqrt{3}+i)(4+3i)$ **29.** $(3+\sqrt{5}i)(3+\sqrt{6}i)$

30. $(2-\sqrt{3}i)(3-\sqrt{2}i)$ **31.** $\dfrac{-3}{i}$ **32.** $\dfrac{7}{i}$ **33.** $\dfrac{5}{4i}$

34. $\dfrac{-3}{2i}$ **35.** $\dfrac{2+i}{-4i}$ **36.** $\dfrac{3-4i}{3i}$ **37.** $\dfrac{-4}{1+2i}$ **38.** $\dfrac{7}{5-2i}$

39. $\dfrac{6}{4-3i}$ **40.** $\dfrac{-8}{6+i}$ **41.** $\dfrac{2i}{5-i}$ **42.** $\dfrac{-4i}{1+3i}$ **43.** $\dfrac{2-i}{2+5i}$

44. $\dfrac{6+i}{3-4i}$ **45.** $\dfrac{2-3i}{-1+5i}$ **46.** $\dfrac{-3+i}{7-2i}$ **47.** $\dfrac{1+4i}{\sqrt{3}+i}$ **48.** $\dfrac{9-2i}{\sqrt{5}+i}$

49. $\dfrac{\sqrt{3}+2i}{\sqrt{3}-2i}$ **50.** $\dfrac{\sqrt{6}-3i}{\sqrt{6}+3i}$

Simplify the following powers of i. Assume k is a positive integer.

51. i^{13} **52.** i^{20} **53.** i^{30} **54.** i^{15} **55.** i^{-3}

56. i^{-5} **57.** i^{4k} **58.** i^{4k+2} **59.** i^{4k+3} **60.** i^{4k+1}

Find the indicated products and simplify.

61. $(x+3i)(x-3i)$ **62.** $(y+5i)(y-5i)$ **63.** $(x+\sqrt{2}i)(x-\sqrt{2}i)$

64. $(2x+\sqrt{7}i)(2x-\sqrt{7}i)$ **65.** $(\sqrt{5}y+2i)(\sqrt{5}y-2i)$ **66.** $(y-\sqrt{3}i)(y+\sqrt{3}i)$

1. $16+24i$
2. $-21+12i$
3. $-7\sqrt{2}+7i$
4. $3+2i\sqrt{3}$
5. $3+12i$
6. $-28-24i$
7. $1-i\sqrt{3}$
8. $-4+2i\sqrt{5}$
9. $3+2i\sqrt{3}$
10. $5\sqrt{2}+10i$
11. $2+8i$
12. $5+44i$
13. $-7-11i$
14. $20+0i$
15. $13+0i$
16. $41+0i$
17. $34+13i$
18. $-3-7i$
19. $-24+70i$
20. $5+12i$
21. $5-i\sqrt{3}$
22. $13+i\sqrt{5}$
23. $23-10i\sqrt{2}$
24. $4+4i\sqrt{7}$
25. $21+0i$
26. $61+0i$
27. $(2+\sqrt{10})+(2\sqrt{2}-\sqrt{5})i$
28. $(8\sqrt{3}-3)+(4+6\sqrt{3})i$
29. $(9-\sqrt{30})+(3\sqrt{5}+3\sqrt{6})i$
30. $(6-\sqrt{6})-(3\sqrt{3}+2\sqrt{2})i$
31. $0+3i$
32. $0-7i$
33. $0-\frac{5}{4}i$

34. $0+\frac{3}{2}i$

35. $-\frac{1}{4}+\frac{1}{2}i$

67. $[(x+2)+6i][(x+2)-6i]$

68. $[(x+1)-\sqrt{8}i][(x+1)+\sqrt{8}i]$

69. $[(y-3)+2i][(y-3)-2i]$

70. $[(x-1)+5i][(x-1)-5i]$

Writing and Thinking About Mathematics

36. $-\frac{4}{3}-i$

37. $-\frac{4}{5}+\frac{8}{5}i$

38. $\frac{35}{29}+\frac{14}{29}i$

39. $\frac{24}{25}+\frac{18}{25}i$

71. Explain why the product of every complex number and its conjugate is a nonnegative real number.

72. Explain why $\sqrt{-4}\cdot\sqrt{-4}\neq 4$. What is the correct value of $\sqrt{-4}\cdot\sqrt{-4}$?

73. What condition is necessary for the conjugate of a complex number, $a + bi$, to be equal to the reciprocal of this number?

HAWKES LEARNING SYSTEMS: INTRODUCTORY & INTERMEDIATE ALGEBRA SOFTWARE

- 9.8 Multiplication and Division with Complex Numbers

40. $-\frac{48}{37}+\frac{8}{37}i$ 41. $-\frac{1}{13}+\frac{5}{13}i$ 42. $-\frac{6}{5}-\frac{2}{5}i$ 43. $-\frac{1}{29}-\frac{12}{29}i$ 44. $\frac{14}{25}+\frac{27}{25}i$ 45. $-\frac{17}{26}-\frac{7}{26}i$ 46. $-\frac{23}{53}+\frac{1}{53}i$
47. $\frac{4+\sqrt{3}}{4}+\left(\frac{4\sqrt{3}-1}{4}\right)i$ 48. $\left(-\frac{1}{3}+\frac{3\sqrt{5}}{2}\right)-\left(\frac{3}{2}+\frac{\sqrt{5}}{3}\right)i$ 49. $-\frac{1}{7}+\frac{4\sqrt{3}}{7}i$ 50. $-\frac{1}{5}-\frac{2\sqrt{6}}{5}i$ 51. $0+i$ 52. $1+0i$
53. $-1+0i$ 54. $0-i$ 55. $0+i$ 56. $0-i$ 57. $1+0i$ 58. $-1+0i$ 59. $0-i$ 60. $0+i$ 61. x^2+9 62. y^2+25
63. x^2+2 64. $4x^2+7$ 65. $5y^2+4$ 66. y^2+3 67. $x^2+4x+40$ 68. x^2+2x+9 69. $y^2-6y+13$
70. $x^2-2x+26$ 71. Given a complex number $(a+bi)$: $(a+bi)(a-bi)=a^2-abi+abi-b^2i^2=a^2+b^2$ which is the sum of squares of real numbers. Thus the product must be a positive real number.
72. The property $\sqrt{a}\sqrt{b}=\sqrt{ab}$ is only true if a and b are positive. The actual value is -4, because $\sqrt{-4}\cdot\sqrt{-4}=2i\cdot 2i=4i^2=-4$. 73. $a^2+b^2=1$

Chapter 9 Index of Key Ideas and Terms

Section 9.1 Roots and Radicals

Perfect Squares pages 682-683

The square of an integer is called a **perfect square**.

Radical Terminology pages 683, 686

The expression $\sqrt[n]{a}$ is called a **radical** or **radical expression**.
The symbol $\sqrt[n]{}$ is called a **radical sign**.
n is called the **index**.
a is called the **radicand**.
(**Note:** If no index is given, the index is 2.)

Square Root page 684

If a is a nonnegative real number, then $\sqrt{a}$ is the **principal square root** of a and $-\sqrt{a}$ is the **negative square root** of a.

Cube Root pages 685-686

If a is a real number, then $\sqrt[3]{a}$ is the **cube root** of a.

Evaluating Radical Expressions Using a Calculator pages 686-688

Section 9.2 Simplifying Radicals

Properties of Square Roots page 690

If a and b are positive real numbers, then

1. $\sqrt{ab} = \sqrt{a}\sqrt{b}$

2. $\sqrt{\frac{a}{b}} = \frac{\sqrt{a}}{\sqrt{b}}$

Simplest Form pages 690, 694

A square root is considered to be in simplest form when the radicand has no perfect square as a factor.
A cube root is considered to be in simplest form when the radicand has no perfect cube as a factor.

Square Root of x^2 page 692

If x is a real number, then $\sqrt{x^2} = |x|$.

Note: If $x \geq 0$ is given, then we can write $\sqrt{x^2} = x$.

Section 9.3 Rational Exponents

Radical Notation and Fractional Exponents page 699

If n is a positive integer, then $\sqrt[n]{a} = a^{\frac{1}{n}}$ (assuming $\sqrt[n]{a}$ is a real number).

Special Notes about the Index n page 699

For the expression $\sqrt[n]{a}$ (or $a^{\frac{1}{n}}$) to be a real number:

1. when a is nonnegative, n can be any index, and
2. when a is negative, n must be odd.
   (If a is negative and n is even, then $\sqrt[n]{a}$ is nonreal.)

Review of the Rules for Exponents page 700

The General Form $a^{\frac{m}{n}}$ page 701

If n is a positive integer, m is any integer, and $a^{\frac{1}{n}}$ is a real number, then

$$a^{\frac{m}{n}} = \left(a^{\frac{1}{n}}\right)^m = \left(a^m\right)^{\frac{1}{n}}.$$

In radical notation:

$$a^{\frac{m}{n}} = \left(\sqrt[n]{a}\right)^m = \sqrt[n]{a^m}.$$

Finding the Value of $a^{\frac{m}{n}}$ with a Graphing Calculator page 704

Section 9.4 Operations with Radicals

Like Radicals page 708

Like radicals have the same index and radicand or they can be simplified so that they have the same index and radicand.

Addition and Subtraction with Radical Expressions page 708

Multiplication with Radical Expressions page 709

Continued on the next page...

Section 9.4 Operations with Radicals (cont.)

To Rationalize a Denominator Containing a Square Root or a Cube Root page 711

1. If the denominator contains a square root, multiply both the numerator and denominator by an expression that will give a denominator with no square roots.
2. If the denominator contains a cube root, multiply both the numerator and denominator by an expression that will give a denominator with no cube roots.

Conjugates page 712

The two expressions $(a-b)$ and $(a+b)$ are called **conjugates** of each other, and their product, $(a-b)(a+b)$, results in the difference of two squares.

To Rationalize a Denominator Containing a Sum or Difference Involving Square Roots page 712

If the denominator of a fraction contains a sum or difference involving a square root, rationalize the denominator by multiplying both the numerator and denominator by the conjugate of the denominator.

1. If the denominator is of the form $a - b$, multiply both the numerator and denominator by $a + b$.
2. If the denominator is of the form $a + b$, multiply both the numerator and denominator by $a - b$.

The new denominator will be the difference of two squares and therefore will not contain a radical term.

Evaluating Radical Expressions with a Graphing Calculator pages 714-715

Section 9.5 Equations with Radicals

Extraneous Solution page 719

An **extraneous solution** is a number that is found when solving an equation but that does not satisfy the original equation. They can be unintentionally introduced by raising both sides of an equation to a power.

Continued on the next page...

Section 9.5 Equations with Radicals (cont.)

Method for Solving Equations with Radicals page 719

1. Isolate one of the radicals on one side of the equation. (An equation may have more than one radical.)
2. Raise both sides of the equation to the power corresponding to the index of the radical.
3. If the equation still contains a radical, repeat Steps 1 and 2.
4. Solve the equation after all the radicals have been eliminated.
5. Be sure to check all possible solutions in the original equation and eliminate any extraneous solutions.

Section 9.6 Functions with Radicals

Radical Function page 727

A **radical function** is a function of the form $y = \sqrt[n]{g(x)}$ in which the radicand contains a variable expression.

The **domain** of such a function depends on the index, n:

1. If n is an even number, the domain is the set of all x such that $g(x) \geq 0$.
2. If n is an odd number, the domain is the set of all real numbers $(-\infty, \infty)$.

Evaluating Radical Functions pages 727-728

Graphing Radical Functions pages 729-730

Using a Graphing Calculator to Graph Radical Functions pages 730-731

Section 9.7 Introduction to Complex Numbers

Complex Numbers page 736

$i = \sqrt{-1}$ and $i^2 = -1$

If a is a positive real number, then $\sqrt{-a} = \sqrt{a} \cdot \sqrt{-1} = \sqrt{a}\, i$.
(To avoid confusion, we sometimes write $i\sqrt{a}$.)

Continued on the next page...

Section 9.7 Introduction to Complex Numbers (cont.)

Standard Form of Complex Numbers page 737

The **standard form** of a **complex number** is $\mathbf{a+bi}$, where a and b are real numbers. a is called the **real part** and b is called the **imaginary part**.

If $b=0$, then $a+bi=a+0i=a$ is a real number.

If $a=0$, then $a+bi=0+bi=bi$ is called a **pure imaginary number** (or an **imaginary number**).

Equality of Complex Numbers page 739

For complex numbers $a+bi$ and $c+di$, if $a+bi=c+di$, then $a=c$ and $b=d$.

Addition and Subtraction with Complex Numbers pages 739-740

For complex numbers $a+bi$ and $c+di$,

$$(a+bi)+(c+di)=(a+c)+(b+d)i$$

and $$(a+bi)-(c+di)=(a-c)+(b-d)i.$$

Section 9.8 Multiplication and Division with Complex Numbers

Multiplication with Complex Numbers page 743

Complex Conjugates page 744

$a+bi$ and $a-bi$ are **complex conjugates** of each other.
The product of two complex conjugates will always be a nonnegative real number: $(a+bi)(a-bi)=a^2+b^2$.

Division with Complex Numbers pages 744-745

To Write a Fraction with Complex Numbers in Standard Form page 745

1. Multiply both the numerator and denominator by the complex conjugate of the denominator.
2. Simplify the resulting products in both the numerator and denominator.
3. Write the simplified result in standard form.

Continued on the next page...

Section 9.8 Multiplication and Division with Complex Numbers (cont.)

Powers of *i*: i^n pages 746-747

$$i^{4n} = (i^4)^n = 1^n = 1$$
$$i^{4n+1} = i^{4n} \cdot i = 1 \cdot i = i$$
$$i^{4n+2} = i^{4n} \cdot i^2 = 1 \cdot (-1) = -1$$
$$i^{4n+3} = i^{4n} \cdot i^3 = 1 \cdot (-i) = -i$$

HAWKES LEARNING SYSTEMS: INTRODUCTORY & INTERMEDIATE ALGEBRA SOFTWARE

- 9.1 Evaluating Radicals
- 9.2 Simplifying Radicals
- 9.3 Rational Exponents
- 9.4a Addition and Subtraction with Radicals
- 9.4b Multiplication with Radicals
- 9.4c Rationalizing Denominators
- 9.5 Equations with Radicals
- 9.6 Functions with Radicals
- 9.7 Complex Numbers
- 9.8 Multiplication and Division with Complex Numbers

Chapter 9 Review

9.1 Roots and Radicals

Simplify the following square roots and cube roots.

1. $\sqrt{36}$ **2.** $\sqrt{196}$ **3.** $\sqrt[3]{8}$ **4.** $\sqrt[3]{729}$

5. $\sqrt{0.0009}$ **6.** $\sqrt[3]{-343}$ **7.** $\sqrt{\frac{144}{49}}$ **8.** $\sqrt[3]{\frac{125}{27}}$

Use your knowledge of square roots and cube roots to determine whether each number is rational, irrational, or nonreal.

9. $\sqrt{-20}$ **10.** $\sqrt{3}$ **11.** $\sqrt{\frac{1}{9}}$ **12.** $-\sqrt{196}$

Use a calculator to find the value (accurate to four decimal places) of each radical expression.

13. $\sqrt{21}$ **14.** $\sqrt{40.5}$ **15.** $-4\sqrt{18}$ **16.** $1+3\sqrt{2}$

9.2 Simplifying Radicals

Simplify each of the radical expressions. Assume that all variables represent positive real numbers.

17. $-\sqrt{225}$ **18.** $\sqrt{9x^3}$ **19.** $\sqrt{8a^4}$ **20.** $\sqrt{50x^3y^2}$

21. $-\sqrt{81x^2y}$ **22.** $\sqrt[3]{40x^4}$ **23.** $\sqrt[3]{81x^5y^7}$ **24.** $\sqrt[3]{54a^4b^2}$

25. $\sqrt{\frac{27x^2}{100}}$ **26.** $\sqrt{\frac{75a^3}{9}}$ **27.** $\sqrt[3]{\frac{3000x^6}{343}}$ **28.** $\sqrt[3]{\frac{8y^{12}}{27x^{15}}}$

9.3 Rational Exponents

Simplify each numerical expression.

29. $81^{\frac{1}{2}}$ **30.** $-36^{-\frac{1}{2}}$ **31.** $225^{-\frac{1}{2}}$ **32.** $\left(\frac{81}{16}\right)^{-\frac{1}{4}}$

33. $\left(\frac{27}{125}\right)^{\frac{2}{3}}$ **34.** $-100^{-\frac{3}{2}}$

Answers:

1. 6
2. 14
3. 2
4. 9
5. 0.03
6. −7
7. $\frac{12}{7}$
8. $\frac{5}{3}$
9. Nonreal
10. Irrational
11. Rational
12. Rational
13. 4.5826
14. 6.3640
15. −16.9706
16. 5.2426
17. −15
18. $3x\sqrt{x}$
19. $2a^2\sqrt{2}$
20. $5xy\sqrt{2x}$
21. $-9x\sqrt{y}$
22. $2x\sqrt[3]{5x}$
23. $3xy^2\sqrt[3]{3x^2y}$
24. $3a\sqrt[3]{2ab^2}$
25. $\frac{3x\sqrt{3}}{10}$
26. $\frac{5a\sqrt{3a}}{3}$
27. $\frac{10x^2\sqrt[3]{3}}{7}$
28. $\frac{2y^4}{3x^5}$
29. 9
30. $-\frac{1}{6}$
31. $\frac{1}{15}$
32. $\frac{2}{3}$
33. $\frac{9}{25}$
34. $-\frac{1}{1000}$

35. 10.9027
36. 82.7037
37. 7.0711 38. 0.2408
39. $27x^2$ 40. $8a^{\frac{3}{2}}$
41. $5y^{\frac{10}{3}}$ 42. $\dfrac{1}{x^{\frac{5}{2}}}$
43. $\dfrac{9a^4b^2}{c^2}$ 44. $\dfrac{1}{2y^{\frac{1}{2}}}$
45. $\sqrt[12]{x^5}$ 46. $2\sqrt[3]{x}$
47. $-4\sqrt{11}$ 48. $5\sqrt{x}$
49. $5\sqrt{2}-26\sqrt{3}$
50. 0 51. -1
52. $7+2\sqrt{10}$
53. $x-y$
54. $8\sqrt{2x}+15x+2$
55. $\dfrac{\sqrt{6}}{4}$
56. $\dfrac{4\sqrt{3b}}{3a}$
57. $\dfrac{\sqrt[3]{100xy^2}}{10y}$
58. $\dfrac{3\sqrt{5}-5}{2}$
59. $\dfrac{5\sqrt{7}-5\sqrt{3}}{4}$
60. $\dfrac{13+3\sqrt{15}}{17}$
61. $\dfrac{6-3x}{4\sqrt{6}+4\sqrt{3x}}$
62. 25.4495
63. -28
64. 0.2185
65. $x=7$ 66. $x=14$
67. $x=68$
68. $x=\pm 9$
69. $x=0$
70. $x=3,5$
71. $x=-5,-4$
72. No solution

Use a graphing calculator to find the value of each numerical expression accurate to 4 decimal places, if necessary.

35. $36^{\frac{2}{3}}$ **36.** $200^{\frac{5}{6}}$ **37.** $\sqrt[4]{2500}$ **38.** $\sqrt[5]{0.00081}$

Simplify each algebraic expression. Assume that all variables represent positive real numbers. Leave the answers in rational exponent form.

39. $\left(3x^{\frac{2}{3}}\right)^3$ **40.** $\left(16a^2\right)^{\frac{3}{4}}$ **41.** $5y^3 \cdot y^{\frac{1}{3}}$ **42.** $\dfrac{x^{\frac{5}{6}} \cdot x^{-\frac{1}{3}}}{x^3}$

43. $\left(\dfrac{27a^6b^3}{c^3}\right)^{\frac{2}{3}}$ **44.** $\dfrac{\left(4x^{-2}y^{\frac{1}{2}}\right)^{-\frac{1}{2}}}{xy^{\frac{1}{4}}}$

Simplify each expression by first changing it into an equivalent expression with rational exponents. Rewrite the answer in simplified radical form.

45. $\dfrac{\sqrt[3]{x}\sqrt[6]{x^5}}{\sqrt[4]{x^3}}$ **46.** $\sqrt{\sqrt[3]{64x^2}}$

9.4 Operations with Radicals

Perform the indicated operations and simplify. Assume that all variables represent positive numbers.

47. $\sqrt{11}-5\sqrt{11}$ **48.** $6\sqrt{x}+\sqrt{x}-2\sqrt{x}$ **49.** $2\sqrt{12}-6\sqrt{75}+\sqrt{50}$

50. $2x\sqrt{y}-\sqrt{4x^2y}$ **51.** $\left(2\sqrt{6}+5\right)\left(2\sqrt{6}-5\right)$ **52.** $\left(\sqrt{5}+\sqrt{2}\right)^2$

53. $\left(\sqrt{x}-\sqrt{y}\right)\left(\sqrt{y}+\sqrt{x}\right)$ **54.** $\left(3\sqrt{x}+\sqrt{2}\right)\left(5\sqrt{x}+\sqrt{2}\right)$

Rationalize the denominator and simplify if possible. Assume that all variables represent positive numbers.

55. $\sqrt{\dfrac{9}{24}}$ **56.** $\dfrac{\sqrt{32b}}{\sqrt{6a^2}}$ **57.** $\dfrac{\sqrt[3]{x^2}}{\sqrt[3]{10xy}}$

58. $\dfrac{\sqrt{20}}{3+\sqrt{5}}$ **59.** $\dfrac{5}{\sqrt{7}+\sqrt{3}}$ **60.** $\dfrac{\sqrt{5}+\sqrt{3}}{2\sqrt{5}-\sqrt{3}}$

61. Rationalize the numerator in the expression $\dfrac{\sqrt{6}-\sqrt{3x}}{4}$ and simplify if possible.

73. $x = 2$ 74. $x = 3$
75. $x = 4$
76. No solution
77. No solution
78. $x = 15$
79. $x = 2, 4$
80. $x = 20$
81. a. 0 b. 5
82. a. $\sqrt[3]{6} \approx 1.8171$
b. $2\sqrt[3]{2} \approx 2.5198$
83. $\left[-\frac{1}{2}, \infty\right)$
84. $(-\infty, \infty)$
85. $\left(-\infty, \frac{3}{2}\right]$
86. $(-\infty, \infty)$
87.
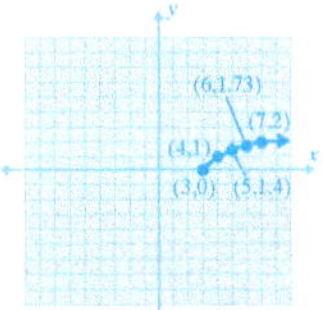

88.
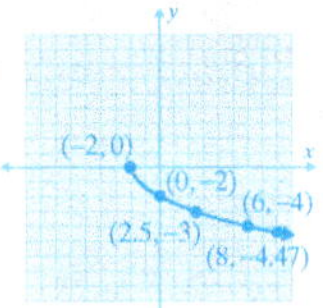

89.
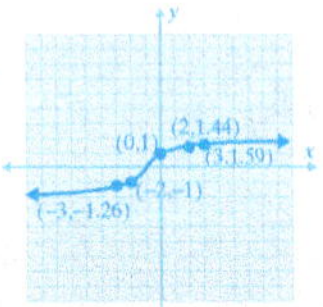

90.
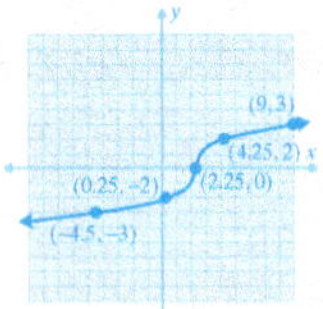

91.
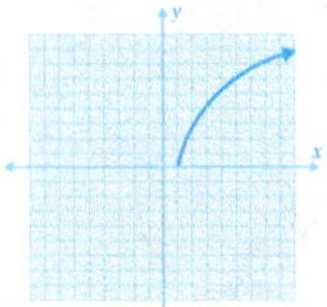

92.
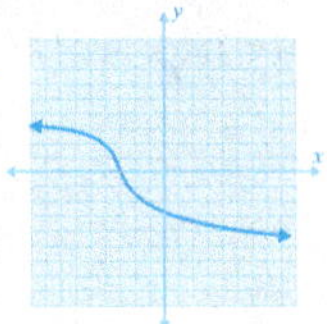

Use a graphing calculator to evaluate each expression. Round answers to 4 decimal places, if necessary.

62. $23+\sqrt{6}$ **63.** $\left(\sqrt{8}+6\right)\left(\sqrt{8}-6\right)$ **64.** $\dfrac{\sqrt{3}}{1+4\sqrt{3}}$

9.5 Equations with Radicals

Solve the following equations. Be sure to check your answers in the original equation.

65. $\sqrt{3x+4}=5$ **66.** $\sqrt{6x-20}=8$ **67.** $\sqrt{x-4}-6=2$

68. $\sqrt{x^2-17}=8$ **69.** $x+3=\sqrt{x+9}$ **70.** $x-3=\sqrt{2x-6}$

71. $\sqrt{x+5}=x+5$ **72.** $\sqrt{2x-7}=3-x$ **73.** $\sqrt{4x+1}=\sqrt{2x+5}$

74. $\sqrt{4-x}=\sqrt{3x-8}$ **75.** $\sqrt{x-3}+\sqrt{x}=3$ **76.** $\sqrt{3x+1}=-4$

77. $\sqrt{2x+3}=2-\sqrt{x-2}$ **78.** $\sqrt{2x-5}=\sqrt{x+1}+1$ **79.** $\sqrt[3]{x(x-6)}=-2$

80. $\sqrt[4]{10x+56}=4$

9.6 Functions with Radicals

Find each function value as indicated and write the answers in both radical notation and decimal notation. (If necessary, round decimal values to 4 decimal places.)

81. Given $g(x)=\sqrt{1-4x}$, find **a.** $g\left(\frac{1}{4}\right)$ and **b.** $g(-6)$.

82. Given $h(x)=\sqrt[3]{6-x}$, find **a.** $h(0)$ and **b.** $h(-10)$.

Use interval notation to indicate the domain of each radical function.

83. $y=\sqrt{2x+1}$ **84.** $f(x)=\sqrt[3]{4x}$ **85.** $y=\sqrt[4]{3-2x}$ **86.** $f(x)=\sqrt[5]{9-x}$

Find and label at least 5 points on the graph of the function and then sketch the graph of the function.

87. $y=\sqrt{x-3}$ **88.** $y=-\sqrt{2x+4}$ **89.** $y=\sqrt[3]{x+1}$ **90.** $y=\sqrt[3]{4x-9}$

Use a graphing calculator to graph each of the functions.

91. $y=3\sqrt{x-1}$ **92.** $f(x)=-2\sqrt[3]{x+3}$ **93.** $y=\sqrt[4]{x+1}$ **94.** $y=\sqrt[5]{x+5}$

9.7 Introduction to Complex Numbers

Find the real part and the imaginary part of each of the complex numbers.

95. $5-2i$ **96.** $-10+\sqrt{3}i$ **97.** $\frac{3}{4}$ **98.** $\frac{3+i}{2}$

Simplify the following radicals.

99. $\sqrt{-36}$ **100.** $\sqrt{-144}$ **101.** $-\sqrt{-72}$ **102.** $2\sqrt{725}$

Solve the equations for x and y.

103. $2x+3i=18-yi+3i$ **104.** $\sqrt{3}-4i=x-3yi$

105. $10i-x+4=14+3i-yi$ **106.** $3+2i+x=8-6i+yi$

Find each sum or difference as indicated.

107. $(3+2i)+(2+6i)$ **108.** $(-7+i)-(-3-i)$

109. $(4-3i)+(\sqrt{2}-i)$ **110.** $(\sqrt{5}+\sqrt{3}i)+(2-\sqrt{3}i)$

111. $(12-4\sqrt{-25})+(3-5\sqrt{-1})$ **112.** $(-3+2i)+(7+i)-(3-2i)$

9.8 Multiplication and Division with Complex Numbers

Perform the indicated operations and write each result in standard form.

113. $4i(3-\sqrt{2}i)$ **114.** $(4+3i)(1-i)$ **115.** $(\sqrt{3}+i)(2\sqrt{3}-i)$

116. $(2+3i)^2$ **117.** $-\frac{8}{5i}$ **118.** $\frac{10}{4-2i}$

119. $\frac{2-i}{3+4i}$ **120.** $\frac{7-2i}{\sqrt{5}-i}$

Simplify the following powers of i.

121. i^{-5} **122.** i^{10}

Find the indicated products and simplify.

123. $(\sqrt{6}x-2i)(\sqrt{6}x+2i)$ **124.** $[(y+4)-4i][(y+4)+4i]$

93.

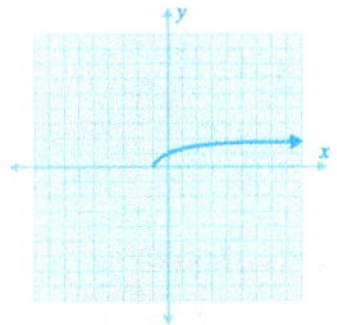

94.

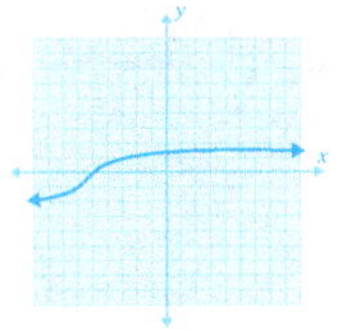

95. Real part is 5, imaginary part is -2

96. Real part is -10, imaginary part is $\sqrt{3}$

97. Real part is $\frac{3}{4}$, imaginary part is 0

98. Real part is $\frac{3}{2}$, imaginary part is $\frac{1}{2}$

99. $6i$

100. $12i$

101. $-6i\sqrt{2}$

102. $10\sqrt{29}$

103. $x=9, y=0$

104. $x=\sqrt{3}, y=\frac{4}{3}$

105. $x=-10, y=-7$

106. $x=5, y=8$

107. $5+8i$

108. $-4+2i$

109. $4+\sqrt{2}-4i$

110. $2+\sqrt{5}$

111. $15-25i$

112. $1+5i$

113. $4\sqrt{2}+12i$

114. $7-i$

115. $7+i\sqrt{3}$

116. $-5+12i$ 117. $0+\frac{8}{5}i$ 118. $2+i$ 119. $\frac{2}{25}-\frac{11}{25}i$ 120. $\frac{2+7\sqrt{5}}{6}+\frac{(7-2\sqrt{5})}{6}i$ 121. $0-i$ 122. $-1+0i$

123. $6x^2+4$ 124. $y^2+8y+32$

Chapter 9 Test

1. $4\sqrt{7}$
2. $2y^2\sqrt{30x}$
3. $2y\sqrt[3]{6x^2y^2}$
4. $\dfrac{7x^6y\sqrt[3]{y}}{2z^2}$
5. $\sqrt[3]{4x^2}$ 6. $2^{\frac{1}{2}}x^{\frac{1}{3}}y^{\frac{2}{3}}$
7. 4 8. $4x^{\frac{7}{6}}$
9. $\dfrac{8y^{\frac{3}{2}}}{x^3}$ 10. $\sqrt[12]{x^{11}}$
11. $17\sqrt{3}$
12. $-xy\sqrt{y}$
13. $5-2\sqrt{6}$
14. $30-7\sqrt{3x}-6x$
15. a. $\dfrac{y\sqrt{10x}}{4x^2}$
    b. $1+\sqrt{x}$
16. $x=14$
17. $x=-4$
18. $x=1$
19. $x=7$
20. a. $\left[-\dfrac{4}{3},\infty\right)$
    b. $(-\infty,\infty)$

Simplify the expressions. Assume that all variables represent positive real numbers.

1. $\sqrt{112}$ **2.** $\sqrt{120xy^4}$ **3.** $\sqrt[3]{48x^2y^5}$ **4.** $\sqrt[3]{\dfrac{343x^{18}y^4}{8z^6}}$

5. Write $(2x)^{\frac{2}{3}}$ in radical notation.

6. Write $\sqrt[6]{8x^2y^4}$ as an equivalent, simplified expression with rational exponents.

Simplify each numerical expression. Assume that all variables represent positive real numbers. Leave the answers in rational exponent form.

7. $(-8)^{\frac{2}{3}}$ **8.** $4x^{\frac{1}{2}}\cdot x^{\frac{2}{3}}$ **9.** $\left(\dfrac{16x^{-4}y}{y^{-1}}\right)^{\frac{3}{4}}$

10. Simplify the expression, $\sqrt[3]{x^2}\cdot\sqrt[4]{x}$, by first changing it into an equivalent expression with rational exponents. Rewrite the answer in simplified radical form.

Perform the indicated operations and simplify. Assume that all variables represent positive real numbers.

11. $2\sqrt{75}+3\sqrt{27}-\sqrt{12}$ **12.** $5x\sqrt{y^3}-2\sqrt{x^2y^3}-4y\sqrt{x^2y}$

13. $\left(\sqrt{3}-\sqrt{2}\right)^2$ **14.** $\left(6+\sqrt{3x}\right)\left(5-2\sqrt{3x}\right)$

15. Rationalize the denominator and simplify, if possible.

a. $\sqrt{\dfrac{5y^2}{8x^3}}$ **b.** $\dfrac{1-x}{1-\sqrt{x}}$

Solve the following equations. Be sure to check your answers in the original equation.

16. $\sqrt{9x-5}-3=8$ **17.** $\sqrt[3]{3x+4}=-2$

18. $\sqrt{x+8}-2=x$ **19.** $\sqrt{5x+1}=1+\sqrt{3x+4}$

20. Find the domain of each of the following radical functions. Write the answer in interval notation.

a. $y=\sqrt{3x+4}$ **b.** $f(x)=\sqrt[3]{2x+5}$

21. Use a graphing calculator to graph each of the following radical functions. Sketch the graph and label 3 points on the graph.

a. $f(x) = -\sqrt{x+3}$ **b.** $y = \sqrt[3]{x-4}$

Find the real part and the imaginary part of each of the complex numbers.

22. $6 - \frac{1}{3}i$ **23.** $\frac{-2+3i}{5}$

24. Solve for x and y: $(2x+3i)-(6+2yi) = 5-3i$.

Perform the indicated operations and write each result in standard form.

25. $(5+8i)+(11-4i)$ **26.** $\left(2+3\sqrt{-4}\right)-\left(7-2\sqrt{-25}\right)$

27. $(4+3i)(2-5i)$ **28.** $(x+2i)(x-2i)$

29. $\frac{2+i}{3+2i}$

30. Simplify the following power of i: i^{23}.

Use a calculator to find the value of each expression accurate to 4 decimal places.

31. $\sqrt[5]{119}$ **32.** $32^{-\frac{3}{5}}$ **33.** $\left(\sqrt{2}+6\right)\left(\sqrt{2}-1\right)$ **34.** $\frac{\sqrt{3}-\sqrt{5}}{\sqrt{7}-\sqrt{10}}$

21. a.

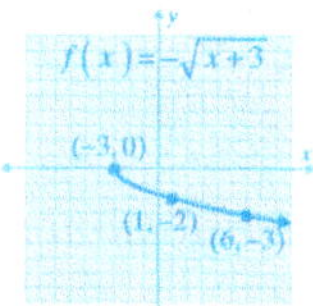

b.

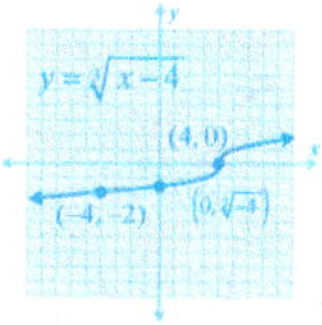

22. Real part is 6, imaginary part is $-\frac{1}{3}$

23. Real part is $-\frac{2}{5}$, imaginary part is $\frac{3}{5}$

24. $x = \frac{11}{2}, y = 3$

25. $16 + 4i$

26. $-5 + 16i$

27. $23 - 14i$

28. $x^2 + 4$

29. $\frac{8}{13} - \frac{1}{13}i$

30. $0 - i$

31. 2.6008

32. 0.1250

33. 3.0711

34. 0.9758

Cumulative Review: Chapters 1 – 9

1. Identity property of addition
2. Distributive property
3. $x = \frac{1}{21}$
4. No solution
5. $n = -\frac{6}{5}, 4$
6. 
$(-\infty, 6)$
7. 
$[0, 2]$
8. 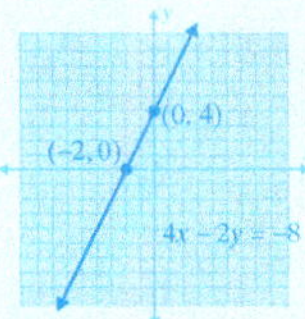

9. 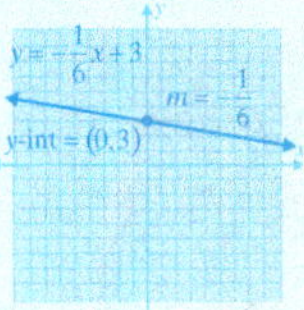

10. $x = 5$
11. $y = -4x - 2$
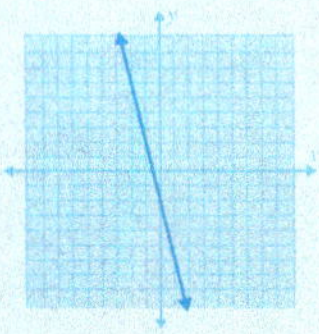
12. a. Function; $D = (-\infty, \infty)$; $R = (-\infty, 0]$
b. Not a function; $D = (-\infty, \infty)$; $R = (-\infty, 5]$
c. Function; $D = [-2\pi, 2\pi]$; $R = [-2, 2]$
13. 1×10^3
14. a. −12 b. 0 c. −168

State the property of real numbers that is illustrated. Assume x is a positive real number.

1. $\sqrt{x} + 0 = \sqrt{x}$ **2.** $7(4-2) = 7 \cdot 4 - 7 \cdot 2$

Solve each of the following equations.

3. $15x + 6(x+1) = 7$ **4.** $4(5x-1) = 10(2x+3)$ **5.** $|5n-7| = 13$

Solve the inequalities and graph the solution sets. Write each solution using interval notation. Assume that x is a real number.

6. $6x - 2 < 4x + 10$ **7.** $2|x-1| + 4 \le 6$

8. Graph the linear equation $4x - 2y = -8$ by locating the y-intercept and x-intercept.

9. Write the equation $x + 6y = 18$ in slope-intercept form. Then find the slope and the y-intercept, and use them to draw the graph.

10. Find an equation for the vertical line through the point $(5, -7)$.

11. Find an equation in standard form for the line passing through the point $\left(\frac{1}{4}, -3\right)$ with slope $m = -4$. Graph the line.

12. The graphs of three curves are shown. Use the vertical line test to determine whether or not each graph represents a function. Then state the domain and range of each graph.

a. 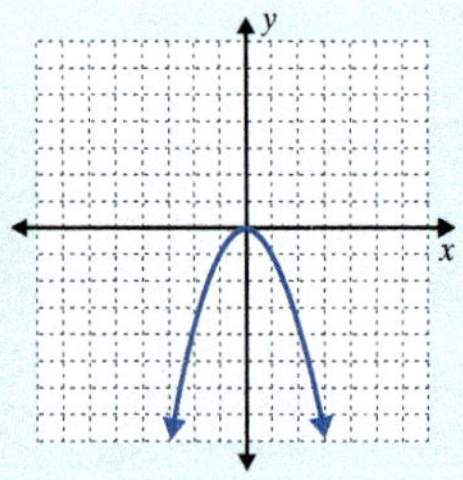

b. 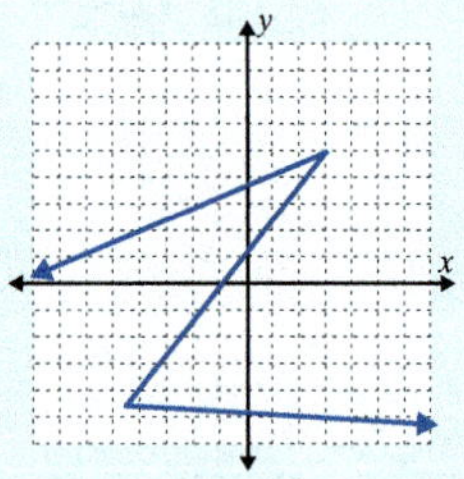

c. 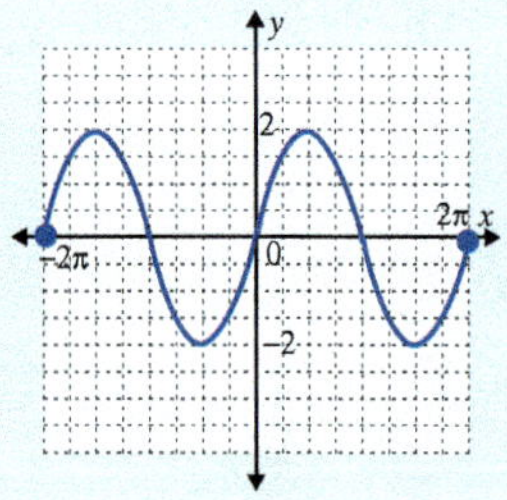

13. Write the expression $\frac{0.008 \times 40000}{320 \times 0.001}$ in scientific notation and simplify.

14. Given $P(x) = x^3 - 8x^2 + 19x - 12$, find **a.** $P(0)$ **b.** $P(4)$ and **c.** $P(-3)$.

Perform the indicated operations.

15. $(x^2 + 7x - 5) - (-2x^3 + 5x^2 - x - 1)$ **16.** $(2x-7)(3x-1)$

17. $(5x+2)(4-x)$

18. $(7x-2)^2$

19. Use the division algorithm to divide and write the answer in the form $Q+\frac{R}{D}$ where the degree of $R <$ the degree of D.

$$\frac{x^3-7x^2+2x-15}{x^2+2x-1}$$

Factor completely.

20. $28+x-2x^2$

21. $64y^4+32y^3+4y^2$

22. $5x^3-320$

23. x^3+4x^2-x-4

Solve each quadratic equation by factoring.

24. $x^2-13x-48=0$

25. $x=2x^2-6$

26. $0=15x^2-11x+2$

27. a. Find an equation that has $x = 4$ and $x = 7$ as roots.

b. Find an equation that has $x = -2$, $x = 12$, and $x = 1$ as roots.

Perform the indicated operations and reduce if possible.

28. $\frac{x+1}{x^2+x-6}+\frac{3x-2}{x^2-2x-15}$

29. $\frac{2x+5}{4x^2-1}-\frac{2-x}{2x^2+7x+3}$

Simplify the complex fractions.

30. $\frac{1-\frac{1}{x^2}}{\frac{2}{x}-\frac{4}{x^2}}$

31. $\frac{x+2-\frac{12}{x+3}}{x-5+\frac{16}{x+3}}$

State the restrictions on the variables and solve each equation.

32. $\frac{3}{x}+\frac{2}{x+5}=\frac{8}{3x}$

33. $\frac{9}{x+7}+\frac{3x}{x^2+4x-21}=\frac{8}{x-3}$

34. Solve the following system of linear equations by graphing both equations and locating the point of intersection.

$$\begin{cases} 3x-2y=7 \\ x+3y=-5 \end{cases}$$

15. $2x^3-4x^2+8x-4$
16. $6x^2-23x+7$
17. $-5x^2+18x+8$
18. $49x^2-28x+4$
19. $x-9+\frac{21x-24}{x^2+2x-1}$
20. $(7+2x)(4-x)$
21. $4y^2(4y+1)^2$
22. $5(x-4)(x^2+4x+16)$
23. $(x+4)(x+1)(x-1)$
24. $(x-16)(x+3)=0$
$x=-3, 16$
25. $(2x+3)(x-2)=0$
$x=-\frac{3}{2}, 2$
26. $(5x-2)(3x-1)=0$
$x=\frac{1}{3}, \frac{2}{5}$
27. a. $x^2-11x+28=0$
b. $x^3-11x^2-14x+24=0$
28. $\frac{4x^2-12x-1}{(x+3)(x-2)(x-5)}$
29. $\frac{4x^2+6x+17}{(2x+1)(2x-1)(x+3)}$
30. $\frac{x^2-1}{2x-4}$
31. $\frac{x+6}{x-1}$
32. $x \neq -5, 0$; $x=-\frac{5}{7}$
33. $x \neq -7, 3$; $x=\frac{83}{4}$
34. $(1, -2)$

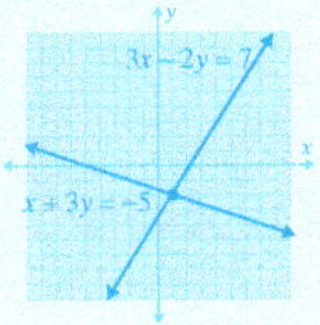

35. $(14, 6)$
36. No solution
37. $(1, 1, 1)$
38. 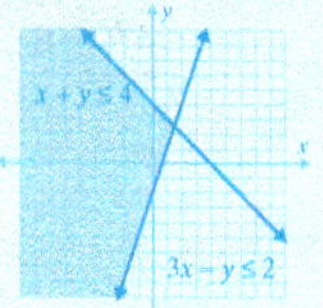

39. $\dfrac{1}{16}$
40. $x^{\frac{7}{3}}$
41. $12\sqrt{2}$
42. $2x^2y^3\sqrt[3]{2y}$
43. $\dfrac{\sqrt{10y}}{2y}$
44. $2\sqrt{30}-11$
45. $x=-1$
46. $x=-1$
47. $(-\infty, 3]$
48. $x=6, y=-5$
49. $2+2i$
50. $14-2i$
51. $-\dfrac{8}{17}-\dfrac{19}{17}i$
52. 1
53. $n=\dfrac{s-a}{d}+1$ or $\dfrac{s-a+d}{d}$
54. $p=\dfrac{A}{1+rt}$

Solve each system by using either the substitution method or the addition method (whichever seems better to you).

35. $\begin{cases} -2y = x-26 \\ y = 2x-22 \end{cases}$ **36.** $\begin{cases} x+y=-5 \\ \frac{1}{2}x+\frac{1}{2}y=\frac{7}{2} \end{cases}$

37. Solve the following system of linear equations using Gaussian elimination. You may use your calculator.

$$\begin{cases} x+y-z=1 \\ 3x-y+z=3 \\ -x+2y+2z=3 \end{cases}$$

38. Solve the following system of inequalities graphically: $\begin{cases} x+y\le 4 \\ 3x-y\le 2 \end{cases}$

Simplify the expressions. Assume that all variables represent positive real numbers.

39. $8^{-\frac{4}{3}}$ **40.** $\left(x^{\frac{1}{2}}\cdot x^{\frac{2}{3}}\right)^2$ **41.** $\sqrt{288}$ **42.** $\sqrt[3]{16x^6y^{10}}$

Rationalize the denominator and simplify if possible.

43. $\dfrac{\sqrt{5}}{\sqrt{2y}}$ **44.** $\dfrac{\sqrt{5}-\sqrt{6}}{\sqrt{5}+\sqrt{6}}$

Solve the radical equations given. Be sure to check your answers in the original equation.

45. $\sqrt{x+10}+1=x+5$ **46.** $\sqrt[3]{x-7}+3=1$

47. Use interval notation to indicate the domain of the radical function $\sqrt[4]{6-2x}$.

48. Solve the equation $x+yi=10+3i-4-8i$ for x and y.

Perform the indicated operations and simplify. Write your answers in standard form.

49. $(3-4i)+(7+7i)-(8+i)$ **50.** $(2+2i)(3-4i)$

51. $\dfrac{4-3i}{1+4i}$ **52.** i^{44}

Solve for the indicated variable.

53. Solve $s=a+(n-1)d$ for n. **54.** Solve $A=p+prt$ for p.

55. a.

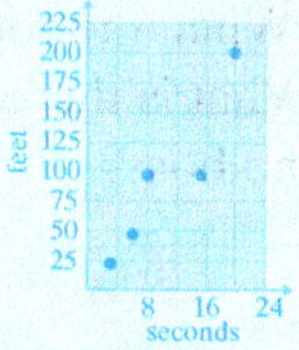

b. $m_1 = 8, m_2 = 24,$
$m_3 = 0, m_4 = \frac{104}{5}$

c. The distance increased by 8 ft/s, then increased by 24 ft/s, then stayed constant, then increased by $\frac{104}{5}$ ft/s.

56. 15, 17

57. $\frac{50}{3}$ in.^{3} or $16.\overline{6}$ in.^{3}

58. 8 ohms

59. 10 mph

60. \$35,000 in 6% and \$15,000 in 10%

61. 7.5 hours

62. 60 pounds of the first type (\$1.25 candy) and 40 pounds of the second type (\$2.50 candy)

63. Burrito price is \$2.60 and taco price is \$1.35

55. Skiing: The following table shows the number of feet that Linda skied downhill from one second to another. (She never skied uphill, but she did stop occasionally.)

Time Elapsed (in seconds)	Distance Traveled (in feet)
3	24
6	48
8	96
15	96
20	200

a. Plot the points indicated in the table.
b. Calculate the slope of the line segments from point to point.
c. Interpret each slope as a rate of change.

56. The sum of the squares of two consecutive positive odd integers is 514. What are the integers?

57. Volume of a gas: V (volume) varies inversely as P (pressure) when a gas is enclosed in a container. In a particular situation, the volume of gas is 25 in.^{3} when a force of 10 lb per in.^{2} is exerted. What would be the volume of gas if a pressure of 15 lb per in.^{2} were to be used?

58. Resistance of a wire: The resistance R (in ohms), in a wire is directly proportional to the length L and inversely proportional to the square of the diameter of the wire. The resistance of a wire 500 ft long with a diameter of 0.01 in. is 20 ohms. What is the resistance of a wire of the same type that is 200 ft long?

59. Boating: Susan traveled 25 miles downstream. In the same length of time, she could have traveled 15 miles upstream. If the speed of the current is 2.5 mph, find the speed of Susan's boat in still water.

60. Investing: Harold has \$50,000 that he wants to invest in two accounts. One pays 6% interest, and the other (at a higher risk) pays 10% interest. If he wants a \$3600 annual return on these two investments, how much should he put into each account?

61. Reporting Sales: Robin can prepare a monthly sales report in 5 hours. If Mac helps her, together they can prepare the report in 3 hours. How long would it take Mac if he worked alone?

62. Selling candy: A grocer plans to make up a special mix of two popular kinds of candy for Halloween. He wants to mix a total of 100 pounds to sell for \$1.75 per pound. Individually, the two types sell for \$1.25 and \$2.50 per pound. How many pounds of each of the two kinds should he put in the mix?

63. Eating out: For lunch, Jason had one burrito and 2 tacos. Matt had 2 burritos and 3 tacos. If Jason spent \$5.30 and Matt spent \$9.25, find the price of one burrito and the price of one taco.

64. 26.5 meters
65. a. $\sqrt{6}$ or 2.45 seconds
b. 96 feet
c. $\sqrt{15}$ or 3.87 seconds
66. $x \approx -1.372, 4.372$

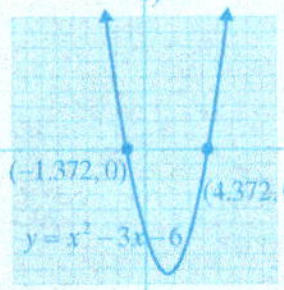

67. $x = 1$

$x \approx -2.303, 1.303$

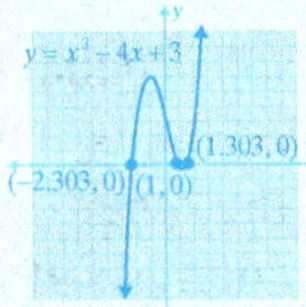

68. $x = -4, 1, 5$

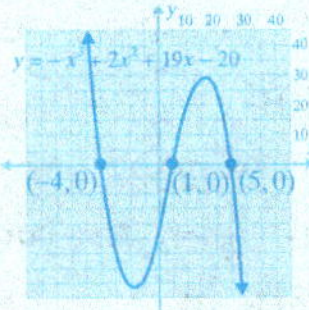

69. 279,936.0000
70. 0.0400
71. 1.3027
72. 6.4552

64. Docking a boat: A boat is being pulled to the shore from a dock. When the rope to the boat is 40 meters long, the boat is 30 meters from the dock. What is the height of the dock (to the nearest tenth of a meter) above the deck of the boat?

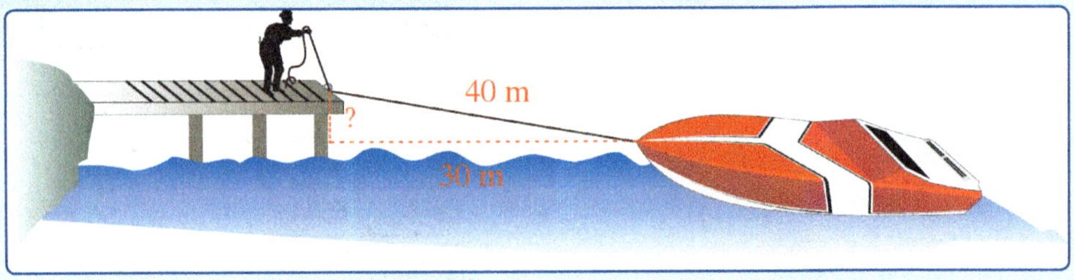

65. Dropping a ball: A ball is dropped from the top of a building that is 240 feet tall. Use the formula $h = -16t^2 + v_0 t + h_0$ to answer the following questions.

a. When will the ball be 144 feet above the ground?

b. How far will the ball have dropped at that time?

c. When will the ball hit the ground?

Use a graphing calculator to first graph each function and then use the CALC *features to find the zeros of each function.*

66. $y = x^2 - 3x - 6$ **67.** $y = x^3 - 4x + 3$ **68.** $y = -x^3 + 2x^2 + 19x - 20$

Use a calculator to find the value of each expression accurate to 4 decimal places.

69. $36^{\frac{7}{2}}$ **70.** $125^{-\frac{2}{3}}$ **71.** $\sqrt{5}\left(\sqrt{21} - 4\right)$ **72.** $\dfrac{13 + \sqrt{12}}{5 - \sqrt{6}}$

CHAPTER 10

Quadratic Equations

Did You Know?

Much of classical algebra has focused on the problem of solving equations, such as the second-degree equation $ax^2+bx+c=0$, which is studied in this chapter. In Section 10.2, you will be introduced to the quadratic formula, a formula that has been known since approximately 2000 B.C. This formula expresses the roots of the quadratic equation in terms of the coefficients a, b, and c. Italian mathematicians during the 16th century also discovered that general cubic equations, in the form $ax^3+bx^2+cx+d=0$, and general fourth-degree equations, in the form $ax^4+bx^3+cx^2+dx+e=0$ could be solved by similar types of formulas. Not many people are aware that these general formulas exist. They can be found in a book of mathematical formulas or in a theory of equations text.

After the discovery of general formulas for the first four cases of polynomial equations (linear, quadratic, cubic, and quartic), it was assumed that a fifth-degree equation also could be solved. However, in the early 1800s, two brilliant young mathematicians proved that the fifth-degree equation was not solvable by algebraic formulas involving the coefficients, as the previous four cases had been. The mathematicians were Niels Henrik Abel (1802 – 1829) and Evariste Galois (1811 – 1832).

Galois

Galois was a fascinating person whose life ended in a romantic duel that may have been arranged by right-wing politicians to eliminate the brilliant young radical. He was involved in the French Revolution and was expelled from school because of his political activity. His brilliance in mathematics was unrecognized because of his youth, his bitterness toward organized science, his radical politics, and the sophistication of his work. The night before his duel, anticipating his death, Galois sat down and wrote out all of the creative mathematics that he could–much of it abstract algebra that he had worked out mentally but not recorded. Included in this work was the unsolvability of fifth-degree equations. Galois' proof that fifth-degree equations could not be solved kept mathematicians from trying to do the impossible and also opened a whole new field in higher mathematics known as **group theory**.

Abel's life ended less dramatically, but also tragically. He died of tuberculosis brought about by poverty. Ironically, he died two days before he was offered a professorship at the University of Berlin. Abel and Galois both left mathematical legacies that kept future mathematicians engaged in developing their ideas up to the present time.

Abel

"Through and through the world is infested with quantity: To talk sense is to talk quantities. It is no use saying the nation is large–How large? It is no use saying that radium is scarce–How scarce? You cannot evade quantity. You may fly to poetry and music, and quantity and number will face you in your rhythms and your octaves."

Alfred North Whitehead (1861 – 1947)

Quadratic equations appear in one form or another in almost every course in mathematics and in many courses in related fields such as business, biology, engineering, and computer science. In this chapter, we will discuss three techniques for solving quadratic equations: factoring, completing the square, and the quadratic formula. Solving quadratic equations by factoring has already been discussed in Section 6.6 and, when possible, is generally considered the method of first choice because it is easiest to apply and is useful in other mathematical situations. However, the **quadratic formula** is very important and should be memorized as it works in all cases. A part of the formula, called the **discriminant**, gives ready information about the nature of the solutions. Additionally, the formula is easy to use in computer programs.

Similarly, the concept of a **function** is present in many aspects of our daily lives. The speed at which you drive your car is a function of how far you depress the accelerator; your energy level is a function of the amount and type of food you eat; your grade in this class is a function of the quality time you spend studying. In this text we quantify these ideas by dealing only with functions involving pairs of real numbers. As we have seen in Chapters 4, 5, and 8, this restriction allows us to analyze functions in terms of their graphs in the Cartesian coordinate system. In Section 10.5, we will be particularly interested in a special category of functions called quadratic functions whose graphs are **parabolas** (curves that, among other things, can be used to describe the paths of projectiles).

10.1 Solving Quadratic Equations

- *Solve quadratic equations by factoring.*
- *Solve quadratic equations using the definition of square root.*
- *Solve quadratic equations by completing the square.*
- *Write equations with given roots.*

Solving Quadratic Equations by Factoring

Not every polynomial can be factored so that the factors have integer coefficients, and not every polynomial equation can be solved by factoring. However, when the solutions of a polynomial equation can be found by factoring, the method depends on the **zero-factor property**, which is restated here for easy reference.

Zero-Factor Property

If the product of two (or more) factors is 0, then at least one of the factors must be 0. That is, if a and b are real numbers, then

if $a \cdot b = 0$, then $a = 0$ or $b = 0$ or both.

Also, as discussed in Section 6.6, polynomial equations of second-degree are called **quadratic equations** and, because these are the equations of interest in this chapter, the definition is restated here.

Quadratic Equations

Quadratic equations are equations that can be written in the form

$\boldsymbol{ax^2 + bx + c = 0}$ where a, b, and c are real numbers and $a \neq 0$.

The procedure for solving quadratic equations by factoring involves making sure that one side of the equation is 0 and then applying the zero-factor property. The following list of steps outlines the procedure.

To Solve a Quadratic Equation by Factoring

1. Add or subtract terms as necessary so that 0 is on one side of the equation and the equation is in the standard form $\boldsymbol{ax^2 + bx + c = 0}$ where a, b, and c are real numbers and $\boldsymbol{a \neq 0}$.
2. Factor completely. (If there are any fractional coefficients, multiply each term by the least common denominator so that all coefficients will be integers.)
3. Set each nonconstant factor equal to 0 and solve each linear equation for the unknown.
4. Check each solution, one at a time, in the original equation.

Example 1: Solving Quadratic Equations by Factoring

Solve the following quadratic equations by factoring.

a. $x^2 - 15x = -50$

Solution: $x^2 - 15x = -50$

$x^2 - 15x + 50 = 0$ Add 50 to both sides. **One side must be 0.**

$(x-5)(x-10) = 0$ Factor the left-hand side.

$x - 5 = 0$ or $x - 10 = 0$ Set each factor equal to 0.

$x = 5$ $\quad$ $x = 10$ Solve each linear equation.

Check: $(5)^2 - 15\cdot(5) \stackrel{?}{=} -50$ $\qquad$ $(10)^2 - 15\cdot(10) \stackrel{?}{=} -50$

$25 - 75 \stackrel{?}{=} -50$ $\qquad$ $100 - 150 \stackrel{?}{=} -50$

$-50 = -50$ $\qquad$ $-50 = -50$

Continued on the next page...

b. $3x^2 - 24x = -48$

Solution: $3x^2 - 24x = -48$

$3x^2 - 24x + 48 = 0$ — Add 48 to both sides. **One side must be 0.**

$3(x^2 - 8x + 16) = 0$ — Factor out the GCF, 3.

$3(x-4)^2 = 0$ — The trinomial is a perfect square.

$x - 4 = 0$ — Two factors are the same.

$x = 4$ — The solution is a **double root**.

Check: $3 \cdot (4)^2 - 24 \cdot (4) \stackrel{?}{=} -48$

$48 - 96 \stackrel{?}{=} -48$

$-48 = -48$

Now that complex numbers have been introduced and discussed, we will see that quadratic equations may have nonreal complex solutions. **In particular, the sum of two squares (previously declared "not factorable" because real factors were implied) can be factored as the product of complex conjugates.** For example,

$$x^2 + 9 = (x + 3i)(x - 3i).$$

As shown in Example 2, such factors can lead to nonreal solutions to a quadratic equation.

Example 2: Quadratic Equations Involving the Sum of Two Squares

Solve the following quadratic equation by factoring: $x^2 + 4 = 0$

Solution: $x^2 + 4 = 0$

$(x + 2i)(x - 2i) = 0$

Note that $x^2 + 4$ is the sum of two squares and can be factored into the product of conjugates of complex numbers.

$x + 2i = 0$ or $x - 2i = 0$

$x = -2i$ $\qquad$ $x = 2i$

Check: $(-2i)^2 + 4 \stackrel{?}{=} 0$ $\qquad$ $(2i)^2 + 4 \stackrel{?}{=} 0$

$4i^2 + 4 \stackrel{?}{=} 0$ $\qquad$ $4i^2 + 4 \stackrel{?}{=} 0$

$-4 + 4 \stackrel{?}{=} 0$ $\qquad$ $-4 + 4 \stackrel{?}{=} 0$

$0 = 0$ $\qquad$ $0 = 0$

Teaching Notes:
You might want to explain how to use fractional exponents in solving equations. This may help your students understand solving higher degree equations later. For example,

$$x^2 = 13$$
$$\left(x^2\right)^{\frac{1}{2}} = 13^{\frac{1}{2}}$$
$$|x| = \sqrt{13}$$
$$x = \pm\sqrt{13}.$$

Using the Definition of Square Root and the Square Root Property

Consider the equation $x^2 = 13.$

Allowing that the variable x might be positive or negative, we use the definition of square root, $\sqrt{x^2} = |x|$.

Taking the square root of both sides of the equation gives: $|x| = \sqrt{13}.$

So, we have two solutions, $x = \sqrt{13}$ or $x = -\sqrt{13}.$

To indicate both solutions, we write $x = \pm\sqrt{13}.$

Similarly, for the equation $(x-3)^2 = 5$

the same process gives $x - 3 = \pm\sqrt{5}.$

This leads to the two equations and the two solutions, as follows:

$$x - 3 = \sqrt{5} \quad \text{or} \quad x - 3 = -\sqrt{5}$$
$$x = 3 + \sqrt{5} \qquad\qquad x = 3 - \sqrt{5}$$

We can write the two solutions in the form $x = 3 \pm \sqrt{5}.$

This discussion shows how the definition of a square root can be used to solve quadratic equations in certain forms. In particular, if one side of the equation is a squared expression and the other side is a constant, we can simply take the square root of both sides. If the constant is negative, then the solutions will involve nonreal numbers. We are using the following **square root property**.

Square Root Property

If $x^2 = c$, then $x = \pm\sqrt{c}$.

If $(x-a)^2 = c$, then $x - a = \pm\sqrt{c}$ $\left(\text{or } x = a \pm \sqrt{c}\right)$.

Note: If c is negative $(c < 0)$, then the solutions will be nonreal.

Example 3: Using the Square Root Property to Solve Quadratic Equations

Teaching Notes:
Early on, remind students that the "±" symbol is an economical way of representing two solutions simultaneously.

Solve the following quadratic equations using the square root property.

a. $(y+4)^2 = 8$

Solution: $(y+4)^2 = 8$ Note that $c > 0$.

$$y+4 = \pm\sqrt{8}$$

$$y = -4 \pm 2\sqrt{2}$$

b. $x^2 = -25$

Solution: $x^2 = -25$ Note that $c < 0$.

$$x = \pm\sqrt{-25}$$

$$x = \pm 5i$$

Completing the Square

Recall that a perfect square trinomial (Section 5.6) is the result of squaring a binomial. Our objective here is to find the third term of a perfect square trinomial when the first two terms are given. This is called **completing the square**. We will find this procedure useful in solving quadratic equations and in developing the quadratic formula.

The following examples illustrate this concept.

Perfect Square Trinomials		Equal Factors		Square of a Binomial
$x^2 - 8x + 16$	=	$(x-4)(x-4)$	=	$(x-4)^2$
$x^2 + 20x + 100$	=	$(x+10)(x+10)$	=	$(x+10)^2$
$x^2 - 9x + \frac{81}{4}$	=	$\left(x-\frac{9}{2}\right)\left(x-\frac{9}{2}\right)$	=	$\left(x-\frac{9}{2}\right)^2$
$\mathbf{x^2 - 2ax + a^2}$	=	$\mathbf{(x-a)(x-a)}$	=	$\mathbf{(x-a)^2}$
$\mathbf{x^2 + 2ax + a^2}$	=	$\mathbf{(x+a)(x+a)}$	=	$\mathbf{(x+a)^2}$

Table 1

The last two examples in Table 1 are in the form of formulas. We see two things in each case:

Teaching Notes:
Some of the *power* of algebra can be illustrated by pointing out how the general representation of the perfect square trinomial, $x^2 + 2ax + a^2$, leads to the procedure for completing the square. This is another example of the advantage in learning to "read" the algebraic form of expressions and equations.

1. The leading coefficient (the coefficient of x^2) is 1.
2. The constant term is the square of $\frac{1}{2}$ of the coefficient of x.

   For example, $\frac{1}{2}(2a) = a$ and the square of this result is the constant a^2.

What constant should be added to $x^2 - 16x$ to get a perfect square trinomial? By following the ideas just discussed, we find that $\frac{1}{2}(-16) = -8$ and $(-8)^2 = 64$. Therefore, to complete the square, we add 64. Thus

$$x^2 - 16x + 64 = (x-8)^2.$$

By adding 64, we have completed the square for $x^2 - 16x$.

Example 4: Completing the Square

Add the constant that will complete the square for each expression, and write the new expression as the square of a binomial.

a. $x^2 + 10x$

Solution: $x^2 + 10x + \underline{\ ?\ } = (\underline{\quad ?\quad})^2$

$\frac{1}{2}(10) = 5$ and $(5)^2 = 25$

Find $\frac{1}{2}$ the coefficient of x and square the result. Add this square constant to complete the square. The resulting trinomial will equal the square of a binomial.

So, add 25: $x^2 + 10x + \underline{25} = (\underline{x+5})^2$

b. $x^2 - 7x$

Solution: $x^2 - 7x + \underline{\ ?\ } = (\underline{\quad ?\quad})^2$

$\frac{1}{2}(-7) = -\frac{7}{2}$ and $\left(-\frac{7}{2}\right)^2 = \frac{49}{4}$

Find $\frac{1}{2}$ the coefficient of x and square the result.

So, add $\frac{49}{4}$: $x^2 - 7x + \underline{\frac{49}{4}} = \left(\underline{x - \frac{7}{2}}\right)^2$

Solving Quadratic Equations by Completing the Square

Now we want to use the process of completing the square to help in solving quadratic equations. This technique involves the following steps.

To Solve a Quadratic Equation by Completing the Square

1. If necessary, divide or multiply both sides of the equation so that the leading coefficient (the coefficient of x^2) is 1.
2. If necessary, isolate the constant term on one side of the equation.
3. Find the constant that completes the square of the polynomial and **add this constant to both sides**. Rewrite the polynomial as the square of a binomial.
4. Use the square root property to find the solutions of the equation.

Example 5: Solving Quadratic Equations by Completing the Square

Solve the following quadratic equations by completing the square.

a. $x^2 - 8x = 25$

Solution: $x^2 - 8x = 25$ The coefficient of x^2 is already 1 and the constant is isolated on one side of the equation.

$x^2 - 8x + 16 = 25 + 16$ Complete the square: $\frac{1}{2}(-8) = -4$ and $(-4)^2 = 16$. Therefore, add 16 to both sides.

$(x-4)^2 = 41$ Factor the polynomial.

$x - 4 = \sqrt{41}$ or $x - 4 = -\sqrt{41}$ Use the square root property.

$x = 4 + \sqrt{41}$ $x = 4 - \sqrt{41}$

There are two real solutions: $4 + \sqrt{41}$ and $4 - \sqrt{41}$.

We write $x = 4 \pm \sqrt{41}$.

b. $3x^2 + 6x - 15 = 0$

Solution: $3x^2 + 6x - 15 = 0$

$\frac{3x^2}{3} + \frac{6x}{3} - \frac{15}{3} = \frac{0}{3}$ Divide each term by 3. **The leading coefficient must be 1**

$x^2 + 2x - 5 = 0$

$x^2 + 2x = 5$ Isolate the constant term.

Teaching Notes:
In Example 5b, we divide by 3 instead of factoring out 3. This process will help in understanding the development of the quadratic formula. This will also eliminate concerns about what to do with 3 if it is a factor.

$x^2+2x+1=5+1$ Complete the square: $\frac{1}{2}(2)=1$ and $1^2=1$. Therefore, add 1 to both sides.

$(x+1)^2=6$ Factor the polynomial.

$x+1=\pm\sqrt{6}$ Use the square root property.

$x=-1\pm\sqrt{6}$

c. $2x^2+2x-7=0$

Solution: $2x^2+2x-7=0$

$x^2+x-\frac{7}{2}=0$ Divide each term by 2 so that the leading coefficient will be 1.

$x^2+x=\frac{7}{2}$ Isolate the constant term.

$x^2+x+\frac{1}{4}=\frac{7}{2}+\frac{1}{4}$ Complete the square: the coefficient of x is 1 and $\frac{1}{2}(1)=\frac{1}{2}$ and $\left(\frac{1}{2}\right)^2=\frac{1}{4}$. Therefore, add $\frac{1}{4}$ to both sides.

$\left(x+\frac{1}{2}\right)^2=\frac{15}{4}$ Factor the polynomial.

$x+\frac{1}{2}=\pm\sqrt{\frac{15}{4}}$ Use the square root property.

$x=-\frac{1}{2}\pm\frac{\sqrt{15}}{2}$

$x=\frac{-1\pm\sqrt{15}}{2}$

d. $x^2-2x+13=0$

Solution: $x^2-2x+13=0$

$x^2-2x=-13$ Isolate the constant term.

$x^2-2x+1=-13+1$ Complete the square: $\frac{1}{2}(-2)=-1$ and $(-1)^2=1$. Therefore, add 1 to both sides.

$(x-1)^2=-12$ Factor the polynomial.

$x-1=\pm\sqrt{-12}$ Use the square root property.

$x-1=\pm i\sqrt{12}=\pm 2i\sqrt{3}$

$x=1\pm 2i\sqrt{3}$ The solutions are nonreal complex conjugates.

Writing Equations with Known Roots

In Section 6.6, we found equations with known roots by setting the product of factors equal to 0 and simplifying. The same method is applied here with roots that are nonreal and roots that involve radicals.

Example 6: Equations with Known Roots

Find polynomial equations that have the given roots.

a. $y = 3 + 2i$ and $y = 3 - 2i$

Solution: $y = 3 + 2i \qquad y = 3 - 2i$

$y - 3 - 2i = 0 \qquad y - 3 + 2i = 0$ Get 0 on one side of each equation.

Set the product of the two factors equal to 0 and simplify.

$$(y-3-2i)(y-3+2i) = 0$$

$$[(y-3)-2i][(y-3)+2i] = 0$$

$$(y-3)^2 - 4i^2 = 0$$

Regroup the terms to represent the product of complex conjugates. This makes the multiplication easier.

$$y^2 - 6y + 9 + 4 = 0$$

Remember, $i^2 = -1$.

$$y^2 - 6y + 13 = 0$$

This equation has two solutions: $y = 3 + 2i$ and $y = 3 - 2i$.

b. $x = 5 - \sqrt{2}$ and $x = 5 + \sqrt{2}$

Solution: $x = 5 - \sqrt{2} \qquad x = 5 + \sqrt{2}$

$x - 5 + \sqrt{2} = 0 \qquad x - 5 - \sqrt{2} = 0$ Get 0 on one side of each equation.

Set the product of the two factors equal to 0 and simplify.

$$(x-5+\sqrt{2})(x-5-\sqrt{2}) = 0$$

$$[(x-5)+\sqrt{2}][(x-5)-\sqrt{2}] = 0$$

Regroup the terms to make the multiplication easier.

$$(x-5)^2 - (\sqrt{2})^2 = 0$$

$$x^2 - 10x + 25 - 2 = 0$$

$$x^2 - 10x + 23 = 0$$

This equation has two solutions: $x = 5 - \sqrt{2}$ and $x = 5 + \sqrt{2}$

c. $x = 3 + i\sqrt{5}$ and $x = 3 - i\sqrt{5}$

Solution: $x = 3 + i\sqrt{5} \qquad x = 3 - i\sqrt{5}$

$x - 3 - i\sqrt{5} = 0 \qquad x - 3 + i\sqrt{5} = 0$ Get 0 on one side of each equation.

1. $x = 3, 15$
2. $y = -\frac{1}{2}, 2$
3. $x = 5, 12$
4. $x = -\frac{3}{7}, 2$
5. $y = -11, 11$
6. $x = -1, 0, 1$
7. $z = -6, 0, 2$
8. $x = 0, 2, 3$
9. $x = -5$
10. $z = -13, 7$

Set the product of the two factors equal to 0 and simplify.

$$\left(x-3-i\sqrt{5}\right)\left(x-3+i\sqrt{5}\right)=0$$

$$\left[(x-3)-i\sqrt{5}\right]\left[(x-3)+i\sqrt{5}\right]=0$$ Regroup the terms to make the multiplication easier.

$$(x-3)^2-\left(i\sqrt{5}\right)^2=0$$

$$x^2-6x+9-i^2\left(\sqrt{5}\right)^2=0$$

$$x^2-6x+9-(-1)(5)=0$$ Remember, $i^2=-1$.

$$x^2-6x+14=0$$ This equation has two solutions: $x=3+i\sqrt{5}$ and $x=3-i\sqrt{5}$

Practice Problems

Solve each of the following quadratic equations by completing the square.

1. $x^2+2x+2=0$ **2.** $3x^2-6x+15=0$ **3.** $x^2-24x+72=0$

4. $2x^2+5x-3=0$ **5.** $x^2-3x+1=0$

6. Find a quadratic equation that has the roots $2+3i$ and $2-3i$.

11. $x = \pm 12$
12. $x = \pm 13$
13. $x = \pm 5i$
14. $x = \pm 9i$
15. $x = \pm 2i\sqrt{6}$
16. $x = \pm 3i\sqrt{2}$

10.1 Exercises

17. $x = -1, 5$
18. $x = -1, 9$
19. $x = \pm\sqrt{5}$
20. $x = \pm 2\sqrt{3}$
21. $x = -3 \pm \sqrt{3}$
22. $x = 1 \pm \sqrt{5}$
23. $x = 3 \pm 2i$
24. $x = -8 \pm 3i$
25. $x = -2 \pm i\sqrt{7}$
26. $x = 5 \pm i\sqrt{10}$
27. $x^2 - 12x + \underline{36} = \underline{(x-6)^2}$

Solve the following equations by factoring.

1. $x^2-18x+45=0$ **2.** $12y^2-18y-12=0$ **3.** $2x^2=34x-120$

4. $7x^2=11x+6$ **5.** $3y^2=363$ **6.** $x^4=x^2$

7. $4z^3+16z^2=48z$ **8.** $5x^3-25x^2=-30x$ **9.** $(x+8)(x+2)=-9$

10. $(z+3)^2=100$

Solve the equations using the square root property.

11. $x^2-144=0$ **12.** $x^2-169=0$ **13.** $x^2+25=0$

Answers to Practice Problems: **1.** $x=-1\pm i$ **2.** $x=1\pm 2i$ **3.** $x=12\pm 6\sqrt{2}$ **4.** $x=-3, \frac{1}{2}$ **5.** $x=\frac{3\pm\sqrt{5}}{2}$
6. $x^2-4x+13=0$

28. $y^2 + 14y + \underline{49} = \left(\underline{y+7}\right)^2$

29. $x^2 + 6x + \underline{9} = \left(\underline{x+3}\right)^2$

30. $x^2 + 8x + \underline{16} = \left(\underline{x+4}\right)^2$

31. $x^2 - 5x + \underline{\frac{25}{4}} = \left(\underline{x - \frac{5}{2}}\right)^2$

32. $x^2 + 7x + \underline{\frac{49}{4}} = \left(\underline{x + \frac{7}{2}}\right)^2$

33. $y^2 + y + \underline{\frac{1}{4}} = \left(\underline{y + \frac{1}{2}}\right)^2$

34. $x^2 + \frac{1}{2}x + \underline{\frac{1}{16}} = \left(\underline{x + \frac{1}{4}}\right)^2$

35. $x^2 + \frac{1}{3}x + \underline{\frac{1}{36}} = \left(\underline{x + \frac{1}{6}}\right)^2$

36. $y^2 + \frac{3}{4}y + \underline{\frac{9}{64}} = \left(\underline{y + \frac{3}{8}}\right)^2$

37. $x = -5, 1$

38. $x = -7, 1$

39. $y = -1 \pm \sqrt{6}$

40. $x = 4 \pm \sqrt{13}$

41. $x = 5 \pm \sqrt{22}$

42. $z = -2 \pm \sqrt{6}$

14. $x^2 + 81 = 0$ **15.** $x^2 + 24 = 0$ **16.** $x^2 + 18 = 0$

17. $(x-2)^2 = 9$ **18.** $(x-4)^2 = 25$ **19.** $x^2 = 5$

20. $x^2 = 12$ **21.** $2(x+3)^2 = 6$ **22.** $3(x-1)^2 = 15$

23. $(x-3)^2 = -4$ **24.** $(x+8)^2 = -9$ **25.** $(x+2)^2 = -7$

26. $(x-5)^2 = -10$

Add the correct constant to complete the square; then factor the trinomial as indicated.

27. $x^2 - 12x + \_\_\_ = (\_\_\_\_)^2$ **28.** $y^2 + 14y + \_\_\_ = (\_\_\_\_)^2$

29. $x^2 + 6x + \_\_\_ = (\_\_\_\_)^2$ **30.** $x^2 + 8x + \_\_\_ = (\_\_\_\_)^2$

31. $x^2 - 5x + \_\_\_ = (\_\_\_\_)^2$ **32.** $x^2 + 7x + \_\_\_ = (\_\_\_\_)^2$

33. $y^2 + y + \_\_\_ = (\_\_\_\_)^2$ **34.** $x^2 + \frac{1}{2}x + \_\_\_ = (\_\_\_\_)^2$

35. $x^2 + \frac{1}{3}x + \_\_\_ = (\_\_\_\_)^2$ **36.** $y^2 + \frac{3}{4}y + \_\_\_ = (\_\_\_\_)^2$

Solve the quadratic equations by completing the square.

37. $x^2 + 4x - 5 = 0$ **38.** $x^2 + 6x - 7 = 0$ **39.** $y^2 + 2y = 5$

40. $x^2 + 3 = 8x$ **41.** $x^2 + 3 = 10x$ **42.** $z^2 + 4z = 2$

43. $x^2 - 6x + 10 = 0$ **44.** $x^2 - 2x + 5 = 0$ **45.** $x^2 + 11 = 12x$

46. $x^2 = 6 - x$ **47.** $y^2 = 10y - 4$ **48.** $x^2 = 3 - 4x$

49. $z^2 + 3z - 5 = 0$ **50.** $x^2 - 5x + 5 = 0$ **51.** $x^2 + x + 2 = 0$

52. $y^2 + 3y + 3 = 0$ **53.** $x^2 + 5x + 2 = 0$ **54.** $4x^2 + 7x + 2 = 0$

55. $3x^2 - 10x + 5 = 0$ **56.** $3y^2 + 5y - 3 = 0$ **57.** $3x^2 + 6x + 18 = 0$

58. $4x^2 + 8x + 16 = 0$ **59.** $2x - 3 = 4x^2$ **60.** $2x + 2 = -6x^2$

61. $5y^2 + 15y + 25 = 0$ **62.** $4x^2 + 20x + 32 = 0$ **63.** $3y^2 = 4 - y$

64. $2x^2 + 4 = -9x$ **65.** $2x^2 - 8x + 4 = 0$ **66.** $3x^2 - 18x + 12 = 0$

43. $x=3\pm i$
44. $x=1\pm 2i$
45. $x=1, 11$
46. $x=-3, 2$
47. $y=5\pm\sqrt{21}$
48. $x=-2\pm\sqrt{7}$
49. $z=\frac{-3\pm\sqrt{29}}{2}$
50. $x=\frac{5\pm\sqrt{5}}{2}$
51. $x=\frac{-1\pm i\sqrt{7}}{2}$
52. $y=\frac{-3\pm i\sqrt{3}}{2}$

Write a quadratic equation with integer coefficients that has the given roots.

67. $x=\sqrt{7}, x=-\sqrt{7}$

68. $x=\sqrt{6}, x=-\sqrt{6}$

69. $x=1+\sqrt{3}, x=1-\sqrt{3}$

70. $z=3+\sqrt{2}, z=3-\sqrt{2}$

71. $y=-2+2\sqrt{5}, y=-2-2\sqrt{5}$

72. $x=1+2\sqrt{3}, x=1-2\sqrt{3}$

73. $x=4i, x=-4i$

74. $x=7i, x=-7i$

75. $y=i\sqrt{6}, y=-i\sqrt{6}$

76. $y=i\sqrt{5}, y=-i\sqrt{5}$

77. $x=2+i, x=2-i$

78. $x=-3+2i, x=-3-2i$

79. $x=1+i\sqrt{2}, x=1-i\sqrt{2}$

80. $x=2+i\sqrt{3}, x=2-i\sqrt{3}$

81. $x=-5+2i\sqrt{6}, x=-5-2i\sqrt{6}$

82. $y=4+3i\sqrt{2}, y=4-3i\sqrt{2}$

Writing and Thinking About Mathematics

53. $x=\frac{-5\pm\sqrt{17}}{2}$
54. $x=\frac{-7\pm\sqrt{17}}{8}$

83. Explain, in your own words, the steps involved in the process of solving a quadratic equation by completing the square.

HAWKES LEARNING SYSTEMS: INTRODUCTORY & INTERMEDIATE ALGEBRA SOFTWARE

- 10.1a Quadratic Equations: The Square Root Method
- 10.1b Quadratic Equations: Completing the Square

55. $x=\frac{5\pm\sqrt{10}}{3}$
56. $y=\frac{-5\pm\sqrt{61}}{6}$
57. $x=-1\pm i\sqrt{5}$
58. $x=-1\pm i\sqrt{3}$
59. $x=\frac{1\pm i\sqrt{11}}{4}$
60. $x=\frac{-1\pm i\sqrt{11}}{6}$
61. $y=\frac{-3\pm i\sqrt{11}}{2}$
62. $x=\frac{-5\pm i\sqrt{7}}{2}$
63. $y=-\frac{4}{3},1$
64. $x=-4,-\frac{1}{2}$
65. $x=2\pm\sqrt{2}$
66. $x=3\pm\sqrt{5}$
67. $x^2-7=0$
68. $x^2-6=0$
69. $x^2-2x-2=0$
70. $z^2-6z+7=0$
71. $y^2+4y-16=0$
72. $x^2-2x-11=0$
73. $x^2+16=0$
74. $x^2+49=0$
75. $y^2+6=0$
76. $y^2+5=0$
77. $x^2-4x+5=0$
78. $x^2+6x+13=0$
79. $x^2-2x+3=0$
80. $x^2-4x+7=0$
81. $x^2+10x+49=0$
82. $y^2-8y+34=0$
83. See page 774.

10.2 The Quadratic Formula: $x = \dfrac{-b \pm \sqrt{b^2 - 4ac}}{2a}$

- Develop the quadratic formula: $x = \dfrac{-b \pm \sqrt{b^2 - 4ac}}{2a}$
- Use the quadratic formula to solve quadratic equations.
- Calculate the discriminant: $b^2 - 4ac$
- Use the discriminant to determine the nature of the solutions (one real, two real, or two nonreal) of a quadratic equation.

Developing the Quadratic Formula: $x = \dfrac{-b \pm \sqrt{b^2 - 4ac}}{2a}$

The **quadratic formula** gives the roots of any quadratic equation in terms of the coefficients a, b, and c of the **general quadratic equation**

$$ax^2 + bx + c = 0.$$

Therefore, if you memorize the quadratic formula, you can solve any quadratic equation by simply substituting the coefficients into the formula. To develop the quadratic formula, we solve the general quadratic equation by **completing the square**.

$ax^2 + bx + c = 0$	The general quadratic equation where $a \neq 0$.
$x^2 + \frac{b}{a}x + \frac{c}{a} = \frac{0}{a}$	Divide each term of the equation by a. The leading coefficient must be 1.
$x^2 + \frac{b}{a}x = -\frac{c}{a}$	Add $-\frac{c}{a}$ to both sides of the equation.
$x^2 + \frac{b}{a}x + \frac{b^2}{4a^2} = \frac{b^2}{4a^2} - \frac{c}{a}$	$\frac{1}{2}\left(\frac{b}{a}\right) = \frac{b}{2a}$ and $\left(\frac{b}{2a}\right)^2 = \frac{b^2}{4a^2}$. Add $\frac{b^2}{4a^2}$ to both sides of the equation.
$\left(x + \frac{b}{2a}\right)^2 = \frac{b^2}{4a^2} - \frac{4ac}{4a^2}$	Factor the left side. $4a^2$ is the common denominator on the right-hand side.
$\left(x + \frac{b}{2a}\right)^2 = \frac{b^2 - 4ac}{4a^2}$	Simplify.
$x + \frac{b}{2a} = \pm\sqrt{\frac{b^2 - 4ac}{4a^2}}$	Use the square root property.
$x + \frac{b}{2a} = \pm\frac{\sqrt{b^2 - 4ac}}{2a}$	Simplify.
$x = -\frac{b}{2a} \pm \frac{\sqrt{b^2 - 4ac}}{2a}$	Solve for x.
$x = \frac{-b \pm \sqrt{b^2 - 4ac}}{2a}$	**This equation is called the quadratic formula.**

NOTES

Note about the coefficient *a*

For convenience and without loss of generality, in the development of the quadratic formula (and in the examples and exercises) the leading coefficient a is positive. If a is a negative number, we can multiply both sides of the equation by -1. This will make the leading coefficient positive without changing any solutions of the original equation.

The Quadratic Formula

For the general quadratic equation

$$ax^2 + bx + c = 0 \qquad \text{where } a \neq 0$$

the solutions are

$$x = \frac{-b \pm \sqrt{b^2 - 4ac}}{2a}.$$

The quadratic formula should be memorized.

Applications of quadratic equations are found in such fields as economics, business, computer science, and chemistry, and in almost all branches of mathematics. Most instructors assume that their students know the quadratic formula and how to apply it. You should recognize that the importance of the quadratic formula lies in the fact that **any** quadratic equation can be solved.

Example 1: The Quadratic Formula

Solve the following quadratic equations using the quadratic formula.

a. $x^2 - 5x + 3 = 0$

Solution: Substitute $a = 1$, $b = -5$, and $c = 3$ into the formula:

$$x = \frac{-b \pm \sqrt{b^2 - 4ac}}{2a} = \frac{-(-5) \pm \sqrt{(-5)^2 - 4 \cdot 1 \cdot 3}}{2 \cdot 1}$$

$$= \frac{5 \pm \sqrt{25 - 12}}{2}$$

$$= \frac{5 \pm \sqrt{13}}{2}$$

Continued on the next page...

b. $7x^2 - 2x + 1 = 0$

Solution: Substitute $a = 7$, $b = -2$, and $c = 1$ into the formula.

$$x = \frac{-b \pm \sqrt{b^2 - 4ac}}{2a} = \frac{-(-2) \pm \sqrt{(-2)^2 - 4 \cdot 7 \cdot 1}}{2 \cdot 7}$$

$$= \frac{2 \pm \sqrt{4 - 28}}{14}$$

$$= \frac{2 \pm \sqrt{-24}}{14}$$

$$= \frac{2 \pm 2i\sqrt{6}}{14}$$

$$= \frac{\cancel{2}\left(1 \pm i\sqrt{6}\right)}{\cancel{2} \cdot 7}$$ Factor and reduce.

$$= \frac{1 \pm i\sqrt{6}}{7}$$ The solutions are nonreal complex conjugates. In standard form: $\frac{1}{7} + \frac{\sqrt{6}}{7}i$ and $\frac{1}{7} - \frac{\sqrt{6}}{7}i$.

c. $\frac{3}{4}x^2 - \frac{1}{2}x = \frac{1}{3}$

Solution: The quadratic formula is easier to use with integer coefficients. Multiply each term by the LCD, 12, so that the coefficients will be integers.

$$12 \cdot \frac{3}{4}x^2 - 12 \cdot \frac{1}{2}x = 12 \cdot \frac{1}{3}$$

$$9x^2 - 6x = 4$$

$$9x^2 - 6x - 4 = 0$$ To apply the formula, one side must be 0.

$$x = \frac{-(-6) \pm \sqrt{(-6)^2 - 4 \cdot 9 \cdot -4}}{2 \cdot 9}$$ Substitute $a = 9$, $b = -6$, and $c = -4$.

$$= \frac{6 \pm \sqrt{36 + 144}}{18}$$

$$= \frac{6 \pm \sqrt{180}}{18} = \frac{6 \pm 6\sqrt{5}}{18}$$

$$= \frac{\cancel{6}\left(1 \pm \sqrt{5}\right)}{\cancel{6} \cdot 3} = \frac{1 \pm \sqrt{5}}{3}$$ Factor and reduce.

NOTES

COMMON ERROR

Many students make a mistake when simplifying fractions by dividing the denominator into only one of the terms in the numerator.

$$\frac{4 + \cancel{2}\sqrt{3}}{\cancel{2}} = 4 + \sqrt{3}$$ **INCORRECT**

The correct method is to divide both terms by the denominator or to factor out a common factor in the numerator and then reduce.

$$\frac{4 + 2\sqrt{3}}{2} = \frac{4}{2} + \frac{2\sqrt{3}}{2} = 2 + \sqrt{3}$$ **CORRECT**

$$\frac{4 + 2\sqrt{3}}{2} = \frac{\cancel{2}\left(2 + \sqrt{3}\right)}{\cancel{2}} = 2 + \sqrt{3}$$ **CORRECT**

In Example 2, the equation is third-degree (a cubic equation) and one of the factors is quadratic. The quadratic formula can be applied to this factor.

Example 2: Cubic Equation

Solve the following cubic equation by factoring and using the quadratic formula.

$$2x^3 - 10x^2 + 6x = 0$$

Solution: $2x^3 - 10x^2 + 6x = 0$

$2x\left(x^2 - 5x + 3\right) = 0$ Factor out $2x$.

$2x = 0$ or $x^2 - 5x + 3 = 0$ Set each factor equal to 0.

$x = 0$ $\quad x = \frac{5 \pm \sqrt{13}}{2}$ Solve each equation. (The quadratic equation was solved in Example 1a using the quadratic formula.)

There are 3 solutions: $0, \frac{5 + \sqrt{13}}{2}$, and $\frac{5 - \sqrt{13}}{2}$.

The Discriminant: $b^2 - 4ac$

The expression $\boldsymbol{b^2 - 4ac}$, the part of the quadratic formula that lies under the radical sign, is called the **discriminant**. The discriminant identifies the number and type of solutions to a quadratic equation. Assuming a, b, and c are all real numbers, there are three possibilities: the discriminant is either positive, negative, or zero.

In Example 1a, the discriminant was positive, $b^2-4ac=(-5)^2-4(1)(3)=13$, and there were two real solutions: $x=\frac{5\pm\sqrt{13}}{2}$. In Example 1b, the discriminant was negative, $b^2-4ac=(-2)^2-4(7)(1)=-24$, and there were two nonreal solutions: $x=\frac{1\pm i\sqrt{6}}{7}$.

The discriminant gives the following information:

Discriminant	Nature of Solutions
$b^2-4ac>0$	Two real solutions
$b^2-4ac=0$	One real solution, $x=\frac{-b\pm 0}{2a}=-\frac{b}{2a}$
$b^2-4ac<0$	Two nonreal solutions

Table 1

In the case where $b^2-4ac=0$, we say $x=-\frac{b}{2a}$ is a **double root**. Additionally, if the discriminant is a perfect square, the equation is factorable.

Example 3: Finding the Discriminant

Find the discriminant and determine the nature of the solutions to each of the following quadratic equations.

a. $3x^2+11x-7=0$

Solution: Substitute $a=3$, $b=11$, and $c=-7$ into the discriminant:

$$b^2-4ac=11^2-4(3)(-7)$$
$$=121+84$$
$$=205>0$$ There are two real solutions.

b. $x^2+6x+9=0$

Solution: Substitute $a=1$, $b=6$, and $c=9$ into the discriminant:

$$b^2-4ac=6^2-4(1)(9)$$
$$=36-36$$
$$=0$$ There is one real solution, a double root.

c. $x^2+1=0$

Solution: Here $b=0$. We could write $x^2+0x+1=0$.

$$b^2-4ac=0^2-4(1)(1)$$
$$=0-4$$
$$=-4$$ There are two nonreal solutions.

Teaching Notes:
You might want to discuss the fact that the discriminant will be 0 when the quadratic expression is a perfect square trinomial, which leads to a double root. (See Examples 3b and 4a.)

Example 4: Using the Discriminant

a. Determine the value(s) for c such that $x^2 + 8x + c = 0$ will have one real solution. (**Hint:** Set the discriminant equal to 0 and solve the equation for c.)

Solution:
$$b^2 - 4ac = 0$$
$$8^2 - 4(1)(c) = 0$$
$$64 - 4c = 0$$
$$-4c = -64$$
$$c = 16$$

Check:
$$x^2 + 8x + (16) = 0$$
$$x^2 + 8x + 16 = 0$$
$$(x+4)^2 = 0$$
$$x = -4$$

There is only one real solution. (Thus -4 is a double root.)

b. Determine the values for a such that $ax^2 - 8x + 4 = 0$ will have two nonreal solutions. (**Hint:** Set the discriminant less than 0 and solve for a.)

Solution:
$$b^2 - 4ac < 0$$
$$(-8)^2 - 4(a)(4) < 0$$
$$64 - 16a < 0$$
$$-16a < -64$$
$$a > 4$$

Thus, if a is any real number greater than 4, the discriminant will be negative and the equation will have two nonreal solutions.

Practice Problems

Solve each of the following quadratic equations using the quadratic formula.

1. $x^2 + 2x - 4 = 0$ **2.** $2x^2 - 3x + 4 = 0$ **3.** $5x^2 - x - 4 = 0$

4. $\frac{1}{4}x^2 - \frac{1}{2}x = -\frac{1}{4}$ **5.** $3x^2 + 5 = 0$

6. Determine the values for c such that $x^2 - 6x + c = 0$ will have two real solutions.

1. 68; Two real solutions
2. 5; Two real solutions
3. 0; One real solution
4. –11; Two nonreal solutions
5. –44; Two nonreal solutions
6. –23; Two nonreal solutions
7. 4; Two real solutions 8. 0; One real solution 9. 19,600; Two real solutions 10. –4356; Two nonreal solutions 11. –11; Two nonreal solutions 12. 49; Two real solutions 13. $x = -2 \pm 2\sqrt{2}$ 14. $x = 3 \pm \sqrt{10}$ 15. $x = -\frac{2}{3}$

10.2 Exercises

Find the discriminant and determine the nature of the solutions of each quadratic equation.

1. $x^2 + 6x - 8 = 0$ **2.** $x^2 + 3x + 1 = 0$ **3.** $x^2 - 8x + 16 = 0$

4. $x^2 + 3x + 5 = 0$ **5.** $4x^2 + 2x + 3 = 0$ **6.** $3x^2 - x + 2 = 0$

7. $5x^2 + 8x + 3 = 0$ **8.** $4x^2 + 12x + 9 = 0$ **9.** $100x^2 - 49 = 0$

10. $9x^2 + 121 = 0$ **11.** $3x^2 + x + 1 = 0$ **12.** $5x^2 - 3x - 2 = 0$

Solve each of the quadratic equations using the quadratic formula.

13. $x^2 + 4x - 4 = 0$ **14.** $x^2 - 6x - 1 = 0$ **15.** $9x^2 + 12x + 4 = 0$

16. $4x^2 - 20x + 25 = 0$ **17.** $x^2 - 2x + 7 = 0$ **18.** $x^2 - 2x + 3 = 0$

19. $2x^2 + 5x - 3 = 0$ **20.** $3x^2 - 7x + 4 = 0$ **21.** $4x^2 + 6x + 1 = 0$

22. $2x^2 - 3x - 1 = 0$ **23.** $4x^2 + 6x + 3 = 0$ **24.** $x^2 - 5x + 7 = 0$

16. $x = \frac{5}{2}$
17. $x = 1 \pm i\sqrt{6}$
18. $x = 1 \pm i\sqrt{2}$
19. $x = -3, \frac{1}{2}$
20. $x = 1, \frac{4}{3}$
21. $x = \frac{-3 \pm \sqrt{5}}{4}$
22. $x = \frac{3 \pm \sqrt{17}}{4}$
23. $x = \frac{-3 \pm i\sqrt{3}}{4}$
24. $x = \frac{5 \pm i\sqrt{3}}{2}$
25. $x = \frac{-3 \pm \sqrt{29}}{2}$
26. $x = \frac{7 \pm \sqrt{61}}{2}$
27. $x = -3, -1$
28. $x = -7$
29. $x = \pm 2i\sqrt{2}$
30. $x = \pm\sqrt{7}$

Answers to Practice Problems: **1.** $x = -1 \pm \sqrt{5}$ **2.** $x = \frac{3 \pm i\sqrt{23}}{4}$ **3.** $x = -\frac{4}{5}, 1$ **4.** $x = 1$ **5.** $x = \frac{\pm i\sqrt{15}}{3}$ **6.** $c < 9$

31. $x = \dfrac{5 \pm \sqrt{17}}{2}$

32. $x = \dfrac{-1 \pm \sqrt{7}}{3}$

33. $x = -\dfrac{1}{4}$

34. $x = -\dfrac{1}{6}, 1$

35. $x = \pm\dfrac{2\sqrt{3}}{3}$

36. $x = \pm\dfrac{3}{2}i$

37. $x = \dfrac{2}{3}$

38. $x = \dfrac{1}{3}$

39. $x = \dfrac{-4 \pm i\sqrt{2}}{2}$

40. $x = \dfrac{3 \pm i\sqrt{3}}{3}$

41. $x = \dfrac{7 \pm i\sqrt{51}}{10}$

42. $x = \dfrac{5 \pm i\sqrt{23}}{8}$

43. $x = -2, \dfrac{5}{3}$

44. $x = -2, -\dfrac{3}{2}$

45. $x = 3 \pm i\sqrt{2}$

46. $x = 4 \pm i\sqrt{2}$

47. $x = -1, 0$

48. $x = -\dfrac{4}{3}, 0$

49. $x = \dfrac{9 \pm \sqrt{65}}{2}, 0$

50. $x = \dfrac{11 \pm \sqrt{133}}{2}, 0$

51. $x = \dfrac{-3 \pm \sqrt{5}}{2}, 0$

52. $x = \dfrac{-5 \pm \sqrt{37}}{4}, 0$

Solve the given equations using any of the techniques discussed for solving quadratic equations: factoring, completing the square, or the quadratic formula.

25. $x^2 + 3x - 5 = 0$

26. $x^2 - 7x - 3 = 0$

27. $x^2 + 4x + 3 = 0$

28. $x^2 + 14x + 49 = 0$

29. $x^2 + 8 = 0$

30. $x^2 - 7 = 0$

31. $x^2 - 5x + 2 = 0$

32. $3x^2 + 2x - 2 = 0$

33. $16x^2 + 8x = -1$

34. $6x^2 = 5x + 1$

35. $3x^2 - 4 = 0$

36. $4x^2 + 9 = 0$

37. $9x^2 - 12x + 4 = 0$

38. $9x^2 - 6x + 1 = 0$

39. $2x^2 = -8x - 9$

40. $3x^2 = 6x - 4$

41. $5x^2 + 5 = 7x$

42. $4x^2 - 5x = -3$

43. $6x^2 + 2x = 20$

44. $10x^2 + 30 = -35x$

45. $3x^2 = 18x - 33$

46. $2x^2 = 16x - 36$

47. $x^2 + 4x = x - 2x^2$

48. $3x^2 + 4x = 0$

49. $x^3 - 9x^2 + 4x = 0$

50. $x^3 - 8x^2 = 3x^2 + 3x$

51. $x^3 + 3x^2 + x = 0$

52. $4x^3 + 10x^2 - 3x = 0$

First multiply each side of the equation by the LCM of the denominators to get integer coefficients and then solve the resulting equation.

53. $3x^2 - 4x + \dfrac{1}{3} = 0$

54. $\dfrac{3}{4}x^2 - 2x + \dfrac{1}{8} = 0$

55. $2x^2 - \dfrac{2}{3}x + \dfrac{2}{9} = 0$

56. $2x^2 + 3x + \dfrac{5}{4} = 0$

57. $\dfrac{1}{2}x^2 - x + \dfrac{3}{4} = 0$

58. $\dfrac{2}{3}x^2 - \dfrac{1}{3}x + \dfrac{1}{2} = 0$

59. $\dfrac{1}{4}x^2 + \dfrac{7}{8}x + \dfrac{1}{2} = 0$

60. $\dfrac{5}{12}x^2 - \dfrac{1}{2}x - \dfrac{1}{4} = 0$

61. Determine the value(s) for c such that $x^2 - 8x + c = 0$ will have two real solutions.

62. Determine the value(s) for c such that $x^2 + 5x + c = 0$ will have two real solutions.

63. Determine the value(s) for c such that $x^2 + 9x + c = 0$ will have one real solution.

64. Determine the value(s) for c such that $x^2 - 7x + c = 0$ will have one real solution.

65. Determine the value(s) for a such that $ax^2 - 6x + 3 = 0$ will have two nonreal solutions.

66. Determine the value(s) for a such that $ax^2 + 4x - 2 = 0$ will have two nonreal solutions.

67. Determine the value(s) for a such that $ax^2 + x - 9 = 0$ will have two real solutions.

68. Determine the value(s) for a such that $ax^2 + 6x + 3 = 0$ will have two real solutions.

69. Determine the value(s) for a such that $ax^2 + 7x + 12 = 0$ will have one real solution.

70. Determine the value(s) for a such that $ax^2 - 2x + 8 = 0$ will have one real solution.

71. Determine the value(s) for c such that $3x^2 + 4x + c = 0$ will have two nonreal solutions.

72. Determine the value(s) for c such that $2x^2 + 3x + c = 0$ will have two nonreal solutions.

Solve the quadratic equations using the quadratic formula and your calculator. Write the solutions accurate to 4 decimal places.

73. $0.02x^2 - 1.26x + 3.14 = 0$

74. $0.5x^2 + 0.07x - 5.6 = 0$

75. $\sqrt{2}x^2 - \sqrt{3}x - \sqrt{5} = 0$

76. $x^2 - 2\sqrt{10}x + 10 = 0$

77. $0.3x^2 + \sqrt{2}x + 0.72 = 0$

78. $\sqrt[3]{4}x^2 - \sqrt[4]{2}x - \sqrt{11} = 0$

79. $x^2 + 2\sqrt{15} - 15 = 0$

80. $0.05x^2 - \sqrt{30} = 0$

Writing and Thinking About Mathematics

81. Find an equation of the form $Ax^4 + Bx^2 + C = 0$ that has the four roots ± 2 and ± 3. Explain how you arrived at this equation.

82. The surface area of a right circular cylinder can be found using the following formula:

$S = 2\pi r^2 + 2\pi rh$, where r is the radius of the cylinder and h is the height.

Estimate the radius of a circular cylinder of height 30 cm and surface area 300 cm^2. Explain how you used your knowledge of quadratic equations.

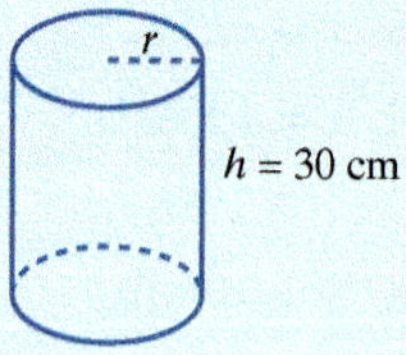

HAWKES LEARNING SYSTEMS: INTRODUCTORY & INTERMEDIATE ALGEBRA SOFTWARE

- 10.2 Quadratic Equations: The Quadratic Formula

53. $x = \frac{2 \pm \sqrt{3}}{3}$ **54.** $x = \frac{8 \pm \sqrt{58}}{6}$ **55.** $x = \frac{1 \pm i\sqrt{3}}{6}$ **56.** $x = \frac{-3 \pm i}{4}$ **57.** $x = \frac{2 \pm i\sqrt{2}}{2}$ **58.** $x = \frac{1 \pm i\sqrt{11}}{4}$ **59.** $x = \frac{-7 \pm \sqrt{17}}{4}$ **60.** $x = \frac{3 \pm 2\sqrt{6}}{5}$ **61.** $c < 16$ **62.** $c < \frac{25}{4}$ **63.** $c = \frac{81}{4}$ **64.** $c = \frac{49}{4}$ **65.** $a > 3$ **66.** $a < -2$ **67.** $a > -\frac{1}{36}$ **68.** $a < 3$ **69.** $a = \frac{49}{48}$ **70.** $a = \frac{1}{8}$ **71.** $c > \frac{4}{3}$ **72.** $c > \frac{9}{8}$ **73.** $x \approx 2.5993, 60.4007$ **74.** $x \approx -3.4174, 3.2774$ **75.** $x \approx -0.7862, 2.0110$ **76.** $x \approx 3.1623$ **77.** $x \approx -4.1334, -0.5806$ **78.** $x \approx -1.1186, 1.8678$ **79.** $x \approx -2.6933, 2.6933$ **80.** $x \approx -10.4664, 10.4664$ **81.** $x^4 - 13x^2 + 36 = 0$; Multiplied $(x-2)(x+2)(x-3)(x+3)$; Answers will vary. **82.** $r \approx 1.5150$ cm; Answers will vary.

10.3 Applications

- *Use quadratic equations to solve applied problems.*

The following strategy for solving word problems, given in Section 2.6 and summarized in Section 6.7, is a valid approach to solving word problems at all levels.

Strategy for Solving Word Problems

1. Understand the problem.
   a. Read the problem carefully. (Read it several times if necessary.)
   b. If it helps, restate the problem in your own words.

2. Devise a plan.
   a. Decide what is asked for; assign a variable to the unknown quantity. Label this variable so you know exactly what it represents.
   b. Draw a diagram or set up a chart whenever possible.
   c. Write an equation that relates the information provided.

3. Carry out the plan.
   a. Study your picture or diagram for insight into the solution.
   b. Solve the equation.

4. Look back over the results.
   a. Does your solution make sense in terms of the wording of the problem?
   b. Check your solution in the equation.

The problems in this section can be solved by setting up quadratic equations and then solving these equations by factoring, completing the square, or using the quadratic formula.

The Pythagorean Theorem

The **Pythagorean theorem** is one of the most interesting and useful ideas in mathematics. We discussed the Pythagorean theorem in Section 6.7 and do so again here because problems with right triangles often generate quadratic equations.

In a **right triangle**, one of the angles is a right angle (measures 90°), and the side opposite this angle (the longest side) is called the **hypotenuse**. The other two sides are called **legs**.

The Pythagorean Theorem

In a right triangle, if c is the length of the hypotenuse and a and b are the lengths of the legs, then

$$c^2 = a^2 + b^2.$$

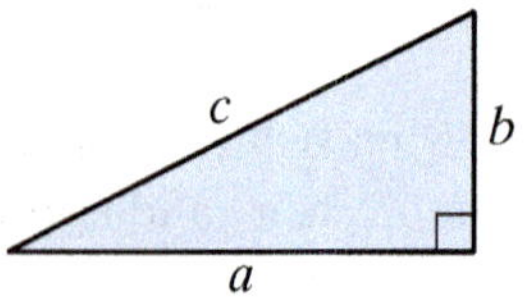

Example 1: The Pythagorean Theorem

Teaching Notes: Some students may tend to believe that all negative solutions are irrelevant or not applicable. Emphasis on the context and interpretation of any application is highly recommended. Students need to understand that algebraic solutions may or may not be related to the problem that generated the original equation.

The length of a rectangular field is 6 meters longer than its width. If a diagonal foot path stretching from one corner of the field to the opposite corner is 30 meters long, what are the dimensions of the field?

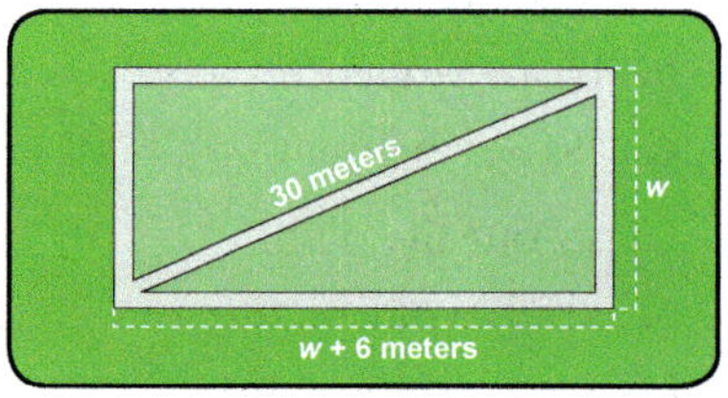

Solution: Let $w = \text{width},$
then $w + 6 = \text{length}.$

$$(w+6)^2 + w^2 = 30^2 \quad \text{Use the Pythagorean theorem.}$$

$$w^2 + 12w + 36 + w^2 = 900$$

$$2w^2 + 12w - 864 = 0$$

$$w^2 + 6w - 432 = 0 \quad \text{Divide both sides by 2.}$$

$$(w+24)(w-18) = 0$$

$$\cancel{w = -24} \quad \text{or} \quad w = 18$$

A negative number does not fit the conditions of the problem as width cannot be negative.

$$w = 18 \text{ meters}$$

$$w + 6 = 24 \text{ meters}$$

The length is 24 meters and the width is 18 meters.

Projectiles

The formula $h = -16t^2 + v_0 t + h_0$ is used in physics and relates to the height of a projectile such as a thrown ball, a bullet, or a rocket.

h = height of object, in feet
t = time the object is in the air, in seconds
v_0 = beginning velocity, in feet per second
h_0 = beginning height ($h_0 = 0$ if the object is initially at ground level.)

Example 2: Projectiles

A bullet is fired straight up from ground level with a muzzle velocity of 320 feet per second.

a. When will the bullet hit the ground?
b. When will the bullet be 1200 feet above the ground?

Solution: In this problem, $v_0 = 320$ feet per second and $h_0 = 0$ feet.

a. The bullet hits the ground when its height h equals 0.

$h = -16t^2 + v_0 t + h_0$

$0 = -16t^2 + 320t + 0$

$0 = t^2 - 20t$ Divide both sides by -16.

$0 = t(t - 20)$ Factor.

$t = 0$ or $t = 20$

The bullet hits the ground in 20 seconds. The solution $t = 0$ confirms the fact that the bullet was fired from the ground.

b. Let $h = 1200$.

$1200 = -16t^2 + 320t$

$0 = -16t^2 + 320t - 1200$

$0 = t^2 - 20t + 75$ Divide both sides by -16.

$0 = (t - 5)(t - 15)$ Factor.

$t = 5$ or $t = 15$

Both solutions are meaningful. The bullet is at a height of 1200 feet twice; once at 5 seconds while going up and once at 15 seconds while coming down.

Geometry

Example 3: Geometry

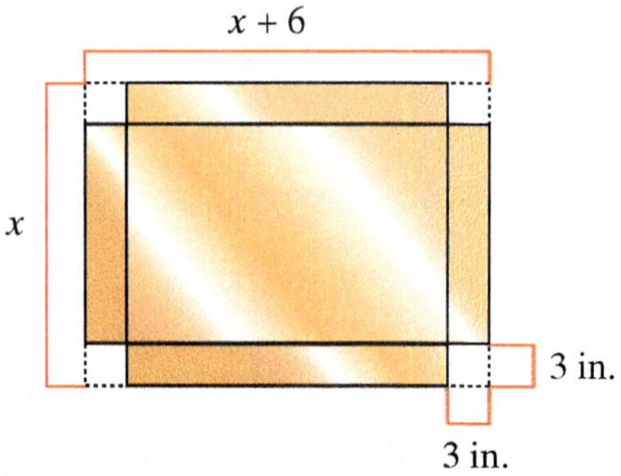

A rectangular sheet of copper was 6 in. longer than it was wide. Three inch by three inch squares were cut from each corner and the sides were folded up to form an open box. If the box has a volume of 336 in.^{3}, what were the dimensions of the original sheet of copper?

Teaching Notes:
As a visual aid to understanding this type of problem, you might have your students take a piece of paper, tear squares from the corners, and fold the paper to make a box. Then take a second piece of paper and tear squares of a different size from the corners to make a box with a different volume. (The sides can be labeled so they can see that the representations of the lengths make sense.) As a further aid, have them calculate the volume of each box.

Solution: Let x = width of copper sheet (before the squares are cut out),
then $x + 6$ = length of copper sheet (before the squares are cut out).

After the box has been folded:

$$3 = \text{height of box}$$

$$(x+6)-3-3 = x = \text{length of box} \qquad \text{3 is subtracted twice.}$$

$$x-3-3 = x-6 = \text{width of box.}$$

The volume of the box is length × width × height: $V = lwh$.

Thus

length × width × height = volume

$$x \cdot (x-6) \cdot 3 = 336$$

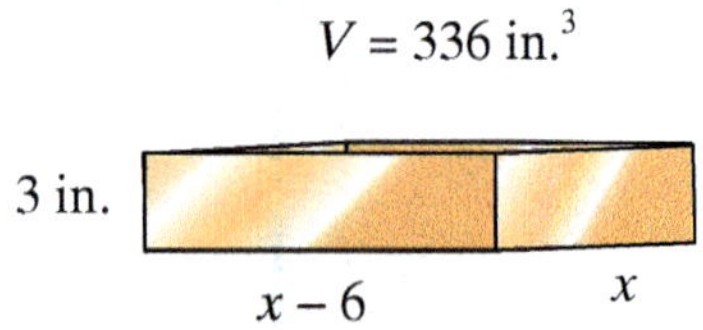

$$3x^2 - 18x = 336$$

$$3x^2 - 18x - 336 = 0$$

$$x^2 - 6x - 112 = 0 \qquad \text{Divide both sides by 3.}$$

$$(x+8)(x-14) = 0$$

$$\cancel{x = -8} \text{ or } x = 14 \qquad \text{Width of the copper sheet}$$

$$x + 6 = 20. \qquad \text{Length of the copper sheet}$$

The length of the sheet was 20 in. and the width was 14 in.

Cost Per Person

In the following example, note that the **cost per person** is found by dividing the total cost by the number of people going to the tournament. The cost per person changes because the total cost remains fixed while the number of people changes.

Example 4: Cost Per Person

The members of a bowling club were going to rent a bus to drive them to a tournament at a total cost of \$2420, which was to be divided equally among the members. At the last minute, two of the members decided to drive their own cars. The cost to the remaining members increased \$11 each. How many members rode the bus?

Solution: Let x = number of club members,
then $x-2$ = number of club members that rode the bus.

$$\underbrace{\text{Final cost per member}} - \underbrace{\text{Initial cost per member}} = \underbrace{\text{Difference in cost per member}}$$

$$\frac{2420}{x-2} - \frac{2420}{x} = 11$$

$$x(x-2)\frac{2420}{x-2} - x(x-2)\frac{2420}{x} = x(x-2)\cdot 11 \qquad \text{LCD is } x(x-2).$$

$$2420x - 2420(x-2) = 11x(x-2)$$

$$2420x - 2420x + 4840 = 11x^2 - 22x$$

$$0 = 11x^2 - 22x - 4840$$

$$0 = x^2 - 2x - 440 \qquad \text{Divide both sides by 11.}$$

$$0 = (x-22)(x+20)$$

$$x = 22 \text{ or } \cancel{x = -20}$$

$$x - 2 = 20$$

−20 does not fit the conditions. That is, the number of people in a club is a positive number.

Check: Final cost per member $= \frac{2420}{20} = \$121$

Initial cost per member $= \frac{2420}{22} = \$110$

$\$121 - \$110 = \$11$ Difference in cost per member

Twenty members rode the bus.

Teaching Notes:
You might need to refer your students back to Section 7.4 and the discussion of equations with rational expressions.

10.3 Exercises

1. 6, 13
2. $-9, 81$ or $8, 64$
3. 4, 14
4. $7, 17$ or $-14, -\frac{17}{2}$
5. $-11, -6$
6. $1+\sqrt{3}$
7. $-5-4\sqrt{3}$
8. $\frac{3-3\sqrt{2}}{2}$ or -0.62
9. $\frac{1+\sqrt{33}}{4}$
10. 7, 15, 46
11. $7\sqrt{2}$ cm, $7\sqrt{2}$ cm
12. $3\sqrt{5}$ meters; $6\sqrt{5}$ meters
13. Mel: 30 mph; John: 40 mph
14. 27 meters
15. Length = 10 feet; Width = 4 feet

1. A positive integer is one more than twice another. Their product is 78. Find the two integers.

2. One number is equal to the square of another. Find the numbers if their sum is 72.

3. Find two positive numbers whose difference is 10 and whose product is 56. (**Hint:** If x is one number, then the other is either $x+10$ or $x-10$. Try both and analyze your answers.)

4. One number is three more than twice a second number. Their product is 119. Find the numbers.

5. The sum of two numbers is –17. Their product is 66. Find the numbers. (**Hint:** If x is one number, then the other is $(-17-x)$.)

6. Find a positive real number such that its square is equal to twice the number increased by 2.

7. Find a negative real number such that the square of the sum of the number and 5 is equal to 48.

8. The square of a negative real number is decreased by 2.25 and the result is equal to 3 times the number. What is the number?

9. Twice the square of a positive real number is equal to 4 more than the number. What is the number?

10. The sum of three positive integers is 68. The second is one more than twice the first. The third is three less than the square of the first. Find the integers.

11. Right triangles: A right triangle has two equal sides. The hypotenuse is 14 cm. Find the length of the sides.

12. Right triangles: The length of one leg of a right triangle is twice the length of the second leg. The hypotenuse is 15 meters. Find the lengths of the two legs.

13. Average speed: Mel and John leave Desert Point at the same time. Mel drives north and John drives east. Mel's average speed is 10 mph slower than John's. At the end of one hour they are 50 miles apart. Find the average speed of each driver.

14. Height of a telephone pole: A telephone pole was broken over at a point $\frac{4}{9}$ of the distance from its base to the top. The top of the pole reached a point on the ground 9 meters from the base of the pole (thus forming a triangle with the ground). What was the original height of the pole?

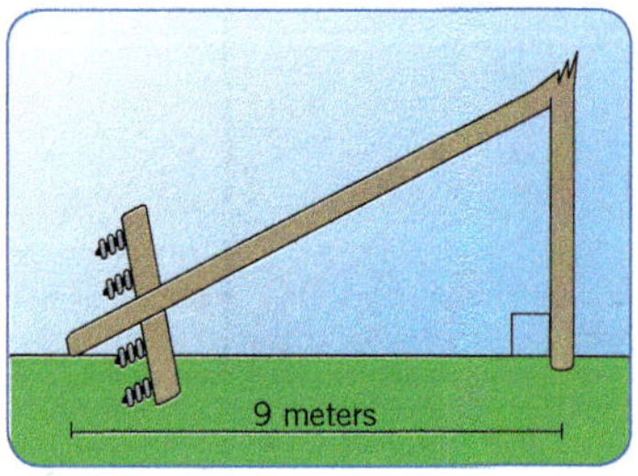

15. Dimensions of a rectangle: The length of a rectangle is 2 feet less than three times the width. If the area of the rectangle is 40 square feet, find the length and width.

16. Length = 8 meters; Width = 5 meters
17. 17 inches × 6 inches
18. Length = 32 cm; Width = 12 cm; Height = 4 cm
19. 70 trees
20. 40 seats
21. 9 floors
22. Length = 26.5 ft; Width = 7.5 ft; Height = 8 ft
23. 3 in.
24. 6 in. thick
25. 5 amperes
26. 2 amperes or 8 amperes

16. Dimensions of a rectangle: A rectangle is 3 meters longer than it is wide. If the width is doubled and the length is decreased by 4 meters, the area is unchanged. Find the original length and width.

17. Dimensions of a rectangle: The area of a rectangle is 102 square inches and the perimeter of the rectangle is 46 inches. Find the dimensions.

18. Dimensions of a box: A rectangular piece of cardboard that is twice as long as it is wide has a small square 4 cm by 4 cm cut from each corner. The edges are then folded up to form an open box with a volume of 1536 cubic cm. What are the dimensions of the box? (See Example 3.)

19. Size of an orchard: An orchard has 2030 trees. The number of trees in each row is 12 more than twice the number of rows. How many trees are in each row?

20. Capacity of an auditorium: A rectangular auditorium seats 960 people. The number of seats in each row is 16 more than the number of rows. Find the number of seats in each row.

21. Size of an apartment complex: An apartment building has the same number of units on each floor. The building has five times as many units per floor as number of floors, and there are 405 units total. How many floors does the building have?

22. Shipping: A large U-Haul truck is 8 ft tall. The length of the truck is 4 ft longer than three times the width. What are the dimensions of the truck if the volume is 1590 ft^3?

23. Dimensions of a frame: A photograph 9 in. wide and 12 in. long is surrounded by a frame of uniform thickness. The area of the frame itself, not including the center, is 162 $in.^2$ Find the thickness of the frame.

24. Art: The Mona Lisa is a famous painting by Leonardo da Vinci. The painting is 30 in. by 21 in. It is surrounded by a frame of uniform thickness whose area (not including the center) is 756 $in.^2$ Find the thickness of the frame.

25. Power of a generator: A 40-volt generator with a resistance of 4 ohms delivers power externally of $40I - 4I^2$ watts, where I is the current measured in amperes. Find the current needed for the generator to deliver 100 watts of power.

26. Power of a generator: Find the current needed for the 40-volt generator in Exercise 25 to deliver 64 watts of power.

27. 12 signs
28. $12
29. a. $307.20
b. 45 cents or 35 cents
30. 12 members
31. 8 workers
32. 64 mph
33. 6 mph
34. 3 hours and 6 hours
35. 4 hours and 12 hours
36. 150 mph

27. Selling signs: Ray operates a small sign-making business. He finds that if he charges x dollars for each sign, he sells $40 - x$ signs per week. What is the least number of signs he can sell to have an income of $336 in one week?

28. Selling picture frames: It costs Mrs. Snow $3 to build a picture frame. She estimates that if she charges x dollars each, she can sell $60 - x$ frames per week. What is the lowest price necessary to make a profit of $432 each week?

29. Selling peanuts: Samuel operates a small peanut stand. He estimates that he can sell 600 bags of peanuts per day if he charges 50¢ for each bag. He determines that he can sell 20 more bags for each 1¢ reduction in price.

a. What would his revenue be if he charged 48¢ per bag?

b. What should he charge in order to have receipts of $315?

30. Rock climbing: The Piton Rock Climbing Club planned a climbing expedition. The total cost was $900, which was to be divided equally among the members going. While practicing, three members fell and were injured so they were unable to go. If the cost per person increased by $15, how many members went on the expedition?

31. Manufacturing: A manufacturing crew needs to assemble 1000 boxes per day, divided equally among the workers. One day, three workers call in sick, and the remaining members each need to assemble 75 more boxes than usual. How many workers are on the manufacturing crew?

32. Traveling by car: Jim traveled 240 miles to a Bridge convention. Later, his wife Ann drove up to meet him. Ann's average speed exceeded Jim's by 4 mph, and the trip took her 15 minutes less time. Find Ann's speed.

33. Boating: In two hours, a motorboat can travel 8 miles down a river and return 4 miles back. If the river flows at a rate of 2 miles per hour, how fast can the boat travel in still water?

34. Product assembly: Two employees together can prepare a large order in 2 hrs. Working alone, one employee takes three hours longer than the other. How long does it take each person working alone?

35. Library processing: Two librarians working together can catalog a new shipment of books in 3 hours. Working alone, one librarian takes 8 hours longer than the other. How long does it take each librarian working alone?

36. Traveling by plane: Fern can fly her plane 240 miles against the wind in the same time it takes her to fly 360 miles with the wind. The speed of the plane in still air is 30 mph more than four times the speed of the wind. Find the speed of the plane in still air.

37. 10 lb of Grade A, 20 lb of Grade B
38. 190 reserved, 150 general
39. a. 6.75 s
b. 3 s, 3.75 s
40. a. 3.25 s
b. 1.25 s, 2 s
41. a. 7 s
b. 2 s, 12 s
42. a. 3.5 s
b. 144 ft
43. 14.14 cm
44. 10.91 in.
45. a. $C = 94.20$ ft, $A = 706.50\text{ ft}^2$
b. $P = 84.85$ ft, $A = 450.00\text{ ft}^2$

37. Selling coffee: A grocer mixes \$9.00 worth of Grade A coffee with \$12.00 worth of Grade B coffee to obtain 30 pounds of a blend. If Grade A costs 30¢ a pound more than Grade B, how many pounds of each were used?

38. Theater: The Andersonville Little Theater Group sold 340 tickets to their spring production. Receipts from the sale of reserved tickets were \$855. Receipts from general admission tickets were \$375. How many of each type of ticket were sold if the cost of a reserved ticket is \$2 more than a general admission ticket?

Use the formula for a projectile $h = -16t^2 + v_0t + h_0$ where t is in seconds, v_0 is in ft/s and h_0 is in feet.

39. Throwing a ball: A ball is thrown vertically from ground level with an initial speed of 108 ft/s.

a. When will the ball hit the ground?

b. When will the ball be 180 ft above the ground?

40. Throwing a ball: After scoring the championship winning touchdown, a football player throws the ball directly upward from the ground with an initial velocity of 52 ft/s.

a. When will the ball strike the ground?

b. When will the ball be 40 ft above the ground?

41. Shooting an arrow: An arrow is shot vertically upward from a platform 40 ft high at a rate of 224 ft/s.

a. When will the arrow be 824 ft above the ground?

b. When will it be 424 ft above the ground?

42. Dropping a stone: A stone is dropped from a platform 196 ft high. (**Hint:** Since the stone is dropped, $v_0 = 0$.)

a. When will it hit the ground?

b. How far has it fallen after 3 seconds?

Use your calculator to find the answers accurate to two decimal places.

43. Geometry: If a triangle is inscribed in a circle so that one side of the triangle is a diameter of the circle, the triangle will be a right triangle (every time). If an isosceles triangle (two sides equal) is inscribed in this manner in a circle with diameter 20 cm, find the length of the two equal sides.

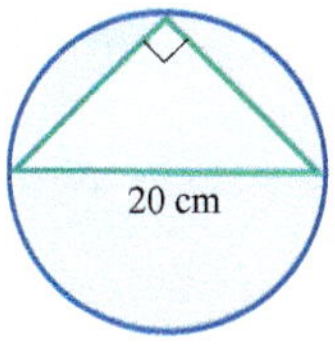

44. Geometry: If a triangle is inscribed in a semicircle such that one side of the triangle is the diameter of a circle with radius 6 in., and one side of the triangle is 5 in., what is the length of the third side? (See Exercise 43.)

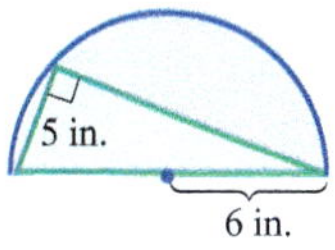

45. Geometry: A square is said to be inscribed in a circle if each corner of the square lies on the circle. (Use $\pi = 3.14$.)

a. Find the circumference and area of a circle with diameter 30 ft.

b. Find the perimeter and area of the square inscribed in the circle.

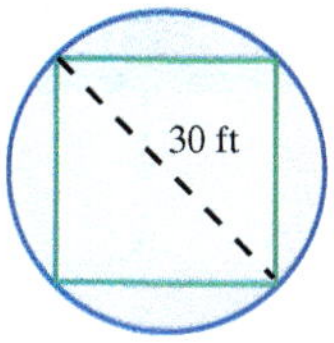

46. a. 127.28 ft
b. 127.28 ft
47. a. No
b. Home plate
c. No
48. 2.24 miles
49. 855.86 feet
50. 8.49 cm
51. See page 780.
52. b^2 must be greater than or equal to $4ac$ to produce a real solution. Also, all the solutions found may not apply to the problem at hand. You must check that each answer makes sense in the context of the problem.

46. Baseball: The shape of a baseball infield is a square with sides 90 ft long.

a. Find the distance from home plate to second base.

b. Find the distance from first base to third base.

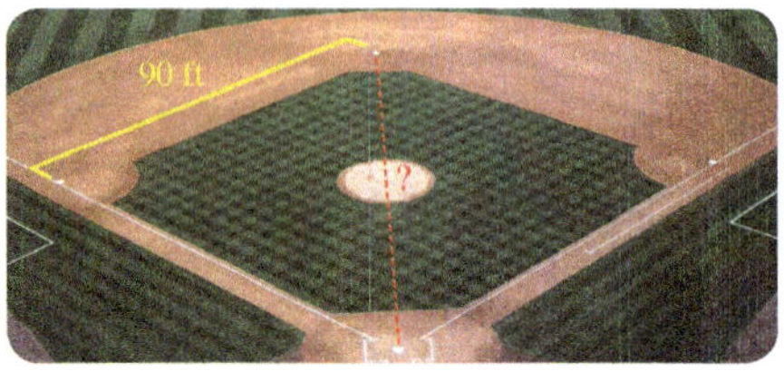

47. Baseball: The distance from home plate to the pitcher's mound for the field in Exercise 46 is 60.5 feet.

a. Is the pitcher's mound exactly halfway between home plate and second base?

b. If not, which base is it closer to, home plate or second base?

c. Do the two diagonals of the square intersect at the pitcher's mound?

48. Flying an airplane: If an airplane passes directly over your head at an altitude of 1 mile, how far is the airplane from your position after it has flown 2 miles farther at the same altitude?

49. Length of a shadow: The GE Building in New York is 850 feet tall (70 stories). At a certain time of day, the building casts a shadow 100 feet long. Find the distance from the top of the building to the tip of the shadow (to the nearest hundredth of a foot).

50. Quilting: To create a square inside a square, a quilting pattern requires four triangular pieces like the one shaded in the figure shown here. If the square in the center measures 12 cm on a side, and the two legs of each triangle are of equal length, how long are the legs of each triangle?

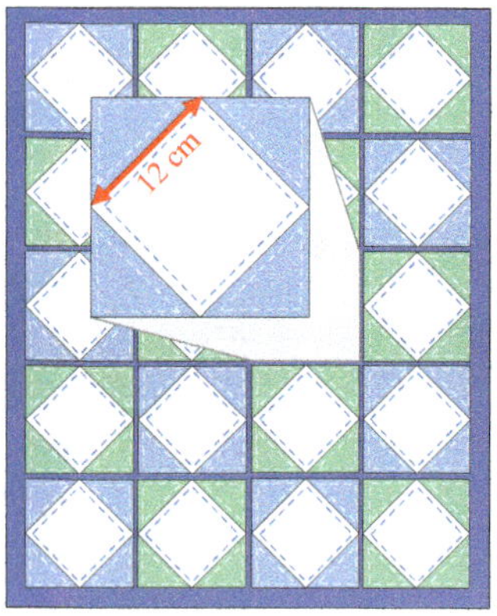

Writing and Thinking About Mathematics

51. Develop the quadratic formula using the technique of completing the square with the general quadratic equation $ax^2 + bx + c = 0$.

52. Suppose that you are to solve an applied problem and the solution leads to a quadratic equation. You decide to use the quadratic formula to solve the equation. Explain what restrictions you must be aware of when you use the formula.

HAWKES LEARNING SYSTEMS: INTRODUCTORY & INTERMEDIATE ALGEBRA SOFTWARE

- 10.3 Applications: Quadratic Equations

10.4 Equations in Quadratic Form

- *Make substitutions that allow equations to be written in quadratic form.*
- *Solve equations that can be written in quadratic form.*
- *Solve equations that contain rational expressions.*
- *Solve higher-degree equations.*

Solving Equations in Quadratic Form

Teaching Notes:
This technique might be introduced to students as a next step in the effort to solve a special family of higher order equations. Point out that solving equations of higher order requires various techniques and that many of these techniques have been around for hundreds of years.

The general quadratic equation is $ax^2 + bx + c = 0$, where $a \neq 0$.

The equations

$$x^4 - 7x^2 + 12 = 0 \qquad \text{and} \qquad x^{\frac{2}{3}} - 4x^{\frac{1}{3}} - 21 = 0$$

are **not** quadratic equations, but they are in **quadratic form** because the degree of the middle term is one-half the degree of the first term. Specifically,

$$\frac{1}{2}(4) = 2 \qquad \text{and} \qquad \frac{1}{2}\left(\frac{2}{3}\right) = \frac{1}{3}.$$

first term exponent (4), middle term exponent (2); first term exponent ($\frac{2}{3}$), middle term exponent ($\frac{1}{3}$)

Equations in quadratic form can be solved using the quadratic formula or by factoring just as if they were quadratic equations. In each case, a substitution can be made to clarify the problem. Follow these suggestions:

Solving Equations in Quadratic Form by Substitution

1. Look at the middle term.
2. Substitute a first-degree variable, such as u, for the variable expression in the middle term.
3. Substitute the square of this variable, u^2, for the variable expression in the first term.
4. Solve the resulting quadratic equation for u.
5. Substitute the results "back" for u in the beginning substitution and solve for the original variable.

The following examples illustrate how such a substitution may help. Study these examples carefully and note the variety of algebraic manipulations used.

Example 1: Using Substitution to Solve Equations in Quadratic Form

Solve the following equations. These equations are in quadratic form and a substitution will reveal the quadratic expression.

Teaching Notes:
You may want to discuss solving Examples 1a – 1d by factoring directly without substitution. This will probably depend on the ability level and mathematical sophistication of your students.

a. $x^4 - 7x^2 + 12 = 0$

Solution:

$x^4 - 7x^2 + 12 = 0$

$u^2 - 7u + 12 = 0$ Substitute $u = x^2$ and $u^2 = x^4$.

$(u-3)(u-4) = 0$ Solve for u by factoring.

$u = 3$ or $u = 4$

$x^2 = 3$ or $x^2 = 4$ Now substitute back x^2 for u.

$x = \pm\sqrt{3}$ $x = \pm 2$ Solve quadratic equations for x.

There are four solutions: $x = -\sqrt{3}, \sqrt{3}, -2$, and 2.

b. $x^{\frac{2}{3}} - 4x^{\frac{1}{3}} - 21 = 0$

Solution:

$x^{\frac{2}{3}} - 4x^{\frac{1}{3}} - 21 = 0$

$u^2 - 4u - 21 = 0$ Let $u = x^{\frac{1}{3}}$ and $u^2 = x^{\frac{2}{3}}$.

$(u-7)(u+3) = 0$ Solve for u by factoring.

$u = 7$ or $u = -3$

$x^{\frac{1}{3}} = 7$ or $x^{\frac{1}{3}} = -3$ Substitute back $x^{\frac{1}{3}}$ for u.

$\left(x^{\frac{1}{3}}\right)^3 = 7^3$ or $\left(x^{\frac{1}{3}}\right)^3 = (-3)^3$ Cube both sides.

$x = 343$ $x = -27$

There are two solutions: $x = -27$ and 343.

c. $x^{-4} - 7x^{-2} + 10 = 0$

Solution:

$x^{-4} - 7x^{-2} + 10 = 0$

$u^2 - 7u + 10 = 0$ Let $u = x^{-2}$ and $u^2 = x^{-4}$.

$(u-2)(u-5) = 0$ Solve for u by factoring.

$u = 2$ or $u = 5$

$x^{-2} = 2$ $x^{-2} = 5$ Substitute back x^{-2} for u.

$\frac{1}{x^2} = 2$ $\frac{1}{x^2} = 5$ Remember, $x^{-2} = \frac{1}{x^2}$.

$x^2 = \frac{1}{2}$ $\quad x^2 = \frac{1}{5}$ Reciprocals

$x = \pm\sqrt{\frac{1}{2}}$ $\quad x = \pm\sqrt{\frac{1}{5}}$

$x = \pm\frac{1}{\sqrt{2}}$ $\quad x = \pm\frac{1}{\sqrt{5}}$

Rationalizing the denominators, we have $x = \pm\frac{\sqrt{2}}{2}, x = \pm\frac{\sqrt{5}}{5}$.

There are four solutions: $x = -\frac{\sqrt{2}}{2}, \frac{\sqrt{2}}{2}, -\frac{\sqrt{5}}{5}$, and $\frac{\sqrt{5}}{5}$.

d. $(x+2)^2 - (x+2) - 12 = 0$

Solution: $(x+2)^2 - (x+2) - 12 = 0$

$u^2 - u - 12 = 0$ Let $u = x + 2$.

$(u-4)(u+3) = 0$ Solve for u by factoring.

$u = 4$ or $u = -3$

$x + 2 = 4$ $\quad x + 2 = -3$ Substitute back $x + 2$ for u.

$x = 2$ $\quad x = -5$

There are two solutions: $x = -5$ and 2.

Solving Equations with Rational Expressions

Equations with rational expressions were first discussed in Section 7.4. In solving equations with rational expressions, first multiply every term on both sides of the equation by the least common multiple (LCM) of the denominators. This will "clear" the equation of fractions. Remember to check the restrictions on the variables. That is, no denominator can have a value of 0. In this section, the **resulting equation will be a quadratic equation**. (A few simple problems of this type already appeared in the application problems in Section 10.3.)

Example 2: Solving Equations with Rational Expressions

Solve the following equation containing rational expressions: $\frac{2}{3x-1} + \frac{1}{x+1} = \frac{x}{x+1}$.

Solution: This equation is not in quadratic form. However, multiplying both sides of the equation by the LCM of the denominators, $(x+1)(3x-1)$, does give a quadratic equation. The restrictions on x are: $x \neq -1, \frac{1}{3}$.

Continued on the next page...

In this case, use the quadratic formula to solve the resulting quadratic equation.

$$\frac{2}{3x-1}+\frac{1}{x+1}=\frac{x}{x+1}$$

$$(x+1)\cancel{(3x-1)}\cdot\frac{2}{\cancel{3x-1}}+\cancel{(x+1)}(3x-1)\cdot\frac{1}{\cancel{x+1}}=\cancel{(x+1)}(3x-1)\cdot\frac{x}{\cancel{x+1}}$$

$$2(x+1)+3x-1=(3x-1)x$$

$$2x+2+3x-1=3x^2-x$$

$$0=3x^2-6x-1$$

$$x=\frac{6\pm\sqrt{(-6)^2-4\cdot3(-1)}}{6}$$

$$=\frac{6\pm\sqrt{48}}{6}$$

$$=\frac{6\pm4\sqrt{3}}{6}$$

$$=\frac{3\pm2\sqrt{3}}{3}$$

Solving Higher-Degree Equations

Example 3: Solving Higher-Degree Equations

Solve the following higher-degree equations.

a. $x^5-16x=0$

Solution: This equation can be solved by factoring and using the square root property.

$x^5-16x=0$

$x(x^4-16)=0$ Factor out the common monomial, x.

$x(x^2+4)(x^2-4)=0$ Factor the difference of two squares.

$x=0$ or $x^2=-4$ or $x^2=4$

$x=\pm2i$ $\quad x=\pm2$

There are five solutions: $x=0,-2i,2i,-2,2$.

1. $x = \pm 2, \pm 3$
2. $x = \pm 2, \pm 5$
3. $x = \pm 2, \pm\sqrt{5}$
4. $y = \pm 3, \pm\sqrt{2}$
5. $y = \pm\sqrt{7}, \pm 2i$
6. $y = \pm\sqrt{3}, \pm 2i$
7. $y = \pm\sqrt{5}, \pm i\sqrt{5}$
8. $x = \pm\sqrt{5}, \pm i\sqrt{5}$
9. $x = 4, \frac{25}{4}$
10. $x = \frac{1}{4}, 1$
11. $x = 1, 4$
12. $y = 1, 9$

b. $x^3 - 27 = 0$

Solution: The polynomial is the difference of two cubes and can be factored. In this case, complex solutions can be found using the quadratic formula.

$$x^3 - 27 = 0$$

$$(x-3)(x^2+3x+9) = 0$$

$$x - 3 = 0 \qquad\qquad x^2 + 3x + 9 = 0$$

$$x = 3$$

Using the quadratic formula:

$$x = \frac{-3 \pm \sqrt{3^2 - 4\cdot 9}}{2} = \frac{-3 \pm \sqrt{-27}}{2} = \frac{-3 \pm 3i\sqrt{3}}{2}$$

There are three solutions: $x = 3, \frac{-3+3i\sqrt{3}}{2}, \frac{-3-3i\sqrt{3}}{2}$.

Practice Problems

Solve the following equations.

1. $x - x^{\frac{1}{2}} - 2 = 0$

2. $x^4 + 16x^2 = -48$

3. $\frac{3(x-2)}{x-1} = \frac{2(x+1)}{x-2} + 2$

4. $x^3 + 8 = 0$

13. $x = -8, \frac{1}{8}$
14. $x = -8, \frac{27}{8}$
15. $x = -27, -8, 0$
16. $x = 0, 49, 64$
17. $x = -17, -2$
18. $x = -2, 3$

10.4 Exercises

19. $x = -\frac{7}{2}, -3$
20. $x = 0, \frac{6}{5}$
21. $x = 0, 8$
22. $x = -5, -1$
23. $x = -1, -\frac{1}{2}$
24. $x = 1, 2$
25. $x = \pm\sqrt{1+i}, \pm\sqrt{1-i}$

Solve the equations.

1. $x^4 - 13x^2 + 36 = 0$

2. $x^4 - 29x^2 + 100 = 0$

3. $x^4 - 9x^2 + 20 = 0$

4. $y^4 - 11y^2 + 18 = 0$

5. $y^4 - 3y^2 - 28 = 0$

6. $y^4 + y^2 - 12 = 0$

7. $y^4 - 25 = 0$

8. $4x^4 - 100 = 0$

9. $2x - 9x^{\frac{1}{2}} + 10 = 0$

10. $2x - 3x^{\frac{1}{2}} + 1 = 0$

11. $x^3 - 9x^{\frac{3}{2}} + 8 = 0$

12. $y^3 - 28y^{\frac{3}{2}} + 27 = 0$

13. $2x^{\frac{2}{3}} + 3x^{\frac{1}{3}} - 2 = 0$

14. $2x^{\frac{2}{3}} + x^{\frac{1}{3}} - 6 = 0$

15. $3x^{\frac{5}{3}} + 15x^{\frac{4}{3}} + 18x = 0$

Answers to Practice Problems: **1.** $x = 4$ **2.** $x = \pm 2i, \pm 2i\sqrt{3}$ **3.** $x = -3 \pm \sqrt{19}$ **4.** $x = -2, 1 \pm i\sqrt{3}$

26. $x = \pm\sqrt{2+i}, \pm\sqrt{2-i}$

27. $x = \pm\sqrt{1+3i}, \pm\sqrt{1-3i}$

28. $x = \pm\sqrt{3+2i}, \pm\sqrt{3-2i}$

29. $x = \pm\sqrt{2+i\sqrt{3}}, \pm\sqrt{2-i\sqrt{3}}$

30. $x = \pm\sqrt{3+i\sqrt{2}}, \pm\sqrt{3-i\sqrt{2}}$

31. $x = \frac{1}{7}, \frac{1}{5}$

32. $z = -\frac{1}{4}, \frac{1}{6}$

33. $x = -\frac{1}{3}, \frac{3}{8}$

34. $x = \frac{1}{2}, \frac{2}{3}$

35. $x = \frac{1}{25}$

36. $y = \frac{1}{4}, 9$

37. $x = \pm\frac{\sqrt{5}}{5}, \pm 1$

38. $x = \pm\frac{\sqrt{6}}{2}, \pm 1$

39. $x = \pm\frac{\sqrt{6}}{2}, \pm\frac{1}{3}i$

40. $x = \pm\frac{1}{2}i, \pm\frac{\sqrt{10}}{5}$

41. $x = -\frac{1}{4}$

42. $x = 9$

43. $x = -4, 1$

44. $x = -\frac{1}{3}, 3$

45. $x = -\frac{1}{2}$

46. $x = -\frac{2}{5}$

16. $2x^2 - 30x^{\frac{3}{2}} + 112x = 0$

17. $(x+7)^2 + 5(x+7) = 50$

18. $(x-1)^2 + (x-1) - 6 = 0$

19. $(2x+3)^2 + 7(2x+3) + 12 = 0$

20. $(5x-4)^2 + 2(5x-4) - 8 = 0$

21. $(x-3)^2 - 2(x-3) - 15 = 0$

22. $(x+4)^2 - 2(x+4) = 3$

23. $(2x+1)^2 + (2x+1) = 0$

24. $(3x-5)^2 + (3x-5) - 2 = 0$

25. $x^4 - 2x^2 + 2 = 0$

26. $x^4 - 4x^2 + 5 = 0$

27. $x^4 - 2x^2 + 10 = 0$

28. $x^4 - 6x^2 + 13 = 0$

29. $x^4 - 4x^2 + 7 = 0$

30. $x^4 - 6x^2 + 11 = 0$

31. $x^{-2} - 12x^{-1} + 35 = 0$

32. $z^{-2} - 2z^{-1} - 24 = 0$

33. $3x^{-2} + x^{-1} - 24 = 0$

34. $2x^{-2} - 7x^{-1} + 6 = 0$

35. $x^{-1} + 5x^{-\frac{1}{2}} - 50 = 0$

36. $3y^{-1} - 7y^{-\frac{1}{2}} + 2 = 0$

37. $x^{-4} - 6x^{-2} + 5 = 0$

38. $3x^{-4} - 5x^{-2} + 2 = 0$

39. $3x^{-4} + 25x^{-2} - 18 = 0$

40. $2x^{-4} + 3x^{-2} - 20 = 0$

41. $\frac{2}{4x-1} + \frac{1}{x+1} = \frac{-x}{x+1}$

42. $\frac{3x-2}{15} - \frac{16-3x}{x+6} = \frac{x+3}{5}$

43. $\frac{2x}{x-4} - \frac{12x}{x^2+x-20} = \frac{x-1}{x+5}$

44. $\frac{x+1}{x+3} + \frac{2x-1}{x-2} = \frac{12x-2}{x^2+x-6}$

45. $\frac{x+5}{3x+2} - \frac{4-2x}{3x^2+8x+4} = \frac{x+4}{x+2}$

46. $\frac{x+5}{3x+4} + \frac{16x^2+5x+6}{3x^2-2x-8} = \frac{4x}{x-2}$

47. $\frac{4x+1}{x-6} - \frac{3x^2-8x+20}{2x^2-13x+6} = \frac{3x+7}{2x-1}$

48. $\frac{3x+2}{x+3} + \frac{22x-31}{x^2-x-12} = \frac{3(x+4)}{x+3}$

49. $\frac{5(x-10)}{x-7} = \frac{2(x+1)}{x-4} + 3$

50. $2 + \frac{2-x}{x+2} = \frac{x-3}{x+5}$

51. $x^5 - 64x = 0$

52. $x^5 = 36x$

53. $8x^3 = 64$

54. $x^3 - 125 = 0$

55. $x^5 + 1000x^2 = 0$

56. $2x^5 + 54x^2 = 0$

Writing and Thinking About Mathematics

47. $x = -7, -\frac{3}{2}$
48. $x = -\frac{3}{4}$
49. $x = \frac{26}{5}$
50. $x = -3$
51. $x = 0, \pm 2\sqrt{2}, \pm 2i\sqrt{2}$
52. $x = 0, \pm\sqrt{6}, \pm i\sqrt{6}$
53. $x = 2, -1 \pm i\sqrt{3}$
54. $x = 5, \frac{-5 \pm 5i\sqrt{3}}{2}$
55. $x = -10, 0, 5 \pm 5i\sqrt{3}$
56. $x = -3, 0, \frac{3 \pm 3i\sqrt{3}}{2}$
57. a. $l^2 - l - 1 = 0$
$l = \frac{1+\sqrt{5}}{2}$ which is the golden ratio
b. 97.08 feet
c. The yellow rectangle is "golden."
58. Using u-substitution where $u = \sqrt{x}$, we get u values of –2, and 3. However, $\sqrt{x}$ never equals –2, so we are left with only one solution, $x = 9$.

57. One of the most studied and interesting visual and numerical concepts in algebra is the **Golden Ratio**. Ancient Greeks thought (and many people still do) that a rectangle was most aesthetically pleasing to the eye if the ratio of its length to its width is the Golden Ratio (about 1.618). In fact, the Parthenon, built by Greeks in the fifth century B.C., utilizes the Golden Ratio. A rectangle is "golden" if its length l and width w satisfy the equation $\frac{l}{w} = \frac{w}{l-w}$.

a. By letting $w = 1$ unit in the equation above, we get the equation $\frac{l}{1} = \frac{1}{l-1}$. Solve this equation for the positive value of l (which is the algebraic expression for the golden ratio).

b. Suppose that an architect is constructing a building with a rectangular front that is to be 60 feet high. About how long should the front be if he wants the appearance of a golden rectangle? (Assume $w = 60$ ft and that you are looking for l.) Round to 2 decimal places.

c. Look at the two rectangles shown here. Which seems most pleasing to your eye? Measure the length and width of each rectangle and see if you chose the golden rectangle.

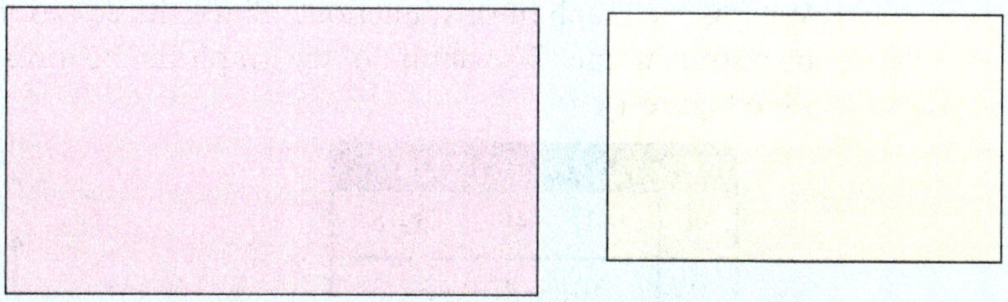

58. Consider the following equation: $x - x^{\frac{1}{2}} - 6 = 0$
In your own words, explain why, even though it is in quadratic form, this equation has only one solution.

HAWKES LEARNING SYSTEMS: INTRODUCTORY & INTERMEDIATE ALGEBRA SOFTWARE

- 10.4 Equations in Quadratic Form

10.5 Graphing Quadratic Functions: Parabolas

- *Graph a parabola (a quadratic function) and determine its vertex, domain, range, line of symmetry, and zeros.*
- *Solve applied problems using quadratic functions and the concepts of maximum and minimum.*

Introduction to Quadratic Functions: $y = ax^2 + bx + c\ (a \neq 0)$

We have studied various types of **functions** and their corresponding graphs: linear functions, polynomial functions, and functions with radicals. Related concepts include **domain**, **range**, and **zeros**. The vertical line test (restated here) can be used to tell whether or not a graph represents a function.

Vertical Line Test

If **any** vertical line intersects a graph in more than one point, then the relation is **not** a function.

In this section, we expand our interest in functions to include a detailed analysis of **quadratic functions**, functions that are represented by quadratic expressions. For example, consider the function

$$y = x^2 - 4x + 3.$$

What is the graph of this function? Since the equation is not linear, the graph will not be a straight line. The nature of the graph can be investigated by plotting several points (See Figure 1).

x	$x^2 - 4x + 3 = y$
-1	$(-1)^2 - 4(-1) + 3 = 8$
0	$(0)^2 - 4(0) + 3 = 3$
$\frac{1}{2}$	$\left(\frac{1}{2}\right)^2 - 4\left(\frac{1}{2}\right) + 3 = \frac{5}{4}$
1	$(1)^2 - 4(1) + 3 = 0$
2	$(2)^2 - 4(2) + 3 = -1$
3	$(3)^2 - 4(3) + 3 = 0$
$\frac{7}{2}$	$\left(\frac{7}{2}\right)^2 - 4\left(\frac{7}{2}\right) + 3 = \frac{5}{4}$
4	$(4)^2 - 4(4) + 3 = 3$
5	$(5)^2 - 4(5) + 3 = 8$

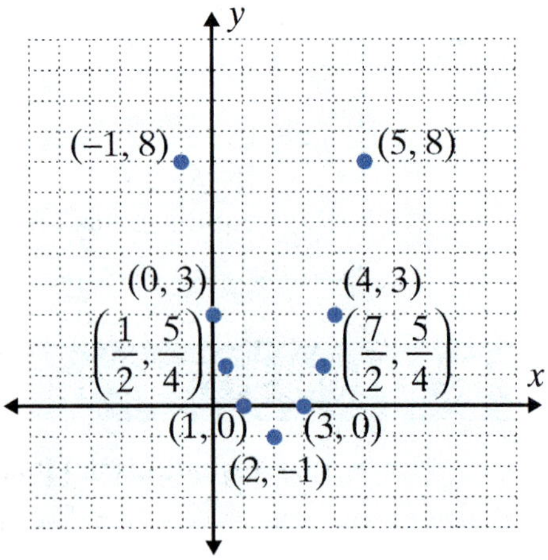

Figure 1

The complete graph of $y=x^2-4x+3$ is shown in Figure 2. The curve is called a **parabola**. The point $(2,-1)$ is the "turning point" of the parabola and is called the **vertex** of the parabola. The line $x=2$ is the **line of symmetry** or **axis of symmetry** for the parabola. That is, the curve is a "mirror image" of itself with respect to the line $x=2$.

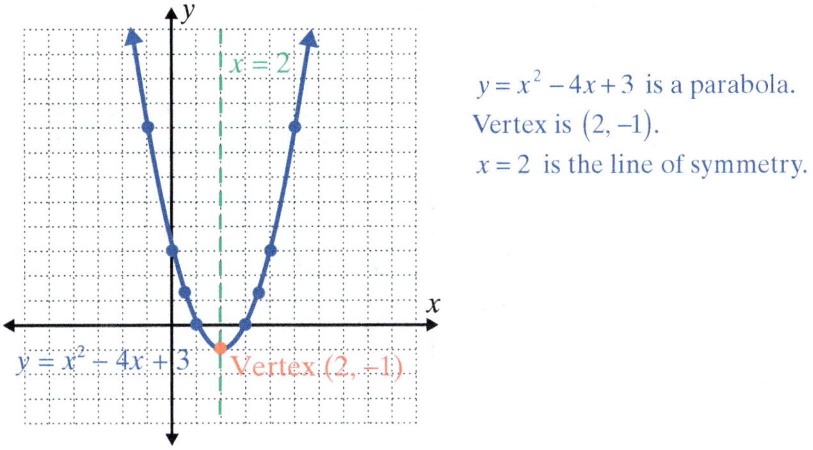

Figure 2

Quadratic Functions

Any function that can be written in the form

$$\boldsymbol{y=ax^2+bx+c}$$

where a, b, and c are real numbers and $a \neq 0$ is a **quadratic function**.

The graph of every quadratic function is a parabola. The position of the parabola, its shape, and whether it "opens upward" or "opens downward" can be determined by investigating the function itself. For convenience, we will refer to parabolas that open up or down as **vertical parabolas**. Parabolas that open left or right will be called **horizontal parabolas**. As we will see in Section 12.2, **horizontal parabolas do not represent functions**.

We will discuss quadratic functions in each of the following five forms where a, b, c, h, and k are constants:

$$y=ax^2$$
$$y=ax^2+k$$
$$y=a(x-h)^2$$
$$y=a(x-h)^2+k$$
$$y=ax^2+bx+c.$$

Functions of the Form $y = ax^2$

For any real number x, $x^2 \geq 0$. So, $ax^2 \geq 0$ if $a > 0$ and $ax^2 \leq 0$ if $a < 0$. This means that the graph of $y = ax^2$ is "above" the x-axis if $a > 0$ and "below" the x-axis if $a < 0$. The **vertex** is at the origin $(0, 0)$ in either of these cases and is the one point where each graph touches (or is tangent to) the x-axis.

For all quadratic functions, the **domain** is the set of all real numbers. That is, x can be replaced by any real number and there will be one corresponding y-value. The **range** of the function $y = ax^2$ depends on the value of a. If $a > 0$, then $y \geq 0$. If $a < 0$, then $y \leq 0$.

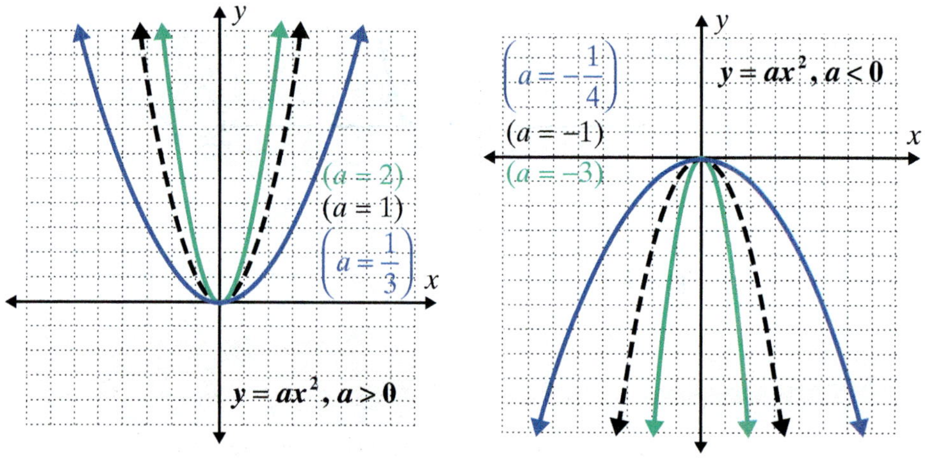

Domain = $\{x \mid x \text{ is any real number}\}$
Range = $\{y \mid y \geq 0\}$

In interval notation:
Domain = $(-\infty, \infty)$, Range = $[0, \infty)$

Domain = $\{x \mid x \text{ is any real number}\}$
Range = $\{y \mid y \leq 0\}$

In interval notation:
Domain = $(-\infty, \infty)$, Range = $(-\infty, 0]$

Figure 3

Figure 3 illustrates the following characteristics of quadratic functions of the form $y = ax^2$:

a. The vertex is at $(0, 0)$.
b. If $a > 0$, the parabola "opens upward."
c. If $a < 0$, the parabola "opens downward."
d. The bigger $|a|$ is, the narrower the opening.
e. The smaller $|a|$ is, the wider the opening.
f. The line $x = 0$ (the y-axis) is the line of symmetry.

Functions of the Form $y = ax^2 + k$

Adding k to ax^2 simply changes each y-value of $y = ax^2$ by k units (increase if k is positive, decrease if k is negative). That is, the graph of $y = ax^2 + k$ can be found by "sliding" or "shifting" the graph of $y = ax^2$ up k units if $k > 0$ or down $|k|$ units if $k < 0$ (Figure 4).

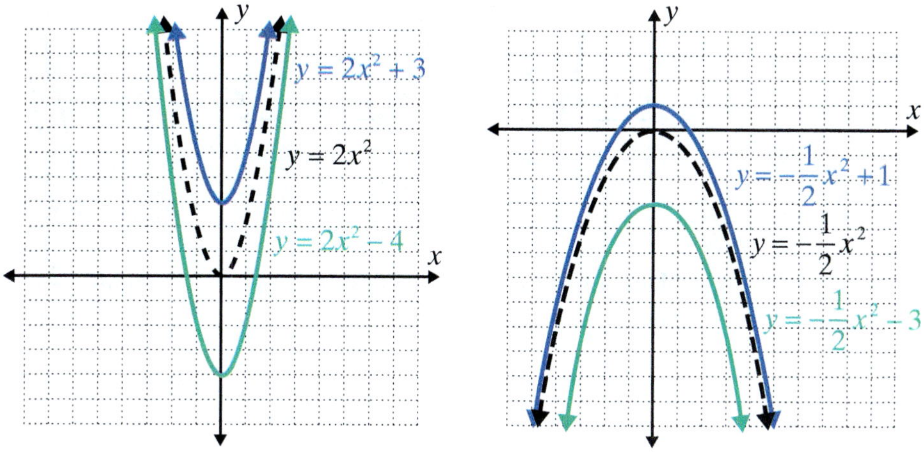

Domain = $\{x \mid x \text{ is any real number}\}$
Range = $\{y \mid y \geq k\}$
In interval notation:
Domain = $(-\infty, \infty)$, Range = $[k, \infty)$

Domain = $\{x \mid x \text{ is any real number}\}$
Range = $\{y \mid y \leq k\}$
In interval notation:
Domain = $(-\infty, \infty)$, Range = $(-\infty, k]$

Figure 4

The vertex of $y = ax^2 + k$ is at the point $(0, k)$. The graph of $y = ax^2 + k$ is a **vertical shift** (or **vertical translation**) of the graph of $y = ax^2$. The line $x = 0$ (the y-axis) is the line of symmetry just as with functions of the form $y = ax^2$.

Functions of the Form $y = a(x - h)^2$

We know that the square of a real number is nonnegative. Therefore, $(x-h)^2 \geq 0$. So, if $a > 0$, then $y = a(x-h)^2 \geq 0$. If $a < 0$, then $y = a(x-h)^2 \leq 0$. Also notice that $ax^2 = 0$ when $x = 0$, and $a(x-h)^2 = 0$ when $x = h$. Thus the vertex is at $(h, 0)$, and the parabola "opens upward" if $a > 0$ and "opens downward" if $a < 0$ (Figure 5).

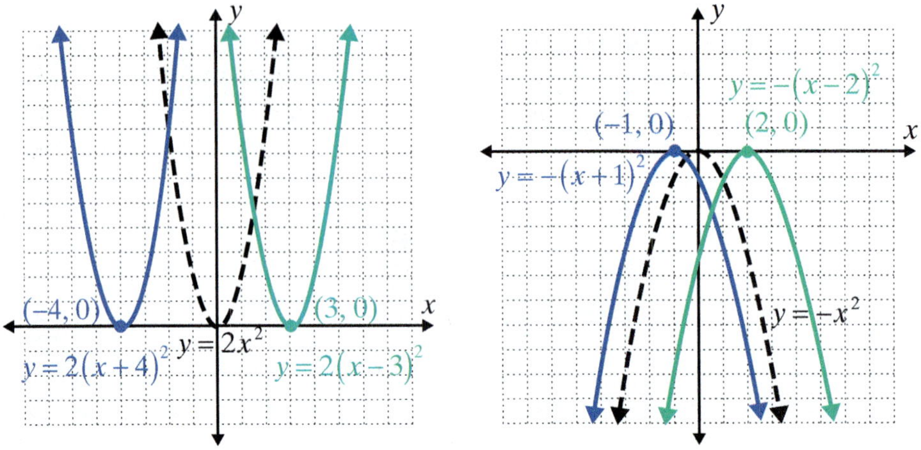

Domain = $\{x \mid x \text{ is any real number}\}$

Range = $\{y \mid y \geq 0\}$

In interval notation:

Domain = $(-\infty, \infty)$, Range = $[0, \infty)$

Domain = $\{x \mid x \text{ is any real number}\}$

Range = $\{y \mid y \leq 0\}$

In interval notation:

Domain = $(-\infty, \infty)$, Range = $(-\infty, 0]$

Figure 5

The graph of $y = a(x-h)^2$ is a **horizontal shift** (or **horizontal translation**) of the graph of $y = ax^2$. The shift is to the right if $h > 0$ and to the left if $h < 0$. As a special comment, note that if $h = -3$, then

$$y = a(x-h)^2$$

gives $y = a(x-(-3))^2$

or $y = a(x+3)^2$. (A shift of 3 units to the left)

Thus if h is negative, the expression $(x-h)^2$ appears with a plus sign. If h is positive, the expression $(x-h)^2$ appears with a minus sign. In either case, the line $\boldsymbol{x = h}$ is the line of symmetry.

Example 1: Quadratic Functions

Find the line of symmetry, the vertex, the domain, and the range of each quadratic function. Then graph the functions, setting up a table of values for x and y as an aid and choosing values of x on each side of the line of symmetry.

a. $y = 2x^2 - 3$

Solution: Line of symmetry is $x = 0$. (The parabola opens upward since a is positive.)

Vertex = $(0,-3)$; Domain = $(-\infty,\infty)$ or $\{x\,|\,x \text{ is any real number}\}$;

Range = $[-3,\infty)$ or $\{y\,|\,y \geq -3\}$

x	y
0	–3
$\frac{1}{2}$	$-\frac{5}{2}$
$-\frac{1}{2}$	$-\frac{5}{2}$
1	–1
–1	–1
2	5
–2	5

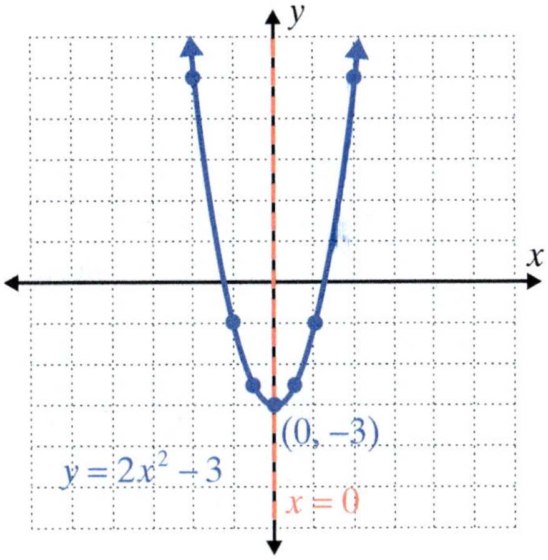

b. $y=-\left(x-\frac{5}{2}\right)^2$

Solution: Line of symmetry is $x=\frac{5}{2}$. (The parabola opens down since a is negative.)

Vertex = $\left(\frac{5}{2},0\right)$; Domain = $(-\infty,\infty)$ or $\{x\,|\,x \text{ is any real number}\}$;

Range = $(-\infty,0]$ or $\{y\,|\,y \leq 0\}$

x	y
$\frac{5}{2}$	0
2	$-\frac{1}{4}$
3	$-\frac{1}{4}$
1	$-\frac{9}{4}$
4	$-\frac{9}{4}$
0	$-\frac{25}{4}$
5	$-\frac{25}{4}$

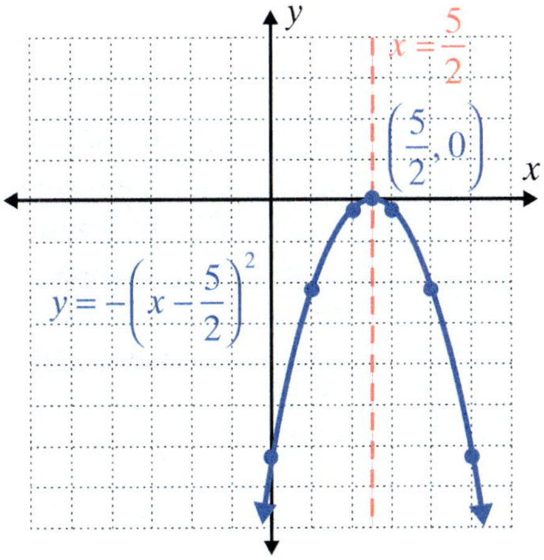

Functions of the Form $y = a(x-h)^2 + k$ and $y = ax^2 + bx + c$

The graphs of equations of the form

$$y = a(x-h)^2 + k$$

combine both the vertical shift of k units and the horizontal shift of h units. The **vertex is at (h, k)**. For example, the graph of the function $y = -2(x-3)^2 + 5$ is a shift of the graph of $y = -2x^2$ up 5 units and to the right 3 units and has its vertex at $(3, 5)$.

The graph of $y = \left(x + \frac{1}{2}\right)^2 - 2$ is the same as the graph of $y = x^2$ but is shifted left $\frac{1}{2}$ unit and down 2 units. The vertex is at $\left(-\frac{1}{2}, -2\right)$ (Figure 6).

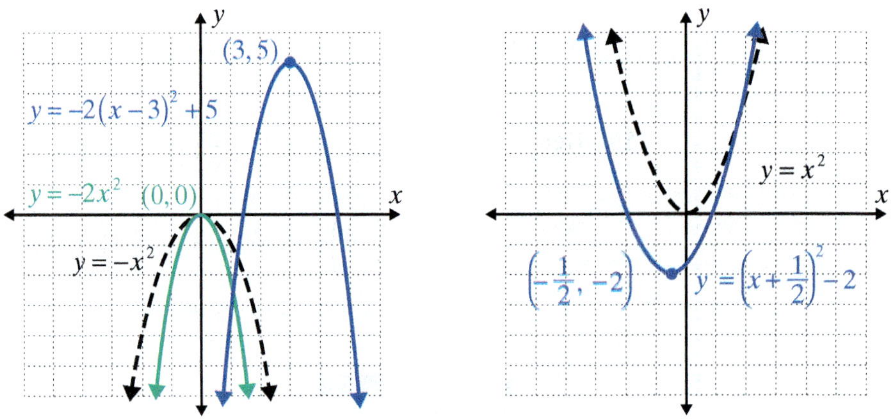

Figure 6

Teaching Notes: This is another good place in this development to make the case for learning to "read" information from an equation so that key characteristics can be known about these functions before (or without) graphing.

The general form of a quadratic function is $y = ax^2 + bx + c$. However, this form does not give as much information about the graph of the corresponding parabola as the form $y = a(x-h)^2 + k$. **Therefore, to easily find the vertex, line of symmetry, and range, and to graph the parabola, we want to change the general form $y = ax^2 + bx + c$ into the form $y = a(x-h)^2 + k$.** This can be accomplished by completing the square using the following technique. (Be aware that you are not solving an equation. You do not "do something" to both sides. You are **changing the form** of a function.) (**Note:** A graphing calculator will give the same graph regardless of the form of the function.)

$y = ax^2 + bx + c$	Write the function.
$= a\left(x^2 + \frac{b}{a}x\right) + c$	Factor a from just the first two terms.
$= a\left(x^2 + \frac{b}{a}x + \frac{b^2}{4a^2} - \frac{b^2}{4a^2}\right) + c$	Complete the square of $x^2 + \frac{b}{a}x$. $\frac{1}{2} \cdot \frac{b}{a} = \frac{b}{2a}$ and $\left(\frac{b}{2a}\right)^2 = \frac{b^2}{4a^2}$. Add and subtract $\frac{b^2}{4a^2}$ inside the parentheses.
$= a\left(x^2 + \frac{b}{a}x + \frac{b^2}{4a^2}\right) - \frac{b^2}{4a} + c$	Multiply $a \cdot \left(\frac{-b^2}{4a^2}\right)$ and write this term outside the parentheses.
$= a\left(x + \frac{b}{2a}\right)^2 + \frac{4ac - b^2}{4a}$	Write the square of the binomial and add the last two terms to get the form $y = a(x-h)^2 + k$.

In terms of the coefficients a, b, and c,

$$x = -\frac{b}{2a} \quad \text{is the } \textbf{line of symmetry}$$

and

$$(h,k) = \left(-\frac{b}{2a}, \frac{4ac - b^2}{4a}\right) \quad \text{is the } \textbf{vertex}. \quad \left(\text{In function notation, } \left(-\frac{b}{2a}, f\left(-\frac{b}{2a}\right)\right).\right)$$

NOTES

Rather than memorize the formula for the coordinates of the vertex, you should just remember that the x-coordinate of the vertex is $x = -\frac{b}{2a}$. Substituting this value for x in the function will give the y-value for the vertex.

Zeros of a Quadratic Function

The points where a parabola crosses the x-axis, if any, are the x-intercepts. This is where $y = 0$. These points are called the **zeros of the function**. We find these points by substituting 0 for y and solving the resulting quadratic equation.

$$y = ax^2 + bx + c \quad \textbf{quadratic function}$$

$$0 = ax^2 + bx + c \quad \textbf{quadratic equation}$$

If the solutions are nonreal complex numbers, then the graph does not cross the x-axis. It is either entirely above the x-axis or entirely below the x-axis.

The following examples illustrate how to apply all our knowledge about quadratic functions.

Example 2: Graphing Quadratic Functions

a. For $y = x^2 - 6x + 1$, find the zeros of the function, the line of symmetry, the vertex, the domain, the range, and graph the parabola.

Solution: $x^2 - 6x + 1 = 0$

$$\begin{aligned} x &= \frac{6 \pm \sqrt{(-6)^2 - 4}}{2} && \text{Quadratic formula} \\ &= \frac{6 \pm \sqrt{32}}{2} \\ &= \frac{6 \pm 4\sqrt{2}}{2} \\ &= 3 \pm 2\sqrt{2} \end{aligned}$$

The zeros are $3 \pm 2\sqrt{2}$.

Change the form of the function for easier graphing.

$$\begin{aligned} y &= x^2 - 6x + 1 \\ &= \left(x^2 - 6x + 9 - 9\right) + 1 && \text{Add } 0 = 9 - 9 \text{ inside the parentheses.} \\ &= \left(x^2 - 6x + 9\right) - 9 + 1 \\ &= (x-3)^2 - 8 \end{aligned}$$

Summary:

Zeros: $3 \pm 2\sqrt{2}$

Line of symmetry is $x = 3$.

Vertex: $(3, -8)$

Domain: $(-\infty, \infty)$ or $\{x \mid x \text{ is any real number}\}$

Range: $[-8, \infty)$ or $\{y \mid y \geq -8\}$

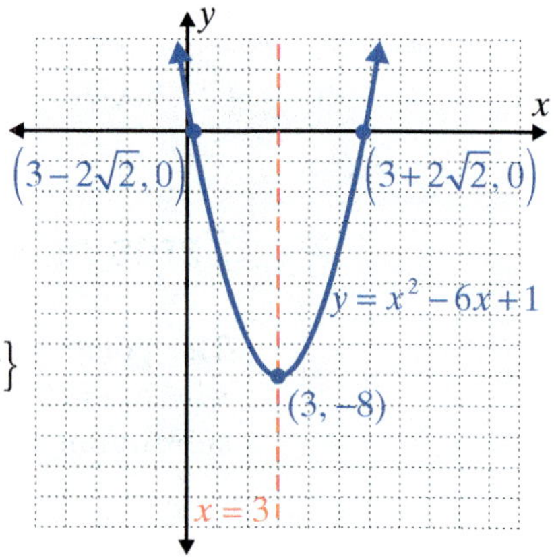

b. For $y=-x^2-4x+2$, find the zeros of the function, the line of symmetry, the vertex, the domain, the range, and graph the parabola.

Solution: $-x^2-4x+2=0$

$$x=\frac{4\pm\sqrt{(-4)^2-4(-1)(2)}}{2(-1)} \qquad \text{Quadratic formula}$$

$$=\frac{4\pm\sqrt{24}}{-2}$$

$$=\frac{4\pm2\sqrt{6}}{-2}$$

$$=-2\pm\sqrt{6}$$

The zeros are $-2\pm\sqrt{6}$.

Change the form of the function for easier graphing.

$y=-x^2-4x+2$

$=-(x^2+4x)+2$ Factor -1 from the first two terms only.

$=-(x^2+4x+4-4)+2$ Add $0=4-4$ inside the parentheses.

$=-(x^2+4x+4)+4+2$ Multiply $-1(-4)$ and put this outside the parentheses.

$=-(x+2)^2+6$

Summary:

Zeros: $-2\pm\sqrt{6}$

Line of symmetry is $x=-2$.

Vertex: $(-2,6)$

Domain: $(-\infty,\infty)$ or $\{x\,|\,x \text{ is any real number}\}$

Range: $(-\infty,6]$ or $\{y\,|\,y\le6\}$

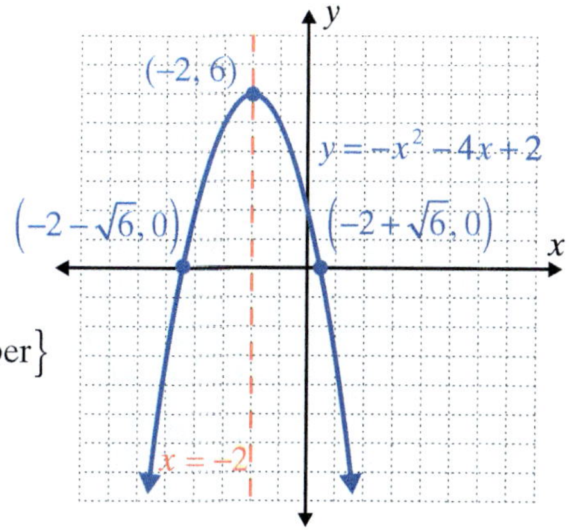

Continued on the next page...

c. For $y = 2x^2 - 6x + 5$, find the zeros of the function, the line of symmetry, the vertex, the domain, the range, and graph the parabola.

Solution: $2x^2 - 6x + 5 = 0$

$$x = \frac{6 \pm \sqrt{(-6)^2 - 4(2)(5)}}{2(2)} \qquad \text{Quadratic formula}$$

$$= \frac{6 \pm \sqrt{-4}}{4}$$

There are **no real zeros** because the discriminant is negative. The graph will not cross the x-axis. Now using another approach, the vertex is at

$$x = -\frac{b}{2a} = -\frac{-6}{2 \cdot 2} = \frac{3}{2}$$

and

$$y = 2\left(\frac{3}{2}\right)^2 - 6\left(\frac{3}{2}\right) + 5$$

$$= \frac{9}{2} - 9 + 5$$

$$= \frac{1}{2}.$$

So we have the vertex at $\left(\frac{3}{2}, \frac{1}{2}\right)$ and the following results:

Summary:

Zeros: No real zeros

Line of symmetry is $x = \frac{3}{2}$.

Vertex: $\left(\frac{3}{2}, \frac{1}{2}\right)$

Domain: $(-\infty, \infty)$ or $\{x \mid x \text{ is any real number}\}$

Range: $\left[\frac{1}{2}, \infty\right)$ or $\left\{y \mid y \geq \frac{1}{2}\right\}$

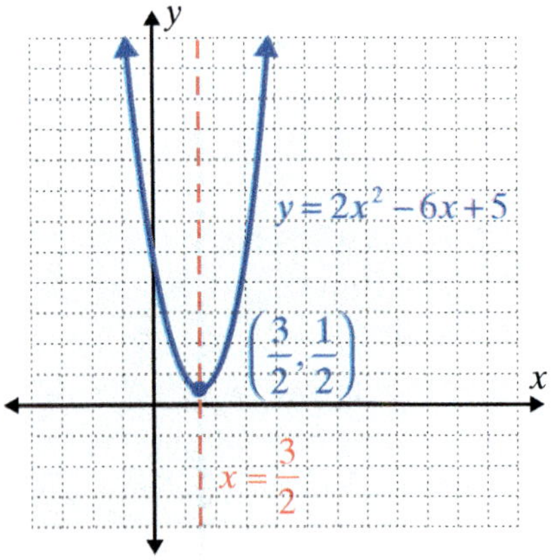

Applications with Maximum and Minimum Values

The vertex of a vertical parabola is either the lowest point or the highest point on the parabola.

Minimum and Maximum Values

For a parabola with its equation in the form $y = a(x-h)^2 + k$,

1. If $a > 0$, then the parabola opens upward and (h, k) is the lowest point and the y-value k is called the **minimum value** of the function.
2. If $a < 0$, then the parabola opens downward and (h, k) is the highest point and y-value k is called the **maximum value** of the function.

If the function is in the general quadratic form $y = ax^2 + bx + c$, then the maximum or minimum value can be found by letting $x = -\frac{b}{2a}$ and solving for y.

The concepts of maximum and minimum values of a function help not only in graphing but also in solving many types of applications. Applications involving quadratic functions are discussed here. Other types of applications are discussed in more advanced courses in mathematics.

Example 3: Minimum and Maximum Values

Teaching Notes:
You might want to introduce your students to the CALC, minimum, and maximum sequence of steps on the TI-84 Plus graphing calculator. Maybe show them these steps at a later time after they have gained the related algebraic skills developed in this section.

a. A sandwich company sells hot dogs at the local baseball stadium for \$3.00 each. On average they sell 2000 hot dogs per game. The company estimates that each time the price is raised by 25¢, they will sell 100 fewer hot dogs. What price should they charge to maximize their revenue (income) per game? What will be the maximum revenue?

Solution: Let x = number of 25¢ increases in price.

Then $3.00 + 0.25x$ = price per hot dog,

and $2000 - 100x$ = number of hot dogs sold.

Revenue(R) = (price per unit)·(number of units sold)

So,

$$R = (3.00 + 0.25x)(2000 - 100x)$$
$$= 6000 - 300x + 500x - 25x^2$$
$$= -25x^2 + 200x + 6000.$$

The revenue is represented by a quadratic function and the maximum revenue occurs at the point where

$$x = -\frac{b}{2a} = -\frac{200}{-50} = 4.$$

Continued on the next page...

1. $x = 0; (0,-4); D = (-\infty, \infty); R = [-4, \infty)$
2. $x = 0; (0,7); D = (-\infty, \infty); R = [7, \infty)$
3. $x = 0; (0,9); D = (-\infty, \infty); R = [9, \infty)$
4. $x = 0; (0,-1); D = (-\infty, \infty); R = [-1, \infty)$
5. $x = 0; (0,1); D = (-\infty, \infty); R = (-\infty, 1]$
6. $x = 0; (0,-6); D = (-\infty, \infty); R = (-\infty, -6]$
7. $x = 0; \left(0, \frac{1}{5}\right); D = (-\infty, \infty); R = \left(-\infty, \frac{1}{5}\right]$
8. $x = 0; \left(0, \frac{7}{8}\right); D = (-\infty, \infty); R = \left[\frac{7}{8}, \infty\right)$

For $x = 4$,

$$\text{price per hot dog} = 3.00 + 0.25(4) = \$4.00,$$

and $$\text{revenue} = (4)(2000 - 100 \cdot 4) = \$6400.$$

Thus the company will make its maximum revenue of $6400 by charging $4 per hot dog.

b. A rancher is going to build three sides of a rectangular corral next to a river. He has 240 feet of fencing and wants to enclose the maximum area possible inside the corral. What are the dimensions of the corral with the maximum area and what is this area?

Solution: Let x = length of one of the two equal sides of the rectangular corral. Then $240 - 2x$ = length of third side of the rectangular corral.

Since area equals length times width, the area, A, of the corral is represented by the quadratic function $A = x(240 - 2x) = 240x - 2x^2$, the maximum area occurs at the point where $x = -\frac{b}{2a} = -\frac{240}{-4} = 60.$

Two sides of the rectangle are 60 feet and the third side is $240 - 2(60) = 120$ feet.

The maximum area possible is $60(120) = 7200$ square feet.

Practice Problems

1. Write the function $y = 2x^2 - 4x + 3$ in the form $y = a(x-h)^2 + k$.
2. Find the zeros of the function $y = x^2 - 7x + 10$.
3. Find the vertex and the range of the function $y = -x^2 + 4x - 5$.

9. $x = -1; (-1,0); D = (-\infty, \infty); R = [0, \infty)$
10. $x = 1; (1,0); D = (-\infty, \infty); R = [0, \infty)$
11. $x = 4; (4,0); D = (-\infty, \infty); R = (-\infty, 0]$
12. $x = -2; (-2,0); D = (-\infty, \infty); R = (-\infty, 0]$
13. $x = -3; (-3,-2); D = (-\infty, \infty); R = [-2, \infty)$
14. $x = 5; (5,1); D = (-\infty, \infty); R = [1, \infty)$
15. $x = -2; (-2,-6); D = (-\infty, \infty); R = [-6, \infty)$

Answers to Practice Problems: **1.** $y = 2(x-1)^2 + 1$ **2.** $x = 2, 5$ **3.** Vertex: $(2,-1)$, Range: $(-\infty, -1]$

10.5 Exercises

For each of the quadratic functions, determine the line of symmetry, the vertex, the domain, and the range.

1. $y=3x^2-4$

2. $y=\frac{2}{3}x^2+7$

3. $y=7x^2+9$

4. $y=5x^2-1$

5. $y=-4x^2+1$

6. $y=-2x^2-6$

7. $y=-\frac{3}{4}x^2+\frac{1}{5}$

8. $y=\frac{5}{3}x^2+\frac{7}{8}$

9. $y=(x+1)^2$

10. $y=(x-1)^2$

11. $y=-\frac{2}{3}(x-4)^2$

12. $y=-5(x+2)^2$

13. $y=2(x+3)^2-2$

14. $y=4(x-5)^2+1$

15. $y=\frac{3}{4}(x+2)^2-6$

16. $y=-2(x+1)^2-4$

17. $y=-\frac{1}{2}\left(x-\frac{3}{2}\right)^2+\frac{7}{2}$

18. $y=-\frac{5}{3}\left(x-\frac{9}{2}\right)^2+\frac{3}{4}$

19. $y=\frac{1}{4}\left(x-\frac{4}{5}\right)^2-\frac{11}{5}$

20. $y=\frac{10}{3}\left(x+\frac{7}{8}\right)^2-\frac{9}{16}$

21. Graph the function $y=x^2$. Then without additional computation, graph the following translations.

a. $y=x^2-2$

b. $y=(x-3)^2$

c. $y=-(x-1)^2$

d. $y=5-(x+1)^2$

22. Graph the function $y=2x^2$. Then without additional computation, graph the following translations.

a. $y=2x^2-3$

b. $y=2(x-4)^2$

c. $y=-2(x+1)^2$

d. $y=-2(x+2)^2-4$

16. $x=-1$; $(-1,-4)$; $D=(-\infty,\infty)$; $R=(-\infty,-4]$

17. $x=\frac{3}{2}$; $\left(\frac{3}{2},\frac{7}{2}\right)$; $D=(-\infty,\infty)$; $R=\left(-\infty,\frac{7}{2}\right]$

18. $x=\frac{9}{2}$; $\left(\frac{9}{2},\frac{3}{4}\right)$; $D=(-\infty,\infty)$; $R=\left(-\infty,\frac{3}{4}\right]$

19. $x=\frac{4}{5}$; $\left(\frac{4}{5},-\frac{11}{5}\right)$; $D=(-\infty,\infty)$; $R=\left[-\frac{11}{5},\infty\right)$

20. $x=-\frac{7}{8}$; $\left(-\frac{7}{8},-\frac{9}{16}\right)$; $D=(-\infty,\infty)$; $R=\left[-\frac{9}{16},\infty\right)$

21. a.
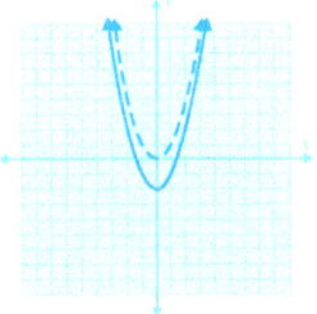

b.
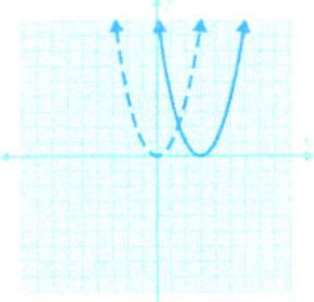

c.
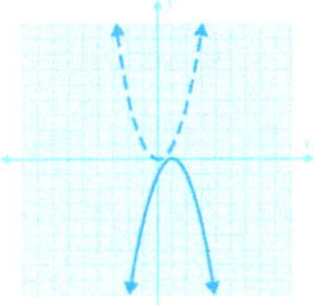

d.
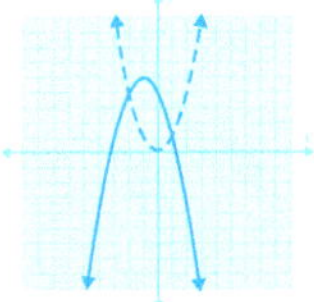

22. a.
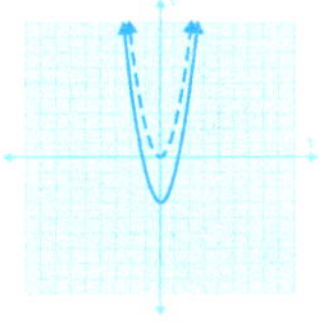

b.
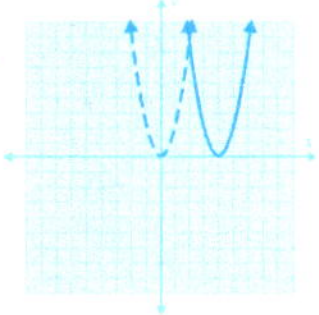

c.
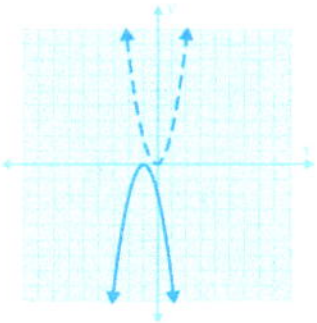

d.
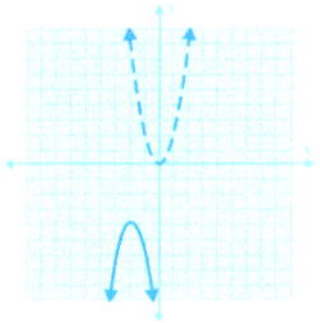

23. a.
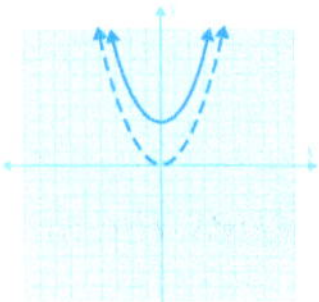
b.
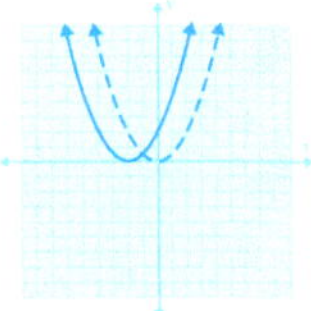
c.
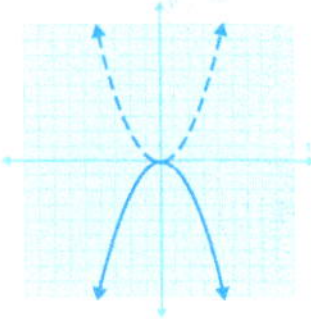
d.
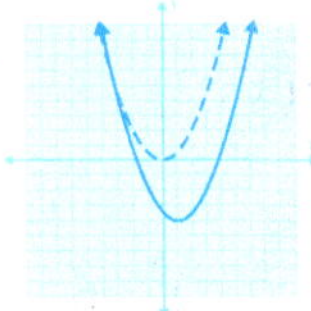
24. a.
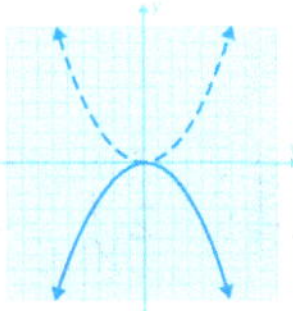
b.
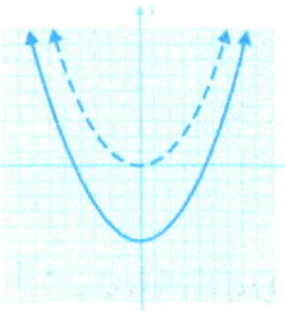
c.
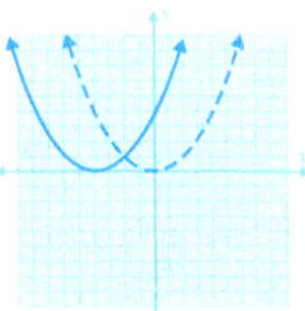
d.
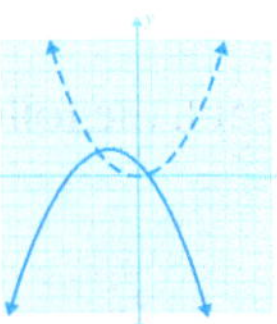

23. Graph the function $y=\frac{1}{2}x^2$. Then without additional computation, graph the following translations.

a. $y=\frac{1}{2}x^2+3$ **b.** $y=\frac{1}{2}(x+2)^2$

c. $y=-\frac{1}{2}x^2$ **d.** $y=\frac{1}{2}(x-1)^2-4$

24. Graph the function $y=\frac{1}{4}x^2$. Then without additional computation, graph the following translations.

a. $y=-\frac{1}{4}x^2$ **b.** $y=\frac{1}{4}x^2-5$

c. $y=\frac{1}{4}(x+4)^2$ **d.** $y=2-\frac{1}{4}(x+2)^2$

Rewrite each of the quadratic functions in the form $y=a(x-h)^2+k$. Find the vertex, range, and zero(s), if any, of each function. Graph the function.

25. $y=2x^2-4x+2$ **26.** $y=-3x^2+12x-12$ **27.** $y=x^2-2x-3$

28. $y=x^2-4x+5$ **29.** $y=x^2+6x+5$ **30.** $y=x^2-8x+12$

31. $y=2x^2-8x+5$ **32.** $y=2x^2-12x+16$ **33.** $y=-3x^2-12x-9$

34. $y=3x^2-6x-1$ **35.** $y=5x^2-10x+8$ **36.** $y=-4x^2+16x-11$

37. $y=-x^2-5x-2$ **38.** $y=x^2+3x-1$ **39.** $y=2x^2+7x+5$

40. $y=2x^2+x-3$

Graph the two given functions and answer the following questions:

a. *Are the graphs the same?*

b. *Do the functions have the same zeros?*

c. *Briefly, discuss your interpretation of the results in parts* **a.** *and* **b.**

41. $\begin{cases} y=x^2-3x-10 \\ y=-x^2+3x+10 \end{cases}$ **42.** $\begin{cases} y=x^2-5x+6 \\ y=-x^2+5x-6 \end{cases}$

43. $\begin{cases} y=2x^2-5x-3 \\ y=-2x^2+5x+3 \end{cases}$ **44.** $\begin{cases} y=-4x^2-15x+4 \\ y=4x^2+15x-4 \end{cases}$

25. $y=2(x-1)^2$; $(1,0)$; $R=[0,\infty)$; Zeros: $x=1$
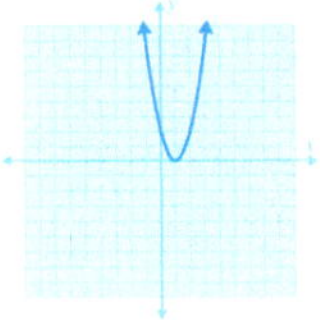

26. $y=-3(x-2)^2$; $(2,0)$; $R=(-\infty,0]$; Zeros: $x=2$
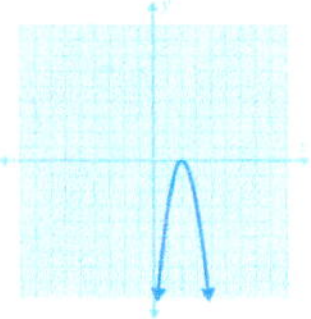

27. $y=(x-1)^2-4$; $(1,-4)$; $R=[-4,\infty)$; Zeros: $x=-1, 3$

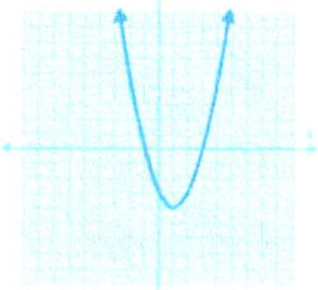

28. $y=(x-2)^2+1$; $(2,1)$; $R=[1,\infty)$; Zeros: none

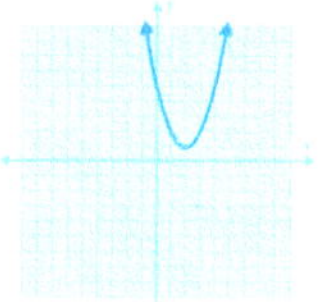

29. $y=(x+3)^2-4$; $(-3,-4)$; $R=[-4,\infty)$; Zeros: $x=-5, -1$

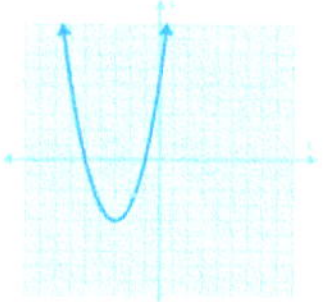

30. $y=(x-4)^2-4$; $(4,-4)$; $R=[-4,\infty)$; Zeros: $x=2, 6$

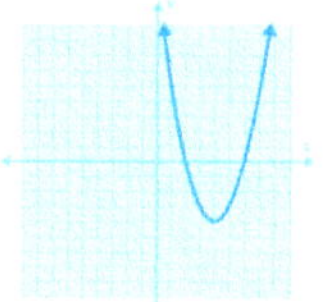

31. $y=2(x-2)^2-3$; $(2,-3)$; $R=[-3,\infty)$; Zeros: $x=\frac{4\pm\sqrt{6}}{2}$

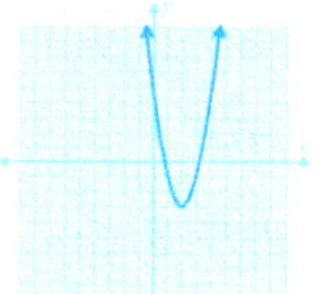

Use the function $h=-16t^2+v_0t+h_0$, where h is the height of the object after time t, v_0 is the initial velocity, and h_0 is the initial height.

45. Throwing a ball: A ball is thrown vertically upward from the ground with an initial velocity of 112 ft/s.

a. When will the ball reach its maximum height?

b. What will be the maximum height?

46. Launching a rocket: A water rocket is launched from the ground and has an initial velocity of 104 ft/s.

a. When will the rocket reach its maximum height?

b. What will be the maximum height?

47. Throwing a stone: A stone is projected vertically upward from a platform that is 20 feet high, at a rate of 160 feet per second.

a. When will the stone reach its maximum height?

b. What will be the maximum height?

48. Throwing a stone: A cannonball is projected vertically upward from a platform that is 32 feet high, at a rate of 128 feet per second.

a. When will the cannonball reach its maximum height?

b. What will be the maximum height?

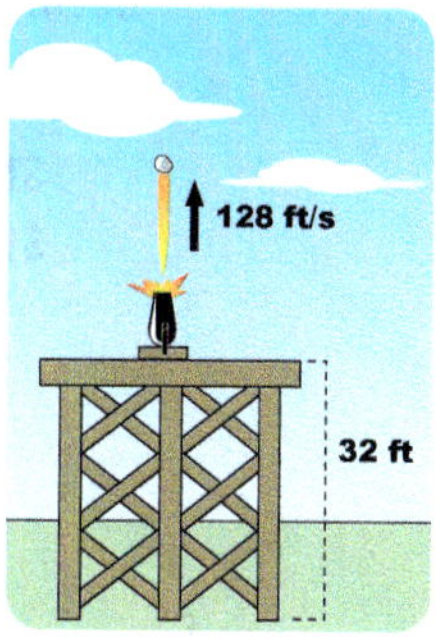

32. $y=2(x-3)^2-2$; $(3,-2)$; $R=[-2,\infty)$; Zeros: $x=2, 4$

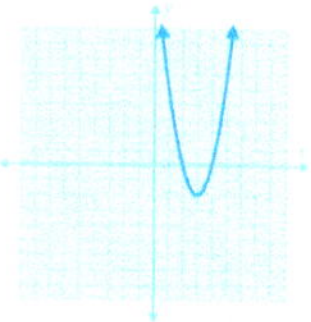

Solve the following application problems.

49. Selling radios: A retailer sells radios. He estimates that by selling them for x dollars each, he will be able to sell $100-x$ radios each month.

a. What price will yield maximum revenue?

b. What will be the maximum revenue?

50. Selling picture frames: Mrs. Richey can sell 72 picture frames each month if she charges $24 each. She estimates that for each $1 increase in price, she will sell 2 fewer frames.

a. Find the price that will yield maximum revenue.

b. What will be the maximum revenue?

51. Selling lamps: A store owner estimates that by charging x dollars each for a certain lamp, he can sell $40-x$ lamps each week. What price will give him maximum sales?

52. Construction: A contractor is to build a brick wall 6 feet high to enclose a rectangular garden. The wall will be on three sides of the rectangle because the fourth side is a building. The owner wants to enclose the maximum area but wants to pay for only 150 feet of wall. What dimensions should the contractor make the garden?

33. $y=-3(x+2)^2+3$; $(-2,3)$; $R=(-\infty,3]$; Zeros: $x=-1, -3$

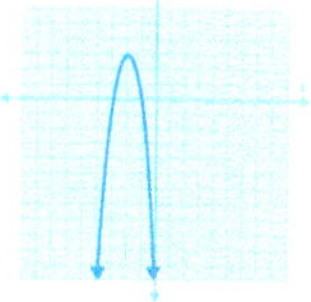

34. $y=3(x-1)^2-4$; $(1,-4)$; $R=[-4,\infty)$; Zeros: $x=\frac{3\pm2\sqrt{3}}{3}$

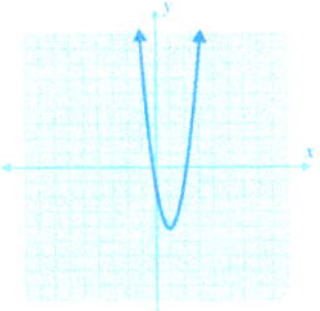

35. $y=5(x-1)^2+3$; $(1,3)$; $R=[3,\infty)$; Zeros: none

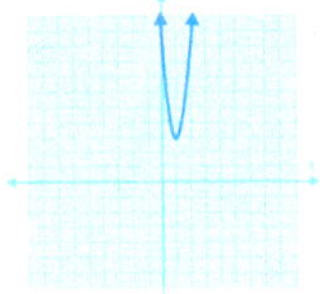

36. $y=-4(x-2)^2+5$; $(2,5)$; $R=(-\infty,5]$; Zeros: $x=\frac{4\pm\sqrt{5}}{2}$

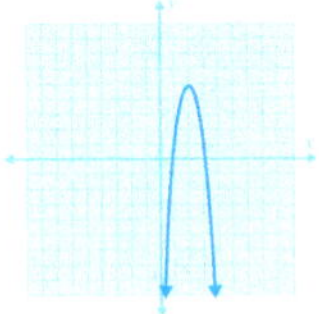

37. $y=-\left(x+\frac{5}{2}\right)^2+\frac{17}{4}$; $\left(-\frac{5}{2},\frac{17}{4}\right)$; $R=\left(-\infty,\frac{17}{4}\right]$; Zeros: $x=\frac{-5\pm\sqrt{17}}{2}$

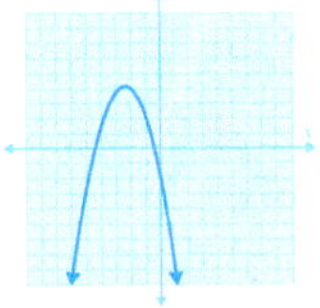

Use the CALC *features of the calculator to find the zeros of the function.* (**Hint:** *Item 2 on the* CALC *menu* `2: zero` *will locate the zeros of the function.) Round answers to four decimal places.*

53. $y=x^2-2x-2$ **54.** $y=3x^2+x-1$ **55.** $y=-2x^2+2x+5$

56. $y=-x^2-2x+7$ **57.** $y=x^2+3x+3$ **58.** $y=-4x^2-x-6$

Use a graphing calculator to graph each function by pressing Y= *and entering the function. Find the coordinates of the* ***maximum*** *as follows. Round answers to four decimal places.*

Step 1: Press CALC (2ND TRACE).
Step 2: Press or choose `4: maximum`.
Step 3: Follow the directions for moving the cursor to `Left Bound?`, `Right Bound?`, and `Guess?`. (Press ENTER each time.)

59. $y=4x-x^2$ **60.** $y=1-2x-x^2$ **61.** $y=-8+4x-x^2$ **62.** $y=3-2x-x^2$

Use a graphing calculator to graph each function by pressing Y= *and entering the function. Find the coordinates of the* ***minimum*** *as follows. Round answers to four decimal places.*

Step 1: Press CALC (2ND TRACE).
Step 2: Press or choose `3: minimum`.
Step 3: Follow the directions for moving the cursor to `Left Bound?`, `Right Bound?`, and `Guess?`. (Press ENTER each time.)

63. $y=x^2-8x+15$ **64.** $y=x^2+10x+22$ **65.** $y=2x^2+4x+3$ **66.** $y=3x^2-6x+5$

Writing and Thinking About Mathematics

67. Discuss the following features of the general quadratic function $y=ax^2+bx+c$.
   a. What type of curve is its graph?
   b. What is the value of x at its vertex?
   c. What is the equation of the line of symmetry?
   d. Does the graph always cross the x-axis? Explain.

68. Discuss the discriminant of the general quadratic equation $ax^2+bx+c=0$ and how the value of the discriminant is related to the graph of the corresponding quadratic function $y=ax^2+bx+c$.

Continued on the next page...

Writing and Thinking About Mathematics (cont.)

69. Discuss the domain and range of a quadratic function in the form $y = a(x-h)^2 + k$.

HAWKES LEARNING SYSTEMS: INTRODUCTORY & INTERMEDIATE ALGEBRA SOFTWARE

- 10.5 Graphing Parabolas

38. $y=\left(x+\frac{3}{2}\right)^2-\frac{13}{4}$; $\left(-\frac{3}{2},-\frac{13}{4}\right)$; $R=\left[-\frac{13}{4},\infty\right)$; Zeros: $x=\frac{-3\pm\sqrt{13}}{2}$

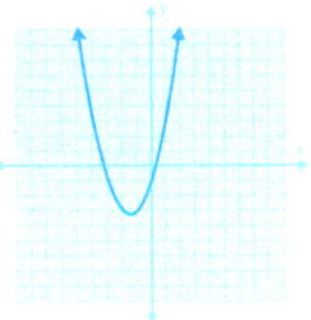

39. $y=2\left(x+\frac{7}{4}\right)^2-\frac{9}{8}$; $\left(-\frac{7}{4},-\frac{9}{8}\right)$; $R=\left[-\frac{9}{8},\infty\right)$; Zeros: $x=-\frac{5}{2},-1$

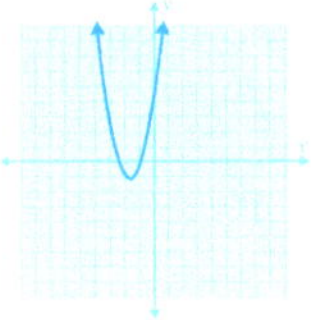

40. $y=2\left(x+\frac{1}{4}\right)^2-\frac{25}{8}$; $\left(-\frac{1}{4},-\frac{25}{8}\right)$; $R=\left[-\frac{25}{8},\infty\right)$; Zeros: $x=-\frac{3}{2},1$

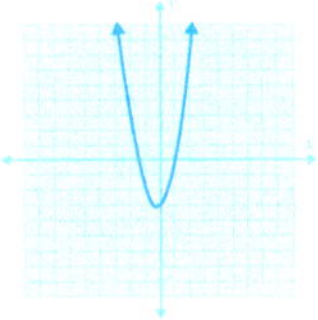

41.

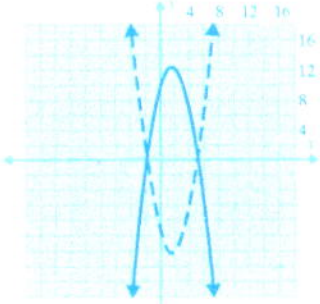

a. No **b.** Yes **c.** One function is a reflection of the other across the x-axis.

42.

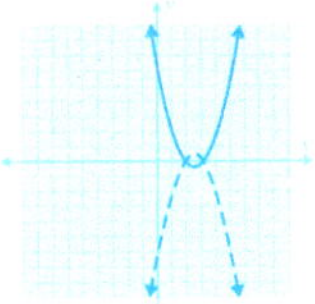

a. No **b.** Yes **c.** One function is a reflection of the other across the x-axis.

43.

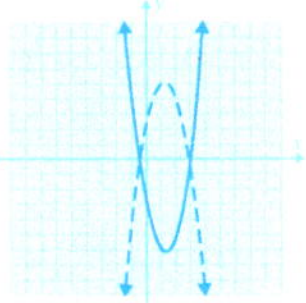

a. No **b.** Yes **c.** One function is a reflection of the other across the x-axis.

44.

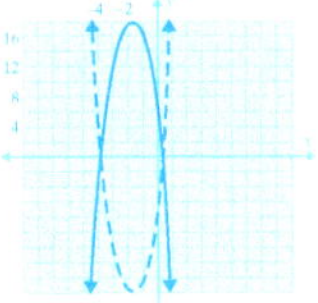

a. No **b.** Yes **c.** One function is a reflection of the other across the x-axis.

45. a. 3.5 s **b.** 196 ft
46. a. 3.25 s **b.** 169 ft
47. a. 5 s **b.** 420 ft
48. a. 4 s **b.** 288 ft
49. a. \$50 **b.** \$2500
50. a. \$30 **b.** \$1800
51. \$20

52. 37.5 ft × 75 ft **53.** Zeros: $x \approx -0.7321, 2.7321$ **54.** Zeros: $x \approx -0.7676, 0.4343$
55. Zeros: $x \approx -1.1583, 2.1583$ **56.** Zeros: $x \approx -3.8284, 1.8284$ **57.** No real zeros **58.** No real zeros

59.

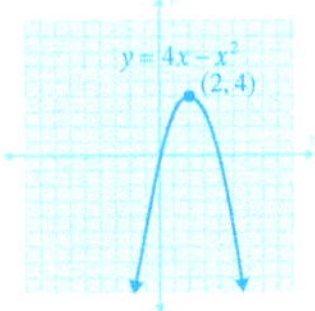

60.

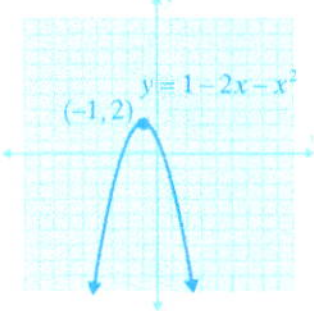

61.

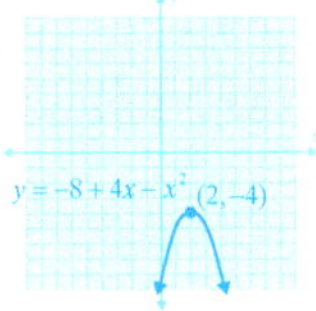

62.

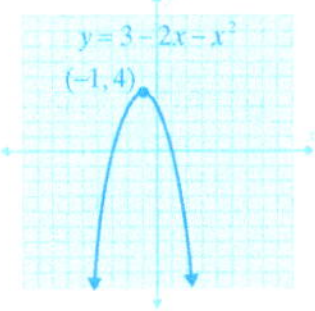

63.

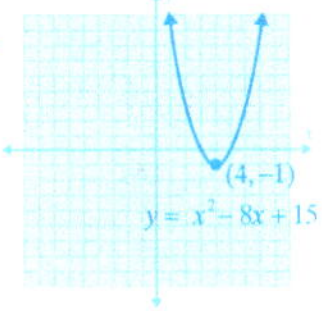

64.

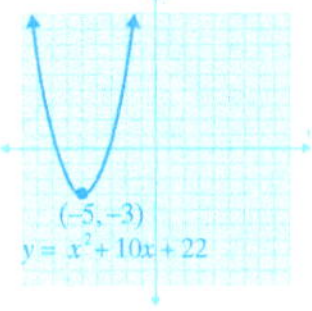

65.

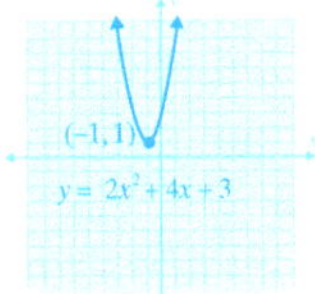

66.

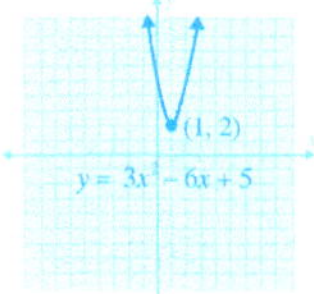

67. a. A parabola **b.** $x=-\frac{b}{2a}$ **c.** $x=-\frac{b}{2a}$
d. No. A graph can be entirely above or below the x-axis.
68. If discriminant > 0, then graph crosses x-axis twice; if discriminant < 0, then graph is entirely above or entirely below the x-axis; if discriminant = 0, then graph touches the x-axis in one point (is tangent to).
69. Domain: $(-\infty,\infty)$; For $a>0$, Range: $[k,\infty)$; For $a<0$, Range: $(-\infty,k]$.

10.6 Solving Quadratic and Rational Inequalities

- *Solve quadratic inequalities.*
- *Solve higher-degree polynomial inequalities.*
- *Solve rational inequalities.*
- *Graph the solutions of inequalities on real number lines.*
- *Use a graphing calculator as an aid in solving quadratic inequalities.*

In this section we will develop algebraic and graphical techniques for solving inequalities with emphasis on **quadratic inequalities** and **rational inequalities**. For example, we will see that inequalities such as

$$x^2 + 3x + 2 \geq 0, \quad x^2 - 2x > 8, \quad x^3 + 4x^2 - 5x < 0 \quad \text{and} \quad \frac{x+3}{x-2} > 0$$

can be solved by factoring or using the quadratic formula and then analyzing the graphs of the corresponding functions. The solution to these inequalities consist of intervals on the real number line.

Teaching Notes:
Many common situations routinely and naturally use the inequality concept. You might want to remind students that common problems dealing with budgets, capacities, sales quotas, profit, fluid pressure, memberships, etc. are more likely to be expressed in terms of "at least," "at most," "not greater than," "not less than," "minimum," and "maximum" than as discrete numbers.

Quadratic Inequalities: Solved Algebraically

The technique of factoring to solve inequalities is based on the simple idea that for values of x on either side of a number a, the sign for an expression of the form $(x-a)$ changes. More specifically:

If $x > a$, then $(x-a)$ is positive.

If $x < a$, then $(x-a)$ is negative.

Graphically,

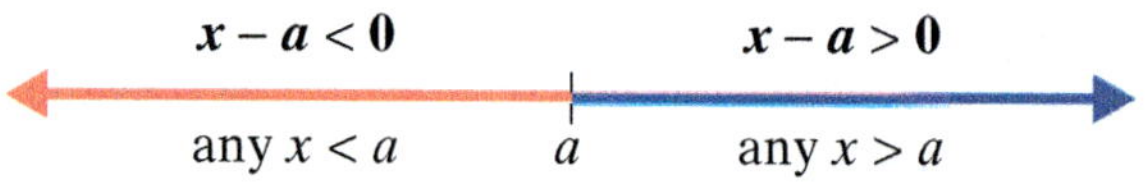

To solve a polynomial inequality algebraically, get 0 on one side of the inequality and then factor the polynomial on the other side. Locate the points where each factor is 0, and then analyze the sign of the polynomial in the intervals on each side of these points.

To Solve a Polynomial Inequality Algebraically

1. Arrange the terms so that one side of the inequality is 0.
2. Factor the polynomial expression, if possible, and find the points where each factor is 0. (Use the quadratic formula, if necessary.)

Continued on the next page...

To Solve a Polynomial Inequality Algebraically (cont.)

3. Mark each of these points on a number line. These are the interval **endpoints**.
4. Test one point from each interval to determine the sign of the polynomial expression for all points in that interval.
5. The solution consists of those intervals where the test points satisfy the original inequality.
6. Mark a bracket for an endpoint that is included and a parenthesis for an endpoint that is not included.

The following examples illustrate this technique.

Example 1: Solving Polynomial Inequalities by Factoring

Solve the following inequalities by factoring and using a number line. Graph the solution set on a number line.

a. $x^2 - 2x > 8$

Solution: $x^2 - 2x > 8$

$x^2 - 2x - 8 > 0$ Add -8 to both sides so that one side is 0.

$(x+2)(x-4) > 0$ Factor.

Set each factor equal to 0 to locate the interval endpoints.

$x + 2 = 0$ $\quad$ $x - 4 = 0$

$x = -2$ $\quad$ $x = 4$

Mark these points on a number line and test one point from each of the intervals formed.

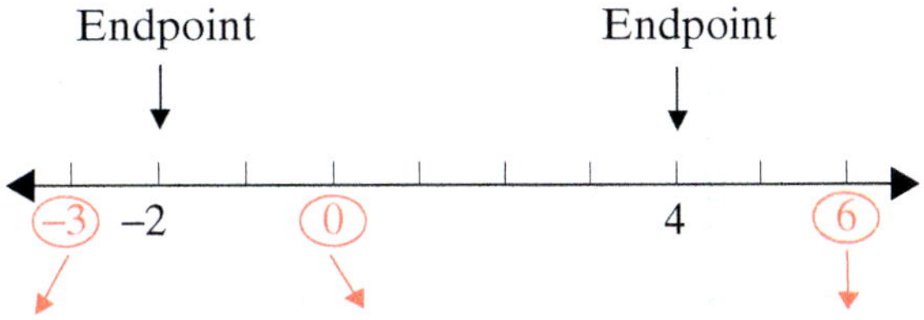

Test $x = -3$:

$(-3+2)(-3-4)$

$= (-1)(-7) = 7 > 0$

This means that $(x+2)(x-4) > 0$ if $x < -2$.

Test $x = 0$:

$(0+2)(0-4)$

$= (2)(-4) = -8 < 0$

This means that $(x+2)(x-4) < 0$ if $-2 < x < 4$.

Test $x = 6$:

$(6+2)(6-4)$

$= (8)(2) = 16 > 0$

This means that $(x+2)(x-4) > 0$ if $x > 4$.

Continued on the next page...

Graphically, the solution set is,

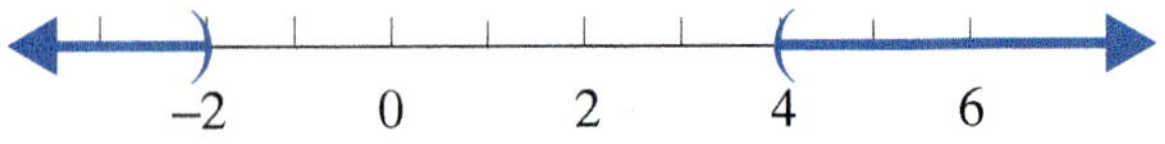

The solution is:

algebraic notation	or	interval notation
$x < -2$ or $x > 4$		$(-\infty, -2) \cup (4, \infty)$

b. $2x^2 + 15 \le 13x$

Solution: $2x^2 + 15 \le 13x$

$2x^2 - 13x + 15 \le 0$ Add $-13x$ to both sides so that one side is 0.

$(2x - 3)(x - 5) \le 0$ Factor.

Set each factor equal to 0 to locate the interval endpoints.

$$2x - 3 = 0 \qquad x - 5 = 0$$

$$x = \frac{3}{2} \qquad x = 5$$

Test one point from each interval formed.

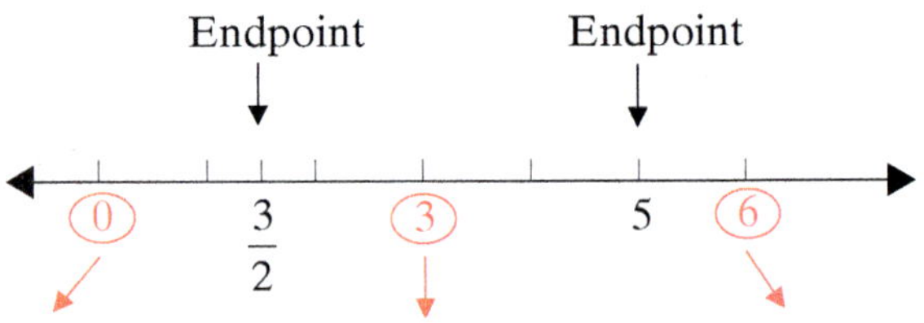

Test $x = 0$:	Test $x = 3$:	Test $x = 6$:
$(2 \cdot 0 - 3)(0 - 5)$	$(2 \cdot 3 - 3)(3 - 5)$	$(2 \cdot 6 - 3)(6 - 5)$
$= (-3)(-5) = 15 > 0$	$= (3)(-2) = -6 < 0$	$= (9)(1) = 9 > 0$
This means that $(2x - 3)(x - 5) > 0$ if $x < \frac{3}{2}$.	This means that $(2x - 3)(x - 5) < 0$ if $\frac{3}{2} < x < 5$.	This means that $(2x - 3)(x - 5) > 0$ if $x > 5$.

The solution includes both endpoints since the inequality ($\le$) includes 0:

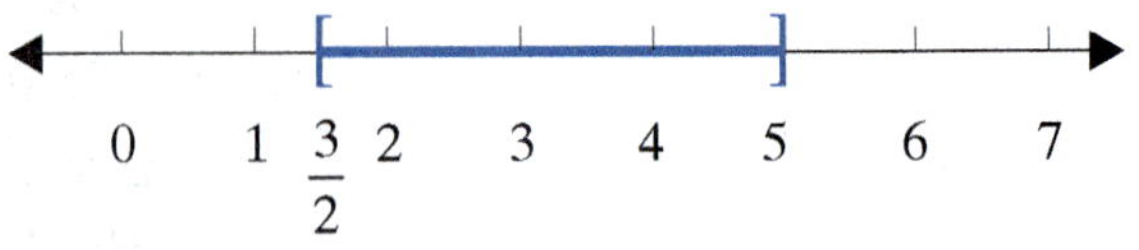

The solution is:

algebraic notation	or	interval notation
$\frac{3}{2} \le x \le 5$		$\left[\frac{3}{2}, 5\right]$

c. $x^3 + 4x^2 - 5x < 0$

Solution:

$$x^3 + 4x^2 - 5x < 0$$
$$x\left(x^2 + 4x - 5\right) < 0$$
$$x(x+5)(x-1) < 0$$

Set each factor equal to 0 to locate the interval endpoints.

$$x = 0 \qquad x + 5 = 0 \qquad x - 1 = 0$$
$$x = -5 \qquad x = 1$$

Test one point from each interval formed.

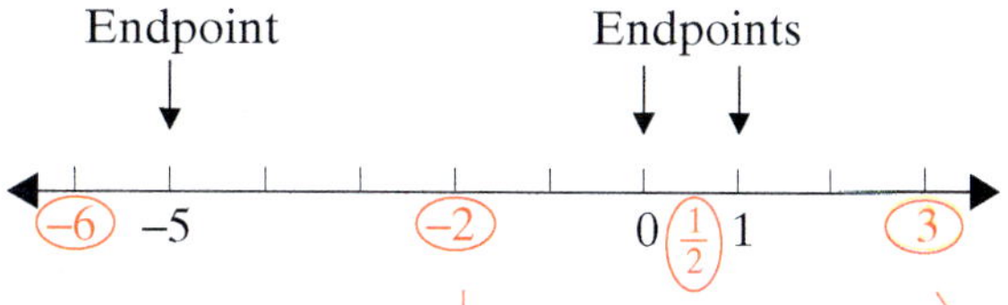

Test $x = -2$:

$$-2(-2+5)(-2-1)$$
$$= -2(3)(-3) = 18 > 0$$

Thus $x(x+5)(x-1) > 0$, if $-5 < x < 0$.

Test $x = 3$:

$$3(3+5)(3-1)$$
$$= 3(8)(2) = 48 > 0$$

Thus $x(x+5)(x-1) > 0$, if $x > 1$.

Test $x = -6$:

$$-6(-6+5)(-6-1)$$
$$= -6(-1)(-7) = -42 < 0$$

Thus $x(x+5)(x-1) < 0$, if $x < -5$.

Test $x = \frac{1}{2}$:

$$\frac{1}{2}\left(\frac{1}{2}+5\right)\left(\frac{1}{2}-1\right)$$
$$= \frac{1}{2}\left(\frac{11}{2}\right)\left(-\frac{1}{2}\right) = -\frac{11}{8} < 0$$

Thus $x(x+5)(x-1) < 0$, if $0 < x < 1$.

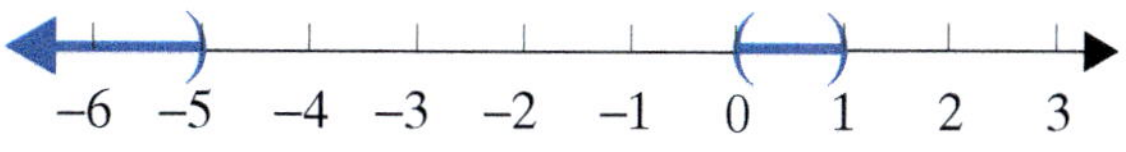

The solution is:

algebraic notation	or	interval notation
$x < -5$ or $0 < x < 1$		$(-\infty, -5) \cup (0, 1)$

Example 2: Solving Polynomial Inequalities Using the Quadratic Formula

Solve the following inequalities using the quadratic formula and a number line. Graph the solution set on a number line.

a. $x^2 - 2x - 1 > 0$

Solution: The quadratic expression $x^2 - 2x - 1$ cannot be factored with integer coefficients. Use the quadratic formula and find the roots of the equation $x^2 - 2x - 1 = 0$. Then use these roots as endpoints for the intervals. The test points can themselves be integers.

$$x^2 - 2x - 1 = 0$$

$$x = \frac{2 \pm \sqrt{(-2)^2 - 4(1)(-1)}}{2(1)} \qquad \text{Use the quadratic formula.}$$

$$x = \frac{2 \pm \sqrt{4+4}}{2}$$

$$x = \frac{2 \pm 2\sqrt{2}}{2} = 1 \pm \sqrt{2}$$

Teaching Notes:
You may want to discuss with your class a strategy for selecting test points.

The endpoints are $x = 1 - \sqrt{2}$ and $x = 1 + \sqrt{2}$.

Test one point from each interval formed.

Note: With a calculator, you can determine that $1 - \sqrt{2} \approx -0.414$ and $1 + \sqrt{2} \approx 2.414$.

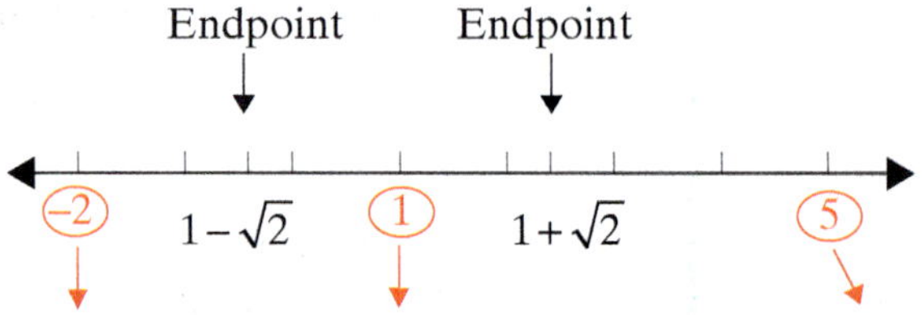

Test $x = -2$:
$(-2)^2 - 2(-2) - 1$
$= 4 + 4 - 1 = 7 > 0$
This means that if $x < 1 - \sqrt{2}$, then $x^2 - 2x - 1 > 0$.

Test $x = 1$:
$1^2 - 2(1) - 1$
$= 1 - 2 - 1 = -2 < 0$
This means that if $1 - \sqrt{2} < x < 1 + \sqrt{2}$, then $x^2 - 2x - 1 < 0$.

Test $x = 5$:
$5^2 - 2(5) - 1$
$= 25 - 10 - 1 = 14 > 0$
This means that if $x > 1 + \sqrt{2}$, then $x^2 - 2x - 1 > 0$.

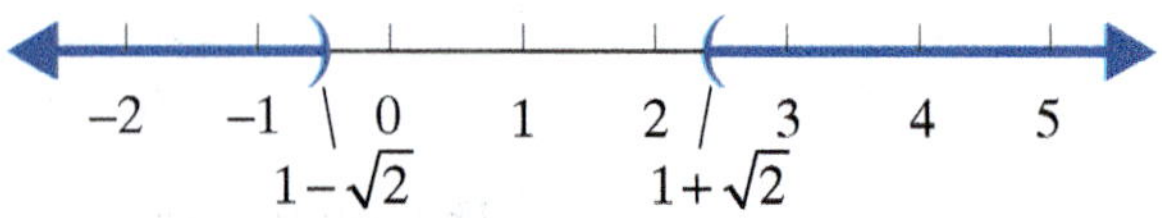

The solution is:

algebraic notation	or	interval notation
$x<1-\sqrt{2}$ or $x>1+\sqrt{2}$		$\left(-\infty,1-\sqrt{2}\right)\cup\left(1+\sqrt{2},\infty\right)$

b. $x^2-2x+13>0$

Solution: To find where $x^2-2x+13>0$, use the quadratic formula:

$$x=\frac{2\pm\sqrt{(-2)^2-4(1)(13)}}{2}=\frac{2\pm\sqrt{-48}}{2}$$

$$=\frac{2\pm4i\sqrt{3}}{2}=1\pm2i\sqrt{3}.$$

Since these values are nonreal, the polynomial is either always positive or always negative for all real values of x. Therefore, we only need to test one point. If that point satisfies the inequality, then the solution is all real numbers. If it does not, then there is no solution. In this example, we test $x=0$ because the polynomial is easy to evaluate for $x=0$.

$$(0)^2-2(0)+13=13>0$$

Since the real number 0 satisfies the inequality, the solution is all real numbers. In interval notation we write $(-\infty,\infty)$, and in algebraic notation we write $\mathbb{R}$. Graphically, the solution set is:

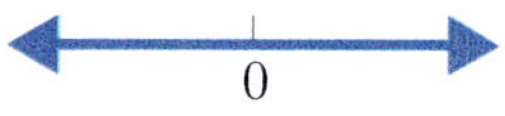

Solving Rational Inequalities

A rational inequality may involve the product or quotient of several first-degree expressions. For example, the inequality

$$\frac{x+3}{x-2}>0$$

involves the two first-degree expressions $x+3$ and $x-2$.

Inequalities of this form can be solved using a similar procedure to that used for quadratic inequalities outlined on the next page.

To Solve a Rational Inequality

1. Simplify the inequality so that one side is 0 and the other side has a single fraction with both the numerator and denominator in factored form.
2. Find the points that cause the factors in the numerator or in the denominator to be 0.
3. Mark each of these points on a number line. These are the interval endpoints.
4. Test one point from each interval to determine the sign of the rational expression for all points in that interval.
5. The solution consists of those intervals where the test points satisfy the original inequality.
6. Mark a bracket for an endpoint that is included and a parenthesis for an endpoint that is not included. Remember that no denominator can be 0.

Example 3: Solving Rational Inequalities

Solve and graph the following rational inequalities.

a. $\frac{x+3}{x-2} > 0$

Solution: Set each factor in the numerator and the denominator equal to 0 to locate the interval endpoints.

$$x+3=0 \qquad x-2=0$$
$$x=-3 \qquad x=2$$

Mark each of these points on a number line and test one point from each of the intervals formed. (Consider these points as endpoints of intervals.)

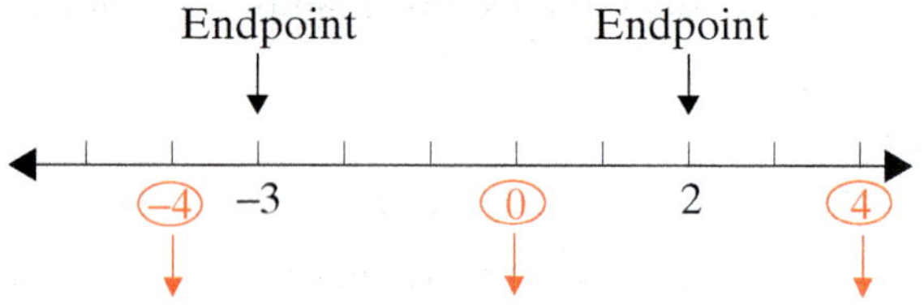

Test $x = -4$:

$$\frac{-4+3}{-4-2} = \frac{-1}{-6} = \frac{1}{6} > 0$$

This means that $\frac{x+3}{x-2} > 0$ if $x < -3$.

Test $x = 0$:

$$\frac{0+3}{0-2} = -\frac{3}{2} < 0$$

This means that $\frac{x+3}{x-2} < 0$ if $-3 < x < 2$.

Test $x = 4$:

$$\frac{4+3}{4-2} = \frac{7}{2} > 0$$

This means that $\frac{x+3}{x-2} > 0$ if $x > 2$.

Graphically, the solution set is:

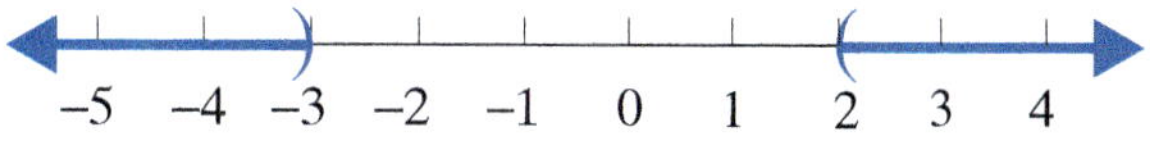

The solution is:

algebraic notation	or	interval notation
$x < -3$ or $x > 2$		$(-\infty, -3) \cup (2, \infty)$

b. $\dfrac{x+3}{x-2} < 0$

Solution: Using the graph and test points from Example 3a, we know that

$$\frac{x+3}{x-2} < 0 \text{ if } -3 < x < 2.$$

Graphically, the solution set is:

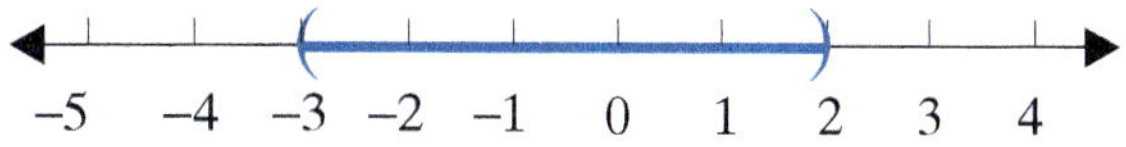

The solution is:

algebraic notation	or	interval notation
$-3 < x < 2$		$(-3, 2)$

Example 4: Solving Rational Inequalities

Solve the following rational inequality.

$$\frac{x+5}{x-4} \geq -1$$

Solution:

$$\frac{x+5}{x-4} + 1 \geq 0$$ One side must be 0.

$$\frac{x+5}{x-4} + \frac{x-4}{x-4} \geq 0$$ Rewrite 1 as a fraction using the LCD, $x-4$, as the denominator.

$$\frac{2x+1}{x-4} \geq 0$$ Simplify to get one fraction. $(x \neq 4)$

Set each linear expression equal to 0 to find the interval endpoints.

$$2x + 1 = 0 \qquad x - 4 = 0$$

$$x = -\frac{1}{2} \qquad x = 4$$

Continued on the next page...

Test a value from each of the intervals:

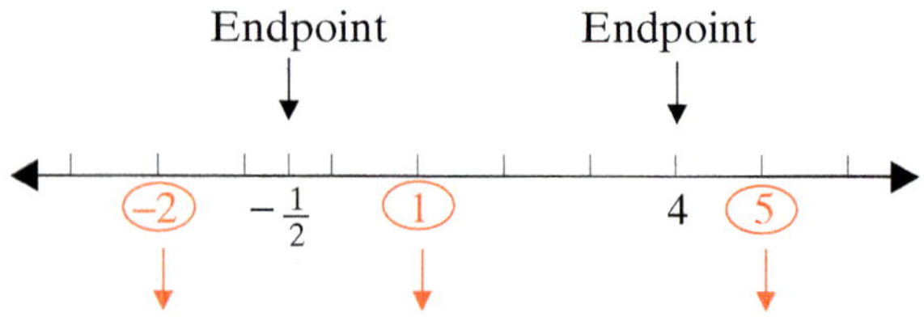

Test $x = -2$:

$$\frac{2(-2)+1}{-2-4} = \frac{-3}{-6} = \frac{1}{2} > 0$$

This means that $\frac{2x+1}{x-4} > 0$ if $x < -\frac{1}{2}$.

Test $x = 1$:

$$\frac{2(1)+1}{1-4} = \frac{3}{-3} = -1 < 0$$

This means that $\frac{2x+1}{x-4} < 0$ if $-\frac{1}{2} < x < 4$.

Test $x = 5$:

$$\frac{2(5)+1}{5-4} = \frac{11}{1} = 11 > 0$$

This means that $\frac{2x+1}{x-4} > 0$ if $x > 4$.

The solution includes the endpoint $-\frac{1}{2}$ since the inequality ($\geq$) includes 0. However, the endpoint 4 is not included in the solution because the rational inequality is undefined (the denominator equals 0) when $x = 4$.

Graphically, the solution set is:

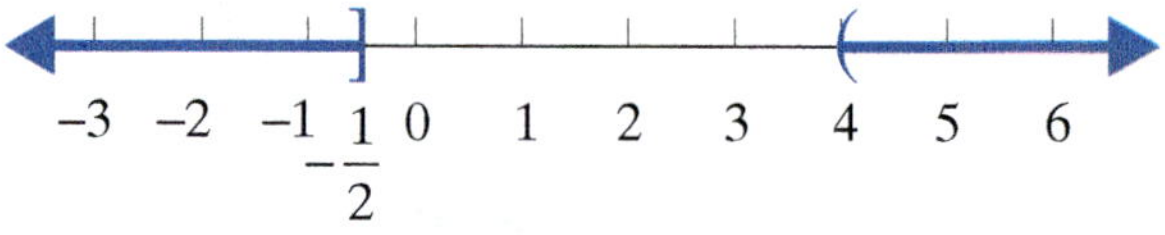

The solution is:

algebraic notation	or	interval notation
$x \leq -\frac{1}{2}$ or $x > 4$		$\left(-\infty, -\frac{1}{2}\right] \cup (4, \infty)$

NOTES

Notice that in the first step, we do **not** multiply by the denominator $x - 4$. The reason is that the variable expression is positive for some values of x and negative for other values of x. Therefore, if we did multiply by $x - 4$, we would not be able to determine whether the inequality should stay as $\geq$ or be reversed to $\leq$.

Solving Quadratic Inequalities Using a Graphing Calculator

To solve quadratic (or other polynomial) inequalities with a TI-84 Plus graphing calculator by graphing the related polynomial function, look for the intervals of x for which the graph of the function is above the x-axis (the y-values will be positive) and where it is below the x-axis (the y-values will be negative). The y-values represent the values of the related polynomial function.

To Solve a Polynomial Inequality Using a Graphing Calculator

1. Arrange the terms so that one side of the inequality is 0.
2. Form a function by letting $y =$ [*the polynomial*], and graph the function. Be sure to set the `WINDOW` so that all of the zeros are easily seen. This may be difficult if large numbers are involved.
3. Use the CALC key (2ND TRACE) and select `2:zero` to find (or approximate) the real zeros of the function, if there are any.
   a. The values of x for which the y-values are above the x-axis satisfy $y > 0$.
   b. The values of x for which the y-values are below the x-axis satisfy $y < 0$.
4. Endpoints of intervals are included if the inequality includes 0, as in $y \geq 0$ or $y \leq 0$.

Example 5: Solving a Quadratic Inequality Using a Graphing Calculator

Solve the following quadratic inequalities using a graphing calculator. Graph the solution set on a real number line.

a. $x^2 - 2x > 8$

Solution: This inequality was solved algebraically in Example 1a. We repeat the solution here using a graphing calculator to show how the two methods are related.

Manipulate the inequality so that one side is 0.

$$x^2 - 2x > 8$$

$$x^2 - 2x - 8 > 0$$

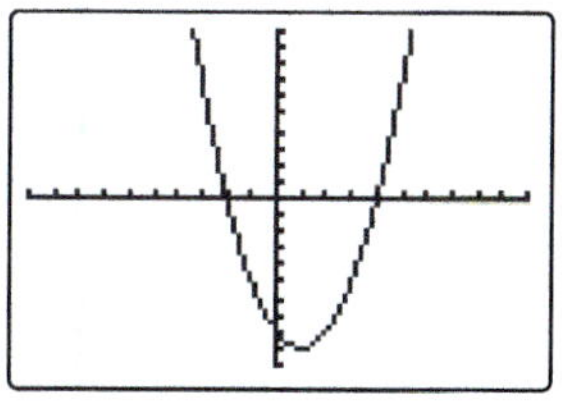

Press Y= and enter the function:

$$y = x^2 - 2x - 8.$$

Press GRAPH to graph the function.

The graph will appear as illustrated above. (In this case the graph is a parabola.)

Find the zeros (or estimate the zeros) as follows:

Step 1: Press CALC (2ND TRACE).

Step 2: Press or choose `2: zero`.

Step 3: Follow the directions for moving the cursor to `Left Bound?`, `Right Bound?`, and `Guess?` for each point. (Press ENTER each time.)

Continued on the next page...

Note: If there are no real zeros, then the graph will be entirely above or entirely below the x-axis. The solution is then dependent on the nature of the inequality. It will be either the empty set (no solution) or the entire set of real numbers $(-\infty, \infty)$.

In this case, the zeros are $x = -2$ and $x = 4$ (just as we found in Example 1a). Because we want to know where $y > 0$, we look at the graph and choose the intervals for x where the curve is above the x-axis. (**Note:** Endpoints are not included because the inequality does not include 0.)

Thus the solution consists of the union of two intervals: $(-\infty, -2) \cup (4, \infty)$.

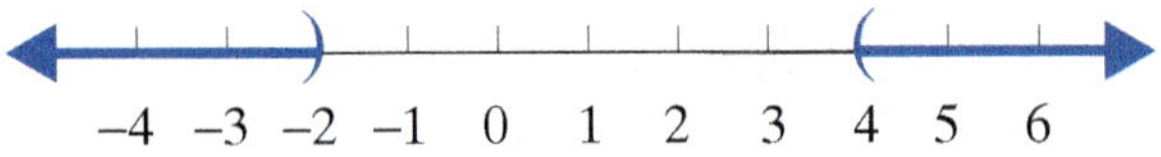

b. $2x^2 + 3x - 10 \le 0$

Solution: Press Y= and enter the function:

$$y = 2x^2 + 3x - 10.$$

Press GRAPH to graph the function.

The graph will appear as illustrated here. (In this case the graph is a parabola.)

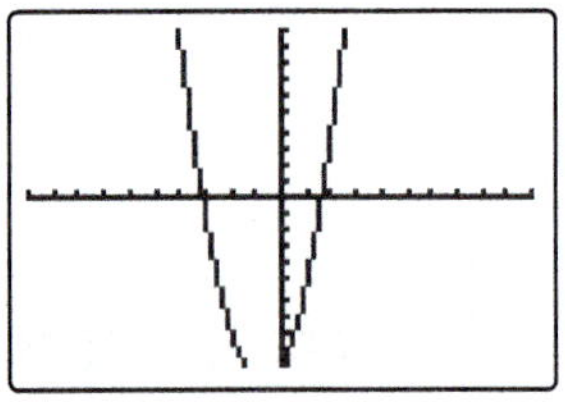

Find the zeros (or estimate the zeros) as follows:

Step 1: Press CALC (2ND TRACE).

Step 2: Press or choose `2: zero`.

Step 3: Follow the directions for moving the cursor to `Left Bound?`, `Right Bound?`, and `Guess?` for each point. (Press ENTER each time.)

In this case the zeros are estimates: $x \approx -3.1085$ and $x \approx 1.6085$.

Because we want to know where $y \le 0$, we look at the graph and find the intervals for x where the curve is below the x-axis and include the endpoints because 0 is included in the inequality.

Thus the solution is the closed interval $[-3.1085, 1.6085]$.

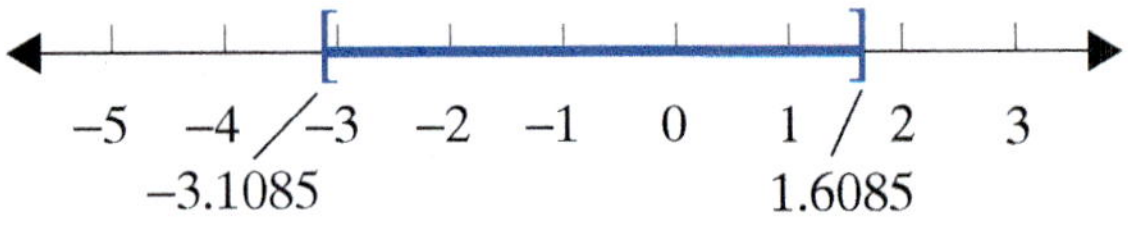

Example 6: Solving a Higher-Degree Polynomial Inequality Using a Graphing Calculator

Solve the following third-degree polynomial inequality using a graphing calculator:

$$x^3 + 2x^2 - 11x - 12 > 0.$$

Solution: Press Y= and enter the function:

$$y = x^3 + 2x^2 - 11x - 12.$$

Press GRAPH to graph the function.

The graph will appear as illustrated here. (In this case the graph is not a parabola.)

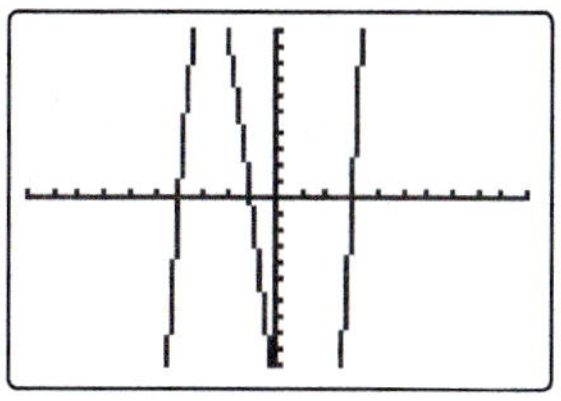

Find the zeros (or estimate the zeros) as follows:

Step 1: Press CALC (2ND TRACE).

Step 2: Press or choose `2: zero`.

Step 3: Follow the directions for moving the cursor to `Left Bound?`, `Right Bound?`, and `Guess?` for each point. (Press ENTER each time.)

In this case there are three zeros: $x = -4$, $x = -1$, and $x = 3$.
Because we want to know where $y > 0$, we look at the graph and find the intervals for x where the curve is above the x-axis. Do not include the endpoints because 0 is not included in the inequality.

Thus the solution is the union of two intervals: $(-4, -1) \cup (3, \infty)$.

1. $(-2, 6)$

2. $(-\infty, -4) \cup (2, \infty)$

3. $\left(-\infty, \frac{2}{3}\right) \cup (5, \infty)$

4. $\left[-1, -\frac{1}{4}\right]$

5. $(-\infty, -7] \cup \left[\frac{5}{2}, \infty\right)$

6. $\left[\frac{3}{5}, 3\right]$

7. $\left[-2, -\frac{1}{3}\right]$
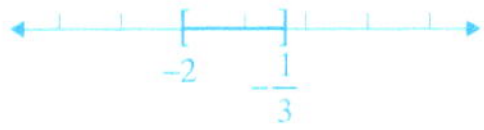

8. $\left(-\infty, \frac{8}{3}\right) \cup (4, \infty)$
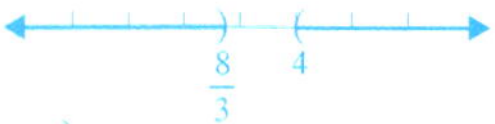

9. $\left(-\infty, -\frac{4}{3}\right) \cup (0, 5)$

10. $(-\infty, -4) \cup \left(-\frac{5}{2}, 1\right)$
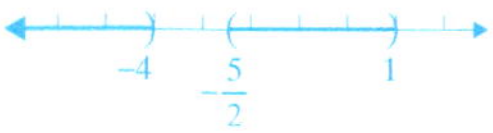

11. $\{-2\}$

12. $(-\infty, -2) \cup \left(\frac{6}{5}, \infty\right)$

13. $\left(-\infty, -\frac{5}{2}\right) \cup (3, \infty)$

14. $\left(-\infty, -\frac{2}{3}\right) \cup \left(\frac{1}{2}, \infty\right)$
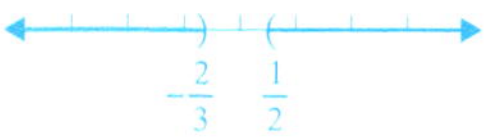

15. $\left(-\frac{1}{4}, \frac{3}{2}\right)$
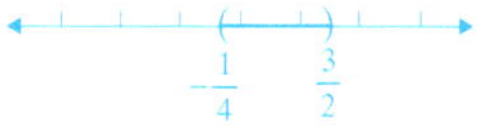

16. $\left(-2, \frac{5}{2}\right)$

17. $\left(-\infty, \frac{1}{2}\right] \cup [2, \infty)$
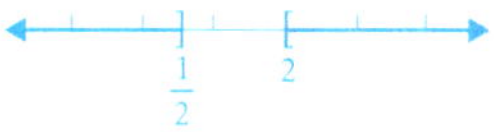

18. $\left(-\frac{3}{5}, 2\right)$

19. $\left(-\frac{2}{3}, -\frac{1}{2}\right)$

Practice Problems

Solve each of the following inequalities algebraically. Graph each solution set on a real number line. Write the answers in interval notation.

1. $(x-5)(x-7)>0$ **2.** $x^2-5x>-4$ **3.** $\dfrac{4+x}{x-2}\le 0$

Use a graphing calculator to solve the following inequalities. Write the answers in interval notation.

4. $2x^2-2x-3\ge 0$ **5.** $x^3-4x+6>0$

20. $\left(-\infty, \frac{1}{3}\right] \cup [3, \infty)$

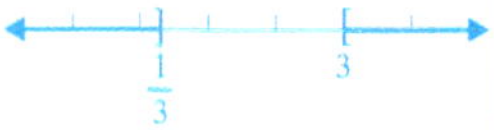

21. $\left(-\infty, \frac{5}{2}\right) \cup \left(\frac{5}{2}, \infty\right)$

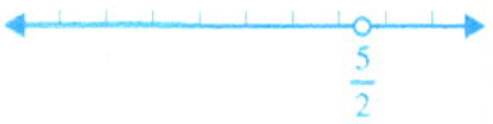

22. $\left[-\frac{2}{3}, \frac{7}{5}\right]$

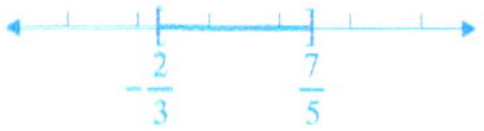

10.6 Exercises

*Solve the quadratic (and higher degree) inequalities algebraically. Write the answers in interval notation, and then graph each solution set on a number line. (**Note:** You may need to use the quadratic formula to find endpoints of intervals.)*

1. $(x-6)(x+2)<0$ **2.** $(x+4)(x-2)>0$ **3.** $(3x-2)(x-5)>0$

4. $(4x+1)(x+1)\le 0$ **5.** $(x+7)(2x-5)\ge 0$ **6.** $(x-3)(5x-3)\le 0$

7. $(3x+1)(x+2)\le 0$ **8.** $(x-4)(3x-8)>0$ **9.** $x(3x+4)(x-5)<0$

10. $(x-1)(x+4)(2x+5)<0$ **11.** $x^2+4x+4\le 0$ **12.** $5x^2+4x-12>0$

13. $2x^2>x+15$ **14.** $6x^2+x>2$ **15.** $8x^2<10x+3$

16. $2x^2<x+10$ **17.** $2x^2-5x+2\ge 0$ **18.** $15y^2-21y-18<0$

19. $6y^2+7y<-2$ **20.** $3x^2+3\ge 10x$ **21.** $4z^2-20z+25>0$

22. $15x^2-11x-14\le 0$ **23.** $8x^2+6x\le 35$ **24.** $7x<6x^2+x^3$

23. $\left[-\frac{5}{2}, \frac{7}{4}\right]$

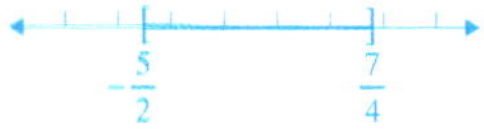

24. $(-7, 0)\cup(1, \infty)$

25. $(-1, 0)\cup(3, \infty)$

26. $(-\infty, 0)$

27. $(0, 1)\cup(4, \infty)$

28. $[0, 1]\cup[3, \infty)$

29. $(-\infty, -1)\cup(4, \infty)$

30. $(-3, 2)$

Answers to Practice Problems:

1. 

$(-\infty, 5)\cup(7, \infty)$

2. 

$(-\infty, 1)\cup(4, \infty)$

3. 

$[-4, 2)$

4. $(-\infty, -0.8229]\cup[1.8229, \infty)$ **5.** $(-2.5251, \infty)$

31. $(-\infty, -2) \cup (-1, 1) \cup (2, \infty)$

32. $(-4, -3) \cup (3, 4)$

33. $[-3, -2] \cup [2, 3]$

34. $(-\infty, -4] \cup [4, \infty)$

35. $(-\infty, -4] \cup [2, \infty)$

36. $\left(-1, \frac{5}{3}\right)$

37. $\left(-\frac{2}{3}, 2\right)$

38. $(-\infty, 0) \cup (1, \infty)$

39. $\left(-\infty, -1-\sqrt{5}\right) \cup \left(-1+\sqrt{5}, \infty\right)$

40. $\left(4-\sqrt{2}, 4+\sqrt{2}\right)$

41. $\left(-\infty, -3-\sqrt{2}\right] \cup \left[-3+\sqrt{2}, \infty\right)$

42. $\left(\frac{-2-\sqrt{10}}{2}, \frac{-2+\sqrt{10}}{2}\right)$

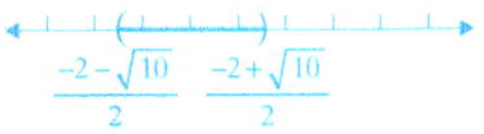

43. $\left(\frac{-5-\sqrt{13}}{6}, \frac{-5+\sqrt{13}}{6}\right)$

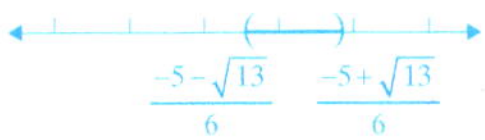

44. $\left(-\infty, -\frac{5}{3}\right] \cup [-1, \infty)$

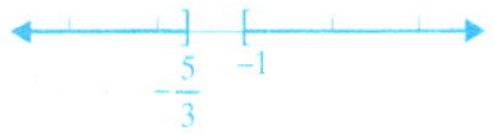

25. $x^3 > 2x^2 + 3x$ **26.** $x^3 < 6x^2 - 9x$ **27.** $x^3 > 5x^2 - 4x$

28. $4x^2 \le x^3 + 3x$ **29.** $(x+2)(x-2) > 3x$ **30.** $(x+4)(x-1) < 2x+2$

31. $x^4 - 5x^2 + 4 > 0$ **32.** $x^4 - 25x^2 + 144 < 0$ **33.** $y^4 - 13y^2 + 36 \le 0$

34. $y^4 - 13y^2 - 48 \ge 0$ **35.** $(x+1)^2 - 9 \ge 0$ **36.** $(3x-1)^2 - 16 < 0$

37. $(2x-3)(3x+2) - (3x+2) < 0$ **38.** $2(x-1)(x-3) > (x-1)(x-6)$

39. $x^2 + 2x - 4 > 0$ **40.** $x^2 - 8x + 14 < 0$ **41.** $x^2 + 6x + 7 \ge 0$

42. $2x^2 + 4x - 3 < 0$ **43.** $3x^2 + 5x + 1 < 0$ **44.** $3x^2 + 8x + 5 \ge 0$

45. $2x^3 \le 7x^2 + 4x$ **46.** $2x^2 > 9x - 8$ **47.** $x^2 - 2x + 2 > 0$

48. $x^2 + 3x + 3 < 0$ **49.** $2x - 1 > 3x^2$ **50.** $6x - 10 < x^2$

The graph of a quadratic function is given. Use the information in the graph to solve the related equations and inequalities in parts **a** – **c**.

51. $y = x^2 - 7x - 10$

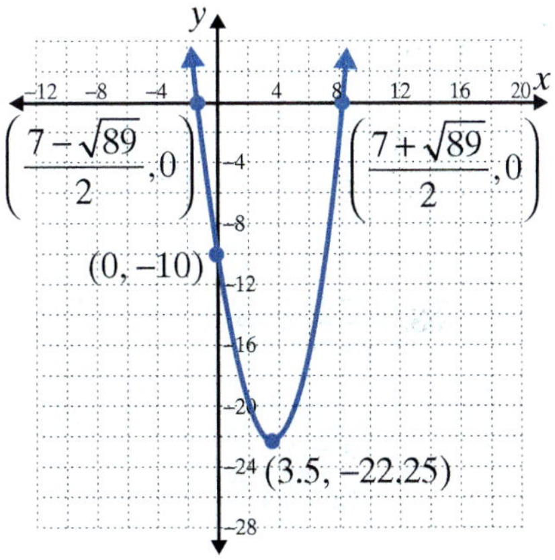

a. $x^2 - 7x - 10 = 0$

b. $x^2 - 7x - 10 > 0$

c. $x^2 - 7x - 10 < 0$

52. $y = x^2 + 5x - 6$

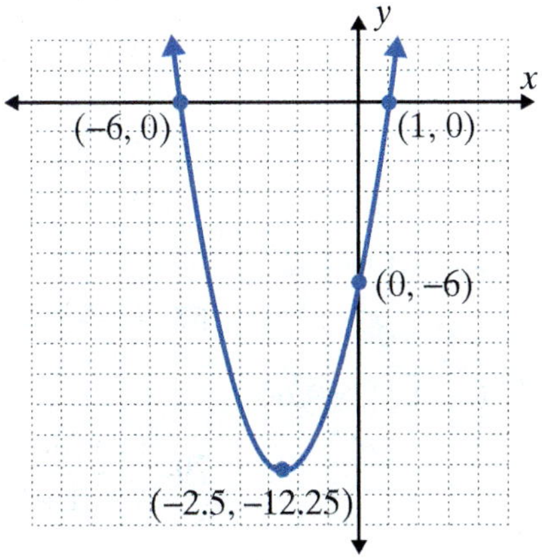

a. $x^2 + 5x - 6 = 0$

b. $x^2 + 5x - 6 > 0$

c. $x^2 + 5x - 6 < 0$

45. $\left(-\infty, -\frac{1}{2}\right] \cup [0, 4]$

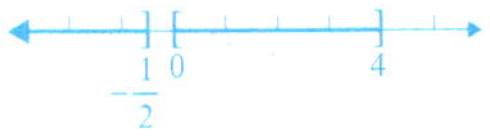

46. $\left(-\infty, \frac{9-\sqrt{17}}{4}\right) \cup \left(\frac{9+\sqrt{17}}{4}, \infty\right)$

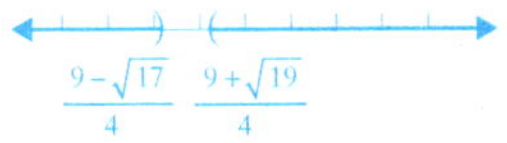

47. $(-\infty, \infty)$

48. $\varnothing$ 49. $\varnothing$

50. $(-\infty, \infty)$

51. a. $x = \frac{7 \pm \sqrt{89}}{2}$
b. $\left(-\infty, \frac{7-\sqrt{89}}{2}\right) \cup \left(\frac{7+\sqrt{89}}{2}, \infty\right)$
c. $\left(\frac{7-\sqrt{89}}{2}, \frac{7+\sqrt{89}}{2}\right)$

52. a. $x = -6, 1$
b. $(-\infty, -6) \cup (1, \infty)$
c. $(-6, 1)$

53. a. $x = -5, 1$ b. $(-5, 1)$
c. $(-\infty, -5) \cup (1, \infty)$

54. a. $x = -1 \pm \sqrt{6}$
b. $\left(-1-\sqrt{6}, -1+\sqrt{6}\right)$
c. $\left(-\infty, -1-\sqrt{6}\right) \cup \left(-1+\sqrt{6}, \infty\right)$

55. $(-\infty, -4] \cup (0, \infty)$

56. $(-\infty, 0] \cup (4, \infty)$

57. $(-\infty, -6)$

53. $y = -x^2 - 4x + 5$

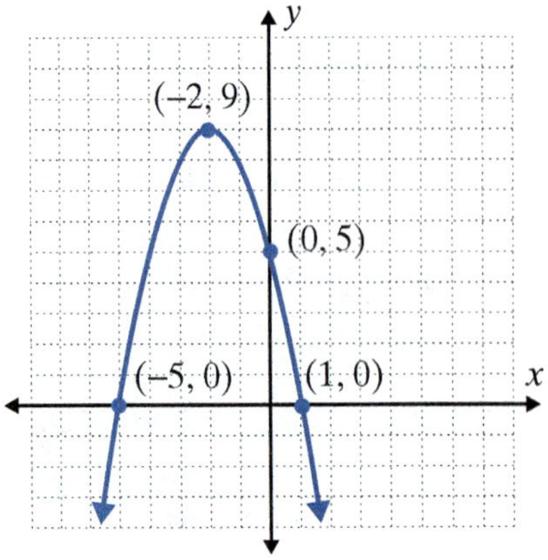

a. $-x^2 - 4x + 5 = 0$

b. $-x^2 - 4x + 5 > 0$

c. $-x^2 - 4x + 5 < 0$

54. $y = -3x^2 - 6x + 15$

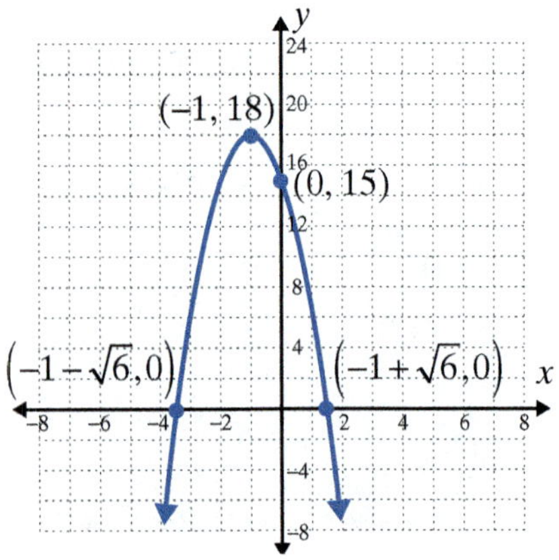

a. $-3x^2 - 6x + 15 = 0$

b. $-3x^2 - 6x + 15 > 0$

c. $-3x^2 - 6x + 15 < 0$

Solve the rational inequalities algebraically. Write the answers in interval notation, and then graph each solution set on a number line.

55. $\frac{x+4}{2x} \geq 0$ **56.** $\frac{x}{x-4} \geq 0$ **57.** $\frac{x+6}{x^2} < 0$ **58.** $\frac{3x^2}{x+1} < 0$

59. $\frac{x+3}{x+9} > 0$ **60.** $\frac{2x+3}{x-4} < 0$ **61.** $\frac{3x-6}{2x-5} < 0$ **62.** $\frac{4-3x}{2x+4} \leq 0$

63. $\frac{x+5}{x-7} \geq 1$ **64.** $\frac{2x+3}{x-1} > 2$ **65.** $\frac{2x+5}{x-4} \leq -3$ **66.** $\frac{3x+2}{4x-1} < 3$

67. $\frac{5-2x}{3x+4} < -1$ **68.** $\frac{8-x}{x+5} < -4$ **69.** $\frac{x(x+4)}{x-3} \leq 0$

70. $\frac{(x+3)(x-2)}{x+1} > 0$ **71.** $\frac{x-5}{x(x+2)} \geq 0$ **72.** $\frac{-(x-3)^2}{(x-1)(x-4)} < 0$

Use a graphing calculator to solve the inequalities. Write the answers in interval notation, and then graph each solution set on a number line. (Estimate endpoints, when necessary, to 4 decimal places.)

73. $x^2 > 10$ **74.** $20 \geq x^2$ **75.** $x^2 - 2.5x + 6.25 < 0$

76. $x^2 + 2x \geq -1$ **77.** $x^3 - 9x < 0$ **78.** $x^3 - 4x^2 + 4x \leq 0$

58. $(-\infty, -1)$

79. $2x^3 - 5x + 4 \geq 0$

80. $x^3 - 4x^2 + 3 < 0$

81. $-x^4 + 6x^2 - 3 > 0$

82. $x^4 - 2x^3 - x^2 - 1 < 0$

Writing and Thinking About Mathematics

59. $(-\infty, -9) \cup (-3, \infty)$

60. $\left(-\frac{3}{2}, 4\right)$
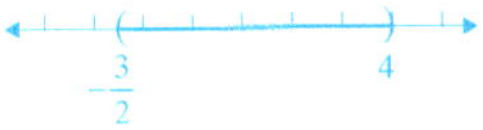

61. $\left(2, \frac{5}{2}\right)$
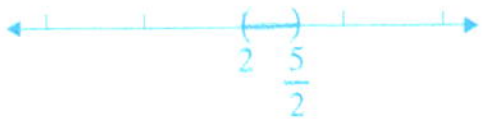

83. Use a graphing calculator to graph the rational function $y = \frac{x^2 + 3x - 4}{x}$.

a. Use the graph to find the solution set for $y > 0$.

b. Use the graph to find the solution set for $y < 0$.

c. Explain the effect of $x = 0$ on the graph and why $x = 0$ is not included in either **a.** or **b.**

84. In your own words, explain why (as in Example 2b), when the quadratic formula gives nonreal values, the quadratic polynomial is either always positive or always negative.

HAWKES LEARNING SYSTEMS: INTRODUCTORY & INTERMEDIATE ALGEBRA SOFTWARE

- 10.6a Solving Quadratic Inequalities
- 10.6b Solving Inequalities with Rational Expressions

62. $(-\infty, -2) \cup \left[\frac{4}{3}, \infty\right)$

63. $(7, \infty)$

64. $(1, \infty)$

65. $\left[\frac{7}{5}, 4\right)$

66. $\left(-\infty, \frac{1}{4}\right) \cup \left(\frac{5}{9}, \infty\right)$
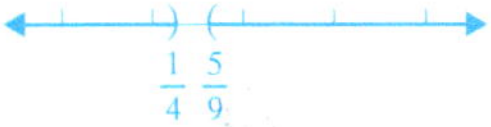

67. $\left(-9, -\frac{4}{3}\right)$

68. $\left(-\frac{28}{3}, -5\right)$
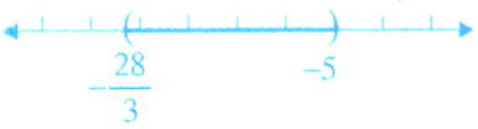

69. $(-\infty, -4] \cup [0, 3)$

70. $(-3, -1) \cup (2, \infty)$

71. $(-2, 0) \cup [5, \infty)$

72. $(-\infty, 1) \cup (4, \infty)$

73. $(-\infty, -3.1623) \cup (3.1623, \infty)$

74. $[-4.4721, 4.4721]$

75. $\varnothing$

76. $(-\infty, \infty)$

77. $(-\infty, -3) \cup (0, 3)$

78. $(-\infty, 0] \cup \{2\}$

79. $[-1.8868, \infty)$

80. $(-\infty, -0.7913) \cup (1, 3.7913)$

81. $(-2.3344, -0.7420) \cup (0.7420, 2.3344)$

82. $(-0.8445, 2.4710)$

83. a. $(-4, 0) \cup (1, \infty)$ b. $(-\infty, -4) \cup (0, 1)$ c. The function is undefined at $x = 0$.

84. Because the polynomial cannot be 0 for any real number, it cannot be negative for some number and positive for another. This corresponds to the graph not crossing the x-axis.

Chapter 10 Index of Key Ideas and Terms

Section 10.1 Solving Quadratic Equations

Review of Solving Quadratic Equations by Factoring pages 768-769

Square Root Property page 771

If $x^2 = c$, then $x = \pm\sqrt{c}$.

If $(x-a)^2 = c$, then $x - a = \pm\sqrt{c}$ (or $x = a \pm \sqrt{c}$).

Note: If c is negative $(c < 0)$, then the solutions will be nonreal.

To Solve a Quadratic Equation by Completing the Square page 774

1. If necessary, divide or multiply both sides of the equation so that the leading coefficient (the coefficient of x^2) is 1.
2. If necessary, isolate the constant term on one side of the equation.
3. Find the constant that completes the square of the polynomial and add this constant to both sides. Rewrite the polynomial as the square of a binomial.
4. Use the square root property to find the solutions of the equation.

Writing Equations with Known Roots page 776

Section 10.2 The Quadratic Formula: $x = \frac{-b \pm \sqrt{b^2 - 4ac}}{2a}$

Quadratic Formula page 781

For the general quadratic equation $ax^2 + bx + c = 0$, where $a \neq 0$, the solutions are $x = \frac{-b \pm \sqrt{b^2 - 4ac}}{2a}$.

Discriminant pages 783-784

The expression $b^2 - 4ac$, the part of the quadratic formula that lies under the radical sign, is called the **discriminant**.

If $b^2 - 4ac > 0$, there are two real solutions.
If $b^2 - 4ac = 0$, there is one real solution.
If $b^2 - 4ac < 0$, there are two nonreal solutions.

Section 10.3 Applications

Section 10.4 Equations in Quadratic Form

1. Look at the middle term.
2. Substitute a first-degree variable, such as u, for the variable expression in the middle term.
3. Substitute the square of this variable, u^2, for the variable expression in the first term.
4. Solve the resulting quadratic equation for u.
5. Substitute the results "back" for u in the beginning substitution and solve for the original variable.

Section 10.5 Graphing Quadratic Functions: Parabolas

Any function that can be written in the form $y = ax^2 + bx + c$, where a, b, and c are real numbers and $a \neq 0$ is a **quadratic function**.

The graph of every quadratic function is a vertical **parabola** (a parabola that opens up or down). Horizontal parabolas (parabolas that open left or right) do not represent functions.

Continued on the next page...

Section 10.5 Graphing Quadratic Functions: Parabolas (cont.)

Forms of Quadratic Functions

$y = ax^2$	page 808
$y = ax^2 + k$	page 809
$y = a(x-h)^2$	pages 809-810
$y = a(x-h)^2 + k$	pages 812-813
$y = ax^2 + bx + c$	pages 812-813

Vertical and Horizontal Shifts (or Translations) pages 809-810

Line of Symmetry page 807

The line $x = -\dfrac{b}{2a}$ is the **line of symmetry** (or **axis of symmetry**) of the graph of the quadratic function $y = ax^2 + bx + c$. The curve is a "mirror image" of itself with respect to the line of symmetry.

Vertex page 807

The **vertex** of the graph of $y = ax^2 + bx + c$ is at the point $\left(-\dfrac{b}{2a}, \dfrac{4ac-b^2}{4a}\right)$. The vertex is the "turning point" of the parabola.

Zeros of a Quadratic Function page 813

The **zeros of a quadratic function** are the points where the parabola crosses the x-axis.

Minimum and Maximum Values page 817

For a parabola with equation of the form $y = a(x-h)^2 + k$,

1. If $a > 0$, then the parabola opens upward and (h,k) is the lowest point and the y-value k is called the **minimum value** of the function.
2. If $a < 0$, then the parabola opens downward and (h,k) is the highest point and the y-value k is called the **maximum value** of the function.

Section 10.6 Solving Quadratic and Rational Inequalities

To Solve a Polynomial Inequality Algebraically pages 824-825

1. Arrange the terms so that one side of the inequality is 0.
2. Factor the polynomial expression, if possible, and find the points where each factor is 0. (Use the quadratic formula, if necessary.)
3. Mark each of these points on a number line. These are the interval endpoints.
4. Test one point from each interval to determine the sign of the polynomial expression for all points in that interval.
5. The solution consists of those intervals where the test points satisfy the original inequality.
6. Mark a bracket for an endpoint that is included and a parenthesis for an endpoint that is not included.

To Solve a Rational Inequality page 830

1. Simplify the inequality so that one side is 0 and the other side has a single fraction with both the numerator and the denominator in factored form.
2. Find the points that cause the factors in the numerator or in the denominator to be 0.
3. Mark each of these points on a number line. These are the interval endpoints.
4. Test one point from each interval to determine the sign of the rational expression for all points in that interval.
5. The solution consists of those intervals where the test points satisfy the original inequality.
6. Mark a bracket for an endpoint that is included and a parenthesis for an endpoint that is not included. Remember that no denominator can be 0.

To Solve a Polynomial Inequality Using a Graphing Calculator pages 832-834

HAWKES LEARNING SYSTEMS: INTRODUCTORY & INTERMEDIATE ALGEBRA SOFTWARE

- 10.1a Quadratic Equations: The Square Root Method
- 10.1b Quadratic Equations: Completing the Square
- 10.2 Quadratic Equations: The Quadratic Formula
- 10.3 Applications: Quadratic Equations
- 10.4 Equations in Quadratic Form
- 10.5 Graphing Parabolas
- 10.6a Solving Quadratic Inequalities
- 10.6b Solving Inequalities with Rational Expressions

Chapter 10 Review

1. $x=-2,7$
2. $x=1,11$
3. $x^2+10x+25=(x+5)^2$
4. $x^2-x+\frac{1}{4}=\left(x-\frac{1}{2}\right)^2$
5. $x=\pm 6i$
6. $x=\pm 10$
7. $x=\pm 2\sqrt{6}$
8. $x=-4\pm\sqrt{10}$
9. $x=3\pm\sqrt{5}$
10. $x=-1\pm 2\sqrt{2}$
11. $x=-1\pm i\sqrt{3}$
12. $x=\frac{1}{2},1$
13. $x=-1,\frac{1}{2}$
14. $x=\frac{3\pm i\sqrt{11}}{2}$
15. $x^2-3=0$
16. $x^2-2x-1=0$
17. $x^2+6=0$
18. $x^2+6x+14=0$
19. -23, Two nonreal solutions
20. 25, Two real solutions
21. 9, Two real solutions
22. 0, One real solution
23. $x=\frac{-1\pm\sqrt{11}}{5}$
24. $x=1\pm\sqrt{5}$
25. $x=\pm\frac{13}{2}i$
26. $x=\frac{-1\pm i\sqrt{11}}{6}$
27. $x=-1,7$
28. $x=-\frac{5}{3},1$
29. $x=-3\pm\sqrt{2}$
30. $x=\frac{-7\pm i\sqrt{23}}{4}$

10.1 Solving Quadratic Equations

Solve the following equations by factoring.

1. $x^2-6x=14-x$ **2.** $2x^2=24x-22$

Add the correct constant to complete the square, then factor the trinomial as indicated.

3. $x^2+10x+\_\_\_\_\_\_=(\_\_\_\_\_\_)^2$ **4.** $x^2-x+\_\_\_\_\_\_=(\_\_\_\_\_\_)^2$

Solve the equations using the square root property.

5. $x^2=-36$ **6.** $x^2-100=0$ **7.** $x^2=24$ **8.** $2(x+4)^2=20$

Solve the quadratic equations by completing the square.

9. $x^2-6x+4=0$ **10.** $x^2+2x=7$ **11.** $3x^2+6x+12=0$

12. $4x^2-6x+2=0$ **13.** $6x^2=3-3x$ **14.** $5x^2-15x+25=0$

Write a quadratic equation with integer coefficients that has the given roots.

15. $x=\sqrt{3}, x=-\sqrt{3}$ **16.** $x=1+\sqrt{2}, x=1-\sqrt{2}$

17. $x=i\sqrt{6}, x=-i\sqrt{6}$ **18.** $x=-3+i\sqrt{5}, x=-3-i\sqrt{5}$

10.2 The Quadratic Formula: $x=\frac{-b\pm\sqrt{b^2-4ac}}{2a}$

Find the discriminant and determine the nature of the solutions of each quadratic equation.

19. $2x^2-3x+4=0$ **20.** $3x^2+x-2=0$

21. $5x^2=3x$ **22.** $x^2-20x+100=0$

Solve each of the quadratic equations using the quadratic formula.

23. $5x^2+2x-2=0$ **24.** $x^2-2x-4=0$ **25.** $4x^2+169=0$ **26.** $3x^2+1=-x$

Solve the given equations using any of the techniques discussed for solving quadratic equations: factoring, completing the square, or the quadratic formula.

27. $x^2-6x-7=0$ **28.** $3x^2+2x=5$ **29.** $x^2+6x+7=0$

31. $x=\frac{3\pm i\sqrt{3}}{4}$
32. $x=0,3\pm2\sqrt{3}$
33. $x=0,\frac{5}{3}$
34. $x=\pm\frac{2}{3}i$
35. $x=\frac{-1\pm\sqrt{31}}{12}$
36. $x=\frac{-5\pm\sqrt{55}}{4}$
37. $c<9$
38. $a>2$
39. 18
40. 8, 21
41. 30 feet, 40 feet
42. 120 seats
43. 9 m × 8 m
44. 1 m
45. 18 members, $110
46. Martin: 15 mph; Milton: 20 mph
47. a. 40 s
b. 1.97 s, 38.03 s
48. Area = 106.76 cm^2; Circumference = 36.62 cm
49. $x=\pm2,\pm\sqrt{6}$
50. $x=\pm\sqrt{2},\pm i\sqrt{6}$
51. $y=-\frac{1}{2},1$
52. $x=64,-\frac{1}{8}$
53. $y=1$
54. $x=-64,-8$
55. $x=-2$
56. $y=7$
57. $x=\pm3\sqrt{i},\pm3i\sqrt{i}$
58. $x=-2,\frac{1}{4}$
59. $x=-8,-2$
60. $x=-10,-\frac{14}{3}$
61. $x=0;(0,-3)$; $D=(-\infty,\infty)$; $R=[-3,\infty)$

30. $2x^2+7x+9=0$

31. $(4x-2)(x-1)=-1$

32. $x^3-5x^2=x^2+3x$

33. $(3x-2)(x-1)=2$

34. $9x^2+4=0$

First multiply each side of the equation by the LCM of the denominator to get integer coefficients and then solve the resulting equation.

35. $4x^2+\frac{2}{3}x-\frac{5}{6}=0$

36. $\frac{2}{5}x^2+x-\frac{3}{4}=0$

37. Determine the value(s) for c such that $x^2+6x+c=0$ will have two real solutions.

38. Determine the value(s) for a such that $ax^2+4x+2=0$ will have two nonreal solutions.

10.3 Applications

39. Find a positive real number such that its square is equal to 144 more than 10 times the number.

40. One integer is five more than twice a second integer. Their product is 168. What are the integers?

41. Right triangles: The length of one leg of a right triangle is 10 feet longer than the length of the other leg. If the hypotenuse is 50 feet long, what are the lengths of the two legs?

42. Football: The seating on one side of a high school football field is in the shape of a rectangle. The capacity of the seating is 1920 fans. The number of seats in each row is 40 less than ten times the number of rows. How many seats are in each row?

43. Area of a rectangle: The area of a rectangle is 72 square meters, and the perimeter of the rectangle is 34 meters. Find the length and the width of the rectangle.

44. Gardening: A garden measures 20 meters by 18 meters. The gardener is going to till strips of equal width along the four sides of the garden to plant a grass border. How wide will the strips be if the non-grassy area of the garden is now 288 square meters?

45. Bridge club: The women in a bridge club are to travel to a bridge tournament by bus for a total cost of $1980 which is to be divided equally among the women. However, two women became ill and could not make the trip. The cost to the remaining members increased $11 each. How many members made the trip? What was the final cost per member?

46. Biking: Martin and Milton leave Denver at the same time. Martin rides his bicycle south and Milton rides his bicycle east. Milton's average speed is 5 mph faster than Martin's. At the end of two hours they are 50 miles apart. Find the average speed of each rider.

62. $x=0;(0,4);$ $D=(-\infty,\infty);$ $R=(-\infty,4]$

63. $x=-4;(-4,0);$ $D=(-\infty,\infty);$ $R=[0,\infty)$

64. $x=0;(0,8);$ $D=(-\infty,\infty);$ $R=(-\infty,8]$

65. $x=-1;(-1,-4);$ $D=(-\infty,\infty);$ $R=[-4,\infty)$

66. $x=\frac{3}{2};\left(\frac{3}{2},-\frac{9}{2}\right);$ $D=(-\infty,\infty);$ $R=\left(-\infty,-\frac{9}{2}\right]$

67. $y=(x-3)^2;(3,0);$ $R=[0,\infty);$ Zeros: $x=3$

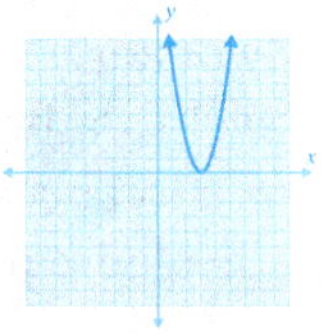

68. $y=(x-1)^2-5;$ $(1,-5);\ R=[-5,\infty);$ Zeros: $x=1\pm\sqrt{5}$

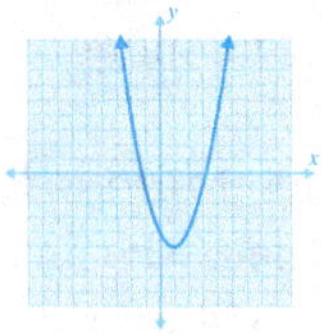

69. $y=(x-2)^2-3;$ $(2,-3);\ R=[-3,\infty);$ Zeros: $x=2\pm\sqrt{3}$

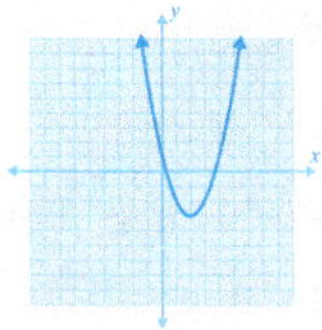

47. Launching a rocket: A rocket is fired straight up from the ground with an initial velocity of 640 ft/s. Using the formula $h=-16t^2+v_0t+h_0$ determine the following:

a. When will the projectile hit the ground?

b. When will the projectile be 1200 ft above the ground? (Round answers to nearest hundredth.)

48. Geometry: A rectangle is inscribed in a circle such that each corner of the rectangle lies on the circle. If the length of the rectangle is 10 cm and the width of the rectangle is 6 cm, what is the area and circumference of the circle? (Use $\pi = 3.14$.) Please round your answer to the nearest hundredth.

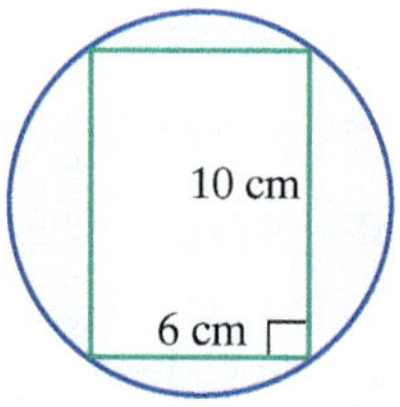

10.4 Equations in Quadratic Form

Solve the equations.

49. $x^4-10x^2+24=0$

50. $x^4+4x^2-12=0$

51. $y^{-2}+y^{-1}-2=0$

52. $2x^{\frac{2}{3}}-7x^{\frac{1}{3}}-4=0$

53. $y^{-1}+6y^{-\frac{1}{2}}-7=0$

54. $x^{\frac{2}{3}}+6x^{\frac{1}{3}}+8=0$

55. $2+\frac{1-x}{x+1}=\frac{x-2}{x+6}$

56. $\frac{y+2}{y-6}+\frac{2y-6}{y+1}=\frac{11y+3}{y+1}$

57. $x^4=-81$

58. $(4x+5)^2-3(4x+5)-18=0$

59. $(x+1)^2+8(x+1)+7=0$

60. $3(x+5)^2+14(x+5)=5$

10.5 Graphing Quadratic Functions: Parabolas

For each of the quadratic functions, determine the line of symmetry, the vertex, the domain, and the range.

61. $y=2x^2-3$

62. $y=-3x^2+4$

63. $y=(x+4)^2$

64. $y=-2x^2+8$

65. $y=(x+1)^2-4$

66. $y=-2\left(x-\frac{3}{2}\right)^2-\frac{9}{2}$

Rewrite each of the quadratic functions in the form $y=a(x-h)^2+k$. Find the vertex, range, and zero(s), if any, of each function. Graph the function.

67. $y=x^2-6x+9$

68. $y=x^2-2x-4$

69. $y=x^2-4x+1$

70. $y=-x^2-6x-2$

71. $y=2x^2+9x+6$

72. $y=x^2+2x+4$

73. Graph the function $y = 3x^2$. Then without additional computation, graph the following translations.

a. $y = 3(x-5)^2$

b. $y = 3(x+1)^2 - 7$

c. $y = -3x^2 + 1$

d. $y = 3(x-3)^2 - 4$

74. Publishing: The profit function (in dollars) for a publisher selling software to accompany an English textbook is given by the function $P(x) = -0.2x^2 + 120x - 3000$ where x is the number of copies of the software sold in one semester.

a. What number of copies of the software should be sold to give the maximum profit?

b. What will be the maximum profit?

75. Rockets: A toy rocket is projected vertically upward from the ground with an initial velocity of 208 ft/s. Using the formula $h = -16t^2 + v_0t + h_0$, answer the following questions.

a. When will the rocket reach its maximum height?

b. What will be the maximum height?

c. When will the rocket hit the ground again?

76. Building a fence: Manuel decides to build a fence to enclose an area for his dogs to play. The area is to be rectangular in shape. If he has 100 feet of fencing, find the dimensions of the rectangle that will give the dogs the maximum area in which to play. Only 3 sides need to be fenced because his house is to be one border of the rectangle.

Use a graphing calculator to graph the functions. Use the CALC *features of the calculator to find the zeros. (Round zeros, when necessary, to four decimal places.)*

77. $y = -2x^2 - 5x + 4$

78. $y = x^2 + 7x + 3$

10.6 Solving Quadratic and Rational Inequalities

Solve the inequalities algebraically, write the answers in interval notation, and then graph each solution set on a real number line.

79. $(x-5)(x+1) < 0$

80. $(x+3)(2x-1) > 0$

81. $2x^2 + x \geq 3$

82. $x^2 + 6x + 9 \leq 0$

83. $3x^3 - 7x^2 - 20x < 0$

84. $4x^2 - 12x + 9 > 0$

70. $y = -(x+3)^2 + 7$; $(-3, 7)$; $R = (-\infty, 7]$; Zeros: $x = -3 \pm \sqrt{7}$

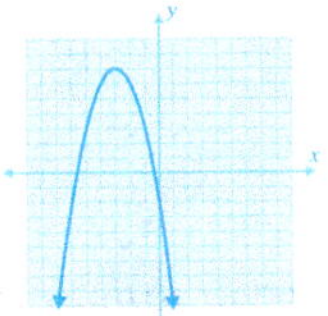

71. $y = 2\left(x + \frac{9}{4}\right)^2 - \frac{33}{8}$; $\left(-\frac{9}{4}, -\frac{33}{8}\right)$; $R = \left[-\frac{33}{8}, \infty\right)$; Zeros: $x = \frac{-9 \pm \sqrt{33}}{4}$

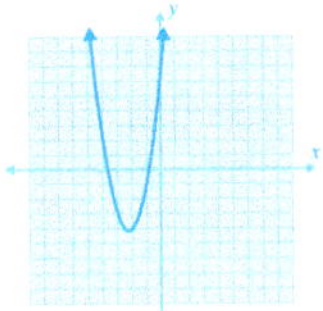

72. $y = (x+1)^2 + 3$; $(-1, 3)$; $R = [3, \infty)$; Zeros: None

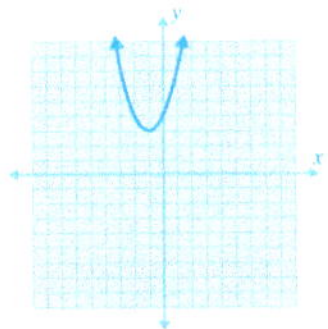

73. a.

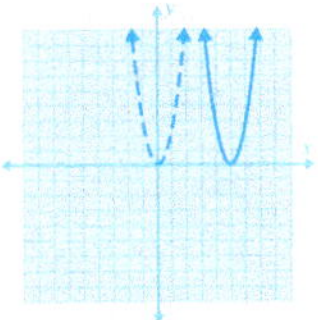

b.

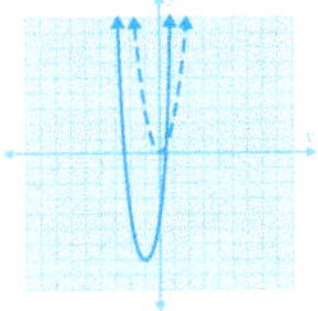

c.

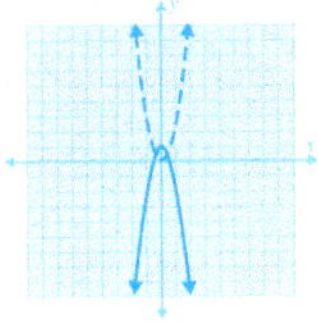

d. 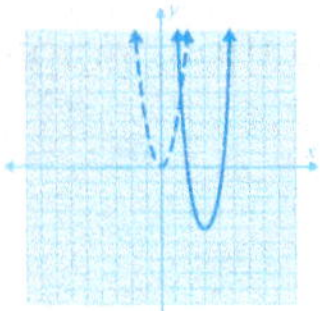

74. a. 300 copies
b. \$15,000

75. a. 6.5 s
b. 676 ft **c.** 13 s

76. Two sides 25 ft, third side 50 ft

77. Zeros: $x \approx -3.1375, 0.6375$

78. Zeros: $x \approx -6.5414, -0.4586$

85. $3x^2 + 1 \geq 4x$ **86.** $y^4 + 2y^2 - 24 > 0$ **87.** $x^2 + 2x - 5 > 0$

88. $4x^2 + 2x - 5 \leq 0$ **89.** $\dfrac{x-6}{2x+1} \leq 0$ **90.** $\dfrac{3x-3}{x-4} > 0$

91. $\dfrac{2x-5}{3-x} < -1$ **92.** $\dfrac{4x+1}{x+2} \geq \dfrac{5}{2}$

Use a graphing calculator to solve the inequalities. Write the answers in interval notation and graph each solution set on a real number line. (When necessary, round values to 4 decimal places.)

93. $x^2 > 8$ **94.** $x^2 - 5x - 6 \leq 0$ **95.** $2x^3 - 6x + 3 \geq 0$

96. $x^3 - 2x^2 + 4 < 0$

79. $(-1, 5)$ 

80. $(-\infty, -3) \cup \left(\frac{1}{2}, \infty\right)$ 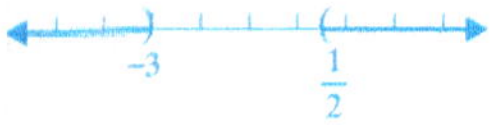

81. $\left(-\infty, -\frac{3}{2}\right] \cup [1, \infty)$ 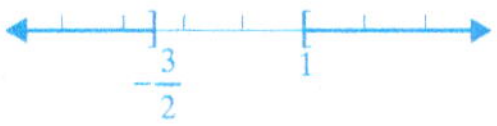

82. $\{-3\}$ 

83. $\left(-\infty, -\frac{5}{3}\right) \cup (0, 4)$ 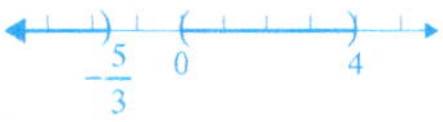

84. $\left(-\infty, \frac{3}{2}\right) \cup \left(\frac{3}{2}, \infty\right)$ 

85. $\left(-\infty, \frac{1}{3}\right] \cup [1, \infty)$ 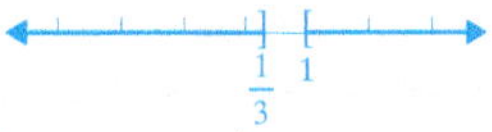

86. $(-\infty, -2) \cup (2, \infty)$ 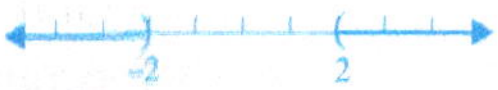

87. $\left(-\infty, -1-\sqrt{6}\right) \cup \left(-1+\sqrt{6}, \infty\right)$ 

88. $\left[\dfrac{-1-\sqrt{21}}{4}, \dfrac{-1+\sqrt{21}}{4}\right]$ 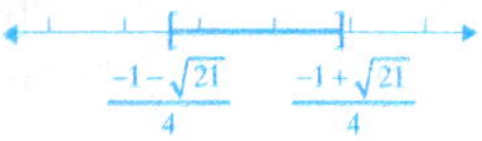

89. $\left(-\frac{1}{2}, 6\right]$ 

90. $(-\infty, 1) \cup (4, \infty)$ 

91. $(-\infty, 2) \cup (3, \infty)$ 

92. $(-\infty, -2) \cup \left[\frac{8}{3}, \infty\right)$ 

93. $(-\infty, -2.8284) \cup (2.8284, \infty)$ 

94. $[-1, 6]$ 

95. $[-1.9422, 0.5579] \cup [1.3844, \infty)$ 

96. $(-\infty, -1.1304)$ 

Chapter 10 Test

1. $x = -4, 2$
2. $x^2 - 30x + \underline{225} = \underline{(x-15)}^2$
3. $x = -2 \pm \sqrt{3}$
4. $x = -\frac{1}{2}, 0$
5. $x^2 - 2x - 4 = 0$
6. 73, Two real solutions
7. $b = \pm 2\sqrt{6}$
8. $x = \frac{-1 \pm i\sqrt{7}}{4}$
9. $x = \frac{3 \pm \sqrt{41}}{4}$
10. $x = \frac{2 \pm i\sqrt{2}}{2}$
11. $x = \pm 1, \pm 3$
12. $x = -1, \frac{3}{2}$
13. $x = 0, 1$
14. $x = 3$; $(3, -1)$; $D = (-\infty, \infty)$; $R = [-1, \infty)$
15. $x = 3$; $(3, -9)$; $D = (-\infty, \infty)$; $R = [-9, \infty)$
16. $y = (x+2)^2 - 10$; $(-2, -10)$; $R = [-10, \infty)$; Zeros: $x = -2 \pm \sqrt{10}$

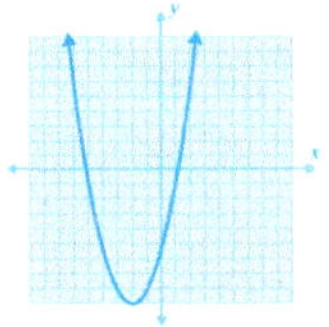

1. Solve the following equation using the square root property: $(x+1)^2 = 9$.

2. Add the correct constant to complete the square; then factor the trinomial as indicated.

$$x^2 - 30x + \_\_\_\_ = (\_\_\_\_\_\_)^2$$

3. Solve the following equation by completing the square: $x^2 + 4x + 1 = 0$

4. Solve the following equation by factoring: $4x^3 = -4x^2 - x$

5. Write a quadratic equation with integer coefficients that has the two numbers $1 \pm \sqrt{5}$ as roots.

6. Find the discriminant of the quadratic equation $4x^2 + 5x - 3 = 0$ and determine the nature of its solutions.

7. Using the discriminant, determine the value(s) for b such that the equation $2x^2 + bx + 3 = 0$ will have exactly one real root.

Solve the given equations using any of the techniques discussed for solving quadratic equations: factoring, completing the square, or the quadratic formula.

8. $2x^2 + x + 1 = 0$

9. $2x^2 - 3x - 4 = 0$

10. $2x^2 + 3 = 4x$

11. $x^4 = 10x^2 - 9$

12. $3x^{-2} + x^{-1} - 2 = 0$

13. $\frac{2x}{x-3} - \frac{2}{x-2} = 1$

For each of the quadratic functions, determine the line of symmetry, the vertex, the domain, and the range.

14. $y = (x-3)^2 - 1$

15. $y = 2(x-3)^2 - 9$

Rewrite each of the quadratic functions in the form $y = a(x-h)^2 + k$. Find the vertex, range, and zero(s) of each function. Graph the function.

16. $y = x^2 + 4x - 6$

17. $y = x^2 - 6x + 8$

Solve the quadratic (and higher degree) inequalities algebraically. Write the answers in interval notation, and then graph each solution set on a number line.

18. $x(x+7)(x-3) \geq 0$

19. $x^2 + 5x + 6 < 0$

Solve the rational inequalities algebraically. Write the answers in interval notation, and then graph each solution set on a number line.

20. $\dfrac{2x+5}{x-3} \geq 0$

21. $\dfrac{x-3}{2x+1} < 2$

22. Use a graphing calculator to solve the following inequality: $x^3 + 3x^2 - 10x > 0$. Write the solution in interval notation, and then graph the solution set on a number line.

23. Right triangles: The length of one leg of a right triangle is one meter less than twice the length of the second leg. The hypotenuse is 17 m. Find the lengths of the two legs.

24. Throwing a ball: A man standing at the edge of a cliff 112 ft above the beach throws a ball straight up into the air with a velocity of 96 ft/s. Use the formula $h = -16t^2 + v_0t + h_0$. (Round answers to the nearest hundredth.)

a. When will the ball hit the beach?

b. When will the ball be 64 ft above the beach?

25. Dimensions of a rectangle: The length of a rectangle is 4 inches longer than the width. If the diagonal is 20 inches long, what are the dimensions of the rectangle?

26. Traveling by car: Sandy made a business trip to a city 200 miles away and then returned home. Her average speed on the return trip was 10 mph less than her average speed going. If her total travel time was 9 hours, what was her average rate in each direction?

27. One number is 10 more than another number. Form a quadratic function that will allow you to find the minimum product of the two numbers. Find the numbers and the minimum product.

28. Maximize area: The perimeter of a rectangle is 22 in. Find the dimensions that will maximize the area.

17. $y = (x-3)^2 - 1$; $(3, -1)$; $R = [-1, \infty)$; Zeros: $x = 2, 4$

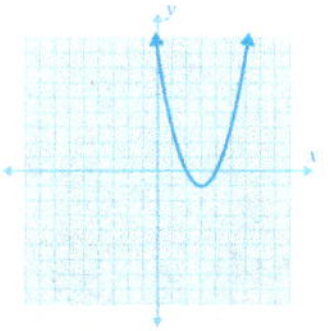

18. $[-7, 0] \cup [3, \infty)$

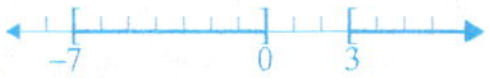

19. $(-3, -2)$

20. $\left(-\infty, -\dfrac{5}{2}\right] \cup (3, \infty)$

21. $\left(-\infty, -\dfrac{5}{3}\right) \cup \left(-\dfrac{1}{2}, \infty\right)$

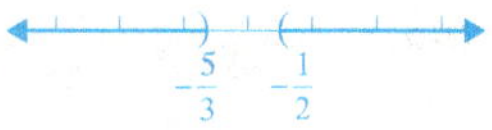

22. $(-5, 0) \cup (2, \infty)$

23. 8 m, 15 m

24. a. $t = 7$ seconds
b. $t = 6.46$ seconds

25. Width = 12 inches; Length = 16 inches 26. Speed returning = 40 mph; Speed going = 50 mph

27. $y = x(x-10)$; $-5, 5$; Minimum product $= -25$ 28. $\dfrac{11}{2}$ in. $\times \dfrac{11}{2}$ in.

Cumulative Review: Chapters 1 – 10

1. $\frac{1}{x^5}$ 2. x^3y^6
3. $\frac{9\sqrt{y}}{x^2}$ 4. $\frac{27x^2}{8y}$
5. $-3x^2y^2\sqrt[3]{y^2}$
6. $2x^2y^3\sqrt[4]{2xy^3}$
7. $\frac{9\sqrt{2}}{2}$ 8. $\frac{4\sqrt{3}}{3}$
9. $5\left(\sqrt{3}-\sqrt{2}\right)$
10. $(x+9)\left(\sqrt{x}-3\right)$
11. $4+2i$ 12. $19+4i$
13. $\frac{-11}{13}+\frac{16}{13}i$
14. $(x-5)(y+2)$
15. $(8x+9)(8x-9)$
16. $2\left(x^2-6y\right)\left(x^4+6x^2y+36y^2\right)$
17. a. -2 b. 37
18. $\frac{9x^2-2x-2}{(x+4)(2x-1)(3x-2)}$
19. $\frac{21x+8}{(x-8)(x+1)(3x-2)}$
20. $\frac{x+3}{3(x-5)}$
21. $\frac{2x^2+3x+1}{2(2x-3)(x-1)}$
22. $x=6$ 23. $x=\frac{7}{8}$
24. $x=-\frac{1}{2},3$
25. $x=-\frac{3}{2},\frac{2}{5}$
26. $x=\frac{-7\pm\sqrt{33}}{8}$
27. $x=-1$
28. $x=-2,\frac{4}{3}$

Simplify the expressions. Assume that all variables are positive.

1. $\left(4x^{-3}\right)\left(2x\right)^{-2}$ **2.** $\frac{x^{-2}y^4}{x^{-5}y^{-2}}$ **3.** $\left(27x^{-3}y^{\frac{3}{4}}\right)^{\frac{2}{3}}$

4. $\left(\frac{9x^{\frac{4}{3}}}{4y^{\frac{2}{3}}}\right)^{\frac{3}{2}}$ **5.** $\sqrt[3]{-27x^6y^8}$ **6.** $\sqrt[4]{32x^9y^{15}}$

7. $\frac{\sqrt{72}}{3}+5\sqrt{\frac{1}{2}}$ **8.** $\frac{1}{2}\sqrt{\frac{4}{3}}+3\sqrt{\frac{1}{3}}$

Rationalize the denominator and simplify each expression.

9. $\frac{5}{\sqrt{2}+\sqrt{3}}$ **10.** $\frac{x^2-81}{3+\sqrt{x}}$

Perform the indicated operations and write the results in the standard form a + bi.

11. $(2+5i)+(2-3i)$ **12.** $(2+5i)(2-3i)$ **13.** $\frac{2+5i}{2-3i}$

Completely factor each polynomial.

14. $xy+2x-5y-10$ **15.** $64x^2-81$ **16.** $2x^6-432y^3$

17. Given $P(x)=x^4-10x^3+20x^2-8x-2$, find **a.** $P(2)$ and **b.** $P(-1)$.

Perform the indicated operations and simplify the results.

18. $\frac{x+1}{2x^2+7x-4}+\frac{3x}{3x^2+10x-8}$ **19.** $\frac{x}{x^2-7x-8}-\frac{3x+1}{3x^2+x-2}$

20. $\frac{x^3+3x^2}{6x^2-36x+30}\cdot\frac{2x^2+2x-4}{x^3+2x^2}$ **21.** $\frac{2x^2+x}{x^2-2x+1}\div\frac{4x^2-6x}{x^2-1}$

Solve the equations.

22. $7(2x-5)=5(x+3)+4$ **23.** $(2x+1)(x-4)=(2x-3)(x+6)$

24. $|3x-2|=|x+4|$ **25.** $10x^2+11x-6=0$ **26.** $4x^2+7x+1=0$

29. $x = -5, -3, 3, 5$

30. $x = -\frac{8}{27}, -1$

31. $x = -\frac{38}{7}$

32. a. -47; No real solution
b. 0; One real solution

33. $4x^2 + 17x - 15 = 0$

34. $x^2 - 2x - 19 = 0$

35. $\left[-\frac{19}{2}, \infty\right)$

36. $\left(-\infty, \frac{7}{5}\right]$

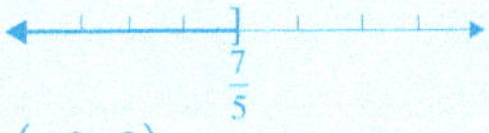

37. $(-2, 3)$

38. $\{4\}$

39. $(-\infty, -4) \cup \left(0, \frac{5}{3}\right)$

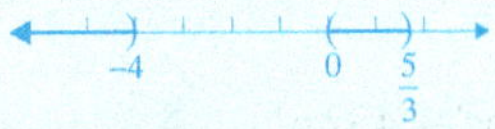

40. $(-2, 6]$

41.

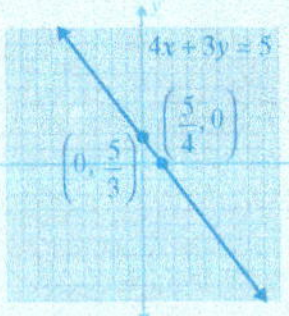

42. $y = x - 7$

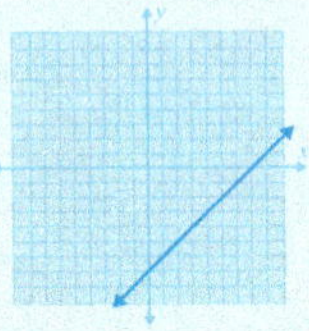

27. $\sqrt{x+5} - 2 = x + 1$ **28.** $8x^{-2} - 2x^{-1} - 3 = 0$ **29.** $x^4 - 34x^2 + 225 = 0$

30. $3x^{\frac{2}{3}} + 5x^{\frac{1}{3}} + 2 = 0$ **31.** $3 + \frac{1-x}{x+2} = \frac{4x+5}{2(x-2)}$

32. For the following equations, find the discriminant and determine the nature of the solutions.

a. $4x^2 - 7x + 6 = 0$ **b.** $x^2 + 10x + 25 = 0$

Find an equation with integer coefficients that has the indicated roots.

33. $x = \frac{3}{4}, x = -5$ **34.** $x = 1 - 2\sqrt{5}, x = 1 + 2\sqrt{5}$

Solve the inequalities algebraically, write the answers in interval notation, and then graph each solution set on a real number line.

35. $4(x+3) - 1 \geq 2(x-4)$ **36.** $\frac{7}{2}x + 3 \leq x + \frac{13}{2}$ **37.** $(x+2)(x-3) < 0$

38. $x^2 - 8x + 16 \leq 0$ **39.** $x(x+4)(3x-5) < 0$ **40.** $\frac{3x-10}{x+2} \leq 1$

41. Graph the equation $4x + 3y = 5$ by locating the x-intercept and the y-intercept.

42. Find an equation in slope-intercept form for the line passing through the points $(-1, -8)$ and $(4, -3)$. Graph the line.

43. Write an equation in standard form for the line parallel to $y = 3x - 4$ that passes through the point $(4, 0)$. Graph both lines.

44. What is the vertical line test for functions? Why does it work?

45. Solve the following system of two linear equations graphically: $\begin{cases} -2x + 3y = -4 \\ x - 2y = 3 \end{cases}$

46. Solve the following system of two linear equations algebraically: $\begin{cases} 3x - 2y = 7 \\ -2x + y = -6 \end{cases}$

47. Use Gaussian elimination to solve the following system: $\begin{cases} x - 3y - z = 1 \\ 2x + y - 2z = -5 \\ 3x - y + 2z = 10 \end{cases}$

48. Solve the following system of linear inequalities graphically: $\begin{cases} x < 4 \\ x - y \geq 3 \end{cases}$

43. $3x - y = 12$

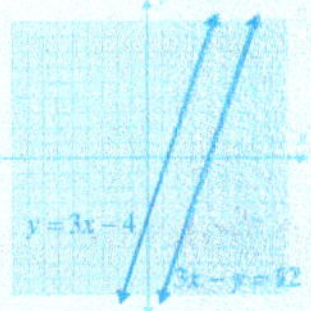

44. If any vertical line intersects the graph of a relation at more than one point, then the relation graphed is not a function. This works because it checks whether any first coordinate appears more than once.

45. $(-1, -2)$

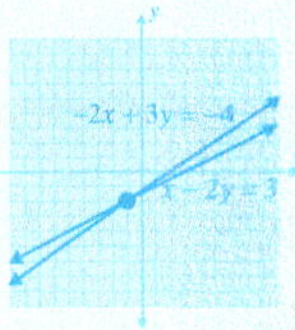

46. $(5, 4)$

47. $(1, -1, 3)$

48.

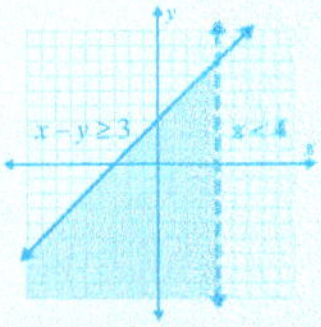

49. $x = -\frac{1}{2}$; $\left(-\frac{1}{2}, -\frac{25}{4}\right)$; $D = (-\infty, \infty)$; $R = \left[-\frac{25}{4}, \infty\right)$; Zeros: $x = -3, 2$

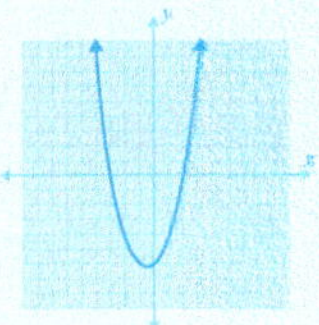

For each of the quadratic functions, determine the line of symmetry, the vertex, the domain, the range, and the zeros. Graph the function.

49. $y = x^2 + x - 6$ **50.** $y = -x^2 + x + 12$ **51.** $y = 2x^2 - 9x - 5$

Solve for the indicated variable.

52. $3x + 2y = 6$ for y **53.** $\frac{3}{4}x + \frac{1}{2}y = 5$ for y

54. Triangles: The base of a triangle is 3 cm more than twice its height. If the area of the triangle is 76 cm^2, find the length of the base and the height of the triangle.

55. Rectangles: Find the dimensions of a rectangle that has an area of 520 m^2 and a perimeter of 92 m.

56. Soapbox Derby: Austin and Jeremy participated in a half mile soapbox car race. If Austin crossed the finish line with a time of 43.2 seconds and Jeremy crossed the finish line in 45.0 seconds, how much faster was Austin going than Jeremy in mph? Please round your answer to the nearest tenth.

57. Find three consecutive even integers such that the square of the first added to the product of the second and third gives a result of 368.

58. Catering: Jennifer and Michael work at a local restaurant and must prepare enough mirepoix (a mixture of onions, celery, and carrots) for a community event. If it takes Michael 3 hours to cut enough mirepoix and 1.25 hours if Michael and Jennifer work together, how long would it take (to the nearest hundredth) for Jennifer to prepare everything by herself?

59. z varies directly as x^2 and inversely as $\sqrt{y}$. If $z = 50$ when $x = 2$ and $y = 25$, find z if $x = 4$ and $y = 100$.

60. Investing: Melissa plans to invest \$10,000 into two accounts, one that returns 9% interest and another that returns 11% interest. If she invests three times as much in the account that returns 11% interest and expects a return of \$1050, how much money does she invest in each account? Please round your answer to the nearest cent.

61. Medicine: A doctor needs to administer two drugs to a patient. The amount of Drug A administered must be three times the amount of Drug B, and the total amount of medication must equal 20 mg. How much of each drug should be given to the patient?

62. Renting cars: A car rental agency rents 200 cars per day at a rate of \$30 per day for each car. For each \$1 increase in the daily rate, the owners have found that they will rent 5 fewer cars per day. What daily rate would give total receipts of \$6125? (**Hint:** Let x = the number of \$1 increases.)

63. Selling golf balls: The profit function (in dollars) for a golf ball manufacturer is given by the function $P(x) = -0.01x^2 + 200x - 5000$ where x is the number of dozens of ball sold in one week.

a. How many dozens of golf balls need to be sold to give the maximum profit in a week?

b. What will be the maximum profit?

c. How many golf balls is this?

50. $x=\frac{1}{2}$; $\left(\frac{1}{2},\frac{49}{4}\right)$; $D=(-\infty,\infty)$; $R=\left(-\infty,\frac{49}{4}\right]$ Zeros: $x=-3, 4$

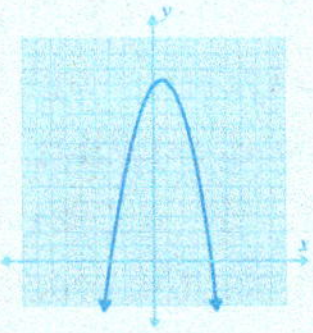

Use a calculator to estimate the value of each number accurate to 4 decimal places.

64. $\sqrt{6}+2\sqrt{3}$ **65.** $\sqrt{2}\left(1+3\sqrt{10}\right)$ **66.** $\dfrac{\sqrt{2}+2}{\sqrt{2}-2}$

Use a graphing calculator and the CALC *and zero features to graph the functions and estimate the x-intercepts to two decimal places.*

67. $f(x)=2x^2-5$ **68.** $y=-x^3+2x^2-1$

69. $g(x)=x^4-x^2+8$ **70.** $h(x)=x^3+3x+2$

Use a graphing calculator to graph each of the functions. State the domain and range of each function.

71. $f(x)=\sqrt{x-3}$ **72.** $g(x)=\sqrt{1-x}$ **73.** $y=-\sqrt{x+1}$

51. $x=\frac{9}{4}$; $\left(\frac{9}{4},-\frac{121}{8}\right)$; $D=(-\infty,\infty)$; $R=\left[-\frac{121}{8},\infty\right)$; Zeros: $x=-\frac{1}{2}, 5$

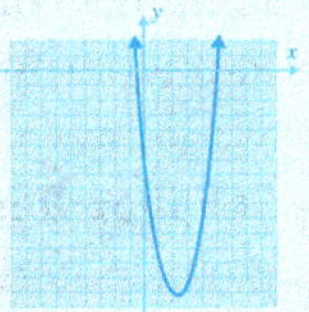

52. $y=-\frac{3}{2}x+3$ **53.** $y=-\frac{3}{2}x+10$ **54.** Length of the base = 19 cm; Height = 8 cm **55.** 20 m × 26 m **56.** 1.7 mph faster **57.** 12, 14, 16 **58.** 2.14 hours **59.** 100 **60.** \$2500 at 9% and \$7500 at 11% **61.** 15 mg of Drug A, 5 mg of Drug B **62.** \$35 **63. a.** 10,000 dozen **b.** \$995,000 **c.** 120,000 golf balls **64.** 5.9136 **65.** 14.8306 **66.** −5.8284

67. x-int: $(-1.58, 0)$ and $(1.58, 0)$

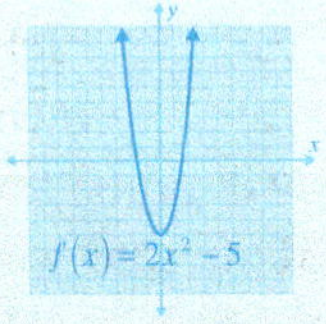

68. x-int: $(-0.62, 0)$, $(1, 0)$ and $(1.62, 0)$

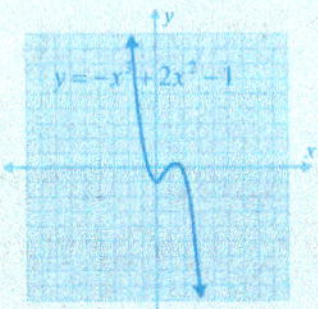

69. x-int: None

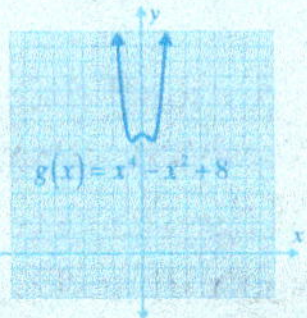

70. x-int: $(-0.60, 0)$

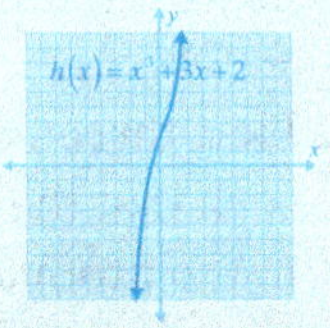

71. $D=[3,\infty)$, $R=[0,\infty)$

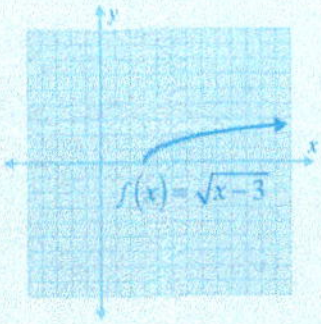

72. $D=(-\infty,1]$, $R=[0,\infty)$

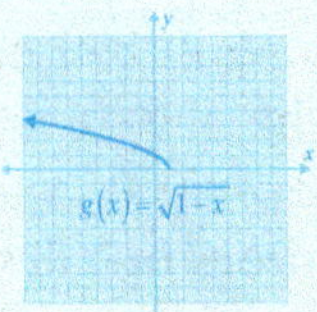

73. $D=[-1,\infty)$, $R=(-\infty,0]$

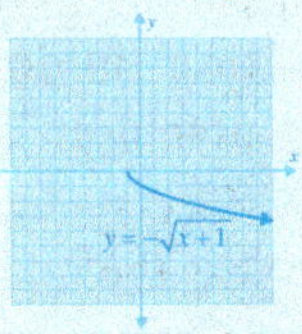

CHAPTER 11

Exponential and Logarithmic Functions

Did You Know?

In this chapter, you will study exponential functions and their related inverses, logarithmic functions. You will also see how the use of logarithms can simplify calculations involving multiplication and division. Although electronic calculators have made calculation with logarithms obsolete, it is still important to study the logarithmic functions because they have many applications other than computing.

Napier

The inventor of logarithms was John Napier (1550 – 1617). Napier was a Scottish nobleman, Laird of Merchiston Castle, a stronghold on the outskirts of the town of Edinburgh. An eccentric, Napier was intensely involved in the political and religious struggles of his day. He had interests in many areas, including mathematics. In 1614, Napier published his "A Description of the Wonderful Law of Logarithms," and thus he is given credit for first publishing and popularizing the idea of logarithms. Napier used a base close to the number e for his system, and natural logarithms (base e) are often called **Napierian logarithms**. Napier soon saw that a base of 10 would be more appropriate for calculations since our decimal number system uses base 10. Napier began work on a base-10 system but was unable to complete it before his death. Henry Briggs (1561 – 1630) completed Napier's work, and base-10 logarithms are often called **Briggsian logarithms** in his honor.

Napier's interest in simplifying calculations was based on the need at that time to do many calculations by hand for astronomical and scientific research. He also invented the forerunner of the slide rule and predicted tanks, submarines, and other advanced war technology. Napier's remarkable ingenuity led the local people to consider him either crazy or a dealer in the black art of magic.

A particularly amusing story is told of Napier's method of identifying which of his servants was stealing from him. He told his servants that his black rooster would identify the thief. Each servant was sent alone into a darkened room to pet the rooster on the back. Napier had coated the back of the rooster with soot, and the guilty servant came out of the room with clean hands.

Napier was a staunch Presbyterian, and he felt that his claim to immortality would be an attack that he had written on the Catholic Church. The scientific community more correctly judged that logarithms would be his greatest contribution.

"The invention of logarithms: 'by shortening the labors doubled the life of the astronomer.'"

Pierre de Laplace (1749 – 1827)

As discussed throughout this text, functions are an important topic in mathematics. Function notation $f(x)$ is particularly helpful in evaluating functions and indicating graphical relationships. Operating algebraically with functions as well as understanding and finding the **composition** and **inverses** of functions rely heavily on function notation. The concepts of composite and inverse functions form the basis of the relationship between logarithmic and exponential functions.

Logarithms are exponents. Traditionally, logarithmic and exponential values were calculated with the extensive use of printed tables and techniques for estimating values not found in the tables. As some of your "older" teachers will tell you, this was a long and detailed process. These tables are no longer printed in textbooks. Now hand-held calculators have programs stored in their electronic memories that calculate the values in these tables with even greater accuracy, and complicated expressions can be evaluated by pressing a few keys.

Of all the topics discussed in algebra, logarithmic functions and exponential functions probably have the most value in terms of applied problems. Learning curves, important in business and education, can be described with logarithmic and exponential functions. Exponential growth and decay are basic concepts in biology and medicine. (Cancer cells grow exponentially and radium decays exponentially.) Computers use logarithmic and exponential concepts in their design and implementation. These concepts are likely to be encountered in almost any field of study.

11.1 Algebra of Functions

- *Find the sum, difference, product, and quotient of two functions.*
- *Graph the sum of two functions.*
- *Use a graphing calculator to graph the sum of two functions.*

If two (or more) functions have the same domain, then we can perform the operations of addition, subtraction, multiplication, and division with these functions. For example, consider the following quadratic functions:

$$f(x) = 2x^2 - 1 \quad \text{and} \quad g(x) = x^2 + 2x - 5.$$

Both have the same domain: $\mathbb{R}$ = all real numbers = $(-\infty, \infty)$. This means that we can

a. choose any value for x from the common domain,
b. evaluate each function for that value of x, and
c. perform operations with those functional values.

The functional values are the *y*-values. For example, if we choose $x = 3$, then

$$f(3) = 2(3)^2 - 1 = 17 \quad \text{and} \quad g(3) = (3)^2 + 2(3) - 5 = 10.$$

Teaching Notes:
As you begin this topic you might want to remind students that most functions they have seen up to this point have been polynomial or rational functions. This chapter introduces them to two related functions that are not based on polynomials.

Now we can easily find the sum and difference

$$f(3)+g(3)=17+10=27 \qquad \text{and} \qquad f(3)-g(3)=17-10=7.$$

However, if we want to find, say $f(5)+g(5)$ and $f(5)-g(5)$, we would need to again evaluate both functions, this time at $x=5$. Another way to find sums and differences of functions is to find the algebraic sum (or difference) of the two expressions. Then these new expressions will allow us to find the sum (or difference) directly for any value of x that is in the original domain of both functions. For example, we find the sum $f+g$

$$\begin{aligned}(f+g)(x)&=f(x)+g(x)\\&=(2x^2-1)+(x^2+2x-5)\\&=3x^2+2x-6.\end{aligned}$$

With this new function, we find directly that

$$(f+g)(3)=3(3)^2+2(3)-6=27.$$

Similarly,

$$\begin{aligned}(f-g)(x)&=f(x)-g(x)\\&=(2x^2-1)-(x^2+2x-5)\\&=x^2-2x+4,\end{aligned}$$

and we have

$$(f-g)(3)=(3)^2-2(3)+4=7.$$

Similar notation is used for the product and quotient of two functions. One important condition is that both functions **must have the same domain**. If not, then the algebraic sums, differences, products, and quotients are **restricted to portions of the domains that are in common**. **Also, in the case of quotients, no denominator can be 0.**

Algebraic Operations with Functions

If $f(x)$ and $g(x)$ represent two functions and x is a value in the **domain of both functions**, then:

1. Sum of two functions: $(f+g)(x)=f(x)+g(x)$
2. Difference of two functions: $(f-g)(x)=f(x)-g(x)$
3. Product of two functions: $(f\cdot g)(x)=f(x)\cdot g(x)$
4. Quotient of two functions: $\left(\dfrac{f}{g}\right)(x)=\dfrac{f(x)}{g(x)}$ where $g(x)\neq 0$.

Example 1: Algebraic Operations with Functions

Let $f(x)=3x^2+x-4$ and $g(x)=x-6$. Find the following functions.

a. $(f+g)(x)$

Solution: $(f+g)(x)=(3x^2+x-4)+(x-6)=3x^2+2x-10$

b. $(f-g)(x)$

Solution: $(f-g)(x)=(3x^2+x-4)-(x-6)=3x^2+2$

c. $(f\cdot g)(x)$

Solution: $(f\cdot g)(x)=(3x^2+x-4)(x-6)=3x^3-17x^2-10x+24$

d. Evaluate each of the functions found in parts **a.**– **c.** at $x=2$.

Solution: Evaluating each of these functions at $x=2$ gives the following results.

$$(f+g)(2)=3(2)^2+2(2)-10=6$$

$$(f-g)(2)=3(2)^2+(2)=14$$

$$(f\cdot g)(2)=3(2)^3-17(2)^2-10(2)+24=-40$$

Example 2: Algebraic Operations with Functions

Let $f(x)=x^2-x$ and $g(x)=2x+1$. Find the following functions.

a. $(f+g)(x)$

Solution: $(f+g)(x)=(x^2-x)+(2x+1)=x^2+x+1$

b. $(g-f)(x)$

Solution: $(g-f)(x)=(2x+1)-(x^2-x)=-x^2+3x+1$

c. $\left(\frac{f}{g}\right)(x)$

Solution: $\left(\frac{f}{g}\right)(x)=\frac{x^2-x}{2x+1}$ where $2x+1\neq 0$ (or $x\neq -\frac{1}{2}$)

d. Evaluate each of the functions found in parts **a.** – **c.** at $x=3$.

Solution: Evaluating each of these functions at $x=3$ gives the following results.

a. $(f+g)(3)=(3)^2+(3)+1=13$

b. $(g-f)(3)=-(3)^2+3(3)+1=1$

c. $\left(\frac{f}{g}\right)(3)=\frac{(3)^2-(3)}{2(3)+1}=\frac{6}{7}$

Note that, except for Example 2c, the domain for all of the functions discussed in Examples 1 and 2 is the set of all real numbers, $(-\infty, \infty)$. In Example 2c we noted that the denominator cannot equal 0. In Example 3 we show how the domain may need to be limited before performing algebra with functions that contain radical expressions.

Example 3: Algebraic Operations with Functions with Limited Domains

Teaching Notes:
You may want to give your students several examples involving radicals.

Let $f(x)=x+5$ and $g(x)=\sqrt{x-2}$. Find the following functions and state the domain of each function.

a. $(f+g)(x)$

Solution: $(f+g)(x)=x+5+\sqrt{x-2}$

The domain of f is the set of all real numbers. However, the domain of the sum is restricted to the domain of g, the radical function. In this case we must have $x-2 \geq 0$. Thus in interval notation, the domain is $[2, \infty)$.

b. $\left(\frac{f}{g}\right)(x)$

Solution: $\left(\frac{f}{g}\right)(x)=\frac{x+5}{\sqrt{x-2}}$

For this function, the denominator cannot be 0, so $x \neq 2$. Therefore, we must have $x-2>0$ and the domain, in interval notation, is $(2, \infty)$.
Note: The domain can become smaller than, but never larger than, the domain of the two original functions.

Graphing the Sum of Two Functions

In this section we discuss how to graph the sum of two functions. (Graphing the difference, product, and quotient can be accomplished in a similar manner.) Remember that, in any case, algebraic operations with functions can be performed only over a common domain.

We begin with two functions f and g that have the same domain and only a finite number of ordered pairs.

$$f=\{(-2,1),(0,4),(3,5)\}$$
$$g=\{(-2,3),(0,1),(3,2)\}$$

Note that the domain of both functions is $\{-2,0,3\}$.

Teaching Notes:
To test their understanding, you might have your students graph the difference of the two functions in Figure 1. Also, you might want to draw several pairs of functions on the board (or on hand-out sheets) and have the students graph the sums, differences, products, and quotients of the functions. Start with a finite number of points. I have found that the students are fascinated with this process and treat it as a fun and rewarding hands-on exercise. It can definitely help in their general understanding about functions and graphs.

Both functions have only three points and are graphed in Figure 1.

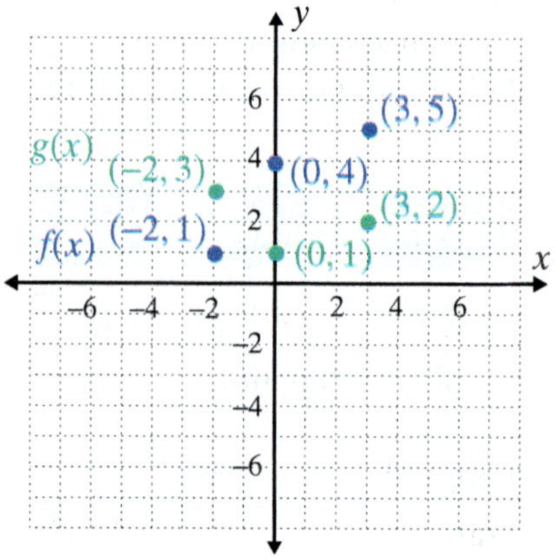

Figure 1

To add the two functions graphically, look at each of the x-values on the x-axis and the points directly above (or below) them and add the corresponding y-values. This process will give new points, and these new points represent a function that is the sum of the two original functions.

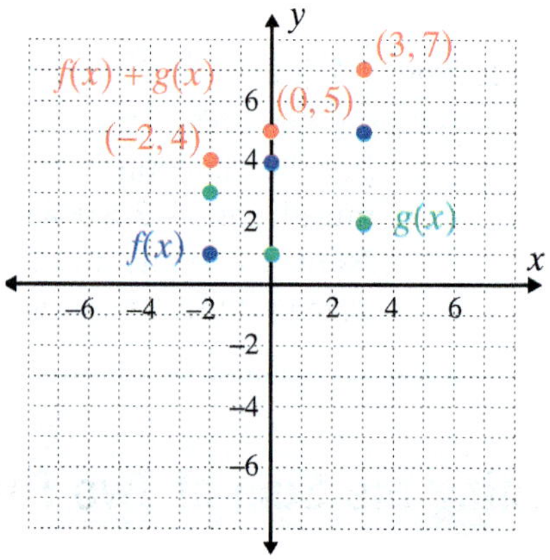

Figure 2

From the points in the Figure 2 we see that

$$\begin{aligned}(f+g)(x) &= f(x)+g(x)\\ &= \{(-2, 1+3), (0, 4+1), (3, 5+2)\}\\ &= \{(-2, 4), (0, 5), (3, 7)\}.\end{aligned}$$

In general, graphing the sum (difference, product, or quotient) of two functions will involve an infinite number of points. Certainly not all of these points can be plotted one at a time. However, by making a table of a few key points and joining these points with smooth curves or line segments, the general nature of the result can be found. In fact, if the two functions consist of line segments, then the sum will also consist of line segments. Table 1 and Figures 3 and 4 illustrate just such a case.

Remember that when operating with functions, the operations are performed with the *y*-values for each value of *x* in the common domain. (Don't mess with the *x*-values!)

Points Illustrated in Figure 3 and Figure 4

x	$f(x)$	$g(x)$	$f(x)+g(x)$
−3	1	−3	$1+(-3)=-2$
−2	1	0	$1+0=1$
−1	1	3	$1+3=4$
0	−2	3	$(-2)+3=1$
1	1	3	$1+3=4$
2	1	4	$1+4=5$
3	0	5	$0+5=5$

Table 1

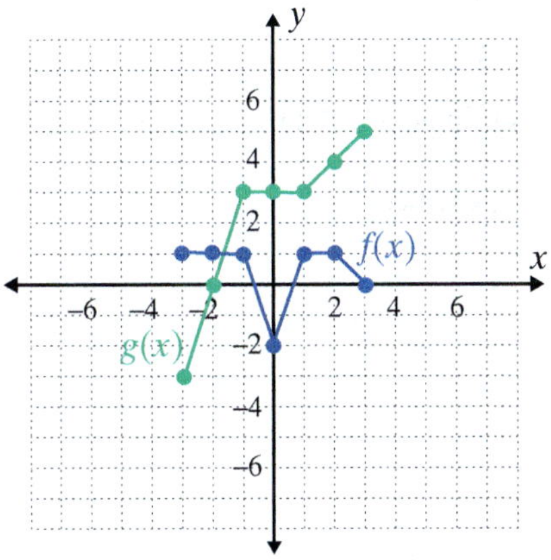

Figure 3

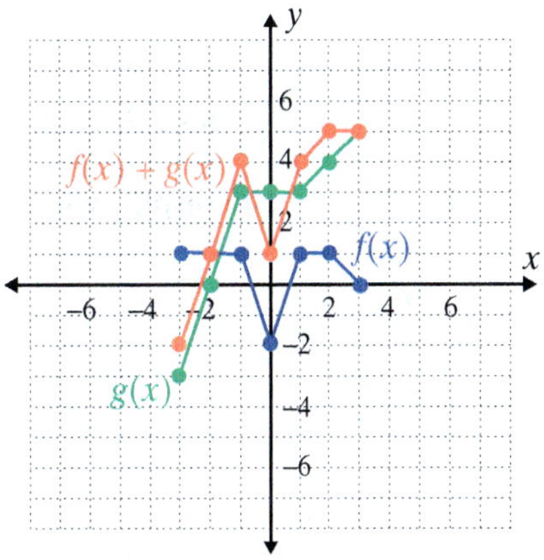

Figure 4

Using a TI-84 Plus Graphing Calculator to Graph the Sum of Two Functions

A graphing calculator can be used to graph the sum (difference, product, or quotient) of two functions. An interesting way to do this is to first press  and enter the functions as Y1 and Y2; then assign Y3 to be the sum of the first two. Figures 5a and 5b show the displays for entering two functions and their graphs.

Using the TI-84 Plus graphing calculator, we enter:

$$\text{Y1} = x^2 - 3 \quad \text{(a parabola)}$$

and

$$\text{Y2} = 5x - 1 \quad \text{(a straight line)}.$$

1. a. $2x-3$
   b. 7
   c. $x^2-3x-10$
   d. $\dfrac{x+2}{x-5},\ x\neq 5$
2. a. $3x+4$
   b. $x-4$
   c. $2x^2+8x$
   d. $\dfrac{2x}{x+4},\ x\neq -4$
3. a. x^2+3x-4
   b. x^2-3x+4
   c. $3x^3-4x^2$
   d. $\dfrac{x^2}{3x-4},\ x\neq \dfrac{4}{3}$
4. a. x^2+x-2
   b. $-x^2+x-4$
   c. x^3-3x^2+x-3
   d. $\dfrac{x-3}{x^2+1}$
5. a. x^2+x-12
   b. x^2-x-6
   c. $x^3-3x^2-9x+27$
   d. $x+3,\ x\neq 3$
6. a. x^2+x-20
   b. x^2-x-30
   c. $x^3+5x^2-25x-125$
   d. $x-5,\ x\neq -5$
7. a. $3x^2+x+2$
   b. x^2+x-2
   c. $2x^4+x^3+4x^2+2x$
   d. $\dfrac{2x^2+x}{x^2+2}$
8. a. x^3+x^2+6x+6
   b. x^3-x^2+6x-6
   c. x^5+12x^3+36x
   d. x

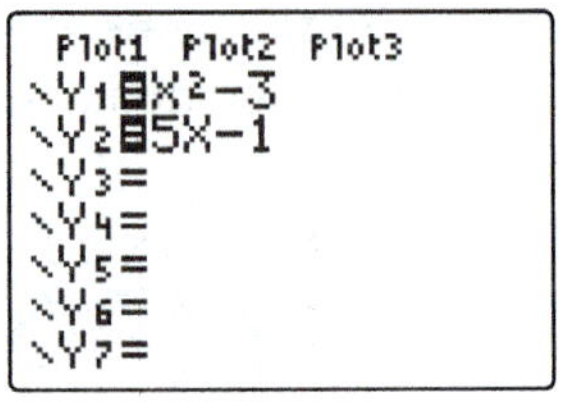

Figure 5a: The two functions entered

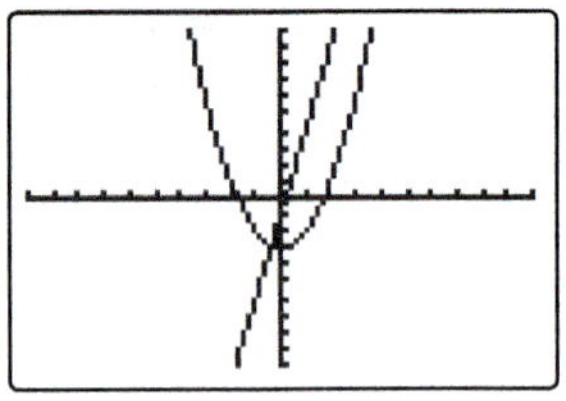

Figure 5b: The two functions graphed

Now we can graph the sum of the two functions, $f(x)+g(x)$, (or Y1 + Y2 on the calculator) by assigning Y3 = Y1 + Y2 as follows. (This saves performing the actual algebraic operations. The calculator does it for us.)

Step 1: Press Y= and select Y3.

Step 2: Press the VARS key.

Step 3: Move the cursor to the Y-VARS at the top of the display.

Step 4: Select 1:FUNCTION... and press ENTER.

Step 5: Select Y1 and press ENTER.

Step 6: Press the + key.

Step 7: Go to Y-VARS again, select 1:FUNCTION... and select Y2.

The display will now appear as follows.

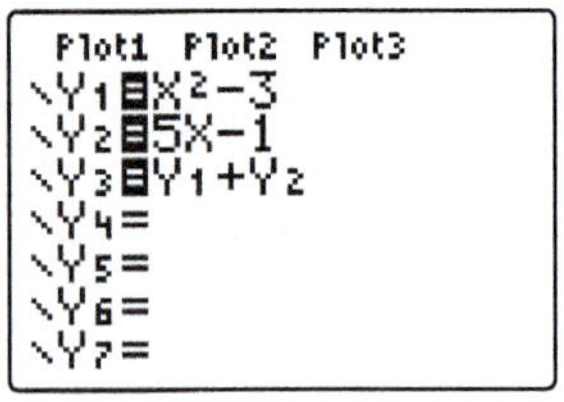

Figure 6

Now press GRAPH and the display will show all three curves, Y1, Y2, and the sum Y3 = Y1 + Y2. You may need to reset the WINDOW to allow a more complete display of all of the graphs.

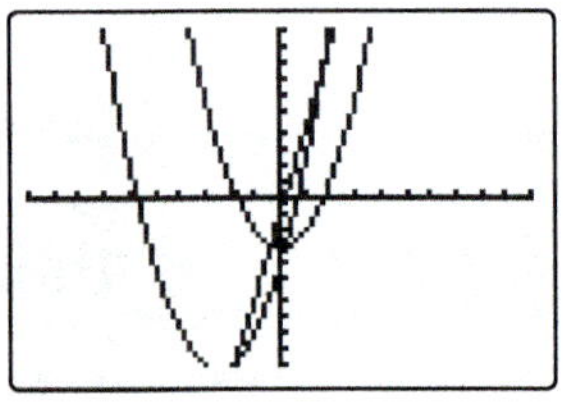

Figure 7

If this graph is somewhat confusing (because three graphs are shown), you can turn off the first two graphs and show just the sum. To turn off a graph, go to the highlighted equal sign, ▀, next to the equation and hit ENTER. You should see the following screens.

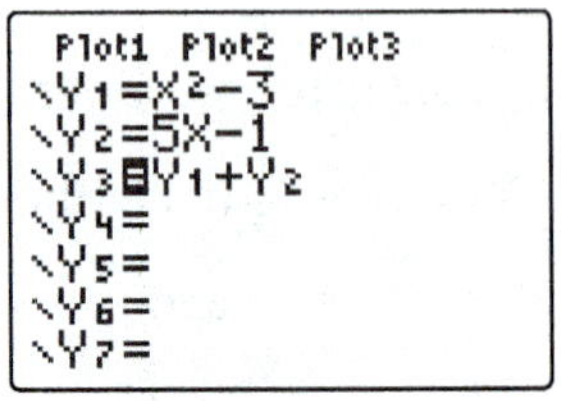

Figure 8a

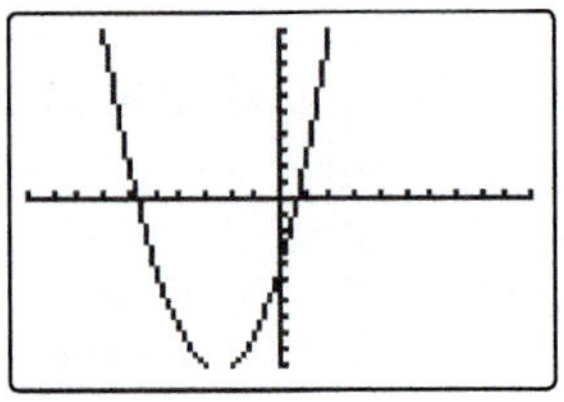

Figure 8b

9. a. $2x^2+2$
b. $8x$
c. x^4-14x^2+1
d. $\dfrac{x^2+4x+1}{x^2-4x+1}$, $x \neq 2 \pm \sqrt{3}$

10. a. x^3-2x^2+6
b. x^3-6
c. $-x^5+x^4+6x^3-6x^2$
d. $\dfrac{x^3-x^2}{6-x^2}$, $x \neq \pm\sqrt{6}$

11. 9 12. 8

13. $-a^2-a-1$

Practice Problems

1. For $f(x)=x^2-2x-3$ and $g(x)=x-3$ find:
   a. $(f+g)(x)$ **b.** $(f-g)(x)$ **c.** $(f \cdot g)(x)$ **d.** $\left(\dfrac{f}{g}\right)(x)$
2. Evaluate each of the functions **a.** – **d.** in Practice Problem 1 for $x = 4$.
3. For $f(x)=2x-5$ and $h(x)=\sqrt{x+6}$,
   a. Find $f(x)+h(x)$ and state the domain of the sum function.
   b. Find $f(x) \cdot h(x)$ and state the domain of the product function.
4. Given $f(x)=\{(-3,2),(-1,1),(0,-1),(2,3)\}$ and $g(x)=\{(-3,4),(-1,2),(0,1),(2,1)\}$ graph the sum of the two functions.

11.1 Exercises

14. $\dfrac{-a+3}{a^2+4}$

15. 27 16. 1.75

17. $\dfrac{8}{5}$ 18. 78

19. -31 20. $\dfrac{4}{5}$

21. $\sqrt{2x-6}+x+4$, $D=[3,\infty)$

22. $x-1$, $D=(-\infty,1)\cup(1,\infty)$

23. $3x^2-19x-14$, $D=(-\infty,\infty)$

24. 2, $D=(-\infty,\infty)$

For the following pairs of functions, find **a.** $(f+g)(x)$, **b.** $(f-g)(x)$, **c.** $(f \cdot g)(x)$, *and* **d.** $\left(\dfrac{f}{g}\right)(x)$

1. $f(x)=x+2,\ g(x)=x-5$
2. $f(x)=2x,\ g(x)=x+4$
3. $f(x)=x^2,\ g(x)=3x-4$
4. $f(x)=x-3,\ g(x)=x^2+1$
5. $f(x)=x^2-9,\ g(x)=x-3$
6. $f(x)=x^2-25,\ g(x)=x+5$
7. $f(x)=2x^2+x,\ g(x)=x^2+2$
8. $f(x)=x^3+6x,\ g(x)=x^2+6$
9. $f(x)=x^2+4x+1,\ g(x)=x^2-4x+1$
10. $f(x)=x^3-x^2,\ g(x)=6-x^2$

Let $f(x)=x^2+4$ *and* $g(x)=-x+3$. *Find the values of the indicated expressions.*

11. $f(2)+g(2)$
12. $f(2) \cdot g(2)$
13. $g(a)-f(a)$
14. $\dfrac{g(a)}{f(a)}$

Answers to Practice Problems: **1. a.** x^2-x-6 **b.** x^2-3x **c.** x^3-5x^2+3x+9
d. $x+1,\ x \neq 3$ (Note the restriction holds even after simplifying.)
2. a. 6 **b.** 4 **c.** 5 **d.** 5 **3. a.** $2x-5+\sqrt{x+6},\ D=[-6,\infty)$
b. $2x\sqrt{x+6}-5\sqrt{x+6},\ D=[-6,\infty)$
4.

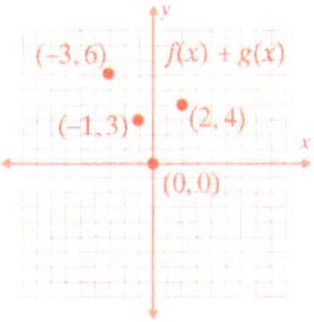

25. $\dfrac{x-5}{\sqrt{x+3}}$, $D=(-3,\infty)$

26. $2x\sqrt{2-x}-8\sqrt{2-x}$, $D=(-\infty,2]$

27. $-3x\sqrt{x-3}$, $D=[3,\infty)$

28. $5-x+\sqrt{4-x}$, $D=(-\infty,4]$

29. $\sqrt[3]{x+3}+\sqrt{5+x}$, $D=[-5,\infty)$

30. $\sqrt{x-1}-\sqrt[3]{2x+1}$, $D=[1,\infty)$

31. a.
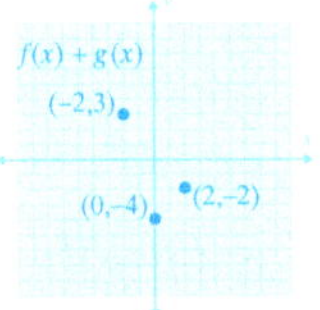

b.
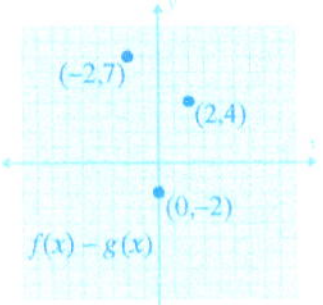

32. a.
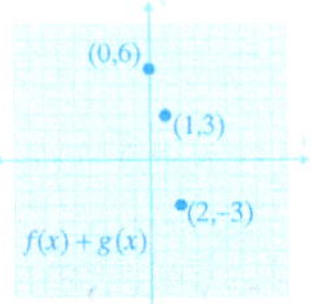

b.
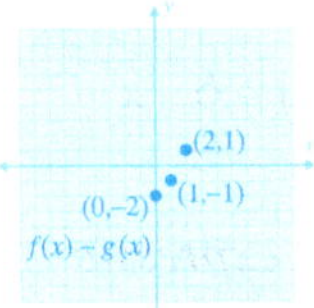

33. a.
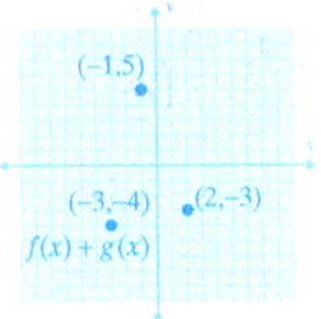

b.
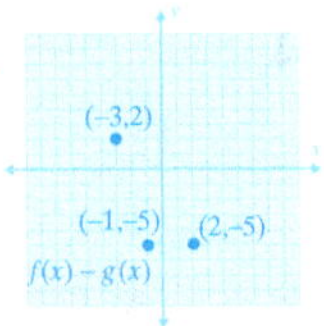

15. $(f+g)(-4)$ **16.** $(f-g)(0.5)$ **17.** $\left(\dfrac{f}{g}\right)(-2)$ **18.** $(f\cdot g)(-3)$

19. $(g-f)(-6)$ **20.** $\left(\dfrac{g}{f}\right)(-1)$

Find the indicated functions and state their domains in interval notation.

21. If $f(x)=\sqrt{2x-6}$ and $g(x)=x+4$, find $(f+g)(x)$.

22. If $f(x)=x^2-2x+1$ and $g(x)=x-1$, find $\left(\dfrac{f}{g}\right)(x)$.

23. Find $f(x)\cdot g(x)$ given that $f(x)=3x+2$ and $g(x)=x-7$.

24. Find $f(x)-g(x)$ given that $f(x)=x^2$ and $g(x)=x^2-2$.

25. For $f(x)=x-5$ and $g(x)=\sqrt{x+3}$, find $\dfrac{f(x)}{g(x)}$.

26. For $f(x)=2x-8$ and $g(x)=\sqrt{2-x}$, find $f(x)\cdot g(x)$.

27. If $f(x)=-\sqrt{x-3}$ and $g(x)=3x$, find $(f\cdot g)(x)$.

28. If $f(x)=-\sqrt{4-x}$ and $g(x)=5-x$, find $(g-f)(x)$.

29. If $f(x)=\sqrt[3]{x+3}$ and $g(x)=\sqrt{5+x}$, find $f(x)+g(x)$.

30. If $f(x)=\sqrt{x-1}$ and $g(x)=\sqrt[3]{2x+1}$, find $f(x)-g(x)$.

*For the following pairs of functions, graph **a.** the sum $(f+g)$ and **b.** the difference $(f-g)$ on two different graphs.*

31. $f=\{(-2,5),(0,-3),(2,1)\}$
$g=\{(-2,-2),(0,-1),(2,-3)\}$

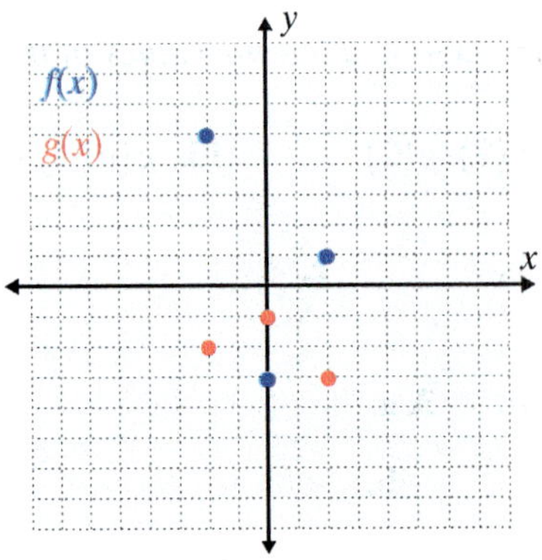

32. $f=\{(0,2),(1,1),(2,-1)\}$
$g=\{(0,4),(1,2),(2,-2)\}$

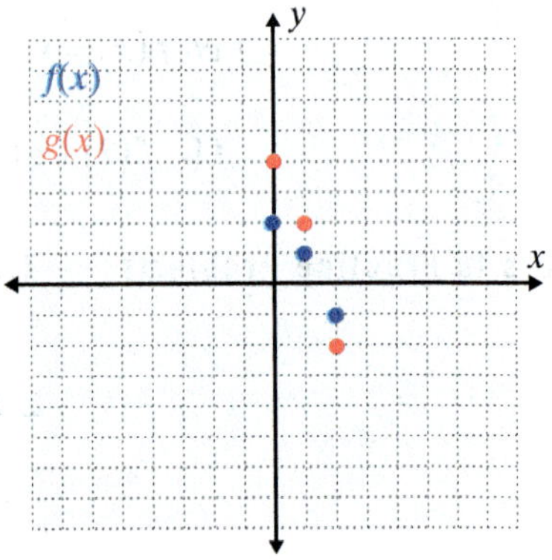

34. a.

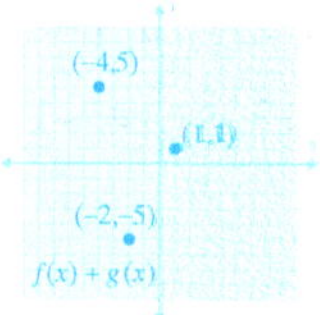

b.

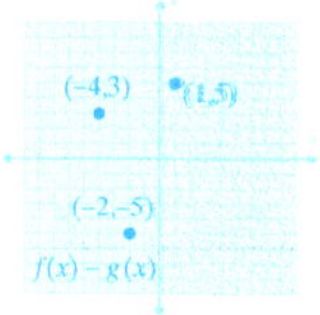

35. a.

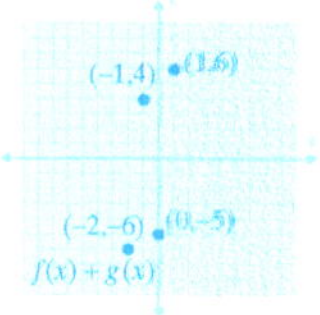

b.

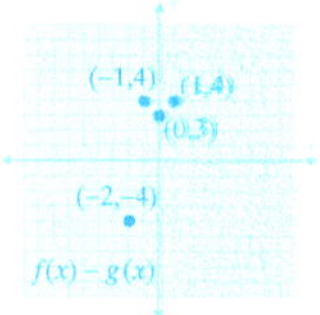

36. a.

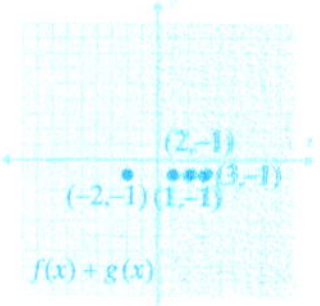

b.

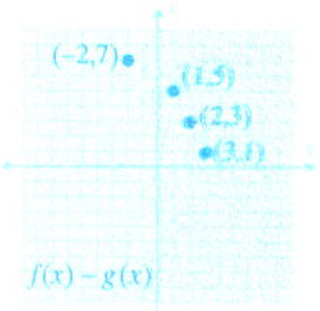

37. a.

b.

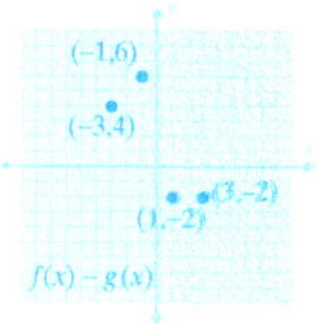

33. $f = \{(-3,-1),(-1,0),(2,-4)\}$
$g = \{(-3,-3),(-1,5),(2,1)\}$

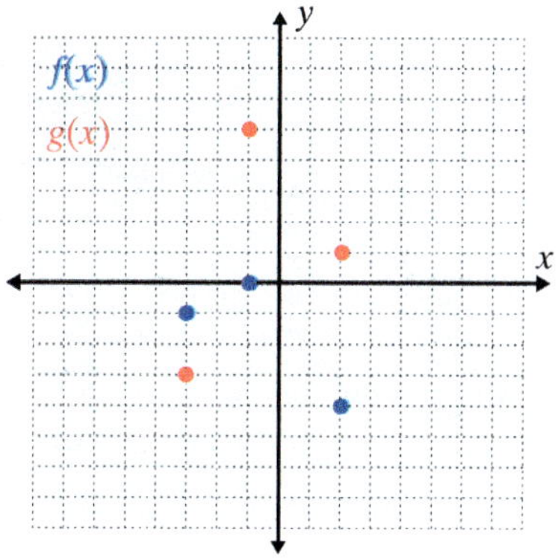

34. $f = \{(-4,4),(-2,-5),(1,3)\}$
$g = \{(-4,1),(-2,0),(1,-2)\}$

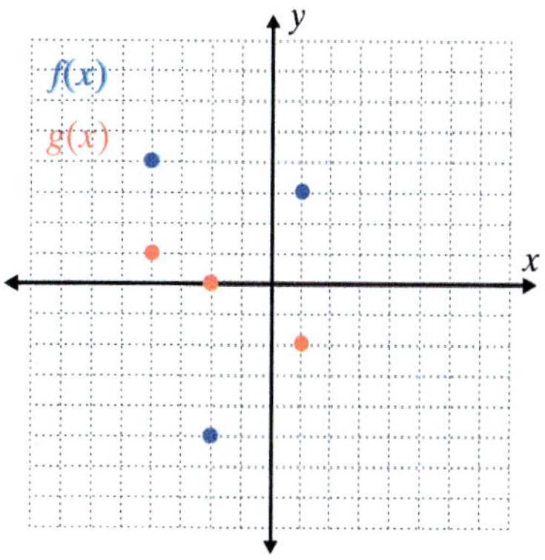

35. $f = \{(-2,-5),(-1,4),(0,-1),(1,5)\}$
$g = \{(-2,-1),(-1,0),(0,-4),(1,1)\}$

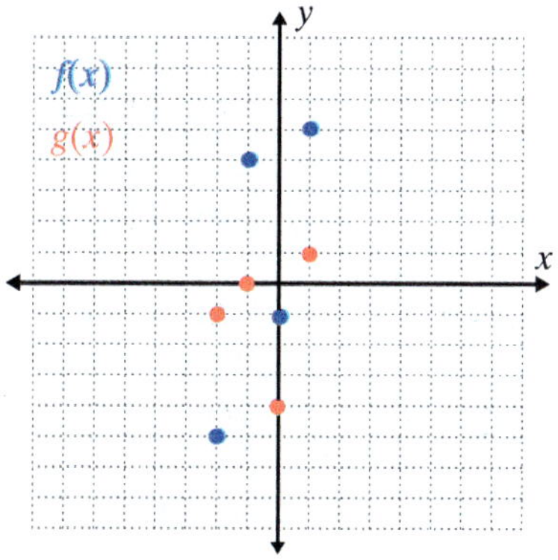

36. $f = \{(-2,3),(1,2),(2,1),(3,0)\}$
$g = \{(-2,-4),(1,-3),(2,-2),(3,-1)\}$

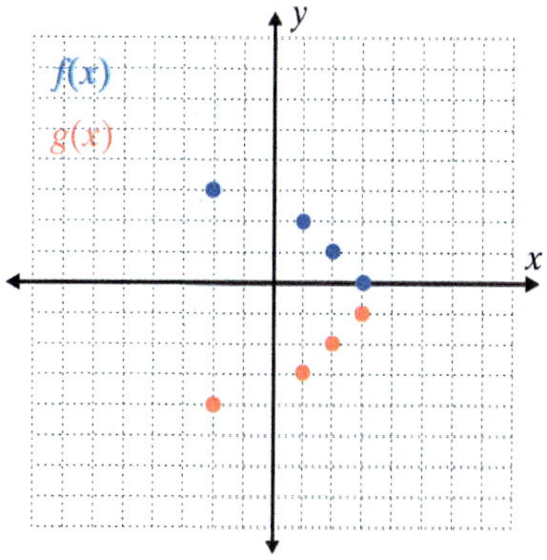

37. $f = \{(-3,1),(-1,2),(1,1),(3,2)\}$
$g = \{(-3,-3),(-1,-4),(1,3),(3,4)\}$

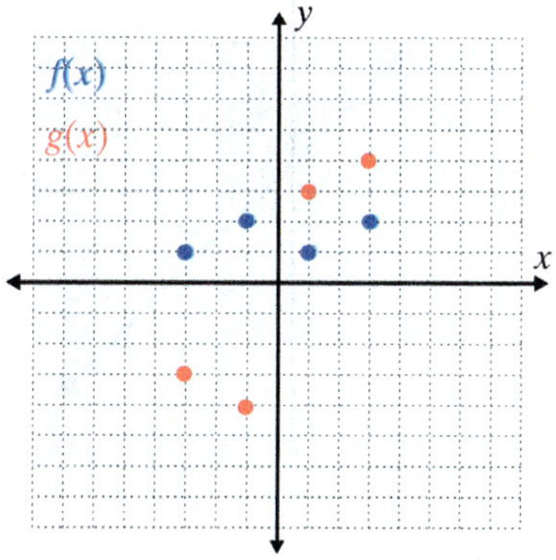

38. $f = \{(-4,6),(-2,0),(0,-2),(3,-3)\}$
$g = \{(-4,0),(-2,-4),(0,1),(3,3)\}$

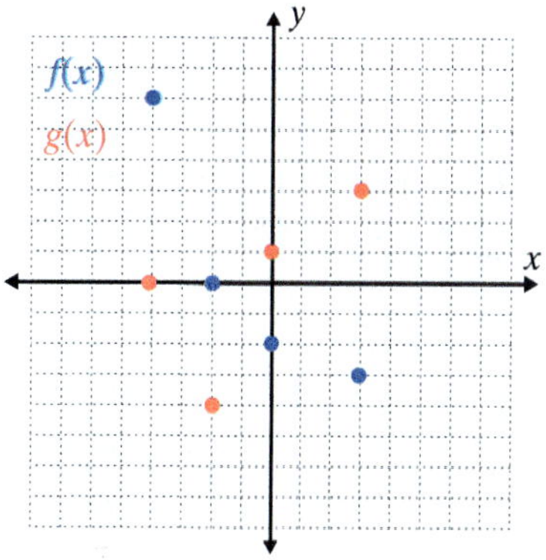

38. a.

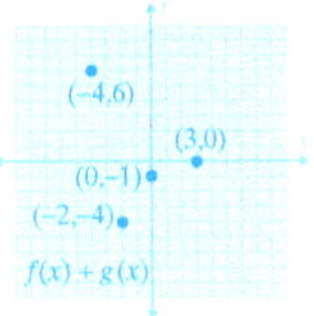

b.

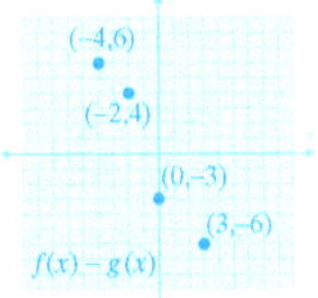

39. a.

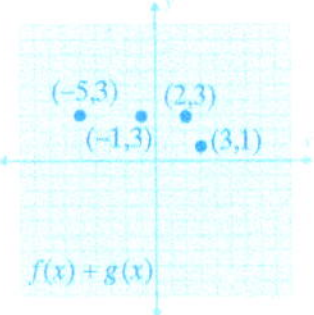

b.

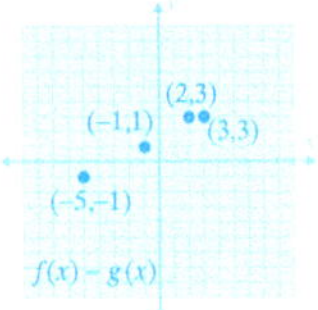

40. a.

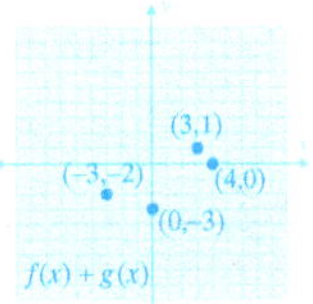

b.

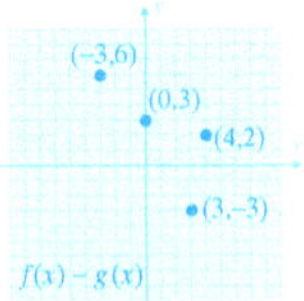

41.

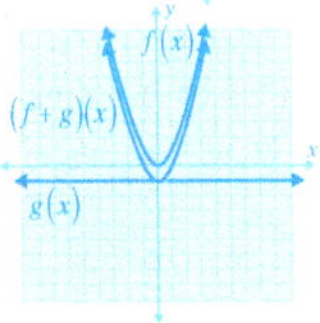

42.

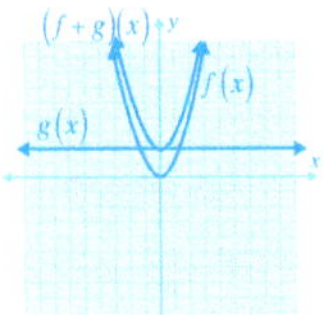

43.

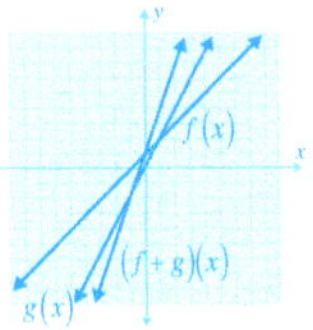

39. $f = \{(-5, 1), (-1, 2), (2, 3), (3, 2)\}$
$g = \{(-5, 2), (-1, 1), (2, 0), (3, -1)\}$

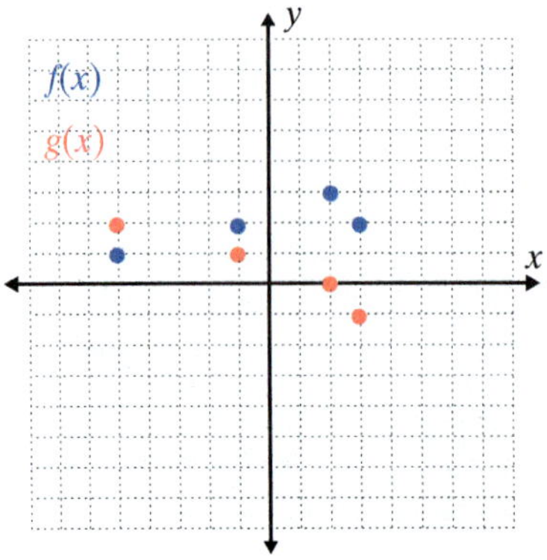

40. $f = \{(-3, 2), (0, 0), (3, -1), (4, 1)\}$
$g = \{(-3, -4), (0, -3), (3, 2), (4, -1)\}$

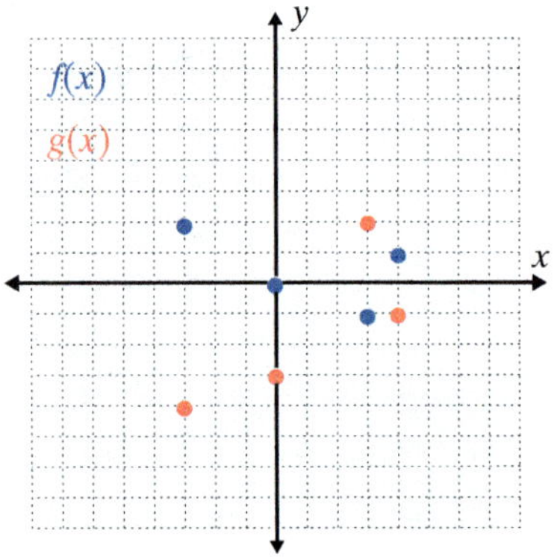

Graph each pair of functions and the sum of these functions on the same set of axes.

41. $f(x) = x^2$ and $g(x) = -1$

42. $f(x) = x^2$ and $g(x) = 2$

43. $f(x) = x + 1$ and $g(x) = 2x$

44. $f(x) = x + 5$ and $g(x) = x - 5$

45. $f(x) = x + 4$ and $g(x) = -x$

46. $f(x) = 2 - x$ and $g(x) = x$

47. $f(x) = x + 1$ and $g(x) = x^2 - 1$

48. $f(x) = x^2 + 2$ and $g(x) = x^2 - 2$

49. $f(x) = \sqrt{x - 6}$ and $g(x) = 2$

50. $f(x) = \sqrt{3 - x}$ and $g(x) = -1$

Use the graph shown here to find the values indicated.

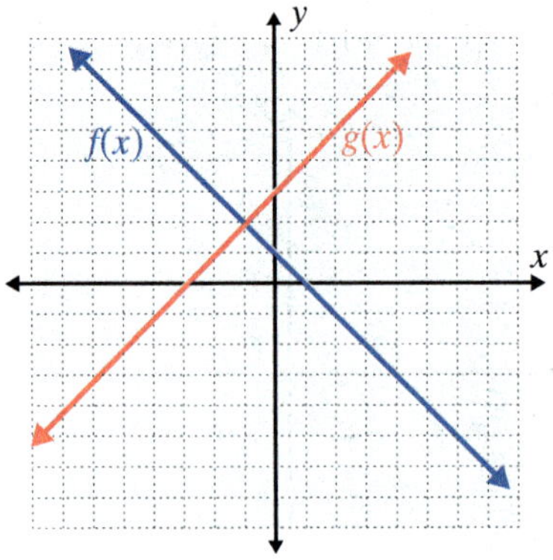

51. $(f + g)(-2)$

52. $(f - g)(2)$

53. $(f \cdot g)(3)$

54. $(g - f)(0)$

55. $\left(\frac{f}{g}\right)(4)$

56. $(g \cdot f)(4)$

The graphs of two functions are given. Graph the sum of these two functions.

57.

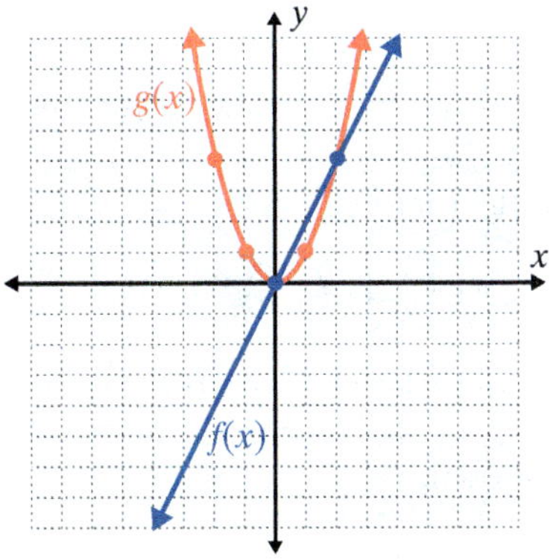

58.

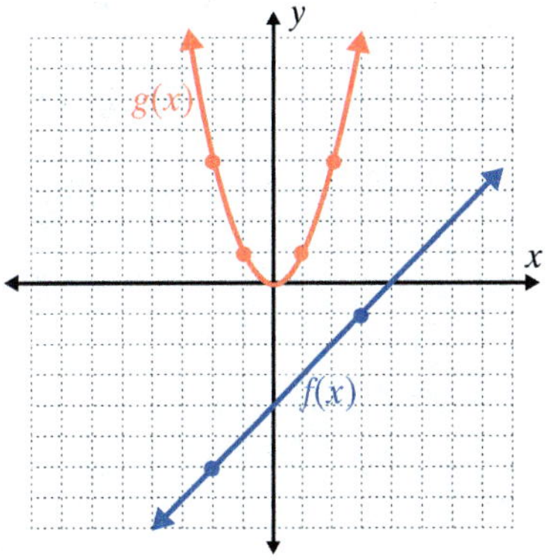

59.

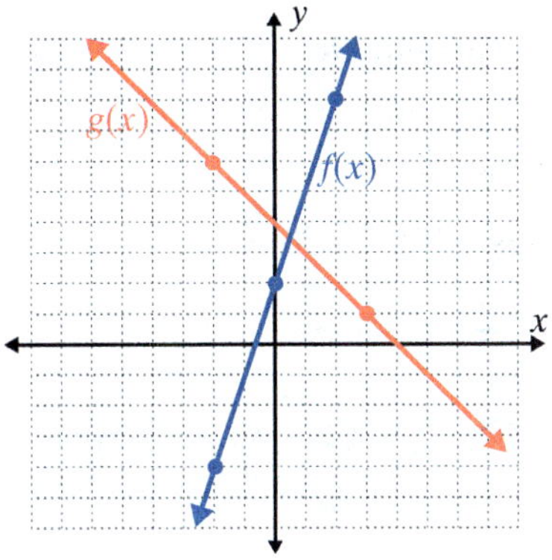

60.

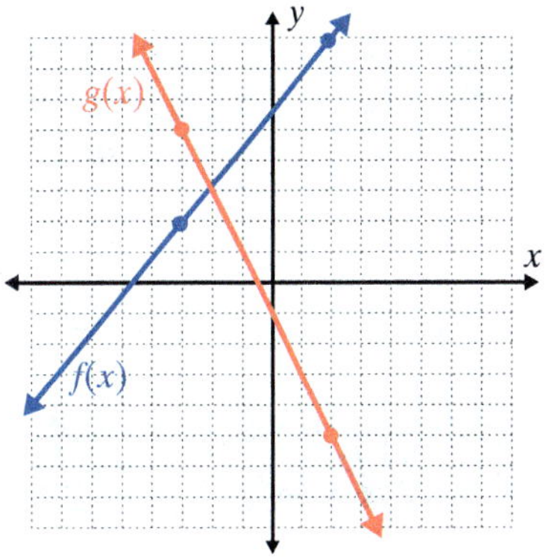

61.

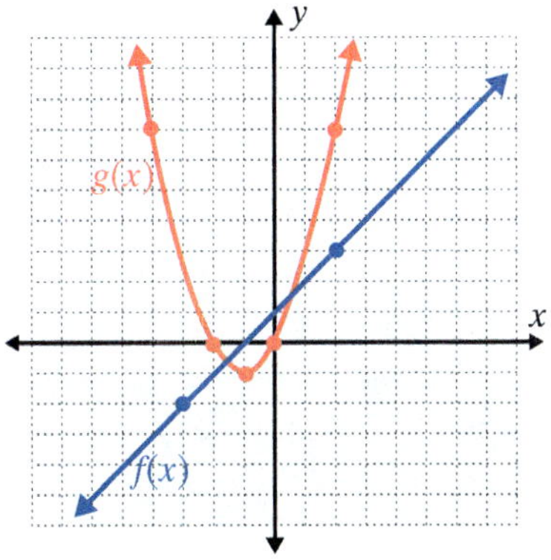

62.

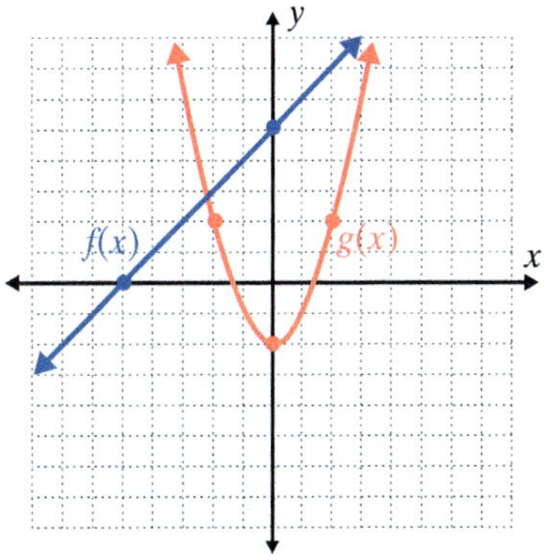

Use a graphing calculator to graph each pair of functions and the sum of these functions on the same set of axes.

63. $f(x)=x^2$ and $h(x)=2x+1$

64. $g(x)=x^2+x$ and $h(x)=3x+4$

65. $f(x)=\sqrt{x+4}$ and $g(x)=-2$

66. $f(x)=-\sqrt{x-1}$ and $g(x)=3$

67. $f(x)=\sqrt[3]{x+5}$ and $h(x)=2x$

68. $h(x)=\sqrt[3]{x-1}$ and $g(x)=x-1$

69. $g(x)=7-x^2$ and $h(x)=x^2-3$

70. $f(x)=x^2+5$ and $g(x)=4-x^2$

44.

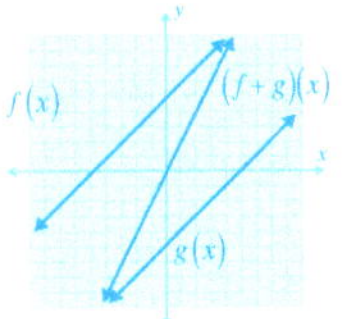

45.

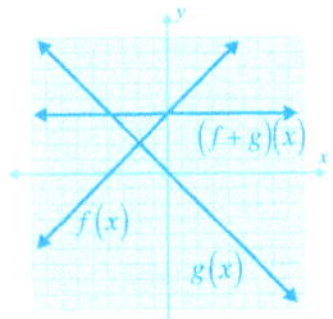

46.

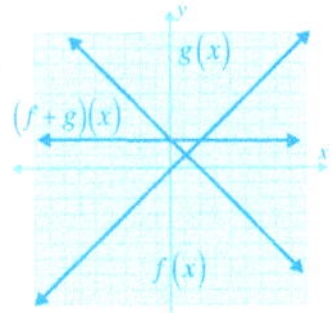

47.

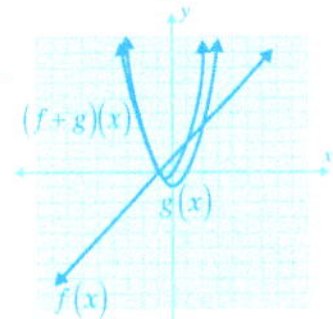

48.

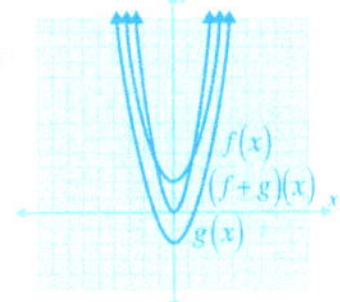

49.

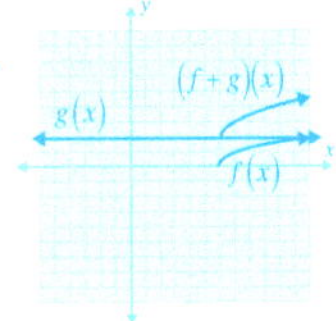

50.

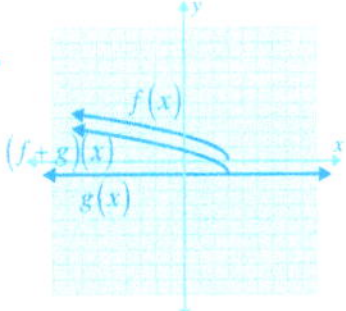

51. 4 52. −6

53. −12 54. 2

55. $-\frac{3}{7}$ 56. −21

57.

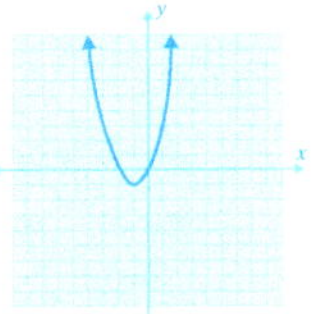

58.

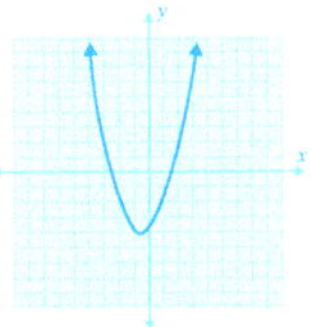

Writing and Thinking About Mathematics

59. 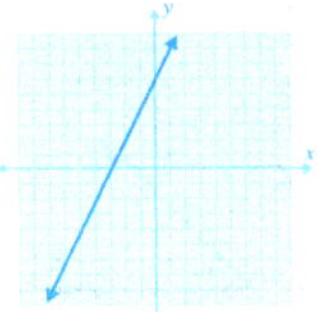

60. 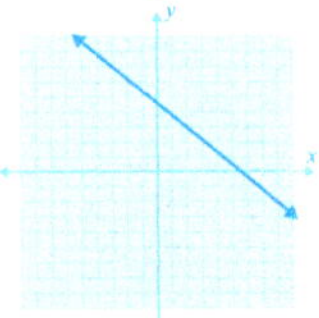

61. 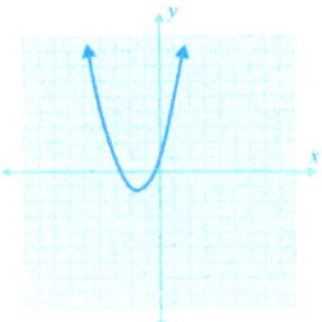

71. Explain why, in general, $(f-g)(x) \neq (g-f)(x)$ if $f(x) \neq g(x)$.

72. Given the two functions f and g,

$$f = \{(-2, 0), (-1, 1), (0, 4), (2, 4), (3, 5), (4, 1)\}$$

$$g = \{(-2, 3), (-1, 4), (0, 1), (2, -1), (3, 2), (4, 6)\}$$

find and graph the following.

a. $f-g$ **b.** $f \cdot g$ **c.** $\dfrac{f}{g}$

73. Use the graphs of the two functions f and g shown in Figure 3 on page 861.

a. Sketch the graph of $f-g$.

b. Sketch the graph of $f \cdot g$.

c. Is $\dfrac{f}{g}$ defined on the entire interval $[-3, 3]$? Briefly discuss your reasoning.

HAWKES LEARNING SYSTEMS: INTRODUCTORY & INTERMEDIATE ALGEBRA SOFTWARE

- 11.1 Algebra of Functions

62. 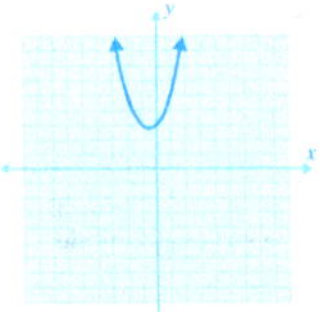

63. 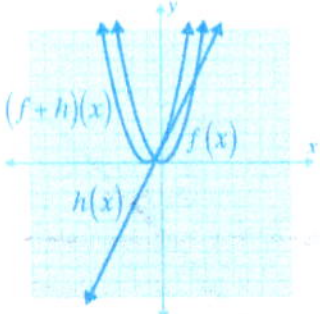

64. 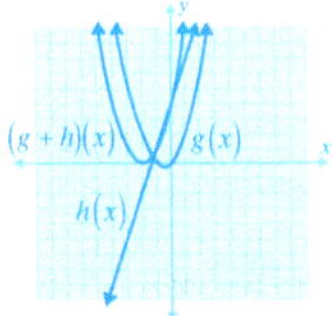

65. 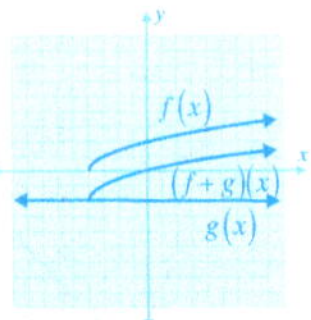

66. 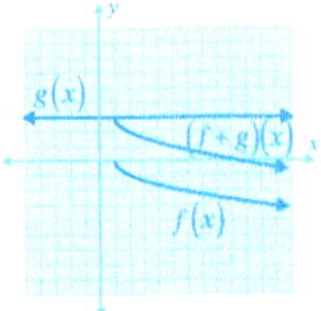

67. 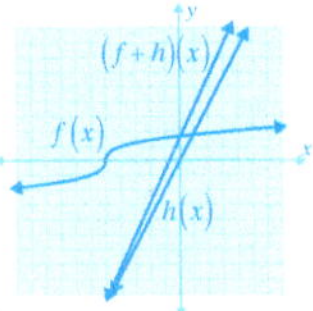

68. 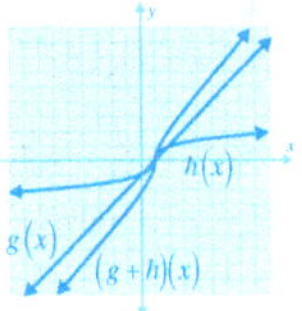

69. 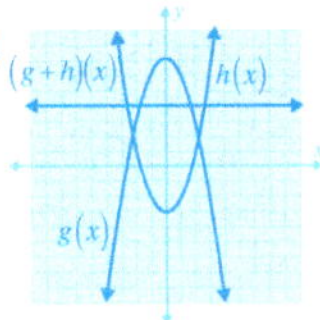

70. 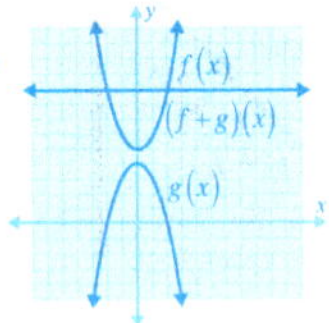

71. In general, subtraction is not commutative. Answers will vary.

72. a. 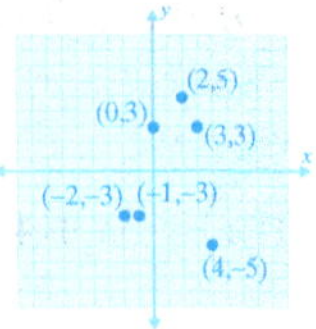

b. 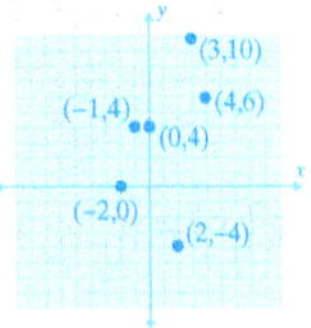

c. 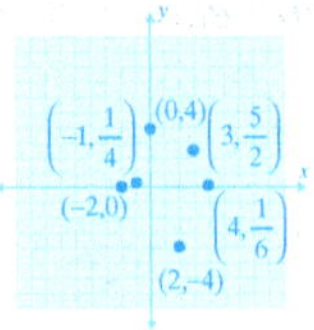

73. a. 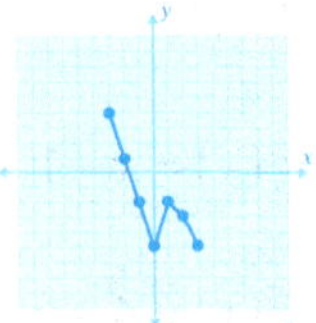

b. 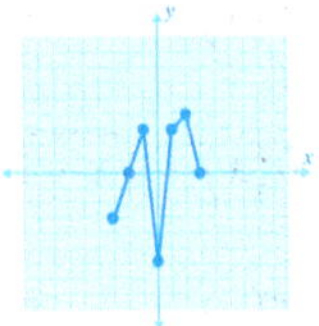

c. $\dfrac{f}{g}$ is undefined at $x=-2$ as $g(-2)=0$.

11.2 Composition of Functions and Inverse Functions

- *Form the* ***composition*** *of two functions.*
- *Determine if a function is one-to-one using the horizontal line test.*
- *Show that two functions are* ***inverses*** *by verifying that* $f(g(x)) = x$ *and* $g(f(x)) = x$.
- *Find the* ***inverse*** *of a one-to-one function.*
- *Graph the inverses of functions by* ***reflecting*** *the graphs of the functions across the line* $y = x$.

Composition of Functions

Teaching Notes:
An overdue question to put to students is whether the sum, difference, product, and quotient of two functions is always a function. You also might want to raise the same question about composition. Is the composition of two functions always a function?

Previously, function notation, $f(x)$, has proven useful in evaluating functions and in indicating arithmetic operations with functions. This same notation is needed in developing the concept of **a function of a function**, called the **composition** of two functions. For example, suppose that

$$g(x) = 2x - 4 \qquad \text{and} \qquad f(x) = x^2 - 4x + 1.$$

Now for $x = 3$, we have

$$g(3) = 2\cdot(3) - 4 = 2 \qquad \text{and} \qquad f(2) = (2)^2 - 4\cdot(2) + 1 = -3.$$

For the **composition** $f(g(3))$ (read "*f* of *g* of 3"), we have

$$f(g(3)) = f(2) = -3.$$

We see that $g(3)$, which equals 2, replaces x in the function $f(x)$.

Teaching Notes:
A diagram can be used to illustrate that composition is a sort of shortcut:

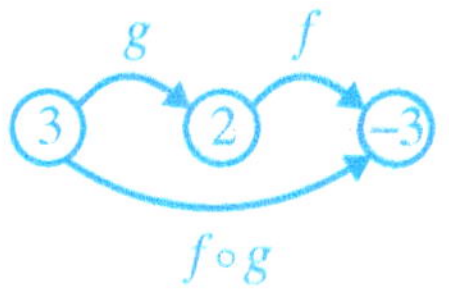

$f \circ g$ takes 3 directly to −3.

More generally, given two functions $f(x)$ and $g(x)$, a new function $f(g(x))$ called the **composition** (or **composite**) of *f* and *g*, is found by substituting the expression for $g(x)$ in place of *x* in the function *f*. In this case the value of $g(x)$ must be in the domain of *f*. Thus for

$$g(x) = 2x - 4 \qquad \text{and} \qquad f(x) = x^2 - 4x + 1,$$

the composition

$$f(g(x)) \qquad \text{(read "}f\text{ of }g\text{ of }x\text{")}$$

is found as follows:

$$f(x) = x^2 - 4x + 1$$

$$f(g(x)) = (g(x))^2 - 4(g(x)) + 1 \qquad \text{Replace the } x \text{ in } f(x) \text{ with } g(x).$$

$$= (2x-4)^2 - 4(2x-4) + 1 \qquad \text{Replace } g(x) \text{ with } 2x-4.$$

$$= 4x^2 - 16x + 16 - 8x + 16 + 1 \qquad \text{Simplify.}$$

$$= 4x^2 - 24x + 33.$$

The composition of *g* and *f* (reversing the order of *f* and *g*) is indicated by

$$g(f(x)) \qquad \text{read} \qquad \text{“}g \text{ of } f \text{ of } x\text{”}$$

and is found by substituting the expression for $f(x)$ in place of *x* in the function *g*. Thus

$$g(x) = 2x - 4$$

$$g(f(x)) = 2(f(x)) - 4 \qquad \text{Replace the } x \text{ in } g(x) \text{ with } f(x).$$

$$= 2(x^2 - 4x + 1) - 4 \qquad \text{Replace } f(x) \text{ with } x^2 - 4x + 1.$$

$$= 2x^2 - 8x + 2 - 4 \qquad \text{Simplify.}$$

$$= 2x^2 - 8x - 2.$$

As we can see with these examples, in general, $f(g(x)) \neq g(f(x))$ and substitutions must be done carefully and accurately. The following definition shows another notation (a small raised circle) often used to indicate the composition of functions.

Composite Function

For two functions *f* and *g*, the **composite function** $f \circ g$ is defined as follows:

$$(f \circ g)(x) = f(g(x)).$$

Domain of $f \circ g$: The domain of $f \circ g$ consists of those values of *x* in the domain of *g* for which $g(x)$ is in the domain of *f*.

Example 1: Compositions

a. Form the compositions $(f \circ g)(x)$ and $(g \circ f)(x)$ if $f(x) = 5x + 2$ and $g(x) = 3x - 7$.

Solution: $(f \circ g)(x) = f(g(x)) = 5 \cdot g(x) + 2 = 5(3x - 7) + 2 = 15x - 33$

$(g \circ f)(x) = g(f(x)) = 3 \cdot f(x) - 7 = 3(5x + 2) - 7 = 15x - 1$

Note: Both $(f \circ g)(x)$ and $(g \circ f)(x)$ are defined for all real numbers.

b. Form the composite functions $(f \circ g)(x)$ and $(g \circ f)(x)$ if $f(x)=\sqrt{x-3}$ and $g(x)=x^2+4$.

Solution: $(f \circ g)(x)=\sqrt{g(x)-3}$

$$=\sqrt{(x^2+4)-3}=\sqrt{x^2+1}$$

Note: For the expression under the radical to be defined, we must have $g(x)\geq 3$. Because $x^2+4\geq 3$ for all real numbers, the domain of $(f \circ g)(x)$ is all real numbers.

$$(g \circ f)(x)=(f(x))^2+4$$

$$=(\sqrt{x-3})^2+4=x-3+4=x+1$$

Note: $(g \circ f)(x)$ is defined only for $x\geq 3$. (Here the domain comes from f before the simplification.)

c. Find $f(g(x))$ and $g(f(x))$ if $f(x)=\sqrt{x+3}$ and $g(x)=2x-5$.

Solution: $f(g(x))=\sqrt{g(x)+3}$

$$=\sqrt{(2x-5)+3}=\sqrt{2x-2}$$

Note: $f(g(x))$ is defined only for $2x-2\geq 0$ or $x\geq 1$.

$$g(f(x))=2(f(x))-5$$

$$=2(\sqrt{x+3})-5=2\sqrt{x+3}-5$$

Note: $g(f(x))$ is defined only for $x\geq -3$.

One-to-One Functions

By the definition of a function, there is only one corresponding y-value for each x-value in a function's domain. Graphically, the **vertical line test** can be used to help determine whether or not a graph represents a function. Now in order to develop the concept of **inverse functions**, we need to study functions that have only one x-value for each y-value in the range. Such functions are said to be **one-to-one functions** (or **1-1 functions**).

Consider the following **functions**:

$$f=\{(1,2),(2,4),(3,6),(4,8),(5,10)\}$$

$$g=\{(-2,6),(0,6),(1,5),(2,4),(4,1)\}.$$

Both sets of ordered pairs are functions because each value of x appears only once. In the function f, each y-value appears only once. But, in function g, the y-value 6 appears twice: in $(-2, 6)$ and $(0, 6)$. Figure 1 illustrates both functions.

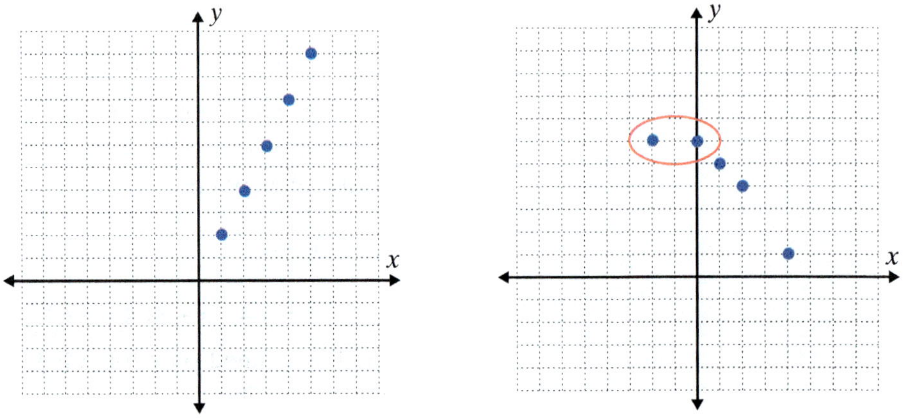

a. Function f is a one-to-one function. **b.** Function g is not a one-to-one function.

Figure 1

One-to-One Functions

A function is a **one-to-one function** (or **1-1 function**) if for each value of y in the range there is only one corresponding value of x in the domain.

Graphically, as illustrated in Figure 1a, if a horizontal line intersects the graph of a function in more than one point then it is **not** one-to-one. This is, in effect, the **horizontal line test**.

Horizontal Line Test

A function is one-to-one if no **horizontal line** intersects the graph of the function at more than one point.

The graphs in Figure 2 (on the next page) illustrate the concept of one-to-one functions. (Note that each graph passes the vertical line test and is indeed a function.)

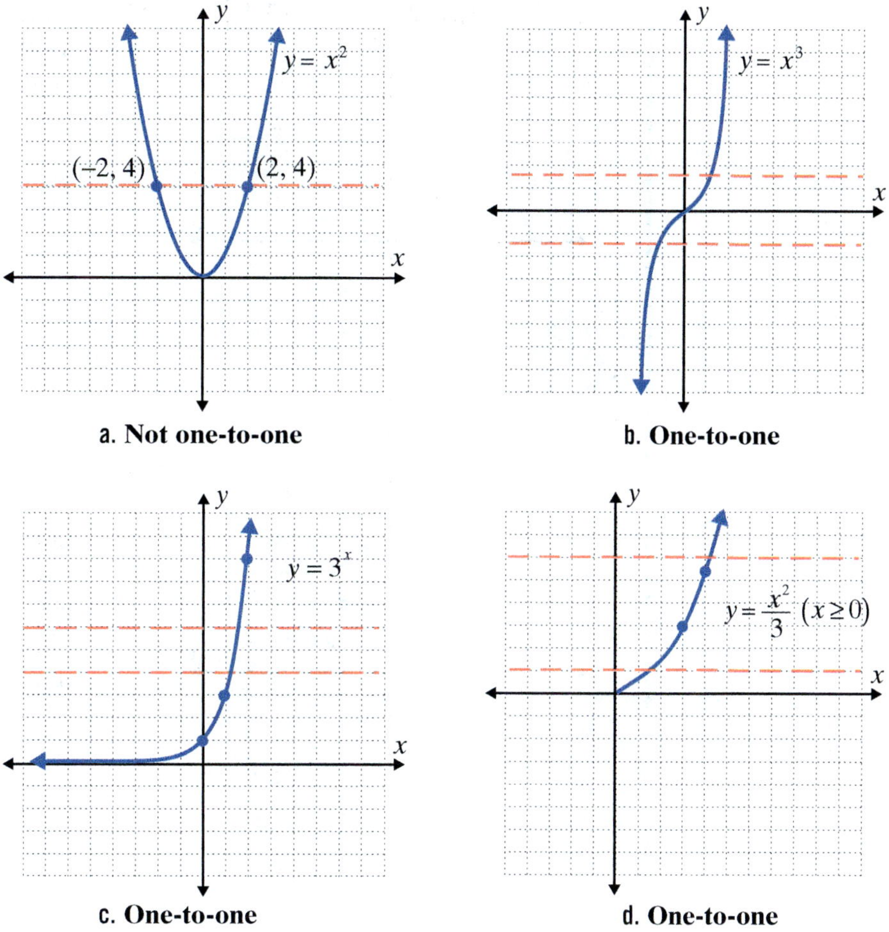

Figure 2

Example 2: One-to-One Functions

Determine whether each function **is** or **is not** one-to-one.

a. $y = \sqrt{x+5}$

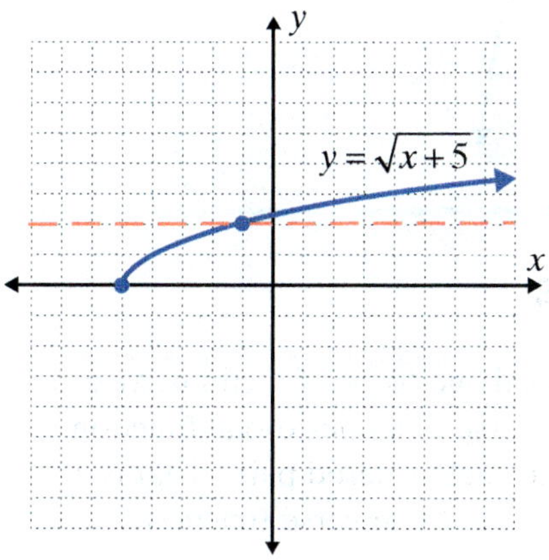

Solution: The horizontal line test shows that this function is one-to-one.

Continued on the next page...

b. $f = \{(-3, 4), (-2, 1), (0, 4), (3, 1)\}$

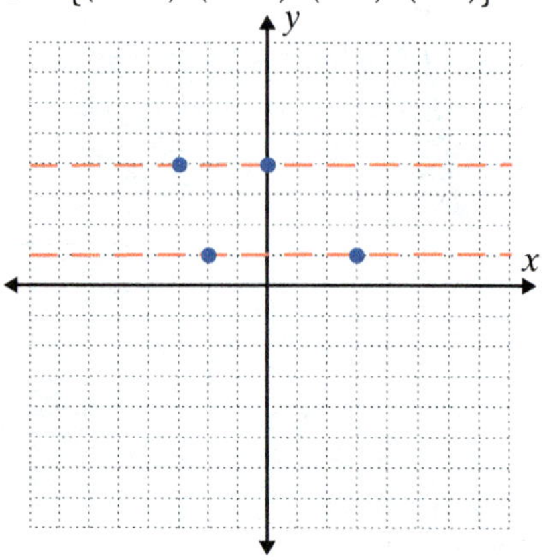

Solution: This function is not one-to-one. Both y-values, 4 and 1, have more than one corresponding x-value.

c. $y = 2x - 1$

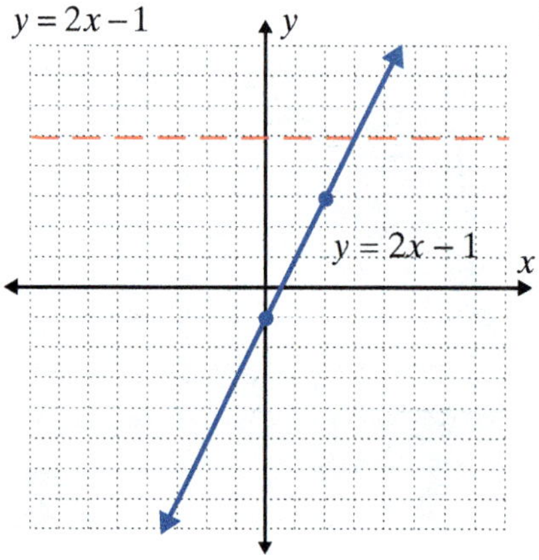

Solution: The graph of the function $y = 2x - 1$ is a straight line. Straight lines that are not vertical and not horizontal represent one-to-one functions. (Vertical lines are not functions in the first place and horizontal lines fail the horizontal line test.)

d. $y = -x^2 + 1$

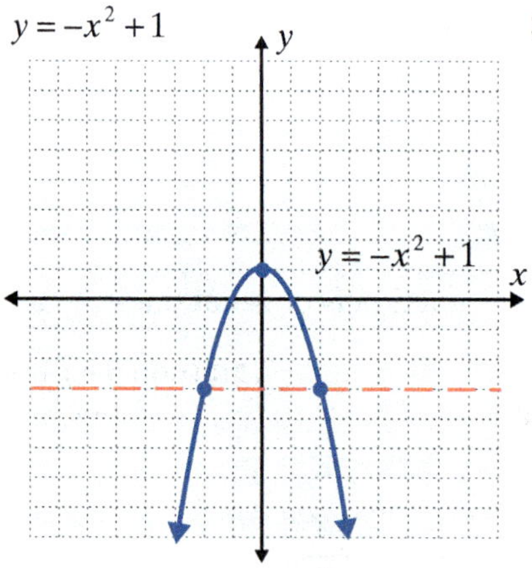

Solution: The graph of the function $y = -x^2 + 1$ is a parabola and the horizontal line test shows that the function is not one-to-one.

Inverse Functions

Now that we have discussed one-to-one functions, we can develop the concept of **inverse functions**. To find the inverse of a one-to-one function represented by a set of ordered pairs, exchange x and y in each ordered pair. That is, if (x, y) is in the original one-to-one function, then (y, x) is in the inverse function.

For example,

$$\text{if } f = \{(-1, 1), (0, 2), (1, 4)\},$$

then interchanging the coordinates in each ordered pair gives

$$g = \{(1, -1), (2, 0), (4, 1)\}.$$

The functions f and g are called **inverses** of each other. If g is the inverse of f we write

$$f^{-1} \text{ (read "}f\text{ inverse") rather than use } g.$$

Thus in this example we can write $f^{-1} = \{(1, -1), (2, 0), (4, 1)\}$.

Inverse Functions

If f is a one-to-one function with ordered pairs of the form (x, y), then its **inverse function**, denoted as f^{-1}, is also a one-to-one function with ordered pairs of the form (y, x).

Only one-to-one functions have inverse functions. If a function is not one-to-one, then interchanging the x- and y-values would yield a relation that is not a function. Graphically, the function must satisfy the horizontal line test. If it did not, then the inverse would not pass the vertical line test and would not be a function.

NOTES

The notation $f^{-1}(x)$ represents the inverse of a one-to-one function. This inverse is a new function in which the x- and y-values have been interchanged. $f^{-1}(x)$ does NOT mean $\dfrac{1}{f(x)}$ because the -1 is NOT an exponent.

The graph of any point (b, a) is the reflection of the point (a, b) across the line $y = x$. Thus the points of the inverse function f^{-1} are reflections of the points of the function f across the line $y = x$. We say that the graphs are **symmetric about the line $y = x$**. Figure 3 illustrates these reflections and the symmetry. **We see that the domain, denoted D_f, and range, denoted R_f, of the two functions are interchanged.**

That is,

$$D_f = R_{f^{-1}} = \{-1, 0, 1\}$$

$$R_f = D_{f^{-1}} = \{1, 2, 4\}.$$

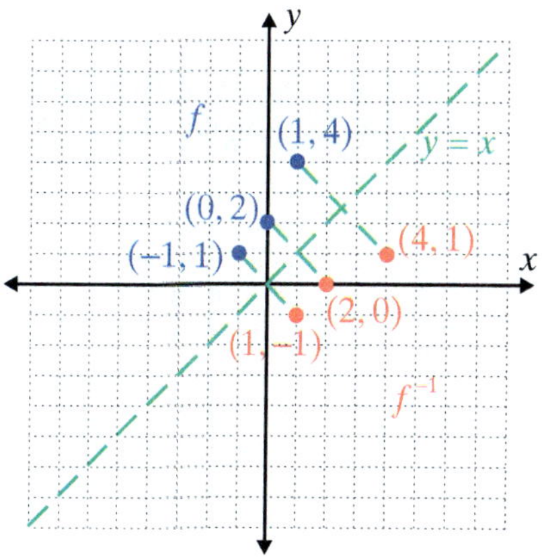

Figure 3

In general, every one-to-one function has an inverse function (or just inverse), and the graph of the inverse function of any one-to-one function f can be found by reflecting the graph of f across the line $y = x$. Figure 4 shows two more illustrations of this concept.

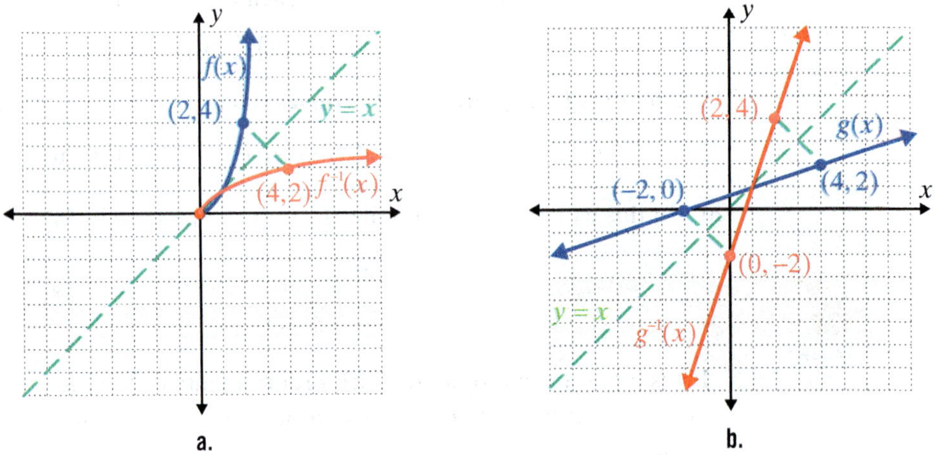

Figure 4

The following definition of inverse functions helps to determine whether or not two functions are inverses of each other.

To Determine whether Two Functions are Inverses

If f and g are one-to-one functions and

$$f(g(x)) = x \quad \text{for all } x \text{ in } D_g, \text{ and}$$

$$g(f(x)) = x \quad \text{for all } x \text{ in } D_f$$

then f and g are **inverse functions**.

That is, $g = f^{-1}$ and $f = g^{-1}$.

Example 3: Inverse Functions

Use the definition of inverse functions to show that f and g are inverse functions.

a. $f(x) = 2x + 6$ and $g(x) = \dfrac{x-6}{2}$.

Solution: $f(x) = 2x + 6$ and $g(x) = \dfrac{x-6}{2}$. The domain of both functions is the set of all real numbers.

We have

$$f(g(x)) = 2 \cdot g(x) + 6$$ Replace the x in $f(x)$ with $g(x)$ and simplify.

$$= 2\left(\frac{x-6}{2}\right) + 6$$

$$= (x-6) + 6$$

$$= x.$$

Also,

$$g(f(x)) = \frac{f(x)-6}{2}$$ Replace the x in $g(x)$ with $f(x)$ and simplify.

$$= \frac{(2x+6)-6}{2}$$

$$= \frac{2x}{2}$$

$$= x.$$

Teaching Notes:
Have students do paper folds along $y = x$ to manually reflect the inverse graph. Onion-skin paper is good for this. Transparent acetates are especially ideal for showing the reflection by tracing the original function and the axes and line $y = x$, and then lifting the image, rotating and repositioning the axes and line $y = x$. The x-axis becomes the y-axis, etc. The inverse graph is now in position.

Therefore, $g = f^{-1}$ and $f = g^{-1}$.

The graph shows that the line $y = x$ is a line of symmetry for the graphs of the inverse functions.

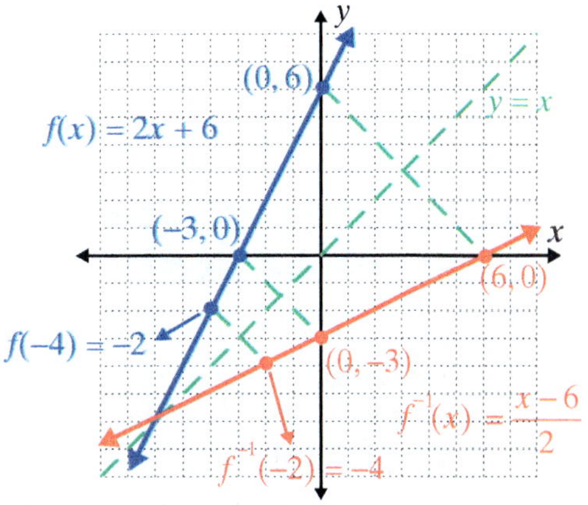

b. $f(x) = \sqrt{x-3}$ and $g(x) = x^2 + 3$ for $x \geq 0$.

Solution: $f(x) = \sqrt{x-3}$ and $g(x) = x^2 + 3$ for $x \geq 0$. The domain of f is the interval $[3, \infty)$ and the domain of g is the interval $[0, \infty)$.

We have

$$f(g(x)) = \sqrt{g(x)-3}$$ Replace the x in $f(x)$ with $g(x)$ and simplify.

$$= \sqrt{(x^2+3)-3}$$

$$= \sqrt{x^2}$$

$$= x \quad \text{for } x \geq 0.$$

Continued on the next page...

Also,

$$\begin{aligned} g(f(x)) &= (f(x))^2 + 3 \\ &= \left(\sqrt{x-3}\right)^2 + 3 \\ &= x - 3 + 3 \\ &= x \qquad \text{for } x \ge 3. \end{aligned}$$

Replace the x in $g(x)$ with $f(x)$ and simplify.

Therefore, $g = f^{-1}$ and $f = g^{-1}$.

The graph shows that the line $y = x$ is a line of symmetry for the graphs of the inverse functions. Note that these graphs are only parts of parabolas and the domains have been restricted accordingly.

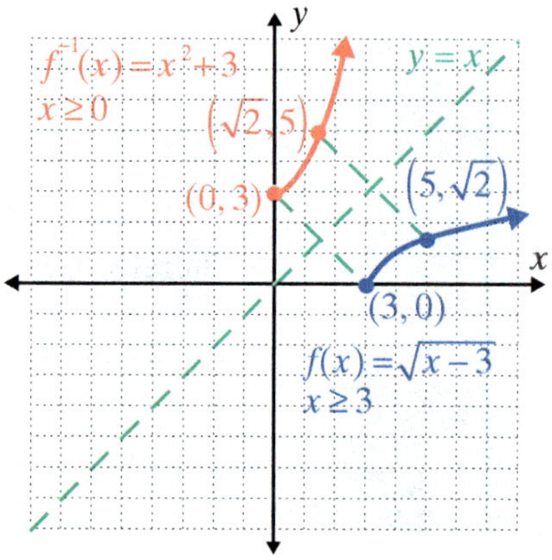

Example 4: Evaluating Compositions of Inverses

Given $f(x) = \sqrt{x+4}$ for $x \ge -4$ and $f^{-1}(x) = x^2 - 4$ for $x \ge 0$, evaluate the indicated compositions.

a. $f(f^{-1}(2))$

Solution: $f^{-1}(2) = (2)^2 - 4 = 0$ so $f(f^{-1}(2)) = f(0) = \sqrt{(0)+4} = \sqrt{4} = 2$

b. $f(f^{-1}(5))$

Solution: $f^{-1}(5) = (5)^2 - 4 = 21$ so $f(f^{-1}(5)) = f(21) = \sqrt{(21)+4} = \sqrt{25} = 5$

c. $f^{-1}(f(-2))$

Solution: $f(-2) = \sqrt{(-2)+4} = \sqrt{2}$ so $f^{-1}(f(-2)) = f^{-1}(\sqrt{2}) = (\sqrt{2})^2 - 4 = 2 - 4 = -2$

d. $f^{-1}(f(-5))$

Solution: $f(-5)$ does not exist because -5 is not in the domain of f. Therefore $f^{-1}(f(-5))$ does not exist.

Finding the Inverse of a One-to-One Function

In Examples 3 and 4 the given functions were inverses of each other. The next question is how to find the inverse of a given one-to-one function. The following procedure shows one method for finding the inverse using the fact that if an ordered pair (x, y) belongs to the function f, then (y, x) belongs to f^{-1}.

To Find the Inverse of a One-to-One Function

1. Let $y = f(x)$. (In effect, substitute y for $f(x)$.)
2. Interchange x and y.
3. In the new equation, solve for y in terms of x.
4. Substitute $f^{-1}(x)$ for y. (This new function is the inverse of f.)

Example 5: Finding the Inverse

a. Find $f^{-1}(x)$ if $f(x) = 5x - 7$.

Solution:

$f(x) = 5x - 7$

$y = 5x - 7$ — Substitute y for $f(x)$.

$x = 5y - 7$ — Interchange x and y.

$x + 7 = 5y$ — Solve for y in terms of x.

$\frac{x+7}{5} = y$

$f^{-1}(x) = \frac{x+7}{5}$ — Substitute $f^{-1}(x)$ for y.

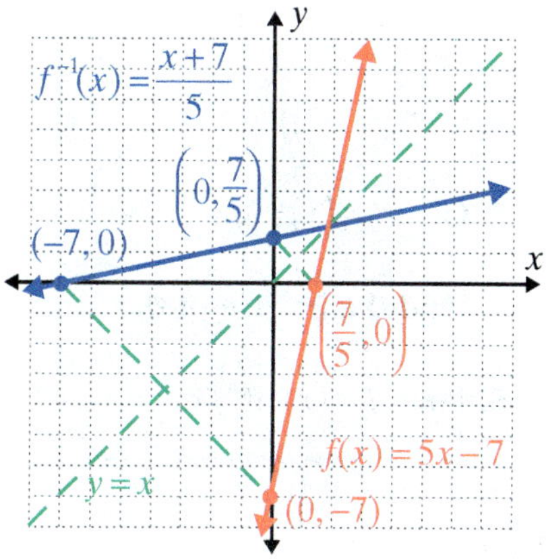

Continued on the next page...

1. a. $f(g(2)) = 14$
   b. $g(f(2)) = \frac{15}{2}$
2. a. $f(g(4)) = \frac{21}{4}$
   b. $g(f(4)) = 5$
3. a. $(f \circ g)(-5) = 49$
   b. $(g \circ f)(-1) = 5$
4. a. $(f \circ g)(3) = 10$
   b. $(g \circ f)(-2) = -1$
5. $f(g(x)) = \sqrt{x^2} = x$
   $g(f(x)) = (\sqrt{x})^2 = x$
6. $f(g(x)) = x$
   $g(f(x)) = x$
7. $f(g(x)) = \sqrt{x-2}$
   $g(f(x)) = \sqrt{x} - 2$
8. $f(g(x)) = \sqrt{x^2 - 9}$
   $g(f(x)) = x - 9$

b. Find $g^{-1}(x)$ if $g(x) = x^2 - 2$ for $x \ge 0$.

Solution:

$g(x) = x^2 - 2$	For $x \ge 0$ (Note that g is one-to-one for $x \ge 0$.)
$y = x^2 - 2$	Substitute y for $g(x)$.
$x = y^2 - 2$	Interchange y and x.
$\pm\sqrt{x+2} = y$	Solve for y in terms of x.
$g^{-1}(x) = \sqrt{x+2}$	Take the positive square root because we must have $y \ge 0$. (The domain of g is $x \ge 0$, so the range of g^{-1} is $y \ge 0$.)

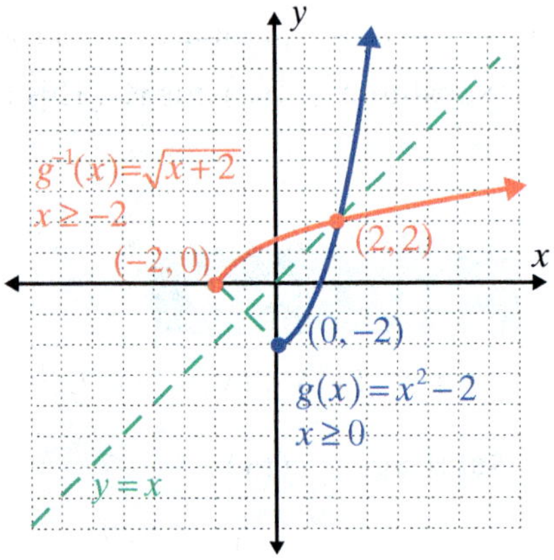

Practice Problems

9. $f(g(x)) = \frac{1}{x^2} - 1$
   $g(f(x)) = \frac{1}{(x-1)^2}$
10. $f(g(x)) = \frac{1}{(x^2+1)^2}$
    $g(f(x)) = \frac{1}{x^4} + 1$
11. $f(g(x)) = x^3 + 3x^2 + 4x + 3$
    $g(f(x)) = x^3 + x + 2$

1. For $f(x) = 2x - 1$ and $g(x) = x^2$, find:

a. $f(g(x))$ **b.** $g(f(x))$

2. Given $f(x) = 3x + 4$ and $f^{-1}(x) = \frac{x-4}{3}$, find the following:

a. $f(f^{-1}(2))$ **b.** $f^{-1}(f(1))$

3. Given $g(x) = \sqrt{x-2}$ and $f(x) = \frac{1}{x-1}$, find $f(g(3))$.

4. Find the inverse of the function $f(x) = 3x - 1$.

5. Find $f^{-1}(x)$ if $f(x) = x^2 - 4$, $x \ge 0$. (Note that f is one-to-one for $x \ge 0$.)

Answers to Practice Problems: **1. a.** $f(g(x)) = 2x^2 - 1$ **b.** $g(f(x)) = 4x^2 - 4x + 1$ **2. a.** 2 **b.** 1 **3.** Undefined **4.** $f^{-1}(x) = \frac{x+1}{3}$ **5.** $f^{-1}(x) = \sqrt{x+4}$, $x \ge -4$

11.2 Exercises

1. $f(x)=3x+5,\ g(x)=\dfrac{x+4}{2}$
Find **a.** $f(g(2))$ and **b.** $g(f(2))$.

2. $f(x)=\dfrac{1}{4}x+1,\ g(x)=6x-7$
Find **a.** $f(g(4))$ and **b.** $g(f(4))$.

3. $f(x)=x^2,\ g(x)=2x+3$
Find **a.** $(f\circ g)(-5)$ and **b.** $(g\circ f)(-1)$.

4. $f(x)=x^2+1,\ g(x)=x-6$
Find **a.** $(f\circ g)(3)$ and **b.** $(g\circ f)(-2)$.

Form the compositions $f(g(x))$ and $g(f(x))$ for each pair of functions.

5. $f(x)=\sqrt{x},\ g(x)=x^2$

6. $f(x)=\dfrac{1}{x},\ g(x)=\dfrac{1}{x}$

7. $f(x)=\sqrt{x},\ g(x)=x-2$

8. $f(x)=\sqrt{x},\ g(x)=x^2-9$

9. $f(x)=x-1,\ g(x)=\dfrac{1}{x^2}$

10. $f(x)=\dfrac{1}{x^2},\ g(x)=x^2+1$

11. $f(x)=x^3+x+1,\ g(x)=x+1$

12. $f(x)=x^3,\ g(x)=2x-1$

13. $f(x)=\dfrac{1}{\sqrt{x}},\ g(x)=x^2$

14. $f(x)=\dfrac{1}{\sqrt{x}},\ g(x)=x^2-4$

15. $f(x)=\dfrac{1}{x},\ g(x)=x^2+7x-8$

16. $f(x)=\dfrac{1}{x+1},\ g(x)=x^2+x-3$

17. $f(x)=x^{3n},\ g(x)=2x-6$

18. $f(x)=x^{\frac{1}{3}},\ g(x)=4x+7$

19. $f(x)=x^3,\ g(x)=\sqrt{x-8}$

20. $f(x)=x^3+1,\ g(x)=\dfrac{1}{x}$

21. For the functions $f(x)=6x-3$ and $g(x)=\dfrac{1}{3}x+3$, find:

a. $f(g(3))$ **b.** $g(f(0))$

c. Does it appear that *f* and *g* are inverses of each other? Explain.

22. For the functions $h(x)=-2x+4$ and $g(x)=\dfrac{4-x}{2}$, find:

a. $h(g(6))$ **b.** $g(h(-4))$

c. Does it appear that *h* and *g* are inverses of each other? Explain.

12. $f(g(x))=(2x-1)^3$
$g(f(x))=2x^3-1$
13. $f(g(x))=\dfrac{1}{x}$
$g(f(x))=\dfrac{1}{x}$
14. $f(g(x))=\dfrac{1}{\sqrt{x^2-4}}$
$g(f(x))=\dfrac{1}{x}-4$
15. $f(g(x))=\dfrac{1}{x^2+7x-8}$
$g(f(x))=\left(\dfrac{1}{x}\right)^2+7\left(\dfrac{1}{x}\right)-8$
16. $f(g(x))=\dfrac{1}{x^2+x-2}$
$g(f(x))=\dfrac{1}{(x+1)^2}+\dfrac{1}{x+1}-3$
17. $f(g(x))=(2x-6)^{3n}$
$g(f(x))=2x^{3n}-6$
18. $f(g(x))=(4x+7)^{\frac{1}{3}}$
$g(f(x))=4x^{\frac{1}{3}}+7$
19. $f(g(x))=(x-8)^{\frac{3}{2}}$
$g(f(x))=\sqrt{x^3-8}$
20. $f(g(x))=\dfrac{1}{x^3}+1$
$g(f(x))=\dfrac{1}{x^3+1}$
21. a. 21 b. 2 c. No
22. a. 6 b. −4 c. Yes
23. a. −9
b. Does not exist
c. $f(4)=\dfrac{1}{9}$ and $\dfrac{1}{9}$ is in the domain of *g*, so $g(f(4))$ is defined. However, $g(2)=-\dfrac{1}{2}$ and $-\dfrac{1}{2}$ is not in the domain of *f*, so $f(g(2))$ is not defined.

23. Given $f(x)=\frac{1}{2x+1}$ and $g(x)=-\frac{1}{x}$, find:

a. $g(f(4))$ **b.** $f(g(2))$

c. Explain the different result in parts **a.** and **b.**

24. Given $f(x)=\sqrt{x-9}$ and $g(x)=x-9$, find:

a. $g(f(109))$ **b.** $f(g(9))$

c. Explain the different result in parts **a.** and **b.**

Show that the given one-to-one functions are inverses of each other. Then graph both functions on the same set of axes and show the line $y=x$ as a dotted line on each graph. (You may use a calculator as an aid in finding the graphs.)

25. $f(x)=3x+1$ and $g(x)=\frac{x-1}{3}$

26. $f(x)=-2x+3$ and $g(x)=\frac{3-x}{2}$

27. $f(x)=\sqrt[3]{x-1}$ and $g(x)=x^3+1$

28. $f(x)=x^3-4$ and $g(x)=\sqrt[3]{x+4}$

29. $f(x)=x^2$ for $x\ge 0$ and $g(x)=\sqrt{x}$

30. $f(x)=\sqrt{x+3}$ and $g(x)=x^2-3$ for $x\ge 0$

31. $f(x)=x^3+2$ and $g(x)=\sqrt[3]{x-2}$

32. $f(x)=\sqrt[5]{x+6}$ and $g(x)=x^5-6$

33. $f(x)=\frac{2}{x}$ and $g(x)=\frac{2}{x}$

34. $f(x)=\frac{3}{x}$ and $g(x)=\frac{3}{x}$

Find the inverse of the given function. Then graph both functions on the same set of axes and show the line $y=x$ as a dotted line on the graph.

35. $f(x)=2x-3$

36. $f(x)=2x-5$

37. $g(x)=x$

38. $g(x)=1-4x$

39. $f(x)=5x+1$

40. $g(x)=-3x+1$

41. $g(x)=\frac{2}{3}x+2$

42. $f(x)=-\frac{1}{2}x-3$

43. $f(x)=-x-2$

44. $f(x)=-2x+4$

45. $f(x)=x^2+1,\ x\ge 0$

46. $f(x)=x^2-1,\ x\ge 0$

47. $f(x)=-\sqrt{x},\ x\ge 0$

48. $f(x)=-\sqrt{x-2},\ x\ge 2$

24. a. 1

b. Does not exist

c. $f(109)=10$ and 10 is in the domain of g, so $g(f(109))$ is defined. However, $g(9)=0$ and 0 is not in the domain of f, so $f(g(9))$ is not defined.

25.

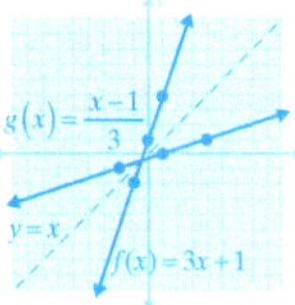

26.

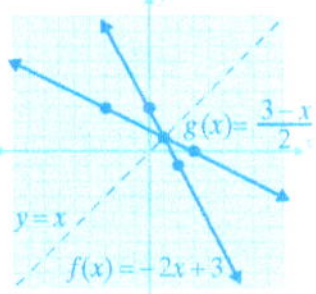

27.

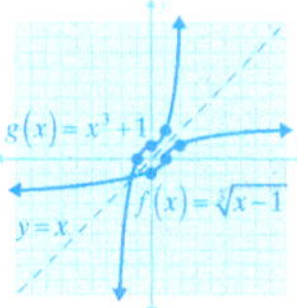

28.

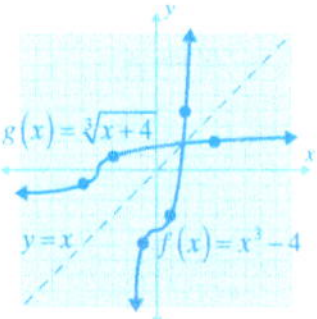

29.

30.

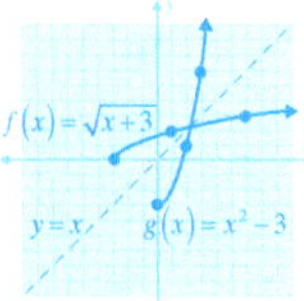

31.

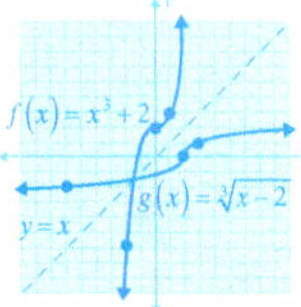

32.

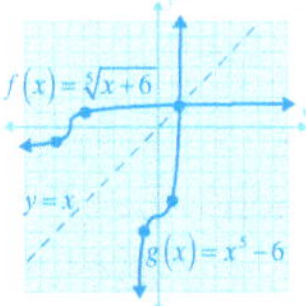

33.

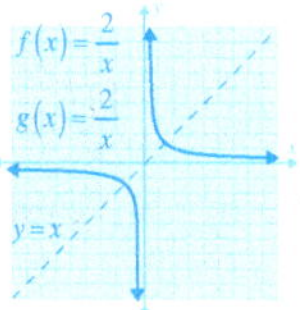

34.

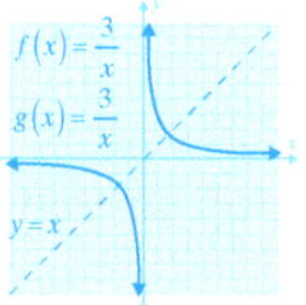

35. $f^{-1}(x)=\frac{x+3}{2}$

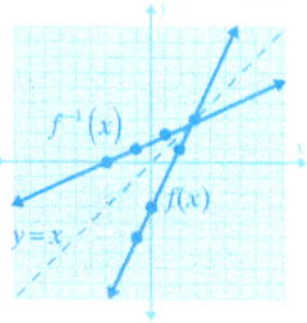

36. $f^{-1}(x)=\frac{x+5}{2}$

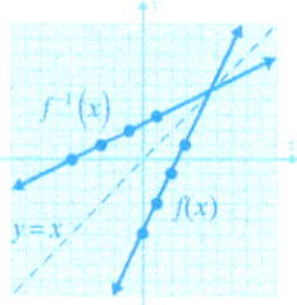

37. $g^{-1}(x) = x$

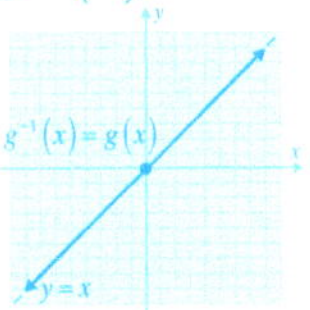

38. $g^{-1}(x) = \frac{1-x}{4}$

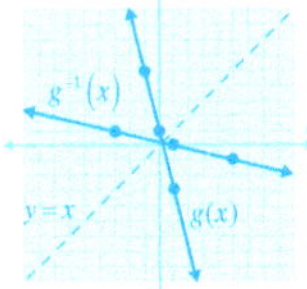

39. $f^{-1}(x) = \frac{x-1}{5}$

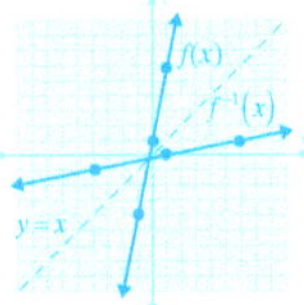

40. $g^{-1}(x) = \frac{1-x}{3}$

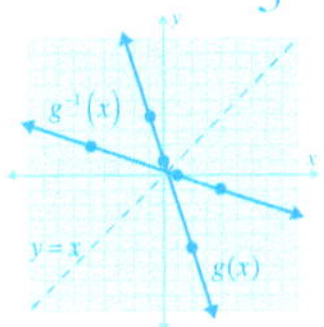

41. $g^{-1}(x) = \frac{3(x-2)}{2}$

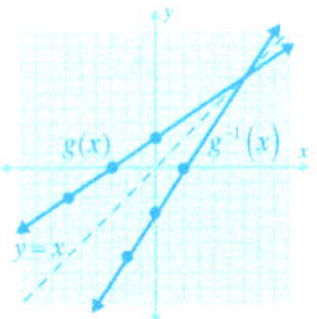

42. $f^{-1}(x) = -2(x+3)$

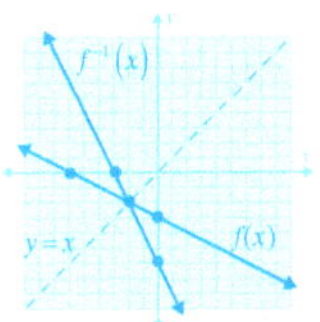

43. $f^{-1}(x) = -x-2$

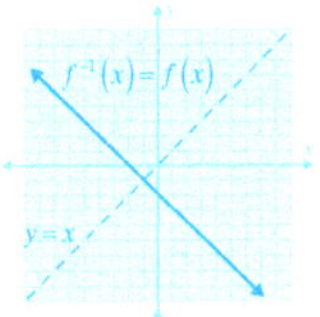

*Using the horizontal line test, determine which of the graphs are graphs of one-to-one functions. If the graph represents a one-to-one function, graph its inverse by reflecting the graph of the function across the line y = x. (**Hint:** If a function is one-to-one, label a few points on the graph and use the fact that the x- and y-coordinates are interchanged on the graph of the inverse.)*

49.

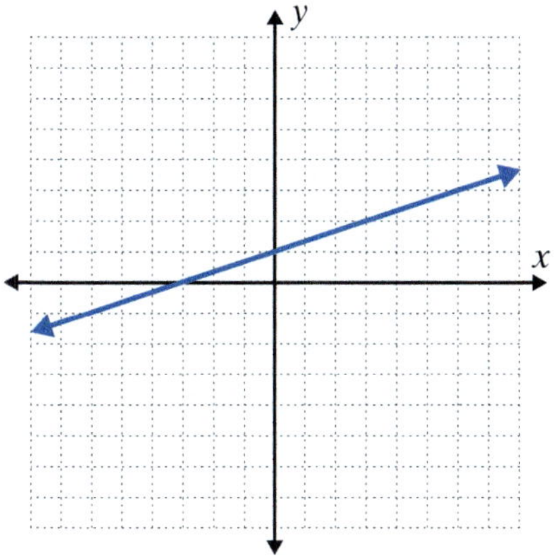

50.

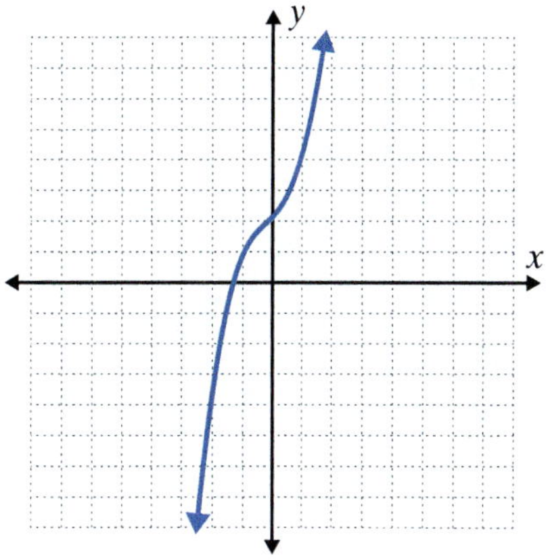

51.

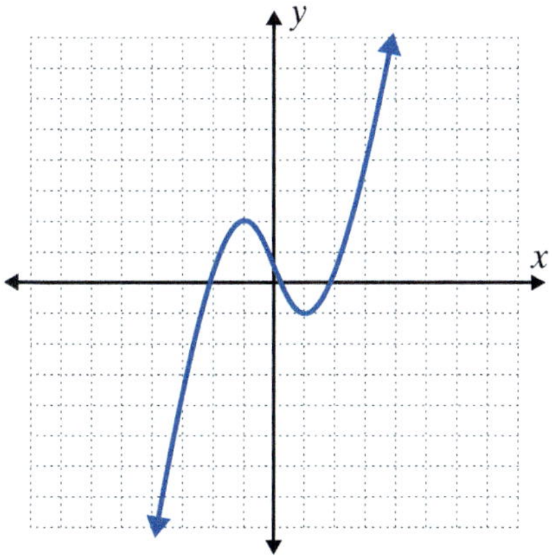

52.

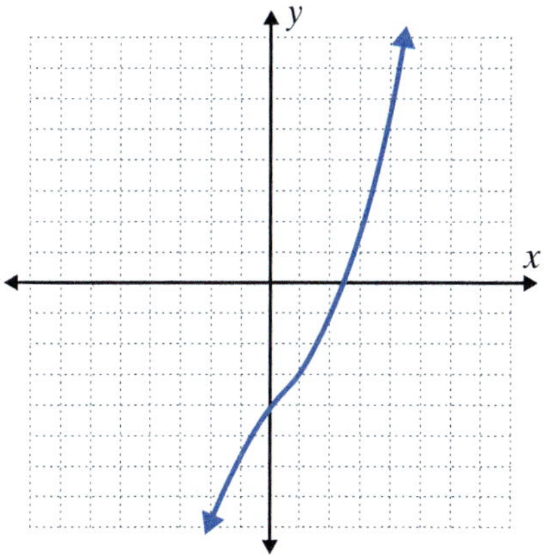

53.

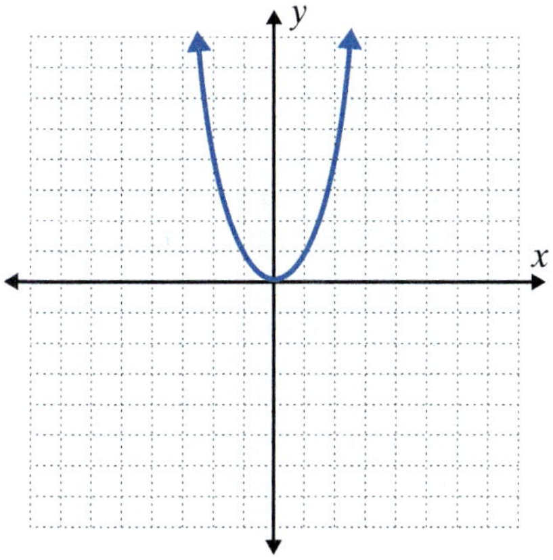

54.

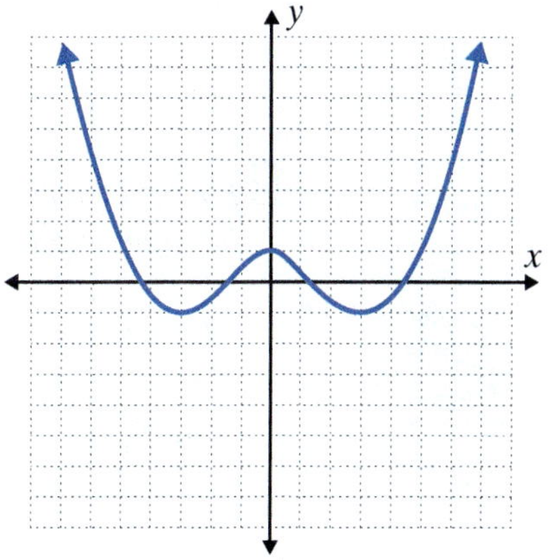

44. $f^{-1}(x)=\frac{4-x}{2}$

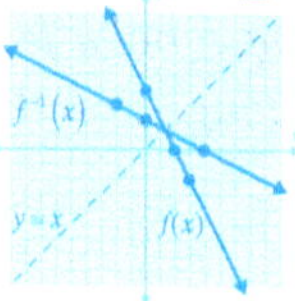

45. $f^{-1}(x)=\sqrt{x-1}$

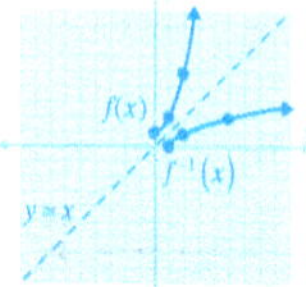

46. $f^{-1}(x)=\sqrt{x+1}$

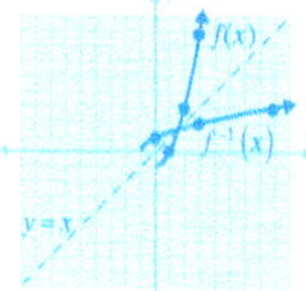

47. $f^{-1}(x)=x^2,\ x\le 0$

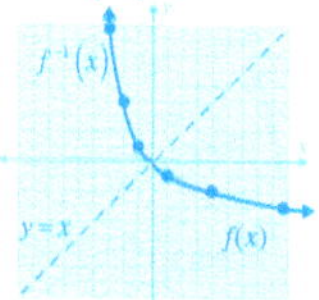

48. $f^{-1}(x)=x^2+2,\ x\le 0$

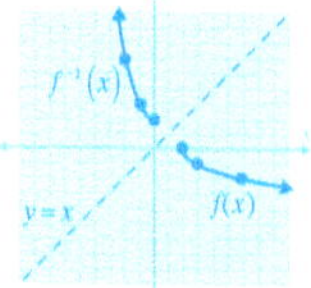

49. One-to-one

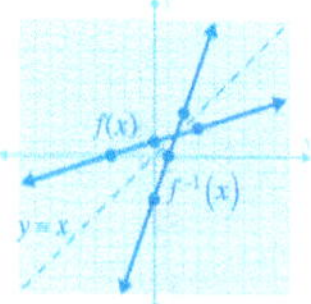

50. One-to-one

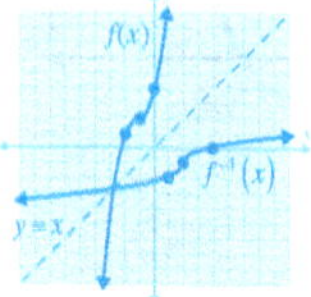

51. Not one-to-one

52. One-to-one

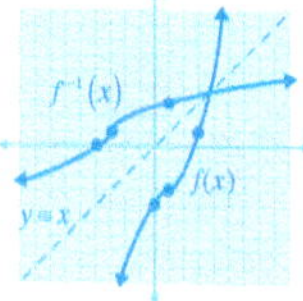

55.

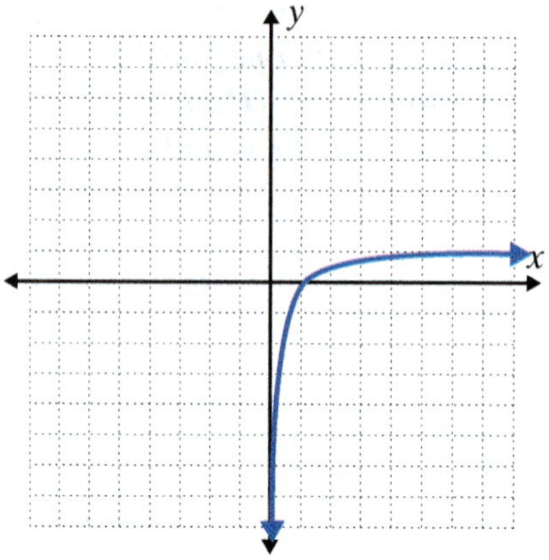

56.

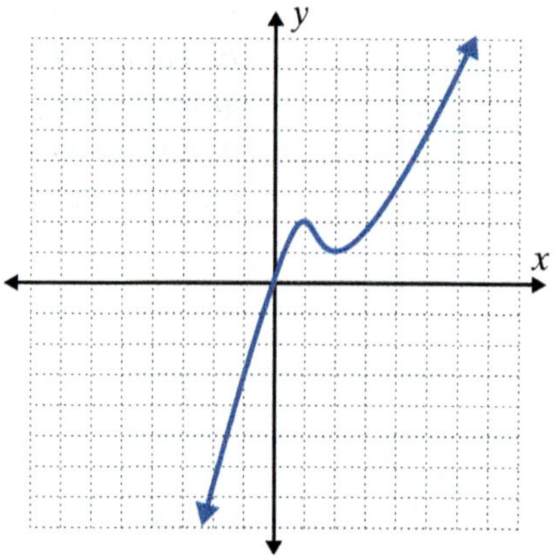

57.

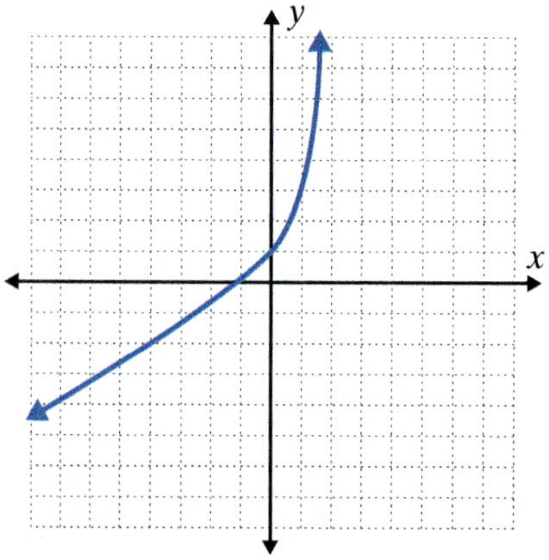

58.

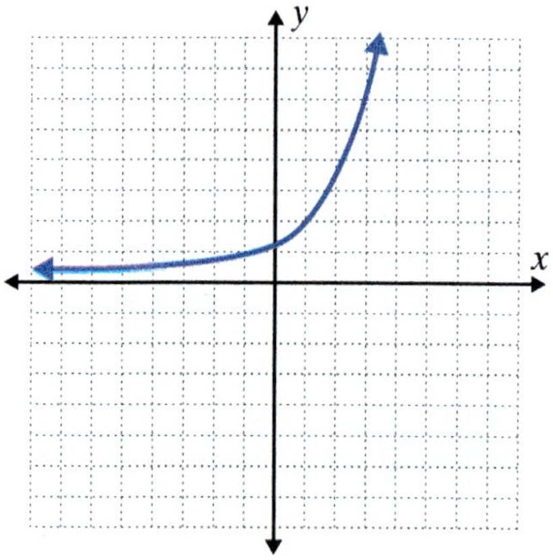

Use a graphing calculator to graph each of the functions and determine which of the functions are one-to-one by inspecting the graph and using the horizontal line test.

59. $f(x)=2x+3$ **60.** $f(x)=7-4x$ **61.** $g(x)=x^2-2$ **62.** $g(x)=9-x^2$

63. $f(x)=4-x^3$ **64.** $f(x)=x^3+2$ **65.** $f(x)=\frac{4}{x}$ **66.** $g(x)=\frac{1}{x}$

67. $g(x)=\sqrt{x-3}$ **68.** $f(x)=\sqrt{x+5}$ **69.** $f(x)=|x+1|$ **70.** $f(x)=|x-5|$

Find the inverse of the given function. Then use a graphing calculator to graph both the function and its inverse. Set the `WINDOW` *so that it is "square."*

71. $f(x)=x^3$ **72.** $f(x)=(x+1)^3$ **73.** $f(x)=\frac{1}{x-3}$

74. $f(x)=\frac{1}{x}$ **75.** $f(x)=x^2,\ x\ge 0$ **76.** $f(x)=x^2+2,\ x\ge 0$

77. $g(x)=x^3+2$ **78.** $g(x)=6-x^3$ **79.** $f(x)=\sqrt{x+5},\ x\ge -5$

80. $g(x)=\sqrt{x-3},\ x\ge 3$ **81.** $f(x)=-x^2+1,\ x\ge 0$ **82.** $g(x)=-x^2-2,\ x\ge 0$

Writing and Thinking About Mathematics

83. Explain, in your own words, why the domains of the two composite functions $f(g(x))$ and $g(f(x))$ might not be the same. Give an example of two functions that illustrate this possibility.

84. Explain briefly why a function must be one-to-one to have an inverse.

HAWKES LEARNING SYSTEMS: INTRODUCTORY & INTERMEDIATE ALGEBRA SOFTWARE

- 11.2 Composition of Functions and Inverse Functions

53. Not one-to-one
54. Not one-to-one
55. One-to-one

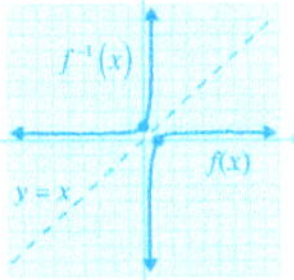

56. Not one-to-one

57. One-to-one

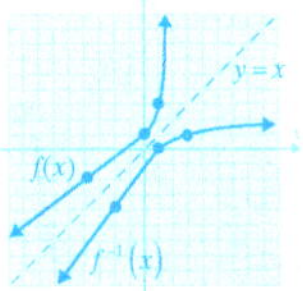

58. One-to-one

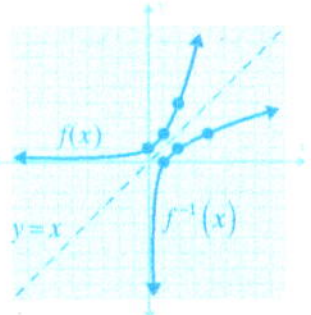

59. One-to-one 60. One-to-one 61. Not one-to-one 62. Not one-to-one 63. One-to-one 64. One-to-one
65. One-to-one 66. One-to-one 67. One-to-one 68. One-to-one 69. Not one-to-one 70. Not one-to-one

71. $f^{-1}(x) = \sqrt[3]{x}$

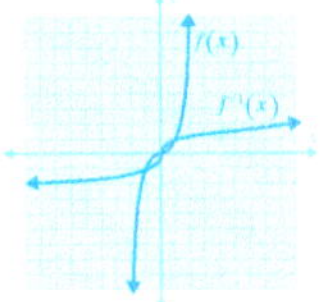

72. $f^{-1}(x) = \sqrt[3]{x} - 1$

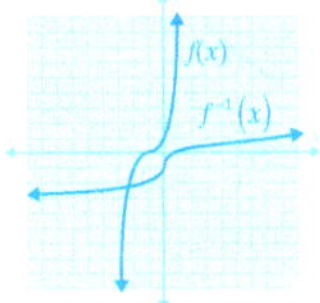

73. $f^{-1}(x) = \dfrac{1}{x} + 3$

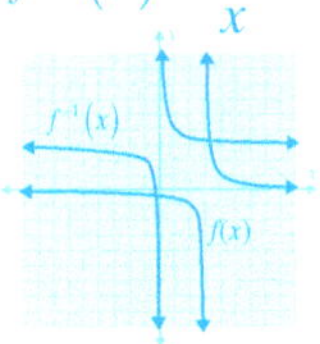

74. $f^{-1}(x) = \dfrac{1}{x}$

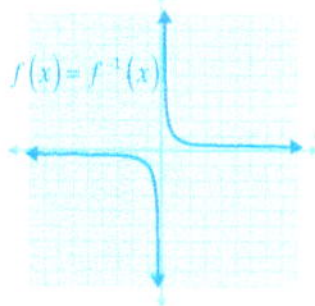

75. $f^{-1}(x) = \sqrt{x}$

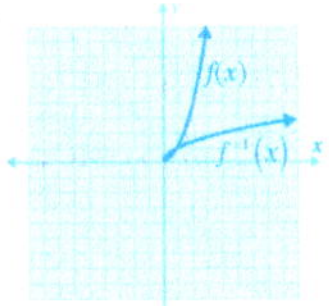

76. $f^{-1}(x) = \sqrt{x-2}$

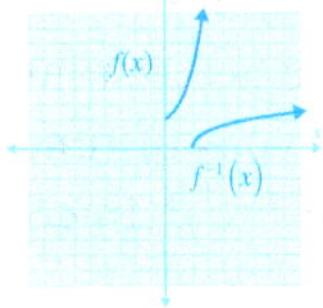

77. $g^{-1}(x) = \sqrt[3]{x-2}$

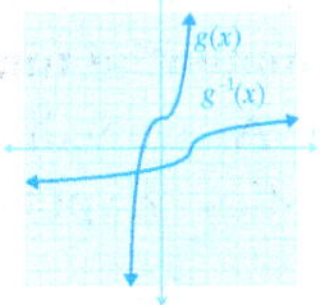

78. $g^{-1}(x) = \sqrt[3]{6-x}$

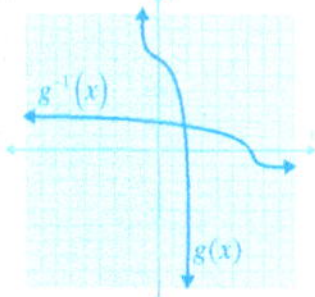

79. $f^{-1}(x) = x^2 - 5,\ x \geq 0$

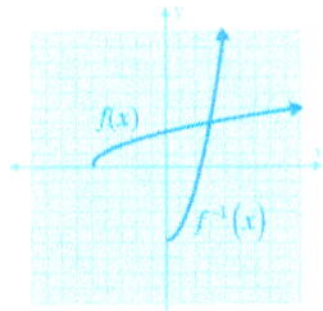

80. $g^{-1}(x) = x^2 + 3, x \geq 0$

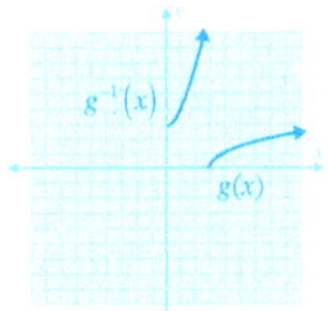

81. $f^{-1}(x) = \sqrt{1-x}$

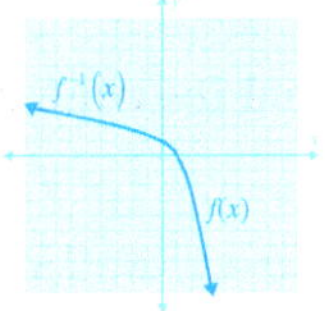

82. $g^{-1}(x) = \sqrt{-2-x}$

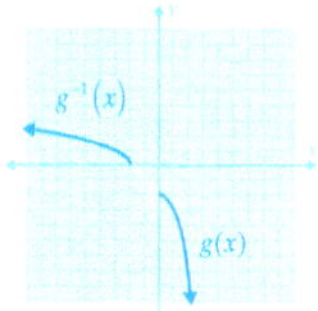

83. The domain of $f(g(x))$ can only include values of x in the domain of g, while $g(f(x))$ can only include values of x in the domain of f. Ex: $f(x) = \sqrt{x}, g(x) = -x$ Answers will vary.

84. In an inverse function, the x- and y-values are interchanged. If a function is not one-to-one, then interchanging the x- and y-values would yield a relation that is not a function.

11.3 Exponential Functions

- *Graph exponential functions.*
- *Solve applied problems involving exponential functions: exponential growth, exponential decay, and compound interest.*

Exponential Functions

You may have read that the population of the world is growing exponentially or studied the exponential growth of bacteria in a biology class. Radioactive materials decay exponentially and never actually disappear. The graph in Figure 1 illustrates that exponential growth has a relatively slow beginning and then builds at an exceedingly rapid rate. This can be extremely important to a doctor trying to curb the growth of "bad" bacteria in a patient.

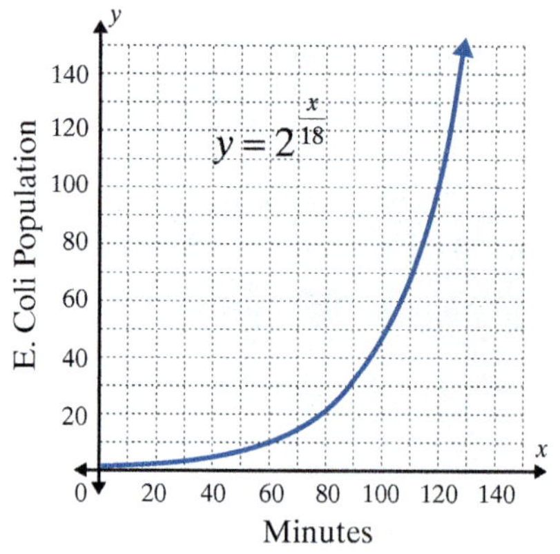

The population of certain strains of E. coli doubles every 18 minutes under optimal conditions.

Figure 1

Quadratic functions have a variable base and a constant exponent, as in $f(x) = x^2$. However, in **exponential functions**, the base is constant and the exponent is a variable, as in $f(x) = 2^x$. As we will see, these two types of functions have major differences in their characteristics. Exponential functions are defined as follows.

Exponential Functions

An **exponential function** is a function of the form

$$\boldsymbol{f(x) = b^x}$$

where $b > 0$, $b \neq 1$, and x is any real number.

Examples of exponential functions are:

$$f(x) = 2^x,\ f(x) = 3^x, \text{ and } y = \left(\frac{1}{3}\right)^x.$$

NOTES

The two conditions $b>0$ and $b \neq 1$ in the definition are important. We must have $b>0$ so that b^x is defined for all real x. For example, we do not consider $y=(-2)^x$ to be an exponential function because $(-2)^{\frac{1}{2}}$ is not a real number. Also, $b \neq 1$. Because the function $y=1^x=1$ for all real x, this function is not considered to be an exponential function.

Exponential Growth

Teaching Note:
At any time in this discussion you may want your students to graph these functions with the TI-84 Plus graphing calculator. Make sure that they understand the limitations on the display in terms of the horizontal asymptote and the appearance of the graph touching the x-axis.

The following table of values and the graphs of the corresponding points give a very good idea of what the graph of the **exponential growth** function $y=2^x$ looks like (see Figure 2a). Because we know that 2^x is defined for all real exponents, points such as $\left(\sqrt{2}, 2^{\sqrt{2}}\right)$, $\left(\pi, 2^{\pi}\right)$, and $\left(\sqrt{5}, 2^{\sqrt{5}}\right)$ are on the graph, and the graph for $f(x)=2^x$ is a smooth curve as shown in Figure 2b.

x	$y=2^x$
3	$2^3=8$
2	$2^2=4$
1	$2^1=2$
$\frac{1}{2}$	$2^{\frac{1}{2}}=\sqrt{2}\approx 1.4142$
0	$2^0=1$
$-\frac{1}{2}$	$2^{-\frac{1}{2}}=\frac{1}{\sqrt{2}}\approx 0.7071$
−1	$2^{-1}=\frac{1}{2}$
−2	$2^{-2}=\frac{1}{2^2}=\frac{1}{4}$
−3	$2^{-3}=\frac{1}{2^3}=\frac{1}{8}$
−4	$2^{(-4)}=\frac{1}{2^4}=\frac{1}{16}$

Domain $=(-\infty, \infty)$
Range $=(0, \infty)$

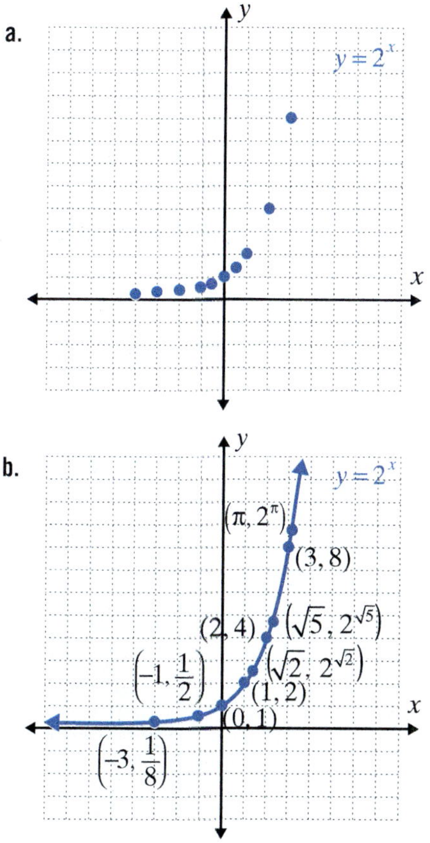

Figure 2

Figure 3 shows a table of values and the graph of the function $y=3^x$. Note that the graphs of $y=2^x$ and $y=3^x$ are quite similar, but that the graph of $y=3^x$ rises faster. That is, the **exponential growth is faster if the base is larger**.

x	$y = 3^x$
2	$3^2 = 9$
1	$3^1 = 3$
0	$3^0 = 1$
−1	$3^{-1} = \frac{1}{3} \approx 0.3333$
−3	$3^{-3} = \frac{1}{27} \approx 0.0370$

Domain = $(-\infty, \infty)$
Range = $(0, \infty)$

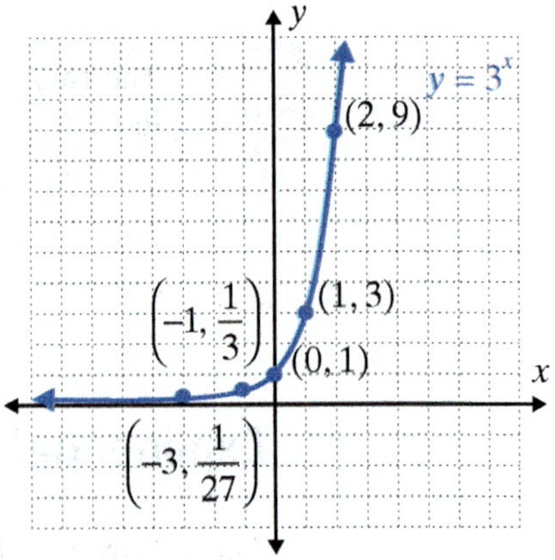

Figure 3

Notice that in both graphs the curves tend to get **very close** to the line $y = 0$ (the x-axis) without ever touching the x-axis. When this happens, the line is called an **asymptote**. If the line is horizontal, as in the cases of exponential growth and (as we will see) exponential decay, the line is called a **horizontal asymptote**. We say that the curve (or function) approaches the line **asymptotically**. In mathematics this phenomenon happens frequently.

Exponential Decay

Now consider the **exponential decay** function $f(x) = \left(\frac{1}{2}\right)^x$. The table and the graph of the corresponding points shown in Figure 4 indicate the nature of the graph of this function.

x	$y = \left(\frac{1}{2}\right)^x = 2^{-x}$
−3	$2^{-(-3)} = 2^3 = 8$
−2	$2^{-(-2)} = 2^2 = 4$
−1	$2^{-(-1)} = 2^1 = 2$
$-\frac{1}{2}$	$2^{-\left(-\frac{1}{2}\right)} = 2^{\frac{1}{2}} = \sqrt{2} \approx 1.4142$
0	$2^{-0} = 2^0 = 1$
$\frac{1}{2}$	$2^{-\frac{1}{2}} = \frac{1}{\sqrt{2}} \approx 0.7071$
1	$2^{-1} = \frac{1}{2}$
2	$2^{-2} = \frac{1}{2^2} = \frac{1}{4}$
3	$2^{-3} = \frac{1}{2^3} = \frac{1}{8}$

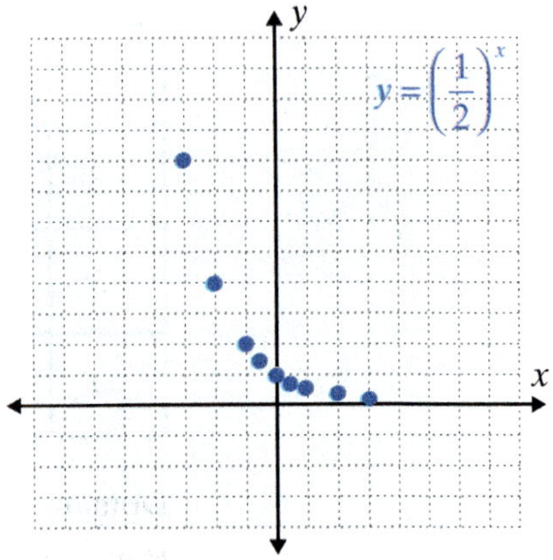

Domain = $(-\infty, \infty)$
Range = $(0, \infty)$

Figure 4

NOTES

Because $\frac{1}{2}=2^{-1}$, $\frac{1}{3}=3^{-1}$, $\frac{1}{4}=4^{-1}$, and so on, for fractions between 0 and 1, we can write an exponential function with a fractional base between 0 and 1 (these are exponential decay functions) in the form of an exponential function with a base greater than 1 and a negative exponent. Thus we write

$$y=\left(\frac{1}{2}\right)^x=\left(2^{-1}\right)^x=2^{-x} \quad \text{and} \quad y=\left(\frac{1}{3}\right)^x=\left(3^{-1}\right)^x=3^{-x}.$$

Figure 5 shows the complete graphs of the two exponential decay functions

$$y=\left(\frac{1}{2}\right)^x=2^{-x} \quad \text{and} \quad y=\left(\frac{1}{3}\right)^x=3^{-x}.$$

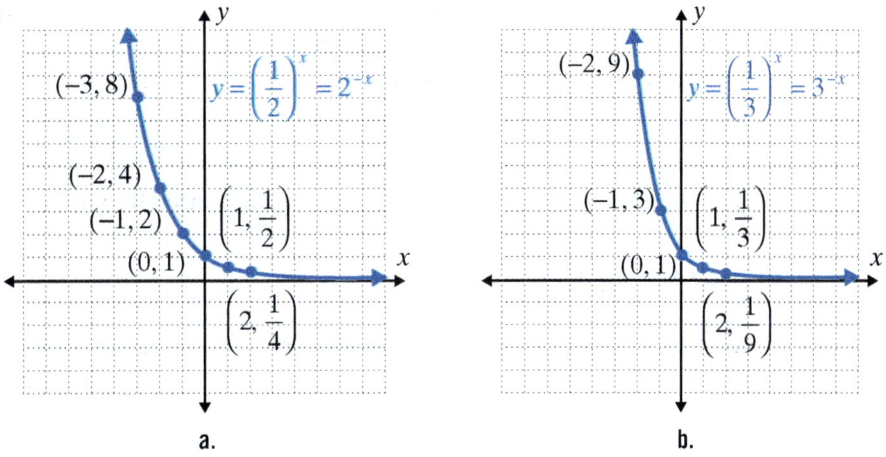

Figure 5

Note again, in Figure 5, that the line $y=0$ (the x-axis) is a **horizontal asymptote** for both curves. For exponential growth the curves approach the asymptote as x moves further in the negative direction as in Figures 2b and 3. For exponential decay, the graph approaches the asymptote as x moves further in the positive direction as in Figure 5.

The following general concepts are helpful in understanding the graphs and the nature of exponential functions, both exponential growth and exponential decay.

General Concepts of Exponential Functions

For $b > 1$:

1. $b^x > 0$
2. b^x increases to the right and is called an **exponential growth function**.
3. $b^0 = 1$, so $(0, 1)$ is the y-intercept.
4. b^x approaches the x-axis for negative values of x. (The x-axis is a horizontal asymptote. **See Figure 3 on page 888.**)

For $0 < b < 1$:

1. $b^x > 0$
2. b^x decreases to the right and is called an **exponential decay function**.
3. $b^0 = 1$, so $(0, 1)$ is the y-intercept.
4. b^x approaches the x-axis for positive values of x. (The x-axis is a horizontal asymptote. **See Figure 5 on page 889.**)

As with all functions, exponential functions can be multiplied by constants, shifted horizontally, and shifted vertically. (**Note:** We did this with parabolas in Chapter 10 and will be more detailed on this topic in Chapter 12.) Thus

$$y = a \cdot b^{r \cdot x}, \quad y = b^{x-h}, \quad \text{and} \quad y = b^x + k$$

and various combinations of these expressions are all exponential functions.

Bacterial Growth

Exponential functions are related to many practical applications, among which are bacterial growth, radioactive decay, compound interest, and light absorption. For example, a bacteria culture kept at a certain temperature may grow according to the exponential function

$$y = y_0 \cdot 2^{0.5t}$$

where t = time in hours and
y_0 = amount of bacteria present when $t = 0$.
(y_0 is called the **initial value** of y.)

Example 1: Exponential Growth

a. A scientist has 10,000 bacteria present when $t = 0$. She knows the bacteria grow according to the function $y = y_0 \cdot 2^{0.5t}$ where t is measured in hours. How many bacteria will be present at the end of one day?

Solution: Substitute $t = 24$ hours and $y_0 = 10{,}000$ into the function.

$$y = 10{,}000 \cdot 2^{0.5(24)} = 10{,}000 \cdot 2^{12}$$
$$= 10{,}000(4096)$$
$$= 40{,}960{,}000$$
$$= 4.096 \times 10^7 \text{ bacteria}$$

You could also use your TI-84 Plus graphing calculator by entering the numbers as shown in the following display. Press ENTER to get the result.

```
10000*2^(0.5*24)
                40960000
```

b. Use the formula for exponential growth, $y = y_0 b^t$, to determine the exponential function that fits the following information: $y_0 = 5000$ bacteria, and there are 135,000 bacteria present after 3 days.

Solution: Use $y = y_0 b^t$ where t is measured in days. Substitute 135,000 for y, 3 for t, and 5000 for y_0, then solve for b.

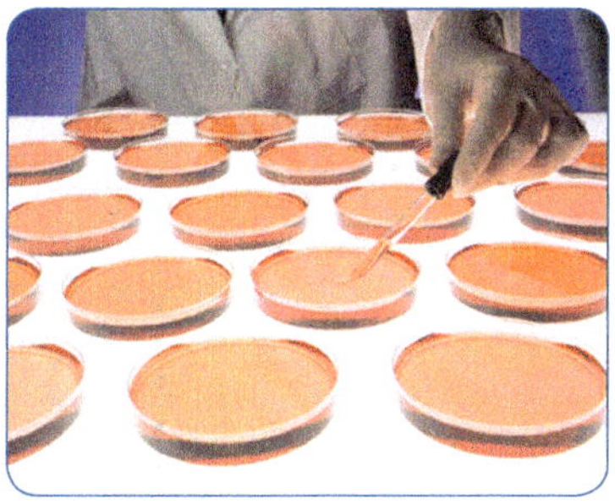

$$135{,}000 = 5000b^3$$

$$27 = b^3$$

$$\sqrt[3]{(27)} = \sqrt[3]{(b^3)}$$

$$3 = b$$

The function is $y = 5000 \cdot 3^t$.

Compound Interest

The topic of compound interest (interest paid on interest) leads to a particularly interesting (and useful) exponential function. The formula $A = P(1+r)^t$ can be used for finding the value (amount) accumulated when a principal P is invested and interest is compounded once a year. If compounding is performed more than once a year, we use the following formula to find A.

Compound Interest

Compound interest on a principal P invested at an annual interest rate r (in decimal form) for t years that is compounded n times per year can be calculated using the following formula:

$$\mathbf{A = P\left(1+\frac{r}{n}\right)^{nt}}$$

where A is the amount accumulated.

Example 2: Compound Interest

a. If P dollars are invested at a rate of interest r (in decimal form) compounded annually (once a year, $n = 1$) for t years, the formula for the amount A becomes $A = P\left(1+\frac{r}{1}\right)^{1\cdot t} = P(1+r)^t$. Find the value of $1000 invested at $r = 6\% = 0.06$ for 3 years.

Solution: We have $P = 1000$, $r = 0.06$, and $t = 3$.

$$A = 1000(1+0.06)^3$$
$$= 1000(1.06)^3$$
$$= 1000(1.191016) \approx \$1191.02$$

Note: To use your calculator to evaluate $(1.06)^3$, enter 1.06, press the ^ key, enter 3, then press the ENTER key.

b. What will be the value of a principal investment of $1000 invested at 6% for 3 years if interest is compounded monthly (12 times per year)?

Solution: Use the formula for compound interest.
We have $P = 1000$, $r = 0.06$, $n = 12$, and $t = 3$.

$$A = 1000\left(1+\frac{0.06}{12}\right)^{12(3)}$$
$$= 1000(1+0.005)^{36}$$
$$= 1000(1.005)^{36}$$
$$= 1000(1.196680524\ldots) \quad \text{Using a calculator}$$
$$\approx \$1196.68$$

c. Find the value of A if $1000 is invested at 6% for 3 years and interest is compounded daily. (**Note:** Banks and savings institutions often use 360 days per year.)

Solution: Use the formula for compound interest.
We have $P = 1000$, $r = 0.06$, $n = 360$, and $t = 3$.

$$A = 1000\left(1+\frac{0.06}{360}\right)^{360(3)}$$
$$= 1000(1.000166666\ldots)^{1080}$$
$$= 1000(1.197199407\ldots) \quad \text{Using a calculator}$$
$$\approx \$1197.20$$

Examples 2a, 2b, and 2c illustrate the effects of compounding interest more frequently over 3 years. The formula gives the results

$$A = \$1191.02 \quad \text{for} \quad n = 1 \text{ (once a year)},$$
$$A = \$1196.68 \quad \text{for} \quad n = 12 \text{ (monthly)},$$
$$A = \$1197.20 \quad \text{for} \quad n = 360 \text{ (daily)}.$$

These numbers might not seem very dramatic, only a difference of \$6.18 for 3 years; but, if you use your calculator and 20 years you will see a difference of \$112.65 for a \$1000 investment. An investment of \$10,000 for 20 years at 9% will show a difference of \$4438.76. The results show that more frequent compounding will result in higher income.

Interest Compounded Continuously

If interest is **compounded continuously** (even faster than every second), then the irrational number e $(e \approx 2.718)$ can be shown to be the base of the corresponding exponential function for calculating interest. Table 1 shows how the expression $\left(1+\frac{1}{n}\right)^n$ changes as n takes on larger and larger values. The number e is the **limit (or limiting value)** of the expression "as n approaches infinity" $(n \to \infty)$ and we write $e = \lim_{n\to\infty}\left(1+\frac{1}{n}\right)^n$. Study the following table to help understand the ideas.

Teaching Notes:
Have students notice how little change is occuring in this sequence of values (3rd column) and that this suggests that there might be a boundary ("limit") to the sequence – and there is – e! Stress that the concept of limit presented in the definition of e is precise, but beyond the scope of this text.

Values of the expression $\left(1+\frac{1}{n}\right)^n$ as n approaches infinity $(n \to \infty)$		
n	$\left(1+\frac{1}{n}\right)$	$\left(1+\frac{1}{n}\right)^n$
1	$\left(1+\frac{1}{1}\right)=2$	$2^1=2$
2	$\left(1+\frac{1}{2}\right)=1.5$	$(1.5)^2=2.25$
5	$\left(1+\frac{1}{5}\right)=1.2$	$(1.2)^5=2.48832$
10	$\left(1+\frac{1}{10}\right)=1.1$	$(1.1)^{10}\approx 2.59374246$
100	$\left(1+\frac{1}{100}\right)=1.01$	$(1.01)^{100}\approx 2.704813829$
1000	$\left(1+\frac{1}{1000}\right)=1.001$	$(1.001)^{1000}\approx 2.716923932$

Continued on the next page...

10,000	$\left(1+\frac{1}{10,000}\right)=1.0001$	$(1.0001)^{10,000} \approx 2.718145927$
100,000	$\left(1+\frac{1}{100,000}\right)=1.00001$	$(1.00001)^{100,000} \approx 2.718268237$
↓		↓
∞		$e = 2.718281828459...$

Table 1

This gives us the following definition of e. (Be aware that these are very sophisticated mathematical concepts and it will take some careful reading to understand how to arrive at e and the formula $A = Pe^{rt}$.)

The Number *e*

The Number *e*

The number e is defined to be

$$e = \mathbf{2.718281828459}\ldots$$

Note: The number e is on the TI-84 Plus calculator above the divide key.

As we know, the formula for compound interest is

$$A = P\left(1+\frac{r}{n}\right)^{nt}.$$

This formula for compound interest takes a different form when interest is **compounded continuously**. The new form involves the number e and is stated below.

Continuously Compounded Interest

Continuously compounded interest on a principal P invested at an annual interest rate r for t years can be calculated using the following formula:

$$\mathbf{A = Pe^{rt}}$$

where A is the amount accumulated.

As illustrated in Example 3, a calculator is needed to use the formula for continuously compounded interest.

Example 3: Calculating Continuously Compounded Interest Using a TI-84 Plus Calculator

Teaching Note:
You might want to discuss interest rates with your students. Point out that interest rates fluctuate with the economy.

Find the value of \$1000 invested at 6% for 3 years if interest is compounded continuously. (In this case, $P = \$1000$, $r = 6\% = 0.06$, and $t = 3$.)

Solution: Press 2ND and LN and e^(will appear on the display.

To find the value of

$$A = Pe^{rt} = 1000e^{0.06 \cdot 3}$$

enter the numbers as shown and press ENTER to get the result.

```
1000*e^(0.06*3)
        1197.217363
```

The entire exponent must be in parentheses.

Thus the value of \$1000 compounded continuously at 6% for 3 years will be \$1197.22. (Note that from Example 2c there is only a 2 cent gain in A when \$1000 is compounded continuously instead of daily at 6% for 3 years.)

Practice Problems

1. Sketch the graph of the exponential function $f(x) = 2 \cdot 3^x$ and label 3 points on the graph.
2. Sketch the graph of the exponential decay function $y = 0.5 \cdot 2^{-x}$ and label 3 points on the graph.
3. Find the value of \$5000 invested at 8% for 10 years if interest is **a.** compounded monthly, **b.** compounded continuously.

Answers to Practice Problems:

1. 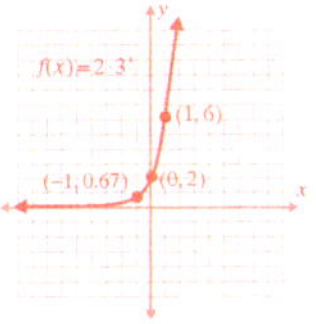

2. 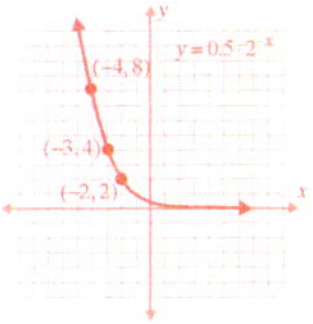

3. a. \$11,098.20 **b.** \$11,127.70

11.3 Exercises

Sketch the graph of each of the exponential functions and label three points on each graph. (Note that some of the graphs are shifts, horizontal or vertical, of the basic exponential functions. These are similar to the shifts performed on parabolas in Chapter 10.)

1. $y = 4^x$ **2.** $y = 5^x$ **3.** $y = \left(\frac{1}{3}\right)^x$ **4.** $y = \left(\frac{1}{5}\right)^x$

5. $y = \left(\frac{2}{3}\right)^x$ **6.** $y = \left(\frac{5}{2}\right)^x$ **7.** $y = \left(\frac{1}{2}\right)^{-x}$ **8.** $y = \left(\frac{3}{4}\right)^{-x}$

9. $y = 2^{x-1}$ **10.** $y = 3^{x+1}$ **11.** $f(x) = 2^x + 1$

12. $f(x) = 3^x - 1$ **13.** $f(x) = -4^{-x}$ **14.** $g(x) = -2^{-x}$

15. $f(x) = 2^{0.5x}$ **16.** $g(x) = 10^{0.5x}$ **17.** $f(x) = 4^{-x} - 1$

18. $g(x) = 10^{-x} - 3$ **19.** $f(x) = 3 \cdot \left(\frac{1}{2}\right)^{x+1}$ **20.** $y = -4 \cdot \left(\frac{1}{3}\right)^{x-1}$

21. If $f(t) = 3 \cdot 4^t$ what is the value of $f(2)$?

22. For $f(x) = 3 \cdot 10^{2x}$, find the value of $f(0.5)$.

23. Use your calculator to find the value (to the nearest hundredth) of $f(2)$ if $f(x) = 27.3 \cdot e^{-0.4x}$.

24. Use your calculator to find the value (to the nearest hundredth) of $f(3)$ if $f(x) = 41.2 \cdot e^{-0.3x}$.

25. Use your calculator to find the value (to the nearest hundredth) of $f(9)$ if $f(t) = 2000 \cdot e^{0.08t}$.

26. Use your calculator to find the value (to the nearest hundredth) of $f(22)$ if $f(t) = 2000 \cdot e^{0.05t}$.

27. Bacteria growth: A biologist knows that in the laboratory, bacteria in a culture grow according to the function $y = y_0 \cdot 5^{0.2t}$, where y_0 is the initial number of bacteria present and t is time measured in hours. How many bacteria will be present in a culture at the end of 5 hours if there were 5000 present initially?

28. Bacteria growth: Referring to Exercise 27, how many bacteria were present initially if at the end of 15 hours, there were 2,500,000 bacteria present?

1.
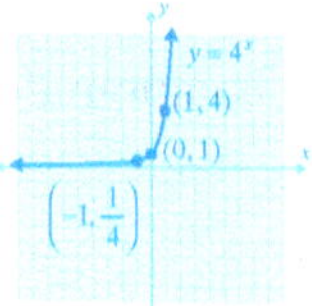

2.
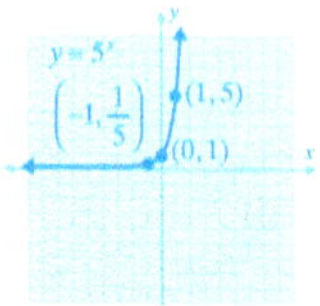

3.
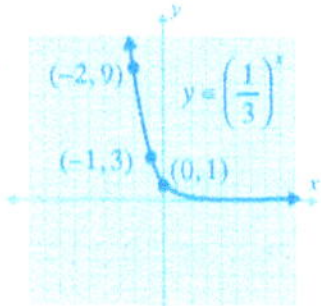

4.
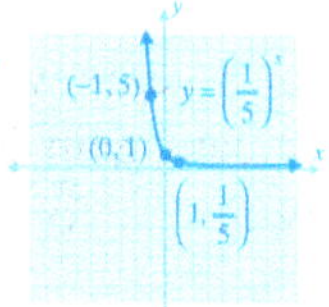

5.
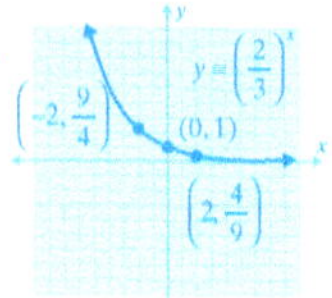

6.
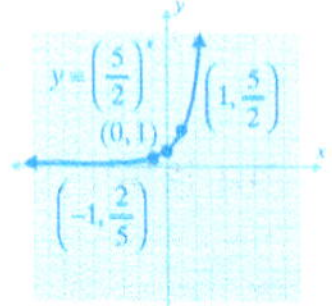

7.
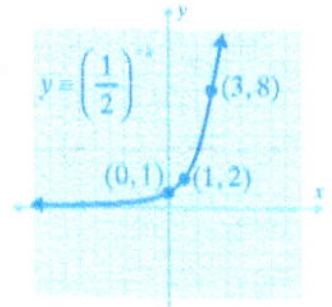

8.
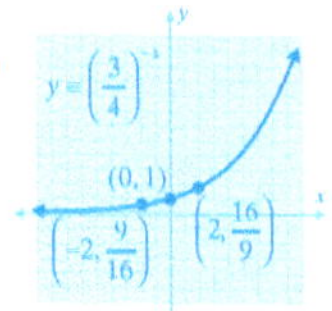

9.
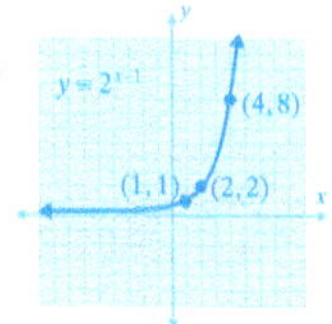

10.
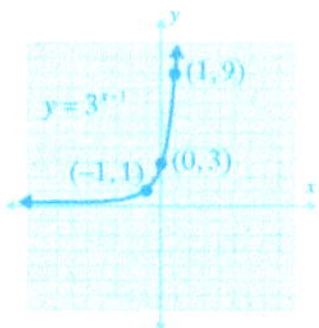

11.
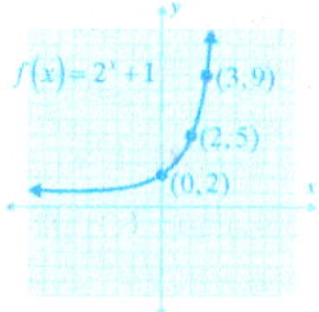

12.
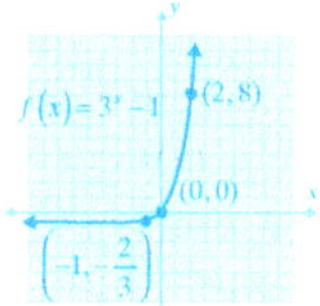

13.
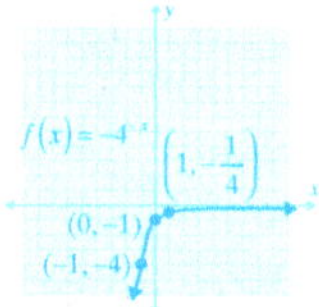

14.
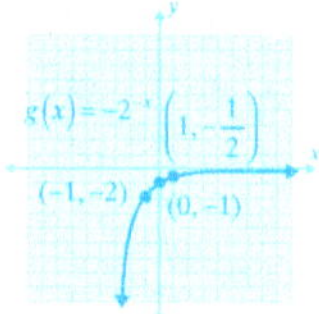

15.
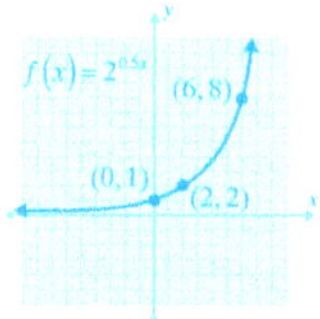

16.
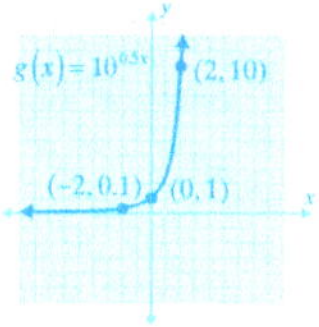

29. Savings accounts: Four thousand dollars is deposited into a savings account with a rate of 8% per year. Find the total amount A on deposit at the end of 5 years if the interest is compounded:

a. annually
b. semiannually
c. quarterly
d. daily (use 360 days)
e. continuously

30. Savings accounts: Find the amount A in a savings account if \$2000 is invested at 7% for 4 years and the interest is compounded:

a. annually
b. semiannually
c. quarterly
d. daily (use 360 days)
e. continuously

31. Investing: Find the value of \$1800 invested at 6% for 3 years if the interest is compounded continuously.

32. Investing: Find the value of \$2500 invested at 5% for 5 years if the interest is compounded continuously.

33. Sales revenue: The revenue function is given by $R(x) = x \cdot p(x)$ dollars, where x is the number of units sold and $p(x)$ is the unit price. If $p(x) = 25(2)^{\frac{-x}{5}}$, find the revenue if 15 units are sold.

34. Sales revenue: Referring to Exercise 33, if $p(x) = 40(3)^{\frac{-x}{6}}$, find the revenue if 12 units are sold.

35. Broadcasting: A radio station knows that during an intense advertising campaign, the number of people N who will hear a commercial is given by $N = A\left(1 - 2^{-0.05t}\right)$, where A is the number of people in the broadcasting area and t is the number of hours the commercial has been run. If there are 500,000 people in the area, how many will hear a commercial during the first 20 hours?

36. Battery reliability: Statistics show that the fractional part of flashlight batteries f that are still good after t hours of use is given by $f = 4^{-0.02t}$. What fractional part of the batteries are still operating after 150 hours of use?

37. Investing: Bethany invested \$45,000 in a retirement fund that earns 8% interest and is compounded continuously, how much money will the account be worth after:

a. 10 years?
b. 20 years?
c. 40 years?

38. Investing: If a principal P is invested at a rate r compounded continuously, the interest earned is given by $I = A - P$.

a. Find the interest earned in 20 years on \$10,000 invested at 10% and compounded continuously?

b. Find the interest earned in 20 years on \$10,000 invested at 5% and compounded continuously.

c. Explain why the interest earned at 5% is not just one-half of the interest earned at 10% in parts **a.** and **b.**

39. Machine value: The value V of a machine at the end of t years is given by $V = C(1-r)^t$, where C is the original cost and r is the rate of depreciation. Find the value of a machine at the end of 4 years if the original cost was \$1200 and $r = 0.20$.

40. Machine value: Referring to Exercise 39, find the value of a machine at the end of 3 years if the original cost was \$2000 and $r = 0.15$.

41. Drug absorption: A cancer patient is given a dose of 50 mg of a particular drug. In five days the amount of the drug in her system is reduced to 1.5625 mg. If the drug decays (or is absorbed) at an exponential rate, find the function that represents the amount of the drug. (**Hint:** Use the formula $y = y_0 b^{-t}$ and solve for b.)

42. Growth of cancer cells: Determine the exponential function that fits the following information concerning exponential growth of cancer cells: $y_0 = 10{,}000$ cancer cells, and there are 160,000 cancer cells present after 4 days. (**Hint:** Use the formula $y = y_0 b^t$ and solve for b.)

43. Use a graphing calculator to graph each of the following functions. In each case the x-axis is a horizontal asymptote.

a. $y = e^x$

b. $y = e^{-x}$

c. $y = e^{-x^2}$

Writing and Thinking About Mathematics

44. Discuss, in your own words, how the graph of each of the following functions is related to the graph of the exponential function $y = b^x$.

a. $y = a \cdot b^x$ **b.** $y = b^{x-h}$ **c.** $y = b^x + k$

45. Discuss, in your own words, the symmetrical relationship of the graphs of the two exponential functions $y = 10^x$ and $y = 10^{-x}$.

46. Discuss, in your own words, the symmetrical relationship of the graphs of the two exponential functions $y = 10^x$ and $y = -10^x$.

Continued on the next page...

17. 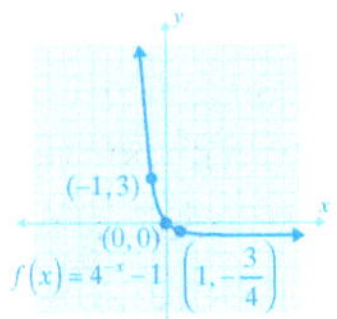

18. 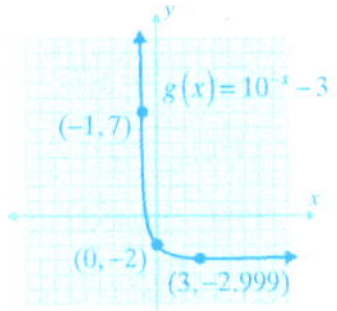

19. 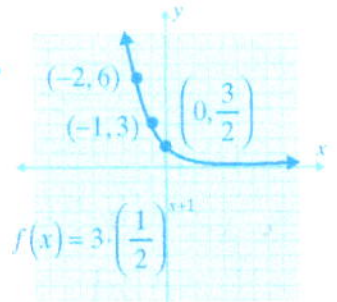

20. 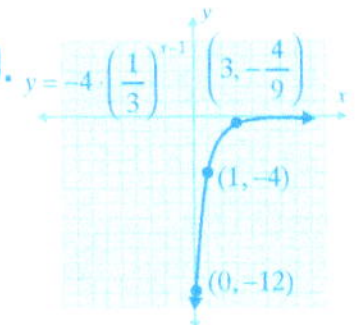

21. 48
22. 30
23. 12.27
24. 16.75
25. 4108.87
26. 6008.33
27. 25,000 bacteria
28. 20,000 bacteria
29. a. \$5877.31 b. \$5920.98 c. \$5943.79 d. \$5967.03 e. \$5967.30
30. a. \$2621.59 b. \$2633.62 c. \$2639.86 d. \$2646.19 e. \$2646.26
31. \$2154.99
32. \$3210.06
33. \$46.88 34. \$53.33 35. 250,000 people 36. $\frac{1}{64}$ 37. a. \$100,149.34 b. \$222,886.46 c. \$1,103,963.86

Writing and Thinking About Mathematics (cont.)

38. a. \$63,890.56
b. \$17,182.82
c. Answers will vary.
39. \$491.52
40. \$1228.25
41. $y = 50(2)^{-t}$
42. $y = 10,000 \cdot 2^t$
43. a.
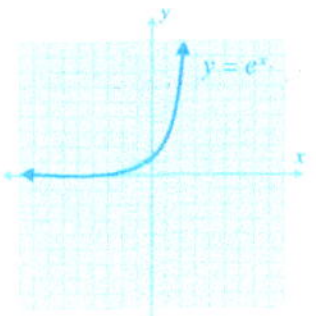

b.
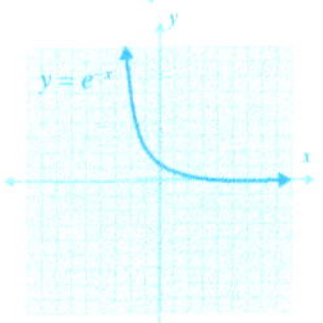

c.
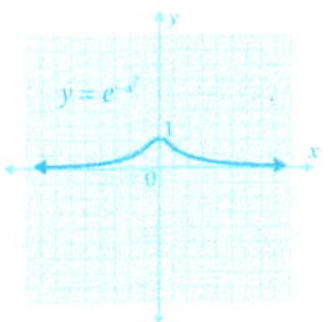

47. The following formula can be used to calculate monthly mortgage payments:

$$A = \frac{P\left(1+\frac{r}{12}\right)^n \cdot \frac{r}{12}}{\left(1+\frac{r}{12}\right)^n - 1}$$

where A = the monthly payment,
P = amount initially borrowed (the mortgage),
r = the annual interest rate (in decimal form), and
n = the total number of monthly payments (12 times the number of years).

With the class divided into teams of 3 or 4 students, each team should complete one table (using different values for r and for P). Discuss the results as a class. Explain what this might mean for you personally.

For annual rate $r =$ _____ and initial mortgage $P =$ ________

Length of Mortgage (in years)	Monthly Payment A	Total Cost of Mortgage n times A
15		
20		
25		
30		

HAWKES LEARNING SYSTEMS: INTRODUCTORY & INTERMEDIATE ALGEBRA SOFTWARE

- 11.3 Exponential Functions

44. a. The graph is stretched vertically by a factor of a.
b. The graph is translated horizontally h units to the right.
c. The graph is translated vertically k units up.
45. The two functions are symmetric about the y-axis.
46. The two functions are symmetric about the x-axis.
47. Answers will vary.

11.4 Logarithmic Functions

- *Write exponential expressions in logarithmic form.*
- *Write logarithmic expressions in exponential form.*
- *Use the basic properties of logarithms to evaluate logarithms.*
- *Solve equations using the definitions of exponential and logarithmic functions.*
- *Graph exponential functions and logarithmic functions on the same set of axes.*

Logarithms

Exponential functions of the form $y = b^x$ are one-to-one functions and, therefore, have inverses. To find the inverse of a function, we interchange x and y in the equation and solve for y. Thus for the function

$$y = b^x,$$

interchanging x and y gives the inverse function

$$x = b^y.$$

Figure 1 shows the graphs of these two functions with $b > 1$. Each function is a reflection of the other across the line $y = x$. Note that the exponential function $y = b^x$ has **the line $y = 0$ (the x-axis) as a horizontal asymptote**, and the inverse function $x = b^y$ has **the line $x = 0$ (the y-axis) as a vertical asymptote.**

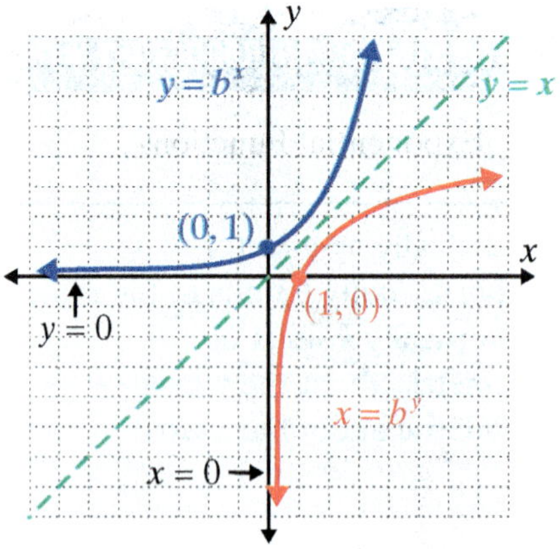

Figure 1

To solve the inverse equation $x = b^y$ for y, mathematicians have simply created a name for y. This name is **logarithm** (abbreviated **log**). This means that **the inverse of an exponential function is a logarithmic function.**

Definition of Logarithm (base b)

For $b > 0$ and $b \neq 1$,

$\boldsymbol{x = b^y}$ is equivalent to $\boldsymbol{y = \log_b x}$.

$\boldsymbol{y = \log_b x}$ is read "y is the logarithm (base b) of x."

A logarithm is an exponent. Example 1 shows how exponential forms and logarithmic forms of equations are related.

Example 1: Translations Between Exponential Form and Logarithmic Form

	Exponential Form		Logarithmic Form	
a.	$2^3 = 8$	$\Leftrightarrow$	$\log_2 8 = 3$	The base is 2. The logarithm is 3.
b.	$2^4 = 16$	$\Leftrightarrow$	$\log_2 16 = 4$	The base is 2. The logarithm is 4.
c.	$10^3 = 1000$	$\Leftrightarrow$	$\log_{10} 1000 = 3$	The base is 10. The logarithm is 3.
d.	$3^0 = 1$	$\Leftrightarrow$	$\log_3 1 = 0$	The base is 3. The logarithm is 0.
e.	$5^{-1} = \frac{1}{5}$	$\Leftrightarrow$	$\log_5 \frac{1}{5} = -1$	The base is 5. The logarithm is –1.

Note that in each case the base of the exponent is the base of the logarithm.

REMEMBER, a logarithm is an exponent. For example,

logarithm ↓ $10^2 = 100$ and logarithm ↓ $\log_{10} 100 = 2$ and logarithm ↓ $100 = 10^{\log_{10} 100}$

are all equivalent. In words,

2 is the **exponent** of the base 10 to get 100 $\left(10^2 = 100\right)$; and

2 is the **logarithm** base 10 of 100 $\left(\log_{10} 100 = 2\right)$.

Basic Properties of Logarithms

We know that **exponents are logarithms** and from our previous knowledge of exponents, we can make the following equivalent statements for logarithms, base b.

$$b^0 = 1 \quad \Leftrightarrow \quad \log_b 1 = 0$$

$$b^1 = b \quad \Leftrightarrow \quad \log_b b = 1$$

Also, directly from the definition of $y = \log_b x$, we can make two more general statements:

$$x = b^{\log_b x} \quad \text{and} \quad \log_b b^x = x.$$

In summary, we have the following four basic properties of logarithms.

Basic Properties of Logarithms

For $b > 0$ and $b \neq 1$,

1. $\mathbf{\log_b 1 = 0}$ Regardless of the base, the logarithm of 1 is 0.
2. $\mathbf{\log_b b = 1}$ The logarithm of the base is always 1.
3. $\mathbf{x = b^{\log_b x}}$ for $\mathbf{x > 0}$
4. $\mathbf{\log_b b^x = x}$

REMEMBER, a logarithm is an exponent.

Example 2: Evaluating Logarithms

Use the four basic properties of logarithms to evaluate the following logarithms.

a. $\log_3 1$

Solution: $\log_3 1 = 0$ By property 1

b. $\log_8 8$

Solution: $\log_8 8 = 1$ By property 2

c. $10^{\log_{10} 20}$

Solution: $10^{\log_{10} 20} = 20$ By property 3

d. $\log_2 32$

Solution: $\log_2 32 = \log_2 2^5$ Write 32 as 2^5 so the base is 2.

$= 5$ By property 4

e. $\log_{10} 0.01$

Solution: $\log_{10} 0.01 = \log_{10} \frac{1}{100}$

$= \log_{10} \frac{1}{10^2}$

$= \log_{10} 10^{-2}$ Write $\frac{1}{10^2}$ as 10^{-2} so the base is 10.

$= -2$ By property 4

Example 3: Using the Exponential Form to Solve Logarithmic Equations

Solve by first changing each equation to exponential form.

a. $\log_{16} x = \frac{3}{4}$

Solution: $\log_{16} x = \frac{3}{4}$

$x = 16^{\frac{3}{4}}$ Write the equation in exponential form and solve for x.

$x = \left(16^{\frac{1}{4}}\right)^3 = 2^3 = 8$

Thus $\log_{16} 8 = \frac{3}{4}$.

b. $\log_4 8 = x$

Solution: $\log_4 8 = x$

$4^x = 8$ Write the equation in exponential form and solve for x.

$\left(2^2\right)^x = 2^3$ Use the common base, 2.

$2^{2x} = 2^3$ The exponents are equal because the bases are the same.

$2x = 3$

$x = \frac{3}{2}$

Thus $\log_4 8 = \frac{3}{2}$.

Graphs of Logarithmic Functions

As illustrated in Figure 2, because logarithmic functions are the inverses of exponential functions, the graphs of logarithmic functions can be found by reflecting the corresponding exponential functions across the line $y = x$. Figure 2a shows how the graphs of $y = 2^x$ and $y = \log_2 x$ are related. Figure 2b shows how the graphs of $y = 10^x$ and $y = \log_{10} x$ are related. Note that in the graphs of both logarithmic functions the values of y are negative when x is between 0 and 1 $(0 < x < 1)$.

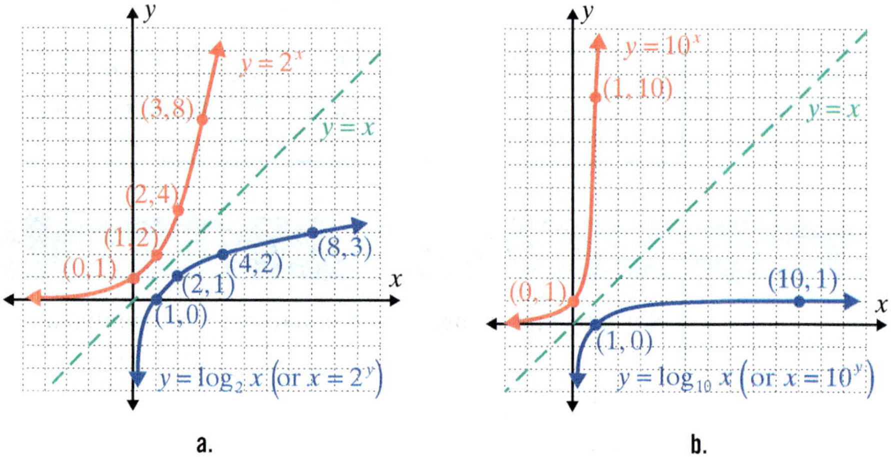

Figure 2

Recall that points on the graphs of inverse functions can be found by reversing the coordinates of ordered pairs. This means that the domain and range of a function and its inverse are interchanged. Thus for exponential functions and logarithmic functions, we have the following.

- For the exponential function $y = b^x$,

  the domain is all real x, and
  the range is all $y > 0$. (The graph is above the x-axis.)
  There is a horizontal asymptote at $y = 0$.

- For the logarithmic function $y = \log_b x$ $(\text{or } x = b^y)$,

  the domain is all $x > 0$, and (The graph is to the right of the y-axis.)
  the range is all real y.
  There is a vertical asymptote at $x = 0$.

1. $\log_7 49 = 2$
2. $\log_3 27 = 3$
3. $\log_5 \frac{1}{25} = -2$
4. $\log_2 \frac{1}{32} = -5$
5. $\log_\pi 1 = 0$
6. $\log_6 1 = 0$
7. $\log_{10} 100 = 2$
8. $\log_{10} 10 = 1$
9. $\log_{10} 23 = k$

Practice Problems

Express the given equation in logarithmic form.

1. $4^2 = 16$ **2.** $5^3 = 125$ **3.** $10^{-1} = \frac{1}{10}$

Express the given equation in exponential form.

4. $\log_2 x = -1$ **5.** $\log_5 x = 2$ **6.** $\log_3 x = 0$

Find the value of each expression.

7. $\log_2 64$ **8.** $\log_3 27$ **9.** $\log_7 7$

11.4 Exercises

10. $\log_4 11.6 = k$
11. $\log_{2/3} \frac{4}{9} = 2$
12. $\log_{3/4} \frac{9}{16} = 2$
13. $3^2 = 9$
14. $5^3 = 125$
15. $9^{\frac{1}{2}} = 3$
16. $b^{\frac{2}{3}} = 4$
17. $7^{-1} = \frac{1}{7}$
18. $\left(\frac{1}{2}\right)^{-3} = 8$
19. $10^{1.74} = N$
20. $2^x = 42.3$
21. $b^4 = 18$
22. $b^{10} = 39$
23. $n^x = y^2$
24. $b^{x^2} = a$
25. $x = 16$ 26. $x = 81$
27. $x = 2$ 28. $x = \frac{3}{2}$
29. $x = -3$ 30. $x = -2$

Express each equation in logarithmic form.

1. $7^2 = 49$ **2.** $3^3 = 27$ **3.** $5^{-2} = \frac{1}{25}$ **4.** $2^{-5} = \frac{1}{32}$

5. $1 = \pi^0$ **6.** $6^0 = 1$ **7.** $10^2 = 100$ **8.** $10^1 = 10$

9. $10^k = 23$ **10.** $4^k = 11.6$ **11.** $\left(\frac{2}{3}\right)^2 = \frac{4}{9}$ **12.** $\left(\frac{3}{4}\right)^2 = \frac{9}{16}$

Express each equation in exponential form.

13. $\log_3 9 = 2$ **14.** $\log_5 125 = 3$ **15.** $\log_9 3 = \frac{1}{2}$ **16.** $\log_b 4 = \frac{2}{3}$

17. $\log_7 \frac{1}{7} = -1$ **18.** $\log_{1/2} 8 = -3$ **19.** $\log_{10} N = 1.74$ **20.** $\log_2 42.3 = x$

21. $\log_b 18 = 4$ **22.** $\log_b 39 = 10$ **23.** $\log_n y^2 = x$ **24.** $\log_b a = x^2$

Solve by first changing each equation to exponential form.

25. $\log_4 x = 2$ **26.** $\log_3 x = 4$ **27.** $\log_{14} 196 = x$ **28.** $\log_{25} 125 = x$

29. $\log_5 \frac{1}{125} = x$ **30.** $\log_3 \frac{1}{9} = x$ **31.** $\log_{36} x = -\frac{1}{2}$ **32.** $\log_{81} x = -\frac{3}{4}$

33. $\log_x 32 = 5$ **34.** $\log_x 121 = 2$ **35.** $\log_8 x = \frac{5}{3}$ **36.** $\log_{16} x = \frac{3}{4}$

37. $\log_8 8^{3.7} = x$ **38.** $\log_{10} 10^{1.52} = x$ **39.** $\log_5 5^{\log_5 25} = x$ **40.** $\log_4 4^{\log_2 8} = x$

Answers to Practice Problems: **1.** $\log_4 16 = 2$ **2.** $\log_5 125 = 3$ **3.** $\log_{10} \frac{1}{10} = -1$ **4.** $x = 2^{-1}$ **5.** $x = 5^2$ **6.** $x = 3^0$ **7.** 6 **8.** 3 **9.** 1

31. $x=\frac{1}{6}$ 32. $x=\frac{1}{27}$
33. $x=2$ 34. $x=11$
35. $x=32$ 36. $x=8$
37. $x=3.7$ 38. $x=1.52$
39. $x=2$ 40. $x=3$

41. 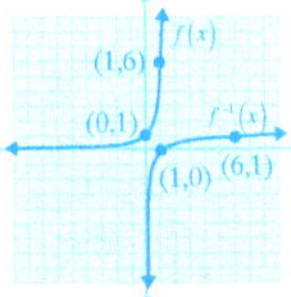

42. 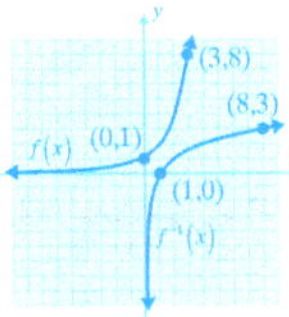

43. 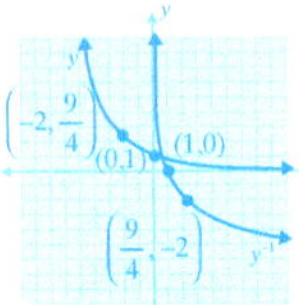

Graph each function and its inverse on the same set of axes. Label two points on each graph.

41. $f(x)=6^x$ **42.** $f(x)=2^x$ **43.** $y=\left(\frac{2}{3}\right)^x$ **44.** $y=\left(\frac{1}{4}\right)^x$

45. $f(x)=\log_4 x$ **46.** $f(x)=\log_5 x$ **47.** $y=\log_{1/2} x$ **48.** $y=\log_{1/3} x$

49. $y=\log_8 x$ **50.** $y=\log_7 x$

51. Consider the function $y=c\left(3^x\right)$ where c is a constant greater than zero. List the following:
a. The domain of the function.
b. The range of the function.
c. Any asymptotes of the graph of the function.
d. Give c two different values and sketch the graphs of both functions.

52. Consider the function $y=c\left(3^{-x}\right)$ where c is a constant greater than zero. List the following:
a. The domain of the function.
b. The range of the function.
c. Any asymptotes of the graph of the function.
d. Give c two different values and sketch the graphs of both functions.

Writing and Thinking About Mathematics

44. 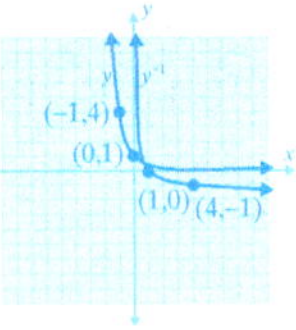

53. Discuss, in your own words, the symmetrical relationship of the graphs of the two functions $y=10^x$ and $y=\log_{10} x$.

54. Discuss, in your own words, the symmetrical relationship of the graphs of the two logarithmic functions $y=\log_{10} x$ and $y=-\log_{10} x$.

HAWKES LEARNING SYSTEMS: INTRODUCTORY & INTERMEDIATE ALGEBRA SOFTWARE

- 11.4 Logarithmic Functions

45. 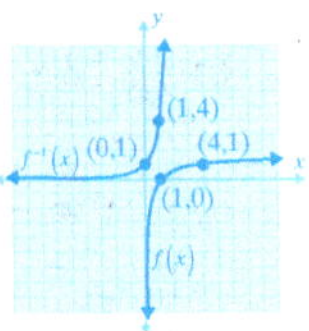

46. 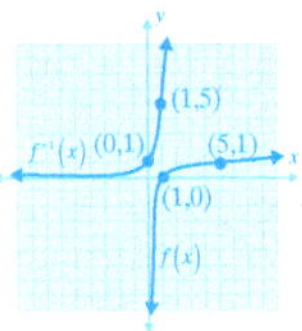

47. 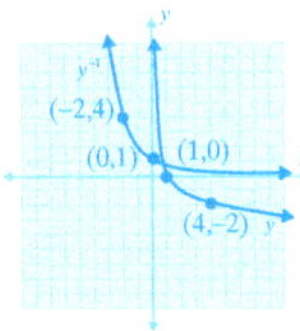

48. 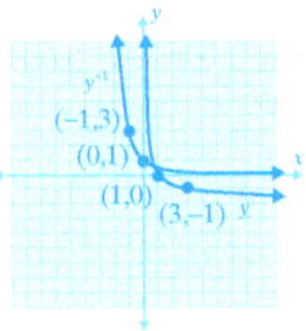

49. 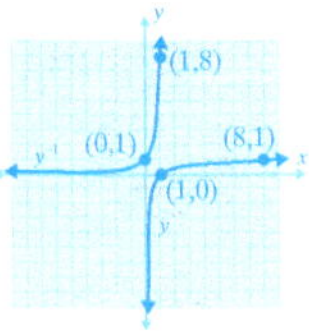

50. 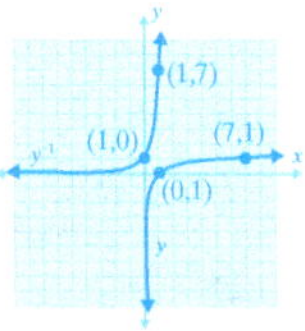

51. a. $D=(-\infty,\infty)$ b. $R=(0,\infty)$ c. $y=0$ d. Answers will vary.
52. a. $D=(-\infty,\infty)$ b. $R=(0,\infty)$ c. $y=0$ d. Answers will vary.
53. The two functions are symmetric about the line $y=x$.
54. The two functions are symmetric about the x-axis.

11.5 Properties of Logarithms

- *Understand the rules for changing the form of logarithmic expressions.*
- *Use the rules for changing the form of logarithmic expressions to evaluate logarithms.*
- *Use the rules to expand logarithmic expressions and to write expanded forms as single logarithms.*

Teaching Notes:
It doesn't hurt to review the product rule for exponents before getting into this section, as the product rule is the same in this context. Similarly, the power rule and quotient rule are the same.

While calculators are effective in giving numerical evaluations, they do not, in general, simplify or solve equations. (You might be aware that there are calculators with computer algebra systems, CAS, that do simplify expressions and solve equations.) In this section, we will discuss several properties (or rules) of logarithms that are helpful in solving equations and simplifying expressions that involve logarithms or exponential functions.

The following four basic properties have been discussed.

For $b > 0$ and $b \neq 1$,

1. $\log_b 1 = 0$

2. $\log_b b = 1$

3. $x = b^{\log_b x}$ for $x > 0$

4. $\log_b b^x = x$.

With these four properties as a basis, we now develop three more properties (or rules) for logarithms: the product rule, the quotient rule, and the power rule.

The Product Rule

Because logarithms are exponents, their properties are similar to those of exponents. In fact, the properties of exponents are used to prove the rules for logarithms. Study the developments carefully and you will see that logarithms are handled just as exponents.

NOTES

Using a calculator to find the values of logarithms is discussed in the next section. To illustrate some of the properties of logarithms in this section, the following base 10 logarithmic values, accurate to 4 decimal places are used.

$$\log_{10} 2 \approx 0.3010 \qquad \log_{10} 3 \approx 0.4771$$
$$\log_{10} 5 \approx 0.6990 \qquad \log_{10} 6 \approx 0.7782$$

Also, because the logarithms are rounded to 4 decimal places, a slight difference may occur in the answers if a calculator is used to find the logarithms in some of the illustrations.

Consider the following analysis.

We know that $6 = 2 \cdot 3.$

Using property 3,

$$10^{\log_{10} 6} = 6, \qquad 10^{\log_{10} 2} = 2, \qquad \text{and} \qquad 10^{\log_{10} 3} = 3.$$

Now using the properties of exponents,

$$\begin{array}{ccccccc} 6 & = & 2 & \cdot & 3 & & \\ \downarrow & & \downarrow & & \downarrow & & \\ 10^{\log_{10} 6} & = & 10^{\log_{10} 2} & \cdot & 10^{\log_{10} 3} & = & 10^{\log_{10} 2 + \log_{10} 3}. \end{array}$$

Equating exponents (with the same base) gives the result:

$$\begin{array}{ccccc} \log_{10} 6 & = & \log_{10} 2 & + & \log_{10} 3 \\ \downarrow & & \downarrow & & \downarrow \\ 0.7782 & \approx & 0.3010 & + & 0.4771. \end{array}$$

Note that the fourth digit is off because of rounding.

This technique can be used to prove the following **product rule for logarithms.**

Product Rule for Logarithms

For $b > 0, b \neq 1$, and $x, y > 0$,

$$\mathbf{\log_b xy = \log_b x + \log_b y}.$$

In words, **the logarithm of a product is equal to the sum of the logarithms of the factors.**

Proof of the Product Rule

$$\underbrace{b^{\log_b(xy)} = xy}_{\text{Property 3}} = \underbrace{x \cdot y = b^{\log_b x} \cdot b^{\log_b y}}_{\text{Property 3 again}} = \underbrace{b^{\log_b x + \log_b y}}_{\text{Add exponents}}$$

Thus

$$b^{\log_b(xy)} = b^{\log_b x + \log_b y}.$$

Equating the exponents gives the product rule for logarithms:

$$\log_b xy = \log_b x + \log_b y.$$

Example 1: Using the Product Rule of Logarithms

Simplify each of the following expressions using the product rule.

a. $\log_{10} 1000$

Solution:
$$\begin{aligned}\log_{10} 1000 &= \log_{10}(10 \cdot 10 \cdot 10)\\ &= \log_{10} 10 + \log_{10} 10 + \log_{10} 10\\ &= 1+1+1\\ &= 3\end{aligned}$$

(We knew this result from another approach: $\log_{10} 1000 = \log_{10} 10^3 = 3$. But the approach shown helps to confirm that the product rule works.)

b. $\log_{10} 30$

Solution:
$$\begin{aligned}\log_{10} 30 &= \log_{10} 10 \cdot 3\\ &= \log_{10} 10 + \log_{10} 3\\ &\approx 1 + 0.4771 = 1.4771\end{aligned}$$

c. $\log_{10} 30$

Solution:
$$\begin{aligned}\log_{10} 30 &= \log_{10}(2 \cdot 3 \cdot 5)\\ &= \log_{10} 2 + \log_{10} 3 + \log_{10} 5\\ &\approx 0.3010 + 0.4771 + 0.6990 = 1.4771\end{aligned}$$

Note that in **b.** and **c.** using different factors of 30 gives the same result.

The Quotient Rule

Now consider the problem of finding the logarithm of a quotient. For example, to find $\log_{10} \frac{3}{2}$, we can proceed as follows:

$$\begin{aligned}10^{\log_{10} \frac{3}{2}} &= \frac{3}{2}\\ &= \frac{10^{\log_{10} 3}}{10^{\log_{10} 2}}\\ &= 10^{\log_{10} 3 - \log_{10} 2}.\end{aligned}$$

Equating exponents gives:

$$\log_{10}\frac{3}{2} = \log_{10}3 - \log_{10}2$$
$$\approx 0.4771 - 0.3010$$
$$= 0.1761.$$

These ideas lead to the following **quotient rule for logarithms**. The proof is left as an exercise for the student.

Quotient Rule for Logarithms

For $b > 0, b \neq 1$, and $x, y > 0$,

$$\log_b \frac{x}{y} = \log_b x - \log_b y.$$

In words, **the logarithm of a quotient is equal to the difference between the logarithm of the numerator and the logarithm of the denominator**.

Example 2: Using the Quotient Rule of Logarithms

Simplify each of the following expressions using the quotient rule.

a. $\log_{10}\frac{1}{2}$

Solution: $\log_{10}\frac{1}{2} = \log_{10}1 - \log_{10}2$
$$= 0 - \log_{10}2 \approx -0.3010$$

b. $\log_2\frac{1}{8}$

Solution: $\log_2\frac{1}{8} = \log_2 1 - \log_2 8$
$$= \log_2 1 - \log_2 2^3$$
$$= 0 - 3 = -3$$

Note: Writing $\frac{1}{8}$ as 2^{-3} produces the same result.

c. $\log_{10}\frac{15}{2}$

Solution: $\log_{10}\frac{15}{2} = \log_{10}15 - \log_{10}2$
$$= \log_{10}(3 \cdot 5) - \log_{10}2$$
$$= \log_{10}3 + \log_{10}5 - \log_{10}2$$
$$\approx 0.4771 + 0.6990 - 0.3010 = 0.8751$$

The Power Rule

The next property of logarithms involves a number raised to a power and the multiplication property of exponents. For example,

$$\begin{aligned} 10^{\log_{10} 2^3} &= 2^3 \\ &= \left(10^{\log_{10} 2}\right)^3 \\ &= 10^{3(\log_{10} 2)}. \quad \text{By the power rule for exponents} \end{aligned}$$

Now equating exponents in the first and last expressions gives

$$\log_{10} 2^3 = 3 \cdot \log_{10} 2.$$

These ideas lead to the following **power rule for logarithms**.

Power Rule for Logarithms

For $b > 0, b \neq 1, x > 0$, and any real number r,

$$\log_b x^r = r \cdot \log_b x.$$

In words, **the logarithm of a number raised to a power is equal to the product of the exponent and the logarithm of the number**.

Proof of the Power Rule for Logarithms

$$\overbrace{b^{\log_b x^r} = x^r = \left(b^{\log_b x}\right)^r}^{\text{Property 3 twice}} = \overbrace{b^{r \cdot \log_b x}}^{\text{Multiply exponents}}$$

Equating the exponents gives the result called the power rule for logarithms:

$$\log_b x^r = r \cdot \log_b x.$$

Example 3: Using the Power Rule

a. $\log_{10} \sqrt{2}$

Solution: $\log_{10} \sqrt{2} = \log_{10} 2^{\frac{1}{2}} = \frac{1}{2} \cdot \log_{10} 2 \approx \frac{1}{2}(0.3010) = 0.1505$

b. $\log_{10} 8$

Solution: $\log_{10} 8 = \log_{10} 2^3 = 3 \cdot \log_{10} 2 \approx 3(0.3010) = 0.9030$

c. $\log_{10} 25$

Solution: $\log_{10} 25 = \log_{10} 5^2 = 2 \cdot \log_{10} 5 \approx 2(0.6990) = 1.3980$

Summary of Properties of Logarithms

For $b > 0, b \neq 1, x, y > 0$, and any real number r,

1. $\log_b 1 = 0$
2. $\log_b b = 1$
3. $x = b^{\log_b x}$
4. $\log_b b^x = x$
5. $\log_b xy = \log_b x + \log_b y$ The product rule
6. $\log_b \frac{x}{y} = \log_b x - \log_b y$ The quotient rule
7. $\log_b x^r = r \cdot \log_b x$ The power rule

Example 4: Using the Properties of Logarithms to Expand Expressions

Use the properties of logarithms to expand each expression as much as possible.

a. $\log_b 2x^3$

Solution: $\log_b 2x^3 = \log_b 2 + \log_b x^3$ Product rule

$= \log_b 2 + 3\log_b x$ Power rule

b. $\log_b \frac{xy^2}{z}$

Solution: $\log_b \frac{xy^2}{z} = \log_b xy^2 - \log_b z$ Quotient rule

$= \log_b x + \log_b y^2 - \log_b z$ Product rule

$= \log_b x + 2\log_b y - \log_b z$ Power rule

c. $\log_b (xy)^{-3}$

Solution: $\log_b (xy)^{-3} = -3\log_b xy$ Power rule

$= -3(\log_b x + \log_b y)$ Product rule

$= -3\log_b x - 3\log_b y$

d. $\log_b \sqrt{3x}$

Solution: $\log_b \sqrt{3x} = \log_b (3x)^{\frac{1}{2}} = \frac{1}{2}\log_b 3x$ Power rule

$= \frac{1}{2}(\log_b 3 + \log_b x)$ Product rule

$= \frac{1}{2}\log_b 3 + \frac{1}{2}\log_b x$

Example 5: Using the Properties of Logarithms to Write Single Logarithm Expressions

Use the properties of logarithms to write each expression as a single logarithm. Note that the bases must be the same when simplifying sums and differences of logarithms.

a. $2\log_b x - 3\log_b y$

Solution: $2\log_b x - 3\log_b y = \log_b x^2 - \log_b y^3$ Power rule

$= \log_b\left(\frac{x^2}{y^3}\right)$ Quotient rule

b. $\frac{1}{2}\log_a 4 - \log_a 5 - \log_a y$

Solution: $\frac{1}{2}\log_a 4 - \log_a 5 - \log_a y$

$= \log_a 4^{\frac{1}{2}} - \log_a 5 - \log_a y$ Power rule

$= \log_a 2 - (\log_a 5 + \log_a y)$ $\left(4^{\frac{1}{2}} = 2\right)$ Factor out –1.

$= \log_a 2 - \log_a 5y$ Product rule

$= \log_a\left(\frac{2}{5y}\right)$ Quotient rule

c. $\log_a(x+1) + \log_a(x-1)$

Solution: $\log_a(x+1) + \log_a(x-1) = \log_a\left[(x+1)(x-1)\right]$ Product rule

$= \log_a\left(x^2 - 1\right)$ Multiply.

d. $\log_b \sqrt{x} + \log_b \sqrt[3]{x}$

Solution: $\log_b \sqrt{x} + \log_b \sqrt[3]{x} = \log_b x^{\frac{1}{2}} + \log_b x^{\frac{1}{3}}$

$= \frac{1}{2}\log_b x + \frac{1}{3}\log_b x$ Power rule

$= \left(\frac{1}{2} + \frac{1}{3}\right)\log_b x$ Distributive property

$= \frac{5}{6}\log_b x$ $\frac{1}{2} + \frac{1}{3} = \frac{3}{6} + \frac{2}{6} = \frac{5}{6}$

Continued on the next page...

OR

$$\log_b \sqrt{x} + \log_b \sqrt[3]{x} = \log_b x^{\frac{1}{2}} + \log_b x^{\frac{1}{3}}$$

$$= \log_b \left(x^{\frac{1}{2}} \cdot x^{\frac{1}{3}}\right) \quad \text{Product rule}$$

$$= \log_b x^{\left(\frac{1}{2}+\frac{1}{3}\right)}$$

$$= \log_b x^{\frac{5}{6}}$$

$$= \frac{5}{6}\log_b x \quad \text{Power rule}$$

NOTES

Common Misunderstandings about Logarithms

There is no logarithmic property for the logarithm of a sum or a difference.

$\log_b(x+y)$ Cannot be simplified

$\log_b(x-y)$ Cannot be simplified

Also,

$\log_b(xy) \neq \log_b x \cdot \log_b y$ The log of a product does not equal the product of the logs.

$\log_b \frac{x}{y} \neq \frac{\log_b x}{\log_b y}$ The log of a quotient does not equal the quotient of the logs.

Practice Problems

1. Write $\log_b(x^3 y)$ as a sum of logarithmic expressions.
2. Write the expression $2\log_b 5 + \log_b x - \log_b 3$ as a single logarithm.
3. Write the expression $2\log_b x - \log_b(x+1)$ as a single logarithm.

Answers to Practice Problems: **1.** $3\log_b x + \log_b y$ **2.** $\log_b\left(\frac{25x}{3}\right)$ **3.** $\log_b\left(\frac{x^2}{x+1}\right)$

11.5 Exercises

1. a. 2 b. 3 c. 5 d. 6
2. a. 12 b. 18 c. 15 d. 2.5
3. 5
4. 2
5. −2
6. −3
7. $\frac{1}{2}$
8. $\frac{3}{2}$
9. 10
10. 17
11. $\sqrt{3}$
12. 5
13. $\log_b 5 + 4\log_b x$
14. $\log_b 3 + 2\log_b x + \log_b y$
15. $\log_b 2 - 3\log_b x + \log_b y$
16. $\log_5 x + 2\log_5 y - \log_5 z$
17. $\log_6 2 + \log_6 x - 3\log_6 y$
18. $\log_3 x + \log_3 y - \log_3 4 - \log_3 z$
19. $2\log_b x - \log_b y - \log_b z$
20. $\log_3 x + 2\log_3 y - 2\log_3 z$
21. $-2\log_5 x - 2\log_5 y$
22. $8\log_b x + 4\log_b y$
23. $\frac{1}{3}\log_6 x + \frac{2}{3}\log_6 y$
24. $\frac{1}{2}\log_5 2 + \frac{3}{2}\log_5 x + \frac{1}{2}\log_5 y$
25. $\frac{1}{2}\log_3 x + \frac{1}{2}\log_3 y - \frac{1}{2}\log_3 z$
26. $\frac{2}{3}\log_6 x - \frac{1}{3}\log_6 y$
27. $\log_5 21 + 2\log_5 x + \frac{2}{3}\log_5 y$

Use the following logarithms (accurate to 4 decimal places) in exercises 1 and 2.

$$\log_{10} 2 \approx 0.3010 \qquad \log_{10} 3 \approx 0.4771$$
$$\log_{10} 5 \approx 0.6990 \qquad \log_{10} 6 \approx 0.7782$$

1. Find the value of the following expressions.

a. $10^{0.3010}$ **b.** $10^{0.4771}$ **c.** $10^{0.6990}$ **d.** $10^{0.7782}$

2. Find the value of the following expressions.

a. $10^{0.3010 + 0.7782}$ **b.** $10^{0.4771 + 0.7782}$ **c.** $10^{0.4771 + 0.6990}$ **d.** $10^{0.6990 - 0.3010}$

Use your knowledge of logarithms and exponents to find the value of each expression.

3. $\log_2 32$ **4.** $\log_3 9$ **5.** $\log_4 \frac{1}{16}$ **6.** $\log_5 \frac{1}{125}$

7. $\log_3 \sqrt{3}$ **8.** $\log_2 \sqrt{8}$ **9.** $5^{\log_5 10}$ **10.** $3^{\log_3 17}$

11. $6^{\log_6 \sqrt{3}}$ **12.** $5^{\log_5 5}$

Use the properties of logarithms to expand each expression as much as possible.

13. $\log_b 5x^4$ **14.** $\log_b 3x^2y$ **15.** $\log_b 2x^{-3}y$ **16.** $\log_5 xy^2z^{-1}$

17. $\log_6 \frac{2x}{y^3}$ **18.** $\log_3 \frac{xy}{4z}$ **19.** $\log_b \frac{x^2}{yz}$ **20.** $\log_3 \frac{xy^2}{z^2}$

21. $\log_5 (xy)^{-2}$ **22.** $\log_b (x^2y)^4$ **23.** $\log_6 \sqrt[3]{xy^2}$ **24.** $\log_5 \sqrt{2x^3y}$

25. $\log_3 \sqrt{\frac{xy}{z}}$ **26.** $\log_6 \sqrt[3]{\frac{x^2}{y}}$ **27.** $\log_5 21x^2y^{\frac{2}{3}}$ **28.** $\log_b 15x^{-\frac{1}{2}}y^{\frac{1}{3}}$

29. $\log_6 \frac{x}{\sqrt{x^3y^5}}$ **30.** $\log_3 \frac{1}{\sqrt{x^4y}}$ **31.** $\log_b \left(\frac{x^3y^2}{z}\right)^{-3}$ **32.** $\log_4 \left(\frac{x^{-\frac{1}{2}}y}{z^2}\right)^{-2}$

Use the properties of logarithms to write each expression as a single logarithm of a single expression.

33. $2\log_b 3 + \log_b x - \log_b 5$ **34.** $\frac{1}{2}\log_b 25 + \log_b 3 - \log_b x$

35. $\log_2 7 + \log_2 9 + 2\log_2 x$ **36.** $\log_5 4 + \log_5 6 + \log_5 y$

37. $2\log_b x + \log_b y$ **38.** $\log_2 x + 3\log_2 y$

28. $\log_b 15 - \frac{1}{2}\log_b x + \frac{1}{3}\log_b y$

29. $-\frac{1}{2}\log_6 x - \frac{5}{2}\log_6 y$

30. $-2\log_3 x - \frac{1}{2}\log_3 y$

31. $-9\log_b x - 6\log_b y + 3\log_b z$

32. $\log_4 x - 2\log_4 y + 4\log_4 z$

33. $\log_b\left(\frac{9x}{5}\right)$

34. $\log_b\left(\frac{15}{x}\right)$

35. $\log_2 63x^2$

36. $\log_5 24y$

37. $\log_b x^2 y$

38. $\log_2 xy^3$

39. $3\log_5 y - \frac{1}{2}\log_5 x$

40. $3\log_{10} x - 2\log_{10} y$

41. $\frac{1}{2}(\log_5 x - \log_5 y)$

42. $\frac{1}{3}(\log_{10} x - 2\log_{10} y)$

43. $\log_2 x - \log_2 y + \log_2 z$

44. $\log_b x - 2\log_b y - 2\log_b z$

45. $\log_b x + 2\log_b y - \frac{1}{2}\log_b z$

46. $-\frac{2}{3}\log_2 x - \frac{1}{3}\log_2 y + \frac{2}{3}\log_2 z$

47. $2\log_5 x + \log_5(2x+1)$

48. $\log_b(3x+1) + 2\log_b x$

49. $\log_2(x-1) + \log_2(x+3)$

50. $\log_{10}(x+3) + \log_{10}(x-3)$

51. $\log_b(x^2-2x-3) - \log_b(x-3)$

52. $\log_2(x-4) - \log_2(x^2-2x-8)$

53. $\log_{10}(x+6) - \log_{10}(2x^2+9x-18)$

54. $\log_5(3x^2+5x-2) - \log_5(3x-1)$

Writing and Thinking About Mathematics

55. Prove the quotient rule for logarithms: For $b > 0, b \neq 1$, and $x, y > 0$,

$$\log_b \frac{x}{y} = \log_b x - \log_b y.$$

56. Prove the following property of logarithms: For $b > 0, b \neq 1$, and $x > 0$,

$$\log_b b^x = x.$$

HAWKES LEARNING SYSTEMS: INTRODUCTORY & INTERMEDIATE ALGEBRA SOFTWARE

- 11.5 Properties of Logarithms

39. $\log_5\left(\frac{y^3}{\sqrt{x}}\right)$ or $\log_5\left(\frac{y^3\sqrt{x}}{x}\right)$ 40. $\log_{10}\left(\frac{x^3}{y^2}\right)$ 41. $\log_5\sqrt{\frac{x}{y}}$ 42. $\log_{10}\sqrt[3]{\frac{x}{y^2}}$ 43. $\log_2\left(\frac{xz}{y}\right)$

44. $\log_b\left(\frac{x}{y^2z^2}\right)$ 45. $\log_b\left(\frac{xy^2}{\sqrt{z}}\right)$ or $\log_b\left(\frac{xy^2\sqrt{z}}{z}\right)$ 46. $\log_2\sqrt[3]{\frac{z^2}{x^2y}}$ 47. $\log_5(2x^3+x^2)$ 48. $\log_b(3x^3+x^2)$

49. $\log_2(x^2+2x-3)$ 50. $\log_{10}(x^2-9)$ 51. $\log_b(x+1)$ 52. $\log_2\left(\frac{1}{x+2}\right)$ 53. $\log_{10}\left(\frac{1}{2x-3}\right)$

54. $\log_5(x+2)$ 55. Answers will vary. 56. Answers will vary.

11.6 Common Logarithms and Natural Logarithms

- *Understand that base 10 logarithms are called* ***common logarithms****.*
- *Use a calculator to find the values of expressions with common logarithms.*
- *Understand that base e logarithms are called* ***natural logarithms****.*
- *Use a calculator to find the values of expressions with natural logarithms.*
- *Use a calculator to find inverse logarithms (called antilogs) for common and for natural logarithms.*

Base 10 logarithms (called **common logarithms**) and base *e* logarithms (called **natural logarithms**) are the two most commonly used logarithms. Common logarithms occur frequently because of their close relationship with the decimal system and are used in chemistry, astronomy, and physical science. Natural logarithms are an important tool in calculus and appear "naturally" in growth and decay problems. Both common and natural logarithms have designated keys on calculators.

Common Logarithms (Base 10 Logarithms)

The abbreviated notation $\log x$ is used to indicate common logarithms (base 10 logarithms). That is, **whenever the base notation is omitted, the base is understood to be 10**:

$$\log_{10} x = \log x.$$

Example 1: Translations Between Exponential Form and Logarithmic Form Using Common Logarithms

	Exponential Form		Logarithmic Form	
a.	$10^4 = 10,000$	$\Leftrightarrow$	$\log 10,000 = 4$	This is a common logarithm that equals 4.
b.	$10^{-2} = 0.01$	$\Leftrightarrow$	$\log 0.01 = -2$	This is a common logarithm that equals –2.
c.	$10^x = 6.3$	$\Leftrightarrow$	$\log 6.3 = x$	This is a common logarithm that equals x.

The TI-84 Plus graphing calculator has the designated key LOG that can be used in the following 3-step process to find the values of common logarithms.

Using a Calculator to Evaluate Common Logarithms (Base 10)

To evaluate a **common logarithm** on a TI-84 Plus graphing calculator:

1. Press the LOG key. `log(` will appear on the display.
2. Enter the number and a right-hand parenthesis,).
3. Press ENTER.

Example 2: Evaluating Common Logarithms Using a TI-84 Plus Graphing Calculator

Use a TI-84 Plus graphing calculator to find the values of the following common logarithms.

a. $\log 200$ **b.** $\log 50{,}000$ **c.** $\log 0.0006$

Solutions: From the display we see the results (accurate to 8 or 9 decimal places).

a. $\log 200 \approx 2.301029996$

b. $\log 50{,}000 \approx 4.698970004$

c. $\log 0.0006 \approx -3.22184875$

```
log(200)
            2.301029996
log(50000)
            4.698970004
log(0.0006)
            -3.22184875
```

Example 2c shows a negative value for the logarithm. This is because 0.0006 is between 0 and 1. **In fact, the logarithm (with base greater than 1) of any number between 0 and 1 will always be negative.** See Figure 1.

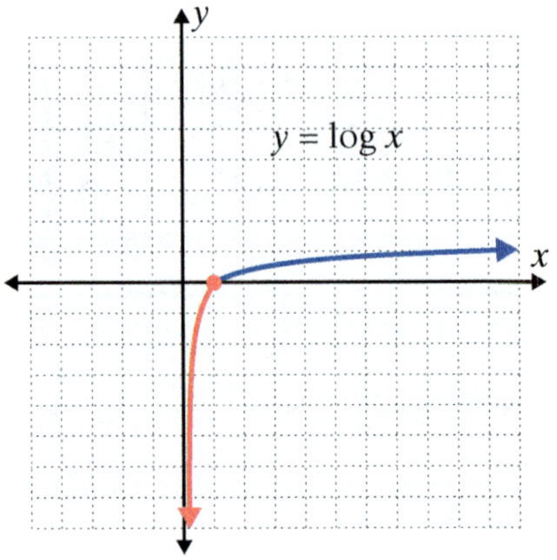

Figure 1

Logarithms are exponents and they may be positive, negative, or 0. For example,

$$10^{-2} = \frac{1}{100} = 0.01 \qquad \text{and} \qquad \log 0.01 = -2.$$

However, there are no logarithms of negative numbers or 0. The domains of logarithmic functions are nonnegative real numbers. The logarithm of a negative number is **undefined**. For example, $\log(-2)$ is undefined. That is, there is no real number x such that $10^x = -2$. If $\log(-2)$ is entered in a calculator, an error message will appear as shown here.

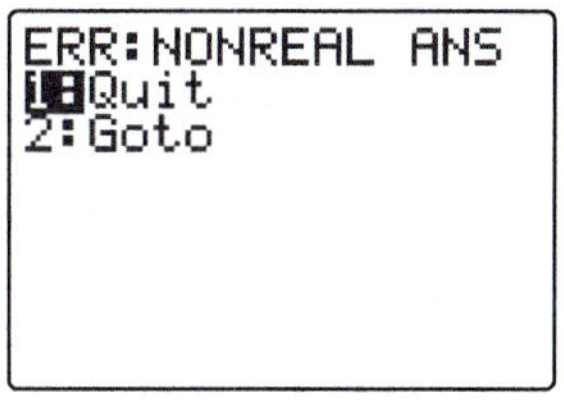

Figure 2

Finding the Inverse of a Logarithm (Base 10)

Because logarithms are exponents, if the value of a common logarithm is known, an exponent is known. Finding the value of the related exponential expression is called finding the **inverse of the logarithm** (or finding the **antilog**). For example,

if $\log x = N$, then $x = 10^N$ and the number x is called the **inverse log of *N***.

Using a TI-84 Plus Graphing Calculator to find the Inverse Log of *N*

Finding the inverse log of N is a three-step process.

Using a Calculator to Find the Inverse Log of *N*

To evaluate the **inverse log of *N*** on a TI-84 Plus graphing calculator:

1. Press 2ND and LOG. The expression 10^(will appear on the display.
2. Enter the value of N and a right-hand parenthesis).
3. Press ENTER.

Example 3: Using a TI-84 Plus Graphing Calculator to find the Inverse Log of *N*

Use a TI–84 Plus graphing calculator to find the **inverse log of *N*** for each expression. (That is, find the value of x.)

a. $\log x = 5$ **b.** $\log x = -3$ **c.** $\log x = 2.4142$ **d.** $\log x = 16.5$

Solutions: Thus the calculator gives,

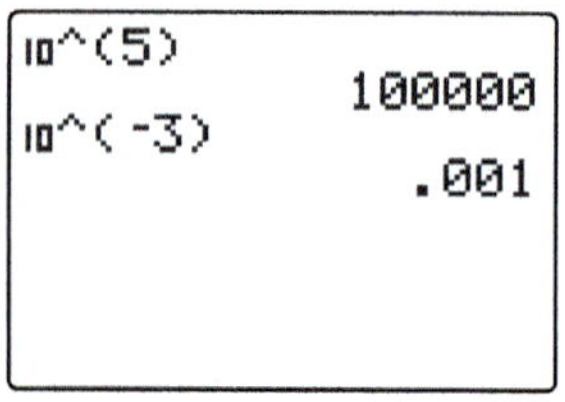

a. $x = 10^5 = 100{,}000$

b. $x = 10^{-3} = 0.001$

c. $x = 10^{2.4142} \approx 259.5374301$

d. $x = 10^{16.5} \approx 3.16227766\text{E}16$

The letter E in the solution is the calculator version of scientific notation. Thus

$$3.16227766\text{E}16 = 3.16227766 \times 10^{16}.$$

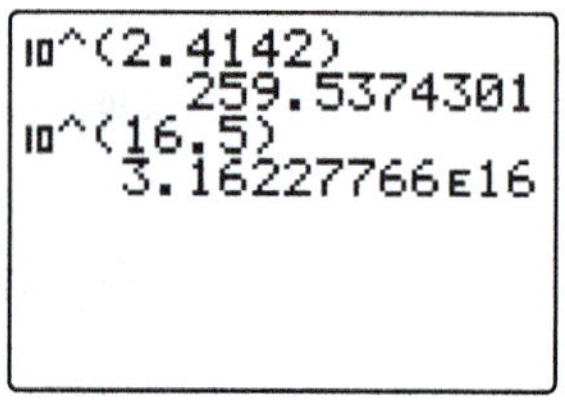

Natural Logarithms (Base *e* Logarithms)

The number $e \approx 2.718281828$ is an irrational number that appears quite naturally in continuously compounded interest. Base e logarithms are called natural logarithms. **The notation for natural logarithms is shortened to ln *x*** (read "L N of x" or "natural log of x"). That is,

$$\log_e x = \ln x.$$

Figure 3 shows the graphs of the two functions $y = e^x$ and its inverse $y = \ln x$.

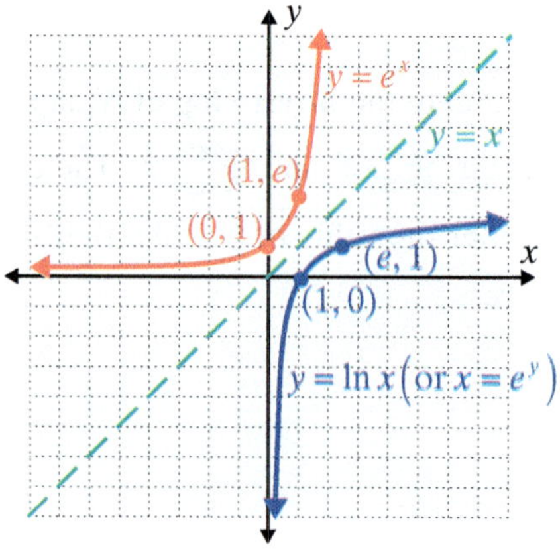

Figure 3

Example 4: Translations Between Exponential Form and Logarithmic Form Using Natural Logarithms

	Exponential Form		Logarithmic Form	
a.	$e^t = 3.21$	$\Leftrightarrow$	$\ln 3.21 = t$	This is a natural logarithm that equals t.
b.	$e^0 = 1$	$\Leftrightarrow$	$\ln 1 = 0$	This is a natural logarithm that equals 0.
c.	$e^2 = n$	$\Leftrightarrow$	$\ln n = 2$	This is a natural logarithm that equals 2.

Using a Calculator to Evaluate Natural Logarithms (Base *e*)

To evaluate a **natural logarithm** on a TI-84 Plus graphing calculator:

1. Press the LN key. `ln(` will appear on the display.
2. Enter the number and a right-hand parenthesis,).
3. Press ENTER.

Example 5: Evaluating Natural Logarithms Using a TI-84 Plus Graphing Calculator

Use a TI-84 Plus graphing calculator to find the values of the following natural logarithms.

a. $\ln 1$ **b.** $\ln 3$ **c.** $\ln 0.02$

Solutions: From the display we see the results (accurate to 9 decimal places).

a. $\ln 1 = 0$

b. $\ln 3 \approx 1.098612289$

c. $\ln 0.02 \approx -3.912023005$

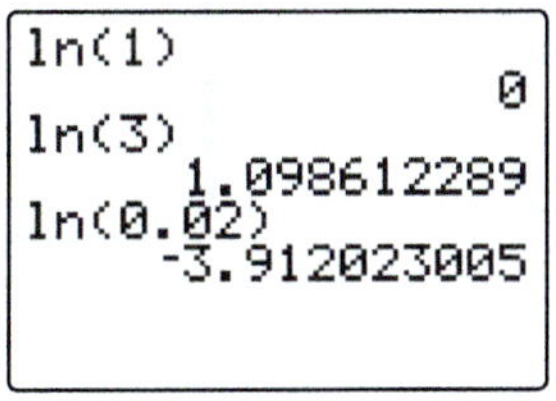

Finding the Inverse of a Natural Logarithm (Base e)

Again, because logarithms are exponents, if we know the value of a natural logarithm, we know an exponent. Finding the value of the related exponential expression is called finding the **inverse of the logarithm**. For example,

if $\ln x = N$, then $x = e^N$ and the number x is called the **inverse ln of *N***.

Using a Calculator to Find the Inverse ln of *N*

To evaluate the **inverse ln of *N*** on a TI-84 Plus graphing calculator:

1. Press 2ND and LN. The expression e^(will appear on the display.
2. Enter the value of N and a right-hand parenthesis).
3. Press ENTER.

Example 6: Using a TI-84 Plus Graphing Calculator to find the Inverse ln of *N*

Use a TI-84 Plus graphing calculator to find the inverse ln of N for each expression. (That is, find the value of x.)

a. $\ln x = 3$ **b.** $\ln x = -1$ **c.** $\ln x = -0.1$ **d.** $\ln x = 50$

Solutions: Thus the calculator gives,

a. $x = e^3 \approx 20.08553692$

b. $x = e^{-1} \approx 0.3678794412$

c. $x = e^{-0.1} \approx 0.904837418$

d. $x = e^{50} \approx 5.184705529\text{E}21$

$= 5.184705529 \times 10^{21}$

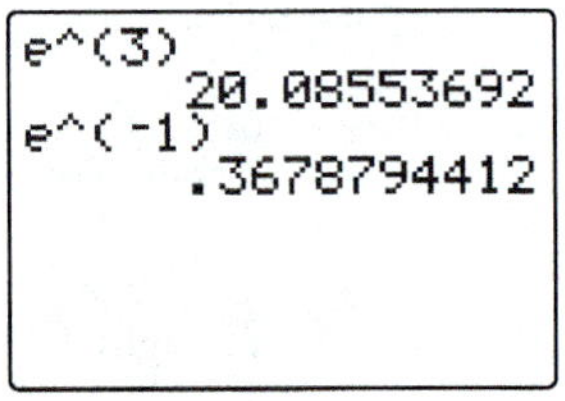

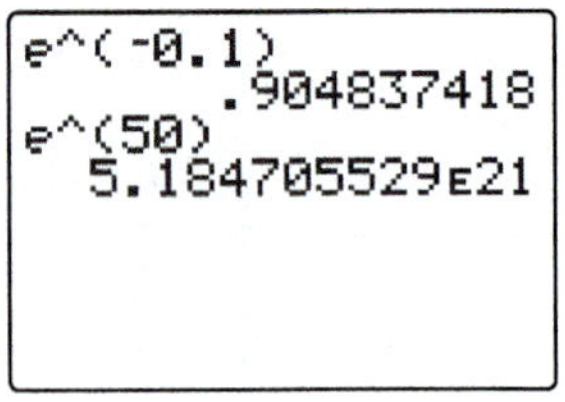

Practice Problems

Express each equation in logarithmic form.

1. $10^1 = 10$ **2.** $e^3 = x$ **3.** $10^{-4} = 0.0001$

Express each equation in exponential form.

4. $\log(0.1) = -1$ **5.** $\log\frac{1}{100} = -2$ **6.** $\ln x = 6$

Use a calculator to evaluate each logarithm (accurate to 4 decimal places).

7. $\log 175$ **8.** $\ln 52$ **9.** $\log(-6.1)$

Use a calculator to find the value of x in each equation (accurate to 4 decimal places).

10. $\ln x = -2$ **11.** $\log x = 0.5$ **12.** $\ln x = 30$

11.6 Exercises

1. $\log x = 1.5$
2. $\log 23 = k$
3. $\log\frac{1}{1000} = -3$
4. $\log 0.0001 = -4$
5. $\ln 27 = x$
6. $\ln 12.4 = k$
7. $\ln 1 = 0$
8. $\ln x = 4$
9. $\log 3.2 = x$
10. $\log x = y$
11. $10^0 = 1$
12. $10^2 = 100$
13. $10^y = 5.4$
14. $e^x = 40.1$
15. $e^{1.54} = x$
16. $10^{25.3} = x$
17. $e^1 = e$

Express each equation in logarithmic form.

1. $10^{1.5} = x$ **2.** $10^k = 23$ **3.** $10^{-3} = \frac{1}{1000}$ **4.** $10^{-4} = 0.0001$

5. $e^x = 27$ **6.** $e^k = 12.4$ **7.** $e^0 = 1$ **8.** $e^4 = x$

9. $10^x = 3.2$ **10.** $10^y = x$

Express each equation in exponential form.

11. $\log 1 = 0$ **12.** $\log 100 = 2$ **13.** $\log 5.4 = y$ **14.** $\ln 40.1 = x$

15. $\ln x = 1.54$ **16.** $\log x = 25.3$ **17.** $\ln e = 1$ **18.** $\log 10 = 1$

19. $\ln x = a$ **20.** $\log a = x$

Use a calculator to evaluate the logarithms accurate to 4 decimal places.

21. log 173 **22.** log 396 **23.** log 88.4 **24.** log 0.0061

Answers to Practice Problems: **1.** $\log 10 = 1$ **2.** $\ln x = 3$ **3.** $\log 0.0001 = -4$ **4.** $10^{-1} = 0.1$ **5.** $10^{-2} = \frac{1}{100}$ **6.** $e^6 = x$ **7.** 2.2430 **8.** 3.9512 **9.** Error (undefined) **10.** 0.1353 **11.** 3.1623 **12.** 1.0686×10^{13}

18. $10^1 = 10$
19. $e^a = x$
20. $10^x = a$
21. 2.2380
22. 2.5977
23. 1.9465
24. −2.2147
25. −1.2418
26. Error (undefined)
27. 3.6243
28. 4.5643
29. Error (undefined)

25. $\log 0.0573$ **26.** $\log(-8.47)$ **27.** $\ln 37.5$ **28.** $\ln 96$

29. $\ln(-14.9)$ **30.** $\ln 157.6$ **31.** $\ln 0.00461$ **32.** $\ln 0.0139$

Use a calculator to find the value of x in each equation accurate to 4 decimal places.

33. $\log x = 2.31$ **34.** $\log x = -3$ **35.** $\log x = -1.7$

36. $\log x = 4.1$ **37.** $2 \log x = -0.038$ **38.** $5 \log x = 9.4$

39. $\ln x = 5.17$ **40.** $\ln x = 4.9$ **41.** $\ln x = -8.3$

42. $\ln x = 6.74$ **43.** $0.2 \ln x = 0.0079$ **44.** $3 \ln x = -0.066$

Writing and Thinking About Mathematics

45. Explain the difference in the meaning of the expressions $\log x$ and $\ln x$.

46. The function $y = \log x$ is defined only for $x > 0$. Discuss the function $y = \log(-x)$. That is, does this function even exist? If it does exist, what is its domain? Sketch its graph and the graph of the function $y = \log x$.

47. What is the domain of the function $y = \ln|x|$? Graph the function.

30. 5.0601
31. −5.3795
32. −4.2759
33. 204.1738
34. 0.0010
35. 0.0120
36. 12,589.2541
37. 0.9572

HAWKES LEARNING SYSTEMS: INTRODUCTORY & INTERMEDIATE ALGEBRA SOFTWARE

- 11.6 Common Logarithms and Natural Logarithms

38. 75.8578 39. 175.9148 40. 134.2898 41. 0.0002 42. 845.5607 43. 1.0403 44. 0.9782
45. $\log x$ is a base 10 logarithm. $\ln x$ is a base e logarithm.
46. The function, $y = \log(-x)$, does exist. Its domain is $(-\infty, 0)$. 47. $D = (-\infty, 0) \cup (0, \infty)$

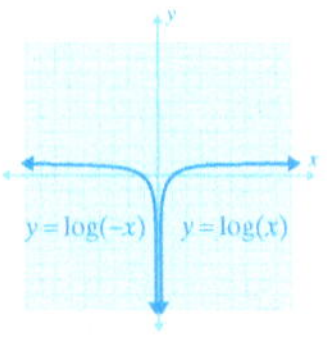

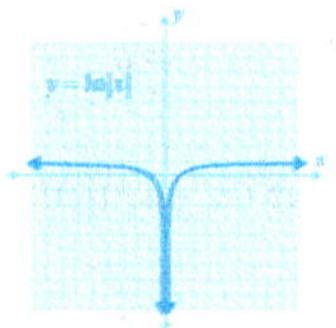

11.7 Logarithmic and Exponential Equations and Change-of-Base

- *Solve exponential equations in which the bases are the same.*
- *Solve exponential equations in which the bases are not the same.*
- *Solve equations with logarithms.*
- *Use the change-of-base formula and a calculator to evaluate logarithmic expressions.*

Solving Exponential Equations with the Same Base

All the properties of exponents that have been discussed are valid for real exponents, not just integers or fractions. For example, $2^{\sqrt{2}} \cdot 2^3 = 2^{\sqrt{2}+3}$ is a valid operation and gives a real number. To be complete in our development of the properties of exponents, the properties for real exponents and positive real bases are stated here.

Properties of Real Exponents

If a and b are positive real numbers and x and y are any real numbers, then:

1. $b^0 = 1$

2. $b^{-x} = \dfrac{1}{b^x}$

3. $b^x \cdot b^y = b^{x+y}$

4. $\dfrac{b^x}{b^y} = b^{x-y}$

5. $\left(b^x\right)^y = b^{xy}$

6. $(ab)^x = a^x b^x$

7. $\left(\dfrac{a}{b}\right)^x = \dfrac{a^x}{b^x}$

The following properties are used in solving equations containing exponents and logarithms.

Properties of Equations with Exponents and Logarithms

For $b > 0$ and $b \neq 1$,

1. If $b^x = b^y$, then $x = y$.

2. If $x = y$, then $b^x = b^y$.

3. If $\log_b x = \log_b y$, then $x = y$ ($x > 0$ and $y > 0$).

4. If $x = y$, then $\log_b x = \log_b y$ ($x > 0$ and $y > 0$).

Example 1: Solving Exponential Equations with the Same Base

Solve each equation for x.

a. $2^{x^2-7} = 2^{6x}$

Solution:

$2^{x^2-7} = 2^{6x}$ Both bases are 2.

$x^2 - 7 = 6x$ The exponents are equal because the bases are the same.

$x^2 - 6x - 7 = 0$ Solve for x.

$(x-7)(x+1) = 0$

$x = 7 \quad \text{or} \quad x = -1$

b. $8^{4-2x} = 4^{x+2}$

Solution:

$8^{4-2x} = 4^{x+2}$ Here the bases are different.

$(2^3)^{4-2x} = (2^2)^{x+2}$ Rewrite both sides so that the bases are the same.

$2^{12-6x} = 2^{2x+4}$ Use the property $(b^x)^y = b^{xy}$.

$12 - 6x = 2x + 4$ The exponents are equal because the bases are the same.

$8 = 8x$ Solve for x.

$1 = x$

Solving Exponential Equations with Different Bases

As Example 1 demonstrated, if the bases are the same in an exponential equation, there is no need to involve logarithms. However, **when the bases are not the same, equations are solved by taking the logarithm of both sides**. (See Property 4.)

Example 2: Solving Exponential Equations with Different Bases

Solve each of the following exponential equations by taking the log (or ln) of both sides or using the definition of logarithm as an exponent.

a. $10^{3x} = 2.1$

Solution: First, take the log of both sides, and then use the definition of logarithm as an exponent as follows.

$$10^{3x} = 2.1$$

$$\log\left(10^{3x}\right) = \log 2.1$$ Take the log of both sides.

$$3x = \log 2.1$$ By the definition of a common logarithm

$$x = \frac{\log 2.1}{3}$$

Use a calculator to find a decimal approximation:

$$x = \frac{\log 2.1}{3} \approx \frac{0.3222}{3} = 0.1074.$$

b. $6^x = 18$

Solution: The base is 6, not 10 or e, but we can solve by taking the **log** of both sides or by taking the **ln** of both sides. The result is the same.

Taking the log of both sides:

$$6^x = 18$$

$$\log 6^x = \log 18$$

$$x \cdot \log 6 = \log 18$$

$$x = \frac{\log 18}{\log 6}$$

Using a calculator,

$$x = \frac{\log 18}{\log 6} \approx \frac{1.2553}{0.7782}$$

$$\approx 1.6131.$$

Taking the ln of both sides:

$$6^x = 18$$

$$\ln 6^x = \ln 18$$

$$x \cdot \ln 6 = \ln 18$$

$$x = \frac{\ln 18}{\ln 6}$$

Using a calculator,

$$x = \frac{\ln 18}{\ln 6} \approx \frac{2.8904}{1.7918}$$

$$\approx 1.6131.$$

c. $5^{2x-1} = 10^x$

Solution:

$$5^{2x-1} = 10^x$$

$$\log 5^{2x-1} = \log 10^x$$ Take the log of both sides.

$$(2x-1)\log 5 = x\log 10$$ Power rule

$$2x \cdot \log 5 - 1 \cdot \log 5 = x \cdot 1$$ $\log 10 = 1$

$$2x\log 5 - x = \log 5$$ Arrange x-terms on one side.

$$x(2\log 5 - 1) = \log 5$$ Factor out the x.

$$x = \frac{\log 5}{2\log 5 - 1}$$

As a decimal approximation,

$$x = \frac{\log 5}{2\log 5 - 1} \approx \frac{0.6990}{2(0.6990) - 1} \approx 1.7563.$$ Using rounded values

Solving Equations with Logarithms

All the various properties of logarithms can be used to solve equations that involve logarithms. **Remember that logarithms are defined only for positive real numbers, so each answer should be checked in the original equation.**

Example 3: Solving Equations with Logarithms

Use the properties of logarithms to solve the following equations.

a. $\log 5x = 3$

Solution: $\log 5x = 3$

$5x = 10^3$ — Definition of a common logarithm

$5x = 1000$

$x = 200$ — Checking will show that the equation is defined at $x = 200$.

b. $\log(x-1) + \log(x-4) = 1$

Solution: $\log(x-1) + \log(x-4) = 1$

$\log[(x-1)(x-4)] = 1$ — Product rule

$(x-1)(x-4) = 10^1$ — Definition of a common logarithm

$x^2 - 5x + 4 = 10$

$x^2 - 5x - 6 = 0$

$(x-6)(x+1) = 0$ — Solve by factoring.

$x = 6$ or ~~$x = -1$~~ — Checking $x = -1$ yields $\log(-1-1) = \log(-2)$, which is undefined.

c. $\log x - \log(x-1) = \log 3$

Solution: $\log x - \log(x-1) = \log 3$

$\log\left(\frac{x}{x-1}\right) = \log 3$ — Quotient rule

$\frac{x}{x-1} = 3$ — If $\log_b x = \log_b y$, then $x = y$.

$x = 3(x-1)$ — Solve for x.

$x = 3x - 3$

$3 = 2x$

$\frac{3}{2} = x$

d. $\ln(x^2 - x - 6) - \ln(x+2) = 2$

Solution:

$$\ln(x^2 - x - 6) - \ln(x+2) = 2$$

$$\ln\left(\frac{x^2 - x - 6}{x+2}\right) = 2 \quad \text{Quotient rule}$$

$$\ln\left(\frac{(x+2)(x-3)}{(x+2)}\right) = 2 \quad \text{Factor the numerator.}$$

$$\ln(x-3) = 2 \quad \text{Simplify.}$$

$$x - 3 = e^2 \quad \text{Change to exponential form with base } e.$$

$$x = 3 + e^2$$

Or, using a calculator, $x = 3 + e^2 \approx 3 + 7.3891 = 10.3891$.

Change-of-Base

Because a calculator can be used to evaluate common logarithms and natural logarithms, most of the examples have been restricted to base 10 or base e expressions. If an equation involves logarithms of other bases, the following discussion shows how to rewrite each logarithm using any base.

Change-of-Base Formula

For $a, b, x > 0$ and $a,\ b \neq 1$,

$$\log_b x = \frac{\log_a x}{\log_a b}.$$

The change-of-base formula can be derived using properties of logarithms as follows.

$$b^{\log_b x} = x \quad \text{Property 3 in Section 11.5}$$

$$\log_a\left(b^{\log_b x}\right) = \log_a x \quad \text{Take the } \log_a \text{ of both sides.}$$

$$\log_b x(\log_a b) = \log_a x \quad \text{By the power rule using } \log_b x \text{ as the exponent } r$$

$$\log_b x = \frac{\log_a x}{\log_a b}. \quad \text{Divide both sides by } \log_a b \text{ to arrive at the change-of-base formula.}$$

Example 4: Change-of-Base

Use the change-of-base formula to evaluate the expressions in parts **a.** and **b.** and to solve the equation in part **c.**

a. $\log_2 3.42$

Solution: This expression can be evaluated using either base 10 or base *e* since both are easily available on a calculator.

$$\log_2 3.42 = \frac{\ln 3.42}{\ln 2} \qquad \text{Change-of-base formula}$$

$$\approx \frac{1.2296}{0.6931} \approx 1.7741 \qquad \text{Using rounded values}$$

(The student can show that $\frac{\log 3.42}{\log 2}$ gives the same result.)

In exponential form: $2^{1.7741} \approx 3.42$

b. $\log_3 0.3333$

Solution: $\log_3 0.3333 = \frac{\log 0.3333}{\log 3}$ Change-of-base formula

$$\approx \frac{-0.4772}{0.4771} \approx -1.0002 \qquad \text{Using rounded values}$$

c. Use the change-of-base formula to find the value of x (accurate to 4 decimal places) in the equation $5^x = 16$.

Solution: Because the base is 5, we can take $\log_5$ of both sides. (This method is not necessary, but it does show how the change-of-base formula can be used.)

$$5^x = 16$$

$$\log_5 5^x = \log_5 16$$

$$x = \log_5 16$$

$$x = \frac{\ln 16}{\ln 5} \qquad \text{Change-of-base formula}$$

$$x \approx \frac{2.7726}{1.6094} \approx 1.7228$$

Practice Problems

Solve each of the following equations.

1. $4^x = 64$

2. $10^x = 64$

3. $2^{3x-1} = 0.1$

4. $15\log x = 45.15$

5. $\ln(x^2+5x+6) - \ln(x+3) = 1$

1. $x = 11$ 2. $x = 5$
3. $x = 9$ 4. $x = 2$
5. $x = \frac{7}{2}$ 6. $x = \frac{7}{12}$
7. $x = \frac{6}{5}$ 8. $x = \frac{8}{3}$
9. $x = \frac{7}{2}$ 10. $x = -2$

11.7 Exercises

Use the properties of exponents and logarithms to solve each of the equations. If necessary, round answers to four decimal places.

1. $2^4 \cdot 2^7 = 2^x$

2. $3^7 \cdot 3^{-2} = 3^x$

3. $\left(3^5\right)^2 = 3^{x+1}$

4. $\left(2^x\right)^3 = 2^{x+4}$

5. $\dfrac{10^4 \cdot 10^{\frac{1}{2}}}{10^x} = 10$

6. $\left(10^2\right)^x = \dfrac{10 \cdot 10^{\frac{2}{3}}}{10^{\frac{1}{2}}}$

7. $2^{5x} = 4^3$

8. $7^{3x} = 49^4$

9. $(25)^x = 5^3 \cdot 5^4$

10. $10^x \cdot 10^8 = 100^3$

11. $8^{x+3} = 2^{x-1}$

12. $100^{2x+1} = 1000^{x-2}$

13. $27^x = 3 \cdot 9^{x-2}$

14. $16^x = 2 \cdot 8^{2x+3}$

15. $2^{3x+5} = 2^{x^2+1}$

16. $10^{x^2+x} = 10^{x+9}$

17. $10^{2x^2+3} = 10^{x+6}$

18. $3^{x^2+5x} = 3^{2x-2}$

19. $\left(3^{x+1}\right)^x = \left(3^{x+3}\right)^2$

20. $\left(10^x\right)^{x+3} = \left(10^{x+2}\right)^{-2}$

21. $3^{x+4} = 9$

22. $2^{5x-8} = 4$

23. $4^{x^2-x} = \left(\dfrac{1}{2}\right)^{5x}$

24. $25^{x^2+x} = \left(\dfrac{1}{5}\right)^{3x}$

25. $5^{2x-x^2} = \dfrac{1}{125}$

26. $10^{x^2-2x} = 1000$

27. $10^{3x} = 140$

28. $10^{2x} = 97$

29. $10^{0.32x} = 253$

30. $10^{-0.48x} = 88.6$

31. $4 \cdot 10^{-0.94x} = 126.2$

32. $3 \cdot 10^{-2.1x} = 83.5$

33. $e^{3x} = 2.1$

34. $e^{4t} = 184$

35. $e^{-0.5x} = 47$

36. $e^{-0.006t} = 50.3$

11. $x = -5$
12. $x = -8$
13. $x = -3$
14. $x = -5$
15. $x = -1, 4$
16. $x = -3, 3$
17. $x = -1, \frac{3}{2}$
18. $x = -2, -1$
19. $x = -2, 3$
20. $x = -4, -1$
21. $x = -2$
22. $x = 2$
23. $x = -\frac{3}{2}, 0$
24. $x = -\frac{5}{2}, 0$
25. $x = -1, 3$
26. $x = -1, 3$
27. $x \approx 0.7154$
28. $x \approx 0.9934$
29. $x \approx 7.5098$
30. $x \approx -4.0572$
31. $x \approx -1.5947$
32. $x \approx -0.6879$
33. $x \approx 0.2473$
34. $t \approx 1.3037$
35. $x \approx -7.7003$
36. $t \approx -653.0008$

Answers to Practice Problems: **1.** $x = 3$ **2.** $x \approx 1.8062$ **3.** $x \approx -0.7740$ **4.** $x = 10^{3.01} \approx 1023.2930$ **5.** $x = e - 2 \approx 0.7183$

37. $t \approx -1.5193$
38. $t \approx 0.9061$
39. $x \approx 3.3219$
40. $x \approx 6.1918$
41. $x \approx -1.4307$
42. $x \approx -0.3944$
43. $x = 1$
44. $x = -\frac{2}{5}$
45. $x \approx -4.1610$
46. $x \approx 1.2203$
47. $x \approx 1.2058$
48. $x \approx -1.6467$
49. $x \approx 1.1292$
50. $x \approx 4.3219$
51. $x \approx 0.9514$
52. $x \approx 0.9741$
53. $x \approx 1.2520$
54. $x \approx 1.5472$
55. $x \approx 25.1189$
56. $x \approx 25{,}118.8643$
57. $x \approx 31.6228$
58. $x \approx 3.1623 \times 10^{7}$
59. $x = 0.0001$
60. $x \approx \pm 5.6234 \times 10^{-21}$
61. $x \approx 4.9530$
62. $x \approx 0.2231$
63. $x \approx \pm 0.3329$
64. $x \approx \pm 1.0798 \times 10^{9}$
65. $x = 3$
66. $x = 5$
67. $x = 100$
68. $x = 10$
69. $x = 6$
70. $x = 3$
71. $x = 20$
72. $x = \frac{25}{2}$
73. $x \approx 3.1893$
74. $x \approx 21.7236$
75. No solution
76. $x = \sqrt{2}$
77. No solution
78. $x \approx 1.1353$
79. $x = 105$
80. No solution

37. $3e^{-0.12t} = 3.6$ **38.** $5e^{2.4t} = 44$ **39.** $2^x = 10$

40. $3^{x-2} = 100$ **41.** $5^{2x} = \frac{1}{100}$ **42.** $7^{3x} = \frac{1}{10}$

43. $5^{1-x} = 1$ **44.** $12^{5x+2} = 1$ **45.** $4^{2x+5} = 0.01$

46. $4^{2-3x} = 0.1$ **47.** $14^{3x-1} = 10^3$ **48.** $12^{2x+7} = 10^4$

49. $7^x = 9$ **50.** $2^x = 20$ **51.** $3^{3x} = 23$

52. $5^{2x} = 23$ **53.** $6^{2x-1} = 14.8$ **54.** $4^{7-3x} = 26.3$

55. $5\log x = 7$ **56.** $3\log x = 13.2$ **57.** $4\log x - 6 = 0$

58. $2\log x - 15 = 0$ **59.** $4\log x^{\frac{1}{2}} + 8 = 0$ **60.** $\frac{2}{3}\log x^{\frac{2}{3}} + 9 = 0$

61. $5\ln x - 8 = 0$ **62.** $2\ln x + 3 = 0$ **63.** $\ln x^2 + 2.2 = 0$

64. $\ln x^2 - 41.6 = 0$ **65.** $\log x + \log 2x = \log 18$

66. $\log(x+4) + \log(x-4) = \log 9$ **67.** $\log x^2 - \log x = 2$

68. $\log x + \log x^2 = 3$ **69.** $\ln(x-3) + \ln x = \ln 18$

70. $\ln(x+5) + \ln(x-1) = \ln 16$ **71.** $\log(x-15) = 2 - \log x$

72. $\log(2x-17) = 2 - \log x$ **73.** $\log(3x-5) + \log(x-1) = 1$

74. $\log(2x-3) + \log(x+3) = 3$ **75.** $\log(x-3) - \log(x+1) = 1$

76. $\ln(x+1) + \ln(x-1) = 0$ **77.** $\log(x^2-9) - \log(x-3) = -2$

78. $\ln(x^2+4x-5) - \ln(x+5) = -2$ **79.** $\log(x^2-4x-5) - \log(x+1) = 2$

80. $\log(x^2-x-12) - \log(x-4) = -2$ **81.** $\ln(x^2-4) = 3 + \ln(x+2)$

82. $\ln(x^2+2x-3) = 1 + \ln(x-1)$ **83.** $\log\sqrt[3]{x^2+2x+20} = \frac{2}{3}$

84. $\log\sqrt{x^2-24} = \frac{3}{2}$

81. $x \approx 22.0855$
82. No solution
83. $x = -10, 8$
84. $x = \pm 32$
85. 2.2619
86. 2.5850
87. 0.3223
88. 1.5342
89. −0.6279
90. −0.0841
91. 2.4391

Use the change-of-base formula to evaluate each of the expressions or solve the equations. Round answers to four decimal places.

85. $\log_3 12$ **86.** $\log_4 36$ **87.** $\log_5 1.68$ **88.** $\log_{11} 39.6$

89. $\log_8 0.271$ **90.** $\log_7 0.849$ **91.** $\log_{15} 739$ **92.** $\log_2 14.2$

93. $\log_{20} 0.0257$ **94.** $\log_9 2.384$ **95.** $2^x = 5$ **96.** $3^{2x} = 10$

97. $9^{2x-1} = 100$ **98.** $5^{x-1} = 30$ **99.** $4^{3-x} = 20$ **100.** $6^{4-3x} = 25$

Writing and Thinking About Mathematics

92. 3.8278
93. −1.2222
94. 0.3954
95. $x \approx 2.3219$
96. $x \approx 1.0480$
97. $x \approx 1.5480$
98. $x \approx 3.1133$
99. $x \approx 0.8390$
100. $x \approx 0.7345$

101. Solve the following equation for x two different ways: $a^{2x-1} = 1$.

102. Rewrite each of the following expressions as products.

a. 5^{x+2} **b.** 3^{x-2}

103. Explain, in your own words, why $7 \cdot 7^x \neq 49^x$ when $x \neq 1$. Show each of the expressions $7 \cdot 7^x$ and 49^x as a single exponential expression with base 7.

HAWKES LEARNING SYSTEMS: INTRODUCTORY & INTERMEDIATE ALGEBRA SOFTWARE

- 11.7 Exponential and Logarithmic Equations

101. Answers will vary. $x = \frac{1}{2}$ 102. a. $5^x \cdot 5^2$ b. $3^x \cdot 3^{-2}$

103. $7 \cdot 7^x = 7^{1+x}$ and $49^x = 7^{2x}$. Since $1 + x \neq 2x$, in general, $7 \cdot 7^x \neq 49^x$. Answers will vary.

11.8 Applications

- *Solve applied problems using logarithmic and exponential equations.*

In Section 11.3 we found that the number e appears in a surprisingly natural way in the formula for continuously compounding interest,

$$A = Pe^{rt},$$

which was developed from the formula for compounding n times per year:

$$A = P\left(1 + \frac{r}{n}\right)^{nt}.$$

There are many formulas that involve exponential functions. A few are shown here and in the exercises.

$A = A_0 e^{-0.04t}$ This is a formula for decomposition of radium where t is in centuries.

$A = A_0 e^{-0.1t}$ This is one formula for skin healing where t is measured in days.

$A = A_0 2^{-\frac{t}{5600}}$ This formula is used for carbon-14 dating to determine the age of fossils where t is measured in years.

$T = Ae^{-kt} + C$ This is Newton's law of cooling where C is the constant temperature of the surrounding medium. The values of A and k depend on the particular object that is cooling.

Example 1: Exponential Growth

Suppose that the formula $y = y_0 e^{0.4t}$ represents the number of bacteria present after t days, where y_0 is the initial number of bacteria. In how many days will the bacteria double in number?

Solution:

$y = y_0 e^{0.4t}$

$2y_0 = y_0 e^{0.4t}$ $2y_0$ is double the initial number present.

$2 = e^{0.4t}$

$\ln 2 = \ln e^{0.4t}$ Take the natural log of both sides.

$\ln 2 = 0.4t \cdot (1)$ Power rule of logarithms; $\ln e = 1$

$t = \frac{\ln 2}{0.4} \approx \frac{0.6931}{0.4}$

$t \approx 1.73$ days

The number of bacteria will double in approximately 1.73 days. Note that this number is completely independent of the number of bacteria initially present. That is, if $y_0 = 10$ or $y_0 = 1000$, the doubling time is the same, namely 1.73 days.

Example 2: Continuously Compounded Interest

If \$1000 is invested at a rate of 6% compounded continuously, in how many years will it grow to \$5000?

Solution:

$$A = Pe^{rt} \quad \text{Formula for continuously compounded interest}$$

$$5000 = 1000e^{0.06t} \quad \text{Substitute } A = 5000,\ P = 1000, \text{ and } r = 0.06.$$

$$5 = e^{0.06t}$$

$$\ln 5 = \ln e^{0.06t} \quad \text{Take the natural log of both sides.}$$

$$\ln 5 = 0.06t \cdot (1) \quad \text{Power rule of logarithms; } \ln e = 1$$

$$t = \frac{\ln 5}{0.06} \approx \frac{1.6094}{0.06}$$

$$t \approx 26.82 \text{ years}$$

\$1000 will grow to \$5000 in approximately 26.82 years.

Example 3: The Richter Scale

a. The magnitude of an earthquake is measured on the **Richter scale** as a logarithm of the intensity of the shock wave. For magnitude R and intensity I, the formula is $R = \log I$. The 1994 earthquake in Northridge, California measured 6.7 on the Richter scale. What was the intensity of this earthquake?

Solution: Substitute 6.7 for R in the formula and solve for I: $6.7 = \log I$

$$I = 10^{6.7}$$

b. The Long Beach earthquake in 1933 measured 6.2 on the Richter scale. How much stronger was the Northridge earthquake than the 1933 Long Beach earthquake?

Solution: The comparative sizes of the quakes can be found by finding the ratio of the intensities. For the Long Beach quake, $6.2 = \log I$ and $I = 10^{6.2}$. Therefore, the ratio of the two intensities is

$$\frac{I \text{ for Northridge}}{I \text{ for Long Beach}} = \frac{10^{6.7}}{10^{6.2}} = 10^{0.5} \approx 3.16.$$

Thus the Northridge earthquake had an intensity about 3.16 times the intensity of the Long Beach earthquake.

Example 4: Half-life of Radium

The **half-life** of a substance is the time needed for the substance to decay to one-half of its original amount. The half-life of radium-226, a common isotope of radium, is 1600 years. If 10 grams are present today, how many grams will remain in 500 years?

Solution: The model for radioactive decay is $y = y_0e^{-kt}$. Since the half-life is 1600 years, if we assume $y_0 = 10$ g, then y would be 5 g after 1600 years. We solve for k as follows.

$$5 = 10e^{-k(1600)} \qquad \text{Substitute } y = 5, y_0 = 10, \text{ and } t = 1600.$$

$$\frac{5}{10} = e^{-1600k}$$

$$\ln 0.5 = \ln e^{-1600k} \qquad \text{Take the natural log of both sides.}$$

$$\ln 0.5 = -1600k \cdot (1) \qquad \text{Power rule for logarithms; } \ln e = 1$$

$$k = \frac{\ln 0.5}{-1600} \approx \frac{-0.6931}{-1600} \approx 0.0004332$$

The model is $y = 10e^{-0.0004332t}$.

Substituting $t = 500$ gives

$$y = 10e^{(-0.0004332)(500)} = 10e^{-0.2166} \approx 10(0.8053) \approx 8.05.$$

Thus there will still be about 8.05 g of the radium-226 remaining after 500 years.

Example 5: Newton's Law of Cooling

Suppose that the room temperature is 70°, and the temperature of a cup of tea is 150° when it is placed on the table. In 5 minutes, the tea cools to 120°. How long will it take for the tea to cool to 100°?

Solution: Using the formula $T = Ae^{-kt} + C$ (Newton's law of cooling), first find A and then k. We know that $C = 70°$ and that $T = 150°$ when $t = 0$.

Find A by substituting these values.

$$150 = Ae^{-k(0)} + 70$$

$$150 = A \cdot (1) + 70 \qquad e^{-k(0)} = e^0 = 1$$

$$80 = A$$

Therefore, the formula can be written as $T = 80e^{-kt} + 70$.

Since $T = 120°$ when $t = 5$, substituting these values allows us to find k.

$$120 = 80e^{-k(5)} + 70$$
$$50 = 80e^{-5k}$$
$$\frac{50}{80} = e^{-5k}$$
$$\ln\frac{5}{8} = \ln e^{-5k} \qquad \text{Take the natural log of both sides.}$$
$$\ln 0.625 = -5k$$
$$k = \frac{\ln 0.625}{-5} \approx \frac{-0.4700}{-5} \approx 0.0940$$

The formula can now be written as $T = 80e^{-0.0940t} + 70$.

With all the constants in the formula known, we can find t when $T = 100°$.

$$100 = 80e^{-0.0940t} + 70$$
$$30 = 80e^{-0.0940t}$$
$$\frac{30}{80} = e^{-0.0940t}$$
$$\ln\frac{3}{8} = \ln e^{-0.0940t} \qquad \text{Take the natural log of both sides.}$$
$$\ln 0.375 = -0.0940t$$
$$t = \frac{\ln 0.375}{-0.0940}$$
$$\approx \frac{-0.9808}{-0.0940} \approx 10.43 \text{ minutes}$$

The tea will cool to 100° in about 10.43 minutes.

11.8 Exercises

Solve the following application problems. If necessary, round answers to two decimal places (unless otherwise specified).

1. $4027.51
2. $4726.34

1. Investing: If Kim invests $2000 at a rate of 7% compounded continuously, what will be her balance after 10 years?

2. Savings accounts: Find the amount of money that will be accumulated in a savings account if $3200 is invested at 6.5% for 6 years and the interest is compounded continuously.

3. 11.55 years
4. 13.86 years
5. $f \approx 0.30$
6. 23.10 hours
7. 1.73 hours
8. 0.05 mg
9. 10.99 days
10. 8166 bees
11. 7.44 hours
12. 12.28 lb per in.^{2}
13. 2350.02 years
14. 15.40 hours

3. Investing: Four thousand dollars is invested at 6% compounded continuously. How long will it take for the balance to be \$8000?

4. Investing: How long does it take \$1000 to double if it is invested at 5% compounded continuously?

5. Battery reliability: The reliability of a certain type of flashlight battery is given by $f = e^{-0.03x}$, where f is the fractional part of the batteries produced that last x hours. What fraction of the batteries produced are good after 40 hours of use?

6. Battery reliability: From Exercise 5, how long will at least one-half of the batteries last?

7. Concentration of a drug: The concentration of a drug in the blood stream is given by $C = C_0 e^{-0.8t}$, where C_0 is the initial dosage and t is the time in hours elapsed after administering the dose. If 20 mg of the drug is given, how much time elapses until 5 mg of the drug remains?

8. Concentration of a drug: Using the formula in Exercise 7, determine the amount of the drug present after 3 hours if 0.60 mg is given.

9. Healing of the skin: One law for skin healing is $A = A_0 e^{-0.1t}$, where A is the number of cm^2 of unhealed area after t days and A_0 is the number of cm^2 of the original wound. Find the number of days needed to reduce the wound to one-third the original size.

10. Beekeeping: A swarm of bees grows according to the formula $P = P_0 e^{0.35t}$, where P_0 is the number present initially and t is the time in days. How many bees will be present in 6 days if there were 1000 present initially? (Round to the nearest integer.)

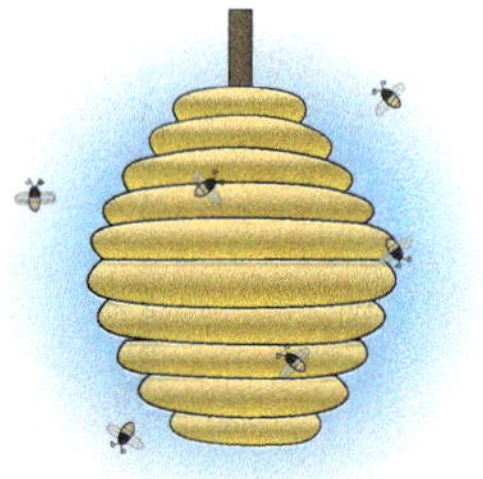

11. Inversion of raw sugar: If inversion of raw sugar is given by $A = A_0 e^{-0.03t}$, where A_0 is the initial amount and t is the time in hours, how long will it take for 1000 lb of raw sugar to be reduced to 800 lb?

12. Atmospheric pressure: Atmospheric pressure P is related to the altitude, h by the formula $P = P_0 e^{-0.00004h}$, where P_0 the pressure at sea level, is approximately 15 lb per in.^{2} Determine the pressure at 5000 in.

13. Radioactive decay: A radioactive substance decays according to $A = A_0 e^{-0.0002t}$, where A_0 is the initial amount and t is the time in years. If $A_0 = 640$ grams, find the time for A to decay to 400 grams.

14. Substance decay: A substance decays according to $A = A_0 e^{-0.045t}$, where t is in hours and A_0 is the initial amount. Determine the half-life of the substance.

15. 2.31 days
16. 4.87 hours
17. 39.65 minutes
18. 8.75 years
19. 7.00 years
20. 9.00 years
21. 5600 years
22. 294.41 days
23. a. 13.86 years
b. 6.93 years
24. a. 27.47 years
b. 13.73 years
25. 100

15. Employee training: An employee is learning to assemble remote-control units. The number of units per day he can assemble after t days of intensive training is given by $N = 80\left(1 - e^{-0.3t}\right)$. How many days of training will be needed before the employee is able to assemble 40 units per day?

16. Lava analysis: A scientist collects a lava sample and measures that its temperature is 1650°. To safely analyze the sample, it must be no warmer than 500°. The scientist stores the sample in a cooling chamber with a temperature of 50° and finds that in 2 hours, the lava has cooled to 1000°. When will the lava sample be safe to analyze?

17. Baking: The temperature of a carrot cake is 350° when it is removed from the oven. The temperature in the room is 72°. In 10 minutes, the cake cools to 280°. How long will it take for the cake to cool to 160°?

18. Investing: How long does it take \$10,000 to double if it is invested at 8% compounded quarterly?

19. Investing: If \$1000 is deposited at 6% compounded monthly, how long before the balance is \$1520?

20. Value of a machine: The value V of a machine at the end of t years is given by $V = C(1-r)^t$, where C is the original cost of the machine and r is the rate of depreciation. A machine that originally cost \$12,000 is now valued at \$3800. How old is the machine if $r = 0.12$?

21. Carbon-14 dating: The formula $A = A_0 2^{-\frac{t}{5600}}$ is used for carbon-14 dating to determine the age of fossils where t is measured in years. Determine the half-life of carbon-14.

22. Half-life of iodine: Radioactive iodine has a half-life of 60 days. If an accident occurs at a nuclear plant and 30 grams of radioactive iodine are present, in how many days will 1 gram be present? (Round k to at least 7 decimal places.)

23. Investing: If a principal P is doubled, then $A = 2P$. Use the formula for continuously compounded interest to find the time it takes the principal to double in value if the rate of interest is **a.** 5% **b.** 10%
(Note that the time for doubling the principal is completely independent of the principal itself.)

24. Investing: If a principal P is tripled, then $A = 3P$. Use the formula for continuously compounded interest to find the time it takes the principal to triple in value if the interest rate is
a. 4% **b.** 8%
(Note that the time for tripling the principal is completely independent of the principal itself.)

25. The Richter scale: The 1906 earthquake in San Francisco measured 8.6 on the Richter scale. In 1971, an earthquake in the San Fernando Valley measured 6.6 on the Richter scale. How many times greater was the 1906 earthquake than the 1971 earthquake? (See Example 3.)

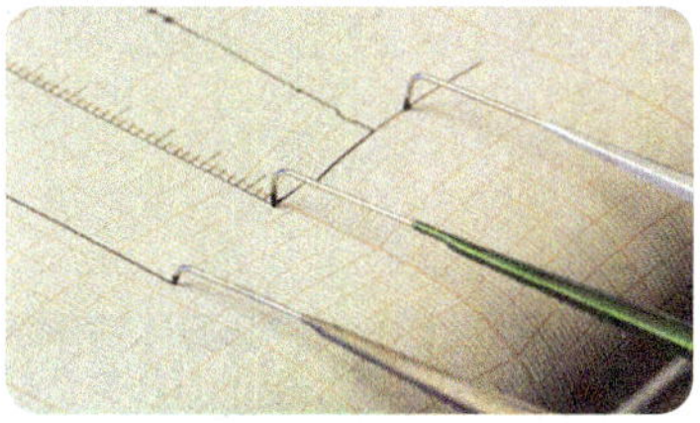

26. 6.31
27. 8.64 million
28. 1083, 1272, 1562
29. 2083, 2437, 2744
30. a. 6.07
b. 3.16×10^{-5}
31. a. 117.95 dB
b. $3.16\times10^{8}I_0$
c. $10^{6}I_0$

26. The Richter scale: In 1985, an earthquake in Mexico measured 8.1 on the Richter scale. How many times greater was this earthquake than the one in Landers, California in 1992 that measured 7.3 on the Richter scale? (See Example 3.)

27. Population growth: Population does not generally grow in a linear fashion. In fact, the population of many species grows exponentially, at least for a limited time. Using the exponential model $y = y_0e^{kt}$ for population growth, estimate the population of a state in 2020 if the population was 5 million in 1990 and 6 million in 2000. (Assume that t is measured in years and $t = 0$ corresponds to 1990.)

28. Fish population growth: Suppose that a lake is stocked with 500 fish, and biologists predict that the population of these fish will be approximated by the function $P(t) = 500\ln(2t+e)$ where t is measured in years. What will the fish population be in 3 years? in 5 years? in 10 years? (Round answers to the nearest integer.)

29. Sales revenue: Sales representatives of a new type of computer predict that sales can be approximated by the function $S(t) = 1000 + 500\ln(3t+e)$ where t is measured in years. What are the predicted sales in 2 years? in 5 years? in 10 years? Round to the nearest integer.

30. pH of a solution: In chemistry, the pH of a solution is a measure of the acidity or alkalinity of a solution. Water has a pH of 7 and, in general, acids have a pH less than 7 and alkaline solutions have a pH greater than 7. The model for pH is $\text{pH} = -\log\left[\text{H}^+\right]$ where $\left[\text{H}^+\right]$ is the hydrogen ion concentration in moles per liter of a solution.

a. Find the pH of a solution with a hydrogen ion concentration of 8.6×10^{-7}.

b. Find the hydrogen ion concentration $\left[\text{H}^+\right]$ of a solution if the pH of the solution is 4.5. Write the answer in scientific notation.

31. Sound levels: A decibel (abbreviated dB) is a unit used to measure the loudness of sound. The decibel level D of a sound of intensity I is measured by comparing it to a barely audible sound of intensity I_0 with the following formula:

$$D = 10\log\left(\frac{I}{I_0}\right).$$

Sounds measuring over 85 dB are not considered safe.

a. Find the decibel level of a rock concert with an intensity of $6.24\times10^{11}I_0$.

b. What is the intensity level of 85 dB?

c. What is the intensity level of 60 dB (normal conversation)?

HAWKES LEARNING SYSTEMS: INTRODUCTORY & INTERMEDIATE ALGEBRA SOFTWARE

- 11.8 Applications: Exponential and Logarithmic Functions

Chapter 11 Index of Key Ideas and Terms

Section 11.1 Algebra of Functions

Algebraic Operations with Functions page 857

If $f(x)$ and $g(x)$ represent two functions and x is a value in the domain of both functions, then:

1. Sum of two functions: $(f+g)(x)=f(x)+g(x)$
2. Difference of two functions: $(f-g)(x)=f(x)-g(x)$
3. Product of two functions: $(f\cdot g)(x)=f(x)\cdot g(x)$
4. Quotient of two functions: $\left(\frac{f}{g}\right)(x)=\frac{f(x)}{g(x)}$ where $g(x)\neq 0$.

Graphing the Sum of Two Functions pages 859-861

Using a Graphing Calculator to Graph the Sum of Two Functions pages 861-862

Section 11.2 Composition of Functions and Inverse Functions

Composite Functions pages 869-870

For two functions f and g, the **composite function** $f\circ g$ is defined as follows: $(f\circ g)(x)=f(g(x))$.
The **domain of $f\circ g$** consists of those values of x in the domain of g for which $g(x)$ is in the domain of f.

One-to-One Functions pages 871-872

A function is a **one-to-one function** (or **1-1 function**) if for each value of y in the range there is only one corresponding value of x in the domain.

Horizontal Line Test page 872

A function is one-to-one if no **horizontal line** intersects the graph of the function at more than one point.

Inverse Functions pages 874-875

If f is a one-to-one function with ordered pairs of the form (x, y), then its **inverse function**, denoted as f^{-1}, is also a one-to-one function with ordered pairs of the form (y, x). (Only one-to-one functions have inverses.)

Continued on the next page...

Section 11.2 Composition of Functions and Inverse Functions (cont.)

To Determine whether Two Functions are Inverses page 876

If f and g are one-to-one functions and

$f(g(x)) = x$ for all x in D_g, and

$g(f(x)) = x$ for all x in D_f

then f and g are **inverse functions**. That is, $g = f^{-1}$ and $f = g^{-1}$.

To Find the Inverse of a One-to-One Function page 879

1. Let $y = f(x)$. (In effect, substitute y for $f(x)$.)
2. Interchange x and y.
3. In the new equation, solve for y in terms of x.
4. Substitute $f^{-1}(x)$ for y. (This new function is the inverse of f.)

Section 11.3 Exponential Functions

Exponential Functions page 886

An **exponential function** is a function of the form $f(x) = b^x$ where $b > 0$, $b \neq 1$, and x is any real number.

Asymptotes page 888

When the graph of a function gets closer and closer to a line (for example the x- or y-axis) without ever touching it, the line is called an **asymptote**.

Exponential Growth and Decay pages 887-890

1. If $b > 1$, then b^x is called an **exponential growth function**. This type of function increases to the right and approaches the x-axis for negative values of x.
2. If $0 < b < 1$, then b^x is called an **exponential decay function**. This type of function decreases to the right and approaches the x-axis for positive values of x.

Compound Interest page 891

Compound interest on a principal P invested at an annual interest rate r (in decimal form) for t years that is compounded n times per year can be calculated using the following formula:

$$A = P\left(1 + \frac{r}{n}\right)^{nt}$$ where A is the amount accumulated.

Continued on the next page...

Section 11.3 Exponential Functions (cont.)

The Number *e* page 894

The number e is defined to be $e = 2.718281828459\ldots$

Continuously Compounded Interest page 894

Continuously compounded interest on a principal P invested at an annual interest rate r for t years, can be calculated using the following formula:

$A = Pe^{rt}$ where A is the amount accumulated.

Section 11.4 Logarithmic Functions

Logarithmic Functions pages 900-901

Logarithmic functions are inverses of exponential functions. For $b > 0$ and $b \neq 1$, $x = b^y$ is equivalent to $y = \log_b x$. (A logarithm is an exponent.)

Basic Properties of Logarithms page 902

For $b > 0$ and $b \neq 1$,

1. $\log_b 1 = 0$
2. $\log_b b = 1$
3. $x = b^{\log_b x}$, for $x > 0$
4. $\log_b b^x = x$

Graphs of Logarithmic Functions page 904

Section 11.5 Properties of Logarithms

Rules for Logarithms pages 907-912

For $b > 0, b \neq 1, x, y > 0$, and any real number r:

1. $\log_b xy = \log_b x + \log_b y$ The product rule
2. $\log_b \dfrac{x}{y} = \log_b x - \log_b y$ The quotient rule
3. $\log_b x^r = r \cdot \log_b x$ The power rule

Section 11.6 Common Logarithms and Natural Logarithms

Common Logarithms — page 917

A **common logarithm** is a base 10 logarithm.

Inverse Logarithms — pages 919, 922

Natural Logarithms — page 920

A **natural logarithm** is a base e logarithm. The notation for natural logarithms is shortened to $\ln x$.

Using a calculator to find:

Common logarithms	page 918
Inverse logarithms	page 919
Natural logarithms	page 921
Inverse natural logarithms	page 922

Section 11.7 Logarithmic and Exponential Equations and Change-of-Base

Properties of Equations with Exponents and Logarithms — page 925

For $b > 0$ and $b \neq 1$,

1. If $b^x = b^y$, then $x = y$.
2. If $x = y$, then $b^x = b^y$.
3. If $\log_b x = \log_b y$, then $x = y$ $(x > 0$ and $y > 0)$.
4. If $x = y$, then $\log_b x = \log_b y$ $(x > 0$ and $y > 0)$.

Change-of-Base Formula — page 929

For $a, b, x > 0$ and $a, b \neq 1$, $\log_b x = \dfrac{\log_a x}{\log_a b}$.

Section 11.8 Applications

Applications

Exponential growth	pages 934-935
Continuously compounded interest	page 935
The Richter scale	page 935
Half-life	page 936
Newton's law of cooling	pages 936-937

HAWKES LEARNING SYSTEMS: INTRODUCTORY & INTERMEDIATE ALGEBRA SOFTWARE

- 11.1 Algebra of Functions
- 11.2 Composition of Functions and Inverse Functions
- 11.3 Exponential Functions
- 11.4 Logarithmic Functions
- 11.5 Properties of Logarithms
- 11.6 Common Logarithms and Natural Logarithms
- 11.7 Exponential and Logarithmic Equations
- 11.8 Applications: Exponential and Logarithmic Functions

Chapter 11 Review

11.1 Algebra of Functions

*For the following pairs of functions, graph **a.** the sum $(f+g)$ and **b.** the difference $(f-g)$ on two different graphs.*

1. $f=\{(-2,3),(-1,4),(0,2),(1,0)\}$

$g=\{(-2,-1),(-1,2),(0,-3),(1,1)\}$

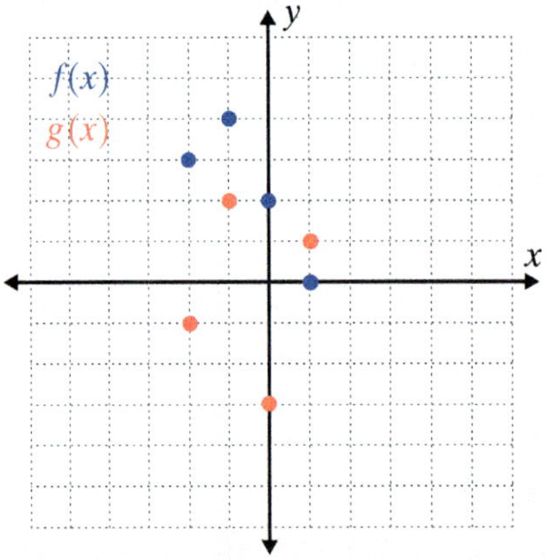

2.

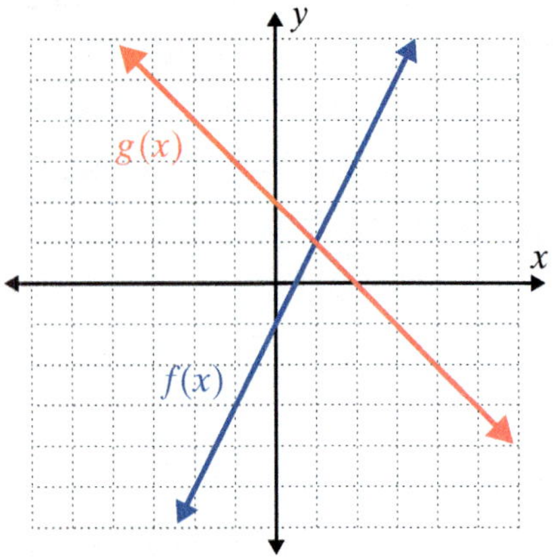

*For the following pairs of functions, find **a.** $(f+g)(x)$ **b.** $(f-g)(x)$ **c.** $(f\cdot g)(x)$ **d.** $\left(\frac{f}{g}\right)(x)$.*

3. $f(x)=2x+3,\ g(x)=x-7$

4. $f(x)=-5x,\ g(x)=x+6$

5. $f(x)=x^2,\ g(x)=x^2-16$

6. $f(x)=2x^2-x,\ g(x)=x^2+5$

Let $f(x)=x^2-4$ and $g(x)=-x+2$. Find the values of the indicated expressions.

7. $f(3)+g(3)$ **8.** $g(1)-f(1)$ **9.** $(f\cdot g)(2)$ **10.** $\left(\frac{f}{g}\right)(3)$

Find the indicated functions and state their domains using interval notation.

11. If $f(x)=\sqrt{x-6}$ and $g(x)=x+2$, find $(f+g)(x)$.

12. For $f(x)=\sqrt{x-1}$ and $g(x)=3x+2$, find $(f\cdot g)(x)$.

Graph each pair of functions and the sum of these functions on the same set of axes.

13. $y=x+3$ and $y=-2x+3$

14. $y=x^2-4$ and $y=x+4$

---

1. a.

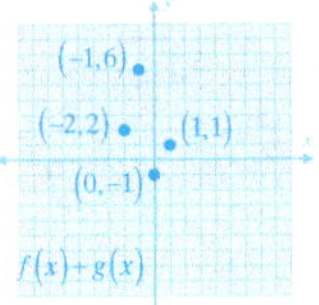

b.

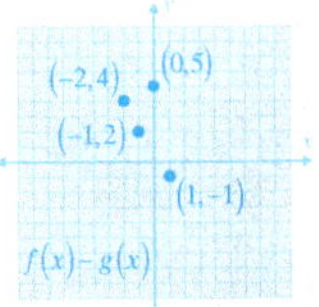

2. a.

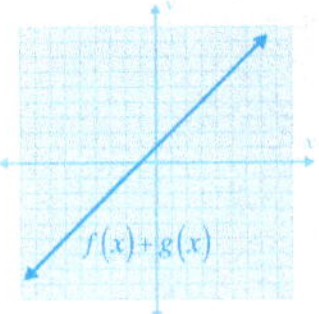

b.

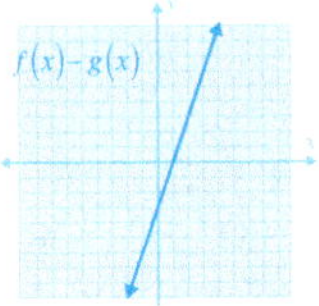

3. a. $3x-4$ b. $x+10$
c. $2x^2-11x-21$
d. $\dfrac{2x+3}{x-7},\ x\neq 7$

4. a. $-4x+6$
b. $-6x-6$
c. $-5x^2-30x$
d. $\dfrac{-5x}{x+6},\ x\neq -6$

5. a. $2x^2-16$ b. 16
c. x^4-16x^2
d. $\dfrac{x^2}{x^2-16},\ x\neq -4, 4$

6. a. $3x^2-x+5$
b. x^2-x-5
c. $2x^4-x^3+10x^2-5x$
d. $\dfrac{2x^2-x}{x^2+5}$

7. 4 8. 4
9. 0 10. −5
11. $\sqrt{x-6}+x+2$, $D=[6,\infty)$

11.2 Composition of Functions and Inverse Functions

15. $f(x)=1-x^2, g(x)=x+5;$ Find **a.** $(f\circ g)(-1)$ and **b.** $(g\circ f)(2)$

Form the compositions $f(g(x))$ *and* $g(f(x))$ *for each pair of functions.*

16. $f(x)=2x+1, g(x)=2x-3$

17. $f(x)=x^2, g(x)=3x-1$

18. $f(x)=\sqrt{x}, g(x)=x^2-4$

19. $f(x)=x^3, g(x)=\frac{1}{x}$

20. $f(x)=\frac{1}{x}, g(x)=\frac{1}{x^2}$

Show that the given one-to-one functions are inverses of each other. Graph both functions on the same set of axes and show the line y = x as a dotted line on each graph. (You may use a calculator as an aid in finding the graphs.)

21. $f(x)=4x-5$ and $g(x)=\frac{x+5}{4}$

22. $f(x)=\sqrt[3]{8-3x}$ and $f(x)=\frac{8-x^3}{3}$

Find the inverse of the given function. Graph both functions on the same set of axes and show the line y = x as a dotted line on the graph.

23. $f(x)=-2x+5$

24. $f(x)=-x^2+2, x\geq 0$

Using the horizontal line test, determine which of the graphs are graphs of one-to-one functions. If the graph represents a one-to-one function, graph its inverse by reflecting the graph of the function across the line y = x.

25.

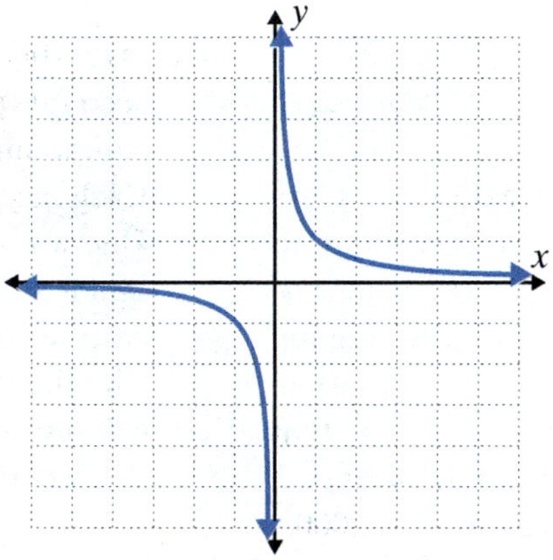

26.

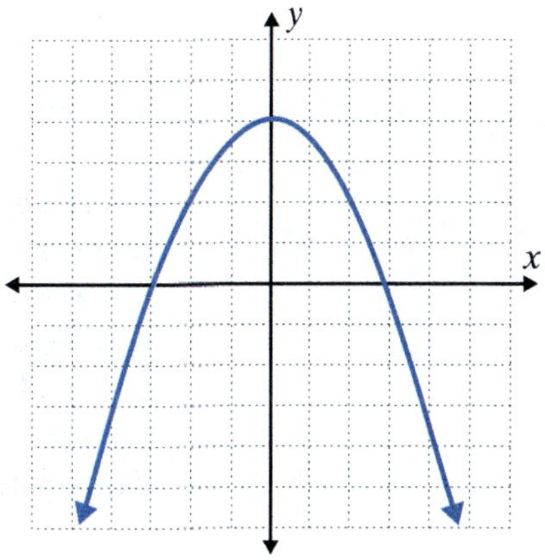

Use a graphing calculator to graph each of the functions and determine which of the functions are one-to-one by inspecting the graph and using the horizontal line test.

27. $f(x)=(x+2)^2, x\geq -2$

28. $f(x)=\frac{1}{x^3}$

---

12. $2\sqrt{x-1}+3x\sqrt{x-1}$, $D=[1,\infty)$

13.

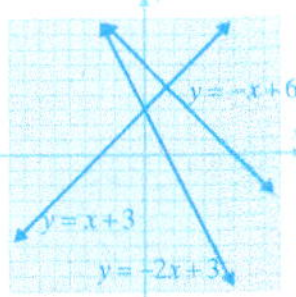

14.

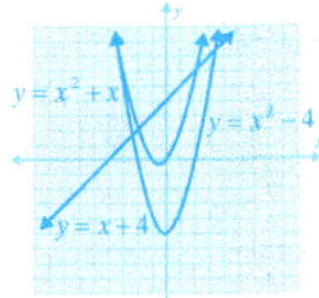

15. a. −15
b. 2

16. $f(g(x))=4x-5$
$g(f(x))=4x-1$

17. $f(g(x))=9x^2-6x+1$
$g(f(x))=3x^2-1$

18. $f(g(x))=\sqrt{x^2-4}$
$g(f(x))=x-4$

19. $f(g(x))=\frac{1}{x^3}$
$g(f(x))=\frac{1}{x^3}$

20. $f(g(x))=x^2$
$g(f(x))=x^2$

21.

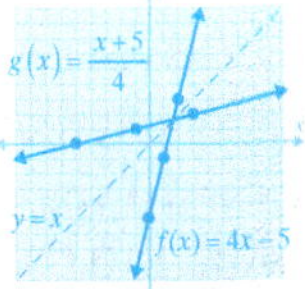

22.

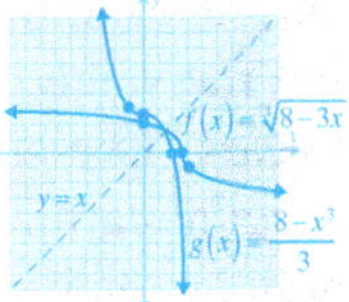

23. $f^{-1}(x)=\frac{5-x}{2}$

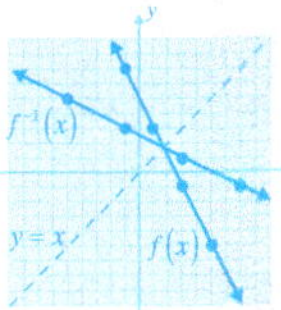

24. $f^{-1}(x) = \sqrt{2-x}$

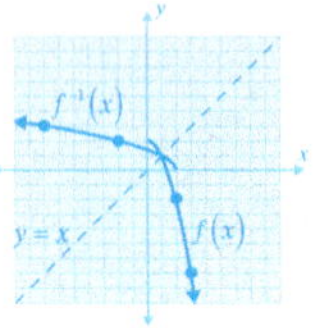

25. One-to-one

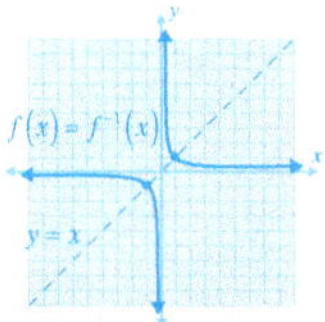

26. Not one-to-one

27. One-to-one

28. One-to-one

29. $f^{-1}(x) = (1+x)^{\frac{1}{3}}$

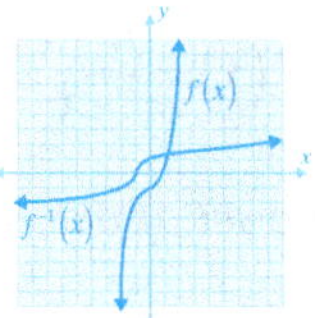

30. $f^{-1}(x) = x^2 + 5,$ $x \geq 0$

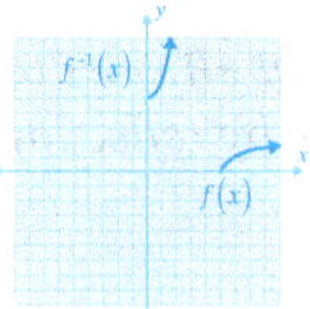

31.

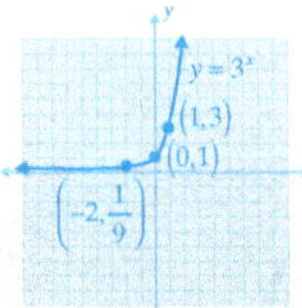

32.

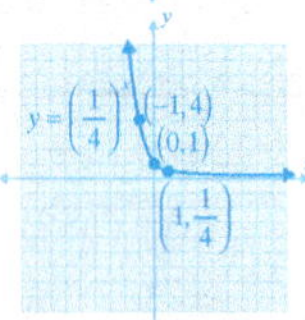

33.

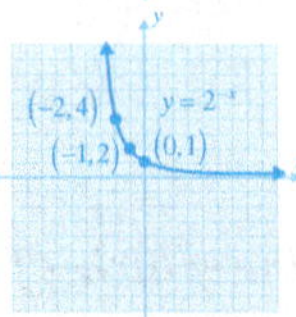

34.

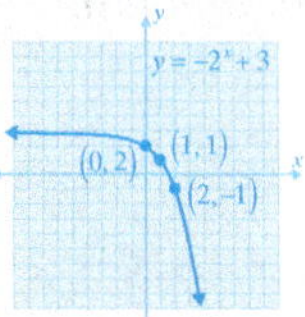

Find the inverse of the given function. Then use a graphing calculator to graph both the function and its inverse. Set the WINDOW *so that it is "square."*

29. $f(x) = x^3 - 1$

30. $f(x) = \sqrt{x-5}, x \geq 5$

11.3 Exponential Functions

Sketch the graph of each of the exponential functions and label three points on each graph.

31. $y = 3^x$

32. $y = \left(\frac{1}{4}\right)^x$

33. $y = 2^{-x}$

34. $y = -2^x + 3$

35. $f(x) = 3^{0.5x}$

36. $g(x) = 2^{x+1}$

37. $h(x) = \frac{1}{2} \cdot 3^{x-1}$

38. $y = \left(\frac{2}{5}\right)^x$

39. Use your calculator to find the value of $f(6)$ if $f(t) = 32.4 \cdot 2^{0.5t}$.

40. If $f(x) = \left(1 + \frac{1}{x}\right)^x$, use your calculator to find the value below rounded to five decimal places.

a. $f(1000)$ **b.** $f(10{,}000)$ **c.** $f(1{,}000{,}000)$

41. Savings accounts: Find the amount A in a savings account of \$5000 invested at 6% for 20 years if the interest is compounded:

a. annually
b. quarterly
c. daily (use 360 days)
d. continuously

42. Automobile depreciation: Cars depreciate in t years according to the formula $V = C(1-r)^t$ where V is the current value, C is the original cost, and r is the rate of depreciation. Find the current value of a car that depreciates at a rate of 15% over 5 years if its original cost was \$40,000.

43. Sales revenue: The revenue function is given by $R(x) = x \cdot p(x)$. If $p(x) = 25\left(2^{-0.025x}\right)$ dollars, where x is the number of units sold, find the revenue if 50 units are sold.

44. Investing: Suppose you have two accounts that are invested so that they compound continuously for 10 years. Each account started with \$10,000. One was earning 4% interest and the other 8% interest. Find the value of each account after 10 years. Explain why the value of the account earning 4% interest is not simply one-half of the account earning 8% interest.

11.4 Logarithmic Functions

Express each equation in logarithmic form.

45. $8^2 = 64$

46. $10^3 = 1000$

47. $2^{-3} = \frac{1}{8}$

48. $\left(\frac{1}{5}\right)^3 = \frac{1}{125}$

Express each equation in exponential form.

49. $\log_2 \frac{1}{2} = -1$ **50.** $\log_5 625 = 4$ **51.** $\log_5 x = -2.5$ **52.** $\log_b x = y$

Solve by first changing each equation to exponential form.

53. $\log_8 x = \frac{1}{3}$ **54.** $\log_{25} 5 = x$ **55.** $\log_{10} 10^{2.35} = x$

56. $\log_2 2^{-5} = x$ **57.** $\log_4 4^{\log_2 64} = x$ **58.** $\log_3 3^{\log_3 27} = x$

Graph each function and its inverse on the same set of axes. Label two points on each graph.

59. $f(x) = 5^x$ **60.** $f(x) = \left(\frac{1}{2}\right)^x$ **61.** $y = \log_6 x$ **62.** $g(x) = \log_{1/4} x$

11.5 Properties of Logarithms

Use your knowledge of logarithms and exponents to find the value of each expression.

63. $\log_4 64$ **64.** $2^{\log_2 \sqrt{10}}$

Use the properties of logarithms to expand each expression as much as possible.

65. $\log_b \left(x^2 y\right)^{-3}$ **66.** $\log_{10} x^2 y^3$ **67.** $\log_b \frac{x}{y^4}$ **68.** $\log_b \left(\frac{x^2}{y^3}\right)^5$

69. $\log_2 \sqrt{\frac{5}{x^2 y^4}}$ **70.** $\log_3 42x^2 y^{-3}$

Use the properties of logarithms to write each expression as a single logarithm of a single expression.

71. $2\log_2 3 + \log_2 x$ **72.** $5\log_3 x - \log_3 y$

73. $3\log_b x - 4\log_b y$ **74.** $\log_4 (4x+1) + 2\log_4 x$

75. $\log_{10}(2x+1) + \log_{10}(x-5)$ **76.** $\log_2(x-3) - \log_2\left(2x^2 - 5x - 3\right) + \frac{1}{2}\log_2 x$

11.6 Common Logarithms and Natural Logarithms

Express each equation in logarithmic form.

77. $10^{-5} = 0.00001$ **78.** $e^{1.2} = x$ **79.** $e^x = 12$ **80.** $10^k = 8.4$

35.
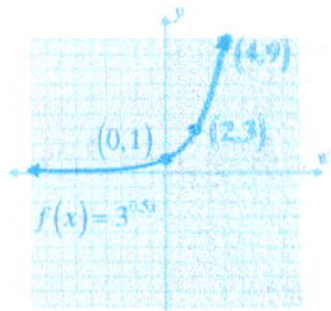

36.
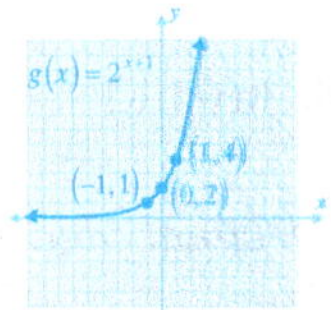

37.
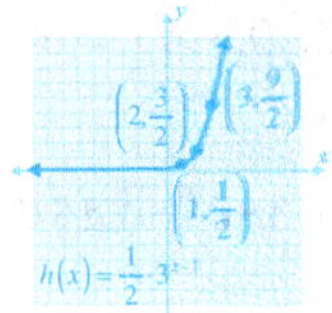

38.
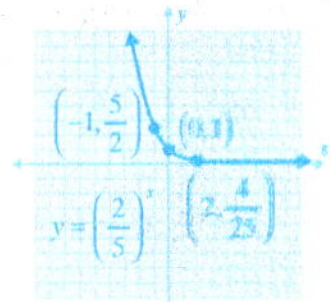

39. 259.2

40. **a.** 2.71692
b. 2.71815
c. 2.71828

41. **a.** \$16,035.68
b. \$16,453.31
c. \$16,598.92
d. \$16,600.58

42. \$17,748.21

43. \$525.56

44. \$14,918.25, \$22,255.41
Answers will vary.

45. $\log_8 64 = 2$

46. $\log_{10} 1000 = 3$

47. $\log_2 \frac{1}{8} = -3$

48. $\log_{1/5} \frac{1}{125} = 3$

49. $2^{-1} = \frac{1}{2}$

50. $5^4 = 625$

51. $5^{-2.5} = x$

52. $b^y = x$

53. $x = 2$ **54.** $x = \frac{1}{2}$

55. $x = 2.35$

56. $x=-5$

57. $x=6$ 58. $x=3$

59.

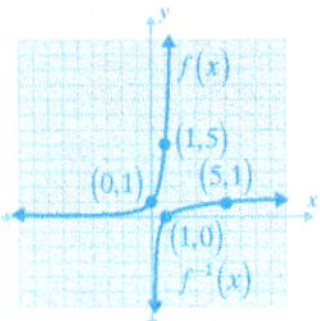

60.

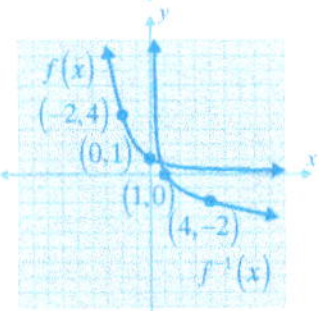

61.

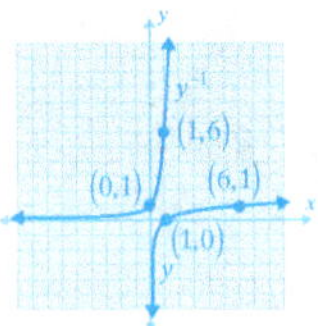

62.

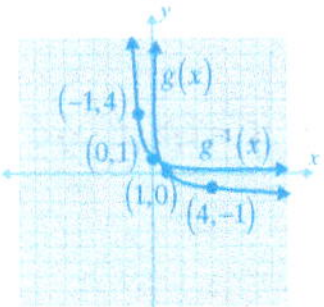

63. 3

64. $\sqrt{10}$

65. $-6\log_b x-3\log_b y$

66. $2\log_{10}x+3\log_{10}y$

67. $\log_b x-4\log_b y$

68. $10\log_b x-15\log_b y$

69. $\frac{1}{2}\log_2 5-\log_2 x-2\log_2 y$

70. $\log_3 42+2\log_3 x-3\log_3 y$

71. $\log_2(9x)$

72. $\log_3\frac{x^5}{y}$

73. $\log_b\left(\frac{x^3}{y^4}\right)$

74. $\log_4\left(4x^3+x^2\right)$

75. $\log_{10}\left(2x^2-9x-5\right)$

76. $\log_2\left(\frac{\sqrt{x}}{2x+1}\right)$

77. $\log 0.00001=-5$

Express each equation in exponential form.

81. $\ln 3=x$ **82.** $\ln x=-1$ **83.** $\log 100{,}000=5$ **84.** $\log x=2.5$

Use a calculator to evaluate the logarithms accurate to four decimal places.

85. $\log 275$ **86.** $\log 84$ **87.** $\ln(-3)$ **88.** $\ln 0.012$

Use a calculator to find the value of x in each equation accurate to four decimal places.

89. $\log x=3.12$ **90.** $5\log x=-4$ **91.** $\ln x=-4.5$ **92.** $0.3\ln x=66$

11.7 Logarithmic and Exponential Equations and Change-of-Base

Use the properties of exponents and logarithms to solve each of the equations. If necessary, round answers to four decimal places.

93. $\ln x^2=\log_3 9$ **94.** $10^{2x-1}=100^2$ **95.** $25^{x^2-3}=5^x$ **96.** $2\log x-6=0$

97. $\ln x+\ln x^2=3$ **98.** $e^{x+5}=36$ **99.** $\ln x^2+5=0$

100. $3^{x+1}=10$ **101.** $\log\sqrt{x^2-25}=2$ **102.** $2\ln x+\ln 4=\ln(4x-1)$

103. $\log(x-3)-2=\log(x+2)$ **104.** $\log(x+11)+\log x=\log 12$

105. $\log(x+3)+\log(x-3)=\log 27$ **106.** $\log\left(x^2-16\right)-\log(x-4)=-1$

Use the change-of-base formula to evaluate each of the expressions or solve the equations. Round answers to four decimal places.

107. $\log_2 12$ **108.** $\log_4 100$ **109.** $2^x=10$ **110.** $4^{x+2}=400$

11.8 Applications

Solve the following application problems. If necessary, round answers to two decimal places.

111. Investing: What will be the value of an investment of $5000 after 20 years if it is invested at 5% and

a. compounded quarterly (4 times a year)?

b. compounded monthly (12 times a year)?

c. compounded continuously?

112. Investing: How long does it take for $10,000 to double when invested at 6% compounded continuously? What would be the doubling time of $5000?

78. $\ln x = 1.2$
79. $\ln 12 = x$
80. $\log 8.4 = k$
81. $e^x = 3$
82. $e^{-1} = x$
83. $10^5 = 100{,}000$
84. $10^{2.5} = x$
85. 2.4393
86. 1.9243
87. Error (undefined)
88. –4.4228
89. 1318.2567
90. 0.1585
91. 0.0111
92. 3.5058×10^{95}
93. $x = \pm e$
94. $x = \frac{5}{2}$
95. $x = -\frac{3}{2}, 2$
96. $x = 1000$
97. $x = e$
98. $x \approx -1.4165$
99. $x \approx \pm 0.0821$
100. $x \approx 1.0959$
101. $x = \pm 5\sqrt{401} \approx \pm 100.1249$
102. $x = \frac{1}{2}$
103. No solution
104. $x = 1$
105. $x = 6$
106. No solution
107. 3.5850
108. 3.3219
109. $x \approx 3.3219$
110. $x \approx 2.3219$
111. a. $13,507.42 b. $13,563.20 c. $13,591.41
112. 11.55 years; 11.55 years

113. Half-life of strontium: Strontium-90 is an isotope of strontium present in radioactive fallout. Find the half-life of strontium-90 using the formula $y = y_0 e^{-0.0239t}$, where t is the time in years.

114. Computers: The value of a computer depreciates at 20% per year. After 3 years, what is the value of a computer that originally cost $2000? (Use the formula $V = C(1-r)^t$.)

115. Cancer cell growth: Certain cancer cells grow according to the formula $y = y_0 e^{0.5t}$ where y_0 is the initial number present and t is measured in days. If $y_0 = 200$, in how many days will the cells number 200,000?

116. Concentration of a drug: The concentration of a drug in the bloodstream is given by $C = 40e^{-0.6t}$ where 40 mg was the initial dose of the drug and t is the time in hours.

a. How much of the drug will be left in the bloodstream after 5 hours?

b. What is the half-life of this drug?

117. Change in temperature: Suppose that the room temperature is 72° and a cup of coffee with temperature 120° is placed in the room. In 6 minutes the temperature of the coffee is 100°. How long will it take for the coffee to cool to 90°? (Use Newton's law of cooling: $T = Ae^{-kt} + C$. C is the constant temperature of the surrounding medium. The values of A and k depend on the particular object that is cooling.)

118. Hydrogen ion concentration: In chemistry, pH is a measure of the hydrogen ion concentration $[H^+]$ of a solution. The pH of a solution is given by the formula $\text{pH} = -\log[H^+]$. Find the hydrogen ion concentration of a solution with pH 6.3.

119. Sound intensity: The decibel level D of a sound of intensity I is measured by comparing it to a barely audible sound of intensity I_0 with the following formula:

$$D = 10\log\left(\frac{I}{I_0}\right).$$

a. Find the decibel level of a lawn mower with intensity of $4.32 \times 10^9 I_0$.

b. What is the intensity level of 140 dB (a loud gun)?

120. The Richter scale: The relationship between the magnitude R and intensity I on the Richter scale is given by the formula $R = \log I$. How much stronger was the 1992 earthquake in Kobe, Japan (7.2 on the Richter scale) than the 1971 earthquake in Los Angeles (6.6 on the Richter scale)?

113. 29 years 114. $1024 115. 13.82 days 116. a. 1.99 mg b. 1.16 hours 117. 10.92 minutes 118. 5.01×10^{-7} 119. a. 96.35 dB b. $10^{14} I_0$ 120. 3.98 times stronger

Chapter 11 Test

For the following pairs of functions, graph **a.** *the sum* $(f+g)$ *and* **b.** *the difference* $(f-g)$ *on two different graphs.*

1. $f=\{(-2,-2),(0,0),(1,1),(2,2)\}$
$g=\{(-2,-5),(0,-1),(1,-2),(2,-5)\}$

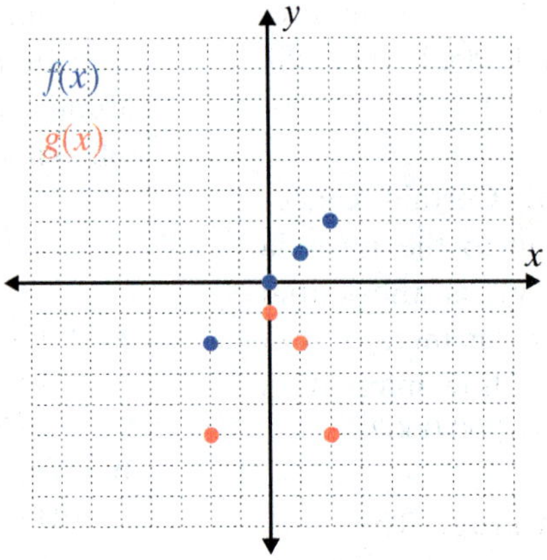

2.

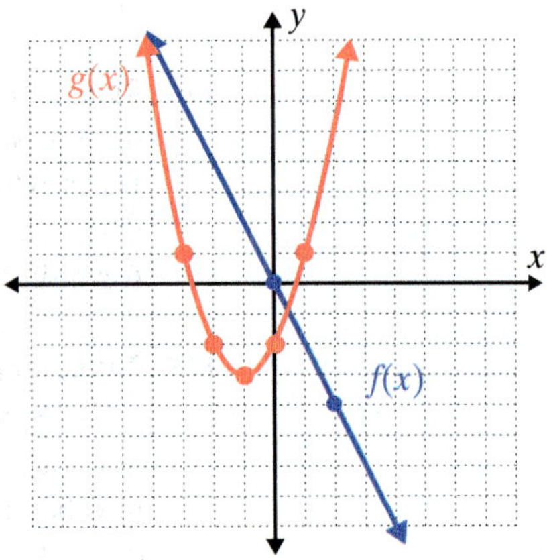

3. Given the two functions, $f(x)=\sqrt{x-3}$ and $g(x)=x^2+1$, find:

a. $(f+g)(x)$ **b.** $(f-g)(x)$ **c.** $(f\cdot g)(x)$ **d.** $\left(\frac{f}{g}\right)(x)$

4. If $f(x)=2x-5$ and $g(x)=3-2x^2$, form the following compositions.

a. $f(g(x))$ **b.** $g(f(x))$

5. Use the horizontal line test to determine whether the graph on the right is a one-to-one function. If the graph represents a one-to-one function, graph its inverse by reflecting the graph of the function across the line $y=x$.

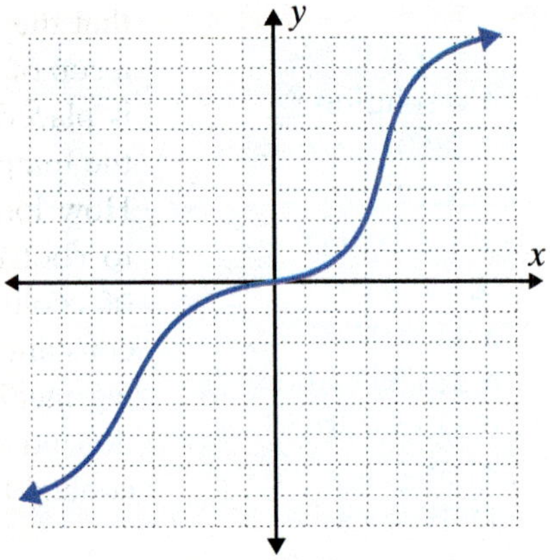

1. a.

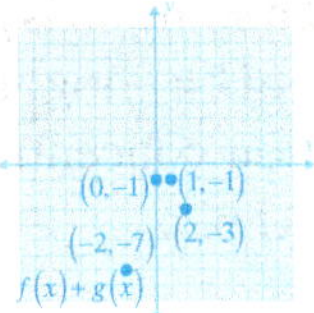

b.

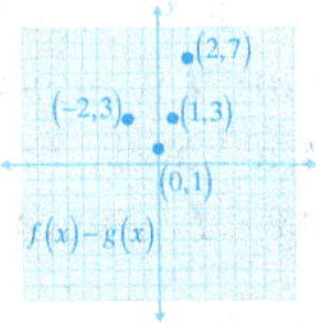

2. a.

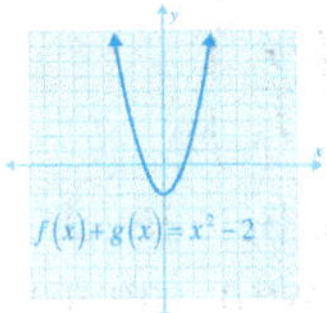

b. $f(x)-g(x)=-(x+2)^2+6$

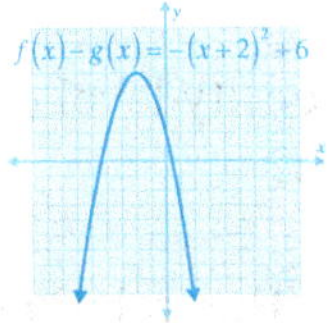

3. a. $\sqrt{x-3}+x^2+1$

b. $\sqrt{x-3}-x^2-1$

c. $\left(\sqrt{x-3}\right)\left(x^2+1\right)$

d. $\dfrac{\sqrt{x-3}}{x^2+1}$

4. a. $-4x^2+1$

b. $-8x^2+40x-47$

5. One-to-one

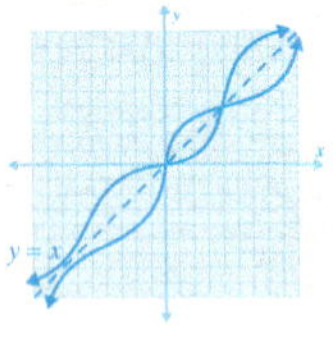

6. a. Not inverses

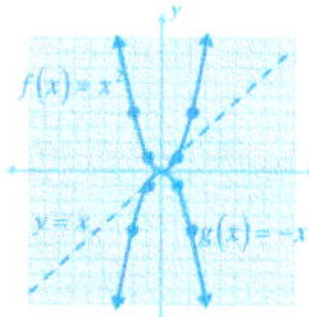

b. Inverses

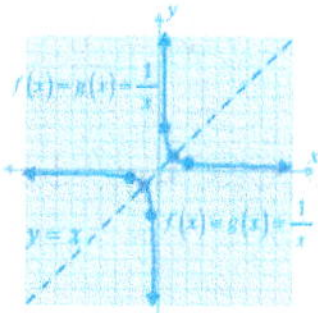

7. $f^{-1}(x)=\dfrac{1}{x+3}-1$

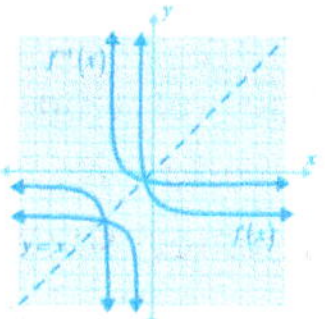

8. a.

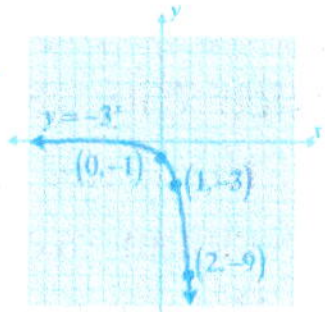

b.

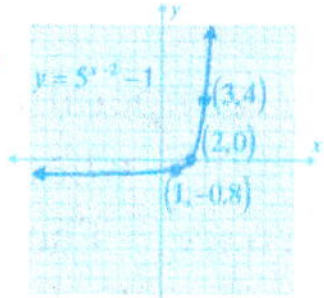

9. a. $\log 100{,}000 = 5$
b. $\log_{1/2} 8 = -3$
c. $x = \ln 15$

10. a. $10^4 = x$
b. $3^{-2} = \dfrac{1}{9}$
c. $4^x = 256$

11. a. $x = 343$ b. $x = \dfrac{3}{2}$

12. $y^{-1} = \log_{1/2} x$

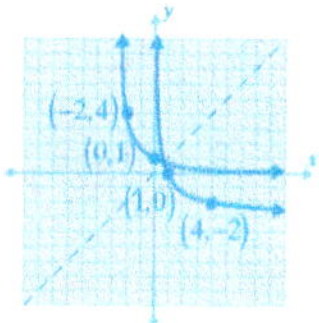

6. Determine, algebraically, whether or not each pair of functions are inverses of each other. Graph both functions on the same set of axes and show the line $y = x$ as a dotted line on each graph.

a. $f(x)=x^2$ and $g(x)=-x^2$ **b.** $f(x)=\dfrac{1}{x}$ and $g(x)=\dfrac{1}{x}$

7. Find $f^{-1}(x)$ if $f(x)=\dfrac{1}{x+1}-3$. Graph both functions on the same set of axes and show the line $y = x$ as a dotted line on the graph.

8. Sketch the graphs of the exponential functions given in the following equations. Label three points on each graph.

a. $y=-3^x$ **b.** $y=5^{x-2}-1$

9. Express the following equations in logarithmic form.

a. $10^5 = 100{,}000$ **b.** $\left(\dfrac{1}{2}\right)^{-3}=8$ **c.** $e^x = 15$

10. Express the following equations in exponential form.

a. $\log x = 4$ **b.** $\log_3 \dfrac{1}{9} = -2$ **c.** $\log_4 256 = x$

11. Solve the following equations by first changing them to exponential form.

a. $\log_7 x = 3$ **b.** $\log_9 27 = x$

12. Find the inverse of the function $y=\left(\dfrac{1}{2}\right)^x$. Graph both functions on the same set of axes and show the line $y = x$ as a dotted line on each graph. Label two points on each graph.

13. Use the properties of logarithms to expand the following expressions as much as possible.

a. $\ln\left(x^2-25\right)$ **b.** $\log\sqrt[3]{\dfrac{x^2}{y}}$

14. Use the properties of logarithms to write each expression as a logarithm of a single expression.

a. $\ln(x+5)+\ln(x-4)$ **b.** $\log\sqrt{x}+\log x^2-\log 5x$

15. Use a calculator to find the value of x accurate to 4 decimal places.

a. $x = \log 579$ **b.** $5 \ln x = 9.35$

13. a. $\ln(x+5) + \ln(x-5)$
b. $\frac{2}{3}\log x - \frac{1}{3}\log y$
14. a. $\ln(x^2+x-20)$
b. $\log\frac{x\sqrt{x}}{5}$
15. a. 2.7627
b. 6.4883
16. $x=-4$
17. $x=-3$
18. $x\approx 0.4518$
19. $x\approx 10.6873$
20. $x\approx 1.7925$
21. No solution
22. $x\approx 21.0855$
23. 1.7965
24. 25,981 bacteria
25. 19.07 years
26. a. 134.29 years
b. 198.04 years
27. 2.00
28. 110.12 minutes

Use the properties of exponents and logarithms to solve each of the equations for x. If necessary, round answers to four decimal places.

16. $7^3 \cdot 7^x = 7^{-1}$

17. $6^{x-1} = 36^{x+1}$

18. $10^{x+2} = 283$

19. $2e^{0.24x} = 26$

20. $4^x = 12$

21. $\log(2x+3) - \log(x+1) = 0$

22. $\ln(x^2+3x-4) - \ln(x+4) = 3$

23. Use the change-of-base formula to evaluate the following expression: $\log_6 25$. Round you answer to four decimal places.

Solve the following application problems. If necessary, round answers to two decimal places (unless otherwise specified).

24. Bacteria growth: A scientist knows that a certain strain of bacteria grows according to the function $y = y_0 \cdot 3^{0.25t}$, where t is a measurement in hours. If she starts a culture with 5000 bacteria, how many will be present after 6 hours? (Round to the nearest integer.)

25. Investing: If \$1000 is invested at 7% compounded continuously, when will the amount reach \$3800?

26. Decomposition: A substance decomposes according to $A = A_0 e^{-0.0035t}$, where t is measured in years and A_0 is the initial amount.

a. How long will it take for 800 grams to decompose to 500 grams?

b. What is the half-life of this substance?

27. The Richter scale: The relationship between the magnitude R and intensity I on the Richter scale is given by the formula $R = \log I$. How much stronger was the 1985 earthquake in Mexico City (8.1 on the Richter scale) than the 2008 earthquake in the Sichuan province of China (7.8 on the Richter scale)?

28. Blacksmithing: The temperature of a piece of iron is 800°F when it is removed from a furnace. The temperature in the room is 80°F. It takes 20 minutes for the iron to cool to 600°F. How long will it take for the iron to cool to 200°F? (Use Newton's law of cooling: $T = Ae^{-kt} + C$. C is the constant temperature of the surrounding medium. The values of A and k depend on the particular object that is cooling.)

Cumulative Review: Chapters 1 – 11

1. 32   2. 7

3. 1   4. $-\dfrac{x+3}{x+1}$

5. $\dfrac{2}{x+4}$ 6. $\dfrac{x-4}{x+1}$

7. $4xy\sqrt{5y}$

8. $\dfrac{2x\sqrt{10y}}{5y}$

9. $\dfrac{x^{\frac{1}{6}}}{y^{\frac{1}{3}}}$ 10. $\dfrac{x^{\frac{1}{2}}}{2y^2}$

11. $x^2y^{\frac{3}{2}}$ 12. $x^{\frac{2}{3}}y$

13. $7x^2-26x-8$

14. $-6y^2-18y+60$

15. $15+6i$

16. 8   17. i

18. $3x^2-4x+1-\dfrac{2x}{x^2+1}$

19. -6 20. 12

21. $3\log_b x+\dfrac{1}{2}\log_b y$

22. $\log_b\left(\dfrac{7x^2}{y}\right)$

23. $x=\dfrac{-5\pm\sqrt{33}}{2}$

24. $x=\dfrac{7}{2}$

25. $x=-3, -2, 2, 3$

26. $x=6$

27. $x=4\pm\sqrt{6}$

28. $x=-0.9212$

29. $x\approx 17.1825$

30. $x=-3$

31. $(-1, 3)$

Find the value of each expression using the rules for order of operations.

1. $8+[5\cdot 6-(9\div 3+3)]$

2. $2^3\div 4+7-10\div 5$

Perform the indicated operations and simplify the results.

3. $\dfrac{2x^2+7x+3}{x^2-3x-18}\cdot\dfrac{x^2-x-30}{2x^2+11x+5}$

4. $\dfrac{9-x^2}{x^2+7x+6}\div\dfrac{x-3}{x+6}$

5. $\dfrac{1}{x-1}+\dfrac{x-6}{x^2+3x-4}$

6. $\dfrac{2x}{2x+3}-\dfrac{7x+12}{2x^2+5x+3}$

Simplify. Assume all variables are positive.

7. $\sqrt{80x^2y^3}$

8. $\dfrac{\sqrt{8x^3y^2}}{\sqrt{5xy^3}}$

9. $\dfrac{x^{\frac{2}{3}}y^{\frac{1}{3}}}{x^{\frac{1}{2}}y^{\frac{2}{3}}}$

10. $\left(x^2y^{-1}\right)^{\frac{1}{2}}\left(4xy^3\right)^{-\frac{1}{2}}$

Change each expression to an equivalent exponential expression and simplify. Assume all variables are positive.

11. $\sqrt{x^4y^3}$

12. $\sqrt[3]{x^2y^3}$

Perform the indicated operations and simplify the results.

13. $(7x+2)(x-4)$

14. $-6(y+5)(y-2)$

15. $\left(6-\sqrt{-4}\right)+\left(9+\sqrt{-64}\right)$

16. $\left(\sqrt{6}-\sqrt{2}i\right)\left(\sqrt{6}+\sqrt{2}i\right)$

17. i^{17}

18. Divide the expression $\dfrac{3x^4-4x^3+4x^2-6x+1}{x^2+1}$ by using the division algorithm. Write the answer in the form $Q+\dfrac{R}{D}$, where the degree of $R <$ the degree of D.

Use your knowledge of logarithms and exponents to find the value of each expression.

19. $\log_2\dfrac{1}{64}$

20. $7^{\log_7 12}$

21. Use the properties of logarithms to expand the expression $\log_b\left(x^3\sqrt{y}\right)$ as much as possible.

22. Use the properties of logarithms to write the expression $2\log_b x-\log_b y+\log_b 7$ as a single logarithm.

32. $(-\infty,-2)\cup\left(\frac{5}{2},\infty\right)$

33. $(-\infty,-1]\cup\left[\frac{3}{2},\infty\right)$

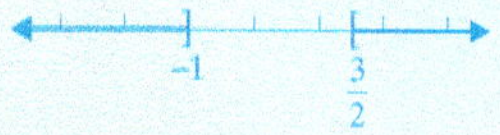

34. $(-9,-3)$

35. $(1,-1,2)$

36. $y=-\frac{2}{3}x+2$

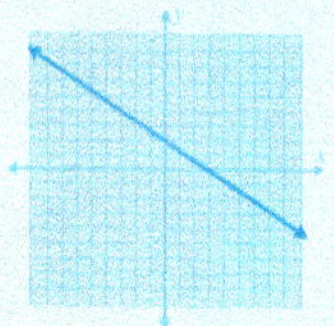

37. a. Parallel
b. No solution
c.

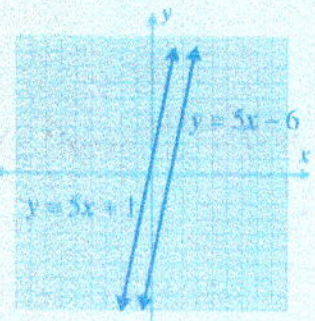

38. a. Perpendicular
b. $(2,1)$
c.

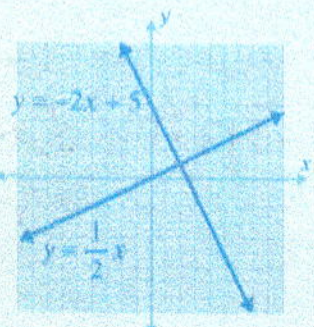

39.

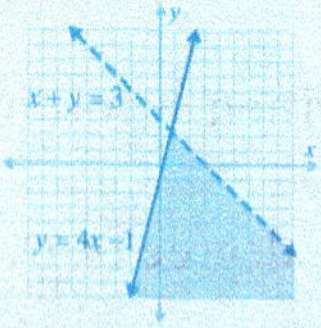

40.

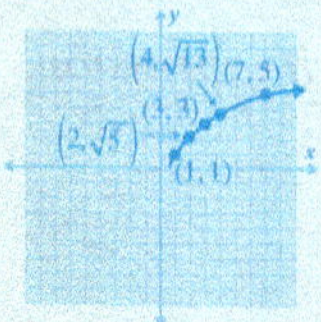

Solve each of the equations. Round answers to four decimal places, if necessary.

23. $x^2+5x-2=0$ **24.** $4x^2-28x+49=0$ **25.** $x^4-13x^2+36=0$

26. $x-2=\sqrt{x+10}$ **27.** $\frac{2}{x-2}+\frac{3}{x-1}=1$ **28.** $6^{3x+5}=55$

29. $\ln(x^2-7x+10)-\ln(x-2)=2.5$

30. Solve the equation $\log_6\frac{1}{216}=x$ by first changing it to exponential form.

Solve the inequalities algebraically and graph each solution set on a real number line. Write the answers using interval notation.

31. $(x-3)(x+1)<0$ **32.** $2x^2-x-10>0$

33. $2x^2-x\geq 3$ **34.** $\left|\frac{1}{2}x+3\right|<\frac{3}{2}$

35. Use the Gaussian elimination method to solve the system of linear equations:

$$\begin{cases} x+y+z=2 \\ 2x+y-z=-1 \\ x+3y+2z=2 \end{cases}$$

36. Find an equation in slope-intercept form for the line passing through the point $(-3,4)$ with slope $m=-\frac{2}{3}$. Graph the line.

For the systems of equations: ***a.*** *Determine whether each pair of lines is parallel, perpendicular, or neither.* ***b.*** *Solve the system of equations.* ***c.*** *Graph both lines.*

37. $\begin{cases} y=5x-6 \\ y=5x+1 \end{cases}$ **38.** $\begin{cases} y=-2x+5 \\ y=\frac{1}{2}x \end{cases}$

39. Solve the following system of two linear inequalities graphically: $\begin{cases} y\leq 4x-1 \\ x+y<3 \end{cases}$.

40. Find and label at least 5 points on the graph of the function $\sqrt{4x-3}$ and then sketch the graph of the function.

Rewrite each of the quadratic functions in the form $y=a(x-h)^2+k$ and find the vertex, range, and zeros of each function. Graph the function.

41. $y=x^2-6x-2$ **42.** $y=2x^2+8x+3$ **43.** $y=x^2-8x+16$ **44.** $y=x^2-2x-6$

41. $y=(x-3)^2-11$;
Vertex: $(3,-11)$;
Range: $[-11,\infty)$;
Zeros: $x=3\pm\sqrt{11}$

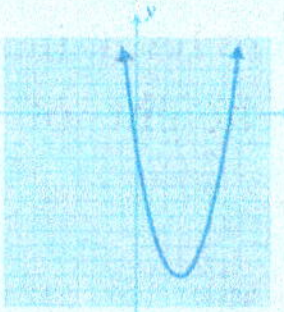

42. $y=2(x+2)^2-5$;
Vertex: $(-2,-5)$;
Range: $[-5,\infty)$;
Zeros:
$x=-2\pm\dfrac{\sqrt{10}}{2}$

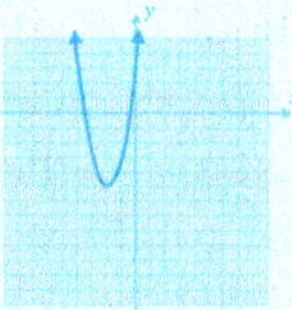

43. $y=(x-4)^2$;
Vertex: $(4,0)$;
Range: $[0,\infty)$;
Zeros: $x=4$

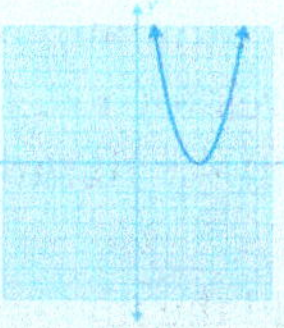

44. $y=(x-1)^2-7$;
Vertex: $(1,-7)$;
Range: $[-7,\infty)$;
Zeros: $x=1\pm\sqrt{7}$

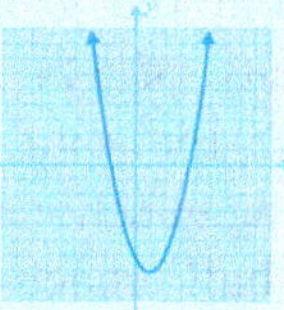

45. The graphs of three curves are shown. Use the vertical line test to determine whether or not each graph represents a function. Then state the domain and range of each graph.

a.

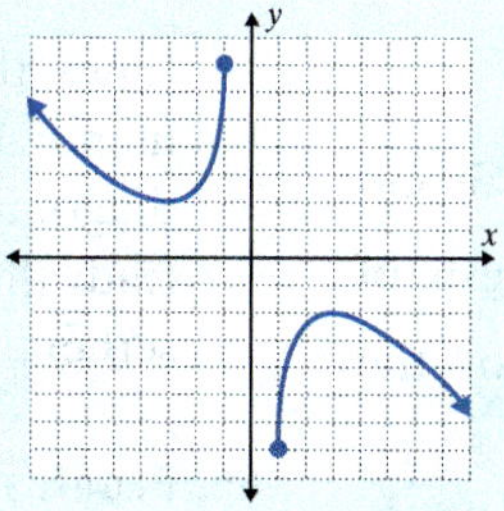

b.

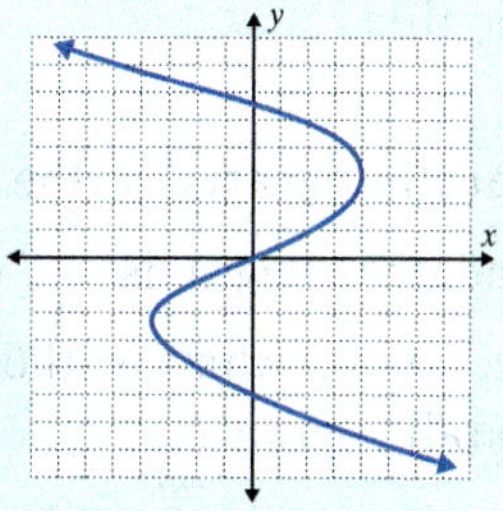

c.

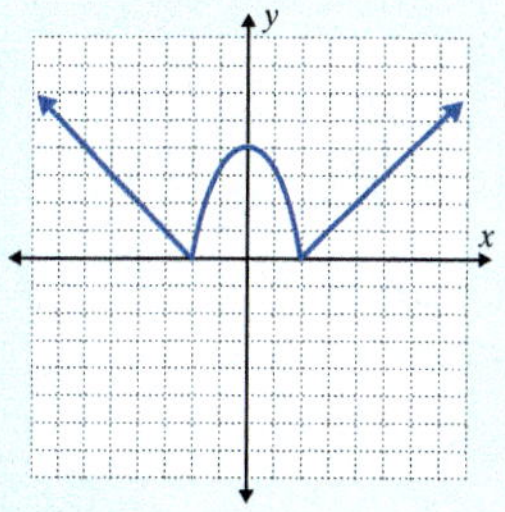

Write the function as a set of ordered pairs for the given equation and domain.

46. $y=x^2-2x+5$
$D=\{-2,-1,0,1,2\}$

47. $y=x^3-5x^2$
$D=\{-2,-1,0,1,2\}$

48. Given the two functions, $f(x)=x^2+x$ and $g(x)=\sqrt{x+2}$, find:

a. $(f+g)(x)$ **b.** $(f-g)(x)$ **c.** $(f\cdot g)(x)$ **d.** $\left(\dfrac{f}{g}\right)(x)$

49. Given the two functions, $f(x)=x^2+3$ for $x\geq 0$ and $g(x)=2x+1$ for $x\geq 0$, find:

a. $f(g(x))$ **b.** $g(f(x))$

Which of the functions are one-to-one functions? If the graph represents a one-to-one function, graph its inverse by reflecting the graph of the function across the line $y=x$.

50.

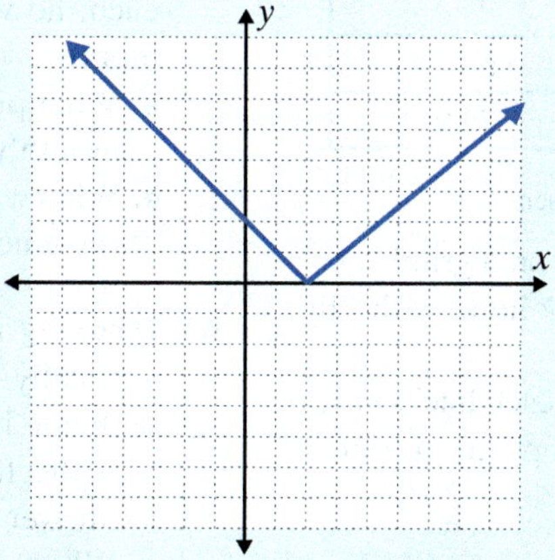

51.

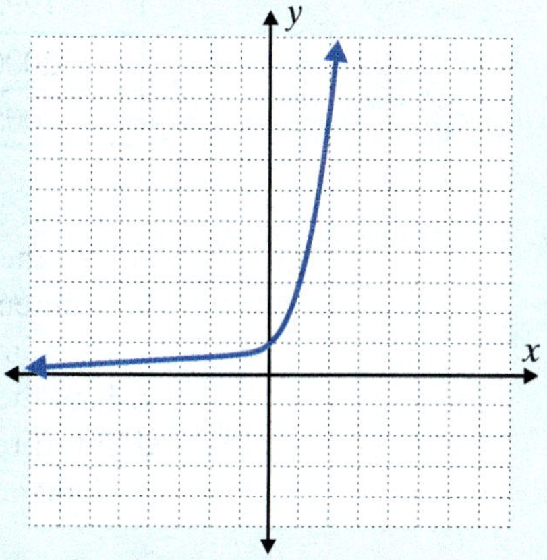

Find the inverse of each function and graph both the function and its inverse on the same set of axes. Include the graph of the line $y=x$ as a dotted line in each graph.

52. $f(x)=x^2-4$ for $x\geq 0$

53. $g(x)=(x-2)^3$

45. a. Function
$D=(-\infty,-1]\cup[1,\infty)$
$R=(-\infty,-2]\cup[2,\infty)$
b. Not a function
$D=(-\infty,\infty)$
$R=(-\infty,\infty)$
c. Function
$D=(-\infty,\infty)$
$R=[0,\infty)$

46. $\{(-2,13),(-1,8),(0,5),(1,4),(2,5)\}$

47. $\{(-2,-28),(-1,-6),(0,0),(1,-4),(2,-12)\}$

48. a. $x^2+x+\sqrt{x+2}$
b. $x^2+x-\sqrt{x+2}$
c. $(x^2+x)\sqrt{x+2}$
d. $\dfrac{x^2+x}{\sqrt{x+2}},\ x>-2$

49. a. $4x^2+4x+4$
b. $2x^2+7$

50. Not a one-to-one function

51. One-to-one function

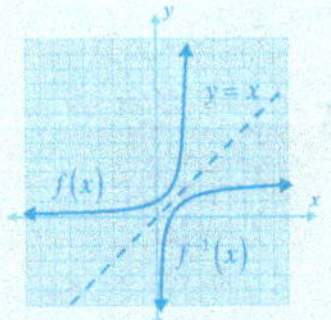

52. $f^{-1}(x)=\sqrt{x+4}$

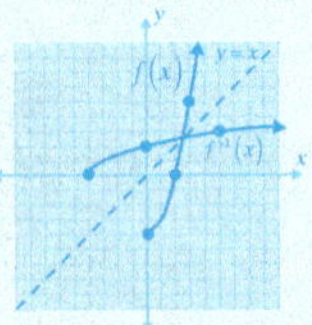

53. $g^{-1}(x)=\sqrt[3]{x}+2$

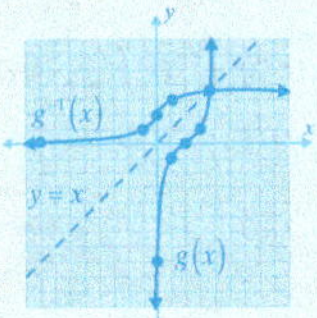

54. Find three consecutive odd integers such that 2 times the second plus 5 times the third is equal to 6 less than the first.

55. Parallelograms: If the base is 30 cm and the height is $\frac{2}{5}$ the length of the base, what is the area of the parallelogram?

56. Cross-country race: Aaryn's first four miles of a five mile cross country race were 6:33 min, 7:09 min, 7:42 min, and 7:21 min. What time does she need to run her last mile if she wants to run an average time of
a. 7:15 minutes per mile?
b. 7:24 minutes per mile?

57. Cable television: The table below shows the number of millions of US households with cable television from 1985-2005.

Year	Households with Cable
1985	36.3
1990	51.9
1995	60.5
2000	68.6
2005	73.9

Source: Nielsen

a. Plot these points on a graph.
b. Connect the points with line segments.
c. Find the slope of each line segment.
d. Interpret the slope as a rate of change.

58. Kayaking: Nadine kayaked 8 miles upstream and then returned. Her average speed on the return trip was 6 mph faster than her speed upstream. If her total travel time was 2.8 hours, find her rate each way.

59. Computer Screens: In order to display images properly, the width of a computer screen must be $\frac{4}{3}$ times the height of the screen. If a manufacturer wants the length of the diagonal of the screen to be 20 inches, find the width and height of the screen.

60. Proofreading: Caleb can proofread a news article in 2 hours working alone. His coworker Tiffany can proofread the same article in $1\frac{1}{2}$ hours working alone. How long will it take if they work together?

61. Publishing books: A book is available in both hardback and paperback. A bookstore sold a total of 43 books during the week. The total receipts were \$297.50. If hardback books sell for \$12.50 and paperbacks sell for \$4.50, how many of each were sold?

62. Selling light fixtures: An electronics store sells a special type of light fixture. The owner estimates that by selling these fixtures for x dollars each, he will sell $1200-x$ fixtures a month.
a. What price will yield the maximum monthly revenue?
b. What will be the maximum monthly revenue?

63. Throwing a ball: A ball is projected vertically upward from the ground with an initial velocity of 144 ft/s. Use the formula $h=-16t^2+v_0t+h_0$ to answer the following questions.
a. When will the ball reach its maximum height?
b. What will be the maximum height?
c. When will the ball be on the ground?

64. Light bulbs: Studies show that the fractional part P of light bulbs that have burned out after t hours of use is given by $P = 1 - 2^{-0.03t}$. What fractional part of the light bulbs have burned out after 100 hours of use?

65. Investing: If P dollars are invested at a rate r (expressed as a decimal), and compounded k times a year, the amount A accumulated at the end of t years is given by $A = P\left(1 + \frac{r}{k}\right)^{kt}$ dollars. Find A if \$500 is invested at 10% compounded quarterly for 2 years.

66. Half-life of radium: Radium decomposes according to $A = A_0 e^{-0.043t}$, where t is measured in centuries and A_0 is the initial amount. Determine the half-life for radium. Round your answer to two decimal places.

67. Friction: When friction is used to stop the motion of a wheel, the velocity may be given by $V = V_0 e^{-0.35t}$, where V_0 is the initial velocity and t is the number of seconds the friction has been applied. How long will it take to slow a wheel from 75 ft/s to 15 ft/s? Round your answer to two decimal places.

Use your calculator to find the value (accurate to four decimal places) of each of the radical expressions.

68. $6 - 5\sqrt{7}$

69. $\dfrac{3 + 2\sqrt{11}}{5}$

Use a calculator to evaluate the logarithms accurate to 4 decimal places.

70. $\log 54.6$

71. $\ln 10000$

Use a graphing calculator to solve the following inequalities. Graph each solution set on a real number line and write the answers in interval notation. (Estimate endpoints when necessary.)

72. $x^2 > 2$

73. $x^2 + 5x - 6 \leq 0$

74. Use a graphing calculator to solve the system of linear equations:

$$\begin{cases} 2x - y = -1 \\ x + 2y = 12 \end{cases}.$$

54. $-5, -3, -1$
55. 360 cm^2
56. 7:30 minutes; 8:15 minutes
57. a. and **b.**

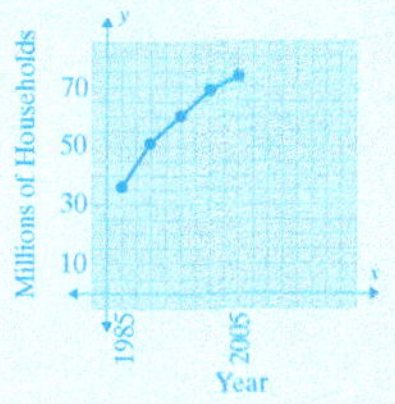

c. 3.12; 1.72; 1.62; 1.06
d. The number of users increased 3.12 million a year from 1985-1990, 1.72 million a year from 1990-1995, 1.62 million a year from 1995-2000, and 1.06 million a year from 2000-2005.
58. 10 mph, 4 mph
59. Width: 16 inches; Height: 12 inches
60. $\frac{6}{7}$ hours
61. 13 hardback; 30 paperback
62. a. \$600 **b.** \$360,000

63. a. 4.5 s **b.** 324 ft **c.** 9 s **64.** $\frac{7}{8}$ **65.** \$609.20 **66.** 16.12 centuries **67.** 4.60 s **68.** -7.2288 **69.** 1.9266

70. 1.7372 **71.** 9.2103 **72.** $(-\infty, -\sqrt{2}) \cup (\sqrt{2}, \infty)$

73. $[-6, 1]$

74. (2, 5)

CHAPTER 12

Conic Sections

Did You Know?

Euler

The mathematician who invented the notation for functions, $f(x)$, was Leonhard Euler (1707 – 1783) of Switzerland. Euler was one of the most prolific mathematical researchers of all time, and he lived during a period in which mathematics was making great progress. He studied mathematics, theology, medicine, astronomy, physics, and oriental languages before he began a career as a court philosopher-mathematician. His professional life was spent at St. Petersburg Academy by invitation of Catherine I of Russia, at the Berlin Academy under Frederick the Great of Prussia, and again at the St. Petersburg Academy under Catherine the Great. The collected works of Euler fill 80 volumes, and for almost 50 years after Euler's death the publications of the St. Petersburg Academy continued to include articles by him.

Euler was blind the last 17 years of his life, but he continued his mathematical research by writing on a large slate and dictating to a secretary. He was responsible for the conventionalization of many mathematical symbols such as $f(x)$ for function notation, i for $\sqrt{-1}$, e for the base of the natural logarithms, π for the ratio of circumference to diameter of a circle, and Σ for the summation symbol.

From the age of 20 to his death, Euler was busy adding to knowledge in every branch of mathematics. He wrote with modern symbolism, and his work in calculus was particularly outstanding. Euler had a rich family life, having fathered 13 children, and he not only contributed to mathematics, but reformed the Russian system of weights and measures, supervised the government pension system in Prussia and the government geographic office in Russia, designed canals, and worked in many areas of physics, including acoustics and optics. It was said of Euler, by the French academician François Arago, that he could calculate without any apparent effort "just as men breathe and eagles sustain themselves in the air."

An interesting story is told about Euler's meeting with the French philosopher Diderot at the Russian court. Diderot had angered the Czarina by his antireligious views, and Euler was called to the court to debate Diderot. Diderot was told that the great mathematician Euler had an algebraic proof that God existed. Euler walked in towards Diderot and said, "Monsieur, $\frac{a+b^n}{n} = x$, therefore God exists, respond." Diderot, who had no understanding of algebra, was unable to respond.

"I have not hesitated in 1900, at the Congress of Mathematicians in Paris, to call the nineteenth century the century of the theory of functions."

Vito Volterra (1860 – 1940)

A circular cone is a three-dimensional figure that can be generated by choosing one point on a line and rotating the line in a circular fashion about this point. (See Figure 1 on page 976.) (You can visualize a cone by holding a meter stick (or a yard stick) with two fingers and rotating the stick so that each end moves in a circle. Note that the cone has a top portion and a bottom portion.) If a plane intersects a cone, the intersections will be curves (or a single point) on the plane. These curves are called **conic sections**. The four conic sections we will discuss in this chapter are circles, ellipses, parabolas, and hyperbolas.

By setting a Cartesian coordinate system on the plane, the conic sections and their properties can be discussed algebraically. The study of geometry with algebraic equations is called analytic geometry. If you continue your studies in mathematics, you will find that analytic geometry can be applied in three dimensions as well as in two dimensions. Related figures would be spheres, ellipsoids (football shapes), and paraboloids (the reflective surfaces of telescopes).

12.1 Translations and Reflections

- *Calculate and understand the difference quotient.*
- *Understand the concepts of horizontal and vertical translations.*
- *Graph translations and reflections of functions. That is, given the graph of a function $y = f(x)$ graph translations and reflections of the form $y = \pm f(x-h)+k$.*

Review of Function Notation: $f(x)$

We have already used function notation $f(x)$ (read "f of x") in Chapter 5 to represent and to evaluate polynomials, in Chapter 9 to represent and evaluate radical functions, and in Chapter 11 in dealing with inverse functions. In this section, function notation is used to indicate a new value or a new function when the single variable x is replaced by an expression such as $(a+1)$ or $(x+h)$. For example,

if $f(x) = 3x+5$, then

$f(2) = 3(2)+5 = 11$ — x is replaced by 2.

$f(a) = 3(a)+5$ — x is replaced by a.

$f(a+1) = 3(a+1)+5$ — x is replaced by $a + 1$.

$f(x+h) = 3(x+h)+5$ — x is replaced by $x + h$.

These substitutions for x relate to formulas and to graphs of functions.

Example 1: Using $f(x)$ Notation

For the function $f(x) = 2x^2 - 4$, find:

a. $f(3)$

Solution: $f(3) = 2(3)^2 - 4 = 2 \cdot 9 - 4 = 14$

b. $f(a)$

Solution: $f(a) = 2(a)^2 - 4 = 2a^2 - 4$

c. $f(a+1)$

Solution: Replace x with $a+1$ and simplify.

$$\begin{aligned} f(a+1) &= 2(a+1)^2 - 4 \\ &= 2(a^2 + 2a + 1) - 4 \\ &= 2a^2 + 4a + 2 - 4 \\ &= 2a^2 + 4a - 2 \end{aligned}$$

The Difference Quotient: $\dfrac{f(x+h) - f(x)}{h}$

The formula

$$\frac{f(x+h) - f(x)}{h}$$

is called the **difference quotient**. As shown in Figure 1, a geometric interpretation of the difference quotient is the **slope** of a line through two points on the graph of a function.

Teaching Notes: You might want to ask your students if they can picture the point Q sliding down the curve, getting closer and closer to P. Then ask what is happening to the secant line and what is happening to h? Give two or three other curves on the board and you can have a lively class discussion.

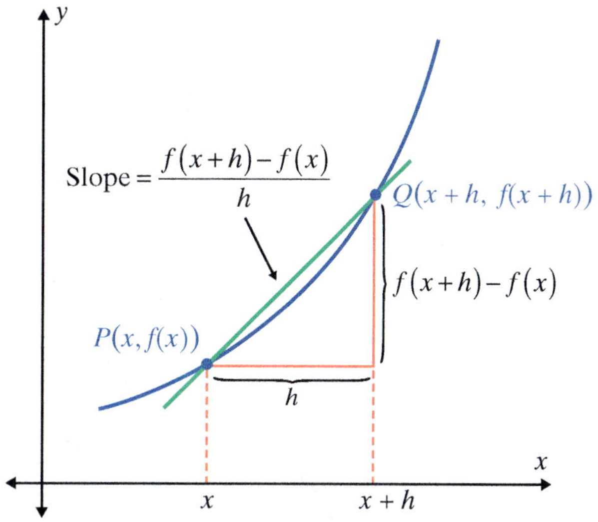

Figure 1

NOTES

The line illustrated in Figure 1 (through two points on the graph) is called a **secant line**.

NOTES

Note carefully that $f(x+h)$ is not the same as $f(x)+h$. For example,

if $\quad f(x) = x^2 + 2x + 3,$

then $\quad f(x+h) = (x+h)^2 + 2(x+h) + 3 = x^2 + 2xh + h^2 + 2x + 2h + 3$

and $\quad f(x) + h = x^2 + 2x + 3 + h.$

Example 2: Finding the Difference Quotient $\frac{f(x+h)-f(x)}{h}$ for a Given Function

a. Find the difference quotient for the function $f(x) = 2 - 6x$.

Solution: $f(x+h) = 2 - 6(x+h)$ and $f(x) = 2 - 6x$

Substituting gives:

$$\frac{f(x+h)-f(x)}{h} = \frac{[2-6(x+h)]-[2-6x]}{h}$$

$$= \frac{2-6x-6h-2+6x}{h}$$

$$= \frac{-6h}{h} = -6.$$

b. Find the difference quotient for the function $f(x) = 2x^2 - 5x$.

Solution: $f(x+h) = 2(x+h)^2 - 5(x+h)$ and $f(x) = 2x^2 - 5x$

Substituting gives:

$$\frac{f(x+h)-f(x)}{h} = \frac{\left[2(x+h)^2 - 5(x+h)\right]-\left[2x^2-5x\right]}{h}$$

$$= \frac{2x^2+4xh+2h^2-5x-5h-2x^2+5x}{h}$$ Expand $(x+h)^2$ and multiply by 2.

$$= \frac{4xh+2h^2-5h}{h}$$

$$= \frac{h(4x+2h-5)}{h}$$ Factor out h.

$$= 4x + 2h - 5.$$

Horizontal and Vertical Translations

In Section 10.5, the graphs of parabolas of the form $y = ax^2$ were discussed. We found that the graph of $y = a(x-h)^2$ is a **horizontal translation** (**horizontal shift**) of h units and the graph of $y = ax^2 + k$ is a **vertical translation** (**vertical shift**) of k units of the graph of $y = ax^2$.

In function notation,

if $f(x) = 2x^2$,

then $f(x-3) = 2(x-3)^2$ is a horizontal translation of 3 units to the right,

and $f(x) - 4 = 2x^2 - 4$ is a vertical translation of 4 units down.

Figure 2 shows the graphs of the three functions and their relationships. Note that the curves themselves are identical, only their positions are changed.

Teaching Notes:
You might ask your students what they think will happen if the leading coefficient 2 is changed.

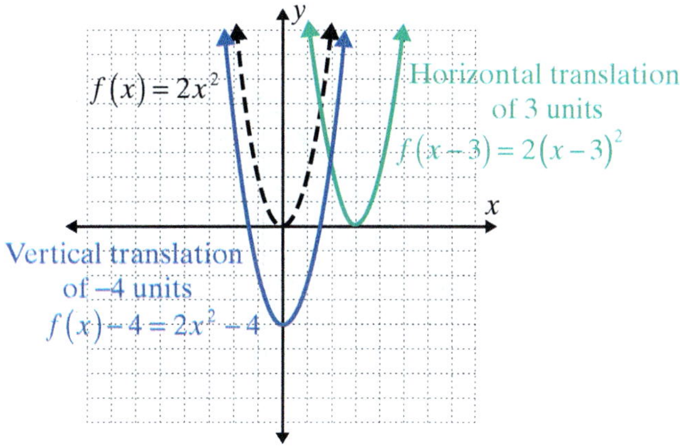

Figure 2

The following general approach can be used to help graph a horizontal and/or vertical translation of any function.

Horizontal and Vertical Translations

Given the graph of $y = f(x)$, the graph of $y = f(x-h)+k$ is

1. a horizontal translation of h units, and
2. a vertical translation of k units

of the graph of $y = f(x)$.

Draw the graph of $y = f(x)$ in relation to (h, k) as if (h, k) were the origin, $(0, 0)$. This new graph will be the graph of $y = f(x-h)+k$.

Another function to use in illustrating translations is the function $f(x)=|x|$. First, we need to know what the graph of $f(x)=|x|$ or $y=|x|$ looks like. The definition of $|x|$ gives

$$f(x)=|x|=\begin{cases} x & \text{if } x \geq 0 \\ -x & \text{if } x < 0 \end{cases}$$

The graph can be analyzed in two pieces.

First Piece:
The graph of $f(x)=x$ is a line, as shown in Figure 3a, but we want only the part where $x \geq 0$, as shown in Figure 3b.

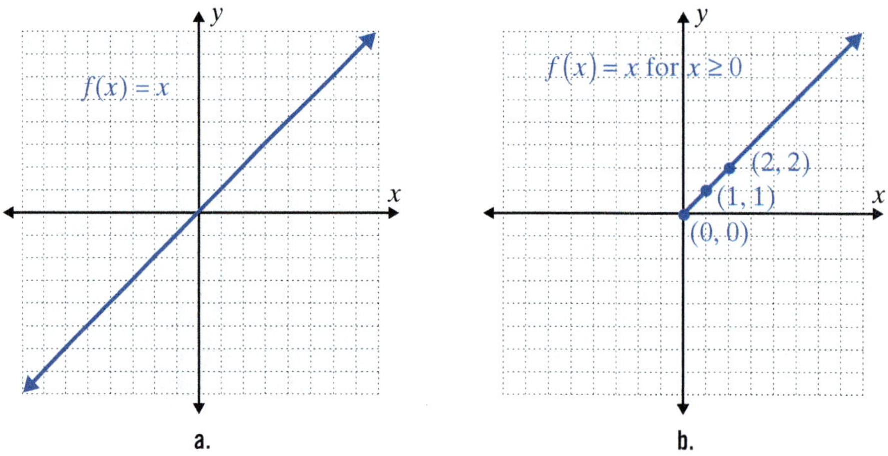

Figure 3

Second Piece:
The graph of $f(x)=-x$ is also a line, as shown in Figure 4a, but we want only the part where $x < 0$, as shown in Figure 4b.

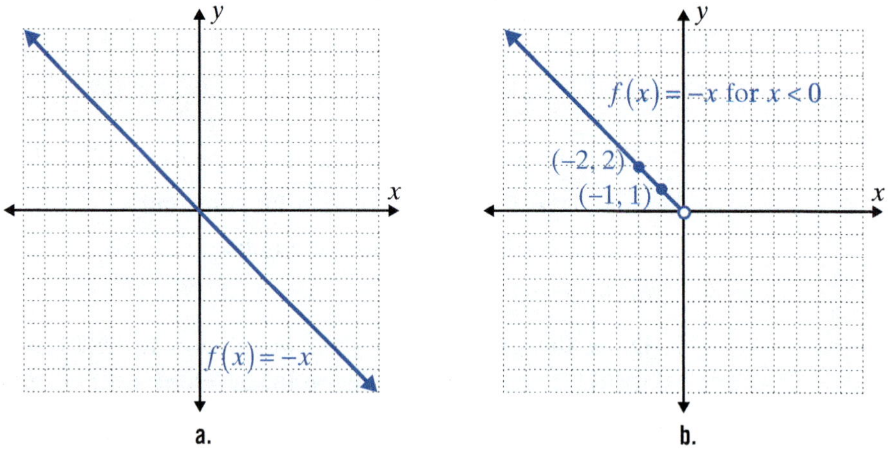

Figure 4

The two graphs in Figures 3b and 4b together give the graph of $f(x)=|x|$, as shown in Figure 5.

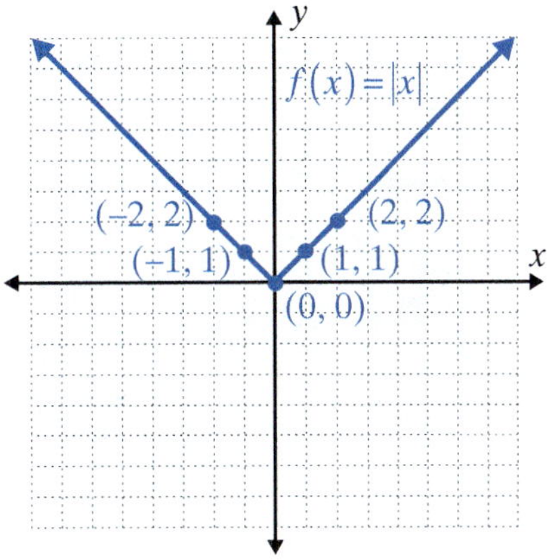

Figure 5

Now, using the graph of the function $f(x)=|x|$ or $y=|x|$, we can examine related horizontal and vertical shifts as in Example 3.

Example 3: Horizontal and Vertical Translations of the Function $y = |x|$

Teaching Notes:
You can explain how points other than $(0,0)$ can be translated by moving 3 units right and 2 units up. For example, referencing Figure 5, we see $(1,1)$ goes to $(4,3)$ and $(-1,1)$ goes to $(2,3)$. This can be done for any point on the graph of the function.

Graph each of the following functions. Use the graph in Figure 5 as a reference.

a. $y=|x-3|+2$

Solution: Here $(h,k)=(3,2)$, so there is a horizontal translation of 3 units right and 2 units up. In effect, $(3,2)$ is now the vertex of the new graph just as $(0,0)$ is the vertex of the original graph. You should check that the points shown on the graph here do indeed satisfy the function.

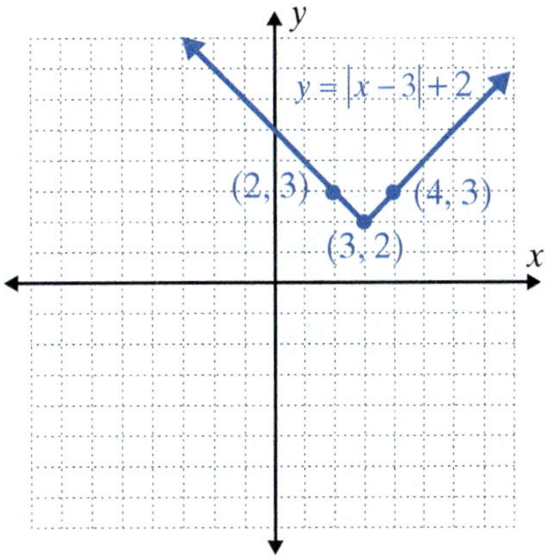

Continued on the next page...

b. $y = |x+4| - 1$

Solution: Here $(h, k) = (-4, -1)$, so the horizontal translation is –4 (4 units left) and the vertical translation is –1 (1 unit down). The effect is that the vertex is now at the point $(-4, -1)$.

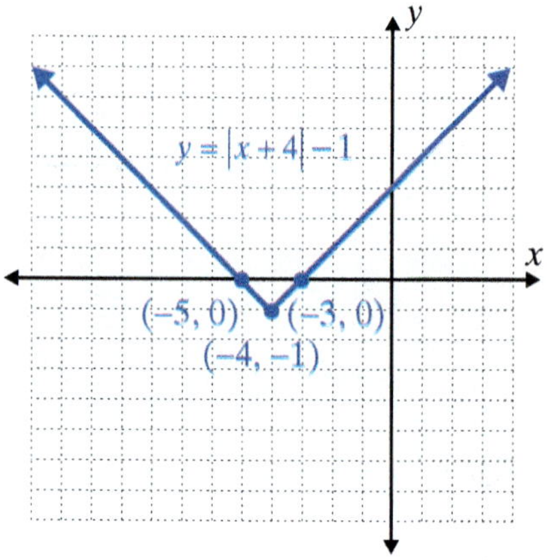

c. $y = |x+2| + 7$

Solution: Here $(h, k) = (-2, 7)$, so the horizontal translation is –2 (2 units left) and the vertical translation is 7 (7 units up). The effect is that the vertex is now at the point $(-2, 7)$.

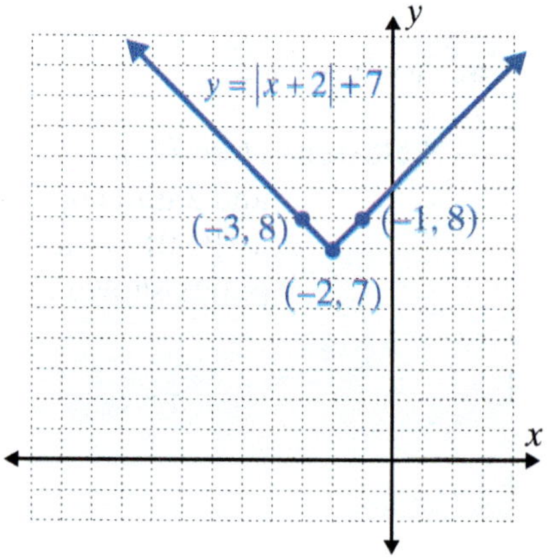

Reflections and Translations

In our discussion of the graphs of $y = ax^2$ and $y = |x|$, the leading coefficient a has not been changed. Changing the coefficient a can have an effect on the shape and direction of the basic function. For example, if a changes from positive to negative, the corresponding graph is reflected across the x-axis. In general, the graph of $y = -f(x)$ is a **reflection across the x-axis** of the graph of $y = f(x)$.

Teaching Notes:
To reinforce this idea, you might consider reviewing the discussions concerning the two functions $y = x^2$ and $y = -x^2$ in Section 10.5. Students should see the similarities.

If we examine the function $y = |x|$ and the graph of the function $y = -|x|$, we see that the graph of each function is the mirror image of the other across the x-axis. The first "opens" upward and the second "opens" downward as illustrated in Figure 6. Both have the same vertex at $(0, 0)$.

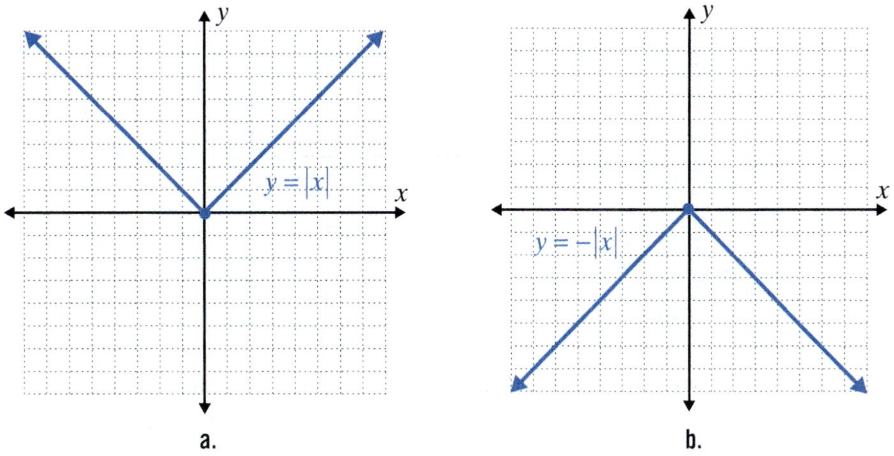

Figure 6

Example 4: Reflections and Translations of the Function $y = |x|$

Graph the function $y = -|x+2| + 5$.

Solution: **The reflection is performed first, followed by the translations.**
Here $(h, k) = (-2, 5)$, and the graph is reflected across the x-axis. We show step-by-step how to "arrive" at the graph. (You may do these steps mentally and graph only the last step.)

Step 1: Graph the reflection $y = -|x|$.

Step 2: Translate the graph 2 units to the left.

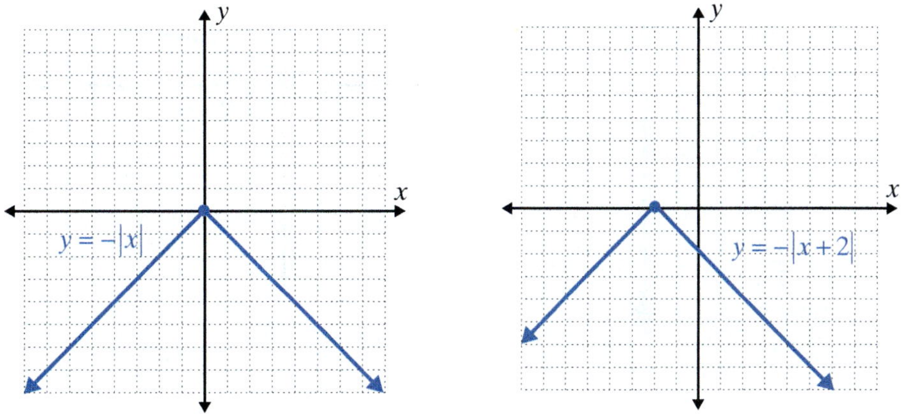

Continued on the next page...

Step 3: Translate the graph 5 units up.

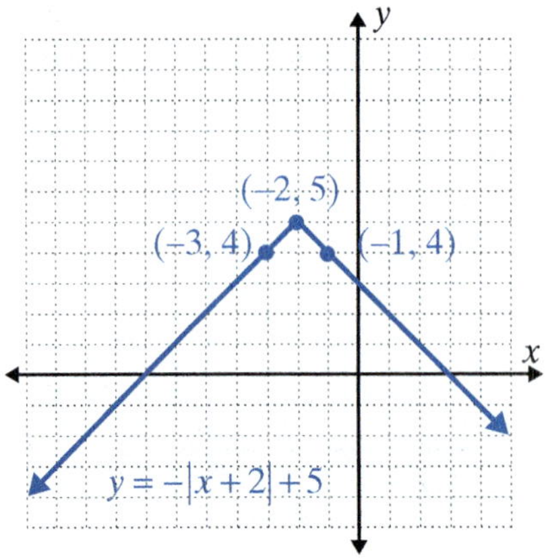

In a translation (horizontal or vertical) the shape of the graph of a function is unchanged. However, translations do change the position of the graph in the coordinate system. Example 5 illustrates two such cases.

Example 5: Translations of Functions with Graphs Given

Teaching Notes:
In graphing reflections and translations, the reflections are performed first. You might have your students do some translations first, followed by the reflections and show them how points on such a graph do not satisfy the new function. One example would be $y=-\sqrt{x-2}+1$ to compare with Example 5a.

a. The graph of $y=\sqrt{x}$ is given. Graph the function $y=\sqrt{x-2}+1$.

Solution: If $y=\sqrt{x}$ is written $y=f(x)$, then $y=\sqrt{x-2}+1$ is the same as $y=f(x-2)+1$. So $(h,k)=(2,1)$, and there is a horizontal translation of 2 units to the right and a vertical translation of 1 unit up.

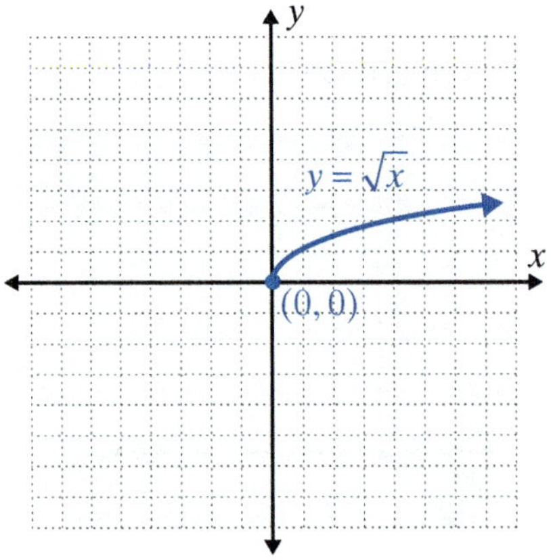

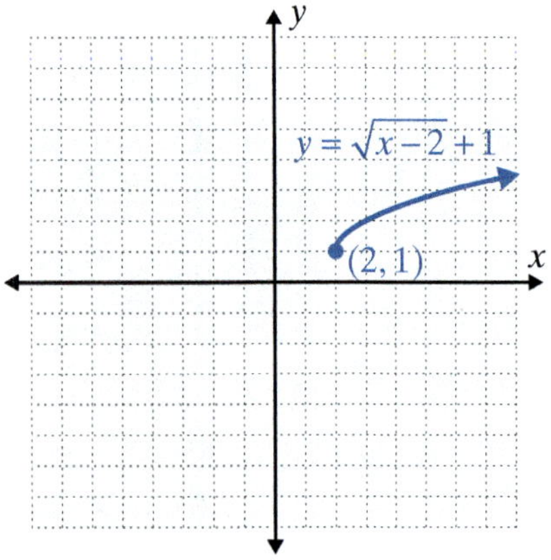

1. a. 0
 b. a^2-6a+5
 c. $x^2+2xh+h^2-4$
 d. $2x+h$
2. a. 0
 b. $1-a^2+2a$
 c. $2-x^2-2xh-h^2$
 d. $-2x-h$
3. a. 0
 b. $2a^2-11a+14$
 c. $2x^2+4xh-3x+2h^2-3h$
 d. $4x+2h-3$
4. a. 44
 b. $3a^2+11a+10$
 c. $3x^2+6xh-x+3h^2-h$
 d. $6x+3h-1$

b. The graph of $y=f(x)$ is given. Graph the function $y=f(x-3)-2$.

Solution: Here $(h,k)=(3,-2)$, so translate the graph horizontally 3 units and vertically −2 units. (Add 3 to each x-value and −2 to each y-value.)

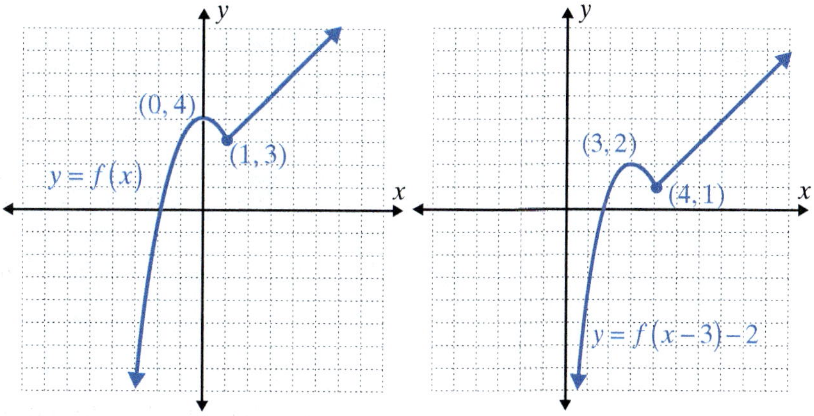

Practice Problems

1. For $f(x)=x^2-5$, find:
 a. $f(0)$ **b.** $f(a)$ **c.** $f(a+2)$
2. If $g(x)=3x+7$, find:
 a. $g(0)$ **b.** $g(x+h)$ **c.** $\dfrac{g(x+h)-g(x)}{h}$

(When evaluating the difference quotient, remember that $g(x+h)\neq g(x)+g(h)$.)

5. 1 6. 2
7. −2 8. 4
9. The results are the coefficients of x.
10. The results equal the slopes of the lines..
11.

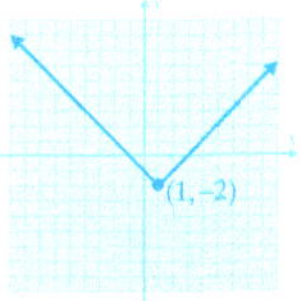

12.1 Exercises

12.

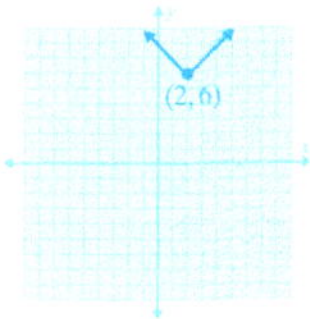

1. For $f(x)=x^2-4$, find:
 a. $f(-2)$
 b. $f(a-3)$
 c. $f(x+h)$
 d. $\dfrac{f(x+h)-f(x)}{h}$

2. For $g(x)=2-x^2$, find:
 a. $g(\sqrt{2})$
 b. $g(a-1)$
 c. $g(x+h)$
 d. $\dfrac{g(x+h)-g(x)}{h}$

Answers to Practice Problems: **1. a.** $f(0)=-5$ **b.** $f(a)=a^2-5$ **c.** $f(a+2)=a^2+4a-1$

2. a. $g(0)=7$ **b.** $g(x+h)=3x+3h+7$ **c.** $\dfrac{g(x+h)-g(x)}{h}=3$

13.

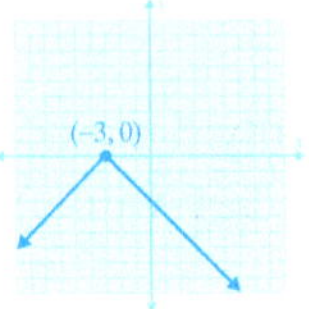

14.

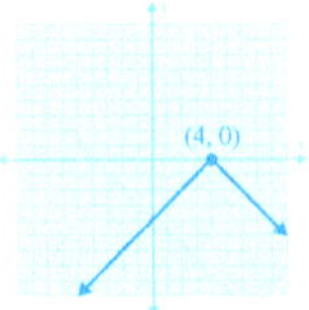

15.

16.

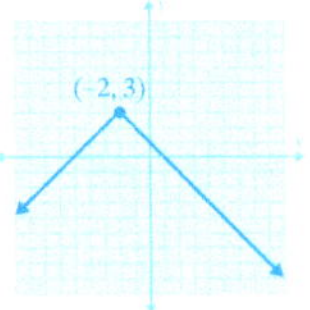

17.

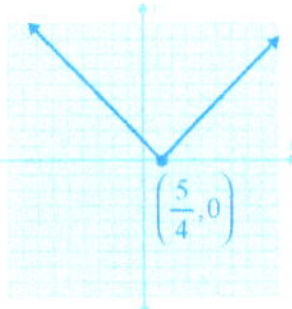

18.

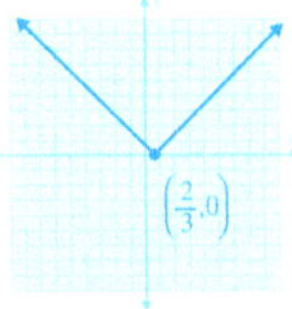

19.

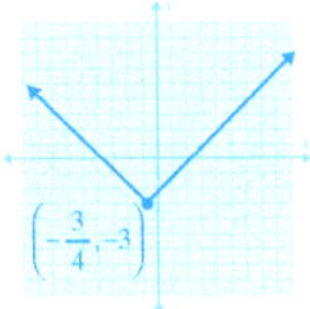

20.

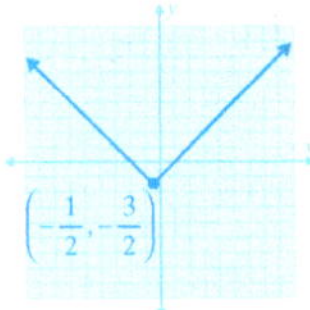

21.

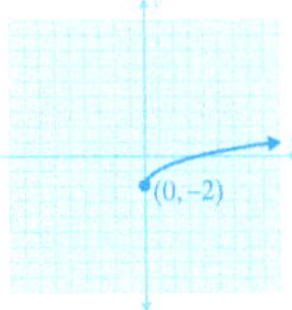

3. For $f(x) = 2x^2 - 3x$, find:

a. $f(0)$

b. $f(a-2)$

c. $f(x+h)$

d. $\dfrac{f(x+h)-f(x)}{h}$

4. For $f(x) = 3x^2 - x$, find:

a. $f(4)$

b. $f(a+2)$

c. $f(x+h)$

d. $\dfrac{f(x+h)-f(x)}{h}$

Find and simplify the difference quotient, $\dfrac{f(x+h)-f(x)}{h}$, *for each function:*

5. $f(x) = x + 7$

6. $f(x) = 2x - 3$

7. $f(x) = 5 - 2x$

8. $f(x) = 4x - 3$

9. What particular information, if any, do you notice about the results in Exercises 5 – 8?

10. Analyze, in your own words, how the results in Exercise 9 relate to the graphs of the functions in relation to the secant line discussion in the text.

The graph of $y = |x|$ *is given along with a few points as aids. Graph the functions using your understanding of reflections and translations with no additional computations.*

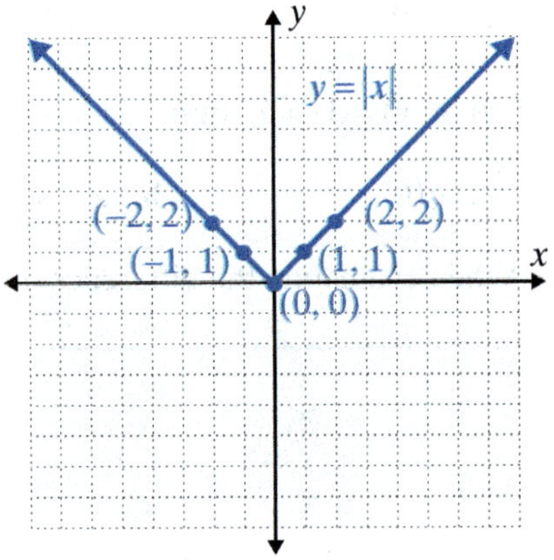

11. $y = |x-1| - 2$

12. $y = |x-2| + 6$

13. $y = -|x+3|$

14. $y = -|x-4|$

15. $y = -|x+5| + 4$

16. $y = -|x+2| + 3$

17. $y = \left|x - \dfrac{5}{4}\right|$

18. $y = \left|x - \dfrac{2}{3}\right|$

19. $y = \left|x + \dfrac{3}{4}\right| - 3$

20. $y = \left|x + \dfrac{1}{2}\right| - \dfrac{3}{2}$

The graph of $y=\sqrt{x}$ is given along with a few points as aids. Graph the functions using your understanding of reflections and translations with no additional computations.

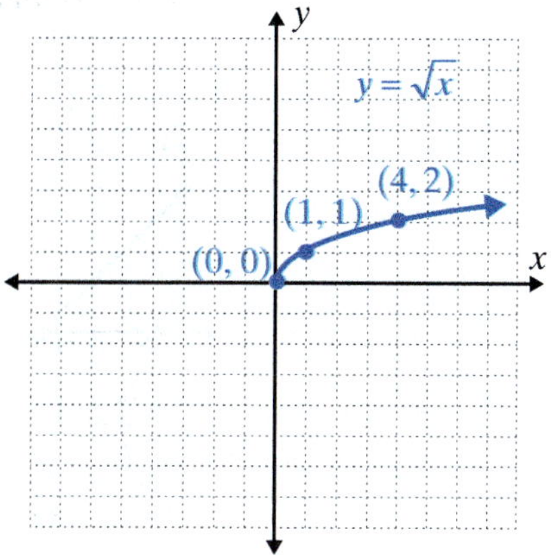

21. $y=\sqrt{x}-2$

22. $y=\sqrt{x}+1$

23. $y=-\sqrt{x+1}$

24. $y=-\sqrt{x-6}$

25. $y=\sqrt{x-4}-3$

26. $y=\sqrt{x-2}-4$

27. $y=\sqrt{x-3}+\frac{1}{2}$

28. $y=\sqrt{x+\frac{3}{2}}+2$

29. $y=5+\sqrt{x+2}$

30. $y=\sqrt{x+4}-3$

Using the graph of $y=x^2$, graph the functions using your understanding of reflections and translations with no additional computations.

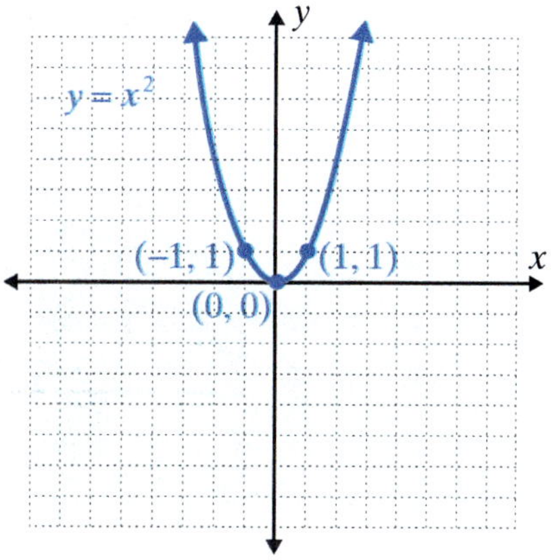

31. $y=x^2-3$

32. $y=x^2+5$

33. $y=(x-1)^2$

34. $y=(x+2)^2$

35. $y=(x-3)^2+1$

36. $y=(x+5)^2-2$

37. $y=(x+1)^2-4$

38. $y=(x-2)^2+3$

39. $y=-(x+4)^2-5$

40. $y=-(x-5)^2+2$

22.
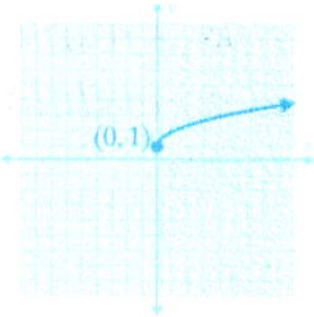

23.
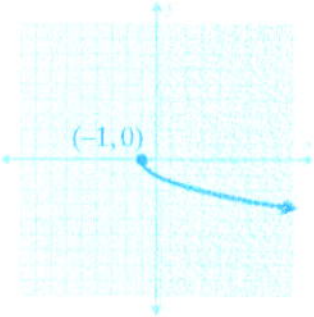

24.
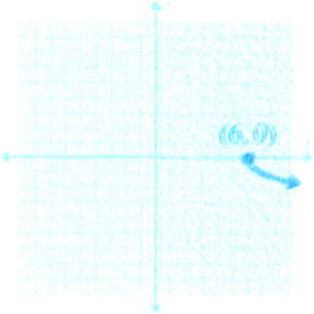

25.
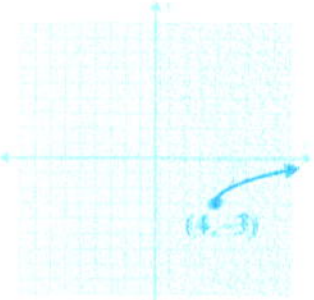

26.
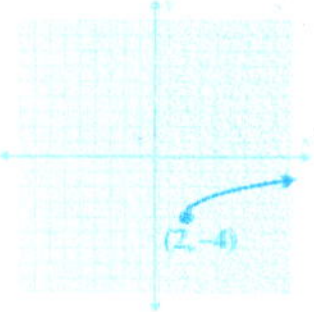

27.
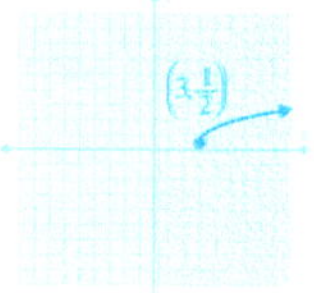

28.
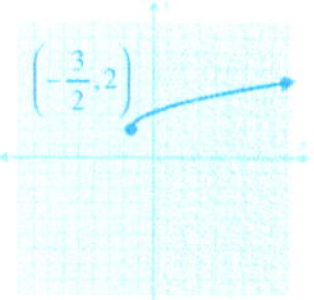

29.
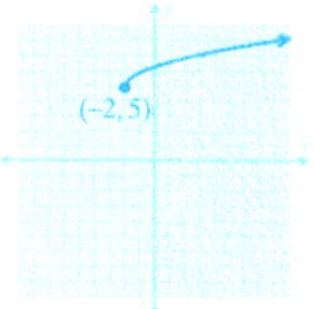

30.
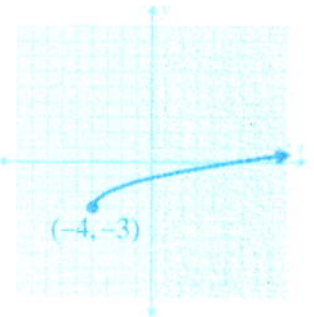

31.
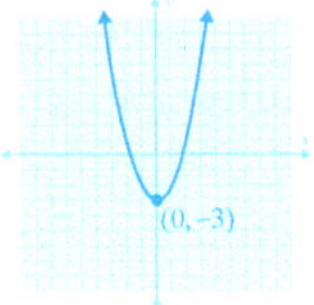

32.
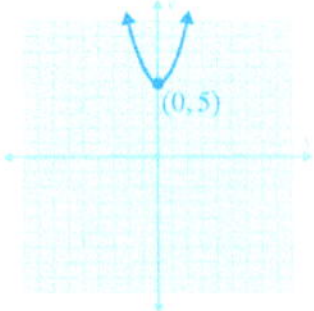

33.
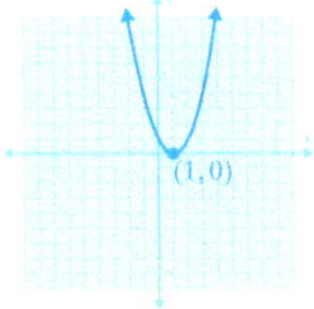

34.
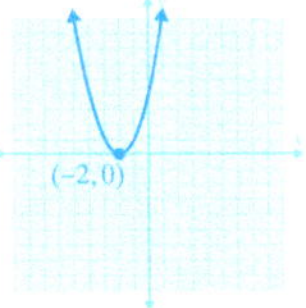

35.
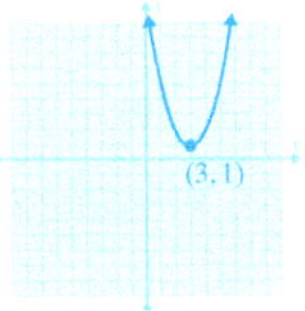

36.
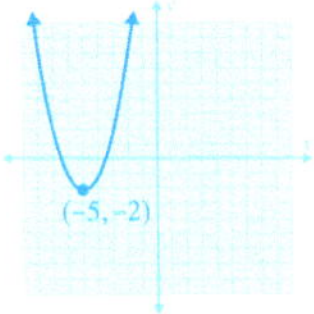

37.
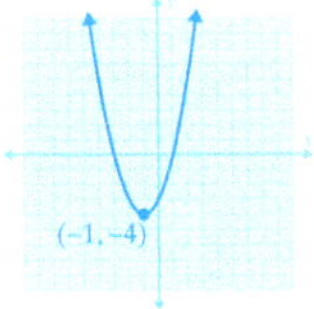

38.
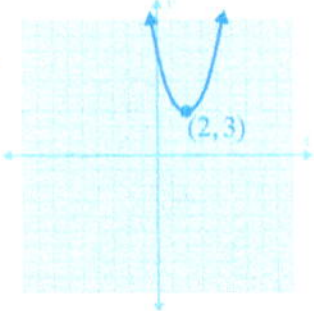

39.
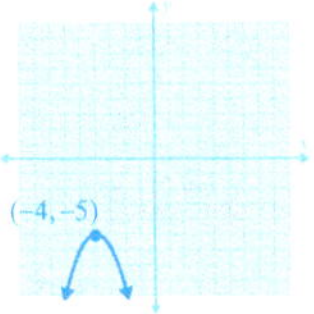

The graph of a function $y = f(x)$ is given with the coordinates of four points. Graph the functions using your understanding of reflections and translations and with no additional computations. Label the new points that correspond to the four labeled points.

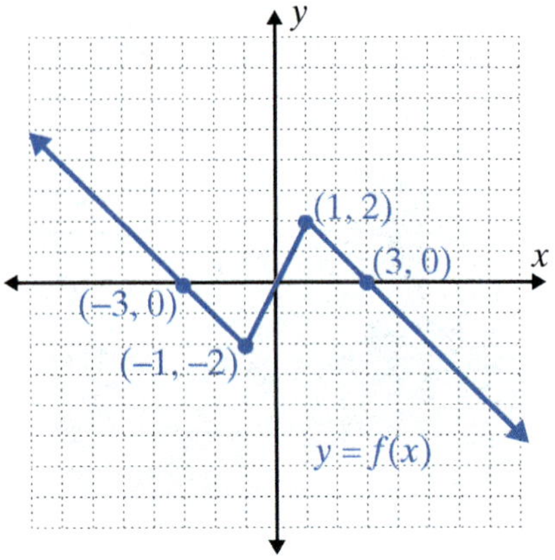

41. $y = f(x) - 1$ **42.** $y = f(x) + 2$ **43.** $y = f(x-3)$

44. $y = f(x+1)$ **45.** $y = -f(x-3)$ **46.** $y = -f(x+1)$

47. $y = f(x+5) + 3$ **48.** $y = f(x-1) + 5$ **49.** $y = f(x+2) - 4$

50. $y = f(x+3) + 2$

The graph of $y = \log(x)$ is given. Graph the functions and state the domain and range of each function.

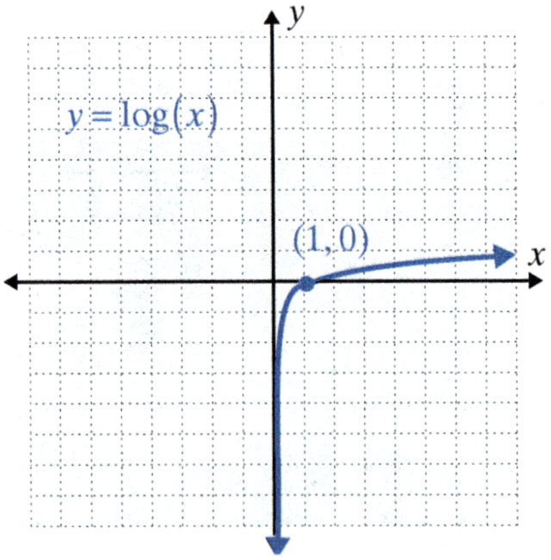

51. $y = \log(x+1)$ **52.** $y = \log(2-x)$ **53.** $y = -\log x$

54. $y = \log(-x)$ **55.** $y = 1 + \log x$ **56.** $y = -3 - \log x$

Use a graphing calculator to graph each pair of functions on the same set of axes.

57. $y = 2x^2$ and $y = -3x^2$

58. $y = 4x^2$ and $y = -x^2$

59. $y = x^2 + 5$ and $y = (x-1)^2$

60. $y = (x+1)^2$ and $y = x^2 - 4$

61. $y = 2(x+3)^2 - 4$ and $y = 2x^2 + 3$

62. $y = -3(x-2)^2 + 1$ and $y = -x^2 + 1$

Writing and Thinking About Mathematics

63. Explain, in your own words, how the graph of the function $y = f(x-h)+k$ represents a horizontal and a vertical shift of the graph of the function $y = f(x)$.

HAWKES LEARNING SYSTEMS: INTRODUCTORY & INTERMEDIATE ALGEBRA SOFTWARE

- 12.1 Translations and Reflections

40.

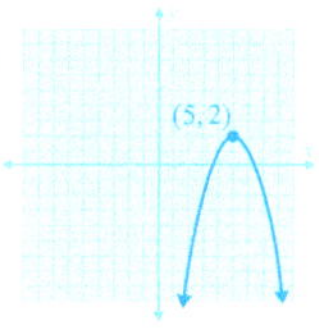

41.

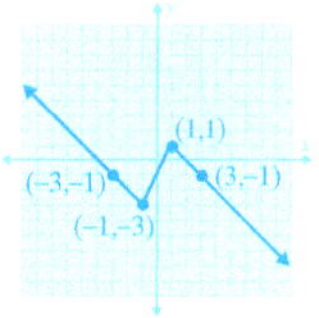

42.

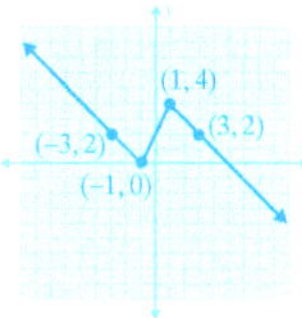

43.

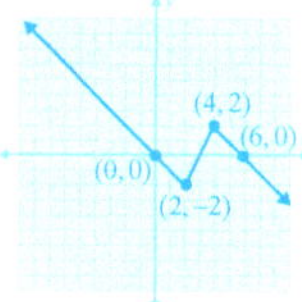

44.

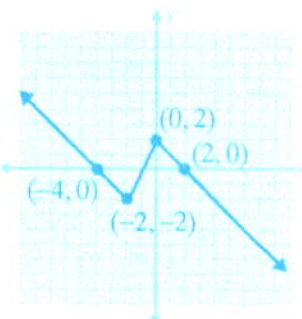

45.

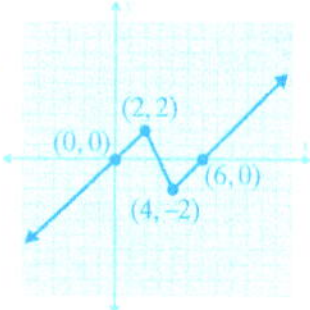

46.

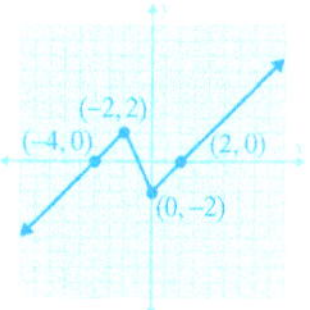

47.

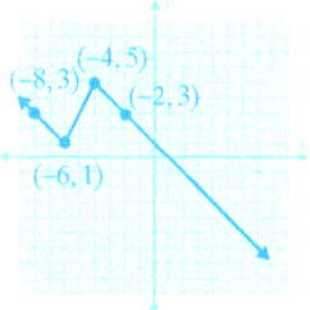

48.

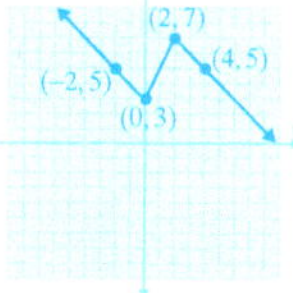

49.

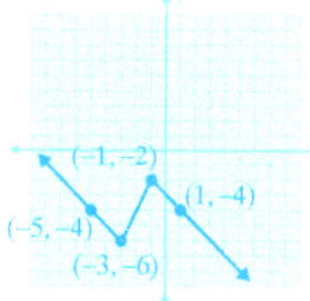

50.

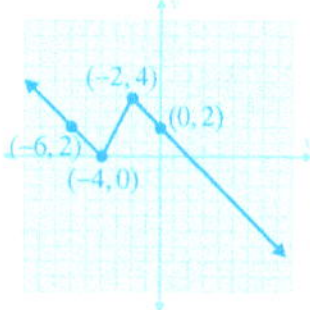

51.

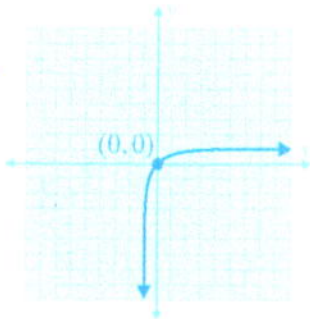

$D = (-1, \infty)$; $R = (-\infty, \infty)$

52.

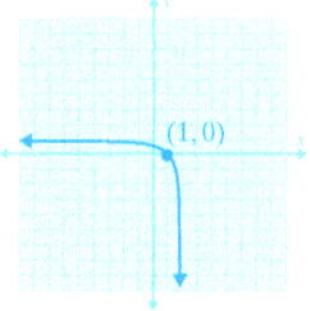

$D = (-\infty, 2)$; $R = (-\infty, \infty)$

53.

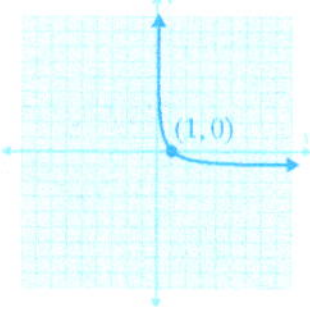

$D = (0, \infty)$; $R = (-\infty, \infty)$

54.

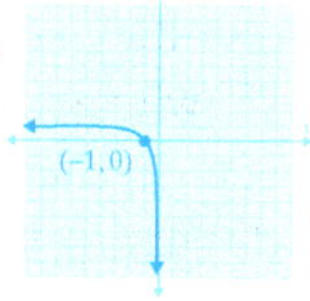

$D = (-\infty, 0)$; $R = (-\infty, \infty)$

55.

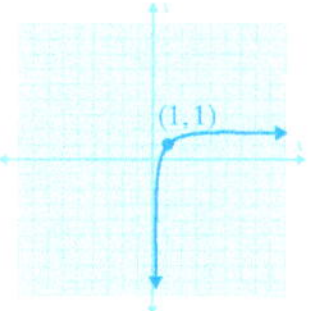

$D = (0, \infty)$; $R = (-\infty, \infty)$

56.

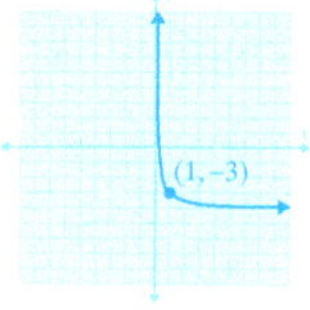

$D = (0, \infty)$; $R = (-\infty, \infty)$

57.

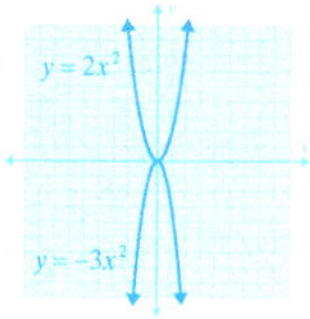

58.

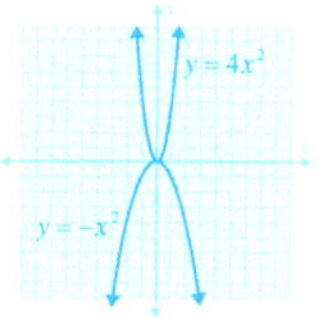

59.

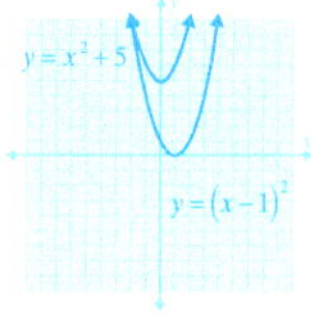

60.

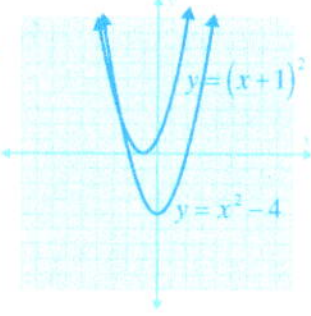

61.

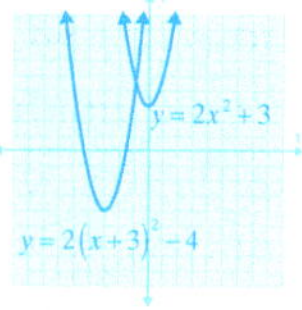

62.

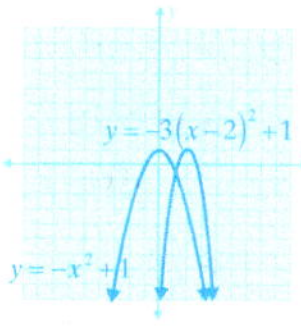

63. The graph of the function $y = f(x-h)+k$ is the graph of the function $y = f(x)$ shifted h units to the right and k units up.

12.2 Parabolas as Conic Sections

- *Graph parabolas that open left or right (horizontal parabolas).*
- *Find the vertices, y-intercepts, and lines of symmetry for horizontal parabolas.*

Conic sections are curves in a plane that are found when the plane intersects a cone. Four such intersections are the circle, ellipse, parabola, and hyperbola, as shown in Figure 1 respectively.

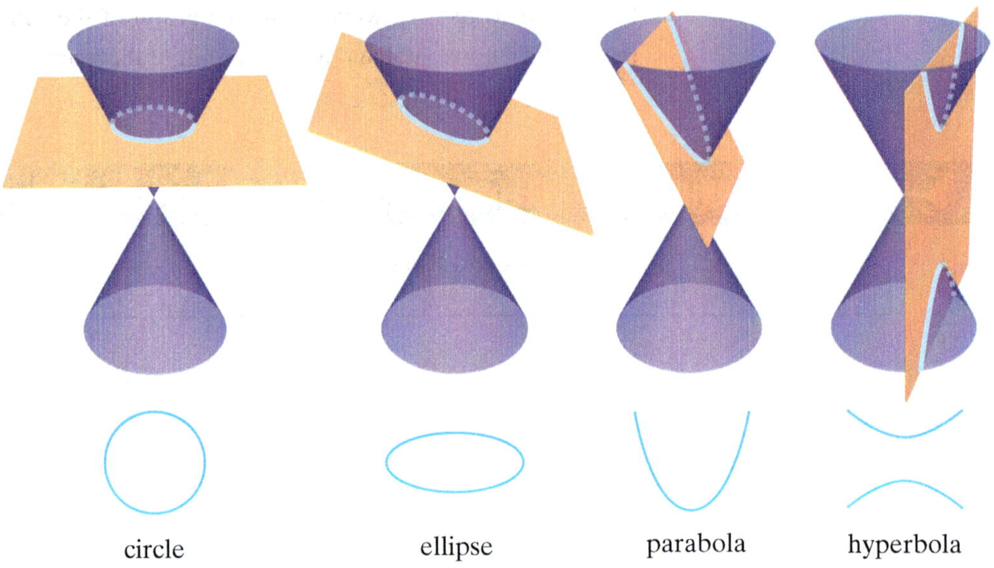

Figure 1

The corresponding equations for these conic sections are called quadratic equations because they are second-degree in x and/or y. Careful examination of these equations can show exactly what type of curve it represents and where the curve is located with respect to a Cartesian coordinate system. The technique is similar to that used in Chapter 4 in discussing lines.

By looking at a linear equation, you can identify

a. the **slope** of the line,
b. the ***y*-intercept** of the line, and
c. **points** on the line.

By looking at a quadratic equation, you will be able to determine if the graph is

a. a **circle** and identify the center and radius,
b. an **ellipse** and identify the center and intercepts,
c. a **parabola** and identify the vertex and line of symmetry, or
d. a **hyperbola** and identify the vertices and asymptotes.

Parabolas

As discussed in Section 10.5, the equations of **quadratic functions** are of the basic form $\boldsymbol{y = ax^2}$, and the corresponding graphs are parabolas. Not all parabolas are functions. Parabolas that open upward or downward are functions, but those that open to the left or to the right are not functions.

The basic form for equations of parabolas that open left or right is $\boldsymbol{x = ay^2}$, and several graphs of equations of this type are shown in Figure 2. As the vertical line test will confirm, these graphs do not represent functions.

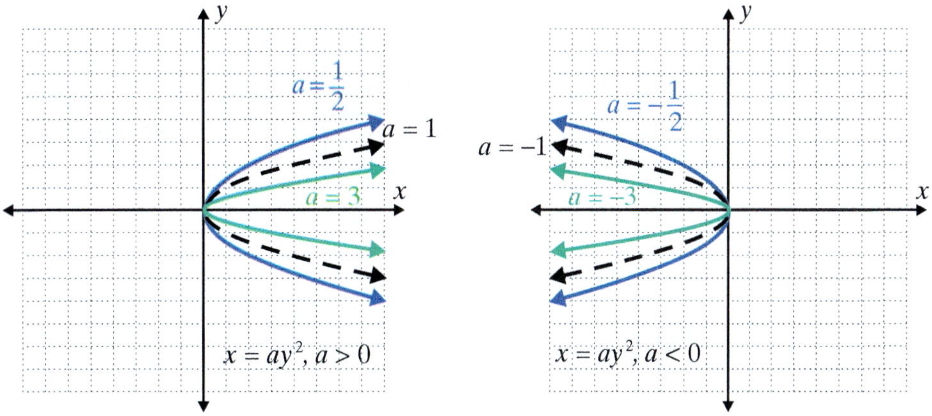

Figure 2

In general, the equations of **vertical parabolas** (parabolas that open upward or downward) can be written in the form

$$y = ax^2 + bx + c \quad \text{or} \quad y = a(x-h)^2 + k \quad \text{where } a \neq 0.$$

The parabolas open down if $a < 0$ or up if $a > 0$ and have their vertex at (h, k). The line $x = h$ is the line of symmetry.

By exchanging the roles of x and y, the equations of **horizontal parabolas** (parabolas that open to the left or right) can be written in the following form.

Equations of Horizontal Parabolas

Equations of horizontal parabolas (parabolas that open to the left or right) can be written in the form

$$\boldsymbol{x = ay^2 + by + c} \quad \text{or} \quad \boldsymbol{x = a(y-k)^2 + h} \quad \text{where } a \neq 0.$$

The parabola opens left if $a < 0$ and right if $a > 0$.

The vertex is at $(\boldsymbol{h}, \boldsymbol{k})$.

The line $\boldsymbol{y = k}$ is the line of symmetry.

In a manner similar to the discussion in Section 12.1, adding h to the right hand side and replacing y with $(y-k)$ in the equation $x=ay^2$ gives the equation $x=a(y-k)^2+h$ whose graph is a horizontal translation of h units and a vertical translation of k units of the graph of $x=ay^2$.

For example, the graph of $x=2(y-3)^2-1$ is shown in Figure 3 with a table of y- and x-values. The vertex is at $(h, k)=(-1, 3)$, and the line of symmetry is $y=3$. The y-values in the table are chosen on each side of the line of symmetry.

y	$2(y-3)^2-1=x$
3	$2(3-3)^2-1=-1$
4	$2(4-3)^2-1=1$
2	$2(2-3)^2-1=1$
5	$2(5-3)^2-1=7$
1	$2(1-3)^2-1=7$

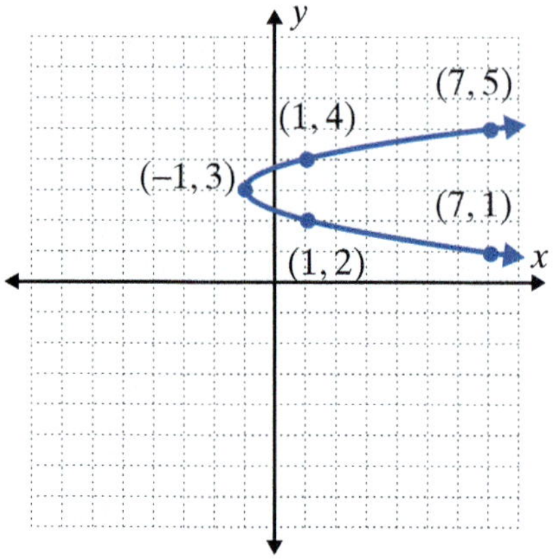

Figure 3

The graph of an equation of the form $x=ay^2+by+c$ can be found by completing the square (as in Section 10.5) and writing the equation in the form

$$x=a(y-k)^2+h.$$

Also, by setting $x=0$ and solving the following quadratic equation

$$0=ay^2+by+c,$$

we can determine the **y-intercepts** (the points, if any, where the graph intersects the **y-axis**).

Example 1: Horizontal Parabolas

a. For $x=y^2-6y+4$, find the vertex, the y-intercepts, and the line of symmetry. Then sketch the graph.

Solution: To find the vertex, complete the square.

$$x=y^2-6y+4$$

$$x=(y^2-6y+9)-9+4$$

$$x=(y-3)^2-5$$

The vertex is at $(-5, 3)$.

Teaching Notes:
You will probably want to remind your students that they can use their calculators to estimate the values of the intercepts when radicals are involved. For example, in 1a they should understand how to determine that $3+\sqrt{5}\approx 5.2$ and $3-\sqrt{5}\approx 0.8$.

Teaching Notes:
In Example 1a you could also use the quadratic formula to find the y-intercept as follows:

$y^2-6y+4=x$

$y^2-6y+4=0$

$y=\frac{6\pm\sqrt{(-6)^2-4\cdot1\cdot4}}{2}$

$y=\frac{6\pm\sqrt{20}}{2}$

$y=3\pm\sqrt{5}.$

To find the y-intercepts, let $x=0$ and use the square root method as follows:

$$\begin{aligned}(y-3)^2-5&=0\\(y-3)^2&=5\\y-3&=\pm\sqrt{5}\\y&=3\pm\sqrt{5}.\end{aligned}$$

Since $a=1$, the parabola has the same shape as $x=y^2$.
Vertex: $(-5,3)$
y-intercepts: $(0,3+\sqrt{5})$ and $(0,3-\sqrt{5})$
Line of symmetry: $y=3$

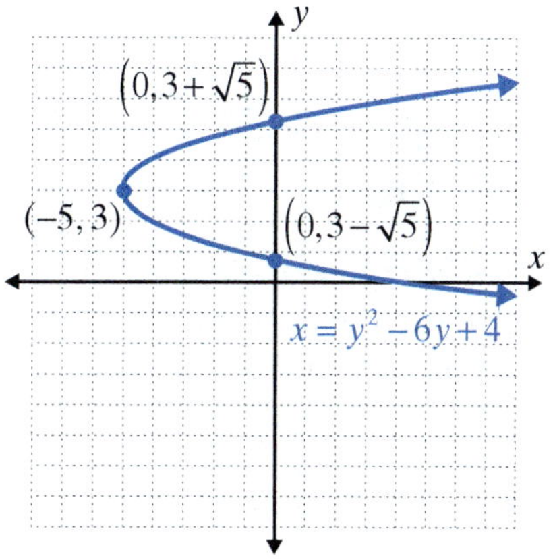

b. For $x=-2y^2-4y+6$, find the vertex, the y-intercepts, and the line of symmetry. Then sketch the graph.

Solution: To find the vertex, complete the square.

$$\begin{aligned}x&=-2y^2-4y+6\\x&=-2(y^2+2y)+6\\x&=-2(y^2+2y+1-1)+6\\x&=-2(y^2+2y+1)+2+6\\x&=-2(y^2+2y+1)+8\\x&=-2(y+1)^2+8\end{aligned}$$

The vertex is at $(8,-1)$.

Continued on the next page...

To find the y-intercepts, let $x = 0$ and factor.

$$-2y^2 - 4y + 6 = 0$$
$$-2(y^2 + 2y - 3) = 0$$
$$-2(y+3)(y-1) = 0$$
$$y = -3 \quad \text{or} \quad y = 1$$

Since $a = -2$, the graph opens to the left and is slightly narrower than $x = y^2$.

Vertex: $(8, -1)$

y-intercepts: $(0, -3)$ and $(0, 1)$

Line of symmetry: $y = -1$

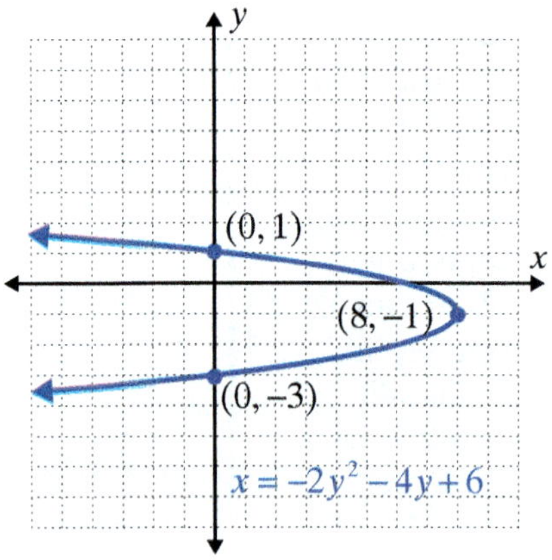

Using a TI-84 Plus Graphing Calculator to Graph Horizontal Parabolas

Horizontal parabolas are not functions and the graphing calculator is designed to graph only functions. Therefore, to graph a horizontal parabola, solve the given equation for y. For example, the graph of the equation $x = y^2 - 2$ is a horizontal parabola opening right with its vertex at $(-2, 0)$. To graph this equation using a graphing calculator, we must first solve for y since equations must be entered with y (to the first power) on the left-hand side. By using the square root property, two functions designated as y_1 and y_2 can be found as follows:

$$x = y^2 - 2$$
$$y^2 = x + 2 \qquad \text{First solve for } y^2.$$
$$\begin{cases} y_1 = \sqrt{x+2} \\ y_2 = -\sqrt{x+2} \end{cases} \qquad \text{Solving for } y \text{ gives two equations that represent two functions.}$$

Graphing these equations individually gives the upper and lower halves of the parabola.

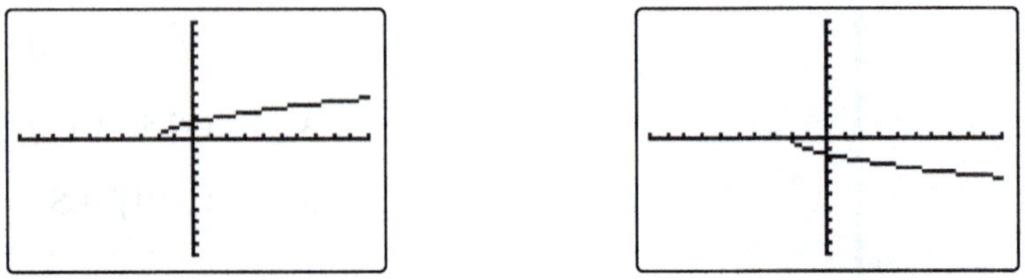

$y_1 = \sqrt{x+2}$ (upper half) $\qquad$ $y_2 = -\sqrt{x+2}$ (lower half)

Figure 4

1. a. $(4, 0)$ b. None
c. $y = 0$

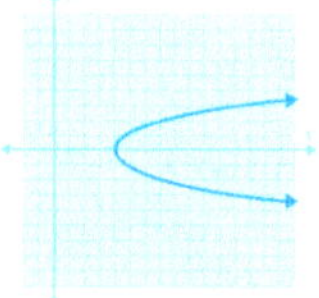

Graphing both halves at the same time gives the entire parabola $x = y^2 - 2$.

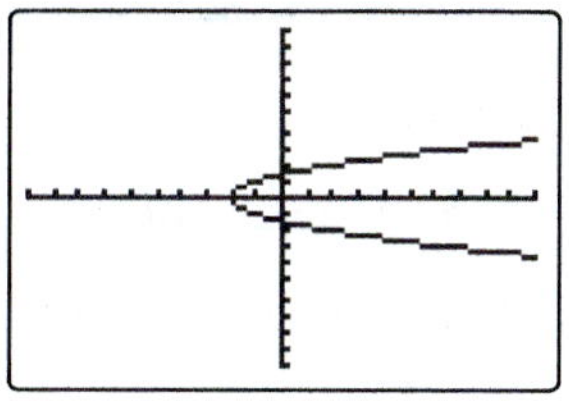

Figure 5

2. a. $(-5, 0)$
b. $(0, \sqrt{5}), (0, -\sqrt{5})$
c. $y = 0$

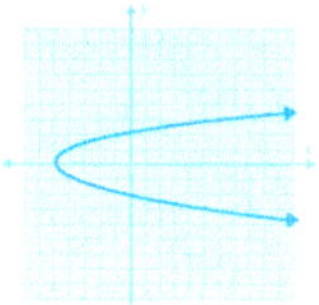

Example 2: Using a Graphing Calculator to Graph Horizontal Parabolas

Use a graphing calculator to graph the horizontal parabola $x = y^2 - 4y + 5$. Find the y-intercepts using the CALC features of the calculator.

Solution: To solve for y, complete the square and use the square root property as follows.

$$x = y^2 - 4y + 5$$

$$y^2 - 4y = x - 5$$

$$y^2 - 4y + 4 = x - 5 + 4 \quad \text{Complete the square.}$$

$$(y-2)^2 = x - 1$$

$$y - 2 = \pm\sqrt{x-1} \quad \text{Use the square root property.}$$

$$\begin{cases} y_1 = 2 + \sqrt{x-1} \\ y_2 = 2 - \sqrt{x-1} \end{cases} \quad \text{Graph both of these equations.}$$

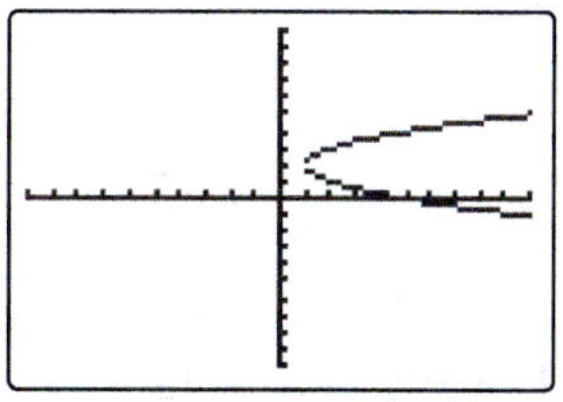

From this graph, we can determine that there are no y-intercepts.

3. a. $(0, -3)$ b. $(0, -3)$
c. $x = 0$

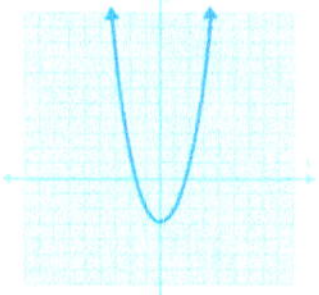

4. a. $(0, 2)$ b. $(0, 2)$
c. $x = 0$

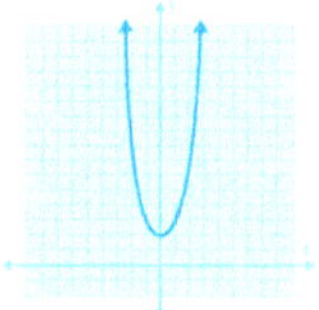

5. a. $(3, 0)$ b. None
c. $y = 0$

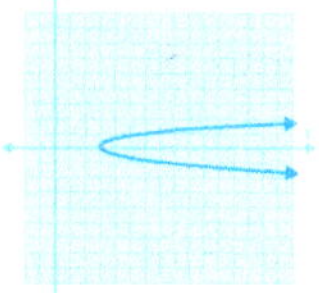

6. a. $(1, 0)$
b. None
c. $y = 0$

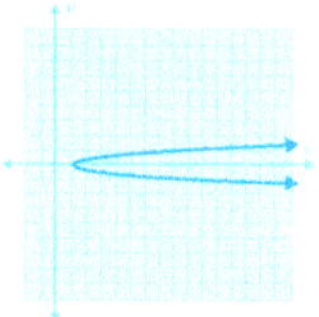

7. a. $(0, 3)$
b. $(0, 3)$
c. $y = 3$

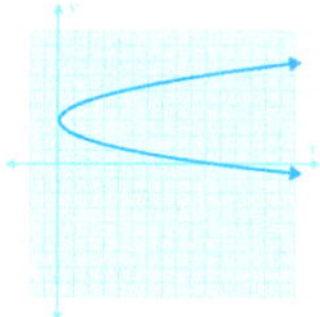

8. a. $(0, 2)$
b. $(0, 2)$
c. $y = 2$

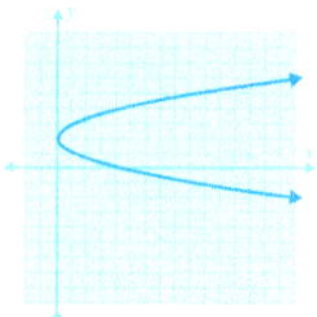

9. a. $(4, -2)$
b. None
c. $y = -2$

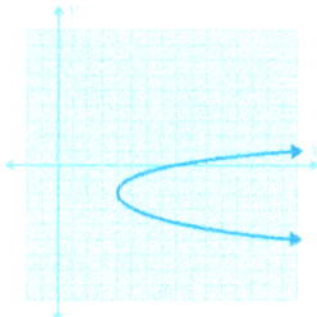

10. a. $(-3, 5)$
b. $(0, 5+\sqrt{3})$, $(0, 5-\sqrt{3})$
c. $y = 5$

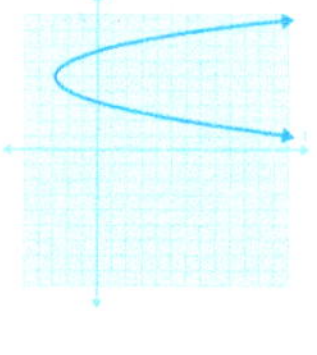

11. a. $(1, -1)$
b. $(0, 0)$
c. $x = 1$

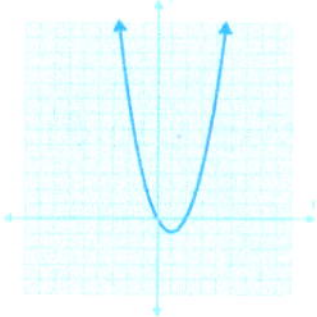

12. a. $(3, 5)$ b. $(0, 14)$
c. $x = 3$

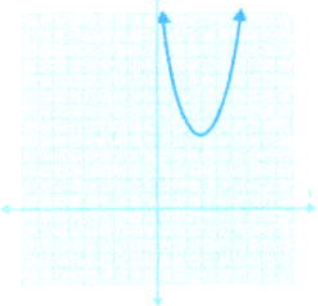

Practice Problems

1. Write the equation $x = -y^2 - 10y - 24$ in the form $x = a(y-k)^2 + h$.
2. Find the vertex, y-intercepts, and line of symmetry for the curve $x = y^2 - 4$.
3. Find the y-intercepts for the curve $x = y^2 + 2y + 2$.

12.2 Exercises

13. a. $(0, -2)$ b. $(0, -2)$
c. $y = -2$

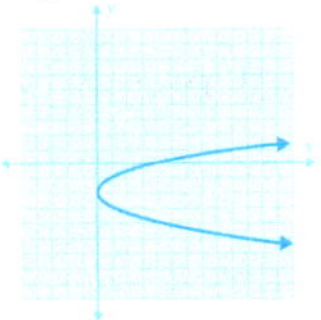

14. a. $(0, 4)$ b. $(0, 4)$
c. $y = 4$

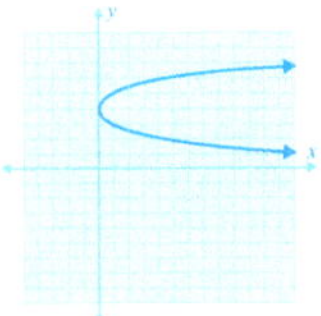

15. a. $(0, 5)$ b. $(0, 5)$
c. $y = 5$

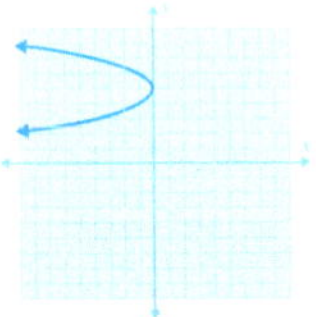

16. a. $(0, -3)$ b. $(0, -3)$
c. $y = -3$

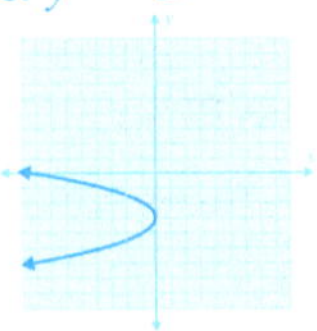

Find **a.** *the vertex,* **b.** *the y-intercepts, and* **c.** *the line of symmetry. Then draw the graph.*

1. $x = y^2 + 4$	**2.** $x = y^2 - 5$	**3.** $y + 3 = x^2$
4. $y - 2 = x^2$	**5.** $x = 2y^2 + 3$	**6.** $x = 3y^2 + 1$
7. $x = (y-3)^2$	**8.** $x = (y-2)^2$	**9.** $x - 4 = (y+2)^2$
10. $x + 3 = (y-5)^2$	**11.** $y + 1 = (x-1)^2$	**12.** $y - 5 = (x-3)^2$
13. $x = y^2 + 4y + 4$	**14.** $x = y^2 - 8y + 16$	**15.** $x = -y^2 + 10y - 25$
16. $x = -y^2 - 6y - 9$	**17.** $y = x^2 + 6x + 5$	**18.** $y = x^2 + 4x + 6$
19. $y = -x^2 - 4x + 5$	**20.** $y = -x^2 + 2x + 5$	**21.** $x = -y^2 + 4y - 3$
22. $x = y^2 + 8y + 12$	**23.** $y = 2x^2 + x - 1$	**24.** $y = -2x^2 + x + 3$
25. $x = -2y^2 + 5y - 2$	**26.** $x = 3y^2 + 5y + 2$	**27.** $x = 3y^2 + 6y - 5$
28. $x = 4y^2 - 4y - 15$	**29.** $y = 4x^2 - 12x + 9$	**30.** $y = -5x^2 + 10x + 2$

Answers to Practice Problems: **1.** $x = -(y+5)^2 + 1$
2. Vertex: $(-4, 0)$; y-intercepts: $(0, 2)$ and $(0, -2)$; Line of symmetry: $y = 0$
3. There are no y-intercepts.

17. a. $(-3,-4)$ b. $(0,5)$
c. $x=-3$

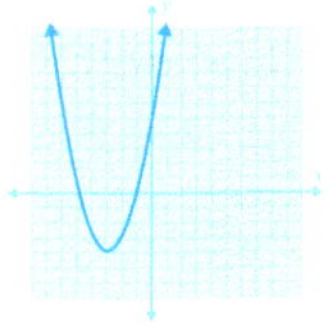

18. a. $(-2,2)$ b. $(0,6)$
c. $x=-2$

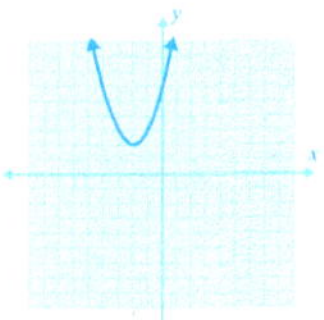

19. a. $(-2,9)$ b. $(0,5)$
c. $x=-2$

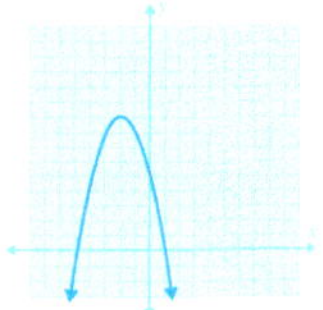

20. a. $(1,6)$ b. $(0,5)$
c. $x=1$

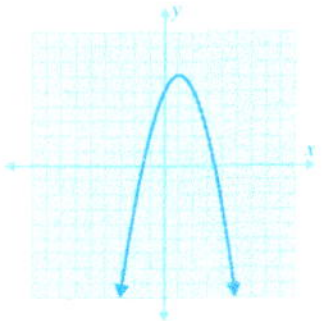

21. a. $(1,2)$
b. $(0,1),(0,3)$
c. $y=2$

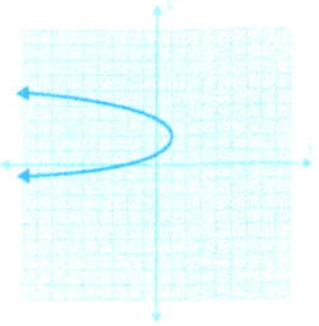

22. a. $(-4,-4)$
b. $(0,-2),(0,-6)$
c. $y=-4$

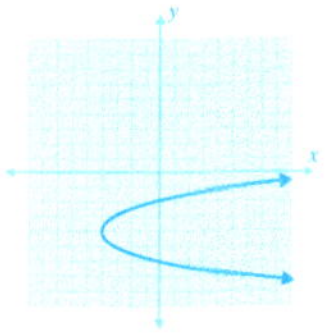

Use a graphing calculator to graph each of the parabolas. Use the trace and zoom features of the calculator to estimate the y-intercepts of the parabola.

31. $x=2y^2-3$ **32.** $x=-3y^2+1$ **33.** $x=-y^2+2y$

34. $x=y^2-5y$ **35.** $x=2y^2+y+1$ **36.** $x=-y^2-4y+1$

37. $x=4y^2+8y-7$ **38.** $x=3y^2+3y+2$ **39.** $x=-2y^2+4y+3$

40. $x=-5y^2-10y-4$

Use your knowledge of parabolas and equations to match the equation with the graph.

41. $x=2(y-3)^2+3$ **42.** $x=-(y+1)^2+5$

43. $x=-y^2-1$ **44.** $x=y^2-6$

a.

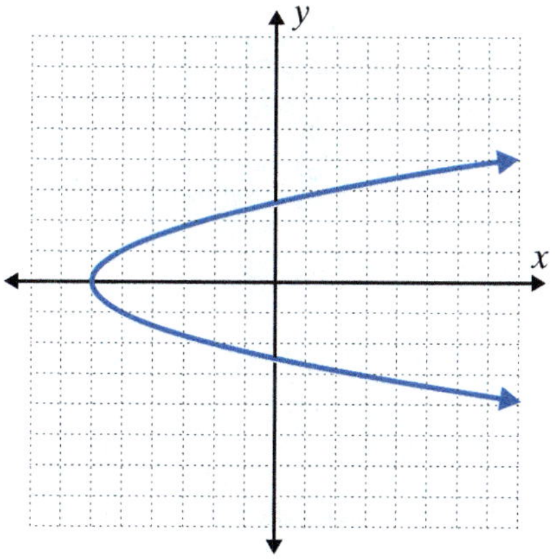

b.

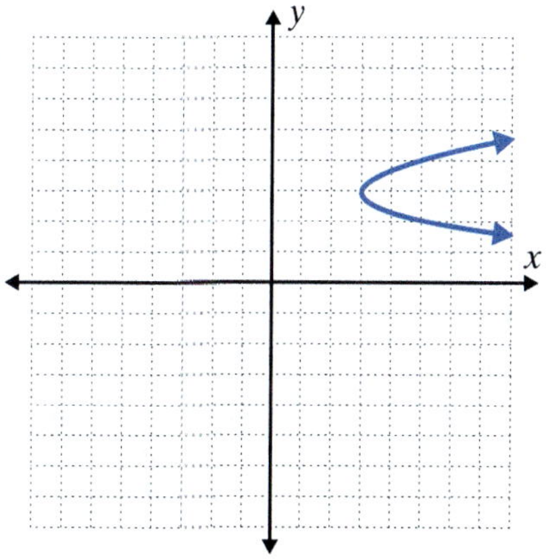

c.

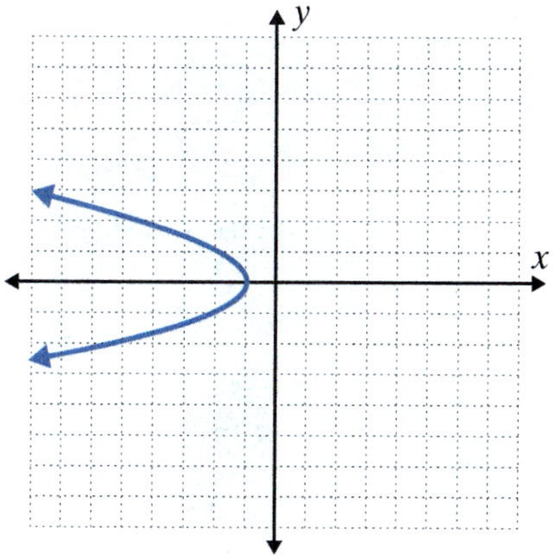

d.

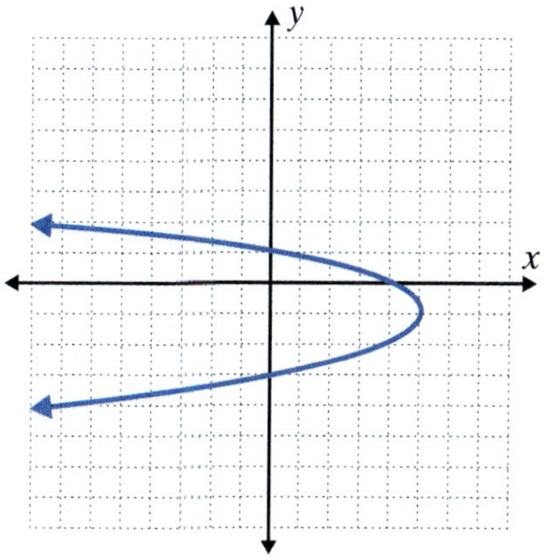

Writing and Thinking About Mathematics

45. For $x = ay^2 + by + c$ we know that the graph of the parabola opens to the right if $a > 0$ and to the left if $a < 0$. Discuss which values of a will cause the parabola to be wider and which will cause it to be narrower than the graph of $x = y^2$.

HAWKES LEARNING SYSTEMS: INTRODUCTORY & INTERMEDIATE ALGEBRA SOFTWARE

- 12.2 Parabolas as Conic Sections

23. a. $\left(-\frac{1}{4}, -\frac{9}{8}\right)$ **b.** $(0,-1)$ **c.** $x = -\frac{1}{4}$

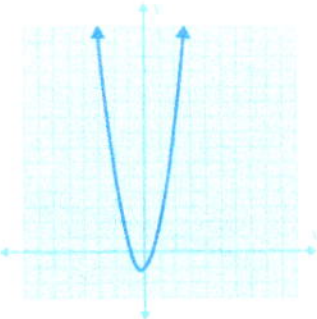

24. a. $\left(\frac{1}{4}, \frac{25}{8}\right)$ **b.** $(0,3)$ **c.** $x = \frac{1}{4}$

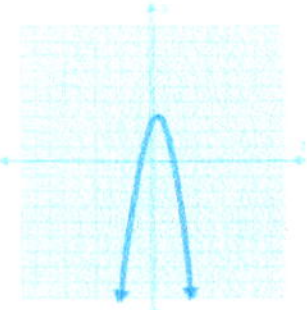

25. a. $\left(\frac{9}{8}, \frac{5}{4}\right)$ **b.** $\left(0, \frac{1}{2}\right), (0,2)$ **c.** $y = \frac{5}{4}$

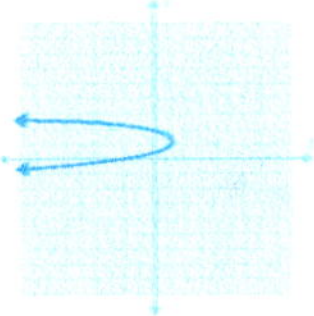

26. a. $\left(-\frac{1}{12}, -\frac{5}{6}\right)$ **b.** $\left(0, -\frac{2}{3}\right), (0,-1)$ **c.** $y = -\frac{5}{6}$

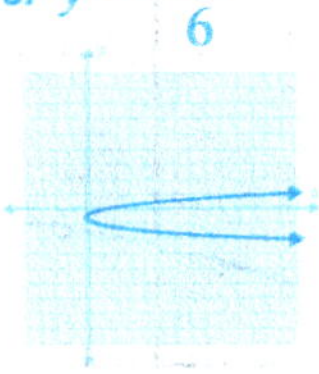

27. a. $(-8,-1)$ **b.** $\left(0, -1 + \frac{2\sqrt{6}}{3}\right), \left(0, -1 - \frac{2\sqrt{6}}{3}\right)$ **c.** $y = -1$

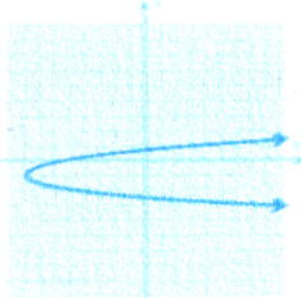

28. a. $\left(-16, \frac{1}{2}\right)$ **b.** $\left(0, \frac{5}{2}\right), \left(0, -\frac{3}{2}\right)$ **c.** $y = \frac{1}{2}$

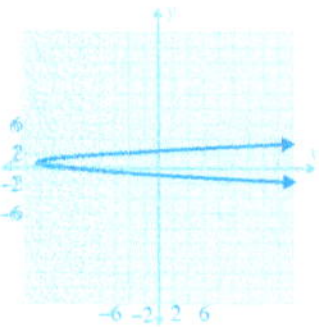

29. a. $\left(\frac{3}{2}, 0\right)$ **b.** $(0, 9)$ **c.** $x = \frac{3}{2}$

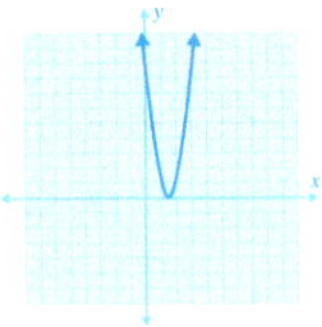

30. a. $(1,7)$ **b.** $(0,2)$ **c.** $x = 1$

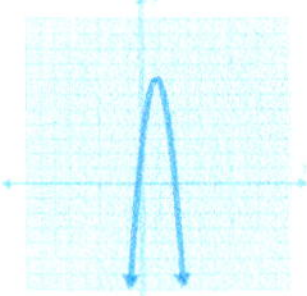

31. $(0, 1.225), (0, -1.225)$

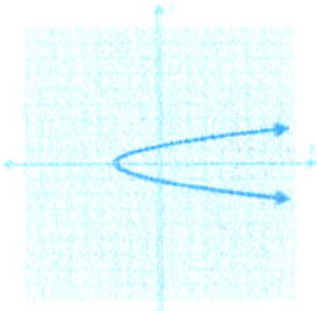

32. $(0, 0.577), (0, -0.577)$

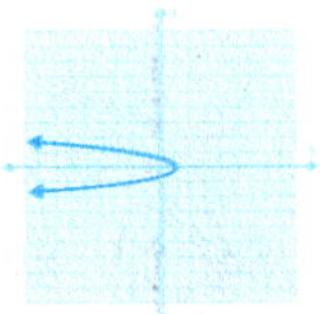

33. $(0, 2), (0, 0)$

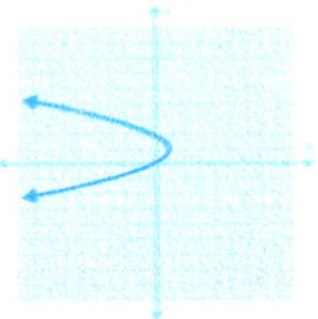

34. $(0, 5), (0, 0)$

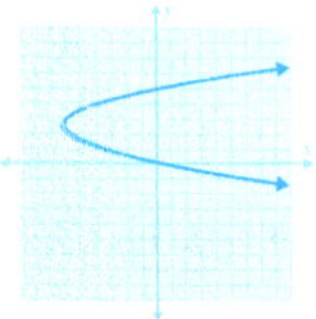

35. No y-intercept

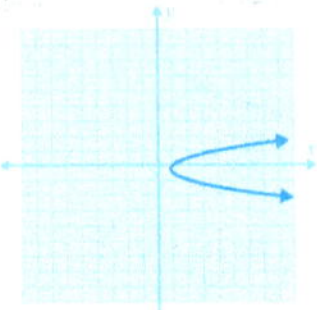

36. $(0, 0.236), (0, -4.236)$

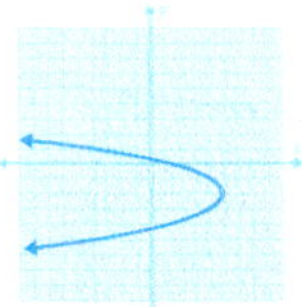

37. $(0, 0.658), (0, -2.658)$

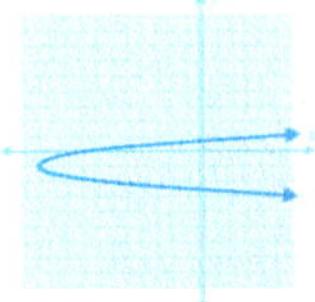

38. No y-intercept

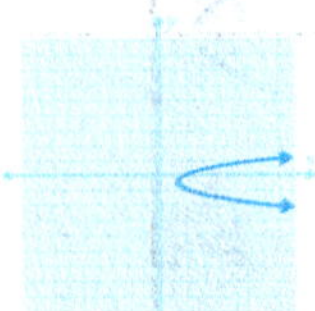

39. $(0, 2.581), (0, -0.581)$

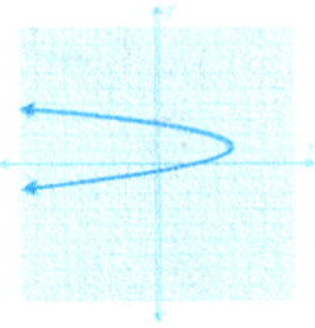

40. $(0, -0.553), (0, -1.447)$

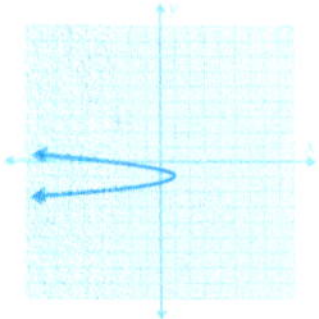

41. b **42.** d **43.** c **44.** a **45.** If $|a| < 1$, then the parabola will be wider than the graph of $x = y^2$. If $|a| > 1$, then the parabola will be narrower than the graph of $x = y^2$.

12.3 Distance Formula and Circles

- *Find the distance between any two points in a plane.*
- *Write the equation of a circle given its center and radius.*
- *Graph circles centered at the point (h, k).*

Distance Between Two Points

The formula for the distance between two points in a plane is used to develop the equations of circles. The **Pythagorean theorem**, previously discussed in Sections 6.7 and 10.3, is the basis for the formula and is repeated here for easy reference. (Remember, the **hypotenuse** is the side opposite the right angle and is the longest side.)

The Pythagorean Theorem

In a right triangle, if c is the length of the hypotenuse and a and b are the lengths of the legs, then

$$c^2 = a^2 + b^2.$$

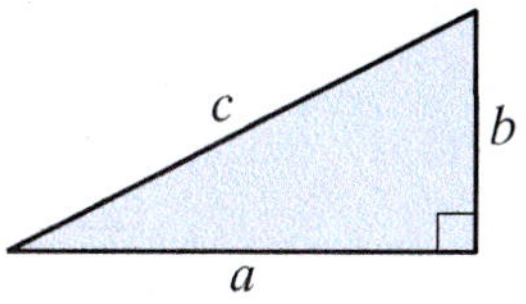

To find the distance between the two points $P(-1, 2)$ and $Q(5, 6)$, as shown in Figure 1a, form a right triangle, as shown in Figure 1b, and find the lengths of the sides a and b. Then using a and b and the Pythagorean theorem, we can find the length of the hypotenuse, which is the distance between the two points.

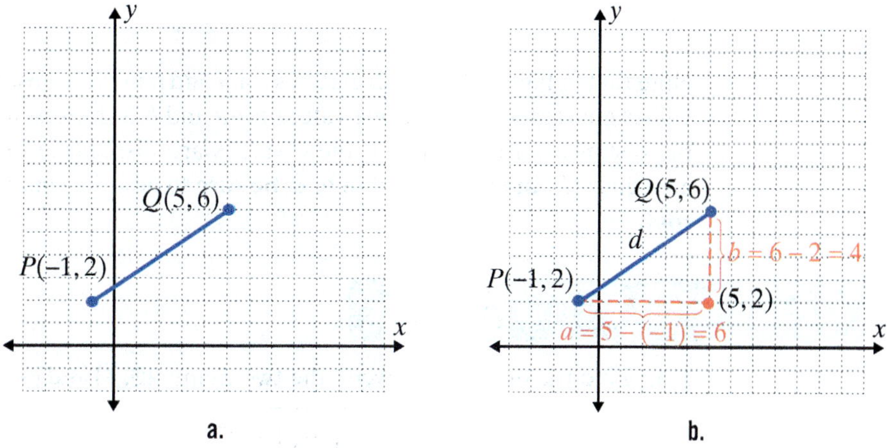

Figure 1

From Figure 1b,

$$d^2 = a^2 + b^2 = (5-(-1))^2 + (6-2)^2 = 6^2 + 4^2 = 36 + 16 = 52.$$

Taking the square root of both sides gives the distance d:

$$d = \sqrt{52} = 2\sqrt{13}. \quad (2\sqrt{13} \approx 7.2111 \text{ estimated with a calculator})$$

In general, for points $P(x_1, y_1)$ and $Q(x_2, y_2)$ in a plane, with $a = |x_2 - x_1|$ and $b = |y_2 - y_1|$, the Pythagorean theorem gives the following distance formula.

The Distance Formula

For two points $P(x_1, y_1)$ and $Q(x_2, y_2)$ in a plane, the distance between the points is

$$\boldsymbol{d = \sqrt{(x_2 - x_1)^2 + (y_2 - y_1)^2}}. \text{ (See Figure 2)}$$

Teaching Notes:
You will probably want to reinforce the fact (with a few examples) that the order of the points (or the order of the subtraction) doesn't matter because of the squaring.

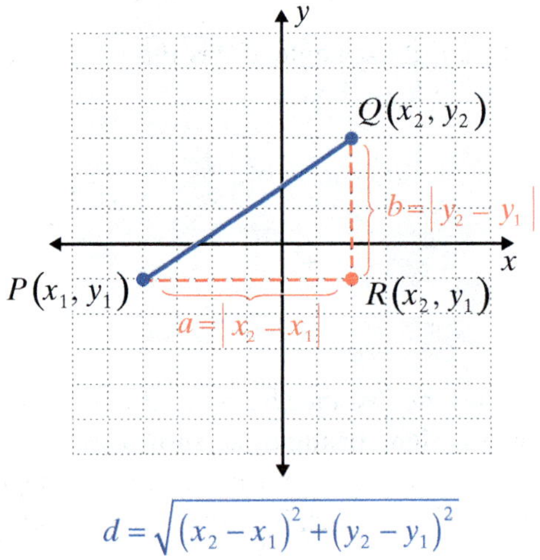

Figure 2

Note that in Figure 2, the calculations for a and b involve absolute values. These absolute values guarantee nonnegative values for a and b to represent the lengths of the legs. In the distance formula, the absolute values are disregarded because $x_2 - x_1$ and $y_2 - y_1$ are squared. **In the calculation of *d*, be sure to add the squares before taking the square root.**

Example 1: The Distance Formula

a. Find the distance between the two points $(3, 4)$ and $(-2, 7)$.

Solution: $d = \sqrt{[3-(-2)]^2 + (4-7)^2}$

$$= \sqrt{5^2 + (-3)^2} = \sqrt{25+9} = \sqrt{34}$$

b. Use the distance formula (3 times) and the Pythagorean theorem to determine whether or not the triangle with vertices at $A(-5,-1)$, $B(2,1)$, and $C(0,7)$ is a right triangle.

Solution: Find the lengths of the three line segments $\overline{AB}$, $\overline{AC}$, and $\overline{BC}$, and decide whether or not the Pythagorean theorem is satisfied.

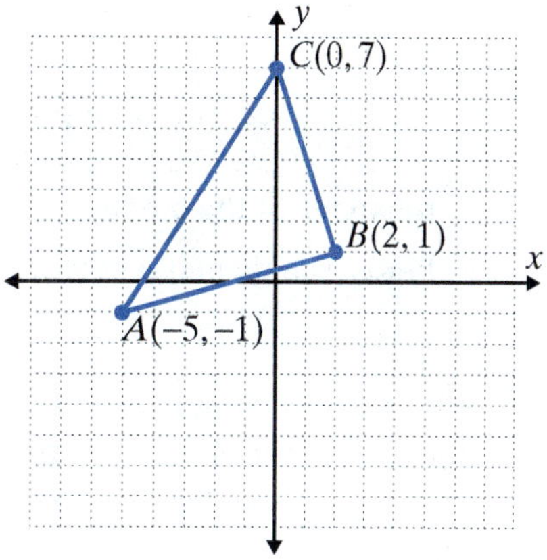

$$AB = \sqrt{(-5-2)^2 + (-1-1)^2} = \sqrt{(-7)^2 + (-2)^2}$$
$$= \sqrt{49+4} = \sqrt{53}$$
$$AC = \sqrt{(-5-0)^2 + (-1-7)^2} = \sqrt{(-5)^2 + (-8)^2}$$
$$= \sqrt{25+64} = \sqrt{89}$$
$$BC = \sqrt{(2-0)^2 + (1-7)^2} = \sqrt{(2)^2 + (-6)^2}$$
$$= \sqrt{4+36} = \sqrt{40}$$

The longest side is $AC = \sqrt{89}$.

The triangle is **not** a right triangle since $(\sqrt{89})^2 \neq (\sqrt{53})^2 + (\sqrt{40})^2$ as $89 \neq 53 + 40$.

Equations of Circles

Circles and the terms related to circles (**center**, **radius**, and **diameter**) are defined as follows.

Circle, Center, Radius, and Diameter

A **circle** is the set of all points in a plane that are a fixed distance from a fixed point.

The fixed point is called the **center** of the circle.

The distance from the center to any point on the circle is called the **radius** of the circle.

The distance from one point on the circle to another point on the circle measured through the center is called the **diameter** of the circle.

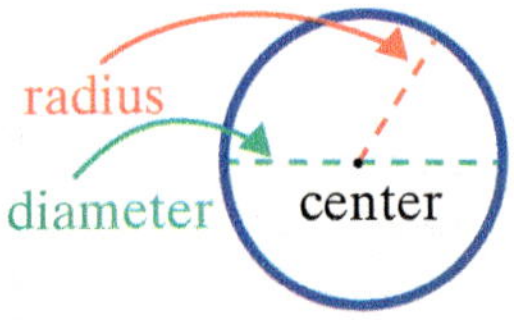

Note: The diameter is twice the length of the radius.

Teaching Notes:
While discussing terms related to circles, you might want to point out the dual use of some terms. "The" radius most often refers to a length, while "a" radius most often refers to a particular line segment. (Similarly with diameter.)

The distance formula is used to find the equation of a circle. For example, to find the equation of the circle with its center at the origin $(0, 0)$ and radius 5, for any point on the circle (x, y) the distance from (x, y) to $(0, 0)$ must be 5. Therefore, using the distance formula,

$$\sqrt{(x_2 - x_1)^2 + (y_2 - y_1)^2} = d$$

$$\sqrt{(x - 0)^2 + (y - 0)^2} = 5$$

$$\sqrt{x^2 + y^2} = 5$$

$$x^2 + y^2 = 25. \qquad \text{Square both sides.}$$

Thus, as shown in Figure 3, all points on the circle satisfy the equation $x^2 + y^2 = 25$.

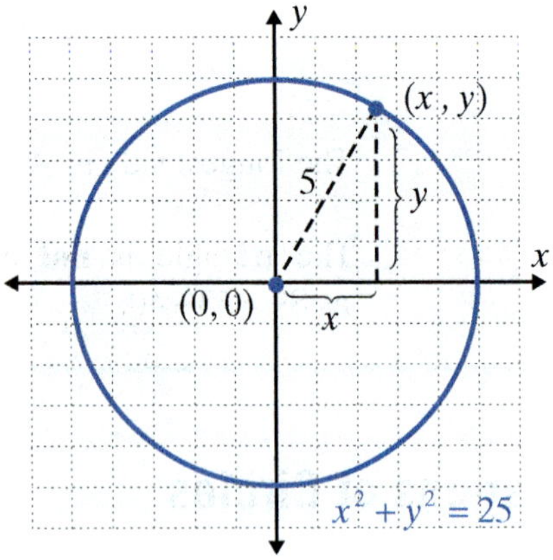

Figure 3

In general, any point (x, y) on a circle with center at (h, k) and radius $r > 0$ must satisfy the equation

$$\sqrt{(x - h)^2 + (y - k)^2} = r.$$

Squaring both sides of this equation gives the **standard form** for the equation of a circle:

$$(x-h)^2+(y-k)^2=r^2.$$

Equation of a Circle

The equation of a circle with radius r and center at (h,k) is

$$\boldsymbol{(x-h)^2+(y-k)^2=r^2}.$$

If the center is at the origin, $(0,0)$, the equation simplifies to

$$\boldsymbol{x^2+y^2=r^2}.$$

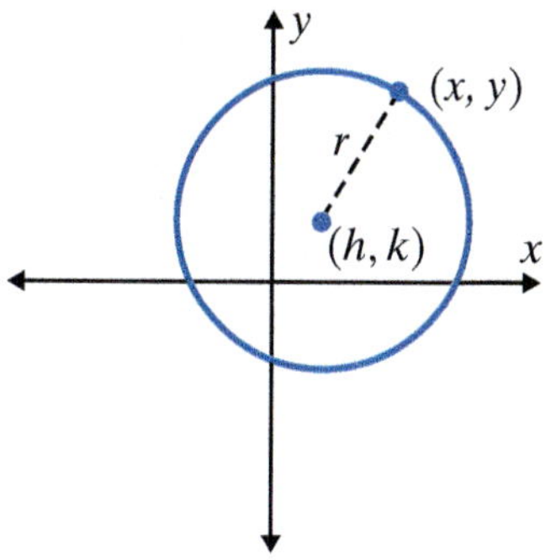

Example 2: Equations of Circles

a. Find the equation of the circle with its center at the origin and radius $\sqrt{3}$. Are the points $(\sqrt{2},1)$ and $(1,2)$ on the circle?

Solution: The equation of the circle is $x^2+y^2=3$.

To determine whether or not the points $(\sqrt{2},1)$ and $(1,2)$ are on the circle, substitute each of these points into the equation.

Substituting $(\sqrt{2},1)$ gives $(\sqrt{2})^2+(1)^2=2+1=3$.

Substituting $(1,2)$ gives $(1)^2+(2)^2=1+4=5\neq 3$.

Therefore, $(\sqrt{2},1)$ is on the circle, but $(1,2)$ is not on the circle.

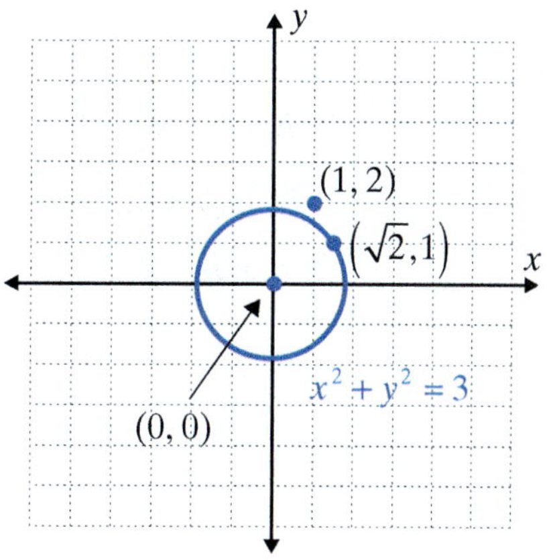

Continued on the next page...

b. Find the equation of the circle with center at $(5, 2)$ and radius 3. Is the point $(5, 5)$ on the circle?

Solution: The equation of the circle is $(x-5)^2+(y-2)^2=9.$

Substituting $(5, 5)$ gives $(5-5)^2+(5-2)^2=0^2+3^2=9.$

Therefore, $(5, 5)$ is on the circle.

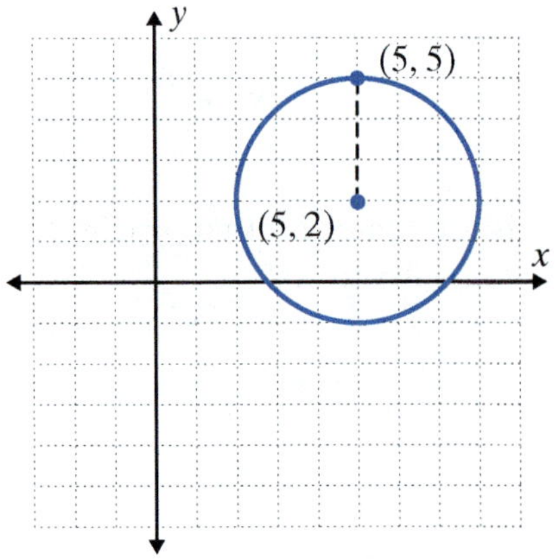

c. Show that $x^2+y^2-8x+2y=0$ represents a circle. Find its center and radius. Then graph the circle.

Solution: Rearrange the terms and complete the square for x^2-8x and $y^2+2y.$

$$x^2+y^2-8x+2y=0$$

$$\left(x^2-8x\right)+\left(y^2+2y\right)=0$$

$$\left(x^2-8x+16\right)+\left(y^2+2y+1\right)=16+1$$ Add 16 and 1 to both sides.

Completes the square

$$(x-4)^2+(y+1)^2=17$$ Standard form for the equation of a circle

$h=4 \quad k=-1 \quad r^2=17$

Center is at $(4, -1)$ and radius $=\sqrt{17}.$

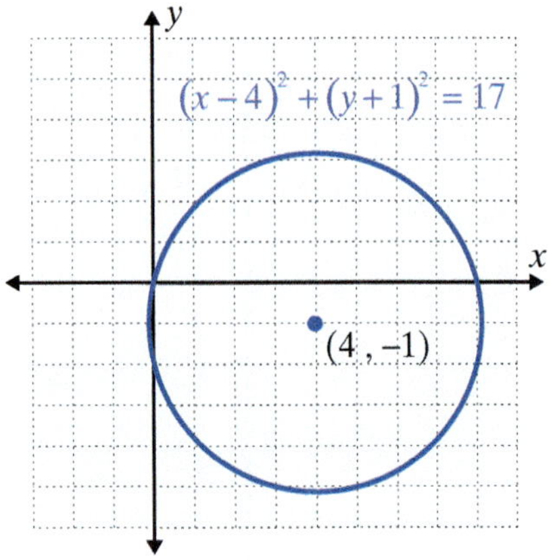

1. 5
2. 13
3. 13
4. 5
5. $\sqrt{29}$
6. $5\sqrt{2}$
7. 3
8. 7
9. $\sqrt{13}$
10. $\sqrt{85}$
11. 17
12. 13
13. $x^2 + y^2 = 16$
14. $x^2 + y^2 = 36$
15. $x^2 + y^2 = 3$
16. $x^2 + y^2 = 7$
17. $x^2 + y^2 = 11$
18. $x^2 + y^2 = 13$
19. $x^2 + y^2 = \frac{4}{9}$
20. $x^2 + y^2 = \frac{49}{16}$
21. $x^2 + (y-2)^2 = 4$
22. $x^2 + (y-5)^2 = 25$
23. $(x-4)^2 + y^2 = 1$
24. $(x+3)^2 + y^2 = 16$
25. $(x+2)^2 + y^2 = 8$
26. $(x-5)^2 + y^2 = 2$
27. $(x-3)^2 + (y-1)^2 = 36$
28. $(x+1)^2 + (y-2)^2 = 25$
29. $(x-3)^2 + (y-5)^2 = 12$
30. $(x-4)^2 + (y+2)^2 = 14$
31. $(x-7)^2 + (y-4)^2 = 10$
32. $(x+3)^2 + (y-2)^2 = 7$

Using a TI-84 Plus Graphing Calculator to Graph Circles

The equation of a circle does not represent a function. The upper half (upper semicircle) and the lower half (lower semicircle) do, however, represent separate functions. Therefore, to graph a circle, solve the equation for two values of y just as we did for horizontal parabolas in Section 12.2. For example, consider the circle with equation $x^2 + y^2 = 4$.

$$x^2 + y^2 = 4$$

$$y^2 = 4 - x^2 \qquad \text{First solve for } y^2.$$

$$\begin{cases} y_1 = \sqrt{4 - x^2} \\ y_2 = -\sqrt{4 - x^2} \end{cases} \qquad \text{Solving for } y \text{ gives two equations that represent two functions.}$$

Graphing both of these functions gives the circle pictured in Figure 4.

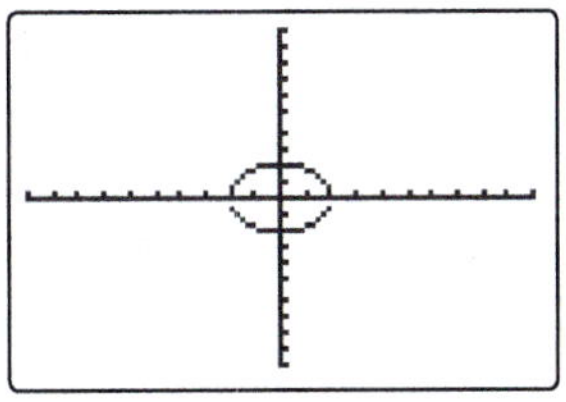

Figure 4

The screen on a TI calculator is rectangular and not a square. The ranges for the standard viewing window are from –10 for both `Xmin` and `Ymin` to 10 for both `Xmax` and `Ymax`. That is, the horizontal scale (x values) and the vertical scale (y values) are the same. Since the rectangular screen is in the approximate ratio of 3 to 2, the graph of a circle using the **standard window** will appear flattened as the circle did in Figure 4.

To get a more realistic picture of a circle press the WINDOW key and set the `Xmin` and `Xmax` values to –6 and 6, respectively. Set the `Ymin` and `Ymax` values to –4 and 4, respectively. Since the numbers 6 and 4 are in the ratio of 3 to 2, the screen is said to show a "square window," and the graphs of y_1 and y_2 will now give a more realistic picture of a circle as shown in Figure 5. Alternatively, pressing the ZOOM key and choosing option `5:ZSquare` automatically "squares" the window.

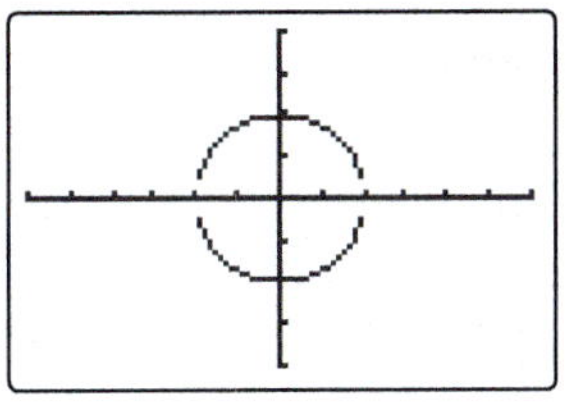

Figure 5

Example 3: Using a Graphing Calculator to Graph Circles

33. $x^2 + y^2 = 9$
Center: $(0,0)$; $r = 3$

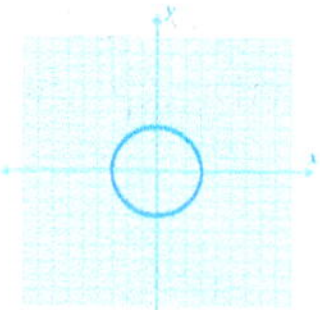

34. $x^2 + y^2 = 16$
Center: $(0,0)$; $r = 4$

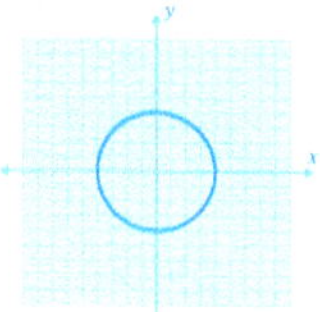

35. $x^2 + y^2 = 49$
Center: $(0,0)$; $r = 7$

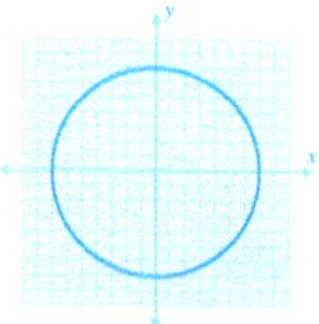

Use a graphing calculator with a "square window" to graph the circle $x^2 + y^2 = 9$.

Solution: Press the WINDOW key and set the values to –6 and 6 for Xmin and Xmax and –4 and 4 for Ymin and Ymax, respectively.

Solving for y^2 gives: $y^2 = 9 - x^2$

Solving for y_1 and y_2 gives: $\begin{cases} y_1 = \sqrt{9 - x^2} \\ y_2 = -\sqrt{9 - x^2} \end{cases}$

Graphing both y_1 and y_2 gives the following graph of the circle.

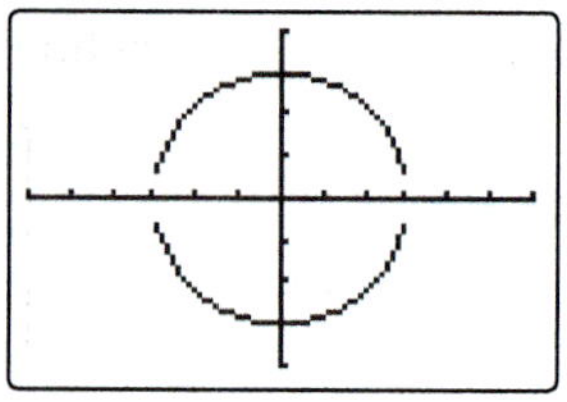

Practice Problems

36. $x^2 + y^2 = 25$
Center: $(0,0)$; $r = 5$

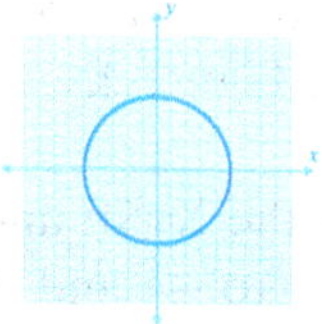

1. Find the distance between the two points $(5, 3)$ and $(-1, -3)$.
2. Find the equation of the circle with center at $(-2, 3)$ and radius 6.
3. Write the equation in standard form and find the center and radius for the circle with equation $x^2 + y^2 + 6y = 7$.

12.3 Exercises

37. $x^2 + y^2 = 18$
Center: $(0,0)$;
$r = 3\sqrt{2}$

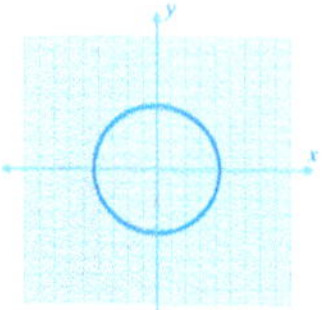

Find the distance between the two given points.

1. $(2, 4), (6, 7)$ **2.** $(1, 0), (6, 12)$ **3.** $(-3, 2), (9, 7)$

4. $(-6, 3), (-2, 0)$ **5.** $(1, 7), (3, 2)$ **6.** $(-2, 1), (3, -4)$

7. $(4, -3), (7, -3)$ **8.** $(-2, 6), (5, 6)$ **9.** $(5, -2), (7, -5)$

Answers to Practice Problems: **1.** $\sqrt{72} = 6\sqrt{2}$ **2.** $(x+2)^2 + (y-3)^2 = 36$
3. $x^2 + (y+3)^2 = 16$; Center: $(0, -3)$; $r = 4$

10. $(6, 4), (8, -5)$ **11.** $(-7, 3), (1, -12)$ **12.** $(3, 8), (-2, -4)$

Find equations for each of the circles.

13. Center $(0, 0)$; $r = 4$ **14.** Center $(0, 0)$; $r = 6$ **15.** Center $(0, 0)$; $r = \sqrt{3}$

16. Center $(0, 0)$; $r = \sqrt{7}$ **17.** Center $(0, 0)$; $r = \sqrt{11}$ **18.** Center $(0, 0)$; $r = \sqrt{13}$

19. Center $(0, 0)$; $r = \frac{2}{3}$ **20.** Center $(0, 0)$; $r = \frac{7}{4}$ **21.** Center $(0, 2)$; $r = 2$

22. Center $(0, 5)$; $r = 5$ **23.** Center $(4, 0)$; $r = 1$ **24.** Center $(-3, 0)$; $r = 4$

25. Center $(-2, 0)$; $r = \sqrt{8}$ **26.** Center $(5, 0)$; $r = \sqrt{2}$ **27.** Center $(3, 1)$; $r = 6$

28. Center $(-1, 2)$; $r = 5$ **29.** Center $(3, 5)$; $r = \sqrt{12}$

30. Center $(4, -2)$; $r = \sqrt{14}$ **31.** Center $(7, 4)$; $r = \sqrt{10}$

32. Center $(-3, 2)$; $r = \sqrt{7}$

Write each of the equations in standard form. Find the center and radius of the circle and then sketch the graph.

33. $x^2 + y^2 = 9$ **34.** $x^2 + y^2 = 16$ **35.** $x^2 = 49 - y^2$

36. $y^2 = 25 - x^2$ **37.** $x^2 + y^2 = 18$ **38.** $x^2 + y^2 = 12$

39. $x^2 + y^2 + 2x = 8$ **40.** $x^2 + y^2 + 6x = 0$ **41.** $x^2 + y^2 - 4y = 0$

42. $x^2 + y^2 - 4x = 12$ **43.** $x^2 + y^2 + 2x + 4y = 11$

44. $x^2 + y^2 + 4x + 4y = 8$ **45.** $x^2 + y^2 - 4x + 10y + 20 = 0$

46. $x^2 + y^2 - 6x - 8y + 9 = 0$ **47.** $x^2 + y^2 - 4x - 6y + 5 = 0$

48. $x^2 + y^2 + 10x - 2y + 14 = 0$

Use the Pythagorean theorem to decide if the triangle determined by the given points is a right triangle.

49. $A(1, -2), B(7, 1), C(5, 5)$ **50.** $A(-5, -1), B(2, 1), C(-1, 6)$

Show that the triangle determined by the given points is an isosceles triangle (has two equal sides).

51. $A(1, 1), B(5, 9), C(9, 5)$ **52.** $A(1, -4), B(3, 2), C(9, 4)$

38. $x^2 + y^2 = 12$
Center: $(0,0)$; $r = 2\sqrt{3}$

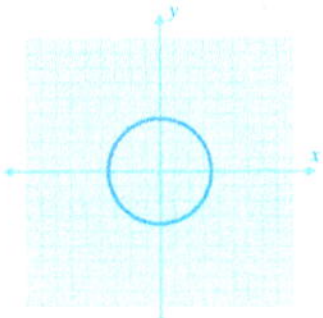

39. $(x+1)^2 + y^2 = 9$
Center: $(-1,0)$; $r = 3$

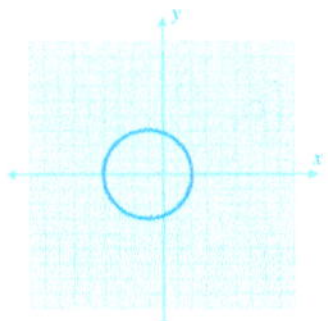

40. $(x+3)^2 + y^2 = 9$
Center: $(-3,0)$; $r = 3$

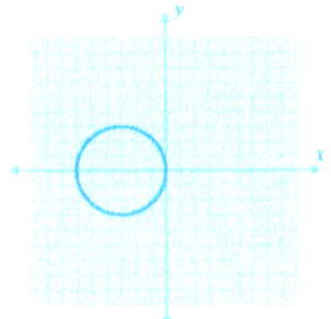

41. $x^2 + (y-2)^2 = 4$
Center: $(0, 2)$; $r = 2$

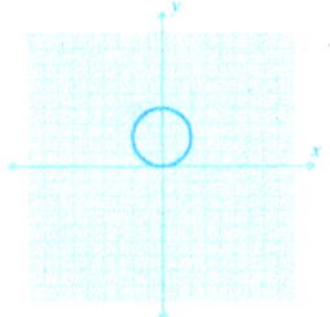

42. $(x-2)^2 + y^2 = 16$
Center: $(2,0)$; $r = 4$

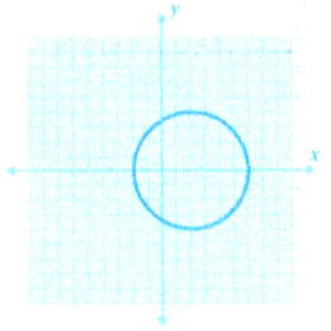

43. $(x+1)^2 + (y+2)^2 = 16$
Center: $(-1, -2)$; $r = 4$

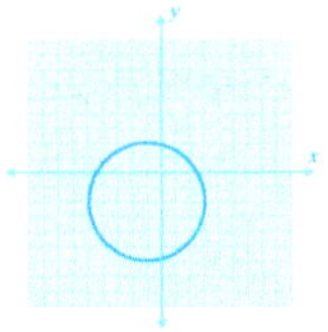

44. $(x+2)^2+(y+2)^2 = 16$
Center: $(-2,-2)$; $r=4$

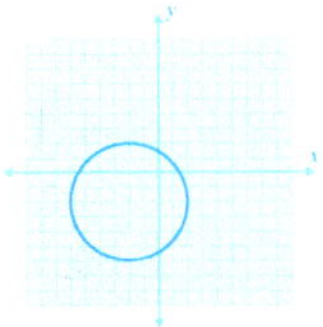

45. $(x-2)^2+(y+5)^2=9$
Center: $(2,-5)$; $r=3$

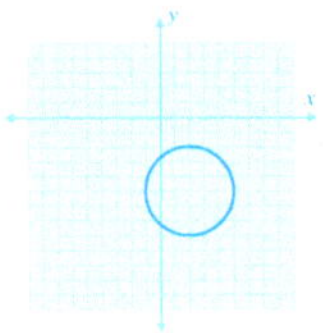

Show that the triangle determined by the given points is an equilateral triangle (all sides equal).

53. $A(1,0), B(3,\sqrt{12}), C(5,0)$

54. $A(0,5), B(0,-3), C(\sqrt{48},1)$

Show that the lengths of the diagonals (AC and BD) of the rectangle ABCD are equal.

55. $A(2,-2), B(2,3), C(8,3), D(8,-2)$

56. $A(-1,1), B(-1,4), C(4,4), D(4,1)$

Find the perimeter of the triangle determined by the given points.

57. $A(-5,0), B(3,4), C(0,0)$

58. $A(-6,-1), B(-3,3), C(6,4)$

Use a graphing calculator to graph the circles. Be sure to set a square window.

59. $x^2+y^2=16$

60. $x^2+y^2=25$

61. $(x+3)^2+y^2=49$

62. $x^2+(y+2)^2=36$

63. $(x-2)^2+(y-5)^2=100$

64. $(x-1)^2+(y+3)^2=64$

Writing and Thinking About Mathematics

46. $(x-3)^2+(y-4)^2 = 16$
Center: $(3,4)$; $r=4$

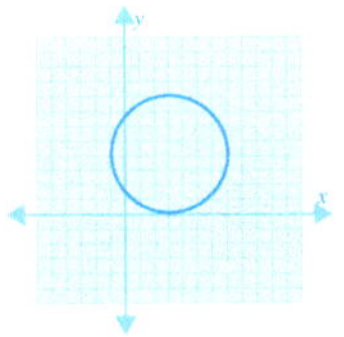

47. $(x-2)^2+(y-3)^2 = 8$
Center: $(2,3)$; $r=2\sqrt{2}$

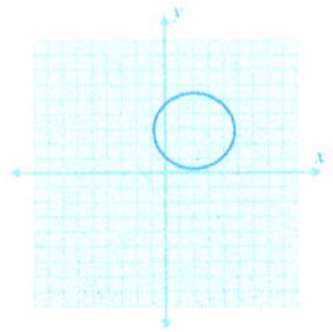

65. For a given line and a point not on the line, a parabola is defined as the set of all points that are the same distance from the point and the line. The point is called the focus and the line is called the directrix. See the figure below.

a. Suppose that (x, y) is any point on a parabola and $(0, p)$ is the focus. Find the distance from (x, y) to the focus.

b. Suppose that (x, y) is any point on the same parabola in part **a.** and the line $y=-p$ is the directrix. Find the distance from (x, y) to the directrix.

c. Show that the equation of the parabola is $x^2=4py$.

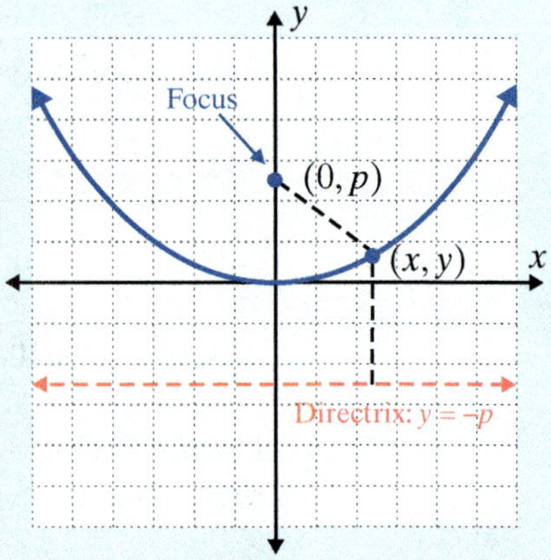

66. Using the equation developed in Exercise 65, find the equation of the parabola with focus at $(0, 2)$ and line $y=-2$ as directrix. Draw the graph.

Continued on the next page...

Writing and Thinking About Mathematics (cont.)

67. For a given line and a point not on the line, a parabola is defined as the set of all points that are the same distance from the point and the line. The point is called the focus and the line is called the directrix. See the figure below.

a. Suppose that (x, y) is any point on a parabola and $(p, 0)$ is the focus. Find the distance from (x, y) to the focus.

b. Suppose that (x, y) is any point on the same parabola in part **a.** and the line $x = -p$ is the directrix. Find the distance from (x, y) to the directrix.

c. Show that the equation of the parabola is $y^2 = 4px$.

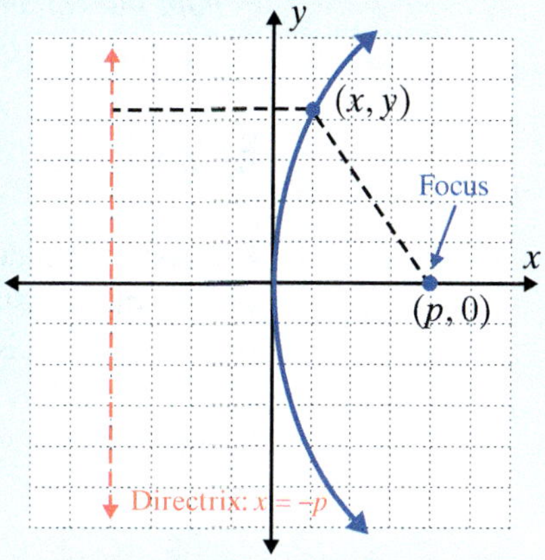

68. Using the equation developed in Exercise 67, find the equation of the parabola with focus at $(-3, 0)$ and line $x = 3$ as directrix. Draw the graph.

48. $(x+5)^2 + (y-1)^2 = 12$
Center: $(-5, 1)$;
$r = 2\sqrt{3}$

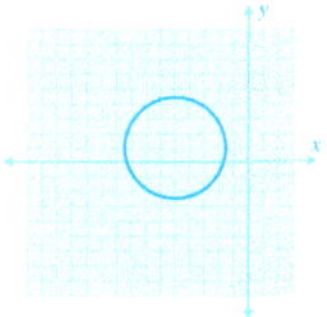

49.
$\overline{AB} = 3\sqrt{5}, \overline{AC} = \sqrt{65}$,
$\overline{BC} = 2\sqrt{5}$,
$(3\sqrt{5})^2 + (2\sqrt{5})^2 = (\sqrt{65})^2$;
Right triangle

50.
$\overline{AB} = \sqrt{53}$, $\overline{AC} = \sqrt{65}$,
$\overline{BC} = \sqrt{34}$,
$(\sqrt{53})^2 + (\sqrt{34})^2 \neq (\sqrt{65})^2$;
Not a right triangle

HAWKES LEARNING SYSTEMS: INTRODUCTORY & INTERMEDIATE ALGEBRA SOFTWARE

- 12.3 The Distance Formula and Circles

51. $\overline{AB} = \overline{AC} = 4\sqrt{5}$ **52.** $\overline{AB} = \overline{BC} = 2\sqrt{10}$ **53.** $\overline{AB} = \overline{AC} = \overline{BC} = 4$ **54.** $\overline{AB} = \overline{AC} = \overline{BC} = 8$

55. $\overline{AC} = \overline{BD} = \sqrt{61}$ **56.** $\overline{AC} = \overline{BD} = \sqrt{34}$ **57.** $10 + 4\sqrt{5}$ **58.** $18 + \sqrt{82}$

59.
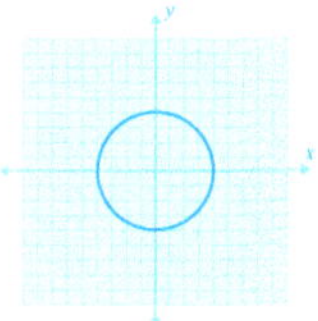

60.
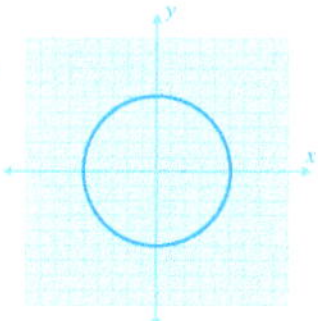

61.
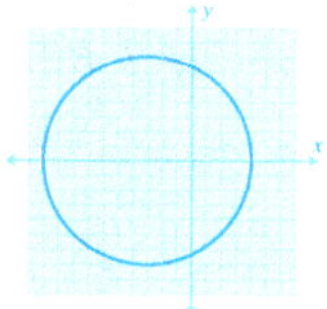

62.
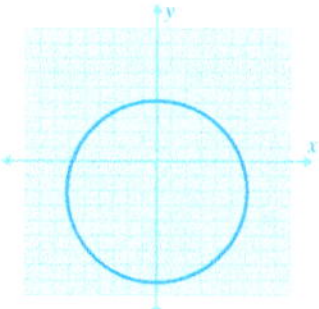

63.
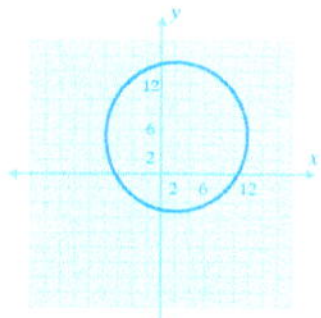

64.
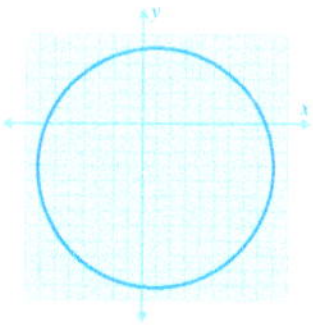

65. **a.** $d = \sqrt{x^2 + (y-p)^2}$ **b.** $d = |y+p|$

c.
$$y + p = \sqrt{x^2 + (y-p)^2}$$
$$(y+p)^2 = x^2 + (y-p)^2$$
$$y^2 + 2py + p^2 = x^2 + y^2 - 2py + p^2$$
$$x^2 = 4py$$

66. $x^2 = 8y$ or $y = \frac{1}{8}x^2$

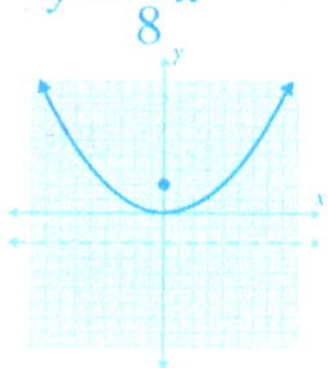

67. **a.** $d = \sqrt{(x-p)^2 + y^2}$ **b.** $d = |x+p|$

c.
$$x + p = \sqrt{(x-p)^2 + y^2}$$
$$(x+p)^2 = (x-p)^2 + y^2$$
$$x^2 + 2px + p^2 = x^2 - 2px + p^2 + y^2$$
$$y^2 = 4px$$

68. $y^2 = -12x$ or $x = -\frac{1}{12}y^2$

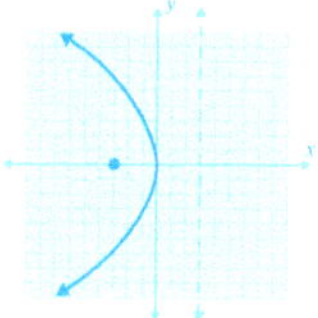

12.4 Ellipses and Hyperbolas

- *Graph* ***ellipses*** *centered at the origin or at another point* (h, k).
- *Graph* ***hyperbolas*** *centered at the origin or at another point* (h, k).
- *Find the equations for the* ***asymptotes*** *of hyperbolas.*

Equations of Ellipses

Ellipses are conic sections that are oval in shape. To draw an ellipse, begin with two fixed points and a distance greater than the distance between the two points. Exercise 45 in the set of exercises is designed to help you understand how points in an elliptical path can be traced by using the following formal definition. You might want to go directly to that exercise now.

Ellipse

An **ellipse** is the set of all points in a plane for which the sum of the distances from two fixed points is constant.

Each of the fixed points is called a **focus** (plural foci).

The **center** of an ellipse is the point midway between the foci.

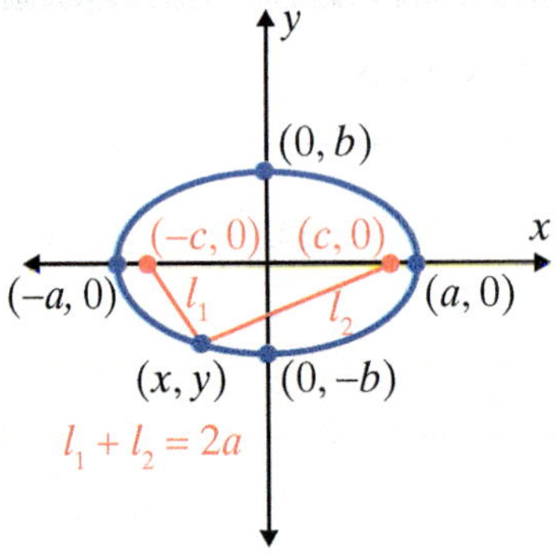

The graph of an ellipse with its center at the origin $(0, 0)$, foci at $(-c, 0)$ and $(c, 0)$, x-intercepts at $(-a, 0)$ and $(a, 0)$, and y-intercepts at $(0, -b)$ and $(0, b)$ (where $a^2 > b^2$).

Ellipses have many practical applications in the sciences, particularly in astronomy. For example, the planets in our solar system (including earth) travel in elliptical orbits (not circular orbits) and the sun is at one of the foci of each ellipse.

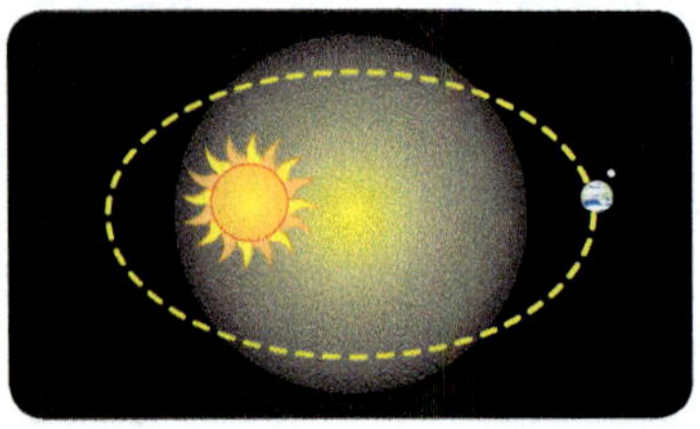

Initially, we will study ellipses with foci either on the x-axis or on the y-axis. Pay close attention to how the constants a, b, and c are used when defining ellipses. Note that the foci are not on the ellipse.

As an example, consider the equation

$$\frac{x^2}{25}+\frac{y^2}{9}=1.$$

Several points that satisfy this equation are given in the following table and then are graphed in Figure 1.

x	y
5	0
–5	0
0	3
0	–3
3	$\frac{12}{5}$
3	$-\frac{12}{5}$
–3	$\frac{12}{5}$
–3	$-\frac{12}{5}$

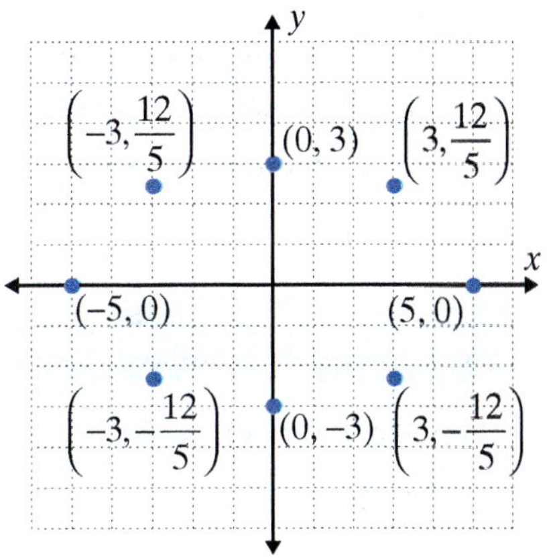

Figure 1

Joining the points in Figure 1 with a smooth curve, produces the ellipse shown in Figure 2. The points $(5, 0)$ and $(-5, 0)$ are the x-intercepts, and the points $(0, 3)$ and $(0, -3)$ are the y-intercepts.

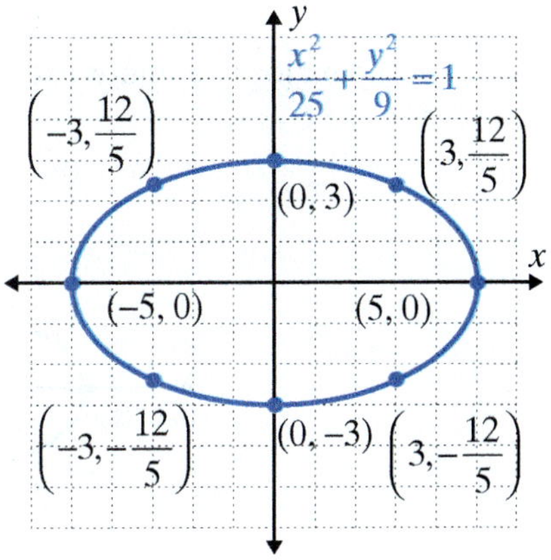

Figure 2

Equation of an Ellipse

Teaching Notes:
You might want to discuss the situation when $a^2 = b^2$ and call a circle a degenerate ellipse. As the two foci move closer together, the ellipse degenerates into a circle with radius r.

The standard form for the equation of an ellipse with its center at the origin is

$$\frac{x^2}{a^2} + \frac{y^2}{b^2} = 1.$$

The points $(a, 0)$ and $(-a, 0)$ are the ***x*-intercepts** (called **vertices**).
The points $(0, b)$ and $(0, -b)$ are the ***y*-intercepts** (called **vertices**).

When $a^2 > b^2$:

1. The segment of length $2a$ joining the x-intercepts is called the **major axis**.
2. The segment of length $2b$ joining the y-intercepts is called the **minor axis**.

When $b^2 > a^2$:

1. The segment of length $2b$ joining the y-intercepts is called the **major axis**.
2. The segment of length $2a$ joining the x-intercepts is called the **minor axis**.

Note: In either case, the foci lie on the major axis.

Example 1: Equation of an Ellipse–Major Axis Horizontal

Graph the ellipse $4x^2 + 16y^2 = 64$.

Solution: First, divide both sides of the given equation by 64 to find the standard form.

$$4x^2 + 16y^2 = 64$$

$$\frac{4x^2}{64} + \frac{16y^2}{64} = \frac{64}{64}$$

$$\frac{x^2}{16} + \frac{y^2}{4} = 1$$

The curve is an ellipse. In this case $a = \sqrt{16} = 4$ and the major axis is $2a = 8$. Also, $b = \sqrt{4} = 2$ and the minor axis is $2b = 4$. The endpoints of the major axis are $(-4, 0)$ and $(4, 0)$. The endpoints of the minor axis are $(0, -2)$ and $(0, 2)$. The major and minor axes intersect at the center of the ellipse, $(0, 0)$.

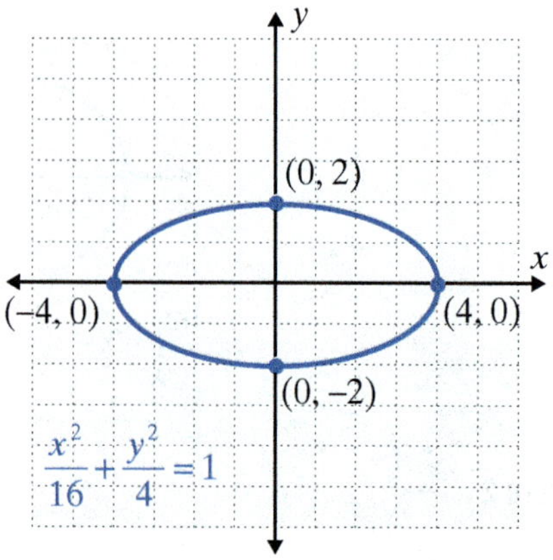

Example 2: Equation of an Ellipse—Major Axis Vertical

Graph the ellipse $\frac{x^2}{1}+\frac{y^2}{9}=1$.

Solution: The equation is in standard form with $a^2=1$ and $b^2=9$.

Because $b^2>a^2$, we know the major axis is vertical. That is, the ellipse is elongated along the y-axis.

The points $(0,-3)$ and $(0,3)$ are the endpoints of the major axis.

The points $(-1,0)$ and $(1,0)$ are the endpoints of the minor axis.

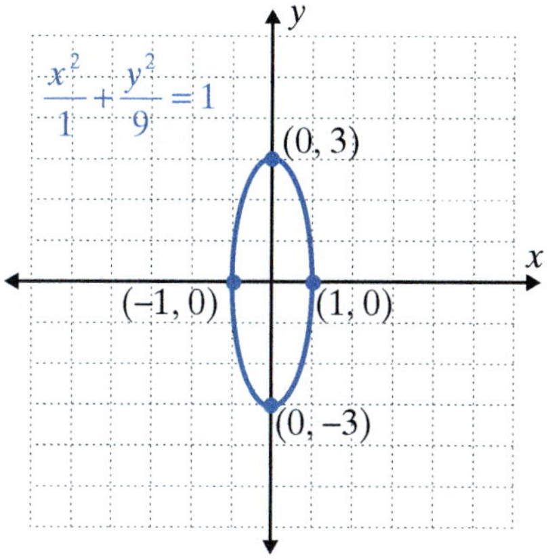

In the equation of an ellipse,

$$\frac{x^2}{a^2}+\frac{y^2}{b^2}=1,$$

the coefficients for x^2 and y^2 are both positive. If one of these coefficients is negative, then the equation represents a **hyperbola**.

Equations of Hyperbolas

Notice how the following definition of a hyperbola differs from the definition of an ellipse. Instead of the sum of the distances from two fixed points, we find the difference of the distances from two fixed points.

Hyperbola

A **hyperbola** is the set of all points in a plane such that the absolute value of the difference of the distances from two fixed points is constant.

Each of the fixed points is called a **focus** (plural foci).

The **center** of a hyperbola is the point midway between the foci.

Continued on next page...

Hyperbola (cont.)

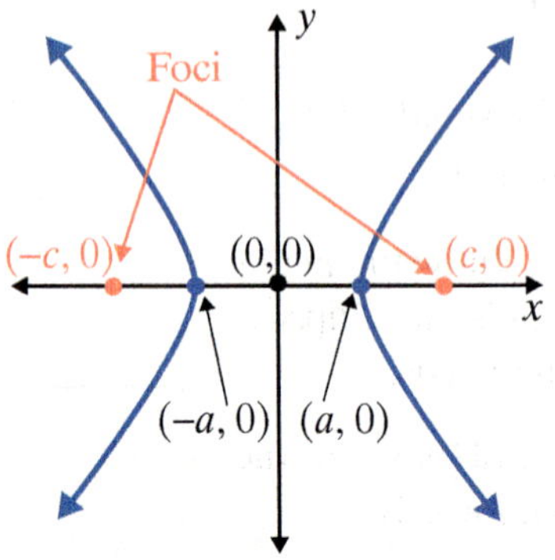

The graph of a hyperbola with its center at the origin $(0, 0)$, foci along the x-axis at $(-c, 0)$ and $(c, 0)$, and x-intercepts at $(-a, 0)$ and $(a, 0)$. There are no y-intercepts.

Several points that satisfy the equation

$$\frac{x^2}{16} - \frac{y^2}{9} = 1$$

and the curves joining these points (a hyperbola) are shown in Figure 3.

x	y
4	0
−4	0
5	$\frac{9}{4}$
5	$-\frac{9}{4}$
−5	$\frac{9}{4}$
−5	$-\frac{9}{4}$

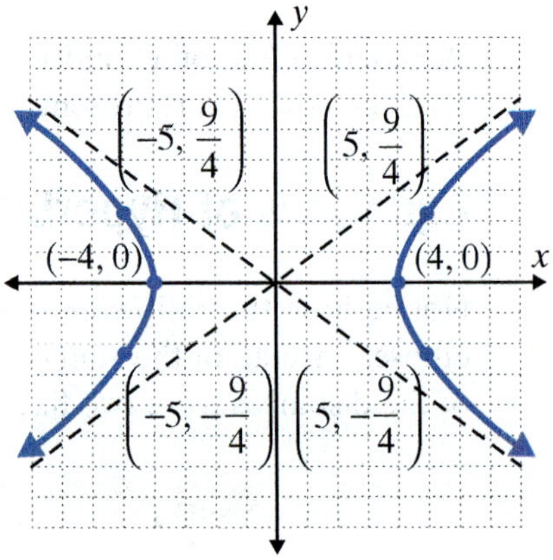

Figure 3

The two dotted lines shown in Figure 3 are called **asymptotes**. (Asymptotes were discussed in Chapter 11 with exponential and logarithmic functions.) These lines are not part of the hyperbola, but they serve as guidelines for sketching the graph. The curve gets closer and closer to these lines without ever touching them. In this figure the equations of the asymptotes are

$$y = \frac{3}{4}x \quad \text{and} \quad y = -\frac{3}{4}x.$$

With the equations of hyperbolas there is a positive term and a negative term. In the standard form for hyperbolas the roles of a and b are that **a^2 is the denominator of the positive term and b^2 is in the denominator of the negative term.** This relationship is very important in determining the equations of the asymptotes. The location of the negative sign determines whether the hyperbola "opens" left and right or "opens" up and down as illustrated in the following discussion.

Equations of Hyperbolas

In general, there are two standard forms for equations of hyperbolas with their centers at the origin.

1. $\dfrac{x^2}{a^2} - \dfrac{y^2}{b^2} = 1$

x-intercepts (vertices) at $(a, 0)$ and $(-a, 0)$

No y-intercepts

Asymptotes: $y = \dfrac{b}{a}x$ and $y = -\dfrac{b}{a}x$

The curves "open" left and right.

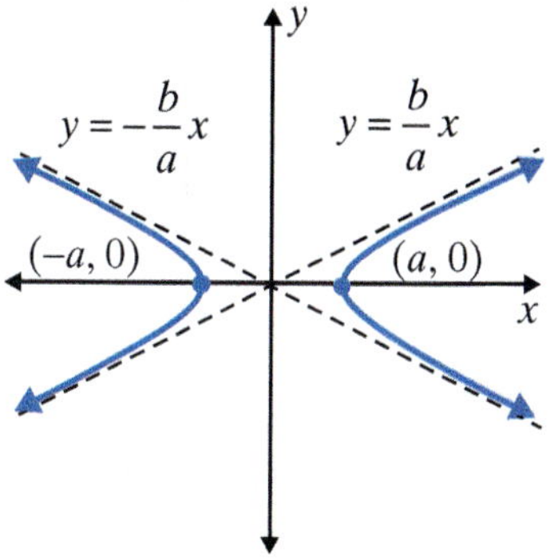

2. $\dfrac{y^2}{a^2} - \dfrac{x^2}{b^2} = 1$

y-intercepts (vertices) at $(0, a)$ and $(0, -a)$

No x-intercepts

Asymptotes: $y = \dfrac{a}{b}x$ and $y = -\dfrac{a}{b}x$

The curves "open" up and down.

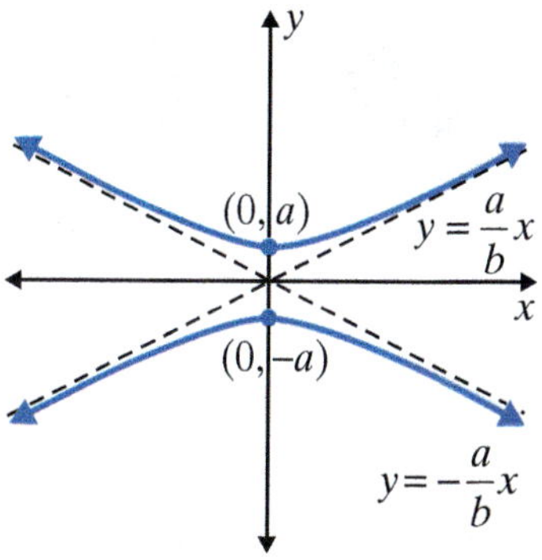

A Geometric Aid for Sketching Asymptotes of Hyperbolas

As a geometric aid in getting the asymptotes in the correct positions, draw a rectangle with sides of lengths $2a$ and $2b$ centered at the origin. The diagonals of this fundamental rectangle are the asymptotes. Study these rectangles in Examples 3 and 4.

Example 3: Hyperbola Opening Left and Right

Graph the hyperbola $x^2 - 4y^2 = 16$.

Solution: Write the equation in standard form by dividing both sides by 16:

$$\frac{x^2}{16} - \frac{y^2}{4} = 1.$$

Here $a^2 = 16$ and $b^2 = 4$. So, using $a = 4$ and $b = 2$, the asymptotes are $y = \frac{2}{4}x = \frac{1}{2}x$ and $y = -\frac{2}{4}x = -\frac{1}{2}x$.

The vertices are $(-4, 0)$ and $(4, 0)$ and the curve opens left and right. (Note that the fundamental rectangle has sides of lengths $2a = 8$ and $2b = 4$.)

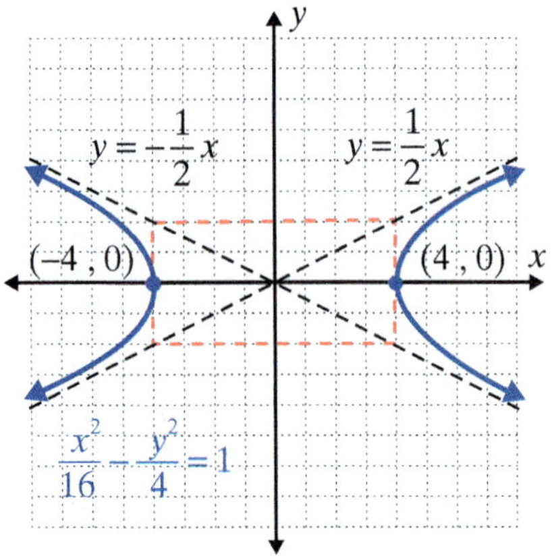

Example 4: Hyperbola Opening Up and Down

Graph the hyperbola $\frac{y^2}{1} - \frac{x^2}{9} = 1$.

Solution: Here $a^2 = 1$ and $b^2 = 9$. So $a = 1$ and $b = 3$ and the asymptotes are $y = \frac{1}{3}x$ and $y = -\frac{1}{3}x$. The vertices are $(0, -1)$ and $(0, 1)$ and the curve opens up and down.
(Note that the fundamental rectangle has sides of lengths $2a = 2$ and $2b = 6$.)

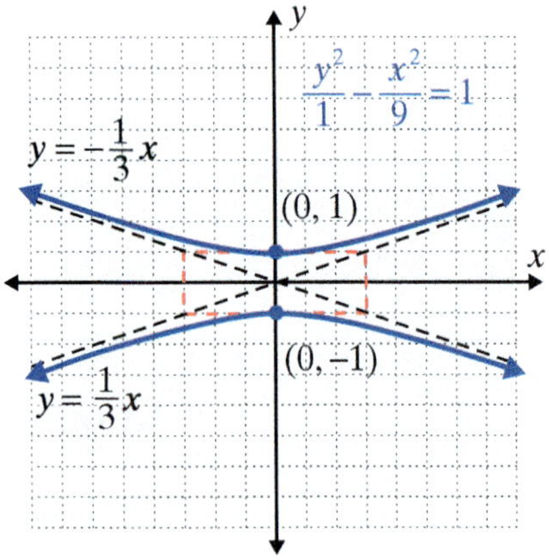

Ellipses and Hyperbolas with Centers at (h, k)

The discussion of translations in Section 12.1 indicates that replacing x with $x - h$ and y with $y - k$ in an equation gives the corresponding graph a horizontal shift of h units and a vertical shift of k units. These ideas were used in the discussion of circles in Section 12.3 for graphs of circles with center at (h, k). $\left[(x-h)^2 + (y-k)^2 = r^2\right]$

The same procedure relates to the equations of ellipses and hyperbolas with centers at (h, k). That is, the graphs of ellipses and hyperbolas can be translated to center at (h, k) by replacing x with $x - h$ and y with $y - k$ in the standard forms. The roles of a^2 and b^2 are the same.

Ellipse with Center at (h, k)

The equation of an ellipse with its center at (h, k) is

$$\frac{(x-h)^2}{a^2} + \frac{(y-k)^2}{b^2} = 1.$$

Note: a and b are distances from (h, k) to the vertices. (See Figure 4)

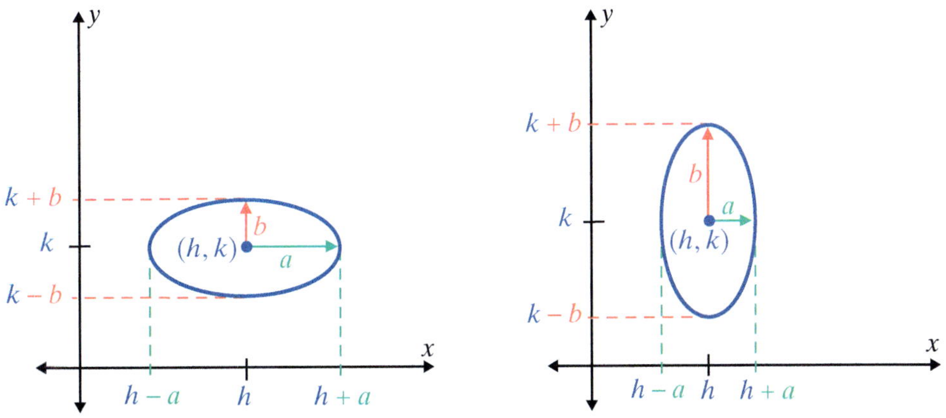

Figure 4

Example 5: Ellipse with Center at (h, k)

Graph the ellipse $\dfrac{(x+2)^2}{16} + \dfrac{(y-1)^2}{9} = 1$.

Solution: The graph of $\dfrac{x^2}{16} + \dfrac{y^2}{9} = 1$ is translated 2 units left and 1 unit up so that the center is at $(-2, 1)$ with $a = 4$ and $b = 3$. The graph is shown here with the center and vertices labeled.

Continued on the next page...

1. $\frac{x^2}{36}+\frac{y^2}{4}=1$

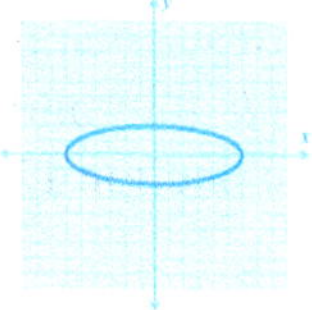

2. $\frac{x^2}{16}+\frac{y^2}{4}=1$

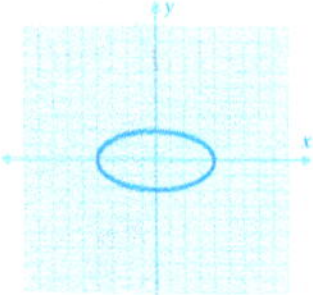

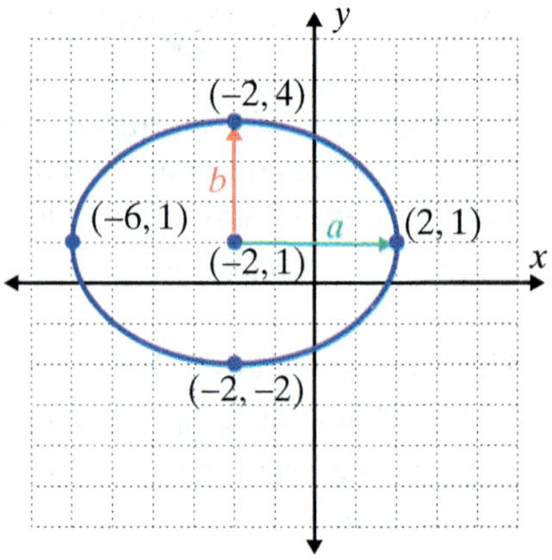

Hyperbola with Center at (h, k)

The equation of a hyperbola with its center at (h, k) is

$$\frac{(x-h)^2}{a^2}-\frac{(y-k)^2}{b^2}=1 \text{ or } \frac{(y-k)^2}{a^2}-\frac{(x-h)^2}{b^2}=1.$$

Note: a and b are used as in the standard form but are measured from (h, k). (See Figure 5)

3. $\frac{x^2}{25}+\frac{y^2}{4}=1$

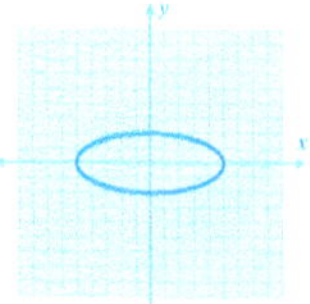

4. $\frac{x^2}{9}+\frac{y^2}{4}=1$

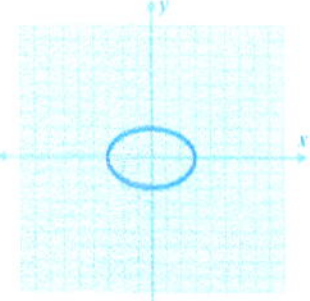

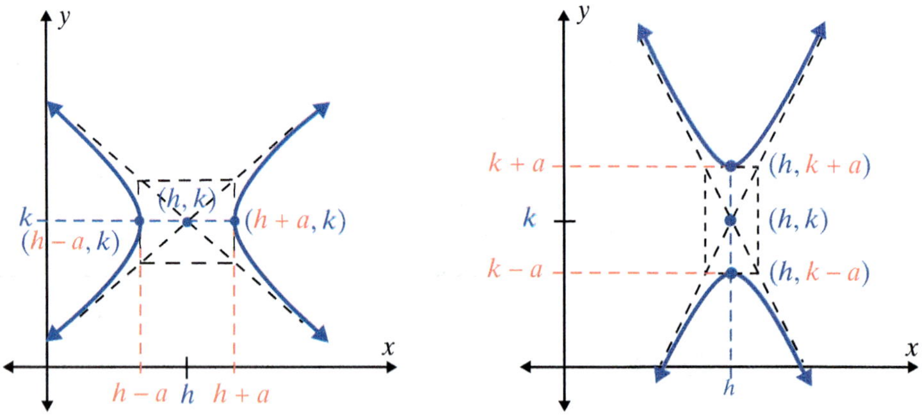

Figure 5

5. $\frac{x^2}{1}+\frac{y^2}{16}=1$

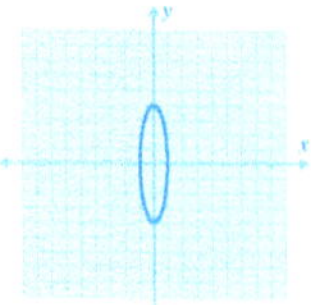

6. $\frac{x^2}{1}+\frac{y^2}{4}=1$

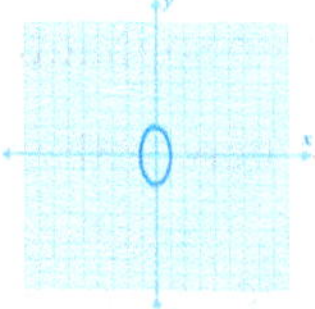

7. $\frac{x^2}{1}-\frac{y^2}{1}=1$

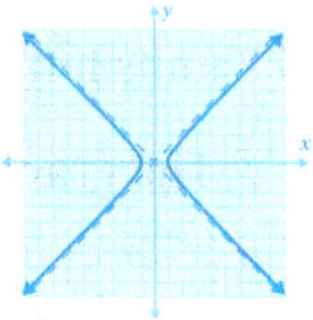

8. $\frac{x^2}{4}-\frac{y^2}{4}=1$

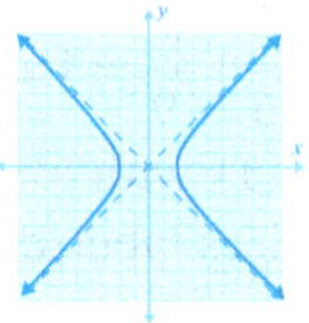

9. $\frac{x^2}{1}-\frac{y^2}{9}=1$

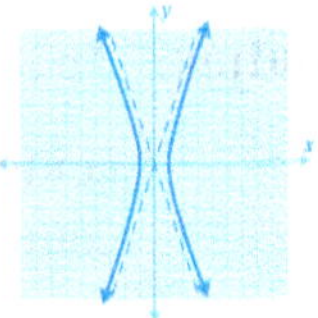

10. $\frac{x^2}{1}-\frac{y^2}{4}=1$

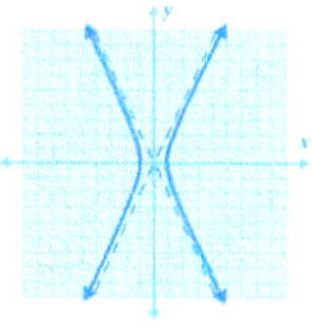

Example 6: Hyperbola with Center at (h, k)

11. $\frac{x^2}{9} - \frac{y^2}{4} = 1$

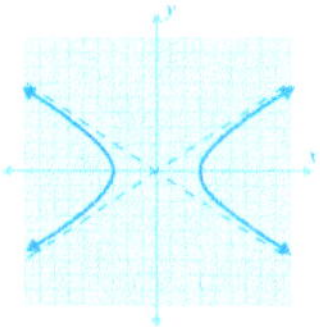

12. $\frac{x^2}{16} - \frac{y^2}{9} = 1$

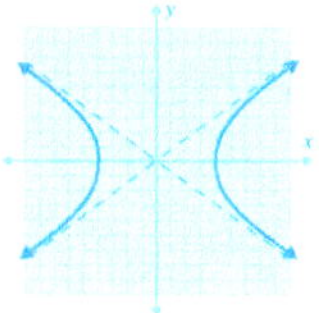

13. $\frac{x^2}{4} + \frac{y^2}{8} = 1$

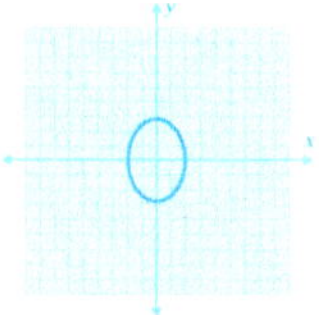

Graph the hyperbola $\frac{(x-3)^2}{25} - \frac{(y+4)^2}{36} = 1$.

Solution: The graph of $\frac{x^2}{25} - \frac{y^2}{36} = 1$ is translated 3 units right and 4 units down so that the center is at $(3, -4)$ with $a = 5$ and $b = 6$. The graph is shown here with the center and vertices labeled.

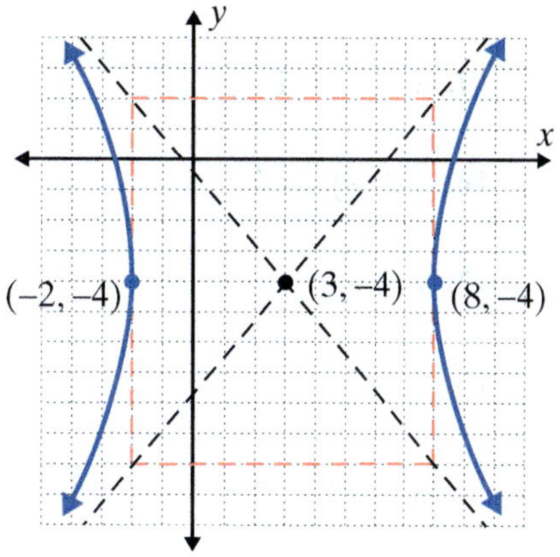

Practice Problems

14. $\frac{x^2}{4} + \frac{y^2}{12} = 1$

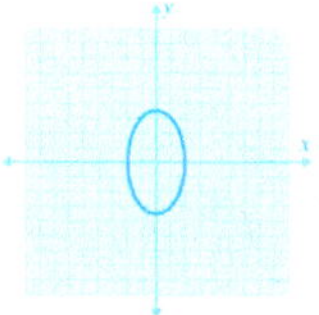

15. $\frac{x^2}{20} + \frac{y^2}{4} = 1$

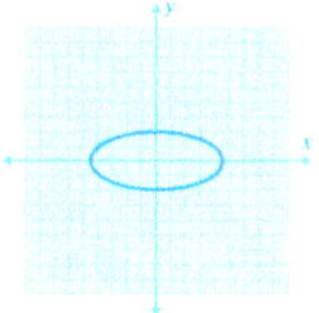

1. Write the equation $2x^2 + 9y^2 = 18$ in standard form. State the length of the major axis and the length of the minor axis.
2. Write the equation $x^2 - 9y^2 = 9$ in standard form. Write the equations of the asymptotes.
3. Graph the ellipse $\frac{(x-2)^2}{4} + \frac{(y+1)^2}{1} = 1$.
4. Graph the hyperbola $\frac{x^2}{16} - \frac{(y-3)^2}{4} = 1$.

Answers to Practice Problems: **1.** $\frac{x^2}{9} + \frac{y^2}{2} = 1$; Major axis: 6, Minor axis: $2\sqrt{2}$ **2.** $\frac{x^2}{9} - \frac{y^2}{1} = 1$; Asymptotes: $y = \frac{1}{3}x$, $y = -\frac{1}{3}x$ **3.**

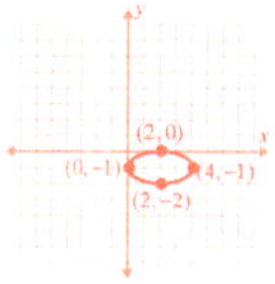

4.

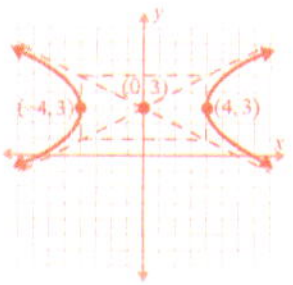

12.4 Exercises

16. $\dfrac{x^2}{28}+\dfrac{y^2}{4}=1$

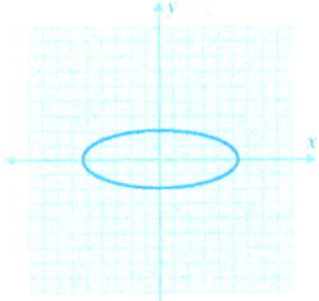

17. $\dfrac{y^2}{9}-\dfrac{x^2}{9}=1$

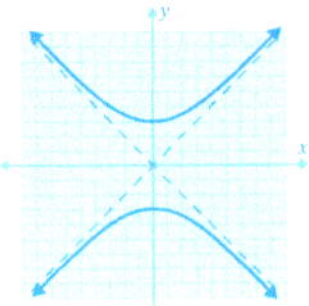

18. $\dfrac{y^2}{16}-\dfrac{x^2}{16}=1$

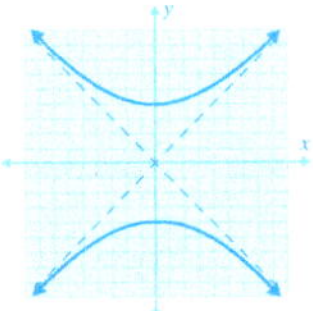

19. $\dfrac{y^2}{8}-\dfrac{x^2}{4}=1$

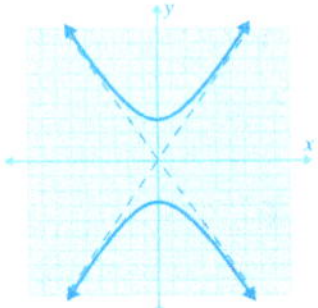

20. $\dfrac{y^2}{12}-\dfrac{x^2}{4}=1$

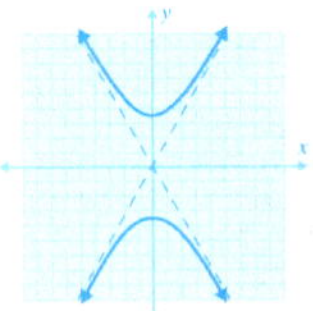

21. $\dfrac{y^2}{18}-\dfrac{x^2}{9}=1$

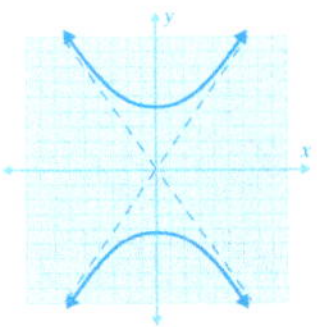

Write each of the equations in standard form. Then sketch the graph. For hyperbolas, graph the asymptotes as well.

1. $x^2+9y^2=36$ **2.** $x^2+4y^2=16$ **3.** $4x^2+25y^2=100$

4. $4x^2+9y^2=36$ **5.** $16x^2+y^2=16$ **6.** $36x^2+9y^2=36$

7. $x^2-y^2=1$ **8.** $x^2-y^2=4$ **9.** $9x^2-y^2=9$

10. $4x^2-y^2=4$ **11.** $4x^2-9y^2=36$ **12.** $9x^2-16y^2=144$

13. $2x^2+y^2=8$ **14.** $3x^2+y^2=12$ **15.** $x^2+5y^2=20$

16. $x^2+7y^2=28$ **17.** $y^2-x^2=9$ **18.** $y^2-x^2=16$

19. $y^2-2x^2=8$ **20.** $y^2-3x^2=12$ **21.** $y^2-2x^2=18$

22. $y^2-5x^2=20$ **23.** $3x^2+2y^2=18$ **24.** $4x^2+3y^2=12$

25. $4x^2+5y^2=20$ **26.** $3x^2+8y^2=48$ **27.** $9y^2-8x^2=72$

28. $4x^2-7y^2=28$ **29.** $3y^2-4x^2=36$ **30.** $3x^2-5y^2=75$

Match the equations with the given graphs.

31. $\dfrac{(x-1)^2}{4}+\dfrac{(y-3)^2}{25}=1$ **32.** $\dfrac{(x+1)^2}{4}+\dfrac{(y+3)^2}{25}=1$ **33.** $\dfrac{(x-1)^2}{25}+\dfrac{(y-3)^2}{4}=1$

34. $\dfrac{(x+1)^2}{25}+\dfrac{(y+3)^2}{4}=1$ **35.** $\dfrac{(x+1)^2}{25}-\dfrac{(y+3)^2}{4}=1$ **36.** $\dfrac{(y+3)^2}{4}-\dfrac{(x+1)^2}{25}=1$

a.

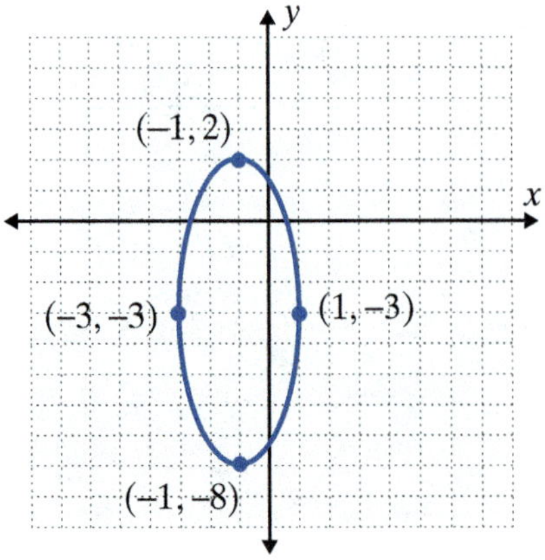

b.

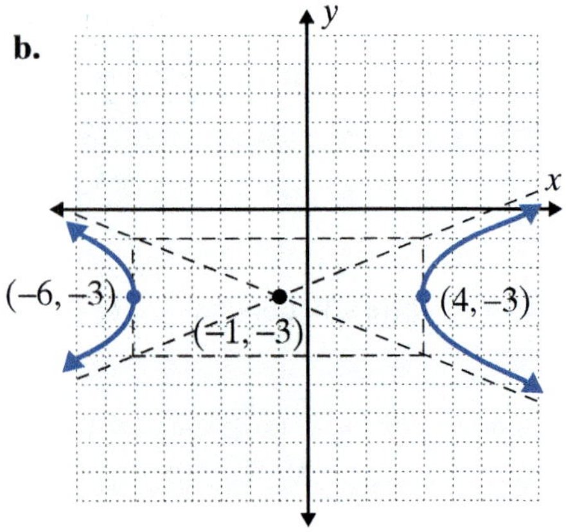

22. $\frac{y^2}{20} - \frac{x^2}{4} = 1$

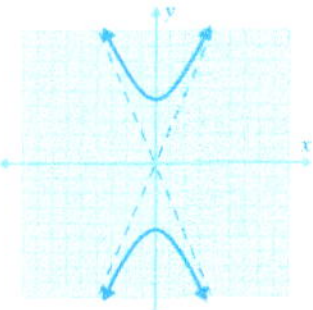

23. $\frac{x^2}{6} + \frac{y^2}{9} = 1$

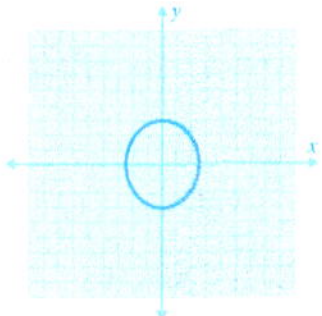

24. $\frac{x^2}{3} + \frac{y^2}{4} = 1$

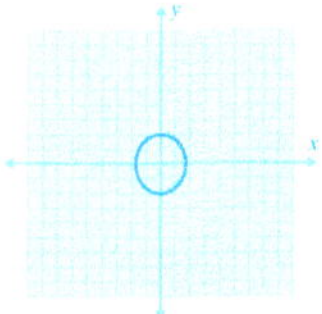

25. $\frac{x^2}{5} + \frac{y^2}{4} = 1$

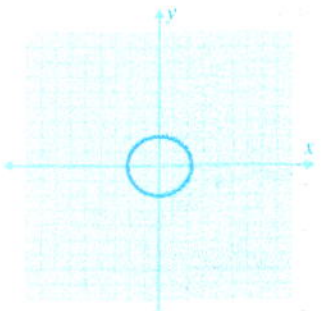

26. $\frac{x^2}{16} + \frac{y^2}{6} = 1$

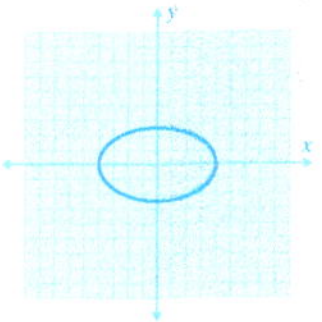

27. $\frac{y^2}{8} - \frac{x^2}{9} = 1$

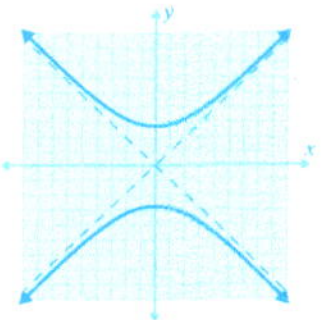

c.

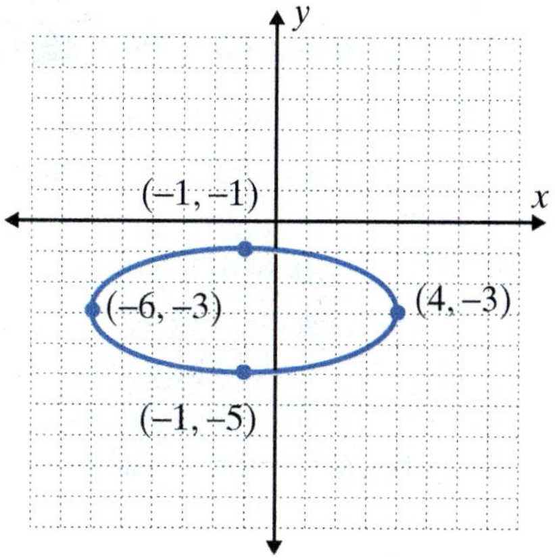

d.

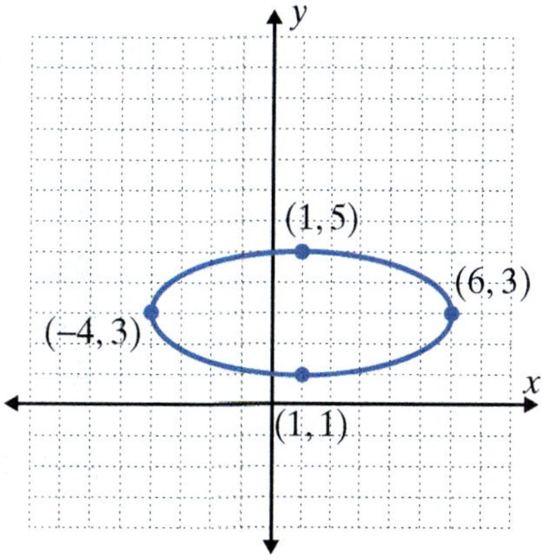

e.

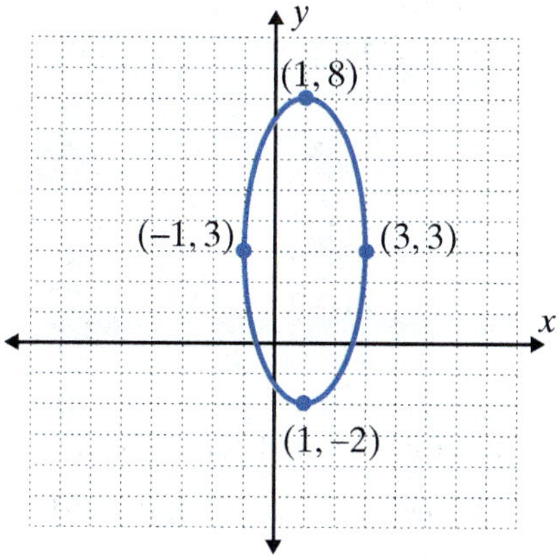

f.

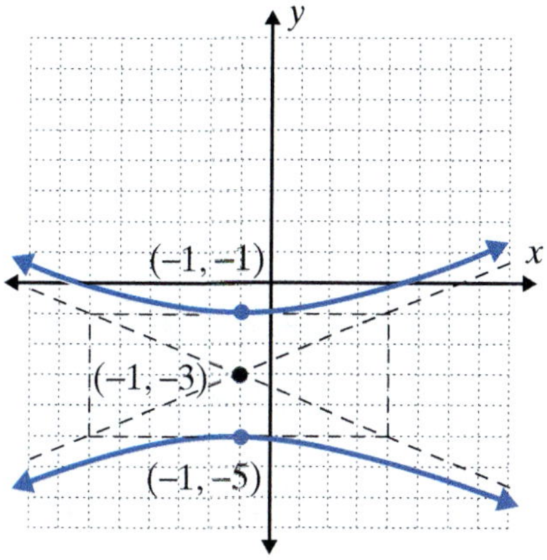

Use your knowledge of translations to graph each of the following equations. These graphs are ellipses and hyperbolas with centers at points other than the origin.

37. $\frac{(x-2)^2}{25} + \frac{(y-1)^2}{9} = 1$ **38.** $\frac{(x+1)^2}{16} + \frac{(y-4)^2}{1} = 1$ **39.** $\frac{(x+5)^2}{1} - \frac{(y+2)^2}{16} = 1$

40. $\frac{(x-4)^2}{9} - \frac{(y-3)^2}{36} = 1$ **41.** $\frac{(x+1)^2}{49} + \frac{(y-6)^2}{100} = 1$ **42.** $\frac{(x+2)^2}{4} + \frac{(y+3)^2}{36} = 1$

43. $\frac{(y-2)^2}{9} - \frac{(x+2)^2}{4} = 1$ **44.** $\frac{(y+5)^2}{25} - \frac{(x-1)^2}{64} = 1$

28. $\frac{x^2}{7} - \frac{y^2}{4} = 1$

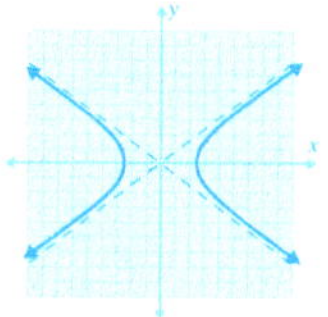

29. $\frac{y^2}{12} - \frac{x^2}{9} = 1$

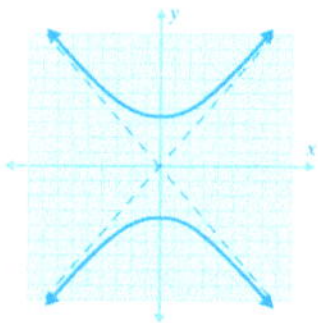

30. $\frac{x^2}{25} - \frac{y^2}{15} = 1$

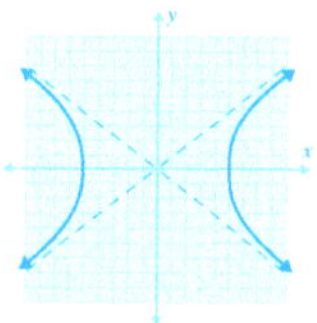

31. e 32. a
33. d 34. c
35. b 36. f

37.

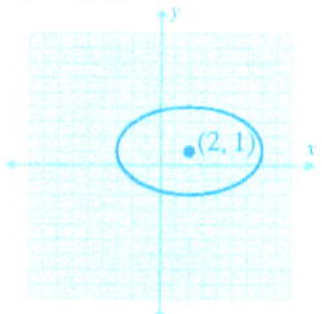

38. 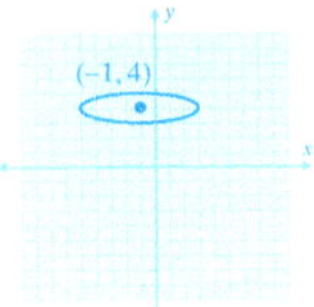

39. 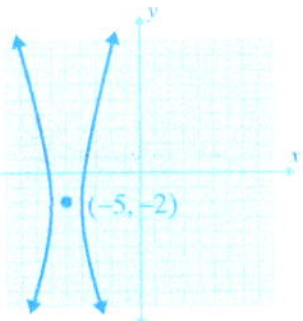

40. 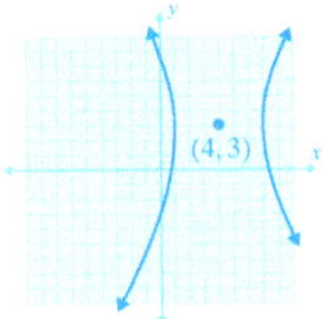

41. 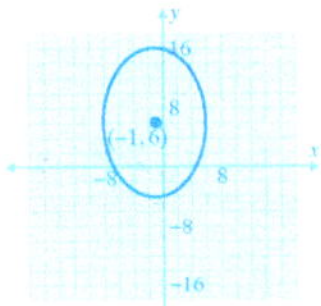

42. 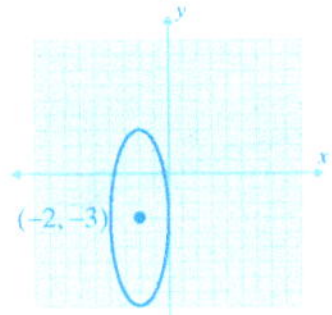

43. 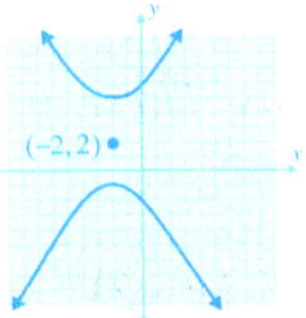

44. 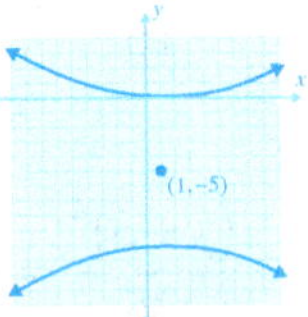

Writing and Thinking About Mathematics

45. The definition of an ellipse is given in the text as follows:
An ellipse is the set of all points in a plane the sum of whose distances from two fixed points is constant.

a. Draw an ellipse by proceeding as follows:

Step 1: Place two thumb tacks in a piece of cardboard.
Step 2: Select a piece of string slightly longer than the distance between the two tacks.
Step 3: Tie the string to each thumb tack and stretch the string taut using a pencil.
Step 4: Use the pencil to trace the path of an ellipse on the cardboard by keeping the string taut. (The length of the string represents the fixed distance from points on the ellipse to the two foci.)

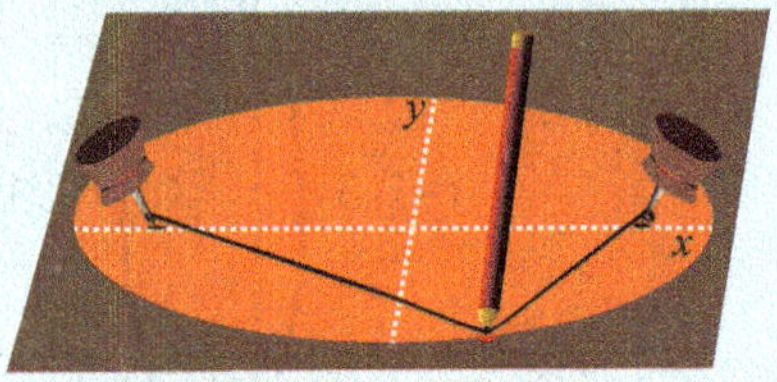

b. Show that the equation of an ellipse with foci at $(-c, 0)$ and $(c, 0)$, center at the origin, and $2a$ as the constant sum of the lengths to the foci can be written in the form $\frac{x^2}{a^2}+\frac{y^2}{a^2-c^2}=1.$

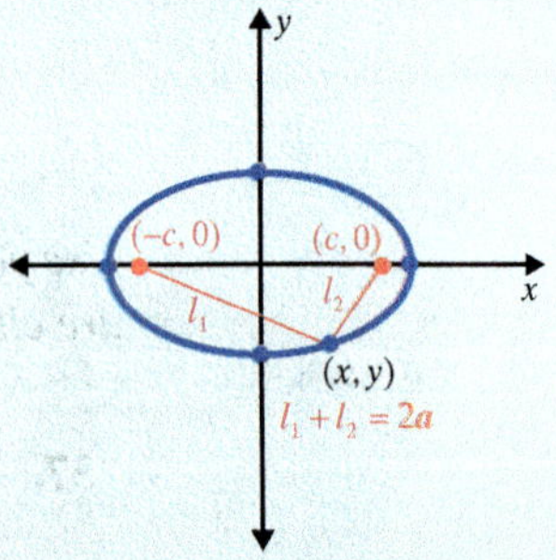

c. In the equation in part **b.**, substitute $b^2=a^2-c^2$ to get the standard form for the equation of an ellipse. Show that the points $(0,-b)$ and $(0,b)$ are the y-intercepts and a is the distance from each y-intercept to a focus.

HAWKES LEARNING SYSTEMS: INTRODUCTORY & INTERMEDIATE ALGEBRA SOFTWARE

- 12.4 Ellipses and Hyperbolas

45. b. Pick an arbitrary point (x, y) located on the ellipse. The sum of the distances from (x, y) to the points $(-c, 0)$ and $(c, 0)$ (the foci) will equal $2a$. Using the distance formula we get the equation $\sqrt{(x+c)^2+(y-0)^2}+\sqrt{(x-c)^2+(y-0)^2}=2a.$ Now to remove the radicals, simplify so that one side of the equation is a single radical and square both sides. Then repeat the previous step and simplify. This gives $\frac{x^2}{a^2}+\frac{y^2}{a^2-c^2}=1.$

c. At the y-intercept $(0, b)$ a right triangle is formed with hypotenuse a and sides b and c. The Pythagorean theorem gives $a^2=b^2+c^2$.

12.5 Nonlinear Systems of Equations

- *Solve systems of equations where one or both equations are nonlinear.*

The equations for the conic sections that we have discussed all have at least one term that is second-degree. These equations are called **quadratic equations.** (Only the equations for parabolas that can be written in the form $y = ax^2 + bx + c$ are **quadratic functions.**) A summary of the equations with their related graphs is shown in Figure 1.

Summary of Equations and Related Graphs

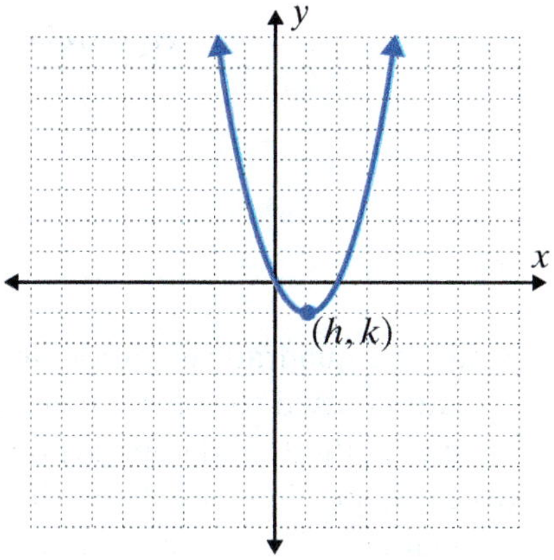

Parabola: $y = a(x-h)^2 + k$
$a > 0$, opens upward
$a < 0$, opens downward

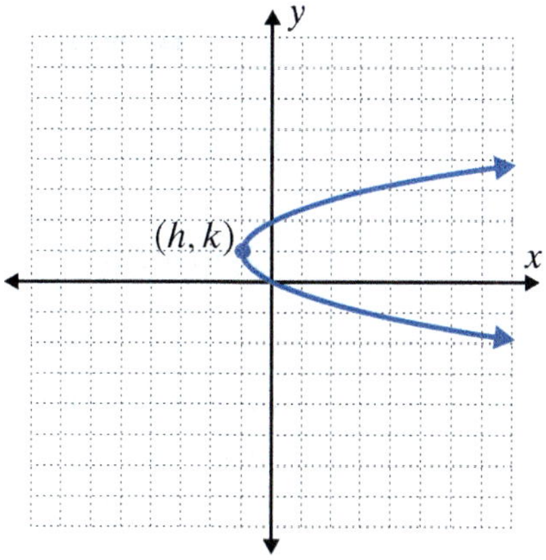

Parabola: $x = a(y-k)^2 + h$
$a > 0$, opens right
$a < 0$, opens left

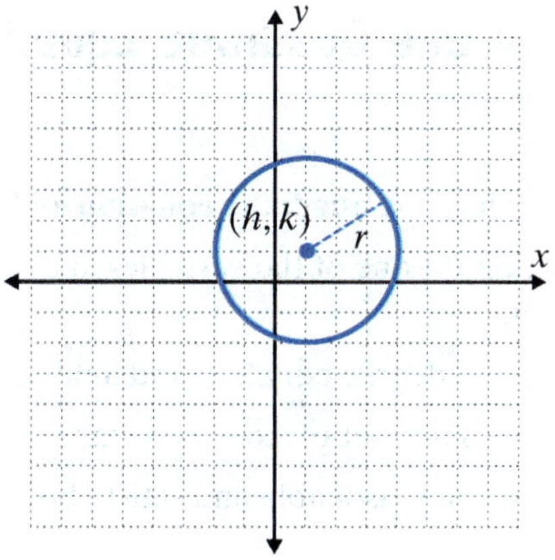

Circle: $(x-h)^2 + (y-k)^2 = r^2$

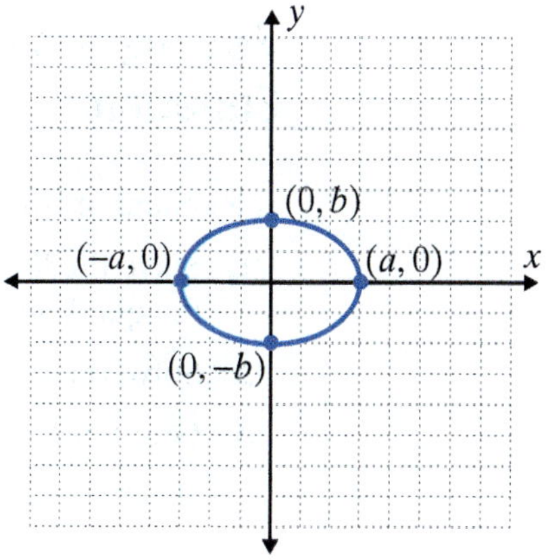

Ellipse: $\frac{x^2}{a^2} + \frac{y^2}{b^2} = 1$

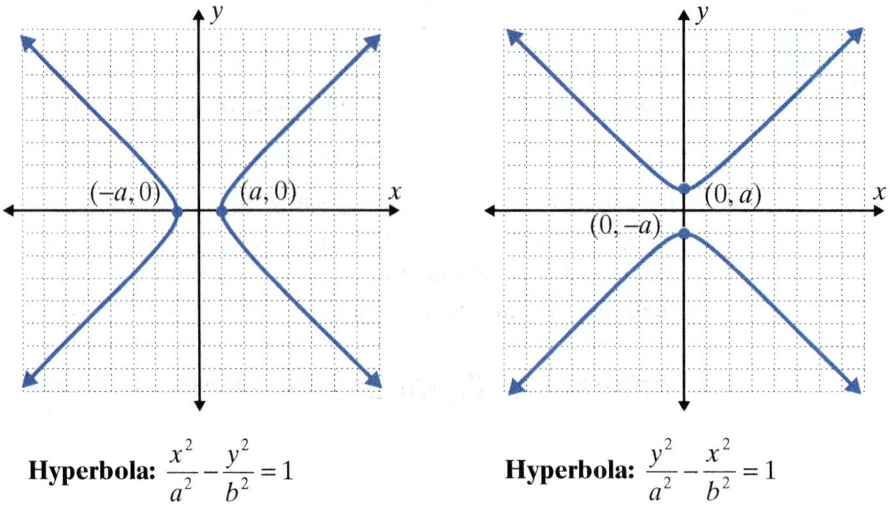

Figure 1

Systems of Equations

If a system of two equations has one quadratic equation and one linear equation, then the method of substitution should be used to solve the system. If the system involves two quadratic equations, then the method used depends on the form of the equations. The following examples show three possible situations. The graphs of the curves are particularly useful for approximating solutions and determining the exact number of solutions.

Examples 1 and 2 illustrate three possibilities and emphasize the value of sketching the graphs to visualize the number of solutions and approximating these solutions.

Solving a System of One Quadratic Equation and One Linear Equation

If a system of two equations has **one quadratic equation and one linear equation**, then:

1. Solve the linear equation for one of the variables and substitute for this variable in the quadratic equation.
2. Solve the resulting second-degree equation and analyze the results.
3. Graph the curves on the same set of axes to visualize the number of solutions and check that the solutions are reasonable and satisfy both equations.

Example 1: A System of One Quadratic and One Linear Equation

Solve the following systems of equations and graph both curves in each system on the same set of axes.

a. A circle and a line: $\begin{cases} x^2 + y^2 = 25 & \text{Circle} \\ x + y = 5 & \text{Line} \end{cases}$

Solution: Solve $x + y = 5$ for y (or x). Then substitute into the other equation.

$$y = 5 - x$$

$$x^2 + (5 - x)^2 = 25$$

$$x^2 + 25 - 10x + x^2 = 25$$

$$2x^2 - 10x = 0$$

$$2x(x - 5) = 0 \qquad \text{Now solve for } x.$$

$$x = 0 \qquad\qquad x = 5$$
$$\text{or}$$
$$y = 5 - 0 = 5 \qquad\qquad y = 5 - 5 = 0$$

The solutions (points of intersection) are $(0, 5)$ and $(5, 0)$.

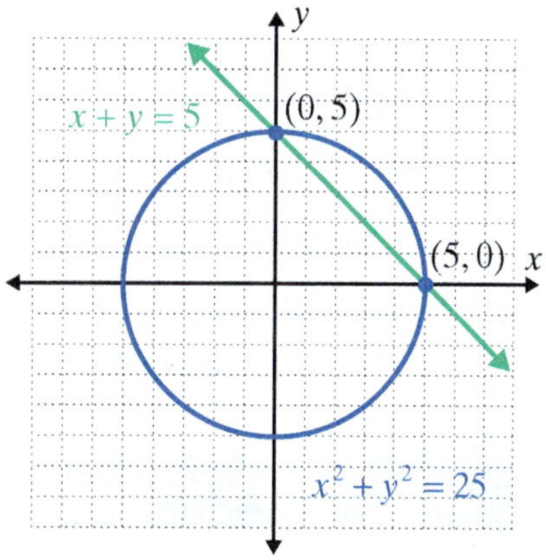

b. A line and a parabola: $\begin{cases} x + y = -7 & \text{Line} \\ y = x^2 - 4x - 5 & \text{Parabola} \end{cases}$

Solution: Solve the linear equation for y (or x), then substitute into the quadratic equation. (In this case, the quadratic equation is already solved for y, so we could have chosen to make the substitution into the linear equation instead.)

$$y = -x - 7$$

$$-x - 7 = x^2 - 4x - 5$$

$$0 = x^2 - 3x + 2$$

$$0 = (x - 2)(x - 1)$$

Continued on the next page...

$$x = 2 \qquad \text{or} \qquad x = 1$$
$$y = -2 - 7 = -9 \qquad\qquad y = -1 - 7 = -8$$

The solutions are $(2, -9)$ and $(1, -8)$.

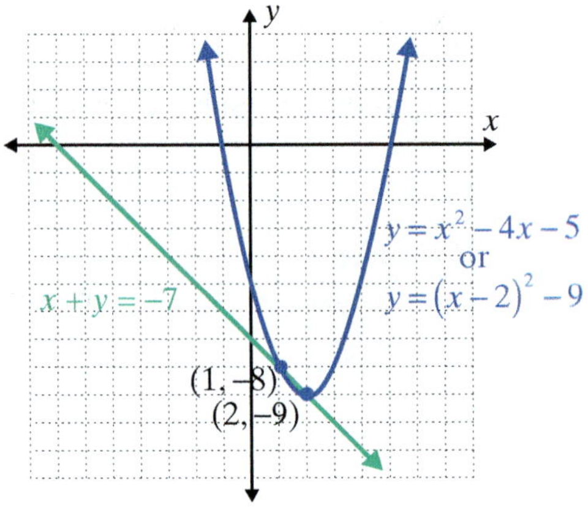

Solving a System of Two Quadratic Equations

If a system of two equations has **two quadratic equations**, then:

1. The method used depends on the form of the equations.
2. Substitution may work or addition may work.
3. Graph the curves on the same set of axes to visualize the number of solutions and check that the solutions are reasonable and satisfy both equations.

Example 2: A System of Two Quadratic Equations

A hyperbola and a circle: $\begin{cases} x^2 - y^2 = 4 & \text{Hyperbola} \\ x^2 + y^2 = 36 & \text{Circle} \end{cases}$

Solution: Here addition will eliminate y^2.

$$\begin{array}{rcl} x^2 - y^2 &=& 4 \\ x^2 + y^2 &=& 36 \\ \hline 2x^2 &=& 40 \\ x^2 &=& 20 \end{array}$$

$$x = \pm\sqrt{20} = \pm 2\sqrt{5}$$

If $x = 2\sqrt{5}$: $\left(2\sqrt{5}\right)^2 + y^2 = 36$

$$20 + y^2 = 36$$
$$y^2 = 16$$
$$y = \pm 4$$

If $x = -2\sqrt{5}$: $\left(-2\sqrt{5}\right)^2 + y^2 = 36$

$$20 + y^2 = 36$$
$$y^2 = 16$$
$$y = \pm 4$$

There are four points of intersection:

$$\left(2\sqrt{5},4\right), \left(2\sqrt{5},-4\right), \left(-2\sqrt{5},4\right), \text{ and } \left(-2\sqrt{5},-4\right).$$

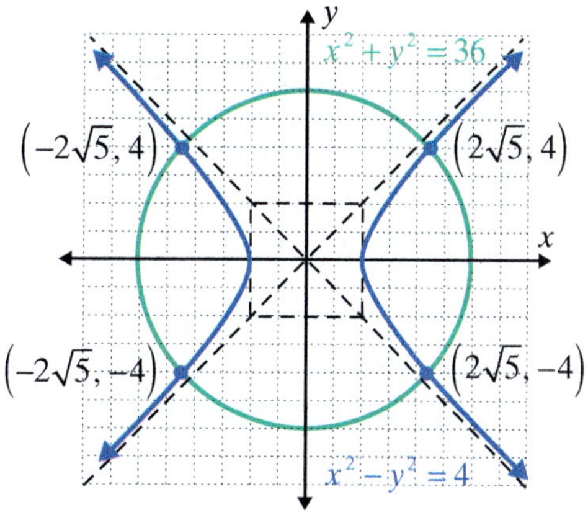

NOTES

In the examples and the exercises the curves do intersect. However, there are many situations where the curves do not intersect. This can be confirmed both algebraically and graphically.

Practice Problems

Solve each of the following systems algebraically and graphically.

1. $\begin{cases} y = x^2 - 4 \\ x - y = 2 \end{cases}$

2. $\begin{cases} x^2 + y^2 = 72 \\ x = y^2 \end{cases}$

Answers to Practice Problems: **1.** $(-1, -3)$ and $(2, 0)$ **2.** $\left(8, 2\sqrt{2}\right)$ and $\left(8, -2\sqrt{2}\right)$

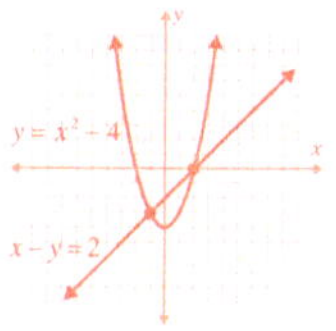

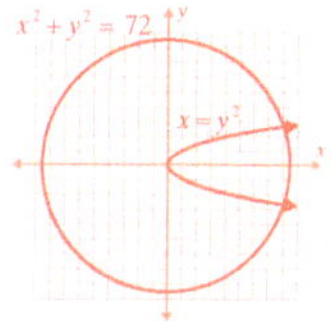

12.5 Exercises

1. $(-3, 10), (1, 2)$

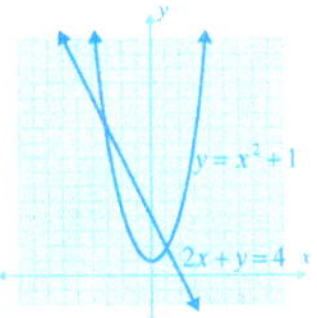

2. $(-2, -1), (3, -6)$

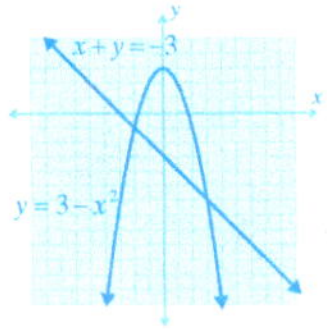

3. $(1, 1), (2, 0)$

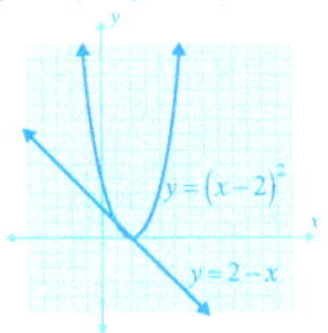

4. $(0, -5), (-5, 0)$

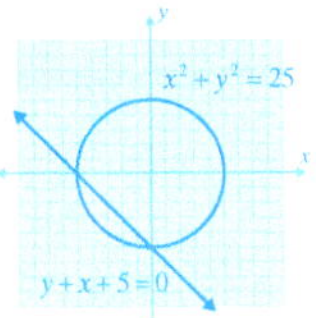

5. $(-2, -4), (4, 2)$

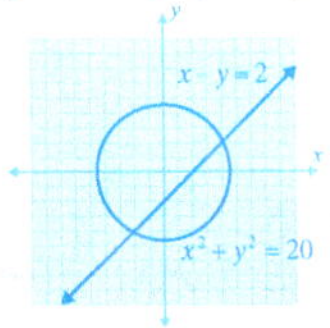

6. $(-5, 3), (5, -3)$

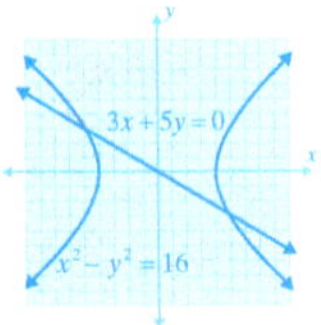

7. $(5, 3)$

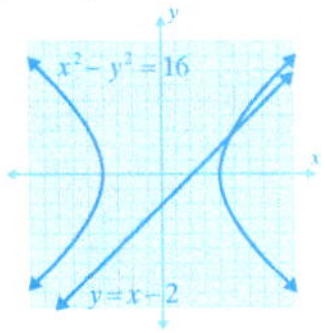

Solve each system of equations. Sketch the graphs.

1. $\begin{cases} y = x^2 + 1 \\ 2x + y = 4 \end{cases}$

2. $\begin{cases} y = 3 - x^2 \\ x + y = -3 \end{cases}$

3. $\begin{cases} y = 2 - x \\ y = (x - 2)^2 \end{cases}$

4. $\begin{cases} x^2 + y^2 = 25 \\ y + x + 5 = 0 \end{cases}$

5. $\begin{cases} x^2 + y^2 = 20 \\ x - y = 2 \end{cases}$

6. $\begin{cases} x^2 - y^2 = 16 \\ 3x + 5y = 0 \end{cases}$

7. $\begin{cases} y = x - 2 \\ x^2 = y^2 + 16 \end{cases}$

8. $\begin{cases} x^2 + 3y^2 = 12 \\ x = 3y \end{cases}$

9. $\begin{cases} x^2 + y^2 = 9 \\ x^2 - y^2 = 9 \end{cases}$

10. $\begin{cases} x^2 + y^2 = 9 \\ x^2 - y + 3 = 0 \end{cases}$

11. $\begin{cases} 4x^2 + y^2 = 25 \\ 3x - y^2 + 3 = 0 \end{cases}$

12. $\begin{cases} x^2 - 4y^2 = 9 \\ x + 2y^2 = 3 \end{cases}$

13. $\begin{cases} x^2 + y^2 + 4x - 2y = 4 \\ x + y = 2 \end{cases}$

14. $\begin{cases} x^2 - y^2 = 9 \\ x^2 + y^2 - 2x - 3 = 0 \end{cases}$

15. $\begin{cases} x^2 - y^2 = 5 \\ x^2 + 4y^2 = 25 \end{cases}$

16. $\begin{cases} 2x^2 - 3y^2 = 6 \\ 2x^2 + y^2 = 22 \end{cases}$

Solve each system of equations.

17. $\begin{cases} x^2 - y^2 = 20 \\ x^2 - 9y = 0 \end{cases}$

18. $\begin{cases} x^2 + 5y^2 = 16 \\ x^2 + y^2 = 4x \end{cases}$

19. $\begin{cases} x^2 + y^2 = 10 \\ x^2 + y^2 - 4y + 2 = 0 \end{cases}$

20. $\begin{cases} x^2 + y^2 = 20 \\ 4x + 8 = y^2 \end{cases}$

21. $\begin{cases} x^2 + y^2 - 4y = 16 \\ x - y = 0 \end{cases}$

22. $\begin{cases} y = x^2 + 2x + 2 \\ 2x + y = 2 \end{cases}$

23. $\begin{cases} 4y + 10x^2 + 7x - 8 = 0 \\ 6x - 8y + 1 = 0 \end{cases}$

24. $\begin{cases} x^2 + y^2 - 4x + 6y = -3 \\ 2x - y - 2 = 0 \end{cases}$

25. $\begin{cases} 2x^2 - y^2 = 7 \\ 2x^2 + y^2 = 29 \end{cases}$

26. $\begin{cases} 4x^2 + y^2 = 11 \\ y = 4x^2 - 9 \end{cases}$

27. $\begin{cases} x^2 - y^2 - 2y = 22 \\ 2x + 5y + 5 = 0 \end{cases}$

28. $\begin{cases} x^2 + y^2 - 6y = 0 \\ 2x^2 - y^2 + 15 = 0 \end{cases}$

29. $\begin{cases} y = x^2 - 2x + 3 \\ y = -x^2 + 2x + 3 \end{cases}$

30. $\begin{cases} y^2 = x^2 - 5 \\ 4x^2 - y^2 = 32 \end{cases}$

8. $(3, 1), (-3, -1)$

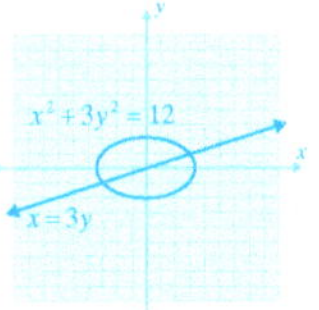

9. $(-3, 0), (3, 0)$

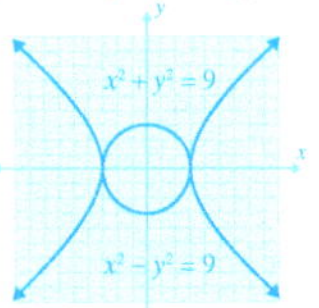

Use a graphing calculator to graph and estimate the solution(s) to each system of equations. If necessary, round values to two decimal places.

31. $\begin{cases} y = x^2 + 3 \\ x + y = 4 \end{cases}$

32. $\begin{cases} y = 1 - x^2 \\ x + y = -4 \end{cases}$

33. $\begin{cases} y = 3 - 2x \\ y = (x-1)^2 \end{cases}$

34. $\begin{cases} x^2 + y^2 = 36 \\ y = x + 5 \end{cases}$

35. $\begin{cases} x^2 + y^2 = 10 \\ x - y = 1 \end{cases}$

36. $\begin{cases} x^2 + y^2 = 4 \\ x^2 - y^2 = 3 \end{cases}$

HAWKES LEARNING SYSTEMS: INTRODUCTORY & INTERMEDIATE ALGEBRA SOFTWARE

- 12.5 Nonlinear Systems of Equations

10. $(0, 3)$

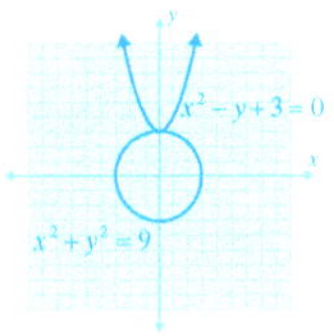

11. $(2, 3), (2, -3)$

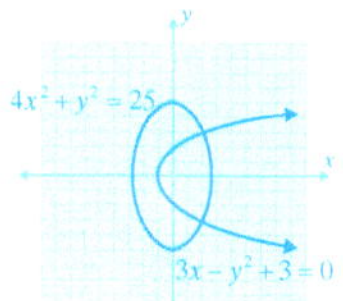

12. $(-5, -2), (-5, 2), (3, 0)$

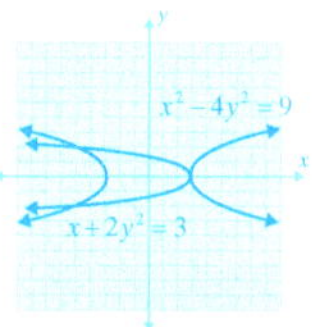

13. $(-2, 4), (1, 1)$

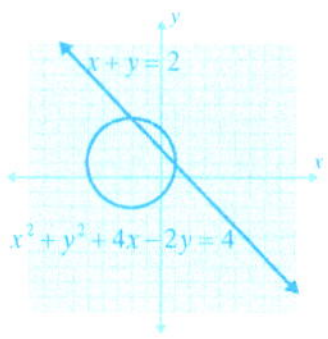

14. $(3, 0)$

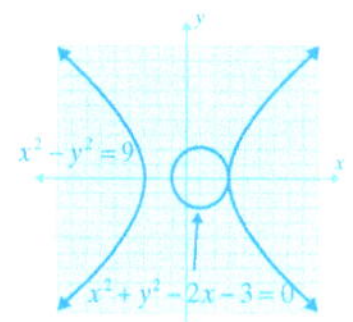

15. $(-3, -2), (3, 2), (-3, 2), (3, -2)$

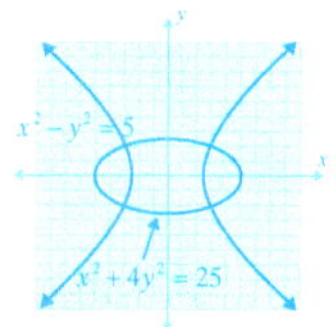

16. $(-3, -2), (3, 2), (-3, 2), (3, -2)$

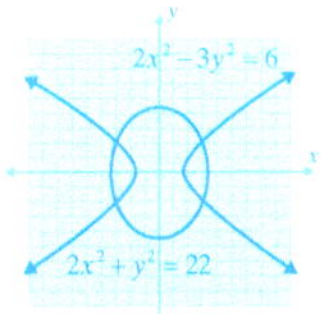

17. $(-3\sqrt{5}, 5), (-6, 4), (3\sqrt{5}, 5), (6, 4)$ **18.** $(1, -\sqrt{3}), (1, \sqrt{3}), (4, 0)$ **19.** $(1, 3), (-1, 3)$

20. $(2, 4), (2, -4)$ **21.** $(4, 4), (-2, -2)$ **22.** $(-4, 10), (0, 2)$ **23.** $\left(-\frac{3}{2}, -1\right), \left(\frac{1}{2}, \frac{1}{2}\right)$

24. $(1, 0), (-1, -4)$ **25.** $(-3, \sqrt{11}), (-3, -\sqrt{11}), (3, \sqrt{11}), (3, -\sqrt{11})$

26. $\left(\frac{\sqrt{7}}{2}, -2\right), \left(-\frac{\sqrt{7}}{2}, -2\right), \left(\frac{\sqrt{10}}{2}, 1\right), \left(-\frac{\sqrt{10}}{2}, 1\right)$ **27.** $(5, -3), (-5, 1)$

28. $(\sqrt{5}, 5), (-\sqrt{5}, 5)$ **29.** $(0, 3), (2, 3)$ **30.** $(3, 2), (3, -2), (-3, 2), (-3, -2)$

31. $(-1.62, 5.62), (0.62, 3.38)$

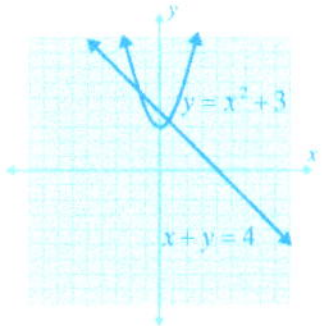

32. $(2.79, -6.79), (-1.79, -2.21)$

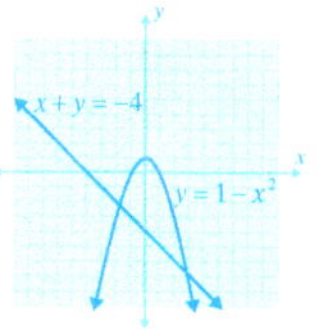

33. $(1.41, 0.17), (-1.41, 5.83)$

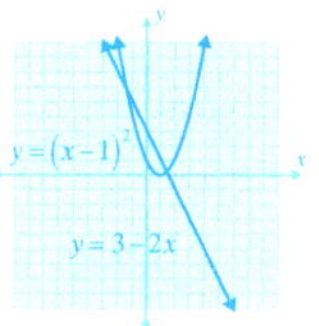

34. $(0.93, 5.93), (-5.93, -0.93)$

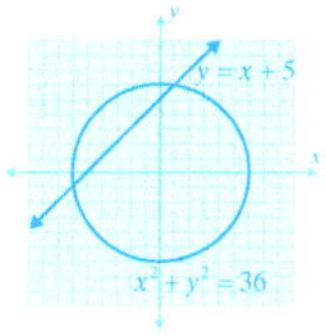

35. $(2.68, 1.68), (-1.68, -2.68)$

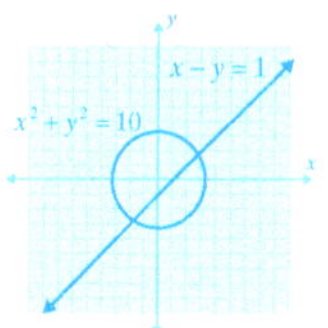

36. $(-1.87, 0.71), (1.87, 0.71), (-1.87, -0.71), (1.87, -0.71)$

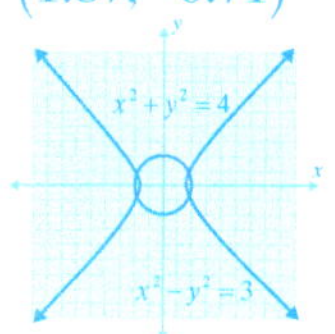

Chapter 12 Index of Key Ideas and Terms

Section 12.1 Translations and Reflections

Difference Quotient page 963

The formula $\frac{f(x+h)-f(x)}{h}$ is called the **difference quotient**.
A geometric interpretation of the difference quotient is the **slope** of a line through two points on the graph of a function.

Horizontal and Vertical Translations page 965

Given the graph of $y=f(x)$, the graph of $y=f(x-h)+k$ is

1. a **horizontal translation** of h units, and
2. a **vertical translation** of k units

of the graph of $y=f(x)$.

Reflections pages 968-969

The graph of $y=-f(x)$ is a **reflection across the *x*-axis** of the graph of $y=f(x)$. This means the graph of $y=-f(x)$ is the mirror image of $y=f(x)$ across the x-axis.

Section 12.2 Parabolas as Conic Sections

Conic Sections page 976

Conic sections are curves in a plane that are found when the plane intersects a cone. Four such sections are the circle, ellipse, parabola, and hyperbola.

Vertical Parabolas page 977

Equations of **vertical parabolas** (parabolas that open up or down) can be written in the form

$y=ax^2+bx+c$ or $y=a(x-h)^2+k$ where $a\neq 0$.

The parabola opens down if $a<0$ and up if $a>0$ and the vertex is at (h,k). The line $x=h$ is the line of symmetry.

Horizontal Parabolas page 977

Equations of **horizontal parabolas** (parabolas that open to the left or right) can be written in the form

$x=ay^2+by+c$ or $x=a(y-k)^2+h$ where $a\neq 0$.

The parabola opens left if $a<0$ and right if $a>0$ and the vertex is at (h,k). The line $y=k$ is the line of symmetry.

Using a Graphing Calculator to Graph Horizontal Parabolas pages 980-981

Section 12.3 Distance Formula and Circles

The Distance Formula page 986

For two points $P(x_1, y_1)$ and $Q(x_2, y_2)$ in a plane, the **distance between the points** is $d = \sqrt{(x_2 - x_1)^2 + (y_2 - y_1)^2}$.

Circles page 988

A **circle** is the set of all points in a plane that are a fixed distance from a fixed point. The fixed point is called the **center** of the circle.

Radius page 988

The distance from the center to any point on a circle is called the **radius** of the circle.

Diameter page 988

The distance from one point on a circle to another point on the circle measured through the center is called the **diameter** of the circle. The diameter is twice the length of the radius.

Equation of a Circle page 989

The equation of a **circle with radius *r* and center at (h, k)** is $(x-h)^2 + (y-k)^2 = r^2$. If the center is at the origin, $(0, 0)$, the equation simplifies to $x^2 + y^2 = r^2$.

Using a Graphing Calculator to Graph Circles page 991

Section 12.4 Ellipses and Hyperbolas

Ellipse page 996

An **ellipse** is the set of all points in a plane for which the sum of the distances from two fixed points is constant. Each of the fixed points is called a **focus** (plural foci). The **center** of an ellipse is the point midway between the foci.

Continued on the next page...

Section 12.4 Ellipses and Hyperbolas (cont.)

Equation of an Ellipse page 998

The standard form for the equation of an **ellipse with its center at the origin** is $\frac{x^2}{a^2}+\frac{y^2}{b^2}=1$.

The points $(a, 0)$ and $(-a, 0)$ are the ***x*-intercepts**.

The points $(0, b)$ and $(0, -b)$ are the ***y*-intercepts**.

When $a^2 > b^2$:

The segment of length $2a$ joining the x-intercepts is called the **major axis**. The segment of length $2b$ joining the y-intercepts is called the **minor axis**.

When $b^2 > a^2$:

The segment of length $2b$ joining the y-intercepts is called the **major axis**. The segment of length $2a$ joining the x-intercepts is called the **minor axis**.

Note: In all cases, the foci lie on the major axis.

Hyperbola pages 999-1000

A **hyperbola** is the set of all points in a plane such that the absolute value of the difference of the distances from two fixed points is constant. Each of the fixed points is called a **focus** (plural foci). The **center** of a hyperbola is the point midway between the foci.

Equations of Hyperbolas page 1001

In general, there are two standard forms for equations of **hyperbolas with their centers at the origin**.

1. $\frac{x^2}{a^2}-\frac{y^2}{b^2}=1$ x-intercepts (vertices) at $(a, 0)$ and $(-a, 0)$; No y-intercepts; Asymptotes: $y=\frac{b}{a}x$ and $y=-\frac{b}{a}x$; The curves "open" left and right.

2. $\frac{y^2}{a^2}-\frac{x^2}{b^2}=1$ y-intercepts (vertices) at $(0, a)$ and $(0, -a)$; No x-intercepts; Asymptotes: $y=\frac{a}{b}x$ and $y=-\frac{a}{b}x$; The curves "open" up and down.

Sketching Asymptotes of Hyberbolas using a Rectangle pages 1001-1002

Continued on the next page....

Section 12.4 Ellipses and Hyperbolas (cont.)

Ellipse with Center at (h, k) page 1003

The equation of an **ellipse with its center at (h, k)** is

$$\frac{(x-h)^2}{a^2}+\frac{(y-k)^2}{b^2}=1.$$

Note: a and b are distances from (h, k) to the vertices.

Hyperbola with Center at (h, k) page 1004

The equation of a **hyperbola with its center at (h, k)** is

$$\frac{(x-h)^2}{a^2}-\frac{(y-k)^2}{b^2}=1 \quad \text{or} \quad \frac{(y-k)^2}{a^2}-\frac{(x-h)^2}{b^2}=1.$$

Note: a and b are used as in the standard form but are measured from (h, k).

Section 12.5 Nonlinear Systems of Equations

Solving a System of One Quadratic Equation and One Linear Equation page 1010

If a system of two equations has **one quadratic** equation and **one linear** equation, then:

1. Solve the linear equation for one of the variables and substitute for this variable in the quadratic equation.
2. Solve the resulting second-degree equation and analyze the results.
3. Graph the curves on the same set of axes to visualize the number of solutions and check that the solutions are reasonable and satisfy both equations.

Solving a System of Two Quadratic Equations page 1012

If a system of two equations has **two quadratic** equations, then:

1. The method used depends on the form of the equations.
2. Substitution may work or addition may work.
3. Graph the curves on the same set of axes to visualize the number of solutions and check that the solutions are reasonable and satisfy both equations.

HAWKES LEARNING SYSTEMS: INTRODUCTORY & INTERMEDIATE ALGEBRA SOFTWARE

- 12.1 Translations and Reflections
- 12.2 Parabolas as Conic Sections
- 12.3 The Distance Formula and Circles
- 12.4 Ellipses and Hyperbolas
- 12.5 Nonlinear Systems of Equations

1. a. -27 **b.** $-5a-12$ **c.** -5 **2. a.** 66 **b.** $2a^2+12a+12$ **c.** $4x+2h$ **3. a.** 19 **b.** a^2+3a+1 **c.** $2x+h-3$

4. 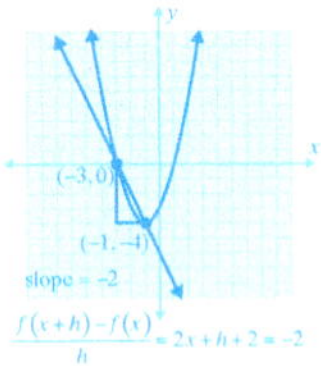

5. 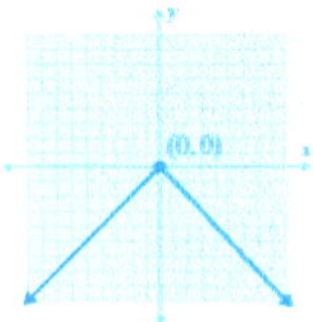

6. 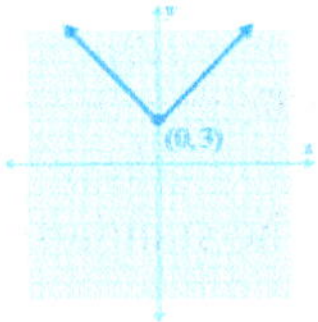

7. 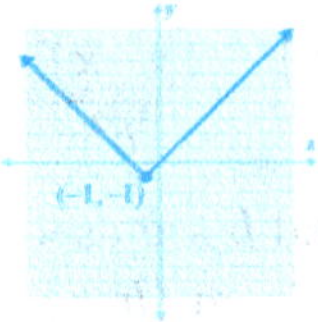

8. 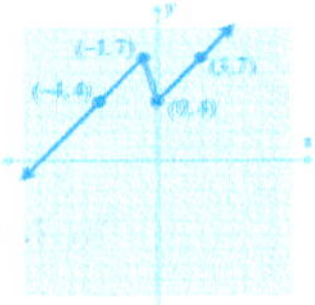

9. 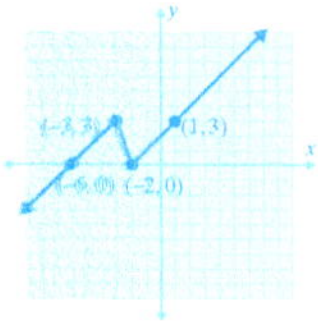

10. 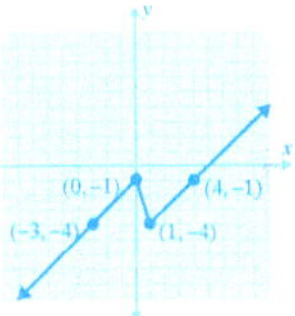

11. 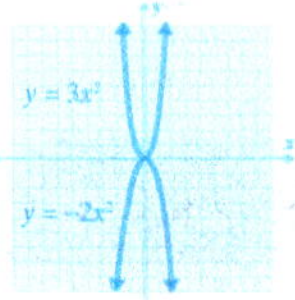

12. 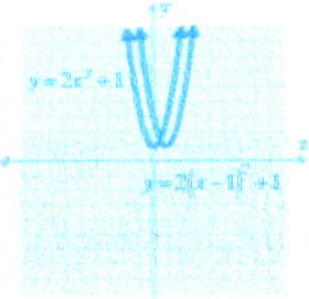

13. a. $(0,0)$ **b.** $(0,0)$ **c.** $y=0$

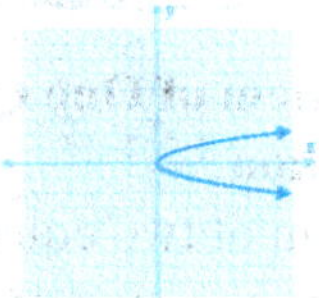

14. a. $(0,0)$ **b.** $(0,0)$ **c.** $y=0$

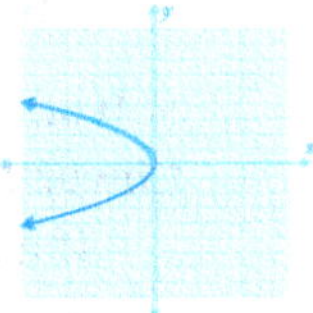

15. a. $(2,1)$ **b.** None **c.** $y=1$

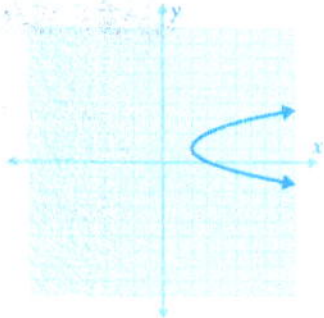

16. a. $(-2,2)$ **b.** $\left(0,2-\sqrt{2}\right)$, $\left(0,2+\sqrt{2}\right)$ **c.** $y=2$

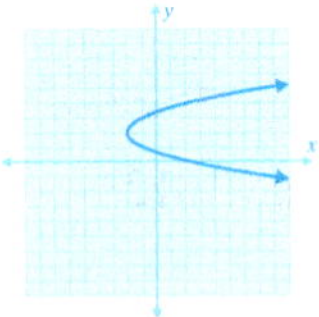

17. a. $(3,-2)$ **b.** None **c.** $y=-2$

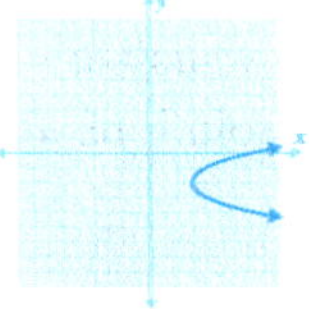

18. a. $(-7,3)$ **b.** $\left(0,3-\sqrt{7}\right)$, $\left(0,3+\sqrt{7}\right)$ **c.** $y=3$

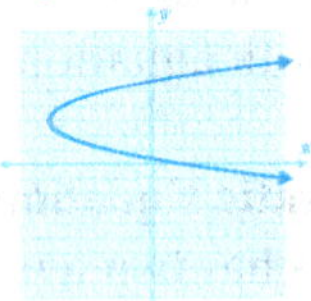

19. a. $\left(\frac{5}{4},\frac{1}{8}\right)$ **b.** $(0,-3)$ **c.** $x=\frac{5}{4}$

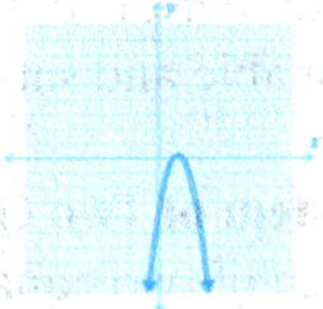

20. a. $(-8,-3)$ **b.** $\left(0,2\sqrt{2}-3\right)$, $\left(0,-2\sqrt{2}-3\right)$ **c.** $y=-3$

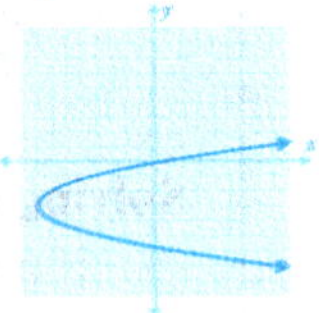

21. $(0,1.7321)$ $(0,-1.7321)$

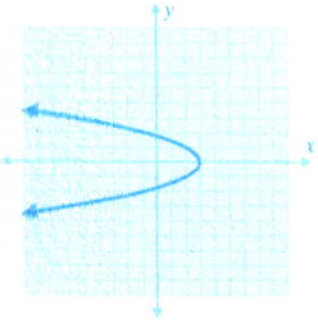

22. $(0,0),(0,1.5)$

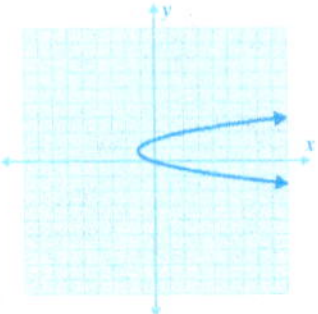

23. 5 **24.** 8 **25.** $\sqrt{97}$ **26.** $3\sqrt{10}$ **27.** $x^2+y^2=8$ **28.** $x^2+y^2=27$ **29.** $(x+2)^2+(y-1)^2=49$ **30.** $(x-5)^2+(y-4)^2=20$ **31.** $(x-4)^2+(y+2)^2=3$ **32.** $x^2+(y-7)^2=16$

33. $x^2+y^2=36$ Center: $(0,0)$; $r=6$

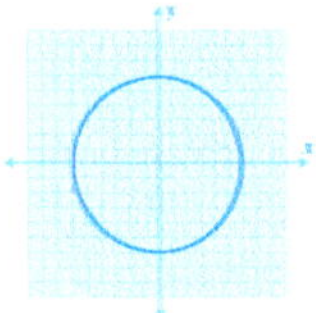

34. $x^2+(y-3)^2=9$ Center: $(0,3)$; $r=3$

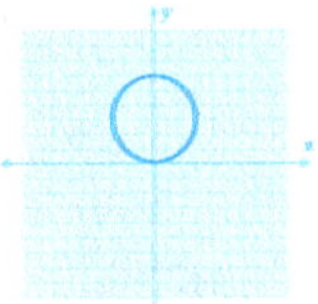

35. $(x+2)^2+(y+1)^2=9$ Center: $(-2,-1)$; $r=3$

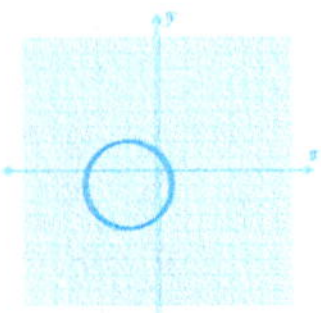

36. $x^2+(y-5)^2=20$ Center: $(0,5)$; $r=2\sqrt{5}$

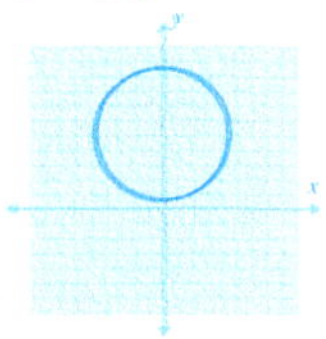

Chapter 12 Review

37. $\overline{AB} = 3\sqrt{2}$, $\overline{AC} = 5\sqrt{2}$, $\overline{BC} = 4\sqrt{2}$, $\left(3\sqrt{2}\right)^2 + \left(4\sqrt{2}\right)^2 = \left(5\sqrt{2}\right)^2$; Right triangle

38. $2\sqrt{29} + 2\sqrt{5} + 6\sqrt{2}$; $\overline{PQ} = 2\sqrt{29}$, $\overline{QR} = 2\sqrt{5}$, $\overline{PR} = 6\sqrt{2}$, $\left(2\sqrt{5}\right)^2 + \left(6\sqrt{2}\right)^2 \neq \left(2\sqrt{29}\right)^2$; Not a right triangle

39. $\dfrac{x^2}{36} + \dfrac{y^2}{9} = 1$

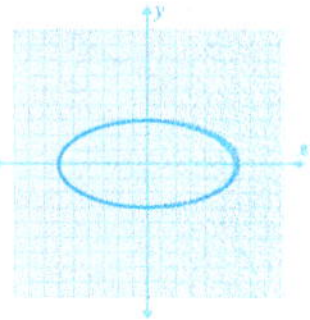

40. $\dfrac{x^2}{4} + \dfrac{y^2}{64} = 1$

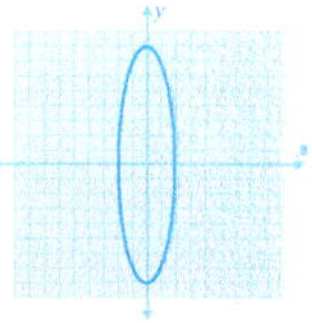

41. $\dfrac{x^2}{81} - \dfrac{y^2}{81} = 1$

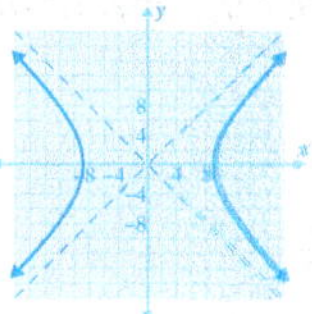

42. $\dfrac{x^2}{9} - \dfrac{y^2}{1} = 1$

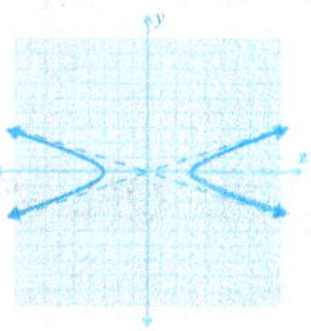

43. $\dfrac{y^2}{9} - \dfrac{x^2}{3} = 1$

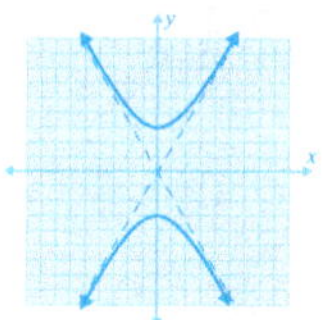

44. $\dfrac{x^2}{10} + \dfrac{y^2}{12} = 1$

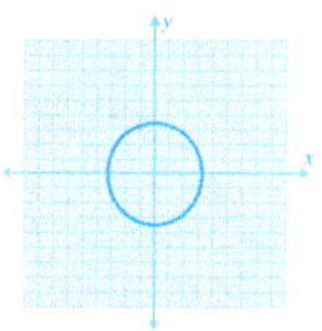

12.1 Translations and Reflections

For each function, find **a.** $f(6)$, **b.** $f(a+3)$ *and* **c.** *the simplified difference quotient* $\dfrac{f(x+h)-f(x)}{h}$.

1. $f(x) = 3 - 5x$

2. $f(x) = 2x^2 - 6$

3. $f(x) = x^2 - 3x + 1$

4. Sketch the curve $y = x^2 + 2x - 3$. Mark two points on the curve: $(-3, 0)$ and $(-1, -4)$. Show that the slope of the line through these two points (the secant line) is equal to the difference quotient where $x = -3$ and $h = 2$.

The graph of $y = |x|$ *is given along with a few points as aids. Graph the functions using your understanding of reflections and translations with no additional computations.*

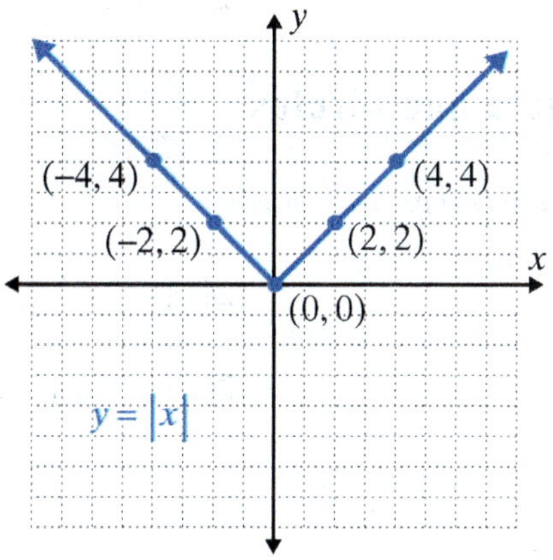

5. $y = -f(x)$

6. $y = f(x) + 3$

7. $y = f(x+1) - 1$

The graph of a function $y = f(x)$ *is given with the coordinates of four points. Graph the functions using your understanding of reflections and translations and with no additional computations. Label the new points that correspond to the four labeled points.*

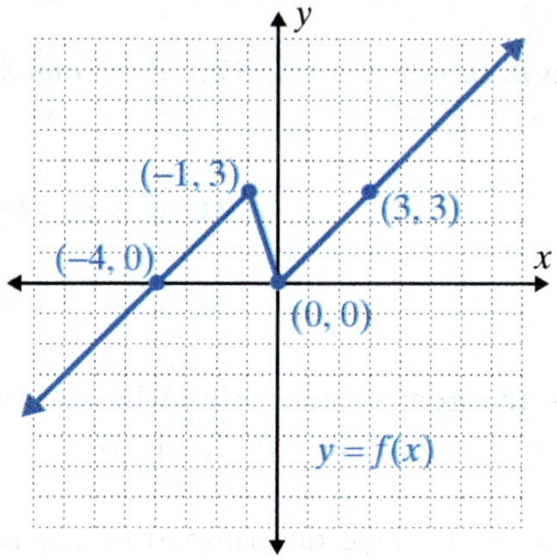

8. $y = f(x) + 4$

9. $y = f(x+2)$

10. $y = f(x-1) - 4$

Use a graphing calculator to graph each pair of functions on the same set of axes.

11. $y = -2x^2$ and $y = 3x^2$

12. $y = 2(x-1)^2 + 1$ and $y = 2x^2 + 1$

12.2 Parabolas as Conic Sections

Find **a.** *the vertex,* **b.** *the y-intercepts, and* **c.** *the line of symmetry. Then draw the graph.*

13. $x = 2y^2$

14. $x = -\frac{1}{2}y^2$

15. $x = (y-1)^2 + 2$

16. $x + 2 = (y-2)^2$

17. $x = (y+2)^2 + 3$

18. $x = y^2 - 6y + 2$

19. $y = -2x^2 + 5x - 3$

20. $x = y^2 + 6y + 1$

Use a graphing calculator to graph each of the parabolas. Use the trace and zoom features of the calculator to estimate the y-intercepts of the parabola.

21. $x = -y^2 + 3$

22. $x = 2y^2 - 3y$

12.3 Distance Formula and Circles

Find the distance between the two given points.

23. $(1, 3), (5, 6)$

24. $(-4, -2), (-4, 6)$

25. $(-3, -4), (1, 5)$

26. $(1, 7), (4, -2)$

Find equations for each of the circles.

27. Center $(0, 0)$; $r = 2\sqrt{2}$

28. Center $(0, 0)$; $r = 3\sqrt{3}$

29. Center $(-2, 1)$; $r = 7$

30. Center $(5, 4)$; $r = 2\sqrt{5}$

31. Center $(4, -2)$; $r = \sqrt{3}$

32. Center $(0, 7)$; $r = 4$

Write each of the equations in standard form. Find the center and radius of the circle and then sketch the graph.

33. $x^2 + y^2 = 36$

34. $x^2 + y^2 - 6y = 0$

35. $x^2 + y^2 + 4x + 2y = 4$

36. $x^2 + y^2 - 10y + 5 = 0$

37. Use the Pythagorean theorem to determine whether or not the triangle with the three given points is a right triangle: $A(-2, -4), B(1, -1), C(-3, 3)$.

38. Find the perimeter of the triangle determined by the points $P(-3, 5)$, $Q(1, -5)$, and $R(3, -1)$. Is this triangle a right triangle? Why or why not?

45. $\frac{x^2}{14} - \frac{y^2}{4} = 1$

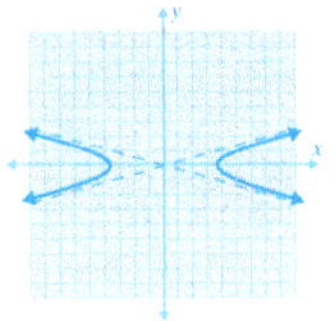

46. $\frac{y^2}{9} - \frac{x^2}{36} = 1$

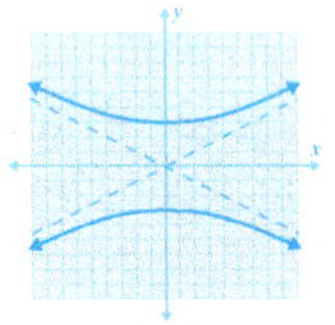

47.

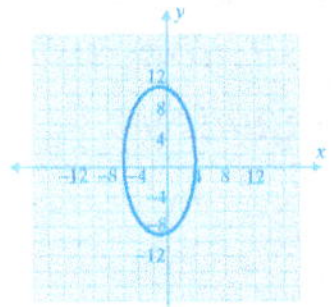

48.

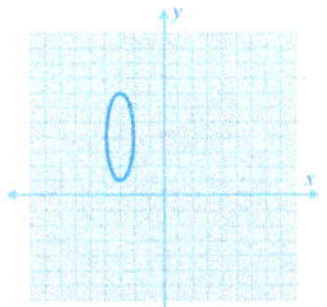

49. $(3, 8), (-2, 3)$

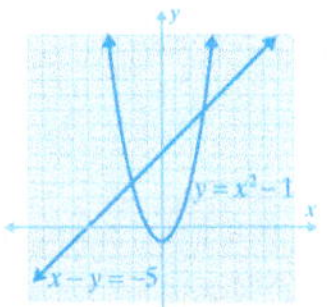

50. $(2, -4), \left(-\frac{5}{2}, -\frac{25}{4}\right)$

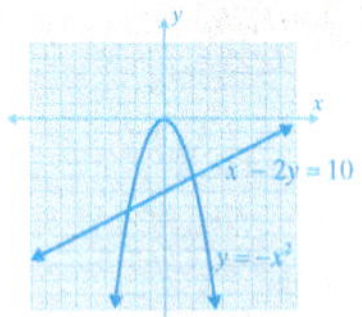

51. $(-2, 0), (0, -2)$

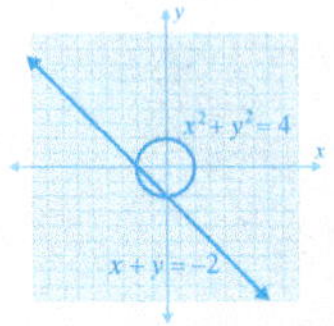

52. $(4, 1), (-4, -1)$

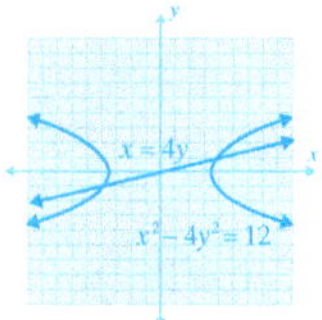

53. $(-4, 0), (4, 0)$

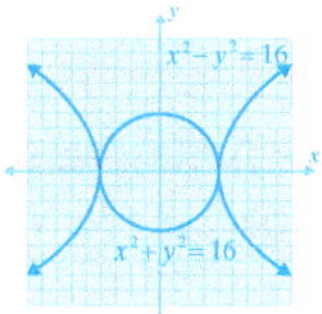

54. $(2, 3), (-2, 3), (-2, -3), (2, -3)$

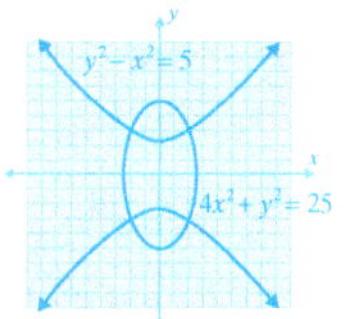

55. $\left(\frac{3+3\sqrt{5}}{2}, \frac{3+3\sqrt{5}}{2}\right), \left(\frac{3-3\sqrt{5}}{2}, \frac{3-3\sqrt{5}}{2}\right)$

56. $(0, 5), (4, 5)$

57. No solution

58. $(1, 3), (1, -3), (-1, 3), (-1, -3)$

59. No solution

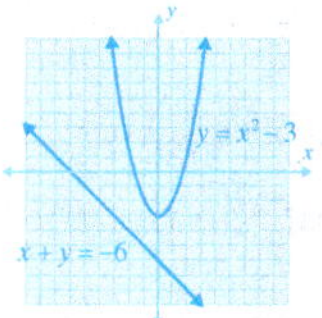

60. $(-1.11, 2.79), (-2.89, -0.79)$

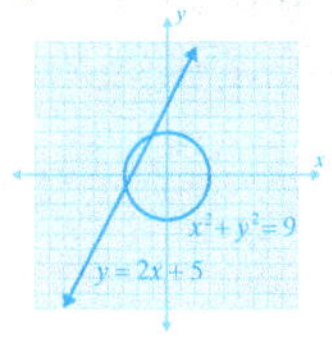

12.4 Ellipses and Hyperbolas

Write each of the equations in standard form. Then sketch the graph. For hyperbolas, graph the asymptotes as well.

39. $x^2 + 4y^2 = 36$ **40.** $16x^2 + y^2 = 64$ **41.** $x^2 - y^2 = 81$ **42.** $x^2 - 9y^2 = 9$

43. $y^2 - 3x^2 = 9$ **44.** $6x^2 + 5y^2 = 60$ **45.** $2x^2 - 7y^2 = 28$ **46.** $4y^2 - x^2 = 36$

Use your knowledge of translations to graph each of the following equations. These graphs are ellipses and hyperbolas with centers at points other than the origin.

47. $\frac{(x+1)^2}{25} + \frac{(y-1)^2}{100} = 1$ **48.** $\frac{(x+3)^2}{1} + \frac{(y-4)^2}{9} = 1$

12.5 Nonlinear Systems of Equations

Solve each system of equations. Sketch the graphs.

49. $\begin{cases} y = x^2 - 1 \\ x - y = -5 \end{cases}$ **50.** $\begin{cases} y = -x^2 \\ x - 2y = 10 \end{cases}$ **51.** $\begin{cases} x^2 + y^2 = 4 \\ x + y = -2 \end{cases}$

52. $\begin{cases} x^2 - 4y^2 = 12 \\ x = 4y \end{cases}$ **53.** $\begin{cases} x^2 + y^2 = 16 \\ x^2 - y^2 = 16 \end{cases}$ **54.** $\begin{cases} y^2 - x^2 = 5 \\ 4x^2 + y^2 = 25 \end{cases}$

Solve each system of equations.

55. $\begin{cases} x^2 + y^2 - 6y = 18 \\ x - y = 0 \end{cases}$ **56.** $\begin{cases} y = x^2 - 4x + 5 \\ y = -x^2 + 4x + 5 \end{cases}$

57. $\begin{cases} x^2 - y^2 = 4 \\ x^2 + y^2 = -2y \end{cases}$ **58.** $\begin{cases} x^2 + y^2 = 10 \\ y^2 - x^2 = 8 \end{cases}$

Use a graphing calculator to graph and estimate the solutions to each system of equations. If necessary, round values to two decimal places.

59. $\begin{cases} y = x^2 - 3 \\ x + y = -6 \end{cases}$ **60.** $\begin{cases} x^2 + y^2 = 9 \\ y = 2x + 5 \end{cases}$

Chapter 12 Test

1\. a. -9 b. $-9-7a$ c. -7 2. a. -6 b. $2a^2+4a-6$ c. $4x+2h+8$

3\.

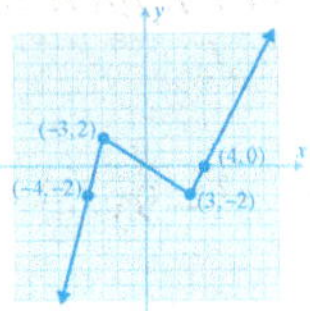

4\.

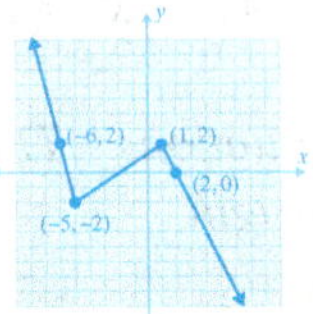

5\.

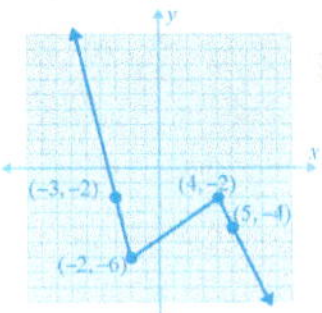

6\. a. $(-5, 0)$

b. $(0, \sqrt{5}), (0, -\sqrt{5})$

c. $y = 0$

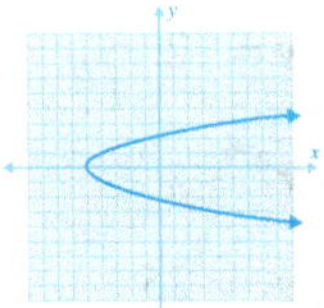

7\. a. $\left(-\frac{25}{4}, -\frac{3}{2}\right)$

b. $(0, -4), (0, 1)$

c. $y = -\frac{3}{2}$

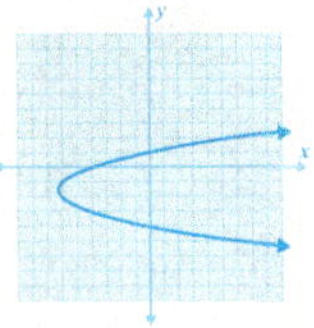

8\. a. $(-5, -5)$

b. $(0, -5+\sqrt{5})$, $(0, -5-\sqrt{5})$

c. $y = -5$

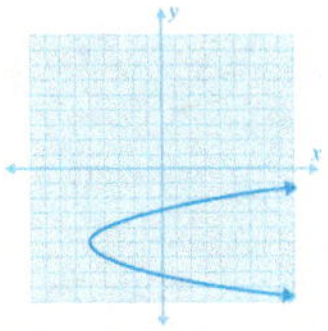

9\. a. $(0, -7)$

b. $(0, -7)$

c. $y = -7$

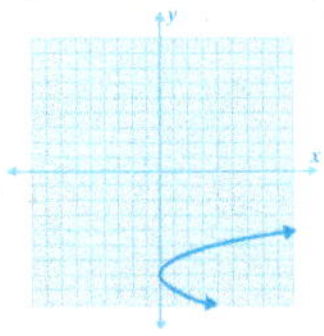

10\. $(0, 1.5447)$, $(0, -1.2947)$

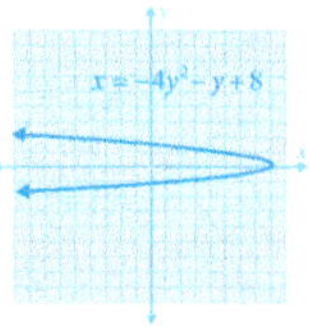

1. Let $f(x) = 5 - 7x$. Find:

a. $f(2)$

b. $f(a+2)$

c. $\dfrac{f(x+h)-f(x)}{h}$

2. Let $f(x) = 2x^2 + 8x$. Find:

a. $f(-3)$

b. $f(a-1)$

c. $\dfrac{f(x+h)-f(x)}{h}$

The graph of $y = g(x)$ is given along with a few points as aids. Graph the functions using your understanding of reflections and translations with no additional computations. Label the new points that correspond to the four labeled points.

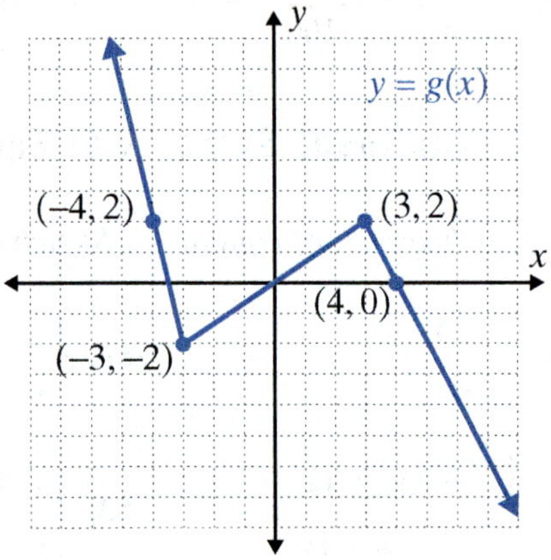

3. $y = -g(x)$

4. $y = g(x+2)$

5. $y = g(x-1) - 4$

*Find **a.** the vertex, **b.** the y-intercepts, and **c.** the line of symmetry. Graph each curve.*

6. $x = y^2 - 5$

7. $x + 4 = y^2 + 3y$

8. $x = y^2 + 10y + 20$

9. $x = (y+7)^2$

10. Use a graphing calculator to graph the parabola $x = -4y^2 + y + 8$. Use the trace and zoom features of the calculator to estimate the y-intercepts of the parabola.

11. Find the distance between the two points $(5, -2)$ and $(-4, 1)$.

12. The triangle determined by the three points $A(-4, -3)$, $B(0, 3)$, and $C(6, -1)$ is a right triangle. Use the distance formula and the Pythagorean theorem to show that this is true.

Find equations for each of the circles.

13. Center $(3, 4)$; $r = 5$

14. Center $(-2, -5)$, $r = 6$

Write each of the equations in standard form. Find the center and radius of the circle and then sketch the graph.

15. $x^2 + y^2 - 2y - 8 = 0$

16. $x^2 + y^2 + 2x - 6y + 10 = 16$

Write each of the equations in standard form. Then sketch the graph. For hyperbolas, graph the asymptotes as well.

17. $x^2 + 4y^2 = 9$

18. $25x^2 + 4y^2 = 100$

19. $9x^2 - 4y^2 = 36$

20. $\dfrac{y^2}{9} - \dfrac{x^2}{16} = 1$

21. Graph the ellipse $\dfrac{(x+1)^2}{36} + \dfrac{(y-2)^2}{9} = 1$. Label the vertices and the center.

22. Graph the given hyperbola and its asymptotes: $\dfrac{(x+1)^2}{16} - \dfrac{y^2}{9} = 1$.

23. State the definition of an ellipse.

Solve each system of equations. Sketch the graphs.

24. $\begin{cases} x^2 + y^2 = 29 \\ x - y = 3 \end{cases}$

25. $\begin{cases} x^2 + 2y^2 = 4 \\ x - y^2 = -2 \end{cases}$

26. $\begin{cases} x^2 + y^2 = 25 \\ x^2 - y^2 = 7 \end{cases}$

27. $\begin{cases} x^2 + y^2 = 25 \\ x + y = 1 \end{cases}$

11. $3\sqrt{10}$

12. $\overline{AB} = 2\sqrt{13}$, $\overline{AC} = 2\sqrt{26}$, $\overline{BC} = 2\sqrt{13}$, $\left(2\sqrt{13}\right)^2 + \left(2\sqrt{13}\right)^2 = \left(2\sqrt{26}\right)^2$

13. $(x-3)^2 + (y-4)^2 = 25$

14. $(x+2)^2 + (y+5)^2 = 36$

15. $x^2 + (y-1)^2 = 9$

Center: $(0,1)$; $r = 3$

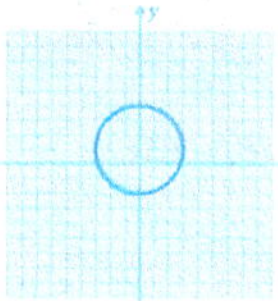

16. $(x+1)^2 + (y-3)^2 = 16$

Center: $(-1,3)$; $r = 4$

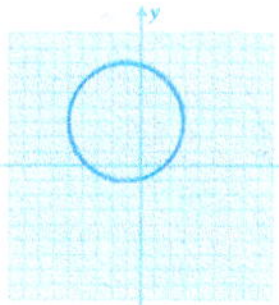

17. $\dfrac{x^2}{9} + \dfrac{y^2}{\left(\frac{9}{4}\right)} = 1$

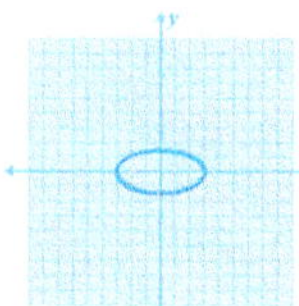

18. $\dfrac{x^2}{4} + \dfrac{y^2}{25} = 1$

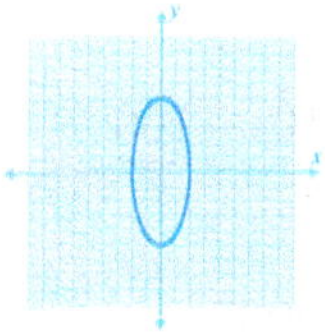

19. $\dfrac{x^2}{4} - \dfrac{y^2}{9} = 1$

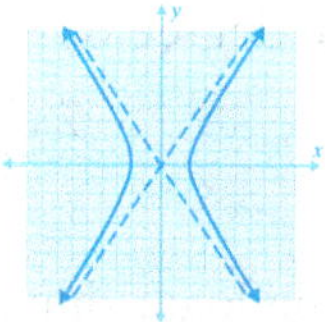

20. $\dfrac{y^2}{9} - \dfrac{x^2}{16} = 1$

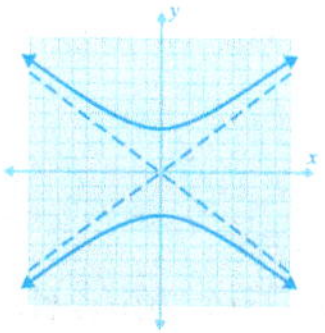

21.

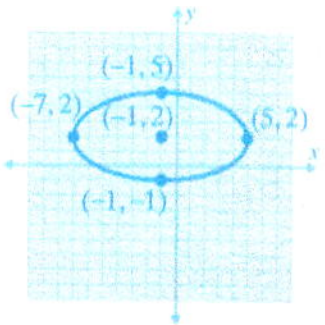

22.

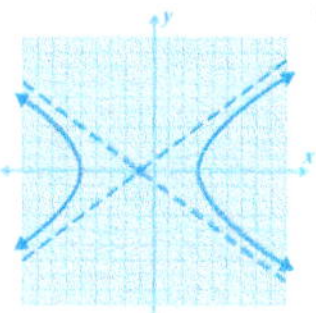

23. An ellipse is the set of all points in a plane for which the sum of the distances from two fixed points is constant.

24. $(-2,-5), (5,2)$

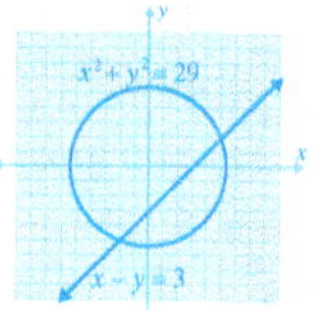

25. $(0,\sqrt{2}), (0,-\sqrt{2}), (-2,0)$

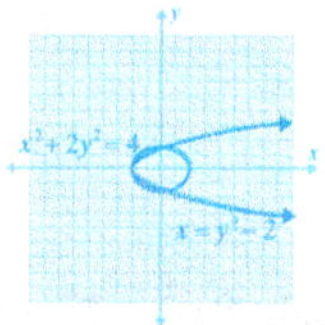

26. $(4,3), (4,-3), (-4,3), (-4,-3)$

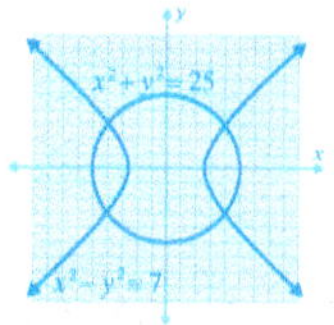

27. $(-3,4), (4,-3)$

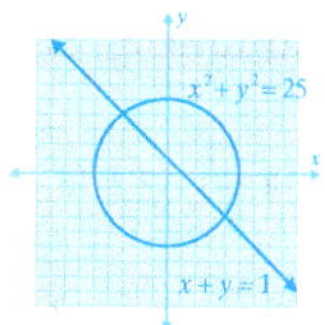

Cumulative Review: Chapters 1 – 12

1. 1
2. $\frac{9x^8}{4y^8}$
3. $5x^{\frac{3}{4}}$
4. $\frac{8y^{\frac{3}{5}}}{x}$
5. $\sqrt[3]{\left(7x^3y\right)^2} = x^2\sqrt[3]{49y^2}$
6. $\left(32x^6y\right)^{\frac{1}{3}} = 2x^2(4y)^{\frac{1}{3}}$
7. $2(x+3)\left(x^2-3x+9\right)$
8. $3y(x+4)(x-4)$
9. $(x-4)\left(x^2+3\right)$
10. $-7x^2+4x+10$
11. $a^2-16a-9$
12. $\frac{2x^2+x+4}{(2x+3)(x-4)(3x-2)}$
13. -1
14. $13\sqrt{3}$
15. $2\sqrt{3x}+10\sqrt{6}$
16. $-3i\sqrt{7}$
17. $\frac{11}{5}-\frac{8}{5}i$
18. $2x^2+2x+20$
19. **a.** 4 **b.** 6 **c.** 0
20. **a.** $\log_b \frac{x^3}{y^2}$ **b.** $\log\frac{10}{y}$
21. 0; One real solution
22. $x = \frac{-1\pm\sqrt{7}}{3}$
23. $x = 1, 9$
24. $x = -\frac{1}{3}, 0, \frac{5}{2}$
25. $x = \frac{3}{2}, -\frac{3}{4}+\frac{3\sqrt{3}}{4}i, -\frac{3}{4}-\frac{3\sqrt{3}}{4}i$
26. $x = \frac{16}{11}, \frac{24}{5}$
27. $x = -9$
28. $x = 2$
29. $x^2-6x+14$

Simplify each of the expressions. Assume all variables represent positive numbers.

1. $\frac{x^{-3}\cdot x}{x^2\cdot x^{-4}}$ **2.** $\left(\frac{2x^{-1}y^2}{3x^3y^{-2}}\right)^{-2}$ **3.** $5x^{\frac{1}{2}}\cdot x^{\frac{1}{4}}$ **4.** $\left(4x^{-\frac{2}{3}}y^{\frac{2}{5}}\right)^{\frac{3}{2}}$

5. Write $\left(7x^3y\right)^{\frac{2}{3}}$ in radical notation.

6. Write $\sqrt[3]{32x^6y}$ in exponential notation.

Completely factor each expression.

7. $2x^3+54$ **8.** $3x^2y-48y$ **9.** $x^3-4x^2+3x-12$

Perform the indicated operations and simplify. Assume that all variables represent positive numbers.

10. $2(8x+5)-\left[13x^2+3x(4-2x)\right]$ **11.** $(a-7)(2a+1)-(a+2)(a+1)$

12. $\frac{x}{2x^2-5x-12}-\frac{x+1}{6x^2+5x-6}$ **13.** $\frac{x^2+2x-3}{x^2+x-2}\div\frac{9-x^2}{x^2-x-6}$

14. $2\sqrt{12}+5\sqrt{108}-7\sqrt{27}$ **15.** $3\sqrt{48x}-2\sqrt{75x}+5\sqrt{24}$

16. $2\sqrt{-28}+\sqrt{-175}-3\sqrt{-112}$ **17.** $\frac{6-i}{2+i}$

18. Divide using the division algorithm. Write the answer in the form $Q+\frac{R}{D}$, where the degree of $R <$ the degree of D.

$$\frac{2x^3-8x^2+10x-100}{x-5}$$

19. Evaluate the following logarithmic expressions.

a. $\log_{10}10{,}000$ **b.** $\log_2 64$ **c.** $\log_6 1$

20. Write each expression as a single logarithm.

a. $3\log_b x-2\log_b y$ **b.** $\frac{1}{3}\log 8+\log 5-\log y$

21. Find the discriminant and determine the nature of the solution(s) of the quadratic equation $9x^2+6x+1=0$.

Solve each of the equations.

22. $3x^2+2x-2=0$ **23.** $(x-5)^2=16$ **24.** $6x^3=13x^2+5x$ **25.** $8x^3-27=0$

26. $|2x-5| = \left|\frac{3}{4}x+1\right|$

27. $\frac{5}{x-3} - \frac{3}{x+2} = \frac{1}{x^2-x-6}$

28. $\sqrt{x+14} - 2 = x$

29. Find an equation with integer coefficients that has $x = 3 + i\sqrt{5}$ and $x = 3 - i\sqrt{5}$ as roots.

30. Solve the following system of linear equations: $\begin{cases} 2x - 3y = 5 \\ -5x + y = 7 \end{cases}$

31. Solve the following system using Gaussian elimination: $\begin{cases} 2x - 3y + z = -4 \\ x + 2y - z = 5 \\ 3x + y + 2z = -5 \end{cases}$

Solve the equations. Use a calculator to find the answers in decimal form (accurate to 4 decimal places).

32. $6^{4-3x} = 6^{x^2}$

33. $5^x = 20$

34. $\log(x-1) + \log(x-5) = 1$

35. $3\ln x - 10 = 0$

Solve the inequalities algebraically. Write the answers using interval notation and then graph each solution set on a number line.

36. $\frac{x^2 - 4x - 5}{x} \geq 0$

37. $|2x-5| \leq 1$

38. $2x^2 + 13x - 7 \leq 0$

39. Solve the following system of linear inequalities graphically: $\begin{cases} 3x + y < 4 \\ x - y \geq 2 \end{cases}$

40. Find an equation in standard form for the line parallel to $5x - 2y = 8$ and passing through $(-2, 3)$.

41. Find an equation in standard form for the line perpendicular to $4x + 3y = 8$ and passing through $(4, -1)$.

42. Find an equation in slope-intercept form for the line passing through the points $\left(6, -\frac{1}{4}\right)$ and $(5, 0)$. Graph the line.

43. Find an equation for the circle with its center at $(-6, -1)$ and a radius of $r = 4$.

Graph each of the equations.

44. $5x + 2y = 8$

45. $y = \left(\frac{5}{4}\right)^{-x}$

46. $2x^2 + 4y^2 = 12$

47. $4x^2 - y^2 = 16$

48. $x^2 + 2x + y^2 - 2y = 4$

30. $(-2, -3)$
31. $\left(\frac{1}{2}, \frac{1}{2}, -\frac{7}{2}\right)$
32. $x = -4, 1$
33. 1.8614 **34.** 6.7417
35. 28.0316
36. $[-1, 0) \cup [5, \infty)$

37. $[2, 3]$

38. $\left[-7, \frac{1}{2}\right]$

39.

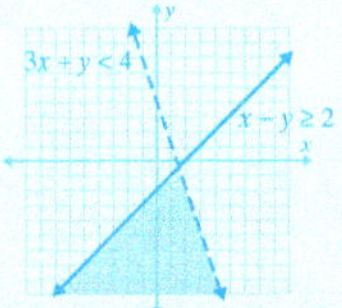

40. $5x - 2y = -16$
41. $3x - 4y = 16$
42. $y = -\frac{1}{4}x + \frac{5}{4}$

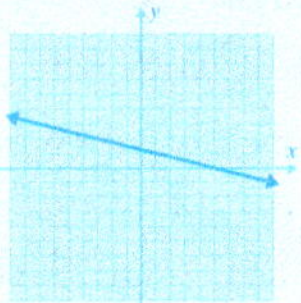

43. $(x+6)^2 + (y+1)^2 = 16$

44.

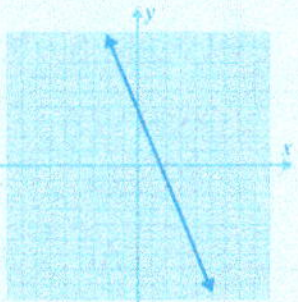

45.

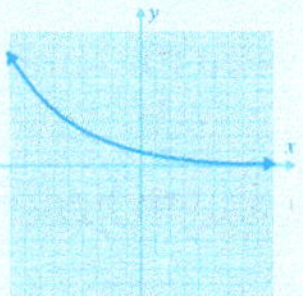

46.

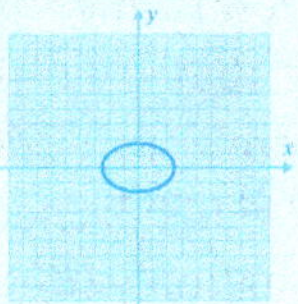

Write the quadratic function in the form $y = a(x-h)^2 + k$. Find the line of symmetry, the vertex, the domain, the range, and the zeros. Graph the function.

49. $y = \frac{1}{2}x^2 - 3$

50. $y = -x^2 + 4x - 4$

51. For the equation $x = -y^2 - 3$, find **a.** the vertex, **b.** the y-intercepts, and **c.** the line of symmetry. Then draw the graph.

52. Given that $g(x) = \sqrt{x}$, graph both $g(x)$ and $h(x) = \sqrt{x-7} + 2$. State the domain and range for each function.

53. Given that $f(x) = 3x - 9$, find $f^{-1}(x)$ and graph both functions.

54. $f(x) = x^2 + 3$ and $g(x) = \sqrt{x-3}$ for $x \geq 3$. **a.** Find $f(g(x))$. **b.** Find $g(f(x))$.
c. Are these two functions inverses of each other? Explain.

55. For $f(x) = 4x - 7$, find: **a.** $f(3)$ **b.** $f(x-1)$ **c.** $f(x) + 3$ **d.** $\frac{f(x+h) - f(x)}{h}$

56. Graph the two functions $y = e^x$ and $y = \ln x$. Label two points on each graph. State which lines are asymptotes for each function.

57. The product of two consecutive positive odd integers is 323. Find the integers.

58. z varies directly as the cube of x and inversely as y. If $z = 10$ when $x = 3$ and $y = 5$, find z if $x = 6$ and $y = 4$.

59. The average of a number and its square root is 21. Find the number.

60. Tallying votes: Bethany and Walter need to count the votes for the school election. If Bethany could count all of the votes in 5 hours working alone, and Walter could count all of the votes in 4 hours working alone, how long will it take to count the votes working together?

61. Investing: Alisha has \$9000 to invest in two different accounts. One account pays interest at the rate of 7%; the other pays at the rate of 8%. If she wants her annual interest to total \$684, how much should she invest at each rate?

62. Path of a ball: The height of a ball projected vertically is given by the function $h = -16t^2 + 80t + 48$ where h is the height in feet and t is the time in seconds.

a. When will the ball reach maximum height?

b. What will be the maximum height?

63. Growth of bacteria: Determine the exponential function that fits the following information: $y_0 = 3000$ bacteria, and there are 243,000 bacteria after 3 days. Use the equation $y = y_0 b^t$ where t is measured in days.

47.

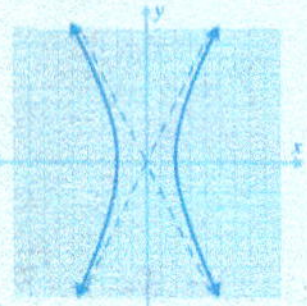

48.

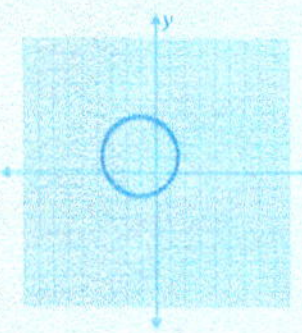

49. $y = \frac{1}{2}x^2 - 3$;
$x = 0$; $(0, -3)$;
$D = (-\infty, \infty)$;
$R = [-3, \infty)$;
Zeros: $x = \pm\sqrt{6}$

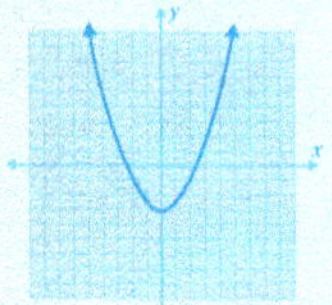

50. $y = -(x-2)^2$;
$x = 2$; $(2, 0)$;
$D = (-\infty, \infty)$;
$R = (-\infty, 0]$;
Zeros: $x = 2$

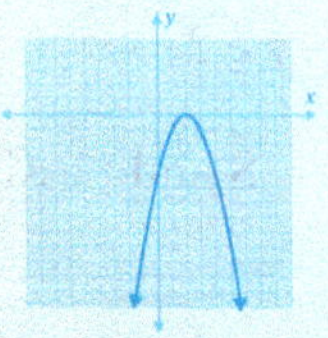

51. a. $(-3, 0)$ **b.** None
c. $y = 0$

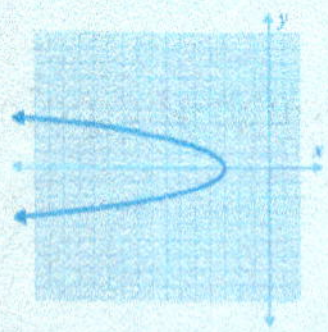

52. For $g(x)$:
$D = [0, \infty)$,
$R = [0, \infty)$
For $h(x)$:
$D = [7, \infty)$,
$R = [2, \infty)$

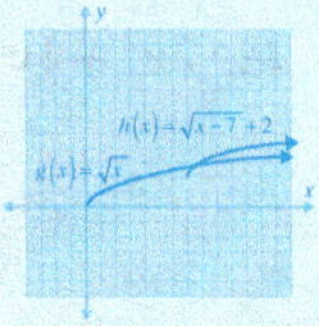

53. $f^{-1}(x) = \frac{x+9}{3}$

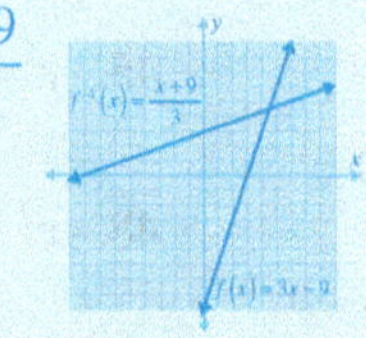

54. a. x b. x
c. Yes. Answers will vary.
55. a. 5 b. $4x-11$
c. $4x-4$ d. 4
56.
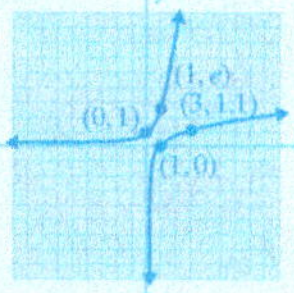

Asymptote for $y=\ln x$ is $x=0$. Asymptote for $y=e^x$ is $y=0$.
57. 17, 19 58. 100
59. 36
60. $\frac{20}{9}$ or $2\frac{2}{9}$ hours
61. \$3600 at 7%; \$5400 at 8%
62. a. $t=2\frac{1}{2}$ seconds
b. 148 ft
63. $y=3000\left(3\sqrt[3]{3}\right)^t$
64. \$2.81
65. $\overline{AB}=\sqrt{65}$, $\overline{BC}=\sqrt{137}$, $\overline{AC}=3\sqrt{2}$; $P=\sqrt{65}+\sqrt{137}+3\sqrt{2}\approx 24.01$
$\left(\sqrt{65}\right)^2+\left(3\sqrt{2}\right)^2\neq\left(\sqrt{137}\right)^2$; Not a right triangle

64. Sales revenue: The revenue function is given by $R(x)=x\cdot p(x)$ dollars, where x is the number of units sold and $p(x)$ is the unit price. If $p(x)=30(2)^{\frac{-x}{3}}$, find the revenue when 24 units are sold.

65. Triangles: Find the perimeter of the triangle determined by the points $A(1,3)$, $B(0,-5)$, and $C(4,6)$. Explain why this triangle is or is not a right triangle.

Use a graphing calculator and the CALC *and zero features to graph the functions and find the x-intercepts. Round zeros to two decimal places.*

66. $y=2x-5$

67. $y=x^2-3$

68. $y=-x^2+5$

69. $y=(x+3)^2-6$

Use a graphing calculator to solve (or estimate the solutions of) the equations and inequalities. Write the solutions to the inequalities in interval notation and find any approximations accurate to two decimal places.

70. $f(x)=x^2-3x-10$

a. $x^2-3x-10=0$

b. $x^2-3x-10>0$

c. $x^2-3x-10<0$

71. $f(x)=-2x^2+3x+5$

a. $-2x^2+3x+5=0$

b. $-2x^2+3x+5>0$

c. $-2x^2+3x+5<0$

Use a graphing calculator to graph both equations and estimate the solutions to the system. If necessary, round values to two decimal places.

72. $\begin{cases} x^2+y^2=35 \\ x+y=4 \end{cases}$

73. $\begin{cases} 3x^2+4y^2=12 \\ x=y^2-3 \end{cases}$

66.
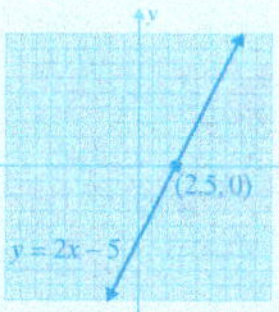

67.
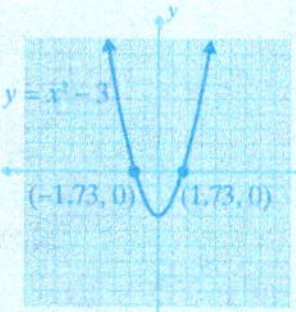

68.
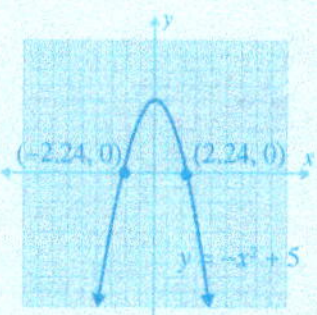

69.
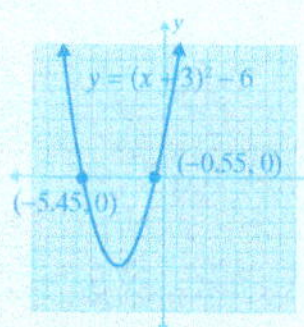

70. a. $x=-2, x=5$ b. $(-\infty,-2)\cup(5,\infty)$ c. $(-2,5)$ 71. a. $x=-1, x=2.5$ b. $(-1,2.5)$ c. $(-\infty,-1)\cup(2.5,\infty)$

72. $(-1.67, 5.67)$, $(5.67,-1.67)$
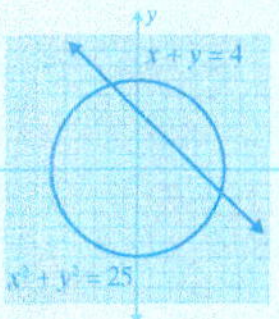

73. $(-1.33, 1.29)$, $(0, 1.73)$, $(-1.33,-1.29)$, $(0,-1.73)$
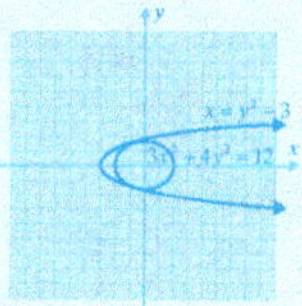

CHAPTER 13

Sequences, Series, and the Binomial Theorem

Did You Know?

One of the outstanding mathematicians of the Middle Ages was Leonardo Fibonacci (Leonardo, son of Bonaccio), also known as Leonardo of Pisa (c. 1170 – 1250). Fibonacci's name is attached to an interesting sequence of numbers 1, 1, 2, 3, 5, 8, 13, . . . , x, y, $x + y$, . . . , the so-called Fibonacci sequence, where each term after the first two is obtained by adding the preceding two terms together. The sequence of numbers arises from a problem found in Fibonacci's writings. How many pairs of rabbits can be produced from a single pair in a year if every month each pair begets a new pair that from the second month on becomes productive? The answer to this odd problem is the sum of the first 12 terms of the Fibonacci sequence. Can you verify this?

The Fibonacci sequence itself has been found to have many beautiful and interesting properties. For example, of mathematical interest is the fact that any two successive terms in the sequence are relatively prime; that is, their greatest common divisor is one. In the world of nature, the terms of the Fibonacci sequence also appear. Spirals formed by natural objects such as centers of daisies, pine cone scales, pineapple scales, and leaves generally have two sets of spirals, one clockwise, one counterclockwise. Each set is made up of a specific number of spirals in each direction, the number of spirals being adjacent terms in the Fibonacci sequence. For example, in pine cone scales, 5 spiral one way and 8 spiral the other; on pineapples, 8 one way and 13 the other.

Fibonacci

Leonardo Fibonacci was best known for the texts he wrote in which he introduced the Hindu-Arabic numeral system. He participated in the mathematical tournaments held at the court of the emperor Frederick I, and he used some of the challenge problems in the book he wrote. Fibonacci traveled widely and became acquainted with the different arithmetic systems in use around the Mediterranean. His most famous text, *Liber Abaci* (1202), combined arithmetic and elementary algebra with an emphasis on commercial applied problems. His attempt to reform and improve the study of mathematics in Europe was not too successful; he seemed to be ahead of his time. But his books did much to introduce Hindu-Arabic notation into Europe, and his sequence has provided interesting problems throughout the history of mathematics. In the United States, a Fibonacci Society exists to study the properties of this mysterious and intriguing sequence of numbers.

"There is no branch of mathematics, however abstract, which may not some day be applied to phenomena of the real world."

Nikolai Ivanovich Lobachevsky
(1792 – 1856)

Chapter 13 provides an introduction to a powerful notation using the Greek letter Σ, capital sigma. With this Σ notation and a few basic properties, we will develop some algebraic formulas related to sums of numbers and, in some cases, even infinite sums. The concept of having the sum of an infinite number of numbers equal to some finite number introduces the idea of limits. Consider adding fractions in the following manner:

$$\frac{1}{2}=\frac{1}{2}; \qquad \frac{1}{2}+\frac{1}{4}=\frac{3}{4}; \qquad \frac{1}{2}+\frac{1}{4}+\frac{1}{8}=\frac{7}{8};$$

$$\frac{1}{2}+\frac{1}{4}+\frac{1}{8}+\frac{1}{16}=\frac{15}{16}; \qquad \frac{1}{2}+\frac{1}{4}+\frac{1}{8}+\frac{1}{16}+\frac{1}{32}=\frac{31}{32}.$$

Continuing to add fractions in this manner, in which the denominators are successive powers of two, will give sums that get closer and closer to 1. We say that the sums "approach" 1, and that 1 is the **limit** of the sum. These fascinating ideas are discussed in Section 13.4 and are fundamental in the development of calculus and higher level mathematics.

The final topic in this chapter, the binomial theorem, has applications in courses in probability and statistics, as well as in more advanced courses in mathematics and computer science.

13.1 Sequences

- *Write several terms of a **sequence** given the formula for its general term.*
- *Find the formula for the general term of a sequence given several terms.*
- *Determine whether a sequence is **increasing**, **decreasing**, or **neither**.*

In mathematics, a **sequence** is a list of numbers that occur in a certain order. Each number in the sequence is called a **term** of the sequence, and a sequence may have a finite number of terms or an infinite number of terms. For example,

2, 4, 6, 8, 10, 12, 14, 16, 18	is a **finite sequence** consisting of positive even integers less than 20.
3, 6, 9, 12, 15, 18, . . .	is an **infinite sequence** consisting of the multiples of 3.

The infinite sequence of the multiples of 3 can be described in the following way.

For any positive integer n, the corresponding number in the list is $3n$. Thus we know that

$3 \cdot 6 = 18$ and 18 is the 6th number in the sequence,
$3 \cdot 7 = 21$ and 21 is the 7th number in the sequence,
$3 \cdot 8 = 24$ and 24 is the 8th number in the sequence, and so on.

Infinite Sequence

An **infinite sequence** (or a **sequence**) is a function that has the positive integers as its domain.

NOTES

A finite sequence will be so indicated. The word sequence, used alone, indicates an infinite sequence.

Consider the function $f(n) = \frac{1}{2^n}$ where n is any positive integer.

For this function,

$$f(1) = \frac{1}{2^1} = \frac{1}{2}$$

$$f(2) = \frac{1}{2^2} = \frac{1}{4}$$

$$f(3) = \frac{1}{2^3} = \frac{1}{8}$$

$$f(4) = \frac{1}{2^4} = \frac{1}{16}$$

$$\vdots$$

$$f(n) = \frac{1}{2^n}$$

$$\vdots$$

Or, using ordered pair notation,

$$f = \left\{\left(1, \frac{1}{2}\right), \left(2, \frac{1}{4}\right), \left(3, \frac{1}{8}\right), \left(4, \frac{1}{16}\right), \ldots, \left(n, \frac{1}{2^n}\right), \ldots\right\}.$$

The **terms** of the sequence are the numbers

$$\frac{1}{2}, \frac{1}{4}, \frac{1}{8}, \frac{1}{16}, \ldots, \frac{1}{2^n}, \ldots$$

Because the order of terms corresponds to the positive integers, it is customary to indicate a sequence by writing only its terms. In general discussions and formulas, a sequence may be indicated with subscript notation as

$$a_1, a_2, a_3, a_4, \ldots, a_n, \ldots$$

The general term a_n is called the ***n*^th^ term** of the sequence. The entire sequence can be denoted by writing the nth term in braces as in $\{a_n\}$. Thus

$$\{a_n\} \quad \text{and} \quad a_1, a_2, a_3, a_4, \ldots, a_n, \ldots$$

are both representations of the sequence with

a_1 as the first term,
a_2 as the second term,
a_3 as the third term,
⋮
a_n as the nth term,
⋮

Example 1: Writing Terms of a Sequence

a. Write the first three terms of the sequence $\left\{\dfrac{n}{n+1}\right\}$.

Solution: $a_1 = \dfrac{1}{1+1} = \dfrac{1}{2}$ $\quad a_2 = \dfrac{2}{2+1} = \dfrac{2}{3}$ $\quad a_3 = \dfrac{3}{3+1} = \dfrac{3}{4}$

b. If $\{b_n\} = \{2n-1\}$, find b_1, b_2, b_3, and b_{50}.

Solution: $b_1 = 2(1) - 1 = 1$ $\quad b_2 = 2(2) - 1 = 3$

$b_3 = 2(3) - 1 = 5$ $\quad b_{50} = 2(50) - 1 = 99$

Example 2: Finding the General Formula of a Sequence

a. Determine a_n if the first five terms of the sequence $\{a_n\}$ are $3, 5, 7, 9, 11$.

Solution: By studying the numbers carefully, we see that they are odd numbers. Odd numbers can be written in the form $2n+1$ or $2n-1$. Because the first term of the sequence is 3, $a_n = 2n+1$.

b. Determine a_n if the first five terms of the sequence $\{a_n\}$ are $0, 3, 8, 15, 24$.

Solution: In this case, study the numbers carefully and look for some pattern (or formula) that seems reasonable for one or two of the numbers. Then after making this educated guess, check to see whether the remaining numbers fit your guess.

Teaching Notes:
There are a variety of examples you can use to have students try to discover a numerical pattern, and thus translate the pattern into an algebraic formula based on "*n*." It is important that students build a confidence level in being able to express (discover) the pattern and then to write a formula that defines it mathematically.

If not, guess again with some basic reasoning for at least one of the positions. In this case you might think as follows:

8 is the 3rd term, $3^2 = 9$ and $9 - 1 = 8$;

15 is the 4th term, $4^2 = 16$ and $16 - 1 = 15$.

So, a good guess seems to be $a_n = n^2 - 1$.

Checking: $a_1 = 1^2 - 1 = 0$

$a_2 = 2^2 - 1 = 3$

$a_3 = 3^2 - 1 = 8$

$a_4 = 4^2 - 1 = 15$

$a_5 = 5^2 - 1 = 24$

Although the formula for a_n may not be obvious, with practice it becomes easier to find.

We see that $a_n = n^2 - 1$ is indeed the correct formula.

Example 3: Application

A pick-up truck sells for \$45,000 new. Each year its value depreciates by 15% of its value for the previous year. What will be its value at the end of each of the next three years?

Solution: Each year the value will be 85% of its value the previous year. (Because it depreciates by 15%, the new value will be 100% – 15% = 85% of its previous value.) The value at the end of each year can be found with the following sequence where $v_0 = \$45{,}000$, the initial value:

Year 1: $v_1 = v_0 \cdot (0.85) = 45{,}000 \cdot (0.85) = \$38{,}250$

Year 2: $v_2 = v_1 \cdot (0.85) = 38{,}250 \cdot (0.85) = \$32{,}512.50$

Year 3: $v_3 = v_2 \cdot (0.85) = 32{,}512.50 \cdot (0.85) \approx \$27{,}635.63$

Alternating Sequence

An **alternating sequence** is a sequence in which the terms alternate in sign.

In an **alternating sequence**, if one term is positive, then the next term is negative. Example 4 illustrates an alternating sequence. Note that alternating sequences generally involve the expression $(-1)^n$ or $(-1)^{n+1}$ as a factor in the numerator or denominator.

Example 4: An Alternating Sequence

Write the first five terms of the sequence in which $a_n = \frac{(-1)^n}{n}$.

Solution: $a_1 = \frac{(-1)^1}{1} = -1$ $\quad a_2 = \frac{(-1)^2}{2} = \frac{1}{2}$ $\quad a_3 = \frac{(-1)^3}{3} = -\frac{1}{3}$

$a_4 = \frac{(-1)^4}{4} = \frac{1}{4}$ $\quad a_5 = \frac{(-1)^5}{5} = -\frac{1}{5}$

Increasing and Decreasing Sequences

In subscript notation, the term a_{n+1} is the term following a_n. This term is found by substituting $n + 1$ for n in the formula for the general term. For example,

$$\text{if } a_n = \frac{1}{3n}, \text{ then } a_{n+1} = \frac{1}{3(n+1)} = \frac{1}{3n+3}.$$

Similarly,

$$\text{if } b_n = n^2, \text{ then } b_{n+1} = (n+1)^2 = n^2 + 2n + 1.$$

If the terms of a sequence grow successively smaller, then the sequence is said to be **decreasing**. If the terms grow successively larger, then the sequence is said to be **increasing**. The following definitions state these ideas algebraically. (Note that a sequence may be **neither** increasing nor decreasing.)

Decreasing Sequence

A sequence $\{a_n\}$ is

decreasing if $a_n > a_{n+1}$ for all n.

(Successive terms become smaller.)

Increasing Sequence

A sequence $\{a_n\}$ is

increasing if $a_n < a_{n+1}$ for all n.

(Successive terms become larger.)

Example 5: A Decreasing Sequence

Teaching Notes: You might have your students substitute several values for n so they can see the numerical pattern. Emphasize, though, that a few examples does not "prove" the pattern.

Show that the sequence $\{a_n\} = \left\{\frac{1}{2^n}\right\}$ is decreasing.

Solution: Write the terms a_n and a_{n+1} in formula form and compare them algebraically:

$$a_n = \frac{1}{2^n} \text{ and } a_{n+1} = \frac{1}{2^{n+1}} = \frac{1}{2 \cdot 2^n}.$$

Comparing denominators we see that $2^n < 2^{n+1}$.

Therefore,

$$\frac{1}{2^n} > \frac{1}{2^{n+1}} \text{ and } a_n > a_{n+1}.$$

Note that the larger denominator gives a smaller fraction.

The sequence $\{a_n\} = \left\{\frac{1}{2^n}\right\}$ is decreasing.

1. $1, 3, 5, 7, 9$
2. $5, 9, 13, 17, 21$
3. $2, \frac{3}{2}, \frac{4}{3}, \frac{5}{4}, \frac{6}{5}$
4. $2, \frac{5}{3}, \frac{3}{2}, \frac{7}{5}, \frac{4}{3}$

Example 6: An Increasing Sequence

Show that the sequence $\{b_n\} = \{n+3\}$ is increasing.

Solution: In formula form: $b_n = n+3$ and $b_{n+1} = (n+1)+3 = n+4$.

Because $n+3 < n+4$, we have $b_n < b_{n+1}$ and the sequence is increasing.

5. $2, 6, 12, 20, 30$
6. $0, -2, -6, -12, -20$
7. $2, 4, 8, 16, 32$
8. $1, 0, -1, 0, 7$
9. $\frac{1}{2}, \frac{1}{4}, \frac{1}{8}, \frac{1}{16}, \frac{1}{32}$
10. $\frac{1}{4}, -\frac{1}{8}, \frac{1}{16}, -\frac{1}{32}, \frac{1}{64}$

Example 7: A Sequence that is Neither Increasing Nor Decreasing

Show that the sequence $\{c_n\} = \{2 + (-1)^n\}$ is neither increasing nor decreasing.

Solution: The first four terms of the sequence are

$$c_1 = 2 + (-1)^1 = 2 - 1 = 1$$

$$c_2 = 2 + (-1)^2 = 2 + 1 = 3 \qquad \text{Here } c_1 < c_2.$$

$$c_3 = 2 + (-1)^3 = 2 - 1 = 1 \qquad \text{Now } c_2 > c_3.$$

$$c_4 = 2 + (-1)^4 = 2 + 1 = 3 \qquad \text{But } c_3 < c_4.$$

From this pattern we see that the sequence is 1, 3, 1, 3, … which is neither increasing nor decreasing.

11. $-2, 5, -10, 17, -26$
12. $3, -9, 27, -81, 243$
13. $-\frac{1}{2}, \frac{2}{3}, -\frac{3}{4}, \frac{4}{5}, -\frac{5}{6}$
14. $-\frac{1}{5}, \frac{1}{7}, -\frac{1}{9}, \frac{1}{11}, -\frac{1}{13}$
15. $1, \frac{1}{3}, \frac{1}{6}, \frac{1}{10}, \frac{1}{15}$
16. $1, \frac{4}{3}, \frac{3}{2}, \frac{8}{5}, \frac{5}{3}$
17. $0, 1, 3, 6, 10$
18. $0, 1, 0, 1, 0$
19. $\{3n-1\}$
20. $\{4n+1\}$

21. $\{n^2\}$ 22. $\{n^2+1\}$
23. $\left\{\frac{1}{2^n}\right\}$ 24. $\left\{\frac{1}{n+2}\right\}$
25. $\{6n\}$ 26. $\{5(2^{n-1})\}$
27. $\{(-1)^{n+1}(2n-1)\}$
28. $\{(-1)^n(4n-1)\}$

Practice Problems

Write the first five terms of each sequence.

1. $\{n^2\}$ **2.** $\{2n+1\}$ **3.** $\left\{\frac{1}{n+1}\right\}$

4. Find a formula for the general term of sequence $-1, 1, 3, 5, 7, \ldots$

5. Determine whether the sequence $\{(-1)^{n+1}\}$ is increasing, decreasing, or neither. Is this sequence an alternating sequence?

13.1 Exercises

29. Increasing; $a_n = n+4$, $a_{n+1} = n+5$; $n+4 < n+5$, so $a_n < a_{n+1}$
30. Decreasing; $a_n = 1-2n$, $a_{n+1} = -1-2n$; $1-2n > -1-2n$, so $a_n > a_{n+1}$
31. Decreasing; $a_n = \frac{1}{n+3}$, $a_{n+1} = \frac{1}{n+4}$; $\frac{1}{n+3} > \frac{1}{n+4}$, so $a_n > a_{n+1}$
32. Decreasing; $a_n = \frac{1}{3^n}$, $a_{n+1} = \frac{1}{3\cdot 3^n}$; $\frac{1}{3^n} > \frac{1}{3\cdot 3^n}$, so $a_n > a_{n+1}$
33. Decreasing; $a_n = \frac{2n+1}{n}$, $a_{n+1} = \frac{2n+3}{n+1}$; $\frac{2n+1}{n} > \frac{2n+3}{n+1}$, so $a_n > a_{n+1}$
34. Increasing; $a_n = \frac{n}{n+1}$, $a_{n+1} = \frac{n+1}{n+2}$; $\frac{n}{n+1} < \frac{n+1}{n+2}$, so $a_n < a_{n+1}$
35. $a_1 = 3$, $a_2 = 1$, and $a_3 = 4$, so $a_1 > a_2$, but $a_2 < a_3$ making the sequence neither inc. nor dec.

Write the first five terms of each of the sequences.

1. $\{2n-1\}$ **2.** $\{4n+1\}$ **3.** $\left\{1+\frac{1}{n}\right\}$

4. $\left\{\frac{n+3}{n+1}\right\}$ **5.** $\{n^2+n\}$ **6.** $\{n-n^2\}$

7. $\{2^n\}$ **8.** $\{2^n-n^2\}$ **9.** $\left\{\left(\frac{1}{2}\right)^n\right\}$

10. $\left\{\left(-\frac{1}{2}\right)^{n+1}\right\}$ **11.** $\{(-1)^n(n^2+1)\}$ **12.** $\{(-1)^{n-1}(3^n)\}$

13. $\left\{(-1)^n\left(\frac{n}{n+1}\right)\right\}$ **14.** $\left\{(-1)^n\left(\frac{1}{2n+3}\right)\right\}$ **15.** $\left\{\frac{2}{n(n+1)}\right\}$

16. $\left\{\frac{2n}{n+1}\right\}$ **17.** $\left\{\frac{n(n-1)}{2}\right\}$ **18.** $\left\{\frac{1+(-1)^n}{2}\right\}$

Find a formula for the general term of each sequence.

19. $2, 5, 8, 11, 14, \ldots$ **20.** $5, 9, 13, 17, 21, \ldots$ **21.** $1, 4, 9, 16, 25, \ldots$

22. $2, 5, 10, 17, 26, \ldots$ **23.** $\frac{1}{2}, \frac{1}{4}, \frac{1}{8}, \frac{1}{16}, \frac{1}{32}, \ldots$ **24.** $\frac{1}{3}, \frac{1}{4}, \frac{1}{5}, \frac{1}{6}, \frac{1}{7}, \ldots$

25. $6, 12, 18, 24, 30, \ldots$ **26.** $5, 10, 20, 40, 80, \ldots$ **27.** $1, -3, 5, -7, 9, \ldots$

28. $-3, 7, -11, 15, -19, \ldots$

For each of the sequences, determine whether it is increasing or decreasing. Justify your answer by comparing a_n with a_{n+1}.

29. $\{n+4\}$ **30.** $\{1-2n\}$ **31.** $\left\{\frac{1}{n+3}\right\}$

Answers to Practice Problems: **1.** 1, 4, 9, 16, 25 **2.** 3, 5, 7, 9, 11 **3.** $\frac{1}{2}, \frac{1}{3}, \frac{1}{4}, \frac{1}{5}, \frac{1}{6}$ **4.** $a_n = 2n-3$ **5.** Neither; Yes

36. $a_1 = \frac{1}{3}$, $a_2 = -\frac{1}{6}$, and $a_3 = \frac{1}{9}$, so $a_1 > a_2$, but $a_2 < a_3$ making the sequence neither inc. nor dec.; Yes, the sequence is alternating.

37. \$28,000 when $n = 1$; \$19,600 when $n = 2$; \$13,720 when $n = 3$; It will be worth \$13,720 after 3 years.

38. 300 when $n = 1$; 900 when $n = 2$; 2700 when $n = 3$; 8100 when $n = 4$; There will be 8100 bacteria after 4 days.

39. 4 m when $n = 1$; 1.6 m when $n = 2$; 0.64 m when $n = 3$; 0.256 m when $n = 4$; It will bounce 0.256 m on the 4th bounce.

32. $\left\{\frac{1}{3^n}\right\}$ **33.** $\left\{\frac{2n+1}{n}\right\}$ **34.** $\left\{\frac{n}{n+1}\right\}$

35. Show that the sequence of digits from the irrational number $\pi = 3.1415926535...$ is neither increasing nor decreasing. (**Note:** There is no formula for a_n.)

36. Show that the sequence $\left\{\frac{(-1)^{n+1}}{3n}\right\}$ is neither increasing nor decreasing. Is this an alternating sequence?

Write the terms of the finite sequences described. Then answer the stated question.

37. Buying a car: A certain automobile costs \$40,000 new and depreciates at a rate of $\frac{3}{10}$ of its current value each year. What will be its value after 3 years?

38. Bacteria growth: A culture of bacteria triples every day. If there were 100 bacteria in the original culture, how many would be present after 4 days?

39. Bouncing a ball: A ball is dropped from a height of 10 meters. Each time it bounces, it rises to $\frac{2}{5}$ of its previous height. How high will it bounce on its fourth bounce?

40. University enrollment: A local university is experiencing a declining enrollment of 6% per year. If the present enrollment is 20,000, what is the projected enrollment after 5 years?

Writing and Thinking About Mathematics

40. 18,800 students when $n = 1$; 17,672 students when $n = 2$; 16,612 students when $n = 3$; 15,615 students when $n = 4$; 14,678 students when $n = 5$; The projected enrollment in 5 years is 14,678 students.

41. The famous Fibonacci sequence is an example of a **recurrence sequence**. That is, each term depends on previous terms.

a. Write the first eight terms of the Fibonacci sequence defined as $F_{n+2} = F_{n+1} + F_n$ where $F_1 = 1$ and $F_2 = 1$.

b. Form the sequence of the differences of successive terms. What do you notice?

HAWKES LEARNING SYSTEMS: INTRODUCTORY & INTERMEDIATE ALGEBRA SOFTWARE

- 13.1 Sequences

41. **a.** 1, 1, 2, 3, 5, 8, 13, 21 **b.** If we call the sequence of differences D_n, then $D_n = F_{n-2}$.

13.2 Sigma Notation

- *Write sums using Σ notation.*
- *Find the values of sums written in Σ notation.*

Teaching Notes:
You might want to make this Σ notation palatable in terms of the need for the economy of expressing a longer number of addends and the convenience when a general form of the term is known. Have students write out several terms of sequences for which the general formula for a term is known – build a feel for the convenience of sigma notation.

A sequence has an infinite number of terms. To find the sum of just a few terms of a sequence means to find the sum of a **finite sequence**. Such a sum is called a **partial sum** of the sequence and can be indicated using **sigma notation** with the Greek letter capital sigma, Σ.

Partial Sums Using Sigma Notation

The nth **partial sum** S_n of the first n terms of a sequence $\{a_n\}$ is

$$\boldsymbol{S_n = \sum_{k=1}^{n} a_k = a_1 + a_2 + a_3 + \ldots + a_n}.$$

k is called the **index of summation**, and k takes the integer values $1, 2, 3, \ldots, n$.

n is the **upper limit of summation**, and 1 is the **lower limit of summation**.

NOTES

As we will see in Section 13.4, sigma notation can be used to indicate the sum of an entire sequence using the symbol for infinity (∞), in place of n. Also, the upper and lower limits of summation can be adjusted to pick out a particular part of the sequence. For example, k can begin with 7 and stop with 10. There may be special times, because of the way formulas are written, when k would begin with 0.

To understand the concept of partial sums, consider the sequence $\left\{\frac{1}{n}\right\}$ and the following partial sums:

$$S_1 = a_1 = \frac{1}{1}$$

$$S_2 = a_1 + a_2 = \frac{1}{1} + \frac{1}{2}$$

$$S_3 = a_1 + a_2 + a_3 = \frac{1}{1} + \frac{1}{2} + \frac{1}{3}$$

$$\vdots$$

$$S_n = a_1 + a_2 + a_3 + \ldots + a_n = \frac{1}{1} + \frac{1}{2} + \frac{1}{3} + \ldots + \frac{1}{n}.$$

In some cases, the lower limit of summation in sigma notation may be an integer other than 1. Also, letters other than k may be used as the index of summation. The lower case letters i, j, k, l, m, and n are commonly used.

For example, the sum of the second through sixth terms of the sequence $\{n^2\}$ can be written in sigma notation as

$$\sum_{i=2}^{6} i^2 = 2^2 + 3^2 + 4^2 + 5^2 + 6^2.$$

If the number of terms is large, then three dots are used to indicate missing terms after a pattern has been established with the first three or four terms. For example,

$$S_{100} = \sum_{k=1}^{100} (k-1) = 0 + 1 + 2 + 3 + \ldots + 99.$$

Example 1: Sigma Notation

Write the indicated sums of the terms and find the value of each sum.

a. $\sum_{k=1}^{4} k^3$

Solution: $\sum_{k=1}^{4} k^3 = 1^3 + 2^3 + 3^3 + 4^3$

$= 1 + 8 + 27 + 64 = 100$

b. $\sum_{k=5}^{9} (-1)^k k$

Solution: $\sum_{k=5}^{9} (-1)^k k = (-1)^5 5 + (-1)^6 6 + (-1)^7 7 + (-1)^8 8 + (-1)^9 9$

$= -5 + 6 - 7 + 8 - 9 = -7$

Example 2: Writing Sums in Sigma Notation

Write each sum using sigma notation.

a. $4 + 9 + 16 + 25$

Solution: Note that each number is a square beginning with 2^2. Two possible forms are the following:

$$4 + 9 + 16 + 25 = \sum_{k=2}^{5} k^2 \quad \text{or} \quad 4 + 9 + 16 + 25 = \sum_{k=1}^{4} (k+1)^2.$$

b. $\frac{5}{2} + \frac{10}{3} + \frac{15}{4} + \frac{20}{5} + \frac{25}{6} + \frac{30}{7}$

Solution: Note that each numerator is a multiple of 5 and each denominator is 1 larger than its position. One possible form is as follows:

$$\frac{5}{2} + \frac{10}{3} + \frac{15}{4} + \frac{20}{5} + \frac{25}{6} + \frac{30}{7} = \sum_{k=1}^{6} \frac{5k}{k+1}.$$

Properties of Σ Notation

The following properties of Σ notation are useful in developing systematic methods for finding sums of certain types of finite and infinite sequences.

Properties of Σ Notation

For sequences $\{a_n\}$ and $\{b_n\}$ and any real number c:

I. $\sum_{k=1}^{n} a_k = \sum_{k=1}^{i} a_k + \sum_{k=i+1}^{n} a_k$ for any i, $1 \le i \le n-1$

II. $\sum_{k=1}^{n} (a_k + b_k) = \sum_{k=1}^{n} a_k + \sum_{k=1}^{n} b_k$

III. $\sum_{k=1}^{n} c a_k = c \sum_{k=1}^{n} a_k$

IV. $\sum_{k=1}^{n} c = nc$

These properties follow directly from the associative, commutative, and distributive properties for sums of real numbers.

I.
$$\begin{aligned}\sum_{k=1}^{n} a_k &= a_1 + a_2 + \ldots + a_i + a_{i+1} + a_{i+2} + \ldots + a_n \\ &= (a_1 + a_2 + \ldots + a_i) + (a_{i+1} + a_{i+2} + \ldots + a_n) \\ &= \sum_{k=1}^{i} a_k + \sum_{k=i+1}^{n} a_k\end{aligned}$$

II.
$$\begin{aligned}\sum_{k=1}^{n} (a_k + b_k) &= (a_1 + b_1) + (a_2 + b_2) + \ldots + (a_n + b_n) \\ &= (a_1 + a_2 + \ldots + a_n) + (b_1 + b_2 + \ldots + b_n) \\ &= \sum_{k=1}^{n} a_k + \sum_{k=1}^{n} b_k\end{aligned}$$

III.
$$\begin{aligned}\sum_{k=1}^{n} c a_k &= c a_1 + c a_2 + \ldots + c a_n \\ &= c(a_1 + a_2 + \ldots + a_n) \\ &= c \sum_{k=1}^{n} a_k\end{aligned}$$

IV.
$$\sum_{k=1}^{n} c = \underbrace{c + c + c + \ldots + c}_{c \text{ appears } n \text{ times}} = nc$$

Example 3: Properties of $\sum$ Notation

a. If $\sum_{k=1}^{7} a_k = 40$ and $\sum_{k=1}^{30} a_k = 75$, find $\sum_{k=8}^{30} a_k$.

Solution: Since $\sum_{k=1}^{7} a_k + \sum_{k=8}^{30} a_k = \sum_{k=1}^{30} a_k$, By property I

then $40 + \sum_{k=8}^{30} a_k = 75$

$$\sum_{k=8}^{30} a_k = 35.$$

b. If $\sum_{k=1}^{50} 3a_k = 600$, find $\sum_{k=1}^{50} a_k$.

Solution: Since $\sum_{k=1}^{50} 3a_k = 3\sum_{k=1}^{50} a_k$, By property III

then $3\sum_{k=1}^{50} a_k = 600$

$$\sum_{k=1}^{50} a_k = 200.$$

Practice Problems

1. Write the indicated sum of terms and find the value of the sum: $\sum_{k=1}^{4} \left(k^2 - 1\right)$.
2. Write the sum $10 + 12 + 14 + 16 + 18$ in Σ notation.
3. $\sum_{k=1}^{5} a_k = 20$ and $\sum_{k=6}^{10} a_k = 30$. Find $\sum_{k=1}^{10} 2a_k$.

Answers to Practice Problems: **1.** $0 + 3 + 8 + 15 = 26$ **2.** $\sum_{k=5}^{9} 2k$ or $\sum_{k=1}^{5} (2k + 8)$ **3.** 100

13.2 Exercises

For each of the sequences given, write out the partial sums S_1, S_2, S_3, and S_4 and evaluate each partial sum.

1. $\{3k-1\}$ **2.** $\{2k+5\}$ **3.** $\left\{\frac{k}{k+1}\right\}$ **4.** $\left\{\frac{k+1}{k}\right\}$

5. $\left\{(-1)^{k-1}k^2\right\}$ **6.** $\left\{(-1)^k k^3\right\}$ **7.** $\left\{\frac{1}{2^k}\right\}$ **8.** $\left\{\left(\frac{2}{3}\right)^k\right\}$

9. $\left\{2k-k^2\right\}$ **10.** $\left\{k^2-k\right\}$

Write the indicated sums in expanded form and evaluate.

11. $\sum_{k=1}^{5} 2k$ **12.** $\sum_{k=1}^{8} 3k$ **13.** $\sum_{k=2}^{6} (k+3)$ **14.** $\sum_{k=9}^{11} (2k+1)$

15. $\sum_{k=2}^{4} \frac{1}{k}$ **16.** $\sum_{k=1}^{3} \frac{1}{2k}$ **17.** $\sum_{k=1}^{3} 2^k$ **18.** $\sum_{k=10}^{15} (-1)^k$

19. $\sum_{k=4}^{8} k^2$ **20.** $\sum_{k=1}^{5} k^3$ **21.** $\sum_{k=3}^{6} (9-2k)$ **22.** $\sum_{k=2}^{7} (4k-1)$

23. $\sum_{k=2}^{5} (-1)^k\left(k^2+k\right)$ **24.** $\sum_{k=1}^{6} (-1)^k\left(k^2-2\right)$ **25.** $\sum_{k=1}^{5} \frac{k}{k+1}$ **26.** $\sum_{k=3}^{5} \frac{k+1}{k^2}$

Write the sums in sigma notation. There may be more than one correct answer.

27. $1+3+5+7+9$ **28.** $4+7+10+13+16$

29. $-1+1+(-1)+1+(-1)$ **30.** $-2+4-8+16-32$

31. $16+25+36+49$ **32.** $8+15+24+35+48$

33. $\frac{1}{8}-\frac{1}{27}+\frac{1}{64}-\frac{1}{125}+\frac{1}{216}$ **34.** $-\frac{1}{8}+\frac{1}{16}-\frac{1}{32}+\frac{1}{64}-\frac{1}{128}$

35. $\frac{4}{5}+\frac{5}{6}+\frac{6}{7}+\ldots+\frac{15}{16}$ **36.** $\frac{6}{25}+\frac{7}{36}+\frac{8}{49}+\frac{9}{64}+\ldots+\frac{13}{144}$

1. $S_1=2, S_2=2+5=7, S_3=2+5+8=15, S_4=2+5+8+11=26$
2. $S_1=7, S_2=7+9=16, S_3=7+9+11=27, S_4=7+9+11+13=40$
3. $S_1=\frac{1}{2}, S_2=\frac{1}{2}+\frac{2}{3}=\frac{7}{6}, S_3=\frac{1}{2}+\frac{2}{3}+\frac{3}{4}=\frac{23}{12}, S_4=\frac{1}{2}+\frac{2}{3}+\frac{3}{4}+\frac{4}{5}=\frac{163}{60}$
4. $S_1=2, S_2=2+\frac{3}{2}=\frac{7}{2}, S_3=2+\frac{3}{2}+\frac{4}{3}=\frac{29}{6}, S_4=2+\frac{3}{2}+\frac{4}{3}+\frac{5}{4}=\frac{73}{12}$
5. $S_1=1, S_2=1+(-4)=-3, S_3=1+(-4)+9=6, S_4=1+(-4)+9+(-16)=-10$
6. $S_1=-1, S_2=-1+8=7, S_3=-1+8+(-27)=-20, S_4=-1+8+(-27)+64=44$
7. $S_1=\frac{1}{2}, S_2=\frac{1}{2}+\frac{1}{4}=\frac{3}{4}, S_3=\frac{1}{2}+\frac{1}{4}+\frac{1}{8}=\frac{7}{8}, S_4=\frac{1}{2}+\frac{1}{4}+\frac{1}{8}+\frac{1}{16}=\frac{15}{16}$
8. $S_1=\frac{2}{3}, S_2=\frac{2}{3}+\frac{4}{9}=\frac{10}{9}, S_3=\frac{2}{3}+\frac{4}{9}+\frac{8}{27}=\frac{38}{27}, S_4=\frac{2}{3}+\frac{4}{9}+\frac{8}{27}+\frac{16}{81}=\frac{130}{81}$
9. $S_1=1, S_2=1+0=1, S_3=1+0+(-3)=-2, S_4=1+0+(-3)+(-8)=-10$
10. $S_1=0, S_2=0+2=2, S_3=0+2+6=8, S_4=0+2+6+12=20$
11. $2+4+6+8+10=30$
12. $3+6+9+12+15+18+21+24=108$
13. $5+6+7+8+9=35$
14. $19+21+23=63$

15. $\frac{1}{2}+\frac{1}{3}+\frac{1}{4}=\frac{13}{12}$

16. $\frac{1}{2}+\frac{1}{4}+\frac{1}{6}=\frac{11}{12}$

17. $2+4+8=14$

18. $1+(-1)+1+(-1)+1+(-1)=0$

19. $16+25+36+49+64=190$

20. $1+8+27+64+125=225$

21. $3+1+(-1)+(-3)=0$

22. $7+11+15+19+23+27=102$

23. $6+(-12)+20+(-30)=-16$

24. $1+2+(-7)+14+(-23)+34=21$

25. $\frac{1}{2}+\frac{2}{3}+\frac{3}{4}+\frac{4}{5}+\frac{5}{6}=\frac{71}{20}$

26. $\frac{4}{9}+\frac{5}{16}+\frac{6}{25}=\frac{3589}{3600}$

27. $\sum_{k=1}^{5}(2k-1)$

28. $\sum_{k=1}^{5}(3k+1)$

29. $\sum_{k=1}^{5}(-1)^k$

30. $\sum_{k=1}^{5}(-1)^k 2^k$

Find the indicated sums.

37. $\sum_{k=1}^{14} a_k = 18$ and $\sum_{k=1}^{14} b_k = 21$. Find $\sum_{k=1}^{14}(a_k + b_k)$.

38. $\sum_{k=1}^{19} a_k = 23$ and $\sum_{k=1}^{19} b_k = 16$. Find $\sum_{k=1}^{19}(a_k - b_k)$.

39. $\sum_{k=1}^{15} a_k = 19$. Find $\sum_{k=1}^{15} 3a_k$.

40. $\sum_{k=1}^{11} a_k = 35$. Find $\sum_{k=1}^{11} 2a_k$.

41. $\sum_{k=1}^{25} a_k = 63$ and $\sum_{k=1}^{11} a_k = 15$. Find $\sum_{k=12}^{25} a_k$.

42. $\sum_{k=1}^{16} a_k = 56$ and $\sum_{k=17}^{40} a_k = 42$. Find $\sum_{k=1}^{40} a_k$.

43. $\sum_{k=13}^{29} a_k = 84$ and $\sum_{k=1}^{29} a_k = 143$. Find $\sum_{k=1}^{12} 5a_k$.

44. $\sum_{k=1}^{27} a_k = 46$ and $\sum_{k=1}^{10} a_k = 122$. Find $\sum_{k=11}^{27} 2a_k$.

45. $\sum_{k=1}^{18} a_k = 41$ and $\sum_{k=1}^{18} b_k = 62$. Find $\sum_{k=1}^{18}(3a_k - 2b_k)$.

46. $\sum_{k=1}^{21} a_k = -68$ and $\sum_{k=1}^{21} b_k = 39$. Find $\sum_{k=1}^{21}(a_k + 2b_k)$.

47. $\sum_{k=1}^{20} b_k = 34$ and $\sum_{k=1}^{20}(2a_k + b_k) = 144$. Find $\sum_{k=1}^{20} a_k$.

48. $\sum_{k=1}^{16} a_k = 28$ and $\sum_{k=1}^{16}(b_k - 3a_k) = -12$. Find $\sum_{k=1}^{16} b_k$.

31. $\sum_{k=4}^{7} k^2$

32. $\sum_{k=3}^{7}(k^2-1)$

33. $\sum_{k=2}^{6}(-1)^k\left(\frac{1}{k^3}\right)$

34. $\sum_{k=3}^{7}(-1)^k\left(\frac{1}{2^k}\right)$

35. $\sum_{k=4}^{15}\frac{k}{k+1}$

36. $\sum_{k=5}^{12}\frac{k+1}{k^2}$

37. 39 38. 7

39. 57 40. 70

41. 48 42. 98

43. 295 44. −152

45. −1 46. 10

47. 55 48. 72

49. $\sum_{k=11}^{15}(-2k)+\sum_{k=1}^{5}3k$

Writing and Thinking About Mathematics

49. Use the sum of two expressions in sigma notation to represent the following sum: $-22+3-24+6-26+9-28+12-30+15$.

HAWKES LEARNING SYSTEMS: INTRODUCTORY & INTERMEDIATE ALGEBRA SOFTWARE

- 13.2 Sigma Notation

13.3 Arithmetic Sequences

- *Determine whether or not a sequence is* ***arithmetic****.*
- *Find the general term for an arithmetic sequence.*
- *Find the specified terms of an arithmetic sequence.*
- *Find the sum of the first n terms of an arithmetic sequence.*

Many types of sequences are studied in higher levels of mathematics. In the next two sections, two types of sequences are discussed: **arithmetic sequences** and **geometric sequences**. In this discussion, sigma notation is used and formulas for finding sums are developed. For arithmetic sequences, sums of only a finite number of terms can be found. For geometric sequences, sums of a finite number of terms and, in some special cases, the sum of an infinite number of terms can be found.

Arithmetic Sequences

The sequences

$$3, 5, 7, 9, 11, 13, \ldots$$

$$4, 5, 6, 7, 8, 9, \ldots$$

$$-2, -5, -8, -11, -14, -17, \ldots$$

all have a common characteristic. This characteristic is that any two consecutive terms in each sequence have the **same difference**.

$$\underbrace{3,\ 5}_{2},\ \underbrace{5,\ 7}_{2},\ \underbrace{7,\ 9}_{2},\ \underbrace{9,\ 11}_{2},\ \underbrace{11,\ 13}_{2}, \ldots \qquad 5-3=2,\ 7-5=2,\ 9-7=2, \text{ and so on.}$$

$$\underbrace{4,\ 5}_{1},\ \underbrace{5,\ 6}_{1},\ \underbrace{6,\ 7}_{1},\ \underbrace{7,\ 8}_{1},\ \underbrace{8,\ 9}_{1}, \ldots \qquad 5-4=1,\ 6-5=1,\ 7-6=1, \text{ and so on.}$$

$$\underbrace{-2,\ -5}_{-3},\ \underbrace{-5,\ -8}_{-3},\ \underbrace{-8,\ -11}_{-3},\ \underbrace{-11,\ -14}_{-3},\ \underbrace{-14,\ -17}_{-3}, \ldots \qquad -5-(-2)=-3,\ -8-(-5)=-3, \text{ and so on.}$$

Such sequences are called **arithmetic sequences** or **arithmetic progressions**.

Arithmetic sequences are closely related to linear functions. To see this relationship we can plot the points of an arithmetic sequence and note that the rise from one point to the next is the constant difference, d. See Figure 1 as an illustration with a positive value for d. Note that the points do indeed lie on a straight line. The slope of the line is d.

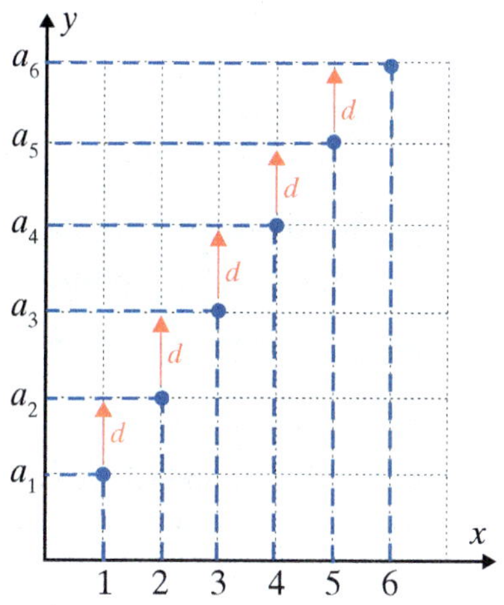

Figure 1

Arithmetic Sequences

A sequence $\{a_n\}$ is called an **arithmetic sequence** (or **arithmetic progression**) if for any natural number k,

$$\boldsymbol{a_{k+1} - a_k = d} \qquad \text{where } d \text{ is a constant.}$$

d is called the **common difference**.

Example 1: An Arithmetic Sequence

Show that the sequence $\{2n-3\}$ is arithmetic by finding d.

Solution: $a_k = 2k-3$ and $a_{k+1} = 2(k+1)-3 = 2k+2-3 = 2k-1$

Therefore, $a_{k+1} - a_k = (2k-1)-(2k-3) = 2k-1-2k+3 = 2.$

So $d = 2$, and the sequence $\{2n-3\}$ **is arithmetic.**

Example 2: A Sequence that is Not Arithmetic

Show that the sequence $\{n^2\}$ is **not** arithmetic.

Solution: Consider a_3, a_2, and a_1 as follows:

$$a_3 = 3^2 = 9,\ a_2 = 2^2 = 4,\ \text{and}\ a_1 = 1^2 = 1.$$

Therefore, $a_3 - a_2 = 9 - 4 = 5$ and $a_2 - a_1 = 4 - 1 = 3$.

So, there is no common difference between successive terms, and the sequence $\{n^2\}$ **is not arithmetic.**

NOTES Note that in Example 2, we needed to show only one case in which the difference between two sets of consecutive terms was not the same. This is called finding a **counterexample**. However, to show that something is true in every case, general formulas, as in Example 1, must be used.

The nth Term of an Arithmetic Sequence

To find a formula for the nth term of an arithmetic sequence, proceed as follows:

$a_1 = a_1$	first term
$a_2 = a_1 + d$	second term
$a_3 = a_2 + d = a_1 + 2d$	third term
$a_4 = a_3 + d = a_1 + 3d$	fourth term
$\vdots$	$\vdots$
$a_n = a_{n-1} + d = a_1 + (n-1)d$	n^{th} term

Teaching Notes:
Working through several examples of arithmetic sequences should remove any "mystery" about the addition of the $(n-1)d$ term.

Having students make absolute sense of formulas, whenever possible, should help them look beyond trying to memorize such equations.

Formula for the nth Term of an Arithmetic Sequence

If $\{a_n\}$ is an arithmetic sequence, then the nth term has the form

$$\boldsymbol{a_n = a_1 + (n-1)d}$$

where d is the common difference between consecutive terms.

Example 3: Finding a Specific Term in an Arithmetic Sequence

a. **Given a_1 and *d*:**
If $a_1 = 5$ and $d = 3$, find a_{16}.

Solution: To find a_{16}, let $n = 16$ in the formula for a_n:

$$\begin{aligned} a_n &= a_1 + (n-1)d \\ a_{16} &= 5 + (16-1)\cdot 3 \\ &= 5 + 15\cdot 3 \\ &= 50 \end{aligned}$$

b. **Given two consecutive terms:**
If the first two terms are –2 and 8, find a_{20}.

Solution: Knowing two consecutive terms, we can find *d*.
Because the sequence is arithmetic, $d = a_2 - a_1 = 8 - (-2) = 10$.
Now using the formula for a_n, we have

$$\begin{aligned} a_n &= a_1 + (n-1)d \\ a_{20} &= -2 + (20-1)\cdot 10 \\ &= -2 + 19\cdot 10 \\ &= 188 \end{aligned}$$

c. **Using a system of equations given two terms:**
If $a_3 = 6$ and $a_{21} = -48$, find a_1 and *d* for the arithmetic sequence.

Solution: Using the formula $a_n = a_1 + (n-1)d$, substituting $n = 3$ and $n = 21$, and solving simultaneous equations, we have

$$\begin{aligned} -48 &= a_1 + 20d \\ 6 &= a_1 + 2d \end{aligned} \qquad \begin{array}{rcrcr} -48 &=& a_1 &+& 20d \\ -6 &=& -a_1 &-& 2d \\ \hline -54 &=& & & 18d \\ -3 &=& & & d. \end{array}$$

Then,

$$\begin{aligned} 6 &= a_1 + 2(-3) \\ 12 &= a_1. \end{aligned}$$

So, $a_1 = 12$ and $d = -3$.

Example 4: Finding n Given Certain Conditions

Given that $\{a_n\}$ is an arithmetic sequence, $a_{10} = -12$, $d = -3$, and $a_n = -72$, find n.

Solution: Because the sequence is arithmetic, the conditions that $a_{10} = -12$ and $d = -3$ can be used to find a_1 as follows:

$$\begin{aligned} a_n &= a_1 + (n-1)d \\ -12 &= a_1 + (10-1)(-3) \\ -12 &= a_1 + (9)(-3) \\ -12 &= a_1 - 27 \\ 15 &= a_1 \end{aligned}$$

Now using the formula again, we can solve for n as follows:

$$\begin{aligned} -72 &= 15 + (n-1)(-3) \\ -72 &= 15 - 3n + 3 \\ -90 &= -3n \\ 30 &= n \end{aligned}$$

Thus –72 is the 30th term in the sequence.

Partial Sums of Arithmetic Sequences

Consider the problem of finding the sum $S = \sum_{k=1}^{6}(4k-1)$. We can, of course, write all the terms and then add them:

$$S = \sum_{k=1}^{6}(4k-1) = 3+7+11+15+19+23 = 78.$$

However, to understand how the general formula is developed, we first write the sum and then write the sum again with the terms in reverse order. Adding vertically gives the same sum six times:

$$\begin{array}{rcrcrcrcrcrcr} S &=& 3 &+& 7 &+& 11 &+& 15 &+& 19 &+& 23 \\ S &=& 23 &+& 19 &+& 15 &+& 11 &+& 7 &+& 3 \\ \hline 2S &=& 26 &+& 26 &+& 26 &+& 26 &+& 26 &+& 26 \\ 2S &=& 6 &\cdot& 26 \\ S &=& 78 \end{array}$$

Using this same procedure with general terms in the subscript notation, the formula for the sum of any finite arithmetic sequence can be developed. Suppose that the n terms are

$$a_1, \quad a_2 = a_1 + d, \quad a_3 = a_1 + 2d, \quad \ldots, \quad a_{n-1} = a_n - d, \quad a_n.$$

Thus, writing the terms in both ascending order and descending order and adding vertically, we have

$$\begin{array}{rcccccccccccc} S &=& a_1 &+& (a_1+d) &+& (a_1+2d) &+ \ldots +& (a_n-2d) &+& (a_n-d) &+& a_n \\ S &=& a_n &+& (a_n-d) &+& (a_n-2d) &+ \ldots +& (a_1+2d) &+& (a_1+d) &+& a_1 \\ \hline 2S &=& (a_1+a_n) &+& (a_1+a_n) &+& (a_1+a_n) &+ \ldots +& (a_1+a_n) &+& (a_1+a_n) &+& (a_1+a_n) \end{array}$$

(a_1+a_n) appears n times

$$2S = n(a_1+a_n)$$

$$S = \frac{n}{2}(a_1+a_n).$$

Partial Sums of Arithmetic Sequences

The nth **partial sum** S_n of the first n terms of an arithmetic sequence $\{a_n\}$ is

$$\boldsymbol{S_n = \sum_{k=1}^{n} a_k = \frac{n}{2}(a_1+a_n).}$$

A special case of an arithmetic sequence is $\{n\}$ and the corresponding sum of the first n terms:

$$\sum_{k=1}^{n} k = 1+2+3+\ldots+n.$$

In this case, n = the number of terms, $a_1 = 1$, and $a_n = n$, so

$$\sum_{k=1}^{n} k = \frac{n}{2}(1+n).$$

Gauss

German mathematician Carl Friedrich Gauss (1777 – 1855) understood and applied this sum at the age of 7 in order to solve an arithmetic problem given to him and his classmates as "busy" work. Gauss probably observed the following pattern when told to find the sum of the whole numbers from 1 to 100:

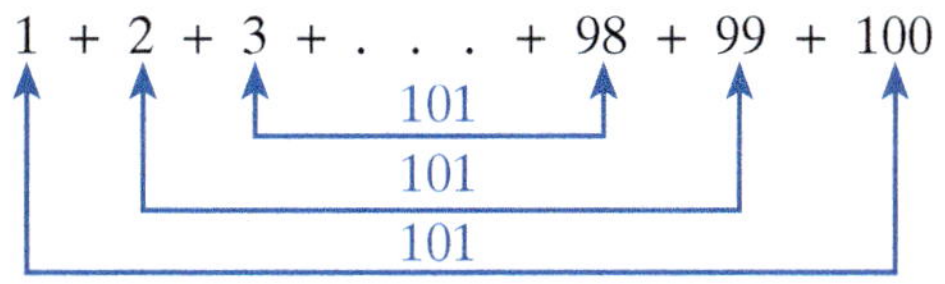

He saw that 101 was a sum 50 times. Thus to find the sum he simply multiplied $101 \cdot 50 = 5050$. Not bad for a 7-year old!

Example 5: Finding Partial Sums of Arithmetic Sequences

First show that the corresponding sequence is an arithmetic sequence by finding $a_{k+1} - a_k = d$. Then find the indicated sum using the formula.

a. $\sum_{k=1}^{75} k = 1+2+3+\ldots+75$

Solution: $a_k = k$ and $a_{k+1} = k+1$

$a_{k+1} - a_k = k+1-k = 1 = d$ and $\{k\}$ is an arithmetic sequence.

$$\sum_{k=1}^{75} k = \frac{75}{2}(1+75) \qquad \text{Here } n = 75,\ a_1 = 1, \text{ and } a_{75} = 75.$$

$$= \frac{75}{2}(76)$$

$$= 2850$$

b. $\sum_{k=1}^{50} 3k = 3+6+9+\ldots+150$

Solution: $a_k = 3k$ and $a_{k+1} = 3(k+1) = 3k+3$

$a_{k+1} - a_k = (3k+3) - 3k = 3 = d$

So, $\{3k\}$ is an arithmetic sequence.

$$\sum_{k=1}^{50} 3k = \frac{50}{2}(3+150) \qquad \text{Here } n = 50,\ a_1 = 3, \text{ and } a_{50} = 150.$$

$$= 25(153)$$

$$= 3825$$

Alternatively, property III of Section 13.2 can be used to find the sum.

$$\sum_{k=1}^{50} 3k = 3\sum_{k=1}^{50} k \qquad \text{Property III}$$

$$= 3 \cdot \frac{50}{2}(1+50) \qquad \text{Here } n = 50,\ a_1 = 1, \text{ and } a_{50} = 50.$$

$$= 3 \cdot 25 \cdot 51$$

$$= 3825$$

c. $\sum_{k=1}^{70} (-2k+5) = 3+1+(-1)+(-3)+\ldots+(-135)$

Solution: $a_k = -2k+5$ and $a_{k+1} = -2(k+1)+5 = -2k+3,$

$a_{k+1} - a_k = (-2k+3) - (-2k+5) = 3-5 = -2 = d$

So, $\{-2k+5\}$ is an arithmetic sequence.

$$\sum_{k=1}^{70}(-2k+5)=\frac{70}{2}[3+(-135)]$$ Here $n=70,\ a_1=3,\ a_{70}=-135.$

$$=35(-132)$$

$$=-4620$$

Alternatively, Properties II, III, and IV of Section 13.2 and the sum of a finite arithmetic sequence from this section can be used to find the sum.

$$\sum_{k=1}^{70}(-2k+5)=\sum_{k=1}^{70}-2k+\sum_{k=1}^{70}5$$ Property II

$$=-2\sum_{k=1}^{70}k+\sum_{k=1}^{70}5$$ Property III

$$=-2\cdot\frac{70}{2}(1+70)+70\cdot5$$ The sum of a finite arithmetic sequence and Property IV

$$=-2\cdot35\cdot71+350$$

$$=-4970+350$$

$$=-4620$$

Example 6: An Application of an Arithmetic Sequence

Teaching Notes:
Example 6 could lead to some interesting class discussion questions. How many years will it take for the salaries to be equal? Will the salaries ever be exactly equal? How many years does it take for the total income to be the same (or approximately the same)? Which job would you prefer and why?

Suppose that you are offered two jobs by the same company. The first job has a starting salary of \$35,000, with a guaranteed raise of \$2000 per year. The second job starts at \$40,000 with a guaranteed raise of \$1200 per year.

a. What would be your salary in the 10th year of each of these jobs?

b. If you were to stay 10 years with the company, which job would pay the most in total salary?

Solution: Since the salary would increase the same amount each year, the yearly salaries form arithmetic sequences and the corresponding formulas for a_{10} and S_{10} can be used.

a. First job: $a_{10}=a_1+(10-1)d=35{,}000+9(2000)=\$53{,}000$

Second job: $a_{10}=a_1+(10-1)d=40{,}000+9(1200)=\$50{,}800$

In 10 years, you would be making a higher salary on the first job.

b. First job: $S_{10}=\frac{10}{2}(35{,}000+53{,}000)=\$440{,}000$

Second job: $S_{10}=\frac{10}{2}(40{,}000+50{,}800)=\$454{,}000$

Over the first 10 years, the second job would pay more in total salary.

Practice Problems

1. Show that the sequence $\{3n+5\}$ is arithmetic by finding d.
2. Find the 40th term of the arithmetic sequence with 1, 6, and 11 as its first three terms.
3. Find $\sum_{k=1}^{50}(3k+5)$.

13.3 Exercises

1. Arithmetic sequence $d=3$; $\{3n-1\}$
2. Arithmetic sequence $d=4$; $\{4n-7\}$
3. Arithmetic sequence $d=-2$; $\{9-2n\}$
4. Arithmetic sequence $d=1$; $\{n+4\}$
5. Not an arithmetic sequence
6. Not an arithmetic sequence
7. Arithmetic sequence $d=-4$; $\{10-4n\}$
8. Arithmetic sequence $d=-5$; $\{9-5n\}$
9. Arithmetic sequence $d=\frac{1}{2}$; $\left\{\frac{n-1}{2}\right\}$
10. Arithmetic sequence $d=\frac{1}{3}$; $\left\{\frac{n+5}{3}\right\}$
11. 1, 3, 5, 7, 9; Arithmetic sequence
12. 3, 2, 1, 0, −1; Arithmetic sequence
13. $\frac{1}{2}, \frac{1}{3}, \frac{1}{4}, \frac{1}{5}, \frac{1}{6}$; Not an arithmetic sequence

Determine which of the sequences are arithmetic. Find the common difference and the nth term for each arithmetic sequence.

1. 2, 5, 8, 11, ... **2.** −3, 1, 5, 9, ... **3.** 7, 5, 3, 1, ... **4.** 5, 6, 7, 8, ...

5. 1, 2, 3, 5, 8, ... **6.** 2, 4, 8, 16, ... **7.** 6, 2, −2, −6, ... **8.** 4, −1, −6, −11, ...

9. $0, \frac{1}{2}, 1, \frac{3}{2}, \ldots$ **10.** $2, \frac{7}{3}, \frac{8}{3}, 3, \ldots$

For each of the following sequences, write the first five terms and then determine whether the sequence is arithmetic.

11. $\{2n-1\}$ **12.** $\{4-n\}$ **13.** $\left\{\frac{1}{n+1}\right\}$ **14.** $\left\{\frac{1}{2n}\right\}$

15. $\left\{n+\frac{n}{2}\right\}$ **16.** $\{5-6n\}$ **17.** $\left\{7-\frac{n}{3}\right\}$ **18.** $\left\{\frac{2}{3}n-\frac{7}{3}\right\}$

19. $\{(-1)^n(3n-2)\}$ **20.** $\{(-1)^{n+1}(2n+1)\}$

Use the given information to find $\{a_n\}$ for each of the arithmetic sequences.

21. $a_1=1,\ d=\frac{2}{3}$ **22.** $a_1=9,\ d=-\frac{1}{3}$ **23.** $a_1=7,\ d=-2$

24. $a_1=-3,\ d=\frac{4}{5}$ **25.** $a_1=10,\ a_3=13$ **26.** $a_1=6,\ a_5=4$

27. $a_{10}=13,\ a_{12}=3$ **28.** $a_5=7,\ a_9=19$ **29.** $a_{13}=60,\ a_{23}=75$

30. $a_{11}=54,\ a_{29}=180$

Answers to Practice Problems: **1.** $d=3$ **2.** $a_{40}=196$ **3.** 4075

14. $\frac{1}{2}, \frac{1}{4}, \frac{1}{6}, \frac{1}{8}, \frac{1}{10}$; Not an arithmetic sequence
15. $\frac{3}{2}, 3, \frac{9}{2}, 6, \frac{15}{2}$; Arithmetic sequence
16. $-1, -7, -13, -19, -25$; Arithmetic sequence
17. $\frac{20}{3}, \frac{19}{3}, 6, \frac{17}{3}, \frac{16}{3}$; Arithmetic sequence
18. $-\frac{5}{3}, -1, -\frac{1}{3}, \frac{1}{3}, 1$; Arithmetic sequence
19. $-1, 4, -7, 10, -13$; Not an arithmetic sequence
20. $3, -5, 7, -9, 11$; Not an arithmetic sequence
21. $\left\{\frac{2n+1}{3}\right\}$
22. $\left\{\frac{28-n}{3}\right\}$
23. $\{9-2n\}$
24. $\left\{\frac{4n-19}{5}\right\}$
25. $\left\{\frac{17+3n}{2}\right\}$
26. $\left\{\frac{13-n}{2}\right\}$
27. $\{63-5n\}$
28. $\{3n-8\}$
29. $\left\{\frac{81+3n}{2}\right\}$
30. $\{7n-23\}$
31. 232 32. −154
33. 44 34. 26
35. 7 36. 12
37. 19 38. 9 39. 154 40. 198 41. 235 42. 324 43. 126 44. −342 45. 231 46. 180 47. $\frac{143}{3}$ 48. 3108 49. 60

Assume $\{a_n\}$ is an arithmetic sequence. Find the indicated quantity.

31. $a_1 = 8,\ a_{11} = 168$. Find a_{15}.

32. $a_1 = 17,\ a_9 = -55$. Find a_{20}.

33. $a_6 = 8,\ a_4 = 2$. Find a_{18}.

34. $a_{16} = 12,\ a_7 = 30$. Find a_9.

35. $a_{13} = 34,\ d = 2,\ a_n = 22$. Find n.

36. $a_4 = 20,\ d = 3,\ a_n = 44$. Find n.

37. $a_{10} = 41,\ d = 4,\ a_n = 77$. Find n.

38. $a_3 = 15,\ d = -\frac{3}{2},\ a_n = 6$. Find n.

Find the indicated sums using the formula for partial sums of arithmetic sequences.

39. $-2 + 0 + 2 + 4 + \ldots + 24$

40. $3 + 6 + 9 + \ldots + 33$

41. $1 + 6 + 11 + 16 + \ldots + 46$

42. $5 + 9 + 13 + 17 + \ldots + 49$

43. $\sum_{k=1}^{9}(3k-1)$

44. $\sum_{k=1}^{12}(4-5k)$

45. $\sum_{k=1}^{11}(4k-3)$

46. $\sum_{k=1}^{10}(2k+7)$

47. $\sum_{k=1}^{13}\left(\frac{2k}{3}-1\right)$

48. $\sum_{k=1}^{28}(8k-5)$

49. $\sum_{k=1}^{9}\left(k+\frac{k}{3}\right)$

50. $\sum_{k=1}^{16}\left(9-\frac{k}{3}\right)$

Find the indicated sums using the properties of sigma notation.

51. If $\sum_{k=1}^{33} a_k = -12$, find $\sum_{k=1}^{33}(5a_k+7)$.

52. If $\sum_{k=1}^{15} a_k = 60$, find $\sum_{k=1}^{15}(-2a_k-5)$.

53. If $\sum_{k=1}^{100}(-3a_k+4) = 700$, find $\sum_{k=1}^{100} a_k$.

54. If $\sum_{k=1}^{50}(2b_k-5) = 32$, find $\sum_{k=1}^{50} b_k$.

55. Construction: On a certain project, a construction company was penalized for taking more than the contractual time to finish the project. The company forfeited $750 the first day, $900 the second day, $1050 the third day, and so on. How many additional days were needed if the total penalty was $12,150?

56. Real estate: It is estimated that a certain piece of property, currently valued at $480,000, will appreciate as follows: $14,000 the first year, $14,500 the second year, $15,000 the third year, and so on. On this basis, what will be the value of the property after 10 years?

50. $\frac{296}{3}$ 51. 171
52. −195 53. −100
54. 141 55. 9 days
56. $642,500
57. 1625 cm
58. 100 blocks
59. 134 seats; 1540 seats
60. $45,000
61. An arithmetic sequence has a constant difference *d*. If the terms alternated being positive and negative, then *d* would also have to alternate being positive and negative. This is a contradiction.

57. **Rungs of a ladder:** The rungs of a ladder decrease uniformly in width from 84 cm to 46 cm. What is the total length of the wood in the rungs if there are 25 of them?

58. **Building blocks:** How many blocks are there in a pile if there are 19 in the first layer, 17 in the second layer, 15 in the third layer, and so on, with only 1 block on the top layer?

59. **Theater:** Theater seats are arranged in semicircles so that there are 6 additional seats in each semicircular "row" moving away from the stage. The first row has 20 seats, and there are 20 rows. How many seats are in the last "row"? How many seats are in the theater?

60. **Student loan repayment:** Samantha accumulated $50,000 in student loans during her four years in college. She has agreed to pay back $1000 the first year and increase the payment by $500 each year thereafter. How much will she have paid back in 12 years?

Writing and Thinking About Mathematics

61. Explain why an alternating sequence (one in which the terms alternate being positive and negative) cannot be an arithmetic sequence.

HAWKES LEARNING SYSTEMS: INTRODUCTORY & INTERMEDIATE ALGEBRA SOFTWARE

- 13.3 Arithmetic Sequences

13.4 Geometric Sequences and Series

- *Determine whether or not a sequence is* ***geometric****.*
- *Find the general term of a geometric sequence.*
- *Find the specified terms of a geometric sequence.*
- *Find the sum of the first n terms of a geometric sequence.*
- *Find the sum of an infinite geometric* ***series****.*

Geometric Sequences

Arithmetic sequences are characterized by having the property that any two consecutive terms have the same difference. **Geometric sequences** are characterized by having the property that **any two consecutive terms are in the same ratio**. That is, if consecutive terms are divided, the ratio will be the same regardless of which two consecutive terms are divided. Consider the three sequences

$$\frac{1}{2}, \frac{1}{4}, \frac{1}{8}, \frac{1}{16}, \frac{1}{32}, \ldots$$

$$3, 9, 27, 81, 243, \ldots$$

$$-9, 3, -1, \frac{1}{3}, -\frac{1}{9}, \ldots$$

As the following patterns show, each of these sequences has a **common ratio** when consecutive terms are divided.

$$\underbrace{\frac{1}{2}, \frac{1}{4}}_{\frac{1}{2}}, \underbrace{\frac{1}{8}}_{\frac{1}{2}}, \underbrace{\frac{1}{16}}_{\frac{1}{2}}, \underbrace{\frac{1}{32}}_{\frac{1}{2}}, \ldots \qquad \frac{\frac{1}{4}}{\frac{1}{2}} = \frac{1}{2}, \frac{\frac{1}{8}}{\frac{1}{4}} = \frac{1}{2}, \frac{\frac{1}{16}}{\frac{1}{8}} = \frac{1}{2}, \text{ and so on.}$$

$$\underbrace{3, 9}_{3}, \underbrace{27}_{3}, \underbrace{81}_{3}, \underbrace{243}_{3}, \ldots \qquad \frac{9}{3} = 3, \frac{27}{9} = 3, \frac{81}{27} = 3, \text{ and so on.}$$

$$\underbrace{-9, 3}_{-\frac{1}{3}}, \underbrace{-1}_{-\frac{1}{3}}, \underbrace{\frac{1}{3}}_{-\frac{1}{3}}, \underbrace{-\frac{1}{9}}_{-\frac{1}{3}}, \ldots \qquad \frac{3}{-9} = -\frac{1}{3}, \frac{-1}{3} = -\frac{1}{3}, \frac{\frac{1}{3}}{-1} = -\frac{1}{3}, \text{ and so on.}$$

Therefore these sequences are **geometric sequences** or **geometric progressions**.

Geometric Sequences

A sequence $\{a_n\}$ is called a **geometric sequence** (or **geometric progression**) if for any positive integer k,

$$\frac{a_{k+1}}{a_k} = r \qquad \text{where } r \text{ is constant and } r \neq 0.$$

r is called the **common ratio**.

Example 1: A Geometric Sequence

Teaching Notes:
You might want to deal with the problem in a more open-ended way after showing these examples. What if we are given a sequence and want to determine if it is geometric?

Show that the sequence $\left\{\frac{1}{3^n}\right\}$ is geometric by finding r.

Solution: $a_k = \frac{1}{3^k}$ and $a_{k+1} = \frac{1}{3^{k+1}}$

$$\frac{a_{k+1}}{a_k} = \frac{\frac{1}{3^{k+1}}}{\frac{1}{3^k}} = \frac{1}{3^{k+1}} \cdot \frac{3^k}{1} = \frac{\cancel{3^k}}{3 \cdot \cancel{3^k}} = \frac{1}{3} = r$$

Example 2: A Sequence that is Not Geometric

Students might be guided through an exploration strategy that includes trying to show there is a common ratio r, but if there is difficulty in resolving the algebra, perhaps they need to change the approach and suspect that no such r-value exists for the given sequence. Then have them choose several terms and test them.

Show that the sequence $\{n^2\}$ is not geometric.

Solution: To show that something is not true, a **counterexample** must be found. In this case, we find two pairs of successive terms and show that they have different ratios.

Consider the ratio $\frac{a_3}{a_2} = \frac{3^2}{2^2} = \frac{9}{4}$ and the ratio $\frac{a_6}{a_5} = \frac{6^2}{5^2} = \frac{36}{25}$.

Because $\frac{9}{4} \neq \frac{36}{25}$ there is no common ratio for successive terms and the sequence $\{n^2\}$ is **not** geometric.

If the first term is a_1 and the common ratio is r, then the geometric sequence can be indicated as follows:

$a_1 = a_1$	$\to$	a_1	first term
$\frac{a_2}{a_1} = r$	$\to$	$a_2 = a_1 r$	second term
$\frac{a_3}{a_2} = r$	$\to$	$a_3 = a_2 r = (a_1 r)r = a_1 r^2$	third term
$\frac{a_4}{a_3} = r$	$\to$	$a_4 = a_3 r = (a_1 r^2)r = a_1 r^3$	fourth term
$\vdots$	$\vdots$		$\vdots$
$\frac{a_n}{a_{n-1}} = r$	$\to$	$a_n = a_{n-1} r = (a_1 r^{n-2})r = a_1 r^{n-1}$	n^{th} term
$\vdots$	$\vdots$		$\vdots$

The Formula for the nth Term of a Geometric Sequence

If $\{a_n\}$ is a geometric sequence, then the nth term has the form

$$\boldsymbol{a_n = a_1 r^{n-1}}$$

where r is the common ratio.

Example 3: Finding a Specified Term of a Geometric Sequence

a. If in a geometric sequence $a_1 = 4$ and $r = -\frac{1}{2}$, find a_8.

Solution: $a_8 = a_1 r^7 = 4\left(-\frac{1}{2}\right)^7 = 2^2\left(-\frac{1}{2^7}\right) = -\frac{1}{2^5} = -\frac{1}{32}$

b. Find the seventh term of the following geometric sequence: $3, \frac{3}{2}, \frac{3}{4}, \ldots$

Solution: Find r using the formula $r = \frac{a_{k+1}}{a_k}$ with $a_1 = 3$ and $a_2 = \frac{3}{2}$.

$$r = \frac{a_2}{a_1} = \frac{\frac{3}{2}}{3} = \frac{3}{2} \cdot \frac{1}{3} = \frac{1}{2}$$

Now the seventh term is $a_7 = a_1 r^{7-1} = 3\left(\frac{1}{2}\right)^6 = \frac{3}{64}$.

Example 4: Finding a_1 and r for a Geometric Sequence

Find a_1 and r for the geometric sequence in which $a_5 = 2$ and $a_7 = 4$.

Solution: Using the formula $a_n = a_1 r^{n-1}$, we get

$$2 = a_1 r^4 \text{ and } 4 = a_1 r^6.$$

Now dividing gives

$$\frac{a_1 r^6}{a_1 r^4} = \frac{4}{2}$$

$$r^2 = 2$$

$$r = \pm\sqrt{2}.$$

Using these values for r and the fact that $a_5 = 2$, we can find a_1.

For $r = \sqrt{2}$:

$$2 = a_1\left(\sqrt{2}\right)^4$$

$$2 = a_1 \cdot 4$$

$$\frac{1}{2} = a_1$$

For $r = -\sqrt{2}$:

$$2 = a_1\left(-\sqrt{2}\right)^4$$

$$2 = a_1 \cdot 4$$

$$\frac{1}{2} = a_1$$

Thus two geometric sequences with $a_5 = 2$ and $a_7 = 4$ exist. In both cases, $a_1 = \frac{1}{2}$. The two possibilities are

$$a_1 = \frac{1}{2} \text{ and } r = \sqrt{2}$$

or $$a_1 = \frac{1}{2} \text{ and } r = -\sqrt{2}.$$

Partial Sum of a Geometric Sequence

The following discussion illustrates the method for finding the formula for the sum of the first n terms of a geometric sequence. Such a sum is called a **partial sum** of the sequence.

The sum of the first 6 terms of the geometric sequence $\left\{\frac{1}{3^n}\right\}$ can be indicated as $S = \sum_{k=1}^{6} \frac{1}{3^k}$.

To find the value of this partial sum, we can write the terms and simply add them:

$$S = \sum_{k=1}^{6} \frac{1}{3^k} = \frac{1}{3} + \frac{1}{3^2} + \frac{1}{3^3} + \frac{1}{3^4} + \frac{1}{3^5} + \frac{1}{3^6}$$

$$= \frac{3^5 + 3^4 + 3^3 + 3^2 + 3 + 1}{3^6} = \frac{364}{729}.$$

Adding a few terms is relatively easy. However, the following procedure will help in understanding the development of a general formula for partial sums. Write the terms in order and then multiply each term by $\frac{1}{3}$ (the common ratio, r). Arranging the terms in a vertical format and then subtracting gives the following results.

$$\begin{array}{rcccccccccccccc} S & = & \frac{1}{3} & + & \frac{1}{3^2} & + & \frac{1}{3^3} & + & \frac{1}{3^4} & + & \frac{1}{3^5} & + & \frac{1}{3^6} & & \\ \frac{1}{3}S & = & & & \frac{1}{3^2} & + & \frac{1}{3^3} & + & \frac{1}{3^4} & + & \frac{1}{3^5} & + & \frac{1}{3^6} & + & \frac{1}{3^7} \\ \hline S - \frac{1}{3}S & = & \frac{1}{3} & - & 0 & - & 0 & - & 0 & - & 0 & - & 0 & - & \frac{1}{3^7} \end{array}$$

$$\left(1 - \frac{1}{3}\right)S = \frac{1}{3} - \frac{1}{3^7} \qquad \text{Factor out the } S.$$

$$S = \frac{\frac{1}{3} - \frac{1}{3^7}}{1 - \frac{1}{3}} = \frac{\frac{1}{3} - \left(\frac{1}{3}\right)^7}{1 - \frac{1}{3}} \qquad \text{Divide by } \left(1 - \frac{1}{3}\right).$$

Teaching Notes:
It might be instructional to point out, if you haven't already, that infinitely repeating decimals (e.g., $0.\overline{3}$) are expandable as an infinite sum of terms having a common ratio. Students' familiarity, at least with the numbers, might help them understand a finite sum from an infinite number of terms.

Using this technique with a general geometric sequence $\{a_1 r^{n-1}\}$ leads to the formula for partial sums of geometric sequences.

Consider the first n terms:

$$a_1,\ a_2 = a_1 r,\ a_3 = a_1 r^2,\ a_4 = a_1 r^3,\ \ldots,\ a_{n-1} = a_1 r^{n-2},\ a_n = a_1 r^{n-1}.$$

Then, as before, the sum can be written

$$\begin{array}{rll} S = & a_1 + a_1 r + a_1 r^2 + \ldots + a_1 r^{n-2} + a_1 r^{n-1} & \\ rS = & \quad\ a_1 r + a_1 r^2 + a_1 r^3 + \ldots + a_1 r^{n-1} + a_1 r^n & \text{Multiply each term by } r. \\ S - rS = & a_1 - 0 - 0 - 0 - \ldots - 0 - 0 - a_1 r^n & \text{Subtract.} \\ \hline (1 - r)S = & a_1\left(1 - r^n\right) & \text{Factor.} \\ S = & \dfrac{a_1\left(1 - r^n\right)}{1 - r}. & \text{Simplify.} \end{array}$$

Partial Sums of Geometric Sequences

The nth **partial sum** S_n of the first n terms of a geometric sequence $\{a_n\}$, is

$$\boldsymbol{S_n = \sum_{k=1}^{n} a_k = \frac{a_1\left(1-r^n\right)}{1-r}}$$ where r is the common ratio and $r \neq 1$.

Example 5: Partial Sums of Geometric Sequences

First show that the corresponding sequence is a geometric sequence by finding $\dfrac{a_{k+1}}{a_k} = r$.

Then find the indicated sum using the formula $\displaystyle\sum_{k=1}^{n} a_k = \frac{a_1\left(1-r^n\right)}{1-r}$.

a. $\displaystyle\sum_{k=1}^{10} \frac{1}{2^k}$

Solution: Represent both a_k and a_{k+1} and find the ratio of these two terms.

$$a_k = \frac{1}{2^k} \text{ and } a_{k+1} = \frac{1}{2^{k+1}}$$

$$\frac{a_{k+1}}{a_k} = \frac{\frac{1}{2^{k+1}}}{\frac{1}{2^k}} = \frac{1}{2^{k+1}} \cdot \frac{2^k}{1} = \frac{1}{2 \cdot \cancel{2^k}} \cdot \frac{\cancel{2^k}}{1} = \frac{1}{2} = r$$

So, $\left\{\dfrac{1}{2^n}\right\}$ is a geometric sequence with $a_1 = \dfrac{1}{2}$ and $r = \dfrac{1}{2}$:

$$\sum_{k=1}^{10} \frac{1}{2^k} = \frac{\frac{1}{2}\left(1-\left(\frac{1}{2}\right)^{10}\right)}{1-\frac{1}{2}} = \frac{\cancel{\frac{1}{2}}\left(1-\frac{1}{1024}\right)}{\cancel{\frac{1}{2}}} = \frac{1023}{1024}.$$

b. $\displaystyle\sum_{k=1}^{5} (-1)^k \cdot 3^{\frac{k}{2}}$

Solution: Represent both a_k and a_{k+1} and find the ratio of these two terms.

$$\frac{a_{k+1}}{a_k} = \frac{(-1)^{k+1} \cdot 3^{\frac{k+1}{2}}}{(-1)^k \cdot 3^{\frac{k}{2}}} = \frac{\cancel{(-1)^k}(-1) \cdot \cancel{3^{\frac{k}{2}}} \cdot 3^{\frac{1}{2}}}{\cancel{(-1)^k} \cdot \cancel{3^{\frac{k}{2}}}}$$

$$= (-1) \cdot 3^{\frac{1}{2}} = -\sqrt{3} = r$$

So, $\left\{(-1)^k \cdot 3^{\frac{k}{2}}\right\}$ is a geometric sequence with $a_1 = -\sqrt{3}$ and $r = -\sqrt{3}$:

$$\sum_{k=1}^{5} (-1)^k \cdot 3^{\frac{k}{2}} = \frac{-\sqrt{3}\left(1-\left(-\sqrt{3}\right)^5\right)}{1-\left(-\sqrt{3}\right)} = \frac{-\sqrt{3}\left(1+9\sqrt{3}\right)}{1+\sqrt{3}}.$$

Example 6: Partial Sums of Geometric Sequences

The parents of a small child decide to deposit \$1000 annually at the first of each year for 20 years for their child's education. If interest is compounded annually at 8%, what will be the value of the deposits after 20 years? (This type of investment is called an **annuity**.)

Solution: The formula for interest compounded annually is $A = P(1+r)^t$ where A is the amount in the account, r is the annual interest rate (in decimal form), and t is the time (in years).

The first deposit of \$1000 will earn interest for 20 years:

$$A_{20} = 1000(1+0.08)^{20} = 1000(1.08)^{20}.$$

The second deposit will earn interest for 19 years:

$$A_{19} = 1000(1.08)^{19}.$$

$$\vdots$$

The last deposit will earn interest for only one year:

$$A_1 = 1000(1.08)^1.$$

The accumulated value of all deposits (plus interest) is the sum of the 20 terms of a geometric sequence.

Value at the end of twenty years:

$$= A_1 + A_2 + \ldots + A_{20} = \sum_{k=1}^{20} 1000(1.08)^k$$

$$= 1000(1.08)^1 + 1000(1.08)^2 + \ldots + 1000(1.08)^{20}$$

$$= \frac{1000(1.08)\left[1-(1.08)^{20}\right]}{1-1.08}$$ Where $a_1 = 1000(1.08)$, $r = 1.08$, and $n = 20$

$$= \frac{1080(1-4.660957)}{-0.08}$$

$$= 49{,}423$$ Rounded to the nearest dollar

Thus the accumulated value of the annuity is \$49,423.

Geometric Series

The indicated sum of all the terms (an infinite number of terms) of a sequence is called an **infinite series** (or simply a **series**). A thorough study of series is a part of calculus. In this text, we will be concerned only with special cases of **geometric series** where $|r| < 1$. The symbol for infinity (∞) is used to indicate that the number of terms is unbounded. The symbol ∞ does not represent a number.

Infinite Series

The indicated sum of all terms of a sequence is called an **infinite series** (or a **series**). For a sequence $\{a_n\}$, the corresponding series can be written as follows:

$$\sum_{k=1}^{\infty}(a_k) = a_1 + a_2 + a_3 + \ldots + a_n + \ldots$$

For geometric sequences in the case where $|r| < 1$, it can be shown, in higher level mathematics, that r^n approaches 0 as n approaches infinity. This does not mean that r^n is ever equal to 0, only that it gets closer and closer to 0 as n becomes larger and larger. In symbols, we write

$$r^n \to 0 \quad \text{as} \quad n \to \infty.$$

Thus we have the following result if $|r| < 1$.

$$S_n = \frac{a_1(1-r^n)}{1-r} \to \frac{a_1(1-0)}{1-r} = \frac{a_1}{1-r} \qquad \text{as } n \to \infty.$$

Sum of an Infinite Geometric Series

If $\{a_n\}$, is a geometric sequence and $|r| < 1$, then the sum of the infinite geometric series is

$$S = \sum_{k=1}^{\infty}(a_k) = a_1 + a_1 r + a_1 r^2 + \ldots = \frac{a_1}{1-r}.$$

Example 7: Geometric Series

Teaching Notes:
You might illustrate, by plotting points on a number line, that the sum of a series is a limiting value and that as more terms are added, the sum gets closer and closer to the limit.

Find the sum of each of the following geometric series.

a. $$\sum_{k=1}^{\infty}\left(\frac{2}{3}\right)^{k-1} = \left(\frac{2}{3}\right)^0 + \left(\frac{2}{3}\right)^1 + \left(\frac{2}{3}\right)^2 + \left(\frac{2}{3}\right)^3 + \ldots$$
$$= 1 + \frac{2}{3} + \frac{4}{9} + \frac{8}{27} + \ldots$$

Solution: Here, $a_1 = 1$ and $r = \frac{2}{3}$. Substitution in the formula yields

$$S = \frac{a_1}{1-r} = \frac{1}{1-\frac{2}{3}} = \frac{1}{\frac{1}{3}} = 3.$$

b. $0.3333\ldots = 0.\overline{3}$ Recall that the bar over the 3 indicates a repeating pattern of digits in the decimal.

Solution: $0.33333\ldots = 0.3 + 0.03 + 0.003 + 0.0003 + 0.00003 + \ldots$

This format shows that the decimal number can be interpreted as a geometric series with $a_1 = 0.3 = \frac{3}{10}$ and $r = 0.1 = \frac{1}{10}$.

Applying the formula gives

$$S = \frac{a_1}{1-r} = \frac{\frac{3}{10}}{1-\frac{1}{10}} = \frac{\frac{3}{10}}{\frac{9}{10}} = \frac{\cancel{3}}{\cancel{10}} \cdot \frac{\cancel{10}}{\cancel{9}_{3}} = \frac{1}{3}.$$

In this way, an infinite repeating decimal can be converted to fraction form:

$$0.33333\ldots = \frac{1}{3}.$$

c. $0.99999\ldots = 0.\overline{9}$

Solution: As shown in Example 7b, the decimal number can be interpreted

$$0.99999\ldots = 0.9 + 0.09 + 0.009 + 0.0009 + \ldots$$

which is a geometric series with $a_1 = 0.9 = \frac{9}{10}$ and $r = 0.1 = \frac{1}{10}$.

Applying the formula gives

$$S = \frac{a_1}{1-r} = \frac{\frac{9}{10}}{1-\frac{1}{10}} = \frac{\frac{9}{10}}{\frac{9}{10}} = \frac{\cancel{9}}{\cancel{10}} \cdot \frac{\cancel{10}}{\cancel{9}} = 1.$$

This very interesting result shows that the infinite decimal notation 0.99999... is just another way of writing 1.

d. $5 - 1 + \frac{1}{5} - \frac{1}{25} + \frac{1}{125} - \frac{1}{625} + \ldots$

Solution: Here, $a_1 = 5$ and $r = -\frac{1}{5}$. (Note that a geometric series that alternates in sign will always have a negative value for r.) Substitution in the formula gives

$$S = \frac{a_1}{1-r} = \frac{5}{1-\left(-\frac{1}{5}\right)} = \frac{5}{1+\frac{1}{5}} = \frac{5}{\frac{6}{5}} = \frac{5}{1} \cdot \frac{5}{6} = \frac{25}{6}.$$

Practice Problems

1. Show that the sequence $\left\{\frac{(-1)^n}{3^n}\right\}$ is geometric by finding r.

2. If in a geometric series $a_1 = 0.1$ and $r = 2$, find a_6.

3. Find the sum $\sum_{k=1}^{5} \frac{1}{2^k}$.

4. Represent the decimal $0.\overline{4}$ as a series using Σ notation, then find the sum.

13.4 Exercises

Determine which sequences are geometric. If the sequence is geometric, find its common ratio and write a formula for the nth term.

1. $2, 4, 6, 8, \ldots$

2. $\frac{1}{12}, \frac{1}{6}, \frac{1}{3}, \frac{2}{3}, \ldots$

3. $3, -\frac{3}{2}, \frac{3}{4}, -\frac{3}{8}, \ldots$

4. $5, 9, 13, 17, \ldots$

5. $\frac{32}{27}, \frac{4}{9}, \frac{1}{6}, \frac{1}{16}, \ldots$

6. $18, 12, 8, \frac{16}{3}, \ldots$

7. $\frac{14}{3}, \frac{2}{3}, \frac{2}{15}, \frac{2}{45}, \ldots$

8. $1, -\frac{2}{3}, \frac{4}{9}, -\frac{8}{27}, \ldots$

9. $48, -12, 3, -\frac{3}{4}, \ldots$

10. $4, -8, 12, -16, \ldots$

For each of the following sequences, write the first four terms and then determine whether the sequence is geometric.

11. $\left\{(-3)^{n+1}\right\}$

12. $\left\{3\left(\frac{2}{5}\right)^n\right\}$

13. $\left\{\frac{2}{3}n\right\}$

14. $\left\{(-1)^{n+1}\left(\frac{2}{7}\right)^n\right\}$

15. $\left\{2\left(-\frac{4}{5}\right)^n\right\}$

16. $\left\{1+\frac{1}{2^n}\right\}$

17. $\left\{3(2)^{\frac{n}{2}}\right\}$

18. $\left\{\frac{n^2+1}{n}\right\}$

19. $\left\{(-1)^{n-1}(0.3)^n\right\}$

20. $\left\{6(10)^{1-n}\right\}$

Use the given information to find $\{a_n\}$ for each of the geometric sequences.

21. $a_1 = 3, r = 2$

22. $a_1 = -2, r = \frac{1}{5}$

23. $a_1 = \frac{1}{3}, r = -\frac{1}{2}$

1. Not a geometric sequence
2. Geometric sequence $r = 2; \left\{\frac{1}{12}\left(2^{n-1}\right)\right\}$
3. Geometric sequence $r = -\frac{1}{2}; \left\{3\left(-\frac{1}{2}\right)^{n-1}\right\}$
4. Not a geometric sequence
5. Geometric sequence $r = \frac{3}{8}; \left\{\frac{32}{27}\left(\frac{3}{8}\right)^{n-1}\right\}$
6. Geometric sequence $r = \frac{2}{3}; \left\{18\left(\frac{2}{3}\right)^{n-1}\right\}$
7. Not a geometric sequence
8. Geometric sequence $r = -\frac{2}{3}; \left\{\left(-\frac{2}{3}\right)^{n-1}\right\}$
9. Geometric sequence $r = -\frac{1}{4}; \left\{48\left(-\frac{1}{4}\right)^{n-1}\right\}$
10. Not a geometric sequence
11. $9, -27, 81, -243$; Geometric sequence
12. $\frac{6}{5}, \frac{12}{25}, \frac{24}{125}, \frac{48}{625}$; Geometric sequence
13. $\frac{2}{3}, \frac{4}{3}, 2, \frac{8}{3}$; Not a geometric sequence

Answers to Practice Problems: **1.** $r = -\frac{1}{3}$ **2.** $a_6 = 3.2$ **3.** $\frac{31}{32}$ **4.** $\sum_{k=1}^{\infty} \frac{4}{10^k}; \frac{4}{9}$

14. $\frac{2}{7}, -\frac{4}{49}, \frac{8}{343}, -\frac{16}{2401}$; Geometric sequence

15. $-\frac{8}{5}, \frac{32}{25}, -\frac{128}{125}, \frac{512}{625}$; Geometric sequence

16. $\frac{3}{2}, \frac{5}{4}, \frac{9}{8}, \frac{17}{16}$; Not a geometric sequence

17. $3\sqrt{2}, 6, 6\sqrt{2}, 12$; Geometric sequence

18. $2, \frac{5}{2}, \frac{10}{3}, \frac{17}{4}$; Not a geometric sequence

19. $0.3, -0.09, 0.027, -0.0081$; Geometric sequence

20. $6, \frac{3}{5}, \frac{3}{50}, \frac{3}{500}$; Geometric sequence

21. $\left\{3(2)^{n-1}\right\}$

22. $\left\{-2\left(\frac{1}{5}\right)^{n-1}\right\}$

23. $\left\{\frac{1}{3}\left(-\frac{1}{2}\right)^{n-1}\right\}$

24. $\left\{5\left(\sqrt{2}\right)^{n-1}\right\}$

25. $\left\{\left(\sqrt{2}\right)^{n-1}\right\}$

26. $\left\{\frac{19}{27}(3)^{n-1}\right\}$

27. $\left\{\frac{1}{3}(3)^{n-1}\right\}$

28. $\left\{10\left(\frac{1}{2}\right)^{n-1}\right\}$

29. $\left\{-5\left(-\frac{3}{4}\right)^{n-1}\right\}$

30. $\left\{2(3)^{n-1}\right\}$ 31. $\frac{1}{4}$ 32. $\frac{5}{16}$ 33. $\frac{32}{9}$ 34. -192 35. $\frac{1}{4}$ or $-\frac{1}{4}$ 36. 768

24. $a_1 = 5, r = \sqrt{2}$ **25.** $a_3 = 2, a_5 = 4, r > 0$ **26.** $a_4 = 19, a_5 = 57$

27. $a_2 = 1, a_4 = 9, r > 0$ **28.** $a_2 = 5, a_5 = \frac{5}{8}$ **29.** $a_3 = -\frac{45}{16}, r = -\frac{3}{4}$

30. $a_4 = 54, r = 3$

Assume $\{a_n\}$ is a geometric sequence. Find the indicated quantity.

31. $a_1 = -32, a_6 = 1$. Find a_8. **32.** $a_1 = 20, a_6 = \frac{5}{8}$. Find a_7.

33. $a_1 = 18, a_7 = \frac{128}{81}$. Find a_5. **34.** $a_1 = -3, a_5 = -48$. Find a_7.

35. $a_3 = \frac{1}{2}, a_7 = \frac{1}{32}$. Find a_4. **36.** $a_5 = 48, a_8 = -384$. Find a_9.

37. $a_1 = -2, r = \frac{2}{3}, a_n = -\frac{16}{27}$. Find n. **38.** $a_1 = \frac{1}{9}, r = \frac{3}{2}, a_n = \frac{27}{32}$. Find n.

Find the indicated sums.

39. $3 + 9 + 27 + 81 + 243$ **40.** $-2 + 4 - 8 + 16$ **41.** $8 + 4 + 2 + \ldots + \frac{1}{64}$

42. $3 + 12 + 48 + \ldots + 3072$ **43.** $\sum_{k=1}^{3} -3\left(\frac{3}{4}\right)^k$ **44.** $\sum_{k=1}^{6}\left(-\frac{5}{3}\right)\left(\frac{1}{2}\right)^k$

45. $\sum_{k=1}^{5}\left(\frac{2}{3}\right)^k$ **46.** $\sum_{k=1}^{6}\left(\frac{1}{3}\right)^k$ **47.** $\sum_{k=4}^{7} 5\left(\frac{1}{2}\right)^k$ **48.** $\sum_{k=3}^{6} -7\left(\frac{3}{2}\right)^k$

49. $\sum_{k=1}^{\infty}\left(\frac{3}{4}\right)^{k-1}$ **50.** $\sum_{k=1}^{\infty}\left(\frac{5}{8}\right)^{k-1}$ **51.** $\sum_{k=1}^{\infty}\left(-\frac{1}{2}\right)^k$ **52.** $\sum_{k=1}^{\infty}\left(-\frac{2}{5}\right)^k$

53. $0.\overline{4}$ **54.** $0.\overline{6}$ **55.** $0.\overline{36}$ **56.** $0.\overline{81}$

57. Trust accounts: When Henry was born, his grandmother deposited \$10,000 in a trust account bearing 5% interest compounded annually for him to use for college expenses when he became 21. How much money was in the account on his 21st birthday?

58. Automobile depreciation: An automobile that costs \$18,500 new depreciates at a rate of 20% of its value each year. What is its value after 4 years?

59. Car repair: The radiator of a truck contains 20 liters of water. 4 liters are drained off and replaced by antifreeze. Then 4 more liters of the mixture are drained off and replaced by antifreeze, and so on. This process is continued until six drain-offs and replacements have been made. How much antifreeze is in the final mixture?

60. Investing: Suppose \$1200 is deposited in a savings account each year for 8 years. If interest is compounded annually at 6%, what would be the value of the account at the end of 8 years?

61. Certificate of deposit: Kathleen buys a \$1000 certificate of deposit each year for 10 years. If the annual interest rate on each CD is 4.5%, what will be the total value of these CDs after 10 years?

62. Decay rate: A substance decays at a rate of $\frac{2}{5}$ of its weight per day. How much of the substance will be present after 4 days if initially there are 500 grams?

63. Bouncing a ball: A ball rebounds to a height that is $\frac{3}{4}$ of its original height. How high will it rise after the fourth bounce if it is dropped from a height of 24 meters?

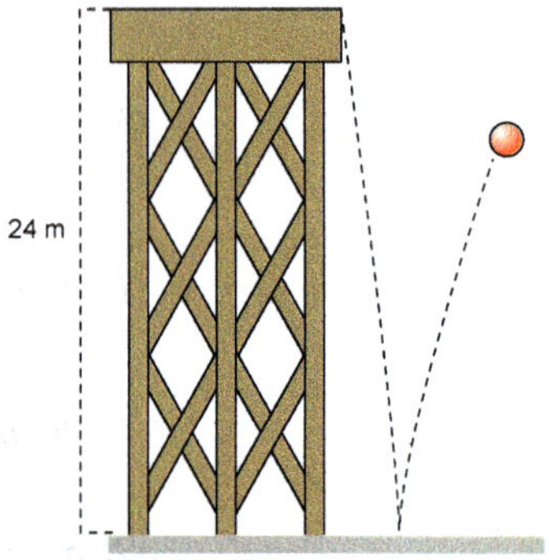

37. $n = 4$ 38. $n = 6$
39. 363 40. 10
41. $\frac{1023}{64}$ 42. 4095
43. $-\frac{333}{64}$ 44. $-\frac{105}{64}$
45. $\frac{422}{243}$ 46. $\frac{364}{729}$
47. $\frac{75}{128}$ 48. $-\frac{12,285}{64}$
49. 4 50. $\frac{8}{3}$
51. $-\frac{1}{3}$ 52. $-\frac{2}{7}$
53. $\frac{4}{9}$ 54. $\frac{2}{3}$
55. $\frac{4}{11}$ 56. $\frac{9}{11}$
57. \$27,859.63
58. \$7577.60
59. 14.76 liters
60. \$12,589.58

Writing and Thinking About Mathematics

64. Graph the first 8 partial sums of each geometric series as points to show how the sum of the series approaches a certain value. Show this value as a horizontal line on the graph.

a. $\sum_{k=1}^{\infty}\left(\frac{1}{2}\right)^{k-1}$ **b.** $\sum_{k=1}^{\infty}\frac{(-1)^{k+1}}{3^k}$

65. Consider the infinite series $4\cdot\sum_{k=1}^{\infty}\frac{(-1)^{k-1}}{2k-1}$. Write out several (at least 10 to 15) of the partial sums and their values until you can identify the number the partial sums "seem" to be approaching. What is this number?

66. Explain why there is no formula for finding the sum of an infinite geometric series when $|r| > 1$.

61. \$12,841.18
62. 64.8 grams
63. 7.59 meters
64. a.

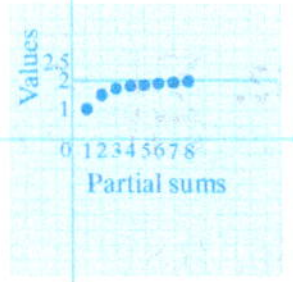

b.

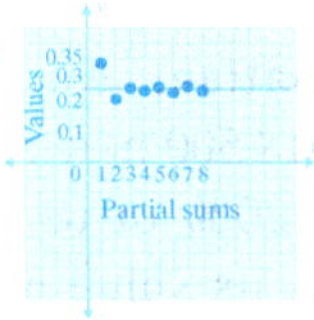

65. π
66. In the formula for S_n, $r^n \to \infty$ if $|r| > 1$.

HAWKES LEARNING SYSTEMS: INTRODUCTORY & INTERMEDIATE ALGEBRA SOFTWARE

- 13.4 Geometric Sequences and Series

13.5 The Binomial Theorem

- *Calculate* ***factorials****.*
- *Expand binomials using the* ***binomial theorem****.*
- *Find specified terms in binomial expressions.*

Factorials

The objective in this section is to develop a formula for writing powers of binomial expressions. This formula is called the **binomial theorem** (or **binomial expansion**). With the binomial theorem products such as

$$(a+b)^3, \quad (x+y)^7, \quad \text{and} \quad (2x+5)^8$$

can be written without having to actually multiply the binomial factors. For example, instead of multiplying three factors as follows,

$$\begin{aligned}(a+b)^3 &= (a+b)(a+b)(a+b)\\ &= (a^2+2ab+b^2)(a+b)\\ &= a^3+2a^2b+ab^2+a^2b+2ab^2+b^3\\ &= a^3+3a^2b+3ab^2+b^3\end{aligned}$$

knowledge of the binomial theorem will lead directly to the final polynomial.

Before discussing the theorem itself, we need to understand the concept of **factorial**. For example, 6! (read "six factorial") represents the product of the positive integers from 6 to 1. Thus

$$6! = 6\cdot5\cdot4\cdot3\cdot2\cdot1 = 720.$$

Similarly,

$$10! = 10\cdot9\cdot8\cdot7\cdot6\cdot5\cdot4\cdot3\cdot2\cdot1 = 3{,}628{,}800.$$

n Factorial ($n!$)

For any positive integer n,

$$\mathbf{n! = n(n-1)(n-2)\ldots(3)(2)(1).}$$

$n!$ is read as "n factorial."

To evaluate an expression, such as $\frac{7!}{6!}$, do **not** evaluate each factorial in the numerator and denominator. Instead, write the factorials as products and reduce the fraction.

$$\frac{7!}{6!}=\frac{7\cdot \cancel{6}\cdot \cancel{5}\cdot \cancel{4}\cdot \cancel{3}\cdot \cancel{2}\cdot \cancel{1}}{\cancel{6}\cdot \cancel{5}\cdot \cancel{4}\cdot \cancel{3}\cdot \cancel{2}\cdot \cancel{1}}=7$$

Note that $n!=(n)(n-1)(n-2)...(3)(2)(1)$

and $(n-1)!=(n-1)(n-2)(n-3)...(3)(2)(1)$.

So $n!=n(n-1)!$

In particular, $\frac{7!}{6!}=\frac{7\cdot (\cancel{6!})}{\cancel{6!}}=7.$

Also, for work with formulas involving factorials, zero factorial is defined to be 1.

0 Factorial (0!)

0! = 1

Using a Calculator to Calculate Factorials

Teaching Notes:
Have your students evaluate 0! with their calculator. This will serve to convince some of your doubting students that 0! = 1.

Factorials can be calculated with the TI-84 Plus calculator by pressing the MATH key and going to the menu under PRB. The fourth item in the list is the factorial symbol, !. For example, 6! can be calculated as follows:

1. Enter 6.
2. Press the MATH key.
3. Go to the PRB heading and press 4. (6! will appear on the display.)
4. Press ENTER and 720 will appear on the display.

Example 1: Factorials

Simplify the following expressions.

a. $\frac{11!}{8!}$

Solution: $\frac{11!}{8!}=\frac{11\cdot 10\cdot 9\cdot (\cancel{8}\cdot \cancel{7}\cdot \cancel{6}\cdot \cancel{5}\cdot \cancel{4}\cdot \cancel{3}\cdot \cancel{2}\cdot \cancel{1})}{(\cancel{8}\cdot \cancel{7}\cdot \cancel{6}\cdot \cancel{5}\cdot \cancel{4}\cdot \cancel{3}\cdot \cancel{2}\cdot \cancel{1})}=990$

or $\frac{11!}{8!}=\frac{11\cdot 10\cdot 9\cdot \cancel{8!}}{\cancel{8!}}=990$

b. $\dfrac{n!}{(n-2)!}$

Solution: $\dfrac{n!}{(n-2)!}=\dfrac{n(n-1)\cancel{(n-2)}!}{\cancel{(n-2)}!}=n(n-1)$

c. $\dfrac{30!}{28!2!}$

Solution: $\dfrac{30!}{28!2!}=\dfrac{\overset{15}{\cancel{30}}\cdot 29\cdot \cancel{28!}}{\cancel{28!}\cdot \cancel{2}\cdot 1}=15\cdot 29=435$

Binomial Coefficients

The expression in Example 1c can be written in the following notation.

$$\binom{30}{2}=\frac{30!}{2!28!} \text{ and } \binom{30}{28}=\frac{30!}{28!2!}$$

Binomial Coefficient $\binom{n}{r}$

For non-negative integers n and r, with $0 \le r \le n$,

$$\binom{n}{r}=\frac{n!}{r!(n-r)!}.$$

Because this quantity appears repeatedly in the binomial theorem, $\binom{n}{r}$ is often called a **binomial coefficient**.

To get a formula for $\binom{n}{n-r}$, we apply the formula for $\binom{n}{r}$ and replace r with $n-r$.

Thus

$$\binom{n}{n-r}=\frac{n!}{(n-r)!(n-(n-r))!}=\frac{n!}{(n-r)!(n-n+r)!}$$

$$=\frac{n!}{(n-r)!r!}=\frac{n!}{r!(n-r)!}=\binom{n}{r}.$$

Thus

$$\binom{n}{n-r}=\binom{n}{r}.$$

Example 2: $\binom{n}{r}$

Evaluate the following.

a. $\binom{8}{2}$ and $\binom{8}{6}$

Solution: $\binom{8}{2} = \dfrac{8!}{2!6!} = \dfrac{\overset{4}{\cancel{8}} \cdot 7 \cdot \cancel{6!}}{\cancel{2} \cdot \cancel{6!}} = 28$ $\quad \binom{8}{6} = \dfrac{8!}{6!2!} = \dfrac{\overset{4}{\cancel{8}} \cdot 7 \cdot \cancel{6!}}{\cancel{6!} \cdot \cancel{2}} = 28$

b. $\binom{17}{0}$

Solution: $\binom{17}{0} = \dfrac{17!}{0!17!} = \dfrac{1}{1} = 1$

The Binomial Theorem

Teaching Notes:
There are several numerical patterns that you could encourage students to discover before moving to the binomial theorem, especially as set up for this example. With the difficult pattern already shown in the example, most students can examine the pattern and write the theorem, if a little guidance is provided. The easier pattern is with the exponents. The other pattern is with the denominator factorials.

The expansions of the binomial $a + b$ from $(a+b)^0$ to $(a+b)^5$ are shown here.

$$(a+b)^0 = 1$$
$$(a+b)^1 = a+b$$
$$(a+b)^2 = a^2 + 2ab + b^2$$
$$(a+b)^3 = a^3 + 3a^2b + 3ab^2 + b^3$$
$$(a+b)^4 = a^4 + 4a^3b + 6a^2b^2 + 4ab^3 + b^4$$
$$(a+b)^5 = a^5 + 5a^4b + 10a^3b^2 + 10a^2b^3 + 5ab^4 + b^5$$

Three patterns are evident.

1. In each case, the **powers of *a* decrease by 1** in each term, and the **powers of *b* increase by 1** in each term.
2. In each term, the sum of the exponents on a and b is equal to the exponent on $(a+b)$.
3. A pattern called Pascal's triangle is formed from the coefficients.

Pascal's Triangle

1

1 1

1 2 1

1 3 3 1

1 4 6 4 1

1 5 10 10 5 1

Pascal

In each case, the first and last coefficients are 1, and the other coefficients are the sum of the two numbers above to the left and above to the right of that coefficient. Thus for $(a+b)^6$, another row of the triangle can be constructed as follows:

1 5 10 10 5 1

1 6 15 20 15 6 1

and

$$(a+b)^6 = a^6 + 6a^5b + 15a^4b^2 + 20a^3b^3 + 15a^2b^4 + 6ab^5 + b^6.$$

Note that the coefficients can be written in factorial notation as follows:

$$\binom{6}{0} = \frac{6!}{0!6!} = 1 \qquad \binom{6}{1} = \frac{6!}{1!5!} = 6 \qquad \binom{6}{2} = \frac{6!}{2!4!} = 15 \qquad \binom{6}{3} = \frac{6!}{3!3!} = 20$$

$$\binom{6}{4} = \frac{6!}{4!2!} = 15 \qquad \binom{6}{5} = \frac{6!}{5!1!} = 6 \qquad \binom{6}{6} = \frac{6!}{6!0!} = 1$$

So, the expansion can be written in the following form:

$$(a+b)^6 = \binom{6}{0}a^6 + \binom{6}{1}a^5b + \binom{6}{2}a^4b^2 + \binom{6}{3}a^3b^3 + \binom{6}{4}a^2b^4 + \binom{6}{5}ab^5 + \binom{6}{6}b^6.$$

This last form is the form used in the statement of the binomial theorem, stated here without proof.

The Binomial Theorem

For real numbers a and b and a nonnegative integer n,

$$(\boldsymbol{a}+\boldsymbol{b})^n = \binom{n}{0}\boldsymbol{a}^n + \binom{n}{1}\boldsymbol{a}^{n-1}\boldsymbol{b} + \binom{n}{2}\boldsymbol{a}^{n-2}\boldsymbol{b}^2 + \ldots + \binom{n}{k}\boldsymbol{a}^{n-k}\boldsymbol{b}^k + \ldots + \binom{n}{n}\boldsymbol{b}^n.$$

In Σ notation, $(\boldsymbol{a}+\boldsymbol{b})^n = \sum_{k=0}^{n}\binom{n}{k}\boldsymbol{a}^{n-k}\boldsymbol{b}^k$.

NOTES

1. There are $n + 1$ terms in $(\boldsymbol{a}+\boldsymbol{b})^n$.
2. In each term of $(\boldsymbol{a}+\boldsymbol{b})^n$, the sum of the exponents of a and b is n.

Example 3: The Binomial Theorem

a. Expand $(x+3)^5$ using the binomial theorem.

Solution:

$$(x+3)^5 = \sum_{k=0}^{5}\binom{5}{k}x^{5-k}3^k$$

$$=\binom{5}{0}x^5+\binom{5}{1}x^4\cdot 3+\binom{5}{2}x^3\cdot 3^2+\binom{5}{3}x^2\cdot 3^3+\binom{5}{4}x\cdot 3^4+\binom{5}{5}3^5$$

$$=\frac{5!}{0!5!}\cdot x^5+\frac{5!}{1!4!}\cdot x^4\cdot 3+\frac{5!}{2!3!}\cdot x^3\cdot 9+\frac{5!}{3!2!}\cdot x^2\cdot 27$$

$$+\frac{5!}{4!1!}\cdot x\cdot 81+\frac{5!}{5!0!}\cdot 243$$

$$=1\cdot x^5+5\cdot x^4\cdot 3+10\cdot x^3\cdot 9+10\cdot x^2\cdot 27+5\cdot x\cdot 81+1\cdot 243$$

$$=x^5+15x^4+90x^3+270x^2+405x+243$$

b. Expand $(y^2-1)^6$ using the binomial theorem.

Solution:

$$(y^2-1)^6 = \sum_{k=0}^{6}\binom{6}{k}(y^2)^{6-k}(-1)^k$$

$$=\binom{6}{0}(y^2)^6+\binom{6}{1}(y^2)^5(-1)^1+\binom{6}{2}(y^2)^4(-1)^2+\binom{6}{3}(y^2)^3(-1)^3$$

$$+\binom{6}{4}(y^2)^2(-1)^4+\binom{6}{5}(y^2)^1(-1)^5+\binom{6}{6}(-1)^6$$

$$=\frac{6!}{0!6!}\cdot y^{12}+\frac{6!}{1!5!}\cdot y^{10}(-1)+\frac{6!}{2!4!}\cdot y^8(+1)+\frac{6!}{3!3!}\cdot y^6(-1)$$

$$+\frac{6!}{4!2!}\cdot y^4(+1)+\frac{6!}{5!1!}\cdot y^2(-1)+\frac{6!}{6!0!}(+1)$$

$$=1\cdot y^{12}-6\cdot y^{10}+15\cdot y^8-20\cdot y^6+15\cdot y^4-6\cdot y^2+1$$

$$=y^{12}-6y^{10}+15y^8-20y^6+15y^4-6y^2+1$$

c. Find the sixth term of the expansion of $\left(2x-\frac{1}{3}\right)^{10}$.

Solution: Since $\left(2x-\frac{1}{3}\right)^{10}=\sum_{k=0}^{10}\binom{10}{k}(2x)^{10-k}\left(-\frac{1}{3}\right)^{k}$, and the sum begins with $k=0$, the sixth term will occur when $k=5$.

$$\binom{10}{5}(2x)^{10-5}\left(-\frac{1}{3}\right)^{5}=\frac{10!}{5!5!}(2x)^{5}\left(-\frac{1}{3}\right)^{5}$$

$$=\overset{28}{\cancel{252}}\cdot 32x^{5}\left(-\frac{1}{\underset{27}{\cancel{243}}}\right)=-\frac{896x^{5}}{27}$$

The sixth term is $-\frac{896x^{5}}{27}$.

d. Find the fourth term of the expansion of $\left(x+\frac{1}{2}y\right)^{8}$.

Solution: $\left(x+\frac{1}{2}y\right)^{8}=\sum_{k=0}^{8}\binom{8}{k}x^{8-k}\left(\frac{1}{2}y\right)^{k}$

The fourth term occurs when $k=3$.

$$\binom{8}{3}x^{8-3}\left(\frac{1}{2}y\right)^{3}=\frac{8!}{3!5!}\cdot x^{5}\cdot\frac{1}{8}y^{3}=\frac{\cancel{8}\cdot 7\cdot\cancel{6}\cdot\cancel{5!}}{\cancel{3}\cdot\cancel{2}\cdot 1\cdot\cancel{5!}}\cdot\frac{1}{\cancel{8}}\cdot x^{5}\cdot y^{3}=7x^{5}y^{3}$$

The fourth term is $7x^{5}y^{3}$.

e. Using the binomial expansion, approximate $(0.99)^{4}$ to the nearest thousandth.

Solution: First rewrite 0.99 in the binomial form $(1-0.01)$ then proceed as follows:

$$(0.99)^{4}=(1-0.01)^{4}=\sum_{k=0}^{4}\binom{4}{k}(1)^{4-k}(-0.01)^{k}$$

$$=\binom{4}{0}\cdot 1^{4}+\binom{4}{1}\cdot 1^{3}\cdot(-0.01)+\binom{4}{2}\cdot 1^{2}\cdot(-0.01)^{2}+\binom{4}{3}\cdot 1\cdot(-0.01)^{3}$$
$$+\binom{4}{4}\cdot(-0.01)^{4}$$

$$=\frac{4!}{0!4!}+\frac{4!}{1!3!}(-0.01)+\frac{4!}{2!2!}(0.0001)+\frac{4!}{3!1!}(-0.000001)$$
$$+\frac{4!}{4!0!}\cdot(0.00000001)$$

$$=1+4(-0.01)+6(0.0001)+4(-0.000001)^{3}+1(0.00000001)$$

$$=1-0.04+0.0006-0.000004+0.00000001$$

$$=0.96059601\approx 0.961$$ To the nearest thousandth

Practice Problems

1. Simplify $\frac{10!}{7!}$.
2. Evaluate $\binom{20}{2}$.
3. Expand $(x+2)^5$ using the binomial theorem.
4. Find the third term of the expansion of $(2x-1)^7$.

13.5 Exercises

1. 56 2. 7920 3. $\frac{1}{15}$
4. 15 5. 4 6. $\frac{1}{30}$
7. $(n-1)!$
8. $n(n-1)(n-2)$
9. $(k+3)(k+2)(k+1)$
10. $\frac{n}{n+2}$
11. 20 12. 5 13. 35
14. 56 15. 1 16. 15
17. $x^7+7x^6y+21x^5y^2+35x^4y^3$
18. $x^{11}+11x^{10}y+55x^9y^2+165x^8y^3$
19. $x^9+9x^8+36x^7+84x^6$
20. $x^{12}+12x^{11}+66x^{10}+220x^9$
21. $x^5+15x^4+90x^3+270x^2$
22. $x^6-12x^5+60x^4-160x^3$
23. $x^6+12x^5y+60x^4y^2+160x^3y^3$

Simplify the following expressions.

1. $\frac{8!}{6!}$ **2.** $\frac{11!}{7!}$ **3.** $\frac{3!8!}{10!}$ **4.** $\frac{5!7!}{8!}$

5. $\frac{5!4!}{6!}$ **6.** $\frac{7!4!}{10!}$ **7.** $\frac{n!}{n}$ **8.** $\frac{n!}{(n-3)!}$

9. $\frac{(k+3)!}{k!}$ **10.** $\frac{n(n+1)!}{(n+2)!}$ **11.** $\binom{6}{3}$ **12.** $\binom{5}{4}$

13. $\binom{7}{3}$ **14.** $\binom{8}{5}$ **15.** $\binom{10}{0}$ **16.** $\binom{6}{2}$

Write the first four terms of the expansions.

17. $(x+y)^7$ **18.** $(x+y)^{11}$ **19.** $(x+1)^9$ **20.** $(x+1)^{12}$

21. $(x+3)^5$ **22.** $(x-2)^6$ **23.** $(x+2y)^6$ **24.** $(x+3y)^5$

25. $(3x-y)^7$ **26.** $(2x-y)^{10}$ **27.** $(x^2-4y)^9$ **28.** $(x^2-2y)^7$

Use the binomial theorem to expand the following expressions.

29. $(x+y)^6$ **30.** $(x+y)^8$ **31.** $(x-1)^7$ **32.** $(x-1)^9$

33. $(3x+y)^5$ **34.** $(2x+y)^6$ **35.** $(x+2y)^4$ **36.** $(x+3y)^5$

Answers to Practice Problems: **1.** 720 **2.** 190 **3.** $x^5+10x^4+40x^3+80x^2+80x+32$ **4.** $672x^5$

37. $(3x-2y)^4$ **38.** $(5x+2y)^3$ **39.** $(x^2+2y)^4$ **40.** $(3x^2-y)^5$

Find the specified term in each of the expressions.

41. $(x-2y)^{10}$, fifth term

42. $(x+3y)^{12}$, third term

43. $(2x+3)^{11}$, fourth term

44. $(4x-1)^9$, seventh term

45. $(5x^2-y^2)^{12}$, tenth term

46. $(2x^2+y^2)^{15}$, eleventh term

Approximate the value of each expression correct to the nearest thousandth.

47. $(1.01)^6$ **48.** $(0.96)^8$ **49.** $(0.97)^7$ **50.** $(1.02)^{10}$

51. $(2.3)^5$ **52.** $(2.8)^6$ **53.** $(0.98)^8$ **54.** $(1.03)^9$

Writing and Thinking About Mathematics

55. Factor the polynomial: $x^4+8x^3+24x^2+32x+16$.

HAWKES LEARNING SYSTEMS: INTRODUCTORY & INTERMEDIATE ALGEBRA SOFTWARE

- 13.5 The Binomial Theorem

24. $x^5+15x^4y+90x^3y^2+270x^2y^3$
25. $2187x^7-5103x^6y+5103x^5y^2-2835x^4y^3$
26. $1024x^{10}-5120x^9y+11{,}520x^8y^2-15{,}360x^7y^3$
27. $x^{18}-36x^{16}y+576x^{14}y^2-5376x^{12}y^3$
28. $x^{14}-14x^{12}y+84x^{10}y^2-280x^8y^3$
29. $x^6+6x^5y+15x^4y^2+20x^3y^3+15x^2y^4+6xy^5+y^6$
30. $x^8+8x^7y+28x^6y^2+56x^5y^3+70x^4y^4+56x^3y^5+28x^2y^6+8xy^7+y^8$
31. $x^7-7x^6+21x^5-35x^4+35x^3-21x^2+7x-1$
32. $x^9-9x^8+36x^7-84x^6+126x^5-126x^4+84x^3-36x^2+9x-1$
33. $243x^5+405x^4y+270x^3y^2+90x^2y^3+15xy^4+y^5$
34. $64x^6+192x^5y+240x^4y^2+160x^3y^3+60x^2y^4+12xy^5+y^6$ 35. $x^4+8x^3y+24x^2y^2+32xy^3+16y^4$
36. $x^5+15x^4y+90x^3y^2+270x^2y^3+405xy^4+243y^5$ 37. $81x^4-216x^3y+216x^2y^2-96xy^3+16y^4$
38. $125x^3+150x^2y+60xy^2+8y^3$ 39. $x^8+8x^6y+24x^4y^2+32x^2y^3+16y^4$
40. $243x^{10}-405x^8y+270x^6y^2-90x^4y^3+15x^2y^4-y^5$ 41. $3360x^6y^4$ 42. $594x^{10}y^2$ 43. $1{,}140{,}480x^8$
44. $5376x^3$ 45. $-27{,}500x^6y^{18}$ 46. $96{,}096x^{10}y^{20}$ 47. 1.062 48. 0.721 49. 0.808 50. 1.219 51. 64.363
52. 481.890 53. 0.851 54. 1.305 55. $(x+2)^4$

Chapter 13 Index of Key Ideas and Terms

Section 13.1 Sequences

Sequences page 1032

A **sequence** is a list of numbers that occur in a certain order.

Terms page 1032

Each number in the sequence is called a **term** of the sequence.

Infinite Sequences page 1033

An **infinite sequence** (or **sequence**) is a function that has the positive integers as its domain.

Alternating Sequences page 1035

An **alternating sequence** is a sequence in which the terms alternate in sign. Alternating sequences generally involve the expression $(-1)^n$ or $(-1)^{n+1}$.

Decreasing Sequences page 1036

A sequence $\{a_n\}$ is **decreasing** if $a_n > a_{n+1}$ for all n.

Increasing Sequences page 1036

A sequence $\{a_n\}$ is **increasing** if $a_n < a_{n+1}$ for all n.

Section 13.2 Sigma Notation

Partial Sum page 1040

The sum of just a few terms of a sequence is called a **partial sum.**

Sigma Notation page 1040

The Greek letter capital sigma Σ can be used to indicate a partial sum.

Partial Sums Using Sigma Notation page 1040

The ***n*^th^ partial sum** S_n of the first n terms of a sequence $\{a_n\}$ is $S_n = \sum_{k=1}^{n} a_k = a_1 + a_2 + a_3 + \ldots + a_n$.

k is called the **index of summation**, and k takes the integer values $1, 2, 3, \ldots, n$. n is the **upper limit of summation**, and 1 is the **lower limit of summation.**

Continued on the next page...

Section 13.2 Sigma Notation (cont.)

Properties of Σ Notation page 1042

For sequences $\{a_n\}$ and $\{b_n\}$ and any real number c:

I. $\sum_{k=1}^{n} a_k = \sum_{k=1}^{i} a_k + \sum_{k=i+1}^{n} a_k$ for any i, $1 \le i \le n-1$

II. $\sum_{k=1}^{n} (a_k + b_k) = \sum_{k=1}^{n} a_k + \sum_{k=1}^{n} b_k$

III. $\sum_{k=1}^{n} ca_k = c\sum_{k=1}^{n} a_k$

IV. $\sum_{k=1}^{n} c = nc$

Section 13.3 Arithmetic Sequences

Arithmetic Sequences pages 1046-1047

A sequence $\{a_n\}$ is called an **arithmetic sequence** (or **arithmetic progression**) if for any natural number k, $a_{k+1} - a_k = d$ where d is a constant. d is called the **common difference**.

n^{th} Term of an Arithmetic Sequence page 1048

If $\{a_n\}$ is an arithmetic sequence, then the nth term has the form $a_n = a_1 + (n-1)d$ where d is the common difference between consecutive terms.

Partial Sums of Arithmetic Sequences pages 1050-1051

The ***n*^th^ partial sum** S_n of the first n terms of an arithmetic sequence $\{a_n\}$ is $S_n = \sum_{k=1}^{n} a_k = \frac{n}{2}(a_1 + a_n)$.

Section 13.4 Geometric Sequences and Series

Geometric Sequences pages 1057-1058

A sequence $\{a_n\}$ is called a **geometric sequence** (or **geometric progression**) if for any positive integer k, $\frac{a_{k+1}}{a_k} = r$ where r is constant and $r \ne 0$. r is called the **common ratio**.

Continued on the next page...

Section 13.4 Geometric Sequences and Series (cont.)

n^{th} Term of a Geometric Sequence page 1059

If $\{a_n\}$ is a geometric sequence, then the nth term has the form $a_n = a_1 r^{n-1}$ where r is the common ratio.

Partial Sums of Geometric Sequences pages 1060-1062

The **n^{th} partial sum** S_n of the first n terms of a geometric sequence $\{a_n\}$ is $S_n = \sum_{k=1}^{n} a_k = \frac{a_1(1-r^n)}{1-r}$ where r is the common ratio and $r \neq 1$.

Infinite Series (or Series) page 1064

The indicated sum of all the terms of a sequence is called an **infinite series** (or a **series**). For a sequence $\{a_n\}$, the corresponding series can be written as follows:

$$\sum_{k=1}^{\infty}(a_k) = a_1 + a_2 + a_3 + \ldots + a_n + \ldots$$

Sum of an Infinite Geometric Series page 1064

If $\{a_n\}$ is a geometric sequence and $|r| < 1$, then the sum of the infinite geometric series is

$$S = \sum_{k=1}^{\infty}(a_k) = a_1 + a_1 r + a_1 r^2 + \ldots = \frac{a_1}{1-r}.$$

Section 13.5 The Binomial Theorem

***n* Factorial $(n!)$** page 1069

For any positive integer n, $n! = n(n-1)(n-2)\ldots(3)(2)(1)$. $n!$ is read as "*n* factorial."

0 Factorial $(0!)$ page 1070

$0! = 1$

Using a Calculator to Calculate Factorials page 1070

Continued on the next page...

Section 13.5 The Binomial Theorem (cont.)

Binomial Coefficient $\binom{n}{r}$ page 1071

For non-negative integers n and r, with $0 \le r \le n$,

$$\binom{n}{r} = \frac{n!}{r!(n-r)!}.$$

Because this quantity appears repeatedly in the binomial theorem, $\binom{n}{r}$ is often called a **binomial coefficient**.

Pascal's Triangle page 1073

The Binomial Theorem page 1073

For real numbers a and b and nonnegative integer n,

$$(a+b)^n = \binom{n}{0}a^n + \binom{n}{1}a^{n-1}b + \binom{n}{2}a^{n-2}b^2 + \ldots + \binom{n}{k}a^{n-k}b^k + \ldots + \binom{n}{n}b^n$$

In Σ notation, $(a+b)^n = \sum_{k=0}^{n}\binom{n}{k}a^{n-k}b^k$.

HAWKES LEARNING SYSTEMS: INTRODUCTORY & INTERMEDIATE ALGEBRA SOFTWARE

- 13.1 Sequences
- 13.2 Sigma Notation
- 13.3 Arithmetic Sequences
- 13.4 Geometric Sequences and Series
- 13.5 The Binomial Theorem

Chapter 13 Review

1. $-1, 1, 3, 5$
2. $4, \frac{5}{2}, 2, \frac{7}{4}$
3. $\frac{1}{2}, \frac{4}{3}, \frac{9}{4}, \frac{16}{5}$
4. $-1, \frac{1}{4}, -\frac{1}{9}, \frac{1}{16}$
5. $7, 5, 7, 5$
6. $-1, -1, -1, -1$
7. $\{5n+5\}$
8. $\{3(-1)^{n+1}\}$
9. $\{n^2+2\}$
10. $\left\{\frac{n+2}{n}\right\}$
11. Increasing
12. Decreasing
13. Increasing
14. Decreasing
15. \$275, \$6750
16. b
17. $S_1 = 5$, $S_2 = 5 + 9 = 14$, $S_3 = 5 + 9 + 13 = 27$, $S_4 = 5 + 9 + 13 + 17 = 44$
18. $S_1 = 3$, $S_2 = 3 + 2 = 5$, $S_3 = 3 + 2 + \frac{5}{3} = \frac{20}{3}$, $S_4 = 3 + 2 + \frac{5}{3} + \frac{3}{2} = \frac{49}{6}$
19. $S_1 = 0$, $S_2 = 0 + 2 = 2$, $S_3 = 0 + 2 + 0 = 2$, $S_4 = 0 + 2 + 0 + 2 = 4$
20. $S_1 = 2$, $S_2 = 2 + 6 = 8$, $S_3 = 2 + 6 + 12 = 20$, $S_4 = 2 + 6 + 12 + 20 = 40$

13.1 Sequences

Write the first four terms of each of the sequences.

1. $\{2n-3\}$ **2.** $\left\{\frac{n+3}{n}\right\}$ **3.** $\left\{\frac{n^2}{n+1}\right\}$

4. $\left\{\frac{(-1)^n}{n^2}\right\}$ **5.** $\{6+(-1)^{n+1}\}$ **6.** $\{(-1)^{2n+1}\}$

Find a formula for the general term of each sequence. There may be more than one correct answer.

7. 10, 15, 20, 25, ... **8.** 3, –3, 3, –3, 3, ...

9. 3, 6, 11, 18, 27, ... **10.** $3, 2, \frac{5}{3}, \frac{6}{4}, \frac{7}{5}, \ldots$

For each of the sequences, determine whether it is increasing or decreasing. Justify your answer by comparing a_n with a_{n+1}.

11. $\{3^n\}$ **12.** $\left\{\frac{1}{4n}\right\}$ **13.** $\left\{\frac{2n}{n+2}\right\}$ **14.** $\{-n^2\}$

15. College tuition: You decide to save for college. Your plan is to save \$100 the first month and then add an extra \$5 each month thereafter for three years. (You save \$100 the first month, \$105 the second month, \$110 the third month, and so on.) How much will you save the last month? What total amount will you save over the three years?

16. Allowances: Your father tells you he will:

a. give you \$10 each day for a month or

b. give you \$0.01 on the first day of the month and double it each day thereafter.

Which should you choose for a 30 day month?

13.2 Sigma Notation

For each of the sequences given, write out the partial sums S_1, S_2, S_3, and S_4 and evaluate each partial sum.

17. $\{4n+1\}$ **18.** $\left\{\frac{n+2}{n}\right\}$ **19.** $\{1+(-1)^n\}$ **20.** $\{n^2+n\}$

21. $5+10+15+20=50$
22. $7+8+9+10+11=45$
23. $5-7+9-11+13=9$
24. $-1+4-9+16-25+36=21$
25. $80+99+120=299$
26. $\frac{1}{3}+\frac{1}{6}+\frac{1}{9}+\frac{1}{12}=\frac{25}{36}$
27. $\sum_{k=1}^{5}(2k+1)$
28. $\sum_{k=3}^{6}\frac{(-1)^{k+1}}{k^3}$
29. $\sum_{k=1}^{6}(-1)^{k+1}(k^2-1)$
30. $\sum_{k=1}^{5}\frac{k+1}{k^2}$
31. 50 32. –56
33. 380 34. 10
35. Arithmetic sequence $d=4;\ \{4n-7\}$
36. Arithmetic sequence $d=\frac{2}{3};\ \left\{\frac{2}{3}n+\frac{5}{3}\right\}$
37. Not an arithmetic sequence
38. Arithmetic sequence $d=7;\ \{7n-8\}$
39. 4, 9, 16, 25, 36; Not an arithmetic sequence
40. $\frac{17}{3},\frac{16}{3},5,\frac{14}{3},\frac{13}{3}$; Arithmetic sequence
41. $\left\{\frac{n+39}{4}\right\}$
42. $\{27-7n\}$
43. $\{2-2n\}$
44. $\{2n-8\}$

Write the indicated sums in expanded form and evaluate.

21. $\sum_{k=1}^{4}5k$ **22.** $\sum_{k=3}^{7}(k+4)$ **23.** $\sum_{k=1}^{5}(-1)^{k+1}(2k+3)$

24. $\sum_{k=1}^{6}(-1)^k k^2$ **25.** $\sum_{k=8}^{10}(k^2+2k)$ **26.** $\sum_{k=1}^{4}\frac{1}{3k}$

Write the sums in sigma notation. There may be more than one correct answer.

27. $3+5+7+9+11$ **28.** $\frac{1}{27}-\frac{1}{64}+\frac{1}{125}-\frac{1}{216}$

29. $0-3+8-15+24-35$ **30.** $2+\frac{3}{4}+\frac{4}{9}+\frac{5}{16}+\frac{6}{25}$

Find the indicated sums.

31. $\sum_{k=1}^{4}a_k=20$ and $\sum_{k=1}^{4}b_k=30$. Find $\sum_{k=1}^{4}(a_k+b_k)$.

32. $\sum_{k=1}^{12}a_k=78$ and $\sum_{k=1}^{12}b_k=134$. Find $\sum_{k=1}^{12}(a_k-b_k)$.

33. $\sum_{k=1}^{10}a_k=150$ and $\sum_{k=11}^{20}a_k=230$. Find $\sum_{k=1}^{20}a_k$.

34. $\sum_{k=1}^{8}b_k=64$ and $\sum_{k=1}^{8}(3a_k+b_k)=94$. Find $\sum_{k=1}^{8}a_k$.

13.3 Arithmetic Sequences

Determine which of the sequences are arithmetic. Find the common difference and the nth term for each arithmetic sequence.

35. –3, 1, 5, 9, ... **36.** $\frac{7}{3}, 3, \frac{11}{3}, \frac{13}{3}, \ldots$ **37.** 12, 6, 3, 1, ... **38.** –1, 6, 13, 20, ...

For each of the following sequences, write the first five terms and then determine whether the sequence is arithmetic.

39. $\{(n+1)^2\}$ **40.** $\left\{6-\frac{n}{3}\right\}$

Use the given information to find $\{a_n\}$ for each of the arithmetic sequences.

41. $a_1=10, d=\frac{1}{4}$ **42.** $a_1=20, a_3=6$ **43.** $a_3=-4, a_7=-12$ **44.** $a_5=2, a_{10}=12$

Assume $\{a_n\}$ is an arithmetic sequence. Find the indicated quantity.

45. $a_5 = 37, a_9 = 57$. Find a_{13}.

46. $a_6 = 18, d = -4, a_n = -38$. Find n.

Find the indicated sums using the formula for partial sums of arithmetic sequences.

47. $-3 + 3 + 9 + 15 + \ldots + 51$

48. $\sum_{k=1}^{6}(2k+3)$

49. $\sum_{k=1}^{10}\left(\frac{k}{2}+4\right)$

50. $\sum_{k=1}^{7}(3k-7)$

Find the indicated sums using the properties of sigma notation.

51. If $\sum_{k=1}^{15} a_k = 120$, find $\sum_{k=1}^{15}(-2a_k + 6)$.

52. If $\sum_{k=1}^{30}(2a_k + 3) = 210$, find $\sum_{k=1}^{30} a_k$.

53. Building blocks: How many building blocks are needed to stack the blocks so that there are 29 blocks in the first layer, 27 blocks in the next layer, 25 blocks in the third layer, and so on with 1 block at the top?

54. Savings account: Jose has decided to deposit part of his paycheck in a savings account each month. The first month he deposited \$200. Each subsequent month he increased the amount he deposited by 5 dollars (\$205 the 2nd month, \$210 the third, etc.). How much will he have deposited after 2 years?

13.4 Geometric Sequences and Series

Determine which sequences are geometric. If the sequence is geometric, find its common ratio and write a formula for the nth term.

55. $12, 6, 3, \frac{3}{2}, \ldots$

56. $3, 6, 9, 12, \ldots$

57. $10, -2, \frac{2}{5}, -\frac{2}{25}, \ldots$

58. $5, -10, 15, -20, \ldots$

For each of the following sequences, write the first four terms and then determine whether the sequence is geometric.

59. $\left\{\frac{(-1)^n}{5^n}\right\}$

60. $\left\{2\left(\frac{2}{3}\right)^n\right\}$

Use the given information to find $\{a_n\}$ for each of the geometric sequences.

61. $a_1 = 7, r = \frac{1}{2}$

62. $a_1 = 4, r = -3$

63. $a_3 = 4, a_4 = -6$

64. $a_2 = \frac{5}{8}, a_5 = 5$

Assume $\{a_n\}$ is a geometric sequence. Find the indicated quantity.

65. $a_1 = -64,\ a_6 = 2$. Find a_8.

66. $a_1 = \frac{5}{8},\ a_6 = 20$. Find a_7.

45. $a_{13} = 77$
46. $n = 20$
47. 240
48. 60 49. $\frac{135}{2}$
50. 35 51. -150
52. 60
53. 225 blocks
54. \$6180
55. Geometric sequence $r = \frac{1}{2}$; $\left\{12\left(\frac{1}{2}\right)^{n-1}\right\}$
56. Not a geometric sequence
57. Geometric sequence $r = -\frac{1}{5}$; $\left\{10\left(-\frac{1}{5}\right)^{n-1}\right\}$
58. Not a geometric sequence
59. $-\frac{1}{5}, \frac{1}{25}, -\frac{1}{125}, \frac{1}{625}$ Geometric sequence
60. $\frac{4}{3}, \frac{8}{9}, \frac{16}{27}, \frac{32}{81}$ Geometric sequence
61. $\left\{7\left(\frac{1}{2}\right)^{n-1}\right\}$
62. $\left\{4(-3)^{n-1}\right\}$
63. $\left\{\frac{16}{9}\left(-\frac{3}{2}\right)^{n-1}\right\}$
64. $\left\{\frac{5}{16}(2)^{n-1}\right\}$
65. $\frac{1}{2}$ 66. 40

67. –48 68. 6
69. 242 70. $\frac{160}{81}$
71. $-\frac{2}{5}$ 72. $\frac{3}{2}$
73. a. $\sum_{k=1}^{\infty}\frac{2}{5}\left(\frac{1}{10}\right)^{k-1}$
b. $\sum_{k=1}^{\infty}\frac{4}{5}\left(\frac{1}{10}\right)^{k-1}$
74. \$110,907.56
75. \$19,966.74
76. 30,240
77. 90 78. $(n-2)!$
79. $(n+3)(n+2)$
80. 56 81. 1
82. $x^8+8x^7y+28x^6y^2+56x^5y^3$
83. $x^9+18x^8+144x^7+672x^6$
84. $1024x^{10}-5120x^9y+11{,}520x^8y^2-15{,}360x^7y^3$
85. $x^{12}-24x^{10}+240x^8-1280x^6$
86. $x^4+20x^3+150x^2+500x+625$
87. $8x^3+36x^2y+54xy^2+27y^3$
88. $x^{10}-15x^8y+90x^6y^2-270x^4y^3+405x^2y^4-243y^5$
89. $x^{10}-10x^9+45x^8-120x^7+210x^6-252x^5+210x^4-120x^3+45x^2-10x+1$
90. $-4032x^4y^5$
91. $-811{,}008x^{10}y^{14}$
92. 1.338
93. 0.630

67. $a_5=-3,\ a_7=-12.$ Find a_9.

68. $a_1=\frac{1}{6},\ r=\frac{3}{2},\ a_n=\frac{81}{64}.$ Find n.

Find the indicated sums.

69. $2+6+\ldots+54+162$

70. $\sum_{k=1}^{4}4\left(\frac{1}{3}\right)^k$

71. $\sum_{k=1}^{\infty}\left(-\frac{2}{3}\right)^k$

72. $\sum_{k=1}^{\infty}\left(\frac{3}{5}\right)^k$

73. Use a geometric series to show the following:

a. $0.\overline{4}\ldots=\frac{4}{9}$

b. $0.\overline{8}\ldots=\frac{8}{9}$

74. Annuities: A professional athlete has a contract for 6 years and decides he might be wise to set up an annuity. He decides to deposit \$15,000 on the first of each year for 6 years. If interest is compounded annually at 6%, what will be the value of the deposits after 6 years?

75. Automobiles: An automobile that costs \$45,000 new depreciates at a rate of 15% of its value each year. What will be its value after 5 years?

13.5 The Binomial Theorem

Simplify the following expressions.

76. $\frac{10!}{5!}$

77. $\frac{7!6!}{8!}$

78. $\frac{n!}{n(n-1)}$

79. $\frac{(n+3)!}{(n+1)!}$

80. $\binom{8}{5}$

81. $\binom{11}{0}$

Write the first four terms of the expansions.

82. $(x+y)^8$

83. $(x+2)^9$

84. $(2x-y)^{10}$

85. $(x^2-4)^6$

Use the binomial theorem to expand the following expressions.

86. $(x+5)^4$

87. $(2x+3y)^3$

88. $(x^2-3y)^5$

89. $(x-1)^{10}$

Find the specified term in each of the expressions.

90. $(x-2y)^9$, sixth term

91. $(4x^2-y^2)^{12}$, eighth term

Approximate the value of each expression correct to the nearest thousandth.

92. $(1.06)^5$

93. $(0.95)^9$

Chapter 13 Test

1. $\left\{\frac{n}{2n+1}\right\}$
2. $0+2+6+12+20=40$
3. $\sum_{k=1}^{4} 3k+4$
4. $\frac{1}{4}, \frac{1}{7}, \frac{1}{10}, \frac{1}{13}, \ldots$; Neither; Decreasing
5. $4, -6, -16, -26, \ldots$; Arithmetic; Decreasing
6. $-\frac{3}{2}, \frac{3}{4}, -\frac{3}{8}, \frac{3}{16}, \ldots$; Geometric; Neither
7. $-5, -2, 3, 10$; Neither; Increasing
8. 26
9. $a_n = 8-2n$
10. $n = 15$
11. 243 12. $\frac{1}{8}$
13. $a_n = \left(\sqrt{3}\right)^{n-2}$
14. $n = 5$
15. $\frac{21}{16}$ 16. 68
17. $-\frac{40}{729}$ 18. $\frac{3}{4}$
19. 74 20. 6
21. $\frac{15}{100} + \frac{15}{10{,}000} + \ldots = \frac{5}{33}$
22. **a.** 180 **b.** 462
23. $32x^5 - 80x^4y + 80x^3y^2 - 40x^2y^3 + 10xy^4 - y^5$

1. Find a formula for the general term of the sequence $\frac{1}{3}, \frac{2}{5}, \frac{3}{7}, \frac{4}{9}, \ldots$

2. Write out and evaluate the partial sum S_5 for the sequence $\{n^2 - n\}$.

3. Write the following sum in Σ notation: $7 + 10 + 13 + 16$.

Write out the first four terms of each sequence. Then determine whether the sequence is arithmetic, geometric, or neither and whether it is increasing, decreasing, or neither.

4. $\left\{\frac{1}{3n+1}\right\}$ **5.** $\{14-10n\}$ **6.** $\left\{3(-2)^{-n}\right\}$ **7.** $\{n^2-6\}$

Assume $\{a_k\}$ is an arithmetic sequence. Find the indicated quantities.

8. $a_1 = 5,\ d = 3$. Find a_8.

9. $a_2 = 4,\ a_7 = -6$. Find a_n.

10. $a_6 = -24, d = -7, a_n = -87$. Find n.

11. $a_4 = 22,\ a_7 = 37$. Find $\sum_{k=1}^{9} a_k$.

Assume $\{a_k\}$ is a geometric sequence. Find the indicated quantities.

12. $a_1 = 8,\ r = \frac{1}{2}$. Find a_7.

13. $a_4 = 3,\ a_6 = 9$. Find a_n.

14. $a_1 = -4, r = \frac{1}{3}, a_n = -\frac{4}{81}$. Find n.

15. $a_2 = \frac{1}{3},\ a_5 = \frac{1}{24}$. Find $\sum_{k=1}^{6} a_k$.

Find each of the sums. Use the formulas for partial and infinite sums where applicable.

16. $\sum_{k=1}^{8}(3k-5)$ **17.** $\sum_{k=3}^{6} 2\left(-\frac{1}{3}\right)^k$ **18.** $\sum_{k=1}^{\infty}\left(\frac{3}{7}\right)^k$

19. If $\sum_{k=1}^{50} a_k = 88$ and $\sum_{k=1}^{19} a_k = 14$, find $\sum_{k=20}^{50} a_k$.

20. $\sum_{k=1}^{14} a_k = 29$ and $\sum_{k=1}^{14} b_k = 52$. Find $\sum_{k=1}^{14}(2a_k - b_k)$

21. Write the decimal number $0.\overline{15}$ in the form of an infinite series and find its sum in the form of a proper fraction.

22. Evaluate **a.** $\frac{6!3!}{4!}$ **b.** $\binom{11}{5}$.

23. Use the binomial theorem to expand $(2x-y)^5$.

24. $5670x^4y^4$
25. $1125x^8y^2$
26. 11 rows of logs; 121 logs in total
27. \$13,050.16

24. Write the fifth term of the expansion of $(x+3y)^8$.

25. Write the third term of the expansion of $(x-5y)^{10}$.

26. Stacking logs: Someone has stacked logs in a pyramid shape. There is one log on the top row and each successive row has 2 more logs than the one before. If the bottom row contains 21 logs, how many rows are there? How many logs have been stacked?

27. Buying a car: A customer intends to buy a new car for \$25,000 and anticipates that it will depreciate at a rate of 15% of its value each year. What will be the value of the car in 4 years when he wants to trade it in for another new car?

Cumulative Review: Chapters 1 – 13

1. $11x+3$
2. $-x^2+3x+5$
3. $(4x+3)(16x^2-12x+9)$
4. $(3x-5)(2x+9)$
5. $(2x+1)(5x+7)(x-2)$
6. $\frac{11x+12}{15}$
7. $\frac{-6x-4}{(x+4)(x-4)(x-1)}$
8. $\frac{1}{x^2}$
9. $\frac{3y^{\frac{1}{3}}}{x^2}$
10. $\frac{2x\sqrt[3]{x}}{3y}$
11. $3\sqrt{3x}$
12. $\frac{-13+11i}{29}$
13. 
$(3, \infty)$
14. 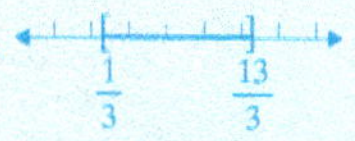
$\left[\frac{1}{3}, \frac{13}{3}\right]$
15. 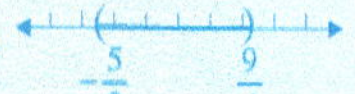
$\left(-\frac{5}{2}, \frac{9}{4}\right)$
16. $x=\frac{26}{7}$
17. $x=\frac{-2\pm i\sqrt{2}}{2}$
18. $x=4$
19. $x=\frac{39}{17}$
20. $x=5$
21. $x=e^{2.4}\approx 11.02$
22. $n=\frac{A-P}{Pi}$

Simplify each expression.

1. $10x-[2x+(13-4x)-(11-3x)]+(2x+5)$

2. $(x^2+5x-2)-(3x^2-5x-3)+(x^2-7x+4)$

Factor completely.

3. $64x^3+27$

4. $6x^2+17x-45$

5. $5x^2(2x+1)-3x(2x+1)-14(2x+1)$

Perform the indicated operations and simplify.

6. $\frac{x+3}{3}+\frac{2x-1}{5}$

7. $\frac{x}{x^2-16}-\frac{x+1}{x^2-5x+4}$

8. $\frac{x^2-9}{x^4+6x^3}\div\frac{x^3-2x^2-3x}{x^2+7x+6}\cdot\frac{x^2}{x+3}$

Simplify each expression. Assume that all variables are positive.

9. $\left(\frac{9x^{-1}y^{\frac{1}{3}}}{x^3y^{-\frac{1}{3}}}\right)^{\frac{1}{2}}$

10. $\sqrt[3]{\frac{8x^4}{27y^3}}$

11. $\sqrt{12x}-\sqrt{75x}+2\sqrt{27x}$

12. $\frac{1+3i}{2-5i}$

Solve the following absolute value inequalities and graph each solution set on a real number line. Write each solution set using interval notation.

13. $6(2x-3)+(x-5)>4(x+1)$

14. $|7-3x|-2\le 4$

15. $8x^2+2x-45<0$

Solve the equations.

16. $4(x-7)+2(3x+2)=3x+2$

17. $2x^2+4x+3=0$

18. $x-\sqrt{x}-2=0$

19. $\frac{1}{2x}+\frac{5}{x+3}=\frac{8}{3x}$

20. $2\sqrt{6-x}=x-3$

21. $5\ln x=12$

22. Solve the formula for n: $P=\frac{A}{1+ni}$

23. $(0, 2, -1)$

24. a. $y = \frac{x+7}{2}$

b. $2x^2 - 5$

c. $2x + 1$

25.

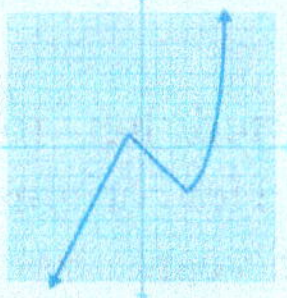

26.

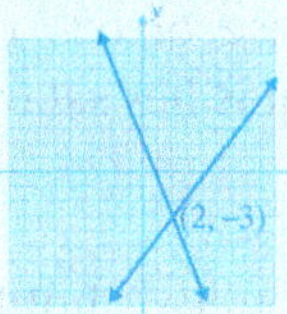

$(2, -3)$

27.

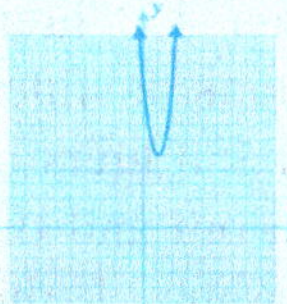

28.

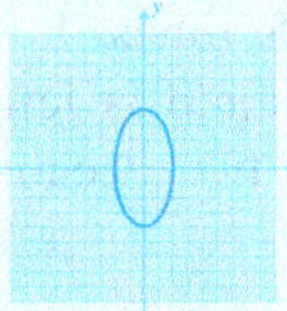

29.

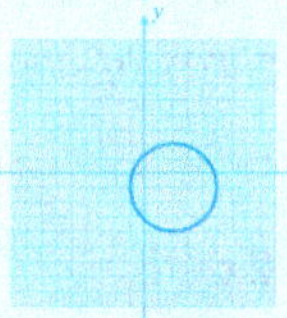

30. a. $a_n = 10 - 2n$

b. -10

31. a. $a_n = 16\left(\frac{1}{2}\right)^{n-1}$

b. $\frac{63}{2}$

32. $x^6 + 12x^5y + 60x^4y^2 + 160x^3y^3 + 240x^2y^4 + 192xy^5 + 64y^6$

23. Solve the following system of equations using the Gaussian elimination method.

$$\begin{cases} 3x + y - 2z = 4 \\ x - 4y - 3z = -5 \\ 2x + 2y + z = 3 \end{cases}$$

24. If $f(x) = 2x - 7$ and $g(x) = x^2 + 1$, find:

a. $f^{-1}(x)$ **b.** $f[g(x)]$ **c.** $g(x+1) - g(x)$

25. The following is a graph of $y = f(x)$. Sketch the graph of $y = f(x-2) - 1$.

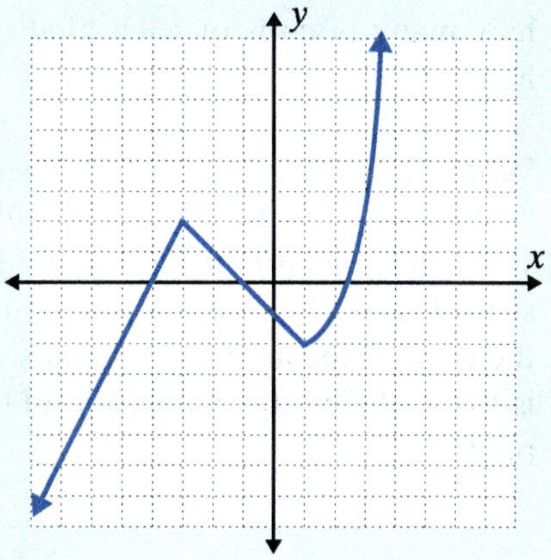

26. Solve the following system by graphing: $\begin{cases} 4x - 3y = 17 \\ 5x + 2y = 4 \end{cases}$

Graph each of the equations.

27. $y = 4x^2 - 8x + 9$ **28.** $\frac{x^2}{4} + \frac{y^2}{16} = 1$ **29.** $x^2 - 4x + y^2 + 2y = 4$

30. If $\{a_n\}$ is an arithmetic sequence where $a_3 = 4$ and $a_8 = -6$,

a. find a_n.

b. find $\sum_{k=1}^{10} a_k$.

31. If $\{a_n\}$ is a geometric sequence where $a_1 = 16$ and $r = \frac{1}{2}$,

a. find a_n.

b. find $\sum_{k=1}^{6} a_k$.

32. Use the binomial theorem to expand $(x + 2y)^6$.

33. 20 mph, 60 mph
34. 16 lb at \$1.40; 4 lb at \$2.60
35. 16 ft by 26 ft
36. 12.56 hours
37. \$16,058.71
38. 443.71 liters
39. $x \approx -0.4746, 1.3953$
40. $x \approx -1.9646, 1.0580$
41. $x \approx 0.2408, 2.7337$
42. $x = 1$
43. $x = -3, 1, 2$

33. Traveling by car: Two cars start together and travel in the same direction, one traveling 3 times as fast as the other. At the end of 3.5 hours, they are 140 miles apart. How fast is each traveling?

34. Grocery stores: A grocer mixes two kinds of nuts. One costs \$1.40 per pound and the other costs \$2.60 per pound. If the mixture weighs 20 pounds and costs \$1.64 per pound, how many pounds of each kind did he use?

35. Swimming pools: A rectangular yard is 20 ft by 30 ft. A rectangular swimming pool is to be built leaving a strip of grass of uniform width around the pool. If the area of the grass strip is 184 sq ft, find the dimensions of the pool.

36. Isotope decomposition: A radioactive isotope decomposes according to $A = A_0 e^{-0.0552t}$, where t is measured in hours. Determine the half-life of the isotope. Round the solution to two decimal places.

37. Job salaries: Joan started her job exactly 5 years ago. Her original salary was \$12,000 per year. Each year she received a raise of 6% of her current salary. What is her salary after this year's raise?

38. Liquid mixtures: A tank holds 1000 liters of a liquid that readily mixes with water. After 150 liters are drained out, the tank is filled by adding water. Then 150 liters of the mixture are drained out and the tank is filled by adding water. If this process is continued 5 times, how much of the original liquid is left? Round the solution to two decimal places.

Use the graphing and CALC *features of a graphing calculator to find the zeros of each function. If necessary, round answers to four decimal places.*

39. $x^4 = 2x + 1$

40. $e^x = -x^2 + 4$

41. $\ln x = x^2 - 2x - 1$

42. $x^3 = 3x^2 - 3x + 1$

43. $x^3 = 7x - 6$

A.1 Decimals and Percents

- *Operate (add, subtract, multiply, and divide) with decimal numbers.*
- *Change decimals to percents.*
- *Change fractions to percents.*

Addition with Decimal Numbers

Addition with decimal numbers can be accomplished by writing the decimal numbers one under the other and keeping the decimal points aligned vertically. In this way, the whole numbers will be added to whole numbers, tenths added to tenths, hundredths to hundredths, and so on. The decimal point in the sum is in line with the decimal points in the addends. Thus, to compute 2.357 + 6.14, we write

$$\begin{array}{r} 2.357 \\ +\ 6.140 \\ \hline 8.497 \end{array}$$

Decimal points are aligned vertically.

0 may be written here to help align digits.

As in the number 6.14, 0's may be written to the right of the last digit in the fractional part to help keep the digits in the correct line. This will not change the value of any number or the sum.

To Add Decimal Numbers

1. Write the addends in a vertical column.
2. Keep the decimal points aligned vertically.
3. Keep digits with the same position value aligned. (Zeros may be filled in as aids.)
4. Add the numbers, just as with whole numbers, keeping the decimal point in the sum aligned with the other decimal points.

Example 1: Adding Decimals

Find the sum: 17 + 4.88 + 50.033 + 0.6

Solution:

$$\begin{array}{r} 17.000 \\ 4.880 \\ 50.033 \\ +\ 0.600 \\ \hline 72.513 \end{array}$$

Align decimal points vertically.
The decimal point is understood to be to the right of 17.

0's are filled in to help keep the digits in line.

sum

Subtraction with Decimal Numbers

To Subtract Decimal Numbers

1. Write the numbers in a vertical column.
2. Keep the decimal points aligned vertically.
3. Keep digits with the same position value aligned. (Zeros may be filled in as aids.)
4. Subtract, just as with whole numbers, keeping the decimal point in the difference aligned with the other decimal points.

Example 2: Subtracting Decimals

Find the difference: 21.715 – 14.823

Solution:

$$\begin{array}{r} 21.715 \\ -\ 14.823 \\ \hline 6.892 \end{array}$$

Example 3: Monetary Arithmetic

At the bookstore, Mrs. Gonzalez bought a text for \$55, art supplies for \$32.50, and computer supplies for \$29.25. If tax was \$9.34, how much change did she receive from a gift certificate worth \$150?

Solution: Find the total of her expenses including tax.

$$\begin{array}{r} \$55.00 \\ 32.50 \\ 29.25 \\ +\ \ 9.34 \\ \hline \$126.09 \end{array} \quad \text{Total}$$

Now subtract the total, \$126.09, from \$150.

$$\begin{array}{r} \$150.00 \\ -\ \ 126.09 \\ \hline \$23.91 \end{array}$$

She received \$23.91 in change.

Multiplication with Decimal Numbers

Decimal numbers are multiplied in the same manner as whole numbers are multiplied with the added concern of the correct placement of the decimal point in the product. Two examples are shown here, in both fraction form and decimal form, to illustrate how the decimal point is to be placed in the product.

Products in Fraction Form	Products in Decimal Form
$\frac{4}{10} \cdot \frac{6}{100} = \frac{24}{1000}$	0.4 ⟵ 1 place; $\times\ 0.06$ ⟵ 2 places; Total of 3 places (thousandths); 0.024
$\frac{5}{1000} \cdot \frac{7}{100} = \frac{35}{100{,}000}$	0.005 ⟵ 3 places; $\times\ 0.07$ ⟵ 2 places; Total of 5 places (hundred-thousandths); 0.00035

The following rule states how to multiply decimal numbers and place the decimal point in the product.

To Multiply Decimal Numbers

1. Multiply the two numbers as if they were whole numbers.
2. Count the total number of places to the right of the decimal points in both numbers being multiplied.
3. Place the decimal point in the product so that the number of places to the right is the same as that found in step 2.

Example 4: Multiplying Decimals

Multiply: 2.435×4.1

Solution:

$$\begin{array}{r l} 2.435 & \longleftarrow \text{3 places} \\ \times \quad 4.1 & \longleftarrow \text{1 place} \\ \hline 2435 & \\ 9740 & \\ \hline 9.9835 & \longleftarrow \text{4 places in the product} \end{array}$$

3 places + 1 place: Total of 4 places (ten-thousandths)

Division with Decimal Numbers

The process of division with decimal numbers is, in effect, the same as division with whole numbers with the added concern of where to place the decimal point in the quotient.

To Divide Decimal Numbers

1. Move the decimal point in the divisor to the right so that the divisor is a whole number.
2. Move the decimal point in the dividend the same number of places to the right.
3. Place the decimal point in the quotient directly above the new decimal point in the dividend.
4. Divide just as with whole numbers: $\text{Divisor}\overline{)\text{Dividend}}$ with the Quotient written above the dividend.

Example 5: Dividing Decimals

Find the quotient: $63.86 \div 6.2$

Solution:

Step 1: Write down the numbers.

$$6.2\overline{)63.86}$$

Step 2: Move both decimal points one place to the right so that the divisor becomes a whole number. Then place the decimal point in the quotient.

$$\begin{array}{r} . \\ 6.2.\overline{)63.8.6} \end{array}$$

← Decimal point in quotient

Step 3: Proceed to divide as with whole numbers.

$$\begin{array}{rr} & 10.3 \\ 62.\!\!\!& \overline{)638.6} \\ & -62 \\ & 18 \\ & -0 \\ & 186 \\ & -186 \\ & 0 \end{array}$$

Divisor → 62.

10.3 ← Quotient

638.6 ← Dividend

0 ← Remainder

If the remainder is eventually 0, as it is in Example 5, then the quotient is a **terminating decimal**. As discussed in Chapter 1, if the remainder is never 0, then the quotient is an **infinite repeating decimal**. That is, the quotient will be a repeating pattern of digits. In division with decimal numbers, some place of accuracy for the quotient is generally agreed to before the division is performed. If the remainder is not 0 by the time this place of accuracy is reached in the quotient, then we divide one more place and round the quotient.

Rounding to the Right of the Decimal Point

Rounding to the right of the decimal point in a decimal number is similar to rounding with whole numbers:

1. Look one digit to the right of the desired place of accuracy.
2. If this digit is 5 or more, raise the digit in the desired place by 1. Otherwise, leave the digit as it is.
3. Drop the remaining digits to the right.

Note: When rounding with whole numbers, the dropped digits must be replaced by 0's.

Example 6: Dividing Decimals and Rounding

Find the quotient $82.3 \div 2.9$ to the nearest tenth.

Solution: Divide until the quotient is in hundredths (one place more than tenths), then round to tenths.

Hundredths (points to the 7 in 28.37)

Read "is approximately" (points to ≈)

$$\begin{array}{r} 28.37 \\ 2.9.\overline{)\,82.3.00} \\ -58 \\ \hline 243 \\ -232 \\ \hline 110 \\ -87 \\ \hline 230 \\ -203 \\ \hline 27 \end{array} \quad \approx 28.4$$

Add 0's as needed.

$82.3 \div 2.9 \approx 28.4$ Accurate to the nearest tenth

Example 7: Calculating Price

The price of a gallon of gas at the pump is \$3.15. If the taxes you pay on each gallon of gas is 0.45 times the original price of a gallon of gas, what is the price of a gallon of gas before taxes (to the nearest penny)?

Solution: To find the price of a gallon of gas before taxes, divide the total price by 1.45.

$$\begin{array}{r} 2.172 \\ 1.45.\overline{)3.15.000} \\ -290 \\ \hline 250 \\ -145 \\ \hline 1050 \\ -1015 \\ \hline 350 \\ -290 \\ \hline 60 \end{array}$$

So, the cost of the gas is about \$2.17 per gallon before taxes.

Understanding Percent

The word **percent** comes from the Latin *per centum*, meaning "per hundred." So, **percent means hundredths**. The symbol % is called the **percent symbol** (or **percent sign**). This symbol can be treated as equivalent to the fraction $\frac{1}{100}$. For example,

$$\frac{25}{100} = 25\left(\frac{1}{100}\right) = 25\% \quad \text{and} \quad \frac{70}{100} = 70\left(\frac{1}{100}\right) = 70\%.$$

In Figure 1, shown here, the large square is partitioned into 100 small squares, and each small square represents 1%, or $1 \cdot \frac{1}{100}$ of the large square. Thus the shaded portion of the large square is

$$\frac{40}{100} = 40 \cdot \frac{1}{100} = 40\%.$$

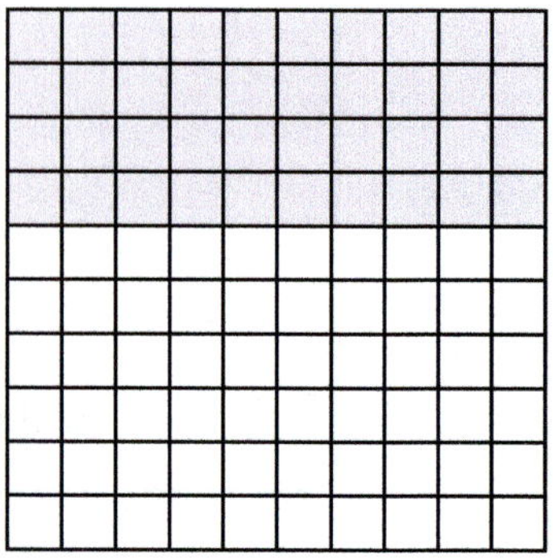

Figure 1

Decimals and Percent

Now the relationship between decimals and percents can be seen by noting the relationship between decimals and fractions with denominator 100. For example,

Decimal Form		Fraction Form		Percent Form
0.33	=	$\frac{33}{100} = 33 \cdot \frac{1}{100}$	=	33%
0.74	=	$\frac{74}{100} = 74 \cdot \frac{1}{100}$	=	74%
0.2	=	$\frac{2}{10} = \frac{20}{100} = 20 \cdot \frac{1}{100}$	=	20%
0.625	=	$\frac{62.5}{100} = 62.5 \cdot \frac{1}{100}$	=	62.5%

By noting the way that the decimal point is moved in these four examples, we can make the change directly (and more efficiently) by using the following rule regardless of how many digits there are in the decimal number.

To Change a Decimal to a Percent

Step 1: Move the decimal point two places to the right.

Step 2: Add the % symbol.

These two steps have the effect of multiplying by 100 and then dividing by 100. Thus the number is not changed. Just the form is changed.

Example 8: Converting Decimals to Percents

Change each decimal to an equivalent percent.

a. 0.254

Solution: $0.254 = 25.4\%$ ← % symbol added on

↖ Decimal point moved two places to the right

b. 0.005

Solution: $0.005 = 0.5\%$ ← % symbol added on

↖ Decimal point moved two places to the right

Note that this value is less than 1%.

Continued on the next page...

c. 1.5

Solution: $1.5 = 150\%$ ← % symbol added on

↖ Decimal point moved two places to the right

Note that this value is more than 100%

d. 0.2

Solution: $0.2 = 20\%$ ← % symbol added on

↖ Decimal point moved two places to the right

To change percents to decimals, the procedure for changing decimals to percents is reversed. For example,

$$38\% = 38\left(\frac{1}{100}\right) = \frac{38}{100} = 0.38.$$

As indicated in the following statement, the same result can be found by noting the placement of the decimal point.

To Change a Percent to a Decimal

Step 1: Move the decimal point two places to the left.

Step 2: Delete the % symbol.

Example 9: Converting Percents to Decimals

Change each percent to an equivalent decimal.

a. 64%

Solution: $64\% = 0.64$ ← % symbol deleted

↑ Understood decimal point (at 64%)

↑ Decimal point moved two places to the left (at 0.64)

b. 16.2%

Solution: $16.2\% = 0.162$

c. 100%

Solution: $100\% = 1.00 = 1$

d. 0.25%

Solution: $0.25\% = 0.0025$ The percent is less than 1%, and the decimal is less than 0.01.

Fractions and Percents

If a fraction has a denominator of 100, the fraction can be changed to a percent by writing the numerator and then writing the % symbol. However, if the denominator is not 100, a more general approach (easily applied with calculators) is to proceed as follows:

To Change a Fraction to a Percent

Step 1: Change the fraction to a decimal.
(Divide the numerator by the denominator.)

Step 2: Change the decimal to a percent.

Example 10: Converting Fractions to Percents

Change $\frac{5}{8}$ to a percent.

Solution: First divide 5 by 8 to get the decimal form. (This can be done with a calculator.)

$$\begin{array}{r} 0.625 \\ 8\overline{)5.000} \\ -\underline{48} \\ 20 \\ -\underline{16} \\ 40 \\ -\underline{40} \\ 0 \end{array}$$

Now change 0.625 to a percent:

$$\frac{5}{8} = 0.625 = 62.5\%$$

Example 11: Calculating Percents

In the U.S. in 1900, there were about 27,000 college graduates (who received bachelor's degrees), of which about 22,000 were men. In 2003, there were about 1,300,000 college graduates, of which about 550,000 were men. Find the percentage of college graduates that were men in each of those years.

Solution: **a.** For 1900, $\frac{22,000}{27,000} \approx 0.8148$ So, about 81.48% of college graduates were men.

b. For 2003, $\frac{550,000}{1,300,000} \approx 0.4231$ So, about 42.31% of college graduates were men.

Appendix 1 Exercises

1. 1.4
2. 12.9
3. 5.76
4. 4.961
5. 1.64
6. 7.93
7. 25.89
8. 1.994
9. 147.385
10. 36.441
11. 19.464
12. 23.55
13. 0.04
14. 1
15. 0.01096
16. 0.6699
17. 5.00
18. 21.34
19. 2.03
20. 17.72
21. \$9691.25
22. \$7.50
23. \$1296
24. \$2355.60
25. 13.3 mph
26. 4.4 yards

Perform the indicated operations. Round any quotient to the nearest hundredth.

1. $0.6+0.4+0.4$ **2.** $7+5.1+0.8$ **3.** $0.79+4.92+0.05$

4. $4.005+0.056+0.9$ **5.** $5.4-3.76$ **6.** $17.83-9.9$

7. $39.6-13.71$ **8.** $55.002-53.008$ **9.** $\begin{array}{r} 57.3 \\ 52.08 \\ +38.005 \\ \hline \end{array}$

10. $\begin{array}{r} 1.007 \\ 30.442 \\ +\ 4.992 \\ \hline \end{array}$ **11.** $\begin{array}{r} 21.007 \\ -\ 1.543 \\ \hline \end{array}$ **12.** $\begin{array}{r} 30. \\ -\ 6.45 \\ \hline \end{array}$

13. $(0.2)(0.2)$ **14.** $8(0.125)$ **15.** $\begin{array}{r} 0.137 \\ \times\ 0.08 \\ \hline \end{array}$ **16.** $\begin{array}{r} 6.09 \\ \times\ 0.11 \\ \hline \end{array}$

17. $28 \div 5.6$ **18.** $35 \div 1.64$ **19.** $2.7\overline{)5.483}$ **20.** $2.54\overline{)45}$

21. Buying a car: Marshall wants to buy a new car for \$28,000. The loan officer at his credit union will lend him \$21,000. He must also pay \$450.50 for a license fee and \$2240.75 for taxes. What amount of cash will he need to buy the car?

22. Cosmetology: Svetlana wants to have a haircut and a manicure. The haircut will cost \$62 and the manicure will cost \$15.50. If she plans to tip the stylist \$15, how much change will she receive from \$100?

23. Electronics: If the sale price of a new flat screen TV is \$1200 and sales tax is 0.08 times the price, what total amount is paid for the television set?

24. Buying a car: To buy a used car, Shane puts down \$300 and makes 18 monthly payments of \$114.20. How much will he pay for the car?

25. Biking: If a bicyclist rode 150.6 miles in 11.3 hours, what was her average speed in miles per hour (to the nearest tenth)?

26. Football: Walter Payton played football for the Chicago Bears for 13 years. In those years he carried the ball 3838 times for a total of 16,726 yards. What was his average yardage per carry (to the nearest tenth)?

27. 0.237
28. 15.85 inches more rain in Charleston; 64.9 inches more snow in Albany
29. a. 77.2 in.
b. 337.68 $in.^2$
30. a. 25.9 cm
b. 28.35 cm^2
31. 3%
32. 5.2%
33. 300%
34. 250%
35. 108%
36. 50%

27. Astronomy: The eccentricity of a planet's orbit is the measure of how much the orbit varies from a perfectly circular pattern. Earth's orbit has an eccentricity of 0.017, and Pluto's orbit has an eccentricity of 0.254. Find the difference between the eccentricity of Pluto's orbit and that of the Earth.

28. Weather: Albany, New York, receives an average rainfall of 35.74 inches and 65.5 inches of snow. Charleston, South Carolina, receives an average rainfall of 51.59 inches and 0.6 inches of snow. On average, Charleston receives how much more rain than Albany? On average, Albany receives how much more snow than Charleston?

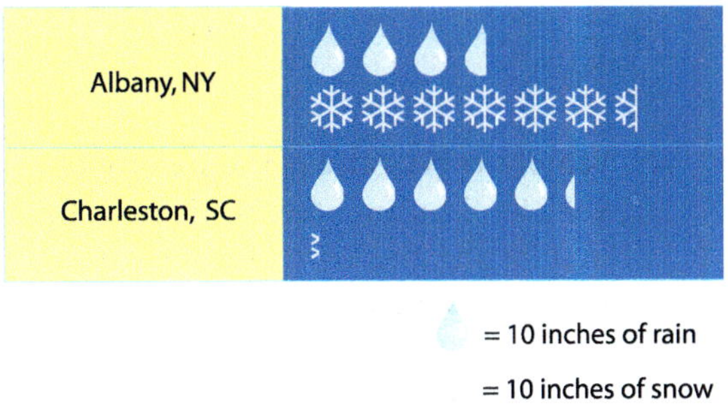

29. Rectangles: The dimensions of a rectangle are shown.
a. Find the perimeter (distance around) of the rectangle.
b. Find the area (length times width) of the rectangle (in square units).

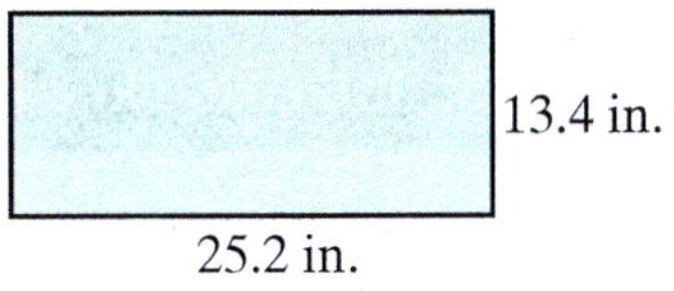

30. Triangles: The dimensions of a triangle are shown.
a. Find the perimeter (distance around) of the triangle.
b. Find the area (base times height divided by 2) of the triangle (in square units).

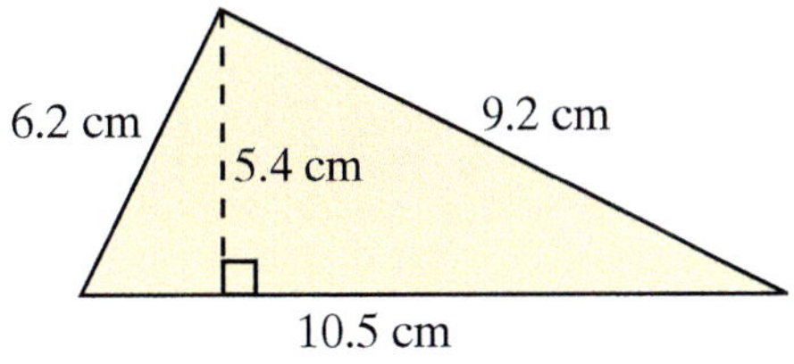

Change the following decimals to percents.

31. 0.03 **32.** 0.052 **33.** 3.0

34. 2.5 **35.** 1.08 **36.** 0.5

37. 0.06
38. 0.11
39. 0.032
40. 0.125
41. 1.2
42. 0.8
43. 15%
44. 30%
45. 96%
46. 25%
47. 162.5%
48. 275%
49. 0.9%
50. 12.91%

Change the following percents to decimals.

37. 6% **38.** 11% **39.** 3.2%

40. 12.5% **41.** 120% **42.** 80%

Change the following fractions and mixed numbers to percents.

43. $\frac{3}{20}$ **44.** $\frac{3}{10}$ **45.** $\frac{24}{25}$

46. $\frac{1}{4}$ **47.** $1\frac{5}{8}$ **48.** $2\frac{3}{4}$

49. Vehicle registration: The state motor vehicle licensing fee is figured by multiplying the cost of your car by 0.009. Change 0.009 to a percent.

50. Population: In the 1990 census, California ranked as the most populated state with about 30,000,000 people. In 2000 California again ranked first with about 33,872,000 people. By about what percent did the population of California grow in the ten years from one census to the other? (Round your answer to two decimal places.)

HAWKES LEARNING SYSTEMS: INTRODUCTORY & INTERMEDIATE ALGEBRA SOFTWARE

- A.1a Decimals and Percents
- A.1b More with Decimals

A.2 Synthetic Division and the Remainder Theorem

- *Divide polynomials by using synthetic division.*

Synthetic Division

In Section 5.7, polynomials are divided by using the division algorithm (long division). In the special case **when the divisor of a rational expression is a first-degree binomial with leading coefficient 1**, long division can be simplified by omitting the variables entirely and writing only certain coefficients. The procedure is called **synthetic division**. The following analysis describes how the procedure works for $\dfrac{5x^3+11x^2-3x+1}{x+3}$. (Note that $x + 3$ is first-degree with leading coefficient 1.)

a. With Variables

$$\begin{array}{rrrrr} & 5x^2 & -4x & +9 & \\ \hline x+3) & 5x^3 & +11x^2 & -3x & +1 \\ & \underline{5x^3} & \underline{+15x^2} & & \\ & & -4x^2 & -3x & \\ & & \underline{-4x^2} & \underline{-12x} & \\ & & & 9x & +1 \\ & & & \underline{9x} & \underline{+27} \\ & & & & -26 \end{array}$$

b. Without Variables

$$\begin{array}{rrrrr} & 5 & -4 & +9 & \\ \hline 1+3) & 5 & +11 & -3 & +1 \\ & \underline{\boxed{5}} & \underline{+15} & & \\ & & -4 & \boxed{-3} & \\ & & \underline{\boxed{-4}} & \underline{-12} & \\ & & & 9 & \boxed{+1} \\ & & & \underline{\boxed{9}} & \underline{+27} \\ & & & & -26 \end{array}$$

The boxed numbers in step **b.** can be omitted since they are repetitions of the numbers directly above them.

c. Boxed numbers omitted

$$\begin{array}{rrrrr} & 5 & -4 & +9 & \\ \hline 1+3) & 5 & +11 & -3 & +1 \\ & & \underline{+15} & & \\ & & -4 & & \\ & & & \underline{-12} & \\ & & & 9 & \\ & & & & \underline{+27} \\ & & & & -26 \end{array}$$

d. Numbers moved up to fill in spaces

$$\begin{array}{rrrrr} & 5 & -4 & +9 & \\ \hline 1+3) & 5 & +11 & -3 & +1 \\ & & +15 & -12 & +27 \\ \hline & & -4 & +9 & -26 \end{array}$$

Next, omit the 1 in the divisor, change +3 to –3, and write the opposites of the boxed numbers (because the quotient coefficient will now be multiplied by –3 instead of +3), as shown in steps **e.** and **f.** This allows the numbers to be added instead of subtracted. The number 5 is written on the bottom line, and the top line is omitted. The quotient and remainder can now be read from the bottom line.

e.
$$\begin{array}{r|rrrr} & 5 & -4 & +9 & \\ \hline 1+3 & 5 & +11 & -3 & +1 \\ & & \boxed{+15} & \boxed{-12} & \boxed{+27} \\ \hline & & -4 & +9 & -26 \end{array}$$

f.
$$\begin{array}{r|rrrr} -3 & 5 & +11 & -3 & +1 \\ & \downarrow & -15 & +12 & -27 \\ \hline & 5 & -4 & +9 & -26 \end{array}$$

This represents

$$5x^2-4x+9+\frac{-26}{x+3}.$$

The numbers on the bottom now represent the coefficients of a polynomial of **one degree less than the dividend**, along with the remainder. The last number to the right is the remainder.

In summary, synthetic division can be accomplished as follows:

1. Write only the coefficients of the dividend and the opposite of the constant in the divisor.

$$\begin{array}{r|rrrr} -3 & 5 & 11 & -3 & 1 \\ \hline & & & & \end{array}$$

2. Rewrite the first coefficient (5) as the first coefficient in the quotient.

$$\begin{array}{r|rrrr} -3 & 5 & 11 & -3 & 1 \\ & \downarrow & & & \\ \hline & 5 & & & \end{array}$$

3. Multiply the coefficient (5) by the constant divisor (–3) and **add** this product (–15) to the second coefficient.

$$\begin{array}{r|rrrr} -3 & 5 & 11 & -3 & 1 \\ & \downarrow & -15 & & \\ \hline & 5 \nearrow & -4 & & \end{array}$$

4. Continue to multiply each new coefficient by the constant divisor and add this product to the next coefficient in the dividend.

$$\begin{array}{r|rrrr} -3 & 5 & 11 & -3 & 1 \\ & \downarrow & -15 & 12 & -27 \\ \hline & 5 \nearrow & -4 \nearrow & 9 \nearrow & -26 \end{array}$$

5. The constants on the bottom line are the coefficients of the quotient and the remainder.

$$\frac{5x^3+11x^2-3x+1}{x+3}=5x^2-4x+9+\frac{-26}{x+3}$$
$$=5x^2-4x+9-\frac{26}{x+3}$$

Example 1: Synthetic Division

Use synthetic division to write each expression in the form $Q+\frac{R}{D}$.

a. $\frac{4x^3+10x^2+11}{x+5}$

Solution:

$\underline{-5}\rfloor$	4	10	0	11
	↓	–20	50	–250
	4	–10	50	–239

Since there is no x-term, 0 is the coefficient. The coefficient is 0 for any missing term.

$$\frac{4x^3+10x^2+11}{x+5}=4x^2-10x+50+\frac{-239}{x+5}$$

$$=4x^2-10x+50-\frac{239}{x+5}$$

b. $\frac{2x^4-x^3-5x^2-2x+7}{x-2}$

Solution:

$\underline{2}\rfloor$	2	–1	–5	–2	7
	↓	4	6	2	0
	2	3	1	0	7

$$\frac{2x^4-x^3-5x^2-2x+7}{x-2}=2x^3+3x^2+x+\frac{7}{x-2}$$

NOTES

Remember that synthetic division is used only when the divisor is a first-degree polynomial of the form $(x+c)$ or $(x-c)$.

The Remainder Theorem

Synthetic division can be used for several purposes, one of which is to find the value of a polynomial for a particular value of x. For example, we know (from Section 5.3) that if

$$P(x)=x^3-5x^2+7x-10,$$

then $$P(2)=2^3-5\cdot2^2+7\cdot2-10=-8.$$

Using synthetic division to divide $x^3-5x^2+7x-10$ by $x-2$ we have

$$\begin{array}{r|rrrr} 2 & 1 & -5 & 7 & -10 \\ & & 2 & -6 & 2 \\ \hline & 1 & -3 & 1 & -8 \end{array} \longleftarrow \text{Remainder}$$

The fact that the remainder is the same as $P(2)$ is not an accident. In fact, as the following theorem states, the remainder when a polynomial is divided by a first-degree factor of the form $(x-c)$ will always be $P(c)$.

The Remainder Theorem

If a polynomial $P(x)$ is divided by $(x-c)$, then the remainder will be $P(c)$.

Proof:

By the division algorithm we know that $\frac{P(x)}{x-c}=Q(x)+\frac{R}{x-c}$ where R is a constant. (Remember that the degree of the remainder must be less than the degree of the divisor.)

Multiplying through by $(x-c)$, we have

$$P(x)=(x-c)\cdot Q(x)+R$$

and substituting $x=c$ gives

$$\begin{aligned} P(c)&=(c-c)\cdot Q(c)+R \\ &=0\cdot Q(c)+R \\ &=0+R \\ &=R. \end{aligned}$$

The proof is complete.

Example 2: The Remainder Theorem and Synthetic Division

a. Use synthetic division to find $P(5)$ given $P(x)=-2x^2+15x-50$.

Solution:

$$\begin{array}{r|rrr} 5 & -2 & 15 & -50 \\ & & -10 & 25 \\ \hline & -2 & 5 & -25 \end{array} \longleftarrow \text{Remainder} = P(5)$$

Thus $P(5)=-25$.

(Checking shows $P(5)=-2\cdot5^2+15\cdot5-50=-50+75-50=-25$.)

1. a. $x-9$
   b. $c=3;\ P(3)=0$
2. a. $x-7$
   b. $c=5;\ P(5)=0$
3. a. $x^2-4x+33-\dfrac{265}{x+8}$
   b. $c=-8;\ P(-8)=-265$
4. a. $x^2-4x-\dfrac{5}{x-2}$
   b. $c=2;\ P(2)=-5$
5. a. $4x^2-6x+9-\dfrac{17}{x+2}$
   b. $c=-2;\ P(-2)=-17$

b. Use synthetic division to find $P(-3)$, given $P(x)=3x^4+10x^3-5x^2+125$.
Note: To evaluate $P(-3)$, think of the divisor in the form $(x+3)=(x-(-3))$. That is, in the form $(x-c)$, $c=-3$.

Solution:

$-3\rfloor$	3	10	−5	0	125	
		−9	−3	24	−72	
	3	1	−8	24	53	← Remainder $=P(-3)$

Thus $P(-3)=53$.

c. Use synthetic division to show that $(x-6)$ is a factor of $P(x)=x^3-14x^2+53x-30$.

Solution:

$6\rfloor$	1	−14	53	−30	
		6	−48	30	
	1	−8	5	0	← Remainder $=P(6)$

Thus the remainder is $P(6)=0$ and $(x-6)$ **is a factor of** $P(x)$.

Note: The coefficients in the quotient tell us that x^2-8x+5 is also a factor of $P(x)$.

Appendix 2 Exercises

6. a. $3x^2+3x+5-\dfrac{10}{x+1}$
   b. $c=-1;\ P(-1)=-10$
7. a. $x^2+7x+55+\dfrac{388}{x-7}$
   b. $c=7;\ P(7)=388$
8. a. $2x^2-8x+25-\dfrac{98}{x+4}$
   b. $c=-4;\ P(-4)=-98$
9. a. $2x^2-2x+6-\dfrac{27}{x+3}$
   b. $c=-3;\ P(-3)=-27$

Divide the following expressions using synthetic division. ***a.*** *Write the answer in the form $Q+\dfrac{R}{D}$ where R is a constant.* ***b.*** *In each exercise, $D=(x-c)$. State the value of c and the value of $P(c)$.*

1. $\dfrac{x^2-12x+27}{x-3}$ **2.** $\dfrac{x^2-12x+35}{x-5}$ **3.** $\dfrac{x^3+4x^2+x-1}{x+8}$

4. $\dfrac{x^3-6x^2+8x-5}{x-2}$ **5.** $\dfrac{4x^3+2x^2-3x+1}{x+2}$ **6.** $\dfrac{3x^3+6x^2+8x-5}{x+1}$

7. $\dfrac{x^3+6x+3}{x-7}$ **8.** $\dfrac{2x^3-7x+2}{x+4}$ **9.** $\dfrac{2x^3+4x^2-9}{x+3}$

10. $\dfrac{4x^3-x^2+13}{x-1}$ **11.** $\dfrac{x^4-3x^3+2x^2-x+2}{x-3}$ **12.** $\dfrac{x^4+x^3-4x^2+x-3}{x+6}$

13. $\dfrac{x^4+2x^2-3x+5}{x-2}$ **14.** $\dfrac{3x^4+2x^3+2x^2+x-1}{x+1}$ **15.** $\dfrac{x^4-x^2+3}{x-\frac{1}{2}}$

16. $\dfrac{x^3+2x^2+1}{x-\frac{2}{3}}$ **17.** $\dfrac{x^5-1}{x-1}$ **18.** $\dfrac{x^5-x^3+x}{x+\frac{1}{2}}$

19. $\dfrac{x^4-2x^3+4}{x+\dfrac{4}{5}}$

20. $\dfrac{x^6+1}{x+1}$

Collaborative Learning Exercise

21. With the class divided into teams of 3 or 4 students, each team should develop answers to the following questions and be prepared to discuss the answers in class.

a. First use long division to divide the polynomial $P(x)=2x^3-8x^2+10x+15$ by $2x-1$.
Then use synthetic division to divide the same polynomial by $x-\dfrac{1}{2}$.
Do the same process with two or three other polynomials and divisors. Next compare the corresponding long and synthetic division answers and explain how the answers are related.

b. Use the results from part **a.** and explain algebraically the relationship of the answers when a polynomial is divided (using long division) by $ax-b$ and (using synthetic division) by $x-\dfrac{b}{a}$.

c. Show how the remainder theorem should be restated if $x-c$ is replaced by $ax-b$.

HAWKES LEARNING SYSTEMS: INTRODUCTORY & INTERMEDIATE ALGEBRA SOFTWARE

- A.2 Synthetic Division

10. a. $4x^2+3x+3+\dfrac{16}{x-1}$ b. $c=1;\ P(1)=16$

11. a. $x^3+2x+5+\dfrac{17}{x-3}$ b. $c=3;\ P(3)=17$

12. a. $x^3-5x^2+26x-155+\dfrac{927}{x+6}$ b. $c=-6;\ P(-6)=927$

13. a. $x^3+2x^2+6x+9+\dfrac{23}{x-2}$ b. $c=2;\ P(2)=23$

14. a. $3x^3-x^2+3x-2+\dfrac{1}{x+1}$ b. $c=-1;\ P(-1)=1$

15. a. $x^3+\dfrac{1}{2}x^2-\dfrac{3}{4}x-\dfrac{3}{8}+\dfrac{45}{16\left(x-\dfrac{1}{2}\right)}$ b. $c=\dfrac{1}{2};\ P\left(\dfrac{1}{2}\right)=\dfrac{45}{16}$

16. a. $x^2+\dfrac{8}{3}x+\dfrac{16}{9}+\dfrac{59}{27\left(x-\dfrac{2}{3}\right)}$ b. $c=\dfrac{2}{3};\ P\left(\dfrac{2}{3}\right)=\dfrac{59}{27}$ 17. a. $x^4+x^3+x^2+x+1$ b. $c=1;\ P(1)=0$

18. a. $x^4-\dfrac{1}{2}x^3-\dfrac{3}{4}x^2+\dfrac{3}{8}x+\dfrac{13}{16}-\dfrac{13}{32\left(x+\dfrac{1}{2}\right)}$ b. $c=-\dfrac{1}{2};\ P\left(-\dfrac{1}{2}\right)=-\dfrac{13}{32}$

19. a. $x^3-\dfrac{14}{5}x^2+\dfrac{56}{25}x-\dfrac{224}{125}+\dfrac{3396}{625\left(x+\dfrac{4}{5}\right)}$ b. $c=-\dfrac{4}{5};\ P\left(-\dfrac{4}{5}\right)=\dfrac{3396}{625}$ 20. a. $x^5-x^4+x^3-x^2+x-1+\dfrac{2}{x+1}$ b. $c=-1;\ P(-1)=2$ 21. a. $x^2-\dfrac{7}{2}x+\dfrac{13}{4}+\dfrac{73}{4(2x-1)};2x^2-7x+\dfrac{13}{2}+\dfrac{73}{4\left(x-\dfrac{1}{2}\right)}$ b. $\dfrac{P(x)}{x-\dfrac{b}{a}}=\dfrac{aP(x)}{ax-b}$

c. If a polynomial $P(x)$ is divided by $(ax-b)$, then the remainder will be $P\left(\dfrac{b}{a}\right)$. Answers will vary.

A.3 Using a Graphing Calculator to Solve Equations

- *Use a TI-84 Plus graphing calculator to solve (or estimate the solutions of) polynomial equations by using one of the following strategies:*
  a. *graph one function and find the zeros of the function, or*
  b. *graph two functions and find the points of intersection of the graphs.*
- *Use a TI-84 Plus graphing calculator to solve absolute value equations and inequalities.*

In Section 4.5, we introduced functions, showed how to use a graphing calculator to graph functions, and listed steps for finding the zeros of a function. **Remember that the zeros of a function are the values of x at the points, if any, where the graph of the function crosses the x-axis.** These are the points where $y = 0$. In this section we will simply expand on these ideas and use the related concepts to solve equations (or to estimate the solutions of equations).

NOTES

When a polynomial is second-degree or higher, the same factor may occur multiple times. This means that the corresponding zero may appear more than once. That is, if a binomial factor is squared, then the corresponding zero is said to be of multiplicity 2. If the factor is cubed, then the corresponding zero is of multiplicity 3, and so on. For example, $P(x) = x^3 - 3x^2 - 24x + 80 = (x+5)(x-4)^2$ and there are technically three zeros, –5, 4, and 4. But 4 appears twice, so 4 is a zero of multiplicity 2 and the only distinct zeros are –5 and 4. As we will see, this has a major effect on the nature of the corresponding graph of the function.

With the calculator as a tool, we can solve (or estimate the solutions of) linear, quadratic, or higher degree polynomial equations. But, we must be careful with the WINDOW settings or the display may not show enough of the graph to illustrate all of the solutions. The following information will help in determining whether or not your WINDOW setting is appropriate.

Zeros of Polynomial Functions

1. Nonconstant linear functions have 1 zero. (The graph crosses the x-axis once.)
2. Quadratic functions have 2 zeros, 1 zero, or none. (The graph crosses the x-axis twice, just touches the x-axis, or doesn't cross at all.)
3. Cubic functions have 3 zeros, 2 zeros or 1 zero. (The graph crosses the x-axis three times, crosses once and just touches once, or crosses just once.)

If you are graphing a cubic function and the display shows only 1 zero (the graph crosses the x-axis only once), you may need a larger WINDOW setting to determine whether or not there are more zeros. If you see 3 zeros, then you know that there are no more and the WINDOW setting is sufficient.

NOTES

Important Note about the Graphs of Polynomial Functions

The graph of every polynomial function is a smooth continuous graph. (That is, there are no holes, jumps from one point to another, or sharp points in the graph of a polynomial function.)

Now there are two basic strategies to solving equations using the graphing calculator. We need to either:

a. graph one function and find the zeros of the function, or
b. graph two functions and find the points of intersection of the graphs.

The following examples illustrate both of these strategies and the steps to use with the graphing calculator.

Example 1: Using One Graph to Solve a Polynomial Equation

Use a graphing calculator to solve the equation $x^3 - 3x^2 = 13x - 15$.

Solution: Manipulate the equation so that one side is 0. Graph the indicated function on the nonzero side. The zeros of this function are the roots of the original equation. (See Section 4.5 for more information about entering and graphing an equation on a TI-84 Plus graphing calculator.)

$$x^3 - 3x^2 = 13x - 15$$

$$x^3 - 3x^2 - 13x + 15 = 0$$

Enter the function as follows:

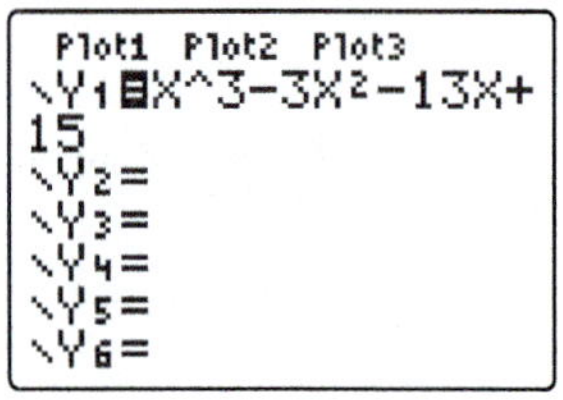

With the standard window the graph will appear as follows:

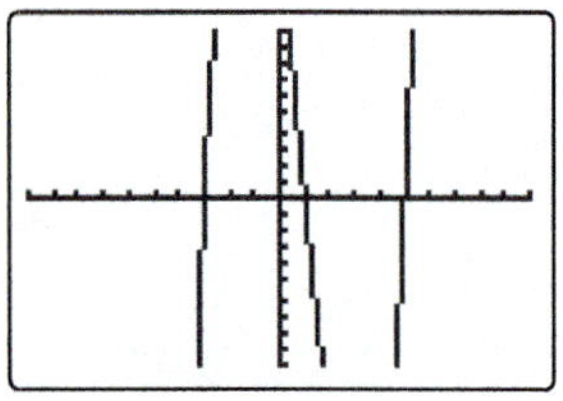

Teaching Notes:
Keep in mind that some students will need instruction in the resetting of the WINDOW limits and scale. Expect to provide some in-class reminders on how and when to reset.

Teaching Notes:
You may want to encourage your students to use the TRACE command to illustrate how the approximations differ from the values found by the 2nd > CALC > `2:zero` sequence of commands.

Note: You may want to increase the y-values on the window to see a more complete graph. This will not change the zeros.

With the 2ND > CALC > `2:zero` sequence of commands you will find the following zeros (and therefore solutions to the equation):

$$x = -3,\ x = 1,\ \text{and } x = 5.$$

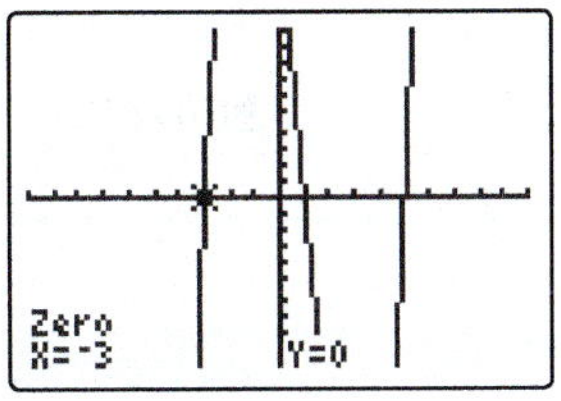

Note: See Section 4.5 for a more in-depth explanation for finding the zero of a function. With the TRACE command you will find only approximations of the zeros.

Example 2: Using Two Graphs to Solve a Polynomial Equation

Solve the polynomial equation $2x^2 = 3x + 1$ using a graphing calculator.

Solution: Graph the function indicated on each side of the equation. Find the points of intersection of these two graphs. The x-values of these points are the roots of the original equation.

Enter the functions as follows:

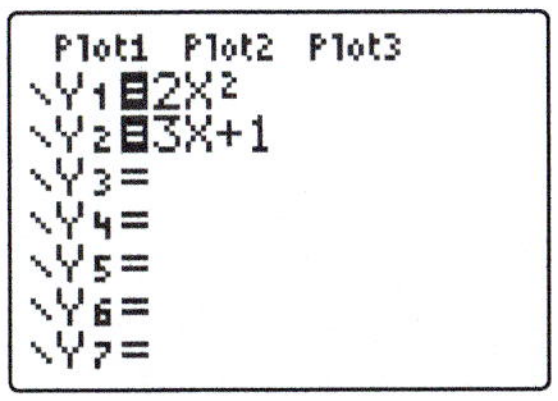

With the standard window the graphs will appear as follows:

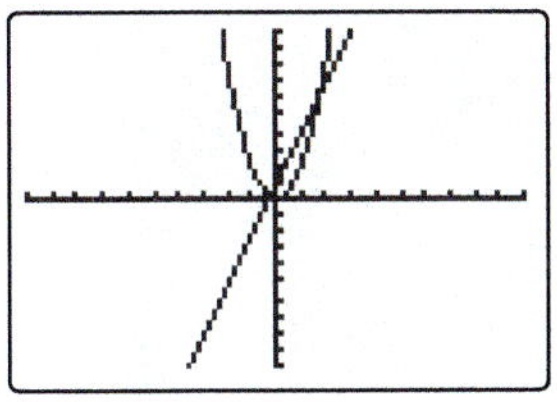

Teaching Notes:
Again, you may want to encourage your students to use the TRACE command so they will understand its limitations.

With the 2ND > CALC > `5:intersect` sequence of commands you will find the following approximate x-values of the points of intersection (and therefore approximate solutions to the equation):

Continued on the next page...

$x \approx -.28$ and $x \approx 1.78$ (accurate to two decimal places).

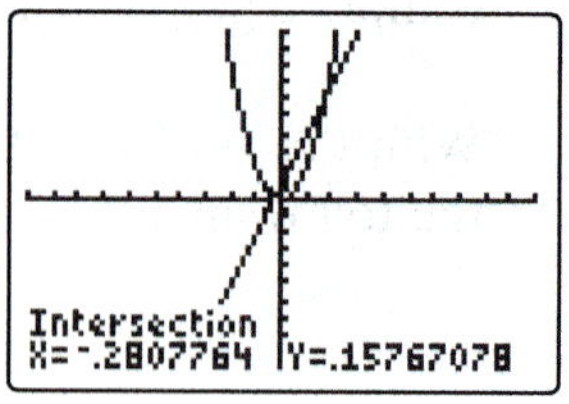

Note: See Section 4.5 for an explanation of the intersect function.

Example 3: Using Two Graphs to Solve an Absolute Value Equation

Solve the equation $|2x-5|=8$ using a graphing calculator.

Solution: Graph the functions indicated on each side of the equation. This includes the constant function. Find the points of intersection of these two graphs. The x-values of these points are the roots of the original equation. Remember that the absolute value command can be found in the MATH > NUM menu.

Enter the functions as follows:

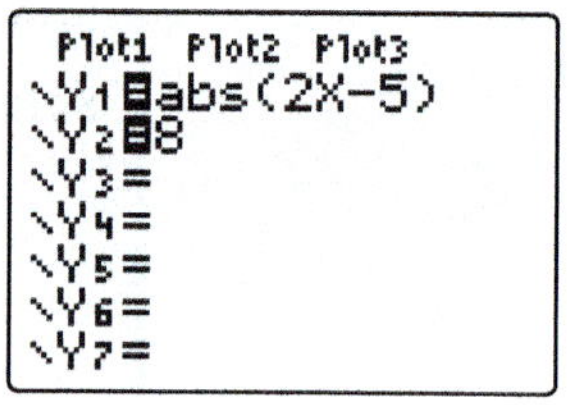

With the standard window the graphs will appear as follows:

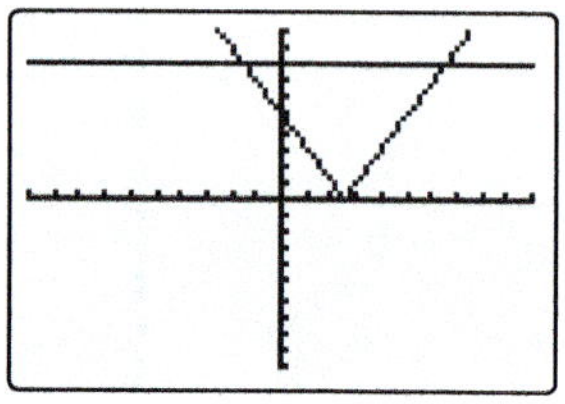

With the 2ND > CALC > 5:intersect sequence of commands you will find the following x-values at the points of intersection (and therefore the solutions to the equation):

$x = -1.5$ and $x = 6.5$

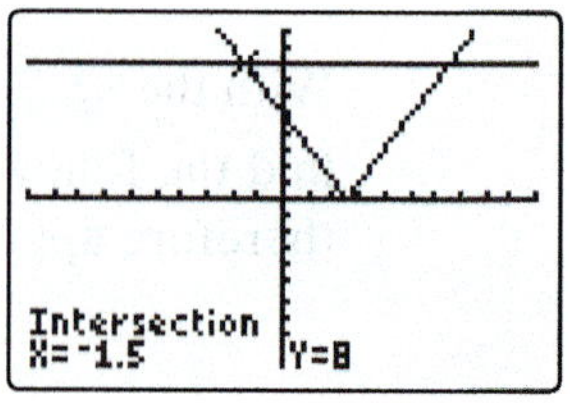

Example 4: Using a Graphing Calculator to Solve Absolute Value Inequalities

Use a graphing calculator to solve the inequalities:

a. $|2x-5|<8$ **b.** $|2x-5|>8$

Both inequalities can be solved by using the graphs from Example 3.

Solution: We change the window for a clearer view: use the interval $[-10, 10]$ for x and the interval $[-1, 15]$ for y. This gives:

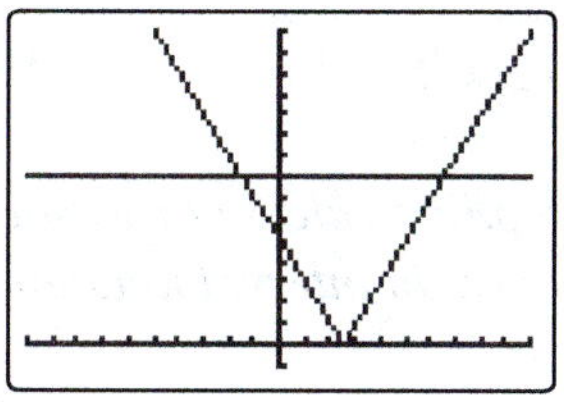

a. From Example 3, we know that the intersections occur at $x = -1.5$ and $x = 6.5$. Looking at the graph we see that the graph of the absolute value is below the line $y = 8$ on the interval $(-1.5, 6.5)$. Thus the interval $(-1.5, 6.5)$ is the solution set for $|2x-5|<8$.

b. Looking at the graph we see that the graph of the absolute value is above the line $y = 8$ on the intervals $(-\infty, -1.5)$ and $(6.5, \infty)$. Thus the solution set is $(-\infty, -1.5)\cup(6.5, \infty)$ for $|2x-5|>8$.

Appendix 3 Exercises

1. $x = -2, 2$
2. $x = -3, 3$
3. $x \approx -1.41, 1.41$
4. $x \approx -3.87, 3.87$
5. $x = -2, 6$
6. $x = -7, 1$
7. $x \approx -4.46, 2.46$
8. $x \approx -1.26, 1.59$
9. $x \approx -1.70, 4.70$
10. No solution
11. No solution
12. $x \approx -2.65, 5.65$
13. $x \approx -1.26, 1.59$
14. $x \approx -1.24$
15. $x \approx -1.77, 0.52, 3.25$
16. $x \approx -2.66, -1.21, 1.87$

Use a graphing calculator to solve the equations. Find any approximations accurate to two decimal places.

1. $x^2-4=0$ **2.** $x^2-9=0$ **3.** $x^2-2=0$

4. $x^2-15=0$ **5.** $x^2-4x=12$ **6.** $x^2+6x=7$

7. $x^2+2x-11=0$ **8.** $3x^2-x-6=0$ **9.** $-x^2+3x+8=0$

10. $-2x^2+4x-5=0$ **11.** $2x^2+x=-2$ **12.** $3x+15=x^2$

13. $3x^2-6=x$ **14.** $x^3=2x^2-5$ **15.** $5x-3=x^3-2x^2$

16. $x^3+2x^2=4x+6$ **17.** $x(x-1)(x-3)=0$

18. $(x-2)(x+1)(x+4)=0$ **19.** $(x+3)(x+1)(x-5)=0$

17. $x = 0, 1, 3$
18. $x = -4, -1, 2$
19. $x = -3, -1, 5$
20. $x = -2, 1, 6$
21. $x \approx 3.40$
22. $x \approx 1.38$
23. $x \approx -3.16, 0, 3.16$
24. $x \approx -1.73, 0, 1.73$
25. $x = -1, 0$
26. $x = -2, 0$
27. $x = -4, 7$
28. $x = -4, 3$
29. $x \approx -1.67, 3$
30. $x = -5, 4.5$
31. $x = -2, 0$
32. $x = 0, 2$
33. $x = 0.5$
34. $x = 3.5$
35. $x = 2, -0.67$
36. $x = -1.25, 0.83$
37. $(-\infty, -6) \cup (6, \infty)$
38. $[-3, 3]$
39. $[2, 4]$

20. $(x+2)(x-1)(x-6) = 0$ **21.** $2x^3 - 8x^2 + 7x - 1 = 9$

22. $3x^3 - x^2 + 4 = 10$ **23.** $x^4 - 10x^2 = 0$ **24.** $x^4 = 3x^2$

25. $x^4 - x^3 + 2x = 0$ **26.** $x^4 - x^2 + 6x = 0$ **27.** $|2x-3| = 11$

28. $|2x+1| = 7$ **29.** $|3x-2| = 7$ **30.** $|4x+1| = 19$

31. $|2x+1| = |x-1|$ **32.** $|5x-4| = |x+4|$ **33.** $|x-3| = |x+2|$

34. $|x-2| = |5-x|$ **35.** $\left|\frac{x}{2}+1\right| = |x|$ **36.** $\left|\frac{x}{5}-1\right| = |x|$

Use a graphing calculator to solve (or estimate the solutions of) the inequalities. Write your answers in interval notation.

37. $|x| > 6$ **38.** $|x| \le 3$ **39.** $|x-3| \le 1$

40. $|x-5| > 2$ **41.** $|2x-3| \ge 4$ **42.** $|3x-8| > 4$

43. $|x+2| - 10 \le 17$ **44.** $|x-4| - 2 > 10$ **45.** $\left|\frac{x}{3}-1\right| < 2$

46. $\left|\frac{x}{4}+3\right| \ge 1$

Writing and Thinking About Mathematics

40. $(-\infty, 3) \cup (7, \infty)$
41. $\left(-\infty, \frac{-1}{2}\right] \cup \left[\frac{7}{2}, \infty\right)$
42. $\left(-\infty, \frac{4}{3}\right) \cup (4, \infty)$

Use a graphing calculator and three graphs to solve the inequality. Write the answer in interval notation. Explain how you might solve this inequality algebraically.

47. $1 \le |x-4| \le 5$

HAWKES LEARNING SYSTEMS: INTRODUCTORY & INTERMEDIATE ALGEBRA SOFTWARE

- A.3 Using a Graphing Calculator to Solve Equations

43. $[-29, 25]$ 44. $(-\infty, -8) \cup (16, \infty)$ 45. $(-3, 9)$ 46. $(-\infty, -16] \cup [-8, \infty)$
47. $[-1, 3] \cup [5, 9]$; You would solve the equations $-1 \le x - 4 \le 5$ and $1 \ge x - 4 \ge -5$. Answers will vary.

A.4 Determinants

- *Evaluate 2 × 2 and 3 × 3 determinants.*
- *Solve equations involving determinants.*

As was discussed in Section 8.7, a rectangular array of numbers is called a matrix, and matrices arise in connection with solving systems of linear equations such as:

$$\begin{cases} a_{11}x + a_{12}y = k_1 \\ a_{21}x + a_{22}y = k_2 \end{cases}$$

The **matrix of the coefficients** (or the **coefficient matrix**) is

$$A = \begin{bmatrix} a_{11} & a_{12} \\ a_{21} & a_{22} \end{bmatrix}.$$

If a matrix is **square** (the number of rows is equal to the number of columns), then the matrix has a number associated with it called the **determinant**. This section covers evaluating determinants and the next section shows how determinants can be used to solve systems of linear equations by using a method called **Cramer's rule**.

Determinant

A **determinant** is a real number associated with a square matrix and is indicated by enclosing the array between two vertical bars. For a matrix A, the corresponding determinant is designated as $\det(A)$ and is read "the determinant of A."

Examples of determinants are:

a. For the matrix, $A = \begin{bmatrix} 3 & 4 \\ 7 & -2 \end{bmatrix}$, $\quad \det(A) = \begin{vmatrix} 3 & 4 \\ 7 & -2 \end{vmatrix}.$

b. For the matrix, $B = \begin{bmatrix} 1 & 6 & -3 \\ 4 & 5 & 5 \\ -1 & -1 & -1 \end{bmatrix}$, $\quad \det(B) = \begin{vmatrix} 1 & 6 & -3 \\ 4 & 5 & 5 \\ -1 & -1 & -1 \end{vmatrix}.$

Example **a.** is a 2 × 2 determinant and has two rows and two columns.
Example **b.** is a 3 × 3 determinant and has three rows and three columns.

$$\begin{array}{l} \text{row 1} \rightarrow \\ \text{row 2} \rightarrow \end{array}\begin{vmatrix} 3 & 4 \\ 7 & -2 \end{vmatrix} \qquad \begin{array}{cc} \text{column 1} & \text{column 2} \\ \downarrow & \downarrow \end{array} \begin{vmatrix} 3 & 4 \\ 7 & -2 \end{vmatrix}$$

$$\begin{array}{l} \text{row 1} \rightarrow \\ \text{row 2} \rightarrow \\ \text{row 3} \rightarrow \end{array}\begin{vmatrix} 1 & 6 & -3 \\ 4 & 5 & 5 \\ -1 & -1 & -1 \end{vmatrix} \qquad \begin{array}{ccc} \text{column 1} & \text{column 2} & \text{column 3} \\ \downarrow & \downarrow & \downarrow \end{array} \begin{vmatrix} 1 & 6 & -3 \\ 4 & 5 & 5 \\ -1 & -1 & -1 \end{vmatrix}$$

A 4×4 determinant has four rows and four columns. A determinant may be of any size $n \times n$ where n is a positive integer and $n \geq 2$. In this text, the discussion will be restricted to 2×2 and 3×3 determinants, and the entries will be real numbers. (Huge matrices and determinants (1000×1000 or larger) are common in industry, and their values are calculated by computers. Even then, someone must understand the algebraic techniques to be able to write the necessary programs.)

Every determinant with real entries has a real value. The method for finding the value of 3×3 determinants involves finding the value of 2×2 determinants. Determinants of larger matrices can be evaluated by using techniques similar to those shown here. Their applications occur in higher mathematics such as linear algebra and differential equations.

Value of a 2×2 Determinant

For the square matrix $A = \begin{bmatrix} a_{11} & a_{12} \\ a_{21} & a_{22} \end{bmatrix}$, $\quad \mathbf{det}(A) = \begin{vmatrix} a_{11} & a_{12} \\ a_{21} & a_{22} \end{vmatrix} = a_{11}a_{22} - a_{21}a_{12}$.

As the definition indicates and the following examples illustrate, the value of a 2×2 determinant is the **product of the numbers in the diagonal containing the term in the first row, first column, minus the product of the numbers in the other diagonal.**

Example 1: 2×2 Determinants

Evaluate the following 2×2 determinants.

a. $\begin{vmatrix} 3 & 4 \\ 7 & -2 \end{vmatrix} = 3(-2) - 7(4) = -6 - 28 = -34$

b. $\begin{vmatrix} -5 & -\frac{1}{2} \\ 6 & 3 \end{vmatrix} = -5(3) - 6\left(-\frac{1}{2}\right) = -15 + 3 = -12$

c. $\begin{vmatrix} 1 & 7 \\ 2 & 14 \end{vmatrix} = 1(14) - 2(7) = 14 - 14 = 0$

One method of evaluating 3 × 3 determinants is called **expanding by minors**. In this method, **one row is chosen** and each entry in that row has a minor. Each minor is found by mentally crossing out both the row and column (shown here in the shaded regions) that contain that entry. The minors of the entries in the first row are illustrated here.

$$\begin{vmatrix} \mathbf{a_{11}} & a_{12} & a_{13} \\ a_{21} & a_{22} & a_{23} \\ a_{31} & a_{32} & a_{33} \end{vmatrix} \longrightarrow \begin{vmatrix} a_{22} & a_{23} \\ a_{32} & a_{33} \end{vmatrix} \longleftarrow \text{minor of } a_{11}$$

$$\begin{vmatrix} a_{11} & \mathbf{a_{12}} & a_{13} \\ a_{21} & a_{22} & a_{23} \\ a_{31} & a_{32} & a_{33} \end{vmatrix} \longrightarrow \begin{vmatrix} a_{21} & a_{23} \\ a_{31} & a_{33} \end{vmatrix} \longleftarrow \text{minor of } a_{12}$$

$$\begin{vmatrix} a_{11} & a_{12} & \mathbf{a_{13}} \\ a_{21} & a_{22} & a_{23} \\ a_{31} & a_{32} & a_{33} \end{vmatrix} \longrightarrow \begin{vmatrix} a_{21} & a_{22} \\ a_{31} & a_{32} \end{vmatrix} \longleftarrow \text{minor of } a_{13}$$

To find the value of a determinant (of any dimension other than 2 × 2), first choose a row (or column) and find the product of each entry in that row (or column) with its corresponding minor. Then the value is determined by adding these products with appropriate adjustments of alternating signs of the minors. We say that the determinant has been expanded by that row (or column). The following illustrates how to find the value of a 3 × 3 determinant by expanding by the first row.

Value of a 3×3 Determinant

For the square matrix $A = \begin{bmatrix} a_{11} & a_{12} & a_{13} \\ a_{21} & a_{22} & a_{23} \\ a_{31} & a_{32} & a_{33} \end{bmatrix}$,

$$\mathbf{det}(\boldsymbol{A}) = \begin{vmatrix} \mathbf{a_{11}} & \mathbf{a_{12}} & \mathbf{a_{13}} \\ a_{21} & a_{22} & a_{23} \\ a_{31} & a_{32} & a_{33} \end{vmatrix} = \mathbf{a_{11}}(\text{minor of } a_{11}) - \mathbf{a_{12}}(\text{minor of } a_{12}) + \mathbf{a_{13}}(\text{minor of } a_{13})$$

$$= \mathbf{a_{11}} \begin{vmatrix} a_{22} & a_{23} \\ a_{32} & a_{33} \end{vmatrix} - \mathbf{a_{12}} \begin{vmatrix} a_{21} & a_{23} \\ a_{31} & a_{33} \end{vmatrix} + \mathbf{a_{13}} \begin{vmatrix} a_{21} & a_{22} \\ a_{31} & a_{32} \end{vmatrix}.$$

NOTES

CAUTION: The negative sign in the middle term of the expansion (representing –1 times a_{12}) is a critical part of the method and is a source of error for many students. **Be careful.**

Teaching Notes:
A good in-class activity is to assign different groups of students a row to use in the determinant (i.e., using different minors) and demonstrate that the value of this determinant is the same, no matter which row is used.

Each minor is multiplied by its corresponding entry and +1 or –1 according to the pattern illustrated in Figure 1. (**Note:** The signs alternate and this pattern can be extended to apply to any $n \times n$ determinant.)

$$\begin{vmatrix} + & - & + \\ - & + & - \\ + & - & + \end{vmatrix}$$

Figure 1

For example, the value of a 3 × 3 determinant can be found by expanding by the minors of the second row as follows:

$$\det(A) = -a_{21}(\text{minor of } a_{21}) + a_{22}(\text{minor of } a_{22}) - a_{23}(\text{minor of } a_{23}).$$

Note the use of the alternating + and – signs from the pattern in Figure 1. You may want to try this for practice with some of the exercises.

NOTES There are methods other than expanding by minors for evaluating 3 × 3 determinants. The advantage of learning to expand by minors is that this method can be used for evaluating higher-order determinants.

Example 2: 3×3 Determinants

Evaluate the following 3 × 3 determinants.

a. $\begin{vmatrix} 5 & 1 & -4 \\ 2 & 6 & 3 \\ 2 & 2 & 1 \end{vmatrix}$

Using Row 1, mentally delete the shaded regions.

Solution:
$$\begin{vmatrix} 5 & 1 & -4 \\ 2 & 6 & 3 \\ 2 & 2 & 1 \end{vmatrix} = (5)\begin{vmatrix} 5 & 1 & -4 \\ 2 & 6 & 3 \\ 2 & 2 & 1 \end{vmatrix} - (1)\begin{vmatrix} 5 & 1 & -4 \\ 2 & 6 & 3 \\ 2 & 2 & 1 \end{vmatrix} + (-4)\begin{vmatrix} 5 & 1 & -4 \\ 2 & 6 & 3 \\ 2 & 2 & 1 \end{vmatrix}$$

$$= 5\begin{vmatrix} 6 & 3 \\ 2 & 1 \end{vmatrix} - 1\begin{vmatrix} 2 & 3 \\ 2 & 1 \end{vmatrix} - 4\begin{vmatrix} 2 & 6 \\ 2 & 2 \end{vmatrix}$$

$$= 5(6 \cdot 1 - 2 \cdot 3) - 1(2 \cdot 1 - 2 \cdot 3) - 4(2 \cdot 2 - 2 \cdot 6)$$

$$= 5(6 - 6) - 1(2 - 6) - 4(4 - 12)$$

$$= 5(0) - 1(-4) - 4(-8)$$

$$= 0 + 4 + 32$$

$$= 36$$

b. $\begin{vmatrix} 6 & -2 & 4 \\ 1 & 7 & 0 \\ -3 & 2 & -1 \end{vmatrix}$

Solution: $\begin{vmatrix} 6 & -2 & 4 \\ 1 & 7 & 0 \\ -3 & 2 & -1 \end{vmatrix} = (6)\begin{vmatrix} 7 & 0 \\ 2 & -1 \end{vmatrix} - (-2)\begin{vmatrix} 1 & 0 \\ -3 & -1 \end{vmatrix} + (4)\begin{vmatrix} 1 & 7 \\ -3 & 2 \end{vmatrix}$

$= 6(-7-0)+2(-1-0)+4(2+21)$

$= -42-2+92$

$= 48$

After some practice, many of these steps can be done mentally.

Example 3: Equations with Determinants

Solve the following equation for x: $\begin{vmatrix} 2 & 3 & 0 \\ 6 & x & 5 \\ 1 & -2 & 9 \end{vmatrix} = 53.$

Solution: First, evaluate the determinant.

$$\begin{vmatrix} 2 & 3 & 0 \\ 6 & x & 5 \\ 1 & -2 & 9 \end{vmatrix} = (2)\begin{vmatrix} x & 5 \\ -2 & 9 \end{vmatrix} - (3)\begin{vmatrix} 6 & 5 \\ 1 & 9 \end{vmatrix} + (0)\begin{vmatrix} 6 & x \\ 1 & -2 \end{vmatrix}$$

$= 2(9x+10)-3(54-5)+0$

$= 2(9x+10)-3(49)$

$= 18x+20-147$

$= 18x-127$

Now solve the equation.

$18x-127 = 53$

$18x = 180$

$x = 10$

The technique of expanding by minors may be used (with appropriate adjustments) to evaluate any $n \times n$ determinant. For example, in a 4×4 determinant, the minors of the entries in a particular row will be 3×3 determinants. Also, there are techniques for simplifying determinants and there are rules for arithmetic with determinants. The general rules governing these operations are discussed in courses like precalculus, finite mathematics, and linear algebra.

Using the TI-84 Plus Graphing Calculator to Evaluate a Determinant

A TI-84 Plus calculator (and other graphing calculators) can be used to find the value of the determinant of a square matrix. The determinant command, `1: det(` is found as the first entry in the `MATRIX / MATH` menu. Example 4 shows, in a step by step format, how to find the determinant of a given 3×3 matrix.

Example 4: Evaluating Determinants with a Calculator

Use a TI-84 Plus calculator to find the value of $\det(A)$ for the following matrix.

$$A = \begin{bmatrix} 2 & 5 & 7 \\ 3 & 1 & 0 \\ 4 & 0 & 3 \end{bmatrix}$$

Solution:

Step 1: Press 2ND > MATRIX, go to the EDIT menu and enter the appropriate dimensions and numbers in the matrix A. The display should appear as follows.

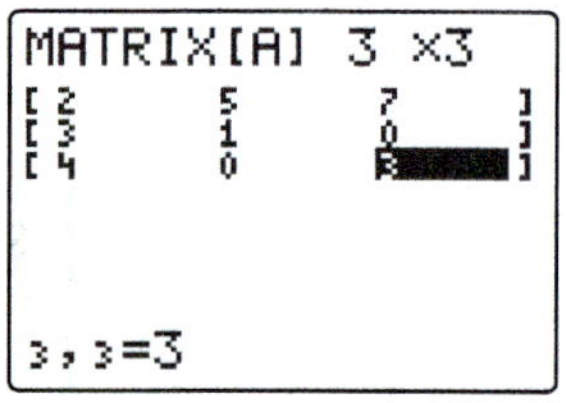

Step 2: Press 2ND > QUIT then 2ND > MATRIX again and go to the MATH menu. On the MATH menu choose `1: det(` and press ENTER. The display should appear as follows.

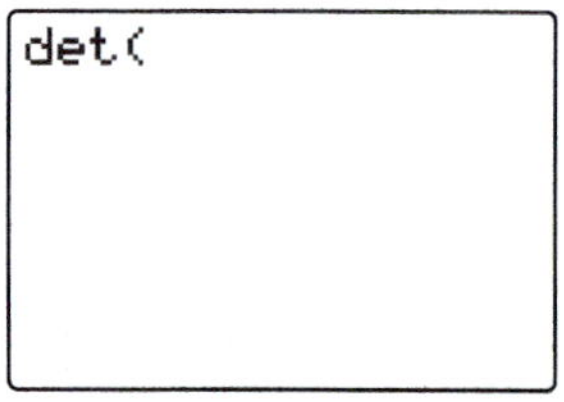

Step 3: Press 2ND > MATRIX again and on the NAMES menu choose `1:[A]3×3` by pressing ENTER. Then type a right parenthesis). The display should appear as follows.

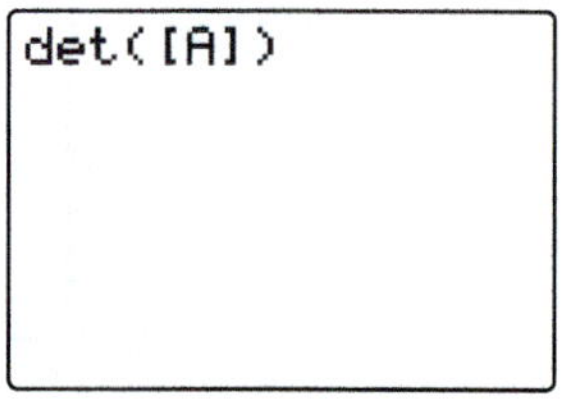

Step 4: Press ENTER and the display should appear as follows with the answer.

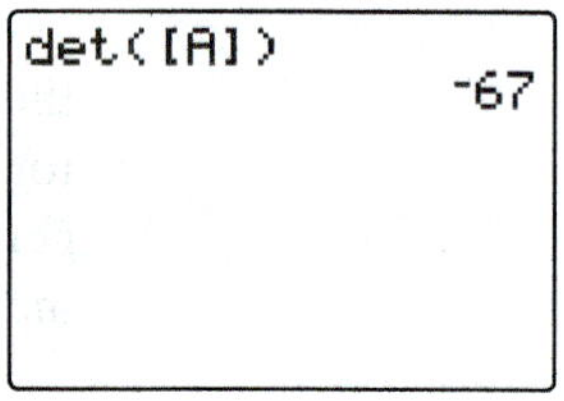

Practice Problems

Evaluate each of the following determinants.

1. $\begin{vmatrix} -3 & 2 \\ 4 & 7 \end{vmatrix}$ **2.** $\begin{vmatrix} 6 & 3 \\ 4 & 2 \end{vmatrix}$ **3.** $\begin{vmatrix} 1 & 4 & 0 \\ 2 & -1 & 5 \\ 0 & 7 & -1 \end{vmatrix}$

Use a graphing calculator to find the value of the determinant.

4. $\begin{vmatrix} 5 & -1 & 3 \\ 0 & 4 & 2 \\ -3 & 1 & 3 \end{vmatrix}$

Use the method for evaluating determinants to solve the equation.

5. $\begin{vmatrix} 3 & 5 \\ 6 & x \end{vmatrix} = 18$

Appendix 4 Exercises

1. −22
2. 11
3. −212
4. −76
5. 11
6. −48
7. 3
8. −17
9. 47
10. 14
11. 36
12. 39
13. −3
14. −23
15. −4
16. 24

In the following exercises the matrix A is given. Find det(A).

1. $A = \begin{bmatrix} 2 & 7 \\ 4 & 3 \end{bmatrix}$ **2.** $A = \begin{bmatrix} 7 & 3 \\ 8 & 5 \end{bmatrix}$

3. $A = \begin{bmatrix} -5 & 2 & 1 \\ 4 & 8 & 0 \\ -2 & 3 & 5 \end{bmatrix}$ **4.** $A = \begin{bmatrix} -6 & 5 & -3 \\ 4 & 0 & -1 \\ -2 & 7 & -2 \end{bmatrix}$

Evaluate the determinants.

5. $\begin{vmatrix} 1 & 3 \\ -2 & 5 \end{vmatrix}$ **6.** $\begin{vmatrix} 7 & 2 \\ 3 & -6 \end{vmatrix}$ **7.** $\begin{vmatrix} 6 & 3 \\ -11 & -5 \end{vmatrix}$ **8.** $\begin{vmatrix} 2 & 3 \\ 3 & -4 \end{vmatrix}$

9. $\begin{vmatrix} 9 & 4 \\ 4 & 7 \end{vmatrix}$ **10.** $\begin{vmatrix} 3 & -4 \\ 8 & -6 \end{vmatrix}$ **11.** $\begin{vmatrix} 0 & -1 & 2 \\ 3 & 5 & -7 \\ -3 & 4 & 1 \end{vmatrix}$ **12.** $\begin{vmatrix} 1 & 0 & -1 \\ -2 & 3 & 5 \\ 6 & -3 & 4 \end{vmatrix}$

13. $\begin{vmatrix} 1 & -1 & 2 \\ -2 & 5 & -7 \\ 6 & 4 & 1 \end{vmatrix}$ **14.** $\begin{vmatrix} 2 & -1 & -3 \\ 5 & 9 & 4 \\ 7 & 6 & -2 \end{vmatrix}$ **15.** $\begin{vmatrix} 2 & 1 & 3 \\ 3 & 4 & 5 \\ 1 & 7 & 2 \end{vmatrix}$ **16.** $\begin{vmatrix} -3 & 2 & 1 \\ 1 & -4 & -1 \\ 2 & 5 & 3 \end{vmatrix}$

Answers to Practice Problems: **1.** −29 **2.** 0 **3.** −26 **4.** 92 **5.** 16

17. −25
18. −36
19. 20
20. −45
21. $x = 7$
22. $x = 9$
23. $x = -7$
24. $x = 2$
25. $x = -3$
26. $\begin{vmatrix} x & y & 1 \\ 3 & 2 & 1 \\ -1 & 4 & 1 \end{vmatrix} = 0$ which simplifies to $x + 2y = 7$
27. $\begin{vmatrix} x & y & 1 \\ -2 & 1 & 1 \\ 5 & 3 & 1 \end{vmatrix} = 0$ which simplifies to $2x - 7y = -11$
28. $\begin{vmatrix} x & y & 1 \\ 4 & -4 & 1 \\ 0 & 6 & 1 \end{vmatrix} = 0$ which simplifies to $5x + 2y = 12$
29. $A = 1$

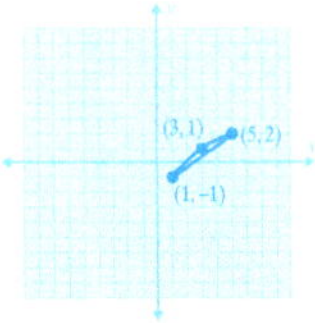

30. $A = \frac{7}{2}$

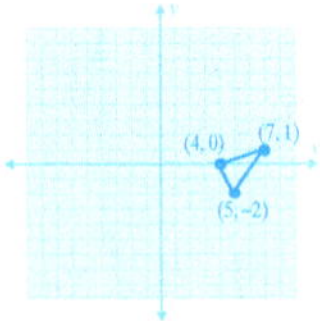

31. $A = \frac{31}{2}$

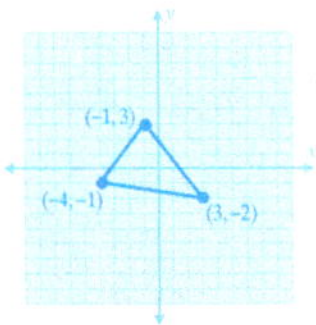

17. $\begin{vmatrix} 2 & 1 & -1 \\ 4 & 3 & 2 \\ 1 & 5 & 5 \end{vmatrix}$ **18.** $\begin{vmatrix} 6 & 7 & 1 \\ 0 & 3 & 3 \\ 4 & 1 & -5 \end{vmatrix}$ **19.** $\begin{vmatrix} 3 & -1 & -1 \\ 2 & 4 & 1 \\ -1 & 1 & 2 \end{vmatrix}$ **20.** $\begin{vmatrix} 2 & 3 & 2 \\ 1 & -1 & 5 \\ 0 & 5 & 1 \end{vmatrix}$

Use the method for evaluating determinants to solve the equations for x.

21. $\begin{vmatrix} 1 & 3 & 4 \\ 2 & x & 3 \\ 1 & 3 & 5 \end{vmatrix} = 1$ **22.** $\begin{vmatrix} -2 & -1 & 1 \\ x & 1 & -1 \\ 4 & 3 & -2 \end{vmatrix} = 7$ **23.** $\begin{vmatrix} 1 & x & x \\ 2 & -2 & 1 \\ -1 & 3 & 2 \end{vmatrix} = 0$

24. $\begin{vmatrix} x & x & 1 \\ 1 & 5 & 0 \\ 0 & 1 & -2 \end{vmatrix} = -15$ **25.** $\begin{vmatrix} 3 & 1 & -2 \\ 1 & x & 4 \\ 2 & x & 0 \end{vmatrix} = 38$

The formula $\begin{vmatrix} x & y & 1 \\ x_1 & y_1 & 1 \\ x_2 & y_2 & 1 \end{vmatrix} = 0$ *is an equation for the line passing through the two points* $P_1(x_1, y_1)$ *and* $P_2(x_2, y_2)$. *Using this formula, find an equation for the line determined by each pair of points given.*

26. $(3, 2), (-1, 4)$ **27.** $(-2, 1), (5, 3)$ **28.** $(4, -4), (0, 6)$

The area of a triangle with vertices $P_1(x_1, y_1)$, $P_2(x_2, y_2)$, *and* $P_3(x_3, y_3)$ *is given by the absolute value of the expression* $\frac{1}{2}\begin{vmatrix} x_1 & y_1 & 1 \\ x_2 & y_2 & 1 \\ x_3 & y_3 & 1 \end{vmatrix}$. *For each of the following sets of points, draw the triangle with the given points as vertices and then find the area of the triangle.*

29. $(3, 1), (1, -1), (5, 2)$ **30.** $(4, 0), (5, -2), (7, 1)$ **31.** $(-1, 3), (-4, -1), (3, -2)$

32. Explain, in your own words, the position of the three points $P_1(x_1, y_1)$, $P_2(x_2, y_2)$, and $P_3(x_3, y_3)$ if the expression $\frac{1}{2}\begin{vmatrix} x_1 & y_1 & 1 \\ x_2 & y_2 & 1 \\ x_3 & y_3 & 1 \end{vmatrix}$ has a value of 0. (**Hint:** Refer to the directions before Exercises 26 – 28.)

32. All three points lie on one line.
33. –25
34. –51.004
35. –33.28

Use a graphing calculator to find the value of the determinant.

33. $\begin{vmatrix} 3 & -4 & 6 \\ 2 & 4 & -1 \\ 7 & 9 & -1 \end{vmatrix}$

34. $\begin{vmatrix} 2.1 & 3.5 & -3.4 \\ 2.6 & 5.0 & 1.2 \\ -1.0 & 3.4 & 6.3 \end{vmatrix}$

35. $\begin{vmatrix} 1.6 & \frac{1}{2} & -5.9 \\ 0.7 & \frac{3}{4} & 1.7 \\ 5.0 & 8.2 & -4.1 \end{vmatrix}$

Writing and Thinking About Mathematics

36. 0; Answers will vary. If the two rows are identical, the determinant is effectively $a_{11}a_{12} - a_{11}a_{12} = 0$.
37. 0; Answers will vary. If you expand by minors using a row of all zeros, each minor will be multiplied by zero resulting in a sum of zero.

36. Suppose that in a 2 × 2 matrix two rows are identical. What will be the value of its determinant? Give two specific examples and a general example to back up your conclusion.

37. Suppose that in a 3 × 3 matrix one row is all 0's. What will be the value of its determinant? Give two specific examples and a general example to back up your conclusion.

38. In each part, give two specific examples and a general example to back up your conclusion.

a. Suppose that in a 2 × 2 determinant two rows (or columns) are switched. How will the value of this new determinant relate to the value of the original determinant?

b. Suppose that in a 3 × 3 determinant two rows (or columns) are switched. How will the value of this new determinant relate to the value of the original determinant?

Hawkes Learning Systems: Introductory & Intermediate Algebra Software

- A.4 Determinants

38. a. Answers will vary. The new determinant will be the negative (the opposite) of the original determinant.
b. Answers will vary. The new determinant will be the negative (the opposite) of the original determinant.

A.5 Cramer's Rule

- *Solve systems of linear equations using Cramer's rule.*

Cramer's rule is a method that uses determinants for solving systems of linear equations. To explain the method and how these determinants are generated, we begin by finding the solution to a system of linear equations by the addition method and do not simplify the fractional answers. We will see that these fractional answers can be represented as determinants.

Consider the following system of linear equations with the equations in standard form:

$$\begin{cases} 2x + 3y = -5 \\ 4x + y = 5 \end{cases}$$

Eliminating y gives:

$$\begin{cases} [1] & (2x + 3y = -5) \\ [-3] & (4x + y = 5) \end{cases}$$

$$\begin{array}{rrr} 1(2x) & +1(3y) & =1(-5) \\ -3(4x) & -3(1y) & =-3(5) \\ \hline \end{array}$$

$$\left[1(2)-3(4)\right]x+\left[1(3)-3(1)\right]y=1(-5)-3(5)$$

$$(-10)x+(0)y=-20$$

$$x=\frac{-20}{-10}$$

Eliminating x gives:

$$\begin{cases} [-4] & 2x + 3y = -5 \\ [2] & 4x + y = 5 \end{cases}$$

$$\begin{array}{rrr} -4(2x) & -4(3y) & =-4(-5) \\ 2(4x) & +2(1y) & =2(5) \\ \hline \end{array}$$

$$\left[-4(2)+2(4)\right]x+\left[-4(3)+2(1)\right]y=-4(-5)+2(5)$$

$$(0)x+(-10)y=30$$

$$y=\frac{30}{-10}$$

Notice that the denominators for both x and y are the same number. This number is the value of the determinant of the coefficient matrix. (Remember that the equations are in standard form.)

$$\text{Determinant of coefficient matrix} = D = \begin{vmatrix} 2 & 3 \\ 4 & 1 \end{vmatrix} = 2\cdot 1 - 4\cdot 3 = -10.$$

In determinant form, the numerator of x, D_x, is found by replacing the x-coefficients with the constant terms:

$$D_x = \begin{vmatrix} -5 & 3 \\ 5 & 1 \end{vmatrix} = (-5)(1) - (5)(3) = -20$$

and the numerator of y, D_y, is found by replacing the y-coefficients with the constant terms:

$$D_y = \begin{vmatrix} 2 & -5 \\ 4 & 5 \end{vmatrix} = 2(5) - 4(-5) = 30.$$

Therefore the values for x and y can be written in fraction form using determinants as follows.

$$x = \frac{D_x}{D} = \frac{-20}{-10} = 2 \qquad \text{and} \qquad y = \frac{D_y}{D} = \frac{30}{-10} = -3$$

The determinant D_x is formed as follows:

1. Form D, the determinant of the coefficients.
2. Replace the coefficients of x with the corresponding constants on the right hand side of the equations.

The determinant D_y is formed as follows:

1. Form D, the determinant of the coefficients.
2. Replace the coefficients of y with the corresponding constants on the right hand side of the equations.

Cramer's rule is stated here only for 2×2 systems (systems of two linear equations in two variables) and 3×3 systems (systems of three linear equations in three variables). However, Cramer's rule applies to all $n \times n$ systems of linear equations.

Cramer's Rule for 2 × 2 Systems

For the system $\begin{cases} a_{11}x + a_{12}y = \boldsymbol{k_1} \\ a_{21}x + a_{22}y = \boldsymbol{k_2} \end{cases}$,

where

$$D = \begin{vmatrix} a_{11} & a_{12} \\ a_{21} & a_{22} \end{vmatrix}, \qquad D_x = \begin{vmatrix} \boldsymbol{k_1} & a_{12} \\ \boldsymbol{k_2} & a_{22} \end{vmatrix}, \qquad \text{and} \qquad D_y = \begin{vmatrix} a_{11} & \boldsymbol{k_1} \\ a_{21} & \boldsymbol{k_2} \end{vmatrix},$$

if $D \neq 0$, then

$$x = \frac{D_x}{D} \quad \text{and} \quad y = \frac{D_y}{D}$$

is the unique solution to the system.

Cramer's Rule for 3 × 3 Systems

For the system $\begin{cases} a_{11}x + a_{12}y + a_{13}z = \boldsymbol{k_1} \\ a_{21}x + a_{22}y + a_{23}z = \boldsymbol{k_2} \\ a_{31}x + a_{32}y + a_{33}z = \boldsymbol{k_3} \end{cases}$,

where

$$D = \begin{vmatrix} a_{11} & a_{12} & a_{13} \\ a_{21} & a_{22} & a_{23} \\ a_{31} & a_{32} & a_{33} \end{vmatrix},$$

$$D_x = \begin{vmatrix} \boldsymbol{k_1} & a_{12} & a_{13} \\ \boldsymbol{k_2} & a_{22} & a_{23} \\ \boldsymbol{k_3} & a_{32} & a_{33} \end{vmatrix}, \quad D_y = \begin{vmatrix} a_{11} & \boldsymbol{k_1} & a_{13} \\ a_{21} & \boldsymbol{k_2} & a_{23} \\ a_{31} & \boldsymbol{k_3} & a_{33} \end{vmatrix}, \quad \text{and} \quad D_z = \begin{vmatrix} a_{11} & a_{12} & \boldsymbol{k_1} \\ a_{21} & a_{22} & \boldsymbol{k_2} \\ a_{31} & a_{32} & \boldsymbol{k_3} \end{vmatrix},$$

if $D \neq 0$, then

$$x = \frac{D_x}{D}, \quad y = \frac{D_y}{D}, \quad \text{and} \quad z = \frac{D_z}{D}$$

is the unique solution to the system.

Cramer's Rule when $D = 0$

If $D = 0$, Cramer's rule cannot be used. In a case where $D = 0$ (either for a 2×2 or a 3×3 matrix), use the algebraic method of elimination or substitution. You will find that the system is either dependent (infinite number of solutions) or inconsistent (no solution).

Example 1: Cramer's Rule

Using Cramer's rule, solve the following systems of linear equations. The solutions are not checked here, but they can be checked by substituting the solutions into all of the equations in the system.

a. $\begin{cases} 2x + y = 3 \\ 3x - 2y = 5 \end{cases}$

Solution: $D = \begin{vmatrix} 2 & 1 \\ 3 & -2 \end{vmatrix} = -7, \qquad D_x = \begin{vmatrix} 3 & 1 \\ 5 & -2 \end{vmatrix} = -11, \qquad D_y = \begin{vmatrix} 2 & 3 \\ 3 & 5 \end{vmatrix} = 1$

$$x = \frac{D_x}{D} = \frac{-11}{-7} = \frac{11}{7} \qquad y = \frac{D_y}{D} = \frac{1}{-7} = -\frac{1}{7}$$

Thus the solution is $\left(\frac{11}{7}, -\frac{1}{7}\right)$.

b. $\begin{cases} 2x + 2y = 8 \\ -x + 3y = -8 \end{cases}$

Solution: $D = \begin{vmatrix} 2 & 2 \\ -1 & 3 \end{vmatrix} = 8, \qquad D_x = \begin{vmatrix} 8 & 2 \\ -8 & 3 \end{vmatrix} = 40, \qquad D_y = \begin{vmatrix} 2 & 8 \\ -1 & -8 \end{vmatrix} = -8$

$$x = \frac{D_x}{D} = \frac{40}{8} = 5 \qquad y = \frac{D_y}{D} = \frac{-8}{8} = -1$$

Thus the solution is $(5, -1)$.

c. $\begin{cases} x + 2y + 3z = 3 \\ 4x + 5y + 6z = 1 \\ 7x + 8y + 9z = 0 \end{cases}$

Solution: $D = \begin{vmatrix} 1 & 2 & 3 \\ 4 & 5 & 6 \\ 7 & 8 & 9 \end{vmatrix} = 1\begin{vmatrix} 5 & 6 \\ 8 & 9 \end{vmatrix} - 2\begin{vmatrix} 4 & 6 \\ 7 & 9 \end{vmatrix} + 3\begin{vmatrix} 4 & 5 \\ 7 & 8 \end{vmatrix}$

$$= 1(-3) - 2(-6) + 3(-3) = 0$$

Because $D = 0$, Cramer's rule cannot be used to solve the system. (You might try to solve the system by using the elimination method to see what happens in this case.)

d. $\begin{cases} x + y + 3z = 7 \\ 2x - y - 3z = -4 \\ 5x - 2y = -5 \end{cases}$

Solution: $D = \begin{vmatrix} 1 & 1 & 3 \\ 2 & -1 & -3 \\ 5 & -2 & 0 \end{vmatrix} = 1\begin{vmatrix} -1 & -3 \\ -2 & 0 \end{vmatrix} - 1\begin{vmatrix} 2 & -3 \\ 5 & 0 \end{vmatrix} + 3\begin{vmatrix} 2 & -1 \\ 5 & -2 \end{vmatrix} = -18$

$$D_x = \begin{vmatrix} 7 & 1 & 3 \\ -4 & -1 & -3 \\ -5 & -2 & 0 \end{vmatrix} = 7\begin{vmatrix} -1 & -3 \\ -2 & 0 \end{vmatrix} - 1\begin{vmatrix} -4 & -3 \\ -5 & 0 \end{vmatrix} + 3\begin{vmatrix} -4 & -1 \\ -5 & -2 \end{vmatrix} = -18$$

$$D_y = \begin{vmatrix} 1 & 7 & 3 \\ 2 & -4 & -3 \\ 5 & -5 & 0 \end{vmatrix} = 1\begin{vmatrix} -4 & -3 \\ -5 & 0 \end{vmatrix} - 7\begin{vmatrix} 2 & -3 \\ 5 & 0 \end{vmatrix} + 3\begin{vmatrix} 2 & -4 \\ 5 & -5 \end{vmatrix} = -90$$

Continued on the next page...

$$D_z = \begin{vmatrix} 1 & 1 & 7 \\ 2 & -1 & -4 \\ 5 & -2 & -5 \end{vmatrix} = 1\begin{vmatrix} -1 & -4 \\ -2 & -5 \end{vmatrix} - 1\begin{vmatrix} 2 & -4 \\ 5 & -5 \end{vmatrix} + 7\begin{vmatrix} 2 & -1 \\ 5 & -2 \end{vmatrix} = -6$$

$$x = \frac{-18}{-18} = 1 \qquad y = \frac{-90}{-18} = 5 \qquad z = \frac{-6}{-18} = \frac{1}{3}$$

Thus the solution is $\left(1, 5, \frac{1}{3}\right)$.

NOTES

The determinants shown in Examples 1c and 1d are expanded by the first row. However, any row or column can be used in the expansion as long as the corresponding adjustments in the + and – signs are used with the minors. This may be particularly useful when a row or column has one or more 0's because multiplication by 0 will always give 0 and this will reduce the time needed for the expansion.

Practice Problems

1. Solve the following system using Cramer's rule.

$$\begin{cases} 2x - y = 11 \\ x + y = -2 \end{cases}$$

2. Find D_x for the following system.

$$\begin{cases} x + 2y + z = 0 \\ 2x + y - 2z = 5 \\ 3x - y + z = -3 \end{cases}$$

Appendix 5 Exercises

Use Cramer's rule to solve the following systems of linear equations, if possible. If the determinant of the coefficient matrix is zero, solve the system using addition or substitution to determine whether the system has no solution or infinitely many solutions.

1. $\begin{cases} 2x - 5y = -7 \\ 3x - 2y = 6 \end{cases}$

2. $\begin{cases} 3x + 5y = 17 \\ x + 3y = 15 \end{cases}$

3. $\begin{cases} 6x - 4y = 5 \\ 3x + 8y = 0 \end{cases}$

4. $\begin{cases} 3x + 4y = 24 \\ 2x + y = 11 \end{cases}$

5. $\begin{cases} 3x + y = 1 \\ -9x - 3y = 2 \end{cases}$

6. $\begin{cases} 4x + 8y = 12 \\ 3x + 6y = 9 \end{cases}$

7. $\begin{cases} 2x + y = 7 \\ 3x + 2y = 10 \end{cases}$

8. $\begin{cases} 5x + 4y = -7 \\ 3x + y = -7 \end{cases}$

1. (4, 3)
2. (−6, 7)
3. $\left(\frac{2}{3}, -\frac{1}{4}\right)$
4. (4, 3)
5. No solution
6. Infinitely many solutions
7. (4, −1)
8. (−3, 2)

Answers to Practice Problems: **1.** (3, −5) **2.** $D_x = 0$

9. $(-1, 0)$
10. $(6, 2)$
11. $\left(-\frac{1}{4}, \frac{3}{2}\right)$
12. $\left(-\frac{1}{4}, -\frac{1}{3}\right)$
13. $\left(\frac{31}{17}, \frac{2}{17}\right)$
14. $\left(\frac{29}{19}, -\frac{6}{19}\right)$
15. $\left(\frac{39}{44}, \frac{41}{44}\right)$
16. $\left(\frac{18}{7}, -\frac{3}{7}\right)$
17. $(-5, -4)$
18. $(-1, 6)$
19. $(3, 7)$
20. $(-4, 2)$
21. $(-2, 1, 3)$
22. $(1, 0, -2)$
23. No solution
24. $\left(-\frac{4}{17}, \frac{7}{17}, \frac{38}{51}\right)$
25. $(4, 2, 6)$
26. $(2, -1, 3)$
27. Infinitely many solutions
28. $(1, -3, 8)$
29. $\left(-\frac{2}{3}, \frac{11}{3}, 2\right)$
30. $\left(\frac{23}{27}, -\frac{5}{27}, -\frac{83}{27}\right)$

9. $\begin{cases} x+8y=-1 \\ 4x-7y=-4 \end{cases}$ **10.** $\begin{cases} 3x-y=16 \\ x+4y=14 \end{cases}$ **11.** $\begin{cases} 12x+4y=3 \\ -10x+3y=7 \end{cases}$ **12.** $\begin{cases} 4x-9y=2 \\ 8x-15y=3 \end{cases}$

13. $\begin{cases} 2x+3y=4 \\ 3x-4y=5 \end{cases}$ **14.** $\begin{cases} 5x+2y=7 \\ 2x-3y=4 \end{cases}$ **15.** $\begin{cases} 7x+3y=9 \\ 4x+8y=11 \end{cases}$ **16.** $\begin{cases} 6x-13y=21 \\ 5x-12y=18 \end{cases}$

17. $\begin{cases} 0.2x+0.1y=-1.4 \\ 0.9x-1.3y=0.7 \end{cases}$ **18.** $\begin{cases} 1.4x+0.5y=1.6 \\ x+0.2y=0.2 \end{cases}$ **19.** $\begin{cases} 1.2x-0.5y=0.1 \\ 0.6x+0.1y=2.5 \end{cases}$

20. $\begin{cases} 0.3x-0.5y=-2.2 \\ 1.1x+1.5y=-1.4 \end{cases}$ **21.** $\begin{cases} x-2y-z=-7 \\ 2x+y+z=0 \\ 3x-5y+8z=13 \end{cases}$ **22.** $\begin{cases} 2x+3y+z=0 \\ 5x+y-2z=9 \\ 10x-5y+3z=4 \end{cases}$

23. $\begin{cases} 5x-4y+z=17 \\ x+y+z=4 \\ -10x+8y-2z=11 \end{cases}$ **24.** $\begin{cases} 9x+10y=2 \\ 2x+6z=4 \\ -3y+3z=1 \end{cases}$ **25.** $\begin{cases} 2x-3y-z=-4 \\ -x+2y+z=6 \\ x-y+2z=14 \end{cases}$

26. $\begin{cases} 2x-3y-z=4 \\ x-2y-z=1 \\ x-y+2z=9 \end{cases}$ **27.** $\begin{cases} 3x+2y+z=5 \\ 2x+y-2z=4 \\ 5x+3y-z=9 \end{cases}$ **28.** $\begin{cases} 8x+3y+2z=15 \\ 3x+5y+z=-4 \\ 2x+3y=-7 \end{cases}$

29. $\begin{cases} 2x-y+3z=1 \\ 5x+2y-z=2 \\ x-2y+5z=2 \end{cases}$ **30.** $\begin{cases} 2x+3y+2z=-5 \\ 2x-2y+z=-1 \\ 5x+y+z=1 \end{cases}$

Set up a system of linear equations that represents the information in each application problem, then solve the system using Cramer's rule.

31. Triangles: The three sides of a triangle are related as follows: the perimeter is 43 feet, the second side is 5 feet more than twice the first side, and the third side is 3 feet less than the sum of the other two sides. Find the lengths of the three sides of the triangle.

32. Nutrition: Joel loves candy bars and ice cream, and they have fat and calories as follows: each candy bar contains 5 grams of fat and 280 calories; each serving of ice cream contains 10 grams of fat and 150 calories. How many candy bars and how many servings of ice cream did he eat the week that he consumed 85 grams of fat and 2300 calories from these two foods?

31. $\begin{cases} x+y+z=43 \\ y=5+2x \\ z=x+y-3 \end{cases}$
6 feet by 17 feet by 20 feet

32. $\begin{cases} 5x+10y=85 \\ 280x+150y=2300 \end{cases}$
5 candy bars and 6 servings of ice cream

33. Investments: A financial advisor has \$6 million to invest for her clients. She chooses, for one month, to invest in mutual funds and technology stocks. If the mutual funds earned 2% and the stocks earned 4% for a total of \$170,000 in earnings for the month, how much money did she invest in each type of investment?

34. Farming: A farmer plants corn, wheat, and soybeans and rotates the planting each year on his 500-acre farm. In one particular year, the profits were: \$120 per acre for corn, \$100 per acre for wheat, and \$80 per acre for soybeans. He planted twice as many acres with corn as with soybeans. How many acres did he plant with each crop the year he made a total profit of \$51,800?

HAWKES LEARNING SYSTEMS: INTRODUCTORY & INTERMEDIATE ALGEBRA SOFTWARE

- A.5 Determinants and Systems of Linear Equations: Cramer's Rule

33. $\begin{cases} x+y=6{,}000{,}000 \\ 0.02x+0.04y=170{,}000 \end{cases}$ \$3,500,000 in mutual funds; \$2,500,000 in stocks

34. $\begin{cases} x+y+z=500 \\ 120x+100y+80z=51{,}800 \\ x=2z \end{cases}$ 180 acres of corn; 230 acres of wheat; 90 acres of soybeans

A.6 Powers, Roots, and Prime Factorizations

Number n	Square n^2	Square Root $\sqrt{n}$	Cube n^3	Cube Root $\sqrt[3]{n}$	Prime Factorization
1	1	1.0000	1	1.0000	
2	4	1.4142	8	1.2599	prime
3	9	1.7321	27	1.4422	prime
4	16	2.0000	64	1.5874	$2 \cdot 2$
5	25	2.2361	125	1.7100	prime
6	36	2.4495	216	1.8171	$2 \cdot 3$
7	49	2.6458	343	1.9129	prime
8	64	2.8284	512	2.0000	$2 \cdot 2 \cdot 2$
9	81	3.0000	729	2.0801	$3 \cdot 3$
10	100	3.1623	1000	2.1544	$2 \cdot 5$
11	121	3.3166	1331	2.2240	prime
12	144	3.4641	1728	2.2894	$2 \cdot 2 \cdot 3$
13	169	3.6056	2197	2.3513	prime
14	196	3.7417	2744	2.4101	$2 \cdot 7$
15	225	3.8730	3375	2.4662	$3 \cdot 5$
16	256	4.0000	4096	2.5198	$2 \cdot 2 \cdot 2 \cdot 2$
17	289	4.1231	4913	2.5713	prime
18	324	4.2426	5832	2.6207	$2 \cdot 3 \cdot 3$
19	361	4.3589	6859	2.6684	prime
20	400	4.4721	8000	2.7144	$2 \cdot 2 \cdot 5$
21	441	4.5826	9261	2.7589	$3 \cdot 7$
22	484	4.6904	10,648	2.8020	$2 \cdot 11$
23	529	4.7958	12,167	2.8439	prime
24	576	4.8990	13,824	2.8845	$2 \cdot 2 \cdot 2 \cdot 3$
25	625	5.0000	15,625	2.9240	$5 \cdot 5$
26	676	5.0990	17,576	2.9625	$2 \cdot 13$
27	729	5.1962	19,683	3.0000	$3 \cdot 3 \cdot 3$
28	784	5.2915	21,952	3.0366	$2 \cdot 2 \cdot 7$
29	841	5.3852	24,389	3.0723	prime
30	900	5.4772	27,000	3.1072	$2 \cdot 3 \cdot 5$
31	961	5.5678	29,791	3.1414	prime
32	1024	5.6569	32,768	3.1748	$2 \cdot 2 \cdot 2 \cdot 2 \cdot 2$
33	1089	5.7446	35,937	3.2075	$3 \cdot 11$

Number n	Square n^2	Square Root $\sqrt{n}$	Cube n^3	Cube Root $\sqrt[3]{n}$	Prime Factorization
34	1156	5.8310	39,304	3.2396	$2 \cdot 17$
35	1225	5.9161	42,875	3.2711	$5 \cdot 7$
36	1296	6.0000	46,656	3.3019	$2 \cdot 2 \cdot 3 \cdot 3$
37	1369	6.0828	50,653	3.3322	prime
38	1444	6.1644	54,872	3.3620	$2 \cdot 19$
39	1521	6.2450	59,319	3.3912	$3 \cdot 13$
40	1600	6.3246	64,000	3.4200	$2 \cdot 2 \cdot 2 \cdot 5$
41	1681	6.4031	68,921	3.4482	prime
42	1764	6.4807	74,088	3.4760	$2 \cdot 3 \cdot 7$
43	1849	6.5574	79,507	3.5034	prime
44	1936	6.6332	85,184	3.5303	$2 \cdot 2 \cdot 11$
45	2025	6.7082	91,125	3.5569	$3 \cdot 3 \cdot 5$
46	2116	6.7823	97,336	3.5830	$2 \cdot 23$
47	2209	6.8557	103,823	3.6088	prime
48	2304	6.9282	110,592	3.6342	$2 \cdot 2 \cdot 2 \cdot 2 \cdot 3$
49	2401	7.0000	117,649	3.6593	$7 \cdot 7$
50	2500	7.0711	125,000	3.6840	$2 \cdot 5 \cdot 5$
51	2601	7.1414	132,651	3.7084	$3 \cdot 17$
52	2704	7.2111	140,608	3.7325	$2 \cdot 2 \cdot 13$
53	2809	7.2801	148,877	3.7563	prime
54	2916	7.3485	157,464	3.7798	$2 \cdot 3 \cdot 3 \cdot 3$
55	3025	7.4162	166,375	3.8030	$5 \cdot 11$
56	3136	7.4833	175,616	3.8259	$2 \cdot 2 \cdot 2 \cdot 7$
57	3249	7.5498	185,193	3.8485	$3 \cdot 19$
58	3364	7.6158	195,112	3.8709	$2 \cdot 29$
59	3481	7.6811	205,379	3.8930	prime
60	3600	7.7460	216,000	3.9149	$2 \cdot 2 \cdot 3 \cdot 5$
61	3721	7.8102	226,981	3.9365	prime
62	3844	7.8740	238,328	3.9579	$2 \cdot 31$
63	3969	7.9373	250,047	3.9791	$3 \cdot 3 \cdot 7$
64	4096	8.0000	262,144	4.0000	$2 \cdot 2 \cdot 2 \cdot 2 \cdot 2 \cdot 2$
65	4225	8.0623	274,625	4.0207	$5 \cdot 13$
66	4356	8.1240	287,496	4.0412	$2 \cdot 3 \cdot 11$
67	4489	8.1854	300,763	4.0615	prime
68	4624	8.2462	314,432	4.0817	$2 \cdot 2 \cdot 17$
69	4761	8.3066	328,509	4.1016	$3 \cdot 23$

Number n	Square n^2	Square Root $\sqrt{n}$	Cube n^3	Cube Root $\sqrt[3]{n}$	Prime Factorization
70	4900	8.3666	343,000	4.1213	$2 \cdot 5 \cdot 7$
71	5041	8.4261	357,911	4.1408	prime
72	5184	8.4853	373,248	4.1602	$2 \cdot 2 \cdot 2 \cdot 3 \cdot 3$
73	5329	8.5440	389,017	4.1793	prime
74	5476	8.6023	405,224	4.1983	$2 \cdot 37$
75	5625	8.6603	421,875	4.2172	$3 \cdot 5 \cdot 5$
76	5776	8.7178	438,976	4.2358	$2 \cdot 2 \cdot 19$
77	5929	8.7750	456,533	4.2543	$7 \cdot 11$
78	6084	8.8318	474,552	4.2727	$2 \cdot 3 \cdot 13$
79	6241	8.8882	493,039	4.2908	prime
80	6400	8.9443	512,000	4.3089	$2 \cdot 2 \cdot 2 \cdot 2 \cdot 5$
81	6561	9.0000	531,441	4.3267	$3 \cdot 3 \cdot 3 \cdot 3$
82	6724	9.0554	551,368	4.3445	$2 \cdot 41$
83	6889	9.1104	571,787	4.3621	prime
84	7056	9.1652	592,704	4.3795	$2 \cdot 2 \cdot 3 \cdot 7$
85	7225	9.2195	614,125	4.3968	$5 \cdot 17$
86	7396	9.2736	636,056	4.4140	$2 \cdot 43$
87	7569	9.3274	658,503	4.4310	$3 \cdot 29$
88	7744	9.3808	681,472	4.4480	$2 \cdot 2 \cdot 2 \cdot 11$
89	7921	9.4340	704,969	4.4647	prime
90	8100	9.4868	729,000	4.4814	$2 \cdot 3 \cdot 3 \cdot 5$
91	8281	9.5394	753,571	4.4979	$7 \cdot 13$
92	8464	9.5917	778,688	4.5144	$2 \cdot 2 \cdot 23$
93	8649	9.6437	804,357	4.5307	$3 \cdot 31$
94	8836	9.6954	830,584	4.5468	$2 \cdot 47$
95	9025	9.7468	857,375	4.5629	$5 \cdot 19$
96	9216	9.7980	884,736	4.5789	$2 \cdot 2 \cdot 2 \cdot 2 \cdot 2 \cdot 3$
97	9409	9.8489	912,673	4.5947	prime
98	9604	9.8995	941,192	4.6104	$2 \cdot 7 \cdot 7$
99	9801	9.9499	970,299	4.6261	$3 \cdot 3 \cdot 11$
100	10,000	10.0000	1,000,000	4.6416	$2 \cdot 2 \cdot 5 \cdot 5$

Answers

Chapter 1

Exercises 1.1, pages 12 - 15

1. 4, 8 **3.** –7, –2, 0, 4, 8 **5.** $-7, -2, -\frac{5}{3}, -1.4, 0, \frac{3}{5}, 4, 5.9, 8$ **7.** 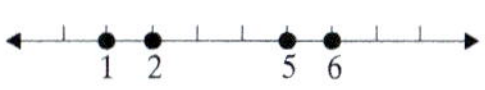 **9.** 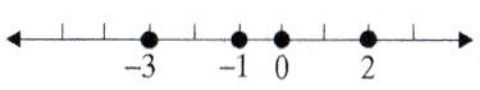

11. 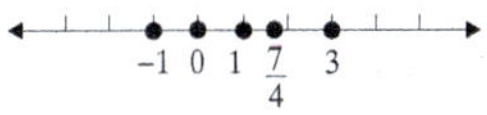 **13.** 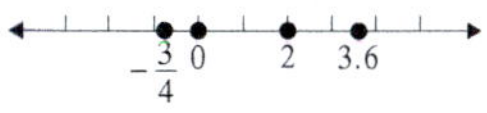 **15.** 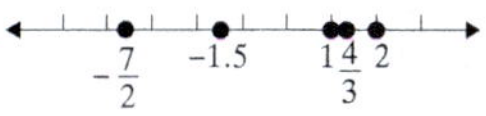 **17.** 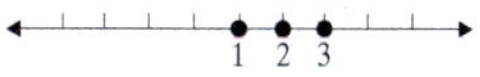

19. 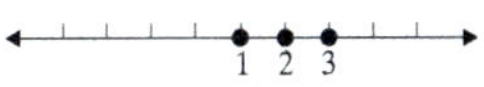 **21.**  **23.** 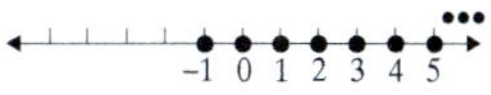 **25.** No solution

27.  **29.** 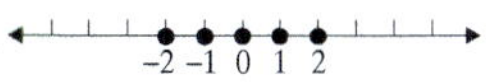 **31.** 10 **33.** –4 **35.** –11.3 **37.** < **39.** > **41.** = **43.** < **45.** > **47.** <

49. = **51.** True **53.** False; $-22 < -16$ **55.** True **57.** True **59.** True **61.** True **63.** False; $-|-3| > -|4|$ **65.** False; $\left|-\frac{5}{2}\right| > 2$ **67.** True

69. False; $-|5| < -|3.1|$ **71.** True **73.**  **75.**  **77.** 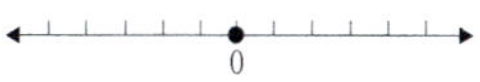

79. No solution **81.**  **83.** 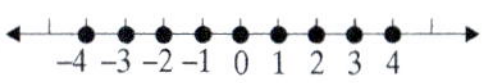 **85.** 

87. 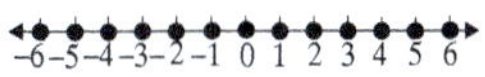 **89.** 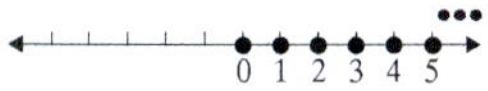 **91.** 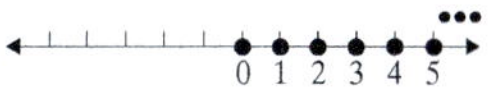 **93.** 46 **95.** –6 **97.** 61.4

99. If y is a negative number then $-y$ represents a positive number. For example, if $y = -2$, then $-y = -(-2) = 2$.

Exercises 1.2, pages 19 - 20

1. 13 **3.** –4 **5.** 0 **7.** 5 **9.** –13 **11.** –8 **13.** –10 **15.** 0 **17.** –2.5 **19.** –29 **21.** –10.1 **23.** –2.5 **25.** 22 **27.** –54 **29.** 40 **31.** –7 **33.** –16 **35.** –26 **37.** 0 **39.** –32 **41.** –5.1 **43.** –83 **45.** 4 **47.** 13.1 **49.** –47 **51.** –42 **53.** –12 **55.** –15 **57.** –6 **59.** 6 **61.** 84 **63.** 3807 **65.** –97,714 **67.** $|0| + |0| = 0$

Exercises 1.3, pages 27 - 30

1. –11 **3.** 6 **5.** –47 **7.** 0 **9.** 5.2 **11.** 5 **13.** –10 **15.** 7 **17.** 3 **19.** –5 **21.** 16 **23.** –13 **25.** 5.9 **27.** –4 **29.** 10 **31.** –15 **33.** –16 **35.** –57 **37.** –5.4 **39.** 1 **41.** –7 **43.** –26 **45.** 11 **47.** –15 **49.** –1 **51.** –3 **53.** –6 **55.** 0 **57.** $-8 < -5$ **59.** $-3 < 3$ **61.** $8 = 8$ **63.** $0 > -27$ **65.** $-23.7 > -24.8$ **67.** –18°F **69.** \$8 **71.** –14,777 ft **73.** 85 years old **75.** $1044 > -39$ **77.** $-15{,}254 > -35{,}090$ **79.** Gained 35 yards **81.** Lost 7 pounds; 203 pounds **83.** Add the opposite of the second number to the first number.

Exercises 1.4, pages 38 - 41

1. –12 **3.** 48 **5.** 56 **7.** –21 **9.** 56 **11.** 26 **13.** –70 **15.** –24 **17.** –288 **19.** 0 **21.** –9 **23.** –7.31 **25.** 4 **27.** –6 **29.** 2 **31.** 0 **33.** –3 **35.** –17 **37.** Undefined **39.** 5 **41.** –0.6 **43.** 20 **45.** –2.3 **47.** False; $(-4)(6) < 3 \cdot 8$ **49.** True **51.** False; $6(-3) = (-14) + (-4)$ **53.** True **55.** True **57.** –12 **59.** \$163 **61.** 26.6 calls **63.** 80 **65.** 78.6 inches **67.** 31.5 hours **69.** 18,902 square miles **71.** –34,459,110 **73.** –2671 **75.** –45.33 **77.** –10.29 **79.** 2.83 **81.** –90.365 **83.** See page 34.

Exercises 1.5, pages 51 - 53

1. 7^4 **3.** $2\cdot 3^2\cdot 11^2$ **5.** $3^3\cdot 7^3$ **7.** 2^4 or 4^2 **9.** 6^2 **11.** 10^3 **13.** A counting number greater than 1 that has exactly two different factors (or divisors): itself and 1. **15.** A counting number with more than two different factors (or divisors). **17.** 13, 17 **19.** 2 **21.** 8, 9 **23.** 4, 17 **25.** 2, 251 **27.** 10, 17 **29.** 4, 111 **31.** $2^2\cdot 13$ **33.** $2^3\cdot 7\cdot 11$ **35.** $2^2\cdot 7\cdot 11$ **37.** 79 is prime. **39.** 17^2 **41.** 5^3 **43.** $2^4\cdot 5^2$ **45.** $2^3\cdot 3\cdot 5$ **47.** $3\cdot 7\cdot 11$ **49.** $2^2\cdot 3^2\cdot 47$ **51. a.** 5, 10, 15, 20, 25, 30, 35, 40, 45, 50 **b.** 6, 12, 18, 24, 30, 36, 42, 48, 54, 60 **c.** 10, 20, 30, 40, 50, 60, 70, 80, 90, 100 **d.** 15, 30, 45, 60, 75, 90, 105, 120, 135, 150 **53.** 105 **55.** 40 **57.** 66 **59.** 140 **61.** 150 **63.** 180 **65.** 196 **67.** 210 **69.** 90 **71.** 100 **73.** 150 **75.** 364 **77.** 2520 **79.** 924 **81.** 7623 **83.** 2880 **85.** 7840 **87.** $40xy$ **89.** $210x^2y^2$ **91.** 53, 59, 61, 67, 71 Answers will vary.

93. a. To be prime, a number has to have exactly two different factors. The number 1 only has one unique factor, itself. **b.** A composite number has more than two different factors. The number 1 only has one unique factor, itself.

Exercises 1.6, pages 61 - 64

1. $\frac{5}{6}=\frac{5}{6}\cdot\frac{8}{8}=\frac{40}{48}$ **3.** $\frac{2}{9}=\frac{2}{9}\cdot\frac{7b}{7b}=\frac{14b}{63b}$ **5.** $-\frac{7}{24}=-\frac{7}{24}\cdot\frac{3x}{3x}=-\frac{21x}{72x}$ **7.** $\frac{2}{5}$ **9.** $\frac{3x}{7}$ **11.** $\frac{3}{25b}$ **13.** $-\frac{2}{3}$ **15.** $-\frac{1}{2x}$ **17.** $-\frac{12z}{35}$ **19.** $\frac{3}{8}$ **21.** 35 **23.** $-\frac{1}{6}$ **25.** $\frac{81}{100}$ **27.** $-\frac{3y}{7}$ **29.** $-\frac{9x}{5}$ **31.** $\frac{2}{3}$ **33.** $-\frac{50}{9}$ **35.** $\frac{2}{5b}$ **37.** $\frac{5y}{3}$ **39.** 0 **41.** Undefined **43.** 22 **45.** $-\frac{1}{10}$ **47.** $\frac{9a}{10b}$ **49.** $\frac{81}{320n}$ **51.** $-\frac{7x}{100z}$ **53.** $\frac{38}{3}$ **55.** $\frac{14}{5}$ **57.** $-\frac{25}{24}$ **59.** 61st floor **61.** 98 pounds **63.** 344 yards **65.** $\frac{5}{8}$ of the student body **67. a.** More than 60 **b.** Less than 60 **c.** 72 passengers **69.** $\frac{1}{18{,}009{,}460}$

71. Division by zero is undefined. For example we could write $0=\frac{0}{1}$. Then the reciprocal would be $\frac{1}{0}$, but this reciprocal is undefined since division by zero is undefined. Thus 0 does not have a reciprocal.

Exercises 1.7, pages 70 - 72

1. $\frac{7}{9}$ **3.** $\frac{3}{7}$ **5.** $\frac{7}{10}$ **7.** $-\frac{7}{20}$ **9.** $\frac{5}{8}$ **11.** 1 **13.** $-\frac{1}{6}$ **15.** $\frac{2}{15}$ **17.** $\frac{17}{36}$ **19.** $\frac{17}{21}$ **21.** $\frac{63}{50}$ **23.** $\frac{4}{7}$ **25.** $\frac{4}{y}$ **27.** $\frac{5}{x}$ **29.** $\frac{16}{15a}$ **31.** $-\frac{2}{5x}$ **33.** $\frac{1}{2y}$ **35.** $\frac{43}{72x}$ **37.** $\frac{5}{6a}$ **39.** $-\frac{11}{12x}$ **41.** 0 **43.** $\frac{25}{24}$ **45.** $\frac{41}{48}$ **47.** $\frac{11}{18}$ of the play **49.** 3 hours **51.** $\frac{9}{10}$ lb **53. a.** $\frac{13}{30}$ **b.** \$1170 **55.** Multiplying the numerator and denominator by the same number is equivalent to multiplying the fraction by 1 and any number multiplied by 1 equals the original number.

Exercises 1.8, pages 78 - 80

1. a. 36 **b.** 16 **3.** −25 **5.** −10 **7.** −45 **9.** −137 **11.** 152 **13.** −189 **15.** 143 **17.** −36 **19.** −10 **21.** $\frac{11}{30}$ **23.** $\frac{1}{24}$ **25.** $\frac{31}{24}$ **27.** $\frac{107}{24}$ **29.** $\frac{41}{32}$ **31.** $\frac{3}{5}$ **33.** $-\frac{341}{30}$ **35.** $-\frac{15}{17}$ **37.** 24 **39.** $\frac{7}{10}$ **41.** $5\frac{1}{12}$ or $\frac{61}{12}$ **43.** $-\frac{13}{64}$ **45.** 42.45 **47.** 67.77 **49.** 15.41 **51.** $\frac{7}{2}$ **53.** $-\frac{23}{21y}$ **55.** $\frac{7}{5}$ **57.** $\frac{13}{36a}$ **59.** $(3^2-9)=0$ and division by 0 is undefined. **61.** It will always be larger. If you square a negative number you get a positive number. All positive numbers are larger than all negative numbers.

Exercises 1.9, pages 84 - 86

1. $3+7$ **3.** $4 \cdot 19$ **5.** $30+48$ **7.** $(2\cdot 3)\cdot x$ **9.** $(3+x)+7$ **11.** 0 **13.** $x+7$ **15.** $2x-24$ **17.** 0
19. Commutative property of addition **21.** Multiplicative identity **23.** Associative property of addition
25. Commutative property of multiplication **27.** Commutative property of multiplication **29.** Multiplicative inverse
31. Additive inverse **33.** Multiplicative identity **35.** Zero-factor law **37.** Associative property of addition
39. $6(11)=66$ and $6\cdot 3+6\cdot 8=66$ **41.** $10(-7)=-70$ and $10\cdot 2-10\cdot 9=-70$ **43.** Commutative property of multiplication; $6\cdot 4=4\cdot 6=24$ **45.** Associative property of addition; $8+(5+(-2))=(8+5)+(-2)=11$ **47.** Distributive property; $5(4+18)=5(4)+90=110$ **49.** Associative property of multiplication; $(6\cdot(-2))\cdot 9=6\cdot(-2\cdot 9)=-108$ **51.** Commutative property of addition; $3+(-34)=(-34)+3=-31$ **53.** Commutative property of addition; $2(3+4)=2(4+3)=14$
55. Commutative property of addition; $5+(4-15)=(4-15)+5=-6$ **57.** Associative property of multiplication; $(3\cdot 4)\cdot 5=3\cdot(4\cdot 5)=60$ **59. a.** Multiply each term that is part of the sum (in the parentheses) by a. **b.** An expression that "distributes addition over multiplication" would look like $a+(bc)=(a+b)(a+c)$, however this is not a true statement. Answers will vary.

Chapter 1 Review, pages 94 - 98

1. a. $\frac{2}{1}$ **b.** $-2, 0, \frac{2}{1}$ **c.** $-9.3, -2, -1.343434..., -\frac{3}{4}, 0, \frac{2}{1}, \frac{10}{3}$ **d.** $\sqrt{2}$ **2.** 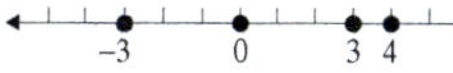

3. 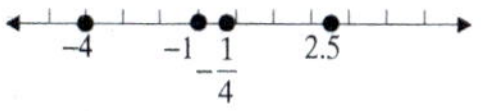

4. 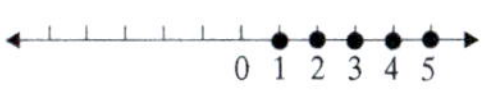

5. 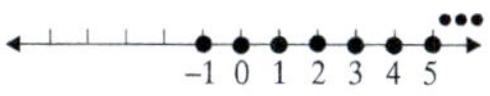

6. 6 **7.** -7.3 **8.** -4.1 **9.** True **10.** True **11.** False; $|-2.3|>0$
12. True **13.** 

14. No solution **15.** 

16. 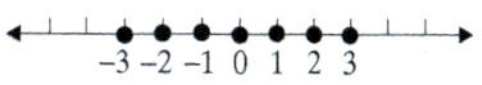

17. 

18. 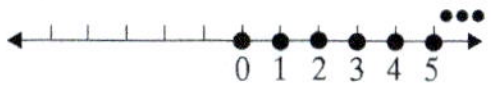

19. 15 **20.** 22 **21.** 5 **22.** -3 **23.** 0 **24.** 0 **25.** -3.4 **26.** 0.7 **27.** -18
28. -6 **29.** 15 **30.** -5 **31.** -35 **32.** -38 **33.** -18 **34.** 4 **35.** 6 **36.** -5.4 **37.** -7 **38.** -2 **39.** -19 **40.** -8 **41.** -8 **42.** -17 **43.** 0.9
44. 0.7 **45.** -4 **46.** -31 **47.** 2.1 **48.** -23.8 **49.** -1 **50.** -41 **51.** -33°F **52.** -6000 feet **53.** -35 **54.** -25.6 **55.** 90 **56.** 91 **57.** 24
58. 0 **59.** 5 **60.** 17 **61.** Undefined **62.** 0 **63.** -0.4 **64.** -40 **65.** 62 **66.** 85.6 **67.** 65.8 inches **68.** 8 hurricanes **69.** $3^3\cdot 5^2$
70. $2^4\cdot 11^2\cdot 13$ **71.** $5^2\cdot 7$ **72.** $2\cdot 7\cdot 11$ **73.** $3\cdot 5\cdot 13$ **74.** $3^2\cdot 5^2\cdot 11$ **75.** 1890 **76.** 120 **77.** 120 **78.** 462 **79.** $270x^2y$
80. $468a^2bc^2$ **81.** $-\frac{3}{16}$ **82.** $-\frac{1}{12}$ **83.** $\frac{2}{5}$ **84.** 1 **85.** $-\frac{1}{5}$ **86.** $\frac{3}{10}$ **87.** $\frac{4}{11}$ **88.** $\frac{7}{12}$ **89.** 12 **90.** $\frac{3}{5}$ was spent; $\frac{2}{5}$ remains
91. $\frac{9}{7}$ **92.** $-\frac{3}{16}$ **93.** $\frac{53}{60}$ **94.** $\frac{7}{24}$ **95.** $\frac{2}{5}$ **96.** $\frac{5}{16}$ **97.** $\frac{7}{8}$ **98.** 0 **99.** $\frac{17}{3a}$ **100.** $-\frac{49}{5x}$ **101.** $\frac{3}{14x}$ **102.** $\frac{5}{12x}$ **103. a.** $\frac{28}{135}$
b. 1120 species **104.** 2790 miles **105.** 88 **106.** 16 **107.** 18 **108.** -4 **109.** 26 **110.** 32 **111.** $\frac{49}{36}$ **112.** $-\frac{9}{2}$ **113.** $-\frac{1}{360}$
114. $\frac{31}{28}$ **115.** $\frac{5}{4x}$ **116.** $\frac{1}{7y}$ **117.** $-\frac{11}{81}$ **118.** $-\frac{9}{8}$ **119.** Commutative property of addition
120. Associative property of addition **121.** Additive identity **122.** Additive inverse
123. Associative property of multiplication **124.** Commutative property of multiplication **125.** Multiplicative identity
126. Multiplicative inverse **127.** Distributive property **128.** Distributive property **129.** 238.43 **130.** -74.65 **131.** $-40{,}473.75$
132. 5.1252

Chapter 1 Test, pages 99 - 100

1. a. $-5, -1, 0$ **b.** $-5, -1, -\frac{1}{3}, 0, \frac{1}{2}, 3\frac{1}{4}, 7.121212...$ **c.** $-\pi$ **d.** All are real numbers **2. a.** True, since for any integer n, $n=\frac{n}{1}$ which is a rational number since the numerator and the denominator are both integers. **b.** False, because rational numbers also include non-integers, such as $\frac{2}{3}$. **3. a.** $<$ **b.** $>$ **c.** $=$ **4. a.** $\{-2,-1,0,1,2\}$ **b.** $\{\ldots, -11, -10, -9, 9, 10, 11, \ldots\}$ **5.** -13

6. 23 **7.** 112 **8.** 0 **9.** –3 **10.** –0.3 **11.** –60 **12.** –22.5 **13.** 21.7 mph **14.** $3^3 \cdot 5 \cdot 7^2$ **15.** $2^4 \cdot 3^2$ **16.** $90x^2y^2$ **17.** $-\frac{9}{2}$ **18.** 8
19. $\frac{9}{10}$ **20.** $-\frac{2}{y}$ **21.** –44 **22.** 19 **23.** $\frac{29}{40}$ **24.** $\frac{17}{90}$ **25.** $-\frac{33}{16}$
26. a. 15 gallons **b.** No, you need \$46.50 to fill the tank so you are short \$6.50. **27.** Additive identity
28. Commutative property of multiplication **29.** Associative property of addition **30.** Zero-factor law

Chapter 2

Exercises 2.1, pages 107 - 109

1. –5, $\frac{1}{6}$, and 8 are like terms; $7x$ and $9x$ are like terms. **3.** $-x^2$ and $2x^2$ are like terms; $5xy$ and $-6xy$ are like terms; $3x^2y$ and $5x^2y$ are like terms. **5.** 24, 8.3, and –6 are like terms; $1.5xyz$, $-1.4xyz$, and xyz are like terms. **7.** 64 **9.** –121 **11.** $15x$ **13.** $3x$ **15.** $-2n$ **17.** $5y^2$ **19.** $34x^2$ **21.** $7x+2$ **23.** $x-4y$ **25.** $13x^2-2y$ **27.** $2n+3$ **29.** $7a-8b$ **31.** $8x+y$ **33.** $2x^2-x$ **35.** $-2n^2+2n-2$ **37.** $3x^2-xy+y^2$ **39.** $2x$ **41.** $-y$ **43. a.** $3x+4$ **b.** 16 **45. a.** $-2x-8$ **b.** –16 **47. a.** $5y+1$ **b.** 16 **49. a.** $-3x-7y$ **b.** –33 **51. a.** $3.6x^2+2$ **b.** 59.6 **53. a.** $3ab+b^2+b^3$ **b.** 6 **55. a.** $3.7x+1.1$ **b.** 15.9 **57. a.** $8a$ **b.** –16 **59. a.** $3b$ **b.** –3 **61. a.** $-x-54$ **b.** –58 **63.** -5^2 is the square of 5 multiplied by –1 by the order of operations; while $(-5)^2$ is the square of –5. ($-5^2=-25$ and $(-5)^2=25$)

Exercises 2.2, pages 114 - 116

1. 4 times a number **3.** 5 more than a number **5.** 7 times the sum of a number and 1.1 **7.** –2 times the difference between a number and 8 **9.** 6 divided by the difference between a number and 1 **11.** 5 times the sum of twice a number and 3 **13.** 3 times a number plus 7; 3 times the sum of a number and 7 **15.** The product of 7 and a number minus 3; 7 times the difference between a number and 3 **17.** $x+6$ **19.** $x-4$ **21.** $\frac{2x}{10}$ **23.** $3x-5$ **25.** $8-2x$ **27.** $20-4.8x$ **29.** $9(x+2)$ **31.** $4(x+1)-13$ **33.** $3(x+6)+8$ **35.** $3(7-x)-4$ **37. a.** $x-6$ **b.** $6-x$ **39. a.** $3x-5$ **b.** $5-3x$ **41.** $24d$ **43.** $3.15x$ **45.** $365y$ **47.** $7t+3$ **49.** $7t+3$ **51.** $20+0.15m$ **53.** $2w+2(2w-3)=6w-6$
55. A phrase whose meaning is not clear or for which there may be two or more interpretations.

Exercises 2.3, pages 124 - 126

1. –2 is a solution **3.** 4 is not a solution **5.** –4 is a solution **7.** –18 is a solution **9.** –28 is a solution **11.** $x=7$ **13.** $y=-4$ **15.** $x=-19$ **17.** $n=37$ **19.** $z=-6$ **21.** $x=5$ **23.** $y=-5.9$ **25.** $x=-1.2$ **27.** $y=\frac{11}{20}$ **29.** $x=9$ **31.** $y=8$ **33.** $x=20$ **35.** $y=10$ **37.** $x=-2$ **39.** $x=12$ **41.** $n=8$ **43.** $y=2.1$ **45.** $x=\frac{20}{9}$ **47.** $x=-13.3$ **49.** $y=-12$ **51.** $x=-\frac{2}{5}$ **53.** $x=-4$ **55.** $x=\frac{5}{8}$ **57.** $n=9.7$ **59.** $x=-4$ **61.** 1945 kanji **63.** 6.25 tons **65.** $x=-50.753$ **67.** $x=-17.214$ **69.** $x=246$ **71.** $x=-153.17$
83. a. Yes; It is stating that 6 + 3 is equal to 9. **b.** No; If we substitute 4 for x, we get the statement 9 = 10, which is not true.

Exercises 2.4, pages 131 - 133

1. $x=-3$ **3.** $x=2$ **5.** $x=2$ **7.** $x=2$ **9.** $y=-1$ **11.** $t=-1$ **13.** $x=-0.12$ **15.** $x=4$ **17.** $x=0$ **19.** $y=0$ **21.** $x=-2$ **23.** $y=-6$ **25.** $n=6$ **27.** $n=8$ **29.** $x=0$ **31.** $x=-7$ **33.** $x=-\frac{1}{8}$ **35.** $x=-\frac{13}{2}$ **37.** $x=-\frac{21}{5}$ **39.** $x=-\frac{8}{15}$ **41.** $y=\frac{28}{5}$ **43.** $x=2$ **45.** $y=\frac{7}{5}$ **47.** $x=-4.5$ **49.** $x=-44$ **51.** $x=2$ **53.** $x=-4$ **55.** $y=0.5$ **57.** $x=1.5$ **59.** $x=0.2$ **61.** 36.5°C **63.** 135 yards **65.** $x=6.1$ **67.** $x=1.12$

Exercises 2.5, pages 139 - 141

1. $x=-5$ **3.** $n=3$ **5.** $y=6$ **7.** $x=3$ **9.** $n=0$ **11.** $y=0$ **13.** $z=-1$ **15.** $y=\frac{1}{5}$ **17.** $x=-3$ **19.** $x=-4$ **21.** $x=-21$ **23.** $y=0$ **25.** $y=1$ **27.** $x=-\frac{3}{2}$ **29.** $x=\frac{1}{4}$ **31.** $x=\frac{3}{17}$ **33.** $x=-\frac{1}{4}$ **35.** $x=\frac{8}{5}$ **37.** $x=\frac{2}{3}$ **39.** $x=6$ **41.** $x=-11$ **43.** $x=\frac{1}{2}$ **45.** $x=-5$ **47.** $n=-1.5$ **49.** $x=0$ **51.** Conditional **53.** Conditional **55.** Contradiction **57.** Identity **59.** Conditional **61.** 1800 square feet **63.** 240 sundaes **65.** $x=-50.21$ **67.** $x=1.067$

Exercises 2.6, pages 148 - 152

1. $x-5=13-x$; 9 **3.** $36=2x+4$; 16 **5.** $7x=2x+35$; 7 **7.** $3x+14=6-x$; -2 **9.** $\frac{2x}{5}=x+6$; -10 **11.** $4(x-5)=x+4$; 8 **13.** $\frac{2x+5}{11}=4-x$; 3 **15.** $2x+3x=4(x+3)$; 12 **17.** $n+(n+2)=60$; 29, 31 **19.** $n+(n+1)+(n+2)=69$; 22, 23, 24 **21.** $n+(n+1)+(n+2)+(n+3)=74$; 17, 18, 19, 20 **23.** $171-n=(n+1)+(n+2)$; 56, 57, 58 **25.** $208-3n=(n+1)+(n+2)+(n+3)-50$; 42, 43, 44, 45 **27.** $n+2(n+2)=4(n+4)-54$; 42, 44, 46 **29.** $4n=(n+2)+(n+4)+44$; 25, 27, 29 **31.** $2n+3(n+2)=2(n+4)+7$; 3, 5, 7 **33.** $c+(c+49.50)=125.74$; Calculator: \$38.12, Textbook: \$87.62 **35.** $2x+90{,}000=310{,}000$; \$110,000 **37.** $(x+56)+x=542$; 243 boys **39.** $50-2x=10.50$; \$19.75 **41.** $(x+68)+x=158$; \$45 million on electric guitars **43.** $20+0.22x=66.20$; 210 miles **45.** $x+x+25{,}000=275{,}000$; Lot: \$125,000, House: \$150,000 **47.** $x+3x-1+2x+5=64$; 10 inches, 29 inches, 25 inches **49.** The difference between five times a number and the number is equal to 8; 2 **51.** Find two consecutive integers whose sum is 33; 16, 17 **53.** Find two consecutive integers such that 3 times the second is 53 more than the first; 25, 26 **55. a.** $n, n+2, n+4, n+6$ **b.** $n, n+2, n+4, n+6$ **c.** Yes; Answers will vary.

Exercises 2.7, pages 159 - 163

1. 91% **3.** 137% **5.** 62.5% **7.** 75% **9.** 0.69 **11.** 1.62 **13.** 0.075 **15.** 0.005 **17.** $\frac{7}{20}$ **19.** $\frac{13}{10}$ **21. a.** 32% **b.** 28% **c.** 12% **23. a.** 38.9% **b.** 19.4% **c.** 66.7% **25.** 61.56 **27.** 40 **29.** 2180 **31.** 125% **33.** 80 **35. a.** 8% **b.** 10% **c.** b **37.** \$85.33 **39.** \$1952.90 **41.** 1.5% **43.** 40 **45. a.** \$700 **b.** 33.33% **47. a.** \$3125 **b.** \$3375 **49. a.** more than \$27 **b.** \$30 **51.** 2000 baskets **53. a.** \$8800 **b.** Sam **c.** Maria

Chapter 2 Review, pages 168 - 171

1. $13x$ **2.** $6y$ **3.** $25x^2$ **4.** $-3a^2$ **5.** $-3x+2$ **6.** $3n+4$ **7.** $5x-34$ **8.** y^2+7y **9.** $7a^2+4a$ **10.** $9y$ **11. a.** $-a^2+4a+7$ **b.** -5

12. a. $17x-17$ **b.** -34 **13. a.** $3.2y^2-0.4y-13.6$ **b.** 36 **14. a.** $5x+5$ **b.** 0 **15. a.** $-2y-26$ **b.** -34 **16. a.** $7y$ **b.** 28 **17.** 3 more than 5 times a number **18.** 5 less than 6 times a number **19.** –3 multiplied by the sum of a number and 2 **20.** –2 multiplied by the sum of a number and 6 **21.** 5 times the difference between a number and 3 **22.** 4 times the difference between a number and 10 **23.** 7 times a number divided by 33 **24.** 50 divided by the product of 6 and a number **25.** $5x+3x$ **26.** $28-6x$ **27.** $72+8(x+2)$ **28.** $2x+3$ **29.** $\frac{22}{x+9}$ **30.** $10x-32$ **31.** $24x+5$ **32.** $0.49y$ **33.** –11 is a solution **34.** –0.28 is a solution **35.** 4.05 is not a solution **36.** –10 is a solution **37.** $x=7$ **38.** $x=14$ **39.** $y=-11$ **40.** $y=-8$ **41.** $n=-8$ **42.** $x=-4$ **43.** $n=10$ **44.** $n=6$ **45.** $y=-18$ **46.** $y=-30$ **47.** $x=18$ **48.** $y=0.4$ **49.** $x=-5$ **50.** $x=-8$ **51.** $y=9$ **52.** $y=-7$ **53.** $x=12$ **54.** $x=-25$ **55.** $y=6$ **56.** $y=-3$ **57.** $n=\frac{17}{9}$ **58.** $n=\frac{2}{5}$ **59.** $x=7$ **60.** $x=28$ **61.** $x=-12.35$ **62.** $x=-8.9$ **63.** $x=12$ **64.** $x=8$ **65.** $y=1.02$ **66.** $y=1.8$ **67.** $n=0$ **68.** $n=0$ **69.** $x=27$ **70.** $x=26$ **71.** $n=-12$ **72.** $n=-15$ **73.** $x=4$ **74.** $x=\frac{27}{5}$ **75.** $y=-2$ **76.** $x=13$ **77.** $x=-\frac{1}{9}$ **78.** $x=\frac{2}{15}$ **79.** $y=-35.6$ **80.** Contradiction **81.** Contradiction **82.** Identity **83.** Identity **84.** Conditional **85.** 19 **86.** 6 **87.** 9 **88.** –9 **89.** 24, 25, 26 **90.** 91, 93, 95 **91.** 10, 12, 14 **92.** 46, 47, 48, 49 **93.** \$245,000 **94.** 12 months **95.** 25 houses **96.** 17 packs **97.** 80% **98.** $\frac{18}{25}$ **99.** 78.2 **100.** 70% **101. a.** 8.4% **b.** 6.25% **c.** a **102.** 1.2% **103.** 18 **104.** \$60.30 **105.** \$21.45; \$281.45 **106.** 6.25%

Chapter 2 Test, pages 172 - 173

1. a. $5x^2+7x$ **b.** 6 **2. a.** $6y-6$ **b.** 12 **3. a.** $5x+2$ **b.** –8 **4. a.** $2y^2+7y$ **b.** 39 **5. a.** The product of 5 and a number increased by 18 **b.** 3 multiplied by the sum of a number and 6 **c.** 42 less the product of 7 and a number **d.** The product of 6 and a number decreased by 11 **6. a.** $6x-3$ **b.** $2(x+5)$ **c.** $2(x+15)-4$ **d.** $\frac{x}{10}+2x$ **7.** $x=-4$ **8.** $y=-29$ **9.** $x=-3$ **10.** $x=-6$ **11.** $x=5$ **12.** $y=0$ **13.** $a=-\frac{9}{8}$ **14.** $x=\frac{3}{2}$ **15.** $x=20$ **16.** $x=-2$ **17.** Identity **18.** Conditional **19.** $(2y+5)+y=-22$; –9, –13 **20.** $3(n+2)=n+(n+4)+27$; $n=25, 27, 29$ **21.** $2n+3(n+1)=83$; 16, 17 **22.** $2(1)+1(3)+2x=21$; 8 2-pt shots **23.** 111.6 **24.** 32% **25.** The \$6000 investment is the better investment since it has a higher percent of profit, 6%, than the \$10,000 investment, 5%. **26.** 3 meters, 4 meters, and 5 meters **27.** \$1875 **28. a.** \$730 **b.** \$605.35

Cumulative Review: Chapters 1-2, pages 174 - 176

1. 20 **2.** –14 **3.** 9 **4.** –15 **5.** –126 **6.** 459 **7.** –39 **8.** 124 **9.** 67 **10.** 23.3 **11.** See page 34. **12.** 108 **13.** $72xy^2$ **14.** Associative property of multiplication **15.** Commutative property of addition **16.** Multiplicative inverse **17.** Additive identity **18. a.** $\{-4, 4\}$ **b.** No solution **19.** True **20.** True **21.** False; $\frac{3}{4}\le|-1|$ **22.** 

23. 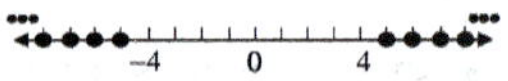

24. 7 **25.** 18 **26.** 189 **27.** 13 **28.** 35 **29.** $\frac{5}{6}$ **30.** $\frac{2}{9}$ **31.** $\frac{6}{5y}$ **32.** $-\frac{1}{36}$ **33.** $\frac{1}{6}$ **34.** 0 **35.** $\frac{49}{40}$ **36.** $-\frac{31}{5}$ **37. a.** $-2y-12$ **b.** –18 **38. a.** $4x+7$ **b.** –1 **39. a.** x^2+13x **b.** –22 **40. a.** $x-7$ **b.** –9 **41.** $13x-5$ **42.** $\frac{a}{6}$ **43. a.** $2x-3$ **b.** $\frac{x}{6}+3x$ **c.** $2(x+10)-13$ **44. a.** The sum of 6 and 4 times a number **b.** 18 times the difference of a number and 5 **45.** $x=-9.8$ **46.** $x=18$ **47.** $y=-3$ **48.** $y=6$ **49.** $x=-1$ **50.** $x=3$ **51.** $x=-11$ **52.** $x=-4$ **53.** $x=\frac{31}{8}$ **54.** $y=\frac{24}{5}$ **55.** $y=\frac{15}{2}$ **56.** $y=0$ **57.** Conditional **58.** Contradiction **59.** Contradiction **60.** Identity **61.** Identity **62.** $-\frac{5}{51}$ **63.** –5 **64.** $\frac{10}{3}$ cups (or $3\frac{1}{3}$ cups) **65.** 12% **66. a.** 1000 Republicans are in favor **b.** 2500 registered voters are in favor **67.** 11

68. 18, 20, 22 **69.** −2, −1, 0, 1 **70.** Hard drive: \$84.60; Battery: \$46.15 **71.** \$506.94 **72.** \$555.80
73. These are equally good investments since the percent of profit, 8%, is the same for each investment.

Chapter 3

Exercises 3.1, pages 183 - 188

1. \$120 **3.** \$10,000 **5. a.** \$183.75 **b.** \$3683.75 **7.** 176 ft/sec **9.** 4 milliliters **11.** \$1030 **13.** 14 rafters **15.** 336 in. or 28 ft **17.** \$1030 **19.** \$76.50 **21.** \$5 million **23.** \$2400 **25.** 7 hours **27.** 1.625 miles per hour **29.** $b = P - a - c$ **31.** $m = \frac{F}{a}$ **33.** $w = \frac{A}{l}$ **35.** $n = \frac{R}{p}$ **37.** $P = A - I$ **39.** $m = 2A - n$ **41.** $t = \frac{I}{Pr}$ **43.** $b = \frac{P-a}{2}$ **45.** $\beta = 180° - \alpha - \gamma$ **47.** $h = \frac{V}{lw}$ **49.** $b = \frac{2A}{h}$ **51.** $\pi = \frac{A}{r^2}$ **53.** $g = \frac{mv^2}{2K}$ **55.** $y = \frac{6-2x}{3}$ **57.** $x = \frac{11-2y}{5}$ **59.** $b = \frac{2A-hc}{h}$ or $b = \frac{2A}{h} - c$ **61.** $x = \frac{8R+36}{3}$ or $x = \frac{8R}{3} + 12$ **63.** $y = -x - 12$ **65.** $C = nt + 9$ **67.** $C = 325n + 5400$
69. a. 0; No, because the numerator will be zero and thus the whole fraction will be equal to zero for all values of s. **b.** $x < 70$ **c.** Answers will vary.

Exercises 3.2, pages 196 - 202

1. e **3.** k **5.** c **7.** d **9.** p **11.** l **13.** m **15.** h **17. a.** 72 mm; **b.** 324 mm^2 **19. a.** 38 in.; **b.** 72 in.^{2} **21. a.** 32 cm; **b.** 44 cm^2 **23.** 108.85 cm^3 **25.** 470.4 ft^3 **27.** 203.47 in.^{3} **29.** 81 ft^2 **31.** 50.24 m **33.** 24.2 m **35.** 2520 ft^2 **37.** 1483.2 in.^{3} **39.** 4186.67 cm^3 **41.** 54.84 in.; 200.52 in.^{2} **43. a.** $P = 2l + 2w$ **b.** $w = \frac{P-2l}{2}$ **c.** 13 feet **45. a.** $A = bh$ **b.** $b = \frac{A}{h}$ **c.** 47 inches **47. a.** $P = a + b + c$ **b.** $a = P - b - c$ **c.** 61 inches **49. a.** $A = \frac{1}{2}h(b+c)$ **b.** $b = \frac{2A}{h} - c$ **c.** 5 feet **51.** $16\frac{1}{8}$ or 16.125 meters **53.** Side $a = 17$ m; Side $b = 11$ m **55.** 15 ft **57.** 40 inches **59.** 5 in. **61.** 9 m, 15 m **63.** 10.5 cm **65.** 5.5 ft **67.** 21 in. **69.** 28 meters **71.** 4.5 cm **73.** 36 cm **75.** $x = 5$ **77.** 52.25 in. **79.** 1326.93 cm^2 **81.** 510.46 ft^3

Exercises 3.3, pages 207 - 213

1. 1.68 mph **3.** 8.75 hours **5.** 3 hours **7.** 60 mph; 300 miles **9.** 7 hours **11.** 36 mph; 60 mph **13.** Day = 56 mph; Night = 69 mph **15.** 4.5 miles **17.** \$14,000 at 5%; \$11,000 at 6% **19.** 4.5% on \$10,000; 5.5% on \$6000 **21.** \$7000 at 6%; \$9000 at 8% **23.** \$24,000 at 4.5%; \$18,000 at 6% **25.** \$600 at 2.5%; \$800 at 4% **27.** \$11,000 at 4%; \$9500 at 5% **29.** \$112.50 **31.** \$60.50 **33.** \$40.50 **35.** 62 min **37.** 6 hours **39.** 71 **41. a.** Not possible **b.** 192 **43. a.** 53.8°F **b.** 44°F **c.** 17°F
45. a. \$24.2 billion **b.** \$26 billion **c.** \$24 billion

Exercises 3.4, pages 225 - 228

1. 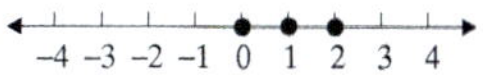

3. 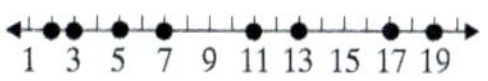

5. 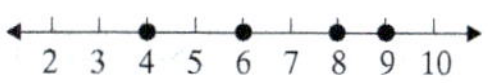

7. No solution **9.** $\{x | 3 \le x < 5\}$ **11.** $\{x | x \ge -2.5\}$ **13. a.** $\{x | -3 \le x \le 1\}$ **b.** $[-3, 1]$ **c.** Closed interval **15. a.** $\{x | -8 < x \le -2\}$ **b.** $(-8, -2]$ **c.** Half-open interval

17. a. $\{x \mid x \le 1\}$ **b.** $(-\infty, 1]$ **c.** Half-open interval **19. a.** $\{x \mid -4 < x < 4\}$ **b.** $(-4, 4)$ **c.** Open interval

21. Half-open interval **23.** Open interval

25. Half-open interval **27.** Closed interval

29. Half-open interval **31.** $(-\infty, 1)$ **33.** $(2, \infty)$

35. $\left(\frac{8}{3}, \infty\right)$ **37.** $[-0.4, \infty)$ **39.** $(-\infty, 2)$

41. $[3, \infty)$ **43.** $(-\infty, 2]$ **45.** $\left(-\frac{1}{3}, \infty\right)$

47. $\left(-\infty, -\frac{11}{2}\right)$ **49.** $[1, \infty)$ **51.** $\left(\frac{9}{2}, \infty\right)$

53. $(-\infty, -0.6)$ **55.** $(6.5, \infty)$ **57.** $(-\infty, -13)$

59. $\left(-\infty, -\frac{10}{9}\right]$ **61.** $[-4, \infty)$ **63.** $(-\infty, 1]$

65. $(-17, \infty)$ **67.** $(-\infty, -30)$ **69.** $\left(-\infty, -\frac{29}{2}\right]$

71. $(-9, 1)$ **73.** $\left[\frac{1}{2}, \frac{3}{2}\right]$ **75.** $[3, 15]$

77. $(-10, -5)$ **79.** $(-2.8, -0.3)$ **81. a.** The student would need a score higher than 102 points, which is not possible. Thus he cannot earn an A in the course. **b.** The student must score at least 192 points to earn an A in the course. **83.** He must sell at least 10 cars. **85.** Answers will vary.

Exercises 3.5, pages 235 - 237

1. $x = -8, 8$ **3.** No solution **5.** $x = -5, -1$ **7.** $x = -\frac{4}{3}, \frac{5}{3}$ **9.** $n = -2, \frac{2}{3}$ **11.** No solution **13.** $x = 2$ **15.** $x = -\frac{3}{2}, 2$ **17.** $x = -\frac{1}{5}, 1$ **19.** $x = -1, \frac{11}{3}$ **21.** $x = -22, 26$ **23.** $x = -8, 4$ **25.** $x = -6, 0$ **27.** $x = -\frac{1}{3}, 3$ **29.** $x = 1$ **31.** $x = -\frac{5}{2}, \frac{3}{4}$ **33.** $x = -\frac{20}{7}, \frac{4}{5}$ **35.** $x = -20, \frac{40}{9}$ **37.** $(-\infty, \infty)$ **39.** $\left[-\frac{4}{5}, \frac{4}{5}\right]$

41. $(-\infty, 1) \cup (5, \infty)$ **43.** $[-10, -2]$

45. $(-\infty, -8] \cup [-2, \infty)$ **47.** $\left(-\infty, -\frac{1}{2}\right] \cup \left[\frac{3}{2}, \infty\right)$

49. No solution **51.** $\left[-2, -\frac{1}{2}\right]$ **53.** $\left(-\frac{5}{3}, -1\right)$

55. $\left(-\infty, -\frac{2}{3}\right] \cup [6, \infty)$ **57.** $[-1, 10]$ **59.** $(-\infty, \infty)$

61. $\left(\frac{1}{2}, \frac{7}{2}\right)$ **63.** $\left(-\infty, -\frac{5}{2}\right) \cup (0, \infty)$ **65.** $\left(-2, -\frac{4}{7}\right)$

67. a. **b.** $|x| \le 10$ **c.** $[-10, 10]$, Closed interval **69. a.** **b.** $|x - 8| > 6$ **c.** $(-\infty, 2) \cup (14, \infty)$ **71. a.** **b.** $|x + 5| < 2$ **c.** $(-7, -3)$, Open interval

Chapter 3 Review, pages 242 - 246

1. $C = 0°$ **2.** 47.5 mph **3.** $y = -2$ **4.** 40 meters **5.** $P = 0.16$ **6.** $\pi = \frac{L}{2rh}$ **7.** $R = P + C$ **8.** $b = \frac{2A}{h}$ **9.** $\alpha = 180° - \beta - \gamma$
10. $g = -\frac{v - v_0}{t}$ **11.** $m = \frac{2gK}{v^2}$ **12.** $y = 6 - 3x$ **13.** $y = 6x - 3$ **14.** $x = \frac{10 + 2y}{5}$ **15.** $x = \frac{4y + 7}{3}$
16. a. $P = 2a + 2b$ **b.** $A = \pi r^2$ **c.** $V = \frac{4}{3}\pi r^3$ **17. a.** 50 cm **b.** 144 cm^2 **18. a.** 38 in. **b.** 80 in.^{2} **19.** 20.2 in. **20.** 530.66 cm^2
21. 27 ft^3 **22.** 468 m^2 **23.** 78.5 in.^{2} **24.** 686.2 cm^3 **25.** $2\frac{2}{3}$ meters **26.** 47 inches **27.** 61 inches **28.** 84 feet by 46 feet
29. 13 cm **30.** 5 yd **31.** 6 meters **32.** $P = 48$ in.; $A = 120$ in.^{2} **33.** $\frac{11}{3}$ hr; $\frac{7}{3}$ hr **34.** 55 mph; 247.5 miles
35. George: 55 mph; Joshua: 50 mph **36.** \$30,000 at 4%; \$10,000 at 6.5% **37.** \$4000 at 6%; \$16,000 at 8% **38.** \$80
39. \$220 **40.** \$220 **41.** 77 points **42. a.** 30.85 million **b.** Aspro Group **c.** 2.7 million **43.** 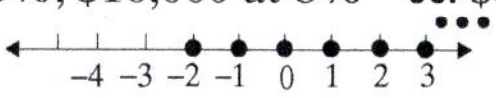

44. 

45. 

46. 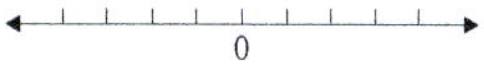

47. 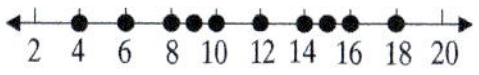

48. $\{x | 3 \le x < 6\}$ **49.** $\{x | -4 < x < 4\}$ **50.** $\{x | x \ge -4.6\}$ **51. a.** $\{x | -4 < x < 2\}$ **b.** $(-4, 2)$ **c.** Open interval
52. a. $\{x | x \ge 3\}$ **b.** $[3, \infty)$ **c.** Half-open interval **53. a.** $\{x | x < 1.5\}$ **b.** $(-\infty, 1.5)$ **c.** Open interval
54. a. $\{x | -2 \le x \le 0\}$ **b.** $[-2, 0]$ **c.** Closed interval **55.** 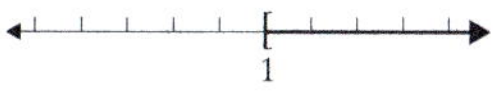

$[1, \infty)$ **56.** 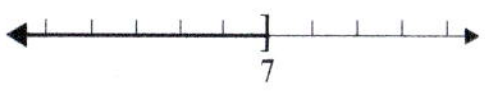

$(-\infty, 7]$
57. 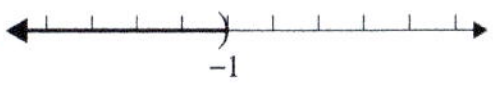

$(-\infty, -1)$ **58.** 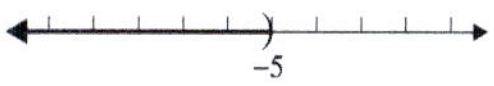

$(-\infty, -5)$ **59.** 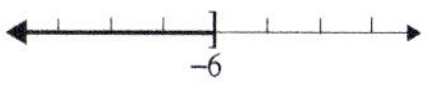

$(-\infty, -6]$
60. 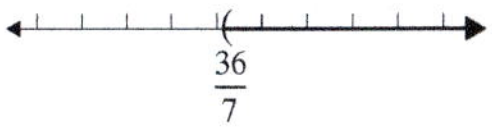

$\left(\frac{36}{7}, \infty\right)$ **61.** 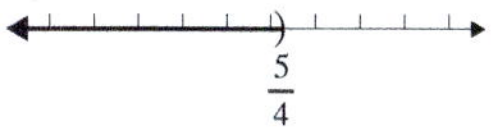

$\left(-\infty, \frac{5}{4}\right)$ **62.** 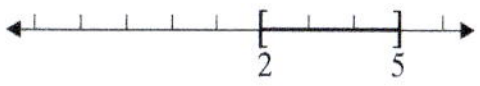

$[2, 5]$
63. 

$(-1, 3)$ **64.** 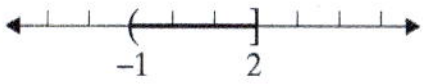

$(-1, 2]$ **65.** 

$\left[-4, \frac{4}{3}\right)$
66. 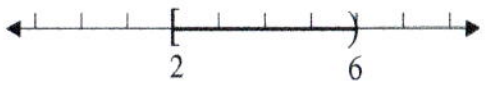

$[2, 6)$ **67.** $x = -3, \frac{7}{3}$ **68.** $x = -10, 4$ **69.** No solution **70.** $x = 3, 11$ **71.** $x = -5$ **72.** $x = 0, 2$
73. 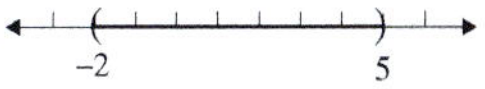

$(-2, 5)$ **74.** 

$[-24, -10]$ **75.** 

$(0, 4)$
76. 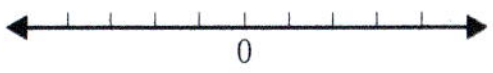

$(-\infty, \infty)$ **77.** 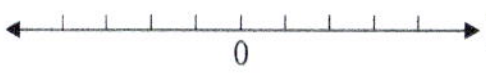

No solution **78.** 

$\left(-\frac{13}{3}, 3\right)$

Chapter 3 Test, pages 247 - 249

1. $m = \frac{N - p}{rt}$ **2.** $y = \frac{7 - 5x}{3}$ **3.** \$29.00 **4. a.** 18 months **b.** \$490 **5. a.** 27.42 ft **b.** 50.13 ft^2 **6.** 4186.67 cm^3 **7.** 200 ft^3
8. 3 meters, 4 meters, and 5 meters **9.** 10 cm **10. a.** $w = 24$ in. **b.** $A = 888$ in.^{2} **11.** \$12,000 at 5%; \$13,000 at 3.5%
12. 4 hours **13.** \$332.50 **14.** 88 **15. a.** 175.05 million **b.** 184.5 million **c.** 128.5 million
16. a. $\{x | x < 5\}$ **b.** $(-\infty, 5)$ **c.** Open interval **17. a.** $\{x | -1 \le x \le 4\}$ **b.** $[-1, 4]$ **c.** Closed interval **18.** $x = -1.9, 0.9$
19. $x = \frac{2}{3}, \frac{8}{3}$ **20.** $x = -\frac{1}{3}, -13$ **21.** 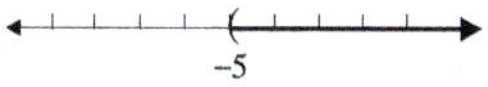

$(-5, \infty)$ **22.** 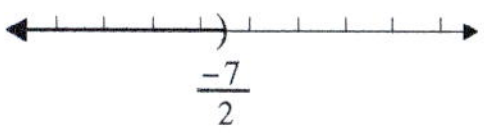

$\left(-\infty, -\frac{7}{2}\right)$
23. 

$(-1, 5)$ **24.** 

$(2, 5)$ **25.** 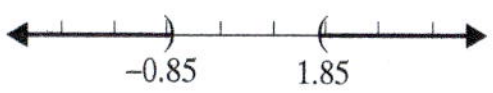

$(-\infty, -0.85) \cup (1.85, \infty)$

Cumulative Review: Chapters 1-3, pages 250 - 253

1. a. 17 **b.** −26 **2.** 21.7 **3.** $x+15$ **4.** $3x+8$ **5.** 540 **6.** $120a^2b^3$ **7.** False; $-15<5$ **8.** True **9.** False; $\frac{7}{8}\geq\frac{7}{10}$ **10.** True **11.** $x=3$ **12.** $x=4$ **13.** $x=\frac{8}{5}$ **14.** $x=-\frac{1}{2}$ **15.** $x=\frac{3}{8}$ **16.** $x=-3$ **17.** Conditional **18.** Identity **19.** Identity **20.** Conditional **21.** $v=\frac{h+16t^2}{t}$ **22.** $r=\frac{A-P}{Pt}$ **23.** $y=-\frac{14-5x}{3}$ **24.** $h=\frac{3V}{\pi r^2}$ **25.** $d=\frac{C}{\pi}$ **26.** $y=\frac{10-3x}{5}$ **27.** $\{x \mid 2<x<5\}$ **28.** $\{x \mid -3\leq x\leq 3\}$ **29.** $\{x \mid x\geq -8.3\}$ **30.** 

31. 

32. 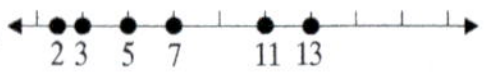

33. 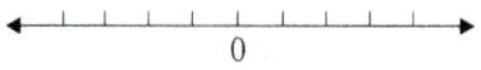

34. 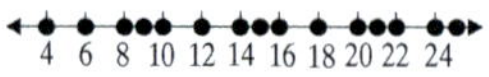

35. 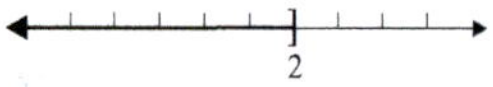

$(-\infty, 2]$ **36.** 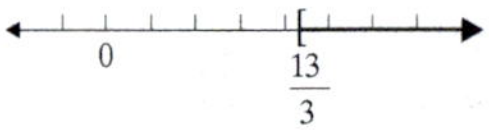

$\left[\frac{13}{3}, \infty\right)$ **37.** 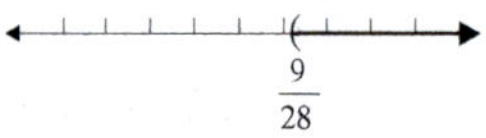

$\left(\frac{9}{28}, \infty\right)$ **38.** 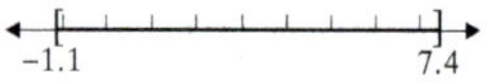

$[-1.1, 7.4]$ **39.** 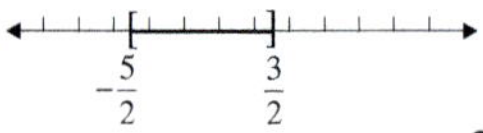

$\left[-\frac{5}{2}, \frac{3}{2}\right]$ **40.** 

$(-6, 2)$ **41.** $x=-3, 3$ **42.** No solution **43.** $x=-19, 1$ **44.** $a=-1.9, 0.9$ **45.** $x=\frac{-21}{2}, \frac{3}{2}$ **46.** $x=-30, 60$ **47.** $x=0, 2$ **48.** $x=0$ **49. a.** $[-3, 3]$ **b.** $(6, 8)$ **c.** $[0.3, 2.7)$
50. 

$(-6, 6)$ **51.** 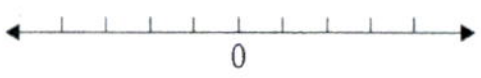

No solution **52.** 

$[-3, 2]$
53. 

$[0, 7]$ **54.** 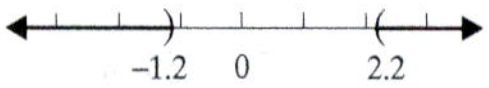

$(-\infty, -1.2)\cup(2.2, \infty)$ **55.** 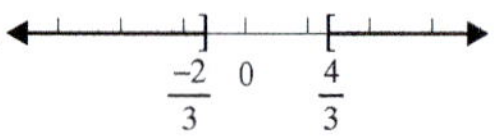

$\left(-\infty, \frac{-2}{3}\right]\cup\left[\frac{4}{3}, \infty\right)$ **56. a.** 45.7 cm **b.** 139.25 cm² **57.** 1177.5 in.³ **58.** 10.71 in. **59.** \$3360 **60.** 20, 22, 24 **61.** 15, 17, 19 **62.** Length = 25 cm, Width = 16 cm **63.** 5 inches, 12 inches, 13 inches **64.** −19 **65.** 1.8 hours (or 1 hour 48 minutes) **66.** \$16,500 at 3%; \$5500 at 4.5% **67.** Less than 140 miles **68.** 87.25 **69.** $86\leq x\leq 100$

Chapter 4

Exercises 4.1, pages 264 - 272

1. $\{A(-5,1), B(-3,3), C(-1,1), D(1,2), E(2,-2)\}$ **3.** $\{A(-3,-2), B(-1,-3), C(-1,3), D(0,0), E(2,1)\}$
5. $\{A(-4,4), B(-3,-4), C(0,-4), D(0,3), E(4,1)\}$ **7.** $\{A(-3,-5), B(-1,4), C(0,-1), D(3,1), E(6,0)\}$
9. $\{A(-5,0), B(-2,2), C(-1,-4), D(0,6), E(2,0)\}$

11. 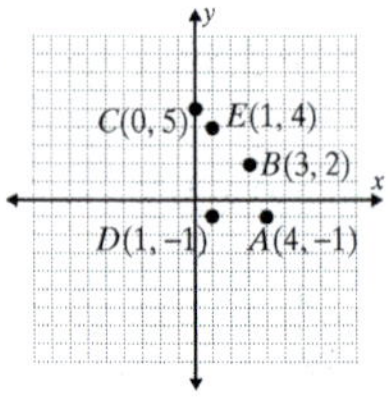

13. 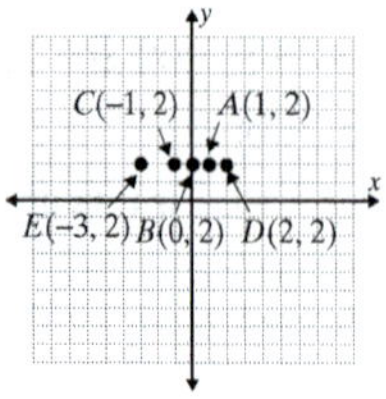

15. 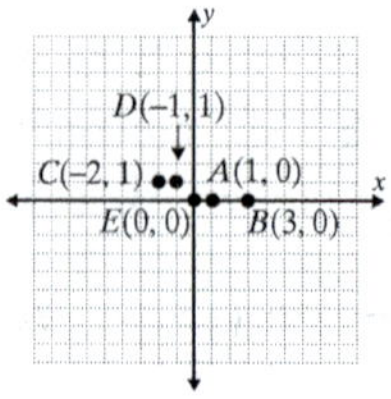

17. 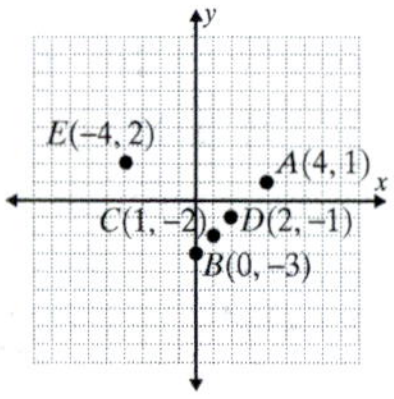

19. 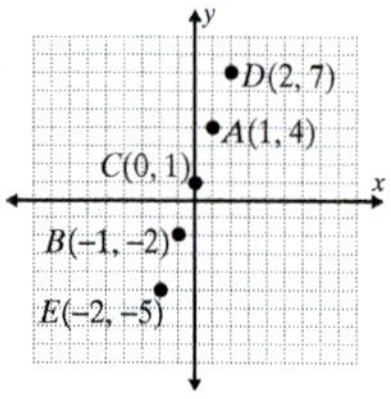

21. 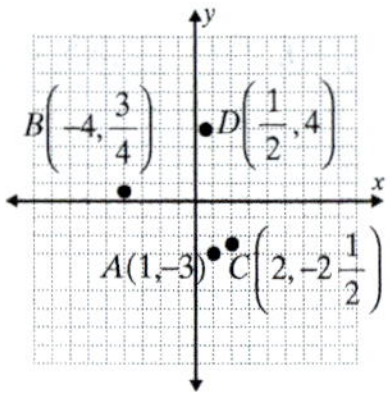

23. 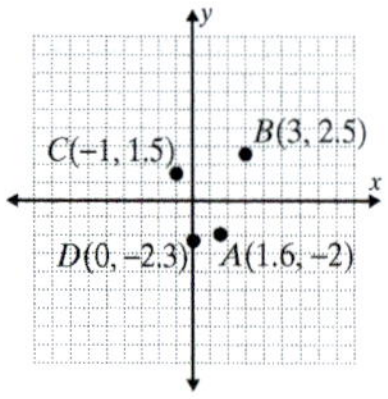

25. b, c, d **27.** a, c **29.** a, c, d **31. a.** $(0,-4)$ **b.** $(2,-2)$ **c.** $(4,0)$ **d.** $(1,-3)$
33. a. $(0,3)$ **b.** $(2,2)$ **c.** $(6,0)$ **d.** $(-2,4)$ **35. a.** $(0,-8)$ **b.** $(1,-4)$ **c.** $(2,0)$ **d.** $(3,4)$

37. **a.** $(0, 2)$ **b.** $\left(-1, \frac{8}{3}\right)$ **c.** $(3, 0)$ **d.** $(6, -2)$ **39.** **a.** $\left(0, -\frac{7}{4}\right)$ **b.** $(1, -1)$ **c.** $\left(\frac{7}{3}, 0\right)$ **d.** $\left(3, \frac{1}{2}\right)$

41. $(0, 0), (-1, -3), (-2, -6), (2, 6)$

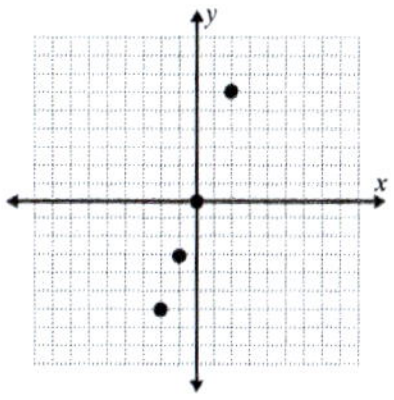

43. $(0, -3), (1, -1), (-2, -7), (3, 3)$

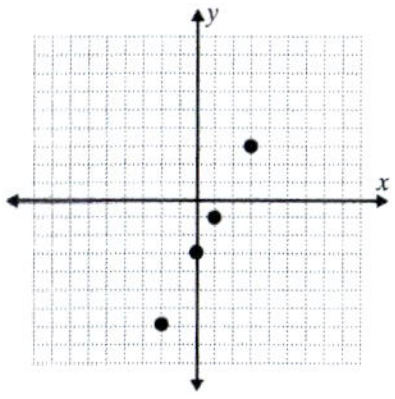

45. $(0, 9), (3, 0), (1, 6), (4, -3)$

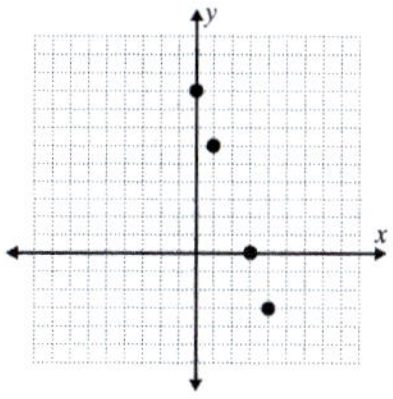

47. $(0, 2), (4, 5), (-4, -1), \left(-1, \frac{5}{4}\right)$

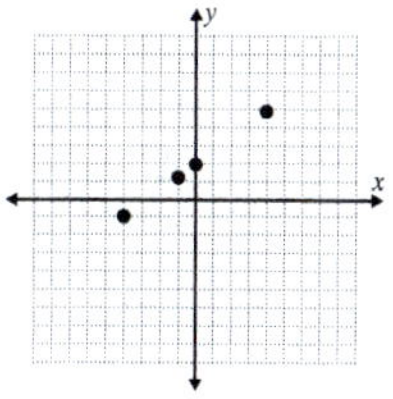

49. $\left(0, -\frac{9}{5}\right), (3, 0), (-2, -3), \left(\frac{4}{3}, -1\right)$

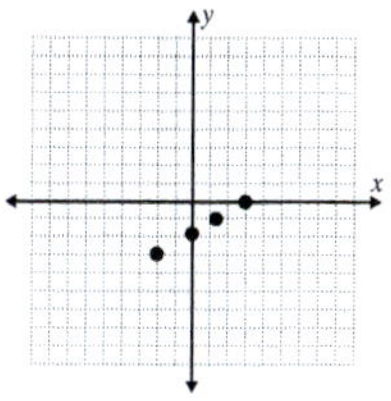

51. $(0, -5), (2, 0), \left(-1, -\frac{15}{2}\right), (4, 5)$

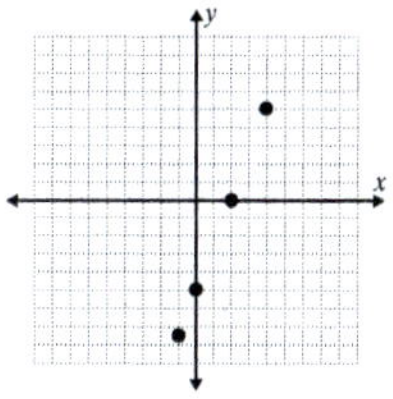

53. $(0, 2), (3.2, 0), (1.92, 0.8), (3.52, -0.2)$

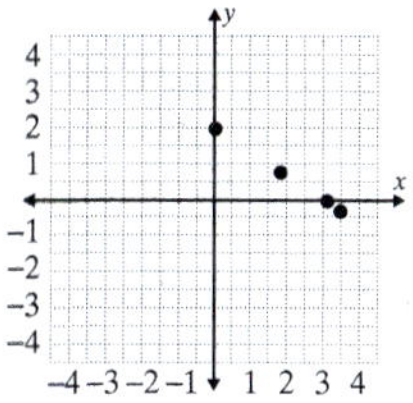

55. For example, $(-3, -3), (0, 1)$, and $(3, 5)$

57. For example, $(-3, -1), (0, 0)$, and $(3, 1)$

59. For example, $(-2, 3), (0, 3)$, and $(4, 3)$

61. For example, $(-4, -4), (0, -3)$, and $(4, -2)$

63. For example, $(-3, -3), (-2, 0)$, and $(-1, 3)$

65. **a.** $(100, 78.02), (200, 156.04), (300, 234.06), (400, 312.08), (500, 390.1)$

b.

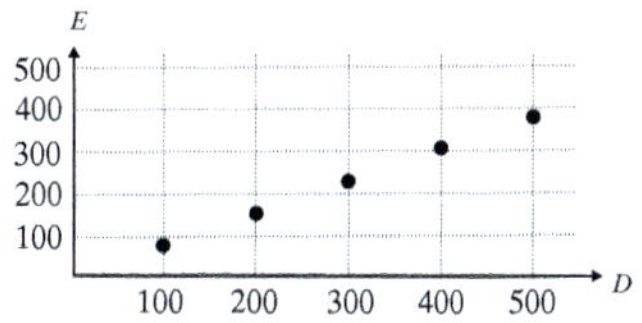

67. **a.** $(1, 16), (2, 64), (3.5, 196), (4, 256), (4.5, 324), (5, 400)$

b.

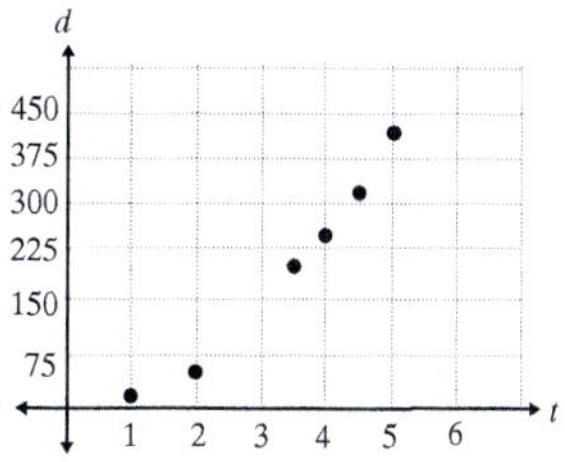

c. The time t is squared in the equation.

69. **a.**

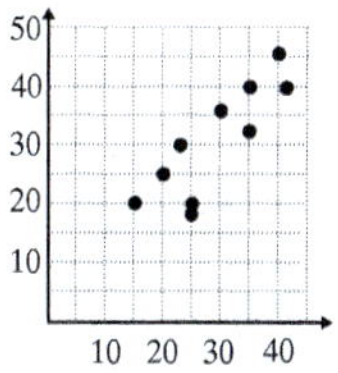

b. Yes, the more sit-ups a person can do it appears the more push-ups he/she can do.

c. 24 sit-ups, 33 sit-ups, 36 sit-ups, 45 sit-ups; Answers will vary.

71. Answers will vary. Not all scatterplots can be used to predict information related to the two variables graphed because not all variables are related.

Exercises 4.2, pages 278 - 280

1. a **3.** d **5.** f

7.

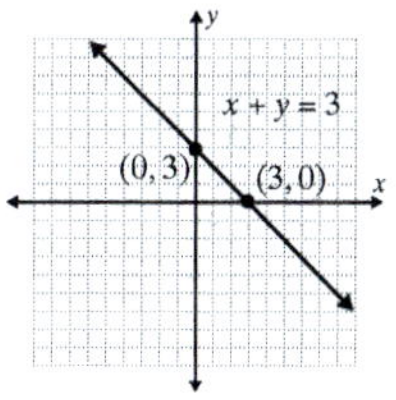

9.

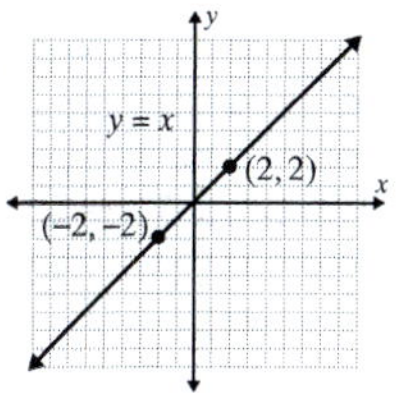

11.

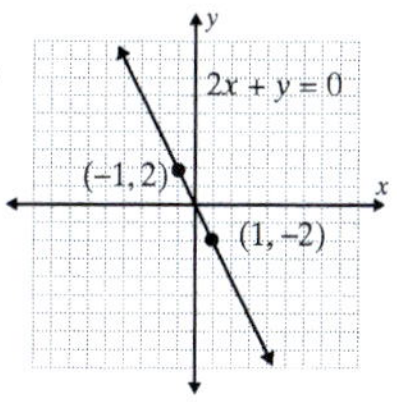

13.

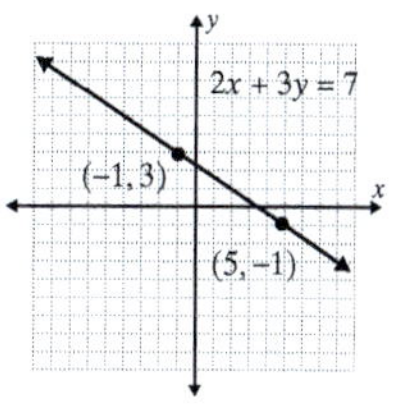

15.

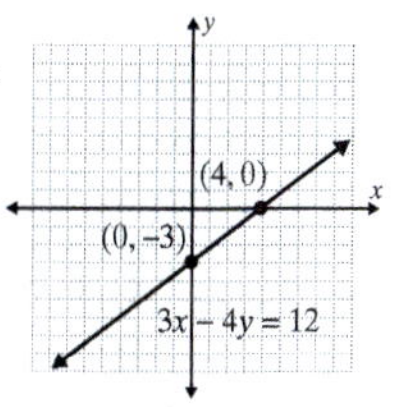

17.
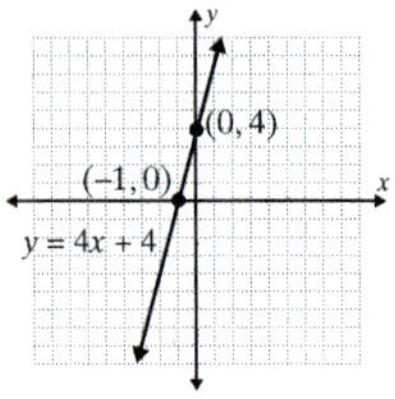

19.
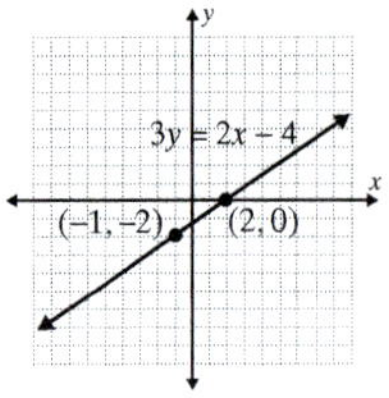

21.
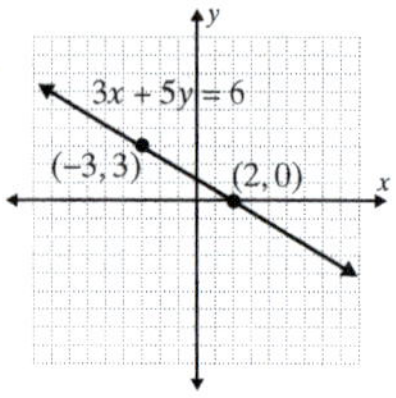

23.
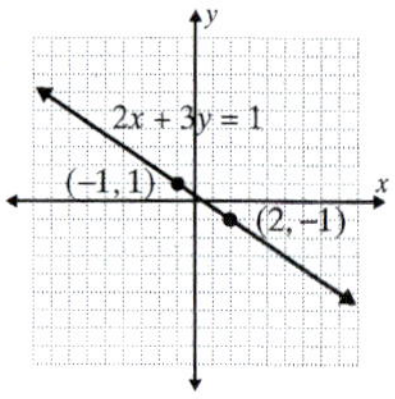

25.
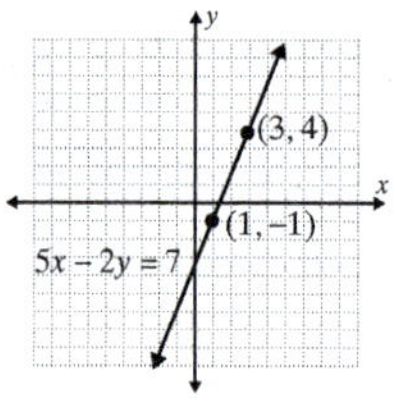

27.
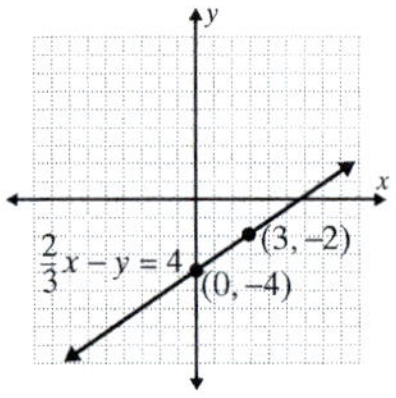

29.
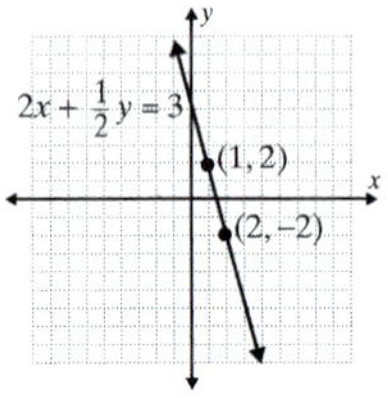

31.
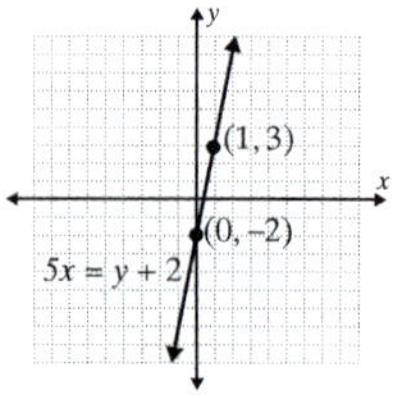

33.
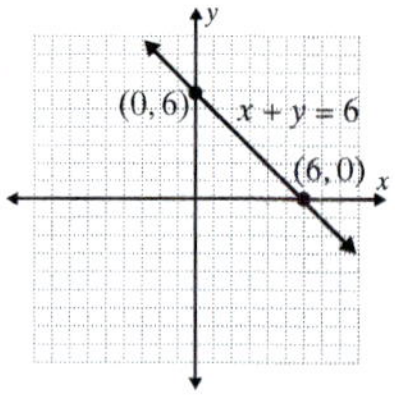

35.
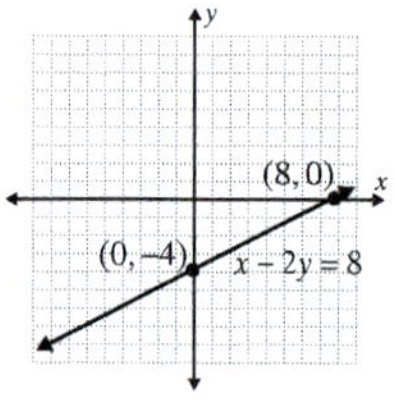

37.
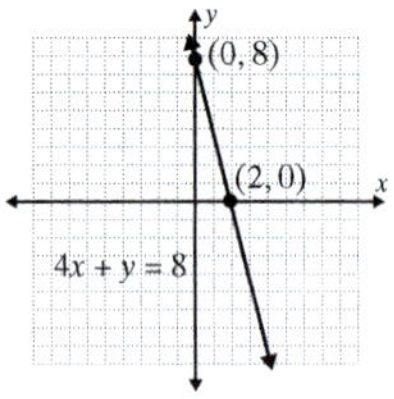

39.
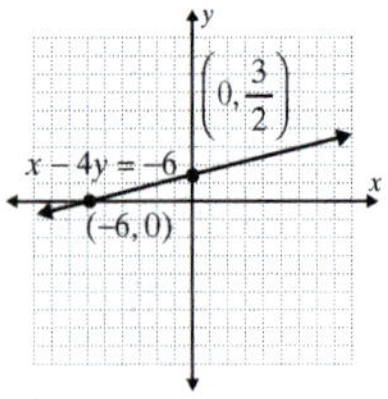

41.
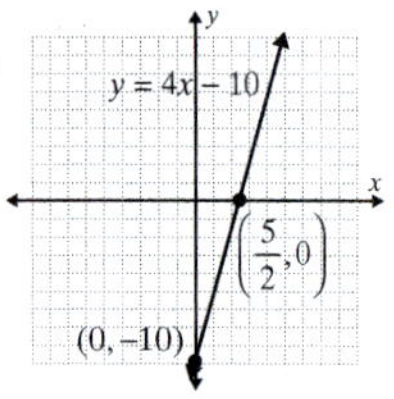

43.
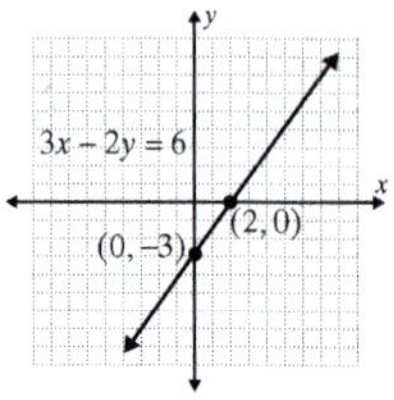

45.
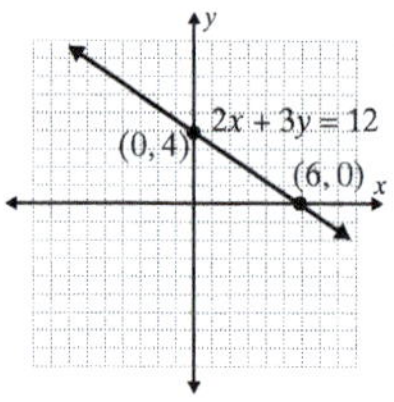

47.
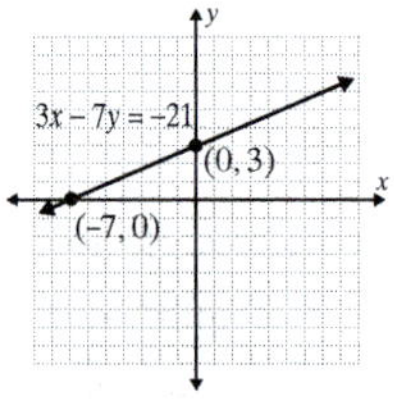

49.
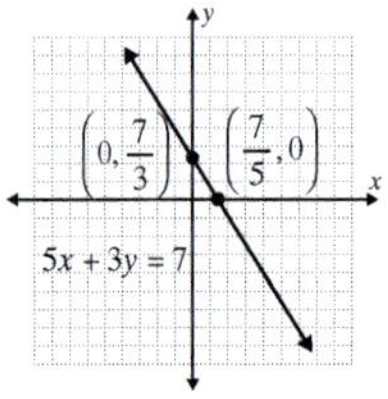

51.
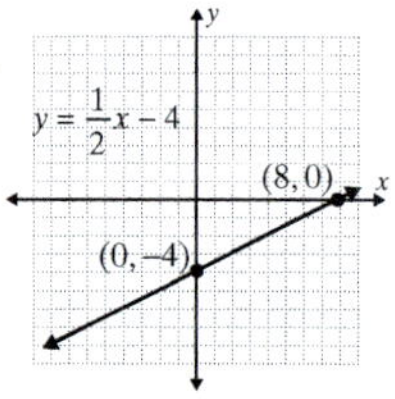

53.
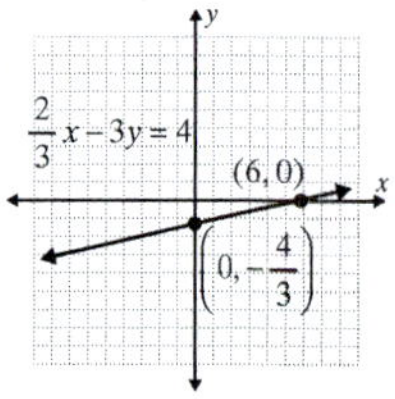

55.
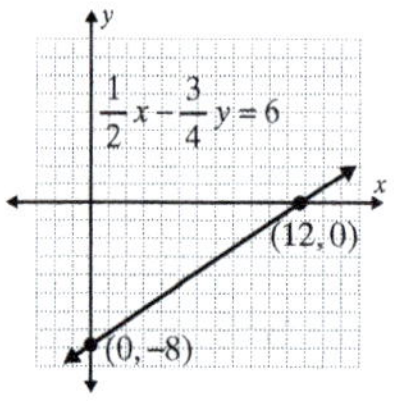

57. Two (unique) points determine a line.

Exercises 4.3, pages 290 - 296

1. $m = 5$ **3.** $m = -\frac{1}{7}$ **5.** $m = 0$ **7.** $m = \frac{1}{2}$ **9.** $m = 2$ **11.** $m = \frac{1}{5}$

13. Horizontal line; $m = 0$

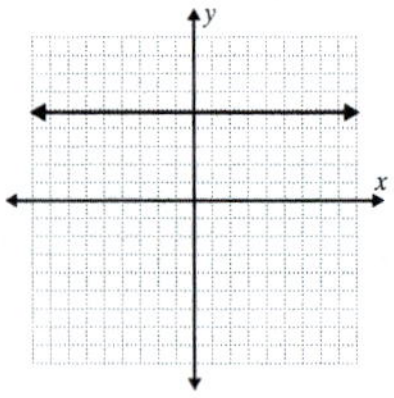

15. Vertical line; m is undefined

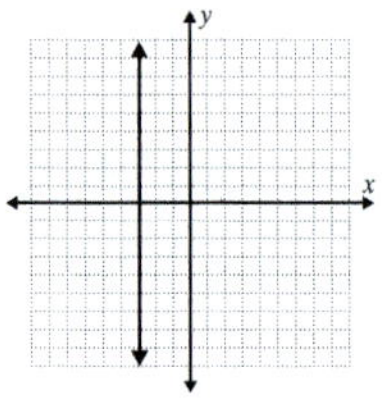

17. Horizontal line; $m = 0$

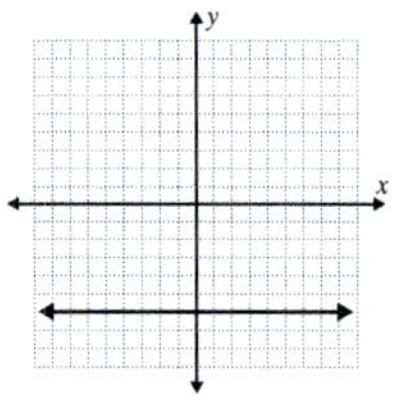

19. Vertical line; m is undefined

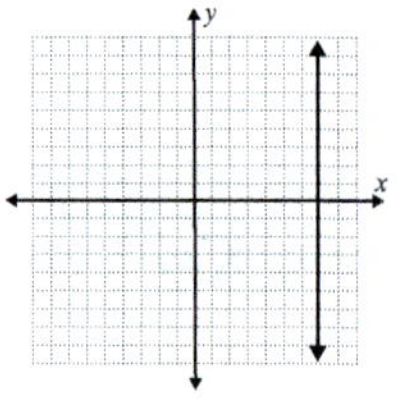

21.
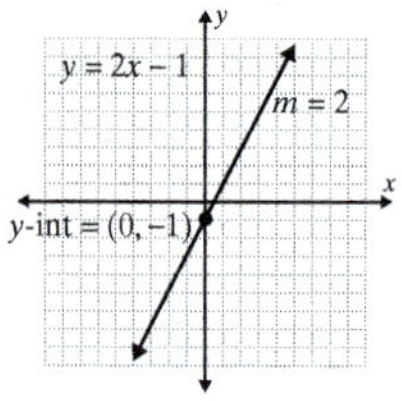

23.
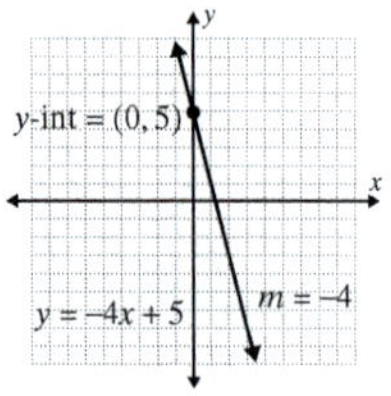

25.
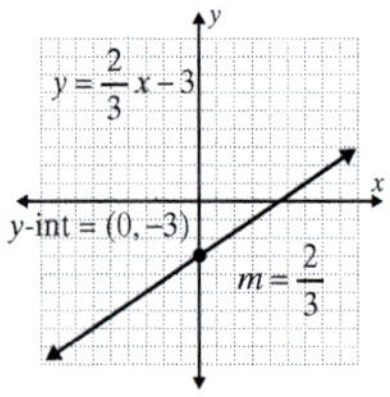

27.
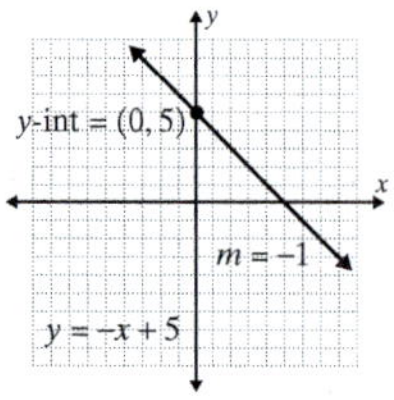

29.
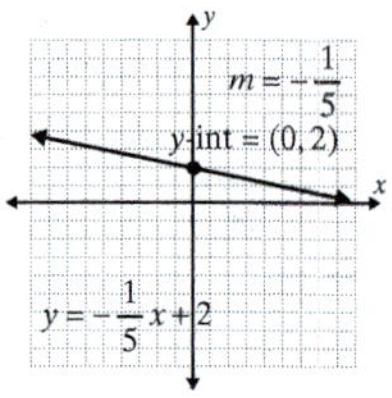

31.
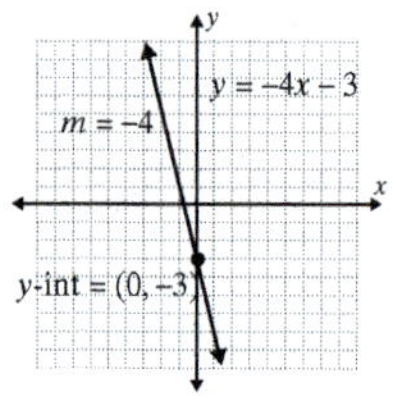

33.
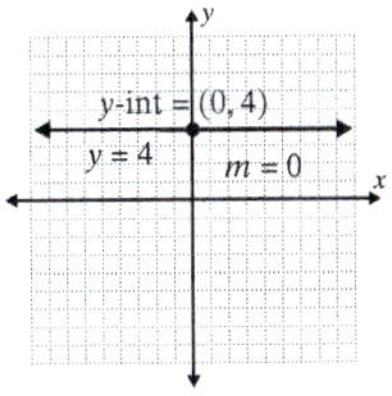

35.
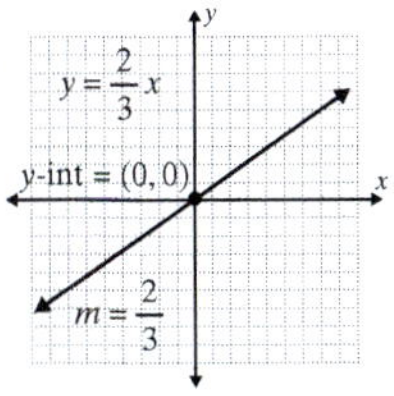

37.
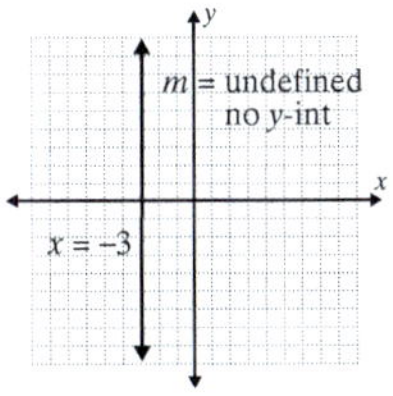

39.
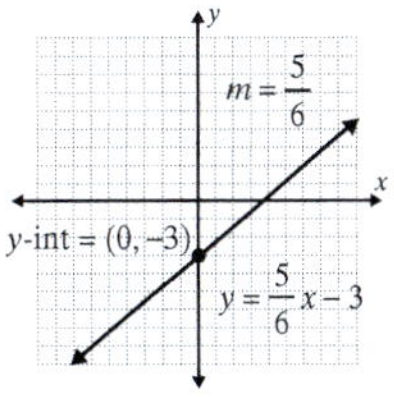

41.
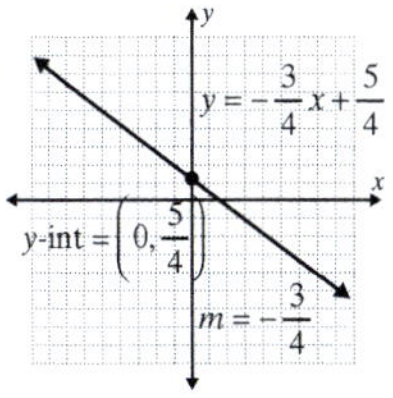

43.
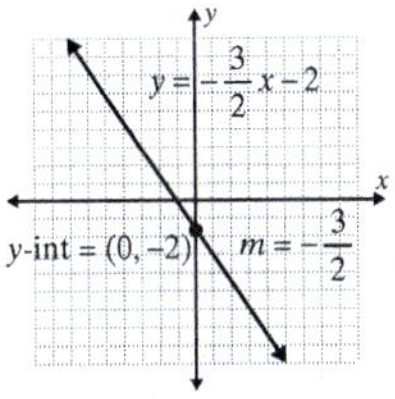

45.
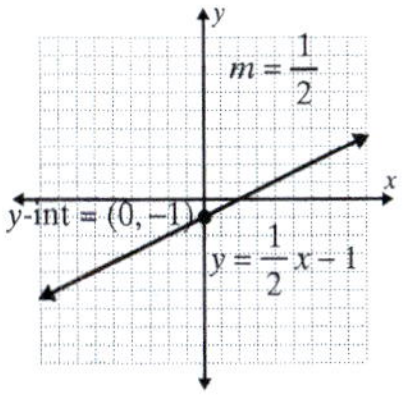

47.
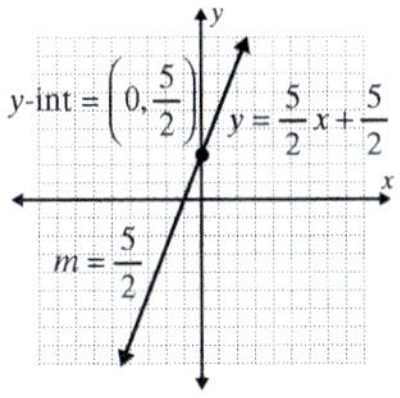

49.
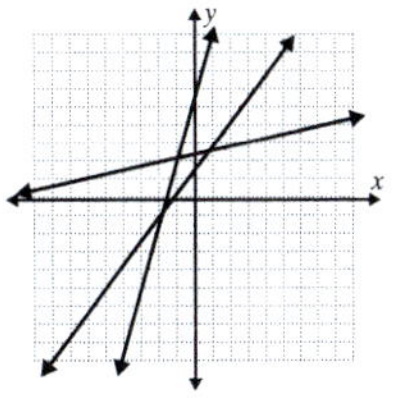

51.
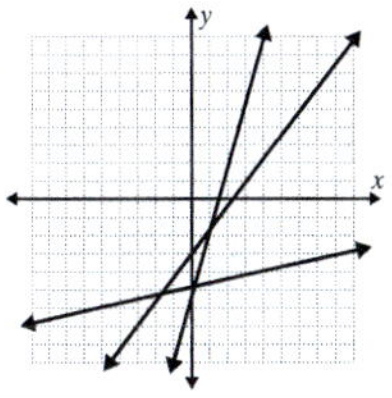

53. $y=-\frac{1}{2}x+3$ **55.** $y=\frac{2}{5}x-3$ **57.** $y=4x-5$ **59.** $y=x-4$ **61.** $y=-\frac{5}{6}x-3$

63. a. $m=\frac{3}{2}$ **b.** $(0,7)$ **c.** $y=\frac{3}{2}x+7$ **65. a.** $m=0$ **b.** $(0,-6)$ **c.** $y=-6$ **67. a.** $m=\frac{1}{2}$ **b.** $(0,-3)$ **c.** $y=\frac{1}{2}x-3$

69. a. $m=-\frac{1}{3}$ **b.** $(0,2)$ **c.** $y=-\frac{1}{3}x+2$ **71.** Yes **73.** No **75.** Yes **77.** \$4000/year

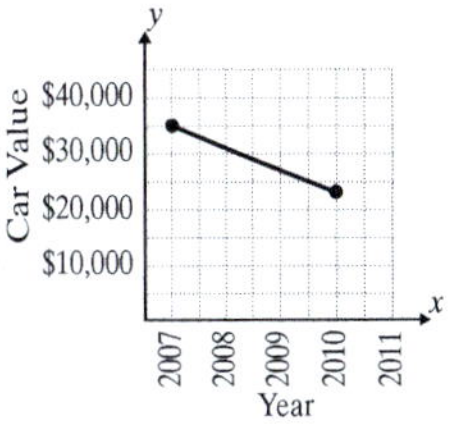

79. a. and **b.**

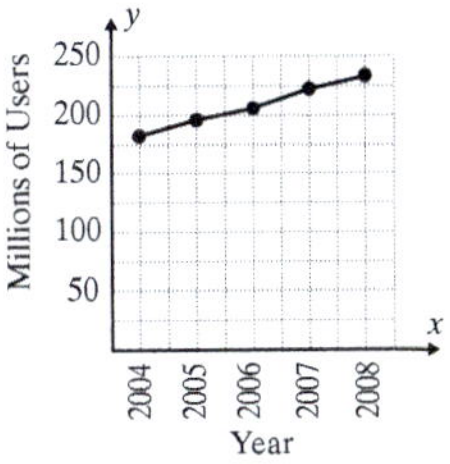

c. 13, 12, 10, 11

d. The number of internet users increased by 13 million people/year from '04-'05, 12 mppy from '05-'06, 10 mppy from '06-'07, and 11 mppy from '07-'08.

81. a. and **b.**

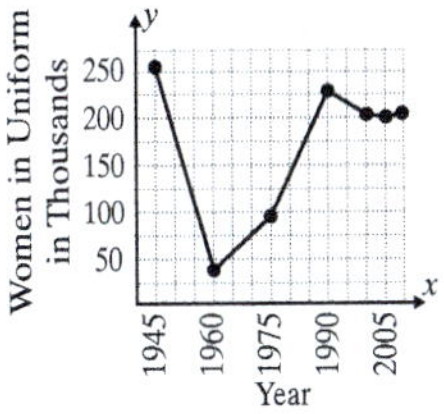

c. −15,647.07; 4354.53; 8676.67; −2441.7; −205.2; 450

d. # of female active duty military personnel decreased by 15,647.07 women/year from 1945-1960; increased by 4354.53 wpy from '60-'75; increased by 8676.67 wpy from '75-'90; decreased by 2441.7 wpy from '90-'00; decreased by 205.2 wpy from '00-'05; and increased by 450 wpy from '05-'09.

83. Answers will vary. **85. a.** The x-axis **b.** The y-axis **87.** A grade of 12% means the slope of the road is 0.12. For every 100 feet of horizontal distance (run) there is 12 feet of vertical distance (rise).

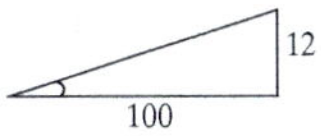

Exercises 4.4, pages 304 - 306

1. a. $m=2$ **b.** $(3,1)$

c.

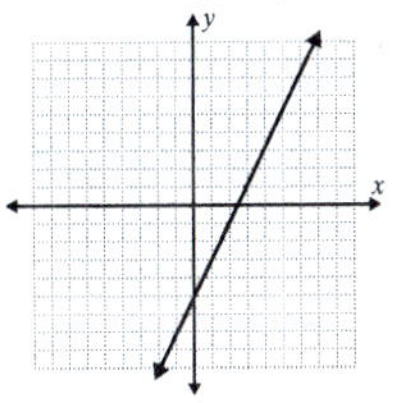

3. a. $m=-5$ **b.** $(0,-2)$

c.

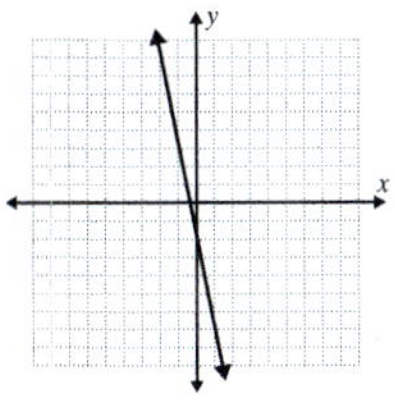

5. a. $m=-\frac{1}{4}$ **b.** $(-2,3)$

c.

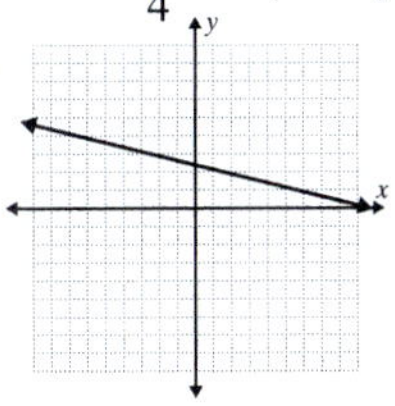

7. $2x+y=-3$

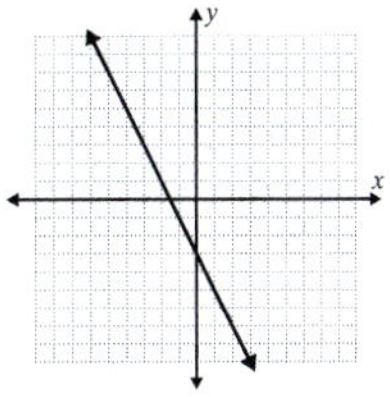

9. $y=-2$

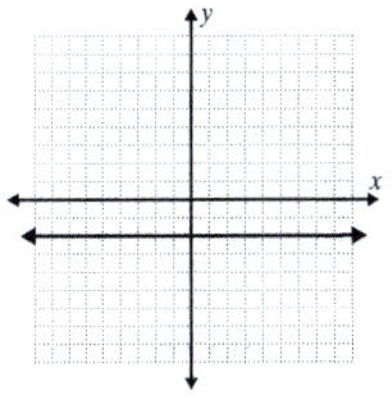

11. $x-2y=-15$

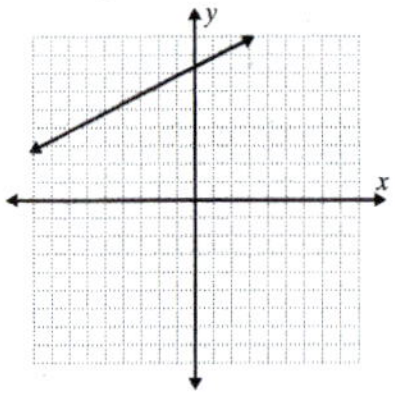

13. $3x-5y=-29$

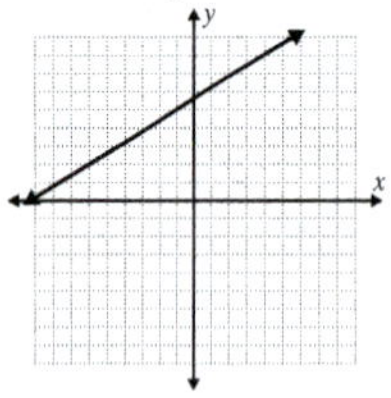

15. $2x-3y=-5$

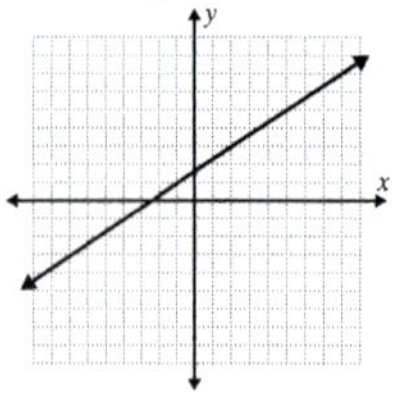

17. $y=\frac{1}{2}x+\frac{9}{2}$ 19. $y=-\frac{1}{7}x+\frac{2}{7}$ 21. $y=-\frac{5}{4}x+2$

23. $y=-5$ 25. $y=-x+4$ 27. $y=6$ 29. $y=7$ 31. $y=7$

33. $y=2x$

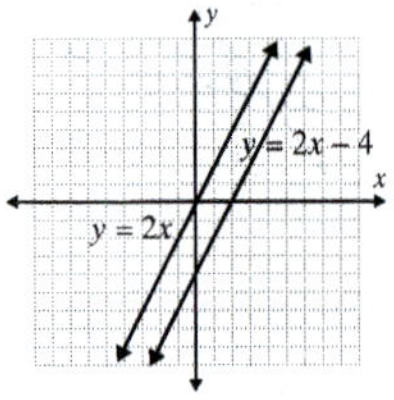

35. $y=5x+2$

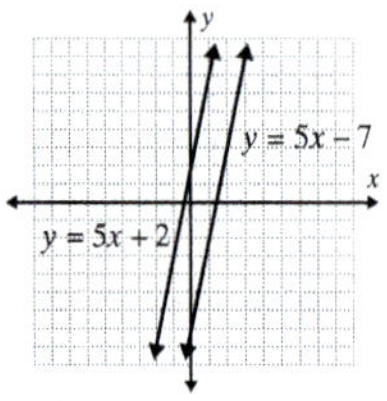

37. $y=-\frac{3}{5}x+\frac{7}{5}$

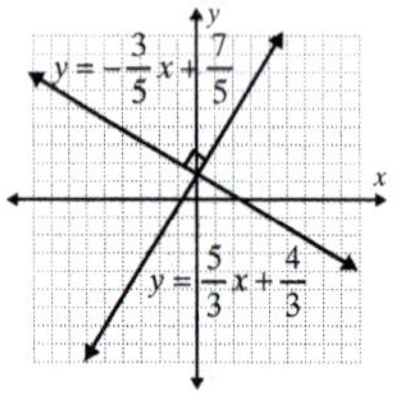

39. $y=-\frac{1}{3}x$ 41. $y=-\frac{1}{2}x-2$

43.

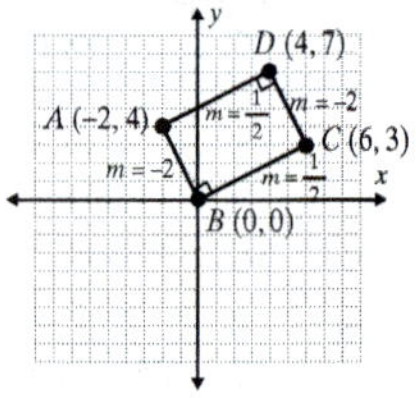

45.

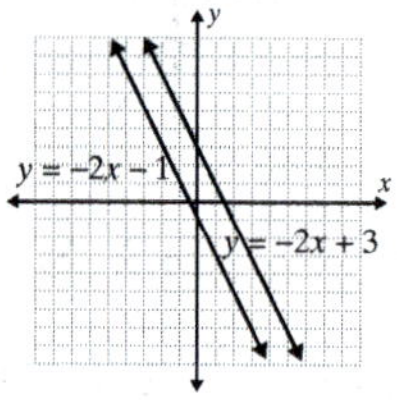

Parallel

47.

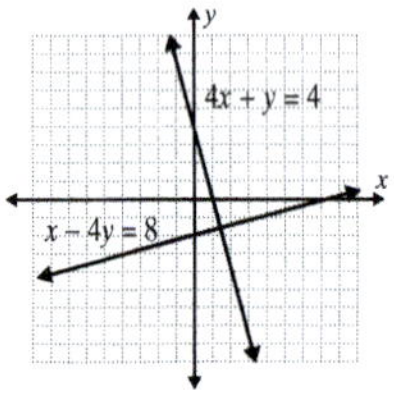

Perpendicular

49.

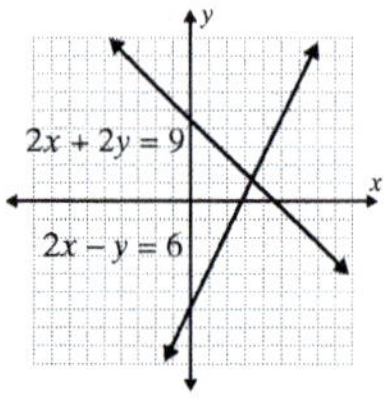

Neither

51. The Americans with Disabilities Act recommends a slope of $\frac{1}{12}$ for business use and a slope of $\frac{1}{6}$ for residential use. Answers will vary.

Exercises 4.5, pages 320 - 325

1. $\{(-4,0),(-1,4),(1,2),(2,5),(6,-3)\}$; $D=\{-4,-1,1,2,6\}$; $R=\{-3,0,2,4,5\}$; Function
3. $\{(-5,-4),(-4,-2),(-2,-2),(1,-2),(2,1)\}$; $D=\{-5,-4,-2,1,2\}$; $R=\{-4,-2,1\}$; Function
5. $\{(-4,-3),(-4,1),(-1,-1),(-1,3),(3,-4)\}$; $D=\{-4,-1,3\}$; $R=\{-4,-3,-1,1,3\}$; Not a function
7. $\{(-5,-5),(-5,3),(0,5),(1,-2),(1,2)\}$; $D=\{-5,0,1\}$; $R=\{-5,-2,2,3,5\}$; Not a function

9. $D=\{-3,0,1,2,4\}$; $R=\{-2,-1,0,5,6\}$; Function

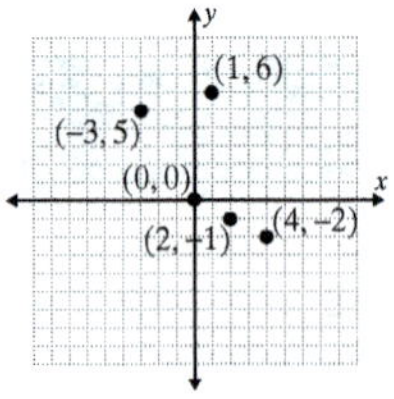

11. $D=\{-4,-3,1,2,3\}$; $R=\{4\}$; Function

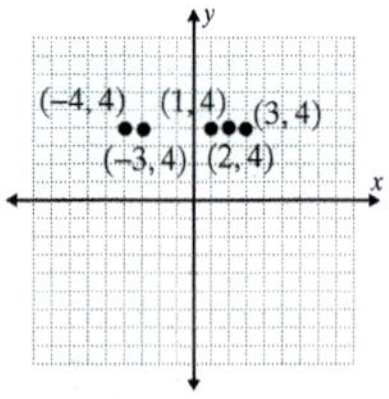

13. $D=\{-3,-1,0,2,3\}$; $R=\{1,2,4,5\}$; Function

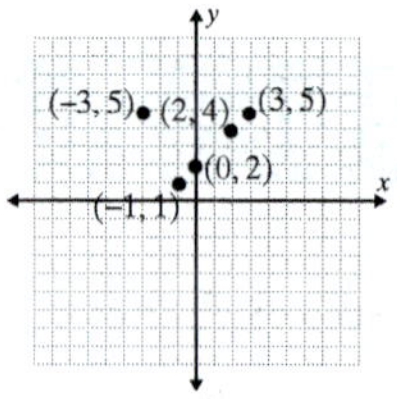

15. $D=\{-1\}$; $R=\{-2,0,2,4,6\}$; Not a function

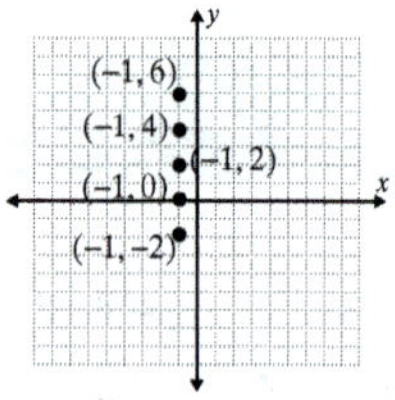

17. Function; $D=(-\infty,\infty)$; $R=[0,\infty)$ 19. Function; $D=(-\infty,\infty)$; $R=(-\infty,\infty)$ 21. Not a function; $D=(-\infty,\infty)$; $R=(-\infty,\infty)$

23. Not a function; $D=(-\infty,\infty)$; $R=(-\infty,\infty)$ 25. Function; $D=[-5,5]$; $R=[-2,2]$

27. Not a function; $D=\{-3\}$; $R=(-\infty,\infty)$ 29. $\left\{(-9,-26),\left(-\frac{1}{3},0\right),(0,1),\left(\frac{4}{3},5\right),(2,7)\right\}$

31. $\{(-2,-11),(-1,-2),(0,1),(1,-2),(2,-11)\}$ **33.** $D=(-\infty,\infty)$ **35.** $D=(-\infty,0)\cup(0,\infty)$ or $x\neq 0$

37. $D=(-\infty,3)\cup(3,\infty)$ or $x\neq 3$ **39. a.** -4 **b.** -16 **c.** -10 **41. a.** 0 **b.** 12 **c.** 56 **43. a.** -3 **b.** 0 **c.** 3

45.
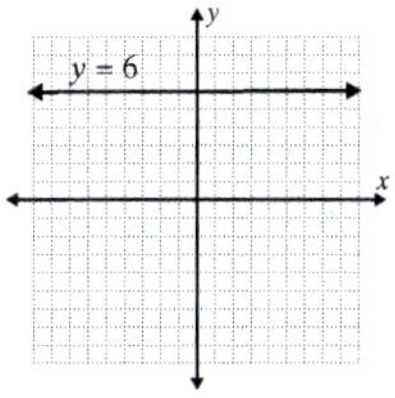

47.
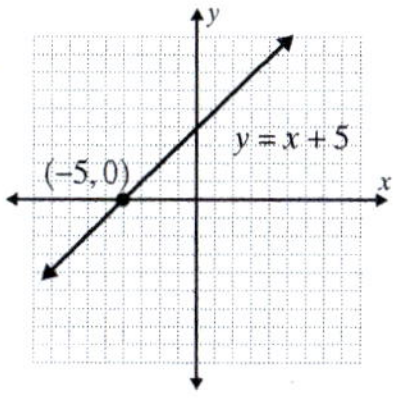

49.
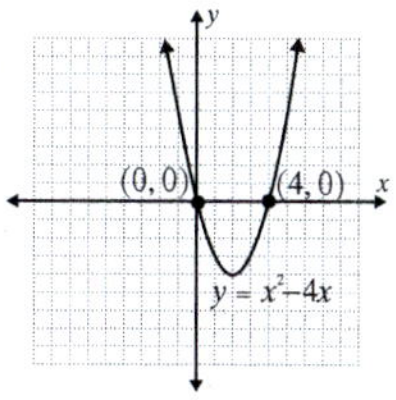

51.
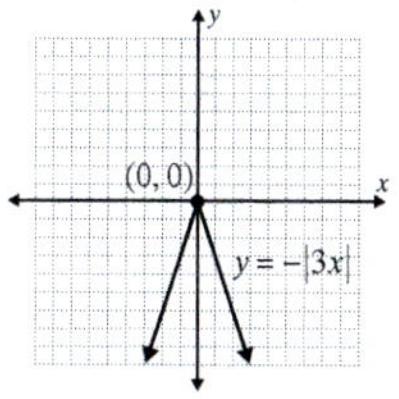

53.
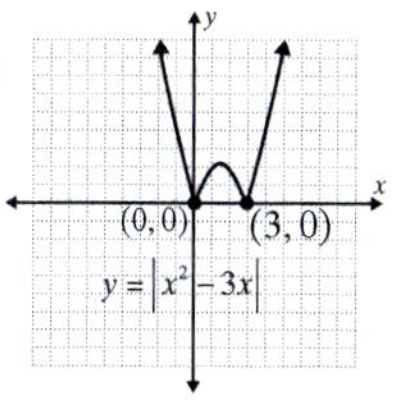

55.
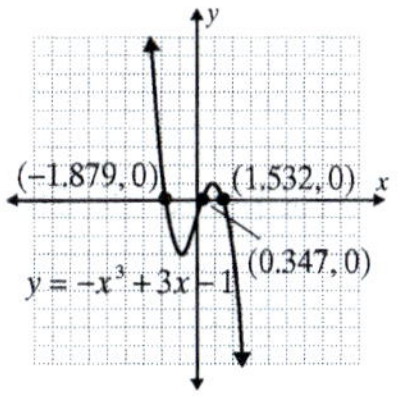

57.
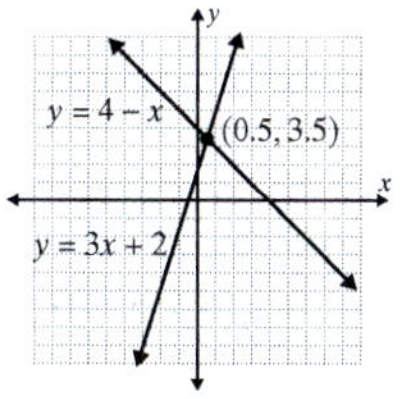

59.
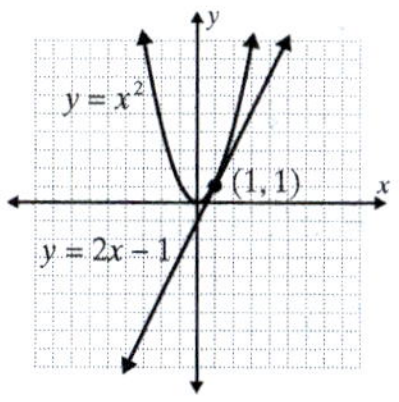

61. y-intercept $=(0,-5)$ (should be $(0,5)$)

63. y-intercept $=(0,-8)$ (should be $(0,-2)$)

65. slope $=-3$ (should be the reciprocal, $-\frac{1}{3}$) and y-intercept $=(0,2)$ (should be $(0,0)$)

Exercises 4.6, pages 331 - 332

1.
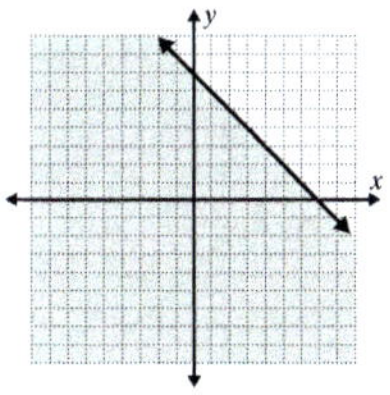

3.
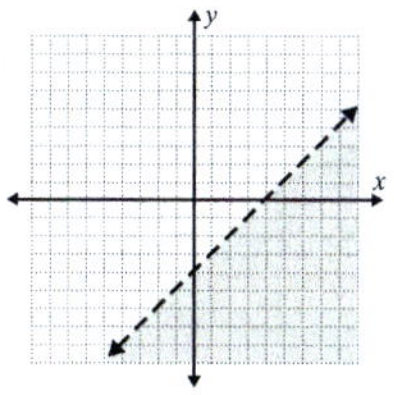

5.
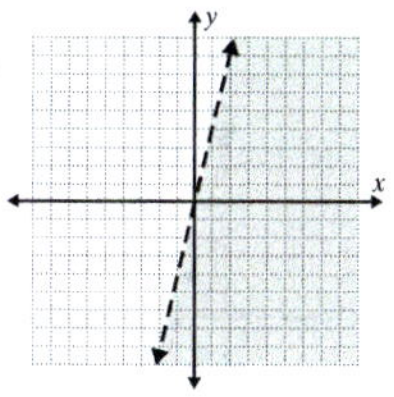

7.
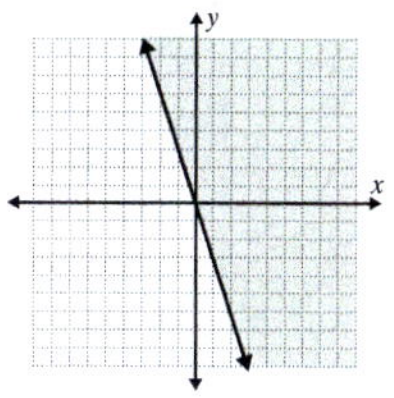

9.
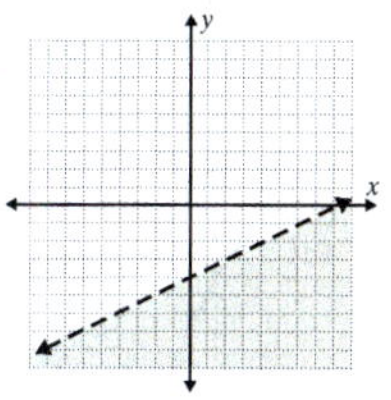

11.
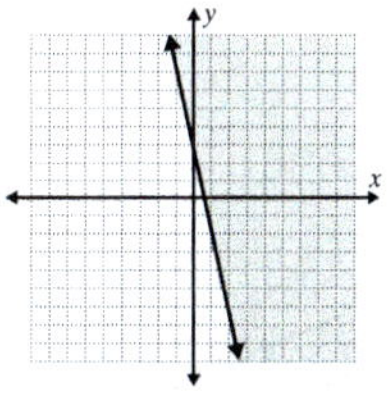

13.
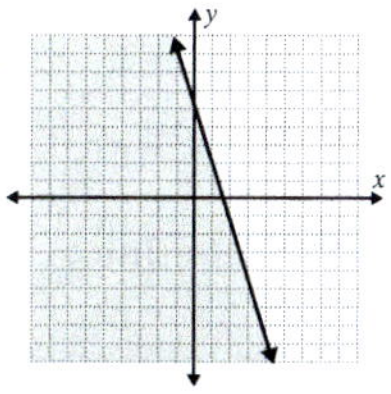

15.
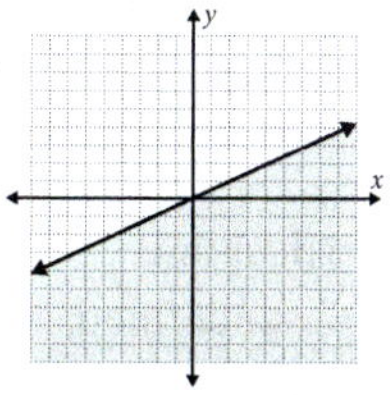

17.
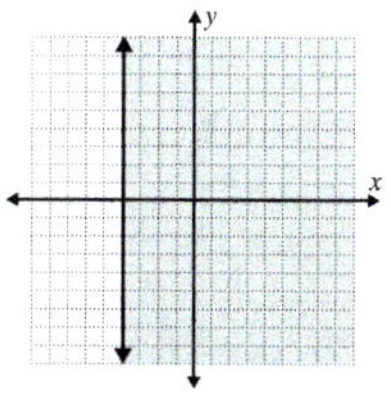

19.
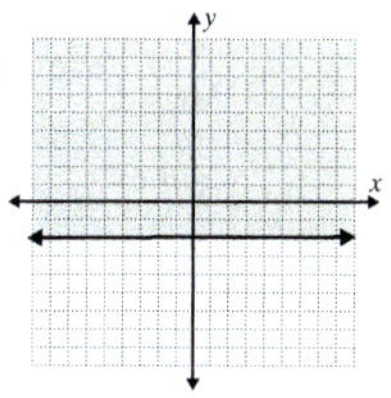

21.
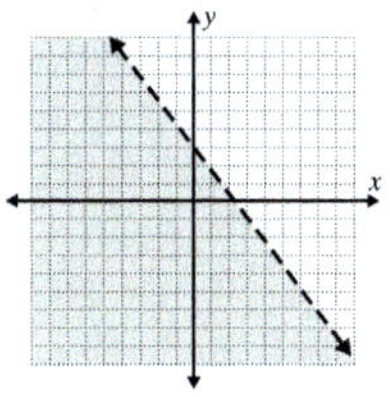

23.
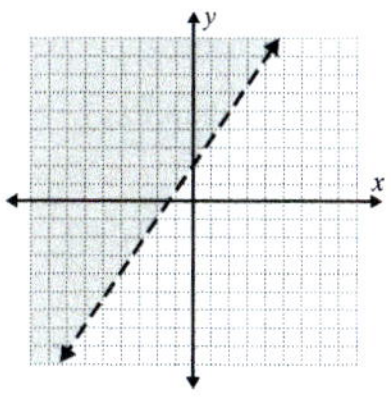

25.
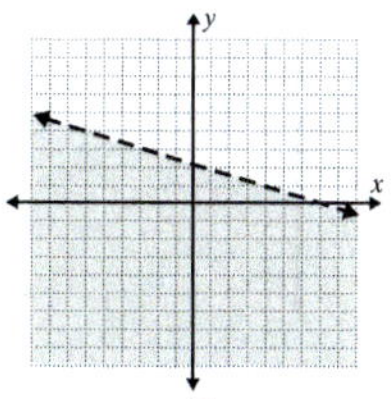

27.
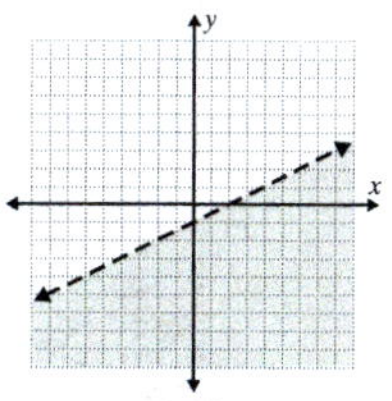

29.
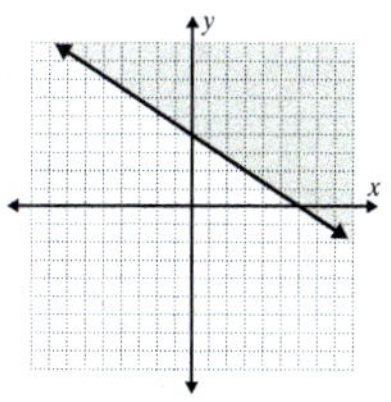

31.
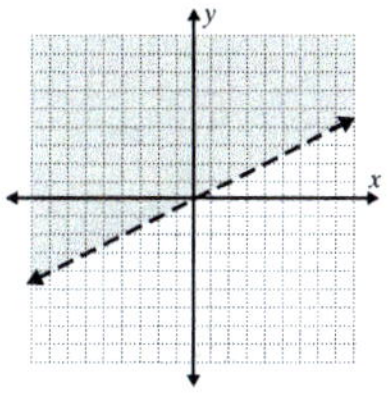

33.
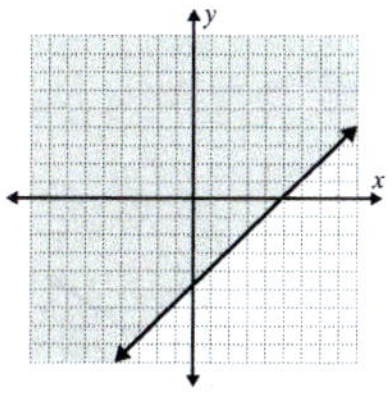

35.
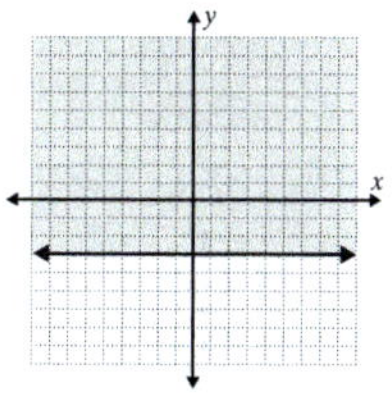

37.
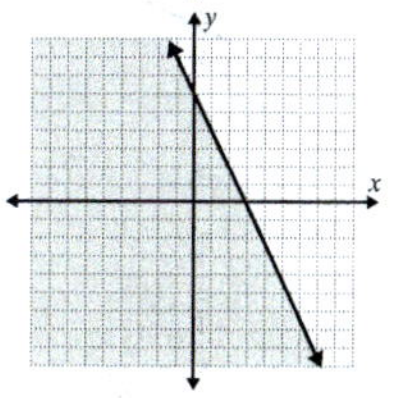

39.
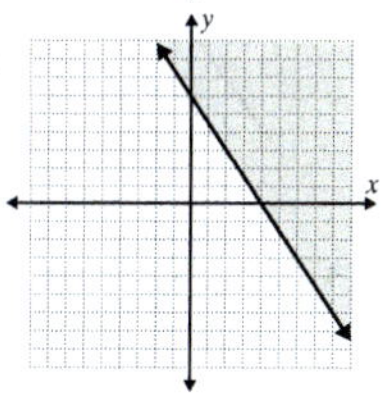

41. Test any point not on the line. If the test point satisfies the inequality, shade the half-plane on that side of the line. Otherwise, shade the other half-plane.

Chapter 4 Review, pages 337 - 343

1. $\{A(-6,-5), B(-5,6), C(1,-3), D(3,6), E(7,-1)\}$ **2.** $\{A(-4,-5), B(-3,1), C(1,2), D(1,-1), E(5,3)\}$

3. $\{A(-7,2), B(-1,7), C(-1,-3), D(3,-1), E(4,-6)\}$ **4.** $\{A(-2,3), B(2,-3), C(3,-4), D(4,-5), E(5,4)\}$

5.

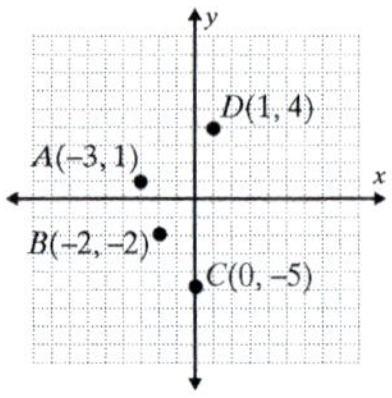

6.

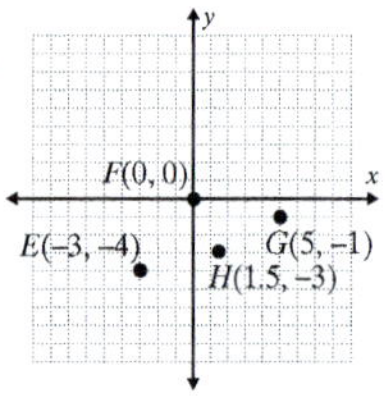

7.

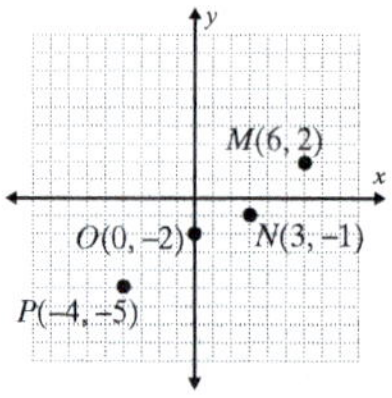

8.

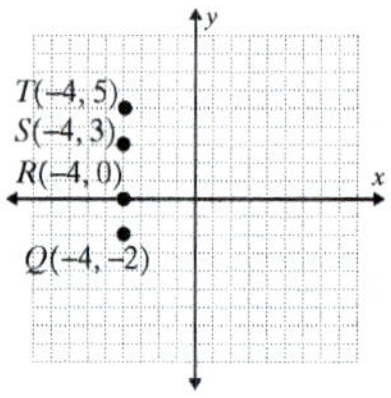

9. None **10.** c

11. a, c, d

12. a, b, c

13. $(0,0), (-1,-5), (-2,-10), (1,5)$

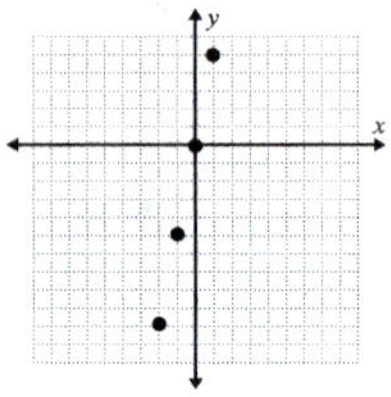

14. $(0,6), (2,0), (-1,9), (4,-6)$

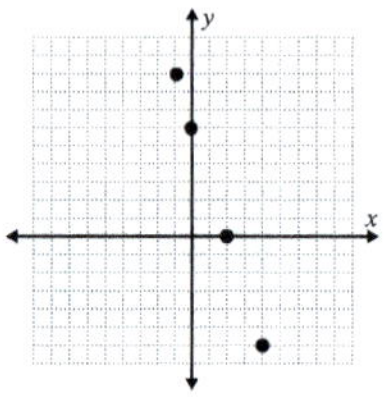

15. $(0,8), (4,0), (3,2), \left(\frac{17}{4}, -\frac{1}{2}\right)$

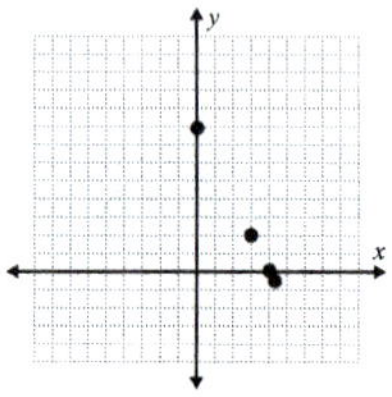

16. $(0,-5), \left(-\frac{4}{3}, -1\right), (-3,4), (-1.5,-0.5)$

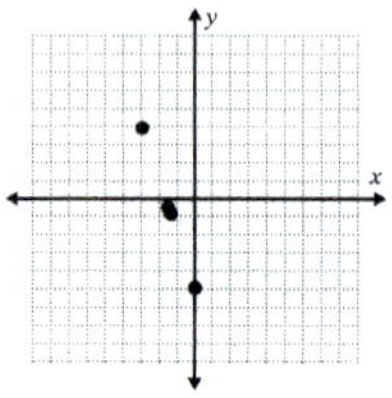

17. c **18.** b **19.** d **20.** a

21.

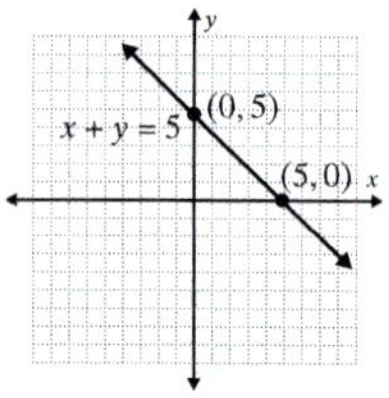

22.

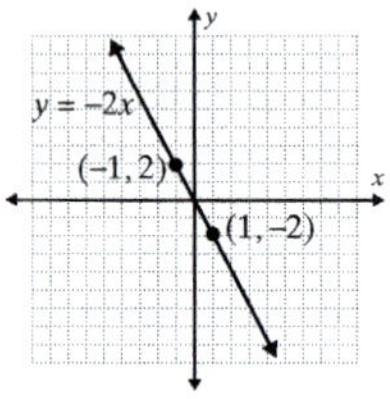

23.

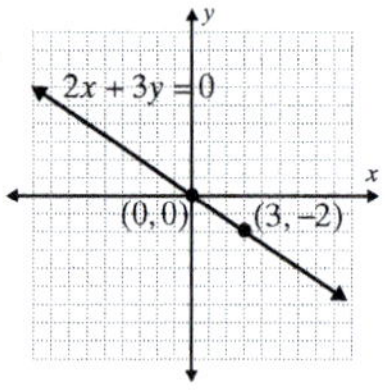

24.

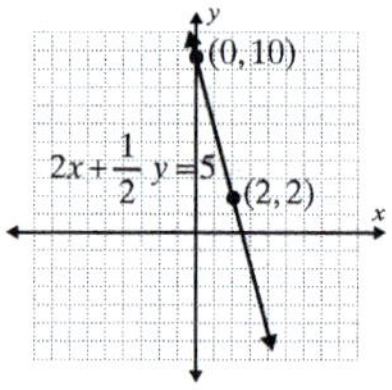

25.

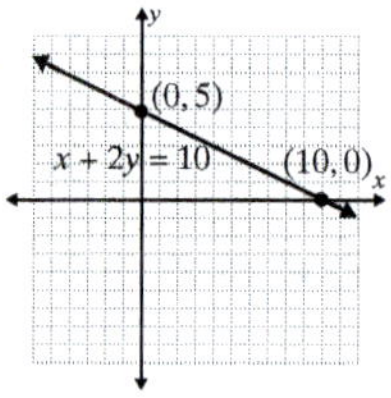

26.

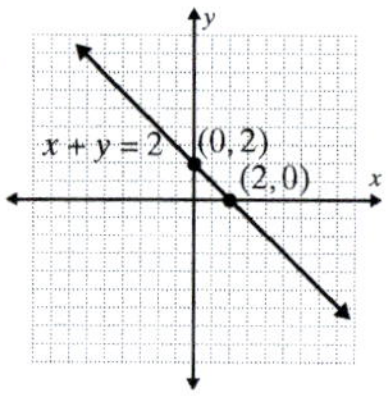

27.

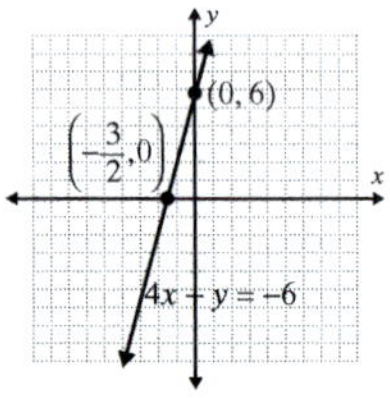

28.

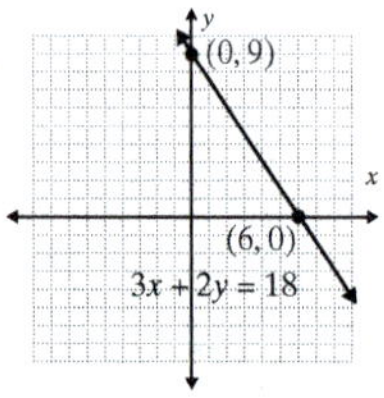

29. $m=-2$ **30.** $m=3$ **31.** $m=4$ **32.** $m=-\frac{1}{2}$

33. Vertical line; m is undefined.

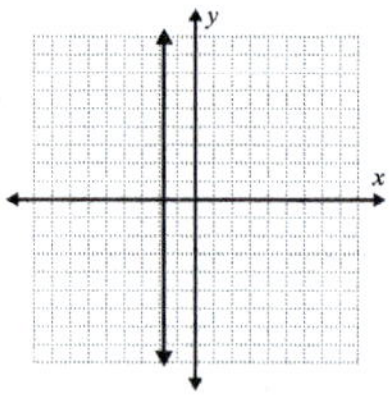

34. Horizontal line; $m=0$

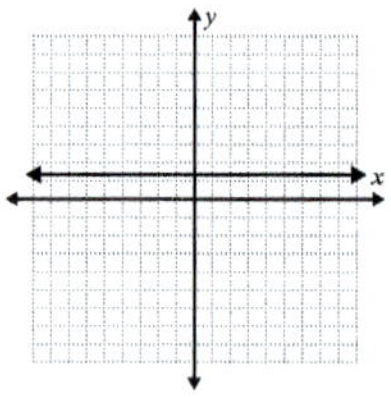

35.

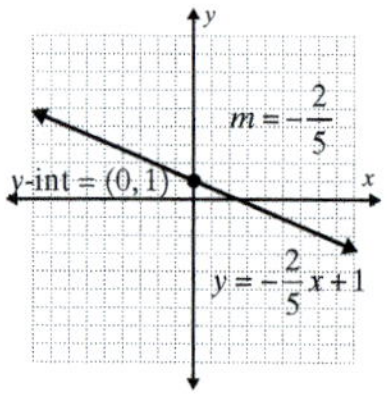

36.

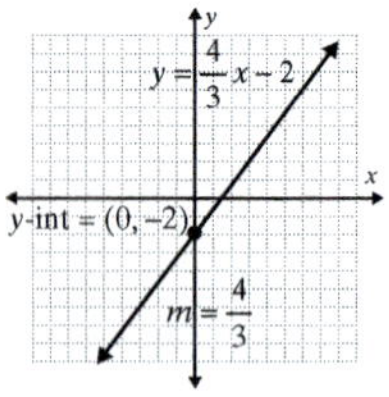

37.

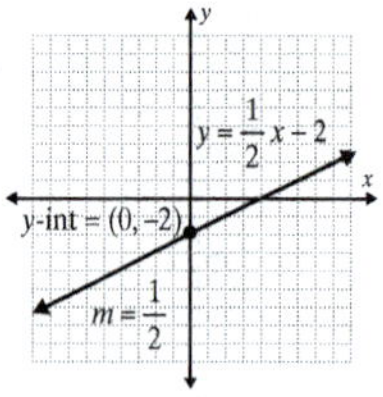

38.

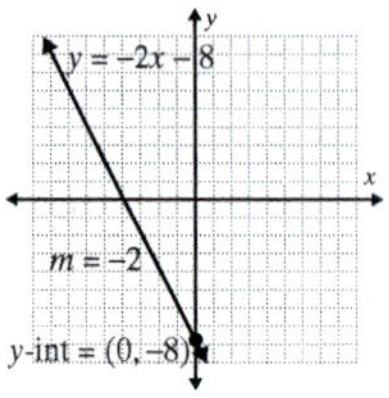

39.

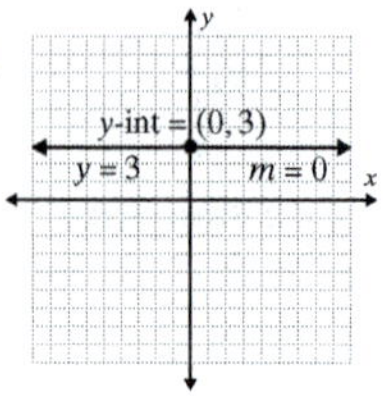

40.

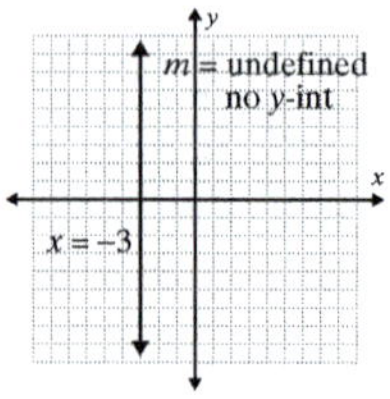

41.

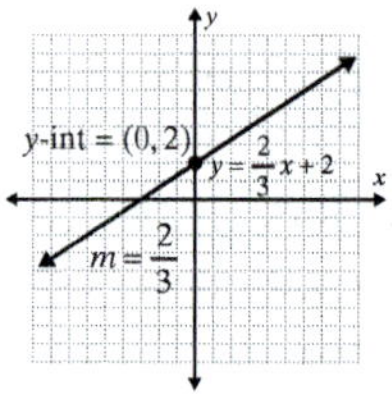

42.

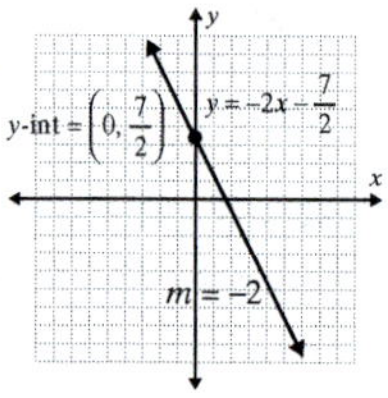

43. a. $m=-\frac{3}{2}$ **b.** $(0,5)$ **c.** $y=-\frac{3}{2}x+5$ **44. a.** m is undefined **b.** No y-intercept **c.** $x=-2$ **45. a.** $m=0$ **b.** $(0,4)$ **c.** $y=4$

46. a. $m=\frac{2}{3}$ **b.** $(0,-3)$ **c.** $y=\frac{2}{3}x-3$

47. a. $m=3$ **b.** $(-1,7)$ **48. a.** $m=-1$ **b.** $(1,-3)$ **49.** $3x+y=-11$ **50.** $5x-2y=19$ **51.** $x=-4$

c.

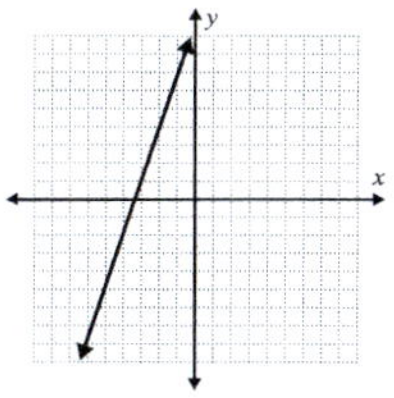

c.

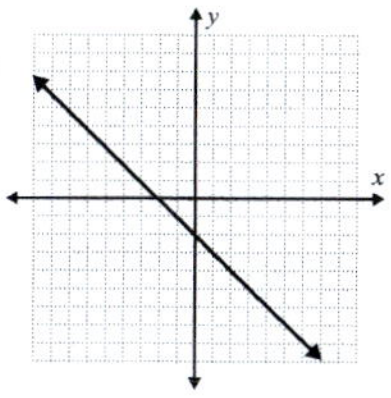

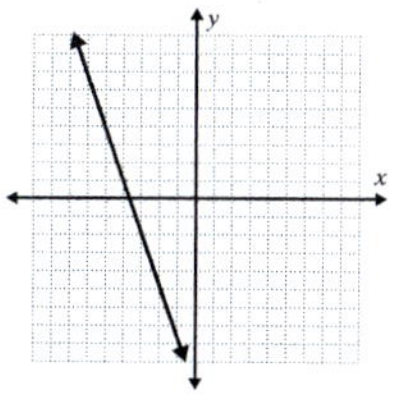

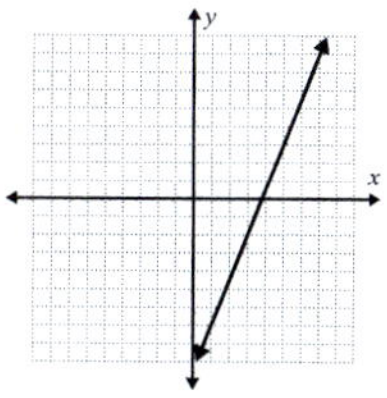

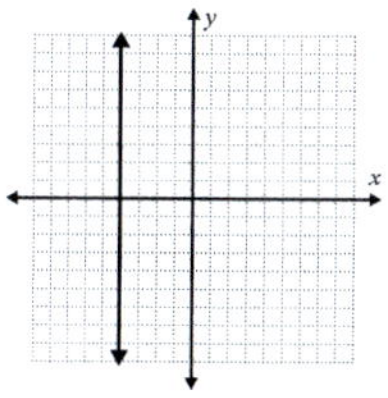

52. $y=-\frac{3}{2}$

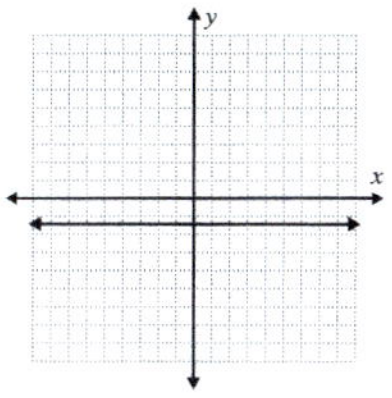

53. $y=-2x-1$ **54.** $y=-\frac{1}{2}x+4$ **55.** $y=-\frac{1}{8}x+\frac{1}{12}$ **56.** $y=\frac{14}{15}x-\frac{1}{2}$ **57.** $y=5$ **58.** $x=5$

59. $y=\frac{1}{4}x-\frac{1}{2}$

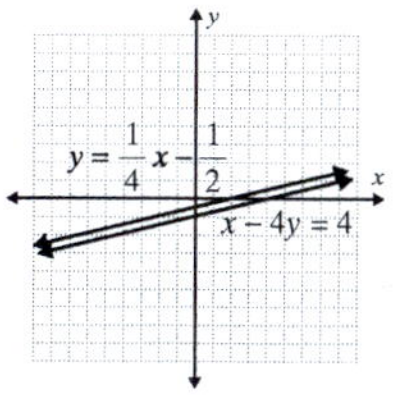

60. $y=-x-3$

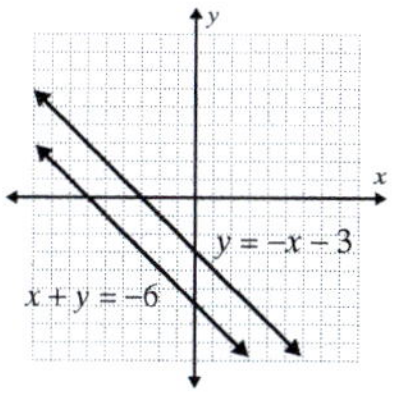

61. $y=\frac{1}{2}x+\frac{1}{2}$

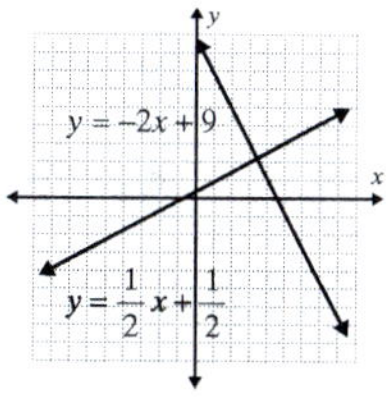

62. $y=2x+6$

63. $y=-\frac{2}{3}x-4$

64. $y=-\frac{1}{3}x+5$

65. Perpendicular

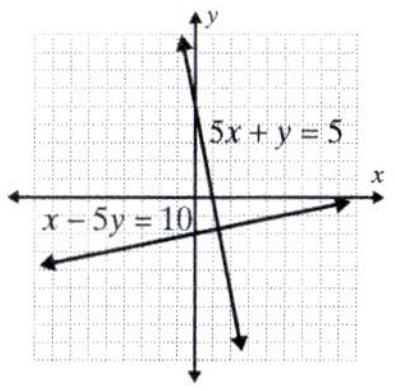

66. Neither

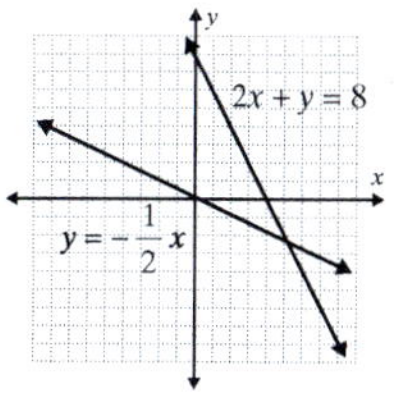

67. $\left\{\begin{matrix}(-5,-1),(-2,3),\\(0,0),\ (4,1),\\(4,-6)\end{matrix}\right\}$

$D=\{-5,-2,0,4\}$;

$R=\{-6,-1,0,1,3\}$;

Not a function

68. $\left\{\begin{matrix}(-5,-3),(-2,0),\\(0,-3),(2,4),\\(6,1)\end{matrix}\right\}$;

$D=\{-5,-2,0,2,6\}$;

$R=\{-3,0,1,4\}$;

Function

69. $D=\{-5,-3,0,1\}$;

$R=\{-2,0,1,2\}$;

Not a function

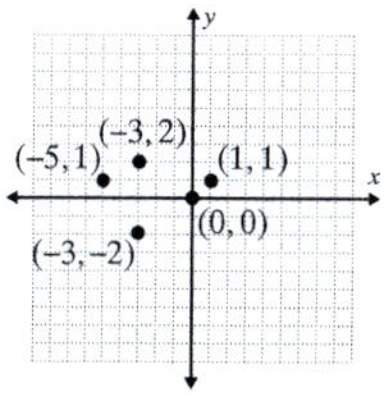

70. $D=\{-4,-3,0,2\}$;

$R=\{-2,0.1,3.2\}$;

Not a function

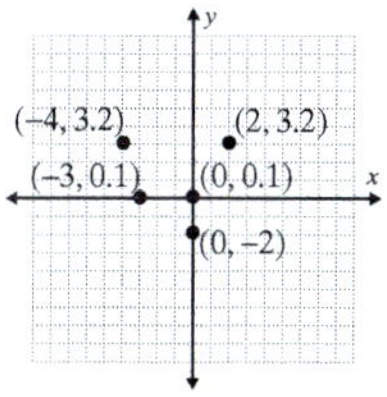

71. Function; $D=(-\infty,\infty)$; $R=(-\infty,\infty)$

72. Not a function; $D=(-\infty,-3]\cup[3,\infty)$; $R=(-\infty,\infty)$ **73.** Function; $D=(-\infty,\infty)$; $R=[-3,3]$

74. Function; $D=(-\infty,\infty)$; $R=(-\infty,4]$ **75. a.** 18 **b.** −17 **c.** −9 **76. a.** 6 **b.** 20 **c.** 6

77.

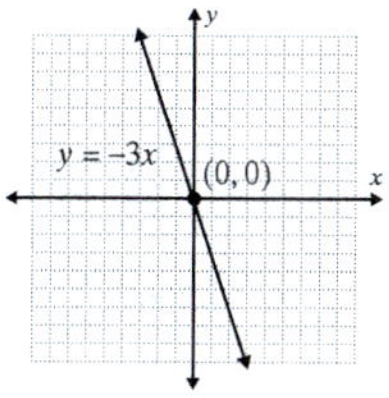

78.

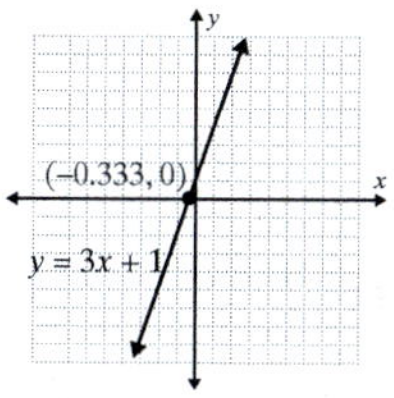

79.

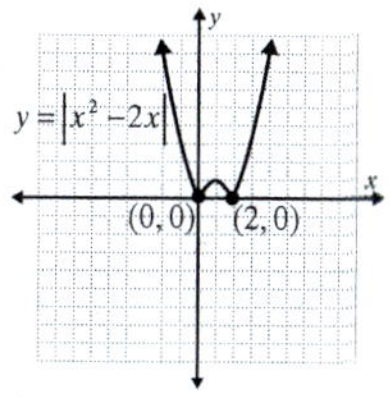

80.

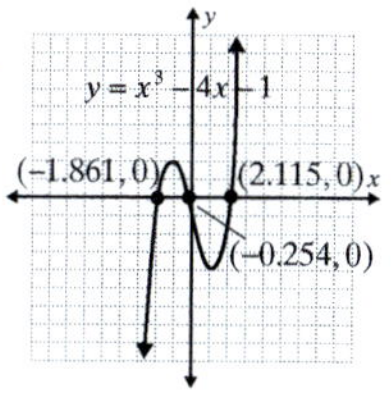

81.

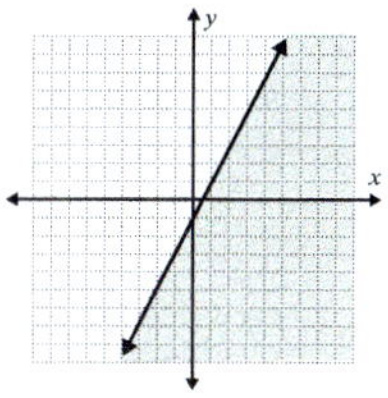

82.

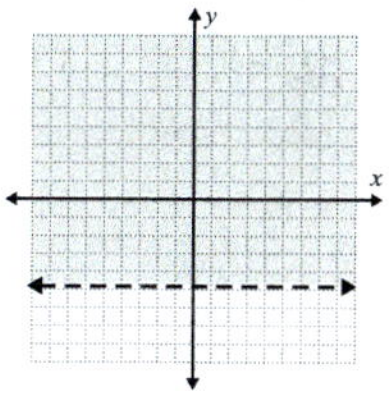

83.

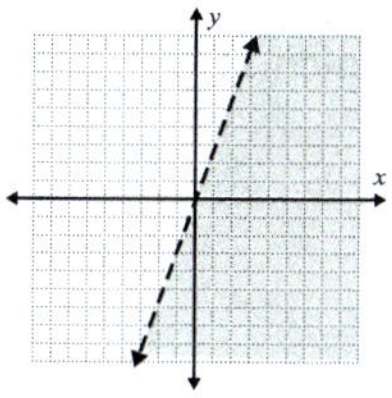

84.

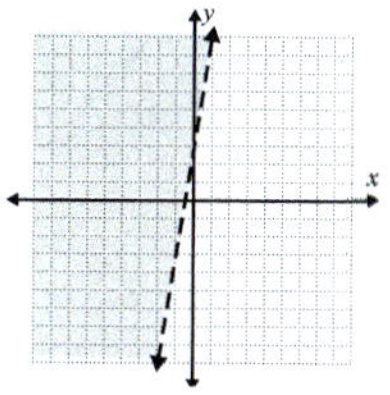

85.

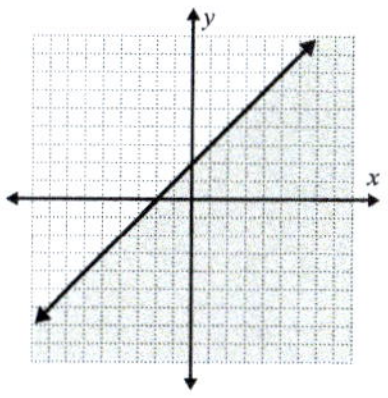

86. 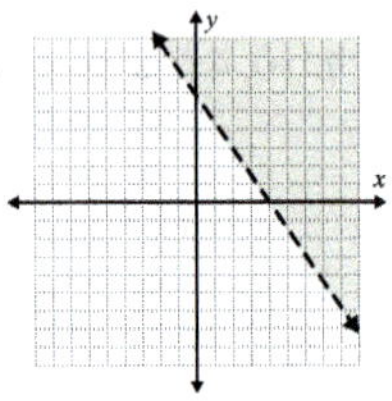

87. 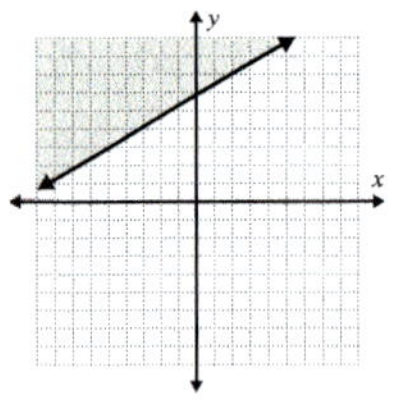

88. 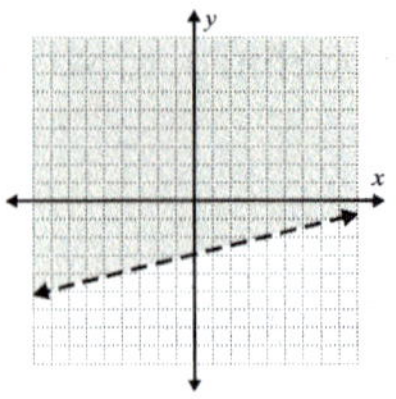

89. 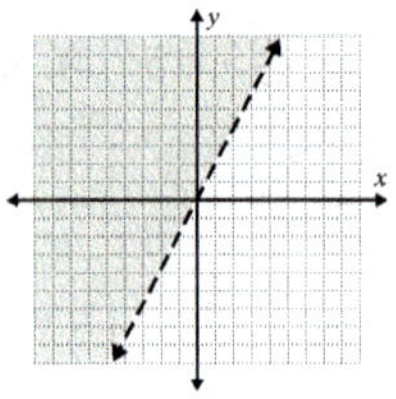

90. 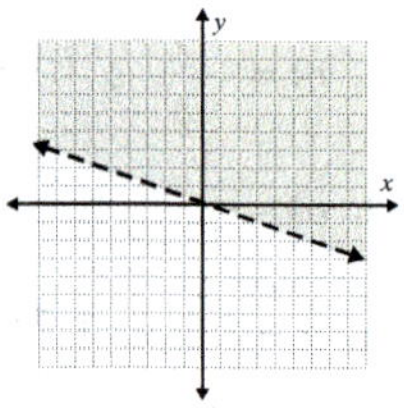

91. 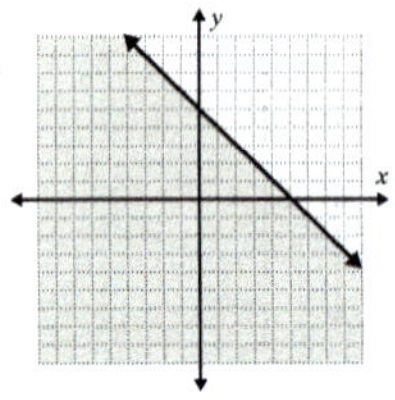

92. 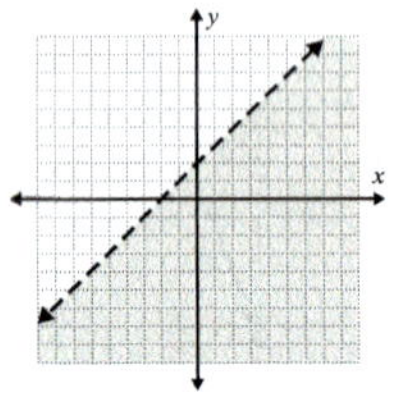

Chapter 4 Test, pages 344 - 346

1. $\{A(-3,2), B(-2,-1), C(0,-3), D(1,2), E(3,4), F(5,0), G(4,-2)\}$

2. 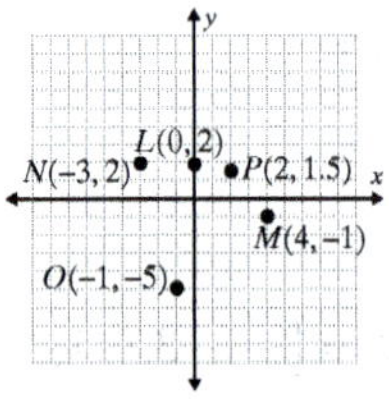

3. a. $(0,2)$ b. $\left(\frac{2}{3},0\right)$ c. $(-2,8)$ d. $(3,-7)$

4. a. $\left(0,-\frac{6}{5}\right)$ b. $(6,0)$ c. $(11,1)$ d. $(-4,-2)$

5. 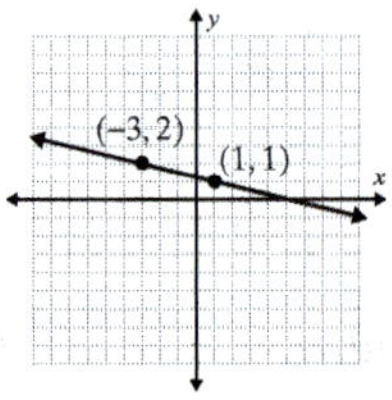

6. 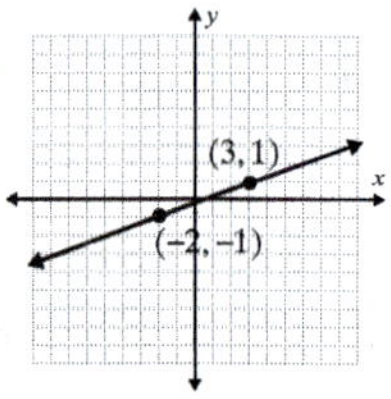

7. 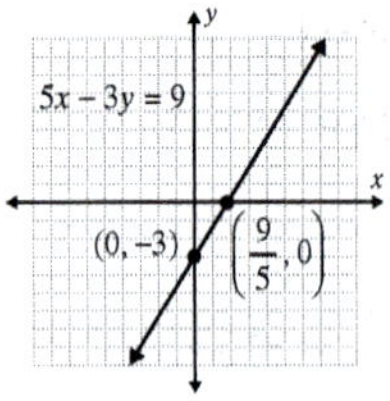

8. 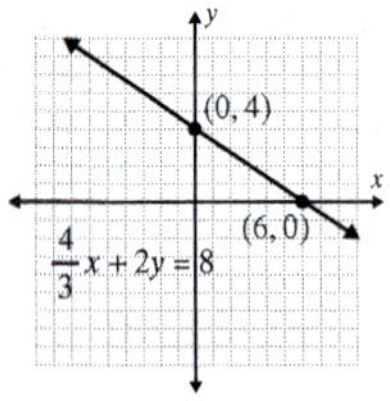

9. $m=\frac{9}{8}$ 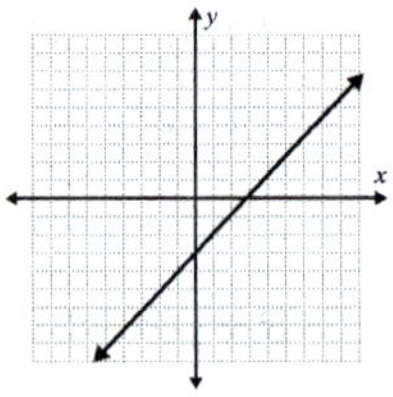

10. $m=-\frac{1}{5}$ 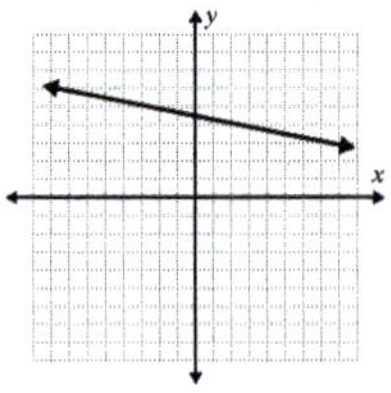

11. 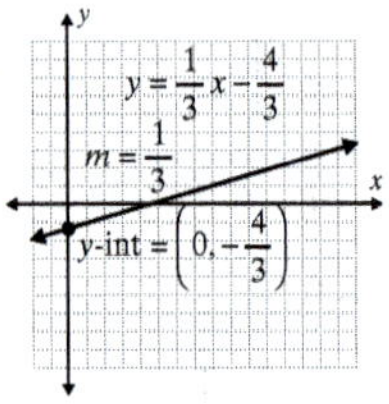

12. 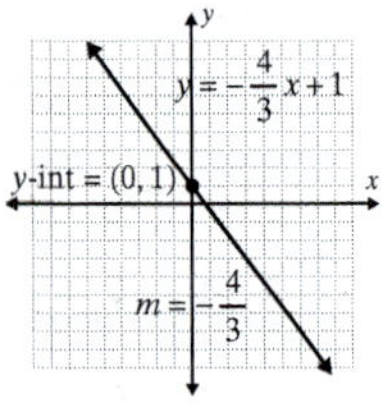

13. a. and b. 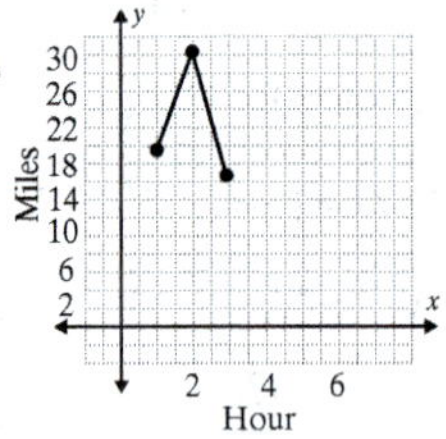

c. 11, −14 d. Sam's speed increased 11 mph from the 1st to the 2nd hour and decreased 14 mph from the 2nd to the 3rd hour.

14. $5x+3y=36$

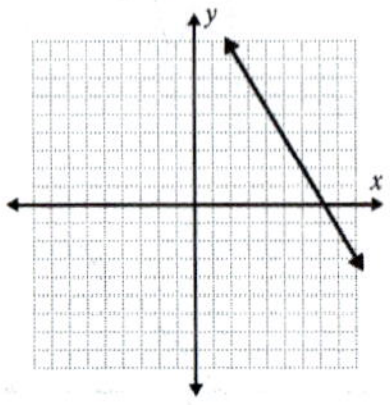

15. $8x+7y=10$

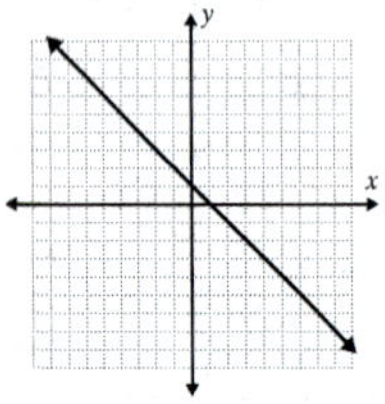

16. $y=6$ 17. $y=-\frac{3}{2}x+7$ 18. $y=-2$ 19. See page 309.

20. $f=\{(-3,2),(-2,-1),(0,-3),(1,2),(3,4),(5,0)\}$;

$D=\{-3,-2,0,1,3,5\}$; $R=\{-3,-1,0,2,4\}$; Function

21. Function; $D=(-\infty,\infty)$; $R=[0,\infty)$ 22. a. 13 b. 5 c. 4

23. a. 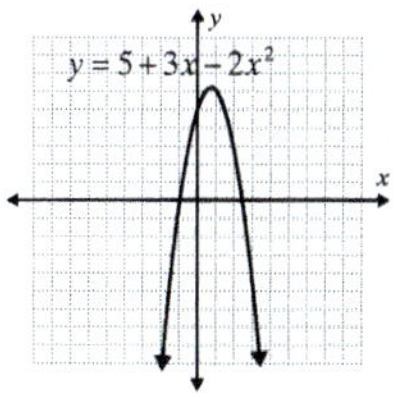

b. $(-1, 0), (2.5, 0)$ **24.** 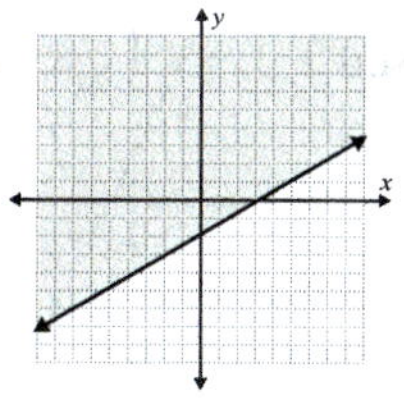
25. 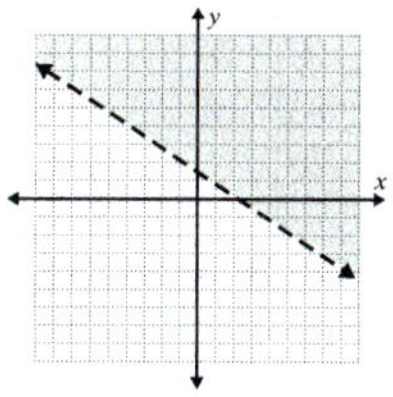
26. 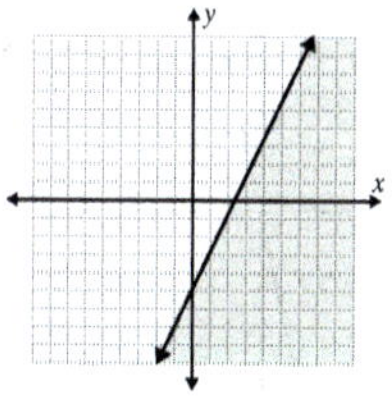

Cumulative Review: Chapters 1-4, pages 347 - 351

1. $93 = 3 \cdot 31$ **2.** $300 = 2^2 \cdot 3 \cdot 5^2$ **3.** $245 = 5 \cdot 7^2$ **4.** 180 **5.** 2100 **6.** $120a^2b^2$ **7.** 13.125 **8.** 83.3 **9. a.** $\{\sqrt{9}\}$ **b.** $\{0, \sqrt{9}\}$
c. $\left\{-10, -\sqrt{25}, -1.6, 0, \frac{1}{5}, \sqrt{9}\right\}$ **d.** $\{-10, -\sqrt{25}, 0, \sqrt{9}\}$ **e.** $\{-\sqrt{7}, \pi, \sqrt{12}\}$ **f.** $\left\{-10, -\sqrt{25}, -1.6, -\sqrt{7}, 0, \frac{1}{5}, \sqrt{9}, \pi, \sqrt{12}\right\}$

10. 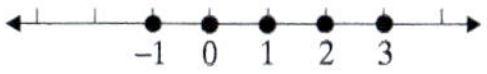

11. 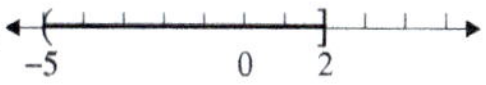

12. Associative property of addition **13.** Additive inverse

14. Multiplicative identity **15.** $x + 15$ **16.** $3x + 8$ **17.** 11 **18.** 22 **19.** −0.8 **20.** $-\frac{1}{12}$ **21.** −40 **22.** $\frac{2}{3}$ **23.** Undefined
24. 0 **25.** −2 **26.** −35 **27.** 26 **28.** −59 **29.** $-2x - 12$ **30.** 0 **31.** $3x^3 - x^2 + 4x - 1$ **32.** $-10x + 4$ **33.** $x = 2$ **34.** $x = -4$
35. $x = -4$ **36.** $x = -8$ **37.** $x = -3.3, 2.3$ **38.** $x = 2, \frac{10}{3}$ **39.** Conditional **40.** Identity **41. a.** $n = 2A - m$ **b.** $f = \frac{\omega}{2\pi}$

42. 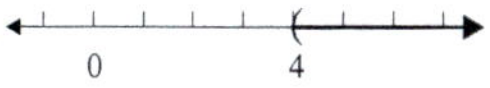

$(4, \infty)$ **43.** 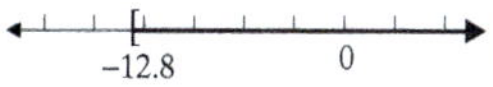

$[-12.8, \infty)$ **44.** 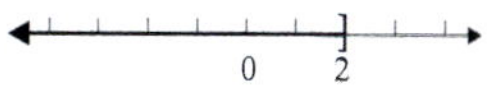

$(-\infty, 2]$

45. 

$(-\infty, -5]$ **46.** 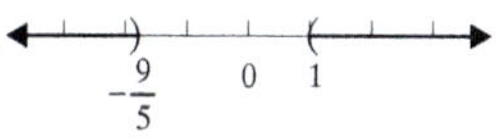

$\left(-\infty, -\frac{9}{5}\right)$ or $(1, \infty)$ **47.** 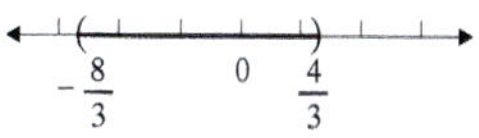

$\left(-\frac{8}{3}, \frac{4}{3}\right)$

48. a. $(0, -4)$ **b.** $(2, 0)$ **c.** $(1, -2)$ **d.** $(3, 2)$ **49. a.** $(0, 2)$ **b.** $(6, 0)$ **c.** $\left(2, \frac{4}{3}\right)$ **d.** $(9, -1)$

50. 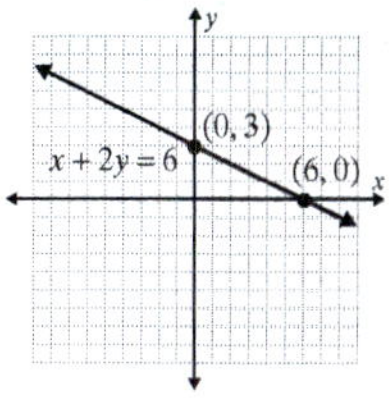

51. 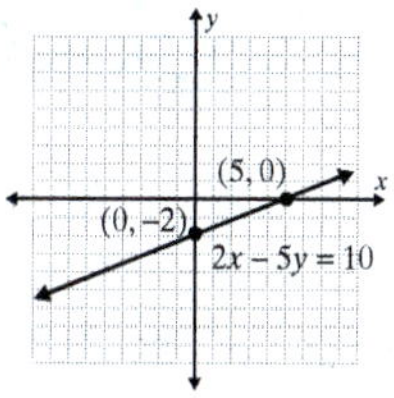

52. 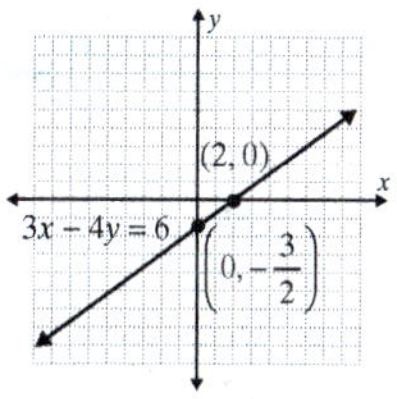

53. 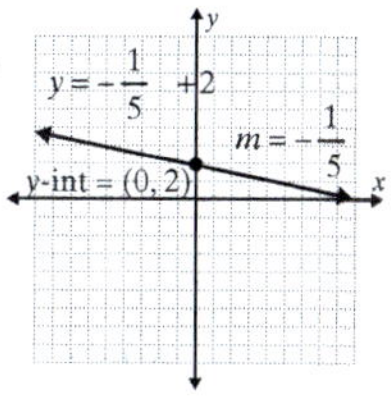

54. 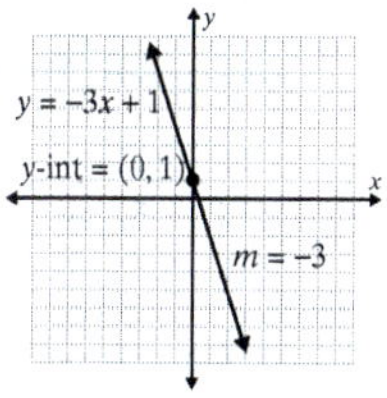

55. 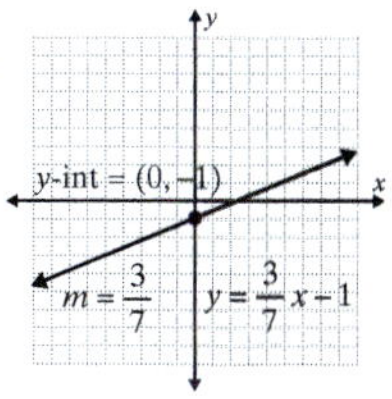

56. $2x - 5y = 17$ **57.** $4x - 3y = -10$ **58.** $2x - y = 0$ **59.** $x = 5$ **60.** $4x + 5y = 15$ **61.** $2x + y = 8$

62. $y = -\frac{3}{2}x + 6$ **63.** $x = 1$ **64.** $y = \frac{3}{4}x - 3$ **65.** $y = -\frac{5}{3}x + 8$

66. $\{(-5, -2), (-3, -2), (-2, 4), (1, -1), (2, 4)\}$; $D = \{-5, -3, -2, 1, 2\}$; $R = \{-2, -1, 4\}$; Function

67. $\{(-3, 1), (-1, -1), (-1, 3), (1, 1), (4, -2), (4, 4)\}$; $D = \{-3, -1, 1, 4\}$; $R = \{-2, -1, 1, 3, 4\}$; Not a function

68. Not a function; $D = (-\infty, 0]$; $R = (-\infty, \infty)$ **69.** Function; $D = (-5, -1] \cup (0, 6]$; $R = \{-4, -2, 2, 4, 5\}$

70. $f(5) = -1$ **71.** $g(3) = 4$ **72.** $F(0) = -10$ **73.** $G(-2) = -32$

74. 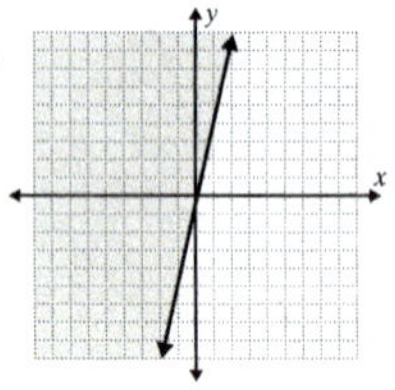

75. 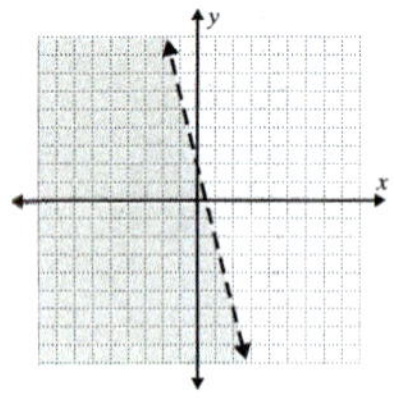

76. Closed 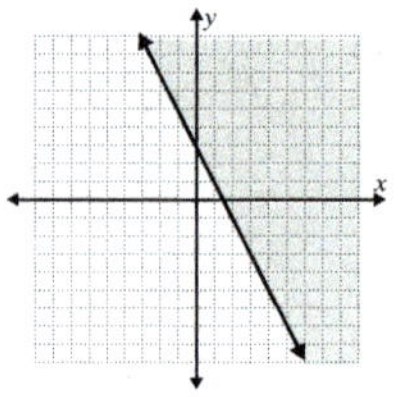

77. Open 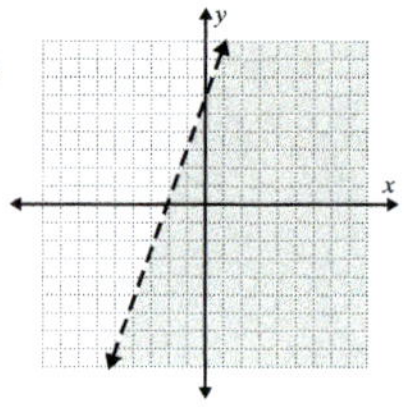

78. 20, 22, 24 **79.** 7 hours **80.** $-\dfrac{1}{3}$ **81.** 6.25% **82.** Length = 25 cm; Width = 16 cm **83.** 84 **84.** 10 miles

85. \$3400 at 4.5%; \$1100 at 7%

86. a. and **b.** 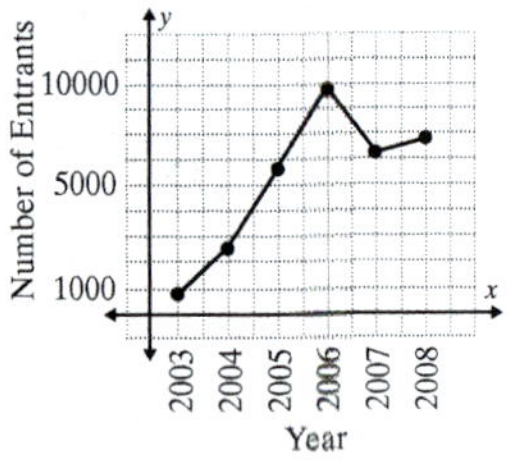

c. 1737; 3043; 3154; –2415; 486 **d.** Number of entrants increased by 1737 people/year from 2003-2004; increased by 3043 people/year from 2004-2005; increased by 3154 people/year from 2005-2006; decreased by 2415 people/year from 2006-2007; increased by 486 people/year from 2007-2008.

87. a. 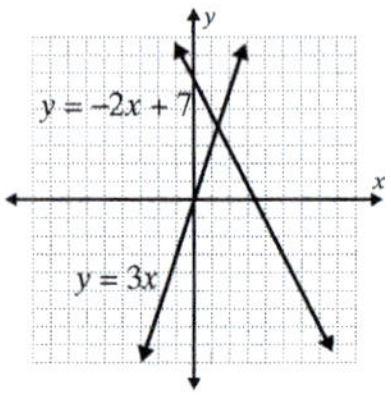

b. $(1.4, 4.2)$

88. a. 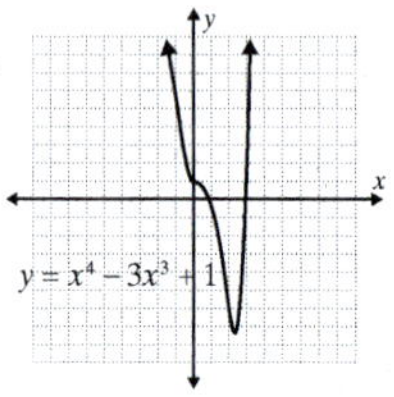

b. $(0.7648, 0)$, $(2.9615, 0)$

Chapter 5

Exercises 5.1, pages 364 - 366

1. 27 **3.** 512 **5.** $\dfrac{1}{3}$ **7.** $\dfrac{1}{25}$ **9.** 16 **11.** 24 **13.** –500 **15.** $\dfrac{3}{8}$ **17.** $-\dfrac{3}{25}$ **19.** x^5 **21.** y^2 **23.** $\dfrac{1}{x^3}$ **25.** $\dfrac{2}{x}$ **27.** $-\dfrac{8}{y^2}$ **29.** $\dfrac{5x^6}{y^4}$

31. 4 **33.** 49 **35.** $\dfrac{1}{10}$ **37.** $\dfrac{1}{8}$ **39.** x^2 **41.** x^2 **43.** x^4 **45.** $\dfrac{1}{x^4}$ **47.** x^6 **49.** x^2 **51.** y^2 **53.** $3x^3$ **55.** x^4 **57.** $36x^3$

59. $-14x^5$ **61.** $-12x^6$ **63.** $4y$ **65.** $3y^2$ **67.** $-2y^2$ **69.** $\dfrac{1}{x^2}$ **71.** 1000 **73.** 1 **75.** $-18x^5y^7$ **77.** $-\dfrac{2y^2}{x}$ **79.** $12a^3b^9c$

81. $-4a^{10}b^3c$ **83.** $\dfrac{5}{x}$ **85.** 1 **87.** 0.390625 **89.** 8875.147264

91. a. $3^2 \cdot 3^2 = 3^4 = 81$, but $6^2 = 36$. **b.** $3^2 \cdot 2^2 = 3 \cdot 3 \cdot 2 \cdot 2 = 6^2$ not 6^4 **c.** $3^2 \cdot 3^2 = 3^4$ not 9^4

Exercises 5.2, pages 377 - 379

1. –81 **3.** –16 **5.** 1,000,000 **7.** $36x^6$ **9.** $-108x^6$ **11.** $\dfrac{5x^2}{y}$ **13.** $-\dfrac{2y^6}{27x^{15}}$ **15.** $\dfrac{27x^3}{y^3}$ **17.** 1 **19.** $4x^4y^4$ **21.** $\dfrac{y^2}{x^2}$ **23.** $\dfrac{1}{3xy^2}$

25. $-\dfrac{x^3y^6}{27}$ **27.** m^2n^4 **29.** $-\dfrac{y^2}{49x^2}$ **31.** $\dfrac{25x^6}{y^2}$ **33.** $\dfrac{x^6}{y^6}$ **35.** $\dfrac{36y^{14}}{x^4}$ **37.** $\dfrac{49y^4}{x^6}$ **39.** $\dfrac{24y^5}{x^7}$ **41.** $\dfrac{4x^3}{3}$ **43.** $\dfrac{16x^{13}}{243y^6}$

45. $96x^2y^4z^4$ **47.** 8.6×10^4 **49.** 3.62×10^{-2} **51.** 1.83×10^7 **53.** 0.042 **55.** 7,560,000 **57.** 851,500,000

59. $(3\times10^2)(1.5\times10^{-4})$; 4.5×10^{-2} **61.** $(3\times10^{-4})(2.5\times10^{-6})$; 7.5×10^{-10} **63.** $\dfrac{3.9\times10^3}{3\times10^{-3}}$; 1.3×10^6 **65.** $\dfrac{1.25\times10^2}{5\times10^4}$; 2.5×10^{-3}

67. $\dfrac{(2\times10^{-2})(3.9\times10^{3})}{1.3\times10^{-2}}$; 6×10^{3} **69.** $\dfrac{(5\times10^{-3})(6.5\times10^{2})(3.3\times10^{0})}{(1.1\times10^{-3})(2.5\times10^{3})}$; 3.9×10^{0} **71.** $\dfrac{(1.4\times10^{-2})(9.22\times10^{2})}{(3.5\times10^{3})(2.0\times10^{6})}$; 1.844×10^{-9}

73. 1.67×10^{-24} grams **75.** 60,000,000,000,000 cells **77.** 5.98×10^{27} grams **79.** 4.0678×10^{16} m **81.** 6.5×10^{-19} grams
83. 3E8 **85.** 8.5E7 **87.** 1E2 **89.** 1.864E2 **91.** See page 370.

Exercises 5.3, pages 384 - 386

1. Monomial **3.** Binomial **5.** Trinomial **7.** Not a polynomial **9.** Binomial
11. $4y$; First-degree monomial; Leading coefficient 4 **13.** x^3+3x^2-2x; Third-degree trinomial; Leading coefficient 1
15. $-2x^2$; Second-degree monomial; Leading coefficient −2 **17.** 0; Monomial of no degree; Leading coefficient 0
19. $6a^5-7a^3-a^2$; Fifth-degree trinomial; Leading coefficient 6 **21.** $2y^3+4y$; Third-degree binomial; Leading coefficient 2
23. 4; Monomial of degree 0; Leading coefficient 4 **25.** $2x^3+3x^2-x+1$; Third-degree polynomial; Leading coefficient 2
27. $4x^4-x^2+4x-10$; Fourth-degree polynomial; Leading coefficient 4 **29.** $9x^3+4x^2+8x$; Third-degree trinomial; Leading coefficient 9 **31.** −16 **33.** −41 **35.** 4379 **37.** 13 **39.** −28 **41.** $3a^4+5a^3-8a^2-9a$ **43.** $3a+11$ **45.** $10a+25$

47. a.
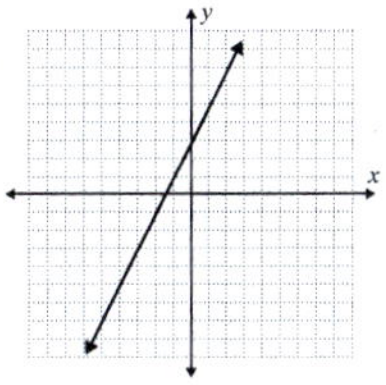
b.
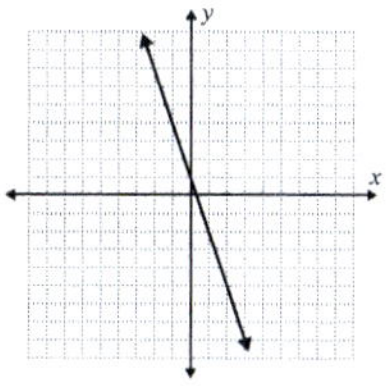
c.
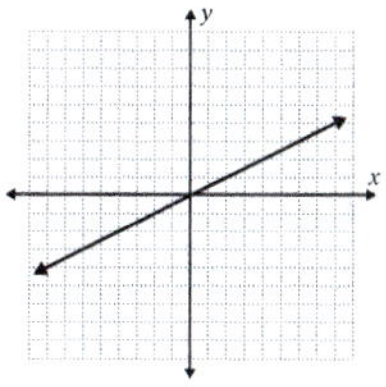

49. a.
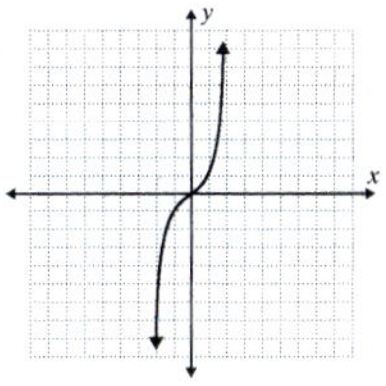
b.
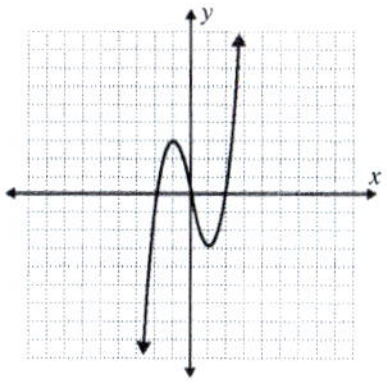
c.
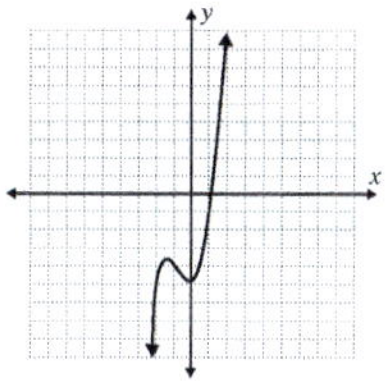

Exercises 5.4, pages 390 - 393

1. $3x^2+7x+2$ **3.** $2x^2+11x-7$ **5.** $3x^2$ **7.** $x^2-5x+17$ **9.** $-x^2+2x-7$ **11.** $2x^3+4x^2-3x-7$ **13.** $-x^2+7x-3$
15. $-x^3+2x^2+3x-4$ **17.** $4x^3+x^2+7$ **19.** $2x^3+11x^2-4x-3$ **21.** x^2+x+6 **23.** $-3x^2-6x-2$ **25.** $3x^2-4x-8$
27. $-x^4-2x^3-16$ **29.** $-4x^4-7x^3-11x^2+5x+13$ **31.** $-4x^2+6x+1$ **33.** $3x^4-7x^3+3x^2-5x+9$ **35.** $6x^2-7x+18$
37. $8x^4+6x^2+15$ **39.** $2x^3+4x^2+3x-10$ **41.** $4x-13$ **43.** $-7x+9$ **45.** $2x+17$ **47.** $3x^3+13x^2+9$ **49.** $8x^2-x-2$
51. $3x-19$ **53.** $x^2-2x+13$ **55.** $7x-3$ **57.** $4x^2+3x+5$ **59.** $7x-18$ **61.** $x^2+11x-8$ **63.** $7x^3-x^2-7$ **65.** Any monomial or algebraic sum of monomials **67.** The largest of the degrees of its terms after like terms have been combined.

Exercises 5.5, pages 397 - 399

1. $-6x^5-15x^3$ **3.** $4x^7-12x^6+4x^5$ **5.** $-y^5+8y-2$ **7.** $-4x^8+8x^7-12x^4$ **9.** $25x^5-5x^4+10x^3$ **11.** $a^7+2a^6-5a^3+a^2$
13. $6x^2-x-2$ **15.** $9a^2-25$ **17.** $-10x^2+39x-14$ **19.** x^3+5x^2+8x+4 **21.** x^2+x-12 **23.** $a^2-2a-48$
25. x^2-3x+2 **27.** $3t^2-3t-60$ **29.** x^3+11x^2+24x **31.** $2x^2-7x-4$ **33.** $6x^2+17x-3$ **35.** $4x^2-9$ **37.** $16x^2+8x+1$

39. y^3+2y^2+y+12 41. $3x^2-8x-35$ 43. $5x^3+6x^2-22x-9$ 45. $2x^4+7x^3+5x^2+x-15$ 47. $3x^2+2x-8$
49. $2x^2+3x-5$ 51. $7x^2-13x-2$ 53. $6x^2-13x-8$ 55. $4x^2+12x+9$ 57. $x^3+3x^2-4x-12$ 59. $4x^2-49$ 61. x^3+1
63. $49a^2-28a+4$ 65. $2x^3+x^2-5x-3$ 67. $x^3+6x^2+11x+6$ 69. a^4-a^2+2a-1 71. $t^4+6t^3+13t^2+12t+4$
73. $2y^2-61$ 75. $3a^2-17a+11$ 77. $-3x-21$ 79. a^2+9a-1 81. Answers will vary.

Exercises 5.6, pages 404 - 407

1. $x^2-14x+49$; Perfect square trinomial 3. $x^2+8x+16$; Perfect square trinomial 5. x^2-9; Difference of two squares
7. x^2-81; Difference of two squares 9. $2x^2+x-3$ 11. $9x^2-24x+16$; Perfect square trinomial
13. $25x^2-4$; Difference of two squares 15. $9x^2-12x+4$; Perfect square trinomial
17. $x^2-16x+64$; Perfect square trinomial 19. $16x^2-25$; Difference of two squares
21. $25x^2-81$; Difference of two squares 23. $x^2-8x+16$; Perfect square trinomial
25. $4x^2-49$; Difference of two squares 27. $10x^4-11x^2-6$ 29. $49x^2+14x+1$; Perfect square trinomial
31. $5x^2+11x+2$ 33. $4x^2+13x-12$ 35. $3x^2-25x+42$ 37. $x^2+10x+25$ 39. x^4-1 41. x^4+6x^2+9
43. x^6-4x^3+4 45. x^4+3x^2-54 47. $x^2-\frac{4}{9}$ 49. $x^2-\frac{9}{16}$ 51. $x^2+\frac{6}{5}x+\frac{9}{25}$ 53. $x^2-\frac{5}{3}x+\frac{25}{36}$ 55. $x^2-\frac{1}{4}x-\frac{1}{8}$
57. $x^2+\frac{5}{6}x+\frac{1}{6}$ 59. $x^2-1.96$ 61. $x^2-5x+6.25$ 63. $x^2-4.6225$ 65. $x^2+2.48x+1.5376$ 67. $2.0164x^2+27.264x+92.16$
69. $129.96x^2-12.25$ 71. $93.24x^2+142.46x-104.04$ 73. a. $A(x)=400-4x^2$ b. $P(x)=4(20-2x)+8x=80$
75. $A(x)=8x+15$ 77. a. $A(x)=150-4x^2$ b. $P(x)=2(10-2x)+2(15-2x)+8x=50$ c. $V(x)=150x-50x^2+4x^3$
79. As indicated in the diagram $(x+5)^2=x^2+2(5x)+5^2$ Answers will vary.

Exercises 5.7, pages 413 - 415

1. y^2-2y+3 3. $2x^2-3x+1$ 5. $10x^3-11x^2+x$ 7. $-4x^2+7x-\frac{5}{2}$ 9. $y^3-\frac{7}{2}y^2-\frac{15}{2}y+4$ 11. $x-6+\frac{4}{x+4}$
13. $3x-4-\frac{7}{2x-1}$ 15. $3x+4+\frac{1}{7x-1}$ 17. $x-9$ 19. $x^2-x-\frac{3}{x-8}$ 21. $4x^2-6x+9-\frac{17}{x+2}$ 23. $x^2+7x+55+\frac{388}{x-7}$
25. $2x^2-9x+18-\frac{30}{x+2}$ 27. $7x^2+2x+1$ 29. $x^2+2x+2-\frac{12}{2x+3}$ 31. $x^2+3x+2-\frac{2}{x-4}$ 33. $2x^2+x-3+\frac{18}{5x+3}$
35. $2x^2-8x+25-\frac{98}{x+4}$ 37. $3x^2+2x-5-\frac{1}{3x-2}$ 39. $3x^2-2x+5$ 41. $x^3+2x+5+\frac{17}{x-3}$ 43. $x^3+2x^2+6x+9+\frac{23}{x-2}$
45. $x^3+\frac{1}{2}x^2-\frac{3}{4}x-\frac{3}{8}+\frac{45}{16\left(x-\frac{1}{2}\right)}$ 47. $3x+5+\frac{x-1}{x^2+2}$ 49. $x^2+x-4+\frac{-8x+17}{x^2+4}$ 51. $2x+3+\frac{6}{3x^2-2x-1}$
53. $3x^2-10x+12+\frac{-x-14}{x^2+x+1}$ 55. $x^2-2x+7+\frac{-17x+14}{x^2+2x-3}$ 57. x^2+3x+9 59. $x^5-x^4+x^3-x^2+x-1$
61. $x^4+x^3+x^2+x+1+\frac{2}{x-1}$ 63. $x^4-\frac{1}{2}x^3-\frac{3}{4}x^2+\frac{3}{8}x+\frac{13}{16}-\frac{13}{32\left(x+\frac{1}{2}\right)}$ 65. $3x^3+4x^2+8x+12$
67. a. 19; $2x^2-4x+2+\frac{19}{x-2}$ b. −5; $2x^2-10x+20-\frac{5}{x+1}$ c. 55; $2x^2+10+\frac{55}{x-4}$ Answers will vary.

Chapter 5 Review, pages 419 - 422

1. 125 **2.** 64 **3.** $-\frac{3}{8}$ **4.** $\frac{5}{9}$ **5.** $\frac{1}{y^3}$ **6.** x^7 **7.** $\frac{1}{y^2}$ **8.** $\frac{4}{x}$ **9.** $-12x^4$ **10.** $-6x^4$ **11.** 1 **12.** $-6a^4b^4c^2$ **13.** $\frac{3a^7}{b^3}$ **14.** $\frac{5}{xy}$ **15.** $-8x^{12}$ **16.** $\frac{17}{x^6}$ **17.** $\frac{x}{y}$ **18.** $9x^2y^4$ **19.** $\frac{9x^6}{y^4}$ **20.** m^6n^9 **21.** $-10x^4y^6$ **22.** $\frac{5}{x^3}$ **23.** $\frac{4}{27a^{14}b^3}$ **24.** $\frac{9}{a^4b^6}$ **25.** 2.93×10^7 **26.** 7.5×10^{-3} **27.** 0.000724 **28.** 94,850,000 **29.** $\frac{5.8\times10^{-3}}{2.9\times10^2}$; 2×10^{-5} **30.** $\frac{2.7\times10^0\cdot2\times10^{-3}\cdot2.5\times10^1}{5.4\times10^1\cdot5\times10^{-4}}$; 5.0×10^0 **31.** $\frac{1.5\times10^{-3}\cdot8.22\times10^2}{4.11\times10^3\cdot3\times10^2}$; 1×10^{-6} **32.** 3.5×10^4 ft **33.** Not a polynomial **34.** Trinomial **35.** Binomial **36.** $6x$; First-degree monomial; Leading coefficient 6 **37.** $6x^2-x$; Second-degree binomial; Leading coefficient 6 **38.** $-10a^3+4a^2+9a$; Third-degree trinomial; Leading coefficient -10 **39.** 5; Monomial of 0 degree; Leading coefficient 5 **40.** $8x^3-6x^2-5$; Third-degree trinomial; Leading coefficient 8 **41.** $4a^3-4a^2-2a+9$; Third-degree polynomial; Leading coefficient 4 **42.** 19 **43.** 14 **44.** 154 **45.** 78 **46.** a^2+6a+9 **47.** $8a-9$ **48.** $4x^2+9x+3$ **49.** $-x^2+x+3$ **50.** $2x^2+19x+4$ **51.** $-5x^2-3x-4$ **52.** $-2x^3+3x^2+3x+7$ **53.** $2x^3+13x^2-x-7$ **54.** $2x^2-x-10$ **55.** $2x^4+7x^3-6x^2+10$ **56.** $6x^2-x$ **57.** $3x^2+2x-2$ **58.** $4x^2-7x+2$ **59.** $10x^3-5x^2+20x+2$ **60.** $5x-18$ **61.** $-x-13$ **62.** $5x^2-x+4$ **63.** $17x-91$ **64.** $5x^2-8x-1$ **65.** $-x^2-11x+12$ **66.** $-6x^5-8x^3$ **67.** $-35a^5+21a^4-7a^3$ **68.** $-6a^2+8a+30$ **69.** x^3+3x^2-6x-8 **70.** x^2-5x-6 **71.** $7t^2-11t-6$ **72.** $9a^2-18a+9$ **73.** $4x^2+19x-5$ **74.** $-4x^2+8x+60$ **75.** $2y^3-5y^2+y+2$ **76.** $3y^3+4y^2-29y-8$ **77.** $x^4+6x^3+12x^2+9x-10$ **78.** $4x^2-14x-3$ **79.** $7x+46$ **80.** $2x^2+5x+3$ **81.** $4x^2+36x+81$; Perfect square trinomial **82.** x^2-169; Difference of two squares **83.** y^2-64; Difference of two squares **84.** $6x^2+25x+25$ **85.** $9x^2-6x+1$; Perfect square trinomial **86.** y^4+2y^2-24 **87.** $9x^2-49$; Difference of two squares **88.** $9x^2-42x+49$; Perfect square trinomial **89.** x^6-100; Difference of two squares **90.** $t^2-\frac{1}{16}$; Difference of two squares **91.** $y^2+\frac{1}{6}y-\frac{1}{3}$ **92.** $2+\frac{8}{y}+\frac{10}{y^2}$ **93.** $5x^2+4x-\frac{6}{x}$ **94.** $\frac{1}{2}x^3+2x^2-3x+1$ **95.** $4y^2+2y-1+\frac{11}{3y}$ **96.** $x+3$ **97.** $8y-14+\frac{47}{y+3}$ **98.** x^2+4x **99.** $4x^2+8x-4-\frac{5}{x-2}$ **100.** $16x^2+20x+25$ **101.** $4x^2-6x+9$ **102.** $3x+4+\frac{-13x-7}{x^2+2}$ **103.** $5x^2-4x+9$ **104.** $x-6+\frac{5x-32}{x^2-2x+3}$ **105.** $x^3+x^2+8x+22+\frac{82}{x-4}$

Chapter 5 Test, pages 423 - 424

1. $-10a^5$ **2.** 1 **3.** $\frac{4x^3}{y^7}$ **4.** $\frac{x}{3y^2}$ **5.** $\frac{x^2}{4y^2}$ **6.** $4x^2y^4$ **7. a.** 135,000 **b.** 0.0000027 **8. a.** $2.5\times10^2\cdot5\times10^5$; 1.25×10^8 **b.** $\frac{6.5\times10^1\cdot1.2\times10^{-2}}{1.5\times10^3}$; 5.2×10^{-4} **9.** $8x^2+3x$; Second-degree binomial; Leading coefficient 8 **10.** $-x^3+3x^2+3x-1$; Third-degree polynomial; Leading coefficient -1 **11.** $5x^5+2x^4-11x+3$; Fifth-degree polynomial; Leading coefficient 5 **12. a.** 20 **b.** -110 **13.** $-3x+2$ **14.** $20x-8$ **15.** $5x^3-x^2+6x+5$ **16.** $7x^4+14x^2+4$ **17.** $15x^7-20x^6+15x^5-40x^4-10x^2$ **18.** $49x^2-9$; Difference of two squares **19.** $16x^2+8x+1$; Perfect square trinomial **20.** $36x^2-60x+25$; Perfect square trinomial **21.** $12x^2+24x-15$ **22.** $6x^3-69x^2+189x$ **23.** $7x^2+7x+14$ **24.** $10x^4+4x^3-15x^2-41x-14$ **25.** $2x+\frac{3}{2}-\frac{3}{x}$ **26.** $\frac{5a}{3}+2a^2+\frac{1}{a}$ **27.** $x-6-\frac{2}{2x+3}$ **28.** $x-9+\frac{15x-12}{x^2+x-3}$ **29.** $2x^2-3x-13$ **30. a.** $A(x)=240-(x^2+3x)=-(x^2+3x-240)$ **b.** $P(x)=64$

Cumulative Review: Chapters 1-5, pages 425 - 428

1. $\frac{34}{9}$ **2.** $-\frac{7}{22}$ **3.** $x+15$ **4.** $3x+8$ **5.** 540 **6.** $120a^2b^3$ **7.** $x=-3$ **8.** $x=\frac{8}{5}$ **9.** $x=-\frac{1}{2}$ **10.** $x=\frac{3}{8}$ **11.** $x=-3$

12. $x=-\frac{20}{7}, 4$ **13.** $x=\frac{y-b}{m}$ **14.** $y=\frac{10-3x}{5}$ **15.** 

$[-6, \infty)$ **16.** 

$(-7, 4)$

17. 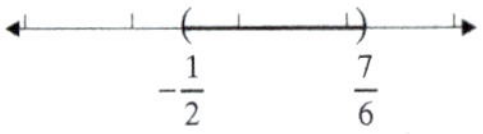

$\left(-\frac{1}{2}, \frac{7}{6}\right)$ **18.** 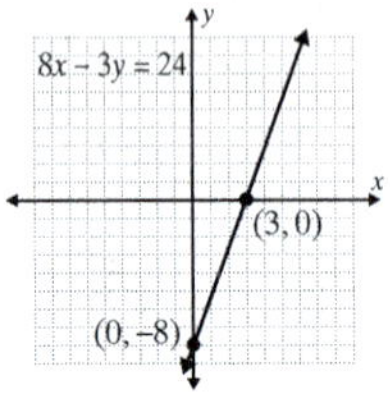

19. $-2x+3y=8$ 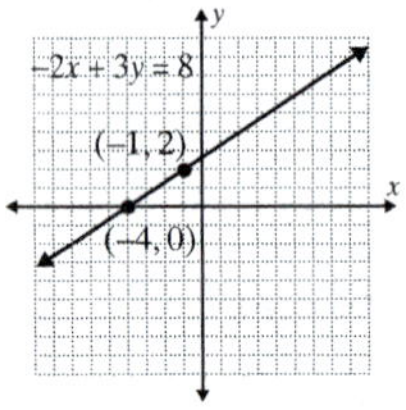

20. $y=-\frac{5}{8}x-\frac{17}{8}$ 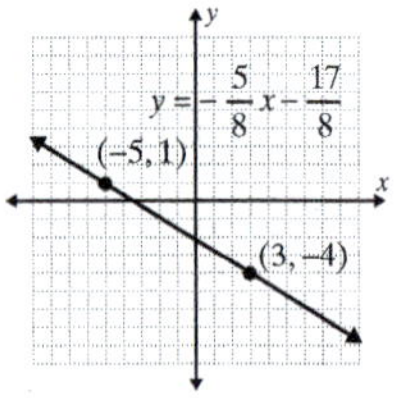

21. $x+6y=42$ 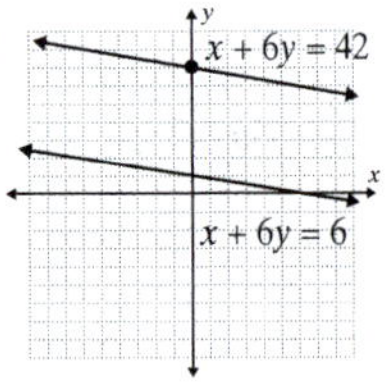

22. a. 22 **b.** 57

23. Open; 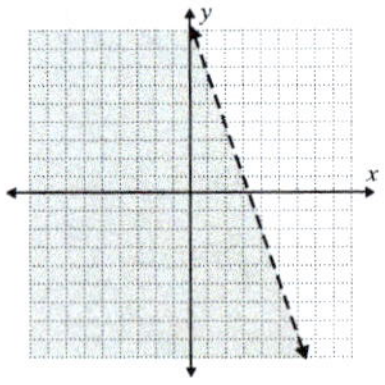

24. Closed; 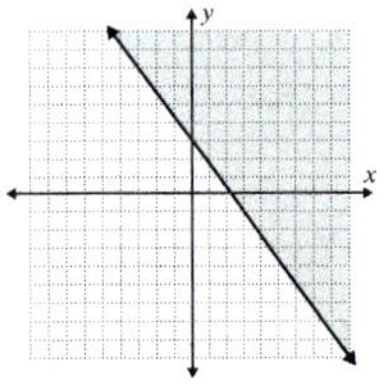

25. $2x$ **26.** $64x^6y^3$ **27.** $\frac{49x^{10}}{y^4}$ **28.** $\frac{36x^4}{y^{10}}$ **29.** $\frac{a^6}{b^4}$ **30.** $\frac{x}{3y^4}$ **31.** 1 **32.** $\frac{x^8}{9y^6}$

33. a. 0.00000028 **b.** 35,100 **34.** $1.5\times10^{-3}\cdot4.2\times10^3; 6.3\times10^0$ **35.** $\frac{8.4\times10^2}{2.1\times10^{-4}}; 4\times10^6$

36. $\frac{5\times10^{-3}\times7.7\times10}{1.1\times10^{-2}\times3.5\times10^3}; 1\times10^{-2}$ **37. a.** $3x^2+8x$ **b.** 51 **c.** $3a^2+8a$

38. a. $-x^3-6x^2+2x-4$ **b.** −79 **c.** $-a^3-6a^2+2a-4$

39. $x-9$; First-degree binomial; Leading coefficient 1 **40.** $x^4-2x^3+4x^2-10x+40$; Fourth-degree polynomial; Leading coefficient 1 **41.** $3x^3+4x^2-7x-3$ **42.** $-x^3+7x^2-6$ **43.** $6x^2-7x+1$ **44.** $-x-17$ **45.** $-3x^3+12x^2-3x$ **46.** $5x^5+10x^4$ **47.** x^2-36 **48.** x^2+x-12 **49.** $9x^2+42x+49$ **50.** $4x^2-1$ **51.** x^4-4x^2+4 **52.** $2x^2-x-36$ **53.** $6x^2+x-12$ **54.** $4x^3-3x^2-x$ **55.** y^3-125 **56.** x^3+27y^3 **57.** $4x-7+\frac{3}{x}$ **58.** $\frac{y^3}{7}-3y+\frac{1}{y}-4y^2$ **59.** $x-2$ **60.** $2x^2-x+3-\frac{2}{x+3}$ **61.** $2x^3-x^2-5x+8-\frac{48}{x^2+2x+3}$ **62.** Function; $D=(-\infty,\infty)$; $R=(-\infty,\infty)$

63. Not a function; $D=(-\infty,\infty)$; $R=[-3,3]$ **64.** 20, 22, 24 **65.** Length = 25 centimeters; Width = 16 centimeters **66.** −19 **67.** 38 feet, 38 feet, and 42 feet **68.** \$35,000 at 8%; \$65,000 at 6% **69.** Length = 105 ft; Width = 90 ft **70.** \$337.50 **71.** 10 miles **72. a.** $A(x)=900-4x^2$ **b.** $P(x)=4(30-2x)+8x=120$

Chapter 6

Exercises 6.1, pages 438 - 440

1. 5 **3.** 8 **5.** 1 **7.** $10x^3$ **9.** $4a^2$ **11.** $13ab$ **13.** $15xy^2z^2$ **15.** x^4 **17.** $-4y$ **19.** $3x^3$ **21.** $2x^2y$ **23.** $m+9$ **25.** $x-6$ **27.** $b+1$ **29.** $3y+4x+1$ **31.** $11(x-11)$ **33.** $4y(4y^2+3)$ **35.** $-3a(2x-3y)$ **37.** $5xy(2x-5)$ **39.** $-2yz(9yz-1)$ **41.** $8(y^2-4y+1)$ **43.** $x(2y^2-3y-1)$ **45.** $4m^2(2x^3-3y+z)$ **47.** $-7x^2z^3(8x^2+14xz+5z^2)$ **49.** $x^4y^2(15+24x^2y^4-32x^3y)$ **51.** $(y+3)(7y^2+2)$ **53.** $(x-4)(3x+1)$ **55.** $(x-2)(4x^3-1)$ **57.** $(2y+3)(10y-7)$

59. $(x-2)(a-b)$ **61.** $(b+c)(x+1)$ **63.** $(x^2+6)(x+3)$ **65.** Not factorable **67.** $(3-b)(x+y)$ **69.** $(y-4)(5x+z)$ **71.** $(z^2+3)(a+1)$ **73.** $(6x+1)(a+2)$ **75.** $(x+1)(y+1)$ **77.** $(2y-7z)(5x-y)$ **79.** $(3x-4u)(y-2v)$ **81.** Not factorable **83.** $(2c-3d)(3a+b)$ **85.** Although both can be factored out of $-3x^2+3$, 3 is greater than -3 making it the greatest common factor.

Exercises 6.2, pages 445 - 447

1. $\{1,15\},\{-1,-15\},\{3,5\},\{-3,-5\}$ **3.** $\{1,20\},\{-1,-20\},\{4,5\},\{-4,-5\},\{2,10\},\{-2,-10\}$ **5.** $\{1,-6\},\{6,-1\},\{2,-3\},\{3,-2\}$ **7.** $\{1,16\},\{-1,-16\},\{4,4\},\{-4,-4\},\{8,2\},\{-8,-2\}$ **9.** $\{1,-10\},\{10,-1\},\{5,-2\},\{2,-5\}$ **11.** 4, 3 **13.** $-7, 2$ **15.** $8, -1$ **17.** $-6, -6$ **19.** $-5, -4$ **21.** $x+1$ **23.** $p-10$ **25.** $a+6$ **27.** $(x-4)(x+3)$ **29.** $(y+6)(y-5)$ **31.** Not factorable **33.** $(x-4)(x-4)$ **35.** $(x+4)(x+3)$ **37.** $(y-1)(y-2)$ **39.** Not factorable **41.** $(x+8)(x-9)$ **43.** $(z-6)(z-9)$ **45.** $x(x+7)(x+3)$ **47.** $5(x-4)(x+3)$ **49.** $10y(y-3)(y+2)$ **51.** $4p^2(p+1)(p+8)$ **53.** $2x^2(x-9)(x+2)$ **55.** $2(x^2-x-36)$ **57.** $2a^2(a-10)(a+6)$ **59.** $3y^3(y-8)(y+1)$ **61.** $x(x-2)(x-8)$ **63.** $5(a^2+2a-6)$ **65.** $20a^2(a+1)(a+1)$ **67.** Base $= x+48$; Height $= x$ **69.** $x+5$ **71.** This is not an error, but the trinomial is not completely factored. The completely factored form of this trinomial is $2(x+2)(x+3)$.

Exercises 6.3, pages 456 - 458

1. $(x+2)(x+3)$ **3.** $(2x-5)(x+1)$ **5.** $(6x+5)(x+1)$ **7.** $-(x-2)(x-1)$ **9.** $(x-5)(x+2)$ **11.** $-(x-14)(x+1)$ **13.** Not factorable **15.** $-x(2x+1)(x-1)$ **17.** $(t-1)(4t+1)$ **19.** $(5a-6)(a+1)$ **21.** $(7x-2)(x+1)$ **23.** $(2x-3)(4x+1)$ **25.** $(3x+4)(3x-5)$ **27.** $2(2x-5)(3x-2)$ **29.** $(3x-1)(x-2)$ **31.** $(3x-1)(3x-1)$ **33.** $(3y+2)(2y+1)$ **35.** $(x-1)(x-45)$ **37.** Not factorable **39.** $2b(4a-3)(a-2)$ **41.** Not factorable **43.** $(4x-1)(4x-1)$ **45.** $(8x-3)(8x-3)$ **47.** $2(3x-5)(x+2)$ **49.** $5(2x+3)(x+2)$ **51.** $-2(9x^2-36x+4)$ **53.** $-15(3y+4)(y-2)$ **55.** $3(2x-5)(2x-5)$ **57.** $3x(2x-1)(x+2)$ **59.** $3x(2x-9)(2x-9)$ **61.** $9xy^3(x^2+x+1)$ **63.** $4xy(3y-4)(4y-3)$ **65.** $7y^2(y-4)(3y-2)$ **67.** If the sign of the constant term is positive, the signs in the factors will both be positive or both be negative. If the sign of the constant term is negative, the sign in one factor will be positive and the sign in the other factor will be negative.

Exercises 6.4, pages 464 - 466

1. $(x-5)(x+5)$ **3.** $(9-y)(9+y)$ **5.** $2(x-8)(x+8)$ **7.** $4(x-2)(x+2)(x^2+4)$ **9.** Not factorable **11.** $(y-8)^2$ **13.** $-4(x-5)(x+5)$ **15.** $(3x-5)(3x+5)$ **17.** $(y-5)^2$ **19.** $(2x-1)^2$ **21.** $(5x+3)^2$ **23.** $(4x-5)^2$ **25.** $4x(x-4)(x+4)$ **27.** $2xy(x+8)^2$ **29.** $(y+3)^2$ **31.** $(x-10)^2$ **33.** $(x^2+5y)^2$ **35.** $(x-5)(x^2+5x+25)$ **37.** $(y+6)(y^2-6y+36)$ **39.** $(x+3y)(x^2-3xy+9y^2)$ **41.** Not factorable **43.** $4(x-2)(x^2+2x+4)$ **45.** $2(3x-y)(9x^2+3xy+y^2)$ **47.** $y(x+y)(x^2-xy+y^2)$ **49.** $x^2y^2(1-y)(1+y+y^2)$ **51.** $3xy(2x+3y)(4x^2-6xy+9y^2)$ **53.** $(x^2-y^3)(x^4+x^2y^3+y^6)$ **55.** $(3x+y^2)(9x^2-3xy^2+y^4)$ **57.** $(2x+y)(4x^2-2xy+y^2)$ **59.** $8(y-1)(y^2+y+1)$ **61.** $(3x-y)(3x+y)$ **63.** $(x-2y)(x+2y)(x^2+4y^2)$ **65.** $(x-y-9)(x-y+9)$ **67.** $(x-y-6)(x-y+6)$ **69.** $(4x+1-y)(4x+1+y)$ **71. a.** x^2-16 **b.**

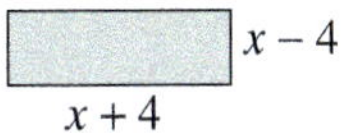

73. a. $xy+xy+x^2+y^2=x^2+2xy+y^2=(x+y)^2$ **b.**

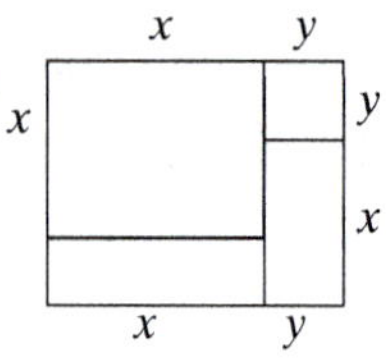

$(x+y)(x+y)=(x+y)^2$

75. For a 3-digit integer: $abc=100a+10b+c=(99+1)a+(9+1)b+c=9(11a+b)+a+b+c$ So, if the sum $(a+b+c)$ is divisible by 3 (or 9), then the number abc will be divisible by 3 (or 9).
For a 4-digit integer: $abcd=1000a+100b+10c+d=(999+1)a+(99+1)b+(9+1)c+d=9(111a+11b+c)+a+b+c+d$ So, if the sum $(a+b+c+d)$ is divisible by 3 (or 9), then the number $abcd$ will be divisible by 3 (or 9).

Exercises 6.5, pages 469 - 470

1. $(m+6)(m+1)$ **3.** $(x+9)(x+2)$ **5.** $(x-10)(x+10)$ **7.** $(m-3)(m+2)$ **9.** Not factorable **11.** $(8a-1)(8a+1)$
13. $(x+5)^2$ **15.** $(x+12)(x-3)$ **17.** $3(a+6)(a-2)$ **19.** $-5(x-6)(x-8)$ **21.** Not factorable **23.** $x(x-6)(x+2)$
25. $-2a(a+8)(a-7)$ **27.** $4x(2x-5)(2x+5)$ **29.** $-(x-5)(3x-2)$ **31.** $(2x-1)(3x-4)$ **33.** $(4m+3)(3m-2)$
35. $2(2x-1)(x-3)$ **37.** $(4x-7)(2x+5)$ **39.** $(5x+6)(4x-9)$ **41.** $-(5x-7)(3x+2)$ **43.** $-(2a-3)(4a-5)$
45. $(4y+5)(5y-4)$ **47.** $(6x-1)(3x-2)$ **49.** $-6(5x-4)(5x+4)$ **51.** $3(4n^2-20n-25)$ **53.** $a(21a^2-13a-2)$
55. $3x(3x-2)(4x+5)$ **57.** $2x(2x-1)(4x-11)$ **59.** $5(24m^2+2m+15)$ **61.** $(y-4)(x+3)$ **63.** $(x+2y)(x-6)$
65. $-(x^2-5)(x-8)$ **67.** $(x+5)(x^2-5x+25)$ **69.** $x^4(y-1)(y^2+y+1)$ **71.** $(2a^2+3b^2)(4a^4-6a^2b^2+9b^4)$
73. $(x^2y-5)(x^4y^2-5x^2y+25)$ **75.** $(x-3)(x+3)(x+7)$ **77.** $(3x+y+6)(3x-y-6)$ **79.** $(y+10+7x)(y+10-7x)$

Exercises 6.6, pages 478 - 479

1. $x=2,3$ **3.** $x=-2,\frac{9}{2}$ **5.** $x=-3$ **7.** $x=-5$ **9.** $x=0,2$ **11.** $x=-6$ **13.** $x=-1,4$ **15.** $x=-3,4$ **17.** $x=-3,0$ **19.** $x=2,4$
21. $x=-4,3$ **23.** $x=-\frac{1}{2},3$ **25.** $x=-\frac{2}{3},2$ **27.** $x=-\frac{1}{2},4$ **29.** $x=-2,\frac{4}{3}$ **31.** $x=\frac{3}{2}$ **33.** $x=0,\frac{8}{5}$ **35.** $x=-2,2$
37. $x=1$ **39.** $x=2$ **41.** $x=-3,3$ **43.** $x=-5,10$ **45.** $x=-6,-2$ **47.** $x=\frac{1}{2}$ **49.** $x=0,2,4$ **51.** $x=-\frac{2}{3},-\frac{1}{2},0$
53. $x=-10,10$ **55.** $x=-5,5$ **57.** $x=-4$ **59.** $x=3$ **61.** $x=-1,3$ **63.** $x=-8,-2$ **65.** $x=-5,2$ **67.** $x=-5,7$ **69.** $x=-6,2$
71. $x=-1,\frac{2}{3}$ **73.** $x=-\frac{3}{2},4$ **75.** $y^2-y-6=0$ **77.** $2x^2+11x+5=0$ **79.** $8x^2-10x+3=0$ **81.** $x^3-x^2-6x=0$
83. $y^3-4y^2-3y+18=0$ **85.** This allows for use of the zero-factor property which says that for the equation to equal zero one of the factors must equal zero. Answers will vary.
87. a. 640 ft; 384 ft **b.** 144 ft; 400 ft **c.** 7 seconds; $0=-16(t+7)(t-7)$

Exercises 6.7, pages 487 - 491

1. $x(x+8)=-16;$ $x=-4$, so the numbers are –4 and 4 **3.** $x^2=7x; x=0,7$ **5.** $x^2+3x=28; x=4$
7. $x(x+7)=78;$ $x=-13, 6;$ so the numbers are –13 and –6 or 13 and 6
9. $(x+6)^2+x^2=260;$ $x=8;$ so the numbers are 8 and 14 **11.** $x+(x+8)^2=124;$ $x=3;$ so the integers are 3 and 11
13. $x(2x-5)=x+56;$ $x=-4$ **15.** $x(x+1)=72;$ $x=8;$ so the integers are 8 and 9

17. $x^2+(x+1)^2=85$; $x=6$; so the integers are 6 and 7 **19.** $x(x+2)=63$; $x=-9,7$; so the integers are –9 and –7 or 7 and 9
21. $4x+(x+1)^2=41$; $x=4$; so the integers are 4 and 5
23. $2x(x+1)=(x+1)(x+2)+88$; $x=10$; so the integers are 10, 11, and 12
25. $6x(x+2)=(x+1)+(x+3)^2$; $x=-2,1$; so the integers are –2, –1, 0, and 1 or 1, 2, 3, and 4
27. $w(2w)=72$; $w=6$; so width is 6 in. and length is 12 in. **29.** $w(4w)=64$; $w=4$; so width is 4 ft and length is 16 ft
31. $l(l-4)=117$; $l=13$; so width is 9 ft and length is 13 ft **33.** $\frac{1}{2}b(b-4)=16$; $b=8$; so base is 8 ft and height is 4 ft
35. $\frac{1}{2}(h+15)h=63$; $h=6$; so base is 21 in. **37.** $w(16-w)=48$; $w=4,12$; so the rectangle is 4 in. by 12 in.
39. $r(r+13)=140$; $r=7$; so there are 7 trees in each row **41.** $r(r+7)=144$; $r=9$; so there are 9 rows
43. $n(n+1675)=8400$; $n=5$, so there are 5 floors **45.** $(w+11)(w+4)=98$; $w=3$; so the rectangle is 3 cm by 10 cm
47. $w(50-2w)=300$; $w=10,15$; so width is 10 ft and length is 30 ft or width is 15 ft and length is 20 ft
49. $h^2+(h-34)^2=(h+2)^2$; $h=48$; so the height of the pole is 48 ft **51.** $h^2+(h-49)^2=(h+1)^2$; $h=60$; so height is 60 ft
53. $l^2+(l-28)^2=(l+8)^2$; $l=60$, so the length of the mat is 60 inches **55.** \$1.50 per pound **57.** \$16 or \$20 per reel
59. $12^2+5^2=13^2$; $20^2+21^2=29^2$; $24^2+7^2=25^2$; $14^2+48^2=50^2$; $60^2+11^2=61^2$

Chapter 6 Review, pages 497 - 500

1. $x-3y+2$ **2.** $2y+3$ **3.** $11(x-2)$ **4.** $-4y(y-7)$ **5.** $4x^2y(4x-3)$ **6.** $a(m^2+11m+25)$ **7.** $-7t^3x(x^3-14tx^2-5t^2)$
8. $-6x^2y^3(xy+y-4)$ **9.** $(a+7)(3a-2)$ **10.** $(x-10)(2a+3b)$ **11.** $(a+c)(x-1)$ **12.** $(1-4y)(x+2z)$ **13.** $(z^2+5)(1+c)$
14. Not factorable **15.** $(a+b)(7a-3)$ **16.** $(1-3y)(x-z)$ **17.** –7, 6 **18.** 3, 9 **19.** $(m+6)(m+1)$ **20.** Not factorable
21. $(n-6)(n-2)$ **22.** $(x-2)(x+5)$ **23.** $(a-10)(a+5)$ **24.** Not factorable **25.** $p(p+2)(p-6)$ **26.** $3(m+2)(m+2)$
27. $2y(y+5)(y+2)$ **28.** $7x^2(x-1)(x+4)$ **29.** $9x^3(x+5)(x+4)$ **30.** $11a(a+1)(a-11)$ **31.** $(a+5b)(a+6b)$
32. $4(y^2-7xy-x^2)$ **33.** $(x+7)(x+5)$ **34.** $-(x-7)(x+3)$ **35.** $-(6x-1)(2x-5)$ **36.** $(3x-2)(4x+3)$
37. $(2y-1)(3y-4)$ **38.** Not factorable **39.** $3(3x+2)(7x-5)$ **40.** $2(2x+5)(4x-7)$ **41.** $-(3x+8)(x-3)$
42. $-(5x-7)(3x+2)$ **43.** Not factorable **44.** Not factorable **45.** $-4(x-10)(x+5)$ **46.** $7x(y-4)(y+6)$
47. $x(3x-2)(6x-1)$ **48.** $(x^3-10)(x^3+10)$ **49.** $(4x-5)(4x+5)$ **50.** Not factorable **51.** $4(y+4)^2$ **52.** Not factorable
53. $2(2x-11)(2x+11)$ **54.** $3x(x-4)(x+4)$ **55.** $-3(x+2)^2$ **56.** $3(4x-y)(4x+y)$ **57.** $(9-4x)(9+4x)$
58. $(2x^3-y^2)(4x^6+2x^3y^2+y^4)$ **59.** $3(x+3)(x^2-3x+9)$ **60.** $4x(x+5)(x^2-5x+25)$ **61.** $(7y-1)(49y^2+7y+1)$
62. $27(2x-1)(4x^2+2x+1)$ **63.** $(z+4)(z-9)$ **64.** $(y+6)(y-7)$ **65.** Not factorable **66.** $4(x+2)(x+3)$
67. $-3(2x-1)(x-5)$ **68.** $-(3x-2)(x+4)$ **69.** $5x(x+2)(2x+3)$ **70.** $(7x+2y)(7x-2y)$ **71.** $2(x-3y)(x^2+3xy+9y^2)$
72. $(3x^2+1)(x-3)$ **73.** $x=-5,0$ **74.** $x=0,6$ **75.** $x=\frac{2}{3},5$ **76.** $y=5,-7$ **77.** $x=-\frac{5}{2},0,\frac{5}{2}$ **78.** $x=-2,0,6$ **79.** $x=-5,5$
80. $y=-\frac{6}{5},0,\frac{6}{5}$ **81.** $x=-5,7$ **82.** $a=-7$ **83.** $x=-6,0$ **84.** $x=-3,13$ **85.** $x=1,7$ **86.** $x=-2,-\frac{3}{2}$ **87.** $x^2+x-20=0$
88. $32x^2+4x-15=0$ **89.** $6x^3+11x^2-2x=0$ **90.** $y^3+7y^2+8y-16=0$ **91.** 13, 10 **92.** 6, –14 or 14, –6 **93.** 9 streets
94. 15×30 yards **95.** 11, 13 **96.** 10, 12 or –8, –6 **97.** 5, 7, 9 **98.** 10, 11, 12, 13 or –7, –6, –5, –4 **99.** 20 inches
100. $l=20$ meters; $w=10$ meters **101.** 4 inches **102.** 20 rows, 30 seats **103.** Length = 21 meters; Width = 20 meters
104. 125 yards

Chapter 6 Test, page 501

1. $7ab^2(4x-3y)$ **2.** $6yz^2(3z-yz+2)$ **3.** $(x-5)(x-4)$ **4.** $-(x+7)^2$ **5.** $(x-5)(y-7)$ **6.** $6(x+1)(x-1)$
7. $2(6x-5)(x+1)$ **8.** $(x+3)(3x-8)$ **9.** $(4x-5y)(4x+5y)$ **10.** $x(x+1)(2x-3)$ **11.** $(2x-3)(3x-2)$
12. $(y+7)(2x-3)$ **13.** Not factorable **14.** $-3x(x^2-2x+2)$ **15.** $10x^3(2y+1)(y+1)$ **16.** $4y^2(x-3y)(x^2+3xy+9y^2)$
17. $x=-1, 8$ **18.** $x=-6, 0$ **19.** $x=-\frac{3}{4}, 5$ **20.** $x=\frac{3}{2}, 4$ **21.** $x^2+5x-24=0$ **22.** $4x^2-4x+1=0$ **23.** 6, 20 or −30, −4
24. Length = 15 centimeters, Width = 11 centimeters **25.** 18, 19 **26.** 3, 12 **27.** 18 cm **28.** $P(x)=4(3x+5)$

Cumulative Review: Chapters 1-6, pages 502 - 506

1. 120 **2.** $168x^2y$ **3.** 6 **4.** −138 **5.** $\frac{55}{48}$ **6.** $\frac{19}{60a}$ **7.** $\frac{3}{10}$ **8.** $\frac{75x}{23}$ **9.** $x=\frac{9}{10}$ **10.** $x=21$ **11.** $x=21$ **12.** $x=-5, 10$

13. Conditional **14.** Identity **15.** Contradiction **16.** 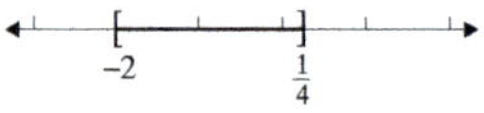

$\left[-2, \frac{1}{4}\right]$ **17.** 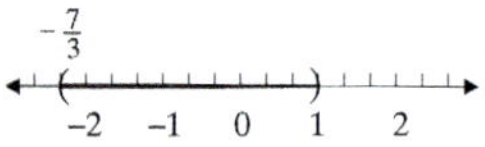

$\left(-\frac{7}{3}, 1\right)$

18. 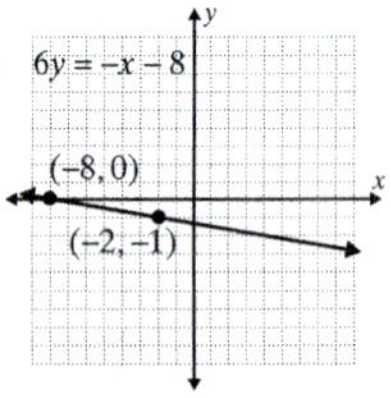

19. 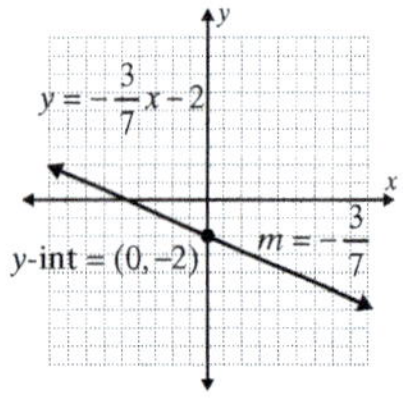

20. $y=7$ 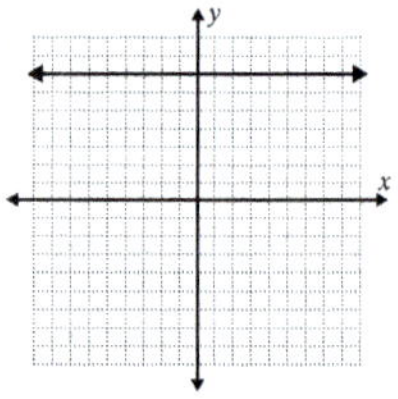

21. $x+y=-2$ 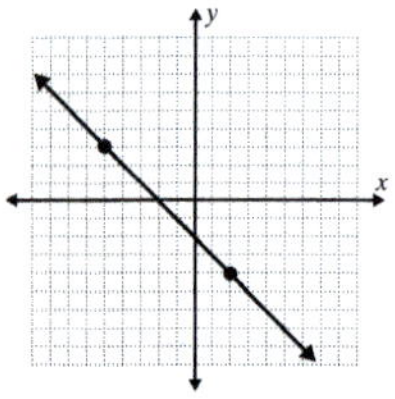

22. $y=3x+4$

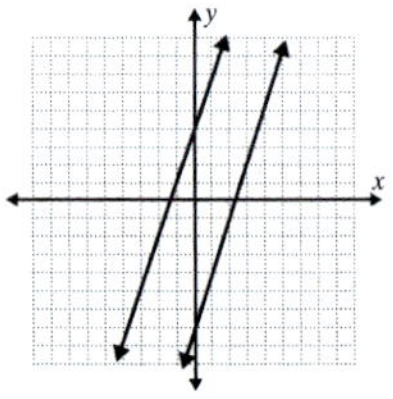

23. $y=-\frac{1}{3}x+\frac{2}{3}$

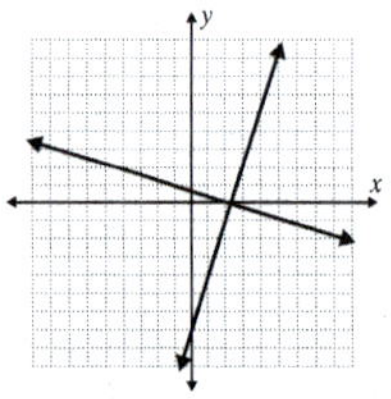

24. a. 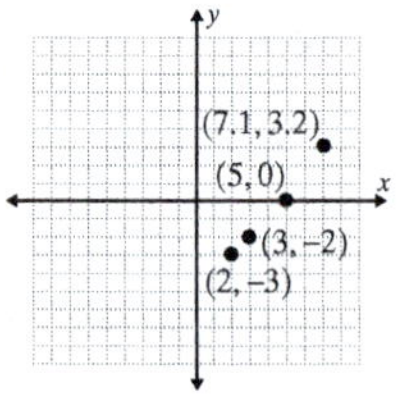

b. $D=\{2, 3, 5, 7.1\}$ **c.** $R=\{-3, -2, 0, 3.2\}$
d. It is a function because each first coordinate (domain) has only one corresponding second coordinate (range).
25. Function; $D=(-\infty, \infty)$; $R=(-\infty, \infty)$
26. Not a function; $D=(-\infty, 4]$; $R=(-\infty, \infty)$
27. a. 58 **b.** 4 **c.** $\frac{11}{4}$

28. 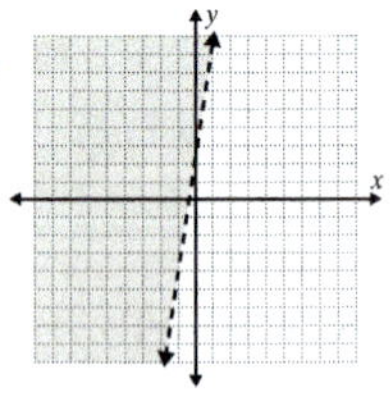

29. 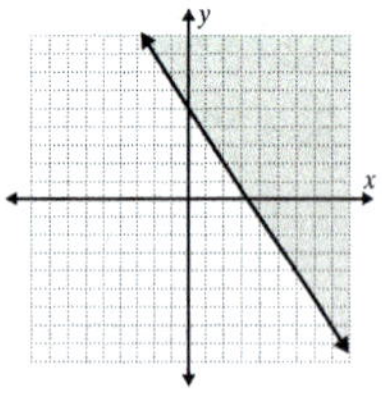

30. $64x^6y^3$ **31.** $-\frac{1}{8x^9y^6}$ **32.** $7x^2y$ **33.** $\frac{36x^4}{y^{10}}$ **34.** $\frac{y^2}{x^3}$ **35.** $\frac{9y^{14}}{x^4}$
36. a. $(5.6\times10^{-7})(3\times10^{-4})$; 1.68×10^{-10}
b. $\frac{(8.1\times10^4)(6.2\times10^3)}{(3\times10^{-3})(2\times10^{-1})}$; 8.37×10^{11}

37. $13x+1$ **38.** $x^2+12x+3$ **39.** $4x^2+5x-8$ **40.** $-4x^2-2x+3$ **41.** $2x^2+x-28$ **42.** $-3x^2-17x+6$
43. $x^2+12x+36$ **44.** $4x^2-28x+49$ **45.** $2x-\frac{5}{4}+\frac{1}{y}$ **46.** $3x+1+\frac{22}{x-2}$ **47.** $4(2x-5)$ **48.** $6(x-16)$
49. $(x+2)(y+3)$ **50.** $(x-2)(a+b)$ **51.** $(x-6)(x-3)$ **52.** $(2x-3)(3x+4)$ **53.** $-4x(3x+4)$ **54.** $8xy(2x-3)$
55. $5x^2(x-5)(x+8)$ **56.** $(2x+1)(2x-1)$ **57.** $3(x+4y)(x-4y)$ **58.** Not factorable **59.** $(x+1)(3x+2)$
60. $2x(x-5)(x-5)$ **61.** $4x(x^2+25)$ **62.** $5(x-3y)(x^2+3xy+9y^2)$ **63.** $x=-1, 7$ **64.** $x=0, 7$ **65.** $x=-6, -2$
66. $x=-4, 7$ **67.** $x=-7, 0$ **68.** $x=-2, 0$ **69.** $x=-\frac{5}{4}, 1$ **70.** $x=-5, 0, 2$ **71.** $x^2+15x+50=0$ **72.** $x^3-17x^2+52x=0$

73. a. 39.7 million tickets **b.** 39.1 million tickets **c.** 12.5 million tickets **74. a.** 13% **b.** 11% **c.** a **75.** 65 mph **76.** \$7500 at 6%; \$2500 at 8% **77.** 13 inches by 17 inches **78.** 4, 11 **79.** 8 and 9 or –8 and –9 **80.** $t = 2.5, 3$ seconds **81.** Length = 6 inches; Width = 4 inches **82.** 5 m

Chapter 7

Exercises 7.1, pages 516 - 518

1. $\frac{3x}{4y}; x \neq 0, y \neq 0$ **3.** $\frac{2x^3}{3y^3}; x \neq 0, y \neq 0$ **5.** $\frac{1}{x-3}; x \neq 0, 3$ **7.** $7; x \neq 2$ **9.** $-\frac{3}{4}; x \neq 3$ **11.** $\frac{2x}{y}; x \neq -\frac{2}{3}, y \neq 0$ **13.** $\frac{x}{x-1}; x \neq -6, 1$ **15.** $\frac{x^2-3x+9}{x-3}; x \neq -3, 3$ **17.** $\frac{x-3}{y-2}; y \neq -2, 2$ **19.** $\frac{x^2+2x+4}{y+5}; x \neq 2, y \neq -5$ **21.** $\frac{ab}{6y}$ **23.** $\frac{8x^2y^3}{15}$ **25.** $\frac{x+3}{x}$ **27.** $\frac{x-1}{x+1}$ **29.** $-\frac{1}{x-8}$ **31.** $\frac{x-2}{x}$ **33.** $\frac{4x+20}{x(x+1)}$ **35.** $\frac{x}{(x+3)(x-1)}$ **37.** $-\frac{x+4}{x(x+1)}$ **39.** $\frac{x+2y}{(x-3y)(x-2y)}$ **41.** $\frac{x-1}{x(2x-1)}$ **43.** $\frac{1}{x+1}$ **45.** $\frac{x+2}{x-2}$ **47.** $\frac{1}{3xy^6}$ **49.** $\frac{6y^7}{x^4}$ **51.** $\frac{x}{12}$ **53.** $\frac{6x+18}{x^2}$ **55.** $\frac{6}{5x}$ **57.** $\frac{3x+1}{x+1}$ **59.** $\frac{x-2}{2x-1}$ **61.** $-\frac{x+4}{x(2x-1)}$ **63.** $\frac{x+1}{x-1}$ **65.** $\frac{6x^3-x^2+1}{x^2(4x-3)(x-1)}$ **67.** $\frac{x^2+4x+4}{x^2(2x-5)}$ **69.** $\frac{x^2-6x+5}{(x-7)(x-2)(x+7)}$ **71.** $\frac{x^2-3x}{(x-1)^2}$ **73.** $\frac{x^2+5x}{2x+1}$ **75.** 1 **77.** $2x-5$ feet **79. a.** A rational expression is an algebraic expression that can be written in the form $\frac{P}{Q}$ where P and Q are polynomials and $Q \neq 0$. **b.** $\frac{x-1}{(x+2)(x-3)}$ Answers will vary. **c.** $\frac{1}{x+5}$ Answers will vary. **81. a.** $x = 4$ **b.** $x = -10, 10$

Exercises 7.2, pages 527 - 528

1. 3 **3.** 2 **5.** 1 **7.** $\frac{2}{x-1}$ **9.** $\frac{14}{7-x}$ **11.** 4 **13.** $\frac{x^2-x+1}{(x+4)(x-3)}$ **15.** $\frac{x-2}{x+2}$ **17.** $\frac{4x+5}{2(7x-2)}$ **19.** $\frac{6x+15}{(x+3)(x-3)}$ **21.** $\frac{x^2-2x+4}{(x+2)(x-1)}$ **23.** $\frac{-x^2-3x-6}{(x+3)(3-x)}$ **25.** $\frac{8x^2+13x-21}{6(x+3)(x-3)}$ **27.** $\frac{3x^2-20x}{(x+6)(x-6)}$ **29.** $\frac{-4x}{x-7}$ **31.** $\frac{4x^2-x-12}{(x+7)(x-4)(x-1)}$ **33.** $\frac{x-6}{(x-10)(x-8)}$ **35.** $\frac{6x}{(x-1)(x-7)}$ **37.** $\frac{4x-19}{(7x+4)(x-1)(x+2)}$ **39.** $\frac{-7x-9}{(4x+3)(x-2)}$ **41.** $\frac{4x^2-41x+3}{(x+4)(x-4)}$ **43.** $\frac{x-4}{2(x-2)}$ **45.** $\frac{x^2-4x-6}{(x+2)(x-2)(x-1)}$ **47.** $\frac{3x^2+26x-3}{(x+7)(x-3)(x+1)}$ **49.** $\frac{6x+2}{(x-1)(x+3)}$ **51.** $\frac{2x^2+x-4}{(x-2)(y+1)(x+1)}$ **53.** $\frac{2x+4xy-15y}{(x+3)(y+2)(x-5)}$ **55.** $\frac{-2x+2}{x^2+x+1}$ **57.** $\frac{2x^2-x-5}{(x-3)(x+3)(x^2+1)}$ **59.** $\frac{5x^3-x^2+6x-4}{(2x+1)(x-1)(x+2)(3x-2)}$ **61.** See page 520. Answers will vary.

Exercises 7.3, pages 534 - 535

1. $\frac{4}{5xy}$ **3.** $\frac{8}{7x^2y}$ **5.** $\frac{2x^2+6x}{2x-1}$ **7.** $\frac{2x-1}{2+3x}$ **9.** $\frac{7}{2(x+2)}$ **11.** $\frac{x}{x-1}$ **13.** $\frac{4x}{3(x+6)}$ **15.** $\frac{7x}{x+2}$ **17.** $\frac{2x+6}{3(x-2)}$ **19.** $\frac{24y+9x}{2(9y-10x)}$ **21.** $\frac{x}{x-1}$ **23.** $\frac{xy}{x+y}$ **25.** $\frac{1}{xy}$ **27.** $\frac{y+x}{y-x}$ **29.** $\frac{3-x}{x}$ **31.** $\frac{x+1}{x+3}$ **33.** $\frac{-1}{x(x+h)}$ **35.** $\frac{-1}{x(x+h)}$

37. $-(x-2y)(x-y)$ **39.** $\frac{2x}{x^2+1}$ **41.** $\frac{(x-3)(x^2-2x+4)}{(x-4)(x-2)(x+1)}$ **43.** $\frac{-5}{x+1}$ **45.** $\frac{29}{4(4x+5)}$ **47.** $\frac{x^2-3x-6}{x(x-1)}$ **49.** $\frac{x^2-4x-2}{(x-4)(x+4)}$
51. a. $\frac{8}{5}$ **b.** 1 **c.** $\frac{x^4+x^3+3x^2+2x+1}{x^3+x^2+2x+1}$

Exercises 7.4, pages 543 - 548

1. $x=7$ **3.** $x\neq 0, 2; x=4$ **5.** $x\neq -3, 4; x=-10$ **7.** $x\neq 0; x=18$ **9.** $x\neq 6; x=-\frac{74}{9}$ **11.** 360 defective computers
13. 28,800 students **15.** 34 miles **17.** 6.8 cups **19.** Width = 3 inches; Length = 7.5 inches **21.** $\overline{LK}=12, \overline{JB}=4$
23. $\overline{AC}=2, \overline{ST}=12$ **25.** $\overline{ST}=8, \overline{TU}=12, \overline{QR}=24$ **27.** $\overline{AP}=\frac{9}{2}$ in., $\overline{PC}=\frac{15}{2}$ in. **29.** $x=\frac{1}{4}$ **31.** $x=6$ **33.** $x=4$
35. $x\neq 0; x=\frac{10}{3}$ **37.** $x\neq 0; x=-\frac{3}{4}$ **39.** $x\neq 0; x=-\frac{3}{16}$ **41.** $x\neq -9, -\frac{1}{4}, 0; x=-2, 1$ **43.** $x\neq \frac{3}{2}, 0, 6; x=\frac{3}{5}, 9$
45. $x\neq \frac{1}{2}, 4; x=-3$ **47.** $x\neq -4, -1; x=2$ **49.** $x\neq -1, \frac{1}{4}; x=\frac{2}{3}$ **51.** $x\neq 2, 3; x=\frac{13}{10}$ **53.** $x\neq -\frac{2}{3}, 2$; No solution
55. $x\neq -1, \frac{1}{3}, \frac{1}{2}; x=\frac{1}{5}$ **57.** $r=\frac{S-a}{S}$ **59.** $s=\frac{x-\overline{x}}{z}$ **61.** $y=m(x-x_1)+y_1$ **63.** $R_{\text{total}}=\frac{R_1R_2}{R_1+R_2}$ **65.** $P=\frac{A}{1+r}$
67. a. $\frac{4x^2+41x-10}{x(x-1)}$ **b.** $x=\frac{1}{4}, 10$ and $x\neq 0, 1$ **69. a.** $\frac{10x-6}{(x-2)(x+2)}$ **b.** $x=\frac{7}{3}$ and $x\neq 2, -2$
71. a. $\frac{x^2-153}{2(x-9)(x+9)}$ **b.** $x=3, -3$ and $x\neq 9, -9$

Exercises 7.5, pages 556 - 559

1. 72, 45 **3.** 9 **5.** $\frac{6}{13}$ **7.** 36, 27 **9.** 7, 12 **11.** 45 shirts **13.** 12 brushes **15.** 1875 miles
17. Person: 505 minutes; Machine: 5.05 minutes **19.** 45 minutes **21.** Beth: 52 mph; Anna: 48 mph
23. Commercial airliner: 300 mph; Private plane: 120 mph **25.** 6 hours **27.** 50 mph **29.** $\frac{9}{4}$ or $2\frac{1}{4}$ hours
31. $\frac{9}{2}$ or $4\frac{1}{2}$ days; 9 days **33.** Boat: 14 mph; Current: 2 mph **35.** 14 kilometers per hour

Exercises 7.6, pages 566 - 571

1. $\frac{7}{3}$ **3.** 2 **5.** $-\frac{32}{9}$ **7.** 36 **9.** 120 **11.** $\frac{56}{3}$ **13.** 40 **15.** 54 **17.** $\frac{48}{5}$ **19.** 27 **21.** 400 feet **23.** \$59.70 **25.** 4.71 feet **27.** 6 m
29. 0.0073 cm **31.** 16,000 lb **33.** 9×10^{-11} N **35.** 15,000 lb **37.** 6400 lb **39.** 1.8 ft^3 **41.** 1700 g per $in.^2$ **43.** 15 ohms
45. 2.56 ohms **47.** 900 lb **49.** 5 ft from the 300 lb weight (or 20 feet from the 75 lb weight)
51. a. When two variables vary directly, an increase in the value of one variable indicates an increase in the other, and the ratio of the two quantities is constant. **b.** When two variables vary inversely, an increase in the value of one variable indicates a decrease in the other, and the product of the two quantities is constant. **c.** Joint variation is when a variable varies directly with more than one other variable. **d.** Combined variation is when a variable varies directly or inversely with more than one variable.

Chapter 7 Review, pages 576 - 580

1. $\frac{3}{16x}; x \neq 0, y \neq 0$ **2.** $-\frac{4}{3}; x \neq 3$ **3.** $\frac{3x+15}{x-5}; x \neq 5$ **4.** $x+4$ **5.** $\frac{2x-1}{x-3};\ x \neq -\frac{5}{3}, 3$ **6.** $\frac{x-1}{x+5};\ x \neq -5, -\frac{3}{2}$ **7.** $\frac{2x}{3(x-2)}$
8. $\frac{10x^2+6x}{5x-3}$ **9.** $\frac{2}{xy^3}$ **10.** $\frac{x}{21}$ **11.** $\frac{x^2-12x+35}{(x-2)(x+3)}$ **12.** $\frac{3x-9}{(x-4)(x+1)(x-2)}$ **13.** $\frac{5x+20}{4x(2x-1)}$ **14.** $\frac{x^2-4x}{(x+1)(x+6)}$
15. $\frac{2x^2-x}{2(2x+1)}$ **16.** $\frac{x^3+9x^2+17x+21}{2(x-3)(x+3)^2}$ **17.** 6 **18.** 8 **19.** $\frac{6x+10}{(x+2)(x-2)}$ **20.** $\frac{x+1}{x-11}$ **21.** $\frac{2x-8}{(x+5)(x-2)}$
22. $\frac{x^4-4x^3-2x^2+1}{(x-1)^2(x+1)^2}$ **23.** $\frac{8x^2+8x+1}{(x+1)(3x+1)(2x+3)}$ **24.** $\frac{11}{2x-3}$ **25.** $\frac{9}{x+4}$ **26.** $\frac{20-x^2}{(x+4)(x+5)}$ **27.** $\frac{x^2}{(x-1)(x^2+x+1)}$
28. $\frac{8x^2+6x+2}{(x-1)(x+3)(3x+1)}$ **29.** $\frac{14y^2}{5x}$ **30.** $\frac{13}{2(5y-1)}$ **31.** $\frac{y+2}{4}$ **32.** $\frac{-2}{(x+2)^2}$ **33.** $\frac{x-2}{x}$ **34.** $\frac{x}{x^2+x+1}$ **35.** $\frac{2x-2}{x^2}$
36. $\frac{4x-12}{x^2-6x-9}$ **37.** $\frac{x}{x-5}$ **38.** $\frac{-2x+10}{x+4}$ **39.** $-\frac{9}{x+2}$ **40.** $\frac{15}{2x}$ **41.** $\frac{20}{3y+2}$ **42.** $\frac{12y+24}{(y+3)^2(y-3)}$ **43.** $x \neq -3, 3; x = 5$
44. $x \neq 0, 3; x = 1, 12$ **45.** 60 gallons **46.** 20 women **47.** 600 times **48.** 40 ft **49.** $\overline{AB} = 8, \overline{FE} = 5$ **50.** $\overline{TU} = 6, \overline{VW} = 7$
51. $x \neq 0; x = -\frac{4}{25}$ **52.** $x \neq -\frac{2}{3}, \frac{4}{7}; x = \frac{8}{53}$ **53.** $x \neq 0, 1; x = \frac{5}{2}$ **54.** $x \neq -3, 3; x = 12$ **55.** $x \neq -3, 4; x = 0, \frac{25}{6}$
56. $x \neq -6; x = -\frac{15}{2}, 4$ **57.** $x_1 = x - \frac{y-y_1}{m}$ **58.** $a_1 = \frac{a_n(1-r)}{1-r^n}$ **59.** 18, 27 **60.** 15 women **61.** $\frac{6}{8}$ **62.** $\frac{17}{7}$ **63.** 21, 27
64. 3 hours; 6 hours **65.** 12 hours **66.** Alice takes 6 hrs; Judy takes 4 hrs **67.** Tyler takes 2 hrs; His son takes 6 hrs
68. 6 mph **69.** 24 mph **70.** Initial rate was 37.5 mph and final rate was 62.5 mph
71. Initial rate was 400 mph and final rate was 430 mph **72.** Raphael takes 1.5 hrs; Matilde takes 3 hrs **73.** 6 **74.** 5
75. $\frac{490}{9}$ **76.** 31,250 **77.** $\frac{100}{7}$ in. **78.** 18.84 in. **79.** 113.04 in.^{2} **80.** 162 rpm **81.** 576 feet **82.** 16.5 ohms **83.** 37.68 cubic feet
84. 5 cups of lemon juice

Chapter 7 Test, pages 581 - 582

1. $\frac{x}{x+4}; x \neq -4, -3$ **2.** $\frac{1}{2x+5}; x \neq -\frac{5}{2}$ **3.** $\frac{x+3}{x+4}$ **4.** $\frac{3x-2}{3x+2}$ **5.** $\frac{-2x^2-13x}{(x+5)(x+2)(x-2)}$ **6.** $\frac{-2x^2-7x+18}{(3x+2)(x-4)(x+1)}$ **7.** $2x^2$
8. $\frac{x^2-7x+1}{(x+3)(x-3)}$ **9.** $\frac{3}{2xy^3}$ **10.** $\frac{-3x}{x-2}$ **11.** $-\frac{1}{xy}$ **12.** $\frac{12x+24}{(x-3)(x+3)^2}$ **13. a.** $\frac{7x+11}{2x(x+1)}$ **b.** $x = \frac{1}{3}$ **14.** $x \neq -4; x = 21$
15. $x \neq 0; x = -\frac{7}{2}$ **16.** $x \neq -4, 1; x = -1$ **17.** $x \neq -2, -1; x = 1$ **18.** $n = \frac{2S}{a_1+a_n}$ **19.** $x = \frac{y-b}{m}$ **20.** $\overline{AC} = 36, \overline{DC} = 24$ **21.** $\frac{2}{7}$
22. 4 hours **23.** Carlos: 42 mph; Mario: 57 mph **24.** 11 mph **25.** $z = 24$ **26.** $z = \frac{400}{9}$ **27.** $\frac{15}{2}$ cm **28.** 5.13 in.^{3}

Cumulative Review: Chapters 1-7, pages 583 - 586

1. 

2. a. $\frac{19}{12}$ **b.** $-\frac{3}{14}$ **c.** $\frac{1}{4}$ **d.** $\frac{1}{12}$ **3.** 5 **4.** $3x^2-5x-28$ **5.** $4x^2-20x+25$ **6.** $6x-23$ **7.** $10x+4$

8. 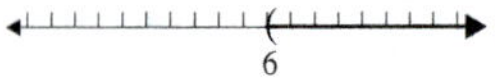

$(6, \infty)$ **9. a.** 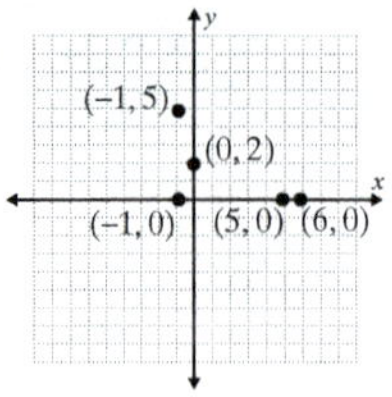

b. $D = \{-1, 0, 5, 6\}$ **c.** $R = \{0, 5, 2\}$ **d.** It is not a function because the x-coordinate −1 has more than one corresponding y-coordinate. **10. a.** 8 **b.** −1 **c.** $-\frac{59}{27}$

11. 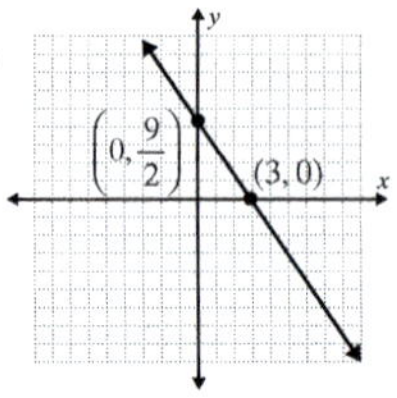

12. $y = \frac{3}{4}x + \frac{11}{2}$;
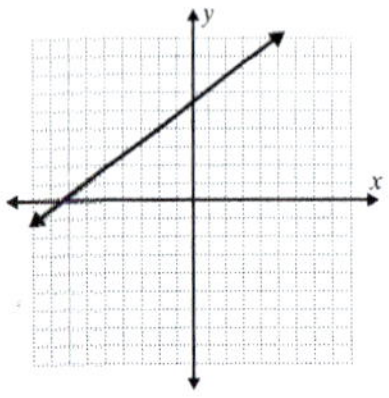
13. $y = \frac{1}{3}x + \frac{2}{3}$;
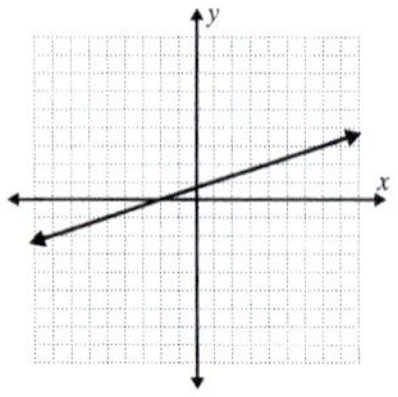
14. $x = -3$;
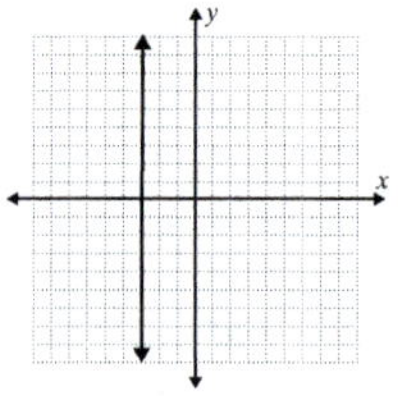
15. $2x + y = -8$;
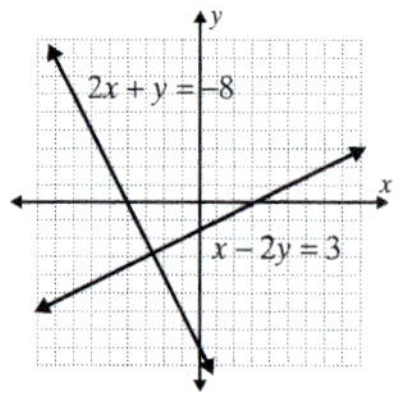

16. Function $D = \{-4, -2, 0, 2, 4\}; R = \{-2\}$ **17.** Function $D = (-\infty, \infty); R = (-\infty, \infty)$ **18.** 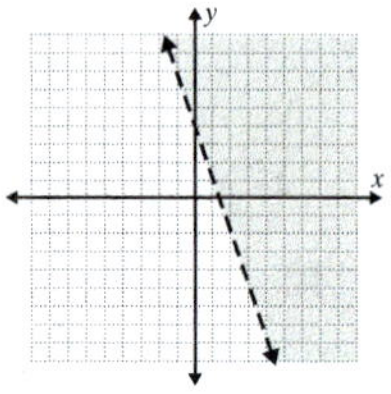
19. $8x^2$
20. $5x^3y^2$ **21.** $(3x + y)(y + 1)$ **22.** $(2x - 5)(2x + 3)$ **23.** $(3x - 2)(2x - 1)$
24. $(2x + 5)(2x + 3)$ **25.** $2(x + 5)(x - 2)$ **26.** $2x(3x + 1)(x - 4)$
27. $(3x^3 + 2y)(3x^3 - 2y)$ **28.** $(2x + 5)(4x^2 - 10x + 25)$ **29.** $x^2 + 3x - 18 = 0$

30. $x^3 + 4x^2 + 3x = 0$ **31.** $\frac{3}{5}x + 2y + \frac{y^2}{x}$ **32.** $1 - 3y + \frac{8}{7}y^2$ **33.** $x - 15 - \frac{1}{x + 1}$ **34.** $2x^2 - x + 3 - \frac{2}{x + 3}$

35. $\frac{4x^2}{x - 3}; x \neq -\frac{1}{2}, 3$ **36.** $x - 3; x \neq -3$ **37.** $\frac{1}{x + 1}; x \neq -1, 0$ **38.** $\frac{x + 5}{2(x - 3)}; x \neq 3$ **39.** $x - y$ **40.** $\frac{x}{3(x + 1)}$ **41.** $\frac{x + 4}{3x}$

42. $\frac{3x^3 + 12x^2 + 12x}{x + 3}$ **43.** $\frac{2x + 1}{x - 1}$ **44.** $\frac{2x^2 + 14x - 8}{(x + 3)(x - 1)(x - 2)}$ **45.** $\frac{2x + 4}{(x - 1)(x + 4)}$ **46.** $\frac{x - 4}{(x + 2)(x - 2)}$ **47.** $\frac{19}{14}$ **48.** $\frac{x + 4}{6x}$

49. $x = -\frac{3}{5}$ **50.** $x = -5, 6$ **51.** $x = -\frac{5}{2}$ **52.** $x = -4, 0, 5$ **53.** $x = 2, 5$ **54.** $x = -\frac{7}{2}, -\frac{3}{10}$ **55.** $x = -45$

56. a. $\frac{9 - 2x}{x(x + 3)}$ **b.** $x = \frac{9}{2}$ **57. a.** $\frac{6x^2 - 16x - 20}{(x - 4)(x + 2)}$ **b.** $x = 7$ **58.** $t = \frac{A - P}{Pr}$ **59.** $P = 30$ in.; $A = 40$ in.^{2} **60.** $\frac{1}{2}$ year

61. She can spend anywhere from 9 to 27 dollars.

62. a. and **b.**
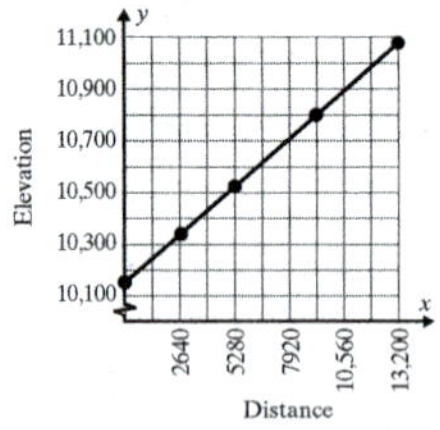

c. Slope = 0.07 for all **d.** For every 100 feet the hikers travel there is a 7 ft change in elevation
63. −6, −5, −4 or 7, 8, 9 **64.** Train's speed is 60 miles per hour; Speed of airplane is 230 mph
65. The father takes $2\frac{2}{3}$ hours and the daughter takes 8 hours **66.** 3 mph **67.** $7\frac{9}{13}$ cm
68. 5.29 minutes

Chapter 8

Exercises 8.1, pages 595 - 598

1. c **3.** a, c **5.** $(0, 2)$ **7.** $(4, 2)$ **9.** $m_1 = -2, b_1 = 3; m_2 = -2, b_2 = \frac{5}{2}$ **11.** $m_1 = \frac{1}{2}, b_1 = 3; m_2 = \frac{1}{2}, b_2 = -\frac{1}{2}$

13. $(-3,-2)$

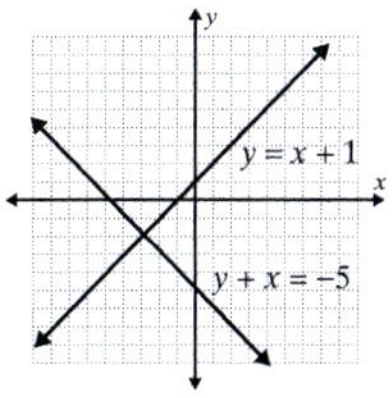

15. $(2,0)$

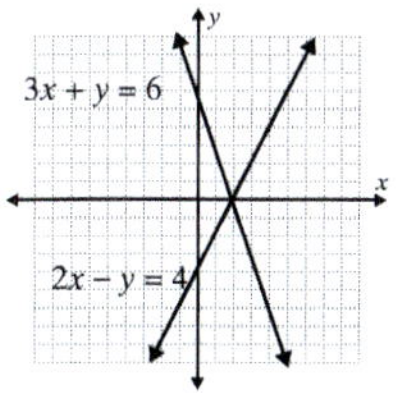

17. $(x,-x+5)$

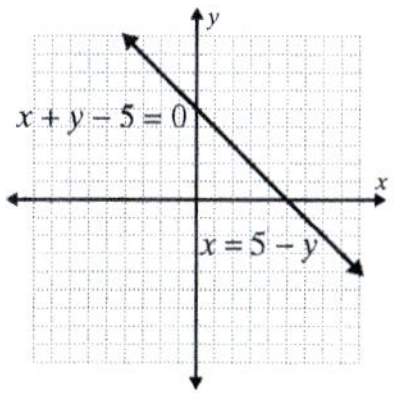

19. No solution

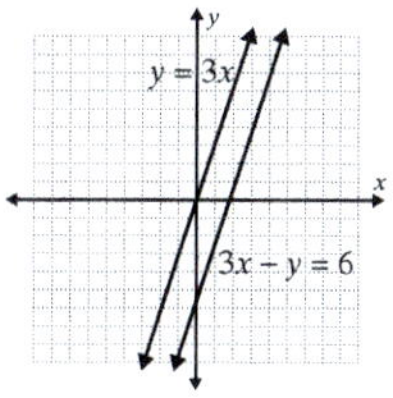

21. $(-3,-8)$

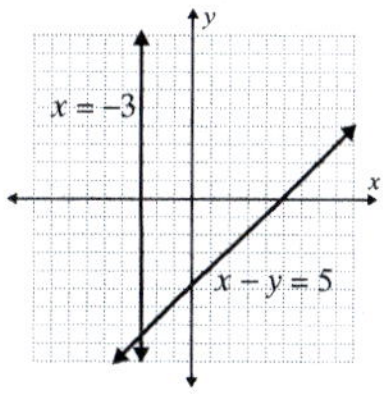

23. $(2,3)$

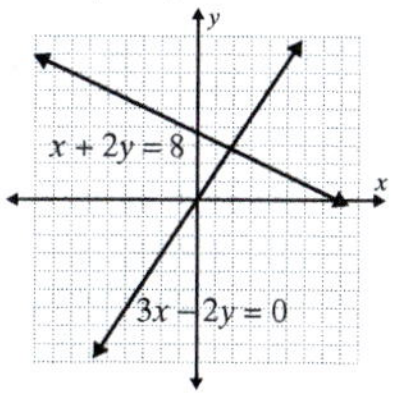

25. $\left(x,\frac{5}{4}x-\frac{5}{4}\right)$

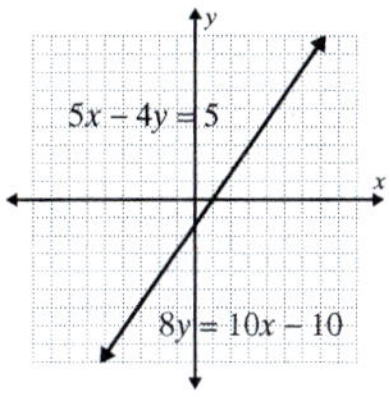

27. $(-1,-1)$

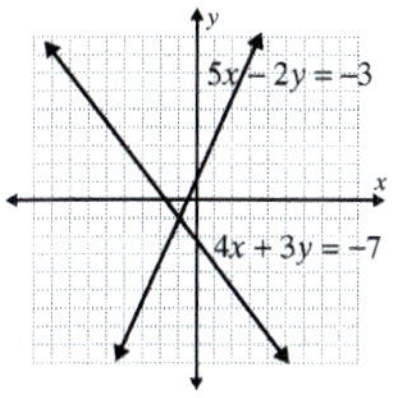

29. $(5,-1)$

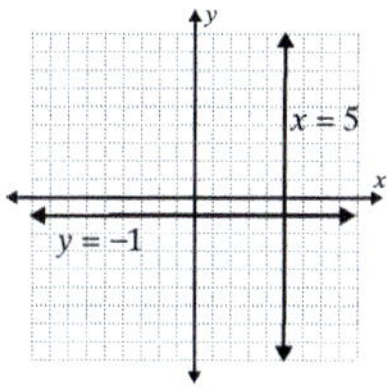

31. No solution

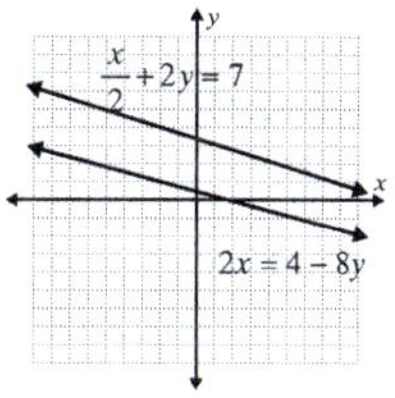

33. $(1,3)$

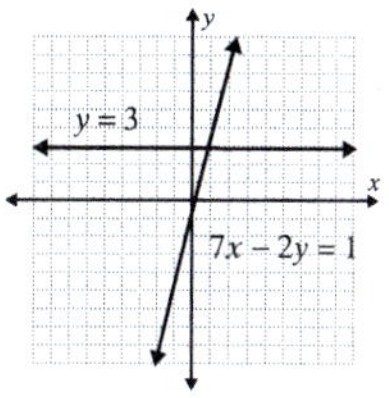

35. $\left(x,\frac{1}{2}x+2\right)$

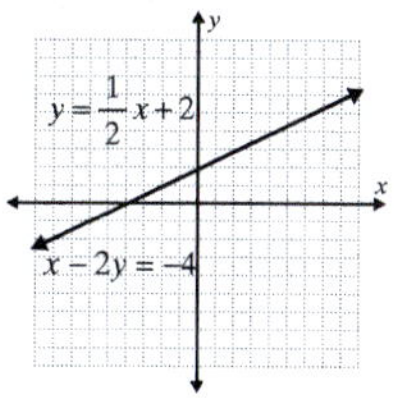

37. $(1,1)$

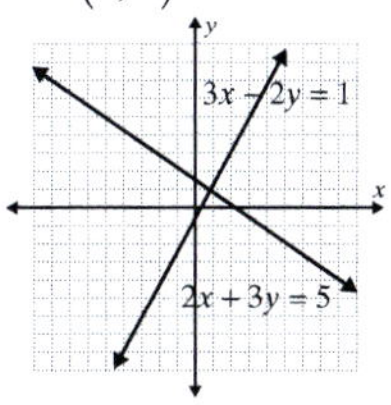

39. No solution

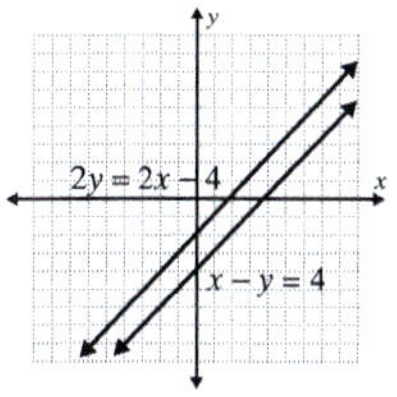

41. $(1,-1)$

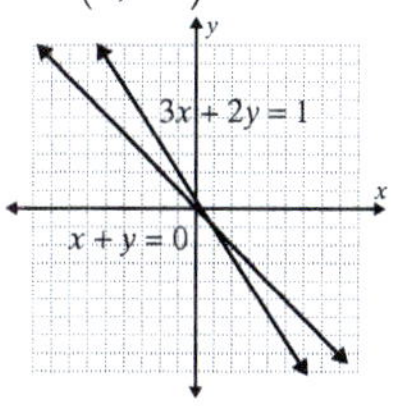

43. $(20,5)$

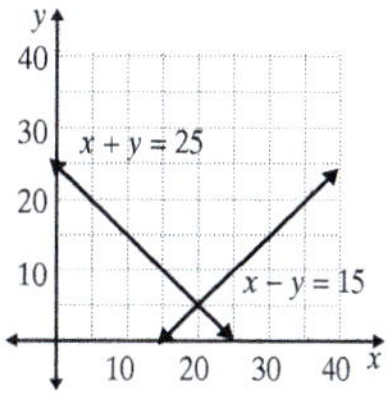

45. 5 gallons of 12%; 10 gallons of 3%

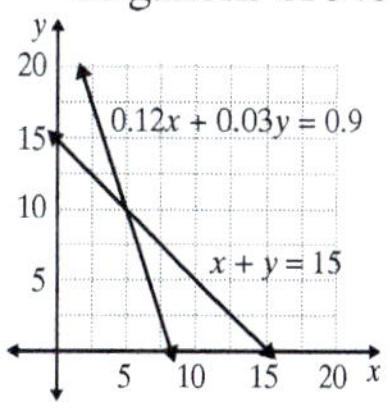

47. $(1,4)$

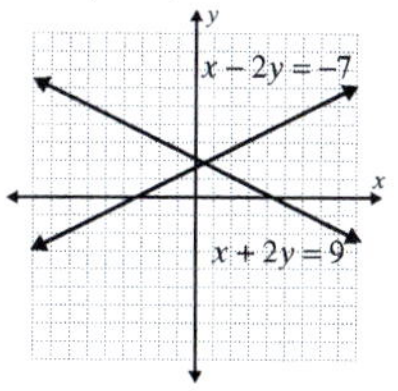

49. $(1.5,2)$

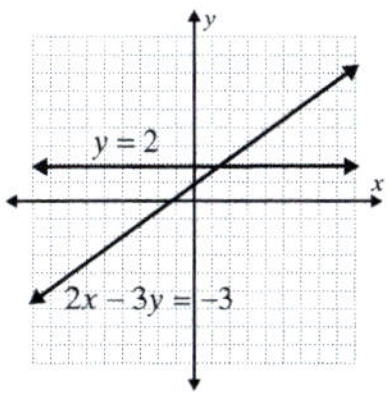

51. $(1.5,-3)$

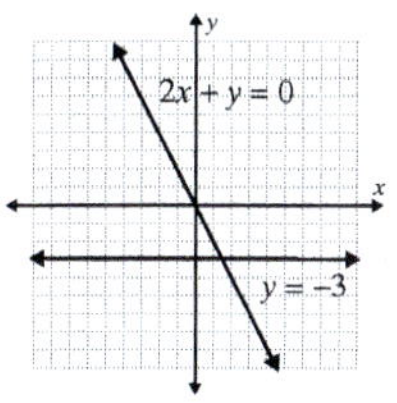

53. $(1.4,2.1)$

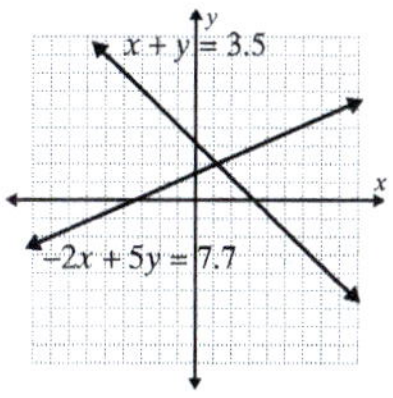

55. The solution to a consistent system of linear equations is a single point, which is easily written as an ordered pair.

Exercises 8.2, pages 603 - 605

1. $(2,4)$ **3.** $(x,3x-7)$ **5.** $(-6,-2)$ **7.** $(4,1)$ **9.** No solution **11.** $(3,2)$ **13.** $(4,-5)$ **15.** $(3,-2)$ **17.** $\left(2,\frac{5}{2}\right)$ **19.** $\left(\frac{1}{2},-4\right)$ **21.** $\left(\frac{7}{2},-\frac{1}{2}\right)$ **23.** $(x,2-3x)$ **25.** $(-2,1)$ **27.** $\left(-\frac{4}{5},-\frac{7}{5}\right)$ **29.** $\left(\frac{11}{7},\frac{8}{7}\right)$ **31.** $\left(x,\frac{1}{6}x+\frac{10}{3}\right)$ **33.** $(3,3)$ **35.** $(10,20)$ **37.** No solution **39.** $\left(x,-\frac{3}{2}x+12\right)$ **41.** $\left(2,\frac{5}{3}\right)$ **43.** $(20,5)$ **45.** 5 gallons of 12%; 10 gallons of 3% **47.** Answers will vary.

Exercises 8.3, pages 611 - 613

1. $(4,3)$ **3.** $\left(1,-\frac{3}{2}\right)$ **5.** No solution **7.** $(x,3-x)$ **9.** $(-2,-3)$ **11.** $(7,5)$ **13.** $(1,-5)$ **15.** $\left(x,\frac{1}{2}x-2\right)$ **17.** $\left(\frac{22}{7},-\frac{2}{7}\right)$

19. $(2,-2)$ **21.** $(x, 2x-4)$ **23.** $(2,-1)$ **25.** $(5,-6)$ **27.** $(3,-1)$ **29.** $(2,4)$ **31.** $(-6,2)$ **33.** No solution **35.** $(20,10)$
37. $\left(2, \frac{10}{9}\right)$ **39.** No solution **41.** $\left(-\frac{45}{7}, \frac{92}{7}\right)$ **43.** $y = 5x - 7; m = 5, b = -7$ **45.** $y = -3; m = 0, b = -3$
47. $y = \frac{1}{2}x + \frac{3}{2}; m = \frac{1}{2}, b = \frac{3}{2}$ **49.** \$4000 at 10%; \$6000 at 6%
51. 40 liters of 30% solution (x), 60 liters of 40% solution (y) **53.** Answers will vary.

Exercises 8.4, pages 619 - 625

1. 23, 33 **3.** 15, 21 **5.** 37, 50 **7.** 80°, 100° **9.** 55°, 55°, 70° **11.** Rate of boat = 10 mph; Rate of current = 2 mph
13. Bolt's speed was 10.32 meters per second and the wind speed was 0.12 meters per second.
15. He traveled $1\frac{1}{2}$ hours at the first rate and 2 hours at the second rate. **17.** Marcos traveled at 40 mph and Cana traveled at 51 mph. **19.** Steve traveled at 28 mph and Tim traveled at 7 mph. **21.** The westbound train was traveling 45 mph and the eastbound train was traveling 40 mph. **23.** He jogged about 12 miles. **25.** 20 nickels and 10 dimes
27. 52 nickels and 130 pennies **29.** Length is 14 meters and width is 8 meters **31.** Length is 40 feet and width is 30 feet
33. 100 yards × 45 yards **35.** Ava is 12 years old and Curt is 4 years old. **37.** 800 adults and 2700 students attended.
39. She bought 10 paperbacks and 5 hardbacks. **41.** 5000 general admission and 7500 reserved tickets were sold.
43. 22 at \$625 and 25 and \$550 **45.** 360 balls or 30 dozen **47.** The store sold 22 of the \$95 jackets and 18 of the \$120 jackets. **49.** One Big Mac costs \$3.58 and one order of medium French Fries costs \$1.79. **51.** They produced 7 of Model X and 10 of Model Y. **53.** The number is 49.

Exercises 8.5, pages 630 - 633

1. \$5500 at 6%; \$3500 at 10% **3.** \$7400 at 5.5%; \$2600 at 6% **5.** \$450 at 8%; \$650 at 10% **7.** \$3500 in each or \$7000 total
9. \$20,000 at 24%; \$11,000 at 18% **11.** \$800 at 5%; \$2100 at 7% **13.** \$8500 at 9%; \$3500 at 11% **15.** \$87,000 in bonds; \$37,000 in certificates **17.** 20 pounds of 20%; 30 pounds of 70% **19.** 20 ounces of 30%; 30 ounces of 20% **21.** 450 pounds of 35%; 1350 pounds of 15% **23.** 20 pounds of 40%; 30 pounds of 15% **25.** 10 g of acid; 20 g of the 40% solution
27. 10 oz of salt; 50 oz of the 4% solution **29.** 2 lb of 72%; 4 lb of 42% **31.** 27 oz of 25%; 9 oz of 45%
33. Answers will vary.

Exercises 8.6, pages 641 - 645

1. $(1,0,1)$ **3.** $(1,2,-1)$ **5.** Infinitely many solutions **7.** $(4,1,1)$ **9.** $(1,2,-1)$ **11.** $(-2,3,1)$ **13.** No solution **15.** $(3,-1,2)$
17. $(2,1,-3)$ **19.** $\left(\frac{1}{2}, \frac{1}{3}, -1\right)$ **21.** 34, 6, 27 **23.** 18 ones, 16 fives, 12 tens **25.** 19 cm, 24 cm, 30 cm
27. 300 main floor, 200 mezzanine, and 80 balcony **29.** 3 lillies, 5 roses, and 8 daisies **31.** \$2.80
33. Savings: \$30,000; Bonds: \$55,000; Stocks: \$15,000 **35.** 3 liters of 10%, 4.5 liters of 30%, 1.5 liters of 40%
37. No. Graphically, the three planes intersect in one point (one solution) or in a line (infinitely many solutions) or they do not have a common intersection (no solution). **39.** $A = 4, B = 2, C = -1$

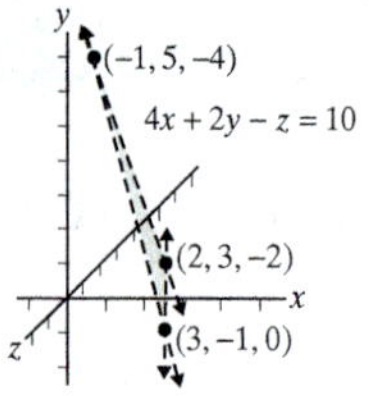

Exercises 8.7, pages 656 - 658

1. $\begin{bmatrix} 2 & 2 \\ 5 & -1 \end{bmatrix}, \left[\begin{array}{cc|c} 2 & 2 & 13 \\ 5 & -1 & 10 \end{array}\right]$ **3.** $\begin{bmatrix} 7 & -2 & 7 \\ -5 & 3 & 0 \\ 0 & 4 & 11 \end{bmatrix}, \left[\begin{array}{ccc|c} 7 & -2 & 7 & 2 \\ -5 & 3 & 0 & 2 \\ 0 & 4 & 11 & 8 \end{array}\right]$ **5.** $\begin{bmatrix} 3 & 1 & -1 & 2 \\ 1 & -1 & 2 & -1 \\ 0 & 2 & 5 & 1 \\ 1 & 3 & 0 & 3 \end{bmatrix}, \left[\begin{array}{cccc|c} 3 & 1 & -1 & 2 & 6 \\ 1 & -1 & 2 & -1 & -8 \\ 0 & 2 & 5 & 1 & 2 \\ 1 & 3 & 0 & 3 & 14 \end{array}\right]$ **7.** $\begin{cases} -3x + 5y = 1 \\ -x + 3y = 2 \end{cases}$

9. $\begin{cases} x + 3y + 4z = 1 \\ 2x - 3y - 2z = 0 \\ x + y = -4 \end{cases}$ **11.** $(-1, 2)$ **13.** $(-1, -1)$ **15.** $(-1, -2, 3)$ **17.** $(1, 0, 1)$ **19.** $(2, 1, -1)$ **21.** $(-2, 9, 1)$ **23.** No solution **25.** Infinitely many solutions **27.** $(1, -3, 2)$ **29.** 52, 40, 77 **31.** Bacon: \$3.09/lb; Eggs: \$4.03/doz; Bread: \$1.40/loaf **33.** $(0, -4)$ **35.** $(2, 1, 7)$ **37.** $(2, -3, 4)$

39. $\left(\frac{13}{12}, \frac{5}{4}, \frac{8}{3}\right)$ **41.** Solving the second equation for z, we can back substitute into the first equation, eliminating z. The result is the equation $x + 5y = 6$ which means the system has an infinite number of solutions.

Exercises 8.8, pages 663 - 664

1.
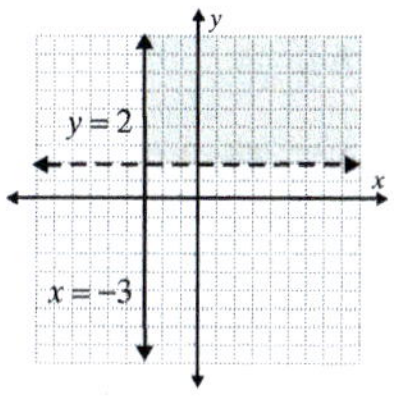

3.
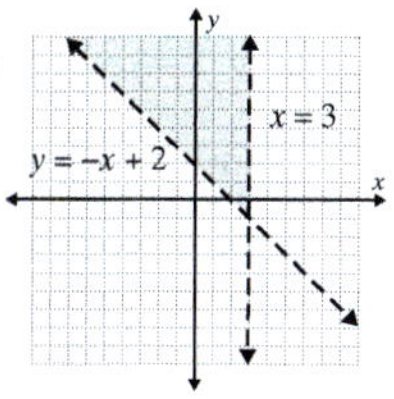

5.
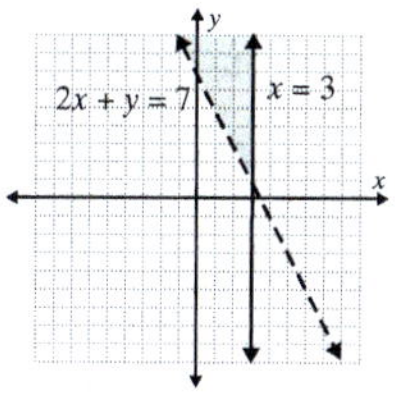

7.
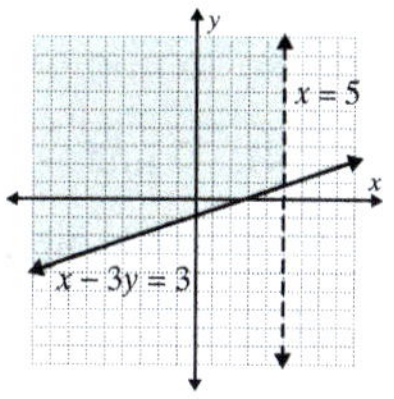

9.
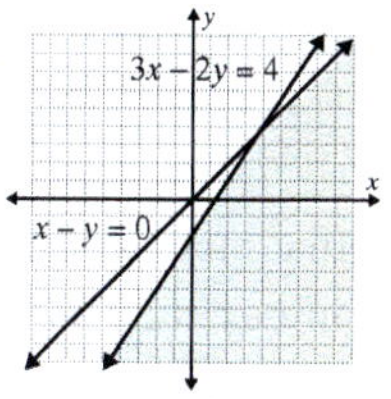

11.
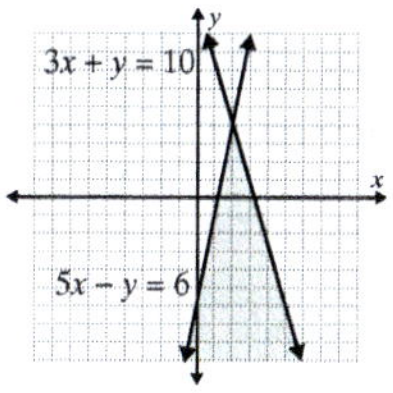

13.
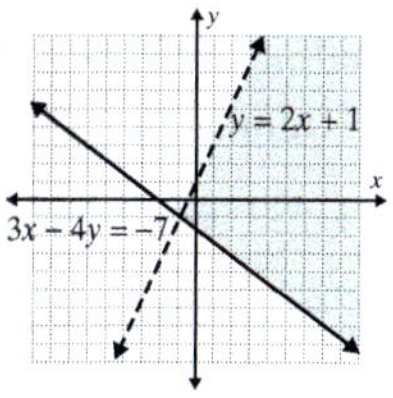

15.
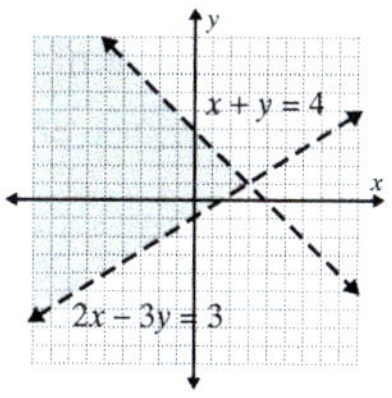

17.
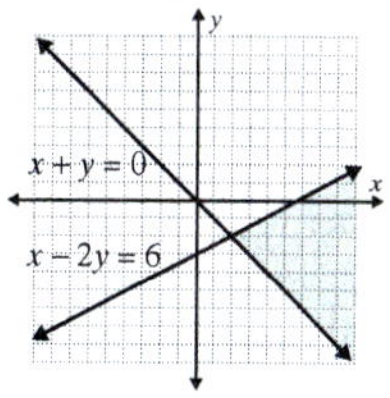

19.
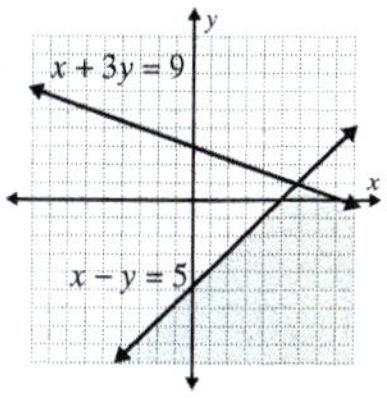

21. No solution

23.
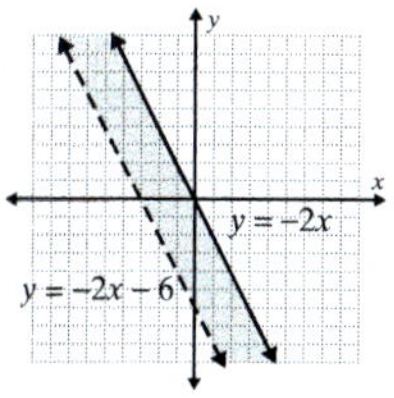

25.
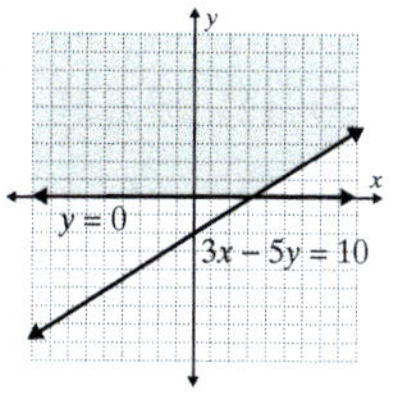

27.
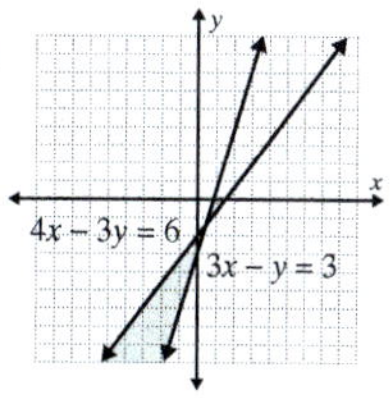

29.
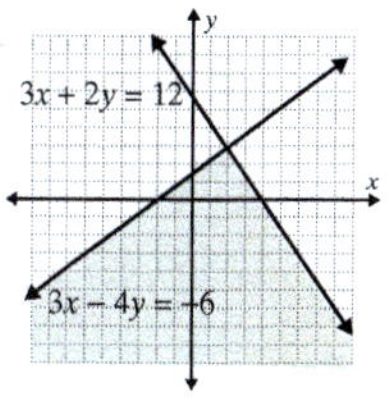

31.
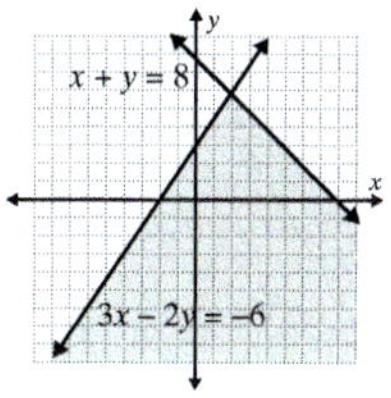

33.
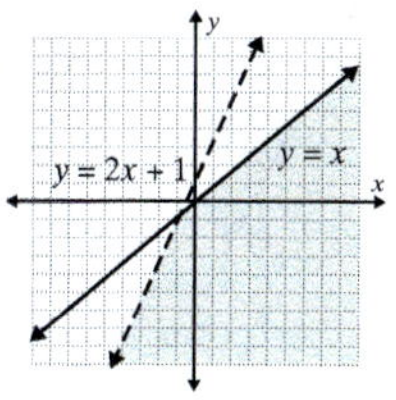

35.
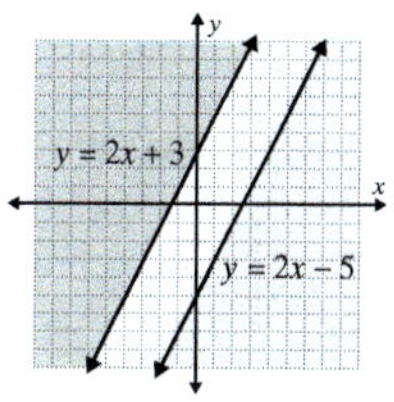

The solutions of the two inequalities do not overlap.

Chapter 8 Review, pages 670 - 674

1. a **2.** b, c, d **3.** $m_1 = 2, b_1 = 0; m_2 = 2, b_2 = 5$ **4.** $m_1 = -3, b_1 = 4; m_2 = -3, b_2 = -\frac{1}{2}$

5. $(1, 3)$

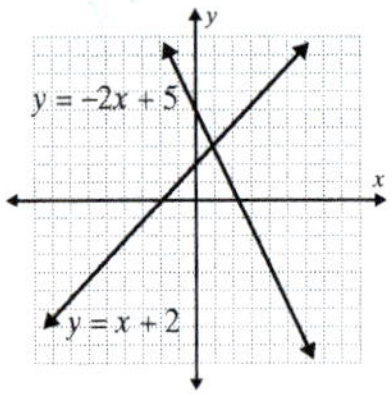

6. No solution

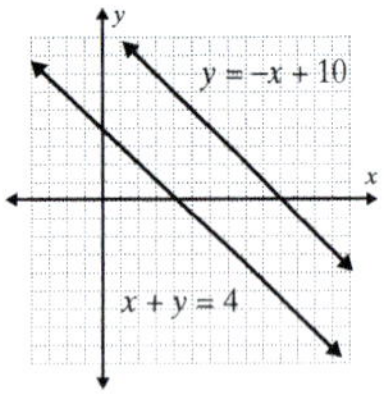

7. $(2, -1)$

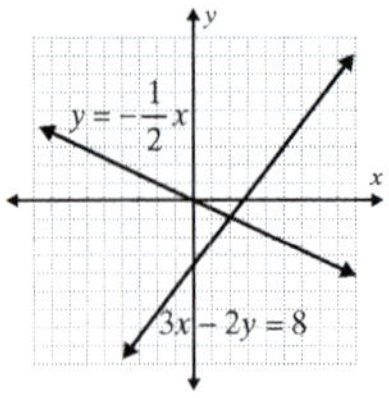

8. $(x, 4-5x)$

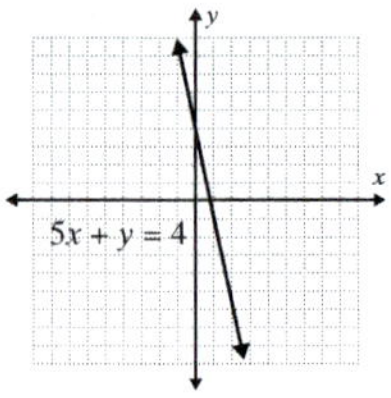

9. No solution

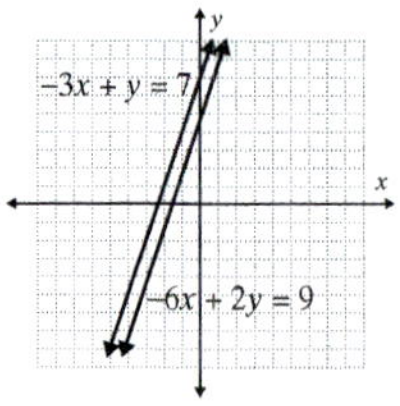

10. $(-1, 4)$

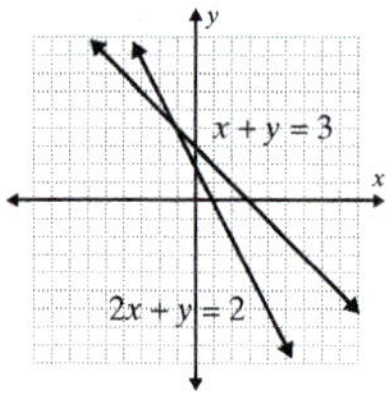

11. $(0, 6)$

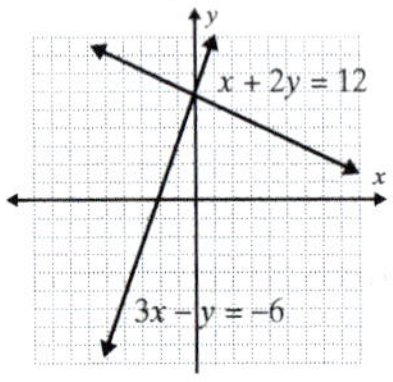

12. $\left(x, -\frac{1}{4}x+\frac{3}{2}\right)$

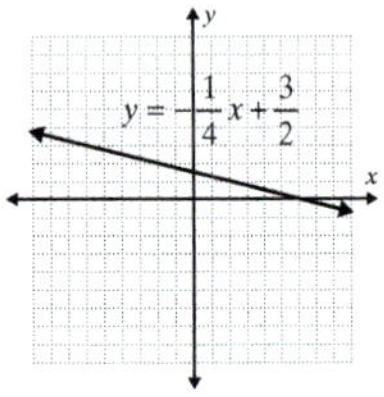

13. $(1.5, 2.5)$

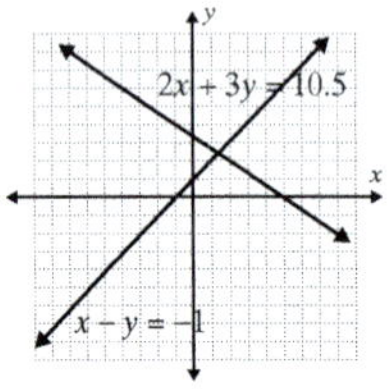

14. $(-2, -1.5)$

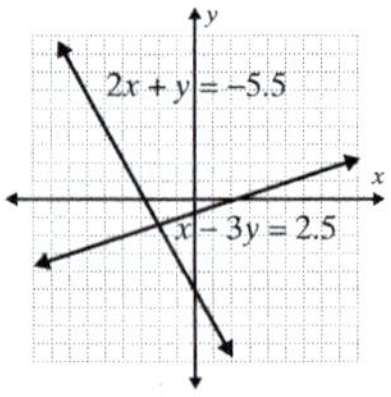

15. $(0, 7)$

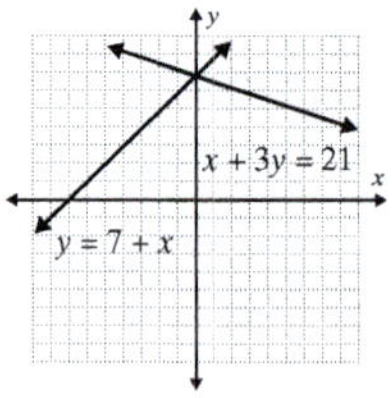

16. $(5.75, 1.45)$

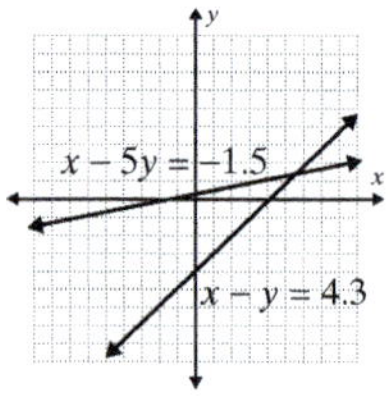

17. $(1, 2)$ **18.** $(x, 4x+1)$ **19.** No solution **20.** No solution **21.** $(x, 5x-6)$ **22.** $\left(\frac{8}{3}, -\frac{1}{3}\right)$ **23.** $(0, -2)$ **24.** $(5, 6)$ **25.** $\left(\frac{3}{4}, \frac{1}{4}\right)$ **26.** No solution **27.** $\left(-\frac{2}{5}, -3\right)$ **28.** $(5, -1)$ **29.** $\left(-\frac{3}{2}, 1\right)$ **30.** $(7, 2)$ **31.** $(2, -1)$ **32.** $\left(\frac{49}{11}, \frac{46}{11}\right)$ **33.** $\left(\frac{3}{7}, \frac{2}{7}\right)$ **34.** $(-1, 6)$ **35.** $(x, 3x+6)$ **36.** No solution **37.** No solution **38.** $\left(x, \frac{1}{3}x+10\right)$ **39.** $y=-4x-1$; $m=-4, b=-1$ **40.** $y=-\frac{3}{8}x+2$; $m=-\frac{3}{8}, b=2$ **41.** −7 and −5 **42.** 5 and 7 **43.** Alice is 11 years old and John is 3 years old. **44.** 2 quarters and 16 dimes **45.** He averaged 2 mph for the first 3.6 hours and 3 mph for the last 2.4 hours. **46.** 36 minutes at 40 mph and 36 minutes at 20 mph **47.** Length = 22.5 meters; width = 17.5 meters **48.** 130°, 50° **49.** They sold 30 of the $110 shirt and 20 of the $65 dollar shirt. **50.** Popcorn: $2.75; Funnel Cake: $4.10 **51.** $15,000 at 6%; $5000 at 8% **52.** $4000 at 8%, $8000 at 5% **53.** $7500 at each rate **54.** $22,000 in each type of investment **55.** 20 gallons of 25% salt; 40 gallons of 40% salt **56.** 40 pounds of 22% fat; 40 pounds of 10% fat **57.** 30 ounces of pure acid, 20 ounces of 10% acid **58.** 80 tons of 20% alloy, 20 tons of 60% alloy **59.** $(3, 1, 6)$ **60.** Infinitely many solutions **61.** $(0, 1, 5)$ **62.** No solution **63.** $\left(3, -4, \frac{1}{2}\right)$ **64.** $\left(2, 0, -\frac{1}{2}\right)$ **65.** 20, 21, and 24 **66.** $10,000 in savings, $15,000 in stocks, and $25,000 in bonds **67.** 8 nickels, 10 dimes, and 16 quarters **68.** Longest side = 20 m; Shortest side = 10 m; Other side = 12 m **69.** $\begin{bmatrix} 2 & -1 \\ 7 & 1 \end{bmatrix}, \left[\begin{array}{cc|c} 2 & -1 & 2 \\ 7 & 1 & 5 \end{array}\right]$ **70.** $\begin{bmatrix} 2 & -1 & 3 \\ 4 & 0 & -1 \\ 1 & -9 & 2 \end{bmatrix}, \left[\begin{array}{ccc|c} 2 & -1 & 3 & 5 \\ 4 & 0 & -1 & 1 \\ 1 & -9 & 2 & 3 \end{array}\right]$ **71.** $(4, -6)$ **72.** $(-4, 5)$ **73.** $(11, 24, 22)$ **74.** $(6, 1, -1)$ **75.** $(1, -10, -6)$ **76.** $(4, -1, -2)$ **77.** $(5, 7)$ **78.** $\left(\frac{17}{11}, \frac{10}{11}, -\frac{16}{11}\right)$

79.

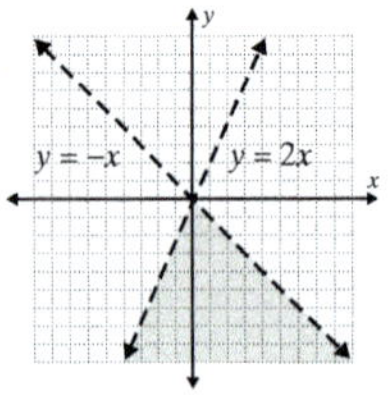

80.

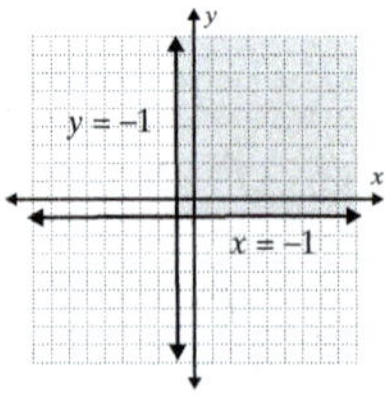

81.

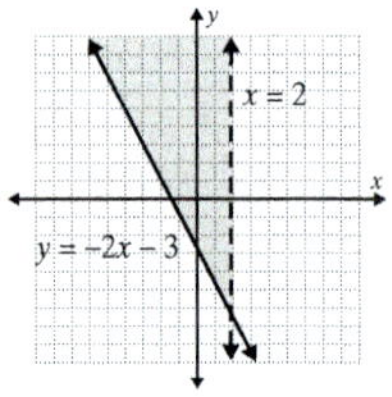

82.

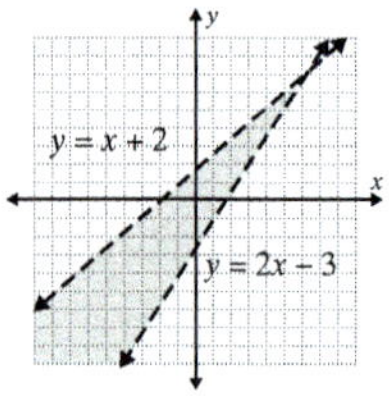

83.

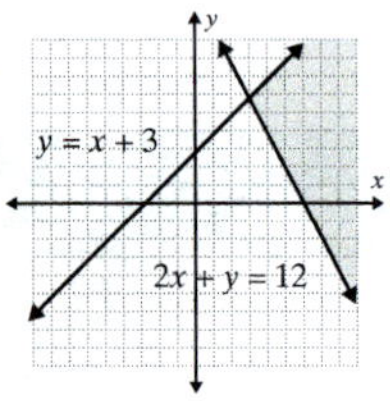

84.

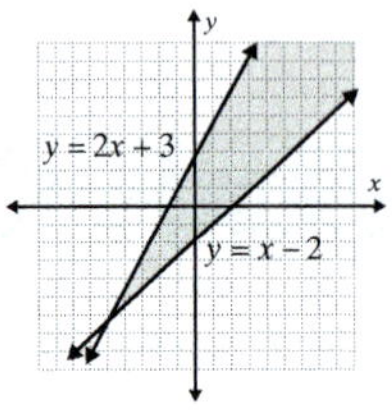

85.

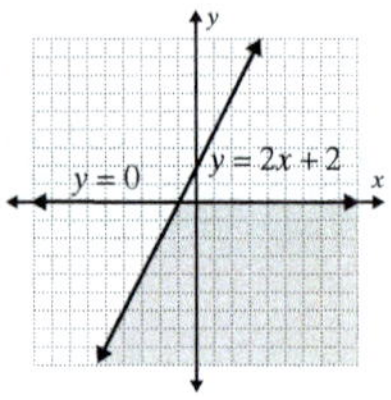

86.

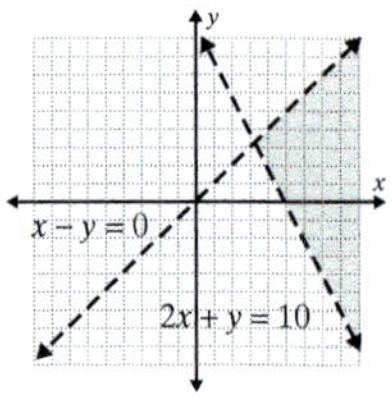

Chapter 8 Test, pages 675 - 676

1. c **2.** $\left(x, -\frac{2}{3}x+3\right)$ **3.** $(4, 1)$ **4.** $(-8, -20)$ **5.** $(-2, 6)$ **6.** No solution **7.** $\left(4, -\frac{1}{2}\right)$ **8.** $(-3, 5)$

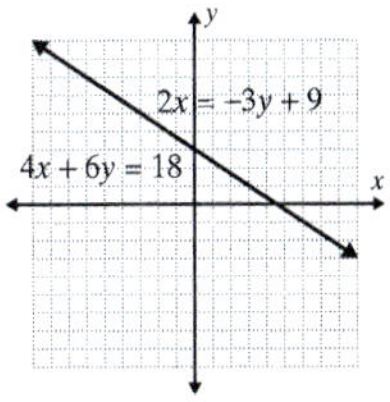

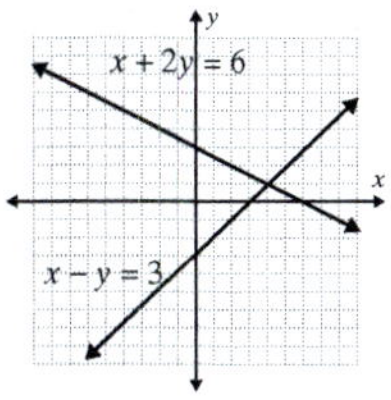

9. No solution **10.** $(3, -6)$ **11.** $\left(x, -\frac{1}{3}x-\frac{2}{3}\right)$ **12.** $(-1, 1, -2)$

13. No solution **14. a.** $\begin{bmatrix} 1 & 2 & -3 \\ 1 & -1 & -1 \\ 1 & 3 & 2 \end{bmatrix}$ $\left[\begin{array}{ccc|c} 1 & 2 & -3 & -11 \\ 1 & -1 & -1 & 2 \\ 1 & 3 & 2 & -4 \end{array}\right]$ **b.** $(1, -3, 2)$

15.

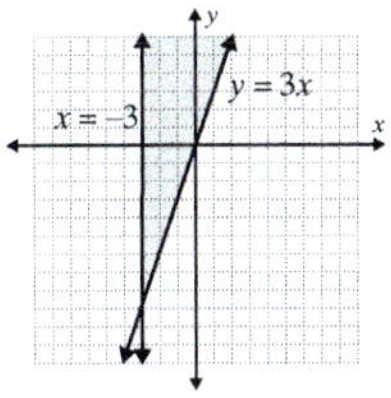

16.

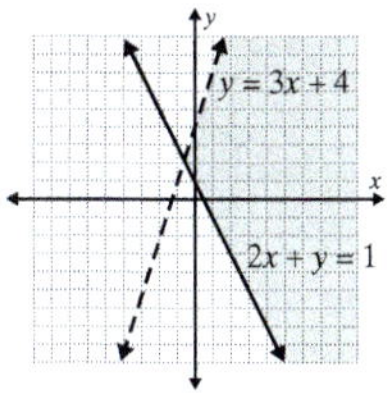

17. Soy latte: \$60.20; Iced coffee: \$18.90

18. Speed of the boat is 14 mph; Speed of the current is 2 mph

19. Pen price: \$0.79; Pencil price: \$0.08 **20.** \$1600 at 8% and \$960 at 6%

21. 17 inches by 13 inches **22.** 1600 lb of 83% and 400 lb of 68%

23. Nickels = 45; Quarters = 60

24. 41¢ stamps: 60; 58¢ stamps: 20; 75¢ stamps: 10

Cumulative Review: Chapters 1-8, pages 677 - 680

1. a. $\left\{-9, 3, \sqrt{25}\right\}$ **b.** $\left\{-9, \frac{1}{4}, 3, \sqrt{25}\right\}$ **c.** $\left\{-\sqrt{7}, \frac{\pi}{2}\right\}$ **d.** $\left\{-9, -\sqrt{7}, \frac{1}{4}, \frac{\pi}{2}, 3, \sqrt{25}\right\}$ **2.** 840 **3.** $36x^2y^3$ **4.** $2(x+3)^2(x-3)$

5. $x = -\frac{9}{8}$ **6.** $x = \frac{29}{11}$ **7.** $x = -\frac{4}{3}, 4$ **8.** $\left(-\infty, -\frac{1}{4}\right]$

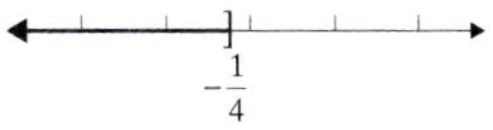

9. $\left(-\infty, -\frac{50}{3}\right)$

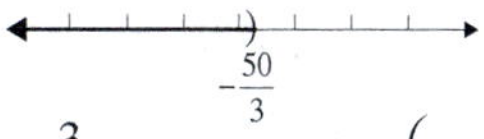

10. $\left[-\frac{8}{3}, 4\right]$

11. m is undefined; No y-intercept; Vertical line **12.** $m = \frac{3}{5}$; y-intercept $= \left(0, -\frac{2}{5}\right)$

13. $m = 4$; y-intercept $= (0, -1)$ **14.** $m = 0$; y-intercept $= \left(0, \frac{3}{2}\right)$; Horizontal line **15.** $2x - y = 10$ **16.** $y = 2x - 6$

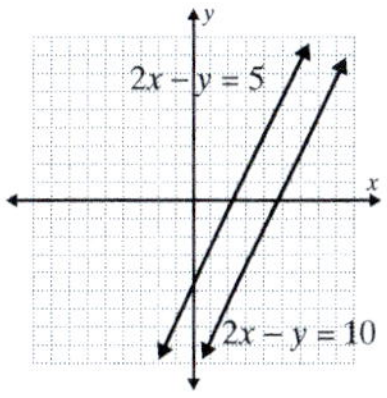

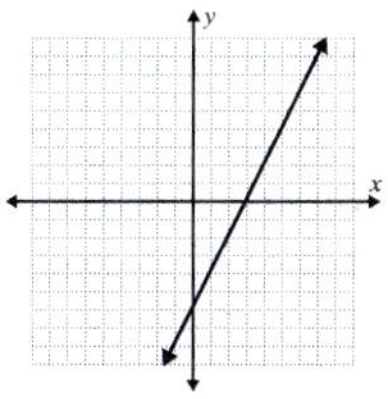

17. a. Not a function; $D = [-5, 5]$; $R = [-3, 3]$

b. Function; $D = \{-7, -3, 0, 5, 6\}$; $R = \{-2, 1, 4\}$ **c.** Function; $D = [-3, 3]$; $R = [0, 6]$

18. $\frac{y^6}{9x^4}$ **19.** $\frac{x^6}{y^6}$ **20.** $\frac{3}{2x^4y}$ **21.** 6.8 **22.** 1.05×10^6 **23.** $(3x+2)(5x+4)$

24. $(x+4)(2x^2+3)$ **25.** $2(2x-3)(4x^2+6x+9)$ **26.** $(2x+1)(x-3)$ **27. a.** −1 **b.** −12 **c.** 15

28. $x = -11$ **29.** $x = -7, 0, 2$ **30.** $x^2 + x - 20 = 0$ **31.** $6x^2 - 7x + 2 = 0$ **32.** $2x^2 + x - 3 + \frac{9}{x+2}$ **33.** $x - 4 + \frac{6x-15}{x^2-6}$

34. $\frac{2x-1}{(x+1)(x-1)}$ **35.** $\frac{x^2+15x-26}{(x-4)(x+2)(2x+3)}$ **36. a.** $x \neq -\frac{1}{2}, -\frac{1}{12}; x = \frac{1}{2}, 2$ **b.** $x \neq -4, 4; x = -2, \frac{2}{3}$ **c.** $x \neq 2, 3; x = \frac{6}{5}$

37. $n = \frac{PV}{RT}$ **38.** $f = \frac{S_1S_2}{S_1+S_2}$ **39.** b, c, d **40.** $(1, 4)$

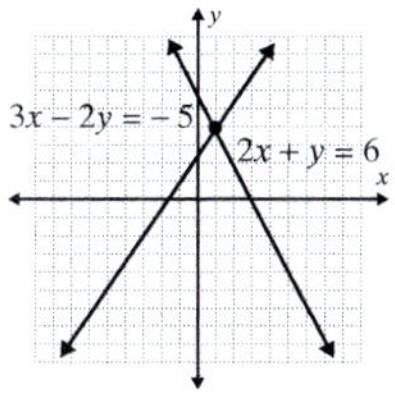

41. $(1, 3)$ **42.** $(-6, 2)$ **43.** $(-1, 4)$

44. No solution **45.** $\left(x, 2 - \frac{1}{5}x\right)$ **46.** $(2, 3)$

47. $(4, 3, 2)$ **48.** $(1, -5, 2)$

49. 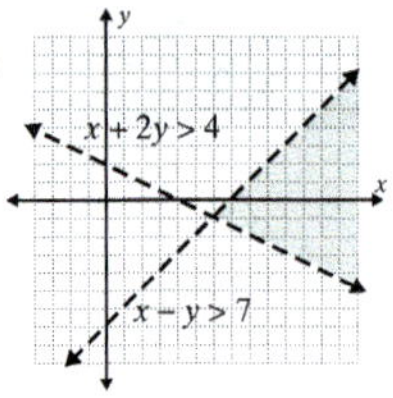

50. 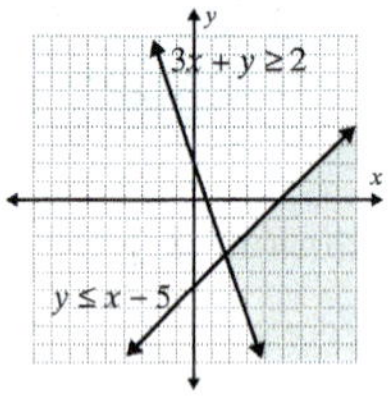

51. a. 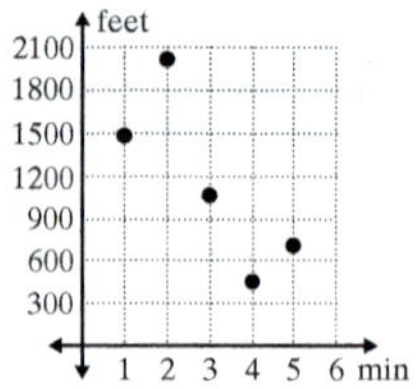

b. 528, –968, –616, 264 **c.** Juan's average speed increased 528 ft/min^2 from minute 1 to 2; decreased 968 ft/min^2 from minute 2 to 3; decreased 616 ft/min^2 from minute 3 to 4; and increased 264 ft/min^2 from minute 4 to 5.

52. 256 feet **53.** 24 ft **54.** 3 and 11 **55.** 120 yd^2 **56.** $\frac{2}{3}$hr or 40 minutes **57.** 5 ft by 9 ft **58.** 3 batches of Choc-O-Nut; 2 batches of Chocolate Krunch **59.** \$4200 at 7%; \$2800 at 8% **60.** Ferris wheel: 5400; Tunnel-of-love; 2250; Tilt-a-whirl: 4350 **61.** $\left(\frac{8}{5}, \frac{27}{10}\right)$ **62.** $(-3, 0, 5)$ **63.** 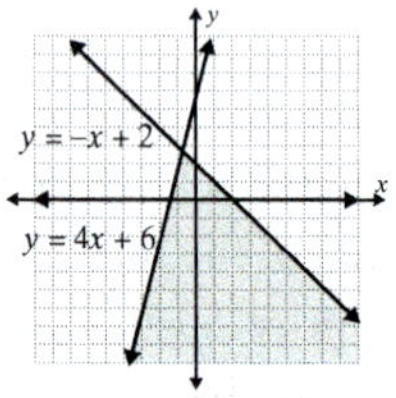

64. 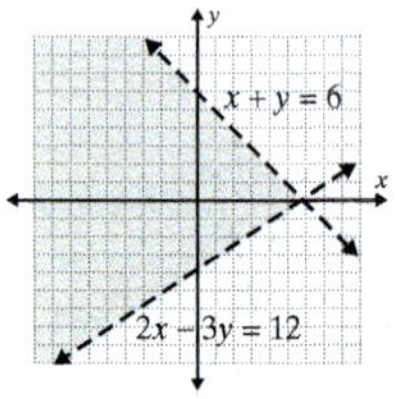

Chapter 9

Exercises 9.1, pages 688 - 689

1. 3 **3.** 9 **5.** 17 **7.** 13 **9.** 1 **11.** 5 **13.** 6 **15.** $\frac{1}{2}$ **17.** $\frac{3}{4}$ **19.** 0.2 **21.** –10 **23.** –0.04 **25.** –3 **27.** –5 **29.** $\frac{3}{5}$
31. $\sqrt{25} < \sqrt{32} < \sqrt{36}$ and $5 < \sqrt{32} < 6$ because $25 < 32 < 36$ or $(5.6569)^2 = 32.00051761$
33. $\sqrt{64} < \sqrt{74} < \sqrt{81}$ and $8 < \sqrt{74} < 9$ because $64 < 74 < 81$ or $(8.6023)^2 = 73.99956529$ **35.** Rational **37.** Rational **39.** Irrational **41.** Nonreal **43.** Rational **45.** Irrational **47.** 6.2450 **49.** 2.4960 **51.** 0.4472 **53.** 8.9443 **55.** –8.2462 **57.** 7.8990 **59.** –5.4641 **61.** There is no real number that when squared results in a negative number.

Exercises 9.2, pages 695 - 697

1. $2\sqrt{3}$ **3.** $12\sqrt{2}$ **5.** $-6\sqrt{2}$ **7.** $-2\sqrt{14}$ **9.** $-5\sqrt{5}$ **11.** $\frac{1}{2}$ **13.** $-\frac{\sqrt{11}}{8}$ **15.** $\frac{2\sqrt{7}}{5}$ **17.** $6x$ **19.** $2x\sqrt{2x}$ **21.** $2x^5y\sqrt{6x}$ **23.** $5xy^3\sqrt{5x}$ **25.** $-3xy\sqrt{2}$ **27.** $2bc\sqrt{3ac}$ **29.** $5x^2y^3z^4\sqrt{3}$ **31.** $\frac{x^2\sqrt{5}}{3}$ **33.** $\frac{4a^2\sqrt{2a}}{9b^8}$ **35.** $\frac{10x^4\sqrt{2}}{17}$ **37.** 6 **39.** $2\sqrt[3]{7}$ **41.** –1 **43.** $-4\sqrt[3]{2}$ **45.** $5x\sqrt[3]{x}$ **47.** $-2x^2\sqrt[3]{x^2}$ **49.** $2a^2b\sqrt[3]{9b}$ **51.** $6x^2y\sqrt[3]{y^2}$ **53.** $2xy^2z^3\sqrt[3]{3x^2y}$ **55.** $\frac{\sqrt[3]{3}}{2}$ **57.** $\frac{5\sqrt[3]{3}}{2}$ **59.** $\frac{5y^4}{3x^2}$ **61.** 7 inches **63.** $\sqrt{6} \approx 2.45$ amperes **65.** 120 volts **67.** When $a < 0$.

Exercises 9.3, pages 705 - 707

1. 3 **3.** $\frac{1}{10}$ **5.** –512 **7.** –4 **9.** Nonreal **11.** $\frac{3}{7}$ **13.** 16 **15.** $-\frac{1}{6}$ **17.** $\frac{5}{2}$ **19.** $\frac{1}{4}$ **21.** $\frac{3}{8}$ **23.** $-\frac{1}{1000}$ **25.** 64 **27.** 8.5499 **29.** 10,000,000 **31.** 99.6055 **33.** 0.0922 **35.** 1.6083 **37.** 0.2236 **39.** 7.7460 **41.** 2.0408 **43.** $8x$ **45.** $\frac{1}{3a^2}$ **47.** $8x^{\frac{5}{2}}$ **49.** $5a^{\frac{13}{6}}$ **51.** $x^{\frac{7}{12}}$ **53.** $x^{\frac{1}{2}}$ **55.** $\frac{1}{a^{\frac{9}{8}}}$ **57.** $a^{\frac{1}{4}}$ **59.** $\frac{a^2}{b^{\frac{6}{5}}}$ **61.** $8x^{\frac{3}{2}}y$ **63.** $\frac{x^{\frac{3}{2}}}{16y^{\frac{2}{5}}}$ **65.** $\frac{x^2y^4}{z^4}$ **67.** $\frac{y^{\frac{3}{2}}z^2}{x}$ **69.** $\frac{8b^{\frac{9}{4}}}{a^3c^3}$ **71.** $x^{\frac{1}{4}}y^{\frac{5}{4}}$ **73.** $\frac{y^{\frac{2}{3}}}{50x^{\frac{4}{3}}}$

75. $\dfrac{b^{\frac{5}{12}}}{a^{\frac{11}{12}}}$ **77.** $\dfrac{3x^{\frac{1}{2}}}{20y^{\frac{2}{3}}}$ **79.** $\sqrt[6]{a^5}$ **81.** $\sqrt[12]{y^7}$ **83.** $\sqrt[30]{x^{11}}$ **85.** $\sqrt[6]{y}$ **87.** $\sqrt[9]{x}$ **89.** $\sqrt[3]{7a}$ **91.** $\sqrt[24]{x}$ **93.** $a^{20}b^5c^{10}$

95. No: $\sqrt[5]{a}\cdot\sqrt{a}=a^{\frac{7}{10}}$; $\sqrt[5]{a^2}=a^{\frac{2}{5}}$; $a^{\frac{7}{10}}\neq a^{\frac{2}{5}}$

Exercises 9.4, pages 716 - 718

1. $-6\sqrt{2}$ **3.** $7\sqrt{11}$ **5.** $-3\sqrt[3]{7x^2}$ **7.** $10\sqrt{3}$ **9.** $-7\sqrt{2}$ **11.** $6\sqrt{2}-\sqrt{3}$ **13.** $8\sqrt{3x}$ **15.** $20x\sqrt{5x}$ **17.** $14\sqrt{5}-3\sqrt{7}$
19. $-7\sqrt[3]{3x^2}-10\sqrt[3]{6x^2}$ **21.** $2x\sqrt{y}-y\sqrt{xy}$ **23.** $x^2y\sqrt{2x}$ **25.** $13+2\sqrt{2}$ **27.** $3x-9\sqrt{3x}+8$ **29.** $2-2\sqrt{7}$ **31.** -3
33. $13+4\sqrt{10}$ **35.** $\sqrt{10}+\sqrt{15}-\sqrt{6}-3$ **37.** $x-2\sqrt{6x}-18$ **39.** 58 **41.** $x+10\sqrt{xy}+25y$ **43.** $4x-9\sqrt{xy}-9y$ **45.** $\dfrac{\sqrt{21}}{3}$
47. $\dfrac{\sqrt[3]{70}}{2}$ **49.** $-\dfrac{\sqrt{6y}}{3y}$ **51.** $\dfrac{2\sqrt{10x}}{5y}$ **53.** $\dfrac{2\sqrt{2y}}{y}$ **55.** $\dfrac{y\sqrt[3]{2x}}{3x}$ **57.** $\dfrac{\sqrt[3]{30a^2b^2}}{5b^2}$ **59.** $\sqrt{2}-1$ **61.** $-\dfrac{\sqrt{15}+4\sqrt{3}}{11}$ **63.** $2\sqrt{3}-2\sqrt{2}$
65. $\dfrac{\sqrt{35}+\sqrt{15}}{4}$ **67.** $\dfrac{\sqrt{6}}{3}$ **69.** $-\dfrac{5+\sqrt{21}}{2}$ **71.** $\dfrac{2x+y\sqrt{x}-y^2}{x-y^2}$ **73.** 4.3397 **75.** 31.6 **77.** −57 **79.** −37.3569 **81.** 0.2831 **83.** 0.3820
85. $5\sqrt{6}+\sqrt{170}\approx 25.29$ ft
87. Multiply both the numerator and the denominator by the conjugate of the denominator. This works because multiplying the denominator by its conjugate results in an expression with no square roots. Answers will vary.

Exercises 9.5, pages 724 - 725

1. $x=3$ **3.** No solution **5.** $x=-3$ **7.** $x=14$ **9.** $x=9$ **11.** No solution **13.** $x=6$ **15.** $x=-4, 1$ **17.** $x=-5, \dfrac{5}{2}$ **19.** $x=-2$
21. $x=2,3$ **23.** $x=2,5$ **25.** $x=-5,5$ **27.** $x=4$ **29.** $x=4$ **31.** $x=3$ **33.** $x=2$ **35.** $x=2$ **37.** $x=7$ **39.** $x=4$ **41.** $x=0$ **43.** $x=5$
45. No solution **47.** $x=4$ **49.** $x=-1,3$ **51.** $x=5$ **53.** $x=2$ **55.** $x=1$ **57.** $x=-4$ **59.** $x=12$ **61.** $x=40$
63. $(a+b)^2=(a+b)(a+b)=a^2+2ab+b^2\neq a^2+b^2$

Exercises 9.6, pages 732 - 734

1. a. $\sqrt{5}\approx 2.2361$ **b.** 3 **c.** $5\sqrt{2}\approx 7.0711$ **d.** 2 **3. a.** 3 **b.** −1 **c.** −2 **d.** $2\sqrt[3]{3}\approx 2.8845$ **5.** $[-8,\infty)$ **7.** $\left(-\infty,\dfrac{1}{2}\right]$ **9.** $(-\infty,\infty)$
11. $[0,\infty)$ **13.** $(-\infty,\infty)$ **15.** E **17.** B **19.** A

21.
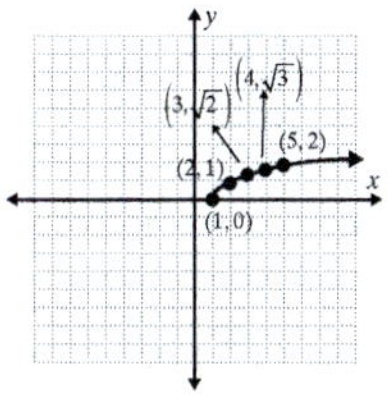

23.
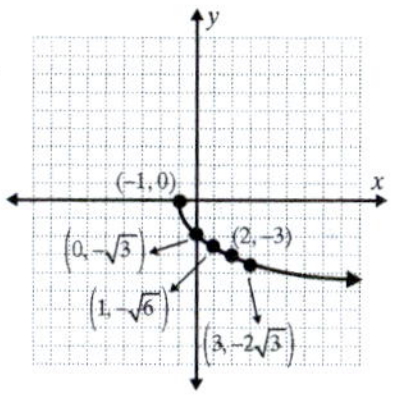

25.
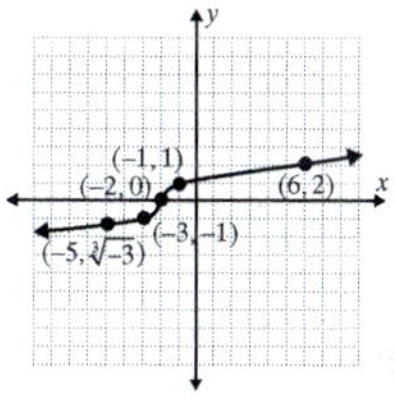

27.
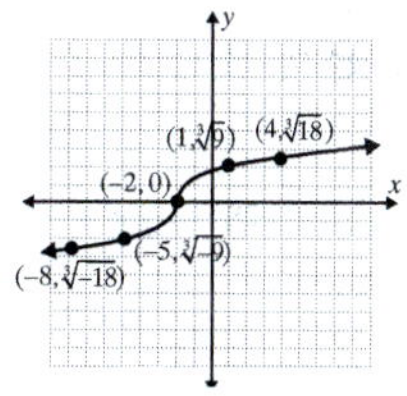

29.
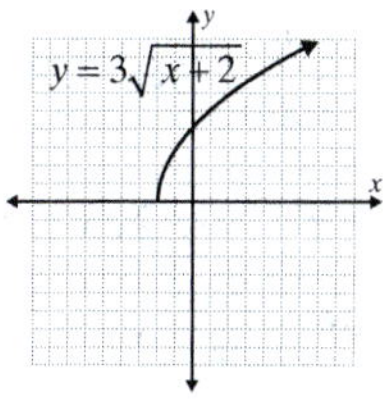

31.
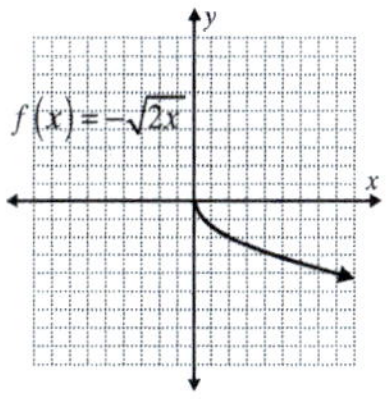

33.
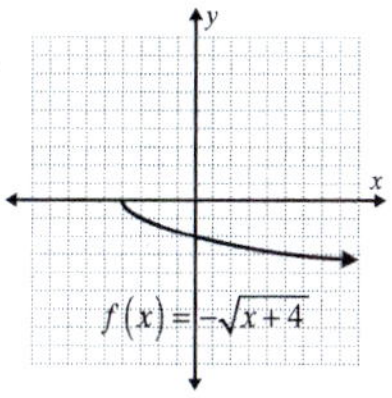

35.
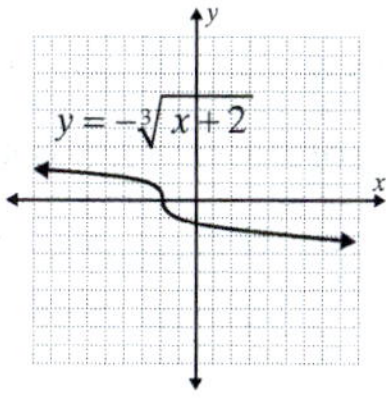

37.
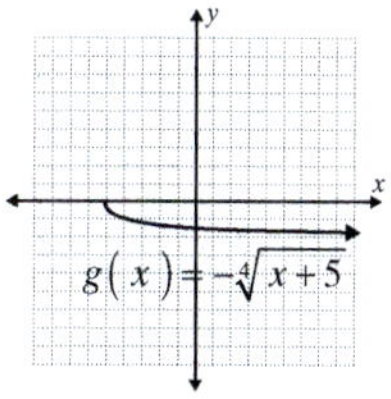

39.
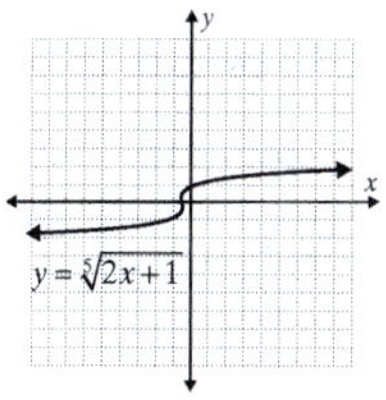

41. a. $\dfrac{1}{\sqrt{3+h}+\sqrt{3}}$ **b.** Slope of the line connecting $(3+h, f(3+h))$ and $(3, f(3))$

c. A line just touching the curve at one point. **d.** $\dfrac{1}{2\sqrt{3}}$; represents the slope of the line tangent to $f(x)$ at $x=3$.

Exercises 9.7, pages 741 - 742

1. Real part is 4, imaginary part is –3 **3.** Real part is –11, imaginary part is $\sqrt{2}$ **5.** Real part is $\frac{3}{8}$, imaginary part is 0 **7.** Real part is $\frac{4}{5}$, imaginary part is $\frac{7}{5}$ **9.** Real part is $\frac{2}{3}$, imaginary part is $\sqrt{17}$ **11.** $7i$ **13.** $-8i$ **15.** $7\sqrt{3}$ **17.** $10i\sqrt{6}$ **19.** $-12i\sqrt{3}$ **21.** $11\sqrt{2}$ **23.** $10i\sqrt{10}$ **25.** $x=6, y=-3$ **27.** $x=-2, y=\sqrt{5}$ **29.** $x=\sqrt{2}-3, y=1$ **31.** $x=2, y=-8$ **33.** $x=2, y=-6$ **35.** $x=3, y=10$ **37.** $x=-\frac{4}{3}, y=-3$ **39.** $6+2i$ **41.** $1+7i$ **43.** $6-6i$ **45.** $14i$ **47.** $(3+\sqrt{5})-6i$ **49.** $5+(\sqrt{6}+1)i$ **51.** $\sqrt{3}-5$ **53.** $-2-5i$ **55.** $11-16i$ **57.** 2 **59.** $3+4i$ **61. a.** Yes **b.** No

Exercises 9.8, pages 748 - 749

1. $16+24i$ **3.** $-7\sqrt{2}+7i$ **5.** $3+12i$ **7.** $1-i\sqrt{3}$ **9.** $3+2i\sqrt{3}$ **11.** $2+8i$ **13.** $-7-11i$ **15.** $13+0i$ **17.** $34+13i$ **19.** $-24+70i$ **21.** $5-i\sqrt{3}$ **23.** $23-10i\sqrt{2}$ **25.** $21+0i$ **27.** $(2+\sqrt{10})+(2\sqrt{2}-\sqrt{5})i$ **29.** $(9-\sqrt{30})+(3\sqrt{5}+3\sqrt{6})i$ **31.** $0+3i$ **33.** $0-\frac{5}{4}i$ **35.** $-\frac{1}{4}+\frac{1}{2}i$ **37.** $-\frac{4}{5}+\frac{8}{5}i$ **39.** $\frac{24}{25}+\frac{18}{25}i$ **41.** $-\frac{1}{13}+\frac{5}{13}i$ **43.** $-\frac{1}{29}-\frac{12}{29}i$ **45.** $-\frac{17}{26}-\frac{7}{26}i$ **47.** $\frac{4+\sqrt{3}}{4}+\left(\frac{4\sqrt{3}-1}{4}\right)i$ **49.** $-\frac{1}{7}+\frac{4\sqrt{3}}{7}i$ **51.** $0+i$ **53.** $-1+0i$ **55.** $0+i$ **57.** $1+0i$ **59.** $0-i$ **61.** x^2+9 **63.** x^2+2 **65.** $5y^2+4$ **67.** $x^2+4x+40$ **69.** $y^2-6y+13$ **71.** Given a complex number $(a+bi)$: $(a+bi)(a-bi)=a^2-abi+abi-b^2i^2=a^2+b^2$ which is the sum of squares of real numbers. Thus the product must be a positive real number. **73.** $a^2+b^2=1$

Chapter 9 Review, pages 756 - 759

1. 6 **2.** 14 **3.** 2 **4.** 9 **5.** 0.03 **6.** –7 **7.** $\frac{12}{7}$ **8.** $\frac{5}{3}$ **9.** Nonreal **10.** Irrational **11.** Rational **12.** Rational **13.** 4.5826 **14.** 6.3640 **15.** –16.9706 **16.** 5.2426 **17.** –15 **18.** $3x\sqrt{x}$ **19.** $2a^2\sqrt{2}$ **20.** $5xy\sqrt{2x}$ **21.** $-9x\sqrt{y}$ **22.** $2x\sqrt[3]{5x}$ **23.** $3xy^2\sqrt[3]{3x^2y}$ **24.** $3a\sqrt[3]{2ab^2}$ **25.** $\frac{3x\sqrt{3}}{10}$ **26.** $\frac{5a\sqrt{3a}}{3}$ **27.** $\frac{10x^2\sqrt[3]{3}}{7}$ **28.** $\frac{2y^4}{3x^5}$ **29.** 9 **30.** $-\frac{1}{6}$ **31.** $\frac{1}{15}$ **32.** $\frac{2}{3}$ **33.** $\frac{9}{25}$ **34.** $-\frac{1}{1000}$ **35.** 10.9027 **36.** 82.7037 **37.** 7.0711 **38.** 0.2408 **39.** $27x^2$ **40.** $8a^{\frac{3}{2}}$ **41.** $5y^{\frac{10}{3}}$ **42.** $\frac{1}{x^{\frac{5}{2}}}$ **43.** $\frac{9a^4b^2}{c^2}$ **44.** $\frac{1}{2y^{\frac{1}{2}}}$ **45.** $\sqrt[12]{x^5}$ **46.** $2\sqrt[3]{x}$ **47.** $-4\sqrt{11}$ **48.** $5\sqrt{x}$ **49.** $5\sqrt{2}-26\sqrt{3}$ **50.** 0 **51.** –1 **52.** $7+2\sqrt{10}$ **53.** $x-y$ **54.** $8\sqrt{2x}+15x+2$ **55.** $\frac{\sqrt{6}}{4}$ **56.** $\frac{4\sqrt{3b}}{3a}$ **57.** $\frac{\sqrt[3]{100xy^2}}{10y}$ **58.** $\frac{3\sqrt{5}-5}{2}$ **59.** $\frac{5\sqrt{7}-5\sqrt{3}}{4}$ **60.** $\frac{13+3\sqrt{15}}{17}$ **61.** $\frac{6-3x}{4\sqrt{6}+4\sqrt{3x}}$ **62.** 25.4495 **63.** –28 **64.** 0.2185 **65.** $x=7$ **66.** $x=14$ **67.** $x=68$ **68.** $x=\pm 9$ **69.** $x=0$ **70.** $x=3,5$ **71.** $x=-5,-4$ **72.** No solution **73.** $x=2$ **74.** $x=3$ **75.** $x=4$ **76.** No solution **77.** No solution **78.** $x=15$ **79.** $x=2,4$ **80.** $x=20$ **81. a.** 0 **b.** 5 **82. a.** $\sqrt[3]{6}\approx 1.8171$ **b.** $2\sqrt[3]{2}\approx 2.5198$ **83.** $\left[-\frac{1}{2},\infty\right)$ **84.** $(-\infty,\infty)$ **85.** $\left(-\infty,\frac{3}{2}\right]$ **86.** $(-\infty,\infty)$

87. 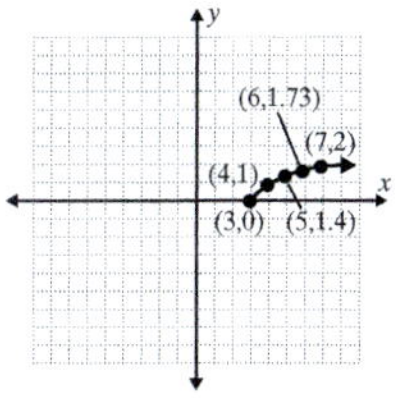

88. 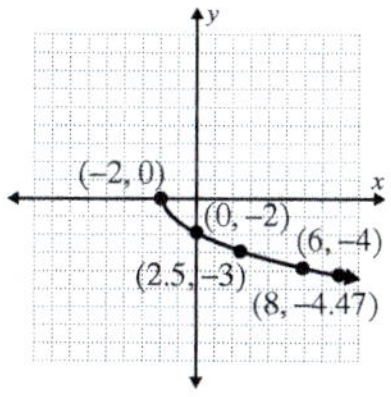

89. 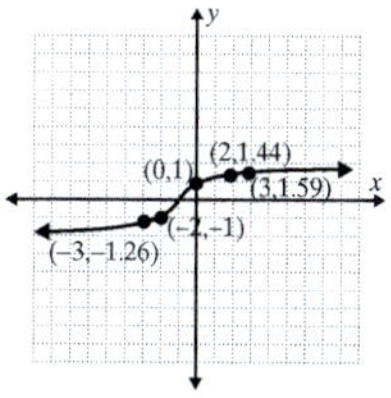

90. 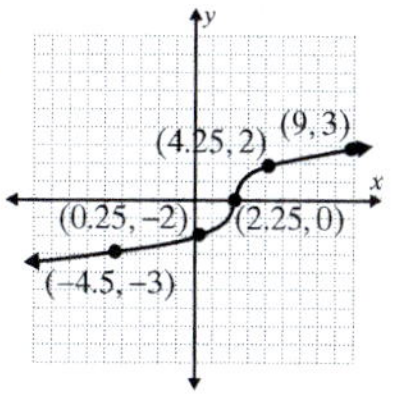

91. 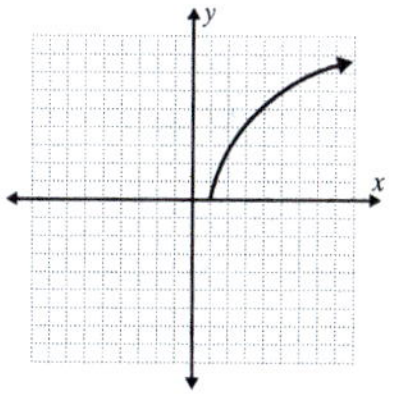

92. 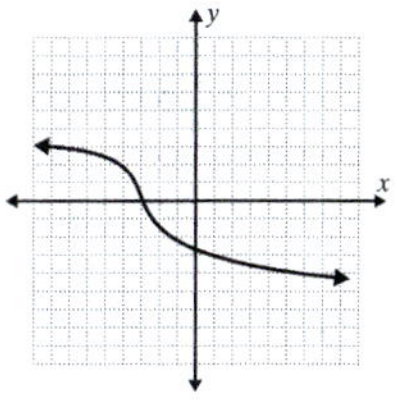

93. 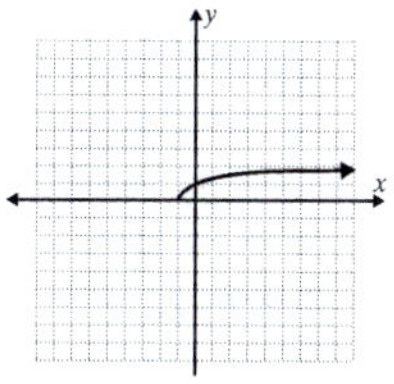

94. 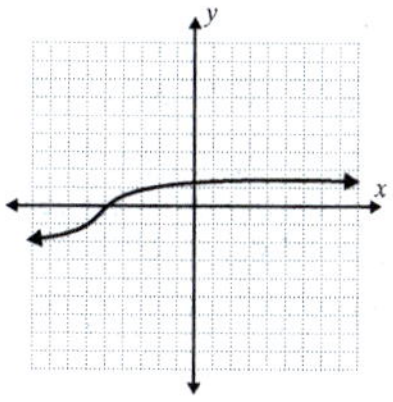

95. Real part is 5, imaginary part is −2 **96.** Real part is −10, imaginary part is $\sqrt{3}$ **97.** Real part is $\frac{3}{4}$, imaginary part is 0 **98.** Real part is $\frac{3}{2}$, imaginary part is $\frac{1}{2}$ **99.** $6i$ **100.** $12i$ **101.** $-6i\sqrt{2}$ **102.** $10\sqrt{29}$ **103.** $x=9, y=0$ **104.** $x=\sqrt{3}, y=\frac{4}{3}$ **105.** $x=-10, y=-7$ **106.** $x=5, y=8$ **107.** $5+8i$ **108.** $-4+2i$ **109.** $4+\sqrt{2}-4i$ **110.** $2+\sqrt{5}$ **111.** $15-25i$ **112.** $1+5i$ **113.** $4\sqrt{2}+12i$ **114.** $7-i$ **115.** $7+i\sqrt{3}$ **116.** $-5+12i$ **117.** $0+\frac{8}{5}i$ **118.** $2+i$ **119.** $\frac{2}{25}-\frac{11}{25}i$ **120.** $\frac{2+7\sqrt{5}}{6}+\left(\frac{7-2\sqrt{5}}{6}\right)i$ **121.** $0-i$ **122.** $-1+0i$ **123.** $6x^2+4$ **124.** $y^2+8y+32$

Chapter 9 Test, pages 760 - 761

1. $4\sqrt{7}$ **2.** $2y^2\sqrt{30x}$ **3.** $2y\sqrt[3]{6x^2y^2}$ **4.** $\frac{7x^6y\sqrt[3]{y}}{2z^2}$ **5.** $\sqrt[3]{4x^2}$ **6.** $2^{\frac{1}{2}}x^{\frac{1}{3}}y^{\frac{2}{3}}$ **7.** 4 **8.** $4x^{\frac{7}{6}}$ **9.** $\frac{8y^{\frac{3}{2}}}{x^3}$ **10.** $\sqrt[12]{x^{11}}$ **11.** $17\sqrt{3}$ **12.** $-xy\sqrt{y}$ **13.** $5-2\sqrt{6}$ **14.** $30-7\sqrt{3x}-6x$ **15. a.** $\frac{y\sqrt{10x}}{4x^2}$ **b.** $1+\sqrt{x}$ **16.** $x=14$ **17.** $x=-4$ **18.** $x=1$ **19.** $x=7$ **20. a.** $\left[-\frac{4}{3}, \infty\right)$ **b.** $(-\infty, \infty)$ **21. a.**

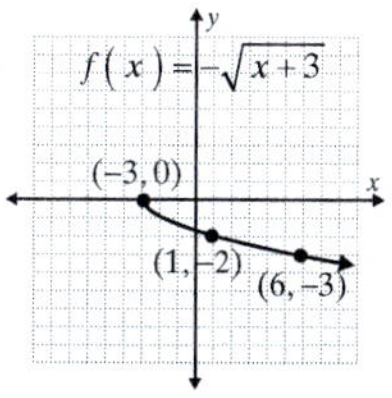

b.

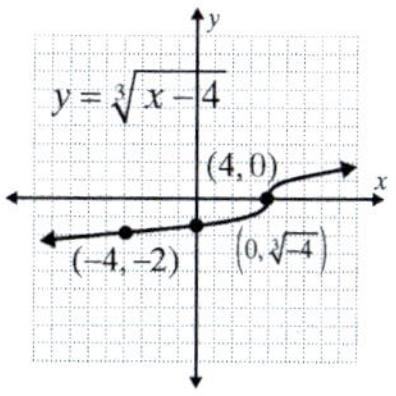

22. Real part is 6, imaginary part is $-\frac{1}{3}$

23. Real part is $-\frac{2}{5}$, imaginary part is $\frac{3}{5}$ **24.** $x=\frac{11}{2}, y=3$ **25.** $16+4i$ **26.** $-5+16i$ **27.** $23-14i$ **28.** x^2+4 **29.** $\frac{8}{13}-\frac{1}{13}i$ **30.** $0-i$ **31.** 2.6008 **32.** 0.1250 **33.** 3.0711 **34.** 0.9758

Cumulative Review: Chapters 1-9, pages 762 - 766

1. Identity property of addition **2.** Distributive property **3.** $x=\frac{1}{21}$ **4.** No solution **5.** $n=-\frac{6}{5}, 4$

6. 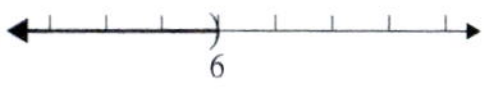

$(-\infty, 6)$ **7.** 

$[0, 2]$

8. 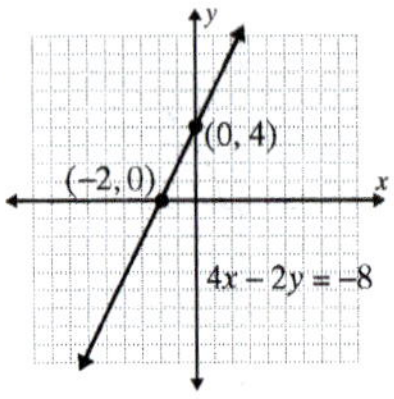

9. 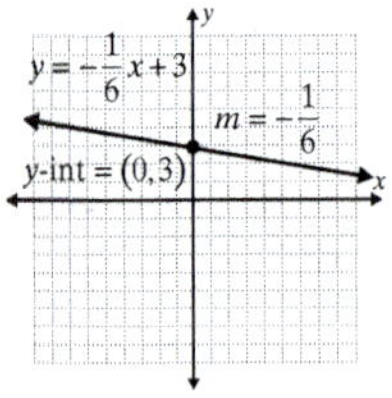

10. $x = 5$ **11.** $y = -4x - 2$

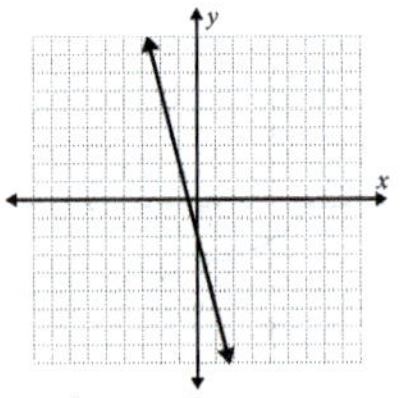

12. a. Function; $D = (-\infty, \infty)$; $R = (-\infty, 0]$ **b.** Not a function; $D = (-\infty, \infty)$; $R = (-\infty, 5]$ **c.** Function; $D = [-2\pi, 2\pi]$; $R = [-2, 2]$ **13.** 1×10^3 **14. a.** -12 **b.** 0 **c.** -168 **15.** $2x^3 - 4x^2 + 8x - 4$ **16.** $6x^2 - 23x + 7$ **17.** $-5x^2 + 18x + 8$ **18.** $49x^2 - 28x + 4$ **19.** $x - 9 + \dfrac{21x - 24}{x^2 + 2x - 1}$ **20.** $(7 + 2x)(4 - x)$ **21.** $4y^2(4y + 1)^2$ **22.** $5(x - 4)(x^2 + 4x + 16)$ **23.** $(x + 4)(x + 1)(x - 1)$ **24.** $(x - 16)(x + 3) = 0$; $x = -3, 16$ **25.** $(2x + 3)(x - 2) = 0$; $x = -\dfrac{3}{2}, 2$ **26.** $(5x - 2)(3x - 1) = 0$; $x = \dfrac{1}{3}, \dfrac{2}{5}$ **27. a.** $x^2 - 11x + 28 = 0$ **b.** $x^3 - 11x^2 - 14x + 24 = 0$ **28.** $\dfrac{4x^2 - 12x - 1}{(x + 3)(x - 2)(x - 5)}$ **29.** $\dfrac{4x^2 + 6x + 17}{(2x + 1)(2x - 1)(x + 3)}$ **30.** $\dfrac{x^2 - 1}{2x - 4}$ **31.** $\dfrac{x + 6}{x - 1}$ **32.** $x \neq -5, 0$; $x = -\dfrac{5}{7}$ **33.** $x \neq -7, 3$; $x = \dfrac{83}{4}$

34. $(1, -2)$ 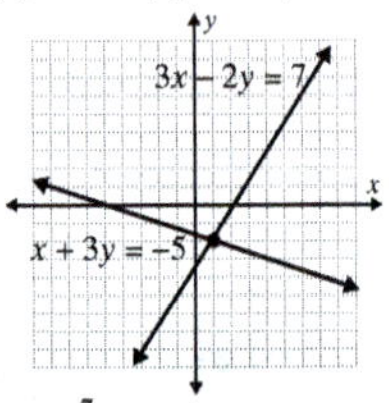

35. $(14, 6)$ **36.** No solution **37.** $(1, 1, 1)$ **38.** 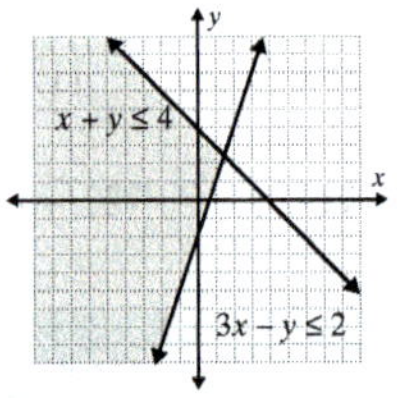

39. $\dfrac{1}{16}$ **40.** $x^{\frac{7}{3}}$ **41.** $12\sqrt{2}$ **42.** $2x^2y^3\sqrt[3]{2y}$ **43.** $\dfrac{\sqrt{10y}}{2y}$ **44.** $2\sqrt{30} - 11$ **45.** $x = -1$ **46.** $x = -1$ **47.** $(-\infty, 3]$ **48.** $x = 6, y = -5$ **49.** $2 + 2i$ **50.** $14 - 2i$ **51.** $-\dfrac{8}{17} - \dfrac{19}{17}i$ **52.** 1 **53.** $n = \dfrac{s - a}{d} + 1$ or $\dfrac{s - a + d}{d}$ **54.** $p = \dfrac{A}{1 + rt}$

55. a. 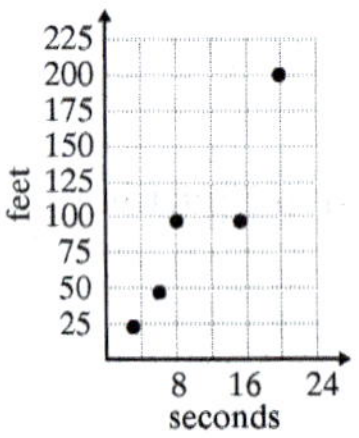

b. $m_1 = 8, m_2 = 24, m_3 = 0, m_4 = \dfrac{104}{5}$ **c.** The distance increased by 8 ft/s, then increased by 24 ft/s, then stayed constant, then increased by $\dfrac{104}{5}$ ft/s. **56.** 15, 17 **57.** $\dfrac{50}{3}$ in.^{3} or $16.\overline{6}$ in.^{3} **58.** 8 ohms **59.** 10 mph **60.** \$35,000 in 6% and \$15,000 in 10% **61.** 7.5 hours **62.** 60 pounds of the first type (\$1.25 candy) and 40 pounds of the second type (\$2.50 candy) **63.** Burrito price is \$2.60 and taco price is \$1.35 **64.** 26.5 meters **65. a.** $\sqrt{6}$ or 2.45 seconds **b.** 96 feet **c.** $\sqrt{15}$ or 3.87 seconds

66. $x \approx -1.372, 4.372$ **67.** $x = 1$, $x \approx -2.303, 1.303$ **68.** $x = -4, 1, 5$ **69.** 279,936.0000 **70.** 0.0400 **71.** 1.3027 **72.** 6.4552

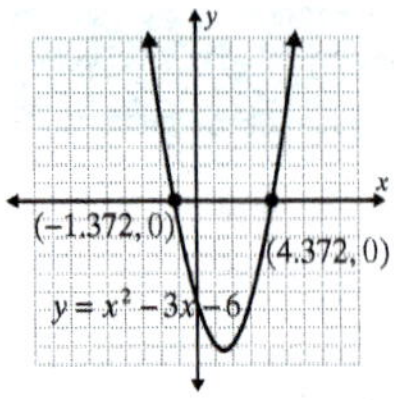

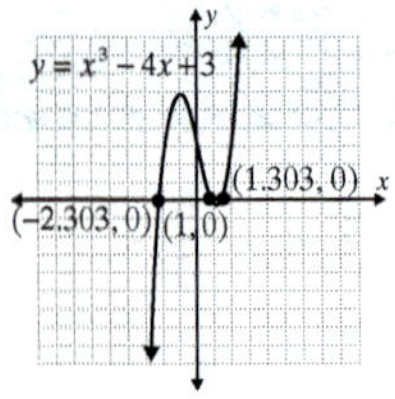

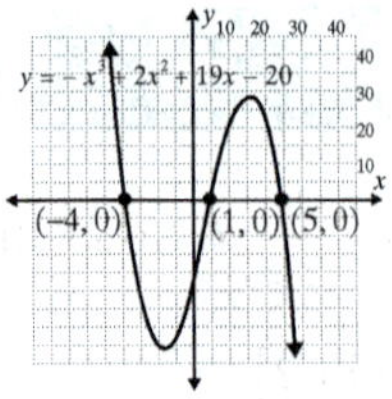

Chapter 10

Exercises 10.1, pages 777 - 779

1. $x = 3, 15$ **3.** $x = 5, 12$ **5.** $y = -11, 11$ **7.** $z = -6, 0, 2$ **9.** $x = -5$ **11.** $x = \pm 12$ **13.** $x = \pm 5i$ **15.** $x = \pm 2i\sqrt{6}$
17. $x = -1, 5$ **19.** $x = \pm\sqrt{5}$ **21.** $x = -3 \pm \sqrt{3}$ **23.** $x = 3 \pm 2i$ **25.** $x = -2 \pm i\sqrt{7}$ **27.** $x^2 - 12x + \underline{36} = \underline{(x-6)}^2$
29. $x^2 + 6x + \underline{9} = \underline{(x+3)}^2$ **31.** $x^2 - 5x + \underline{\frac{25}{4}} = \underline{\left(x - \frac{5}{2}\right)}^2$ **33.** $y^2 + y + \underline{\frac{1}{4}} = \underline{\left(y + \frac{1}{2}\right)}^2$ **35.** $x^2 + \frac{1}{3}x + \underline{\frac{1}{36}} = \underline{\left(x + \frac{1}{6}\right)}^2$
37. $x = -5, 1$ **39.** $y = -1 \pm \sqrt{6}$ **41.** $x = 5 \pm \sqrt{22}$ **43.** $x = 3 \pm i$ **45.** $x = 1, 11$ **47.** $y = 5 \pm \sqrt{21}$ **49.** $z = \frac{-3 \pm \sqrt{29}}{2}$
51. $x = \frac{-1 \pm i\sqrt{7}}{2}$ **53.** $x = \frac{-5 \pm \sqrt{17}}{2}$ **55.** $x = \frac{5 \pm \sqrt{10}}{3}$ **57.** $x = -1 \pm i\sqrt{5}$ **59.** $x = \frac{1 \pm i\sqrt{11}}{4}$ **61.** $y = \frac{-3 \pm i\sqrt{11}}{2}$
63. $y = -\frac{4}{3}, 1$ **65.** $x = 2 \pm \sqrt{2}$ **67.** $x^2 - 7 = 0$ **69.** $x^2 - 2x - 2 = 0$ **71.** $y^2 + 4y - 16 = 0$ **73.** $x^2 + 16 = 0$ **75.** $y^2 + 6 = 0$
77. $x^2 - 4x + 5 = 0$ **79.** $x^2 - 2x + 3 = 0$ **81.** $x^2 + 10x + 49 = 0$ **83.** See page 774.

Exercises 10.2, pages 786 - 788

1. 68; Two real solutions **3.** 0; One real solution **5.** –44; Two nonreal solutions **7.** 4; Two real solutions
9. 19,600; Two real solutions **11.** –11; Two nonreal solutions **13.** $x = -2 \pm 2\sqrt{2}$ **15.** $x = -\frac{2}{3}$ **17.** $x = 1 \pm i\sqrt{6}$
19. $x = -3, \frac{1}{2}$ **21.** $x = \frac{-3 \pm \sqrt{5}}{4}$ **23.** $x = \frac{-3 \pm i\sqrt{3}}{4}$ **25.** $x = \frac{-3 \pm \sqrt{29}}{2}$ **27.** $x = -3, -1$ **29.** $x = \pm 2i\sqrt{2}$ **31.** $x = \frac{5 \pm \sqrt{17}}{2}$
33. $x = -\frac{1}{4}$ **35.** $x = \pm\frac{2\sqrt{3}}{3}$ **37.** $x = \frac{2}{3}$ **39.** $x = \frac{-4 \pm i\sqrt{2}}{2}$ **41.** $x = \frac{7 \pm i\sqrt{51}}{10}$ **43.** $x = -2, \frac{5}{3}$ **45.** $x = 3 \pm i\sqrt{2}$
47. $x = -1, 0$ **49.** $x = \frac{9 \pm \sqrt{65}}{2}, 0$ **51.** $x = \frac{-3 \pm \sqrt{5}}{2}, 0$ **53.** $x = \frac{2 \pm \sqrt{3}}{3}$ **55.** $x = \frac{1 \pm i\sqrt{3}}{6}$ **57.** $x = \frac{2 \pm i\sqrt{2}}{2}$
59. $x = \frac{-7 \pm \sqrt{17}}{4}$ **61.** $c < 16$ **63.** $c = \frac{81}{4}$ **65.** $a > 3$ **67.** $a > -\frac{1}{36}$ **69.** $a = \frac{49}{48}$ **71.** $c > \frac{4}{3}$ **73.** $x \approx 2.5993, 60.4007$
75. $x \approx -0.7862, 2.0110$ **77.** $x \approx -4.1334, -0.5806$ **79.** $x \approx -2.6933, 2.6933$
81. $x^4 - 13x^2 + 36 = 0$; Multiplied $(x-2)(x+2)(x-3)(x+3)$; Answers will vary.

Exercises 10.3, pages 794 - 798

1. 6, 13 **3.** 4, 14 **5.** –11, –6 **7.** $-5 - 4\sqrt{3}$ **9.** $\frac{1 + \sqrt{33}}{4}$ **11.** $7\sqrt{2}$ cm, $7\sqrt{2}$ cm **13.** Mel: 30 mph; John: 40 mph
15. Length = 10 feet; Width = 4 feet **17.** 17 inches × 6 inches **19.** 70 trees **21.** 9 floors **23.** 3 in. **25.** 5 amperes
27. 12 signs **29. a.** $307.20 **b.** 45 cents or 35 cents **31.** 8 workers **33.** 6 mph **35.** 4 hours and 12 hours
37. 10 lb of Grade A, 20 lb of Grade B **39. a.** 6.75 s **b.** 3 s, 3.75 s **41. a.** 7 s **b.** 2 s, 12 s **43.** 14.14 cm
45. a. $C = 94.20$ ft, $A = 706.50$ ft^2 **b.** $P = 84.85$ ft, $A = 450.00$ ft^2 **47. a.** No **b.** Home plate **c.** No **49.** 855.86 feet
51. See page 780.

Exercises 10.4, pages 803 - 805

1. $x=\pm2,\pm3$ **3.** $x=\pm2,\pm\sqrt{5}$ **5.** $y=\pm\sqrt{7},\pm2i$ **7.** $y=\pm\sqrt{5},\pm i\sqrt{5}$ **9.** $x=4,\frac{25}{4}$ **11.** $x=1,4$ **13.** $x=-8,\frac{1}{8}$
15. $x=-27,-8,0$ **17.** $x=-17,-2$ **19.** $x=-\frac{7}{2},-3$ **21.** $x=0,8$ **23.** $x=-1,-\frac{1}{2}$ **25.** $x=\pm\sqrt{1+i},\pm\sqrt{1-i}$
27. $x=\pm\sqrt{1+3i},\pm\sqrt{1-3i}$ **29.** $x=\pm\sqrt{2+i\sqrt{3}},\pm\sqrt{2-i\sqrt{3}}$ **31.** $x=\frac{1}{7},\frac{1}{5}$ **33.** $x=-\frac{1}{3},\frac{3}{8}$ **35.** $x=\frac{1}{25}$ **37.** $x=\pm\frac{\sqrt{5}}{5},\pm1$
39. $x=\pm\frac{\sqrt{6}}{2},\pm\frac{1}{3}i$ **41.** $x=-\frac{1}{4}$ **43.** $x=-4,1$ **45.** $x=-\frac{1}{2}$ **47.** $x=-7,-\frac{3}{2}$ **49.** $x=\frac{26}{5}$ **51.** $x=0,\pm2\sqrt{2},\pm2i\sqrt{2}$
53. $x=2,-1\pm i\sqrt{3}$ **55.** $x=-10,0,5\pm5i\sqrt{3}$ **57. a.** $l^2-l-1=0$; $l=\frac{1+\sqrt{5}}{2}$ which is the golden ratio **b.** 97.08 feet
c. The yellow rectangle is "golden."

Exercises 10.5, pages 819 - 823

1. $x=0;(0,-4);D=(-\infty,\infty);R=[-4,\infty)$ **3.** $x=0;(0,9);D=(-\infty,\infty);R=[9,\infty)$ **5.** $x=0;(0,1);D=(-\infty,\infty);R=(-\infty,1]$
7. $x=0;\left(0,\frac{1}{5}\right);D=(-\infty,\infty);R=\left(-\infty,\frac{1}{5}\right]$ **9.** $x=-1;(-1,0);D=(-\infty,\infty);R=[0,\infty)$ **11.** $x=4;(4,0);D=(-\infty,\infty);R=(-\infty,0]$
13. $x=-3;(-3,-2);D=(-\infty,\infty);R=[-2,\infty)$ **15.** $x=-2;(-2,-6);D=(-\infty,\infty);R=[-6,\infty)$
17. $x=\frac{3}{2};\left(\frac{3}{2},\frac{7}{2}\right);D=(-\infty,\infty);R=\left(-\infty,\frac{7}{2}\right]$ **19.** $x=\frac{4}{5};\left(\frac{4}{5},-\frac{11}{5}\right);D=(-\infty,\infty);R=\left[-\frac{11}{5},\infty\right)$

21. a.
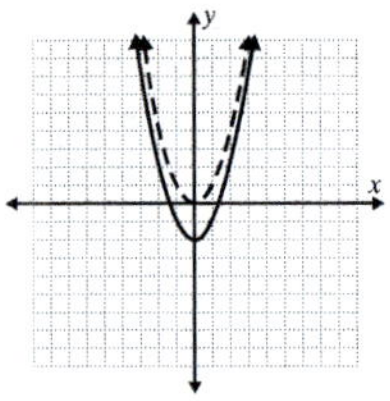
b.
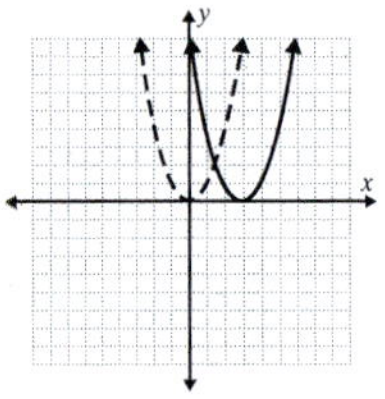
c.
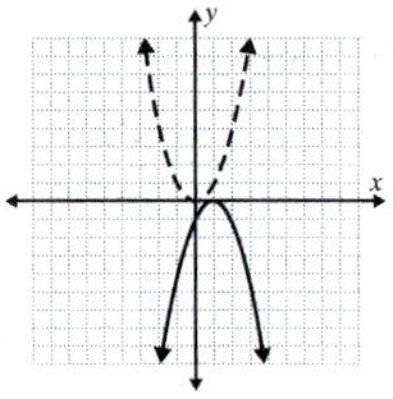
d.
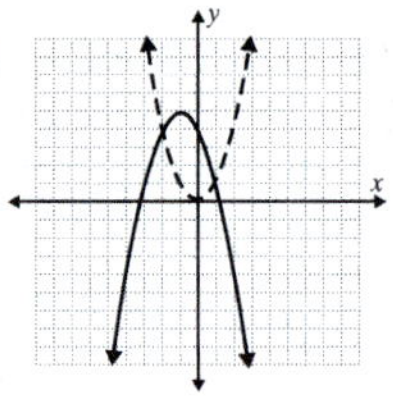

23. a.
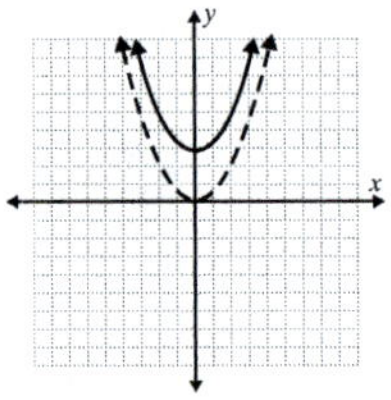
b.
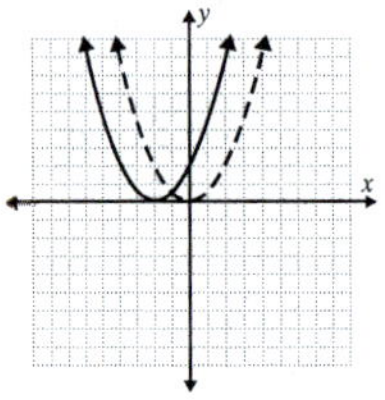
c.
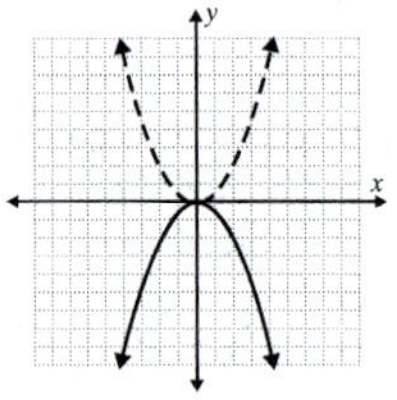
d.
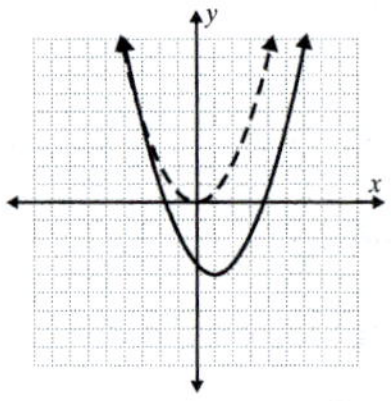

25. $y=2(x-1)^2$; $(1,0)$; $R=[0,\infty)$; Zeros : $x=1$
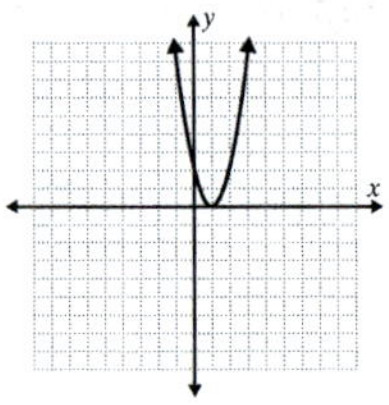

27. $y=(x-1)^2-4$; $(1,-4)$; $R=[-4,\infty)$; Zeros: $x=-1,3$
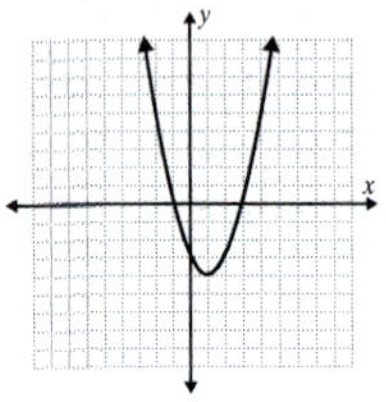

29. $y=(x+3)^2-4$; $(-3,-4)$; $R=[-4,\infty)$; Zeros: $x=-5,-1$
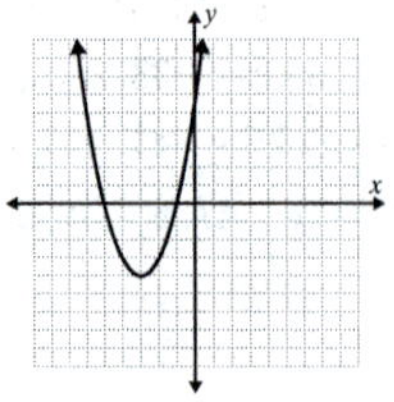

31. $y=2(x-2)^2-3$; $(2,-3)$; $R=[-3,\infty)$; Zeros: $x=\frac{4\pm\sqrt{6}}{2}$
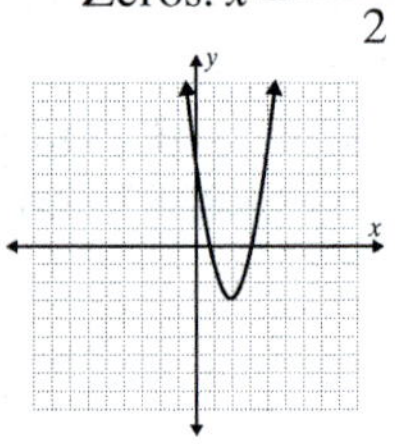

33. $y=-3(x+2)^2+3$; $(-2,3)$; $R=(-\infty,3]$; Zeros: $x=-1,-3$
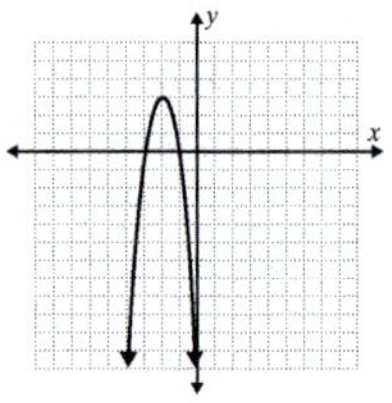

35. $y=5(x-1)^2+3$; $(1, 3)$; $R=[3, \infty)$; Zeros: none

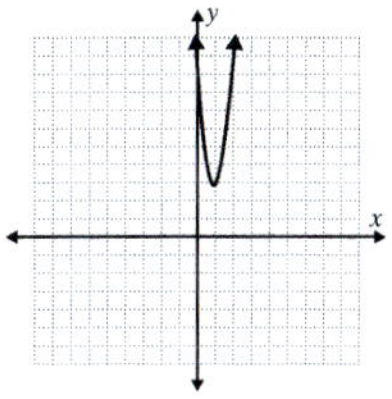

37. $y=-\left(x+\frac{5}{2}\right)^2+\frac{17}{4}$; $\left(-\frac{5}{2}, \frac{17}{4}\right)$; $R=\left(-\infty, \frac{17}{4}\right]$; Zeros: $x=\frac{-5\pm\sqrt{17}}{2}$

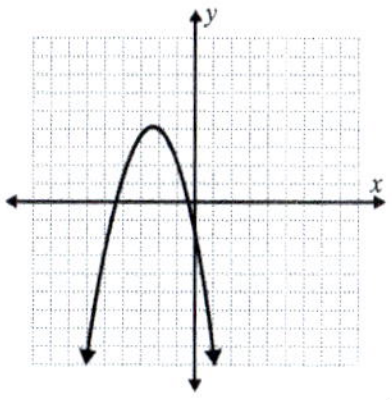

39. $y=2\left(x+\frac{7}{4}\right)^2-\frac{9}{8}$; $\left(-\frac{7}{4}, -\frac{9}{8}\right)$; $R=\left[-\frac{9}{8}, \infty\right)$; Zeros: $x=-\frac{5}{2}, -1$

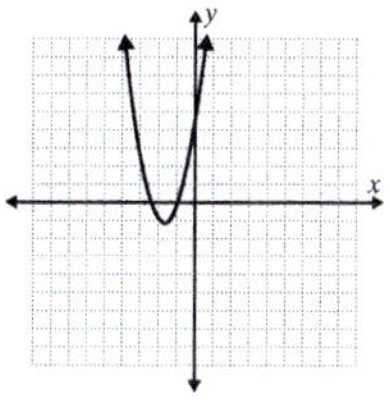

41.

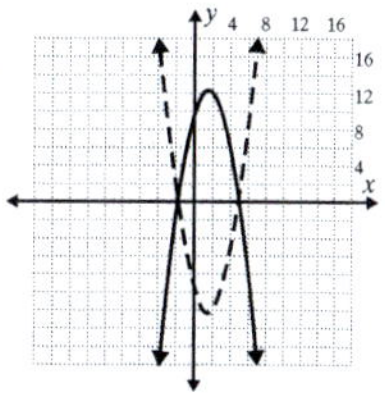

a. No **b.** Yes **c.** One function is a reflection of the other across the x-axis.

43.

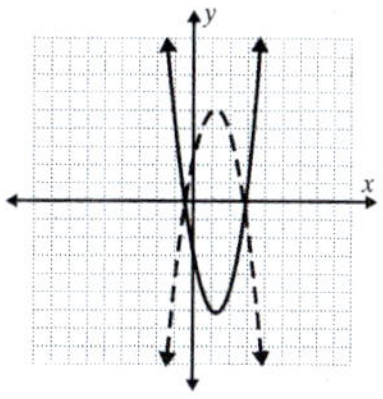

a. No **b.** Yes **c.** One function is a reflection of the other across the x-axis.

45. a. 3.5 s **b.** 196 ft **47. a.** 5 s **b.** 420 ft **49. a.** \$50 **b.** \$2500 **51.** \$20 **53.** Zeros: $x\approx -0.7321, 2.7321$
55. Zeros: $x\approx -1.1583, 2.1583$ **57.** No real zeros

59.

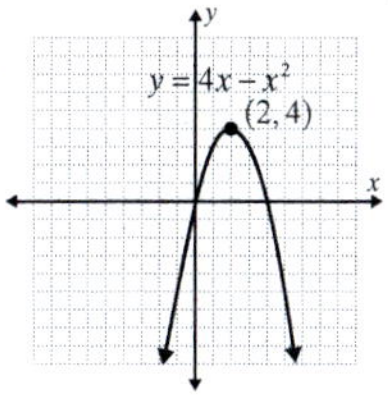

61.

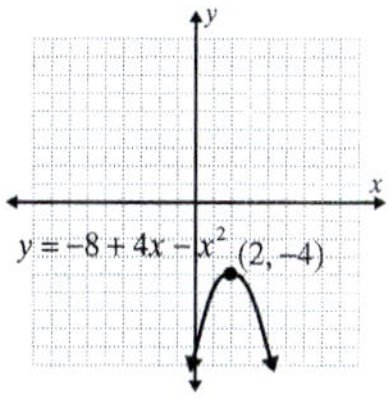

63.

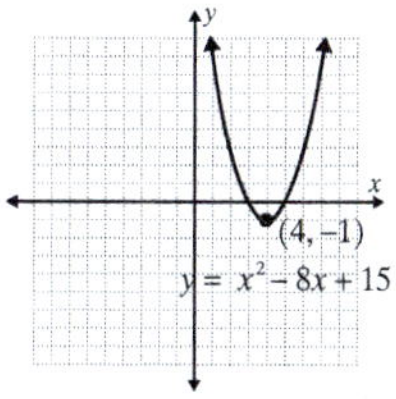

65.

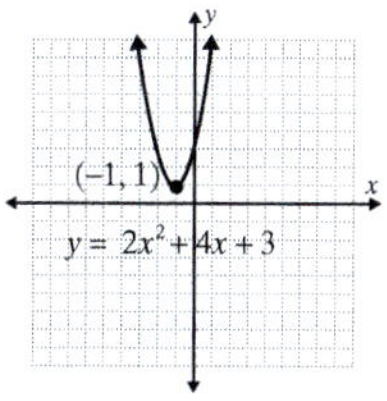

67. a. A parabola **b.** $x=-\frac{b}{2a}$ **c.** $x=-\frac{b}{2a}$ **d.** No. A graph can be entirely above or below the x-axis.

69. Domain: $(-\infty, \infty)$; For $a>0$, Range: $[k, \infty)$; For $a<0$, Range: $(-\infty, k]$.

Exercises 10.6, pages 836 - 839

1. $(-2, 6)$

3. $\left(-\infty, \frac{2}{3}\right)\cup(5, \infty)$

5. $(-\infty, -7]\cup\left[\frac{5}{2}, \infty\right)$

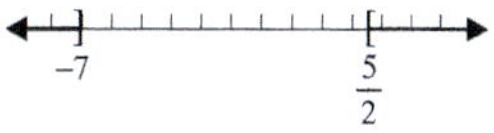

7. $\left[-2, -\frac{1}{3}\right]$

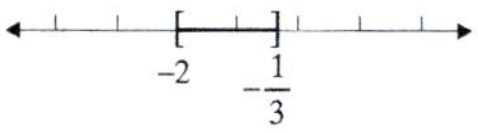

9. $\left(-\infty, -\frac{4}{3}\right)\cup(0, 5)$

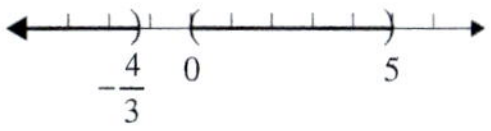

11. $\{-2\}$

13. $\left(-\infty, -\frac{5}{2}\right)\cup(3, \infty)$

15. $\left(-\frac{1}{4}, \frac{3}{2}\right)$

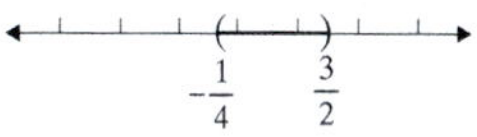

17. $\left(-\infty, \frac{1}{2}\right]\cup[2, \infty)$

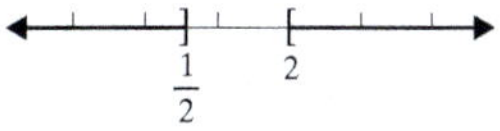

19. $\left(-\frac{2}{3}, -\frac{1}{2}\right)$

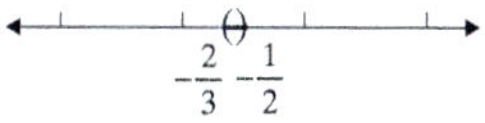

21. $\left(-\infty, \frac{5}{2}\right)\cup\left(\frac{5}{2}, \infty\right)$

23. $\left[-\frac{5}{2}, \frac{7}{4}\right]$

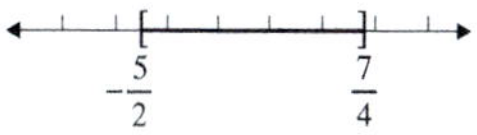

25. $(-1, 0)\cup(3, \infty)$

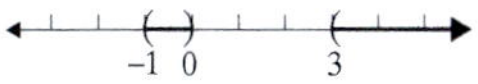

27. $(0, 1)\cup(4, \infty)$

29. $(-\infty, -1)\cup(4, \infty)$

31. $(-\infty, -2)\cup(-1, 1)\cup(2, \infty)$

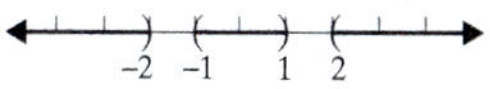

33. $[-3,-2]\cup[2,3]$

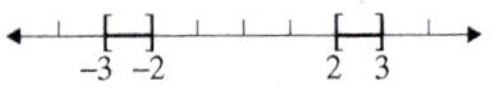

35. $(-\infty,-4]\cup[2,\infty)$

37. $\left(-\dfrac{2}{3},2\right)$

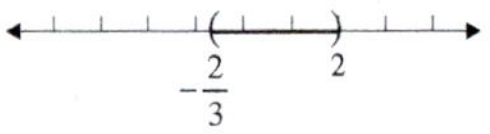

39. $\left(-\infty,-1-\sqrt{5}\right)\cup\left(-1+\sqrt{5},\infty\right)$

41. $\left(-\infty,-3-\sqrt{2}\right]\cup\left[-3+\sqrt{2},\infty\right)$

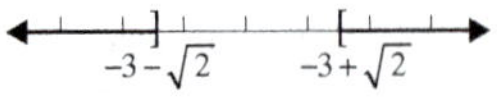

43. $\left(\dfrac{-5-\sqrt{13}}{6},\dfrac{-5+\sqrt{13}}{6}\right)$

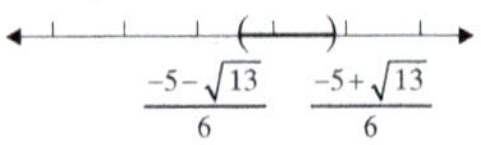

45. $\left(-\infty,-\dfrac{1}{2}\right]\cup[0,4]$

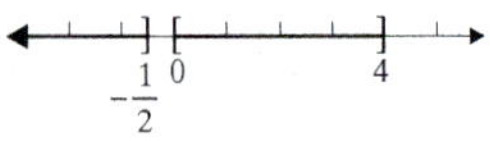

47. $(-\infty,\infty)$

49. $\varnothing$ **51. a.** $x=\dfrac{7\pm\sqrt{89}}{2}$ **b.** $\left(-\infty,\dfrac{7-\sqrt{89}}{2}\right)\cup\left(\dfrac{7+\sqrt{89}}{2},\infty\right)$ **c.** $\left(\dfrac{7-\sqrt{89}}{2},\dfrac{7+\sqrt{89}}{2}\right)$

53. a. $x=-5,1$ **b.** $(-5,1)$ **c.** $(-\infty,-5)\cup(1,\infty)$

55. $(-\infty,-4]\cup(0,\infty)$

57. $(-\infty,-6)$

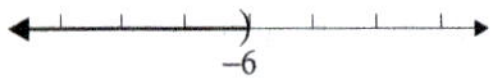

59. $(-\infty,-9)\cup(-3,\infty)$

61. $\left(2,\dfrac{5}{2}\right)$

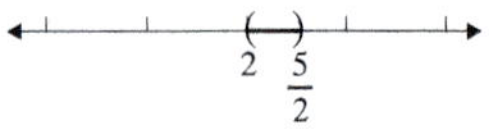

63. $(7,\infty)$

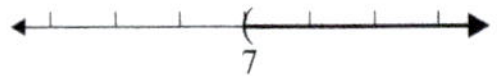

65. $\left[\dfrac{7}{5},4\right)$

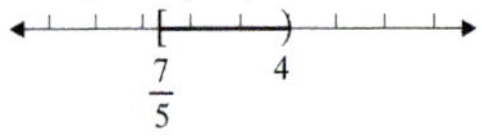

67. $\left(-9,-\dfrac{4}{3}\right)$

69. $(-\infty,-4]\cup[0,3)$

71. $(-2,0)\cup[5,\infty)$

73. $(-\infty,-3.1623)\cup(3.1623,\infty)$

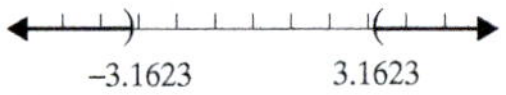

75. $\varnothing$

77. $(-\infty,-3)\cup(0,3)$

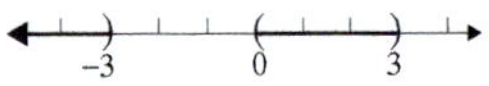

79. $[-1.8868,\infty)$

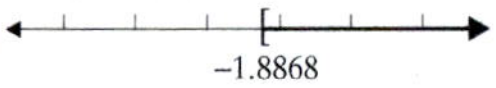

81. $(-2.3344,-0.7420)\cup(0.7420,2.3344)$

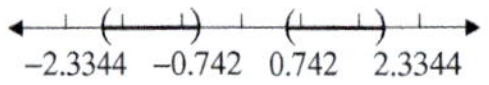

83. a. $(-4,0)\cup(1,\infty)$ **b.** $(-\infty,-4)\cup(0,1)$ **c.** The function is undefined at $x=0$.

Chapter 10 Review, pages 844 - 848

1. $x=-2,7$ **2.** $x=1,11$ **3.** $x^2+10x+\underline{25}=(\underline{x+5})^2$ **4.** $x^2-x+\underline{\dfrac{1}{4}}=\left(\underline{x-\dfrac{1}{2}}\right)^2$ **5.** $x=\pm 6i$ **6.** $x=\pm 10$ **7.** $x=\pm 2\sqrt{6}$

8. $x=-4\pm\sqrt{10}$ **9.** $x=3\pm\sqrt{5}$ **10.** $x=-1\pm 2\sqrt{2}$ **11.** $x=-1\pm i\sqrt{3}$ **12.** $x=\dfrac{1}{2},1$ **13.** $x=-1,\dfrac{1}{2}$ **14.** $x=\dfrac{3\pm i\sqrt{11}}{2}$

15. $x^2-3=0$ **16.** $x^2-2x-1=0$ **17.** $x^2+6=0$ **18.** $x^2+6x+14=0$ **19.** –23, Two nonreal solutions

20. 25, Two real solutions **21.** 9, Two real solutions **22.** 0, One real solution **23.** $x=\dfrac{-1\pm\sqrt{11}}{5}$ **24.** $x=1\pm\sqrt{5}$

25. $x=\pm\dfrac{13}{2}i$ **26.** $x=\dfrac{-1\pm i\sqrt{11}}{6}$ **27.** $x=-1,7$ **28.** $x=-\dfrac{5}{3},1$ **29.** $x=-3\pm\sqrt{2}$ **30.** $x=\dfrac{-7\pm i\sqrt{23}}{4}$ **31.** $x=\dfrac{3\pm i\sqrt{3}}{4}$

32. $x=0,3\pm 2\sqrt{3}$ **33.** $x=0,\dfrac{5}{3}$ **34.** $x=\pm\dfrac{2}{3}i$ **35.** $x=\dfrac{-1\pm\sqrt{31}}{12}$ **36.** $x=\dfrac{-5\pm\sqrt{55}}{4}$ **37.** $c<9$ **38.** $a>2$ **39.** 18 **40.** 8, 21

41. 30 feet, 40 feet **42.** 120 seats **43.** 9 m × 8 m **44.** 1 m **45.** 18 members, \$110 **46.** Martin: 15 mph; Milton: 20 mph

47. a. 40 s **b.** 1.97 s, 38.03 s **48.** Area = 106.76 cm^2; Circumference = 36.62 cm **49.** $x=\pm 2,\pm\sqrt{6}$ **50.** $x=\pm\sqrt{2},\pm i\sqrt{6}$

51. $y=-\frac{1}{2},1$ **52.** $x=64,-\frac{1}{8}$ **53.** $y=1$ **54.** $x=-64,-8$ **55.** $x=-2$ **56.** $y=7$ **57.** $x=\pm3\sqrt{i},\pm3i\sqrt{i}$ **58.** $x=-2,\frac{1}{4}$

59. $x=-8,-2$ **60.** $x=-10,-\frac{14}{3}$ **61.** $x=0;(0,-3);D=(-\infty,\infty);R=[-3,\infty)$ **62.** $x=0;(0,4);D=(-\infty,\infty);R=(-\infty,4]$

63. $x=-4;(-4,0);D=(-\infty,\infty);R=[0,\infty)$ **64.** $x=0;(0,8);D=(-\infty,\infty);R=(-\infty,8]$

65. $x=-1;(-1,-4);D=(-\infty,\infty);R=[-4,\infty)$ **66.** $x=\frac{3}{2};\left(\frac{3}{2},-\frac{9}{2}\right);D=(-\infty,\infty);R=\left(-\infty,-\frac{9}{2}\right]$

67. $y=(x-3)^2;(3,0);$
$R=[0,\infty);$
Zeros: $x=3$

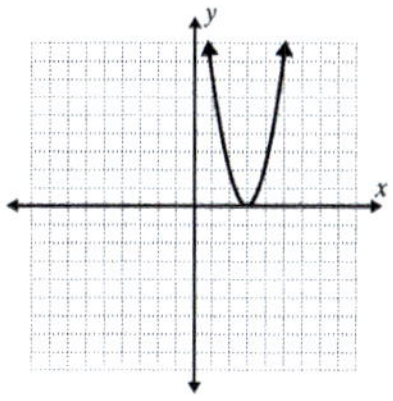

68. $y=(x-1)^2-5;(1,-5);$
$R=[-5,\infty);$
Zeros: $x=1\pm\sqrt{5}$

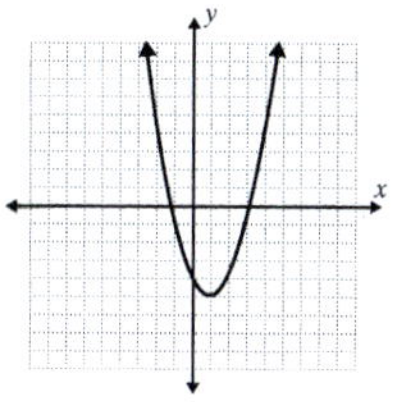

69. $y=(x-2)^2-3;(2,-3);$
$R=[-3,\infty);$
Zeros: $x=2\pm\sqrt{3}$

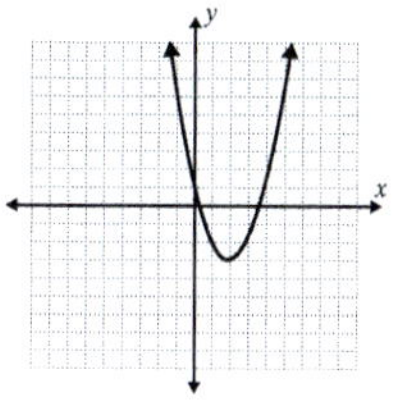

70. $y=-(x+3)^2+7;(-3,7);$
$R=(-\infty,7];$
Zeros: $x=-3\pm\sqrt{7}$

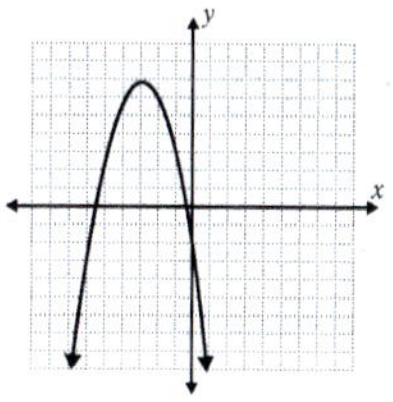

71. $y=2\left(x+\frac{9}{4}\right)^2-\frac{33}{8};\left(-\frac{9}{4},-\frac{33}{8}\right);$
$R=\left[-\frac{33}{8},\infty\right);$ Zeros: $x=\frac{-9\pm\sqrt{33}}{4}$

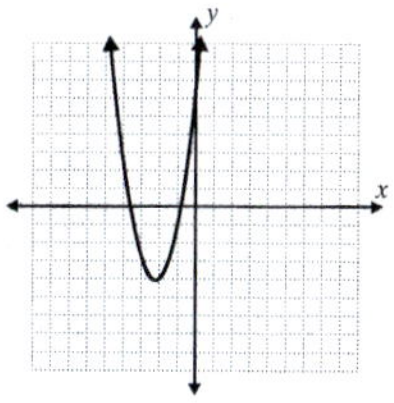

72. $y=(x+1)^2+3;(-1,3);$
$R=[3,\infty);$ Zeros: None

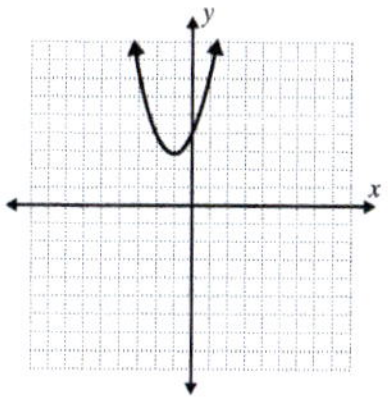

73. a.

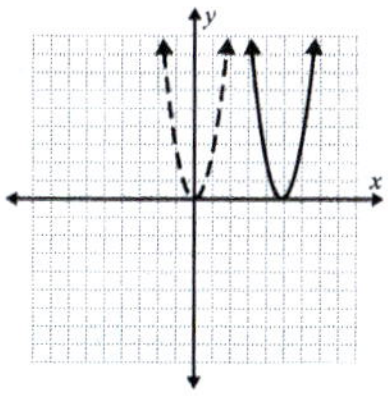

b.

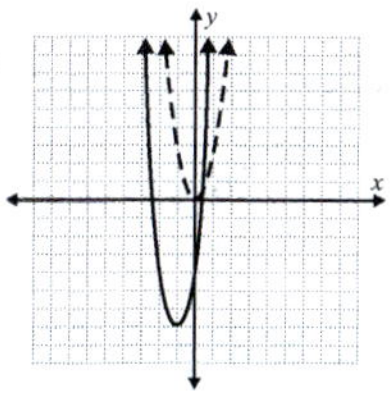

c.

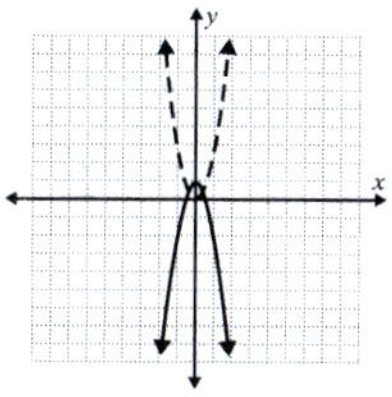

d.

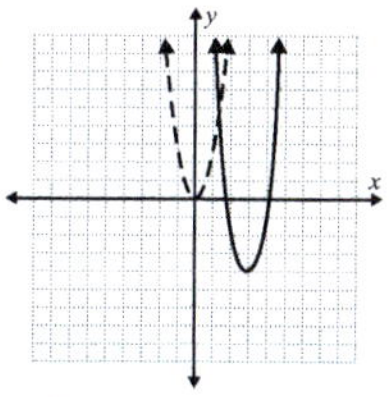

74. a. 300 copies **b.** $15,000
75. a. 6.5 s **b.** 676 ft **c.** 13 s
76. Two sides 25 ft, third side 50 ft
77. Zeros: $x\approx-3.1375, 0.6375$
78. Zeros: $x\approx-6.5414,-0.4586$

79. $(-1,5)$

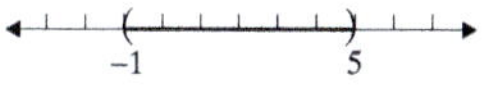

80. $(-\infty,-3)\cup\left(\frac{1}{2},\infty\right)$

81. $\left(-\infty,-\frac{3}{2}\right]\cup[1,\infty)$

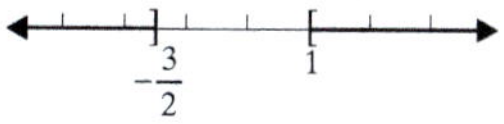

82. $\{-3\}$

83. $\left(-\infty,-\frac{5}{3}\right)\cup(0,4)$

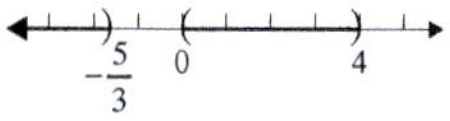

84. $\left(-\infty,\frac{3}{2}\right)\cup\left(\frac{3}{2},\infty\right)$

85. $\left(-\infty,\frac{1}{3}\right]\cup[1,\infty)$

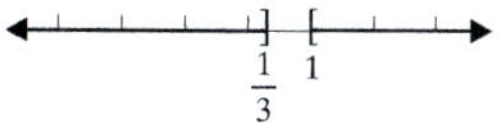

86. $(-\infty,-2)\cup(2,\infty)$

87. $(-\infty,-1-\sqrt{6})\cup(-1+\sqrt{6},\infty)$

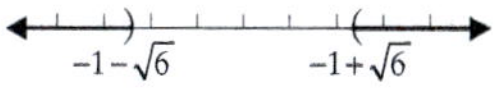

88. $\left[\frac{-1-\sqrt{21}}{4},\frac{-1+\sqrt{21}}{4}\right]$

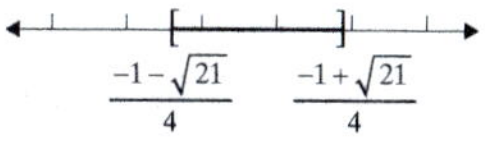

89. $\left(-\frac{1}{2},6\right]$

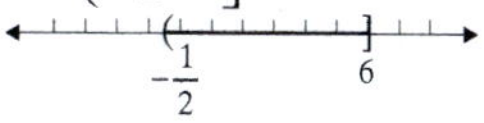

90. $(-\infty,1)\cup(4,\infty)$

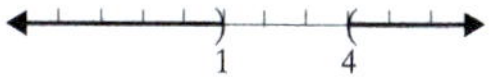

91. $(-\infty,2)\cup(3,\infty)$

92. $(-\infty,-2)\cup\left[\frac{8}{3},\infty\right)$

93. $(-\infty,-2.8284)\cup(2.8284,\infty)$

94. $[-1,6]$

95. $[-1.9422, 0.5579] \cup [1.3844, \infty)$

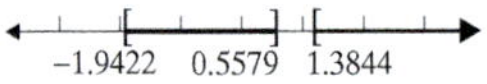

96. $(-\infty, -1.1304)$

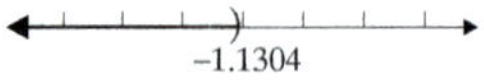

Chapter 10 Test, pages 849 - 850

1. $x=-4, 2$ **2. a.** $x^2-30x+\underline{225}=(\underline{x-15})^2$ **3.** $x=-2\pm\sqrt{3}$ **4.** $x=-\frac{1}{2}, 0$ **5.** $x^2-2x-4=0$ **6.** 73, Two real solutions

7. $b=\pm 2\sqrt{6}$ **8.** $x=\frac{-1\pm i\sqrt{7}}{4}$ **9.** $x=\frac{3\pm\sqrt{41}}{4}$ **10.** $x=\frac{2\pm i\sqrt{2}}{2}$ **11.** $x=\pm 1, \pm 3$ **12.** $x=-1, \frac{3}{2}$ **13.** $x=0, 1$

14. $x=3; (3,-1); D=(-\infty,\infty); R=[-1,\infty)$ **15.** $x=3; (3,-9); D=(-\infty,\infty); R=[-9,\infty)$

16. $y=(x+2)^2-10; (-2,-10)$; $R=[-10,\infty)$; Zeros: $x=-2\pm\sqrt{10}$

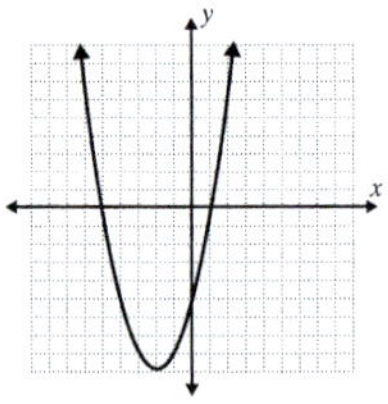

17. $y=(x-3)^2-1; (3,-1)$; $R=[-1,\infty)$; Zeros: $x=2, 4$

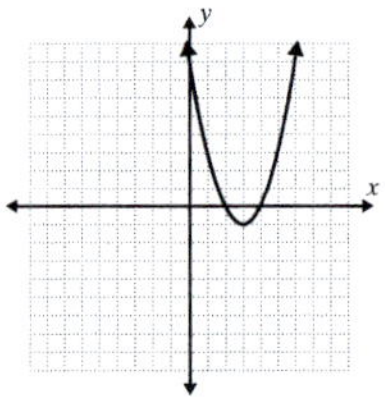

18. $[-7,0]\cup[3,\infty)$

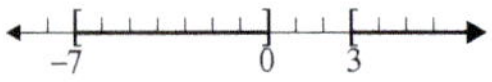

19. $(-3,-2)$

20. $\left(-\infty,-\frac{5}{2}\right]\cup(3,\infty)$

21. $\left(-\infty,-\frac{5}{3}\right)\cup\left(-\frac{1}{2},\infty\right)$

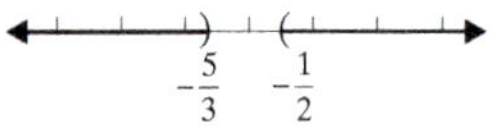

22. $(-5,0)\cup(2,\infty)$

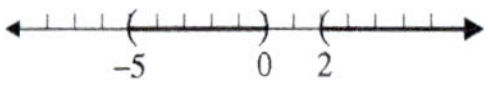

23. 8 m, 15 m

24. a. $t=7$ seconds **b.** $t=6.46$ seconds **25.** Width = 12 inches; Length = 16 inches **26.** Speed returning = 40 mph; Speed going = 50 mph **27.** $y=x(x-10)$; $-5, 5$; Minimum product $=-25$ **28.** $\frac{11}{2}$ in. $\times\frac{11}{2}$ in.

Cumulative Review: Chapters 1-10, pages 851 - 854

1. $\frac{1}{x^5}$ **2.** x^3y^6 **3.** $\frac{9\sqrt{y}}{x^2}$ **4.** $\frac{27x^2}{8y}$ **5.** $-3x^2y^2\sqrt[3]{y^2}$ **6.** $2x^2y^3\sqrt[4]{2xy^3}$ **7.** $\frac{9\sqrt{2}}{2}$ **8.** $\frac{4\sqrt{3}}{3}$ **9.** $5(\sqrt{3}-\sqrt{2})$

10. $(x+9)(\sqrt{x}-3)$ **11.** $4+2i$ **12.** $19+4i$ **13.** $\frac{-11+16i}{13}$ **14.** $(x-5)(y+2)$ **15.** $(8x+9)(8x-9)$

16. $2(x^2-6y)(x^4+6x^2y+36y^2)$ **17. a.** -2 **b.** 37 **18.** $\frac{9x^2-2x-2}{(x+4)(2x-1)(3x-2)}$ **19.** $\frac{21x+8}{(x-8)(x+1)(3x-2)}$ **20.** $\frac{x+3}{3(x-5)}$

21. $\frac{2x^2+3x+1}{2(2x-3)(x-1)}$ **22.** $x=6$ **23.** $x=\frac{7}{8}$ **24.** $x=-\frac{1}{2}, 3$ **25.** $x=-\frac{3}{2}, \frac{2}{5}$ **26.** $x=\frac{-7\pm\sqrt{33}}{8}$ **27.** $x=-1$ **28.** $x=-2, \frac{4}{3}$

29. $x=-5,-3,3,5$ **30.** $x=-\frac{8}{27}, -1$ **31.** $x=-\frac{38}{7}$ **32. a.** -47; No real solution **b.** 0; One real solution

33. $4x^2+17x-15=0$ **34.** $x^2-2x-19=0$ **35.** $\left[-\frac{19}{2},\infty\right)$

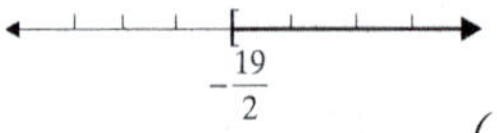

36. $\left(-\infty,\frac{7}{5}\right]$

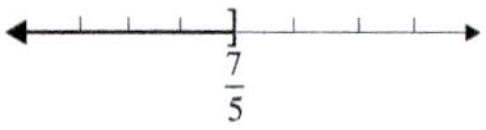

37. $(-2,3)$

38. $\{4\}$

39. $(-\infty,-4)\cup\left(0,\frac{5}{3}\right)$

40. $(-2,6]$

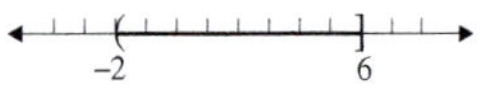

41.

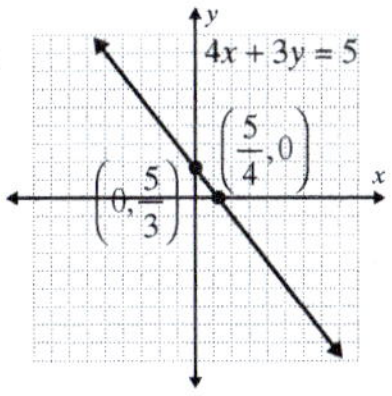

42. $y = x - 7$

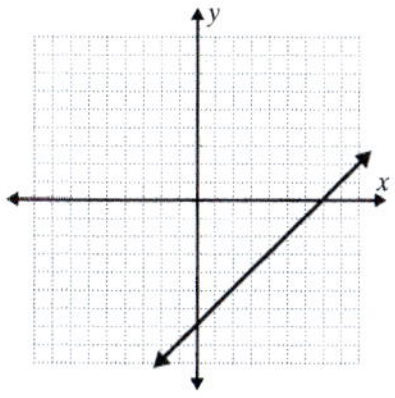

43. $3x - y = 12$

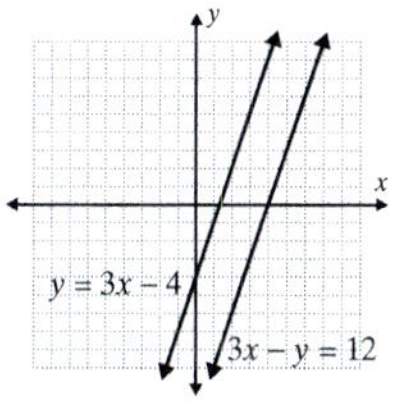

44. If any vertical line intersects the graph of a relation at more than one point, then the relation graphed is not a function. This works because it checks whether any first coordinate appears more than once.

45. $(-1, -2)$

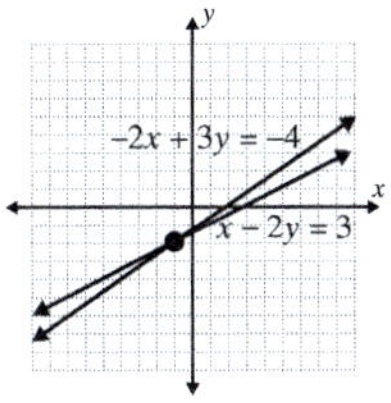

46. $(5, 4)$

47. $(1, -1, 3)$

48.

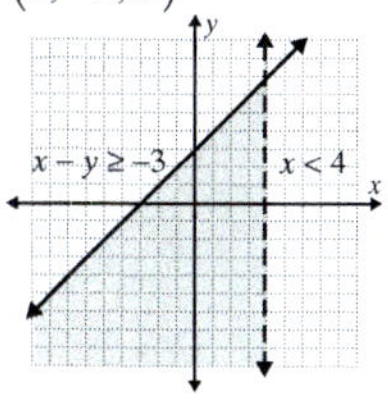

49. $x = -\frac{1}{2}; \left(-\frac{1}{2}, -\frac{25}{4}\right)$; $D = (-\infty, \infty)$; $R = \left[-\frac{25}{4}, \infty\right)$; Zeros: $x = -3, 2$

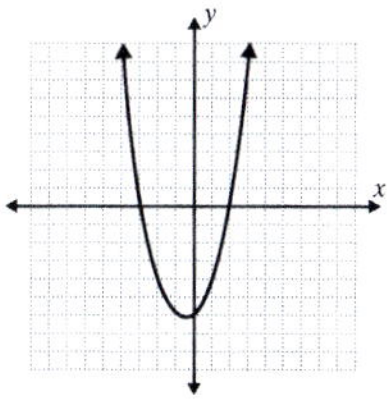

50. $x = \frac{1}{2}; \left(\frac{1}{2}, \frac{49}{4}\right)$; $D = (-\infty, \infty)$; $R = \left(-\infty, \frac{49}{4}\right]$; Zeros: $x = -3, 4$

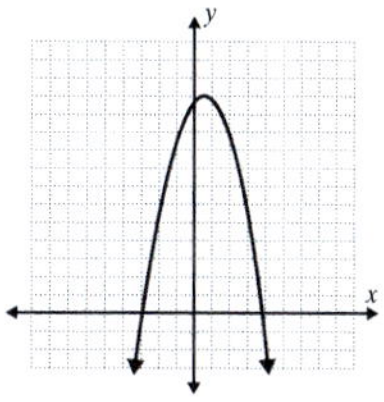

51. $x = \frac{9}{4}; \left(\frac{9}{4}, -\frac{121}{8}\right)$; $D = (-\infty, \infty)$; $R = \left[-\frac{121}{8}, \infty\right)$; Zeros: $x = -\frac{1}{2}, 5$

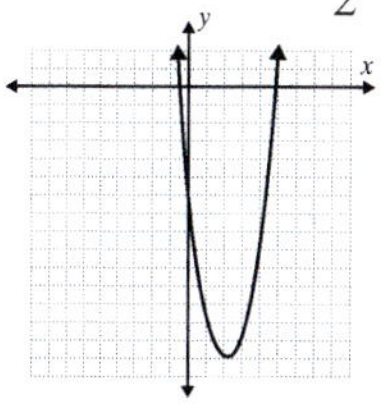

52. $y = -\frac{3}{2}x + 3$ **53.** $y = -\frac{3}{2}x + 10$

54. Length of the base = 19 cm; Height = 8 cm

55. 20 m × 26 m **56.** 1.7 mph faster **57.** 12, 14, 16

58. 2.14 hours **59.** 100 **60.** \$2500 at 9% and \$7500 at 11% **61.** 15 mg of Drug A, 5 mg of Drug B **62.** \$35

63. a. 10,000 dozen **b.** \$995,000 **c.** 120,000 golf balls **64.** 5.9136 **65.** 14.8306 **66.** –5.8284

67. x-int: $(-1.58, 0)$ and $(1.58, 0)$

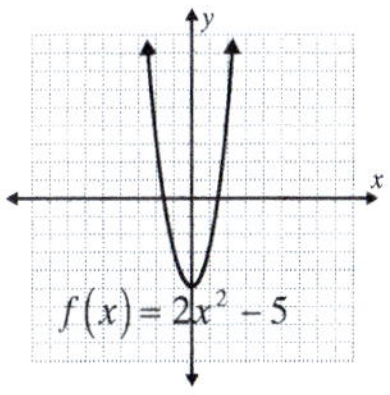

68. x-int: $(-0.62, 0)$, $(1, 0)$ and $(1.62, 0)$

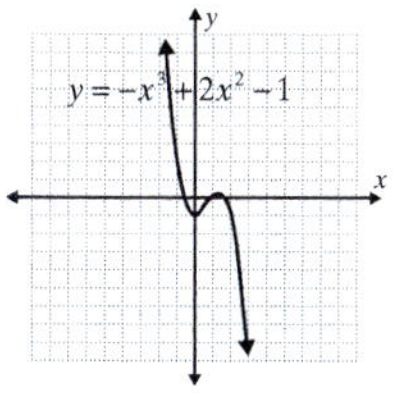

69. x-int: None

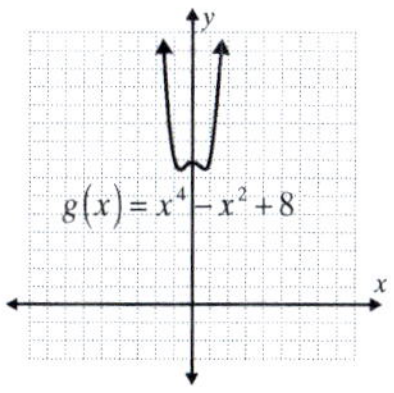

70. x-int: $(-0.60, 0)$

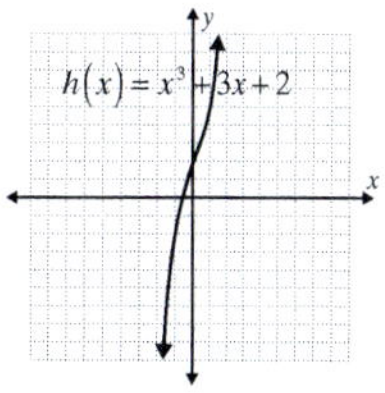

71. $D = [3, \infty)$; $R = [0, \infty)$

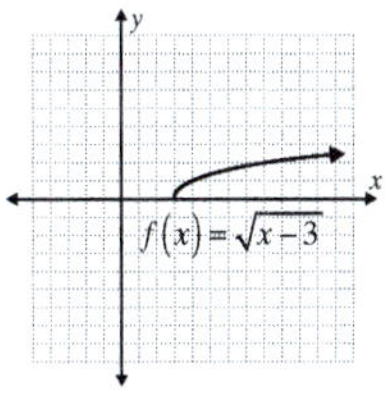

72. $D = (-\infty, 1]$; $R = [0, \infty)$

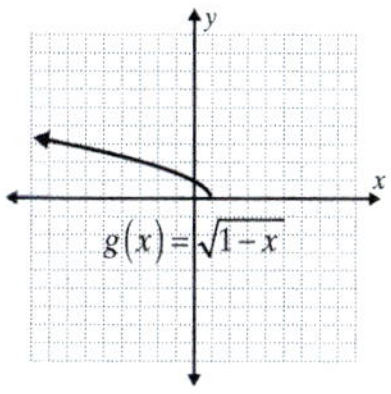

73. $D = [-1, \infty)$; $R = (-\infty, 0]$

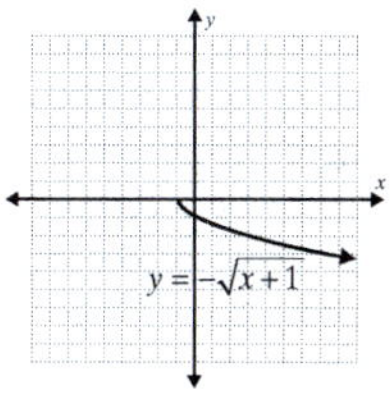

Chapter 11

Exercises 11.1, pages 863 - 868

1. a. $2x - 3$ **b.** 7 **c.** $x^2 - 3x - 10$ **d.** $\frac{x+2}{x-5}, x \neq 5$ **3. a.** $x^2 + 3x - 4$ **b.** $x^2 - 3x + 4$ **c.** $3x^3 - 4x^2$ **d.** $\frac{x^2}{3x-4}, x \neq \frac{4}{3}$

5. a. x^2+x-12 **b.** x^2-x-6 **c.** $x^3-3x^2-9x+27$ **d.** $x+3, x\neq 3$ **7. a.** $3x^2+x+2$ **b.** x^2+x-2 **c.** $2x^4+x^3+4x^2+2x$ **d.** $\dfrac{2x^2+x}{x^2+2}$ **9. a.** $2x^2+2$ **b.** $8x$ **c.** x^4-14x^2+1 **d.** $\dfrac{x^2+4x+1}{x^2-4x+1}, x\neq 2\pm\sqrt{3}$ **11.** 9 **13.** $-a^2-a-1$ **15.** 27 **17.** $\dfrac{8}{5}$ **19.** -31 **21.** $\sqrt{2x-6}+x+4, D=[3,\infty)$ **23.** $3x^2-19x-14, D=(-\infty,\infty)$ **25.** $\dfrac{x-5}{\sqrt{x+3}}, D=(-3,\infty)$ **27.** $-3x\sqrt{x-3}, D=[3,\infty)$ **29.** $\sqrt[3]{x+3}+\sqrt{5+x}, D=[-5,\infty)$

31. a.
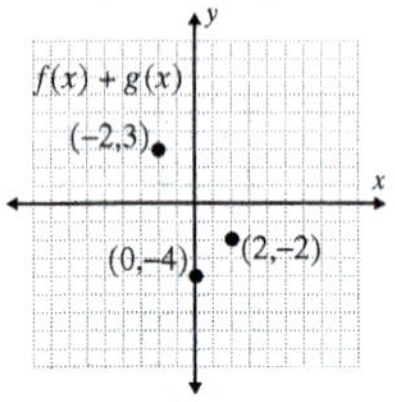

b.
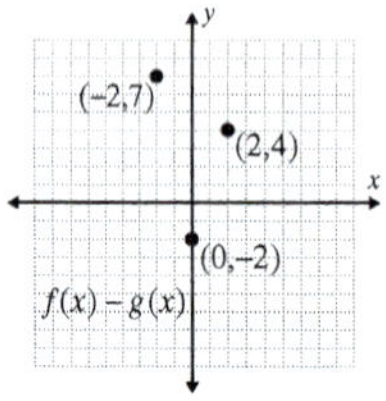

33. a.
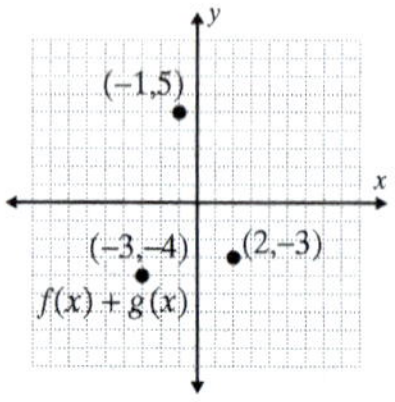

b.
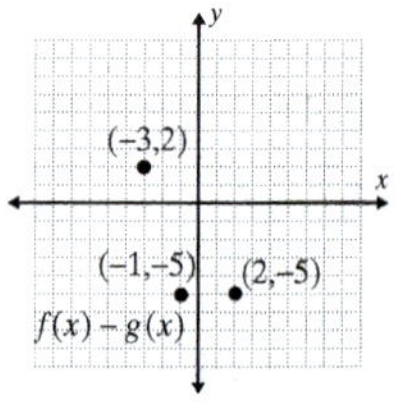

35. a.
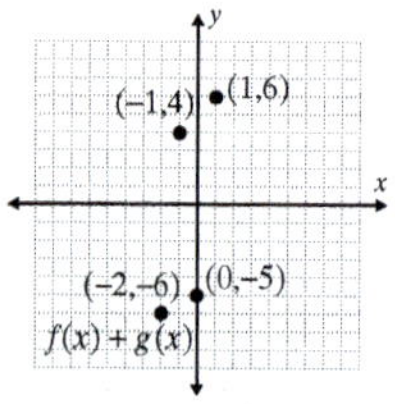

b.
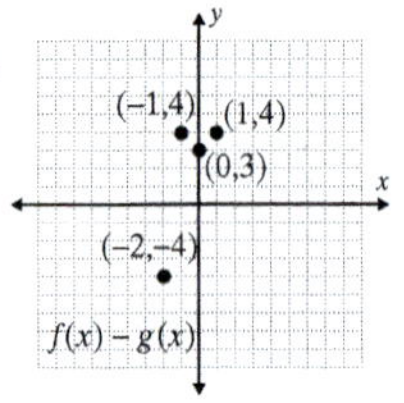

37. a.
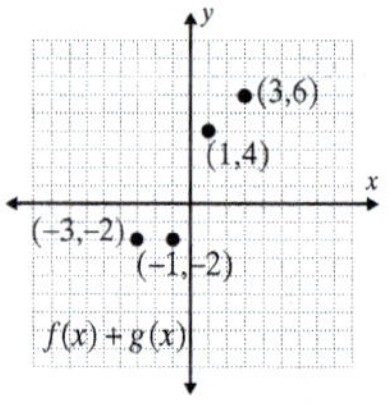

b.
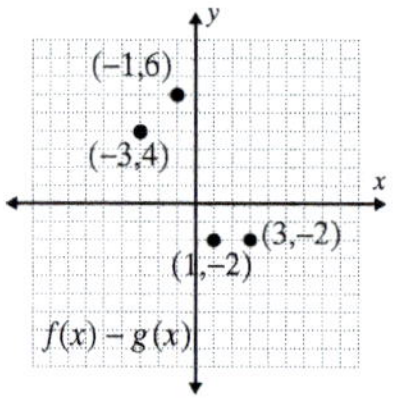

39. a.
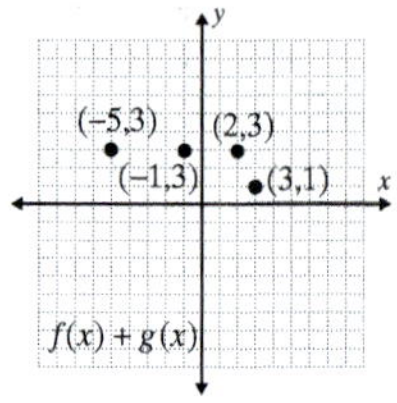

b.
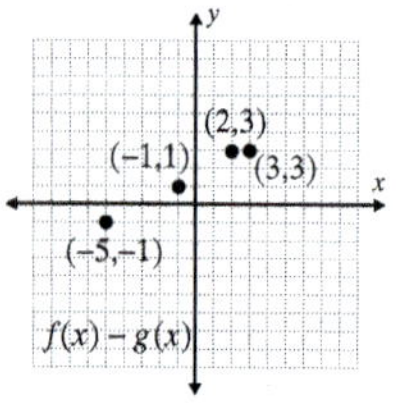

41.
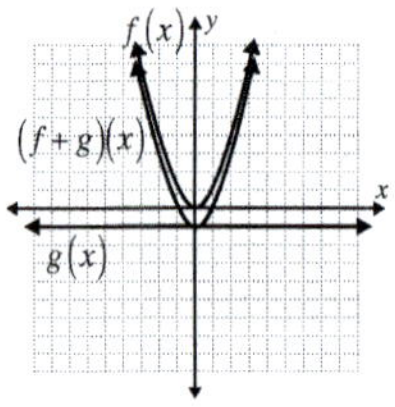

43.
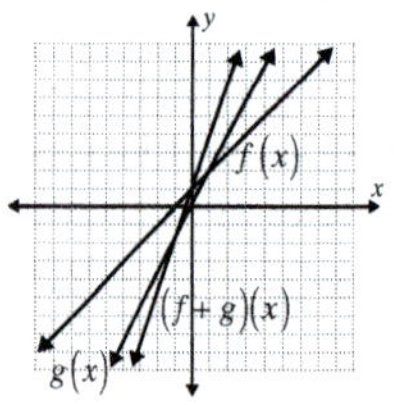

45.
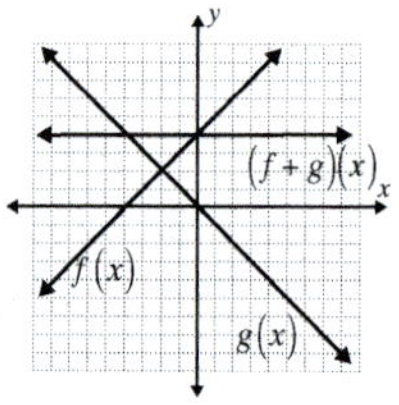

47.
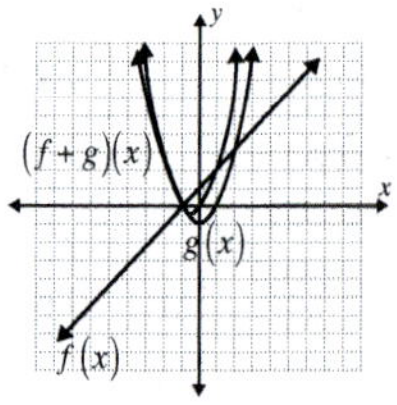

49.
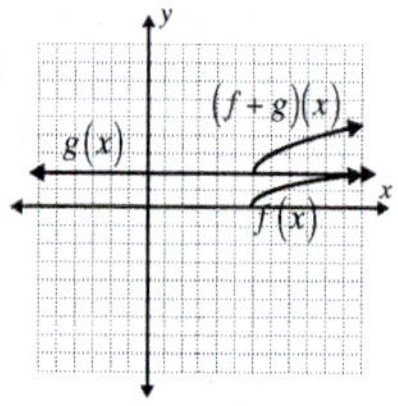

51. 4 **53.** -12 **55.** $-\dfrac{3}{7}$

57.
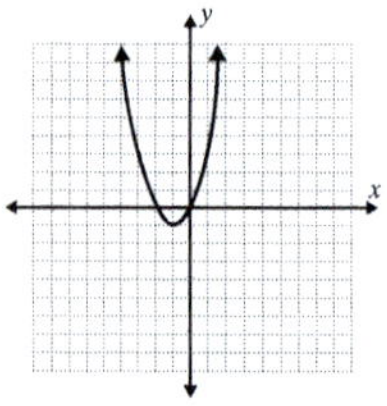
59.
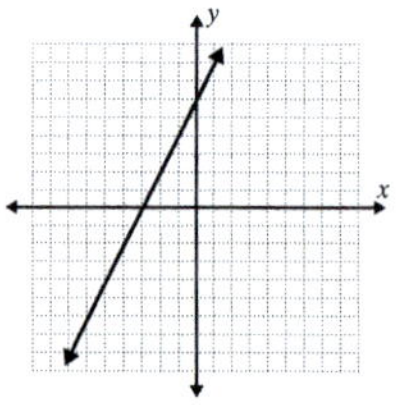
61.
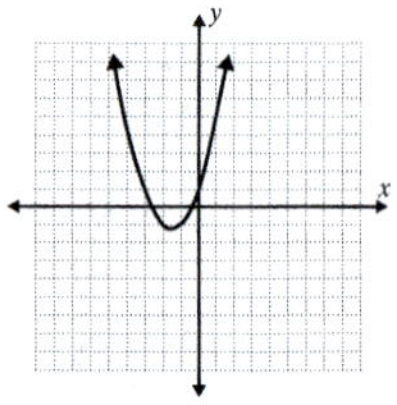
63.
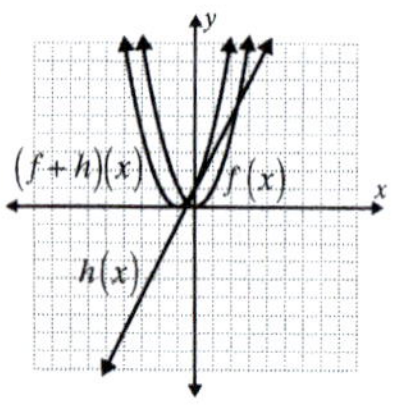

65.
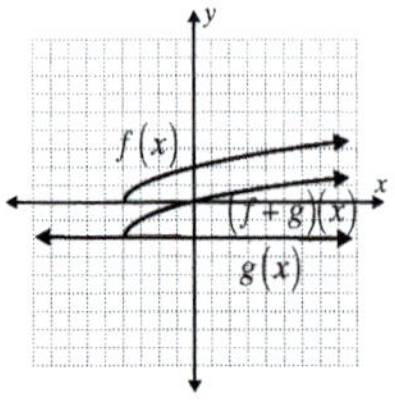

67.
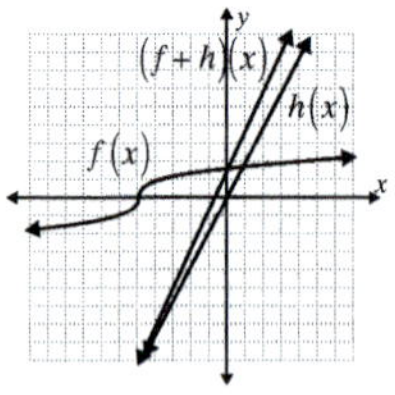

69.
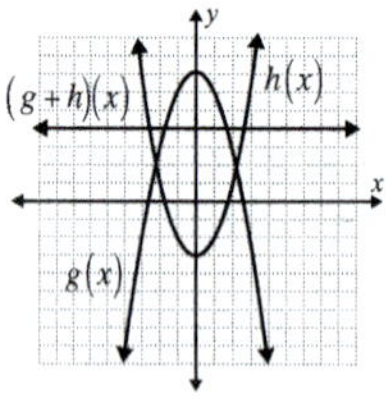

71. In general, subtraction is not commutative. Answers will vary.

73. a.
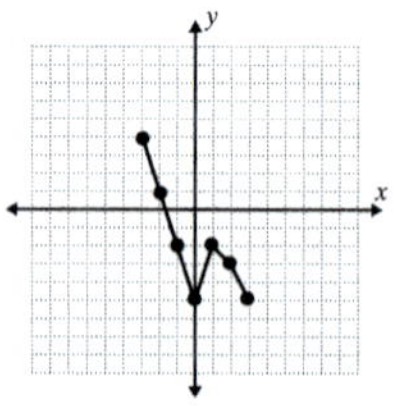
b.
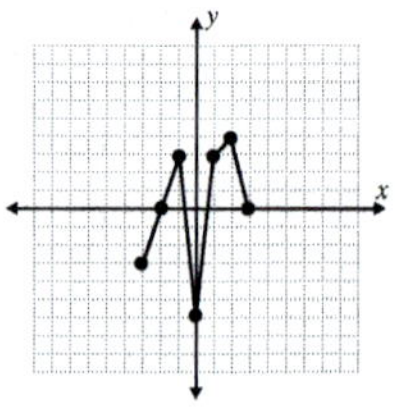
c. $\dfrac{f}{g}$ is undefined at $x=-2$ as $g(-2)=0$.

Exercises 11.2, pages 881 - 885

1. a. $f(g(2))=14$ **b.** $g(f(2))=\dfrac{15}{2}$ **3. a.** $(f\circ g)(-5)=49$ **b.** $(g\circ f)(-1)=5$ **5.** $f(g(x))=\sqrt{x^2}=x, g(f(x))=(\sqrt{x})^2=x$

7. $f(g(x))=\sqrt{x-2},\ g(f(x))=\sqrt{x}-2$ **9.** $f(g(x))=\dfrac{1}{x^2}-1,\ g(f(x))=\dfrac{1}{(x-1)^2}$

11. $f(g(x))=x^3+3x^2+4x+3,\ g(f(x))=x^3+x+2$ **13.** $f(g(x))=\dfrac{1}{x},\ g(f(x))=\dfrac{1}{x}$

15. $f(g(x))=\dfrac{1}{x^2+7x-8},\ g(f(x))=\left(\dfrac{1}{x}\right)^2+7\left(\dfrac{1}{x}\right)-8$ **17.** $f(g(x))=(2x-6)^{3n},\ g(f(x))=2x^{3n}-6$

19. $f(g(x))=(x-8)^{\frac{3}{2}},\ g(f(x))=\sqrt{x^3-8}$ **21. a.** 21 **b.** 2 **c.** No **23. a.** –9 **b.** Does not exist **c.** $f(4)=\frac{1}{9}$ and $\frac{1}{9}$ is in the domain of *g*, so $g(f(4))$ is defined. However, $g(2)=-\frac{1}{2}$ and $-\frac{1}{2}$ is not in the domain of *f*, so $f(g(2))$ is not defined.

25.

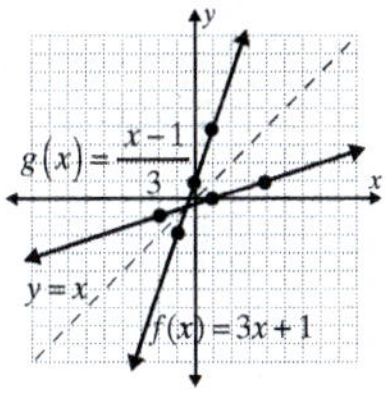

27.

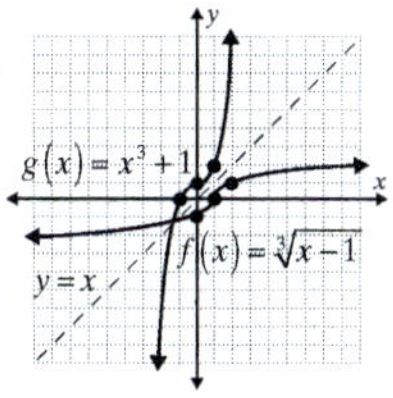

29.

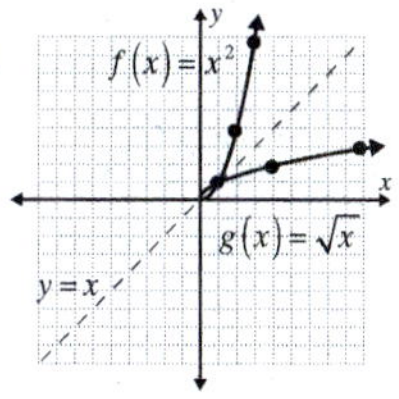

31.

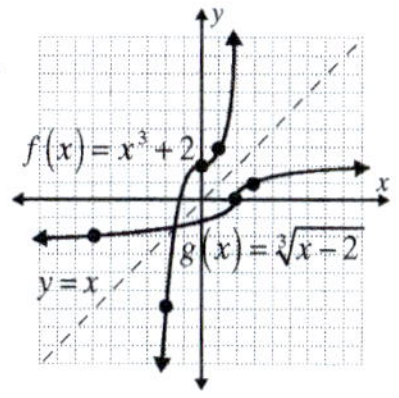

33.

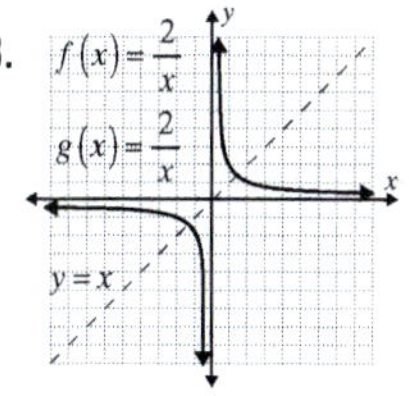

35. $f^{-1}(x)=\dfrac{x+3}{2}$

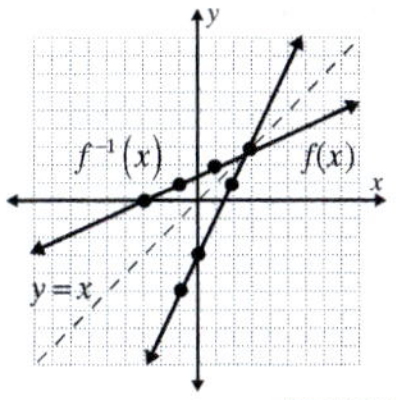

37. $g^{-1}(x)=x$

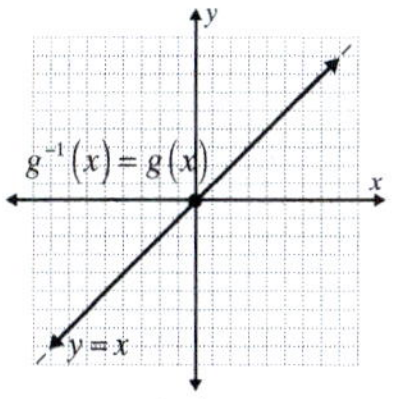

39. $f^{-1}(x)=\dfrac{x-1}{5}$

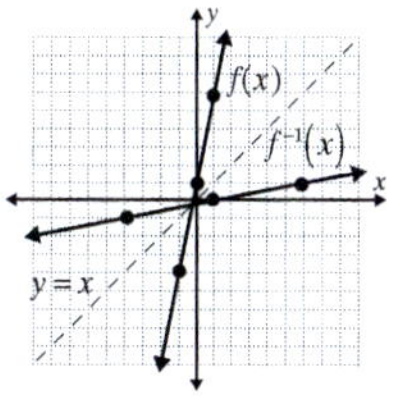

41. $g^{-1}(x)=\dfrac{3(x-2)}{2}$

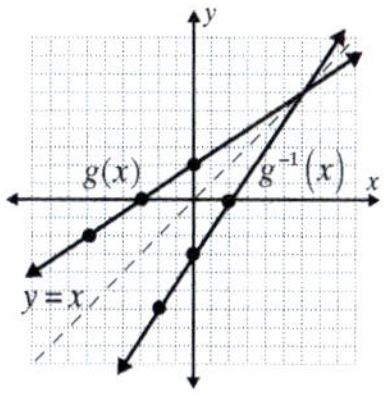

43. $f^{-1}(x)=-x-2$

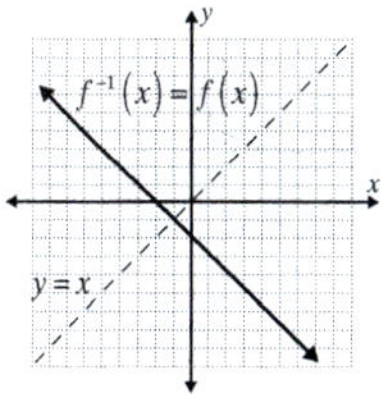

45. $f^{-1}(x)=\sqrt{x-1}$

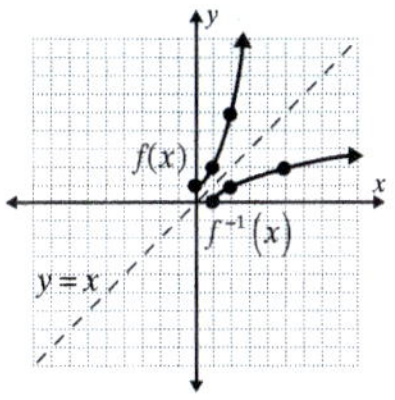

47. $f^{-1}(x)=x^2,\ x\le 0$

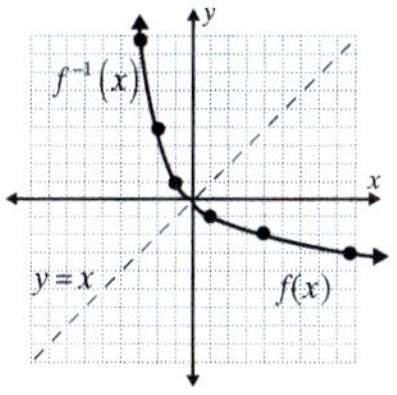

49. One-to-one

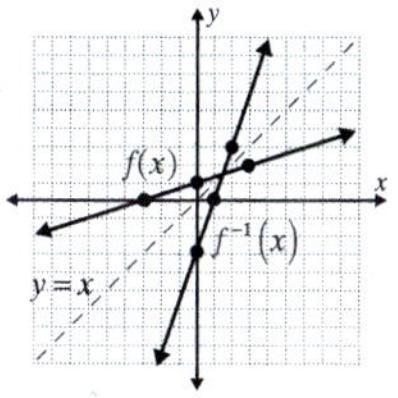

51. Not one-to-one

53. Not one-to-one

55. One-to-one

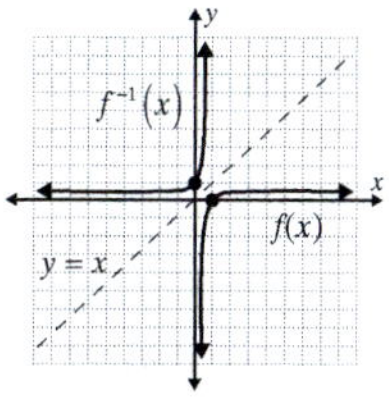

57. One-to-one

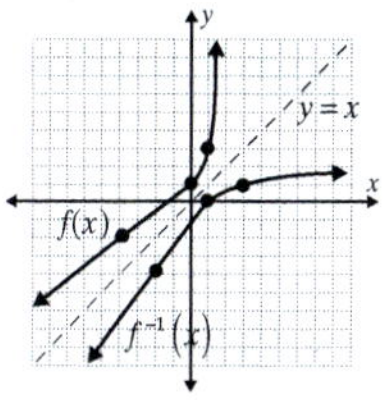

59. One-to-one **61.** Not one-to-one **63.** One-to-one **65.** One-to-one **67.** One-to-one

69. Not one-to-one

71. $f^{-1}(x)=\sqrt[3]{x}$

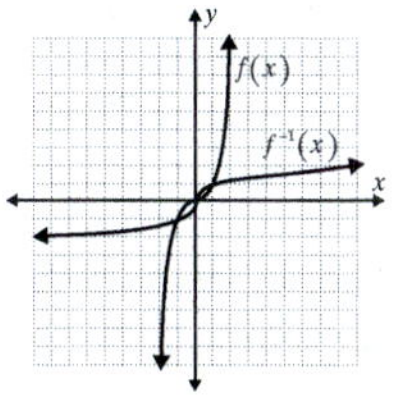

73. $f^{-1}(x)=\dfrac{1}{x}+3$

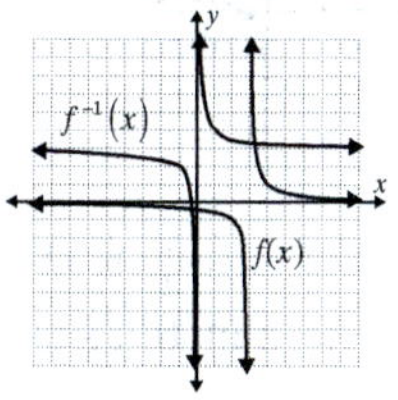

75. $f^{-1}(x)=\sqrt{x}$

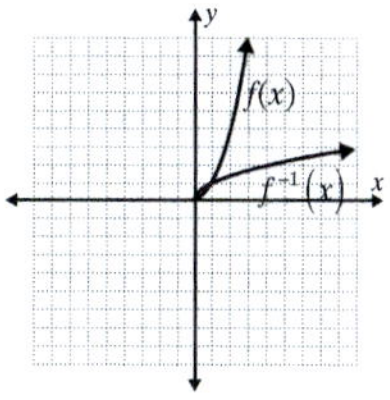

77. $g^{-1}(x)=\sqrt[3]{x-2}$

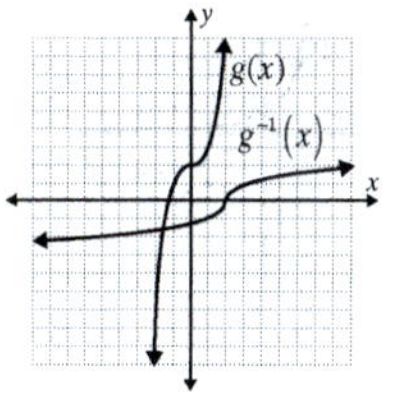

79. $f^{-1}(x)=x^2-5,\ x\ge 0$

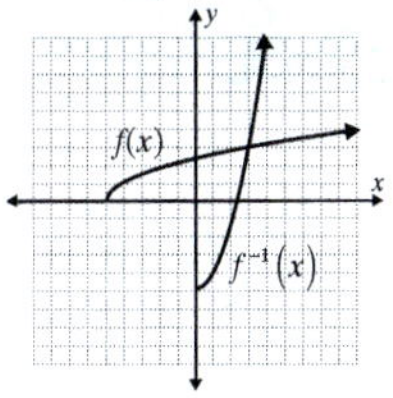

81. $f^{-1}(x)=\sqrt{1-x}$

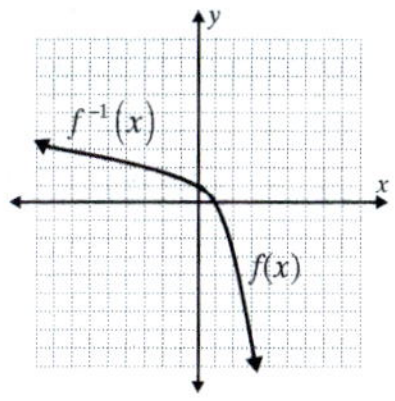

83. The domain of $f(g(x))$ can only include values of *x* in the domain of *g*, while $g(f(x))$ can only include values of *x* in the domain of *f*. Ex: $f(x)=\sqrt{x}, g(x)=-x$ Answers will vary.

Exercises 11.3, pages 896 - 899

1. 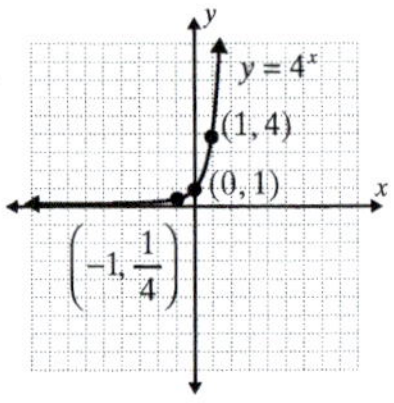

3. 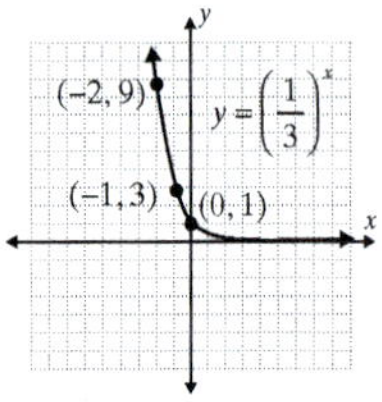

5. 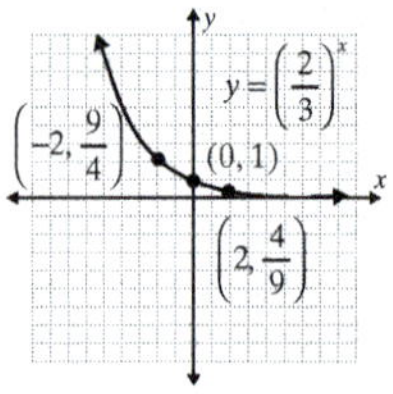

7. 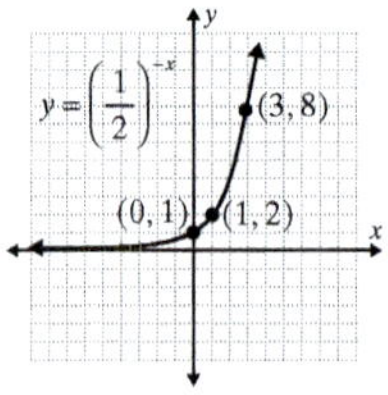

9. 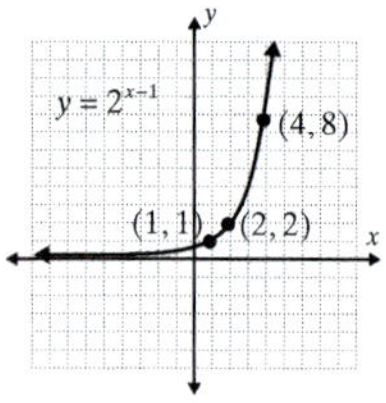

11. 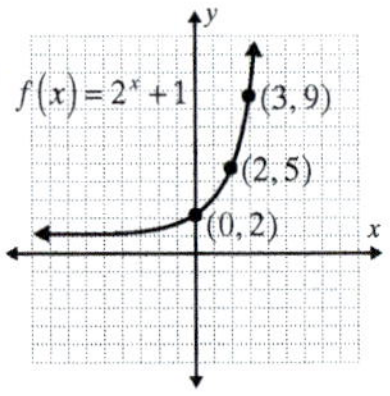

13. 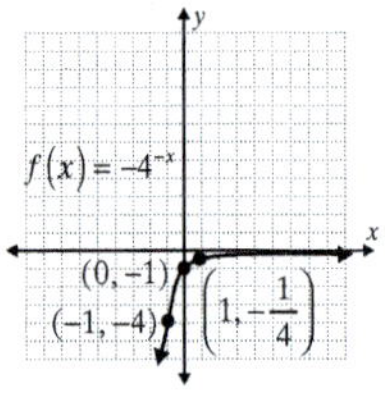

15. 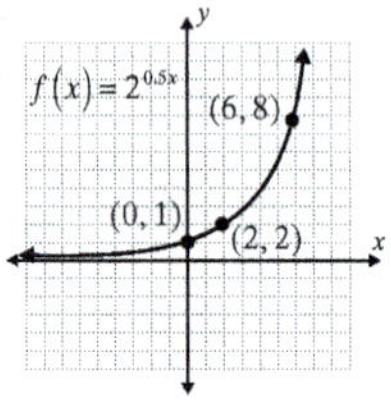

17. 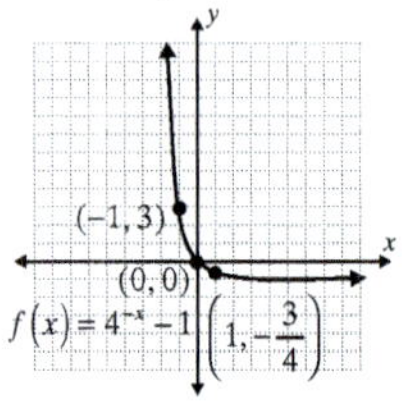

19. 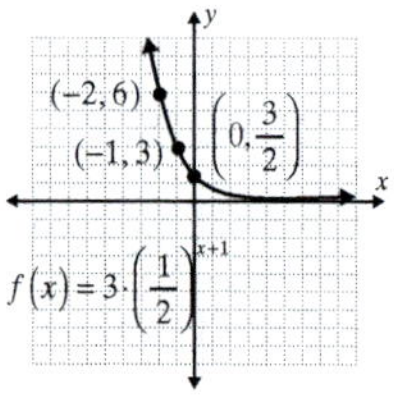

21. 48 **23.** 12.27 **25.** 4108.87 **27.** 25,000 bacteria **29. a.** \$5877.31 **b.** \$5920.98 **c.** \$5943.79 **d.** \$5967.03 **e.** \$5967.30

31. \$2154.99 **33.** \$46.88 **35.** 250,000 people **37. a.** \$100,149.34 **b.** \$222,886.46 **c.** \$1,103,963.86 **39.** \$491.52 **41.** $y = 50(2)^{-t}$

43. a. 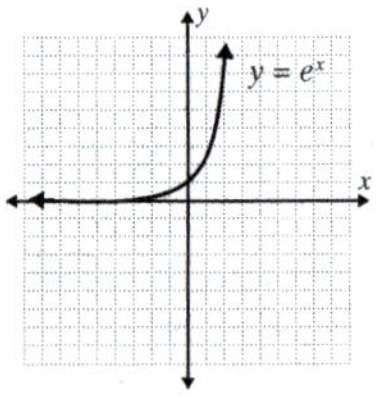

b. 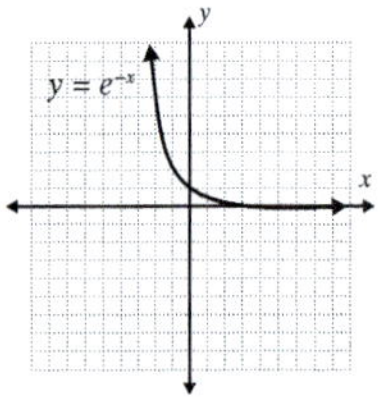

c. 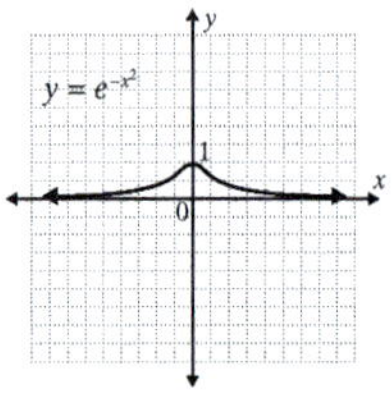

45. The two functions are symmetric about the y-axis.

47. Answers will vary.

Exercises 11.4, pages 905 - 906

1. $\log_7 49 = 2$ **3.** $\log_5 \frac{1}{25} = -2$ **5.** $\log_\pi 1 = 0$ **7.** $\log_{10} 100 = 2$ **9.** $\log_{10} 23 = k$ **11.** $\log_{2/3} \frac{4}{9} = 2$ **13.** $3^2 = 9$

15. $9^{\frac{1}{2}} = 3$ **17.** $7^{-1} = \frac{1}{7}$ **19.** $10^{1.74} = N$ **21.** $b^4 = 18$ **23.** $n^x = y^2$ **25.** $x = 16$ **27.** $x = 2$ **29.** $x = -3$ **31.** $x = \frac{1}{6}$

33. $x = 2$ **35.** $x = 32$ **37.** $x = 3.7$ **39.** $x = 2$

41. 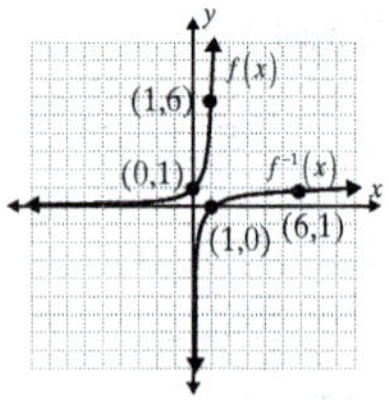

43. 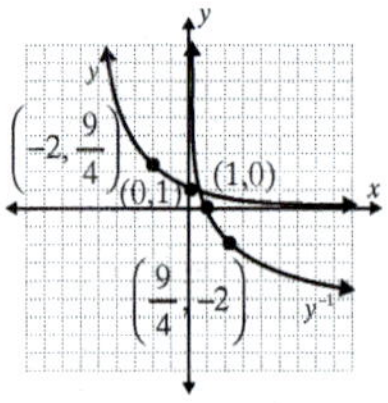

45. 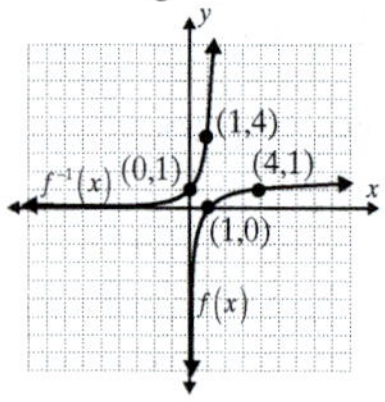

47. 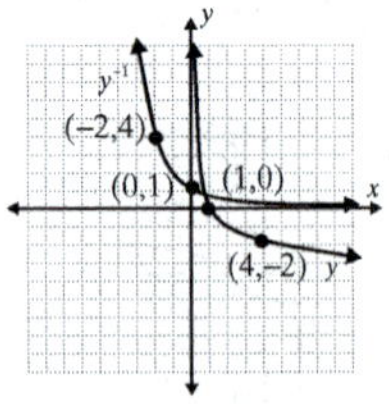

49. 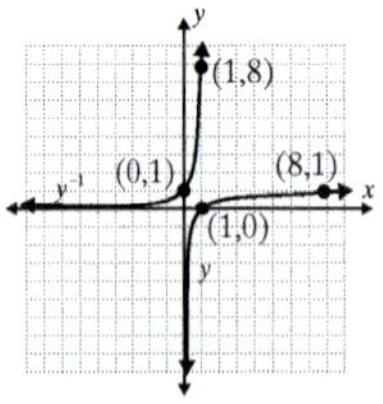

51. a. $D = (-\infty, \infty)$ **b.** $R = (0, \infty)$ **c.** $y = 0$ **d.** Answers will vary.

53. The two functions are symmetric about the line $y = x$.

Exercises 11.5, pages 915 - 916

1. a. 2 **b.** 3 **c.** 5 **d.** 6 **3.** 5 **5.** −2 **7.** $\frac{1}{2}$ **9.** 10 **11.** $\sqrt{3}$ **13.** $\log_b 5 + 4\log_b x$ **15.** $\log_b 2 - 3\log_b x + \log_b y$

17. $\log_6 2+\log_6 x-3\log_6 y$ **19.** $2\log_b x-\log_b y-\log_b z$ **21.** $-2\log_5 x-2\log_5 y$ **23.** $\frac{1}{3}\log_6 x+\frac{2}{3}\log_6 y$

25. $\frac{1}{2}\log_3 x+\frac{1}{2}\log_3 y-\frac{1}{2}\log_3 z$ **27.** $\log_5 21+2\log_5 x+\frac{2}{3}\log_5 y$ **29.** $-\frac{1}{2}\log_6 x-\frac{5}{2}\log_6 y$ **31.** $-9\log_b x-6\log_b y+3\log_b z$

33. $\log_b\left(\frac{9x}{5}\right)$ **35.** $\log_2 63x^2$ **37.** $\log_b x^2 y$ **39.** $\log_5\left(\frac{y^3}{\sqrt{x}}\right)$ or $\log_5\left(\frac{y^3\sqrt{x}}{x}\right)$ **41.** $\log_5\sqrt{\frac{x}{y}}$ **43.** $\log_2\left(\frac{xz}{y}\right)$

45. $\log_b\left(\frac{xy^2}{\sqrt{z}}\right)$ or $\log_b\left(\frac{xy^2\sqrt{z}}{z}\right)$ **47.** $\log_5\left(2x^3+x^2\right)$ **49.** $\log_2\left(x^2+2x-3\right)$ **51.** $\log_b(x+1)$ **53.** $\log_{10}\left(\frac{1}{2x-3}\right)$

55. Answers will vary.

Exercises 11.6, pages 923 - 924

1. $\log x = 1.5$ **3.** $\log\frac{1}{1000}=-3$ **5.** $\ln 27 = x$ **7.** $\ln 1 = 0$ **9.** $\log 3.2 = x$ **11.** $10^0 = 1$ **13.** $10^y = 5.4$ **15.** $e^{1.54} = x$

17. $e^1 = e$ **19.** $e^a = x$ **21.** 2.2380 **23.** 1.9465 **25.** −1.2418 **27.** 3.6243 **29.** Error (undefined) **31.** −5.3795 **33.** 204.1738

35. 0.0120 **37.** 0.9572 **39.** 175.9148 **41.** 0.0002 **43.** 1.0403 **45.** $\log x$ is a base 10 logarithm. $\ln x$ is a base e logarithm.

47. $D=(-\infty,0)\cup(0,\infty)$

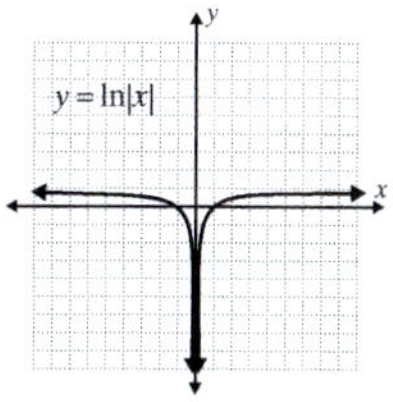

Exercises 11.7, pages 931 - 933

1. $x = 11$ **3.** $x = 9$ **5.** $x=\frac{7}{2}$ **7.** $x=\frac{6}{5}$ **9.** $x=\frac{7}{2}$ **11.** $x=-5$ **13.** $x=-3$ **15.** $x=-1, 4$ **17.** $x=-1, \frac{3}{2}$ **19.** $x=-2, 3$

21. $x=-2$ **23.** $x=-\frac{3}{2}, 0$ **25.** $x=-1, 3$ **27.** $x\approx 0.7154$ **29.** $x\approx 7.5098$ **31.** $x\approx -1.5947$ **33.** $x\approx 0.2473$ **35.** $x\approx -7.7003$

37. $t\approx -1.5193$ **39.** $x\approx 3.3219$ **41.** $x\approx -1.4307$ **43.** $x=1$ **45.** $x\approx -4.1610$ **47.** $x\approx 1.2058$ **49.** $x\approx 1.1292$

51. $x\approx 0.9514$ **53.** $x\approx 1.2520$ **55.** $x\approx 25.1189$ **57.** $x\approx 31.6228$ **59.** $x=0.0001$ **61.** $x\approx 4.9530$ **63.** $x\approx \pm 0.3329$

65. $x=3$ **67.** $x=100$ **69.** $x=6$ **71.** $x=20$ **73.** $x\approx 3.1893$ **75.** No solution **77.** No solution **79.** $x=105$ **81.** $x\approx 22.0855$

83. $x=-10, 8$ **85.** 2.2619 **87.** 0.3223 **89.** −0.6279 **91.** 2.4391 **93.** −1.2222 **95.** $x\approx 2.3219$ **97.** $x\approx 1.5480$ **99.** $x\approx 0.8390$

101. Answers will vary. $x=\frac{1}{2}$ **103.** $7\cdot 7^x = 7^{1+x}$ and $49^x = 7^{2x}$. Since $1+x\neq 2x$, in general, $7\cdot 7^x\neq 49^x$. Answers will vary.

Exercises 11.8, pages 937 - 940

1. \$4027.51 **3.** 11.55 years **5.** $f\approx 0.30$ **7.** 1.73 hours **9.** 10.99 days **11.** 7.44 hours **13.** 2350.02 years **15.** 2.31 days **17.** 39.65 minutes **19.** 7.00 years **21.** 5600 years **23. a.** 13.86 years **b.** 6.93 years **25.** 100 **27.** 8.64 million **29.** 2083, 2437, 2744 **31. a.** 117.95 dB **b.** $3.16\times 10^8 I_0$ **c.** $10^6 I_0$

Chapter 11 Review, pages 946 - 951

1. a.

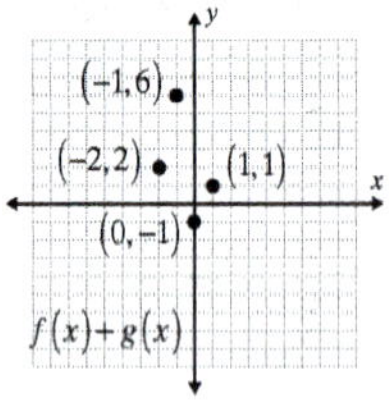

b.

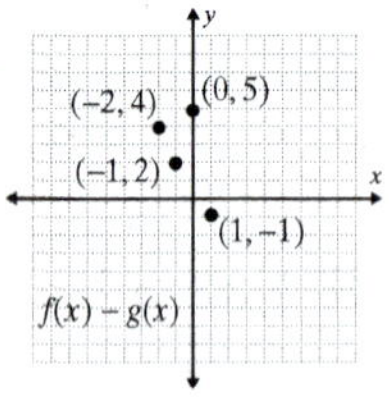

2. a.

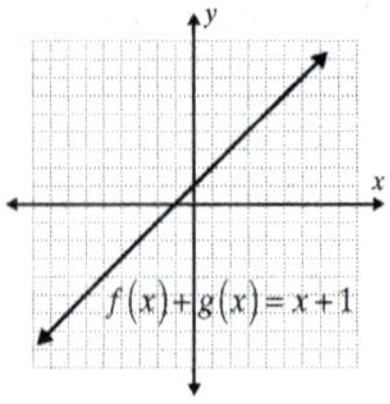

b.

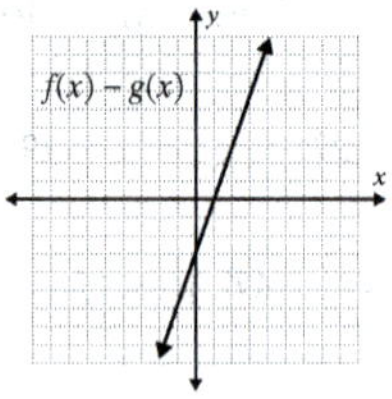

3. a. $3x-4$ **b.** $x+10$ **c.** $2x^2-11x-21$ **d.** $\dfrac{2x+3}{x-7}, x \neq 7$ **4. a.** $-4x+6$ **b.** $-6x-6$ **c.** $-5x^2-30x$ **d.** $\dfrac{-5x}{x+6}, x \neq -6$

5. a. $2x^2-16$ **b.** 16 **c.** x^4-16x^2 **d.** $\dfrac{x^2}{x^2-16}, x \neq -4, 4$ **6. a.** $3x^2-x+5$ **b.** x^2-x-5 **c.** $2x^4-x^3+10x^2-5x$

d. $\dfrac{2x^2-x}{x^2+5}$ **7.** 4 **8.** 4 **9.** 0 **10.** −5 **11.** $\sqrt{x-6}+x+2; D=[6,\infty)$ **12.** $2\sqrt{x-1}+3x\sqrt{x-1}; D=[1,\infty)$

13.

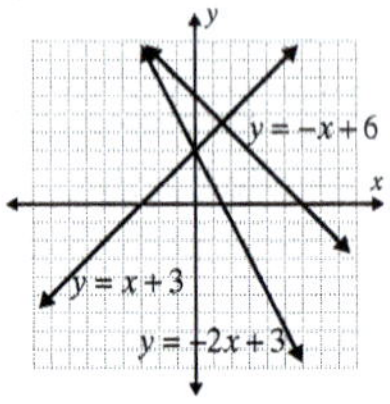

14.

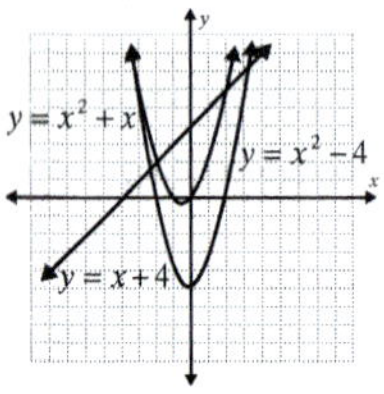

15. a. −15 **b.** 2 **16.** $f(g(x))=4x-5; g(f(x))=4x-1$

17. $f(g(x))=9x^2-6x+1; g(f(x))=3x^2-1$

18. $f(g(x))=\sqrt{x^2-4}; g(f(x))=x-4$ **19.** $f(g(x))=\dfrac{1}{x^3}, g(f(x))=\dfrac{1}{x^3}$

20. $f(g(x))=x^2, g(f(x))=x^2$

21.

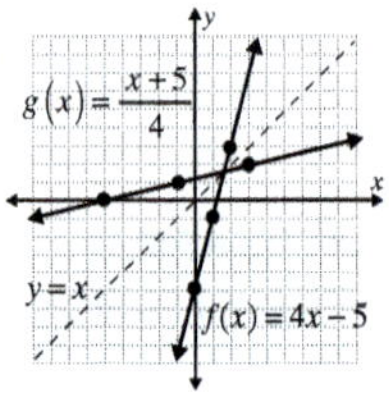

22.

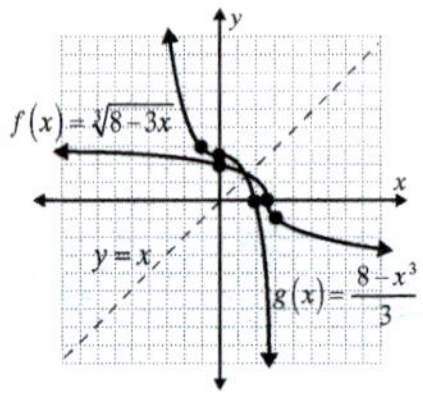

23. $f^{-1}(x)=\dfrac{5-x}{2}$

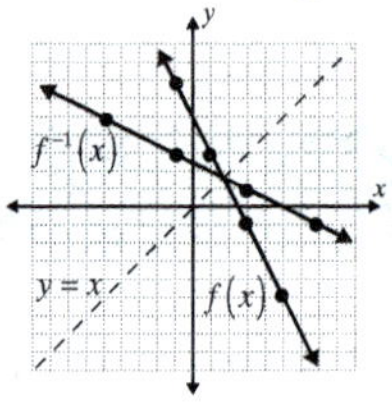

24. $f^{-1}(x)=\sqrt{2-x}$

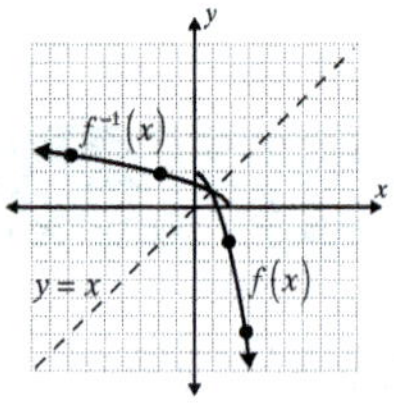

25. One-to-one

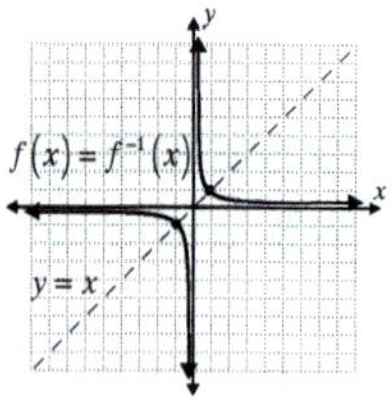

26. Not one-to-one
27. One-to-one
28. One-to-one

29. $f^{-1}(x)=(1+x)^{\frac{1}{3}}$

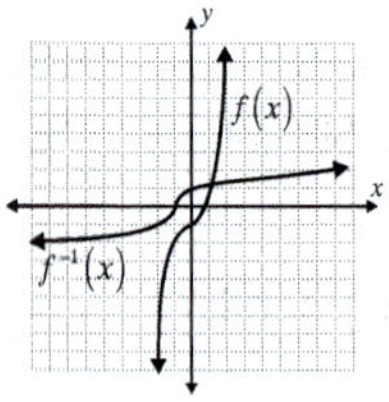

30. $f^{-1}(x)=x^2+5,$ $x \geq 0$

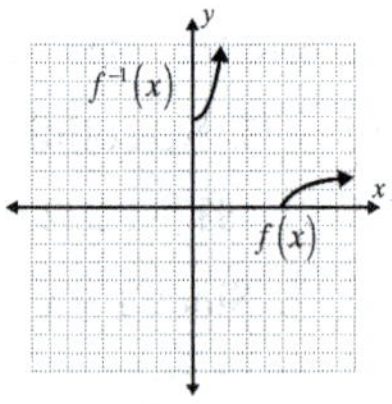

31.

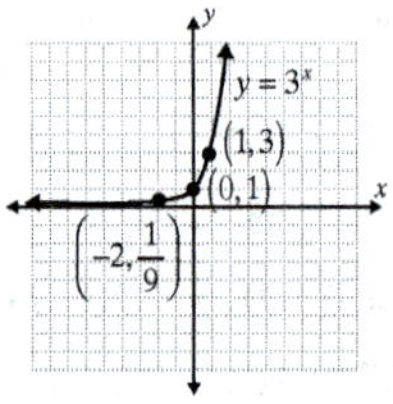

32.

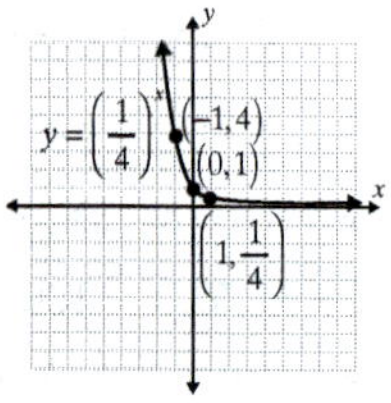

33.

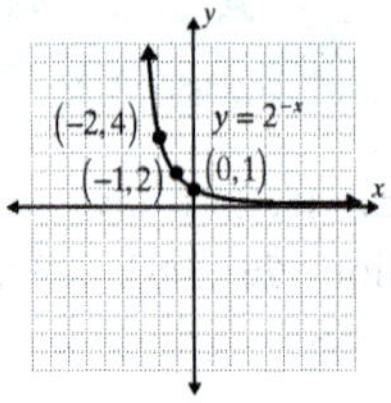

34.

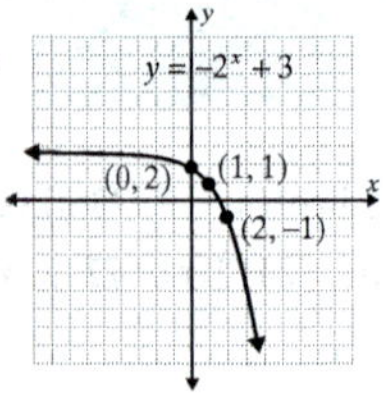

35.

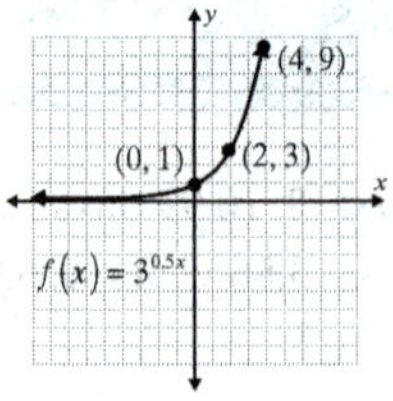

36.

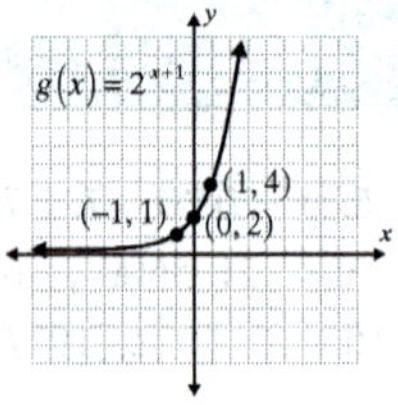

37.

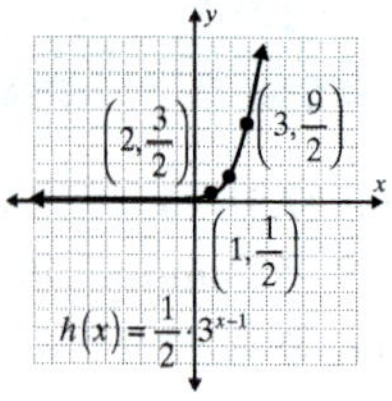

38. 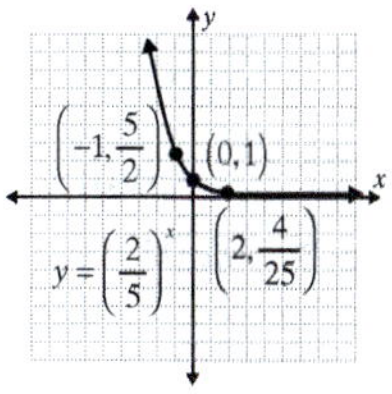

39. 259.2 40. a. 2.71692 b. 2.71815 c. 2.71828 41. a. \$16,035.68 b. \$16,453.31 c. \$16,598.92 d. \$16,600.58 42. \$17,748.21 43. \$525.56 44. \$14,918.25, \$22,255.41 Answers will vary. 45. $\log_8 64 = 2$ 46. $\log_{10} 1000 = 3$ 47. $\log_2 \frac{1}{8} = -3$ 48. $\log_{1/5} \frac{1}{125} = 3$ 49. $2^{-1} = \frac{1}{2}$ 50. $5^4 = 625$ 51. $5^{-2.5} = x$ 52. $b^y = x$ 53. $x = 2$ 54. $x = \frac{1}{2}$ 55. $x = 2.35$ 56. $x = -5$ 57. $x = 6$ 58. $x = 3$

59. 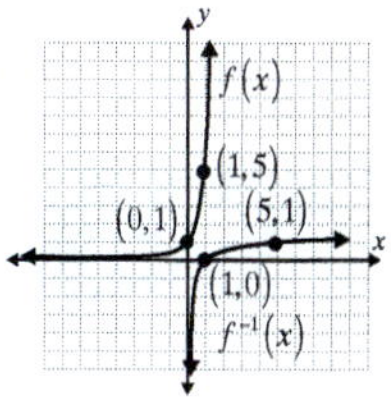

60. 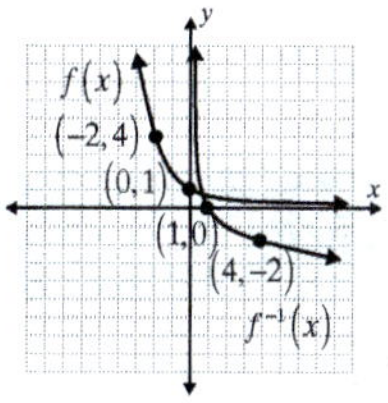

61. 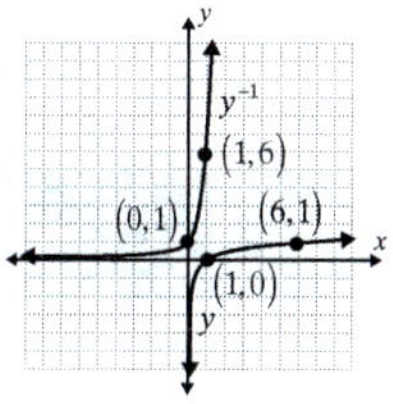

62. 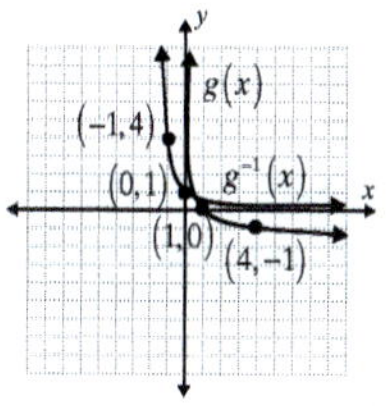

63. 3 64. $\sqrt{10}$ 65. $-6\log_b x - 3\log_b y$ 66. $2\log_{10} x + 3\log_{10} y$ 67. $\log_b x - 4\log_b y$ 68. $10\log_b x - 15\log_b y$ 69. $\frac{1}{2}\log_2 5 - \log_2 x - 2\log_2 y$ 70. $\log_3 42 + 2\log_3 x - 3\log_3 y$ 71. $\log_2(9x)$ 72. $\log_3 \frac{x^5}{y}$ 73. $\log_b\left(\frac{x^3}{y^4}\right)$ 74. $\log_4\left(4x^3 + x^2\right)$ 75. $\log_{10}\left(2x^2 - 9x - 5\right)$ 76. $\log_2\left(\frac{\sqrt{x}}{2x+1}\right)$ 77. $\log 0.00001 = -5$ 78. $\ln x = 1.2$ 79. $\ln 12 = x$ 80. $\log 8.4 = k$ 81. $e^x = 3$ 82. $e^{-1} = x$ 83. $10^5 = 100{,}000$ 84. $10^{2.5} = x$ 85. 2.4393 86. 1.9243 87. Error (undefined) 88. −4.4228 89. 1318.2567 90. 0.1585 91. 0.0111 92. 3.5058×10^{95} 93. $x = \pm e$ 94. $x = \frac{5}{2}$ 95. $x = -\frac{3}{2}, 2$ 96. $x = 1000$ 97. $x = e$ 98. $x \approx -1.4165$ 99. $x \approx \pm 0.0821$ 100. $x \approx 1.0959$ 101. $x = \pm 5\sqrt{401} \approx \pm 100.1249$ 102. $x = \frac{1}{2}$ 103. No solution 104. $x = 1$ 105. $x = 6$ 106. No solution 107. 3.5850 108. 3.3219 109. $x \approx 3.3219$ 110. $x \approx 2.3219$ 111. a. \$13,507.42 b. \$13,563.20 c. \$13,591.41 112. 11.55 years; 11.55 years 113. 29 years 114. \$1024 115. 13.82 days 116. a. 1.99 mg b. 1.16 hours 117. 10.92 minutes 118. 5.01×10^{-7} 119. a. 96.35 dB b. $10^{14} I_0$ 120. 3.98 times stronger

Chapter 11 Test, pages 952 - 954

1. a. 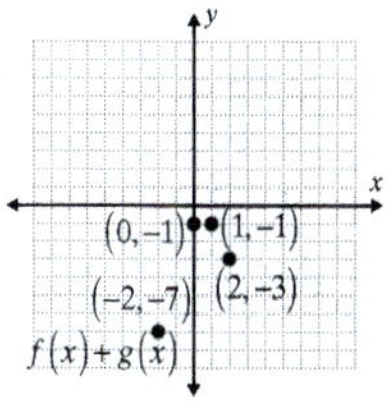

b. 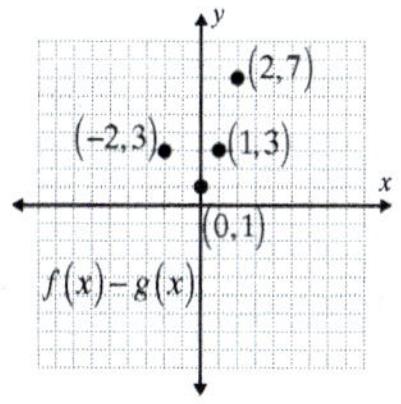

2. a. 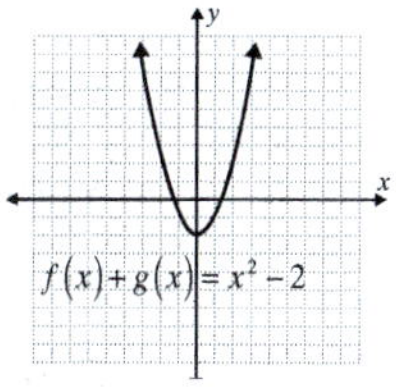

b. 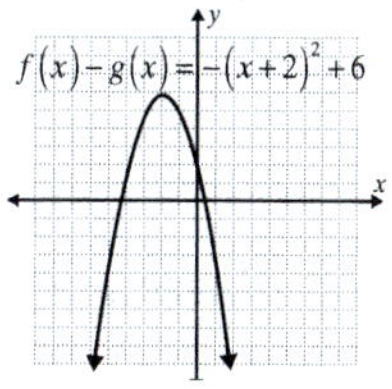

3. a. $\sqrt{x-3} + x^2 + 1$ b. $\sqrt{x-3} - x^2 - 1$ c. $\left(\sqrt{x-3}\right)\left(x^2+1\right)$ d. $\frac{\sqrt{x-3}}{x^2+1}$

4. a. $-4x^2 + 1$ b. $-8x^2 + 40x - 47$

5. One-to-one 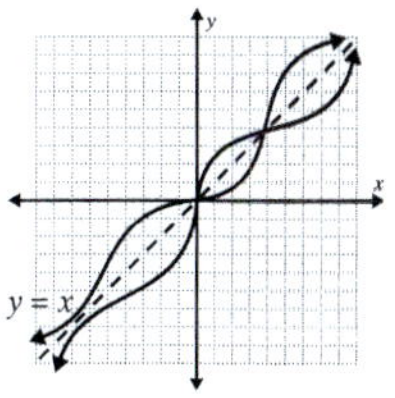

6. a. Not inverses 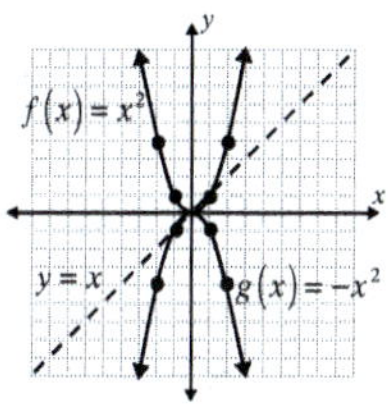

b. Inverses 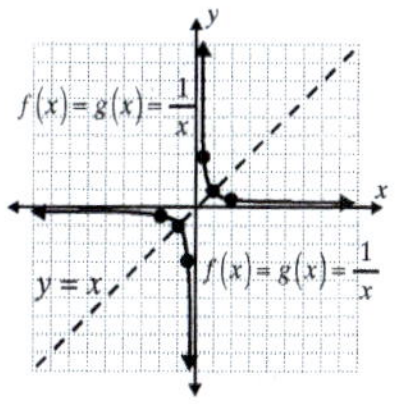

7. $f^{-1}(x) = \frac{1}{x+3} - 1$ 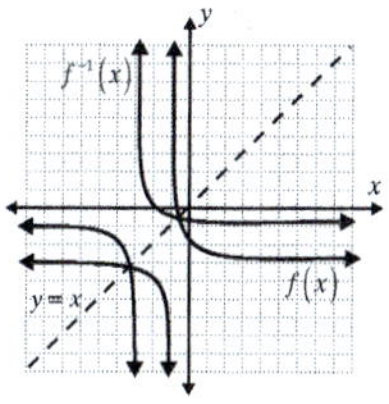

8. a. 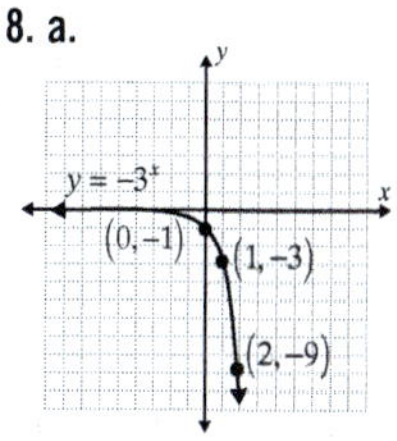

b. 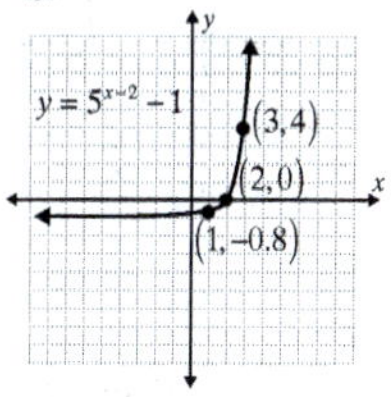

9. a. $\log 100{,}000 = 5$ b. $\log_{1/2} 8 = -3$ c. $x = \ln 15$ 10. a. $10^4 = x$ b. $3^{-2} = \frac{1}{9}$ c. $4^x = 256$ 11. a. $x = 343$ b. $x = \frac{3}{2}$

12. $y^{-1} = \log_{1/2} x$

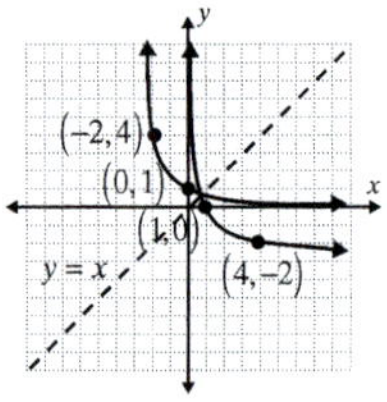

13. a. $\ln(x+5) + \ln(x-5)$ **b.** $\frac{2}{3}\log x - \frac{1}{3}\log y$ **14. a.** $\ln(x^2 + x - 20)$ **b.** $\log \frac{x\sqrt{x}}{5}$ **15. a.** 2.7627 **b.** 6.4883 **16.** $x = -4$ **17.** $x = -3$ **18.** $x \approx 0.4518$ **19.** $x \approx 10.6873$ **20.** $x \approx 1.7925$ **21.** No solution **22.** $x \approx 21.0855$ **23.** 1.7965 **24.** 25,981 bacteria **25.** 19.07 years **26. a.** 134.29 years **b.** 198.04 years **27.** 2.00 **28.** 110.12 minutes

Cumulative Review: Chapters 1-11, pages 955 - 959

1. 32 **2.** 7 **3.** 1 **4.** $-\frac{x+3}{x+1}$ **5.** $\frac{2}{x+4}$ **6.** $\frac{x-4}{x+1}$ **7.** $4xy\sqrt{5y}$ **8.** $\frac{2x\sqrt{10y}}{5y}$ **9.** $\frac{x^{\frac{1}{6}}}{y^{\frac{1}{3}}}$ **10.** $\frac{x^{\frac{1}{2}}}{2y^2}$ **11.** $x^2 y^{\frac{3}{2}}$ **12.** $x^{\frac{2}{3}} y$

13. $7x^2 - 26x - 8$ **14.** $-6y^2 - 18y + 60$ **15.** $15 + 6i$ **16.** 8 **17.** i **18.** $3x^2 - 4x + 1 - \frac{2x}{x^2+1}$ **19.** −6 **20.** 12 **21.** $3\log_b x + \frac{1}{2}\log_b y$

22. $\log_b\left(\frac{7x^2}{y}\right)$ **23.** $x = \frac{-5 \pm \sqrt{33}}{2}$ **24.** $x = \frac{7}{2}$ **25.** $x = -3, -2, 2, 3$ **26.** $x = 6$ **27.** $x = 4 \pm \sqrt{6}$ **28.** $x = -0.9212$

29. $x \approx 17.1825$ **30.** $x = -3$ **31.** $(-1, 3)$

32. $(-\infty, -2) \cup \left(\frac{5}{2}, \infty\right)$

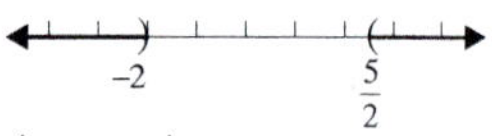

33. $(-\infty, -1] \cup \left[\frac{3}{2}, \infty\right)$

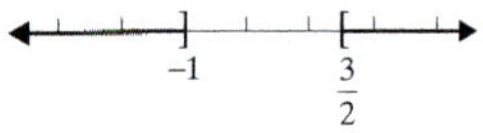

34. $(-9, -3)$

35. $(1, -1, 2)$

36. $y = -\frac{2}{3}x + 2$

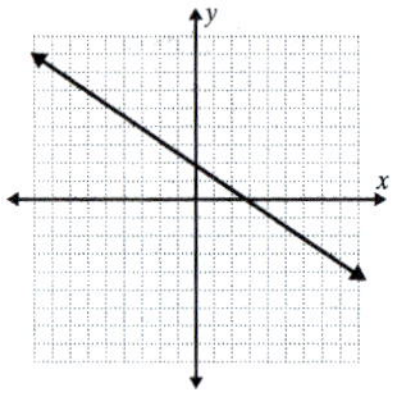

37. a. Parallel
b. No solution
c.

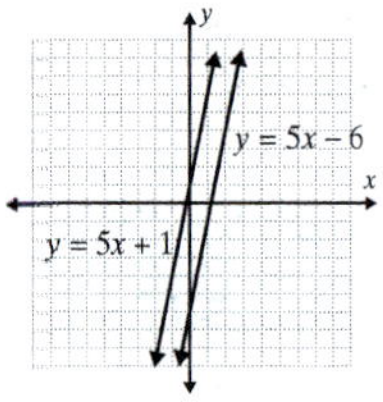

38. a. Perpendicular
b. $(2, 1)$
c.

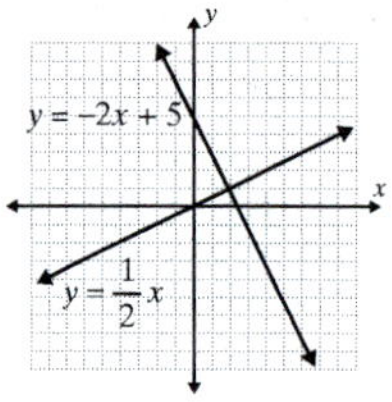

39.

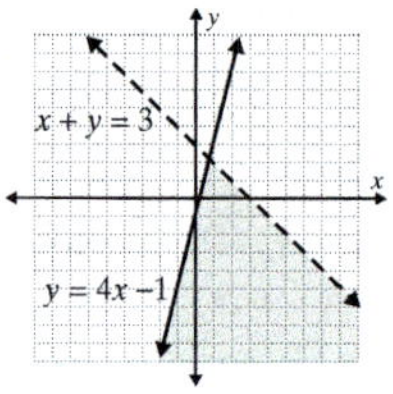

40.

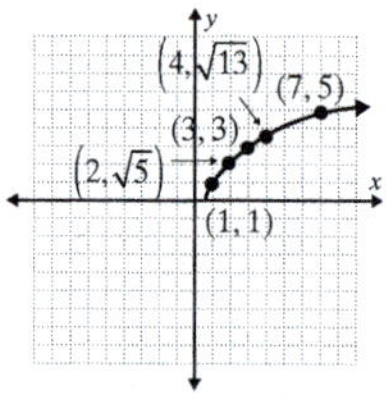

41. $y = (x-3)^2 - 11$;
Vertex: $(3, -11)$;
Range: $[-11, \infty)$;
Zeros: $x = 3 \pm \sqrt{11}$

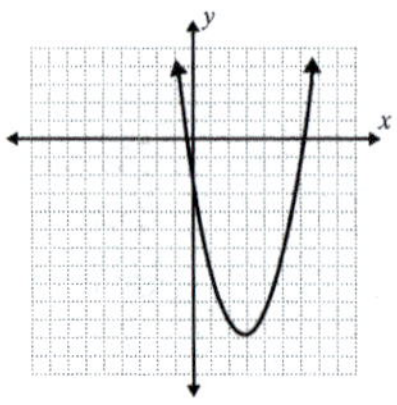

42. $y = 2(x+2)^2 - 5$;
Vertex: $(-2, -5)$;
Range: $[-5, \infty)$;
Zeros: $x = -2 \pm \frac{\sqrt{10}}{2}$

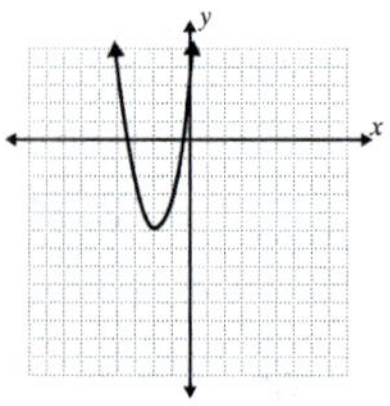

43. $y = (x-4)^2$;
Vertex: $(4, 0)$;
Range: $[0, \infty)$;
Zeros: $x = 4$

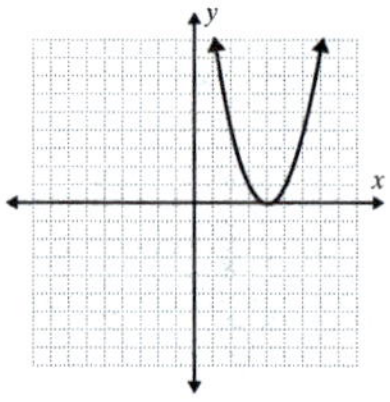

44. $y = (x-1)^2 - 7$;
Vertex: $(1, -7)$;
Range: $[-7, \infty)$;
Zeros: $x = 1 \pm \sqrt{7}$

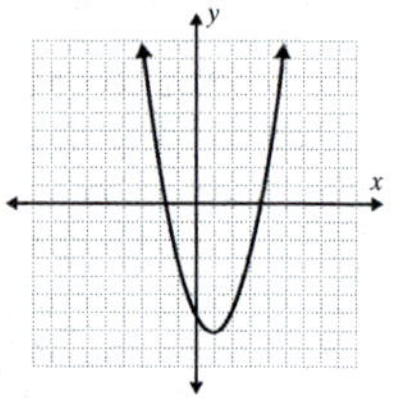

45. a. Function; $D = (-\infty, -1] \cup [1, \infty)$; $R = (-\infty, -2] \cup [2, \infty)$ **b.** Not a function; $D = (-\infty, \infty)$; $R = (-\infty, \infty)$ **c.** Function; $D = (-\infty, \infty)$; $R = [0, \infty)$ **46.** $\{(-2, 13), (-1, 8), (0, 5), (1, 4), (2, 5)\}$ **47.** $\{(-2, -28), (-1, -6), (0, 0), (1, -4), (2, -12)\}$

48. a. $x^2 + x + \sqrt{x+2}$ **b.** $x^2 + x - \sqrt{x+2}$ **c.** $(x^2 + x)\sqrt{x+2}$ **d.** $\frac{x^2 + x}{\sqrt{x+2}}, x > -2$ **49. a.** $4x^2 + 4x + 4$ **b.** $2x^2 + 7$

50. Not a one-to-one function

51. One-to-one function

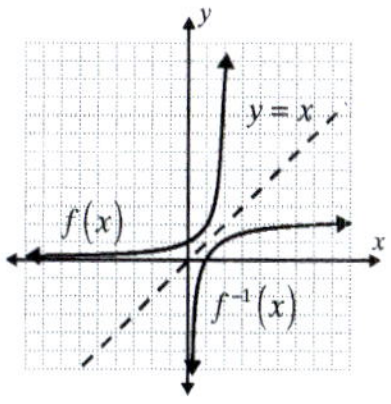

52. $f^{-1}(x)=\sqrt{x+4}$

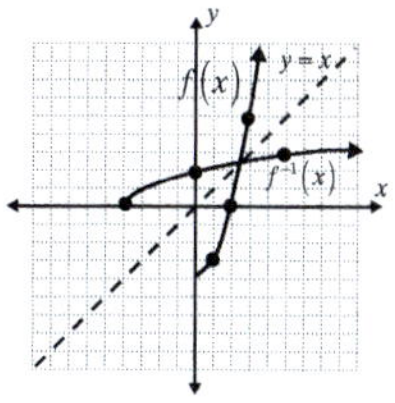

53. $g^{-1}(x)=\sqrt[3]{x}+2$

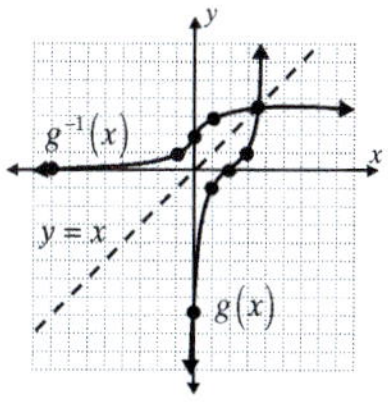

54. –5, –3, –1 **55.** 360 cm^2 **56.** 7:30 minutes; 8:15 minutes

57. a. and **b.**

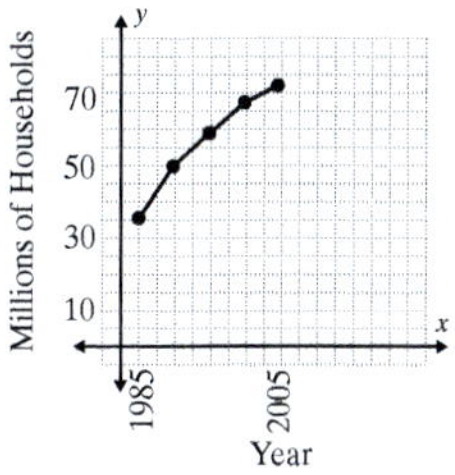

c. 3.12; 1.72; 1.62; 1.06 **d.** The number of users increased 3.12 million a year from 1985-1990, 1.72 million a year from 1990-1995, 1.62 million a year from 1995-2000, and 1.06 million a year from 2000-2005.

58. 10 mph, 4 mph **59.** Width: 16 inches; Height: 12 inches **60.** $\frac{6}{7}$ hours

61. 13 hardback; 30 paperback **62. a.** \$600 **b.** \$360,000 **63. a.** 4.5 s **b.** 324 ft **c.** 9 s

64. $\frac{7}{8}$ **65.** \$609.20 **66.** 16.12 centuries **67.** 4.60 s **68.** –7.2288 **69.** 1.9266 **70.** 1.7372

71. 9.2103 **72.** $\left(-\infty,-\sqrt{2}\right)\cup\left(\sqrt{2},\infty\right)$ 

73. $[-6,1]$ 

74. $(2,5)$

Chapter 12

Exercises 12.1, pages 971 - 975

1. a. 0 **b.** a^2-6a+5 **c.** $x^2+2xh+h^2-4$ **d.** $2x+h$ **3. a.** 0 **b.** $2a^2-11a+14$ **c.** $2x^2+4xh-3x+2h^2-3h$ **d.** $4x+2h-3$

5. 1 **7.** –2 **9.** The results are the coefficients of x.

11.
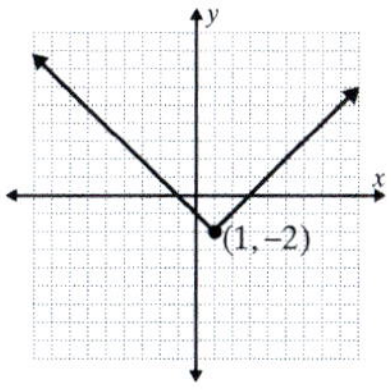

13.
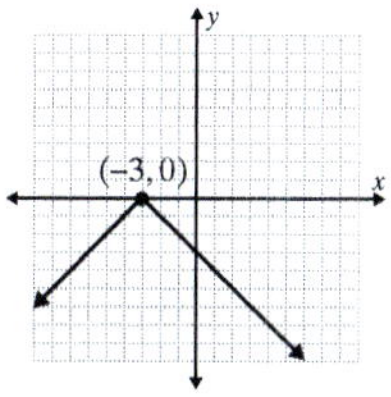

15.
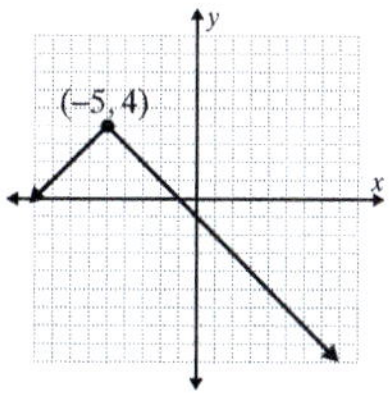

17.
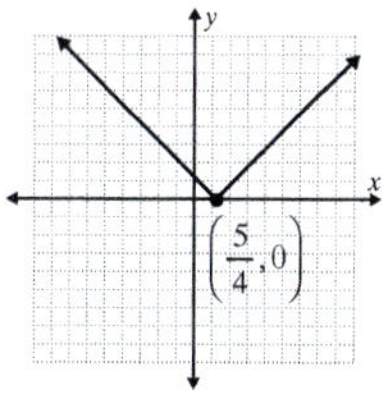

19.
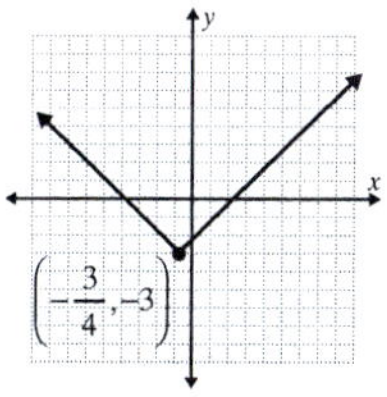

21.
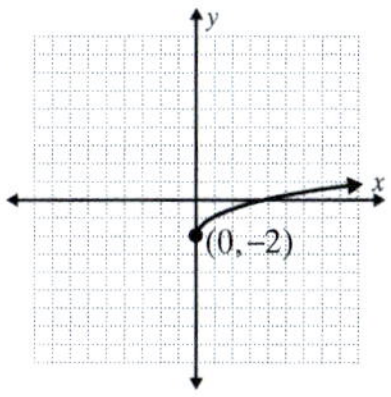

23.
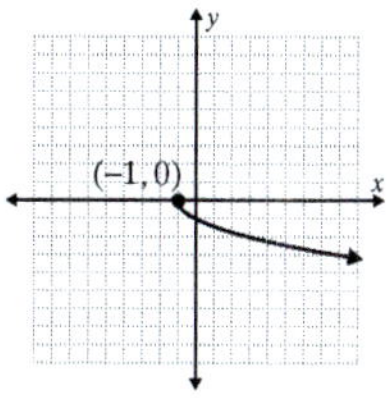

25.
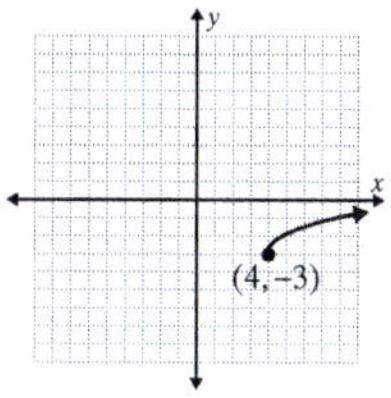

27.
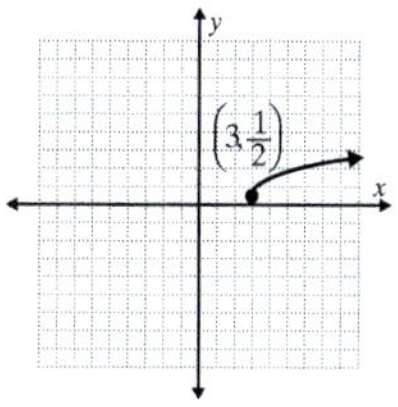

29.
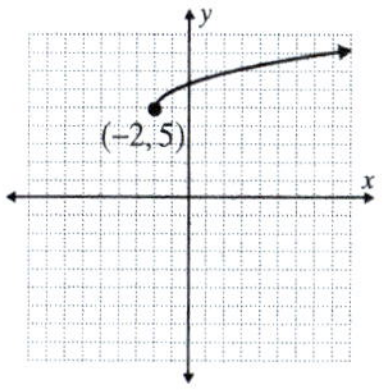

31.
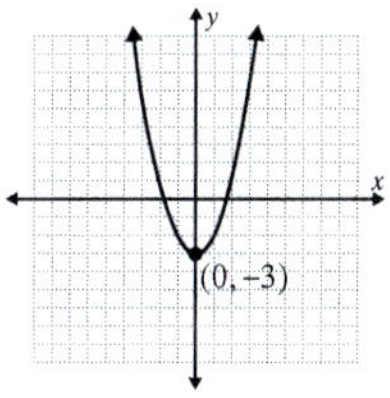

33.
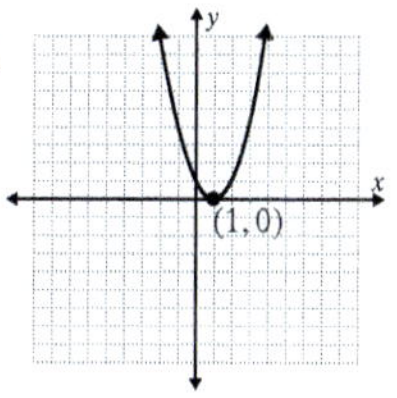

35.
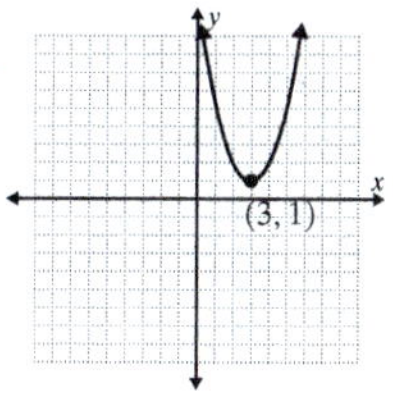

37.
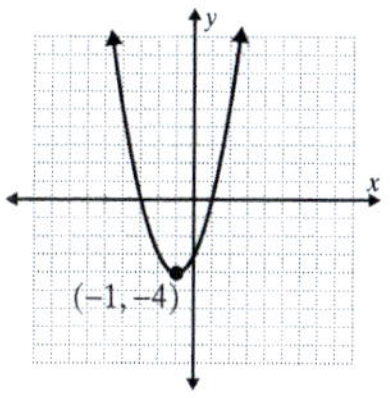

39.
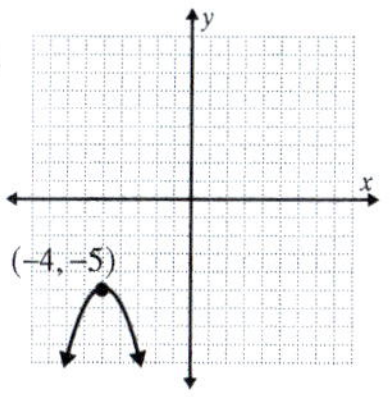

41. 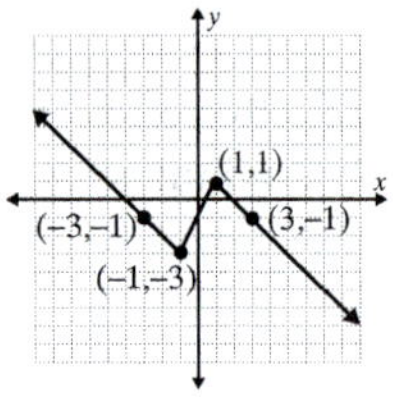

43. 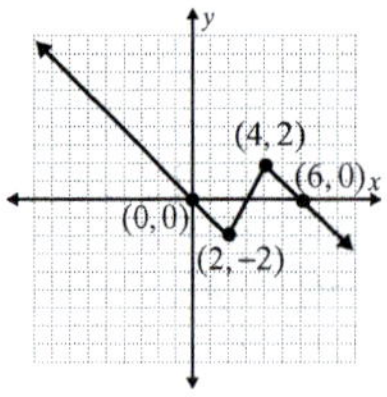

45. 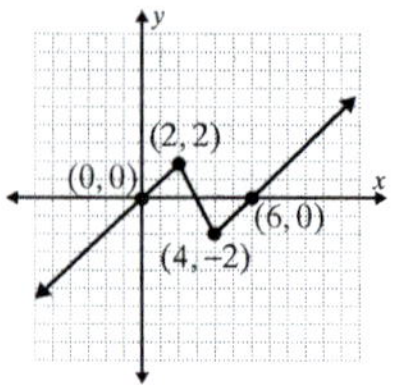

47. 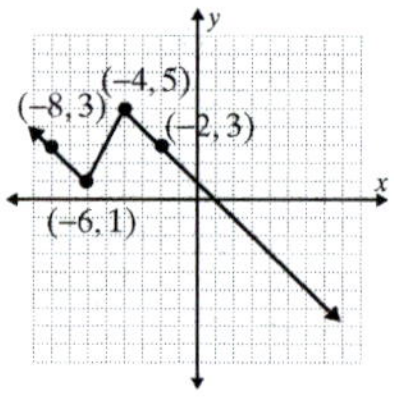

49. 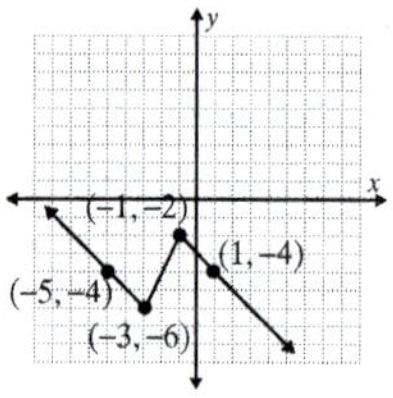

51. 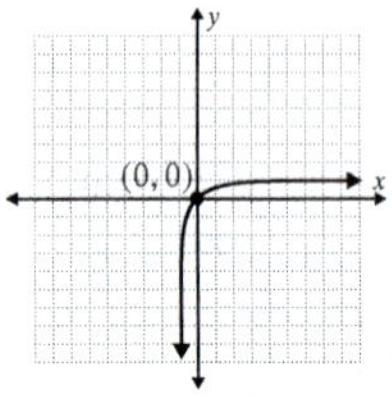

$D = (-1, \infty)$;

$R = (-\infty, \infty)$

53. 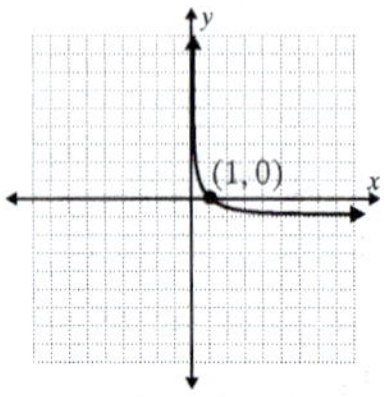

$D = (0, \infty)$;

$R = (-\infty, \infty)$

55. 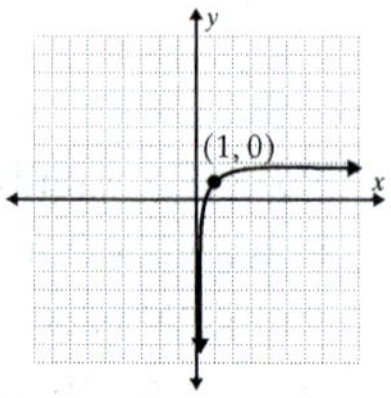

$D = (0, \infty)$;

$R = (-\infty, \infty)$

57. 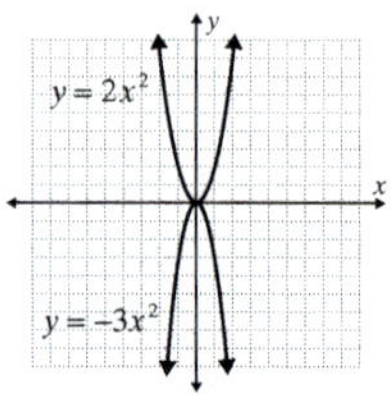

59. 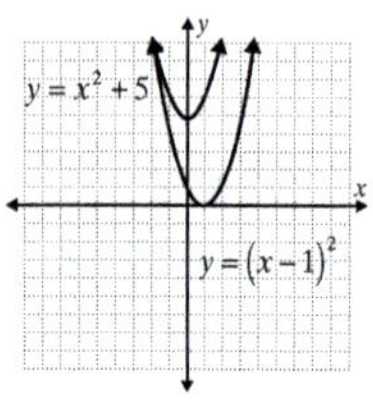

61. 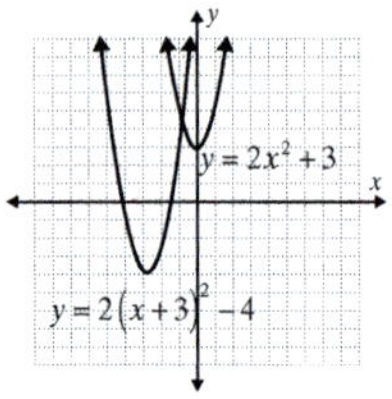

63. The graph of the function $y = f(x-h)+k$ is the graph of the function $y = f(x)$ shifted h units to the right and k units up.

Exercises 12.2, pages 982 - 984

1. a. $(4, 0)$
b. None
c. $y = 0$

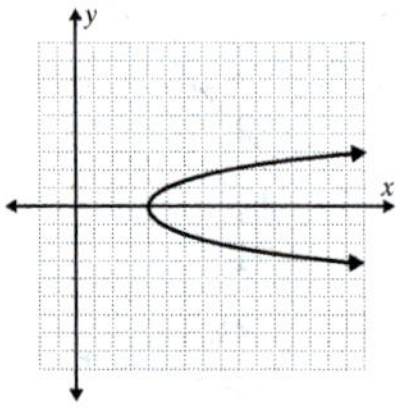

3. a. $(0, -3)$
b. $(0, -3)$
c. $x = 0$

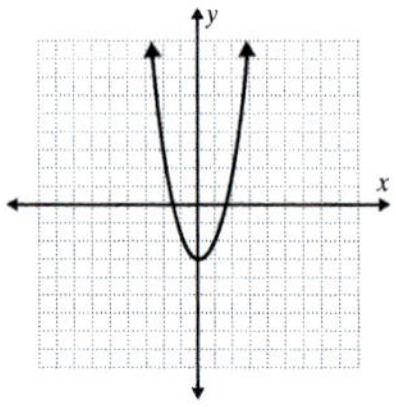

5. a. $(3, 0)$
b. None
c. $y = 0$

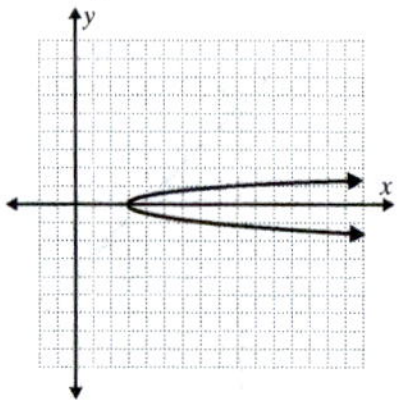

7. a. $(0, 3)$
b. $(0, 3)$
c. $y = 3$

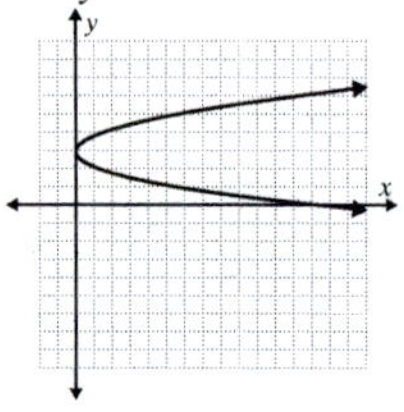

9. a. $(4, -2)$
b. None
c. $y = -2$

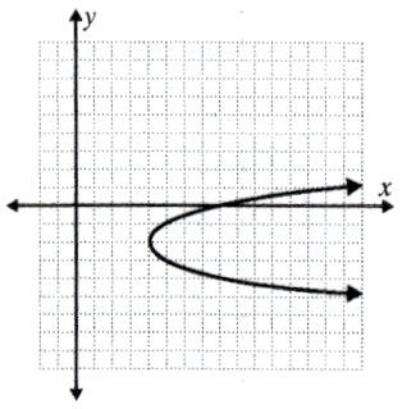

11. a. $(1, -1)$
b. $(0, 0)$
c. $x = 1$

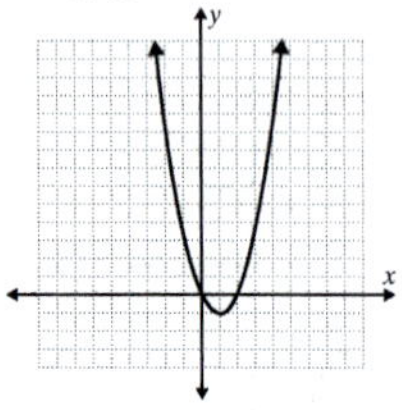

13. a. $(0, -2)$
b. $(0, -2)$
c. $y = -2$

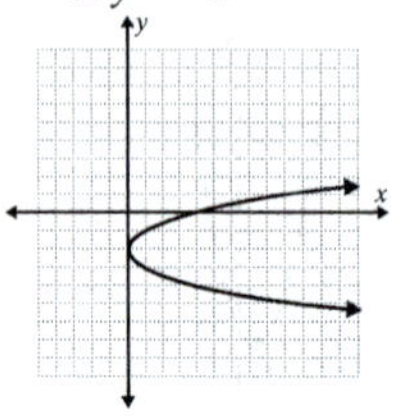

15. a. $(0, 5)$
b. $(0, 5)$
c. $y = 5$

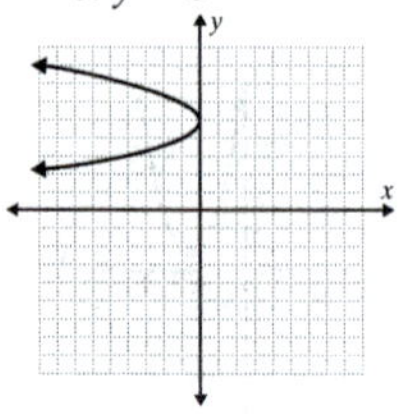

17. a. $(-3, -4)$
b. $(0, 5)$
c. $x = -3$

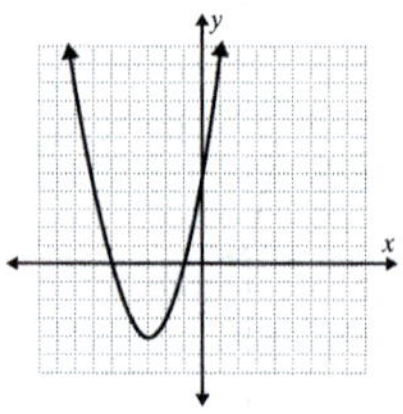

19. a. $(-2, 9)$
b. $(0, 5)$
c. $x = -2$

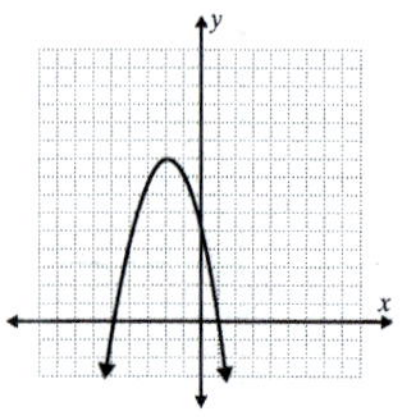

21. a. $(1, 2)$ **b.** $(0, 1), (0, 3)$ **c.** $y = 2$

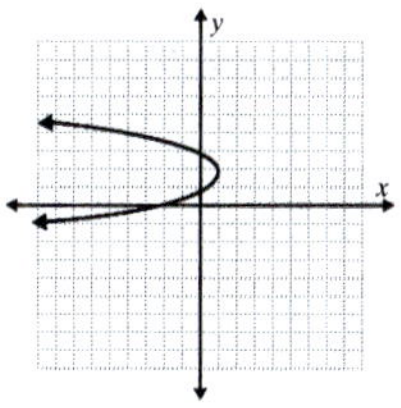

23. a. $\left(-\frac{1}{4}, -\frac{9}{8}\right)$ **b.** $(0, -1)$ **c.** $x = -\frac{1}{4}$

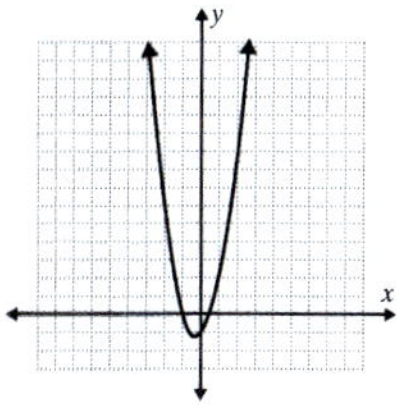

25. a. $\left(\frac{9}{8}, \frac{5}{4}\right)$ **b.** $\left(0, \frac{1}{2}\right), (0, 2)$ **c.** $y = \frac{5}{4}$

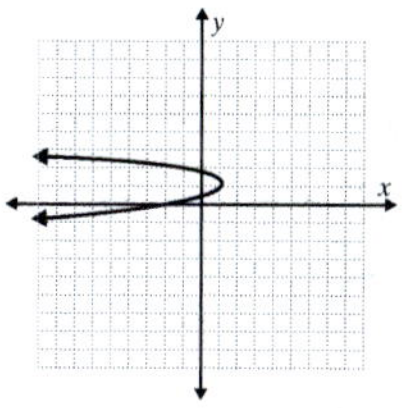

27. a. $(-8, -1)$ **b.** $\left(0, -1 + \frac{2\sqrt{6}}{3}\right)$, $\left(0, -1 - \frac{2\sqrt{6}}{3}\right)$ **c.** $y = -1$

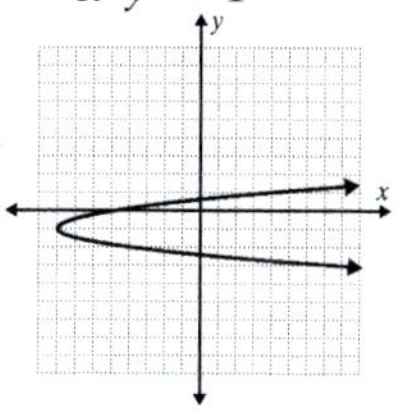

29. a. $\left(\frac{3}{2}, 0\right)$ **b.** $(0, 9)$ **c.** $x = \frac{3}{2}$

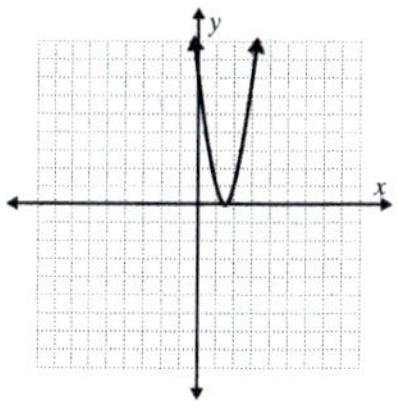

31. $(0, 1.225), (0, -1.225)$

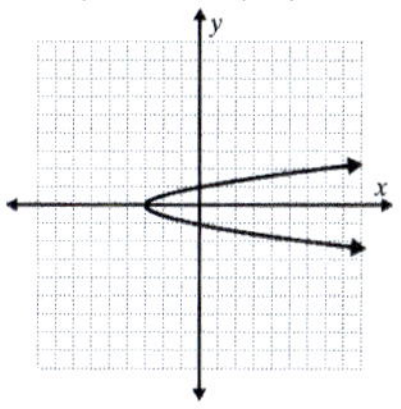

33. $(0, 2), (0, 0)$

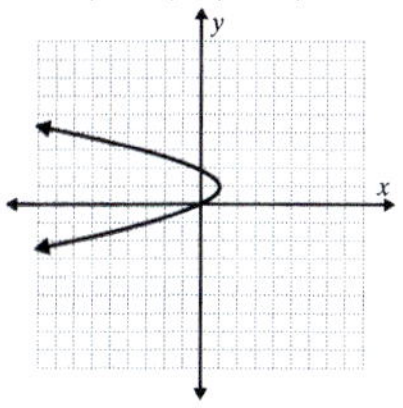

35. No y-intercept

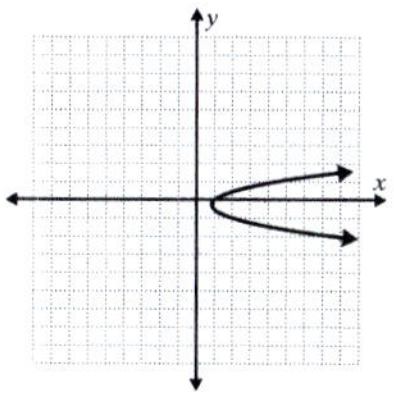

37. $(0, 0.658), (0, -2.658)$

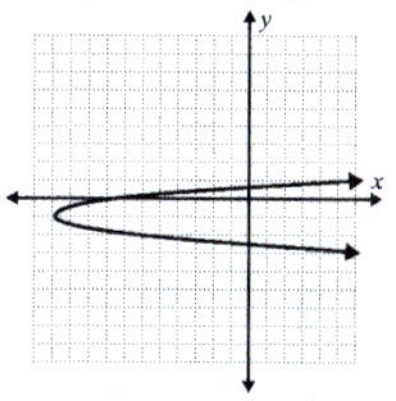

39. $(0, 2.581), (0, -0.581)$

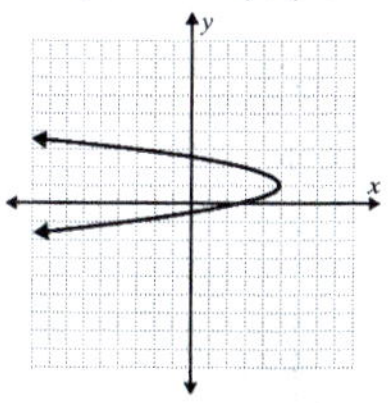

41. b **43.** c **45.** If $|a| < 1$, then the parabola will be wider than the graph of $x = y^2$. If $|a| > 1$, then the parabola will be narrower than the graph of $x = y^2$.

Exercises 12.3, pages 992 - 995

1. 5 **3.** 13 **5.** $\sqrt{29}$ **7.** 3 **9.** $\sqrt{13}$ **11.** 17 **13.** $x^2 + y^2 = 16$ **15.** $x^2 + y^2 = 3$ **17.** $x^2 + y^2 = 11$ **19.** $x^2 + y^2 = \frac{4}{9}$ **21.** $x^2 + (y-2)^2 = 4$ **23.** $(x-4)^2 + y^2 = 1$ **25.** $(x+2)^2 + y^2 = 8$ **27.** $(x-3)^2 + (y-1)^2 = 36$ **29.** $(x-3)^2 + (y-5)^2 = 12$ **31.** $(x-7)^2 + (y-4)^2 = 10$

33. $x^2 + y^2 = 9$
Center: $(0,0)$; $r = 3$

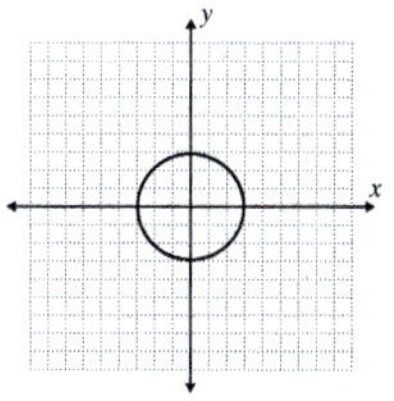

35. $x^2 + y^2 = 49$
Center: $(0,0)$; $r = 7$

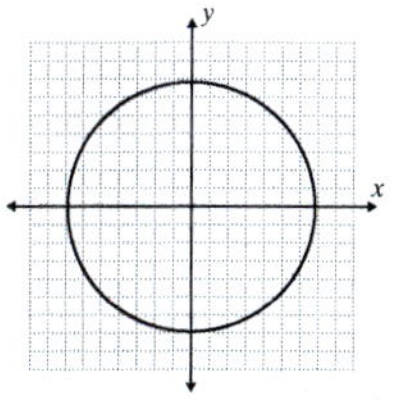

37. $x^2 + y^2 = 18$
Center: $(0,0)$; $r = 3\sqrt{2}$

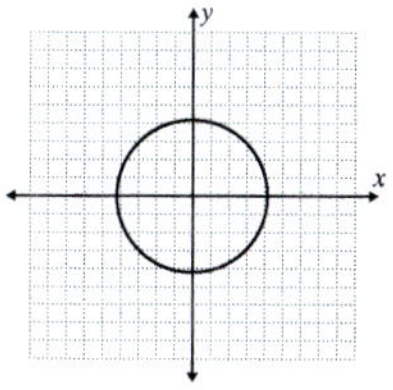

39. $(x+1)^2 + y^2 = 9$
Center: $(-1,0)$; $r = 3$

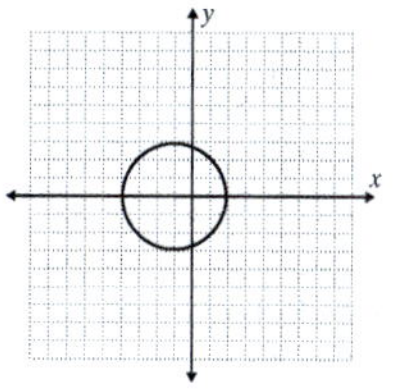

41. $x^2 + (y-2)^2 = 4$
Center: $(0, 2)$; $r = 2$

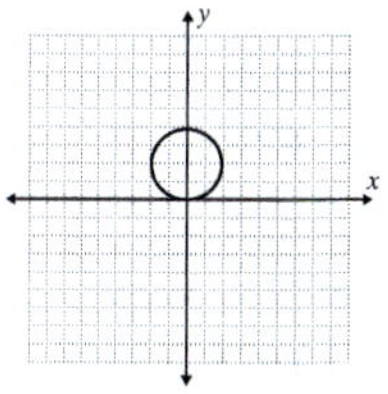

43. $(x+1)^2 + (y+2)^2 = 16$
Center: $(-1, -2)$; $r = 4$

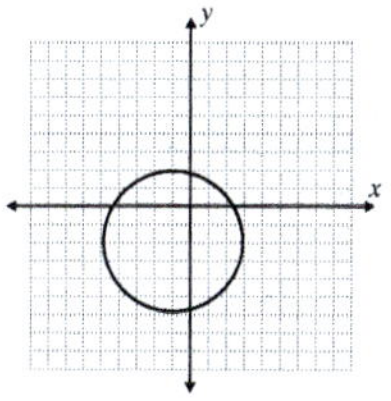

45. $(x-2)^2 + (y+5)^2 = 9$
Center: $(2, -5)$; $r = 3$

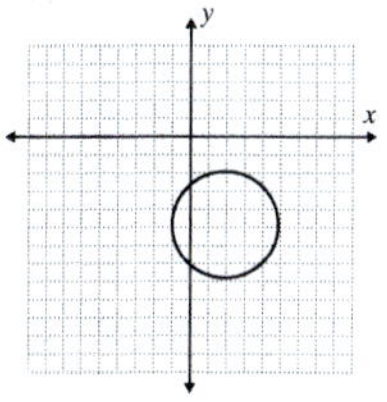

47. $(x-2)^2 + (y-3)^2 = 8$
Center: $(2,3)$; $r = 2\sqrt{2}$

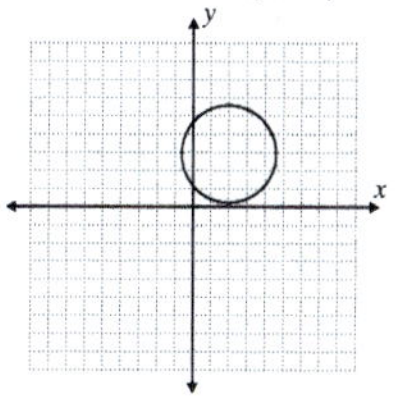

49. $\overline{AB} = 3\sqrt{5}$, $\overline{AC} = \sqrt{65}$, $\overline{BC} = 2\sqrt{5}$,
$(3\sqrt{5})^2 + (2\sqrt{5})^2 = (\sqrt{65})^2$;
Right triangle

51. $\overline{AB} = \overline{AC} = 4\sqrt{5}$

53. $\overline{AB} = \overline{AC} = \overline{BC} = 4$

55. $\overline{AC} = \overline{BD} = \sqrt{61}$

57. $10 + 4\sqrt{5}$

59. 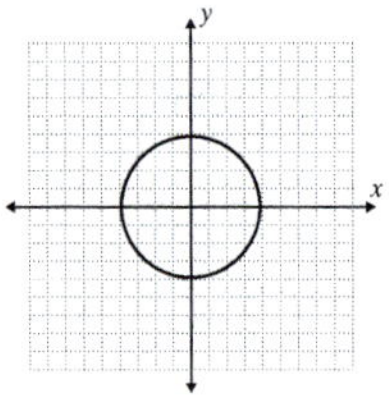

61. 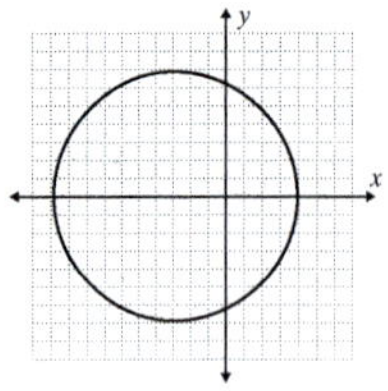

63. 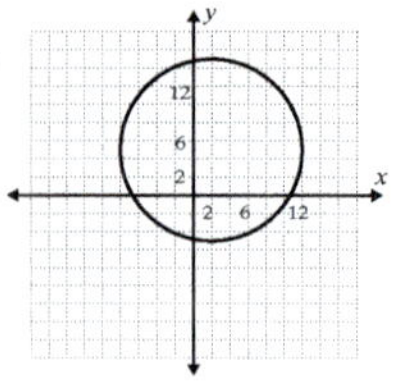

65. a. $d=\sqrt{x^2+(y-p)^2}$ **b.** $d=|y+p|$

c.
$$y+p=\sqrt{x^2+(y-p)^2}$$
$$(y+p)^2=x^2+(y-p)^2$$
$$y^2+2py+p^2=x^2+y^2-2py+p^2$$
$$x^2=4py$$

67. a. $d=\sqrt{(x-p)^2+y^2}$ **b.** $d=|x+p|$ **c.**
$$x+p=\sqrt{(x-p)^2+y^2}$$
$$(x+p)^2=(x-p)^2+y^2$$
$$x^2+2px+p^2=x^2-2px+p^2+y^2$$
$$y^2=4px$$

Exercises 12.4, pages 1006 - 1008

1. $\frac{x^2}{36}+\frac{y^2}{4}=1$

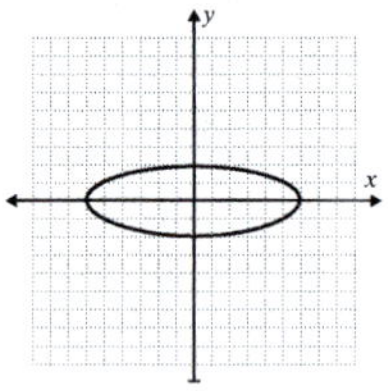

3. $\frac{x^2}{25}+\frac{y^2}{4}=1$

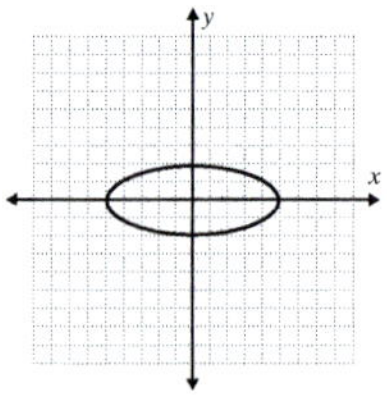

5. $\frac{x^2}{1}+\frac{y^2}{16}=1$

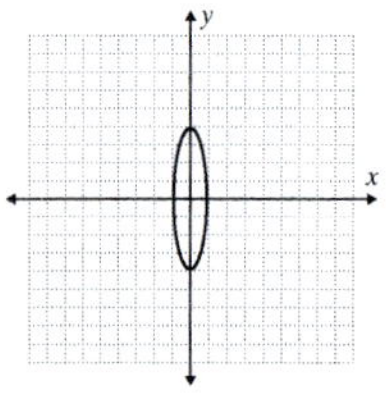

7. $\frac{x^2}{1}-\frac{y^2}{1}=1$

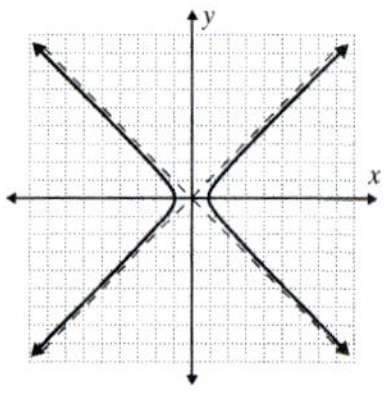

9. $\frac{x^2}{1}-\frac{y^2}{9}=1$

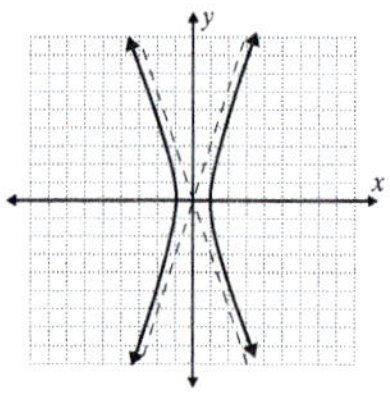

11. $\frac{x^2}{9}-\frac{y^2}{4}=1$

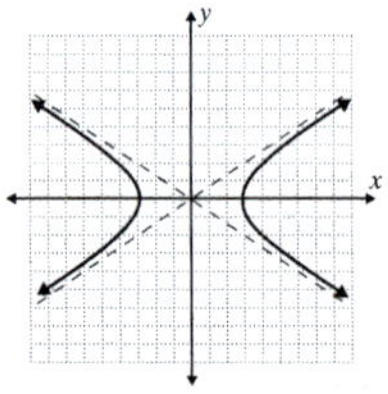

13. $\frac{x^2}{4}+\frac{y^2}{8}=1$

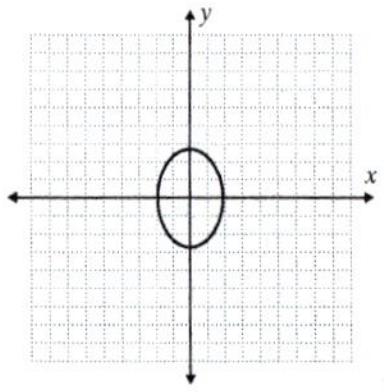

15. $\frac{x^2}{20}+\frac{y^2}{4}=1$

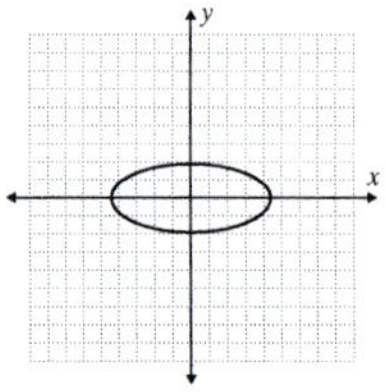

17. $\frac{y^2}{9}-\frac{x^2}{9}=1$

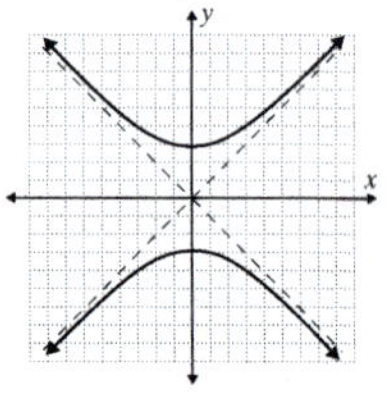

19. $\frac{y^2}{8}-\frac{x^2}{4}=1$

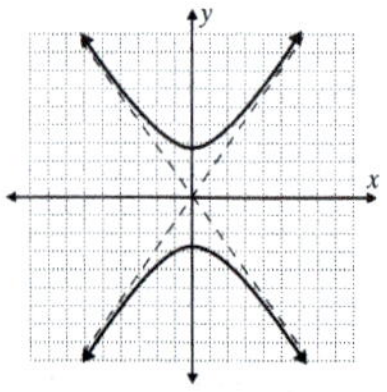

21. $\frac{y^2}{18}-\frac{x^2}{9}=1$

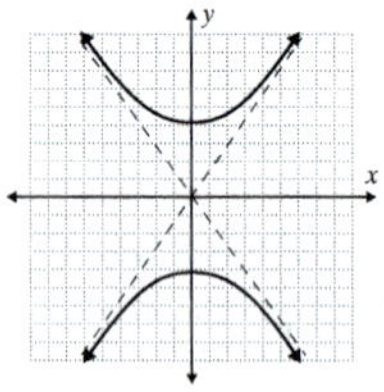

23. $\frac{x^2}{6}+\frac{y^2}{9}=1$

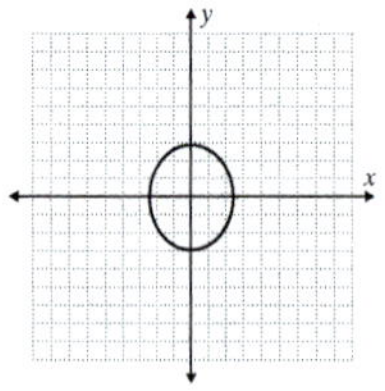

25. $\frac{x^2}{5}+\frac{y^2}{4}=1$

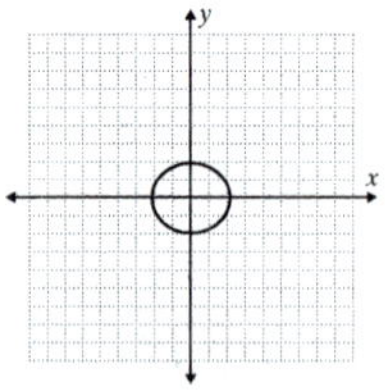

27. $\frac{y^2}{8}-\frac{x^2}{9}=1$

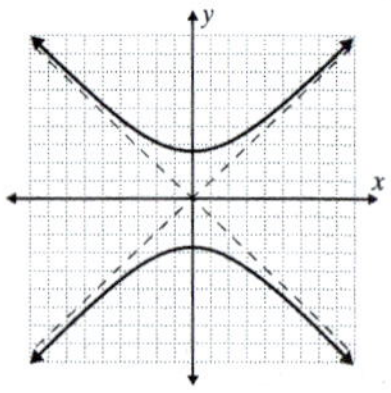

29. $\frac{y^2}{12}-\frac{x^2}{9}=1$

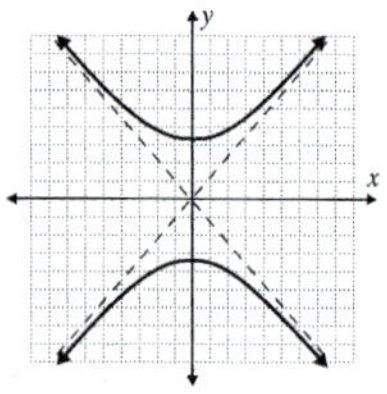

31. e **33.** d **35.** b **37.**

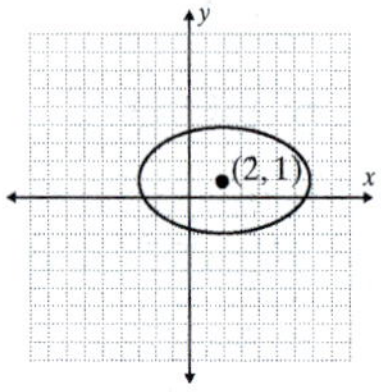

39.

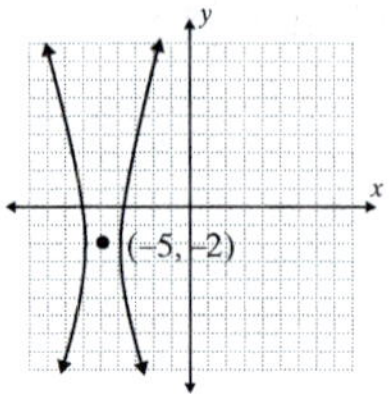

41.

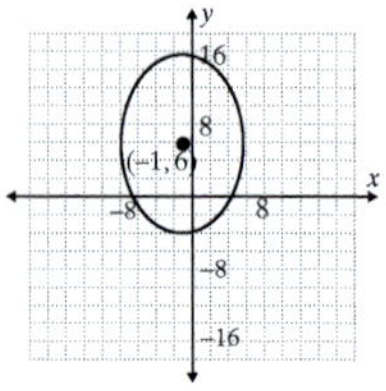

43.

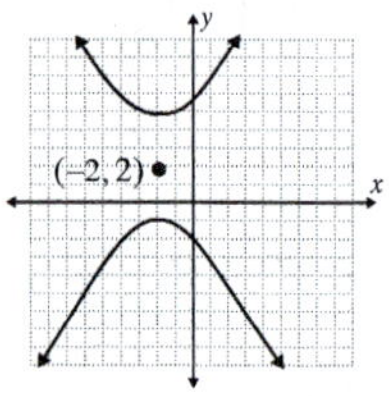

45. b. Pick an arbitrary point (x, y) located on the ellipse. The sum of the distances from (x, y) to the points $(-c, 0)$ and $(c, 0)$ (the foci) will equal $2a$. Using the distance formula we get the equation $\sqrt{(x+c)^2+(y-0)^2}+\sqrt{(x-c)^2+(y-0)^2}=2a$. Now to remove the radicals, simplify so that one side of the equation is a single radical and square both sides. Then repeat the previous step and simplify. This gives $\frac{x^2}{a^2}+\frac{y^2}{a^2-c^2}=1$. **c.** At the y-intercept $(0, b)$ a right triangle is formed with hypotenuse a and sides b and c. The Pythagorean theorem gives $a^2=b^2+c^2$.

Exercises 12.5, pages 1014 - 1015

1. $(-3, 10), (1, 2)$

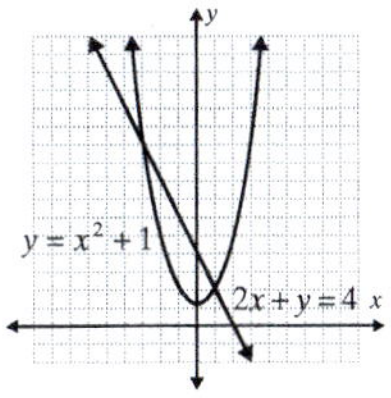

3. $(1, 1), (2, 0)$

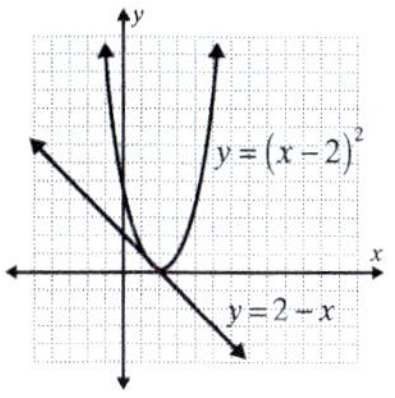

5. $(-2, -4), (4, 2)$

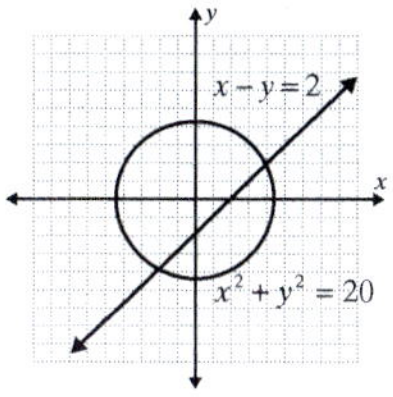

7. $(5, 3)$

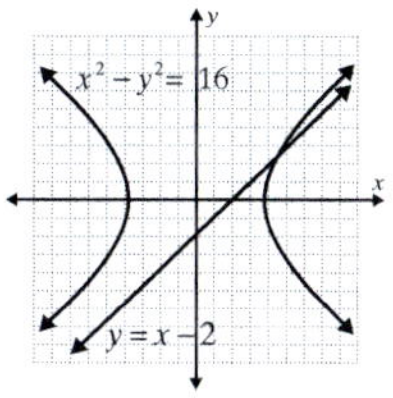

9. $(-3, 0), (3, 0)$

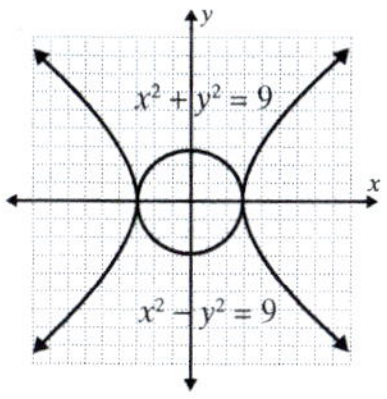

11. $(2, 3), (2, -3)$

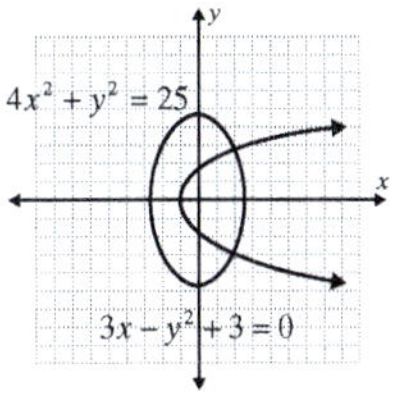

13. $(-2, 4), (1, 1)$

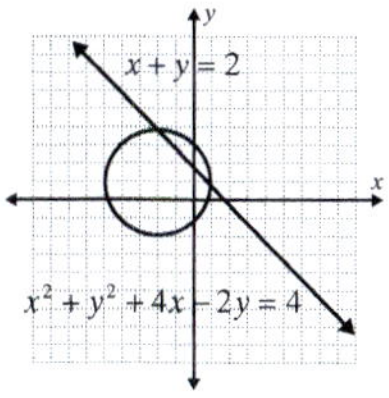

15. $(-3, -2), (3, 2), (-3, 2), (3, -2)$

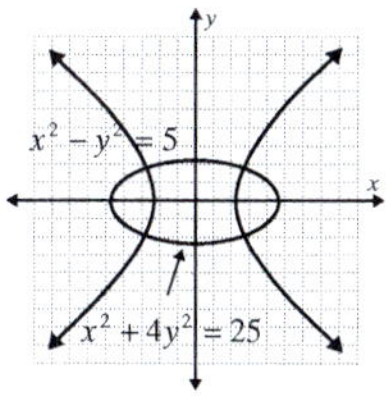

17. $(-3\sqrt{5}, 5), (-6, 4), (3\sqrt{5}, 5), (6, 4)$
19. $(1, 3), (-1, 3)$ **21.** $(4, 4), (-2, -2)$
23. $\left(-\frac{3}{2}, -1\right), \left(\frac{1}{2}, \frac{1}{2}\right)$
25. $(-3, \sqrt{11}), (-3, -\sqrt{11}), (3, \sqrt{11}), (3, -\sqrt{11})$
27. $(5, -3), (-5, 1)$ **29.** $(0, 3), (2, 3)$

31. $(-1.62, 5.62), (0.62, 3.38)$

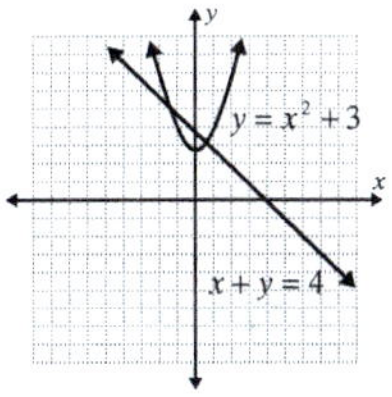

33. $(1.41, 0.17), (-1.41, 5.83)$

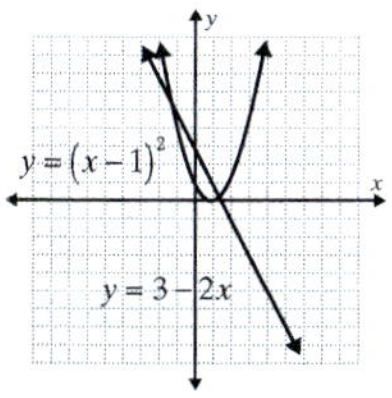

35. $(2.68, 1.68), (-1.68, -2.68)$

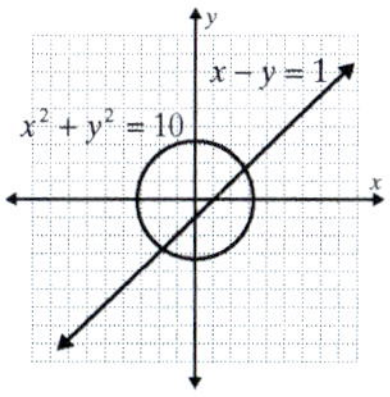

Chapter 12 Review, pages 1021 - 1023

1. a. -27 **b.** $-5a-12$ **c.** -5 **2. a.** 66 **b.** $2a^2+12a+12$ **c.** $4x+2h$ **3. a.** 19 **b.** a^2+3a+1 **c.** $2x+h-3$

4.

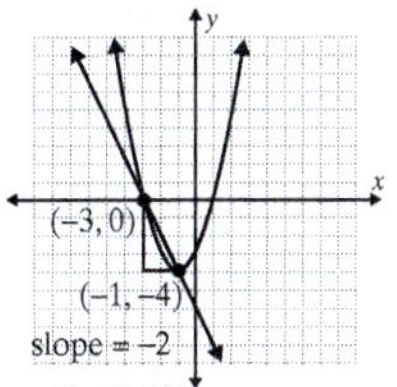

$\frac{f(x+h)-f(x)}{h}=2x+h+2=-2$

5.

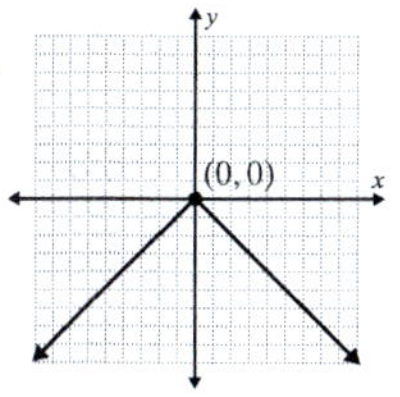

6.

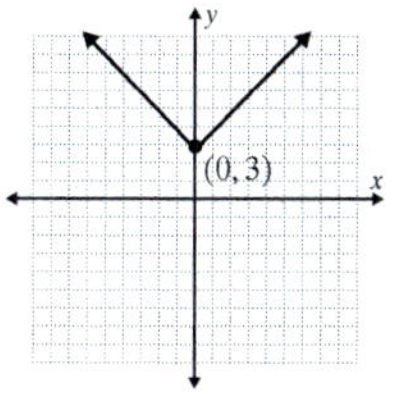

7.

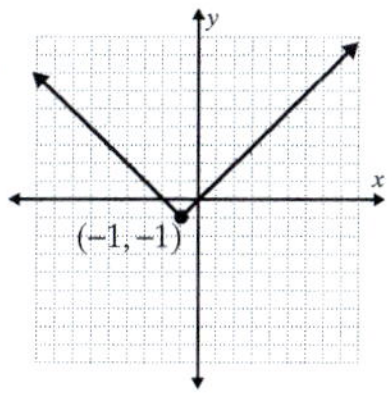

8.

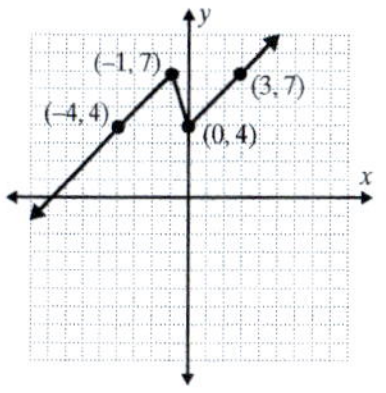

9.

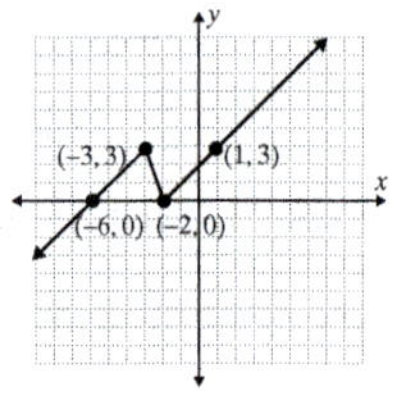

10.

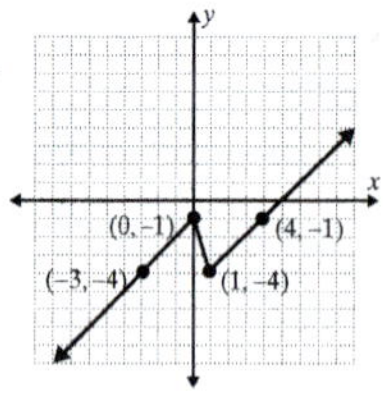

11.

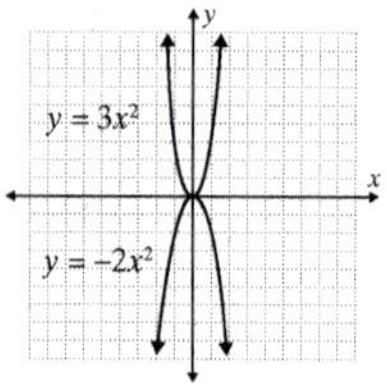

12.

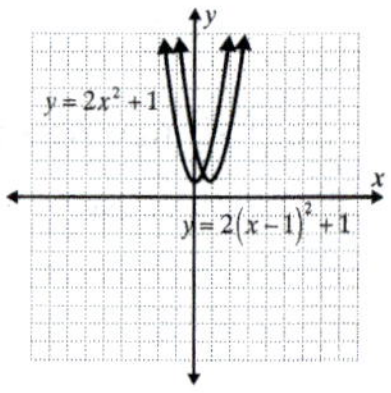

13. a. $(0, 0)$
b. $(0, 0)$
c. $y = 0$

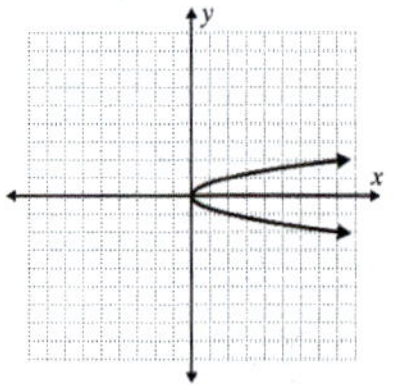

14. a. $(0, 0)$
b. $(0, 0)$
c. $y = 0$

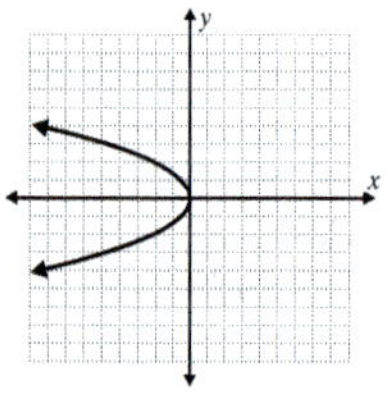

15. a. $(2, 1)$
b. None
c. $y = 1$

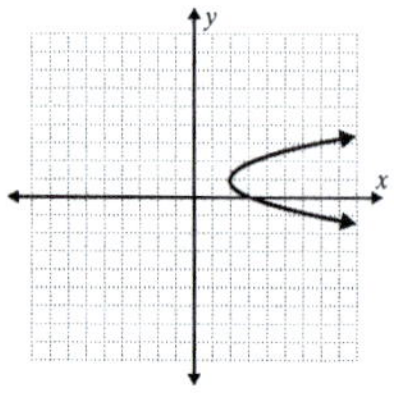

16. a. $(-2, 2)$
b. $\left(0, 2-\sqrt{2}\right)$, $\left(0, 2+\sqrt{2}\right)$
c. $y = 2$

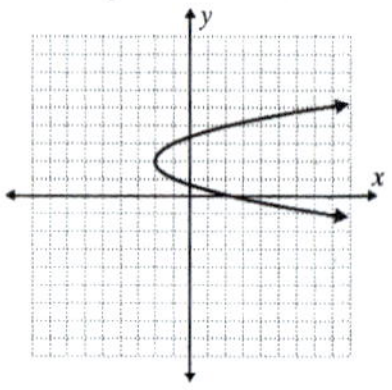

17. a. $(3, -2)$
b. None
c. $y = -2$

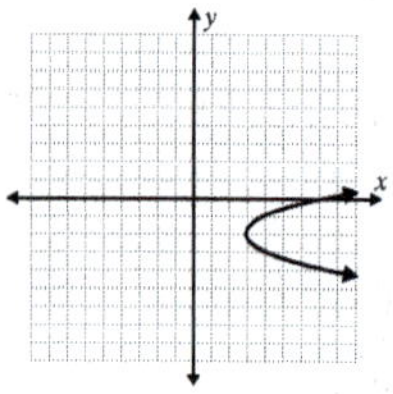

18. a. $(-7, 3)$
b. $\left(0, 3-\sqrt{7}\right)$, $\left(0, 3+\sqrt{7}\right)$
c. $y = 3$

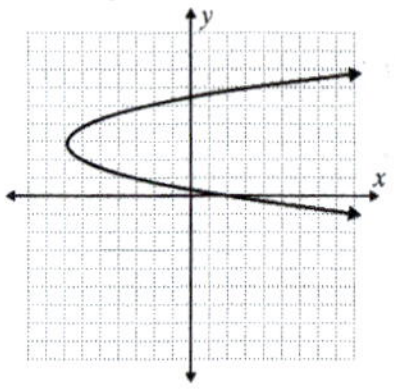

19. a. $\left(\frac{5}{4}, \frac{1}{8}\right)$
b. $(0, -3)$
c. $x = \frac{5}{4}$

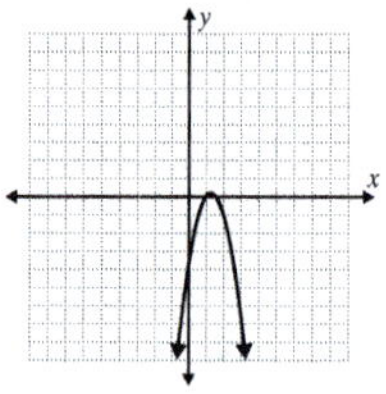

20. a. $(-8, -3)$
b. $\left(0, 2\sqrt{2}-3\right)$, $\left(0, -2\sqrt{2}-3\right)$
c. $y = -3$

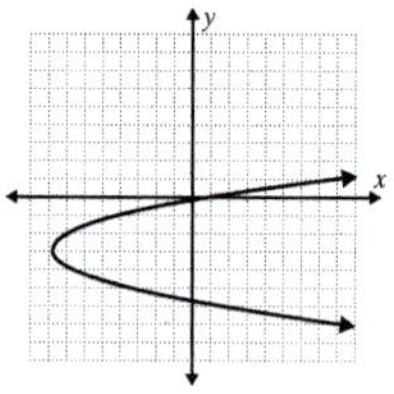

21. $(0, 1.7321)$, $(0, -1.7321)$

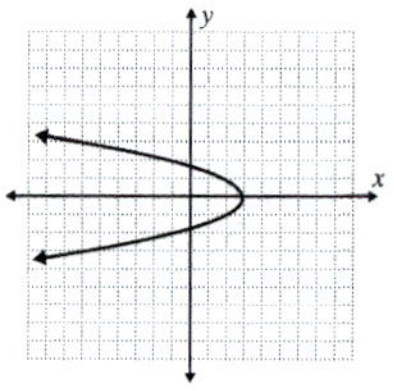

22. $(0, 0), (0, 1.5)$

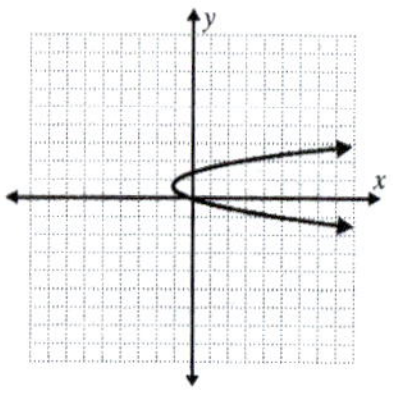

23. 5 **24.** 8 **25.** $\sqrt{97}$ **26.** $3\sqrt{10}$ **27.** $x^2 + y^2 = 8$

28. $x^2 + y^2 = 27$ **29.** $(x+2)^2 + (y-1)^2 = 49$ **30.** $(x-5)^2 + (y-4)^2 = 20$ **31.** $(x-4)^2 + (y+2)^2 = 3$ **32.** $x^2 + (y-7)^2 = 16$

33. $x^2 + y^2 = 36$
Center: $(0,0)$;
$r = 6$

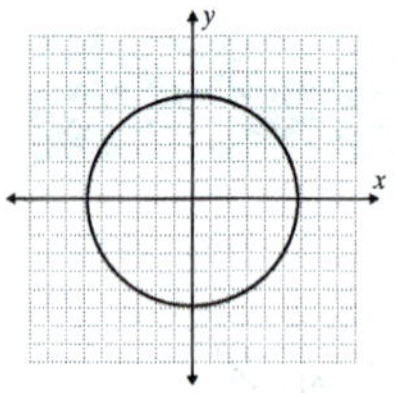

34. $x^2 + (y-3)^2 = 9$
Center: $(0,3)$;
$r = 3$

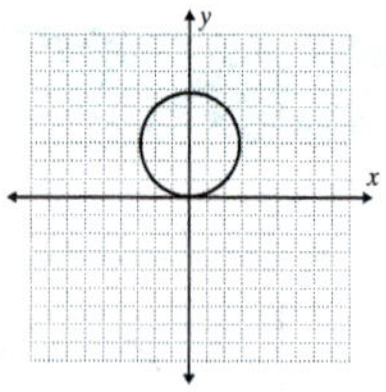

35. $(x+2)^2 + (y+1)^2 = 9$
Center: $(-2,-1)$;
$r = 3$

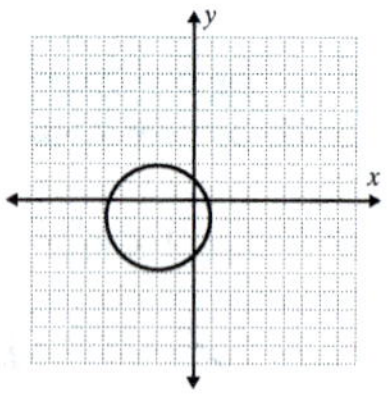

36. $x^2 + (y-5)^2 = 20$
Center: $(0,5)$;
$r = 2\sqrt{5}$

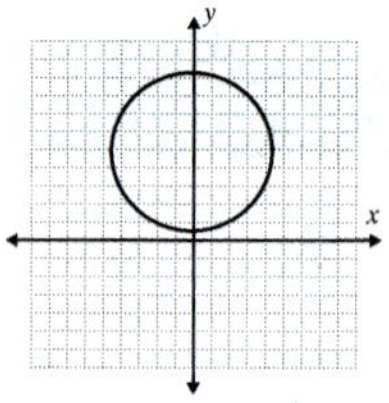

37. $\overline{AB} = 3\sqrt{2}$,
$\overline{AC} = 5\sqrt{2}$,
$\overline{BC} = 4\sqrt{2}$;
$\left(3\sqrt{2}\right)^2 + \left(4\sqrt{2}\right)^2 = \left(5\sqrt{2}\right)^2$;
Right triangle

38. $2\sqrt{29} + 2\sqrt{5} + 6\sqrt{2}$; $\overline{PQ} = 2\sqrt{29}$, $\overline{QR} = 2\sqrt{5}$, $\overline{PR} = 6\sqrt{2}$; $\left(2\sqrt{5}\right)^2 + \left(6\sqrt{2}\right)^2 \neq \left(2\sqrt{29}\right)^2$; Not a right triangle

39. $\frac{x^2}{36}+\frac{y^2}{9}=1$

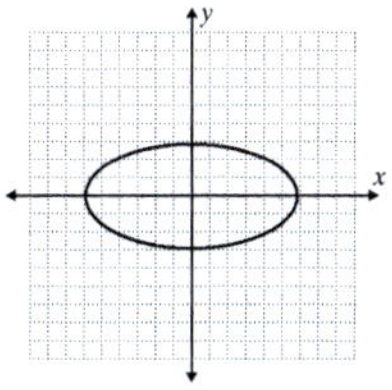

40. $\frac{x^2}{4}+\frac{y^2}{64}=1$

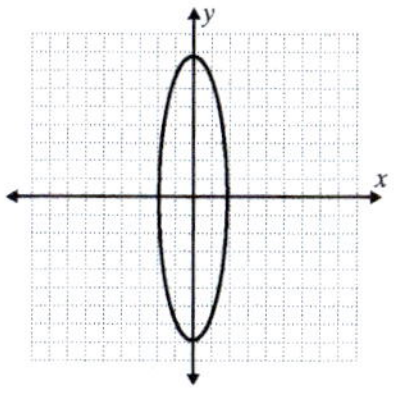

41. $\frac{x^2}{81}-\frac{y^2}{81}=1$

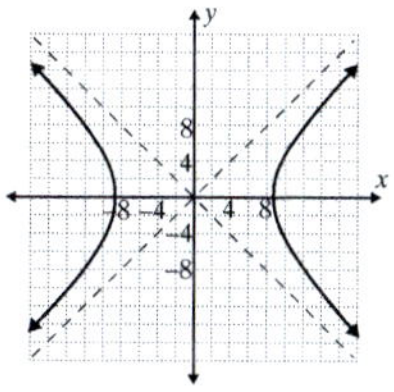

42. $\frac{x^2}{9}-\frac{y^2}{1}=1$

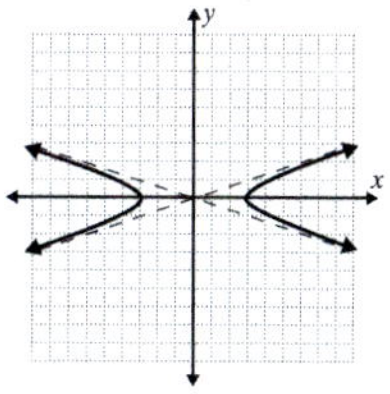

43. $\frac{y^2}{9}-\frac{x^2}{3}=1$

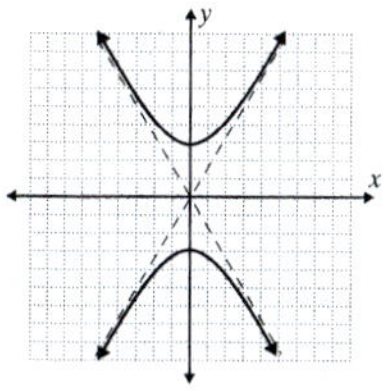

44. $\frac{x^2}{10}+\frac{y^2}{12}=1$

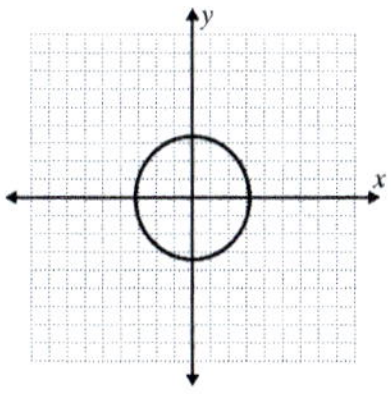

45. $\frac{x^2}{14}-\frac{y^2}{4}=1$

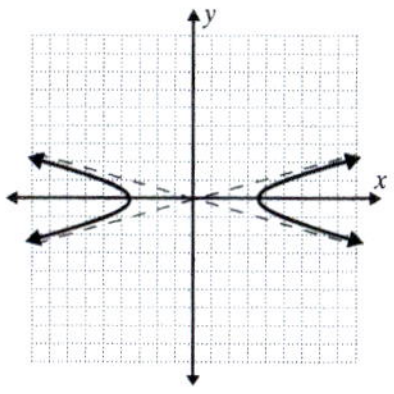

46. $\frac{y^2}{9}-\frac{x^2}{36}=1$

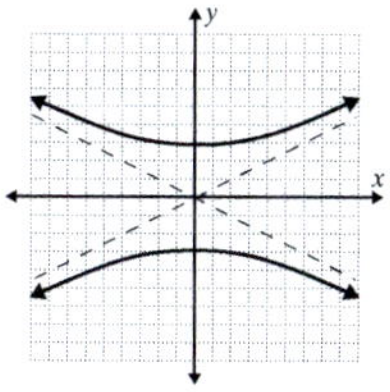

47.

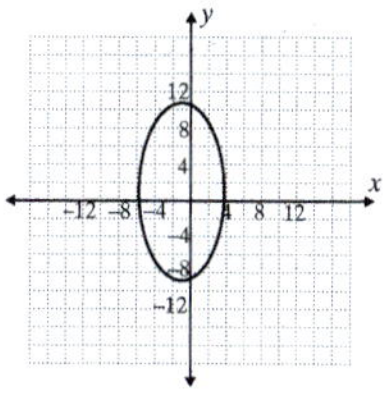

48.

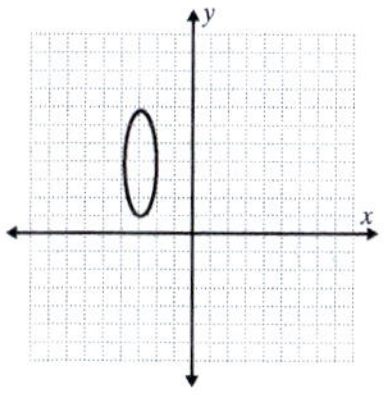

49. $(3, 8), (-2, 3)$

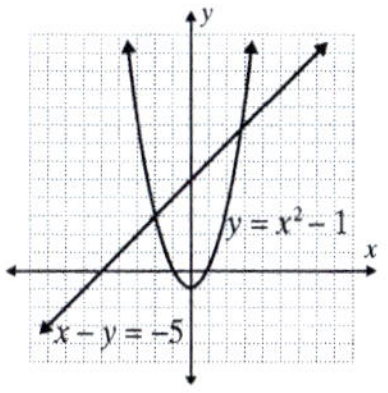

50. $(2, -4), \left(-\frac{5}{2}, -\frac{25}{4}\right)$

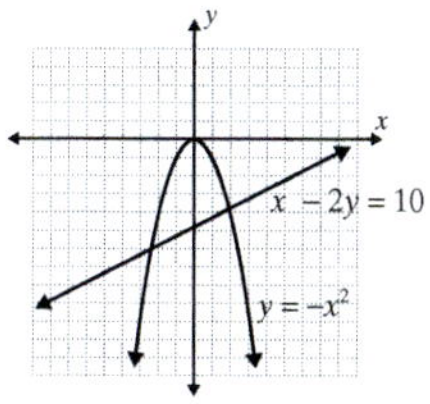

51. $(-2, 0), (0, -2)$

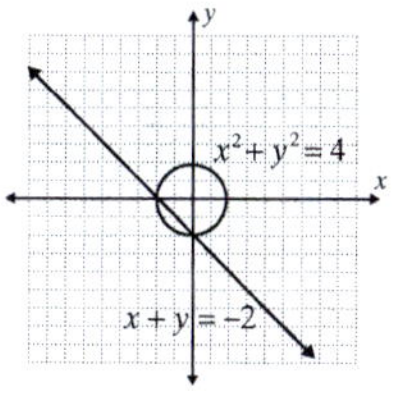

52. $(4, 1), (-4, -1)$

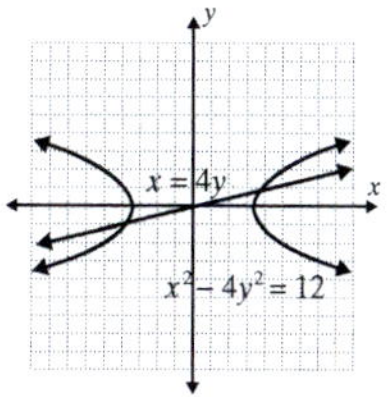

53. $(-4, 0), (4, 0)$

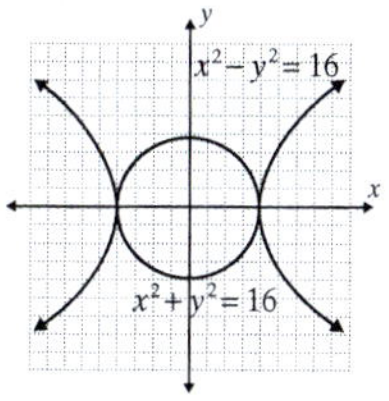

54. $(2, 3), (-2, 3), (-2, -3), (2, -3)$

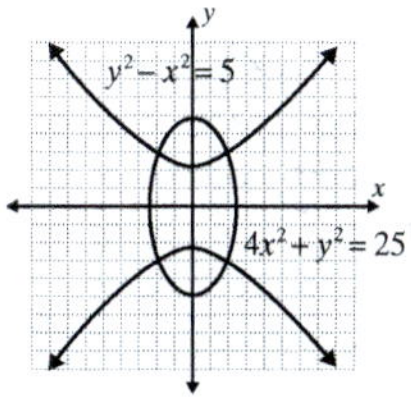

55. $\left(\frac{3+3\sqrt{5}}{2}, \frac{3+3\sqrt{5}}{2}\right)$, $\left(\frac{3-3\sqrt{5}}{2}, \frac{3-3\sqrt{5}}{2}\right)$

56. $(0, 5), (4, 5)$

57. No solution

58. $(1, 3), (1, -3), (-1, 3), (-1, -3)$

59. No solution

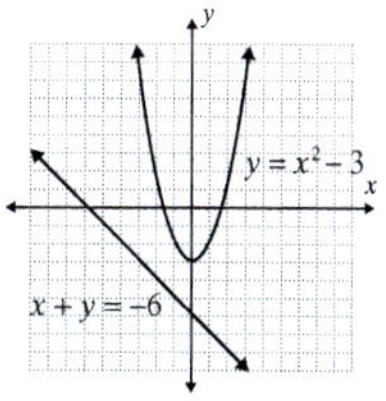

60. $(-1.11, 2.79), (-2.89, -0.79)$

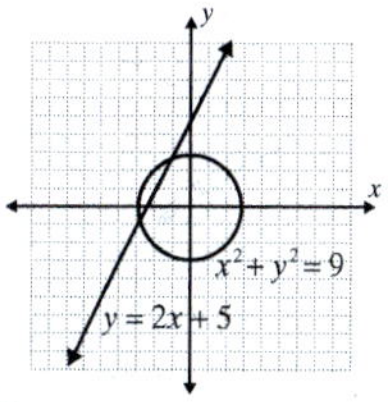

Chapter 12 Test, pages 1024 - 1025

1. a. -9
   b. $-9-7a$
   c. -7

2. a. -6
   b. $2a^2+4a-6$
   c. $4x+2h+8$

3.

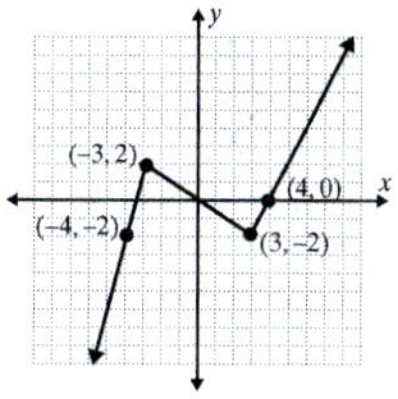

4.

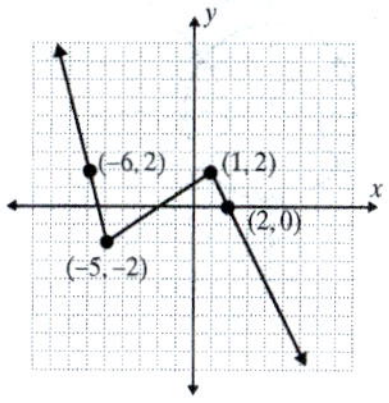

5.

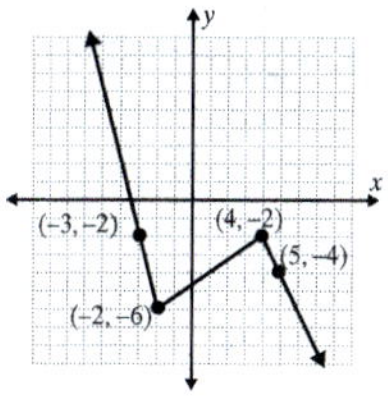

6. a. $(-5, 0)$

b. $(0, \sqrt{5}), (0, -\sqrt{5})$

c. $y = 0$

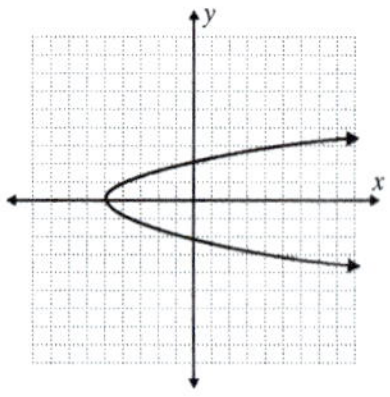

7. a. $\left(-\frac{25}{4}, -\frac{3}{2}\right)$

b. $(0, -4), (0, 1)$

c. $y = -\frac{3}{2}$

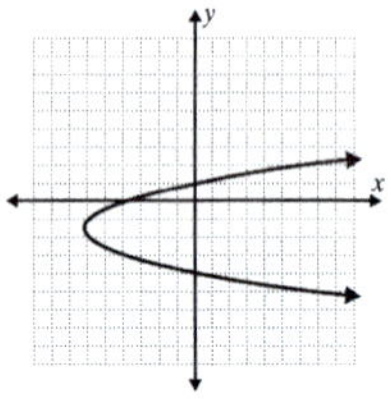

8. a. $(-5, -5)$

b. $(0, -5+\sqrt{5})$, $(0, -5-\sqrt{5})$

c. $y = -5$

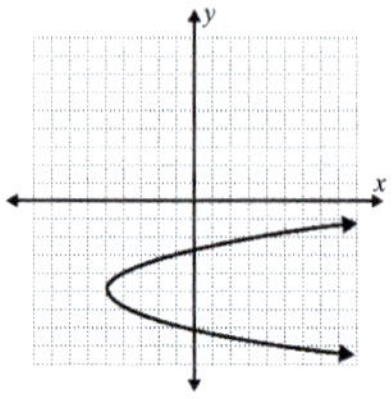

9. a. $(0, -7)$

b. $(0, -7)$

c. $y = -7$

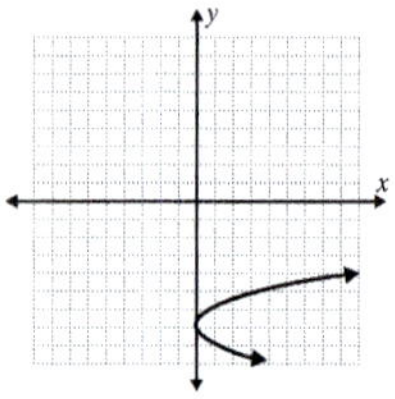

10. $(0, 1.5447)$, $(0, -1.2947)$

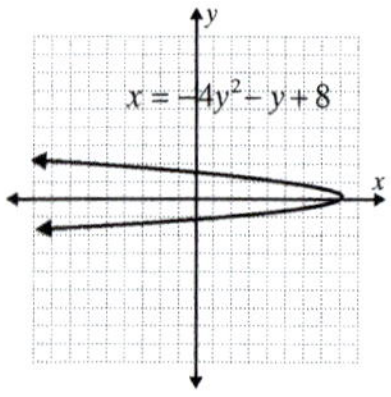

11. $3\sqrt{10}$ **12.** $\overline{AB} = 2\sqrt{13}, \overline{AC} = 2\sqrt{26}, \overline{BC} = 2\sqrt{13}; \left(2\sqrt{13}\right)^2 + \left(2\sqrt{13}\right)^2 = \left(2\sqrt{26}\right)^2$ **13.** $(x-3)^2 + (y-4)^2 = 25$

14. $(x+2)^2 + (y+5)^2 = 36$

15. $x^2 + (y-1)^2 = 9$

Center: $(0, 1)$; $r = 3$

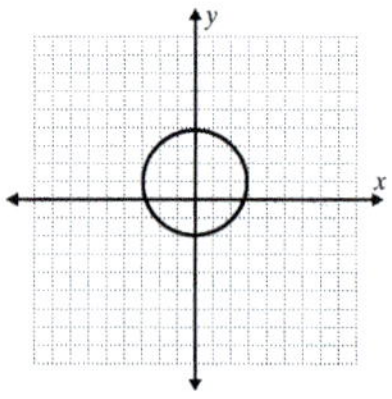

16. $(x+1)^2 + (y-3)^2 = 16$

Center: $(-1, 3)$; $r = 4$

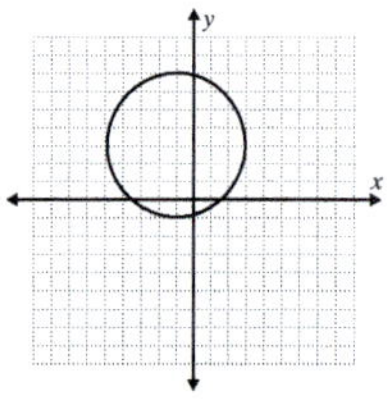

17. $\frac{x^2}{9} + \frac{y^2}{\left(\frac{9}{4}\right)} = 1$

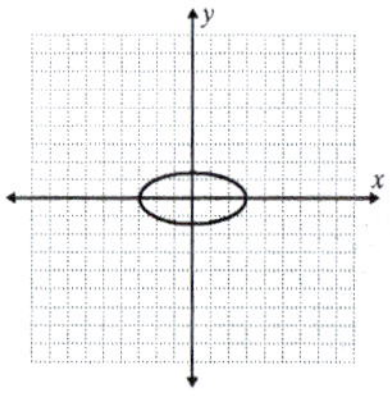

18. $\frac{x^2}{4} + \frac{y^2}{25} = 1$

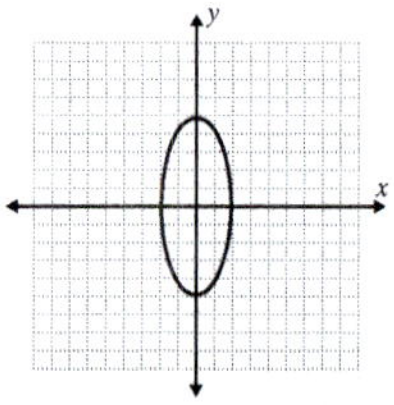

19. $\frac{x^2}{4} - \frac{y^2}{9} = 1$

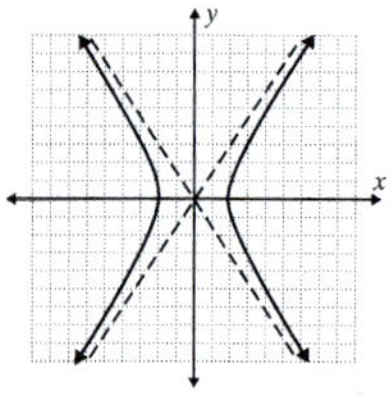

20. $\frac{y^2}{9} - \frac{x^2}{16} = 1$

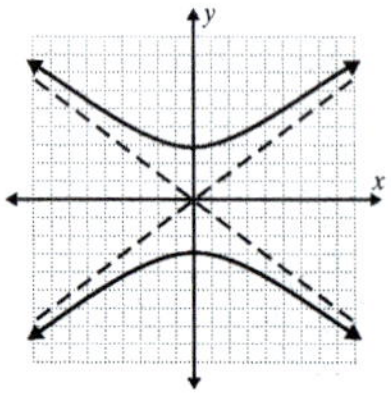

21.

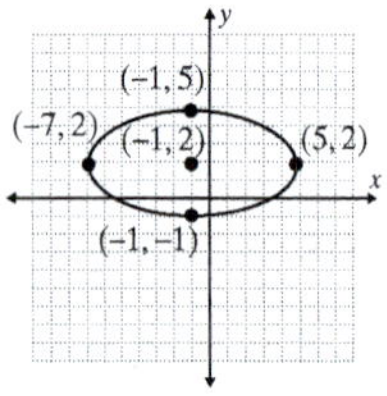

22.

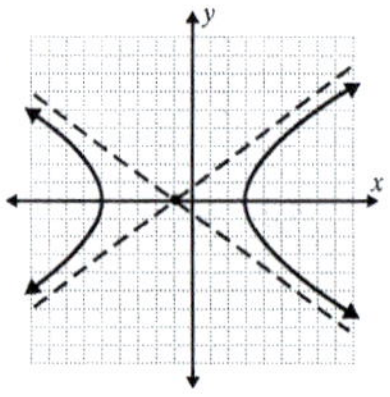

23. An ellipse is the set of all points in a plane for which the sum of the distances from two fixed points is constant.

24. $(-2, -5), (5, 2)$

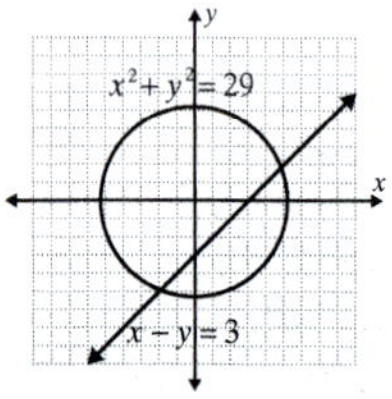

25. $(0, \sqrt{2}), (0, -\sqrt{2})$, $(-2, 0)$

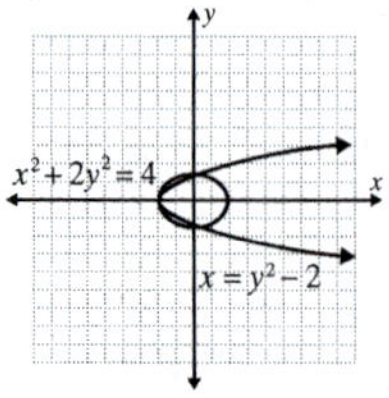

26. $(4, 3), (4, -3)$, $(-4, 3), (-4, -3)$

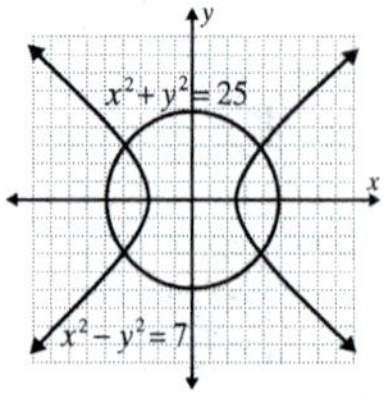

27. $(-3, 4), (4, -3)$

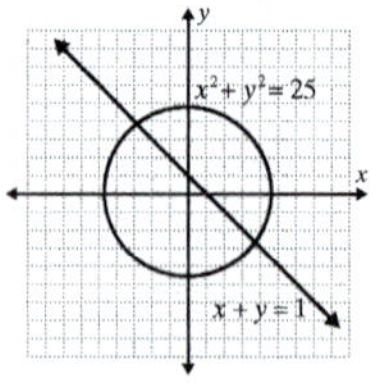

Cumulative Review: Chapters 1-12, pages 1026 - 1029

1. 1 **2.** $\frac{9x^8}{4y^8}$ **3.** $5x^{\frac{3}{4}}$ **4.** $\frac{8y^{\frac{3}{5}}}{x}$ **5.** $\sqrt[3]{(7x^3y)^2} = x^2\sqrt[3]{49y^2}$ **6.** $(32x^6y)^{\frac{1}{3}} = 2x^2(4y)^{\frac{1}{3}}$ **7.** $2(x+3)(x^2-3x+9)$

8. $3y(x+4)(x-4)$ **9.** $(x-4)(x^2+3)$ **10.** $-7x^2+4x+10$ **11.** $a^2-16a-9$ **12.** $\frac{2x^2+x+4}{(2x+3)(x-4)(3x-2)}$ **13.** -1 **14.** $13\sqrt{3}$

15. $2\sqrt{3x}+10\sqrt{6}$ **16.** $-3i\sqrt{7}$ **17.** $\frac{11}{5}-\frac{8}{5}i$ **18.** $2x^2+2x+20$ **19. a.** 4 **b.** 6 **c.** 0 **20. a.** $\log_b \frac{x^3}{y^2}$ **b.** $\log \frac{10}{y}$

21. 0; One real solution **22.** $x = \frac{-1 \pm \sqrt{7}}{3}$ **23.** $x = 1, 9$ **24.** $x = -\frac{1}{3}, 0, \frac{5}{2}$ **25.** $x = \frac{3}{2}, -\frac{3}{4}+\frac{3\sqrt{3}}{4}i, -\frac{3}{4}-\frac{3\sqrt{3}}{4}i$ **26.** $x = \frac{16}{11}, \frac{24}{5}$

27. $x=-9$ **28.** $x=2$ **29.** $x^2-6x+14$ **30.** $(-2,-3)$ **31.** $\left(\frac{1}{2}, \frac{1}{2}, -\frac{7}{2}\right)$ **32.** $x=-4, 1$ **33.** 1.8614 **34.** 6.7417 **35.** 28.0316

36. $[-1,0)\cup[5,\infty)$ 

37. $[2,3]$ 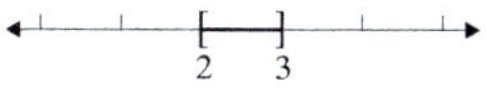

38. $\left[-7, \frac{1}{2}\right]$ 

39. 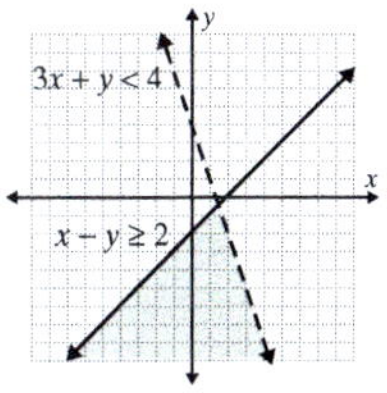

40. $5x-2y=-16$ **41.** $3x-4y=16$ **42.** $y=-\frac{1}{4}x+\frac{5}{4}$ 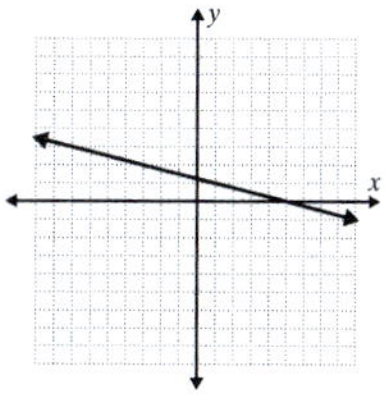

43. $(x+6)^2+(y+1)^2=16$

44. 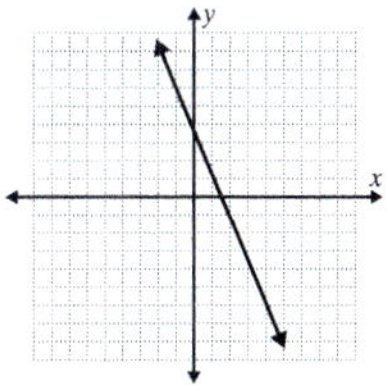

45. 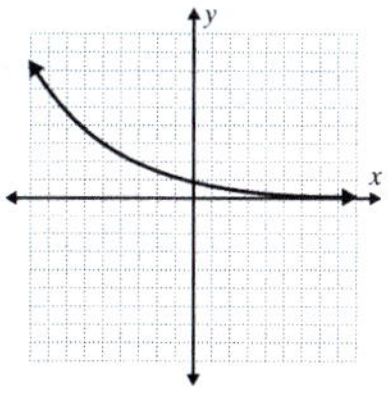

46. 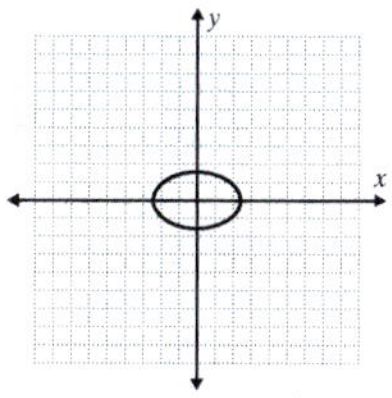

47. 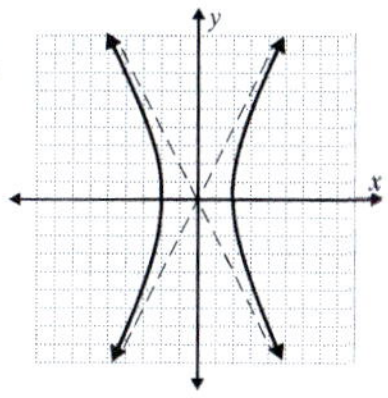

48. 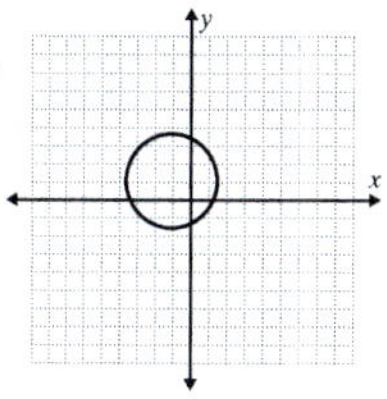

49. $y=\frac{1}{2}x^2-3$;
$x=0$; $(0,-3)$;
$D=(-\infty,\infty)$;
$R=[-3,\infty)$;
Zeros: $x=\pm\sqrt{6}$

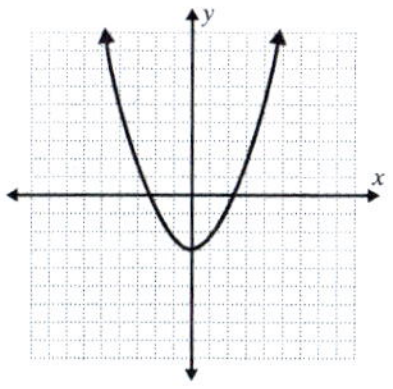

50. $y=-(x-2)^2$;
$x=2$; $(2,0)$;
$D=(-\infty,\infty)$;
$R=(-\infty,0]$;
Zeros: $x=2$

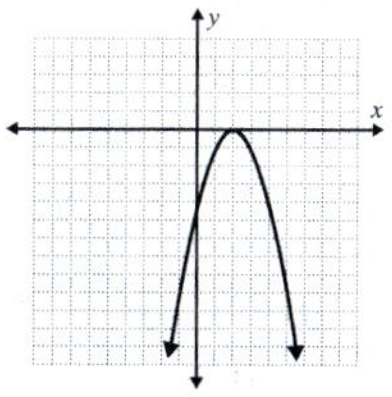

51. a. $(-3,0)$ **b.** None
c. $y=0$

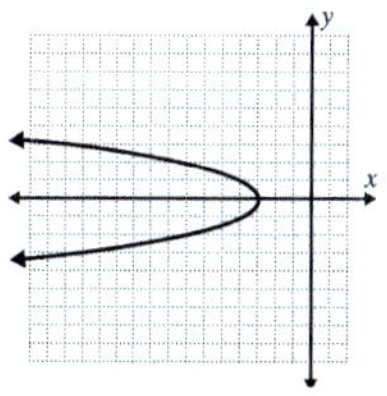

52. For $g(x)$: $D=[0,\infty)$,
$R=[0,\infty)$;
For $h(x)$: $D=[7,\infty)$,
$R=[2,\infty)$

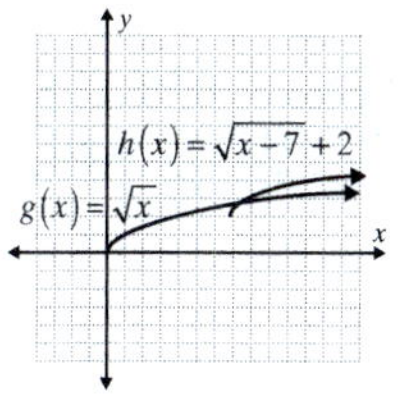

53. $f^{-1}(x)=\frac{x+9}{3}$

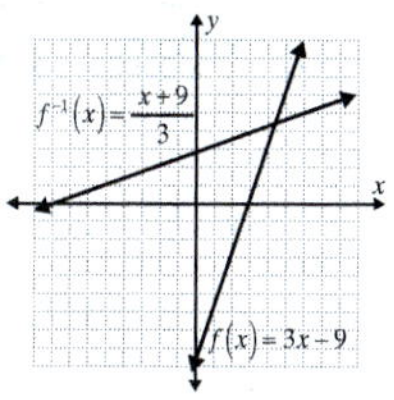

54. a. x **b.** x **c.** Yes. Answers will vary.

55. a. 5 **b.** $4x-11$ **c.** $4x-4$ **d.** 4

56. 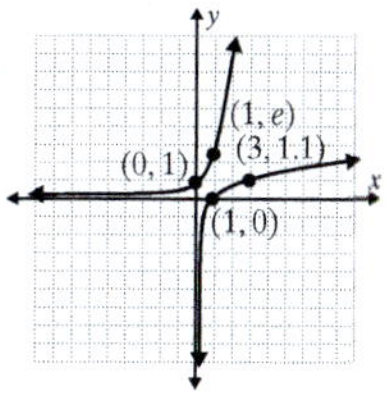

Asymptote for $y=\ln x$ is $x=0$. Asymptote for $y=e^x$ is $y=0$.

57. 17, 19 **58.** 100 **59.** 36 **60.** $\frac{20}{9}$ or $2\frac{2}{9}$ hours **61.** \$3600 at 7%; \$5400 at 8%

62. a. $t = 2\frac{1}{2}$ seconds **b.** 148 ft **63.** $y = 3000\left(3\sqrt[3]{3}\right)^t$ **64.** $2.81

65. $\overline{AB} = \sqrt{65}, \overline{BC} = \sqrt{137}, \overline{AC} = 3\sqrt{2}; P = \sqrt{65} + \sqrt{137} + 3\sqrt{2} \approx 24.01; \left(\sqrt{65}\right)^2 + \left(3\sqrt{2}\right)^2 \neq \left(\sqrt{137}\right)^2$; Not a right triangle

66. 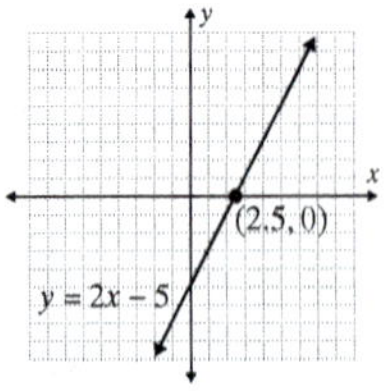

67. 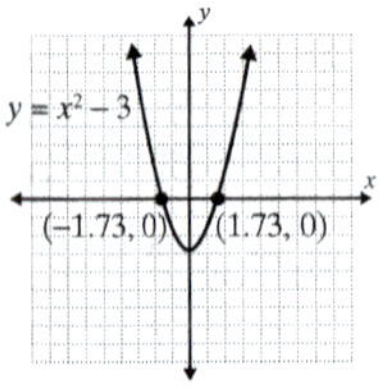

68. 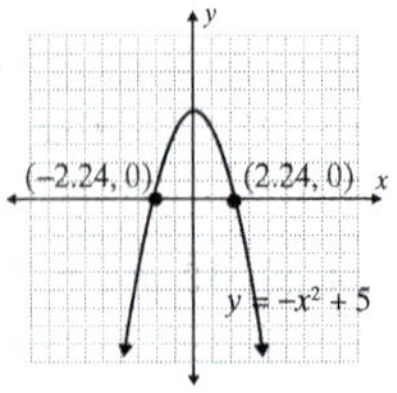

69. 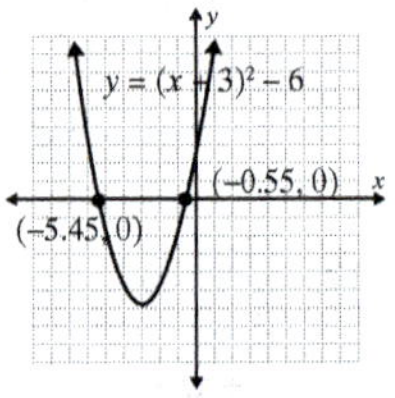

70. a. $x = -2, x = 5$ **b.** $(-\infty, 2) \cup (5, \infty)$ **c.** $(-2, 5)$ **71. a.** $x = -1, x = 2.5$ **b.** $(-1, 2.5)$ **c.** $(-\infty, -1) \cup (2.5, \infty)$

72. $(-1.67, 5.67)$, $(5.67, -1.67)$ 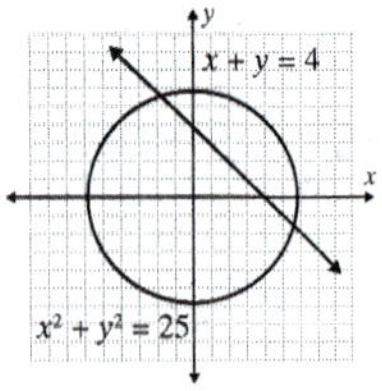

73. $(-1.33, 1.29), (0, 1.73)$, $(-1.33, -1.29), (0, -1.73)$ 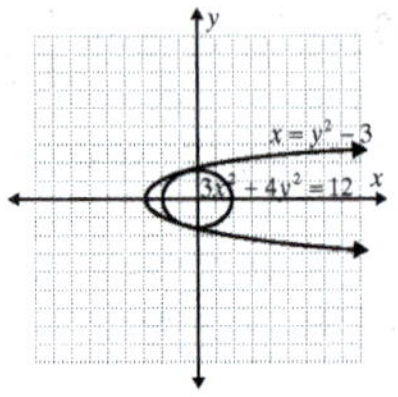

Chapter 13

Exercises 13.1, pages 1038 - 1039

1. 1, 3, 5, 7, 9 **3.** $2, \frac{3}{2}, \frac{4}{3}, \frac{5}{4}, \frac{6}{5}$ **5.** 2, 6, 12, 20, 30 **7.** 2, 4, 8, 16, 32 **9.** $\frac{1}{2}, \frac{1}{4}, \frac{1}{8}, \frac{1}{16}, \frac{1}{32}$ **11.** −2, 5, −10, 17, −26 **13.** $-\frac{1}{2}, \frac{2}{3}, -\frac{3}{4}, \frac{4}{5}, -\frac{5}{6}$ **15.** $1, \frac{1}{3}, \frac{1}{6}, \frac{1}{10}, \frac{1}{15}$ **17.** 0, 1, 3, 6, 10 **19.** $\{3n - 1\}$ **21.** $\{n^2\}$ **23.** $\left\{\frac{1}{2^n}\right\}$ **25.** $\{6n\}$ **27.** $\left\{(-1)^{n+1}(2n-1)\right\}$ **29.** Increasing; $a_n = n + 4, a_{n+1} = n + 5; n + 4 < n + 5$, so $a_n < a_{n+1}$ **31.** Decreasing; $a_n = \frac{1}{n+3}$, $a_{n+1} = \frac{1}{n+4}; \frac{1}{n+3} > \frac{1}{n+4}$, so $a_n > a_{n+1}$ **33.** Decreasing; $a_n = \frac{2n+1}{n}, a_{n+1} = \frac{2n+3}{n+1}; \frac{2n+1}{n} > \frac{2n+3}{n+1}$, so $a_n > a_{n+1}$
35. $a_1 = 3, a_2 = 1$, and $a_3 = 4$, so $a_1 > a_2$, but $a_2 < a_3$ making the sequence neither inc. nor dec. **37.** $28,000 when n = 1; $19,600 when n = 2; $13,720 when n = 3; it will be worth $13,720 after 3 years.
39. 4 m when n = 1; 1.6 m when n = 2; 0.64 m when n = 3; 0.256 m when n = 4; it will bounce 0.256 m on the 4th bounce
41. a. 1, 1, 2, 3, 5, 8, 13, 21 **b.** If we call the sequence of differences D_n, then $D_n = F_{n-2}$.

Exercises 13.2, pages 1044 - 1045

1. $S_1 = 2, S_2 = 2 + 5 = 7, S_3 = 2 + 5 + 8 = 15, S_4 = 2 + 5 + 8 + 11 = 26$
3. $S_1 = \frac{1}{2}, S_2 = \frac{1}{2} + \frac{2}{3} = \frac{7}{6}, S_3 = \frac{1}{2} + \frac{2}{3} + \frac{3}{4} = \frac{23}{12}, \quad S_4 = \frac{1}{2} + \frac{2}{3} + \frac{3}{4} + \frac{4}{5} = \frac{163}{60}$
5. $S_1 = 1, S_2 = 1 + (-4) = -3, S_3 = 1 + (-4) + 9 = 6, S_4 = 1 + (-4) + 9 + (-16) = -10$
7. $S_1 = \frac{1}{2}, S_2 = \frac{1}{2} + \frac{1}{4} = \frac{3}{4}, S_3 = \frac{1}{2} + \frac{1}{4} + \frac{1}{8} = \frac{7}{8}, S_4 = \frac{1}{2} + \frac{1}{4} + \frac{1}{8} + \frac{1}{16} = \frac{15}{16}$
9. $S_1 = 1, S_2 = 1 + 0 = 1, S_3 = 1 + 0 + (-3) = -2, S_4 = 1 + 0 + (-3) + (-8) = -10$ **11.** $2 + 4 + 6 + 8 + 10 = 30$

13. $5+6+7+8+9=35$ **15.** $\frac{1}{2}+\frac{1}{3}+\frac{1}{4}=\frac{13}{12}$ **17.** $2+4+8=14$ **19.** $16+25+36+49+64=190$
21. $3+1+(-1)+(-3)=0$ **23.** $6+(-12)+20+(-30)=-16$ **25.** $\frac{1}{2}+\frac{2}{3}+\frac{3}{4}+\frac{4}{5}+\frac{5}{6}=\frac{71}{20}$ **27.** $\sum_{k=1}^{5}(2k-1)$ **29.** $\sum_{k=1}^{5}(-1)^k$
31. $\sum_{k=4}^{7}k^2$ **33.** $\sum_{k=2}^{6}(-1)^k\left(\frac{1}{k^3}\right)$ **35.** $\sum_{k=4}^{15}\frac{k}{k+1}$ **37.** 39 **39.** 57 **41.** 48 **43.** 295 **45.** –1 **47.** 55 **49.** $\sum_{k=11}^{15}(-2k)+\sum_{k=1}^{5}3k$

Exercises 13.3, pages 1054 - 1056

1. Arithmetic sequence $d=3$; $\{3n-1\}$ **3.** Arithmetic sequence $d=-2$; $\{9-2n\}$ **5.** Not an arithmetic sequence
7. Arithmetic sequence $d=-4$; $\{10-4n\}$ **9.** Arithmetic sequence $d=\frac{1}{2}$; $\left\{\frac{n-1}{2}\right\}$ **11.** 1, 3, 5, 7, 9; Arithmetic sequence
13. $\frac{1}{2},\frac{1}{3},\frac{1}{4},\frac{1}{5},\frac{1}{6}$; Not an arithmetic sequence **15.** $\frac{3}{2},3,\frac{9}{2},6,\frac{15}{2}$; Arithmetic sequence
17. $\frac{20}{3},\frac{19}{3},6,\frac{17}{3},\frac{16}{3}$; Arithmetic sequence **19.** –1, 4, –7, 10, –13; Not an arithmetic sequence **21.** $\left\{\frac{2n+1}{3}\right\}$ **23.** $\{9-2n\}$
25. $\left\{\frac{17+3n}{2}\right\}$ **27.** $\{63-5n\}$ **29.** $\left\{\frac{81+3n}{2}\right\}$ **31.** 232 **33.** 44 **35.** 7 **37.** 19 **39.** 154 **41.** 235 **43.** 126 **45.** 231 **47.** $\frac{143}{3}$
49. 60 **51.** 171 **53.** –100 **55.** 9 days **57.** 1625 cm **59.** 134 seats; 1540 seats **61.** An arithmetic sequence has a constant difference d. If the terms alternated being positive and negative, then d would also have to alternate being positive and negative. This is a contradiction.

Exercises 13.4, pages 1066 - 1068

1. Not a geometric sequence **3.** Geometric sequence $r=-\frac{1}{2}$; $\left\{3\left(-\frac{1}{2}\right)^{n-1}\right\}$ **5.** Geometric sequence $r=\frac{3}{8}$; $\left\{\frac{32}{27}\left(\frac{3}{8}\right)^{n-1}\right\}$
7. Not a geometric sequence **9.** Geometric sequence $r=-\frac{1}{4}$; $\left\{48\left(-\frac{1}{4}\right)^{n-1}\right\}$ **11.** 9, –27, 81, –243; Geometric sequence
13. $\frac{2}{3},\frac{4}{3},2,\frac{8}{3}$; Not a geometric sequence **15.** $-\frac{8}{5},\frac{32}{25},-\frac{128}{125},\frac{512}{625}$; Geometric sequence
17. $3\sqrt{2},6,6\sqrt{2},12$; Geometric sequence **19.** 0.3, –0.09, 0.027, –0.0081; Geometric sequence **21.** $\left\{3(2)^{n-1}\right\}$
23. $\left\{\frac{1}{3}\left(-\frac{1}{2}\right)^{n-1}\right\}$ **25.** $\left\{\left(\sqrt{2}\right)^{n-1}\right\}$ **27.** $\left\{\frac{1}{3}(3)^{n-1}\right\}$ **29.** $\left\{-5\left(-\frac{3}{4}\right)^{n-1}\right\}$ **31.** $\frac{1}{4}$ **33.** $\frac{32}{9}$ **35.** $\frac{1}{4}$ or $-\frac{1}{4}$ **37.** $n=4$ **39.** 363
41. $\frac{1023}{64}$ **43.** $-\frac{333}{64}$ **45.** $\frac{422}{243}$ **47.** $\frac{75}{128}$ **49.** 4 **51.** $-\frac{1}{3}$ **53.** $\frac{4}{9}$ **55.** $\frac{4}{11}$ **57.** $27,859.63 **59.** 14.76 liters
61. $12,841.18 **63.** 7.59 meters **65.** π

Exercises 13.5, pages 1076 - 1077

1. 56 **3.** $\frac{1}{15}$ **5.** 4 **7.** $(n-1)!$ **9.** $(k+3)(k+2)(k+1)$ **11.** 20 **13.** 35 **15.** 1 **17.** $x^7+7x^6y+21x^5y^2+35x^4y^3$
19. $x^9+9x^8+36x^7+84x^6$ **21.** $x^5+15x^4+90x^3+270x^2$ **23.** $x^6+12x^5y+60x^4y^2+160x^3y^3$
25. $2187x^7-5103x^6y+5103x^5y^2-2835x^4y^3$ **27.** $x^{18}-36x^{16}y+576x^{14}y^2-5376x^{12}y^3$

29. $x^6+6x^5y+15x^4y^2+20x^3y^3+15x^2y^4+6xy^5+y^6$ **31.** $x^7-7x^6+21x^5-35x^4+35x^3-21x^2+7x-1$
33. $243x^5+405x^4y+270x^3y^2+90x^2y^3+15xy^4+y^5$ **35.** $x^4+8x^3y+24x^2y^2+32xy^3+16y^4$
37. $81x^4-216x^3y+216x^2y^2-96xy^3+16y^4$ **39.** $x^8+8x^6y+24x^4y^2+32x^2y^3+16y^4$ **41.** $3360x^6y^4$ **43.** $1,140,480x^8$
45. $-27,500x^6y^{18}$ **47.** 1.062 **49.** 0.808 **51.** 64.363 **53.** 0.851 **55.** $(x+2)^4$

Chapter 13 Review, pages 1082 - 1085

1. $-1, 1, 3, 5$ **2.** $4, \frac{5}{2}, 2, \frac{7}{4}$ **3.** $\frac{1}{2}, \frac{4}{3}, \frac{9}{4}, \frac{16}{5}$ **4.** $-1, \frac{1}{4}, -\frac{1}{9}, \frac{1}{16}$ **5.** 7, 5, 7, 5 **6.** $-1, -1, -1, -1$ **7.** $\{5n+5\}$ **8.** $\{3(-1)^{n+1}\}$
9. $\{n^2+2\}$ **10.** $\left\{\frac{n+2}{n}\right\}$ **11.** Increasing **12.** Decreasing **13.** Increasing **14.** Decreasing **15.** \$275, \$6750 **16.** b
17. $S_1=5, S_2=5+9=14, S_3=5+9+13=27, S_4=5+9+13+17=44$
18. $S_1=3, S_2=3+2=5,\ S_3=3+2+\frac{5}{3}=\frac{20}{3}, S_4=3+2+\frac{5}{3}+\frac{3}{2}=\frac{49}{6}$
19. $S_1=0, S_2=0+2=2, S_3=0+2+0=2, S_4=0+2+0+2=4$
20. $S_1=2, S_2=2+6=8, S_3=2+6+12=20, S_4=2+6+12+20=40$ **21.** $5+10+15+20=50$ **22.** $7+8+9+10+11=45$
23. $5-7+9-11+13=9$ **24.** $-1+4-9+16-25+36=21$ **25.** $80+99+120=299$ **26.** $\frac{1}{3}+\frac{1}{6}+\frac{1}{9}+\frac{1}{12}=\frac{25}{36}$
27. $\sum_{k=1}^{5}(2k+1)$ **28.** $\sum_{k=3}^{6}\frac{(-1)^{k+1}}{k^3}$ **29.** $\sum_{k=1}^{6}(-1)^{k+1}(k^2-1)$ **30.** $\sum_{k=1}^{5}\frac{k+1}{k^2}$ **31.** 50 **32.** -56 **33.** 380 **34.** 10
35. Arithmetic sequence $d=4; \{4n-7\}$ **36.** Arithmetic sequence $d=\frac{2}{3}; \left\{\frac{2}{3}n+\frac{5}{3}\right\}$ **37.** Not an arithmetic sequence
38. Arithmetic sequence $d=7; \{7n-8\}$ **39.** 4, 9, 16, 25, 36; Not an arithmetic sequence
40. $\frac{17}{3}, \frac{16}{3}, 5, \frac{14}{3}, \frac{13}{3}$; Arithmetic sequence **41.** $\left\{\frac{n+39}{4}\right\}$ **42.** $\{27-7n\}$ **43.** $\{2-2n\}$ **44.** $\{2n-8\}$ **45.** $a_{13}=77$
46. $n=20$ **47.** 240 **48.** 60 **49.** $\frac{135}{2}$ **50.** 35 **51.** -150 **52.** 60 **53.** 225 blocks **54.** \$6180
55. Geometric sequence $r=\frac{1}{2}; \left\{12\left(\frac{1}{2}\right)^{n-1}\right\}$ **56.** Not a geometric sequence **57.** Geometric sequence $r=-\frac{1}{5}; \left\{10\left(-\frac{1}{5}\right)^{n-1}\right\}$
58. Not a geometric sequence **59.** $-\frac{1}{5}, \frac{1}{25}, -\frac{1}{125}, \frac{1}{625}$ Geometric sequence **60.** $\frac{4}{3}, \frac{8}{9}, \frac{16}{27}, \frac{32}{81}$ Geometric sequence
61. $\left\{7\left(\frac{1}{2}\right)^{n-1}\right\}$ **62.** $\{4(-3)^{n-1}\}$ **63.** $\left\{\frac{16}{9}\left(-\frac{3}{2}\right)^{n-1}\right\}$ **64.** $\left\{\frac{5}{16}(2)^{n-1}\right\}$ **65.** $\frac{1}{2}$ **66.** 40 **67.** -48 **68.** 6 **69.** 242 **70.** $\frac{160}{81}$ **71.** $-\frac{2}{5}$
72. $\frac{3}{2}$ **73. a.** $\sum_{k=1}^{\infty}\frac{2}{5}\left(\frac{1}{10}\right)^{k-1}$ **b.** $\sum_{k=1}^{\infty}\frac{4}{5}\left(\frac{1}{10}\right)^{k-1}$ **74.** \$110,907.56 **75.** \$19,966.74 **76.** 30,240 **77.** 90 **78.** $(n-2)!$
79. $(n+3)(n+2)$ **80.** 56 **81.** 1 **82.** $x^8+8x^7y+28x^6y^2+56x^5y^3$ **83.** $x^9+18x^8+144x^7+672x^6$
84. $1024x^{10}-5120x^9y+11520x^8y^2-15360x^7y^3$ **85.** $x^{12}-24x^{10}+240x^8-1280x^6$
86. $x^4+20x^3+150x^2+500x+625$ **87.** $8x^3+36x^2y+54xy^2+27y^3$ **88.** $x^{10}-15x^8y+90x^6y^2-270x^4y^3+405x^2y^4-243y^5$
89. $x^{10}-10x^9+45x^8-120x^7+210x^6-252x^5+210x^4-120x^3+45x^2-10x+1$ **90.** $-4032x^4y^5$ **91.** $-811,008x^{10}y^{14}$
92. 1.338 **93.** 0.630

Chapter 13 Test, pages 1086 - 1087

1. $\left\{\frac{n}{2n+1}\right\}$ **2.** $0+2+6+12+20=40$ **3.** $\sum_{k=1}^{4} 3k+4$ **4.** $\frac{1}{4},\frac{1}{7},\frac{1}{10},\frac{1}{13},\ldots$; Neither; Decreasing

5. 4, –6, –16, – 26,...; Arithmetic; Decreasing **6.** $-\frac{3}{2},\frac{3}{4},-\frac{3}{8},\frac{3}{16},\ldots$; Geometric; Neither **7.** –5, –2, 3, 10; Neither; Increasing

8. 26 **9.** $a_n=8-2n$ **10.** $n=15$ **11.** 243 **12.** $\frac{1}{8}$ **13.** $a_n=\left(\sqrt{3}\right)^{n-2}$ **14.** $n=5$ **15.** $\frac{21}{16}$ **16.** 68 **17.** $-\frac{40}{729}$ **18.** $\frac{3}{4}$

19. 74 **20.** 6 **21.** $\frac{15}{100}+\frac{15}{10,000}+\ldots=\frac{5}{33}$ **22. a.** 180 **b.** 462 **23.** $32x^5-80x^4y+80x^3y^2-40x^2y^3+10xy^4-y^5$

24. $5670x^4y^4$ **25.** $1125x^8y^2$ **26.** 11 rows of logs; 121 logs in total **27.** $13,050.16

Cumulative Review: Chapters 1-13, pages 1088 - 1090

1. $11x+3$ **2.** $-x^2+3x+5$ **3.** $(4x+3)(16x^2-12x+9)$ **4.** $(3x-5)(2x+9)$ **5.** $(2x+1)(5x+7)(x-2)$ **6.** $\frac{11x+12}{15}$

7. $\frac{-6x-4}{(x+4)(x-4)(x-1)}$ **8.** $\frac{1}{x^2}$ **9.** $\frac{3y^{\frac{1}{3}}}{x^2}$ **10.** $\frac{2x\sqrt[3]{x}}{3y}$ **11.** $3\sqrt{3x}$ **12.** $\frac{-13+11i}{29}$ **13.** 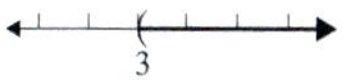

$(3,\infty)$

14. 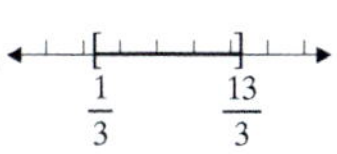

$\left[\frac{1}{3},\frac{13}{3}\right]$ **15.** 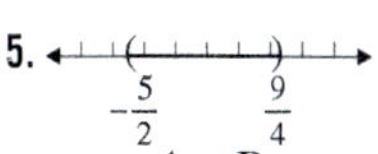

$\left(-\frac{5}{2},\frac{9}{4}\right)$ **16.** $x=\frac{26}{7}$ **17.** $x=\frac{-2\pm i\sqrt{2}}{2}$ **18.** $x=4$ **19.** $x=\frac{39}{17}$

20. $x=5$ **21.** $x=e^{2.4}\approx 11.02$ **22.** $n=\frac{A-P}{Pi}$ **23.** $(0,2,-1)$ **24. a.** $y=\frac{x+7}{2}$ **b.** $2x^2-5$ **c.** $2x+1$ **25.** 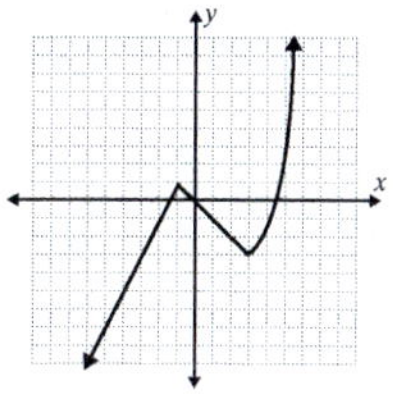

26. 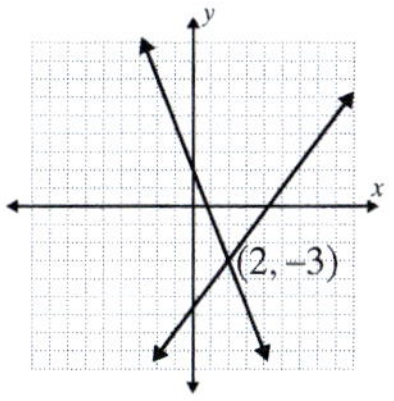

(2, –3) **27.** 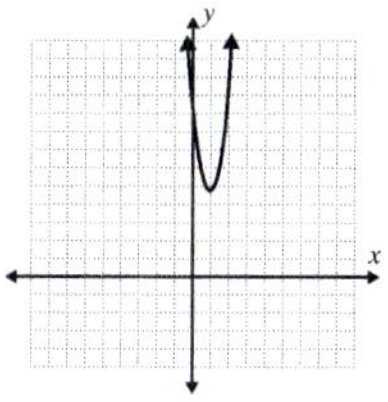 **28.** 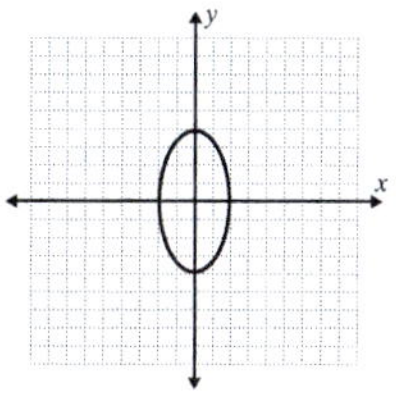 **29.** 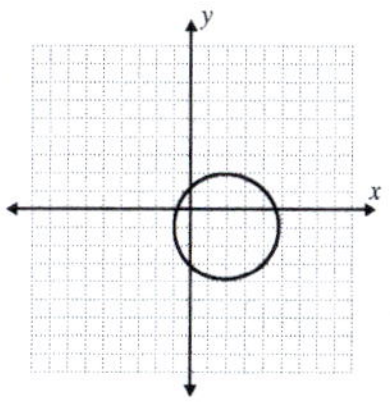

30. a. $a_n=10-2n$ **b.** –10 **31. a.** $a_n=16\left(\frac{1}{2}\right)^{n-1}$ **b.** $\frac{63}{2}$ **32.** $x^6+12x^5y+60x^4y^2+160x^3y^3+240x^2y^4+192xy^5+64y^6$

33. 20 mph, 60 mph **34.** 16 lb at $1.40; 4 lb at $2.60 **35.** 16 ft by 26 ft **36.** 12.56 hours **37.** $16,058.71 **38.** 443.71 liters **39.** $x\approx -0.4746, 1.3953$ **40.** $x\approx -1.9646, 1.0580$ **41.** $x\approx 0.2408, 2.7337$ **42.** $x=1$ **43.** $x=-3,1,2$

Appendix

Section A.1, pages 1100 - 1102

1. 1.4 **3.** 5.76 **5.** 1.64 **7.** 25.89 **9.** 147.385 **11.** 19.464 **13.** 0.04 **15.** 0.01096 **17.** 5.00 **19.** 2.03 **21.** $9691.25 **23.** $1296 **25.** 13.3 mph **27.** 0.237 **29. a.** 77.2 in. **b.** 337.68 in.^{2} **31.** 3% **33.** 300% **35.** 108% **37.** 0.06 **39.** 0.032 **41.** 1.2 **43.** 15% **45.** 96% **47.** 162.5% **49.** 0.9%

Section A.2, pages 1107 - 1108

1. a. $x-9$ **b.** $c=3;\ P(3)=0$ **3. a.** $x^2-4x+33-\dfrac{265}{x+8}$ **b.** $c=-8;\ P(-8)=-265$ **5. a.** $4x^2-6x+9-\dfrac{17}{x+2}$ **b.** $c=-2;\ P(-2)=-17$ **7. a.** $x^2+7x+55+\dfrac{388}{x-7}$ **b.** $c=7;\ P(7)=388$ **9. a.** $2x^2-2x+6-\dfrac{27}{x+3}$ **b.** $c=-3;\ P(-3)=-27$ **11. a.** $x^3+2x+5+\dfrac{17}{x-3}$ **b.** $c=3;\ P(3)=17$ **13. a.** $x^3+2x^2+6x+9+\dfrac{23}{x-2}$ **b.** $c=2;\ P(2)=23$

15. a. $x^3+\dfrac{1}{2}x^2-\dfrac{3}{4}x-\dfrac{3}{8}+\dfrac{45}{16\left(x-\dfrac{1}{2}\right)}$ **b.** $c=\dfrac{1}{2};\ P\left(\dfrac{1}{2}\right)=\dfrac{45}{16}$ **17. a.** $x^4+x^3+x^2+x+1$ **b.** $c=1;\ P(1)=0$

19. a. $x^3-\dfrac{14}{5}x^2+\dfrac{56}{25}x-\dfrac{224}{125}+\dfrac{3396}{625\left(x+\dfrac{4}{5}\right)}$ **b.** $c=-\dfrac{4}{5};\ P\left(-\dfrac{4}{5}\right)=\dfrac{3396}{625}$

21. a. $x^2-\dfrac{7}{2}x+\dfrac{13}{4}+\dfrac{73}{4(2x-1)};\ 2x^2-7x+\dfrac{13}{2}+\dfrac{73}{4\left(x-\dfrac{1}{2}\right)}$ **b.** $\dfrac{P(x)}{x-\dfrac{b}{a}}=\dfrac{aP(x)}{ax-b}$ **c.** If a polynomial $P(x)$ is divided by $(ax-b)$, then the remainder will be $P\left(\dfrac{a}{b}\right)$. Answers will vary.

Exercises A.3, pages 1113 - 1114

1. $x=-2,2$ **3.** $x\approx-1.41,1.41$ **5.** $x=-2,6$ **7.** $x\approx-4.46,2.46$ **9.** $x\approx-1.70,4.70$ **11.** No solution **13.** $x\approx-1.26,1.59$ **15.** $x\approx-1.77,0.52,3.25$ **17.** $x=0,1,3$ **19.** $x=-3,-1,5$ **21.** $x\approx3.40$ **23.** $x\approx-3.16,0,3.16$ **25.** $x=-1,0$ **27.** $x=-4,7$ **29.** $x\approx-1.67,3$ **31.** $x=-2,0$ **33.** $x=0.5$ **35.** $x=2,-0.67$ **37.** $(-\infty,-6)\cup(6,\infty)$ **39.** $[2,4]$ **41.** $\left(-\infty,\dfrac{-1}{2}\right]\cup\left[\dfrac{7}{2},\infty\right)$ **43.** $[-29,25]$ **45.** $(-3,9)$

47. $[-1,3]\cup[5,9]$ You would solve the equations $-1\le x-4\le5$ and $1\ge x-4\ge-5$. Answers will vary.

Section A.4, pages 1121 - 1123

1. −22 **3.** −212 **5.** 11 **7.** 3 **9.** 47 **11.** 36 **13.** −3 **15.** −4 **17.** −25 **19.** 20 **21.** $x=7$ **23.** $x=-7$ **25.** $x=-3$

27. $\begin{vmatrix} x & y & 1 \\ -2 & 1 & 1 \\ 5 & 3 & 1 \end{vmatrix}=0$ which simplifies to $2x-7y=-11$

29. $A=1$

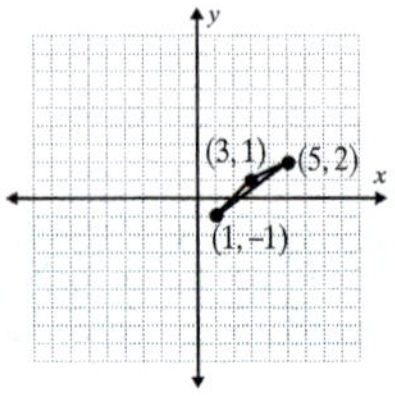

31. $A=\dfrac{31}{2}$

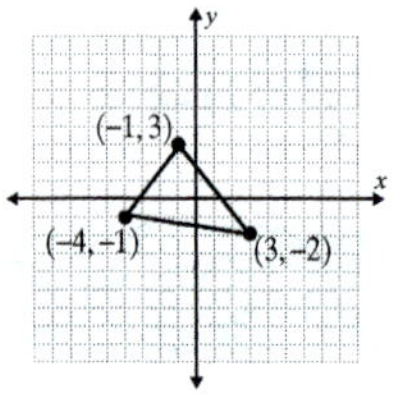

33. −25 **35.** −33.28

37. 0; Answers will vary. If you expand by minors using a row of all zeros, each minor will be multiplied by zero resulting in a sum of zero.

Section A.5, pages 1128 - 1130

1. $(4, 3)$ **3.** $\left(\frac{2}{3}, -\frac{1}{4}\right)$ **5.** No solution **7.** $(4, -1)$ **9.** $(-1, 0)$ **11.** $\left(-\frac{1}{4}, \frac{3}{2}\right)$ **13.** $\left(\frac{31}{17}, \frac{2}{17}\right)$ **15.** $\left(\frac{39}{44}, \frac{41}{44}\right)$ **17.** $(-5, -4)$

19. $(3, 7)$ **21.** $(-2, 1, 3)$ **23.** No solution **25.** $(4, 2, 6)$ **27.** Infinitely many solutions **29.** $\left(-\frac{2}{3}, \frac{11}{3}, 2\right)$

31. $\begin{cases} x + y + z = 43 \\ y = 5 + 2x \\ z = x + y - 3 \end{cases}$ 6 feet by 17 feet by 20 feet **33.** $\begin{cases} x + y = 6{,}000{,}000 \\ 0.02x + 0.04y = 170{,}000 \end{cases}$ \$3,500,000 in mutual funds; \$2,500,000 in stocks

Index

A

B

E

F

G

H

I

Q

R

S

T

U

V

W

X

Y

Z

CHAPTER 8 Systems of Linear Equations (cont.)

Matrices:

System of Linear Equations

$$\begin{cases} a_1x+b_1y=c_1 \\ a_2x+b_2y=c_2 \end{cases}$$

Coefficient Matrix

$$\begin{bmatrix} a_1 & b_1 \\ a_2 & b_2 \end{bmatrix}$$

Augmented Matrix

$$\left[\begin{array}{cc|c} a_1 & b_1 & c_1 \\ a_2 & b_2 & c_2 \end{array}\right]$$

Elementary Row Operations:

1. Interchange two rows.
2. Multiply a row by a nonzero constant.
3. Add a multiple of a row to another row.

Upper Triangular Form and Row Echelon Form:

$$\begin{bmatrix} a_{11} & a_{12} & a_{13} \\ 0 & a_{22} & a_{23} \\ 0 & 0 & a_{33} \end{bmatrix}$$

A matrix is in **upper triangular form** if its entries in the lower left triangular region are all 0's. If a_{11}, a_{22}, and a_{33} (the entries along the main diagonal) all equal 1 when the matrix is in upper triangular form, then the matrix is also in **row echelon form** (or **ref**).

Gaussian Elimination:

1. Write the augmented matrix for the system.
2. Use elementary row operations to transform the matrix into row echelon form.
3. Solve the corresponding system of equations by using back substitution.

CHAPTER 9 Roots, Radicals, and Complex Numbers

Radical Terminology:

The symbol $\sqrt{\ }$ is called a **radical sign.**
The number under the radical sign is called the **radicand**.
The complete expression, such as $\sqrt{64}$, is called a **radical** or **radical expression**.

Square Roots:

If a is a nonnegative real number, then $\sqrt{a}$ is the **principal square root** of a, and $-\sqrt{a}$ is the **negative square root** of a.

Properties of Square Roots:

If a and b are positive real numbers, then

1. $\sqrt{ab}=\sqrt{a}\sqrt{b}$
2. $\sqrt{\dfrac{a}{b}}=\dfrac{\sqrt{a}}{\sqrt{b}}$

The Square Root of x^2:

If x is a real number, then $\sqrt{x^2}=|x|$ and if $x \geq 0$, then we can write $\sqrt{x^2}=x$.

Cube Roots:

If a is a real number, then $\sqrt[3]{a}$ is the **cube root** of a.

Properties of Radicals:

If n is a positive integer, m is any integer, and $\sqrt[n]{a}$ is a real number then

1. $\sqrt[n]{a}=a^{\frac{1}{n}}$
2. $a^{\frac{m}{n}}=\left(a^{\frac{1}{n}}\right)^m=\left(a^m\right)^{\frac{1}{n}}$ or, in radical notation, $a^{\frac{m}{n}}=\left(\sqrt[n]{a}\right)^m=\sqrt[n]{a^m}$

Conjugates:

The two expressions $(a-b)$ and $(a+b)$ are called **conjugates** of each other, and their product, $(a-b)(a+b)$, results in the difference of two squares.

Definition of *i*:

$i=\sqrt{-1}$ and $i^2=\left(\sqrt{-1}\right)^2=-1$

If a is positive real number, $\sqrt{-a}=\sqrt{a}\cdot\sqrt{-1}=\sqrt{a}\,i=i\sqrt{a}$.

Complex Numbers:

The **standard form** of a **complex number** is $\boldsymbol{a+bi}$ where a and b are real numbers. a is called the **real part** and b is called the **imaginary part**.

CHAPTER 10 Quadratic Equations

Square Root Property:

If $x^2=c$, then $x=\pm\sqrt{c}$.
If $(x-a)^2=c$, then $x-a=\pm\sqrt{c}$ $\left(\text{or } x=a\pm\sqrt{c}\right)$.

Completing the Square:

To complete the square, find the third term of a perfect square trinomial when the first two terms are given. The trinomial should have the following characteristics:

1. The leading coefficient (the coefficient of x^2) is 1.
2. The constant term is the square of $\frac{1}{2}$ of the coefficient of x.

Quadratic Formula:

For the general quadratic equation $ax^2+bx+c=0$, where $a\neq 0$, the solutions are

$$x=\frac{-b\pm\sqrt{b^2-4ac}}{2a}.$$

Discriminant:

The expression b^2-4ac, the part of the quadratic formula that lies under the radical sign, is called the **discriminant**.
If $b^2-4ac>0 \rightarrow$ There are two real solutions.
If $b^2-4ac=0 \rightarrow$ There is one real solution, $x=-\dfrac{b}{2a}$.
If $b^2-4ac<0 \rightarrow$ There are two nonreal solutions.

Parabolas:

Parabolas of the form $y=ax^2+bx+c$ $(a\neq 0)$:

If $a>0$, the parabola opens upward.
If $a<0$, the parabola opens downward.
The bigger $|a|$ is, the narrower the opening.
The smaller $|a|$ is, the wider the opening.

Vertex: $\left(-\dfrac{b}{2a}, \dfrac{4ac-b^2}{4a}\right)$

Line of Symmetry: $x=-\dfrac{b}{2a}$

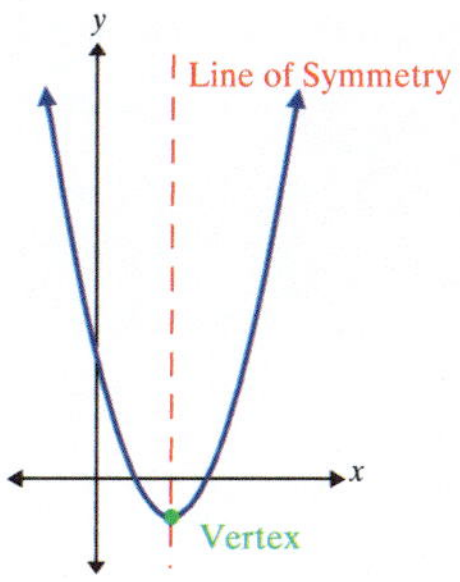

Parabolas of the form $y=a(x-h)^2+k$:

Vertex: (h,k)
Line of Symmetry: $x=h$
The graph is a horizontal shift of h units and a vertical shift of k units of the graph of $y=ax^2$.

Minimum and Maximum Values:

For a parabola with its equation given in the form $y=a(x-h)^2+k$:

1. If $a>0$, then the parabola opens upward and (h,k) is the lowest point, and the y-value is called the **minimum value** of the function.
2. If $a<0$, then the parabola opens downward and (h,k) is the highest point, and the y-value is called the **maximum value** of the function.

CHAPTER 11 Exponential and Logarithmic Functions

Algebraic Operations with Functions:

For functions $f(x)$ and $g(x)$ where x is in the domain of both functions,

$(f+g)(x)=f(x)+g(x)$

$(f-g)(x)=f(x)-g(x)$

$(f\cdot g)(x)=f(x)\cdot g(x)$

$\left(\frac{f}{g}\right)(x)=\frac{f(x)}{g(x)},\ g(x)\neq 0$

Composite Functions:

For two functions f and g, $(f\circ g)(x)=f(g(x))$.

Domain of $f\circ g$: The domain of $f\circ g$ consists of those values of x in the domain of g for which $g(x)$ is in the domain of f.

One-to-One Functions:

A function is a **one-to-one function** if for each value of y in the range there is only one corresponding value of x in the domain.

Horizontal Line Test:

A function is one-to-one if no **horizontal line** intersects the graph of the function at more than one point.

Inverse Functions:

If f is a one-to-one function with ordered pairs of the form (x, y), then its **inverse function**, denoted as f^{-1}, is also a one-to-one function with ordered pairs of the form (y, x).

If f and g are one-to-one functions and $f(g(x))=x$ for all x in D_g and $g(f(x))=x$ for all x in D_f, then f and g are **inverse functions**.

Exponential Functions:

An **exponential function** is a function of the form $f(x)=b^x$ where $b>0, b\neq 1$, and x is any real number.

General Concepts of Exponential Functions:

For $b>1$:

1. $b^x>0$
2. b^x increases to the right and is called an **exponential growth function**.
3. $b^0=1$, so $(0,1)$ is the y-intercept.
4. b^x approaches the x-axis for negative values of x. (The x-axis is a horizontal asymptote.)

For $0<b<1$:

1. $b^x>0$
2. b^x decreases to the right and is called an **exponential decay function**.
3. $b^0=1$, so $(0,1)$ is the y-intercept.
4. b^x approaches the x-axis for positive values of x. (The x-axis is a horizontal asymptote.)

Compound Interest:

Compound interest on a principal P invested at an annual interest rate r (in decimal form) for t years that is compounded n times per year can be calculated using the following formula:

$$A=P\left(1+\frac{r}{n}\right)^{nt}$$

where A is the amount accumulated.

The Number *e*:

The number e is defined to be $e=2.718281828459\ldots$

Continuously Compounded Interest:

Continuously compounded interest on a principal P invested at an annual interest rate r for t years can be calculated using the following formula:

$$A=Pe^{rt}$$

where A is the amount accumulated.

Logarithms:

For $b>0$ and $b\neq 1$, $x=b^y$ is equivalent to $y=\log_b x$.

Properties of Logarithms:

For $b, x, y>0, b\neq 1$, and any real number r:

1. $\log_b 1=0$
2. $\log_b b=1$
3. $x=b^{\log_b x}$
4. $\log_b b^x=x$
5. $\log_b(xy)=\log_b x+\log_b y$ The product rule
6. $\log_b \frac{x}{y}=\log_b x-\log_b y$ The quotient rule
7. $\log_b x^r=r\cdot\log_b x$ The power rule

Properties of Equations with Exponents and Logarithms:

For $b>0$ and $b\neq 1$:

1. If $b^x=b^y$, then $x=y$.
2. If $x=y$, then $b^x=b^y$.
3. If $\log_b x=\log_b y$, then $x=y$ $(x>0$ and $y>0)$.
4. If $x=y$, then $\log_b x=\log_b y$ $(x>0$ and $y>0)$.

Change-of-Base Formula:

For $a, b, x>0$ and $a, b\neq 1$, $\log_b x=\frac{\log_a x}{\log_a b}$.

CHAPTER 12 Conic Sections

Horizontal and Vertical Translations:

Given the graph of $y=f(x)$, the graph of $y=f(x-h)+k$ is

1. a horizontal translation by h units, and
2. a vertical translation by k units.

of the graph of $y=f(x)$.

Reflections:

The graph of $y=-f(x)$ is a **reflection across the x-axis** of the graph of $y=f(x)$. This means the graph of $y=-f(x)$ is the mirror image of $y=f(x)$ across the x-axis.

Horizontal Parabolas:

Parabolas of the form $x=ay^2+by+c$ or $x=a(y-k)^2+h,\ a\neq 0$:

If $a>0$, the parabola opens right.

If $a<0$, the parabola opens left.

Vertex: (h, k)

Line of Symmetry: $y=k$

Distance Formula (Distance Between Two Points):

For two points, $P(x_1, y_1)$ and $Q(x_2, y_2)$, in a plane, the distance between the points is $d=\sqrt{(x_2-x_1)^2+(y_2-y_1)^2}$.

Circles:

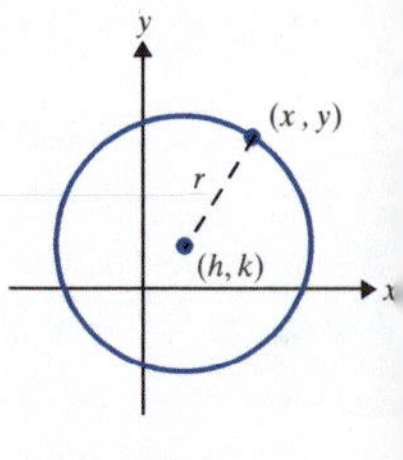

A **circle** is the set of all points in a plane that are a fixed distance from a fixed point.

The fixed point is called the **center** of the circle.

The distance from the center to any point on the circle is called the **radius** of the circle.

The distance from one point on the circle to another point on the circle measured through the center is the **diameter** of the circle.

Standard form for the equation of a circle: $(x-h)^2+(y-k)^2=r^2$

The radius is r and the center is at (h, k).